W9-BRM-058

General Microbiology

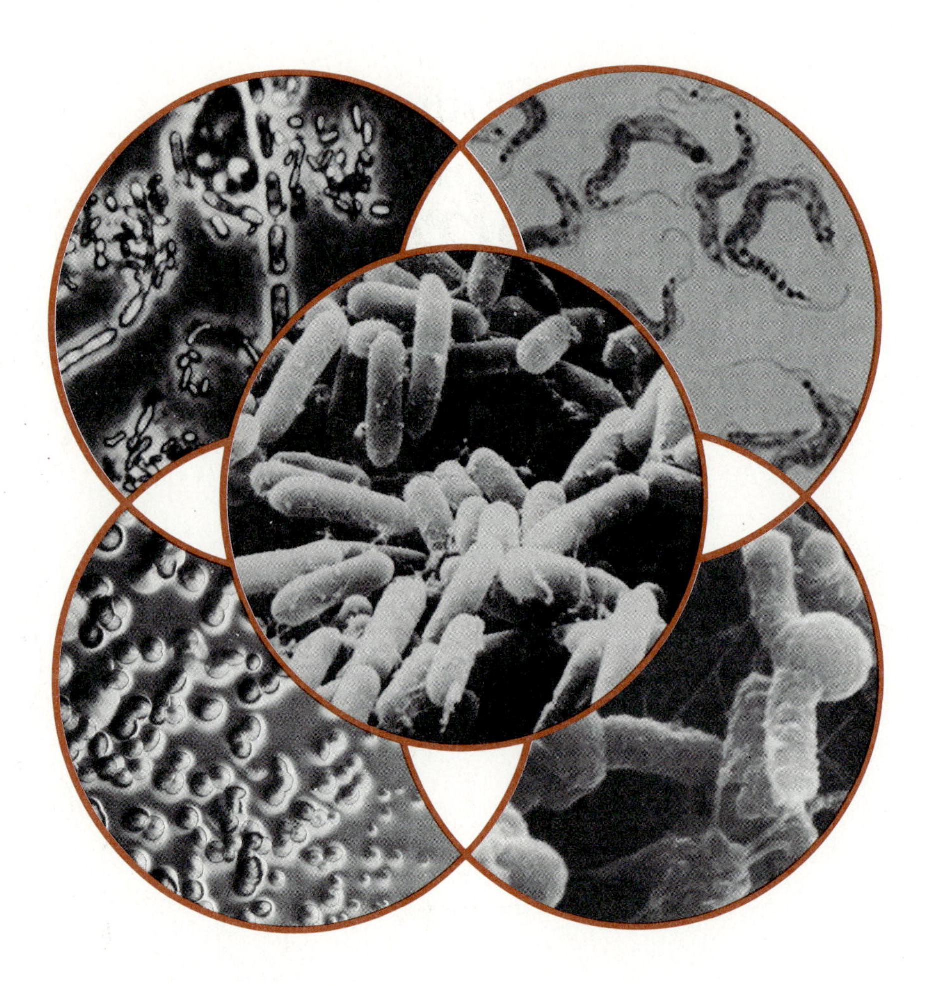

Center	**Salmonella bacteria** Magnification 12,000:1 SEM Color Cause diarrhea.
Top right	**Trypanosoma gambensi** Magnification 2000:1 Causes sleeping sickness.
Bottom right	**E-Coli under the influence of antibiotica** Magnification 8000:1 SEM Color The antibiotica prevent building of a new cell wall; in the picture the spherical part of the dividing bacteria is actually without cell wall.
Bottom left	**E-Coli** Magnification 40:1 Photographed in Schlierenoptics (relief effect).
Top left	**Sewage bacteria** Magnification 1200:1 Photographed in phase contrast; these bacteria actually clean polluted water by feeding on on it.

Cover photos by Manfred Kage/Peter Arnold Inc.

General Microbiology

Robert F. Boyd, PhD
Wirtz, Virginia

Second Edition

With 860 illustrations and 40 color plates

Original artwork by Ruth Steinberger

Times Mirror/Mosby College Publishing

ST. LOUIS • TORONTO • SANTA CLARA

Editor David Kendric Brake
Developmental Editor Kathy Sedovic
Project Manager Mark Spann
Production Editor Carol Sullivan Wiseman
Design John R. Rokusek

Second Edition

A division of The C.V. Mosby Company
11830 Westline Industrial Drive, St. Louis, Missouri 63146

Printed in the United States of America

Library of Congress Cataloging-in-Publication Data

Boyd, Robert F.
General microbiology.

Includes bibliographies and index.
1. Microbiology. I. Title. [DNLM: 1. Microbiology.
QW 4 B789g]
QR41.2.B69 1988 576 87-18066

ISBN 0-8016-1291-8

Contributors

Martin Alexander, PhD

Department of Agronomy, Cornell University, Ithaca, New York

Gordon Carter, PhD

College of Veterinary Medicine, Virginia Polytechnic and State University, Blacksburg, Virginia

Glenn Chambliss, PhD

Department of Bacteriology, University of Wisconsin—Madison, Madison, Wisconsin

Robert Deibel, PhD

Department of Bacteriology, University of Wisconsin—Madison, Madison, Wisconsin

Bryan Hoerl, PhD

Department of Microbiology, School of Dentistry, Marquette University, Milwaukee, Wisconsin

Gary Hooper, PhD

Department of Plant Pathology, Virginia Polytechnic and State University, Blacksburg, Virginia

F.L. Singleton, PhD

Department of Biological Sciences, Old Dominion University, Norfolk, Virginia

Preface

General Microbiology was written with the intent of providing students with a basic understanding of the biology of microorganisms and their importance to the scheme of life on this planet. No other area of science has caught the public attention as microbiology. Recombinant DNA technology and related technologies have demonstrated how microorganisms can be used to solve biological problems that only a few years ago were considered to be part of science fiction. In addition, the appearance of acquired immune deficiency syndrome (AIDS) and other sexually transmitted diseases has made us aware that infectious disease is *still* an important part of microbiology. To fully comprehend these developments in microbiology, a basic understanding of biochemistry, genetics, and immunology is required.

APPROACH

Since the first edition of this book, the explosion of information in applied and basic research involving microorganisms has made textbook writing a hazardous occupation. The challenge is not only what to discuss but how to discuss it in a meaningful way. The easiest course is to be purely descriptive and comprehensive (the marketplace is loaded with such books). This approach might be suitable to students for whom microbiology represents their first and/or final biology course. I do not, however, find this approach suitable for students who are biology majors or others who consider biology, or some aspect of biology, as part of their future occupation. My approach in this text is to discuss the concepts or the "how" and "why" of microbiology. This road is laid with dangers, because it requires some knowledge of biochemistry and genetics, topics that are not always easily understood by the average student. My experiences have shown me that the conceptual or mechanistic approach best prepares a student for problem solving. This approach was used for the first edition, but it suffered from a basic problem: some discussions were so long that the student could not see the forest (concept) through the clutter of trees. The second edition of *General Microbiology* has gone through three drafts—each critiqued by several instructors at universities in the United States and Canada. In addition, all line drawings and other illustrations were critiqued separately to ensure that there were no inadvertent errors. The consequence of these critiques has been a revision in which 70% of the book, including line drawings and other illustrations, has been reorganized and rewritten. The end result is a book with a "nuts and bolts" approach that is current, crisp, and well suited for students in the biological sciences, allied health, agriculture, and veterinary science areas.

ORGANIZATION AND NEW COVERAGE

General Microbiology is divided into seven parts. Part I, "Introduction to the Microbial World," consists of seven chapters, beginning with the History of Microbiology. This chapter discusses the historical aspects of microbiology and its importance to humankind now and in the future. Chapter 2 is "Biochemical Background for Microbiology," which reviews the biochemistry of macromolecules and also includes a discussion of enzymes. In response to reviewer suggestions, coverage of inorganic chemistry has been omitted. The remaining chapters in Part I are devoted to microscopic technique (Chapter 3) and the structure and function of prokaryotes and eukaryotes (Chapters 4, 5, 6, and 7). Chapter 4, "Prokaryotes and Eukaryotes: A Comparison" is unique to the second edition, because it presents the student with a comprehensive view of the similarities and differences between eukaryotes and prokaryotes. This chapter prepares the student for the more detailed discussions of structure and function of microorganisms that appear in Chapters 5, 6, and 7.

Part II, "Microbial Genetics," consists of four chapters (8-11) devoted primarily to prokaryotic systems. These chapters have been greatly expanded in terms of illustrations to clarify specific concepts. The second edition does not contain a separate chapter on eukaryotic genetics. Instead, specific aspects of eukaryotic genetics are compared with prokaryotic systems in the appropriate chapter topics. The second edition also contains a separate chapter on mutation. Chapter 11, "Recombination, Gene Transfer, Genetic Engineering, and Recombinant DNA," has been expanded to accentuate the increasing importance of research in genetic engineering. Genetic engineering is not left to die in this chapter. Instead, aspects of this technology as it relates to agriculture and industry are discussed later in the text (Chapters 20 and 22, respectively).

Part III, "Microbial Metabolism and Growth," consists of four chapters beginning with "Energy Production in Microorganisms" (Chapter 12). The remaining chapters discuss the metabolic processes of organisms, including nutrition, biosynthesis, and metabolic control (Chapter 13) and microbial growth and factors affecting growth (Chapter 14). These chapters rely heavily on summary tables and illustrations to clarify and review basic metabolic concepts. One of the organizational changes in Part III is the inclusion of "Chemotherapy, Sterilization and Disinfection" (Chapter 15), which was part of the infectious disease section in the first edition.

Part IV, "Viruses," consists of two chapters. Chapter 16, "Basic Properties of Viruses," discusses the general characteristics of bacterial, animal, plant, and fungal viruses. Included in this chapter is a discussion of viroids and satellites. Chapter 17, "Viral Replication and Tumor Development," has been almost completely revised and now contains detailed discussions on the mechanisms of virus infection and replication. In addition, considerable changes have been made in the discussion of viruses and their relationship to cancer.

Part V, "Environmental and Applied Microbiology," has been revised in both organization and content. The biogeochemical cycles that were divided between three chapters in the first edition are now together in Chapter 18, "Concepts of Microbial Ecology." The various interactions between microorganisms in the environment originally in the soil microbiology chapter are now discussed in Chapter 18. An additional content change for Chapter 18 has been the inclusion of the subject of metal extraction by microorganisms. The major changes in Chapter 19, "Aquatic Microbiology," have been the expanded discussion of sewage treatment and the inclusion of a detailed section on water treatment. The first edition chapter, "Soil Microbiology," is now part of a chapter called "Agricultural Microbiology" (Chapter 20). This chapter includes a discussion of soil microorganisms and their interactions with plants and animals. Consequently, the first edition chapters called "Infectious Diseases of Plants" and "Infectious Diseases of Animals" have now been eliminated and some of their material is part of Chapter 20. In Chapter 21, "Food Microbiology," the new material includes a discussion of oriental food processes and a more comprehensive discussion of food-borne disease. The major change in Chapter 22, "Industrial Microbiology," has been the inclusion of techniques for improvement of yeast strains used in industry.

Part VI, "Infection and Immunity," consists of four chapters, beginning with "Host-parasite Interaction" (Chapter 23). This chapter has been rewritten and updated to include more material on the recent discoveries of the molecular relationships between host and microorganism. The first immunology chapter, "Factors and Functions of the Immune System" (Chapter 24), now contains a detailed discussion of antibody fine structure, plus a detailed discussion of vaccines and the role of genetic engineering. In addition, the tests used to detect antigen and antibody, which was a separate chapter in the first edition, are now part of Chapter 24. In the chapter, "Immune Disorders" (Chapter 25), there is now an expanded and updated discussion of AIDS.

Part VII, "Microorganisms and Infectious Disease," has undergone considerable organizational and content changes. The chapter on epidemiology (Chapter 26) now consists of a discussion of community and hospital-associated diseases. Hospital-associated diseases was a separate chapter in the first edition. The chapters on microbial diseases have less discussion of the clinical aspects of infection and greater elaboration on the nature of the infectious agent and the molecular biology of the infectious-disease process. In addition, many of the photographs characterizing the clinical manifestations of disease have been deleted.

In response to reviewer recommendations, separate chapters on bacterial, viral, and sexually transmitted diseases are not included in the second edition. Chapter 27 now contains a discussion of bacterial and viral diseases, including those transmitted sexually. Fungal and animal parasite diseases are now part of a single chapter (Chapter 28) in the second edition. The discussion of helminths (worms) and arthropods has been deleted from the animal parasite disease section for the second edition.

There are two appendices. One consists of a review

of mathematical terms used throughout the text. The second appendix is devoted to an expanded discussion of taxonomic techniques and classification. Included in this appendix is the most recent Bergey's (ninth edition) classification scheme with the appropriate page reference for the bacterial group discussed in the text.

PEDAGOGICAL AIDS

General Microbiology contains a variety of learning aids designed to assist instructors and enhance student comprehension.

Chapter Outline—*New to this edition,* each chapter begins with an outline of the chapter, enabling the student to preview the content and direction of the chapter.

Key Terms—Key terms appear in boldface and are defined in the text the first time they appear. The most important terms are also included in a revised and expanded glossary at the end of the text for quick reference by the student.

Concept Checks—Following each major topic within a chapter, a *new and unique* learning aid referred to as a "Concept Check" is introduced to the student. Each Concept Check includes three or more questions to test student recall of the material. This feature will assist the student in identifying concepts that are not being retained before continuing to the next section of the text.

Points of Interest—Another feature *new to this edition* of *General Microbiology* is the boxed Point of Interest sections. Many chapters contain one or more of these boxed essays, which discuss applied or unusual aspects of microbial research or elaborate on the interesting characteristics of specific groups of microorganisms. This unique feature was specifically designed to engage student interest in and discussion of the material.

Chapter Summaries—Concluding each chapter a numbered summary provides a quick review of the major points in the chapter.

Self-study Quiz—A list of objective questions has been added at the end of each chapter to be used by students in testing their retention of the material. The answers to these questions are provided at the end of the text.

Selected Readings—A brief list of books and journal articles related to the chapter topic is provided for further reading at the end of each chapter.

Appendices—A review of mathematical terms that appear within the text is provided in Appendix A. To aid the student in locating the discussions of some specific groups of bacteria, a Bergey classification scheme is presented in Appendix B. The appropriate reference page in the text is conveniently included next to each group.

Illustrations and Tables—To clearly illustrate basic concepts for students, many *new* two-color illustrations are provided in this second edition, particularly in those chapters related to biochemistry, genetics, and metabolism. In addition, summary tables are used more extensively to assist the student in reviewing large amounts of information.

SUPPLEMENTS

The following supplements are available to adopters of *General Microbiology*.

Instructor's Manual and Test Bank, prepared by Mary Burke, PhD, of Oregon State University, includes chapter overviews and outlines, teaching objectives, lists of key terms, extensive lecture outlines, a list of pertinent audiovisual aids for each chapter, and questions for further consideration. The 2000-item testbank offers 50-75 questions of varying formats (matching, completion, short answer) for each chapter; questions are coded for each chapter category and level of difficulty, as well as text page–number references.

Overhead Transparency Acetates—A selection of 50 two-color acetates representing key illustrations from the book is provided free of charge to adopters of the text. Chosen for their pedagogical value, these acetates will prove a valuable part of your classroom presentation.

ACKNOWLEDGEMENTS

The many changes in the second edition are a result of the input of many reviewers through several drafts of the manuscript. Their suggestions and contributions to the development of this text are greatly appreciated.

W. Peter Astin, PhD
Queens University

W. Murray Bain, PhD
University of Maine at Orono

Larry Barton, PhD
University of New Mexico

R. L. Bernstein, PhD
San Francisco State University

Mary Burke, PhD
Oregon State University

Anthony Catena, PhD
San Francisco State University

June Chadwick, PhD
Queens University

Wendy Champness, PhD
Michigan State University

Donald Deters, PhD
University of Texas at Austin

Jacquelin Dudley, PhD
University of Texas at Austin

Richard Fleming, PhD
California State University at Los Angeles

Robert Gessner, PhD
Western Illinois University

Robyn Hillam
South Dakota State University

Thomas Kerr, PhD
University of Georgia

Ken Keudell, PhD
Western Illinois University

Kathleen Killick, PhD
St. John's University

A. Kropinski
Queens University

Joseph Layne, PhD
Memphis State University

Ron Lyric, PhD
University of Manitoba

William O'Dell, PhD
University of Nebraska at Omaha

Shelley Payne, PhD
University of Texas

Dan Portnoy, PhD
Washington, D.C.

Fred Rosenberg, PhD
Northeastern University

Robert Sjogren, PhD
University of Vermont

James Struble, PhD
North Dakota State University

James Urban, PhD
Kansas State University

Carl Westby, PhD
South Dakota State University

Helen Westfall, PhD
South Dakota State University

Gary Wilson, PhD
Texas A & M University

A special debt of thanks is due those individuals who allowed me to use their photomicrographs, tables, and graphs. I am grateful to Ruth Steinberger for the artwork and to my editor at Times Mirror/Mosby College Publishing, David Brake, and Kathy Sedovic, developmental editor, and to the production editor, Carol Wiseman, who worked tirelessly to get this book into print. Finally, I would like to thank the members of my family, Beverly, T. J., and Sasha Boyd for their support and encouragement.

Contents

Detailed contents

PART II
Microbial genetics

PART I

Introduction to the microbial world

CHAPTER

1

History of microbiology

Chapter Outline

Introduction

Microbiology is a science that is involved with microscopic forms of life. The adage, "What you can't see won't hurt you" must have been coined by someone who had no knowledge of bacteria or other microorganisms. We are now aware that microorganisms can be detrimental, as well as beneficial, to the well-being of all life forms. Microbiology today is a very diverse area of study, encompassing many other branches of science. As early as the late 1800s the renowned bacteriologist, Theobald Smith, stated "Bacteria are not an end in themselves but knowledge concerning them contributes to the solution of higher problems."

The purposes of this chapter are to describe how microorganisms came to be recognized as "life forms," to demonstrate how the study of microorganisms evolved into a science, and to briefly describe the impact that microbiology has on human affairs.

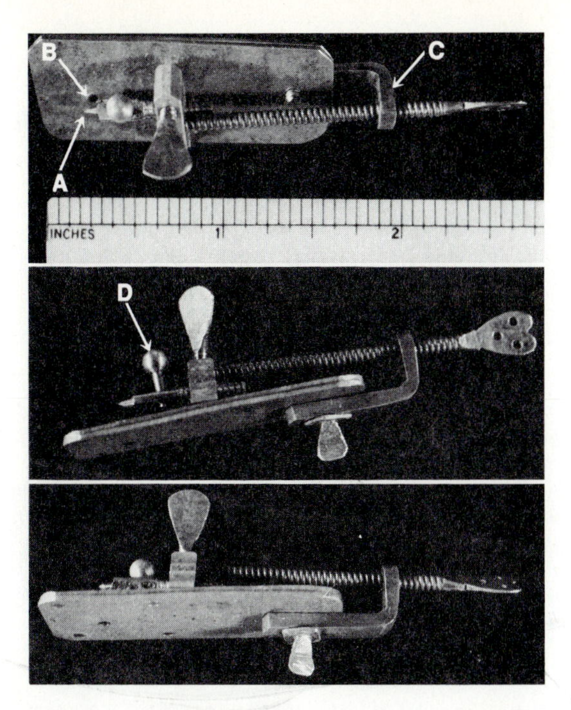

Figure 1-1 *Actual-size replica of Leeuwenhoek microscope from Leyden, Holland, showing various views of instrument.* **A,** *Pin on which object is placed for viewing;* **B,** *lens;* **C,** *screw for coarse adjustment;* **D,** *fine adjustment.*

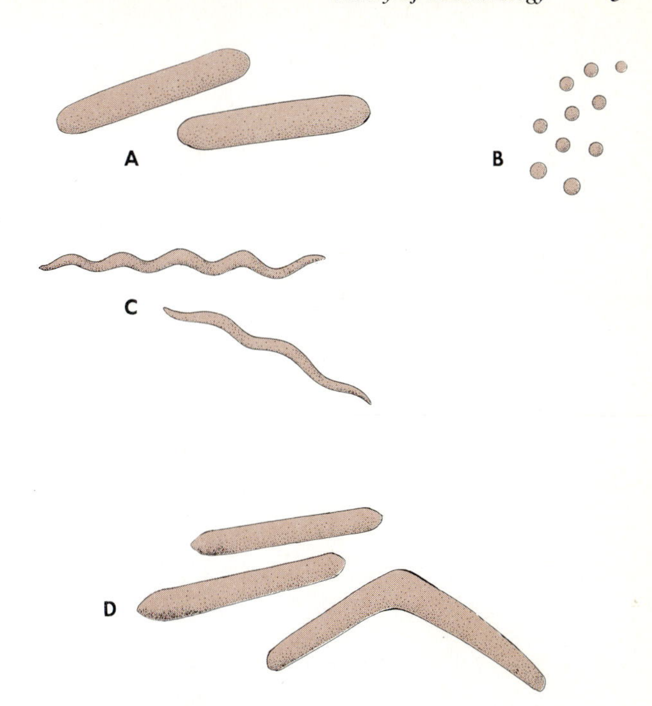

Figure 1-2 *Leeuwenhoek's figures of bacteria from human mouth.* **A,** *Rod-shaped (bacilli).* **B,** *Spherical (cocci).* **C,** *Spiral-shaped (spirochetes).* **D,** *Cigar-shaped rods.*

DISCOVERY OF SMALL "ANIMALCULES"

Leeuwenhoek and primitive microscopes

One of the major obstacles, in early times, to the discovery of microorganisms was an inability to see them. Nearly all microorganisms (bacteria, protozoa, primitive fungi, and microscopic algae), as well as viruses, cannot be seen with the naked eye and require some degree of magnification. Anton van Leeuwenhoek (circa 1685) is thought to be the original discoverer of bacteria. He observed, using primitive lenses (Figure 1-1), various forms of life that he characteristically described as little "animalcules." He observed these microscopic forms in such diverse environments as pond water, scrapings from teeth, pepper infusions, and even the discharges from his own bowels and reported his findings to other members of the scientific community. We can conclude from Leeuwenhoek's drawings that he had observed one-celled animals, called protozoa, as well as variously shaped bacteria (Figure 1-2). His findings were confirmed shortly thereafter, although 150 years passed before any serious investigations into the nature of microorganisms were begun. One of the reasons for this was that lenses used before 1820 produced images with spherical and chromatic aberrations (fuzzy images resulting from the inability of the lens to focus light rays). These aberrations were corrected, and this led to the development of the compound microscope. Later, in 1866 Ernst Abbe developed the technique for using oil immersion as a medium for the transmission of light rays from the specimen to the lens. Immersion oil prevented loss of light and increased the clarity of the images produced.

Soon after Leeuwenhoek's discovery many investigators began to wonder about the origin of these "animalcules." Two schools of thought prevailed. Some believed that animalcules arose spontaneously from organic matter, whereas others believed that they arose from the "seeds" or germs of the animalcules.

THE ORIGIN OF MICROBIAL LIFE

The theory of spontaneous generation

The debate concerning the origin of life is perhaps the single most important factor that led to the development of microbiology as a science and paved the way for the discovery of microorganisms as part of the biological world. Extremely bizarre ideas about the origin of life have been recorded through history. The origin of larger animals and plants that could be easily observed was obvious. The origin of microorganisms that were ordinarily hidden remained a mystery, and, consequently, mysterious forces were conjured up, even by renowned scientists. It was a common belief that decay-

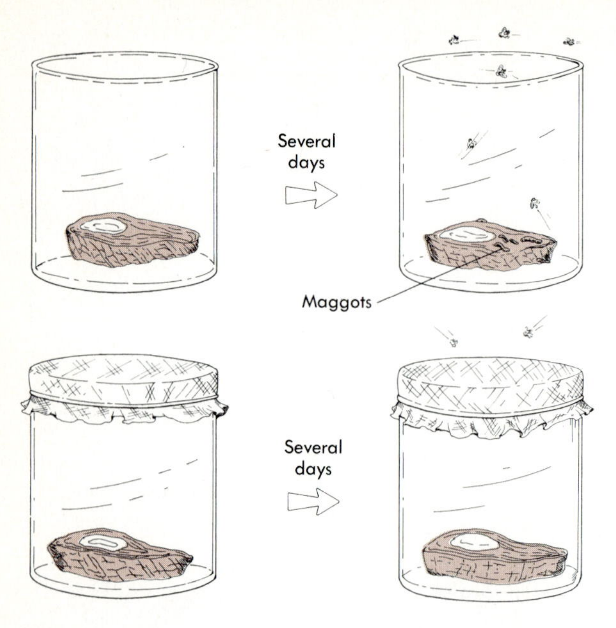

***Figure 1-3** Redi's experiment disproving spontaneous generation. Many believed that meat exposed to air would give rise to maggots. To disprove this theory Redi placed gauze over jar containing meat and left it exposed to air for several days. No maggots developed on meat in covered jar.*

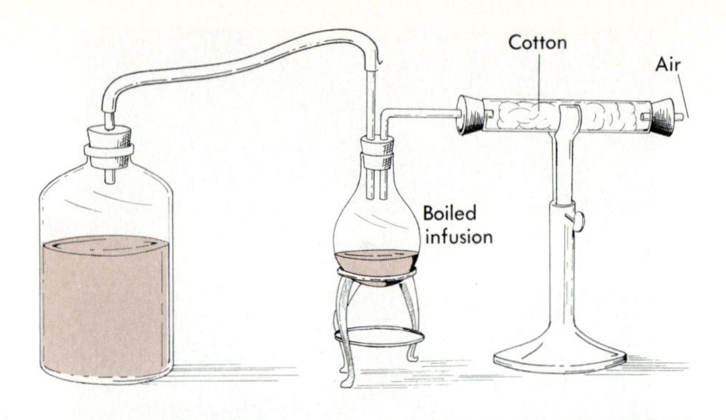

***Figure 1-4** Schröder and von Dusch experiment that helped disprove spontaneous generation. Aspirating bottle drew air into infusion that was boiled previously to destroy any living microorganisms. Cotton in tube trapped any microorganisms or dust in air being passed into infusion. Under these conditions infusions remained clear and free of microorganisms, thus demonstrating that living things could not generate spontaneously from nonliving matter.*

ing meat generated maggots and that lice and fleas arose from sweat. The famous chemist Helmont (1652) gave his formula for the production of mice: "place some dirty rags together with a few grains of wheat or a piece of cheese in a dark place and in a few days they will be transformed into mice."

These ideas of spontaneous generation lasted into the nineteenth century, even though experiments performed by scientists for nearly 300 years had suggested the theory to be false. These attempts to disprove spontaneous generation led to the observation of bacteria and to the demonstration of their association with disease.

The Italian physician, Francisco Redi (1626-1697), was one of the first scientists who sought to disprove spontaneous generation. Redi put pieces of meat in an open vessel and covered it with fine gauze to protect the meat from flies (Figure 1-3). Another jar was prepared in the same manner except the jar was left uncovered. In both cases the meat decayed but only in the uncovered vessel did maggots appear and flies emerge. Redi's experiments convinced the supporters of spontaneous generation that higher forms of life could not arise by nonbiological methods but, as we shall see later, they still firmly believed that smaller forms of life arose spontaneously.

In 1749, Joseph Needham reported that boiled mutton broth became turbid when left to stand overnight in a flask, although the broth was initially clear. When Needham observed small samples of the broth under the microscope, he noted the "animalcules" observed by Leeuwenhoek. From these observations Needham declared that bacteria arose spontaneously from the dead organic matter in the broth. This observation is not too farfetched because many bacteria can divide in less than 30 minutes. Another Italian, Abbe Spallanzani, performed experiments similar to Needham's except that the flasks containing the broth were boiled for several hours and sealed so no air could enter. No matter how long the sealed flasks were left to stand, there was no turbidity in the flasks. If the flasks were opened to the air, the broth became cloudy. When Spallanzani observed a diluted portion of the broth under the microscope, he saw only a few bacterial cells. As he continued to observe the bacteria he saw them elongate and divide, and within an hour the drop of broth was covered with bacteria. Undaunted by Spallanzani's experiments the proponents of spontaneous generation created new theories to support their views, which in turn stimulated the opposition to plan their own experiments. Spallanzani's detractors proposed that by sealing the flask the "vital force" necessary for spontaneous generation had been excluded. The consequence of this controversy was the accumulation of a large body of information about microorganisms.

Theodor Schwann and Franz Schulze improved on Spallanzani's experiments by allowing air to enter the flasks only after it had passed through a solution of sulfuric acid and potassium hydroxide. Those who supported spontaneous generation then suggested that air was a "vitalizing power" that had been destroyed on

Figure 1-5 *Louis Pasteur (1822-1895).*

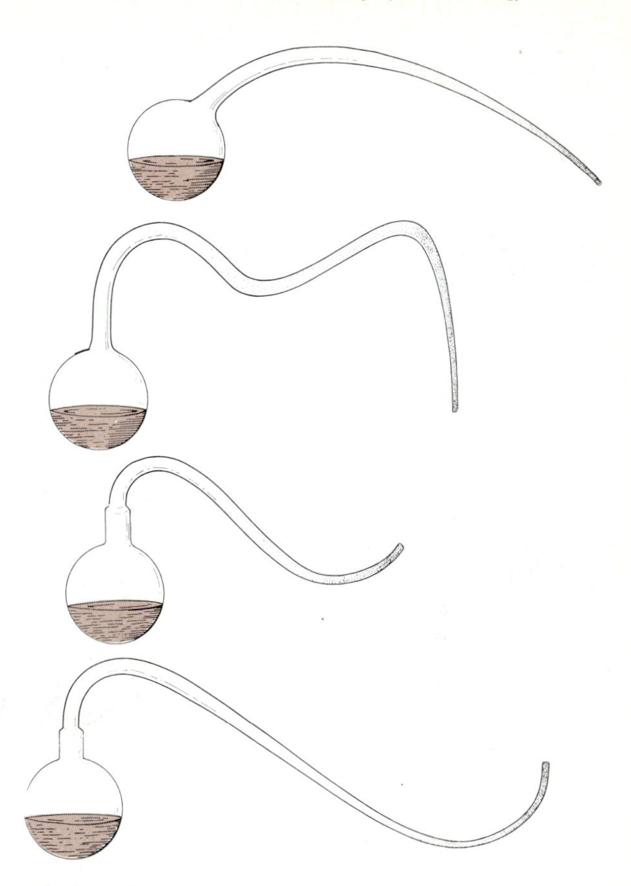

Figure 1-6 ***Pasteur's swan-necked flasks. Strippling in necks represents microorganisms unable to get into sterile growth medium. Shape of flask permits air to enter, but microorganisms cannot.***

passing through the chemicals used by Schwann and Schulze. In 1854, Schröder and von Dusch countered the objections to Schwann and Schulze's experiment by filtering the air through cotton wool before it entered the flasks (Figure 1-4). The experiments that would finally disprove spontaneous generation were those designed by Louis Pasteur, the French chemist who is called the **Father of Microbiology** (Figure 1-5), and John Tyndall, an English physicist (1820-1893).

Pasteur's and Tyndall's experiments

Pasteur repeated the experiments of Schröder and von Dusch. The flasks were incubated at 25° to 30° C for several months with no visible growth evident in any of the flasks. If a piece of the cotton plug used to filter air was added to one of the flasks, growth occurred in 24 to 48 hours. Examination of the cotton plug before it had been added to the broth revealed spores and various microbial forms. Later Pasteur used flasks in which the neck of the flask had been drawn out and bent down so that air could enter but microorganisms could not ascend into the flask (Figure 1-6). Again, incubation over several months produced no visible growth. Growth occurred if the neck of the flask were broken or if the flask were tipped and the infusion allowed to make contact with the exposed parts of the neck and then returned to the flask. Pasteur had conclusively shown from these experiments and others that growth in infusions were the result of microbes in the air (Pasteur called them organized corpuscles) and not from organic matter that had been "vitalized." Pasteur concluded his experiments with this statement: "When one incubates the corpuscles and the amorphous debris with which they are associated into liquids which have been subjected to boiling and which would have remained unchanged in previously heated air if this inoculation had not been performed, one observes the appearance in this liquid of exactly the same organisms which they develop in the open air."

In 1876, Tyndall supplemented Pasteur's experiments and proved that microorganisms are carried on dust by showing that dust is the source of contamination of fluids exposed to air. He observed that open tubes of boiled broth remained free of bacteria when enclosed

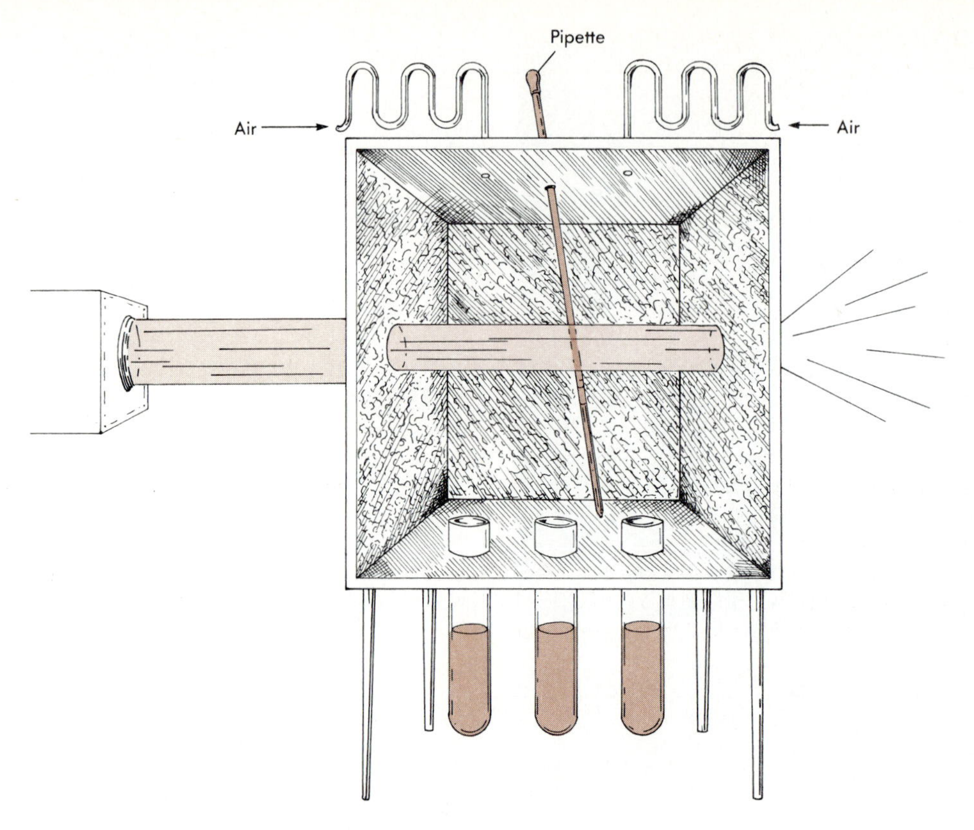

Figure 1-7 *Tyndall's dust-free chamber with front panel removed to show contents. Inside of chamber is coated with glycerin, a sticky substance that will catch dust particles. Light is shown through chamber until dust particles settle out. Tubes were filled with broth using pipette. Open tubes were sterilized by heating in oil bath. Air passed through convoluted tubing and any particles in air would be filtered out before reaching inner chamber. Broth in tubes remained free of microorganisms for indefinite period.*

in a dust-free chamber through which a beam of light could be passed (Figure 1-7).

Fermentation studies

One of the factors that spurred further experimentation in microbiology was the interest in fermentation. Fermented drinks and foods have been known for many centuries. The mechanism of fermentation was believed by most scientists to be a chemical process, although Cagniard de la Tour demonstrated in 1838 that a microbial agent, yeast, was responsible. Again, experiments designed by proponents of both theories provided substantial information on microbial metabolism. Diseases of wine made in France attracted Pasteur's attention. He studied the fermentation process and observed that yeasts produced alcohol by breaking down the sugar, resulting in palatable wines. Spoiled wines were a consequence of products released by rod-shaped bacteria that had entered the fermentation vats through the air. The growth of these rod-shaped bacteria showed that heating the wine to a particular temperature could destroy the unwanted bacteria without affecting the taste of the wine. This procedure, called **pasteurization,** is used today in the treatment of milk and certain alcoholic beverages.

Concept Check

1. What is spontaneous generation?
2. When and how were microorganisms first observed?
3. Describe the nature of the experiments, including Pasteur's, that helped disprove spontaneous generation.

MICROORGANISMS AS INFECTIOUS AGENTS

Infectious diseases and the gods

Diseases caused by microorganisms (infectious diseases) were as commonplace centuries ago as they are today. However, early civilizations were unaware of the origin of disease. History has shown that when answers to scientific questions are not available, humankind often invents answers that somehow involve a god, and who can question the gods? Thus it should not seem peculiar to us that earlier civilizations regarded disease as being the curse of gods and witches. Even today certain forms of disaster, such as hurricanes and volcanic eruptions, are legally termed "acts of God." The gods and superstition remained the ruling forces in medicine for most of civilization until the nineteenth century.

One of the first to separate medicine from the supernatural was Hippocrates (circa 460 BC), the Greek physician. Hippocrates believed disease was part of the order of nature and could be cured by natural means. He did not refer to biological agents as the cause of infectious disease but instead referred to *miasmas* (which is Greek for pollution) that were present in the air. The notion that a living agent could be the cause of infectious disease would not become a viable hypothesis until centuries later.

EARLY EXPERIMENTS

As early as 1546, Fracastorius (1478-1553), a physician, expressed the belief that invisible living organisms can cause disease. Anton von Plenciz (1705-1786), also a physician, published articles expressing a similar belief that living agents were the cause of disease. However, it was not until the first half of the nineteenth century that reports circulated implicating microorganisms in various disease conditions. For example, a disease of silkworms, called pebrine, was believed to have been caused by a microorganism.

In the nineteenth century silk was a major industry in France, and pebrine had spread to Spain, through Turkey, Syria, and China. At the request of the French government, Pasteur in 1849 investigated the problem and tried to determine the cause of the disease. Pasteur worked on the problem for 5 years and discovered that the silkworms were afflicted with infectious disease caused by single-celled protozoa. He provided the industry with the means of preventing the diseases through the recognition of infected moths. The concept that microorganisms were the cause of infectious disease (the germ theory of disease) was difficult for many scientists to accept, and it spurred further research.

During his investigations of silkworm disease, Pasteur received a letter from Joseph Lister. Lister, an English physician, stated that Pasteur's ideas on fermentation and the causes of bitter wine led him to believe that infections in humans arose from the same cause. Lister demonstrated that boiling material dipped in carbolic acid before applying to the skin prevented microbial agents from causing wound infections. Boiling had been recognized as a technique for destroying cells but had not been used on surgical instruments because many in

Point of Interest

IGNAZ SEMMELWEISS AND CHILDBIRTH FEVER

Pregnant women in Europe in the 1800s who could not afford a private physician or midwife came to "lying-in" hospitals for childbirth. The principal ward contained large numbers of beds placed in alcoves along each side. Pregnant women unable to find a bed would lie on straw spread on the floor, some would sit on wooden benches, and others would crouch in corners. Floors and partitions were washed once a month. Ceilings were seldom washed. Pregnant women who became ill were transferred to isolation rooms regardless of the nature of their illness, whether it was diarrhea, diphtheria, bronchitis, or measles. These were the conditions observed by Ignaz Semmelweiss (circa 1845), a Hungarian physician. Semmelweiss also found that patients in labor laid on filthy sheets that stank of decomposed blood and uterine discharge. No connection was made between filth and disease. He discovered that the head nurse had accepted the bedding as clean. To his dismay Semmelweiss later found that the laundry contractor had accepted the contract at an especially low rate and that the head nurse had been bribed. Amidst similar conditions in Paris, the number of patients handled at the lying-in hospital from 1861 to 1864 was 9886, of whom 1,226 died. Most of these deaths occurred among mothers following childbirth, hence the name childbirth, or puerperal, fever.

Semmelweiss meticulously sought to discover the reasons for childbirth fever in his own Vienna hospital. He observed that three times as many deaths occurred when women were examined by medical students than by midwives. He discovered that medical students had performed autopsies before examining postpartum mothers and that even washing with soap and water was not sufficient to remove contaminating particles. Semmelweiss required medical students to wash their hands in chloride of lime before making examinations. At that time in his division there were 120 deaths in 1000 births but after instituting his procedures the deaths fell to 12 in 1000 births.

Figure 1-8 *Robert Koch (1843-1910).*

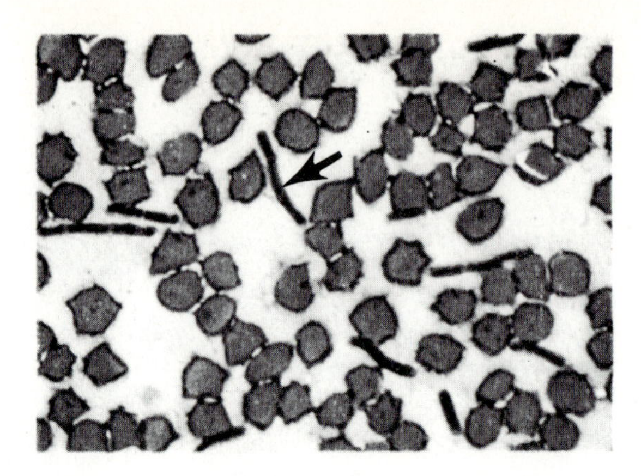

Figure 1-9 *Appearance in blood of the rod-shaped bacterium causing anthrax* (arrow). *Actually several bacteria shown are attached to one another.*

the scientific community in the early 1800s had not accepted the concept that microorganisms were the agents of disease. Oliver Wendell Holmes, an American physician, published an article in 1843 on the nature of an infection called puerperal sepsis,* which afflicts mothers following childbirth. Holmes reported that it was safer to give birth to a child at home than in the hospital where physician handling contributed to disease. Semmelweiss, a Hungarian physician, was ridiculed for his insistence that physicians wash their hands before handling postpartum (after childbirth) women in the hospital ward (see Point of Interest).

These early investigations eventually led to **aseptic technique**, which is the removal of viable microorganisms from a particular environment. It is hard to imagine surgery with contaminated hands and instruments or drinking water that has been untreated or unfiltered. Asepsis is also important in the study of microorganisms. Aseptic techniques permit the investigator to use a growth medium devoid of unwanted microorganisms. Thus if a single cell of a particular microbial species is inoculated into an uncontaminated medium, a population of cells can be produced that is made up of only one microbial species **(pure culture technique)**. Only with a large population of a single species can the biochemistry, genetics, physiology, or other characteristics of that species be effectively studied. These cannot be studied if the microorganism is growing in a medium contaminated by other microorganisms.

* *Sepsis* is derived from the Greek, *sepsis,* which means decay. It is used to indicate the presence of pathogenic microorganisms or their products. Asepsis means without sepsis. Antisepsis means removal of organisms.

Koch's experiments

In the 1840s Jacob Henle made some important contributions to the germ theory of disease by stating that for a microbe to be considered the cause of disease it must be consistently found in association with the disease and isolated from the disease. Robert Koch, a German physician (Figure 1-8), carried these ideas further and was the first to establish the relationship between an individual bacterial species and a specific disease. Koch's experiments proved that differences exist between bacteria that cause disease and that each species has a constant set of characteristics. Koch began his experiments with the agent of anthrax *(Bacillus anthracis)*. Anthrax is a particularly devastating disease of cattle that is capable of destroying entire herds. The anthrax bacillus had been microscopically observed by others as early as 1850. Davaine in the 1860s had observed the rod-shaped organism in the blood of anthrax victims (Figure 1-9), but others could not repeat his experiments and refuted his findings. Koch in 1876 elaborated on the life cycle of the anthrax bacillus and demonstrated how it formed resistant endospores in the soil, thus shedding some light on the epidemiological aspects of the disease. In 1878 Koch published a paper on the role of microorganisms in wound infections. He showed that six different infections in mice were caused by six different bacteria that differed morphologically and biologically from one another. In 1881 he proposed four postulates, **Koch's postulates**, that would prove whether or not an infectious agent is the cause of a disease:

1. The causative agent of the disease must be present in all cases of the disease and must be absent from healthy animals.
2. The agent of disease can be isolated from the diseased animal and can be cultivated in pure culture

Table 1-1 *Discoverers of some of the major pathogens during the Golden Age of Microbiology (1879-1900)*

Disease	Causative agent	Discoverer	Date
Gonorrhea	*Neisseria gonorrhoeae*	Albert Neisser	1879
Malaria	Species of *Plasmodium*	Charles Laveran; Ronald Ross described life cycle of microbes	1880
Typhoid fever	*Salmonella typhi*	Karl Eberth	1880
Pneumonia	*Streptococcus pneumoniae*	Independent isolation by Louis Pasteur and George Sternberg	1881
Tuberculosis	*Mycobacterium tuberculosis*	Robert Koch	1882
Cholera	*Vibrio cholerae*	Robert Koch	1883
Diphtheria	*Corynebacterium diphtheriae*	Edwin Klebs and Friedrich Loeffler made initial isolation, but Emile Roux and Alexander Yersin established importance of toxin in 1888	1883
Tetanus	*Clostridium tetani*	Arthur Nicolaier	1884
Diarrhea	*Escherichia coli*	Theodor Escherich	1885
Meningitis	*Neisseria meningitidis*	Anton Weichselbaum	1887
Undulant fever	*Brucella* species	David Bruce	1887
Gas gangrene	*Clostridium perfringens*	William Welch and G.H. Nuttall	1892
Plague	*Yersinia pestis*	Shibasaburo Kitasato and Alexander Yersin	1894
Botulism	*Clostridium botulinum*	Emile van Ermengem	1896
Dysentery	*Shigella dysenteriae*	Kiyoshi Shiga	1898

(a population of one species of microorganism) in the laboratory.

3. The disease can be reproduced by inoculating a portion of the pure culture into healthy animals.
4. The agent of disease can be reisolated from the infected animal and again cultivated in the laboratory.

Koch's postulates, with few exceptions, can now be demonstrated for all bacterial diseases. They cannot be specifically applied to viruses or rickettsia, which cannot be grown in artificial media. In addition, in some viral diseases the agent may not be directly demonstrable.

GOLDEN AGE OF MICROBIOLOGY

The results of Koch's experiments initiated a period of investigation called the Golden Age of Microbiology, in which the causes of many diseases were determined. These are outlined in Table 1-1. This golden age would have been slow to develop if it had not been for specific contributions made by Koch and other investigators. Koch's postulates can be proved only if the organism causing disease can be cultivated in the laboratory. In other words, we must have a pure culture. Koch observed that exposure of a potato slice or gelatin to the air resulted in the formation of discrete colonies on its surface and that each colony was made up of a population of cells that had apparently arisen from a single cell. He later demonstrated that if a small population of cells could be thinned out on the surface of a solid nutrient surface, single cells could be deposited that would give rise to individual colonies.

Other important contributions were made that involved the growth medium for bacteria. Potato slices and gelatin are not ideal media for bacterial growth because (1) many microorganisms cannot grow on them because of a lack of specific nutrients, (2) many microorganisms produce enzymes that can liquefy gelatin, and (3) gelatin liquefies at 37° C. The German physician, Walter Hesse, who was interested in the bacteriology of air, used gelatin to trap microorganisms from the air, but the gelatin melted at high temperatures. In 1881 Hesse's wife suggested the use of **agar-agar**, which she had used for many years in preparing fruits and jellies. Agar-agar was solid at the usual temperatures of incubation for bacterial cultures; it was heat stable; and it was resistant to microbial attack. This observation was reported to

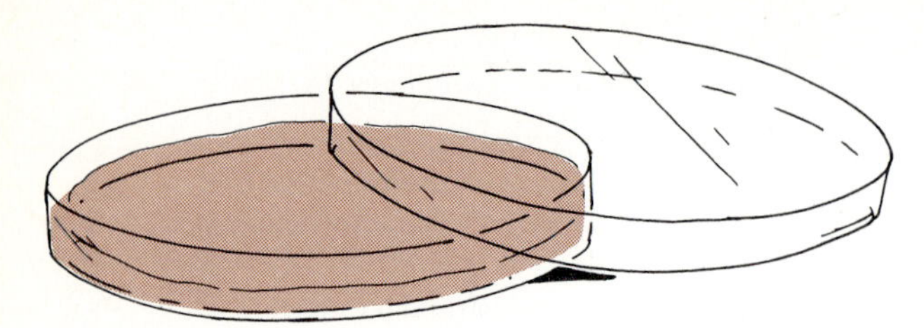

Figure 1-10 Petri dish. Petri dish consists of bottom plate that holds the semisolid medium and cover plate that helps prevent contamination of medium from air.

Koch, who immediately adapted the new medium for the cultivation of the agent of tuberculosis. Friedrich Loeffler, who was an associate of Koch, devised a nutrient medium in 1881 that was composed of meat extract and digests of protein. This medium is the basis of a **nutrient broth** that is still used today. The addition of agar-agar to the nutrient medium would now provide a simple means for isolating and cultivating microorganisms. This is called **nutrient agar**. These observations coupled with the design of the **Petri dish** in 1877, which holds semi-solid media, provided bacteriologists with an important tool for isolating bacteria (Figure 1-10).

Koch also contributed to the more rapid identification of microorganisms through the development of a microbial stain. He developed a technique for staining the bacillus that is the cause of tuberculosis. Koch's pioneering work stimulated other workers to continue the search for new stains. Among these other workers were Paul Ehrlich, Franz Ziehl, Friedrich Neelsen, and Hans Christian Gram, whose names are associated with stains that are routinely used today in microbiological laboratories.

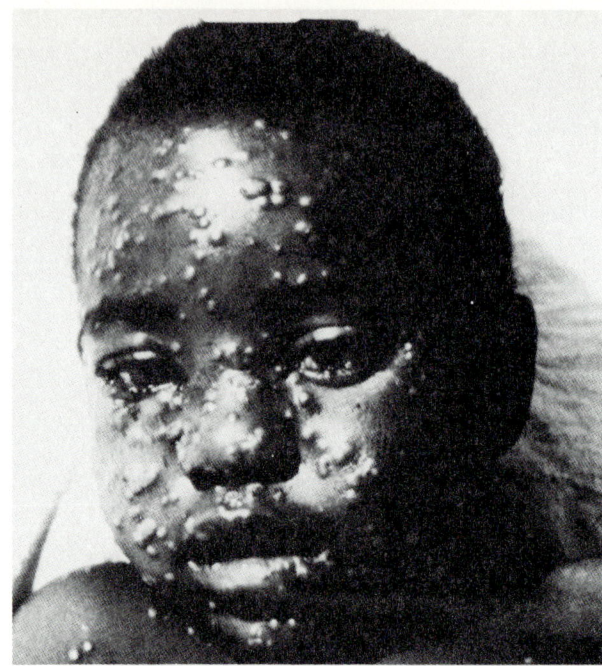

Figure 1-11 Smallpox lesions on face of African boy.

Concept Check

1. What were the preliminary experiments that led to the theory that microorganisms could be infectious agents?
2. Outline Koch's postulates and explain their importance.
3. Describe the contributions made by the following investigators in the field of microbiology: Koch, Pasteur, Frau Hesse, Semmelweiss, Roux, and Yersin.
4. What is meant by the term *pure culture?*

MICROBIOLOGY IN THE TWENTIETH CENTURY

Immunology

History of immunology. Although the cause of many diseases had been discovered during the Golden Age of Microbiology, more practical minds also sought to prevent or cure disease. The study of resistance to infection is call **immunology**. The concept of immunity began with the Chinese and their methods of preventing smallpox. This concept was also revealed in the eighteenth century in a letter written by Lady Mary Wortley Montagu from France in 1779. She described smallpox parties in which an old woman, carrying a nutshell full of matter from smallpox sores, asked each individual what veins they would like to have opened. The old lady put some of the smallpox matter on the head of a pin and then scratched the designated vein. Lady Montagu stated that no one had died from this operation and they did not acquire smallpox during later epidemics of the disease.

Edward Jenner, an English physician, in the 1790s observed that horses suffered from a disease called "grease," which resembled the human disease smallpox. Horse attendants who applied dressings to the diseased horses often were asked to milk cows. When this occurred the disease was often transmitted to the cows and from the cows to the milkmaids who milked them. The disease, which they called **cowpox**, appeared as pustules (elevation of the skin containing pus) on the hands of the milkmaids after coming in contact with the

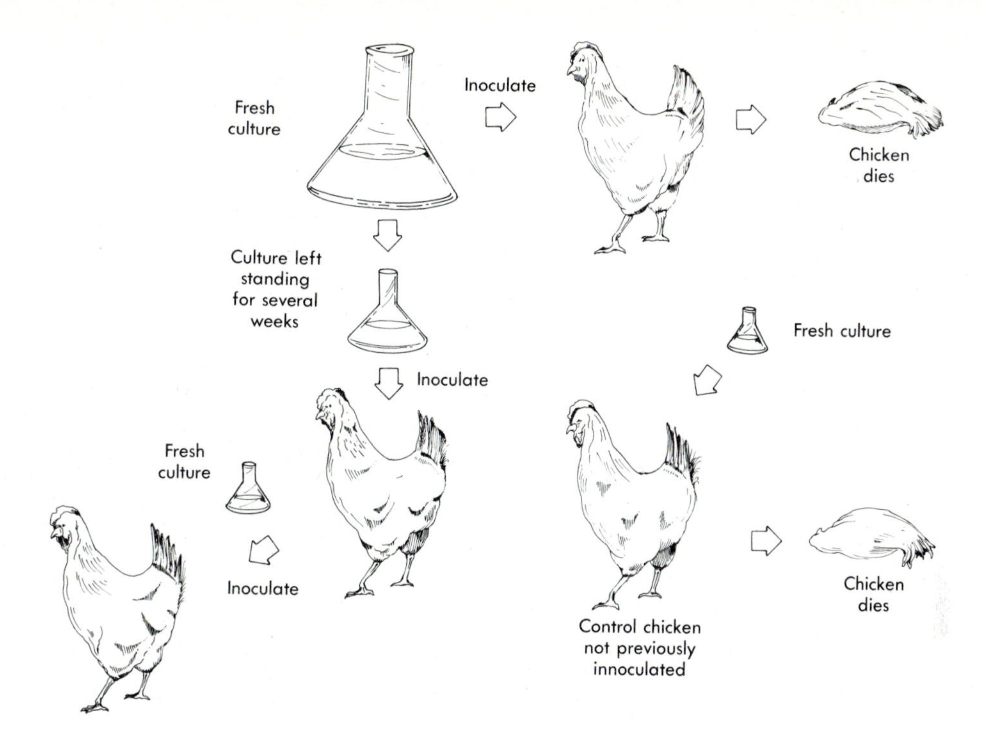

Figure 1-12 *Pasteur's accidental discovery of principle of immunization. Fresh culture of chicken cholera bacteria injected into chicken causes it to die. When culture had been left standing for several weeks and then used to inject into chicken, chicken survived and remained healthy. If chicken inoculated with old culture was injected with fresh culture of bacteria, it remained healthy. However, if chicken that had not been previously inoculated was inoculated with fresh culture, it died. Thus culture that had been left standing was not capable of causing disease but was capable of inducing immunity to disease.*

diseased animals. These symptoms were followed by a fever that lasted from 3 to 4 days. What surprised Jenner was the fact that the milkmaids infected with cowpox were immune to smallpox either from exposure to individuals with smallpox or from direct inoculation of smallpox matter into their skin. Jenner seized on this observation and developed a technique for introducing the infectious material from milkmaids cowpox sores into healthy individuals. He showed that this procedure could protect individuals for life from the ravages of smallpox (Figure 1-11).

An accidental discovery that opened the way for the use of vaccines was made by Pasteur in 1875. Pasteur was studying chicken cholera and noticed that the bacteria grown in broth did not remain viable unless fresh broth was frequently added (Figure 1-12). A few drops of fresh broth containing the microorganisms and placed on bread invariably induced the disease and killed the chickens. Pasteur by chance fed chickens with bacteria that had remained for several days in the same broth. When these chickens were fed virulent bacteria that had been cultivated in fresh broth, they did not even become sick. The chickens had apparently acquired immunity from the weakened (attenuated) bacteria. Pasteur having been influenced by Jenner's work coined the term **vaccination** (vacca means cow in Latin) in honor of Jenner. Pasteur also applied the principle of vaccination to the prevention of anthrax in sheep. Up to this point Pasteur's work had benefited the wine industry, the silk industry, and the sheep industry but his work was not complete because he desperately wanted to apply his discoveries to the control of human disease. He chose to study the disease rabies, which occurs primarily in animals but can be transmitted to humans. Rabies was more prevalent in those days, because of the absence of animal vaccines, and a bite from a rabid animal was invariably fatal. The agent of rabies, a virus, was never isolated by Pasteur; however, he cultivated it in the nervous system of rabbits. After the animals died the spinal cords were removed and dried—a process that weakened the virus. The dried spinal cords were ground up and when injected into live rabbits produced an immunity to rabies. Pasteur applied his vaccine to a 9-year-old Alsatian boy, Joseph Meister, who had been bitten by a rabid dog. The boy recovered and later became the gatekeeper at the Pasteur Institute.

Pasteur's success stimulated other scientists to seek methods for preventing disease. Application of his principles led, for example, to the discovery of diphtheria antitoxin. Diphtheria is caused by the bacterium, *Corynebacterium diphtheriae,* which causes an infection in the throat and releases a potent toxin that passes into the blood and affects the heart and nervous system. Nearly 50% of the children who contracted diphtheria during this period died from the disease. Emile Roux and Alexander Yersin established the importance of the diphtheria toxin; this led to the discovery that persons who recover from diphtheria do so because their bodies produce an **antitoxin** that neutralizes the toxin. Emil von Behring and Shibasaburo Kitasato showed that antitoxin can be produced in horses if the horses are injected with gradually increasing amounts of toxin. The antitoxin is produced in the blood of horses and can be recovered from the serum of these animals. Antitoxin was administered to humans in 1895, and within 5 years the death (mortality) rate had dropped to 12%.

Paul Ehrlich, an associate of Koch, and Elie Metchnikoff, a student of Pasteur, made important contributions with their ideas of the mechanisms of resistance to infection. Ehrlich was convinced that resistance was a result of soluble substances in the blood (now called **antibodies**), whereas Metchnikoff believed that white blood cells could digest foreign invaders such as bacteria—a process called **phagocytosis** and the cells that perform the function, **phagocytes.** Both men were correct because soluble substances called antibodies (**humoral immunity**), as well as phagocytes (**cellular immunity**), are part of the host's defense against infections. They shared the Nobel prize in 1908 for their contributions to this branch of microbiology.

Immunology in the twentieth century. Immunity from its earliest beginnings was considered a problem in chemistry, and this is even more evident today. One of the first contributions to the biochemical nature of the immune response in the twentieth century came from the laboratory of Karl Landsteiner (1868-1943). Landsteiner demonstrated that small chemical groups could be attached to existing antigens (large chemical groups that can stimulate the formation of antibodies) and could impart specificity in the immune response. Landsteiner's work on specificity led to the discovery of the four major blood groups: A, B, AB, and O. His work in cooperation with Alexander Weiner in 1940 led to the discovery of the Rh antigen system.

Modern theories of antibody formation are in large part a result of the studies of Felix Haurowitz and others in the 1930s. They suggested that antigens were held firm by the antibody-forming cells and thus served as a template for the synthesis of antibody. Later, in 1959 Burnet expanded on these ideas and conceived the **clonal selection theory**. This theory states that cells are endowed with the genetic capacity to produce a specific antibody. The antigen, by combining with a specific cell, causes that cell to divide and to produce significant antibody. Thus antigen, according to Burnet, acts as a trigger for proliferation and not as a template for specificity,

Point of Interest

THE DISCOVERY OF PENICILLIN

Alexander Fleming, a Scottish bacteriologist, had been working in 1928 with variants of the bacterial genus, *Staphylococcus.* Frequent examination of the colonies of these bacteria resulted in their exposure to air and thus contamination. He noticed that around one large contaminating colony the staphylococcal colonies became transparent, indicating that the bacterial cells had died. Fleming did not throw away the contaminated plates but instead isolated the fungus called *Penicillium* and cultivated it in broth. He found that the fungus secreted a substance in the broth that had the ability to inhibit the growth of or kill many common pathogenic bacteria. Several species of *Penicillium* plus other fungal species were tested for antibacterial activity but only one strain of *Penicillium* produced the antibiotic, **penicillin.** What intrigued Fleming was that penicillin not only killed bacteria but also was nontoxic to white blood cells. After years of developing methods for purification and standardization of penicillin, Florey and his colleagues at Oxford purified the drug in 1938. They demonstrated its amazing therapeutic properties and their information was taken to the United States where the antibiotic was produced on a large scale. By D-Day in 1944 there was enough penicillin for every wounded man who would need it.

It should be noted that the antagonistic action of the substance produced by *Penicillium* had been observed as early as 1876 by Tyndall. However, the techniques for cultivation, as well as the chemical methods for drug concentration, had not been refined until the 1930s. The chance observation by Fleming and the hard labor of Florey and his colleagues are a testimony to the declaration by the American physicist Joseph Henry when he said "The seeds of great discoveries are constantly floating around us, but they only take root in minds well prepared to receive them."

as proposed by Haurowitz. Peter Medawar and Burnet in the years 1946 to 1960 applied the clonal selection theories to the characteristics of immunological tolerance; that is, one does not make antibodies to one's own antigens. Burnet and Medawar shared the Nobel price in 1960.

One of the more recent landmarks in immunity was the discovery of **interferon** by Isaacs and Lindenmann in 1957. Interferons are substances that are excreted by cells harboring viruses. The interferons migrate to adjacent cells and protect them from infection by the virus. Interferons exert their effect for only a short period and are unlike antibodies, which may afford protection for many years.

Chemotherapy

Chemotherapy is the treatment of disease in which a chemical is specifically targeted for a microbial agent or a specific tissue. For many centuries herbs were used to cure infectious disease. Unfortunately herbs exerted no curative powers, but neither did they do any harm. In the late 1400s a practitioner of medicine named Paracelsus rebuked those who used herbs and instead substituted specific mineral substances. He was firmly convinced that there was a specific remedy for each disease. During this period syphilis was a new disease and Paracelsus prescribed mercury as a curative agent. The advocates of herbs, not to be outdone by Paracelsus, prescribed imported herbs, such as sarsaparilla and sassafras, but with little success. Mercury did effect a cure for many afflicted with syphilis and became so popular that vagabond quacks were prescribing the drug. Unfortunately it was being prescribed in extremely large doses and resulted in mercury poisoning causing loss of teeth and hair and creating permanent health problems. A simple method for determining the value of drugs in the treatment of disease did not emerge until the nineteenth century. Ehrlich's studies of resistance to infectious agents made him believe that microorganisms could be affected by chemicals in a very specific way and the drug would not harm the host. He called these specific chemicals "magic bullets." Ehrlich sought to find a cure for syphilis and spent several years developing an organic arsenical compound, **salvarsan**, or **606**. The figure represents the number of experiments required to develop this arsenic compound. Salvarsan, which was marketed in 1906, was most effective as a cure for syphilis during the early stages of the disease. Even today there is no cure for the final stages of syphilis. Research directed at finding drugs for the treatment of other diseases soon followed but it was not until 1932 that another drug would be discovered that could control a bacterial disease. Domagk demonstrated the effects of a dye called prontosil during the treatment of infections in mice. Later it was discovered that prontosil is broken down in the body into a compound called **sulfanilamide** (a member of a large group of compounds called sulfonamides). Sulfanilamide was later mass-produced and used in the treatment of meningitis, gonorrhea, and other diseases.

Penicillin was described years before sulfonamides but it was not concentrated for practical use until 1940 (see Point of Interest). The discovery that microorganisms, such as fungi, can produce antimicrobial substances of therapeutic value was soon followed by the isolation of other antimicrobial substances from microorganisms. The term **antibiotic** was used to designate these antimicrobial substances. Dubos in 1939 isolated the antibiotic **tyrothricin** from the soil bacterium, *Bacillus brevis,* and Waksman in 1943 described the antibiotic **streptomycin**, which is produced by the bacterium, *Streptomyces griseus.* Since 1943 many other drugs have been marketed as chemotherapeutic agents. Some of these are produced synthetically in the laboratory, others are antibiotics isolated from microorganisms cultivated in the laboratory, and others are semisynthetic compounds in which chemical groups have been added to the basic antibiotic nucleus.

Virology

Viruses are submicroscopic agents of disease that cannot exist on their own and require a living host. **Virus** is a Latin word that means a slimy liquid, poison, or venom. During the history of microbiology virus was used to denote any infective agent of disease. When the causes of many bacterial diseases were discovered there remained a group of diseases whose cause could not be determined using Koch's postulates. These agents of disease were not visible with the ordinary microscope of those times and could not be cultivated on any synthetic laboratory medium. Beijerinck in 1898 discovered that the agent of tobacco mosaic disease passed through bacteriological filters (porcelain filters were used to trap bacteria). The filtered fluid was capable of causing disease when applied to healthy tobacco plants. The first animal virus was isolated by Loeffler and Frosch in 1898 from animals with foot-and-mouth disease. The causative agent was also filterable and could not be considered a bacterium. Other animal viruses were soon discovered (Figure 1-13). In 1915 Twort discovered that even bacteria could become infected and could be killed by viruses (Figure 1-14). Bacterial viruses were called **bacteriophage**, which means bacterial eaters. d'Herelle in 1922 showed that bacterial viruses reproduced and thus wrongly concluded that they must be living microorganisms.

In 1935 Stanley crystallized tobacco mosaic virus (TMV) and demonstrated that it contained protein. In the period from 1936 to 1940 several different experiments showed that nucleic acid was also a component

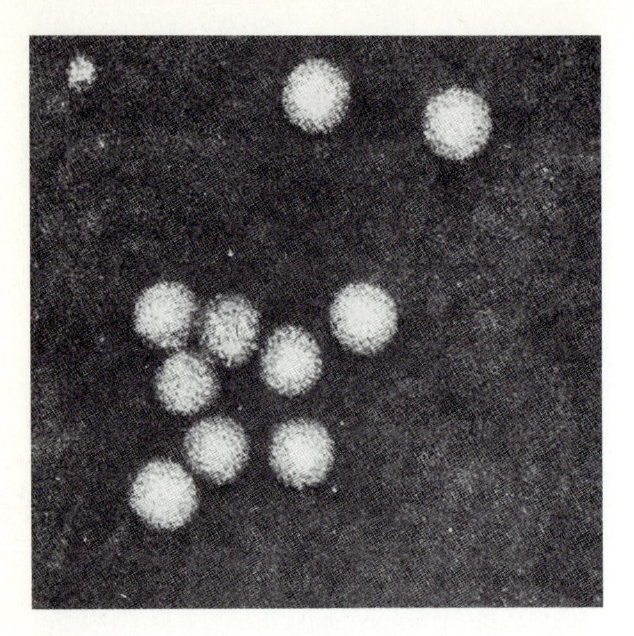

Figure 1-13 *Electron micrograph ($\times$29,500) of animal virus (poliovirus).*

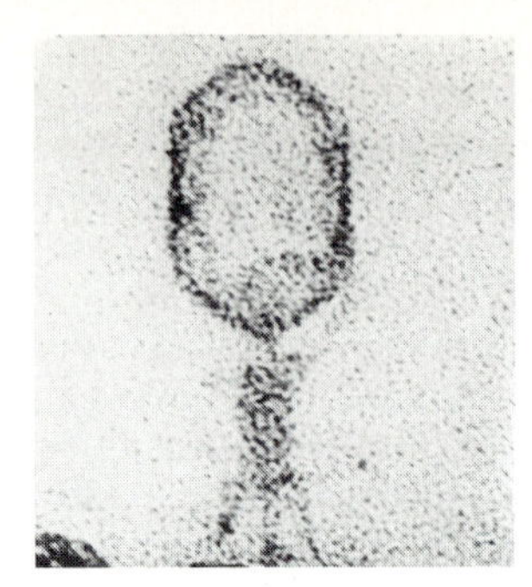

Figure 1-14 *Electron micrograph of bacterial virus similar to one that Twort characterized before development of electron microscope.*

of viruses. Stanley's achievement initiated a new concept in biology—biological properties, formerly thought to be exclusively associated with the cellular level of organization, were found to be characteristic of a purified protein. Thus scientists were beginning to consider viruses as different from microorganisms. In the 1950s Delbruck and others investigated the life cycle of the bacterial virus. They found that the biochemical machinery of the infected cell is diverted to making more virus particles. The virus, although not a microorganism, was capable of performing a cellular function—**replication**. Fraenkel-Conrat, Gierer, and Schramm later demonstrated that the viral particle consists of a protein coat that surrounds a core of nucleic acid. They separated the nucleic acid from the protein coat of a virus and demonstrated the former's infectious properties and its ability to initiate replication.

The emergence of the electron microscope in the 1930s and the development of new staining techniques enabled investigators to characterize viruses based on their shape and size. Until this time the only virus characteristic that could be recognized was based on the type of host the virus infected. One of the single most important events that enabled the study of viruses to be propelled to the forefront of scientific investigation was the development of **cell culture**. Biologists, such as Puck, Earle, and Eagle, performed the techniques of cultivating mammalian cells in the laboratory. This development enabled the investigator to forego cultivating virus in the whole animal, from whom virus concentration was impractical or impossible. The era of molecular virology was now a reality. Virus protein and nucleic acid material could now be chemically and physically analyzed. Cell culture is currently the basic biological material for research in all fields of biology, including genetics, biochemistry, and endocrinology.

ERA OF GENETICS AND MOLECULAR BIOLOGY

The discovery that deoxyribonucleic acid **(DNA)** was the genetic material represented the first step in the development of molecular biology. Griffith in 1927 demonstrated that the trait for producing an extracellular polymer called the capsule could be transferred from one bacterial species to another member of the species that did not possess the trait. The transfer resulted in the conversion of the recipient bacterium from noncapsule producing to capsule producing. What was so amazing about the experiment was that the transfer involved dead bacteria that had possessed the trait and live bacteria that did not possess the trait. Later Avery and Hotchkiss prepared extracts from the dead bacteria to see which of the chemical components was responsible for this amazing transformation process. Several years of work resulted in the discovery that the chemical compound responsible for the capsule-producing trait was DNA. These and other experiments clearly demonstrated that genes are made of DNA.

Scientists had assumed that DNA was a self-copying molecule but there had been no precise suggestions as to how the process would occur. Then in 1953, Watson and Crick, using only the experimental results of others, constructed a scale model of DNA. Their model showed the DNA molecule to consist of two chains of nucleotides that were bonded to each other by hydrogen bonds between specific bases on the nucleotide chains. From x-ray diffraction studies this speculative model was later demonstrated by Wilkins and others to be correct. The Watson and Crick model illustrated the mechanism by which DNA duplicated itself. A few years later, through further model building and experimental results, the basic mechanism of DNA duplication was worked out. This work provided tremendous impetus to the study of molecular genetics.

Molecular genetics is a science that investigates the structure and function of nucleic acids using biochemical, biophysical, and genetic techniques. The birth of bacterial genetics began in 1943 with the work of Luria and Delbruck. They published an article that provided a model for experimental design and protocol. Their unambiguous results provided the impetus for other scientists to become involved in the emerging field of bacterial genetics. Later in the 1950s and 1960s, Jacob and Monod made significant contributions to the understanding of molecular genetics. They demonstrated how bacteria are able to control the expression of their genetic material. The discovery that bacteria have genetic mechanisms similar to those of higher organisms provided a simple genetic model from which experimental results could be extrapolated to more advanced forms of life. A revolutionary discovery by Temin in the 1970s provided new avenues of approach to molecular genetics. Temin discovered that some viruses carry an enzyme that makes a DNA copy from ribonucleic acid (RNA). This novel enzyme was found to be carried by those RNA viruses that are known to cause tumors. In 1975 Temin, Baltimore, and Dulbecco were awarded the Nobel prize for their contributions to the understanding of the relationship between viruses and certain tumors.

The advances made in the field of molecular genetics brought together many branches of science. Manipulation of microorganisms and their genetic material has tremendous consequences for humankind—now and in the future.

Concept Check

1. What specific contributions were made by Pasteur, Ehrlich, and Metchnikoff to the field of immunology?
2. What is Ehrlich's "magic bullet" theory?
3. Describe how one of the first antibiotics, penicillin, was discovered.
4. Explain why viruses are not considered microorganisms and why early scientists had difficulty in establishing them as infectious agents.

IMPACT OF MICROBIOLOGY ON HUMAN AFFAIRS

Until the twentieth century, microorganisms, as agents of disease, kept the world's population in check. Cholera, the plague, influenza, and diphtheria decimated populations. It is an established fact that most of the casualties of war resulted from typhoid, typhus, and var-

Point of Interest

WOMEN MICROBIOLOGISTS

Because of societal prejudices, women in the 1800s and early 1900s were not encouraged to study science. Although there are more women microbiologists today (35% of the American Society for Microbiology are women) than in 1914 (approximately 10%) or than in 1890 (less than 1%), womens' contributions to microbiology are sometimes ignored.

Most often, women in historical biographies are noted as the wives of famous men. For example, Fannie Hesse (discoverer of the use of agar), Mary Bruce, and Grace Coleridge Franklin are mentioned as the wives of Walter Hesse (a student of Koch), David Bruce (discoverer of *Brucella melitensis*), and Percy Franklin (a pioneer in water microbiology). However, two women, Lydia Rabinowitsch and Guiseppina Cattani, are exceptions to the rule. Rabinowitsch worked with the tuberculosis bacillus and made a number of discoveries about the pathogenicity and serology of the bacterium. Cattani worked with the tetanus bacillus and, with a colleague, grew the bacterium anaerobically to produce the first antitoxin used to cure tetanus.

From the 1930s through the 1950s women microbiologists made significant discoveries in pathogenic microbiology. For example, Alice Evans contributed to the field of dairy microbiology and our knowledge of the disease, brucellosis. Rebecca Lancefield devised the classification system used to differentiate hemolytic streptococci. In 1950, microbiologist Elizabeth Hazen and biochemist Rachel Brown discovered nystatin. Nystatin was the first antifungal antibiotic, and it proved especially effective against yeasts like *Candida albicans*. Royalties for the discovery of nystatin were placed in the Brown-Hazen Fund for the support of scientific work, and by the time the patent expired in 1974, over 13 million dollars had been generated.

During the 1950s the embryonic field of molecular biology began. When Watson, Crick, and Wilkins were given the Nobel prize in 1962 for proposing the double helix as the structure of the DNA molecule, it was based heavily on the x-ray crystalography work of Rosaland Franklin, who died of cancer in 1958. Unfortunately, the Nobel Prize is never awarded posthumously.

The 1970s and 1980s have seen the awarding of the Nobel prize to two women scientists who, although not traditional microbiologists, were working in microbiological areas: Rosaland Yalow for her work in developing the radioimmunoassay, and Barbara McClintock for the discovery of transposons.

ious gastrointestinal diseases. Today many of these diseases are controlled by vaccination or antibiotics. The world's population is expected to double to 10 billion by the year 2000, and our ability to control infectious disease and thus increase longevity has contributed significantly to this increase in population. The expansion of the population has put constraints on many resources taken for granted. Resources, such as coal, natural gas, sulfur, and copper, many of which were formed by microbial activities, are being depleted. Top soil is being lost through deforestation and overfarming, thus requiring the use of chemically produced fertilizers. Air and water supplies are becoming polluted. New animal and plant diseases are hindering progress in supplying sufficient protein for the world's population. Technology removed many burdens from our lives but has placed considerable stress on our environment. Social and political, as well as scientific decisions will have an impact on the future of our world. Microbiology is one of those sciences that will greatly influence the quality of our lives. Two recent advances in microbiology demonstrate this influence on human endeavors.

Genetic engineering

One of the most important advances in molecular biology is in the field of **recombinant DNA technology.** Recombinant DNA techniques permit selective removal of specific segments of DNA or genes from one organism for integration into the DNA of totally unrelated or related organisms. Thus the organism can be "genetically engineered." Genetic engineering has great potential in the following areas:

1. Vaccine development. There are no suitable vaccines for such diseases as acquired immune deficiency syndrome (AIDS), syphilis, gonorrhea, and others because the gene or genes associated with the disease process have not been identified or isolated. Isolation of specific genes by genetic engineering will permit investigators to produce vaccines that will be cheaper and result in a purer product.
2. Noninfectious human disease. Disorders, such as diabetes and growth hormone deficiency, are being treated with products synthesized in genetically engineered organisms. Insulin, for example, was originally recovered from pig pancreas. It is now possible to remove the animal gene or synthetically make the human gene and insert either into the DNA of a bacterial cell. Human insulin produced in the bacterial cell prevents the allergic reactions that are associated with animal insulin. Scientists are now exploring the possibility of using viruses to carry functional human genes and insert them into the cells of humans with defective genes. Replacement gene therapy has been performed successfully in animals.
3. Agriculture. Techniques are being created that introduce genes into plants to develop pest resistance, upgrade protein content, and fix nitrogen from the air into amino acids, the component of protein. The latter would reduce the need for excessive amounts of nitrogen-containing fertilizers.
4. Industry. Several chemicals (antibiotics, citric acid, and ethanol) are produced industrially by microorganisms. Recombinant DNA technology is being used to enhance production levels of these compounds.
5. Food source. Microorganisms are being used as a source of protein for animal feed and in some areas of the world as a source of food for humans. Recombinant DNA technology may help increase the growth rate or protein content of these microorganisms.-

Monoclonal antibodies

Another important advance in molecular biology is **monoclonal antibody production.** In 1975 Kohler and Milstein were able to fuse a skin cancer cell from a mouse to an antibody-producing white blood cell. This was an important development for two important reasons. First, the antibodies produced by this hybrid are of only one type **(monoclonal)**, which is in sharp contrast to the complex mixtures of antibodies found in normal sera. Second, the hybridoma tissue culture cells can be frozen and used later to inject into animals to produce tumors. Large amounts of homogenous antibody can then be obtained from the affected animal. Monoclonal antibodies have been generated against a great many antigens, but human monoclonal antibodies were not developed until 1980, when Olsson and Kaplan fused a human myeloma cell to immune spleen cells. It is hoped that by using monoclonal antibodies, we will now be able to study more precisely the structure and function of proteins produced by viruses during infection. With this knowledge we will be able to construct inhibitors that can be used for therapy against viral diseases. Monoclonal antibodies directed against specific virus components will also aid in their identification in the clinical laboratory, which today is a time-consuming process. Monoclonal antibodies will probably replace horse and rabbit antibodies that are presently used to treat such diseases as tetanus and transplant rejection. This would prevent allergic reactions that occur with second injections of animal antibodies. It was also suggested that immunization with human monoclonal antibodies may become the preferred treatment for drug overdoses and some diseases caused by microorganisms.

Microbiology and the future

The future of microbiology for the enhancement of human life is limitless. Many economic and social problems may be solved when the microbe is involved. Scientists have been successful in the genetic manipulation

Table 1-2 *Major contributors to microbiology as a science up to the twentieth century*

Contributor	Contribution	Date(s)
Anton van Leeuwenhoek	First to observe microorganisms with primitive microscope.	1685
Francisco Redi	Demonstrated that animals do not arise spontaneously from dead organic matter.	1660
Abbe Spallanzani	One of the first to demonstrate that heated broth, in the absence of air, did not support spontaneous generation.	1770
Schröder and von Dusch	Demonstrated that broth heated in the presence of filtered air did not support spontaneous generation.	1854
John Tyndall	Demonstrated that open tubes of broth remained free of bacteria if air was dust free.	1860
Louis Pasteur	His experiments put theory of spontaneous generation to rest (1861). Contributed to understanding of fermentation (1858). Technique for selective destruction of microorganisms (pasteurization) (1866). Study of diseases of wine (1866) and silkworms (1868). Attenuated vaccines for anthrax (1881) and chicken cholera. Immunization against rabies (1885).	1855-1890s
Joseph Lister	Contributed to concept of aseptic technique and pure culture concept.	1870s
Robert Koch	Developed postulates for proving the cause of infectious disease (1884). Observed anthrax bacilli (1876). Developed solid culture media (1882). Discovered organisms causing tuberculosis (1882).	1870s to 1890s
Paul Ehrlich	Formulated humoral theory of resistance. Developed new staining techniques. Developed first chemotherapeutic agent.	1890s to 1900
Elie Metchnikoff	Formulated cellular theory of resistance.	1890s
Emil von Behring	Developed method for producing immunity by using antitoxin against diphtheria.	1890s

Concept Check

1. How has the study of microorganisms contributed to an understanding of higher forms of life?
2. Discuss the specific impact of microbiology on human endeavors, particularly the implications of genetic engineering.

of animal and plant genes, but what about the application of such techniques to humans? What are the ethical issues, for example, if sex selection becomes possible for humans? Any innovative diagnostic or therapeutic techniques should be understood by the public so that wise and informed decisions concerning their application can be made. A biological revolution based on DNA and genetic manipulation has arrived, and it is important that we remain informed of its progress. One purpose of this book is to make this kind of information available.

SUMMARY

1. Two events that propelled microbiology into a major science were the experiments refuting spontaneous generation and the proof that microorganisms cause disease. Table 1-2 lists some of the individuals who made significant contributions to the development of microbiology before the twentieth century.
2. Microbiology has developed rapidly, especially in the last 30 years. Immunology, virology, and chemotherapy and associated fields have developed due primarily to advances in biochemistry and genetics (molecular genetics). Watson and Crick in 1953 initiated the field of molecular genetics when they described the helical structure of DNA.
3. Two of the most important advances in microbiology have occurred since 1973: the construction of hybridomas and techniques in genetic engineering. They are leading to developments in the prevention and cure of infectious, as well as noninfectious, diseases and the solution of environmental problems confronting humankind.

Self-study Quiz

Multiple choice

1. Which of the following diseases was used as the basis for Koch's postulates?
   A. Tuberculosis
   B. Syphilis
   C. Smallpox
   D. Bacterial meningitis
   E. Anthrax
2. Paul Ehrlich discovered which of the antimicrobial agents to be effective against syphilis?
   A. Penicillin
   B. Streptomycin
   C. Salvarsan
   D. Mercury
   E. None of the above
3. Which of the following is recognized as Paul Ehrlich's major contribution to immunology?
   A. Discovery of Rh antigens
   B. Humoral theory of immunity
   C. Vaccine for tuberculosis
   D. Development of monoclonal antibodies
   E. Cellular theory of immunity
4. Which of the following characteristics suggest that viruses are not living microorganisms?
   A. Viruses pass through porcelain filters.
   B. Viruses cannot be cultivated on bacteriological media.
   C. Vaccines cannot be made for viruses.
   D. Viruses cannot be observed with any kind of microscope.
   E. None of the above.
5. What is the major stumbling block in applying Koch's postulates to viral disease?
   A. Viruses cannot be cultivated in artificial media.
   B. Viruses cannot be isolated in pure culture.
   C. Viruses cannot be attenuated.
   D. Viruses are too small to be seen.
6. Which of the following was not a result of the Golden Age of Microbiology?
   A. Use of agar as a solidifying agent for media
   B. Discovery of the causative agent of cholera
   C. Discovery of the simple microscope
   D. Description of phagocytosis

Matching

1. ________ Griffith
2. ________ Kohler and Milstein
3. ________ Watson and Crick
4. ________ Jacob and Monod
5. ________ Alexander Fleming
6. ________ Ignaz Semmelweiss
7. ________ Francisco Redi
8. ________ Edward Jenner
9. ________ Frau Hesse

A. Recognized relationship between dirty hands and childbirth fever.
B. Discovered streptomycin.
C. Described helix of DNA.
D. Recognized relationship between smallpox and cowpox.
E. Suggested agar-agar as a solidifier of bacteriologic media.
F. Demonstrated how bacteria can control the expression of genes.
G. Formed hybridoma cells.
H. Discovered penicillin.
I. Showed that naked DNA can transform a bacterium.
J. Showed that animals do not arise by spontaneous generation.

Thought questions

1. It has been asserted that Frau Hesse's suggestion to use agar as a solidifying agent was one of the most significant breakthroughs in the history of medical microbiology. What arguments can you point out to support this assertion?
2. List as many activities as you can think of that you have engaged in since rising this morning that have microbiological implications.

SELECTED READINGS

Books

Brock, T.D. *Milestones in microbiology.* American Society for Microbiology, Washington, D.C., 1975.

Bulloch W. *The history of bacteriology.* Oxford University Press, London, 1938.

Clark, P.F. *Pioneer microbiologists of America.* University of Wisconsin Press, Madison, Wis., 1961.

Dixon, B. *Magnificent microbes,* Atheneum Publishers, New York, 1976.

Dobell, C. (Ed.) *Anton van Leeuwenhoek and his "little animals."* Dover Publications, Inc., New York, 1960.

Dubos, R.J. *Louis Pasteur: free lance of science.* Little, Brown & Co., Boston, 1976.

Koprowski, H., and Plotkin, S. (Eds.). *World's debt to Pasteur.* Alan R. Liss, Inc., New York, 1985.

Lechevalier, H.A., and Solotorovsky, M. *Three centuries of microbiology.* Dover Publications, Inc., New York, 1974.

Reid R. *Microbes and men.* Saturday Review Press, New York, 1975.

Journal Articles

Bardell, D. The roles of the sense of taste and clean teeth in the discovery of bacteria by Antoni van Leeuwenhoek. *Microbiol. Rev.* **47:**121, 1983.

Bibel, D.J. William Bulloch's pioneer women of microbiology. *ASM News* **51**(7):328, 1985.

Gröschel, D.H.M. The etiology of tuberculosis: a tribute to Robert Koch on the centenary of his discovery of the tubercle bacillus. *ASM News* **48**(6):248, 1982.

Howard, D.H. Friedrich Loeffler and his history of bacteriology. *ASM News* **48**(7):297, 1982.

CHAPTER 2

Biochemical background for microbiology

Chapter Outline

Introduction

Biochemistry is the study of living matter, which is composed of both inorganic and organic compounds. The sciences of biochemistry and microbiology have become so intertwined that they are practically inseparable. Using biochemical techniques, scientists are probing even more deeply into the detailed mechanisms that permit certain molecules to behave in specific ways. Much of what we know about biochemical mechanisms has been learned from microbial systems. The molecular architecture of biological components is also being unraveled through biochemical techniques. Any understanding of what makes the microorganism and other biological systems "tick" cannot be fully appreciated unless some fundamental concepts of biochemistry are understood.

The purpose of this chapter is to review the basic characteristics of the inorganic elements that make up biological molecules and to outline the characteristics of those organic molecules that are associated with cell structure and function.

Throughout this chapter and the remainder of the book various metric units will be used. The most frequently used metric units and their conversions are listed in the box on p. 21.

INTERCONVERSION SCALE FOR COMMON METRIC UNITS

The numerical difference between the various metric units is 10 or a factor of 10. For example, the difference between an angstrom (Å) and a nanometer (nm) is a factor of 10 but between an Å and a micrometer (μm), the factor is 10,000. When going down the conversion scale as indicated, divide by 10 or a factor of 10, but multiply when going up the conversion scale. For example, to convert 1 μm to angstroms multiply by 10,000; thus 1 μm = 10,000 Å. To convert 1 nm to micrometers divide by 1000; thus, 1 nm = 0.001 μm.

↓ Divide by 10 or factor of 10		↑ Multiply by 10 or factor of 10
	Angstrom (Å) = 10^{-10} meters	
	Nanometer (nm) = 10^{-9} meters	
	Micrometer (μm) = 10^{-6} meters	
	Millimeter (mm) = 10^{-3} meters	
	Centimeter (cm) = 10^{-2} meters	
	Meter (m) = 1	

BUILDING BLOCKS: THE ELEMENTS

Organisms are chemical machines that take elements from the environment, which exist in various molecular forms, and convert them into molecules that enable cells to grow and divide. Over 90% of the atoms of an organism are composed of carbon, nitrogen, phosphorous, oxygen, and hydrogen. Other elements, such as sulfur, calcium, potassium, sodium, magnesium, and chlorine, make up the majority of the remaining 10%. The role of these elements in microorganisms is illustrated in Table 2-1. The major elements: carbon, nitrogen, phosphorous, oxygen, and hydrogen are discussed in detail here.

Carbon

Carbon contains four electrons in its outer shell, which can be represented as being equidistant from each other (Figure 2-1). Carbon therefore contains four binding sites that can be occupied by other elements, in particular, hydrogen, oxygen, and carbon atoms. Most organic molecules are made up primarily of two or more carbon atoms. The maximum number of covalent bonds between two carbon atoms is three (Figure 2-2). Single bonds between carbon atoms permit rotation about the bond axis, and this allows other attached atoms or groups to lie in different planes. Double and triple bonds between carbon atoms reduces the distance between them and does not permit free rotation around the bonds. Hydrogen atoms, for example, attached to these inflexible carbons lie in the same plane (Figure 2-3, *A*). Long-chained molecules, in which there is free

Table 2-1 *Function of the major* and minor elements present in microbial systems*

Element	Function
Major elements	
Carbon (C)	Major component of organic compounds, used as carbon and energy source.
Nitrogen (N)	Major component of organic compounds. Inorganic forms, such as nitrate (NO_3) and nitrite (NO_2), used as electron acceptors in respiration.
Sulfur (S)	A component of amino acids methionine and cysteine. Sulfur-sulfur bonds help stabilize proteins. Some inorganic forms can be used as sources of energy or as electron acceptors in respiration
Phosphorous (P)	A component of nucleic acids and the energy molecule, ATP.
Oxygen (O)	A major component of most organic molecules. An electron acceptor during respiration in air.
Hydrogen (H)	Major component of all organic matter. Some bacteria can use it as an energy source.
Minor elements	
Magnesium (Mg)	Required for enzyme activity. Helps stabilize negatively charged nucleic acid molecules.
Calcium (Ca)	Required during bacterial spore formation. Plays a role in microbial movement.
Potassium (K)	Important in ribosome function. Used by some bacteria to maintain osmotic balance.
Iron (Fe)	Important in electron transport proteins, cytochromes. Influences disease potential of bacteria.
Manganese (Mn)	Can substitute for magnesium in affecting enzyme activity.
Sodium (Na)	Used by some bacteria to maintain osmotic balance. Required by some marine microorganisms.
Zinc (Zn)	Required for activity of some enzymes.
Chlorine (Cl)	Used by some bacteria to maintain osmotic balance.
Silicon (Si)	Required by some algae and diatoms.

*The term *major* is used here to denote quantitative importance.

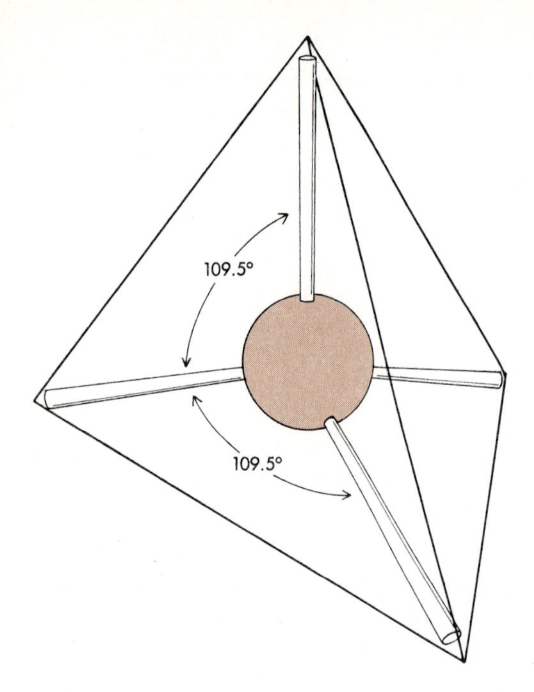

Figure 2-1 *Tetrahedral arrangement of atoms on carbon (colored). Four single bonds are same length and are spaced 109.5° apart.*

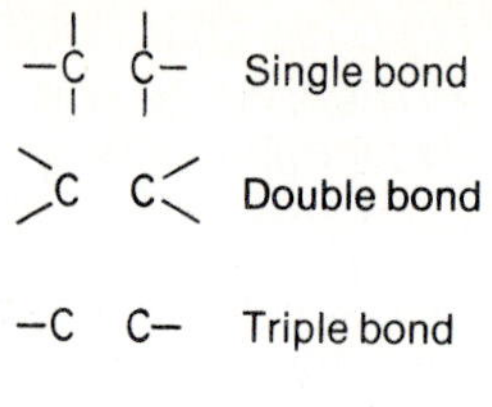

−C C− Single bond

>C C< Double bond

−C C− Triple bond

Figure 2-2 *Carbon-carbon bonds.*

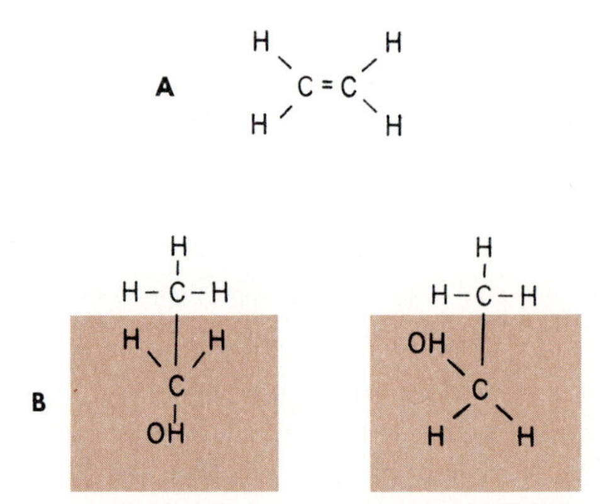

Figure 2-3 *Effect of carbon-carbon bond on conformation of molecule.* ***A,*** ***Double bond between carbon atoms does not permit free rotation, and attached hydrogen atoms all lie in same plane.*** ***B,*** ***Single bond between carbon atoms permits rotation, and different conformations of molecule can be obtained.***

rotation about the chemical bond, can assume a variety of different shapes or conformations (Figure 2-3, *B*).

The tetrahedral arrangement of the bonds around the carbon atom imparts another important characteristic on some organic molecules. A carbon atom with four different atoms or groups attached to it is said to be **asymmetric.** An asymmetric carbon atom can exist in two isomeric forms. Structural isomers are compounds having the same molecular formula but different spatial arrangements of their atoms (Figure 2-4). Compounds that possess asymmetric atoms can rotate plane-polarized light and are considered optically active. The instrument used to measure optical activity is called a polarimeter. Passing plane-polarized light through a solution containing an optically active compound will result in light being rotated clockwise or counterclockwise as it approaches the viewer's eye. Compounds that rotate the light to the right or clockwise are said to be **dextrorotary** (+), whereas those that rotate the light to the left or counterclockwise are said to be **levorotatory** (−). Isomers play a very important role in biological systems because only certain ones are recognized by those enzymes normally found in the cell. The most common groups of molecules found in biological systems that exhibit isomeric properties are amino acids and sugars. An interesting example of how this knowledge is being applied is in the case of commercial sweeteners. Refined sugars are optically active and when ingested are recognized by salivary enzymes that hydrolyze them to smaller metabolic products. Refined sugars are being tested in which the optical activity has been reversed and the sugars are not recognized by human enzymes. The sweet taste is still present, but the sugar cannot be metabolized, an important consideration for diabetics, for those having problems with weight, and for those (many of us) suffering from tooth decay.

Oxygen

Oxygen is a highly reactive molecule that is used by biological systems for the combustion of organic compounds and the release of energy. Molecular oxygen (O_2) is not ordinarily found free in the environment because of its highly reactive nature and must be continually supplied to us through the process of photosynthesis. Oxygen forms more compounds with carbon than any other element; some of these are discussed later.

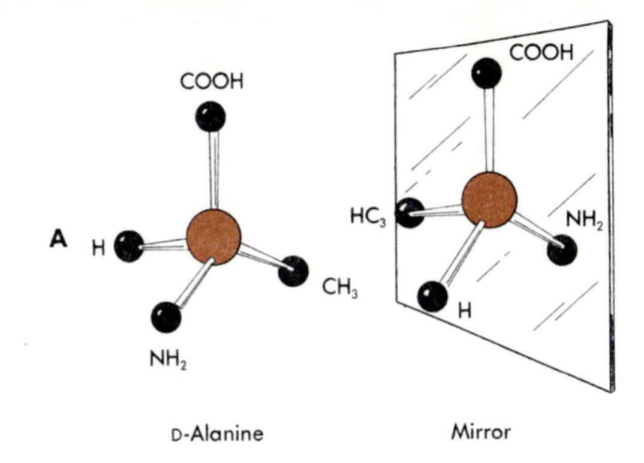

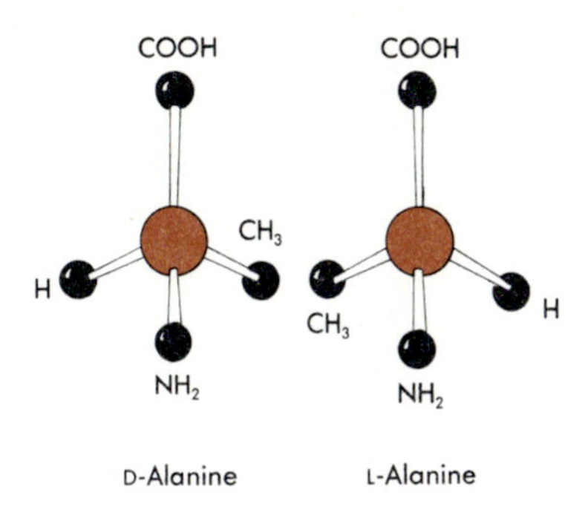

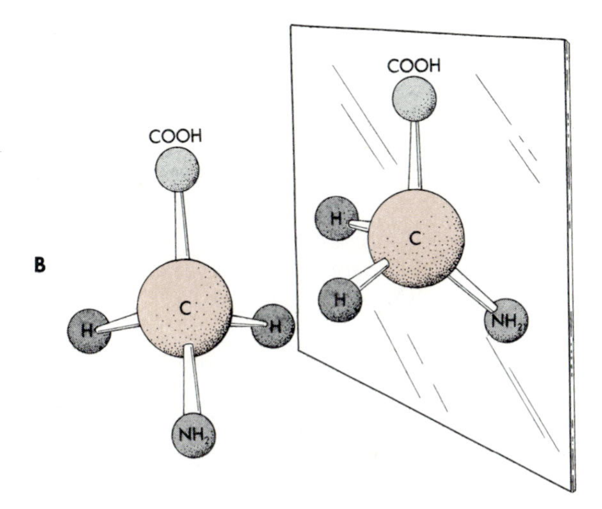

Figure 2-4 *Asymmetrical and symmetrical carbon compounds.* ***A,*** *Asymmetrical compound. The carbon atom has four different groups: H, COOH, CH_3, and NH_2. Compound here is amino acid alanine. Two arrangements of alanine that are mirror images but are not superimposable are illustrated on right and are designated D- and L-alanine.* ***B,*** *Symmetrical compound. Only three dissimilar groups attached to carbon. Two hydrogens make this compound symmetrical, and mirror images are superimposable. Compound here is amino acid glycine.*

Hydrogen

Hydrogen, the lightest element, can share its solitary electron with many other elements. Hydrogen bonds are important in maintaining the shape and structure of macromolecules such as proteins and nucleic acids (see Proteins and Nucleic Acids).

Organic compounds of carbon, hydrogen, and oxygen

The bonds between carbon atoms, as discussed previously, can be single, double, or triple, leaving three, two, or one electron(s), respectively, available for bonding with other elements. Only the single and double bonds are common in biological systems. The covalent bond between carbon atoms is so strong that carbon chains of great length, or rings of carbon, can be produced to provide an endless number of molecular possibilities.

Hydrocarbons. Chains containing only carbon and hydrogen are referred to as hydrocarbons. Those hydrocarbons in which there are only single bonds between adjacent carbon atoms are referred to as **saturated** (Figure 2-5, *A*). **Unsaturated** hydrocarbons (Figure 2-5, *B*) are those in which a double bond exists between any of the carbons in the chain. Many hydrocarbons are linear molecules, but some are also cyclic. The arrangement of six carbon atoms into an unsaturated ring is characteristic of a group of molecules called **aromatic** (having an aroma) compounds. This six-carbon arrangement produces the benzene ring (Figure 2-5, *C*), in which three double bonds contribute to the molecule's unsaturation. The cyclic nature of benzene prevents free rotation about the carbon bonds, and the molecule is planar. The heterocyclic compounds (Figure 2-5, *D*), which contain additional atoms of nitrogen or sulfur, are another important group of hydrocarbons.

Various oxygen-containing groups are found on hydrocarbons, which can be divided into a number of classes, the most important of which are alcohols, aldehydes, ketones, esters, acids, and anhydrides. Each of these classes contains groups that are particularly reactive and enter into many biochemical reactions.

Alcohols. Alcohols are hydroxy derivatives of hydro-

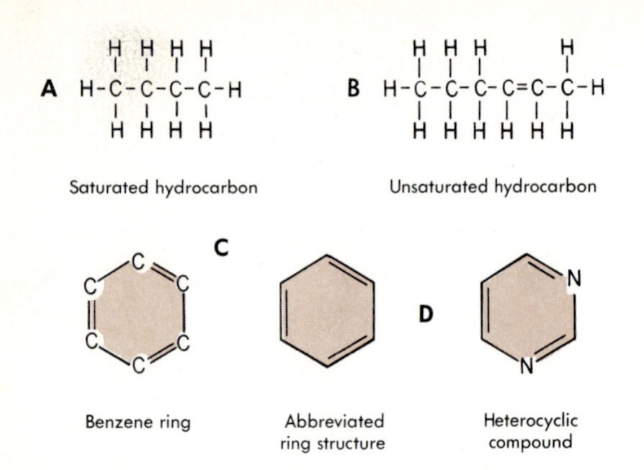

Figure 2-5 *Structure of various hydrocarbons.* ***A,*** *Saturated;* ***B,*** *unsaturated;* ***C,*** *benzene rings;* ***D,*** *heterocyclic.*

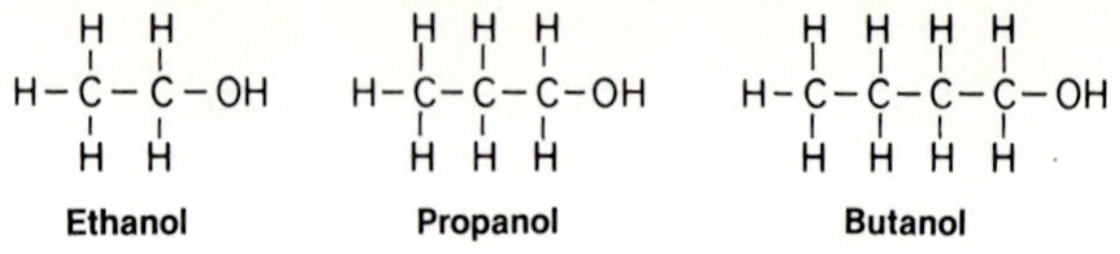

Ethanol **Propanol** **Butanol**

Figure 2-6 *Structure of some important alcohols.*

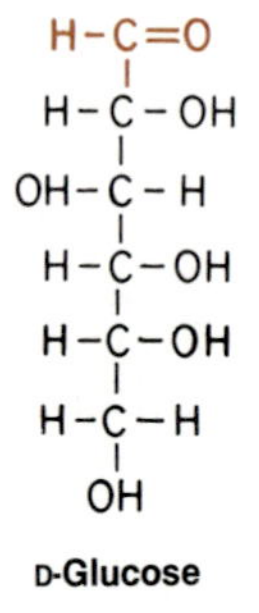

D-Glucose

Figure 2-7 *Structure of D-glucose. Aldehyde group is color.*

carbons and possess a hydroxyl (OH) group. Alcohols, such as ethanol, propanol, and butanol (Figure 2-6), are produced as by-products of microbial metabolism. Some of them are of obvious importance in the beverage industry. Other oxygen-containing groups, such as aldehydes, ketones, acids, and esters, are the result of oxidation of alcohols; discussions follow.

Aldehydes. Aldehydes contain a carbonyl group (C=O) and possess at least one hydrogen atom (H—C=O). Many sugars, such as glucose (Figure 2-7) and ribose, contain an aldehyde group that is very reactive and can bond to other molecules.

Ketones. Ketones are carbonyl groups in which there is no hydrogen atom attached ($\text{—C—}\overset{\text{O}}{\overset{\|}{\text{C}}}\text{—C—}$). Pyruvic acid, for example, contains a keto group ($\text{H}_3\text{C—}\overset{\text{O}}{\overset{\|}{\text{C}}}\text{—COOH}$) and is a very important product in the fermentation of sugars. Carbonyl oxygen groups can be replaced by amino groups (NH_2), and many metabolic acids, containing a keto group, can be converted to amino acids in this manner.

Esters. Esters are produced from the reaction of alcohols with acids, and they contain the following: $\text{R—}\overset{\text{O}}{\overset{\|}{\text{C}}}\text{—O—R}_1$ in which R stands for the rest of the molecule. Many lipids, for example, are esters or possess an ester linkage. The formation of a simple ester is described in Figure 2-8. The ester linkage is also important in the formation of nucleotide chains in nucleic acids such as deoxyribonucleic acid (DNA).

Ethers. Ethers have the general formula of R—O—R, in which two groups are attached to oxygen. The R groups may be identical or different. The ethers can be produced in the laboratory by a dehydration reaction involving two molecules of an alcohol:

$$\text{R—OH} + \text{HO—R} \xrightarrow{\text{H}_2\text{SO}_4} \text{R—O—R} + \text{H}_2\text{O}$$

The ether linkage has some important implications in the bacteria. A group of bacteria recently classified as the archaebacteria possess ether linkage in their membrane lipids instead of the ester linkage found in all other bacteria. The importance of the ether linkage is discussed in Chapter 14.

Carboxylic acids. The carboxylic acid group (COOH) on the hydrocarbon consists of a carbonyl group (C=O) and a hydroxyl group (OH). Carboxylic acids are weak acids and may exist as monocarboxylic, dicarboxylic, or tricarboxylic acids (Figure 2-9). Many of these organic acids are produced as end products during metabolism or as intermediates that can be used in the biosynthesis of amino acids, which are the building blocks of proteins. Other organic acids are long-chained fatty acids that are components of lipids.

Anhydrides. The anhydride linkage is formed by the elimination of water from two molecules of acid (Figure 2-10). The acids may be organic or inorganic. One important aspect of the anhydride linkage is that it releases a great deal of energy when the bond is broken. The symbol (~) denotes the high-energy bond (see Phosphorus).

Nitrogen

Nitrogen, which contains five electrons in its outermost shell, can exist in various states of oxidation ranging

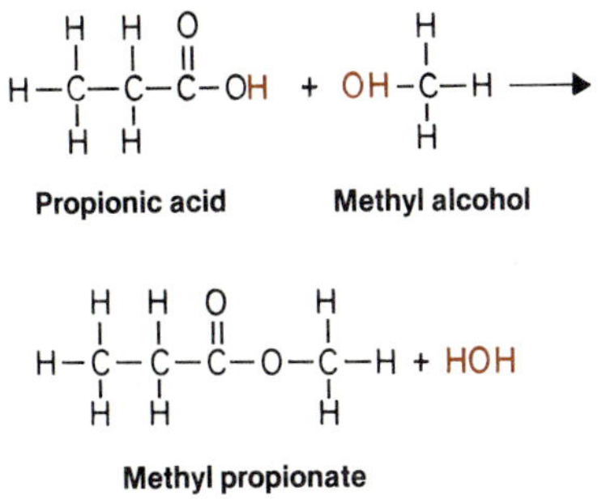

Propionic acid + Methyl alcohol → Methyl propionate + HOH

Figure 2-8 *Formation of simple ester.*

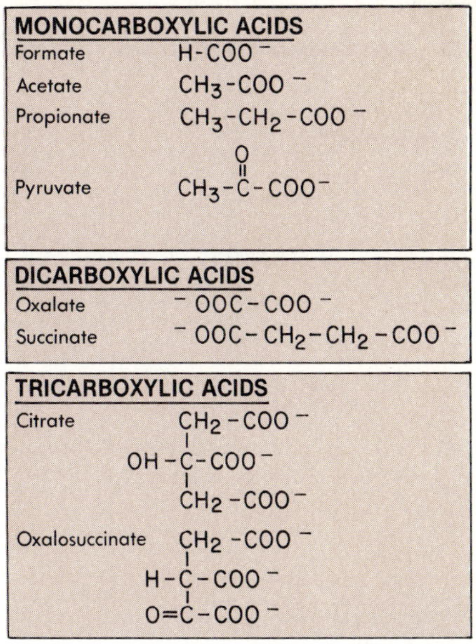

MONOCARBOXYLIC ACIDS	
Formate	$H-COO^-$
Acetate	CH_3-COO^-
Propionate	$CH_3-CH_2-COO^-$
Pyruvate	$CH_3-C(=O)-COO^-$

DICARBOXYLIC ACIDS	
Oxalate	$^-OOC-COO^-$
Succinate	$^-OOC-CH_2-CH_2-COO^-$

TRICARBOXYLIC ACIDS	
Citrate	CH_2-COO^- / $OH-C-COO^-$ / CH_2-COO^-
Oxalosuccinate	CH_2-COO^- / $H-C-COO^-$ / $O=C-COO^-$

Figure 2-9 *Important monocarboxylic, dicarboxylic, and tricarboxylic acids formed in biological systems.*

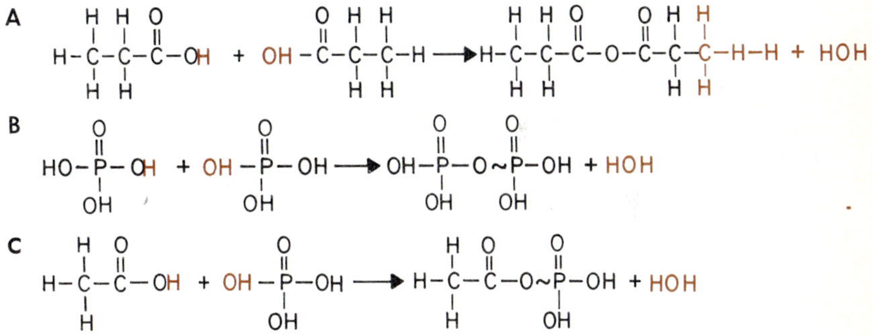

Figure 2-10 *Anhydride formation between organic acids (propionic acids),* ***A;*** *inorganic acids (phosphoric acids),* ***B;*** *and organic (propionic acid) and inorganic acids (phosphoric acid),* ***C;*** *~ = high energy bond.*

from −3 to +3. It can substitute very readily for hydrogen on organic compounds to form an amino group ($—NH_2$), which helps to give polarity to the molecule. Nitrogen is available to microorganisms in many inorganic and organic forms. Important inorganic forms are ammonium (NH_4^+), nitrate (NO_3^-), nitrite (NO_2^-), and nitrogen gas (N_2). Organic nitrogen compounds include amino acids, nucleotides, and some of the macromolecules from which they are totally or partially derived, such as proteins and nucleic acids. One of the most important basic organic nitrogen compounds is the amino acid, which possesses at least two ionizable groups—an amino group ($—NH_2$) and a carboxyl group ($—COOH$). Under most physiological conditions the amino group is protonated and the carboxyl group ionized. Amino acids have the following basic formula:

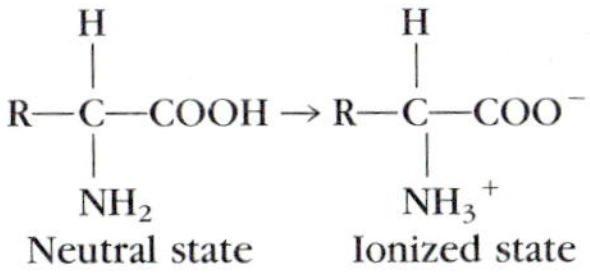

$$\begin{array}{c} H \\ | \\ R-C-COOH \\ | \\ NH_2 \end{array} \rightarrow \begin{array}{c} H \\ | \\ R-C-COO^- \\ | \\ NH_3^+ \end{array}$$

Neutral state — Ionized state

Amino acids: the building blocks of proteins. All amino acids possess a central carbon atom (the alpha carbon) to which is attached the NH_2 group, the COOH group, and a hydrogen atom. The distinctive chemical properties of amino acids are governed by the chemical nature of the atom or group of atoms that make up the fourth bond of the alpha carbon, the R group. There are twenty different amino acids in biological systems, and they can be divided into different classes based on the chemical nature of the R group (Table 2-2). The alpha

Table 2-2 *Twenty amino acids found universally in proteins*

Side-chain (R group) characteristic	Amino acid	Chemical structure
Aliphatic, nonpolar	Glycine	$H-C(H)(NH_3^+)-C\,OO^-$
	Alanine	$CH_3-C(H)(NH_3^+)-C\,OO^-$
	Valine	$(CH_3)_2CH-C(H)(NH_3^+)-C\,OO^-$
	Leucine	$(CH_3)_2CH-CH_2-C(H)(NH_3^+)-C\,OO^-$
	Isoleucine	$CH_3-CH_2-CH(CH_3)-C(H)(NH_3^+)-C\,OO^-$
Alcoholic, aliphatic, and aromatic	Serine	$HO-CH_2-C(H)(NH_3^+)-C\,OO^-$
	Threonine	$CH_3-C(H)(OH)-C(H)(NH_3^+)-C\,OO^-$
	Tyrosine	$HO-C_6H_4-CH_2-C(H)(NH_3^+)-C\,OO^-$
	Phenylalanine	$C_6H_5-CH_2-C(H)(NH_3^+)-C\,OO^-$
	Tryptophan	(indole ring: C=CH, N) $-CH_2-C(H)(NH_3^+)-C\,OO^-$

Table 2-2—cont'd

Side-chain (R group) characteristics	Amino acid	Chemical structure
Carboxylic (acidic)	Aspartic	$^{-}OO\backslash C-CH_2-C(H)(NH_3^+)-C/OO^-$
	Glutamic	$^{-}OO\backslash C-CH_2-CH_2-C(H)(NH_3^+)-C/OO^-$
Amine bases (basic)	Lysine	$NH_3^+-CH_2-CH_2-CH_2-CH_2-C(H)(NH_3^+)-C/OO^-$
	Arginine	$NH_2-C(=NH_2^+)-NH-CH_2-CH_2-CH_2-C(H)(NH_3^+)-C/OO^-$
	Histidine	$HC{=}C(-NH-C(H){=}HN^+-)-CH_2-C(H)(NH_3^+)-C/OO^-$
Sulfur containing	Cysteine	$HS-CH_2-C(H)(NH_3^+)-C/OO^-$
	Methionine	$CH_3-S-CH_2-CH_2-C(H)(NH_3^+)-C/OO^-$
Amides	Asparagine	$O{=}C(NH_2)-CH_2-C(H)(NH_3^+)-C/OO^-$
	Glutamine	$O{=}C(NH_2)-CH_2-CH_2-C(H)(NH_3^+)-C/OO^-$
Imino	Proline	$CH_2-CH_2-CH(-C/OO^-)-NH_2^+-CH_2$ (ring)

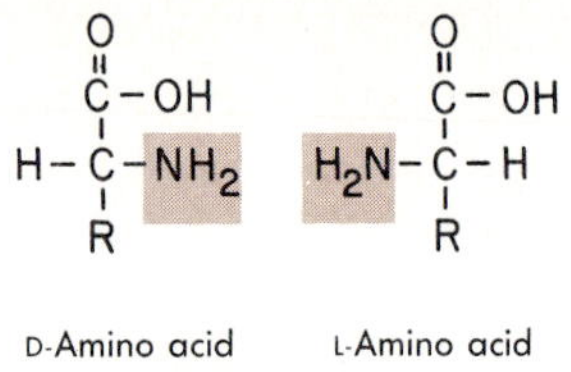

$$\begin{array}{cc} \begin{array}{c} O \\ \| \\ C-OH \\ | \\ H-C-NH_2 \\ | \\ R \end{array} & \begin{array}{c} O \\ \| \\ C-OH \\ | \\ H_2N-C-H \\ | \\ R \end{array} \\ \text{D-Amino acid} & \text{L-Amino acid} \end{array}$$

Figure 2-11 *Arrangement of amino group in* D- *and* L- *amino acids.*

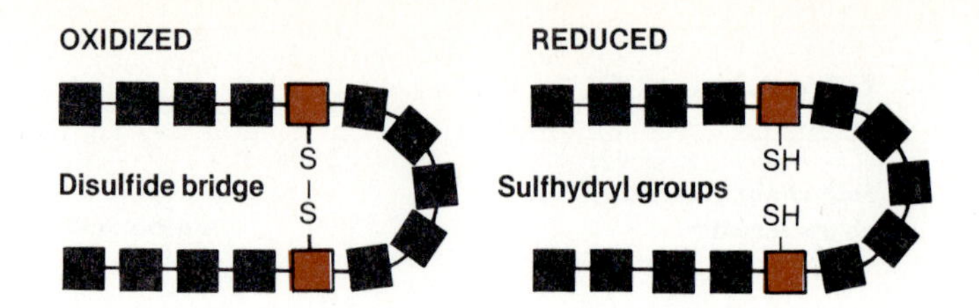

Figure 2-12 *Polypeptide containing cysteine in oxidized state and reduced state. Each box represents amino acid; colored boxes are cysteine.*

carbon is asymmetric for all amino acids except glycine.

The amino acids are divided into **D**- and **L**-families based on the position of the NH_2 group on the alpha carbon. When the COOH group is written at the top and the carbon chain downward, the amino acid belongs to the D-series if the NH_2 group is on the right (Figure 2-11). Conversely, if the NH_2 group is on the left side, the amino acid belongs to the L-series. Most amino acids in biological systems belong to the L-series but some amino acids found in the cell wall of bacteria belong to the D-series. Only a few D-amino acids are found in bacterial components.

In aqueous solution of neutral pH, amino acids are not undissociated, that is, they exist as **dipolar** ions having positively charged ($—NH_3^+$) and negatively charged ($—COO^-$) poles. Thus the amino acid can act as a proton donor (acid):

$$\begin{array}{c} H \\ | \\ R—C—COO^- \\ | \\ NH_3^+ \end{array} \rightleftharpoons \begin{array}{c} H \\ | \\ R—C—COO^- \\ | \\ NH_2 \end{array} + H^+$$

or as a proton acceptor (base):

$$\begin{array}{c} H \\ | \\ R—C—COO \\ | \\ NH^{3+} \end{array} + H^+ \rightleftharpoons \begin{array}{c} H \\ | \\ R—C—COOH \\ | \\ NH_3^+ \end{array}$$

The acid-base properties of some amino acids will be affected by those R groups that are also ionizable. Proteins are composed of amino acids; therefore they also display acid-base properties and this accounts for their buffering properties.

Sulfur

Sulfur is similar to oxygen chemically, but only a few organic compounds in biological systems contain sulfur. Some of the most important of these are the amino acids cysteine and methionine (Table 2-2), which are found in some proteins. Sulfur in organic compounds is in the form of **sulfhydryl** groups (—SH). The sulfhydryl group is particularly important in the enzyme molecule because many enzymes maintain their activity only when they possess free sulfhydryl groups. Oxidation of sulfhydryl groups inactivates many enzymes:

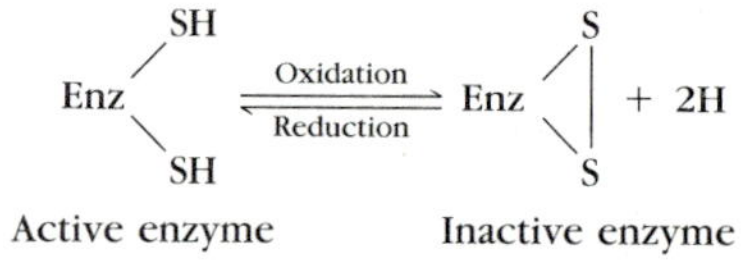

$$\underset{\text{Active enzyme}}{\text{Enz}\begin{array}{l} \diagup SH \\ \diagdown SH \end{array}} \underset{\text{Reduction}}{\overset{\text{Oxidation}}{\rightleftharpoons}} \underset{\text{Inactive enzyme}}{\text{Enz}\begin{array}{l} \diagup S \\ \quad | \\ \diagdown S \end{array}} + 2H$$

Sulfhydryl groups are also important in the structure of many proteins where cross-linkages within a polypeptide as well as between polypeptides contribute to chain folding. Oxidation of free sulfhydryl groups in the polypeptide strands leads to the formation of a disulfide (S—S) group (Figure 2-12).

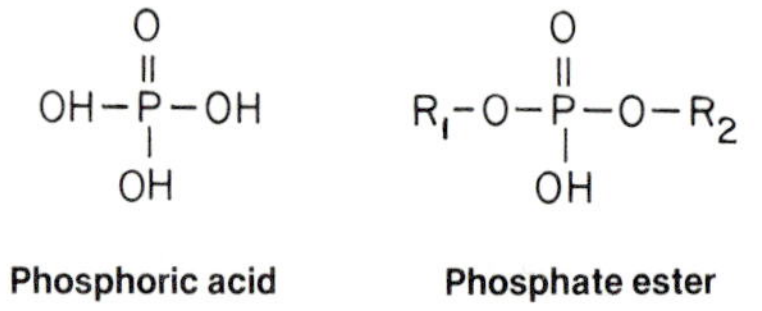

$$\begin{array}{cc} \begin{array}{c} O \\ \| \\ OH-P-OH \\ | \\ OH \end{array} & \begin{array}{c} O \\ \| \\ R_1-O-P-O-R_2 \\ | \\ OH \end{array} \\ \textbf{Phosphoric acid} & \textbf{Phosphate ester} \end{array}$$

Figure 2-13 *Structure of phosphoric acid (H_3PO_4). Replacement of two hydrogen atoms with organic groups results in formation of phosphate ester.*

Phosphorus

Phosphorus is an important element in biochemistry, since it takes part in many reactions in which energy is either expended or produced. Phosphorus occurs as derivatives of phosphoric acid (H_3PO_4), which is depicted in Figure 2-13. One or two of the hydrogen atoms in the phosphoric acid molecule may be replaced by organic radicals to produce phosphate esters. One important function of the phosphate ester is to act as a bridge linking the various nucleoside residues of nucleic acids, such as **deoxyribonucleic acid (DNA)** and **ribonucleic acid (RNA)**. Derivatives of phosphoric acid are also capable of interacting to form anhydrides (Figure 2-10), linkages that possess high energy. One of the most important organic molecules possessing the anhydride linkage is adenosine triphosphate (ATP), a molecule that

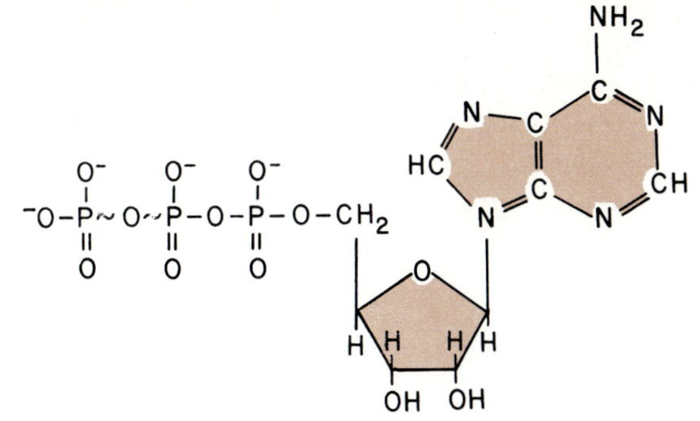

Figure 2-14 *Structure of adenosine triphosphate (ATP) molecule.*

Concept Check

1. What are the characteristics of carbon that make it ideal for the backbone material of living matter?
2. What are the basic formulas of the following chemical groups: alcohol, aldehyde, ketone, ester, ether, carboxylic acid, and anhydride? Give an example in which each of those groups appear in biological material.
3. What are the basic inorganic and organic forms of nitrogen, phosphorus, and sulfur that are available to microorganisms as nutrients?

is involved in energy exchange (Figure 2-14). Hydrolysis of the energy-rich phosphoanhydride bond occurs easily because of the repulsion between negative charges on the phosphate groups in the ionized state. The energy release is available for many biological activities, such as movement, metabolism, and other activities requiring energy.

Phosphate is also a component of other molecules in the cell such as the following:

1. The **phospholipids** that make up the cytoplasmic membrane of all cells, including bacteria. Phospholipids are also associated with the bacterial cell wall of a specific group of bacteria called gram-negative bacteria. In gram-positive bacteria, phosphates are part of the wall component called **teichoic acid**. The negative charge of cell wall phosphates contributes to the overall negative charge of the bacterial cell surface.
2. The **coenzymes**, nicotinamide adenine dinucleotide (NAD), nicotinamide adenine dinucleotide phosphate (NADP), flavin adenine dinucleotide (FAD), and others.
3. Sugars, such as glucose, are phosphorylated during their metabolism in the cell. Some of these **sugar-phosphates** will yield a considerable amount of energy because of the anhydride linkage of the phosphate group.

Table 2-3 *Principal monosaccharides found in microbial systems*

No. of carbons in monosaccharide	Name of monosaccharide
3	Glyceraldehyde
4	Erythrose
5	Ribose, arabinose, xylose, deoxyribose, xylulose
6	Glucose, mannose, galactose, fructose
7	Sedoheptulose

CHARACTERISTICS OF THE MAJOR BIOLOGICAL MOLECULES

Polymers, the large organic molecules of the cell associated with cell structure, are composed of smaller, yet similar individual units called **monomers**. The polymers and their monomeric units are:

POLYMERS	MONOMERS
Proteins	Amino acids
Nucleic acids (DNA or RNA)	Nucleotides
Complex carbohydrates	Monosaccharides

Some **lipids** are also large organic molecules; however, they are composites consisting of two units: **glycerol** and **fatty acids**.

Carbohydrates

Carbohydrates are so called because their general formula can be represented as $C_n(H_2O)_n$. The most important example is glucose, where n = 6, or $C_6H_{12}O_6$. Chemically, carbohydrates may be defined as carbonyl derivatives of polyhydric alcohols. The smallest carbohydrates are called **monosaccharides**, which can be linked together to produce more complex molecules called **polysaccharides**.

Monosaccharides. The most important monosaccharides in biological systems consist of three to seven carbons and can be divided into classes called, respectively, trioses, tetroses, pentoses, hexoses, and heptoses (Table 2-3). Each of these classes can be represented by more than one member, with each member having different

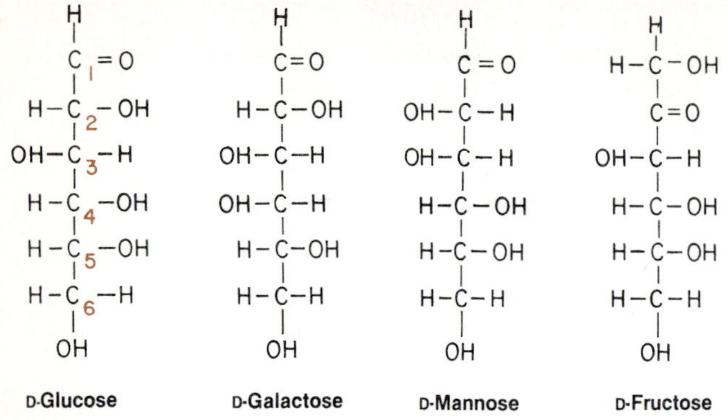

D-Glucose D-Galactose D-Mannose D-Fructose

Linear form

Ring form, all carbons shown

Figure 2-15 *Structure of some monosaccharides having general formula $C_6H_{12}O_6$ (hexoses). Numbering system for carbons is indicated in color.*

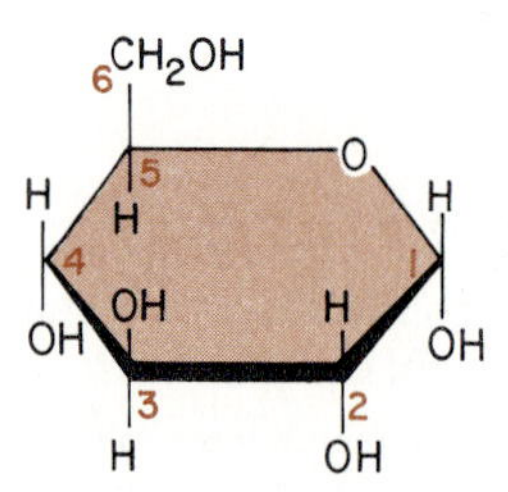

Abbreviated ring form

Figure 2-16 *Linear and chain forms representing sugar D-glucose.*

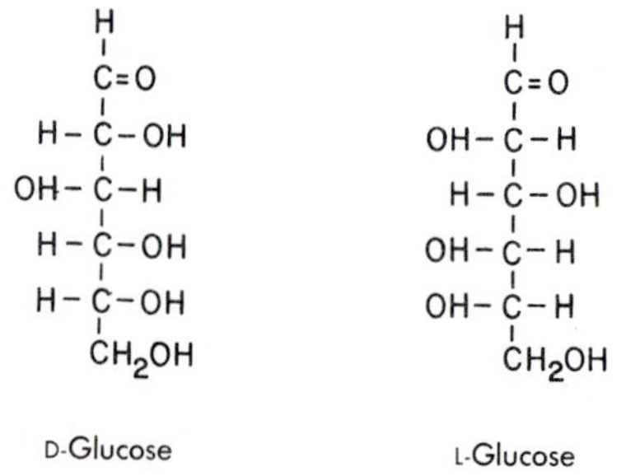

chemical properties based on the positions of their hydroxyl groups. The most important of the monosaccharides quantitatively are the trioses, pentoses, and hexoses. Many different hexoses have the general formula $C_6H_{12}O_6$, but the position of their hydroxyl groups differ (Figure 2-15). The carbonyl group can appear either at carbon 1 or carbon 2 of the monosaccharide. The monosaccharide is an aldehyde, if the carbonyl group is on carbon 1, and such monosaccharides are referred to as **aldoses**. The monosaccharide is a ketone if the carbonyl group is at carbon 2, and such monosaccharides are called **ketoses**. Glucose, for example, is an aldose, but fructose is a ketose (Figure 2-15). The five carbon sugars have the general formula $C_5H_{10}O_5$ and two of them, ribose and deoxyribose, are especially important as components of the nucleic acids to be described later.

The monosaccharides are often depicted in a linear chain form, but in solution they are usually in the ring form. Ring formation is commonly the result of interaction between the highly reactive carbonyl group and the hydroxyl group of the next to last carbon atom. An example of a six-membered monosaccharide ring structure is depicted in Figure 2-16.

Isomerism is a characteristic of sugars. Glucose, fructose, and galactose are structural **isomers**, that is, they have the same empirical formula, $C_6H_{12}O_6$, but have different structural formulas and exhibit distinct chemical properties. Carbon 2 through 5 for galactose and carbons 3 through 5 for fructose are bonded to 4 different groups and are thus asymmetric. Each of the monosaccharides therefore has optical activity and when synthesized in the laboratory a 50-50 mixture of mirror images is produced. The stereoisomers of monosaccharides, as well as all other optically active compounds, can all be related to a standard reference compound, glyceraldehyde, which has a single asymmetric carbon. Because

D-Glucose L-Glucose

Figure 2-17 *D and L configurations of glucose.*

monosaccharides have more than one asymmetric carbon the prefixes **D** and **L** are used to refer to the three-dimensional configuration around the asymmetric carbon farthest from the carbonyl carbon (Figure 2-17). The D and L designations have nothing to do with the direction of rotation of plane-polarized light. The hydroxyl group on carbon 5 of hexoses is always to the right in D-sugars and to the left in L-sugars. Biological systems for the most part use only D-sugars.

Disaccharides. Carbohydrates composed of two molecules of hexose-monosaccharides are called **disaccharides** (oligosaccharides have from two to ten mono-

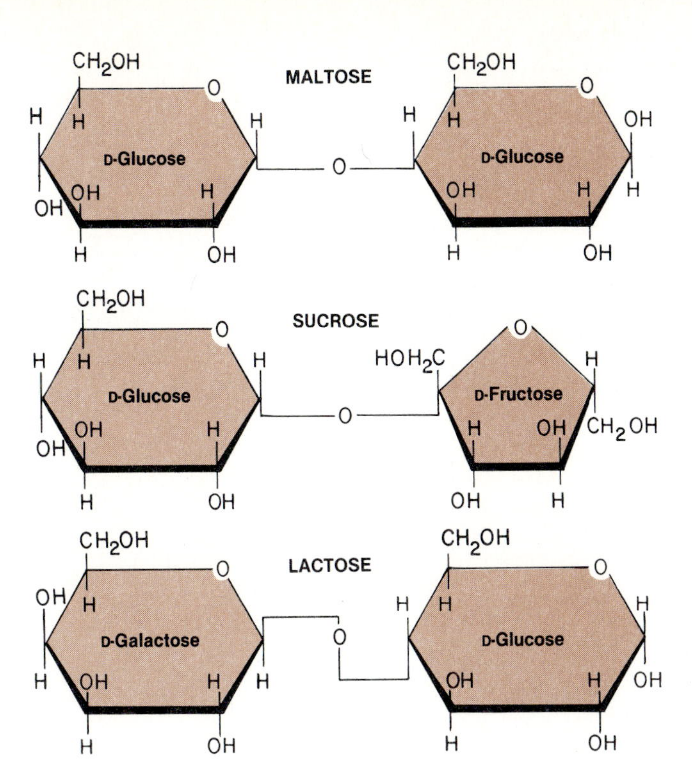

Figure 2-18 *Structure of disaccharides: maltose, sucrose, and lactose.*

saccharide units.) The most frequently encountered disaccharides are lactose, sucrose, and maltose (Figure 2-18), which have the same molecular formula: $C_{12}H_{22}O_{11}$. All are formed by the union of two monosaccharides with the elimination of water. **Lactose** is composed of glucose and galactose; **sucrose**, of glucose and fructose; and **maltose**, of glucose and glucose. These disaccharides are found normally in nature and can be hydrolyzed into their respective monosaccharides by many groups of microorganisms. Note that each disaccharide contains a molecule of glucose, the carbohydrate most readily catabolized by microorganisms.

Polysaccharides. Polysaccharides consist of many units of monosaccharides linked together to produce high molecular weight compounds ranging in size from 10,000 to 4,000,000. Some of the major polysaccharides are starch, pectin, cellulose, glycogen, and dextran, each of which consists of glucose units linked in such a way that each can be chemically distinguished from one another. Most linkages are between carbon 1 of one glucose unit and either carbon 4 (1-4 linkage), carbon 6 (1-6 linkage), or, infrequently carbon 3 of the adjacent glucose unit. Multiple linkages within the monosaccharide subunits result in extensive branching of the polysaccharide (Figure 2-19). Microorganisms, as well as plants, use polysaccharides to make cell walls and as fuel reserves.

Many polysaccharides are composed of monosaccharides and noncarbohydrate subunits, such as peptides (two or more amino acids joined by a chemical linkage called the peptide bond), proteins, and lipids. The carbohydrate component of many of these hybrid molecules is an amino sugar in which an amino group replaces a hydroxyl group at carbon 2. The addition of acetyl and lactic acid groups to one type of amino sugar produces ***N*-acetylglucosamine** and ***N*-acetylmuramic** acid (Figure 2-20). They are particularly important because they form the backbone of bacterial cell walls called **peptidoglycan** and are discussed in Chapter 5. Chitin is a component of fungal cell walls and is composed of *N*-acetylglucosamine.

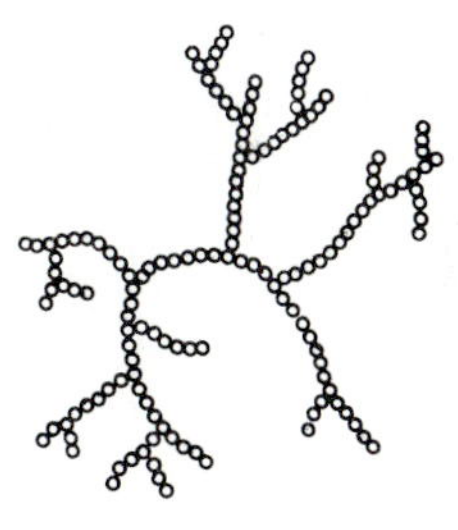

Figure 2-19 *Branching in polysaccharides such as glycogen. Each circle represents glucose unit.*

Proteins associated with carbohydrates are called **glycoproteins** and can be found in most organisms. Examples of glycoproteins are blood group antigens, cer-

```
  H-C=O                              H-C=O
    |                                  |
  H-C-NH-C-CH3             COOH      H-C-NH-C-CH3
    |    ||                 |          |    ||
 OH-C-H  O               H-C--------O-C-H   O
    |                       |          |
  H-C-OH                   CH3       H-C-OH
    |                                  |
  H-C-OH                             H-C-OH
    |                                  |
   CH2OH                              CH2OH
```

N-Acetylglucosamine **N-Acetylmuramic acid**

Figure 2-20 *Structure of amino sugars* N-*acetylglucosamine and* N-*acetylmuramic acid, found in cell wall of bacteria.*

```
    H
    |
  H-C-O-R
    |
  H-C-O-R
    |     O
    |     ||
  H-C-O-P-O-CH2-CH2-NH3+
    |   |
    H   O-
```

Phospholipid (phosphotidylethanolamine)

Figure 2-22 *Structure of compound lipid.* R = *fatty acid chain.*

```
    H
    |
  H-C-OH                H
    |                   |
  H-C-OH   +   3  H-C-(CH2)n -COOH  -->
    |                   |
  H-C-OH                H
    |
    H
```

Glycerol + **Fatty acid** →

```
    H      O             H
    |      ||            |
  H-C-O -C -(CH2)n - C-H
    |                    |
    |                    H
    |      O             H
    |      ||            |
  H-C-O-C- (CH2)n -C-H  +  3HOH
    |                    |
    |                    H
    |      O             H
    |      ||            |
  H-C-O -C - (CH2)n -C-H
    |                    |
    H                    H
```

Triglyceride (fat)

Figure 2-21 *Triglyceride formation, or reaction between alcohol and fatty acid.*

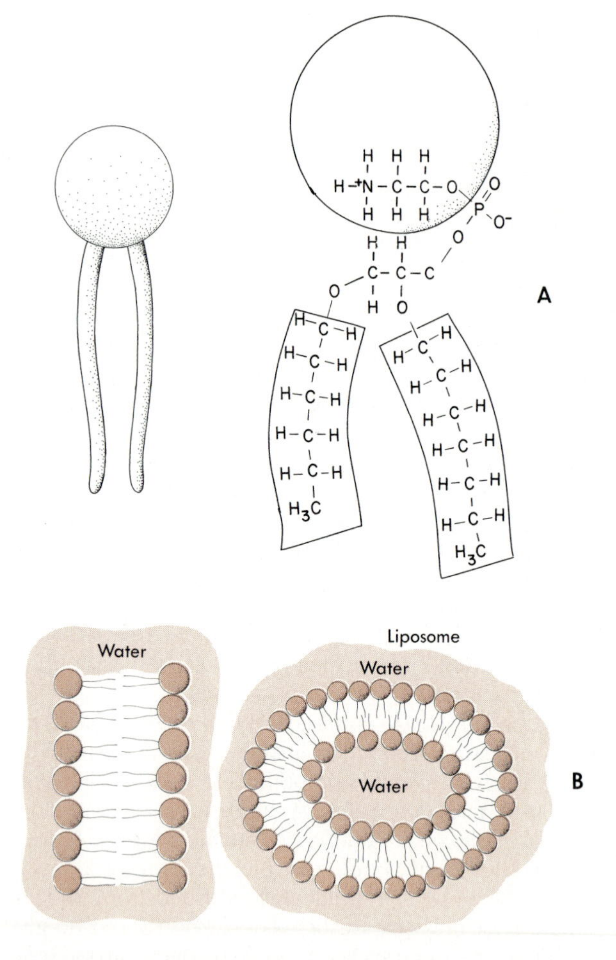

Figure 2-23 *Representations of phospholipid molecule.* ***A,*** *Diagrammatic representation of hydrophilic head and fatty acid tail of molecule, showing how the chemical groups are arranged in head and tail* (right). ***B,*** *Phospholipid arrangement in water. Phospholipid bilayer with polar (hydrophilic) heads of each phospholipid in contact with water and hydrophobic tails contacting each other to exclude water* (left). *Closed phospholipid bilayer, called liposome, that separates two aqueous compartments* (right).

tain hormones, the antiviral substance interferon, and most cell surfaces where the glycoprotein consists primarily of polysaccharide plus small amounts of protein. Certain viruses contain a glycoprotein envelope that contributes to the disease process. The glycoprotein component of some viruses, such as the influenza virus, has been used in the preparation of vaccines.

Lipopolysaccharides are also important macromolecules in biological systems, particularly bacteria, and they are discussed with the lipids.

Lipids

Lipids (fats and fatlike substances) are a very diverse group of organic compounds. They are generally classi-

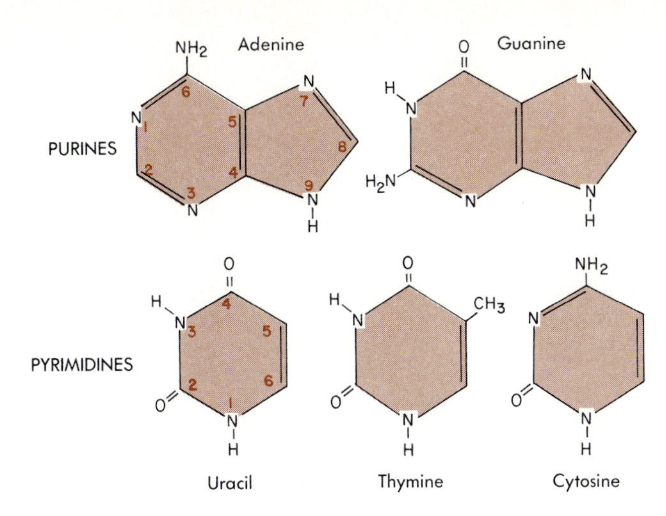

Figure 2-24 *Structure of purine and pyrimidine bases. Numbering of atoms is in color.*

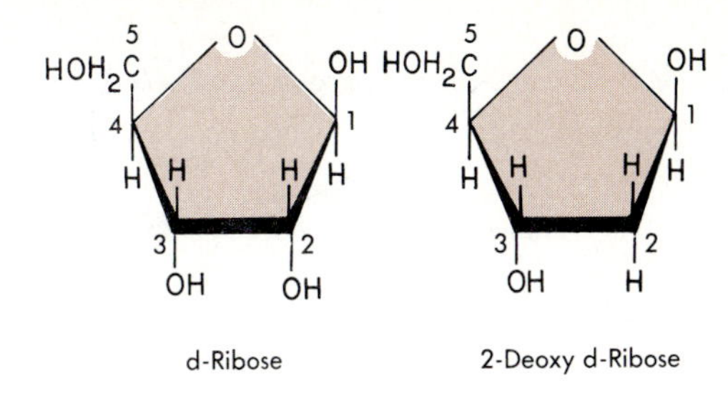

Figure 2-25 *Structure of nucleic acid sugars* D*-ribose and 2-deoxy-ribose. Numbering of atoms is in color.*

fied as esters or potential esters of fatty acids that are soluble in nonpolar solvents (ether, chloroform, benzene, etc.) but insoluble in water. They may be further classified as simple or compound. **Simple lipids** are esters of fatty acids and alcohol. The **triglycerides** are the most important class of simple lipids (the other class is made up of waxes, which are more important in plants and animals) and are esters of glycerol and fatty acids. Glycerol is a derivative of the simplest sugar, glyceraldehyde, and, during lipid formation in the cell, each hydroxyl group is esterified to a fatty acid (Figure 2-21) that may be 10 to 20 carbons in length. Each fatty acid may be saturated or unsaturated. Simple lipids function primarily as a source of energy because oxidation of the hydrocarbon-rich fatty acid chains releases intermediates that lead to the synthesis of ATP.

Compound lipids are also esters of fatty acids with alcohol but contain other groups as well, such as phosphoric acid (Figure 2-22), sulfate, or nitrogen bases. Some compound lipids are also attached to macromolecules such as proteins (**lipoproteins**) or polysaccharides (**lipopolysaccharides**). Lipopolysaccharide (LPS) is a component of the cell wall of certain bacteria (gram-negative types) and contributes to their toxicity during the disease process. Compound lipids function primarily as structural units within cell structures: the cell membrane and cell envelope (structures outside cell membrane). The compound lipids found in the microbial cell are primarily phospholipids and lipoprotein.

Phospholipids are the most common membrane lipid. They consist of a hydrophilic (high affinity for water) head that is made up of a phosphate group linked to another residue, such as ethanolamine, choline, or glycerol, and two hydrophobic (low affinity for water) tails each of which is a fatty acid (Figure 2-23, *A*). When phospholipids are suspended in a watery medium, they have a tendency to arrange themselves into a bilayer. The molecules in both layers are arranged in such a way that the hydrophilic heads face water on both sides of the bilayer (Figure 2-23, *B*). Polar lipids can also spontaneously form bilayers that separate two water phases by extending inward to form a continuous inner core (Figure 2-23, *B*). These closed bilayers are called **liposomes.** The phospholipid bilayer is especially suited for biological membranes because the hydrocarbon interior makes the membrane impermeable to nutrient molecules, such as amino acids, sugars and larger molecules, such as proteins and nucleic acids, and small ions. Physiologically the bilayer is a pliable structure having the viscosity of olive oil. Cytoplasmic membrane structure is discussed in greater detail in Chapter 5.

Nucleic acids

Nucleic acids are components of all living cells, as well as acellular viruses. They are large complex molecules that, on the basis of chemical differences, can be divided into two types: DNA and RNA. Each type is made up of two classes of nitrogen bases, called **purines** and **pyrimidines**, plus **pentose phosphates**.

The structure of the purine and pyrimidine bases is illustrated in Figure 2-24. The purines adenine and guanine are both found in DNA and RNA. The pyrimidines thymine and cytosine appear in DNA, and cytosine and uracil are present in RNA. Both groups of nitrogen bases possess keto and amino groups that, as we will see, can take part in hydrogen bonding. The carbohydrate components of the nucleic acids are D-ribose and D-2-deoxyribose. The latter sugar has a hydrogen in place of a hydroxyl group on carbon 2 (Figure 2-25). **Nucleosides** are compounds that contain the pentose sugar bonded to a purine or pyrimidine base and are referred to as ribonucleosides or deoxyribonucleosides (Figure 2-26). They are more specifically named according to the base attached to them (Table 2-4). Attachment of phosphate either to carbon 5 or to carbon 3 of the ribose or deoxyribose moiety produces what is called

Table 2-4 *Nomenclature of some nucleosides and nucleotides*

Base	Nucleoside		Nucleotide	
	Ribose	**Deoxyribose**	**Ribose**	**Deoxyribose**
Adenine	Adenosine	Deoxyadenosine	5′-Adenylic acid or adenosine 5′-monophosphate	5′-Deoxyadenylic acid or deoxyadenosine 5′-monophosphate
Cytosine	Cytidine	Deoxycytidine	5′-Cytidylic acid or cytidine 5′-monophosphate	5′-Deoxycytidylic acid or deoxycytidine 5′-monosphosphate
Guanine	Guanosine	Deoxyguanosine	5′-Guanylic acid or guanosine 5′-monophosphate	5′-Deoxyguanylic acid or deoxyguanosine 5′-monophosphate
Uracil	Uridine	Deoxyuridine	5′-Uridylic acid or uridine 5′-monophosphate	5′-Deoxyuridylic acid or deoxyuridine 5′-monophosphate
Thymine	Thymidine	Deoxythymidine	′5-Thymidylic acid or thymidine 5′-monophosphate	5′-Deoxythymidylic acid or deoxythymidine 5′-monophosphate

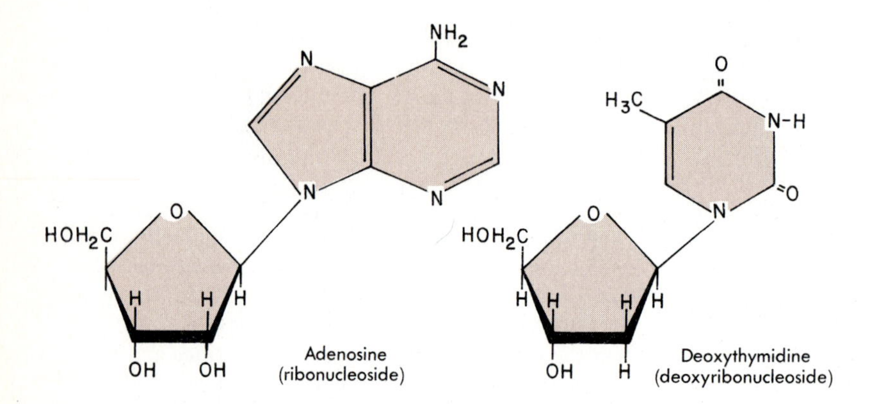

Figure 2-26 *Structure of two nucleosides: adenosine and deoxythymidine.*

nucleoside phosphates (Table 2-4). More than one phosphate may be attached to a nucleoside (a nucleoside diphosphate or nucleoside triphosphate), but not if the nucleoside is part of the internal structure of the nucleic acid molecule. The combination of sugar plus phosphate plus base is called a **nucleotide**. The arrangement of purines, pyrimidines, and sugar phosphates contributes to their primary, as well as secondary, structure.

Deoxyribonucleic acid

Primary structure. The DNA molecule typically is composed of two strands, each of which consists of individual nucleosides linked together via a phosphate ester linkage between the carbon 5′-hydroxyl of one nucleoside and the carbon 3′-hydroxyl of another nucleoside (Figure 2-27). The nucleosides commonly number from 1×10^4 to 1×10^8, depending on the species in which they are found. The molecular weight of the DNA molecule may range from 1×10^5 to 2×10^9.

Secondary structure. Most DNA molecules (except those in some viruses) are double stranded, whereas RNA is single stranded (except in a few viruses that possess double-stranded RNA). X-ray diffraction studies in the 1950s revealed that the two chains of DNA are twisted around a central axis to form a right-handed double helix or B form (B-DNA) (Figure 2-28). There

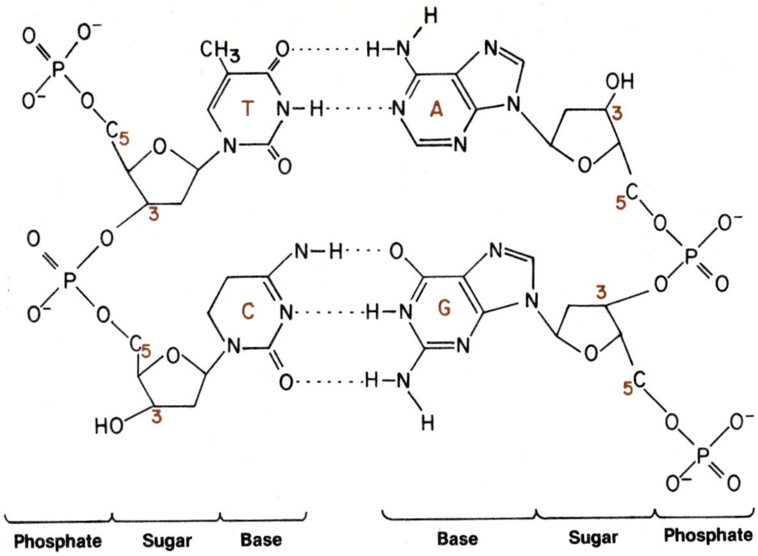

Phosphate Sugar Base Base Sugar Phosphate

Figure 2-27 *Base pair arrangements between adjacent nucleotides on portion of DNA molecule. Note that strands are antiparallel and sugar groups on right chain appear to be upside down (strand on right is 3′ to 5′, whereas strand on left runs 5′ to 3′ from top to bottom). Three hydrogen bonds bind guanine with cytosine, whereas two hydrogen bonds bind adenine with thymine.*

are 10 base pairs for each turn of the helix, and each base pair is separated by 0.34 nm. The backbone of the chain, which consists of alternating deoxyribose and negatively charged phosphate groups, is hydrophilic and faces the outside or surrounding water. The hydrophobic purine and pyrimidine bases are stacked inside the helix so that they are perpendicular to the long axis of the helix. The purine and pyrimidine bases of one strand pair with the bases of the other strand in a very specific way. A purine on one strand pairs with a pyrimidine on the adjacent strand, that is, adenine (A) pairs with thymine (T) and guanine (G) pairs with cytosine (C). This is called **complementary base pairing**, and because of it the molar ratio of A to T and G to C in the DNA molecule is 1:1. This complementarity exists because the chains are antiparallel and one chain of the DNA appears to be upside down (Figure 2-27). The base at the top of the chain on the left in Figure 2-27 contains a free, unlinked 5′-phosphate, and the complementary base on the opposite chain contains a free 3′-hydroxyl. The chain on the left therefore runs in the 5′ to 3′ direction, whereas the chain on the right runs from 3′ to 5′, going from top to bottom; that is, it is antiparallel.

The spatial relationship between the two strands of the DNA molecule produces a major and minor groove. The **major groove** has a diameter and depth that can serve as a site for binding of proteins. Proteins possess groups of amino acids arranged in a conformation whose diameter is comparable to the diameter of the major groove. Thus amino acids in the protein make contact with the edges of the base pairs in the major groove. The implications of these interactions are discussed in Chapter 9. X-ray diffraction studies have now revealed a new form of DNA called Z-DNA (see Point of Interest) whose conformation differs from that of B-DNA.

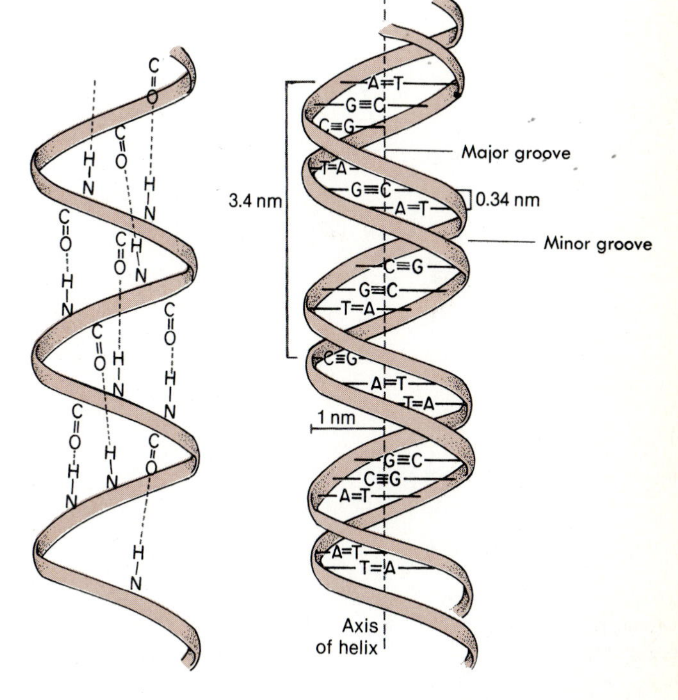

Figure 2-28 *DNA helix (B-DNA) showing distance between adjacent nucleotides and distance between each turn of helix. This DNA configuration produces large major groove and small minor groove.*

Ribonucleic acid

Primary and secondary structures. The primary structure of RNA differs from DNA: the sugar in RNA is D-ribose, uracil is substituted for thymine, and RNA is single stranded (except some viruses).

The secondary structure (bonding between noncontiguous bases that defines the conformation of the molecule) of RNA is known for a few molecules. The identified RNA molecules have helical domains that in turn can apparently interact to provide tertiary structure. The coiling of the RNA brings about the random matching of complementary regions (Figure 2-29), thereby producing double helical (duplex) regions.

The RNA found in nature may be of three different functional types: **transfer RNA (tRNA)**, **messenger RNA (mRNA)**, and **ribosomal RNA (rRNA)**.

1. Transfer RNA is a low molecular weight molecule

Point of Interest

Z-DNA

In 1979 Alexander Rich and colleagues discovered a new DNA conformation while performing x-ray diffraction studies on synthetic DNA consisting of alternating guanine and cytosine. Their synthetic molecule, unlike B-DNA, was left-handed in its twist and its backbone zigzagged down the molecule. They called this molecule Z-DNA. Z-DNA has 12 base pairs per helical turn with each base separated by 0.372 nm. Each turn of the helix is separated by 4.46 nm; thus Z-DNA is slimmer and possesses a larger number of base pairs per helical turn than B-DNA. The bases in Z-DNA are also turned around or flipped over. Z-DNA is not as stable as B-DNA because the negatively charged phosphate groups on opposite strands are closer together than in B-DNA and this leads to a greater electrostatic repulsion. There are situations in which Z-DNA is more stable. Laboratory studies have demonstrated that Z-DNA forms from sequences that contain alternating purines and pyrimidines, such as CGCGCG but not alternating AT sequences. Even with the proper sequences, Z-DNA will convert to the B form because the latter structure requires less energy to maintain. It now appears that adding methyl groups to the carbon 5 position of cytosine-quanine sequences, as well as supercoiling (circular form of the DNA in which it twists to form coils) of the DNA, helps maintain and stabilize Z-DNA. Supercoiling experiments have demonstrated that Z-DNA segments can be formed in the DNA molecule even though the remainder of the DNA molecule is in the B form. The identification of Z-DNA regions is possible because Z-DNA, unlike B-DNA, can stimulate the formation of antibodies. These antibodies have been useful in identifying Z-DNA in many biological systems. What are the biological implications of Z-DNA? The conformational differences between Z-DNA and B-DNA may provide different recognition sites for proteins that regulate gene activity. Several proteins have been isolated from cells that are specific for Z-DNA but not B-DNA. Some of these proteins not only bind specifically to Z-DNA but can flip the B-DNA into its Z conformation. A test-tube model has also provided some evidence that Z-DNA is generated at sites where recombination is most likely to take place. The enzyme that promotes pairing between DNA molecules involved in the recombination process has a tremendous affinity for the Z-DNA conformation compared with the B-DNA conformation. However, there is no direct evidence for the in vivo existence of Z-DNA or for its physiological function.

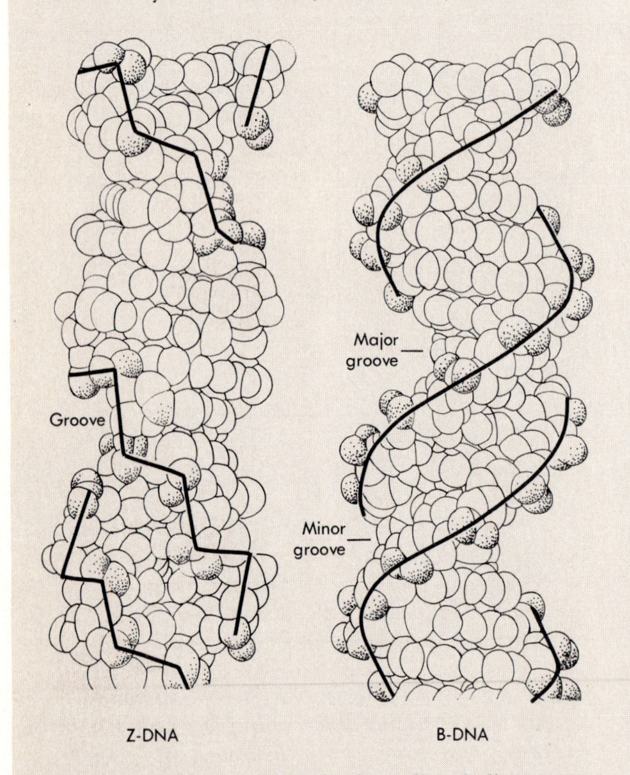

Comparison of Z-DNA and B-DNA. Heavy lines indicate sugar-phosphate backbone of DNAs. The zigzag path of Z-DNA is easily discernible as compared with smooth path of sugar-phosphate backbone of B-DNA. Minor groove of Z-DNA is very deep and penetrates to helix axis, whereas major and minor grooves of B-DNA are relatively shallow.

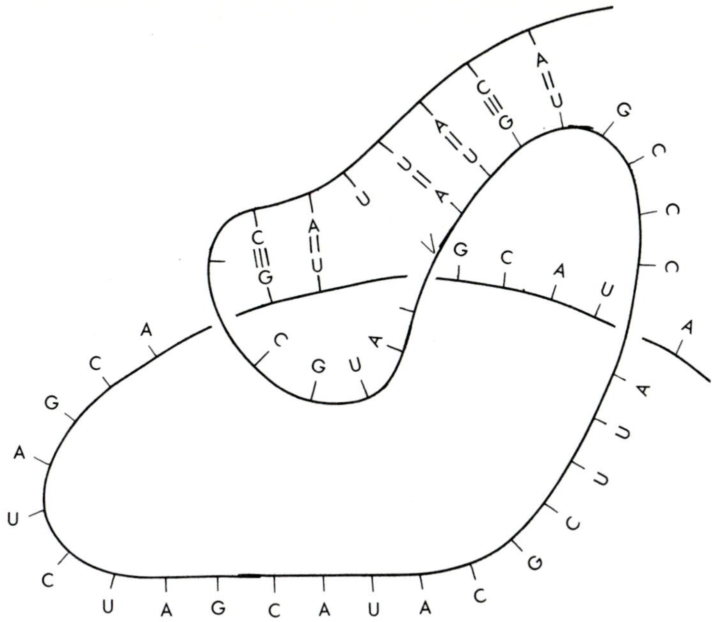

Figure 2-29 *Single-stranded RNA molecule in which folding brings about random pairing between complementary base pairs.*

containing 70 to 80 nucleotides. Its function in the cell is to carry amino acids that ultimately are used in the process of protein synthesis.

2. Ribosomal RNA is a component of those structural units in the cell called **ribosomes.** The ribosome, in addition to its rRNA component, also contains proteins. Ribosomes are the sites of protein synthesis.
3. Messenger RNA is a molecule that is complementary to one of the DNA strands. Its formation is the result of a transcription of the DNA's genetic information. The information in this mRNA will eventually be translated to a protein.

Proteins

Proteins are high molecular weight polymers consisting of amino acids linked by peptide bonds. Different polypeptides may contain nearly the same number of amino acids, but their activities may vary considerably. It is the number, as well as the sequence, of amino acids in the polypeptide that determines the latter's chemical properties and, hence, function in the cell. The number of totally different polypeptides that could be produced from an average polypeptide containing 200 amino acids would be 20^{200}, assuming that the number of different amino acids is 20. The order in which the amino acids appear in the polypeptide is a function of the hereditary material of the cell. The mechanism by which DNA is translated into specific amino acid sequences is discussed in Chapter 9.

Proteins can be divided into two types based on their composition and solubility. There are the **simple proteins,** which are composed only of amino acids, and the **conjugated proteins,** which contain a nonprotein or prosthetic group. The prosthetic group may be phosphate (phosphoproteins), carbohydrate (glycoproteins), lipid (lipoproteins), nucleic acid (nucleoproteins), iron or zinc (metalloproteins), and heme (hemoproteins). Proteins may also be divided into groups based on their function, for example, structural, enzymatic, hormonal, or viral. Most of our discussion will be concerned with enzymatic proteins.

Protein structure. Proteins fold up and assemble into specific structures based on the interaction of their respective amino acids. This structural organization can be divided into four levels: primary, secondary, tertiary, and quaternary.

Primary structure. The sequence of amino acids in a polypeptide chain or chains of a protein is the basis for its primary structure. The bond linking the successive amino acid residues is called the **peptide bond** (Figure 2-30). The backbone of the polypeptide chain remains the same for all polypeptides. Only the R groups projecting from the backbone are different in the various polypeptides. The R groups determine the tertiary and quaternary structure of the protein. The amino acid residue at the end of the polypeptide containing a free alpha (α) amino group is called the **amino terminal.** The amino acid residue at the opposite end that has a free carboxyl group is called the **carboxyl terminal.** By convention the sequence of amino acid residues is named from the amino terminal of the polypeptide. Many proteins have been totally sequenced including insulin, a

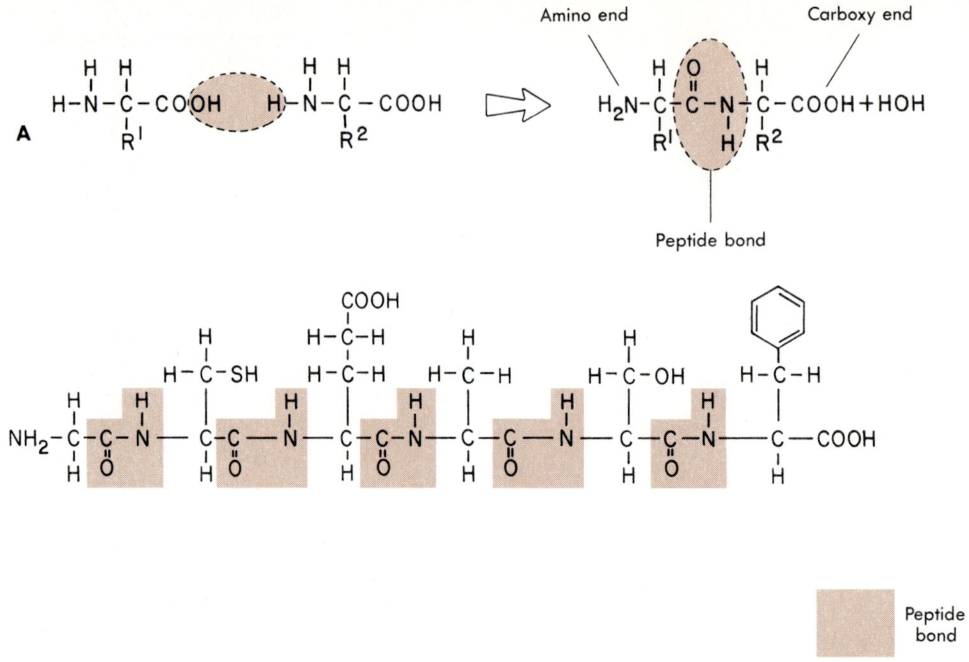

Amino end

Carboxy end

A

Peptide bond

B

Peptide bond

Figure 2-30 *Primary structure of proteins. **A,** Peptide bond formation. Reaction involves amino group on one amino acid and carboxyl group of another amino acid. By convention amino end is on left and free carboxyl on right. **B,** Structure of peptide chain containing six amino acid residues. Note the R groups projecting from backbone of peptide chain. Amino acids beginning from amino end of the chain to carboxyl end are glycine, cysteine, glutamic acid, alanine, serine, and phenylalanine.*

polypeptide that has 51 amino acid residues, hemoglobin that has 4 polypeptide chains totalling 574 amino acid residues, and cytochrome c, a single polypeptide chain with approximately 100 amino acid residues, depending on the organism in which it is found. Proteins that perform the same function in different species have homologous amino acid sequences. Sequence analysis of proteins, such as cytochrome c, from different species has greatly elucidated the ancestral relationship, especially among vertebrates and even bacteria.

Secondary structure. Proteins in solution assume a particular conformation or three-dimensional structure. Secondary structure is a consequence of the hydrogen bonding between the C═O and N—H groups of different peptide bonds. Two types of secondary structure are possible:

1. In the **pleated sheet** arrangement there is hydrogen bonding between two polypeptides to produce a sheetlike arrangement (Figure 2-31). Collagen is an example of the pleated sheet arrangement and is characteristic of fibrous proteins, which function as support structures (bone, muscle) in the cell.
2. The **α-helix,** which is the result of intramolecular hydrogen bonding, is the most stable conformation for the polypeptide. Hydrogen bonding occurs between the C═O of one amino acid and the N—H of a peptide bond three amino acids down the chain. Once all of the hydrogen bonds be-

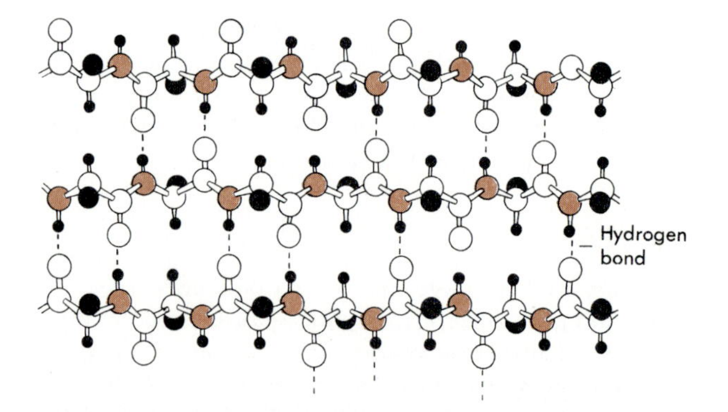

Figure 2-31 *Example of pleated sheet secondary structure of certain fibrous proteins. Note that three polypeptides are involved with hydrogen bond cross links between adjacent chains.*

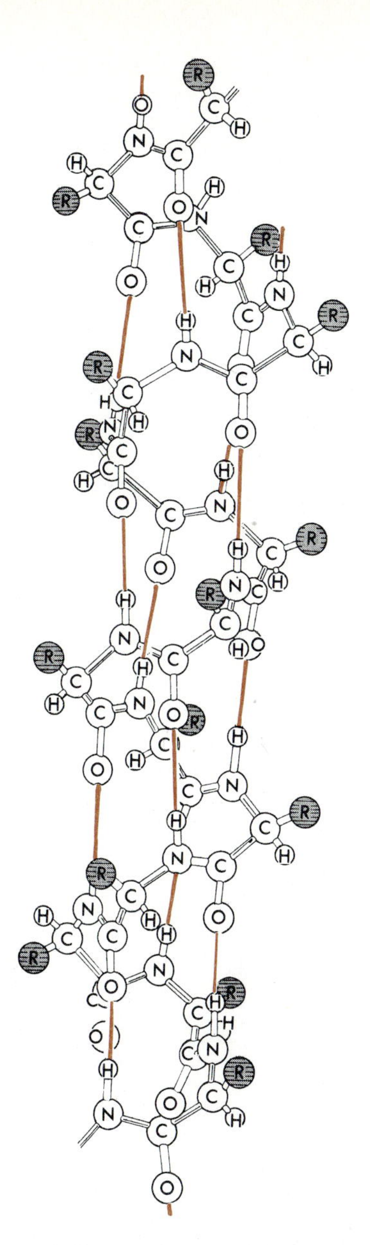

Figure 2-32 *Model of α-helical polypeptide chain. Backbone of chain consists of repeating sequences* C, C, N. R *represents side groups of the different amino acids. The colored lines represent hydrogen bonds that stabilize the helix, that is, hydrogen bond involving one amino acid and another three amino acids down chain.*

tween peptide units have been made, the polypeptide assumes the conformation of the α-helix (Figure 2-32). The α-helix gives the appearance of a tightly coiled spring in which the R groups project away from the center of the molecule. Bends can appear in the helix when the α-helical structure is disrupted. Bends appear at those positions occupied by proline or hydroxyproline, the only amino acids in which there is no hydrogen available for bonding once the peptide has been produced.

Tertiary structure. The α-helical coils of the polypeptide can fold into various shapes depending on the interaction of various R groups and the number of proline or hydroxyproline residues. The spatial arrangement of the helical coils is called the tertiary structure. The R group interactions that contribute to the folding are:

1. Disulfide bonds. The disulfide bonds are the only covalent bonds involved in tertiary structure. They can bring together amino acids of the polypeptide that are normally far apart in terms of their position in the polypeptide chain (Figure 2-33).
2. Hydrogen bonds. Hydrogen bonding may occur between various groups: the carboxyl groups of glutamic or aspartic acid and the hydroxyl group of tyrosine.
3. Salt bonds. Ionizable groups such as NH_2 and COOH interact to produce a salt bridge, which is an electrostatic interaction.
4. Hydrophobic and hydrophilic interactions. The R groups projecting from the polypeptide chain are either hydrophilic or hydrophobic and can influence tertiary structure. Their interactions serve to arrange the polypeptide into the shape of a sphere, with the hydrophobic groups pushed together in the center of the sphere away from water (Figure 2-33) and the hydrophilic groups projecting away from the sphere and closest to water molecules in solution.

The folding of the polypeptide is responsible for the globular or spherical shape that is characteristic of most proteins of biological origin, such as the enzymes. The folded conformation of globular proteins is responsible for their biological activities. When these proteins are heated or exposed to extremes of pH, they undergo **denaturation** and lose their biological activities. Denaturation does not affect the covalent backbone structure of polypeptides but does change the conformation or tertiary structure.

Proteins that consist of more than 200 amino acid residues appear to consist of two or more spherical structural groups called **domains**. When the polypeptide is folded there is the appearance of stiff polypeptide segments interspersed with tight bends (Figure 2-34). These domains play a major role in the binding of some proteins to nucleic acids. These types of interactions are responsible for the control exerted by certain proteins on the expression of nucleic acid information (see Chapter 9).

Quaternary structure. Many proteins consist of more than one polypeptide, and the manner in which these polypeptides interact with one another is called the quaternary structure. The R group interactions that

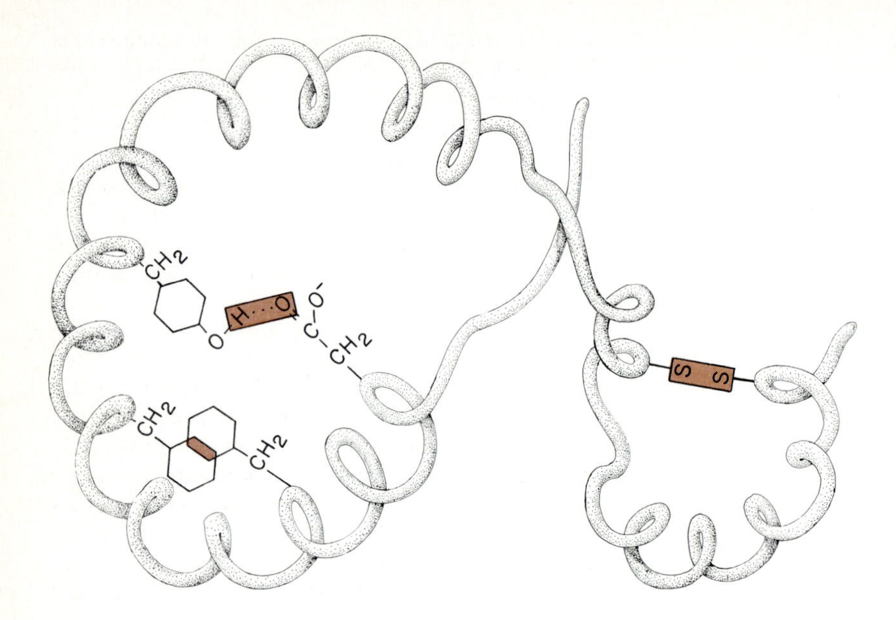

Figure 2-33 *Secondary and tertiary structure of protein.*

Concept Check

1. Describe the composition of the major biological molecules, proteins, nucleic acids, complex carbohydrates, and lipids and the role they play in the cell.
2. Explain the basic design of cell membranes.
3. What is meant by the primary and secondary structure of DNA and RNA?
4. What is meant by the primary, secondary, tertiary, and quaternary structure of proteins?

contribute to intrachain tertiary structure are also important in interchain quaternary conformation. Disulfide bridges, salt bonds, and ionic bonds all contribute to the quaternary structure. The biological function of proteins, particularly enzymes, is dependent on the interaction of the polypeptide chains.

ENZYMES

Inorganic and organic elements can enter into reactions that result in the formation of compounds, but most require extremes of temperature, pH, pressure, or radiation. These conditions would be detrimental to living cells, such as microorganisms. Chemical reactions occurring in the cell at normal temperatures in the absence of catalysts would not take place or would take

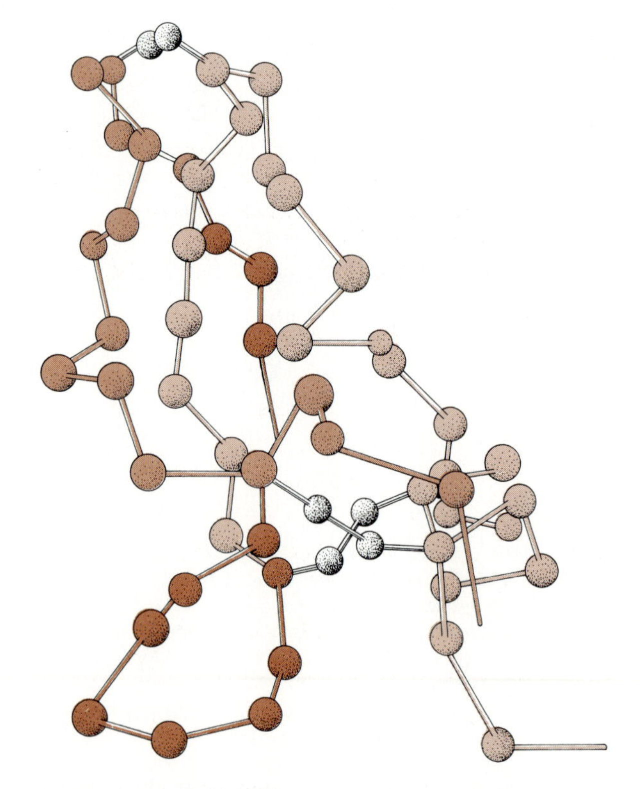

Figure 2-34 *Ball-and-stick model of polypeptide showing domains. Polypeptide backbone within domain may be irregular but generally runs straight course across domain and then turns and continues in relatively direct path to other side. Domains are indicated by different colored shadings. Stippled uncolored balls represent disulfide bonds.*

place so slowly that before the cell could grow and reproduce, it would die of old age. To speed up these biochemical reactions, nature has provided the cell with enzymes, which act as **catalysts.** Catalysts modify the speed of a chemical reaction without being used up or appearing as one of the reaction products. It is no wonder that much of the information locked up in the hereditary material of all living cells codes for the synthesis of enzymes.

Characteristics

All enzymes are globular proteins with molecular weights of from 10,000 to over 2,000,000. Their activity as catalysts is associated with their three-dimensional shape, which creates a site for binding of substrate, called the **active site.**

Active site. The active site of the enzyme consists of a group of amino acids that are involved in making contact with the substrate. The amino acids at the site may not be contiguous and can be from widely separated areas of the polypeptide. They may even be from different polypeptides. The active site is accessible to the substrate and once the latter is bound causes a conformational change in the enzyme. This conformational change brings the amino acids associated with the active site in closer proximity to the substrate. The conformational change in the enzyme brought about by the substrate is a reversible process. The enzyme can be used again and again.

Changes in the conformation of an enzyme, whether caused by high temperature, variations in pH, or other chemical or physical means, can lead to alterations at the active site and loss of catalytic activity.

Enzyme specificity. The extent of specificity between enzyme and substrate can vary widely among different enzymes. Some enzymes exhibit steric specificity. Recall that D-sugars and L-amino acids are the most predominant types of stereoisomers in nature. The ability of enzymes to distinguish these differences has resulted in a hypothesis suggesting that there are multiple sites (a minimum of three) for attachment of substrate to enzyme (Figure 2-35). Other enzymes recognize particular groups (amino, carboxyl, ester, peptide, etc.), and any alterations in nonspecific areas of the substrate may have no effect, or minor effects, on enzyme activity. Most enzymes exhibit a specificity that is dependent not only on the catalyzed groups of the substrate but on other regions of the substrate as well. These accessory regions may be required not only for binding to the enzyme but for proper orientation of the substrate at the active site.

Enzyme cofactors. Many enzymes require additional inorganic or organic components, referred to as **cofactors,** in order to exhibit catalytic activity. The most important of the cofactors are the organic **coenzymes,** or **vitamins,** without which no catalytic activity occurs (Table 2-5). Most coenzymes are loosely attached to the enzyme, but some organic groups are covalently bound and are called prosthetic groups. Inorganic cofactors include metal ions, which are usually bound by weak bonds (seldom covalent) to specific amino acids at the active site. Some important metal cofactors are magnesium, manganese, cobalt, iron, and molybdenum. When the enzyme protein, called the **apoenzyme,** is combined

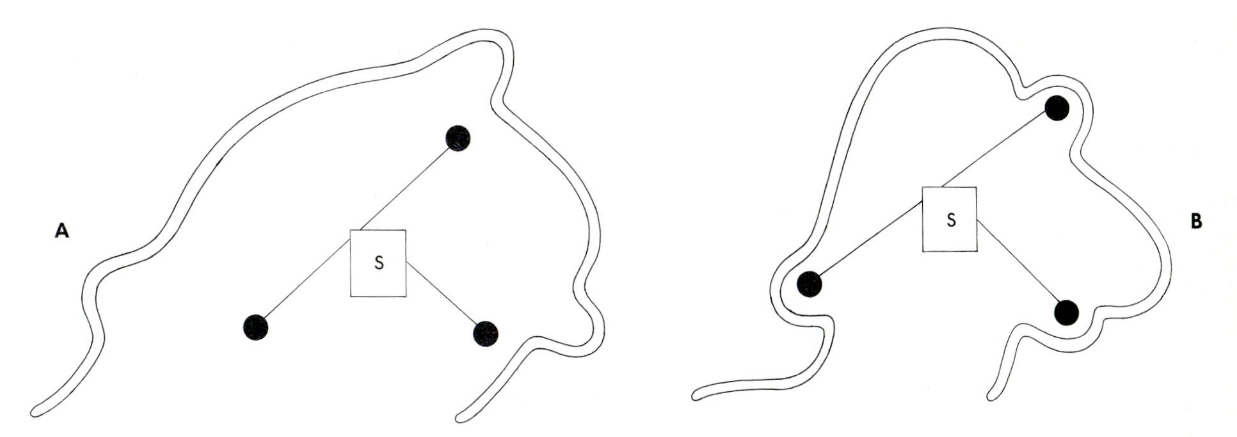

Figure 2-35 *Multiple site arrangement for attachment of substrate to active site of enzyme.* ***A,*** *Substrate approaches active site of enzyme.* ***B,*** *Substrate interacts with amino acids at active site and causes conformational changes in enzyme.*

Table 2-5 *Vitamins and their coenzyme forms*

Vitamin	Coenzyme form
Riboflavin (vitamin B_2)	Flavin adenine dinucleotide (FAD) and its reduced form ($FADH_2$); flavin mononucleotide (FMN) and its reduced form $FMNH_2$
Nicotinic acid (niacin)	Nicotinamide adenine dinucleotide (NAD) and its reduced form ($NADH_2$); nicotinamide adenine dinucleotide phosphate (NADP) and its reduced form ($NADPH_2$)
Thiamine (vitamin B_1)	Thiamine pyrophosphate (TPP)
Pyriodoxine (vitamin B_6)	Pyridoxal phosphate; pyridoxamine phosphate
Folic acid	Tetrahydrofolic acid
Biotin	Biotin
Pantothenic acid	Coenzyme A

with its cofactor, the active complex is called a **holoenzyme.**

Classification. Enzymes, for many years, were given names that did not depict the type of reaction in which they were involved. A commission on nomenclature in 1961 declared that biochemical reactions were catalyzed by six general types of enzymes. Various classes, within each type, which represented more specific enzyme activities, were also introduced. The names of all enzymes end in *ase.* Listed below are the six major groups of enzymes (Table 2-6).

1. **Oxidoreductases** are involved in oxidation-reduction reactions. Some of the enzymes of this group remove hydrogen from a reactant and pass it to another reactant. These are called **dehydrogenases.** Other enzymes of this group, like the oxidases, incorporate molecular oxygen into reactants.
2. **Transferases** transfer one group of atoms from one molecule to another. They may transfer an amino group, acetyl group, or phosphate group.
3. **Hydrolases** bring about the hydrolysis of molecules—such as proteins, carbohydrates, lipids and so on—into their respective monomeric units. Protein hydrolases, for example, cleave peptide bonds and release amino acids.
4. **Lyases** break covalent bonds and remove groups from a compound other than by hydrolysis. The removal of a carboxyl group or a water molecule from a compound is an example of the activity of lyases.
5. **Isomerases** bring about the rearrangement of groups of atoms within a molecule. The conversion of a D-amino acid to an L-amino acid is an example of the activity of isomerases.
6. **Ligases** (sometimes called **synthetases**) join two molecules together, and during the reaction there is the breakdown of ATP or a related energy molecule. One type of ligase binds an amino acid to a tRNA molecule.

General mechanism of enzyme activity

Some chemical reactions proceed spontaneously with the release of **free energy**, but most chemical reactions in biological systems do not proceed spontaneously. The greater the amount of free energy that is released, the more likely a chemical reaction will occur. Free energy, which can be measured in calories, is the form of energy that can perform work in the cell at constant temperature and pressure. Even though a chemical reaction may take place with the release of free energy, the rate at which it is released will depend on other factors, for example, enzymes. Look at the following chemical reaction:

$$A + B \rightleftharpoons C + D + \text{free energy}$$

For free energy to be released from this chemical reaction the molecules of reactant A must interact with molecules of reactant B with sufficient energy to yield products C and D. The magnitude of this energy, the **energy of activation**, influences the rate of the reaction. If the activating energy in this reaction were in the form of an increase in temperature, molecules of A would collide with molecules of B and products C and D would be formed and free energy would be released. The higher the temperature applied to the reaction (within limits) the greater the fraction of activated molecules, the greater the number of collisions between A and B, and the greater the number of molecules of product formed (Figure 2-36). Biological systems can survive only within a limited temperature range, and the temperatures required to activate the reactants is so high that it would destroy the cell. Enzymes act as catalysts and in so doing reduce the activating energy required to promote molecular collisions (Figure 2-37). Enzymes increase the rate at which a chemical reaction proceeds by acting as a site for the interaction of reactants (we should now

Table 2-6 *Classes of enzymes and examples of the reactions they catalyze**

Class of enzyme	Example and trivial name	Systematic name	Reaction catalyzed
Oxidoreductase	D-Lactate dehydrogenase	D-Lactate: NAD oxidoreductase	D-Lactate + NAD → Pyruvate + $NADH_2$
Transferase	Acetate kinase	ATP: acetate phosphotransferase	ATP + Acetate → ADP + Acetylphosphate
Hydrolase	Lipase	Glycerol ester hydrolase	Triglyceride + H_2O → Diglyceride + Fatty acid
Lyase	Oxalate decarboxylase	Oxalate carboxylyase	Oxalate → Formate + CO_2
Isomerase	Alanine racemase	Alanine racemase	L-Alanine → D-Alanine
Ligase	Methionyl tRNA synthetase	L-Methionine: tRNA ligase (AMP)	ATP + L-Methionine + tRNA → AMP + PP + L-Methionyl tRNA

*Examples are those characteristically found in microbial cells.

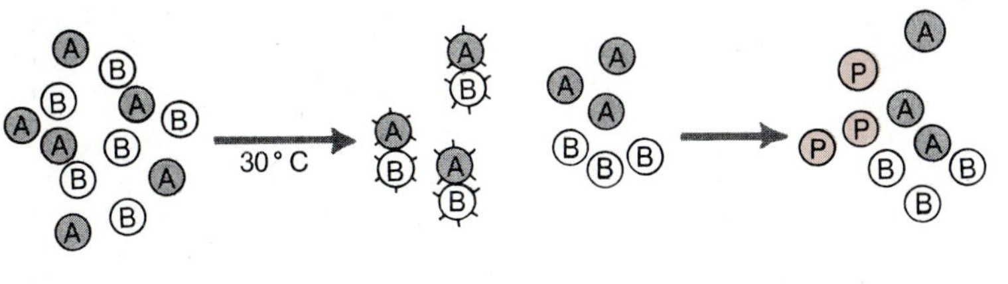

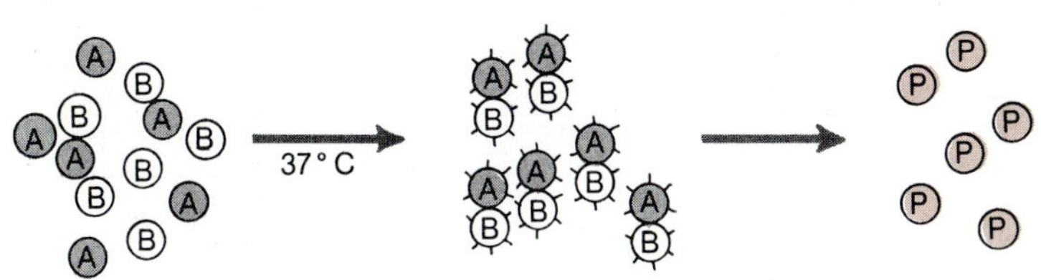

Figure 2-36 *Effect of temperature on chemical reactions. Two reactants,* A *and* B, *collide at temperatures of 30° C and 37° C. Only three collisions per unit time occur at 30° C, but this number doubles at 37° C; therefore product* (P) *also doubles.*

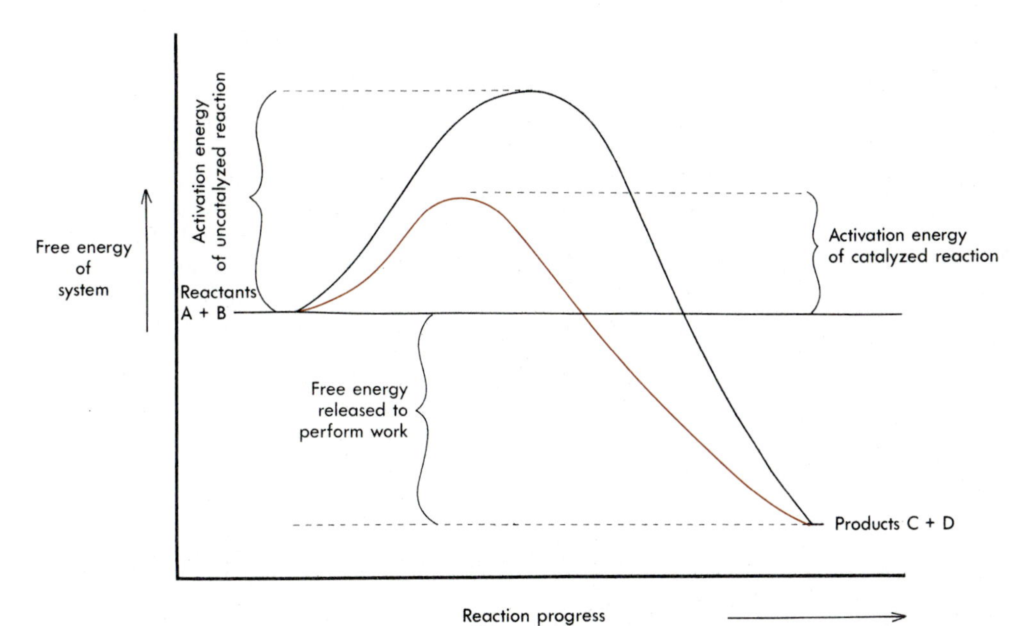

Figure 2-37 *Comparison of activation energy for catalyzed and uncatalyzed reaction: A + B → C + D. Amount of energy required to activate uncatalyzed reaction is greater than for catalyzed reaction. Free energy released from catalyzed and uncatalyzed reactions is same; only amount of energy required to activate them differs.*

call the reactants the **substrates** for the enzyme). Enzymes bring the substrates together in the form of an enzyme-substrate complex from which products will be formed.

Specific mechanisms of enzyme activity

The binding of substrate to the active site is the result of the specific conformation of the polypeptide and the charges of the R groups of some of its amino acids. In addition, some of the metal ions that are required as cofactors also aid in the binding of substrate to enzyme. The enzyme catalyzed reaction can be represented as:

E	+	S ⟶	ES ⟶	E	+	P
Enzyme		Substrate	Enzyme-substrate complex	Enzyme		Product

Most chemical reactions do not go to completion, and in a reaction in which substrates A and B are converted to products C and D, an equilibrium is established; that is, the reaction is reversible. This can be expressed as:

$$A + B \xrightarrow{k_1} C + D \qquad \text{Forward reaction}$$

$$C + D \xrightarrow{k_{-1}} A + B \qquad \text{Reverse reaction}$$

where k_1 and k_{-1} are the rate constants for the forward and reverse reactions, respectively. When the two reactions are combined, we obtain:

$$A + B \underset{k_{-1}}{\overset{k_1}{\rightleftharpoons}} C + D$$

Applying this equation to that of an enzyme-catalyzed reaction, we obtain:

$$E + S \underset{k_{-1}}{\overset{k_1}{\rightleftharpoons}} ES \underset{k_{-2}}{\overset{k_2}{\rightleftharpoons}} P + E$$

Thus at equilibrium the energy released in the forward reaction equals the energy put into the system to reverse the reaction. Enzymes affect the rate at which equilibrium is established, and more free energy in the cell is realized when the forward reaction is favored.

Factors affecting enzyme activity

Enzyme concentration. To determine the effect of enzyme concentration on a reaction, the substrate must be saturating. In other words, the rate or velocity of the reaction is independent of the substrate (a zero order reaction or rate = k), and the product produced per unit of time is related directly to the enzyme concentration. Figure 2-38 illustrates that the formation of product is linear with time, and as the time doubles, so does the

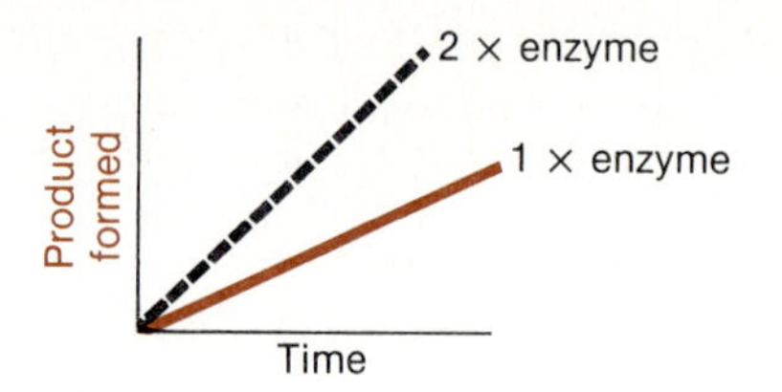

Figure 2-38 *Effect of enzyme concentration on activity when substrate is in excess.*

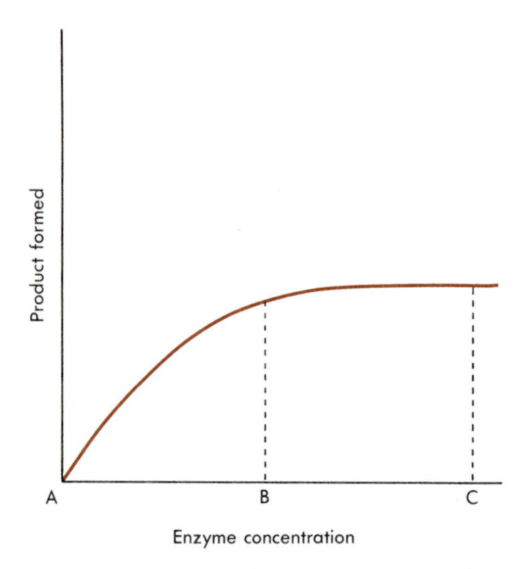

Figure 2-39 *Enzyme activity vs. enzyme concentration. Enzyme activity is proportional to enzyme concentration between points* A *and* B, *indicating excess substrate in reaction. Enzyme activity is not proportional to enzyme concentration between points* B *and* C *because substrate has become limiting. Product formed is expressed as* μM *per unit time.*

concentration of product. In addition when the concentration of enzyme is doubled, the product also doubles. Depletion of the substrate results in a loss of zero order rates, and there is no longer a direct proportionality between enzyme concentration and enzyme activity. Figure 2-39 illustrates that between points *A* and *B* there is proportionality between enzyme concentration and product formed, but this does not occur between points *B* and *C*. What has happened between *B* and *C* is that only some of the active sites of the enzyme are saturated with substrate; consequently, less product is formed. This type of reaction is called a **first-order reaction** because the rate is proportional to the first power of the substrate concentration (rate = k[S]). Enzyme activity is best determined when the enzyme is the only limiting factor.

Substrate concentration. It can be demonstrated that if the enzyme concentration is kept constant and the substrate concentration is increased gradually during a reaction, a point will be reached where the velocity will be maximum. Any further increase in substrate beyond this point has no effect on the velocity of the reaction (Figure 2-40, *A*). Thus at **maximum velocity (V_{max})** all of the enzyme is tied up as an enzyme-substrate complex. Michaelis derived equations to calculate enzyme activity. He proposed the **Michaelis constant**, or **K_m**, which is defined as the substrate concentration at one-half maximum velocity. K_m can be expressed as:

$$K_m = \frac{k_1 + k_2}{k_{-1}} = \frac{[S]\,V_{max}}{2}$$

The K_m is actually a dissociation constant for the enzyme-substrate complex. The faster the ES complex is converted to E and P, the less chance there is for ES to be converted to E + S in the reverse reaction. The reaction velocity for an enzyme-catalyzed reaction, based on K_m and substrate concentration, can be expressed as:

$$V_t = \frac{V_{max}\,[S]}{K_m + [S]}$$

where v_t equals the velocity of the reaction at any time during the reaction.

Michaelis constants have been determined for most of the commonly used enzymes. They are very descriptive and tell the experimenter some important characteristics of the enzyme. A small K_m (10^{-6} to 10^{-7} M) implies that only small amounts of substrate are necessary to saturate the enzyme, and maximum velocities are obtained at low substrate concentrations. A high K_m (10^{-2} to 10^{-3} M) indicates that large amounts of substrate are necessary to obtain maximum velocities.

Temperature and pH. An increase in temperature, within certain limits, increases the rate of enzyme-catalyzed reactions. A 10° C rise in temperature results approximately in a two-fold increase in activity. The optimum temperature for most enzymes is between 40° and 50° C. Above these temperatures the reaction rate decreases because of denaturation (Figure 2-40, *B*). Most enzyme determinations are carried out at temperatures between 25° and 37° C. If carried out at higher temperatures, many reactions proceed so quickly that accurate measurements cannot be made.

Enzymes are also affected by the hydrogen ion concentration, and the optimum pH may vary considerably from one enzyme to another. The enzyme pepsin, for example, has a pH optimum of 1.5, whereas pancreatic lipase is most active at a pH near 8.0. On either side of the optimum pH, enzyme activity decreases (Figure 2-40, *C*). An extremely high or low pH denatures most enzymes.

Inhibitors. Enzyme inhibitors are substances that alter catalytic activity. They may be of two types: competitive or noncompetitive.

Competitive inhibitors. Competitive inhibitors are compounds that are similar to the normal substrate. They bind to the enzyme and prevent access of normal substrate to the enzyme (Figure 2-41, *B*). The inhibitor

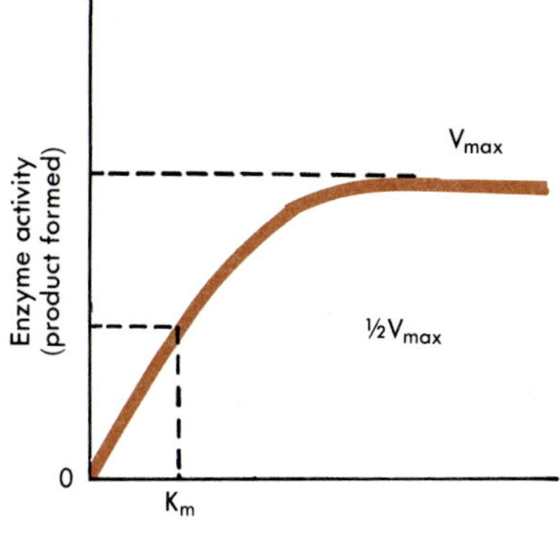

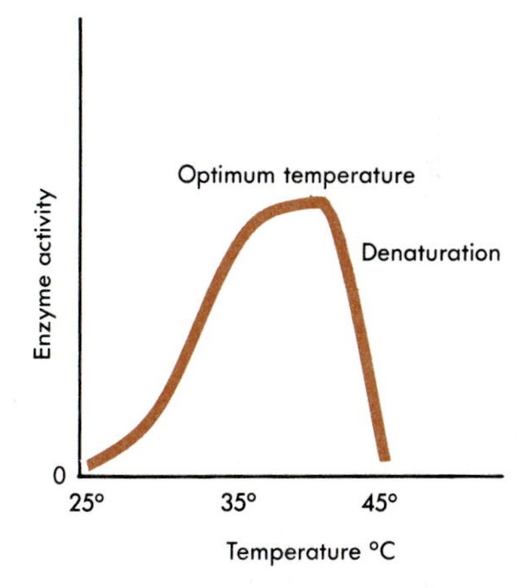

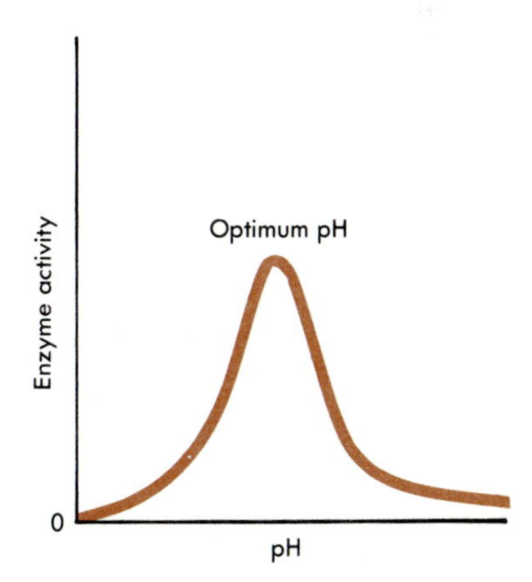

Figure 2-40 *Effect of various factors on velocity of enzyme-catalyzed reaction.*

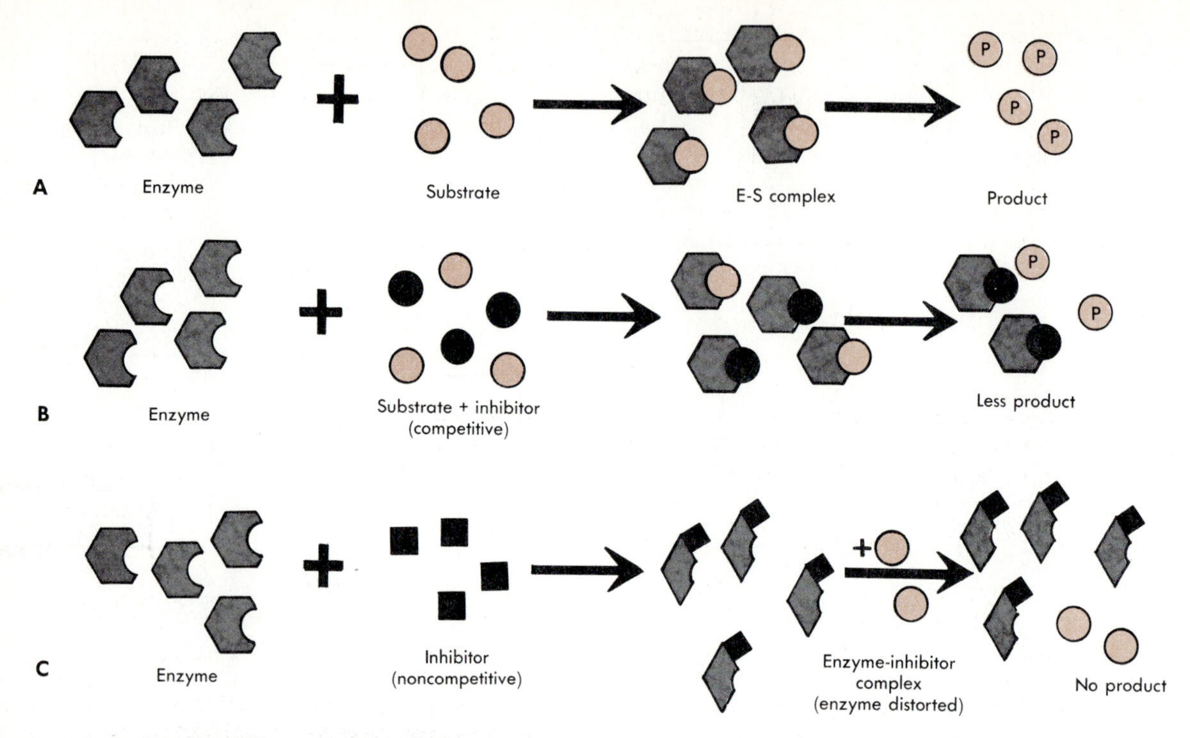

Figure 2-41 *Effect of inhibitors on enzyme activity.* ***A,*** *No inhibitors.* ***B,*** *Competitive inhibitors.* ***C,*** *Noncompetitive inhibitors.*

is not converted to a product. The velocity of the enzyme-catalyzed reaction will be influenced by the concentration of inhibitor, the concentration of substrate, and the relative affinities of the substrate and inhibitor for the enzyme. Enzyme activity may be abolished if the inhibitor is bound very tightly to the active site. It is possible to reverse the effect of the competitive inhibitor by increasing the substrate concentration, if the inhibitor is not bound tightly to the enzyme. The mechanism of competitive inhibition has been used with certain drugs in the treatment of infectious disease. Sulfa drugs, for example, mimic the normal microbial substrate paraaminobenzoic acid, which is required for folic acid synthesis by disease-causing microorganism (see Chapter 15).

Noncompetitive inhibitors. Noncompetitive inhibition is a reversible process in which the inhibitor combines with the enzyme at a point other than the active site (Figure 2-41, *C*). This interaction changes the conformation of the enzyme, and the active site is so distorted that it cannot be recognized by its normal substrate. The velocity of this reaction is influenced by the inhibitor concentration. If enough inhibitor is added so that all of the enzyme is complexed with it, there will be no normal enzyme activity, and addition of substrate will have no effect.

Protein interactions with other molecules

Proteins serve many functions in the cell: structural proteins, enzymes, repressors, immunoglobulins, membrane transport proteins, toxins, and motility proteins. Their activity in nearly all cases is related to their interaction with other molecules. The molecules with which a protein interacts are **ligands.** In most instances a single polypeptide will have a single binding site for a specific ligand. The specificity of binding is determined by the relative binding affinities between closely related ligands. For oligomeric proteins, which may contain from 4 to 32 polypeptides, the ligand-binding sites may be identical, having the same intrinsic binding properties, or there may be different sites with different affinities.

Binding of a ligand to one site can increase or decrease the affinities of other sites, that is, there can be either positive or negative cooperation. When the binding of one ligand molecule affects the binding of further molecules of the same ligand the interactions are **homotropic.** When the binding of one ligand molecule affects the binding of different ligands to other sites the interactions are **heterotropic.** The ligands that bind to proteins may be as small as ions or prosthetic groups, such as NAD, or the ligand may be another similar or different protein molecule.

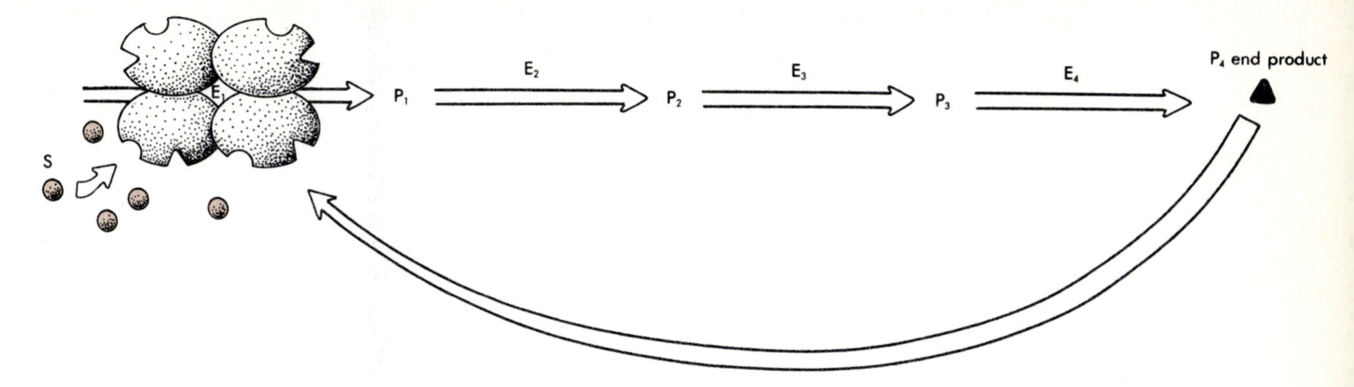

Figure 2-42 *Position of allosteric enzyme in metabolic pathway. Allosteric enzyme consists of subunits with sites for binding substrate (●), as well as regulatory or effector molecules (▲). Allosteric enzyme catalyzes first reaction in pathway, which is converted to product P_1 that is further metabolized to final product P_4. P_4 (▲) acts as effector molecule and can bind to allosteric enzyme to cause conformational change and prevent binding of substrate. The effector therefore shuts down synthesis of P_4, the end product.*

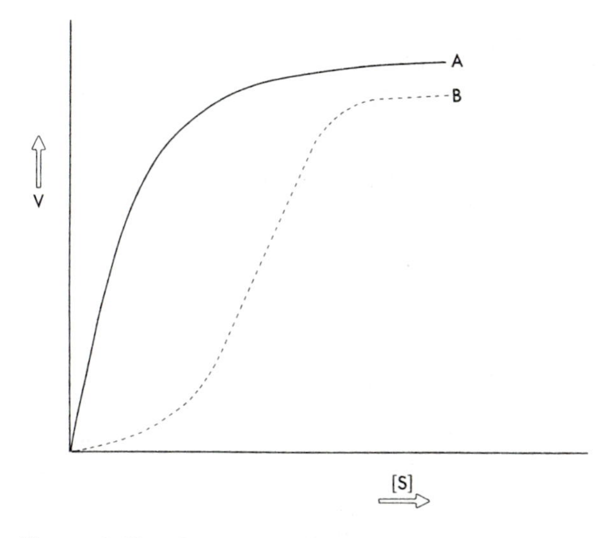

Figure 2-43 *Comparison of kinetics of enzyme reactions catalyzed by nonregulatory* (A) *and regulatory* (B) *enzymes. Note sigmoidal nature of regulatory enzyme at low substrate concentrations indicating the cooperative effects of substrate.*

Concept Check

1. Define enzyme, and explain how enzymes reduce the energy of activation of a chemical reaction.
2. What are the factors that affect enzyme activity and how do they affect it?
3. Explain the relationship of K_m to V_{max} and the value of knowing the K_m.
4. Distinguish between competitive and noncompetitive inhibition.
5. What are allosteric enzymes and how are they beneficial to the cell?

Allosteric proteins. Some oligomeric proteins have regulatory binding sites that are distinct from their catalytic sites. The term **allosteric**, which means another site, has been used to characterize these regulatory proteins. The regulatory sites bind ligands (also called effectors or modifiers) that will enhance (positive effector) the binding of substrate to the catalytic site or reduce (negative effector) substrate binding. Allosteric enzymes can usually be found at the beginning of a long metabolic pathway (Figure 2-42). The end product of the pathway acts as a negative effector. When the end product is not used by the cell, it accumulates and feeds back on the first enzyme in the pathway. The end product binds to the enzyme and alters its conformation, thereby reducing its affinity for substrate. The rate at which the first substrate in the pathway is degraded thus influences the rate of all other reactions in the pathway.

The activity curves of regulatory enzymes are different from those of nonregulatory enzymes. In reactions catalyzed by nonregulatory enzymes the velocity of the reaction increases linearly at low substrate concentrations, but in reactions catalyzed by regulatory enzymes the velocity curve is sigmoidal (Figure 2-43). The reason for the sigmoidal curve is that when one substrate molecule binds to a polypeptide of the oligomeric protein at low substrate concentrations, it changes the conformation of the protein and enhances the binding of other substrate molecules (Figure 2-44). As more substrate becomes available the binding sites are made more accessible to substrate and the curve assumes a linear (at low

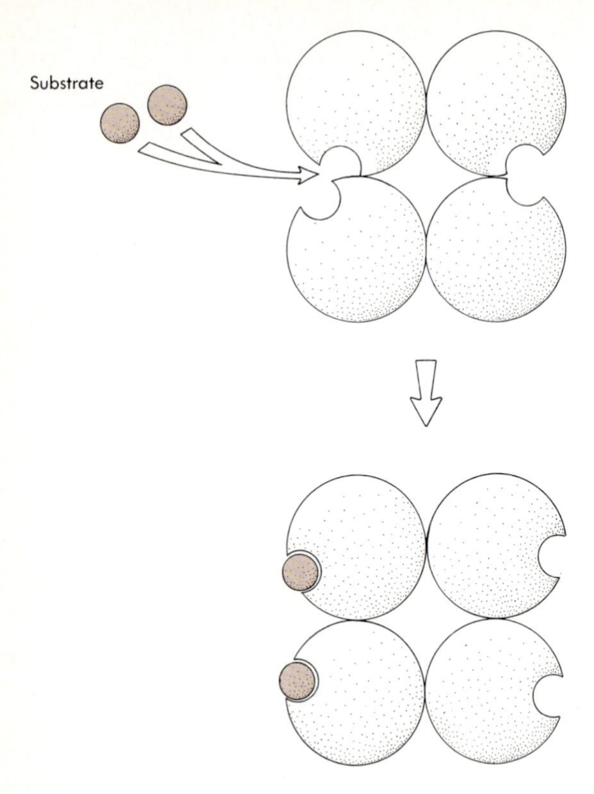

Figure 2-44 *Effect of substrate on regulatory enzyme. Binding of first substrate molecule changes conformation of enzyme so that future substrate molecules are more easily bound.*

substrate concentrations) and then a hyperbolic (at high substrate concentrations) character. The effector molecules are usually low molecular weight compounds but may even include the substrate itself. Effector molecules usually exhibit a negative effect by reducing the activity of an enzyme. The negative effector binds to a subunit and changes the conformation of the enzyme so that substrate-binding sites become inaccessible to substrate. As more negative effector becomes bound the activity of the enzyme is decreased. Enzyme activity can be restored only when excess substrate becomes available. There are some enzymes in which the effector stimulates the activity of the enzyme.

SUMMARY

1. Carbon, nitrogen, phosphorus, oxygen, and hydrogen make up over 90% of the elements in organic matter. Carbon is the backbone of all organic matter, and its four bonds enable it to interact with elements (C, H, N, O, P, and S) to produce a variety of compounds (Table 2-7).
2. Nitrogen, sulfur, and phosphorus are also found in specific organic molecules. Nitrogen is a component of amino acids, proteins, and nucleic acids. Sulfur is primarily a component of the amino acids cysteine and methionine and plays a role in enzyme activity. Phosphorus is associated with nucleic acids and the energy molecule, ATP.
3. The major biological molecules are proteins, complex carbohydrates, nucleic acids, and lipids. Proteins function primarily as catalysts (enzymes) but are also involved in motility and regulation and serve structural roles as well. Complex carbohydrates may be structural but many are also sources of energy. Nucleic acids, such as DNA and RNA, are primarily informational molecules. Lipids are important in the structure of biological membranes.
4. As catalysts, enzymes are affected by many factors: substrate concentration, cofactors and coenzymes, pH, temperature, and inhibitors. Inhibitors may be competitive by mimicking the substrate or they may be noncompetitive by binding to sites other than the catalytic site.
5. Proteins can interact with other molecules both inorganic and organic that will affect activities, such as motility, regulation, transport, and catalysis. Some oligomeric proteins have regulatory binding sites that are distinct from the active site. They are allosteric proteins. Allosteric proteins are characteristically found as controlling agents in the first reaction of a metabolic sequence where several enzymes are involved.

Table 2-7 *Bonding characteristics of the major elements in organic compounds*

Element	Element covalently bonded to	Structure of bond	Type or class of compound
Carbon	Carbon	—C—C— (single bond); —C=C—; —C≡C—	Linear hydrocarbon
	Carbon	six-membered carbon ring (C—C—C—C—C—C)	Cyclic hydrocarbon
	Carbon plus nitrogen or sulfur	six-membered ring of four C and two N (C=N, C=C, C=N double bonds)	Heterocyclic hydrocarbon
Oxygen	Carbon and hydrogen	—C—OH	Alcohol
		H—C=O	Aldehyde
		—C—C(=O)—C—	Ketone
		R_2—C(=O)—O—R_1	Ester
		—C(=O)—OH	Carboxylic acid
		R—C(=O)—O—C(=O)—R	Anhydride
Nitrogen	Carbon and hydrogen	N—C—; N=C—; N≡C—; —C—NH_2	Amines (C—NH_2) important in amino acid formation, i.e.: R—CH(NH_2)—COOH
Sulfur	Carbon and hydrogen	—C—S—H	Sulfhydryl (SH) group found in compounds called thiols
	Sulfur	S—S	Disulfide bond formed between sulfur-containing amino acids
Phosphorus	Oxygen	—P(=O)—O—P(=O)—	Anhydride

Self-study Quiz

Multiple choice

1. Phosphorus would not be found in which of the following?
   A. Nucleotides
   B. Phospholipids
   C. Amino acids
   D. RNA
2. Which of the following types of RNA associate with proteins in the cell to provide the site for protein synthesis?
   A. Viral RNA
   B. rRNA
   C. tRNA
   D. mRNA
3. Oxidation of free sulfyhydryl groups between sulfur-containing amino acids in polypeptide strands has what effect?
   A. Formation of disulfide bridges, which contributes to chain folding and protein conformation
   B. Release of free sulfur groups, which poison the cell
   C. Formation of sulfuric acid, which changes the pH of the enzyme
   D. Breakage of the peptide bond and the release of free amino acids
4. Which of the following would be found in RNA but not DNA?
   A. Adenine
   B. Guanine
   C. Uracil
   D. Thymine
5. Which of the following statements about enzyme-catalyzed reactions is not true?
   A. The fact that the enzyme is broken down during the reaction supplies the energy to drive an enzyme-catalyzed reaction.
   B. Enzymes reduce the activating energy required to promote molecular collision.
   C. Enzymes increase the rate at which a chemical reaction proceeds by acting as a site for the interaction of the substrates.
   D. Enzymes bring the substrates together in the form of an enzyme-substrate complex.
6. An enzyme that catalyzes the joining of nucleotides to a DNA molecule belongs to which of the following classes?
   A. Lyase
   B. Isomerase
   C. Ligase
   D. Transferase
   E. None of the above

Completion

1. The substrate concentration at one-half V_{max} is called the ________________.
2. Increasing the substrate concentration is a means of reversing the effect of ________________ inhibitors on enzyme activity.
3. Enzymes or other proteins that possess regulatory sites distinct from the active site for binding effector molecules are called ________________ proteins.
4. The term ________________ is used to describe a process in which heat causes a change in the conformation of a polymer such as protein.
5. The only symmetric amino acid is ________________.

Matching

Match the functional group structure in the right-hand column with its name from the list in the left-hand column.

1. Hydroxyl	a. $R{-}C({=}O){-}NH_2$
2. Ketone	b. $R{-}C({=}O){-}O{-}R$
3. Aldehyde	c. $R{-}O{-}R$
4. Ether	d. $R{-}C({=}O){-}C$
5. Carboxylic acid	e. $R{-}C({=}O){-}H$
6. Carbonyl	f. $R{-}C({=}O){-}OH$
7. Amino	g. $-OH$
8. Ester	h. $-C{=}O$
9. Sulfhydryl	i. $R{-}O{-}P({=}O)(OH){-}O{-}R$
10. Phosphate ester	j. $R{-}SH$

Thought questions

1. Many bacteria produce enzymes that are capable of hydrolizing proteins in their environment to provide amino acid nutrients for the growth of the bacterial cell. These hydrolytic enzymes, called proteases, are synthesized as inactive proenzymes and are activated to active protein-digesting enzymes only on release from the bacterial cell. This avoids what potential problem for the cell? Part of the structure of the bacterial cell wall is composed of short peptide sequences; however, some of the amino acids in this recurring peptide structure are D-amino acids. What advantage is that?
2. Detergents and quaternary ammonium compounds act to disrupt the orientation of hydrophobic/hydrophilic molecules. What parts of the bacterial cell would be most susceptible to disruption by these agents?
3. Certain poisons are toxic because they are competitive inhibitors of an essential cellular enzyme. Other poisons are noncompetitive inhibitors. It should be easier to reverse the effects of which type of poisoning? How would you do that?
4. Explain the chemical basis of the alteration of egg white (a protein known as ovalbumin) from viscous liquid to opaque solid on heating.
5. The structure and activity of functional proteins is determined by the sequence and nature of the constituent amino acids. The cell sometimes makes mistakes in the insertion of the correct amino acid at a given point in the sequence. Predict how the following alterations might affect the function of an enzyme:
   a. Substitution of an uncharged amino acid for a charged amino acid involved in the active site.
   b. Insertion of many more proline molecules into the chain.
   c. Substitution of hydrophobic amino acids for hydrophilic amino acids.

SELECTED READINGS

Books

Christensen, H.N., and Palmer, G.A. *Enzyme kinetics: a learning program for students of biological and medical sciences.* W.B. Saunders Co., Philadelphia, 1967.

Cohen, A. *Handbook of cellular chemistry* (2nd ed.). The C.V. Mosby Co., St. Louis, 1979.

Conn, E.E., and Stumpf, P.K. *Outline of biochemistry* (5th ed.). John Wiley & Sons, Inc., New York, 1987.

Davies, J., and Littlewood, B.S. *Elementary biochemistry: an introduction to the chemistry of living cells.* Prentice-Hall, Inc., Englewood Cliffs, N.J., 1979.

Debey, H.J. *Introduction to the chemistry of life: biochemistry* (2nd ed.). Addison-Wesley Publishing Co., Inc., Reading, Mass., 1976.

Dickerson, R.E., and Geis, I. *The structure and action of proteins.* Harper & Row, Publishers, Inc., New York, 1969.

Lehninger, A.L. *Principles of biochemistry.* Worth, New York, 1982.

Speakman, J.C., *Molecules,* McGraw-Hill Book Co., New York, 1966.

Stryer, L., *Biochemistry* (2nd ed.). W.H. Freeman & Co., Publishers, San Francisco, 1981.

Suttie, J.W. *Introduction to biochemistry* (2nd ed.). Holt, Rinehart & Winston, New York, 1977.

Wold, F. *Macromolecules: structure and function.* Prentice-Hall, Inc., Englewood Cliffs, N.J., 1971.

Yudkin, M., and Offord, R. *Comprehensible biochemistry.* Longman Group Ltd., London, 1973.

Journal articles

Dickerson, R.E. The structure and history of an ancient protein. *Sci. Am.* **226**(4):58, 1972.

Doolittle, R.F. Proteins. *Sci. Am.* **253**(4):88, 1985.

Felsenfeld, G. DNA. *Sci. Am.* **253**(4):58, 1985.

Frieden, E. The chemical elements of life. *Sci. Am.* **227**(1):52, 1972.

Rich, A., Nordheim, A., and Wang, H.J. The chemistry and biology of Z-DNA. *Annu. Rev. Biochem.* 53:791, 1984.

Sharon, N. Carbohydrates. *Sci. Am.* **243**(1):90, 1980.

CHAPTER 3

Introduction to microscopic technique

Chapter Outline

- Brightfield Microscopy
  - Magnification
  - Resolving Power
  - Contrast (Staining)
- Darkfield Microscopy
- Phase-contrast Microscopy
- Fluorescent Microscopy
- Electron Microscopy

Introduction

It is difficult to study any living entity if you are unable to see it. Microscopes enable us to see microorganisms, and it should not be a surprise to learn that the initial advances in microbiology as a science paralleled the development of microscopy. Construction of simple microscopes in the nineteenth century, accompanied by staining techniques, allowed scientists to identify external microbial characteristics. Such identification potential permitted differentiation of many microbial species and led to the correlation of microorganisms with specific diseases. Advances in microscopy led to the development of more sophisticated instruments, such as the electron microscope. The electron microscope enabled the microbiologist to examine the structural detail of microorganisms. When electron microscopy was supplemented with specific biochemical testing procedures, the microbiologist gained insight into the relationship between microbial structure and function.

One purpose of this chapter is to describe the functional properties of the microscope, that is, how to magnify the object, how to achieve maximum resolution, and how to provide sufficient contrast. This chapter then describes how microscopy, when supplemented with certain staining techniques, can be used to obtain information about the microbial cell.

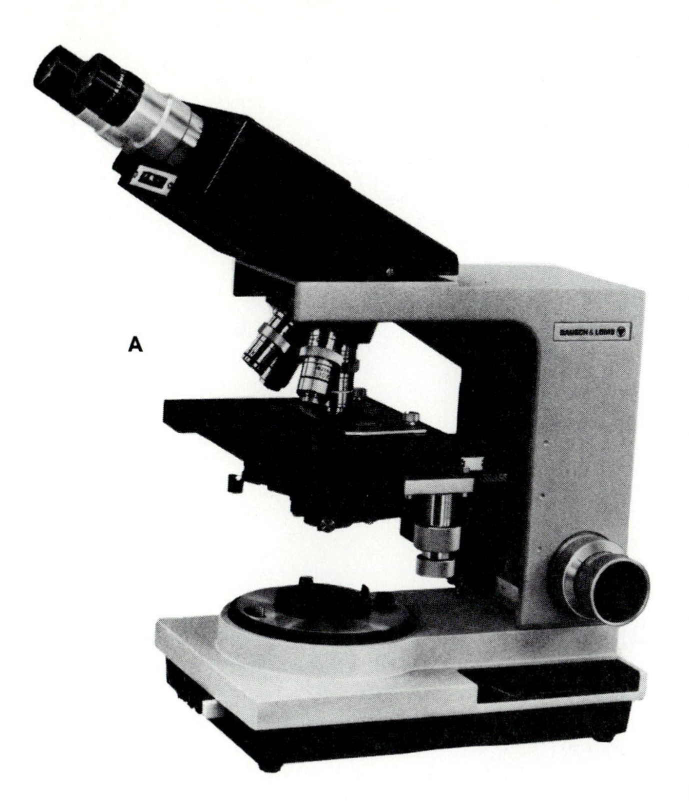

BRIGHTFIELD MICROSCOPY

The basic microscopic system used in the microbiological laboratory is the **compound brightfield microscope** (Figure 3-1). This system is called *compound* because it possesses at least two lens systems and *brightfield* because visible light is passed directly through the lenses until it reaches the observer's eye; thus the specimen appears in a bright background. Our discussion begins with an explanation of the optical properties associated with brightfield microscopy.

Magnification

Magnification of an image is obtained in the compound microscope by two lens systems: the **objective lens**, which is closest to the specimen to be examined, and the **ocular (eyepiece) lens.** The objective lens magnifies the specimen and produces a **real image** (Figure 3-1, *B*). The real image is projected through the microscope to the ocular lens, which magnifies the real image and produces an image on the observer's retinal membrane. This image is then processed and the brain creates an imagined image called the **virtual image.** This image appears to be at the level of the specimen. Most compound microscopes are equipped with three objective lenses attached to a nosepiece: the low-power lens or

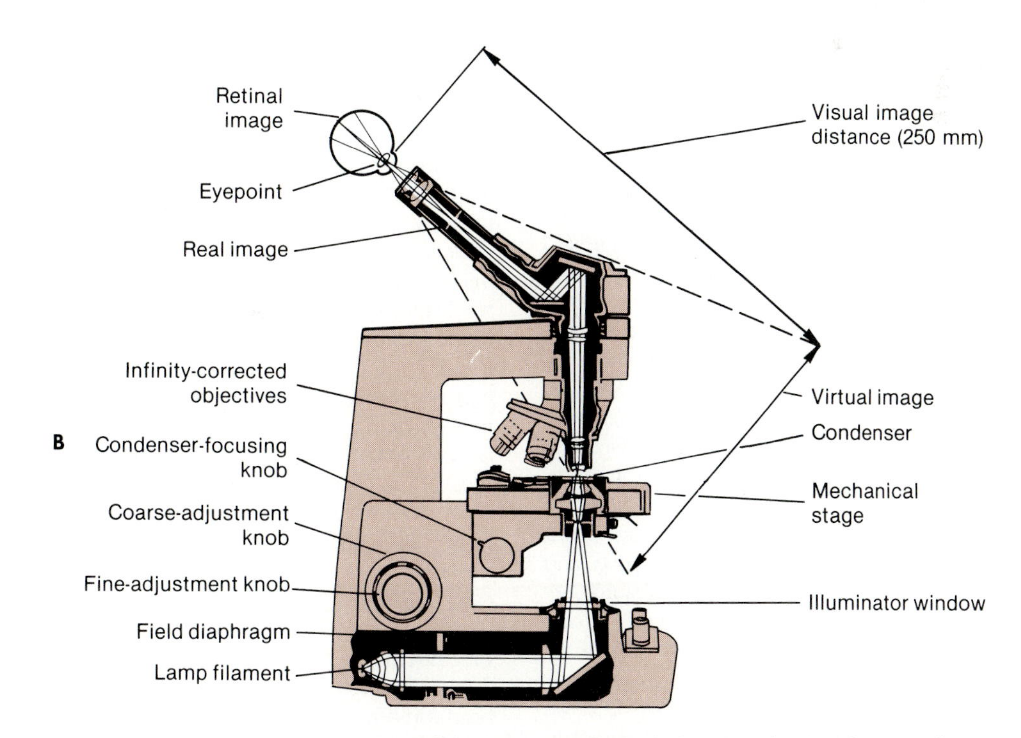

Figure 3-1 ***A,*** *Compound brightfield microscope.* ***B,*** *Optical and mechanical features of compound microscope.*

10× magnification, the high-power lens or 45× magnification, and the oil immersion lens or 100× magnification. The total magnification for the lens system is obtained by multiplying the magnification of the objective lens and the magnification of the ocular lens, which is usually 10× magnification. Thus the maximum magnification obtained by using the oil immersion lens is approximately 1000×.

Resolving power

One might think that by increasing the magnification power of the objective lens, a magnification greater than 1000× could be achieved with maximum clarity. Such is not the case, since all lenses have imperfections. The objective lens magnifies the real image, but no lens is capable of producing a perfect image, and lens errors (aberrations) are an inherent part of the optical system. Magnification is of little value unless the enlarged image possesses increased detail and thus clarity. Image clarity is associated with the term **resolving power**, which is the ability of the lens to distinguish two adjacent objects at a particular distance apart. Resolving power is best explained by assuming that you are microscopically observing two objects. How close can you bring the two objects together and still see them as separate entities? When they are brought too close together, they will be observed as a single object. Resolving power (RP) is dependent on the wavelength (λ) of light used in the optical system and the numerical aperture (NA) according to the following formula:

$$RP = \frac{\lambda}{2 \times NA}$$

The **numerical aperture** is a function of the angle of the maximum cone of light that can enter the objective and the refractive index* of the medium between the specimen and the lens. The medium may be air or oil (Figure 3-2). The greater the numerical aperture, the greater the resolving power. Manipulation of the condenser and iris diaphragm control the amount of light reaching the specimen. The condenser must be at its highest position to ensure that all the light rays focus on the specimen. A reduction in the amount of light reaching the specimen reduces the effective numerical aperture and consequently reduces the resolution. Light from the specimen is collected by the objective and transmitted to the ocular lens. Since there is a space between the specimen and the objective lens, light passing through this space at an angle is lost as a result of refraction because the refractive indices of glass and air are different. When immersion oil is placed between the specimen and the objective lens, more light is gathered by the objective lens because the refractive indices of glass and oil are comparable and light is not bent away from the objective lens. Thus the maximum numerical aperture can be achieved.

* *Refractive index* is related to the change in direction of light as it passes at an angle from air into a more dense medium such as a liquid.

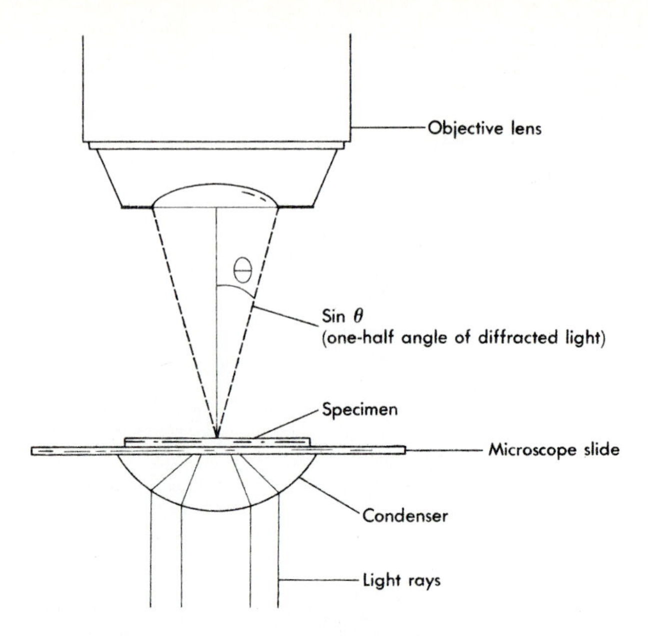

***Figure 3-2** Numerical aperture determination of objective lens. NA = n sin θ, where n is refractive index of medium (can be oil or air) and sin θ is one-half angle created by light passing through condenser and specimen and transmitted to objective.*

Compound microscopes employ a tungsten filament lamp, which generates wavelengths in the visible spectrum. The shorter the wavelength, the greater the resolving power according to the preceding formula. A blue filter is inserted between the light source and the slide, because the most energetic wavelengths in the visible spectrum are in the blue-violet range (Figure 3-3). This filter permits the transmission of blue-violet wavelengths and the absorption of the longer (yellow, green, and red) wavelengths. The resolving power of the compound microscope is about 0.2 μm when oil immersion is used and the numerical aperture is maximum. This means that two objects can be distinguished as two separate entities when they are separated by 0.2 μm or more but not if by less than 0.2 μm.

Contrast (staining)

Brightfield microscopy can be used to observe both stained and unstained bacteria. One of the advantages of brightfield microscopy is that unstained microorganisms can be observed in a living condition. However, unstained bacteria are transparent and it is difficult to observe them unless there is a reduction in the amount of

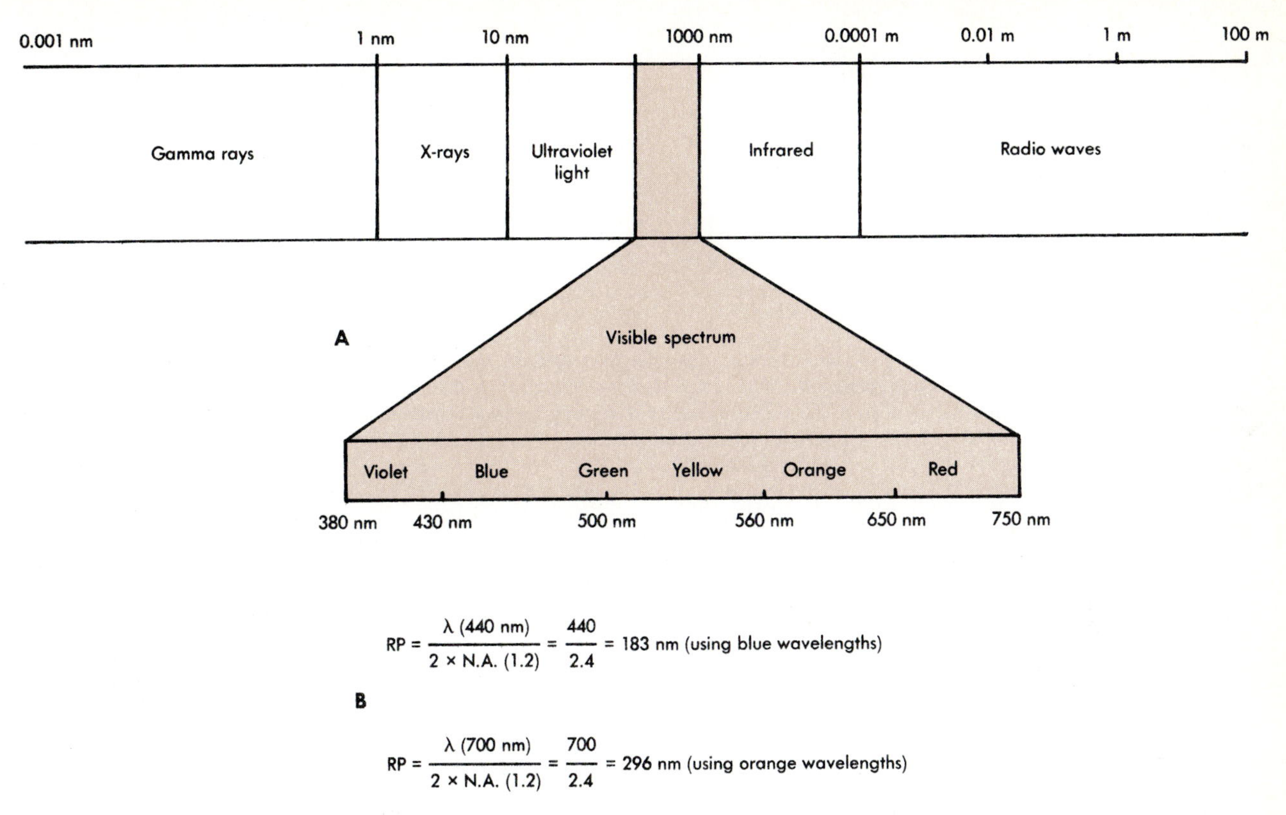

$$RP = \frac{\lambda\ (440\ nm)}{2 \times N.A.\ (1.2)} = \frac{440}{2.4} = 183\ nm\ \text{(using blue wavelengths)}$$

$$RP = \frac{\lambda\ (700\ nm)}{2 \times N.A.\ (1.2)} = \frac{700}{2.4} = 296\ nm\ \text{(using orange wavelengths)}$$

Figure 3-3 *Effects of different wavelengths of light on resolving power.* ***A,*** *Spectrum of ultraviolet, visible, and infrared regions.* ***B,*** *Resolving power at two different wavelengths.*

light used to illuminate the specimen. Compound microscopes have an iris diaphragm, located between the condenser and the light source, which can be adjusted to reduce illumination. Even this procedure does not always produce sufficient contrast between the background and the microbial cell. Staining the microbial cell can make it visible even at magnifications lower than 1000×, but only at approximately 1000× is there adequate resolution. Microbiological stains are called **dyes**. A dye is an organic compound containing a **chromophore** group, which imparts color to the compound, and an **auxochrome**, which is a dissociable group (one that gives up or takes a proton to possess a negative [−] or positive [+] electrical charge) that binds to the substrate and intensifies the color. The bacterial cell contains many negatively charged groups (carboxyl, for example), and only basic dyes, which contain positively charged ionizable groups, will bind to them (Figure 3-4). The more commonly used basic dyes are methylene blue, crystal violet, safranin, and malachite green. These stains may be used individually to stain the cell, and the staining procedure is called **simple staining**.

More complex staining procedures can be used to stain individual components of the cell, such as flagella and endospores, but these procedures are not routinely used. **Differential staining** procedures use more than one stain to distinguish differences in the chemical composition of the bacterial cell. Two of the most important differential staining techniques are the **Gram stain** and the **acid-fast stain.**

The Gram stain is the single most important procedure in the identification of most bacteria in the clinical laboratory. Gram staining divides the bacterial world into groups that are either **gram-positive** or **gram-negative.** A smear of the organisms is prepared on a glass slide in the gram-staining procedure. The smear is stained with a primary stain called crystal violet. The excess stain is washed off with water and a solution of Gram's iodine is applied to the smear. The iodine acts as a mordant and fixes the crystal violet to the cell wall. The smear is then destained by adding a few drops of an alcohol-acetone solution. Gram-positive bacteria can resist decolorization, but gram-negative bacteria are destained, and crystal violet is removed from the cell. A secondary stain, safranin (red), is applied to the smear. Gram-positive cells that have already been stained with crystal violet appear blue or violet (Plate 3). If the organ-

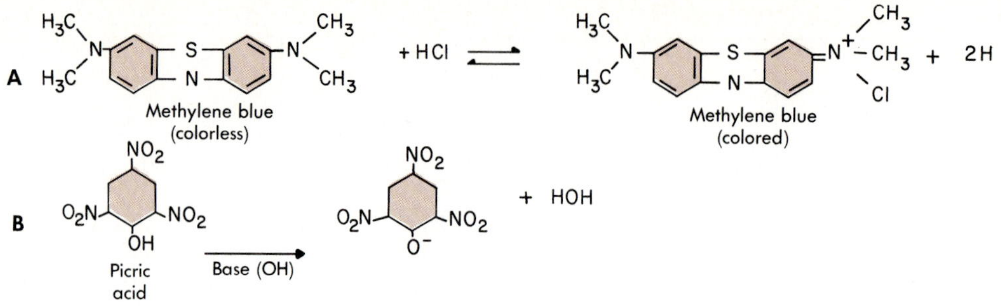

A H_3C, H_3C N S N CH_3 N CH_3 + HCl ⇌ H_3C N H_3C S N N^+ CH_3 CH_3 Cl + 2H

Methylene blue (colorless) — Methylene blue (colored)

B NO_2 O_2N NO_2 OH Picric acid — Base (OH) → NO_2 O_2N NO_2 O^- + HOH

Figure 3-4 *Structure of selected acidic and basic dyes.* ***A,*** *Basic dye contains dissociable group called auxochrome that forms salts with acids. Methylene blue salt is blue in solution and nitrogen of auxochrome is pentavalent.* ***B,*** *Acid dye ionizes to impart negative charge to molecule. Picric acid is example of acidic dye because of dissociable hydroxyl (OH) group.*

isms are gram-negative, they take up the secondary stain and appear red (Plate 1).

The acid-fast stain, like the Gram stain, is a differential stain, but its use is more limited. The acid-fast stain is used to distinguish species of *Mycobacterium* (agents of tuberculosis and leprosy), which are acid fast. The primary stain is carbol-fuchsin (red). After the cells have been stained, a decolorizing solution made up of acid and alcohol is added to the smear. The secondary stain, methylene blue, is applied after the colorization step. Mycobacterial species contain as much as three times the amount of lipid as gram-negative species, and, unlike gram-negative bacteria, they can resist decolorization. Acid-fast organisms are not stained by the secondary stain but remain red or pink, whereas non-acid-fast cells take up the secondary stain and appear blue (Plate 2).

Sometimes it is not necessary to stain the bacterial cell to visualize it. There is a technique called **negative staining** in which the background is filled with particles of dye that do not stain the bacterial cell. This procedure makes the background appear dark, while the bacterial cells remain clear and transparent. India ink or nigrosin dyes are frequently used to "stain" the background around cells that produce a polysaccharide or polypeptide substance called a capsule (Figure 3-5). Nigrosin, for example, is an acidic dye (negative charge) that does not bind to the capsule or the bacterial cell, which are also negatively charged. One of the advantages of "negative staining" is that living microbial cells can be observed by brightfield microscopy. Dyes that stain the cell usually kill it.

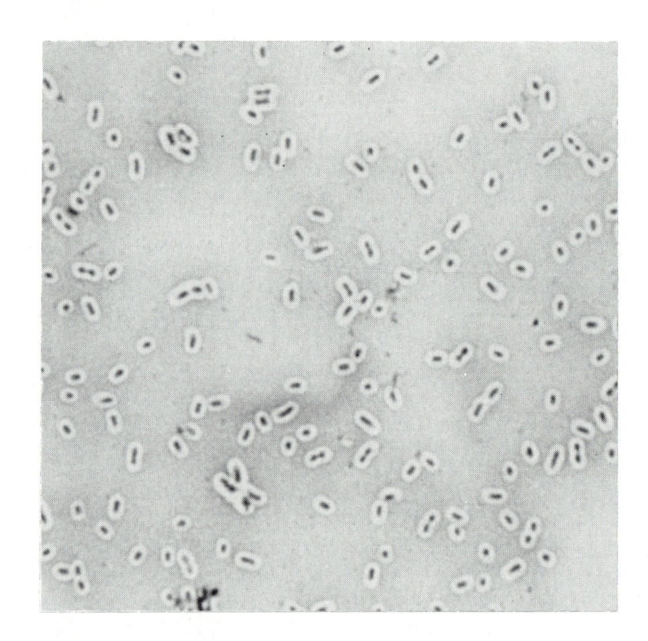

Figure 3-5 *Negative staining (×500) using congo red to "stain" background and observe capsule. Bacterial cells (*Pasteurella *species) are rod-shaped and dark and surrounded by white halo (capsule).*

DARKFIELD MICROSCOPY

Some bacteria, such as certain spirochetes, have a diameter (0.1 to 0.15 μm) that is less than the resolving power of the brightfield microscope and are difficult to observe. Thin transparent objects, like bacteria, can be more easily observed when the background is dark (Plate 3). The compound microscope can be fitted with a special condenser called a darkfield condenser, which has a numerical aperture greater than the objective and blocks all direct light from reaching the objective. The condenser allows only angled light to strike the specimen. This light is not seen unless the specimen diffracts it into the objective lens. The cells in the specimen appear light against a dark background because of light scattering. There is no increase in resolution, since the wavelengths and numerical aperture used are the same as those in brightfield microscopy.

widely used fluorochromes for immunofluore
fluorescein isothiocyanate and lissamine r
Fluorescein emits a green color (517 nm), wh
damine emits a yellow color (580 nm). Fo
when an animal is injected with an antigen, s
rabies virus, antibodies are produced that a
for that virus. The virus-specific antibodies ca
ered from the animal and conjugated with
chrome. Later, it may be necessary to deter
animal has been infected by the rabies virus.
gator removes a specimen of nerve tissue fro
animal, or animal suspected of being rabid,
fluorochrome-coated rabies antibodies. La
bodies will bind and coat the virus if the
does contain the rabies virus, and when ob

Concept Check

1. Describe some characteristics that distinguish types of microscopes: darkfield, phase-contrast, and brightfield.
2. How can a microorganism be identified in a bo fluorescent microscopy?

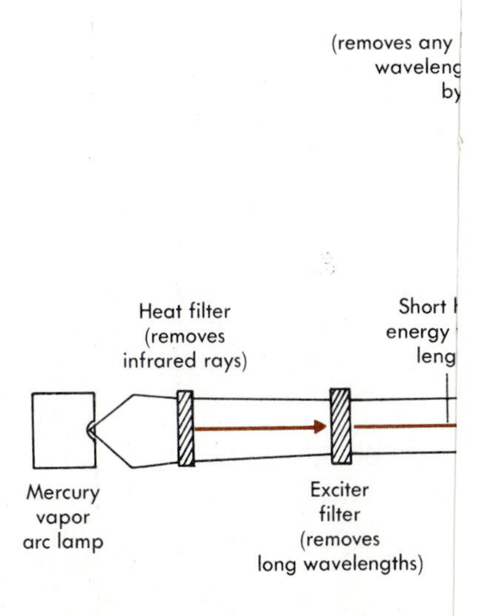

PHASE-CONTRAST MICROSCOPY

As we have already noted, the transparency of whole cells makes them difficult to distinguish from the background. The same difficulty is encountered in trying to differentiate organelles or other dense components within a single cell. The phase-contrast microscope consists of special condensers and objectives that enable one to increase the contrast between the transparent components in the cell by exploiting differences in their densities. When light strikes the transparent cell, some of the light rays, called **direct rays**, pass through and are unchanged in terms of amplitude and phase. Some components in the cell have different densities and the light striking them will emerge with their wavelengths retarded **(indirect rays)** by one fourth. The amplitude of direct and indirect rays when brought into phase is the sum of the two waves, and this gives the specimen the appearance of being bright against a dark background (Figure 3-6). If the direct and indirect rays are made to be out of phase (one-half wavelength off), their amplitudes cancel each other out and the object appears dark against a bright background. Phase plates are present in the objective lens that will either retard (positive-phase plate) or increase (negative-phase plate) the phase of the diffracted light relative to the direct light. The technique is called bright-phase microscopy when the image appears brighter than the background and darkphase microscopy when the image is darker than the background. An example of the optical system of the phase-contrast microscope can be seen in Figure 3-7.

Concept Check

1. Explain resolving power and the factors that influence it.
2. Describe the path that light follows in a compound microscope from when it strikes the object to be observed, until it reaches the eye of the observer.
3. Distinguish between simple and differential stains and give examples of each.
4. Distinguish between basic and acidic dyes and explain why only certain types can be used to stain microorganisms, such as bacteria.

FLUORESCENT MICROSCOPY

Fluorescent microscopy is characterized by the self-luminous property of the treated specimen. Certain molecules when struck by light waves emit light of a longer wavelength. The electromagnetic wave striking the molecule imparts energy and causes the electrons of the molecule to oscillate. The energy in these oscillations is released by the molecule in the form of heat or light

Point of Interest

HOW DOES THE GRAM STAIN WORK?

When Hans Christian Gram developed the gram stain (p. 55) in 1884, it was to reveal bacteria in diseased tissues. It soon became apparent that this was a valuable stain to distinguish between different types of bacteria called gram positive and gram negative. It was a long time before the actual difference between the gram-positive and gram-negative bacteria was shown to be a difference in cell wall composition. However, now that the composition of the cells walls of gram-positive and negative bacteria is known, a hypothesis to explain how the stain works has been developed. Gram-positive bacteria have a cell wall composed of a thick layer of peptidoglycan (p. 104) whereas gram-negative bacteria have a thin layer of peptidoglycan and an additional outer membrane layer. In both cases the crystal violet forms an insoluble complex with the iodine (CV-I) inside the cells. When treated with alcohol, the plasma membrane is dissolved but the thick peptidoglycan layer of the gram-positive cell wall is dehydrated causing the pores of the peptidoglycan layer to close, thus preventing the CV-I complex from leaving the cell. Thus these cells retain the purple color. In the case of the gram-negative bacteria, the alcohol dissolves the plasma and outer membrane layer and the thin peptidoglycan layer is too thin to retain the CV-I complex so the purple color is washed away during alcohol decolorization. These gram negative cells then pick up the red safarin counterstain.

Thus the gram staining property is not due to a direct chemical reaction but is due to the physical structure of the cell wall. Nonbacterial cells such as yeast that have a thick cell wall of totally different composition also stain gram-positive. Also while gram-negative bacteria always stain gram negative, the ability to stain gram positive can change during the growth cycle. Older cells tend to lose the ability to stain gram positive and thus the results of the stain are most accurate if the stain is performed on young actively growing cultures.

This stain continues to be valuable today for this easy stain tells a lot about the cell wall structure of the bacteria which is related to other important differences such as susceptibility to various antibiotics and chemicals.

and can be used to initiate chemical reaction
energized molecule releases its energy in the
light, whose wavelength is greater than the
wave, the compound is called a **fluorochrome.**
chromes, therefore, exhibit the property of
cence. Some of the important fluorochromes a
mine O, acridine orange, fluorescein, and rhc
Microbiological material can be examined by c
with fluorochromes and using the fluorescen
scope, which differs considerably from the br
microscope.

The fluorescent microscope uses a mercu
lamp instead of the incandescent lamp usec
brightfield microscope. The mercury lamp p
more energetic waves (ultraviolet, violet, blue
are needed to impart a sufficient amount of e
the fluorochromes. Two light filters are placed
the mercury lamp and the fluorochrome-coat
men (Figure 3-8). They serve to remove th
wavelengths generated by the mercury lamp a
mit the shorter wavelengths. A darkfield con
used to produce a dark background, which pro
best contrast. The condenser also deflects mo
shorter wavelengths, such as ultraviolet, which

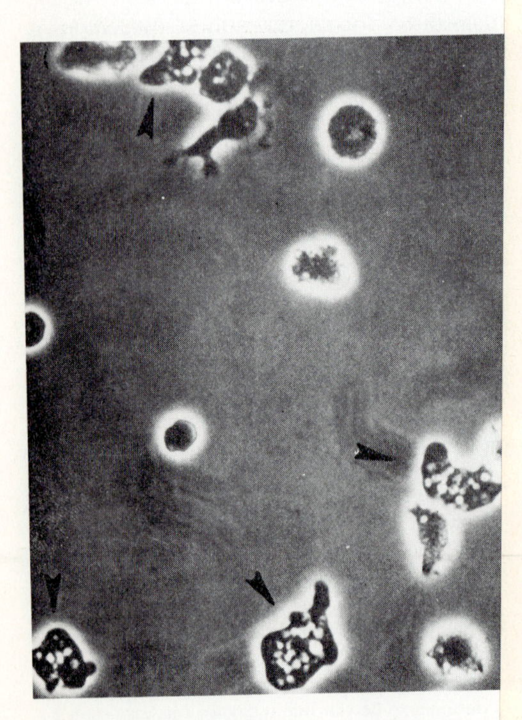

Figure 3-6 *Phase-contrast microscopy. Phase co smear of cerebrospinal fluid of patient (×2500). A cellular material appears to have halo. Arrow poin parasitic amebas, which contain many vacuoles a distinguished from more spherical red blood cells.*

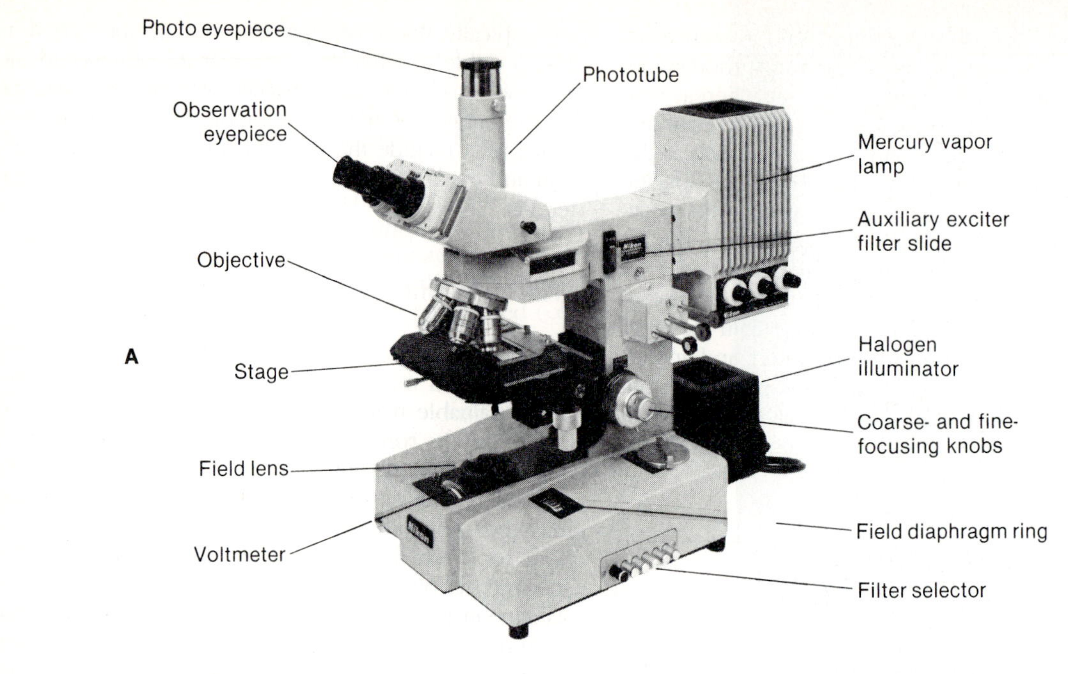

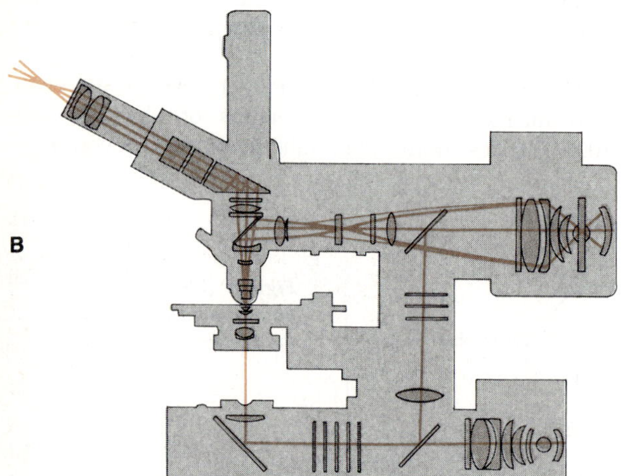

Figure 3-9 *Mechanical and optical features of fluorescent microscope.* ***A,*** *Mechanical features of Nikon fluorescent microscope.* ***B,*** *Optical properties of Nikon fluorescent microscope.*

already learned, resolving power is related to the wavelength. A tungsten filament is placed in a vacuum and heated to emit electrons that are accelerated and passed through an evacuated cylinder until they reach a condenser system. The condenser system consists of **electromagnetic coils** and is analogous to the lens system of the brightfield microscope. The first condenser system directs the electrons to the specimen. Below the specimen is a second magnetic coil, which acts as an objective and collects the electrons that have passed through and around the specimen. The electrons are passed to a third magnetic coil, which enlarges the image produced by the passage of electrons through the specimen and focuses the image on a **fluorescent screen** that can be flipped up to take a photograph. A final coil system acts as an eyepiece. The image can be recorded either on a plate or on a 35 mm roll of film. Direct screen magnifications of 1,000,000× are possible. Most large electron microscopes go up to magnifications of 500,000×, whereas most small ones go up to magnifications of 100,000×. Figure 3-12 compares the lens system of an electron microscope with the brightfield microscope.

There are two types of electron microscopes: **transmission electron microscopes** (TEMs) and **scanning electron microscopes** (SEMs). The electrons generated in the tungsten filament in TEMs pass through the specimen and onto a photographic plate. Using the highest accelerated voltages, specimens up to 2 μm in thickness

dehydrated in alcohol or acetone, and finally impregnated with polymers such as epoxy resins. The resin-embedded specimen is then cut into thin sections. The specimen can be prestained with a heavy metal such as uranyl acetate, lead citrate, or lead hydroxide, to increase contrast.

The last technique for specimen preparation is called **replication** and is used when the object cannot be brought directly into the electron microscope. A replica can be produced in place of the original specimen. Replication electron microscopy uses a technique called **freeze-fracturing**, which is important in the investigation of biological structures, particularly membrane structure. The specimen is first frozen to immobilize the molecules. The frozen specimen is placed in a vacuum and a cooled microtome knife is used to splinter off parts of the specimen, producing a fracture plane. Structures not in the fracture plane are exposed, and a replica of the exposed surfaces is made (Figure 3-14). The replica is made by first covering the exposed surface with a fine layer of platinum and then reinforcing it with a layer of carbon. The replica is liberated by digesting away the adhering organic material. Figure 3-15 demonstrates the image produced in a microbial cell subjected to a freeze-fracturing technique.

The principal advantage of scanning electron microscopy is the depth of field that can be observed. SEMs

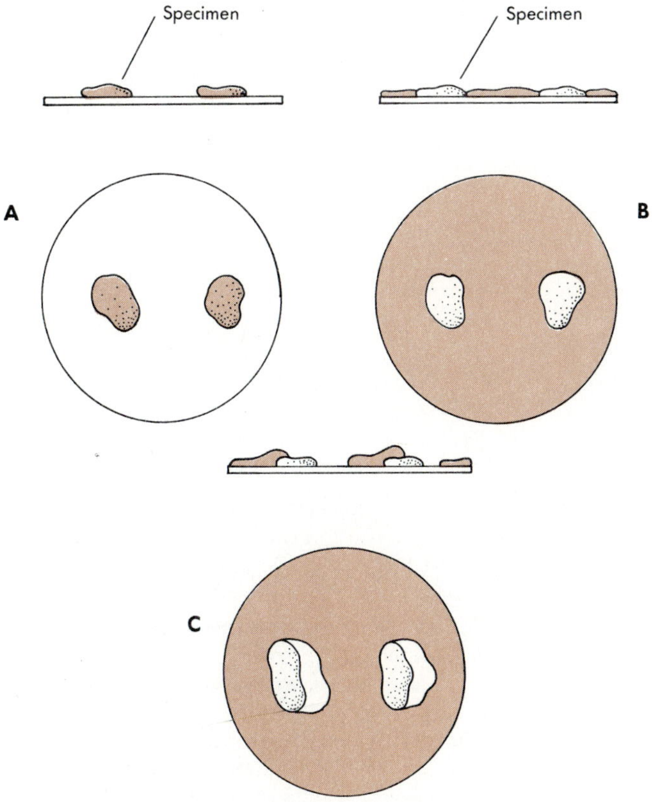

Figure 3-13 *Different types of whole mount preparations using heavy metals.* ***A,*** *Positive staining—specimen is stained.* ***B,*** *Negative staining—background is stained.* ***C,*** *Heavy metal shadow casting (unidirectional)—one side stained.*

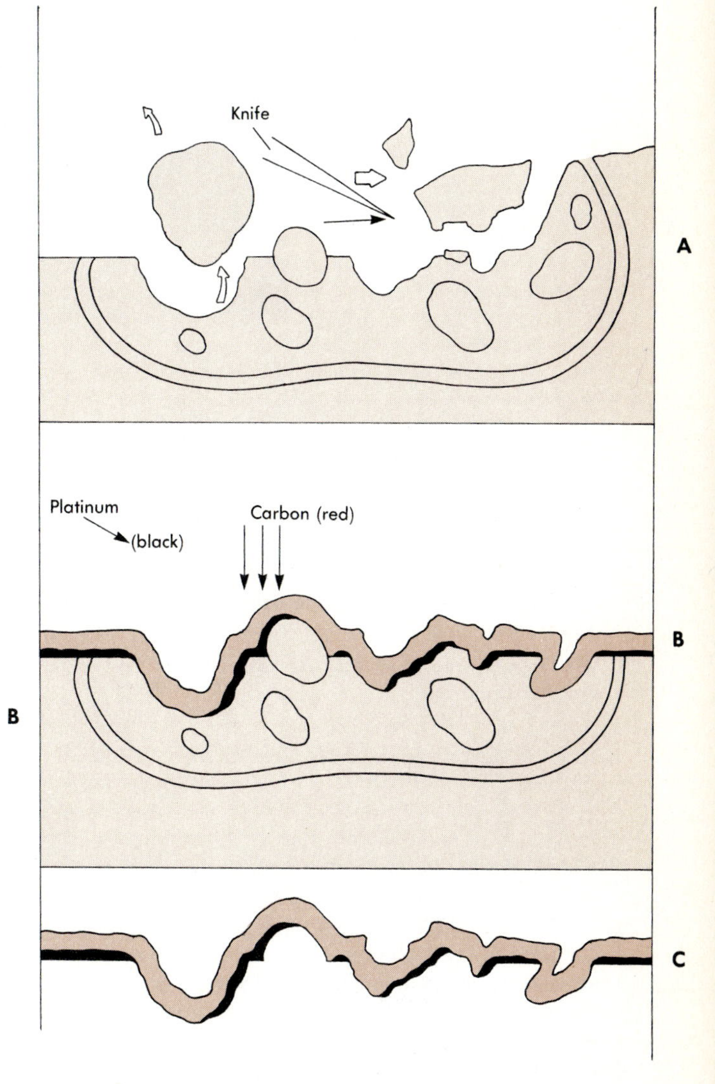

Figure 3-14 *Replication electron microscopy. Replica is made by freeze-fracture technique.* ***A,*** *Frozen specimen (−100° C) is fractured with a liquid nitrogen-cooled (−196° C) knife. Procedure splinters off parts of specimen.* ***B,*** *Replica is made by evaporating platinum atoms onto specimen at 30 to 45 degree angle. This produces platinum layer 1 to 2 nm thick (black). Depressions and protrusions therefore receive one-sided coating of platinum. Platinum replica is reinforced by depositing large layer of carbon about 20 Nm thick in direction perpendicular to fracture plane (dark color).* ***C,*** *Coated specimen is thawed and organic material is removed by treatment with sulfuric acid and bleach. This leaves replica in place of original specimen.*

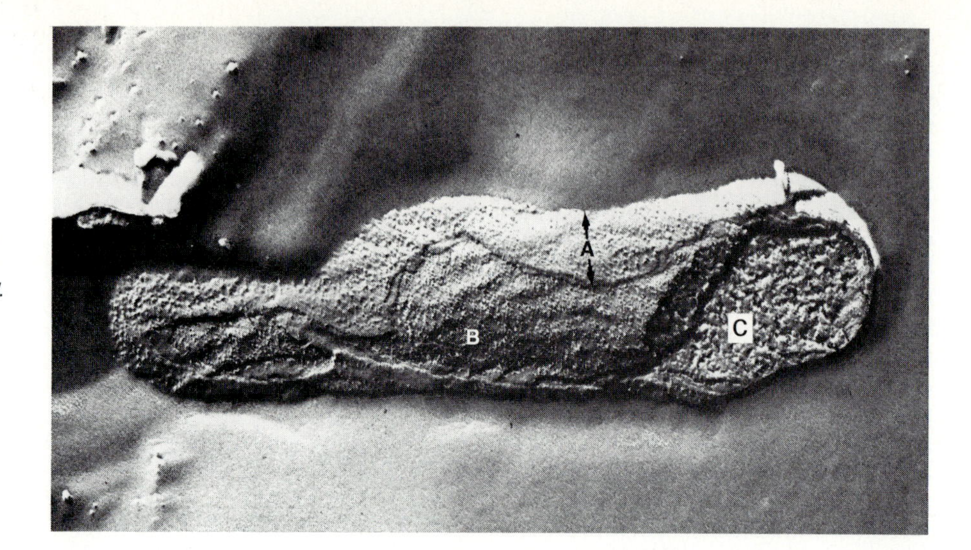

Figure 3-15 *Transmission electron micrograph of replica of freeze-fractured preparation of* Escherichia coli. ***A,*** *Cell wall.* ***B,*** *Cytoplasmic membrane.* ***C,*** *Cytoplasm.*

Concept Check

1. What are the advantages and disadvantages of electron microscopy compared with brightfield microscopy?
2. How can heavy metals be used to stain an object in electron microscopy?
3. What is freeze-fracturing? What are its benefits and how is the process carried out?

provide a three-dimensional image as opposed to the two-dimensional TEM image. However, only the surface structures can be observed by SEM and this is a distinct disadvantage. The specimen, for examination by scanning electron microscopy, is coated with a heavy metal and then scanned with a focused beam of electrons. When the beam of electrons strikes the specimen, secondary electrons are released, collected, and used to produce an image on a cathode ray tube. Magnifications of up to 100,000 × are possible with a resolution of 0.02 μm. An example of the three-dimensional effect of scanning electron microscopy is illustrated in Figure 3-16.

The student might consider microscopy to be more appropriate for the laboratory section of a microbiology course. It is true that microscopic identification of microorganisms is often necessary in determining the cause of infectious diseases or the source of pollution. Keep in mind, however, that our understanding of the biology of microorganisms is directly associated with the correlations that have been made between biochem-

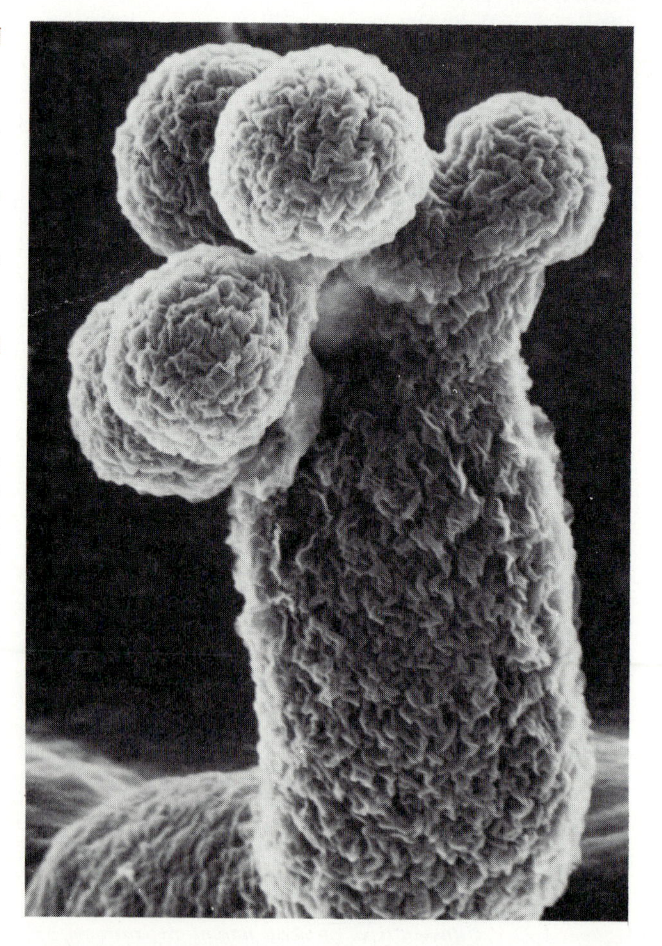

Figure 3-16 *Scanning electron micrograph of fruiting body of species of* Myxobacteria. *Note detail, as well as three-dimensional aspects. Bar = 20 μm.*

Table 3-1 *Characteristics of the various types of microscopes*

Type	Resolving power	Useful magnification	Advantages	Disadvantages
Brightfield	0.2 μm	1000	Can observe morphology of microorganisms, such as bacteria, protozoa, fungi, and algae in living (unstained) and nonliving (stained) state.	Cannot resolve organisms less than 0.2 μm, such as spirochetes and viruses.
Darkfield	0.2 μm	1000	Background is dark, and unstained organisms can be seen. Useful for examining spirochetes.	Slightly more difficult to operate than brightfield.
Phase contrast	0.2 μm	1000	Can observe dense structures in living prokaryotic and eukaryotic microorganisms.	Requires trained personnel.
Fluorescent	0.2 μm	1000	Bright dye attached to organism. Primarily a diagnostic technique (immunofluorescence) to detect organisms in cells, tissue, and clinical specimens.	Training required in specimen preparation and microscope operation.
Transmission electron microscope (TEM)	0.0005 μm	200,000	Specimen can be viewed on screen. Tremendous resolution. Allows examination of cellular ultrastructure, as well as viruses.	Specimen is nonliving. Image is two dimensional.
Scanning electron microscope (SEM)	0.02 μm	10,000	Specimen can be viewed on screen. Three-dimensional view of specimen. Useful in examining surface structure of cells and viruses.	Specimen is nonliving. Resolution limited compared with TEM.

ical properties and structure, as revealed by microscopy. This correlation becomes evident in the next chapters, which describe the relationship between structure and function in the microbial cell.

The characteristics of the various microscopes discussed are outlined in Table 3-1.

SUMMARY

1. The basic laboratory microscope is the compound brightfield microscope. It is equipped with objective lenses and can obtain magnifications of 100× to 1000×, and its resolving power reaches 0.2 μm. The resolving power or image clarity, is defined as:

$$\text{Resolving power (RP)} = \frac{\text{Wavelength of light } (\lambda)}{\text{Numerical aperture (NA)} \times 2}$$

2. When a specimen is stained for microscopic visualization, the technique is called positive staining. If only the background is stained, the technique is called negative staining.
3. The objective lenses of the darkfield microscope receive only diffracted light from the specimen, thus the specimen appears bright against a dark background. The resolution obtained with darkfield microscopy is the same as that with the brightfield microscope.
4. In phase-contrast microscopy, differences in the density between bacterial components and the cytoplasm can be exploited through the use of special lenses. Certain intracellular components can be resolved using phase contrast.
5. In fluorescent microscopy the biological material is coated with a fluorescing compound called a fluorochrome. Immunofluorescence, in which the antibody is tagged with a fluorochrome, is the most widely used fluorescent technique. The fluorescing antibody will bind to specific antigens. Antigens can also be tagged if one wishes to detect specific antibodies in a specimen.

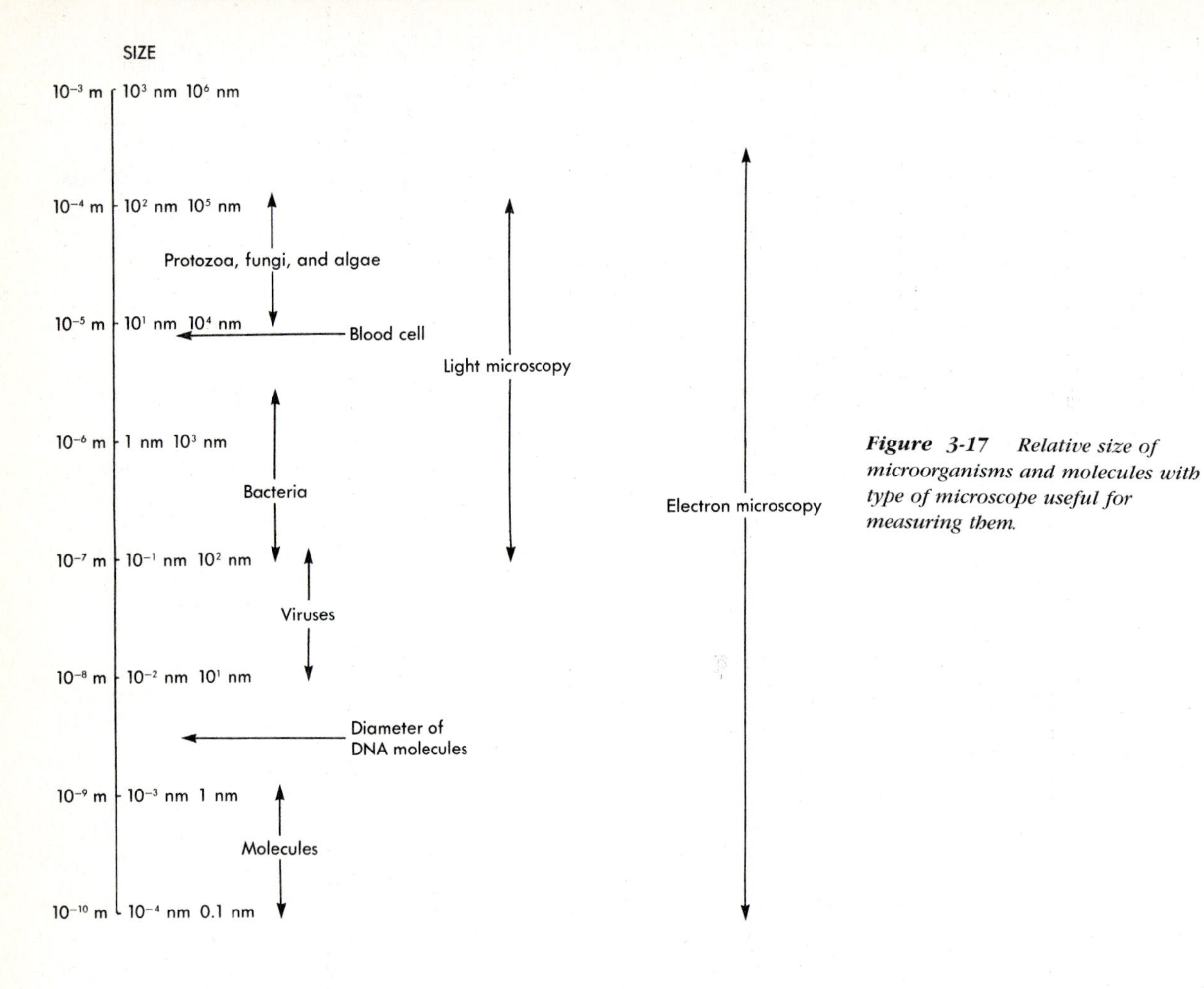

Figure 3-17 *Relative size of microorganisms and molecules with type of microscope useful for measuring them.*

6. The electron microscope uses electrons instead of light waves, and electromagnetic coils instead of lenses. Magnifications of close to 1,000,000× can be obtained with a resolving power of 0.0005 μm. Two types of electron microscopes are available: transmission electron microscopes (TEMs) and scanning electron microscopes (SEMs). In transmission electron microscopy, electrons pass through the specimen and are diffracted to different degrees depending on the mass of the atoms in the cellular component. Specimens for transmission electron microscopy are nonliving and must be prepared by staining or freeze-fracturing. Scanning electron microscopy provides a three-dimensional image in which the specimen (also nonliving) is coated with a heavy metal and is then scanned with a beam of electrons. Magnifications for SEMs approximate 100,000×. The size range of the various microorganisms discussed, as well as the type of microscopy used in the study of each, is illustrated in Figure 3-17.

Self-study Quiz

Multiple choice

1. Which of the following would lead to the improvement of the resolving power of a lens system?
   A. Decreasing the wavelength of light
   B. Increasing the wavelength of light
   C. Increasing the number of lenses in the system
   D. Decreasing the numerical aperture
2. Why are virtually all microbiological dyes known as basic dyes?
   A. Because the basic nature of the molecule is needed for the molecule to penetrate the cell wall.
   B. Because the bacterial cell has an excess of negatively charged groups and basic dyes contain positively charged ionizable groups.
   C. Because basic dyes are needed to neutralize the aminos and nucleic acid.
   D. Because no other type of dye will bind to the cell.
3. Which of the following statements concerning fluorescent microscopy is true?
   A. The specimen will appear dark against a light background.
   B. The fluorescent microscope is similar to a brightfield microscope, with the only difference being the use of fluorescent dyes.
   C. The fluorescent microscope uses a different wavelength of light and different filters than the brightfield microscope.
   D. The fluorescent microscope uses either phase-positive or phase-negative plates; therefore the specimen may appear either lighter or darker than its background.
4. The extracellular polymeric substances surrounding the bacterial cell is most easily observed by which of the following?
   A. Transmission electron microscopy
   B. Negative staining and observation with brightfield microscope
   C. Freeze-etching and observation with transmission electron microscopy
   D. Phase-contrast microscopy
   E. Fluorescent microscopy

Matching

Match the microscope that best fits the particular laboratory situation.

1. ________Transmission electron
2. ________Darkfield
3. ________Phase-contrast
4. ________Brightfield
5. ________Scanning electron
6. ________Fluorescent

A. Unstained preparation of bacteria for determining movement
B. Observing the spiral-shaped organism that causes syphilis
C. Three-demensional structure of a cluster of bacteria
D. The presence of virus in tissue
E. Observing the dense intracellular spore in a bacterial cell
F. Freeze-fracture of cytoplasmic membrane

Thought questions

1. Will it be possible someday to have unlimited magnification in microscopic systems? Explain why or why not.
2. When Leeuwenhoek first observed microorganisms with his simple lenses he described the organisms as small animalcules. Why do you suppose he referred to them as animals and not plants?

SELECTED READINGS

Books

Barer, R. Microscopy. In Bourne, G.H. (Ed.). *Cytology and cell physiology.* Academic Press, Inc., New York, 1964.

Glauert, A.M. (Ed). *Practical methods in electron microscopy.* North-Holland Publishing Co., Amsterdam, 1975.

James, J. *Light microscopic techniques in biology and medicine.* Martinus Nijoff Publishers, Medical Division, Amsterdam, 1976.

Kopp, F. *Electron microscopy.* Scientific Publication Division, Carolina Biological Supply Co., Burlington, N.C., 1981.

Lickfield, K.G. Transmission electron microscopy of bacteria. In Norris, J.R. (Ed). *Methods in microbiology* (Vol. 9). Academic Press, Inc., Ltd., London, 1976.

Quesnel, L.B. Microscopy and microtomy. In Norris, J.R., and Ribbons, D.W. (Eds). *Methods in microbiology* (Vol. 5a). Academic Press, Inc., Ltd., London, 1971.

Slayter, E.M. *Optical methods in biology.* Wiley-Interscience, New York, 1970.

Smith, R.F. *Microscopy and photomicrography.* Appleton-Century-Crofts, New York, 1982.

Wilson, M.B. *The science and art of basic microscopy.* American Society for Medical Technology, Bellaire, Tex., 1976.

Journal articles

Costerton, J.W. The role of the electron microscope in the elucidation of bacterial structure and function. *Annu. Rev. Microbiol.* **33**:459, 1979.

CHAPTER

Prokaryotes and eukaryotes: a comparison

Chapter Outline

Introduction

The earliest and simplest types of living organisms to inhabit the earth were the bacteria. All bacteria are called **prokaryotes** because they possess genetic information that is not enclosed by a membrane—the nuclear membrane. The term prokaryote is derived from the Greek, *protos* meaning primitive and *karyon* meaning nucleus. Prokaryotes evolved over millions of years into more complex organisms or **eukaryotes** (from the Greek, *eu* meaning true and *karyon* meaning nucleus), that is, the organism possessed a nuclear membrane that surrounded the genetic information.

Throughout evolution a tremendous number of changes in size, structure, metabolism, movement, and reproduction have occurred among prokaryotes, as well as eukaryotes. Because of these many changes it has been necessary to arrange organisms into specific groups and to assign them names. This chapter will therefore have two functions; first, to indicate how organisms are classified and named and second, to briefly describe the basic similarities and differences that exist between prokaryotes and eukaryotes.

CLASSIFICATION AND NOMENCLATURE OF PROKARYOTES AND EUKARYOTES

Classification is that branch of the science of **taxonomy** that is used to arrange organisms into specific groups called **taxa** based on a similarity of characteristics. **Nomenclature** is also a branch of taxonomy in which names are assigned to a taxonomic group. The characteristics that enable one to name and classify organisms are derived from laboratory tests. These tests are used to determine the biochemical, morphological, physiological, genetic, and immunological properties of the organism. What this all boils down to is that we must name an organism based on a given set of characteristics that has the same meaning to all scientists. Thus when the name of an organism appears in the literature, certain characteristics are immediately brought to mind. Most of the approaches used to classify microorganisms are based on genetic characteristics and are discussed in Chapters 8 to 11 and listed in Appendix B.

The lack of sophisticated microscopes and biochemical techniques prevented any scientists from recognizing the differences that existed between organisms. Organisms were originally separated into two kingdoms, *Plant* and *Animal.* In the late 1800s scientists microscopically identified microorganisms and noted that they possessed traits that could not be attributed to either plant or animal. They proposed a third kingdom, called *Protista,* to encompass single-celled organisms. The discovery that some single-celled organisms lacked a nuclear membrane led to the division of organisms into prokaryotes and eukaryotes.

The many differences, as well as similarities, that exist between eukaryotes and prokaryotes (Table 4-1) have resulted in various classification schemes. One recent (1978) scheme proposed by Whittaker and Margulis still divides the living world into prokaryotes and eukaryotes but also includes five kingdoms to embrace all biological groups. The scheme is as follows:

- Superkingdom: Prokaryotae
  - Kingdom: Monera (Bacteria)
- Superkingdom: Eukaryotae
  - Kingdom: Protista
    - Branch: Protophyta (plantlike, for example, primitive algae)
    - Branch: Protomycota (funguslike, for example, slime molds)
    - Branch: Protozoa (animal-like, for example, protozoans)
  - Kingdom: Fungi (unicellular yeasts and multicellular molds)
  - Kingdom: Plantae (complex algae and plants such as mosses and ferns)
  - Kingdom: Animalia (multicellular vertebrate and invertebrate animals)

This classification illustrates that animals and plants also have a eukaryotic cellular organization just like the microscopic protozoa, fungi, and algae.

The species

The **species** is the basic unit of classification. The definition of a species in eukaryotes is based on more solid scientific data than it is in prokaryotes. Morphology, reproduction, and fossil records can be used to characterize eukaryotic species, but these are of little value in prokaryotic classification. Bacteria are not as morphologically distinctive as are eukaryotes, which can be observed with the naked eye. Many higher eukaryotes exhibit only a sexual stage in their reproductive cycle; thus certain animals can be included in a species if they are part of a group that interbreeds and cannot interbreed with other groups. This species definition does not apply to bacteria, which reproduce asexually. Fossil records are numerous for plants and animals, but only a few fossil records exist for bacteria. Classification based on fossil records or inferred evolutionary relationships is called **phylogeny**; it is not particularly useful for comparing prokaryotes.

A **bacterial species** is defined as a collection of **strains** that have certain common characteristics that differentiate them from other strains. A strain represents a number of descendants derived initially from a single colony using pure culture techniques. Thus a number of strains, although possessing certain similar characteristics, may differ in one or more properties. The strains of a species may be given special designation in the form of letters or numbers, for example, *E. coli* K10. One strain of the species is regarded as the **type strain.** Type strains are maintained by specific agencies referred to as **culture collections.** Type strains can be obtained from these agencies and used as a reference for comparison with newly isolated organisms. In the United States this function is performed by the American Type Culture Collection.

Taxonomists often disagree as to which characteristics should be considered the most important in separating and delineating the bacterial species. Consequently, many changes in classification occur as new microbial features are elucidated and others are given less importance. The characteristics that have been used traditionally to determine the degree of similarity between organisms include morphological, biochemical, ecological, and serological data. The importance of each characteristic is and will always be a matter of debate. Ideally one strives to have a natural classification that is based on evolutionary (phylogenetic) trends as well as biological relationships. There is very little the taxonomist can do about evolutionary trends because bacterial microfossils discovered in Australia and elsewhere have

Table 4-1 *A comparison of eukaryotic and prokaryotic characteristics and structures*

Characteristic or structure	Eukaryote	Prokaryote
Nucleus	Present; DNA is in form of chromosomes enclosed in nucleus with nuclear membrane	Absent; DNA dispersed in cell into nuclear areas (nucleoids) but no nuclear membrane
Nucleolus	Present	Absent
DNA	More than one chromosome; each chromosome believed to be composed of single DNA molecule; DNA complexes with proteins (histones); chloroplasts and mitochondria contain their own DNA, which resembles prokaryotic DNA; plasmid DNA uncommon	Single chromosome made up of only one molecule of DNA; histones absent; extrachromosomal component, called plasmid DNA, is common
Cell wall	Present in most algae and fungi; absent in protozoa; absence of peptidoglycan layer	Present in all groups except mycoplasmas; peptidoglycan layer present except in mycoplasmas and archaebacteria
Cytoplasmic membrane	Contains sterols	Sterols absent except in *Mycoplasma*
Cytoplasmic streaming	Present	Absent
Mitosis and meiosis	Present	Absent
Ribosomes	Larger than prokaryotes; sedimentation coefficient* of 80S; attached to membranous system called endoplasmic reticulum; ribosomes similar to prokaryotes found in mitochondria and chloroplasts	Sedimentation coefficient* of 70S; exist independently in cytoplasm
Respiration	Associated with organelle called a mitochondrion	Associated with cytoplasmic membrane
Photosynthesis	Present; associated with membranes called thylakoids that are enclosed within an organelle called a chloroplast	Present; some associated with thylakoids, others not; none associated with organelle
Motility	Flagella (cilia) or ameboid movement; flagellum more complex than prokaryote; some gliders	Flagella, axial filament, and some gliders; flagellum has simple construction
Reproduction	Sexual or asexual; conjugation part of reproductive process; sporulation used as reproductive process	Asexual (binary fission); conjugation rare and not part of reproduction; sporulation used for survival value and not as reproductive process except in *Streptomyces*
Size and metabolism	Usually larger than prokaryotes (average 20 μm); metabolism slower than prokaryote	Smaller then eukaryote (average 1 to 6 μm) and have greater ratio of surface area to volume; metabolism faster than eukaryote
Cellular interactions	Have complex cell aggregations with coordinated activities	Most of those that produce colonies show no cooperation between cells
Endosymbiosis (see p. 93)	Nearly always the host in such relationships	Never the host in such relationships
Habitat	Almost exclusively aerobic (presence of air) environments	Aerobic and anaerobic (absence of air) environments

*See Point of Interest (p. 88) for discussion of sedimentation.

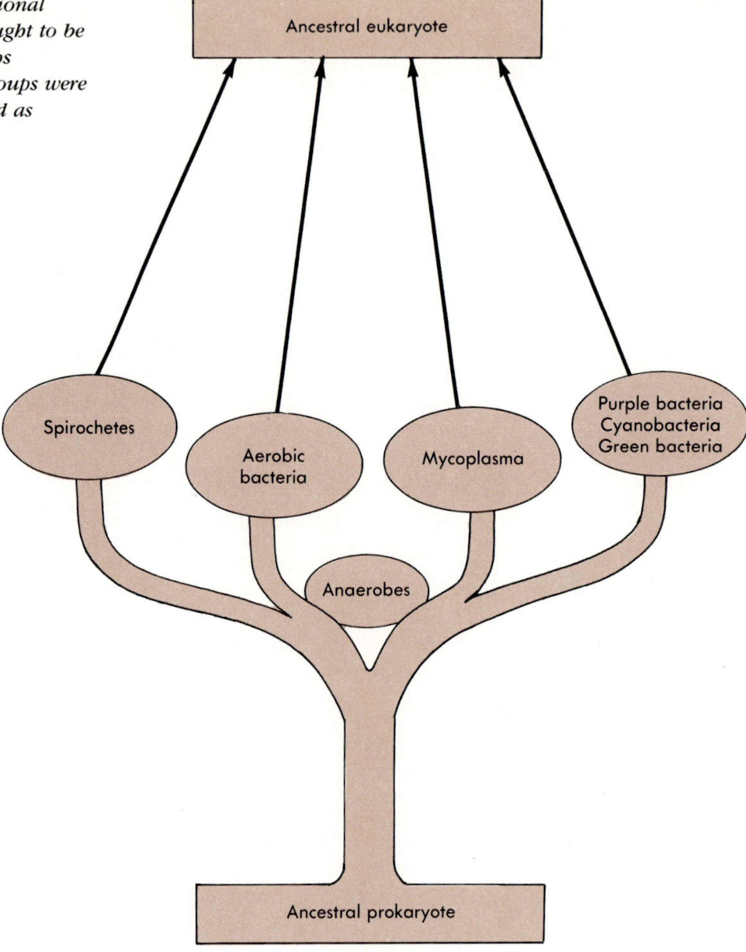

Figure 4-1 *Phylogenetic tree based on conventional taxonomic schemes. Ancestral prokaryote was thought to be anaerobe that evolved into various bacterial groups (spirochetes, aerobes, cyanobacteria, etc.). Some groups were endowed with characteristics that eventually served as prototype for eukaryotic functions.*

revealed only the age of bacteria (approximately 3.5 billion years) and general shape but little else. The history of the bacterial cell is recorded in the proteins and nucleotide sequences of DNA and RNA, and their analysis has provided some important discoveries (see Appendix B).

Groups of related species are placed in **genera**, and related genera are grouped into **families**. Related families are placed into **orders**, orders into **classes**, classes into **phyla**, and phyla into **kingdoms**. This categorization applies to both prokaryotes and eukaryotes. The species in eukaryotes, however, is more precisely defined because of phylogeny and because morphological characteristics can be used to distinguish one species from another. Some of the categories such as families or phyla may be further subdivided into less frequently employed categories such as superfamilies and subphyla, respectively. The category, **subdivision**, is frequently used in the classification of fungi, protozoa, and algae. The subdivision category is equivalent to phylum, that is, it is a collection of similar classes.

Naming an organism (nomenclature)

The method for naming an organism is based on the binomial system of nomenclature first used by Linneaus in the eighteenth century. It employs two words that are given Latin endings to describe a species. The first word indicates the **genus** (pl. genera) with the first letter of the genus capitalized. A genus can be defined as a group of species with a specific set of properties. The second word is not capitalized and indicates the **species** of the genus. Both genus and species are italicized, as in *Bacillus anthracis.* The genus and species names are very descriptive and designate certain characteristics to the reader that may be based on the organism's metabolism, shape, or colony color. Genera such as *Bacillus* (rod), *Streptococcus* (chain of spherical cells), and *Spirillum* (spiral) indicate shape, whereas species names such as *aureus* (gold) and *albus* (white) indicate the color that develops when the organism forms a colony. Not all genus-species nomenclature is so descriptive, and some organisms are named according to the individual who isolated them or the geographic site from which they were isolated. Theodor Escherich, a German microbiologist, for example, first described the organism we now call *Escherichia coli.*

The classification of bacteria has undergone constant revision. Currently the most extensive classification scheme for the bacteria is **Bergey's Manual of Systematic Bacteriology**, which first appeared in 1923. The current ninth edition places the bacteria in the kingdom Prokaryotae, which is separated into four divisions: *Gracilicutes* for gram-negative–type cell walls, *Firmicutes* for gram-positive–type cell wall, *Tenericutes* for organisms lacking a cell wall, and *Mendosicutes* for bacteria having walls lacking peptidoglycan. For convenience the bacteria are divided into a number of **sections** on the basis of staining reaction, spore formation, oxygen requirements, and metabolic patterns. Each section is further divided at the species level. Bergey's ninth edition is in four volumes; however, only two volumes are currently available. Approved lists of species can be determined by consulting the *International Journal of Systematic Bacteriology.* The currently established and proposed sections of *Bergey's Manual* can be examined in Appendix B. The classification schemes for the fungi, algae, and protozoa are briefly outlined in Chapter 6.

Prokaryotes, eukaryotes, and their descendants

It is now believed that the first line of prokaryotes was derived from an ancestral prototype and that eukaryotes in turn evolved from prokaryotes (Figure 4-1). The first line of prokaryotes was believed to have been anaerobic cells (anaerobes can grow only in the absence of oxygen) that derived their energy from fermentation. A group of prokaryotes has been discovered that differs from all other prokaryotes in the structure and biochemistry of certain large molecules. For example, the cell wall of this group lacks a true peptidoglycan, a component found in the cell walls of all other prokaryotes that have cell walls. This was an unusual finding because today's microorganisms are for the most part replicas based on an early theme. These unusual prokaryotes are found in habitats where environmental extremes exist, for example, some are found in habitats with high temperatures and elevated acid conditions while others are found where high salt concentrations exclude all other forms of life or in sediments that are totally devoid of oxygen. These prokaryotes have been placed into a group called the **Archaebacteria.** Carl Woese and others, who examined the characteristics of these organisms, believe that the archaebacteria may have diverged from the original "predecessor" of all life before the evolution of other bacteria. Woese and others now believe that the cells present today were derived from at least three primary lines of descent: **archaebacteria, urkaryotes,** and **eubacteria** (Figure 4-2). Most recognized bacterial species belong to the eubacteria.

The evolutionary transition from prokaryote to eukaryote. As we have all learned from occasional news items, anthropologists have recovered fossils that are helping them to determine the line of apes that gave rise to humans and their ancestors. The search still continues, with new fossils being recovered each year. Similarly, microbiologists continue to search for the organisms that represent a link between prokaryotes and eukaryotes. The task is much more difficult for the microbiologist than the anthropologist because there are relatively few bacterial fossil records and those that are available reveal little other than relative size and shape. Historical "footprints" available by looking at biochemical characteristics of present-day microorganisms (see Appendix B) as well as the discovery of some unusual photosynthetic bacteria have provided some clues to the ancestors of microscopic eukaryotes. Photosynthesis is a property that exists in some prokaryotes and eukaryotes. One of the distinctions between prokaryotic and eukaryotic photosynthesizers is the presence of a distinct photosynthetic organelle, the **chloroplast**, in eukaryotes and its absence in prokaryotes. The photosynthetic apparatus in prokaryotes is part of a diverse and sometimes scattered membrane system that is associated with the cytoplasmic membrane. In addition, there are biochemical differences in the primary pigment molecules, chlorophylls, and accessory pigment molecules such as carotenoids and others. How then did the photosynthetic system of prokaryotes evolve into the eukaryotic chloroplast? It is currently believed that photosynthetic bacteria such as the **cyanobacteria** were engulfed by a prototype of a nonphotosynthetic eukaryotic cell. This resulted in a relationship (symbiosis) in which the photosynthetic membranes of the engulfed bacterium evolved into a chloroplast. The chloroplasts of eukaryotes are biochemically distinct and can be divided into three groups based on the type of chlorophylls and accessory pigments: (1) green algae and plants, (2) red algae, and (3) brown algae and diatoms. The photosynthetic systems of three bacteria show similarities to the chloroplasts of these eukaryotic groups (Table 4-2). Two of these bacteria, *Heliobacterium chloron* and *Prochloron didemni,* have only recently been isolated. The remaining bacteria are the cyanobacteria, a group of organisms sometimes referred to as blue-green bacteria (see Point of Interest on p. 76). Fossil evidence indicates that cyanobacteria were probably one of the predominant forms of life during the Precambrian period. They possess chlorophyll *a*, which is characteristic of chloroplasts, and they also release oxygen, a characteristic associated with all eukaryotic photosynthesizers but absent from most prokaryotes. The cyanobacteria possess accessory pigments, called **phycobilins,** that are identical to those of the red algae. *Prochloron* was discovered in 1975 on the surface of certain tropical and subtropical

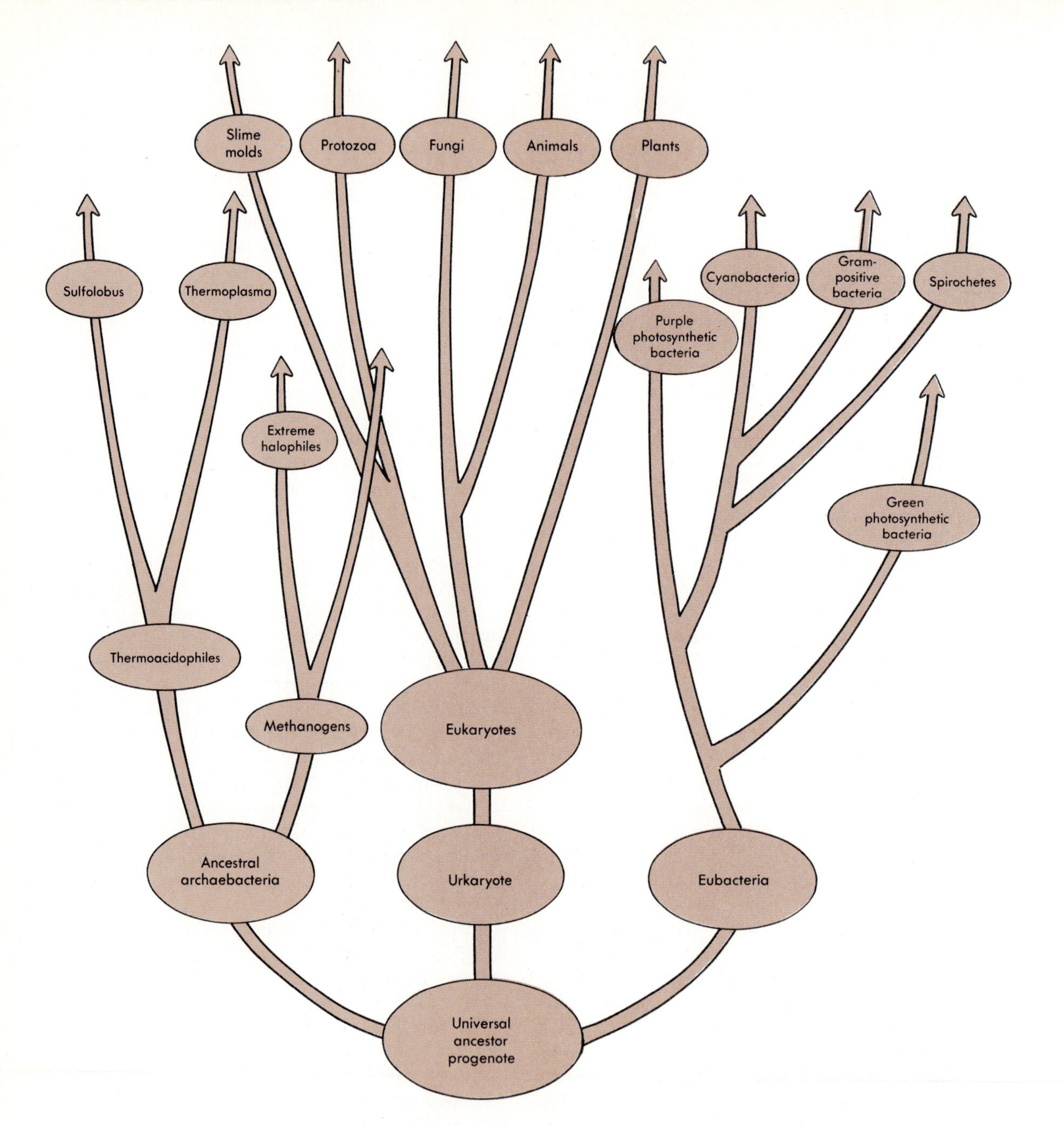

Figure 4-2 *Major lines of prokaryotic descent derived in part from 16S rRNA analysis.*

marine invertebrates (Figure 4-3). *Prochloron* is morphologically similar to cyanobacteria and also releases oxygen during photosynthesis. *Prochloron* possesses chlorphylls *a* and *b,* which is characteristic of the green algae—the latter being the predecessor to our present-day plants. *Prochloron* appears to be a transition organism between cyanobacteria and green algae. *Heliobacterium* (Figure 4-4) was discovered in 1983 and is a rod-shaped photosynthesizer that grows only in the absence of oxygen. This organism, since it can photosynthesize in the absence of oxygen, could have functioned in the early primordial environment, which was initially devoid of oxygen. *Heliobacterium* possesses chlorophylls resembling chlorophylls *a* and *c,* which are present in the golden-brown algae and thus may have given rise to this eukaryote.

Table 4-2 *Characteristics of photosynthetic prokaryotes that are believed to have given rise to specific photosynthetic eukaryotes*

Bacterial photosynthesizer	Photosynthetic characteristics	Characteristics of the eukaryote that prokaryote most closely resembles
Cyanobacteria species	Evolves oxygen; produces chlorophyll *a;* contains accessory pigments called phycobilins	Red algae that evolves oxygen; produces chlorophyll *a* and *d;* produces phycobilins
Prochloron didemni	Evolves oxygen; produces chlorophylls *a* and *b;* does not produce phycobilins	Green algae that evolves oxygen; produces chlorophylls *a* and *b;* does not produce phycobilins but does produce carotenes and xanthophylls
Heliobacterium chloron	Does not evolve oxygen; produces chlorophylls *a* and *c;* does not produce phycobilins	Golden brown algae that evolves oxygen; produces chlorophylls *a* and *c;* does not produce phycobilins but does produce carotenes and xanthophylls

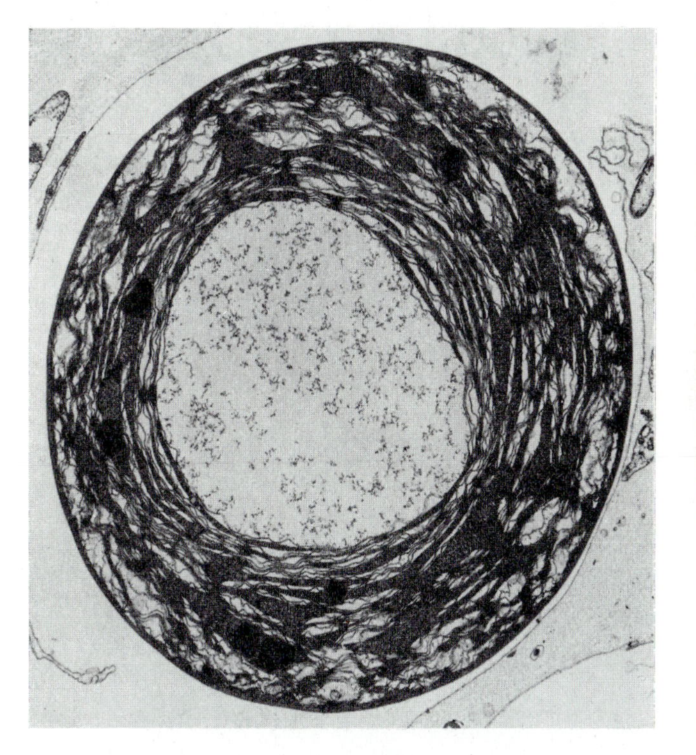

Figure 4-3 *Electron micrograph of* Prochloron.

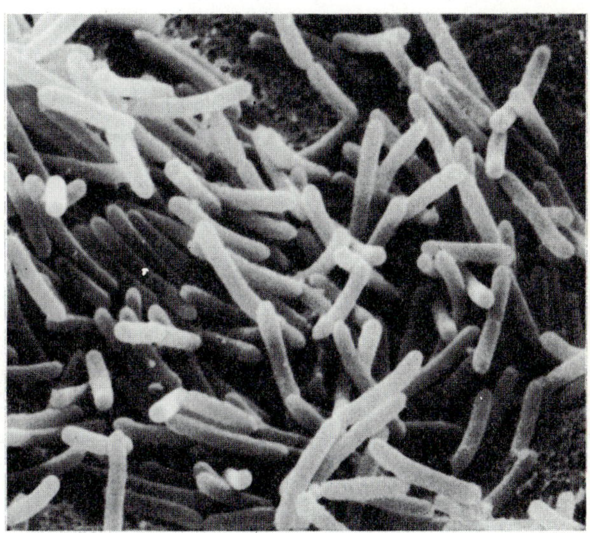

Figure 4-4 *Electron micrograph of* Heliobacterium chlorium.

The mitochondrion, which is present in eukaryotes but absent from prokaryotes, is believed to have evolved from the ingestion of a prokaryote by a prototype of the eukaryotic cell. Two reasons for this suggestion relate to function as well as size. The respiratory activities of the eukaryotic mitochondrion are duplicated by enzymes present in the cytoplasmic membrane of prokaryotes. In addition, the mitochondrion and many small bacteria are similar in size: 1 to 2 μm in length and 1.5 μm in diameter. It is hypothesized that if the ingested prokaryote possessed well-developed respiratory activities it evolved into a mitochondrion in the host cell. The bacteria believed to be the ancestor of the mitochondrion are the purple bacteria.

Point of Interest

THE CYANOBACTERIA

The cyanobacteria are phototrophic (use light as energy source) prokaryotes that were originally called blue-green algae. The older terminology was not without justification, since many cyanobacterial species are the size of some algae and all cyanobacteria possess chlorophyll *a*, a pigment associated with oxygen production in photosynthetic eukaryotes. Cyanobacteria, however, possess a gram-negative cell wall containing peptidoglycan and do not possess a nuclear membrane.

Various groups of cyanobacteria exist in marine and freshwater environments; some inhabit hot sulfur springs, and many are common to all types of soil. They range in size from the smaller unicellular rods and cocci to the filamentous types (figure, top right). The unicellular types divide either by splitting in two (binary fission) or by budding. The filamentous types multiply, by breaking into smaller units of cells called **hormongonia.** Some, but not all, species exhibit a gliding type of motility that is associated with a slime layer. Cyanobacteria produce extracellular products such as anatoxin, a compound similar in structure to cocaine. A compound called **geosmin** is also produced by cyanobacteria and is responsible for the earthy odor associated with drinking water contaminated by these organisms. Geosmin is also produced by other soil bacteria.

The cytoplasmic membrane is the source of the photosynthetic membranes called **thylakoids,** which traverse the cytoplasm of the cell (figure, middle right). Proteins called **phycobiloproteins** are attached to the outer surface of the thylakoids in aggregates called **phycobilisomes.** Phycobilisomes are composed of light-harvesting pigments involved in energy transfer. One of the unusual characteristics of the cyanobacteria is their ability to simultaneously carry out the processes photosynthesis and nitrogen fixation. Photosynthesis in cyanobacteria results in the release of oxygen, whereas nitrogen fixation can occur only in the absence of oxygen. These two processess can be carried out simultaneously in cyanobacteria because of a cellular differentiation process in which actively dividing (vegetative) cells give rise to a specialized cell called a **heterocyst** (bottom right). After differentiation takes place, the actively dividing cells of the filament are involved in photosynthetic reactions leading to the evolution of oxygen, whereas the heterocysts engage only in the fixation of nitrogen and photosynthetic ATP production. The mechanism by which the cyanobacteria can accomplish two apparently mutually exclusive functions is discussed in Chapter 9.

Spores called **akinetes** can also be found in the multicellular filaments of cyanobacteria. They are similar in appearance to vegetative cells but are larger (figure, bottom right). In most of their physiological processes akinetes are similar to vegetative cells. Akinetes, although more resistant to environmental factors than vegetative cells, do not approach the degree of resistance of bacterial endospores. Akinetes do not fix nitrogen but do carry out photosynthetic processes like the actively dividing cells.

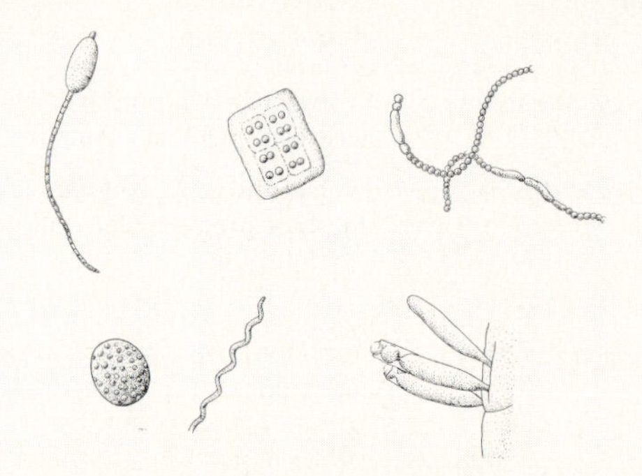

Morphological types of cyanobacteria.

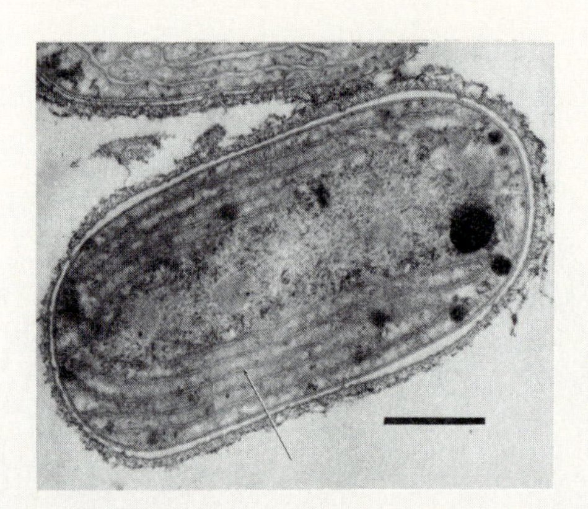

Electron micrograph of cyanobacterium demonstrating arrangement of thylakoids (arrow) *in cytoplasm.*

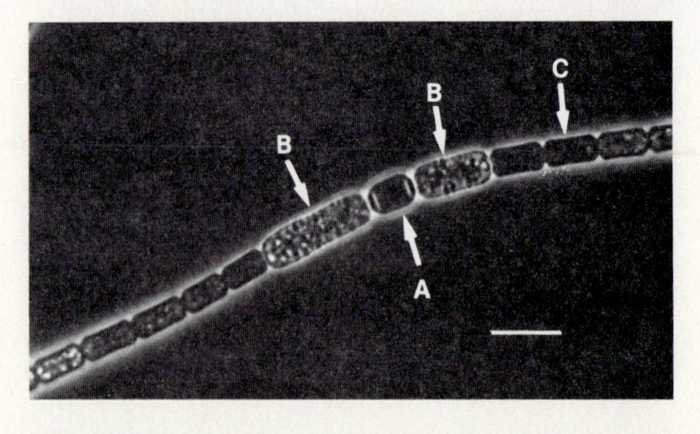

Phase-contrast photomicrograph of differentiated cells within filamentous cyanobacteria. Bar = 10 μm. ***A,*** *Heterocyst.* ***B,*** *Akinete.* ***C,*** *Vegetative cell.*

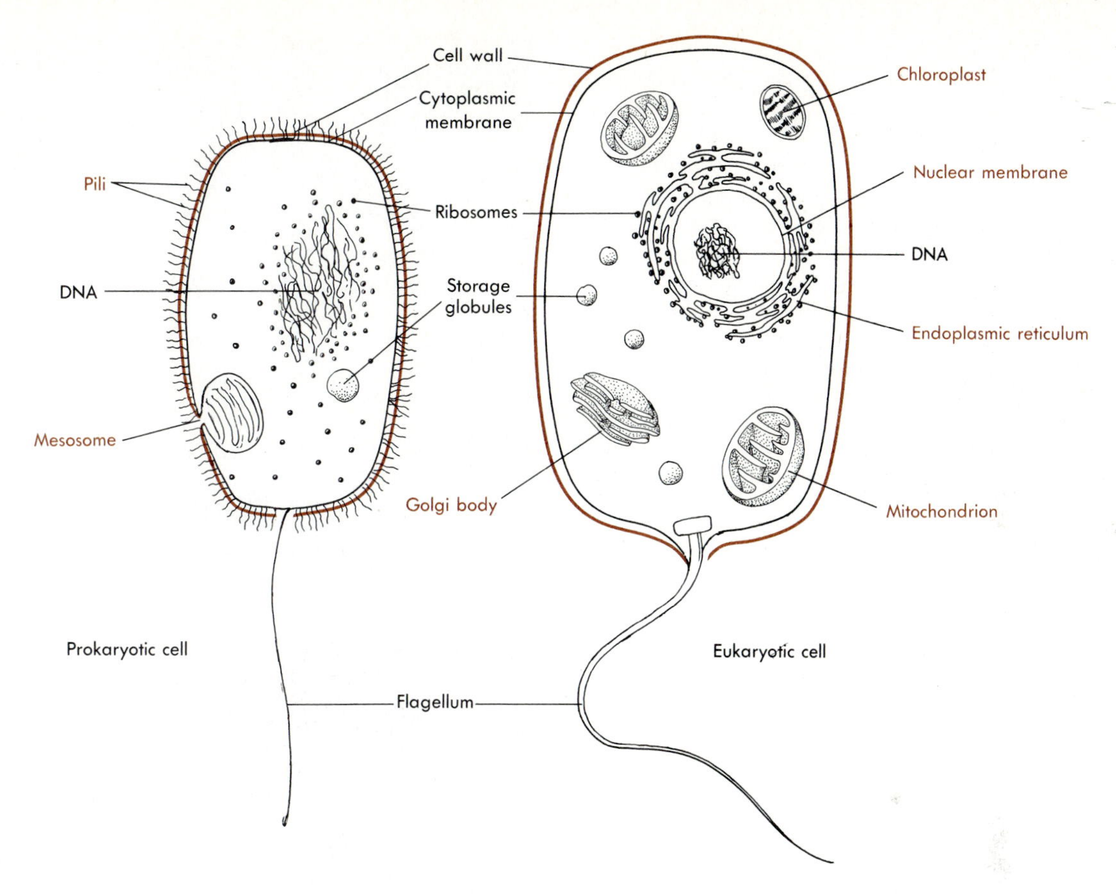

Figure 4-5 *Basic morphological features of prokaryotes and eukaryotes. Not all structures that are present in various members of cell types are indicated. Structures that are comparable among two groups are labeled in black; structures that are unique to group are color.*

Concept Check

1. Define the terms genus, species, and type strain as they apply to microorganisms.
2. What evidence exists that eukaryotes are derived from prokaryotes?
3. What are five basic differences between prokaryotes and eukaryotes?

GENERAL FEATURES OF PROKARYOTES AND EUKARYOTES

Figure 4-5 demonstrates the basic morphological features of prokaryotic and eukaryotic cells. The figures are in no way complete, but they will help the student understand the relative position and shape of the most important cell structures.

Size

Prokaryotes. On an average bacteria are between 1 and 6 μm in length and 0.2 and 1.5 μm in diameter. There are some bacteria such as the mycoplasma (Figure 4-6) that are only 0.2 μm in length and some spiral-shaped bacteria that reach 400 to 500 μm in length, but they are exceptions.

Eukaryotes. Most eukaryotic microorganisms average 20 μm in length, which includes the nearly spherical algae and yeasts. Some protozoa, such as the giant ameba, are 1 to 5 mm (1000 to 5000 nm) long (Figure 4-7), whereas other ameba are only 4 to 5 μm in length. The relative sizes of various prokaryotes and eukaryotes is depicted in Figure 4-8.

The considerable size difference between bacteria and the more advanced eukaryotes is significant. The rate of metabolism in an organism increases with an increase in the surface/volume ratio, and in prokaryotes this ratio is much greater than in eukaryotes. The surface/volume ratio for prokaryotes, such as spherical bacteria with a diameter of 0.5 μm, is approximately 100,000, whereas in an organism the size of a hen's egg, whose diameter may be 1.5 inches, the surface/volume ratio is between 1 and 2. The bacterium, therefore, with its proportionately increased surface area, can effect a rapid exchange of nutrients and other materials with the surrounding environment. A bacterium such as *E. coli,* a common inhabitant of the intestinal tract, can degrade over 1000 times its own weight of glucose within an hour at 37° C. It would take a human one half a lifetime to degrade glucose that amounted to 1000 times his own weight. The prokaryotic cell not only experiences a faster rate of metabolism than eukaryotes but also divides more rapidly than the eukaryotic cell. Eukaryotes compensate for their small surface/volume ratio by compartmentalization. This compartmentalization results from the organization of the internal membrane system, the endoplasmic reticulum.

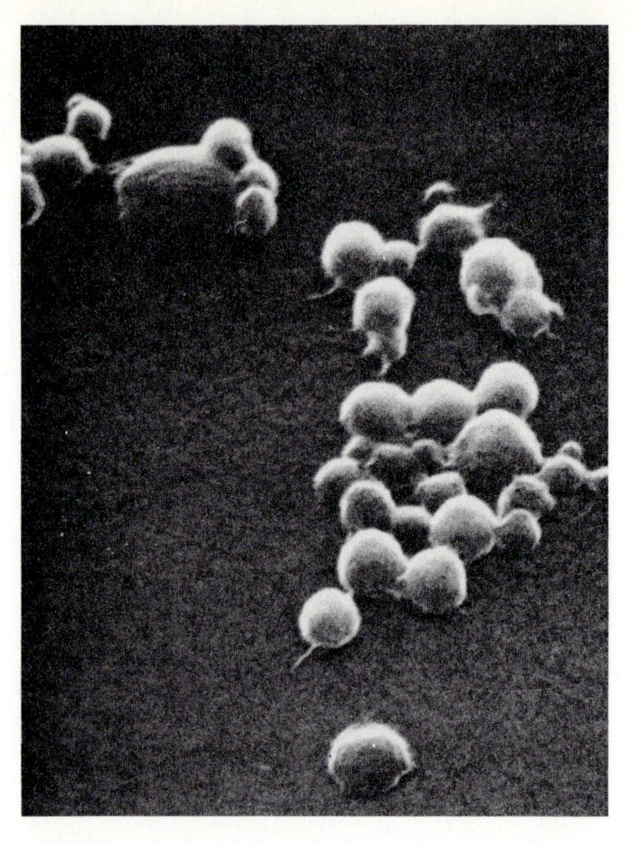

Figure 4-6 *Scanning electron micrograph (× 10,000) of* Mycoplasma *species.*

Shape

Prokaryotes. Bacteria can be separated into four basic shapes: spherical, rod, curved or spiral, and square, with variations in most (Figure 4-9).

Spherical bacteria. Spherical bacteria are referred to as **cocci** (sing. **coccus**), which, depending on the spe-

Figure 4-7 *Photomicrograph of giant ameba. Note arrow points to another protozoan,* Paramecium, *that has been ingested.*

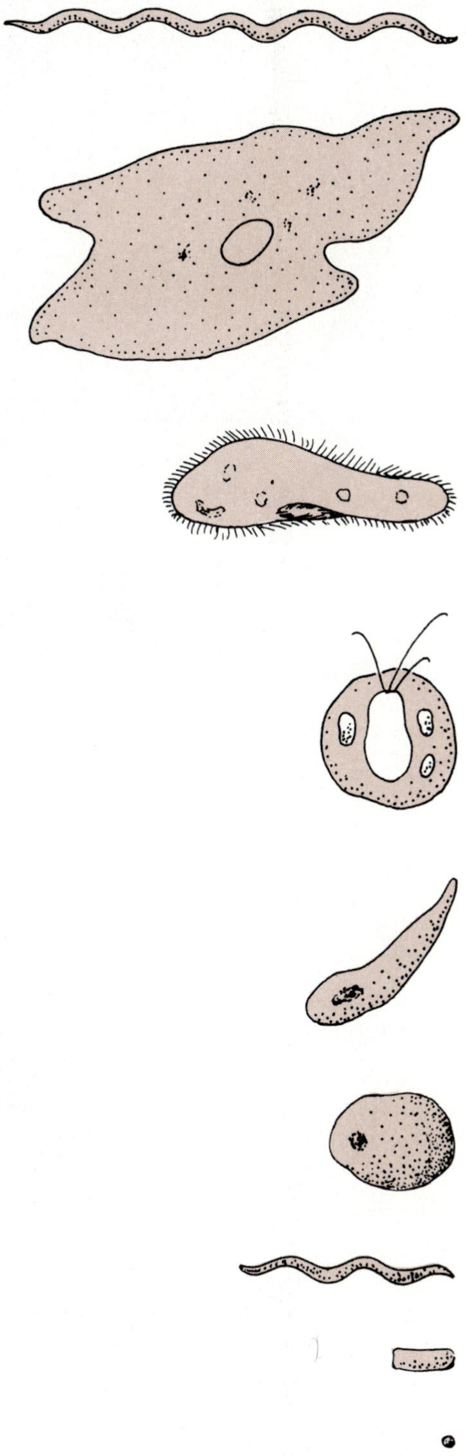

MICROBIAL SHAPE	GENUS AND/OR SPECIES OR TYPE	SIZE (LENGTH)	MICROBIAL GROUP
	Saprosphira	500 μm	Bacterium
	Giant amoeba	1 mm	Protozoa
	Paramecium	300 μm	Protozoa
	Chlamydomonas	25 μm	Alga
	Malaria parasite	15 μm	Protozoa
	Yeast Saccharomyces cerevisiae	10 μm	Fungus
	Treponema pallidum	10 μm	Bacterium
	Escherichia coli	3 μm	Bacterium
	Mycoplasma	0.2 μm	Bacterium

Figure 4-8 *Relative size of various unicellular prokaryotic and eukaryotic microorganisms.*

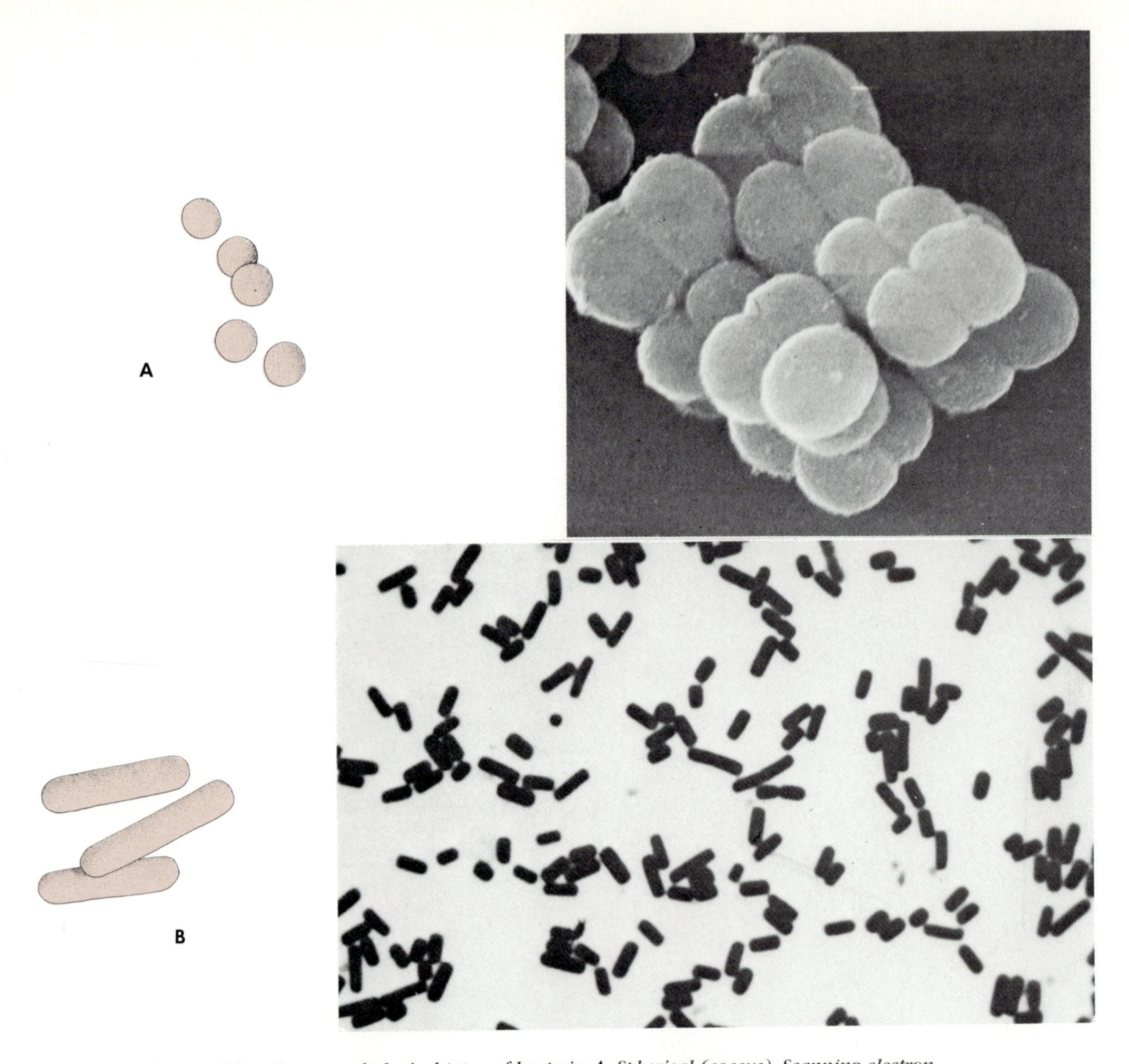

Figure 4-9 *Four morphological types of bacteria.* ***A,*** *Spherical (coccus). Scanning electron micrograph is* Staphylococcus aureus. ***B,*** *Rod (bacillus). Light micrograph is of* Clostridium perfringens. *(× 2560.)*

cies, exist in different arrangements depending on the plane of division during reproduction (Figure 4-10). Division in one plane results in the formation of either a pair of cells, called **diplococci**, or a chain of cells, called **streptococci**. Cells of **diplococci** split before the next cell division, preventing the streptococcal arrangement. Division in two planes results in the formation of **tetrads**. Division in three planes produces cubical masses that appear as groups of eight called **sarcinae** or as clusters of cells called **staphylococci**.

Rod-shaped bacteria. Rod-shaped organisms are called **bacilli** (sing. bacillus). The term **bacillus** is used here as a morphological characteristic, but it can also be used to indicate a taxonomic group: the genus *Bacillus*. Rod-shaped bacteria exhibit considerable variation in length and diameter. The length and diameter of some rods is so similar that they are difficult to distinguish from cocci. The ends of rods also vary in shape and may be flat, rounded, bifurcated (divided into two branches), or cigar shaped (Figure 4-11). Cell division in most rod-shaped bacteria results in the formation of two similar daughter cells, but this may be subject to variations. The actinomycetes actually form branches and develop into filaments **(hyphae)** that become so extensive that large mats of filaments, called **mycelia** (sing. mycelium), are formed (Figure 4-12). This type of growth is also characteristic of the fungi that the actinomycetes resemble in many other ways.

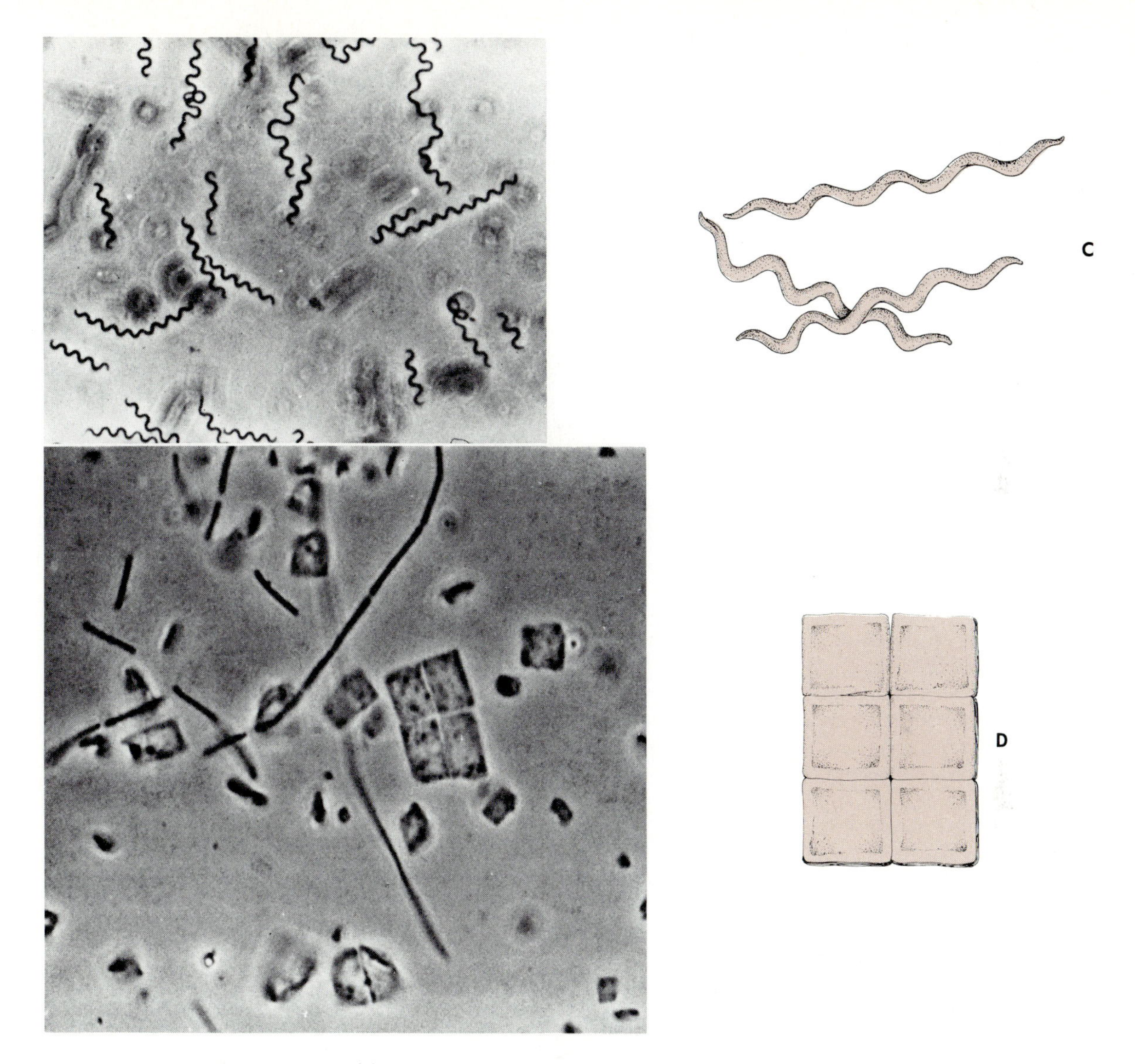

Figure 4-9 cont'd *C, Spiral-shaped bacteria. Phase-contrast micrograph is of* Treponema *species. (×2000.)* ***D,*** *Square-shaped bacteria. Micrograph is phase contrast. (×1600.)*

Spiral-shaped bacteria. Some rod-shaped organisms are curved or exist in the shape of a helix (Figure 4-9, *C*). The rods of the spiral bacteria may be flexible or rigid, depending on the species. Motility is typical of most members of this group and is characterized by a corkscrew movement (for example, the spirochetes).

Square bacteria. One of the oddities of the microbial world is the square bacteria. These organisms were originally discovered in 1981 by Walsby on the shores of the Red Sea (Figure 4-9, *D*). The sides of these halophilic (salt-loving) organisms range from 2 to 4 μm and are 0.25 μm thick.

Pleomorphism. Pleomorphism can be defined as the existence of an organism in different morphological forms. Environmental factors influence the size and sometimes the shape of bacteria. Size variations, which are more readily detected in rod-shaped microorganisms than in spherical cells, are associated with changes in the biosynthesis and growth of the cell wall. Rod-shaped cells, cultured in media in which the cells are dividing at a rapid rate (for example, 20 minutes), are two to three times larger than cells that are dividing at lower rates (for example, every 90 minutes). Some small rod-shaped cells in nutrient-poor media are barely distinguishable from cocci; hence the term **coccobacillus** is used to describe them. This variation in size is related to the growth-supporting ability of the medium as well as the ambient temperature for growth. An excellent example of pleomorphism is observed in the mycoplasmas that do not possess cell walls. Under optimal growth conditions these organisms are filamentous, but under less than ideal conditions they appear donut-shaped (Figure 4-13).

Abnormal or aberrant forms of rod-shaped bacteria can be produced when the growth medium becomes deprived of nutrients or a growth inhibitor is present in the medium. Large filaments may be produced by rod-shaped bacteria when the growth medium contains an inhibitor of growth such as penicillin. Penicillin in low concentrations permits the cell to grow, and because the drug affects cell wall synthesis, the cell wall does not maintain its normal shape; thus filamentous extensions are produced (Figure 4-14).

Eukaryotes

Fungi. There is no basic shape for eukaryotes, but some generalizations can be made. Most fungi, like the bacterial actinomycetes, are filamentous and exhibit extensive branching. They are referred to as **molds**. Sometimes the hyphal mass is loosely arranged into a cottony structure, as in bread molds (Figure 4-15). Other times

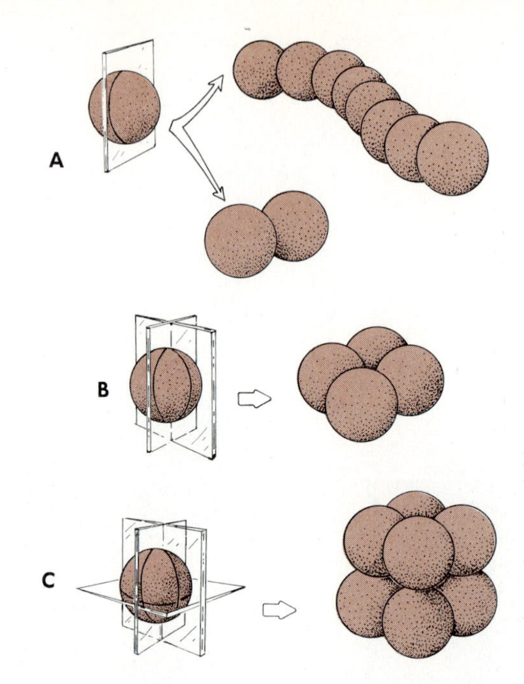

Figure 4-10 *Various arrangements of cocci depending on plane of cell division.* ***A,*** *Division in one plane produces diplococci or streptococci.* ***B,*** *Division in two planes produces tetrad.* ***C,*** *Division in three planes produces group of eight (sarcina) or clusters (staphylococcus).*

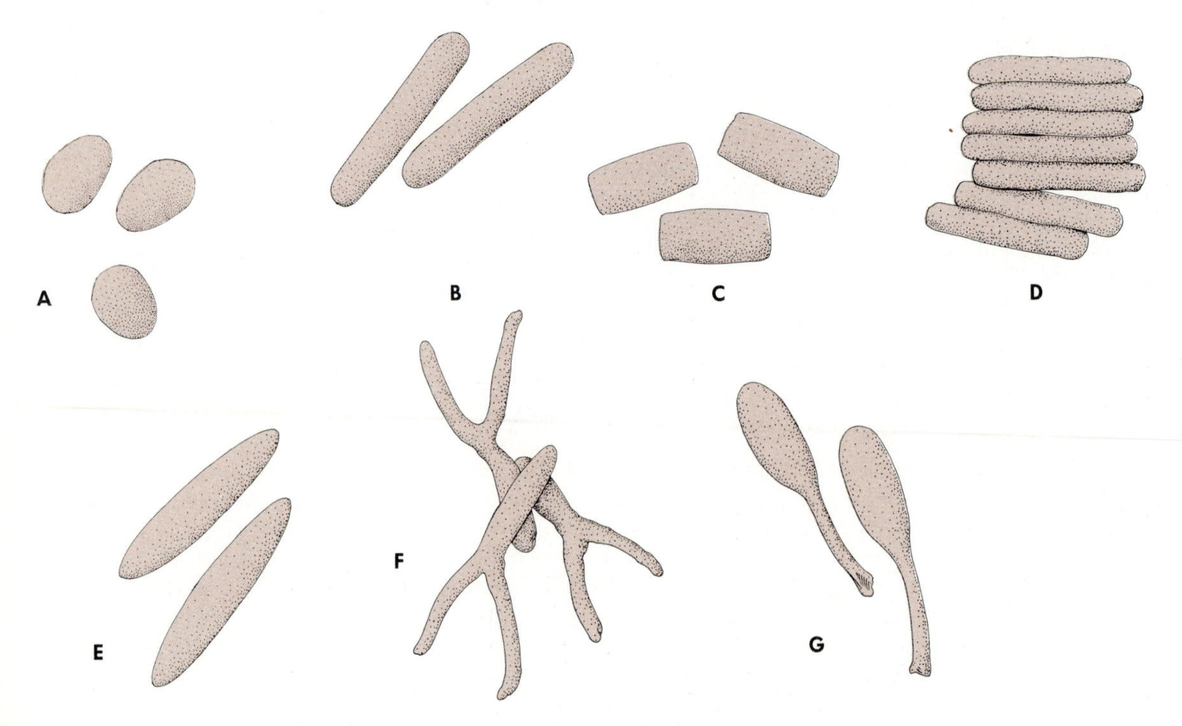

Figure 4-11 *Variations in morphology and arrangement of rod-shaped bacteria.* ***A,*** *Coccobacillus.* ***B,*** *Rods with rounded ends.* ***C,*** *Rods with flat ends.* ***D,*** *Rods stacked in palisade arrangement.* ***E,*** *Cigar-shaped rods.* ***F,*** *Bifurcated rods.* ***G,*** *Rods with stalks.*

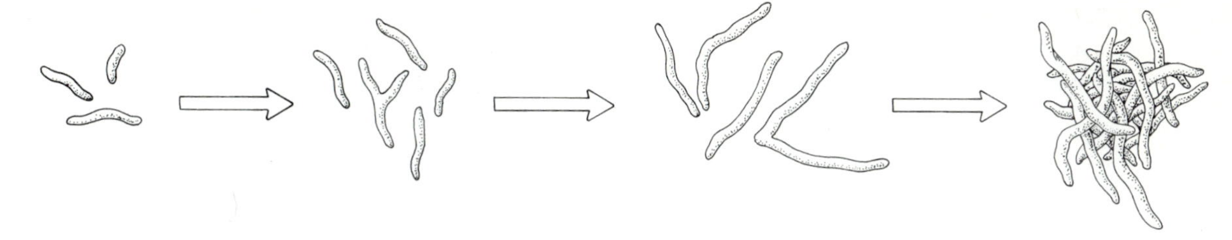

Figure 4-12 *Actinomycete, a filamentous bacterium, is initially a rod-shaped cell that forms filaments (hyphae) that develop into large mats or mycelia.*

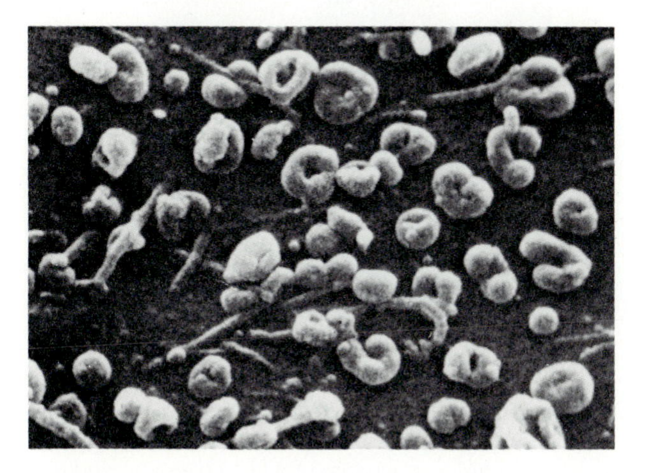

Figure 4-13 *Pleomorphic properties of* Mycoplasma. *Electron micrograph shows donut shape of majority of cells but filamentous forms are also present.*

the mycelial mass is compact, hard, and leathery, as in mushrooms or bracket fungi (Figure 4-16). Some fungi called **yeasts** remain unicellular; however, there are fungal species that can exist in either a spherical or filamentous state, depending on environmental conditions. This property is called **dimorphism** and is characteristic of those fungi that cause disease.

Protozoa. Protozoa show a great variation in shape; some are spherical, some are wormlike, and others, because of the flexibility of their cytoplasmic membrane or because of secreted surface structures, can assume a variety of shapes (Figure 4-17).

Algae. Algae also show variation in shape. The more primitive forms are unicellular and are frequently found as plankton on the surface of inland waters or the ocean. The more advanced forms are usually multicellular, existing as filaments or two-dimensional organisms whose cells are arranged into sheets (Figure 4-18).

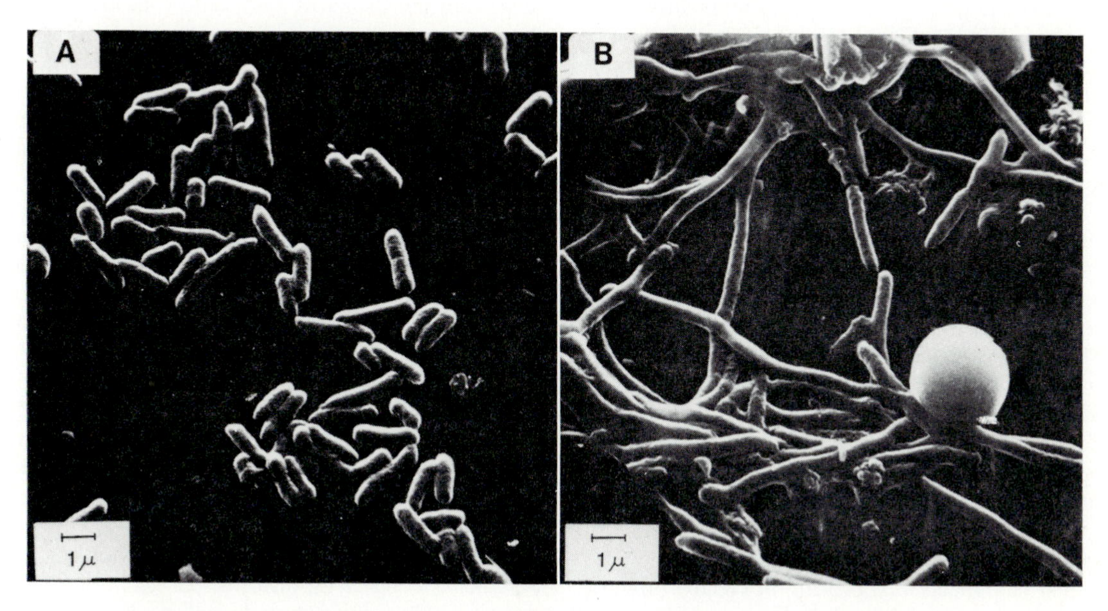

Figure 4-14 *Effects of penicillin derivative (ampicillin) on growth and morphology of rod-shaped bacterium* Escherichia coli. ***A,*** *Untreated control.* ***B,*** *Treatment with ampicillin. Note formation of long filaments.*

Figure 4-15 *Mold. On suitable sustrate, such as commercial jelly, molds can form colony or mass of loosely arranged hyphae.*

Figure 4-16 *Mushroom.*

Structures protecting the cytoplasmic contents (cell wall and cytoplasmic membrane)

Cell wall in prokaryotes. All bacteria except the mycoplasmas possess cell walls that protect the cell and determine to a great extent the cell's shape. The cell walls of nearly all bacteria, except the mycoplasmas and archaebacteria, possess a cell wall polymer component called **peptidoglycan.** Most bacteria have been divided into two groups based on the chemical components that lie outside this peptidoglycan layer: **gram-positive** and **gram-negative** bacteria. This division is based on the gram-staining characteristics of the cell wall components, that is, the effect of a decolorizing solution of alcohol on the crystal violet-iodine complex (see Chapter 3). Wall porosity is believed to be an important factor in determining the ability of cell wall components to retain or not retain the crystal violet–iodine complex, but other factors may also be important. Porosity may be influenced by the cell wall lipid concentration, which is considerably higher in gram-negative bacteria than in gram-positive bacteria. Lipids are disrupted and extracted by alcohol, which alters membrane permeability. Porosity is also influenced by the thickness of the peptidoglycan, which in gram-negative bacteria may exist as a monomolecular layer, but which in gram-positive bacteria may be several layers thick. The relative composition and structure of gram-positive and gram-negative walls are illustrated in Figure 4-19. (Detailed descriptions are given in chapter 5.)

Bristlelike appendages, appearing in electron micrographs to emanate from the cell wall of many bacteria, are called **pili** (sing. **pilus**, see Figure 4-5). The cytoplasmic membrane is the actual origin of these structures. They are not associated with movement and are characteristic of prokaryotes but not eukaryotes. One type of pili, **common pili** (formerly called fimbriae), are found in great numbers on the surface of the bacterial cell. They are primarily associated with the ability of the bacterium to adhere to different surfaces. The other pili, called **sex pili**, are not only less numerous but are also wider than common pili. Sex pili are characteristic of specific groups of bacteria that engage in a conjugation process in which a copy of DNA is transferred to a recipient bacterium.

Cell wall in eukaryotes. Only fungi and algae of the eukaryotic microorganisms possess a cell wall. The composition of their cell walls is radically different from that of bacteria. The cell wall of fungi is made up of primarily polysaccharides (70% to 80% of dry weight) and protein making up most of the remainder. The principal polysaccharide is **chitin** (a polymer of *N*-acetylglucosamine), or **cellulose** (a polymer of glucose) (Figure 4-19). The cell wall of most algae is composed primarily of cellulose with a thin outer layer of pectin, a polysaccharide containing galactose, arabinose, and galacturonic acid. Some algal cell walls may also contain inorganic substances such as calcium carbonate, iron, or silicon. The walls of diatoms, for example, are composed of 96% silicon dioxide (SiO_2).

Cytoplasmic membrane in prokaryotes. The cytoplasmic membrane of prokaryotes as well as eukaryotes

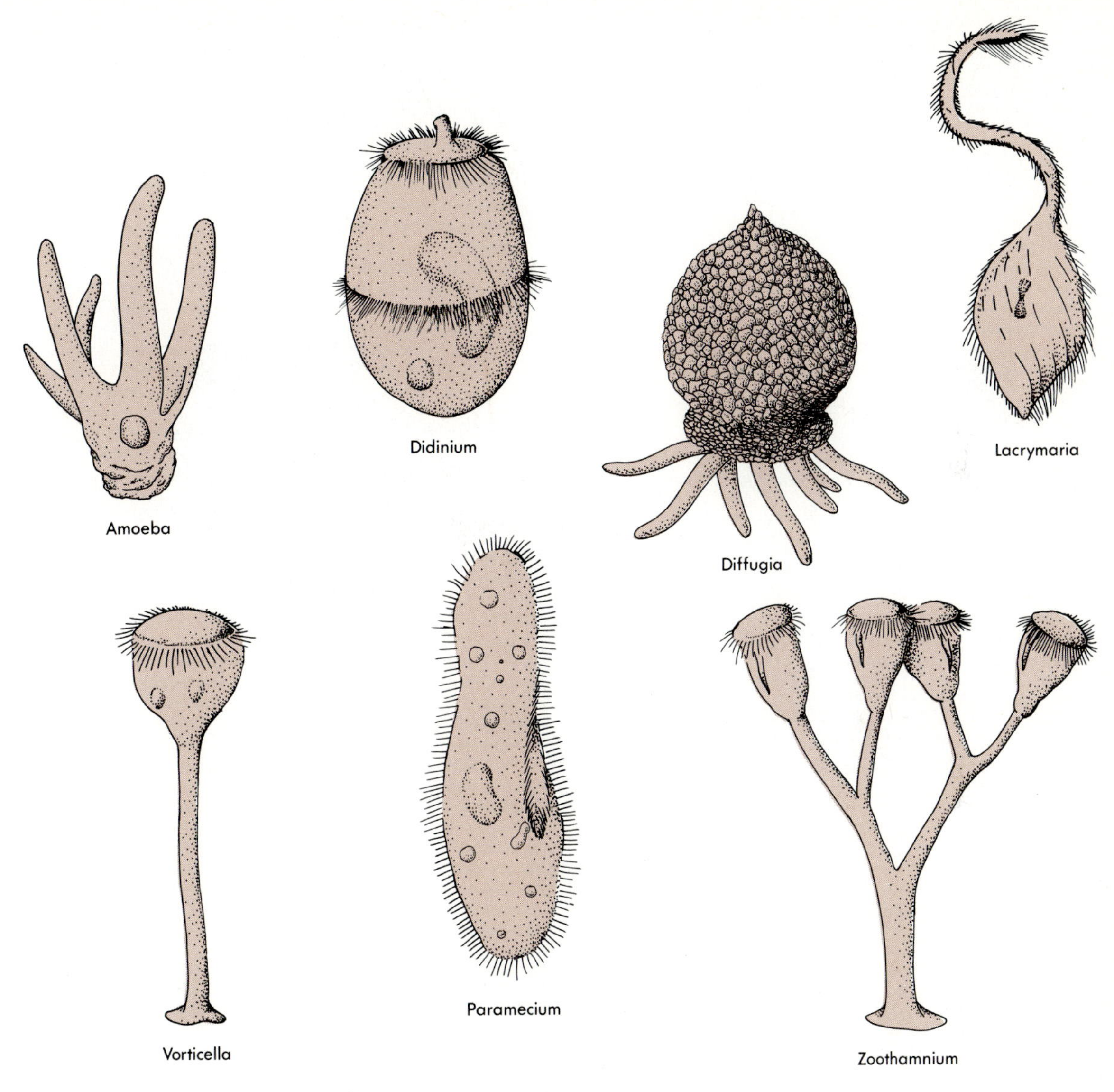

Figure 4-17 *Shapes of some selected protozoa.*

is a three-layered structure in which a central hydrophobic fatty acid area, containing embedded proteins, is bounded on either side by hydrophilic phospholipid molecules (see Chapter 2). This type of membrane structure is called a **unit membrane**. The principal chemical difference between prokaryotic and eukaryotic membranes is the presence of sterols in the eukaryotic membrane and their absence from those bacteria possessing cell walls. Only the wall-less bacteria, mycoplasma, contain sterols in their cytoplasmic membrane. Sterols help to stabilize the membrane and give it some support, since it is the only structure separating the cytoplasmic contents from the environment.

The cytoplasmic membrane in prokaryotes and eukaryotes functions primarily in the transport of nutrients into the cell and waste products out of the cell. In bacteria the cytoplasmic membrane also serves as an anchor for the attachment of DNA during the replication process. The bacterial cytoplasmic membrane possesses enzymes that function in energy production (respiration) as well as cell wall synthesis.

The bacterial cytoplasmic membrane is bounded to the interior of the cell wall, but sometimes when electron micrographs are examined the membrane appears to be invaginated into the cytoplasm of the cell forming an organelle-like structure referred to as a **mesosome** (Figure 4-5). Mesosomes may be the site for attachment of DNA during replication or the site for cell wall syn-

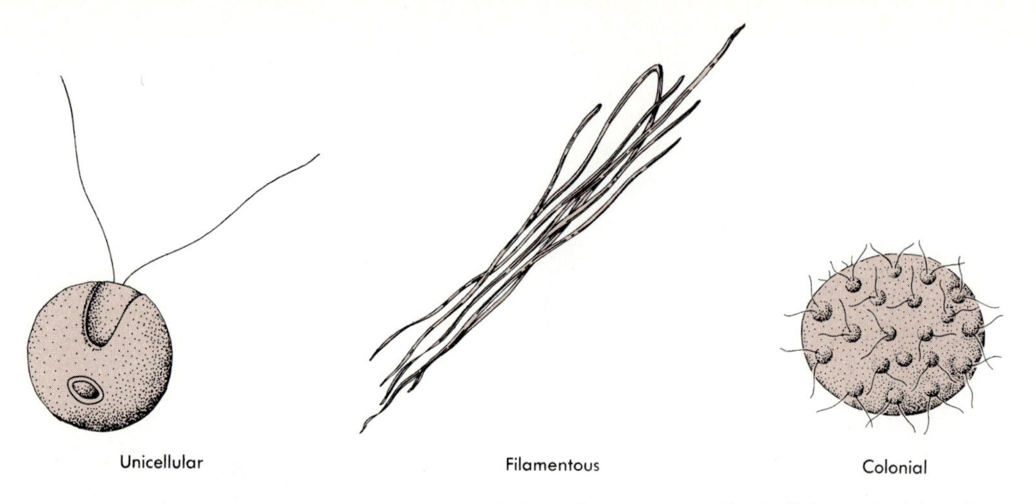

Figure 4-18 *Representation of various morphological arrangements (unicellular, colonial, and filamentous) of algae.*

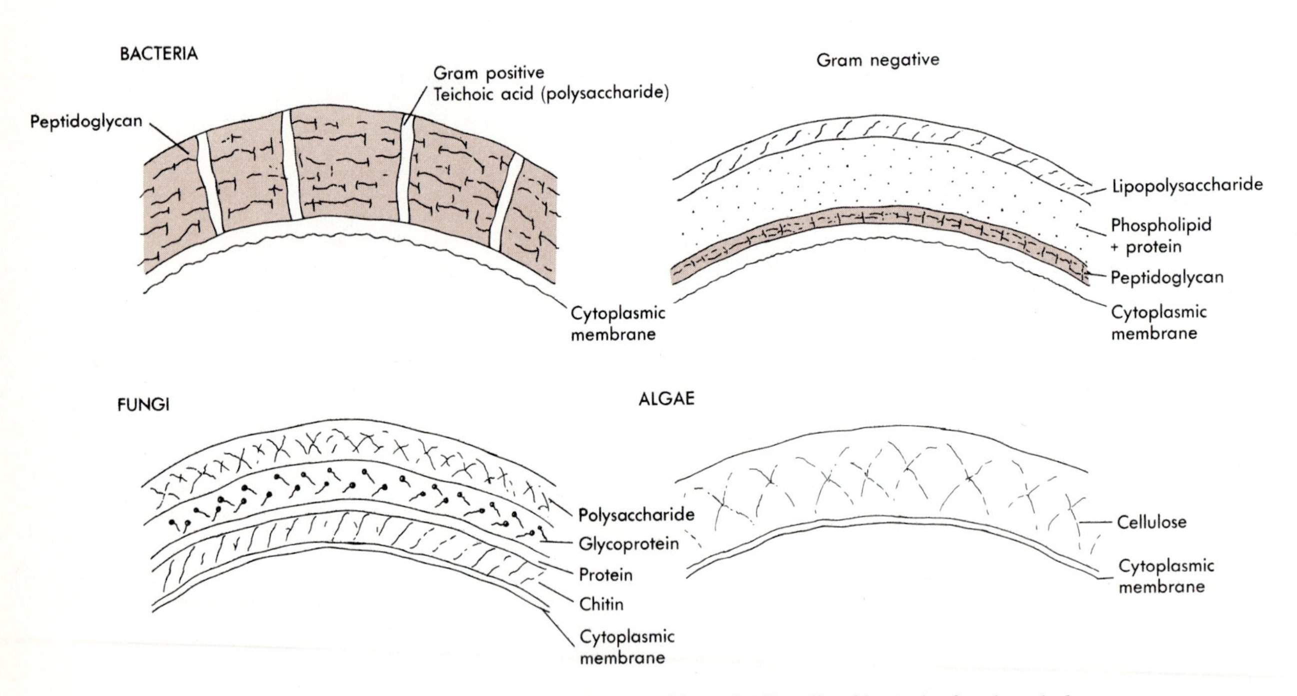

Figure 4-19 *Comparative structure and composition of cell walls of bacteria, fungi, and algae.*

thesis during cell division. The photosynthetic bacteria also possess unusual membrane organization in which the membrane assumes various forms in the cytoplasm to increase the surface area and accommodate the photosynthetic apparatus (Figure 4-20). These specialized photosynthetic membranes are attached to or are continuous with the cytoplasmic membrane. The purple bacteria and cyanobacteria possess the photosynthetic membrane system, called **thylakoids**, that is continuous with the cytoplasmic membrane. The green bacteria possess a photosynthetic membrane system, called **chlorosomes**, that are attached to the cytoplasmic membrane.

Cytoplasmic membrane in eukaryotes. The organization of the cytoplasmic membrane in eukaryotes is more complex than in prokaryotes. All eukaryotes possess an internal membrane system called the **endoplasmic reticulum (ER)** that is continuous with the cytoplasmic membrane (also called **plasma membrane**). The ER fills the cytoplasm and carries out three impor-

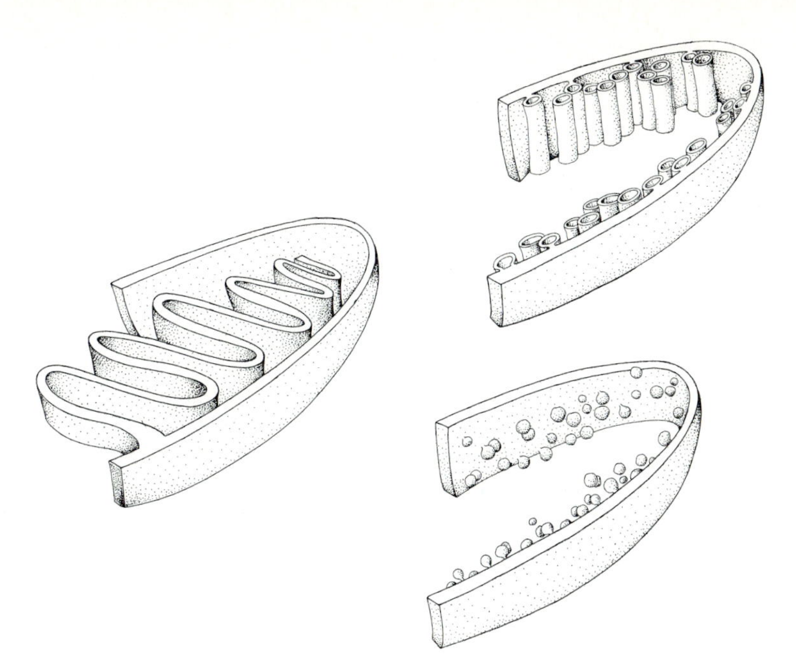

Figure 4-20 *Examples of arrangement of photosynthetic membranes in bacteria.*

tant functions: (1) It provides a channel for transport of molecules between various regions of the cell. Some of these molecules, such as proteins, are synthesized on ribosomes attached to the ER. (2) It is the site for enzymes that engage in carbohydrate and lipid synthesis. (3) It forms subcompartments in the cell that help to separate certain cellular functions, for example, degradative from biosynthetic functions. Most of the ER (Figure 4-5) is a highly flexible membrane structure that is constantly undergoing degradation and synthesis to suit the particular needs of the cell.

The protozoa are organisms that lack a cell wall, and their cytoplasmic membrane must suffice as a structural support. Although sterols do provide some ridigity to the cytoplasmic membrane, there are other modifications that are inherent in this group of prokaryotes. In some species a layer of organic material, called the **pellicle**, lies external to the cytoplasmic membrane. In still other protozoa protective coverings lie over the pellicle. These coverings are loose fitting and are composed of an organic matrix impregnated with inorganic substances such as calcium carbonate. For example, the foraminifers have flourished in different geological eras, and their deposits consist of calcium carbonate shells that serve as markers of their existence.

Concept Check

1. Give the minimal and maximal size dimensions of prokaryotic and eukaryotic microorganisms and the effect size has on the organism.
2. Describe the basic shapes of prokaryotes and eukaryotes and the terminology used to identify prokaryotic shapes.
3. Why is shape not useful in distinguishing one bacterial species from another?
4. What is the difference in cell wall composition between prokaryotes and eukaryotes, and how can the prokaryotic cell wall be used to subdivide bacteria?
5. What are the similarities and differences that exist between the prokaryotic and eukaryotic cell membrane? What variations exist in the cytoplasmic membrane of prokaryotes?

The cytoplasm and its contents

Prokaryotes. The organization of the cytoplasm in prokaryotes is simple compared to eukaryotes (Figure 4-5). The fluid portion of the cytoplasm consists of soluble enzymes and various inorganic and organic compounds. When the prokaryotic cell is examined in electron micrographs, the cytoplasm appears to be very granular because of the widely distributed ribosomes (Figure 4-21). The prokaryotic ribosome is smaller than its eukaryotic counterpart and has a sedimentation constant of 70S.* The genetic information of the cell is contained in an area called the **nuclear area** or **nucleoid**. These terms are used to distinguish it from the eukaryotic nucleus, which is surrounded by a nuclear membrane. The nucleoid consists of a single, circular, double-stranded DNA molecule that is not attached to any proteins. The bacterial cell may have from one to four nucleoids, each consisting of a single chromosome and containing information identical to the DNA in other nucleoids (Figure 4-22).

Nearly all bacteria so far discovered also possess DNA molecules that are considered extrachromosomal; that is, they are not associated with nucleoid DNA. These circular DNA molecules are called **plasmids** and they carry information not essential for survival but do provide specific advantages to the bacterium (for example, resistance to antibiotics). Plasmids are not associated with higher eukaryotes but they have been found in some lower eukaryotes such as yeasts.

* S stands for Svedberg units. They represent the relative sedimentation constant of a component being centrifuged at very high speeds. The greater the molecular weight or density of a component, the faster it sediments in a high-viscosity suspending medium, and the larger the S value (see Point of Interest).

Electron micrographs also reveal that in some bacteria large bodies of insoluble material are suspended in the cytoplasm. These bodies are called **inclusion bodies** and often consist of higher molecular weight polymers such as glycogen or β-polyhydroxybutyrate granules or inorganic molecules such as granules of sulfur or phosphate. Many inclusions are carbon or energy storage forms for the cell, but they can have other functions.

Eukaryotes. The eukaryotic cytoplasm and its contents are complex in organization (Figure 4-23). The fluid portion, like that of prokaryotes, contains various soluble enzymes and organic and inorganic molecules. Ribosomes can be found free in the cytoplasm of eukaryotes as well as attached to the endoplasmic reticulum. The eukaryotic ribosome has a sedimentation constant of 80S. Suspended in the cytoplasm are various organelles enclosed within membranes, most of which are derived from the endoplasmic reticulum. These organelles include the **nucleus, chloroplast, mitochondrion**, and **Golgi body**.

The nucleus is the largest organelle in the eukaryotic

Point of Interest

SEDIMENTATION AND THE ULTRACENTRIFUGE

Separation of various bacterial components can be accomplished in special centrifuges called preparative ultracentrifuges (right). One technique, called **density gradient centrifugation,** permits one to separate particles or molecules, such as ribosomes or enzymes, based on their size and shape. The material to be sedimented, for example, a solution of ribosomes, is layered over the top of a tube containing an inert soluble material, such as sucrose, that is at a very high concentration. When the tube is spun in the centrifuge at a high rate of speed (present-day centrifuges can rotate at speeds of up to 80,000 rpm), a density gradient of sucrose is formed. The rate at which the bacterial components in the solution sediment depends on their size and shape and is expressed as its sedimentation coefficient, or *s* value. Sedimentation coefficients are in units of seconds, and because they are so small, are expressed in **Svedberg** units (S) where $1S = 1 \times 10^{-13}$ second.

A very good example of the influence of shape on the sedimentation coefficient is in the case of ribosomes. The bacterial ribosome is composed of two units that can be separated. One unit has a sedimentation value of 50S, while the other unit has a sedimentation value of 30S. When the two components are combined and sedimented together, the shape of the ribosome complex is different and the sedimentation of the ribosome complex turns out to be 70S, not the 80S to be expected if one were combining the sedimentation rates of the individual ribosome units.

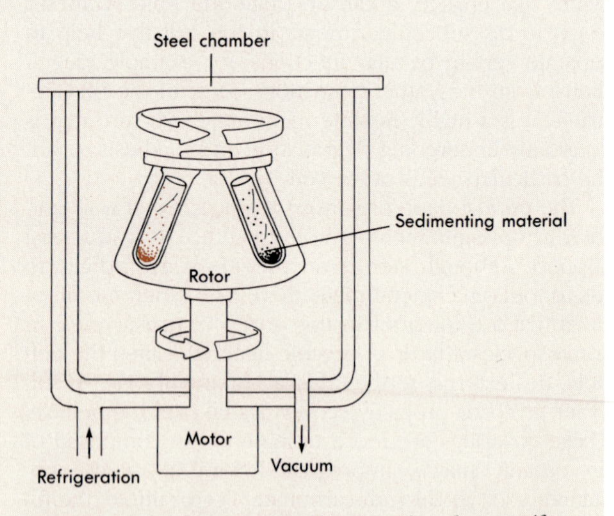

Diagrammatic representation of preparative ultracentrifuge. Material to be sedimented is placed in tubes that are inserted into cylindrical holes in metal rotor. Rotation of rotor generates centrifugal forces that cause particles in tube to sediment. Unit is refrigerated (4° C), which prevents heat damage to sample material. In addition, vacuum is produced in system that reduces friction.

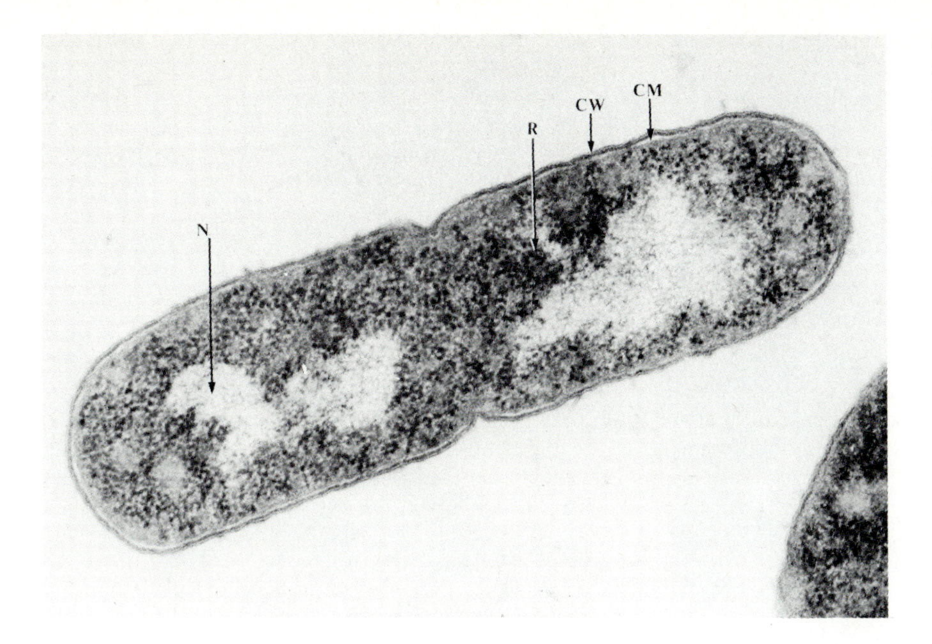

Figure 4-21 *Ribosomes. Electron micrograph of* Escherichia coli *(×60,000) demonstrating ribosomes* (R) *with some concentrated around nucleoids* (N). *(*CM, *Cytoplasmic membrane;* CW, *cell wall).*

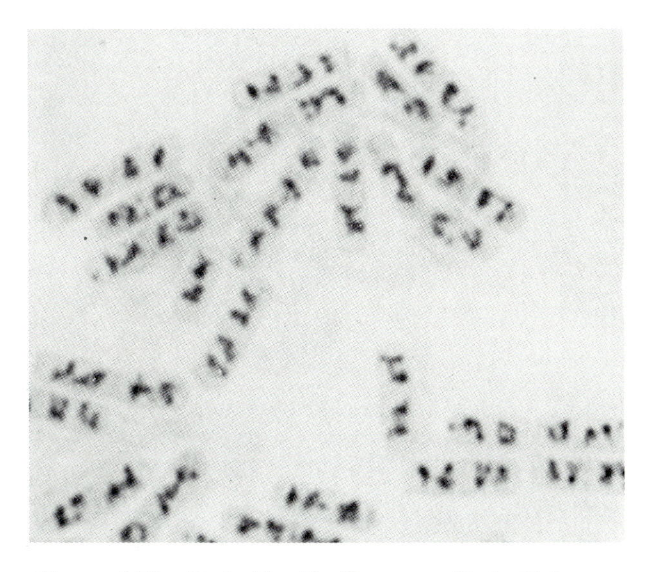

Figure 4-22 *Nucleoids of* Bacillus cereus. *Nucleoids have been fixed in osmium tetraoxide and stained with Giemsa.*

cell and contains the genetic information (Figure 4-23). The genetic information is double-stranded linear DNA complexed with proteins called **histones** to form compact units or **chromosomes.** Located within the nucleus is a granular region called the **nucleolus** (Figure 4-23). The nucleolus engages in the synthesis of ribosomal RNA and is the site of ribosome assembly in the cell. The chloroplast is associated with photosynthetic organisms and among eukaryotic microorganisms is found primarily in the algae. The mitochondrion (Figure 4-23) is found in all eukaryotes. Its task is to generate energy, a property that in prokaryotes is carried out by enzymes in the cytoplasmic membrane. Both mitochondria and chloroplasts contain DNA and ribosomes, and both organelles are self-replicating. The DNA of the organelle is required for their replication, but most of the necessary proteins are still encoded in nuclear DNA. The DNA of these organelles is naked (no histones) and is more closely related to prokaryotic DNA. Their ribosomes are also more closely related to prokaryotic ribosomes. These characteristics support the hypothesis that both organelles may have arisen by a process in which a bacterium was ingested by the predecessor of the eukaryotic cell. The Golgi bodies are flattened stacks of membranes derived from the endoplasmic reticulum (Figure 4-24). Their primary function is in the collection, packaging, and distribution of molecules produced on the endoplasmic reticulum. In addition to the cytoplasmic organelles, one can also find various storage forms of molecules in the form of polymers of glycogen or lipid.

The cytoplasm of some eukaryotes has been observed to show a characteristic movement referred to as **cytoplasmic streaming.** Cytoplasmic streaming is associated with cellular fluids that exhibit a low viscosity, a property not found in prokaryotic cells. Cytoplasmic streaming enables nutrients to be more quickly and evenly distributed throughout the cell. Simple diffusion (movement of molecules from a point of high concentration to an area of low concentration) is not an efficient process for the distribution of molecules and com-

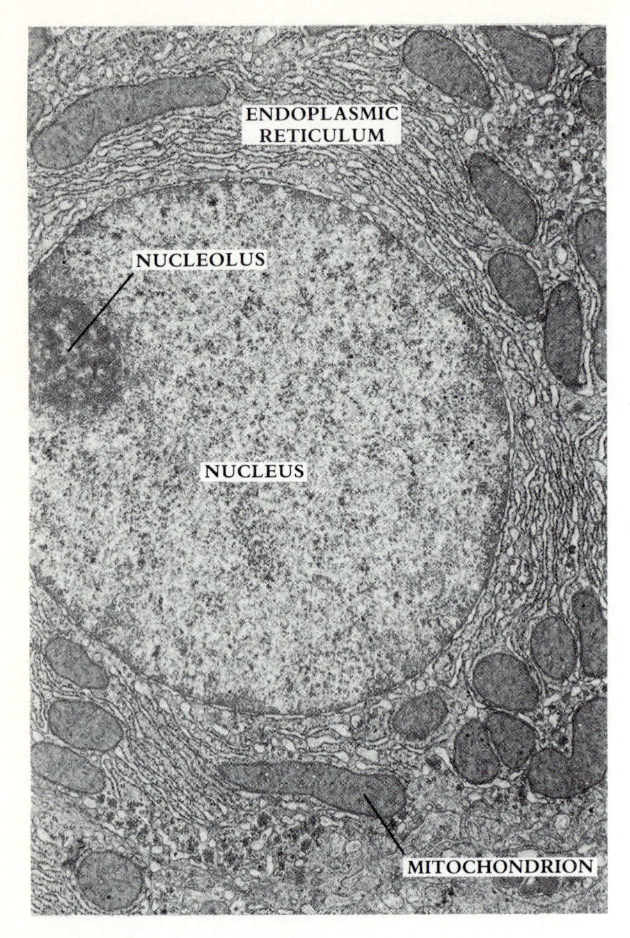

Figure 4-23 *Cross-section of eukaryotic cell showing cell nucleus, nucleolus, endoplasmic reticulum, and mitochondrion.*

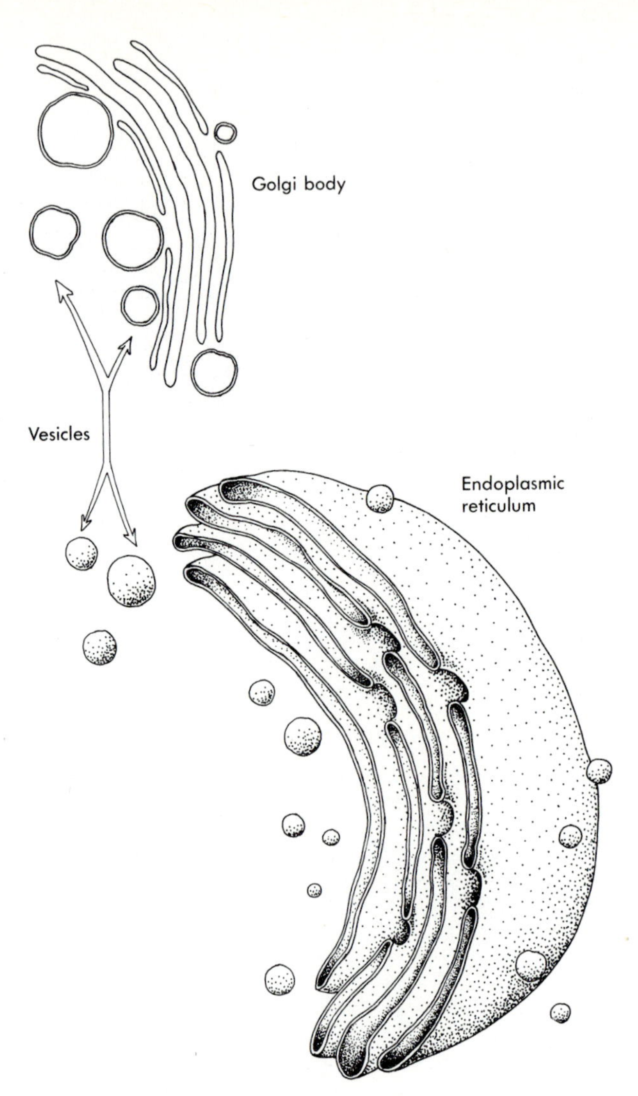

Figure 4-24 *Representation of Golgi body in cross-section, demonstrating its saclike structure. Golgi body in electron micrograph* (upper) *and* (lower) *artist's concept. Spherical structures are vesicles, which are formed by pinching off ends of Golgi sacs. Vesicles contain enzymes and other substances derived from endoplasmic reticulum.*

ponents in the eukaryotic cell because of the large size of the eukaryotic cell. Even with the help of cytoplasmic streaming, which may reach 1 mm/sec, nutrient distribution is not as efficient as in prokaryotes. Eukaryotes have compensated for this deficiency by exhibiting reduced metabolic rates and by internal compartmentalization. The fastest growing eukaryote may have a generation time (time for the cell to divide) of 1 hour, whereas the fastest prokaryote *(Vibrio natriegens)* may divide 6 times in 1 hour.

Metabolism and nutrition—prokaryotes and eukaryotes

The first organisms were believed to have evolved on a primitive earth's surface in which organic molecules (derived originally from the transformation of inorganic molecules) were present in an atmosphere devoid of oxygen. These primitive forms of life were prokaryotic cells whose metabolism was based on obtaining **carbon** from the breakdown of organic material. This type of metabolism is called **heterotrophic**, and the organisms carrying it out are called **heterotrophs**. In addition, these heterotrophs also obtained their **energy** from organic carbon molecules.

There must have been a population explosion of these heterotrophic organisms until the rate of consumption of organic molecules exceeded their rate of replenishment. The products of heterotrophic metabolism are hydrogen, carbon dioxide, and various acids

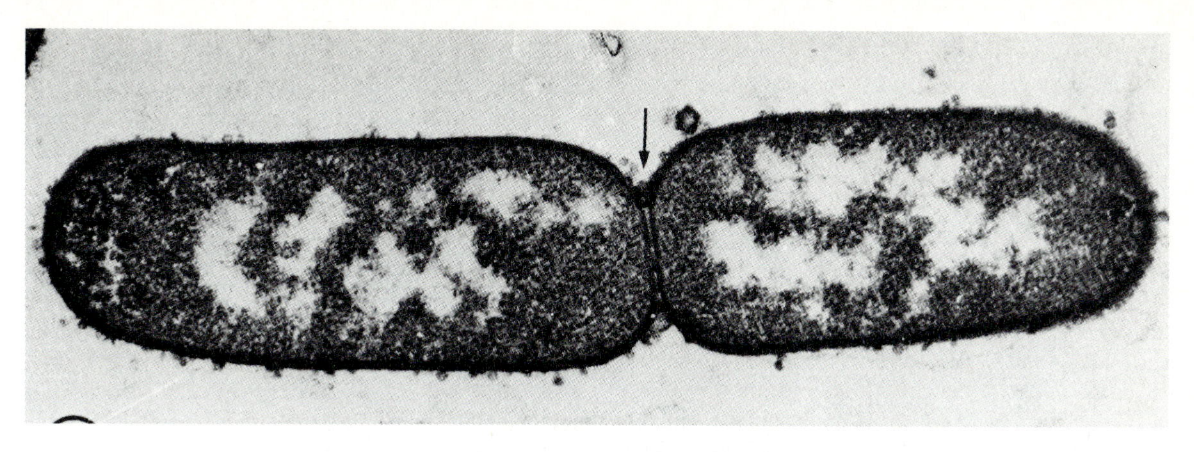

Figure 4-25 *Division in rod-shaped bacterial cell. Division is perpendicular to long axis of cell* (arrow). *Point of indentation where cell divides is transverse septum.*

such as acetic acid plus other energy-depleted molecules. These breakdown products were not suitable as carbon and energy sources for the existing microorganisms. This scenario resulted in a metabolic evolution and the formation of other metabolic types. Any newly evolved microorganisms would have to produce cellular organic matter from the byproducts released by the original microorganisms. These byproducts in turn would eventually become depleted unless another metabolic selection process developed. Eventually microorganisms evolved that obtained their energy from light **(phototrophs)** and converted inorganic carbon into organic molecules **(autotrophs)**. Such organisms are called **photoautotrophs.** Photoautotrophs released oxygen into the atmosphere, and this enabled the early prokaryotes to evolve into cellular systems with greater metabolic diversity and to exploit many ecological niches in which oxygen was present.

Bacteria can obtain their energy and carbon in several ways. Carbon may be obtained from inorganic carbon dioxide or from organic molecules. Energy can be obtained from organic or inorganic molecules (for example, iron, hydrogen, or sulfur compounds), or it may be extracted from light for conversion into chemical energy. Eukaryotes can obtain their carbon from inorganic carbon dioxide or from organic molecules and can obtain energy from organic molecules just like bacteria. However, eukaryotes cannot extract energy from inorganic molecules. The inability to extract energy from inorganic sources and their almost absolute requirement for oxygen has placed severe restrictions on the types of habitat eukaryotes can colonize. Bacteria can be found in hot springs where only inorganic sources of energy are present and where temperatures may exceed 80° C while others can be found near thermal vents located several miles deep on the ocean floor where temperatures exceed 260° C and where oxygen is absent. Other bacteria can be found beneath several meters of ice in the Antarctic. Some bacteria such as the archaebacteria are present in environments that will not support the growth of other prokaryotes or eukaryotes. This type of environmental exploitation is not characteristic of most eukaryotes, which for the most part are confined to areas where oxygen is present and where sunlight or organic sources of energy are available.

Reproduction

Prokaryotes. Sexual reproduction, a characteristic of more advanced forms of life, is absent in bacteria. **Binary fission**, an asexual process, is the most common method of bacterial reproduction and is characterized by an approximately equal separation of the cell into two daughter cells. Division is perpendicular to the long axis of the cell in rods and spherical cells (Figure 4-25) and is preceded by intercalary* growth in which only a small portion of new cell wall is synthesized. The daughter cells, especially of spherical bacteria, do not always separate, and depending on the plane of division, different morphological arrangements are produced (see Figure 4-10). Each daughter cell receives a single unpaired chromosome, that is, the cell is **haploid.**

Budding is another method of asexual reproduction but is relatively infrequent in bacteria, except for a few species (Figure 4-26). It is more common in the eukaryotic yeasts, where it is usually polar, with one end of the parent cell bulging and producing a spherical daughter body that separates from the parent. Budding may appear to be similar to binary fission, but there is a difference. Budding is considered a reproductive process

* Intercalary, inserted between components.

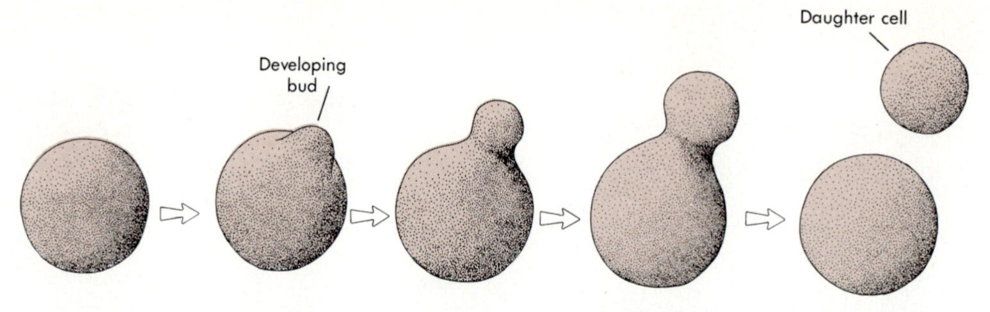

Figure 4-26. *Budding in prokaryotes. Daughter cell or bud appears as protuberance on mother cell. Bud enlarges until it constricts and separates from mother cell. Released daughter cell is smaller than mother cell.*

only if the daughter cell is smaller than the parent cell and all or most of the bud-cell wall and cytoplasmic material are newly synthesized.

Sporulation (formation of spores) is a technique used by several groups of bacteria to protect themselves from environmental stress when nutrients become depleted in their environment. The genera *Clostridium* and *Bacillus* use this technique to produce a single intracellular body form, called an **endospore**, that is unusually resistant to heat, radiation, and chemicals. Sporulation, however, can also be used as a reproductive device by certain members of the genus *Streptomyces*. *Streptomyces* produces filaments during growth which are multinucleated. As the filamentous mass ages, some of the aerial filaments develop a series of cross walls which separates the filaments into individual cellular units. Each unit contains a nucleus and will develop into a spore (Figure 4-27). When the spore reaches a suitable environment it will germinate into a new individual. The genus *Nocardia,* which is related to *Streptomyces,* also grows by producing filaments, but the filament separates by **fragmentation** into smaller units, each capable of independent growth, and no sporulation is involved.

Although sexual reproduction is absent from bacteria, some species possess mating types that do make physical contact. Following contact genetic information passes only from one cell (donor) to a recipient but there is no mutual exchange. This **conjugation** act, therefore, is not a reproductive process but it does enable some bacteria to acquire new traits.

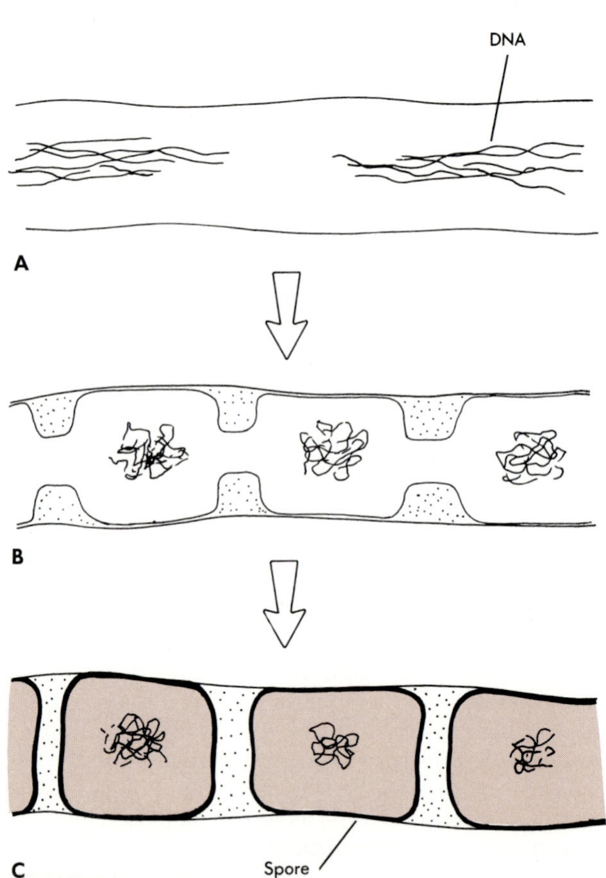

Figure 4-27 *Spore formation in* Streptomyces. ***A,*** *Aerial filament (hypha) contains molecules of DNA in dispersed form.* ***B,*** *Series of cross walls begin to form, separating filament into individual units with molecule of DNA.* ***C,*** *Each unit in filament is spore.*

Eukaryotes. Sexual reproduction is a characteristic that separates eukaryotes from prokaryotes. Sexual processes result in the formation of cells that possess two complete sets of genetic information **(diploid)**. The diploid state results from the fusion of male and female gametes (sex cells), each of which possesses only one set (haploid) of genetic information. The diploid state appears very fleetingly in many eukaryotic microorganisms, that is, most of their life cycle is spent in the haploid state. The more complex eukaryotic microorganisms as well as higher eukaryotes, such as plants and animals, maintain a permanent diploid state, with the haploid state appearing only during formation of gametes. Details of specific sexual processes in eukaryotes will be discussed in Chapter 6.

The most frequent type of reproductive process among the microbial eukaryotes is asexual. **Binary fis-**

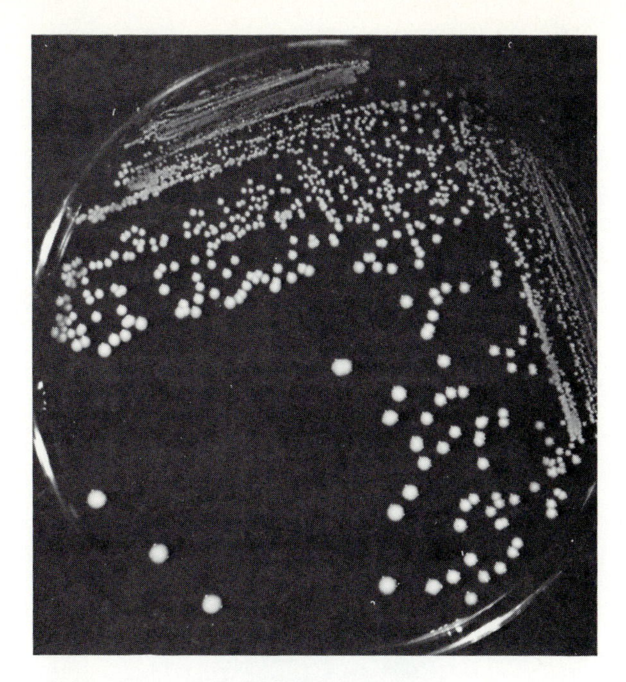

Figure 4-28 *Bacterial colonies on agar surface. Colonies appear as they would to unaided eye.*

sion is the most common form of asexual reproduction in unicellular algae and protozoa, but in the fungi the **spore** is the primary reproductive unit. Fungal spores can be produced asexually as well as sexually; however, asexual spore formation is the most common. **Budding** is a frequent form of reproduction among the nearly spherical yeast cells. **Fragmentation** is also a characteristic type of reproduction in the filamentous algal species. There are variations on these asexual processes, and they will be discussed with the specific eukaryotic group in Chapter 6.

Cellular interactions

Prokaryotes, for the most part, do not exhibit cell-to-cell interactions. Many bacteria form colonies when they reproduce on or in a substratum (Figure 4-28), but these represent random accumulations of cells that do not enter into a cooperative state. Bacteria within the colonies are also competing for existing nutrients and do not differentiate into groups with specific activities, which is characteristic of multicellular organisms. Eukaryotic microorganisms produce more complex colonies in which there is coordination of activities, and it is this colonial coordination of cells that has permitted eukaryotes to evolve into larger multicellular life forms. There are some prokaryotes, however, that produce colonies in which some of the cells exhibit specific roles. The rod-shaped cells of the actinomycetes, for example, develop into long filaments, some of which penetrate the substratum to seek nutrients, whereas other filaments rise above the substratum and produce aerial spores, a cell type that can be released to the surface to produce individual colonies.

Symbiosis. Most environmental niches contain various groups of organisms that interact to create a variety of relationships called **symbioses.** Microorganisms can interact with plants and animals as well as with other microorganisms to create **endosymbiotic** or **ectosymbiotic** relationships. Microorganisms that exhibit endosymbiosis grow within the host cell, whereas those microorganisms that remain attached but outside the host cell demonstrate ectosymbiosis. Ectosymbiotic or endosymbiotic relationships can be categorized as **mutualistic** if both partners benefit from the association; **commensalistic** if one partner is benefited and one is unaffected; and **parasitic** if one organism is harmed and the other is benefited. The most widely studied and clearly defined types of symbioses have been those existing between microorganisms and plants or animals. Several species of bacteria, fungi, and protozoa, for example, are the frequent cause of disease in plants, and animals, including humans. There are instances where one bacterium preys on another bacterium, but these are very rare.

Bacteria can interact with other microorganisms, but this type of relationship has been difficult to evaluate in their natural environment, and most of what we know comes from laboratory studies. Many bacteria, because of their small size, are the prey for eukaryotic microorganisms, and laboratory studies have demonstrated that the ingested prokaryote occasionally survives in its host and becomes an endosymbiont. The eukaryotic microorganism is always the host when endosymbiosis occurs, and this is believed to be the result of the greater flexibility of its cytoplasmic membrane. The cytoplasmic membrane of eukaryotic microorganisms can surround and engulf the prokaryote, but the prokaryotic cytoplasmic membrane of wall-less bacteria is inflexible, and large particles, such as microorganisms, cannot penetrate into the cytoplasm. For those bacteria that possess a cell wall, the wall is a natural barrier to the ingestion process.

One example of endosymbiosis is that existing between bacteria and some protozoa. The protozoa harbor a bacterium that produces a toxic substance called the "killer" trait (Figure 4-29). The toxin is not lethal to its host but will kill those strains of the same protozoa that do not possess the killer factor. This type of association provides the protozoa, containing the killer factor, with a competitive edge in an environment containing several strains of the species. The endosymbiont in turn is provided nutrients from the protozoa. Another frequent microbial association is that existing between an alga or cyanobacterium and a fungus. This association is called a **lichen** and is characterized by the penetration of specialized fungal hyphae into the photosynthetic partner (Figure 4-30). Some of the compounds synthesized by the alga are used as nutrients by the fungus. The fungus, which quantitatively makes up most of the lichen, provides its algal partner with an environment that protects it from wind, heat, and dessication. Lichens are ubiquitous and can be found in varied habitats such as the Arctic and Antarctic and solidified lava flows in Hawaii. Many are found on the bark of trees or on rocks. The reindeer moss is a lichen and provides some winter food for caribou and reindeer in Europe and North America.

Microorganisms also develop mutualistic relationships with higher eukaroytes such as plants and animals; these are outlined in Table 4-3. More detailed discussions of these relationships will appear later in this book.

Movement (motility)

Prokaryotes. Independent movement is not necessary for the survival of the microbial cell, but it does provide it with certain advantages. Movement directed to nutrients or away from toxic substances is of obvious value to the organism that must compete with other microorganisms in its environment. Movement to or away from a chemical stimulus is called **chemotaxis**. Motility is usually associated with single cells, but this need not always be the case. Some bacteria, for example, use motility to form aggregates of cells that swarm as a single unit. Swarming may be a mechanism for feeding or forming a colonial structure that protects some of the

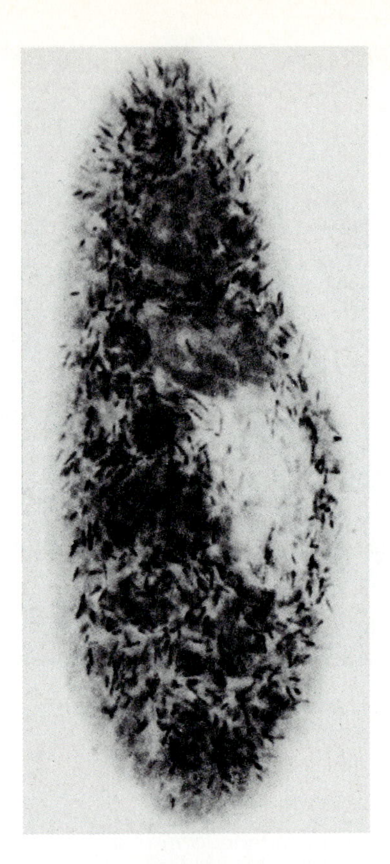

Figure 4-29 *Phase-contrast photomicrograph of protozoa,* Paramecium *species. (×500.) Protozoa contains many endosymbionts (bacteria) that appear as dark-stained rods.*

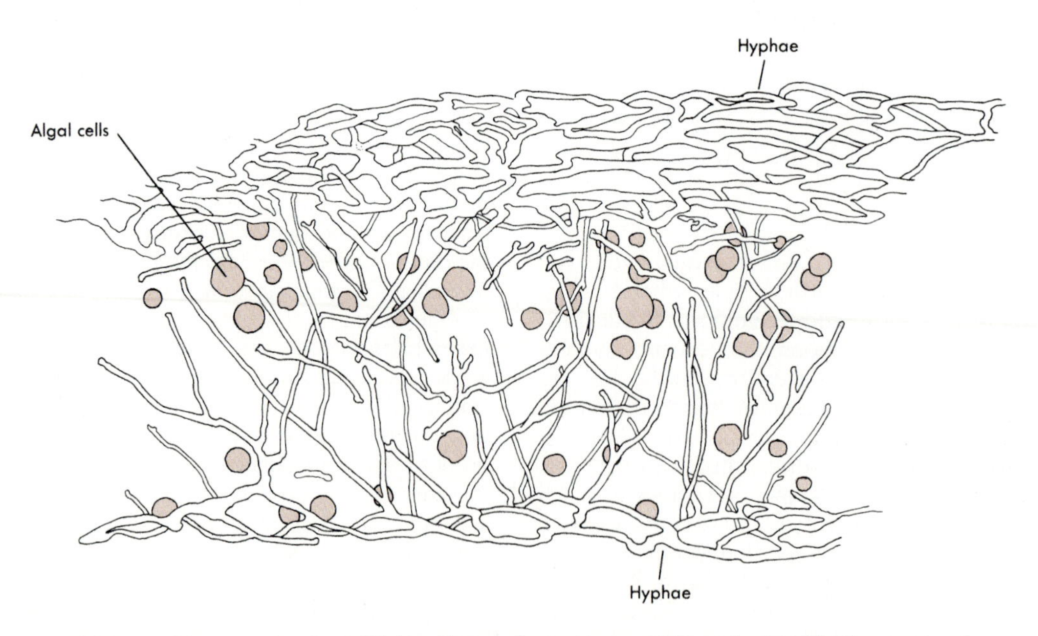

Figure 4-30 *Cross-section of lichen. Fungus forms upper and lower layers of hyphae, some of which penetrate into central area that contains algal cells.*

Table 4-3 *Mutualistic relationships between microorganisms and animals or plants*

Eukaryotic host	Site of microbial attachment	Microorganism involved	Benefit provided by microorganism	Benefit provided by host
Human	Skin	Primarily bacteria	Prevent colonization by disease-producing organisms	Sebaceous gland supplies nutrients such as salts, amino acids, and lipids
	Oral cavity and upper respiratory tract	Primarily bacteria but some yeasts	Prevent colonization by disease-producing organisms	Nutrients
	Intestinal tract	Primarily bacteria but some yeasts	Release products that inhibit colonization by other bacteria; produce vitamins (B_{12}, thiamin, etc.)	Nutrients
Ruminant animals (goat, sheep, cow, etc.)*	Rumen	Primarily bacteria but some protozoa	Digest polymers such as pectin, cellulose, and starch; products of polymer digestion are organic acids used by animal as source of energy; synthesize vitamins and amino acids for animal use	Nutrients
Plant				
Leguminous (pea, bean, clover, alfalfa)	Root	Bacterium *(Rhizobium)*	Fixes atmospheric nitrogen into ammonia, which can be used for amino acid synthesis	Supplies nutrients in form of carbohydrates; provides compound that protects nitrogen-fixing enzyme
Trees (forest)	Root; hyphae penetrate cortical cells (endomycorrhizae) or lie between cortical cells (ectomycorrhizae)	Fungus	Provides greater surface area for nutrient absorption by root	Supplies nutrients in form of carbohydrates

*Ruminant animals possess a special digestive compartment called the **rumen** where plant material is degraded.

individuals of the colony from the extremes of the environment. Motility by most organisms is provided by appendages called **flagella**.

Flagella. The flagellum (Figure 4-5) is the principal organelle associated with movement in bacteria. It is a long, helical, unbranched filament ranging in length from 10 to 20 μm, which makes it several times longer than the average bacterial cell. It is so thin (0.2 μm), however, that it cannot be observed unstained with the compound microscope. Flagella may be arranged on the cell according to four basic types (Figure 4-31): **peritrichous** (flagella cover the entire cell surface), **monotrichous** (a single polar flagellum), **lophotrichous** (two or more flagella at one pole of the cell), and **amphitrichous** (tufts of flagella at both poles of the cell). The arrangement of flagella on an organism is not always constant. Recent studies have shown that qualitative changes in nutrients in a medium can cause variations in the flagellar arrangement.

Flagella are better suited for movement through relatively nonviscous liquids. When the viscosity is increased, flagellar locomotion is severely impeded because the medium offers resistance to flagellar movement. Some flagellated species such as *Pseudomonas aeruginosa,* which is 3 to 4 μm in length, can move 37 cell lengths/second. This is a remarkable rate, since horses move about 6 lengths/second.

Flagella-like filaments called **axial filaments**, which do not project from the surface of the cell but are wrapped about the body of the cell, are characteristic of the spirally shaped bacteria called **spirochetes** (Figure 4-32). Spirochete locomotion is not hampered when the

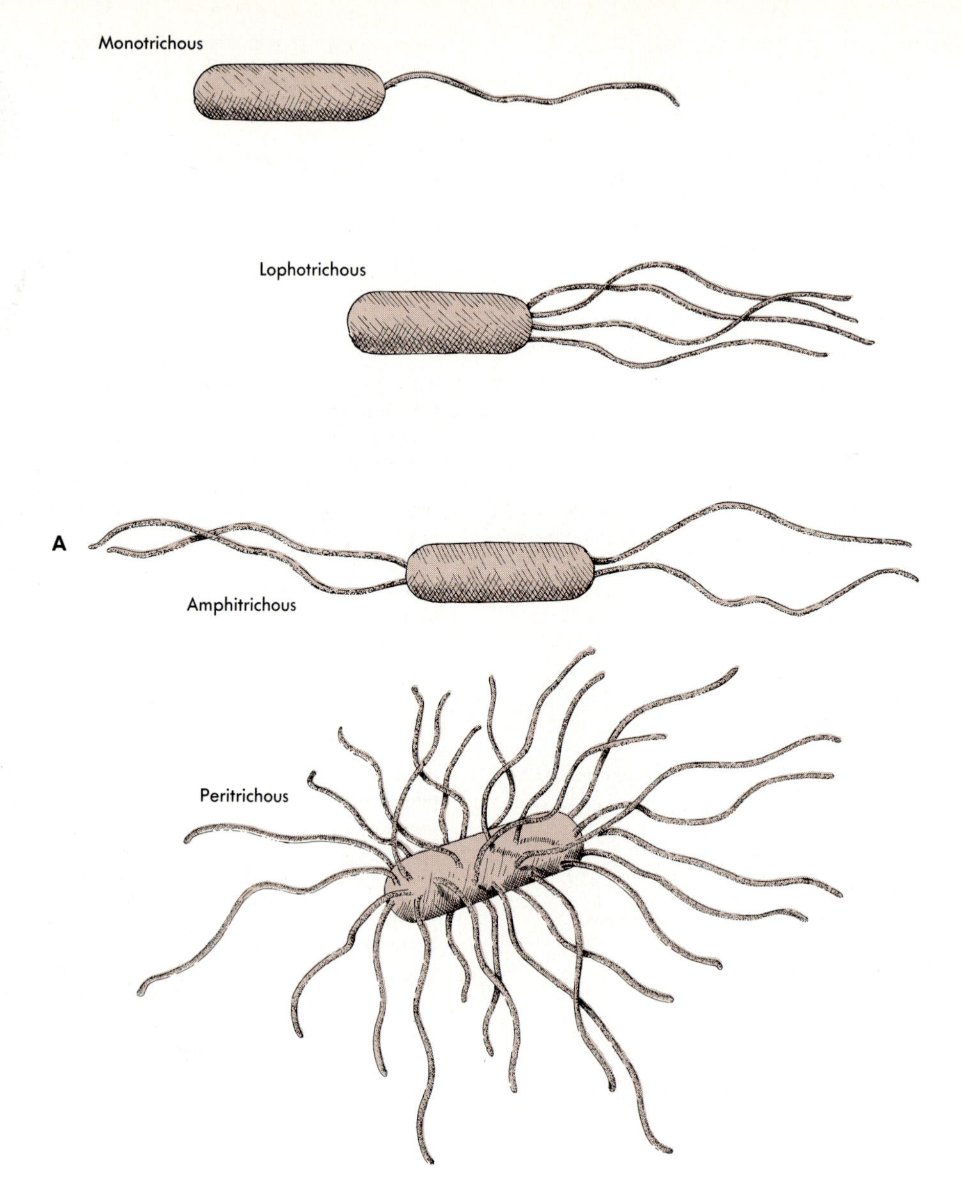

Figure 4-31 ***A, Flagellar types. B to D, Electron micrographs of some representative flagellar types.***

viscosity of the medium is increased, and in many instances the velocity actually increases. This is probably because of the location of the axial fibers, which are beneath the outer sheath of the cell. The axial fibers are therefore not in direct contact with the extracellular environment. The velocity of the spirochetes depends on the viscosity of the medium and is somewhere between 20 and 30 μm/second.

Gliding. Some prokaryotes, especially cyanobacteria and myxobacteria, show a gliding type of motility over solid or semisolid surfaces. Gliding motility may be defined as that type of movement in which (1) obvious locomotion organelles are absent, (2) there is no visible wriggling or contraction, and (3) the change of shape is restricted to bending. The only characteristic common to all of the gliders is the secretion of an extracellular **slime layer** outside the cell wall. Gliding may be associated with single cells or a filament of cells (trichome*). The velocity of movement is affected by moisture levels, temperature, and the nature of the substratum. Some single-cell gliders move only 1 to 2 μm/minute, whereas others may move as fast as 150 μm/minute. Trichomes of some cyanobacteria move as rapidly as 600 μm/min-

* Trichome is a filament of cells surrounded by a slime layer.

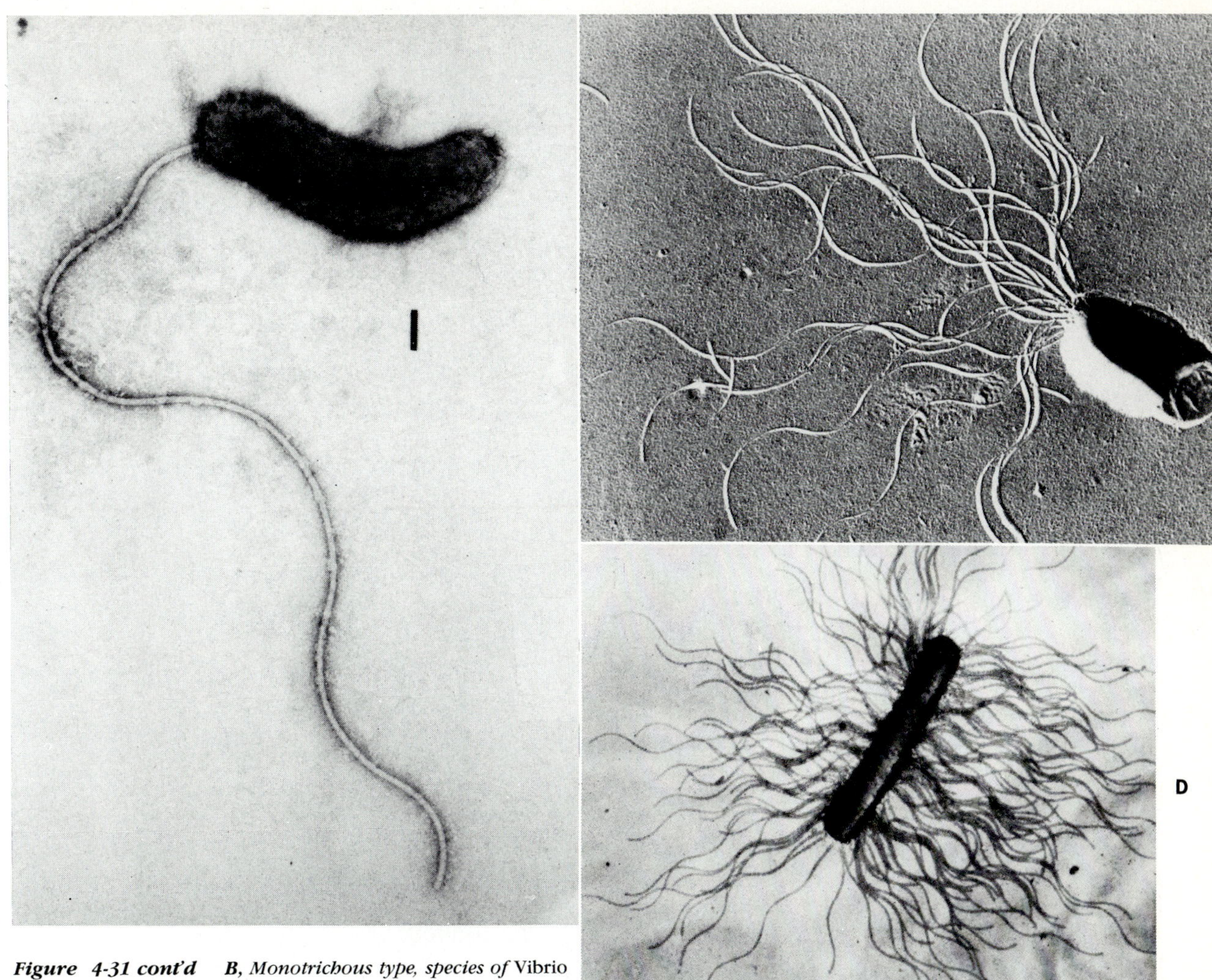

Figure 4-31 cont'd ***B,** Monotrichous type, species of* Vibrio *(×24,000).* ***C,** Lophotrichous type, species of* Pseudomonas *(×30,000).* ***D,** Peritrichous type, species of* Salmonella.

ute. None of these velocities compares with the rapid movements of flagellated bacteria. The exact mechanism(s) for gliding movement is as yet unknown.

Eukaryotes

Flagella and cilia. The flagellum of eukaryotes is of more complex construction than the prokaryotic flagellum. The term **cilium** is equivalent to the term **flagellum** with respect to structure and function. Cilia, however, are shorter and more numerous than flagella. Eukaryotic flagella may reach lengths of 250 μm, whereas cilia are usually no longer than 5 to 10 μm in length. Smaller eukaryotes such as microscopic algae often possess one or more flagella, while larger eukaryotes such as certain protozoa may have several hundred cilia covering the entire surface of the organism (Figure 4-33). Like prokaryotes, eukaryotes use movement to perform functions other than responses to nutrients. Many eukaryotes use motility to aggregate with members of the same species. The aggregation process often culminates in the formation of a colony of cells that undergoes differentiation, with different cell types performing specific functions. The colonial state is of survival value to the species. Some eukaryotes show no motility in the adult stage, but their gametes (sex cells) are motile, particularly in the presence of chemical attractants. This ensures that there will be rapid zygote for-

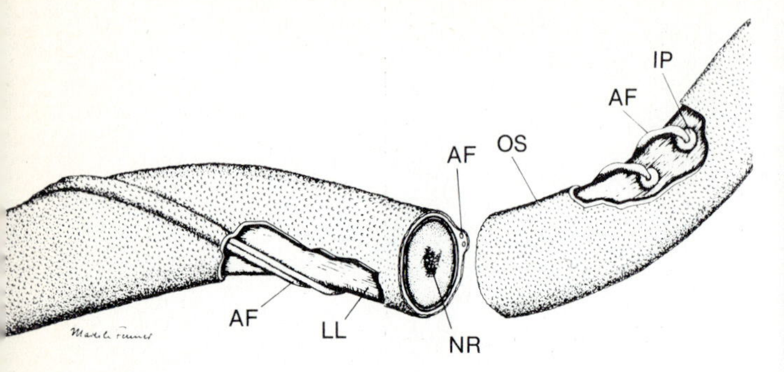

Figure 4-32 *Relationship of flagella-like fibers (axial filaments) in spirochetes to other cell components. Axial filaments* (AF) *are located between outer sheath* (OS) *and lipoprotein layer* (LL) *of cell wall. Axial filaments are inserted into protoplasmic cylinder by insertion pores* (IP). NR, *Nuclear region.*

Figure 4-33 *Scanning electron micrograph of ciliate* Paramecium caudatum. *Organism has been preserved in state of movement showing arrangement of cilia.*

mation before the cells succumb to environmental influences.

Cytoplasmic streaming. Amebas use the process of cytoplasmic streaming as a means of movement over a solid surface. The cytoplasm of amebas contains a large core of fluid called the **endoplasm**. Within the endoplasm are located cellular organelles (Figure 4-34). The endoplasm is surrounded by a transparent gel called the **ectoplasm**. Movement in amebas is believed to occur by contraction of the ectoplasm, which squeezes the endoplasm, causing it to stream. The loose cytoplasmic membrane projects foward into elongations called **pseudopodia**. The endoplasm streams into the pseudopods, and the ectoplasm becomes solid at its tip. In other parts of the cell the ectoplasm changes into a more fluid endoplasm and streams toward the pseudopod, causing it to extend further and thus create movement. Some evidence suggests that actin fibers (like those in mammalian muscle) are present and are capable of contraction, thus promoting cytoplasmic streaming.

Concept Check

1. What differences exist in terms of the type of cytoplasmic components found in eukaryotes as compared with prokaryotes?
2. What differences exist in the organization and location of the genetic material of prokaryotes and eukaryotes?
3. What are the similarities or differences in the way prokaryotes and eukaryotes obtain their energy from metabolic processes?
4. Describe the various types of asexual reproduction in prokaryotes. In eukaryotes.
5. Define the terms used to explain how bacteria interact with other bacteria or bacteria with eukaryotes.
6. Describe the various types of movement found in prokaryotes and eukaryotes.

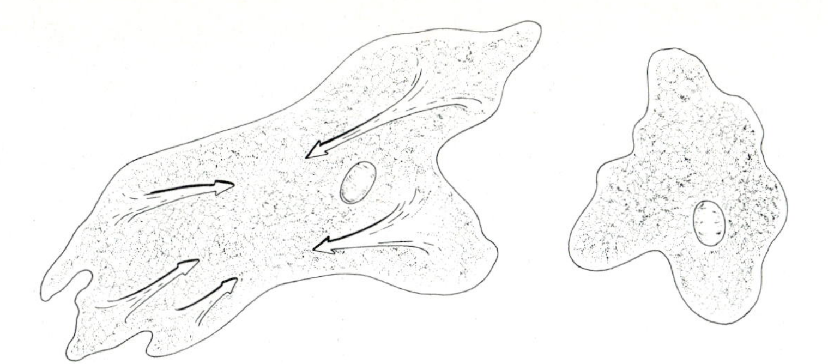

Figure 4-34 *Cytoplasmic streaming in ameba.*

SUMMARY

1. The biological world is separated into two groups—prokaryotes and eukaryotes—based primarily on the presence of a nuclear membrane in eukaryotes and its absence in prokaryotes. Microorganisms belong to both groups and this has resulted in several classification schemes. Bacteria are now placed in the kingdom *Monera*, while eukaryotes belong to the kingdom *Protista*.
2. The species is the basic unit of classification for all organisms. For bacteria a species represents a collection of strains that have certain common characteristics. Species are placed into genera, and therefore every organism is identified by its genus and species. The most widely accepted reference text for the various bacterial species is *Bergey's Manual of Systematic Bacteriology*.
3. Most bacteria are believed to have been derived from a common ancestral prototype; however, the discovery of an unusual group of bacteria called Archaebacteria suggests that the latter may have diverged before the evolution of true bacteria.
4. The evolution of prokaryotes to eukaryotes may have originated with the photosynthetic bacteria. The similarities between the photosynthetic machinery and respiratory mechanisms of some prokaryotes and eukaryotes suggest an evolutionary link between the two groups.
5. The general features of prokaryotes and eukaryotes, including characteristics and structures, are outlined in Table 4-1.

Point of Interest

ORIGIN OF THE EUKARYOTIC CELL

The question of how life evolved on earth interests almost everyone. Most scientists (but not all) believe that life on earth began on earth. It is generally thought that the first life forms on earth were the simple prokaryotic cells, appearing around 3 billion years ago. These prokaryotic cells were photosynthetic cyanobacteria, or blue-green algae and bacteria. Three main theories of how these prokaryatic cells evolved into more complex eukaryotic cells are given below.

The first theory is the serial endosymbiosis theory, or the exogenous theory. Evidence suggests that eukaryotic organelles such as mitochondria, chloroplasts, and microtubular structures arose by the symbiotic merging of two organisms. For example, an aerobic prokaryote may have invaded an anaerobic prokaryote to form a eukaryotic heterotroph (and eventually the aerobic prokaryote lost its independence and became the mitochondria). Algae evolved when this eukaryotic heterotroph merged with a cyanobacterium (and the cyanobacterium lost its independence and became the chloroplast). Of course, all plants evolved from the algae. Finally, spirochetes joined with a eukaryote and became the first protozoans, from which all animals evolved.

In contrast to the exogenous approach, the autogenous theory proposes that the eukaryotic cell evolved by progressive compartmentalization. The complex organelles of the eukaryote, such as the nuclear membrane, golgi bodies, vacuoles, and endoplasmic reticulum, evolved originally from a prokaryotic cell through a series of mutations. This theory suggests that all life forms arose from an ancestrial prokaryote along a single line of descent.

The last theory of the origin of the eukaryotic cell is called the fungal theory. According to this hypothesis, fungi is believed to be the first eukaryote, rising from the prokaryotic cell (the bacteria). All other eukaryotic cells then evolved from the primitive fungal cell.

Although it is still uncertain which of the theories is true, the exogenous theory is the most exciting, because it provides an explanation of evolutionary process based on symbiosis and the merger of two organisms to make a totally new one.

Self-study Quiz

Identification

In the blank space insert the letter *P* if the indicated characteristic is found primarily in prokaroytes, *E* for eukaroytes, or *both* if the characteristic is found consistently in both groups.

1. ________ Photosynthesis
2. ________ Flagella
3. ________ Pili
4. ________ Plasmids
5. ________ Reproduction by sporulation
6. ________ Sterols in cytoplasmic membrane
7. ________ Greatest ratio of surface area to volume
8. ________ Sexual reproduction
9. ________ Nuclear membrane
10. ________ Confined to aerobic habitats

Multiple choice

1. Bacteria are classified as members of the kingdom:
   A. Protista
   B. Protozoa
   C. Monera
   D. Protophyta
   E. Protomycota
2. A bacterial strain is defined as:
   A. A collection of species having the same characteristics.
   B. A collection of species that are similar in biochemical characteristics.
   C. A number of descendants derived from a single colony using pure culture techniques.
   D. A number of descendants originally isolated from the same ecological habitat.
   E. A collection of species that exhibit the same method of reproduction.
3. Which of the following characteristics separate the archaebacteria from the true bacteria?
   A. Only archaebacteria have sterols in their cytoplasmic membranes.
   B. Only archaebacteria have sterols in their cell walls.
   C. Only archaebacteria are found in habitats where the temperature exceeds 50° C.
   D. Archaebacteria lack a true peptidoglycan layer in the cell wall.
   E. Archaebacteria exhibit photosynthesis.
4. The group of bacteria that uses sporulation as a means of reproduction is:
   A. Archaebacteria
   B. *Nocardia*
   C. *Streptococcus*
   D. *Staphylococcus*
   E. *Streptomyces*
5. Dimorphism is characteristic of:
   A. Algae
   B. True bacteria
   C. Fungi
   D. Archaebacteria
   E. Protozoa
6. Which of the following is characteristic of gram-negative bacteria?
   A. Have a cell wall peptidoglycan that is quantitatively less than in gram-positive cells
   B. Cell wall crystal violet iodine complex decolorized by alcohol in the gram stain
   C. Possess cell wall lipids that are not found in gram-positive bacteria
   D. Stain red following decolorization in gram stain
   E. All of the above are characteristic of gram-negative bacteria
7. The layer of organic material that lies external to the cytoplasmic membrane of the protozoa is called the:
   A. Mesosome
   B. Pellicle
   C. Peduncle
   D. Protoplasmic membrane
   E. Plasmid
8. The lichen represents a relationship best described as:
   A. Dimorphism
   B. Parasitism
   C. Heterotrophism
   D. Mutualism
   E. Phototropism
9. Axial filaments are characteristic of which of the following prokaryotes?
   A. Archaebacteria
   B. Spirochetes
   C. *Clostridium*
   D. *Bacillus*
   E. *Cyanobacterium*

Thought questions

1. Considering that fungi are found primarily in the soil, what is the value of dimorphism?
2. What is the value of pleomorphism to rod-shaped bacteria in an environment such as the soil? Would the same reasons apply to an organism whose habitat is the intestinal tract?

SELECTED READINGS

Books

Bull, A.T., and Slater, J.H. (Ed.) *Microbial interactions and communities*, Academic Press, Inc., New York, 1982.

Carlisle, M.J., Collins, J.F., and Moseley, B.E.B. (Eds.) *Molecular and cellular aspects of microbial evolution: thirty-second symposium of the society for General Microbiology,* Cambridge University Press, New York, 1981.

Freifelder, D., *Molecular biology: a comprehensive introduction to prokaryotes and eukaryotes,* Science Books International, Boston, 1983.

Goldman, R. (Ed.) *Cell motility*, Cold Spring Harbor Laboratory, Cold Spring Harbor, New York, 1976.

Gooda, G.W., Lloyd, D., and Trinici, A.P.J. (Eds.) *The eukaryotic microbial cell: the thirtieth symposium of the Society for General Microbiology,* Cambridge University Press, New York, 1980.

Hazelbauer, G.L. (Ed.) *Taxis and behavior,* Chapman and Hall, New York, 1978.

Starr, M.P., and others (Eds.) *The prokaryotes,* Springer-Verlag New York, Inc., New York, 1981.

Journal articles

Burchard, R.P. Gliding motility of prokaryotes: ultrastructure, physiology, and genetics. *Annu. Rev. Microbiol.* **35**:497, 1981.

Gray, M.W. The bacterial ancestry of plastids and mitochondria. *Bioscience* **33**:693, 1983.

Schwartz, R.M., and Dayhoff, M.O. Origins of prokaryotes, eukaryotes, mitochondria and chloroplasts, *Science* **199**:395, 1978.

Stackenbrandt, E., and Woese, C.R. The evolution of prokaryotes. *Symp. Soc. Gen. Microbiol.* **32**:1, 1981.

Woese, C.R. Archaebacteria. *Sci. Am.* **244**:98, 1981.

CHAPTER 5

Prokaryotic ultrastructure and function

Chapter Outline

Introduction

The term **ultrastructure** in cell biology implies that a structure is composed of individual units, which are arranged to provide shape and impart specific chemical and physical properties to that structure. The structures that make up a microbial cell provide it with certain basic requirements that enable it to survive. The requirements for cell survival are (1) a protective structure surrounding the cytoplasmic contents, (2) a site and mechanism for the storage and replication of genetic information, respectively, (3) a mechanism for synthesis of cellular components, (4) a mechanism for generating energy, and, for some, (5) a mechanism for movement.

This chapter discusses the ultrastructure of the various components that make up a bacterial cell and shows how the specific design of bacterial cell structures is related to their function in the cell. The electron micrograph in Figure 5-1 illustrates the basic bacterial structures.

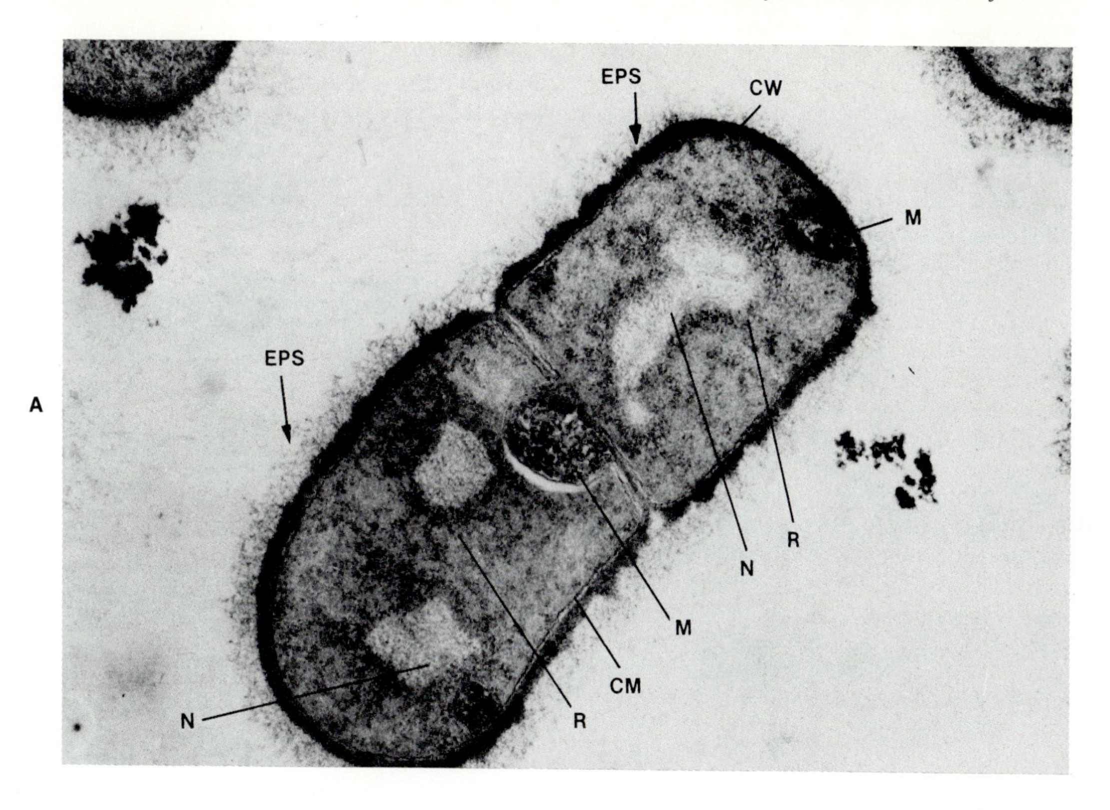

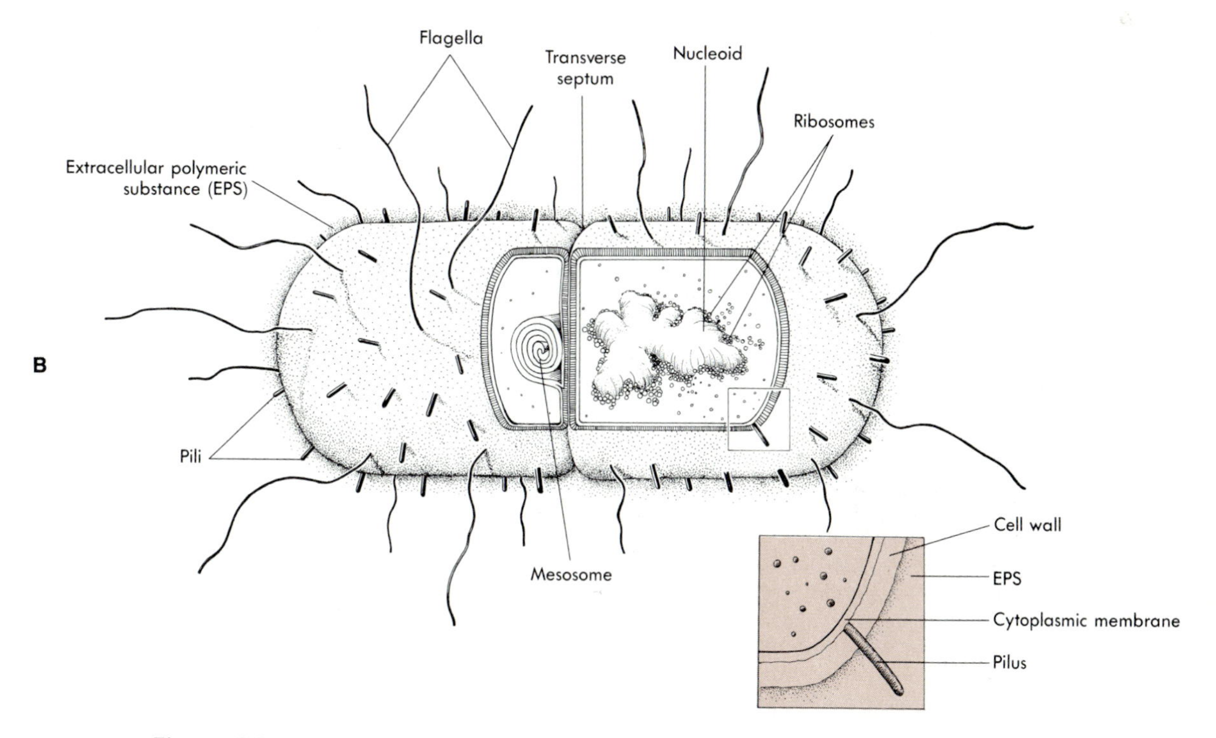

Figure 5-1 ***A,*** *Electron micrograph of typical bacterial cell demonstrating some important structures.* ***B,*** *Diagrammatic illustration of structures found in typical bacterial cell.* CW, *Cell wall;* M, *mesosome;* N, *nucleoid;* R, *ribosomes;* EPS, *extracellular polymeric substance.*

CELL WALL

The cell wall represents the first structural component of the bacterial cell that reacts with the external environment. The fact that bacteria can be found in extremely harsh environments, such as the high temperatures and acidity of hot sulfur springs, the high alkalinity of areas such as the Dead Sea, and the arid areas of deserts, suggests that the wall is a formidable structure. The cell wall has a basic unit of design that has apparently evolved various modifications to accommodate the harsh environments to which some bacteria are exposed. Some of these modifications have been detected in the group now designated archaebacteria. We will examine the basic design and some of the modifications of the two major groups of bacteria that have been separated based on the staining characteristics of the cell wall: the gram-positive and gram-negative bacteria.

Gram-positive cell wall types

Peptidoglycan. The cell wall of gram-positive bacteria is of simple design compared with the gram-negative cell wall, but both possess the basic unit called **peptidoglycan** (also called mucopeptide) layer (Figure 5-2). The peptidoglycan forms a corset that covers the surface of the cell. It may account for 15% to 50% of the dry weight of the gram-positive cell wall, depending on the species as well as environmental conditions. The peptid-

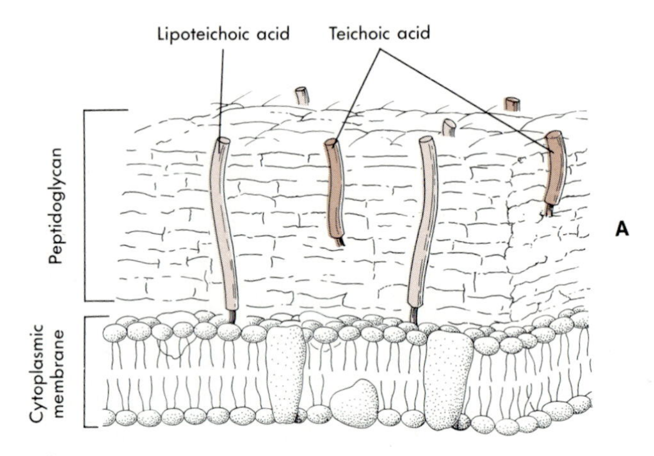

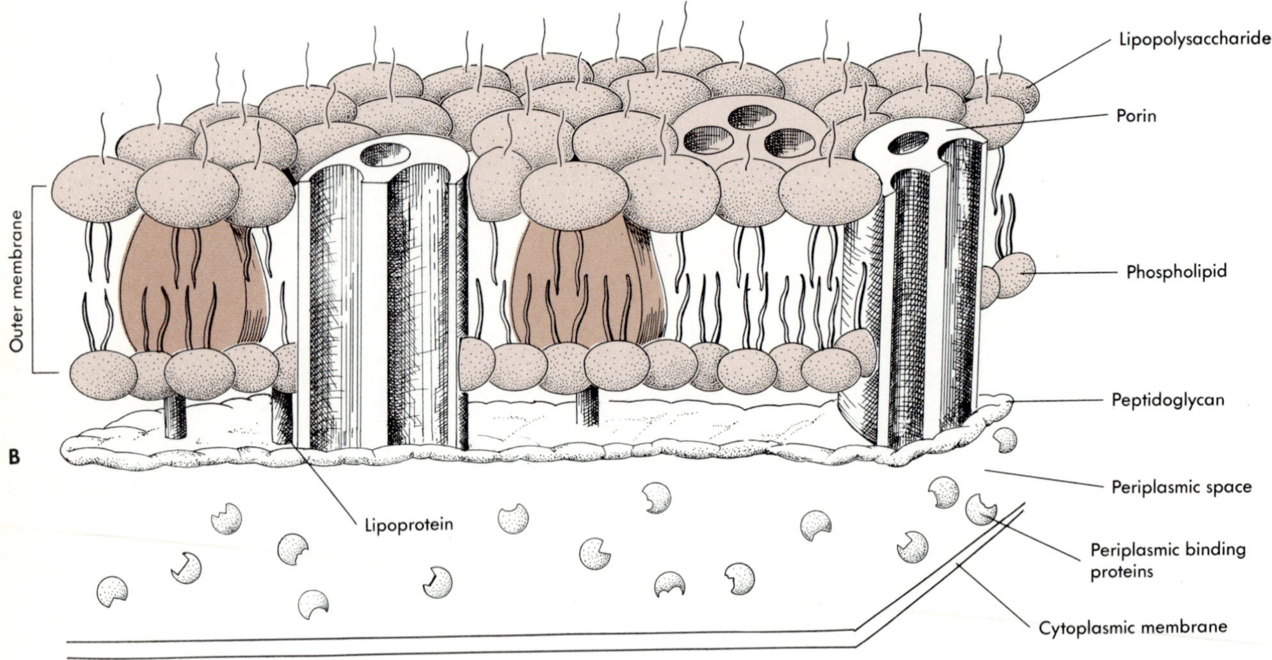

Figure 5-2 *Diagrammatic representation of basic design of gram-positive and gram-negative bacterial cell wall.* ***A,*** *Gram-positive cell wall consists of thick multilayered peptidoglycan, within which are interspersed acidic polysaccharides called teichoic acids. Teichoic acids can be replaced by teichuronic acids and are linked to peptidoglycan. Lipoteichoic acids are also part of some cell walls and are linked to cytoplasmic membrane. (Cytoplasmic membrane not drawn to scale.)* ***B,*** *Gram-negative cell wall contains very thin monolayer of peptidoglycan. Outside peptidoglycan is outer membrane that consists of sheet of lipopolysaccharide molecules in which are embedded proteins involved in transport, as well as lipoproteins, which are covalently attached to peptidoglycan. Periplasmic space separates cytoplasmic membrane from outer membrane and is site of several proteins involved in transport, as well as outer membrane synthesis (not drawn to scale).*

oglycan of gram-negative species, however, accounts for less than 5% of the dry weight of the cell wall.

The glycan portion of the peptidoglycan is a network of linear strands of connected disaccharides (may be 20 to 100 or more disaccharides long) that consist of the repeating units *N*-**acetylglucosamine** (NAG) and *N*-**acetylmuramic acid** (NAM). The strands are linked randomly to adjacent strands by tetrapeptide branches that are covalently attached to the NAM residues by the carboxyl of the lactic acid group (Figure 5-3). In other words, a tetrapeptide branch on one glycan strand is attached to a tetrapeptide in an adjacent strand. The range of amino acids in the tetrapeptide from different species of bacteria is large and varied. Most tetrapeptides contain D-alanine, L-alanine, D-glutamate, and L-lysine (Figure 5-3, *B*). Occasionally diaminopimelic acid (DAP) is substituted for lysine (Figure 5-4). In most gram-negative bacterial species the amino acid arrangement, beginning at the amino end of the tetrapeptide, is often L-alanine, D-glutamate, L-lysine, and D-alanine. The attachment between the tetrapeptide of one strand and the tetrapeptide of the adjacent strand is brought about by **interpeptide bridges**. The reaction that links glycan strands by interpeptide bridges is called **transpeptidation**. It is the transpeptidation reaction that is affected by antibiotics such as penicillin and cephalosporin (Figure 5-3, *B*). The interpeptide bridge may be direct and consist of a peptide bond between the amino acids on the tetrapeptide branches, or the bridge may consist of short peptides that link the tetrapeptides. For example, in *Staphylococcus aureus* a pentaglycine bridge links the tetrapeptides of adjacent glycan strands (Figure 5-3, *B*). Not all tetrapeptide bridges are linked by interpeptide bridges, and in some gram-negative species as much as

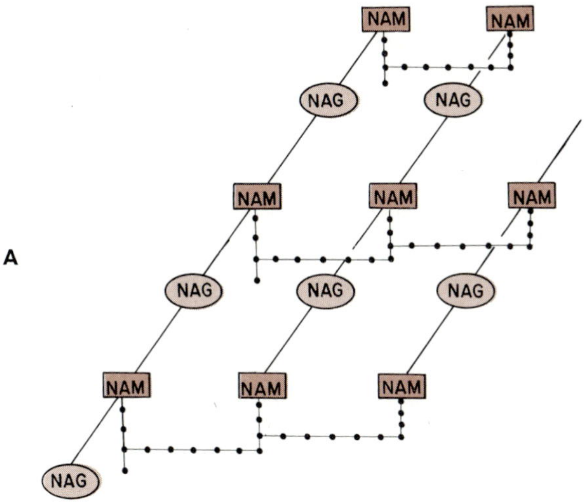

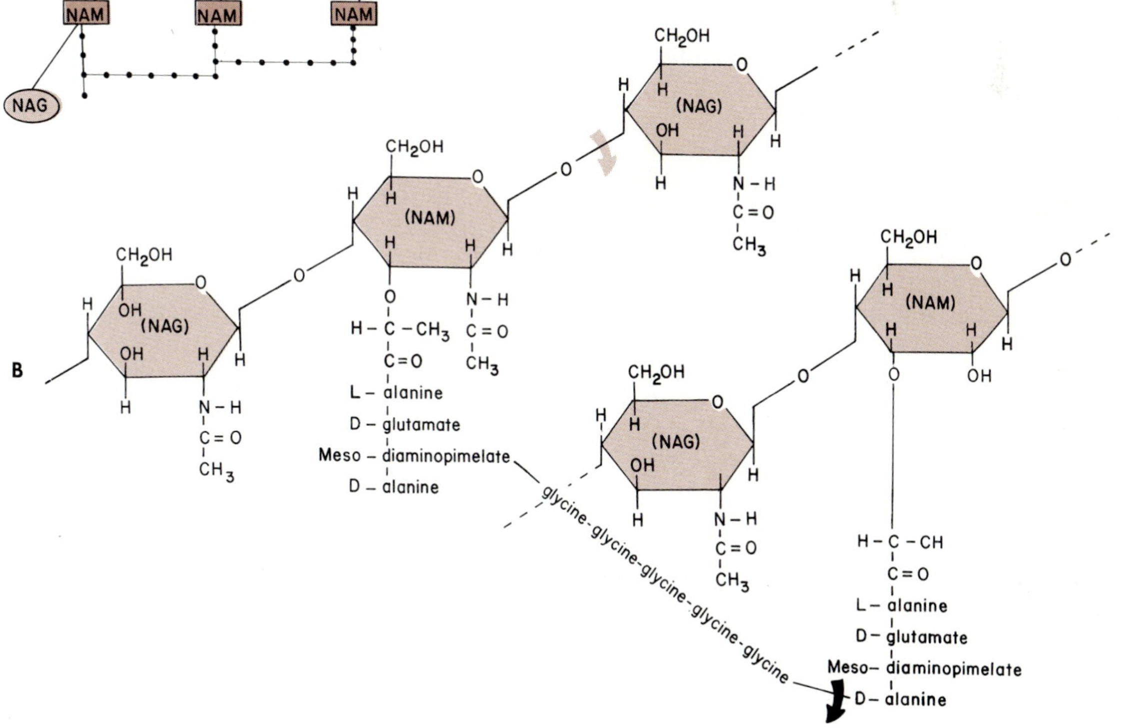

Figure 5-3 *Basic design of gram-positive peptidoglycan layer* (Staphylococcus aureus). ***A,*** *Diagrammatic relationship of peptidoglycan components. Each (●) represents amino acid residue.* NAG, N-*acetylglucosamine;* NAM, *N-acetylmuramic acid.* ***B,*** *Molecular representation of peptidoglycan layer. Colored arrow points to site of action of lysozyme. Black arrow points to site where penicillin prevents linkage of peptide groups on adjacent peptidolgycan strands.*

50% or more of the tetrapeptide branches are not cross-linked. For comparison, *S. aureus* has only 7% to 8% of the bridges that are not cross-linked. The degree of cross-linkage is directly proportional to the rigidity of the cell wall. Secondary polymers, such as teichoic acid and teichuronic acid, may also be covalently linked to NAM in the peptidoglycan; these are discussed in the next section.

Teichoic acid. Teichoic acids are acidic polysaccharides consisting of repeating units of glycerol or ribotol that are linked by phosphate esters and have other sugars and D-alanine attached (Figure 5-5). The relationship of teichoic acids to the peptidoglycan layer has not been entirely clear in many gram-positive species (Figure 5-2, *A*). Teichoic acid in *Bacillus subtilis* is 20 to 30 residues in length and makes up as much as 50% of the dry weight of the cell wall. One half of the teichoic acid is intertwined within the peptidoglycan matrix and is covalently bonded to the C 6′-hydroxyl group of NAM. Some of the teichoic acid is also exposed on the surface and is associated with the outer face of the peptidoglycan. Surface teichoic acid forms part of the receptor for various bacteriophages. In addition, teichoic acid of pathogenic species of *Streptococcus* and *Staphylococcus* is important for attachment to epithelial surfaces. Teichuronic acid (*N*-acetylgalactosamine plus glucuronic acid linked as a disaccharide) contains no phosphorus and is a secondary polymer found in species of *Bacillus*. Teichuronic acid can replace teichoic acid when the phosphate level in the medium is very low.

Lipoteichoic acid may also be found in the cell walls of some species of *Lactobacillus*. Lipoteichoic acid has its polyglycerol phosphate chains extending into the wall, with some reaching the cell surface while the glycolipid is inserted into the cytoplasmic membrane. Lipoteichoic acids in some bacteria therefore provide a physical bridge binding the cell wall to the cytoplasmic membrane.

Function of the gram-positive wall components. The peptidoglycan layer, with its cross-linking and interpeptide bridges, gives structural rigidity to the wall, but other components also play a role. Both teichoic acid and teichuronic acid contribute to the negative charge associated with the surface of the gram-positive cell wall. Both polymers bind divalent cations, such as magnesium, which allows the wall to become more compact (less electronegative repulsion) but does not alter the porosity of the wall. When mutants of *B. subtilis* are produced that are deficient in teichoic acid or teichuronic acid, the cells exhibit coccoid shapes but return to their normal rod shape under conditions that permit insertion of these polymers. This suggests that teichoic acid and teichuronic acid lend structural support to the peptidoglycan. Teichoic acid and teichuronic acid are also involved in the control of **autolysins**, enzymes that hydrolyze the cell wall. Controlled hydrolysis of cell wall components is necessary in the normal

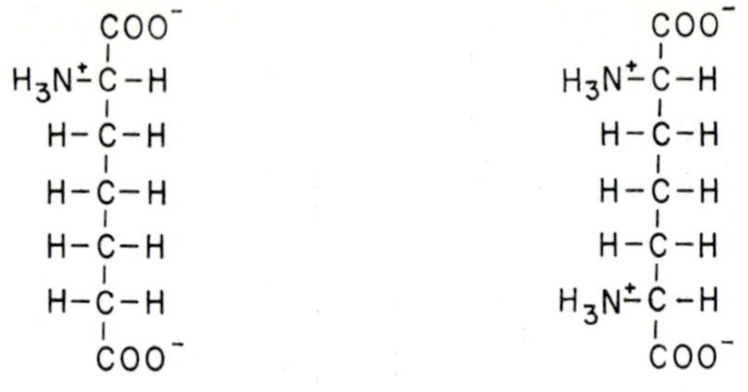

Lysine	Diaminopimelate
COO^-	COO^-
H_3N^+–C–H	H_3N^+–C–H
H–C–H	H–C–H
H–C–H	H–C–H
H–C–H	H–C–H
H–C–H	H_3N^+–C–H
COO^-	COO^-

Figure 5-4 *Structure of lysine and diaminopimelic acid.*

growth cycle because old cell wall must be cleaved to allow for the insertion of new cell wall components. Autolysins may be totally destructive to the cell, particularly when a cell ages. The destructive effects are seen when a broth culture of gram-positive bacteria is centrifuged. If the pelleted cells are left at room temperature, lysis of the cells begins rapidly. Autolytic activity is held in check only at lower temperatures (0° to 5° C).

Gram-negative cell wall types

The cell wall of gram-negative bacteria is a complex unit consisting of an **outer membrane** and an underlying peptidoglycan, which may be loosely or strongly adherent to the cytoplasmic membrane. The peptidoglycan is also part of an area called the **periplasmic space**, which is adjacent to the cytoplasmic membrane. Diaminopimelic acid is found in the peptidoglycan of all gram-negative bacteria but not in all gram-positive bacteria. The structural arrangement of the gram-negative cell wall is illustrated in Figure 5-2, *B*; the basic design of the gram-negative peptidoglycan layer is illustrated in Figure 5-6.

Outer membrane. The outer membrane consists of 20% to 25% **phospholipid**, 30% **lipopolysacharide (LPS)**, and 45% to 50% protein. The outer face of this membrane is composed of LPS and proteins, and the inner portion is primarily phospholipid. This arrangement creates a lipid bilayer in which the fatty acid chains are arranged at right angles to the plane of the membrane, thus forming a hydrophobic (water-repelling) interior. The polysaccharide portion of the LPS produces a hydrophilic (water-loving) exterior. Proteins are located in the outer membrane and include the matrix proteins and lipoproteins. The matrix protein forms a lattice that is bound to the peptidoglycan by ionic bonds. Some of the matrix proteins are arranged in such a way as to form channels or pores **(porins)** for the passive transfer of solutes (Figure 5-2, *B*). Proteins other than porins are also located in the outer membrane and are involved in solute binding and transport. Some of these proteins form channels for the transport of substrates such as certain sugars, nucleosides, amino acids, and iron compounds. Lipoproteins are smaller than matrix proteins and are the most abundant proteins (7.2×10^5 per cell). One third of the lipoprotein is covalently bonded by an amino group to every tenth or twelfth meso-diaminopimelic acid of the peptidoglycan, whereas two thirds of the lipoprotein remains free in the outer membrane. The covalently bound lipoprotein provides a bridge holding the outer membrane to the peptidoglycan.

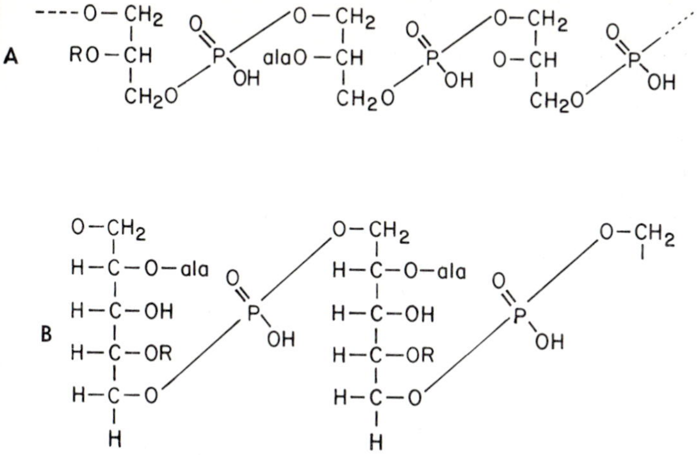

Functions. The polysaccharide portion of the LPS is composed of a core of sugars that is common to one group of bacteria. There are sugar side chains, however, that vary from one species to another within the bacterial groups (Figure 5-7). These side chains are antigenic, since they stimulate the formation of immunoglobulins (antibodies) when injected into an animal, and are called **O antigens.** Unless an extracellular polymeric substance is present on the cell, the sugar side chains represent the outermost component of the outer membrane. The O antigens in some bacterial species protect the cell from digestion (phagocytosis) by white blood cells in the vertebrate host. The resistance is due to the composition of the sugar side chains. Outside of the vertebrate host, these same bacteria are apparently protected from digestion by free-living amebas, which also employ the phagocystic process.

The LPS is often referred to as **endotoxin,** and some types are capable of inducing a physiological response such as fever in animal hosts, including humans. The

Figure 5-5 *Two types of teichoic acid in gram-positive bacterial cell wall.* ***A,*** *Glycerol teichoic acid.* ***B,*** *Ribitol teichoic acid.* ala, *Alanine;* R, *different organic groups.*

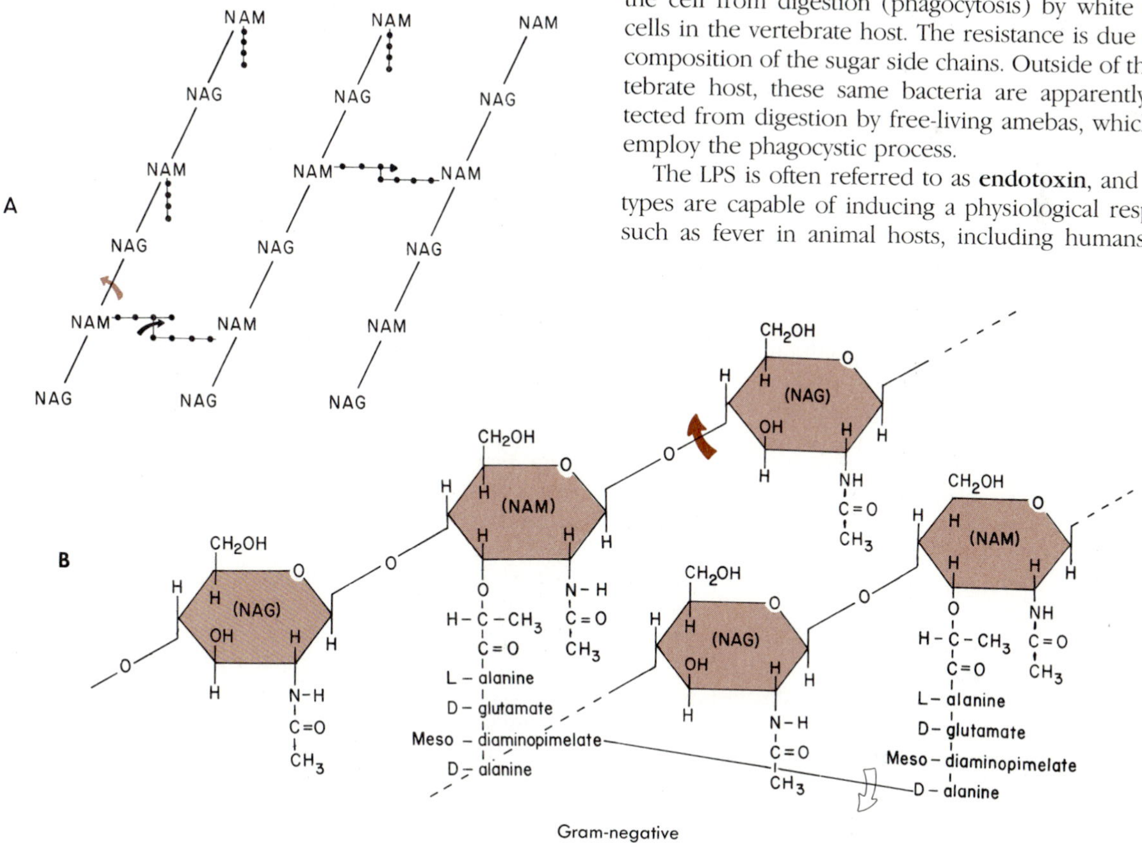

Figure 5-6 *Design of gram-negative peptidoglycan layer.* ***A,*** *Diagrammatic representation. Each (●) is equivalent to amino acid.* ***B,*** *Molecular representation of gram-negative peptidoglycan. Colored arrow points to site of lysozyme. Outline arrow points to site where penicillin prevents linkage of peptide groups on adjacent peptidoglycan strands.* NAG, *N-acetylglucosamine;* NAM, *N-acetylmuramic acid.*

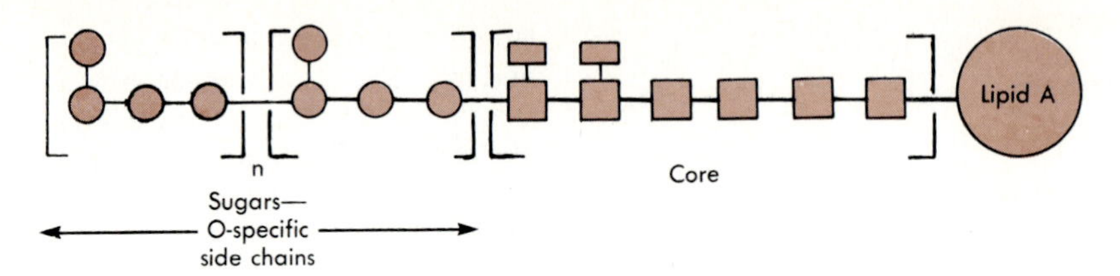

Figure 5-7 *Structure of LPS in gram-negative bacterial cell walls. Lipid A, toxic component, is bound to core of sugars (□—□—□—□) whose composition remains relatively stable within a bacterial group, such as genus* Salmonella. *Core is bound to variable region of sugars called O-specific side chain. Serological difference between* Salmonella *species are related to differences in sugar composition of these O-specific side chains, which are complex.* N, *repeating unit.*

toxic component of LPS is the lipid portion, which is called **lipid A.** Lipid A is covalently bound to the core of the polysaccharide (Figure 5-7).

The outer membrane acts as a barrier to the penetration of certain molecules by the very nature of its construction and chemical composition. Certain dyes such as crystal violet, antibiotics such as erythromycin and actinomycin, heavy metals and bile salts do not readily penetrate the gram-negative cell wall but do penetrate the gram-positive cell. The mechanism of transport of molecules across the cell wall remains highly speculative. The tightly clustered hydrophobic fatty acid side chains of the LPS are believed to slow down the penetration process, particularly of hydrophobic molecules. Once hydrophobic molecules penetrate the LPS, they easily pass through the phospholipid bilayer. This relative impermeability to hydrophobic compounds may explain the large number of gram-negative bacteria present in the intestinal tract, where bile salts are continuously present. The outer membrane thus offers the gram-negative cell a selective advantage. Hydrophilic molecules that are small enough are thought to pass through the outer membrane by diffusion through the **porins.** The porins presumably exclude hydrophobic compounds. The porin proteins appear to be of three types. The first type provides an aqueous channel that permits passive nonspecific diffusion of solutes across the membrane. Selectivity is based on the size of the solute in relation to the radius of the pore. A second class of porin proteins is involved in the transport of oligosaccharides and nucleosides. A third type of porin functions in the uptake of solute molecules of relatively large size, and this type may actually interact directly with solute. All porin proteins span the outer membrane, with the exterior polypeptide chains serving as receptor sites for bacterial virus attachment and bacteriocin binding.* The polypeptide chains on the inner surface of the porin channel make direct contact with the periplasmic space.

Peptidoglycan. The peptidoglycan of gram-negative bacteria performs the same function as the peptidoglycan of gram-positive bacteria. The gram-negative peptidoglycan, however, differs in amount and location from the gram-positive component. The peptidoglycan accounts for 5% to 10% of the dry weight of the cell wall. The tetra-peptide linkage of most gram-negative types is of the direct type, such as appears in *Escherichia coli* (see Figure 5-6, *A*). The attachment of the peptidoglycan to the outer membrane has already been discussed.

Periplasmic space. The periplasmic space separates the outer membrane from the cytoplasmic membrane (see Figure 5-2, *B*). It contains all of the peripherally attached proteins of the inner face of the outer membrane and outer face of the cytoplasmic membrane. The periplasm contains several enzymes whose functions are discussed below.

Functions. The periplasm is in equilibrium with the cytoplasm of the cell; that is, it is isosmotic. If the osmotic pressure outside the cell is increased, the periplasmic space also enlarges and thus acts as an osmotic buffer zone between the cytoplasm and its external environment.

The periplasm is associated with three distinct functions. First, there are enzymes in the periplasm that are degradative and help to break down molecules into forms that can penetrate the cytoplasmic membrane. These enzymes include proteases, lipases, and nucleases, and they appear to be attached to the outer surface of the cytoplasmic membrane. Second, several enzymes are present in the periplasm of some drug-resistant species that are capable of modifying antibiotics, thus rendering them harmless to the bacterium. The enzymes beta-lactamases, for example, inactivate beta-lactam antibiotics such as the penicillins and cephalosporins. Enzymes capable of inactivating streptomycin, kanamycin, and gentamicin are also present in the periplasm of some species. These latter antibiotics are normally very effective against gram-negative bacteria, but the presence of inactivating or modifying enzymes renders the bacteria resistant to their effects. Third, specific solute-binding proteins are present in the periplasm. These binding proteins may represent the first contact

*Bacteriocins are chemical compounds produced by bacteria that inhibit or kill closely related species.

```
NAG—T—NAG
    |
   Glu
    |
   Ala
    |
   Lys———Glu
    |     |
   Glu   Lys———Glu
    |     |
   Ala   Ala
          |
         Glu
          |
     NAG—T—NAG
```

Figure 5-8 *Arrangement of peptidoglycan in* Methanobacterium thermoautotrophicum, member of archaebacteria. All amino acids belong to the L-series. NAG, *N-acetylglucosamine; T,* N-*acetyltalosaminuronic acid;* Glu, *glutamic acid;* Ala, *alanine;* L, *lysine.*

of a solute with a transport protein and thus facilitate transfer of the solute through the cytoplasmic membrane. Some periplasmic binding proteins interact with outer membrane proteins and thus represent the second contact of the solute as it is transported into the cell. The periplasmic proteins that are specific for the transport of galactose, maltose, and ribose play a dual function because they are receptors in the chemotactic response of organisms (see chemotaxis, p. 122). Mechanisms of solute transport are discussed later in the chapter.

Cell walls of the archaebacteria. The archaebacteria have atypical peptidoglycans or related cell wall structures and provide an interesting insight into wall modifications. They do not possess muramic acid and have only L-amino acids in the cell wall. *Methanobacterium thermoautrophicum* contains an aminohexuronic acid called *N*-acetyltalosaminuronic acid, which substitutes for NAM (Figure 5-8). *Halococcus morrhuae* contains *N*-glycylglucosamine in its glycan, an example of the amino acid glycine being substituted for an amino group on NAG. The glycine residues are connected by peptide bonds to sugars in the cell wall. *Methanococcus vannielii* contains neither a rigid cell wall nor sheath but exhibits an outer protein layer consisting of subunits that disintegrate very easily. One interesting observation from these studies is that many of the species show a gram-negative reaction that is based not on chemical characteristics but on the thickness of the cell wall. Those with thick cell walls are gram-positive, whereas those with thin cell walls are gram-negative.

Cell wall and osmosis. In addition to its function in maintaining the shape of the cell, the cell wall is a dynamic structure that responds to its environment. The concentration of solute outside the cell for many bacteria in their natural environment is hypotonic (the concentration of salts is lower outside the cell than inside the cell). Under such circumstances water tends to flow into the cell to compensate for this concentration difference, a process called **osmosis.** In a hypotonic medium water rushes into the cell and causes the cytoplasmic membrane to expand. The cell wall also expands but does not burst. It acts as a cushion for the expanding cytoplasmic membrane. The protective effect of the cell wall can be demonstrated if a bacterial culture is placed in a hypotonic medium containing lysozyme, an enzyme that can effect the removal of the cell wall. Lysozyme is found naturally in tears, saliva, and other body fluids. It can cause the breakdown of the cell wall by cleaving the covalent linkage between NAM and NAG (Figure 5-3). As the cell wall is being removed or weakened in a hypotonic medium, water rushes into the cell, causing the unprotected cytoplasmic membrane to expand and eventually burst; that is, the cell lyses. The removal of the bacterial cell wall need not result in cell lysis if the medium contains solutes in such a concentration that osmotic forces outside the cell balance osmotic forces inside the cell. Under such circumstances a spherically shaped wall-less cell, called a **protoplast**, is formed (Figure 5-9). Protoplasts are spherical whether the original cell was spherical or rod shaped. Protoplasts are prepared primarily from gram-positive cells because the entire cell wall can be easily removed. Treatment of gram-negative cells usually results in some of the cell wall, such as the outer membrane, still attached to the cell; these cells are called **spheroplasts.** If the solute concentration outside the protoplast is higher than that inside the cell, water will flow out of the cell (dehydration), and it will collapse, a process called **plasmolysis.** Many bacteria are inhibited by high concentrations of solutes such as salt and sugars, and this property is used in the curing of various meats and in the preparation of jellies, respectively.

Bacteria such as *E. coli* have evolved mechanisms for living in a world deficient in water. They can prevent dehydration and loss of water by producing protective molecules (osmoprotectant molecules) that balance the osmotic strength of the cytoplasm with that of the environment. A class of genes in *E. coli* called OSM (osmotic tolerance) genes govern the production of osmoprotectant molecules. Some of the osmoprotectant molecules include glycine betaine, choline, and proline betaine (Figure 5-10). One of the striking features of these molecules is their extreme solubility in water and their osmoprotectant value. Apparently when the osmotic strength of the environment increases the OSM genes act as an osmometer and overproduce compounds such as glycine betaine. These compounds accumulate in the cytoplasm to a level 10^5 times the level in the environment. In addition to their osmotic balancing properties, osmoprotectant molecules may interact with protein

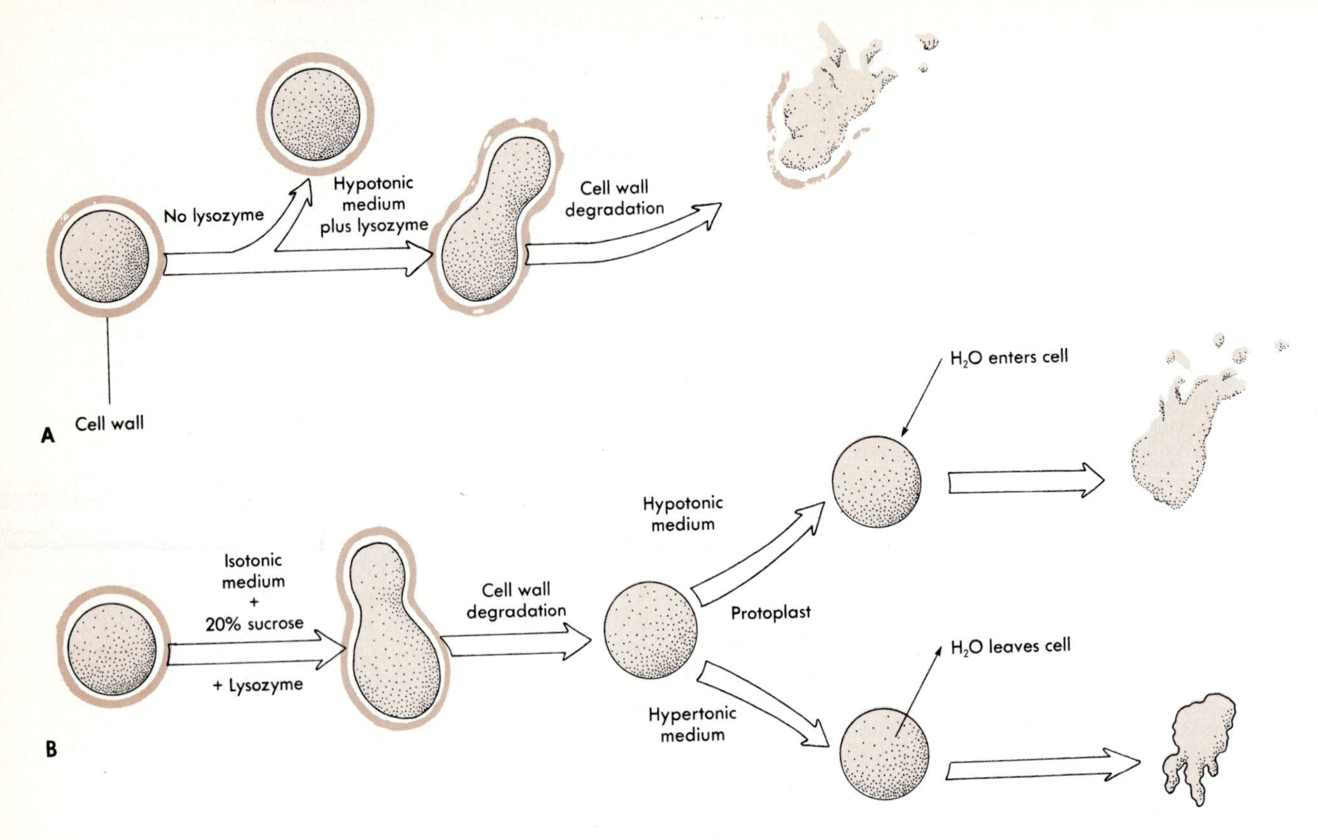

Figure 5-9 *Effect of wall lysing agents on bacterial cell in media of different osmotic pressures.* ***A,*** *Hypotonic media in presence and absence of lysozyme.* ***B,*** *Isotonic medium containing 20% sucrose plus lysozyme. Protoplast formed will burst if put into hypotonic medium or collapse if put into hypertonic medium.*

molecules and increase their water-binding properties. Such a property would protect proteins during severe stress.

$(CH_3)_3\ N^+\ CH_2\ COO^-$ Glycine betaine

$(CH_3)_3\ N^+\ CH_2\ CH_2\ OH$ Choline

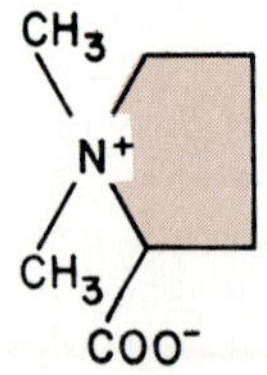

CH_3, N^+, CH_3, COO^- Proline betaine

Figure 5-10 *Structure of osmoprotectant molecules glycine betaine, choline, and proline betaine.*

WALL-LESS BACTERIA

Wall-less bacteria can be divided into two groups: naturally occurring wall-less bacteria called **mycoplasmas,** and **L-forms,** which are derived from gram-negative and gram-positive bacteria. Mycoplasmas are the smallest free-living organisms capable of self-replication. They possess no peptidoglycan layer, and the cell cytoplasm is bounded only by a phospholipid bilayer with associated proteins. The stability of the mycoplasmas is related to their small size and the high concentrations of cholesterol or other sterols that are incorporated into the cytoplasmic membrane (see Point of Interest).

L-forms were originally described at the Lister Institute in London, hence their name. L-forms are not capable of self-replication. They can be produced from any gram-positive or gram-negative bacterium using the appropriate cell-wall–degrading agents, such as lysozyme, or inhibitors of cell wall synthesis, such as penicillin. Once isolated, the L-form may be stable or unstable. Stable L-forms maintain their wall-less state in the absence of wall-lysing agents. Naturally occurring L-forms are believed to represent a technique for survival. Under unfavorable conditions bacteria may lose much

Point of Interest

MYCOPLASMA: WALL-LESS BACTERIA

The term **mycoplasma** is a trivial name for all members of a class of bacteria called Mollicutes. Several important genera in this group are *Mycoplasma, Ureaplasma, Acholeplasma, Spiroplasma,* and *Thermoplasma.* They are pleomorphic organisms ranging in size from 0.2 to 0.3 μm. They are primarily intracellular parasites and can be found in plants, animals, humans, and insects. One species, *Thermoplasma acidophilum,* can be found in highly acidic coal refuse piles where internal temperatures of 55° C provide them with an optimal temperature for growth. Sterols are present in the cytoplasmic membrane of all mycoplasmas except *Acholeplasma* and *Thermoplasma.* Even though sterols enable the cell membrane to be relatively rigid, the mycoplasmas still are osmotically sensitive and cultivation in laboratory media has been accomplished with only a few species.

Motility among the mycoplasmas, except for *Spiroplasma,* appears to be of a gliding nature and is believed to be associated with specialized polar structures such as stalks, blebs, or tapered tips (figure on left). These structures apparently help the organism to attach to surfaces and to push forward in a gliding movement. *Spiroplasma* organisms, which appear primarily as short helical filaments (figure on right), exhibit the same type of motility as the spirochetes. *Spiroplasma,* however, lack flagella and have no cell wall or axial filaments. Membrane-associated fibrils similar to the actin proteins of muscle have been detected in *Spiroplasma,* and they may be involved in the motility of this group.

Few species of *Mycoplasma* are agents of disease in humans. *M. pneumoniae* is a cause of minor pneumonia, but most mycoplasmas are more frequently found as agents of disease in cattle and plants. *M. mycoides* is the cause of contagious bovine pleuropneumonia, a major disease of cattle in parts of Africa and Asia. *Mycoplasma* inhabits the phloem of plants and is transmitted by leaf hoppers who naturally feed on phloem. Many of the plant mycoplasmas have not been completely characterized because of difficulties in cultivation and have been referred to as mycoplasma-like organisms (MLOs). MLOs are associated with the yellow diseases of a large variety of plants (see Chapter 20).

Ureaplasma is a common inhabitant of the vagina and cervix. It is believed to be a possible cause of a sexually transmitted disease in which the urethra becomes inflamed.

Mycoplasmas are important research tools because of their limited genetic information, minimal biochemical activity, and reduced size. They possess a minimal set of structures that are necessary for replication: cell membrane, DNA, and ribosomes. It has not been determined if on the evolutionary scale they are at the bottom of the prokaryotic ladder or represent a more developed prokaryotic cell that during evolution lost some genetic information.

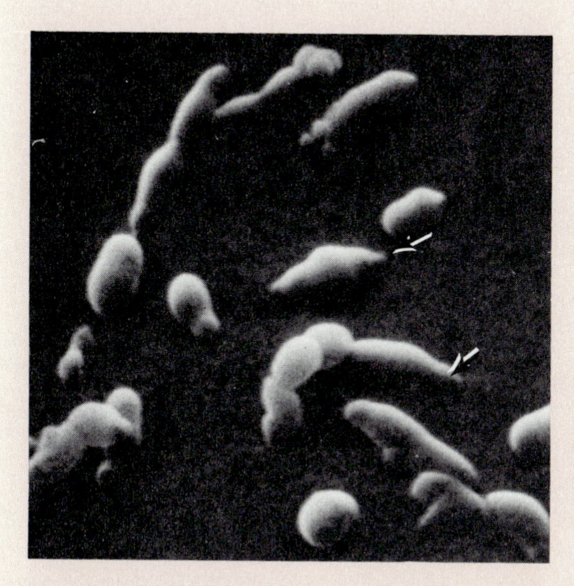

Electron micrograph of Mycoplasma gallisepticum *shows* blebs (arrows) *involved in motility.*

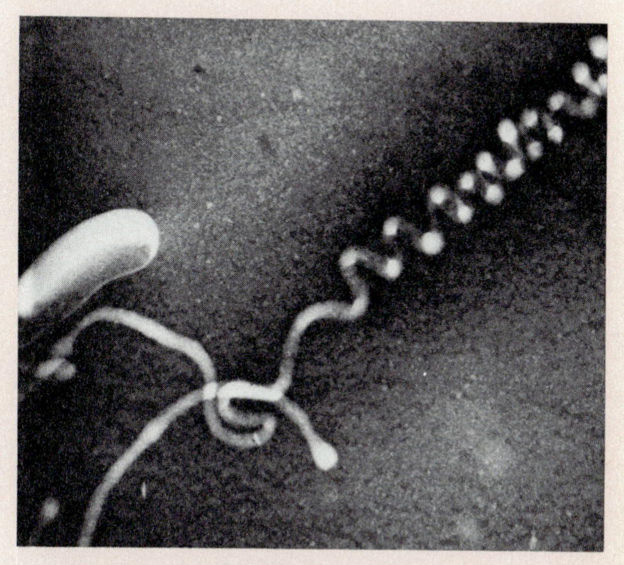

Electron micrograph of Spiroplasma citri, *agent of plant disease. (×34,500.) Note spiral nature of bacterium.*

of their cell wall by the induction of autolytic enzymes, and only when conditions are favorable for normal growth and reproduction will the cell revert back to its natural state. Bacteria that infect a host, for example, may be exposed to antibiotics such as penicillin or to immune responses by the host that result in destruction of the cell wall. If left exposed to these agents, the bacterial population would cease to exist, but in the absence of a cell wall these agents have no effect and the survival of the microbial cell is ensured.

CYTOPLASMIC MEMBRANE

The cytoplasmic membrane separates the cytoplasm of the cell from its external environment. Damage to the membrane usually results in the loss of important cytoplasmic constituents, and the cell dies. The cytoplasmic membrane of the bacterial cell is a prototype for other biological membranes. This universal membrane structure is referred to as a **unit membrane**, which distinguishes it from complex membranes. The currently devised evolutionary scheme proposes that prokaryotes preceded eukaryotes, and it is not surprising to find that the cytoplasmic membrane of prokaryotes carries out

Concept Check

1. What are the differences in cell wall composition of gram-positive and gram-negative bacteria?
2. Describe the chemical composition and function of the various cell wall components found in bacteria.
3. Describe how the bacterial cell wall responds to changes in concentration of solutes in its environment.
4. What are the different types of wall-less bacteria and how do they arise?

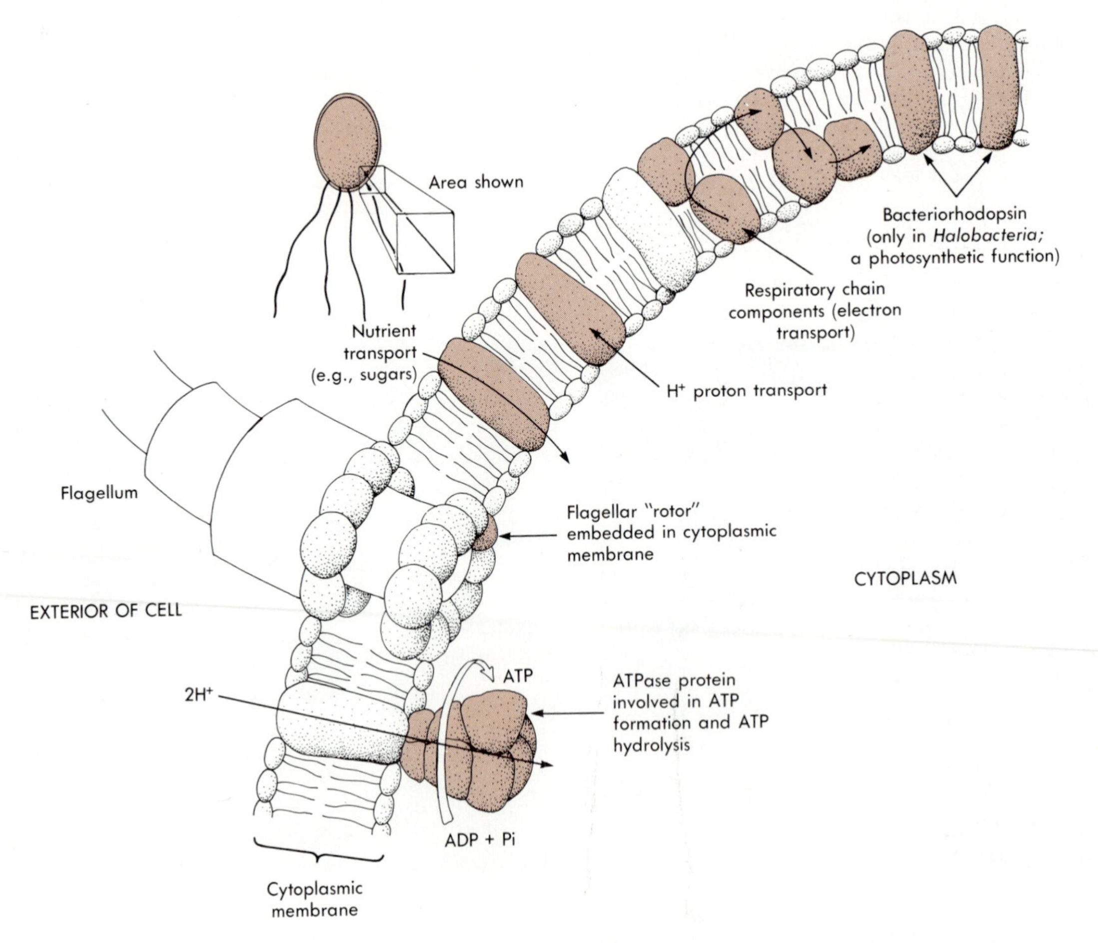

Figure 5-11 *Diagrammatic representation of cytoplasmic membrane and protein components that can be found there.*

many of the functional roles of the membranes surrounding the organelles of eukaryotic cells.

Structure

The cytoplasmic membrane is chemically composed of 40% lipid and 60% protein. Phospholipids represent the major lipid components, and two of the principal ones found in bacteria are phosphatidylethanolamine and phosphatidylglycerol (see Chapter 2). Lipoteichoic acids are also present in the cytoplasmic membrane of some prokaryotes. The fatty acid side chains of microbial lipids vary in length, but most are between 16 and 18 carbon units long. The lipid composition of the prokaryotic membrane may vary considerably from that of the eukaryotic membrane. The chemical and physical properties of membrane lipids, particularly their fatty acid side chains, permit some prokaryotes to survive in environments where there are chemical and/or physical extremes. As we described in Chapter 2, the phospholipid bilayer arrangement is characteristic of membranes whose function is to act as a selective barrier to various types of molecules. The phospholipids form the matrix of the cytoplasmic membrane, but it is the proteins that carry out its principal functions (Figure 5-11). The positioning of the proteins in the phospholipid bilayer is dependent on their hydrophilic and hydrophobic domains. The hydrophilic parts of the protein appear oriented adjacent to the membrane, whereas the hydrophobic portions are buried in the lipid bilayer. Some proteins are large enough to span the bilayer, whereas others are distributed at various levels. There appears to be no covalent bonding between the proteins and lipid; thus the membrane demonstrates fluidity and some freedom of movement.

Proteins not only lend structural integrity to the cytoplasmic membrane but perform many other functions as well: the "rotor" (Figure 5-11) mechanism of the bacterial flagellum is attached to the cytoplasmic membrane; one of the subunits of the enzyme ATPase involved in ATP formation and hydrolysis spans the cytoplasmic membrane; the electron transport components are arranged across the cytoplasmic membrane; some proteins form channels for transport of solutes into the cytoplasm; and one protein is involved in photosynthesis (see Point of Interest).

Functions

The functions of the cytoplasmic membrane can be divided into those activities associated with solute transport, energy formation, biosynthesis, and secretion.

Solute transport. The cytoplasmic membrane consists of highly charged phosphate groups and hydrophobic fatty acid chains that are highly selective to molecular transport. This specific selectivity is based on size, as well as electrostatic charge of the solute.

Located within the cytoplasmic membrane are transport proteins, which exist in many forms. They are designed to transfer a specific class of chemical compounds into the cytoplasm of the cell. Most solute molecules are transported into the microbial cell by three mechanisms: **active transport** systems, **facilitated diffusion**, or **group translocation.** A **simple passive transport** system in which solute molecules such as oxygen and water, pass across the cytoplasmic membrane from an area of high solute concentration to an area of low solute concentration does operate in microbial cells. However, this type of passive diffusion is very slow and nonspecific. The molecular size, shape, charge, and solubility of the solute on either side of the membrane are limiting factors to simple passive diffusion. Under most environmental conditions the concentration of solute outside the cell is less than that inside the cell. The survival of the microbial cell depends on moving growth-requiring nutrients from a point of low concentration (the environment) to a point of high concentration (inside the cell). This ability to concentrate nutrients against a concentration gradient provides the cell with a better chance for survival when such nutrients are not plentiful in the environment. **Active transport** allows the accumulation of nutrients in an unmodified form against a concentration gradient (Figure 5-12). Active transport systems are of two types: simple and complex, both requiring an input of energy. **Simple active transport systems** involve a single cytoplasmic membrane protein that is energized by ion gradients (see Chapter 12). Simple sugars such as lactose, and unbranched amino acids utilize only a single transport protein. **Complex active transport** systems require three or more proteins to operate. The complex transport systems are called "periplasmic-binding proteins" * or shock-sensitive binding proteins because one of their components is located in the periplasmic space and can be released by osmotic shock. Binding proteins demonstrate solute specificity and permit rapid transfer of solutes into the cell. During the transport of a solute the periplasmic binding proteins interact with binding proteins in the cytoplasmic membrane, which results in the formation of a pore for transport of solute into the cytoplasm (Figure 5-13). Branched chain amino acids and sugars such as galactose, maltose, arabinose, and ribose use more than one transport protein.

Molecules can also be transported on carriers without the expenditure of energy in a process called **facilitated diffusion.** Facilitated diffusion does not concentrate solutes in the cell like active transport but does aid in the rapid equilibration of solutes across the cytoplasmic membrane. The only carbohydrate transferred by facilitated diffusion in the systems so far examined is glycerol. **Group translocation** is also a mechanism using

* Binding proteins are sometimes referred to as permeases.

carriers. This mechanism is unique in that the solute is altered chemically during transport on the carrier. The best-characterized group translocation system is the phosphoenolpyruvate-dependent sugar phosphotransferase system, in which the solute becomes phosphorylated during its transfer across the cytoplasmic membrane (Figure 5-14). Several sugars, including fructose, mannitol, and rhamnose, are transported by this mechanism.

It is important to realize that some substrates may be transported by more than one mechanism in a single bacterial species. In addition, the transport mechanism used by one bacterial species for transport of a specific substrate may not operate in another species.

Energy production. Most energy in nonphotosynthetic eukaryotes is produced in organelles called mitochondria or chloroplasts. There are no such organelles in prokaryotes, and some energy production is carried out in the cytoplasmic membrane that contains the appropriate enzymes (see Chapter 12).

Biosynthesis. Many enzymes and enzyme systems attached to the cytoplasmic membrane are involved in biosynthetic activities, including the following:

1. Biosynthesis of cell membrane lipid components
2. Biosynthesis of cell wall polymers such as peptidoglycan and teichoic acids
3. Biosynthesis of polysaccharides employed as extracellular polymeric substances
4. Biosynthesis of LPS that makes up part of the outer membrane of gram-negative bacteria

Secretion. The cytoplasmic membrane is involved in the release of extracellular proteins that include various toxins, bacteriocins, and enzymes produced in the cytoplasm of the cell. These products are important to the survival of the microbial species. Many of the enzymes, for example, are used to hydrolyze nontransportable

POINT OF INTEREST

BACTERIORHODOPSIN: A SPECIAL GLOBULAR PROTEIN

We are only beginning to understand the structure and function of the globular proteins that make up the cytoplasmic membrane. One of the most widely studied and best understood globular proteins is a pigment called **bacteriorhodopsin.** This protein is found in the cytoplasmic membrane of the bacterium *Halobacterium halobium,* a member of the Archaebacteria. Halobacteria are found in environments where the concentration of salt (usually NaCl) is at or near saturation. This organism can grow in the presence or absence of oxygen. Under anaerobic conditions (oxygen is absent), halobacteria rely on photosynthesis for growth and only under these conditions is bacteriorhodopsin synthesized. Pigment formation continues until it occupies approximately 50% to 80% of the membrane surface where it is organized into patches called the "purple membrane." Bacteriorhodopsin is the only protein in the purple membrane and 70% of its amino acids are hydrophobic; these are distributed in clusters. Each protein contains seven alpha helical domains that are arranged back and forth through the membrane (see figure). The helices are packed together to form a globular protein. The carboxy terminus of the polypeptide is on the inside of the membrane, while the amino terminus is on the outside of the membrane. The light-absorbing component of bacteriorhodopsin is the chromophore retinal, which is similar to vitamin A. Retinal is covalently bonded to a lysine residue in the polypeptide and is actively engaged in capturing photons of light. The energy from the captured light is used to pump protons across the membrane, creating an electrochemical gradient that is used in the synthesis of ATP as well as performing other functions (see Chapter 12 for a further discussion of proton pumps).

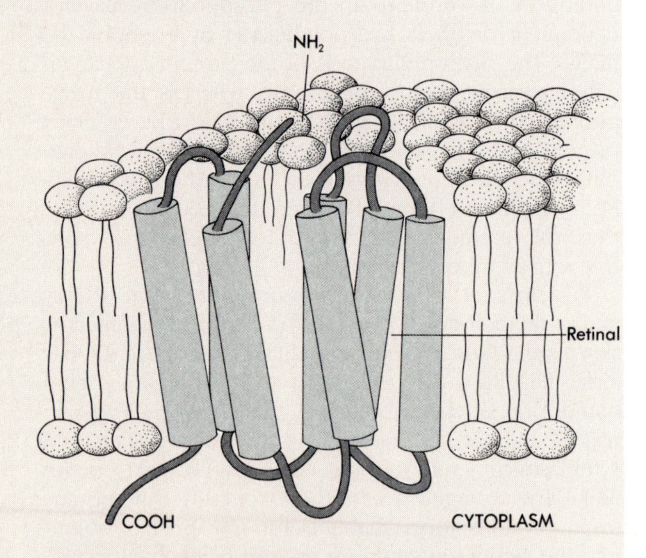

Model demonstrating position of bacteriorhodopsin in cytoplasmic membrane. Bacteriorhodopsin is made up of seven hydrophobic alpha helices (represented as cylinders) embedded in cytoplasmic membrane. Amino and carboxyl ends of protein are indicated. Each alpha helix is joined to other by hydrophobic sequences, which are found on either side of membrane. Pigment molecule called retinal is attached to one of alpha helices. Retinal captures light energy and triggers bacteriorhodopsin to pump protons across membrane. Process initiates special kind of photosynthesis in certain halophilic (salt-loving) bacteria.

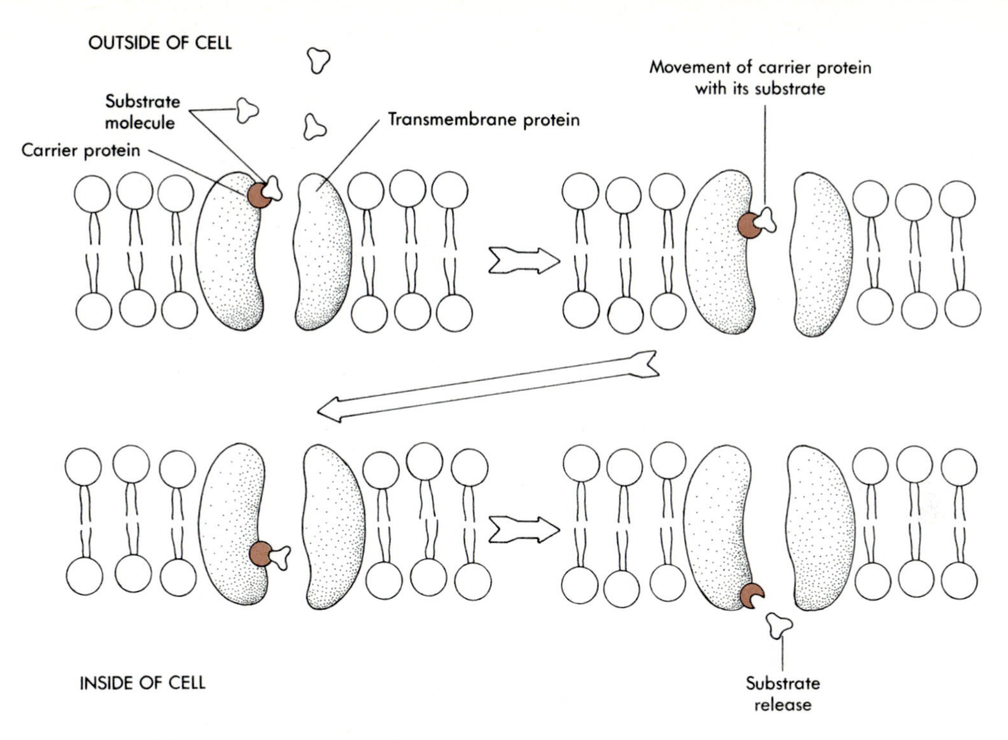

Figure 5-12 *Basic mechanism for active transport system. Substrate combines with carrier protein that transports substrate across cytoplasmic membrane unchanged and via a transmembrane protein. Energy required for transport is not indicated in this diagram.*

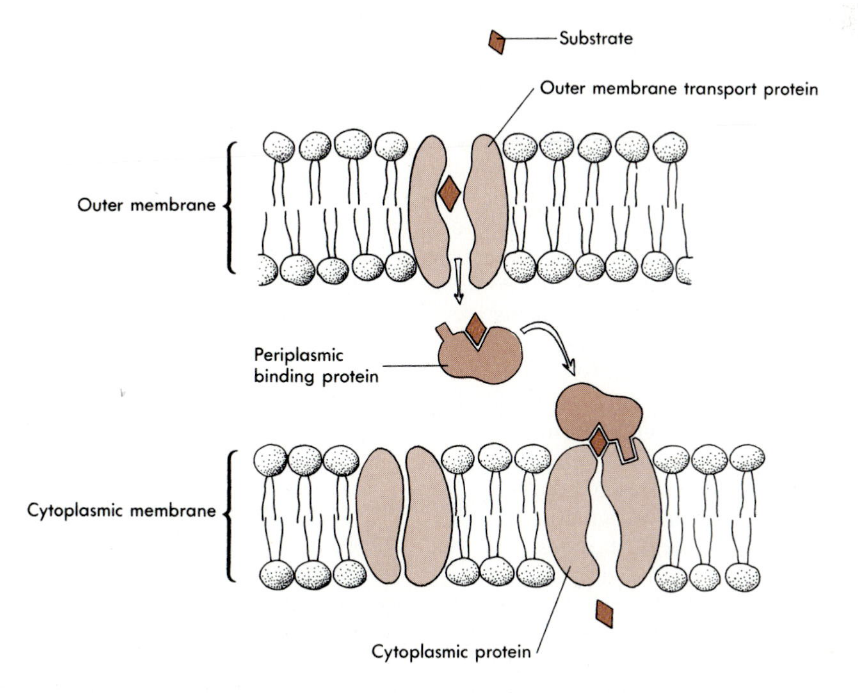

Figure 5-13 *Complex active transport system using gram-negative cell as example. Substrate is transported through outer membrane of cell wall via transport protein embedded in cell wall. When substrate reaches periplasmic space it interacts with periplasmic binding protein. Periplasmic binding protein and its bound substrate are specifically recognized by cytoplasmic membrane protein and substrate is transported into cytoplasm of cell.*

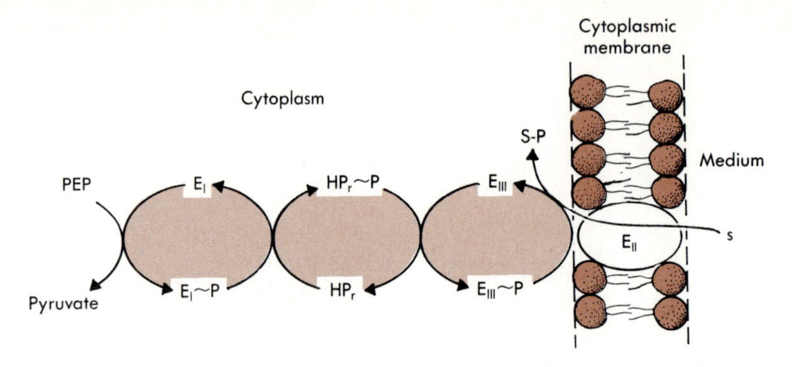

Figure 5-14 *Group translocation (phosphoenolpyruvate-dependent sugar phosphotransferase system). Phosphoenolpyruvate (PEP) transfers its high-energy phosphate to enzyme I (E_I) and then to heat-stable protein* (HP_r)*. High-energy phosphate on HP_r is transferred to enzyme III* (E_{III})*, which in turn interacts with sugar and enzyme II* (E_{II}) *to facilitate transfer of sugar* (S) *across cytoplasmic membrane into cytoplasm of cell as phosphorylated derivative (S-P).*

macromolecules, such as proteins, polysaccharides, and lipids, into lower molecular weight units for transport from the environment into the cytoplasm of the cell. Some enzymes are used in the construction of the cell wall. Bacteriocins released by the bacterial cell are able to inhibit the growth of organisms that compete with them for nutrients. Bacteriocins are specific in that they inhibit only related species of bacteria.

During infection of mammalian hosts some bacteria produce toxins that, when released, inhibit the defense mechanisms of the host (see Chapter 23). Most of the secreted molecules appear to be too large or hydrophilic to pass through the cytoplasmic membrane of the bacterium. A model called the **signal hypothesis** has been proposed, based originally on protein secretion in eukaryotes, to explain protein export in bacteria. In both eukaryotes and prokaryotes proteins are synthesized in precursor forms that are later made active by the removal of amino-terminal amino acids. These terminal sequences consist of 15 to 30 amino acids (referred to as the **signal** or **leader sequences**). They comprise a hydrophobic sequence of at least 11 amino acid residues preceded by a short stretch of positively charged amino acid residues at the amino terminus. The current version of the signal hypothesis proposes that as soon as the signal is made on the ribosome during protein synthesis, it interacts with a recognition particle in the cytoplasm (Figure 5-15). The ribosome-protein-recognition particle complex interacts with a cytoplasmic-associated protein (docking protein). Synthesis of the protein continues, and the polypeptide chain is extruded through a protein channel in the cytoplasmic membrane. A processing enzyme (peptidase) removes the signal as it passes through the cytoplasmic membrane. Once the signal is removed the externalized polypeptide will fold into its normal configuration and become active.

Mesosomes

Mesosomes, whose function is still in doubt and which many scientists regard as artifacts, are discussed in Chapter 4.

Concept Check

1. **Describe the chemical composition of the cytoplasmic membrane and how its components are organized.**
2. **What are the potential functions of the cytoplasmic membrane?**
3. **Describe the ways in which solutes can be transported across the cytoplasmic membrane into the cytoplasm.**
4. **How is the cytoplasmic membrane involved in the export of proteins (signal hypothesis)?**

STRUCTURES LYING OUTSIDE THE CELL WALL

Extracellular polymeric substances (EPS)

Adjacent to the cell wall are located polymers, called extracellular polymeric substances (EPS), that provide certain advantages to the cells that possess them. Included in this group are **S-layers** and **capsules** and **slime layers**.

S-layer. S-layers, or subunit surface layers, are not found in all bacteria but when present may represent the outermost structure of the cell. They are layers of protein or glycoprotein that are linked to each other as well as to the underlying cell wall (Figure 5-16). The function of the S-layer has not been unequivocally de-

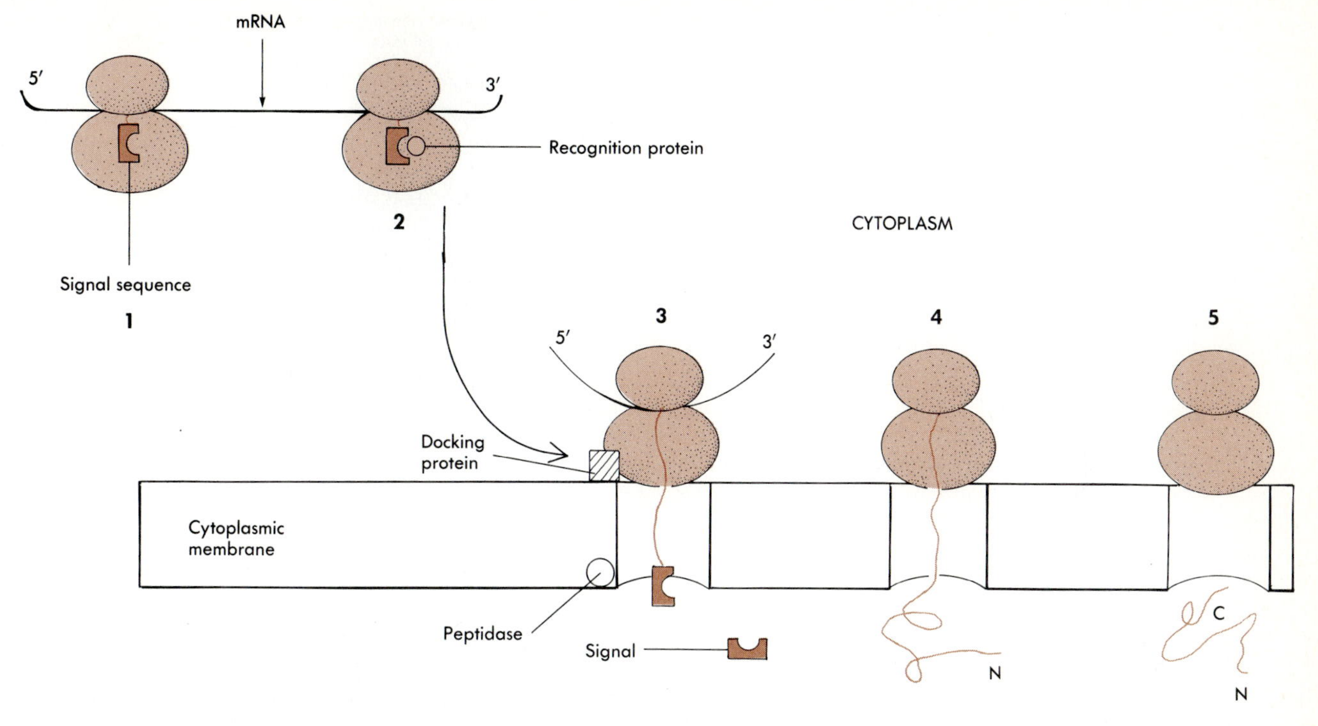

Figure 5-15 *Model to explain signal hypothesis for protein export. (1) Export protein partially synthesized with signal sequence. (2) Ribosome-protein complex interacts with recognition protein in cytoplasm blocking further protein synthesis. (3) Ribosome with export protein and recognition protein interact with docking protein on cytoplasmic membrane and synthesis of export protein continues with translocation of export protein through pore in cytoplasmic membrane. (4) Peptidase in cytoplasmic membrane removes signal sequence. (5) Export protein is totally synthesized and folded into proper configuration.* N, *N-terminal amino acid residue;* C, *carboxyl terminal amino acid residue.*

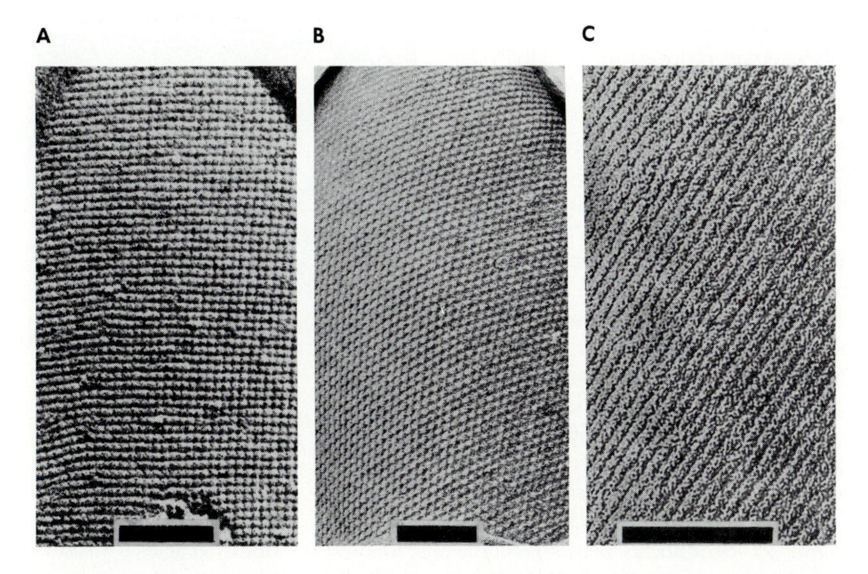

Figure 5-16 *Electron micrography of replicas of freeze-etched preparations of S-layer of three different species of bacteria.* ***A,*** *Square S-layer morphology.* ***B,*** *hexagonal S-layer morphology.* ***C,*** *oblique S-layer morphology. Bar = 100 nm.*

termined but is probably a protective device for the cell. In some archaebacteria in which a true peptidoglycan layer is absent, the S-layer acts as a supporting framework for the cell. The S-layer in some species of *Bacillus* appears to act as a shield against degradative enzymes such as lysozyme.

Capsules and slime layers. Capsules and slime layers are polymers surrounding the cell wall of many bacteria. Capsules are firmly adherent to the cell wall and exhibit structural integrity (Figure 5-17). Slime layers are amorphous masses released by the cell and do not show the structural definition of capsules.

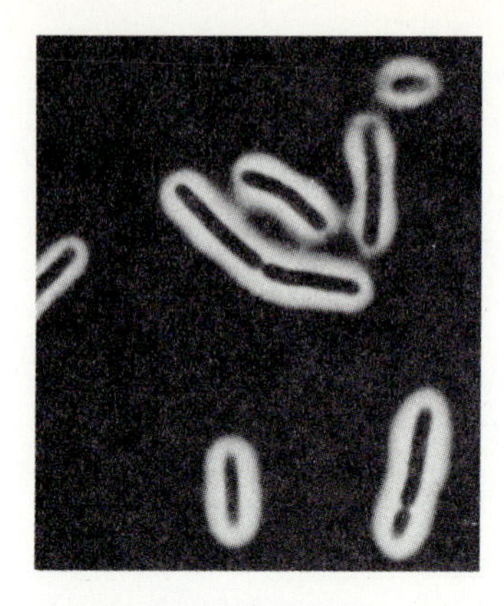

Figure 5-17 India ink preparation demonstrating bacterial capsule. Capsule appears as refractive halo around bacillus.

Structure

Most capsules or slime layers are either homopolymers or heteropolymers of a wide variety of monosaccharides, such as hexoses, uronic acid, and amino sugars, as well as other constitutents. The chemical composition of the capsule or slime layer of some bacterial species is described in Table 5-1. Some strains of a species will produce the same extracellular polymer, whereas strains of some species, such as *Streptococcus pneumoniae,* may produce different types. The consequences of such a characteristic are discussed later. Some capsules are made of protein; for example, *Bacillus anthracis* produces a polyglutamic acid.

Functions and properties

Adherence. The EPS is a very hydrated and sticky substance and therefore adheres to various substrates or is adhered to by the same or other species in its particular habitat. Some groups of microorganisms, such as the myxobacteria, use capsules or slime layers for aggregation when nutrients become reduced in the environment. Many species in aquatic habitats use them to form aggregates of cells of the same or different species. Capsules or slime layers may afford the cell protection and permit the cell(s) to attach to a specific substrate in order to remain sessile in fast-moving streams and/or obtain nutrients from the substrate. Once the capsule or slime layer is formed, bacteria can divide and form a microcolony of many cells (Figure 5-18). The capsule or slime layer acts like a resin exchange column that can retain some ions and molecules and let other molecules pass through. This permits the microcolony to be protected from heavy metals, adsorption by bacterial viruses, and the action of antibiotics and bacteriocins produced by other microorganisms.

Figure 5-18 *Mucoid appearance of colonies on agar surface produced by bacterial cells that release EPS.*

The capsule or slime layer is used as a recognition site for the attachment of nitrogen-fixing bacteria, such as species of *Rhizobium,* to leguminous plants (Figure 5-19). Plant surface proteins or glycoproteins called **lectins** act as the binding site for the encapsulated bacteria.

The bacterium *Streptococcus mutans,* known to cause tooth decay (caries), produces an extracellular slime layer composed of a water-soluble glucose polymer called **dextran** (mutan) (Figure 5-20). The polymer enables the organism to bind to the enamel surface and

Table 5-1 *Chemical composition of the extracellular polymeric substances of some bacterial species*

Bacterium	Chemical composition
Acinetobacter calcoaceticus	Glucose plus glucuronic acid
Azotobacter vinelandii	Alginate
Bacillus anthracis	Polyglutamic acid
Erwinia amylovora	Polysaccharide
Leuconostoc mesenteroides	Dextran*-dextran-sucrose
Rhizobium meliloti	Glucose plus galactopyruvate
Ruminococcus albus	Glycoprotein
Streptococcus mutans	Dextran plus other glucan
Streptococcus (group B)	Galactose plus *N*-acetylglucosamine or sialic acid plus galactose, heptose, glucose, glucosamine, and mannose
Streptococcus pneumoniae	Various oligosaccharides
Yersinia pestis	Protein plus glycoprotein containing oligomeric galactan

*Dextrans are polymers of glucose containing predominantly α, 1-6 linkages.

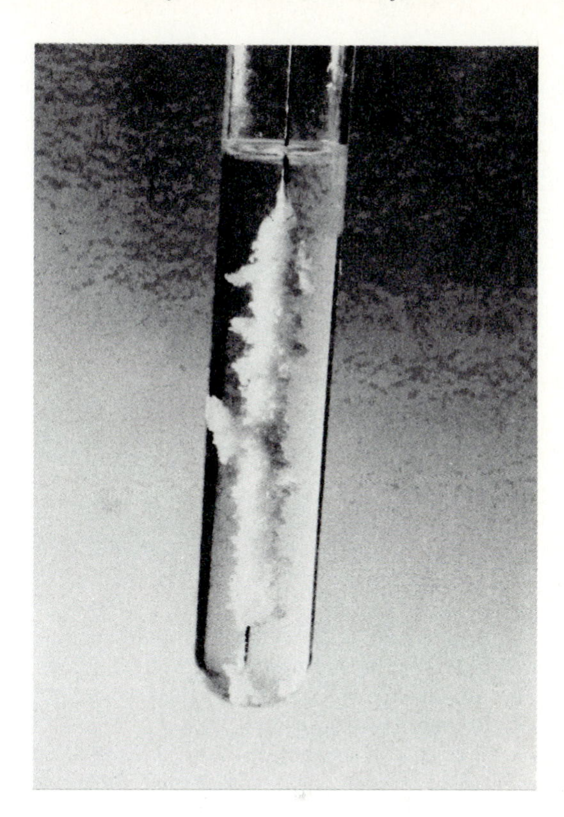

Figure 5-20 Streptococcus mutans *grown in culture tube containing synthetic medium plus sucrose. Wire in tube is covered with sticky white polysaccharide (dextran) produced by growing microorganisms.*

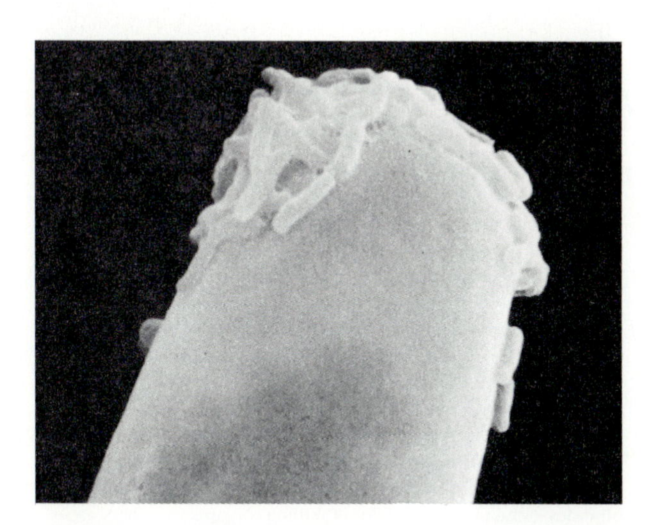

Figure 5-19 *Adherence of nitrogen-fixing bacteria (species of* Rhizobium*) to clover root hair.*

also serves as the site for accumulation of other microbial species. As the number of *S. mutans* cells increases in the microcolony, and when sweets containing sucrose are ingested, fermentation results in excessive acid production, which reduces the pH level to a point where demineralization of enamel takes place. Adherence of microorganisms to various substrates has also created problems in industry (see Point of Interest).

Virulence. The disease-producing potential (virulence) of several microorganisms is related to the production of capsules or slime (Plates 4 and 5). Some protect the cell from phagocytosis (digestion by phagocytes) during the infection process. The capsule or slime layer can be a factor in the spread of the infectious agent because of the latter's ability to multiply within the phagocyte. *Streptococcus pneumoniae,* the most common cause of pneumonia, is virulent only when it produces a capsule, and over 100 capsular types have been described. Mutants of *S. pneumoniae* unable to produce capsules are readily digested by phagocytes and are no longer virulent. The capsule or slime layer produced by certain species of human pathogens can also inhibit the transport of certain molecules into the bacterial cell. Species of *Pseudomonas,* for example, that are normally sensitive to certain antibiotics become resistant when, through genetic changes, they develop the capacity to produce extracellular polymers.

Vaccines can be produced from some EPS-producing bacteria, such as *S. pneumoniae,* because many capsular components are strongly antigenic. Vaccines stimulate the formation of antibodies but do not cause disease and thus make the host resistant to future infections by the same species. Many capsular types of *S. pneumoniae* cause disease; therefore the vaccine used must be composed of more than one type of capsule (that is, it must be polyvalent).

Movement. The production of slime is a characteristic of all gliding bacteria (see Chapter 4).

FLAGELLA

Structure

The flagellum in most bacteria is composed of a **filament, hook,** and **basal structure** (Figure 5-21). This organization is also characteristic of the axial filaments of spirochetes. The filament is made up of a single type of protein subunit called **flagellin,** whose amino acid composition differs slightly from one species to another. The protein subunits in the filament are arranged in a helical fashion, leaving it with a central hole. Rigorous conditions such as heat or low pH can lead to disaggregation of the filament into its individual subunits. The subunits can be made to reaggregate to produce flagella-like filaments under suitable laboratory conditions. Aggregation occurs in the absence of enzymes and is similar to the assembly of protein subunits that make up the outer structure of viruses. If the filament is sheared off at the surface of the bacterial cell, regeneration takes place by the addition of subunits to the distal end. The rate of growth of the flagellin filament under ideal conditions is 0.2 μm/minute. Apparently flagellin travels along the base of the flagellum to the tip during the regeneration process. Most filaments are naked, but in some genera, such as *Vibrio, Pseudomonas, Bdellovibrio,* and one species of *Rhizobium,* a sheath of protein covers the filament. The sheath appears to be an extention of the outer membrane of the cell wall.

The hook is attached to the base of the filament and is composed of a single type of polypeptide that differs from the subunits of the filament.

The basal structure is the most complex component of the flagellum and consists of several different polypeptides. There are ringlike structures associated with the basal structure: four rings in gram-negative bacteria and two rings in gram-positive bacteria. In *E. coli,* a gram-negative bacterium, the two rings immediately adjacent to the hook are embedded in the outer membrane (Figure 5-21) and remain stationary. The two other rings are associated with the peptidoglycan layer and the cytoplasmic membrane. The ring embedded in the peptidoglycan layer remains stationary and in mechanical terms is referred to as the **stator.** The stator is

Point of Interest

EXTRACELLULAR POLYMERIC SUBSTANCES: A STICKY SUBJECT

Microbial adhesion to animate or inanimate surfaces can be mediated by polysaccharide capsules or slime and protein pili. These adherence polymers are collectively called **adhesins.** Microbial adhesin plays an important role in the infectious disease process, but it is also an important factor in various technologies. Microorganisms tend to adhere to any surface, and the layer they produce is called a **biofilm.** Biofilms have created many problems in the oil industry and in other situations in which pipelines are used. Biofilms of anaerobic bacteria in metal pipes can result in the formation of iron-sulfide particles. These particles are corrosive and cause perforations in the metal pipe. The leakage of oil from these corroded pipes is not only of economic importance but is also a cause of environmental pollution. The formation of biofilms also reduces pipe diameters, resulting in a need for increased pressures and thereby placing greater demands on energy sources. The adhesion of microorganisms also causes problems in the oil recovery process. Hydraulic fluid systems used to lubricate valves in oil and gas drilling rigs provide an ideal environment for the formation of biofilms. These biofilms, if left untreated, can cause serious fouling problems.

The capsules or slime layers produced by bacteria in biofilms enables the aggregated bacteria to accumulate nutrients, a property useful in sewage treatment. Dissolved organic matter in sewage is trapped in the biofilm and converted to inorganic compounds, a process called **mineralization.** This process is discussed further in Chapter 18.

The problems associated with biofilms has led to the search for chemicals capable of dispersing microorganisms. Microorganisms in biofilms are more resistant to commercial biocides (chemicals capable of killing organisms) than are their free-floating counterparts. Chlorine compounds, such as hypochlorite, have been shown to be capable of breaking down slime layers and dispersing bacteria, but some biocides are also dangerous to the environment and require constant monitoring. Biocides specific for the organism causing the biofilm are currently under investigation to circumvent this problem.

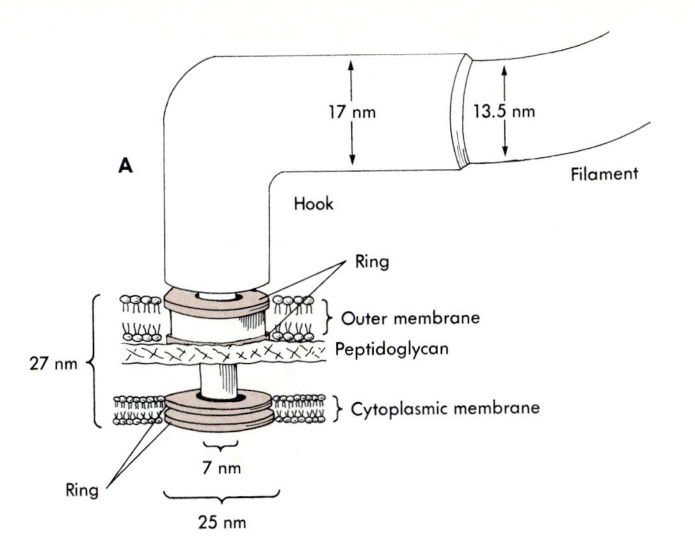

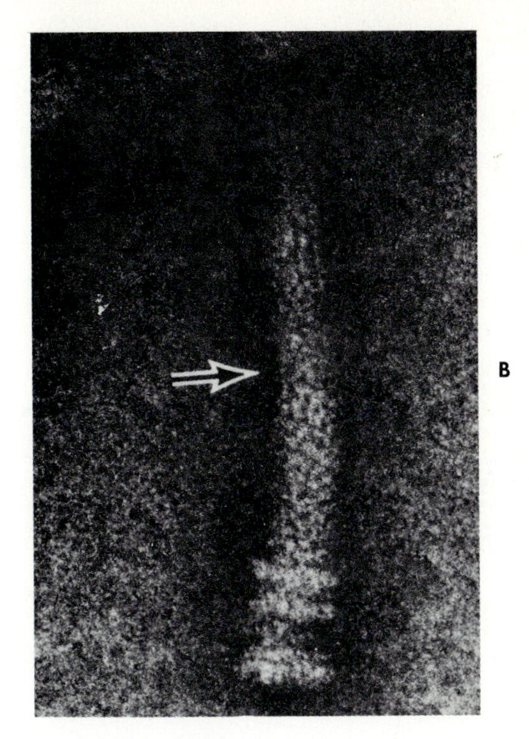

Figure 5-21 *Structure of prokaryotic flagellum.* ***A,*** *Model of gram-negative flagellum showing relationship of rings to outer membrane and cytoplasmic membrane. Numbers represent dimensions in nanometers.* ***B,*** *Electron micrograph of flagellum from gram-negative bacterium. Arrow points to junction between hook and filament.*

in direct contact with the most distal ring, which is attached to the cytoplasmic membrane and is capable of rotation. The most distal ring acts like a "**rotor**" and forms a direct link to the hook and filament by means of a protein rod. The basal structure thus acts like a motor.

Mechanism of flagellar motion

Bacterial flagellar movement is accomplished by the rotation of the filament, which is mediated by the "rotor" embedded in the cytoplasmic membrane. Movement of the rotor is achieved by the energy generated when protons move across the cytoplasmic membrane (proton movement and energy generation are discussed in Chapter 12). The best way to visualize flagellar movement is to imagine yourself sitting inside the cell facing the rotor. If a handle were attached to the rotor, you could move the rotating parts to the right or left. In this way a spiral wave travels down the flagellum and a rotary motion is achieved. Rotary motion does not occur during the movement of eukaryotic flagella; instead movement is oarlike (see Chapter 7). Experiments have revealed that when latex beads are attached to the flagellum of a bacterial cell, the beads can be seen to rotate about the axis of the flagellum during movement while the body of the cell rotates in a different direction. Motile bacteria, such as the peritrichously flagellated *E. coli,* swim forward in almost a smooth straight-line fashion, with each run being interrupted by jiggling motions called **twiddles** or **tumbles**. Following the short tumbling period, the bacterial cell begins a new run but in a random direction. These changes in direction are also accompanied by a change in the direction of rotation of the flagella. Smooth swimming is characterized by **counterclockwise** rotation of the flagella, whereas tumbles are clockwise.

The clockwise rotation of the flagella is from the vantage point looking along the flagellum toward the cell. In *E. coli* the flagella form a coordinated bundle during smooth swimming (Figure 5-22) that enables the cell to move and rotate in one direction. The flagellar bundle (or flagellum in polarly flagellated cells) pushes the cell forward like a ship's propeller. A change of direction or tumbling results in clockwise rotation and the flagellar bundle flies apart (Figure 5-22, *B*). Not all bacteria tumble to change direction. Some monoflagellated bacteria "back up" after a period of smooth swimming and then change direction.

The mechanism by which spirochetes move is not completely understood. They move in a screwlike motion when observed in vivo, and many investigators believe they rotate about their longitudinal axes. One mechanism that has been proposed suggests that the rotation of axial fibers in one direction causes the rotation of the protoplasmic cylinder in the opposite direction. The outer sheath, which is believed to be flexible, rotates in a direction opposite that of the protoplasmic cylinder.

Taxes and tropisms. Microorganisms have sensory

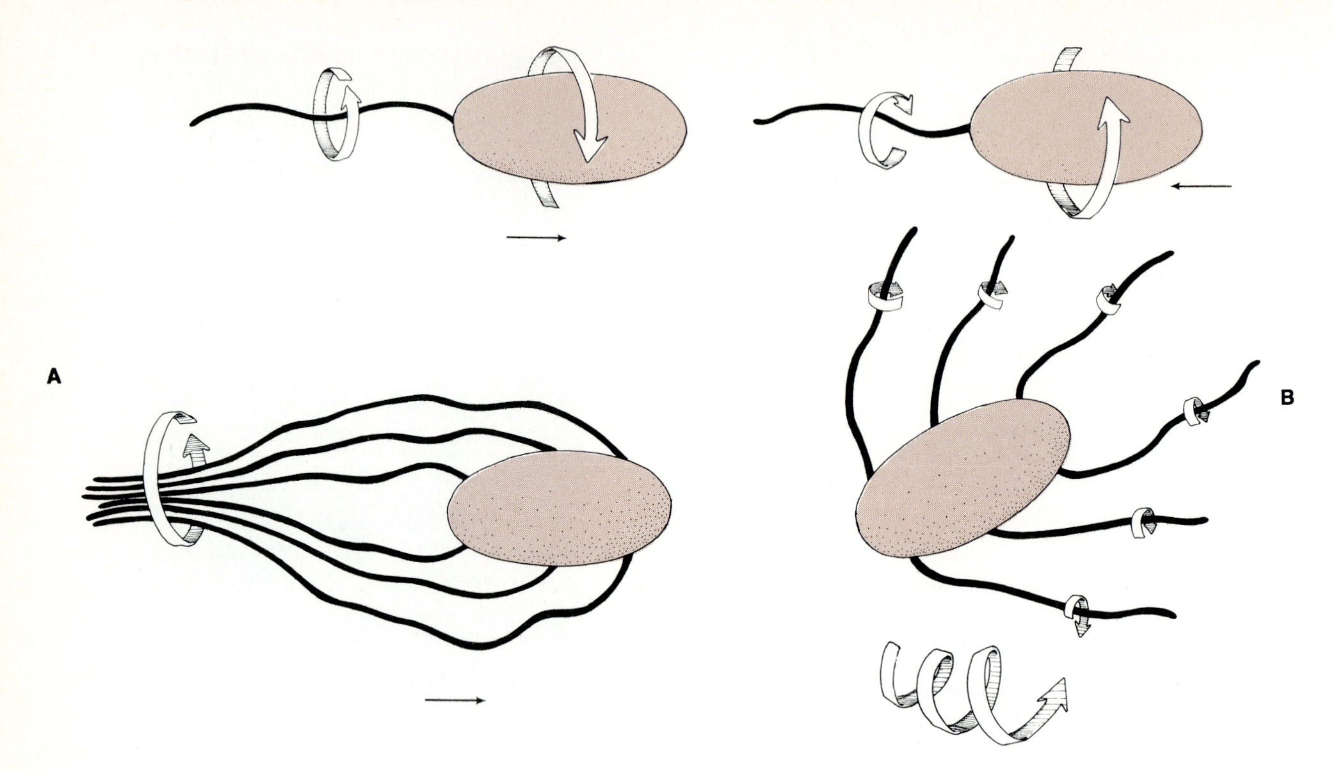

Figure 5-22 *Flagellar rotation and movement.* ***A,*** *Response of cell with polar flagellum. On left is normal body rotation in which single flagellum rotates counterclockwise and cell moves forward (period of smooth swimming) with flagellum trailing cell. On right, single flagellum shows clockwise rotation. Direction of movement (period of tumbling) is with flagellum ahead of cell.* ***B,*** *Response of cell with peritrichous flagellation. On left, counterclockwise rotation (period of smooth swimming) causes flagella to form bundle behind cell. On right, clockwise rotation (period of tumbling) causes individual flagella to fly apart.*

receptors that enable them to respond to stimuli of a chemical or physical nature. These responses may be in the form of taxes and tropisms. **Tactic** responses are movements by cells that are motile, and **tropisms** are growth orientations of organisms that are normally attached to a surface and are unable to demonstrate locomotion. Tropisms are more characteristic of eukaryotic microorganisms. The tactic response in prokaryotes may be due to stimuli such as light, heat, electricity, and chemical agents. The most widely studied responses are those resulting from the presence of chemical agents, that is, chemotactic responses.

Chemotaxis. Chemotaxis is a biological phenomenon that is common to the microbial world as well as higher organisms. It is a property that has been demonstrated to be important in controlling many behavioral responses, including sexual activity, feeding, predatory-prey relationships, and aggregation. Bacteria may be attracted to chemicals **(positive chemotaxis)** or repelled by chemicals **(negative chemotaxis)**. Chemotaxis can be demonstrated by creating a concentration gradient, and one technique is to use a capillary tube con-

Figure 5-23 *Demonstration of bacterial chemotaxis.* ***A,*** *Bacteria evenly dispersed in flask.* ***B,*** *Capillary tube with chemical attractant is added to culture medium. Bacteria are seen to aggregate around and within capillary tube.*

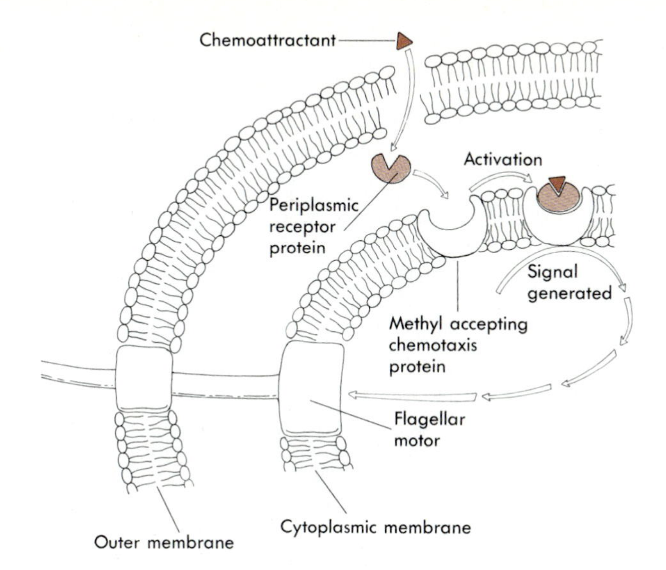

Figure 5-24 *Model for chemotaxis. Chemoattractant (amino acid, for example) interacts with periplasmic receptor protein. Periplasmic receptor protein-chemoattractant complex then interacts with methyl-accepting chemotaxis protein (MCP) in cytoplasmic membrane and activates the MCP. Activated MCP generates signal (may be enzymatic reaction) that causes flagellar motor to rotate counterclockwise, thus enabling bacterium to swim towards chemoattractant. For convenience, only one group of periplasmic or methyl-accepting chemotaxis proteins have been indicated.*

taining a specific chemical. If the capillary tube contains an attractant and is placed in a suspension of bacteria, the bacteria will soon be found in high concentrations around the capillary opening and within the capillary tube (Figure 5-23). By removing the capillary tube at intervals of time, it can be seen that the number of bacteria in the capillary tube increases with exposure time. If a repellent is added to the bacterial suspension, there will be a rapid increase in the number of bacteria within the capillary tube where the concentration of repellent is least. Because chemotaxis is associated with flagellar movement, any factors that affect flagellar activity will also affect chemotaxis. Such factors include pH, temperature, and heavy metal concentrations.

How does the cell respond in its movement toward an attractant or away from a repellent? In a medium in which there is no chemical gradient (the concentration of the chemical in the surrounding environment is equal at any one site), a bacterium will spend equal time swimming and tumbling. Any change in direction of movement is dependent on Brownian collisions (bombardment of the cell by water molecules). Once a chemical is added to the medium a gradient (high concentration down to a low concentration) is formed, and the cell responds in one of two ways. If the chemical is an attractant, the microbial cell moves to the area where the concentration of attractant is highest. This movement is characterized by less tumbling and longer periods of smooth swimming. The cell, in the presence of a chemical that acts as a repellent, swims away from the highest concentration of repellent to the area of lowest concentration of repellent. Again the tumbling is decreased. In both of the foregoing instances there is specific orientation and progress to or away from the stimulus.

Microorganisms may be attracted to amino acids, sugars, oxygen, vitamins, and inorganic ions; however, it is not necessary that these attractants be actually metabolized by the cell. It has been observed in some instances that the attractant can become a repellent as the concentration of the chemical increases or if other activities in the cell change. Aerobic microorganisms, for example, can be attracted to a high concentration of oxygen (aerotaxis), but when respiration slows down because of a lack of nutrients in the medium, oxygen acts as a repellent. Negative chemotaxis is most frequently observed with chemicals that produce extremes of pH, that is, pH values between 2 and 4, and 10 and 12. Such responses to stimuli are of obvious selective advantage to the microorganism because they provide the cell with a mechanism for avoiding lethal or deleterious regions of its environment.

Mechanism of chemotaxis. The sensory response of bacteria to various chemicals is associated with **chemoreceptor proteins** located in the cell. When an attractant is present, it binds to a specific periplasmic receptor protein. The receptor then interacts with one of three cytoplasmic membrane proteins (Figure 5-24). This interaction results in the release of an intracellular mediator molecule that causes the flagellar motor to continue to rotate counterclockwise and maintain smooth swimming. This adaptation in chemotaxis is believed to be due to **methylation** of cytoplasmic membrane proteins. When attractant is present, cytoplasmic membrane proteins are methylated; when attractant is absent, the proteins are demethylated. Methylation is associated with smooth swimming, while demethylation results in tumbling of the bacterial cell. The rotary motor of the flagellum, which is embedded in the cytoplasmic membrane, is driven by a proton (H^+) gradient across this membrane. The information from the methylation process is believed to be processed by a tumble regulator that is the actual "switch" controlling the direction of flagellar rotation. How this process works is not known, but in *Bacillus subtilis* calcium ion concentration appears to play a role. At low cytoplasmic calcium concentrations the cells swim smoothly; at slightly higher concentrations the organism exhibits mixed smooth swimming and tumbling; and at high levels of calcium there is continuous tumbling. Calcium also plays a role in eukaryotic movement (see Chapter 14).

Swarming. Motile species of *Proteus* exhibit a response on the surface of agar called **swarming**. The re-

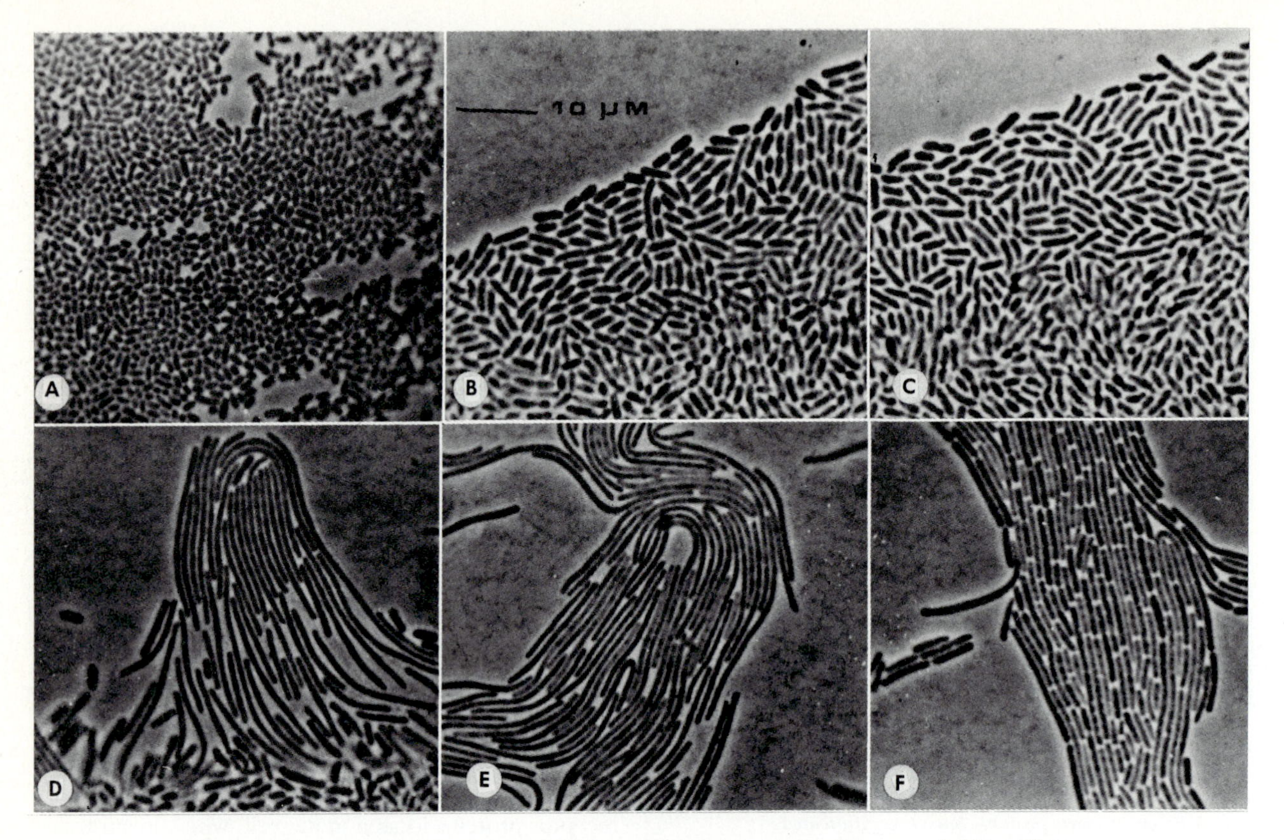

Figure 5-25 *Swarming phenomenon. Phase-contrast micrographs of* Proteus mirabilis *within colony.* ***A,*** *Cells are initially coccoid.* ***B*** *and* ***C,*** *Cells at edge of colony begin to enlarge.* ***D*** *and* ***E,*** *Cells at edge of colony move out into agar in large groups of long swarm cells.* ***F,*** *Swarm cells begin to divide and form short rods as they consolidate.*

sponse is observed when cells are inoculated near the center of the plate. The cells undergo division and form colonies, but later the cells on the perimeter of the colony change morphologically and swarm to fresh areas. The process will continue until the entire surface is covered with microbial growth. The mechanism controlling this reponse is not understood, but it is believed to be a chemotactic response. The most interesting aspect of this phenomenon is the morphological changes that occur (Figure 5-25).

Proteus normally possesses a polar flagellum but as a colony develops, cells at the edge elongate and become peritrichously flagellated. The elongated (swarm) cells then move out from the colony in larger and larger groups, and later the swarm cells consolidate and undergo normal growth and division to produce short cells. As the colony enlarges, short cells at the edge of the colony again elongate, and the whole process is repeated until the plate is covered with zones of light and heavy growth (Figure 5-26).

PILI

Pili (sing. *pilus*), as stated in Chapter 4, may be divided into common pili and sex pili. **Common pili** are for the most part found on gram-negative bacteria, particularly members of the family Enterobacteriaceae (Figure 5-27). Appendages similar to common pili have been observed on some gram-positive bacteria, such as species of *Streptococcus* (Figure 5-28) and *Corynebacterium,* but sex pili have not been found in gram-positive bacteria.

Common pili, which appear to arise from the cytoplasmic membrane, may cover the entire bacterial surface. A cell may carry from one to more than 100 pili, which can vary in length from 0.2 to 20 μm, but a few species have pili located only at the poles. Common pilus production is controlled by genes on the chromosome, in contrast to sex pili, which are governed by genes located on extrachromosomal DNA called plasmids. Although common pili do confer certain properties on the cell, their significance in terms of selective advantage over cells without them has been difficult to

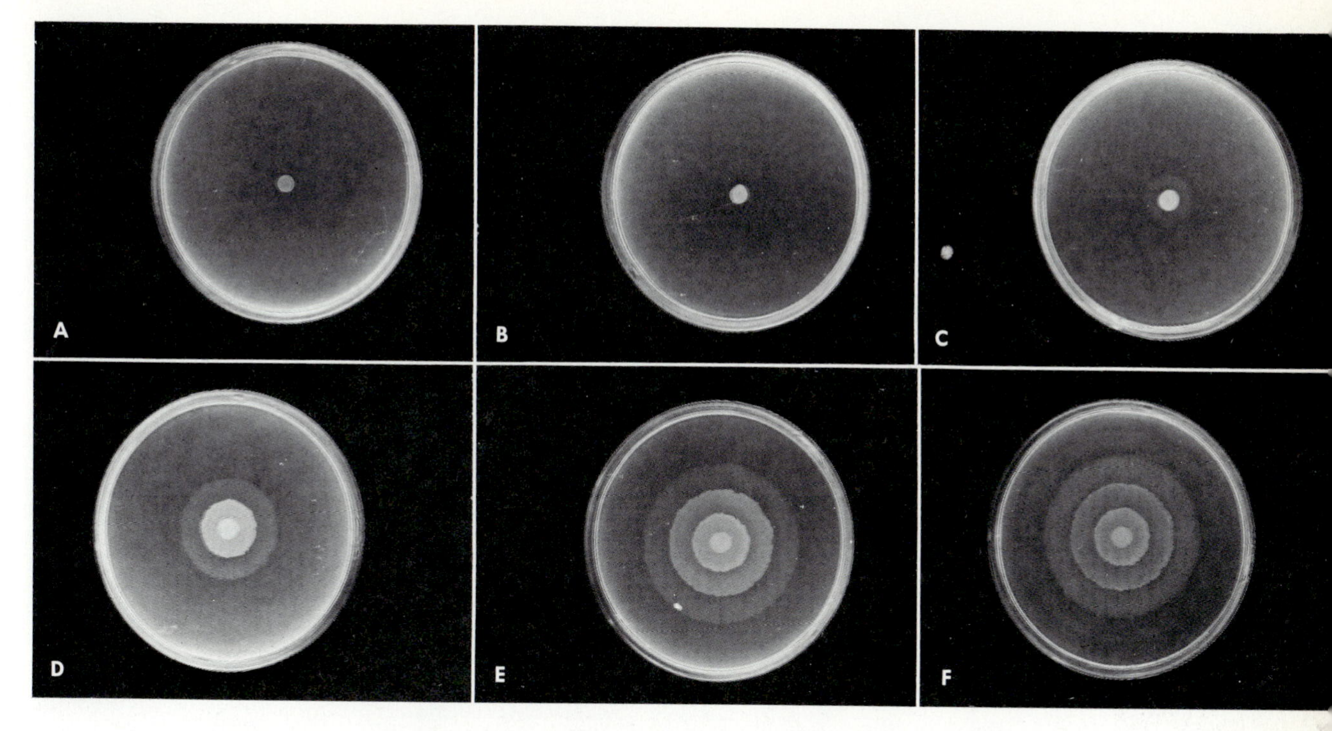

Figure 5-26 *Appearance of agar plate demonstrating swarming of* Proteus mirabilis *after 2 hours.* ***A****; 3 hours,* ***B****; 4 hours,* ***C****; 8 hours,* ***D****; 12 hours,* ***E****; and 16 hours,* ***F*** *(see text for details).*

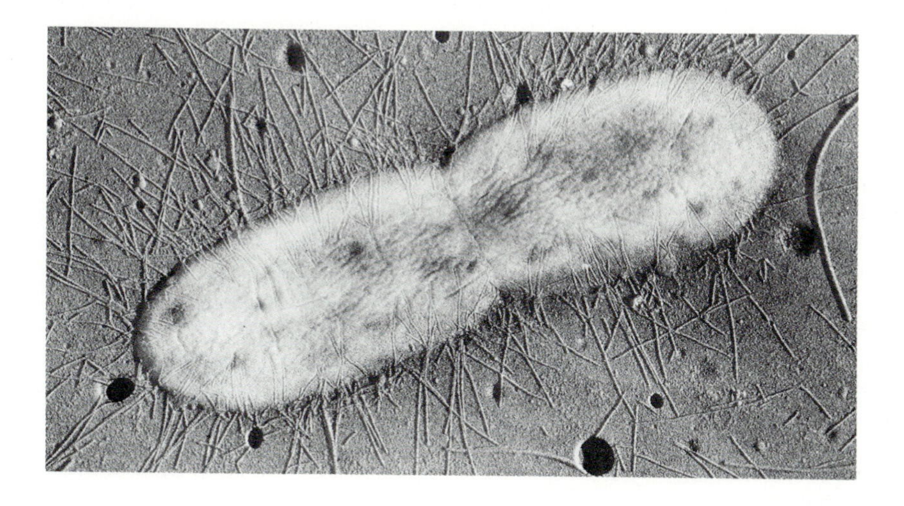

Figure 5-27 *Electron micrograph of piliated gram-negative* (E. coli) *bacterium.*

evaluate. Some of the suggested properties are as follows:

1. **Antigenicity.** Pili are antigentic, and their presence on the bacterial cell surface makes them readily accessible to those immune products of mammalian systems that engage in antibody formation. Common pilus formation by the causative agent of gonorrhoeae *(Neisseria gonorrhoeae)* has prompted an attempt at producing a vaccine by using purified pilus protein.
2. **Adherence.** Pili confer adhesive properties to the cell and enable it to bind to epithelial surfaces such as those found on the respiratory, intestinal, and genitourinary tracts of animals and humans.

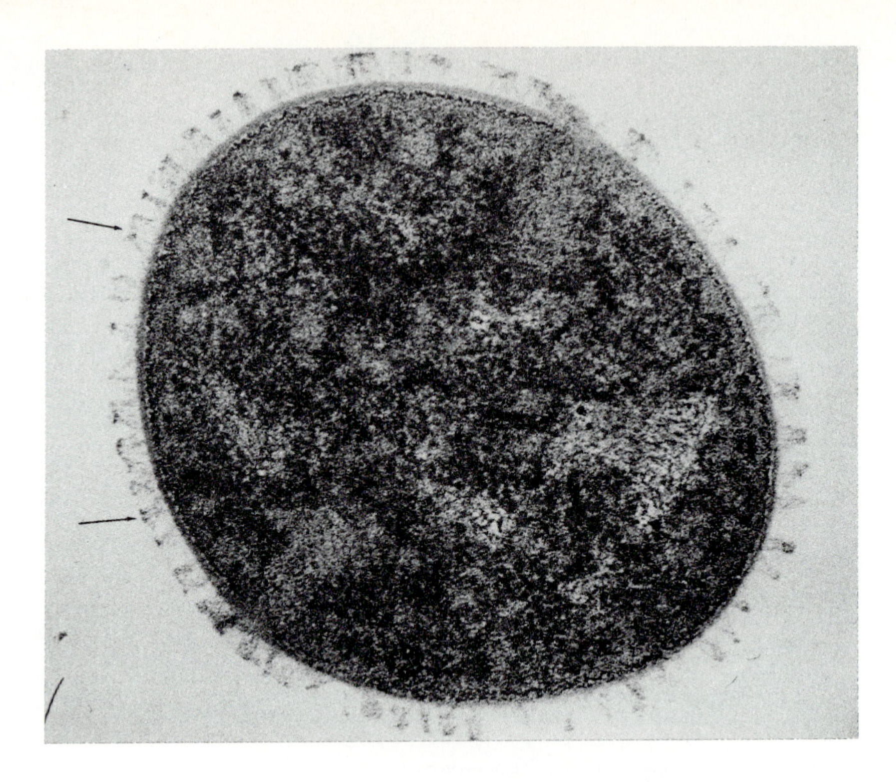

Figure 5-28 *Electron micrograph of pili-like appendages* (arrows) *on surface of gram-positive bacterium* (Streptococcus pyogenes). *(× 140,000.)*

Attachment permits microbial cells to replicate and increase to higher numbers than cells that do not have mechanisms for attachment (see Chapter 23).

3. **Aggregation.** Pili cause certain bacteria to aggregate and form pellicles or thin layers of cells packed together on a surface of a liquid medium. This characteristic may facilitate oxygen delivery to the cell in oxygen-poor or oxygen-depleted areas.

Sex pili are appendages on the cell surface that are involved in the transfer of nucleic acid from one cell to another (Figure 5-29). The sexual preference may be between species of the same genus or between species from different but related gram-negative genera. The DNA transferred may be totally chromosomal, viral, or plasmid (extrachromosomal DNA). The information for the production of sex pili is found on a special plasmid referred to as the fertility factor or F^+ factor. The cell producing the sex pilus is called the male or donor cell. Although sex pili are essential for the mating process in bacteria (the process is called **conjugation**), their role in DNA transfer is still unclear.

Sex pili are hollow cylindrical filaments 8 nm in diameter with a central axial hole of 2 nm. Knobs have been observed occasionally on one end of the pilus. Sex pili extend from 1 to 2 μm from the surface of the cell, and their average number per cell ranges from 1 to 3. The sex pilus can act as an organ of attachment for certain RNA bacterial viruses that attach along its entire length. This characteristic helps to distinguish the pilus from other appendages on the cell surface. Filamentous single-stranded DNA bacterial viruses can also attach to the sex pilus at its tip before invading the bacterium and lysing it.

Concept Check

1. **Explain the chemical composition and possible functions of the structures that lie outside the cell wall of bacteria.**
2. **Explain how flagella rotate when bacteria are attracted to a chemical or repulsed by it.**
3. **Describe a currently held hypothesis for bacterial chemotaxis.**

THE CYTOPLASM AND ITS CONTENTS

The cytoplasm of the cell is that aqueous solution which is bounded by a cytoplasmic membrane and within which are located various insoluble inclusions. Some of these cytoplasmic components are absolutely essential to all cells, while others are specialized structures that are found only in a few prokaryotes.

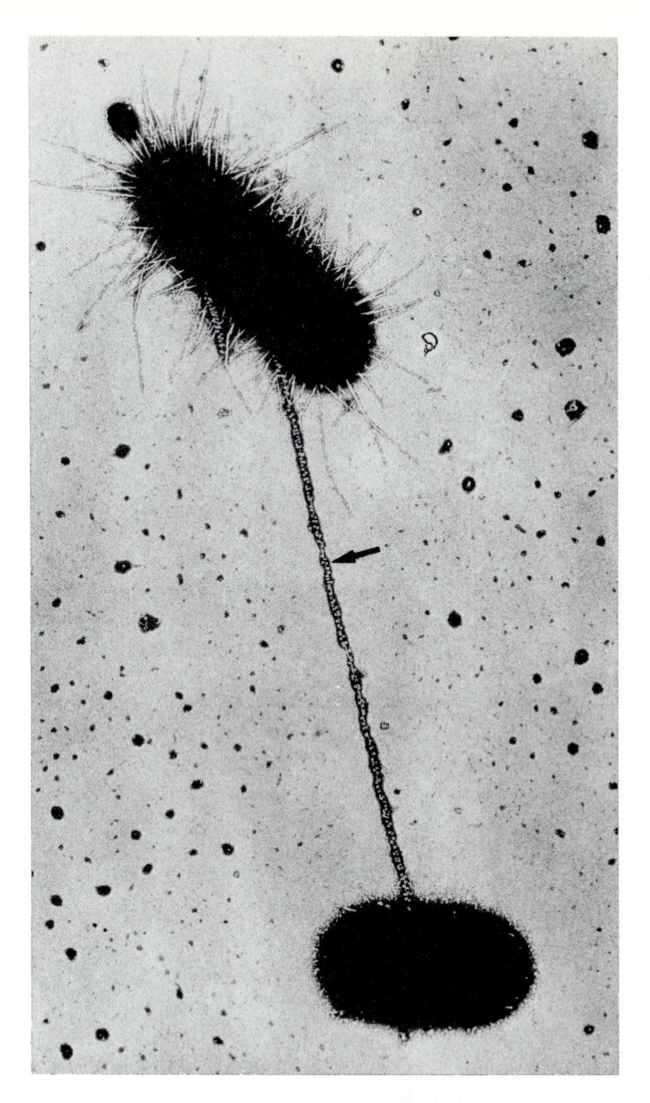

Figure 5-29 ***Electron micrograph demonstrating sex pilus in*** E. coli. ***Upper bacterial cell is covered with common pili, but there is only one sex pilus*** (arrow), ***which in this photo is covered with special viral particles. Cell with sex pilus is male, or donor, and cell in lower portion of photograph is female, or recipient.***

Essential cytoplasmic components

Ribosomes. Ribosomes are cytoplasmic particles that engage in protein synthesis. They can be seen with the electron microscope when the cell is carefully prepared and stained (see Figure 4-21). The prokaryotic ribosome, which is composed of 60% RNA and 40% protein, has a sedimentation constant of **70S** and is composed of two subunits: **30S** and **50S**. Bacterial ribosomes are composed of approximately 55 different protein molecules and three different ribonucleic acid molecules. When these individual components are incubated under appropriate conditions they spontaneously reform into their original structure. The 70S ribosomes are stable in solutions with relatively high magnesium concentrations, but when the magnesium concentration is reduced, the two subunits can be disassociated. As the growth rate of a cell increases, there is also an increase in the number of ribosomes. The structure and biosynthesis of the ribosome differs considerably between prokaryotes and eukaryotes, even though ribosomal function and the overall mechanics of protein synthesis are the same. This disparity is revealed indirectly when their sensitivities to certain antibiotics are compared. Streptomycin, for example, binds to prokaryotic ribosomes and inhibits protein synthesis but has no effect on eukaryotic ribosomes. This difference indicates a possible independent evolutionary origin of the ribosome or ribosomal proteins for the two groups. The mechanisms of protein synthesis are discussed in Chapter 9.

Nucleoids. One of the most distinctive differences between eukaryotic and prokaryotic cells is the absence of a nuclear membrane in prokaryotic cells. The nuclear material, or DNA, occupies a position in the cell called the **nucleoid** region (see Figure 5-1). Under optimal growth conditions as many as four macromolecules of DNA may be present in the cell, each representing a nucleoid region. Nucleoid regions are easily observed in electron micrographs, as well as in phase-contrast microscopy (Figure 4-22). Each nucleoid is a bacterial chromosome or single circular double-stranded DNA molecule.

Specialized cytoplasmic components

Gas vacuoles. Gas vacuoles are hollow cavities made up of protein **vesicles** and are found (Figure 5-30) in many aquatic prokaryotes, such as the various photosynthetic bacteria and halobacteria. The protein subunits are assembled in the shape of a cylinder, producing a hollow space within the protein clusters. Gas enters the vesicles by passive diffusion through pores in the protein, but water cannot penetrate. The inner surface of the vesicle appears to be hydrophobic, and this prevents the entrance of water. Gas vesicles, which have no counterpart in eukaryotes, are not bounded by a unit membrane, which is characteristic of other organelles. These vesicles provide buoyancy, and this keeps the organism at a depth where nutrients, oxygen, light, etc., are most suitable for microbial growth. Once the vesicles collapse, they cannot be reinflated, and new vesicles must be produced to replace them.

Nearly all gas vesicle–containing bacteria are aquatic, and few possess flagella. Aquatic organisms that do not possess gas vacuoles are flagellated. Thus the ability of aquatic microorganisms to position themselves in water where various chemical gradients exist may be accom-

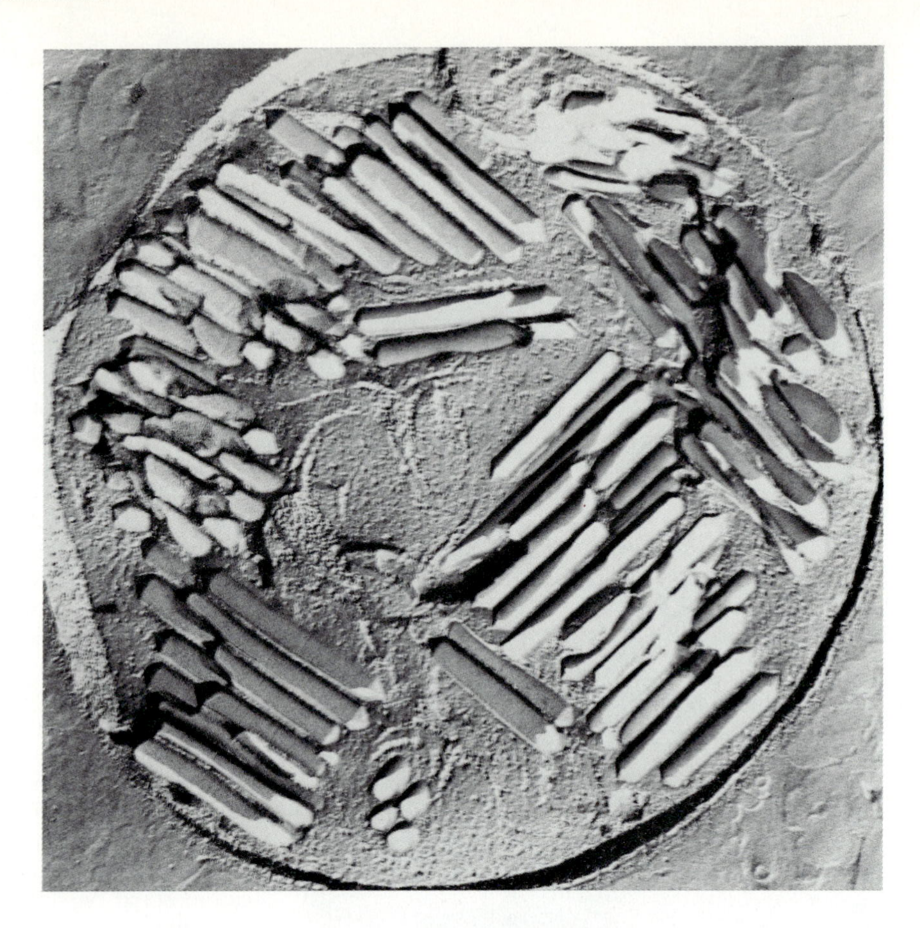

Figure 5-30 *Gas vacuoles. Freeze-etched preparation (electron micrograph) showing cross section of gas vacuoles in cyanobacterium.*

plished by either motility and chemotaxis or by the property of buoyancy associated with gas vacuoles.

Glycogen granules. Glucose is stored by many bacterial cells, in the form of a polymer called glycogen, under conditions of carbon excess and certain nutrient limitations such as nitrogen depletion. The accumulated glycogen (Figure 5-31) forms inclusions that in most bacteria is surrounded by a membrane. The glycogen reserve is later hydrolyzed to glucose when nutrients become available.

Polyphosphate granules. Inorganic phosphate accumulates in some bacterial cells during maximum growth. The phosphates are converted to polyphosphates, which form inclusions in the cell called **metachromatic granules** (also called volutin or Babés-Ernst bodies). Their only function is as a reserve of phosphate, which could be used in DNA or RNA synthesis. They are useful in the identification of corynebacteria.

Poly-β-hydroxybutyrate granules. Poly β-hydroxybutyrate (PHB) granules are lipid storage components accumulated by many bacteria, including the cyanobacteria. They have also been observed to accumulate in sporulating cells of *Clostridium botulinum* type E, one of the causative agents of a type of food poisoning called botulism. PHB granules act as cellular reserves of carbon and energy or energy alone. When excess carbon and energy sources are present in the medium, they may account for up to 50% of the dry weight of the cell. The granules are surrounded by a nonunit membrane that is believed to contain the enzymes for PHB synthesis (Figure 5-32).

Sulfur granules. Certain bacteria contain inorganic sulfur in the form of inclusions (Figure 5-33). Sulfur granules are surrounded by a nonunit membrane that is believed to be proteinaceous. Sulfur granules are produced by the nonphotosynthetic *Beggiatoa* during the oxidation of hydrogen sulfide, the organism's energy source, and by the purple bacteria such as species of *Chromatium.* Intracellular sulfur can be oxidized to sulfate to produce energy when hydrogen sulfide becomes depleted as an energy source. Sulfur oxidation is discussed in Chapter 12.

Magnetosomes. In 1975, bacteria collected from marine marsh muds in Cape Cod, Massachusetts, were observed to be magnetotactic. When observed microscopically (without the involvement of light), the microorganisms moved to one side of a drop of mud and

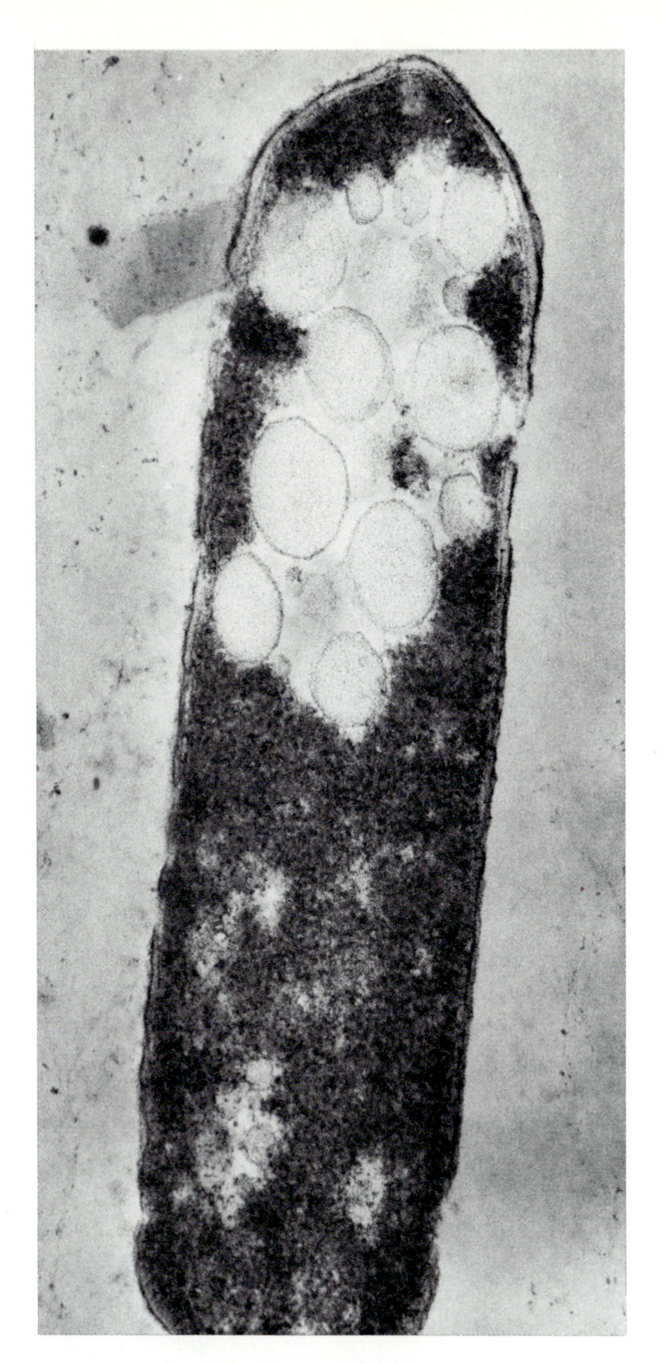

Figure 5-31 *Electron micrograph of glycogen granules from* Clostridium pasteurianum. *Cell is damaged, revealing membranous coats surrounding granules.*

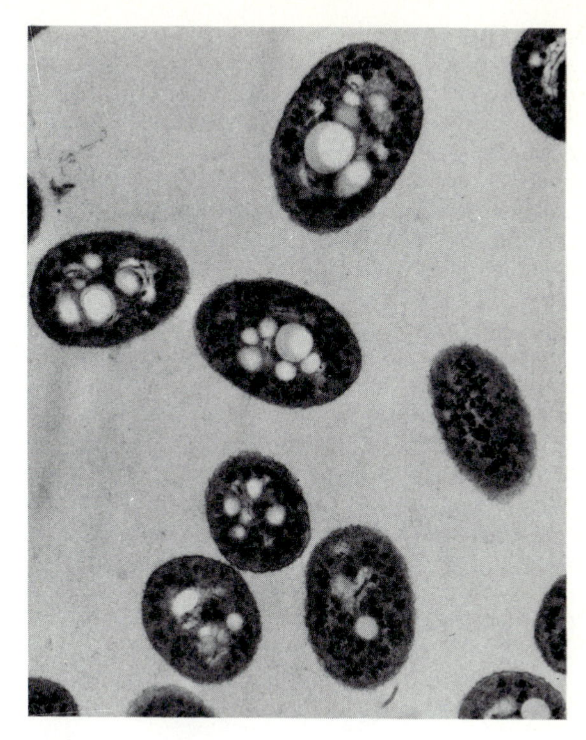

Figure 5-32 *Electron micrograph of thin section of* Ferrobacillus ferrooxidans *showing poly β-hydroxybutyrate (PHB) granules. (×20,500.)*

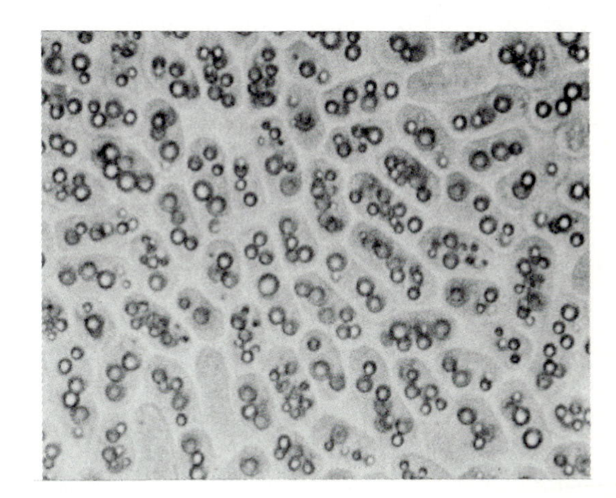

Figure 5-33 *Sulfur inclusions. Photomicrograph of sulfur granules in species of* Chromatium.

changed direction only when magnets were moved in their vicinity. Ultrastructural studies of these organisms revealed the presence of a chain of electron-dense particles (Figure 5-34), each of which was surrounded by a nonunit protein membrane. These particles are rich in cellular iron resembling magnetite and are now referred to as magnetosomes.

Magnetotaxis means movement along the earth's magnetic lines of force. These force lines point down as well as north and south. Organisms exhibiting magnetotaxis are motile and can exist in the presence of little or no oxygen. It has been sugggested that magnetotaxes serve to direct organisms downward toward sediments and anaerobic areas favorable to their growth.

Chlorosomes. Chlorosomes are the light-harvesting structures found in the photosynthetic green bacteria (*Chlorobiaceae* and *Chloroflexaceae)* and contain bacteriochlorophylls c, d, and e. Chlorosomes are vesicular structures that lie immediately beneath the cytoplasmic membrane (Figure 5-35) and are not found in other photosynthetic bacteria. Each vesicle is surrounded by an electron-dense nonunit membrane.

Carboxysomes. Carboxysomes are polyhedral or hexagonal bodies (Figure 5-36) that are found in a wide variety of prokaryotes, particularly the cyanobacteria, the nitrifying bacteria, and the thiobacilli. Carboxysomes, which are surrounded by a nonunit membrane, contain a mass of particles that have been identified as the enzyme **ribulose 1, 5-diphosphate carboxylase**; hence the name carboxysomes. This enzyme is used for carbon dioxide fixation (the organisms possessing it use carbon dioxide as their sole carbon source). It has not been determined if the carboxysomes are active in carbon dioxide fixation or if they are merely a storage area for the enzyme.

BACTERIAL ENDOSPORES (DIFFERENTIATION PROCESS)

Differentiation in the biological world implies that during an organism's cycle of development or growth some of the cells take on specialized functions; that is, they become differentiated. Differentiation is a property generally associated with eukaryotic organisms or those

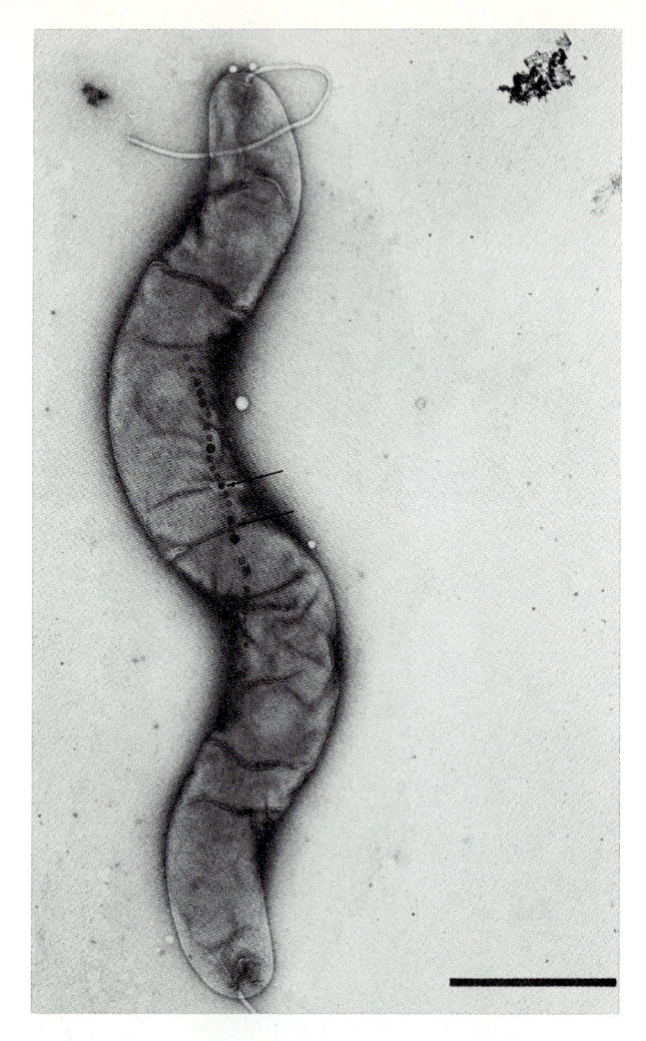

Figure 5-34 *Magnetosomes. Electron micrograph of magnetotactic spirillum showing chain of magnetosomes* (arrows). *Bar = 1 μm.*

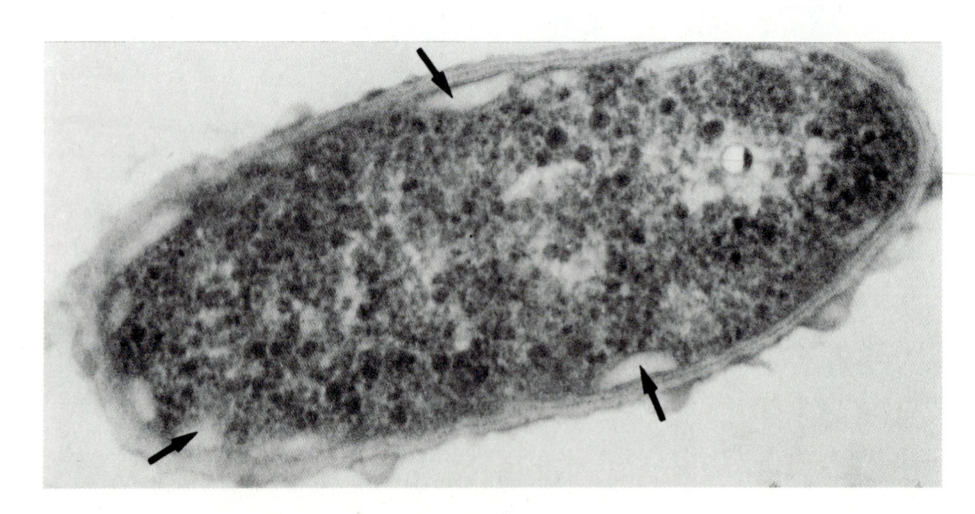

Figure 5-35 *Chlorosomes. Electron micrograph of chlorosomes* (arrows) *in species of green bacteria. (×90,000.)*

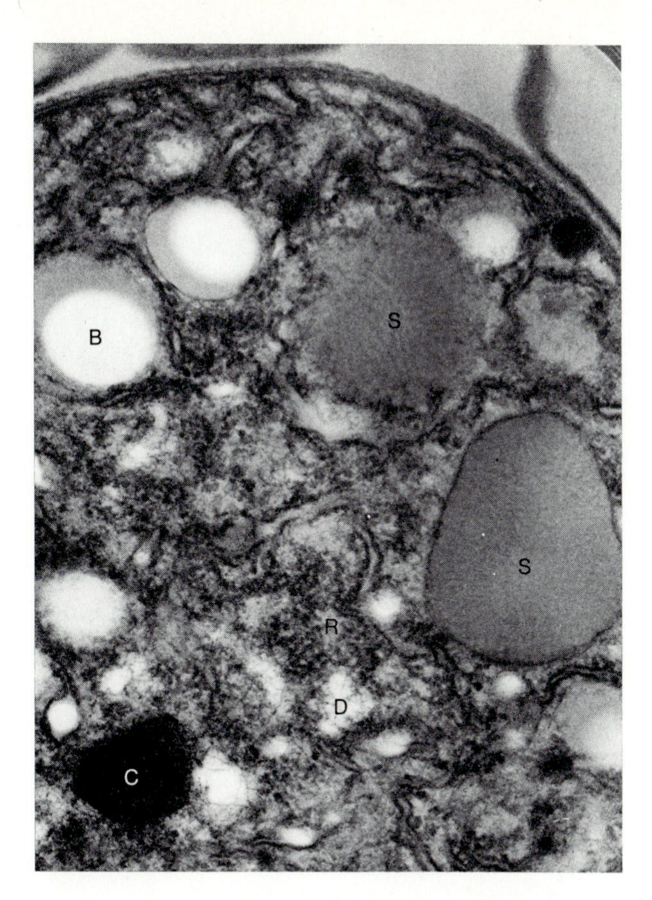

Figure 5-36 *Carboxysomes. Electron micrograph of cyanobacterial cell illustrating carboxysome.* C, *Carboxysome;* S, *structural granules;* B, *poly β-hydroxybutyrate granule;* T, *thylakoid;* R, *ribosomes;* D, *DNA.*

possessing more than one cell type. One type of differentiation process in prokaryotes is spore formation, which is a response to detrimental environmental conditions. Spore formation does not take place in all bacterial species. Most species that are indigenous to animals and humans do not form spores, since they inhabit areas of the host that are most suitable for their growth and development and where dramatic changes in their environment are unlikely to occur. Bacterial spore formation is more often associated with those species that inhabit the soil, such as species of *Bacillus, Clostridium, Sporosarcina, Thermoactinomyces, Actinobifida,* and *Streptomyces.* All of these except *Streptomyces* produce endospores. *Streptomyces* species produce spores in specialized multinucleated hyphal filaments that are resistant to dessication but in no way compare with the resistance of endospores to both chemical and physical agents. Spores are a mode of reproduction in *Streptomyces* but not in others.

Endospore formation (sporulation process)

Endospore formation occurs most frequently when there is a depletion of an essential metabolite in the environment. Nutrient depletion triggers biochemical as well as morphological changes in the cell, during which time vegetative growth and cell division cease and a spore (endospore) develops in the cytoplasm of the cell (the vegetative part of the cell is called the **sporangium**). The efficiency of sporulation appears to depend on cell density and a proteinlike factor released by individual bacterial cells. When the mature spore is released from the sporangium, metabolic processes are almost imperceptible, and it is able to resist heat, radiation, and various chemical agents. The endospore, therefore, represents the dormant stage of the vegetative cell. Once environmental conditions are again suitable for vegetative growth, a previously silent mechanism is triggered, which results in germination and the establishment of vegetative growth. The biochemical and morphological events that occur during sporulation are described later. They occur in stages and coincide with those of Figure 5-37. It should be mentioned that many of the biochemical processes that are triggered by the sporulation process may or may not be specific for the process:

Stage I. It is believed that DNA replication is one of the earliest sporulation events. During this period the bacterial chromosome changes from a compact coiled unit into a filamentous structure called the **axial filament.** During this period there is the release of antibiotics by species of *Bacillus.* The role if any of these antibiotics in sporulation is not known. There are changes in certain enzymes during this period such as the increase in enzymes involved in energy production. Adequate levels of ATP are required to maintain the sporulation process. Enzymes such as serine protease are also released during this period, but its role in sporulation is not known. Highly phosphorylated guanosine is also found in high concentration during sporulation, but the nature of its involvement is also not known.

Stage II. This is the stage where the first morphological change takes place. Invagination of the cytoplasmic membrane and formation of a membrane septum occur near one pole of the cell. This event partitions the sporangium into two compartments

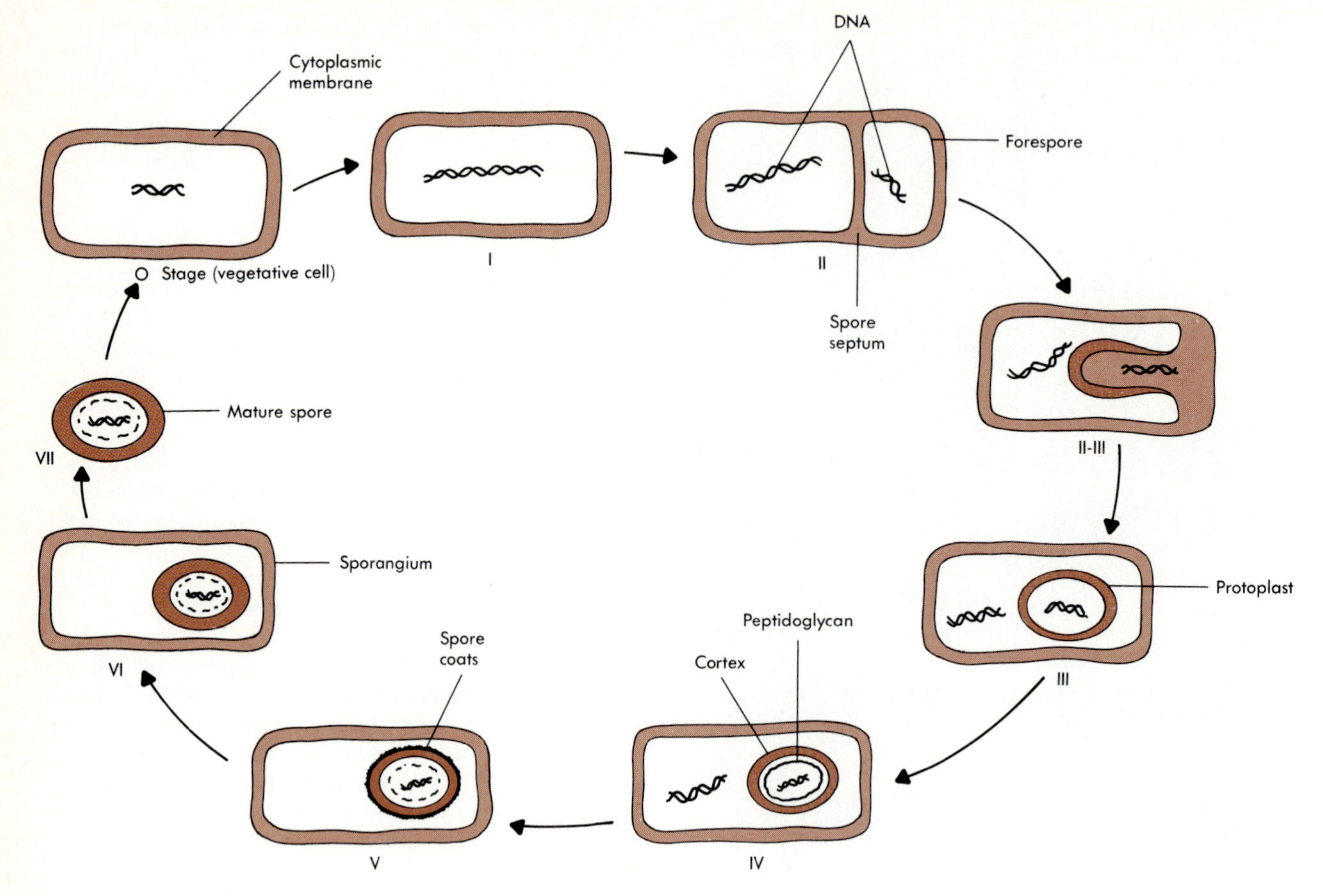

Figure 5-37 *Stages in formation of endospore (see text for details).*

of unequal size, that is, the larger **mother cell** and the smaller **forespore**. It is the forespore that will eventually develop into an endospore. The partitioning of the sporangium also results in the separation of DNA molecules into the mother cell and forespore.

Stage III. During this stage the mother cell septum membrane engulfs the forespore in a second membrane layer. This process results in the pinching off of the forespore as a free protoplast in the mother cell. The engulfment process is easily detected because of the appearance of the enzyme alkaline phosphatase. The role of this enzyme in sporulation is not known.

Stage IV. The area between the two membranes of the forespore protoplast is filled with a cell wall–like material called the **cortex**. The cortex is a peptidoglycan but of different composition than the peptidoglycan of the vegetative cell wall. During this stage calcium is taken up by the forespore and dipicolinic acid, which is not present in vegative cells, is synthesized in the forespore. There is a 1:1 ratio of dipicolinic acid to calcium. Calcium binds (chelates) to dipicolinic acid to form calcium dipicolinate, which accounts for up to 15% of the dry weight of the mature spore. A separate layer of peptidoglycan is formed underneath the cortex. This is the "primordial cell wall," which will become the cell wall of the germinating spore. Enzymes for cortex formation are believed to be associated with the mother cell, while the enzymes for primordial cell wall are associated with the protoplast. Stage IV can be monitored because of the presence of the enzyme glucose dehydrogenase, whose function in sporulation is not known.

Stage V. Protein coats containing high levels of cysteine are deposited on the membrane surrounding the cortex. These spore coats are electron dense

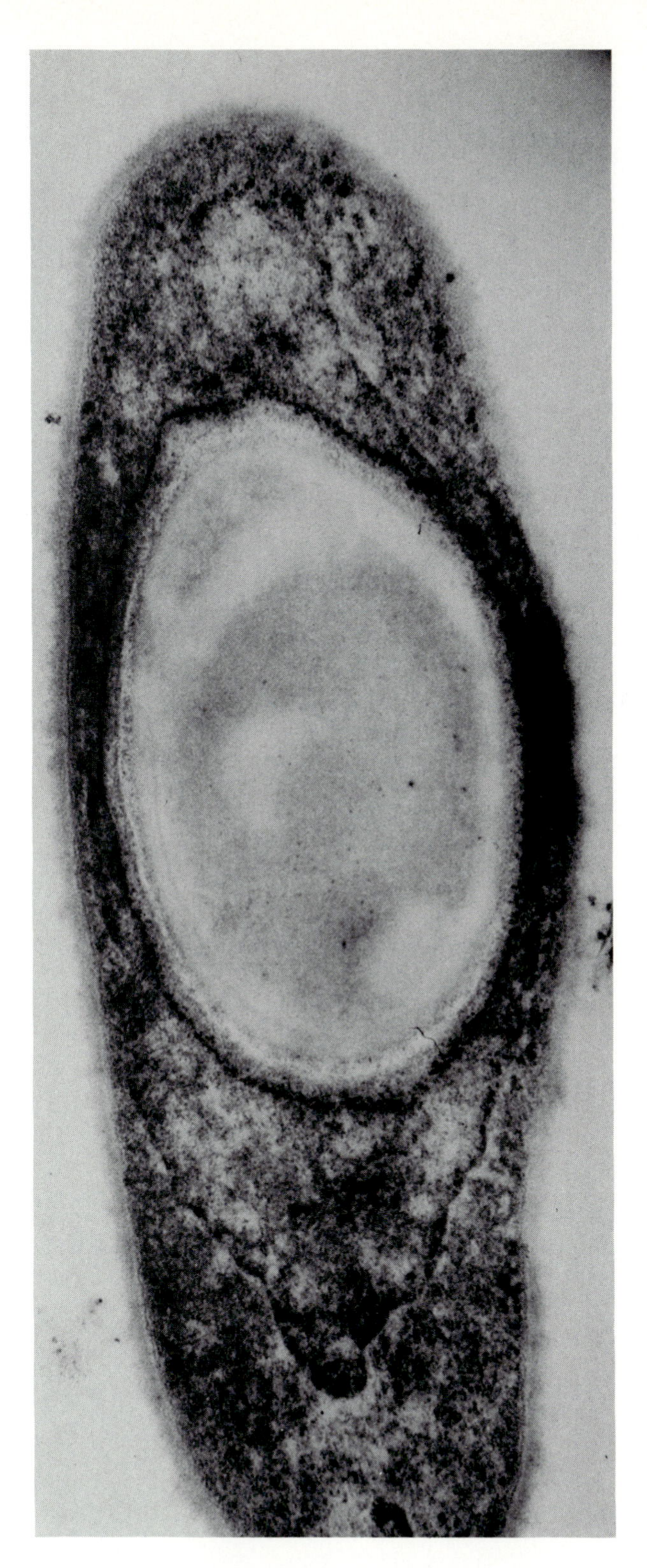

Figure 5-38 *Electron micrograph of ultrathin section of* Bacillus cereus. *(×76,250.) Endospore is white refractile body in center of vegetative cell.*

when viewed by the electron microscope and give the appearance of a white refractile body (Figure 5-38).

Stage VI. Maturation of the spore occurs during this stage. At the end of this period the developing endospore is resistant to heat, organic solvents, radiation, and lysozyme.

Stage VII. The sporangium may lyse, releasing the mature spore into the environment.

Although sporulation is described in stages, there is no rigid separation, because many of the activities overlap into other stages. For example, during the entire process water is slowly removed from the spore.

Mechanism of spore resistance and its implications

Organisms are sensitive to chemicals because the chemicals cause the hydrolysis of important proteins (enzymes) or nucleic acids. Spores therefore are more resistant to chemicals because of the absence of free water, which is needed for hydrolysis.

Heat resistance of spores appears to be associated with three components: (1) an **inherent molecular component**, (2) **dehydration**, and (3) **mineralization**. The inherent molecular component is related to the evolved temperature optimum for the spore-forming organism. In other words, thermophiles produce the most heat-resistant spores, while psychrophiles produce the least resistant spores. The adaptation to heat is probably due to the inherent stabilization of macromolecules in the spore. In regards the second component, dehydration, there is evidence that there is an unequal distribution of water in the spore. The core of the spore is relatively dehydrated while the cortex contains most of the spore-associated water. The degree of dehydration of the core therefore determines the degree of heat resistance. Mineralization, the third component, is a major extrinsic factor affecting dehydration. For example, the concentration of an element such as calcium controls the degree of dehydration. It is postulated that minerals stabilize spores by replacement of water in polymers. For example, calcium dipicolinate has been shown to intercalate in the DNA molecule, thereby providing protection from hydrolysis.

The mechanism of resistance to ultraviolet (UV) radiation has not been determined. It has been suggested that a special class of proteins is synthesized in the developing spore; these proteins interact, with the nucleic acids, changing their conformation and making them resistant to UV light. During the germination process these proteins are degraded, making the DNA susceptible to UV light.

In addition to their survival value in harsh environments, spores also present problems in the medical community, especially spores of the genera *Bacillus* and

Clostridium. For example, *Bacillus anthracis* is responsible for a highly fatal disease in cattle and other herbivores called **anthrax**; *Clostridium botulinum* is the etiological agent of one type of food poisoning called **botulism**; *C. tetani* is the etiological agent of tetanus; and *C. perfringens* is one of the etiological agents of **gas gangrene** and **food poisoning.** These organisms in their spore state exhibit resistance to boiling, antibiotics, and many antiseptics. Vegetative cells of bacteria and yeasts, for example, are killed in 5 to 10 minutes at 80° C. Studies have shown that it takes 330 minutes of boiling to kill botulinum spores, 90 minutes to kill tetanus spores, and 30 minutes to kill the spores of the gas gangrene organisms. They all can be destroyed by autoclaving, that is, steam at 121° C under 15 pounds pressure/square inch for 15 minutes. Wounds or abrasions as well as food contaminated by some of these soil-residing spores, offer environments most suitable for rapid germination and growth. They all produce deadly toxins in their vegetative states, but in one species, *C. perfringens,* the toxin responsible for food poisoning is released when sporulation occurs in the intestinal tract (see Chapters 21 and 27).

Spore-forming species, particularly those of *Bacillus* and *Streptomyces,* are responsible for the synthesis of most of our antibiotics. They are discussed in Chapter 15.

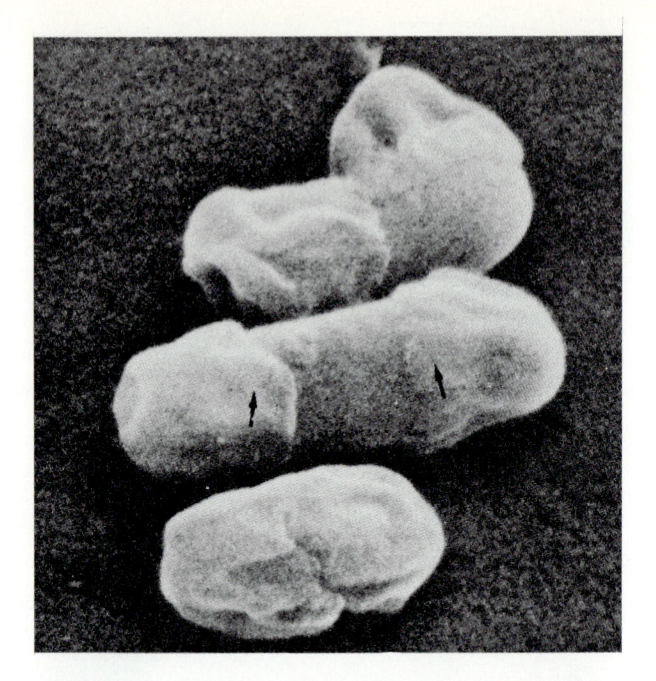

Figure 5-39 *Scanning electron micrograph of spores in process of germination and breakage of spore coat* (arrows). *(×30,000.)*

Concept Check

1. What are the essential and nonessential components found in the microbial cell? What is their chemical composition and function in the cell?
2. What are the metabolic and physical changes that occur in the microbial cell during endospore formation?

Germination

Germination, or the conversion of the spore to the vegetative state, is divided into three stages: **activation, initiation,** and **outgrowth.** Spores do not germinate efficiently unless they are exposed to activation conditions such as mild heat treatment. Initiation, unlike activation, is an irreversible process and involves the hydration of the spore. Once hydration begins (the process takes less than 1 minute), the resistance properties of the spore are lost. Initiation may be induced by exposure of activated spores to such compounds as L-alanine or a mixture of glucose, fructose, asparagine, and potassium chloride. Outgrowth represents the actual morphological and physiological changes that convert the spore into a vegetative cell (Figure 5-39). The process usually takes more than 1 hour and represents a period of macromolecular synthesis. The protein synthesis that occurs during this period uses amino acids derived from the degradation of a family of proteins called **acid-soluble proteins.** These acid-soluble proteins are believed to be the same ones that provide spore-associated nucleic acids with resistance to ultraviolet light.

SUMMARY

1. The prokaryotic cell, that is, bacterial cell, possesses a variety of structures that have chemical and functional characteristics that distinguish them from eukaryotic cells. Most species of bacteria have been divided into two groups based on the staining characterists of their cell wall: gram-positive and gram-negative. The characteristics of the major structures found in the prokaryotic cell are outlined in Table 5-2.
2. Some bacterial species, primarily members of the genera *Bacillus* and *Clostridium,* produce endospores. The formation of a spore is not a reproductive process, as in *Streptomyces* and some eukaryotes, but merely a mechanism used by the vegetative cell to survive stringent environmental conditions. During sporulation there are biochemical and morphological changes that distinguish it from the vegetative cell. During sporulation there is a depletion of water and the formation of a dense wall-like material around a DNA molecule. The spore contains high concentrations of calcium and dipicolinic acid, which are not found in the vegetative cell. The spore is unusually resistant to heat and radiation as well as various chemical and physical conditions that might be detrimental to the vegetative cell. Once environmental conditions are appropriate, spores germinate and return to the vegetative state.

Table 5-2 *Characteristics of prokaryotic structures and inclusions*

Structure or inclusion	Chemical composition	Function	Comments
Cell wall	Peptidoglycan is basic structural unit, making up 15% to 50% of dry weight of gram-positive cell wall but only 5% to 15% of gram-negative cell wall; gram-negative cell has outer memrane made up of lipopolysaccharide (LPS), protein, and phospholipid; gram-positive cell has teichoic acid	Structural support to cell; also acts as osmotic barrier to certain molecules	There are natural wall-less bacteria called mycoplasmas and artificially induced wall-less types called L-forms; archaebacteria have modified walls
Cytoplasmic membrane	Consists of phospholipid bilayer with proteins embedded in matrix	Abuts against cell wall but may invaginate in some bacteria; associated with activities such as solute transport, respiratory activity, secretion of molecules, and biosynthesis of cell membrane and cell wall components; internal membrane system (thylakoids) associated with photosynthetic activities in phótosynthetic bacteria	Mesosomes may be cytoplasmic extensions or artifacts
Extracellular polymeric substances (EPS) may be adherent to cell (capsule, S-layer) or an amorphous mass (slime)	S-layers are protein or glycoprotein; capsules and slime layers are usually polysaccharide with occasional amino sugars.	S-layers may be protective; capsules and slime layers protect some cells from phagocytosis and may be used for adherence. Slime layer associated with gliding motility	Vaccines are produced from some encapsulated species such as *Streptococcus pneumoniae*

Table cont'd

Table 5-2—cont'd

Structure or inclusion	Chemical composition	Function	Comments
Flagella	Protein	Appendages extending from surface of cell; involved in locomotion toward or away from chemical or physical stimuli	Spirochetes have flagella-like filaments called axial fibers that wrap around body of cell and beneath a sheath
Pili	Protein	Hairlike appendages found on surface of cell; sex pili are important in sexual conjugation; common pili may be useful in adherence to substrates or for aggregation	Found primarily on gram-negative bacteria
Gas vacuole	Protein	Provides buoyancy to organism	Found primarily in aquatic bacteria
Nucleoid	DNA	Contains hereditary information	Nucleoids increase with increase in growth rate; as many as four nucleoids may appear in cytoplasm; each nucleoid is a copy of the same circular chromosome
Ribosomes	RNA and protein	Protein synthesis	Sedimentation constant for prokaryotic ribosome is 70S; that is, it contains 50S and 30S subunits
Magnetosomes	Iron (magnetite?)	Orientation in environment	Found in marine marsh muds
Chlorobium vesicles	Bacteriochlorophylls c, d, and e	Light harvesting	—
Carboxysomes	Ribulose 1,5-diphosphate carboxylase	Carbon dioxide fixation or storage area for enzyme	—
Granules	Polyphosphate, poly-β-hydroxybutyrate (PHB), sulfur, and glycogen	Storage products used primarily as energy, or energy and carbon sources	Type of storage product depends on species as well as nature of medium on which they are growing

Self-study Quiz

Matching

Match the prokaryotic structure or chemical component with the appropriate function(s) in the right column

1. ________ Teichoic acid
2. ________ Porins
3. ________ Periplasm
4. ________ Mesosome
5. ________ S-layer
6. ________ Capsule
7. ________ Pili
8. ________ Magnetosomes
9. ________ Chlorosome
10. ________ Peptidoglycan

A. Lends structural support and rigidity to cell wall.
B. Aids in attachment to epithelial surfaces.
C. The photosynthetic apparatus of green bacteria.
D. Receptor for some bacteriophages.
E. Acts as an osmotic buffer zone between cytoplasm and environment.
F. A site for DNA attachment during replication.
G. Transport protein (passive transport of solutes).
H. Attachment to various surfaces.
I. Enables cell to move to anaerobic areas favorable for growth.
J. Enables many organisms possessing them to resist phagocytosis.
K. Contains transport-binding proteins.
L. Storage vesicle of the cell.

Multiple choice

1. Which of the following treatments would result in the formation of protoplasts?
   A. Cells treated with lysozyme in a hypotonic solution
   B. Cells exposed to penicillin in distilled water
   C. Cells treated with lysozyme in a high-solute concentration medium
   D. Endospores placed in a suitable nutrient medium under proper growth conditions
2. Penicillin interferes with which of the following reactions?
   A. Methylation-demethylation of cytoplasmic membrane protein
   B. Synthesis of NAG-NAM in the cell cytoplasm
   C. Action of autolysins to cleave sites in the cell wall for addition of cell wall subunits
   D. Transpeptidation between peptide tails of adjacent peptidoglycan strands
3. Where does energy production occur in prokaryotes?
   A. Cytoplasmic membrane
   B. Mitochondria
   C. Polyphosphate granules
   D. Periplasmic space
4. A mutant strain of bacteria that has lost the ability to synthesize the following structure might show some variations in its cultural characteristics, but it would still be able to survive under normal environmental conditions in nature.
   A. Cell wall
   B. Pili
   C. Ribosome
   D. Cytoplasmic membrane
5. What structure facilitates the binding of *Streptococcus mutans* to the enamel surface of teeth and initiates development of dental caries?
   A. Pili
   B. Cell wall
   C. Extracellular polymeric substance
   D. S-layer
6. Which of the following processes could be classified as a differentiation process?
   A. Development of magnetosomes
   B. Utilization of bacteriochlorophylls in photosynthetic bacteria
   C. Formation of bacterial endospores
   D. Bacterial conjugation

7. If you formed antibodies against the 0 antigen of certain bacterial strains, where would the antibody bind to the bacterial cell?
   A. Nucleoid
   B. Ribosome
   C. Capsule
   D. Outer membrane of gram-negative cell wall
8. Which of the following statements is most descriptive of bacterial flagellar movement?
   A. The flagellum moves by means of contraction and expansion of the filament within the hook.
   B. Motility results from rotation of the flagellum by rotation of the basal body anchored in the cytoplasmic membrane.
   C. The bacterial flagellum is whipped back and forth, setting up waves of coordinated motion in the filament.
   D. By contraction and expansion of the cytoplasmic membrane the cell "rows" itself through the medium.
9. The compounds glycine, betaine, choline, and proline are associated with which of the following bacterial properties?
   A. Attachment
   B. Movement
   C. Osmotic tolerance
   D. Spore formation
10. Which of the following is not found in the cytoplasmic membrane of bacteria?
   A. Cytochromes
   B. Flagellar ring
   C. Protein (ATPase) associated with ATP formation
   D. Porins

Completion

1. The gram reaction is based on the structure of the ________________.
2. The slow movement of solute from an area of high concentration to an area of low concentration is called ________________.
3. The process that permits the movement of nutrients into the cell against a concentration gradient is called ________________.
4. The "signal" associated with secretory proteins is composed of mainly ________________amino acids.
5. Most capsules and slime layers of bacteria are composed of ________________.
6. Smooth swimming in motile bacteria is characterized by ________________rotation of the flagella.
7. The bacterial structure that enables certain gram-negative organisms to engage in conjugation or DNA transfer is called the ________________.
8. The lipid storage component of certain bacteria is called ________________.
9. The peptidoglycan-containing component of the spore is called the ________________.
10. The bacterial structure that provides buoyancy for aquatic bacteria is called ________________.

Thought questions

1. The basic structure of the cell wall of bacteria is composed of the glycan NAG-NAM backbone with the tetrapeptide tail composed of four amino acids attached to the glycan backbone. One of the amino acids is diaminopimelic acid (DAP), which is unique to the prokaryote cell wall. Another unique feature is that two of the amino acids are found in the D-amino acid configuration, whereas all other amino acids are in the L-configuration.
   a. Would you expect that the cell wall tetrapeptide is synthesized by the same mechanism as other cell protein? Explain.
   b. Can you propose any advantage to the bacterial cell in having unique amino acids in its cell wall structure?
2. The dye triphenyl tetrazolium chloride (TTC) possesses the property of being taken up by bacterial cells; when coupled with enzymatic oxydation-reduction reactions within the cell, its reduced form is insoluble and dark red. When intact bacterial cells reduce TTC, red crystals form near the sites of concentration of the respiratory oxidative enzyme system within the cell. If growing cells of *Bacillus megaterium* are allowed to take up TTC and then the cells are observed under the microscope, where would you expect to see the red crystals concentrated in the cells?
3. You isolated a strain of *Bacillus cereus* that was incapable of synthesizing dipicolinic acid (DPA).
   a. What observable difference, if any, would there be in the vegetative cells of the culture?
   b. What observable difference, if any, would there be in the endospores formed in the culture?
   c. What difference, if any, would there be in the heat resistance of the culture?
4. Design an experiment to demonstrate bacterial chemotaxis.

SELECTED READINGS

Books

Freifelder, D. *Molecular biology: a comprehensive introduction to prokaryotes and eukaryotes*. Science Books International, Boston, 1983.

Hazelbauer, G.L. (Ed.). *Taxis and behavior*. Chapman & Hall, Ltd., London, 1978.

Inouye, M. (Ed.). *Bacterial outer membrane: biogenesis and function*, John Wiley & Sons, Inc., New York, 1979.

Mandelstam, J., and McQuillen, K. *Biochemistry of bacterial growth (3rd ed)*. John Wiley & Sons, Inc., New York, 1982.

Reiss, J.L. (Ed.). *Microbial interactions*. Chapman & Hall, Ltd., London, 1977.

Rose, A.H. *Chemical microbiology (3rd ed.)* Plenum Publishing Corp., New York, 1976.

Rogers, H.J., Perkins, H.R., and Ward, J.B. *Microbial cell walls and membranes*. Chapman & Hall, Ltd., London, 1980.

Rogers, H.J. *Bacterial cell structure. Aspects of Microbiology No. 6*, American Society for Microbiology, Washington, D.C., 1983.

Rosen, B.P. (Ed.). *Bacterial transport*, Marcel Dekker, Inc., New York, 1978.

Stanier, R.Y., Rogers, H.J., and Ward, J.B. (Eds.). *Relations between structure and function in the prokaryotic cell: twenty-eighth symposium of the Society for General Microbiology*, Cambridge University Press, New York, 1978.

Sutherland, I.W. *Surface carbohydrates of the prokaryotic cell*. Academic Press, Ltd., London, 1977.

Journal articles

Adler, J. The sensing of chemicals by bacteria. *Sci. Am.* **234**:40, 1976.

Beveridge, T.J. Ultrastructure, chemistry, and function of the bacterial cell wall. *Int. Rev. Cytol.* **72**:229, 1981.

Bove, J.M. Wall-less prokaryotes of plants. *Annu. Rev. Phytopathol.* **22**:36, 1984.

Boyd, A., and Simon, M. Bacterial chemotaxis. *Annu. Rev. Physiol.* **44**:401, 1982.

Burchard, R.P. Gliding motility of prokaryotes: ultrastructure, physiology and genetics. *Annu. Rev. Microbiol.* **35**:497, 1981.

Canale-Parola, E. Motility and chemotaxis of spirochetes. *Annu. Rev. Microbiol.* **32**:9, 1977.

Dawes, E.A., and Senior, P.J. The role and regulation of energy reserve polymers in microorganisms. *Adv. Microb. Physiol.* **10**:135, 1973.

Dills, S.S., et al. Carbohydrate transport in bacteria. *Microbiol. Rev.* **44**:385, 1980.

Ferris, F.G., and Beveridge, T.J. Functions of bacterial cell surface structures. *Bioscience* **35**:172, 1985.

Irvin, J.W.T., and Cheng, K.J. The bacterial glycocalyx in nature and disease. *Annu. Rev. Microbiol.* **35**:299, 1981.

Keynan, A. The transformation of bacterial endospores into vegetative cells. *Symp. Soc. Gen. Microbiol.* **23**:85, 1973.

Klemm, P. Fimbrial adhesins of *Escherichia coli*. *Rev. Infect. Dis* **7**(3):321, 1985.

Kurland, C.G. Structure and function of the bacterial ribosome. *Annu. Rev. Microbiol.* **46**:173, 1977.

Lodish, H., and Rothman, J.E. The assembly of cell membranes. *Sci. Am.* **240**:48, 1979.

Nikado, H., and Vaara, M. Molecular basis of bacterial outer membrane permeability. *Microbiol. Rev.* **49**:1, 1985.

Oliver, D. Protein secretion in *Escherichia coli*. *Annu. Rev. Microbiol.* **39**:615, 1985.

Papahadjopoulos, D. Liposomes and their uses in biology and medicine. *Ann. N.Y. Acad. Sci.* **308**:1, 1978.

Postma, P.W., and Leninger, J.W. Phosphoenolpyruvate: carbohydrate phosphotransferase system of bacteria. *Microbiol. Rev.* **49**:232, 1985.

Razin, S. The mycoplasmas. *Microbiol. Rev.* **42**:414, 1978.

Salton, M.R.J. Structure and function of bacterial plasma membranes. *Symp. Soc. Gen. Microbiol.* **28**:201, 1978.

Shockman, G.D., and Barrett, J.F. Structure, function, and assembly of cell walls of gram-positive bacteria. *Annu. Rev. Microbiol.* **37**:501, 1983.

Silverman, M., and Simon, M.I. Bacterial flagella. *Annu. Rev. Microbiol.* **31**:397, 1977.

Sleytr, U.B., and Messner, P. Crystalline surface layers on bacteria. *Annu. Rev. Microbiol.* **37**:311, 1983.

Ward, J.B. Teichoic and teichuronic acids. *Microbiol. Rev.* **45**:211, 1981.

CHAPTER

Characteristics of eukaryotes

Chapter Outline

Introduction

The eukaryotic microorganisms consist of three groups: the **Fungi**, the **Protozoa**, and the **Algae**. Eukaryotes carry out the same biological processes as prokaryotes, such as energy metabolism, nutrient transport, motility and replication, and transcription and translation of genetic information, but they carry out these processes in different ways and often with more complex structures than do prokaryotes. Most eukaryotes require more time for reproduction and are also more difficult to cultivate than prokaryotes. However, studies on eukaryotic morphology, genetics, metabolism, and physiology have important implications for humans, because we too are eukaryotes. More research is now being directed toward understanding the role of eukaryotic microorganisms in the fields of agriculture, geochemical activities, and pollution control. Such studies may be necessary for the survival of humankind.

The purpose of this chapter is to compare the algae, fungi, and protozoa's morphology, nutrition, reproduction, classification, role in nature, and importance to humans.

Table 6-1 *Characteristics of the subdivisions of the true fungi (Eumycota)*

Subdivision	General characteristics	Importance
Mastigomycotina	Flagellated. Many unicellular species. Found mainly in water or moist habitats.	Freshwater plant and animal parasites. Soil parasites cause plant diseases, such as powdery mildew and damping-off disease of seedlings.
Zygomycotina	Sexual reproduction results in formation of thick-walled zygospore.	Soil and dung fungi. A few parasites of humans. Bread mold. Some insect parasites.
Ascomycotina	All produce spore-holding saclike structure called the ascus. Yeasts included in this group.	Associated with food spoilage and plant diseases. Several parasites of humans. Yeasts used in industry. Some edible mushrooms.
Basidiomycotina	Spores borne externally on small base called a basidium.	Most important decomposers of wood (lignin). Important associations with the roots of plants (mycorrhizae). Some edible mushrooms. Many plant parasites (rusts).
Deuteromycotina	All lack described sexual stages. Asexual conidia formation most important method of reproduction.	Important decomposers on land and in water. Most human fungal diseases caused by this group.

CHARACTERISTICS OF FUNGI

Fungi can be found in most habitats throughout the world, although they are more common in warm, humid climates. Fungi play important roles as parasites, symbionts, and **saprobes**, which exist on dead organic matter and prevent the accumulation of organic matter in the environment.

Fossil records for fungi are meager; therefore an evolutionary history for this group is incomplete. Morphological and physiological characteristics are used for classification purposes, but environmental factors play such a major role in determining fungal form and physiology that classification has been difficult and frustrating. Fungi were originally classified as plants but because of their nonphotosynthetic nutrition and their cell structure, fungi are now placed in a separate kingdom: the Fungi, or Myceteae. Most mycologists (those who study fungi) have separated the kingdom into two divisions: the **Myxomycota**, which are slime molds, and the **Eumycota**, which are considered true fungi. The true fungi are separated into five subdivisions as outlined in Table 6-1.

Morphology

As discussed in Chapter 4 the basic unit of structure for most fungi is the **hypha**, (pl. **hyphae**), except for yeasts and members of the subdivision Mastigomycotina, which are single cells and more or less spherical.

Hyphal filaments exhibit diverse structure and function. Hyphae can penetrate substrates, such as the soil, to obtain nutrients, whereas other hyphae become aerial and develop structures necessary for reproduction. Fungi that parasitize hosts, such as plant cells, use a specialized hyphal filament called a **haustorium** (pl. haustoria). Haustoria function as absorptive structures that extract nutrients from the host (Figure 6-1). This ability to parasitize plants has been used commercially as a means for biological control. Fungal plant pathogens, **mycoherbicides**, are being used to kill or inhibit the growth of weeds, such as jointvetch, in the rice fields of Arkansas, Louisiana, and Mississippi and water hyacinthe in many southern states.

Hyphae also show some variation in internal structure. The hyphae of some fungi possess cross walls called **septa** (sing. septum) that separate one segment from another (Figure 6-2). Some septa contain pores through which the cytoplasm, nuclei, and organelles can pass freely. This characteristic results in multinucleated cells. Some septa contain pores through which the cytoplasm flows freely but nuclei and organelles are unable to move from one compartment to another, thus each compartment in the hyphal mass remains uninuclear. **Coenocytic** describes a hyphal mass in which the cytoplasm flows freely from one compartment to another.

Nutrition

Fungi lack chlorophyll and therefore are unable to produce their own food like algae and plants. Fungi live primarily on dead organic matter, using it as a source of

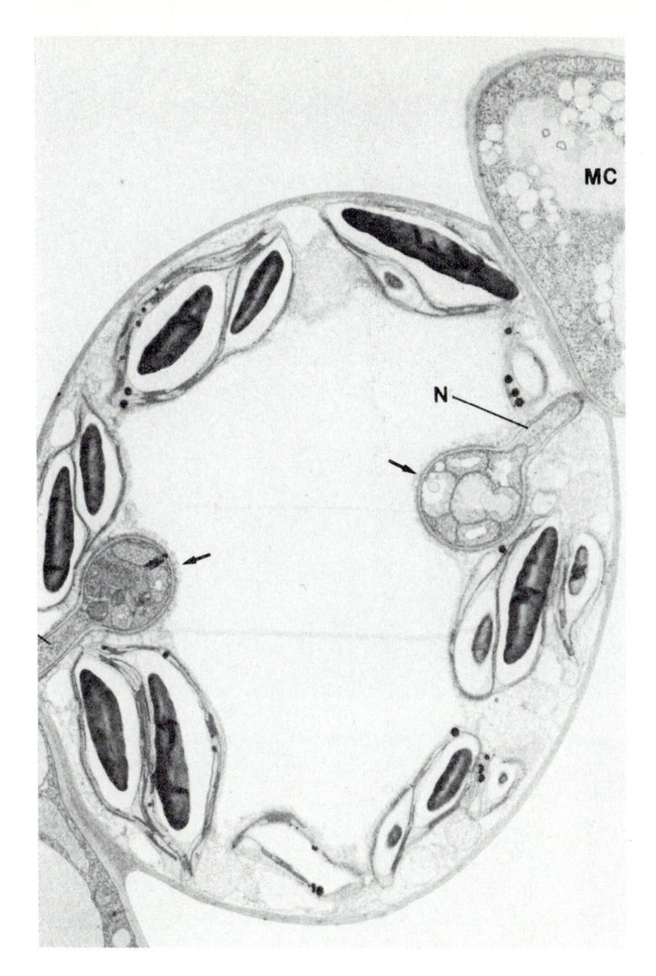

Figure 6-1 *Haustoria. Section through plant cell showing presence of haustorial bodies* (arrows) *connected by neck* (N) *to their respective mother cells* (MC).

carbon and energy; thus their nutrition is **heterotrophic**. Heterotrophism is greatly facilitated by the ability of fungi to produce many extracellular enzymes that degrade macromolecules associated with plants, animals, soil, or fabrics found in the environment. Several fungal species are parasitic and obtain their nutrients from a living host. Some fungal species can exist as saprobes or parasites. For example, many fungi that infect humans or animals live in the soil where they exist totally as filamentous saprobes. Occasionally, when fungi infect animals or humans, they exist as yeastlike parasites. This ability to exist in two different morphological forms is called **dimorphism**. The filamentous or mycelial state provides the fungus with an increased surface area for obtaining nutrients that may be in short supply in the soil. Inside the host the fungal species encounters an environment that may be hostile. Since the temperature in the host is 37° C, a temperature considerably higher than the soil (20° C), many scientists believe that the

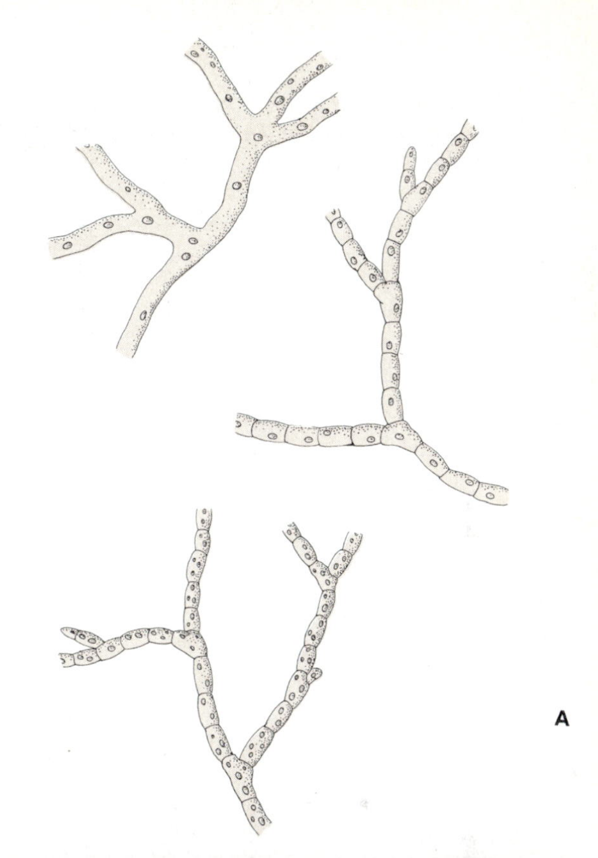

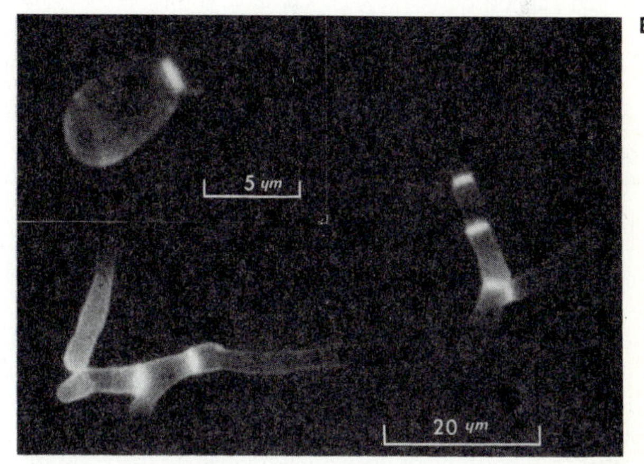

Figure 6-2 *Fungal hyphae.* ***A,*** *Diagrammatic illustration. Hyphae, top, contain no septa and there is freedom of movement of cytoplasm and nuclei or other organelles. Hyphae in center have septa that do not permit movement of nuclei or organelles but do permit movement of cytoplasm. Hyphae, bottom, possess septa but they are porous enough to permit free mingling of nuclei thus more than one nucleus in each segment.* ***B,*** *Photomicrograph of* Saccharomyces pombe *cells in which septa have been stained with fluorescent dye.* Upper left, *single septum following division of single cell.* Lower *shows septa with several branches having been formed. (Reduced 50%.)*

parasitic, yeastlike state is a response to elevated temperature.

Many fungi that parasitize plants are filamentous in the soil and remain so even after they infect their hosts. Hyphae penetrate the tough outer coat of the plant (see Figure 6-1). Once inside the plant, fungal enzymes separate host cells and permit extensive branching of haustoria between them. The haustoria penetrate the cell wall but not the cytoplasmic membrane, which is pushed aside to produce a pocket where the haustoria can develop and absorb nutrients.The myxomycota are capable of actively engulfing their food by phagocytosis, which is a process that is usually associated with the protozoa. These fungi lack rigid cell walls and are called slime molds. They have presented scientists with a taxonomic dilemma because of their morphological and nutritional variations.

Most fungi are easily cultivated in the laboratory, requiring only a carbon source such as glucose and mineral salts. Generally they grow best in acid conditions (pH between 5 and 6) with high humidity and at temperatures between 20° C to 25° C. Differentiation of the yeast and mycelial phases of dimorphic fungi can be accomplished by growing the organisms at different temperatures or by altering nutritional conditions. The mycelial phase, which develops at room temperature (20° C), is usually represented as a fuzzy colony.* This colony produces aerial spores, whereas the yeast phase, which can be detected at 37° C, is granular or mucoid and produces no aerial spores (Plates 6 and 7).

Reproduction

Reproduction in fungi may occur by sexual or asexual means. Asexual reproduction by budding (Figure 6-3) or fission is characteristic of a few species, such as the yeasts. The primary reproductive unit in fungi is the **spore**, which may be produced asexually or sexually. Asexually, spores may be produced within specialized structures or they may be produced externally on modified hyphae. **Sporangiospores**, for example, are produced within a globe-shaped structure called a **sporangium**, which is borne at the apex of a modified hypha called the **sporangiophore** (Figure 6-4, *A*). In contrast, **conidia** are spores that arise externally on a modified hypha called a **conidiophore** (Figure 6-4, *B*). The size, shape, color, and number of conidia and conidiophores are useful in the identification of fungal species. Asexual spores may arise directly from the mycelium and there are different types (Figure 6-5). The asexual reproductive process begins with the formation of a protuberance called a **germ tube** (Figure 6-6), which elongates and forms a hyphal filament. The hypha grows and produces a mycelium from which new spore-bearing structures may be produced, then the cycle is repeated.

* Fungi that form mycelia and appear as cottony growths are sometimes called **molds**.

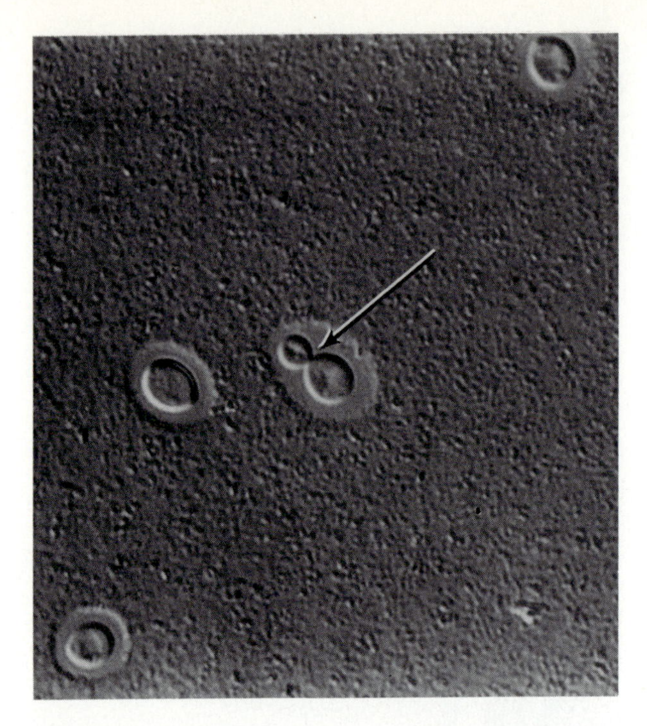

Figure 6-3 *Budding. Differential interference phase-contrast microscopy of yeast cell in process of budding* (arrow).

Sexual reproduction as a means of spore formation can be divided into four stages: (1) union of compatible sex cells **(gametes)** or organs, (2) mingling of the cytoplasms of the two cell types **(plasmogamy)**, (3) fusion of nuclei of the cell types **(karyogamy)**, and (4) **meiosis**. Genetic units are exchanged during meiosis, and haploid spores are produced. The names given to the types of spores produced depends on the manner in which they are produced; these will be discussed later.

Fungal sporulation is usually triggered by a depletion of nutrients. Other factors, such as light or chemical products in the environment, can also induce spore formation; however, little is known of the specific mechanisms involved. Many spores are produced during sporulation and, if left to germinate at one site on a substrate, they would compete for nutrients. It is to the advantage of the species that spores are dispersed at some distance from the parent to reduce such competition. The spores, or in some species, the spore-bearing structures (sporangia), are violently released from the mycelium when a pressure change occurs within the spore sac. The triggering condition may be light, temperature, or a physical jolt, and the released spores may travel from a few centimeters to 1 or 2 m. Some spores can be disseminated by air currents, animals, or water

Concept Check

1. What are the different ways in which hyphae may function in fungi?
2. In what ways is the nutrition of fungi similar to bacteria? Different?
3. List the different types of fungal spores and the manner in which they are produced.

A

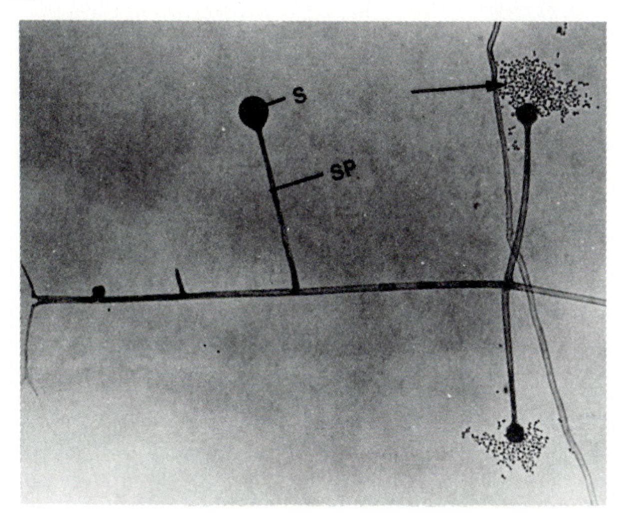

B

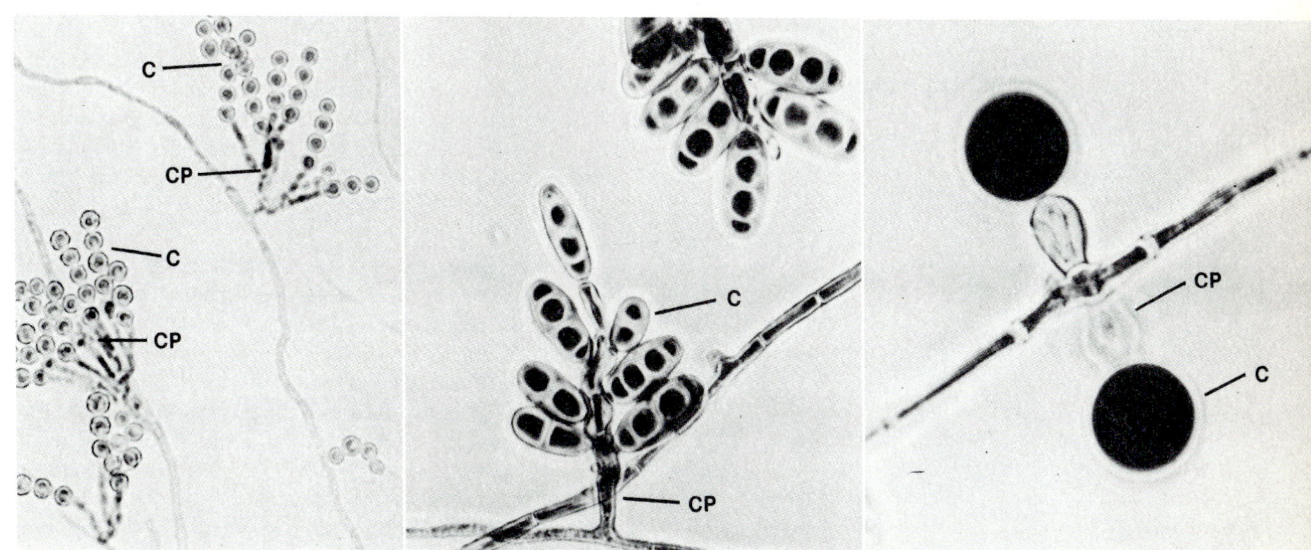

Figure 6-4 *Specialized fungal structures that bear spores.* ***A,*** *Sporangium* (S) *and sporangiosphore* (SP), *the modified hyphal filament (arrow points to sporangiospores).* ***B,*** *Conidiophore* (Cp) *and arrangement of conidia* (C) *that they bear. Some Conidia are multicelled* (middle).

vectors.* Germination of the spores will occur if environmental conditions of temperature, humidity, and nutrients are adequate.

Characteristics of the major subdivisions of fungi

Mastigomycotina. This subdivision is made up of four classes: Chytridiomycetes, Plasmodiophoromycetes, Oomycetes, and Hyphochytridiomycetes. The Mastigomycotina produce motile cells that posses a single flagellum—a characteristic that separates them from nearly all other fungi. Many orders within this subdivision are made up of members that are aquatic; however, several are soil molds. The order Peronosporales, for example, is made up of several genera, such as *Phytophthora, Pythium,* and *Albugo,* that are parasites of higher plants. They are of great economic importance because they are responsible for damping-off disease, which causes wilting of seedlings; downy mildew of crop leaves in which the sporangia and sporangiophores cover the leaf surface (see Point of Interest); and potato blight. Potato blight caused the widespread famines in Ireland during the 1840s.

The order Saprolegniales in the class Oomycetes contains genera that are common water molds. Species of *Saprolegnia* can become parasites on the external mucus of fish. The fungus produces enzymes that can

* The importance of water in spore dispersal is easily recognized by rose growers. A fungus disease called black spot attacks the rose leaves and causes them to turn yellow and fall off. The infection process will continue the following spring if the dead leaves are not removed from the ground surrounding the plant because raindrops splashing on the dead leaves cause the dispersal of spores, which can attack healthy leaves.

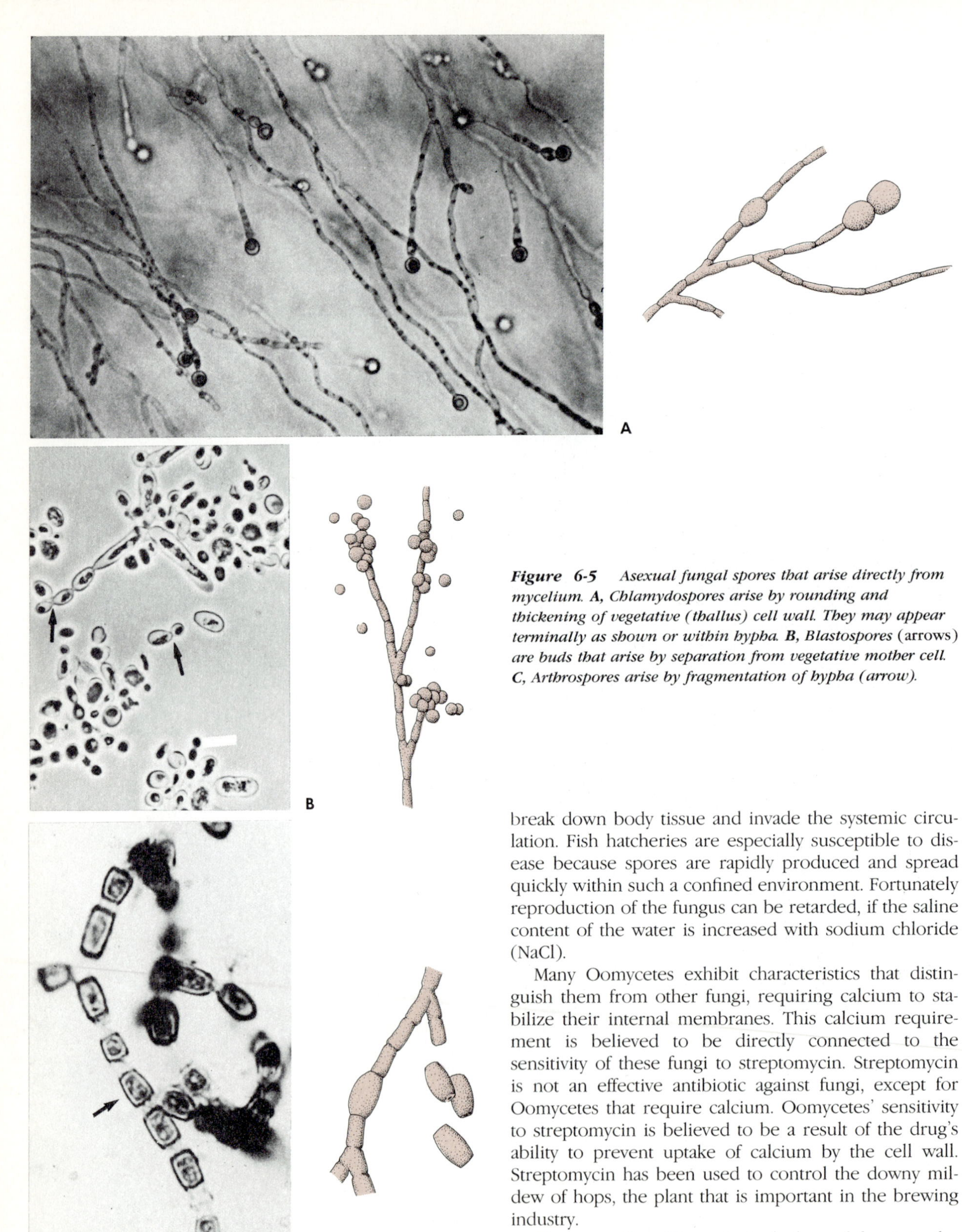

Figure 6-5 *Asexual fungal spores that arise directly from mycelium.* ***A,*** *Chlamydospores arise by rounding and thickening of vegetative (thallus) cell wall. They may appear terminally as shown or within hypha.* ***B,*** *Blastospores* (arrows) *are buds that arise by separation from vegetative mother cell.* ***C,*** *Arthrospores arise by fragmentation of hypha (arrow).*

break down body tissue and invade the systemic circulation. Fish hatcheries are especially susceptible to disease because spores are rapidly produced and spread quickly within such a confined environment. Fortunately reproduction of the fungus can be retarded, if the saline content of the water is increased with sodium chloride (NaCl).

Many Oomycetes exhibit characteristics that distinguish them from other fungi, requiring calcium to stabilize their internal membranes. This calcium requirement is believed to be directly connected to the sensitivity of these fungi to streptomycin. Streptomycin is not an effective antibiotic against fungi, except for Oomycetes that require calcium. Oomycetes' sensitivity to streptomycin is believed to be a result of the drug's ability to prevent uptake of calcium by the cell wall. Streptomycin has been used to control the downy mildew of hops, the plant that is important in the brewing industry.

Zygomycotina. The principal class of fungi in the subdivision Zygomycotina is Zygomycetes, which is

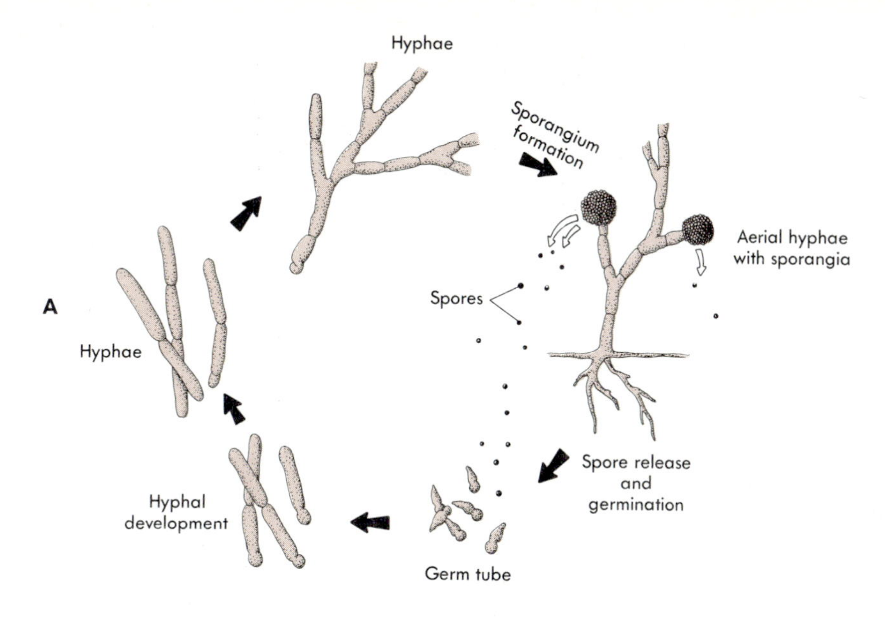

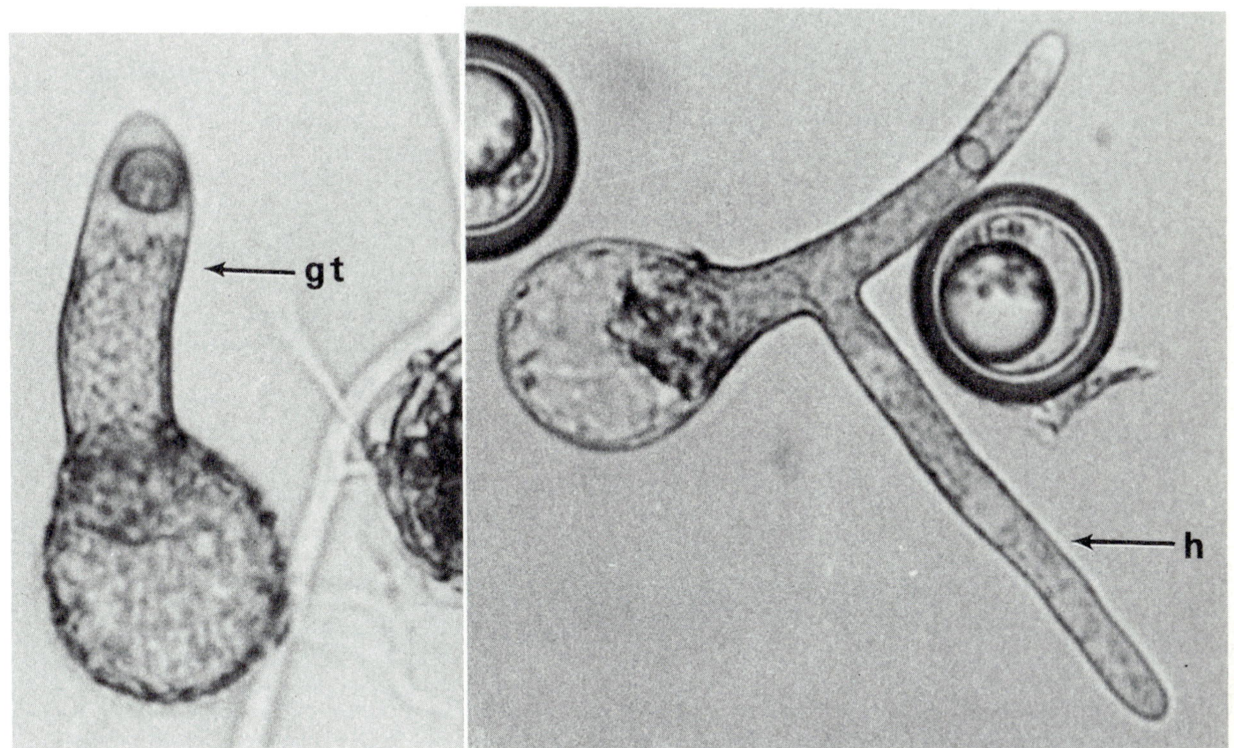

Figure 6-6 ***A, Diagrammatic illustration of germ tube formation and hyphal development. B and C, Photomicrographs*** (top) ***show germ tube*** (gt) ***formation followed by development of hyphal branches*** (h).

composed of three orders: Mucorales, Entomophthorales, and the Zoopagales. Many members of the Zygomycotina have economic value (Table 6-2).

The Mucorales are terrestrial fungi that can be isolated from soil, dung, and other decaying matter. Some species can cause severe infections in animals and plants. Asexual spore production is usually associated with the formation of a sporangium (Plate 8), whereas

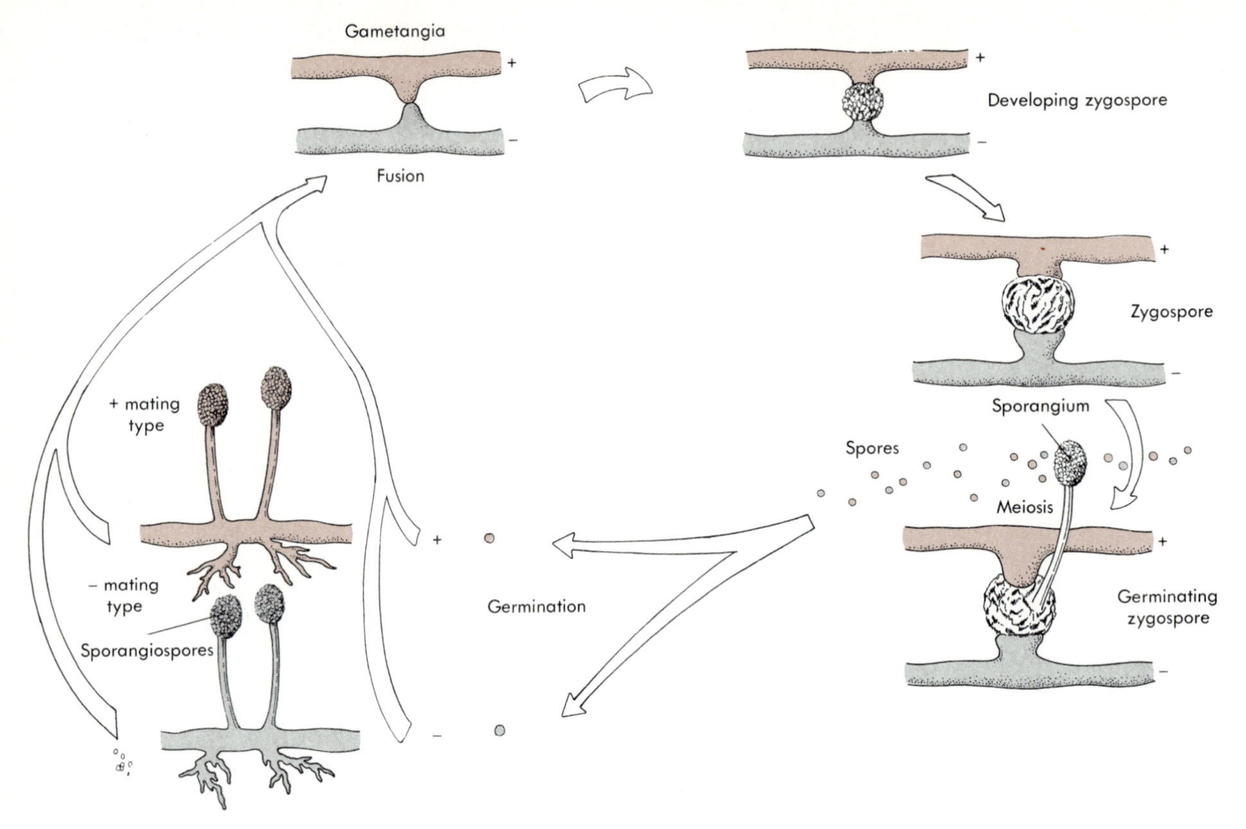

Figure 6-7 *Life cycle of* Rhizopus stolonifer. *Asexual reproduction* (left) *occurs when sporangia produce multinucleate sporangiospores, which germinate to form haploid hypha. Sexual reproduction is initiated by formation of gametangia that are isolated from the parent (+ and −) hypha. Gametangia fuse forming zygospore that goes through period of development including meiosis. Zygospore germination results in formation of sporangium and life cycle is completed.*

Point of Interest

DOWNY MILDEW OF GRAPES

In the mid-1800s the grape vines of many French winemakers were being damaged by root aphids. Since American grape vines were known to be resistant to root aphids the French grafted the American vine to the French vine. The grafted vine was resistant to root aphids but unfortunately the American vine carried the downy mildew fungus *(Plasmopara viticola).* The American vine was resistant to this fungus but the French vine was not and foilage destruction became rampant. The destruction became so widespread in the late 1800s that the economy of France was being severely affected.

In 1882, a French professor, Pierre Millardet, noticed that grapes along a road outside the University of Bordeaux were not covered by the downy mildew. He also noticed that the grapes had an unusual bluish tinge. Puzzled by this chance observation the professor made inquiries and discovered that the farmer had sprayed the vines with some left over chemicals that had imparted the blue tinge. The farmer's object was to prevent passersby from eating the grapes near the roadside. The farmer had not only discouraged passersby from eating the grapes but had also prevented the downy mildew from attacking the grape foilage. The chemicals used by the farmer were copper sulfate and calcium oxide. Millardet worked out the correct proportions of chemicals that would kill the fungus but not adversely affect the plant since copper can be toxic to plants. Millardet's mixture, which is called **Bordeaux Mixture,** was one of the first inorganic fungicides. Although it is still used today, organic fungicides have replaced it for most commercial uses.

Table 6-2 *Beneficial and detrimental properties of Zygomycotina*

Species	Economic value
Beneficial properties	
Mucor rouxii	Commercial production of ethyl alcohol via degradation of starch to glucose.
Mucor and *Rhizopus* sp.	Commercial production of organic acids—gluconic, citric, succinic, and others.
Mucor and *Rhizopus* sp.	Commercial transformation of steroids into active compounds.
Mucor ramannianus	Commercial production of the antibiotic, fusidic acid.
Rhizopus oligosporus	Commercial production of the Oriental food, tempeh, produced by fermentation of soybean or coconut.
Mucor sp.	Fermentation of soybean curd to an Oriental food product called sufu.
Mucor mucedo	Aging of meat (tenderizer).
Endogone and *Glomus* sp.	Mycorrhizae on trees and plants, such as Juniper, rose, citrus, orchids, and Sequoia.
Many species	Dung decomposition, breakdown of sewage sludge, and soil improvement.
Detrimental properties	
Rhizopus and *Mucor* sp.	Damage to stored fruits and vegetables, baked goods, textiles, fabrics, and paper products.
Mucor, Rhizopus, and *Absidia* sp.	Disease in humans compromised by underlying conditions, such as diabetes.
Absidia sp.	Tuberculosis-like disease in pigs. Respiratory infections in birds and sheep.

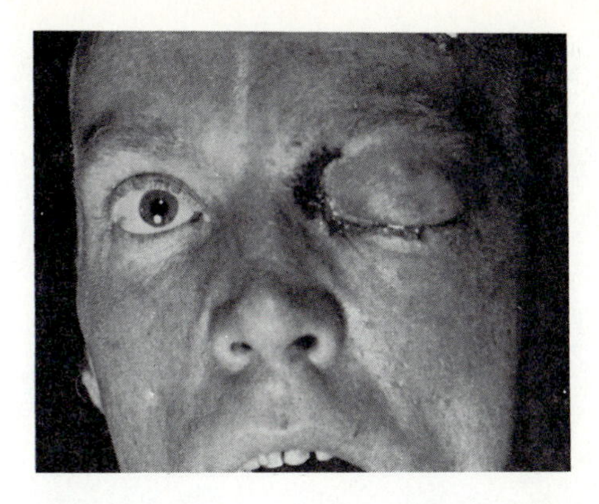

Figure 6-8 *Mucormycosis. Fungal disease caused by members of genera* Rhizopus, *and* Mucor. *Sporangiospores are inhaled and germinate in nasal region with hyphae spreading to orbit of eye.*

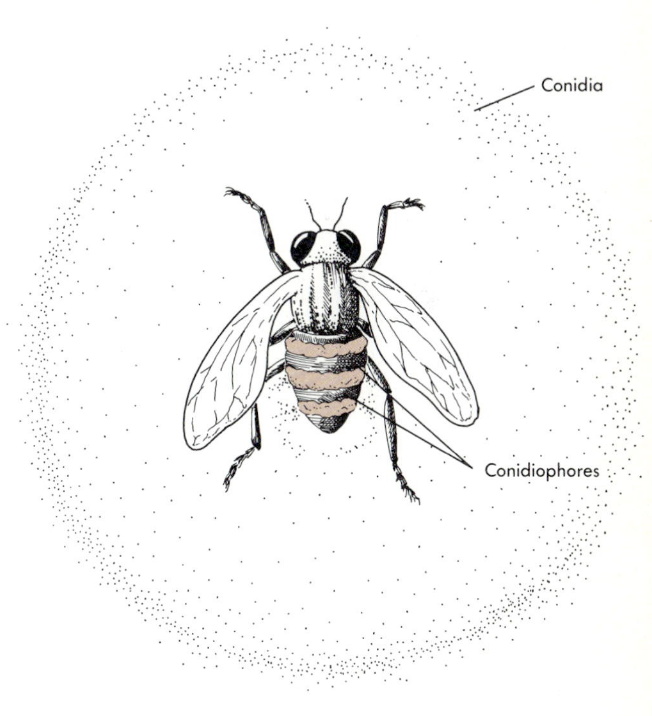

Figure 6-9 *Conidia covering housefly following penetration of conidiophores through abdominal cavity.*

sexual reproduction is initiated by the formation of sex organs that fuse and form a zygospore (thus the subdivision's name). The sexual cycle of the bread mold, *Rhizopus stolonifer,* is illustrated in Figure 6-7. Species of *Rhizopus,* as well as *Mucor,* are important human pathogens (Figure 6-8 and Table 6-2).

Many Entomophthorales are parasites of insects. The fungus grows in the abdominal cavity of a fly, for example, until all the organs are digested and the mycelium occupies the entire abdominal cavity. Later conidiophores are produced that penetrate between abdominal segments to the exterior and release their spores (Figure 6-9).

The majority of Zoopagales are microscopic fungi that parasitize microscopic organisms, such as protozoans, rotifers, and nematodes. They consume the body contents of their victims and produce a sticky substance that coats the mycelium. Once the microscopic organ-

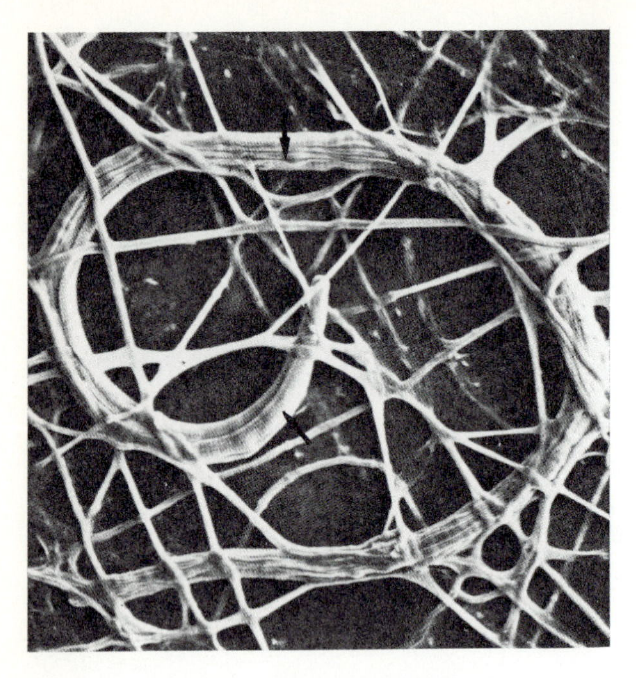

Figure 6-10 *Scanning electron micrograph of nematode* (arrows) *entrapment by adhesive network of filaments produced by a member of Zoopagales. (×1900.)*

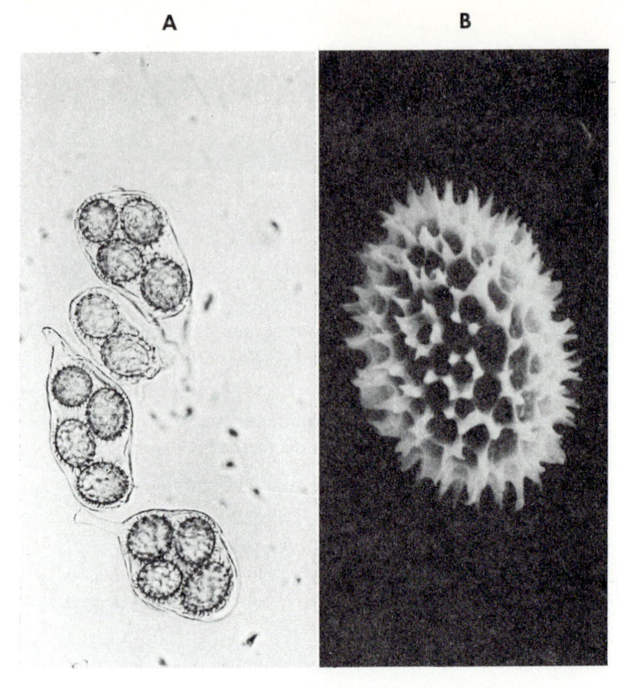

Figure 6-11 *Ascopores.* ***A,*** *Photomicrograph of asci containing ascospores. (×360.)* ***B,*** *Scanning electron micrograph of ascospore.*

ism comes in contact with the fungus, the two stick together. The hyphae penetrate the host and digest the internal contents (Figure 6-10).

Ascomycotina. The members of this subdivision are called **ascomycetes** and include both unicellular and filamentous forms. Most yeasts belong to the Ascomycotina. The one characteristic that distinguishes this group from all others is the **ascus**—a sac that holds the sexually produced **ascospores.** Ascospores are usually produced as four or multiples of four (Figure 6-11).

Filamentous forms reproduce sexually and asexually. Asexual reproduction is primarily by the formation of conidia. Sexual reproduction is characterized by meiosis in the ascus. The asci produced are not free on the body of the fungus but are enclosed in a fruiting body called an **ascocarp.** The ascocarp is tissue composed of several different cell types that have arisen from modified hyphae. The organization and structure of some of these fruiting bodies (Figure 6-12) are important for the separation of various groups.

Yeasts can also reproduce either asexually or sexually. Asexual reproduction is by budding or fission. Budding takes place by the formation and enlargement of a protuberance, called the **bud,** on the surface of the mother or parent cell (Figure 6-3). When the bud reaches a certain size (but always smaller than the mother cell), it separates from the parent. The molecular mechanisms of this process are described in the Point of Interest box. Fission is superficially similar to the bacterial reproductive process. The yeast cell undergoes nuclear division, swells, and then divides into two cells of equal size. Sometimes when multiplication is occurring rapidly, cells divide without dislodging from the parent. Consequently the cells appear in chains or filaments and are referred to as **pseudohyphae** (Figure 6-13). Sexual reproduction in yeasts is a sporulation process characterized by the union of vegetative cells, the product of which develops into an ascus that contains ascospores.

Despite its simple life cycle the yeast *Saccharomyces cerevisiae* exists in three distinct cell types. Haploid cell types designated *a* cells and α cells are able to fuse and form the third, or diploid, cell type *a*/α (Figure 6-14). The cell type *a* and α are referred to as opposite mating types because mating can only occur between *a* and α and not between *a* and *a* or α and α mating types. Diploid cells, under conditions of starvation, undergo meiosis and sporulation to produce an ascus containing four ascospores: two *a* spores and two α spores. The initial mating process is controlled by the reciprocal action of diffusible peptides secreted by each of the opposite mating types. These peptides are called **pheromones.** The *a* mating type produces the pheromone called *a* factor, whereas the α mating type produces α factor. The pheromone of one mating type is able to influence the opposite mating type cell in such a way

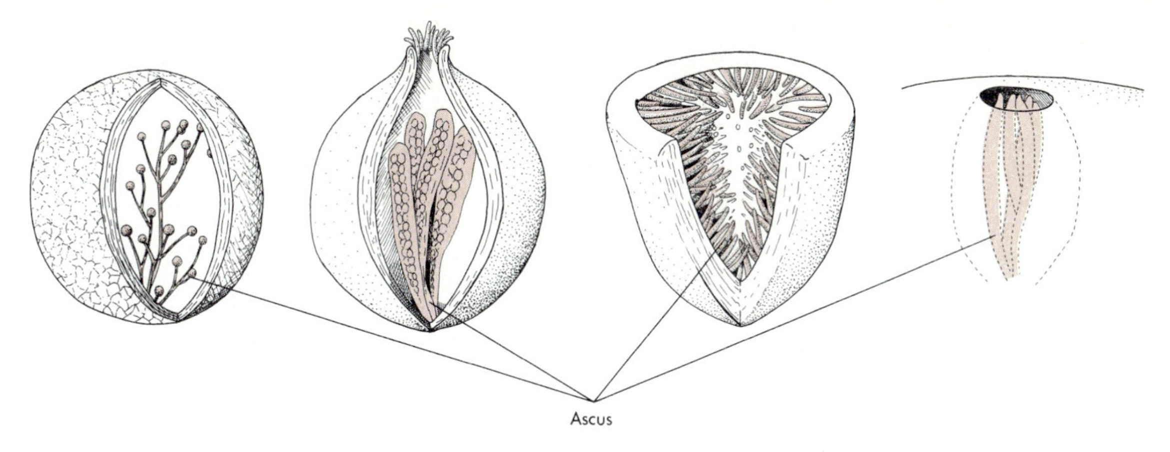

Figure 6-12 *Organization and structure of some ascocarps (fruiting bodies). Asci of different types are seen within ascocarp.*

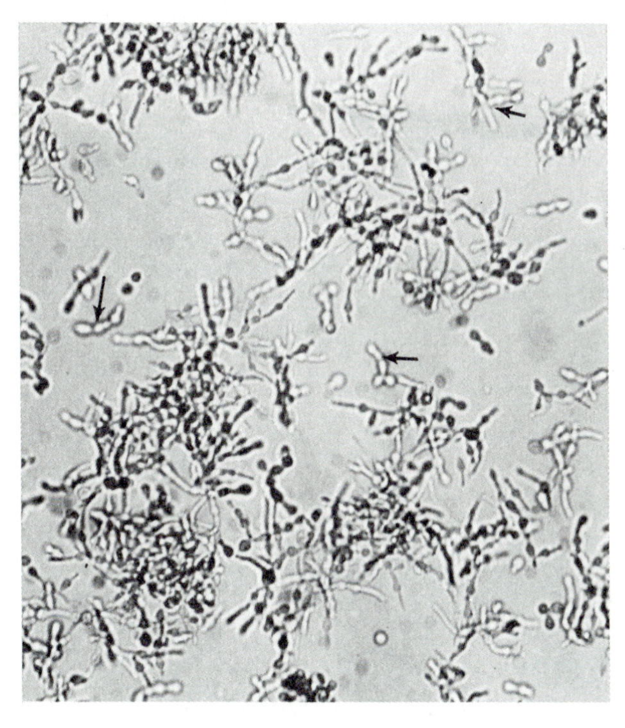

Figure 6-13 *Pseudohyphae in yeast cells. Filaments* (arrows) *are not true hyphae. (True hyphae have parallel walls with no invaginations at septa.)*

that it differentiates into a form that is characteristic of a sex cell or gamete. Thus the *a* cell type is differentiated into a sex cell only by the action of α pheromone and vice versa. Yeast pheromones are similar to the hormones that are characteristic of higher eukaryotes, including humans.

Saccharomyces species, such as *S. cerevisiae,* have many industrial applications, based on their physiology (saccharomyces means "sugar fungus"). Many free-living yeasts can be cultivated in a medium containing only inorganic salts and an organic carbon source, such as glucose. Domesticated yeasts, however, often cannot synthesize all of their vitamins from a single carbon source and thus the vitamin must be supplied in the growth medium. Malt extract, which is produced from sprouted barley, can provide these nutritional requirements. Malt extract is used as the substrate for initiation of fermentation in beermaking. Yeast species carry out vigorous fermentations in which glucose is converted, in the absence of air, to ethanol and carbon dioxide. Ethanol is the desired product in the brewing, winemaking and liquid fuel industries, whereas in the baking industry carbon dioxide is the desired product because it causes bread to rise. Yeasts can also grow in the presence of air. The production of baker's yeast, for example, is accomplished by growing yeast in the presence of air. This process favors the conversion of glucose to carbon dioxide and water and discourages the formation of ethanol.

Some yeasts are **osmophilic**; that is, they thrive in environments in which the concentration of salts or other solutes is extremely high. Osmophilic species such as *S. rouxii* are used in the production of Oriental soy sauce (see Chapter 21). These species are also important spoilage organisms in foods that have high concentrations of sugar or salt, such as honey, jellies, jams, or salted meats.

Yeasts are now being used for the production of nonyeast products. Genetic engineering studies have demonstrated that foreign genes can be inserted into the yeast cell to produce materials such as insulin and pituitary growth hormone. Many additional nonyeast materials can probably be produced by genetic engineering

Point of Interest

BUDDING AND CELL WALL SYNTHESIS IN YEAST

Yeast cell wall is composed primarily of two classes of polysaccharides: polymers of mannose **(mannan)** covalently linked to peptides and polymers of glucose **(glucans).** Both polymers occur in approximately equal proportions in the cell wall. **Chitin,** a polymer of *N*-acetylglucosamine, is a minor component of yeast cell walls but is a major component of the cell walls of other fungi. Over 90% of the chitin in the yeast cell wall is found in the region of bud formation. The glucose and mannose polymers appear to be intertwined, giving the cell wall the appearance of a cross-linked structure. The external portion of the cell wall appears to be primarily mannan. A net of microfibrils is attached to the inner layer of the cell wall and is composed of beta (1→3) glucan. If the microfibrils are removed, the cell wall is easily disrupted by enzymatic treatment. Microfibrils appear to function as a supporting framework for the cell wall.

The bud or future daughter cell first appears as a protuberance on the surface of the yeast cell (figure below). Bud formation is usually centered at the poles of the cell or the areas of maximum curvature. The cell wall of the developing bud is synthesized de novo and is not derived from the mother cell. Glucan and mannan synthesis occur continuously during bud development. Incorporation of new cell wall material at first occurs uniformly over the surface of the developing bud. However, during subsequent bud development, incorporation of cell wall components becomes localized at the bud apex (that portion distal to the mother cell). New cell wall is integrated into older material but the exact mechanism of integration is not known. The most likely mechanism would involve lysis of preexisting cell wall followed by insertion of new cell wall. The bud and its wall grow to a size that is smaller than the mother cell. At the constriction separating mother cell from developing bud there is a ring of chitin. The plasma membrane invaginates at the junction between the two dividing cells, and at the same time the chitin ring grows centripetally until the primary septum is formed.

Glucan synthesis is catalyzed by the enzyme glucan synthetase, which is located on the inner surface of the plasma membrane. Uridine diphosphate (UDP)-glucose* is the substrate for the enzyme and gives rise to a beta (1→3)-linked product. The active enzyme is believed to be carried in vesicles to the site of bud formation. When the bud has reached its maximum size the enzyme is somehow inactivated. Guanidine diphosphate mannose* (GDP-mannose) is the precursor for all the mannosyl groups in mannan. The oligosaccharide is transferred to a protein probably in the endoplasmic reticulum. The glycoprotein product is believed to be processed in the Golgi and released in secretory vesicles. The secretory vesicles would then pass through the plasma membrane and deposit glycoprotein for incorporation into cell wall.

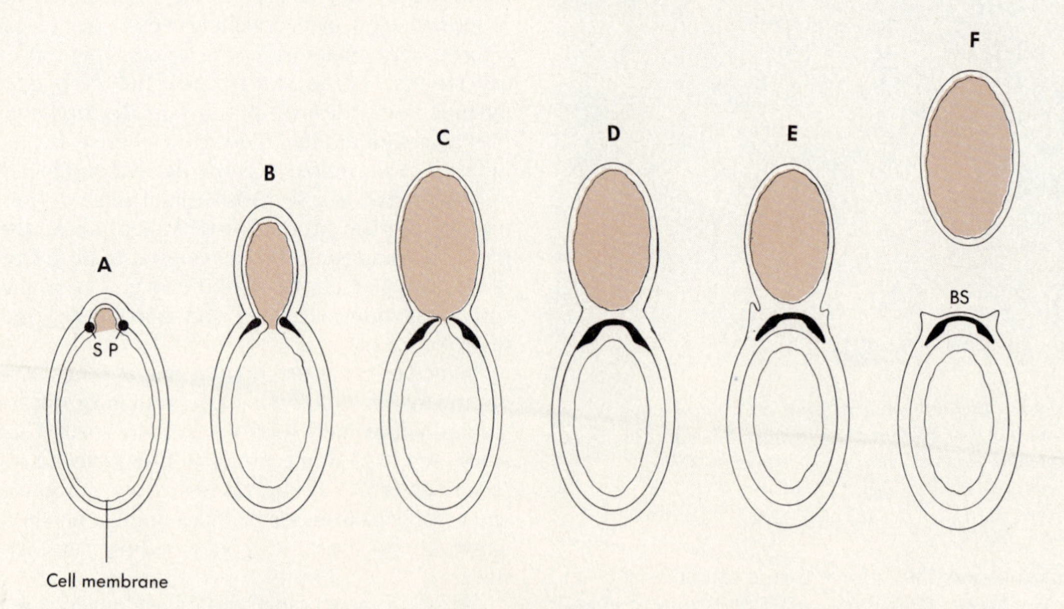

Cell cycle of budding yeast. ***A,*** *Bud appears as protuberance near pole of cell.* SP *indicates site of septum formation.* ***B*** *to* ***E,*** *Septum increases with localization of chitin (filled in black).* ***F,*** *Separation of daughter cell from mother cell leaves bud scar* (BS).

*UDP-glucose and GDP-mannose act as glucose and mannose donors, respectively, in the enzymatic conversion of these sugars into polymers.

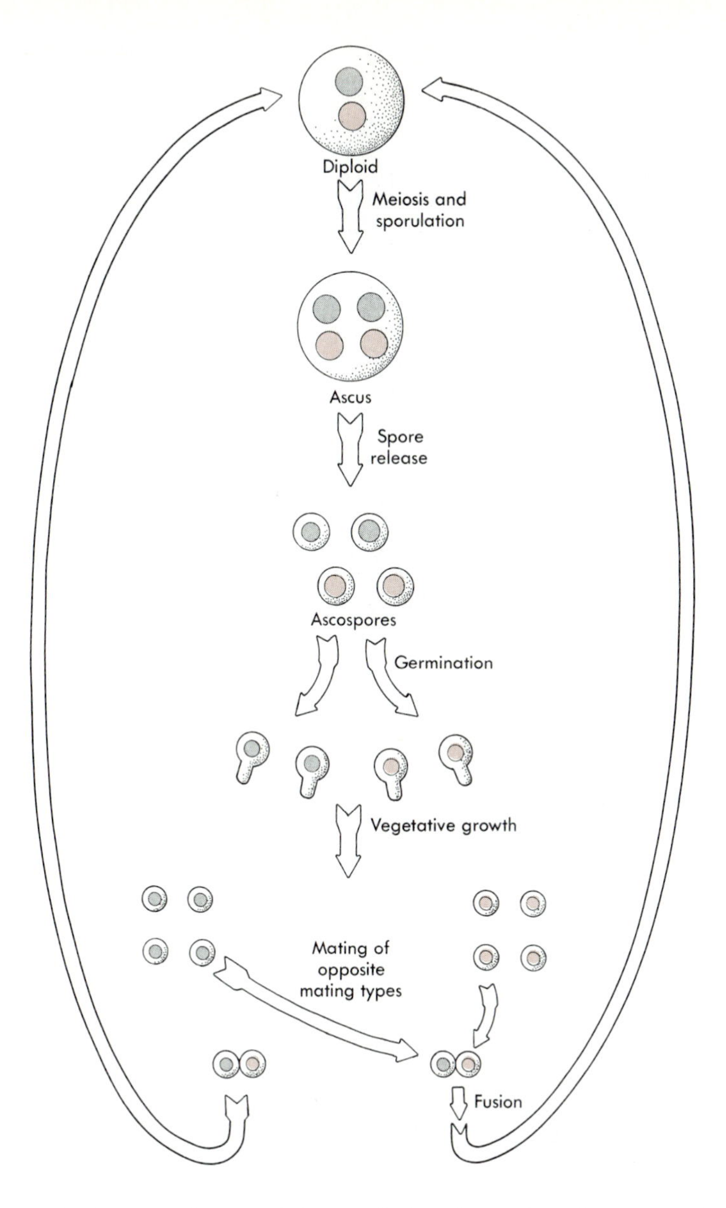

Figure 6-14 *Life cycle of typical yeast cell.*

techniques, and research is now being directed to that task. Genetic improvement in yeast for industrial applications is discussed in Chapter 22.

Some of the beneficial and detrimental properties of yeasts and other ascomycetes are outlined below:

1. *Saccharomyces cerevisiae*, because of its fermentative properties, is used in the baking, brewery, and fuel industries, producing alcohol in breweries and carbon dioxide for the raising of dough. The organism is also an important research tool for the study of biological phenomena. It is easy to cultivate and exhibits sexual and asexual cycles. It is being used to clone genes in genetic engineering studies.
2. *Penicillium roquefortii* is used in the production of roquefort cheese.
3. *Penicillium chrysogenum* is the source of the antibiotic penicillin.
4. *Taphrina deformans* is a parasite of vascular plants that causes leaf curl and witches'-broom (characteristic appearance of the foliage) in such plants as peach and almond trees (Plate 9).
5. *Histoplasma capsulatum* can cause a serious, sometimes fatal, disease called histoplasmosis.
6. *Trichophyton* species (Plate 10) are one of the principal causes of skin diseases called dermatomycoses, for example, athlete's foot (Figure 6-15).
7. *Aspergillus flavus* is responsible for the formation of a group of toxins called **aflatoxins** that have been implicated in liver cancer, particularly among Asians. This species grows on stored grains and foods, such as peanuts and rice.

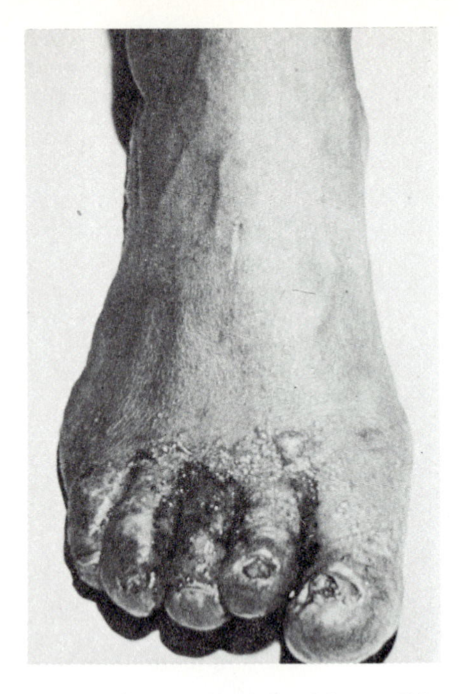

Figure 6-15 *Athlete's foot (tinea pedis).*

Figure 6-16 *Morchella.*

8. *Claviceps purpurea* parasitizes the rye plant and is a cause of **ergot**. Cattle grazing on grasses contaminated by this fungus are prone to disease, and cows may abort their calves. The improper cleaning of grain in earlier times often resulted in the contamination of flour and was the cause of disease symptoms characterized by convulsions and often death. The compounds causing ergot are alkaloids. Semisynthetic ergot alkaloids have been used for a variety of purposes: induction of labor, prevention of postpartum hemorrhage during childbirth, and suppression of postpartum lactation. The work on LSD is an outgrowth of studies on ergot alkaloids.
9. Truffles are underground fruiting bodies of certain ascomycetes. They are not only extremely delicious but expensive (approximately $200 an ounce). Pigs or dogs are used to snuff them out. Truffles are usually found associated with the roots of the oak and hazelnut trees.
10. Morels, which belong to the genus *Morchella,* are also edible; unlike truffles, the fruiting body is above ground (Figure 6-16).
11. *Neurospora crassa,* a red filamentous species, and others in the genus *Neurospora* produce eight ascospores arranged in a linear fashion in the ascus. The spores can be cultured so that individual genetic and observable traits can be analyzed (Figure 6-17). Thus the organism is an important tool for meiosis analysis.

Figure 6-17 ***Photomicrograph of eight-ascospore arrangement in asci of genus*** Neurospora.

12. *Aspergillus* species cause aspergillosis, a pulmonary disease in humans, and a pulmonary disease in young chicks called brooder's pneumonia.
13. *Ceratocystis ulmi* causes Dutch elm disease—a disease that destroyed nearly all the American elms between 1930 and 1970.

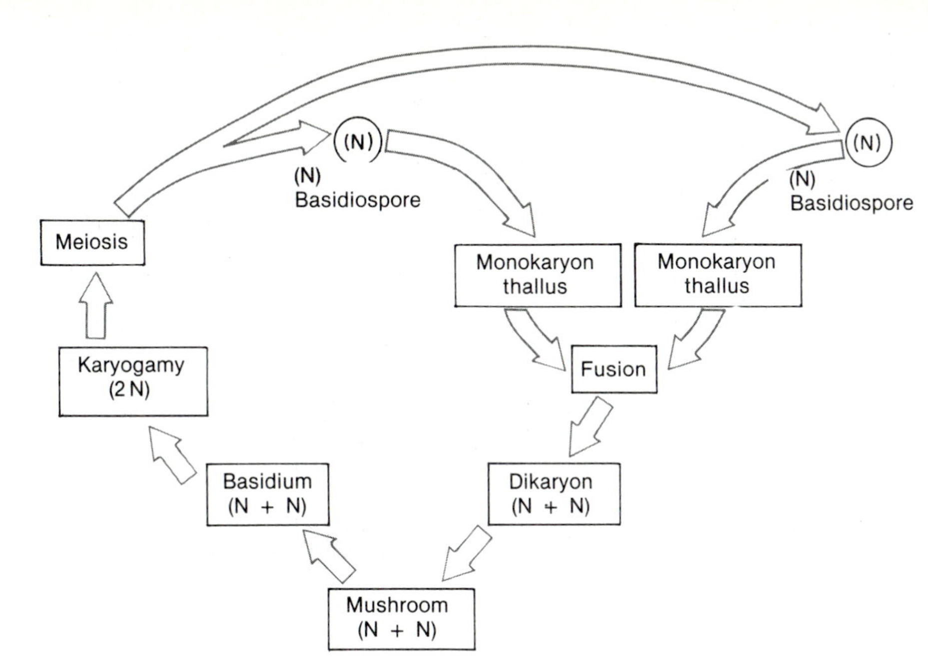

Figure 6-18 *Sexual reproductive cycle of Basidiomycotina.* N, *Haploid genetic state;* 2N, *diploid genetic state.* Monokaryon, *single genetic nucleus;* dikaryon, *two genetic nuclei.*

Deuteromycotina (fungi imperfecti). The Deuteromycotina are a heterogenous group of fungi that are ubiquitous in the environment. They are called imperfect fungi because they all lack a classical sexual cycle. Many species have been reclassified as Ascomycotina or Basidiomycotina because of more recently discovered sexual cycles. Reproduction is usually by hyphal fragmentation or conidia formation. Many deuteromycetes play an important role as parasites but others also have great commercial value. Some of the beneficial and detrimental properties of the Deuteromycotina are as follows:

1. *Fusarium, Alternaria, Verticillium,* and *Sclerotium* species cause such diseases as wilt, leafspot, and white rot in various vegetables and fruits.
2. *Candida albicans* is part of the normal flora of humans. It causes disease in patients stressed by antibiotics or steroid therapy.
3. *Alternaria* species grow on dying plant material. Inhalation of spores can cause "farmer's lung."
4. *Cladosporium* species can decompose or stain paper in various stages of the manufacturing process.
5. *Fusarium* species produce mycotoxins when growing on stored grains or grain products. Ingestion of moldy grains, such as corn, can cause toxicosis (mycotoxicosis) in cattle and humans.
6. *Coccidioides immitis* causes coccidioidomycosis in humans who inhale spores found in dry, hot climates, such as the southwestern United States.
7. *Blastomyces dermatitidis* causes blastomycosis.
8. *Cryptococcus neoformans* causes cryptococcosis.
9. *Cephalosporium* species produce antibiotics called cephalosporins.
10. *Botrytis cinerea* grows on ripening grapes. Presence of the fungus causes grapes to increase their sugar content and reduces the acid content of resulting wine.
11. *Candida utilis* can grow in wastes from the paper industry. It is rich in proteins, vitamins, and fats and has been used as a source of food (single-cell protein).
12. *Trichoderma reesei* is used for commercial production of enzymes that hydrolyze cellulose. These enzymes (cellulases) convert cellulose to glucose that can be fermented by other microorganisms to ethanol and other organic compounds.
13. Many species are important in the degradation of natural and synthetic organic molecules in the environment.

Basidiomycotina. The Basidiomycotina are the most familiar to humans because its members include the edible and inedible mushrooms (Plates 11 and 12), puffballs, stinkhorns, rusts and smuts, and many species that attack lumber. This group differs from all others because of the **basidium**, a swollen-shaped cell in which sexually produced spores (**basidiospores**) are produced. Asexual reproduction processes, such as hyphal fragmentation and conidia or chlamydospore formation, do occur in the Basidiomycotina but they play a lesser role than in other fungal subdivisions.

The mycelium, which can be observed on wet, dead leaves, bark, or other organic matter, penetrates its host to obtain nutrients. The sexual reproductive cycle in mycelial development is relatively simple in most Basidiomycetes (Figure 6-18).

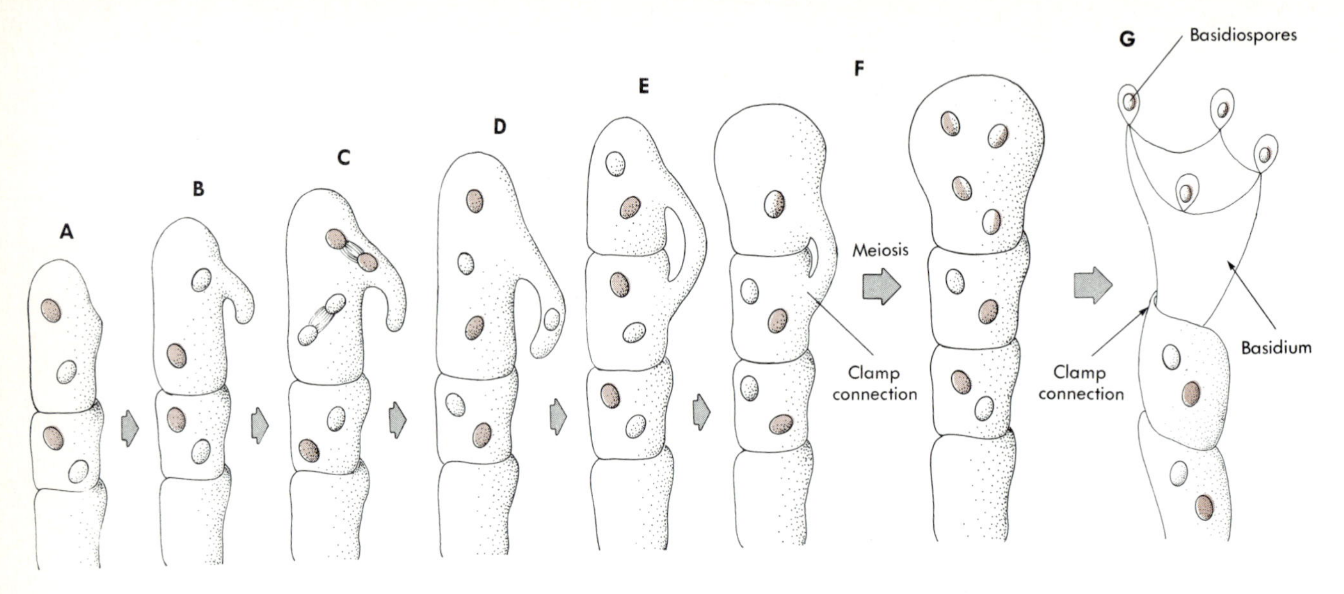

Figure 6-19 *Clamp connection and basidium formation.* ***A*** *to* ***E,*** *Terminal cell in dikaryotic hypha undergoes mitosis. Process ensures that each new cell will contain one light and one dark nucleus.* ***F,*** *Nuclei fuse, producing diploid nucleus that undergoes meiosis* ***(G).*** *Meiosis produces four haploid nuclei that form basidiospores in mature basidium.*

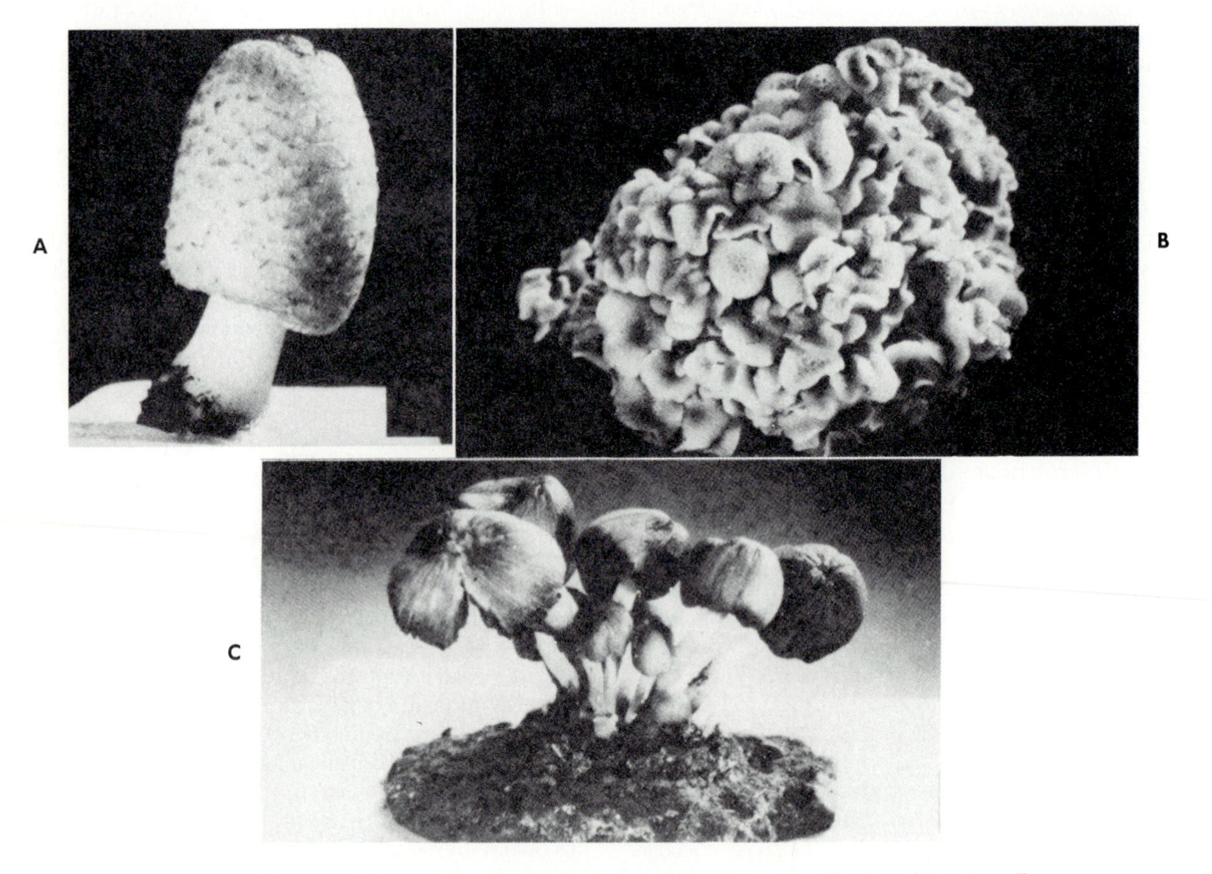

Figure 6-20 *Basidiocarps (fruiting bodies) of some mushrooms.* ***A,*** *Species of* Coprinus. ***B,*** *Compound basidiocarp of species of* Polyporus. ***C,*** *Species of* Coprinus.

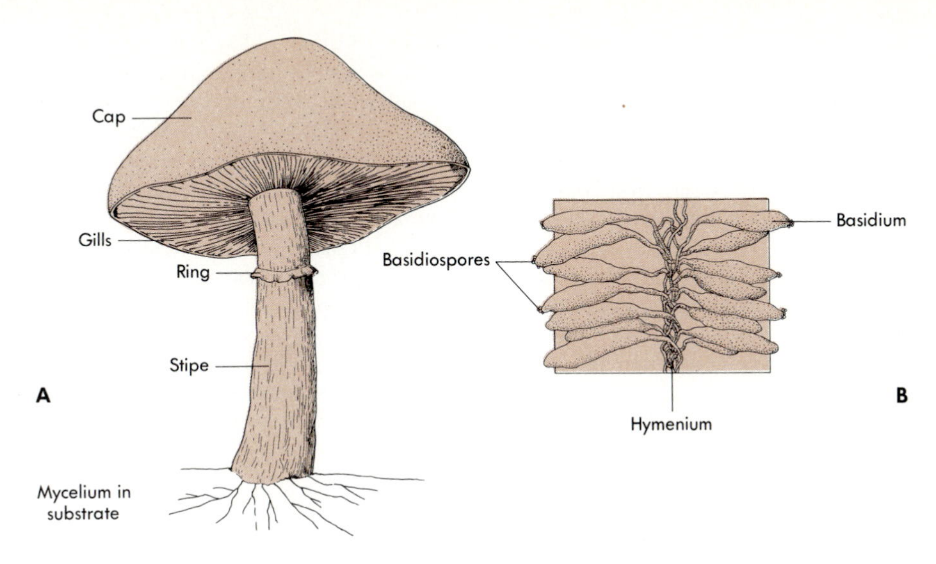

Figure 6-21 ***Morphological characteristics of basidiocarp of mushroom. A, Mature basidiocarp. B, Vertical section through gill demonstrating structure of hymenium with its basidia and basidiospores.***

Genetically different hyphae (called **monokaryons** because they contain only one kind of genetic nucleus) fuse and produce a **dikaryon.** The nuclei derived from the two hyphae do not fuse, and the characteristics of the dikaryon are determined by the two sets of haploid genes. The dikaryotic hyphae grow and form a mycelium that attaches to their normal substrate: bark, grain, soil, etc. An unusual characteristic of mycelial growth is the formation of structures called **clamps** (Figure 6-19). Clamps represent a mechanism for ensuring that each cell produced from division remains a dikaryon and does not receive two identical nuclei. A clamp is produced near one of the nuclei when the tip of a hyphal branch has finished its growth and a septum is to be formed. The nuclei divide, and a septum is formed, resulting in four nuclei: one is in the clamp, one is in the subapical region, and two are in the apical portion of the hypha. The clamp grows backward and fuses with the subapical region to which the clamp nucleus has now migrated. Nuclear fusion in the apical portion of hypha followed by meiosis results in four haploid nuclei. These nuclei form the four basidospores of the apical hypha, which develops into a basidium (Figure 6-19).

A **basidiocarp** (the macroscopic spore-bearing structure) and **basidium** or a basidium only will develop from the mycelium. The basidiocarp, the fruiting body in the Basidiomycetes, may vary in size and shape (Figure 6-20). Mushrooms, puffballs, and others are examples of basidiocarps of the fungi producing them. The basidiocarp bears basidia that are arranged in layers called **hymenia** (Figure 6-21) and may cover the entire surface or only a portion of the basiodiocarp. The arrangement and microscopic features of the hymenia are used to classify the various Basidiomycetes.

One of the most important contributions of the basidiomycetes is their association with the roots of various plants (see Point of Interest). Some of the beneficial, as well as detrimental, properties of the basidiomycetes are outlined in the box on p. 159.

Slime molds

Slime molds are viscous organisms whose taxonomic relationship is not clear. This taxonomic uncertainty is a result of the fact that slime molds exhibit fungal and protozoal qualities. They are nonphotosynthetic and produce fruiting bodies, which are characteristics associated with fungi. They pursue and engulf their food and lack a cell wall, which are characteristics resembling the protozoa. Most investigators consider them to be fungi. Whenever mycologists isolate true fungi from the environment, slime molds are frequently found in the same ecological niche. The fruiting body in many groups of slime molds is the direct result of cellular aggregation, and this differentiation process is characteristic of the prokaryotic myxobacteria. There are several taxonomic divisions of the slime molds, but this discussion will center only on the most interesting groups.

Cellular slime molds. Cellular slime molds consist of individual nonflagellated cells called **myxamebae.** One species, *Dictyostelium discoideum,* has been extensively studied because of the nature of its differentiation process (Figure 6-22). The principal vegetative unit is an ameba that feeds on bacteria and divides by mitosis forming a large population of cells. Each cell behaves independently of other cells if the food supply is adequate. When the food supply becomes depleted, one or more of the myxameba release a chemical that diffuses away and stimulates other myxameba to release the same chemical signal; this causes the cells to aggregate (Figure 6-23). As the myxameba aggregate they secrete another chemical that causes them to adhere to each other, forming long streams. The aggregation of the myxameba is caused by the chemical, cyclic adenosine monophosphate (cAMP). Aggregation eventually results in the formation of a mass of cells (usually 100,000 to 200,000) called a **slug**. The slug migrates for short distances up to 2 mm/hour in response to light (phototropism), as well as temperature (thermotropism). The multicellular slug differentiates into a fruiting body called a **sorocarp**. The sorocarp consists of a stalk, at the tip of which individual cells give rise to spores. The stalk of the sorocarp may reach a length of 1 cm. The spores in the stalk synthesize cellulose that the spores use to surround themselves for protection until conditions are favorable for germination. Sexual reproduction has been demonstrated to exist in some species.

Acellular or true slime molds. The propagative unit of the true slime molds is a dormant, spherical, thick-walled spore that is often covered with spines. The first stage in the life cycle of this group is characterized by

Point of Interest

MYCORRHIZAE

The associations between fungi and plant roots are called **mycorrhizae.** Fungal hyphae invade the root tissue where the transfer of water, inorganic nutrients, and carbohydrates may occur in both directions. Both partners appear to benefit from the association and neither is harmed. Mycorrhizal relationships are common in woody and herbaceous plants. There are three forms of mycorrhizae: ectotrophic, endotrophic, and ectendotrophic. **Ectotrophic** mycorrhizae are widespread in temperate forest trees (pine, beech, and birch) and most are Basidiomycetes. Ectotrophic mycorrhizae form a mycelial mantle around the root surface (Figure 6-B). Hyphae invade the root cortex by secreting enzymes (pectases) that breakdown pectin, the plant cell polymer. The hyphae grow between the cortical cells (intercellularly) of the plant. Hyphae on the root surface penetrate the soil and thus an avenue between soil and plant cortex is established. The fungus absorbs carbohydrates (usually in the form of sucrose) and converts them into fungal sugars, such as trehalose and mannitol. Because most ectotrophic fungi cannot degrade soil cellulose and lignin, the structural polymers of plants, they must rely on plant carbohydrates as a carbon source.

Endomycorrhizae appear similar to ectomycorrhizae except that the fungus grows internally within the cortical cells of the plant. This type of association is obligatory in plants, such as orchids, where dependence appears to be due to a lack of plant enzymes required for metabolism of storage products used for orchid growth. Endomycorrhizae are found on most cultivated plants and some trees. The **ectendomycorrhizae** are characterized by hyphae that penetrate into and between cortical cells of the plant, and they may or may not have a fungal mantle covering the roots. Benefits to the plant from mycorrhizal relationships include the following:

1. Possible compensation for loss of roots caused by disease.
2. Increase in the availability of mineral elements in infertile soils. Phosphorus, for example, appears to be concentrated by the fungus and then slowly released to the plant. Mycorrhizae also enhance the uptake of iron, sulfate, and possibly manganese and zinc.
3. Deterrence of the infection of feeder roots by soil-borne pathogens by reduction of the level of free carbohydrates and other nutrients in root exudate that normally stimulate pathogen activity. This may be accomplished by establishing a physical barrier to penetration or by releasing inhibitory compounds. Some ectomycorrhizae provide protection by producing antibiotics. Certain endomycorrhizae increase the arginine content in the roots of tobacco plants, which suppresses chlamydospore formation by some plant pathogens.
4. Reduction of the amount of susceptible plant tissue to infection.

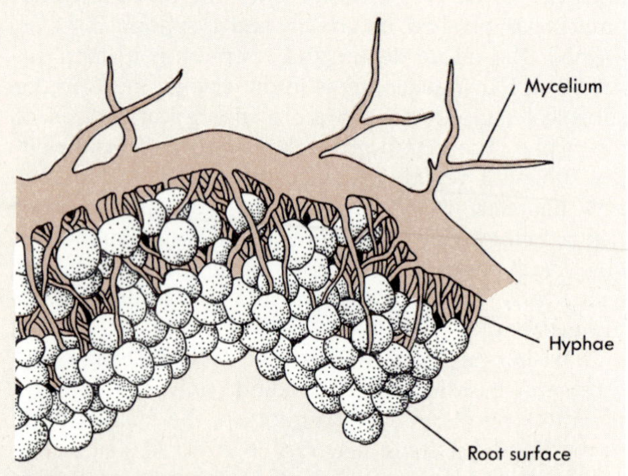

Ectrophic mycorrhiza. Hyphae penetrate between cells of root with fungal mycelium forming mantle over root surfaces.

spore germination and the release of a uninucleate ameboid cell, which becomes flagellated (Figure 6-24, *A*). The second stage involves the formation of a **plasmodium**, a multinucleated, wall-less structure arising from one ameba or from the fusion of ameboid units (Figure 6-24, *B*). The plasmodium is the primary vegetative phase and may vary in size from a few hundred micrometers to several square meters. The plasmodia are often highly pigmented structures, usually yellow or white, but colors of red and black have been reported. The plasmodium feeds by ingesting microscopic organisms, such as bacteria and protozoa, but later ceases to feed and forms fruiting bodies. The fruiting bodies (sporangia) are usually one of two types: a sporangium in which the stalk is hollow or one in which the interior of the stalk is fibrous. The life cycle is completed by cleavage of the multinucleated protoplasm into uninucleate segments, which acquire a thick wall and become spores (Plates 13 to 15).

Net slime molds. Net slime molds are aquatic, and some are known to parasitize algae and plants. The basic cellular unit is the **spindle cell**, which is nonflagellated and possesses indentations on the periphery. These indentations correspond to special organelles called **sae-**

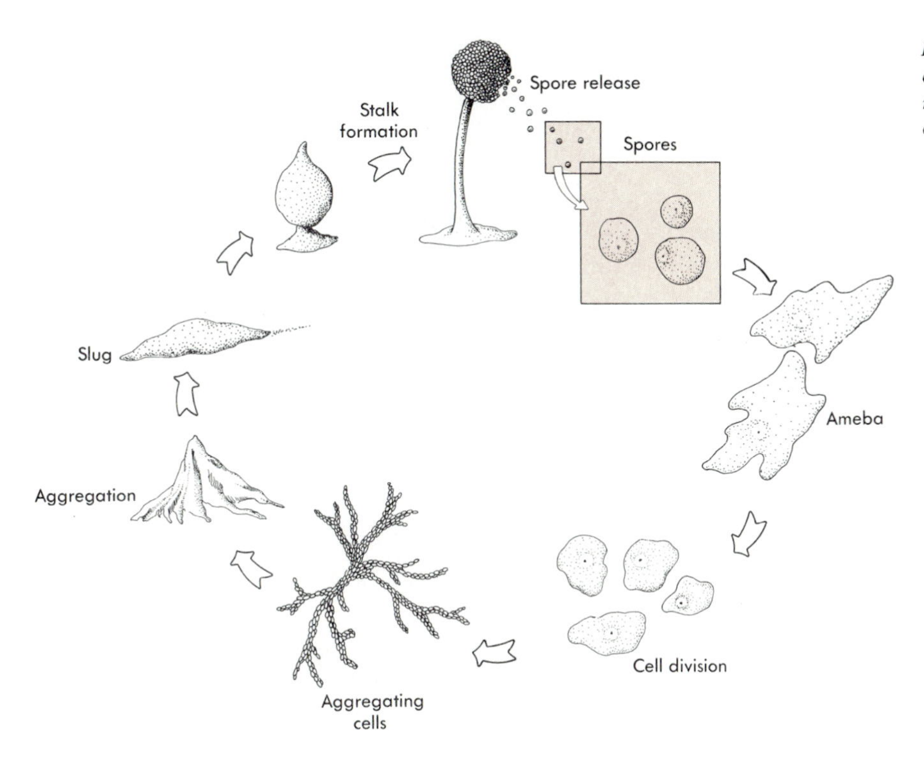

Figure 6-22 *Morphological changes in life cycle of slime mold* Dictyostelium discoideum.

BENEFICIAL AND DETRIMENTAL PROPERTIES OF BASIDIOMYCOTINA

BENEFICIAL

1. Several species form mycorrhizal relationships with several species of trees, including pine, hickory, oak, and birch.
2. Basidiomycetes are the most important decomposers of the macromolecules cellulose and lignin, which are found in plants. Therefore they are important decomposers of various forest litter.
3. Mushrooms for human consumption are primarily members of *Agaricus* species.

DETRIMENTAL

1. Plant parasites, such as wheat rust, corn smut, white pine blister rust, and cedar apple rust.
2. Poisonous mushrooms, such as *Amantia* species.
3. *Serpula lacrymans* and other fungi cause dry rot of timbers.
4. Species of *Schizophyllum, Stereum,* and *Fomes* cause rotting of the heartwood (dead interior wood of the tree) and sapwood (live outer wood) of standing trees.

Figure 6-23 *Aggregation in cellular slime mold* Dictyostelium discoideum. ***A,*** *Starved ameboid cells begin to elongate (×200).* ***B,*** *Streams of elongated cells merge toward center (×200).* ***C,*** *Lower magnification of aggregating amebas forming discrete colonies (×10).*

genogens, from which extrude membranous sacs that eventually coalesce to form a membrane (**extoplasmic sheath**) around the spindle cells (Figure 6-25). The membranous sheath extends ahead of the spindle cells and covers potential food sources. The sheath serves not only to capture prey such as bacteria and yeasts but also to aid the spindle cells in locomotion, which is characterized by a gliding movement. The spindle cells migrate to the food source when prey is captured, but if no food is available, the net begins to spread in all directions as if hunting. Under adverse conditions the vegetative state is followed by cell aggregation and sporangium formation. Flagellated zoospores, which develop from the sporangium, become spindle cells.

Concept Check

1. Name the major subdivisions of fungi and give examples of their importance in the environment or their commercial value.
2. Describe the asexual and sexual life cycle of yeast.
3. Describe some of the diseases associated with the subdivision Deuteromycotina.
4. What are mycorrhizae and what is their value?
5. What is the differentiation process that is characteristic of cellular slime molds?

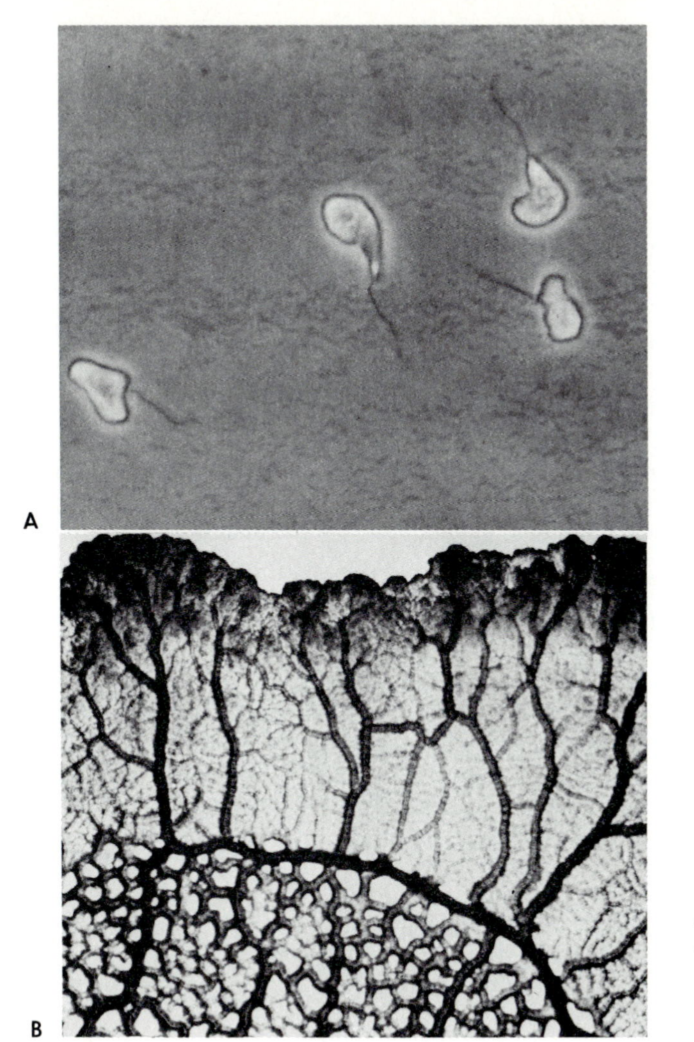

Figure 6-24 *True slime molds. **A,** Flagellated ameboid cells of genus* Physarum. ***B,*** *Plasmodium of genus* Physarum.

Figure 6-25 *Ectoplasm sheath and spindle cells of net slime mold.*

CHARACTERISTICS OF THE PROTOZOA

The term **protozoa** signifies unicellular microscopic animals, a term that probably does not mean much to the average reader until he or she examines a drop of pond water with a microscope. That microscopic world is filled with small unicellular animals whose structural clarity and movements through the watery medium remind one of a ballet. It is a mesmerizing experience that one never forgets and is always willing to repeat.

The study of diseases caused by animal parasites, such as the protozoa, is called **parasitology**. The importance of animal parasites in causing disease cannot be underestimated. The majority of human and animal disease in underdeveloped countries is a result of animal parasites. Malaria, for example, affects 300 million people worldwide and causes 2 to 4 million deaths each year. Over a billion people are infected with another animal parasite, the parasitic worm. Although parasitic worms do not cause as many deaths as the protozoa, they can cause various states of debilitation and loss of energy in the host. In the United States, a protozoan called *Eimeria tenella* is responsible for the loss of over 40 million dollars each year in the poultry industry. The organism can infect a flock and cause up to 100% mortality. In Africa over 4 million square miles of potential agricultural land will not support cattle because they invariably become infected by a protozoan of the genus *Trypanosoma*. Mortality from infection is near 100%.

Protozoa also play some role in the degradation of sewage through their oxidative metabolic activities and abililty to remove suspended bacteria. Protozoa are found in the digestive tract of ruminants but their role is not clear. It is believed that they are a major source of lipid for the host animal. Protozoa are also excellent tools for research because they are easily cultivated in the laboratory. They reproduce asexually, and clones are

Table 6-3 *Modification of some surface structures in the Protozoa*

Surface feature	Characteristics	Function
Pellicle	A thick covering over the cell membrane and believed to be derived from it	Protection or support
Shell (test)	A capsular-like covering secreted by the organism—some are like skeletons and contain calcium carbonate; others secrete a glue-like material that can be mixed with sand to produce a matrix	Protection; sometimes used to attach to a substrate
Mucocyst	Sacs that lie beneath but pressed against pellicle; discharge mucus to the exterior	Protective device by aiding in cyst development; aids phagocytosis of particles
Cytosome	An opening through which food is ingested—may be round or slitlike	Feeding
Oral groove	Indentation of pellicle of ciliates; the cilia around the oral groove appear to fuse and give appearance of undulating membrane	Mechanism for concentrating food
Cytopharynx	Region associated with cytosome in which food passes before being pinched off in a vacuole	Mechanism for concentrating digested food
Contractile vacuole	Fixed or variably positioned membrane-bound vacuole; cytoplasmic fluid fills vacuole; when pushed against the cytoplasmic membrane, fluid is expelled	Osmoregulatory device; may also excrete products such as ammonia and urea
Food vacuoles	Membrane-bound vesicles that contain food ingested through cytosome or food acquired at cell surface by invagination of cell membrane and formation of vacuole; vacuole will acquire enzymes to digest food and then move through cytoplasm	Food digestion
Cytopyge	Fixed opening through which digested food is eliminated	Waste elimination
Extrusomes	Filamentous structures whose function varies; there are four types:	
	TRICHOCYSTS: occur in ciliates and flagellates; located beneath pellicle, they can eject filament	Function in doubt but may be protective response to irritants
	HAPTOCYSTS: small sacs located at the tips of tentacles of aquatic sessile species; when tentacle touches a prey, the haptocyst penetrates pellicle and binds prey	To capture prey
	TOXICYSTS: fibers concentrated around the cytosome	To capture and paralyze prey
	EJECTOSOMES: coiled refractile bodies located beneath the pellicle of one class of protozoa	When activated, believed to propel organism away from danger
Uroid	Posterior protrusion of certain amebas	Believed to be associated with pseudopodial activity or movement

easily generated with the same genetic profile as any desired "parent." The effects of environmental factors, such as radiation, are easily analyzed in these organisms.

The protozoa are classified in the Kingdom Protista, but they are difficult in many ways to distinguish from other eukaryotes. For our purposes we are only concerned with four groups, based on their mechanism of locomotion.

Sarcomastigophora—locomotion by flagella or ameboid movement

Sporozoa—spore-forming protozoa, no locomotor organelles, but movement by gliding in some species

Myxospora—spores that contain a filament, vegetative cells use ameboid movement

Ciliophora—locomotion by cilia

The subdivisions presented above are an artificial and simplified scheme; more recent and detailed classification schemes can be found in textbooks of protozoology.

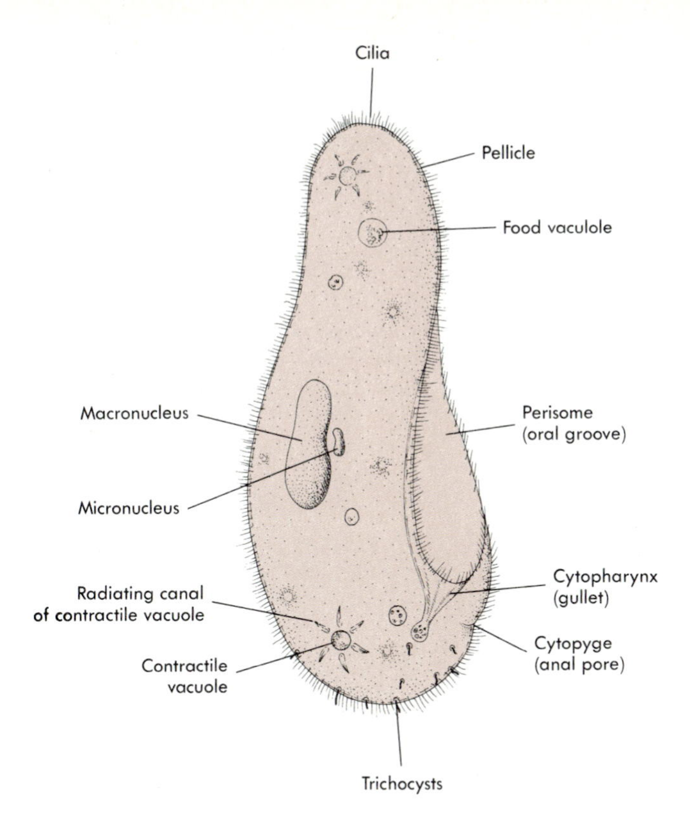

Figure 6-26 *Structural features of protozoan* (Paramecium).

Morphology

Chapter 4 demonstrated that the protozoa show a tremendous variation in size (4 nm to 5 mm) and shape (see Figure 4-17). They show many subcellular specializations and modifications that enable them to survive in their environment (Table 6-3 and Figure 6-26). This group, you will recall, does not possess a cell wall.

Nutrition

Protozoa can be found as free-living species in the soil and in both fresh and marine waters, but some species are parasitic to invertebrates and vertebrates. Free-living protozoa require some level of water for survival, such as ditches, bogs, moist soil, lakes, or oceans. Many protozoa form cellular protective structures called **cysts** that can be dispersed by air currents, insects, and birds, if water becomes unavailable. Bacteria are an important protozoan food source, and their presence often determines the relative numbers of protozoa that are present in an environment. Most protozoa live at the edges of or on the bottom of shallow bodies of water, where decaying vegetation, temperature, and oxygen levels are adequate for their survival.

Unlike bacteria and fungi, protozoa are not primary decomposers. They are heterotrophic and can ingest organic debris or feed on other organisms, such as bacteria, algae, and nematodes, as well as other protozoa. Food may be taken into the cell by (1) passing through permanent structures such as the mouth or cytosome of some flagellates and ciliates, (2) dissimilation either by invagination of the cell membrane at various sites on the body surface as in ameboid protozoa, or (3) diffusion through the cell membrane. Food ingested by amebas (Figure 6-27) is surrounded by a membrane to produce a **food vacuole**, in which digestive enzymes break down the food. The food vacuole circulates freely in the cytoplasm. The ciliates and some flagellates have a permanent indentation at the cell surface called the **oral groove** that leads to the **cytosome**, an area for digestion of ingested food. Food taken into the oral groove passes through the **cytopharynx** before it is pinched off into a food vacuole in the cytosome. Undigested material left in food vacuoles is eliminated at a site at the cell surface called the **cytopyge**. Freshwater protozoa, because they live in an environment that is hypotonic, continually take water into the cell to maintain an osmotic balance. The cell might burst if it were not for the **contractile vacuole**, which maintains water balance by pumping out water after the cell has reached a critical size.

Most free-living protozoa are aerobic. Anaerobic forms are found primarily as parasites, such as those species that inhabit the intestinal tracts of vertebrates and invertebrates.

Reproduction

Asexual reproduction. Asexual reproduction occurs in all protozoa and is the most frequent means of propagation. **Binary fission** is the most common method of asexual reproduction, but other mechanisms including budding and multiple fission are available to some species.

Binary fission. Binary fission in flagellates is longitudinal (Figure 6-28, *A*). The basal body of the flagellum and the nucleus divide during binary fission, and when the division is complete, each daughter cell receives a nucleus as well as a flagellum. Binary fission in ciliates is transverse and occurs at right angles to the long axis of the cell (Figure 6-28, *C*).

Budding. Budding occurs more frequently in the protozoa that are sessile and anchored to stones, algae, or detritus (disintegrated materials). A portion of the adult constricts and leaves the parent cell during the budding process.

Multiple fission (schizogony). Multiple fission is a process in which the nucleus within a cell divides repeatedly to produce as many as several thousand nuclei. Each nucleus becomes surrounded by a small portion of cytoplasm and breaks away from the multinucleated

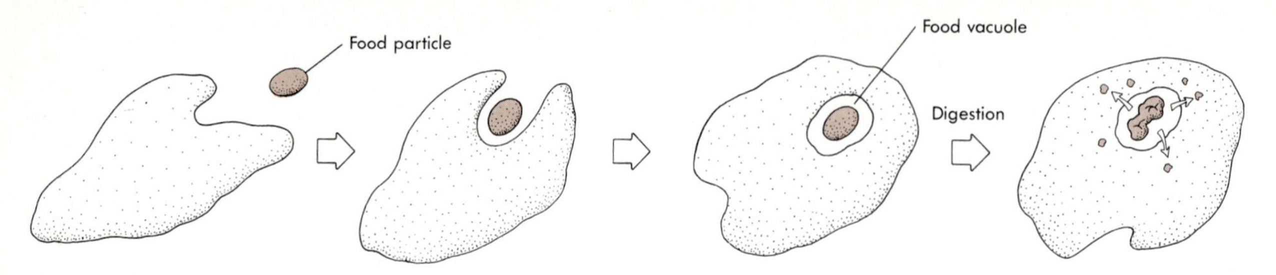

Figure 6-27 *Capture of food by ameba. Food particles become enclosed by membrane that produces food vacuole. Enzymes in food vacuole degrade food particles into molecules that diffuse to rest of cell.*

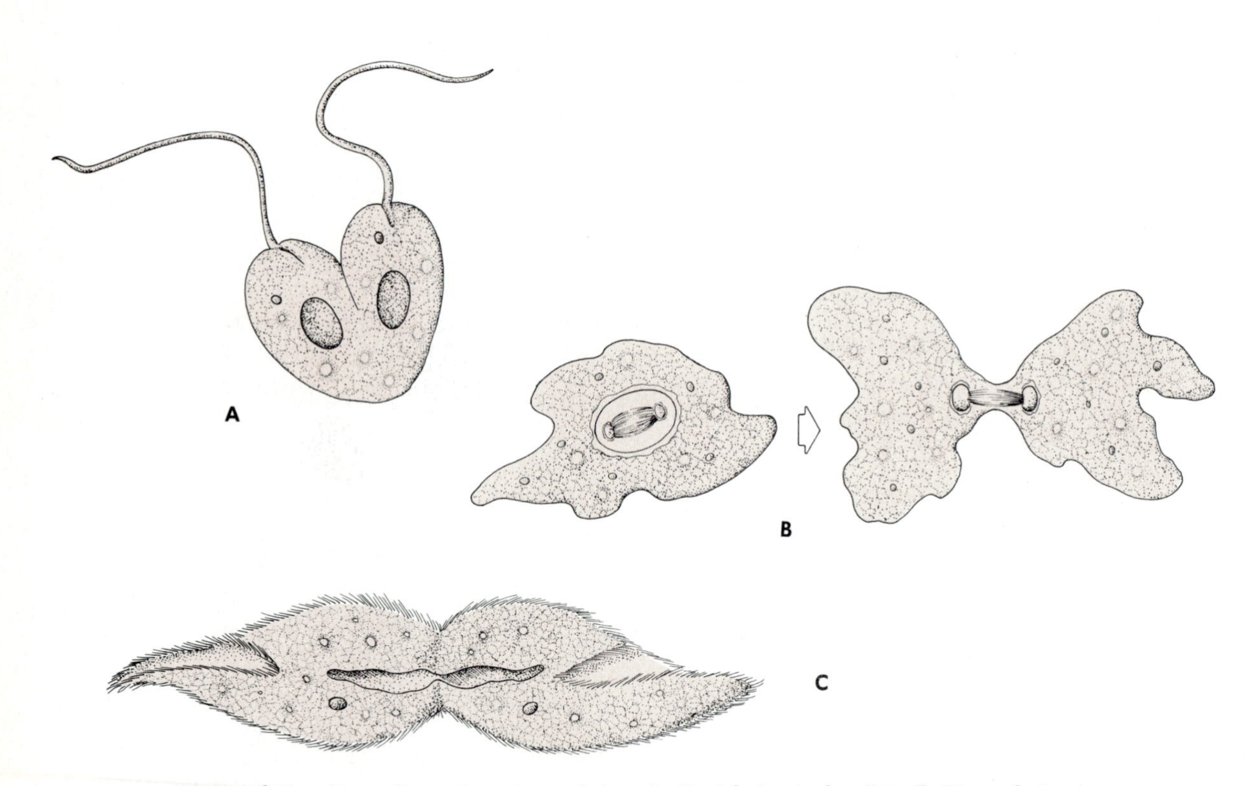

Figure 6-28 *Binary fission in protozoa.* ***A,*** *Longitudinal fission in flagellate.* ***B,*** *Binary fission in ameba.* ***C,*** *Transverse fission in ciliate.*

mass as an individual uninucleated cell. This type of process is characteristic of the sporozoa that are parasites of animals, including humans (Figure 6-29). The high fever and concomitant chills that appear periodically in malaria are a result of multiple fission in the red blood cells. Thousands of these uninucleated forms are released when the infected red blood cell lyses and the body responds to them by an increase in temperature.

Sexual reproduction. Sexual reproduction appears during the life cycle of most protozoa except in those species that are parasitic; however, sexual reproduction is influenced primarily by environmental conditions. Sexual reproductive patterns vary in the protozoa as they do in the fungi. The vegetative cell of some species spends the majority of its life cycle in the haploid state; in others, it is the diploid cell that predominates. Finally, there are those species that show an equivalent alternation between the haploid and diploid states.

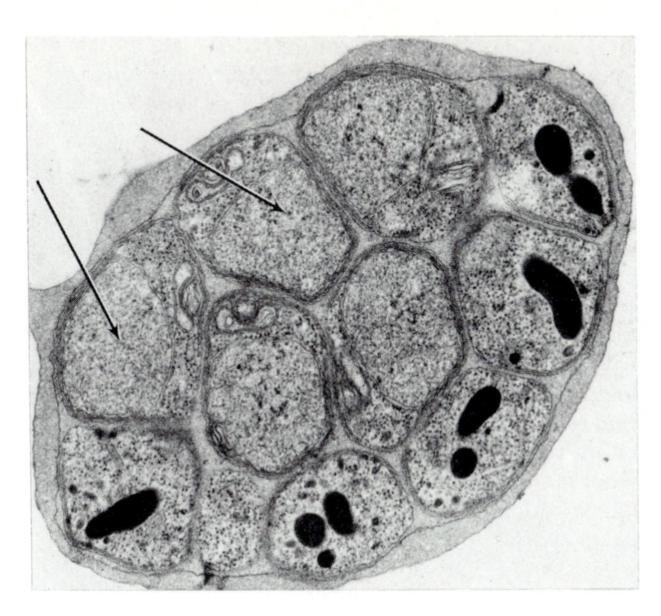

Figure 6-29 *Example of multiple fission in infected red blood cell. Several merozoites are seen* (arrow). *(Reduced 70%.)*

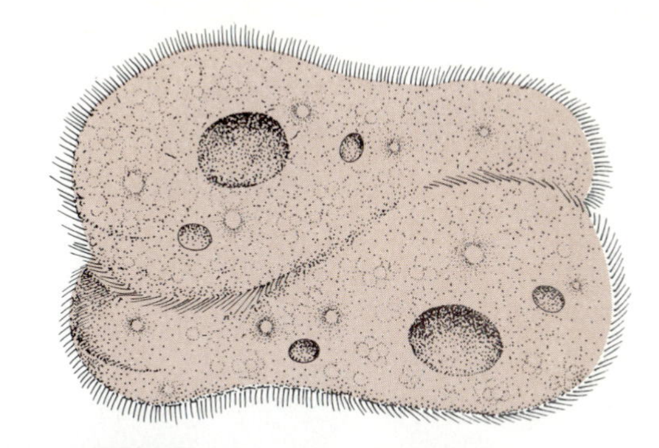

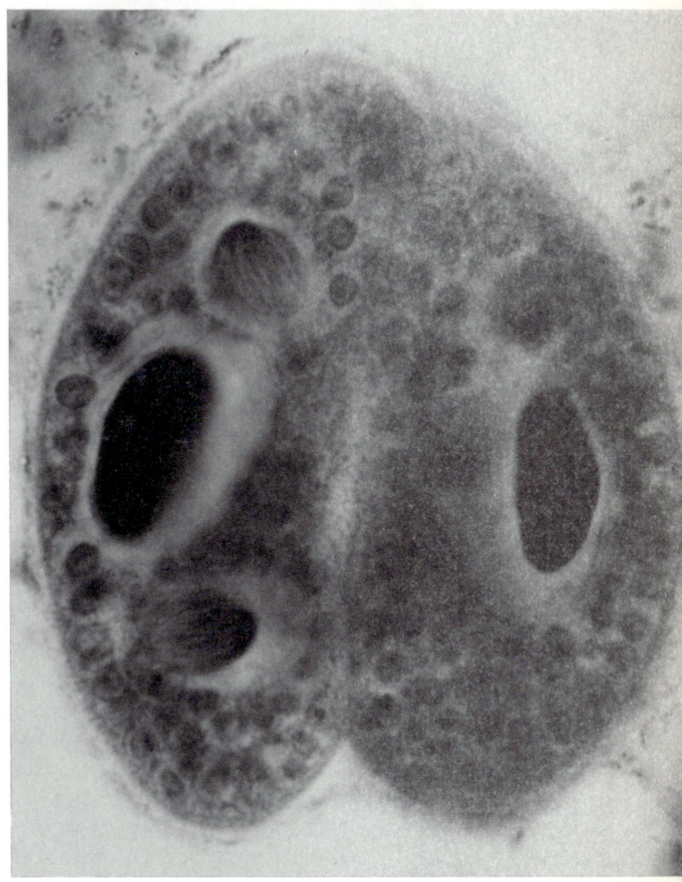

Figure 6-30 *Conjugation in ciliates. Photomicrograph and line drawing of conjugating* Paramecium *species.*

Basically two types of sexual reproduction occur in protozoa: **syngamy** and **conjugation**. Syngamy involves the fusion of free-swimming gametes that vary in size and morphology. Conjugation occurs only in ciliates, and during the process two parent cells, called **conjugants**, initially bind to each other at their anterior ends and in the area of the oral groove (Figure 6-30). A bridge is produced after the binding process that permits cytoplasmic flow between the conjugants.

Cysts

Many protozoa, under appropriate conditions, can reduce their size, eliminate wastes and water, store food reserves, and develop a thick cuticle, all of which results in the formation of a **cyst**. The cyst is usually a protective device for the cell, but it can also function in a reproductive manner, or as an agent of transmission. Cyst formation (encystment) can be stimulated by a number of environmental conditions including evaporation, pH changes, nutrient depletion, and overcrowding. These factors initiate encystment for the purpose of protection of the genetic material in the cell. The nucleus within the cyst divides into four and sometimes eight nuclei in some species—a typical reproductive function. Some species are parasitic, however, and the cyst also becomes a unit of transmission. The amebas causing dysentery in humans, for example, will encyst while in the host and will become infective only when eliminated in the feces and ingested by a suitable host. Emergence from the encysted state (excystment) is usually stimulated by conditions that favor vegetative growth. Excystment is the result of digestion of the cyst wall by enzymes present in the gut of the host. Some cysts have a preformed hole or plug that is closed during encystment but can be dissolved during excystment, releasing a vegetative cell.

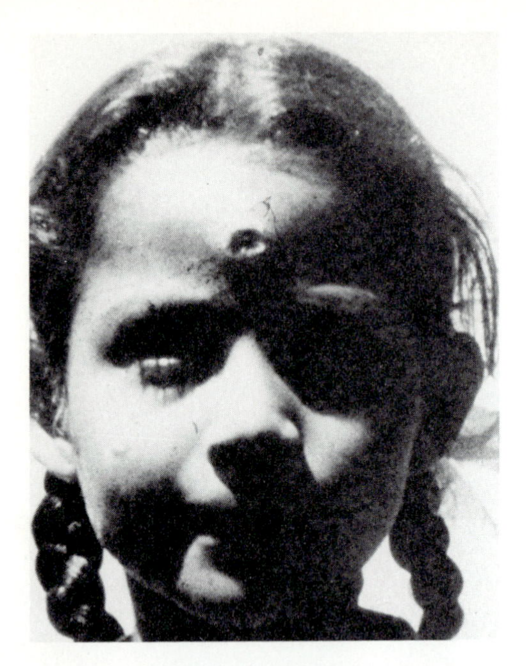

Figure 6-31 *Cutaneous lesion (cutaneous leishmaniasis) caused by* Leishmania *species.*

Characteristics of the major groups of protozoa

Sarcomastigophora. Members of the Sarcomastigophora group include not only flagellates and ameboid forms but species that can utilize both methods of movement. Flagella, when present, may be single or many and can be used not only for movement but for feeding, as a sensory organ, and for surface cleaning, attachment, and gamete locomotion.

Flagellates are heterotrophic in their nutrition. Some species are found in water as solitary or colonial forms, whereas others are commensals or parasites of vertebrates and invertebrate hosts such as snails, insects, mollusks, nematodes, and humans. Some diseases of humans are caused by protozoans transmitted by insect vectors. For example, leishmaniasis (Figure 6-31) is caused by a protozoan transmitted to humans by sandflies. The sandfly acquires the disease agent by feeding on an infected human and transmits it to other humans via biting. The genera of flagellates that are parasitic to humans are *Giardia, Trichomonas, Leishmania,* and *Trypanosoma*. Certain species of *Trichomonas* can also infect poultry and wild birds, causing interference with feeding and sometimes death.

The ameboid members of the Sarcomastigophora utilize the pseudopodial movements of the cytoplasm not only for movement but to capture food. Amebas feed on bacteria, algae diatoms, and ciliated and flagellated protozoa. They are found in both fresh and marine waters, and some parasitic forms can be found in the intestinal tract of both vertebrates and invertebrates. *Entameba gingivalis* is a commensal in the oral cavity of humans, whereas *Entameba coli* is part of the intestinal microflora. *Entameba histolytica*, however, can cause dysentery (amebic dysentery) in humans. The life cycle of this organism is illustrated in Figure 6-32. The pseudopodia produced by ameboid forms have characteristics that are useful in classification (Figure 6-33). Some species secrete a **shell (test)**, or skeleton, that is also used for identification purposes (Figure 6-34). The foraminifera, which produce calcium carbonate shells that are useful in acquiring geological data, are almost exclusively marine organisms that move slowly along the ocean floor or float in the water. They produce shells of various shapes and exhibit colors whose hues depend on the mineral content of the ocean environment. Some of the larger species contain algal symbionts.

Sporozoa. Sporozoa do not have flagella (except for some gametes), and movement is usually accomplished by flexing the body, which creates an undulating movement. The sporozoa are all **parasites** and infect nearly all species of the animal kingdom. They are responsible for large economic losses in cattle and poultry, as well as serious infections of wildlife. Human diseases, such as malaria and toxoplasmosis, are also caused by members of this group (Table 6-4). Malaria is discussed in Chapter 28. Toxoplasmosis is an important human disease because domestic cats are the source of cysts that are infective for humans (Figure 6-35). Sandboxes, as well as soil in the yard, are areas where infected cat feces are deposited. Fecal contamination of food, water, or hands can lead to infections in humans. The disease in humans is not particularly debilitating but the organism can be transmitted from an infected pregnant woman to the fetus in whom infections are disastrous. Fetal infections may result in stillbirths or serious involvement of the central nervous system, eyes, and viscera. A special sporozoan is *Pneumocystis carinii*. This organism appears to be present in the lungs of most humans and can infect many kinds of mammals. It causes severe pneumonia in those whose immune system has been depressed, such as AIDS patients (see p. 756).

Nearly all members of Sporozoa exhibit a spore-forming stage called **oocyst**. Sporozoa life cycles usually involve three phases.

1. The infective stage of sporozoa is called a **sporozoite**, which invades the host. Some sporozoites, which develop in the mosquito and are transmitted to humans via a bite, undergo multiple fission in the human host, producing uninucleate cells called **merozoites** (See Figure 6-29).
2. Merozoites differentiate into gametes (male microgametocyte and female macrogametocyte).
3. **Gametocytes** fuse to produce a zygote. Meiosis in

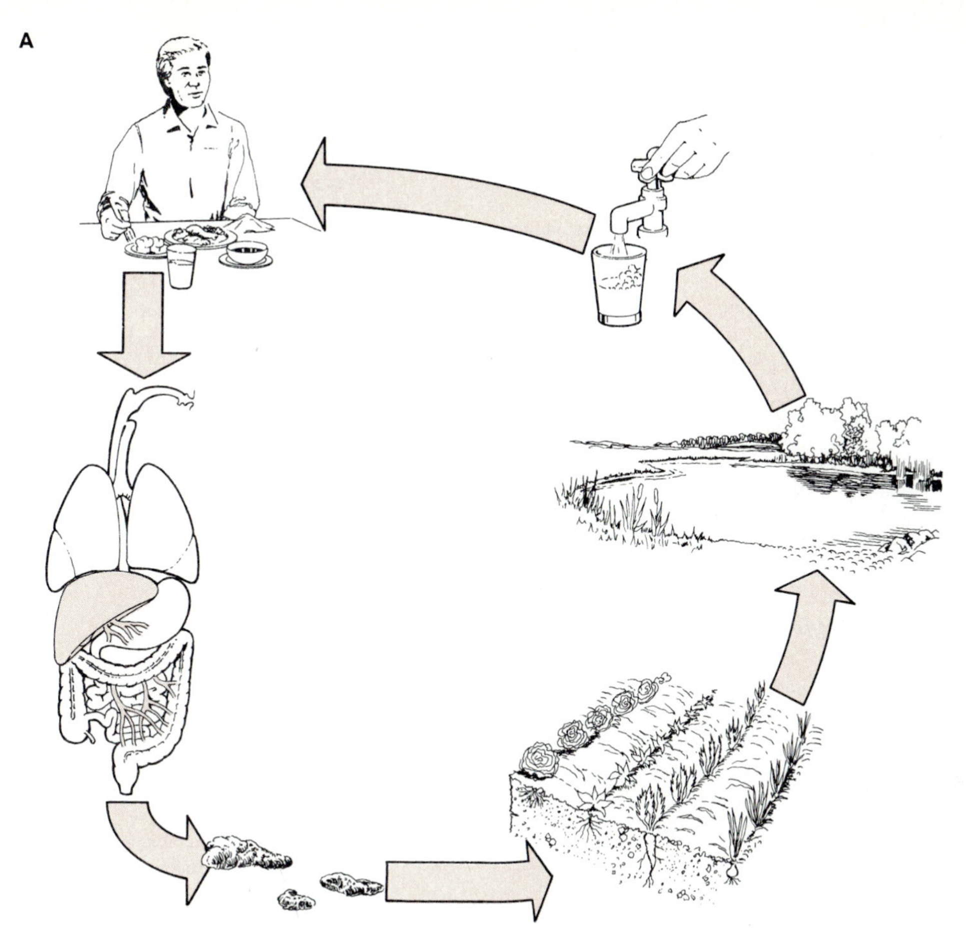

Figure 6-32 Life cycle of Entamoeba histolytica. Food and water contaminated with infective cysts are ingested. Cysts develop into trophozoites in intestinal tract, which can enter circulation. Cysts are also produced in intestinal tract and these are passed in feces to contaminate water and food until ingested by humans.

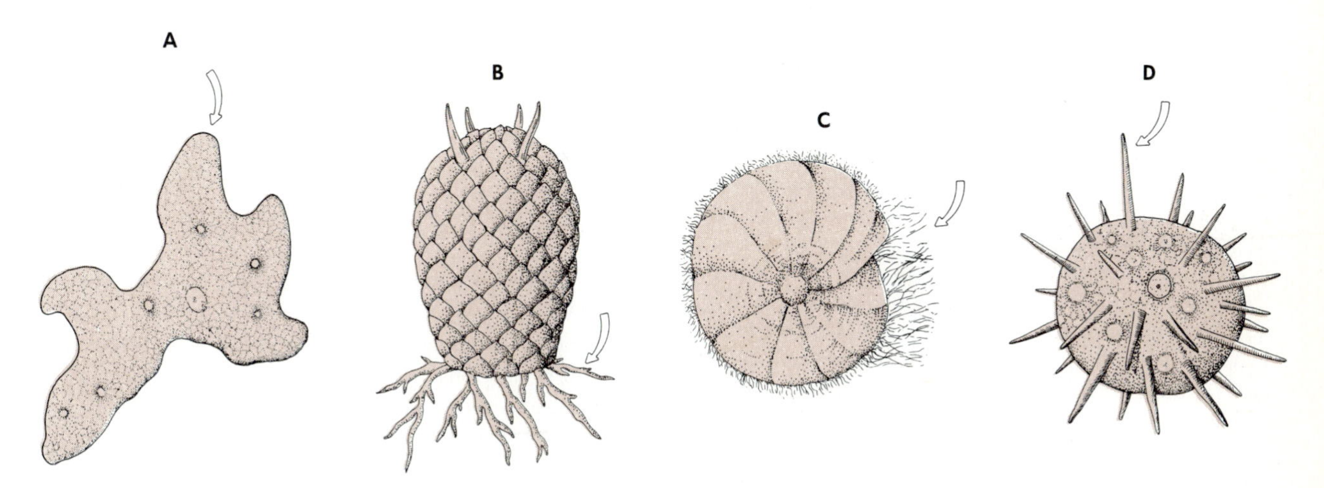

Figure 6-33 Pseudopodia (arrows) produced by ameboid forms of protozoa. **A,** Lobopodia—blunt and rounded as in amebas. **B,** Filopodia—slender and tapered to a point. **C,** Myxopodia—slender pseudopodia anastamose to form mesh. **D,** Axopodia—axial filaments about which cytoplasm flows.

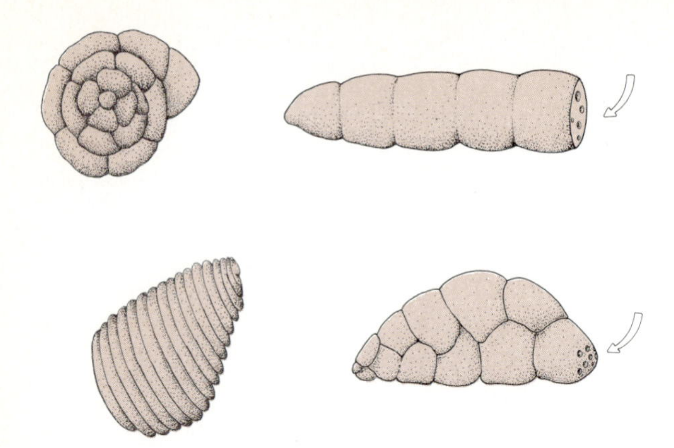

Figure 6-34 *Shells (tests) of some foraminifera. Shells contain pores at tips* (arrows) *through which cytoplasmic extensions flow.*

the zygote results in the formation of oocysts, which engage in multiple fissions and the formation of many uninucleate sporozoites (a repeat of the cycle).

Myxospora. Myxospora are parasites of invertebrates and lower vertebrates, especially amphibians, fish, mollusks, and insects. In some instances there is no cure for infection, and the animals must be destroyed. Spores are produced within cells that are lodged in various tissues and organs of the host. The spores possess unique structures called **polar capsules** (Figure 6-36), which contain filaments that are either coiled within the capsule or extruding from the cell. After the host ingests these spores, the polar filaments are extruded and used to attach to the host tissue, such as the epithelium of the intestinal tract. During this period of attachment the ameboid unit within the spore, called a **spiroplasm**, is released, and it migrates to specific organs where it develops into a multinucleated unit. One preferred site is the cartilage of the head and spine of fish. Infection results in an interference with coordinated swimming, and the fish often falls victim to predators. If the fish survives, deformities appear in the head (Figure 6-37).

Ciliophora. Ciliophora possess both simple and complex ciliary organelles whose distribution and organization on the cell have been useful for taxonomic purposes. Cilia completely cover some species and function together in a concerted manner for movement. Cilia in some species may be more concentrated near the oral groove where their movement aids in the feeding process. Some cilia fuse to form an undulating membrane that is used to generate a current of water through the oral cavity, bringing particles near the cytosome. Examples of ciliated species are illustrated in Figure 6-38.

Ciliophora are unique in that they possess two different nuclei: a **macronucleus** and a **micronucleus**. Micronuclei, which are small and spherical, can give rise to

Table 6-4 *Members of the Sporozoa and the hosts they infect*

Species	Infected host
Invertebrate	
Monocystis lumbrici	Earth worm
Gregarina polymorpha	Meal worm
Haemogregarina stepanowi	Leech
Vertebrate	
Eimeria tenella	Chicken
Toxoplasma gondii	Rodents, pigs, many mammals, and humans (toxoplasmosis)
Sarcocystis sp.	Herbivores, such as cattle
Levineia sp.	Dogs and cats
Plasmodium sp.	Humans (malaria)
Haemoproteus sp.	Birds and reptiles
Leucocytozoon simondi	Ducks, geese, and swans
Babesia bigemina	Cattle (babesiosis or Texas red-water fever)
Babesia microti	Deer mice and various mammals; human infection called Nantucket fever
Theileria parva	Cattle, zebu, and Cape buffalo (East Coast fever)
Isospora belli	Many vertebrates, including humans
Pneumocystis carinii	Human infection (pneumonia) in those whose immune system is depressed (AIDS patients)
Cryptosporidium sp.	Farm and domestic animals and humans (cryptosporidosis)

Concept Check

1. Explain how the nutrition of protozoa differs from that of fungi and bacteria. What structures are involved?
2. Explain the protozoan reproductive processes of multiple fission and conjugation.
3. What are the four major groups of protozoa and their beneficial and detrimental contributions to the environment?
4. What is unique about the nucleus of ciliophora in the sexual cycle?

macronuclei. Micronuclei are diploid and contain genetic information that can be exchanged during sexual conjugation. Macronuclei, which are large and variable in shape, are polyploid (containing several sets of chromosomes) and function in the control of metabolic processes in the cell. The micronucleus undergoes mitotic division when ciliates divide by binary fission, but the macronucleus merely elongates and is halved as the cell divides.

The ciliates can be found in a variety of watery habitats including bogs, fresh and marine waters, and polluted environments. Those ciliates that form cysts are found primarily in soil environments. Ciliophora swim freely in their environment, but a few are sessile and remain attached to substrates. Only a few species are parasitic to marine or freshwater animals, and only one species (*Balantidium coli*) is parasitic to humans. Humans who are in contact with swine are more likely to acquire disease by this organism. Humans are infected by ingestion of cysts released by infected animals. The ciliate excysts in the cecum or colon and begins to feed on the intestinal epithelium. An ulcer develops that can lead to hemorrhage and perforation of the intestine.

Figure 6-35 *Epidemiology of toxoplasmosis. Domestic cats pass cysts in feces, which are ingested by humans and various animals. Flies, cockroaches, and other hosts can disseminate cysts in environment. Soiled hands in mouth and contaminated food and water lead to infection. Insufficiently cooked meat from domestic animals containing asexual stages of parasite also results in infection.* Toxoplasma gondii *is transmitted to offspring through placenta during pregnancy.*

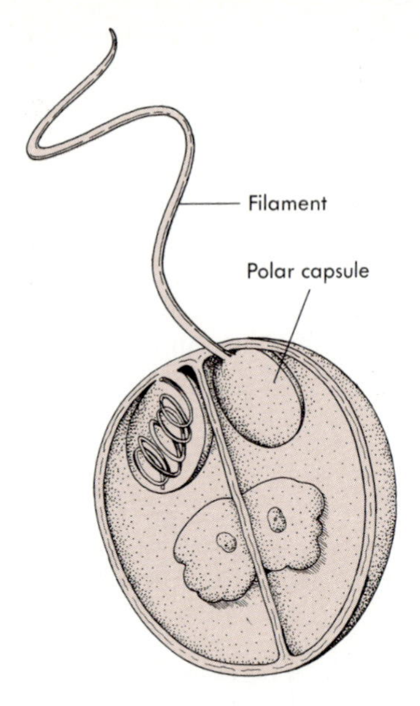

Figure 6-36 *Polar capsules found in spores of Myxospora. Note one filament is extruded from polar capsule.*

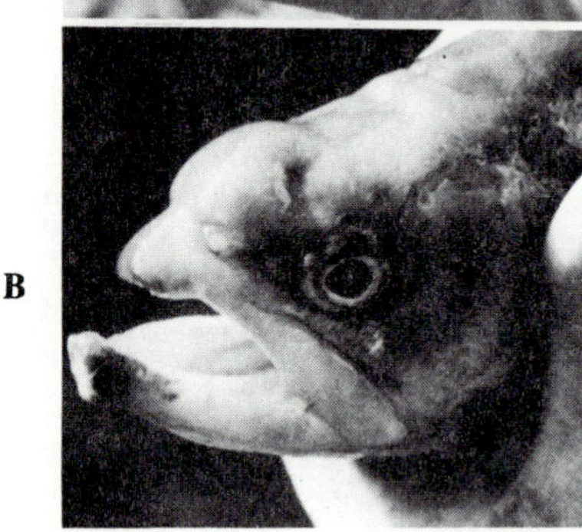

Figure 6-37 *Axial skeleton deformities in living rainbow trout that has recovered from infection by Myxospora.* ***A,*** *Note bulging eyes and curvature of spinal column.* ***B,*** *Note deformity on cranium.*

CHARACTERISTICS OF THE ALGAE

The study of algae is called **phycology** (Gr. *phykos,* seaweed). The algae are a diversified group of organisms that are difficult to categorize. Algae are primarily nonvascular plantlike organisms that possess chlorophyll *a* and whose reproductive structures are relatively simple. There are exceptions to this definition that are discussed later.

The principal role of algae in the environment is **producer**; that is, they use light energy to convert carbon dioxide and water into chemical energy in the form of carbohydrates $(CH_2O)_n$. A simplified equation for this photosynthetic process is:

$$CO_2 + H_2O \xrightarrow{\text{light}} (CH_2O)_n + O_2$$

The energy rich molecules found in the algae can be transferred to grazing organisms (**consumers**) on land and in the sea. Oxygen is a byproduct of photosynthesis and thus algae play an important role in supplying this gas to those organisms that require it in their life cycle.

Algae are found in all aquatic environments, including soil, when sufficient moisture and light are present. Most algae are free living but some form symbioses with plants, animals, protozoa, and fungi. The algal requirement for light precludes their development in areas without light; however, there are some exceptions to this rule.

The algae are classified primarily according to the pigments of the organism. They all contain chlorophyll *a* but other pigments in the cell may mask the green color with which we associate plant chlorophyll. A classification scheme is provided in Table 6–5 but our attention is devoted exclusively to the unicellular members of the Euglenophyta, Chlorophyta, Bacillariophyta, and Pyrrophyta.

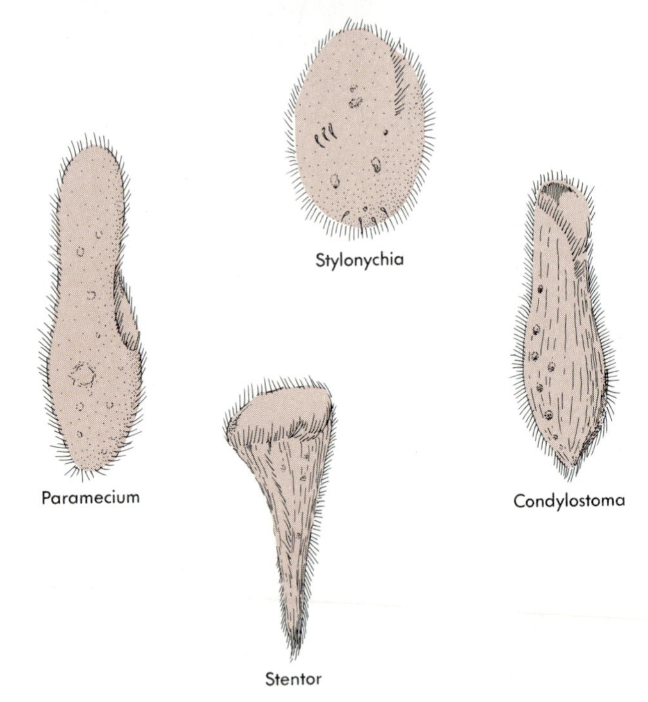

Figure 6-38 *Examples of ciliates.*

Table 6-5 *Major divisions of the algae and their characteristics*

Characteristics	Euglenophyta	Chlorophyta	Bacillariophyta	Pyrrophyta	Phaeophyta	Rhodophyta
Common name	Euglenoids	Green algae	Diatoms	Dinoflagellates	Brown algae	Red algae
Morphology	Most unicellular	Unicellular, colonial, multicellular, and filamentous	Most unicellular	Unicellular	Multicellular	Most multicellular
Chlorophylls	a and b	a and b	a and c	a and c	a and c	a
Other pigments	Carotenoids, xanthophylls	Carotenoids, xanthophylls	Fucoxanthin	Carotenoids, peridinin	Carotenoids, fucoxanthin	Phycobilins, carotenoids, xanthophylls
Cell covering	Pellicle	Cellulose when wall present	Silicon dioxide, protein, and carbohydrate	None, or cellulose thecal plate under cytoplasmic membrane	Cellulose and algenic acid	Cellulose, or mannan microfibrils embedded in mucilage such as agars or car rageenans, both composed of galactose; some deposit $CaCO_3$ in walls
Storage products	Paramylon and fats	Starch in pyrenoid	Chrysolaminarin	Starch and complex oils	Laminarin	Floridean starch
Other characteristics	Colorless species common, freshwater	90% freshwater species	Half marine and other half fresh water; important producers of oxygen	Primarily plankton in oceans; some parasites of animals, symbionts of marine animals (corals); some cause blooms as "red tide"	Most marine; commercial source of alginates used in foods	Most species marine; commercial source of agar for microbiology laboratories and for foods; carrageenans used as gels in ice cream and puddings

Morphology

Algae range in size from 3 to 20 μm in diameter for unicellular types to the giant multicellular kelp that may reach a length of over 200 feet. Unicellular types are common among the primitive algae and some will divide by mitosis without complete separation, forming a multicellular colonial type (Figure 6-39) of organism.

Flagellar motility is especially common among the unicellular algae and occurs in a spiraling manner. Anywhere from one to four flagella may be present on the cell. Those species that do not have motile adult stages usually produce motile gametes. Fusion of motile gametes in sessile forms is characterized by resorption of the flagella, but later, when the adult produces gametes, the flagella become reassembled. Structurally, the flagella of algae are typical of other eukaryotic flagella (see Chapter 7).

Diatoms (Bacillariophyta) are without flagella (except for male gametes of certain marine species), but gliding movement is still possible in some species because of mucous secretions. They have been observed to move at a rate of 2 to 14 μm/sec. This type of movement is similar to that found in certain bacteria, but the mechanism is controversial.

Algae, like most other eukaryotes, possess the usual intracellular structures but the chloroplast is the most distinctive structure in the cell. The membranes (thylakoids), which make up the chloroplast, contain various pigments that are required for the photosynthetic process. These pigments are also useful for differentiating the algae (Table 6-5). In addition the polymer storage products, which are synthesized as a consequence of the photosynthetic process, are also useful in differentiating algae. These polymers are composed of glucose in var-

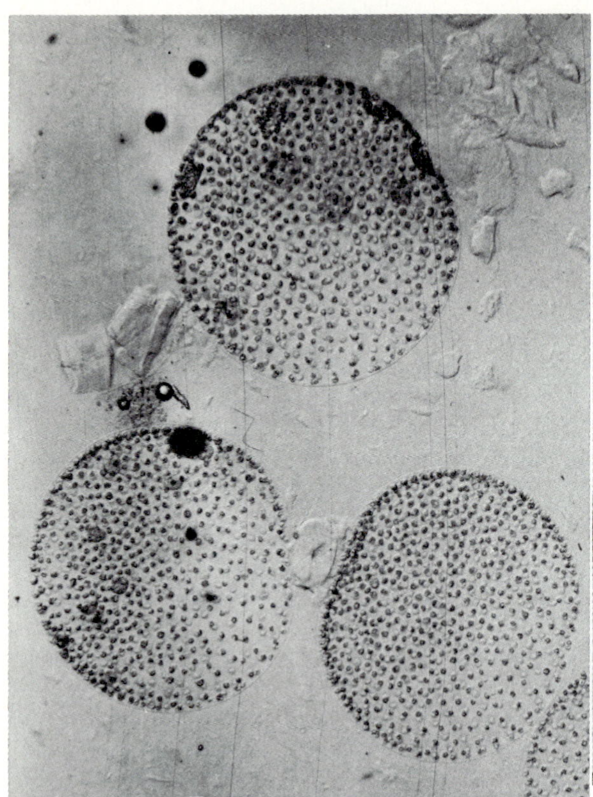

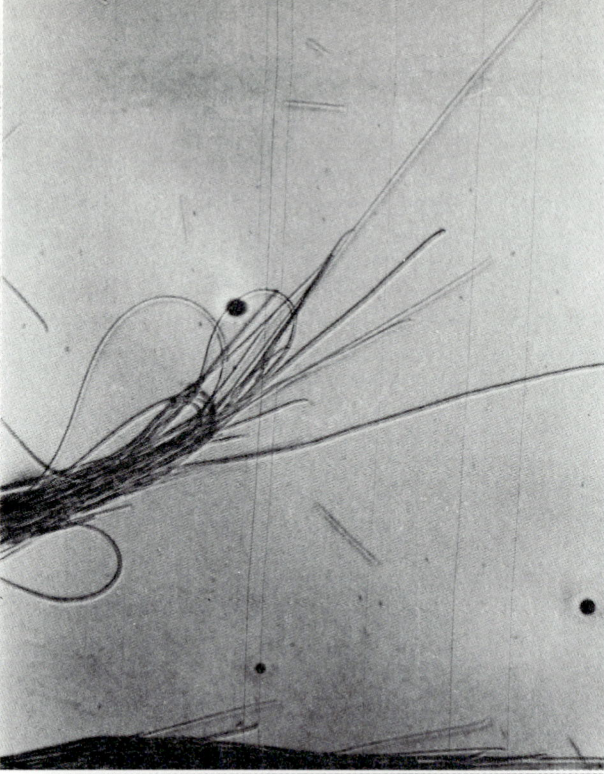

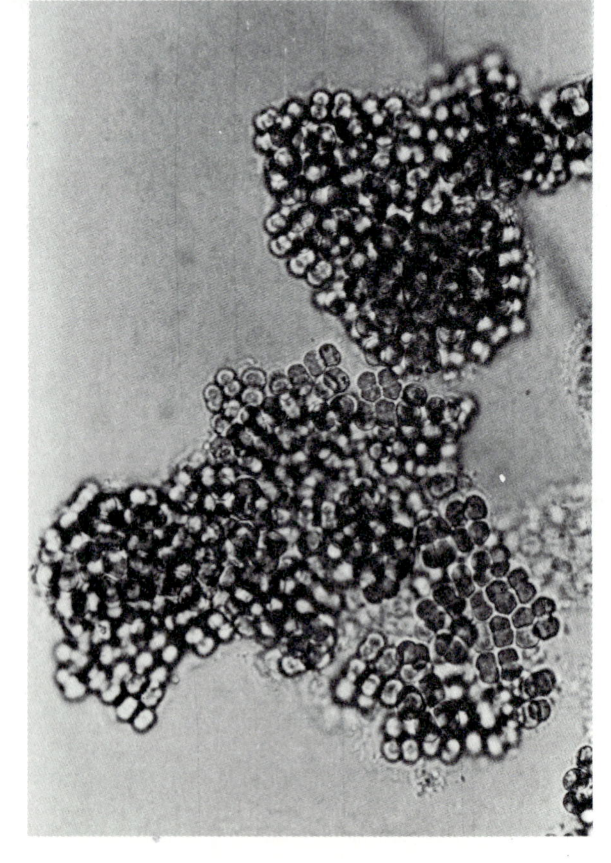

Figure 6-39 *Morphological types of microscopic algae.* ***A,*** *Colonial* (Volvox) *(×33).* ***B,*** *Filamentous* (Oscillatoria) *(×33).* ***C,*** *Sheets (*Merismopedia) *(×33).*

ious linear arrangements; for example, starch consists of glucose in an α-1-4, 1-6 arrangement, while paramylon consists of a β-1,3 glucose arrangement. Starch granules are formed on the surface of a dense region in the chloroplast called the **pyrenoid**.

Nutrition

As stated, most algae are characterized by their ability to use carbon dioxide as a carbon source and light as an energy source. There are species of algae that, although photosynthetic and autotrophic, still require certain preformed organic compounds, such as amino acids, vitamins, and other nutrients, to survive. There are even some colorless species that rely entirely on preformed organic molecules for energy. Some species of *Euglena,* although phototrophic, lose most of their chloroplasts when grown in the dark for long periods of time and are capable of a heterotrophic existence. Chloroplast formation is possible when the organisms are placed again in the light. Finally, some heterotrophic species behave like protozoa and ingest other living organisms (**phagotrophs**), especially other algae.

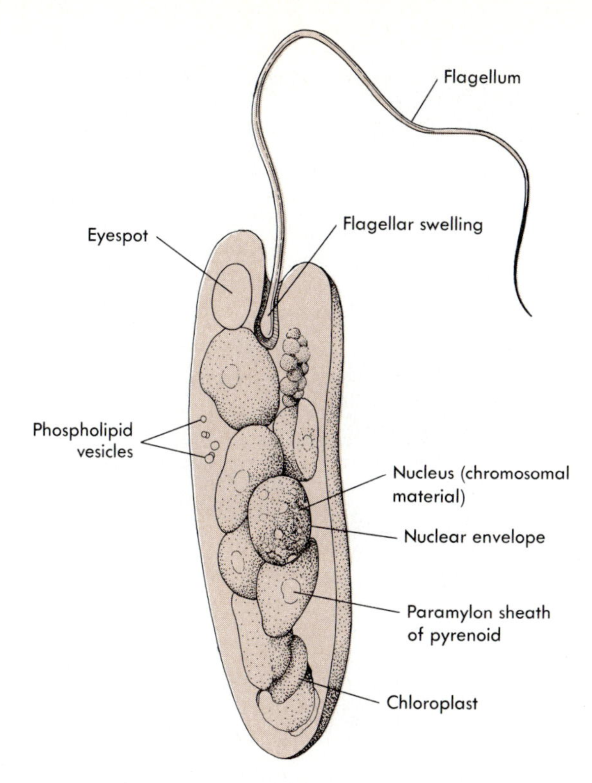

Figure 6-40 *Various structures in* Euglena, *a member of Euglenophyta.*

Reproduction

Both sexual and asexual reproduction occur in the algae, but some groups have no apparent sexual mechanisms. Asexual reproduction in unicellular types may be by binary fission as in bacteria in which division is followed by longitudinal division of the cytoplasm. Many multicellular forms of algae reproduce by fragmentation.

In most instances during sexual reproduction the male and female gametes are motile, and sexual union occurs by random collision. Gamete collision in one species results in a cohesive reaction between their respective flagella, and this is soon followed by nuclear fusion. In other groups only the male gamete is motile, and it is attracted to the nonmotile female gamete by chemotactic mechanisms. For example, in the colonial algae called *Volvox,* a chemical factor is produced in the male colony that causes asexual colonies to produce male and female gametes. The chemotactic factor is produced in response to nutrient depletion and decreased light intensity.

Characteristics of the unicellular algal groups

Euglenophyta. Members of the division Euglenophyta are called **euglenoids.** They are an unusual group because they exhibit both animal and plant characteristics. They are unicellular and lack a cell wall but possess a tough flexible pellicle located underneath the cytoplasmic membrane. They contain chlorophyll *a* and *b* but none are capable of a total photosynthetic nutrition and all species require some preformed organic nutrients. Most algal species require at least one vitamin, and for this reason some species have been used to quantitate vitamins in solutions. A few species of euglenoids are phagotrophs and obtain their nutrients by ingesting other organisms, usually algae. The euglenoids are found primarily in fresh water where high concentrations of organic material are found, such as in farm ponds, but some species are marine.

Encystment is characteristic of the euglenoids and is the result of mucilage production by muciferous glands. A wall-like material is also produced during encystment that surrounds the cyst and protects the internal protoplasm. The germination of cysts in ditches and ponds is responsible for the unmistakable green and red blooms observed after heavy rains. Euglenoids possess an eyespot (Figure 6-40) that contains concentrations of red or orange pigments. The eyespot appears to act as a shading device for the major light receptor in the cell, which is the flagellar swelling located near the base of the flagellum. The eyespot apparently controls the direction in which the organism moves. Euglenoids move toward light (positive phototaxis) of low intensity and away from light (negative phototaxis) of high intensity. These phototactic responses are particularly evident in those euglenoids found in tidal mud flats. The organisms respond negatively to light at high tide and burrow into the mud, whereas at low tide they respond positively and migrate to the surface.

One species of *Euglena (Euglena sanguinea)* produces a red pigment, **astaxanthin**, which colors the entire organism when the light intensity is high. Large populations of these organisms are present in water high in organic nutrients and are responsible for the appearance of its red color. It has been suspected that the biblical plague in which the Nile River turned blood red may have been caused by this organism.

One of the heterotrophic euglenoids lives in the gut of dragonfly nymphs in the winter, causing their posterior to appear green. In the spring the algae leave the dragonfly rectum as free-swimming cells.

Euglenoids have some economic significance. A few species dominate some sewage oxidation ponds where they degrade organic materials. When large populations of euglenoids become associated with community water supplies, they impart a fishy odor. Euglenoids are an important laboratory tool for research on nucleic acids, chloroplasts, and mitochondria. Only one species, *Euglena gracilis,* has been successfully cultivated in pure culture, and, because of its plant and animal characteristics, is an ideal research tool. *E. gracilis,* because of its

requirement for vitamin B_{12}, has been used for vitamin assays, whereas other euglenoids because of their specific nutrient requirements have been used in various nutritional studies.

Chlorophyta. The Chlorophyta, or green algae, are for the most part unicellular organisms, but colonial forms, such as *Volvox,* multicellular forms such as *Ulva,* and flamentous types such as *Oedogonium* are also part of this group. They are more frequently found in shallow waters, on tree trunks and sheltered sides of buildings, and in snow, ice, and soil. Some species also appear in marine environments. In other words, any environmental niche that has inorganic nutrients, is moist, and receives sufficient light will support their existence. They are not usually found in deep water because their pigments, chlorophyll *a* and *b,* absorb the red (650 nm) and blue (450 nm) wavelengths that do not penetrate deeply into waters that are cloudy. In clearer tropical waters light can penetrate more deeply, and some green algae are found at depths of up to 100 to 200 m.

The unicellular green algae are one of the predominant phytoplankton found in fresh waters and thus are the major primary producers. The dense mats of algae found in confined bodies of water such as ponds and fish tanks are green algae. Some species of *Chlamydomonas* are found in the snows of some mountains in North America; one species gives a red color to the snowfields. An unusual relationship exists between an alga and the sea slug. Slugs graze on species of *Codium* and *Caulerpa* and their chloroplasts are retained in the digestive system. The chloroplasts apparently continue to photosynthesize and the photosynthetic product is used by the slug. Several species of *Chlorella* have also evolved a relationship with protozoa, such as the *Paramecium,* sea anemones, various marine flatworms, and others. This association usually results in the animal being slightly green in color.

Some unicellular species, such as species of *Chlorella* and *Chlamydomonas,* are used in sewage treatment facilities. The algae help bacterial decomposition by providing oxygen, and they also recover mineral nutrients from the sewage that would otherwise be lost in the effluent. The excess algae can be used to feed livestock. Humans, however, cannot digest *Chlorella* well and its use as a food supplement is not widespread. The Japanese and Chinese use species of *Chlorella, Ulva,* and others in soups and for flavoring meat and other commercial products. The green algae are of little medical interest, although a few species parasitize the leaves of certain plants such as *Magnolia* and *Rhodedendron* and *Thea* (tea leaves). Species of *Protheca,* an inhabitant of soil, can cause a dermatologic disease in humans called **protothecosis.** Occasionally severe disease can occur in cattle who ingest large numbers of these algae.

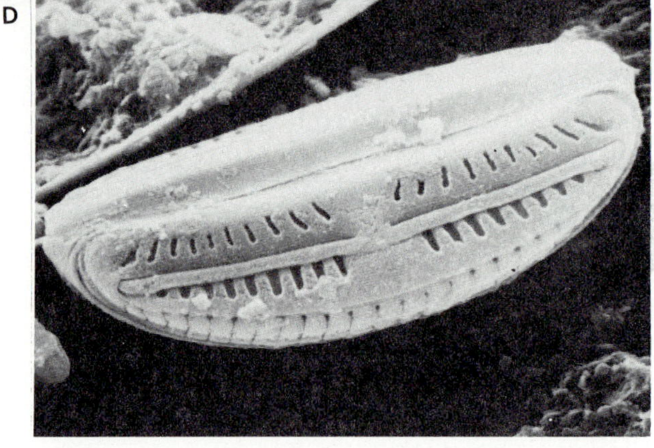

Figure 6-41 *Morphological types of diatoms.* **A,** Navicula scutiloides. *(×2200.)* **B,** Surirella *species. (×500.)* **C,** Coloneis lewissii. *(×1200.)* **D.** Amphora *species. (×5100.)*

The green algae have been the most widely studied because of their size and ease of cultivation in the laboratory. They also have been studied because they are believed to be the direct ancestors of green land plants. The characteristics they have in common with the higher plants are (1) cellulose in the cell wall, (2) chlo-

Figure 6-42 *Scanning electron micrograph of centric diatom* (Cyclotella meneghiniana) *demonstrating frustules* (F). *Note marginal spines* (arrows) *and ornamentation of mantle* (M).

rophyll *a* and *b,* (3) starch as a storage product, and (4) similar chloroplast structure.

Bacillariophyta. The Bacillariophyta are primarily unicellular and are characterized by the presence of large quantities of the brown pigment fucoxanthin, a xanthophyll. The **diatoms** are perhaps the most important class in this division and appear in a variety of striking shapes (Figure 6-41). They are important as primary producers in marine environments. The siliceous cell walls of diatoms are composed of two overlapping halves called **frustules**, which give the diatom the appearance of a pill box or Petri dish (Figure 6-42).

The diatoms are divided into two groups based on the symmetry of their cell walls; radial symmetry and bilateral symmetry. In motile diatoms with bilateral symmetry a furrow called a **raphe** extends between the poles of the cell (Figure 6-43). The secretion of a mucilagenous substance through the raphe is believed to be involved in a gliding movement. The sticky substance attaches to the substrate and during locomotion leaves behind a slime trail.

The peculiar nature of the cell wall of diatoms dictates that the cell undergo an unusual division process. The overlapping valves during mitosis come apart, and each produces a new bottom (Figure 6-44). Thus, the bottom part of the parent cell becomes the top part of the daughter cell. Successive division results in a population of cells, many of which are considerably smaller than the original parent cell. It appears that when a critical minimal size is reached, a specialized growth spore is formed from the union of two gametes in a sexual process. The growth spore is called an **auxospore**, which, following germination, produces a diatom whose size is the same as the original parent.

Diatoms are a major component of freshwater and marine phytoplankton. As primary producers they must remain near the surface in a zone that receives sufficient light (photic zone). Diatoms have a cytoplasmic density

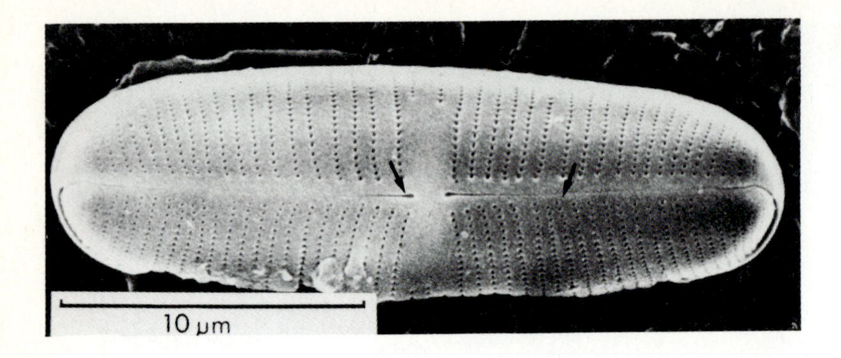

***Figure 6-43** Scanning electron micrograph of motile diatom demonstrating furrow called raphe* (arrows), *which extends between two poles.*

of 1.03 to 1.10 g/cc, which makes them more dense than sea water, yet they are still capable of buoyancy. Buoyancy appears to be provided by a large vacuole that excludes heavier ions, and thus its contents are at a lower density than sea water. This type of mechanism does not work in fresh water, but some colonial diatoms are capable of flotation because of a gelatinous sac that surrounds the aggregate.

The most important contribution of diatoms is their role as primary producers in fresh and marine waters where they act as food for the various aquatic animals. When the diatom dies, the silicon-containing walls are deposited in the sediment of marine and fresh waters. These deposits are called **diatomaceous earth** (Figure 6-45). Geological changes have raised these deposits above the surface of seas so they can be quarried. Certain diatoms have also been used as markers for geological deposits to identify potential oil deposits and gas-bearing strata. Diatomaceous earth is used primarily for industrial processes, such as the filtration of fluids (wine, beer, and antibiotics). They can also be used as abrasives in toothpaste and metal polishes.

Diatoms vary considerably in their response to environmental conditions and because of this have been used to determine the quality of water with respect to such factors as alkalinity, acidity, temperature, and pollution. Certain species, for example, are present only at certain alkalinities, and by finding which species is present, an estimate of the alkalinity of water can be obtained.

Pyrrophyta. The Pyrrophyta, or dinoflagellates, have both plant and animal characteristics. They are usually motile organisms found in both fresh and marine waters. The cells are usually flattened and have a transverse constriction around the equator of the cell called the **girdle** (Figure 6-46). Flagella are inserted in the girdle; one circles the cell and is flat, and the other trails the cell.

The cell covering (**thecal plate**) of the dinoflagellates is a cytoplasmic membrane beneath which lies a layer of flattened vesicles. The vesicles contain plates of cellulose that give the cell covering its characteristic structure. The plates may be unornamental, or they may contain a series of cellulose plates that overlap and are held together at their joints. The plates appear as horns or wings. Few species have a true cell wall.

Pyrrophyta are second only to the diatoms as primary producers. Blooms of certain species *(Gonyaulax* and *Gymnodinium)* are responsible for environmental hazards, such as the **red tide**. Several factors are believed to control blooms, including temperature, salinity, pH, illumination, and trace elements in the water. Initially, the bloom gives the water a yellow color, which is followed by brown and then red coloration. These color changes are caused by changes in pigment concentration within the cells. As many as 10^8 cells can be found per liter of water with blooms extending over several kilometers. A toxin produced by some of the species is often released as the cells break up during passage through the gills of fish. The toxin adversely affects fish, and as many as 50 million have been killed from one bloom in the Gulf of Mexico. Paralytic shellfish poisoning in humans occurs when shellfish are eaten that have filtered a toxic species of dinoflagellate out of the water. The toxin concentrates in the tissues of the shellfish. When shellfish are eaten by humans the toxin is released and blocks nerve transmission. Over 220 human fatalities from respiratory paralysis have been recorded following shellfish poisoning. The toxin in some dinoflagellates is heat stable and can remain within the shellfish for up to a year after the algal bloom has disappeared.

Certain dinoflagellates form relationships with animal coral. The coral provides the alga with protection, shade, carbon dioxide from its own respiratory processes, and waste products containing nitrogen and phosphorous. The alga in turn provides the coral with oxygen, waste removal, and carbohydrates. The coral relies heavily on this association because the areas where coral is found are nutrient poor. The coral is restricted to shallow waters where light can penetrate and be used by the algal cells.

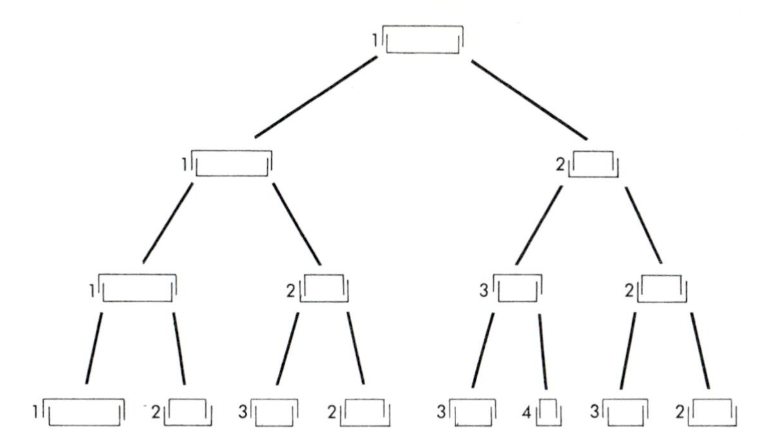

Figure 6-44 Diagrammatic representation of progressive diminution in average size of cell frustules in any diatom population. Each top of pillbox synthesizes new bottom and each bottom synthesizes new bottom (thus becomes top of new cell). Over extended period of time population cell size has diminished.

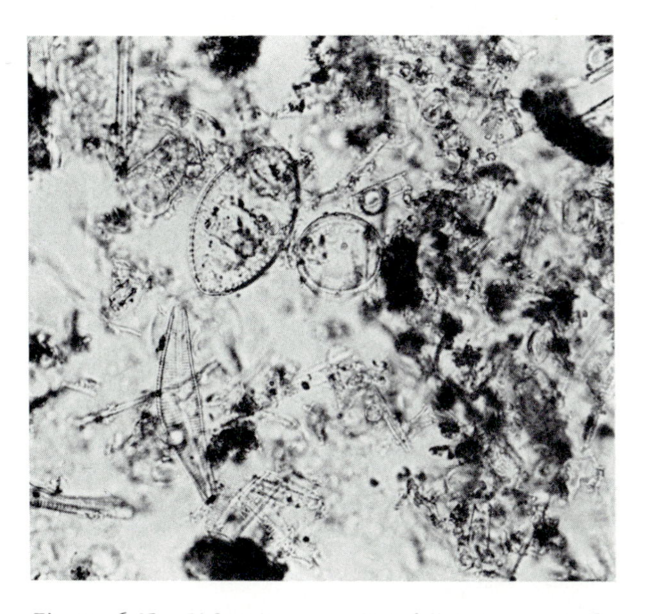

Figure 6-45 Light microscope view of diatomaceous earth mounted in water. Individual diatoms are visible. (×250.)

Figure 6-46 Scanning electron micrograph of dinoflagellate. Girdle (G) surrounds cell, which is covered with thecal plates (TP). Transverse flagellum (TF) is inserted in girdle; posterior flagellum (P) is partially seen.

Concept Check

1. What characteristics separate the algae from all other eukaryotes?
2. What morphological characteristic of the euglenoids makes them particularly distinctive? What is the economic value of euglenoids?
3. What are the major subdivisions of unicellular algae, what is their role in nature or their commercial value, and where are they found in nature?

SUMMARY

A brief comparison of the algae, fungi, and protozoa is presented in Table 6-6.

Fungi

1. The hypha is the vegetative unit of those fungi referred to as filamentous, whereas the nearly spherical fungi are called yeasts. A few species are dimorphic and can exist in either state.

Table 6-6 *Comparison of eukaryotic microorganisms*

Characteristic	Fungi	Protozoa	Algae
Cell wall	Present (chitin)	Absent, although in some, cytoplasmic membrane forms thick pellicle	Present (cellulose)
Motility	Primarily nonmotile except for flagellated sex cells	Many flagellated and ciliated species	Many flagellated species
Nutrition	Primarily dead organic matter	Ingest living or dead organic matter	Use energy of sun (photosynthesis) to produce organic matter
Reproduction	Asexual by spore formation is the most frequent form	Asexual by binary fission the most frequent form	Both asexual and sexual are frequent forms
Symbiosis	Lichen (alga and fungus) frequent type. Fungi and plant roots (mycorrhizae) also a frequent type	Protozoa and alga—a frequent type	Lichen most frequent type
Ecological significance	Decomposition of organic matter in soil; mycorrhizae important in plant nutrition	Control levels of other microorganisms by feeding on them	Source of food for other organisms; supplies oxygen to atmosphere
Parasitism and disease	Infect many plants, animals, and humans to cause disease	Cause many fatal diseases in humans and animals	Not parasitic but some species produce toxins, which when ingested by invertebrates and humans can cause death
Economic importance	Source of important compounds, including enzymes, antibiotics, acids, and alcohol; food source (truffles); production of beer and wine; fermentation of foods	A few species important in sewage treatment	Algal blooms detrimental to water supplies; source of food

2. The spore is the basic unit of reproduction and can be reproduced asexually or sexually. Asexual spores may arise directly from hypha or they may be produced on specialized structures.
3. Fungi are separated into two divisions: true fungi and slime molds. The true fungi are further separated into five subdivisions. The Zygomycotina are terrestrial species, many of which parasitize plants, animals, and humans. The Mastigomycotina, which includes aquatic and terrestrial species, possess a flagellum that distinguishes them from all other fungi. The Ascomycotina are characterized by a spore-bearing sac called the ascus. The Deuteromycotina have no apparent sexual stage. Most of the fungi pathogenic to humans are included in this group. The Basidiomycotina are characterized by a spore-bearing structure called the basidium, which in some species is enclosed by a fruiting body, the basidiocarp. The slime molds possess protozoan and fungal qualities.
4. Fungi perform an important function in the environment: they degrade organic matter. Many fungi are parasites for a variety of animals, plants, and humans. Several fungi have industrial importance in the production of antibiotics, acids, and alcohols.

Protozoa

1. Protozoa are without cell walls and have evolved various surface structures to support the cytoplasmic membrane. The plasticity of the cytoplasmic membrane has permitted their differentiation into various morphological types.
2. The most common form of reproduction is binary fission. Sexual reproduction does occur but is influenced by environmental factors. Two major types of sexual reproduction occur: conjugation, which is characteristic of ciliated protozoa and is carried out by the union of two parental cells; and syngamy, which is the fusion of free-swimming gametes. Cysts are part of the sexual and asexual reproductive cycles. They can serve in a reproductive capacity or be used as a unit of transmission.
3. The major subdivisions of protozoa include the Sarcomastigophora, which are either ameboid or flagellated. The Sporozoa are all parasites whose infective stage is called a sporozoite. The Myxospora are parasites of fish, amphibians, and insects. The Ciliophora are characterized by cilia covering all or parts of the organism. They possess two nuclei: a macronucleus and a micronucleus that control metabolism and propagation, respectively.
4. Protozoa are found in most aquatic environments, including sewage treatment facilities. They make up part of the microbial flora of the intestinal tracts of ruminants where they are believed to supply lipids to the host animal. Several species cause life-threatening diseases, such as malaria.

Algae

1. Algae are photosynthetic aquatic organisms, most of which possess thick cell walls. They are primary producers and serve as the first link in the food chain.
2. Most algae are motile by flagella but some (diatoms) secrete a slime layer that provides some movement.
3. Both sexual and asexual reproduction occur in algae. Asexual reproduction in unicellular forms is primarily by binary fission.
4. Algae are often endosymbionts with other organisms and supply nutrients in the form of photosynthetically-produced carbohydrates.
5. The principal unicellular algae belong to the following groups:

   The Euglenophyta lack a cell wall and are primarily freshwater organisms. They give color to algal blooms.

   The Chlorophyta are primarily freshwater species. They are predominant phytoplankton in fresh water. They are believed to be direct ancestors of the green land plants.

   The Bacillariophyta are the primary producers in marine environments. Their cell wall is composed almost entirely of silicon dioxide.

   The Pyrrophyta are characterized by the modifications that occur in their cytoplasmic membrane. They are important as the cause of red tide and shellfish poisoning in humans.

Self-study Quiz

Completion

1. The fungal subdivision that contains the majority of human pathogens is ____________________.
2. The modified hyphal structure that is used to parasitize plant cells is called a ________________.
3. The cell covering of dinoflagellates is called the ______.
4. The ability of fungi to exist in either a yeast or filamentous state is called ________________.
5. The asexual reproductive process in protozoa in which the cell divides to produce large numbers of nuclei, which with a small amount of cytoplasm break off to produce individual uninucleate cells, is called _______.
6. Organisms that live off dead organic matter are referred to as ________________.

Multiple choice

1. Which of the following structures is found in both prokaryotes and eukaryotes?
   A. Mitochondria
   B. Golgi
   C. Chloroplasts
   D. Ribosomes
2. Which of the following characteristics is considered typical of protozoa?
   A. Possess cell walls
   B. Carry out photosynthesis
   C. Produce cysts that function as protective and reproductive devices
   D. Produce germ tubes
3. Which of the following would most likely leave a fossil?
   A. *Escherichia coli*
   B. Ascomycetes
   C. Diatoms
   D. Sporozoa
4. What coloration would you expect the water to take on during a bloom caused by Chlorophyta algae?
   A. Red
   B. Blue
   C. Green
   D. Brown
5. The mats of filamentous growth associated with molds are called?
   A. Saegenogens
   B. Coenobium
   C. Germ tubes
   D. Mycelia
6. Which of the following is a sexual mechanism of reproduction in protozoa?
   A. Conjugation
   B. Binary fission
   C. Fragmentation
   D. Schizogony

Identification

For the following microbial characteristics, structures, or diseases indicate that group of eukaryotes with which they are associated. Mark *A*, for algae; *P*, for protozoa; and *F*, for fungi.

1. ________ Sporangiospores
2. ________ Cell wall chitin
3. ________ Pellicle
4. ________ Rarely cause of human disease
5. ________ Flagellated members
6. ________ Source of the antibiotic penicillin
7. ________ Pyrenoid
8. ________ Basidium
9. ________ Mycorrhizae
10. ________ Slime molds
11. ________ Budding
12. ________ Malaria
13. ________ Athlete's foot

Thought questions

1. If you were given a sample of pond water, what procedures would you use to isolate protozoa, fungi, and algae? What criteria would you use in the identification of each group?
2. What possible explanation might you give for the inability of algae to infect humans?

SELECTED READINGS

Books

Alexopolous, C.J., and Mims, C.W. *Introductory mycology* (3rd ed.). John Wiley & Sons, Inc., New York, 1980.

Ashworth, J.M., and Dee, J. *The biology of slime molds.* Edward Arnold (Publishers) Ltd., London, 1975.

Baker, J.R. *The biology of the parasitic protozoa.* Edward Arnold (Publishers), Ltd., London, 1982.

Deacon, J.W. *Introduction to modern mycology* (3rd ed.). Blackwell Scientific Publications, Oxford, 1984.

Goff, L.J. *Algal Symbiosis.* Cambridge University Press, Cambridge, 1983.

Goodaa, G.W., Lloyd, D., and Trinici, A.P.J. (Eds.). *The eukaryotic microbial cell.* In the 30th Symposium of the Society for General Microbiology, Cambridge University Press, Cambridge, 1980.

Ingold, C.T. *The biology of fungi.* Hutchinson, London, 1984.

Krier, J.P. *Parasitic protozoa.* Academic Press, Inc., New York, 1978.

Kumar, H.D., and Singh, H.N. *A textbook on algae.* Macmillan Publishing Co., Inc. New York, 1979.

Loomis, W.F. (ed.). *The development of dictyostelium discoideum.* Academic Press, Inc., New York, 1982.

Martin, G.W., Alexopolous, C.J., and Farr, M.L. *The genera of myxomycetes.* University of Iowa Press, Iowa City, 1983.

Moore-Landecker, E. *Fundamentals of the fungi* (2nd ed.). Prentice-Hall, Englewood Cliffs, N. J., 1982.

Pritchard, H.N., and Bradt, P. *Biology of nonvascular plants.* Times Mirror/Mosby College Publishing, St. Louis, 1984.

Ross, I.K. *Biology of the fungi.* McGraw-Hill Book Co., New York, 1979.

Round, F.E. *The biology of the algae* (2nd ed.). Edward Arnold (Publishers) Ltd., London, 1973.

Schmidt, G.D., and Roberts, L.S. *Foundations of parasitology* (3rd ed.). Times Mirror/Mosby College Publishing, St. Louis, 1985.

Sleigh, M.A. *The biology of the protozoa.* American Elsevier, New York, 1973.

Trainor, F.R. *Introductory phycology.* John Wiley & Sons, Inc., New York, 1978.

Webster, J. *Introduction to fungi.* Cambridge University Press, Cambridge, London, 1977.

CHAPTER 7

Eukaryotic ultrastructure and function

Chapter Outline

Introduction

Eukaryotic microorganisms are more recent inhabitants of the earth than prokaryotes and are believed to have evolved from prokaryotes somewhere between 1.2 and 2.1 billion years ago. Their formation is believed to have coincided with the evolution of aerobic ecosystems. The structural changes or modifications that occurred during the evolution of the eukaryotic cell were often accompanied by variations in function that this chapter will try to elucidate.

The primary purpose of this chapter is twofold: first, to describe chemically those structures that are unique to eukaryotes (primarily fungi and protozoa) or have similar counterparts in prokaryotic cells, and second, to assess the functional significance of these structural changes or modifications. Figure 4-1 gives some morphological perspective of the similarities and differences between prokaryotic and eukaryotic cells.

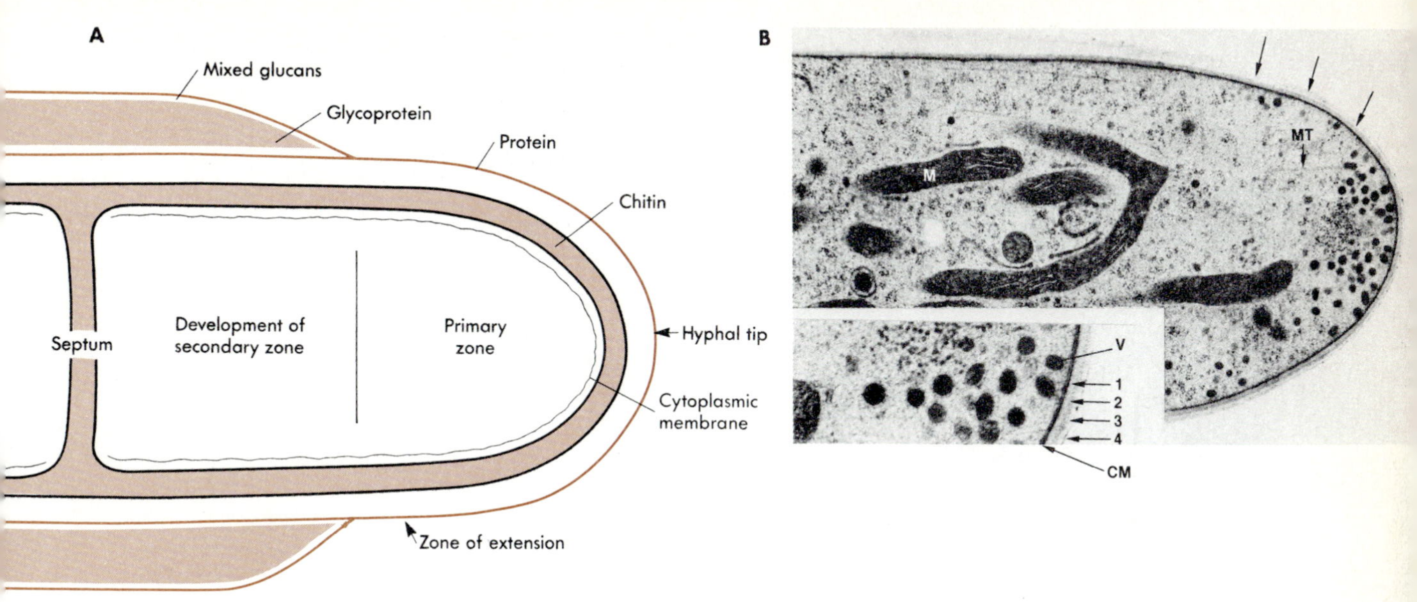

Figure 7-1 *Fungal cell wall structure.* ***A,*** *Hyphal cell wall tip. Wall increases in thickness farther away from tip of hypha that is, in secondary zone. Tip grows in length only in zone of extension or primary zone.* ***B,*** *Electron micrograph of hyphal tip cell. Lower photograph is enlargement of boxed area in larger photograph.* MT, *Microtubule;* M, *mitochondrion;* V, *vesicles;* CM, *cytoplasmic membrane. Arrows in upper photograph point to outer cell wall layer. Note four layers in cell wall of insert, constituting primary and secondary layers.*

CELL WALL

The cell wall in eukaryotic microorganisms is a feature of the fungi and the algae. Chemically the cell wall of fungi differs considerably from the cell wall of bacteria although both function in the same way. The hypha, which represents the unit of fungal growth, has a cell wall whose appearance varies with age (Figure 7-1). The growing point or tip of the hypha reveals a primary cell wall that is a smooth-structured unit showing little demarcation. Older areas of the hypha contain not only a primary wall but a secondary layer of structureless material. The **primary wall**, which is adjacent to the cytoplasmic membrane, lends structural support to the cell. As described in Chapter 4, polysaccharides make up nearly 80% of the dry weight of the fungal cell wall. They appear as crystalline **microfibrils** (Figure 7-2) that are believed to have special orientation, depending on their location in the wall. A thin layer of protein overlaps the polysaccharide and is believed to be covalently attached to it. The **secondary wall**, which increases in thickness with distance from the tip, is composed of layers of glycoprotein and mixtures of glucans. There is no peptidoglycan in the fungal cell wall and thus no transpeptidation reaction. This accounts for the inability of bacterial cell wall inhibitors such as penicillin to have any effect on fungal growth. Large molecules can pass into and out of the fungal cell wall, indicating less compactness than in prokaryotic cells. It is not known

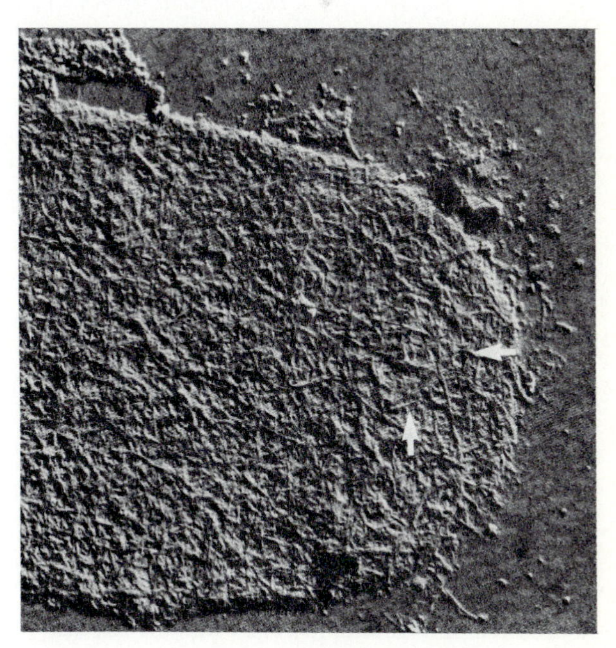

Figure 7-2 *Electron micrograph of tip of hypha demonstrating thin microfibrils* (arrow).

whether permanent pores exist in the cell wall as avenues for transport or whether pores are formed only when contact is made between specific cell wall components and the penetrating molecule.

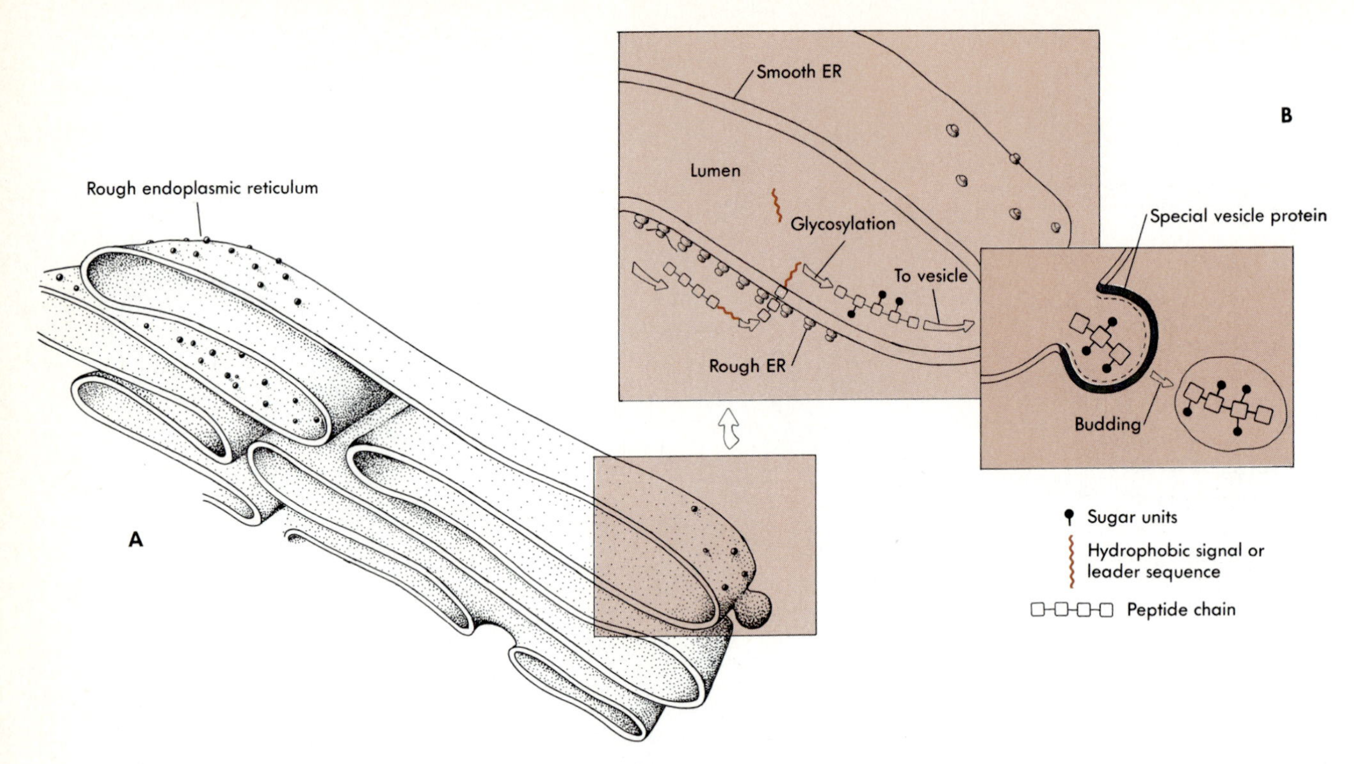

Figure 7-3 ***A,*** *Diagrammatic representation of endoplasmic reticulum* (ER). *ER consists of flattened sacs (cisternae) whose outer surface (rough ER) is studded with ribosomes. Internal space formed by convoluted sacs is lumen and is continuous with rough ER. Luminal side of ER is devoid of ribosomes and is called smooth ER.* ***B,*** *Mechanism of protein transport and modification in endoplasmic reticulum. Polypeptide is synthesized with signal (〰) that enables it to pass through inner membrane and enter lumen. Signal is removed after movement through lumen. Polypeptide is then glycosylated with sugars, such as mannose and glucose derivatives, in smooth ER. Protein is then packaged into vesicle that requires special surface protein for budding from ER. The budded vesicle will be transported to Golgi.*

CYTOPLASMIC MEMBRANE AND ENDOPLASMIC RETICULUM

There are some chemical differences between prokaryotic and eukaryotic cytoplasmic membranes even though both are unit membranes. One of the major differences between them is the presence of **sterols** in the membrane of eukaryotes and the absence of sterols in prokaryotes (except mycoplasmas and some others). Sterols stabilize the membrane, a function that is readily noted when one compares the instability of laboratory-induced wall-less bacteria (L-forms) with the stability of mycoplasmas in isotonic media.

A network of unit membranes, the **endoplasmic reticulum (ER)**, is present in all eukaryotes. The ER is a highly convoluted structure appearing as a network of interconnecting flattened sacs that is continuous with the cytoplasmic membrane, as well as the nuclear membrane (see Figure 4-23). The ER has a lipid bilayer arrangement and, like the cytoplasmic membrane, has transport proteins embedded in it. The continuity of the cytoplasmic membrane with the ER is somewhat analogous to the membranes of the photosynthetic bacteria (see Figure 4-20). The internal space enclosed by the convoluted nature of the ER is the ER **lumen**. The lumen is separated from the **cytosol** (a term used to refer to the cytoplasm without its suspended organelles) by a single membrane. There are two types of ER, **rough** and **smooth,** but they are continuous with each other. The rough portion is defined as the cytoplasmic side, which is studded with ribosomes, whereas the smooth side is the luminal side, which is devoid of ribosomes (Figure 7-3, *A*).

Rough ER with its attached ribosomes is associated with the synthesis of proteins to be secreted from the cell or to be transported to other organelles (Figure 7-3, *B*). The amino terminal end of these proteins possesses a **leader** or **signal** sequence of hydrophobic amino acids that assist the protein in its transport

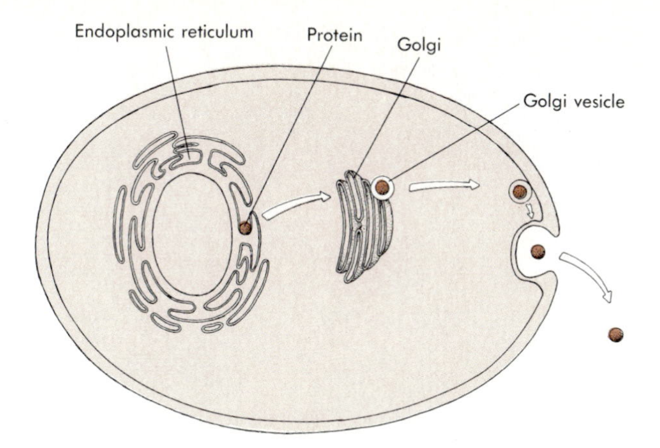

Figure 7-4 *Protein or lipid synthesized in ER is transported to Golgi for further processing. Transported molecule is surrounded by Golgi membrane to form vesicle. When vesicle is released from Golgi, it fuses with target organelle or in this instance cell membrane from which it is secreted from cell.*

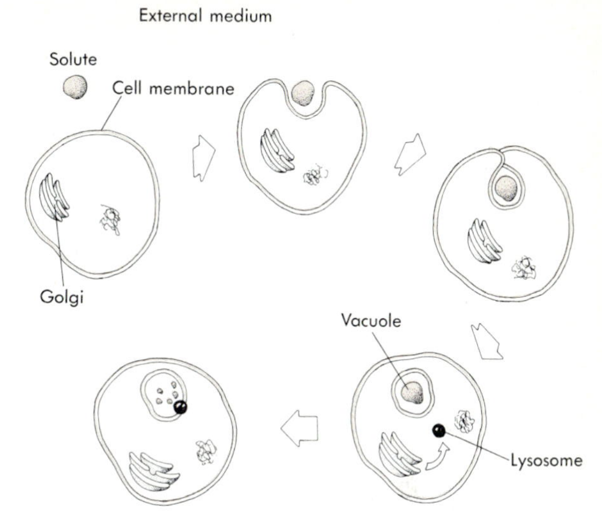

Figure 7-5 *Mechanism of pinocytosis as exemplifed in protozoa. Solute particle is engulfed by cytoplasmic membrane. Vacuole, consisting of detached cytoplasmic membrane plus solute moves into cytoplasm. Lyosome, produced by Golgi apparatus and containing hydrolytic enzymes, moves into cytoplasm to engage vacuole. Lysosome fuses with vacuole (see Figure 7-6), releasing enzymes that break down solute.*

through the lipid bilayer of the ER to the lumen (see p. 187). Most proteins synthesized on rough ER will also have sugar groups added to them—a process called **glycosylation**. The addition of sugar groups is a major function of the rough ER and is believed to take place on the luminal side of this internal membrane system. Once the glycosylated protein reaches the lumen, it is packaged and channeled to the Golgi body where further modification of the molecule occurs. Not all proteins synthesized on the rough ER pass directly into the lumen. Some are retained in the ER membrane, while others are thought to be selectively transported to other organelles.

The smooth ER contains enzymes that are associated with the synthesis of lipids, such as phospholipids and cholesterol. These lipids are eventually transported to various organelles where they become part of the membrane system. The smooth ER also possesses enzymes that are involved in the oxidation of drugs, a process called **detoxification**.

Once proteins or lipids are synthesized in the ER, most are transported directly to the Golgi body by budding of the ER into specific transport vesicles (Figure 7-4). These vesicles will fuse with the membrane of their target organelle and the orientation of lipid and protein in the vesicle membrane will be retained on the target membrane.

Free ribosomes are found in the cytosol and, unlike the ribosomes embedded in the ER, are associated with synthesis of primarily enzymatic proteins.

Two of the principal functions of the cytoplasmic membrane in both eukaryotes and prokaryotes are transport of nutrients into the cell and excretion of wastes or other products from the cell into the environment. These functions are shared by both, but the mechanism differs in those eukaryotic cells that are wall-less—for example, the protozoa. The cytoplasmic membrane in prokaryotic cells is a much more selective barrier to the passage of nutrients into the cell. The cytoplasmic membrane in those eukaryotes without walls is relatively pliable and provides the cell with unique methods for nutrient and waste transport: pinocytosis, phagocytosis, and exocytosis. **Pinocytosis** is a transport process in which fluids or suspensions of material from the external environment are delivered across the cytoplasmic membrane in a closed membrane produced at the cell surface (Figure 7-5). **Phagocytosis** is a transport process in which the material being transported is particulate and not in solution. Pinocytosis and phagocytosis are referred to as **endocytoses** (sing. endocytosis). A more lengthy discussion of phagocytosis, as carried out by cells of the immune system in humans, is presented in Chapter 23.

Endocytosis does not entail the indiscriminate uptake of molecules from the environment. It is a process that is induced by the chemical nature of the solute or solution in the cell's environment. There is also a limit to the number of endocytic activities that can take place within any period of time. Since the cytoplasmic membrane of the cell is used to produce a closed structure around the solute during endocytosis, new cytoplasmic membrane must of necessity be regenerated. It is believed that the number of endocytoses is controlled by the amount of cytoplasmic membrane precursor available in the cell. Endocytosis provides the wall-less eukaryotes with a means of capturing their "meals" in a concentrated particulate form. This is a selective advantage for the eukaryotic cell over the prokaryotic cell because eukaryotes do not have to produce extracellular enzymes to break down large molecules or food particles that cannot penetrate the cytoplasmic membrane, a necessity for most prokaryotic cells.

Endocytosis also explains the presence of endosymbionts. Prokaryotes are often the prey for eukaryotes; for example, protozoa phagocytose bacteria as a source of food. Sometimes the engulfed prokaryote survives in the host, and the host may even benefit from the association. These relationships are discussed in Chapter 4.

The extrusion of materials from within the cell to the environment also takes place in eukaryotes in a process that is the reverse of endocytosis called **exocytosis.** Exocytosis occurs by fusion of intracellular vesicles with the cytoplasmic membrane. The products are released at the cell surface.

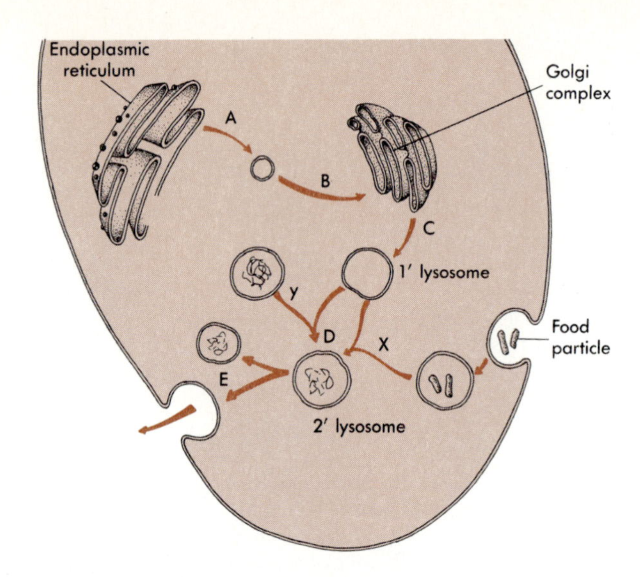

***Figure* 7-6** *Relationship of lysosomes to phagocytic or pinocytic capture of food particles.* A, *Vesicle produced by endoplasmic reticulum contains hydrolytic enzymes.* B *and* C, *Vesicle interacts with Golgi and enzymes become concentrated to produce primary (1′) lysosome.* D, *Lysosome fuses with vacuole or vesicle-containing food particles* (X) *or cellular debris (Y) to produce secondary (2′) lysosome.* E, *Particles or debris in secondary lysosome may be expelled at surface or some debris, such as cellular components, may remain in secondary lysosomes.*

VACUOLES, VESICLES, AND LYSOSOMES

The Golgi body and ER are involved in the formation of membranous structures called vacuoles, vesicles, and lysosomes. These three membranous structures can be defined as separate entities, but their functions in the cell are interrelated. For example, a cell that captures its food by endocytosis has the food particles surrounded by a closed membranous structure. If phagocytosis is the mechanism of capture, the membranous structure is called a **vacuole**; if pinocytosis is the mechanism of capture, the membranous structure is called a **vesicle.** Vesicles may also contain endogenous cellular components such as mitochondria and ribosomes that undergo self-digestion. The digestion of exogenous or endogenous material occurs in a series of events initiated first by the formation of hydrolytic enzymes on the endoplasmic reticulum. The enzymes are then packaged within a membranous unit, produced by the Golgi body, called a **lysosome.** Cellular constituents would be hydrolyzed by the lysosomal enzymes if they were not packaged in these individual membranous sacs. Fusion of the lysosomal membrane with that of the particle-containing vacuole or vesicle results in the formation of a single membranous structure called the **secondary lysosome,** where digestion of the captured particles occurs. Digested material may pass directly through the lysosome membrane and back into the cytosol, whereas undigested material or particles may remain within the cell or be expelled at the cell surface (Figure 7-6). A lysosome can therefore be considered a specialized vesicle.

Vesicle formation at the surface of the cell is also a mechanism for transport of viruses into susceptible animal cells. Once the virus makes contact with the surface of the host cell it becomes enclosed in the vesicle and enters the cytoplasm where the vesicle membrane is later dissolved (Figure 7-7). Vesicles may also be involved in certain biosynthetic reactions within the cell—for example, during fungal growth in which the hyphal tips are actively engaged in cell wall synthesis and tip elongation. It is believed that vesicles, formed in the cytosol, carry various enzymes and polysaccharides that are involved in cell wall synthesis. The vesicles fuse with the cytoplasmic membrane and release their contents at the hyphal tip.

GOLGI BODIES

The Golgi body is a complex membranous network found exclusively in eukaryotes, but not all eukaryotes possess them. The Golgi is organized into six or more flattened stacks of membranes (Figure 7-8). Each membrane-enclosed sac is called a **cisterna** (pl. cisternae), which carries out specialized functions involved in the processing of ER proteins and lipids. The region of the cisternae receiving glycoproteins or lipids from the ER is called the **cis** face. The opposite end of the Golgi body stack is called the **trans** face, that is, the region

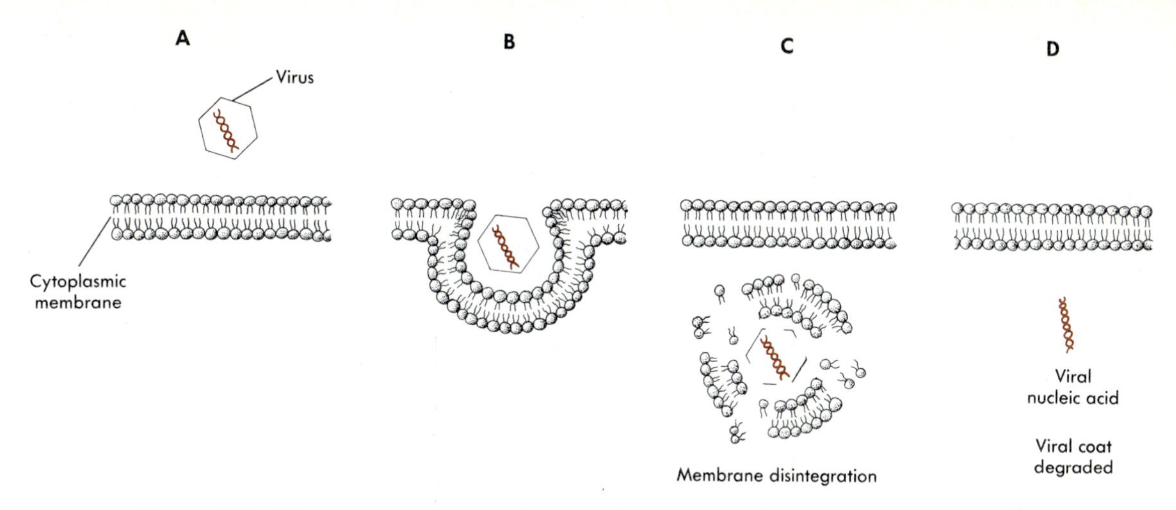

Figure 7-7 *Virus penetration into cell by vesicle formation.* A, *Binding of virus to cytoplasmic membrane* B, *Invagination of cytoplasmic membrane about virus particle.* C, *Vesicle, which contains virus, is pinched off from cytoplasmic membrane.* D, *Vesicle membrane is degraded and viral coat is removed, releasing viral nucleic acid into cytoplasm.*

containing the processed glycoprotein or lipid.

Molecules synthesized on the ER are transported to the Golgi via transport vesicles containing a piece of reticular membrane. The vesicle is covered with a protein that appears to be required for budding from the ER (Figure 7-3, *B*). Once the vesicle buds from the ER the coat protein is shed. The vesicle moves to the cis face of the Golgi body where it fuses with the cisternae (Figure 7-9). As glycoproteins are passed through the Golgi body stacks, they are chemically modified by enzymes located within specific cisternae. There appears to be three specific areas of modification—those located near the cis face, those in the middle of the stack (**medial face**), and those at the trans face. This modification process is believed to occur in the following manner:

1. All glycoproteins arriving at the cis face have the same identical oligosaccharide chains consisting of six carbon sugars, such as mannose and glucose derivatives. The mannose units are removed in two enzymatic steps by **mannosidase.** Other proteins, such as those destined for lysosomes, are only phosphorylated. Phosphorylation apparently makes the oligosaccharide chain resistant to enzymatic attack during transport through the remaining Golgi body stacks. The partially modified glycoproteins are packaged into a vesicle, which is transported to the medial face.
2. The processed glycoprotein arriving at the medial face is further modified by ***N-acetylglucosaminyl transferase,*** which adds *N*-acetylglucosamine to the glycoprotein. A vesicle formed on the medial face transports the glycoprotein to the trans face.
3. Glycoproteins arriving at the trans face are further modified by the addition of galactose (**galactose transferase**) and sialic acid (**sialic acid transferase**) to the molecule. Once modification is finished at the trans face the proteins are sorted out and packaged into vesicles—some to be transported to lysosomes, others to the plasma membrane, and others as secretory granules that are

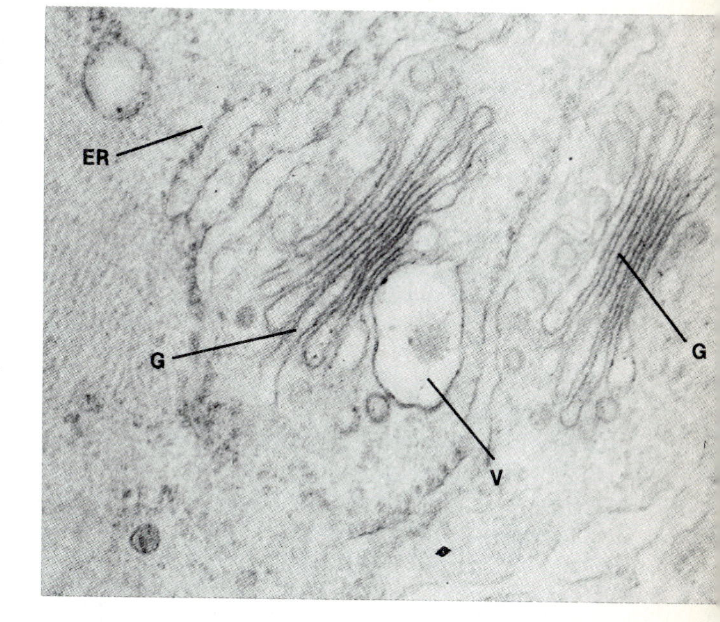

Figure 7-8 *Golgi complex. Electron micrograph of algal cell showing Golgi complex with its flattened sacs (cisternae).* G, *Golgi complex;* V, *vesicle containing granular material;* ER, *endoplasmic reticulum.*

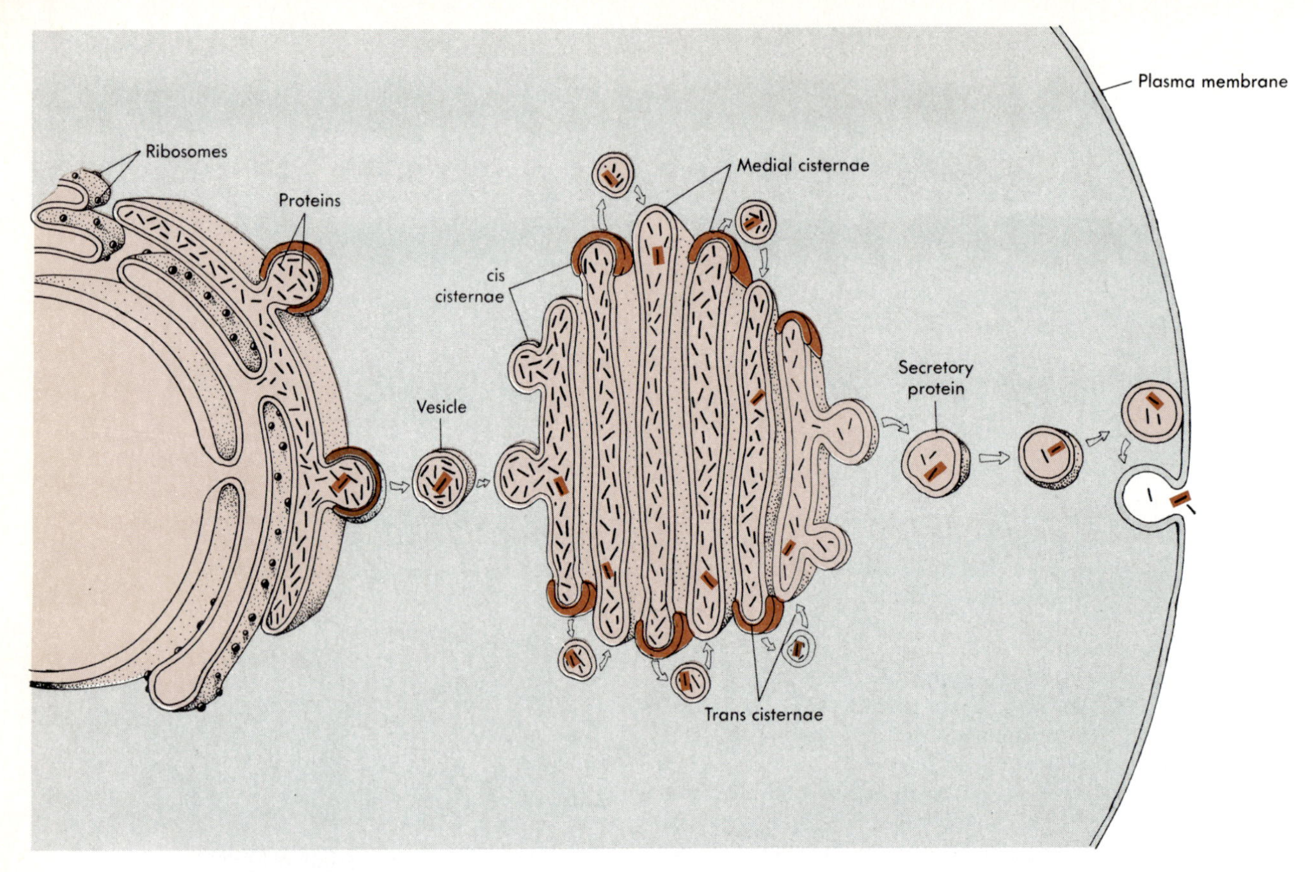

Figure 7-9 *Golgi transport process. Proteins synthesized on rough ER interact with smooth ER membrane to form vesicles. Vesicle becomes covered with protein that enables it to bud. As vesicle buds, surface protein is removed. Vesicle migrates to Golgi and fuses with cis cisternae and then is processed. Vesicle migrates to medial and finally trans cisternae for further processing. Transported proteins (■) are sorted and packaged for delivery to specific targets in cell and in this instance to cell membrane for secretion from cell.*

released in response to an external signal (Figure 7-9). Thus the Golgi body acts like a distilling apparatus that takes crude unfinished proteins and refines them to a product ready for delivery to various cellular compartments.

Concept Check

1. Describe the location of the various components of the fungal cell wall and explain their chemical composition.
2. Explain the basic organization of the ER and its function in the cell.
3. What are the different ways in which the plasma membrane is involved in nutrient transport?
4. What are the steps involved in the modification of proteins when they leave the ER and enter the Golgi body?

MITOCHONDRIA

We have indicated so far that the eukaryotic cell is a compartmentalized unit in which certain cell functions are restricted to specific areas of the cell. Thus the substrates and products of metabolic reactions are confined within membrane-enclosed structures, or organelles. One of these organelles, the **mitochondrion**, is found in all eukaryotic cells except anaerobic protozoa. Mitochondria are themselves compartmentalized, and it is this compartmentalization that is associated with specific biochemical functions. The mitochondrion is a complex organization of enzymes whose activity is associated with aerobic respiration, that is, aerobic energy production. Structurally, the mitochondrion is a double-membrane organelle.

The two membranes, which are lipid bilayers, enclose two compartments: the internal matrix and the narrow intermembrane space (Figure 7-10). The outer membrane possesses several transport proteins that form channels for transport of molecules of 10,000 molecular

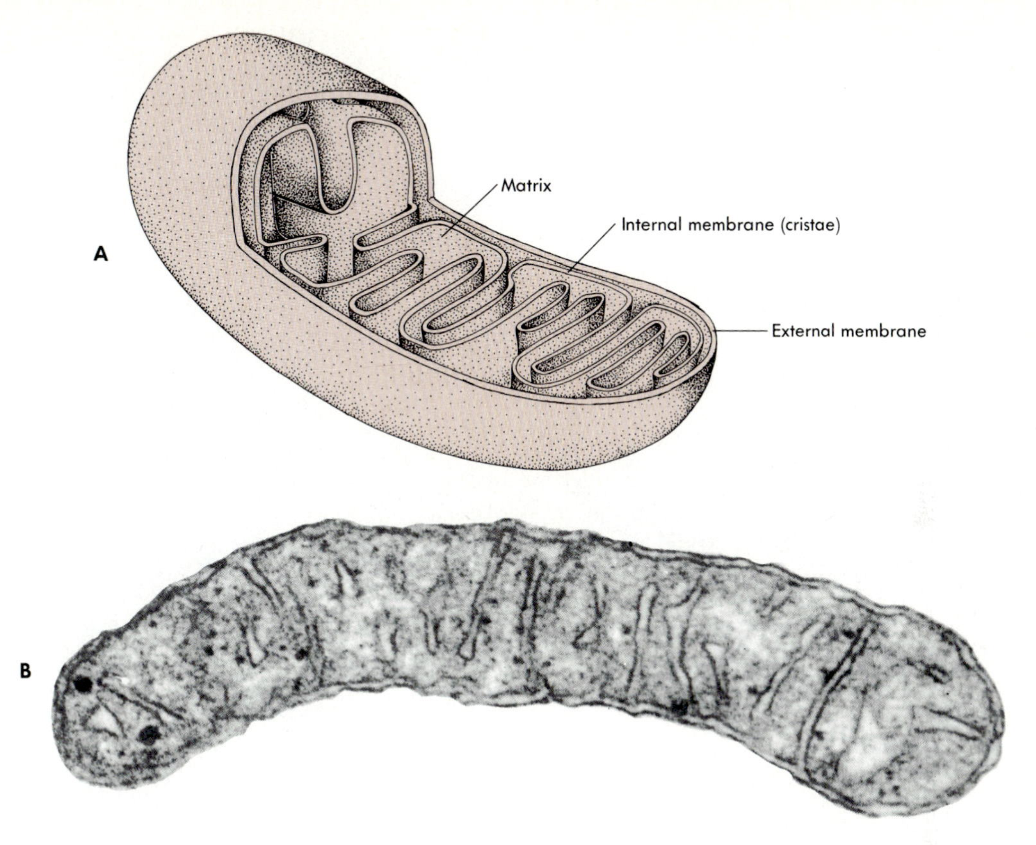

Figure 7-10 *Mitochondrion.* ***A,*** *Three-dimensional illustration of mitochondrion.* ***B,*** *Electron micrograph of mitochondrion in thin section.*

weight or less. Other proteins in the outer membrane are enzymes involved in lipid modification for metabolism in the matrix. The intermembrane space contains enzymes that phosphorylate nucleotides other than ATP.

The inner membrane is organized into a number of folds called **cristae**, which increase the total surface areas of the mitochondria. The three major proteins in the inner membrane are: (1) those involved in electron transport reactions in the respiratory chain, (2) those involved in ATP synthesis, and (3) those that control the transport of metabolites into and out of the matrix. The matrix contains many different enzymes, including those that are required for the oxidation of pyruvate and molecules of the citric acid cycle (Krebs cycle) that are associated with electron transport.

The mitochondrion can multiply by division and possesses its own genetic system—that is, DNA, ribosomes, and other machinery associated with DNA replication and translation. Not all of the genes necessary for replication and other functions are found in mitochondrial DNA. Some are found in the nuclear DNA. Mitochondrial DNA is much smaller than nuclear DNA and ranges in molecular weight from 9×10^6 to 7×10^7, depending on the microorganism. Mitochondrial DNA is also naked and not associated with histones as is nuclear DNA. The ribosomes in the mitochondrion are more closely related to prokaryotic ribosomes than the ribosomes in the eukaryotic cell cytosol. These latter characteristics have suggested that bacteria parasitizing eukaryotes lost their cell wall and evolved as membranous organelles with their own genetic system but incapable of synthesizing cell wall. During evolution the eukaryotes became dependent on the endosymbiont for survival and vice versa.

CHLOROPLASTS

Chloroplasts are organelles found in those eukaryotes that carry out photosynthetic reactions. Eukaryotic photosynthesis results in the evolution of oxygen. Photosynthesis can also occur in prokaryotes, but the chloroplast as an organelle is found only in eukaryotic cells. The generalized structure of a chloroplast (Figure 7-11, *A*),

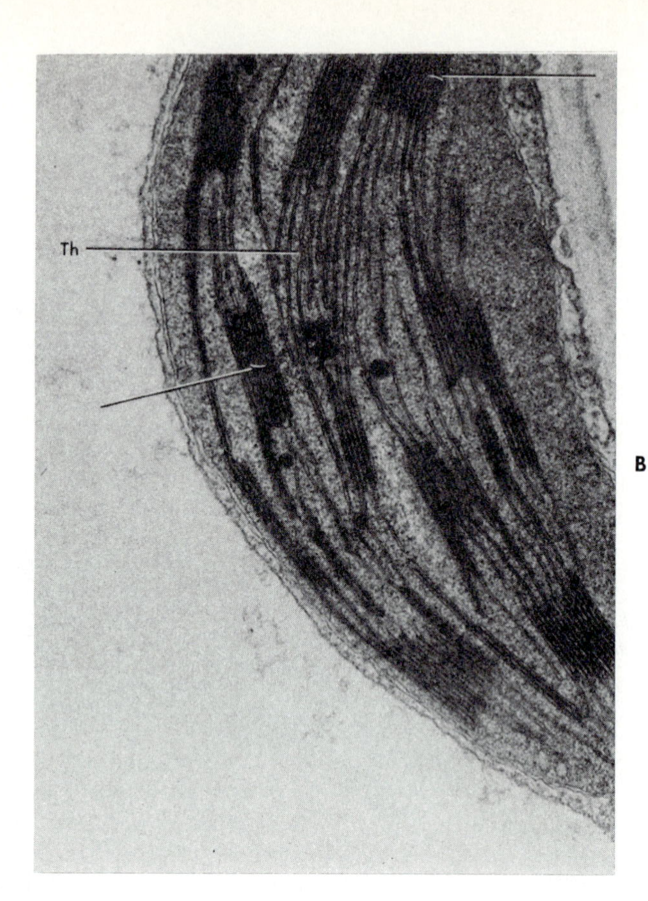

Figure 7-11 *Chloroplast.* ***A,*** *Diagrammatic illustration of chloroplast.* ***B,*** *Electron micrograph of plant chloroplast. (Reduced 70%.) Th, thylakoid. Arrows point to grana.*

like a mitochondrion, possesses an external envelope and internal membrane system. The outer envelope consists of an outer and inner membrane. The outer membrane of the envelope is permeable to molecules of molecular weight less than 1000, whereas the inner membrane is passively permeable only to small molecules, such as water, ammonia, and carbon dioxide. Only in chloroplasts does a rough ER completely envelop the organelle and isolate it from the cell's cytoplasm. The chloroplast, is usually pressed against the nucleus, and the outer membrane of the nuclear envelope is continuous with the sheath of ER surrounding the chloroplast. The internal homogenous matrix of the chloroplast is called the **stroma** in which are embeded the system of internal membranes called the **thylakoids** (Figure 7-11, *A*). Thylakoids may be free within the stroma of the chloroplast, or they may be stacked into units called **grana**, such as those in the Chlorophyta. The thylakoids are made up of chlorophyll and other pigments associated with photosynthesis, as well as glycolipids, phospholipids, and proteins. The conversion of light energy to ATP production occurs within the thylakoid membrane. The stroma is the site of reduction of carbon into organic materials. Chloroplasts, like mitochondria, contain DNA and ribosomes and other protein-synthesizing components.

NUCLEUS AND DNA

The genetic material in eukaryotes is enclosed in a membranous structure called the **nucleus**. The nuclear membrane surrounding the nuclear material is composed of two unit membrane layers, with a perinuclear space between. (Figure 7-12). The outer nuclear membrane is continuous with the ER and often contains ribosomes on its surface. The presence of two bilipid membranes in the nucleus presents special problems in

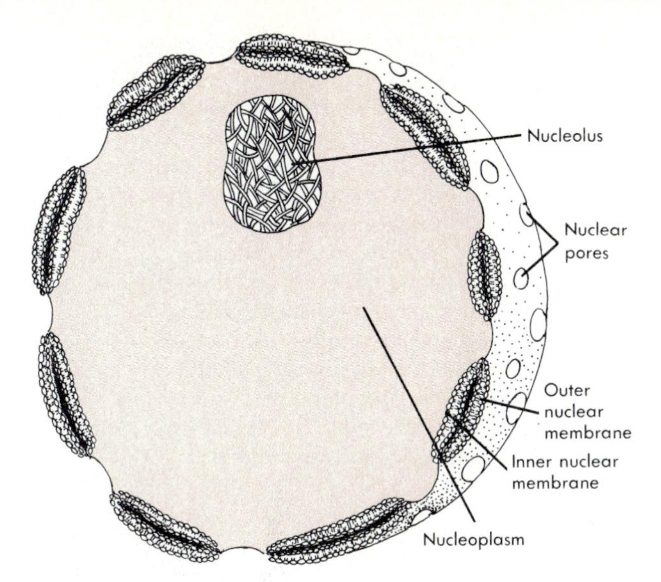

Figure 7-12 *Diagrammatic illustration of nucleus showing double membrane and muclear pores.*

Figure 7-13 *Nuclear pores* (NP). *Electron micrograph of freeze-etched preparation of plant cell showing studded appearance of nuclear pores.*

the transport of larger molecules between the cytosol and nucleus. This problem has been overcome by the presence of **nuclear pores** (Figure 7-13), which are widely distributed over the membrane surface. The nuclear pore represents the site where the nuclear outer membrane is connected to the inner membrane. Protein molecules with molecular weights larger than 60,000 appear to be unable to penetrate the nuclear pores. Therefore large particles, such as mature ribosomes, cannot pass through the pores and protein synthesis is restricted to the cytoplasm. Molecules with molecular weights of 17,000 to 30,000 can apparently pass through the pores with little difficulty.

The **nucleolus** (Figure 7-14) is a granular-appearing area within the nucleus that is specifically engaged in ribosomal RNA synthesis. Ribosomal proteins from the cytoplasm enter the nucleus where the smaller ribosomal RNA-protein subunits are assembled. The smaller ribosomal subunits then pass into the cytoplasm where they are mobilized into the larger mature ribosome.

Eukaryotic DNA organization and structure

The DNA of eukaryotic cells, which is a linear molecule and not circular like that of prokaryotes, is coiled into compact units called **chromosomes**. Eukaryotes, unlike prokaryotes, possess more than one chromosome, and in each chromosome there is one DNA molecule. The DNA in the chromosome is much longer than the chromosome length because it is folded many times into a compact unit. Intimately associated with the chromosome are various proteins. These proteins are divided

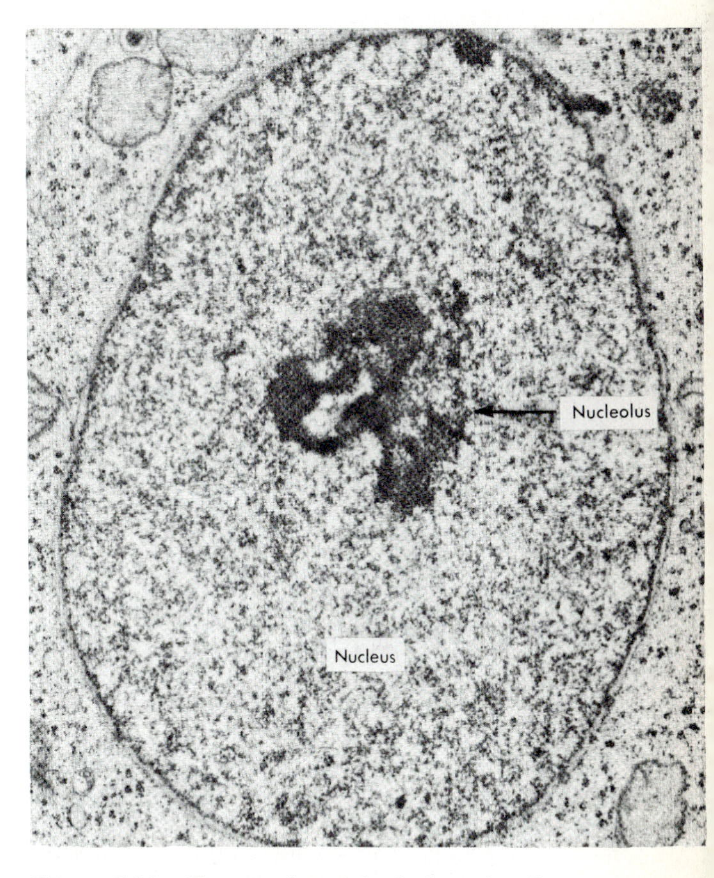

Figure 7-14 *Electron micrograph of eukaryotic cell showing nucleus and intensely granular nucleolus. Arrows point to nuclear pores.*

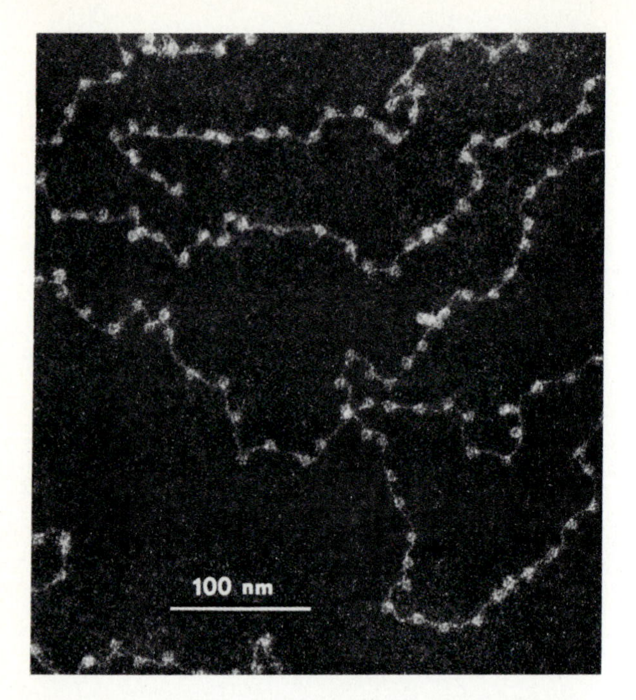

Figure 7-15 *Nucleosomes. Darkfield electron micrograph of edge of chicken erythrocyte nucleus demonstrating beaded nucleosomes associated with eukaryotic chromosome.*

into two types: histone and nonhistone. Histones are rich in basic amino acids, such as lysine. Basic amino acids help to neutralize the negatively charged phosphate groups on the DNA molecule and thus have a stabilizing effect. DNA associated with histones is called **chromatin.** The arrangement of histones on the DNA is such that it folds the DNA into a pattern that gives the appearance of beads in electron micrographs (Figure 7-15). These repeating nucleoprotein beads are called **nucleosomes.** Each nucleosome bead consists of a core of histone protein around which a double-stranded DNA segment of approximately 140 base pairs is wound. Each nucleosome bead is separated from the next by a spacer of 40 to 60 base pairs of DNA (Figure 7-16). The nucleosome is made up of 8 histone molecules (two each of the histones H2A, H2B, H3, and H4). The regularity of the nucleosome repeating unit is believed to be due to a histone protein called HI. Histone HI molecules are located at the regions where DNA enters and exits from the core particle. The HI protein is made up of a central globular region linked to amino and carboxyl terminal "arms." The ability of the arms to make contact with adjacent nucleosomes is believed to be responsible for nucleosome spacing.

Some of the histone proteins are known to undergo modification, whereas others show no modification. Modification may involve a reversible acetylation process involving the amino acid lysine. Lysine is a positively charged amino acid, resulting from the presence of a free amino group (Figure 7-17, *A*). When this amino acid is acetylated, the free amino group is tied up and the positive charge is removed. Acetylation-deacetylation reactions may therefore be associated with a genetic control mechanism. A second type of modification may involve the phosphorylation of the amino serine. Serine phosphorylation results in the addition of negative charges to the histone molecule and thus may also represent a genetic control mechanism (Figure 7-17, *B*).

The most abundant nonhistone proteins associated with eukaryotic DNA are those called high mobility group (HMG) because they are small and highly charged. They are believed to be regulatory proteins but their mechanism of action is not known.

MICROTUBULES

Microtubules are rigid hollow filaments found in all eukaryotic cells. They are usually long and between 150 and 250 Å wide. Most of the microtubules so far isolated and chemically analyzed are composed of protein subunits called **tubulin** and arranged in a helical manner (Figure 7-18). Once tubulin is disaggregated, the protein subunits are capable of self-assembly, a process that is also characteristic of the protein subunits of flagella and viruses. Microtubules are known to be associated with motility because they are components of flagella. Microtubules are also involved in nuclear division because they form part of the mitotic spindle (Figure 7-19). Because of their location in the cytoplasm and their apparent mobility, it has been suggested that microtubules may affect cell shape in wall-less eukaryotes, such as the amebas, and that they may be involved in the transport of material for biosynthesis. The association of microtubules with movement is discussed in the next section.

CILIA, FLAGELLA, AND CELLULAR MOVEMENT

Movement created by eukaryotic cilia is brought about by the bending of a ciliary core called the **axoneme,** which has a constant diameter of approximately 0.2 μm in all species. Each axoneme contains microtubules that are derived from a **basal body** located at a point just inside the cytoplasmic membrane from where the cilium protrudes. The microtubules of the axoneme are in the 9 + 2 arrangement, that is, there is a circle of 9 microtubular pairs surrounding 2 central pairs (Figure 7-20). Attached to one microtubule of each doublet are protein linkages called **dynein arms.** The nine doublet microtubules on the periphery appear to be linked to each other by structural units called **nexin links.** Spokes are present that appear to bond each doublet microtubule to the two central microtubules. Near the axis of

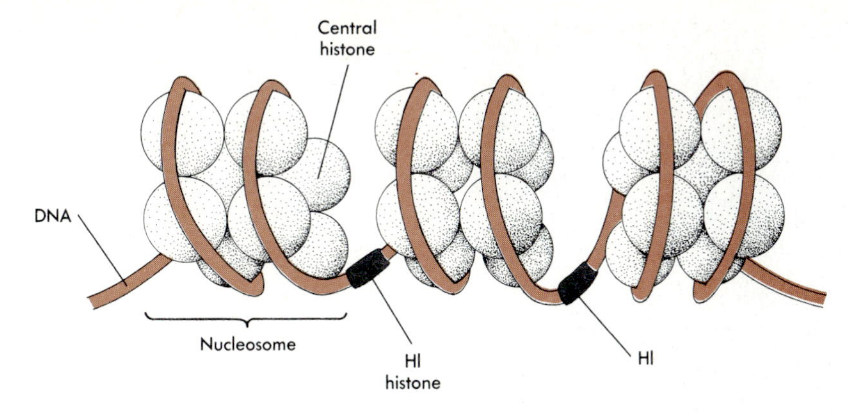

Figure 7-16 *Nucleosomes. DNA molecule is wound twice around aggregates of histone proteins to form nucleosome core. Nucleosome cores are connected to each other by stretch of linker DNA approximately 60 base pairs in length. One histone molecule called HI is located just outside of nucleosome core and is believed to be simultaneously associated with both ends of DNA as it enters and exits from core.*

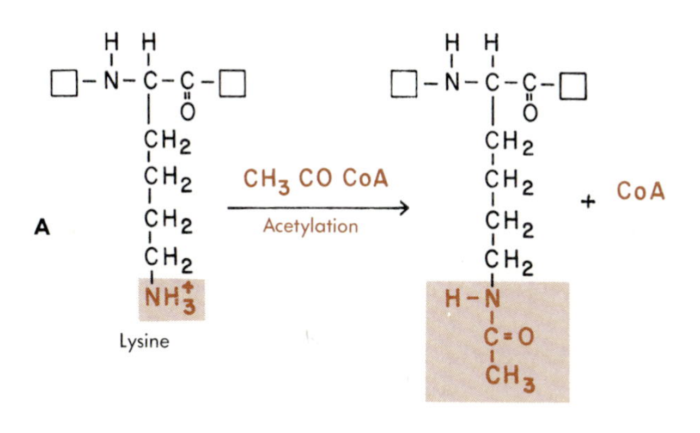

Figure 7-17 *Modification reactions involving acetylation and phosphorylation of histone amino acids.* ***A,*** *Acetylation of lysine.* ***B,*** *Phosphorylation of serine. Uncolored boxes represent polypeptide backbone.*

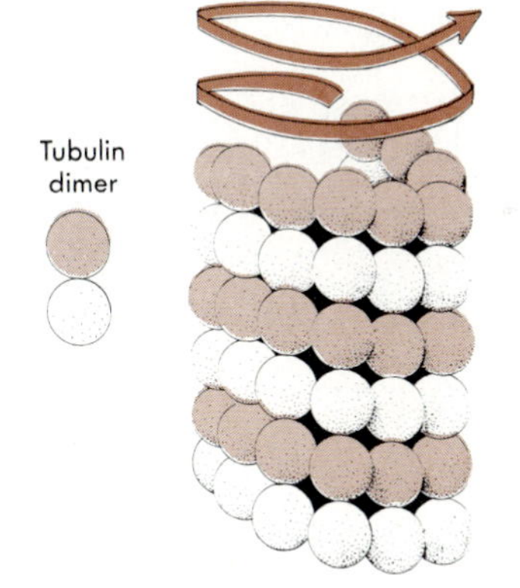

Figure 7-18 *Microtubules. Microtubules consist of a protein called tubulin, which is arranged in helical fashion. Tubulin subunits form dimers, which attach to growing chain.*

the cilium, projections from each central microtubule form a series of rings called the central sheath. Surrounding the entire microtubular system is a membrane that is an extension of the cytoplasmic membrane of the cell. Recall that a membrane is not found around most prokaryotic flagella. The basal body contains the basic nine microtubular pairs; however, it does not contain the two central microtubules. If a cilium is broken off at the cell surface, the basal body serves as a center for synthesis of new ciliary material.

Mechanism of movement

We know from laboratory observations that eukaryotic ciliary or flagellar movement, unlike flagellar movement in prokaryotes, is oarlike or undulating. The oarlike motion results from a bending at the base of the cilium causing the tip to produce a circular arc. Undulating movements are generally observed when a cell possesses a single flagellum, whereas the oarlike movements are observed when many cilia are present on the cell. The mechanism used by eukaryotes to cause the cilium to bend is not understood. However, enough has been learned to project a basic mechanism for movement.

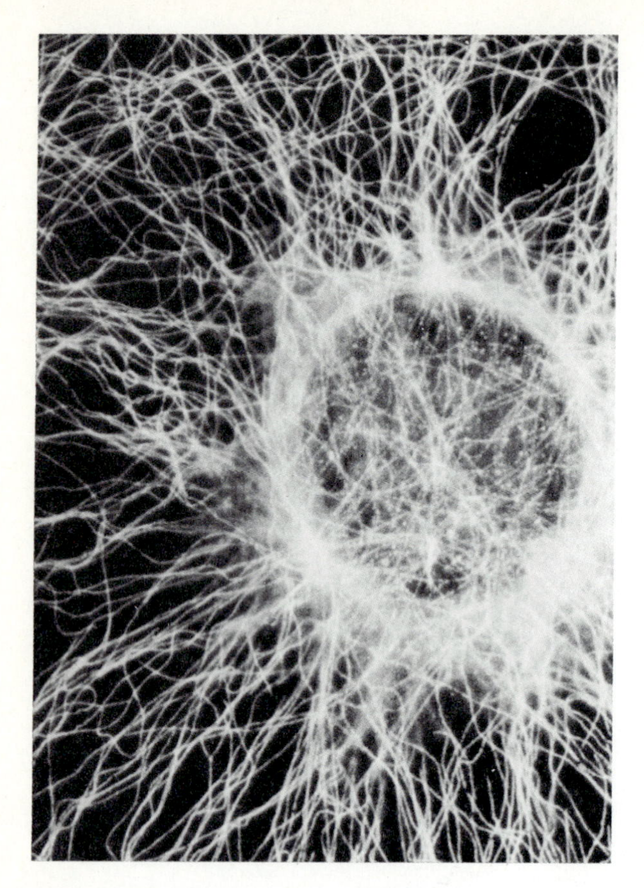

Figure 7-19 ***Photomicrograph of tubulin fibers surrounding nucleus of mouse cell.***

Concept Check

1. Describe the organization of the mitochondrion and chloroplast, and explain the function of the various compartments.
2. Explain the structure of the nucleus and the importance of nuclear pores?
3. What is the structure of histones and how are they arranged on the DNA?
4. How does the structure of eukaryotic flagella differ from prokaryotic flagella?
5. How does eukaryotic flagellar movement differ from that in prokaryotes?
6. Create a table listing the similarities and differences between prokaryotes and eukaryotes.

Cilia can be removed from the cell and will continue to bend as long as ATP and magnesium or calcium ions are present. When treated briefly with low concentrations of a protease and in the presence of ATP and magnesium or calcium ions, the cilium does not bend but the axoneme elongates to seven to eight times its original length. This type of protease treatment is known to destroy the nexin links but not the dynein arms (Figure 7-21). The elongation is believed to be a result of the sliding of doublet microtubules. When dynein arms are selectively removed from the cilium, movement ceases, and, when purified dynein is restored, movement is initiated. It is believed that in the intact cilium, dynein arms interact with adjacent doublets in such a way as to cause bending of the cilium. Although the dynein arms do not extend all the way from one doublet to another, they will make contact when ATP is being used up. The force generated by the dynein arms results from the fact that they possess ATPase activity. Contact between dynein arms and a neighboring doublet activates ATP hydrolysis and generates a sliding force between the microtubules. The doublets, because they are held in the cilium by nexin cross links, will bend when there is doublet sliding (Figure 7-22). The action of the dynein arms appears to be controlled by other proteins. The dynein arms are normally inactive until specific signals, perhaps generated by the radial spokes or inner sheath, are encountered. Once the signals are transmitted the dynein arms are then activated into an organized pattern of response.

For example, the control of ciliary movement in the protozoa *Paramecium* is believed to be associated with bioelectric properties of the cytoplasmic membrane. When the anterior end of a *Paramecium* collides with an obstacle, there is ciliary reversal and backward movement. This is called the "avoiding reaction" and is similar to the tumbling reaction in bacteria. The avoiding reaction is due to calcium ion (Ca^{2+}) influx into the cytoplasm and depolarization of the cytoplasmic membrane. When the calcium concentration is kept below 10^{-7}M normal swimming occurs but when the calcium concentration is raised above 10^{-6}M ciliary reversal occurs. The *Paramecium,* soon after the avoiding reaction, uses a calcium pump in the cytoplasmic membrane to force calcium ion out of the cytoplasm to a level below 10^{-6}M, and forward swimming resumes. Mechanical stimulation of the posterior end of *Paramecium* is associated with an influx of potassium (K^{+}) and hyperpolarization of the cytoplasmic membrane.

Chemotaxis in eukaryotes

Chemotaxis is a property that is found in eukaryotic microorganisms as well as prokaryotes. The chemotactic response in several eukaryotic groups is involved in the aggregation of single cells into multicellular bodies that undergo differentiation processes. One of the more

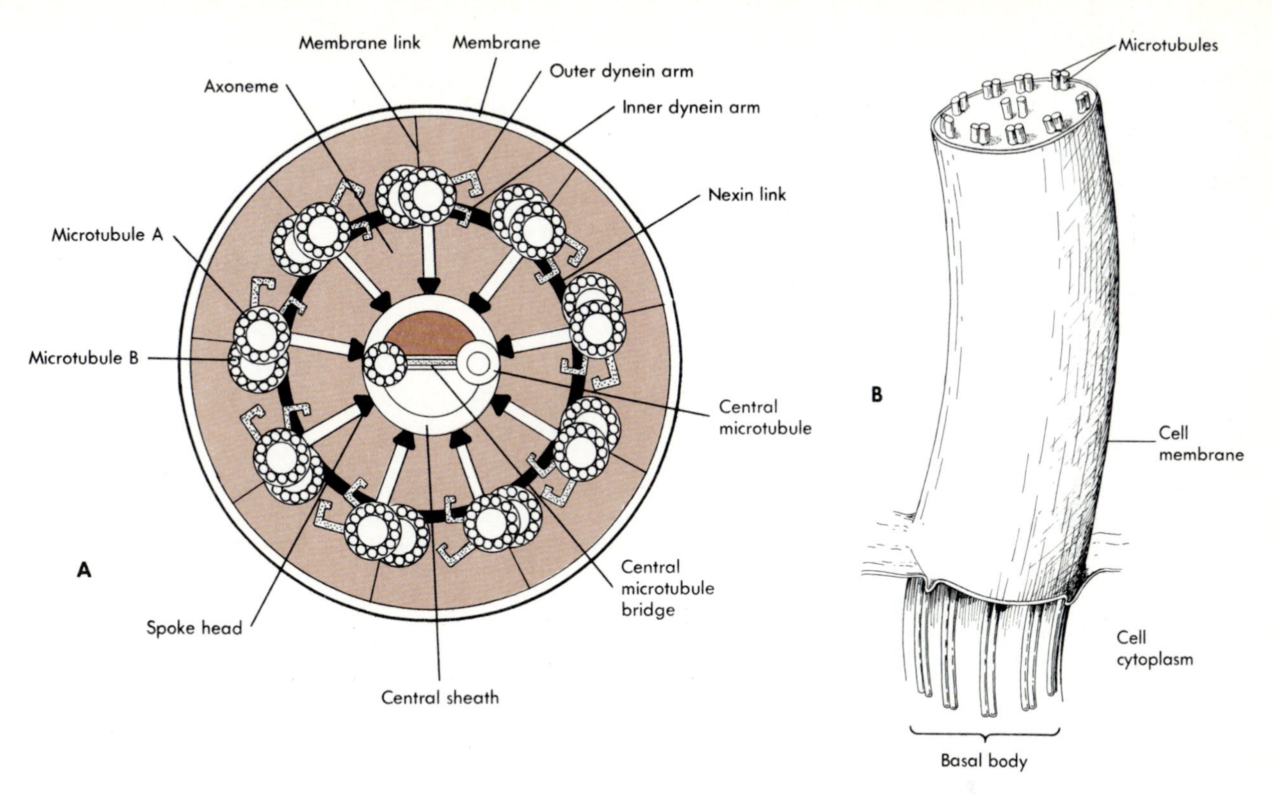

Figure 7-20 *Cilium ultrastructure.* ***A,*** *Cross section of typical cilium. Two central microtubules are surrounded by nine doublet microtubules.* ***B,*** *Side view of eukaryotic cilium showing basal body, which lies in cytoplasm.*

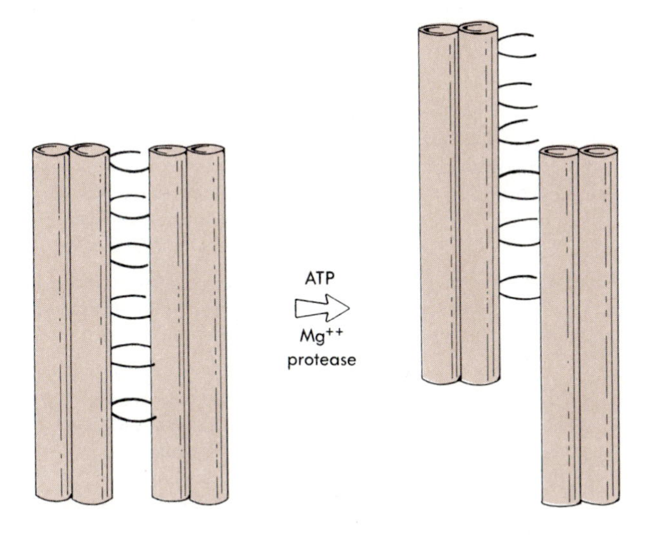

Figure 7-21 *Sliding of doublet microtubules as a result of pushing action of dynein arms of one doublet when nexin links are removed by protease.*

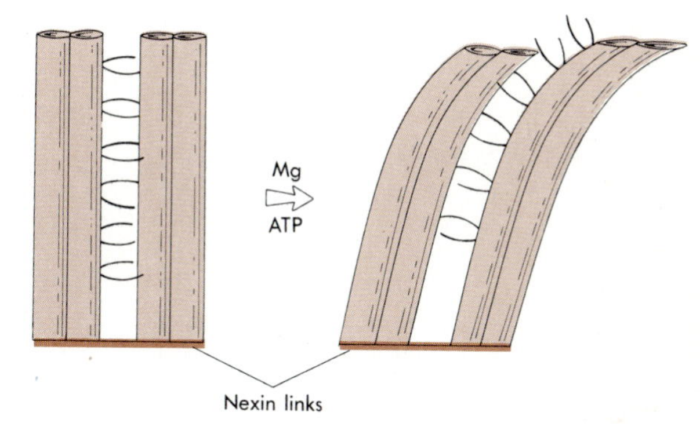

Figure 7-22 *Bending of microtubules when nexin links are left intact and ATP hydrolysis occurs.*

widely studied eukaryotic chemotactic responses is in the cellular slime mold *Dictyostelium discoideum,* discussed in Chapter 6.

SUMMARY

Eukaryotes are structurally and functionally more complex than prokaryotes. Some of these differences are outlined in Table 4-1. Listed in Table 7-1 are the characteristics of the basic structures of eukaryotes. Additional characteristics and structures are unique to a particular eukaryotic group, but these are discussed in detail and at more length in Chapter 6, characteristics of eukaryotes.

Table 7-1 *Characteristics of eukaryotic structures and inclusions*

Structure or inclusion	Chemical structure	Function	Comments
Cell wall	In green algae, primarily cellulose; in other algae, silica, $CaCO_3$, and other polysaccharides in addition to cellulose; fungal cell wall made up of 80% polysaccharide and 20% protein	Gives structural support and shape to the cell	Protozoa have no cell wall; fungal wall is more porous than in prokaryotes
Cytoplasmic membrane	A unit membrane, but in addition contains sterols	Nutrient transport; in protozoa the membrane can capture food (endocytosis)	
Endoplasmic reticulum (ER)	A unit membrane with a rough portion containing ribosomes and a smooth portion that does not contain ribosomes	Permeation of cytosol, smooth portion functions to produce lipids and structures such as vesicles; rough ER engages in protein synthesis and protein and lipid transport	Proteins synthesized are primarily secretory or membrane
Mitochondria	Membranous structure containing many enzymes and electron transport components; also contains some DNA and ribosomes	Fatty acid oxidation, electron transport, and ATP production	Independent unit capable of self-duplication inside cell only
Chloroplast	Membranous system with an inner membrane organized into thylakoids containing glycolipid, phospholipid, and protein; also contains pigments such as chlorophyll and some DNA	Conversion of light energy into ATP and reduction of carbon dioxide to organic molecules	Independent unit capable of self-duplication inside cell only
Golgi body	Membranous organization of sacs	Synthesis and formation of lysosomes; may be a site for refinement of protein synthesized on ER	Probably modifies proteins by acetylation, sulfation, and adding or removing sugars
Microtubules	Protein units called tubulin	Movement and reproduction; make up structural components of flagella and mitotic spindle	May also be involved in intracellular transport and in maintaining shape of wall-less protozoa; movement of chromosomes
Flagella or cilia	Both made up of protein microtubules; also contain ATPase and ATP	Movement associated with feeding, sexual activity, and aggregation; cilia shorter and more numerous on cell than flagella	Flagella or cilia more complex structurally than prokaryotic flagella
Nucleus	A porous unit membrane structure containing DNA; membrane continuous with endoplasmic reticulum	Contains hereditary information for cell; nucleolus involved in production of ribosomal RNA	DNA in nucleus associated with proteins called histones; organized into units called chromosomes

Self-study Quiz

Completion

Identify the structure in the eukaryotic cell associated with the following function or chemical component.

1. ________________ ATP formation
2. ________________ Chitin
3. ________________ Tubulin
4. ________________ Protein synthesis
5. ________________ Sterols
6. ________________ Photosynthesis
7. ________________ Histones
8. ________________ Cytochromes
9. ________________ Protein glycosylation
10. ________________ Compartmentalization of hereditary material

Multiple choice

1. The fungal cell wall is not sensitive to penicillin because it does not contain:
   A. Cholesterol
   B. Chitin
   C. Peptidoglycan
   D. Teichoic acid
2. The smooth endoplasmic reticulum is associated with which of the following:
   A. Synthesis of secretory proteins
   B. Synthesis of enzymatic proteins
   C. Synthesis of ribosomes for rough endoplasmic reticulum
   D. Synthesis of lipids
3. Extrusion of undigested material to the outside of the cell is called:
   A. Pinocytosis
   B. Phagocytosis
   C. Biotransformation
   D. Exocytosis
4. Glycoproteins reaching the Golgi body have which of the following sugars initially removed:
   A. Sialic acid
   B. Mannose
   C. Glucose
   D. *N*-acetylglucosamine
5. Lysosomes are composed of which of the following materials:
   A. Hydrolytic enzymes
   B. Enzymes associated with protein synthesis
   C. Enzymes associated with lysosome membrane synthesis
   D. Processed glycoprotein from the Golgi body
6. The stacked units of thylakoids found in the chloroplasts of plants are called:
   A. Stroma
   B. Grana
   C. Dynein
   D. Chlorosomes
7. Cytoplasmic streaming is not associated with which of the following:
   A. Prokaryotes
   B. Nutrient distribution in the cell
   C. Eukaryotic metabolism
   D. Ameboid movement
8. The ciliary structure believed to be required for bending of the cilium is:
   A. Basal body
   B. Dynein arms
   C. Nexin links
   D. Radial spokes
9. Where in the cell would you expect to find the greatest amount of protein synthesis?
   A. Smooth endoplasmic reticulum
   B. Rough endoplasmic reticulum
   C. Golgi
   D. Nucleolus
10. The principal function of the nucleolus is:
   A. Synthesis of ribosomal protein
   B. Synthesis of cytoplasmic protein
   C. Synthesis of ribosomal RNA
   D. Synthesis of mitochondrial proteins

Thought questions

1. What might be the explanation for mitochondrial ribosomes resembling bacterial ribosomes?
2. Why would a developmental biologist be interested in the organism *Dictyostelium discoideum?*

SELECTED READINGS

Books

Alberts, B., et al. *Molecular biology of the cell.* Garland Publishing, Inc., New York, 1983.

Carlile, M.J. (Ed.) *Primitive sensory and communication systems: the taxes and tropisms of microorganisms and cells.* Academic Press, Inc., New York, 1975.

Cole, G.T. Architecture and chemistry of the cell walls of higher fungi. In Schlessinger, D. (Ed.). *Microbiology 1981.* American Society for Microbiology, Washington, D.C.

Dyson R.D. *Essentials of cell biology* (2nd ed.). Allyn & Bacon, Boston, 1978.

Goldman, R. (Ed.). *Cell motility.* Cold Spring Harbor Laboratory, Cold Spring Harbor, N.Y., 1976.

Goodaa, G.W., Lloyd, D., and Trinci, A.P.J. (Eds.). *The eukaryotic microbial cell: thirtieth symposium of the Society for General Microbiology,* Cambridge University Press, New York, 1980.

Hoch, J.A., and Setlow, P. (Eds.). *Molecular biology and microbial differentiation.* American Society for Microbiology, Washington, D.C., 1984.

Peberdy, J.F. *Developmental microbiology.* Blackie and Son, Ltd., Glasgow, 1980.

Reissig, J.L. (Ed.). *Microbial interactions.* Chapman and Hall, Ltd., London, 1977.

Journal articles

Calvalier-Smith, T. The origin and early evolution of the eukaryotic cell. *Symp. Soc. Gen. Microbiol.* **32**:33, 1981.

Farkas, V. Biosynthesis of cell walls of fungi. *Microbiol. Rev.* **43**:117, 1979.

Rothman, J.E. The Golgi apparatus. *Science* **213**:1212, 1981.

Satir, P. How cilia move. *Sci. Am.* **243**:44, 1980.

Schwartz, R.M., and Dayhoff, M.O. Origins of prokaryotes, eukaryotes, mitochondria, and chloroplasts. *Science* **199**:395, 1978.

Wood, U.G. The structure and function of eukaryotic ribosomes. *Annu. Rev. Biochem.* **48**:719, 1979.

PART

II

Microbial genetics

CHAPTER

8

DNA replication, plasmids, and other DNA elements

Chapter Outline

Introduction

Genetics is the science that deals with the mechanisms by which organisms pass traits from one generation to another. Although considerable differences exist in the traits expressed by various groups of organisms, the nature of hereditary material and the chemical and physical factors that affect it are the same. Microorganisms, especially bacteria, are used to study genetic mechanisms because they multiply rapidly, are easily cultivated in the laboratory, are not burdened by complex cellular interactions as are multicellular organisms, are haploid, and have genetic material that is not surrounded by a nuclear membrane. Genetic studies with prokaryotic organisms make it possible to analyze genetic mechanisms in higher organisms.

This chapter describes the chemical and physical mechanisms used by bacteria and eukaryotes to replicate their DNA before cell division and the characteristics of plasmid DNA, which is not part of the cell's chromosomal DNA, and how it confers special properties on the bacterial cell. Also discussed is how certain small elements of DNA, transposons and insertion sequences, can move from one position on the chromosome to another position on the same or different chromosome to alter genetic traits.

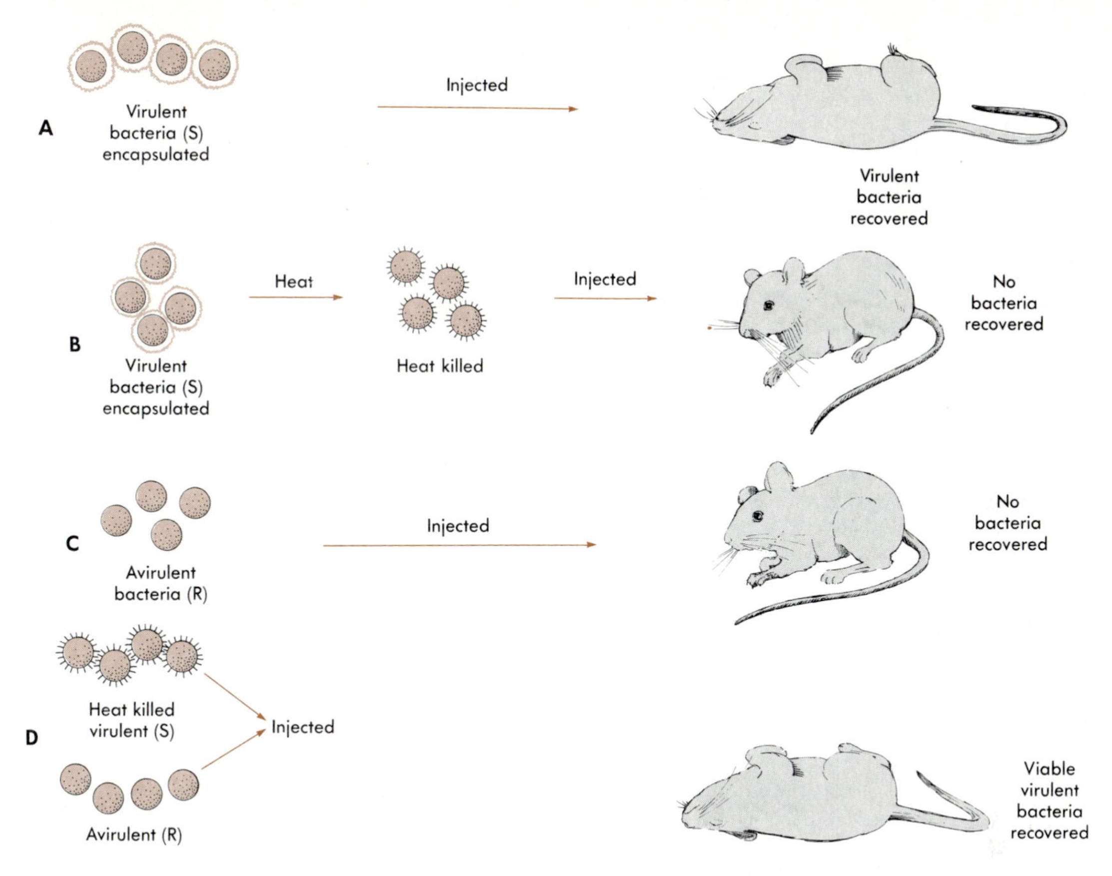

Figure 8-1 *Transformation experiment. **A,** Virulent encapsulated* Streptococcus pneumoniae *cells (S type) are injected into mouse. Mouse dies, and virulent bacteria recovered. **B,** Virulent encapsulated cells of the organism are heat killed and injected into mouse. Mouse survives and no bacteria are recovered. **C,** Avirulent cells (R type) of the organism contain no capsule and are injected into mouse. Mouse survives and no bacteria are recovered (nonspecific immune cells of host destroy the bacteria). **D,** Experiments **B** and **C** are combined by injecting heat-killed virulent plus live avirulent bacteria into mouse. Mouse dies, and viable virulent bacteria are recovered, indicating that avirulent cells were somehow transformed into virulent types.*

DNA REPLICATION

Nature of the hereditary material

The nature of hereditary material remained a mystery until a series of experiments, beginning in 1928, revealed important genetic mechanisms. Each experiment, in its own way, provided vital information that would eventually help us understand the replication process.

Griffith in 1928 showed that the disease-producing ability of *Streptococcus pneumoniae* was due to the formation of a capsule. Encapsulated strains produce a smooth colony on laboratory media and are referred to as *S* strains. Strains of *S. pneumoniae* that do not produce a capsule form rough colonies and are referred to as *R* strains. Laboratory mice die of septicemia when injected with a smooth strain but are unaffected by injected R strains (Figure 8-1). In later experiments Griffith used a combination of live R bacteria and heat-killed S bacteria to inject into mice. He noted that not only did the mice die, but smooth strains could be isolated from their blood. Griffith reasoned that something from the dead S strains had converted the R strains into smooth strains—a process called **transformation**.

Avery and others in 1944 more clearly established that the **transforming principle** was DNA. They isolated DNA from S strains of *S. pneumoniae* and demonstrated in vitro that the nucleic acid could transform an R strain into an S strain (Figure 8-2). Furthermore, the isolated DNA fraction, when incubated in the presence of a DNA-hydrolyzing enzyme called deoxyribonuclease (DNase), lost its transforming properties.

Chemical analysis of the nucleic acid by Chargaff and others in the 1950s revealed a composition of phosphate, sugar, and purine and pyrimidine bases. They also established that there were equal amounts of purines and pyrimidines in the DNA. Shortly thereafter, Wilkins and coworkers, using x-ray diffraction techniques, noted that DNA formed a helix. This prompted Watson and Crick to propose a model for the DNA molecule. Chemical and physical properties of this molecule are discussed in Chapter 2.

Hershey and Chase in 1952 demonstrated the genetic nature of DNA with experiments involving bacterial viruses. They studied the viral infection process in the bacterium *Escherichia coli* by labeling the viral protein and nucleic acid with radioactive isotopes. Viral particles can be labeled by allowing them to replicate in bacteria in which the bacterial growth medium contains radioactive sulfur (^{35}S) and phosphorus (^{32}P). Radioactive sulfur is incorporated preferentially into viral protein, and radioactive phosphorus is incorporated into viral DNA (Figure 8-3). Unlabeled bacteria were infected with the doubly labeled virus. A few minutes after the adsorption process, the suspension of infected bacteria was agitated in a blender. This procedure removed the radioactive viral protein coat from the bacterial surface but did not interfere with the production of new viral particles, because viral DNA penetrated the cell. DNA was considered the hereditary material and thought to be responsible for the production of new viral particles.

The hereditary material is usually a double-stranded DNA molecule, although variations do exist in the viruses. For example, in some viruses the hereditary material is RNA, existing either as a single strand or double-stranded molecule. Few viruses possess single-stranded DNA instead of double-stranded. Chemical and physical properties of the hereditary material are the same de-

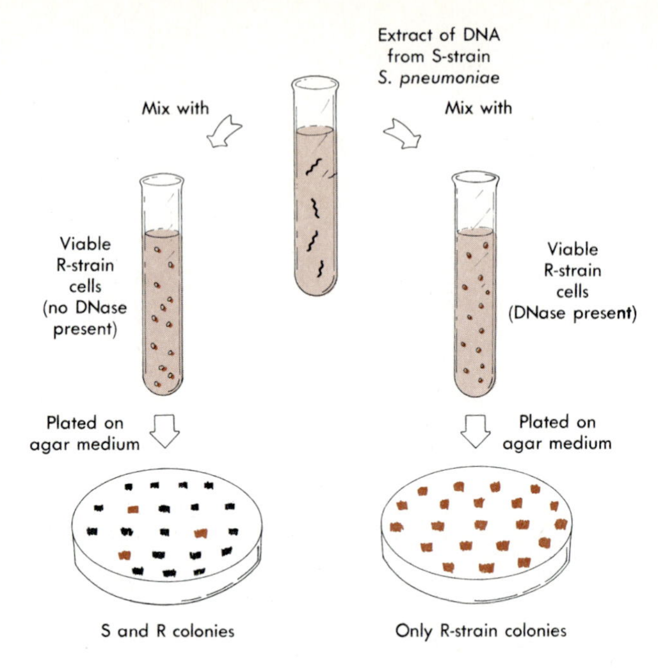

Figure 8-2 *In vitro transformation process. DNA extracted from smooth* (S) *strain (virulent) of* S. pneumoniae *and mixed with viable rough* (R) *strain cells (avirulent) with and without DNase. In absence of DNase, some R strain cells convert to S strain cells, but in presence, only R strain cells appear following cultivation. Transforming factor hydrolyzed in DNase presence.*

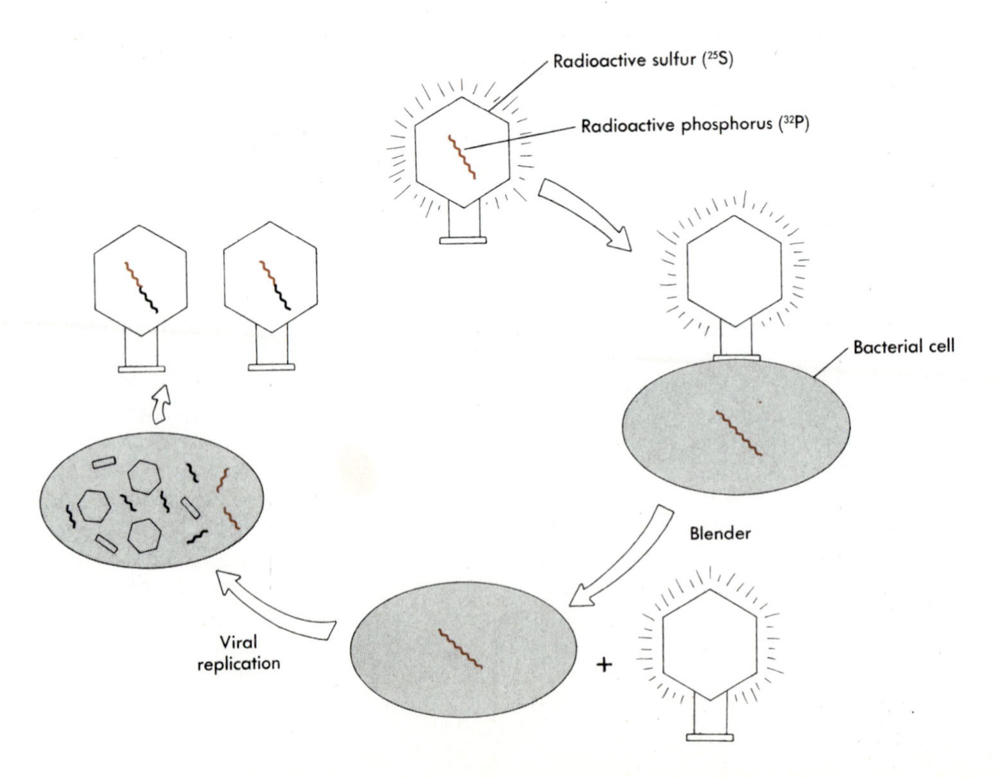

Figure 8-3
Experiment used to prove that only DNA and not protein of virus enter bacterial cell following infection. Agitation in blender removes radioactive sulfur from cell, but radioactive phosphorous, which is part of viral DNA (color), remains inside cell. Newly produced viral particles contain radioactive phosphorous but not radioactive sulfur. Uncolored DNA is nonradioactive. See text for details.

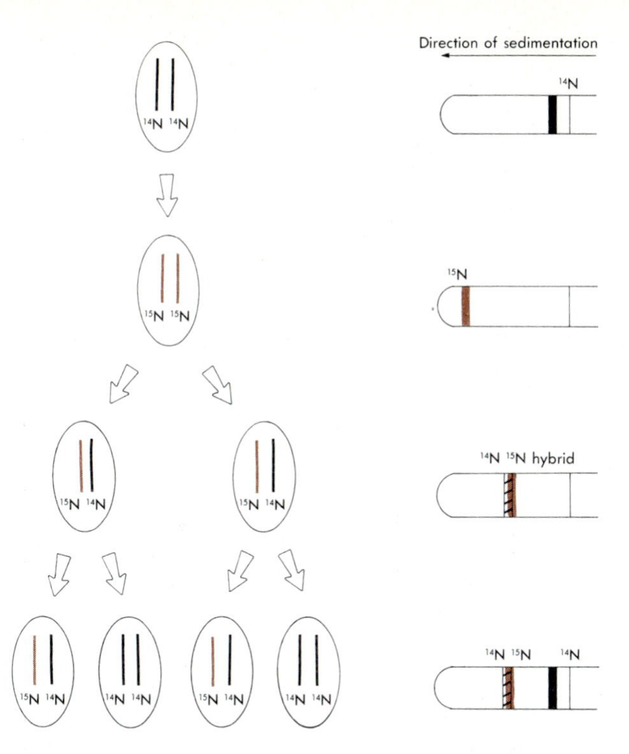

Figure 8-4 *Schematic representation of Meselson-Stahl experiment demonstrating semiconservative replication.* E. coli *cells containing light nitrogen (^{14}N) are grown for several generations in medium containing heavy nitrogen (^{15}N). Heavy cells are transferred to medium containing only light nitrogen and are then harvested after one generation (one doubling) and then harvested after subsequent generations. DNA was extracted from cells and subjected to centrifugation in cesium chloride. All cells of first generation contain hybrid DNA in which one strand contains ^{14}N and one ^{15}N. Population of cells harvested after two generations is of two types. One half contains hybrid DNA and the other half contains only light DNA. Positions of DNA following density gradient centrifugation are shown on the right.*

spite variations, and they all share the same properties or functions in the cell, which include the ability:

1. To replicate and transfer hereditary molecules to progeny during cell division. A unit of replication is a **replicon**, that is, it contains a replication origin and is capable of autonomous replication, for example, a bacterial or viral chromosome or a plasmid (see p. 206).
2. To mutate (see Chapter 10).
3. To act as a blueprint for the expression of nucleic acid information into functional proteins (enzymes) or molecules that regulate metabolic processes (see Chapter 9).
4. To transfer wholly as chromosomes or in part as genes or gene clusters to other cell types (outside cell division) and engage in recombination with other nucleic acid molecules (see Chapter 11).

DNA REPLICATION MECHANISMS

Semiconservative replication

What mechanisms in the cell ensure that genetic determinants will be passed faithfully from one generation to another? Watson and Crick suggested that during replication the chains of the DNA molecule separate and that each acts as a template from which a complementary chain can be produced. The evidence that strand separation did occur came from the laboratories of Meselson and Stahl in 1958. They grew *E. coli* in a medium containing heavy nitrogen (^{15}N) for several generations until all of the DNA contained the heavy isotope. The cells were collected and then transferred to a medium containing light nitrogen (^{14}N). Some cells were allowed to divide just once and were then removed from the medium, and others were removed after two or more generations. DNA was extracted from the cells and centrifuged in cesium chloride. DNA containing the heavier isotope had a greater density and sedimented faster, DNA containing light nitrogen did not sediment as rapidly and remained near the top of the centrifuge tube. DNA isolated after the cell population had doubled was shown to be intermediate in density between DNA containing all light nitrogen and DNA containing all heavy nitrogen. This demonstrated that the parent molecule does separate and act as a template for the synthesis of a complementary strand (Figure 8-4). They concluded that replication of DNA is semiconservative; that is, one parent DNA strand is conserved during the replication process, whereas its complementary strand is newly synthesized.

DNA polymerase and the replication process

Meselson and Stahl confirmed the theory of strand separation postulated by Watson and Crick but gave no idea as to mechanism(s) involved or enzymes required. Much of the enzymatic details of replication were discovered by Kornberg and his colleagues in the 1950s. They found that in vitro DNA synthesis required the following components:

1. Mixture of deoxyribonucleoside 5′-triphosphates: dATP, dGTP, dCTP, and dTTP. (Omission of any resulted in no replication.)
2. Magnesium ion (Mg^{++}). (Magnesium stabilizes DNA by binding to negatively charged phosphate groups.)
3. High molecular weight DNA. (DNA acts as **primer**, or growing point, for addition of nucleotides and **template**; that is, the product formed is complementary to the original DNA molecule.)
4. DNA polymerase.

The DNA polymerase enzyme isolated by Kornberg required a free 3′ hydroxyl on one of the DNA strands of the primer template and a region of single strandedness along the other strand. This meant that the direction of DNA synthesis was only in the 5′ to 3′ direction (Figure

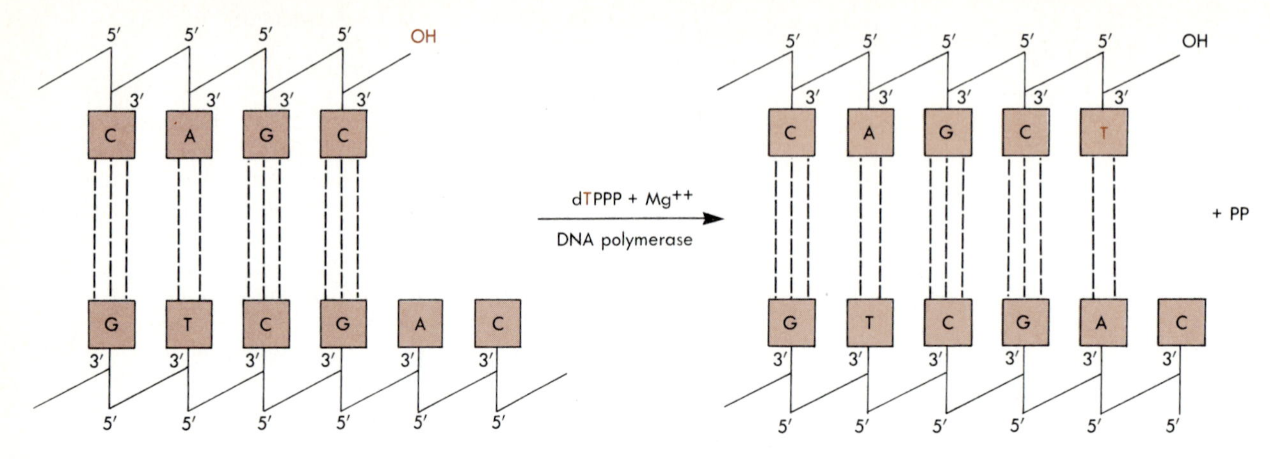

Figure 8-5 *Requirements for in vitro DNA synthesis. Double-stranded DNA required for synthesis, with one strand containing free 3′ hydroxyl* (OH) *group for continued synthesis. Through action of DNA polymerase, pyrophosphate* (PP) *is cleaved from deoxythymidine triphosphate* (dTPPP) *and nucleotide* dTMP *is added to 3′ hydroxyl group on growing chain.*

8-5). The DNA used as template could be from sources other than *E. coli*. Every experiment of Kornberg's using *E. coli* DNA or other DNA showed that the ratios of each nucleotide in the template were equivalent to those in the product. For example, if the template contained adenine, thymine, guanine, and cytosine, nucleotides in the ratio 0.64 to 0.66 to 1.34 to 1.35, the same ratio would appear in the product. Thus the Kornberg enzyme could catalyze the self-copying of DNA and 3′ ends along each strand could serve as growing points. Since Kornberg's discovery of DNA polymerase a number of DNA polymerases were discovered: three in *E. coli* and four in eukaryotes. Their precise functions are not entirely clear but in *E. coli*, **DNA polymerase I** is believed to be a repair enzyme, whereas **DNA polymerase III** is the major enzyme involved in replication. **DNA polymerase II** may also be a repair enzyme. Most of the prokaryotic DNA polymerases, in addition to 5′ to 3′ polymerase activity, also exhibit **exonuclease** activity that may be in the 3′ to 5′ or 5′ to 3′ direction, that is, nucleotides are removed from the 3′ or 5′ ends, respectively, of polynucleotides. In this capacity they act as repair enzymes (Figure 8-6). DNA repair is discussed in Chapter 10.

Structure of DNA in the cell

DNA is a right-handed helix (looking down the axis of DNA, each strand follows a clockwise path), which exists in the bacterial cell as a circle (Figure 8-7, *A*), that is, the two ends of a linear DNA molecule are brought together and joined. The contour length of a circular DNA molecule is 1100 μm, and because *E. coli* has a diameter of 1 to 2 μm, the DNA must be packaged in a highly folded or coiled configuration. DNA can be twisted around its axis to produce **supercoils** (Figure 8-7, *B*). Twisting a rubber band about itself is a close analogy. Many proteins or enzymes will not bind to DNA unless it is supercoiled. The type of supercoiling necessary for DNA functions, such as strand separation and replication, is called **negative supercoiling**, that is, the DNA is twisted in a direction opposite to the clockwise turns of the right-handed helix. DNA with negative supercoils is **unwound** and occurs because one of the DNA strands is broken by an enzyme, **DNA gyrase**. Negative supercoiling may be converted into strand separation by DNA gyrase (Figure 8-7, *C*). This enzyme belongs to a special class called **topoisomerases** that convert DNA into different topological configurations. There are two types: **type I isomerase** that can cut single strands of DNA and **type II**, such as gyrase, that cuts two DNA strands simultaneously. Enzymes and the replication process are discussed later.

DNA replication (bidirectional vs. unidirectional)

In 1962 Cairns used autoradiography to show how DNA synthesis occurs in the microbial cell. In the experi-

Concept Check

1. Which experiments helped confirm the theory that DNA was the genetic material of the cell?
2. Describe the experiment that demonstrated the semiconservative nature of DNA replication in bacteria.
3. What is meant by negative supercoiling and what is its importance in DNA replication?
4. What are the chemical components that are needed to carry out DNA synthesis in the test tube?

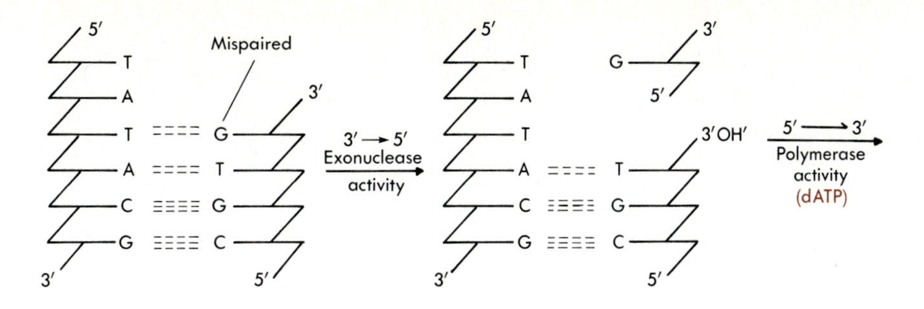

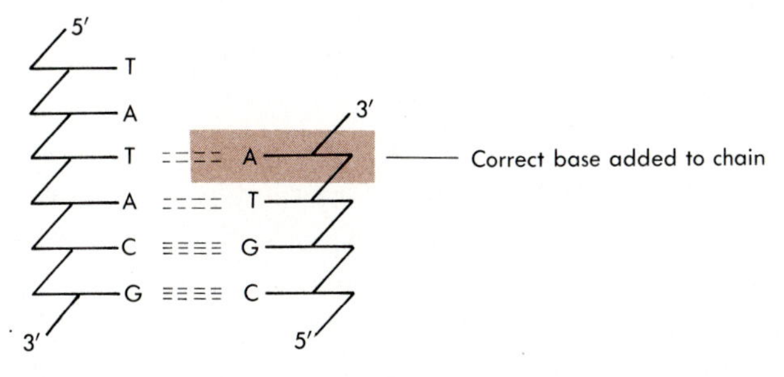

Figure 8-6 Exonuclease and polymerase activity of DNA polymerase. During replication mistake is made and incorrect nucleotide (GTP) is added to primer strand. Replication is halted, DNA polymerase recognizes error, and its activity is shifted to that of exonuclease, whereby mispaired nucleotide is removed. DNA polymerase now exhibits polymerase activity and correct nucleotide (ATP) is covalently attached to primer strand. Estimated error rate is between 10^{-8} to 10^{-11} per base pair synthesized. (100 million to 100 billion base pairs are synthesized before error.)

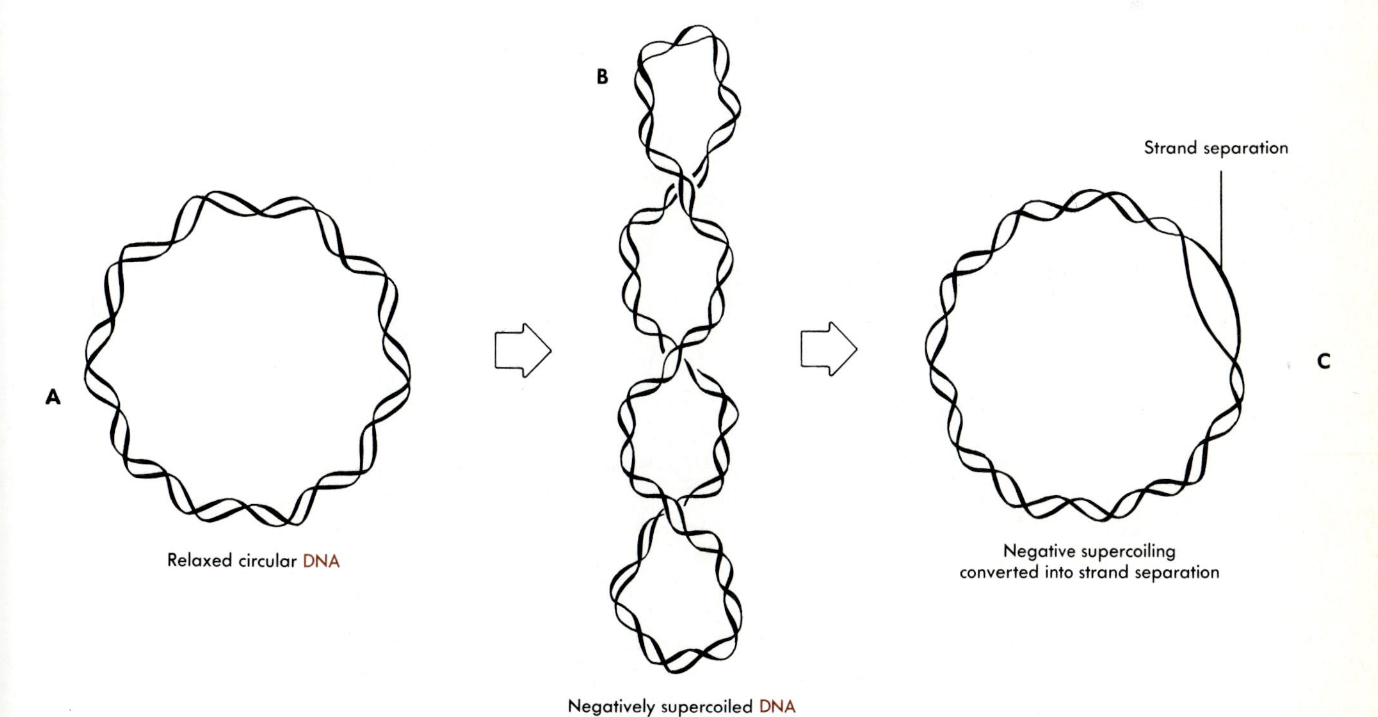

Figure 8-7 Conformational forms of DNA. **A,** Circular relaxed form of double-stranded DNA that can be converted into negatively supercoiled form **(B)** by action of DNA gyrase. Negative supercoiling can be transformed into strand separation **(C)**, a requirement for DNA replication.

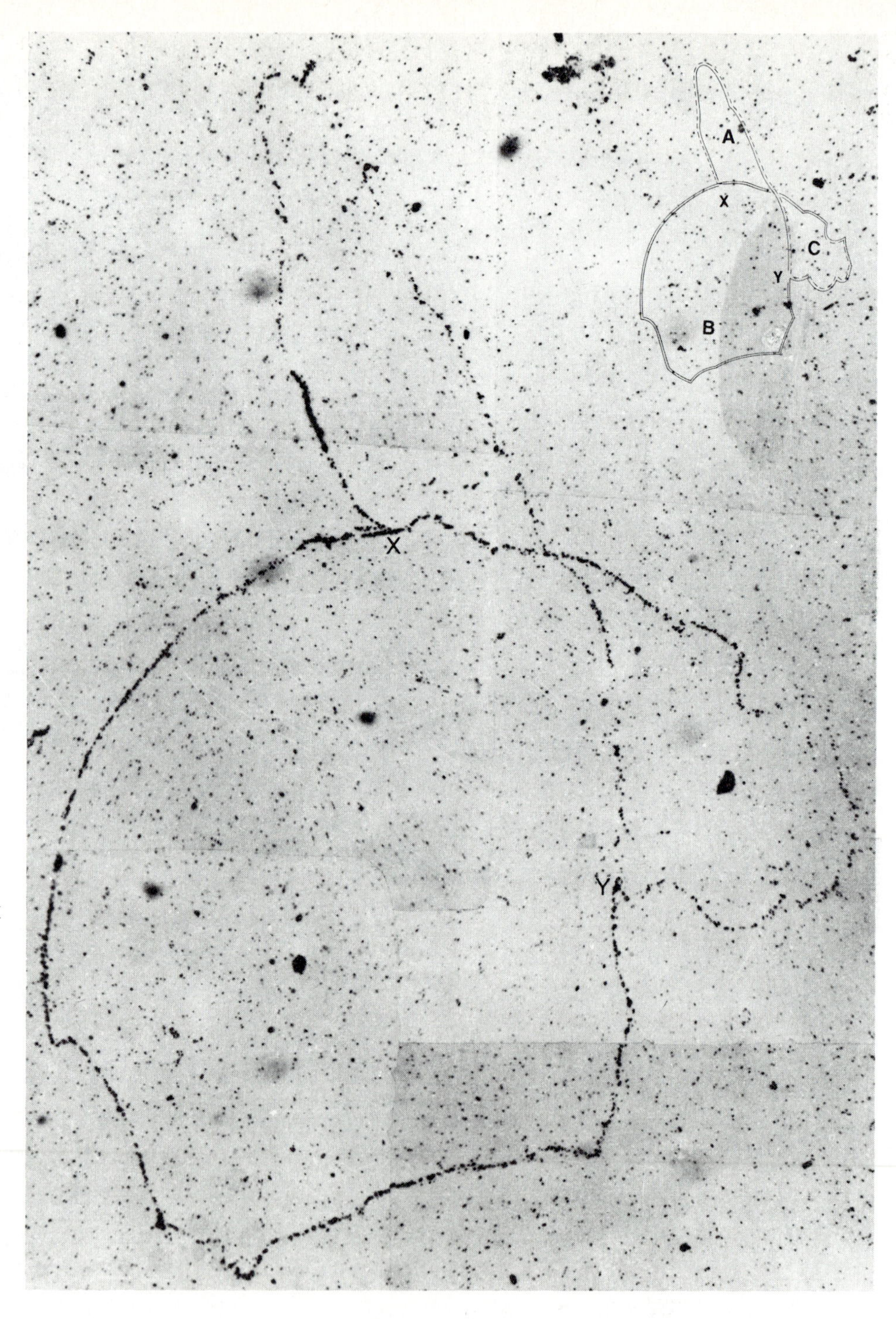

Figure 8-8
Autoradiograph technique demonstrating duplication of E. coli *chromosome. Tritiated thymidine is revealed as line of dark grains in photographic emulsion. Interpretation of autoradiograph is based on varying densities of grains. Insert is interpretive drawing showing doubly labeled DNA as two solid lines; singly labeled DNA is solid line and dashed line. Duplication began at* X *with newly labeled strands extending as far as* Y. *Loop* B *exhibits twice grain density of loop* A *(both strands contain tritiated thymidine). Loops have completed second round of replication in thymidine. Section* C *has replicated only once. Originally believed either* X *or* Y *was replication fork and other was "swivel" that served as axis of rotation for unwinding DNA molecule. Swivel would represent terminus. Unidirectional model was later demonstrated in error and replication was actually bidirectional, that is, both* X *and* Y *were replication forks.*

ment, bacteria are grown in a medium containing tritiated (^3H) thymidine. The DNA, which is labeled, is gently isolated from the bacteria and covered with a photographic emulsion on a microscopic slide. The slide is developed, and observation with a light microscope reveals a pattern of silver grains that outlines the structure of the chromosome. Interpretation of the autoradiograph (Figure 8-8) is based on the varying densities of the lines of grains. Cairns' interpretation showed that semiconservative replication occurred at a site called the **origin** and proceeded in one direction (unidirectional replication) around the chromosome.

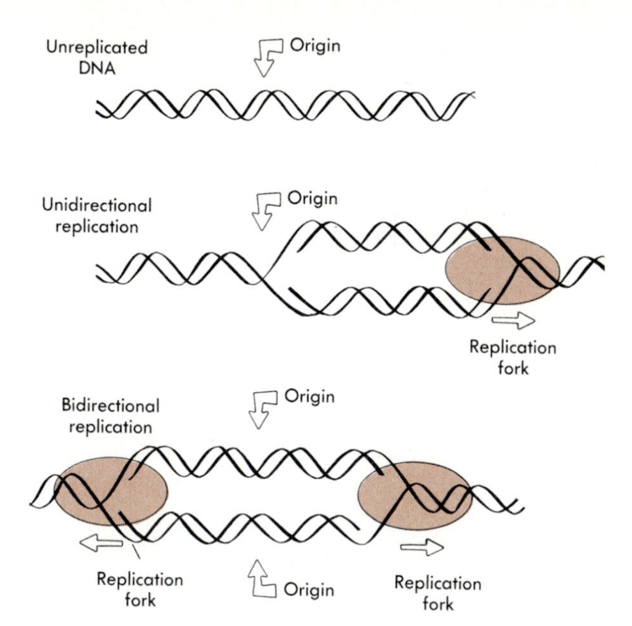

Figure 8-9 *Comparison of unidirectional and bidirectional replication. One replication fork in unidirectional replication and two replicating forks in bidirectional replication.*

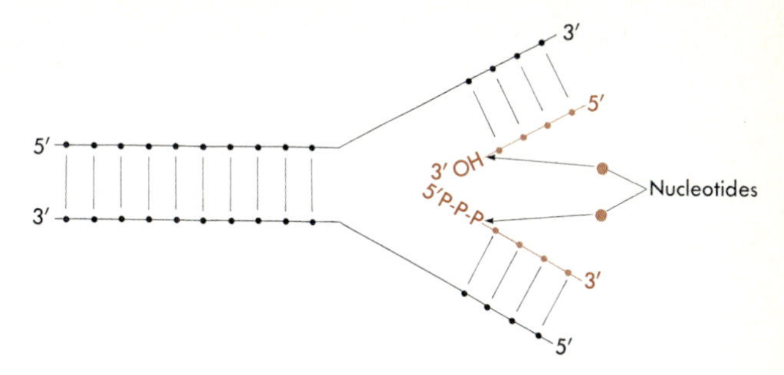

Figure 8-10 *Simplest method, but incorrect one, for DNA replication. Arrows indicate sites (3′-OH and 5′-phosphate) where nucleotides would have to be added for replication to continue in same direction.*

The site of replication is shaped like a fork and referred to as a **replicating fork** (see *X* and *Y*, Figure 8-8). Another fork-shaped site on the chromosome was believed to be a swivel for unwinding the DNA. Subsequent experiments showed that both forks are replicating forks and that each **Y**-shaped fork moves in an opposite direction around the circular chromosome (bidirectional replication) until a **termination site** 180 degrees from the origin is reached. Bidirectional replication occurs in organisms and most viruses but it is not universal. Some bacterial viruses with a circular chromosome replicate unidirectionally (Figure 8-9).

Discontinuous DNA synthesis. Although resolving a fundamental question, the Cairns' experiment raised two others. First, can DNA synthesis take place on both DNA strands if they are antiparallel and if the DNA polymerase enzyme catalyzes DNA synthesis only in the 5′ to 3′ direction? For synthesis of both strands to occur in the same direction the cell needs two DNA polymerases; one requires a nucleotide containing a free 3′ hydroxyl and the other requires a free 5′ phosphate (Figure 8-10). Second, if the strands unwind at one end, the opposite end must rotate continuously. A closed circle characteristic of the chromosome seems to prevent this necessary type of rotation.

In the early 1970s Okazaki and others tried to resolve the problem of how DNA synthesis could proceed on both strands of the DNA from a fixed point. They added radioactive DNA precursors to cultures of *E. coli* and isolated DNA from the culture at various times before one round of replication took place. DNA was extracted and centrifuged in a sucrose gradient and analyzed for radioactivity. They demonstrated that at early times during the replication process most of the radioactivity was associated with low molecular weight DNA, whereas at later times in the replication process the radioactivity was predominantly part of high molecular weight DNA. They concluded that short pieces of DNA are synthesized early (the short pieces are now referred to as **Okazaki fragments**) and later are covalently bonded to produce high molecular weight DNA.

Okazaki's results demonstrate that DNA synthesis on one strand is **discontinuous**, while DNA synthesis on the other strand is **continuous**. Both strands are synthesized in the 5′ to 3′ direction but the strand that is being extended overall in the 3′ to 5′ direction grows by the synthesis of Okazaki fragments, which are subsequently joined together by a polynucleotide ligase. Thus DNA synthesis is continuous for the strand that grows overall in the 5′ to 3′ direction (also called the **leading strand**) and is discontinuous for the strand growing overall in the 3′ to 5′ direction (also called the **lagging strand**) (Figure 8-11). The replication process in *E. coli* and the proteins involved are as follows:

1. Replication events begin at the origin where the two strands of DNA have separated. The replicating proteins are assembled into a complex referred to as the **primosome**. The protein complex can then follow the replicating fork during the synthesis of DNA.
2. One of the DNA strands at the origin is nicked, thus exposing one end of a single strand. Before Okazaki fragments can be synthesized on the lagging strand a special polymerizing enzyme, **RNA primase**,* is needed. RNA primase, unlike DNA polymerase, can start a polynucleotide chain by

* Two enzymes synthesize RNA primer molecules. *RNA primase* is associated with lagging strand synthesis, whereas *RNA polymerase* is involved in leading strand synthesis.

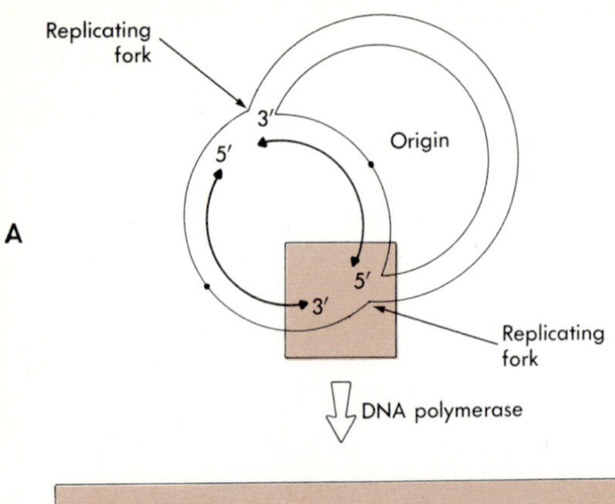

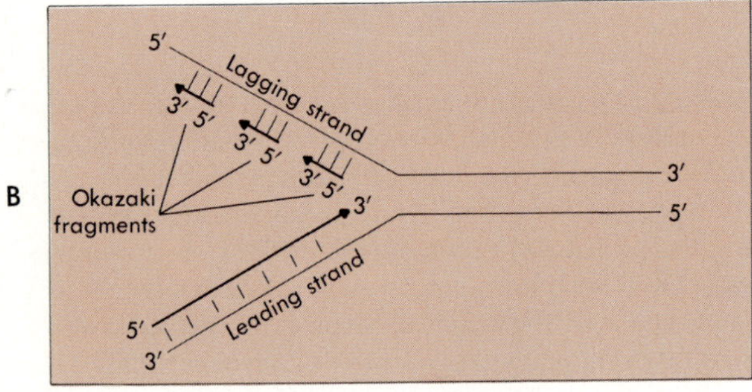

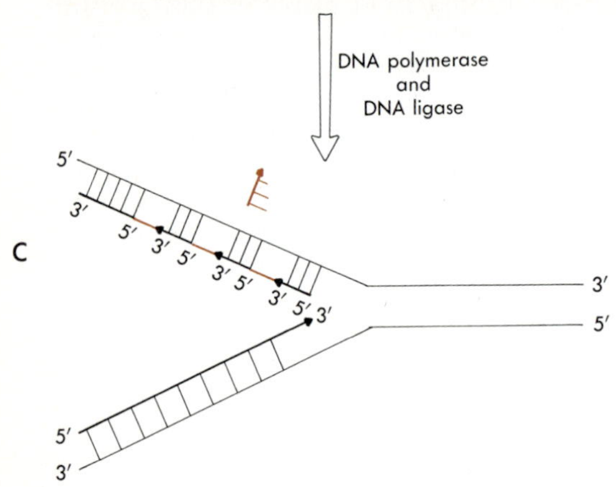

Figure 8-11 *Bidirectional replication coordinated with autoradiographic analysis.* ***A,*** *Autoradiographic analysis shows both newly synthesized DNA chains are extended overall in same direction at each replication fork.* ***B,*** *Analysis of replication fork shows DNA synthesis to be discontinuous on lagging strand but continuous on leading strand.* ***C,*** *As DNA synthesis continues short Okazaki fragments on lagging strand are joined by DNA ligase.*

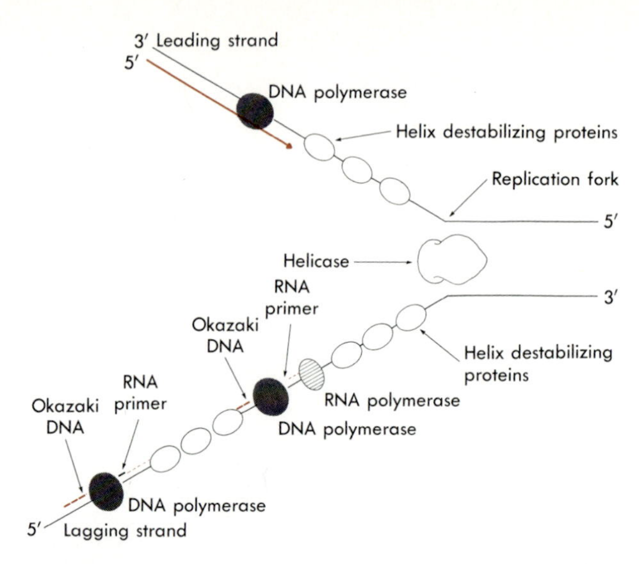

Figure 8-12 *Detailed scheme for discontinuous replication of DNA. DNA helix is opened at replication origin producing replicating fork. DNA helicase helps to unwind DNA. Helix destabilizing proteins maintain DNA in unwound state ahead of replicating fork. Leading strand provides template for continuous synthesis of DNA via DNA polymerase III. Short pieces of RNA synthesized on lagging strand by RNA primase serve as primers for addition of DNA called Okazaki fragments. Later RNA primers will be removed and deoxyribonucleotides added in their place by DNA polymerase I. Okazaki fragments will be linked to produce complete DNA strand.*

joining together two nucleoside triphosphates. RNA primase uses ribonucleotide triphosphates to synthesize RNA primers that are approximately 10 nucleotides long. Primers are made at intervals on the lagging strand. Later, DNA polymerase III is used to elongate the RNA primers, thus producing Okazaki fragments. The elongation by DNA polymerase III continues until it runs into the 5′ end of the previous DNA fragment. The RNA primers are removed by repair enzymes (polymerase I) and are replaced with DNA. The enzyme, **DNA ligase**, joins the 3′ hydroxyl end of the new DNA fragment to the 5′ phosphate end of the previous one (Figure 8-11), and the daughter strand complimentary to the lagging strand is synthesized.

The rate of DNA synthesis in *E. coli* is rather amazing. The *E. coli* chromosome has approximately 3.8×10^6 base pairs, and the cell divides on an average of 30 to 40 minutes in most media. This means that to duplicate the chromosome the DNA must polymerize at an overall rate of 1584 base pairs per second (2400 seconds if cell division time is 40 minutes). Because DNA replication in *E. coli* is bidirectional, each replicating fork polymerizes at a rate of 792 base pairs per second.

There are two possibilities for initiation of leading-strand synthesis. One, a primer RNA similar to one used for the lagging strand, is initiated but once the initial primer is made there is continu-

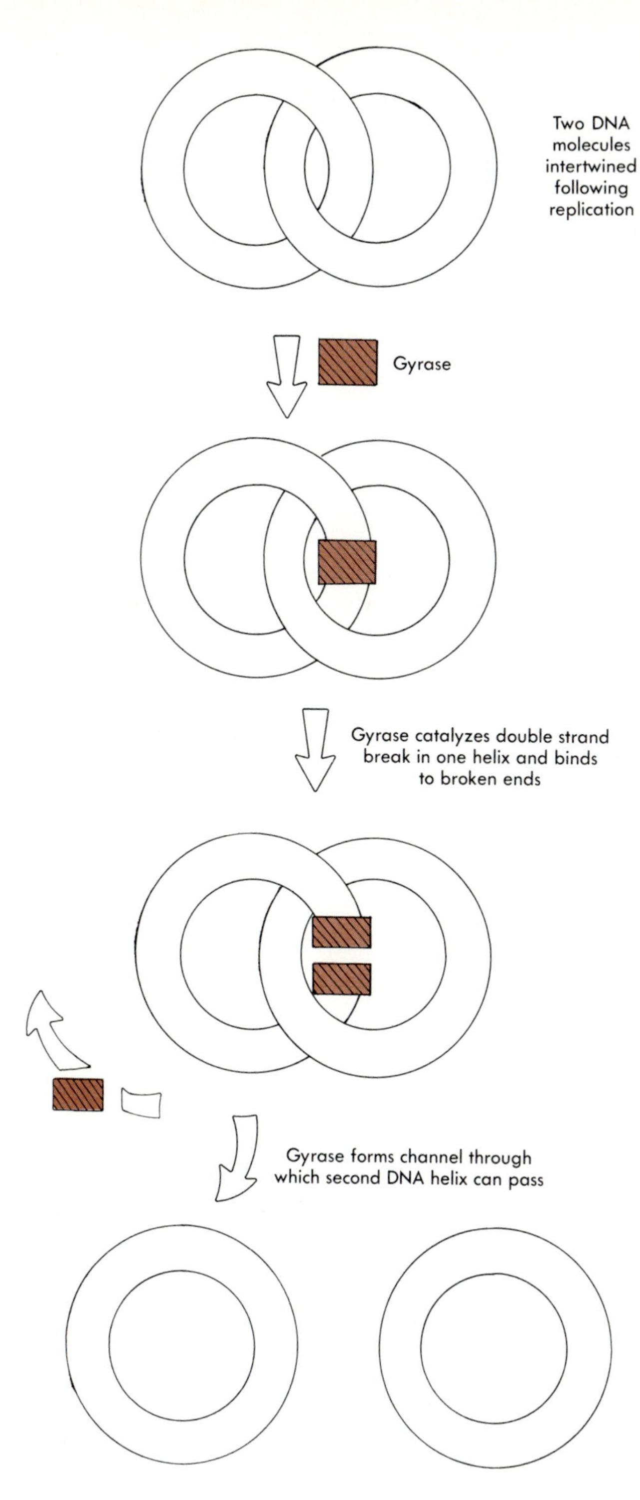
Two DNA molecules intertwined following replication
Gyrase
Gyrase catalyzes double strand break in one helix and binds to broken ends
Gyrase forms channel through which second DNA helix can pass

Figure 8-13 *Model showing how topoisomerase type II enzymes (gyrases) permit passage of one DNA duplex through another.*

ous elongation and no further RNA primers are required. Second, a nick is introduced into the parental strand whose 3′ hydroxyl is used as a primer. At this time the first proposal is favored.

3. The synthesis of DNA is not as straightforward as implied in the previous discussion. The DNA molecule, once it is opened at the origin, must be continually unwound in front of the replication fork to expose the template strand. In addition, the unwound DNA must maintain that state, a process that requires energy. Three types of proteins are involved in these processes: helix destabilizing proteins, DNA helicase, and topoisomerase (Figure 8-12). **DNA helicases** are enzymes that participate in the unwinding of DNA just ahead of the replication fork. They exhibit DNA dependent ATPase activity, thus energy is provided for their activity by the hydrolysis of ATP. Once the DNA is unwound, **helix destabilizing proteins** bind to the DNA to keep the strands separated. Type I topoisomerase acts as a swivel that relieves the torsional strain placed on the DNA ahead of the replication fork. If this strain were not relieved, fork movement would cease.

DNA gyrases appear to be important once replication has terminated. DNA gyrases can break both strands of DNA at once and attach covalently to the free 5′ end of each strand. This double-strand breaking facility appears to be important for the separation of interlocked daughter chromosomes (Figure 8-13). Gyrases may also be involved in preventing some of the tangling problems encountered when DNA is being replicated and unwound. The primary function of gyrases is to keep DNA in a negatively supercoiled form.

Rolling circle mechanism of DNA replication. In the rolling circle process of DNA replication (Figure 8-14) one of the DNA strands of an intact circle is cut by an endonuclease to produce a linear strand with 5′ and 3′ ends. The other strand remains as an intact circle. The 3′ end of the cut strand acts as a primer for chain elongation via DNA polymerase, using the closed intact circle as a template. DNA synthesis is completed when the intact single-stranded circle has revolved 360 degrees and all DNA complementary to the intact strand has been synthesized. The displaced portion of the replicating strand serves as a template for the synthesis of its complement. One advantage of this technique is that many copies of DNA can be produced (**concatameric** describes copies of DNA that are joined together) as long as the intact strand (the rolling circle) revolves. The linear copies produced are covalently attached to each other but they can be separated by enzymes. Each separated DNA molecule can circularize by way of single-stranded overlaps that come together. The rolling circle mechanism of DNA replication can be used by some bacterial viruses to generate concatameric molecules required during the packaging of DNA into individual virus particles.

DNA replication in eukaryotes. Eukaryotes contain many times the amount of DNA as prokaryotes, and this DNA is packaged into several chromosomes. Each chromosome is present in two or more copies. In addition, eukaryotic DNA is wrapped around histone octamers (see Chapter 7). Eukaryotic chromosomes contain repetitive DNA sequences of up to 300 nucleotides in length. As many as 100,000 copies of these repeated sequences may be found in a single cell. Some of these repetitive sequences constitute transposable elements in higher eukaryotes (see p. 216). Except for ribosomal RNA genes, repetitive sequences are not present in prokaryotes.

Once eukaryotic DNA synthesis begins, it will continue until all of the DNA is replicated. Replication of eukaryotic DNA, like the replication of prokaryotic DNA, occurs in a semiconservative manner. Eukaryotic DNA replication begins at the replication origin where a replication fork is produced. As yet, we do not know what type of sequence constitutes the origin. One suggestion is that it may be a member of the repetitive sequence family. The genome of prokaryotes is small enough that only one or two replication forks are needed to duplicate the DNA in the division cycle. Such is not the case for eukaryotes in which the DNA is 40 to 50 times longer than prokaryotic DNA. As many as 50 to 100 replication forks may be produced on each chromosome during the division cycle. Most replication forks appear in pairs with the two forks of a pair moving away from each other to produce a typical replication bubble (Figure 8-15). The evidence suggests that DNA synthesis is discontinuous, like that in prokaryotes, but the evidence for RNA primers initiating the replication process is still unsubstantiated in many eukaryotic systems. The Oka-

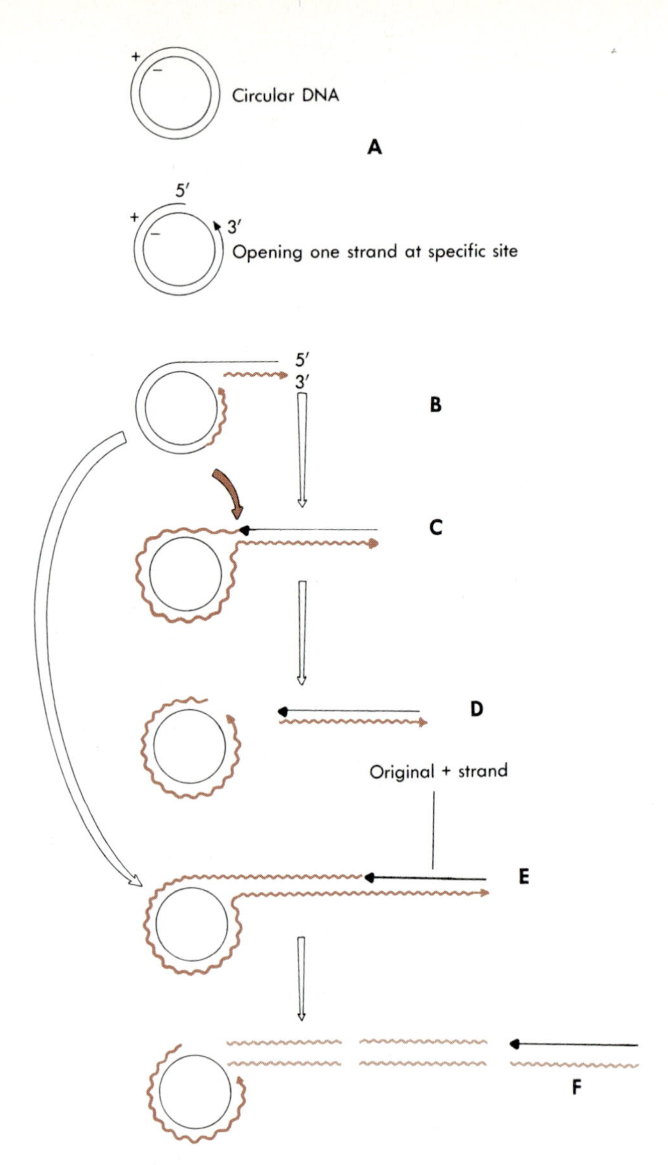

Figure 8-14 *Rolling circle mechanism of DNA synthesis.* ***A,*** *Double-stranded DNA molecule with positive (+) and negative (−) strands. Positive strand nicked by endonuclease.* ***B,*** *Intact negative strand begins to roll and 5′ end of positive strand is displaced (attached to membrane of cell). DNA synthesis on positive strand is begun as negative strand rolls with nucleotides added to 3′ end of positive strand (color). Displaced 5′ end of positive strand now acts as template and complementary DNA strand is synthesized (color).* ***C,*** *As negative strand has completed 360 degree turn, template strands are fully copied. Enzyme breaks single strands* (color arrow). ***D,*** *Two molecules of DNA, one circular and one linear have been synthesized.* ***E,*** *Several 360 degree rolls of negative strand (as in* ***B****) have been completed and ligase connects newly synthesized linear strands into concatamer.* ***F,*** *A site-specific nuclease cleaves concatamer into specific lengths of DNA.*

Concept Check

1. What is the difference between unidirectional and bidirectional replication? What is a replicating fork?
2. Why does DNA synthesis have to be discontinuous on one of the strands of DNA?
3. What was Okazaki's contribution to the mechanism of DNA synthesis?
4. What are the major enzymes involved in DNA strand separation and replication, and what is their function?
5. Why is the "rolling circle" a useful mechanism of DNA replication for viruses?
6. What is the most important difference between eukaryotic and prokaryotic DNA replication?

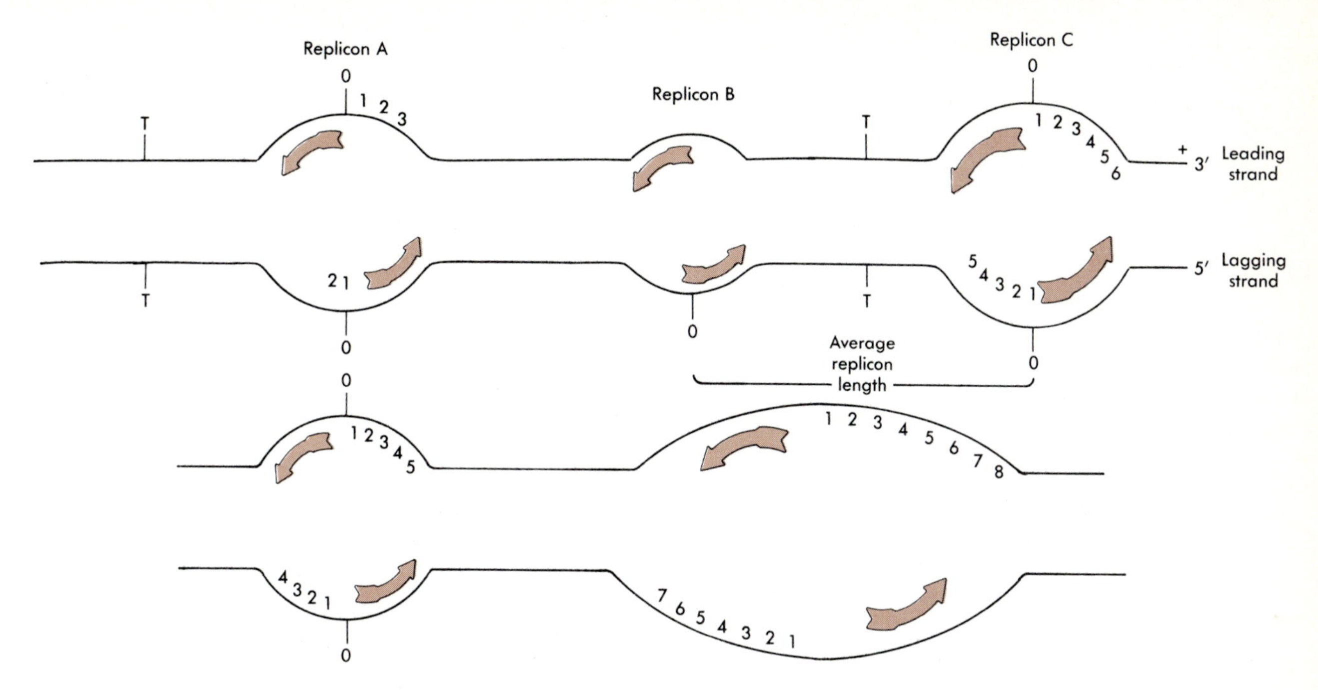

Figure 8-15 *Eukaryotic DNA replication. Eukaryotic DNA has several replication origins on single chromosome. Each replication unit has origin and two termination sites. Such replication units are called "replicons" and distance between each measures about 40,000 bases in lower eukaryotes, such as yeasts. As replication proceeds in each replicon, bubble is produced that enlarges until next replicon is met. Large arrows indicate direction of DNA synthesis, while numbers indicate relative number of nucleotides that have been added to complementary newly synthesized strand. Note distinction in number of nucleotides added to leading and lagging strands. When termination sites in adjacent replicons have been reached, they fuse and produce even larger bubble. Eventually all replicons will fuse and daughter strand DNA synthesis will be complete.*

zaki fragments produced during DNA synthesis in some eukaryotes are much smaller (10 to 60 nucleotides) than in prokaryotes (1000 to 2000 nucleotides).

PLASMIDS

Characteristics

Some of the genetic material in an uninfected microbial cell is extrachromosomal, that is, it is not part of the genome of the cell. These independent pieces of DNA are called **plasmids**. They are closed circles of double-stranded DNA having molecular weights of 10^6 to 10^8 (for comparison, the genome of *E. coli* has a molecular weight of 2×10^9) and are passed to daughter cells following cell division. Plasmids can also be "cured" or removed from the cell when the latter is subjected to certain environmental stresses, such as changes in temperature, the presence of certain dyes, or nutrient depletion. Plasmids may be lost naturally from the cell when other organisms in the population without plasmids outreplicate those that possess them. Plasmids are not indispensable to the cell, but they may confer on the cell certain selective advantages over other organisms (Table 8-1). Some plasmids, for example, carry information for the degradation of certain substrates, and antibiotic resistance plasmids confer resistance to specific antibiotics that may be present in the microorganism's environment. Other observable properties carried by plasmids are described later. The first plasmid to be discovered was one that conferred the ability to transfer itself to a recipient cell in a mating like process called conjugation. (See Chapter 11 for details of this process.) It was called the **sex factor**, fertility factor, or F factor. Therefore the organism possessing the sex factor (also called the donor cell) is capable of transferring a copy of the sex factor to a recipient.

Plasmid replication can occur at two different times: (1) when the bacterial cell divides (plasmid DNA also divides, ensuring that both daughter cells receive a plasmid) and (2) during the conjugation process, when a replicated molecule can enter the recipient cell. It appears that during plasmid replication the DNA attaches to the cytoplasmic membrane and uses the same enzymes and machinery that are used in the replication of chromosomal DNA. Replication may be unidirectional (one replicating fork) or bidirectional (two replicating forks), depending on the type of plasmid.

Some plasmids are present in the cell in low copy numbers (1 to 10), whereas others are in high numbers (10 to 100). The **copy number** is controlled by the rate of initiation of DNA synthesis. The plasmid replicates until the proper copy number is reached. It is believed that a plasmid encoded inhibitor affects the initiation rate for plasmid DNA synthesis and thus controls copy number.

Many plasmids have the ability to confer the property of fertility **(conjugative)**, whereas other plasmids are **nonconjugative** and cannot effect their own transfer. Antibiotic-resistant plasmids in gram-positive bacteria such as the staphylococci cannot be transferred by a conjugation process, because conjugation does not exist or at least has not been discovered in most gram-positive bacteria.* The plasmids in the gram-positive staphylococci can only be transferred during a process called transduction, which involves a viral agent. (See Chapter 11 for details of this process.) Nonconjugative plasmids in gram-negative bacteria can be transferred only if the cell also contains conjugative plasmids. The transfer factor on one plasmid can therefore effect the transfer of nonconjugative plasmids, a process called **plasmid mobilization**.

Bacteria may contain a number of different plasmid types. Enterobacteriaceae, such as *E. coli,* possess one to two conjugative plasmids per chromosomal DNA but can also carry 10 to 15 nonconjugative plasmids per chromosomal DNA. When two or more different plasmids are stably inherited, they are said to be **compatible**. Other plasmids are considered **incompatible** when, after a few cell divisions, one of the plasmid types is lost. The molecular mechanisms controlling this incompatibility are not known but may be due to genetic factors controlling plasmid replication and segregation at cell division.

Table 8-1 *Some important plasmid-coded functions in microorganisms*

Function	Example
Resistance	Antibiotics: penicillin, streptomycin, chloramphenicol, erythromycin, kanamycin, and tetracycline Bacteriocins Heavy metals: mercury, lead, cadmium, nickel, and arsenic Ultraviolet light Bacteriophage
Gene transfer	Sex pili DNA mobilization Sensitivity to male-specific phage
Catabolism	Sugars, such as lactose and sucrose Opine* degradation (*Agrobacterium,* for example) Degradation of hydrocarbons, such as camphor, octane, naphthalene, and salicylate (*Pseudomonas* species)
Antagonism	Antibiotic production in soil Bacteriocin production Toxins: plant toxins and animal toxins, such as enterotoxins (affects intestinal tract) produced by *E. coli;* hemolysins (lyse red blood cells) produced by *Streptococcus* species; exfoliative toxin (causes peeling of epidermis) produced by *Staphylococcus aureus* Killer plasmid in yeast
Interactions between organisms	Tumor inducing *(Agrobacterium)* Nitrogen fixation *(Rhizobium)* Opine synthesis *(Agrobacterium)* Adhesion to plant *(Erwinia)* and animal surfaces *(Neisseria)* Nodulation in plants *(Rhizobium)* Plant hormone *(Pseudomonas syringae)*
Others	Restriction endonuclease DNA polymerase DNA methylase for cytosine Iron uptake

*Opines are compounds that are derivatives of amino acids. They are not normally produced by plant tissue but are produced following infection by *Agrobacterium* and transfer of a plasmid called Ti (tumor inducing).

Classes

The characteristics conferred on the host by plasmids are extensive and some of them are indicated below.

Sex factor plasmids. As previously discussed, the sex factor (F factor) is important for plasmid transfer to recipient cells. Plasmids like the sex factor are capable of either integrating into the host chromosome or remaining independent of the host chromosome. An integrated sex factor enables the cell to mobilize the transfer of the bacterial chromosome during the conjugation process (see Chapter 11).

R plasmids. Antibiotic resistance in many microorganisms is due to the presence of plasmids that contain specific information for the synthesis of enzymes that inactivate specific antibiotics. These are called resistance or **R factor** plasmids. R factors were first discovered in

* Conjugation occurs in *Bacillus subtilis* and in some species of *Streptococcus* and *Streptomyces.* Conjugation plasmids coding for antibiotic resistance have been found in *Streptococcus faecalis.*

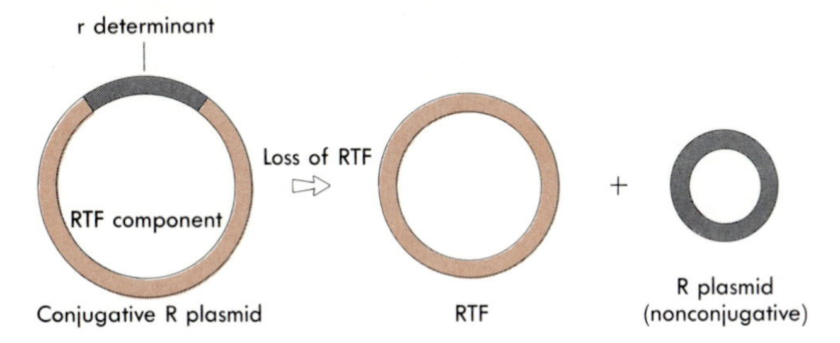

Figure 8-16 *Composition of conjugative and nonconjugative plasmid.*

Japan during an epidemic of bacillary dysentery. It was discovered that a strain of *Shigella dysenteriae* had acquired resistance to four different antibiotics and that these determinants could spread rapidly in a population of related gram-negative species that were initially sensitive to the same antibiotics. R factors are composed of two components: the **resistance (r) determinant** and the **resistance transfer factor (RTF)**. The RTF is needed for the transfer of the determinants. The RTF contains information for pilus formation, a requirement for conjugal DNA transfer in gram-negative bacteria. Some resistance determinants do not possess the RTF segment, therefore the bacteria carrying them are unable to transfer determinants to other bacteria during conjugation (Figure 8-16). Both r and RTF are capable of autonomous replication in their independent states (existing on separate pieces of DNA) and can integrate themselves into other extrachromosomal elements as well as chromosomal units.

Conjugative R plasmids conferring resistance to antibiotics such as kanamycin, chloramphenicol, tetracycline, penicillin, and ampicillin, as well as others, have been discovered in several virulent strains of bacteria (see Chapter 15).

Col plasmids. Many bacteria produce proteins, called **bacteriocins**, that are released into the environment and are lethal to related species of bacteria. Most are produced in gram-negative bacteria such as *E. coli,* in which the bacteriocin is called **colicin** and the plasmid encoding them is called **Col plasmid**. Col plasmids are divided into two groups: those that are conjugative and those that are nonconjugative. Some of the conjugative Col plasmids may also code for drug resistance or some virulence factor. Some colicins are a complex of two proteins: one that is active in killing sensitive cells and the other, which confers immunity. Every Col plasmid confers immunity on the host to the colicin for which it encodes. This immunity for some colicins is a result of a small protein that binds tightly to the active region of the colicin, thus temporarily inactivating it. By some unexplained mechanism the binding protein is removed by the bacterium affected by the colicin. The bacterial cell, however, may be susceptible to other types of colicins. All colicins bind to specific receptors on the cell surface. The specific mechanism by which many colicins act to kill sensitive cells has not been clarified. Some colicins, such as E1, achieve their toxic effects by collapsing the membrane potential gradient and nutrients cannot be transported into the cell. The bacterial cell starves to death. Colicin E2 is an endonuclease that can cause single- or double-strand breaks in the bacterial genome. Colicin E3 is a ribosomal nuclease that cleaves a single phosphodiester bond of the 16S RNA of the 30S ribosome.

The effectiveness of colicins in offering a selective advantage to cells in their natural environment is unclear. It is not known, for example, if colicin is produced in significant quantities in the intestinal tract, where *E. coli* normally resides, to control the population of other gram-negative species. Other bacteriocins produced by gram-negative bacteria are pesticin *(Yersinia pestis)* and pyocin *(Pseudomonas aeruginosa).*

Only a small percentage of gram-positive bacteria, such as the staphylococci (staphylococcin), streptococci (streptococcin), and *Clostridium perfringens* (perfringocin), are known to produce bacteriocins.

Mercury and other heavy metal ion–resistant plasmids. Some bacterial species, such as *Staphylococcus aureus* and species of *Pseudomonas,* are resistant to certain heavy metals, such as mercury, arsenic, cadmium, and lead. Mercury resistance has received much scientific and public attention. Inorganic mercury compounds, which are nontoxic, can be converted by anaerobic bacteria into methylated organic compounds that possess considerable neurotoxic activity for humans. Inorganic mercury resistance in microorganisms is due to the enzyme mercuric reductase, which reduces inorganic mercury according to the following reaction:

$$Hg^{2+} \xrightarrow{\text{Mercuric reductase}} Hg^{0}$$

Organic mercury resistance is due to hydrolases that separate mercury from the organic moiety.

Arsenic exists in the environment at levels 50 to 100 times that of mercury. Arsenic may exist in the arsenite (+3) state in which it acts as a sulfhydryl (SH) poison or it can exist in the arsenate (+5) state. Arsenate can be toxic in two ways. Arsenate is an analogue of phosphate:

$$O^{-}\text{—}\underset{\underset{^{-}O}{|}}{\overset{\overset{O^{-}}{|}}{As}}\text{=}O \qquad {}^{-}O\text{—}\underset{\underset{^{-}O}{|}}{\overset{\overset{O^{-}}{|}}{P}}\text{=}O$$

and when present affects the affinity of phosphate for its transport system. This can result in phosphate starvation. Arsenate also affects cellular kinases. For example, when

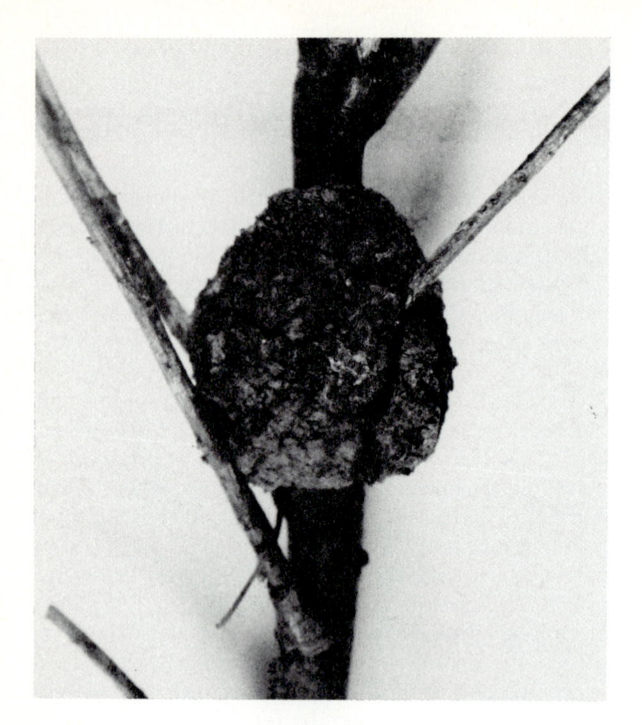

Figure 8-17 *Characteristic crown gall observed on stem of willow tree.*

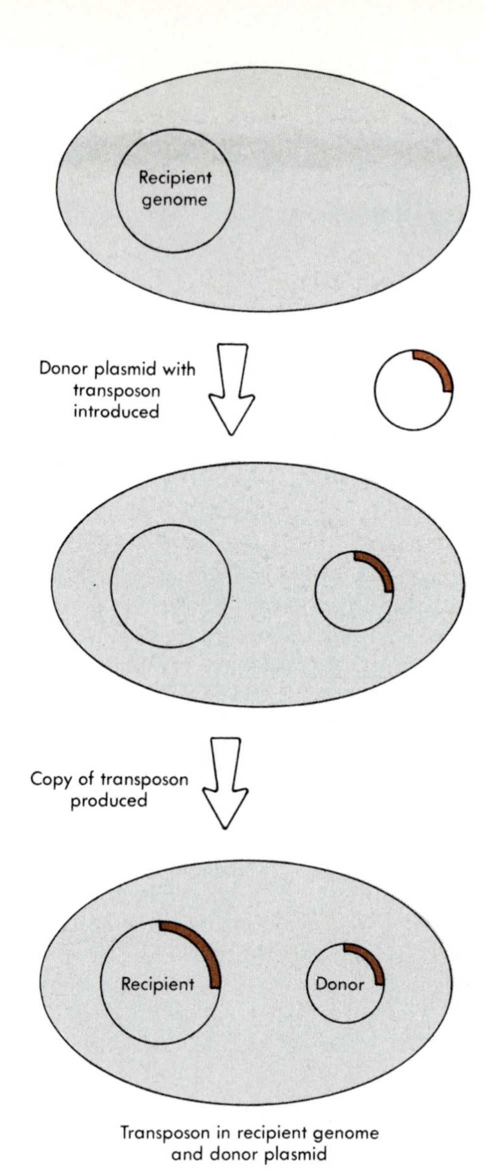

Figure 8-18 *Transposition involving donor and recipient. Bacterium containing plasmid acts as target for transposon-containing plasmid. Transposon makes copy of itself that is inserted into target plasmid. Thus recipient plasmid has transposon, and donor plasmid retains original transposon.*

arsenate is present, sugars transported into the bacterial cell become arsenylated instead of phosphorylated. Arsenylated sugars hydrolyze rapidly before they can be converted to energy compounds, such as ATP, and thus energy is lost to the cell. Resistance to arsenic is found on the plasmids of gram-negative and gram-positive bacteria. Resistance to arsenate is shown to be possible by alteration of the chromosomal genes encoding phosphate transport. Thus arsenate is unable to bind to phosphate transport proteins. The mechanism of resistance to arsenite is not yet known.

Plasmids that confer resistance to metal ions are selected in those environments containing high levels of the metal. Plasmid-associated resistance to silver, for example, has been found in those bacteria isolated from sludge produced by industrial processing of used photographic film.

Plasmids and catabolic activity. The ability of certain bacterial species, particularly species of *Pseudomonas,* to extract carbon and energy from unusual organic compounds is determined by plasmids. For example, some species can catabolize over 100 different compounds, and the enzymes responsible for the breakdown of some of them are encoded in plasmids. Plasmid-encoded enzymes can break down such products as naphthalene, xylene, camphor, and octane into metabolites, such as pyruvate and acetate.

Plasmids are used in the field of genetic engineering for the creation of microorganisms with special properties. Chakrabarty, for example, constructed an organism that could degrade 75% of the components making up crude petroleum. This constructed organism has been patented and may someday play a role in the clearing of oil spills. How plasmids are used in genetic engineering is discussed in Chapter 11.

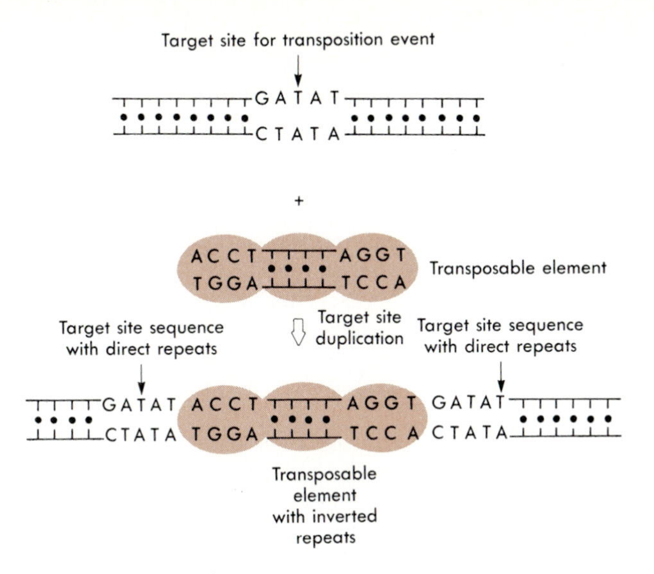

Figure 8-19 *Insertion of transposable element into genome at specific site (target site). Transposable element possesses central DNA sequence flanked by inverted repeats (color). During transposition target site duplicates and insertion of transposable element results in transposon flanked by identical sequences (direct repeats).*

Plasmids and plant hyperplasias. Crown gall is one type of cancerous disease of plants caused by a bacterium, *Agrobacterium tumefaciens* (Figure 8-17). The information for tumor (hyperplasia) development is carried on the bacterial plasmid. The evidence suggesting that the bacterium can transfer genetic information to the plant has come from in vitro studies. The excised cancerous tissue, which can be subcultured in the laboratory without the addition of plant hormones, exhibits tumor-specific properties. (More detailed information on crown gall appears in Chapter 20). Gall or tumor formation also occurs on oleander plant stems infected with the bacterium *Pseudomonas savastanoi*. The organism produces two enzymes involved in the synthesis of the plant growth compound, indoleacetic acid (IAA). The genes coding for the two enzymes are located on a plasmid carried by the bacterium. Mutants of *P. savastanoi* lacking IAA synthetic capabilities fail to induce gall formation, and those with increased capacity for IAA synthesis induce more severe symptoms in the plant.

Virulence plasmids for mammals. Plasmids in various bacteria carry information that contributes to their virulence during infection of mammals, including humans. Some of this plasmid information codes for (1) toxin production, for example, the enterotoxin produced by *E. coli* and (2) the production of cell surface components that permit certain bacteria to adhere to epithelial surfaces, for example, the surface of *E. coli* strains that cause diarrhea in humans and animals. Other virulence plasmids are discussed in Chapter 23.

Concept Check

1. What differentiates plasmid DNA from chromosomal DNA?
2. How does a conjugative plasmid differ from a nonconjugative one?
3. What are the different types of plasmids found in bacterial cells, and what are their function?
4. How do arsenic compounds affect the bacterial cell? What is the mechanism?

INSERTION SEQUENCES AND TRANSPOSONS

Insertion sequences (IS) and **transposons** are small elements of DNA that are capable of movement from one position in the DNA to another position in the DNA. This movement may be from one site on the chromosome to another site on the same chromosome, or the DNA element may move from one chromosome to another. Because of their mobility these DNA elements have been called **transposable genetic elements**. All transposable elements contain the information for their own transposition. The smallest transposable elements contain only the information for their own transposition but larger transposable elements carry additional genetic markers. During the transposition process an additional copy of the transposable element is produced, thus a copy remains at the target site, whereas the original copy remains at the donor site (Figure 8-18). Transposition thus provides the organism with a mechanism for DNA rearrangement. The frequency of transposition is comparable to the spontaneous mutation rate that occurs in bacteria, that is, 10^{-5} to 10^{-7} per generation. The mechanisms that characterize mutations caused by transposition are discussed in Chapter 10.

Insertion sequences

Insertion sequences are basically a class of transposons. They are usually smaller than transposons and have a length of 800 to 1400 base pairs compared to 2000 base pairs for transposons. Insertion sequences do not carry easily detectable genetic markers other than the information for their own transposition. The termini of all IS elements share a common characteristic—they all possess short **inverted repeats** that are 15 to 40 base pairs long. In other words, as we proceed from both ends of the IS element toward the center the same sequence is encountered (Figure 8-19). The target sites for insertion of transposable elements are usually not random and some sites seem to be preferred over others. Each transposable element has a broad range of specificities—

some appear to recognize certain nucleotides, whereas bacteriophage Mu* can insert almost anywhere. One characteristic of the insertion process is that there is duplication of a short nucleotide sequence, the **target sequence**, in the recipient DNA molecule. This results in the transposable element being flanked by repeated nucleotides called **direct repeats**, that is, they are in the same orientation (Figure 8-19). For any transposable element the target-site sequence is different for each insertion site.

Transposons

Transposons carry other genetic markers, in addition to the information for transposition. The most common transposon-carried genes confer antibiotic resistance. Other genetic markers can be carried by transposons, including genes for lactose fermentation, raffinose metabolism, enterotoxin formation in *E. coli* and heavy metal (cadmium, mercury, and arsenic) resistance.

There are basically two types of transposons (excluding the IS elements). **Class I** has a genetic marker flanked by primarily two copies of an IS element as inverted repeats (IR) and less commonly direct repeats (Figure 8-20). The IS element in this class provide the transposition function. **Class II** consists of a sequence that is flanked by IR but not IS elements. Between the IR are genes encoding the transposition function, as well as other markers, such as antibiotic resistance (Figure 8-20). Several plasmids in a wide variety of bacterial species are known to carry Class II transposons. The **Tn3 transposon** family belongs to the Class II transposons. Tn3 contains three genes: *A, R,* and *bla* (Figure 8-21). Two of these genes, *A* and *R,* are involved in a transposition process that differs from other transpositions. The *bla* gene encodes a protein called beta-lactamase that destroys the activity of the antibiotic ampicillin which, like other penicillins, contains a beta-lactam ring structure (see Chapter 15).

Transposition in Tn3. The transposition of Tn3 involves the formation of an intermediate or **cointegrate** (Figure 8-22). The cointegrate represents the fusion in the cell of two plasmids, one of which contains a transposon. The enzyme **transposase**, the product encoded by the *A* gene, is responsible for cointegrate formation. The final step in transposition involves a process called **resolution**. This means that the cointegrate can be resolved and the two plasmids can return to their independent status. The enzyme responsible for this is **resolvase**, the product encoded by the *R* gene. Resolvase catalyzes a site-specific exchange at the *res* sites. This recombination event results in the exchange of genetic information between two copies of Tn3. Thus two plasmids are formed and both possess Tn3 transposons (Figure 8-22).

Transposons: genetic tools. Transposons are exciting tools for the geneticist because they can be used as cloning vehicles, that is, genes of various origins can be inserted into the transposons of plasmids and the latter can be introduced into bacterial cells. Copies of the engineered transposons are produced in bacteria and the gene products can be purified. These techniques are discussed in Chapter 11.

* The term *Mu* was given to the virus because of its ability to cause mutations from random insertions into the host genome.

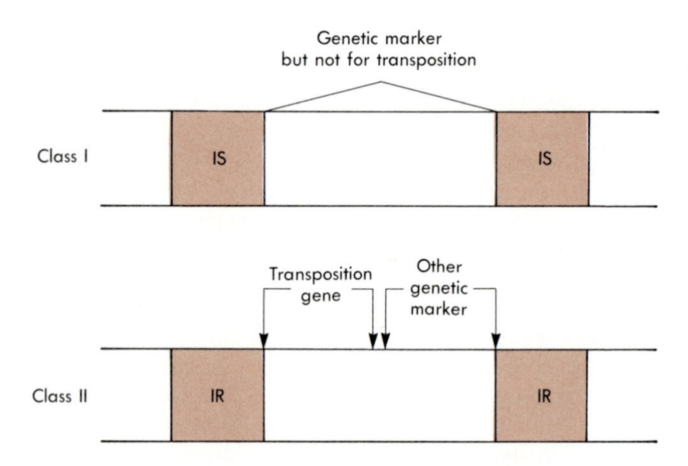

Figure 8-20 *Types of transposons. Representation of transposons as Class I and Class II. Transposition function in Class I provided by IS elements flanking internal genetic marker.*

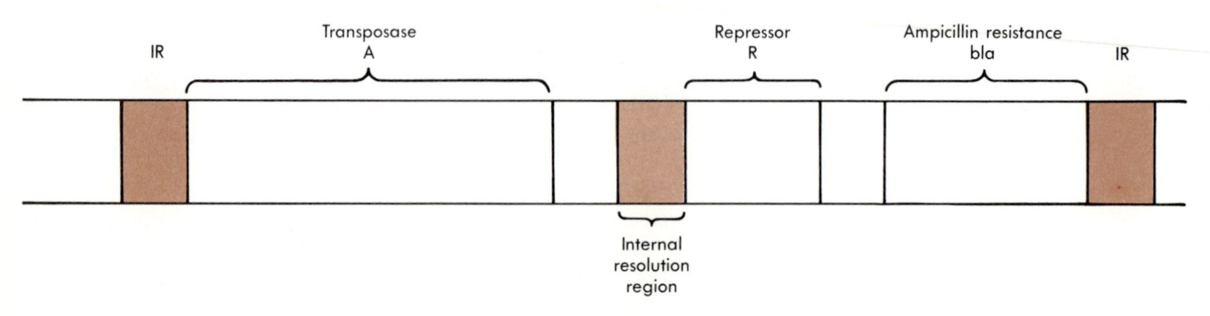

Figure 8-21 *Arrangement of genes in Tn3 transposon.* IR, *Inverted repeat.*

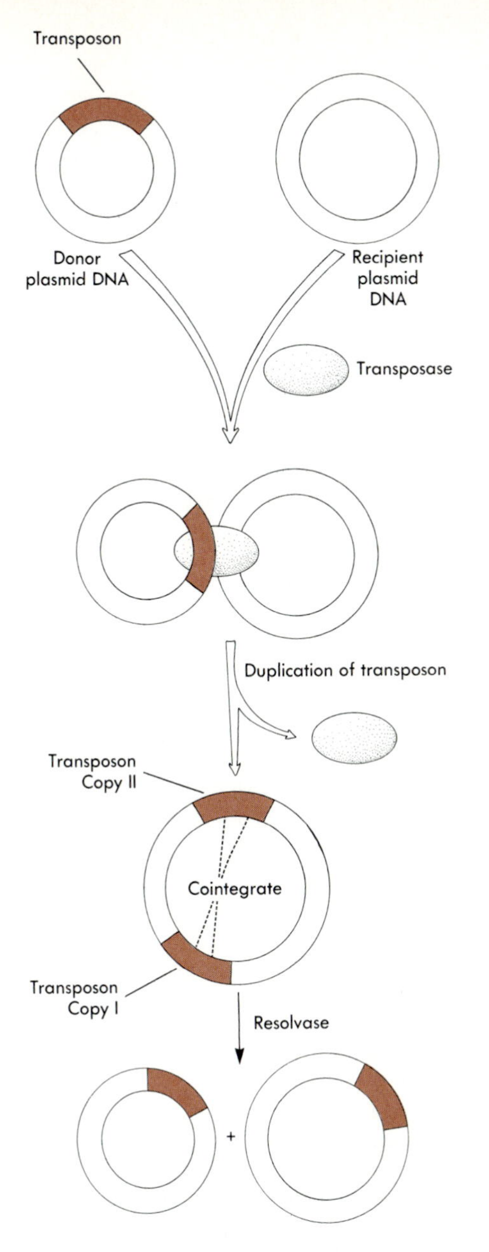

Figure 8-22 *Mechanism for transposition in Tn3. Enzyme transposase mediates binding of donor plasmid containing transposon Tn3 to recipient plasmid containing no transposon. Tn3 transposon duplicates and is followed by fusion of plasmids producing cointegrate containing two copies of transposon. Cointegrate is resolved into two independent plasmids by enzyme resolvase. During resolution there is genetic exchange between copies of transposon at res sites. After strand exchange, breaks are sealed by DNA ligase. Each plasmid now contains recombinant Tn3 transposon.*

SUMMARY

1. Bacterial cell DNA engages in a number of processes. These processes or characteristics include duplication and transfer of duplicated material to daughter cells, mutating, being translated into proteins, and being transferred totally or in part to other cell types with the possibility of recombination.
2. In the bacterial cell DNA is circular. DNA duplication occurs in a semiconservative manner, resulting in two DNA molecules, each containing one parental strand and one strand of newly synthesized DNA. DNA synthesis requires a double-strand DNA molecule, with one strand acting as a template and the other as a primer stand. Also required is DNA polymerase, magnesium ion, and all the deoxyribonucleotides.
3. DNA replication begins at a site called the origin and proceeds in both directions on a supercoiled circular DNA molecule. On one strand DNA synthesis is discontinuous, but is continuous on the opposite strand. During DNA replication the circular strand is nicked with an enzyme and is then unwound and stabilized through the aid of various proteins. As the DNA is unwound, new DNA is synthesized on both strands until the termination site on the DNA is reached, and a duplicate DNA is produced. In some viruses DNA is replicated in one direction by a rolling circle mechanism. DNA replication in eukaryotes is probably discontinuous like prokaryotic replication.
4. Plasmids (extrachromosomal DNA) appear in many bacterial cells. Extrachromosomal DNA is similar to chromosomal DNA in structure but plasmid genes are dispensable, smaller, and subject to loss from the cell by environmental manipulation. Plasmids may be conjugative or nonconjugative, depending on the presence or absence, respectively, of a sex (F) factor. Plasmids are not essential to the survival of the cell but offer properties that make it more competitive in its environment, such as the ability to conjugate with related species, resistance to antimicrobials, and increased catabolic activity. Also, plasmids carry virulence factors that cause disease in plants and animals.
5. Insertion sequences and transposons are mobile genetic elements of DNA that can insert themselves into chromosomal DNA. Insertion sequences contain only enough information for transposition but transposons possess additional markers, such as the genes encoding antibiotic resistance. Transposons show great potential for genetic engineering experiments.

Concept Check

1. Differentiate between IS elements and transposons.
2. Describe the mechanism by which IS elements are transposed from one site to another on the chromosome.
3. What is meant by cointegrate, and how is it produced?

Self-study Quiz

Identification

Identify the enzymes or other proteins involved in each of the following steps of DNA replication:

A. Negative supercoiling ________________.
B. Synthesis of short RNA primers ________________.
C. Final attachment of the phosphodiester bonds between DNA fragments ________________.
D. Growth of DNA chain in the 5′ to 3′ direction ______.
E. Relieves torsional strain of DNA ahead of the replication fork ________________.
F. Helps keep DNA strands separated ________________.
G. Removes RNA primers ________________.

Multiple choice

1. The term *concatameric* refers to:
   A. Type of DNA replication.
   B. Copies of DNA that are joined together.
   C. Type of plasmid.
   D. Type of excision process by the virus Mu.
2. The primary difference between IS elements and transposons is:
   A. IS elements are always longer than transposons.
   B. IS elements contain only information for mobilization while transposons possess additional information.
   C. IS elements contain inverted terminal repeats and transposons do not.
   D. IS elements can insert into plasmids; transposons cannot.
3. The term cointegrate refers to:
   A. Union of a plasmid with the bacterial genome.
   B. Union of an insertion sequence with a target site on the same chromosome.
   C. The product resulting from the union of two replicons each containing a transposon
   D. The product resulting from the union of a replicon containing a transposable element and a replicon lacking one.
4. The enzyme resolvase performs which of the following functions?
   A. Synthesis of RNA primers
   B. Initiates unwinding of DNA
   C. Catalyzes genetic exchange between copies of Tn3
   D. Negative supercoiling
5. The function of transposase is to:
   A. Produce staggered cuts in the recipient DNA.
   B. Catalyzes recombination between copies of the transposon Tn3.
   C. Duplicate the flanking sequences on IS elements.
   D. Join the transposable element to the single-stranded ends of the recipient DNA.
6. Which of the following inactivates the transforming principle isolated by Griffith and Avery?
   A. Proteinase
   B. ATPase
   C. DNase
   D. RNase
7. Which of the following statements is not true concerning DNA polymerase I?
   A. It synthesizes the growing strand in the 5′ to 3′ direction.
   B. Its function is to synthesize the RNA primer needed for chain elongation.
   C. It serves to remove the RNA primer sequence and fill the gaps with DNA.
   D. It cannot serve as a polymerase unless all of the deoxyribonucleoside 5′ triphosphates are present (ATP, GTP, CTP, TTP).
8. Which of the following staements is most descriptive of plasmids?
   A. The bacterial genome when it is carrying genetic information that is not normally a trait of the species.
   B. A virus capable of infecting bacteria.
   C. A form of single-stranded, linear DNA not found attached to the bacterial genome that allows the bacterial cell to adapt to its environment.
   D. A piece of double-stranded, circular DNA that is not part of the bacterial genome.
9. Which of the following is responsible for plasmid transfer and conferring the conjugative donor state on the cell?
   A. Col plasmid
   B. Sex factor plasmid
   C. Resistance factor plasmid
   D. Mercury ion resistance plasmid
10. What is the state of the plasmid after a few divisions when one of the plasmid types is lost?
    A. Cointegration
    B. Resolution
    C. Incompatibility
    C. Plasmid mobilization

SELECTED READINGS

Books

Birge, E.A. *Bacterial and bacteriophage genetics.* Springer-Verlag, Inc., New York, 1981.

Freifelder, D. *Molecular biology* (2nd ed.). Jones and Bartlett Publishers, Inc., Boston, 1987.

Hardy, K. *Bacterial plasmids* (2nd ed.): aspects of microbiology (No. 4). American Society for Microbiology, Washington, D.C., 1986.

Kornberg, A. *DNA synthesis.* W.H. Freeman & Co., Publishers, Inc., San Francisco, 1974.

Lewin, B. *Genes.* John Wiley & Sons, Inc., New York, 1983.

Scaife, J., Leach, D., and Galizzi, A. (Eds.). *Genetics of bacteria.* Academic Press, Inc., New York, 1985.

Shapiro, J. (Ed.). *Mobile Genetic Elements,* Academic Press, Inc., New York, 1983.

Suzuki, D., Griffiths, A.J.F., Miller, J.H., and Lewontin, R.C. *Introduction to genetic analysis* (3rd ed.). W.H. Freeman & Co., Publishers, Inc., San Francisco, 1986.

Watson, J., Hopkins, N ., Roberts, J., Steitz, J., and Weiner, A. *Molecular biology of the gene* (4th ed.), Benjamin/Cummings Publishing, Inc. Menlo Park, California, 1987.

Watson, J., and Tooze, J. *The DNA story.* W.H. Freeman & Co., Publishers, Inc., San Francisco, 1981.

Journal Articles

Bauer, W.R., Crick, F.H.C., and White, J.H. Supercoiled DNA. *Sci. Am.* **243**:118, 1980.

Cohen, S.N., and Shapiro, J.A. Transposable genetic elements. *Sci. Am.* **242**:40, 1980.

Challberg, M.D., and Kelly, T.J. Eukaryotic replication: viral and plasmid model systems. *Annu. Rev. Biochem.* **51**:901, 1982.

Drlica, K. Biology of bacterial DNA topoisomerases. *Microbiol. Rev.* **48**:273, 1984.

Kleckner, N. Transposable elements in prokaryotes. *Annu. Rev. Gen.* **15**:341, 1981.

Konisky, N. Colicins and other bacteriocins with established modes of action. *Annu. Rev. Microbiol.* **36**:125, 1982.

Novick, R.P. Plasmids. *Sci Am.* **243**:102, 1980.

Syvanen, M. The evolutionary implications of mobile genetic elements. *Annu. Rev. Gen.* **18**:271, 1984.

CHAPTER 9

Gene expression and its control

Chapter Outline

Introduction

The previous chapter showed how DNA functions as a molecule of replication. It is now time to discuss how DNA acts as an informational molecule. Beadle and Tatum in the 1940s demonstrated that genes control sequentially related metabolic reactions and that one gene codes for one protein. Later investigations showed that enzyme function is under the control of specific genes. We now know that all proteins are synthesized on structural units in the cytoplasm called ribosomes and a ribonucleic acid molecule called messenger RNA (mRNA) carries the information for synthesis of the protein. The normal flow of information in the cell is as follows:

$$\text{DNA} \rightarrow \text{RNA} \rightarrow \text{Protein}$$

Major questions left unanswered are how the information on each gene is translated to a polypeptide and how each gene is able to exert control over the vast number of proteins in the cell.

This chapter describes how certain sequences of purine and pyrimidine bases in the DNA molecule can be transcribed into information specifying the amino acids, how transcribed information can be translated into a polypeptide, and the mechanisms used by the cell to control the rate at which polypeptides are synthesized.

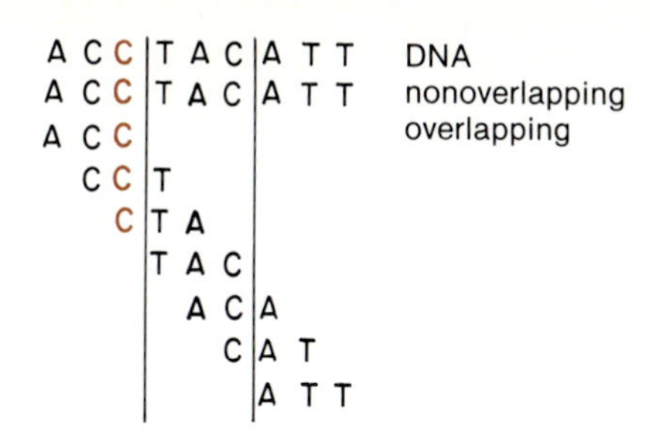

Figure 9-1 *Effect of mutational event involving single base* (color) *in genetic systems that possess either overlapping or nonoverlapping code. Note that in overlapping code, mutation in one base will cause changes in three triplet sequences.*

GENETIC CODE

Characteristics

The genetic code is triplet, nonoverlapping, commaless, degenerate, and universal. Each of these characteristics will now be discussed.

Triplet code. It was apparent that the conversion of nucleotide information into amino acids could take place only if the DNA contained coded information. One of the first considerations in deciphering a code is to evaluate the components of the system. The number of different nucleotides in the DNA is 4, and the number of amino acids to be coded is 20. The simplest code is one in which a combination of nucleotides stands for one amino acid. A two letter code gives only 16 possibilities, that is, 4^2, and leaves us 4 short of the 20 code words required. A three-letter code gives 64 possibilities, that is, 4^3, which is far in excess of what is needed. Experiments described later confirmed this three-letter code.

Nonoverlapping code. The nucleotide sequences on the DNA molecule are obviously very long. The question to be answered is whether the code words should be read in an overlapping or a nonoverlapping manner (Figure 9-1). A nonoverlapping code had been suggested from earlier studies with the hemoglobin molecule. Hemoglobin from normal adults was compared with hemoglobin from those with a type of sickle cell anemia. The hemoglobin in patients with sickle cell anemia is altered in such a way that it results in an increase in the rate of destruction of erythrocytes, which leads to hemolytic anemia. Analysis of the polypeptides from these two groups showed that the only difference between them is a single amino acid on one of the polypeptides; for example, in one type of sickle cell anemia, valine, an uncharged (nonpolar) amino acid is found in place of the normal glutamic acid, a charged (polar) amino acid. This single amino acid substitution is enough to change the conformation of the hemoglobin molecule and make it function abnormally. It was reasoned that a mutational event in an overlapping code would produce three new code words and hence three amino acid changes in the polypeptide, but a mutation in a nonoverlapping code would produce only a single amino acid change.

Commaless code. Can the code be read from any point, or is it read from one end to the other? Experiments performed on T4 bacteriophage determined the manner in which the code should be read. The T4 virus was treated with proflavine (a mutagen that can produce deletions and insertions) to produce mutants referred to as rII, which cannot grow on a particular strain of *Escherichia coli.* The rII mutants were treated with proflavine, which reversed the effects of the first mutation and enabled the virus to grow in the *E. coli* strain. The ability of the revertants to grow on *E. coli* is best understood by examining Figure 9-2. The deletion or addition of a base causes a change in the reading frame of the code when we assume a triplet nonoverlapping code. (The reading frame is a segment of RNA corresponding to the DNA segment that codes for a polypeptide.) The addition or deletion of a base changes the reading frame of the message (mRNA) from the point of the addition or deletion. These code changes result in changes in the amino acids of the polypeptide for which the gene codes. The rII revertants therefore represent double mutants in which the addition or deletion of a base restores part or all of the reading frame, depending on its location (Figure 9-2, *C*) in the message. For example, if the second mutation is close to the first mutation, only a few amino acids will be changed in the polypeptide because most of the original reading frame has been restored. A second mutation located some distance from the first also restores the reading frame but will leave a large part of the altered message unchanged. The polypeptide produced under the latter circumstance will probably be nonfunctional. Experiments showed that if the number of nucleotides deleted or added to the gene equaled the **codon number*** (or multiple thereof), some activity could be restored in the polypeptide. These results demonstrated that the code contained no punctuation, such as commas, and that the nucleotides were used in successive nonoverlapping groups of probably three. With no punctuation to indicate the beginning of one codon or the end of the next the reading frame must therefore be correctly established at the beginning of the message.

The code is degenerate. The above-mentioned experiments provided valuable information as to the nature of the code but told us nothing about the specific code words. Remember, there are 64 code words for 20

* The codon number refers to the base sequence (in mRNA or single-stranded DNA from which it is derived) that determines which of the 20 amino acids will be inserted at a particular position in a polypeptide chain.

amino acids. Scientists had proposed that the code was actually **degenerate**; that is, there might be more than one code word for a single amino acid. This means that there could exist a maximum of three code words for each amino acid, with four left over. The solution to this coding problem was made possible by an in vitro protein-synthesizing system developed by Nirenberg in 1961. The cell-free system consisted of ribosomes, transfer RNAs (tRNAs),* amino acids, energy sources, protein-synthesizing enzymes (aminoacyl tRNA synthetases),* and cofactors. Protein could be synthesized in this system only when mRNA is added. A series of experiments were set up in which reaction mixtures contained a different radioactive amino acid. The first experiment used synthetic messages containing one base, for example, polyuridylic acid (poly U) or polyadenylic acid (poly A). The polypeptides produced contained polyphenylalanine or polylysine, respectively; that is, the code word for phenylalanine was UUU, and the code word for lysine, AAA,† assuming a triplet code. These results prompted further experiments using synthetic mRNAs in which different triplets could be produced. Difficulties were encountered in distinguishing the arrangement of codons on a large synthetic message. These difficulties were solved by the development of the tRNA-binding technique. The basis of the technique is that ribosomes and mRNA bind to cellulose nitrate filters. More important, tRNAs, with their attached amino acids, also bind to the ribosome mRNA complex as long as they are specific for a particular triplet on the message. A filter with bound poly (U) mRNA and ribosomes binds only those tRNAs carrying the radioactive amino acid phenylalanine, whereas all other tRNAs pass through the filter. Experiments were designed so that trinucleotide mRNAs, such as UAC, CCA, and AGA, could be added to an extract of *E. coli* that had lost the ability to produce protein. Following suitable incubation the extract was examined to determine which aminoacyl tRNA bound to the ribosome-mRNA complex. These studies led to the complete deciphering of the code (Table 9-1). Examination of the code words reveals that there are 61 code words for the 20 amino acids **(degeneracy)**. Three of the codons—UAA, UGA, and UAG—represent no amino acids and are called **nonsense**.

The code is universal. The amino acid code words are identical in all species of animals, plants, bacteria, and viruses thus tested. This implies that the genetic code had been fixed before the divergence of all life forms. A mutation that changes a codon assignment could cause several mistakes in the protein and this could be lethal. Studies in mitochondria, which have their own ribosomes, tRNA, and mRNAs, demonstrated that some of the amino acid codons are not identical to those of the standard code-word dictionary. For example, yeast mitochondria use AUA as a code word for methionine rather than isoleucine. Many scientists believe that because mitochondria produce only a limited number of proteins they can tolerate such changes. Recent findings, however, indicate that the notion of universality in the code may have to be reevaluated. The codon UAA is normally considered a termination or nonsense codon; however, in certain protozoa it codes for glutamine. In certain *Mycoplasma* species the codon UGA, normally a termination codon, codes for tryptophan. How or when such changes occurred is still a matter of speculation.

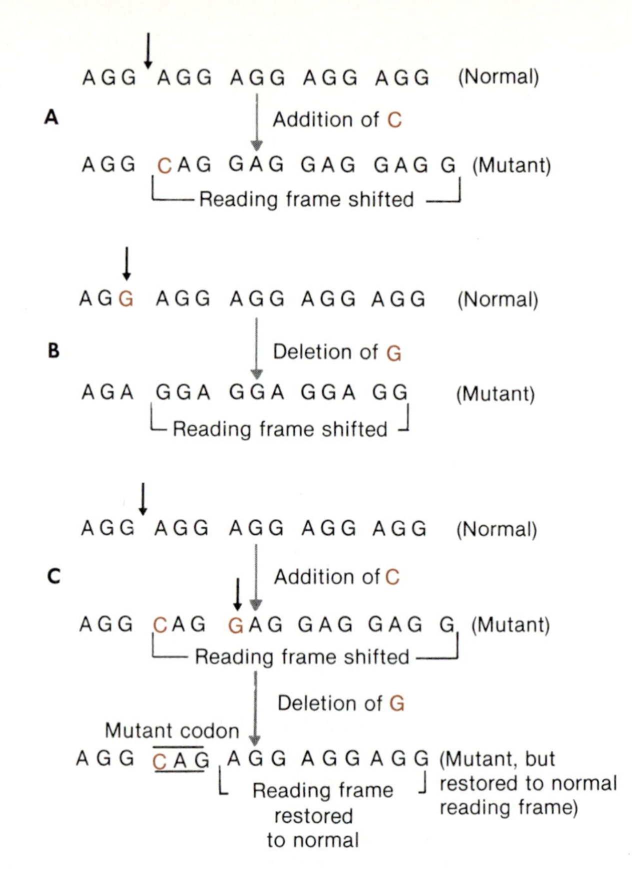

Figure 9-2 *Effect of deletion or insertion of bases in DNA frameshifts. Information encoded on gene is read in sequence from specific point. Code words are in form of base pair triplets that are read from left to right. When base is inserted* (color), *reading frames of all base pair triplets downstream from site of insertion are altered.* ***A,*** *Addition of base;* ***B,*** *deletion of base;* ***C,*** *addition of base followed by deletion of base to restore reading frame.*

* Transfer RNAs carry amino acids. The charging of the amino acid to a tRNA molecule is catalyzed by aminoacyl tRNA synthetases. Transfer RNAs are discussed at more length on p. 227.

† Artificial mRNA polymers are poor stimulators of polypeptide synthesis unless the first codon is AUG. The AUG codon is called the initiating codon and is important for ribosome binding on the mRNA (see p. 231). Nirenberg's experiments worked only because artificial mRNAs in the presence of high concentrations of magnesium ions are not dependent on the initiating codon.

Table 9-1 *The genetic code*

First letter	Second letter: U	C	A	G	Third letter
U	Phe	Ser	Tyr	Cys	U
	Phe	Ser	Tyr	Cys	C
	Leu	Ser	Nonsense	Nonsense	A
	Leu	Ser	Nonsense	Trp	G
C	Leu	Pro	His	Arg	U
	Leu	Pro	His	Arg	C
	Leu	Pro	Gln	Arg	A
	Leu	Pro	Gln	Arg	G
A	Ileu	Thr	Asn	Ser	U
	Ileu	Thr	Asn	Ser	C
	Ileu	Thr	Lys	Arg	A
	[Met]*	Thr	Lys	Arg	G
G	Val	Ala	Asp	Gly	U
	Val	Ala	Asp	Gly	C
	Val	Ala	Glu	Gly	A
	Val	Ala	Glu	Gly	G

*Boxed codon represents most common one for initiation. Nonsense refers to "stop" codons, that is, the reading of the message terminates at the "stop" codon.

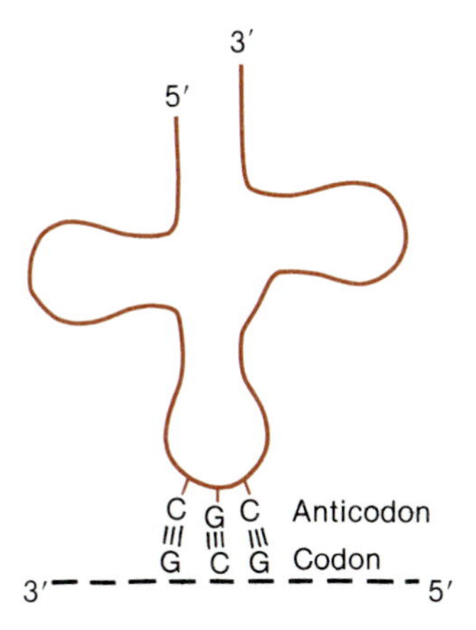

Figure 9-3 *Binding between mRNA codon and tRNA anticodon.*

Inosine Adenine

Inosine Uracil

Inosine Cytosine

Figure 9-4 *Pairing properties of base inosine, which is found in tRNA molecule. Inosine can pair with three bases, because it can rotate slightly on its sugar-phosphate linkage, thus producing "wobble."*

Wobble hypothesis

Studies on tRNA, at the time of the decoding experiments, indicated that an amino acid may be bound to more than one tRNA species. Does that mean that there are 61 different tRNA molecules in the cell, each one recognizing a different codon? Each tRNA has a triplet, called the **anticodon**, that is complementary to the **codon** on the mRNA (Figure 9-3). Nirenberg's experiments with triplet messages demonstrated that several tRNA species can carry the same amino acid and that some different tRNA species recognize the same codon. One proposal formulated to explain the range of binding activities for tRNA is called the wobble hypothesis. This proposal envisages that only the first two letters of the code, except the code words for serine, arginine, and leucine, specify the amino acid. Serine, arginine, and leucine have six possible codons that vary not only in the third letter but also in the first and second letter (Table 9-1). For example, CCU, CCA, CCG, and CCC code for proline. According to the wobble hypothesis, wobble can make possible the formation of more than one type of codon-anticodon interaction at the third codon position. This flexibility in hydrogen bonding is believed to be due to the minimal steric constraints at the 5' end of the anticodon, where there is a turn in the molecule (see Figure 9-11) and the appearance of minor bases, such as inosine at the 5' end of the anticodon. Inosine, for example, can pair with adenine, uracil, and cystosine (Figure 9-4). The range of base interactions at the third codon position are illustrated in Figure 9-5.

PROTEIN SYNTHESIS IN PROKARYOTES

The preceding discussion of the genetic code gave us some concept as to the nature of the gene. A **gene** can be defined as a segment of DNA (in some viruses the hereditary material is RNA) in which a sequence of nucleotides is transcribed into an RNA molecule, which in turn is translated into a sequence of amino acids in a polypeptide chain. Not all DNA sequences are transcribed into mRNA; some are transcribed into ribosomal RNA (rRNA) and tRNA. In the following paragraphs the manner in which genetic information is transcribed and translated in the cytoplasm of the prokaryotic cell is discussed.

Transcription process

Transcription of messenger RNA. Information on the DNA molecule is first **transcribed** into an RNA molecule called **messenger RNA (mRNA, tRNA,** or **r-RNA).** Recall that the nucleotides in DNA and RNA are the same except that uracil substitutes for thymine in the RNA molecule (see Chapter 2).

A strand of DNA in the gene serves as a template for the synthesis of mRNA. The enzyme that directs the synthesis of mRNA from the DNA template is a DNA-dependent RNA polymerase (Figure 9-6). Unlike the eukaryotic cell, the prokaryotic cell uses this same enzyme to produce all of its RNAs. The reaction catalyzed by RNA polymerase involves the transfer of a ribonucleoside monophosphate to the 3′hydroxyl terminus of a growing RNA chain using a ribonucleoside triphosphate as substrate (Figure 9-7). RNA initiation, unlike DNA initiation, does not require a primer. The transcription process is an exact one because of the manner in which hydrogen bonding occurs between complementary bases. The size of the mRNA is a function of the size of the polypeptides for which it codes. The mRNAs of prokaryotes are with few exceptions very unstable molecules and are degraded by ribonucleases in 2 minutes or less following their transcription. Messenger RNA instability is actually an invaluable property to the cell. The enzymatic needs of the cell vary considerably, depending on nutritional sources and environmental conditions. Genes are turned on and off to meet these needs. A stable mRNA, producing more protein, after the gene controlling its transcription has been turned off, causes a waste of energy, but an unstable mRNA allows the cell to respond rapidly to environmental changes.

Messenger RNA that is synthesized from the DNA template of the gene in the 5' to 3' direction (Figure 9-6) requires initiation, as well as termination, signals. The major regulatory unit of transcription is a segment of DNA called the **promoter.** The promoter precedes the **startpoint** or the base pair that corresponds to the first

Anticodon (5′ end)	Codon (3′ end)
Inosine	U C A
Guanine	U C
Adenine	U
Cytosine	G
Uracil	G A

Figure 9-5 *Base pair possibilities for third (5′) base of anticodon with base of its codon.*

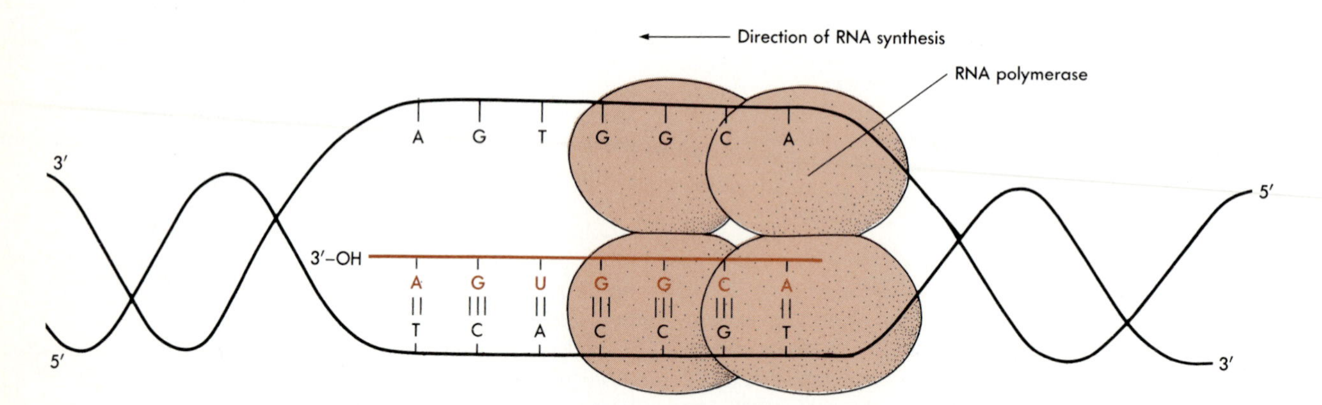

Figure 9-6 *Transcription of RNA from DNA double helix. DNA unwinds, exposing both strands, but only one (template strand) is transcribed. RNA polymerase synthesizes mRNA in 5′ to 3′ direction, therefore DNA strand is being transcribed in 3′ to 5′ direction.*

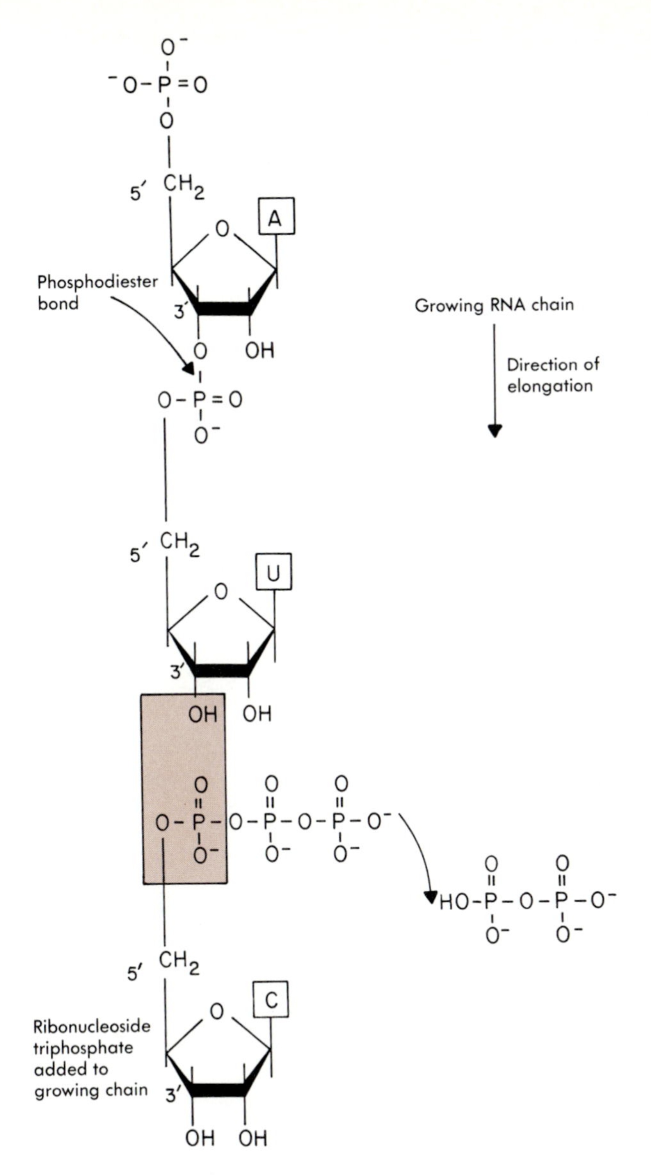

Figure 9-7 *Mechanism of RNA chain elongation by RNA polymerase. Growing RNA chain acts as primer to which will be added ribonucleoside triphosphate. Shaded area represents components involved in elongation reaction—3′ OH of growing chain attacks 5′ PO_4 of incoming ribonucleoside triphosphate. A 3′ = 5′ phosphodiester bond is formed, resulting in release of pyrophosphate. New phosphodiester bond between U and C is same as that between A and U in figure.*

nucleotide incorporated into the transcript by RNA polymerase. Pribnow discovered an unusual characteristic of nucleotide sequences in the promoter. These sequences consist of seven nucleotides having the general form of TATAATG or variants of this and now referred to as the **Pribnow box.** These sequences act as signals that are believed to be important for binding of RNA polymerase and its conversion to a form capable of initiating mRNA synthesis (Figure 9-8). The RNA polymerase holoenzyme consists of subunits of two alpha (α_1 and α_2), one beta (β), one beta prime (β'), and sigma (σ). The **sigma factor** is not bound as tightly as other subunits and readily dissociates from the complex leaving behind a β, β^1, α_1, α_2 enzyme referred to as the **core enzyme** (Figure 9-9). The sigma subunit is especially important to core enzyme specificity because in its absence the enzyme would bind to random sites on the DNA. Sigma is a recycled element whose function is to aid the polymerase in finding and binding to promoter sites for initiation of mRNA synthesis. Each RNA polymerase holoenzyme recognizes two sites on the promoter that are located 10 base pairs and 35 base pairs **upstream** from the startpoint (that is, before the startpoint). They are referred to as **−10** and **−35** sequence and are highly conserved regions found in most prokaryotes so far examined. Although RNA polymerase recognizes specific nucleotide sequences on the promoter, the promoter nucleotides are not transcribed. The region for unwinding the DNA and initiation reactions are at the right end of the Pribnow box and extend just beyond the startpoint, that is, −9 to +3 (+ refers to base pairs after the startpoint or **downstream**). RNA polymerase catalyzes the bonding of the first two incoming ribonucleotides, which begins the mRNA, and produces a triphosphate tail at the 5′ end. Ribonucleotides are then added one at a time in the 5′ to 3′ direction until the last nucleotide is added, which represents the **termination site.**

Transcription termination. Termination of transcription is necessary to stop further RNA elongation and to release the RNA transcript. Termination sequences may be of two types; those that are **self-terminating** in that they rely only on the DNA base sequence and those that require a special protein called **rho.** The signals for self-termination are believed to be sites rich in GC and include a stretch of several consecutive uridine residues. Rho is a protein that possesses RNA-dependent nucleoside triphosphate phosphorylase activity. It binds to specific termination sites, such as those rich

Concept Check

1. What are the advantages of having a genetic code that is triplet, commaless, degenerate, and universal?
2. What are the ingredients that are necessary to carry out the synthesis of polyphenylalanine in a test tube?
3. What is the transcription process? What elements on the DNA strand are involved and how is RNA polymerase involved?

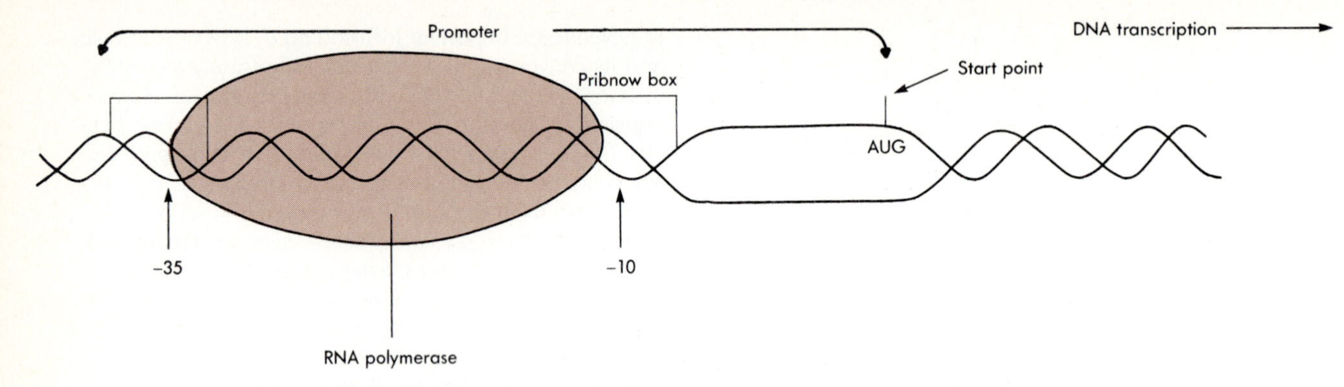

Figure 9-8 *Basic structure of promoter. Two areas of RNA polymerase recognition are within Pribnow box centered at −10 (upstream from startpoint) and −35 sequence. Unwinding of DNA occurs within Pribnow box and extends just beyond startpoint. DNA transcription begins at startpoint whose coding sequence is AUG.*

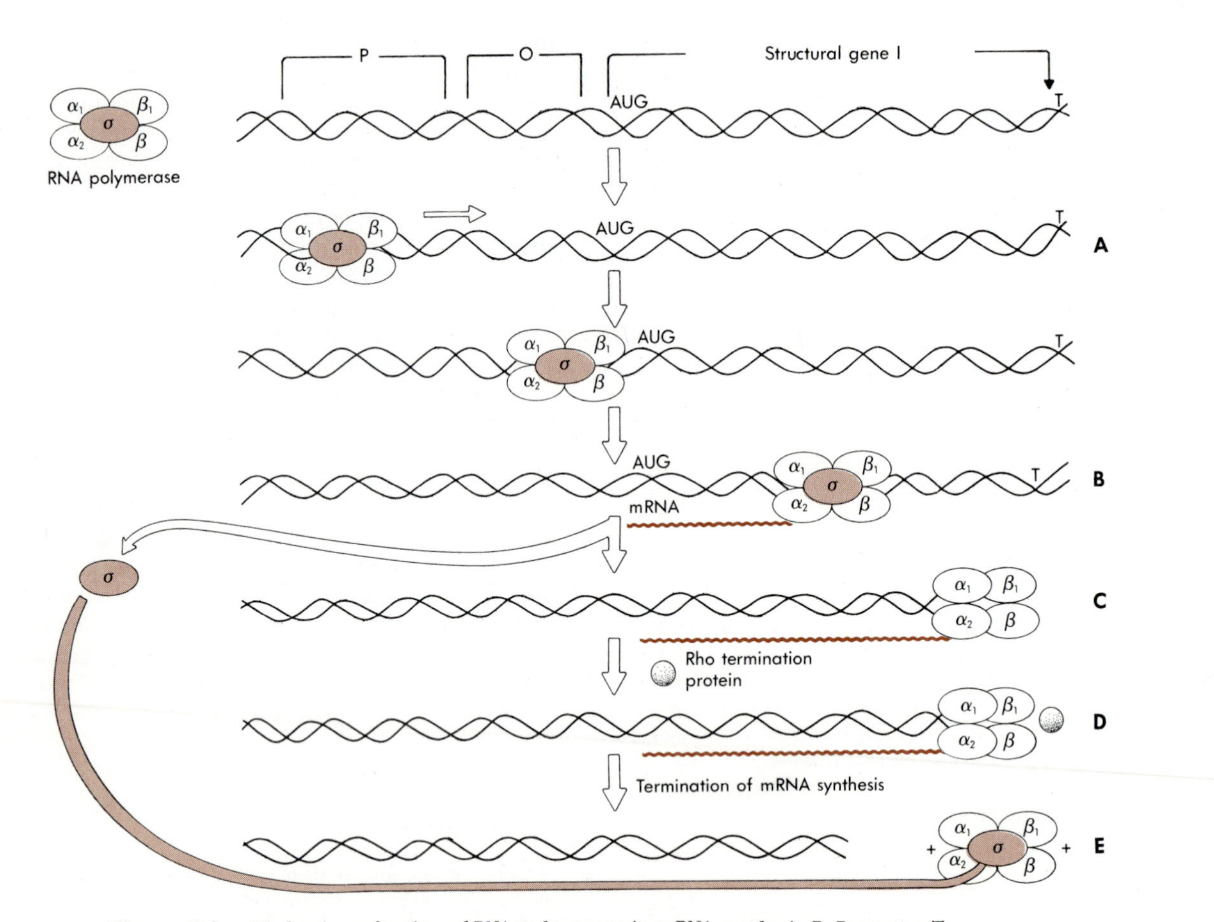

Figure 9-9 *Mechanism of action of RNA polymerase in mRNA synthesis.* P, *Promoter;* T, *termination site.* **A,** *RNA polymerase binds to promoter region of template strand of gene I and moves to start site (AUG).* **B,** *Sigma subunit of RNA polymerase is released. Release occurs after about 8 nucleotides have been put down in growing mRNA.* **C,** *RNA polymerase (core enzyme) catalyzes further transcription of gene I into mRNA molecule.* **D,** *Rho protein causes termination of mRNA synthesis.* **E,** *Core enzyme is released from DNA strand and combines with recycled sigma factor, and mRNA molecule is completely synthesized.*

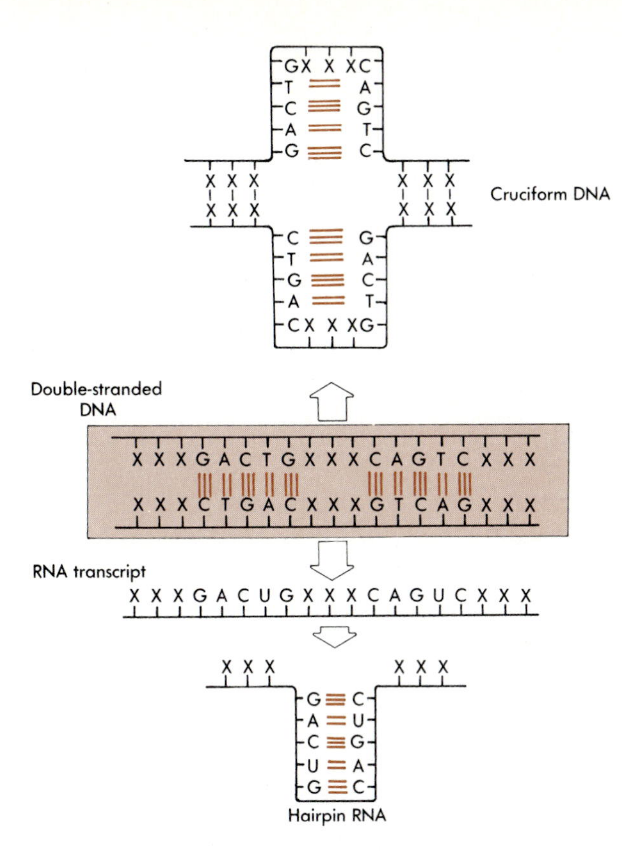

Figure 9-10 *Palindromic sequences consist of adjacent inverted repeats that generate secondary structures, such as hairpins in RNA and cruciforms (cross-shaped) in DNA.*

in cytidine, and then begins to cause hydrolysis of nucleoside triphosphates that are essential to release of the mRNA transcript.

Both types of transcription terminators possess **palindromic sequences**. A palindrome is a sequence of double-stranded DNA that is the same when either strand is read with a particular orientation, for example:

5'G G T A C C 3'
3' C C A T G G 5'

The presence of palindromes in the DNA can result in the formation of secondary structures, such as **hairpins** in RNA and **cruciforms** in DNA (Figure 9-10). It is believed that hairpins in the RNA product cause RNA polymerase to slow down or pause during transcription. Once the pause is initiated the final step in termination depends on the type of termination site. How these features lead to termination is not clearly understood.

How RNA polymerase is able to distinguish which strand of the double-stranded DNA molecule is to be transcribed has not been elucidated. It might be reasoned that because the DNA strands have opposite polarity, the enzyme can distinguish this difference. Unfortunately, experiments have shown that sometimes both strands can be read. For example, the area coding for the enzymes involved in histidine biosynthesis can be read on one strand, and the area coding for enzymes involved in lactose catabolism can be read from another strand. Under no circumstances, however, is a single region transcribed from both strands.

Transcription of transfer RNA and ribosomal RNA. The smallest RNA molecules in the cell, **transfer RNAs (tRNAs)**, are single stranded and contain from 70 to 80 nucleotides. Each carries a specific amino acid in protein synthesis. There is at least one specific tRNA molecule for each of the 20 amino acids found in the cell. The tRNA molecule contains several functional areas that are associated with protein synthesis. One end of the molecule always contains the sequence CCA, whereas G is at the other end of the molecule. The specific amino acid carried by the tRNA is attached to the adenine of the CCA end of the molecule. A second site consists of three bases called the anticodon, which is complementary to a triplet sequence on the mRNA (this specificity is discussed later). A third site is associated with binding to the ribosome during protein synthesis. A fourth site engages the enzyme (aminoacyl tRNA synthetase), which covalently attaches the amino acid to the CCA end of the molecule. The tRNA exhibits a helical structure, even though it is single stranded. Several tRNAs have been subjected to nucleotide analysis, and they are believed to exhibit a secondary structure similar to that of a "cloverleaf" (Figure 9-11). Other hydrogen bonding is believed to bend the cloverleaf into a tertiary structure that has a more stable conformation (Figure 9-11, *B*). Transfer RNAs contain several "abnormal" bases, such as pseudouracil, 5,6-dihydrouracil, and dimethylguanine. The normal bases are produced on the tRNA during transcription from the DNA, and the abnormal bases are produced (for example, methylation of guanine) after transcription—a process called **post-transcriptional modification.** The modification of certain bases on the tRNA may protect the molecule from degradation by RNA nucleases in the cell. Transfer RNA molecules, unlike mRNA, are used repeatedly during protein synthesis and may require this modification process to resist digestion.

Ribosomal RNA (rRNA) and proteins are components of ribosomes. The rRNA found in prokaryotes consists of 23S, 16S, and 5S molecules. The 23S and 5S components combine with a large number of proteins to produce a 50S ribosomal subunit, and the 16S rRNA combines with other proteins to produce a 30S ribosomal subunit. The 50S subunit has approximately 3330 nucleotides, whereas the 30S subunit has about half that number. The 50S and 30S subunits combine to produce the typical 70S prokaryotic ribosome. Similar rRNA subunits in eukaryotic cells combine with proteins to produce 60S and 40S ribosomal subunits, which combine

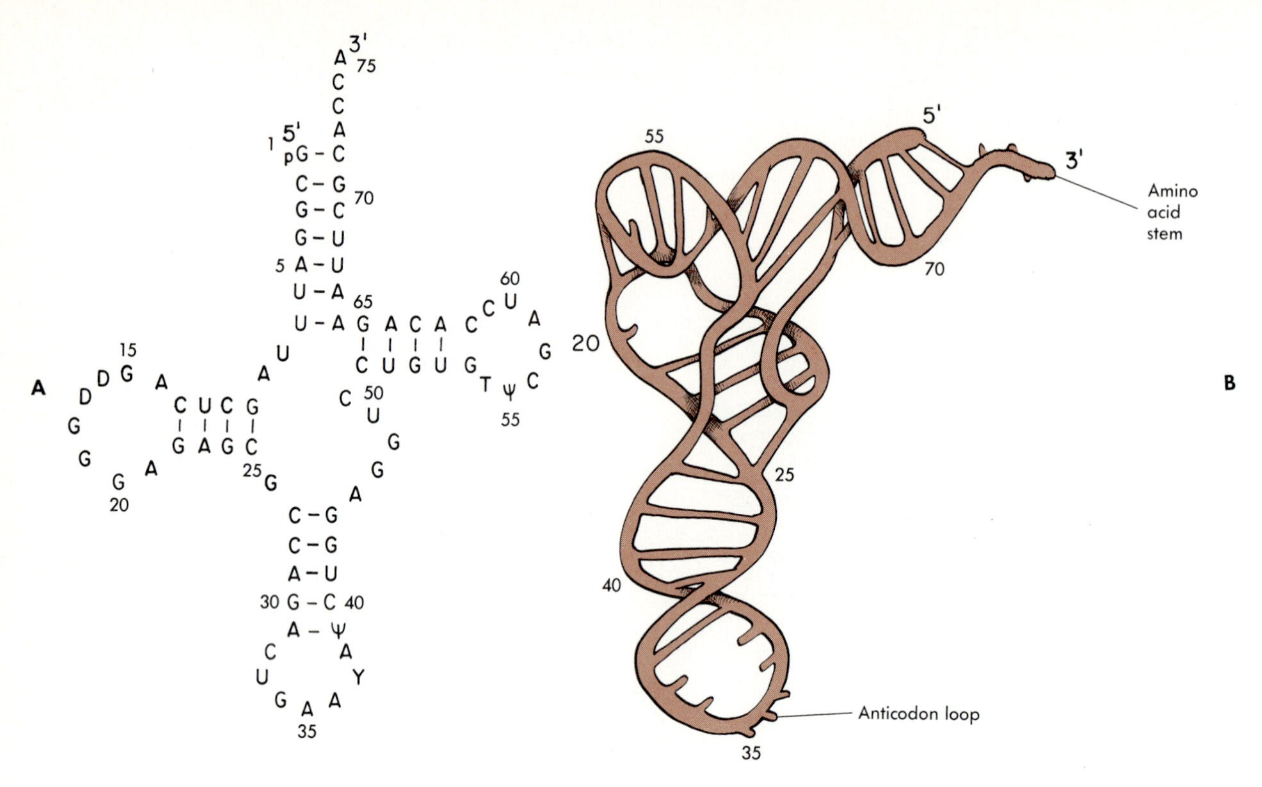

Figure 9-11 *Yeast phenylalanyl tRNA.* ***A,*** *Primary structure of phenylalanyl tRNA demonstrating nucleotide positions.* ***B,*** *Tertiary structure of phenylalanyl tRNA. Ribose phosphate backbone is drawn as continuous cylinder with various bars to show tertiary interactions. Numbers indicate nucleotide positions starting at 5′ end.*

to form the typical 80S ribosome (Figure 9-12). There are many copies of rRNA gene sets on the prokaryotic chromosome, and these are arranged as tandem repeats for the 16S, 23S, and 5S rRNA sequences. Each repeat unit is initially transcribed as a precursor 30S molecule that is later cleaved by a ribonuclease (RNase) into 16S, 23S, and 5S subunits (Figure 9-13). Once the ribosomal proteins are assembled, the rRNAs interact with them in such a way that they become buried and protected from nuclease activity.

Translation process (mechanics of protein synthesis)

Once the sequences of bases* on the mRNA have been transcribed, the molecule is ready for translation into a specific sequence of amino acids that make up the polypeptide chain. A sequence of three nucleotides on the mRNA, called the codon, codes for a single amino acid in the polypeptide chain. (The characteristics of the genetic code are discussed earlier.) This conversion process is called **protein synthesis**.

Charging of amino acids in preparation for protein synthesis. Amino acids do not attach directly to the mRNA. Amino acids are first charged and attached to small carrier molecules called tRNA and are thus termed **aminoacyl tRNAs**. Aminoacyl tRNAs float free in the cytoplasm and are waiting to be engaged in protein synthesis. The tRNA molecules in a starving cell are devoid of their passenger amino acids, but in rapidly growing cells, where amino acids are plentiful and protein synthesis is at a maximum, they are fully charged.

Each amino acid is carried by a specific tRNA molecule. This specificity is maintained because of enzymes that recognize only certain amino acids and specific tRNA molecules. These enzymes are called **aminoacyl tRNA synthetases**, and each possesses a site for attachment of the amino acids and a site for the specific tRNA. Thus each enzyme's name corresponds to the particular amino acid for which it is specific. The activation of

* The size of the mRNA molecules varies considerably. Small mRNA molecules have about 150 nucleotides. Larger mRNA molecules average in size from 2000-3000 nucleotides.

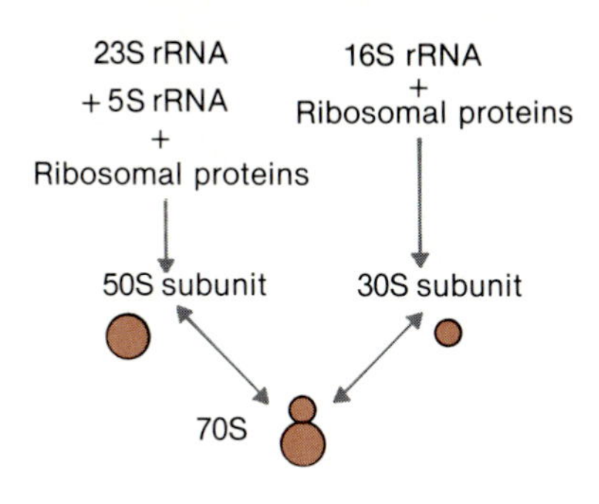

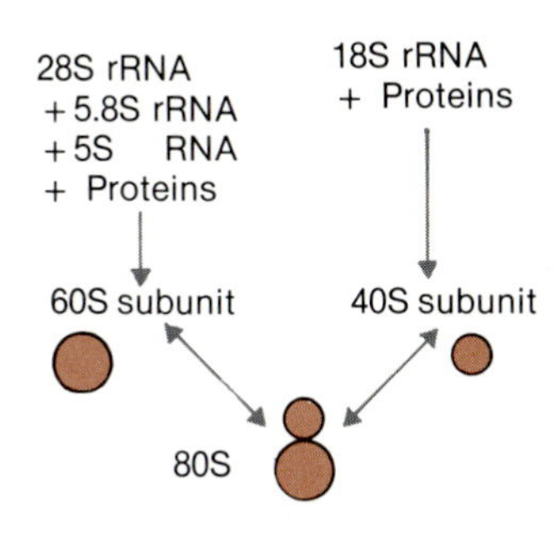

Figure 9-12 *Subunit association scheme for prokaryotic and eukaryotic ribosomes.*

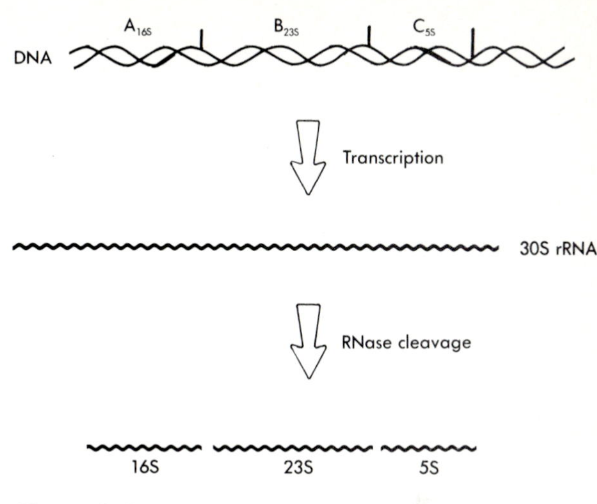

Figure 9-13 *Proposed scheme for transcription of ribosomal genes. Genes for rRNAs indicated in illustration are 16S* (A), *23S* (B), *and 5S* (C), *and they are linked together at various sites on chromosome. Three genes are transcribed into single 30S rRNA precursor, which is later cleaved by RNase into 16S, 23S, and 5S rRNA.*

amino acids and tRNA binding occur in the following manner:

$$\text{Amino acid} + \text{ATP} \underset{\text{synthetase}}{\overset{\text{Aminoacyl tRNA}}{\rightleftharpoons}} \text{Aminoacyl AMP} + \text{PP}$$

$$\text{Aminoacyl AMP} + \text{tRNA} \underset{\text{synthetase}}{\overset{\text{Aminoacyl tRNA}}{\rightleftharpoons}} \text{Aminoacyl tRNA} + \text{AMP}$$

The activated amino acid, which is covalently attached by its carboxyl group to the CCA end of a tRNA molecule, now possesses energy for the formation of a peptide bond in polypeptide synthesis. On each tRNA molecule, as mentioned earlier, is a triplet sequence that is complementary to the triplet code on the mRNA.

Translation of messenger RNA. The translation of mRNA occurs on the ribosomes that contain sites for the binding of mRNA, aminoacyl tRNA, and the growing polypeptide chain. Just as each mRNA sentence has a beginning and an end, so does the polypeptide coded by the mRNA. The polypeptide is synthesized in a step-wise manner, one amino acid at a time, beginning from the amino end (the first amino acid has a free amino group) and terminating with an amino acid containing a free carboxyl end. The synthesis of the polypeptide can be divided into the following steps.

Initiation. In prokaryotes protein synthesis is initiated when DNA transcription is initiated (Figure 9-14). This type of arrangement would be impossible in eukaryotes where a nuclear membrane separates the ribosomes from the chromosomes. The initiation of polypeptide synthesis in all prokaryotes begins with the binding of a tRNA that carries **N-formylmethionine (fMet-tRNA$_f$)** (Figure 9-15). The fMet-tRNA$_f$ carries an anticodon that recognizes the initial mRNA codon, AUG or GUG. The formyl unit attached to the amino group of methionine prevents the amino acid from engaging in peptide bond formation and therefore will never appear in the interior of the polypeptide. The formyl group is subsequently removed by an enzyme, leaving a polypeptide that begins with methionine. Sometimes the entire fMet is cleaved from the polypeptide, and an amino acid other than methionine is located at the amino end.

One of the first prerequisites for protein synthesis is the formation of the **70S ribosome initiation complex.** This process can be broken down into two steps: (1) formation of a 30S ribosome initiation complex and (2) formation of the 70S initiation complex (Figure 9-16).

1. The 30S ribosome binds three proteins called **initiation factors** (IF-1,IF-2, and IF-3). This is followed by the binding of the initiator fMet-tRNA$_f$, GTP, and mRNA. (The mechanism of recognition of the initiation on the mRNA is discussed later.) The formation of the 30S complex can be depicted as follows:

$$\text{fMet-tRNA}_f + \text{30S·IF-1·IF-2·IF-3} + \text{mRNA} + \text{GTP} \rightarrow [\text{30S·IF-1·IF-2·mRNA·GTP·fMet-tRNA}_f] + \text{IF-3}$$

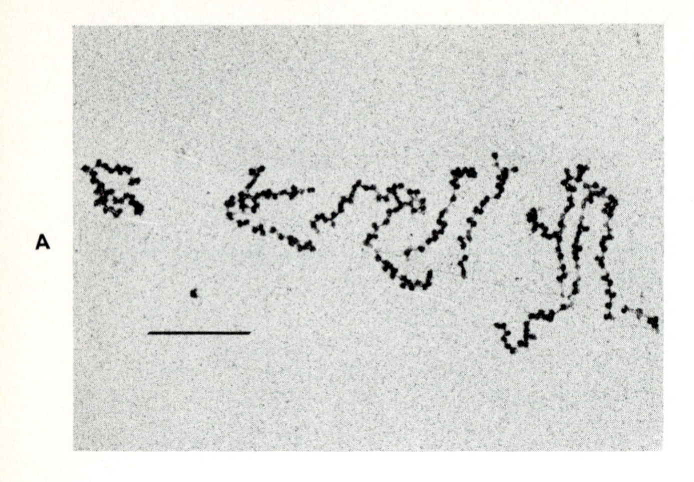

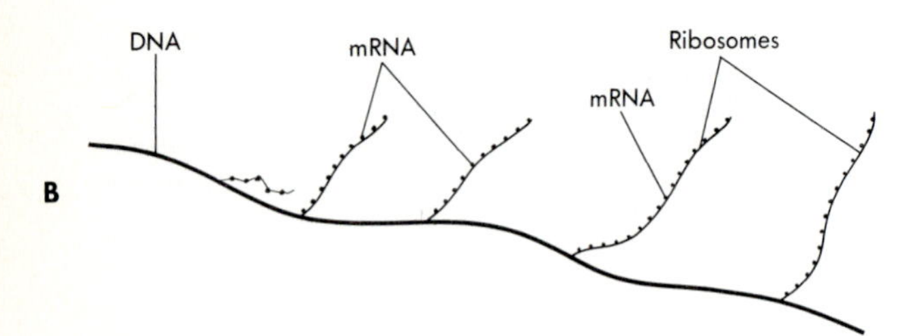

Figure 9-14 *Translation of mRNA in prokaryotic cells. Translation apparently begins as soon as mRNA is being transcribed on DNA.* ***A,*** *Electron micrograph demonstrates single strand of* E. coli *DNA to which are attached mRNA molecules in various stages of translation. To each mRNA are attached ribosomes (dark granules) that increase in number from left to right (the direction of translation).* ***B,*** *Diagrammatic illustration of electron micrograph. Bar = 500 nm.*

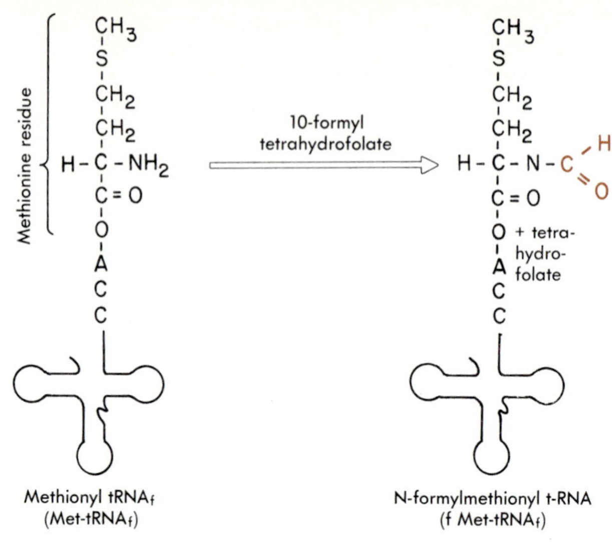

Figure 9-15 *Generation of* N-*Formylmethionyl-tRNA with formyl tetrahydrofolate as cofactor. Formyl group indicated in color.*

GTP in this reaction acts as a steric factor for the complex and prevents fMet-$tRNA_f$ from participating in subsequent peptide bond formation.

2. Addition of the 50S ribosome to the 30S initiation complex results in the formation of the 70S initiation complex. This reaction is accompanied by the release of the remaining initiation factors and hydrolysis of GTP. The reaction can be written as follows:

$$[30S \cdot mRNA \cdot IF\text{-}2 \cdot IF\text{-}1 \cdot fMet\text{-}tRNA_f \cdot GTP] + 50S \rightarrow$$

$$[mRNA \cdot 70S \cdot fMet\text{-}tRNA_f] + IF\text{-}2 + IF\text{-}1 + GDP + P_i$$

The energy from GTP hydrolysis is used to release GDP from the ribosomal complex and in so doing changes the configuration of the complex so that fMet-$tRNA_f$ can participate in peptide bond formation.

Messenger RNA that is to be transcribed may contain information for one polypeptide (monocistronic)* or for more than one polypeptide (polycistronic). Each mRNA

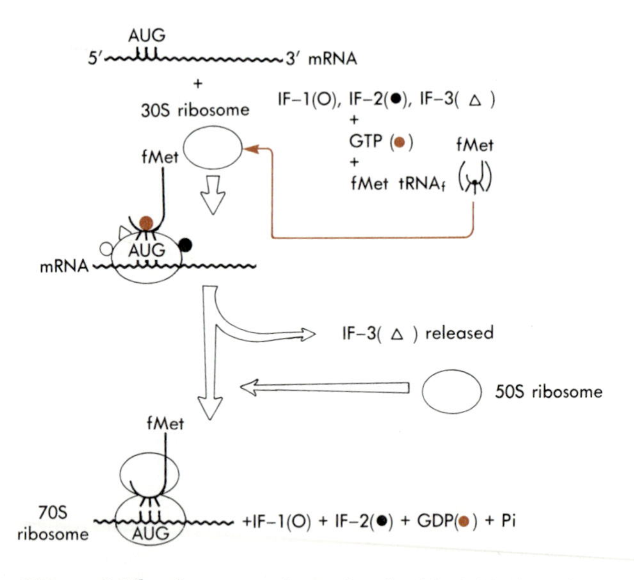

Figure 9-16 *Summary of steps involved in initiation of protein synthesis.* IF, *Initiation factor;* GTP, *guanosine triphosphate;* fMet-tRNA, *formylmethionyl tRNA. Several IFs, GTP, and fMet-tRNA bind to 30S ribosome that attaches to mRNA startpoint* (AUG). *IF-3 is then released and this is followed by binding of 50S ribosome and formation of 70S complex.*

* A cistron is the same as a gene—a functional unit that has been recognized through mutant testing.

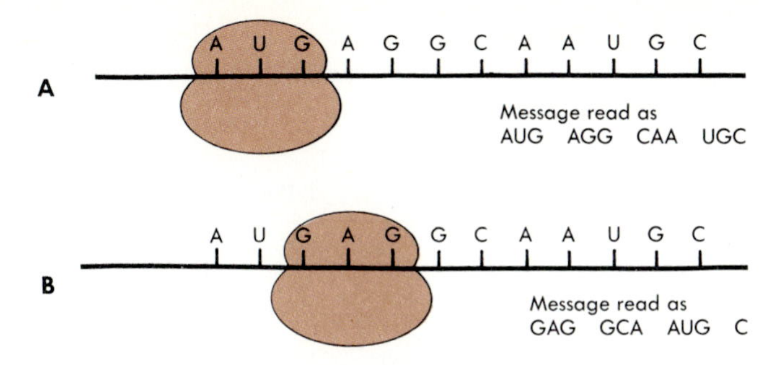

Figure 9-17 *Effect of incorrect orientation of ribosome on mRNA.* ***A,*** *Correct orientation;* ***B,*** *incorrect orientation.*

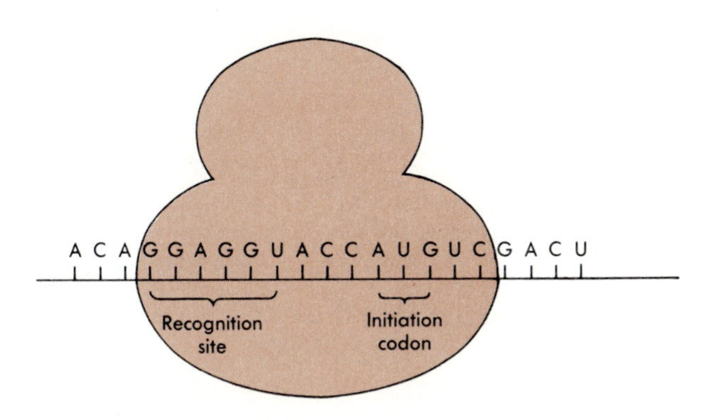

Figure 9-18 *Ribosome binding to mRNA involves recognition of Shine-Dalgarno sequences that precede initiation codon AUG.*

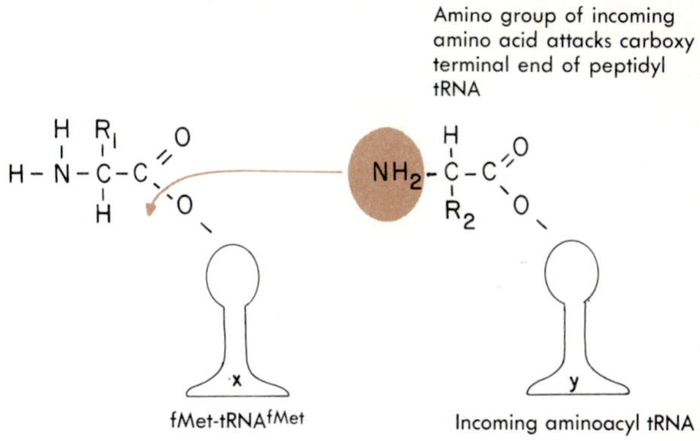

Figure 9-19 *Formation of peptide bond by addition of the incoming amino acid* (y) *to acyl terminal end of the peptidyl tRNA* (x). R, *R group of amino acid.*

coding for a polypeptide begins with a sequence that must be correctly oriented with the 30S subunit so that the message is read completely and in the correct reading frame (Figure 9-17). These initiating sequences on the mRNA are called **ribosome-binding sites.**

There is a ribosome-binding site for each independently synthesized polypeptide product. Shine and Dalgarno and Steitz and associates showed that virtually all ribosome-binding sites have an AGGAGGU (3′) sequence or closely related sequence at a similar position on the 5′ side (upstream) of the initiation codon. This sequence base pairs to a pyrimidine-rich region at the 3′ end of 16S ribosomal RNA chains. These sequences are called **Shine-Dalgarno sequences**, and they position the initiating codon AUG, so that it can bind to the anticodon of initiator tRNA (Figure 9-18).

Elongation. The formation of an active 70S initiation complex is followed by the addition of amino acids and the formation of peptides—the elongation step. Two sites exist on the 70S ribosome for binding of aminoacyl tRNA: the **A site** (refers to aminoacyl-tRNA binding site) and the **P site** (refers to peptidyl-tRNA binding site). The P site is initially bound by *N*-formylmethionyl $tRNA^{fMet}$. Formyl methionyl tRNA is the only incoming aminoacyl tRNA that can bind to the P site. All other incoming aminoacyl tRNAs will bind to the A site. The anticodon of the aminoacyl tRNA molecules hydrogen bond with its complementary mRNA codon with the subsequent hydrolysis of GTP.

In the next step of elongation the first peptide bond is formed between the amino acids whose tRNAs are located on the P and A sites. This occurs by transfer of the *N*-formylmethionine acyl group from its tRNA to the amino group of the incoming amino acid at the A site. The enzyme catalyzing this step is called peptidyl transferase (Figure 9-19), which is bound to the 50S ribosome. The energy for peptide bond formation is provided by ATP during amino acid activation, that is, during attachment of the amino acid to its tRNA. Consequently a dipeptidyl tRNA is formed on the A site with an uncharged (carries no amino acid) $tRNA^{fMet}$ attached to the P site. The uncharged $tRNA^{fMet}$ will enter the pool of free tRNAs in the cytoplasm. Further additions of amino acids to the growing peptide chain require a translocation process.

Translocation. The growing peptidyl tRNA chain formed during the elongation step is initially at the A site and must be translocated to the P site. The translo-

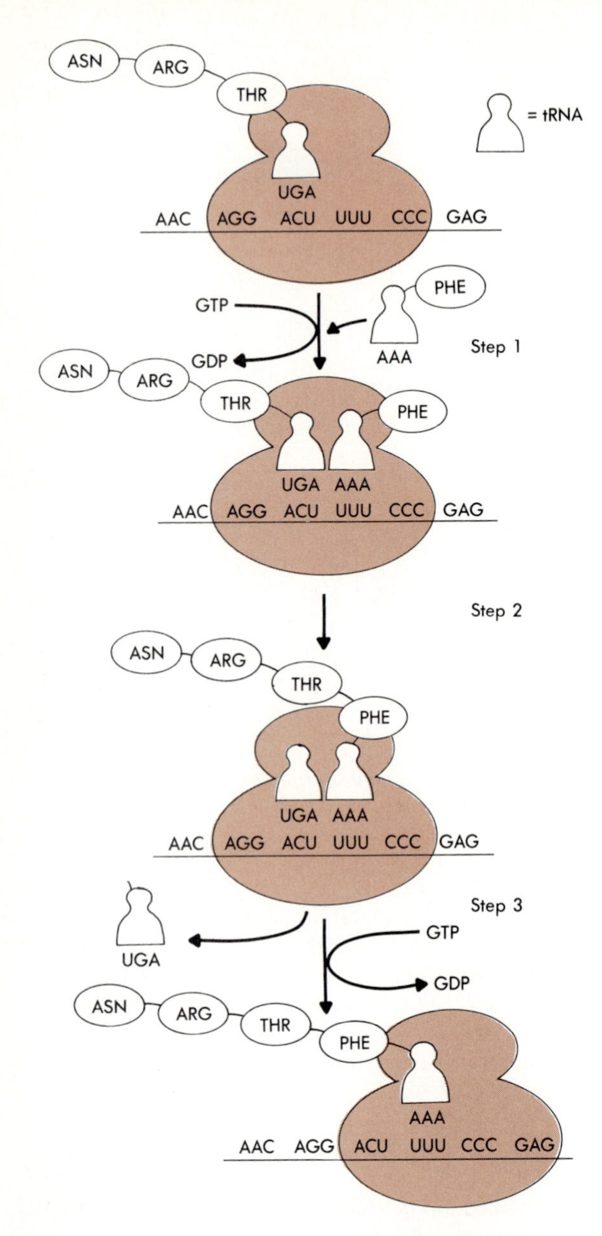

Figure 9-20 *Summary of elongation and translocation steps in protein synthesis. Step 1, incoming phenylalanyl tRNA binds to A site. Step 2, peptide bond formation between phenylalanine and threonine. Peptidyl tRNA is on A site. Step 3, translocation of peptidyl tRNA to P site and release of threonyl tRNA into cytoplasm. Note translocation associated with movement of ribosome three nucleotides along mRNA chain.*

cation process occurs as the ribosome moves three nucleotides to the next codon along the mRNA. This movement frees the A site and enables it to accept the next aminoacyl tRNA. The translocation process requires a protein elongation factor (EF-G) and energy provided by the hydrolysis of GTP. The end of translocation signals the start of the cycle and addition of the next amino acid to the polypeptide. Figure 9-20 summarizes the elongation and translocation steps by assuming that the P site contains peptidyl tRNA three amino acids in length (tripeptidyl tRNA).

Table 9-2 *Comparison of the transcription process in eukaryotes and prokaryotes*

Characteristic	Prokaryote	Eukaryote
Ribosome	70S	80S
Initiation factors	3	10
Initiator tRNA	Formyl methionyl tRNA	Methionyl tRNA
mRNA	Less stable, primarily polycistronic, no cap structures, size correlates with genome size	More stable, monocistronic, presence of cap structures, size larger than expected for genome size
Promoter signal	Pribnow box	Goldberg-Hogness box and enhancer
Termination site	Defined: rho dependent and self-termination	Not defined, but some genes have a polyadenylation signal

Termination. Each mRNA has a termination or "stop" signal that does not recognize any aminoacyl tRNA anticodon. Termination codons are UAA, UAG, and UGA. A protein called **release factor** (two in *E. coli,* RF 1 and RF 2) binds to the "stop" codons and interferes with the normal activity of the ribosome-bound peptidyl transferase. Peptidyl transferase now behaves in a hydrolytic manner; the bond holding the carboxyl end of the polypeptide to the tRNA is hydrolyzed. This results in the release of the polypeptide, as well as the last tRNA, into the cytoplasm. In addition, the 70S ribosome dissociates into its 30S and 50S subunits, thus making them ready for initiation of new polypeptide synthesis. The synthesized polypeptide begins to assume a three-dimensional structure even before it is released from the ribosome.

Other aspects of prokaryotic protein synthesis. The mRNA of prokaryotic cells will have several ribosomes bound to it during protein synthesis, and this complex is called a **polyribosome** (polysome). This means that more than one polypeptide can be made from a single mRNA before the latter is degraded by nucleases. At any one time on the mRNA the polypeptide appearing on each ribosome will vary in length, depending on the dis-

Concept Check

1. What are palindromic sequences, and what is their relationship to termination of mRNA synthesis?
2. What are the functional sites on the tRNA molecule?
3. What are the steps involved in synthesis of a protein?
4. What are the steps involved in the formation of the 70S ribosome complex and initiation of protein synthesis?
5. How is the growing polypeptide chain elongated, one amino acid at a time?

tance of the ribosome from the 5′ end (initiation site) of the mRNA (Figure 9-21).

PROTEIN SYNTHESIS IN EUKARYOTES

Transcription process

The transcription of eukaryotic DNA, like that of prokaryotic DNA, requires RNA polymerases. Eukaryotic systems, however, possess three types of RNA polymerase. **RNA polymerase I** is found in the nucleolus and is involved in the transcription of the genes for 28S, 5.8S, and 18S rRNA. **RNA polymerase** II is associated with mRNA synthesis, and **RNA polymerase III** is involved in the transcription of tRNA and 5S rRNA. Eukaryotic RNA polymerases, like those in prokaryotes, begin and end transcription at specific sites on the chromosome. The transcription process in eukaryotes follows the same basic pattern as occurs in prokaryotes. However, some aspects of the eukaryotic process are different (Table 9-2).

A region similar to the promoter in prokaryotes is found in eukaryotes and is called the 5′-flanking region. This region contains the **Goldberg-Hogness box** or **TATA box** and consists of the sequences TAT*X*A*X*, where X is either T or A. The TATA box is found about 30 base pairs upstream from the point of initial gene transcription. Deletion of the TATA sequence does not abolish in vivo transcription of the eukaryotic gene. This characteristic is unlike that in prokaryotes where deletion of the Pribnow box eliminates prokaryotic gene transcription. A second regulatory region appears anywhere from 50 to 500 nucleotides upstream from the transcription initiation site. These regulatory sequences or **enhancer elements** increase the frequency with which RNA polymerase initiates transcription. Deletion of the enhancer greatly reduces transcription. We are unsure if the enhancer is an individual element or merely part of the promoter. Enhancers can be removed from one genetic element and inserted into another so as to increase the expression of specific genes. Enhancer transfer is a valuable tool in genetic engineering (see Chapter 11).

Eukaryotic mRNA is more stable than prokaryotic mRNA. This stability is believed to be a result of modification steps immediately following transcription. One modification system is called **capping**. The 5′ ends of nearly all eukaryotic mRNAs are capped with a guanosine that protects the next to last (penultimate) base (Figure 9-22). The 5′ cap is unusual because it is methylated and is attached to the penultimate base by a unique triphosphate linkage. The penultimate base is also methylated. Methylation renders the bases polar and thus capable of interaction with charged molecules. Eukaryotic mRNA is also modified by the addition of (poly A) sequences **polyadenylation** to the 3′ end of the molecule, which is near but not at the transcription termination site. The length of the poly A sequences may range from 30 to 200 bases. Their function is not understood at this time, but they may be a recognition signal during translation, or they may be a binding site for some cytoplasmic component. A schematic diagram of a typical eukaryotic transcription unit is illustrated in Figure 9-23.

Messenger RNA processing

In prokaryotes the immediate product of transcription, the **primary transcript**, is mRNA but this is not the case in eukaryotes. Eukaryotes must process the primary transcript and transform it into a translatable mRNA (mature mRNA). This RNA processing involves modification (previously discussed) and excision of untranslatable sequences. The eukaryotic genome contains many repetitive sequences that do not code for any information. As much as 70% of the nucleotide sequences in the eukaryotic primary transcript are lost before the latter's conversion to translatable mRNA. Experiments have confirmed that noncoding sequences are copied from regions on the gene called **introns**. The translatable or coding sequences are called **exons**. Eukaryotic mRNA contains coding sequences that have been spliced together following excision of the noncoding regions (Figure 9-24). Introns have been divided into two groups based on apparently different mechanisms of splicing. Group I consists of mRNA, rRNA, and tRNA introns and also includes those RNAs from mitochondria and chloroplasts, which contain their own DNA, RNA, and ribosomes). Group II introns are associated with RNAs from fungal mitochondria and plant chloroplasts.

Precise excision and splicing or pre-mRNA or the primary transcript appears to be related to recognition sequences located at the intron-exon junction. The 5′ and 3′ boundaries of most introns consist of dinucleotides, *GU* and *AG,* respectively. These dinucleotides are part of

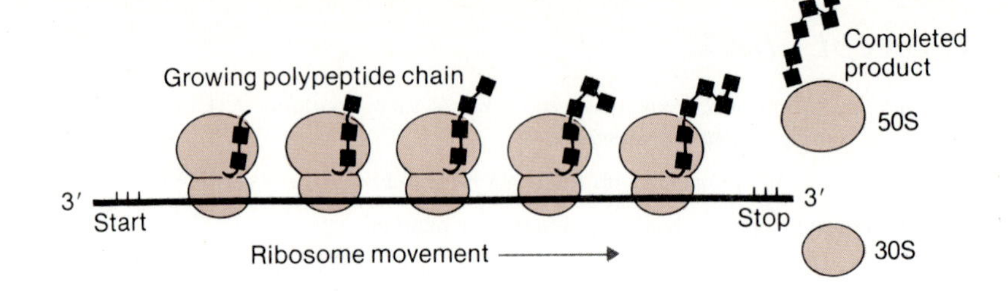

Figure 9-21 *Polysome engaged in protein synthesis.*

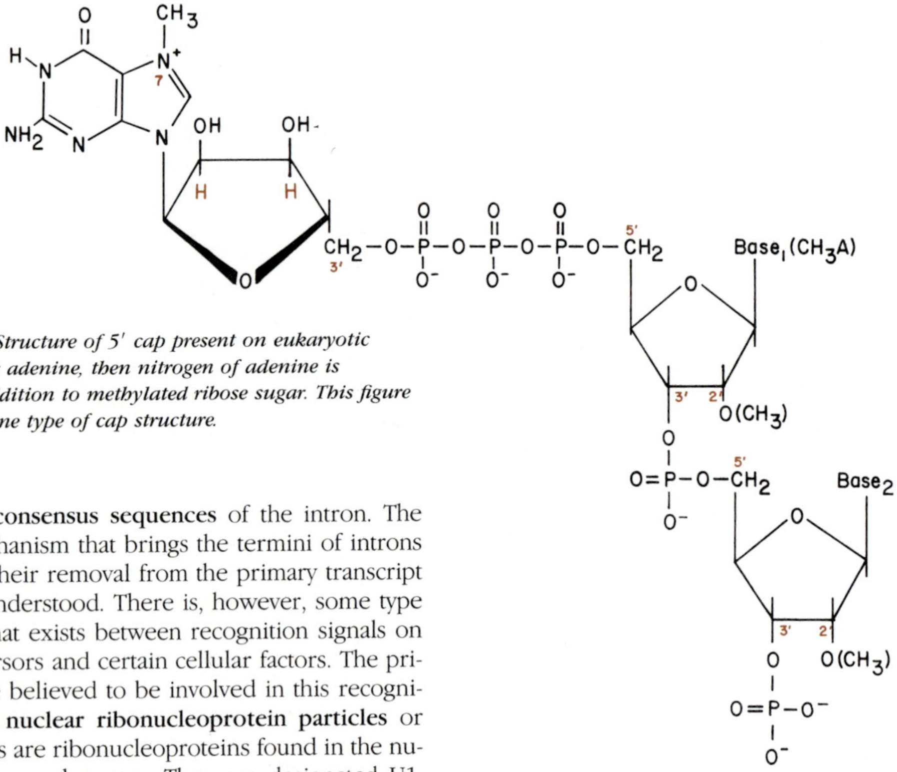

Figure 9-22 *Structure of 5′ cap present on eukaryotic mRNAs. If base is adenine, then nitrogen of adenine is methylated in addition to methylated ribose sugar. This figure represents only one type of cap structure.*

conserved or **consensus sequences** of the intron. The enzymatic mechanism that brings the termini of introns together after their removal from the primary transcript is not totally understood. There is, however, some type of specificity that exists between recognition signals on the RNA precursors and certain cellular factors. The primary candidate believed to be involved in this recognition are **small nuclear ribonucleoprotein particles** or **snRNPs.** snRNPs are ribonucleoproteins found in the nuclei of all higher eukaryotes. They are designated U1, U2, etc. and each contains a unique RNA molecule. It is believed that U1 snRNP is complementary to and can bind to the 5′ site, whereas U5 is complementary to and binds to the 3′ site:

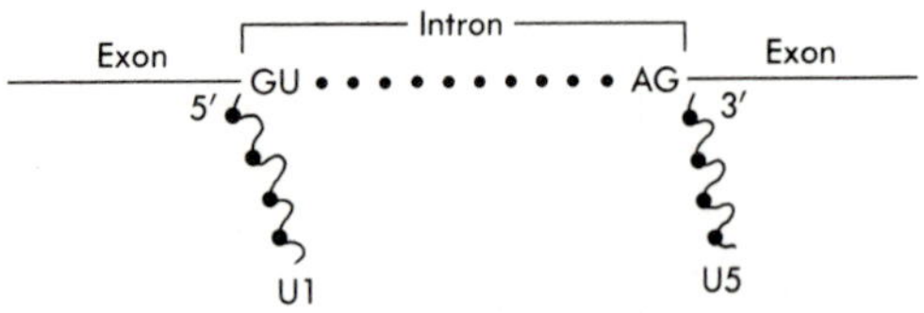

Cleavage at the 5′ site occurs and is accompanied by binding of the 5′ end of the intron to the 3′ end to form a lariat. In the final stage the 3′ splice site is cleaved and the exons are ligated, releasing the intron as a **lariat.** This model is believed to be the most common mechanism for splicing or pre-mRNA; however, there also exists a self-splicing mechanism that involves other RNA transcripts (see Point of Interest).

The splicing process offers a mechanism for producing different mRNAs from the same DNA sequences. For example, if a DNA segment possesses five exons, one mRNA could be produced by splicing all exons, whereas another mRNA could be produced by splicing only three exons. Whether a five-exon or three-exon mRNA is produced would depend on whether the splicing signal in the intron is used during the processing (Figure 9-25). This characteristic of eukaryotic genomes provides a degree of genetic flexibility for all eukaryotes, including those viruses that infect eukaryotes.

Transcription does not terminate at the poly A site

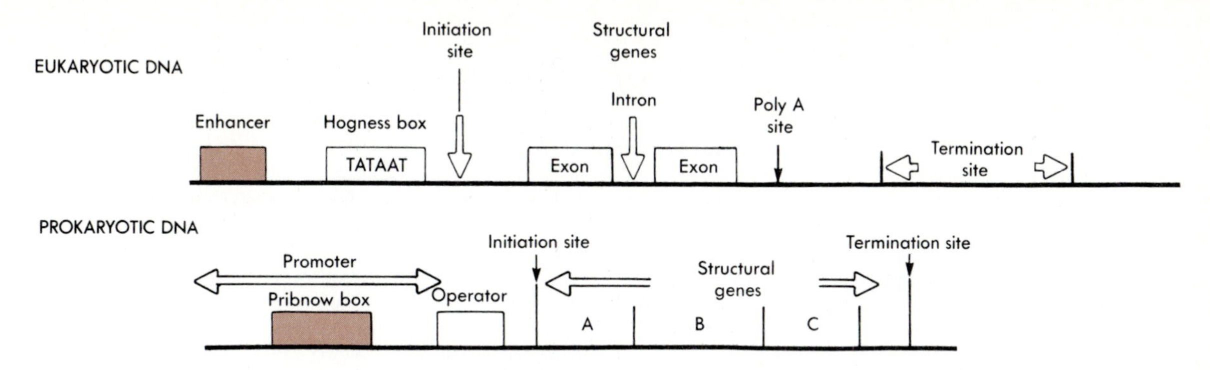

Figure 9-23 *Comparison of sites on eukaryotic and prokaryotic DNA that are involved in expression and regulation of transcription.*

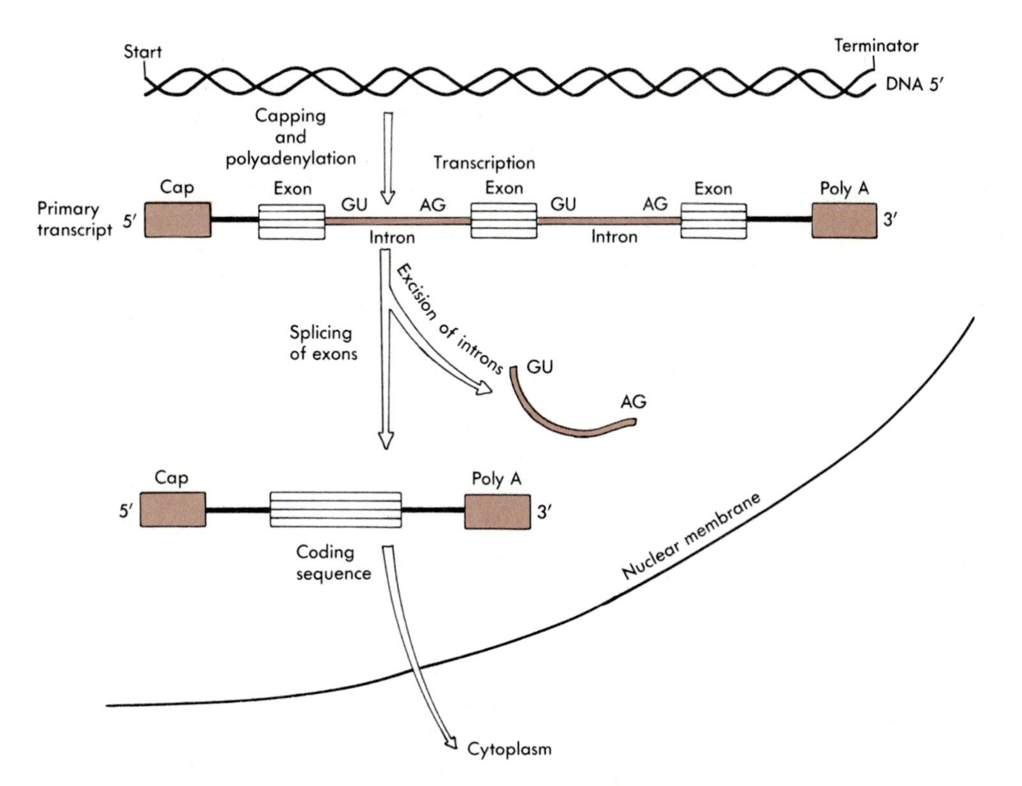

Figure 9-24 *Processing of nuclear RNA into cytoplasmic mRNA.*

but at some distance downstream. The reason for this is not clear. Once the nuclear RNA transcript has been processed the mRNA that is produced leaves the nucleus and migrates into the cytoplasm. Movement of mRNA, as well as rRNA and tRNA, through the nuclear membrane is not associated with prokaryotes. The 5′ capping unit of the monocistronic mRNA is bound by an 80S ribosome in the cytoplasm, and, with the appropriate initiation factors, translation begins (see Table 9-2). Recall that mRNA translation in prokaryotic cells begins even before the mRNA is completely transcribed from the DNA. This difference in translation may be associated with the differences in mRNA stability between eukaryotes and prokaryotes.

Point of Interest

RNA SPLICING—IS RNA AN ENZYME?

Experiments with splicing reactions involving rRNA in the protozoa *Tetrahymena* revealed that enzymatic activity for splicing in some introns is contained within the primary RNA transcript. The RNA of *Tetrahymena* is involved not only in self-splicing reactions but in other related reactions as well. The reactions that result in an intron removing itself from *Tetrahymena* rRNA (figure on left) are as follows:

1. A free guanosine or guanosine triphosphate possesses a hydroxyl group that attacks the phosphate at the 5′ end of the intron. This results in the breakage of the phosphodiester bond linking the left exon to the intron. The attacking guanosine becomes attached to the 5′ end of the intron and the left exon now has a free hydroxyl group.
2. The hydroxyl group on the left exon is free to attack the 3′ end of the intron. When this occurs, the exons become joined (ligated) and the intron is released.
3. Similar reactions to those above enable the intron to form a circle, and during this process, several nucleotides are released.

The activity of the intron is brought about only in a folded state, just as the three-dimensional structure of a protein is essential for enzymatic activity. Folding of the intron creates double-stranded regions that act as sites for binding of guanosine. The intron-binding site for the free guanosine consists of a nucleotide sequence, GGAGGG. These nucleotides pair with 6 pyrimidines: CUCUCU at the 5′ end of the exon. The folded state of the intron also activates the phosphate groups at each reaction site, thus favoring cleavage at those sites by hydroxyl ion. How the folding activates the phosphate at the splice sites is not known. However, it is known that mutations preventing the intron from forming normal doube-stranded regions hinder the capacity of the intron to splice itself out of the rRNA primary transcript.

Some self-splicing systems do not require guanosine. Instead, one of several hydroxyl groups on the intron attacks the 5′ splice site (figure on right). The 5′ end of the intron is not joined to the 3′ end but to a point a short distance away; this produces a branched-looped structure or **lariat.** The branching of the loop is created by the formation of an unusual 2′ to 5′ phosphodiester bond. When the exons are ligated the lariat is released. The lariat is the product of the splicing of nuclear pre-mRNA, as well as the Group II splicing process.

This self-splicing process implies that RNA may have preceded protein in early evolution. Thus RNA could have replicated without the assistance of enzymatic proteins.

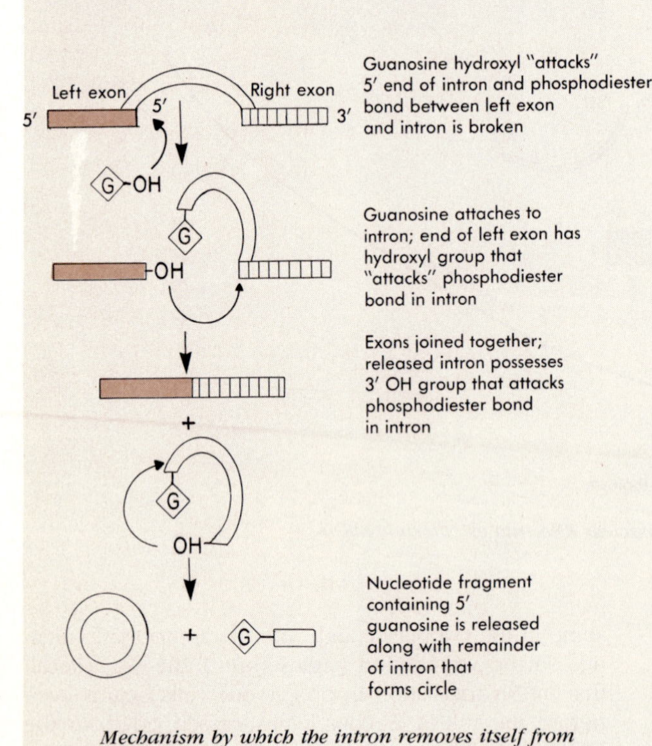
Left exon
Right exon
5′
5′
3′
Guanosine hydroxyl "attacks" 5′ end of intron and phosphodiester bond between left exon and intron is broken
G-OH
G
OH
Guanosine attaches to intron; end of left exon has hydroxyl group that "attacks" phosphodiester bond in intron
Exons joined together; released intron possesses 3′ OH group that attacks phosphodiester bond in intron
+
G
OH
+
G
Nucleotide fragment containing 5′ guanosine is released along with remainder of intron that forms circle

Mechanism by which the intron removes itself from Tetrahymena *ribosomal RNA.*

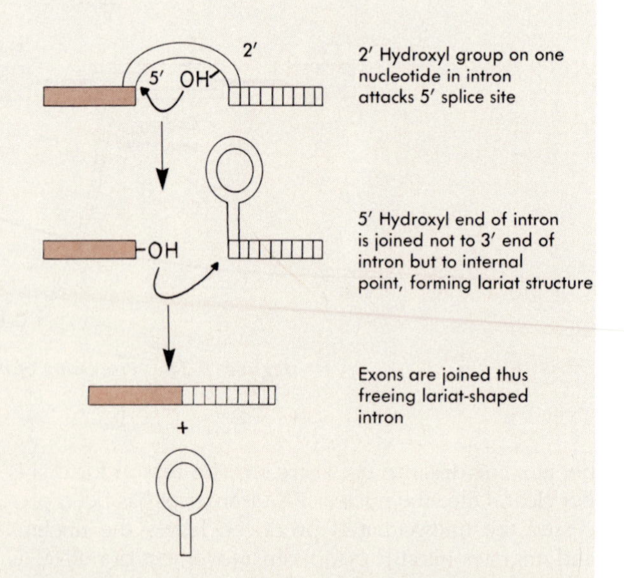
2′
5′
OH
2′ Hydroxyl group on one nucleotide in intron attacks 5′ splice site
OH
5′ Hydroxyl end of intron is joined not to 3′ end of intron but to internal point, forming lariat structure
Exons are joined thus freeing lariat-shaped intron
+

Mechanism of lariat formation during self-splicing.

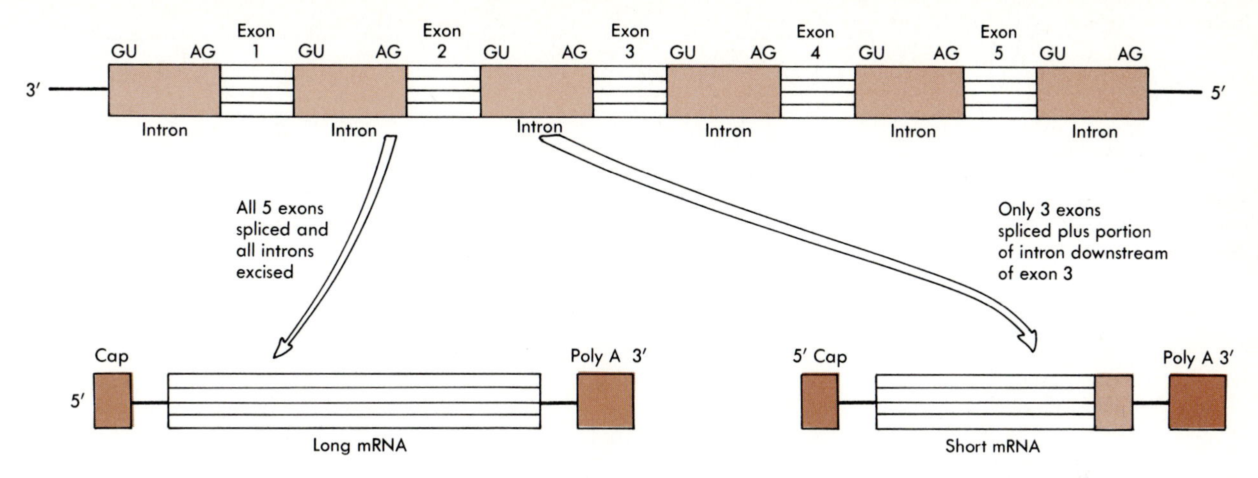

Figure 9-25 *Splicing technique in which RNA transcript can be processed to produce different proteins.*

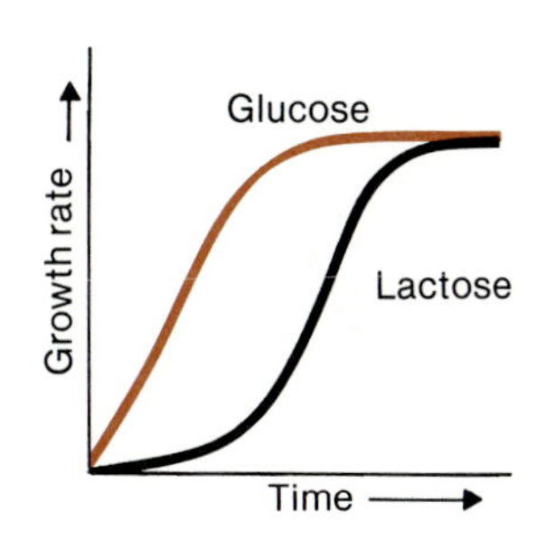

Figure 9-26 *Graphic description of enzyme induction. Bacteria* (E. coli) *grown on glucose show no lag in growth, but cells grown on lactose do show growth lag. Once enzymes for lactose catabolism have been synthesized during lag period, lactose-grown cells have same rate of growth as glucose-grown cells.*

Concept Check

1. What factors are required in the translocation process during protein synthesis?
2. Illustrate by a schematic diagram the essential components of a typical protein-coding transcriptional unit for prokaryotes. Eukaryotes.
3. What is the sequence of steps involved in the processing of eukaryotic mRNA?
4. What is the mechanism behind "self-splicing" of introns?
5. What are the similarities and differences of the transcription and translation processes of eukaryotes and prokaryotes?

Chain elongation and termination

The elongation process and chain termination are bascially the same for eukaryotes and prokaryotes. The protein elongation factors in eukaryotes are similar to those in prokaryotes, even though they are not interchangeable, and some termination codons operate in both groups. Information about the actual site of termination and the process involved is lacking.

REGULATION OF PROTEIN SYNTHESIS IN PROKARYOTES

The availability of nutrients, as well as many other environmental factors, determines what enzymes are required by the cell for adequate growth. The synthesis of all the potentially usable enzymes would be a waste of energy; therefore regulatory mechanisms have evolved in biological systems. This regulation is achieved by control mechanisms that act at the level of the gene and control the synthesis of enzymes, or that act at the level of the enzyme and control the activity of already-synthesized enzymes. Control of enzyme activity is discussed in Chapter 2. The following discussion is concerned with the genetic control of enzyme synthesis.

Induction and repression (negative control)

The phenomenon of enzyme **induction** was recognized long before the mechanism of protein synthesis had been elucidated. Investigators in the 1940s noted the microbial cells growing on a carbon source, such as glucose, could adapt to growth on other carbon sources, such as lactose, when glucose became exhausted in the medium. An initial lag in growth was observed when lactose was added to the glucose-exhausted medium, but later the rate of growth was equal to that of cells grown with only glucose present in the medium (Figure 9-26). The lag period was recognized as the period for synthesis of enzymes that had been absent during growth on glucose; consequently, the enzymes for lac-

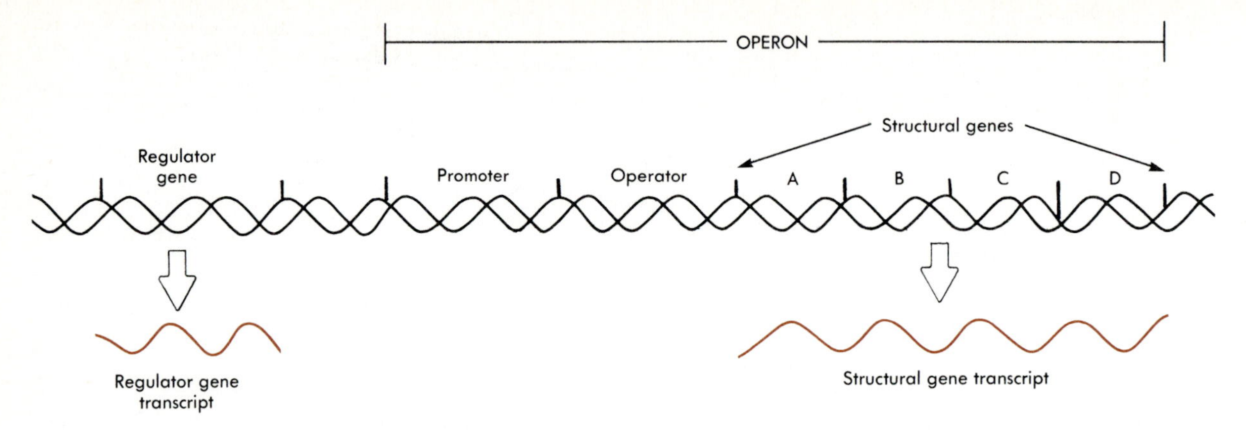

Figure 9-27 *Organizational features of operon. Only regulator gene and structural genes produce functional mRNA transcripts. Although regulator gene controls operon, it is not part of operon.*

tose catabolism (degradation) were inducible. Small molecules (such as lactose in the example) that cause the production of enzymes capable of metabolizing them are **inducers.**

Sometimes the cell must shut down the synthesis of enzymes. For example, in *E. coli* the amino acid histidine is synthesized by several endogenous enzymes. If histidine is added to the medium, however, these enzymes are no longer required and their continued production is halted. This type of response is called **repression.** Small molecules (such as histidine in the example) that prevent the production of enzymes capable of synthesizing them are called **corepressors.** Both inducers and corepressors are highly specific, although closely related molecules can often substitute for them.

Another interesting phenomenon was observed when cells growing on lactose exhausted their carbon source and were transferred to media containing glucose. There was no lag in the growth rate following this transfer. This meant that the enzymes for glucose catabolism are present in the cell whether or not glucose is present in the medium; that is, the enzymes for glucose catabolism are **constitutive,*** whereas **inducible** enzymes are produced only in the presence of their substrate. Later investigations revealed that one of the proteins induced by the substrate lactose is a carrier protein. Carrier proteins are responsible for the transport of various nutrient molecules across the cytoplasmic membrane and play a role in microbial metabolism. The lag period observed in enzyme induction experiments during growth on lactose, for example, was found to be a result of a lag in production of carrier proteins, as well as enzymes that catabolize lactose. How does the inducer molecule control the synthesis of enzymes involved in its own metabolism? The answer to this question came from genetic studies on *E. coli* and the brilliant work of two French investigators: Jacob and Monod, who studied the induction of enzymes involved in the catabolism of lactose. Lactose as a disaccharide is not metabolized and must first be broken down into its respective monosaccharide units: glucose and galactose. Jacob and Monod showed that enzyme induction is under genetic control because the enzymes for lactose catabolism can be produced constitutively through mutation.

The experiments of Jacob and Monod resulted in the formulation of the operon theory of control. An **operon** is a contiguous cluster of genes whose expression involves a particular trait (for example, the enzymes involved in the biosynthesis of histidine or the enzymes involved in the catabolism of lactose) and is in turn under the control of an operator and a repressor. Organizationally, the operon consists of **structural genes,** an **operator,** and a **promoter** region. An associated gene, which is located outside the operon, is the **repressor gene** (Figure 9-27). The characteristics of these genetic units as they apply to the lactose operon are discussed below. The organizational features of the lactose operon are illustrated in Figure 9-28.

Structural genes. The lactose operon possesses three structural genes: the gene for β-galactosidase (Z), which cleaves lactose into glucose and galactose; the carrier transport gene (Y), which transports lactose across the cytoplasmic membrane; and the transacetylase gene (A), whose function is as yet not fully understood.

The structural genes are transcribed as a single polycistronic mRNA (Figure 9-21). Ribosomes translate the message from the 5′ end of the gene and stop at the termination site of the Z gene and then continue to

* Constitutive enzymes are produced in constant amounts in the cell whether their substrates are present or not. Inducible enzymes vary in their concentration in the cell. They are present in trace amounts when the substrate is absent but their concentration increases greatly when their substrates are present in the medium.

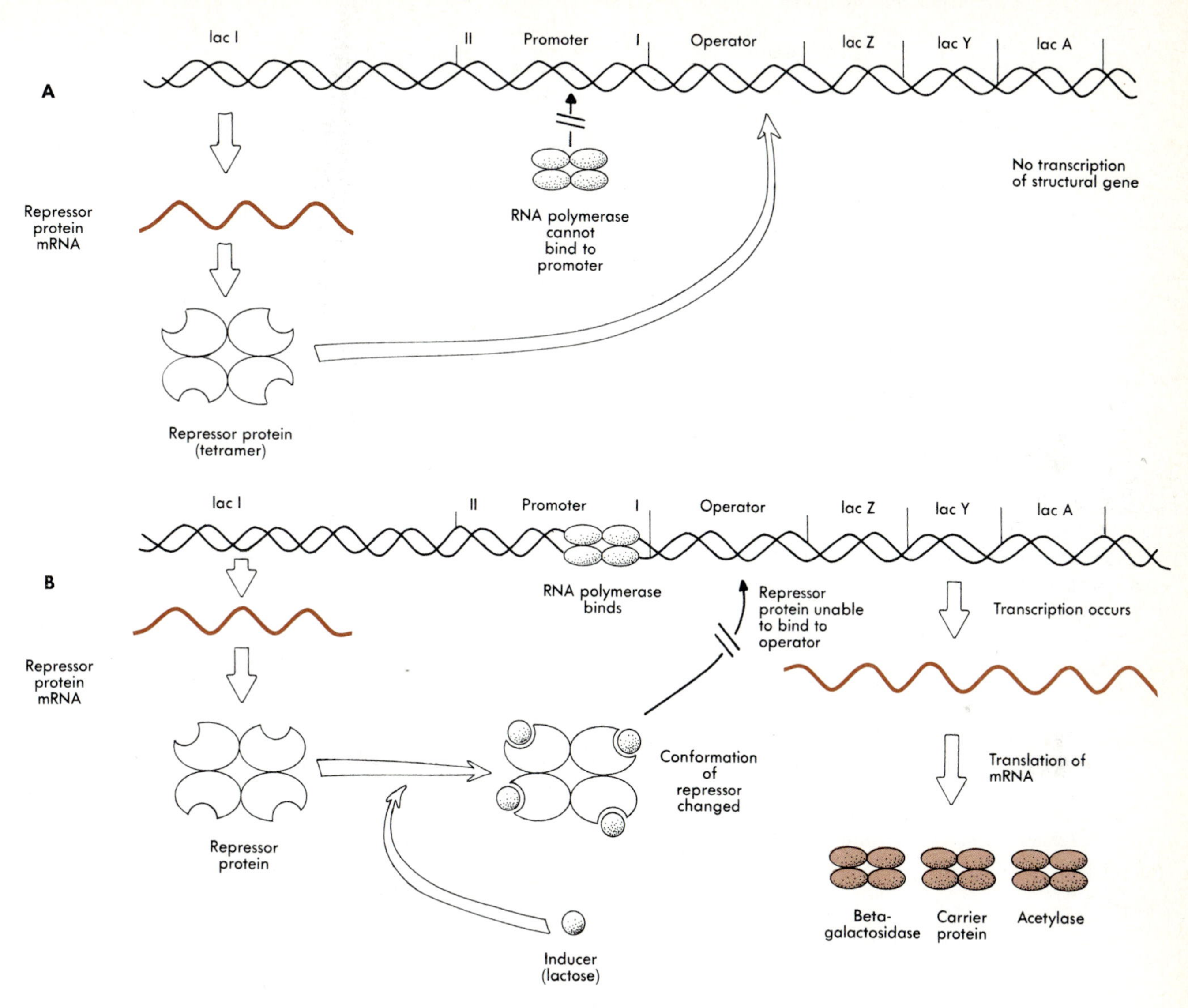

Figure 9-28 *Organization of lactose operon.* ***A,*** *Condition of operon in absence of inducer lactose. Operon operates negatively and genes for lactose catabolism cannot be transcribed. Binding of repressor (encoded in lactose I gene) to operator inhibits binding of RNA polymerase to promoter.* ***B,*** *Condition of operon in presence of lactose. Inducer lactose binds to repressor and changes its confirmation so it is unable to bind to operator. RNA polymerase is free to bind to promoter and structural genes for lactose catabolism can be transcribed resulting in formation of β-galactosidase, carrier protein, and acetylase.* Lac, *Lactose.*

translate the subsequent genes in the same manner. Ribosomes cannot initiate translation at the start regions of the Y or A genes, because only the 5′ end of the mRNA contains the "correct" initiation sequence.

Operator. The operator is a sequence of nucleotides positioned between the promoter and the structural genes. The operator is not transcribed into a mRNA or other functional nucleic acid molecule. The function of the operator is to control the expression of the structural genes by binding the repressor in the absence of inducer (lactose* in the lactose operon) and preventing transcription of the structural genes. Mutations in the operator can result in the inability of repressor to bind to the operator, and a permanent constitutive state can be obtained. Any condition or mutation that results in either a faulty operator or a faulty repressor (reduced affinity for operator) creates the constitutive state.

* Lactose is converted to allolactose, the inducer of the lactose operon.

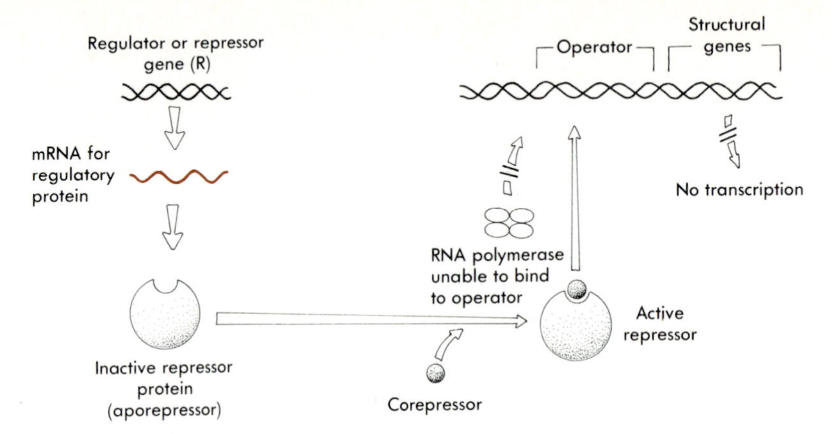

Figure 9-29 *Model demonstrating basic mechanism of repression.*

Promoter. The promoter region lies upstream of the operator and like the operator does not produce an mRNA or functional nucleic acid. The promoter is the central regulatory element of the transcription process. It contains signals that direct proper binding of RNA polymerase, which is activated to initiate RNA transcription. The currently accepted mechanism for promoter recognition by proteins is discussed in the Point of Interest box on p. 241. Mutations involving the promoter region prevent transcription because of an inability to bind RNA polymerase, and this prevents transcription of structural genes. Thus a condition of repression can exist without the help of a repressor molecule. The promoter is also capable of binding the protein, catabolite receptor protein (CRP), which is discussed in the section on repression.

Repressor. The repressor gene (**R gene** or I gene in lactose operon) is not contiguous with the operon that it controls. The repressor gene is transcribed into an mRNA that codes for the repressor protein. The lactose repressor protein is made up of four identical subunits, that is, it exists as a tetramer and is subject to conformational change like other regulatory proteins. The repressor (tetramer) in the absence of inducer exists in a conformational state that binds strongly to the operator region. This complex prevents the binding of RNA polymerase to the promoter region because the repressor overlaps the site on the promoter where RNA polymerase would normally bind. Likewise, if RNA polymerase were first bound to the promoter it would prevent binding of repressor to the operator. Once repressor is bound transcription of structural genes cannot take place. Mutation in the repressor gene results in the formation of a repressor molecule with binding properties, ranging from a state of increased binding to a state of inability to bind to the operator. The various binding properties are the result of changes in the conformation of the molecule.

The induction mechanism for the lactose operon works in the following way:

1. The repressor molecule is synthesized constitutively in the cell and in the absence of an inducer (lactose) is free to bind to the operator. The transcription of structural genes Z, Y, and A is prevented, and no lactose-metabolizing enzymes are synthesized. Thus the lactose operon is under negative control.
2. Lactose, when present in the medium, is converted to allolactose and binds to the repressor molecule forming a complex that has reduced affinity for the operator. The operon is free to bind RNA polymerase at the promoter, and the structural genes are transcribed into mRNA (Figure 9-28). The mRNA is translated into the enzymes involved in lactose metabolism.

Repression

Operons, in addition to being inducible, are also repressible. As stated previously, **repressible operons** function only in the absence of a small molecule called the corepressor. Repression can be achieved when a corepressor activates a repressor protein (Figure 9-29). Negative control via repression is characteristic of operons that control the synthesis of enzymes involved in biosynthesis, such as enzymes required in the synthesis of amino acids. Many microbial cells are capable of synthesizing all or most of their amino acids from a simple carbon source, such as glucose, and thus do not require amino acids in the growth medium. In other words the enzymes involved in amino acid biosynthesis are produced constitutively by the microorganism. If one of the amino acids, for example, histidine, is added to the growth medium that already contains glucose, the level of enzymes associated with histidine synthesis is greatly reduced, whereas the enzymes involved in the synthesis of amino acids other than histidine are produced at normal levels. Because histidine is the endproduct this type of negative control is called **endproduct repression**. The endproduct of the biosynthetic pathway is the **corepressor**, and it combines with an inactive repressor molecule called the **aporepressor**. The corepressor may be the free amino acid or the aminoacyl tRNA form of the amino acid. The union of corepressor and aporepressor produces an active repressor complex that binds to the operator and represses transcription of the structural genes of the operon.

One of the most thoroughly studied repressible systems is the tryptophan (Trp) operon (Figure 9-30). When tryptophan is in excess (more than is needed for protein synthesis at the particular growth rate), the Trp operon is repressed because the corepressor-repressor

Point of Interest

INTERACTIONS BETWEEN DNA AND PROTEINS

Only recently, we began to understand the molecular nature of the binding process between certain proteins and specific DNA sequences. What are the recognition factors that enable proteins, such as cyclic AMP (cAMP), catabolite receptor protein (CRP), RNA polymerase, or lactose repressor molecules to bind to specific sites on the DNA? X-ray analysis of the structure of proteins and DNA reveal that macromolecules, in addition to differences in size and tertiary structure, possess different **domains.** Domains represent specific chemical groups brought together by the conformation of the molecule. With the help of X-ray crystallography, genetic studies, biochemical studies, and model building, it is possible to interpret the binding that occurs between DNA and protein domains. A protein, for example, may have an amino *(N)* terminal domain and a carboxyl *(C)* terminal domain. Each domain could consist of various alpha helices interrupted by turns of beta sheets. The domains of the DNA consist of nucleotide sequences that make up the area in and around a major or minor groove. The currently accepted hypothesis for protein-DNA recognition is that base pair sequences in the grooves of the DNA act as hydrogen bond donors, and acceptors that recognize a complimentary cluster of hydrogen donors, and acceptors that make up the protein domain. The DNA-binding proteins studied have alpha helices interrupted by turns that recognize B-DNA, the right-handed form that is found in solution, although Z-DNA (see Chapter 2) may also be involved. DNA-binding proteins, such as CRP, use the alpha helices to make most of their contacts between side chains and bases in the major groove (figure on right). Adjacent alpha helices are probably used to properly orient the protein along the DNA. Hydrogen bonding between the nucleic acid bases and the amino acids of the protein determine specificity but the stability of the binding is probably because of other interactions, such as electrostatic interactions between hydrophobic amino acids and the methyl group of thymine, which juts out from the backbone into the major groove. It is believed that regulatory proteins bind strongly to the promoter in at least one major groove. Some regulatory proteins apparently enhance the formation of a complex between RNA polymerase and the promoter site, and this ensures that transcription will be initiated. (See Catabolite Repression).

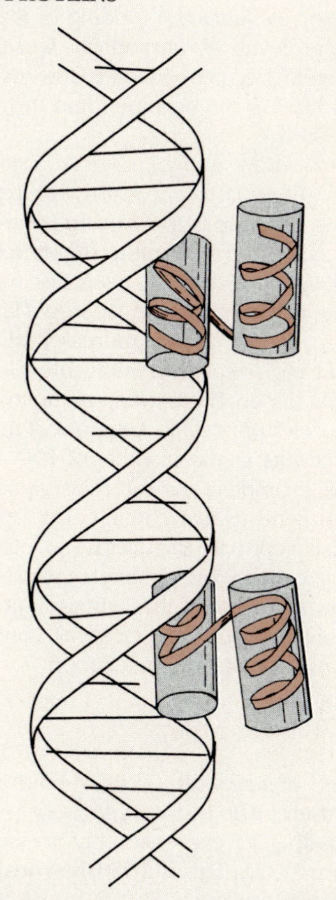

Interaction of binding proteins with DNA. One example in which alpha helices of protein make contact with DNA. Polypeptide alpha helices are represented as cylinders in illustration. Contacts between DNA and protein may involve hydrogen bonding between amino acid residues (for example, hydroxyl groups of tyrosine or threonine, amino group of lysine or guanidinium group of arginine) and nucleic acid bases.

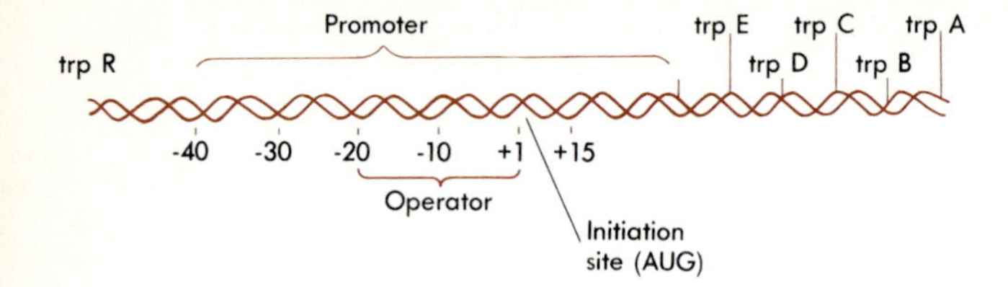

Figure 9-30 *Organizational features of tryptophan operon. Note operator for* trp *genes lies within promoter. TrpR is repressor gene.*

protein complex is bound to the operator. The operator for the tryptophan genes lies within the promoter (Figure 9-30) and, as in the inducible lactose operon, binding of repressor excludes the binding of RNA polymerase. When the level of tryptophan is reduced, the repressor protein becomes inactive and RNA polymerase is able to bind to the promoter and the Trp operon becomes activated.

Multiple operators and repression. Examination of repression in the galactose operon has revealed that more than one operator is involved in control. The typical operator is located adjacent to the structural genes, but a second operator has also been discovered within the first gene of the galactose operon (Figure 9-31). What seems to happen is that repressor binds to both operators, and each repressor molecule binds to each other to form a dimer. The binding of the two repressor molecules causes a physical looping out of the promoter region. This results in the inability of RNA polymerase to bind to the promoter, and the transcription of galactose structural genes is prevented. Other operator sites had been discovered in the lactose structural genes; however, their capacity for binding repressor was often 100 times less than that of the primary operator. Such studies indicate that our knowledge of control mechanisms is still not completely understood.

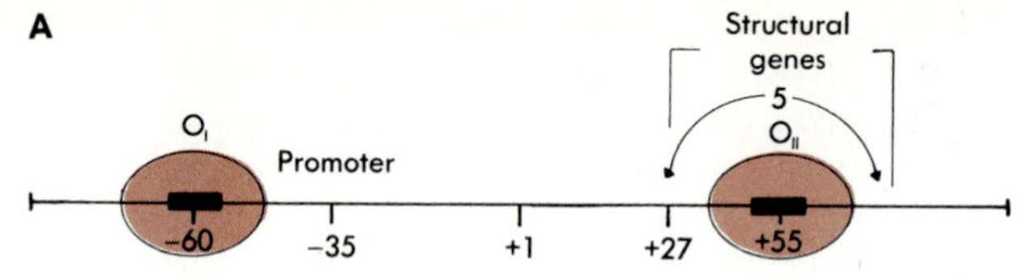

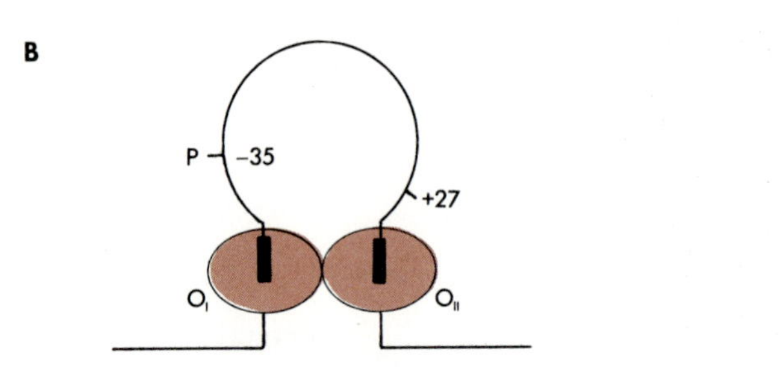

Figure 9-31 *Proposed mechanism for repressor-operator interaction in galactose operon.* ***A,*** *Partial genetic map of galactose operon. Numbers indicate distance, in base pairs, from start site of transcription initiation. Two operators, O_I and O_{II}, are involved.* ***B,*** *Looping out on galactose operon, O_I is located upstream (−60) from transcript start site, while O_{II} is located downstream (+55). Repressor dimer binds simultaneously to both O_I and O_{II} and causes looping out of DNA. Occupation of both operators by repressor molecule blocks rapid access of RNA polymerase to promoter, which has been looped out.*

Concept Check

1. **Diagram the components of the operon. What are the functions of each component?**
2. **How can regulatory proteins bind to specific areas on the DNA molecule?**
3. **What is the difference between an inducible and repressible operon?**

Positive control of the operon

Positive control of genes can occur when the regulatory protein is present, that is, the regulatory protein does not interfere with gene expression but is essential for its activity (Figure 9-32). This shows that positively controlled promoters are not fully functional in the presence of RNA polymerase alone and require activation. Positively controlled systems use a **regulator (R) gene** that encodes the **activator.** The activator controls the activity of the promoter of at least one gene. One example of this type of control is called **catabolite repression.**

Catabolite repression. When *E. coli* is grown in a medium containing glucose and lactose, glucose is preferentially metabolized and the enzymes for lactose metabolism are temporarily repressed until the glucose is exhausted. The regulator protein involved in this phenomenon is an activator called the **catabolite receptor protein (CRP)**—the product of the R gene. The events that are associated with catabolite repression are:

1. Glucose, which is transported into the cell by transport proteins, blocks the activity of the enzyme adenylcyclase. Adenylcyclase is involved in the conversion of AMP to cyclic AMP (cAMP). Catabolite repression is mediated by the intracellular concentration of cAMP.
2. CRP is active only in the presence of cAMP. As mentioned previously, the lactose promoter possesses two protein binding sites: one for RNA polymerase and one for the CRP (Figure 9-33).
3. When glucose is present, CRP is inactive and the

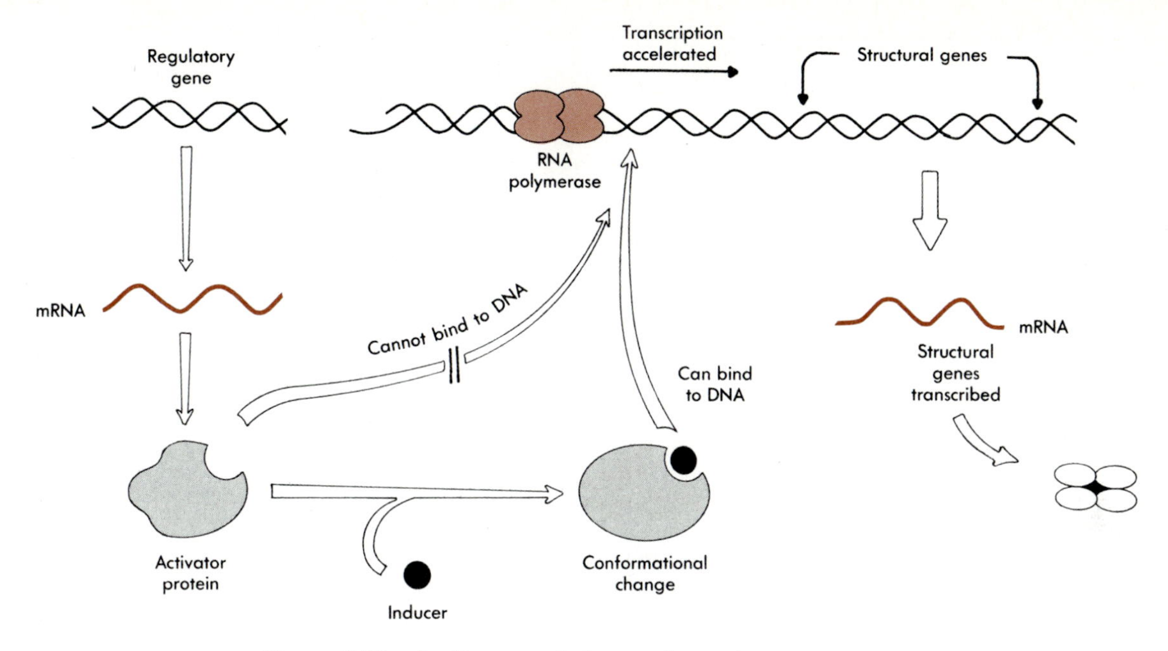

Figure 9-32 *Positive control of operon by regulatory gene product.*

enzymes for lactose catabolism are repressed. This effect is sometimes called the **glucose effect**. When glucose is absent, cAMP levels increase and the CRP becomes activated.

4. The binding of cAMP causes a conformational change in the CRP and allows the latter to bind to the promoter at a site near the RNA polymerase binding site. CRP is a dimer that can apparently interact with RNA polymerase and increase lactose transcription about 50-fold. CRP is also associated with other operons including the arabinose and galactose operons.

Catabolite repression is an important energy-saving device for cells growing in the presence of two carbon sources such as glucose and lactose. Glucose is much more easily metabolized than lactose, and with the lactose operon repressed, the enzymes for which it codes are not produced and energy is saved. Many catabolic enzymes in *E. coli* and other enteric bacteria are sensitive to glucose repression.

Other types of positive control. The activator (CRP) in catabolite repression is an example of a direct control mechanism that allows RNA polymerase to initiate transcription at specific promoters. However, other mechanisms of positive control do exist and considerable research is being directed at their elucidation. In one of these the activator is a factor that replaces one of the subunits of RNA polymerase, thus altering its promoter-recognition specificity. For example, a number of different sigma factors, both bacterial and viral coded, have been discovered. In bacteria, transcriptional specificity is altered in sporulating cells of *Bacillus subtilis.* The RNA polymerase of **B. subtilis** during vegetative growth has the same structure as that of *E. coli,* a nonsporulating bacterium. In sporulating *B. subtilis* there are different sigma factors of the enzyme that vary significantly from the sigma factor of vegetative cells. During sporulation the presence of new sigma factors enables RNA polymerase to recognize some genes (those involved in the sporulation process) and to ignore those associated with the vegetative state. The affinity of the sigma factor for core RNA polymerase therefore affects the expression of certain genes.

Another type of positive control could involve the synthesis of an entirely new RNA polymerase. Such a mechanism exists for some viruses when they invade a cell and must divert transcription of cellular genes to transcription of viral genes.

Attenuation: a control mechanism without a regulatory protein

Apparently repression is not the only way that the microbial cell can regulate the Trp operon. Mutant analysis showed that even if a functional repressor is lacking, the cell can still respond to tryptophan starvation by increasing the rate of tryptophan mRNA synthesis. In addition, some mutants with deletions in the transcribed region of the operon have a sixfold increase in expression of the remaining genes of the operon. Repressor control is not affected by these deletions, suggesting that the deleted region is distinct from the operator but still in-

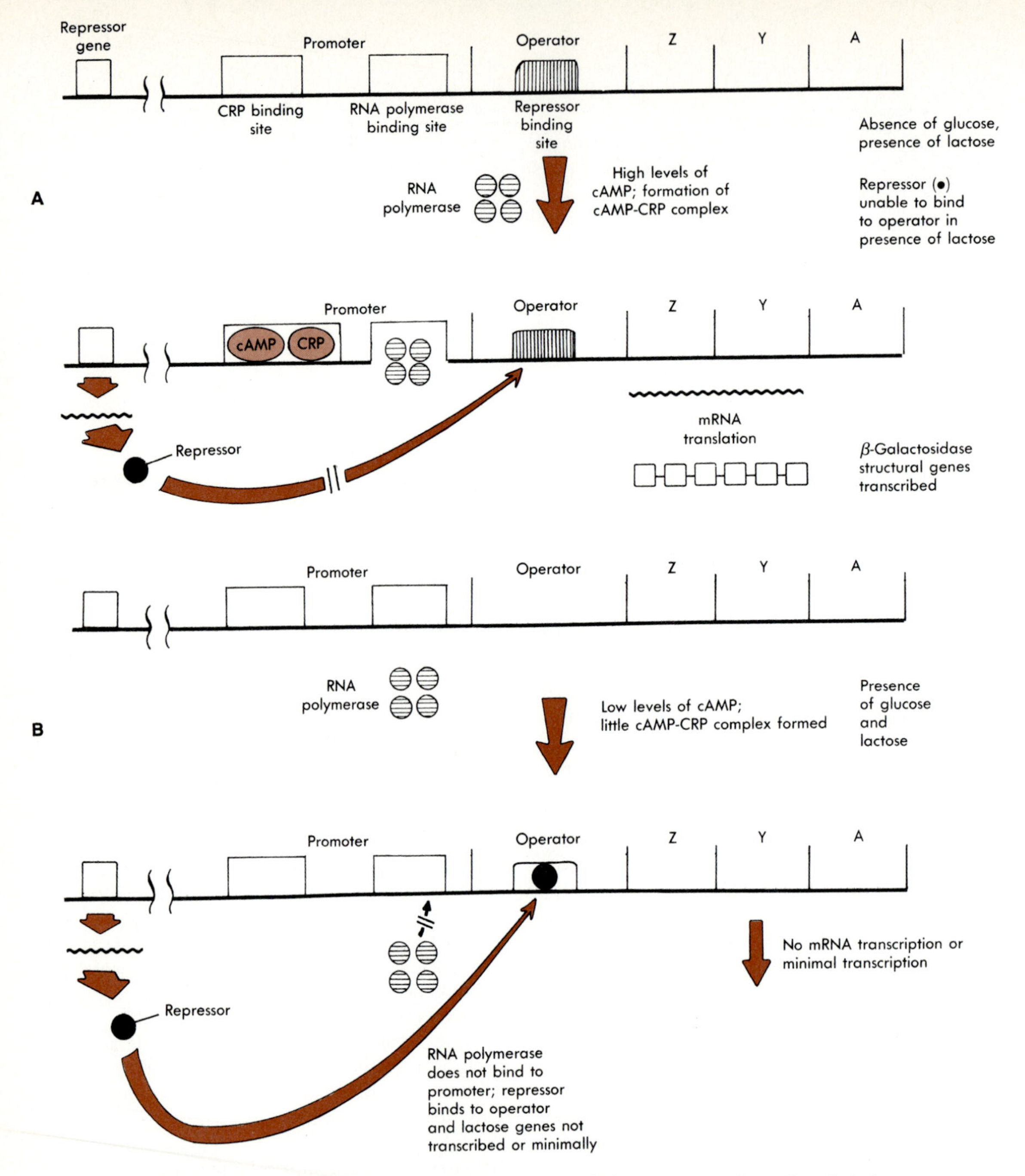

Figure 9-33 *Catabolite repression in lactose operon.* ***A,*** *Lactose present in medium but glucose absent; thus cAMP concentration is increased and acts as inducer of CRP converting it to active state. CRP interacts with RNA polymerase and increases enzyme activity so that level of lactose structural gene-encoded proteins is increased 50-fold (50× concentration).* ***B,*** *Glucose and lactose present in medium, cAMP concentration is greatly reduced. RNA polymerase activity is minimal; thus transcription of lactose structural genes is also minimal (1× concentration).*

volved in control of the operon. The regulatory site is the **attenuator**, and it controls events that occur after RNA polymerase leaves the promoter. Attenuation produces a finer control of biosynthetic enzyme synthesis by promoting the premature termination of transcription. This regulatory mechanism is affected by the concentration of the particular amino acid. Six amino acid biosynthetic operons including the Trp operon are controlled by attenuation (Figure 9-34). The details of this mechanism can be found in the listings in the selected readings.

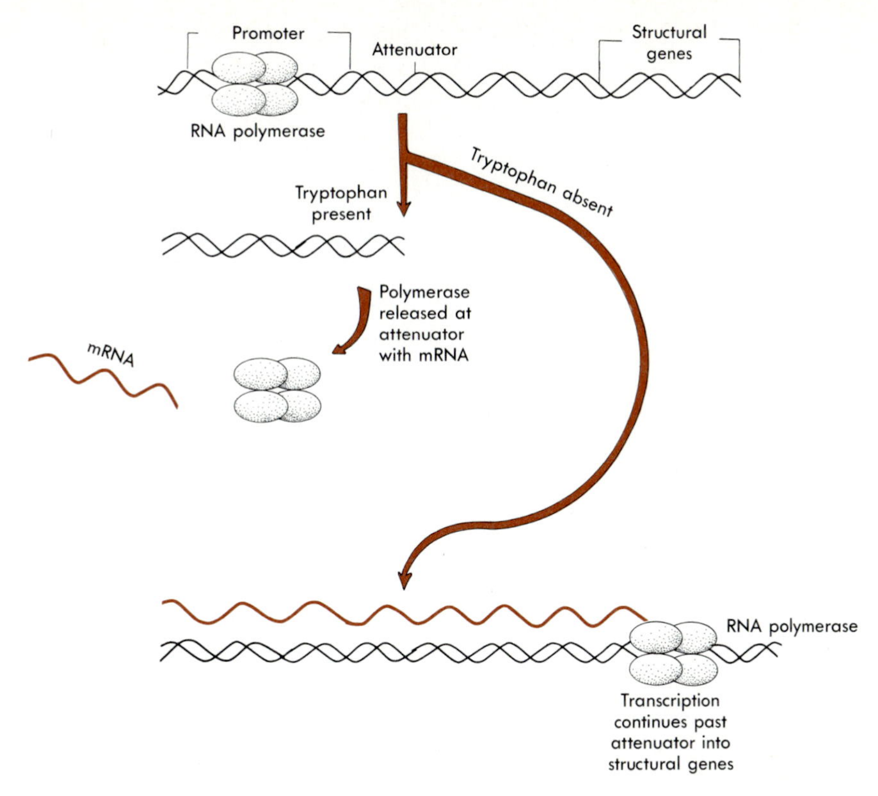

Figure 9-34 *Basic scheme of attenuation using tryptophan operon as example. RNA polymerase binds to promoter and begins synthesis of mRNA. In presence of tryptophan RNA polymerase can go only as far as attenuator at which point it drops off and releases shortened mRNA transcript. In absence of tryptophan RNA polymerase transcribes beyond attenuator into structural genes (details of this mechanism are described in literature listed in selected readings).*

Regulons

Some genetic systems have several operons that are under the control of a single regulatory protein called **regulons.** The arginine biosynthetic pathway, for example, has several operons scattered on the chromosome (Figure 9-35). Each operon is controlled by a separate operator, which in turn is controlled by a single repressor protein. This implies that each operator possesses common base sequences that recognize the repressor. Because the genes of a regulon have separate operators the synthesis of their products is controlled noncoordinately, that is, the amount produced by the activity of one gene may vary from that of another gene.

Control at the level of translation

Our discussion of protein synthesis regulation up to now has centered on mechanisms that affect transcription of the gene. There are ways in which the cell might control protein synthesis at the level of translation. The products of some polycistronic mRNAs are produced in stoichiometrically equal amounts. The lactose operon contains three different genes whose coded products are not produced in equal amounts. The enzymes coded by the Z, Y, and A genes occur in the relative proportions of 10:5:2. Two possible mechanisms may account for the unequal production of enzymes coded by a polycistronic mRNA:

1. Nucleotide sequences (intercistronic) exist between the termination signal of the Z gene and the initiation signal of the Y gene, as well as between the termination signal of the Y gene and the initiation signal of the A gene. Some ribosomes translating the Z gene could fall off the mRNA before reaching the Y gene. Proportionately more ribosomes would drop off during their intercistronic movement from the Y gene to the A gene.
2. A second mechanism for translation control might involve the use of different codons for the same amino acid. For example, the codon for leucine in gene Z might be different from that in gene Y. Disproportionate amounts of enzyme could be produced if the concentration of one charged tRNA · Leu was different from another.

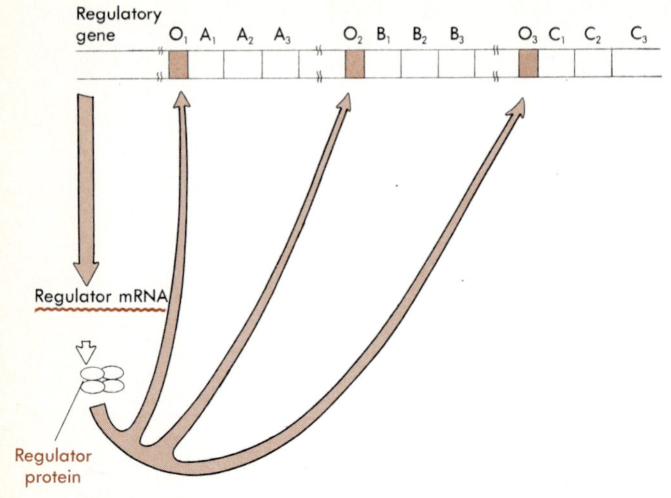

Figure 9-35 *Model of regulon. Genes A, B, and C are scattered on chromosome (noncontiguous nature of three genes is indicated by broken line). Each gene is under control of different operator but each operator is controlled by single regulator protein (repressor protein).*

Posttranslational control

Proteins, once their synthesis has been completed on ribosomes, sometimes require modification before they are able to function in the cell. The removal of the formyl or formyl-methionine group at the amino end of the protein is a typical posttranslational response in bacteria. Proteins required in the periplasm and in the outer membrane of gram-negative bacteria or proteins secreted into the medium also undergo modification. These proteins possess special signal sequences at their amino ends. These signal sequences, which are recognized by the cytoplasmic membrane, facilitate the movement of the protein to its final destination. The signal sequence must be cleaved by proteolytic enzymes before the protein becomes functional (see Chapter 5).

Concept Check

1. What is the distinction between negative and positive control of the operon?
2. Explain the process of catabolite repression.
3. What are the various types of transcriptional control?
4. What is the mechanism of attenuation?
5. In what ways can protein synthesis be controlled?

SUMMARY

1. The hereditary information of the cell is present in the DNA in the form of genes. This DNA information is translated primarily into enzymatic proteins, which impart to the cell certain characteristics. In addition, some areas of the DNA can act as a template for the formation of transfer RNA (tRNA) and ribosomal RNA (rRNA).
2. The genetic code was found to be a triplet code in which most triplets code for an amino acid. There are 64 code words for the 20 biologically available amino acids; therefore an amino acid may be coded for by more than one triplet (the code is degenerate).
3. The code, which is nonoverlapping and contains no punctuation, is read from a fixed point on the message. The addition or deletion of a base in the message produces a shift in the reading frame, which in turn results in changes in the amino acids of the polypeptide.
4. The code is universal. The amino acid code words are identical (with a few exceptions) in all species of cells so far tested.
5. There are 20 molecular species of tRNA, but several can carry more than one amino acid. The wobble hypothesis explains this characteristic and assumes that a rigid specificity exists in the first two letters of the codon-anticodon interaction, whereas the third-letter interactions show more flexibility (wobble).
6. The information in the DNA molecule during protein synthesis is transcribed into messenger RNA, following the binding of RNA polymerase to the promoter site on the DNA. The mRNA is translated into proteins in a series of events: initiation, elongation, translocation, and termination.
7. The cell must regulate the synthesis of certain proteins not always required by the cell under certain environmental conditions. Control mechanisms are basically negative and positive.
8. Inducible and repressible operons demonstrate negative control because a regulatory protein turns off the activity of the operon. Positively controlled operons require a regulatory protein that activates them, such as occurs in catabolite repression.
9. Some operons can be controlled without a regulatory protein. In the process called attenuation a regulatory site on the DNA called the attenuator determines whether RNA polymerase can or cannot transcribe the structural genes.
10. Control may also exist at the level of translation, resulting in different amounts of structural gene products being synthesized.

Self-study Quiz

Multiple choice

1. The Pribnow box is:
   A. A sequence of bases in the operator that are important for transcription.
   B. A sequence of bases in the mRNA transcript that recognize ribosomes.
   C. The termination site on eukaryotic mRNA.
   D. A sequence of bases on the promoter region of DNA that help to bind RNA polymerase.
2. The sequences -10 and -35 refer to:
   A. Sequences in the promoter that recognize the repressor molecule.
   B. The sequences on mRNA that recognize rho termination factor.
   C. Highly conserved sequences on the promoter that recognize RNA polymerase.
   D. Highly conserved sequences on the operator that recognize the repressor.
3. Which of the following is a palindrome?
   A. GTACTT
   CATGAA
   B. GGGCCC
   CCCGGG
   C. GGACCC
   CCTGGG
   D. GATGAT
   CTACTA
4. The Goldberg-Hogness box is:
   A. The eukaryotic version of the prokaryotic Pribnow box.
   B. The sequences on the eukaryotic operator that bind regulatory proteins.
   C. The site on eukaryotic mRNA recognized as a termination site.
   D. The sequence of bases that precede each intron in eukaryotic DNA.
5. What initiation factor, in addition to GTP, initiation factors, 30S ribosome, 50S ribosome, and mRNA is needed to initiate protein synthesis in prokaryotes?
   A. Sigma factor
   B. fMet-$tRNA_f$
   C. Peptidyl transferase
   D. RNA polymerase core enzyme
6. Messenger RNA synthesis is initiated when the RNA polymerase core enzyme binds to a sequence of nucleotides called the :
   A. Promoter.
   B. Terminator.
   C. 30S subunit.
   D. Structural gene.
7. On which molecule would you find the anticodon sequence of nucleotides?
   A. Sigma factor
   B. tRNA
   C. mRNA
   D. DNA
8. Which of the following is not part of an operon?
   A. Promoter
   B. Structural genes
   C. Operator
   D. Regulatory gene
9. Which of the following would be composed of the longest sequence of nucleotides?
   A. RNA polymerase core enzyme
   B. An operon
   C. tRNA
   D. Polycistronic mRNA
10. Which of the following steps in protein synthesis requires ATP to provide the energy for the process to occur?
    A. Transcription
    B. Elongation
    C. Amino acid activation for attachment to tRNA
    D. Translocation

Completion

1. The triplet of nucleotide bases on mRNA that codes for a single amino acid is called a ________________.
2. A triplet of nucleotide bases on mRNA that does not specify an amino acid is called ________________.
3. The process whereby mRNA is synthesized from the DNA template is called ________________.
4. Shift of the mRNA relative to the ribosome to expose a new A site codon is called ________________.
5. mRNA of prokaryotic cells with several ribosomes bound to it is called ________________.
6. The process in which the preferential metabolism of glucose occurs while the lactose operon is repressed is called ________________.
7. Enzymes that are synthesized by the cell whether or not the enzyme substrates are present in the environment are called ________________.
8. The component of RNA polymerase whose function is in the selection of the correct binding site on DNA is called ________________.

Matching

1. Capping ________
2. Corepressor ________
3. Positive control ____
4. Negative control ____
5. Aporepressor ______
6. Attenuator ________
7. Enhancer elements __
8. Exon ________

A. Translatable RNA information.
B. Nontranslatable RNA.
C. The repressor in its inactive configuration.
D. Control site affected by configuration of mRNA transcript.
E. A control process in which regulatory protein activates the operon.
F. Control process in which the regulatory protein turns off the operon.
G. Eukaryotic DNA sequence that increases promoter activity.
H. The endproduct of a biosynthetic pathway that activates the repressor.
I. The process in which a unique nucleotide is added to the eukaryotic mRNA transcript.

Thought question

1. For your research project you chose to work on an obscure enzyme of *Bacillus subtilis* that breaks down arabinose. The enzyme is inducible and appears in the cell in response to arabinose in the medium. In the course of your research you isolate a mutant strain in which your enzyme appears to be constitutive.
   (a) Explain how this phenomenon might occur.
   (b) What advantages is there to the cell for the arabinose pathway to be inducible rather than constitutive?

SELECTED READINGS

Books

Freifelder, D. *Molecular biology* (2nd ed.). Jones and Bartlett Publishers, Inc., Boston, 1987.

Hunt, T., Prentis, S., and Tooze, J. (Eds.). *DNA makes RNA makes protein*. Elsevier, Amsterdam, 1983.

Watson, J. *Molecular biology of the gene* (4th ed.). Benjamin/Cummings Publishing, Inc. Menlo Park, Calif. 1987.

Journal Articles

Crombrugghe, B., Busby, S., and Buc, H. Cyclic AMP receptor protein: role in transcription activation. *Science* **224**:831, 1984.

Darnell, J.E., Jr. The processing of RNA. *Sci. Am.* **249**:90, 1983.

Darnell, J.E., Jr. RNA. *Sci. Am.* **253**:68, 1985.

Dickson, R.C., et al. Genetic regulation of the lac control region. *Science* **187**:27, 1975.

Garen, A. Sense and nonsense in the genetic code. *Science* **160**:149, 1968.

Gottesman, S. Bacterial regulation: global regulatory networks. *Annu. Rev. Genet.* **18**:415, 1984.

Jacob, F., and Monod, J. Genetic regulatory mechanisms in the synthesis of proteins. *J. Mol. Biol.* **3**:318, 1961.

Maniatis, T., and Ptashne, M. DNA-operator repressor systems. *Sci. Am.* **234**:64, 1976.

Nirenberg, M., and Leder, R. RNA code words and protein synthesis. *Science* **145**:1399, 1964.

Raibaud, O., and Schwartz, M. Positive control of transcription initiation in bacteria. *Annu. Rev. Genet.* **18**:173, 1984.

Reznikoff, W.S., Seigele, D.A., Cowling, D.W., and Gross, C.A. The regulation of transcription initiation in bacteria. *Annu. Rev. Genet.* **19**:355, 1985.

Takeda, Y, Ohlendorf, D.H., Anderson, W.F., and Matthews, B.W. DNA binding proteins. *Science* **221**:1020, 1983.

Umbarger, H.E. Feedback control by endproduct inhibition. *Cold Spring Harbor Symp. Soc. Quant. Biol.* **26**:301, 1961.

Yanofsky, C. Attenuation in the control of expression of bacterial operons. *Nature* **289**:751, 1981.

CHAPTER

10

Mutation

Chapter Outline

- Characteristics of the Mutation Process
  - Classes of Mutants and Phenotypic Effects of Mutation
  - Spontaneous vs. Induced Mutations
  - Forward, Back, and Suppressor Mutations
- Molecular Basis of Mutation
  - Base Pair Substitutions
  - Deletions and Insertions
  - Transposons, IS Elements, and Mutation
- Mutation and the Genetic Code
  - Missense Mutations
  - Nonsense Mutations
  - Reading Frame Mutations
  - Suppressor Mutations
- DNA Repair Systems
  - Repair of DNA Damaged by Ultraviolet Light
  - Repair of DNA Other than Those for Ultraviolet Damage
- Mutant Isolation and Detection
  - Enrichment
  - Detection
- Practical Application of Microbial Mutants
  - Indicators of Mutagens
  - Vaccine Production
- Mutation and Microbial Evolution

Introduction

Previous chapters show that the DNA molecule is faithfully reproduced from one generation to the next but errors are sometimes made. Alterations in the chemical and/or physical makeup of the DNA molecule are called **mutations**. The chemical or physical agents causing the mutations are called **mutagens** (Table 10-1). Mutations are the source of genetic variety among organisms and without them adaptation to environmental changes would not be possible. Organisms therefore possess the inherent tendency to produce heritable changes from one generation to another.

The most important advances in the study of mutation have been achieved using microorganisms such as *Escherichia coli*. We are beginning to understand the molecular mechanisms of the mutation process. Advances in sequencing of DNA, as well as recombinant DNA technology, enable scientists to analyze the mutational process in much greater detail than possible with earlier classical approaches. The study of mutational events is important because of the variety of mutagens in our environment. Now that the link between mutation and cancer is established, the mechanisms involved in mutation are even more important.

This chapter describes the molecular mechanisms of mutation, the effect of mutation on the genetic code, how mutations are repaired, and some practical applications of microbial mutants.

Table 10-1 *Characteristics of the common microbial mutagens*

	Mutagen	Mechanism of action	Effect of mutation
Physical	Ultraviolet light	Causes dimerization between pyrimidine residues on DNA	Inhibits replication; repair may lead to deletion of base
	X-rays	Causes formation of free radicals such as hydroxyl, as well as superoxide ion	Oxidation and breakage of DNA chains
Chemical	5-Bromouracil	Analogue of thymine that in its rare "enol" form pairs with guanine	Causes transitions* from A:T base pairs to G:C bases pairs, and vice versa
	2-Aminopurine	Analogue of adenine that in its rare imino form bonds to cytosine; in natural state can also bind to cytosine	Causes transitions from A:T base pairs to G:C base pairs, and vice versa
	Nitrous acid	Oxidatively deaminates bases such as adenine, cytosine, and guanine	Causes mispairing and transitions from A:T base pairs to G:C base pairs, and vice versa
	Hydroxylamine	Causes deamination of cytosine	Transistions from G:C base pairs to A:T base pairs only
	Nitrogen mustard	Alkylating agent that can act as binding component between purines on opposite strands of DNA	Causes removal of bases that can be later replaced by any of the four bases; thus transitions and transversions† are possible
	Nitrosoguanidine	Like nitrogen mustard causes interstrand linkage of purine bases; causes alkylation	Similar to nitrogen mustard
	Ethylethanesulfonate	Alkylating agent like nitrogen mustard	Causes removal of purines, especially guanine
	Acridines	Insert themselves between bases on DNA strand	Depending on whether DNA strand is parental or is being synthesized, addition or deletion, respectively, will occur
Biological	Insertion sequences	Transposable from site to site on chromosome	Alter expression of genes into which they insert themselves as well as genes near insertion site
	Transposons	Same as insertion sequence	Same as insertion sequences except that transposons carry a specific trait that is linked to genes where they insert on chromosome

*Mutation in which one pyrimidine is substituted by the other, or in which one purine is substituted for the other.
†Mutation in which a purine is replaced by a pyrimidine or vice versa.

CHARACTERISTICS OF THE MUTATION PROCESS

Classes of mutants and phenotypic effects of mutations

In biology the organism isolated from nature that has not been exposed to a mutagen in the laboratory is called the **wild type**. The progeny of the wild type in which the mutation has been expressed are **mutants**. The mutant may produce some recognizable change in a trait, that is, a change in its phenotype, or the change may be so slight or minor that only a special biochemical or biophysical process can detect it. A mutation in a gene will result in a new form or **allele** of that gene.

There are several classes of phenotypic mutants, and the most important are described in Table 10-2. One of the most useful types of mutation for analyzing biologi-

Table 10-2 *Major phenotypic classes of bacterial mutants*

Class	Description of defect
Auxotrophs	Defect in biosynthetic pathway; mutant is unable to grow unless its growth medium is supplemented with amino acid, purine or pyrimidine base, or vitamin.
Carbon source	Defect in pathway that metabolizes the organism's carbon source; for example, mutation in a step in glycolysis prevents mutant from growing in presence of any product that is normally synthesized before mutation.
Nitrogen source	Defect in metabolism of nitrogen to ammonia; mutant unable to grow in presence of any nitrogen product before mutational step.
Cryptic mutants	Possess cytoplasmic potential for particular function but are deficient in some other function and potential is not realized. For example, a bacterium unable to metabolize lactose may possess cytoplasmic enzymes for lactose metabolism but is deficient in a transport process that normally concentrates lactose in cytoplasm
Temperature sensitive	Cannot synthesize a product at one temperature (restrictive temperature) but can synthesize a product at another temperature (permissive temperature. May be cold sensitive, lower temperature (usually 20° C) is restrictive and higher temperature (usually 37° C) is permissive; or heat sensitive, higher temperatures are restrictive (usually 42° C) but lower temperatures (usually 37° C) are permissive
Drug resistant	Exhibit tolerance to various drugs, such as antibiotics.
Morphological	Exhibit some change in morphology, such as inability to produce flagella, pili, capsules, etc.

cal processes is called the **conditional mutant**. This means that under one set of conditions the mutant will not function (the **restrictive** set of conditions) but in another environment the mutant will function (the **permissive** set of conditions). Some of the most common types of conditional mutants are the temperature sensitive mutants, suppressor-sensitive mutants, and the auxotrophic class of mutants.

1. *Temperature senstive mutants.* Temperature sensitive **(ts)** mutants produce enough product to perform a function at one temperature but are unable to produce a sufficient amount of product or fully perform a function at another temperature. Most times the mutation involves the secondary, tertiary, or quaternary structure of a protein. The protein or product of mutation is produced at both temperatures but at the restrictive temperature, changes in conformation prevent the protein from becoming active.
2. *Suppresor-sensitive mutants.* Suppressor mutants are unable to function unless a second mutation or factor, the **suppressor**, is present. The suppressor corrects or compensates for the defect in phenotype caused by the original suppressor-sensitive mutation. The special types of suppressors are discussed on (p. 260).
3. *Auxotrophic mutants.* **Auxotrophic** mutants are unable to synthesize one or more growth factors, such as amino acids, purines, pyrimidines, or vitamins. They are derived from "wild type" strains, called **prototrophs**, that can synthesize the specific growth factor. An auxotroph unable to synthesize histidine, for example, is designated His^-, whereas the prototroph is designated His^+. (The minus sign denotes a defective or mutant copy or allele of a gene, and a plus sign indicates a normal copy of the gene.) As indicated in Table 10-2, auxotrophs have defects in enzymes involved in the metabolism of compounds that serve as nitrogen or carbon sources. Sometimes the mutant synthesizes the necessary enzymes for metabolism but the carbon or nitrogen source cannot permeate the cell surface because of a defective transport protein. Such mutants are called **cryptic**.

Spontaneous vs. induced mutations

Mutations may be spontaneous or induced. Spontaneous mutations occur in the absence of human intervention. They may be caused by mistakes during DNA replication or by exposure of the organism to natural extracellular influences, such as radiation or chemical agents in the environment. Induced mutations are caused by deliberate exposure of the organism to natural extracellular influences, such as radiation or chemical agents in the environment (see Point of Interest, p. 257).

Mutations are rare events, and the rate at which they appear in a bacterial cell ranges from 1×10^{-4} to 1×10^{-10} per generation for a particular gene. In other words, one bacterial cell in 10,000 or one in 10 billion may show a mutation for a specific gene. Mutation rates are generally defined as the probability of mutation per cell per generation. The mutation rate varies for each characteristic examined and also varies for the same characteristic between species or even between strains of the same species. Some areas of DNA are more susceptible to the same mutational event and are called **hot**

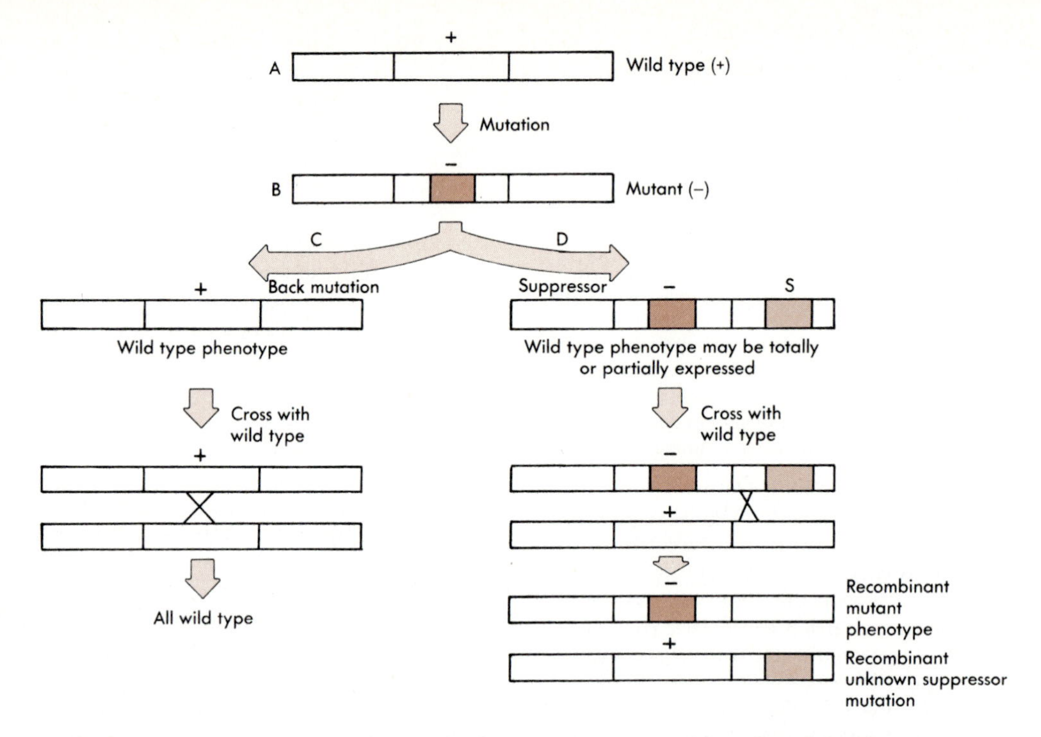

Figure 10-1 *Relationship of forward, back, and suppressor mutations.* ***A*** *to* ***C****, Forward mutation produces mutant (−). Back mutation at same site as forward mutation produces wild type phenotype (+). When revertant crossed with wild type, all progeny will be wild type.* ***A*** *to* ***B*** *to* ***D****, Mutant phenotype can be reversed by suppressor mutation (S) at another site on genome. When suppressor mutation is at site relatively distant from site of forward mutation, some revertants will produce mutant phenotype when backcrossed with wild type.*

spots. Thus, when mutants of a bacterial species are isolated, genetic analysis may reveal that many of them occur at a single base on the chromosome. Mutations involving a single base are called **point mutations.** The mutation rate for a particular mutagen at a specific site depends on the nature of the base at the 5′ side of the affected base. It appears that repair systems used by the organism to correct the mutation differ in their efficiency because of the presence of specific bases near the site of mutation. The mutation rate for a particular characteristic will also vary depending on the type of mutagen used. The spontaneous mutation rate for penicillin sensitivity (Pen^s) might be 1×10^{-10} but in the presence of a chemical or physical agent the rate could increase to $1 \times 10.^{-4}$

The error rate in the DNA replication system is estimated at 10^{-8} to 10^{-11} per base that is synthesized (that is, an error will occur once for every 10^8 to 10^{11} base pairs synthesized). This accuracy is based on the activity of DNA polymerase, as well as repair enzymes, present in both prokaryotic and eukaryotic cells. Most DNA replication errors are corrected or are silent; that is, no visible change can be detected in the phenotype.

The spontaneous mutation rate in some *E. coli* strains is several times greater than in most strains of the species. Closer examination of these strains has revealed the presence of **mutator genes**, which are mutant genes that increase the rate of mutation. For example, one mutator gene in *E. coli* produces a thermolabile DNA polymerase that inserts incorrect nucleotides during replication at a higher rate than wild type enzyme.

Forward, back, and suppressor mutations

Some mutations remain stable and are transferred to future generations. A mutation from the wild type, such as His^+ to the auxotroph His^- is a **forward mutation** but a mutation could reverse the direction and go from His^- to His^+ and thus restore the function of histidine synthesis. The chances for the forward mutation from His^+ to His^- and a loss of function are high, because there are nine enzymes involved in the synthesis of histidine and a mutation in any one could result in a loss of function. In order to restore function (His^- to His^+) the mutation must reverse the effects of the original mutation. These reversed mutations are **back mutations** or **reversions** and usually are less frequent than the forward mutation. Back mutations that occur by alteration

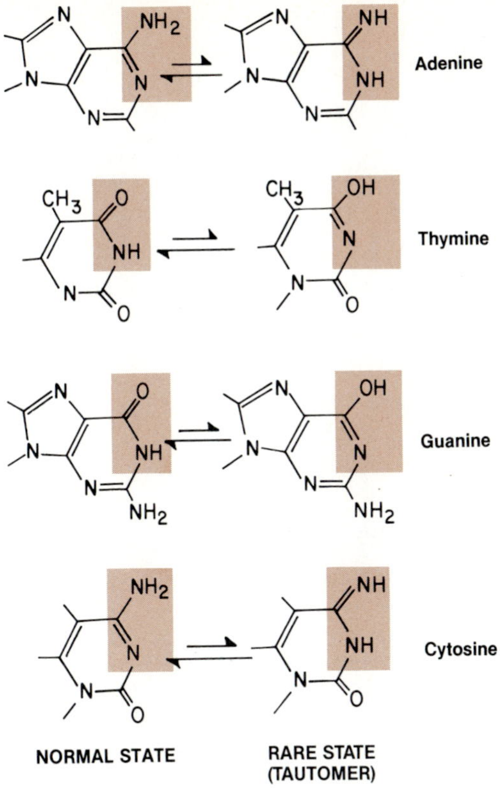

Figure 10-2 *Tautomeric shifts of purine and pyrimidine bases to rare molecular state.*

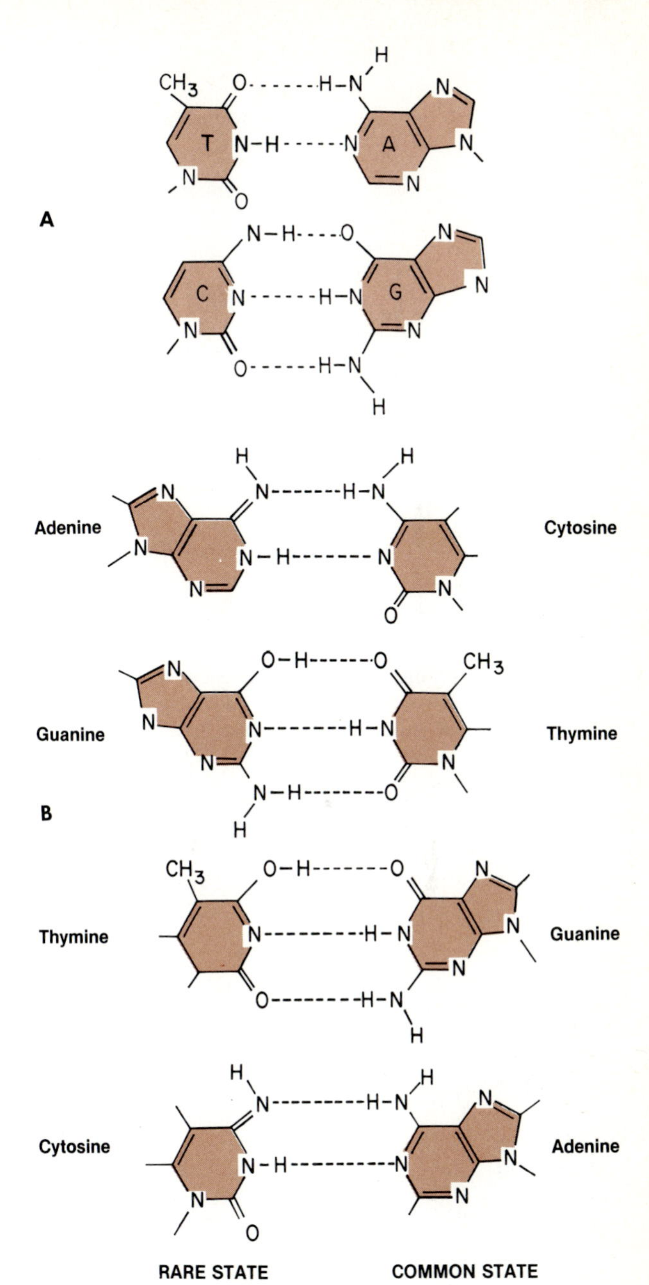

Figure 10-3 *Abnormal base pairing resulting from formation of tautomers.* ***A,*** *Normal base pairing between purines and pyrimidines in DNA molecule.* ***B,*** *Abnormal base pairing between purines and pyrimidines. All of "forbidden" base pairings in illustration are called transitions.*

of the genetic material at a site distant from the original mutation are **suppressor mutations**. Suppressor mutations may occur at a site in the same gene as the original forward mutation, in different genes, or even on different chromosomes. The relationship of forward, back, and suppressor mutations is illustrated in Figure 10-1. A more detailed discussion of suppressor mutations appears on p. 260.

MOLECULAR BASIS OF MUTATION

Base pair substitutions

Spontaneous base pair substitutions. Watson and Crick had predicted that if the two strands of DNA unwound and acted as a template in the replication process, errors could be made as the complementary strands were being synthesized. Such errors can take place because "abnormal" base pairing is possible between bases on the DNA molecule. Abnormal base pairing is a consequence of rearrangements in the distribution of electrons on the purine and pyrimidine bases. These arrangements, called **tautomeric shifts**, result in the formation of **tautomers.** Tautomers are compounds that exist in equilibrium but differ in the arrangement of hydrogen atoms and electrons (Figure 10-2). The normal hydrogen bonding between purine and pyrimidine bases is associated with those forms that possess amino (NH_2) and keto (C═O) groups. Tautomers are formed on rare occasions when the hydrogen bonding involves imino (C═NH) and enol (C—OH) groups (Figure 10-3) and normal purine-to-pyrimidine pairing is not possible. The replacement of a purine or pyrimidine on a

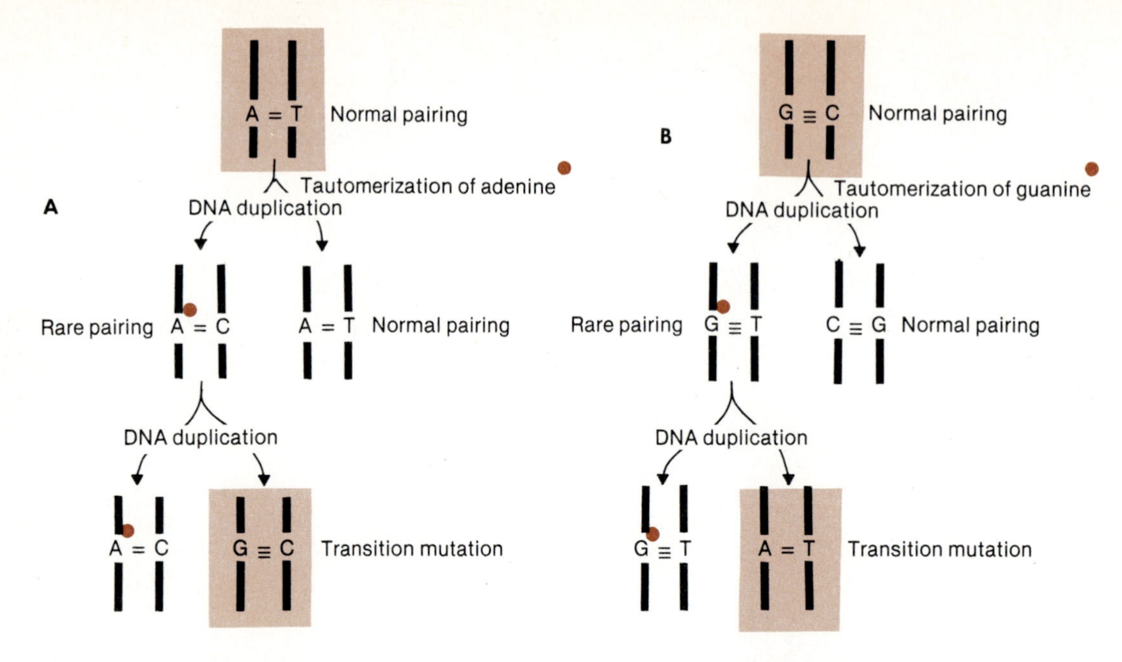

Figure 10-4 *Consequences of tautomerization involving bases adenine and guanine.* ***A,*** *Tautomerization of adenine and transition from A:T to G:C base pairing.* ***B,*** *Tautomerization of guanine and transition from G:C to A:T base pairing. Tautomers that gave rise to abnormal pairing will not persist, therefore in second round of duplication they will demonstrate normal pairing.*

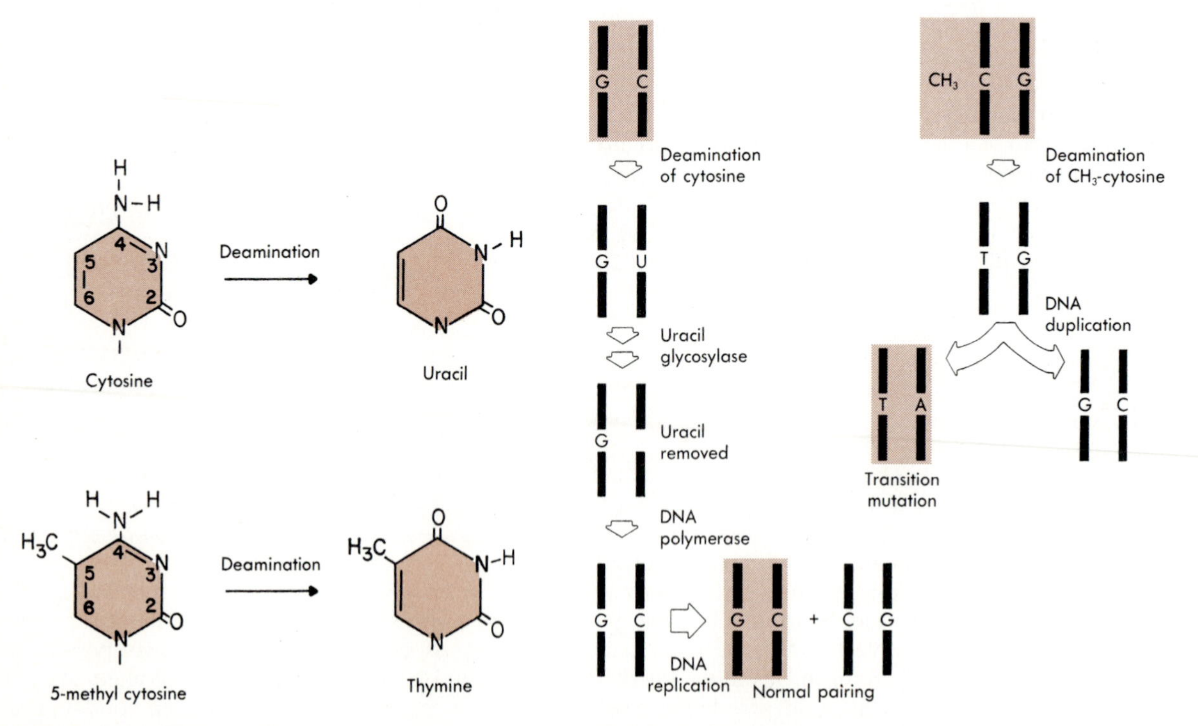

Figure 10-5 *Deamination of cytosine and 5-methylcytosine produces uracil and thymine, respectively.*

Figure 10-6 *Comparison of the effects of deamination of cytosine and 5-methylcytosine.*

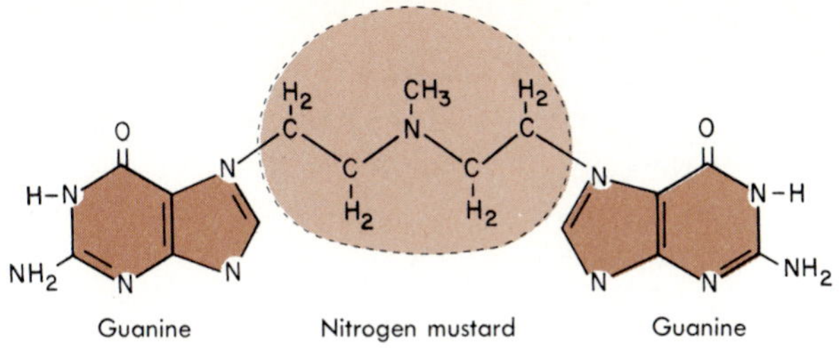

Figure 10-7 *Nitrogen mustard* (color) *alkylating adjacent guanines.*

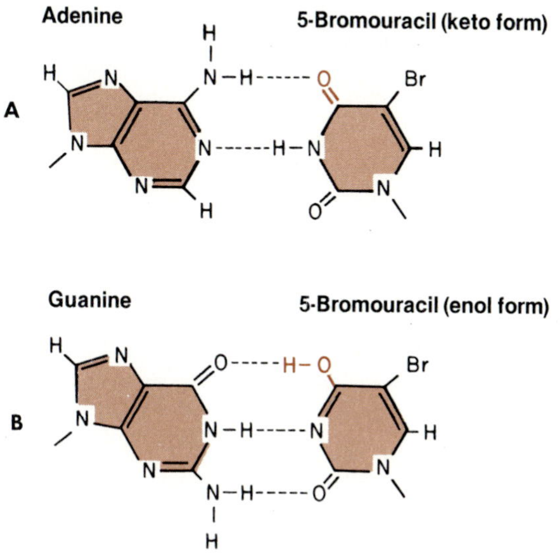

Figure 10-8 ***A,*** *Pairing properties of 5-Bu in its normal keto state.* ***B,*** *Abnormal enol state.*

DNA strand by a different purine or pyrimidine, respectively, is a **transition.** For example, adenine, which normally pairs with thymine, can pair with cytosine when adenine is in its rare imino state. The primary consequence of the initial pairing mistake is the formation of a DNA molecule, following replication, that contains a "forbidden" base pair; that is, there is a transition from A:T to G:C or G:C to A:T (Figure 10-4).

A major cause of spontaneous point mutations in *E. coli* is associated with a modified base, 5-methylcytosine. Modification by addition of methyl groups is generated by enzymes called **methylases,** a protective device by microorganisms. In the lactose I gene of *E. coli* the hot spots for spontaneous mutation occur in sequences where 5-methylcytosine is present. Strains of *E. coli* unable to use a methylase do not show these hot spots. The reason for this peculiarity is that 5-methylcytosine has a tendency to spontaneously deaminate with the replacement of an amino group by a keto group (Fig. 10-

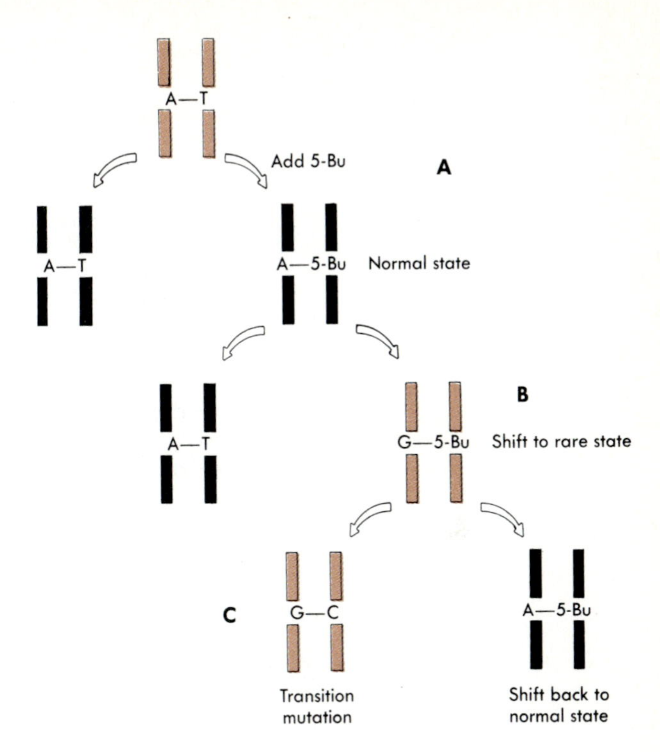

Figure 10-9 *How 5-Bu causes transitions from A:T to G:C.* ***A,*** *5-Bu acts like thymine in its normal state and is base paired with adenine.* ***B,*** *During duplication of DNA, 5-Bu shifts to rare enol form and behaves like cytosine and therefore pairs with guanine.* ***C,*** *During next round of replication, guanine on template strand pairs with cytosine of complementary strand.*

5). This replacement process results in the formation of thymine, thus creating at the next replication a G:C to A:T transition. Unmethylated cytosine can spontaneously deaminate to uracil. However, in *E. coli* an enzyme, *N*-uracil DNA-glycosylase, can remove uracil residues from DNA before it can pair with adenine at the next replication (Figure 10-6).

Induced base pair substitutions. Three of the more widely used chemical mutagens that cause transitions are 5-bromouracil, nitrous acid, and alkylating agents, such as nitrogen mustard (Figure 10-7) and ethylmethane sulfonate. **5-Bromouracil (5-Bu)** is an analogue of thymine and has a tendency to ionize more frequently than thymine. 5-Bu in its normal keto state pairs with adenine, just as its analogue thymine does, but in its rare enol state it pairs with guanine (Figure 10-8). The transition from A:T to G:C or G:C to A:T, following incorporation of 5-Bu into one of the DNA strands, appears after two duplications (Figure 10-9).

Nitrous acid (HNO_2), unlike 5-Bu, can exert its mutagenic effects on nonreplicating DNA. Nitrous acid deaminates amino groups from purines and pyrimidines and causes the substitution of hydroxyl (OH) groups for

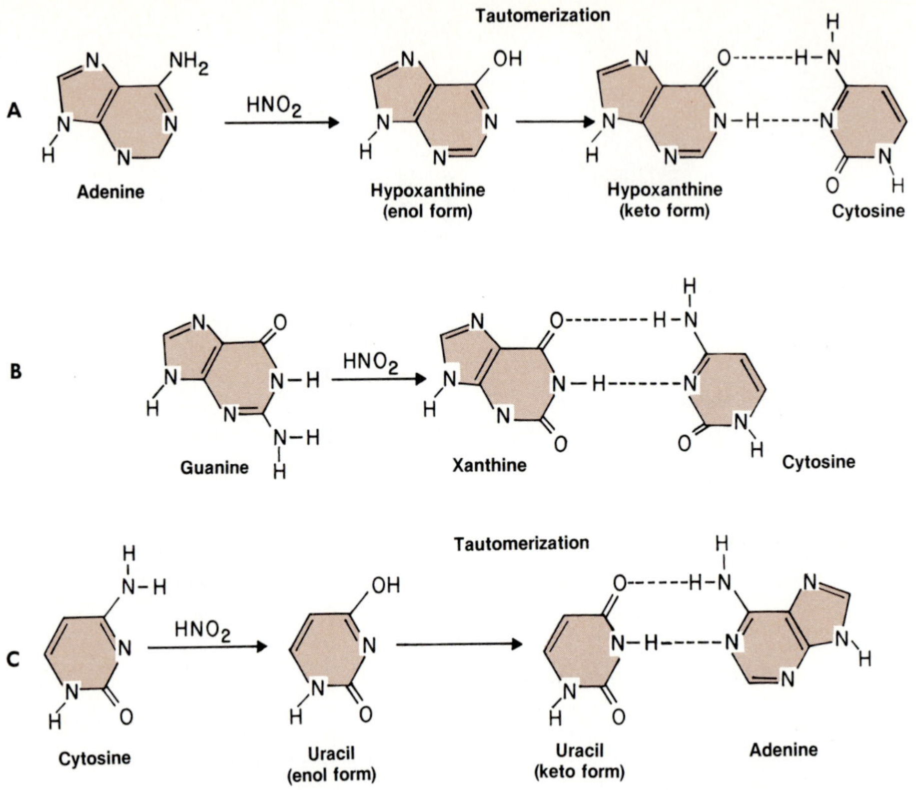

Figure 10-10 *Effects of nitrous acid on purine and pyrimidine bases.* ***A,*** *Nitrous acid deaminates adenine to hypoxanthine, which can pair with cytosine, thus causing transition mutation.* ***B,*** *Deamination of guanine with nitrous acid produces xanthine, which pairs with cytosine, and no base pair substitution occurs.* ***C,*** *Deamination of cytosine produces uracil, which pairs with adenine, therefore causing transition mutation.*

amino groups. Adenine and cytosine appear to be the most sensitive to the action of nitrous acid. Deamination of adenine leads to the formation of hypoxanthine, which has hydrogen bonding specificities for cytosine (Figure 10-10), whereas deamination of cytosine leads to the formation of uracil, which has hydrogen-bonding specificities for adenine. One of the characteristics of nitrous acid is its ability to cause transitions in both directions:

$$A:T \leftrightarrows G:C$$

Alkylating agents, such as ethylethane sulfonate ($CH_3CH_2SO_3CH_2CH_3$) and ethylmethane sulfonate ($CH_3CH_2SO_3CH_3$), possess ethyl groups that can be donated to nitrogen in the 7 position of purine bases, particularly guanine. Alkylation can cause hydrolysis of the base-sugar linkage and result in the release of the base from the DNA molecule. This creates a gap in the DNA chain where the base had formerly been present. Any base can be inserted into the gap (Figure 10-11) during the replication process. The inserted bases can be the same as the normal one, or they can be a different base. A purine or pyrimidine replacement may be in the nature of a transition, but the replacement of purine for a pyrimidine or a pyrimidine for a purine can also take place. Such replacements are called **transversions.**

Alkylation is also known to occur at the 0^6 position of guanine when chemical agents such as *N*-methyl-*N*-nitrosourea ($NO—NCH_3—CO—NH_2$) are used. Donation of a methyl group produces methyl guanine and on replication frequent incorporation of thymine instead of cytosine residues occurs. This tendency can lead to the accumulation of transition mutations.

Deletions and insertions

Acridines are dyes, such as proflavin, that resemble purine bases (Figure 10-12). They are able to insert (intercalate) themselves into the DNA molecule and create gaps between normal bases. This can eventually lead to the formation of deletions or insertions. Deletions represent a loss of nucleotide bases from a gene, whereas

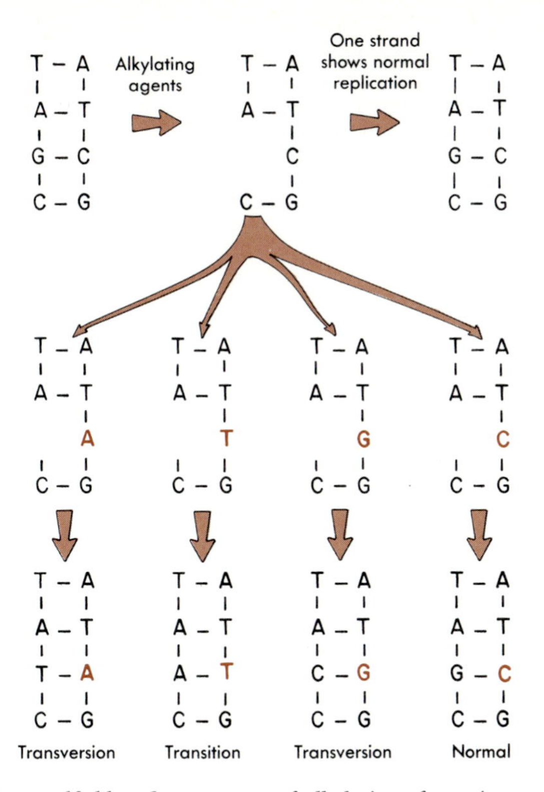

Figure 10-11 *Consequences of alkylation of guanine. Removal of guanine by alkylation creates gap in one strand of DNA. At next replication unaffected strand gives rise to normal double-stranded DNA molecules. Strand with gap can give rise to complementary strand with one of four possible bases. Only one out of four DNA molecules will be normal; the rest will be mutant.*

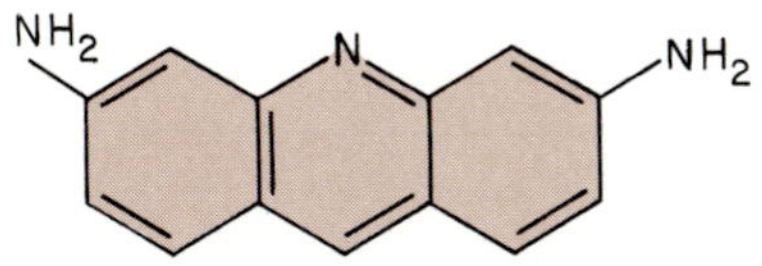

Figure 10-12 *Structure of proflavin.*

insertions occur when nucleotide bases are added to the DNA of a gene. The inserted or deleted DNA can amount to a single base or many bases. Because of the effect of these mutations on genetic activity, they are referred to as **frameshift mutations.** Frameshift mutations represent the addition or deletion of any number of bases not divisible by three (the genetic code is a triplet base sequence) and are among the most frequent unrepaired mutations. The term **frameshift** means that the reading frame in the gene is changed beyond the point of mutation (Figure 10-13). Consequently, the information downstream from the frameshift is not read in its original sequence and will not make sense.

Acridines can be intercalated into the template strand and when the DNA polymerase encounters it during complementary strand formation a new base is inserted to pair with it. This results in an insertion mutation during the next round of replication. Acridines can also be intercalated into the complementary strand as it is being synthesized. This means that the acridine will mask the base in the template strand and no base will pair with it. A deletion is produced during the next round of DNA replication. Deletions occur less frequently than insertions.

Spontaneous frameshifts resulting from additions and deletions have also been observed in microbial systems such as the lactose I gene of *E. coli.* The frameshifts appear to predominate at short sequence repeats that may be tandem, for example, CTGG CTGG CTGG, or they may be separated by several hundred base pairs. This observation led to the formulation of a **slipped mispairing** model for frameshift mutations. In this scheme, single strand breaks during replication allow one strand to slip and loop out one or more base pairs. Slippage on the growing daughter strand will generate additions, whereas slippage on the template strand will produce deletions (Figure 10-14). This model could also apply to frameshift mutagens, such as acridines; however, other ways can account for deletions, such as those generated by normal host recombination systems.

Point of Interest

LAMARCK'S IDEAS ABOUT INDUCED MUTATIONS

The French zoologist Jean Baptiste de Lamarck in the early 19th century argued that species are not immutable. This concept was used by Darwin in his theory of evolution. However, Lamarck believed that 'mutations' occurred because of use or disuse of an organ, that new organs developed because of need and most importantly that organisms could inherit acquired characteristics. Thus he argued that animal and plant species should be modified and in the course of many generations these acquired characteristics would be transmitted to their offspring. This view of 'mutations' was accepted by some into the 20th century. It was not until the nature of DNA and its role in inheritance was proven that this concept was discarded.

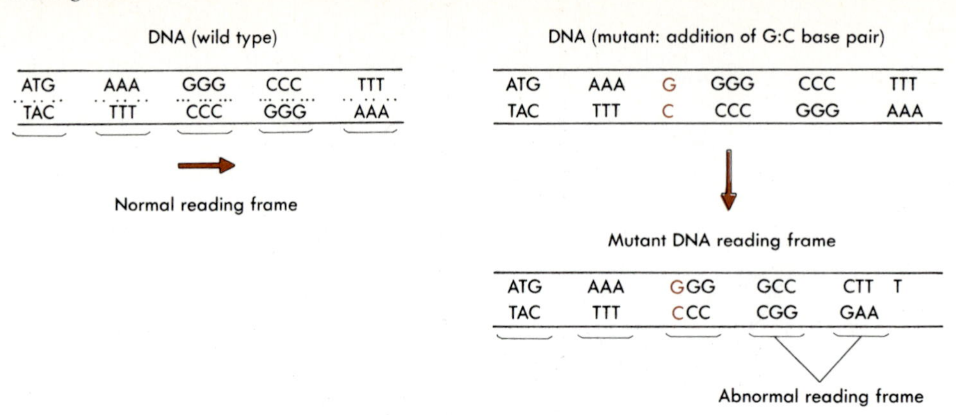

Figure 10-13 *Example of frameshift mutation resulting from addition of G:C base pair to the DNA. Reading frame downstream of addition is out of its normal sequence.*

X-ray induced mutations. DNA is the most radiosensitive component of cells. X-rays can cause mutations by directly affecting the DNA molecule through the absorption of energy. X-ray photons can cause deletions in the DNA backbone by breaking both chains at opposing sites. X-rays can also indirectly affect the DNA molecule when their energies are absorbed by small molecules in the surrounding medium. Various free radicals, such as HO_2 and organic peroxides, can react with the DNA and chemically alter it (see Chapter 14).

Ultraviolet light–induced mutations. Ultraviolet light can generate different types of mutations: base substitutions, frameshifts, large deletions, and duplications. Ultraviolet light is readily absorbed by molecules, such as nucleic acids, whose constitutents, such as purine and pyrmidine bases, possess alternating single and double bonds. Following absorption of ultaviolet light the excited bases on the DNA interact with adjoining bases on the same strand to form dimers (Figure 10-15). Dimers slow down DNA synthesis but do not absolutely block it. Ultraviolet light also activates the cell's repair system (see DNA Repair section) to remove the dimers but sometimes during the repair process errors are made that result in mutations. Thus dimer formation is merely a premutational stage that initiates the events that are the actual cause of the mutation. A more detailed discussion of the effect of ultraviolet light on growth can be found in Chapter 14.

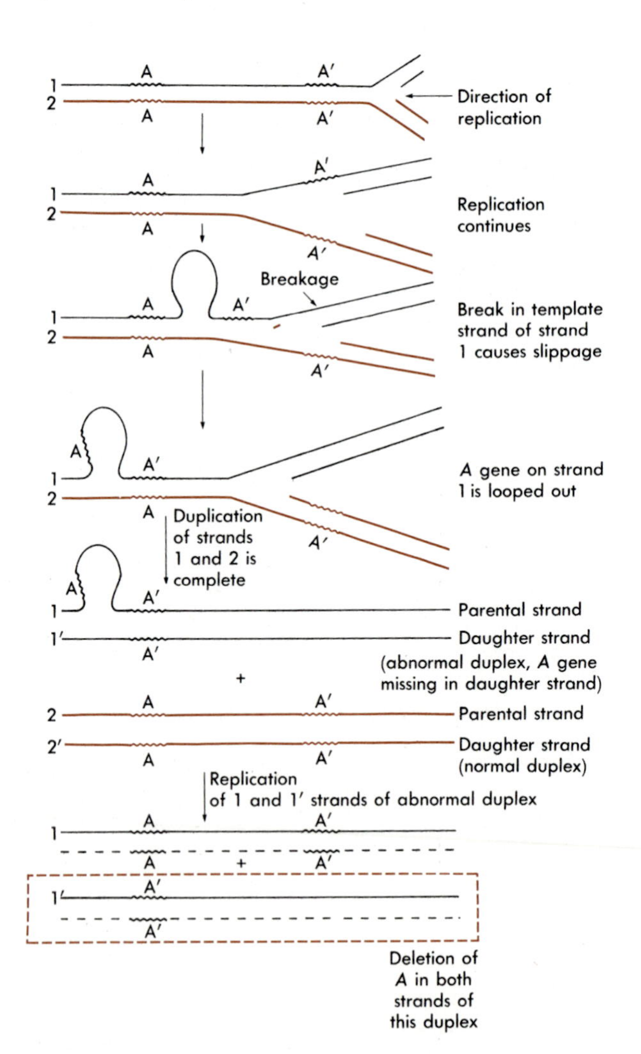

Figure 10-14 *Postulated mechanism for deletion formation by slipped mispairing. Homologous sequences A and A′ are indicated.*

Transposons, IS elements, and mutation

For many years geneticists believed that point mutations were the principle means of genetic change. We are now aware that long sequences of DNA material called transposons and insertion sequences (IS) are inserted naturally into the microbial genome (see Chapter 8). Transposons and IS elements behave like mutator genes and are now believed to be the principle cause of spontaneous mutations. Many transposable elements carry

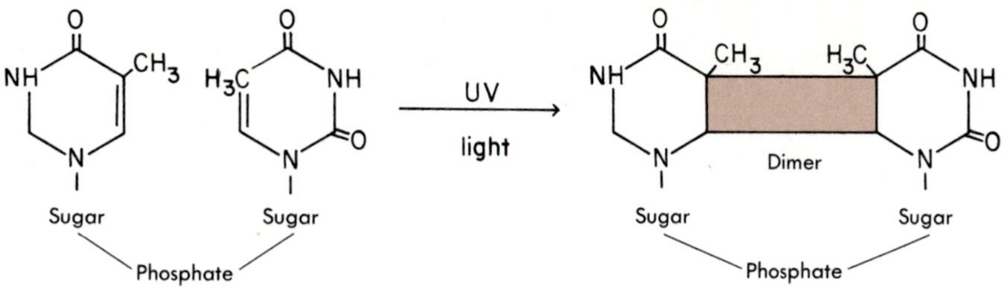

UV light

Dimer

Sugar Sugar Phosphate

Figure 10-15 *Thymine dimer formation. Adjacent thymine residues (can be intrachain or interchain) when exposed to UV light form a dimer.*

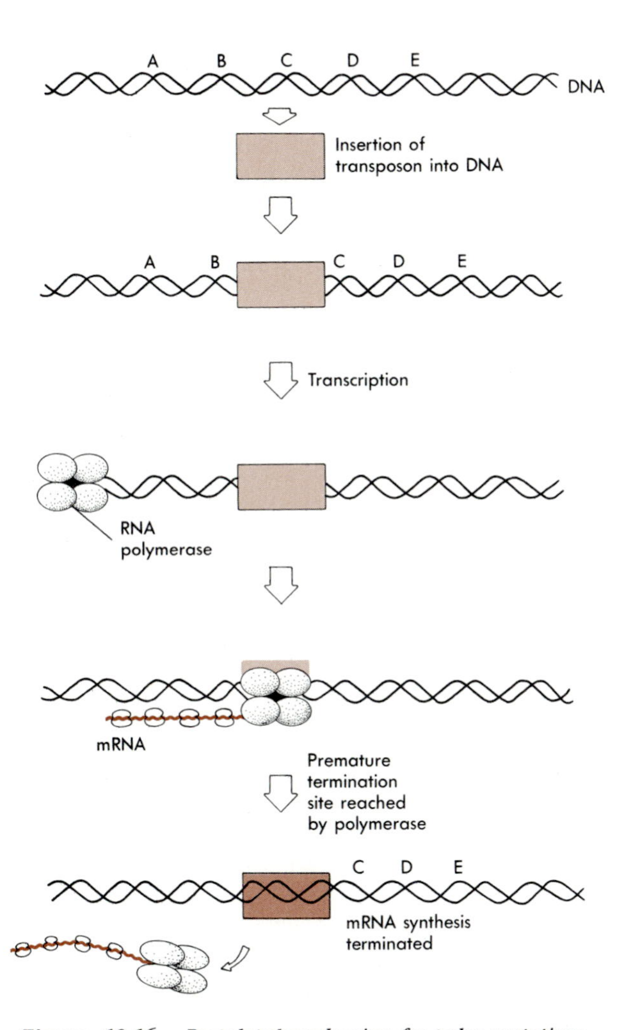

Figure 10-16 *Postulated mechanism for polar mutations. Transposon carrying termination site is inserted into gene B of bacterial genome. Gene A and part of gene B are transcribed by RNA polymerase. However, premature termination site of inserted transposon causes RNA polymerase to discontinue transcription. The synthesized mRNA is incomplete and genes downstream or distal to inserted transposon are not transcribed.*

Concept Check

1. How are temperature sensitive, suppressor, and auxotrophic mutants used in experimental genetic research?
2. Why is the reversion rate for many cellular characteristics so much higher, for example, 10^{-4}, than the forward mutation, for example, 10^{-10}, for the same characteristic?
3. Discuss the molecular basis of spontaneous and induced base pair substitutions. What is believed to be the reason for the increased number of spontaneous mutations at specific sites on the genome?
4. Explain the mechanism of mutations caused by nitrous acid, alkylating agents, acridines, 5-Bu, ultraviolet light, and x-rays?

transcriptional "stop" or terminating signals, and when they are inserted into a gene the continuity of transcription downstream from the insertion is interrupted. When transposable elements are inserted into an operon the expression of all genes downstream from the insertion is greatly reduced. This downstream effect is the basis of **polar mutations** which is the type of mutation caused by these elements. It is believed that the premature rho-dependent termination of transcription prevents synthesis of mRNA for genes downstream from the premature termination site (Figure 10-16).

MUTATION AND THE GENETIC CODE

Four types of mutations alter the activity of polypeptides: missense, nonsense, reading frame, and suppressor mutations. Their ability to affect the phenotype of the cell depends on their location in the gene.

Missense mutations

Mutations that involve a single base pair substitution in the DNA may result in the substitution of one amino

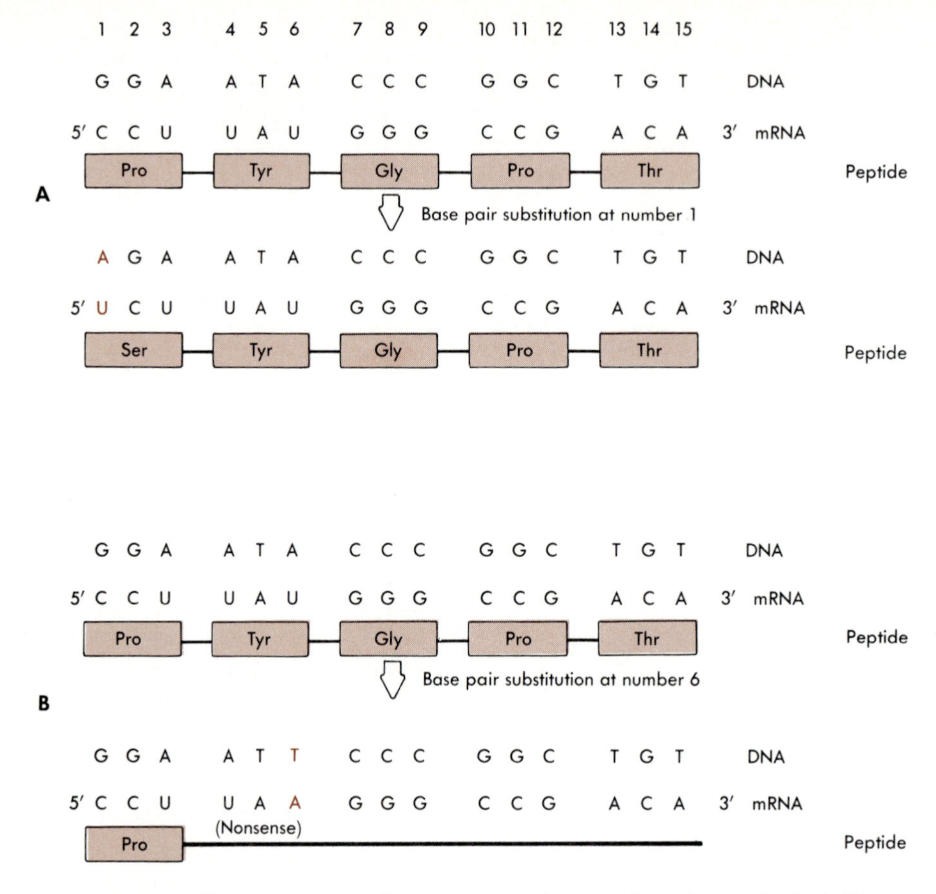

Figure 10-17 *Effect of missense and nonsense mutation on polypeptide synthesis.* ***A,*** *Missense mutation; substitution of adenine for guanine at position 1.* ***B,*** *Nonsense mutation; base pair substitution at position 6, resulting in premature termination.*

acid for another in the polypeptide **missense.** Occasionally the codon change is such that because of degeneracy the replacement amino acid is the same as the original amino acid in the polypeptide. This is referred to as a **sense codon. Missense codons** often alter the activity of the polypeptide because of the type of substitution. Substitution of polar for nonpolar amino acids or vice versa and replacement of cysteine by another amino acid are important causes of altered activity. Such amino acid substitutions result in changes in the conformation of the polypeptide and hence its activity (Figure 10-17, *A*).

Nonsense mutations

Nonsense codons do not specify for any amino acids. They are recognized by ribosomes as termination signals. Ribosomes, when they meet the termination codon, fall off the message, and this results in the formation of a polypeptide that is shorter than normal (Figure 10-17, *B*). Nonsense mutations located near the end of the message may not result in a loss of polypeptide activity, but the closer the mutation is to the 5′ end (the origin of the message), the greater the loss of polypeptide activity.

Reading frame mutations

Reading frame or frameshift mutations were discussed earlier in the section on deletions and insertions.

Suppressor mutations

Suppressor mutations, as stated on p. 253, represent secondary mutations that antagonize or suppress the original mutation. Suppressor mutations may be **intragenic**, that is, the second mutation occurs in the gene near the first mutation; or they may be **extragenic**, that is, the second mutation occurs at a site outside the gene in which the first mutation occurred. Each type of suppression results in the restoration of some polypep-

	Normal codon	Nonsense triplet
Serine	UCG	UAG
	UCA	UAA
Tyrosine	UAU	UAA
	UAC	UAG
Glutamic acid	GAA	UAA
	GAG	UAG
Lysine	AAG	UAG
Glutamine	CAG	UAG
Tryptophan	UGG	UAG
Leucine	UUG	UAG
Tryptophan	UGG	UGA

Figure 10-18 *Base changes in those amino acid codons that result in formation of nonsense or stop sequence. Note that most nonsense triplets are UAG* (amber).

tide activity (also called the **pseudowild type**) but seldom in restoration of full activity. Suppression can be broken down into several types: nonsense suppression, frameshift suppression, missense suppression, ribosomal suppression, and polarity suppression. The most common is nonsense suppression. Consult the readings at the end of the chapter for selections on the other types of suppression.

Mutations causing chain termination can arise in several ways. Single base changes in codons, such as CAG, AAG, GAG, UCG, UGG, UAC, UUG, UGG, and UAA (Figure 10-18), can give rise to the nonsense triplet, **UAG**. Mutants that are defective because of the presence of this mutation are **amber mutants**. When amber mutations occur within a gene there is a premature interruption in translation because there is no tRNA that recognizes UAG. The end result is a mutant protein with little or no activity.

The phenotypic change brought about by amber mutants can be rectified either partially or nearly completely by a second mutation, an extragenic suppressor mutation. The **amber suppressor mutation** involves an alteration of the gene coding for tRNAs. The suppressor mutation enables the anticodon of the tRNA to acquire the ability to recognize the nonsense triplet and provide an amino acid for insertion at the nonsense or terminating site. Thus the anticodon of the tRNA is changed so that it makes sense and not nonsense. The amino acids inserted at the site of the chain-terminating codon are the same ones that can be altered to form a UAG site. For example, if a mutation in a structural gene, (such as β-galactosidase gene) changes a UUG (leucine) codon to UAG, there is premature termination of the mRNA during translation. A second mutation (suppressor) occurs that involves a tRNA molecule, such as leucyl tRNA. The suppressor mutation alters the leucyl tRNA anticodon and enables it to recognize UAG. Now leucine occupies the UAG codon site. In this example, the suppressor restores the original activity of the protein because leucine replaced a mutated leucine codon. This does not always occur because tRNAs other than leucyl tRNA can suppress amber mutations. For example, if lysyl tRNA recognizes UUG then lysine will be inserted in place of leucine. The chemical nature of lysine (positively charged) could interfere with the folding of the polypeptide or affect the nature of the active site; this could severely affect polypeptide activity.

Other nonsense triplets, such as **UGA (opal)** and **UAA (ochre)**, also have their suppressors (Figure 10-18).

Suppressor mutations are important genetic tools because they enable the investigator to analyze newly isolated mutants and to study the nature of the genetic lesion. For example, the study of the structure-function relationships of the lactose repressor is made possible by using suppressor mutations and knowing the position of nonsense codons in the lactose I gene. By suppressing each of the nonsense mutations, different amino acids are incorporated into specific positions in the repressor polypeptide. Thus a large number of mutant repressor molecules are available for analysis. Each mutant repressor molecule is tested for activity (binding properties, tetramer formation, enzyme repression, and so on). In this way, specific amino acid domains in the repressor can be associated with specific functions.

Concept Check

1. **What is the distinction between a transposon and insertion sequence element? What types of mutations can they cause?**
2. **What is the effect of missense, nonsense, and suppressor mutations on the genetic code?**
3. **How can suppressor mutations revert nonsense mutations?**

DNA REPAIR SYSTEMS

To ensure that DNA is replicated with high fidelity the cell evolved intricate enzymatic mechanisms to correct mistakes that are made. The pathways of DNA repair in an organism, such as *E. coli,* are similar to those that exist in mammalian systems. When *E. coli* is exposed to chemical or physical agents that damage DNA and/or prevent replication, DNA repair proteins are synthesized. These repair proteins are part of two regulatory networks called the **SOS response** and the **adaptive response network** (see Point of Interest). The activities of these networks in some instances are responsible for

Point of Interest

REGULATION OF DNA REPAIR SYSTEMS

Two types of inducible repair networks are at work when bacterial DNA is damaged: the SOS network and the adaptive response network. A variety of diverse physiological phenomena occur in the cell when DNA is damaged or DNA replication is inhibited. The SOS (or international distress signal) regulatory network is induced primarily by ultraviolet light. When a cell's DNA is damaged, an inducing signal is generated that activates specific genes that in turn control the expression of other genes (figure below). The expression of repair genes, for example, genes for excision repair or postreplication repair in the SOS system, is controlled by the products of two genes, *rec A* and *lex A.* The Lex A protein acts as a repressor of every SOS gene. When cells are exposed to damaging ultraviolet irradiation, there is activation of the Rec A protein. Rec A protein, which in addition to its involvement in recombination also possesses proteolytic activity, interacts with the Lex A protein and cleaves it. This proteolytic cleavage inactivates Lex A protein, and SOS genes are derepressed, that is, activated. Once the DNA is repaired the inducing signal is eliminated, and the SOS genes are repressed.

The adaptive response network controls the induction of repair processes when DNA is damaged by methylating and ethylating agents. The gene controlling the repair enzymes of this network is the *ada* gene. There are at least two repair enzymes known to exist in this system, and each is controlled by genes that are regulated by the *Ada* protein. One enzyme, 3-methyladenine-DNA glycosylase II, initiates excision repair of methylated and ethylated bases, whereas the other enzyme, O^6-alkylguanine-DNA alkyltransferase, can directly remove methyl and ethyl groups from affected bases without breaking the DNA. The molecular mechanism by which the *Ada* protein exerts its regulatory effects has not been elucidated.

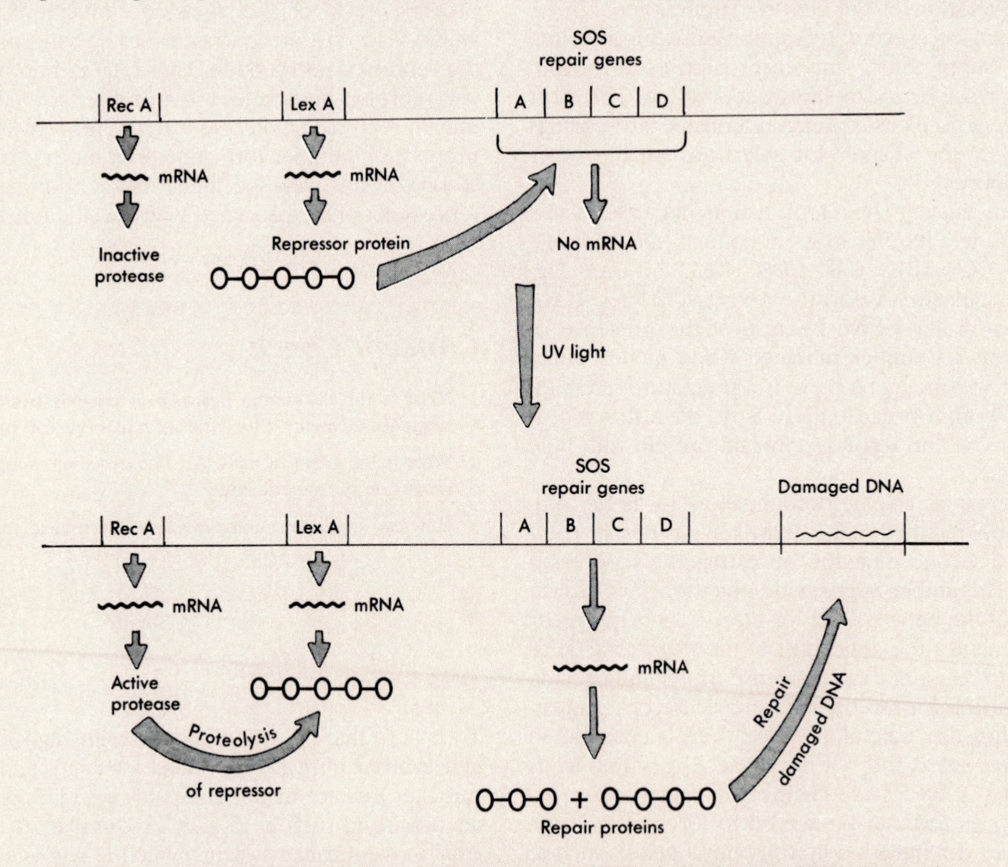

SOS response of bacterial cell to ultraviolet light damage of DNA. ***A,*** *In absence of DNA damage by ultraviolet light, SOS genes are repressed by* Lex A *repressor.* ***B,*** *When DNA damaged by ultraviolet light,* rec A *protease is activated. Protease inactivates* Lex A *product until damage is repaired. Note during SOS repair, nucleotides can be inserted incorrectly into damaged portion and may lead to mutation.*

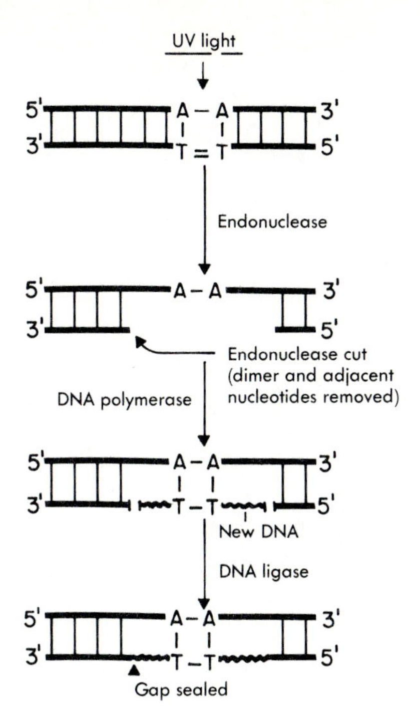

Figure 10-19 *Excision repair of DNA.*

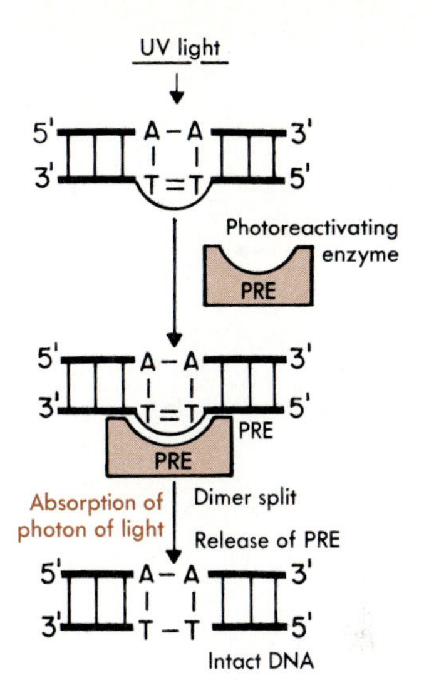

Figure 10-20 *Photoreactivation repair of DNA.*

mutations. Thus the cell takes an active role in mutating its DNA. Mutagenesis by ultraviolet light and various chemicals is therefore not a passive process. The following section details the major pathways of DNA repair.

Repair of DNA damaged by ultraviolet light

Excision repair. **Excision repair** is a frequently used mechanism for repairing DNA damaged by ultraviolet light. The process occurs in the dark and is specific for lesions on single strands. Such sites are recognized for repair because the dimer produced by ultraviolet light causes a distortion in the helix. The process can be broken down into the following steps:

1. An incision is made at or near the lesion (Figure 10-19). For example, the thymine dimer caused by ultraviolet light is cut on the 5′ side of the dimer by an endonuclease.
2. A second enzyme excises the fragment of DNA containing the damage. This includes the damaged base, as well as adjacent bases (12 to 13 bases are often involved).
3. The complementary strand is used to direct the resynthesis of DNA that was removed by using DNA polymerase I.
4. The sugar-phosphate linkage is sealed by DNA ligase.

Photoreactivation repair. **Photoreactivation repair** is accomplished through the activity of a photoreactivating enzyme (PRE) produced by the microorganism. PRE, which binds to the dimer in the dark, becomes activated and splits the dimer (Figure 10-20) when the organism is exposed to certain wavelengths in the visible spectrum (310 to 600 nm). PRE is also called **photolyase.**

Daughter strand–gap repair. When an ultraviolet-damaged DNA strand containing a dimer serves as a template for synthesis of the daughter strand, gaps appear. These gaps appear in the daughter strand because replication is blocked at the site of the dimer and resumes at some site past the lesion. *E. coli* has a strategy for repairing these gaps called **daughter strand–gap repair** or **postreplication repair** (Figure 10-21). During the repair process the gaps are filled and the discontinuous strands are joined. The mechanism involves an exchange of strands, or recombination. The process attempts to replicate past certain lesions but does not remove the original lesions themselves. Once strand exchange has occurred lesions can be removed by excision repair. Daughter strand–gap repair is a mechanism that permits the cell to temporarily tolerate the presence of ultraviolet light damage.

Error prone–excision repair. Despite the fact that ultraviolet light damage to DNA can be repaired by several mechanisms, mutations can still arise because of an **error prone–excision repair** process. One such postulated mechanism is illustrated in Figure 10-22.

Repair of DNA other than those for ultraviolet light damage

DNA glycosylases. DNA glycosylases are found universally in organisms. They catalyze the cleavage of base-

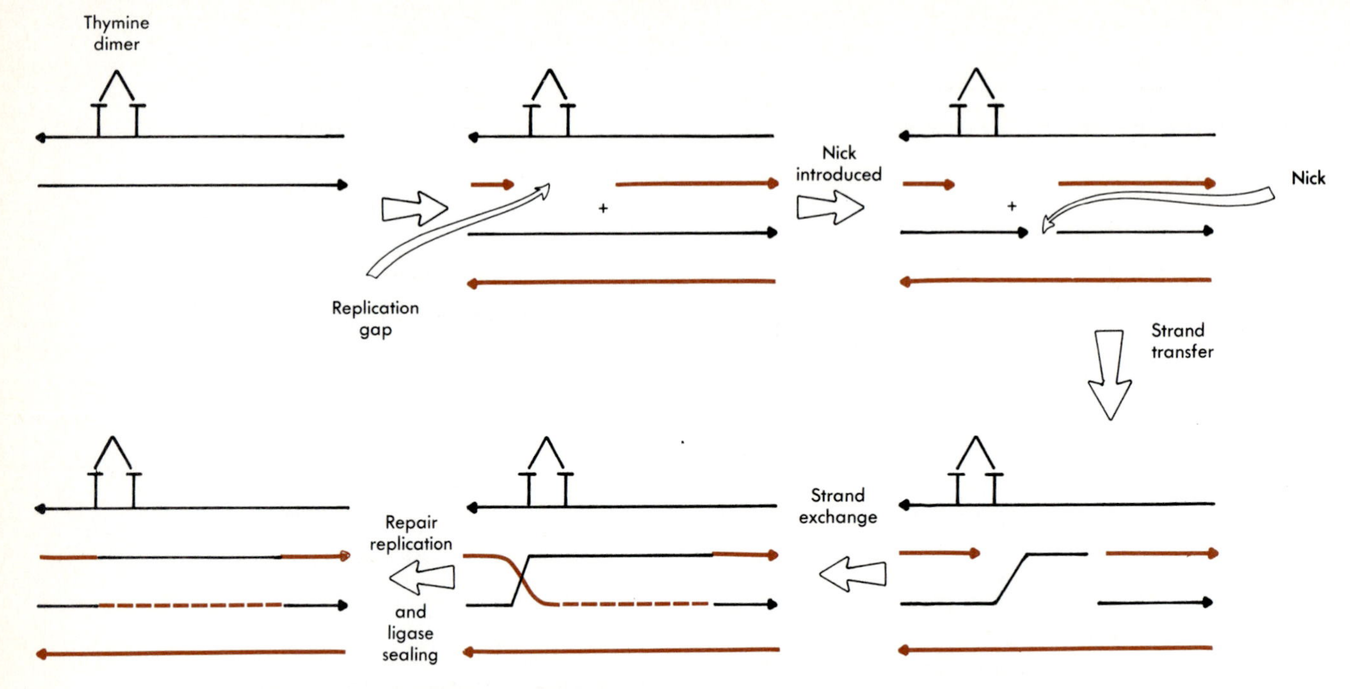

Figure 10-21 *Postulated mechanism for daughter strand–gap repair. Double-stranded DNA with dimer on one strand undergoes replication. One of the daughter DNA duplexes* (color) *contains gap in one of strands. Nick is introduced into undamaged DNA duplex and strand transfer is initiated. Strands are exchanged, and repair enzymes fill the gaps. Ligase seals final sugar-phosphate gaps but dimer still present.*

sugar bonds in the DNA (Figure 10-23). Most glycosylases prefer double-stranded DNA and each has a narrow substrate specificity. Because there is a glycosylase for almost every altered base, damage to any base results in their quick removal. This leaves the site with a missing base (apurinic or apyrimidinic) that is filled in with the help of DNA polymerase I and sealed with DNA ligase. One of the most important glycosylases is *N*-uracil-DNA-glycosylase. Figure 10-23 illustrates the mechanism by which this enzyme operates.

Endonucleases. A variety of endonucleases act as repair enzymes. One of these was described earlier during the discussion of repair of DNA damaged by ultraviolet light. When apurinic or apyrimidinic sites on the DNA are generated, for example, by glycosylases, endonucleases initiate the correction process. There are two classes of endonucleases: those that cleave the 3′ side of the missing base and those that cleave the 5′ side of the missing base. In either of these two cases, a 5′ phosphate and 3′ hydroxyl group are produced at the cleavage site. Endonucleases show a broad specificity and thus are able to recognize a variety of different types of damaged DNAs.

Transmethylases. Transmethylases recognize bases that have been methylated or ethylated following exposure to alkylation agents, such as *N*-methyl-*N*-nitrosourea. One of the major mutagenic DNA lesions produced by exposure to alkylating agents is 0^6-methylguanine. The transmethylase for 0^6-methylguanine is not involved in an excision process. Instead, the enzyme catalyzes the transfer of the methyl group to one of its own cysteine residues. Thus the products of the reaction are S-methylcysteine in the protein and unsubstituted guanine in the DNA.

MUTANT ISOLATION AND DETECTION

Most mutations are recessive. When a recessive mutation has been introduced into the microbial cell, it must be grown for a sufficient period so the mutant allele can be expressed. The time period for expression varies, depending on the type of mutant sought, as well as other factors. For example, a culture of *E. coli* cells under ideal growth conditions possesses four nucleoids. A mutant allele in just one of the DNA molecules is not ex-

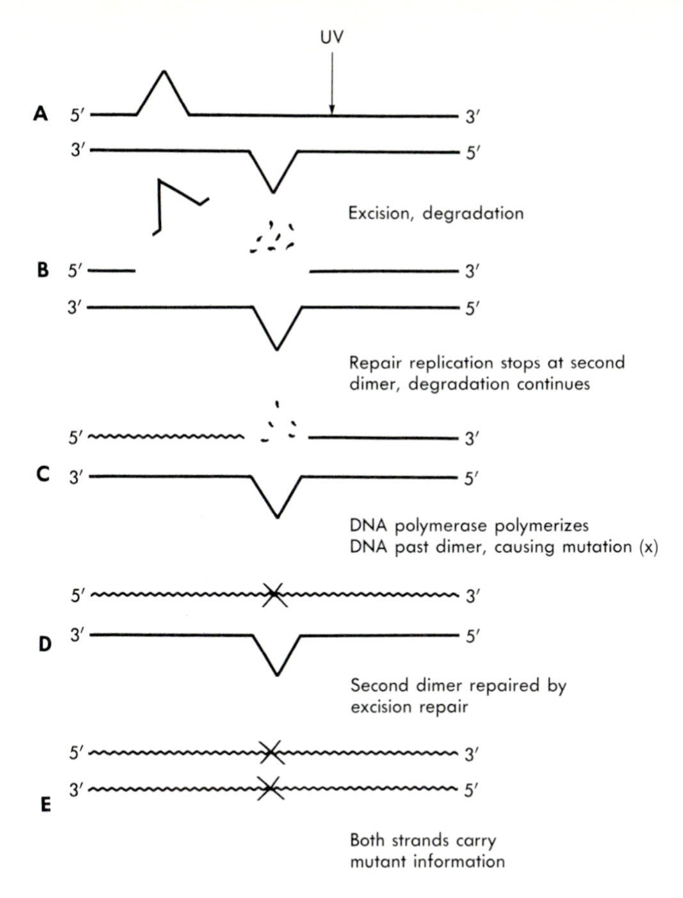

Figure 10-22 *Model for mutation via error-prone excision repair mechanism.* ***A,*** *Dimers (⩘) on both strands of DNA.* ***B,*** *Excision and degradation of dimer and adjacent nucleotides on one strand.* ***C,*** *DNA polymerase repair stops at site opposite second dimer, and degradation of original dimer strand continues.* ***D,*** *Special DNA polymerase activity is induced that polymerizes DNA past dimer, thereby causing mutation (X).* ***E,*** *Second dimer is repaired by excision repair mechanism, and both strands contain mutant information.*

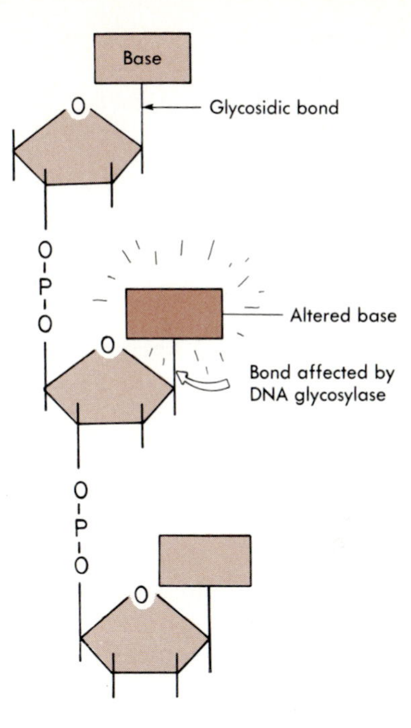

Figure 10-23 *Schematic diagram demonstrating site of DNA molecule recognized by DNA glycosylases.*

pressed for at least two generations (Figure 10-24), that is, until the DNA of the daughter cell possesses only mutant alleles. If the recessive mutation involves the loss of a gene product that is normally present in the cytoplasm of the parent, for example, an enzyme involved in carbon metabolism, then many generations may elapse before the product is diluted out (Figure 10-25).

Enrichment

Direct selection. Once a few mutants have been produced in a population, it is necessary to allow them to grow and increase their numbers. This process is called enrichment, and one type available is direct selection. This procedure is useful for mutants that are resistant to a particular chemical, physical, or biological agent. The investigator creates a growth condition that favors growth of the mutant and inhibits the wild-type or parent. For example, when selecting a mutant resistant to a particular metal, the mutagenized culture is plated on a medium containing the metal. Only the metal-resistant cells form colonies, whereas the metal-sensitive or wild type cells will be unable to grow or will be killed.

Counterselection is another technique for enrichment. In this procedure a growth condition is selected that kills the wild type or parent cells. One counterselection technique uses penicillin in the growth medium for enriching auxotrophs. Penicillin acts on the bacterial cell wall of only growing cells by preventing the formation of cross-links between peptidoglycan strands. This weakens the cell wall, and osmotic forces cause the wall to burst. In a medium supporting the growth of parental but not mutant cells, penicillin kills only the parental cells. Surviving mutant cells can be selected after the

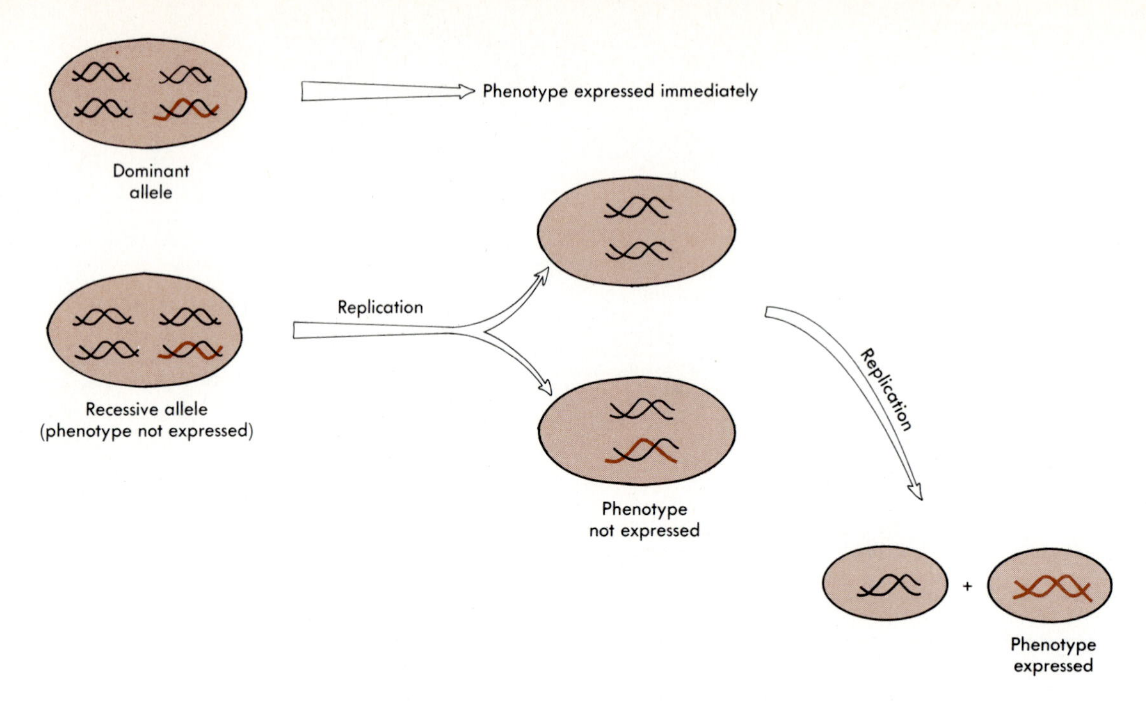

Figure 10-24 *Phenotypic expression in cells with dominant and recessive alleles. Each initial cell has 4 nucleoids.*

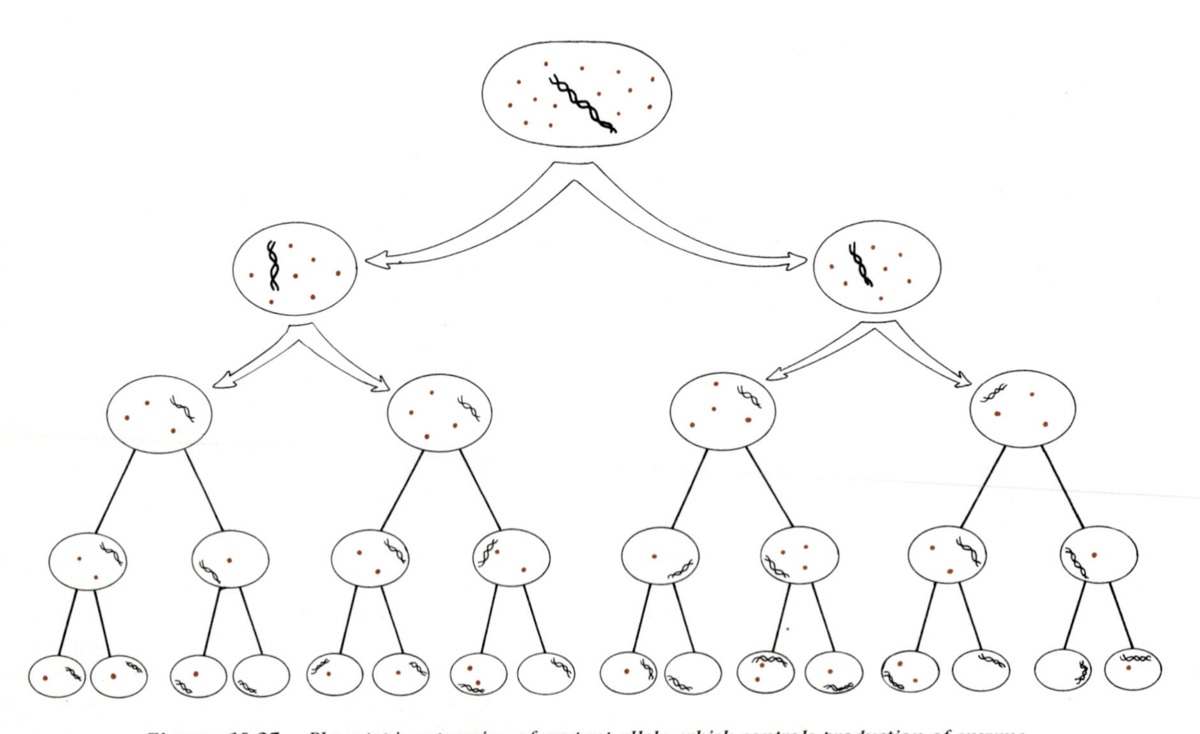

Figure 10-25 *Phenotypic expression of mutant allele, which controls production of enzyme involved in carbon metabolism. Red dots represent enzyme molecules present in cell at time of mutant formation. Several generations may be required before enzyme is diluted out and mutant phenotype (no new enzyme synthesis) is expressed.*

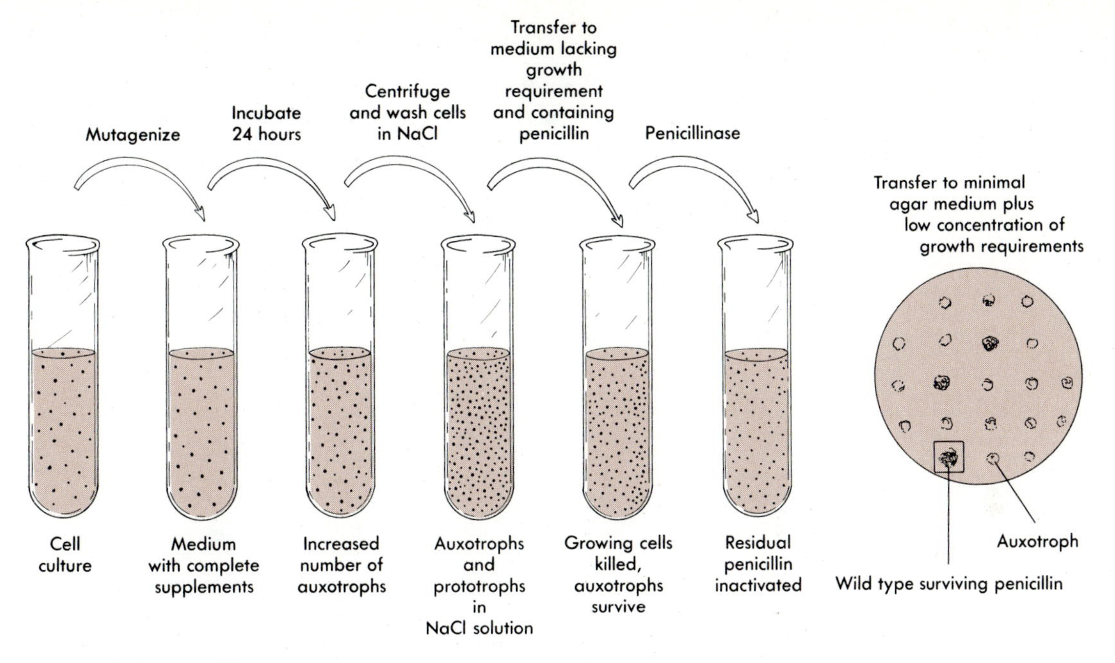

Figure 10-26 *Penicillin counterselection technique for isolation of auxotrophs. Large population of cells is exposed to mutagen and then is grown in nutritionally supplemented medium. Culture is incubated for 24 hours to increase number of auxotrophs. Culture is centrifuged and washed with sodium chloride to remove any traces of medium. Washed cells, containing auxotrophs and prototrophs, are transferred to medium that is deficient in auxotrophic requirement and also contains penicillin. Growing cells are killed by penicillin, but auxotrophs survive. Penicillinase is later added to destroy any residual penicillin. Culture is spread on agar surface containing minimal amounts of auxotrophic requirement. Mutants will produce small colonies on medium, whereas any wild-type cells that survived penicillin treatment will be much larger.*

penicillin has been inactivated with penicillinase (Figure 10-26).

Physical separation can be used to enrich certain mutants. For example, mutants unable to grow as large as the wild type can be separated via filtration or in some instances by centrifugation.

Detection

Once the mutant cell is isolated and enriched, it is usually necessary to apply a test that specifically identifies the mutant. A simple test for auxotrophic mutants removes a portion of the colony and tests for its phenotype. For example, if the mutant selected requires histidine (His^-) for growth, the mutant cells can be introduced into media that lack histidine and media containing histidine. Growth in the presence of histidine but not in its absence indicates the histidine auxotroph. A widely used detection device is **replica plating** (Figure 10-27). In this procedure colonies on an agar surface can be screened to determine which is mutant. The plate is pressed face down over a block covered with a velveteen pad. The fibers on the pad pick up samples of the colonies. The impregnated pad is used like a stamp and is pressed on the surface of agar plates that will either support or not support the growth of the mutant. Because the inoculation procedure is the same for all replica plates, the mutant colony can be easily detected.

PRACTICAL APPLICATIONS OF MICROBIAL MUTANTS

Microorganisms have been used for many years in industrial processes. Production of antibiotics, amino acids, enzymes, ethanol, and single-cell protein are some of the major industrial uses of microorganisms. Strain improvement, in each microbial-related process to increase the yield of microbial products, is usually obtained through mutation. The uses of microbial mutants are discussed in more detail in other chapters. Let us look at two additional industrial uses for microbial mutants that have some very modern practical applications.

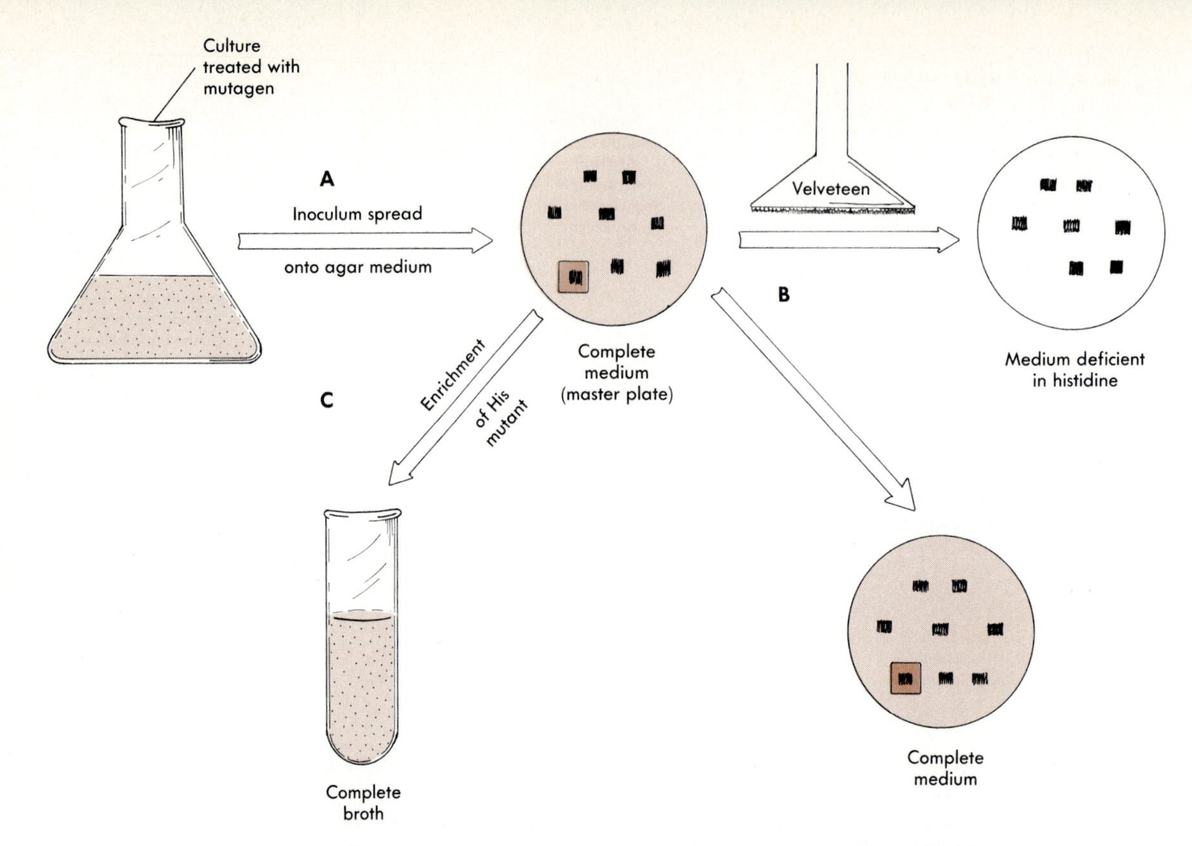

Figure 10-27 *Replica plating technique for isolation of auxotroph requiring histidine.* ***A,*** *Culture is treated with mutagen, such as ultraviolet light. An inoculum of mutagenized culture is spread on solid medium supporting growth of auxotrophs and prototrophs (master plate). Normally auxotroph cannot be identified on master plate; however, for convenience we have preidentified it.* ***B,*** *Velveteen pad is used like stamp and pressed on surface of master plate. Fibers of pad pick up samples of all colonies. Impregnated pad is pressed on surface of solid medium containing complete medium, as well as one deficient in histidine. Comparison of two plates, reproduced from master plate, indicate absence of colony in lower left corner of plate in which histidine is absent. Cells in corresponding colony on complete medium or master plate are thus unable to synthesize histidine (His$^-$). His auxotrophs from this colony can be cultivated in pure culture* **(C)** *and exact nature of mutation can be determined.*

Indicators of mutagens

Industrial chemicals have found their way into our food and water supplies, and some have been implicated as mutagens that may be also carcinogenic (causing malignant tumor formation). One technique used to detect the mutagenic potential of these chemicals is to test them on preformed mutants of *Salmonella typhimurium* that have an absolute requirement for the amino acid histidine. The histidine auxotrophs have a characteristic low frequency of reversion that can be increased when another mutation is introduced, for example, by a chemical compound. This is the theoretical basis of a test called the **Ames mutagenicity test.** The tester strain (the histidine auxotroph) has three other modifications that make it more sensitive and suitable for the test:

1. It is made more sensitive to DNA damage by eliminating its ability to perform DNA repair following mutation.
2. Increased error-prone repair is introduced into the strain that makes DNA more susceptible to replicating errors.
3. A mutation is introduced that makes the envelope of the bacterium more permeable to chemical agents.

The Ames test is carried out by adding the chemical to be tested to a rat liver extract (Figure 10-28). The reason for the rat liver extract is that many chemicals do not become mutagenic for animals until they have been converted to other metabolites by enzymes present in the liver extract. This conversion process in the intact animal usually takes place in the liver in the process called **biotransformation.** The mixture is plated on a solid nutrient medium deficient in histidine. After 48 hours the plates are examined for any revertants (only organisms that can synthesize histidine will be able to

Figure 10-28 *Ames test for identifying potential carcinogens. Tester bacteria containing histidine mutation are mixed with rat liver extract. Mixture is spread on agar medium deficient in histidine so that confluent layer of bacteria covers surface. Specific amount of potential chemical mutagen is soaked into filter disk and dried and then placed on surface of plate. During 48-hour incubation period, chemical in filter disk will diffuse outward with highest concentration nearest disk. If chemical gives rise to mutations, histidine mutants will revert to wild type and give rise to individual colonies. Control plate without mutagen indicates number of spontaneous mutations that caused reversion to wild type. Mutagenesis indicated by increased reversion frequency above control plate. Test can also be performed by mixing bacteria, chemical mutagen, and liver extract and then plating.*

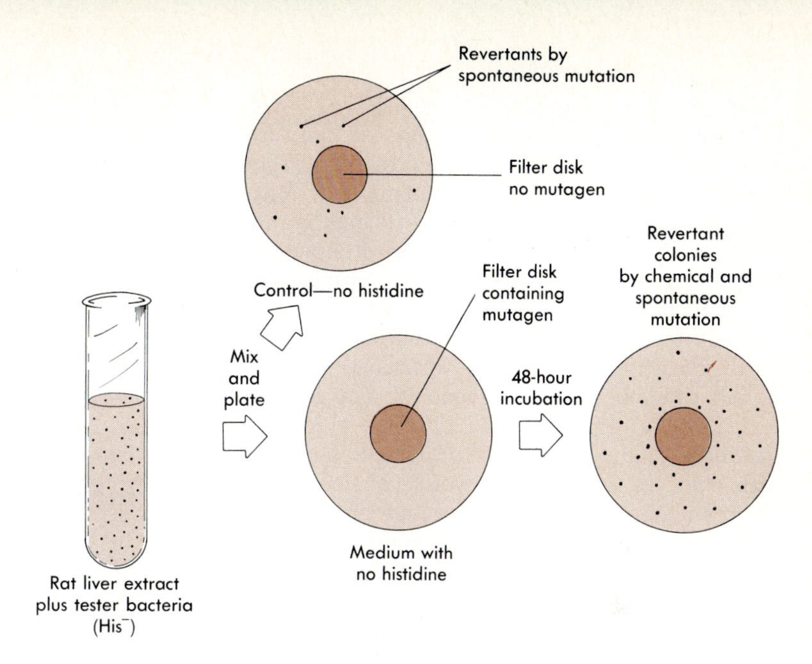

grow). The frequency of reversion to the wild type in the presence of the mutagenic agent is compared with the reversion frequency in the absence of the mutagen. The number of colonies per mole of mutagen is a quantitative estimate of the mutagenic potential of the chemical. False-negative results can arise because the mutagen may not restore the specific histidine function and may thus go undetected. This type of problem can be prevented by using other tester bacteria. A very high reversion frequency in the presence of the mutagen indicates that the chemical agent may be carcinogenic, and further tests are initiated using laboratory animals. It may be necessary to consider removal of the chemical from public or industrial use if the animal tests are positive.

Vaccine production

Vaccines are made up of either living or dead microorganisms. They induce immunity to infection by the same microorganism but do not cause disease. Mutants are used to reduce the disease-producing qualities of some living microorganisms that are to be used as vaccines. One type of vaccine uses temperature-sensitive organisms, that is, the mutants are able to grow at one temperature but not another. For example, a vaccine to prevent respiratory tract damage following vaccination has been developed. Respiratory tract damage can occur in those areas where host defense mechanisms are less effective. The temperature-sensitive strain to be used as the vaccinating agent will not grow at 37° C, the temperature of the lower respiratory tract where tissue damage could occur, but will grow in the upper respiratory tract, where the temperature is cooler and respiratory tissue damage will not occur.

MUTATION AND MICROBIAL EVOLUTION

Sexual mechanisms of reproduction in eukaryotic systems permit variations in the genotype. Genetic variation in prokaryotes is believed to occur primarily through spontaneous mutation and selection of those mutants with survival advantages. The microbial world is supplied with an almost inexhaustible supply of genetic possibilities because of the short generation time and randomness of mutation. Spontaneous mutations in the laboratory, for example, have been detected for practically every gene locus. The mutation rate depends on the gene locus involved. It is speculated that during evolution the first microorganisms possessed very few genes and that the products of gene expression—the enzymes—demonstrated a broad substrate specificity, which enabled them to act as catalysts on a wide range of related substrates. Greater enzyme specificity or specialization evolved later. For example, a gene could be duplicated with alterations occurring in only one copy. Several naturally occurring duplications are known in *E. coli*, in which there are at least seven copies of the genes that code for rRNA. The duplicated genes may be in tandem, or they may appear at different sites on the chromosome. A mechanism for tandem duplication could involve a reciprocal exchange (recombination) between identical or nearly identical genetic sequences located at different sites on the chromosome (Figure 10-29). The mutated copy could produce an inactive product while the parent gene retained its original product specificity. The inactive gene, through other mutations or reversions, could express a product with the same specificities as the original unmutated gene, or it could possess new substrate specificities. The new gene would enable the cell to have a selective advantage over other cells, provided that the substrate for the new enzyme was present in the environment.

Another mechanism for acquiring new cellular functions may be associated with the activities of different genera and species in a microbial community. Microbial communities are better equipped to handle novel substrates that are introduced into the environment than a single species. It is believed that one of the microbial members possesses an enzyme that recognizes the new substrate and at the very least modifies it. The modified product may be recognized by other enzymes produced by other members of the community. The gene coding for the novel enzyme could be transferred to other members of the community. This new gene can be maintained as long as the novel substrate stays in the environment and offers a selective advantage to the microorganisms. We already know that the maintenance in humans of bacteria with plasmid-encoded antibiotic resistance genes is coupled with the continued presence of the antibiotic. Once the antibiotic is no longer taken by the patient, the antibiotic-resistant strains are supplanted by the antibiotic-sensitive strains. Experiments in the laboratory have also demonstrated that when a community of different microorganisms is permitted to grow in a medium containing degradable carbon and nitrogen sources plus a nondegradable carbon source, for example, a herbicide, one of the organisms in the community through mutation may acquire the ability to convert the herbicide into a product that can be degraded by other microorganisms.

It can be expected that we will continue to challenge microorganisms with new chemicals that make their way into the environment. It is hoped and expected that this challenge will be met because of the ability of microorganisms to produce a multitude of genetic changes during evolution.

The role of spontaneous mutation in evolution has been questioned because of the presence of plasmids, as well as the knowledge we have gained from a better understanding of the functional role of chromosome organization. The F factor plasmid, which is responsible for sexual contact and genetic transfer between related

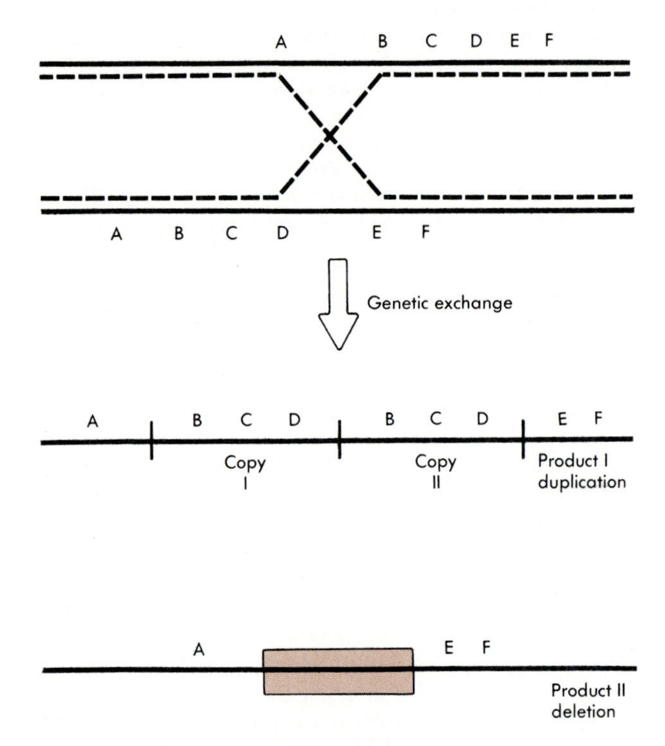

Figure 10-29 *Genetic exchange between two DNA molecules resulting in tandem duplication of genetic sequence* B, C, *and* D. *Solid lines indicate double-stranded DNA. Dashed lines represent reciprocal recombination event.*

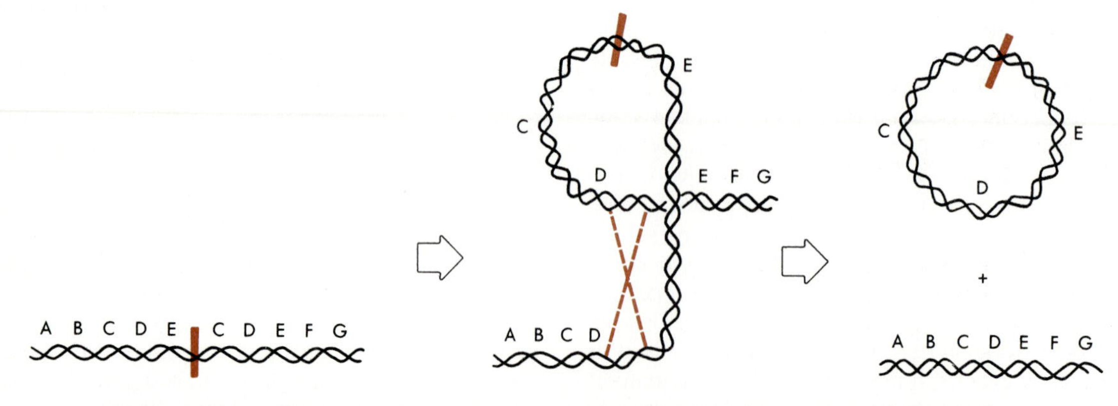

Figure 10-30 *Hypothetical mechanism for formation of plasmid from chromosome containing tandem sequence of genes. Solid line is double-stranded DNA. Tandem sequence is CDE. Single-crossover (exchange) event could produce unduplicated chromosome (ABCDEFG) and DNA circle containing genes CDE. Colored line indicates point on chromosome where tandem genes are fused.*

species, is believed to play a minor role in evolution via the recombination process between chromosomes. Investigations into other types of plasmids has revealed that factors other than chromosomal recombination may play a vital role in evolution. Transposable genetic elements have been discovered in several locations outside of structural genes on both chromosomes and plasmids. What this means is that genetic exchange between transposable genetic elements can occur when these elements are present on plasmids and chromosomes and that structural genes may be carried with them. This type of genetic exchange can occur within the cell or between cells if the mating process is involved. In addition, a method for the formation of plasmids has been suggested. We already know that gene duplication occurs in prokaryotes as well as eukaryotes. Duplicated genes located in tandem on the chromosome are also unstable and create an avenue for genetic exchange, that is, reciprocal crossover between the homologous genes. This can lead to the formation of plasmids (Figure 10-30) and genes on these plasmids may contain transposable elements, as well as transfer potential. Plasmids can be transferred to other species and inserted into any replicon: viral, plasmid, or chromosomal DNA. Proponents of this theory believe that plasmids play a major and not a minor role in evolution among bacteria.

Concept Check

1. **Explain the mechanism by which ultraviolet light damage can be repaired. What mechanisms regulate DNA repair systems?**
2. **How do DNA glycosylases, endonucleases, and transmethylases repair genetic damage?**
3. **What are the techniques by which you would select, enrich, and detect an auxotrophic mutant for histidine?**
4. **What is the Ames test and what is its experimental and/or commercial value?**
5. **How has mutation resulted in the evolution of new bacterial characteristics?**

SUMMARY

1. Mutations are alterations in the DNA molecule that give rise to genetic variety among organisms. Important microbial mutants for analyzing biological processes are conditional mutants: temperature-sensitive, suppressor, and auxotrophic mutants.
2. Mutations may be spontaneous, that is, they arise without human intervention, or induced, that is, they arise from deliberate exposure to mutagenic agents. The rate of mutation is related to the type of mutagen, as well as the microbial species.
3. Conversion of wild type to mutant phenotype is called forward mutation, but the process can be reversed. Reversion involving the original mutation site is back mutation, whereas reversion involving a site distant from the original mutation site is called suppressor mutation.
4. Mutations may involve base pair substitutions and can be spontaneous or induced. Insertions and deletions may arise from agents such as alkylating compounds and acridines, and physical agents, such as ultraviolet light and x-rays. Biological mutagens, such as transposons and insertion sequence elements, are believed to be the principle cause of spontaneous mutations. They are responsible for mutations called polar mutations.
5. Four types of mutations affect the genetic code and alter polypeptide activity. They include missense mutations, nonsense mutations, reading frame mutations, and suppressor mutations.
6. DNA can be repaired by a variety of enzymes that belong to two regulatory networks: the SOS response and adaptive response networks. The repair pathways can be divided into two major groups: those that repair DNA damaged by ultraviolet light and those that repair damage by chemical agents. Ultraviolet light damage can be repaired by processes such as excision repair, photoreactivation repair, and daughter strand–gap repair. Error prone–excision repair appears to be responsible for mutagenesis following ultraviolet light damage. Enzymes involved in repairing damage by chemical agents include DNA glycosylases, endonucleases, and transmethylases.
7. Once a culture has been mutagenized it is necessary to isolate, enrich, and detect the mutant. Enrichment of mutants can occur by direct or indirect methods. One of the most widely used is the indirect method of counterselection in which a growth condition is used that kills wild type cells but doesn't affect the mutant. An important technique for detecting mutants is replica plating.
8. Microbial mutants used in industrial processes involve the production of antibiotics, vaccines, amino acids, and enzymes and are used to detect potential environmental carcinogens.
9. Mutation has played an important role in the evolution of microorganisms, particularly in respect to their ability to metabolize a wide variety of organic substrates. It has been suggested that some novel metabolic activities could be handled in a sequential manner by two or more members of the microbial community. The metabolic potential of one member could be transferred to other members of the community either in the form of chromosomal or extrachromosomal DNA.

Self-study Quiz

Multiple choice

1. Which of the following would have the most serious outcome for a bacterial cell?
   A. Deletion of a nucleotide near the beginning of the structural gene DNA sequence that codes for a constitutive enzyme
   B. Addition of a nucleotide at the 5′ end of the structural gene DNA sequence that codes for a constitutive enzyme
   C. Substitution of thymine for guanine at the sixth nucleotide of the structural gene DNA sequence that codes for an inducible enzyme
   D. An addition of a nucleotide between the fifteenth and sixteenth nucleotides of the structural gene DNA sequence that codes for a repressible enzyme
2. Mutants unable to synthesize growth factors, such as amino acids, are called:
   A. Prototrophs.
   B. Phototrophs.
   C. Mutotrophs.
   D. Auxotrophs.
3. Most suppressor mutations are associated with genes concerned with the formation of:
   A. DNA polymerase.
   B. Transfer RNA.
   C. RNA polymerase.
   D. DNA ligase.
4. Which of the following mutagenic agents is most frequently associated with deletions?
   A. Nitrous acid
   B. 5-Bromouracil
   C. Alkylating agents
   D. Acridines
5. Given the following spontaneous mutation rates for different genes within the same species, which mutation would occur more frequently than the others?
   A. 1×10^{-6}
   B. 1×10^{-10}
   C. No mutation occurs unless it has been created by a mutagen, and in that case the rates would be the same for all genes.
   D. Spontaneous mutation rates do not vary for different genes within a species.
6. Which of the following repair mechanisms does *not* repair the original lesion?
   A. Excision repair
   B. Daughter strand–gap repair
   C. Photoreactivation repair
   D. Transmethylation
7. Which of the following would most likely produce an induced base pair substitution type of genotypic change?
   A. Ultraviolet radiation
   B. X-ray radiation
   C. Nitrous acid
   D. Acridines
8. Which of the following would be responsible for the appearance of spontaneous mutations?
   A. Introduction of a plasmid into a cell
   B. Introduction of insertion sequences into the bacterial genome
   C. Growth in a medium containing penicillin
   D. Replica plating
9. What is the Ames test designed to do?
   A. Detect the mutagenic effects of environmental chemicals
   B. Distinguish auxotrophs from prototrophs
   C. Isolate antibiotic-resistant organisms
   D. Isolate biochemically-deficient auxotrophs that are carrying certain plasmids
10. Which of the following repair systems or enzymes does not cleave DNA?
    A. Transmethylase for 0^6-methylguanine
    B. Endonuclease
    C. DNA glycosylase
    D. Photolyase

Completion

1. Areas on the DNA that are more susceptible to mutagens than other sites are called ________________.
2. Compounds that exist in equilibrium but differ in the arrangement of their hydrogen atoms and electrons are called ________________.
3. The replacement of one purine for another on a DNA strand is called a ________________.
4. The deamination of cytosine results in the formation of ________________, which can pair with ____________.
5. Chemical compounds that can donate their ethyl or methyl groups to certain bases on the DNA are called ________________ agents.
6. The "slipped mispairing" model is used to explain the formation of ________________ mutations.
7. A codon change in which one amino acid is substituted for another in the polypeptide is called a ____________ mutation.
8. Mutants that are defective because of the presence of an abnormal UAG triplet are called ________________ mutants.
9. The regulatory network of the *E. coli* cell that is induced by ultraviolet light damage is called ________________.
10. In the Ames test the liver extract is used to bring about the ________________ of potentially carcinogenic substances.

Thought questions

1. Why is it not possible to induce nonsense mutations by treating wild-type strains with mutagens that cause only AT to GC or TA to CG transitions in the DNA?
2. Why are frameshift mutations more likely to result in proteins lacking normal function than nonsense mutations?

SELECTED READINGS

Books

Friedberg, E.C., *DNA repair.* W.H. Freeman and Co., San Francisco, 1985.

Hanawalt, P.C., Friedberg, E.C., and Fox, C.F. *DNA repair mechanisms.* Academic Press, Inc., New York, 1978.

Hurst, A., and Nasim, A. (Eds.). *Repairable lesions in microorganisms.* Academic Press, Inc., New York, 1984.

Watson, J., Hopkins, N., Roberts, J., Steitz, J., and Weiner, A. *Molecular biology of the gene* (4th ed.). Benjamin/Cummings Publishing, Inc., Menlo Park, Calif., 1987.

Journal Articles

Ames, B.W. Identifying environmental chemicals causing mutations and cancer. *Science* **204**:587, 1979.

Devoret, R. Bacterial tests for potential carcinogens. *Sci. Am.* **244**:40, 1979.

Drake, J.W., and Baltz, R.H. The biochemistry of mutagenesis. *Annu. Rev. Biochem.* **45**:11, 1976.

Lederberg, J., and Lederberg, E.M. Replica plating and indirect selection of bacterial mutants. *J. Bacteriol.* **63**:399, 1962.

Perlman, D.A., et al. Molecular models for DNA damaged by photoreaction. *Science* **227**:1304, 1985.

Streisinger, G., et al. Frameshift mutations and the genetic code. *Cold Spring Harbor Symp. Quant. Biol.* **31**:77, 1966.

Walker, G.C. Mutagenesis and inducible responses to deoxyribonucleic acid damage in *Escherichia coli. Microbiol. Rev.* **48**:60, 1984.

Witkin, E.W. Ultraviolet mutagenesis and inducible DNA repair in *Escherichia coli. Microbiol. Rev.* **40**:869, 1976.

CHAPTER 11

Recombination, gene transfer, genetic engineering, and recombinant DNA

Chapter Outline

- Recombination
  - General Recombination
  - Site-Specific Recombination
  - Illegitimate Recombination
- Gene Transfer
  - Transformation
  - Transduction
  - Conjugation
- Genetic Engineering and Recombinant DNA
  - Requirements for Forming and Cloning Recombinant DNA
  - Significance of Recombinant DNA Research
  - Benefits vs. Fears of Genetic Engineering

Introduction

To understand evolution, the student must have some knowledge of genetic systems and the processes of mutation and recombination. Mutation ensures that variability will occur in the gene, and genetic recombination guarantees that a number of different gene combinations will be produced. Any significant complexity in organisms could not have arisen without these processes. Recombination in bacteria, in which some of the genetic traits of two separate cells have been exchanged, occurs as the result of natural processes referred to as **transformation**, **transduction**, and **conjugation**. These processes of gene transfer are being used in the laboratory to genetically engineer microorganisms.

This chapter describes the basic mechanics of the recombination process and the enzymes involved, discusses the ways in which genes can be transferred from various donors to a recipient cell, and outlines how microbiologists are able to use natural gene transfer mechanisms to produce recombinant DNA; that is, to genetically engineer microorganisms.

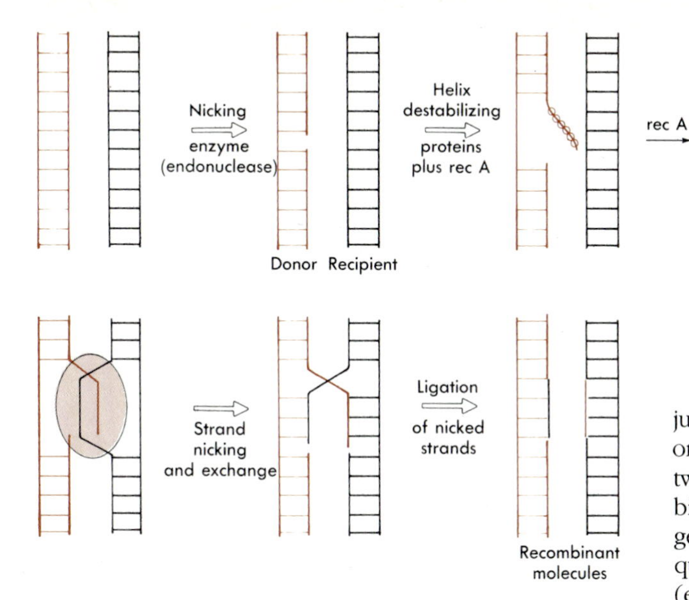

Figure 11-1 *Mechanism of recombination and cross-strand exchange of homologous DNA segments. Nicking enzyme (endonuclease) frees single strand in one DNA duplex. For convenience first DNA to be nicked is called donor strand. Helix destabilizing proteins plus Rec A protein form clusters along single strand. Rec A protein holds apart strands of recipient DNA molecule and promotes interaction between single strands of donor and incoming strand of recipient DNA duplex. Recipient DNA is nicked by enzyme, thus freeing homologous strand for exchange with donor strand. Nicked strands are ligated and exchange is complete.*

RECOMBINATION

The term **recombination** in bacterial genetics has undergone some changes in the past 25 years. Originally the term was used to denote the transfer of DNA from one cell to another and the establishment of some of this genetic information in the recipient cell. Today, recombination is thought of as a reassortment of nucleotide sequences within the DNA molecule, a process that may take two forms. First, the reassortment may result from the rearrangement of pieces of DNA derived from two parent molecules (**general recombination**). The recombination could involve chromosomal DNA from both a donor cell and a recipient cell, plasmid DNA and plasmid DNA, viral DNA and chromosomal DNA, and plasmid DNA and chromosomal DNA. The interactions of these molecules may produce a single molecule derived from each parent molecule or two molecules derived in part from each parent molecule. Second, the recombination does not necessarily involve two DNA molecules in the exchange. For example, the rearrangement of nucleotides may occur on a single DNA molecule. Deletions, inversions, duplications, or transpositions may lead to DNA rearrangements, and these are discussed later. The modes of genetic recombination may be divided into the following types: **general, site specific**, and **illegitimate**.

General recombination

General recombination occurs more frequently than other types of recombination because of the homology that exists between the nucleotide sequences of the two DNA molecules. General recombination is common in processes such as transformation, transduction, and conjugation where the DNA molecules are from the same or similar species. Binding of homologous areas between two DNA molecules usually affects a rapid recombinational event but requires the products of specific genes called **rec genes**. Mutual exchange of DNA requires the formation of single-stranded DNA. An enzyme (endonuclease) is produced in the cell that nicks the strands of the double-stranded DNA molecules, causing them to unwind and form single strands. The mechanism for general recombination is discussed below.

Model for general recombination. One protein that plays a pivotal role in the recombination process is **Rec A**. This protein acts as a DNA-dependent ATPase that promotes homologous pairing of a single strand of DNA with a linear double-stranded DNA molecule. Rec A protein also cleaves certain repressors and is involved in derepressing certain genes, such as those involved in DNA repair (see Chapter 10). Rec A protein is unable to pair fully double-stranded molecules, but it will pair one DNA duplex with a variety of single-stranded or partially single-stranded substrates; for example, circular single-stranded DNA, or duplex DNA with single-stranded ends. This demonstrates the importance of single-stranded DNA in the recombination process. The homologous pairing activities of Rec A protein can be divided into three phases (Figure 11-1):

1. Rec A and apparently helix-destabilizing proteins bind to the single-stranded DNA formed by the nicking enzyme. These protein complexes form clusters along the sugar-phosphate backbone of the DNA and thus promote a fully stretched-out segment of single-stranded DNA. The complexes promote unwinding of DNA and ensure that hydrogen bonding between complementary segments can occur.
2. Rec A holds apart the strands of the recipient DNA molecule, thus providing a mechanism for pairing the incoming single strand with its complement. This pairing, or **synapsis**, involves hydrogen bonding between bases. Note that nicking (unless one of the strands is circular) and single strand formation must occur on both of the DNA segments that are to be exchanged.

3. Strand exchange occurs and involves two strands from each of the two double-stranded DNA molecules. The exchange requires energy, which is provided by the hydrolysis of ATP, a function of Rec A. Rec A drives a strand exchange in the 5′ to 3′ direction with respect to the displaced strand. The protein can push an exchange past thymine dimers or even deletions and insertions. The two crossing strands are cut to terminate any further pairing. The exchanged pieces of DNA are then ligated to produce two recombinant DNA molecules.

Site-specific recombination

Site-specific recombination differs from general recombination in that no Rec A protein is required and only limited homology exists between the participating molecules. Instead, a recombination enzyme recognizes special nucleotide sequences on one or both of the participating DNA molecules—hence the term **site-specific recombination**. There are two types of site-specific recombination. One is nonreplicative; that is, no replication of DNA occurs. The other type initiates transposition of genetic elements, uses the enzyme **transposase**, and requires DNA replication.

Nonreplicative site-specific recombination. The most studied example of nonreplicative site specific recombination involves the integration and excision of bacteriophage lambda (λ). **Lambda** is a virus that can integrate itself into the host chromosome, which is **lysogeny**, or initiate its own replication and kill or lyse the bacterial cell. Viruses that integrate into the host genome exist in a **provirus** state. In the case of bacterial viruses this state is called the **prophage*** state. Lambda possesses three elements important to these processes: an attachment site (integration site), called *att P,* and two genes, *int* and *xis,* whose products are associated with integration and excision, respectively (Figure 11-2).

Lambda can integrate into the bacterial chromosome because of recombination between the viral attachment site *(att P)* and a comparable attachment site on the bacterial chromosome, the *att B* site. Integration is now believed to occur in the following way:

1. DNA gyrase of bacterial origin supercoils the DNA, allowing the integration reaction to occur.
2. The integration gene product, **integrase** *(int)*, binds extensively to the attachment sites *att P* and *att B.* The attachment sites are several hundred base pairs in length, but only a 15 base-pair region, the **core**, shows homology:

   —GCTTTTTTATACTAA—
   —CGAAAAAATATGATT—

   The integrase exhibits topoisomerase activity and nicks DNA within the *att* core sequences.

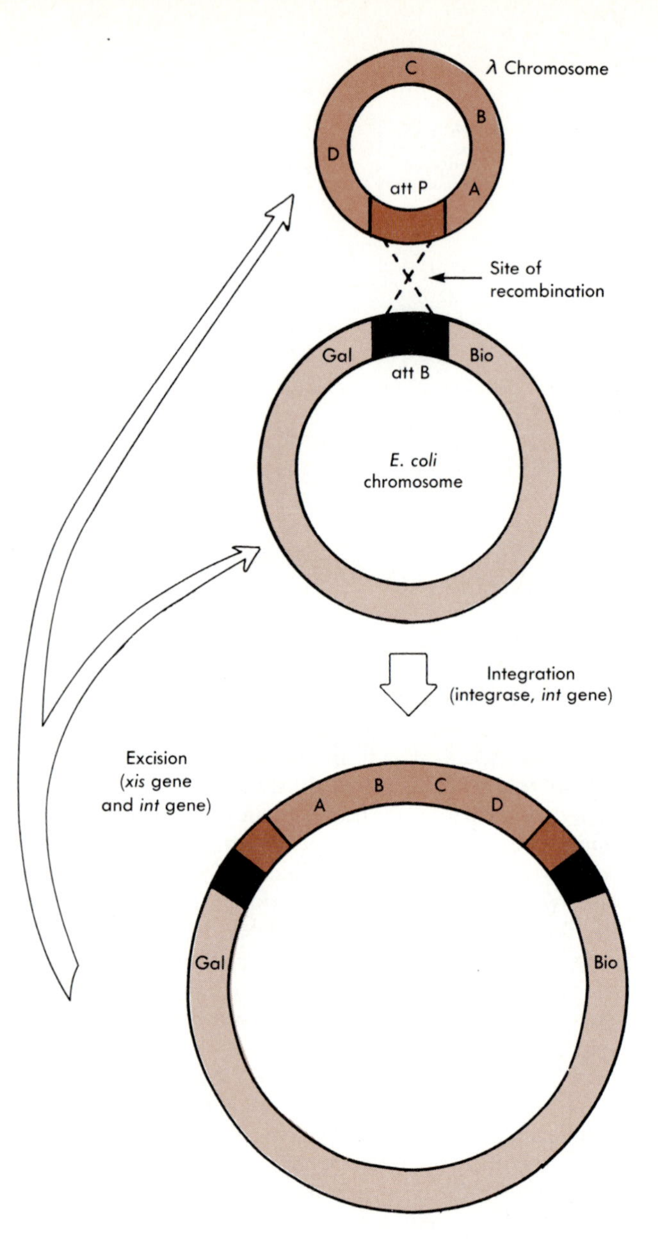

Figure 11-2 *Site-specific recombination. Integration of bacteriophage lambda (λ) into chromosome of* E. coli. *Att P and* att B *represent attachment sites for phage and bacterial chromosome, respectively.* Gal *and* bio *represent genes for galactose and biotin metabolism, respectively. Recombination between* att P *and* att B *sites results in integration of lambda. The process requires activity of integrase* (int) *gene. The excision of lambda requires activity of excisase gene* (xis) *plus integrase.*

3. The breakage of DNA at the core permits recombination to occur. This results in the integration of the lambda chromosome into the bacterial DNA, with the regions flanking the viral DNA containing

* Bacteriophage that can enter a prophage state are **temperate,** and the host carrying a prophage is called **lysogenized** or **lysogenic.**

attachment sites that are part bacterial and part viral in origin.

The excision of virus from the bacterial chromosome can also take place and is basically a reversal of the integration process. Excision requires the products of the *int* and excisase *(xis)* genes; however, the function of the *xis* gene is not clear. Reciprocal crossover occurs between the attachment sites that flank the integrated viral genome, and the products of this event are a fully circular viral chromosome and an intact bacterial chromosome. This excision process is necessary if the virus is to replicate and produce hundreds of new progeny.

Replicative site-specific recombination. Replicative site-specific recombination is characteristic of transposition processes involving bacteriophage Mu and transposons, discussed earlier in Chapter 8.

Another example of this type of recombination is observed in the flagellated bacterial species, *Salmonella typhimurium* (see Point of Interest).

Concept Check

1. Diagrammatically illustrate the mechanism of general recombination. What is the function(s) of the Rec A protein?
2. Explain the mechanism of integration and excision of lambda on the bacterial genome.
3. What are the various types of genetic recombination? Can you give examples of where they occur?

Point of Interest

SITE-SPECIFIC RECOMBINATION: A MECHANISM FOR *SALMONELLA* TO EVADE THE IMMUNE SYSTEM

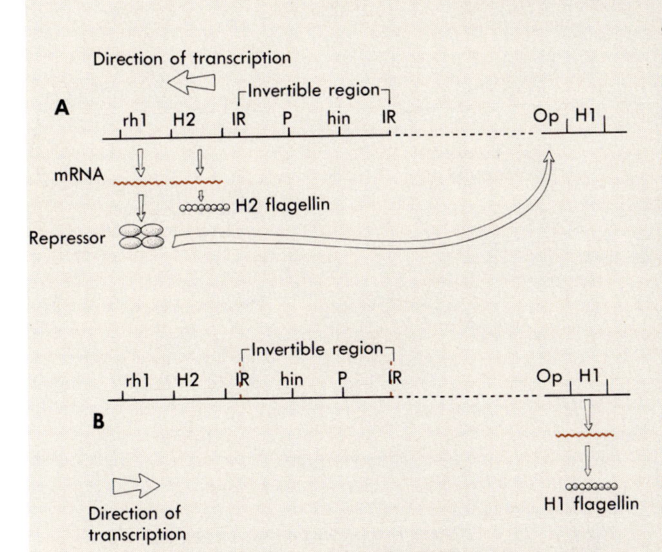

Inversion process associated with flagellin phase variation in Salmonella. ***A,*** *Configuration of genome when* H2 *gene is expressed and converted to H2 flagellin. Note promoter is upstream from* hin *gene. In this orientation the* rhl *gene is also expressed; it acts as a repressor of the* H1 *gene by binding to the operator* (O_p). ***B,*** *Configuration of genome when* H1 *gene is expressed. Product of* hin *gene causes inversion of invertible region, and promoter (P) now lies downstream from* hin. *In this configuration transcription directed toward* H1 *gene and H1 flagellin is produced.*

Salmonella can synthesize one type of flagellin (flagellar protein), H2, and then switch to synthesis of another type of flagellin, H1. This switching process is called **phase variation.** *Salmonella* species are agents of disease in animals and humans, and by changing their type of flagellar protein can temporarily evade the host's antibodies.

The region of the genome associated with flagellin synthesis is outlined in figure on the left. In the region is a sequence of bases flanked by 14 base-pair inverted repeats. Located between the inverted repeats (the **invertible region**) is a promoter that controls the transcription of two genes, *H2* and *rh1*. The *H2* gene is associated with H2 flagellin synthesis, whereas the *rh1* gene is a repressor that controls expression of the *H1* gene, which encodes H1 flagellin. The *H1* gene is a considerable distance from the invertible region. The key element to flagellin switching is the gene *hin*, whose product, the Hin protein, can cause the inversion of the invertible region by catalyzing site-specific recombination between the inverted repeats. Basically, the Hin protein causes a change in the direction of transcription—the promoter is turned around on the genome. This inversion property affects the activity of the repressor *rh1*. In one orientation *rh1* represses the *H1* gene, hence H2 flagellar protein is favored. In the opposite orientation transcription is directed away from the *H2* and *rh1* genes, thus *H1* gene activity is expressed and H1 flagellar protein is synthesized. What has been discovered is that a sequence homology exists between *hin* and other genes that are associated with site-specific recombination, such as the *R* gene of the Tn3 transposon (see Chapter 8). It seems possible that all these controlling elements have some common evolutionary relationship.

Illegitimate recombination

Recombination can take place at very low frequencies at random sites on the chromosome, with no apparent homology between the nucleotide sequences and no apparent specific sites for recombination. This type of recombination is called **illegitimate**. Duplications and deletions (other than those caused by transposons or insertion sequences) are caused by this type of recombination; however, little is known of the mechanisms involved.

GENE TRANSFER

Recombination in bacteria is facilitated by mechanisms that transfer DNA from various donor sources into a recipient cell. The fate of the transferred DNA depends on its capacity to be taken up by the host cell, its stability in the host cell, and its homology with the host chromosome. There are three mechanisms for gene transfer: transformation, transduction, and conjugation.

Transformation

Transformation was discussed earlier as a microbial mechanism that implicated DNA as the genetic material (see Chapter 8). The following discussion explores the mechanics of the transformation process and its significance in recombination. **Transformation** is a process in which free DNA is taken up by a cell, resulting in a genotypic change in the recipient. Not all cells can be transformed, but those that can are considered **competent**. The composition of the medium in which the transformation is to take place, as well as the phase of the bacterial cell cycle, influences the transformation process. Most transformation experiments have been performed with gram-positive species of *Streptococcus* and *Bacillus*, both of which exhibit similarities in the process. Experiments with streptococci have shown that the entire culture becomes competent when the cell concentration is from 10^7 to 10^8 cells/ml. This response is caused by the release of a small activating protein called the **competence factor**. The competence factor, which increases as the culture density increases, has been isolated and shown to induce competence in noncompetent cells. The competence factor interacts with a cell surface receptor, and the interaction induces the cell to produce several competence-specific proteins, one of which is an autolysin. The autolysin acts on the cell wall, exposing a membrane-associated DNA-binding protein and a second protein, an endonuclease. The function of these proteins and others are examined below in transformations involving gram-positive bacteria (Figure 11-3).

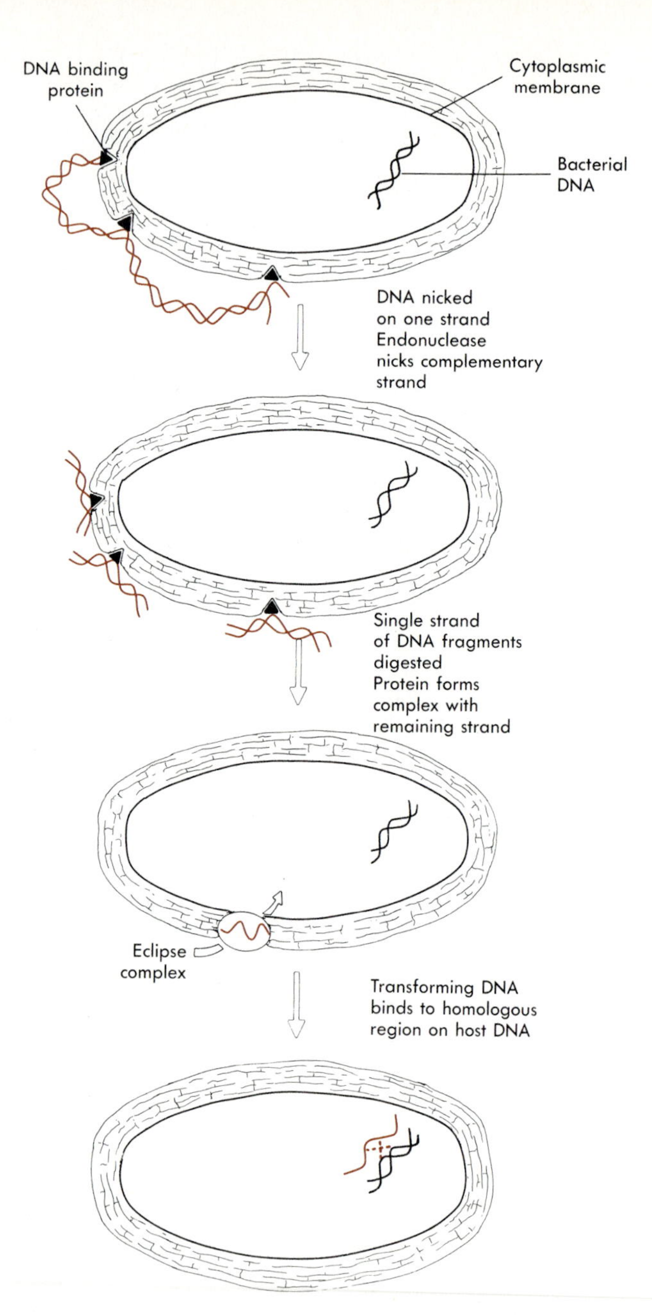

Figure 11-3 *Transformation process in gram-positive bacteria. Several sites exist on cell surface for binding of single DNA duplex. Strand on DNA is nicked at intervals, and complementary strand is nicked at opposite sites. This results in several small fragments of double-stranded DNA that attach to the cell surface. Exonuclease digests away one strand of each DNA duplex. Remaining single strands are complexed with protein (eclipse complex) and are transported into cytoplasm. Transported single strands of DNA binds to homologous region on host DNA, where gene exchange occurs.*

1. Donor DNA, probably derived from bacterial cells that have lysed in the culture, is bound at several sites to the cell surface by a **DNA-binding protein**. The binding protein nicks the DNA on one strand at intervals of approximately 6000 to 8000 base pairs.
2. A membrane-associated **endonuclease** causes breaks in the DNA at sites opposite those nicked by the DNA-binding protein. This causes the formation of many fragments of double-stranded DNA, some of which will be released from the cell surface, and some of which will become irreversibly bound and then taken up into the cell.
3. An **exonuclease** acts on the double-stranded DNA fragments, digesting away one of the strands. Only single-stranded DNA can enter the cell, but either strand may be involved.
4. The single-stranded DNA is complexed with a protein **(the eclipse complex)**, which protects the DNA from DNase digestion.
5. The eclipse complex is transported across the cytoplasmic membrane into the cytoplasm, where the single-stranded DNA binds to a homologous region on the recipient genome.
6. The single strand of the donor physically displaces a homologous strand on the recipient. The integrated segment measures up to approximately 30,000 base pairs in length. The frequency of integration (proportion of recipients that are transformed) depends in large part on the degree of homology between the donor and recipient DNA.

Transformation can also occur naturally in gram-negative bacteria such as *Neisseria gonorrhoeae* and some *Haemophilus* species. The process is similar to what occurs in gram-positive bacteria, but differences exist, such as:

1. Gram-negative bacteria do not have competence factors. Competence is internally regulated and is induced by factors such as those affecting growth.
2. In gram-negative bacteria the DNA adsorbed by the cell is only from related strains. The transforming DNA of genera such as *Haemophilus* appear to require specific sequences before they can be adsorbed to sites on the cell surface. This specificity resides in an 11 base-pair sequence; there are approximately 600 copies of this sequence in the genome.
3. Both gram-negative and gram-positive bacteria require double-stranded DNA for binding to the cell surface. However, in gram-negative bacteria, double-stranded DNA is not degraded by an exonuclease, but instead is transported into the cell in double-stranded form. During integration only one strand of the double-stranded molecule is incorporated into the recipient genome. The other donor strand and the displaced recipient strand are then degraded.

Transformation can be induced by artifical means in many bacteria that do not possess natural mechanisms, such as *E. coli*. For example, treating *E. coli* with calcium chloride in the cold can induce transformation. The transformation process is most successful when the DNA is a **replicon**; that is, an autonomous replicating unit such as a plasmid or a viral genome. Linear bits of chromosome DNA apparently must be hydrolyzed by intracellular nucleases before they can be integrated into host DNA. Artificially induced transformation has important implications in recombinant DNA research and will be discussed later. As we will see in the next section, other methods are available for the uptake of replicons into the cell. One of these, protoplast formation, has important implications in industrial microbiology (see Chapter 22).

At least two genes can be involved in the natural process of transforming DNA. The actual number of genes involved is based on the length of the transforming DNA. If the genes (for example, A and B) are closely linked, there is a likelihood that transformation produced by a single DNA molecule at the A locus would also produce transformation at the B locus; that is, **cotransformation** occurs. If the genes on the DNA being used for the transformation experiment are widely separated, then they will be carried on different transforming segments. The linkage of genes can be established by determining the frequency of cotransformation as compared with the frequencies of single marker transformations. Transformation can be used to determine close linkage or genetic fine structure, but it is not useful for primary localization of genes on the bacterial map. One of the handicaps in using transformation for genetic studies is that, under the best of circumstances, only a fraction of the population becomes competent. In addition, the efficiency of transformation varies greatly depending on the genetic markers transformed.

Transduction

Transduction is a process in which bacterial DNA is transferred from one cell to another with the help of a virus. It was discovered in 1952 by Zinder and Lederberg, whose experiments were designed to demonstrate that transfer of genetic material between strains of *Salmonella* does not require cell-to-cell contact (up until this time only conjugation and transformation were known as mechanisms of gene transfer). They used a U tube (Figure 11-4) in which a glass filter separated the two arms of the tube. The glass filter was porous, but the pores were too small for bacteria to pass through them. A minimal medium that would not support the growth of biosynthetic mutants was placed in the tube.

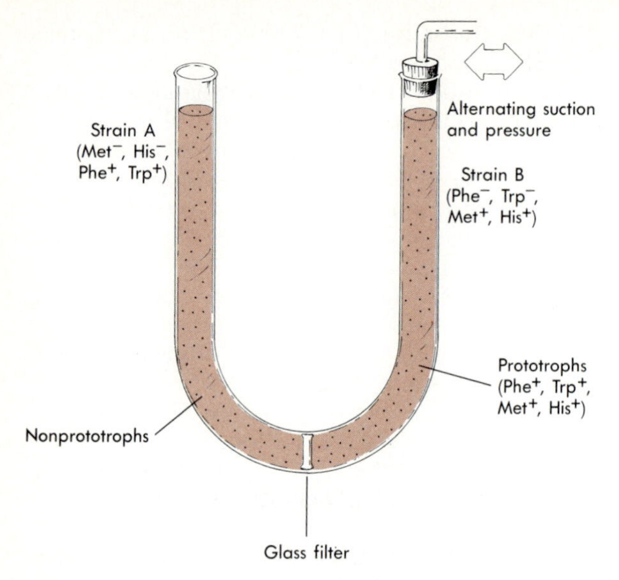

Figure 11-4 *U tube used to demonstrate genetic transfer by virus of bacterial genes (transduction). Bacterial strain on left cannot synthesize methionine* (Met^-) *or histidine* (His^-), *and bacterial strain on right cannot synthesize phenylalanine* (Phe^-) *or tryptophan* (Trp^-). *Glass filter prevents bacterial transfer but not viral transfer. Alternating suction and pressure is used to transfer media between arms of U tube. Following incubation bacterial genes are transferred on virus from left side of tube to right side. Auxotrophic strain on right is converted to prototroph, i.e.,* Phe^+, Trp^+, Met^+, His^+. *See details of experiment.*

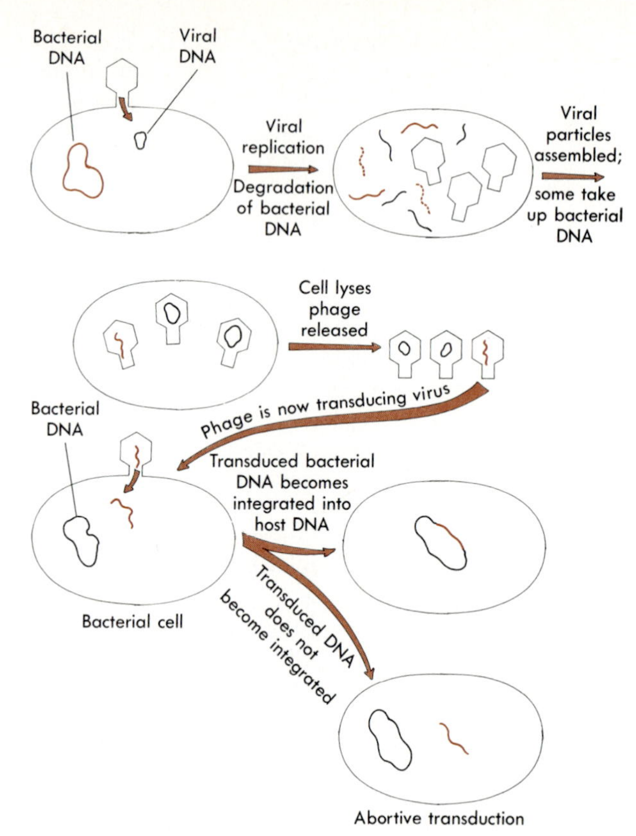

Figure 11-5 *Mechanism of generalized transduction.*

One strain of a double amino acid auxotroph (a mutant defective in the synthesis of two amino acids) was placed in one arm of the tube, and another auxotroph requiring two different amino acids was placed in the other arm. Wild-type bacteria (prototrophs) appeared on only the right side of the glass filter, suggesting that a filterable agent* had passed from the left arm into the right arm.

The explanation for this discovery was that the bacterial strain on the left harbored a virus that could transfer bacterial genes into the right arm of the U tube. The virus was produced only when both bacterial strains shared the same growth medium. The strain on the left, when grown in isolation from the strain on the right, did not produce the virus, which was the filterable agent. The process has been called transduction and is believed to occur in the following way. The strain on the right in Figure 11-4 carries a temperate virus, which integrates into the bacterial chromosome. Occasionally the temperate virus is excised from the bacterial chromosome and begins to replicate (vegetative virus), producing progeny viral particles that may lyse the cell. The released phage particles move across the glass filter and infect and lyse the bacterial cells on the left, which are not lysogenic. Infrequently, some of the phage particles incorporate bacterial DNA during its excision from the bacterial genome and its packaging into the virus particle. The phage particles move back across the glass filter and lysogenize the bacterial cells on the right. Wild-type bacterial recombinants appear on the right side if the viruses lysogenizing them are carrying the necessary wild-type bacterial genes. Wild-type genes cannot appear on the left because the cells are nonlysogenic and are lysed. Lysogeny is discussed at length in Chapter 17.

There are two types of transduction processes: generalized and specialized.

Generalized transduction. The term **generalized** implies that any gene has a chance of being packaged and transduced. The virus that engages in generalized transduction carries primarily bacterial DNA. This was the type of transduction observed by Zinder and Lederberg in 1952. The process is carried out by preparing a high concentration (titer) of phage obtained by lytic infection of a bacterial strain (Figure 11-5). Some of the viral particles produced in the cell incorporate bacterial DNA instead of viral DNA, and these represent the trans-

* The agent was not naked DNA or RNA, because neither DNase nor RNase treatment interfered with the transduction. Thus transformation was eliminated as a possibility.

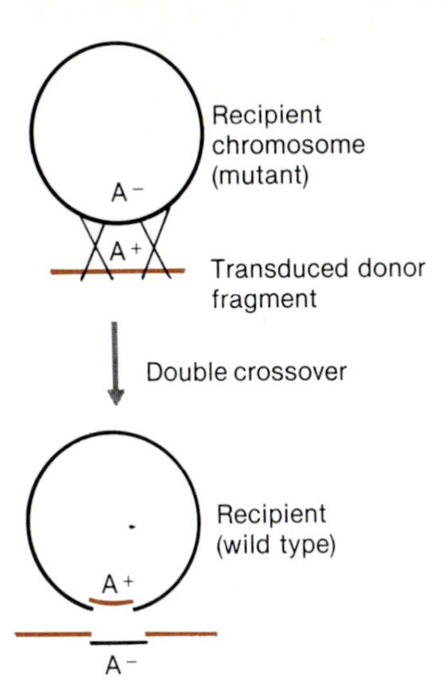

Figure 11-6 *Recombination during generalized transduction. Reciprocal recombinational event (double crossover) results in exchange of homologous genes and formation of recipient cell containing wild-type gene.*

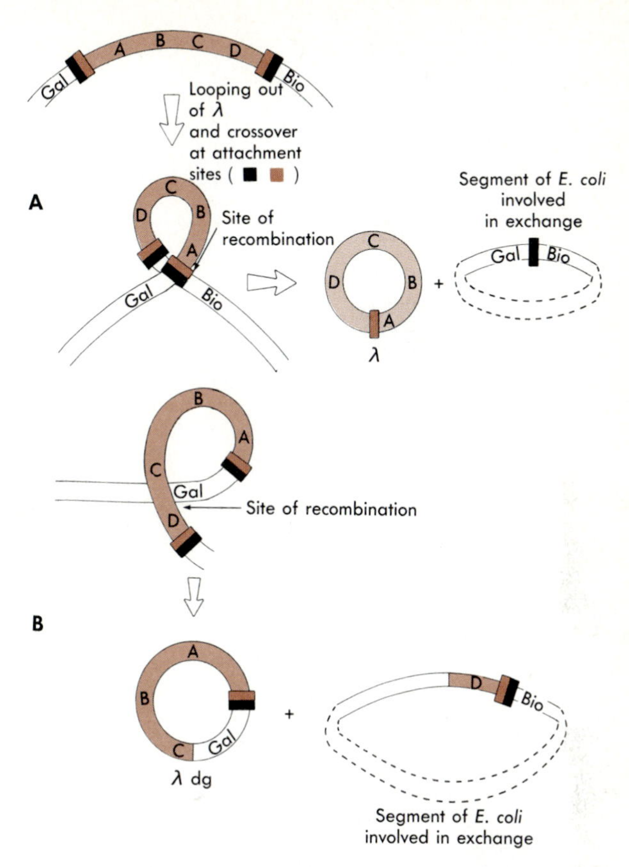

Figure 11-7 *Normal and abnormal excision of phage lambda* (color) *from* E. coli *chromosome.* ***A,*** *Normal excision results in formation of original genomes.* ***B,*** *Abnormal excision results in formation of specialized transducing particle, called* λdg. *During this event excision does not occur between attachment sites of two chromosomes, but between* gal *region of* E. coli *genome and region between* C *and* D *genes of lambda genome. Excised lambda chromosome will contain part of* E. coli *chromosome at* gal *region, whereas host chromosome* (E. coli) *will contain part of* D *gene of lambda chromosome.*

ducing particles. Once they are released from the bacterial cell, the transducing particles can infect susceptible bacterial cells but cannot lyse them, because they do not contain viral information. A recombinational event can occur between transduced and recipient DNA. The recombination (Figure 11-6) involves incorporating a double-stranded segment of transducing DNA in exchange for recipient bacterial genes (this differs from transformation, in which a single-stranded DNA is incorporated into the bacterial chromosome). Abortive transduction can occur if there is no integration of transduced DNA into the recipient DNA. As the bacterial population multiplies, only one cell carries the transduced DNA (see Figure 11-5).

Specialized transduction. Specialized transduction involves the transfer of specific bacterial genes that are located near the viral integration site. A variety of temperate viruses can take part in specialized transduction, including the virus lambda, which integrates into the bacterial chromosome near the genes that control galactose and biotin metabolism (see Figure 11-2). Lambda, which has been most widely studied, rarely is excised from the bacterial chromosome but can be induced to deintegrate by ultraviolet (UV) light. Abnormal excision, involving crossover events between lambda and the bacterial chromosome, releases a transducing particle that carries the galactose or biotin genes (Figure 11-7). The excised transducing particle also lacks many viral genes because they are left behind during the excision and crossover events. The transducing particles are usually defective and cannot become lytic, but some are infectious. Defective lambda particles carrying the galactose or biotin genes are denoted λ-d Gal (**λ*dg***) or λ-d Bio (**λ*db***). When λdg particles are used to infect an *E. coli* cell that contains a defective galactose gene, the λdg engages in a crossover event with the galactose region of the recipient (Figure 11-8). The result is the formation of a partial diploid *(gal^+/gal^-)*. It is called a *gal^+/gal^-* **heterogenote** and contains a gal^+ **exogenote** (donor DNA fragment) and a *gal^-* **endogenote** (recipient chromosome). *Gal^+* will be the dominant phenotype; that is, the cell will be able to use galactose as a carbon source. Different temperate phages, because of their ability to integrate at different sites on the bacterial chromosome, can transduce specific bacterial genes and are used to map the bacterial chromosome.

Transduction and gene mapping. Transduction is a suitable method for determining the order of genes, as

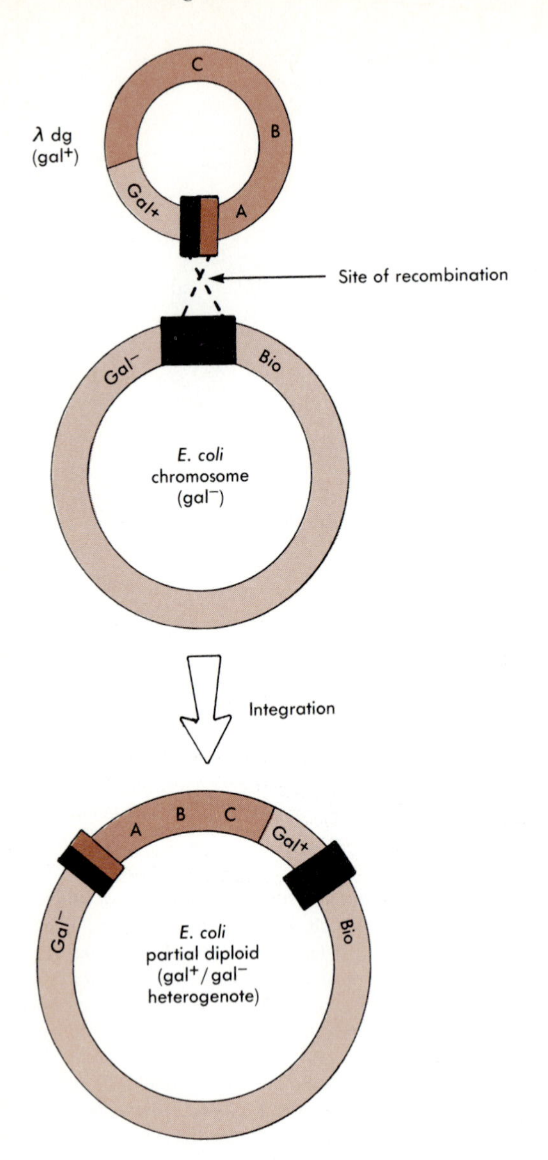

Figure 11-8 *Integration of* λdg *chromosome, carrying wild type galactose gene* (gal$^+$), *into recipient chromosome carrying a defective galactose gene* (gal$^-$).

well as the relative distances between genes, when two or more genes are incorporated into the transducing viral particle. Specialized temperate phages are not used in mapping experiments because of their different attachment *(att)* sites. Phages such as lambda are of little use in mapping because only genes adjacent to the *att* site are picked up. Generalized transducing phages, such as P1 or P22 do not need to integrate to generate transducing particles, these particles contain only host DNA, and any gene may be transduced. Generalized transducing phages are useful in genetic mapping experiments. The phages are used to transduce the recipient bacterium, which is mutant for the selected markers. The order of transduced genes can best be determined by examining three genes, two of which may or may not be cotransduced with a third gene (the frequency of cotransduction decreases as the distance between markers increases). For example, if we used the genetic markers for threonine synthesis *(thr+)*, leucine synthesis *(leu+)*, and the ability to ferment arabinose *(ara+)*, the arrangement of genes on the chromosome could be determined (Table 11-1). Results of such an experiment might show that when arabinose use is selected in the transductants, 75% also carry the leucine gene, but the threonine gene does not appear. The arabinose gene is therefore linked more closely to the leucine gene than to the threonine gene. Consequently, the gene order could be *ara-leu-thr* or *leu-ara-thr*. When the threonine-leucine transductants are sought, 85% of the time the arabinose gene is also costransduced, suggesting that the gene order is *leu-ara-thr*. (Cotransduction is not 100% because of quadruple crossovers). Wild-type leucine and threonine transductants would transduce very infrequently if the arabinose gene were between leucine and threonine.

Table 11-1 *Cotransduction frequencies for markers in P_1 phage involving the cross ara^+ thr^+ leu^+ with ara^- thr^- leu^-*

Selected marker	Appearance of unselected marker
ara^+	75% = leu^+
	5% = thr^+
thr^+ leu^+	85% = ara^+

More accurate measurements of the linkage order between two genes can be obtained by performing reciprocal crosses in which each parent is used alternatively as donor and recipient. Such crosses are set up by growing the transducing phage on each mutant strain of bacteria. Let us assume that we are examining three genes (A, B, and C) and we do not know if the gene order is CBA or BCA (Figure 11-9). Various crossovers between the mutant donor and recipient DNA molecules must occur to obtain the wild type in the recipient cell.

Conjugation

Conjugation in bacteria is a mechanism for gene transfer that requires cell-to-cell contact. The original experiments used to demonstrate conjugation were designed

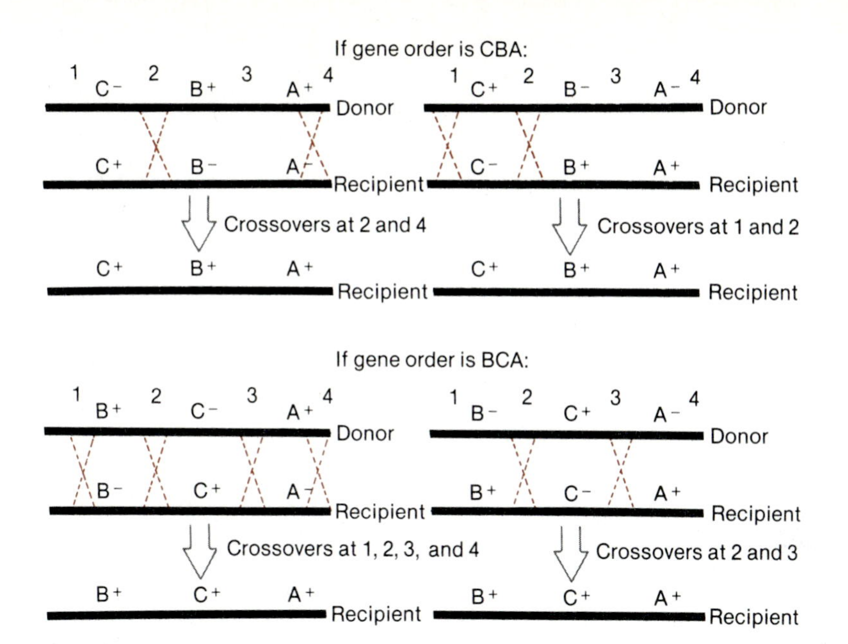

Figure 11-9 *Determination of gene order through transduction experiments by using reciprocal crosses. Postulated gene order is CBA or BCA. Number of crossovers required to obtain prototrophs (C^+, B^+, A^+) for each postulated gene order is indicated. If gene order were BCA, four crossover events would be required in one of reciprocal crosses, which is very infrequent. If gene order were CBA, ratio of recombinants in two crosses of upper figure would be approximately one. Ratio of recombinants for reciprocal crosses in lower figure would be much less than one because of infrequency with which a quadruple crossover can occur.*

Concept Check

1. **Compare and contrast the mechanisms of transformation, transduction, and conjugation.**
2. **Describe the transformation process as it occurs in gram-positive bacteria. How does it differ in gram-negative bacteria?**
3. **Explain the differences between generalized and specialized transduction.**
4. **Explain why all phages are not suitable for gene mapping experiments.**

by Lederberg and Tatum in 1947. Their discovery was based on the formation of prototrophs of *E. coli* from a mixture of cells that consisted of two kinds of auxotrophs. The auxotrophs can be represented as $A^+B^+C^+D^-E^-F^-$ and/or $A^-B^-C^-D^+E^+F^+$ (Figure 11-10). The auxotrophs were cultivated together overnight in a complete medium (supporting the growth of mutants). The cells were centrifuged, washed, and then plated on a minimal medium (supporting only the growth of prototrophs). One prototroph for every 1×10^7 cells plated appeared on the minimal agar. (No prototrophs appeared when each of the auxotrophs was plated on a minimal medium.*) A **U** tube similar to the one used in transduction experiments (see Figure 11-4) was used to show that mutant strains placed on either side of the glass filter produced no prototrophs unless the filter was removed and cell contact took place. Lederberg and Tatum originally believed that the two bacterial cells entering the conjugation process played equal roles, but this was discredited by Hayes' discovery of the sex factor.

Sex factor and conjugation. Hayes in the early 1950s discovered that treatment of one of the Lederberg and Tatum auxotrophs (strain A) with streptomycin before the conjugation process took place had no effect on the number of prototrophs produced. However, similar treatment of the other auxotrophic strain (strain B) prevented recombination. This experiment suggested that recombination takes place in only one direction and that one cell (strain A) acts as a **donor** and the other (strain B) acts as a **recipient.** In other words, in the experiment the donor cell could be inhibited in its growth but still enter into the recombination process, but the recipient cell, which was sensitive to streptomycin, could not grow and so was unable to produce recombinant colonies on an agar surface. Further experiments involving strain A and strain B cells showed that some failed to produce recombinants. Crossing one of these strains with the corresponding partner of known ferility showed that strain A had become defective and was no longer acting as a donor. The defective donor could be made fertile if it were crossed with a normal donor strain. Hayes proposed that the donor strain contained an infectious particle, or **sex factor**, that could be trans-

* The reason that no reversions occurred is that the probability of a reversion to the wild type for a cell containing three mutations is the product of the individual probabilities. For example, if the reversion frequency for each of three mutations is 1×10^{-6}, then the reversion frequency for a cell containing three mutations will be 1×10^{-18} (1 cell in 1×10^{18} will be a prototroph).

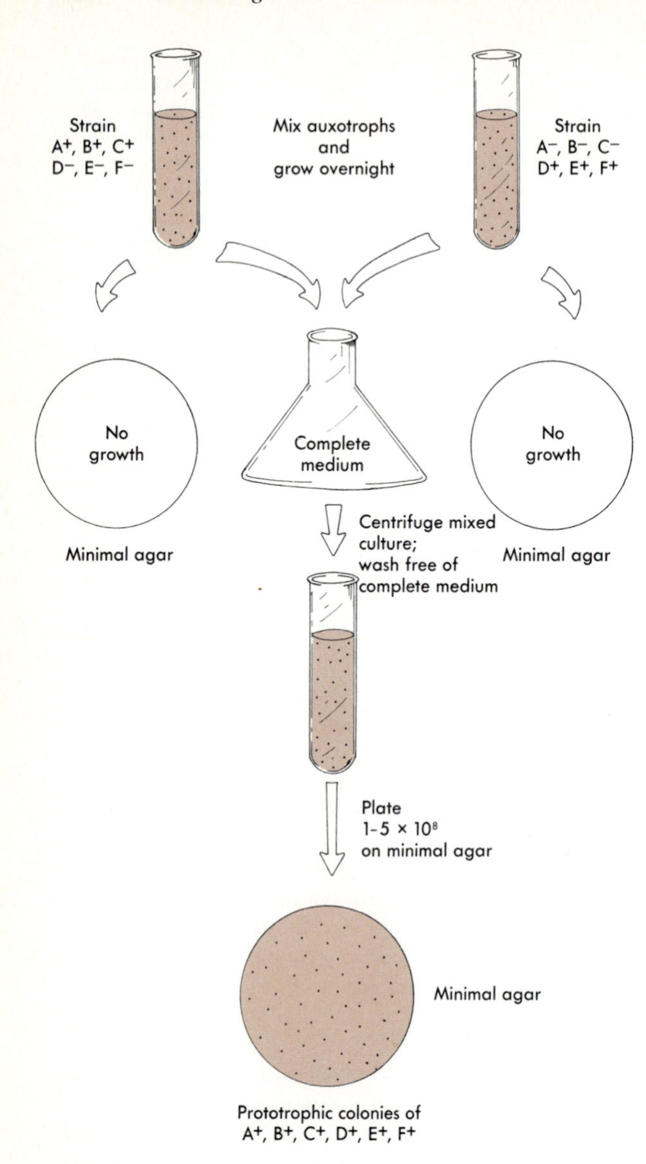

Figure 11-10 *Lederberg and Tatum experiment demonstrating conjugation between two auxotrophic strains of* E. coli *K12.*

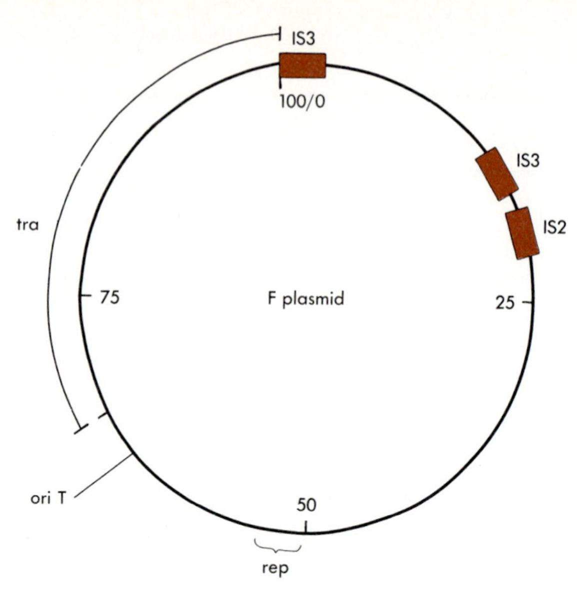

Figure 11-11 *Map of F, or sex factor, plasmid. Inside circle are kilobase coordinates—the distance in thousands of bases. Abbreviations are* IS3 *and* IS2, *insertion sequence elements;* tra, *genes associated with transfer of F factor;* oriT, *origin of transfer; and* rep, *genes associated with replication.*

ferred during the conjugation process. Cells containing the sex factor could therefore donate genes to a recipient. Cells that possess the sex factor are called **F^+**; those without the sex factor are called **F^-**. This means that $F^+ \times F^-$ crosses are fertile, whereas $F^- \times F^-$ crosses are infertile.

Extraction of DNA from cells containing the sex factor reveals a circular, double-stranded DNA molecule whose molecular weight is approximately 5×10^7. The sex factor can exist independently of the host chromosome as a plasmid, or it may be integrated into the host chromosome.

The F factor was the first plasmid discovered and happened to possess the genes necessary for conjugation, but did not possess any phenotypic traits. Since the F factor was first discovered many conjugative plasmids have been found in other gram-negative bacteria. The phenotypic traits coded by these conjugative plasmids include resistance to several heavy metals and antibiotics, toxin production, colicin production, metabolism of certain carbon sources such as toluene, and resistance to ultraviolet light (see Chapter 8).

The F factor codes for about 22 genes, and one-third of these are associated with the conjugative process (the conjugative genes are transfer, or *tra,* genes) (Figure 11-11). Conjugation can be divided into two steps: formation of mating pairs, and DNA transfer (Figure 11-12). The **sex pilus** plays an important role by recognizing the recipient cell in the mating process. Twelve genes are required for pilus synthesis, and one of these codes for the pilus protein, called **pilin**. The pilus makes contact with the recipient cell, then contracts to bring the two cell surfaces closer together. The close contact between the cells apparently triggers unwinding of DNA and strand transfer. The DNA of the F factor is unwound by a plasmid-encoded helicase. The process of DNA transfer begins with the nicking of one strand of the donor plasmid at the origin *(oriT).* The nicked strand is transferred to the recipient, beginning with the 5′ end. A strand complementary to the transferred strand is synthesized in the recipient by host-specified enzymes, in-

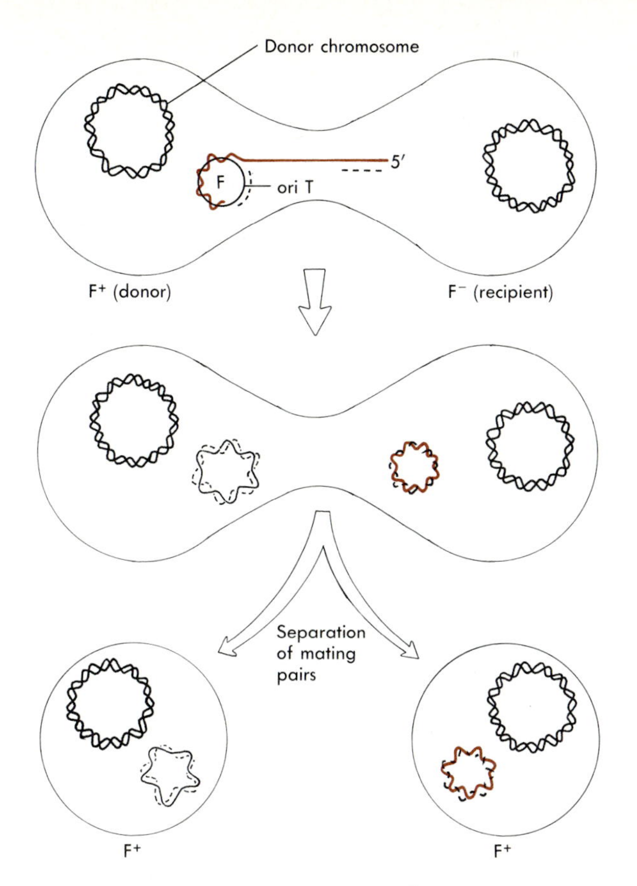

Figure 11-12 *Conjugation between F^+ and F^-. Donor strand nicked by enzyme at origin* (ori T) *and strand transferred at 5′ end. Displaced strand acts as template when it enters recipient, and a complementary strand is synthesized. In addition, remaining donor strand also acts as template and another complementary strand is synthesized. Donor and recipient separate and each contains copy of F factor.*

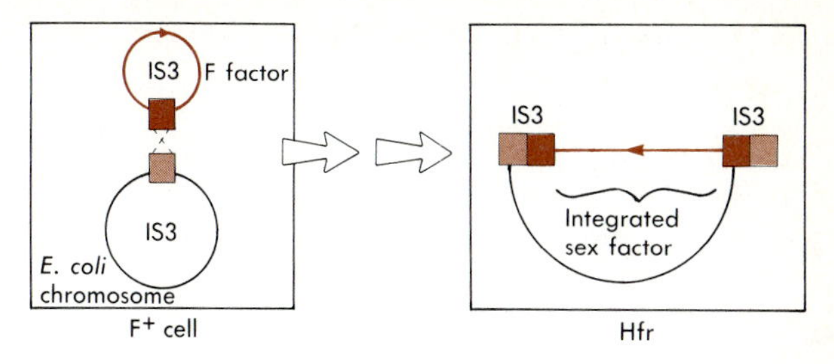

Figure 11-13 *Integration of sex factor into host chromosome, producing Hfr. Site for integration occurs at same location as IS elements. Thus homology between IS elements of sex factor and* E. coli *chromosome results in integrated sex factor being flanked by identical IS elements. Arrow on sex factor indicates site of origin and direction of transfer of genes during conjugation.*

cluding DNA polymerase III. The strand remaining in the donor also acts as a template for the synthesis of a complementary strand. The result is the formation of a circular F factor in the recipient cell (or conversion of an F^- cell to an F^+ cell).

The F factor also encodes genes to prevent the donor strand from acting as a recipient. In other words, an interaction between two F^+ cells would be unproductive. The genes that prevent this reaction encode proteins that are located on the inner and outer surface of the cytoplasmic membrane.

The F factor replication mechanism is patterned after the rolling circle model described in Chapter 8. The transfer process takes only a few minutes, so that in a large population of F^- cells and only a few F^+cells, the entire population can become F^+ after only a few cell divisions. Only one recipient in a million that has been converted to F^+ shows recombination of chromosomal genes.

Sex factor and the integrated state. Hayes and others discovered a strain of *E. coli* in which recombinants appeared 100 to 1000 times more frequently than in the Lederberg and Tatum strains (recombinants appeared once for every 10,000 to 20,000 cells examined). This "super male" strain was referred to as **Hfr** (high frequency of recombination). The sex factor in Hfr bacteria is integrated into the chromosome in a rare recombinational event between F^+ DNA molecule and the bacterial chromosome. The recombination involves a single crossover and takes place at different locations on the chromosome depending on the strain (there is an insertion sequence recognition between the sex factor and specific chromosomal sites (Figure 11-13). The F factor can also be deintegrated, converting the Hfr to the F^+ state. The mating process between an Hfr and an F^- is the same as in $F^+ \times F^-$ matings. The major difference is that **chromosomal DNA** is transferred from the donor to the recipient in an Hfr × F^- mating. The transfer can be interrupted by breakage of the chromosome; therefore the number of recombinants detected will be dependent on the amount of chromosome transferred (Figure 11-14). All bacteria of one Hfr strain transfer their chromosomal genes in a particular sequence, such as ABCDEFG. Different strains may transfer their genes in a different sequence, for example, CDEFG or FGHI. Some strains transfer their genes in reverse order, such as HGFEDC or ZYXW. However, the relationship of the various genes to one another remains the same. Consequently, there is an origin and termination site on the bacterial chromosome during the conjugation process. For different Hfr strains the sex factor is inserted at different sites, which gives rise to unique origins of transfer. The first segment of genes attached to the origin, which enters the recipient, are some F factor genes that do not control pilus formation or the donor state. The genes of the last chromosomal segment to enter are also F factor genes that do control the donor state (see Figure 11-14). It is easy to understand why few F^- cells in Hfr × F^- matings are converted to the Hfr state.

Mapping of the chromosome by conjugation. The observation that each Hfr strain tested transfers its chromosomes from a specific site offered an opportunity for gene mapping. Conjugation can be used for coarse mapping (position of genes mapped in terms of time units), whereas transduction and transformation are used for fine-structure mapping (determining linkage and gene order.) Conjugation mapping, unlike transformation or transduction mapping, is based on time of entry into the recipient and not on recombination frequencies. Jacob and Wollman devised the interrupted mating technique as a method for mapping genes by conjugation. Their experiment was set up in the following way:

1. The genetic cross involved an Hfr strain that had the genotype thr^+, leu^+, azi^R, $T1^R$, lac^+, gal^+, and str^S; that is, they could synthesize threonine and leucine, were resistant to sodium azide and *T1* bacteriophage, could use lactose and galactose, and were sensitive to streptomycin and a F^- strain that was thr^-, leu^-, azi^S, $T1^S$, lac^-, gal^-, and str^r; that is, they could not synthesize threonine or leucine, were sensitive to sodium azide and *Tl* bacteriophage, could not use lactose or galactose, and were resistant to streptomycin.
2. The Hfr and F^- cells were mixed in a nutrient medium for about 5 minutes to promote conjugation, and then were diluted to prevent the formation of new mating pairs. Samples of the mating pairs, usually taken at 1-minute intervals, were agitated in a blender to break the conjugation bridge and interrupt chromosome transfer. A large number of recipient cells therefore contained various lengths of donor chromosome. Because the F^- cells were resistant to streptomycin but the Hfr cells were sensitive to the drug, only F^- recombinants (not Hfr cells) were recovered.
3. Following conjugation the cells were placed on a medium containing the antibiotic streptomycin but lacking the amino acids threonine and leucine. On this type of medium only recombinant cells carrying the leu^+ and thr^+ genes of the Hfr parent cell and the str^r gene of the F^- cell would grow.
4. Colonies produced by thr^+, leu^+, str^r recombinants were replica plated (see chapter 10) on different selective media to determine which donor genes were present. The media included sodium azide, to score resistance or sensitivity to sodium azide; *T1* bacteriophage, to score resistance or sensitivity to *T1*; lactose, to score metabolism of lactose; and galactose, to score metabolism of galactose.
5. The experiment showed that when conjugation was interrupted before 8 minutes elapsed, no thr^+, leu^+, or str^r recombinants were formed (Figure 11-15). After 8 minutes the genetic markers were transferred in a specific temporal sequence. A small genetic map was constructed from this experiment and read: *origin, thr, leu, azi, T1, lac,* and *gal.*

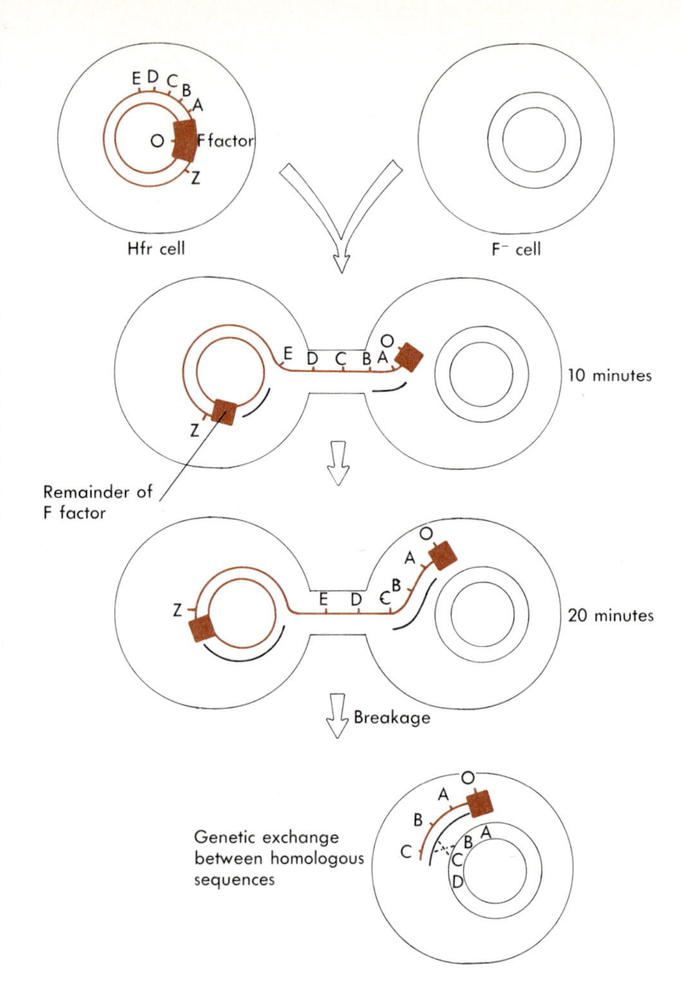

Figure 11-14 *Interrupted mating experiment between an Hfr cell and F^- cell. Hfr cell transfers chromosome at origin, which is in middle of F factor; thus only part of F factor transferred (colored block). Transmission of chromosome (in our example) continues for 20 minutes, at which time conjugation process is halted. As donor chromosome enters F^- cell a complementary strand is synthesized. Remaining donor strand also acts as template for synthesis of complementary strand. In the F^- cell, strand exchange can occur between donor and recipient homologous sequences.*

The more distant the gene is from the origin, the more difficult it is to obtain recombinants because of chromosome breakage. This problem was rectified by

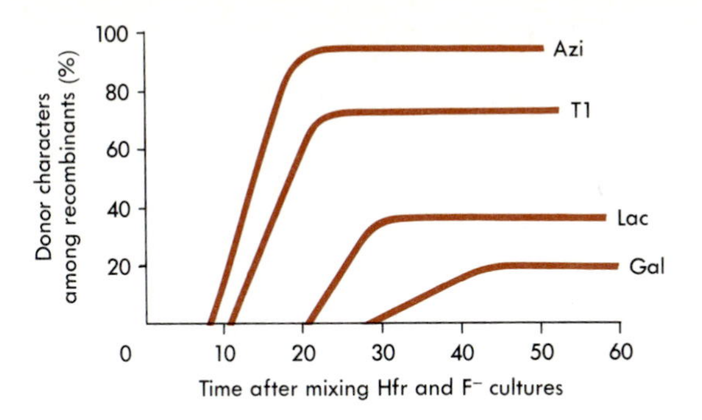

Figure 11-15 *Kinetics of inheritance of unselected markers after conjugation between prototrophic Hfr strain (*thr^+, leu^+, azi^R, Tl^R; lac^+, gal^+, *and* str^S*) and* F^- *cell (*thr^-, leu^-; azi^S, Tl^S; lac^-, gal^-, *and* Str^R*). Conjugation is interrupted by agitation in blender. Selected markers among recombinants are* thr^+ *and* leu^+*, which appear 8 to 8½ minutes after conjugation. Appearance of unselected markers is indicated in graph.* R, *Resistant;* S, *sensitive;* thr, *threonine;* leu, *leucine;* azi, *sodium azide;* Tl, Tl *bacteriophage;* lac, *lactose;* gal, *galactose; and* str, *streptomycin.*

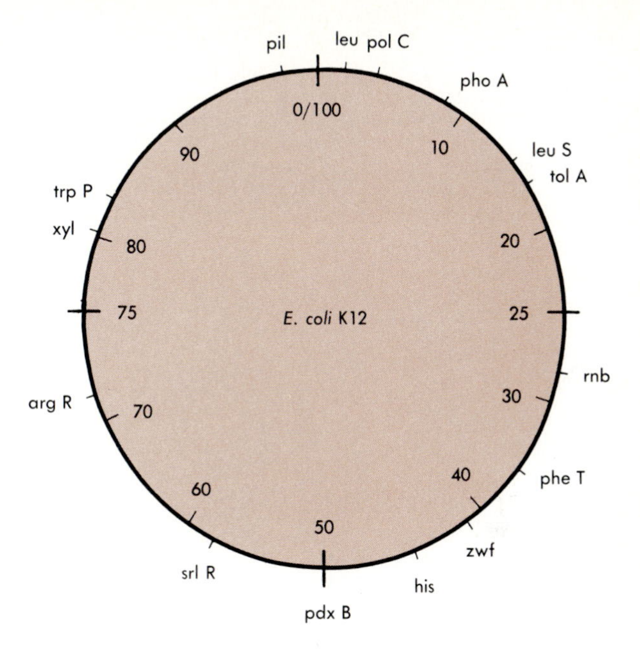

Figure 11-16 *Circular map of* E. coli *K12 showing a few of the genetic markers obtained from conjugation experiments involving several Hfr strains. Abbreviations are* leu, *leucine;* pol C, *DNA polymerase III;* pho A, *alkaline phosphatase;* leu S, *leucyl tRNA synthetase;;* tol A, *tolerance to colicins;* rnb, *ribonuclease;* phe T, *phenylalanyl tRNA synthetase;* zwf, *glucose 6-phosphate dehydrogenase;* his, *histidine;* pdx B, *pyrimidine requirements;* srl R, *sorbitol regulation;* arg R, *arginine regulation;* xyl, D*-xylose use;* trp P, *tryptophan permease; and* pil, *pili.*

Concept Check

1. Explain the process of conjugation in terms of the mechanism of formation of mating pairs and DNA transfer.
2. Explain the characteristics of cells carrying the sex factor in the integrated vs. nonintegrated state.
3. Which methods of gene transfer are suitable for coarse gene mapping and fine-structure gene mapping? Why?

using different Hfr strains whose origins are all different. The results of these experiments has permitted a rather detailed genetic map of *E. coli* and some other species. The *E. coli* map is divided into 100 one-minute divisions (100 minutes is the time it takes for transfer of the entire chrosomsome when the conjugation process takes place at 37° C), with the threonine gene indicating the approximate site of the origin. A modified version of this map is presented in Figure 11-16.

Conjugation in gram-positive bacteria. Only a few species of gram-positive bacteria are known to be capable of conjugation. Most of the studies to date have involved species of *Streptococcus*. Conjugation and chromosomal transfer apparently do occur in gram-positive bacteria, but not at a very high frequency. However, plasmid transfer by conjugation does occur at a high frequency, and the mechanism used is unique for bacteria. *Streptococcus faecalis* can transfer some of its plasmids to recipient cells at a very high frequency (as great as 10^{-1} per donor), but others are transferred poorly (less than 10^{-6} per donor). In those systems in which there is efficient plasmid transfer, the recipient cells release small peptides, or **sex pheromones.*** The use of sex pheromones by higher organisms (fungi and higher eukaryotes) had been recognized earlier, but they had not previously been found in bacteria. When these peptides make contact with the donor cell, the latter synthesizes surface proteins that facilitate the formation of aggregates of donor-recipient cells. The pheromones also induce the transfer of the plasmid to the recipient cell. The mechanism of plasmid transfer is not known.

GENETIC ENGINEERING AND RECOMBINANT DNA

Our discussion of recombination so far has involved the rearrangement or reciprocal exchange of genetic mate-

* Sex pheromones are a class of hormones or hormonelike substances secreted by an individual of a species to attract another individual of the same species.

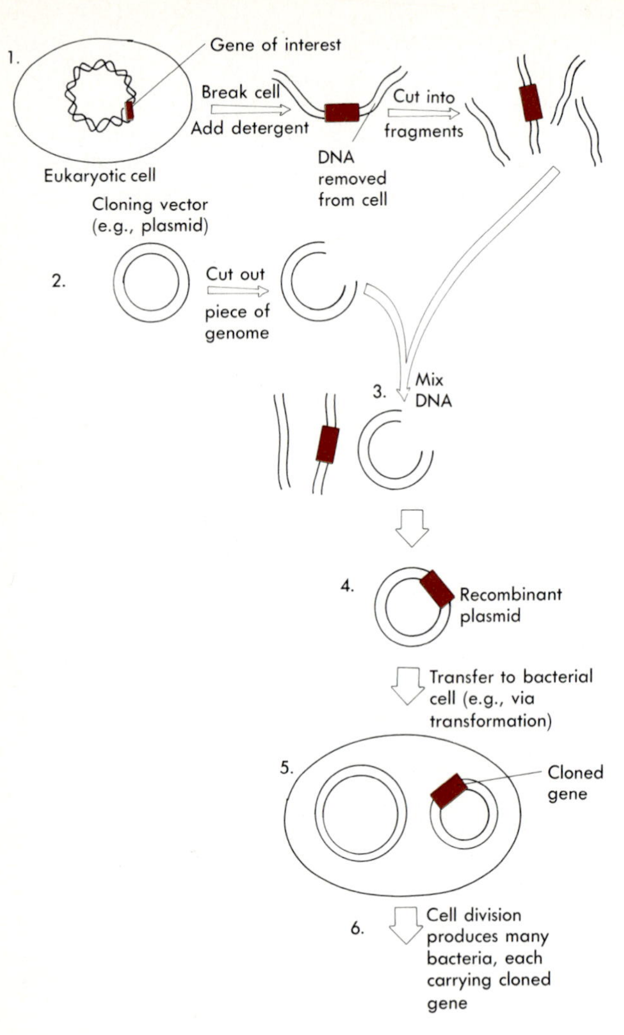

Figure 11-17 *Basic steps performed in gene cloning experiment. 1. Eukaryotic cells are broken apart to release DNA. DNA is cut into fragments of varying length by restriction endonucleases. Gene of interest is in color. 2. Plasmid cloning vector is treated with enzyme to remove fragment of DNA. 3. Fragmented eukaryotic DNA and plasmid DNA are mixed. 4. Eukaryotic gene is integrated into plasmid to produce recombinant molecule. 5. Plasmid is transferred to bacterial cell via transformation process. 6. Bacterial cell can divide, and copies of cloned gene will be carried by all descendants.*

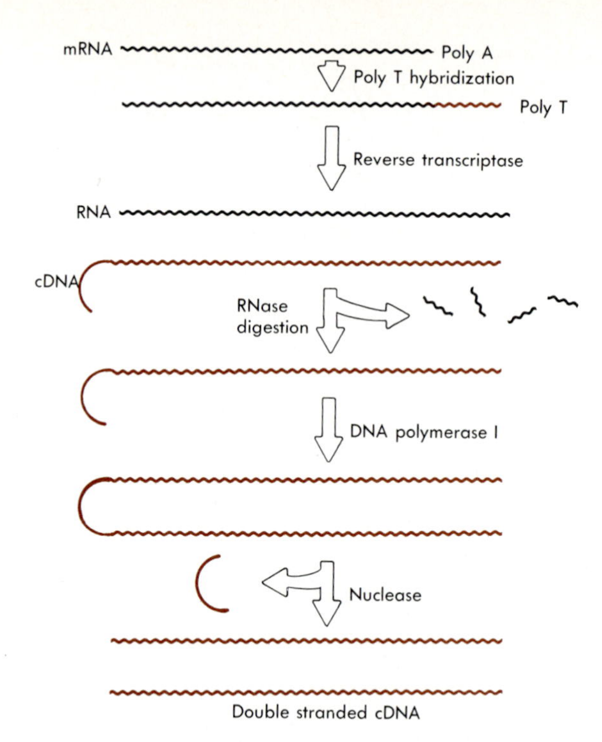

Figure 11-18 *Mechanism for synthesis of double-stranded complementary DNA (cDNA) from messenger RNA (mRNA). The mRNA (uncolored) normally carries a poly A tail (eukaryotic mRNA). An oligo-dT (deoxythymidine) chain is used to hybridize with poly A tail. Oligo-dT chain will serve as primer for polymerase, reverse transcriptase; complementary DNA strand is synthesized. RNA strand is then digested away with RNase. Note complementary DNA has hairpin loop at its 3′ end. Hairpin loop will serve as primer for DNA polymerase I during formation of double-stranded DNA. Nuclease is used to remove hairpin, which results in formation of double-stranded complementary DNA molecule.*

rial between the same or related species of bacteria. Recently developed experimental procedures now permit scientists not only to fuse genes but also to insert genes from any eukaryotic or prokaryotic source into a bacterial replicon using phage or plasmid DNA. It is now possible to obtain millions of copies of a specific region of DNA by placing the DNA inside a cell and allowing the cell to multiply. This process, **gene cloning**, is one form of recombinant DNA technology, or **genetic engineering**. The potential uses of genetic engineering range from replacing defective genes in humans suffering from inherited genetic disorders to developing plants with special traits.

Before discussing gene cloning procedures in detail, let us examine the basic steps that are required (Figure 11-17).

1. The cells from which the DNA is to be obtained are sheared in a blender and then treated with a detergent to release DNA.
2. DNA is purified by spooling it out on a glass rod from the broken cell mixture.
3. Specific genes of interest are spliced out of the DNA by enzymes called **restriction endonucleases**.
4. The genes of interest are spliced into a DNA molecule called the **cloning vehicle** or **vector**. The cloning vector, for example, a plasmid, can be used to penetrate the cell wall of the microbial cell.

5. The cloning vehicle, which is now a recombinant DNA molecule, is transferred into a host cell (usually bacteria or yeast) via transformation.
6. The host cell multiplies, and each member of the clone carries a new piece of genetic information.

Requirements for forming and cloning recombinant DNA

Source of the foreign gene. The source of the gene to be cloned will dictate the methodology used in its isolation (see Point of Interest). Genes in eukaryotic organisms are often quite complex because they contain base sequences, or intervening **(introns)** sequences, that are not involved in determining the amino acid sequence of the protein encoded by the gene. (see Chapter 9). When the gene is transcribed, the resultant RNA molecule is processed by special nucleases that delete specific portions of the RNA molecule. Thus the mature messenger RNA (mRNA) is shorter than the original gene from which it was transcribed. If the entire gene were cloned in a bacterium that lacked the special nucleases, the mRNA from that gene would not be processed, consequently, the protein encoded by the unprocessed mRNA would be different from that found in the eukaryotic cell and probably would not be functional. Therefore it is best to start with the mRNA for the eukaryotic protein of interest. A copy of DNA can be made from the mRNA by the enzyme reverse transcriptase. The complementary copy is **cDNA.** The hybrid RNA-DNA produced by reverse transcriptase can be used as a template to make a DNA strand complementary to cDNA (the RNA of the hybrid can be removed by RNase) (Figure 11-18). The double-stranded DNA product contains the exact eukaryotic coding information; hence mRNA transcribed from it will contain only protein-encoded information.

Bacterial genes do not contain introns and can be cloned directly without first isolating mRNA and making DNA from it.

Cloning strategy. There are two basic approaches to cloning genes. In the first approach, many laboratories, using recombinant techniques, develop **gene libraries,** or **banks.** A gene library is a collection of recombinant molecules that contain all of the DNA sequences in a specific genome (for example, bacteria, yeasts, and humans). Thus the laboratory will have available hybrid-plasmid–containing bacteria in which each segment of the DNA source (for example, yeast) will be represented at least once. This technique eliminates having to go through a complete cloning procedure each time a new DNA segment is needed. Each vector can be cloned and then screened for the gene it carries.

The second approach is to purify a specific gene from all the fragments generated by endonuclease digestion. The cloning procedure will become simpler and faster if the desired gene can be enriched. Relatively

Point of Interest

SYNTHESIZING GENES

Another method available in genetic engineering is to actually synthesize the gene wanted and combine it with a vector. How does one go about synthesizing a gene? The first thing one needs to know is the amino acid sequence of the protein for which one wishes to synthesize a gene. Machines have been developed that will determine the sequence of protein from as little as a few micrograms of protein. Once one has determined the amino acid sequence of a protein, a nucleotide sequence of the gene can be deduced.

Gene machines are available that can synthesize a piece of DNA of approximately 40 nucleotides and newer machines can actually synthesize genes of 1000 nucleotides. These machines are fully automated; a researcher types in the genetic code for the protein's amino acid sequence and the machine will synthesize the correct piece of DNA. The machine sequentially adds the correct nucleotide at the correct time. A column consisting of silica beads binds the first nucleotide which has a protective agent attached to its 5' phosphate group so that it will not react with another nucleotide of the same type. Once the first nucleotide is bound, the column is washed of any free nucleotides and the protective agent is removed from the 5' end of the first nucleotide. The second nucleotide is then added which again has a protective agent on its 5' phosphate. The 5' phosphate of the first nucleotide will react with the 3' position of the second nucleotide. The protective agent on the 5' phosphate of the 2nd nucleotide is then removed after free nucleotides are washed from the column and the third nucleotide is added. This is continued until the desired piece of DNA is synthesized.

The 'synthesized gene' is then combined with a vector and cloned in a host cell so that large amounts of the protein can be made. Because no promoter is attached to the gene, an expression factor must be present in the vector to allow proper expression of the gene. For many genes that one wishes to clone this is the easiest way to obtain the correct piece of DNA so it can be added to a host cell. This technique is also used to synthesize small pieces of synthetic nucleic acid which can be used as probes to detect the host cell that has picked up the correct piece of recombinant DNA formed by one of other previously discussed methods.

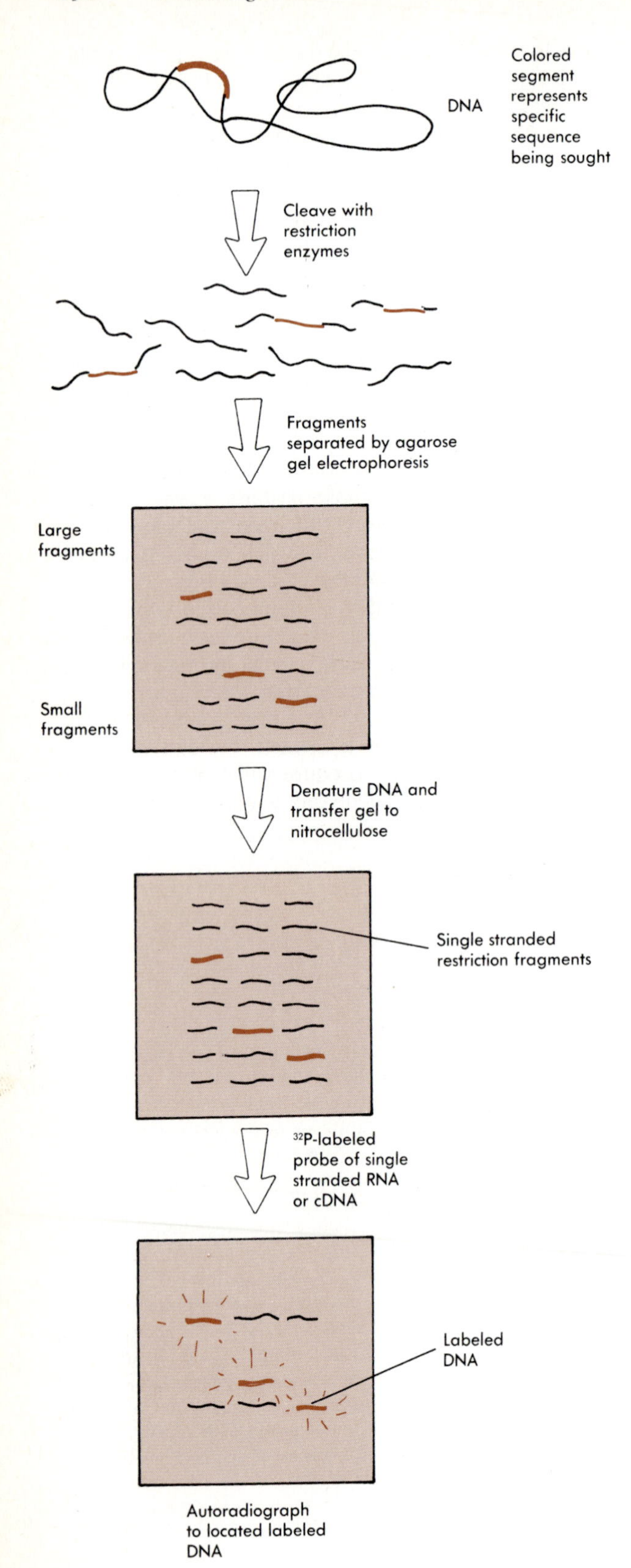

few enrichment methods are available, and most are based on size fractionation procedures such as sucrose density gradient (see Chapter 4) or gel electrophoresis. Monitoring the fractionated DNAs can be performed by the Southern blotting technique. In **Southern blotting** (Figure 11-19) the isolated DNA, cleaved by restriction enzymes, is separated into fragments by gel electrophoresis. The fragments are denatured, and single strands are immobilized on nitrocellulose (only single-stranded DNA sticks to nitrocellulose). The single strands of DNA will stick to the nitrocellulose in a pattern that mirrors their positions on the elctrophoretic gel. Unlabeled single-stranded DNA from an unrelated source is used to saturate the remaining sites on the nitrocellulose. ^{32}P-labeled purified DNA or RNA that is specific for the gene being sought is used to hybridize with the single-stranded DNA. The position of the bands that anneal to the ^{32}P-labeled probe can be detected by autoradiography. An alternate method can be used if a purified gene product is available. Part of the amino acid sequence of the protein must be determined, and from this the nucleotide sequence of the gene can be determined and synthesized. Usually only 20 to 30 nucleotides (6 to 10 amino acids) are needed. The synthetic gene fragment is radioactively labeled and used as a probe.

Need for a vector. A gene or specific genetic sequence being transferred to another cell must first be inserted into a molecule (vector) that will permit its transfer. The vector should have the following properties:

1. The vector should replicate autonomously.
2. Appropriate vector cleavage sites, preferably those with single cleavage sites, should exist for as many endonucleases as possible. In this way foreign DNA can be inserted without disrupting any essential function, such as replication.
3. The vector should be able to enter the host cell relatively easily. Usually small DNA molecules (plasmids and viruses) have higher transformation rates than do larger DNA molecules.
4. The vector should express a physical marker, such as antibiotic resistance, that is useful for selection.

Two types of vectors that have these properties are **plasmids** and **viruses**. Plasmids exist in the cell in certain copy numbers. Some plasmids may have 15 copies per

Figure 11-19 *Southern blotting technique for enriching a cloned gene. DNA is treated with restriction endonuclease. Colored segment represents specific sequence being sought. Fragments are separated by agarose gel electrophoresis, which separates by size. DNA is denatured, and single-stranded fragments are immobilized on nitrocellulose filter. Single-stranded DNA from unrelated source saturates unfilled sites on nitrocellulase. Fragments on filter can be hybridized to suitable labeled probe of ^{32}P complementary DNA (cDNA) or to a RNA specific for DNA fragment. Autoradiography allows for detection of labeled probe and its hybridized fragment.*

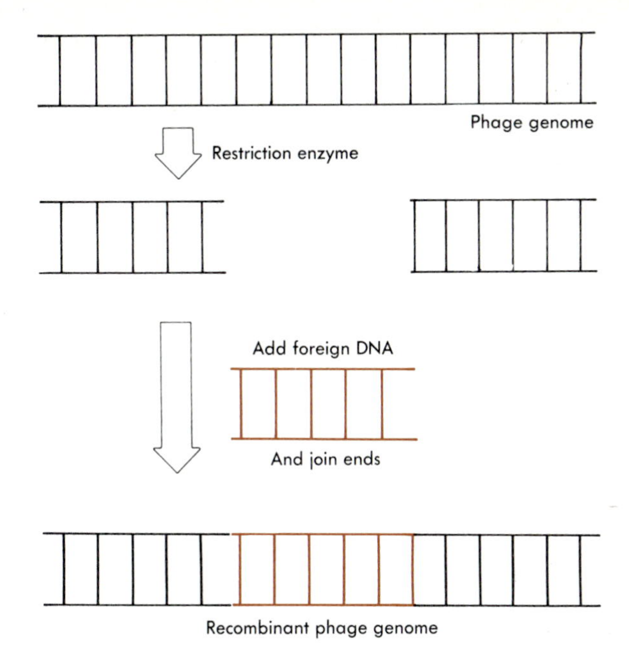

Figure 11-20 ***Phage vector can be cut into two fragments by single site-restriction enzyme. Foreign DNA can be inserted into ends of fragments to produce recombinant phage molecule that can be used to infect bacteria.***

host chromosome, whereas others are at levels of 1 to 5 copies per host chromosome. In addition, some multicopy plasmids accumulate in large numbers (75 to 100) when bacteria stop growing. These types are often used as cloning vectors. Manipulating the copy number can be used to increase the yield of protein specified by the cloned gene. Many cloning experiments use plasmids that carry resistance to two antibiotics. Resistance to one antibiotic is used to identify bacteria carrying the plasmid. The other resistance determinatn is used to distinguish recombinant from parental plasmids. Sometimes the plasmid vector is not totally suitable for the intended cloning experiment, and changes are made in the vector. One of the most widely used plasmid cloning vectors is called *pBR 322*. This vector was developed from a natural plasmid by removing nonessential DNA segments and by inserting selectable drug resistance markers. Also, cleavage sites at unique locations on the plasmid were added.

Most plasmids used today are restricted to *E. coli* and closely related enteric bacteria such as *Salmonella, Proteus,* and *Serratia*. It is important to remember that some plasmids are incompatible with others and will not be stable when maintained in the same cell. Some plasmids can be maintained in almost any gram-negative bacterial species, including strains of economic and medical importance such as *Agrobacterium tumefaciens, Pseudomonas, Rhizobium, Azotobacter,* and *Neisseria* species. *A. tumefaciens* carries a plasmid called T1 that induces tumors in dicotyledonous plants by transferring part of the plasmid DNA into the plant cell, where it integrates into the plant chromosomal DNA. This DNA transfer system possibly can be exploited for the genetic engineering of higher plants (see Chapter 20). The cloning and manipulation of eukaryotic genes in prokaryotic systems such as *E. coli* is at best impractical. The fundamental mechanisms of gene expression differ markedly between prokaryotes and eukaryotes. Some of the problems associated with the eukaryotic system are: (1) the mRNA may contain introns (see Chapter 9); (2) the eukaryotic promoter may not be recognized by bacterial RNA polymerase; (3) often, the eukaryotic gene product must be processed; and (4) eukaryotic proteins are often hydrolyzed by bacterial proteases. The intron problem can be eliminated by using cDNA. The promoter problem can be avoided by coupling the gene to a lambda or *lac* promoter. Performing the processing *in vitro* sometimes rectifies the processing problem. The protease problem can be handled by mutation of the bacterial gene. The best eukaryotic host cell candidate at present is the yeast, *Saccharomyces cerevisiae*. Yeast cells are 50 times larger than bacteria, easy to grow, and easy to separate from the medium in which they are cultivated. They also have a relatively thick cell wall that is resistant to relatively high concentrations of byproducts such as alcohol. Yeast is being actively pursued for the enzymatic conversion of cellulose to alcohol (see Chapter 22).

Phages provide another vector system. Phage DNA is usually a linear molecule, and a single site-restriction enzyme cut can generate two fragments (Figure 11-20). Foreign DNA can be joined to the two fragments to produce a recombinant molecule. The phage is then allowed to proceed through the lytic cycle to produce more virus, with each virus carrying foreign DNA. The length of foreign DNA that can be inserted into the phage genome is limited because the amount of phage DNA that can be packaged into the phage particle is limited; that is, there is a minimum and maximum length requirement. This problem can be remedied by removing some of the nonessential phage DNA and replacing it with larger fragments of foreign DNA. This approach has led to the formation of cosmids, which have the advantages of plasmids and phages. **Cosmids** are plasmids containing antibiotic resistance genes as selective markers and unique restriction sites for cloning. In addition, they contain the DNA sequence from phage lambda that is required for packaging DNA into the phage. This sequence, called the cohesive end site, or **cos site**, is recognized and acted on by the lambda packaging system (Figure 11-21). As discussed previously, the viral genome is replicated by the rolling circle method, which produces a long chain of viral genomes, or concatamer. Lambda enzymes chop the concatamer into individual

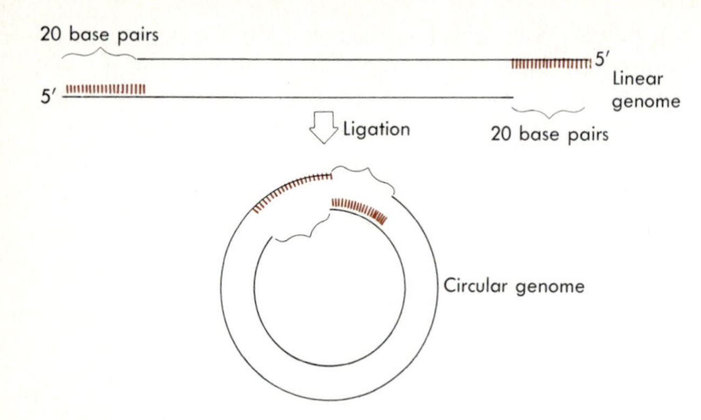

Figure 11-21 *Linear lambda genome consists of cohesive ends at 5' phosphate termini* (color). *Ligation can join cohesive ends by complementary-base pairing, thus producing circular molecule.*

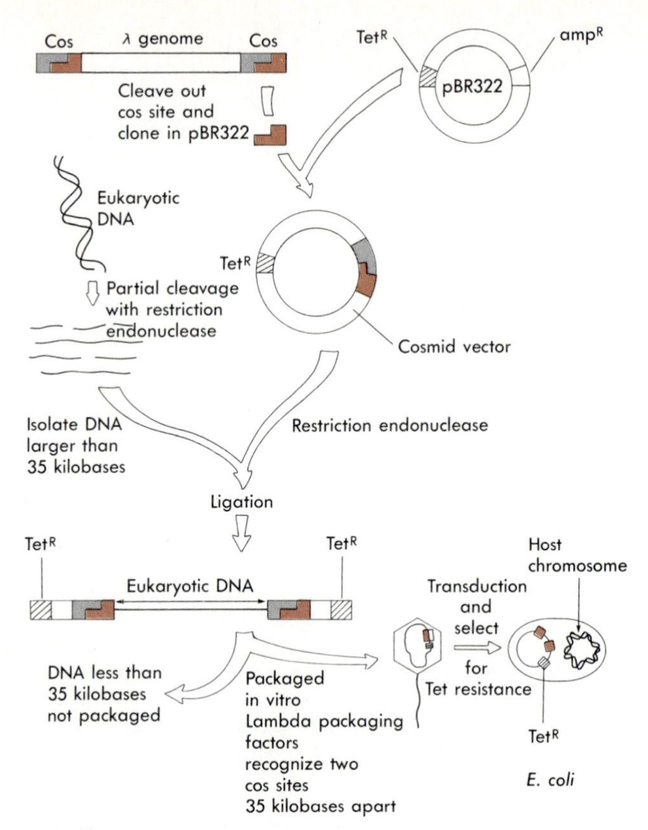

Figure 11-22 *Use of cosmids in cloning process (see text for details).*

genome units at the cos sites, which are at the end of the phage genome. Thus the cos sites are all that is necessary to package DNA into phage particles. The cos sites are cloned into the ampicillin resistance gene of the plasmid *pBR 322*, leaving the tetracycline resistance gene intact (Figure 11-22). Eukaryotic DNA is cleaved with restriction enzymes to produce large fragments, which are then ligated to *pBR 322* plasmid that contains the cos sites. Packaging of the cosmid with its eukaryotic DNA does not require an *in vivo* process. Instead, a lambda packaging extract is added to the cosmids. Enzymes in the extract will recognize the cos sites and cleave them, producing large and small fragments of eukaryotic DNA. Only the larger fragments of eukaryotic DNA will be packaged (those containing 35 to 45 kilobases, the approximate length of the missing lambda genome). The phage particles can be used to infect *E. coli* and are selected for tetracycline resistance. The advantage of cosmids is that DNA fragments up to 45 kilobases can be cloned. Only DNA up to 25 kilobases in length can be cloned using plasmids or lambda. The efficiency of cloning with cosmids has not yet equalled that of plasmid or lambda phage cloning. Cosmids have been useful for constructing gene libraries from *E. coli*, yeast, *Drosophila*, and humans.

Some animal tumor viruses such as simian virus 40 (SV40) and retroviruses are also candidates as vectors. SV40 makes an ideal vector because of our detailed understanding of its organization and the availability of a complete nucleotide sequence. However, the small size of the virus and restrictions on the size of foreign DNA that can be inserted make its use limited. Retroviruses have the following properties, which make them ideally suited for gene transfer into mammalian cells: (1) the viral genome, which is RNA, is efficiently transmitted to recipient cells and is integrated into the host chromosome as DNA; (2) large amounts of DNA can be packaged into the viral genome, and (3) retroviruses have a wide host range. One drawback is their potential for causing malignancies in certain situations (see chapter 17).

Concept Check

1. What are the necessary steps in gene cloning?
2. What is the Southern blotting technique?
3. Why do plasmids, used as vectors in genetic engineering, carry two genes resistant to antibiotics?
4. What is a cosmid? How are they produced and what is their value?

One of the major discoveries that enhanced genetic engineering studies was the isolation of certain enzymes, the **restriction endonucleases**. Bacteria can protect themselves from invasion by viruses or other foreign DNA molecules by using restriction enzymes. Restriction endonucleases can cleave the foreign DNA and leave it with no function or structural integrity. The fact that the DNA of bacterial cells is left intact during growth and that many viruses are still capable of successfully infecting bacteria indicates that there are mechanisms operating that can protect certain DNA molecules, even foreign ones. DNA can be protected from endonuclease digestion by methylation of the purine and pyrimidine bases that the endonuclease normally selects for cleavage.

Restriction enzymes are divided into types I, II, and III. **Type II** cleaves DNA at or very close to a defined recognition sequence. This is the reason why type II restriction endonucleases are used for genetic engineering studies. Most type II enzymes recognize and act on a particular tetranucleotide or hexanucleotide sequence. Many of the recognition sites contain a two-fold axis of symmetry. For example, in the sequence below, rotation 180 degees around the dot (axis of symmetry) leaves the sequence unchanged. Thus the sequences are palindromic, or contain inverted repeats (also see Chapter 8).

5′-G-T-C-Py-Pu-C-A-G-3′

●

3′-C-A-G-Pu-Py-G-T-C-5′

Another feature of restriction enzymes is the position of bonds that are cleaved. Some restriction enzymes cut the DNA in such a way as to produce termini that are fully base paired (flushed ends). In the following example, *Hae* III restriction enzyme produces:

↓
5′—G p G p C p C—3′
3′—C p C p G p G—5′
↑

Other enzymes produce staggered cuts. For example, *Eco*RI enzyme produces termini whose structures are:

$5'_{p}$. . . .G $_{p}A_{p}A_{p}T_{p}T_{p}C$. . . .$3'_{OH}$
$3'_{OH}$. . . .$C_{p}T_{p}T_{p}A_{p}A_{p}$ G. . . .$5'_{p}$

The complementary ends of the fragment are cohesive ends. Consequently, any *Eco*RI-produced termini can base pair with any other *Eco*RI-produced end. The recognition and cleavage sites of some restriction enzymes are outlined in Table 11-2.

The number of cleavage sites in DNA depends on the source of the DNA as well as the restriction enzyme used. For example, the genome of SV40 contains only one cleavage site for *Eco*RI endonuclease, therefore cleavage of the viral genome produces a linear molecule with cohesive ends (Figure 11-23). The restriction enzyme *Hae*III produces over 50 cleavage sites on the lambda chromosome, thus yielding many fragments.

Restriction enzymes have other uses besides those associated with creating recombinants for commercial applications. Those other uses include:

1. Counting gene copies. It is known that some genes in a genome are not affected by certain restriction enzymes. When these enzymes digest the entire genome, they leave fragments containing the unaffectable gene intact (assuming more than one copy exists). Application of the Southern blotting technique using a specific mRNA for the gene in question will reveal the approximate, if not actual, copy number.
2. Restriction mapping and sequencing. Small genomes such as plasmids, bacteriophages, and viruses can be digested by restriction enzymes. The fragments left are electrophoresed through agarose and their molecular weights determined. The fragments can be aligned by analyzing what has been digested. After the fragments have been positioned they can be sequenced. Normally a length of 100 bases can be sequenced on a single gel.
3. Detecting a genetic defect. Some genetic diseases are caused by simple mutation of a structural gene. If mutations either remove or produce a restriction enzyme site, then the site serves as a marker.

The formation of recombinant molecules using restriction enzymes may involve eukaryotic as well as prokaryotic DNA. Therefore it is possible to insert eukaryotic genes into either prokaryotic or eukaryotic vectors. This synthetic application of restriction enzymes would proceed in the following way. Foreign DNA containing many genes and plasmid DNA are treated with the same restriction enzyme and are then mixed together. Some of the foreign genes will form hydrogen bonds with the plasmid DNA. The enzyme DNA ligase is used to zipper up (produce covalent bonds between the phosphate and sugar groups) the plasmid and its foreign genes, thus producing recombinant DNA molecules. The plasmid being derived from more than one source, is often referred to as a **chimeric plasmid** (the chimera, a mythological creature, was part goat, lion, and serpent). At this point we have produced a number of plasmids containing different foreign genes.

But a problem occurs when double-stranded DNA to be cloned is synthesized from mRNA. This DNA possesses ends that are not complementary to those of the cloning vector, thus it cannot be incorporated as such into the vector. This problem can be avoided in many ways, one of which uses short pieces of double-stranded DNA that contain the cleavage sites for the restriction enzyme. These short pieces are called **synthetic linkers**, and they can be chemically attached to each end of the piece of DNA that is to be cloned. When DNA with synthetic linkers is treated with specific endonuclease, cohesive ends are generated by the enzyme cutting at the cleavage sites contained in the linkers.

Once the chimeric plasmid has been constructed, the trick is to put it into a viable host so that this new information can be expressed.

Table 11-2 *Recognition and cleavage sites of some important restriction endonucleases*

Enzyme	Source organism	Recognition site*
*Eco*RI	*Escherichia coli*	5′-G↓A-A-T-T-C- ● -C-T-T-A-A↑G-5′
*Eco*RII	*E. coli*	5′-↓C-C-T-G-G- ● -G-G-A-C-C↑-5′ (printed: 5′-G-C-C-T-G-G-C- / -C-G-G-A-C-C-G-5′)
*Hind*II	*Hemophilus influenzae*	5′-G-T-Py↓Pu-A-C- ● -C-A-Pu↑Py-T-G-5′
*Hind*III	*H. influenzae*	5′-A↓A-G-C-T-T- ● -T-T-C-G-A↑A-5′
*Hae*III	*H. aegyptius*	5′-G-G↓C-C- ● -C-C↑G-G-5′
*Hpa*II	*H. parainfluenzae*	5′-C↓C-G-G- ● -G-G-C↑C-5′
*Bam*I	*Bacillus amyloliquefaciens*	5′-G↓G-A-T-C-C- ● -C-C-T-A-G↑G-5′
*Bgl*II	*B. globiggi*	5′-A↓G-A-T-C-T- ● -T-C-T-A-G↑A-5′

*Dot indicates axis of symmetry.

Inserting vector foreign recombinant DNA into the host cell. The chimeric plasmids can be inserted into bacteria (although other recipients could theoretically be used) in a transformation process* (Figure 11-24). The uptake of plasmid DNA into the bacterial cell is fa-

*Introduction of viral DNA into a recipient cell is **transfection**.

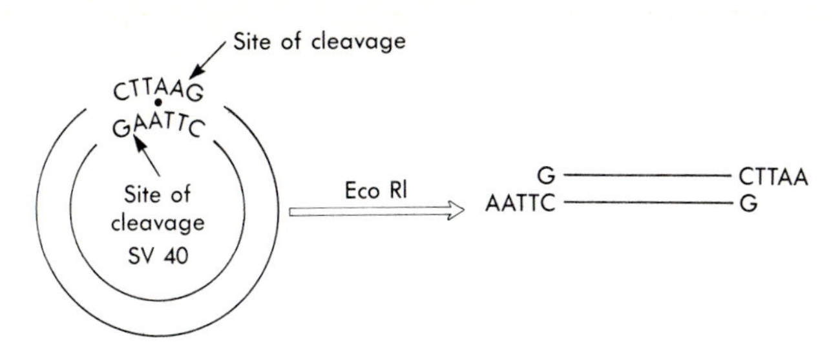

Figure 11-23 *Cleavage of SV40 DNA by the restriction enzyme* EcoRI *produces a linear DNA with cohesive ends. Black dot indicates point of symmetry.*

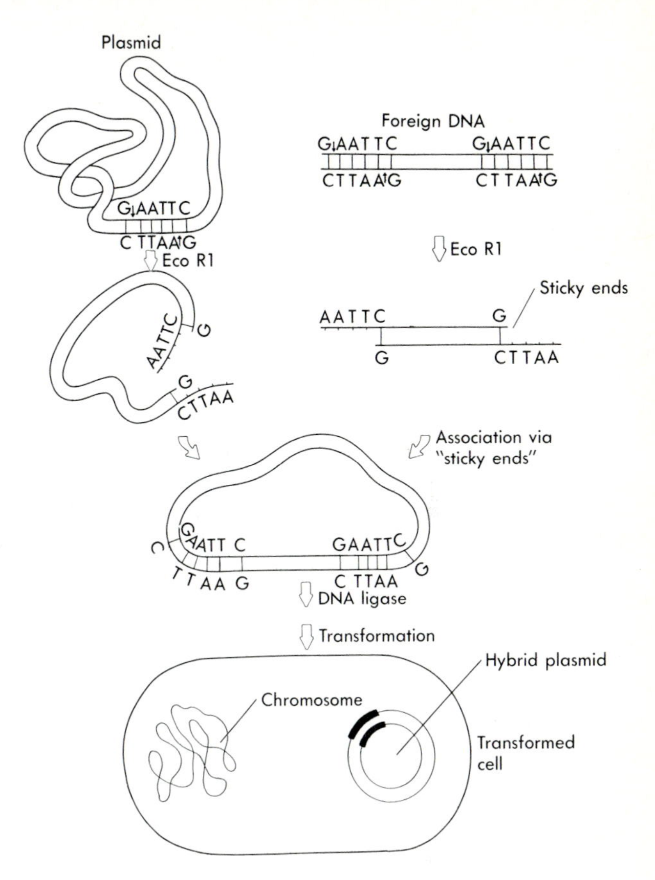

Figure 11-24 *Recombinant DNA experiment. Plasmid and foreign DNA are both treated with* EcoRI *restriction endonuclease. Fragments of both DNAs are produced with complementary sticky ends. Foreign gene with sticky ends will be incorporated into plasmid DNA. Gaps in plasmid are sealed with DNA ligase and circular molecule is produced. Recombinant DNA plasmid is then used to transform bacterial cell. Transformed cell can be cloned, then DNA information can be translated.*

cilitated by the presence of calcium chloride in the medium. Once the foreign genes have been transferred to the host bacterial cell, the cells are plated and cloned on an agar medium. We now have a number of bacterial cells containing recombinant DNA plasmids, and this presents us with a problem: selecting the specific recombinant.

Selecting recombinants. The population of cells containing recombinant DNA will be quite numerous, and some of these cells will have nonrecombinant plasmids that circularized spontaneously following endonuclease digestion and were subsequently ligated. Others will be recombinants that the experimenter will not want. How does the experimenter select a clone in which cells carry the specific gene that he wishes to study? First, he must select all of the recombinants. One technique is to initially use a plasmid that contains two drug-resistant markers, one of which has a specific restriction endonuclease site located within it. Three types of cells will be produced following the formation of recombinant DNA and the transformation process if we use the antibiotic-resistant markers: (1) those cells in which no plasmids appear and that are sensitive to both antibiotics; (2) those cells in which the foreign gene is inserted into the site originally occupied by the antibiotic-resistant marker (these recombinant cells will be sensitive to one of the antibiotics [Figure 11-25] and resistant to the other); and (3) those cells containing the plasmid but no recombinant DNA (these are resistant to both antibiotics). The second group will be evaluated for study.

Selecting cloned genes. Now that we have identified all the bacteria carrying recombinant DNA we have to identify those cells carrying the specific recombinant plasmid. If the gene being sought codes for an enzyme, colonies can be assayed for its presence, but identification is more difficult if the gene codes for a nonenzymatic protein. One technique uses immunological principles, in which antibodies to the protein are used to detect the presence of the protein in cloned cells. A second technique employs a radioactive probe (Figure 11-26). The experiment is carried out by replica plating all the clones obtained from the agar plate onto a nitrocellulose filter. The pattern of cells on the filter will mimic those on the agar plate. The cells are then exposed to a chemical (such as sodium hydroxide) that lyses the cells and denatures DNA. Denatured plasmid DNA remains attached to the filter following a baking process. The filter is placed in a solution containing a radioactive probe that consists of either ^{32}P-labeled mRNA specific for the gene, or a complementary DNA, either of which will hybridize with the denatured DNA. The filter is washed to remove unhybridized material, and a photographic emulsion is placed over the filter. Sequences homolo-

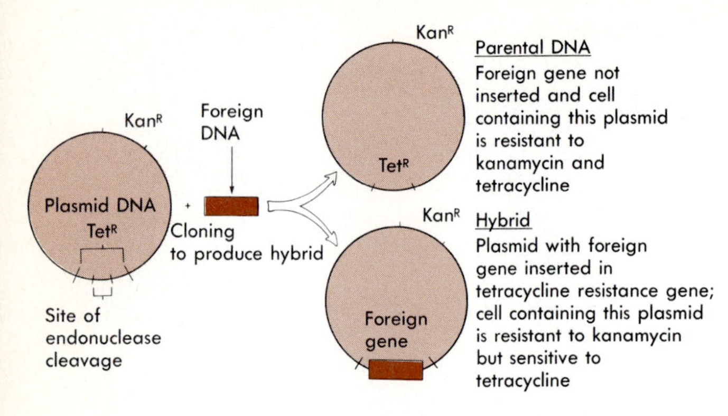

***Figure 11-25** Technique for selecting recombinants using antibiotic-resistant markers. Abbreviations are* Kan^R, *kanamycin resistance; and* Tet^R, *tetracycline resistance.*

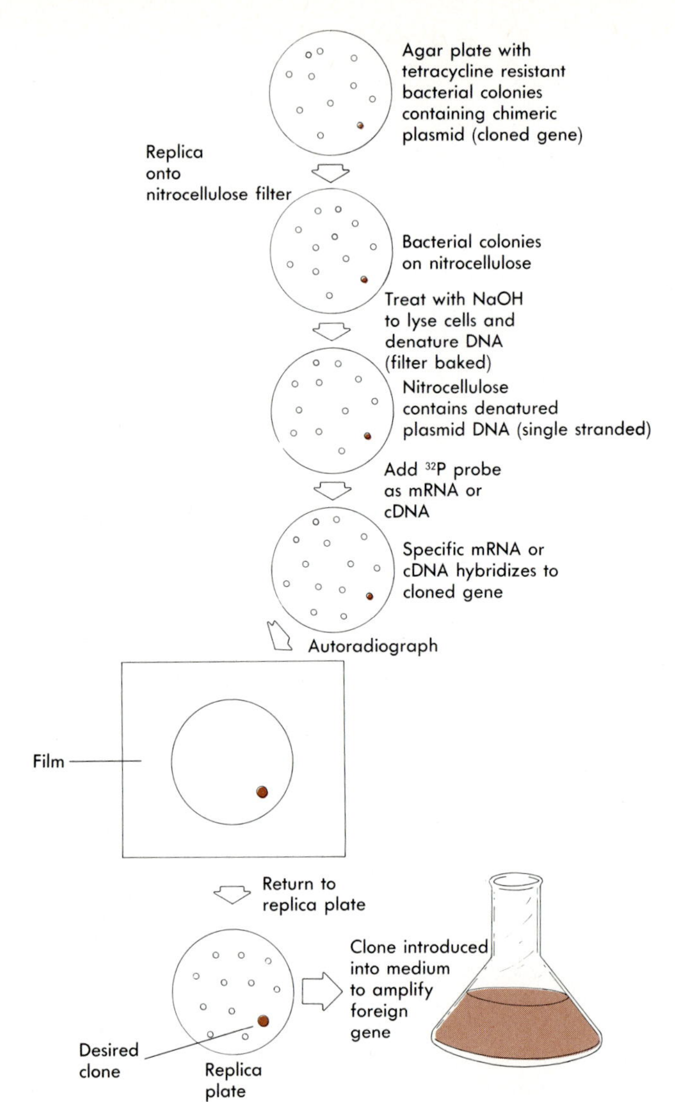

***Figure 11-26** Technique for locating a specific chimeric plasmid (cloned gene) in a bacterial colony. Colony with cloned gene has been preidentified and is in color.*

gous to the probe will hybridize with it and be detected by autoradiography. The position of the positive colony on the replica filter can then be compared with the master plate, and that colony can be recovered for further experimentation.

Radiolabeled DNA probes are unstable and have a limited shelf-life, therefore ^{32}P has a short half-life (14 days). In addition, the use of radiolabeled materials requires specially trained personnel and expensive disposal procedures. A nonradioactive technology has been developed in which biotin is attached to deoxyuridine triphosphate (dUTP). The **biotinylated derivative** substitutes for thymidine triphosphate in DNA-labeling reactions and hybridizes normally to complementary DNA. Hybridized biotinylated probes can be easily detected because of their strong affinity to the protein streptavidin. Streptavidin is coupled to a signal-generating system that catalyzes the synthesis of a colored product or fluorescent molecule. For example, the enzyme, horseradish peroxidase, has been complexed with streptavidin. When the proper substrate is added, a colored product is formed. The biotinylated probe has a shelf-life of 2 years or more.

Significance of recombinant DNA research

Recombinant DNA research has many applications, the most important of which is genetic analysis. Eukaryotic and prokaryotic genomes alike have been analyzed to determine the function of certain genes, as well as to map entire genomes such as those of certain viruses. One fundamental advantage of examining eukaryotic genes in a prokaryotic environment is that the genes are free of their eukaryotic regulatory elements. Thus gene function is more easily ascertained. However, one problem is that proper folding of some eukaryotic proteins is hampered in a prokaryotic cytoplasm (other problems are discussed on p. 291).

Recombinant DNA research also has great potential for humankind. It has been used for the purposes listed below, some of which will be referred to again in later chapters (see Chapters 20, 22, and 24).

1. Research is exploring formation of improved vaccines. By isolating specific microbial components rather than using the whole organism, it will be possible to develop vaccines that (1) are easier to prepare, (2) will not cause serious side effects, and (3) will induce an effective immune response in the host. A genetically engineered vaccine is now available for foot-and-mouth disease of cattle, and a vaccine for human viral hepatitis B has been approved.

Concept Check

1. What is the function of restriction endonucleases in the cell, and what is their commercial value? What types of cuts do they make in the DNA?
2. How are recombinants selected during gene cloning, and how can specific cloned genes be recognized?
3. What are the potential benefits of recombinant DNA technology?

To boost the immune respone of the host after exposure to various microbial components scientists have been testing vaccinia virus as a vector. Vaccinia virus, which was used to protect humans against smallpox before the disease was eradicated, is a safe virus whose DNA has a large capacity. Work has already begun on using vaccinia to express proteins that would protect against rabies and hepatitis B. Modified vaccinia virus so far appears to be very promising in protecting animals against such diseases as ovine foot rot, bovine virus diarrhea, and liver fluke disease—commercially important diseases that affect sheep and other livestock.

2. Scientists are studying the formation of proteins not normally synthesized by microorganisms. Mammalian genes coding for insulin, somatostatin (a hypothalmic hormone that controls the release of several hormones from the pituitary), and human growth hormone have been inserted as recombinant plasmids in *E. coli.* These genetically engineered products are currently on the market. Interferon is also a protein, produced normally by the body in respone to viral infections, that has been shown to restrict the growth of certain tumors and hairy cell leukemia. Interferon is produced in very small amounts in cell cultures but is produced in greater quantities in bacterial cells carrying the gene for interferon synthesis.
3. Microorganisms for agricultural use are being investigated. It may be possible to create plants resistant to a wide variety of pests by splicing foreign genes into plant chromosomes. This innovation will lead to decreased use of chemical fungicides and insecticides, most of which are toxic to humans. Genetic engineering is also being considered to increase legume yields. The microorganism to be studied is *Rhizobium,* which naturally infects leguminous plants. *Rhizobium* fixes nitrogen and supplies it to the plant, and the plant supplies the bacteria with nutrients. The normal nitrogen fixation process wastes energy by releasing hydrogen gas. Insertion of a gene coding for the enzyme hydrogenase into species of *Rhizobium* results in the breakdown of hydrogen into free protons (H^+) and electrons (e^-) that drive the reactions of photosynthesis and conserve energy.

   Useful additions to the plant genome may eventually be accomplished by using the bacterium *Agrobacterium tumefaciens. A. tumefaciens* causes crown gall tumors in plants by injecting a plasmid (T1) into the nuclear genome of the plant (see Chapter 20).

   Expression of bacterial genes in plants has also been accomplished. A bacterial gene was introduced into a tobacco plant and enabled the plant to show increased resistance to the herbicide glyphosate. This will help farmers to rid their fields of weeds without harming their crop plants (see Chapter 20).
4. Viruses are currently being investigated for the purpose of transferring genes into research animals to cure disease. It is predicted that in a few years it will be technically feasible to treat humans with certain rare inherited disorders using this technique of virus transfer (see Point of Interest).
5. Research includes developing a technique for producing site-specific mutagenesis. Mutations ordinarily occur randomly in the bacterial genome or within a particular gene, and it is difficult to produce specific mutations in the laboratory. Isolation of genes from clones offers an opportunity to produce point mutations (mutations involving a single base pair). The isolated gene can be treated in the laboratory with specific agents and then reinserted into the host cell and exchanged for the wild-type gene. This type of mutagenesis has tremendous application in industrial microbiology, where an increase in production of special compounds, such as amino acids, can be controlled by genetic means rather than random selection of mutants. In addition, important determinants, such as antibiotic resistance genes and virulence factors, can be analyzed in pathogenic organisms.

Benefits vs. fears of genetic engineering

The possible benefits of recombinant DNA research are innumerable, but it is possible to produce organisms with less than desirable attributes. Resistance to all known antibiotics, and toxin production in normally nonpathogenic microrganisms that are indigenous to humans, are just two examples of potentially harmful products of this research. Considerable controversy surrounded initial recombinant DNA research, and strict guidelines were instituted but have subsequently been relaxed.

Derivatives of the *E. coli* strain K12 have been developed with several mutational defects that ensure their own destruction outside the laboratory. Some of these mutations make the bacterium (1) very sensitive to ultraviolet light, (2) incapable of conjugating with other enteric organisms, (3) resistant to transducing phages, (4) very sensitive to bile salts and detergents, and (5) unable to replicate except under controlled laboratory conditions. But even though the fears of using *E. coli* in recombinant DNA research are no longer prevalent, genetic engineers are still searching for other organisms to manipulate. *Bacillus subtilis,* for example, a free-living soil bacterium, is considered to be a safer organism. This organism, unlike *E. coli,* ordinarily secretes its products, has no surface toxins, and is not infectious. But considerable research needs to be performed, elucidating the genetic control mechanisms of *B. subtilis,* before it can be substituted for *E. coli.*

SUMMARY

1. Nucleotide sequences in microbial systems can be reassorted, and this often (but not always) involves an exchange or recombinational event between two DNA molecules. Recombination takes place by one of three methods: general, site specific, and illegitimate. The differences between them are related primarily to the homology (relatedness) of the interacting DNA molecules.
2. General recombination is the most common type of recombination and involves considerable homology between two DNA molecules. In site-specific recombination there is homology only at certain sites on the interacting DNA molecules. Illegitimate recombination involves no apparent homology between the interacting sites of recombination.
3. Recombination can occur when genetic information is transferred from one cell to another in one of three processes: transformation, transduction, or conjugation.
4. In transformation, naked DNA (no viable cell is required) is taken up by a cell, and part of the DNA is incorporated into the recipient DNA.
5. In transduction, viral DNA is used to transfer bacterial DNA from one cell to another, resulting in an exchange of bacterial DNAs in the recipient cell. Sometimes the DNA transferred is mostly viral (specialized transduction), but usually it is bacterial (generalized transduction).

Point of Interest

VIRUSES AND GENE THERAPY

Most viruses basically consist of genetic material wrapped in a protective coat of protein. They require the metabolic processes and machinery of the host to replicate and survive. Many viruses infecting a cell direct the synthesis of more virus, which results in the lysis of the cell. The viruses being used for gene therapy are a special type of virus (called tumor or retroviruses) that, after infection, insert their viral genes into the host chromosome. Scientists engineered the virus so that the tumor-causing genes are removed and replaced with a functional version of an animal gene that is defective in the host.

One of the first experiments using such recombinant viruses was performed on mouse bone marrow cells. If one type of bone marrow cells, called stem cells, integrates the functional gene carried on the recombinant, then all the blood cells derived from the stem cells will carry the gene. This is possible because stem cells divide indefinitely. The mouse experiment was successful, and now scientists are sure that the same sort of success can be obtained in humans. Through a series of experiments, scientists have been able to construct a virus carrying functional genes that can be used to infect defective human cells, such as bone marrow. In human terms, the transfer of functional genes into stem cells will eventually be important in the treatment of hemolytic diseases such as sickle cell anemia, in which blood cell production is adversely affected. One of the problems confronting molecular biologists is the inability to control the fate of the genes that are transferred; that is, the site where the genes will be integrated. Inherited anemias are caused by defective globin genes, and, unfortunately, these types of genes are precisely regulated (regulation ensures the correct proportion of beta-globin and alpha-globin in the cell). Probably the best candidates for gene therapy are those diseases in which the genes are not tightly regulated and are expressed in many types of cells. Diseases such as Lesch-Nyhan syndrome might be influenced by gene therapy. People with Lesch-Nyhan syndrome are deficient in the gene that encodes hypoxanthine-guanine phosphoribosyl transferase (HGPRT). People with this untreatable disease are characterized by mental retardation, cerebral palsy, self-mutilation, and an accumulation of uric acid in the body. An additional reason that this disease may be amenable to gene therapy is that the cells carrying the missing gene have a competitive advantage over cells not carrying the gene.

Another way to introduce genes into animals is to use a retrovirus to infect embryos. It is hoped that the gene is expressed throughout the life of the animal, not just in the embryo.

Many scientific and ethical issues have to be addressed before treating humans. Any disastrous gene therapy attempt could arouse public outcry and seriously curb further development in the field.

6. In conjugation, a cell-to-cell interaction occurs with the transfer of DNA from a male donor (F^+) to a female recipient (F^-). The sex factor may be the only information transferred if the factor exists as a plasmid outside the chromosomal DNA. If the sex factor is part of the chromosome, all or part of the chromosomal DNA can be transferred to engage in recombination.
7. Several discoveries have led to the new science of genetic engineering, in which genes from a variety of sources can be cloned in a host cell that is either prokaryotic or eukaryotic. The genes are carried on vectors such as plasmids or bacteriophages and are removed from their sources by restriction enzymes. The restriction enzymes are also used to insert the genes into the vector. The vector and its new genes are then taken up by a bacterial cell in the process of transformation. The transferred genes within the new host act like plasmids and can be expressed and transferred to other bacterial progeny.

Self-study Quiz

Multiple choice

1. Which of the following characteristics is common to bacteriophage Mu, transposons, and *Salmonella typhimurium?*
   A. All three are viruses.
   B. All three can take part in transduction.
   C. All three have demonstrated replicative site-specific recombination.
   D. All three can act as cloning vectors.
2. Which of the following does *not* take place during transformation involving gram-positive bacteria?
   A. Double-stranded DNA is bound to the recipient's cell surface.
   B. DNA is transported into the cytoplasm in double-stranded form.
   C. Competence factors are produced by the culture.
   D. An eclipse complex protects single-stranded DNA during transport into the cytoplasm.
3. The sex factor is integrated into the bacterial chromosome in which type of cell?
   A. F^-
   B. F^+
   C. Hfr
   D. All donor cells
4. Transformation would be inhibited by which of the following?
   A. Prevention of cell-to-cell contact
   B. DNase
   C. RNase
   D. Competence factor
5. The transfer of the largest segment of bacterial chromosome information would be expected in which of the following?
   A. Conjugation of Hfr $\times$ F^-
   B. Abortive transduction
   C. Conjugation of F^+ $\times$ F^-
   D. Transformation
6. Long sequences of homologous DNA and rec genes are involved in which of the following?
   A. General recombination
   B. Site-specific recombination
   C. Transposition
   D. Illegitimate recombination
7. Which of the following mechanisms of gene transfer is least helpful in general mapping of the bacterial chromosome?
   A. Specialized transduction
   B. Conjugation
   C. Generalized transduction
   D. Transformation
8. Which of the following is most descriptive of a chimeric plasmid?
   A. Extrachromosomal element of DNA derived from more than one source
   B. The plasmid when it is integrated into the bacterial chromosome
   C. The DNA of a plasmid after it has been cleaved by a restriction endonuclease
   D. Another name for insertion sequence
9. Which of the following is not characteristic of the restriction endonuclease?
   A. Cleaves specific sites on both strands of the DNA molecule
   B. All cleave at the same palindromic sequence of nucleotides
   C. Can cleave both eukaryotic and prokaryotic DNA
   D. Are produced by various bacterial species
10. Why are nuclear DNA sequences of eukaryotic genes inappropriate to serve as a source of recombinant DNA?
    A. They are not based on the same genetic code as prokaryotic organisms
    B. They cannot be cleaved by restriction endonucleases
    C. They require synthetic linkers
    D. They contain intron base sequences that do not code for the amino acid sequence of the gene protein

Completion

1. The vehicle used to carry a gene for cloning is called a (an) ________________.
2. Plasmids containing antibiotic resistance genes as well as the cohesive end sites of lambda phage are called ________________.
3. The protein that promotes homologous pairing of a single strand with a double-stranded molecule of DNA during general recombination is called ________________.
4. During an Hfr × F^- conjugation the last genes that can enter the F^- cells are ________________.
5. The process in which a bacterial virus integrates into the host genome is called ________________.
6. When cloned DNA is not complementary to the DNA of the cloning vector, the problem can be solved by attaching ________________ to the cloned DNA.
7. The high frequency recombinant strain of bacteria in which the sex factor is integrated into the bacterial chromosome is called ________________.
8. The uptake of free DNA by a cell and its acquisition of new traits is called ________________.
9. The process of transfer of genetic information from one cell to another by cell-to-cell contact is called ________________.
10. The process in which temperate virus can transfer specific bacterial genes that are located near the viral integration site is called ________________.

Thought questions

1. List the methods and briefly describe a strategy for determining the gene map if you wanted to map the genes of a bacterium.
2. If recombinant DNA research and bacterial gene manipulation were allowed to be carried out indiscriminately and without careful control, what problems would you anticipate regarding the classification methods for bacterial species?
3. The following is a partial map of *E. coli* genes. Locate the point of integration of the sex factor in two different Hfr strains of *E. coli*.

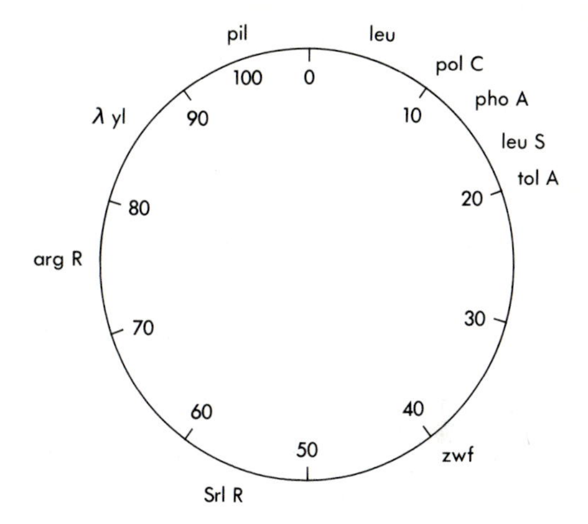

a. Hfr strain 1 when crossed with an F^- strain transferred the following markers in the sequence:

polC-3 min	pil-10 min	stlR-50 min
leu-5 min	argR-36 min	

b. Hfr strain 2 when crossed with an F^- strain transferred the following markers:

leu-8 min	phoA-15 min	tol A-23 min
polC-10 min	leuS-21 min	

Indicate on the above map where the sex factor is situated in the two strains and the direction of transfer.

SELECTED READINGS

Books

Booth, I.R., and Higging, C.F. (Eds.). *Regulation of Gene Expression—25 Years On*. Society for General Microbiology Symposium 39, Cambridge University Press, New York, 1986.

Freifelder, D. *Molecular Biology* (2nd ed.), Jones and Bartlett Publishers, Inc., Boston, 1987.

Hawkins, J.D. *Gene Structure and Expression*. Cambridge University Press, New York, 1985.

Hendrix, R.W., Roberts, J.W., Stahl, F.W., and Weisberg, R.A. (Eds.) *Lambda II*. Cold Spring Harbor Laboratory, Cold Spring Harbor, N. Y., 1983.

Watson, J., Tooze, J., and Kurtz, D.T. *Recombinant DNA—A Short Course*. Scientific American Books, New York, 1984.

Wilson, J.H. (Ed.) *Genetic Recombination*. Benjamin-Cummings Publishing Co., Menlo Park, Cal., 1985.

Journal Articles

Grindley, N.D.F., and Reed, R.R. Transpositional recombination in prokaryotes. *Annu. Rev. Biochem*. **54**:863, 1985.

Ippen-Ihler, K.A., and Minkley, E.G., Jr. The conjugation system of F1, the fertility factor of *Escherichia coli*. *Annu. Rev. Genet*. **20**:593, 1986.

Kucherlapati, R.S. Introduction of purified genes into mammalian cells. *ASM News* **50**:49, 1984.

Morse, M.L., Lederberg, E.M., and Lederberg, J. Transduction in *Escherichia coli* K-12. *Genetics* **41**:142, 1956.

Nash, H.A. Integration and excision of bacteriophage lambda: the mechanism of conservative site specific recombination. *Annu. Rev. Genet*. **15**:143, 1981.

Stahl, F.W. Genetic recombination. *Sci. Am*. **256**(2):90, 1987.

Stewart, G.J., and Carlson, C.A. The role of natural transformation. *Annu. Rev. Microbiol*. **40**:211, 1986.

Willets, N., and Wilkins, B. Processing of plasmid DNA during bacterial conjugation. *Microbiol. Rev*. **48**:24, 1984.

Vane, J., and Cuatrecasas, P. Genetic engineering and pharmaceuticals. *Nature* **312**:303, 1984.

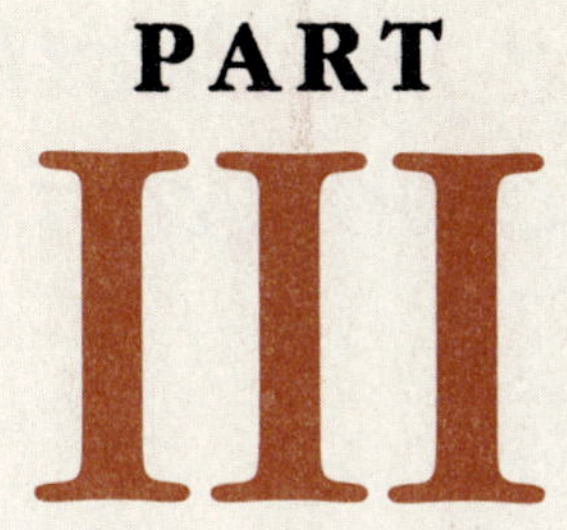

Part III: *Microbial metabolism and growth*

CHAPTER 12

Energy production in microorganisms

Chapter Outline

Introduction

Energy, which can be defined as the capacity to do work, may exist in various states, such as potential, radiant, thermal, and electrical. Survival in the biological world requires energy for certain cellular functions—transport of solutes across the cytoplasmic membrane, movement, and biosynthesis. The energy to perform these functions can be obtained by the cell in two ways. First, organisms called **phototrophs** convert the radiant energy from sunlight into chemical energy. Second, some organisms called **chemotrophs** transfer energy from preformed organic or inorganic molecules into specific energy molecules.

This chapter's functions are to describe how the laws of thermodynamics apply to an organism's capacity to derive and use energy from metabolic processes; second, to describe the different techniques that a microorganism can call on to obtain energy from chemical reactions; and third, to describe how energy is produced by two nutritional types of microorganisms—those that obtain their carbon for biosynthetic processes from organic compounds (**heterotrophs**) and those that obtain their carbon for biosynthesis from carbon dioxide (**autotrophs**).

Table 12-1 *Heats of combustion (ΔH_c) of some cellular fuels*

Fuel	Molecular weight	ΔH_c (cal/mole)	Caloric value (kcal/g)
Carbohydrate			
Glucose	180	−673,000	3.74
$CH_3—CH(OH)—COOH$ Lactic acid	90	−326,000	3.62
Lipids			
$CH_3(CH_2)_{14}COOH$ Palmitic acid	256	−2,380,000	9.30
Amino acid			
$H—CH(NH_2)—COOH$ Glycine	75	−234,000	3.12

Kcal, kilocalories.

LAWS OF ENERGY TRANSFORMATION (THERMODYNAMICS)

Energy transformations, which involve the production or expenditure of energy, are the basis of the science of thermodynamics. Every chemical or physical event that involves an exchange of energy between the system (such as a chemical reaction) and the environment outside the system is governed by two laws of thermodynamics. First, the total energy of the universe remains constant, or the energy gained by one reaction is lost by another, and vice versa. Second, only part of the energy released in a reaction can be used for work. The usable energy obtained from a chemical reaction is **free energy,** and as it declines there is an increase in the amount of nonusable energy, or **entropy**. Reactions in which two compounds, A and B, are converted to products C and D can be described thermodynamically, because there is a change in their energy contents. Gibbs explained the changes in the energy content of such a system with the following formula:

$$\Delta G = \Delta H - T\Delta S$$

Let us examine each of the components of the Gibbs formula: ΔG refers to the change (Δ) in free energy of a system or the energy that can be obtained from a chemical reaction to do work. ΔG is the difference in free energy between products and reactants, or

$$\Delta G = (G_C + G_D) - (G_A + G_B)$$

The sign and magnitude of ΔG are influenced by the values of ΔH and $T\Delta S$. In most instances the ΔH is nearly equal to the ΔG, and the two terms are often used interchangeably. ΔG^0 is the standard free energy change when the initial concentration of all components is 1.0 M, the pH is 7.0, and the temperature is 25° C. Every chemical reaction has a standard free energy change or ΔG^0.

ΔH refers to the heat that is transferred between the reaction system and its surroundings, or the **enthalpy** change. The enthalpy change can be determined from the characteristic heats of combustion (ΔH_c) of organic molecules when the latter are burned in the presence of oxygen (Table 12-1). The heat given up to the surroundings when the organic molecule is burned in oxygen is expressed in calories (a **calorie** is the amount of energy required to raise the temperature of 1 g of water from 14.5° to 15.5° C). The ΔH is equal to the difference between the sum of the ΔH of the products and the sum of the ΔH of the reactants, or ΔH (products) − ΔH (reactants). Let us use the reversible reaction A + B⇌C + D and assume three different conditions in which the ΔH is different from products and reactants:

A + B	⇌ C + D	ΔH FOR REACTION
Reaction 1		
ΔH = 1000 cal/mole	ΔH = 900 cal/mole	+100 cal/mole
Reaction 2		
ΔH = 900 cal/mole	ΔH = 1000 cal/mole	−100 cal/mole
Reaction 3		
ΔH = 900 cal/mole	ΔH = 900 cal/mole	0 cal/mole

The positive ΔH for reaction 1 means that for the reaction to occur energy will have to be put into the system, that is, the reaction is **endergonic**. The negative ΔH of reaction 2 means that heat is given off, that is, the reaction is **exergonic**. Reaction 3 in which the ΔH is 0 represents a state of equilibrium, and no net chemical change takes place.

In TΔS, T represents the absolute temperature expressed in degrees Kelvin (subtract 273° from degrees Kelvin to obtain degrees Celsius) of the reaction, and ΔS is the change in entropy. Entropy is a measure of the randomness of a system or simply the energy in a system that tends to go downhill and become randomized. TΔS represents that part of the total energy change that is lost to the system in the form of random molecular motions and is therefore not available to do work. In most biochemical reactions in a cell the TΔS is usually small in comparison with the quantities for ΔG and ΔH, which are negative in those reactions that yield energy. The oxidation of glucose, for example, by molecular oxygen can be represented in the following way:

$$\text{Glucose} + 6O_2 \rightarrow 6CO_2 + 6H_2O$$

where

ΔG° = − 686,000 cal/mole (−686 kcal/mole or −2870 kJ/mole*
ΔH = −673,000 cal/mole (−673 kcal/mole or −2815 kJ/mole
ΔS = +43.6 cal/mole-degree or 182 J/mole-degree

When we insert these values into the equation ΔG° = ΔH − TΔS we obtain

* kJ = kilojoules. There are 4.184 kj in 1 kcal. The joule has been recommended as the international unit of energy: however, the calorie is still observed in the literature.

$$\Delta G^\circ\ (-686{,}000 \text{ cal/mole}) = \Delta H\ (-673{,}000, \text{cal/mole}) - T\Delta S\,[298^\circ(+43.6 \text{ cal/mole-degree})]$$

or

$$-686{,}000 \text{ cal/mole} = -685{,}993 \text{ cal/mole}$$

THERMODYNAMICS AND METABOLISM

Metabolism is the sum total of all the chemical reactions that occur in the cell and, like other chemical reactions, is governed by the laws of thermodynamics. Thousands of metabolic reactions occur in the cell; some are exergonic and others endergonic. Exergonic reactions are those processes that release free energy, for example, oxidations involving carbohydrates, fats, and proteins, as well as hydrolytic reactions involving energy-rich molecules such as ATP. Endergonic reactions are processes that require the addition of energy, for example, flagellar movement, nutrient transport, and reactions that are involved in the synthesis of amino acids, lipids, and carbohydrates.

A chemical reaction will reach a final equilibrium point, that is, a point at which no further chemical change takes place in the reaction, and the rate of conversion of reactants A and B to products C and D will be balanced by the rate of conversion of C and D to A and B. This can be expressed as an equilibrium constant, K_{eq} or:

$$K_{eq} = \frac{\text{Product of concentrations of products}}{\text{Product of concentrations of reactants}} = \frac{[C][D]}{[A][B]}$$

K_{eq} is a function of the free energy change of the components of the reaction and is related to it in the following equation:

$$\Delta G^{\circ\prime} = -RT \ln K_{eq}$$

where $\Delta G^{\circ\prime}$ is the standard free energy in which 1 mole of reactant is converted to 1 mole of product under standard conditions of temperature and pressure. R is the gas constant (1.987 cal/mole-degree), T is the absolute temperature, and $\ln K_{eq}$ is the natural log of the equilibrium constant. If, for example, the equilibrium constant is high (1000), it means that more product has been released, the reaction tends to go to completion, and the free energy change is negative. If the equilibrium constant is low (0.001), it means that little product is formed, the reaction does not go far to the right, and the free energy change is positive. If the equilibrium constant is low, converting 1 mole of reactant to 1 mole of product requires that energy be put into the system. High-energy compounds, such as ATP, are responsible for supplying most of the energy required for ender-

Table 12-2 *Energy-rich compounds found in microbial cells*

Compound	General formula for energy linkage	$\Delta G°$ (cal/mole)
Phosphoenolpyruvate	$R-C(=CH_2)-O\sim P(=O)(O^-)-O^-$	−14,800
Adenosine triphosphate (ATP)	$R-O-P(=O)(O^-)-O\sim P(=O)(O^-)-O\sim P(=O)(O^-)-O^-$	−7300 (for terminal and adjacent phosphate)
Acetyl CoA	$R-C(=O)\sim SH$	−10,500
Acetyl phosphate	$R-C(=O)-O\sim P(=O)(O^-)-O^-$	−10,100

gonic reactions. Hydrolysis of the terminal phosphate of ATP releases approximately −7300 cal/mole of free energy. Other energy-rich compounds that may be found in microbial cells are described in Table 12-2.

The $\Delta G°'$ of metabolic reactions is a thermodynamic concept and refers only to the potential of the reaction occurring. It gives no information about the rate. Even though a reaction has a high negative free energy and proceeds spontaneously, it may take years for the reaction to reach its final equilibrium point. The rate at which the reactants are converted to products is influenced by enzymes discussed in Chapter 2.

HOW CELLS DERIVE ENERGY FROM CHEMICAL REACTIONS

An actively growing cell carries out hundreds of chemical reactions that require energy in the form of ATP. The mechanisms that enable the cell to trap energy in the ATP molecule are discussed below.

Oxidation and reduction

The chemical bonds of inorganic, as well as organic, molecules possess potential energy that can be extracted by living organisms. Whether these potential energy sources can be used depends on the microbial species. The sources of energy and the sources of carbon used by microorganisms varies among bacteria, and this allows us to devise a nutritional classification scheme (Table 12-3). Energy is extracted from chemicals in a process called **oxidation**. Oxidation was originally intended to characterize inorganic reactions involving the element oxygen. In terms of atomic structure, however, oxidation means the transfer of electrons from one atom to another. Electrons may be transferred directly or they may be transferred in the form of hydrogen atoms to another atom (remember that the hydrogen atom consists of a proton [H^+] and an electron [e^-]. In oxidations involving organic molecules electrons are transferred in the form of hydrogen atoms, and the process is referred to as **dehydrogenation**. In the oxidation described here a hydrogen donor AH_2 transfers its electrons in the form of hydrogen atoms to the electron acceptor B:

$$AH_2 + B \rightarrow BH_2 + A$$

In any metabolic reaction in which a molecule is oxidized, another molecule is also reduced and in the preceding example B becomes reduced to BH_2. Depending on the nature of the reaction oxidations may be the addition of oxygen, loss of electrons, or loss of hydrogen atoms. Conversely, reduction means the loss of oxygen, gain of electrons, or gain of hydrogen. One example of a typical oxidation in bacteria, as well as other forms of life, involves the conversion of lactic acid to pyruvic acid. If this reaction is carried out in a test tube with the appropriate enzyme, the reaction could be expressed:

Table 12-3 *Nutritional classification of microorganisms based on carbon and energy sources*

Nutritional type	Carbon source	Energy source	Examples
Chemoautotroph*	Carbon dioxide	Chemicals	Hydrogen bacteria, colorless sulfur bacteria, iron bacteria, nitrifying bacteria, CO oxidizing
Chemoheterotroph	Organic compounds	Chemicals	Animals (including humans), most bacteria, fungi, and protozoa
Photoautotroph	Carbon dioxide	Light	Green plants, most algae, purple and green bacteria
Photoheterotroph	Organic compounds	Light	Some cyanobacteria, some algae, some purple and green bacteria

*Also called chemolithotrophs. All chemoautotrophs are chemolithotrophs but not all lithotrophs are autotrophic because some can use organic carbon.

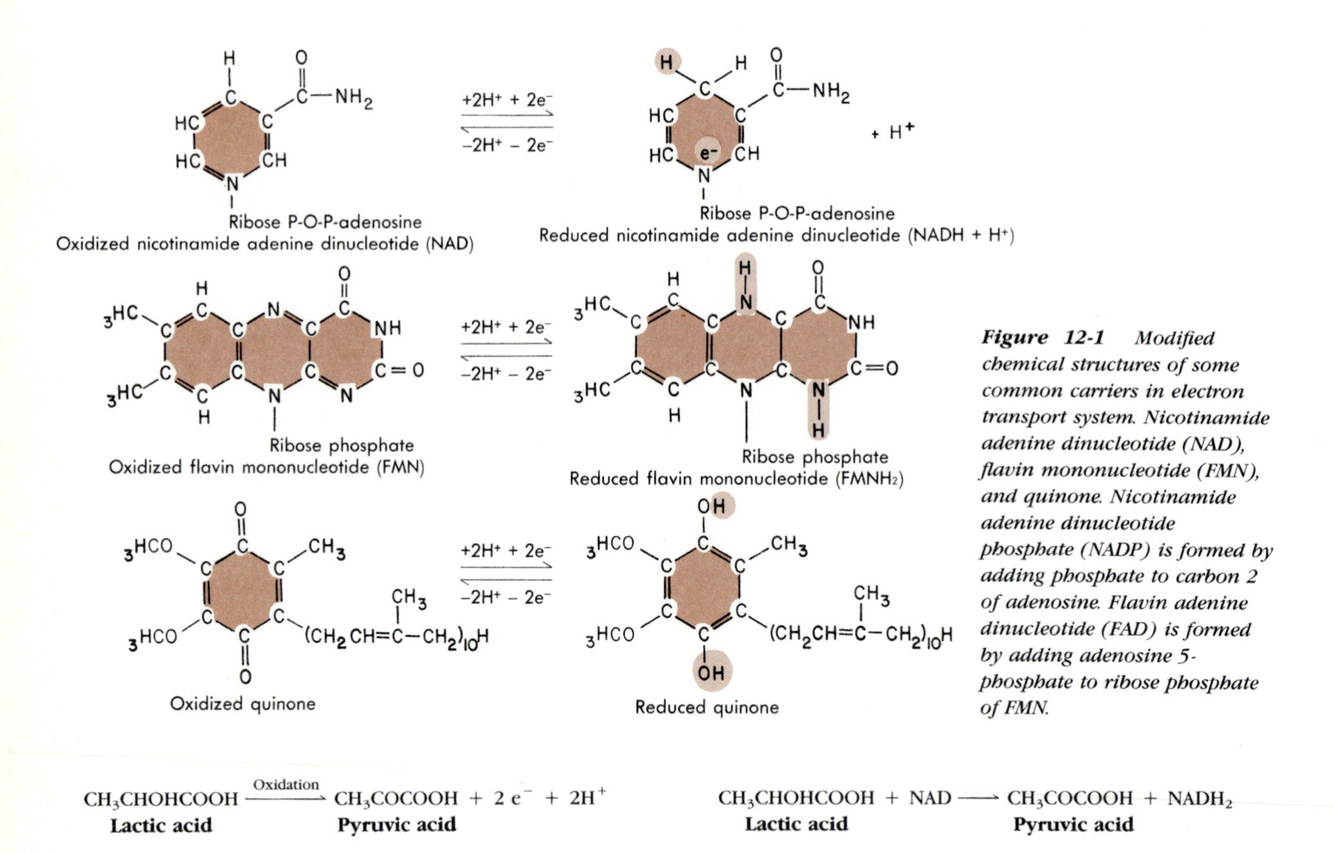

Figure 12-1 *Modified chemical structures of some common carriers in electron transport system. Nicotinamide adenine dinucleotide (NAD), flavin mononucleotide (FMN), and quinone. Nicotinamide adenine dinucleotide phosphate (NADP) is formed by adding phosphate to carbon 2 of adenosine. Flavin adenine dinucleotide (FAD) is formed by adding adenosine 5-phosphate to ribose phosphate of FMN.*

$$\underset{\textbf{Lactic acid}}{CH_3CHOHCOOH} \xrightarrow{\text{Oxidation}} \underset{\textbf{Pyruvic acid}}{CH_3COCOOH} + 2\,e^- + 2H^+$$

This oxidation releases two hydrogen ions and two electrons with the formation of pyruvic acid; however, the reduction process is not indicated. When this oxidation process is expressed in terms of reduction and oxidation, the hydrogen atoms released are used to reduce a carrier molecule, such as nicotinamide adenine dinucleotide (NAD), which is one of the principal hydrogen carriers in the cell (Figure 12-1). The oxidation of lactic acid in the cell is therefore best illustrated in the following way:

$$\underset{\textbf{Lactic acid}}{CH_3CHOHCOOH} + NAD \longrightarrow \underset{\textbf{Pyruvic acid}}{CH_3COCOOH} + NADH_2$$

As we shall see, it is during oxidation reduction reactions that energy is released for the eventual formation of ATP. There are three biological processes in which ATP is produced: **substrate phosphorylation**, **electron transport phosphorylation**, and **photophosphorylation**.

Mechanisms of ATP formation

The oxidation of food material by the cell releases energy that is used to produce ATP. ATP is produced in a process called phosphorylation in which inorganic phosphate is enzymatically added to adenosine diphosphate (ADP):

$$\text{ADP + Inorganic phosphate (iP)} \xrightarrow[\text{oxidation of nutrients}]{\text{Energy from}} \text{ATP}$$

The energy released from the oxidation of nutrients in the cell is stored by incorporating it into the phosphoanhydride bond of the ATP molecule. When energy is required for various cellular processes the ATP molecule can serve as a source of this energy because of its high free energy of hydrolysis. The ATP molecule on hydrolysis has a ΔG^0 of -7.3 kcal/mole as does the ADP molecule:

$$\text{ATP} + H_2O \rightleftharpoons \text{ADP} + \text{iP} \quad \Delta G^0 = -7.3 \text{ kcal/mole}$$
$$\text{ADP} + H_2O \rightleftharpoons \text{AMP} + \text{iP} \quad \Delta G^0 = -7.3 \text{ kcal/mole}$$

Adenosine monophosphate (AMP) is considered a "low energy" molecule because its hydrolysis yields:

$$\text{AMP} + H_2O \rightleftharpoons \text{Adenosine} + \text{iP} \quad \Delta G^0 = -3.4 \text{ kcal/mole}$$

Before we consider the three processes in which ATP is formed in the cell we must realize a fundamental characteristic of cellular oxidation. When food material is completely oxidized the final products are carbon dioxide and water. This oxidation does not occur all at once but takes place in a series of reactions in which the components of the hydrogen atom are transferred between one molecule and another.

Concept Check

1. **What is the function of each component of Gibb's formula?**
2. **What is the equation that relates the equilibrium constant to the free energy change of the components of a chemical reaction? What does a high or low equilibrium constant indicate?**
3. **Why does oxidation result in the release of energy?**
4. **What are the major energy molecules of the cell? Why do they possess this energy?**

Substrate phosphorylation

Some intermediate compounds produced during the oxidation of organic compounds carry phosphate groups that react directly or indirectly with ADP to produce ATP. This type of phosphorylation is called **substrate level phosphorylation** because the high-energy phosphate group of the substrate is transferred directly to ADP. One example of this occurs during the oxidation of glucose (see p. 324) in which one of the high-energy phosphate groups of phosphoenolpyruvate (PEP) is used to phosphorylate ADP:

$$\text{PEP} + \text{ADP} \rightarrow \text{Pyruvate} + \text{ATP}$$

Substrate level phosphorylation is not the most efficient mechanism in the cell for obtaining energy.

Electron transport phosphorylation (oxidative phosphorylation)

Hydrogen atoms released during oxidation are picked up by electron carrier molecules, such as **nicotinamide adenine dinucleotide (NAD), nicotinamide adenine dinucleotide phosphate (NADP), flavine adenine dinucleotide (FAD)**, or **flavin mononucleotide (FMN)**. The carriers become reduced by accepting two protons and two electrons. Actually NAD and NADP accept only one hydrogen atom at a time, but for convenience they are represented as carrying two hydrogens (Figure 12-1):

$$2\text{H} + \text{NAD or NADP, FAD, or FMN} \rightarrow \text{NADH}_2\text{, NADPH}_2\text{, FADH}_2\text{, or FMNH}_2$$

The pair of electrons in the reduced carrier molecules are at a particular energy level and their removal can result in the release of a large quantity of energy. A sudden burst of energy usually results in the loss of a substantial fraction of this energy as heat, which could be harmful to the cell. If reduced NAD were directly oxidized by oxygen, the amount of free energy released would be:

$$\text{NADH} + H^+ + \tfrac{1}{2}O_2 \rightarrow \text{NAD} + H_2O$$
$$\Delta G^0 = -52{,}440 \text{ cal/mole}$$

The 52,440 calories could not be captured all at once by a single ADP molecule. The cell has evolved a mechanism for liberating only small quantities of energy in several distinct steps. Electrons are transferred along a chain of components called the **electron transport system** (also called **respiratory chain**), which is located in the cytoplasmic membrane of bacteria. The respiratory chain in eukaryotes is located in the mitochondrion. Each of the components in the chain is alternately reduced and oxidized as electrons are transported until a final acceptor molecule is reached (Figure 12-2). This final electron acceptor is oxygen if the organism grows in the presence of air or it may be an inorganic molecule other than oxygen, such as nitrate (NO_3^-), or an organic molecule if the organism grows in the absence of air.

Each of the electron donors in the respiratory chain

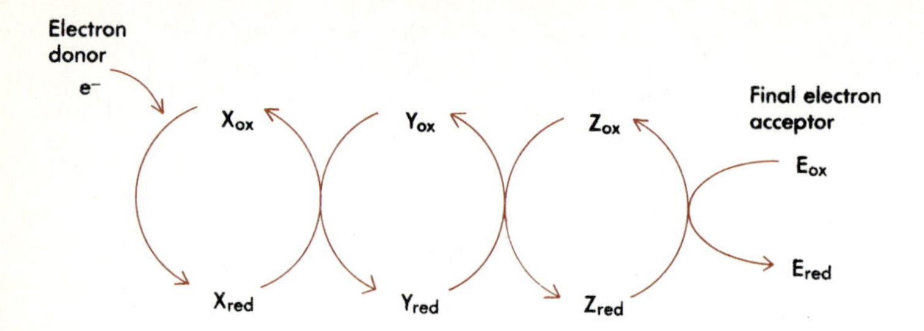

Figure 12-2 *Basic scheme used to demonstrate electron transfer in which there is cyclic oxidation and reduction of components X, Y, and Z.*

has a characteristic standard oxidation-reduction potential (E_o', expressed in volts), which represents their relative electronegativity (Table 12-4). Electrons tend to travel from a more negative carrier to a more positive one, and the decline in free energy between adjacent respiratory components is small enough that energy is released in small increments. The amount of energy released is related to the magnitude of difference in the oxidation-reduction potentials of the adjacent respiratory chain components.

The electron transport system consists of protein and nonprotein components that have oxidation-reduction potentials between those of NAD and oxygen (Figure 12-3). The protein components are flavoproteins, iron-sulfur (FeS) proteins, and cytochromes, whereas the nonprotein component is a quinone. **Flavoproteins** are enzymes (dehydrogenases) composed of a protein component and a yellow component that gives rise to the name "flavo." The yellow constituent is FAD or flavin mononucleotide FMN. The primary dehydrogenases include NADH dehydrogenase, succinate dehydrogenase, formate dehydrogenase, lactate dehydrogenase, and others. Not all of them will be present in any one cell type. In addition, the presence and concentration of the different dehydrogenases are regulated by the growth conditions, such as the presence or absence of oxygen or the presence of specific carbon sources (lactate, succinate, formate, etc.) in the medium. **Iron-sulfur proteins** are nonheme proteins associated with dehydrogenases, such as NADH dehydrogenase and succinate dehydrogenase. They accept only electrons and transfer them to the next component in the respiratory chain. **Quinones** (Q) are lipid-soluble molecules belonging to a class of compounds called **isoprenoids** (see figure in Point of Interest on p. 319). Quinones can accept hydrogen atoms. **Cytochromes** are iron-containing electron carriers called heme proteins. (The hemoglobin of blood, for example, is a heme protein.) There are several types of cytochromes with two to four of them appearing sequentially in an electron transport chain. Each cytochrome receives electrons from electron carriers and donates them to the next cytochrome in the sequence. Electrons received by cytochromes reduce the iron component from its ferric state to the ferrous state.

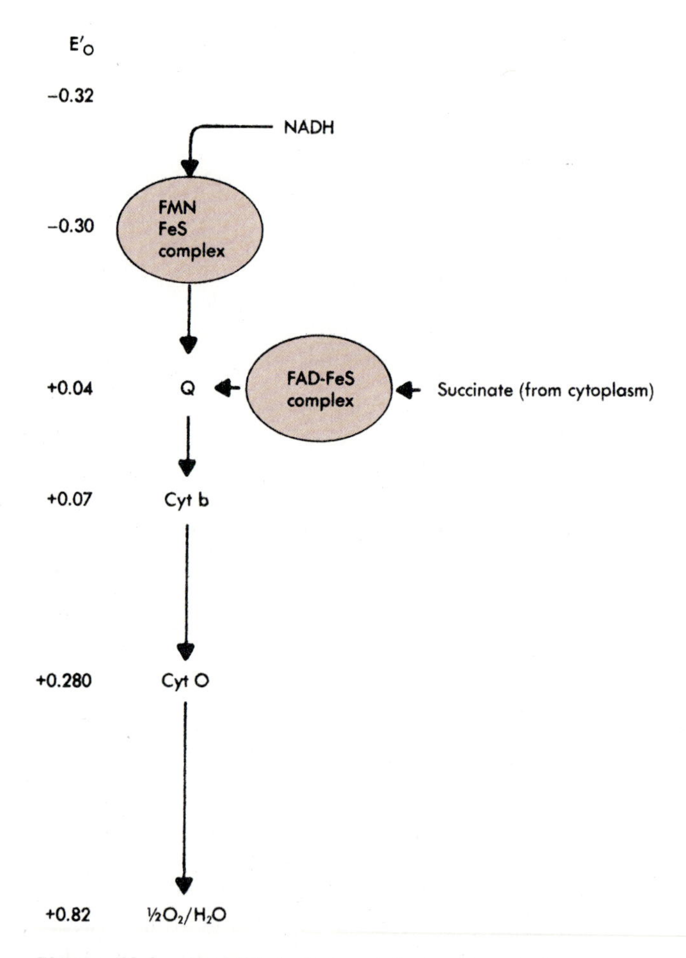

Figure 12-3 *Simplified scheme of electron transport components associated with aerobic respiration in bacteria* (E. coli). *Oxidation-reduction potentials (E_o') of electron transport components* (left). *FMN-FeS complex represents flavoprotein or NADH dehydrogenase. FAD-FeS complex represents flavoprotein or succinate dehydrogenase. Substrates other than succinate, such as D-lactate and L-alpha glycerophosphate, may enter at quinone (Q) under aerobic conditions. In bacteria cytochromes are not always linear and branched pathways exist. It should be pointed out that scheme presented here is abbreviated. At least two b type cytochromes have been purified from cytoplasmic membrane of* E. coli *grown under aerobic conditions. In addition, more than one quinone may also be involved.*

Table 12-4 *The oxidation-reduction potential ($E^{o\prime}$, expressed in volts) of some redox pairs in microbial systems*

Redox pairs	E_o' or oxidation-reduction potential (volts)
CO_2/formate	−0.432
H^+/H_2	−0.420
NAD/NADH	−0.320
S^o/HS^-	−0.280
FAD/FADH	−0.22
FMN/FMNH (flavoprotein)	−0.19
SO_3^-/S	−0.11
Fumarate/succinate	+0.03
Ubiquinone ox/ubiquinone red	+0.04
Cytochrome b ox/cytochrome b red	+0.07
Cytochrome c_1 ox/cytochrome c_1 red	+0.23
Cytochrome o ox/cytochrome o red	+0.280
Cytochrome a_3 ox/cytochrome a_3 red	+0.385
NO_3^-/NO_2^-	+0.43
Fe^{+++}/Fe^{++}	+0.770
O_2/H_2O	+0.82

The numerical values for the oxidation-reduction potentials reflect the reduction potential relative to the $H^+ + e^- \rightarrow \frac{1}{2} H_2$ half reaction, which is −0.42 volts at pH 7.0. A complete oxidation-reduction reaction is composed of two half reactions, one in which a substance gains electrons and one in which a substance loses electrons. The E^o, therefore, represents the voltage required to remove an electron from a substance. One can calculate the free energy change that occurs during the transfer of a pair of electrons between two redox pairs when the $E^{o\prime}$ values are known. The equation representing this change is $\Delta G^{o\prime} = nF \Delta E^{o\prime}$, where $\Delta G^{o\prime}$ is the standard free energy at pH 7.0; n is the number of electrons transferred; F is a faraday, a constant equal to 23,000 cal/volt; and $\Delta E^{o\prime}$ is the difference in volts between the two redox pairs. For example, the free energy potential between NAD/NADH and O_2/H_2O during the transfer of electrons from NAD to oxygen follows.

$$\begin{aligned}\Delta G^{o\prime} &= -2 \times 23{,}000 \text{ cal/volt} \times (-0.320) - (+0.82) \\ &= -46{,}000 \text{ cal/volt} \times 1.14 \text{ volts} \\ &= -52{,}440 \text{ cal/mole}\end{aligned}$$

$$Fe^{+++}(\text{ferric}) \xrightarrow{e^-} Fe^{++}\ (\text{ferrous})$$

We shall discuss the electron transport process as it occurs in bacterial systems in which oxygen is the final electron acceptor (Figure 12-4). The redox carriers in bacteria are of the same basic type as those found in the mitochondria of eukaryotes except for some minor variations in their composition and number. The following transport scheme is that found in *E. coli*:

1. Many dehydrogenases in the cell transfer hydrogen atoms (reducing equivalents) from their substrates to NAD. The reducing equivalents carried by NAD are transferred to the prosthetic group (FMN) of the flavoprotein (or flavin-linked dehydrogenase):

$$NADH_2 + FMN \xrightarrow[\text{dehydrogenase}]{\text{NADH}} FMNH_2 + NAD$$

The iron atoms of the FeS-protein that are associated with NADH dehydrogenase undergo their $Fe^{3+} = Fe^{2+}$ cycles and thus are specifically involved in electron transfer.

2. The reduced NADH dehydrogenase of step one donates its reducing equivalents via the FeS centers to the next respiratory chain component, quinone:

$$FMNH_2 + Q \rightarrow QH_2 + FMN$$

The free energy released during the transfer of reducing equivalents from NAD to quinone is sufficient to couple it to the synthesis of one ATP molecule. Succinate dehydrogenase can also feed into the quinone, thereby providing a means for another substrate, with a different $E^{o\prime}$ and coenzyme and prosthetic group specificity, to be channeled into the respiratory chain (Figure 12-3). Succinate dehydrogenase contains a FeS center and has FAD covalently linked to it. Succinate is an intermediate in the tricarboxylic acid (TCA) cycle, a metabolic process that generates reducing equivalents (see p. 331). Succinate is oxidized to fumarate with the release of reducing equivalents. The oxidation-reduction potential for the fumarate/succinate pair is +0.03 volts (Table 12-4). Electrons transferred from the fumarate/succinate couple to the FMN/FMNH couple (−0.19 volts) require energy because the transfer is uphill, that is, the FMN/FMNH couple is less oxidized than the fumarate/succinate couple. The natural flow of electrons is to a component in the respiratory chain that is more oxidized than the fumarate/succinate couple, which in this case is the quinone.

3. The reducing equivalents captured by quinone are transferred to oxidized cytochrome *b**, which becomes reduced. The hydrogen ions are released outside the cytoplasmic membrane:

$$QH_2 + \text{Cytochrome } b_{ox} \rightarrow \text{Cytochrome } b_{red} + H_2 + Q$$

4. The reducing equivalents in reduced cytochrome b are transferred to oxidized cytochrome o.

* More than one type of cytochrome b may be involved during the transport of electrons from Q to oxygen.

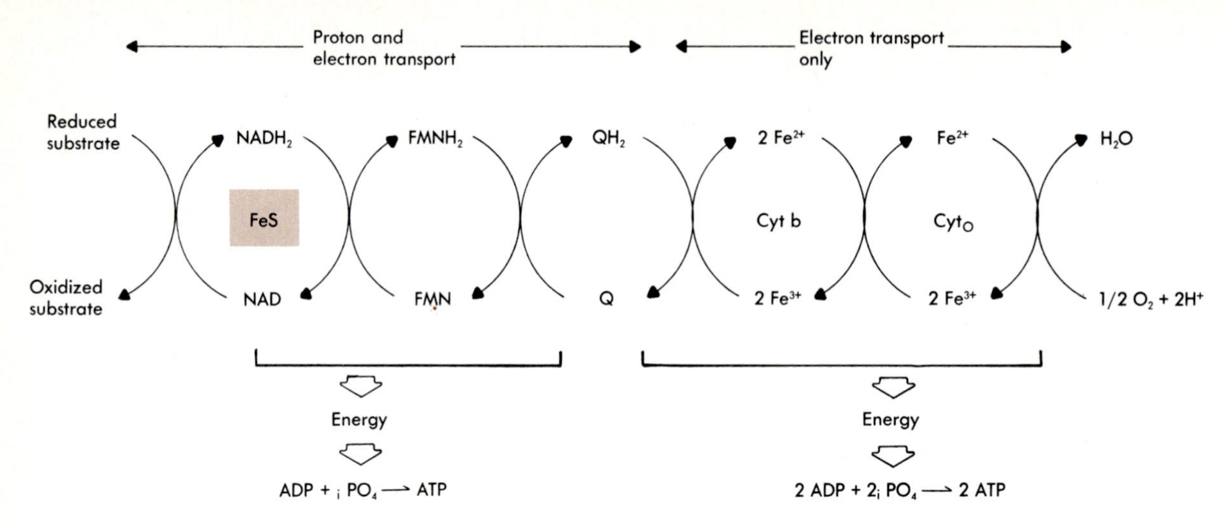

Figure 12-4 *Electron transport system and sites of energy formation in bacteria. During oxidation of substrate, hydrogens (protons) and concomitant electrons are removed and transferred along a series of transport molecules. Protons are carried only as far as quinone, whereas electrons are carried along entire respiratory chain. Sufficient energy (oxidation-reduction potential) is generated at points along chain, which is coupled to oxidative phosphorylation and formation of ATP. In final step, deenergized electrons and hydrogens are combined with oxygen to produce water. Some reduced substrates such as succinate may transfer their protons to flavoprotein component of chain, in which cases only two ATP will be produced. Mechanism for coupling of electron transport to ATP formation is illustrated in Figure 12-7.*

$$\text{Cytochrome } b_{red} + \text{Cytochrome } o_{ox} \rightarrow \text{Cytochrome } b_{ox} + \text{Cytochrome } o_{red}$$

During the transfer of electrons from QH_2 to the cytochromes b and o there is sufficient energy released that it results in the formation of two molecules of ATP. Three ATPS are the most that can be synthesized during the transfer of reducing equivalents through the respiratory chain. In the case where succinate is the substrate (succinate is more oxidized than glucose, the common energy source of the cell), reducing equivalents are transferred to FAD and then to quinone. The NAD step is bypassed and thus only two ATPs are produced. The direct oxidation of NAD by molecular oxygen results in the release of −52,440 cal/mole. Since the free energy of hydrolysis of a single molecule of ATP is −7300 cal/mole, the efficiency of energy conserved by the respiratory chain can be calculated as follows:

$$3(-7300\text{cal/mole}) = -21{,}900 \text{ cal/mole}$$

therefore

$$-21{,}900 \text{ cal/mole divided by } -52{,}440 \text{ cal/mole} = 42\%$$

5. The electrons in reduced cytochrome o (also called **cytochrome oxidase**) are transferred to the final electron acceptor, oxygen. The electrons are now at their lowest energy state and, in combination with the hydrogen ions released during oxidation, form water. **Water**, therefore, represents one of the products resulting from the complete oxidation of organic molecules in the presence of oxygen—the other is **carbon dioxide** and its formation will be discussed shortly.

Some type of electron transport system is found in all organisms, except fermenters that are strict anaerobes. The phosphorylation process discussed previously in which oxygen is the final electron acceptor during the

Concept Check

1. **How can substrate phosphorylation, electron transport phosphorylation, and photophosphorylation be distinguished from one another?**
2. **How does electron transport in the respiratory chain result in energy release?**
3. **Concerning the respiratory chain in bacteria, list in sequence the protein and nonprotein components and whether they transport electrons or protons. What are the sites where energy is released to produce ATP?**

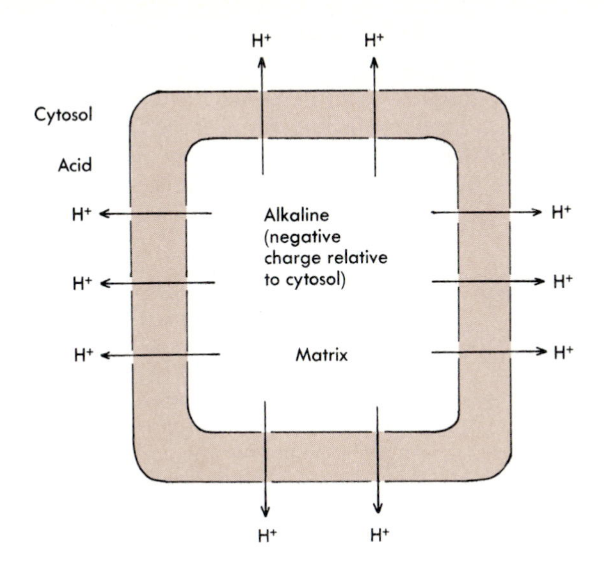

Figure 12-5 *Model demonstrating electrochemical effect when protons are pumped from mitochondrial matrix into cytosol. Comparable to protons being pumped from cytoplasm of bacterial cell to the exterior of the cell.*

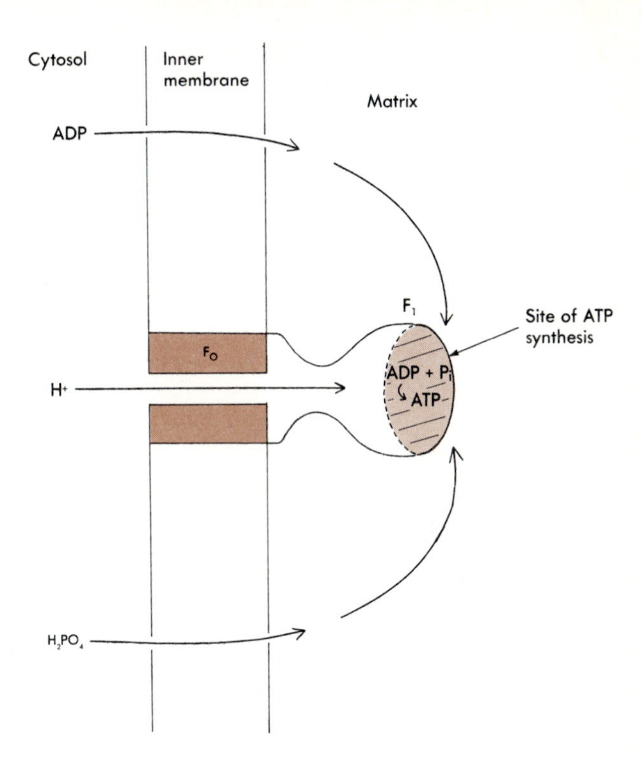

Figure 12-6 *Model showing location of ATPase protein complex of membrane (inner membrane of mitochondrion as shown in illustration or cytoplasmic membrane of bacteria). Flow of protons through ATPase controlled by F_o subunit. Protons induce formation of ATP in F_1 region.*

transport of electrons is called oxidative phosphorylation. Since oxygen is not always the final electron acceptor in bacteria, the process is more appropriately called **electron transport phosphorylation**.

Oxidative phosphorylation and chemiosmosis. The electrons removed from organic substrates can be combined with oxygen, and the energy generated can be used to phosphorylate ADP and phosphate to ATP. How electron transport was coupled to ATP formation remained a mystery for many years. Mitchell in 1961 proposed a mechanism based on the impermeability of the mitochondrial membrane to protons (hydrogen ions), as well as the arrangement of the respiratory or electron carriers across the membrane. This mechanism is called **chemiosmosis** and can also be applied to prokaryotic systems, that is, the cytoplasmic membrane of bacteria. Basically this mechanism states that as electrons pass down the respiratory chain the energy released is used to pump protons across the mitochondrial membrane or cytoplasmic membrane in bacteria. The movement of protons produces two effects: one, protons are moved from inside the mitochondrion (matrix) to the outside. This movement produces a positively charged area outside the mitochondrion and a negatively charged area inside the mitochondrion (Figure 12-5). Thus a **voltage gradient** is generated. Second, proton movement produces a **pH gradient** in which the area outside the membrane is more acid than inside the membrane. The sum of the electrical and pH potentials is the **proton motive force** (PMF), which can be measured in millivolts. Thus the electrochemical gradient that is produced during electron transport contains potential energy.

The hydrogen ions that are translocated across the membrane during electron transport flow back into the mitochondrion through a specific channel or pore created by a protein called **ATPase.** ATPase is a complex protein that spans the membrane (cytoplasmic membrane in bacteria, mitochondrial membrane in animals, and chloroplast membrane in plants). ATPase is composed of three functionally different units called F_o,F_1, and I. The **F_o protein** functions to transfer protons across the lipid bilayer of the membrane. The **F_1 protein** is involved in a catalytic step involving the phosphorylation of ADP and phosphate to ATP (Figure 12-6). The **I protein** appears to be a regulatory protein that prevents premature hydrolysis of ATP once it is formed. As with all respiratory transport components, ATPase functions are reversible, that is, ATP can be hydrolyzed to ADP and inorganic phosphate with the release of energy.

The manner in which the electron transport process pumps protons (H^+) from the matrix of the mitochondrion to the exterior occurs by a looping mechanism, as suggested by Mitchell. The carriers of hydrogen and electrons alternate in the membrane to form loops (Figure 12-7). The system works in the following way. Oxidation of a substrate releases electrons, which are transported as hydrogen atoms by a carrier (for example, quinone) located at the inner surface of the membrane. An electron carrier (for example, the cytochromes) facing the outer surface of the membrane becomes reduced by the hydrogen carrier. This loop reaction or oxidation-reduction results in the release of two protons ($2H^+$). Each loop, consisting of a hydrogen carrier and electron carrier, translocates $2H^+$ from the matrix to the outside of the mitochondrion. The number of redox

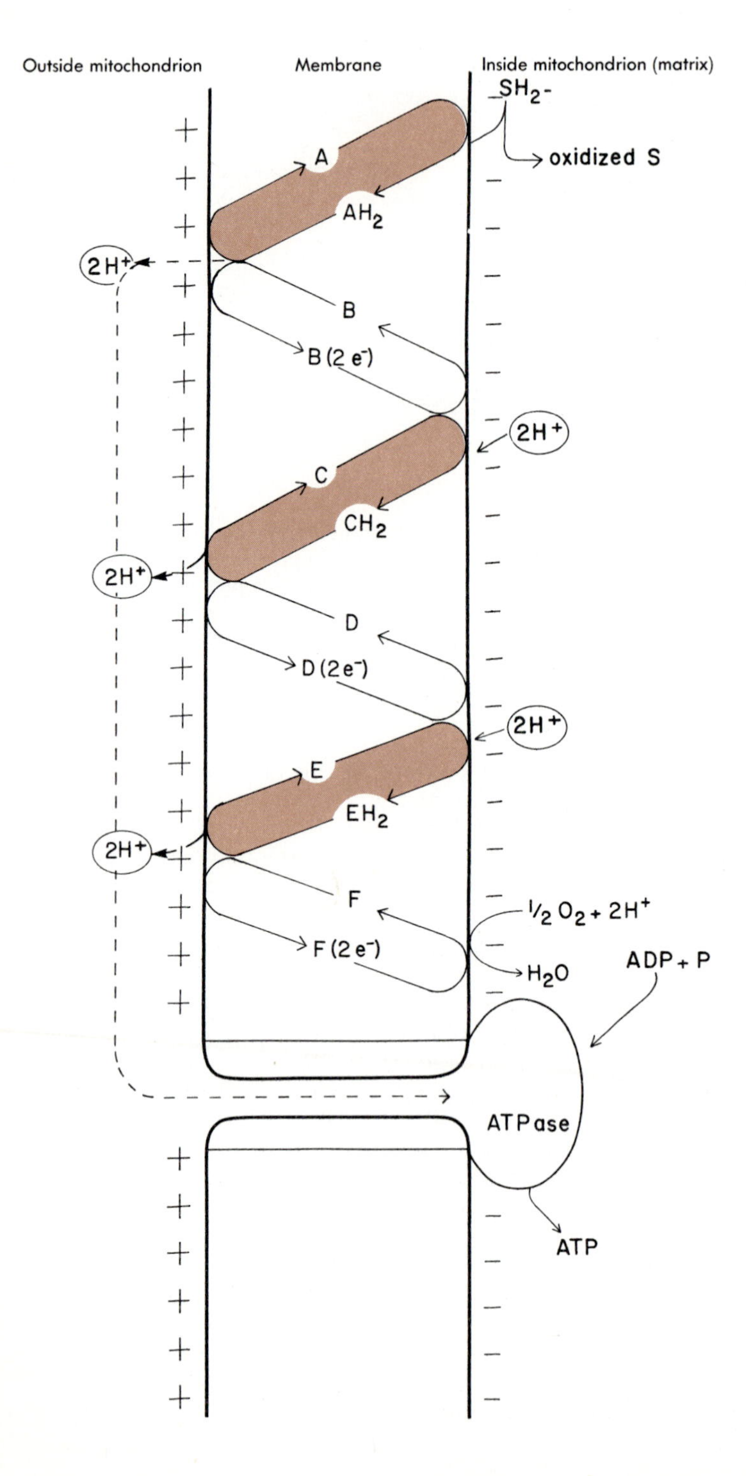

Figure 12-7 *Model for loop mechanism proposed by Mitchell to explain chemiosmosis. Hydrogen* (color) *and electron (no color) carriers alternate to form three loops. Oxidation of substrate (SH_2) releases electrons that are transferred as H atoms, for example, AH_2. At outer surface of membrane electron carrier* (B) *accepts electrons from AH_2 and hydrogen is released as protons ($2H^+$). Electrons are transported back to matrix and engage second loop (C plus D) and so on, until final electron acceptor, oxygen, is reached. Protons released during translocation can reenter membrane through pore in ATPase protein and ATP can by synthesized.*

loops was first suggested to be three, but so far only two have been uncovered. Regardless, after protons are translocated they reenter the membrane through the ATPase pore. For each molecule of ATP synthesized two to four protons must pass through the pore.

Proton motive force is used not only to generate ATP but is required for other cellular functions, such as active transport of certain ions, sugars, and amino acids. In bacteria additional processes, such as flagellar movement and sensing mechanisms involving oxygen and light also rely on the proton motive force.

Coupling of proton transport to ATP synthesis and hydrolysis. The movement of protons across the mitochondrial membrane, chloroplast membrane, and cytoplasmic membrane of bacteria, which creates a transmembrane potential, is used to drive the synthesis of ATP. The mechanism that couples these two events is still a matter of debate, but certain experimental findings led to some important conclusions. Evidence suggests that ATP processing by F_oF_1 subunits of ATPase does not involve a phosphoenzyme intermediate. It is estimated that ATP bond cleavage occurs without much free energy change (the only energy-requiring steps would be substrate binding and product release) and for this to occur the binding-proteins of ATPase must exist in at least two conformational states: (1) a state of loosely bound substrate and (2) a state of tightly bound substrate. The conformational changes are believed to be induced by membrane energization and the flow of H^+ through the ATPase pore. It is believed that there is a gating mechanism on the F_1 subunits adjacent to the membrane that controls the flow of H^+ to the catalytic subunits (Figure 12-8). When the membrane potential is sufficiently high (150 to 400 mV) for ATP synthesis then the flux of H^+ through the ATPase pore is increased. How protons participate in catalysis is not understood. The most widely accepted hypothesis is that proposed by P.D. Boyer called the **alternating catalytic mechanism** (Figure 12-9). He suggests that ATP is bound tightly at a site on the F_1 protein and that as membrane energization increases, a conformational change occurs that leads to the tighter binding of ADP and PO_4 to site one and the relaxation of binding at the ATP site (site two). As ATP is formed at site one, preformed ATP is released at site two, which now binds ADP and PO_4. This cycle is repeated until ATP is released from site one.

Photophosphorylation

Photophosphorylation (**photosynthetic phosphorylation**) is a process in which light energy is converted to chemical energy in the form of ATP. Photophosphoryla-

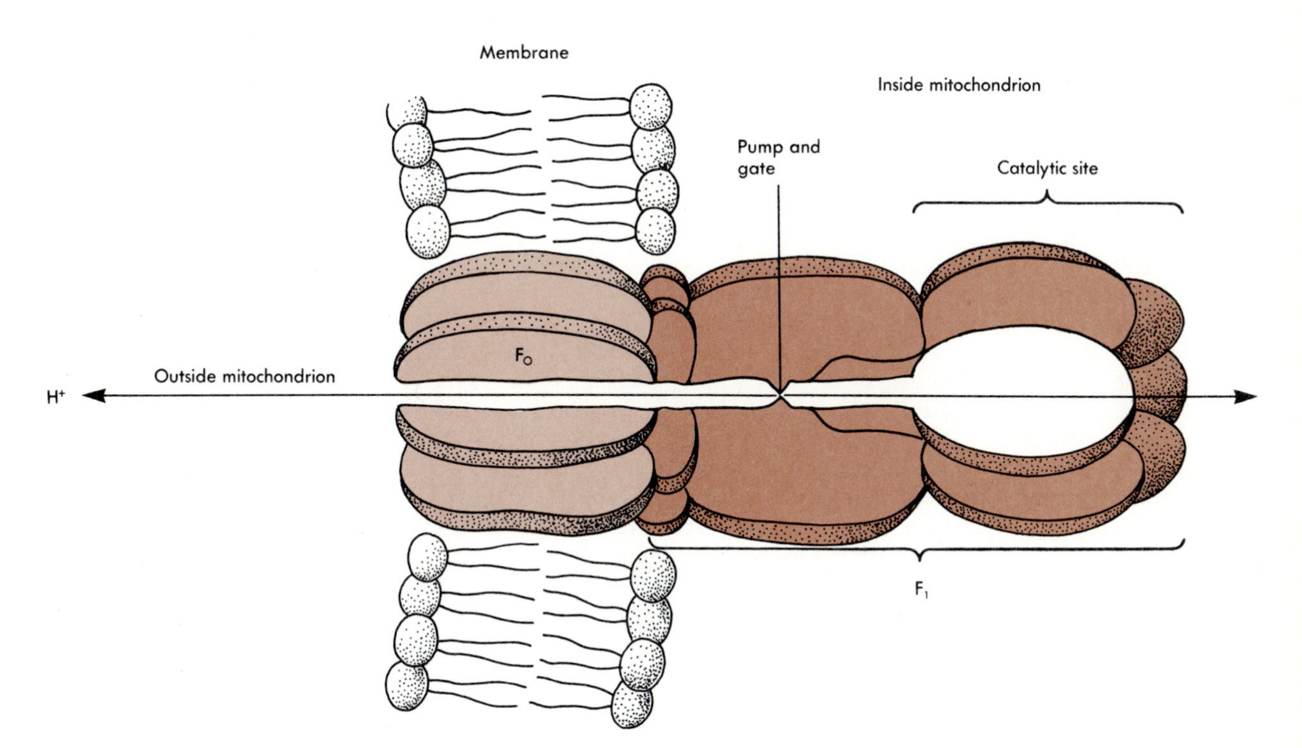

Figure 12-8 *Diagrammatic representation of organization of F_1 and F_o in ATPase protein. F_1 subunits possess ATPase activity and also act as proton pump and gate while F_o appears to function as proton channel. Depending on growth circumstances F_1F_o complex may eject or inject protons. Mitochondrial membrane is comparable to cytoplasmic membrane in bacteria.*

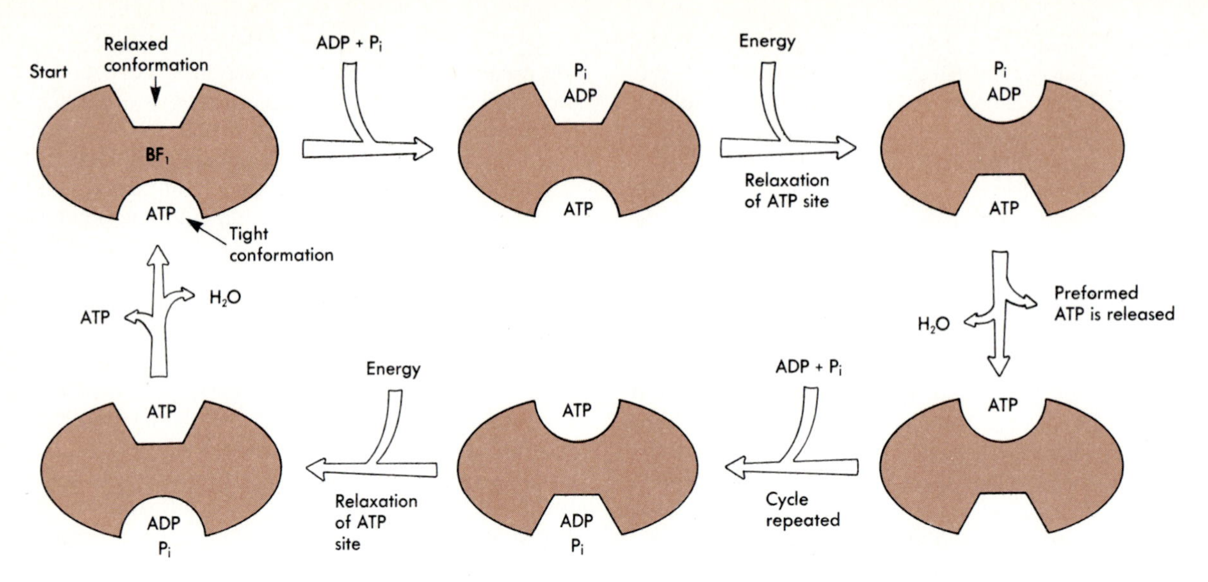

Figure 12-9 *Alternating catalytic site mechanism of ATP synthesis proposed by Boyer (see text for details).*

tion is but one phase of photosynthesis. The ATP that is generated by this process is used to convert carbon dioxide to sugars. The biosynthetic phase of photosynthesis is discussed in Chapter 13, p. 348.

The basic feature of photophosphorylation is that a quantum of light, called a **photon**, has sufficient energy to excite chlorophyll pigment molecules. The excited pigment molecules release high-energy electrons, which are captured by electron carriers. During electron transport, energy is released and coupled to the phosphorylation of ADP and PO_4 to ATP.

Photophosphorylation occurs in both eukaryotes and prokaryotes (Table 12-5). The cyanobacteria, prochlorophyta, purple bacteria, and green bacteria are all photosynthetic prokaryotic organisms, whereas plants and algae represent the eukaryotic kingdom. Photosynthetic organisms can be divided into two classes: those that produce oxygen (**oxygenic**) and those that do not produce oxygen (**anoxygenic**). All photosynthetic eukaryotes produce oxygen, but only the cyanobacteria and prochlorophyta among the prokaryotes produces oxygen. The evolutionary significance of the similarity of photosynthesis between the oxygenic eukaryotes and prokaryotes is discussed elsewhere (see Chapter 4).

Let us now examine in more detail the mechanism by which light energy is used to energize electrons and how this energy is converted to chemical energy in the form of ATP.

Light-harvesting structures. **Chloroplasts** and their system of thylakoid membranes are the light-harvesting structures in plants and algae. The chloroplast organelle is absent from prokaryotes, but an organized **thylakoid membrane** system is present in cyanobacteria. Located within these light-harvesting structures are pigments, such as chlorophyll a, and related molecules, such as bacteriochlorophylls and carotenes. They all act like antennae and absorb light energy (see Table 12-6). The kind and content of the chlorophylls and carotenes determine the color of the organism and the wavelength of light that can be used by that organism. Despite the importance of chlorophyll a and the carotenes as light-harvesting pigments, as much as 50% of the light-harvesting capacity in cyanobacteria is associated with a family of colored proteins called **phycobiliproteins.** These proteins are assembled into aggregates, called **phycobilisomes,** which are attached in regular arrays to the outer surface of the thylakoids. The major biliproteins are the red phycoerythrin (maximum absorption at 565 nm), blue phycocyanin (maximum absorption at 620 nm), and allophycocyanin (maximum absorption at 650 nm). Phycobilisomes are composed of a triangular

Concept Check

1. **What is the relationship between proton movement and ATP formation?**
2. **Explain the mechanism in which protons are pumped from the cytoplasm of the bacterial cell or mitochondrial matrix to the exterior of the cell or mitochondrion, respectively.**
3. **Briefly outline the mechanism by which photophosphorylation results in ATP formation.**

Table 12-5 *Characteristics of the photosynthetic system in eukaryotic and prokaryotic microorganisms*

		Prokaryotes			
Characteristic	**Eukaryotes**	**Procholorophyta**	**Cyanobacteria**	**Purple bacteria**	**Green bacteria**
Substructure containing photosynthetic machinery	Chloroplast	Stacked thylakoids	Thylakoids—phycobilisomes are protein aggregates attached to thylakoids	Intracytoplasmic membrane system continuous with cytoplasmic membrane	Chlorosomes, which are appressed to cytoplasmic membrane
Photosynthetic pigment	Chlorophyll a and b	Chlorophyll a and b	Chlorophyll a	Bacteriochlorophylls a and b	Bacteriochlorophylls a and c, d or e
Oxygen production	Yes	Yes	Yes	No	No
Electron donors	H_20	H_20	H_20; some can use H_2S	H_2S, S_0, H_2, S_20_3, organic compounds	H_2S, S_o, H_2, S_20_3
Photosystem I	Present	Present	Present	Present	Present
Photosystem II	Present	Present	Present	Absent	Absent

Table 12-6 *Spectroscopic properties of purified chlorophylls in ether solution*

Absorption maxima (wavelengths in nm; relative peak heights [highest = 100] in parentheses)							
Chlorophyll a			430 (100)		615 (13)	662 (77)	
Bchl c			428 (100)		622 (29)	660 (63)	
Bchl d			424 (100)		608 (17)	654 (61)	
Bchl e			458 (100)		593 (12)	647 (32)	
Bchl a	358 (100)	391 (68)		577 (29)			773 (126)
Bchl b	368 (100)	407 (87)		582 (30)			795 (96)

core of allophycocyanin surrounded by peripheral rods of phycocyanin and/or phycoerythrin (Figure 12-10). The characteristic absorbance of the biliproteins in the visible range is due to the open-chain tetrapyrrole prosthetic groups covalently attached to these proteins (Figure 12-11). Incident light in the 550 to 660 nm range that strikes the bacterium is absorbed by these pigments and the energy is transferred to the major light-harvesting pigment chlorophyll a.

Light absorption and energy capture. Chlorophyll a (Figure 12-12) is a light-harvesting pigment found in plants and oxygenic prokaryotes. It is made up of closed pyrrole rings, which surround a central magnesium atom, and have alternating single and double bonds that enable the molecule to absorb light. A hydrophobic side chain anchors the chlorophyll to the thylakoid membrane. Carotenoids, such as beta carotene, are accessory pigments that absorb light but at wavelengths different from the principal light-absorbing pigments. They assist in transferring energy to the chlorophyll reaction center (see Point of Interest).

The light-absorbing pigments are arranged in clusters referred to as **photosystems**. There are two photosystems: **photosystem I**, which is maximally excited by light at wavelengths above 700 nm, and **photosystem II**, which is maximally excited by light at wavelengths below 680 nm. Oxygenic species possess both photosystems, whereas anoxygenic species contain only photo-

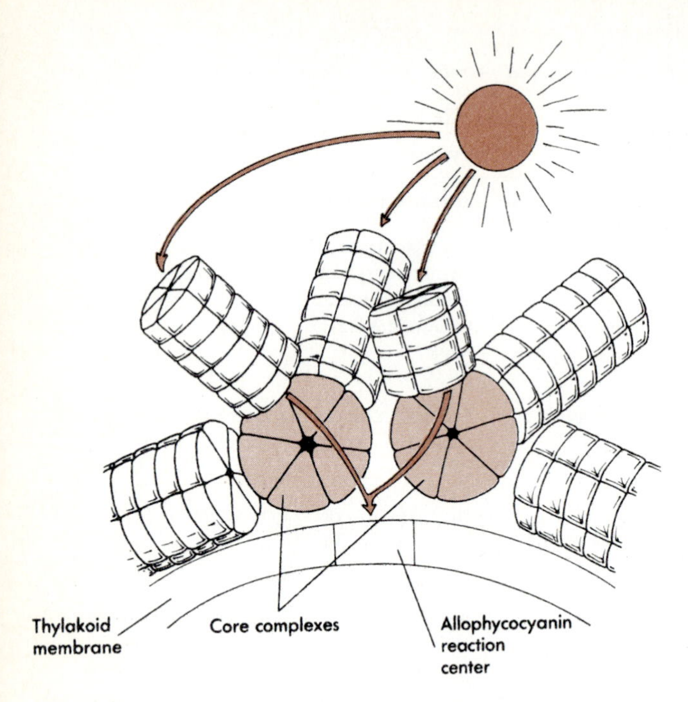

Figure 12-10 *Schematic model of intact phycobilisome. There are hexameric phycocyanin complexes that are attached to core cylinders* (color). *Light energy absorbed by complex is channeled to reaction center in thylakoid membrane.*

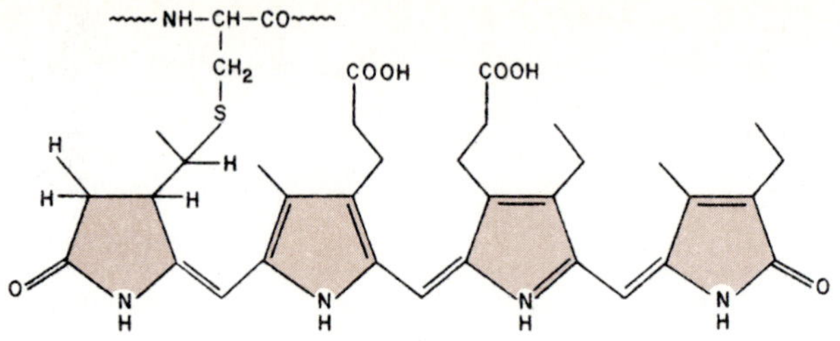

Figure 12-11 *Structure of open chained biliprotein prosthetic group. Wavy line indicates polypeptide of protein.*

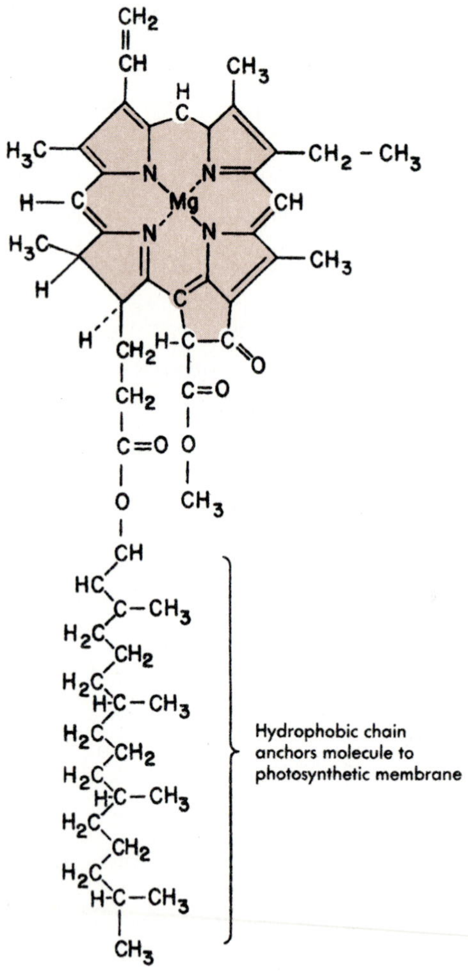

Hydrophobic chain anchors molecule to photosynthetic membrane

Figure 12-12 *Structure of bacteriochlorophyll a.*

system I. Two hundred or more pigment molecules may be present in the photosystem with each molecule capable of absorbing light. Only one pigment molecule, however, is capable of converting light energy into chemical energy, and it is combined with a protein to form what is called a **reaction center**.

When a photon of light strikes a light-harvesting molecule, the latter is energized to an excited state and an electron is raised to a high energy level (Figure 12-13). The energy absorbed by an electron is called an **exciton.** This energy might normally be lost again as heat or light (fluorescence) if the electron returned to its original stable condition or ground state. Instead the packet of energy migrates through the highly organized chlorophyll molecules of the photosynthetic apparatus to the **reaction center** of the photosystem (Figure 12-14). The energy that is trapped in the reaction center is transferred to an electron. This highly energized electron is released from the reaction center to be accepted by electron carriers.

Electron flow in oxygenic photosynthesis. The ejection of the highly energized electron from the reaction center leaves it in an oxidized state and thus an electron void or electron hole is produced. How is this electron hole filled? This question can best be answered by examining the characteristics of the photosystems of oxygenic organisms. These two systems cooperate to generate electron flow from an electron donor, H_20 (anoxygenic organisms use electron donors other than H_20—see Table 12-5). The two photosystems form a zigzag pattern (**Z scheme**) outlined in Figure 12-15.

Photosystem I maximally absorbs light at wavelengths greater than 700 nm. The highly energized electron re-

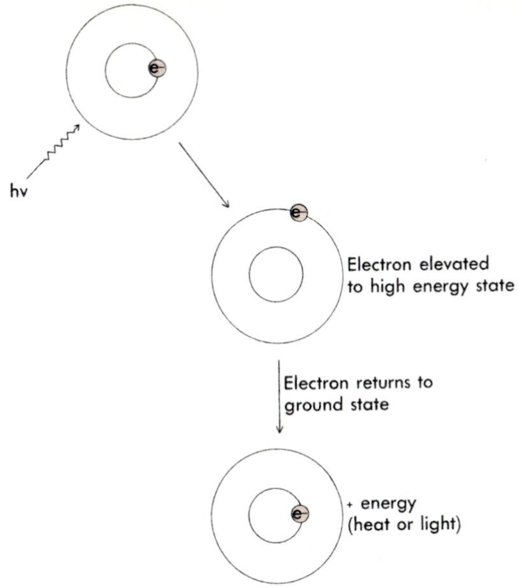

Figure 12-13 *Atom excited by light (hv) results in electron jumping to higher energy level. Excited electron quickly returns to ground state with release of energy as heat or light (fluorescence) unless energy can be trapped by other molecules.*

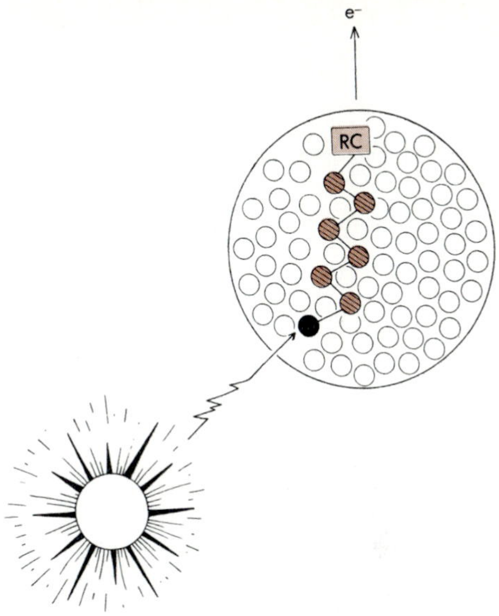

Figure 12-14 *Model of photosystem. Light energy is absorbed by pigment molecule and energy is transferred from pigment molecule to pigment molecule until reaction center (RC) is reached. RC pigment molecule converts exciton energy (energy absorbed by electron) into electron flow.*

Point of Interest

CAROTENOIDS

Carotenoids are open-chained and cyclic hydrocarbon molecules whose precursors are isoprenoids (*A*, below). They are among the oldest chemicals on earth and have been found in fossil remains of photosynthetic and nonphotosynthetic bacteria. Carotenoids play an accessory role in transferring energy to acceptor chlorophyll molecules during photosynthesis. Their efficiency in energy transfer is much lower than that of the bacteriochlorophylls and they appear to have other roles that may be more important in the cell. Carotenoid synthesis in the microbial cell can lead to the formation of sterols, but they are not a major product. Carotenoids and their derivatives function in the cytoplasmic membrane. They are believed to stabilize the bilayer arrangement of the cytoplasmic membrane, and this in turn can affect permeability, nutrient transport, and osmotic stability. Carotenoids appear to be more important as photoprotective devices. Carotenoids are able to protect other molecules against light-induced destruction in the presence of molecular oxygen. Studies with carotenoid-deficient mutants of photosynthetic bacteria show that bacteriochlorophyll can be destroyed in the presence of light and air—a process called **photooxidation**. Several membrane-associated molecules are now known to be protected by carotenoids from photooxidation, including succinic dehydrogenase, NADH oxidase, ATPase, quinones, and respiratory chain components. The protective effects of carotenoids is directly associated with the number of conjugated double bonds in the molecule. At least nine conjugative double bonds must be present for the carotenoid to be effective.

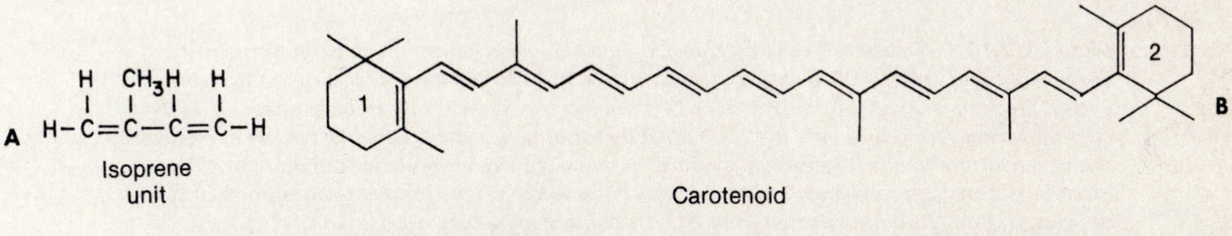

Isoprenoids. ***A,*** *Basic isoprene unit.* ***B,*** *Structure of carotenoid in which two rings (1 and 2) are connected by chain of repeating isoprene units.*

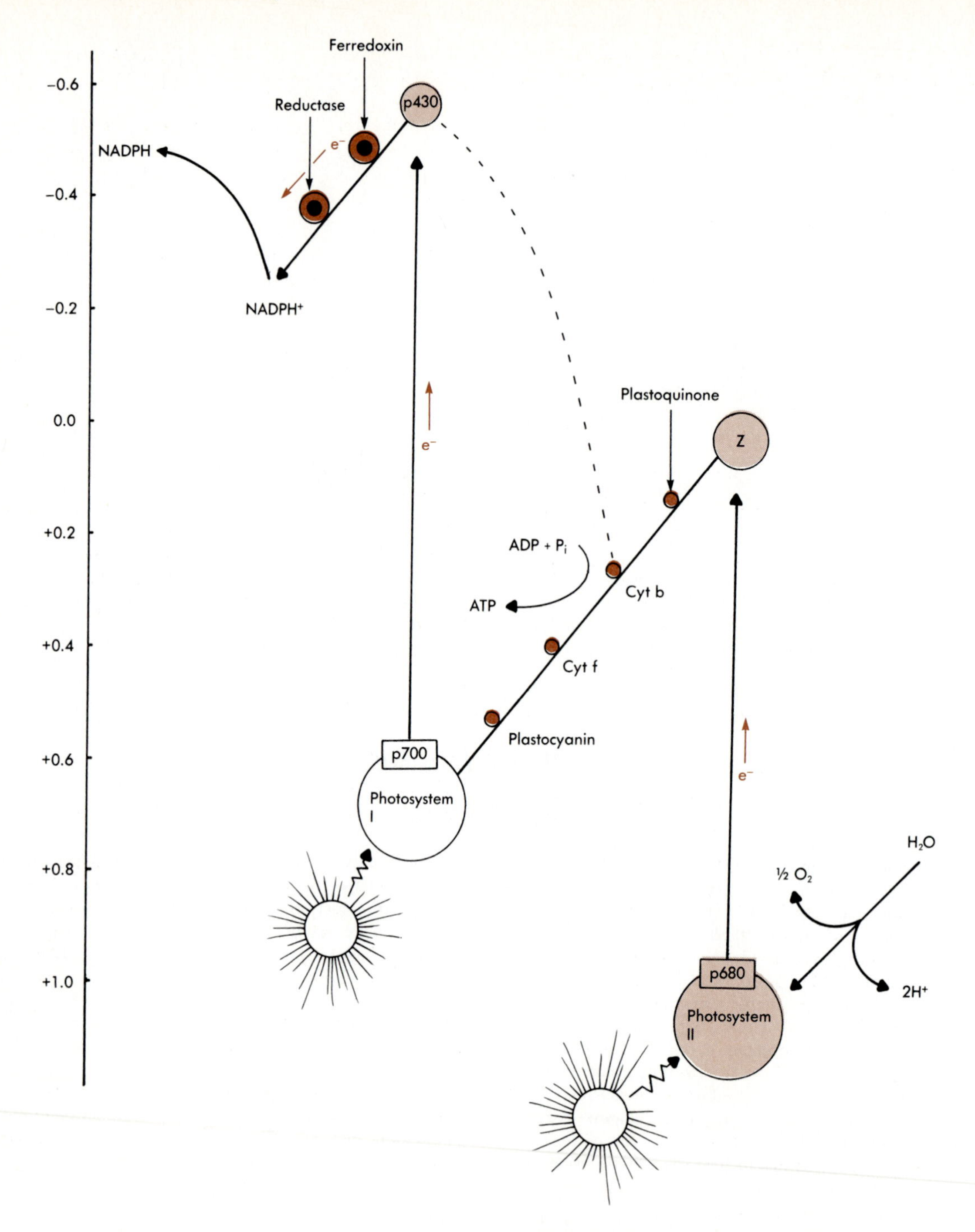

Figure 12-15 *Z scheme of electron transfer. Two photosystems are involved in transfer of electrons from H_2O to NADP. Energy for elevating electrons is provided by absorption of light energy. Electron is boosted by energy released from reaction center P680 in photosystem II to the electron carrier Z (electron hole in P_{680} is filled by splitting of water). Downhill travel of electron through cytochromes and plastocyanin results in release of free energy and formation of ATP. Electron is then captured by photosystem I, which has electron hole following absorption of light by reaction P700. Electron released from P700 is boosted to higher energy level (P430) by absorption of light. Electron flows downhill through carriers such as ferredoxin and reductase and finally reduces NADP to $NADPH^+$. Pathway is noncyclic but can become cyclic (------) when electron at P430 returns to cytochrome b. Under these circumstances NADP is not reduced but ATP is synthesized and only photosystem I is used.*

leased by the reaction center, designated P700, flows through a series of electron carriers to NADP, which is reduced to NADPH. The electron hole in the reaction center is filled with an electron released from the reaction center of photosystem II, designated P680, after absorbing wavelengths at 680 nm. This now leaves the reaction center of photosystem II with an electron hole. This hole is filled by an electron that is released by the photolysis of water. Water is split into electrons, hydrogen ions (H^+), and molecular oxygen according to the following equation:

$$2H_20 \xrightarrow{\text{light}} 4H^+ + 4e^- + O_2$$

In Figure 12-15 components of the photosystems are aligned with their corresponding oxidation-reduction potentials. In both photosystems a quantum of light is used to energize an electron and raise it from a positive oxidation-reduction potential to a negative one. Once raised, the electron flows "downhill" on electron carriers with the release of free energy, and in the process ATP and reducing power in the form of NADPH are generated. The Z scheme demonstrates how electrons released from water, which has a positive oxidation-reduction potential, can be used to reduce NADP. The energy for electron movement is provided by the quanta of light absorbed by photosystems I and II.

The first electron carrier in photosystem I is designated P430, an iron-sulfur protein. This carrier can release the electron downhill to be captured by the second carrier, ferredoxin. The third carrier is called ferredoxin-NADP oxidioreductase, which transfers an electron from ferredoxin to NADP:

$$2F_{red}^{++} + 2H^+ + NADP \longrightarrow 2F_{ox}^{+++} + NADPH + H^+$$

The first electron carrier in photosystem II is a carrier of which little is known; it is designated Z. Electrons are then transferred downhill to plastoquinone, two cytochromes, and plastocyanin. During electron transfer between the cytochromes, ATP is produced by the coupling of free energy to the phosphorylation of ADP to ATP. Apparently one ATP is produced per pair of electrons passing down the chain. Plastocyanin, which is a copper protein, donates its electron to fill the electron hole that was left in reaction center P700 of photosystem I. The electron hole left in photosystem II is filled by electrons removed from water by a mechanism that is not totally understood.

Cyclic vs. noncyclic electron flow. The Z pathway described in Figure 12-15 represents a undirectional flow of electrons from water to NADP. The Z pathway, however, can become cyclic when the energized electron at the first electron carrier P430 of photosystem I returns to fill the electron hole in reaction center P700. This shunt pathway involves the electron carriers in photosystem II; thus ATP is formed but NADPH and oxygen are not produced. This cyclic pathway is believed to be used when the cell has an ample supply of NADPH but requires additional ATP.

Photosynthetic phosphorylation resembles oxidative phosphorylation. The electron transport components of photosynthetic organisms are arranged or organized on the thylakoid membrane in such a way that the flow of electrons results in the pumping of protons from the stroma of the chloroplast across the thylakoid membrane to the inner compartment. The protons return through an ATPase located in the thylakoid membrane, and ATP is synthesized. Thus the chemiosmotic mechanism originally discovered in mitochondria also applies to the chloroplast.

Photophosphorylation in anoxygenic organisms. Anoxygenic photosynthetic organisms, such as the purple and green bacteria, possess only photosystem I and utilize a cyclic electron transfer system. The electron donors for photosynthesis are molecules other than water (see Table 12-5). In addition, these organisms possess pigment molecules referred to as bacteriochlorophylls. Many species have photosynthetic units in which up to 1500 molecules of bacteriochlorophyll are present per cell. This enormous number of "antennae" enables many species to grow in relatively weak light. In the green bacteria the antennae pigments are located in the chlorosomes, the ovoidlike membranous structures appressed to the cytoplasmic membrane. In the purple bacteria the whole photosynthetic apparatus is located within the intracytoplasmic membranes, which are continuous with the cytoplasmic membrane (see Chapter 5).

Many of the reactions involving electron donors and the formation of NADPH have not been elucidated in the purple and green bacteria. A postulated scheme for electron transport in the purple bacteria is represented in Figure 12-16. The longest wavelength constituent is a bacteriochlorophyll designated P870. P870 passes its energy to a reaction center (a bacteriopheophytin molecule termed Bph). The energized electron from the reaction center is transferred to a quinone-iron complex and thence to cytochrome b, an iron-sulfur protein, and cytochrome c. Because the quinone-iron complex is not sufficiently electronegative to reduce NADP, cyclic electron transfer generates a potential across the membrane, which in turn drives **reversed electron transfer** from succinate to NAD (see Table 12-4 for oxidation-reduction potentials).

The photosynthetic electron transfer system appears to be organized asymmetrically across the intracytoplasmic membrane in purple bacteria. Just as in nonphotosynthetic bacteria, there is also sufficient evidence to show that the cyclic transfer system is compatible with the basic mechanism of chemiosmosis found in mitochondria, chloroplasts, and nonphotosynthetic bacteria.

Photosynthesis and bacteriorhodopsin. Bacteriorhodopsin is a photosynthetic protein pigment found only in the Halobacteria (see p. 321). Bacteriorhodopsin

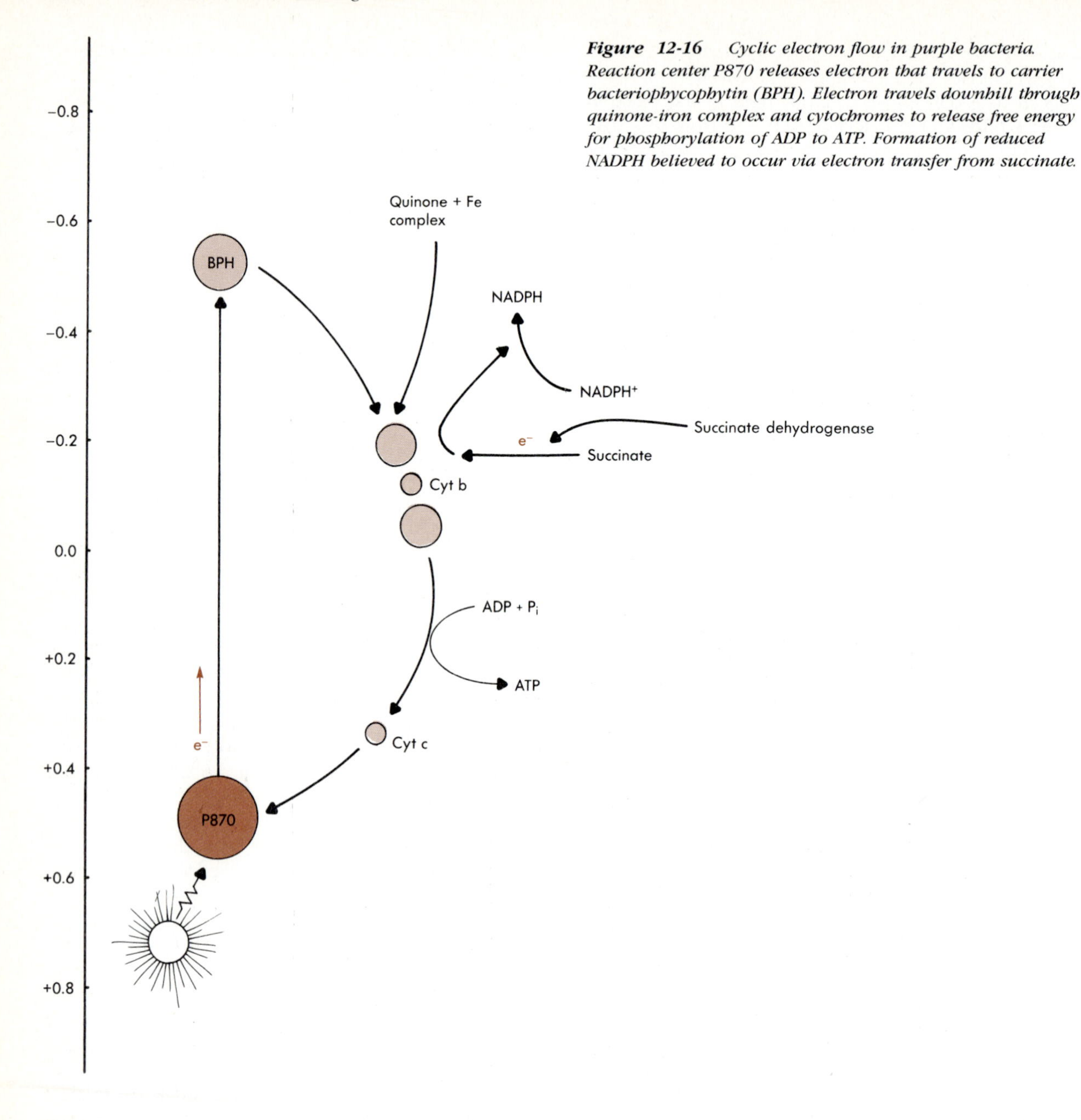

Figure 12-16 *Cyclic electron flow in purple bacteria. Reaction center P870 releases electron that travels to carrier bacteriophycophytin (BPH). Electron travels downhill through quinone-iron complex and cytochromes to release free energy for phosphorylation of ADP to ATP. Formation of reduced NADPH believed to occur via electron transfer from succinate.*

forms aggregates in the cytoplasmic membrane that resemble purple patches (see p. 114). **Retinal**, which is vitamin A aldehyde, is the prosthetic group of bacteriorhodopsin. Retinal is identical to the light-absorbing pigment found in the retina of the vertebrate eye. When retinal is activated by a photon of light, the conformation of bacteriorhodopsin is changed and protons (H^+) from inside the cell are transferred across the cytoplasmic membrane to the outside of the cell. This proton transfer results in the formation of ATP. This is the simplest mode of proton translocation because the release or use of protons by substrate molecules is not involved.

Ion exchange is also affected by the proton motive force in the *Halobacteria*. *Halobacteria* live in environments of high salt concentration (4M Na^+ and 0.03M K^+) and must maintain a salt balance in the cell (1.3M Na^+ and 3M K^+). During illumination, sodium is extruded from the cell and potassium is taken up by the cell until K^+ reaches molar concentrations in the cytoplasm. The extrusion of sodium and uptake of potassium are coupled to the uptake of protons resulting from the proton motive force. The membrane potential drives potassium into the cell, and the result is an approximate 1:1 exchange of sodium for potassium (Fig-

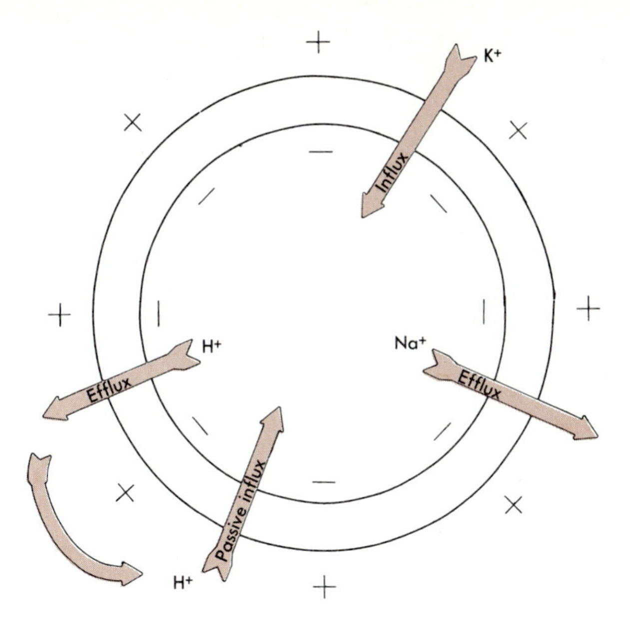

Figure 12-17 *Scheme for light driven ion exchange in Halobacteria. Pumping of protons out of cell and passive influx of protons into cell are balanced by mechanism that allows sodium to be transported out of cell as protons enter cell. Process results in positive charge outside cell and negative charge inside cell.*

ure 12-17). A recently discovered retinal pigment called **halorhodopsin** is a light-driven anion pump specific for the halide chlorine. Experiments on the purified pigment indicate that chlorine influences the absorption maximum of the pigment and helps stabilize it. Thus on excitation by light, bacteriorhodopsin pumps protons out of the cell and halorhodopsin pumps chloride into the cell, an action that creates a membrane potential.

Concept Check

1. What contributions do photosystems I and II make to electron transport, and in what types of organisms are they found?
2. What are the end products of cyclic and noncyclic electron flow?
3. How does photophosphorylation in anoxygenic organisms differ from that in oxygenic organisms?
4. Explain the mechanism by which proton transfer is coupled to ATP synthesis in *Halobacteria?*

ENERGY PRODUCTION DURING OXIDATION OF ORGANIC MOLECULES

In the previous discussions we have seen the mechanisms used by organisms to produce energy in the form of ATP. In the following discussion we will be concerned with those metabolic pathways in which ATP is generated by the oxidation of organic compounds.

Metabolic pathways

Metabolic pathways are characterized by a sequence of chemical reactions, each of which is catalyzed by a specific enzyme. The end product of one enzyme reaction is the substrate for the following enzyme (Figure 12-18). Each of the products in the metabolic pathway is referred to as a **metabolite**. Each metabolite becomes chemically altered by the addition, removal, or rearrangement of chemical groups until a final product is formed. These end products may be excreted by the cell, or they may be used as substrates in other metabolic reactions. Metabolic reactions may be divided into two types: **anabolic** and **catabolic**. Anabolic pathways (biosynthesis) use energy to build cellular components, and they will be discussed in a subsequent chapter. Catabolic reactions are represented by a sequence of reactions whose function is to degrade or oxidize molecules and release energy. Since energy-producing reactions involve an oxidation and reduction, each pathway can be described in terms of the electron (hydrogen) donors and final electron acceptors. There are three major pathways for the extraction of energy from organic molecules: fermentation, respiration, and anaerobic respiration.

Fermentation

Fermentation* is a process in which an organic donor is oxidized but oxygen is not involved and the final electron acceptor is another organic molecule. Fermentation is believed to have been the principal method for extracting energy by the first organisms that inhabited the planet. The atmosphere in the primordial environment probably consisted of carbon dioxide, methane, argon, nitrogen, and water vapor but no oxygen. If oxygen had been present, all of the organic sources of energy would have been completely oxidized to carbon dioxide and water. Fermentation is an incomplete oxidation process in which the end products are mostly organic molecules, such as lactic acid or pyruvic acid, that possess extractable energy. Fermentation therefore does not provide a mechanism for extracting all of the available

* Fermentation is defined in a strict sense. *Fermentation* to the industrial microbiologist means any process by which an organism produces a desired product (see Chapter 22).

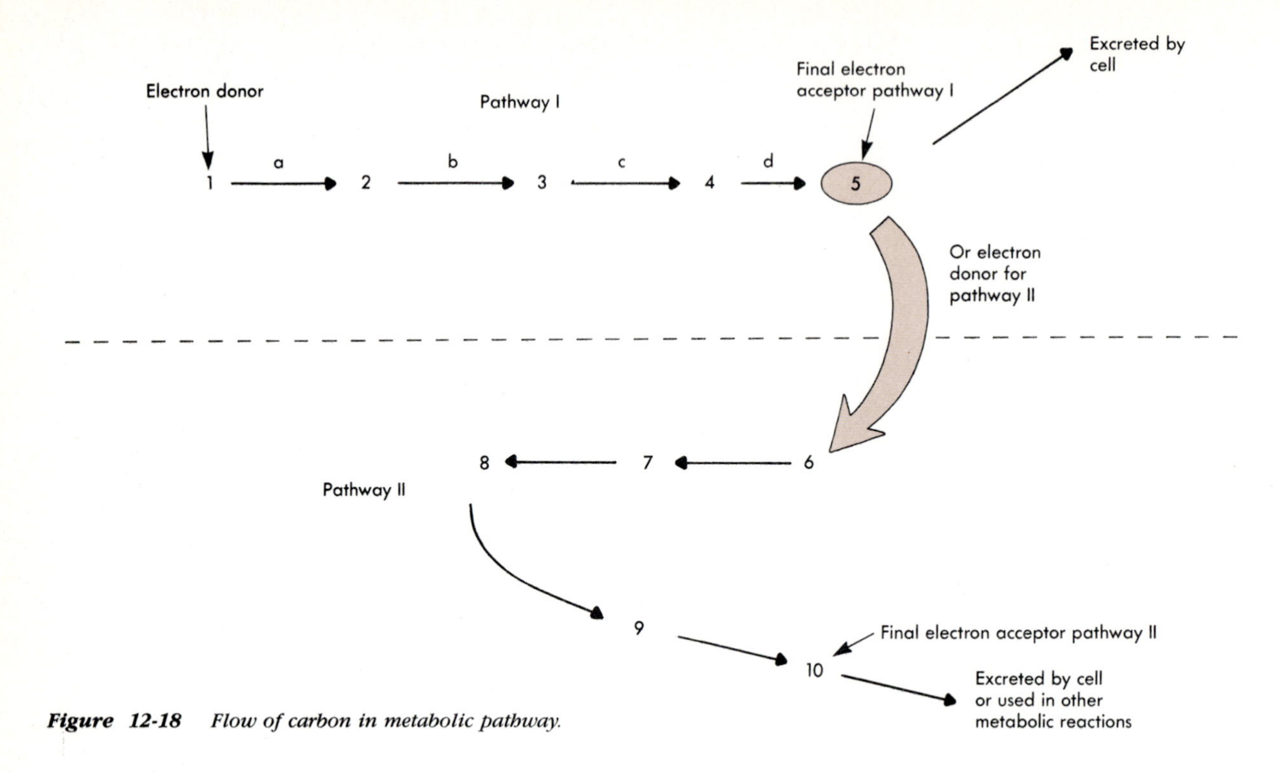

Figure 12-18 *Flow of carbon in metabolic pathway.*

energy from an organic molecule. Organisms that can obtain their energy only in the complete absence of oxygen are called **obligate anaerobes** (see Table 12-7 for the differentiation of organisms based on their oxygen requirements). For obligate anaerobes, fermentation via substrate level phosphorylation is the only mechanism for obtaining energy.

Glycolysis-A fermentation involving glucose. The most common fermentation process in organisms is called **glycolysis,** in which glucose is the source of electrons. Glycolysis is also called the **Embden-Meyerhoff-Parnas** (EMP) pathway (after the men who made the most significant contributions to its discovery). Even though oxygen is not involved in any glycolytic reactions, glycolysis can be used by organisms that grow in the presence of oxygen and therefore can be used in fermentation and respiration. The final electron acceptor in glycolysis is a three-carbon unit called **pyruvate** (Figure 12-19), which can be reduced to lactate and other products (Figure 12-20). Energy in the form of ATP is produced in glycolysis by substrate phosphorylation (see p. 309). For simplicity's sake, glycolysis can be divided into two phases:

Phase I. Glucose proceeds through two phosphorylation steps in which ATP is the source of phosphate. This priming phase actually involves five reactions and ends with the cleavage of glucose into two carbon units: **glyceraldehyde-3-phosphate** and **dihydroxyacetone phosphate.** Only glyceraldehyde-3-phosphate can be

Table 12-7 *Differentiation of microorganisms based on their oxygen requirements*

Group	Oxygen requirement
Aerobes	
Obligate	Grow only in presence of oxygen at approximately one atmosphere of air or 20% oxygen
Anaerobes	
Obligate	Grow only in absence of oxygen; oxygen is lethal
Facultative	Grow best in presence of one atmosphere of air but can also grow in absence of oxygen
Aerotolerant	Are tolerant to presence of oxygen and can grow in presence or absence of oxygen; oxygen not a terminal electron acceptor in energy formation
Microaerophiles	Grow in presence of oxygen but at concentrations less than 20%

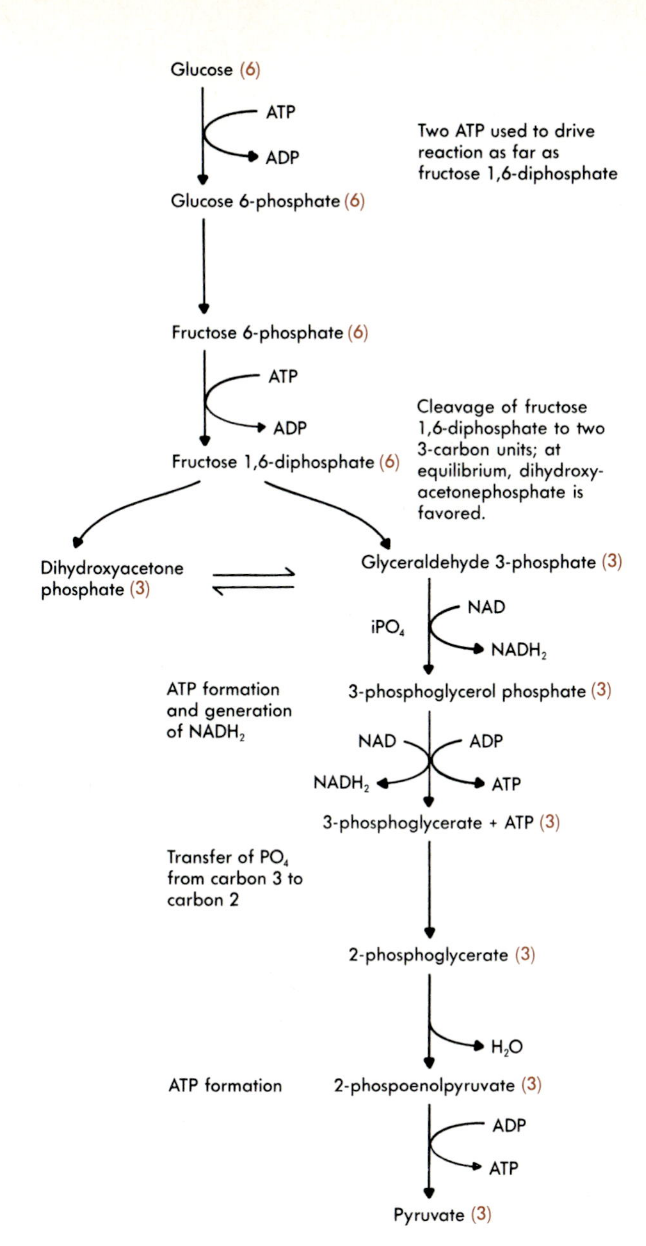

Figure 12-19 *Sequence of reactions occurring in glycolysis or the Embden-Meyerhoff-Parnas (EMP) pathway. Numbers beside compounds indicate number of carbons in molecule. Pyruvate, product of glycolysis under anaerobic conditions, is converted to lactate resulting in regeneration of NAD. Under aerobic conditions pyruvate is oxidized to acetyl coenzyme A whose further metabolism is discussed on p. 331 (see Figure 12-25).*

converted to glycolytic products, and thus dihydroxyacetone phosphate is enzymatically converted to glyceraldehyde-3-phosphate. Therefore, two molecules of glyceraldehyde-3-phosphate will proceed through the remaining reactions.

Phase II. High-energy phosphates are generated in two separate reactions, and these are coupled to the formation of ATP (Figure 12-21). In addition, during the conversion of glyceraldehyde-3-phosphate to 1,3-diphosphoglycerate, an oxidation has taken place in which two hydrogen atoms are transferred to NAD. Overall in phase II, two molecules of glyceraldehyde-3-phosphate are converted to two molecules of pyruvate, and in the process four molecules of ATP are produced. The reduced NAD may follow one of two paths: If the organism carries out glycolysis, in the absence of oxygen, reduced NAD is used to reduce pyruvate to lactate. This process regenerates NAD for future use in glycolysis, and thereby precluding the need for synthesis of more NAD.

$$\underset{\text{Pyruvate}}{CH_3COCOO^-} + NADH_2 \rightarrow \underset{\text{Lactate}}{CH_3{-}\overset{\overset{\displaystyle H}{|}}{\underset{\underset{\displaystyle OH}{|}}{C}}{-}COO^-} + NAD$$

If the organism carries out glycolysis in the presence of oxygen, reduced NAD transports electrons to the electron transport chain, where additional ATP may be produced by electron transport phosphorylation (discussed later). In the overall process of glycolysis, two ATP molecules are gained (two were used to prime the pathway and four were generated), and reduced NAD is produced.

Some obligate anaerobes can obtain their energy only by substrate phosphorylation. These organisms do not possess electron transfer components or proton pumping mechanisms. They hydrolyze some of the ATP produced by substrate phosphorylation by the reverse activity of membrane ATPase (recall that in electron transport phosphorylation ATPase was used in the synthesis of ATP). In this way these anaerobic species can energize the cytoplasmic membrane for such activities as solute transport and flagellar movement.

Products of pyruvate metabolism. Pyruvate is the final electron acceptor in glycolysis, but pyruvate can be metabolized to other products, depending on the microbial species or group. Fermentations are often classified according to the major end products (Figure 12-20). For example, in lactic acid fermentation, lactic acid is the principal product when glucose is fermented by species of *Streptococcus* and *Lactobacillus*. Alcoholic fermentation is characterized by the formation of ethanol by species of yeast, while propionic acid fermentation results in the formation of propionic acid by *Propionibacterium* species. Some end products of pyruvate metabolism are of commercial value, and many industries are

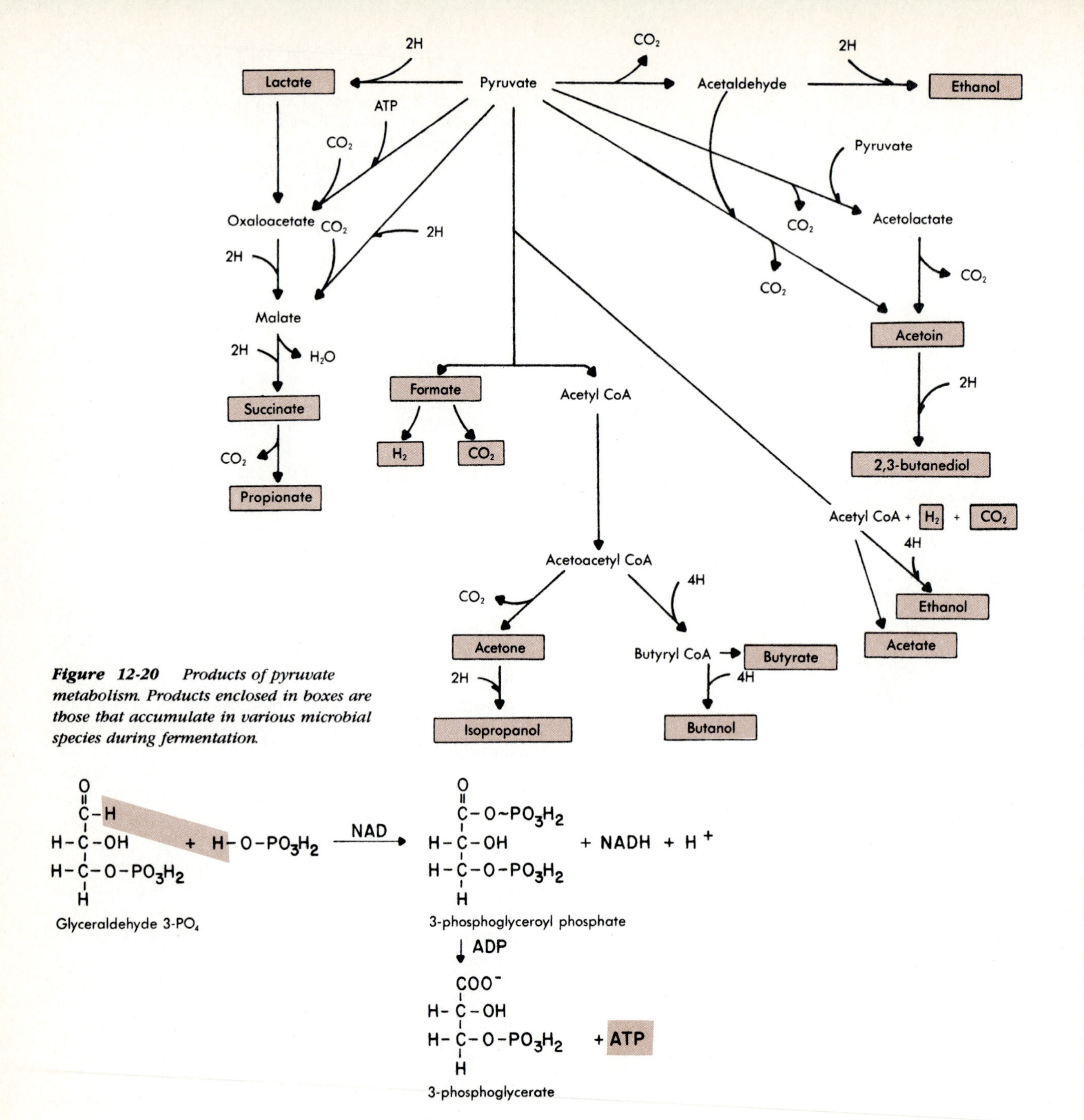

Figure 12-20 *Products of pyruvate metabolism. Products enclosed in boxes are those that accumulate in various microbial species during fermentation.*

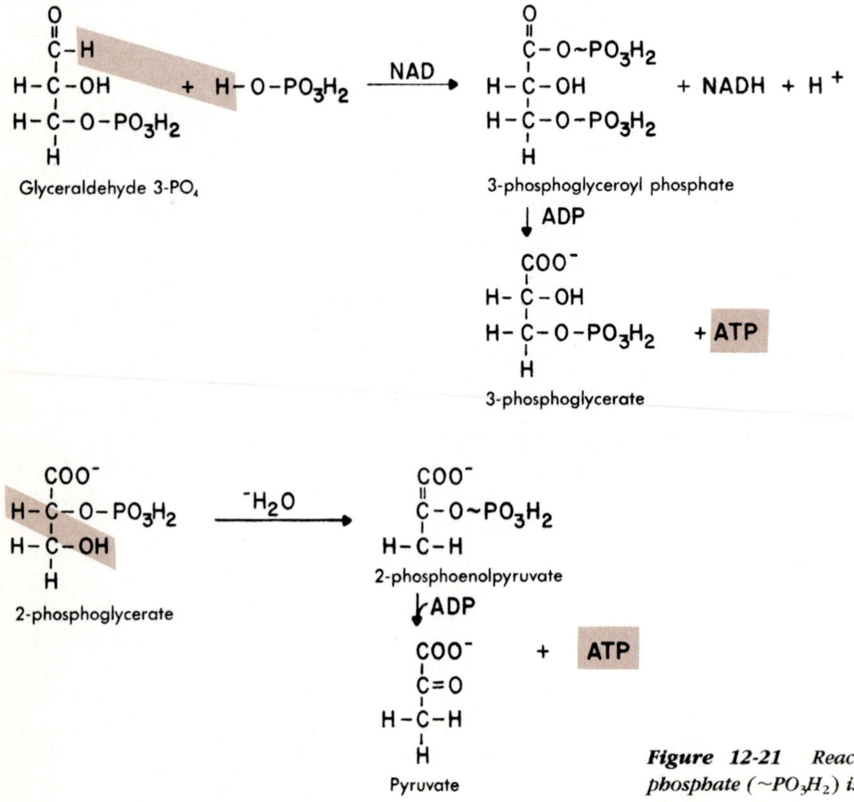

2-phosphoglycerate $\xrightarrow{-H_2O}$ 2-phosphoenolpyruvate

2-phosphoenolpyruvate $\xrightarrow{ADP}$ Pyruvate + ATP

Figure 12-21 *Reactions in glycolysis in which high energy phosphate ($\sim PO_3H_2$) is produced.*

devoted to their development (see Chapter 22).

Aerobic microorganisms convert pyruvate to acetyl coenzyme A (acetyl CoA), a compound whose oxidation will provide energy for the cell (Figure 12-19). This aspect of metabolism will be discussed later.

It is no accident that there are variations in the fermentation end products released by microorganisms. Some of them have been shown to inhibit growth of other species capable of inhabiting the same ecological niche. Anaerobes in the intestinal tract produce fatty acids, such as butyric and propionic acids, that prevent the colonization of pathogenic species, such as *Salmonella* and *Shigella,* which can cause gastroenteritis, dysentery, and typhoid fever. There are organisms in the oral cavity, such as *Veillonella,* that do not possess enzymes for the fermentation of glucose but can metabolize the products of glycolysis. These products include such compounds as lactate and acetate, both of which are readily supplied via fermentation of glucose by other microorganisms. There are also organisms whose growth is severely restricted because of the products of their own metabolism. In anaerobic environments, for example, macromolecules and other substrates are degraded to fatty acids, ethanol, and hydrogen, which can inhibit the growth of microorganisms producing them. Some anaerobes (organism I), however, can catabolize a product such as ethanol to acetate and hydrogen:

$$\textbf{Organism I} \quad 2\ CH_3CH_2OH + 2\ H_2O \rightarrow 2\ CH_3COOH + 4\ H_2$$

The evolution of H_2 from NADH (NADH + H+ $\rightarrow$ NAD + H_2) is a thermodynamically unfavorable reaction in which the standard free energy is +4.3 kcal/mole under standard conditions (one atmosphere of hydrogen, 25° C, and pH 7.0). The accumulation of hydrogen gas results in a reduction in the amount of substrate (ethanol) that can be metabolized and consequently causes a reduction in growth. If hydrogen gas is removed rapidly from the environment, the free energy change becomes negative and hydrogen can be evolved from NADH. A species of *Methanobacterium* (organism II) supplies this function by removing hydrogen gas and producing methane according to the following reaction:

$$\textbf{Organism II} \quad 4\ H_2 + CO_2 \rightarrow \underset{\textbf{Methane}}{CH_4} + 2\ H_2O$$

Thus organism II is necessary for the successful development of organism I in the environment.

Fermentation of organic molecules other than glucose. Many organisms are equipped to ferment organic compounds other than glucose. A variety of sugars, amino acids, purines, pyrimidines, and other molecules can be fermented. These oxidations as well as others will be discussed later in the chapter.

Entry of sugars into the glycolytic pathway

It might appear that for an organism to use the glycolytic pathway glucose must be present. Most microorganisms in their natural environment do not have glucose readily available to them. Polysaccharides, disaccharides, trisaccharides, and other complex carbohydrates are sometimes present, and most are characterized by the presence of glucose as one of their covalently bonded constituents. Some polysaccharides, such as starch and cellulose, have molecular weights of several hundred thousand and are composed of glucose units linked to each other in specific arrangements. The disaccharides maltose, sucrose, and lactose, as well as certain trisaccharides, also contain glucose as one of their components. Complex polysaccharides cannot penetrate the cytoplasmic membrane, but many microorganisms produce inducible* extracellular enzymes that can hydrolyze the polysaccharide and release the individual glucose or other monosaccharide units so they can be transported into the cell. Some microbes also have the enzymatic potential to hydrolyze disaccharides and trisaccharides found in the environment and release the individual monosaccharides. Monosaccharides such as galactose (a component of lactose), mannose, and fructose (a component of sucrose) can be converted into glycolytic intermediates by phosphorylation and by rearrangement of their hydroxyl groups (Figure 12-22).

Other pathways of glucose oxidation

There are variations of glycolysis in which different intermediates are produced, and pyruvate is not always the final electron acceptor. These pathways represent alternative routes for glucose catabolism, but in other species they may be the sole pathway of fermentation. These pathways include those discussed here.

Pentose phosphate pathway. The pentose phosphate pathway is another pathway for the metabolism of glucose. It provides the cell with pentoses, such as ribose 5-phosphate, that are required for nucleic acid synthesis and also provides much of the NADPH required for biosynthetic reactions in the cell. There are many reactions and enzymes involved in the pentose pathway but two of the pivotal enzymes are called **transketolase** and **transaldolase**. It is their catalytic activity that is most responsible for this pathway being cyclic and for generating so many diverse sugars (Figure 12-23). The cycle can be divided into three basic sets of reactions:

I. Glucose is converted to glucose-6-phosphate just as in glycolysis, but unlike glycolysis oxidation

* Inducible enzymes are those produced by an organism in response to its substrate. In the absence of its substrate inducible enzymes are not synthesized.

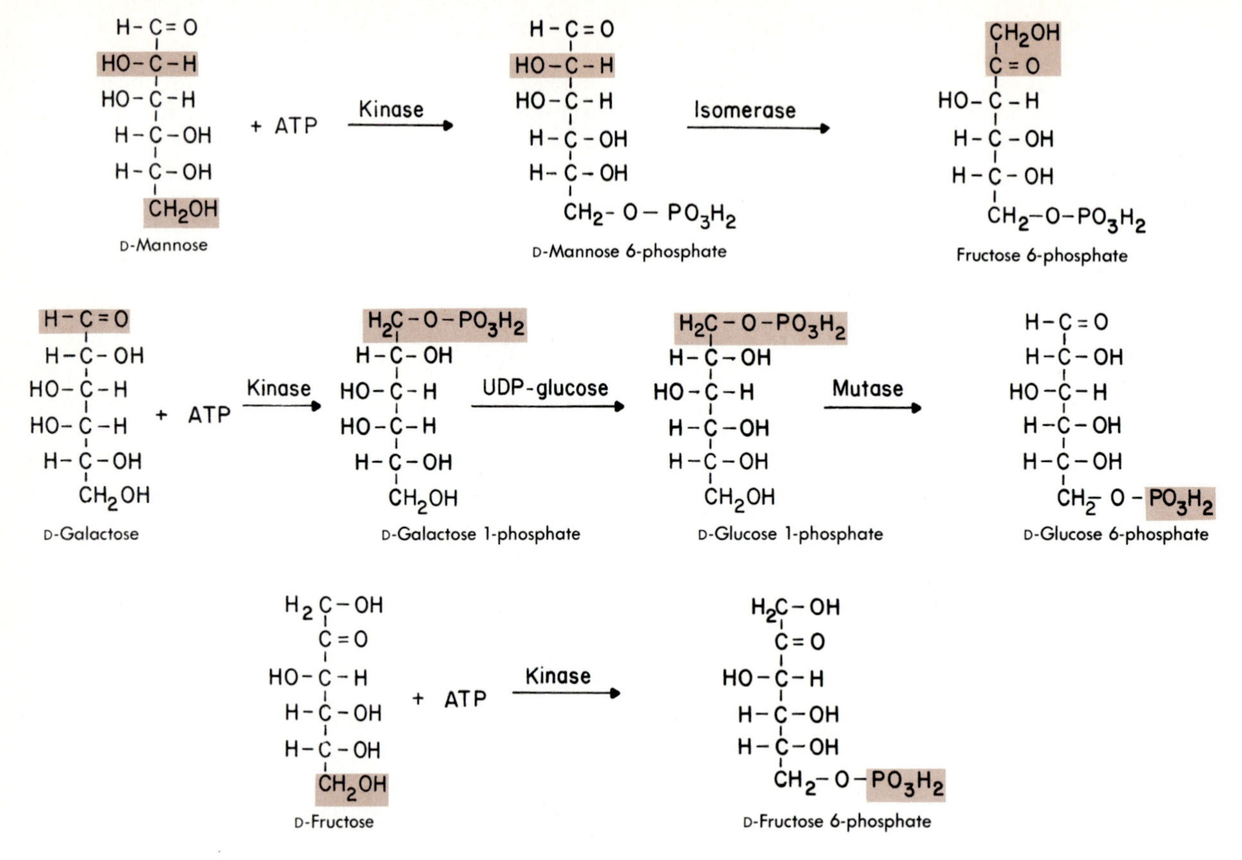

Figure 12-22 *Reactions permitting entry of carbohydrates other than glucose into glycolytic pathway. Boxed areas represent groups that are enzymatically rearranged.*

and decarboxylation steps occur, resulting in the formation of the pentose ribulose-5-phosphate. Ribulose-5-phosphate by epimerization (sugars differing only in the steric configuration of one specific carbon atom are called epimers) forms ribose-5-phosphate and xylulose-5-phosphate.

II. Ribose-5-phosphate and xylulose-5-phosphate are starting points for a reaction involving the enzyme transketolase (transfer of the glycoaldehyde group, —CH_2OH—CO—) and transaldolase (transfer of a 3-carbon —CH_2OH—CO—CHOH). Transketolase transfers a two-carbon keto group from xylulose-5-phosphate to ribose 5-phosphate, producing a seven-carbon sugar sedoheptulose-7-phosphate and glyceraldehyde-3-phosphate. A transaldolase then catalyzes the transfer of the three-carbon unit of sedoheptulose-7-phosphate to the three-carbon glyceraldehyde-3-phosphate to form fructose-6-phosphate and erythrose-4-phosphate.

III. Transketolase catalyzes the interconversion of erythrose-4-phosphate and xylulose-5-phosphate to form fructose-6-phosphate and glyceraldehyde-3-phosphate. These two products can enter glycolysis, and therefore glucose-6-phosphate can be produced. Catabolically glyceraldehyde-3-phosphate can be converted to pyruvate and then to acetyl CoA for inclusion into the Krebs cycle, which provides a source of reducing equivalents for ATP production.

The passage of six molecules of glucose through the cycle results in the complete oxidation of one molecule of glucose-6-phosphate to carbon dioxide and the reduction of 12 NADP to 12 NADPH. In addition to the formation of pentose and NADPH, the cycle also generates erythrose-4-phosphate, an important four-carbon sugar that is a precursor to the amino acids tryptophan, tyrosine, and phenylalanine. In addition, erythrose and sedoheptulose are used in the biosynthesis of cell walls.

Entner Doudoroff (ED) pathway. Glucose is converted in the ED pathway to a six-carbon unit called 2-keto-3-deoxy-6-phosphogluconate, which is later cleaved by an enzyme called aldolase into pyruvate and glyceraldehyde 3-phosphate (Figure 12-24). Glyceraldehyde 3-phosphate produced in the ED pathway can be used in the EMP pathway for conversion to pyruvate and hence

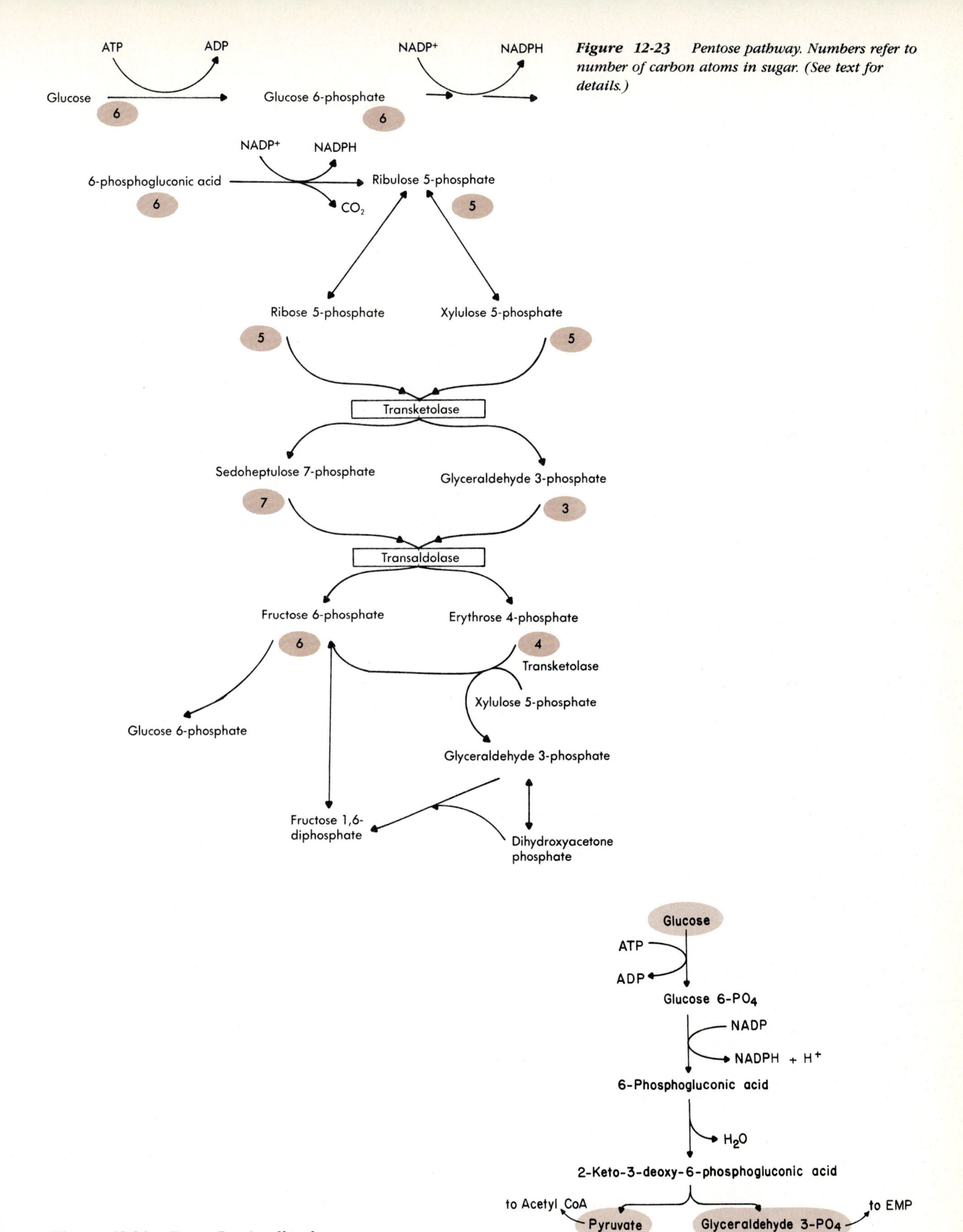

Figure 12-23 *Pentose pathway. Numbers refer to number of carbon atoms in sugar. (See text for details.)*

Figure 12-24 *Entner-Doudoroff pathway.*

to acetyl CoA. The ED pathway was first discovered in aerobic organisms—strict aerobes and facultative anaerobes. Strict aerobes appear to exclusively use the ED pathway, while facultative anaerobes use both the ED and EMP pathways. Strict anaerobes have also been found that have an ED pathway. Only one ATP is gained in the ED pathway, and thus the use of this pathway by strict anaerobes would appear to be an inefficient way to metabolize glucose unless the organism had a defective EMP pathway.

Point of Interest

RUMEN METABOLISM—A SPECIALIZED FERMENTATION PROCESS

Plant material in ruminants (sheep, cattle, goats, etc.) is indigestible by mammalian enzymes, and microorganisms (figure, right) are the primary digesters of food. The rumen stomach, unlike the nonruminant stomach, is divided into compartments, the first of which acts like a large fermentation vessel that in cattle may hold up to 100 liters. The first chamber is kept at alkaline pH and a temperature of 39° C, which is ideal for microbial growth. The oxidation-reduction potential in the rumen is approximately −30 mV; therefore all the indigenous microorganisms engage in anaerobic metabolism. The principal microorganisms found in the rumen are primarily gram-negative rodlike bacteria, but some protozoa are also present.

The diet of rumens consists primarily of plant material that is made up of cellulose, hemicellulose, and pectin but little protein, fats, or digestible carbohydrates. Fibrolytic microorganisms, which make up only 5% of the rumen population, produce extracellular enzymes that break down plant polysaccharides into soluble sugars such as glucose. Some of these decomposers include species of *Bacteroides, Ruminococcus, Clostridium,* and *Streptococcus.* A large percentage of the soluble sugars is utilized by nonfibrolytic microorganisms as a source of energy, and thus the population numbers are maintained in the rumen. The products of rumen fermentation include gases such as hydrogen, methane, and carbon dioxide plus a variety of organic acids such as acetic, propionic, formic, succinic, and lactic. Gases produced in the rumen are eliminated by constant belching. The organic acids pass through the rumen epithelium and are used as energy sources by the animal. Sugars are the principal energy source for most mammals. Thus the rumen is an example of a symbiotic relationship in which both microorganism and animal benefit.

Examination of the rumen contents reveals a constant proportion of propionic acid, acetic acid, butyric acid, and methane but little hydrogen gas, formic acid, lactic acid, or succinic acids. These compounds are rapidly metabolized according to the following reactions by some of the indicated microbial species:

1. HCOOH (formic acid) → $H_2 + CO_2$ *(Ruminococcus albus)*
2. $4H_2 + CO_2 \rightarrow CH_4 + 2H_20$ *(Methanobacterium ruminantium)*
3. $CH_3CHOHCOOH$ (lactic acid) → CH_3COOH (Acetic acid) or CH_3CH_2COOH (propionic acid) or $CH_3CH_2CH_2COOH$ (butyric acid)

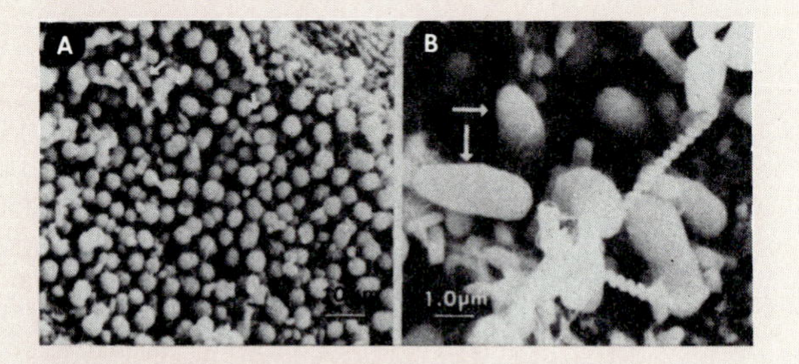

Rumen bacteria in lamb. ***A,*** *Coccus is predominant morphological type. Some short rods* (arrow) *are also present.* ***B,*** *Spiral-shaped organisms are present in low numbers. Large rods* (arrow) *present in depressed areas of rough topography.*

4. $COOHCH_2CH_2COOH$ (succinic acid) → CH_3CH_2COOH (propionic acid) + CO_2

In the absence of methanogens the products of metabolism in the rumen are hydrogen, carbon dioxide, formic acid, and succinic acid. In the presence of methanogens the products are primarily carbon dioxide, methane, and unusually low levels of hydrogen. Methanogens perform an important role by capturing electrons produced during fermentation and converting the excess hydrogen, produced by nonmethanognic bacteria, to methane. By the use of electrons for the formation of hydrogen and methane, pyruvate is converted to acetyl CoA, and energy is generated in the form of ATP. If the electrons generated during fermentation were not used to produce hydrogen and methane, pyruvate would be reduced to ethanol and lactate, and considerable potential energy would be lost.

Most of the rumen bacteria can use ammonium (NH_4) as a source of nitrogen; thus inorganic nitrogen is converted to amino acids and thence to protein in the microbial cell. Microbial cells pass out of the rumen together with undigested plant material, such as lignin, into other stomach compartments. These micro-organisms are digested by host enzymes such as proteases; thus the microorganism, and not protein from the ingested food, serves as a source of nitrogen for the ruminant. Since ammonium is a source of nitrogen for the ruminant bacteria, urea (NH_2—CO—NH_2), which is normally a waste product, can be used as a supplemental nitrogen source for the ruminant.

Figure 12-25 *Tricarboxylic acid (TCA) cycle.*

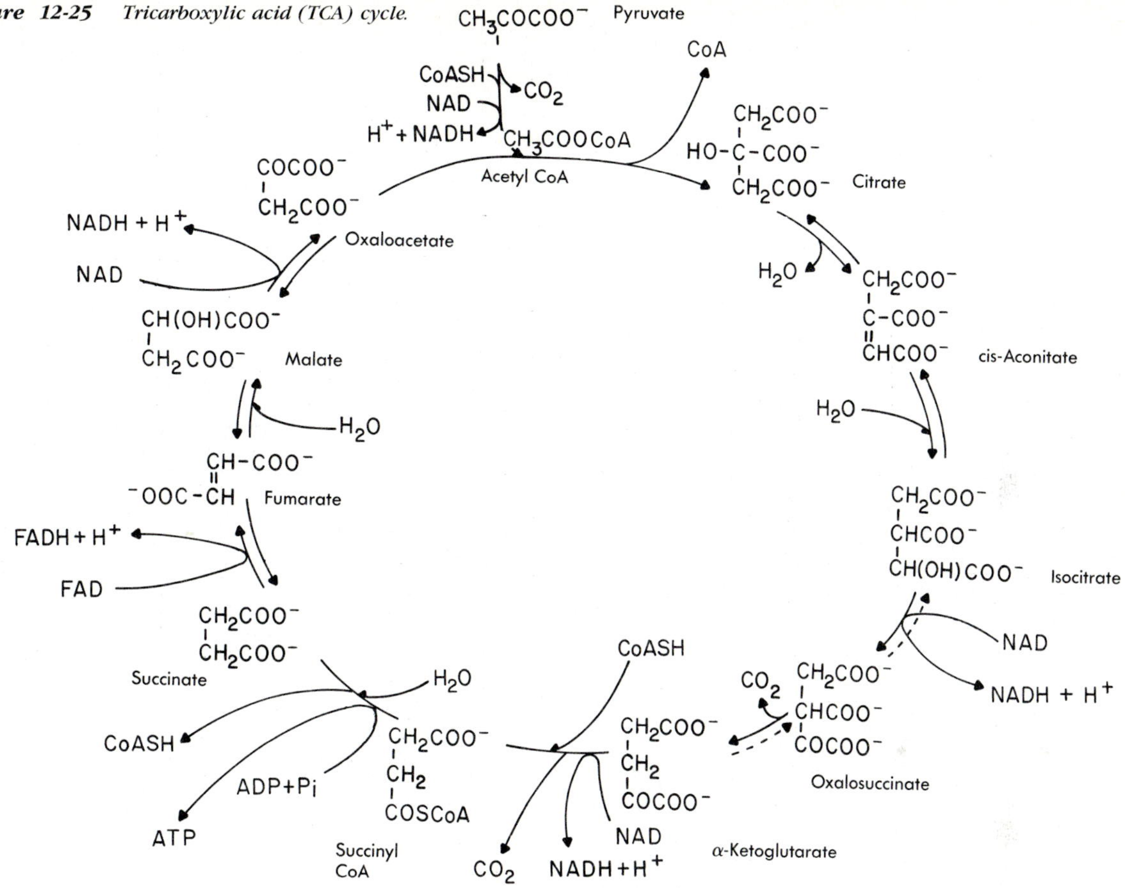

Respiration

Respiration is defined as a process in which the electron donor, either an organic (could be glucose) or inorganic compound, is oxidized and the final electron acceptor is oxygen. It is speculated that oxygen accumulated in the primordial atmosphere only after the appearance of oxygenic photosynthetic microorganisms and that heterotrophs later evolved the machinery for using oxygen as a final electron acceptor. The potential for energy release during respiration is greater than in fermentation because the organic electron donor is completely oxidized to carbon dioxide and water with the release of many electron pairs:

$$\text{Hydrogenated substrate (e.g., glucose)} \xrightarrow{\text{Dehydrogenation}} 2H^+ + 2e^- + (CO_2 \text{ and } H_2O \text{ release})$$

The electrons that are removed from organic fuels are captured primarily by NAD and then transported along the electron transport chain where the release of free energy is coupled to the phosphorylation of ADP to ATP. Electron transport phosphorylation was discussed earlier (p. 309). Our immediate concern is to determine how organic substrates are completely oxidized to carbon dioxide and water.

Tricarboxylic acid (TCA) cycle. The TCA cycle (also called **citric acid** or **Krebs cycle**) can be viewed as a pathway for the oxidation of acetate derived from the fermentation glucose, but acetate can also be derived from the catabolism of proteins and lipids. The TCA cycle is responsible for the release of a large number of electrons and protons and produces compounds that can be used as intermediates for biosynthesis (see Chapter 13). The complete TCA cycle is summarized in Figure 12-25, some of the major reactions are discussed here.

1. Before initiation of the cycle, pyruvate, produced in glycolysis, is oxidized to acetate with the release of carbon dioxide and two hydrogen atoms. The acetate is actually in an activated state as part of the coenzyme A (CoA) molecule (Figure 12-26). CoA is a carrier of acetyl groups, and the bond between the acetate and the sulfhydryl (SH) group of CoA is a high-energy linkage. Hydrolysis of the

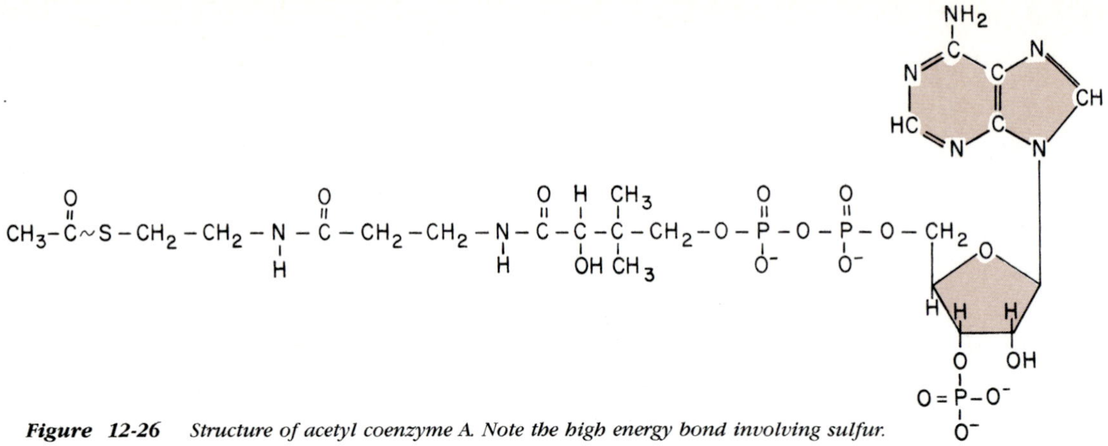

Figure 12-26 *Structure of acetyl coenzyme A. Note the high energy bond involving sulfur.*

SH-acetate linkage results in the release of more free energy than is obtained from the hydrolysis of the terminal phosphate of ATP (Table 12-2). Hydrogen ions and their electrons are picked up by specific carriers such as NAD (nicotinamide adenine dinucleotide).

2. The acetyl group of acetyl CoA is transferred enzymatically to oxaloacetate (OAA), a four-carbon unit, to produce **citrate**, a six-carbon compound. It is this reaction that initiates the TCA cycle and regenerates CoA.
3. Citrate, in a series of reactions, loses a carbon atom as carbon dioxide and two hydrogen atoms to produce a five-carbon intermediate called **α-ketoglutarate (α-kg)**. NAD is the carrier of the hydrogen atoms and electrons.
4. α-kg loses a carbon atom as carbon dioxide to form **succinate**, a four-carbon intermediate, and two electrons and hydrogen ions are released and trapped by NAD. Actually during this conversion process succinyl-CoA is an intermediate. The energy of the thioester of succinyl-CoA is used to synthesize ATP from ADP and inorganic phosphate. This is an example of substrate-level phosphorylation.
5. Succinate is oxidized to **fumarate**, a four-carbon intermediate, with the release of hydrogen, which is captured by flavine adenine dinucleotide (FAD). Fumarate is hydrated to become **malate.**
6. Malate, a four-carbon compound, is oxidized to the four-carbon **oxaloacetate** with the release of hydrogen, which is captured by NAD.

In summary, respiration may be described as occurring in three stages:

Stage I. Two acetyl groups are produced by the oxidation of one glucose (acetyl groups can also be produced by the oxidation of amino acids, or fatty acids).

Stage II. Acetyl groups are channeled into the Krebs cycle by condensation with oxaloacetate (OAA). The acetyl groups are enzymatically degraded to carbon dioxide with the release of energy-rich hydrogen atoms.

Stage III. Hydrogen atoms are separated into protons (H^+) and energy-rich electrons. The electrons are transferred along the electron transport chain until the final electron acceptor, oxygen, is reduced to water, that is:

$$O_2 + H^+ + NADH \rightarrow H_2O + NAD$$

Energy profile: respiration vs. fermentation

During glycolysis there is a net gain of two ATP via substrate phosphorylation. However, reduced NAD is also produced, which during respiration is not used to reduce pyruvate to lactate or some other reduced end product. Instead, the reduced NAD enters the electron transport chain where three ATPs are produced for every pair of electrons transported through it. The other electron carrier, FAD, transports its electrons to the electron transport chain, but only two ATPs are produced per pair of electrons transported. The complete oxidation of glucose during respiration results in the formation of 38 molecules of ATP. Thus the complete oxidation of glucose produces considerably more ATPs than can be produced by the fermentation of glucose. The reactions in both cycles from which ATP can be realized are outlined in Table 12-8.

Although fermentation might appear to be a wasteful process, in habitats where several species are present the process is useful. The products of fermentation still possess potential energy, and when they are excreted, can provide a source of energy to other microorganisms.

Anaerobic respiration. Anaerobic respiration is a

Table 12-8 *Total number of ATPs gained from the complete oxidation of glucose*

Step in oxidation		**No. of ATP**
Glucose oxidized to pyruvate (substrate phosphorylation)		2
Glyceraldehyde 3-phosphate ⇌ 1,3-diphosphoglycerate + 2 NADH (to electron transport system)		6
Pyruvate → acetyl CoA + 2 NADH (to electron transport system)		6
TCA cycle	6 NADH	18
TCA cycle	2 FADH	4
2 succinyl-CoA → 2 succinate + 2 ATP* (substrate phosphorylation)		2
	TOTAL	38

*Overall reaction involves GTP:

$$\text{Succinyl CoA} + \text{GDP} \xrightarrow{\text{iPO}_4} \text{Succinate} + \text{GTP}$$
$$\text{GTP} + \text{ADP} \longrightarrow \text{GDP} + \text{ATP}$$

process in which the electron donor is an inorganic or organic molecule and the final electron acceptor is usually an inorganic molecule other than oxygen or occasionally an organic molecule. The mechanism of electron transport and its coupling to ATP formation is basically the same in nonfermentative anaerobes as it is in aerobes but there are some variations in the anaerobic system:

1. The final electron acceptor is not oxygen but is some other inorganic molecule, such as nitrate (NO_3^-), sulfate ($SO_4^=$), nitrite (NO_2^-), or an organic molecule, such as fumarate. Table 12-9 lists the various groups of bacteria that utilize these electron acceptors during anaerobic respiration.
2. The terminal enzyme component in electron transport in anaerobes is not cytochrome oxidase but enzymes such as **reductases** that catalyze the reduction of nitrate to nitrite and nitrite to ammonia, sulfate to hydrogen sulfide, and nitrite to nitric oxide (NO). Other reductases, such as nitric oxide reductase and nitrous oxide reductase, complete denitrification. One of the most frequently used inorganic electron acceptors is **nitrate,** whereas **fumarate** is a more common organic electron acceptor. The synthesis of terminal reductases is repressed and their activities inhibited by oxygen during anaerobic respiration. This mechanism is especially important in facultative anaerobes that switch from anaerobic to aerobic growth when oxygen becomes available. Nitrate reductase has been studied extensively in *Escherichia coli.* When this organism is grown anaerobically with nitrate, nitrate reductase can account for up to 20% of the protein in the cytoplasmic membrane. During anaerobic growth, formate is one of the principal products of metabolism, and this is accompanied by the synthesis of formate dehydrogenase. Formate and nitrate together make up a simple respiratory circuit. **Formate dehydrogenase** is a molybdenum-selenium–containing protein

Table 12-9 *Examples of electron acceptors used by various genera of bacteria during anaerobic respiration*

Genera	**Electron acceptor**	**Reduced product**
Escherichia, Klebsiella and all other Enterobacteriaceae	Nitrate (NO_3^-) NO_2^-	Nitrite (NO_2^-) Ammonia (NH_3)
Bacillus, Pseudomonas, Alcaligenes	NO_3^- NO_2^-	NO_2^- Nitrous oxide (N_2O) or N_2
Vibrio, Escherichia, Streptococcus, Bacteroides, Desulfovibrio, Propionibacterium, Haemophilus, and *Bacillus*	Fumarate ($^-OOC{-}CH{=}CH{-}COO^-$)	Succinate ($^-OOC{-}CH_2{-}CH_2{-}COO^-$)
Desulfotomaculum, Desulfovibrio, Desulfuromonas	Sulfate ($SO_4^=$) Elemental sulfur (S^o)	Sulfide (H_2S) H_2S
Methanobacterium	CO_2	Methane (CH_4)
Bacillus, Pseudomonas	Ferric iron (Fe^{3+})	Ferrous iron (Fe^{2+})

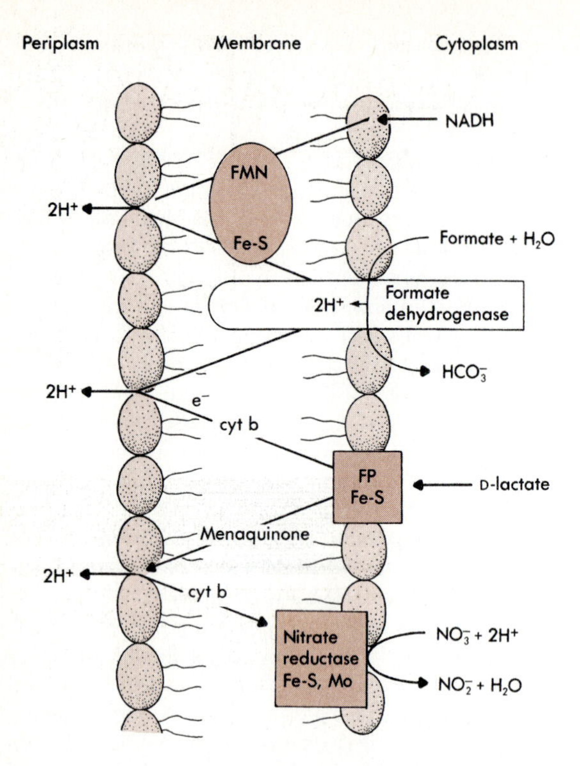

Figure 12-27 *Proposed pathway for anaerobic respiration in* E. coli *with nitrate as terminal electron acceptor and NADH as electon donor. During anaerobic growth, formate is produced and formate dehydrogenase is induced. When lactate is the carbon source, pathway begins at FP-FeS complex. There are two types of cytochrome b involved in this scheme.* FP, *Flavoprotein;* FeS, *iron-sulfur protein;* Mo, *molybdenum.*

that also contains a *b*-type cytochrome and is similar to nitrate reductase. Formate dehydrogenase spans the cytoplasmic membrane and is believed to pump protons across it. **Nitrate reductase** possesses a molybdenum-containing cofactor as well as FeS centers and cytochrome b activity. There is some controversy as to the number and kind of transport proteins involved in this system, but a proposed system is illustrated in Figure 12-27.

3. Less energy is realized during electron transport under anaerobic conditions. The difference in the oxidation-reduction potential between the donor and acceptor (for example, between NADH and nitrate; see Table 12-4) is not sufficient to generate three molecules of ATP. Depending on the donor and acceptor, the number of ATP molecules generated during electron transport under anaerobic conditions is usually between one and two. The rate of growth of anaerobes is, therefore, considerably slower than that of aerobes. A facultative anaerobe, for example, under anaerobic conditions might require 48 hours of optimal growth conditions to produce a population of cells that under aerobic conditions could be established in 16 hours or less.

Concept Check

1. **What is the significance of the variety of end products produced during fermentation?**
2. **How can simple sugars, other than glucose, as well as more complex carbohydrates, be made available for glycolysis?**
3. **What are the basic differences between respiration and fermentation? What metabolic pathways are associated with them?**
4. **What is the function of the pentose phosphate pathway, and what is the role of transaldolase and transketolase?**
5. **What is the function of the TCA cycle? Compare the number of ATPs produced by the TCA cycle with the number produced during glycolysis.**

ENERGY PRODUCTION FROM ORGANIC SUBSTRATES OTHER THAN CARBOHYDRATES

One might get the impression after reading this chapter that heterotrophic microorganisms can use only carbohydrates as organic energy sources. It is true that the principal pathway involved in energy production is the EMP pathway, where glucose and other sugar intermediates are oxidized. Glucose or other simple sugars are not primary sources of energy for many microorganisms in their environment. Instead, they must extract energy from other organic compounds that might be present, such as amino acids, purines or pyrimidines, fatty acids, and a variety of hydrocarbons. These compounds are either channeled directly into energy-producing pathways or oxidized to products that can be channeled. Only some amino acids, purines, and pyrimidines, and some organic acids can be fermented. Molecules such as fatty acids, or hydrocarbons are too reduced to be fermented. Fermentable compounds are those that are neither too oxidized or reduced. If substrates are too oxidized, for example, acetate, not enough energy can be realized from their further oxidation, and maximum growth of an organism is severely hampered. If substrates are too reduced, they will be unable to serve as metabolic electron acceptors. Some of the pathways used by the cell to derive energy from organic compounds other than carbohydrates follow.

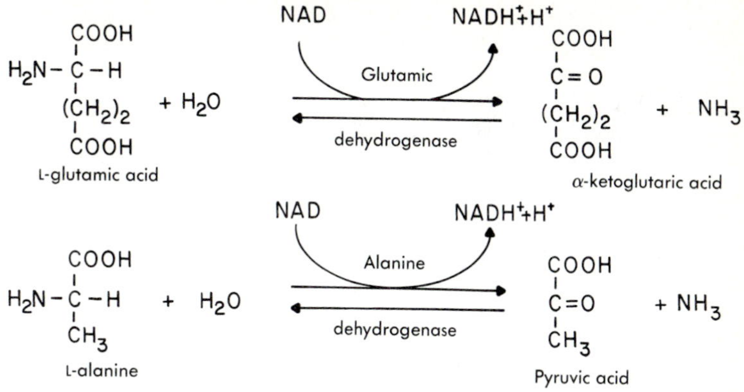

Figure 12-28 *Examples of oxidative deamination of amino acids that result in formation of keto acids.*

$$\underset{\text{L-glutamic acid}}{HOOC-(CH_2)_2-CH(NH_2)-COOH} + H_2O \underset{\text{dehydrogenase}}{\overset{\text{Glutamic}}{\rightleftharpoons}} \underset{\alpha\text{-ketoglutaric acid}}{HOOC-(CH_2)_2-C(=O)-COOH} + NH_3 \quad (NAD \rightarrow NADH^+ + H^+)$$

$$\underset{\text{L-alanine}}{CH_3-CH(NH_2)-COOH} + H_2O \underset{\text{dehydrogenase}}{\overset{\text{Alanine}}{\rightleftharpoons}} \underset{\text{Pyruvic acid}}{CH_3-C(=O)-COOH} + NH_3 \quad (NAD \rightarrow NADH^+ + H^+)$$

Amino acid oxidation or hydrolysis

Amino acids may be free in the environment, or they may be the products of protein hydrolysis by proteases excreted by microorganisms. Amino acids can be oxidatively deaminated to keto acids, which can directly enter the TCA cycle (Figure 12-28). Other amino acids may be reductively deaminated to produce saturated fatty acids, for example:

$$R—CH(NH_2)—COOH \rightarrow R—CH_2—COOH + NH_3$$

A very few microbes can obtain their energy directly from the hydrolysis of an amino acid such as arginine according to the following reactions:

$$\text{Arginine} \rightarrow \text{Citrulline} + NH_3$$

$$\text{Citrulline} + Pi + ADP \rightarrow \text{Ornithine} + CO_2 + NH_3 + ATP$$

The hydrolysis of arginine is a significant source of energy for some fermentative bacteria but plays a minor role in aerobic energy production.

One of the characteristics of the anaerobic clostridia and some other anaerobes is that they obtain energy from the fermentation of amino acids. Some clostridia are unable to ferment sugars and therefore rely on amino acid fermentation as a means of extracting energy. Many of the clostridia that ferment amino acids are also proteolytic, for example, *Clostridium perfringens* and *C. histolyticum*, which are causative agents of gas gangrene in animals. Proteolysis of muscle by these organisms and subsequent amino acid fermentation results in the formation of foul-smelling products characteristic of disease. One of the most important discoveries associated with microbial fermentation was recorded in 1934 by Stickland, who observed that some clostridia ferment single amino acids poorly but readily degrade certain pairs of amino acids. The amino acid pairs are coupled in oxidation-reduction reactions that are referred to as **Stickland reactions**. The oxidation reactions are similar to those catalyzed by aerobic organisms except that oxygen is absent. The oxidations include oxidative deaminations or transaminations, but the reduction reactions are distinctive because a suitable electron acceptor must be generated from the amino acids that the anaerobe metabolizes. The major electron acceptors are various amino acids, keto acids, protons, or unsaturated acids or their CoA thiolesters. The simplest example of a Stickland reaction involves two amino acids, glycine and alanine, fermented by *Clostridium sporogenes*. The overall reaction is as follows:

$$\underset{\textbf{Alanine}}{CH_3CHNH_2COOH} + 2\ \underset{\textbf{Glycine}}{CH_2NH_2COOH} + ADP + P_i \rightarrow 3\ NH_3 + 2\ CH_3COOH + CO_2 + ATP$$

in which glycine is the electron acceptor and alanine the electron donor. The various reactions in this fermentation process are shown in Figure 12-29.

Purine and pyrimidine oxidation

The hydrolysis of nucleic acids results in the release of nucleosides and nucleotides. The purines and pyrimidines can be converted to a variety of products, including glycine, carbon dioxide, formate, aspartate, alanine, propionic acid, and isobutyric acid. Many of these compounds can be used as precursors in the biosynthesis of new purine and pyrimidine bases (see p. 370) and can also be channeled into TCA intermediates.

Lipid and fatty acid oxidation

Lipid degradation is not a common mechanism among microorganisms for obtaining their energy, because lipids are not always a frequent commodity in the environment. Lipid degradation is often associated with microorganisms that are parasites of animals. Staphylococci, for example, inhabit the human skin and break down

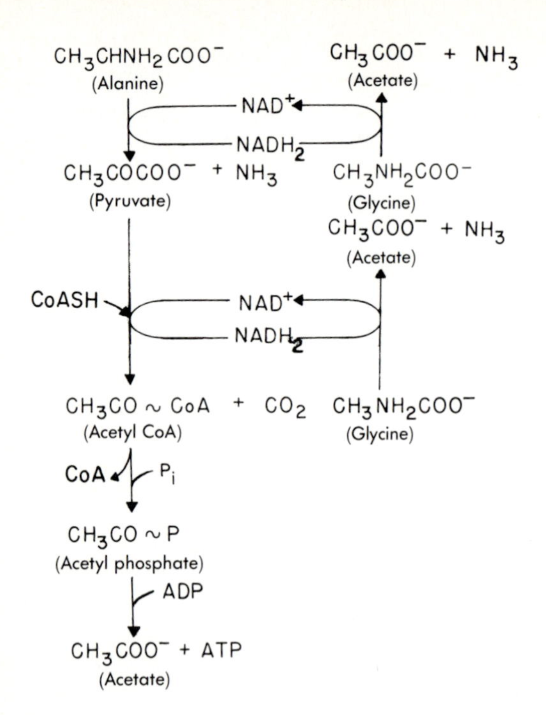

Figure 12-29 *Stickland reactions. Example of amino acids acting as electron donors (alanine in example) and electron acceptors (glycine).*

surface lipids to fatty acids, which may be assimilated directly by the cell into lipids, or they may be further oxidized to acetyl CoA units in a process called **beta-oxidation**. Fatty acid catabolism usually occurs when the chain length is more than eight carbons. The fatty acid is first converted to a CoA ester by acyl CoA synthetase. The ester is then oxidized in the beta position:

$$\underset{\substack{\uparrow \\ \textbf{Beta}}}{RCH_2CH_2COCoA} \xrightarrow{\text{oxidation}} RCH{=}CHCOCoA$$

Subsequently, acetyl CoA is released (after addition of water across the double bond and oxidation of beta carbon to keto), leaving behind a CoA ester of the fatty acid shortened by two carbons. Even-numbered fatty acids yield only acetyl CoA units, whereas odd-numbered fatty acids yield acetyl CoA units plus propionyl CoA. Reduced NAD and FAD are generated during fatty acid oxidation and can be transported to the respiratory chain to generate energy. Acetyl CoA can also enter the glyoxylate cycle, which functions in the conservation of TCA intermediates (see Chapter 13).

Hydrocarbon oxidation

The ability to hydrolyze hydrocarbons, that is, **alkanes** and **alkenes**, is found among several species of microorganisms. This type of reaction has recently drawn considerable attention. Microorganisms have been genetically manipulated to produce enzymes that rapidly degrade petroleum products and related hydrocarbons. They are being considered for use in oil spills so that degradation processes can be speeded up and environmental recovery enhanced. The oxidation of alkanes with 10 to 18 carbons by microorganisms takes place according to the following reactions:

$$\underset{\textbf{Alkane}}{R{-}(CH_2)_nCH_3} \xrightarrow{O_2} \underset{\textbf{Alcohol}}{R{-}(CH_2)_nCH_2OH} \rightarrow$$
$$\underset{\textbf{Aldehyde}}{R{-}(CH_2)_nCHO} \rightarrow \underset{\textbf{Fatty acid}}{R{-}(CH_2)_nCOOH}$$

One of the major problems for microorganisms that degrade hydrocarbons and use them for carbon and energy sources is that hydrocarbons are deficient in nitrogen, phosphorus, and oxygen. Since most natural aquatic environments are deficient in usable nitrogen and phosphorus, the latter nutrients must be added to the system. One technique that is being considered to eliminate the need for additions of nitrogen is to manipulate the microorganism so that it is genetically capable of fixing nitrogen.

The manufacture of many toxic synthetic hydrocarbons and their subsequent release into the environment are causing considerable ecological problems. Bacteria and fungi can metabolize many diverse hydrocarbons, but a few, such as pesticides and herbicides, are unusually recalcitrant. The chlorinated insecticides, such as DDT, aldrin, and chlordane, are among the most resistant to microbial degradation* and can persist in the soil in an unaltered state for up to 5 years. The metabolism of some of these recalcitrant molecules can be speeded up if an additional organic energy source, which can be readily degraded, is also present. This phenomenon is called **cometabolism**.

Organisms that cannot generate their own energy

Some microorganisms are obligate intracellular parasites unable to survive outside of a living host. Many of these parasites require a variety of different nutrients or enzymes that can be supplied only by the host cell. The parasite, in some instances, is unable to produce any energy, and two groups of parasites fall into this category: chlamydiae, which are gram-negative bacteria, and viruses. Chlamydiae possess RNA, DNA, and ribosomes but lack certain key enzymes that lead to the synthesis of certain amino acids and nucleosides. In addition, they

* Recent genetic engineering experiments have demonstrated that a gene coding for the enzymatic removal of chlorine can be cloned in bacteria. The dechlorinated compounds are more readily metabolized by microorganisms normally present in the environment.

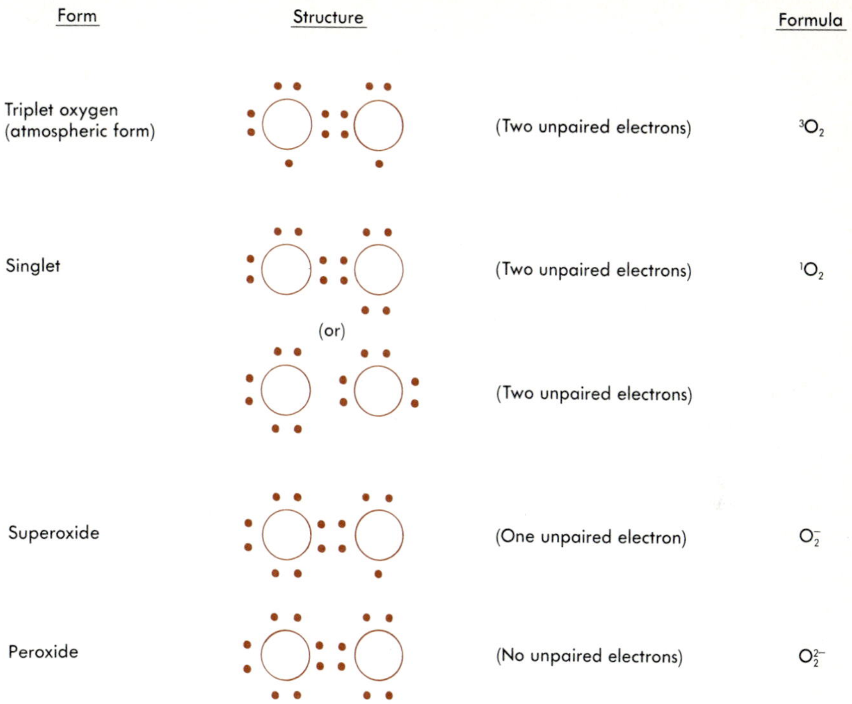

Figure 12-30 *Different molecular forms of oxygen. Not included is hydroxyl radical (·OH) that possesses one unpaired electron and is produced during reduction of oxygen to water. Singlet oxygen is not produced during microbial reduction of oxygen to water but can be produced by reaction of visible light with triplet oxygen in mammalian cells as in white blood cells.*

Form	Structure		Formula
Triplet oxygen (atmospheric form)		(Two unpaired electrons)	$^3O_2$
Singlet		(Two unpaired electrons)	$^1O_2$
	(or)	(Two unpaired electrons)	
Superoxide		(One unpaired electron)	O_2^-
Peroxide		(No unpaired electrons)	O_2^{2-}

also require the mitochondria of the host cell to supply energy. These pathogens are permeable to ATP, a characteristic that is not associated with other pathogens. Viruses contain substantially less genetic information than eukaryotes or prokaryotes, and the enzymes necessary for generating energy are missing. Viruses can produce new virus particles only as long as the host's energy-generating systems are intact. Once the energy-producing components of the host have been destroyed, no new viral particles can be synthesized.

THE ROLES OF OXYGEN IN METABOLISM

Oxygen plays an especially important role as the terminal electron acceptor in aerobic respiration. Facultative anaerobes and microaerophiles can use oxygen as a terminal electron acceptor for energy production, but the growth of microaerophiles and even some aerobes can be inhibited, depending on the concentration of oxygen. Oxygen can exist in different molecular forms (Figure 12-30). Oxygen is lethal to some obligate anaerobes, but in others oxygen may be tolerated but no growth occurs. Aerotolerant anaerobes such as lactic acid bateria, however, can grow in the presence of oxygen.

Oxygen and aerobic bacteria

Cells that use oxygen as a terminal electron acceptor possess cytochrome oxidase or a comparable oxidase to catalyze the final transfer of electrons and the subsequent formation of water. Nearly all aerobes possess additional enzymes that use oxygen as an electron acceptor, but products other than water are formed and no energy is produced. Two of these products, **hydrogen peroxide** and **superoxide**, are toxic to the cell. Most cells synthesize **flavoprotein oxidases** that catalyze the divalent (a two-electron transfer) reduction of oxygen with the formation of hydrogen peroxide (H_2O_2):

$$H_2 + O_2 \xrightarrow[\text{oxidase}]{\text{flavoprotein}} H_2O_2$$

Microorganisms produce two kinds of enzymes that can break down hydrogen peroxide: **catalase** and **peroxidase**. Catalase catalyzes the following reaction:

$$2\ H_2O_2 \xrightarrow{\text{catalase}} 2\ H_2O + O_2$$

Peroxidases catalyze the reduction of hydrogen peroxide to water through the oxidation of organic compounds.

Oxygen can also accept a single electron (univalent reduction) from reduced flavines and be converted to the superoxide anion, O_2^-, as well as other intermediates, such as the hydroxyl radical. The sequence of univalent reductions of oxygen can be represented in the following way:

$$O_2 \xrightarrow{e^-} O_2^{\overline{\cdot}} \xrightarrow{e^- + 2H^+} H_2O_2 \xrightarrow{e^- + H^+} \cdot OH \xrightarrow{e^- + H^+} H_2O$$

Superoxide anion **Hydrogen peroxide** **Hydroxyl radical**

The superoxide radical is more stable than the hydroxyl radical, but the latter is believed to be the intermediate that is most toxic to the cell. Where superoxide and hydrogen peroxide are generated simultaneously, the possibility of hydroxyl radical production exists. The microbial cell prevents the buildup of superoxide and other intermediates through the synthesis of the enzyme **superoxide dismutase,** which removes the toxic anion according to the following reaction:

$$2H^+ + O_2^- + O_2^- \xrightarrow[\text{dismutase}]{\text{superoxide}} H_2O_2 + O_2$$

Hydrogen peroxide is not as deleterious as once suspected, and in most aerobes catalase or peroxidase can handle any excess of this compound. Nearly all aerobic cells possess superoxide dismutase, but in some microbial species excess oxygen may inhibit the enzyme. Key enzymes important in energy formation may be inhibited by elevated concentrations of superoxide, and this may be the reason that microaerophiles cannot handle oxygen at partial pressures greater than 0.2 atmospheres.

The formation of superoxide dismutase indicates that in addition to the divalent reduction of oxygen via the usual respiratory enzymes, such as cytochrome oxidase, some univalent reductions do occur. A cell, for example, might reduce 95% of the available oxygen to water, but 5% could be converted to the superoxide radical.

Facultative anaerobes have the luxury of growing in the presence or absence of oxygen. Respiratory enzymes in facultative anaerobes are induced by oxygen when the gas is present in the environment, but anaerobic energy-forming systems are inhibited. The enzyme lactate dehydrogenase, for example, which catalyzes the conversion of pyruvate to lactate during fermentation, is not only inhibited by oxygen, but its synthesis is also repressed. Yet, in aerotolerant anaerobes such as lactic acid bacteria, oxygen does not inhibit the enzyme.

Oxygen and anaerobic bacteria

Obligate anaerobes are not able to grow in the presence of air, but they do show different sensitivities to oxygen. Some anaerobic bacterial populations are almost completely killed when exposed to oxygen for 3 minutes, whereas others remain viable even though they cannot grow. Indifference to oxygen by some anaerobes is the result of their inability to reduce oxygen; therefore toxic intermediates are not produced. In other anaerobes the concentration of oxygen is the critical factor. Obligate anaerobes are often found associated with facultative anaerobes in areas where oxygen is present. Facultative anaerobes can remove oxygen from the environment by using it in aerobic respiration, thus making the environment suitable for the growth of obligate anaerobes. Studies with oxygen-tolerant anaerobes have demonstrated that a positive correlation exists between oxygen tolerance and the content of superoxide dismutase in the cell. It is now clear that microbial cells, whether aerobes or anaerobes, rely on scavenger systems such as catalase, peroxidase, and superoxide dismutase to remove toxic intermediates produced by the reduction of oxygen.

Concept Check

1. **What are the products of oxidation of amino acids, purines, and pyrimidines, fatty acids, and hydrocarbons? What metabolic pathways or chemical components will those oxidation products enter?**
2. **What are some of the oxygen derivatives produced during electron transport and reduction of oxygen, and what is their effect on aerobic and anaerobic bacteria?**

ENERGY PRODUCTION DURING OXIDATION OF INORGANIC MOLECULES (CHEMOAUTOTROPHS)

Inorganic molecules can serve as a source of energy in those microorganisms referred to as **chemoautotrophs.** Chemoautotrophy is unique to bacteria. The term **chemolithotroph** is sometimes used in place of chemoautotrophs. Lithotroph means "rock eating" and implies an existence on minerals. These organisms, like the photosynthetic organisms, use carbon dioxide as a source of carbon.

Chemoautotrophs, according to present-day view, represent a more complex group of microorganisms than was once supposed. Scientists had previously considered them to be among the most primitive prokaryotes. This was an understandable conclusion, since they have an unusually slow rate of growth and extract their energy and carbon from inorganic sources. Reduced organic compounds were believed to be abundant during the anaerobic period of the earth's history. The fermentative bacteria that were present during this period flourished abundantly, and when the organic material diminished, there began a competition for energy sources. Reduced inorganic compounds, which were undoubtedly present in large quantities during this period because of volcanic activity, represented alternative

Table 12-10 ***Principal inorganic energy sources used by chemoautotrophs (chemolithotrophs)***

Energy source (electron donor)	Electron acceptor	Metabolic products	Physiological group	Bacterial species
H_2	O_2	H_2O	Hydrogen bacteria	*Hydrogenomonas* sp.
S	O_2	H_2SO_4	Sulfur-oxidizing bacteria	*Thiobacillus* spp., *Beggiatoa*
H_2S	O_2	H_2SO_4	Sulfur-oxidizing bacteria	*Thiobacillus* spp., *Beggiatoa*
S	NO_3	H_2SO_4, N_2	Sulfur-oxidizing bacteria	*Thiobacillus denitrificans*
Fe^{++}	O_2	Fe^{+++}	Iron-oxidizing bacteria	*Thiobacillus ferrooxidans*
H_2	SO_4	S^{--}, SO_3^{--}, H_2O	Dissimilatory sulfate reducers	*Desulfovibrio* spp.*
NO_2	O_2	NO_3^-, H_2O	Nitrifying bacteria	*Nitrobacter* spp.
NH_4^+	O_2	NO_2^-, H_2O	Nitrifying bacteria	*Nitrosomonas* spp.
CO	O_2	CO_2, H_2O	Carboxydotrophic bacteria	*Pseudomonas* spp., *Clostridium* spp.

*Can also use certain organics, such as lactic acid, as energy source.

energy sources, and therefore autotrophic microorganisms had the best chance of survival.

Most chemoautotrophs can survive in an entirely inorganic environment as long as carbon dioxide and oxygen are available. The growth of many species is hindered by the presence of organic compounds, but some species can use organic substrates, in a limited way, as carbon and energy sources. Chemoautotrophs are found in many diverse habitats in the biosphere—soil and fresh and marine waters—and are engaged in the cycling of elements in the biosphere (these activities are discussed in Chapters 18 to 20). The principal inorganic compounds used as energy sources by this group of microorganisms are listed in Table 12-10.

Inorganic substrates oxidized by chemoautotrophs

Hydrogen. Several species of bacteria in several genera use the oxidation of hydrogen to generate energy as ATP and couple this energy to the fixation of carbon dioxide (see p. 348). Hydrogen is used by the hydrogen bacteria in two types of reactions: (1) a reaction that ultimately ends in the generation of ATP through electron transport and oxidative phosphorylation:

$$H_2 + \frac{1}{2} O_2 \rightarrow H_2O + \text{Energy from electron transport}$$

and (2) a reductive process to generate NADH:

$$H_2 + NAD \rightarrow NADH + H^+$$

The NADH is used to reduce NADP, which in turn is used in biosynthetic reactions and in chemoautotrophs to fix carbon dioxide and produce organic compounds.

Iron. Only in the organism *Thiobacillus ferrooxidans* has iron oxidation been studied in any detail. The oxidation reaction that ultimately results in the production of energy follows:

$$4\ FeSO_4 + O_2 + 2\ H_2SO_4 \rightarrow 2\ Fe_2(SO_4)_3 + 2\ H_2O$$

The sulfuric acid required in the oxidation of $FeSO_4$ is generated by sulfur-oxidizing bacteria. Many of the iron oxidizers can also obtain their energy from the oxidation of sulfide in such compounds as the sulfides of copper, lead, nickel, zinc, and cobalt. The importance of the sulfur oxidizers in biotechnology will be discussed in Chapter 18.

Ammonia and nitrite. The nitrifying bacteria oxidize only two inorganic nitrogen compounds, either ammonium (NH_4^+) or nitrite (NO_2^-). Ammonium is produced in the environment following mineralization of organic nitrogen—a process called **ammonification**. Nitrite is found in the environment often as the result of the microbial oxidation of ammonia. In anaerobic environments nitrite occurs as a result of anaerobic respiration (nitrate reduction). Ammonium and nitrite oxidation results in the production of energy according to the following reactions:

$$NH_4^+ + 1\frac{1}{2}\,O_2 \rightarrow NO_2^- + H_2O + 2\,H^+$$

$$NO_2^- + \frac{1}{2}\,O_2 \rightarrow NO_3^-$$

Nitrification in soils and water plays an important role in agriculture and water quality; these aspects are discussed in Chapters 18 and 19.

Sulfur. Several types of reduced sulfur compounds found in the environment may be oxidized by various species of bacteria. Species of *Thiobacillus,* for example, can oxidize the following sulfur compounds:

$$S_8 + 12\,O_2 + 8\,H_2O \rightarrow 8\,H_2SO_4$$

$$H_2S + 2\,O_2 \rightarrow H_2SO_4$$

$$Na_2S_2O_3 + 2\,O_2 + H_2O \rightarrow Na_2SO_4 + H_2SO_4$$

Sulfur is polymerized in its elemental form and is represented as S_8 (Figure 12-31). The sulfide and thiosulfate states of sulfur are represented as H_2S and S_2O_3, respectively. The interrelationships of these three sulfur states during oxidation and energy formation are outlined in Figure 12-32.

Single carbon substrates (methylotrophs). Methylotrophs are microorganisms that gain their energy from the oxidation of one-carbon (C1) compounds other than carbon dioxide (see Point of Interest on p. 342). These carbon compounds include **formaldehyde (HCHO), methanol (CH_3OH), methane (CH_4)**, and other related compounds. The methylotrophs, even though they oxidize organic compounds, exhibit properties similar to the autotrophs and have, therefore, been included here. The oxidation of one-carbon compounds results in the formation of carbon dioxide or formaldehyde. The methylotrophs, unlike the autotrophs, do not reduce atmospheric carbon dioxide to cell material*—methylotrophs require a compound more reduced than carbon dioxide, for example, formaldehyde or a mixture of formaldehyde and carbon dioxide. The overall scheme for the oxidation of one-carbon compounds is presented in Figure 12-33. The oxidation reactions for the principal carbon substrates are outlined later.

Methane is found in coal and oil deposits and is associated with the anaerobic decomposition of organic matter, sewage sludge digestors, sediments and muds of various aquatic environments, and the intestinal tract of animals and humans (see Point of Interest on p. 341). Methanol is a degradation product of several naturally occurring substances, such as pectins (polymer of galacturonic acid), which are found especially in fruits. The oxidation of methane and methanol proceeds via reactions outlined in Figure 12-33. Their oxidation results in the formation of formaldehyde, which is used to produce cellular organic material.

Formaldehyde as an energy source is oxidized primarily by a route involving two reactions catalyzed by separate enzymes. Formate is the product of the first reaction and is later cleaved to carbon dioxide according to the following reactions:

* Other microorganisms using one-carbon compounds convert carbon dioxide to usuable organic compounds in the cell via the Calvin cycle (see p. 348).

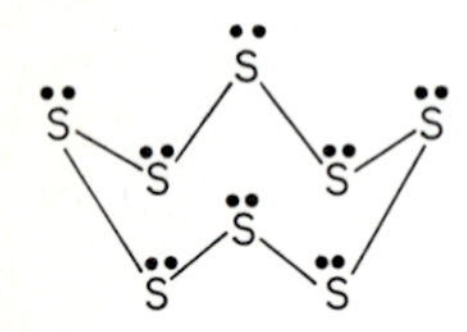

Figure 12-31 *Molecular forms of elemental sulfur.*

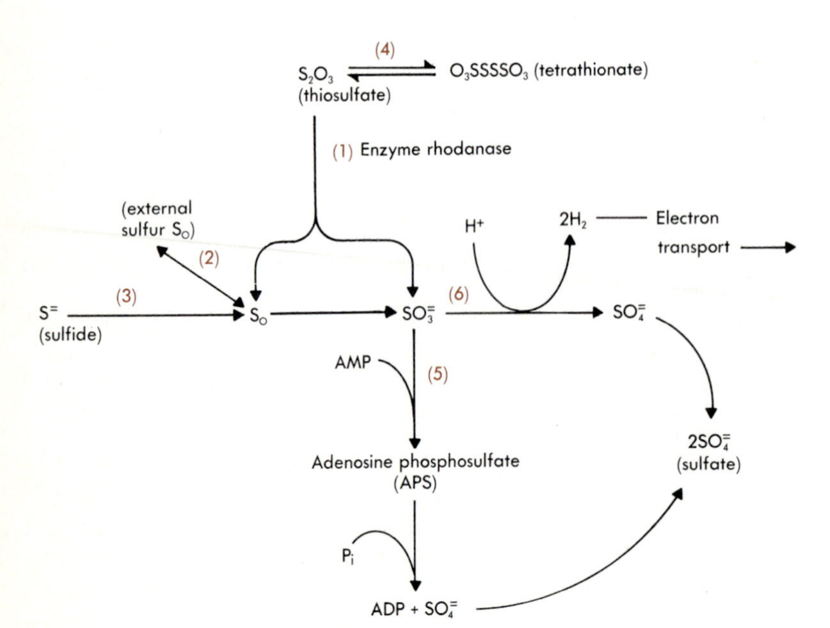

Figure 12-32 *Steps in oxidation of different sulfur compounds. Thiosulfate, sulfide, and elemental sulfur are oxidized in reactions 1, 2, and 3, respectively. Thiosulfate is cleaved to S_o and SO_3 by enzyme rhodanase. Sulfide is oxidized to S_o. Reaction 4 indicates that under certain conditions of growth tetrathionate can also be produced or used. Tetrathionate is excreted into medium or can be further oxidized. Sulfite in reaction 5 is oxidized by adenosine phosphosulfate reductase, which ultimately results in formation of ADP and finally ATP production in electron transport system. Sulfite in reaction 6 can be oxidized to sulfate by sulfite-cytochrome c-oxidoreductase.*

$$HCHO + X + H_2O \rightarrow HCOO^- + XH_2$$

$$HCOO^- + NAD \rightarrow CO_2 + NADH + H^+$$

Methylamines are substrates for many bacterial species. These compounds are found as degradation products of various plants and animal tissue. Oxidation of methylamines occurs in a series of reactions in which successive carbon-nitrogen bonds are cleaved and methyl groups are released as formaldehyde. The oxidation of a typical methylamine is outlined in Figure 12-34.

Carbon monoxide–oxidizing bacteria (also called **carboxydobacteria** or **carboxydotrophs**) use carbon monoxide aerobically as an energy and carbon source. These organisms represent a ubiquitous group that have been isolated from polluted streams, soils of junkyards, gardens, and sewage treatment plants. They convert carbon monoxide to carbon dioxide and are able to assimilate parts of the carbon dioxide formed into organic matter in the Calvin cycle (see Chapter 13). The carboxydotrophic bacteria possess a terminal cytochrome oxidase that is insensitive to carbon monoxide. The cytochrome oxidase of all other aerobic bacteria is sensitive to carbon monoxide, and thus the carboxydotrophic bacteria occupy a selective niche in the environment.

Mechanism of energy production in chemoautotrophs

Chemoautotrophs use the conventional methods of electron transport and oxidative phosphorylation and, therefore, possess flavoproteins, quinones, and cytochromes.

Figure 12-33 *Pathway for oxidation of methane (CH_4) and methanol (CH_3OH), as well as other one carbon substrates. Formaldehyde formation can lead to C3 compounds and thence to TCA cycle.*

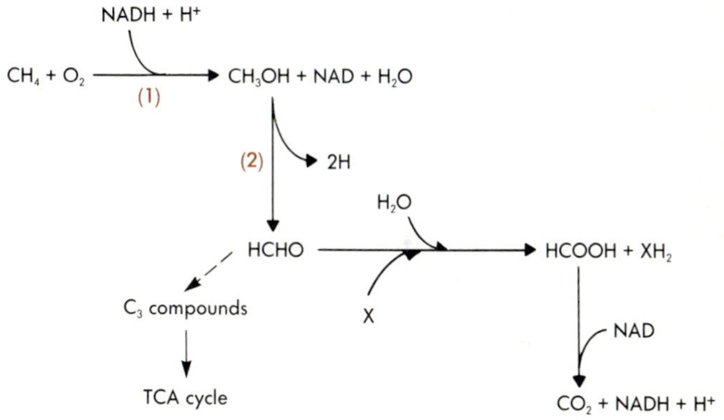

$CH_4 + O_2 \xrightarrow[(1)]{NADH + H^+} CH_3OH + NAD + H_2O$

$CH_3OH \xrightarrow{(2)} HCHO + 2H$

$HCHO + H_2O + X \rightarrow HCOOH + XH_2$

$HCHO \rightarrow C_3$ compounds $\rightarrow$ TCA cycle

$HCOOH + NAD \rightarrow CO_2 + NADH + H^+$

$$(CH_3)_3NH^+ + \text{Flavoprotein} + H_2O \longrightarrow (CH_3)_2NH_2^+ + HCHO + \text{Reduced flavoprotein}$$

Figure 12-34 *Oxidation of methylamine.*

Point of Interest

METHANE-BASED SYMBIOSIS BETWEEN ANIMAL AND INTRACELLULAR BACTERIA

Researchers in 1986 discovered a marine mollusk (mussel) in the Gulf of Mexico that derives its carbon from methane gas. The mussels, which were found at depths of 1800 feet, consumed methane and oxygen at an extremely high rate. When various tissues of the mussel were examined, only the gills showed significant methane consumption, and much of the carbon from methane was retained throughout the animal as organic carbon. The question to be answered was "What is responsible for this high degree of oxidation in the gills?" Electron microscopy of gill tissue revealed the presence of abundant intracellular coccoid bacteria in vacuoles within the gills. In many of the bacteria there were large stacks of intractytoplasmic membranes characteristic of certain methanotrophs that are known to derive their organic carbon from oxidation of methane. In addition, the endosymbionts are also found near the surface of the gills, thus facilitating the acquisition of dissolved methane from seawater. Thus there is an apparent symbiosis in which the organic carbon of the mussel is derived from methane gas consumed by intracellular bacteria.

Electrons generated during the oxidation of fuels are important for ATP formation, but they are also important in the reduction of NAD to NADH. Reduced NAD is used to reduce NADP, which is important in the biosynthesis of the cell's metabolites, The hydrogen bacteria couple hydrogen oxidation to the reduction of NAD, and the entire electron transport chain is available for electron transport and generation of three molecules of ATP. The oxidation-reduction potentials of the energy substrates (Table 12-4) for other chemoautotrophs is higher (more positive) than NAD, and oxidation of the energy source cannot be directly coupled to the reduction of NAD. The oxidation-reduction potential of the Fe^{+++}/Fe^{++} couple ($E^{\circ\prime}$ = + 0.770 volt), for example, is too high to drive the reaction in which NAD is reduced (the E_o' for the $NAD^+/NADH$ couple is − 0.320 volt). The chemolithotrophs compensate for this characteristic by using "uphill" reactions that are just the reverse of the normal electron flow in the respiratory chain. Oxidation of iron, for example, releases electrons that probably enter the respiratory chain at cytochrome c. The energy gained in their oxidation is used to drive the uphill reaction that terminates in the reduction of NAD or NADP. The oxidation of ammonia and nitrite is also an example of where the oxidation-reduction potentials for the reactions are sufficiently positive (E_o' for NO_2^-/NH_3 = + 0.44 and NO_3^-/NO_2 = + 0.35 to + 0.43) that reverse electron flow is required to generate reduced NAD (Figure 12-35). Thus most chemolithotrophs, because of reverse electron flow, do not produce large amounts of energy for growth. They all grow slowly, and doubling rates of up to 6 hours are obtained and only then when their energy source is in excess.

Concept Check

1. What inorganic elements serve as energy sources for some bacteria, and what are their oxidation products?
2. What single carbon compounds serve as energy sources?
3. How do chemoautotrophs compensate for the fact that the oxidation reduction potential of some of their energy sources are considerably higher (more positive) than NAD?

Point of Interest

SINGLE CELL PROTEIN

The term **single cell protein (SCP)** was first used at a conference on food production in 1966 to overcome the unpleasant connotation associated with eating bacteria, algae, or fungi. With the growth of world population and a decrease in available energy sources SCP seems to be a logical source of food, for it can be produced rapidly and with much less land, water, and personnel than are now needed to produce food. In addition, inexpensive nutrients could be used to grow SCP, and microbes have a high conversion of nutrients to protein.

The principal organisms that have been studied as sources of SCP are yeast and methylotrophs, organisms that grow on 1 C compounds. Yeasts have been grown on a number of waste products, such as paper plant wastes and whey, especially in the Soviet Union. Methylotrophs have a higher conversion of nutrients to protein and thus have some advantage over yeast.

When this research was originally begun in the 1960s, petroleum was inexpensive and methanol for the growth of the methylotrophs was obtained from petroleum. Now more inexpensive sources of methanol are used. Processes have been developed that successfully grow the organism *Methylophilus methylotrophus* in large aerated tanks, using methanol as substrate. Methanol has the advantage of being free of impurities, soluble in water, and readily separated from the end product. Using this method, 0.5 gram of *M. methylotrophus* can be obtained from 1 gm of methanol. The bacteria is 70% crude protein.

At this point SCP has mainly been proposed as a supplement to animal feed. Europeans have been the main developers of the processes because soybeans, the most common high protein feed for animals, do not grow well there. In 1980, a British company began producing Pruteen (*M. methylotrophus*) commercially, but stopped production after several years, becasue it could not compete with the price of soybean and fish meal. Thus, while the biological and technical problems of growing SCP for food have been solved, the ecomonic problems remain.

Also, although SCP may be accepted as a supplement to animal feed, work is still going on to find the best microbes and best conditions to produce SCP that can be fed directly to humans. The problems here are widespread lack of acceptance because the product is not necessarily pleasing to the taste and may cause gastrointestinal upset. Gastrointestinal upset is caused by the lipospolysaccharide outer membrane of gram-negative organisms. The nucleic acid content of bacteria is also high, which contributes to gout and kidney stones. So while some day in the future bacteria may be a common food, much work still needs to be done to reach that goal.

Table 12-11 *Comparison of fermentation, respiration, and anaerobic respiration*

Characteristic	Fermentation	Respiration	Anaerobic respiration
Carbon and electron source	Organic compound	Organic compound	Organic compound
Final carbon oxidation product	Fermentation product such as lactic acid	CO_2	CO_2
Final electron acceptor	Organic molecule such as pyruvate	O_2	Inorganic, such as NO_3^-, $SO_4^=$, NO_2^-, Fe^{+++}, or organic, such as fumarate
Principal mode of ATP synthesis	Substrate level phosphorylation	Electron transport phosphorylation	Electron transport phosphorylation

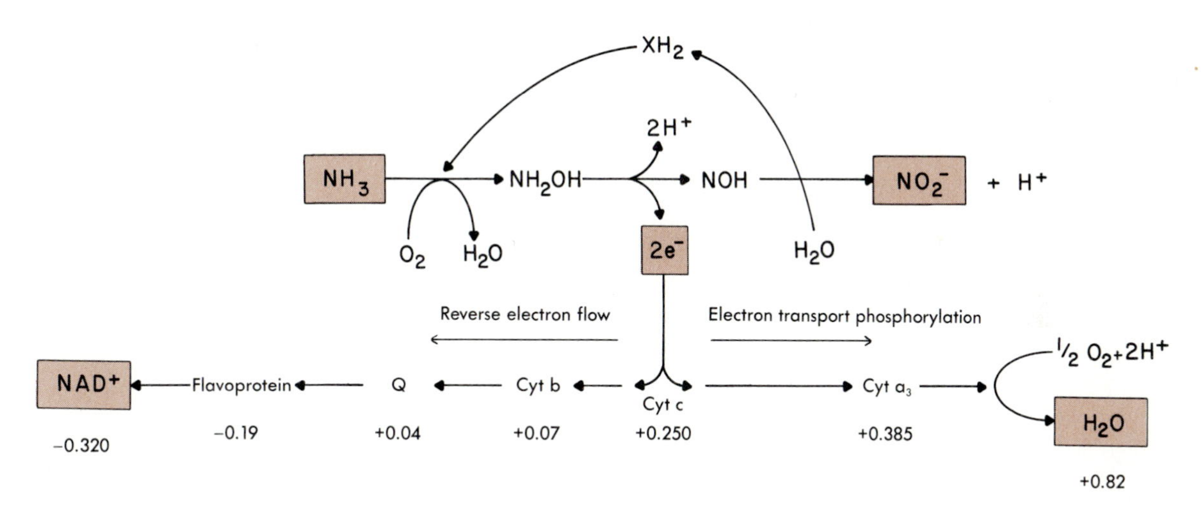

Figure 12-35 *Oxidation of ammonia (NH_3). Ammonia is oxidized to hydroxylamine (NH_2OH) and thence to unstable intermediate (NOH) and finally to nitrite (NO_2). Oxidation of NH_3 to NH_2OH is believed to involve cytoplasmic cytochrome P460* (X). *Little is known of spatial organization of chain, however, E_o of redox couple in ammonia oxidation is such that electrons enter respiratory chain at cytochrome c. Thus transport phosphorylation yields probably only one ATP. Reversed electron flow is required to reduce NAD and energy for reaction is supplied by ATP hydrolysis.*

SUMMARY

1. Microorganisms obtain their energy for metabolism by either converting sunlight into chemical energy (phototrophs) or extracting energy from preformed organic or inorganic chemicals (chemotrophs).
2. Oxidation-reduction reactions are the basis of all energy-producing reactions. ATP is the energy molecule of the cell, and it can be produced by three different mechanisms: substrate phosphorylation, electron transport phosphorylation, and photophosphorylation. Each mechanism ultimately involves the phosphorylation of ADP to ATP.
3. In organisms with electron transfer components the transfer of electrons generates energy that is used to produce an electrochemical gradient (chemiosmosis) across the membrane (cytoplasmic, mitochondrial, or chloroplast). This gradient causes protons to pass from inside the membrane to outside the membrane. The reverse flow of protons through ATP synthetase pores, which spans the membrane, results in the synthesis of ATP.
4. Some bacteria (photosynthetic) can use light energy to generate ATP. A quantum of light is used to excite chlorophyll pigments, which release high energy

electrons. These electrons are captured by carriers and, during electron transport, energy is released that is coupled to the phosphorylation of ADP and phosphate to ATP.

5. Energy can be extracted from organic substrates in one of three ways: fermentation, respiration, and anaerobic respiration. These metabolic processes are compared in Table 12-11.
6. Glycolysis is a special type of fermentation in which glucose is the electron donor. Glucose can also be oxidized through other pathways such as the pentose phosphate pathway or Entner Doudoroff pathway. The pentose phosphate pathway is especially important because it supplies reduced NADP for biosynthesis as well as pentose sugars for nucleic acid synthesis.
7. Pyruvate, one of the major products of glucose fermentation or respiration, can also be converted to a variety of products, including many different organic acids, alcohols, gases and water. Microbial species show a variation in the types of products derived from pyruvate, and these products are often a clue to their identity.
8. Oxygen not only plays an important role as a final electron acceptor in energy production in aerobic microorganisms but can also influence cells that do not use it for energy production. Oxygen is often reduced to intermediates such as H_2O_2 and the superoxide radical that are toxic to anaerobes. Superoxide produced in aerobes is removed by the enzyme superoxide dismutase, which is generally absent in anaerobes.
9. Although carbohydrates are the most frequently metabolized organic compounds, other molecules such as amino acids, purines and pyrimidines, lipids and fatty acids, and hydrocarbons can also be used as energy sources.
10. Energy from the oxidation of inorganic compounds is characteristic of the group called the chemoautotrophs. This group extracts both carbon and energy from inorganic sources. The carbon source is carbon dioxide. The energy sources include hydrogen, inorganic sulfur compounds, iron, ammonia, and nitrite. These bacteria use the electron transport chain, but growth is usually slow, and the number of ATP produced per electron pair is less than that obtained by organisms that use organic substrates.
11. One-carbon compounds such as methane, methanol, and carbon monoxide can also be metabolized by methylotrophs to produce energy.

Self-study Quiz

Completion

1. The usable energy obtained from a chemical reaction is called ________________.
2. Flavoproteins are enzymes that are described as ______ ________________.
3. The chemical composition of quinones is __________ ________________.
4. The light-harvesting structures bound to the thylakoid membranes of the cyanobacteria are called _________ ________________.
5. The fermentation process in which glucose is converted to pyruvate is called the ________________.
6. The products of the Entner Doudoroff pathway are ________________ and ________________.
7. The enzyme that prevents a buildup of hydrogen peroxide in the cell is ________________.

Multiple choice

1. Which of the following reactions or processes is considered exergonic?
   A. ATP hydrolysis
   B. Synthesis of tubulin
   C. Synthesis of glucose from CO_2 and water
   D. Phosphorylation of ADP to ATP
2. The type of enzyme that catalyzes the conversion of oxygen to water in aerobic respiration is called a(an):
   A. Transferase
   B. Oxidase
   C. Dehydrogenase
   D. ATPase

3. Facultative anaerobes grow more slowly under anaerobic conditions than under aerobic conditions because:
   A. Electron transport phosphorylation is lacking
   B. Only substrate phosphorylation is operative
   C. The carbon source is incompletely oxidized
   D. There is excessive buildup of superoxide in the cell
4. The pathway that leads to the formation of reduced NADP and ribose 5-phosphate is:
   A. Entner-Doudoroff
   B. Embden-Meyerhoff-Parnas
   C. Pentose phosphate
   D. Tricarboxylic acid
5. Which of the following compounds does not contain a *high*-energy phosphate?
   A. Phosphoenolpyruvate
   B. 3-Phosphoglycerol phosphate
   C. Glucose 6-phosphate
   D. Adenosine triphosphate
6. The number of ATP molecules produced per electron pair during aerobic respiration with succinate as a carbon source is:
   A. 1
   B. 2
   C. 3
   D. 0
7. The terminal enzyme component in electron transport phosphorylation during growth on inorganic molecules such as nitrate is a(an):
   A. Oxidase
   B. Transferase
   C. Reductase
   D. Dehydrogenase
8. The divalent reduction of oxygen results in the formation of:
   A. Superoxide
   B. Hydroxyl radical
   C. Singlet oxygen
   D. Hydrogen peroxide
9. Which of the following growth conditions would be considered chemolithotrophic?
   A. CO_2, O_2, Fe^{++}
   B. Glucose, O_2
   C. Succinate, nitrate, O_2
   D. CO_2, nitrate, Fe^{++}
10. Which of the following sources of energy by chemolithotrophs would not require reverse electron flow to generate reduced NAD?
    A. NO_3
    B. H_2
    C. Fe^{++}
    D. S^o

Matching

Match the chemical component or process in the left column with its appropriate function in the right column.

1. ________ Bacteriorhodopsin
2. ________ Cometabolism
3. ________ Substrate phosphorylation
4. ________ Proton motive force
5. ________ Beta oxidation
6. ________ Respiration
7. ________ Chemiosmosis
8. ________ Tetrapyrroles
9. ________ Dehydrogenation

A. Light-harvesting pigment in cyanobacteria
B. Oxidation in which electrons are removed from an organic compound
C. Component of chlorophyll molecules that absorbs light
D. Light-harvesting pigment found in halobacteria
E. A process that generates oxygen
F. A process in which the movement of protons across the membrane generates a force that can be converted to chemical energy
G. The process in which light energy is converted to chemical energy
H. A component of phycobilisomes
I. A process in which the electron donor is organic or inorganic and the electron acceptor is oxygen
J. The sum of the electrical and pH potentials across the membrane
K. The process in which the energy of an organically bound phosphate is used to phosphorylate ADP
L. Process in which acetyl CoA can be generated
M. Process in which a more degradable energy source is used to speed up the metabolism of a metabolically recalcitrant molecule
N. Oxidation process in which superoxide is produced

Thought questions

1. Many of the end products of the metabolic pathways of microorganisms are important in the desired flavoring of foods and beverages and also in the spoilage of the character and flavor of the foods. Fermenting of foods and beverages then becomes a careful balancing of conditions to favor the expression of desirable metabolic pathways. Briefly describe the desirable metabolic pathways that would be encouraged in the following food products and how the pathways involved could be controlled.
   A. Dry, still grape wine
   B. Sweet, sparkling wine
   C. Wine vinegar
   D. Swiss cheese (whose flavor comes from propionates and whose holes come from gas formation)
2. If you found an antibiotic substance that selectively could cause the inhibition of the membrane-bound ATPase complex, predict what effect the substance would have on the following functions:
   A. ATP production by oxidative phosphorylation
   B. Oxidation of glucose to CO_2 and H_2O
   C. Operation of the TCA cycle
   D. ATP formation by substrate phosphorylation
3. Arsenate ($HAsO_4^{2}$) is a molecule known to work by uncoupling the phosphorylation event occurring at the oxidation of glyceraldehyde 3-phosphate. This occurs because the enzyme catalyzing the reaction, glyceraldehyde 3-phosphate dehydrogenase, utilizes arsenate instead of inorganic phosphate. The compound formed, glycerate-1-arseno-3-phosphate, spontaneously decomposes to glycerate 3-phosphate and free arsenate.
   A. Which pathway is influenced by this reaction?
   B. What are the consequences in the cell exposed to arsenate?
   C. Would this compound be equally toxic for all living systems? Explain.

SELECTED READINGS

Books

Doelle, H.W. *Bacterial metabolism* (2nd ed.). Academic Press, Inc., New York, 1975.

Gottschalk, G. *Bacterial metabolism* (2nd ed.). Springer-Verlag New York, Inc., New York, 1985.

Jones, C.W. *Bacterial respiration and photosynthesis. Aspects of Microbiology #5,* American Society for Microbiology, Washington, D.C., 1982.

Knowles, C.J. *Diversity of bacterial respiratory systems.* CRC Press, Inc., Boca Raton, Fla., 1980.

Mandelstam, J., and McQuillen, K. *Biochemistry of bacterial growth* (3rd ed.). Holsted Press, New York, 1982.

Moat, A.G. *Microbial physiology.* John Wiley & Sons, New York, 1979.

Rose, A.H. *Chemical microbiology* (3rd ed.). Plenum Publishing Corp., New York, 1976.

Weinberg, E.D. (Ed.). *Microorganisms and minerals.* Marcel Dekker, Inc., New York, 1977.

Journal articles

Amzel, L.M., and Pedersen, P.L. Proton ATPases: structure and function. *Annu. Rev. Biochem.* **52**:801, 1983.

Beaman, L., and Beaman, B.L. The role of oxygen and its derivatives in microbial pathogenesis and host defenses. *Ann. Rev. Microbiol.* **38**:27, 1984.

Colby, J., Dalton, H., and Whittenbury, R. Biological and biochemical aspects of microbial growth on C_1 compounds. *Ann. Rev. Microbiol.* **33**:481, 1979.

Glazer, A.N. Comparative biochemistry of photosynthetic light-harvesting systems. *Ann. Rev. Biochem.* **52**:125, 1983.

Higgins, I.J., et al. Methane-oxidizing microorganisms. *Microbiol. Rev.* **45**:556, 1981.

Hinkle, P.C., and McCarty, R.E. How cells make ATP. *Sci. Am.* **238**:104, 1978.

Hooper, A.B., and DiSpirito, A.A. In bacteria which grow on simple reductants, generation of a proton gradient involves extracytoplasmic oxidation of substrate. *Microbiol. Rev.* **49**(2):140, 1985.

Ingledew, W.J., and Poole, R.K. The respiratory chain of *Escherichia coli. Microbiol. Rev.* **48**:222, 1984.

Klass, D.L. Methane from anaerobic fermentation. *Sci.* **223**:1021, 1984.

McCarty, R.E. H^+-ATPases in oxidative and photosynthetic phosphorylation. *Bioscience* **35**(1):27, 1985.

Meyer, O., and Schlegel, H.G. Biology of aerobic carbon monoxide-oxidizing bacteria. *Annu. Rev. Microbiol.* **37**:277, 1983.

Miller, T.L., and Wolin, M.J. Interspecies hydrogen transfer: 15 years later. *ASM News* **48**(12):561, 1982.

Mitchell, P. Coupling of phosphorylation to electron and hydrogen transfer by a chemi-osmotic type of mechanism. *Nature* **191**:144, 1961.

Morris, J.G. The physiology of obligate anaerobiosis. *Adv. Microb. Physiol.* **12**:169, 1975.

Osterhelt, P. Light driven proton pumping in *Halobacteria. Bioscience* **35**(1):18, 1985.

Prince, R.G. Redox-driven proton gradients. *Bioscience* **35**(1):22, 1985.

Smith, A.J., and Hoare, D.S. Specialist phototrophs, lithtrophs, and methylotrophs: a unity among a diversity of procaryotes. *Microbiol. Rev.* **41**:419, 1977.

Stoeckenius, W., and Bogomolni, R.A. Bacteriorhodopsin and related pigments of the halobacteria. *Annu. Rev. Biochem.* **51**:587, 1982.

Taylor, R.F. Bacterial triterpenoids. *Microbiol. Rev.* **48**:181, 1984.

CHAPTER 13

Nutrition, biosynthesis, and metabolic control

Chapter Outline

Introduction

You have already learned that biological energy is derived from the oxidation of inorganic or organic molecules or from the sun. The cell uses this energy to perform specific functions, one of which is the biosynthesis of new cellular components. The materials that are used to construct new cellular components are called **nutrients**. The nutrients required for growth are those inorganic or organic compounds that supply the major elements (carbon, hydrogen, oxygen, sulfur, phosphorus, nitrogen, etc.) to the cell. Elements that are not major components of building block material but are still necessary for growth are called **micronutrients** and include such minerals as magnesium, iron, and manganese. Some nutrients may actually be organic molecules that are directly incorporated into building block materials or used as energy sources or cofactors for some enzymes. The cell has evolved various types of metabolic pathways to construct organic molecules. To efficiently use the various nutrients for biosynthesis the cell is equipped with a variety of control mechanisms.

The purposes of this chapter are threefold: (1) to outline the elements required for growth and the various ways they are incorporated into organic molecules, the building blocks (amino acids, fatty acids, sugars, etc.); (2) to discuss the pathways used by the cell to construct the building blocks of organic matter; and (3) to elaborate on the different control mechanisms used by the cell to prevent oversynthesis of building block material.

CHEMICAL COMPOSITION OF THE BACTERIAL CELL

Before we discuss in detail the chemical requirements for growth and metabolism of the bacterial cell, let us look at the chemical composition of the bacterial cell. Table 13-1 lists the percentages of various chemical elements found in the cell, while Table 13-2 illustrates the total cell weight percentages of important molecules that make up the bacterial cell. Table 13-2 demonstrates that, except for water, macromolecules represent the greatest percentage of total cell weight. Many cellular metabolic processes understandably concern the synthesis of macromolecules. These macromolecules become incorporated into various cell structures and components. In addition, macromolecules are broken down into smaller units as sources of energy or as precursors in the synthesis of other macromolecules. Table 13-3 outlines the chemical composition of an *Escherichia coli* cell when grown under ideal laboratory conditions. In their natural environment most bacteria are subjected to limited nutrients; therefore Table 13-3 must be viewed from that perspective. The size, morphology, and chemical composition of the bacterial cell can vary considerably, depending on environment and the genetic makeup of the organism. An organism that lives in the intestinal tract might have a difficult time surviving in an aquatic environment, where nutrient concentrations are low. In the aquatic environment the bacterial cell, provided it possessed the necessary genetic information, might convert nutrients into storage forms such as polyhydroxybutyrate, but in the intestinal tract such storage products might not be necessary. The effect of nutrient limitation on the growth of microorganisms is discussed in Chapter 14.

NUTRITION

Inorganic nutrients and metabolism

Carbon. Carbon exists in many different chemical forms in nature. It may appear as inorganic carbon, such as carbonate or gaseous carbon dioxide, or as a component of simple or complex organic compounds, such as the various hydrocarbons abundant in the environment. Some microorganisms can use carbon dioxide from the atmosphere or bicarbonate to synthesize more complex organic compounds. These microbes are called **autotrophs** and include the photosynthetic bacteria. Eukaryotic autotrophs include plants and algae.

Autotrophs and carbon. The energy used for autotrophic biosynthesis may come from the oxidation of inorganic molecules (**chemoautotrophy**) or may be derived from the sun's energy (**photoautotrophy**). In the following discussion we will describe how autotrophs are able to incorporate inorganic carbon into an organic molecule.

Table 13-1 *Major elements that make up the organic matter of the typical bacterial cell**

Element	Dry weight percentage
Carbon	50
Oxygen	20
Nitrogen	15
Hydrogen	8
Phosphorus	3
Sulfur	1

*The function of these major elements as well as minor elements were discussed in Chapter 2 (see Table 2-1).

The overall reaction of photosynthesis in photoautotrophs can be described by the following equation:

$$2H_2A + CO_2 \xrightarrow{\text{Light energy}} (CH_2O) + H_2O + 2A$$

CH_2O in the equation represents the carbohydrate into which carbon dioxide has been incorporated. **A** represents the oxidized form of H_2A, which happens to be oxygen for the eukaryotes and oxygen-producing prokaryotes. H_2A in the equation is the electron donor, which is water for the cyanobacteria, algae, and green plants. However, the purple and green bacteria replace water with other reductants, such as H_2S, S^{2-}, or $S_2O_3^{2-}$; for example:

$$2H_2S + CO_2 \xrightarrow{\text{Light energy}} (CH_2O) + H_2O + 2S$$

Some species of purple bacteria can also use organic acids as a source of electrons:

$$2 \text{ lactate} + CO_2 \xrightarrow{\text{Light energy}} (CH_2O) + H_2O + 2 \text{ pyruvate}$$

Photosynthesis is a process involving two types of reactions: **light reactions** and **dark reactions.** Light reactions are concerned with the absorption of light by pigment molecules, and in the process electrons are transferred and ATP, NADPH, and oxygen are produced. These reactions are discussed in detail in Chapter 12. Dark reactions, or those occurring in the absence of light, take the ATP and NADPH generated during light reactions to fix carbon dioxide and produce carbohydrates such as glucose. The dark reactions, part of the **Calvin-Benson cycle**, can also occur during periods of light and will be discussed next.

Ribulosebisphosphate carboxylase is the key enzyme in the Calvin-Benson cycle and catalyzes the initial step

*Table 13-2 Composition and function of the chemical components of a bacterial cell**

Chemical	Total cell weight (%)	Function
Water	70	Hydrolysis of organic molecules; solute transport; source of electrons for some photosynthetic organisms
Inorganic ions (K^+, Na^+, Mg^{2+}, PO_4^{4-}, SO_4^{2-}, etc.)	1	Active components of many enzymes and proteins; Mg^{2+} in chlorophyll; Fe in cytochromes, etc.
Sugar and precursors	3	Major source of energy; component of cell wall, capsule, and slime layers; component of DNA and RNA
Amino acids and precursors	0.4	Precursors of proteins; source of energy; component of cell wall
Nucleotides and precursors	0.4	Precursors of nucleic acids; source of energy as ATP, coenzymes
Lipids and precursors	2	Components of gram-negative cell wall and cytoplasmic membrane; source of energy
Other small molecules (quinones, products of catabolism such as lactic acid, etc.)	0.2	
Macromolecules		
Proteins	15	Components of cytoplasmic membrane, cell wall, pili, flagella, and ribosomes; most are enzymes
Nucleic Acids		
DNA	1	Primary genetic components of cell; components of ribosomes, mRNA, and tRNA that functions in protein synthesis; some viruses have RNA as genetic material
RNA	6	

*These values are characteristic of *E. coli* grown in a salts medium with glucose as the source of carbon and energy—a minimal growth medium. These values could vary considerably for some molecules, depending on the microorganisms and the chemical and physical conditions of growth.

in carbon dioxide fixation. The enzyme simultaneously catalyzes the insertion of carbon dioxide (carboxylation) into ribulose-1,5-diphosphate and the cleavage of an unstable intermediate into two molecules of 3-phosphoglycerate (Figure 13-1). Ribulosebisphosphate carboxylase is only one of several enzymes involved in the Calvin-Benson cycle. When the cycle is examined more closely, many of its reactions can be seen to mimic the reactions in the pentose phosphate pathway. One of the major distinctions between these two pathways is that in the Calvin-Benson cycle NADPH is consumed while in the pentose phosphate pathway NADPH is generated. Ribulosebisphosphate carboxylase is the most abundant protein in the world. It is estimated that worldwide there is some 40 million tons of it and that it annually fixes approximately 10^{11} tons of CO_2. The enzyme, which is structurally and functionally similar in most prokaryotes and eukaryotes is being studied extensively, hoping to improve the efficiency of photosynthesis. Results from these studies are important, if attempts are made to genetically engineer crop efficiency. The en-

Table 13-3 Chemical composition of an E. coli *cell grown under ideal environmental conditions*

Macromolecule	Approximate total dry weight (%)
Protein	55
RNA	20.5
DNA	3
Lipid	9
Lipopolysaccharide	3.4
Peptidoglycan	2.5
Glycogen	2.5
TOTAL	95.9
Inorganic ions and metabolites	4.1

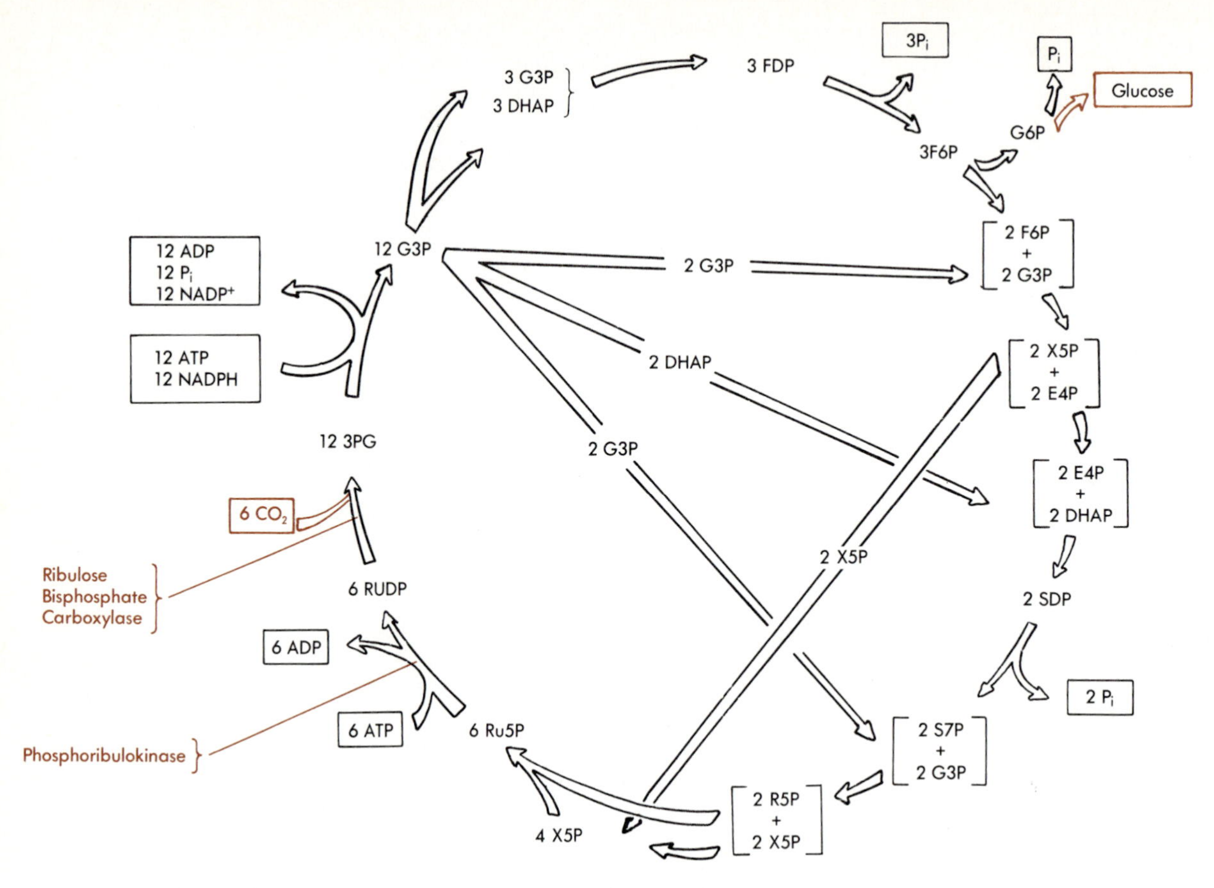

Figure 13-1 *Calvin-Benson cycle for conversion of carbon dioxide into glucose during photosynthesis. Carbon dioxide and glucose are in colored boxes.* 3PG, *3-Phosphoglycerate;* G3P, *glyceraldehyde 3-phosphate;* DHAP, *dihydroxyacetone phosphate;* FDP, *fructose 1,6-diphosphate;* G6P, *glucose 6-phosphate;* E4P, *erythrose 4-phosphate;* X5P, *xylulose 5-phosphate;* SDP, *sedoheptulose 1,7-diphosphate;* S7P, *sedoheptulose 7-phosphate;* R5P, *ribose 5-phosphate;* Ru5P, *ribulose 5-phosphate;* RuDP, *ribulose 1,5-diphosphate.*

zyme does not work at maximum efficiency because it also has oxygenase activity. If the oxygenase activity can be reduced, the carboxylation activity may be increased (both activities occur at the same site on the enzyme). It is composed of eight large catalytic subunits and eight smaller subunits. The enzyme is activated by carbon dioxide and divalent metal ions such as magnesium. The second most important enzyme in the Calvin-Benson cycle is **phosphoribulokinase**. This enzyme catalyzes the conversion of ribulose-5-phosphate to ribulose-1,5-diphosphate. Phosphoribulokinase, therefore, is the enzyme that closes the cycle.

To convert carbon dioxide to glucose requires six molecules of carbon dioxide. The overall equation can be written as follows:

$$6CO_2 + 18\ ATP + 12\ H_2O + 12NADPH + 12H^+ \rightarrow C_6H_{12}O_6 + 18\ P_i + 18ADP + 12NADP^+$$

Because six molecules of ribulose-1,5-diphosphate are used in the fixation process and six molecules are regenerated, it is not necessary to include them in the overall equation; they would cancel out.

The key intermediate in the Calvin-Benson cycle is **glyceraldehyde-3-phosphate**, which is also the key intermediate in glycolysis. Glyceraldehyde-3-phosphate may serve as a precursor to amino acids or fatty acids; some is converted to fructose-6-phosphate and glucose-6-phosphate for conversion to glucose, while some is used to regenerate ribulose diphosphate. Both chemoautotrophs and photoautotrophs use the same enzymatic mechanisms for CO_2 fixation. The green and methanogenic bacteria may fix carbon dioxide by a pathway distinct from the Calvin-Benson cycle, since some of the key enzymes in the latter cycle have not been detected. Organisms that possess a Calvin-Benson cycle also have an alternative pathway that produces

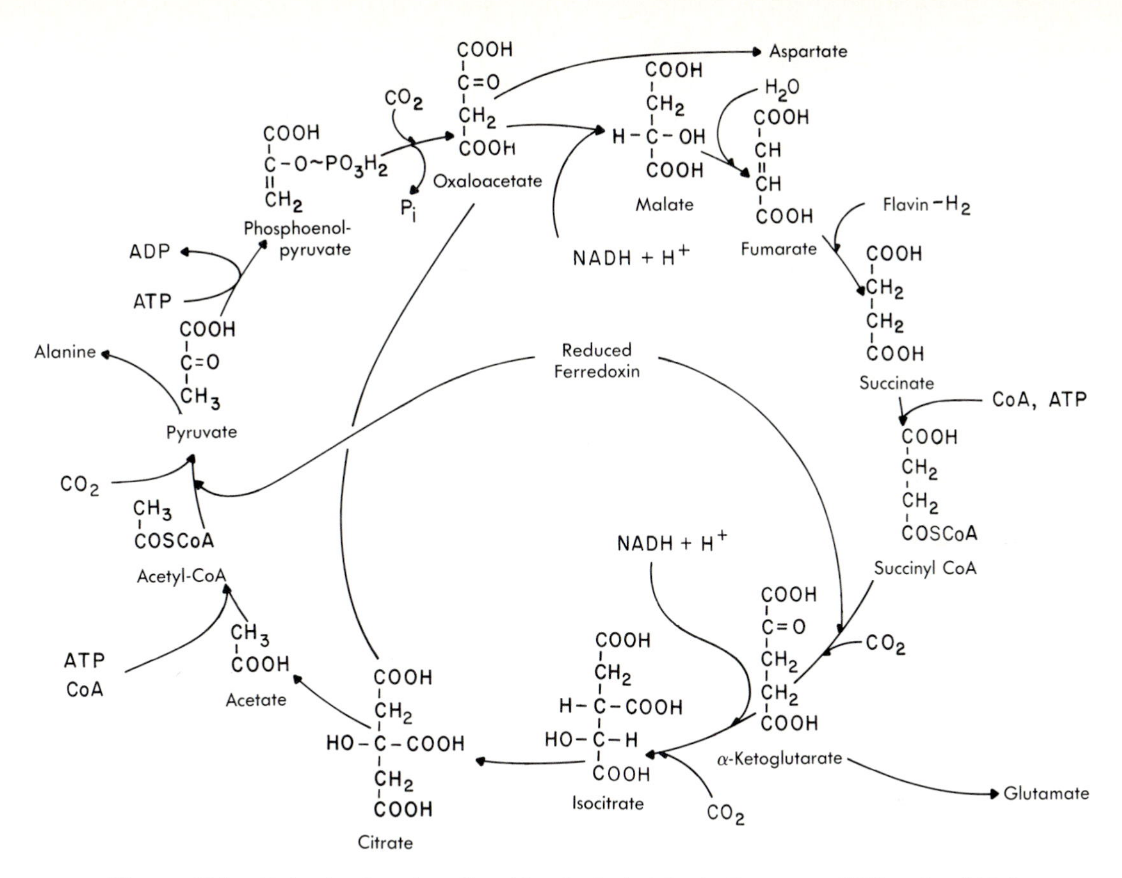

Figure 13-2 *Reductive C_4-carboxylic acid cycle. Cycle is basically reverse of TCA cycle, which is oxidative cycle. Two enzyme systems that are different in C_4 cycle are ferredoxin-dependent and include pyruvate synthetase (pyruvate oxidase is in TCA cycle) and α-ketoglutarate synthetase (α-ketoglutarate oxidase is in TCA cycle).*

four carbon intermediates from CO_2 fixation. This pathway is called the **C_4-carboxylic acid cycle**. The cycle (Figure 13-2) resembles the tricarboxylic acid (TCA) cycle; the major distinction is the presence of two ferredoxin-dependent enzymes—**pyruvate synthetase** and **alpha-ketoglutarate synthetase**—that are absent from the TCA cycle. Examination of the C_4 cycle reveals that it is important for capturing CO_2 and supplying amino acid intermediates, but it does not result in sugar synthesis.

Heterotrophs and carbon. Those microorganisms that require an organic source of carbon are called heterotrophs. Heterotrophism is characteristic of most eukaryotes, including humans. The ability to use one or more organic carbon sources is determined by the enzymatic machinery of the microorganism. Some microorganisms are limited to one or a few carbon sources, whereas others, such as species of *Pseudomonas,* have the potential to catabolize up to 100 different carbon sources. Each carbon source must be transported across the cytoplasmic membrane and enzymatically metabolized in the cytoplasm. Many carbon sources, such as low molecular weight compounds (monosaccharides and disaccharides), are transported directly into the microbial cell. Some carbon sources, such as the polymers of glucose (cellulose, starch, etc.) have molecular weights of 1×10^6 or more and cannot be transported into the cell as such but are hydrolyzed outside the cell (Table 13-4). Microbial cells can produce extracellular enzymes that degrade these polymers into smaller units that are transportable across the cytoplasmic membrane. Some eukaryotes, such as the protozoa, can use the mechanism of phagocytosis to transport polymers into the cell. Most carbohydrate polymers, except for those such as agar and cellulose, are more easily hydrolyzed by microorganisms. Other polymers such as lipids, proteins, and nucleic acids are also potential sources of carbon skeletons and can be hydrolyzed into monomers by lipases, proteases, and nucleases, respectively. Hydrolytic enzymes play a special role in the breakdown of organic material by fungi and bacteria in the soil and water and thus have ecological significance.

Table 13-4 Principal carbohydrate polymers found in nature and degradable by microorganisms

Polymer	Chemical composition	Source in nature	Degrading enzyme	Microorganism degrading the polymer
Starch	Polymer of D-glucose (α-1,4 linkage)	Plants	Amylase	Many bacteria and fungi
Cellulose	Polymer of D-glucose (β-1,4 linkage)	Plants	Cellulase	Several groups of microorganisms, particularly those in the stomach of ruminants such as cows
Pectin	Polymer of methyl D-galacturonic acid	Plants	Pectinase	Many microorganisms especially bacterial species of the Enterobacteriaceae
Agar	Polymer of D- and L-galactose	Marine algae	Agarase	Degradable by only a few microorganisms
Glycogen	Polymer of D-glucose (α-1,4, linkage)	Animal storage product	Amylase	Many microorganisms
Chitin	Polymer of *N*-acetyl glucosamine	Shells of invertebrates such as lobsters and exoskeletons of insects	Chitinase	Many bacteria and fungi; bacterial species; *Streptomyces* are most active decomposers
Hyaluronic acid	Polymer of *N*-acetyl glucosamine and glucuronic acid	Connective tissue of animals	Hyaluronidase	Primarily bacterial species; *Clostridia*
Paramylon	Polymer of glucose (β-1,3 linkage)	Energy storage product of certain algae	β-1,3-Glucanase	Many microorganisms
Laminarin	Polymer of glucose (β-1,3 linkage)	Energy storage product of certain algae	β-1,3-Glucanase	Many microorganisms

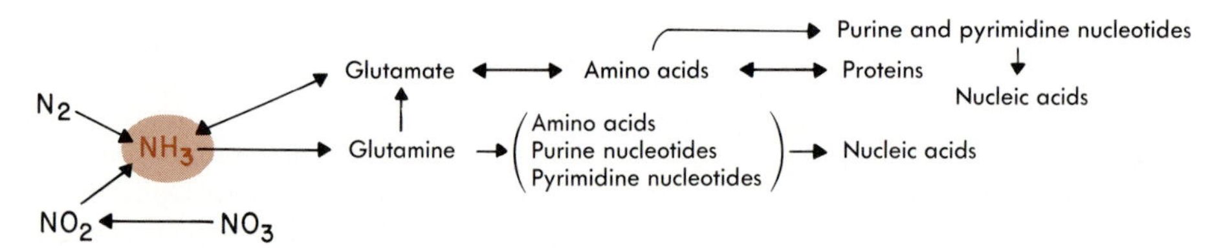

Figure 13-3 *Pathways of nitrogen metabolism. Ammonia is basic nitrogen compound from which organic nitrogen compounds will be synthesized.*

Nitrogen. Nitrogen can exist in several valence states and as a consequence can appear inorganically in reduced (NH_3) or oxidized (NO_3^-, NO_2^-) forms. Nitrogen also exists in an elemental form (N_2) in the atmosphere. Inorganic nitrogen can be converted by many bacteria into organic nitrogen compounds such as amino acids, nucleotides, and other nitrogen-containing molecules (Figure 13-3). Still other bacteria require preformed amino acids or nucleotides in their diet.

Nitrogen gas. Nitrogen gas has a valence of 0 and must be reduced to NH_3 before it can be incorporated directly into cellular organic compounds (the amino group of amino acids). Nitrogen gas, which makes up 78% of the atmosphere, can be used by several bacteria in a process called **nitrogen fixation**. Nitrogen fixation can occur in the atmosphere (lightning discharges, for example), but most occurs in biological processes. Biological nitrogen fixation is associated primarily with certain agricultural plants in which bacteria in a symbiotic or free-living state (Table 13-5) fix atmospheric nitrogen and produce NH_3 for use by the plant. The fixation of nitrogen is a complex process that is catalyzed by the enzyme **nitrogenase**. Nitrogenase is an iron-sulfur protein complex made up of two units referred to as **com-**

Table 13-5 *Examples of nitrogen-fixing microorganisms**

Relationship to host	Organisms
Free living (nonsymbiotic)	
Aerobic	*Azotobacter*
	Beijerinckia
	Azomonas
	Aquaspirillum peregrinum
	Cyanobacteria (heterocystous species)
	Gloeocapsa (coccoid cyanobacterium)
Microaerophilic	*Xanthobacter autotrophicum*
	Azospirillum lipoferum
	Thiobacillus ferrooxidans
	Alcaligenes latus
	Many nonheterocystous cyanobacteria
	Rhizobium japonicum
	Arthrobacter species
Facultative	*Klebsiella pneumoniae*
	Klebsiella aerogenes
	Klebsiella oxytoca
	Enterobacter species
	Citrobacter freundii
	Bacillus polymyxa
	Bacillus macerans
Anaerobic	*Clostridium* species
	Photosynthetic species of *Chromatium, Chlorobium,* and *Rhodospirillum*
Symbiotic	*Rhizobium* species (nodulate leguminous plants such as peas, soybeans, and clover)
	Certain actinomycetes (certain nodulate plants such as *Comptonia* and *Frankia* [symbiont on alder trees])
	Anabaena azollae (fixes nitrogen within leaf pores of the water fern *Azolla*)

*No eukaryotes are known to fix nitrogen.

ponent I and **component II**. The nitrogenases from different organisms are similar, and component I from one genus can interact with component II of another genus to produce enzyme activity. The characteristics of these components are as follows:

1. Component I. Component I is the larger of the two components and consists of two molybdenum atoms and 28 to 34 nonheme iron atoms. It functions in the binding as well as reduction of substrate (N_2). Note that substrates other than N_2 can bind to component I and be reduced, for example, acetylene (C_2H_2), protons, and cyanide (CN). Acetylene reduction is used to determine nitrogenase activity.
2. Component II. Component II contains no molybdenum but does possess four nonheme iron atoms. Its function is to supply electrons, one at a time, to component I. It also binds ATP and magnesium.

Six electrons are required for the reduction of nitrogen, but the sequence of the steps has not been determined. The overall reaction is:

$$6\,H^+ + 6\,e^- + N{\equiv}N \rightarrow 2\,NH_3$$

The reduction of the triple-bonded nitrogen (or C≡C in acetylene or C≡N in cyanide formation) requires energy that is released from the hydrolysis of ATP. The total ATP requirement in vivo varies from one genus of bacteria to another. From 6 to 15 ATPs are required for species of *Azotobacter,* but as many as 20 ATPs are required for nitrogen fixation by *Clostridium pasteurianum.* ATP binds to component II and is not hydrolyzed until component II transfers an electron to component I.

Electrons for nitrogen fixation may be derived from photosynthetic reactions, such as occur in the cyanobacteria, or from carbon substrates such as pyruvate, as in *Clostridium pasteurianum.* The electron carriers appear to be ferredoxin (Figure 13-4) or NADPH.

Both components I and II of the nitrogenase enzyme are inhibited by oxygen. This presents no special problem for the strict anaerobes, such as *Clostridium pasteurianum,* which grows only in the absence of oxygen. Facultative anaerobes, such as species of *Klebsiella,* fix nitrogen only under anaerobic conditions. Some organisms that carry out growth under conditions in which oxygen is present have evolved systems that still permit nitrogen fixation. *Azotobacter* species, for example, are strict aerobes, but their respiratory activity is so high that oxygen is quickly reduced to water. The rapid reduction of oxygen occurs as long as carbon substrates are plentiful and electrons and protons can be generated. When carbon substrates become limited, the oxygen available during respiration is not quickly reduced and could potentially inhibit nitrogenase activity. *Azoto-*

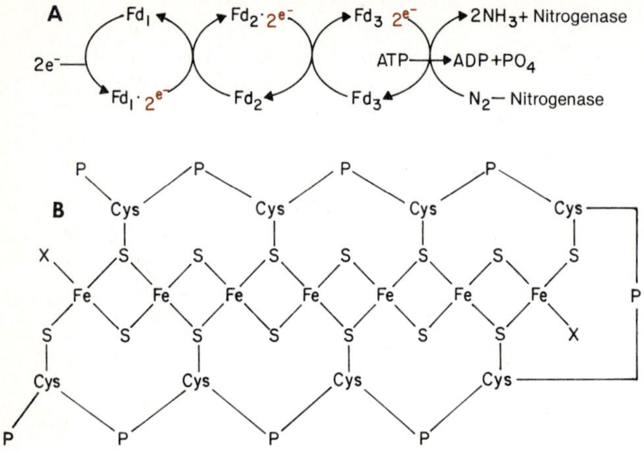

Figure 13-4 *Nitrogen fixation.* ***A,*** *Nitrogen fixation in bacteria. Two electrons ($2e^-$) generated by electron source are picked up and transferred by three different ferredoxin components (Fd_1, Fd_2 and Fd_3) of nitrogenase enzyme system. Hydrogen atoms (not shown) are also believed to be carried on intermediates bound to enzyme system. Energy for fixation of nitrogen comes from ATP hydrolysis.* ***B,*** *Partial structure of ferredoxin molecule.* Fe, *Iron;* S, *sulfur;* Cys, *cysteine;* P, *segment of protein chain.*

bacter, however, can cope with this condition because it produces another iron-sulfur protein that complexes with nitrogenase and protects it from inactivation. This new protein prevents nitrogenase from fixing nitrogen, but as soon as the respiratory activity of the organism increases, it dissociates from the nitrogenase and activity is restored. Oxygen lability is also a potential problem for some of the photosynthetic cyanobacteria that can fix nitrogen. They exist in an environment of oxygen and also produce it from photosynthetic reactions. They have adapted to the problem in the following way.

Cyanobacteria produce **heterocysts** that enable them to have active nitrogenase under aerobic as well as anaerobic conditions. Heterocysts are thick-walled refractive structures interspersed among the vegetative cells of the filamentous species (see Point of Interest, Chapter 4). Vegetative cells are produced only when the concentration of nitrogenous compounds such as NH_3 and NO_3 are at sufficient levels in the environment. When these compounds fall below a critical level, nitrogen fixation is the only mechanism for obtaining combined nitrogen, and the cell responds by producing heterocysts. The vegetative cells that lie on either side of the heterocyst supply nutrients (fixed carbon such as maltose plus ATP,

Point of Interest

NITRIFICATION AND DENITRIFICATION EFFECTS ON THE AVAILABILITY OF NITROGEN IN THE SOIL

Farmers normally fertilize soil for nitrogen with ammonia. Ammonia is a form of nitrogen that all plants can assimilate; however, under aerobic conditions ammonia is volatile or is tightly bound to clay particles in the soil. Under aerobic conditions, ammonia is rapidly converted to nitrates by chemoautotrophic bacteria that use the oxidation of ammonia as a source of energy (p. 340). This process, called **nitrification**, is usually considered beneficial because nitrate is more water soluble and thus more readily available to plants and is the form in which nitrogen is usually assimilated by plants.

Under conditions of heavy rainfall, nitrates because of their solubility in water, are readily leached from the soil and accumulate in groundwater and rivers. High concentrations of nitrates in water can be dangerous. In the anaerobic environment of babies' intestines, nitrate is converted to nitrite, which combines with hemoglobin to form methemoglobin that is incapable of carrying oxygen. In addition to the danger to humans, nitrates carried by rivers are deposited in the oceans and thus nitrogen is effectively removed from land environments on earth.

Also, with a high buildup of nitrates in the soil, **denitrification** can occur in anaerobic pockets or under waterlogged conditions. Denitrification is a form of anaerobic respiration in which normally aerobic bacteria use nitrate in place of oxygen as the final electron acceptor and convert nitrate to nitrogen gas (p. 356) resulting in a net loss of nitrogen to the soil. This is an unfavorable reaction from the standpoint of soil fertility. When this process was originally discovered, it was thought that all nitrogen would eventually be lost from the soil by this process until it was discovered that there were other bacteria that could fix nitrogen into the soil (p. 352). This process is no longer considered entirely detrimental because denitrification helps to remove nitrates from rivers and to prevent the nitrates from being deposited in the oceans.

Neither of these reactions can be looked on as being beneficial or detrimental by themselves. They are both reactions in the nitrogen cycle that have managed to keep the supply of nitrogen available to life on earth. However, man has affected this cycle by the increased use of nitrogen fertilizers on soil and may have to adjust these practices to bring the cycle back into balance. Amazingly, the cycle still does remain somewhat in balance because we find that there are competing reactions in the soil in which nitrates and nitrites are reduced back to ammonia in the soil by other anaerobic bacteria.

and NADPH) to the heterocyst. The septum separating the vegetative cells from the heterocyst become very small, and only small pores called microplasmadesmata connect their cytoplasms. Through these pores nutrients are carried from the vegetative cell to the heterocyst (Figure 13-5). The glycolipid envelope of the heterocyst provides a barrier to the entrance of oxygen. Studies with mutants have shown that cells deficient in this glycolipid envelope are more sensitive to oxygen than normal cells. The heterocysts possess photosystem 1 activity and generate large amounts of the ATP required for nitrogen fixation, but they do not contain photosystem II activity, which generates oxygen. The NH_4^+ produced in heterocysts flows through microplasmadesmata into neighboring vegetative cells.

Nitrogen fixation in leguminous plants. Species of bacteria belonging to the genus *Rhizobium* exist in a symbiotic relationship with leguminous plants (peas, alfalfa, clover, peanuts, and beans) in which nitrogen is fixed in root nodules. Rhizobium species in the soil are attracted to the roots of legumes by root exudates. The specific recognition between bacterium and root involves the interaction of a glycoprotein, or **lectin**, on the root hair and a polysaccharide on the bacterial surface. There is evidence that other factors may be involved in the recognition process. The bacteria then form a firm attachment to the root surface. The first visible response of the host to the bacterium is a deformation or curling on the part of the root hair. The deformation process appears to be a prerequisite for infection to occur. Penetration of the root hair cell wall begins, and the plant responds by depositing cell wall–like material around the bacteria, forming a tubular structure called the **infection thread** (Figure 13-6). The infection thread passes into neighboring cells. Bacteria are released from the infection thread into cortical cells and are enclosed in a membrane of host origin. The cortical cells are induced to increase their rate of cell division, and the bacteria differentiate into larger bodies called **bacteroids**. The result of this process is an organized structure in the root called a **nodule** in which bacteroids are capable of nitrogen fixation. These bacteria have evolved a mechanism for protecting the nitrogenase enzyme from inactivation by oxygen. Surrounding the bacteroids are large quantities of **leghemoglobin**, whose function is to bind oxygen and facilitate its diffusion to rhizobia while preventing free oxygen from inactivating nitrogenase. The heme portion of leghemoglobin is contributed by the bacteria, and the globin portion is coded for by the plant. Bacteria in the nodules fix nitrogen, and the ammonia produced is used by the plant to synthesize amino acids. Sugars synthesized by the plant are supplied to the bacteroids, thus completing the symbiosis. Thus soil deficient in fixed nitrogen can support the growth of leguminous plants because the bacterial symbiont takes nitrogen from the atmosphere and not from

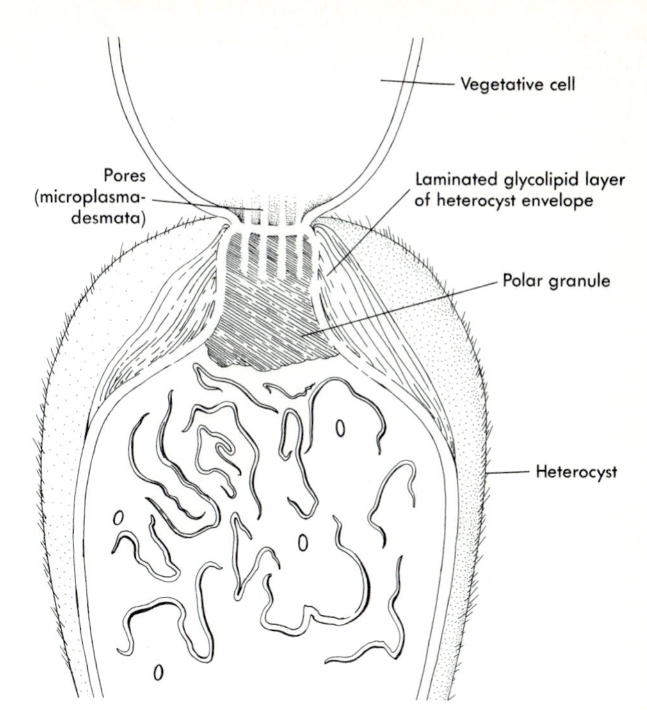

Figure 13-5 *Cytoplasmic association of heterocyst and adjacent vegetative cell. Cytoplasm of heterocyst and vegetative cells are connected by small pores (microplasmadesmata).*

the soil. Leguminous plants can be plowed under to replace lost nitrogen when the soil has been depleted of nitrogen by nonleguminous plants.

Nitrogen fixation and industry. The industrial conversion of gaseous nitrogen to ammonia for use as a fertilizer is expensive and difficult because of the need for extremes of temperature and pressure. A great deal of research has been devoted to understanding the enzymatic mechanism of nitrogen fixation in the microbial cell. The development of an industrial catalyst similar in activity to biological systems could be an important energy-saving device in the production of commercial fertilizers. A better understanding of the genetics of nitrogenase, however, would be even more beneficial. The genes for nitrogen-fixing microorganisms are transferable by conjugation and can be manipulated on plasmids. The transfer of nitrogen-fixing genes through recombinant DNA technology to other microorganisms or to plants may someday increase crop yields and help to rejuvenate grazing lands. There are inherent problems in this type of genetic transfer because the nitrogenase enzyme complex, when synthesized in the recipient organism, requires protection from oxygen. In addition, large quantities of ATP will be required to fix nitrogen in the recipient organism. See Chapter 18 for further discussion of nitrogen fixation.

Nitrate. The nitrogen in nitrate has a valence of +5 and to be assimilated is reduced to −3, the valence of

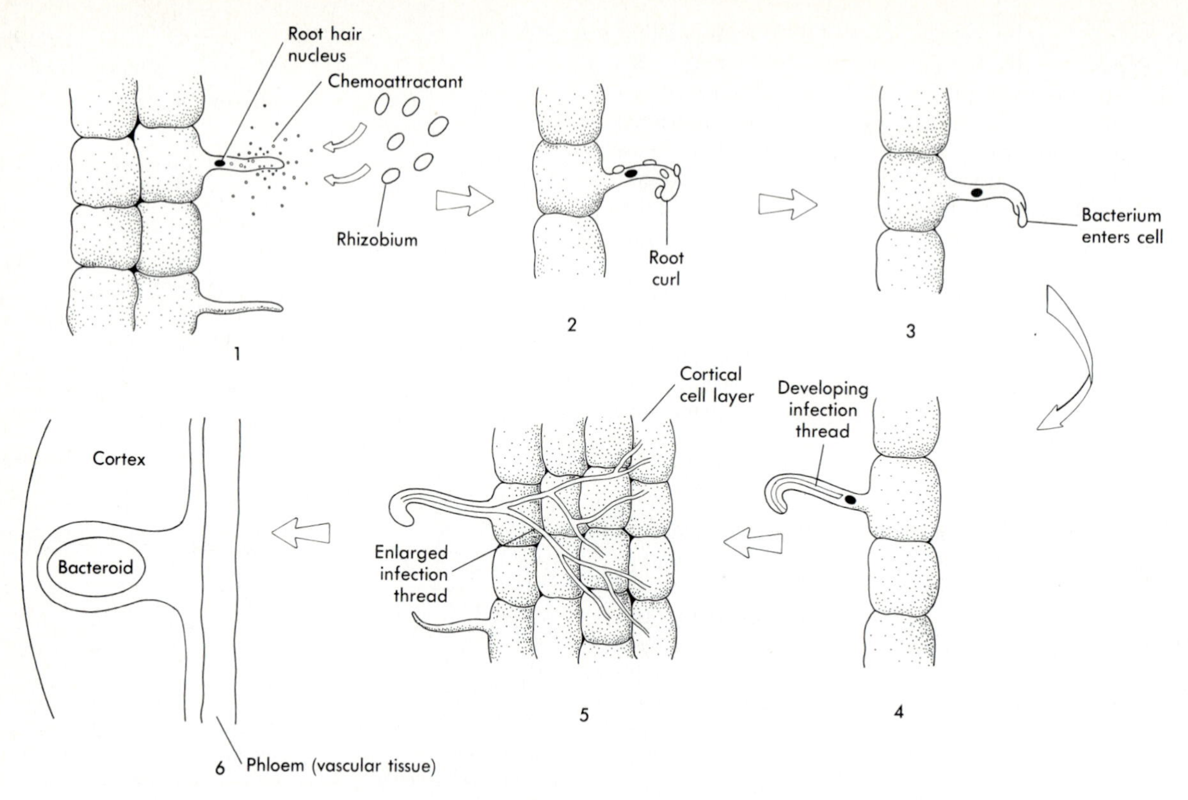

Figure 13-6 *Mechanism of nodule formation in root hairs of leguminous plants. Root hairs release chemical that attracts* Rhizobium *to epidermal surface. While attached to root surface, tip of root curls and this is apparently required for bacteria to penetrate. Bacteria then form infection thread that permits them to penetrate epidermal cell layer until cortical layer of cells is reached. Nucleus of plant cell is in touch with tip of infection thread. In cortex, bacteria are released from infection thread enclosed in membrane of host origin. Cortical cells are induced to divide and enclosed bacteria differentiate into larger body called bacteroid. Thus root possesses organized nodule in which infected cells contain bacteroids capable of dinitrogen (N_2) fixation.*

nitrogen in NH_3. The assimilatory process is accomplished through the activity of two enzymes—nitrate reductase and nitrite reductase—with electrons supplied by $NADPH_2$. The reaction is not coupled to ATP formation.

$$NO_3^- \xrightarrow[\text{reductase}]{\text{Nitrate}} NO_2^- \xrightarrow[\text{reductase}]{\text{Nitrite}} NH_3$$

The ammonia can be assimilated as described previously (see Figure 13-3). Nitrite, which is toxic, is seldom excreted as such and is rapidly reduced to ammonia. Nitrites have been used for many years as preservatives for certain meats and meat products such as bacon and frankfurters. Concern over their use has arisen because nitrites are converted in the body to nitrosamines, which have been described as carcinogenic in animals (see Chapter 21).

Nitrate can also be used by some microorganisms under anaerobic conditions as a terminal electron acceptor in ATP formation (denitrification or dissimilatory nitrate reduction). All denitrifying bacteria can substitute nitrite for nitrate with molecular nitrogen (N_2) as the usual major end product. This is discussed in Chapter 12. Denitrification in the soil is detrimental because of the loss of nitrogen. Farmers can reduce this loss by tilling the soil and providing aeration (see Point of Interest, p. 354).

Ammonia. Ammonia (NH_3) is the most efficiently used nitrogen source for microorganisms because it can be incorporated directly into carbon skeletons to produce amino acids. The concentration of ammonia in the environment, as well as the level of energy sources, determines its fate. The scheme of the bacterial cell is to use ammonia immediately, if biosynthesis of amino acids and other nitrogen-containing material is required, or to store it in the form of amino acids for future use when biosynthesis is repressed. A **glutamine synthetase/glutamate synthase** enzyme system (Figure 13-7) exists in the bacterial cell to control the assimilation of nitrogen. When adequate concentrations of ammonia are available to the cell, glutamate dehydrogenase catalyzes the assimilation of nitrogen into glutamate, and some of

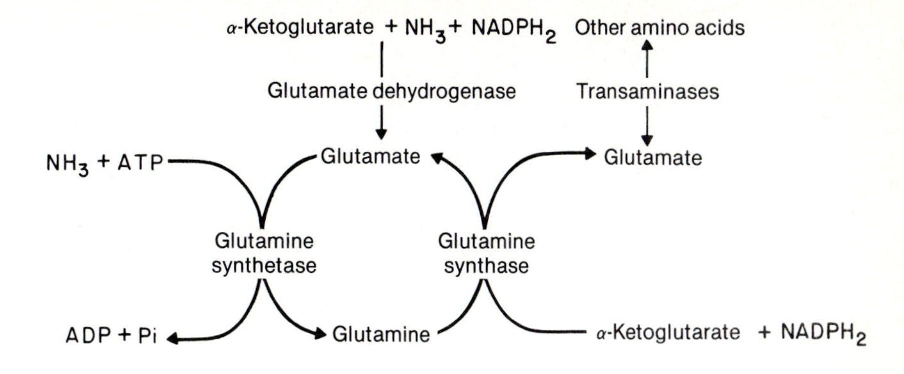

Figure 13-7 *Mechanism used by cell to control assimilation of nitrogen: glutamine synthetase/ glutamate synthase enzyme system. See text for details.*

the glutamate is converted into glutamine by glutamine synthetase. Under limited ammonia availability all of the nitrogen is incorporated into glutamine via the glutamine synthetase reaction and then much of this is converted to glutamate via glutamine synthase. Basically, the nitrogen in the ammonia molecule under conditions of biosynthesis will become part of the glutamate molecule according to the following reaction catalyzed by glutamate dehydrogenase.

$$\underset{\alpha\text{-Ketoglutarate}}{\begin{array}{c}COO^-\\|\\C{=}O\\|\\H{-}C{-}H\\|\\H{-}C{-}H\\|\\COO^-\end{array}} + \underset{\text{Ammonia}}{NH_3} \xrightarrow{NADPH_2} \underset{\text{Glutamate}}{\begin{array}{c}COO^-\\|\\H{-}C{-}NH_2\\|\\H{-}C{-}H\\|\\H{-}C{-}H\\|\\COO^-\end{array}} + H_2O + NADP$$

Glutamate provides the amino (NH_2) group for most other amino acids by participating in reactions called **transaminations**, in which an amino group is transferred:

$$\underset{\text{Pyruvate}}{\begin{array}{c}O\\ \|\\ CH_3{-}C{-}COO^-\end{array}} + \underset{\text{Glutamate}}{\begin{array}{c}COO^-\\|\\H{-}C{-}NH_2\\|\\H{-}C{-}H\\|\\H{-}C{-}H\\|\\COO^-\end{array}} \xrightarrow{\text{Transaminase}} \underset{\text{Alanine}}{\begin{array}{c}CH_3\\|\\H{-}C{-}NH_2\\|\\COOH\end{array}} + \underset{\alpha\text{-Ketoglutarate}}{\begin{array}{c}COO^-\\|\\C{=}O\\|\\H{-}C{-}H\\|\\H{-}C{-}H\\|\\COO^-\end{array}}$$

The glutamine produced in the cell supplies nitrogen to a number of metabolites such as the amino sugars and nucleotides, and the amino acids histidine and tryptophan.

Sulfur. Sulfur is a component of two amino acids, methionine and cysteine, as well as the vitamins biotin and thiamin. Sulfur appears in the environment primarily as inorganic sulfate ($SO_4^=$) which can be assimilated by microorganisms and reduced, in a series of reactions, to hydrogen sulfide (H_2S). Inorganic sulfur can become part of organic molecules in the following reaction:

$$\underset{\text{Hydrogen sulfide}}{H_2S} + \underset{\text{Serine}}{\begin{array}{c}NH_2\\|\\CH_2OH{-}C{-}COOH\\|\\H\end{array}} \rightarrow \underset{\text{Cysteine}}{\begin{array}{c}SH\quad NH_2\\|\qquad|\\CH_2{-}C{-}COOH\\|\\H\end{array}} + \underset{\text{Water}}{H_2O}$$

In some organisms cysteine is produced via the intermediate *O*-acetylserine. Cysteine helps maintain the tertiary structure of proteins through the formation of sulfur-sulfur bonds.

The formation of hydrogen sulfide occurs in the following steps:

$$\text{ATP} + \text{Sulfate} \xrightarrow{\text{ATP sulfurylase}} \text{Adenosine } 5'\text{-phosphosulfate (APS)} + PP_i$$

Concept Check

1. What is the importance of ribulosebisphosphate carboxylase?
2. What forms of nitrogen can be used in microbial biosynthesis?
3. What is the mechanism of nitrogen fixation by bacteria and its significance with respect to *Rhizobium* species in the soil?
4. What is the mechanism by which nitrate is assimilated into amino acids?

$$ATP + APS \xrightarrow{\text{APS phosphokinase}} \text{Adenosine 3' phosphate 5'-phosphosulfate (PAPS)} + ADP$$

$$2\ RSH\ (\text{Thioredoxin reduced}) + PAPS \xrightarrow{\text{PAPS reductase}} SO_3^{=} + AMP\ 3'\text{-phosphate} + RSSR\ (\text{Thioredoxin oxidized})$$

$$SO_3^{=} + 3\ NADPH_2 \xrightarrow{\text{Sulfite reductase}} H_2S + 3\ NADP$$

Inorganic sulfur is also an important component of certain proteins such as ferredoxin and of oxidation-reduction enzymes, where it behaves as a cofactor. The inorganic sulfur in these proteins is usually associated with other cofactors: iron-sulfur, as in ferredoxins, hydrogenases, and succinate dehydrogenase; and iron-sulfur molybdenum, as in nitrogenases and nitrate reductase. Many iron-sulfur enzymes possess extremely reduced redox centers and are consequently involved in oxidation-reduction reactions in which the reduction potentials are between 0 and −550 mV.

Sulfate can also be used as a terminal electron acceptor, and this function is discussed in Chapter 12. A more detailed discussion of the transformation of sulfur by soil and water microorganisms is discussed in Chapter 18.

Phosphorus. Phosphorus appears in organic molecules primarily as a component of nucleotides, such as ATP, which is important as a carrier of energy and phosphate and as a constituent of nucleic acids. Some microorganisms transport phosphate into the cell and accumulate it in the form of granules called **metachromatic granules** or **volutin**. The microbial cell assimilates inorganic phosphate into organic compounds through different reactions, including the following reaction:

$$\underset{\text{Glyceraldehyde-3-phosphate}}{\begin{array}{c} CHO \\ | \\ H{-}C{-}OH \\ | \\ H{-}C{-}O{-}PO_3H_2 \\ | \\ H \end{array}} + \underset{\text{Inorganic phosphate}}{iPO_4} \xrightarrow{NAD \rightarrow NADH_2} \underset{\text{3-Phosphoglyceroyl phosphate}}{\begin{array}{c} CO{-}PO_3H_2 \\ | \\ H{-}C{-}OH \\ | \\ H{-}C{-}O{-}PO_3H_2 \\ | \\ H \end{array}}$$

Media in which microorganisms, especially bacteria, are cultivated is prepared with salts of phosphate, such as sodium phosphate and potassium phosphate. The phosphates serve not only as a source of phosphorus but also as a buffering agent for the medium, since they are capable of binding one, two, or three hydrogen atoms:

$$Na_3PO_4 \xrightarrow{H^+} Na_2HPO_4 \xrightarrow{H^+} NaH_2PO_4 \xrightarrow{H^+} H_3PO_4$$

Phosphorylation of proteins is also an important process in the cell, but many of its function(s) are not completely understood. Possible functions include protein inactivation, protein activation, a regulatory device, and a protective device against protease digestion. One particularly important phosphorylated protein is the heat-stable protein (HPr) required for the first step in the phosphorylation of all phosphotransferase system carbohydrates. This phosphorylation process occurs at the expense of phosphoenolpyruvate and is called **group translocation** (see Chapter 5).

Oxygen. Oxygen is the final electron acceptor during aerobic respiration and is converted to nontoxic as well as toxic intermediates (see Chapter 12). In addition to its role as a terminal electron acceptor in respiration, dioxygen is a component of cellular materials such as organic compounds, water, and inorganic ions. The oxygen atoms of most cellular material are derived from supplied nutrients, carbon dioxide, and water, since biosynthesis is primarily an anaerobic process. In aerobically grown *E. coli* less than 0.1% of cell material is derived from atmospheric oxygen. An exception to this occurs when the organism uses an oxygenase-dependent reaction to initiate the breakdown of a substance for growth and energy. For example, organisms that use methane (CH_4) as a carbon and energy source use an oxygenase in the following reaction:

$$\underset{\text{Methane}}{CH_4} + O_2 \xrightarrow[\text{monooxygenase}]{\text{Methane}} \underset{\text{Methanol}}{CH_3OH} + H_2O$$

Up to 40% of the oxygen in the cellular material of methane-oxidizing bacteria is derived from atmospheric oxygen.

Even though most biosynthetic routes are anaerobic, some biosynthetic products can be produced only by oxygen-dependent routes. For example, the conversion of squalene to a sterol (lanosterol) in certain yeasts such as *Saccharomyces cerevisiae* requires oxygen. Consequently, if these organisms are grown anaerobically, sterols must be supplied in the culture medium; otherwise they are limited in their growth.

Many oxygen-dependent enzymes found in microorganisms serve in metabolic pathways that are alternatives to equivalent anaerobic pathways. In this way the organism is able to survive under aerobic or anaerobic conditions.

Oxygen is also a metabolic regulator, since many enzymes (for example, nitrogenase and hydrogenase) are sensitive to its effects or the effects of intermediates formed during its reduction.

Hydrogen. Hydrogen makes up 8% of the total cell dry weight and is a major constituent of inorganic and organic molecules. The hydrogen bond (energy of 3 to 7 kcal/mole compared to the covalent bond, which has a bond energy of 50 to 110 kcal/mole) is weak enough to be broken but strong enough to hold two atoms together, such as the bonds between hydrogen and oxygen or hydrogen and nitrogen, in molecules. Although the individual hydrogen bond is weak, when many of them are present in large macromolecules the secondary structure of the molecule is maintained. For example, in a polypeptide the hydrogen bonds between the carbonyl group of one residue and the imino group of the fourth residue along the chain hold the alpha helix together. The two chains of the DNA molecule are held together because of the many hydrogen bonds between purine and pyrimidine residues on opposite chains.

The removal of hydrogen atoms from organic or inorganic molecules is important in the generation of energy during electron transport. In addition, molecular hydrogen (H_2) is a source of energy for the chemoautotrophic hydrogen bacteria (see Chapter 12).

Water. The principal component of living matter, at least quantitatively, in the microbial cell is water. Water, which makes up from 70% to 80% of the weight of the cell, acts as a solvent for many molecules and carries them across the cytoplasmic membrane in a hydrated form. Water, in addition to its inert activity as a carrier, is also involved in most biochemical reactions and, because of its polar nature and capacity to bind other polar molecules, influences the shape and architecture of many biological polymers, especially proteins. More detailed discussions of water and water activity are presented in Chapters 2 and 21, respectively.

Mineral metabolism

The crust of the earth is composed of many minerals that are important in the life of the microbial cell. These elements include magnesium, calcium, manganese, zinc, iron, selenium, potassium, sodium, cobalt, copper, and molybdenum. Some of these minerals are major nutrients and are required in high concentration in all cells, whereas others are required only by certain species. Minor nutrients are those that are used by cells at low concentrations, and laboratory media will often contain sufficient impurities to satisfy the cell's needs.

Magnesium. Magnesium is a divalent cation that plays an important role in metabolism. All cells require magnesium and actively accumulate it by specific energy-dependent transport systems. Most of the magnesium in the cell is associated with the nucleic acids, especially ribosomes; the magnesium neutralizes the negatively charged phosphate groups and thus acts to stabilize ribosomes. Magnesium deficiency can lead to filament formation in ordinarily unicellular forms, to ribosome degradation, and to a decrease in nucleic acid synthesis. Magnesium also binds to the negatively charged teichoic acids, which are believed to act as a repository for magnesium ions in gram-positive cells. Magnesium is a cofactor for many enzymes and is also a component of the chlorophyll molecule.

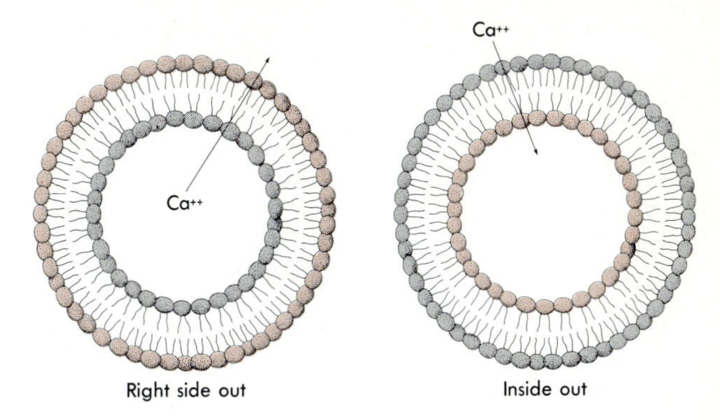

Figure 13-8 *Direction of movement of calcium in membranes prepared right side out and inside out.*

Calcium. The calcium level in microbial cells is very low compared with its concentration in the external environment. Calcium represents a special case because it is secreted during metabolically active processes and is not actively accumulated like nearly all other minerals. Lowering the temperature or using inhibitors of energy formation causes microbial cells to accumulate calcium, thus indicating a carrier-mediated transport system.

The reverse orientation of calcium transport can be demonstrated by a very unusual technique of preparing microbial cell membranes with different orientations. In other words, the cell membrane can be prepared in a normal way in which the outer part of the membrane remains outside (right side out), or it can be prepared abnormally, in which the membrane is turned inside out (Figure 13-8). The right-side-out orientation causes calcium to be secreted, but in the inside-out orientation calcium is accumulated.

The only instance where calcium has been observed to be accumulated by bacterial cells is during the sporulation process by members of the genus *Bacillus.* A carrier is believed to be involved, but it is not known if it is the same one used for secretion of calcium by nonsporulating cells. Calcium plays a role in the chemotactic behavior of *Bacillus subtilis* (see Chapter 5). It is also a cofactor for some enzymes such as proteinases.

Calcium appears to play some specific roles in eukaryotic cells. It may be important for the aggregation associated with the differentiation process in the slime mold *Dictyostelium discoideum.* Calcium increases dramatically in cells during slug migration and stalk and spore formation. Calcium is also associated with ciliary movement in protozoa such as *Paramecium.* Changes in calcium concentration appear to be associated with changes in direction of movement of the protozoa. When the organism meets an obstacle and must swim backward, the calcium concentration increases in the cell; for the organism to swim forward, calcium is pumped out of the cell.

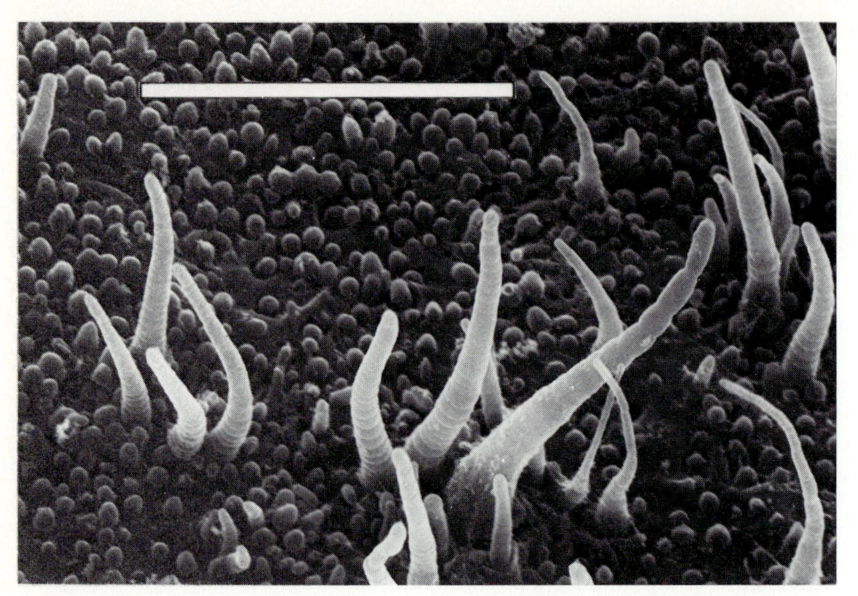

Figure 13-9 *Scanning electron micrograph of microorganisms encrusted with metal deposits on ocean floor. Microorganisms are attached to mussel beds in vicinity of thermal submarine vents at Galapagos rift at depth of over 7600 feet. Spherical cells are encrusted with manganese and/or iron and are interspersed with clusters of filamentous forms. Hydrogen sulfide is source of energy for these bacteria, which are sole source of food for animal populations at these extreme depths. Bar = 10 μm.*

Manganese. Manganese is required by microbial cells in trace amounts, and its functions are largely unknown. It can serve as an alternate co-factor for magnesium during magnesium deficiency, particularly in magnesium-requiring enzymes.

Microbial oxidation of manganese occurs among many soil and marine species. Nodules of ferromanganese (Figure 13-9) are produced on the ocean floor, presumably through the activity of microbial enzymes, but the exact mechanism is not known. Soil bacteria and fungi reduce insoluble salts of manganese (Mn^{4+}) to soluble manganese (Mn^{2+}) compounds. This reduction process, which is characteristic of some bacteria that inhabit the roots of some plants, makes manganese available in a soluble form. The sheathed bacteria, such as *Leptothrix* and *Sphaerotilus,* oxidize manganese (Mn^{2+}) and accumulate it within the sheath as deposits of manganese oxides.

Manganese is required for the synthesis of secondary metabolites such as antibiotics. Peptide antibiotics produced industrially by species of *Bacillus*, for example, cannot be synthesized in the absence of manganese. Peptide antibiotics, which are produced during sporulation, are believed to act as carriers of calcium and thus may help the cell to accumulate this divalent cation.

Zinc. Zinc is required only in trace amounts (0.05 to 0.1 μM) by the microbial cell. It is essential for the activity of several enzymes, including DNA and RNA polymerase. Zinc is also essential for the synthesis of many secondary metabolites produced by microorganisms that are used industrially and medically. For example, the commercial production of bacterial proteases requires that zinc be present in the medium.

Iron. The earth was originally enveloped in an anaerobic atmosphere, and one of its primary elements, iron, existed in the ferrous state (Fe^{++}). The appearance of oxygen from photosynthetic processes resulted in the oxidation of much of the ferrous iron to the ferric (Fe^{+++}) state. Oxidation produces iron complexes that make the mineral unavailable for biological transport in the ionic form. Microbial cells have evolved techniques to compensate for this transport deficiency. Ferric iron is made available to the cell through the production of metal-binding (chelator) molecules called **siderophores**. The iron is bound to these transport molecules via an oxygen, sulfur, or nitrogen atom. These atoms have a high potential negative charge and thus have strong affinities for ferric iron, which has a high potential positive charge.

There are three major classes of natural iron transport molecules: hydroxamic acids, catechols (derivatives of 2,3-dihydroxybenzoic acid), and citrate. Three identifiable hydroxamates are ferrichrome, mycobactin, and ferrioxamine B (Figure 13-10). Mycobactin is found in the mycobacteria, which have a high lipid content in their envelope. Mycobactin, interestingly enough, has the greatest lipid solubility among the siderophores. An important catechol derivative is enterobactin (Figure 13-11), which is produced by members of the Enterobacteriaceae. Specific siderochromes are produced by different genera of bacteria, but some microorganisms can use the chelating agents produced by microorganisms from other genera (see Point of Interest).

Iron is a metal required by all bacteria, and its major role is in electron/hydrogen movement. Iron is a component of the hemin molecule that makes up part of the cytochromes and thus functions in the transport of electrons by being alternatively reduced and oxidized. Other heme proteins in which iron is a component are catalase and peroxidase, two enzymes that pro-

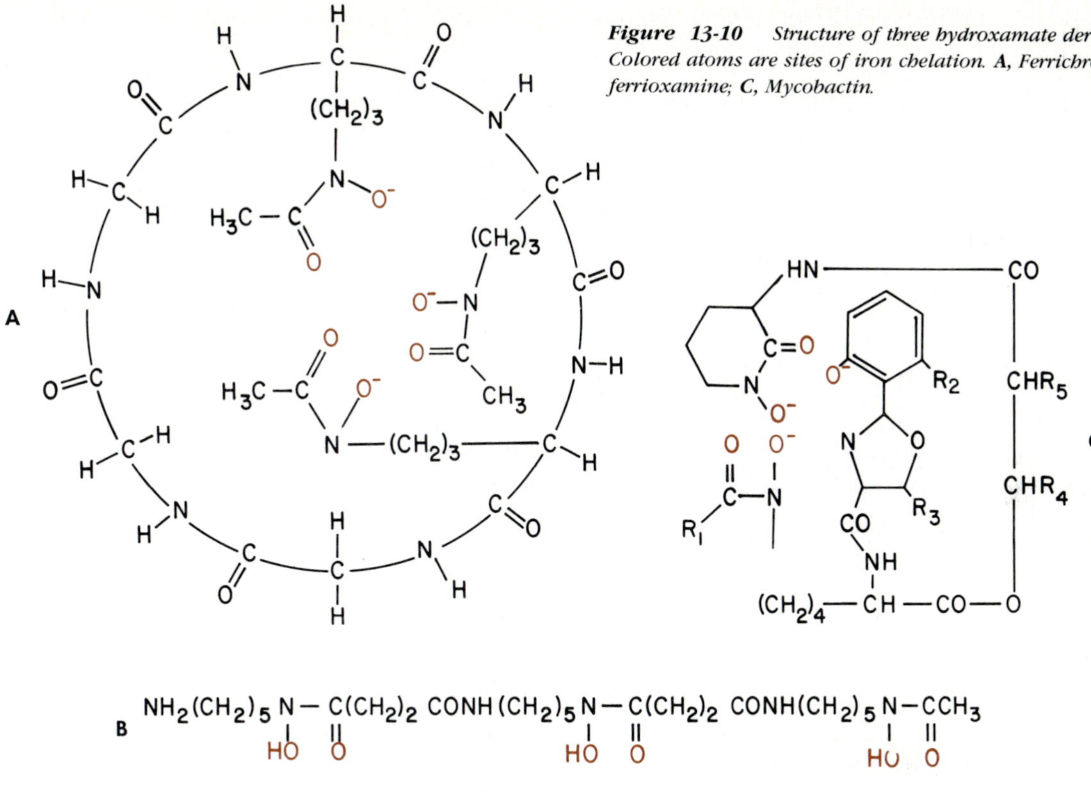

Figure 13-10 *Structure of three hydroxamate derivatives. Colored atoms are sites of iron chelation.* ***A,*** *Ferrichrome;* ***B,*** *ferrioxamine;* ***C,*** *Mycobactin.*

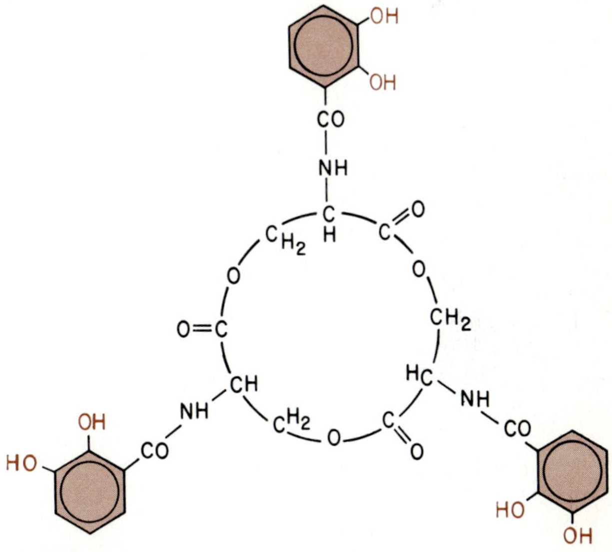

tect the cell from the deleterious effects of hydrogen peroxide generated during aerobic respiration. Nonheme iron proteins involved in electron transport are hydrogenase, flavoproteins, ferredoxin, and other FeS proteins. Iron is also believed to play an important role in the host's response to invading microbial pathogens. Nearly all vertebrates respond to the infection process by altering their metabolic processes in such a way as to deprive the invading pathogen of iron. Two vertebrate chelators of iron are **transferrin** and **lactoferrin**. Transferrin is found predominately in blood and lymph, while lactoferrin is found primarily in external secretions, such as tears, milk, saliva, and seminal fluid. Iron-binding proteins such as transferrin increase in concentration at the site of infection and are believed to behave as bacteriostatic agents by siphoning away iron that could be used for growth of the invading parasite.

Toxin production in *Corynebacterium diphtheriae*, the causative agent of diphtheria, is controlled by the concentration of iron, and this aspect of the host-parasite relationship is discussed in Chapter 23.

Figure 13-11 *Structure of siderophore enterobactin. Atoms in color represent sites for chelation of iron.*

Potassium. Potassium is required by all microbial species tested to date and is especially important in ribosome function. Cells deprived of potassium or mutants unable to transport potassium into the cell are unable to carry out protein synthesis. Potassium plays a crucial role in the *Halobacteria* that live in environments where the salt (NaCl) concentration is very high (see p. 411). Potassium is also a cofactor for some enzymes.

Other minerals. Molybdenum, selenium, cobalt, and

copper are minor nutrients that play very specific roles in the function of some microorganisms. Molybdenum plays a role in nitrogen metabolism, since it is an essential component of ferredoxins, nitrogenase, and nitrate reductase. Cobalt is part of the vitamin B_{12} molecule, and copper is a component of cytochrome oxidase and is involved in electron transport during aerobic respiration.

Specific organic nutrients (growth factors)

Some microorganisms are unable to convert inorganic or organic nutrients from the environment into organic metabolites. Consequently, these organisms require supplements of specific organic products, called **growth factors.** Growth factors include vitamins, amino acids, and nucleotides. Vitamins are especially important because of their role in enzymatic reactions. They form the coenzyme portion of many enzymes and have specific functions, which are described in Table 13-6. Many microorganisms isolated from their natural environment and cultivated in the laboratory require growth factors. Microorganisms in their natural environment may have these requirements supplied by other organisms that inhabit the same ecological niche. Microorganisms that normally colonize animals and humans often have growth factor requirements when cultivated in the laboratory. Scientists have suggested that during evolution these microorganisms could produce all of the necessary metabolites from simple carbon sources; however, once they became permanent colonizers of mammals, who could supply preformed metabolites, biosynthetically deficient mutants were selected. One especially important exception is *E. coli*, which inhabits the intestinal

Point of Interest

IRON TRANSPORT

The problems associated with getting iron into the cell have resulted in the evolution of a very elaborate transport system in bacteria. The process occurs in the following way (see figure): Once a siderophore has been released from the bacterial cell, it complexes with an iron molecule. The siderophore is too large to readily pass through cytoplasmic membrane pores, but a membrane receptor specific for the siderophore is available for rapid transport (these siderophore receptors are also used by bacteriophages when they attach to the bacterial cells). The siderophore-iron complex either passes into the cytoplasm, releasing iron that has been reduced to the ferrous state, or only the iron is released into the cytoplasm with the siderophore remaining outside the cell. The released ferrous iron is used for various iron-requiring enzymes and proteins. Iron transport into the cell occurs when there is iron limitation. The system can also be inhibited when the iron requirement of the cell has been satisfied. Thus the genes controlling iron synthesis operate in bacterial cells and, under conditions of excess iron, combine with a protein molecule called a repressor that switches off the system.

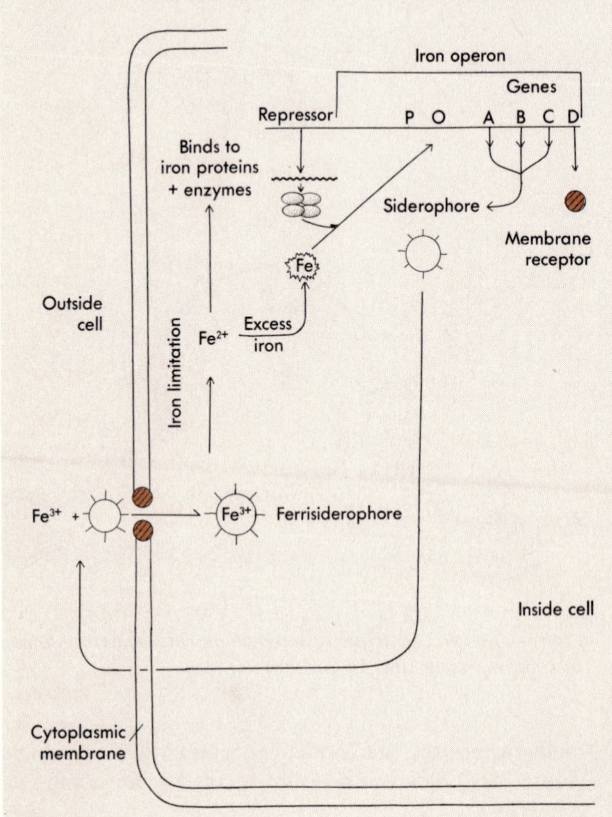

Model for iron transport in bacteria. Three genes (A, B, *and* C) *control synthesis of siderophore and they are active under conditions of low iron stress. Fourth gene codes for membrane receptor of iron complex. Siderophore passes through cytoplasmic membrane and complexes with iron (ferric) outside cell (for simplicity we have shown only one possible mechanism for transport of iron-siderophore complex). Ferrisiderophore complex is reduced in cytoplasm with release of ferrous iron to be used for iron-containing enzymes and proteins. Under iron excess, ferrous iron complexes with repressor protein and binds to genetic sequence that inhibits activity of genes that code for synthesis of siderophore.* P, *Promotor;* O, *operator.*

Table 13-6 *Vitamins and their function in microbial metabolism*

Vitamin	Function
Thiamin	As thiamin pyrophosphate, a coenzyme involved in decarboxylation reactions
Riboflavin	In combined form such as flavin adenine dinucleotide (FAD), a coenzyme important in oxidation-reduction reactions
Niacin	In combined form such as nicotinamide adenine dinucleotide (NAD) or its phosphorylated derivative (NADP), a coenzyme in oxidation-reduction reactions
Pantothenic acid	Part of the coenzyme A (CoA) molecule; functions as a carrier of acyl groups; precursor of prosthetic group of acyl carrier protein (ACP) involved in lipid metabolism
Pyridoxal	In phosphorylated form, functions as a coenzyme; important in amino acid metabolism, particularly group transfer reactions involving the amino group
Biotin	Coenzyme important as a carrier of carbon dioxide
Folic acid	In reduced form (tetrahydrofolic acid), acts as a carrier of formyl groups; especially important in purine nucleotide biosynthesis
Vitamin B_{12}	Coenzyme functioning as a carrier of alkyl groups
Vitamin K	Precursor of menaquinone, which is involved in electron transport

Concept Check

1. How is inorganic sulfur incorporated into organic matter, and in what type of cellular components is sulfur found?
2. Describe the reaction in which inorganic phosphorus is incorporated into organic molecules. What is the role of phosphate?
3. What functions does hydrogen serve in organic molecules?
4. What are the functions of oxygen in the microbial cell, and how it is assimilated into organic molecules?
5. What are the functions of the following minerals: manganese, magnesium, calcium, zinc, iron, and potassium? How is iron transported into the cell?

Table 13-7 *Chemical composition of two kinds of synthetic media**

Medium	mg/L
Medium A	
K_2HPO_4	7000
KH_2PO_4	3000
Sodium citrate	500
$MgSO_4$	100
$(NH_4)_2SO_4$	1000
Glucose	5000
Distilled or deionized water	1 L
Medium B	
Cysteine	15
Sodium glutamate	500
$(NH_4)_2SO_4$	1200
Sodium acetate	10,000
K_2HPO_4	7000
NaH_2PO_4	1200
$MgSO_4$	400
Folic acid	0.005
Biotin	0.0025
Para-aminobenzoic acid	0.1
Thiamin	0.5
Riboflavin	0.5
Pyridoxal	1.0
Pantothenate	0.5
Nicotinic acid	1.0
Glucose	10,000
Distilled or deionized water	1 L

*Medium A is a basal salts minimal medium used to cultivate such enteric organisms as *E. coli*. Medium B is a more complex medium used to cultivate some species of oral streptococci.

tract of animals. Most strains of this species are capable of growth on a simple defined medium (Table 13-7) and can synthesize all of their organic requirements from inorganic salts and a single organic carbon source such as glucose. Compare this with the requirement of a species of *Streptococcus* isolated from the oral cavity (Table 13-7). The ability of *E. coli* to synthesize all of its organic needs is one of the reasons why this organism has been used extensively in genetic and metabolic studies.

BIOSYNTHESIS

How energy is extracted from inorganic and organic compounds and how this energy is used for movement and nutrient transport is demonstrated in earlier chapters. Energy is also used to convert the basic nutrients, discussed earlier in this chapter, into organic compounds that make up the structural and enzymatic components of the cell. What organic compounds can be synthesized by each cell is determined by the genetic content of the cell and by what enzymes are potentially inducible. This biosynthetic capacity may be the result of natural evolutionary processes, or it may be due to unnatural laboratory manipulation, such as genetic engineering, in which foreign genes are introduced into the cell.

Biosynthetic capacity

The biosynthetic potential of microorganisms is as variable as any other property that can be measured. We have already discussed the fact that some microorganisms are able to take inorganic nutrients and assimilate them into organic molecules whereas other organisms require external supplies of many organic compounds in order to grow. Biosynthesis is a continual process in cells, and whether the cell has a short (20-minute) or long (24-hour) generation time, biosynthesis is required to construct new cell material and to replenish certain molecules in the cell that have been degraded. Macromolecules, such as proteins and nucleic acids, which are continually being degraded during the cell cycle, may have a short or long half-life. Some messenger RNA (mRNA) molecules in bacteria have a half-life of 60 to 120 seconds, whereas in eukaryotic cells the mRNA half-life may be up to 20 minutes. Biosynthesis occurs at a faster rate in prokaryotic cells (mycobacteria are exceptions) than in eukaryotic cells—a property that is especially important when antimicrobials are administered to humans suffering from infectious diseases.

One of the most significant properties of both eukaryotic and prokaryotic cells is that the pathways leading to the biosynthesis of a central metabolite, such as an amino acid, are invariably the same or similar. This indicates that little change has occurred after the evolutionary divergence of these cellular types. This is one of the reasons why research of biosynthetic processes in microorganisms has been so helpful in understanding human biosynthetic processes.

Requirements in biosynthesis

In the discussion that follows we assume that the microorganism involved in biosynthesis is capable of synthesizing all of its organic matter using glucose as the sole source of carbon and energy, for example, *E. coli.*

All biosynthetic processes have very basic requirements and conform to specific patterns of organization. Most biosynthetic reactions require **energy**, and they are invariably coupled to the hydrolysis of ATP. The energy from ATP hydrolysis can therefore be used to produce covalent bonds between two molecules, as in the following:

$$A + B \xrightarrow[\text{hydrolysis}]{\text{ATP}} A\text{—}B$$

Biosynthesis also requires **reducing power.** Recall that reducing power in the form of NADPH is generated during the pentose phosphate pathway and during photophosphorylation. **NADPH** is used specifically in biosynthetic reactions to supply hydrogen for the reduction of carbon (NADH is used primarily for catabolic or oxidative processes).

During biosynthesis larger molecules are formed by the addition of smaller activated units to a preexisting larger molecule. This building-up process usually takes place in one of two ways:

1. Coenzymes are used as carriers of small units. Acetyl coenzyme A (acetyl CoA), for example, acts as a carrier of acetyl groups. In the TCA cycle the acetyl group is condensed with oxaloacetate to form citrate (see p. 331). The coenzymes that act as carriers of carbon units (or hydrogen) are outlined in Table 13-8.
2. Some chains of molecules (polymers) do not require carriers for their buildup but instead are formed by the condensation or addition of similar molecules (Figure 13-12). The synthesis of the bacterial cell wall or the synthesis of macromolecules such as DNA or RNA occurs by the transfer of an activated monomer to a growing polymer chain. The addition of a monomer to a growing chain involves three basic steps: (1) The monomer that is to be added to the growing polymer is activated, for example, amino acids are activated by conversion to aminoacyl tRNA. (2) Each of the monomers will also carry a high-energy bond (~) that will be used for the addition of itself or the next monomer to the growing chain. One exception to this rule is in the case of glycogen synthesis in bacteria where a monomer in the form of ADP-glucose is added on to a growing polymer and ADP is released without a high-energy bond being used. (3) The addition of the monomer to the growing chain involves the loss of water (by contrast, the breakdown or polymers involves the addition of water, that is, hydrolysis).

The synthesis of nucleic acids and proteins is intimately associated with genetic processes;

Table 13-8 *Coenzymes involved as carriers in biosynthetic reactions*

Coenzyme	Group transferred	Biosynthesis involved in
Biotin	CO_2 (carboxy)	Fatty acids
Thiamin pyrophosphate	CH_3-CHO (acetaldehyde)	Sugars
Coenzyme A	CH_3-CO (acetyl)	Lipids
$NADPH_2$	H_2 (hydrogen)	All biosyntheses
Uridine diphosphate glucose (UDP-glucose)	Glucose	Disaccharides and polysaccharides (glycogen)
5-Adenosylmethionine	CH_3 (methyl)	RNA, DNA, and fatty acid methylation
Transfer RNA (tRNA)	$COCHNH_2R$ (amino acid as aminoacyl group)	Proteins
Folic acid	HCHO (formate)	Purine nucleotides

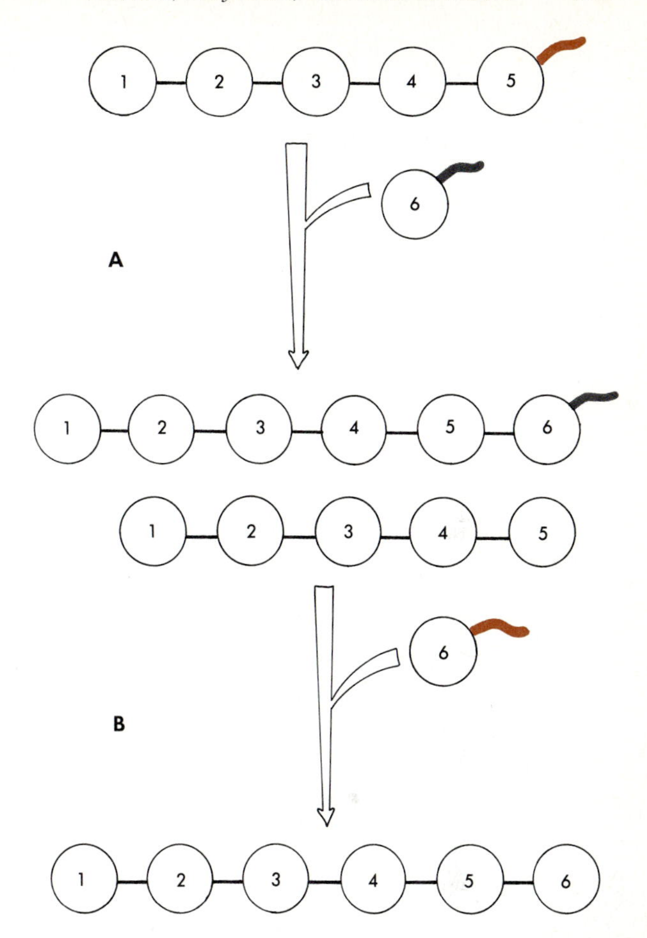

Figure 13-12 *Two mechanisms for synthesis of polymers. **A,** Last monomer on growing chain has high energy bond (color) that will be used for addition of next monomer. This type of reaction is characteristic in synthesis of proteins and fatty acids. **B,** Monomer on growing chain does not contain high energy bond. Incoming monomer (6) will carry energy for its own addition. This type of reaction occurs in synthesis of RNA, DNA, and some polysaccharides.*

therefore it is discussed in detail in Chapters 11 and 12. Cell wall synthesis, because it is affected by several antibiotics, is discussed in Chapter 15.

Biosynthetic pathways

Catabolic pathways are intimately connected to biosynthetic pathways and provide the necessary precursors for fatty acids, amino acids, purines and pyrimidines, and various sugars. Figure 13-13 illustrates how these metabolic systems are integrated. This figure should be referred to throughout the remainder of the discussion of biosynthesis.

Biosynthesis of amino acids. We discussed earlier how nitrogen in all of its inorganic forms is converted to ammonia, which in turn is incorporated into organic compounds to produce amino acids. Four different mechanisms are available to microorganisms for the assimilation of ammonia. The reactions are outlined in Figure 13-14. One of the reactions involves a system devised to aid the cell in ammonia assimilation when high or low concentrations of the compound are present in the medium, as discussed earlier in the chapter.

One of the principal functions of the TCA cycle is to produce intermediates that act as precursors to the amino acids. The amino acids have been divided into families based on their precursor types (Table 13-9). Oxaloacetate and pyruvate provide pathways for the synthesis of eight amino acids (Figure 13-13). Only L-lysine, of all the amino acids, is synthesized via two different pathways (Figure 13-15). This is an example of the type of divergence that is believed to have accompanied the evolutionary split between prokaryotes and eukaryotes. L-Lysine in prokaryotes is synthesized from the amino acid L-aspartate, whereas in eukaryotic fungi and most algae L-glutamate is the primary metabolite from which L-lysine is produced. The aspartate pathway provides two

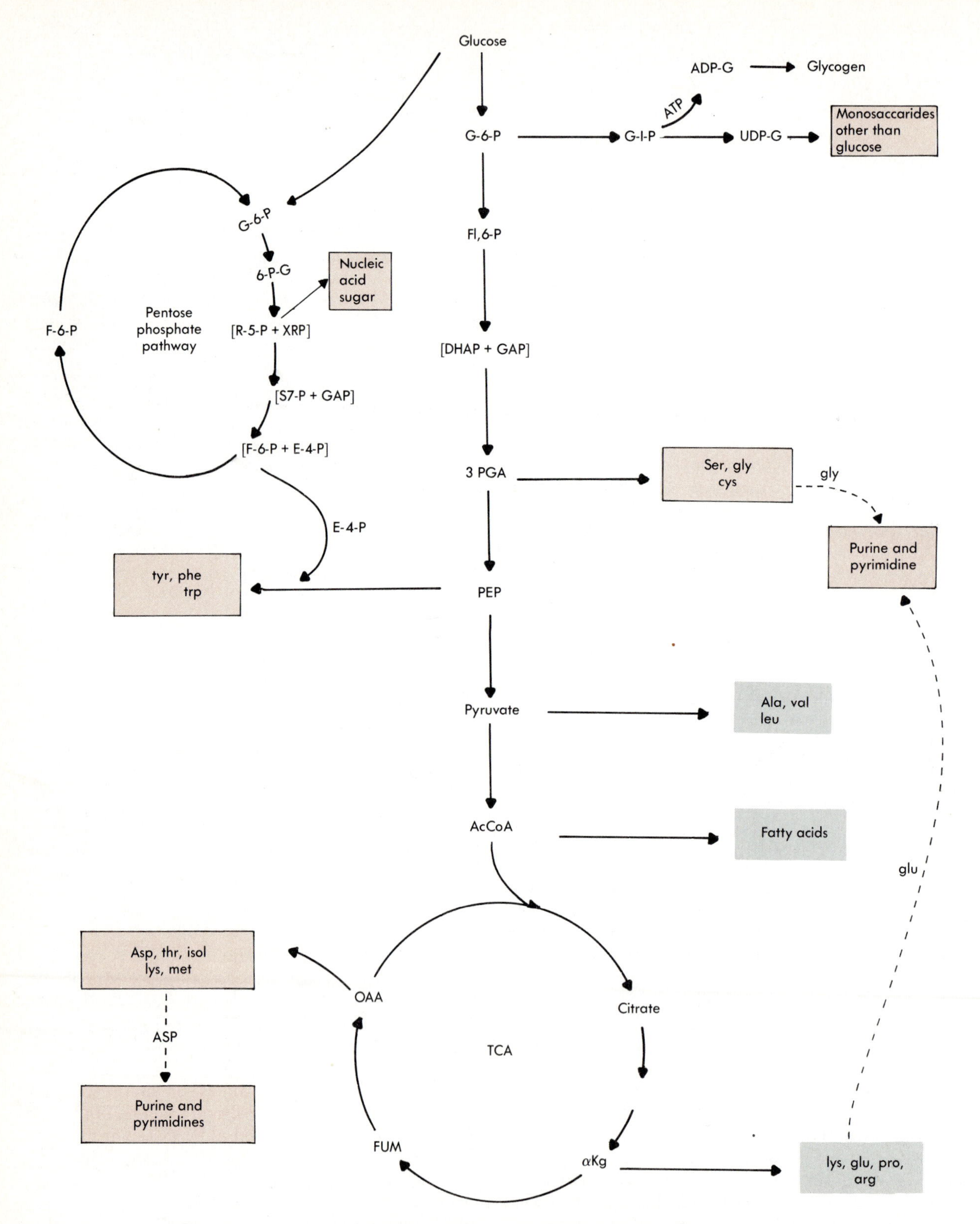

Figure 13-13 *Summary of pathways that provide precursors in biosynthesis. Broken arrow indicates amino acids, which supply carbon and nitrogen for purine and pyrimidines.*

Table 13-9 *Metabolic intermediates serving as precursors in the biosynthesis of amino acids*

Precursor(s)	Amino acid end product(s)
Pyruvate	Alanine, valine, leucine
Oxaloacetate	Aspartate, threonine, isoleucine, lysine, methionine
α-Ketoglutarate	Lysine, glutamate, proline, arginine
3-Phosphoglycerate	Serine, glycine, cysteine, cystine
Erythrose phosphate and phosphoenolpyruvate	Tyrosine, phenylalanine, typtophan
Pentose phosphate	Histidine

1. Alpha-ketoglutarate + NH_4^+ + $NADPH_2$ $\xrightleftharpoons[\text{dehydrogenase}]{\text{L-glutamate}}$ L-glutamate + NADP + H_2O + H^+

2. Fumarate + NH_4^+ $\xrightleftharpoons[\text{deaminase}]{\text{aspartate}}$ L-aspartate + H^+

3. Pyruvate + NH_4^+ + $NADPH_2$ $\xrightleftharpoons[\text{dehydrogenase}]{\text{alanine}}$ L-alanine + NADP

4. L-glutamate + NH_4^+ + ATP $\xrightarrow[\text{synthetase}]{\text{glutamine}}$ L-glutamine + ADP + $_iPO_4$

alpha-ketoglutarate + L-glutamine + $NADPH_2$ $\xrightarrow[\text{synthase}]{\text{glutamate}}$ 2-L-glutamate + NADP

Net reaction: alpha-ketoglutarate + NH_4^+ + ATP + $NADPH_2$ $\rightarrow$ L-glutamate + ADP + $_iPO_4$ + NADP

Figure 13-14 *Four reactions in bacterial cell in which ammonia can be assimilated to produce amino acids.*

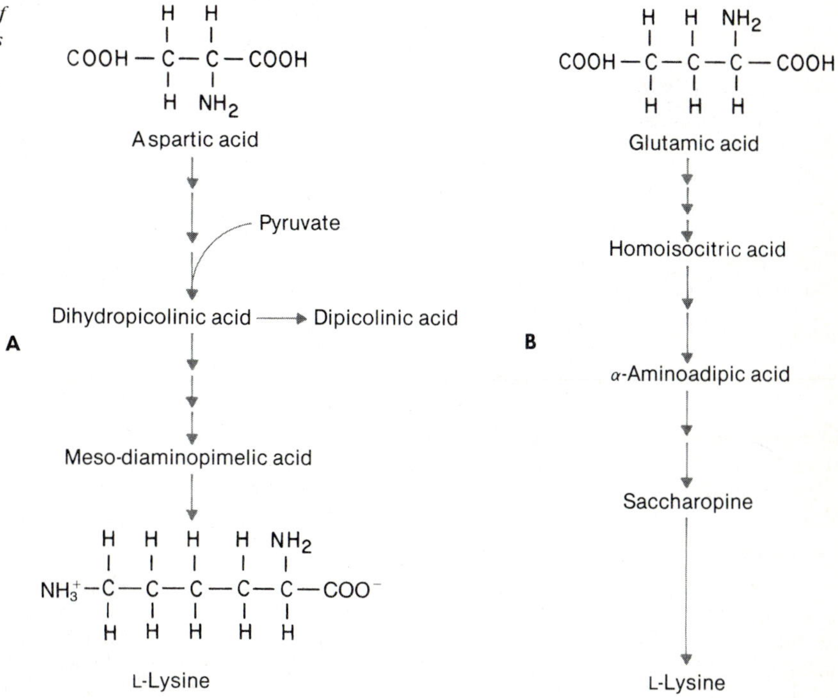

A

$COOH-CH_2-CH(NH_2)-COOH$

Aspartic acid

↓ ↓ (+ Pyruvate)

Dihydropicolinic acid → Dipicolinic acid

↓ ↓ ↓

Meso-diaminopimelic acid

↓

$NH_3^+-CH_2-CH_2-CH_2-CH_2-CH(NH_2)-COO^-$

L-Lysine

B

$COOH-CH_2-CH_2-CH(NH_2)-COOH$

Glutamic acid

↓ ↓ ↓

Homoisocitric acid

↓ ↓

α-Aminoadipic acid

↓ ↓

Saccharopine

↓

L-Lysine

Figure 13-15 *Alternative pathways of lysine biosynthesis (not all intermediates are indicated).* ***A,*** *Pathway used in bacteria.* ***B,*** *Pathways used by fungi and most algae.*

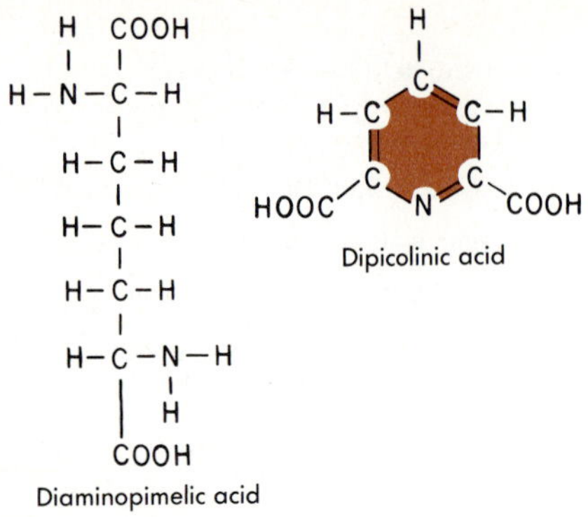

Diaminopimelic acid

Dipicolinic acid

Figure 13-16 *Structure of diaminopimelic acid (DAP) and dipicolinic acid.*

essential intermediates that are an important part of the bacterial life cycle: diaminopimelic acid (DAP) and dipicolinic acid (Figure 13-16). DAP is a component in the peptidoglycan layer of some bacterial cell walls, and dipicolinic acid is a component of the cortex of the bacterial endospore. Neither of these intermediates is used by eukaryotic cells and consequently they are not produced.

Some microorganisms cannot synthesize all of the necessary amino acids, and they must be supplied preformed in the medium, either as free amino acids or in the form of proteins. Many microbial species are genetically equipped to produce **extracellular proteases** that hydrolyze proteins into amino acid residues. The clostridia, which are capable of causing gas gangrene in humans, produce proteases that degrade muscle protein during the infection process. The amino acids released by hydrolysis are transported across the cytoplasmic membrane to be fermented and to engage in protein synthesis. The virulence of pathogenic species of *Neisseria* and *Streptococcus* appears to be related to the production of specific proteases. *Neisseria gonorrhoeae* (the causative agent of gonorrhea) owes part of its virulence to its ability to adhere to epithelial cells of the urogenital tract. Epithelial secretions in the urogenital tract (or other secretory surfaces) contain secretory antibodies, called secretory IgA (sIgA), that can prevent the attachment of pathogenic species to the epithelial surface. *N. gonorrhoeae,* however, produces a protease that cleaves specific sIgA molecules and inactivates them—a process that permits attachment of the virulent organisms to the host. Nonpathogenic species of *Neisseria* apparently do not produce the protease.

Most of the amino acids in nature are L-isomers, and cellular enzymes recognize only that form. There are some instances, for example, in the bacterial cell wall, where the D-isomer of an amino acid exists (see Chapter 15). Conversion of the L-isomer to the D-isomer is catalyzed by enzymes called **racemases.** Without racemases a microorganism would have to be supplied preformed D-amino acids, which are minor forms in nature.

Amino acids not only contribute to the formation of high molecular weight proteins but are used in other biosynthetic processes. Amino acids are used to produce small peptides that make up the bacterial cell wall. These peptides lend stability to the cell wall when they are linked to peptides on adjacent peptidoglycan strands (see Chapter 5). Amino acids also contribute their carbon skeletons and the nitrogen of their amino groups to the formation of purines and pyrimidines (see Figure 13-13). Amino acids are precursors for some vitamins, coenzymes, and antibiotics produced by bacteria.

Biosynthesis of lipids and fatty acids. Lipids play a very important role in the construction of microbial membranes. We have already discussed in previous chapters the importance of cytoplasmic membrane lipids in terms of solute transport, excretion of proteins, and stability to environmental extremes. The membrane lipids also provide a protective device surrounding the protein coat of some viruses that obtain their membrane when released from the infected cell (see Chapters 16 and 17).

Microbial lipids. Lipids are not considered true macromolecules because they are not formed from monomers that are covalently linked to one another. Lipids are complex molecules, however, that are uniquely important to the structure of the cellular membrane and the transport processes with which they are associated.

The principal lipids found in the microbial membrane are **glycerophospholipids.** Their basic structure is illustrated in Figure 13-17. The fatty acid components of microbial lipids are most frequently chains containing 15 to 19 carbons. The chains are usually straight, but occasionally they are cyclic. Any unsaturation of the molecule is usually at carbon 2 of the fatty acid. Unsaturated fatty acids are associated with psychrophiles, in whom fluidity of the membrane can be maintained at low temperatures because of the lower melting point of unsaturated fatty acids (see Chapter 14). Glycolipids, which are also present in bacteria, contain hexoses as the sugar moiety. Glycolipids make up part of the cell wall–outer membrane of gram-negative bacteria and the cytoplasmic membrane of all bacteria. Glycoplipids are also important in the cell envelope of the mycobacteria (some species cause tuberculosis and leprosy), in which the fatty acid side chains are **mycolic acid.** These glycolipids are believed to provide the mycobacteria with their disease-producing potential (see Chapter 27). The basic structure of some glycolipids is illustrated in Figure 13-17.

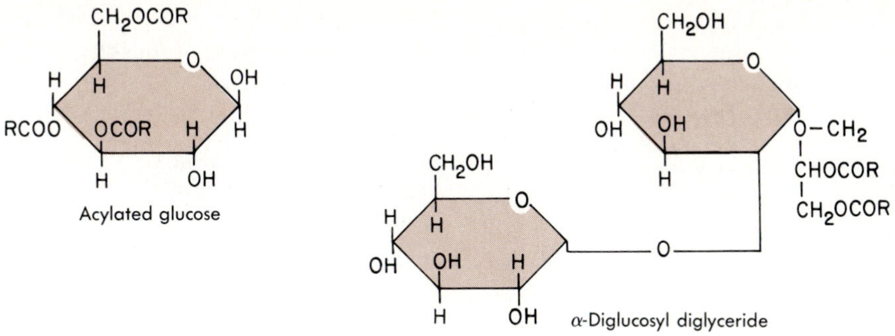

CH_2OCOR H H OH RCOO OCOR H H H OH

Acylated glucose

CH_2OH H H OH OH H $O-CH_2$ $CHOCOR$ CH_2OCOR **A**

CH_2OH H H OH OH H H OH

α-Diglucosyl diglyceride

B

$$\begin{array}{l} H_2C-O-C(=O)-R \\ R-C(=O)-O-CH \\ H_2C-O-P(=O)(O^-)-O-X \end{array}$$

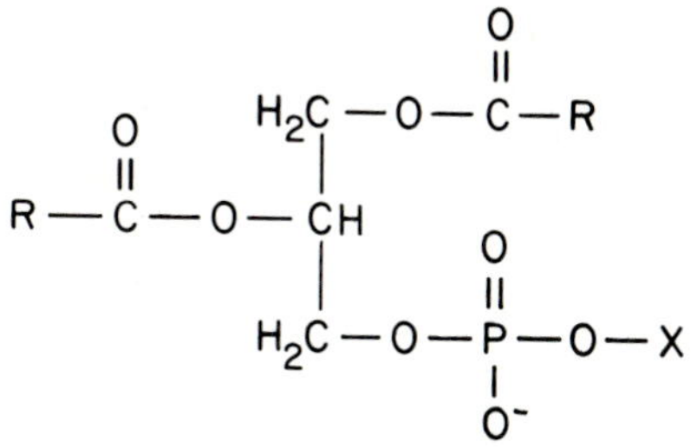

Structure of X	Phospholipid	Microorganism
$-H$	Phosphatidic acid	Bacteria
$-CH_2-CH(OH)-CH_2OH$	Phosphatidylglycerol	Bacteria
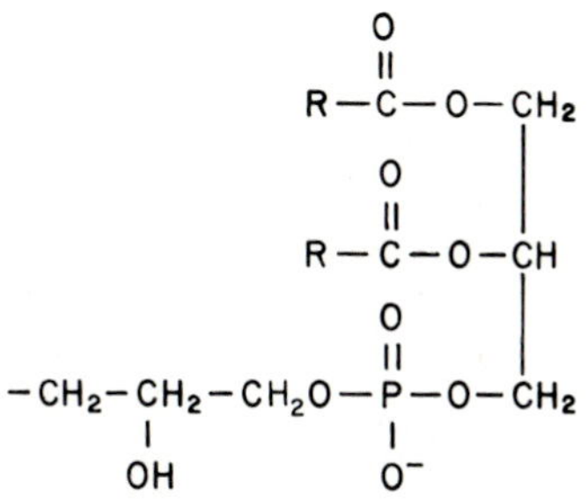 $-CH_2-CH(OH)-CH_2O-P(=O)(O^-)-O-CH_2-CH(-O-C(=O)-R)-CH_2-O-C(=O)-R$	Diphosphatidylglycerol (cardiolipin)	Bacteria
$-CH_2-CH(NH_2)-COOH$	Phosphatidylserine	Bacteria
$-CH_2-CH_2-NH_2$	Phosphatidylethanolamine	Bacteria
$-CH_2-CH_2-N^+(CH_3)_3$	Phosphatidylcholine (lecithin)	Plants, animals, fungi
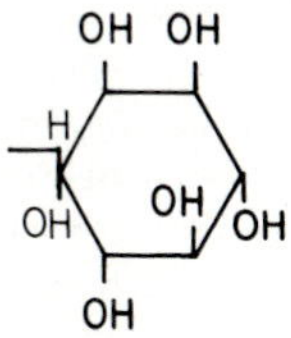 Inositol ring (OH OH, H, OH, OH, OH, OH)	Phosphatidylinositol	Mycobacteria

Figure 13-17 *Microbial lipids.* ***A,*** *Two types of glycolipids found in bacteria.* ***B,*** *Basis structure of microbial glycerophospholipids.* R, *Fatty acid constitutents.*

Biosynthesis of fatty acids. Two components in the microbial cell are especially important during the formation of saturated fatty acids: **acetyl CoA** and **acyl carrier protein (ACP)**. Very simply, fatty acid biosynthesis involves the repeated addition of two carbons in the form of acetyl groups to a growing hydrocarbon chain. During the addition of these two carbon units the acyl carrier protein binds to the growing fatty acid moiety. Seven steps are required to extend the length of the fatty acid chain, and these are repeated until the final chain length is attained. The synthesis of long-chain, even-numbered fatty acids is referred to as the **malonyl CoA pathway**, in which acetyl CoA acts as a primer molecule to which are added two carbon units at a time from malonyl CoA ($COOH\text{-}CH_2\text{-}CO\text{-}CoA$). The seven steps involved in the synthesis of even-numbered fatty acids are illustrated in Figure 13-18. The overall reaction for the biosynthesis of a fatty acid containing 18 carbons (stearic acid) would be:

$$(CH_3\text{-}CO\text{-}CoA)_8 + COOH\text{-}CH_2\text{-}CO\text{-}S\text{-}CoA + 14\ NADPH_2 \rightarrow CH_3(CH_2)_{16}CO\text{-}SCoA + 8\ CoASH + 14\ NADP + 8\ H_2O + CO_2$$

The synthesis of odd-numbered fatty acid chains in some bacteria can take place by using valeral CoA (a five-carbon compound). Unsaturated fatty acids such as those containing one, two, or three double bonds are also components of microbial lipids. Their synthesis occurs by one of two mechanisms: one that requires aerobic conditions and one that requires anaerobic conditions. The aerobic system is characterized by double-bond insertion between carbon 9 and carbon 10, whereas in the anaerobic system, which occurs in most bacteria, dehydration and double-bond formation occur at carbon 10.

Biosynthesis of glycerophospholipids. The basic components of glycerophospholipids are glycerol phosphate, which is an intermediate in the glycolytic pathway, and fatty acids. Carbons 1 and 2 of glycerol phosphate are esterified with fatty acids to produce phosphatidic acid. The sequence of reactions by which **phosphatidic acid** is formed and converted to other phospholipids is illustrated in Figure 13-19.

Table 13-10 *Origin of the atoms of purine and pyrimidine bases**

Atom and position	Purine (molecule derived from)	Pyrimidine (molecule derived from)
N-1	Aspartate	Aspartate
C-2	Formate	Carbon dioxide
N-3	Glutamine	Ammonia
C-4	Glycine	Aspartate
C-5	Glycine	Aspartate
C-6	Carbon dioxide	Aspartate
N-7	Glycine	—
C-8	Formate	—
N-9	Glutamine	—

*The atom numbers and positions are illustrated in Figure 13-20.

Biosynthesis of purines and pyrimidines. The structure of purine and pyrimidine bases is illustrated in Chapter 2. These cyclic compounds are made up of elements derived from various metabolic intermediates (aspartate, bicarbonate, ammonia, glycine, etc.) The structure and numbering of the atoms in the purine and pyrimidine molecules are illustrated in Figure 13-20. The origin of these atoms is described in Table 13-10.

Biosynthesis of purines. The basic purine ribonucleotide molecule, inosine 5-phosphate (inosinic acid), is synthesized in a series of reactions outlined in Figure 13-21. **Inosinic acid** serves as precursor to two of the major purines: guanylic acid (GMP) and adenylic acid (AMP). The pathways leading to the synthesis of AMP and GMP are outlined in Figure 13-22.

Biosynthesis of pyrimidines. The pyrimidines—thymine, cytosine, and uracil—are synthesized in a series of reactions in which the initial precursor molecules are

1 $ACP\text{-}SH + CH_3CO\text{-}S\text{-}CoA \rightleftharpoons CH_3CO\text{-}S\text{-}ACP + CoASH$

2 $COOH\text{-}CH_2CO\text{-}S\text{-}CoA + ACP\text{-}SH \rightleftharpoons COOH\text{-}CH_2CO\text{-}SACP + CoASH$

3 $CH_3\text{-}CO\text{-}S\text{-}ACP + COOH\text{-}CH_2COS\text{-}ACP \rightleftharpoons CH_3\text{-}CO\text{-}CH_2CO\text{-}S\text{-}ACP + HS\text{-}ACP + CO_2$

4 $CH_3\text{-}CO\text{-}CH_2CO\text{-}S\text{-}ACP + NADPH + H^+ \rightleftharpoons CH_3\text{-}CHOH\text{-}CH_2CO\text{-}S\text{-}ACP + NADP^+$

5 $CH_3\text{-}CHOH\text{-}CH_2CO\text{-}S\text{-}ACP \rightleftharpoons CH_3\text{-}CH{=}CH\text{-}CO\text{-}S\text{-}ACP + H_2O$

6 $CH_3\text{-}CH{=}CH\text{-}CO\text{-}S\text{-}ACP + NADPH + H^+ \rightleftharpoons CH_3\text{-}CH_2\text{-}CH_2CO\text{-}S\text{-}ACP + NADP^+$

7 $CH_3CH_2\text{-}CH_2CO\text{-}S\text{-}ACP + CoASH \rightleftharpoons CH_3\text{-}CH_2\text{-}CH_2CO\text{-}S\text{-}CoA + ACP\text{-}SH$

Figure 13-18 *Seven steps required in biosynthesis of fatty acids. Reactions 1 and 2 are involved in formation of acetyl-ACP (acetyl-acyl carrier protein) and malonyl-ACP, respectively. In reaction 3, acetyl-ACP provides first two carbons of fatty acid. Remaining reactions involve reduction processes to produce butyryl-ACP.*

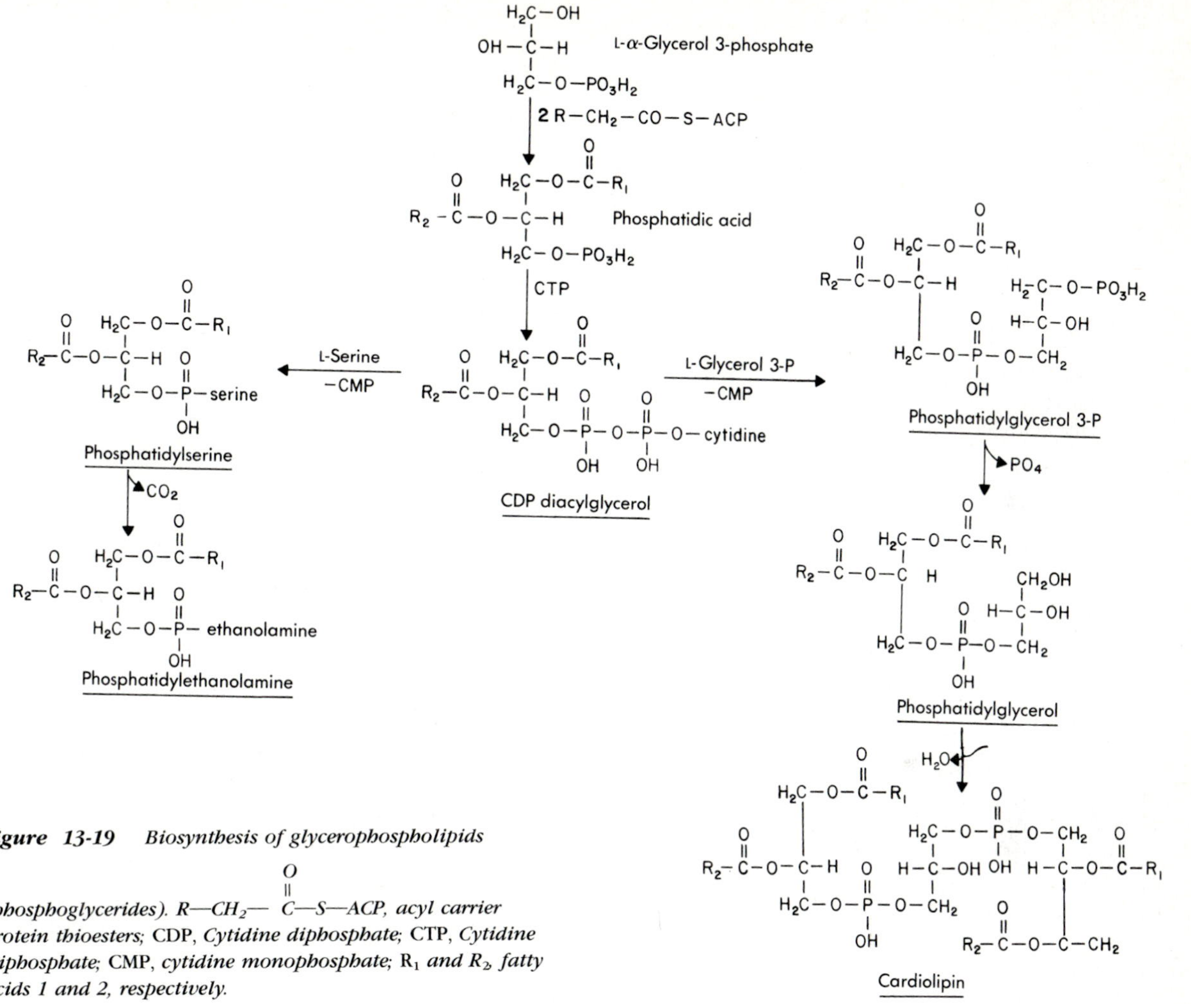

Figure 13-19 *Biosynthesis of glycerophospholipids (phosphoglycerides).* $R—CH_2—\overset{O}{\overset{\|}{C}}—S—ACP$, *acyl carrier protein thioesters;* CDP, *Cytidine diphosphate;* CTP, *Cytidine triphosphate;* CMP, *cytidine monophosphate;* R_1 *and* R_2, *fatty acids 1 and 2, respectively.*

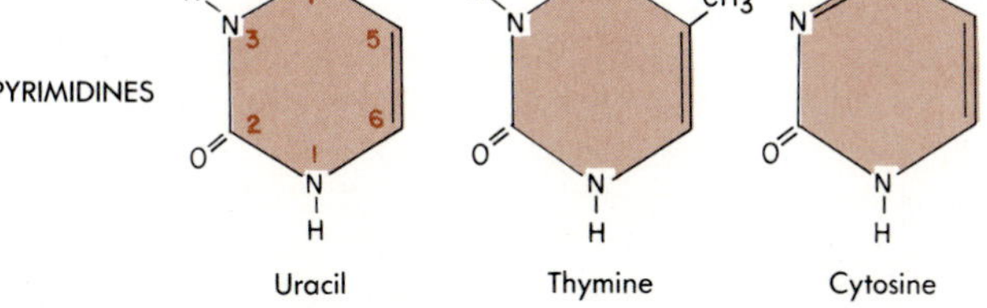

Figure 13-20 *Position and atom number in purine and pyrimidine bases.*

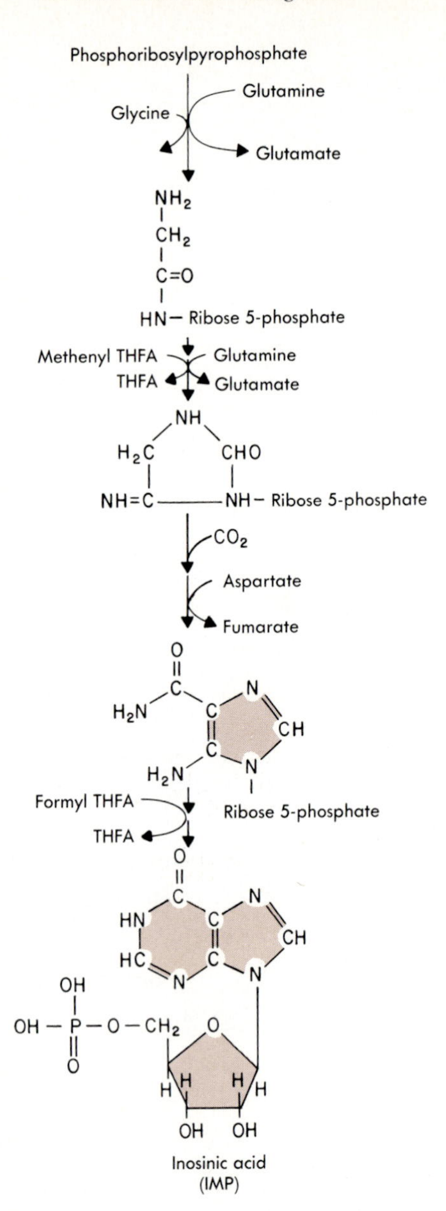

Figure 13-21 *Biosynthesis of inosinic acid. Several intermediates have been omitted. Note importance of amino acids in pathway.* THFA, *tetrahydrofolate.*

Figure 13-22 *Biosynthesis of purines adenylic acid and guanylic acid from inosinic acid.*

aspartate and **carbamoyl phosphate** (Figure 13-23).

Most cells can synthesize purine and pyrimidine nucleotides de novo (from basic nutrients in the cell), but preformed bases and nucleosides in some instances supplied in the growth medium can be converted directly into ribonucleotides, or they may be converted to other ribonucleotide bases if the organism lacks an enzyme in the de novo pathway. Some microorganisms can convert free bases such as adenine to a nucleotide by reacting them with phosphoribosylpyrophosphate (PRPP):

$$\text{Adenine} + \text{PRPP} \rightarrow \text{AMP} + \text{PP}_i$$

Biosynthesis of carbohydrates and polysaccharides. Carbohydrates, in addition to their role in energy production, are also important as structural components of the cell. Capsules, the cell wall, and the cytoplasmic membrane are composed of carbohydrates that appear either as oligosaccharides or as polysaccharides. Monosaccharides can be activated by nucleoside triphosphates such as CTP, UTP, and ATP to produce sugar nucleotides, which are capable of entering into several biosyn-

Figure 13-23 *Biosynthesis of pyrimidines; uridylic acid, cytidylic acid, and thymidylic acid. Since thymidylic acid is a deoxyribonucleotide, uridylic acid, a ribonucleotide, is first converted to the deoxy form.* THFA, tetrahydrofolate.

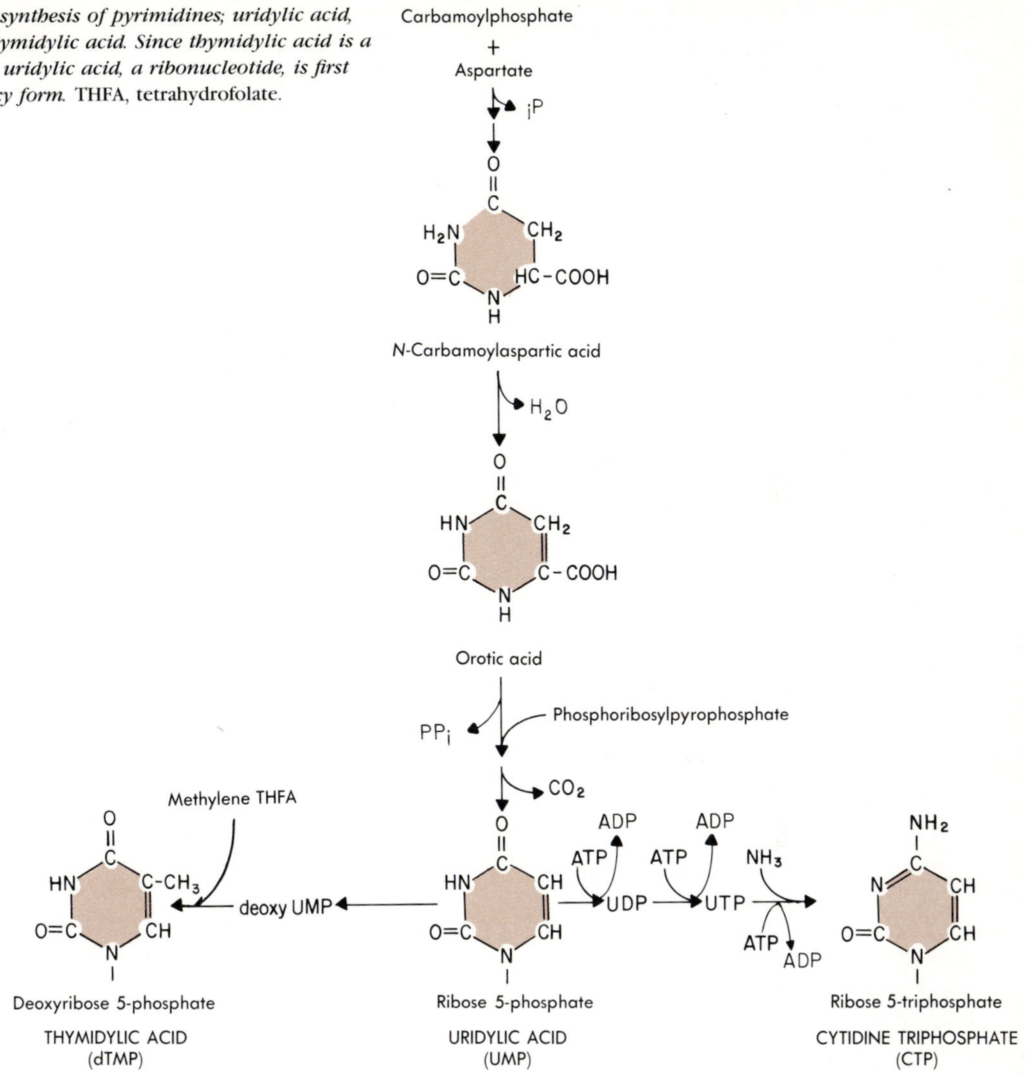

thetic processes. Glucose, for example, can become activated and ready for participation in biosynthetic reactions in the following type of reaction:

$$\textbf{Glucose-1-phosphate + UTP} \rightarrow \textbf{UDP-glucose + PP}_i$$

The major biosynthetic reactions in which sugar nucleotides participate are the formation of monosaccharides other than glucose and polysaccharide synthesis.

Biosynthesis of monosaccharides other than glucose. Glucose can serve as a precursor to other monosaccharides. Galactose, glucuronic acid, *N*-acetylglucosamine, arabinose, and other sugar derivatives can be produced from sugar nucleotides such as UDP-glucose. Many of these derivatives are important in cell wall peptidoglycan formation. (See Chapter 15 for discussion of cell wall synthesis.)

Biosynthesis of polysaccharides. A variety of polysaccharides, either linear or branched, are synthesized by microbial cells. The most common one is a high molecular weight polymer called **glycogen**, which acts as an energy storage form of glucose. Glycogen synthesis occurs under conditions where there is excess glucose and a deficiency in a specific nutrient required for biosynthesis. The cell, instead of hydrolyzing glucose to produce more energy as ATP, will convert the glucose into glycogen. The conversion of glucose to glycogen includes the following reactions:

1. D-Glucose + ATP → D-glucose-6-phosphate + ADP
2. Glucose-6-phosphate ⇌ Glucose-1-phosphate
3. Glucose-1-phosphate + ATP → ADP-glucose + PP_i
4. ADP-glucose + Glucose primer(n) → $(\text{Glucose})_{n+1}$ + ADP

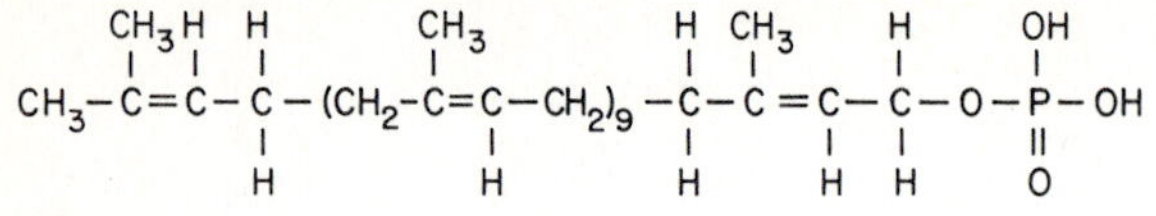

```
     CH3H  H         CH3            H CH3     H        OH
      |  |  |         |             |  |      |         |
CH3–C=C–C–(CH2–C=C–CH2)9–C–C=C–C–O–P–OH
            |         |             |      |  |         ||
            H         H             H      H  H         O
```

Figure 13-24 *Structure of lipid carrier polyisoprenoid (phosphorylated derivative).*

During the synthesis of some polysaccharides (such as in cell wall formation) the sugar nucleotide is not transferred directly to the growing polymer but is first linked to a lipid. The lipid has been identified as a **C_{55} polyisoprenoid** (Figure 13-24), which must be phosphorylated before it accepts a glycosyl (sugar) unit. The C_{55} lipid carrier is involved in the biosynthesis of polysaccharides, such as the lipopolysaccharide of gram-negative bacteria, and the teichoic acids of the gram-positive bacteria, and the peptidoglycan of the bacterial cell wall.

Gluconeogenesis. Glucose can be synthesized by the reversal of most of the reactions of glycolysis—a process called **gluconeogenesis**. Many microorganisms in their natural environment seldom encounter free extracellular glucose and must therefore synthesize it from other metabolites such as those in the TCA cycle. This means that a TCA intermediate can be converted to glucose only by the reversal of glycolysis. During the reversal process, however, the conversion of pyruvate to phosphoenolpyruvate is irreversible, and the cell bypasses the reaction by converting the TCA intermediate to phosphoenolpyruvate in the following reaction:

$$\text{Oxaloacetate} + \text{ATP} \rightarrow \text{Phosphoenolpyruvate} + CO_2 + \text{ADP}$$

Phosphoenolpyruvate is easily converted to glucose by the reversal of glycolysis. The reactions between phosphoenolpyruvate and fructose-1, 6-diphosphate are reversible. The enzyme phosphofructokinase catalyzes the irreversible phosphorylation of fructose-6-phosphate to fructose-1, 6-diphosphate and therefore cannot participate in the uphill reaction. This reaction is bypassed by a second enzyme, fructose-1,6-diphosphatase, which carries out the irreversible hydrolysis of the 1-phosphate group to produce fructose-6-phosphate.

Concept Check

1. **How are polymers produced from monomers?**
2. **Describe the reactions in which ammonia is assimilated into amino acids. What amino acids are derived from TCA intermediates?**
3. **Briefly, how are fatty acids synthesized?**
4. **Describe a mechanism by which a TCA intermediate such as oxaloacetate can be converted to glucose.**
5. **What are the products of catabolic pathways that serve as links to biosynthetic pathways?**

METABOLIC CONTROL

Up to this point we have observed that the primary function of glycolysis is to provide energy, which can be used in the biosynthesis of metabolites whose precursors are formed primarily in the TCA cycle. The microbial cell is often confronted with metabolic conditions where either energy is not available and biosynthesis must be discontinued, or excess sources of energy are available but certain nutrients are unavailable for biosynthesis. Control mechanisms are available to the cell that enable it to conserve energy sources when they are present and help to prevent the biosynthesis of unneeded metabolites.

Control mechanisms

We learned (Chapter 9) that one of the mechanisms used by the cell to control metabolic reactions is of genetic origin and that enzymes controlling certain reactions in the cell are subject to induction or repression, that is, their synthesis can be turned on or off. This genetic control affects the synthesis of enzymes but has no effect on enzymes that are already present in the cell. A second major control system has evolved in organisms that controls the activity of preformed enzymes.* Every chemical reaction in the cell is catalyzed by an enzyme, but at strategic places in many metabolic pathways there are special enzymes that control the amount of product produced in the pathway. These enzymes are called **allosteric** (means other site) or **regulatory enzymes** (see Chapter 2).

Control in biosynthetic pathways

Feedback inhibition. Regulatory enzymes are usually found at the first reaction of a biosynthetic pathway, but not always. The reactions involved in the synthesis of the amino acid L-isoleucine is a typical example of a pathway controlled by a mechanism called **feedback inhibition**. In this type of inhibition the end products of a pathway inhibit the activity of an early (usually first) en-

* A third but less-understood mechanism of control is called enzyme inactivation and is discussed on p. 377.

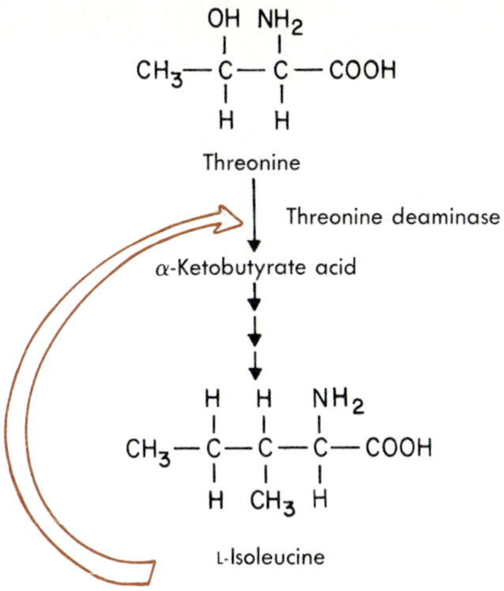

Threonine

Threonine deaminase

α-Ketobutyrate acid

L-Isoleucine

Figure 13-25 *Feedback inhibition in biosynthesis of* L-*isoleucine.*

zyme in the pathway. The biosynthesis of L-isoleucine is characterized by the formation of several intermediates between the precursor threonine and the final product, L-isoleucine (Figure 13-25). The first enzyme in the pathway is **threonine deaminase**, which is the regulatory enzyme. The allosteric effector molecule is L-isoleucine, which in high concentration inhibits the activity of threonine deaminase. Thus the product of threonine deaminase, as well as other products in the pathway, are not formed. If a bacterial cell such as *E. coli* is grown in a medium containing mineral salts plus glucose as a carbon source, all of the amino acids, including L-isoleucine, are synthesized de novo. If we were now to add excess L-isoleucine to the growth medium, the cell would have no need to synthesize L-isoleucine, and threonine deaminase would be inhibited. There are several variations of feedback inhibition, particularly when there are branched pathways leading to the synthesis of more than one product and where certain reactions are shared.

Feedback inhibition in branched amino acid pathways. In a branched pathway a common precursor gives rise to several end products through a series of enzymatic reactions. One of the best examples of a branched amino acid pathway is that involving the aspartate family of amino acids (Figure 13-26). Aspartic acid is the common precursor for the amino acids lysine, methionine, threonine, and isoleucine. In the initial step the enzyme **aspartokinase** catalyzes the phosphorylation of aspartate. The phosphorylated aspartate is subsequently reduced and dephosphorylated to aspartate-beta-semialdehyde by **aspartate-beta-semialdehyde dehydrogenase**. The third enzyme in the sequence, **homoserine dehydrogenase**, transforms aspartate-beta-semialdehyde to homoserine, which serves as a precursor for methionine, threonine, and isoleucine. Aspartate-beta-semialdehyde is also a starting product for lysine biosynthesis. The amino acid cysteine, although not a

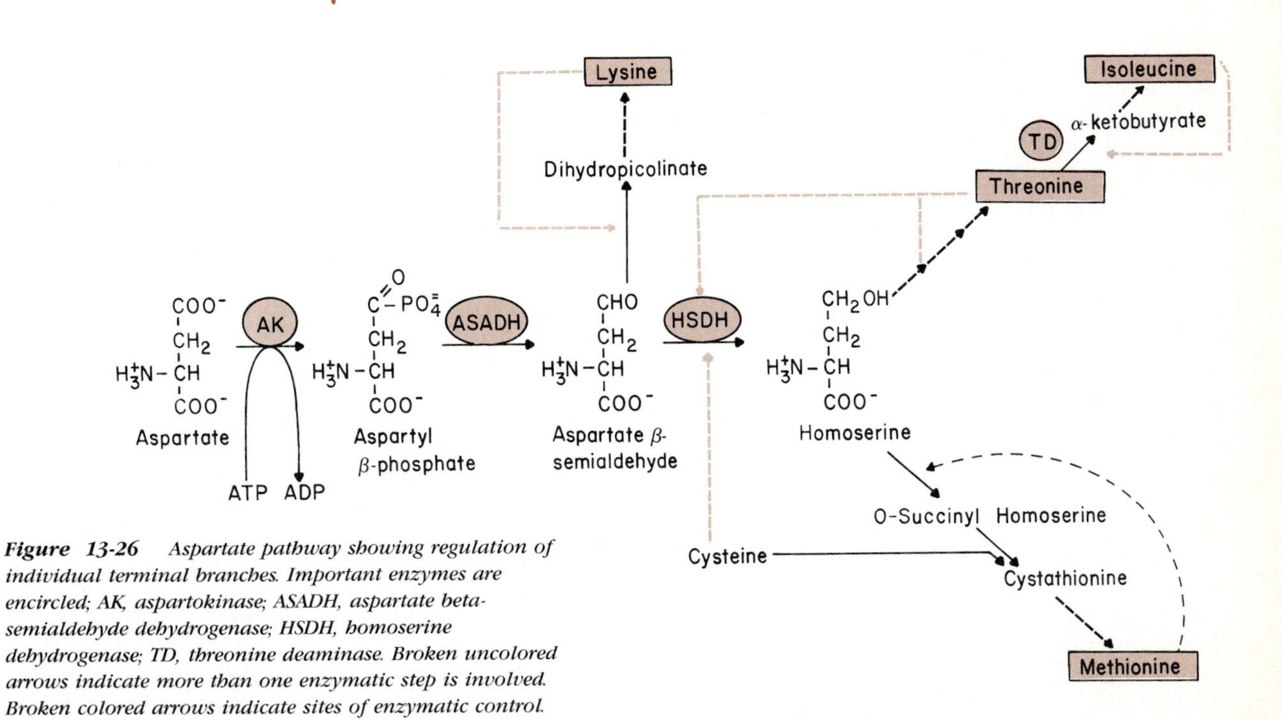

Figure 13-26 *Aspartate pathway showing regulation of individual terminal branches. Important enzymes are encircled; AK, aspartokinase; ASADH, aspartate beta-semialdehyde dehydrogenase; HSDH, homoserine dehydrogenase; TD, threonine deaminase. Broken uncolored arrows indicate more than one enzymatic step is involved. Broken colored arrows indicate sites of enzymatic control.*

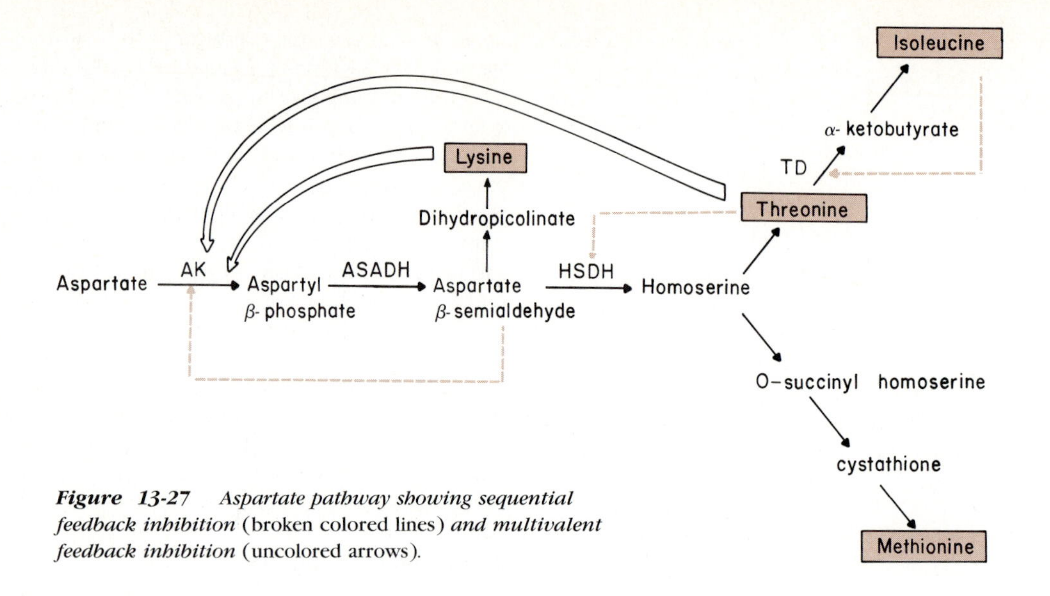

Figure 13-27 *Aspartate pathway showing sequential feedback inhibition* (broken colored lines) *and multivalent feedback inhibition* (uncolored arrows).

direct intermediate in the aspartate pathway, does donate its three-carbon backbone for the formation of methionine.

Two enzymes—aspartokinase and aspartate-beta-semialdehyde dehydrogenase—are common to the synthesis of four amino acids, while homoserine dehydrogenase is shared for the synthesis of three amino acids. Homoserine and aspartate-beta-semialdehyde are crucial intermediates and their concentrations must be maintained at proper levels in the cell. If one particular end product is in excess, its synthesis can be reduced by control of the linear part of the sequence. For example, lysine could regulate its own synthesis by feeding back on the enzyme that catalyzes the conversion of aspartate-beta-semialdehyde to dihydropicolinate (Figure 13-26). However, this will not affect the other three amino acids. The mechanisms for controlling those reactions common to all the amino acid end products are discussed next.

Amino acid regulatory enzymes may be affected by a single product or by more than one product. Let us look at some examples of single product control. If aspartokinase activity were regulated solely by lysine, then aspartate-beta-semialdehyde would not be produced and the formation of the other amino acids would be prevented. To circumvent this problem the bacterium has evolved three separate aspartokinases: one sensitive to threonine, one sensitive to methionine, and one sensitive to lysine. Enzymes catalyzing the same reaction but sufficiently different to be distinct proteins are called **isoenzymes.** In addition to the aspartokinase isoenzymes there are also homoserine dehydrogenase isoenzymes: one controlled by methionine and the other threonine. An alternative to control by isoenzymes is control by sequential feedback inhibition. In this type of regulation the end product controls the first enzyme of the terminal branch (Figure 13-27). This results in the accumulation of the branch point intermediate, which, if it is in excess, regulates earlier enzymatic steps. For example, overproduction of isoleucine inhibits the enzyme threonine deaminase and this eventually causes the accumulation of threonine. Excess threonine inhibits the activity of homoserine dehydrogenase, and this leads to the accumulation of aspartate-beta-semialdehyde, which in turn slows down its own production by decelerating the formation of aspartyl phosphate. The concentration of aspartate-beta-semialdehyde, however, does not fall below a "critical" level (the level needed for synthesis of lysine and methionine) because aspartate-beta-semialdehyde is a competitive inhibitor for aspartate and ATP on the active site of aspartokinase, and threonine is a competitive inhibitor for aspartate-beta-semialdehyde on the active site of homoserine dehydrogenase.

From the types of mechanism discussed it is clear that a safeguard must exist for maintaining an adequate pool of intermediate metabolites after the blocked reaction to provide normal synthesis of the remaining end products. One safeguard relies on the concerted action of two end products and is called **multivalent feedback inhibition** (Figure 13-27). For example, in some bacteria early enzymes in biosynthetic pathways are not affected by any single end product but are affected when two of the end products are present simultaneously. In some bacteria aspartokinase, for example, is unaffected by lysine alone, but in the presence of excess lysine plus threonine, enzyme activity is reduced by as much as 75%. Thus residual activity of this enzyme provides some level of intermediates beyond the block. Control of regulatory enzymes can be brought about by genetic mechanisms. Thus the concentration as well as activity of regulatory enzymes can be mediated by end products of biosynthetic pathways. Repression of enzyme synthesis is discussed in Chapter 9.

Inactivation of enzymes. A third but not entirely understood mechanism of control involves the inactivation of enzymes. The term inactivation does not mean that enzymes are degraded to their constituent amino acids. Inactivation may be due to a physical or conformational change in the enzyme resulting from the binding of low molecular weight molecules such as NH_4^+, a chemical modification of the enzyme, or a cleavage of a peptide bond in the enzyme. One type of modification, such as the phosphorylation of proteins, is mentioned earlier in the chapter. Other types of modification may include glucosylation, acetylation or deacetylation, and methylation. Glucosylation and/or methylation of DNA is a mechansism for protecting bacteriophage DNA from enzymatic degradation by host enzymes (see Chapter 17). Most examples of enzyme inactivation occur during shifts in carbon or nitrogen metabolism. An organism on one substrate, for example, may exhaust its supply and have to use other substrates. One of the most extensively studied inactivation systems occurs in the organism *Rhodopseudomonas gelatinosa* during its growth on citrate under anaerobic conditions. Growth on citrate results in the catabolism of the molecule to acetate, but when citrate is exhausted, the organism must shift to citrate synthesis. Citrate-catabolizing enzymes that are already present in the cell are prevented from engaging in citrate synthesis because the first enzyme in the catabolic pathway, citrate lyase, is inactivated by another enzyme, *S*-acetylcitrate lyase deacetylase. The inactivating enzyme removes an acetyl group from a sulfhydryl group on citrate lyase. Should citrate later accumulate in the cell, the inactivated citrate lyase can be reactivated by an acetylation process.

Control in catabolic pathways. Regulatory enzymes are also a feature of catabolic systems. Since catabolism is involved in energy formation, it should not seem unusual that molecules such as ATP and AMP are effector molecules. One of the principal control sites in glycolysis is in the reaction:

$$\text{Fructose-6-phosphate} + \text{ATP} \rightleftharpoons \text{Fructose-1,6-diphosphate} + \text{ADP}$$

The forward reaction, which is catalyzed by the enzyme phosphofructokinase, is inhibited by high concentrations of ATP but stimulated by ADP or AMP. In other words, if the cell has excess ATP (the ATP/AMP + ADP ratio is high), then the forward reaction, in which more ATP will eventually be produced, is unnecessary. High concentrations of AMP or ADP are signals to the cell that ATP has been used up in biosynthetic reactions and more is needed; hence the forward reaction is stimulated. The reverse reaction, which is catalyzed by fructose diphosphatase, is stimulated by ATP and inhibited by AMP or ADP. Stimulating the reverse reaction prevents the forward reaction from proceeding and vice versa. Inhibition of enzyme activity is the only means of control in glycolysis, since all of the enzymes are synthesized constitutively (under any environmental condition) and are not subject to genetic control.

The effect ATP produces on phosphofructokinase is characteristic of facultative organisms when sufficient oxygen is present in the cell. Respiration, not fermentation, is the principal means of ATP formation. Increased levels of ATP inhibit glycolysis, producing fewer acids such as lactic acid and more ATP using less glucose. This effect was first observed by Pasteur in fermenting yeast and is known as the **Pasteur effect**.

Anaplerotic pathways

A third class of reactions occurring in the cell, which are neither catabolic or biosynthetic, is called **anaplerotic**. Anaplerotic pathways function to replenish intermediates that have been drained from cyclic processes such as acetylation of oxaloacetate to citrate in the TCA cycle. At maximum activity the TCA cycle acts primarily as a source of intermediates for biosynthesis and secondarily as a supplier of reducing equivalents for energy production. Replenishing TCA intermediates can take place in two different ways, depending on the organism's carbon source:

1. If the organism is growing on glucose, intermediates of glucose can be carboxylated:

$$\text{Pyruvate} + CO_2 + \text{ATP} \rightarrow \text{Oxaloacetate} + \text{ADP} + \text{iP}$$

or

$$\text{Phosphoenolpyruvate} + CO_2 \rightarrow \text{Oxaloacetate} + \text{iP}$$

Concept Check

1. Describe those reactions in glycolysis that are controlled by allosteric enzymes.
2. Explain the mechanism by which amino acid biosynthesis can be controlled.
3. What are the types of molecules that serve to control enzymatic reactions in catabolic pathways?
4. Explain the mechanism by which microorganisms are able to replenish those carbons lost during the operation of the TCA cycle.

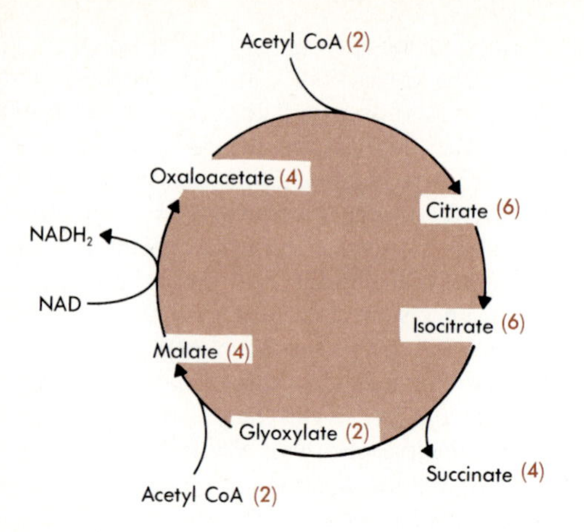

Figure 13-28 *Glyoxylate cycle. Number of carbons in each intermediate is indicated in color. Compare with TCA cycle.*

2. If the organism is growing on a two-carbon source such as acetate, the glyoxylate cycle provides a means of replenishment (Figure 13-28). The glyoxylate cycle is an abbreviated version of the TCA cycle, but unlike the TCA cycle possesses enzymes that conserve carbon and do not lose it as carbon dioxide.

SUMMARY

1. Nutrients are those elements or compounds that must be supplied to the cell for its normal growth and development. Inorganic or organic forms of carbon, hydrogen, sulfur, oxygen, and phosphorus represent the major nutrient sources, whereas various minerals, such as magnesium, manganese, iron, and others, are considered minor nutrients.
2. Most microorganisms, except for autotrophs, require organic carbon. Autotrophs use carbon dioxide as a source of carbon in a process called carbon dioxide fixation. The key enzyme in this process is ribulosebisphosphate carboxylase, which catalyzes only one reaction in a series of reactions called the Calvin-Benson cycle. The key intermediate produced in the cycle is glyceraldehyde-3-phosphate, which serves key roles in glucose synthesis as well as synthesis of amino and fatty acids.
3. Nitrogen may appear in many forms, including amino acids, proteins, and purine and pyrimidine bases, but inorganic nitrogen as ammonia and nitrate is the principal source of nitrogen for conversion into organic material in the cell. Some bacteria are capable of fixing nitrogen from the air.
4. Sulfur and phosphorus as components of organic molecules are most readily available as nutrients to the cell in the form of inorganic sulfate and phosphate, respectively.

5. Minor nutrients are usually cofactors for enzymes or are involved in energy reactions. Growth factors are cellular metabolites that cannot be synthesized from nutrients supplied in the environment and must be added to the medium. They include vitamins, amino acids, and purine and pyrimidine bases.
6. Saprophytes from the soil can synthesize nearly all organic metabolites from a single carbon source. Parasites, however, have lost much of their biosynthetic potential and when cultivated outside the host often require a multitude of organic metabolites.
7. Amino acid synthesis is usually dependent on the incorporation of ammonia into a keto acid. Precursors to the amino acids are therefore derived primarily from the TCA cycle. The pathways leading to the synthesis of amino acids in both eukaryotes and prokaryotes are similar except for L-lysine.
8. The principal lipids synthesized in the microbial cell are glycerophospholipids and glycolipids. The fatty acid components of lipids are synthesized through the repeated addition of two carbons, in the form of an acetyl group, to a growing hydrocarbon chain. The two carbon units are transferred by a carrier called the acyl carrier protein (ACP).
9. Purine and pyrimidine nucleotides are synthesized from precursor molecules, particularly amino acids such as aspartic acid, glutamine, and glycine. Inosinic acid is the intermediate in the formation of the purines guanylic and adenylic acid. In pyrimidine biosynthesis uridylic acid is produced, which then serves as an intermediate in the formation of thymidylic acid and cytidylic acid.
10. Carbohydrate or polysaccharide biosynthesis requires activation of monomeric sugars into sugar nucleotides, using CTP, UTP, or ATP. During some polysaccharide syntheses the sugar nucleotides are transferred directly to a lipid carrier such as C_{55} isoprenoid before incorporation into the growing polysaccharide chain. When glucose or another carbohydrate is not available to the cell for incorporation into cell components, glucose can be synthesized by the reversal of glycolysis.
11. Metabolism can be controlled by allosteric enzymes situated at specific points in biosynthetic and catabolic pathways. In linear biosynthetic pathways the end product controls the first enzyme in the pathway, but in branched pathways other mechanisms exist to ensure synthesis of other end products.
12. Allosteric enzymes present in catabolic pathways are controlled by energy compounds such as ATP, ADP, and AMP, which act as inhibitors and activators. A less well understood mechanism of enzyme control is called enzyme inactivation.
13. Anaplerotic reactions are neither catabolic or anabolic. They function to replenish intermediates that have been drained from metabolic cycles such as the TCA cycle. The glyoxylate cycle is an example of an anaplerotic pathway.

Self-study Quiz

Multiple choice

1. Acetyl CoA and acyl carrier protein are precursors for which of the following biosynthetic pathways?
   A. Amino acid
   B. Purine
   C. Pyrimidine
   D. Fatty acid
2. A transaminase catalyzes which of the following reactions:
   A. Pyruvate + L-Glutamate $\rightleftharpoons$ L-Alanine + Alpha-ketoglutarate
   B. Pyruvate + $NADH_2$ + NH_3 $\rightleftharpoons$ L-Alanine + NAD + H_20 + H^+
   C. Alpha-ketoglutarate + NH_3 + $NADPH_2$ $\rightleftharpoons$ L-Glutamate + NADP + H_20 + H^+
   D. L-Glutamate + NH_3 + ATP $\rightleftharpoons$ L-Glutamine + ADP + Pi
3. Which of the following is *not* an anabolic pathway?
   A. Gluconeogenesis
   B. Calvin-Benson cycle
   C. C_4-carboxylic acid cycle
   D. Glycolysis
4. Which of the following molecules would most likely be an effector for an allosteric enzyme in glycolysis?
   A. Alanine
   B. Adenosine monophosphate
   C. Aspartate
   D. CO_2
5. The most important enzyme in the Calvin Benson cycle is:
   A. Phosphofructokinase
   B. Glyceraldehyde-3-phosphate dehydrogenase
   C. Ribulosebisphosphate carboxylase
   D. Ribulose-1, 5-diphosphatase
6. The primary function of the glyoxylate cycle is to:
   A. Replenish glucose
   B. Provide reducing power in the form of NADPH
   C. Replenish TCA intermediates lost by growth on acetate
   D. Store nitrogen in the form of glutamine
7. The ability of certain bacteria such as cyanobacteria to fix nitrogen even under aerobic conditions is due to the presence of a structure called a(n):
   A. Bacteroid
   B. Heterocyst
   C. Infection thread
   D. Endospore

Completion

1. Enzymes that catalyze the same reaction in the cell but are considered sufficiently different proteins are called ________________.
2. The type of inhibition in which an enzyme is affected by the simultaneous presence of two or more end products is called ________________.
3. An enzyme that would be responsible for the conversion of L-amino acids to D-amino acids is called a(n) ________________.
4. The conversion of nitrogen gas to a biologically available nitrogen within the cell is called ________________.
5. The metal-binding molecules produced by microorganisms that aid in scavenging iron from the environment are called ________________.

Matching

In the left column are various coenzymes, elements, or ions. Match them with their functions described in the right column.

1. ________ Biotin
2. ________ Calcium
3. ________ Iron
4. ________ Coenzyme A
5. ________ Magnesium
6. ________ Folic acid
7. ________ Phosphate

A. Transfer of acetyl group
B. Transfer of formyl group
C. Component of chlorophyll molecule
D. Required during endospore formation
E. Transfer of carboxy (CO_2) group
F. Transfer of methyl group
G. Important buffering ion
H. Important nutrient of the hydrogen bacteria
I. Precursor of menaquinone
J. Important component of the cytochromes

Thought questions

Design an experiment:

A. To identify which enzyme in an enzyme series is inhibited by feedback inhibition of the end product
B. To determine whether the inhibitory control mechanism is feedback inhibition or enzyme inactivation
C. To determine the sensitivity of enzymes in catabolic pathways to the ATP/AMP + ADP ratio

SELECTED READINGS

Books

Bothe, H., and Trebst, A. *Biology of inorganic nitrogen and sulfur.* Springer-Verlag, New York, Inc., New York, 1981.

Doelle, H.W. *Bacterial metabolism* (2nd ed.). Academic Press, Inc., New York, 1975.

Gottschalk, G. *Bacterial metabolism.* (2nd ed.) Springer-Verlag, New York, Inc., New York, 1985.

Ingraham, J.L., Maale, O., and Neidhardt, F.C. *Growth of the bacterial cell,* Sinaver, Sunderland, Mass., 1983.

Jones, C.W. *Bacterial respiration and photosynthesis. Aspects of microbiology (# 5).* American Society for Microbiology, Washington, D.C., 1982.

Mandelstam, J., and McQuillen, K. *Biochemistry of bacterial growth* (3rd ed.). Halsted Press, New York, 1982

Moat, A.G. *Microbial physiology.* John Wiley & Sons, Inc., New York, 1979.

Weinberg, E.D. (Ed.). *Microorganisms and minerals.* Marcel Dekker, Inc., New York, 1977.

Journal articles

Biebel, H., and Pfenning, N. Growth of sulfate-reducing bacteria with sulfur as electron acceptor. *Arch. Microbiol.* **112:**115, 1977.

Knowles, C.J. Microbial metabolic regulation by adenine nucleotide pools. *Symp. Soc. Gen. Microbiol.* **27:**241, 1977.

Magasanik, B. Genetic control of nitrogen assimilation. *Annu. Rev. Genet.* **16:**135, 1982.

Monod, J. The growth of bacterial cultures. *Annu. Rev. Microbiol.* **3:**371, 1949.

Monod, J., Changeux, J.P., and Jacob, F. Allosteric proteins and cellular control systems. *J. Mol. Biol.* **6:**306, 1963.

Morris, J.G. The physiology of obligate anaerobiosis. *Adv. Microb. Physiol.* **12:**169, 1976.

Nierlich, D.P., and Magasanik, B. Regulation of purine ribonucleotide synthesis by end product inhibition. *J. Biol. Chem.* **240:**358, 1965.

Preiss, J. Bacterial glycogen synthesis and its regulation. *Annu. Rev. Microbiol.* **38:**419, 1984.

Sanwal, B.D. Allosteric controls of amphibolic pathways in bacteria. *Microbiol Rev.* **34:**20, 1970.

Switzer, R.L. The inactivation of microbial enzymes in vivo. *Annu. Rev. Microbiol.* **31:**135, 1977.

Umbarger, H.E. Amino acid biosynthesis and its regulation. *Annu. Rev. Biochem.* **47:**533, 1978.

CHAPTER 14

Microbial growth and factors affecting growth

Chapter Outline

Introduction

Under ideal conditions growth may be defined as the orderly increase in all of the components of the cell. Ideal growth conditions are the exception and not the rule for most microorganisms in their natural environment. Physical growth processes, which depend on biochemical reactions, are usually observed as an increase in the size of an individual or in the number of individuals in a population. The biochemical and physical events associated with bacterial growth are integrated into a cycle that terminates with cell division. Many factors determine whether a cell will grow and divide: the availability of carbon and energy sources, the ability of DNA to replicate, the size and mass of the cell, and the effect of chemical and physical agents in the environment.

This chapter outlines the stages of the bacterial cell cycle and how various cellular processes are integrated with cell division, describes the mathematics of bacterial growth under ideal as well as less than ideal conditions, briefly outlines the mechanisms used by bacteria to control their rate of growth, and discusses the laboratory techniques used by the microbiologist to measure bacterial growth and the growth cycle of a eukaryote, such as a yeast cell, and how it is coordinated with cell division. Finally, the chapter describes those factors, both chemical and physical, that affect the growth of microorganisms.

GROWTH IN PROKARYOTES

Methods for studying the cell cycle

Single cell studies. The bacterial cell, like all other cells, goes through a cycle of events that end in cell division. These events can be examined in numerous ways. First, the physical growth of a cell (its increase in size) and its subsequent division can be visualized by direct microscopic observation of living cells (wet mount), for example, by time-lapse photography. Single-cell studies are generally of limited value primarily because the bacterial cell is of small size. In addition, we cannot yet detect at a micro level, the many biochemical changes that occur in the cell. Most studies of the bacterial cell cycle involve large cell populations.

Synchronized cultures. Ordinarily, when one cultures a population of bacteria in the laboratory, the cells will be in various stages of growth. Some cells will be dividing, others dying, and others in various stages of the growth cycle. One could not accurately determine, for example, the rate of RNA synthesis in a cell if in the population of cells some were synthesizing RNA and others were not. For this reason, many population studies utilize synchronized cultures so that most cells are in the same stage of the growth process.

Synchronized cultures may be obtained by an induction process in which the population is subjected to environmental manipulation or by a selection process that is based on the physical state of the organism. The whole purpose is to obtain a culture in which all of the cells divide at the same time. In practice this is never achieved because age differences in cells cause them to divide at different times.

Environmental manipulations. Several chemical or physical techniques are used to obtain a synchronous culture. The technique to be employed is often related to the type of organism used in the experiment. One method uses **temperature variations** during growth; for example, a culture may be grown at 42° C for a period of time and then at 28° C. Repetition of these temperature changes in some way coordinates the population as long as the length of time at each temperature closely follows the generation time (time required for a cell to divide) of the organism or the population of cells. Photosynthetic microorganisms can be synchronized by subjecting a culture to **alternating cycles of light and dark exposure**. If one has a species of bacteria that produces spores, **germination** can provide some synchrony for a short period of time. A frequently used technique is to **starve** cells by placing them in a medium in which one of the nutrients is in short supply. In other words, the concentration of nutrient in the medium will be enough to support the growth of only a specific number of cells. Once that population number has been reached, there is not enough nutrient for future generations. Presumably, during the starvation period the cell population approaches a uniform state in the cell cycle. Later the cells are placed in a nutritionally complete medium where they will initiate growth at approximately the same stage in the cell cycle.

Selection techniques. Selection techniques are of two types, and both are related to the physical aspects of cell division that is the **size** of the organism. Immediately following cell division bacterial cells are smallest. In a nutritionally complete medium the cell will enlarge as it grows (note that this enlargement is barely visible in spherical bacterial cells). One selection method uses a **membrane filter** (Figure 14-1, *A*). The asynchronous culture is poured over a filter on which the cells adhere but whose porosity prevents the bacterial cells from passing through. The filter is inverted, and fresh medium is added to the filter. The cells on the filter will divide, with one of the daughter cells adhering to the filter and the other daughter cell being eluted. The eluted cells represent newly divided cells. The second synchronization technique uses **density centrifugation.** An asynchronous culture is placed in a centrifuge tube containing a nonmetabolizable solute such as sucrose (30% solution) and is centrifuged for a specific period of time. A density gradient of sucrose is produced during centrifugation that increases with the distance from the axis of rotation; that is, the density is greatest at the bottom of the centrifuge tube (Figure 14-1, *B*). Bacteria will also form a gradient when centrifuged in this medium, with the larger and heavier population of cells at the bottom of the centrifuge tube and the smaller (recently divided) cells near the top. This same technique is also used to separate viruses, macromolecules, and other particulate matter (see Point of Interest, Chapter 4).

Once the microbial population has been synchronized, we are ready to investigate the cycle of events that is associated with the growth and division of the microbial cell.

Bacterial cell cycle

A number of biochemical and physical events occur in the bacterial cell from the time it is derived from a parent cell until it divides into two daughter cells. This period of time, called the bacterial cell cycle, includes such events as (1) growth, or the coordinated synthesis of macromolecules as well as cellular components in the cell; (2) septum formation; and (3) division, or the process in which each daughter cell receives a copy of DNA and half the cytoplasmic contents of the parent cell. Many individual aspects of cell cycle control have been investigated, but as yet they have not been pieced together into an integrated scheme of events. The most significant results have come from studies that have at-

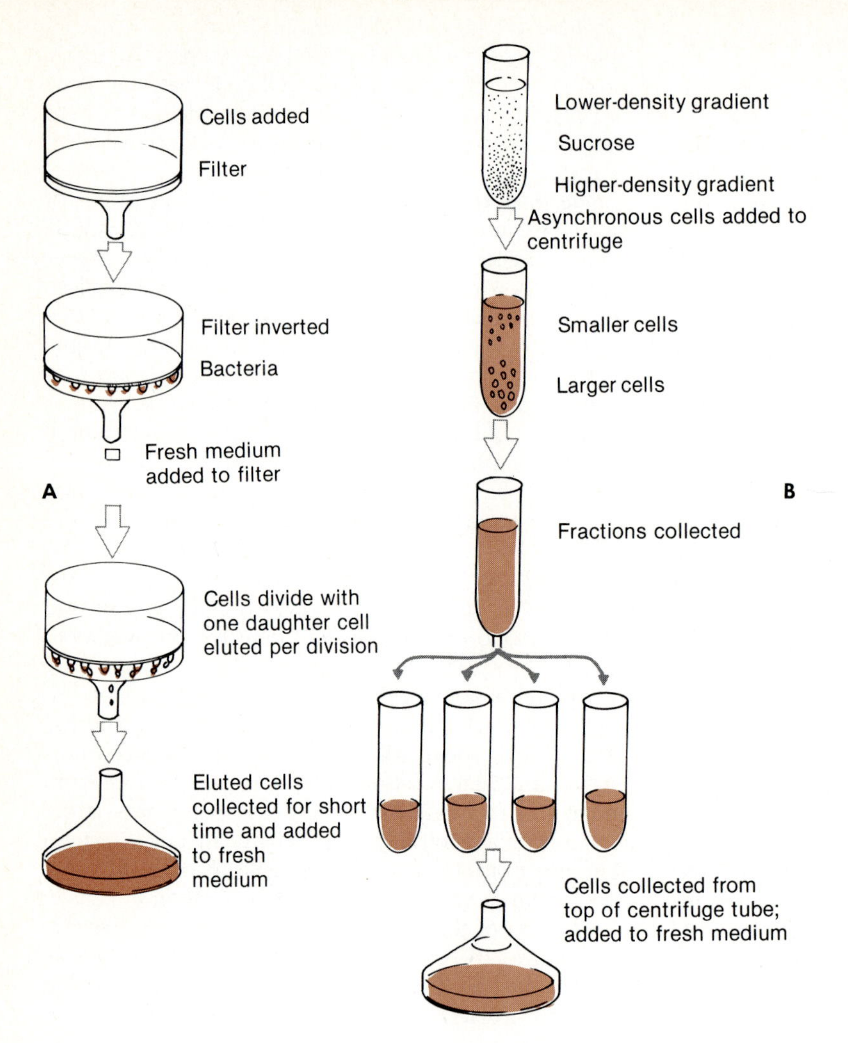

Figure 14-1 *Selection techniques for obtaining synchronized cultures.* ***A,*** *Membrane filter technique.* ***B,*** *Centrifugation technique.*

tempted to correlate DNA synthesis and cell envelope (cell wall and cytoplasmic membrane) synthesis with cell division.

DNA replication. Studies have demonstrated that if DNA synthesis is interrupted, cell division is also interrupted; however, if cell division is interrupted by an antimicrobial, such as penicillin, DNA synthesis still continues. This indicates that DNA replication is somehow involved in the cell division process.

Chapter 8 explains that the bacterial chromosome is a circular molecule in which replication begins at a specific site called the **origin** (Figure 14-2). Once initiated, replication becomes bidirectional and proceeds in opposite directions from the origin via two replication forks until they meet at the terminus. The actual point of replication is called a **replication fork.** The end result is the formation of two DNA duplexes, each duplex containing one parental strand of DNA and one newly synthesized DNA strand. We will now see how DNA replication and the growth cycle are integrated.

The time it takes for a bacterial cell such as *Escherichia coli* to divide can be controlled by adjusting those factors that affect growth, such as nutrients, pH, and oxygen tension, *E. coli* grown under ideal conditions in a medium rich in nutrients at a temperature of 37° C will divide every 20 to 25 minutes or less. The generation time is 60 minutes if the same medium is used, but the temperature of incubation is 30° C. The reason for mentioning this is that in *E. coli,* where the cell cycle has been most extensively studied, DNA replicates once every 40 minutes (provided that the cell divides in a time period between 25 and 70 minutes). The generation time for most bacteria is between 40 and 60 minutes. The question that probably comes to mind is, how can a cell that is dividing every 20 to 30 minutes synthesize a complete molecule of DNA if DNA replicates

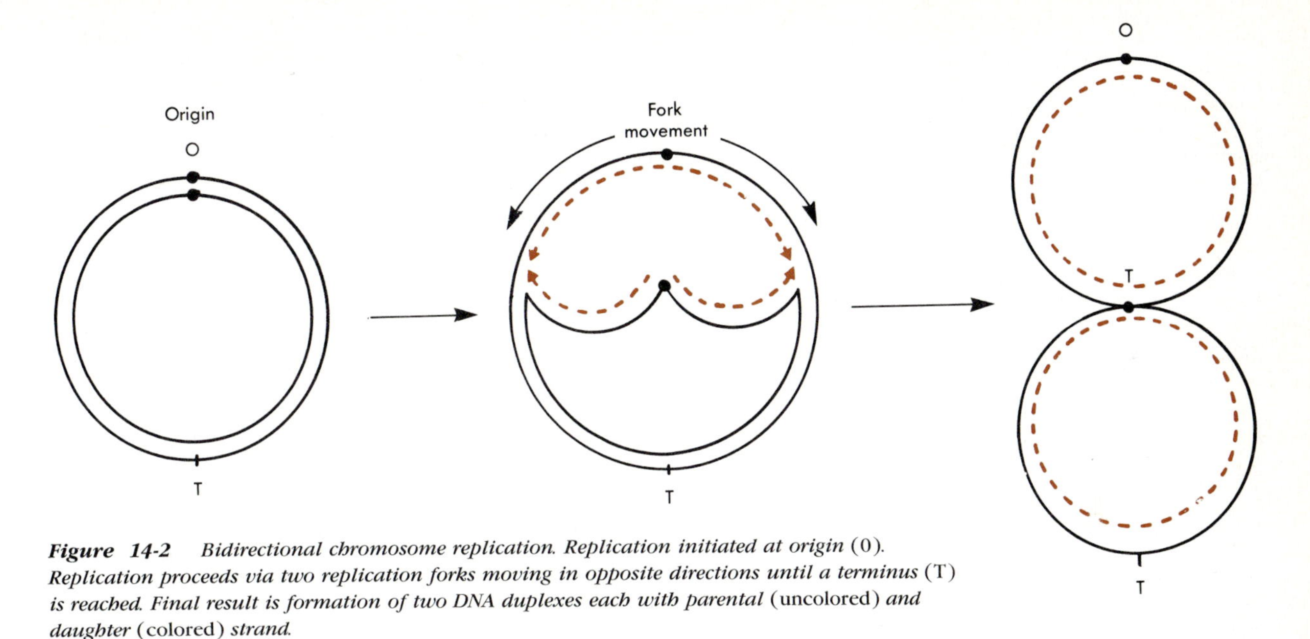

Figure 14-2 *Bidirectional chromosome replication. Replication initiated at origin* (0). *Replication proceeds via two replication forks moving in opposite directions until a terminus* (T) *is reached. Final result is formation of two DNA duplexes each with parental* (uncolored) *and daughter* (colored) *strand.*

every 40 minutes? To understand this apparent paradox let us examine the cell cycle of *E. coli* in more detail.

The cell cycle can be divided into three periods: I, C, and D.

1. The *I,* or initiation, period is devoted to preparing the cell for chromosome replication. In slowly dividing cells this period is very long; in fast-growing cells (division times of 60 minutes or less), the I period is short or even absent.
2. The *C* period, or chromosome replication event, has a fixed time course of 40 minutes in *E. coli* when the doubling time is between 20 to 70 minutes. This indicates that the rate of replication is independent of the rate of growth.
3. The *D,* or division, period represents the time between termination of chromosome replication and cell division. In *E. coli* the D period lasts for 20 minutes. Thus the C and D periods together equal 60 minutes in cells with doubling times of 20 to 70 minutes.

The paradox between the fixed replication time of 40 minutes (plus 20-minute D period) in cells with doubling times of 20 to 30 minutes was resolved with the discovery of multiple replication forks. This means that a new round of replication can occur before the previous one has been completed. For example, in a growing cell with a doubling time of 40 minutes the initiation of chromosome replication will have occurred 20 minutes into the previous cell cycle (Figure 14-3).

Chromosome replication and cell division. The rate of replication is governed by the frequency of chromosome initiation. It is believed that initiation occurs when the cell has reached a critical mass (M_i or cell initiation mass) (Figure 14-4). Concomitantly the concentration of an initiation protein is believed to parallel the increased cell mass. When the initiation protein has reached a critical level, chromosome replication is triggered. Chromosome replication ends with the synthesis of termination proteins (Figure 14-4), and this initiates the D period. Termination proteins are believed to interact with cell division proteins (septum precursor proteins, for example), and this initiates the septum formation process, or invagination of the cytoplasmic membrane and cell wall. The D period is apparently divided into a phase of extensive macromolecule synthesis followed by an event committed to cell division.

Cell envelope and cell division in cocci. The cell envelope, which is composed of the cell wall and cytoplasmic membrane, is intimately associated with growth and cell division. Studies of only the cell wall, however, have yielded sufficient data to obtain a clear picture of how cell surfaces respond during the cell cycle.

Cytoplasmic membrane. The cytoplasmic membrane is believed to play a role in the regulation of DNA initiation as well as in the segregation of replicated DNA. The **replicon hypothesis** proposed by Jacob and others assumed that at an early stage in the cell cycle one strand of the parental DNA is attached to a special membrane site, which signals the start of DNA replication. The other parental strand remains attached to a previously synthesized membrane appendage. The membrane attachment site of the first strand grows, and in conjunction with the region between the two attachment

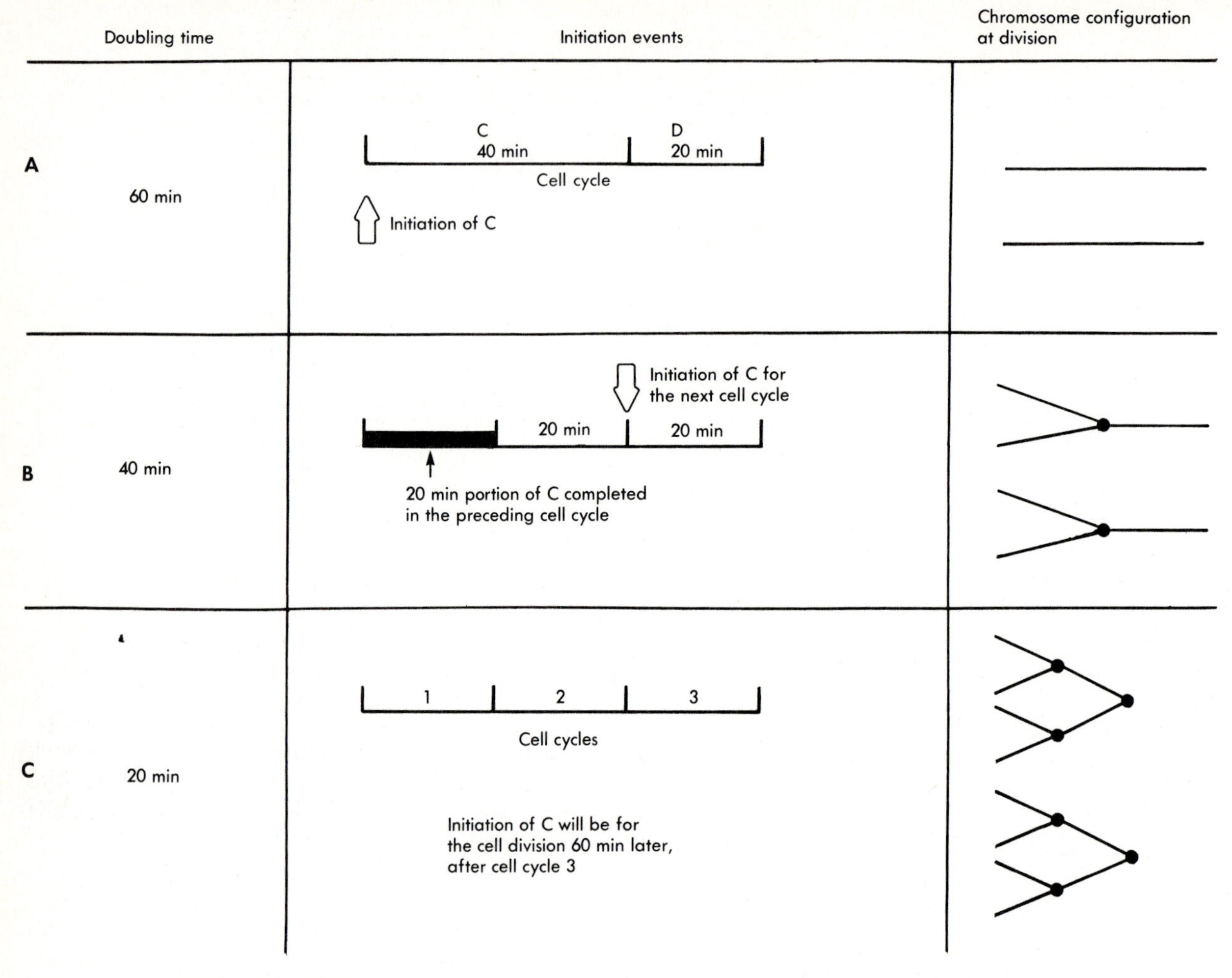

Figure 14-3 *Initiation of chromosome replication in* E. coli *cells growing with doubling time of 60, 40, or 20 minutes.* ***A,*** *Initiation of C (chromosome replication) period occurs at or just after cell division. Single parallel lines represent replicated circular chromosome at end of D or division period.* ***B,*** *Initiation of C occurs 20 minutes into previous cell cycle, that is, at beginning of D. Solid circles represent replication forks, which at division are already halfway round each chromosome.* ***C,*** *Initiation of C is for cell division 60 minutes later (*C, *40 minutes +* D, *20 minutes), which in this case occurs three generations later.*

points, as it grows the replicating genomes become segregated (see Figure 14-5). Most of the evidence for this type of mechanism comes from biochemical studies. Whether or not the membrane attachment sites coincide with the mesosome is still a matter of speculation, since many scientists consider mesosomes to be artifacts.

Cell wall. Experiments have shown that cell wall synthesis does not occur in the same way in all species of bacteria. The clearest picture to date of cell wall synthesis during the cell cycle has been obtained with those organisms that divide in one plane, such as the streptococci. Streptococci can be grown in a medium that contains an antibody capable of reacting with cell wall antigen. The antibody is first labeled with a fluorescent molecule, called a fluorochrome, thus making it observable microscopically. Once the streptococci have been labeled and the entire surface appears to fluoresce, the microorganisms are removed from the medium and inoculated into a medium containing no fluorescent antibody. As the population of streptococci grows, newly synthesized cell wall will not contain fluorescent antibody. The results of such an experiment are illustrated in Figure 14-6. It can be seen from this figure that new cell wall is laid down centripetally, so that the new hemispheres for each of the eventual daughter cells are back to back. Figure 14-7 is a model demonstrating how cell wall synthesis is organized. The model illustrates the following:

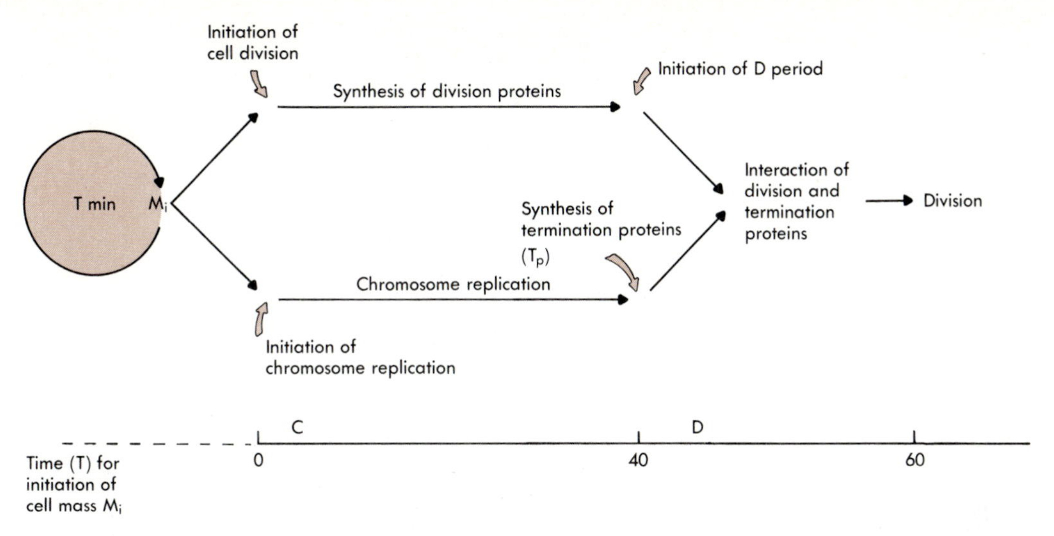

Figure 14-4 *Events that are believed to occur during* E. coli *division cycle. Cell reaches critical cell mass* (Mi), *which initiates chromosome replication and synthesis of division proteins. DNA replication time is 40 minutes and ends with synthesis of termination proteins* (Tp). *Interaction of termination proteins and division proteins initiates cell division.*

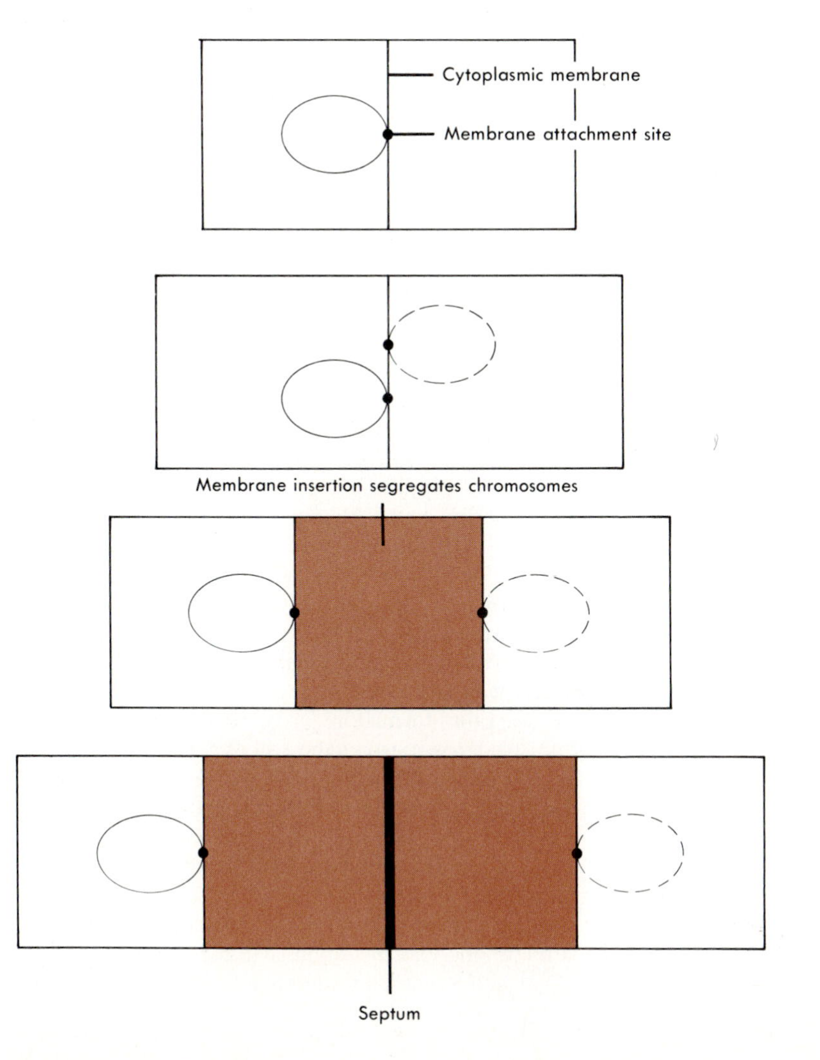

Figure 14-5 *Replicon model for segregation of DNA by the membrane of* E. coli. *Single chromosome is attached to membrane at one site. Replication of chromosome results in formation of daughter strand* (dotted line) *and duplication of membrane attachment sites. Eventually membrane and cell wall duplication will segregate chromosomes resulting in each cell, following division, possessing copy of genome. Cell division will occur at site called septum.*

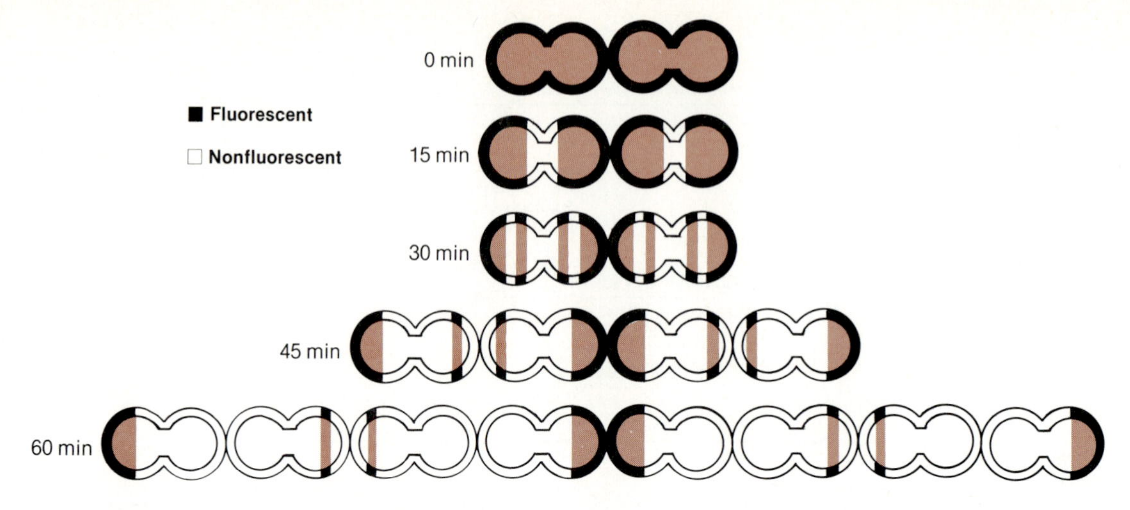

Figure 14-6 *Cell wall growth in* Streptococcus pyogenes. *Bacteria are grown in medium containing fluorescent antibody that attaches to cell wall. Fluorescent cells are then transferred to medium containing no fluorescent antibody. Times indicate period of incubation in nonfluorescent medium.*

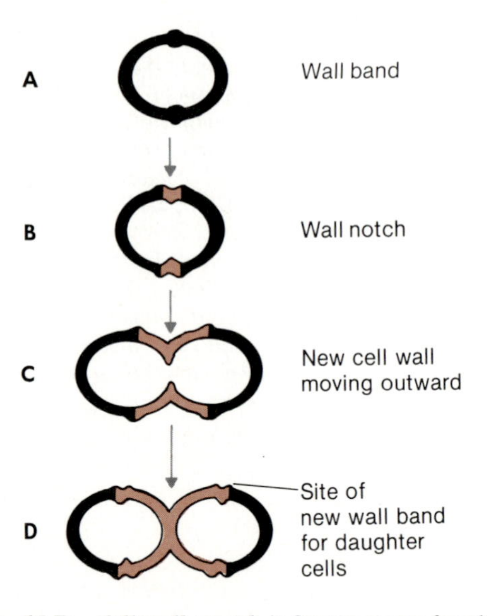

Figure 14-7 *Cell wall growth in* Streptococcus faecalis. ***A,*** *Raised area on cell envelope, called wall band, appears.* ***B,*** *Wall band invaginates and forms wall notch, which becomes septum. In* ***B*** *through* ***D,*** *new cell wall is synthesized and moves outward from wall notch.* ***D,*** *New cell wall bands are being initiated at center of each daughter coccus even before cell divides.*

1. New cell wall synthesis begins at areas of the cell envelope, called **wall bands**, that have been produced during the previous cell cycle (Figure 14-7). A membrane forms beneath each wall band, which represents the membrane attachment site. Except for a middle band at the cell equator, the right and left hemispheres of the cell show no new cell wall growth.
2. Cell wall precursors are added below the wall band, and as they are added, the membrane invaginates and forms a **wall notch**. The wall notch will eventually become a **septum**, or point of cell division. The addition of new cell wall precursors requires certain cell wall lytic enzymes to break down preexisting peptidoglycan.
3. The cytoplasm becomes completely divided as new cell wall precursors are added centripetally, and the cell is cleaved into two daughter cells by the septum. Each daughter cell possesses a wall band that represents cell wall material produced during two different cell generations.

Cell envelope and cell division in rod-shaped bacteria. Experiments to identify cell wall growing points in rod-shaped bacteria have been difficult to evaluate because of conflicting results. Rod-shaped organisms lengthen by insertion of new wall at multiple sites along the side walls (Figure 14-8). New cell wall material is inserted with the existing wall, and at the same time new cell wall is formed by ingrowth of the wall at the point of septum formation. Division takes place after the septum has grown across the cell to form two daughter cells. Cell wall material at the caps or ends of the future daughter cells is conserved during the whole process, and no new cell wall material appears to be deposited in these areas. Thus the process in bacilli appears to be similar to that in streptococci.

Unequal cell division. Cell division in some mutant bacteria is unequal, resulting in the formation of one daughter cell containing a nucleoid and one small anu-

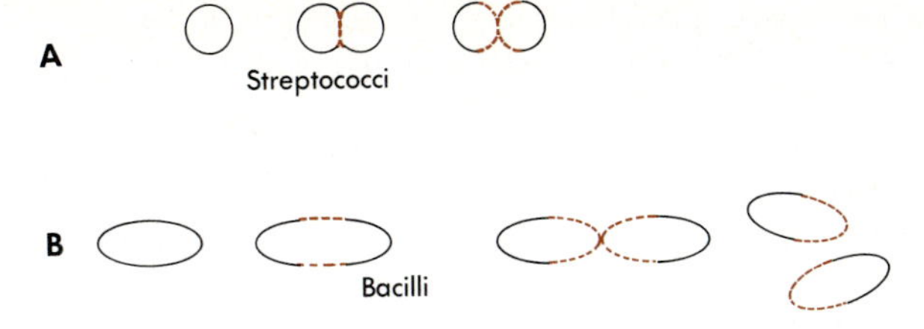

Figure 14-8 *Cell wall synthesis (---) and cell division in **A,** Streptococci. **B,** Bacilli. Note in bacilli new cell wall material becomes integrated with existing wall but some wall (like that of streptococci) is formed by ingrowth of wall midway along cell. Division takes place after septum has grown across cell to form two daughter cells.*

cleated cell called a **minicell.** Minicells have been used as a research tool because they lack genetic chromosomal material. Minicells can be used as vehicles for the insertion of plasmids coding for specific proteins. Minicells are currently being evaluated for the preparation of certain vaccines. *E. coli,* for example, causes diarrhea in calves because of certain surface antigens. The idea behind the vaccine is to insert a plasmid coding for the surface antigen into a minicell-producing strain of *E. coli.* Minicells derived from the cell line produce the surface antigen and will provide a safe vaccine free from products (for example, endotoxins) coded by chromosomal genes. Minicells are also being evaluated as vehicles for the synthesis of specific industrial products, such as enzymes, specialized proteins, and therapeutic agents.

Concept Check

1. What is meant by cell culture synchronization, and in which ways can it be accomplished?
2. How can a bacterial cell divide every 20 minutes if the chromosome replicates every 40 minutes?
3. Briefly, what is the difference in cell wall growth of spherical vs. rod-shaped bacteria?

Population growth

Large populations of cells are routinely used in laboratory experiments in which microbial growth and related processes are measured. A small inoculum of cells is usually introduced into a culture medium, and after several hours of growth overnight, the cells are harvested. Microbial populations may be prepared in one of two ways: batch culture or continuous culture.

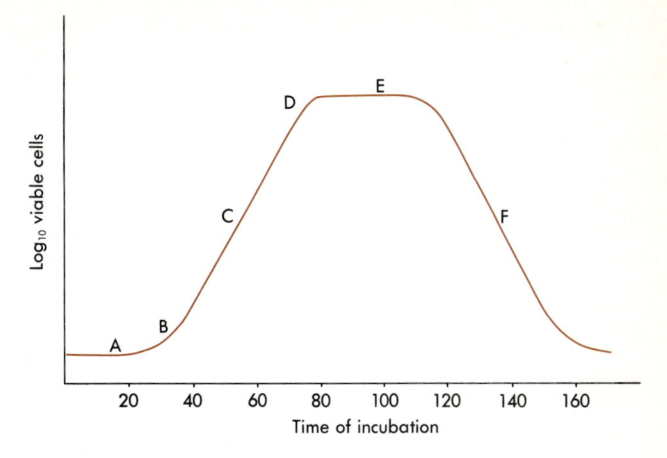

Figure 14-9 *Growth curve.* A, *Lag phase,* B, *acceleration phase;* C, *logarithmic phase;* D, *deceleration phase;* E, *stationary phase;* F, *Death or decline phase.*

Batch culture. The rate of growth of a cell population will be seen to vary with time following the inoculation of a small number of cells into the culture medium. In other words, from the time of inoculation of cells into the medium until growth ceases, the cell population will go through a number of phases that can be expressed as a growth curve. The growth curve can be separated into the following stages (Figure 14-9).

Lag phase. Following inoculation into the medium, microorganisms usually require a period of time to adjust to their new environment. There is no lag if the cells to be used for the inoculum were grown in the same medium under the same environmental conditions. Any lag time refers only to a lag in cell division and not to other aspects of growth. The cells, for example, increase in size, and there is synthesis of enzymes to ensure that nutrients in the medium are metabolized. There is an increase in the synthesis of macromolecules such as RNA and protein, but DNA synthesis remains constant. The length of the lag phase is related to the physiological and genetic state of the species at the time of inoculation into the culture medium. Cells that are nutritionally starved before their inoculation into a medium demonstrate a long lag phase, but if the original inoculum of cells contains an abundance of enzymes from previous growth in the same type of medium, the lag phase will be short. Organisms that do not have the necessary inducible enzymes exhibit an indefinite lag phase until induction of the enzymes has been completed (see Chapter 9 for detailed information on inducible enzymes). At the end of the lag phase a very short period of acceleration occurs, in which some of the cells are entering a period of cell division but not yet at a constant rate.

Logarithmic phase. The logarithmic phase (also called **log** or **exponential phase**) represents that period of time in which cell division is maintained at a constant and maximum rate under the environmental conditions present. *E. coli,* for example, in one type of medium might divide every 30 minutes in the log phase, but if it is grown in a less-rich medium or at a different temperature, the division time could be 60 minutes or longer. The growth rate, or doubling time, remains constant for any organism during the log phase regardless of the medium. The culture in the log phase is said to be in **balanced growth.** Balanced growth also means that if the cell population doubles, so do all of the components of the cell, such as DNA, RNA, and protein. This does not mean that if a population of cells were examined at a specific time the cell mass in each cell would be the same. The population during exponential growth expresses the average of individual cell cycle events. Some cells can be expected to be in each of an infinite number of quantitative increments of cell mass in each phase of the cell cycle. The exponential culture will therefore display an average mass. Cultures in a closed system, in which no nutrients enter the culture flask and no waste products are removed, reach a point where environmental factors slow down the rate of growth. These factors might include overcrowding, nutrient depletion, toxin production, and pH changes due to acid formation. This slowing-down process is referred to as a **deceleration phase** and precedes the next stage, called the **stationary phase.**

Stationary phase. The number of viable microorganisms in the stationary phase remains constant. The conditions that initiate the deceleration phase are accentuated in the stationary phase. This is a period of **unbalanced growth,** and cellular components and macromolecules are synthesized at different rates. When nutrient depletion is the cause of unbalanced growth, the cells that divide produce smaller cells. Stationary phase cells are always smaller than exponential phase cells. Viable cells in the stationary phase are producing more acids and waste products, which further inhibit cell division. Some dying cells lyse and release autolytic enzymes, such as proteases, nucleases, and lipases, that break down macromolecules into their respective building blocks, which can be used by those cells that have survived. The stationary phase can be short or very long, depending on the bacterial species.

Death or decline phase. The rate at which a population dies varies with the microbial species involved. The rapid lysis of some microbial species, such as *Neisseria* species, during this period is brought about by a self-destruct mechanism that is characterized by the induction of autolytic enzymes that degrade the cell envelope. These autolytic enzymes are the same enzymes that are necessary for growth, because they allow the insertion of newly formed peptidoglycan into the cell wall. Obviously, a balance might be maintained in the cell between cell wall lytic action and cell wall synthesis. Studies with *Streptococcus pneumoniae* indicate that the nature of the lipoteichoic acid in the cell wall is an important factor in control. Apparently, when the lipoteichoic acid is in an acylated state, the cell wall hydrolytic enzyme (β-1,4-*N*-acetylmuramide glycan hydrolase) is inhibited. The deacylated form does not inhibit the enzyme, suggesting that a deacylase enzyme may be the controlling factor. Other mechanisms must also exist for cells without lipoteichoic acid in the cell envelope.

The rate of death can be exponential, but sometimes it is not. It is believed that if the cell requires only one target site to be affected in order to cause death, then the death curve will be exponential. If two or more sites are required, then a nonexponential rate of death takes place in the population.

Continuous culture. One of the disadvantages of batch culture is that one cannot independently control those factors that decrease the growth rate, such as nutrient supply and waste removal, although air supply and pH can be controlled. A technique for overcoming this problem involves the use of an open-culture system called **continuous culture.** The primary purpose of continuous culture is to maintain a population of cells at a constant growth rate in the exponential phase of growth for an indefinite period of time. The devices used to produce continuous culture are of two types: **chemostats** (also called bactogens) and **turbidostats.** Before we discuss the implications of continuous culture, let us examine the mathematics of bacterial growth.

MATHEMATICS OF BACTERIAL GROWTH

Bacterial cells divide in geometric or exponential fashion, that is, 1-2-4-8-16-32, etc. The population therefore increases as an exponent of 2 or 2^n where n equals the number of doublings or generations. For example, if we were to start with 1 cell, the number of cells after 3 generations or doublings would be 2^3 or $2 \times 2 \times 2$ or 8. The actual number of cells measured under laboratory conditions is much larger, often between 1×10^5 to 1×10^9 cells per milliliter of suspension. Suppose, for example, we have two sample cultures that were collected at two different times during the growth cycle. Further suppose that we have a method (to be discussed later) for determining the number of cells in the sample. First, we could either plot the number of microorganisms as a function of time or plot the logarithm of the number of cells as a function of time (Figure 14-10). The disadvantage of plotting the arithmetic number of cells versus time is that at higher cell concentrations the curve is nearly impossible to evaluate. The logarithm of the number of organisms is usually plotted as the logarithm to the base 10 (obtained by multiplying the loga-

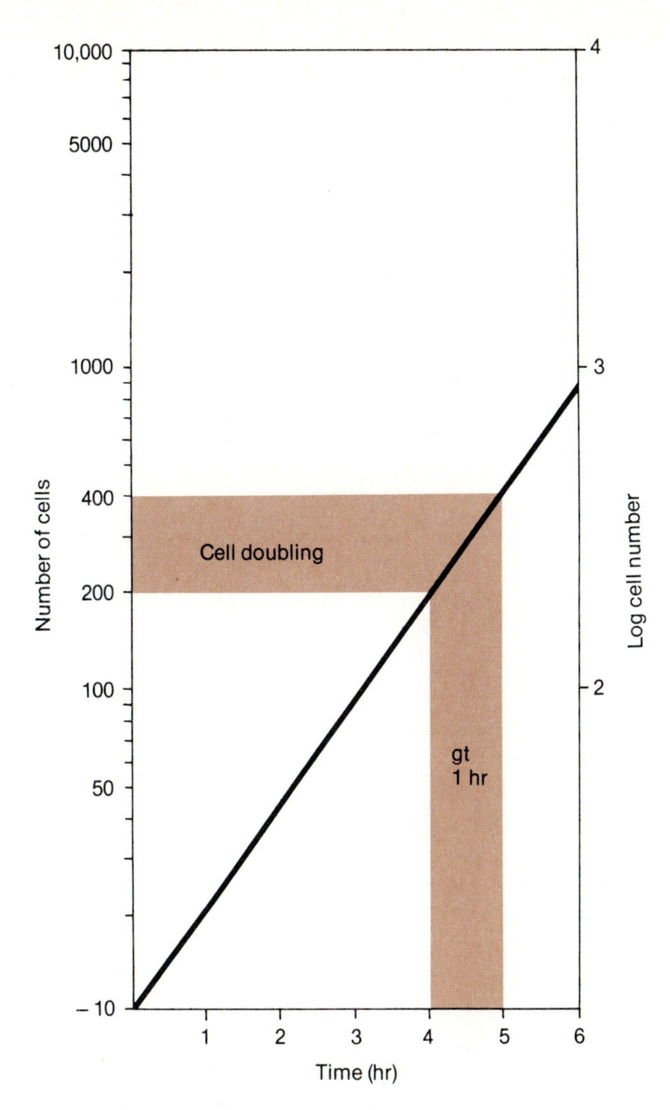

Figure 14-10 *Graph depicting log number of cells as function of time when cultivated in laboratory growth medium.* gt, *Generation time.*

rithm to the base 2 of a number by 0.3010). The number of generations (n) can be determined by using the following formula:

$$n = \log_2 N_1 - \log_2 N_0$$

or

$$\frac{\log_{10} N_1 - \log_{10} N_0}{\log_{10} 2}$$

where N_1 and N_0 are the number of organisms at times T_1 and T_0, respectively. If in our example N_0 equals 100 organisms and N_1 equals 800 organisms, then the equation reads:

$$n = \frac{\log_{10}800 - \log_{10}100}{\log_{10}2} = \frac{2.903 - 2.0}{0.301} = 3 \text{ generations}$$

Because the generation or doubling time (gt) is equal to the time interval between T_1 and T_0 (assume it to be 90 minutes) divided by the number of generations, then:

$$gt = \frac{T_1 - T_0}{n} = \frac{90 \text{ min}}{3} = 30 \text{ min}$$

From the previous equations one can also determine the specific growth rate constant (k) for any culture during unrestricted growth. During this time the rate of increase of cells is proportional to the number of cells present at any particular time. The rate of increase is not influenced by the concentration of any particular nutrient in the medium but only by the overall composition or growth-supporting ability of the medium. In mathematical terms the growth rate is expressed as follows:

$$kN = \frac{dN}{dT}$$

where $\frac{dN}{dT}$ is the change in a measurable parameter of the cell, which may include cell number or cell mass, for example, between two time periods. The symbol *k* represents the specific growth rate constant. When the equation is converted to logarithms we obtain:

$$k = \frac{\ln_2 N_1 - \ln_2 N_0}{T_1 - T_0} \text{ or } \frac{2.303(\log_{10}N_1 - \log_{10}N_0)}{T_1 - T_0}$$

In practice, samples of cells are removed from a culture at various times during exponential growth and cell numbers are determined (Figure 14-11). When these values are plotted, a straight line is obtained whose slope can be used to determine the value of k. The dimensions of k are reciprocal hours or per hour. The growth rate constant will be the same during exponential growth regardless of the component (except for a reserve product such as glycogen that increases when other components are not increasing) measured (cell mass, cell number, protein, carbon, etc.). However, it can be influenced by environmental conditions, including oxygen tension, pH, temperature, and type of growth medium. The growth rate constant provides the researcher with a valuable tool for comparison between different microbial species provided standard environmental conditions are maintained.

When the culture has doubled, then the generation time (doubling time) can be expressed as follows:

$$gt = \frac{\ln_2}{k} = \frac{0.693}{k}$$

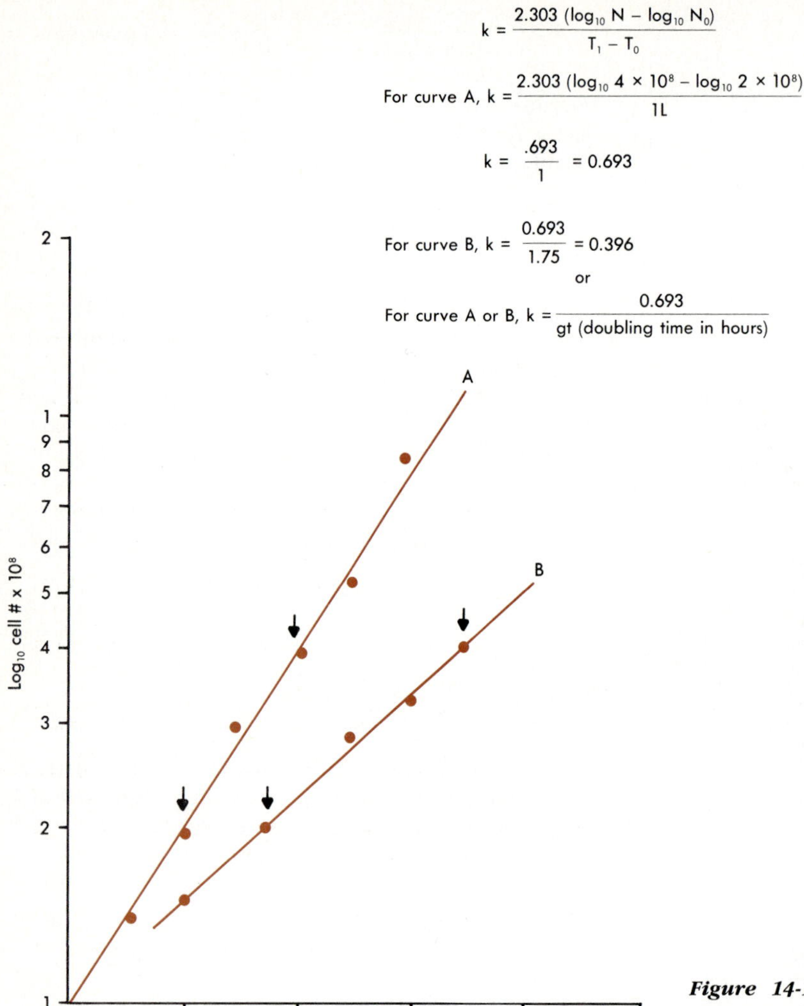

$k = \frac{2.303\ (\log_{10} N - \log_{10} N_0)}{T_1 - T_0}$
For curve A, $k = \frac{2.303\ (\log_{10} 4 \times 10^8 - \log_{10} 2 \times 10^8)}{1L}$
$k = \frac{.693}{1} = 0.693$
For curve B, $k = \frac{0.693}{1.75} = 0.396$
or
For curve A or B, $k = \frac{0.693}{\text{gt (doubling time in hours)}}$
A
B
Log_{10} cell # x 10^8
Time (hours)

Figure 14-11 *Graph depicting specific growth rate constants of two cultures with different generation times* (gt)*. Arrows on each plot indicate doubling time intervals.*

The reciprocal of the doubling time is called μ and represents the number of doublings per hour:

$$\mu = \frac{1}{gt} = \frac{k}{0.693}$$

Chemostats. The simple chemostat (Figure 14-12) consists of a vessel in which the culture is grown. Devices are connected to the growth vessel that control the amount of air for aerobes and fresh medium to be added to the culture, as well as the overflow of medium containing cells and cell products. The culture will reach a constant volume and cell density when an equilibrium between inflow of fresh medium and outflow of medium is reached. All of the constituents of the medium in a chemostat except one are in excess. The deficient nutrient will support the growth of only a limited number of microorganisms; therefore the density of the culture in a chemostat is controlled by nutrient limitation. The carbon source, for example, glucose, may be used as the limiting nutrient, but other nutrients such as nitrogen, magnesium, sulfate, or phosphorus may also be used. Nutrient limitation is a frequent condition of microorganisms in their natural environment, especially those found in soil and water. Chemostats, therefore, provide a means of duplicating in the laboratory the natural condition of microorganisms.

By controlling the rate at which nutrients are added to the chemostat, the rate of bacterial growth (k) may also be controlled. The rate of nutrient addition is referred to as the dilution rate (D) and is equal to the volume of nutrient added per hour expressed as a fraction of the volume of the vessel or turnover per hour. *D,* therefore, not only accounts for the rate of nutrient

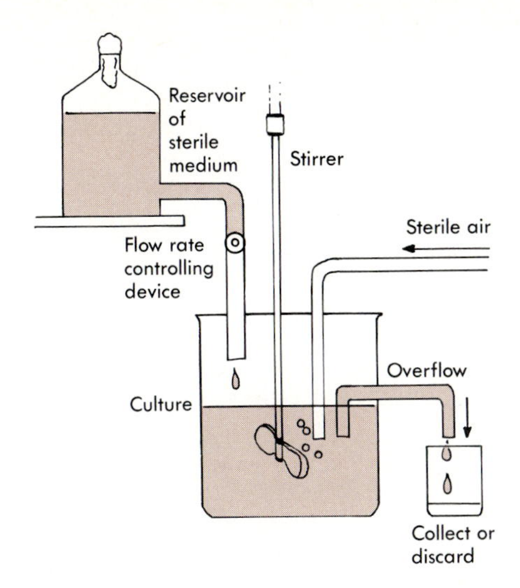

Figure 14-12 *Chemostat.*

addition but also for loss of cells from the chemostat. The dilution rate can be expressed as follows:

$$Dx = \frac{dx}{dt}$$

where $\frac{dx}{dt}$ is the change in number of bacteria between two time periods. If the specific growth rate constant (k) is also expressed:

$$kx = \frac{dx}{dt}$$

then the dilution rate equals the growth rate constant. Any changes in cell concentration can be defined as follows:

$$dx/dt = kx - Dx$$

meaning that a change in cell concentration equals growth of bacteria minus dilution of bacteria. Thus the growth rate of a culture in the chemostat is determined by the flow rate (culture volumes/hour) and the volume of the culture vessel, that is, k = D at steady state or equilibrium. At low dilution rates (D_0) the concentration of cells in a continuous culture system is at its highest (Figure 14-13). A very gradual decline in bacterial concentration and a gradual increase in substrate concentration occur when the dilution rate begins to increase. A point is reached, during this increase in dilution rate (D_c), where the rate at which bacteria are diluted out exceeds the maximum growth of the culture and there is rapid decline in the concentration of bacteria. In other words, more bacteria are washed out of the culture vessel than are produced by cell division.

The chemostat is a stable system because it is capable of self-correction. If there is an increase or decrease in

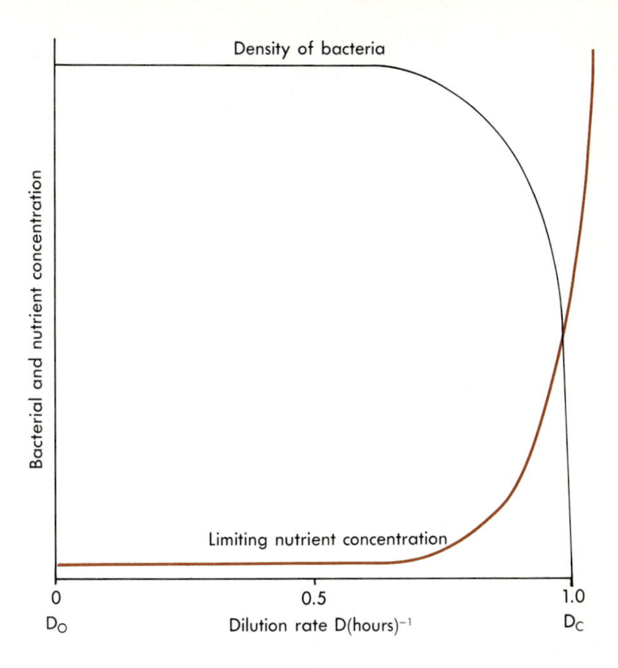

Figure 14-13 ***Relationship between cell density, limited nutrient concentration, and dilution rate in continuous culture system. See text for details.***

the rate of addition of medium (which is accompanied by loss of culture via washout), the system corrects itself by a decrease or increase in the growth rate of the culture. This assumes that the rate at which medium is being added does not exceed the maximum rate of growth that the organism is able to attain under conditions of unrestricted growth in the same medium and under the same environmental conditions. Otherwise the culture is likely to wash out of the vessel. The chemostat works best in those regions of the growth curve in which the rate of growth varies with the concentration of the limited nutrient.

Turbidostats. Cell density in turbidostats is controlled by a device that increases or decreases the flow of culture medium into the growth vessel. There is no limiting nutrient. A photoelectric device is attached to the turbidostat that measures cell density or the relative turbidity of the culture. Any deviations in culture density are passed on as a signal to a pump that increases or decreases the flow of growth medium into the growth vessel.

Continuous cultures have played a very important role in studies of microbial physiology and have been used extensively in industry. The effects of nutrient limitation have revealed many important features of cell composition, enzyme production, and growth, that could not have been obtained with batch cultures. Continuous cultures have been used to analyze mutation rates of organisms, and many of the regulatory devices in microbial systems have been amenable to study through this technique. The results of these studies have led to a greater understanding of control mechanisms in animal systems. The technique of continuous culture has also been applied in industry to the production of single-celled proteins. Microbes are being considered as a food source because of their ability to produce a high yield of protein when fed certain carbon compounds. Large-scale single-cell protein processes use carbon sources as a limiting nutrient in continuous culture (see Chapter 22 and Point of Interest, Chapter 12).

Concept Check

1. What are the various stages of the bacterial growth cycle and the physiological changes that occur during each stage?
2. Distinguish between balanced and unbalanced growth. How can one maintain balanced growth?
3. What is the mathematical equation expressing the growth rate? Of what value is the growth rate to the microbiologist?
4. What factors control the growth rate of a culture during continuous growth?

MICROBIAL RESPONSES TO NUTRIENT LIMITATION

Bacteria express only a portion of their genome under any set of environmental conditions. Much of the genome is not expressed until a set of circumstances induces the bacterial cell to express some portion of it. Microorganisms that do not possess the genetic potential to respond to environmental change will invariably die. Natural ecosystems are frequently limited in nutrient availability because of the activities of its indigenous species. A microbial cell, therefore, is continually confronted with a stiuation in which nutrient uptake is against a concentration gradient, that is, the concentration of nutrient outside the cell is less than it is inside the cell. The growth rate of a microorganism remains constant over a wide range of external nutrient concentrations because of active transport systems. When growth rate is plotted against nutrient concentration, we obtain a hyperbolic curve similar to the one obtained when enzyme activity is compared to enzyme concentration (Figure 14-14). The growth rate falls when nutrients become limiting. Nutrient limitation not only involves the basic elements, such as carbon, nitrogen, sulfur, phosphorus, potassium, and magnesium, but may also include gases such as oxygen and carbon dioxide. In addition, other factors, such as pH, light, and tempera-

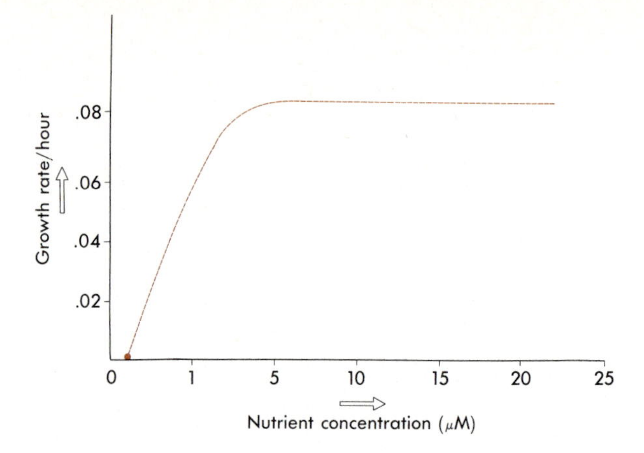

Figure 14-14 *Effect of nutrient concentration on specific growth rate.*

ture, play a role in how an organism responds to its environment. Nutrient limitation, however, is the most likely factor to affect microbial proliferation. The heterogenous nature of natural ecosystems makes it difficult to assess the response of a single microbial species within a population. It is established practice, therefore, to cultivate specific microbial species in vitro, and this is best accomplished using flow-controlled continuous culture.

Microorganisms can respond to nutrient stress by structural changes, that is, altering certain macromolecular components of the cell, its ultrastructure, or its morphology, or by using control mechanisms to adjust metabolic activities within the cell.

Structural changes

Microorganisms can modulate the rate of synthesis of all macromolecular components so as to allow more of certain cell materials to be produced from a given quantity of the growth-limiting nutrient. Most microorganisms, when placed under stringent conditions that limit the growth rate, show a decrease in content of DNA and protein. However, these stringent conditions result in a marked differential increase in RNA; this is associated with an increase in ribosome content above that expected. Even though some of the excess ribosomes may not be functioning at a low growth rate, when environmental conditions are more suitable for an increase in the rate of growth the cell may respond immediately with an increase in the rate of protein and RNA synthesis (see p. 399 for further details).

Many bacteria respond to nutrient limitation by reducing their size (Figure 14-15). This type of response increases the surface-to-volume ratio of the cell, thus enhancing the uptake of nutrients. Rod-shaped soil micro-

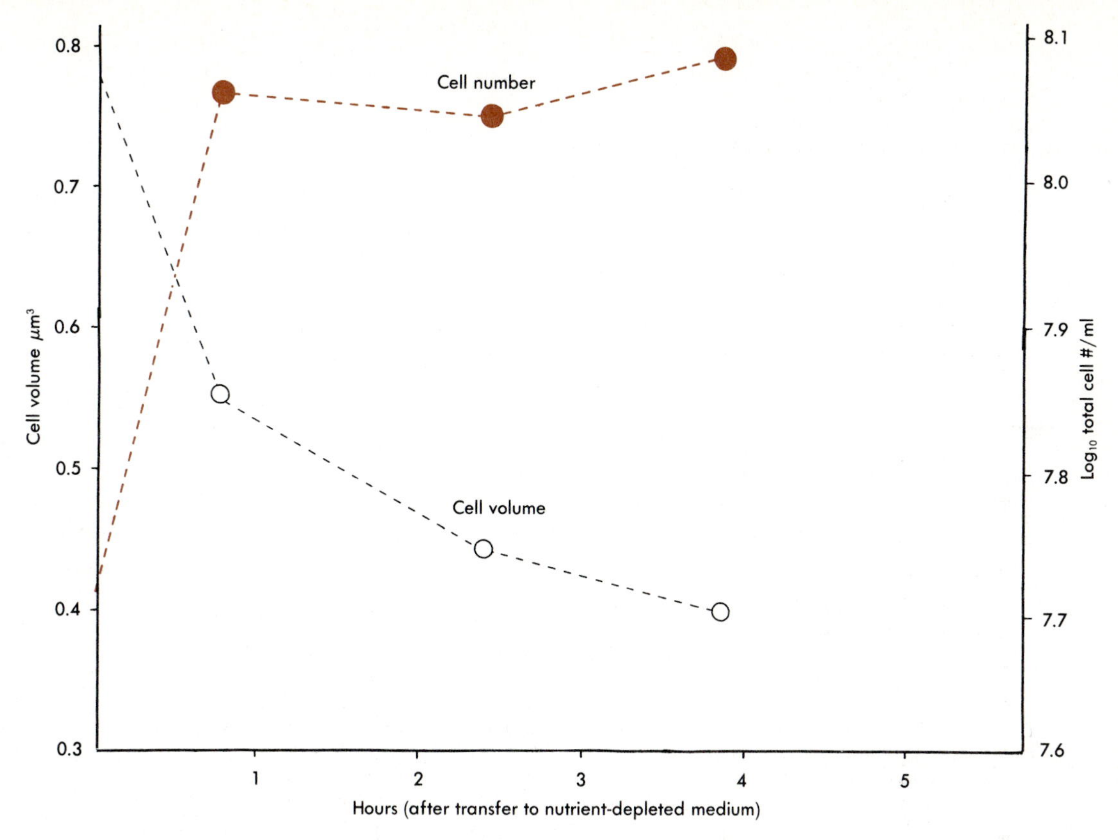

Figure 14-15 *Changes in cell volume (○) and cell number (●) of* Vibrio *species following transfer from nutritionally rich medium to starvation medium. Bacteria respond to starvation by fragmenting, which initially causes increase in viable cell number and is accompanied by continuous cell size reduction.*

organisms, such as *Arthrobacter* species, have been shown to respond to nutrient limitation by assuming a coccoid morphology—a form that predominates in the soil. This morphologic diversity may be one reason that *Arthrobacter* represents the most numerous group of bacteria in the soil. Some soil microorganisms are able to respond to nutrient depletion by forming endospores. The ability to sporulate enables the microbial species to survive long periods of starvation and other environmental stresses (see Chapter 5 for discussion of endospore formation).

Metabolic changes

Microorganisms possess control devices that allow them to fluctuate their intracellular metabolic activities in response to changes in the environment. These responses may involve the following:

1. Many microorganisms synthesize higher levels of the enzymes normally associated with the uptake of a particular nutrient or they are able to induce the synthesis of an alternative uptake mechanism. For example, the presence of a carbon source such as lactose may induce the cell to synthesize enzymes that are required in its uptake and catabolism. In the absence of lactose, expression of the relevant genes is blocked by a repressor protein molecule. This process of induction is a universal characteristic of microorganisms and is discussed at more length in Chapter 9.

In natural environments microorganisms often encounter mixtures of nutrients, for example, two or more carbon sources, with one or both of them limiting. Batch culture studies have demonstrated that microorganisms adapt themselves in such a way that the substrate supporting the highest growth rate is used preferentially. During the period when the first substrate is being preferentially used, synthesis of enzymes that could metabolize the second substrate is inhibited. Only when the first substrate has been completely exhausted is the second substrate utilized. Immediately following the exhaustion of the first substrate there is a lag in growth during which time those enzymes required for

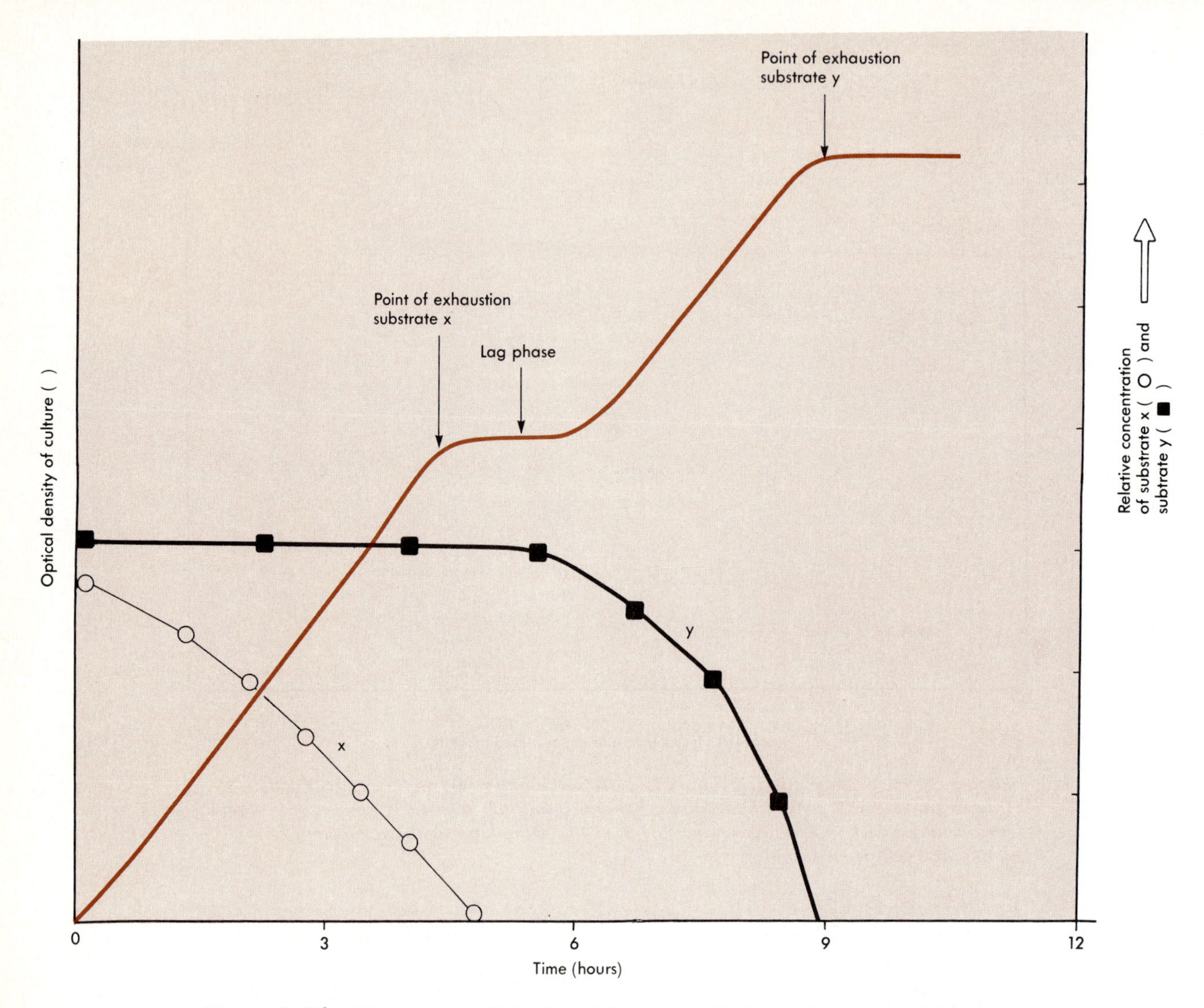

Figure 14-16 *Diagrammatic illustration of diauxic growth of microbial species. Culture medium contains two carbon sources, x and y. Inoculum for this culture was first pregrown on substrate x. Leveling-off period after exhaustion of x is time required for synthesis of enzymes involved in metabolism of substrate y.*

the preferential uptake and utilization of the second substrate are synthesized. When this type of growth is graphed, a **diauxic** curve (growth consisting of two phases between which is a lag phase) is obtained (Figure 14-16).

2. Many microorganisms can modulate the rate of uptake and metabolism of a nutrient. The uptake of glucose by *Klebsiella aerogenes,* for example, is carried out by a phosphotransferase system. Under conditions of glucose insufficiency the activity (maximum velocity or V_{max}) of this enzyme system is increased, but under conditions of glucose sufficiency the activity of the enzyme is decreased.

3. Some microorganisms can use alternative pathways so as to circumvent the restrictions placed on the cell by the limiting nutrient. Glycerol can be metabolized by *K. aerogenes* using two different enzymes (Figure 14-17). One of the enzymes, glycerol kinase, has a low K_m (1×10^{-6}M), that is, the enzyme shows maximal activity when the concentration of glycerol is very low in the environment. The second enzyme, glycerol dehydrogenase, has a high K_m (1×10^{-2}M), that is, the enzyme requires high concentrations of glycerol to show maximal activity. Under aerobic conditions and glycerol excess there is little kinase activity but increased dehydrogenase activity. When glycerol is limiting, kinase

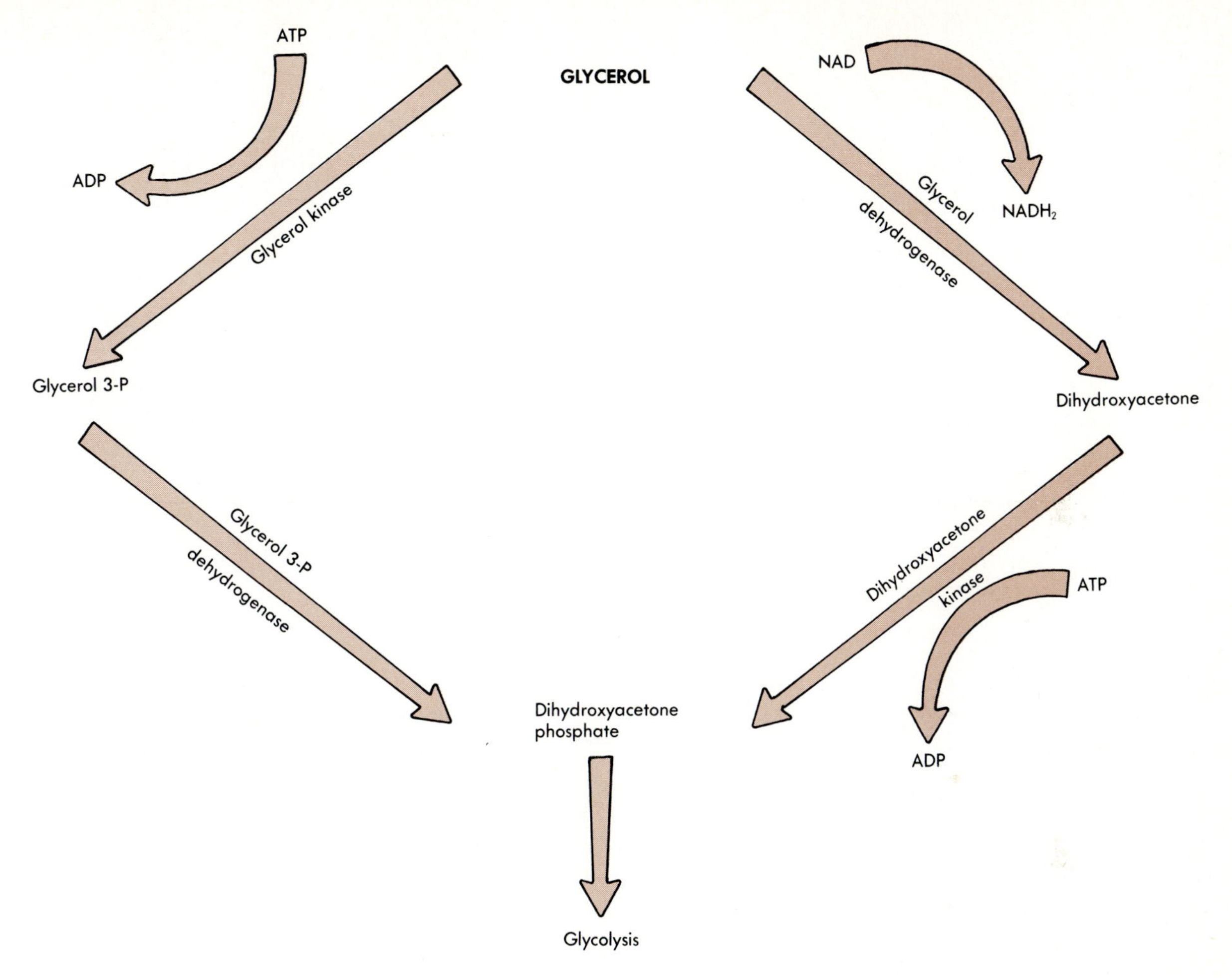

Figure 14-17 *Pathways of glycerol metabolism in* Klebsiella aerogenes. *Pathway used depends on glycerol concentration in environment. At low concentrations of glycerol the glycerol kinase pathway is used because enzyme is most active at those concentrations. When glycerol concentration is high glycerol, dehydrogenase shows maximal activity and this pathway is used by organism.*

activity predominates. Thus the cell can utilize an enzyme that has a high affinity (low K_m) for substrate. This allows the organism to sustain a higher rate of metabolism of the limited nutrient.

GROWTH YIELDS AND EFFICIENCY

Studies with chemoheterotrophs in which the carbon and energy source were the same revealed that the mass of cells (mg of cells) produced per unit (mg) of limiting nutrient is a constant, the **growth yield *(Y)*.** The growth yield can be mathematically expressed as follows:

$$Y = \frac{N - N_0}{C_0 - C}$$

where N is the dry weight per milliliter of cells at a point during growth just before the stationary phase and N_0 is the dry weight per milliliter of cells just after inoculation. C_0 is the initial concentration of nutrient and C is the concentration of nutrient when growth stops. If we were to graph the growth obtained from separate experiments using different concentrations of a limiting nutrient, such as glucose, we would find that the growth yield was independent of the initial substrate concentration (Figure 14-18). It is possible to take any nutrient and find the value of Y. We might, for example, wish to determine the Y values for carbohydrates such as fructose, lactose, mannose, etc., using a particular microbial species. Once the Y value has been determined, it is possible to ascertain the concentration of nutrient in an unknown medium that is complete except for the limit-

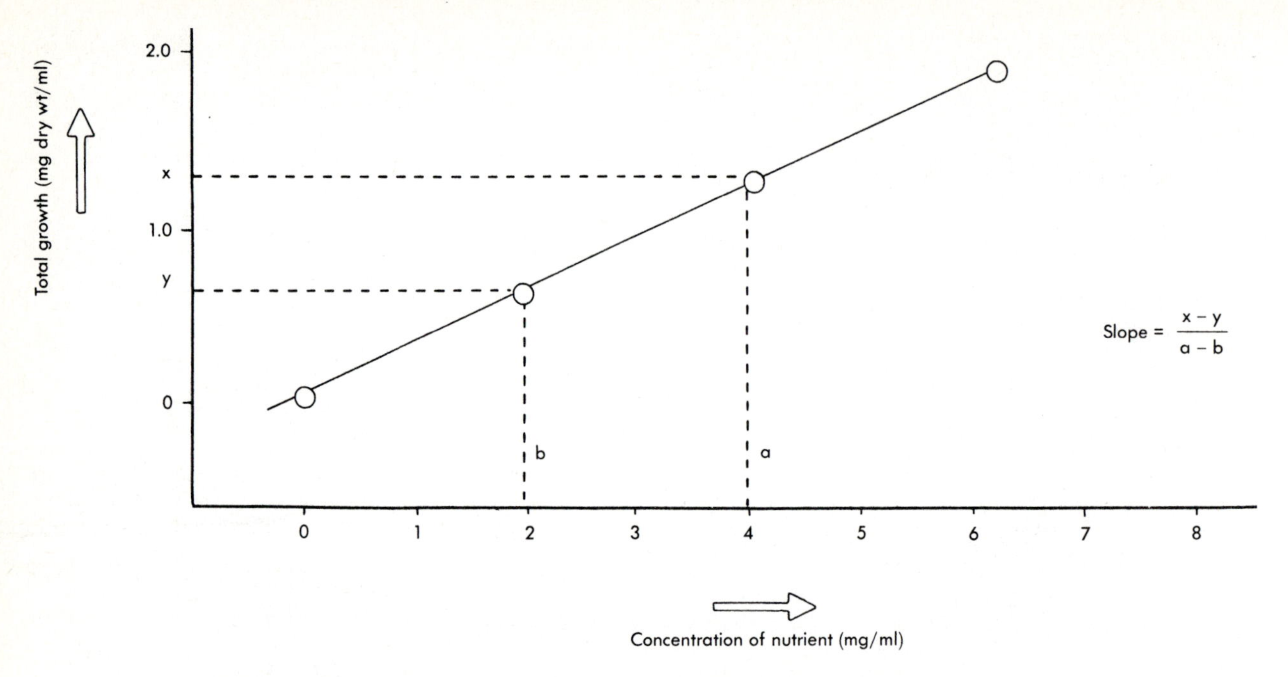

Figure 14-18 *Relationship between total growth of organism and initial concentration of limiting nutrient. Organism is grown in synthetic medium with nutrient serving as source of carbon and energy. Growth yield* Y *is obtained from slope of curve.*

ing nutrient. This type of determination is called a **bioassay** (see Point of Interest, p. 401). For example, the concentration of nutrients such as vitamins or amino acids can be determined by using a microorganism that requires the nutrient in question. Once the amount of growth in the culture has been determined, the concentration of the nutrient can be determined by plugging the value into the growth yield equation. In addition to bioassays, growth yield values are extremely important to the industrial microbiologist, who is continually in search of nutrient resources that will be most efficiently and economically metabolized during product development.

The efficiency with which microorganisms convert the carbon of a limiting nutrient, such as glucose, into cell substance varies between different microbial species. These differences for the most part reflect the relative ability of the organisms to produce ATP. After all, we know that different catabolic pathways exist for the production of ATP and not all of them are equally efficient. This disparity in ATP formation was originally discovered in the early 1950s by Gunsalus and others. They cultivated two different bacterial species under identical environmental conditions in a complex medium to which glucose had been added. The complex medium provided the bulk of the cell carbon while the glucose was used almost exclusively as a source of energy. The glucose carbons were converted to lactic acid. The yield (mg dry weight cells per mg glucose fermented) for one of the organisms, *Leuconostoc mesenteroides,* was substantially lower than that of the second organism, *Streptococcus faecalis.* It was discovered that *S. faecalis* fermented glucose by the Embden-Meyerhofff-Parnas (EMP) pathway, which yielded 2 moles of ATP per mole of glucose fermented. *L. mesenteroides* fermented glucose utilizing a pathway in which only 1 mole of ATP was produced per mole of glucose fermented. Further growth studies using a variety of nutrients as energy sources revealed that within certain limits a linear relationship existed between the dry weight of organisms produced and the concentration of the energy source. This relationship was expressed as the $\mathbf{Y_{ATP}}$. It was later postulated that an average of 10.5 gm dry weight of cells resulted from anaerobic growth of fermentative bacteria under ideal conditions at the expense of each mole of ATP generated, that is, the $Y_{ATP} = 10.5$. Calculations of Y_{ATP} are complicated because not all ATP is used for growth; some energy is used for maintenance, nutrient transport, motility, and polymer storage. Actual Y_{ATP} measurements are restricted to studies on cultures that produce ATP via fermentation because the ATP yield per mole of substrate is known only for such conditions.

Maintenance energy

Growth yield determinations, which were initially developed using batch cultures, were eventually calculated using chemostat culture techniques. The steady state

A

B

C

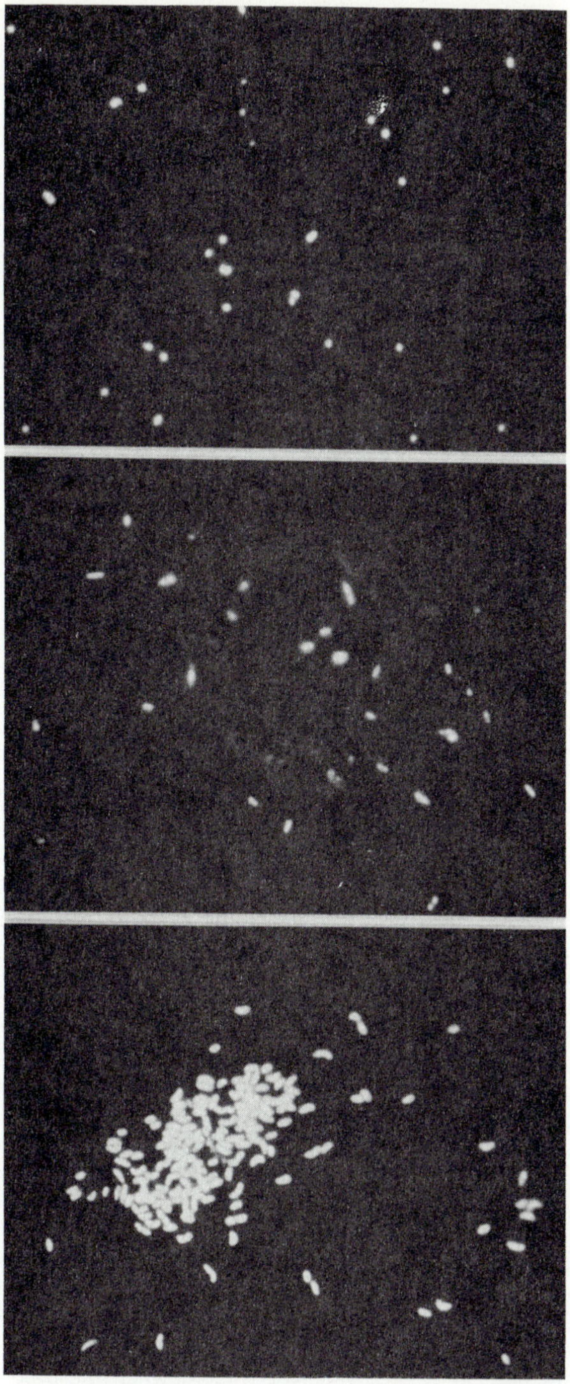

Figure 14-19 *Effect of nutrition on size of rod-shaped bacteria. Nutrient-starved* Vibrio cholerae *cells were stained with fluorescent dye and observed by epifluorescence microscopy.* ***A,*** *Cells before addition of nutrient.* ***B,*** *Cells two hours after nutrient addition.* ***C,*** *Cells 24 hours after nutrient addition. Note that cells have doubled in size as compared to* ***A.***

conditions of the chemostat permitted investigators to grow organisms at a variety of predetermined submaximal rates by setting different dilution (D) rates in an environment that did not vary with time. These studies showed that, when the limiting nutrient is an energy source, growth will cease at low dilution rates. What this means is that a certain amount of energy, beyond that needed for growth, is required to maintain the cell. This maintenance energy is required to sustain solute gradients, osmotic regulation, and motility.

REGULATION OF BACTERIAL GROWTH

Growth can be considered a manifestation of an organism's ability to produce macromolecules such as proteins. Some proteins, for example, are catalytic and thus control many thousands of chemical reactions that occur in the cell. The protein-forming system, specifically the ribosome, is also central to the control of growth. One of the techniques used by microbiologists to study the regulation of bacterial growth is to take a culture (for example, in a chemostat) and speed up or slow down the rate of growth. The speeding up of the growth rate, called a **shift up** experiment, can be carried out by adding richer nutrients to the growth medium. Conversely, the growth rate can be slowed down, called a **shift down** experiment, by transferring a washed culture that had been previously cultivated in a rich medium and transferring it to a nutritionally poorer medium. Measurements of DNA, RNA, protein, cell size, and cell mass can be made during shift experiments to determine their relationship to the growth rate. Two important conclusions drawn from these studies are as follows:

1. Cells growing at faster growth rates are larger than cells grown at slower growth rates. The average cell size just before division during slow growth may be one fourth the average size of a cell just before division during fast growth (Figure 14-19).
2. Cells growing at the faster growth rates are very rich in RNA as expressed by their RNA:protein or RNA:DNA ratios. Nearly all of this RNA increase results from ribosomal RNA (Figure 14-20). Under optimal conditions, over 40% of the total dry weight of the cell will be ribosomes.

Ribosome content in slow-growing and fast-growing cells

The data from some of the earlier studies on bacterial growth made it appear that ribosome concentration was the single factor determining growth rate. Those studies showed that the cell appears to produce just enough ribosomes so that the specific growth rate can be maintained. Unfortunately, later studies with cells growing at

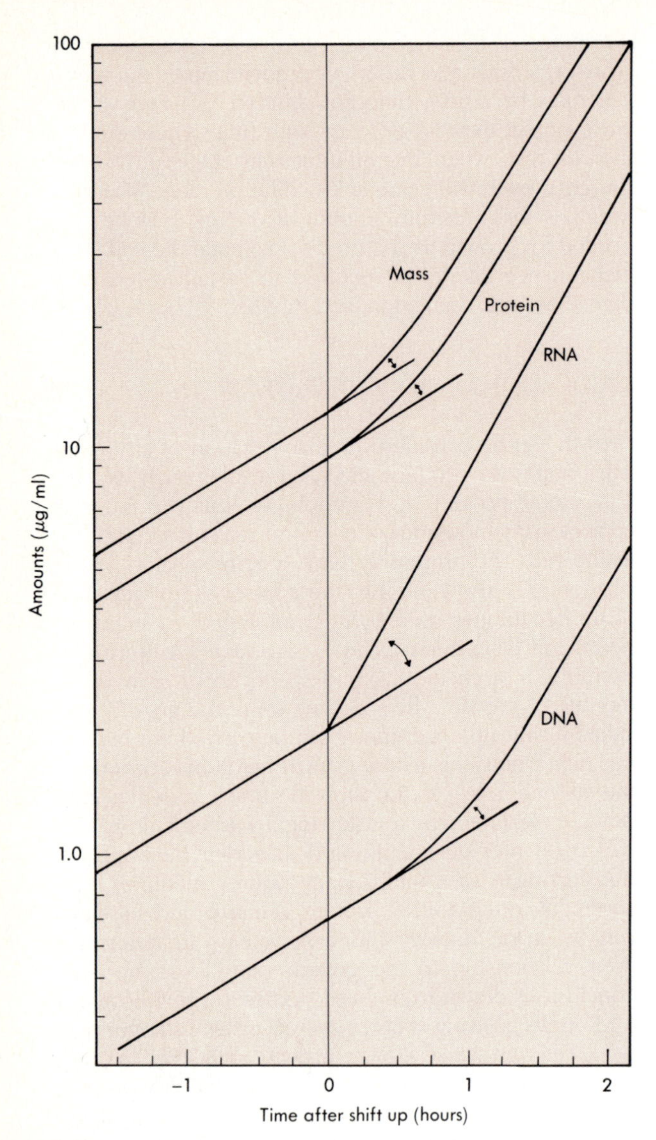

Figure 14-20 *Relative increases in RNA, protein, DNA, and mass of a bacterial species following "shift up." Organisms first grown in minimal medium in which doubling time is long, for example, 90 minutes. Rates of macromolecular synthesis before shift up are indicated. "Preshift up" rates are extended beyond point of shift to demonstrate relative rate of increase of macromolecular components in rich medium. At time of "shift up" culture was washed to remove minimal medium and cells were added to rich medium, which shortened doubling time, for example, to 30 minutes. RNA is seen to initially increase at rate higher than other components in cell.*

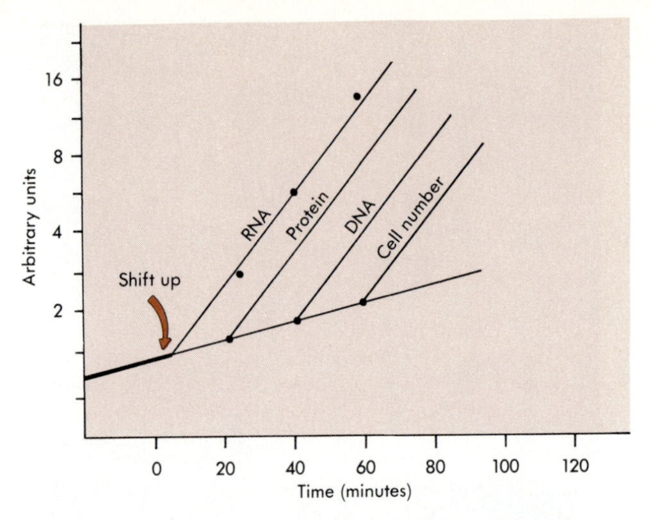

Figure 14-21 *Representation of shift up. Culture growing in minimal medium (doubling time is 60 minutes) is shifted to rich medium (doubling time is 20 minutes). Straight line before shift up indicates relative increment of various cellular components. After shift up, times at which molecular components have changed to faster growth rate are indicated. RNA synthesis accelerates to new growth rate almost immediately followed by protein and DNA synthesis. Cell division continues at original growth rate until cells have acquired larger size characteristic of faster growth rate.*

extremely slow growth rates did not support this conclusion. Slowly growing cells contain an excess of ribosomes, over what would be expected, some of which are nonfunctioning. If a culture of slowly growing cells is subjected to a shift up, there is an immediate increase in the rate of RNA and protein synthesis (Figure 14-21). This occurs because the previously nonfunctioning ribosomes are activated. During the shift up the nonfunctioning ribosomes become immediately engaged in protein synthesis. (NOTE: the protein chain elongation rate of each functioning ribosome at a particular temperature is the same regardless of the cell's growth rate). Thus slowly growing cells have reserve informational capacity (messenger RNA) that enables them to take advantage of conditions that favor an increase in the rate of growth. In rapidly growing cells all the ribosomes are utilized at peak efficiency. Thus, when these cells are shifted to an even higher growth rate, ribosomes are synthesized to accommodate the increase. In rapidly growing cells the concentration of ribosomes increases approximately in proportion to the growth rate.

In a rich medium a bacterial cell, such as *E. coli* will not have to synthesize some of the enzymes required for many biosynthetic processes. A rich medium con-

Point of Interest

MICROBIOLOGICAL ASSAY

Microorganisms are useful for the measurement or assay of small amounts of vitamins, amino acids, and other growth factors. One needs to have a culture that has a requirement for the chemical to be assayed. Lactobacillus is used in many microbiological assays because it has multiple growth requirements. Culture medium is prepared that contains all the growth requirements of the bacterium except for the chemical that is going to be assayed. The growth factor to be measured is added at different concentrations to different preparations of the culture medium. The relationship between the amount of growth factor added and the amount of growth obtained in the culture is determined (figure at right). At low concentrations, there normally is a linear relationship between the amount of growth and the concentration of growth factor. After such a relationship has been determined, the concentration of growth factor in a sample can be measured by determining the amount of growth that results and referring to the standard graph to determine the concentration of the growth factor.

Microbiological assay originally determined amino acid concentrations. Chemical methods do not distinguish between the D and L optical isomers of an amino acid. Most vitamins are assayed by chemical means today, but vitamin B12 and biotin are still assayed microbiologically. Also, microbiological assay is still used in foods and pharmaceuticals when other substances could interfere with chemical tests.

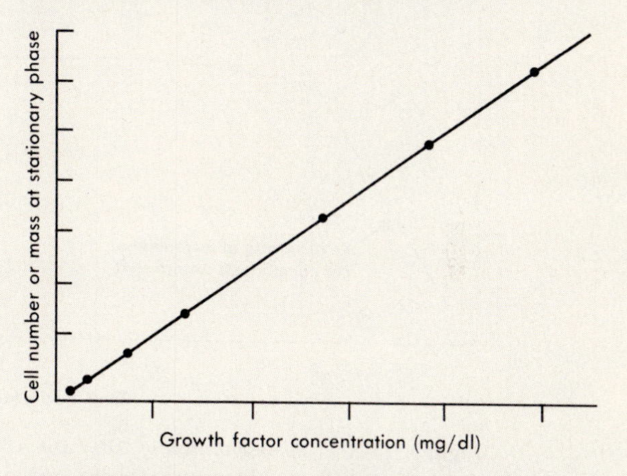

Microbiological assay of growth factors. Maximum growth of bacterial culture is directly proportional to growth factor concentration in range tested.

tains preformed amino acids, purines, pyrimidines, and vitamins, and thus their synthesis in the cell is repressed. If we take a fast-growing culture and transfer it to a nutritionally poor environment (for example, a medium containing only a single carbon and energy source such as succinate), there is a lag in the increase of cell mass (Figure 14-22). The bacterial cell must adjust to the new medium, and during this period there is no net synthesis of RNA and protein. In other words, synthesis of any new enzyme and RNA during this transition period results from the amino acids and RNA supplied by the degradation of proteins and ribosomes made at the fast growth rate. DNA synthesis and cell division still occur during the transition period, but the cells become smaller and their content of RNA decreases. After a period of time the needed biosynthetic enzymes are synthesized, and net protein and RNA synthesis will resume at the new and slower growth rate.

LABORATORY METHODS FOR DETERMINATION OF GROWTH

There are many instances in research, industry, and medicine where the number of microorganisms in a sample must be determined. Determining the potability of water requires that specific bacterial counts be performed daily in water treatment facilities. Many food products are subjected to microbial counts to determine their suitability for human consumption. Bacterial counts are sometimes required by the physician to evaluate the necessity for antimicrobial therapy. Whatever the measurement of growth that is used, it is important to distinguish between total counts and viable counts. Total-count measurements include dead as well as viable cells, whereas viable counts include only those cells that are alive and capable of growth. Both physical and chemical methods are used to determine bacterial growth and to distinguish between total and viable counts.

Concept Check

1. In morphological and physiological ways, how does a microorganism respond to nutrient limitation?
2. What is the mathematical expression for growth yield? What is the value of knowing the growth yield?
3. What is the meaning of shift up and shift down experiments? What do they reveal about microbial growth?

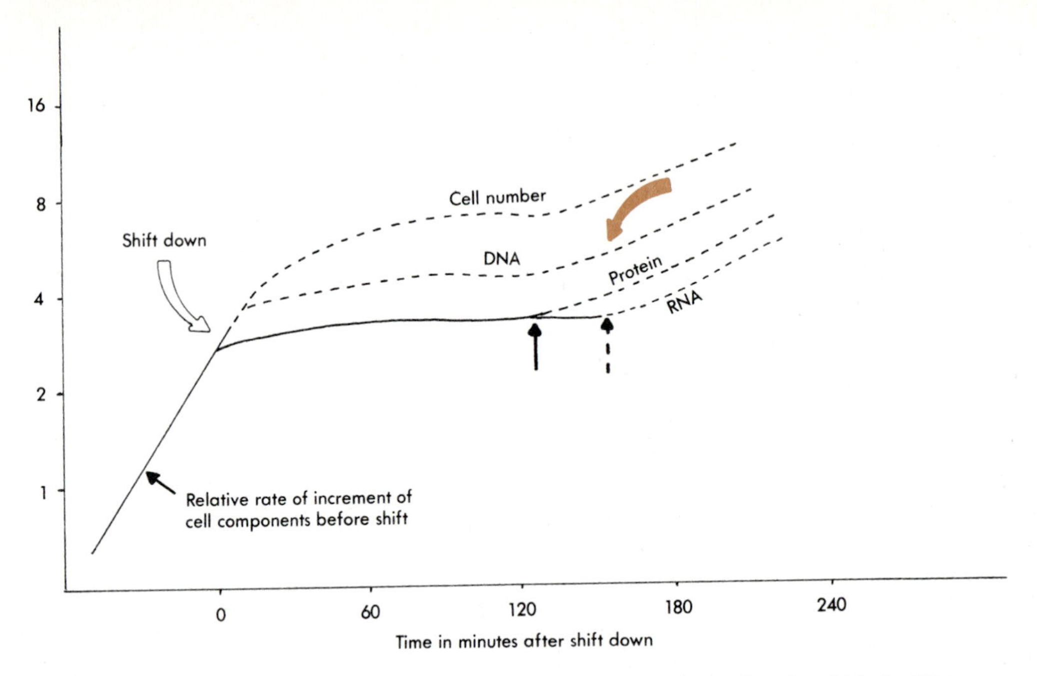

Figure 14-22 *Representation of "shift down." Culture growing in medium in which doubling time of organisms is 30 minutes is transferred to medium in which doubling time of organisms will be increased to 60 minutes or more. Straight line before shift down represents relative increments of various cellular components before shift. After "shift down" net RNA and protein synthesis are resumed only after more than two hours* (black arrows) *lag is due to turnover of RNA that accumulated at faster growth rates). Cell division and DNA synthesis continue for some time resulting in smaller cells but new rate is not reached until almost three hours after shift* (colored arrow).

Cell count

Microscopy. Microorganisms can be counted directly by using a specially etched microscopic slide. A grid is engraved on a microscopic slide containing a pattern of squares (Figure 14-23). Each small square on the grid is 50 μm on each side. On either side of the grid are raised edges so that when a drop of fluid is placed on the grid and covered with a coverslip, the underside of the coverslip is exactly 20 μm above the grid; thus a known volume of fluid is enclosed on the grid. The bacteria are counted in the squares when observed microscopically, and this figure is multiplied by a volume figure (derived from the size of the grid squares and the height above them). The cell count will have to be multiplied by a dilution factor if the sample was originally diluted. The engraved grid used for this technique is called the **Petroff-Hausser counter**. One cannot, however, differentiate living from nonviable cells using this technique. To use cell count microscopy, there must be at least 10^6 cells per milliliter in the sample.

Colony count. A colony count is the most effective way to distinguish between viable and nonviable cells in a bacterial suspension. This is an indirect technique in which a sample of bacteria is diluted to an appropriate number of cells that can be spread on the surface of agar in a Petri dish. Each cell plated on the agar surface, following suitable incubation, will give rise to discrete aggregates of cells called colonies. The number of colonies that can be counted and that is of statistical significance is between 30 and 300 on a standard glass Petri dish. A bacterial suspension that is faintly turbid to the eye will contain approximately 1×10^7 bacteria per milliliter and will require dilution to obtain a measurable number of colonies (Figure 14-24). Once diluted, an aliquot of the suspension is spread on the surface of an agar plate and often incubated for 18 to 24 hours between 35° and 37° C. (Some bacterial species require longer times and different temperatures of incubation.) Most colonies will be visible after this period of time. An alternative method, called the **pour tube** technique, can also be used. A 0.1 to 1.0 ml sample of the diluted bacterial suspension is placed in molten (42° C) agar, and then the mixture is poured into the bottom of a Petri dish. Once the agar has hardened and after suitable incubation, colonies will be produced not only on the surface but also within the agar unless a strict obligate

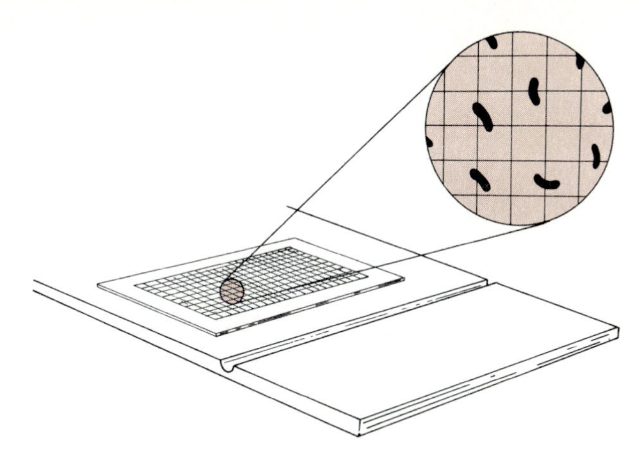

Figure 14-23 *Representation of etched grid found on slide chamber of Petroff-Hauser counting chamber. Dimensions of each small square are 0.05 mm × 0.05 mm × 20 μm, or volume of 50,000 cubic μm. Bacteria are counted microscopically in large square, and this figure is multiplied by volume figure times original dilution factor if original sample was diluted.*

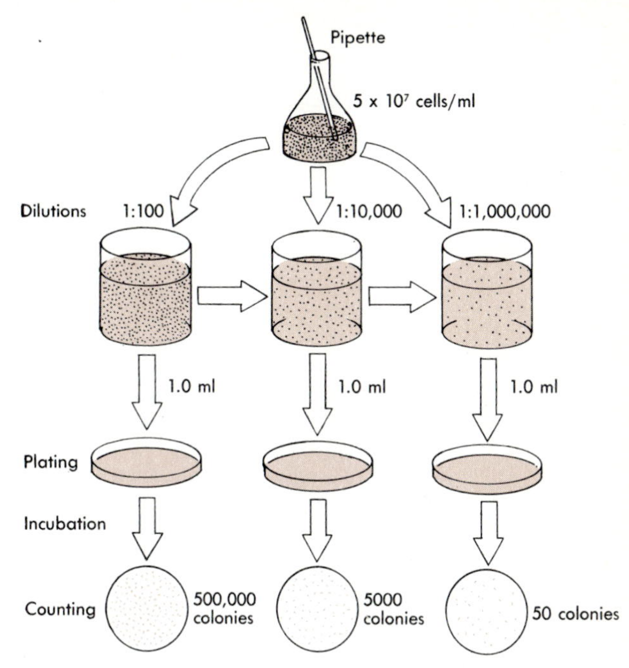

Figure 14-24 *Dilution technique for determining number of viable microorganisms (bacteria) in culture. Dilutions, 100 fold or more when necessary, are used to reduce cell number in original culture. For simplicity sake, we have indicated only final dilution. Under ordinary laboratory conditions all of vessels and dishes would be covered to prevent contamination from air.*

aerobe is plated. Obligate aerobes are found only on the surface of the agar.

Electrical resistance. Microbial cells are poor conductors of electricity and offer resistance to current flow. The **Coulter counter** is a device in which a suspension of cells is made to pass through an opening on either side of which are electrodes. A microbial cell appearing between the electrodes causes the current to be impeded, and there is a drop in voltage. This voltage drop is recorded as a pulse, the height of which is proportional to the volume of the cell. Cell number and cell size can therefore be determined using this technique. No distinctions, however, can be made between viable and nonviable cells, and contaminating dust particles or fibers in the bacterial suspension can be a source of error.

Cell density

Photometric devices are the most widely used methods for determination of microbial growth. **Photometry** is a technique that employs a beam of light to estimate bacterial populations. A beam of light focused on the bacterial suspension (Figure 14-25) is scattered once it strikes the cell wall. The undeviated light can be measured by a technique called **turbidometry**, and scattered light is measured by a technique called **nephelometry**. Turbidometry is used almost exclusively over nephelometry for growth measurements.

The instruments used to measure light absorbed by a microbial population are called **photometers** or **spectrophotometers**. The theory behind the use of these instruments is related to the Beer-Lambert law for solutes, which states that there is a straight-line relationship between absorbance (also called optical density or OD) of a solution and the concentration of the solute).* A bacterial suspension can be considered the solute (but in reality is a suspension of particles), and therefore a linear relationship exists beweeen its optical density and cell number per milliliter. Measurements are carried out by first placing the bacterial suspension in a tube of uniform-bore diameter. The tube is inserted into the spectrophotometer, in which a beam of light strikes the suspension, and the light passing through the suspension is measured. The intensity of the light passing through the suspension will be governed by the concentration of bacteria in the suspension, the intensity of light, the wavelength of light, and the diameter of the tube containing the bacterial suspension. A photodetector mea-

* The optical density is a function of the original intensity of the light beam (I_o) and the intensity of the beam (I) passing through the solution, or:

$$\text{optical density} = \log\left(\frac{I_o}{I}\right)$$

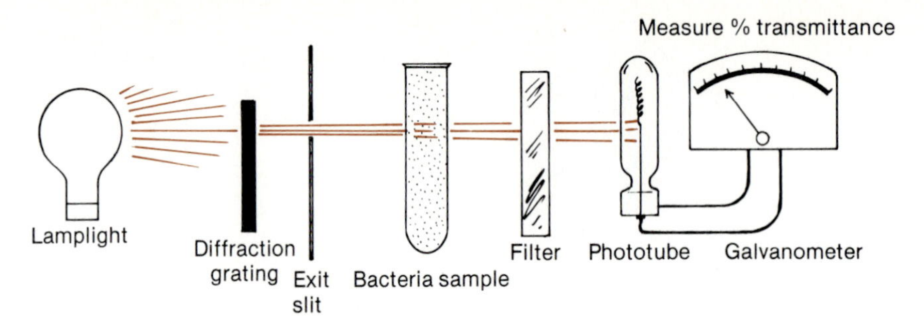

Figure 14-25 *Spectrophotometer used to determine turbidity of bacterial suspension. White light generated by tungsten filament strikes diffracting grating and produces spectrum of wavelengths. Only wavelengths that fall on exit slit go through to strike sample (wavelengths are adjustable). Light that passes through sample strikes phototube, which activates galvanometer. The more light transmitted, the less concentrated the sample, whereas the less light transmitted, the greater the turbidity of concentration or cells in suspension.*

sures the intensity of light passing through the bacterial suspension and translates it on a meter or chart as percent transmittance (%T) and/or absorbance (A).

The values of absorbance and/or percent transmittance can be correlated with the cell number by using additional procedures, provided that one uses the same species of bacteria for experiments over a long period of time. This can be accomplished by doing a growth curve for the bacterial species in question. The samples taken during growth should be subjected to cell counts as well as optical density measurements (Figure 14-26). This type of correlation will be of considerable value in future experiments because any optical density measurement can be extrapolated on the growth curve. Cell size influences the results obtained by turbidometry; therefore the same organism must be used in the same or similar growth medium, and samples should be taken during the exponential phase of growth for the data to be of value.

Cell mass

As pointed out previously, one can determine the rate of growth during the exponential phase by measuring other parameters such as RNA, DNA, and protein content, as well as cell number. One can also measure cell mass by determining the dry or wet weight of a population of bacterial cells. To determine the dry weight, the cell suspension (the sample should contain approximately 1×10^{11} cells) is centrifuged and washed several times with distilled water and recentrifuged. The sample is placed in a dish and dried in an oven at 100° C or in a vacuum at 80° C until a constant weight is achieved. Wet-weight determinations are made by collecting the bacterial cells on a filter that has been previously dried to a constant weight. Extracellular water is removed when the membrane and bacteria are dried in a vacuum at 40° C.

The most consistent relationship between turbidity and cell density is obtained when correlations are made between optical density and cell number, since cell size influences optical density. Determining cell mass is a time-consuming process and is used only under special research conditions, such as those in which the culture forms diplococci, forms filaments, or clumps excessively.

Other methods

Chemical methods. Several chemical components of the cell can be analyzed by colorimetric means. One can measure RNA, DNA, or protein by treating a washed suspension of cells with agents that release these molecules from the cell and hydrolyze them. The addition of a specific reagent produces a color complex with one of the macromolecular components whose intensity can be measured with a spectrophotometer. Although seldom used as a technique for determining growth, colorimetric methods have certain advantages in that fewer cells are needed and the test is not as time consuming as some other techniques.

Estimations of the efficiency of cellular growth can also be made by measuring the total ATP produced. The ability of an organism to convert the carbons of a fermentable carbohydrate, such as glucose, to cellular carbon is a function of the organism's efficiency at producing ATP. The total ATP obtained from the fermentation of carbohydrates is known for many microorganisms and is relatively constant when total cell mass is measured. When the ATP yield is compared with the amount of carbohydrate fermented, this equivalency may disappear. This difference is related to the different pathways that some microorganisms use to catabolize a carbon source (see Chapter 12). Unknown catabolic pathways can be evaluated from known ATP yields.

Metabolic uptake. When environmental conditions

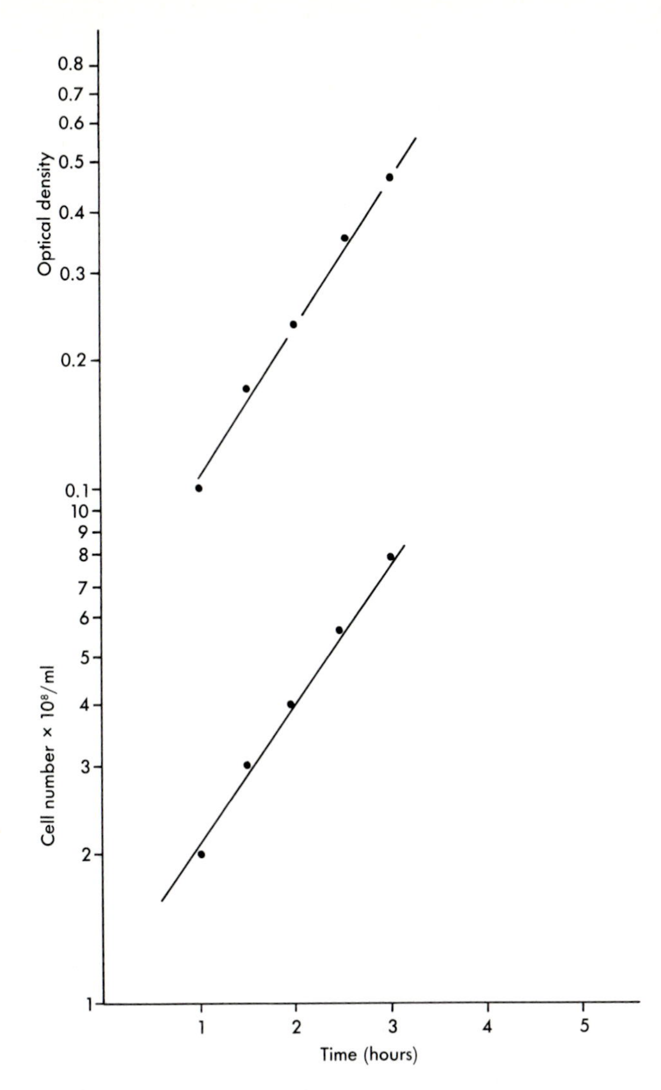

Figure 14-26 *Camparison of optical density and cell counts during exponential growth. Note that slope of both curves is same indicating rate of increment increase for both cell-growth functions is same.*

are standardized, growth can be related to the oxygen consumed by an organism or to the excretion or uptake of certain metabolites, such as fermentable sugars. One can also measure the amount of acid produced from the fermentation of sugars, and this can be related to the growth yield, defined as the milligrams of dry weight per mole of acid produced.

GROWTH IN EUKARYOTES

Eukaryotic cell cycle

One of the major differences between eukaryotic and prokaryotic cell cycles is the pattern of DNA synthesis. DNA synthesis in eukaryotes occurs only periodically and does not occupy the entire cell cycle, as it does in prokaryotes. In addition, the mitochondria and chloroplasts of eukaryotes also contain DNA with functional genes. There are more morphological events associated with eukaryotic growth, and this tends to make the eukaryotic cell cycle more complex than the cycle of events in prokaryotes.

Stages in the eukaryotic cell cycle. The eukaryotic cell cycle can be divided into four stages: G_1, S, G_2, and **M**. The duration of each cycle, which varies from species to species, may take from 2 hours to several days and is controlled by the growth conditions, particularly nutrient supply. Table 14-1 reviews the stages in the cell cycle and describes some of the cellular events that occur in them.

Cell cycle in a single-celled eukaryote. Our understanding of the eukaryotic cell cycle events, particularly their control, is very meager. Most of what we know about cell cycle events in eukaryotes has been revealed from studies with single-celled microorganisms, such as *Saccharomyces cerevisiae,* a yeast that divides by budding and *Schizosaccharomyces pombe,* which divides by fission.

The cell cycle in *S. cerevisiae* is best explained by starting with an unbudded cell in the G_1 period (Figure 14-27) and following the events described below:

1. The G_1 period prepares the cell for DNA synthesis. Most of the GI period is thought to be involved in accumulation of cellular mass. Several biochemical events occur during this period, and a point is reached where the cell is irreversibly committed to initiate DNA synthesis and complete cell division. The period is terminated by the initiation of DNA synthesis, the duplication of the **spindle plaque** (centrioles), and the emergence of the bud. The spindle plaque is comparable to the centrioles of higher eukaryotes and is concerned with microtubule and spindle formation. The DNA division cycle is initiated only when the cell has reached a critical size. The faster the growth rate, the shorter the time required to attain this critical size.
2. The S period occupies about 25% of the cell cycle and is characterized by the duplication of the chromosomes. The spindle plaques produced in G_1 separate to produce a complete spindle, and the bud that was formed in G_1 continues to enlarge.
3. The G_2 period is characterized by migration of the nucleus to the neck (area separating mother cell and bud). The nucleus undergoes its first stage of nuclear division and spindle elongation such that one nucleus is in the mother cell and the other is located in the bud.
4. The M period is represented by further spindle

Table 14-1 *Characteristics of the stages in the eukaryotic cell cycle*

Stage	Percentage of cycle*	Characteristics
G_1	Up to 50%	Most RNA and protein synthesis occur in this period; dry mass increase is greatest in G_1
S	From 20% to 25%	DNA duplication; DNA content doubles
G_2	Up to 25%	Very little synthetic activity except for precursors for mitotic spindle and precursors to components involved in cytoplasmic division (cytokinesis)
M	Approximately 5%	Period of mitosis; practically no synthesis; spindle is visible; period terminates with nuclear division and separation of daughter cell from parent

*Applies only to the simplest of eukaryotic microorganisms, such as the single-celled fungi.

elongation as well as the completion of the remaining phases of nuclear division. At the end of the M period there is cell membrane separation (cytokinesis) and cell wall separation. Finally, the daughter cells separate, and two unbudded cells are produced. The division of a budding yeast cell is asymmetric, yielding two cells, one of which is the older parent and one of which is the new daughter cell. Once the bud is approximately the same size as the mother cell, cell separation occurs.

Cell cycle control in *Saccharomyces cerevisiae*

One of the reasons for the increasing interest in cell cycle control in yeasts is their relevance to the control of cell proliferation, especially in cancerous growth. In addition, yeasts play an important role in industry and any knowledge concerning their growth may lead to their use at maximum efficiency. The use of temperature-sensitive mutants, called **cell division cycle (cdc) mutants**, has contributed extensively to our understanding of the cell cycle in *S. cerevisiae.* Such mutants divide normally at the permissive temperature (23° C), but at restrictive temperatures (36° C) all cells in the culture will arrest at a particular stage of the cell cycle. Many of these mutants have been shown to be defective in a single nuclear gene. They have been used to characterize the relationship between various cell cycle events and the influence of each gene product on cell division. Studies with such mutants have revealed that the individual events that make up the cell division process are organized into a group of interlocking, dependent pathways (Figure 14-28). The events converge before the G_1 period and then diverge at the end of the G_1 period. Control of cell proliferation in *S. cerevisiae* appears to be affected by the regulation of a single event in the G_1 period, which has been defined by the cdc 28 mutation. This point has been termed **start.** For cells to proceed beyond "start," certain conditions, such as the presence of adequate nutrients and the attainment of critical size,

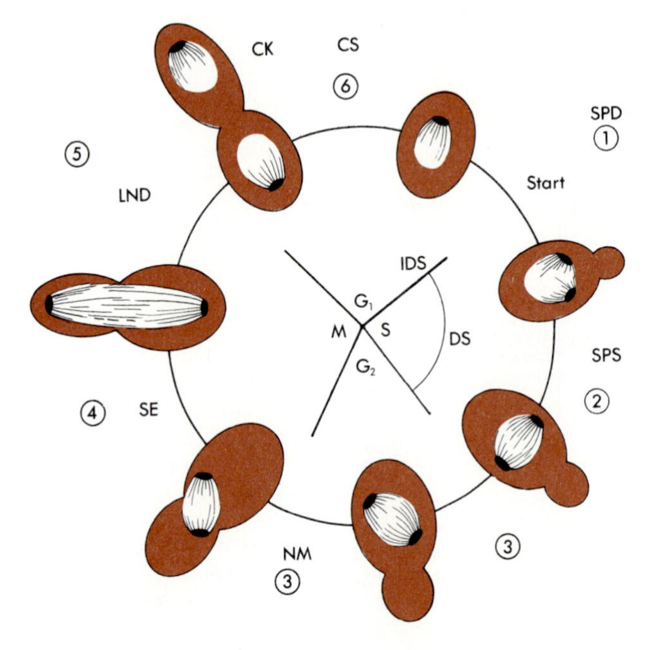

Figure 14-27 *Cell division cycle in the yeast* Saccharomyces cerevisiae. 1, *Mitotic spindle duplicates and bud emerges. This is beginning of "start" period.* 2, *Spindle separates and bud enlarges.* 3, *Nuclear migration.* 4, *Spindle elongates and nuclear division initiated.* 5, *Nuclear division complete; bud approaches size of parent.* 6, *Cytokinesis and cell separation.* SPD, *Spindle pole duplication;* SPS, *spindle pole separation;* IDS, *initiation of DNA synthesis;* DS, *DNA synthesis;* NM, *nuclear migration;* SE, *spindle elongation;* LND, *late nuclear division;* CK, *cytokinesis;* CS, *cell separation.*

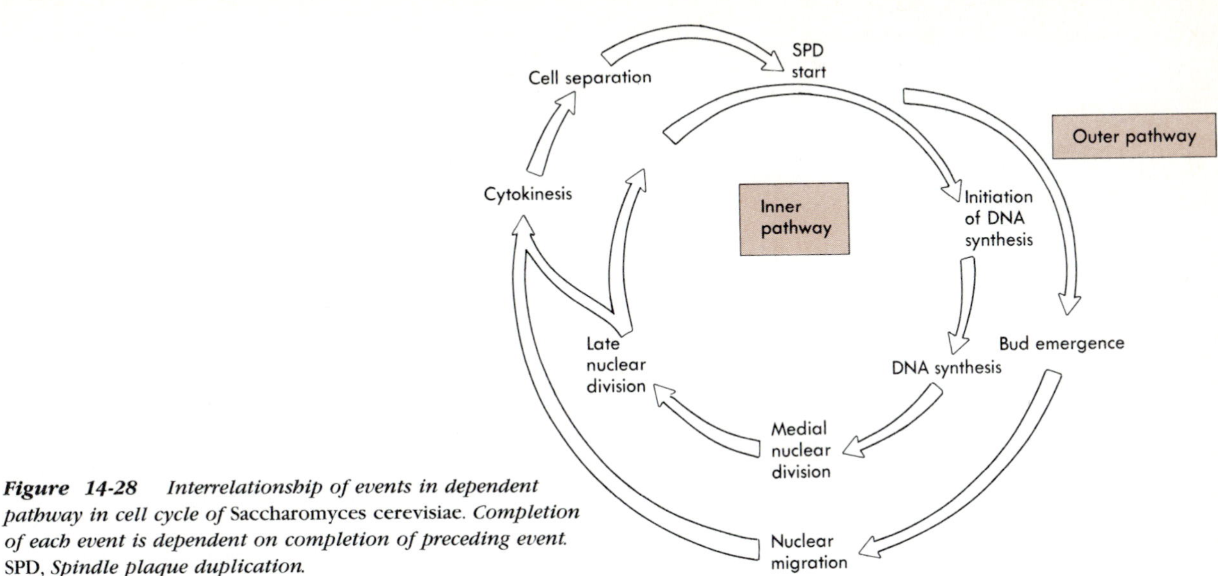

Figure 14-28 *Interrelationship of events in dependent pathway in cell cycle of* Saccharomyces cerevisiae. *Completion of each event is dependent on completion of preceding event.* SPD, *Spindle plaque duplication.*

must be fulfilled. We will now discuss some of the experimental results that suggest the existence of this type of control.

Pheromones and the cell cycle

Unbudded cells of *S. cerevisiae* under appropriate conditions may conjugate to form diploid cells. The conjugates are of opposite mating types referred to as *a* and α. When haploid cells of opposite mating type are mixed together, the proportion of budded cells decreases rapidly and that of unbudded cells increases. This characteristic is brought about by the action of hormonelike molecules called **pheromones**—*a* cells produce *a* factor pheromone and α cells produce α factor pheromone. The *a* pheromone arrests α cells in the G_1 period. Similarly, the α factor pheromone arrests *a* cells in the G_1 period. Examination of the arrested cells showed that in both instances the spindle pole is unduplicated. In addition, the start mutation acted at nearly the same point in the G_1 period as the pheromones. Further genetic studies indicated that pheromones apparently require the product of the "start" gene and without it the "start" event cannot be completed. Thus spindle duplication during sexual conjugation cannot be initiated until pheromone and a "start" product are present. Recent studies on the "start" product indicate that it is a polypeptide that has protein kinase activity. The role of protein kinases in cancer caused by viruses is discussed in Chapter 17.

Nutrition, cell size, and the cell cycle

Yeast cultures have been examined at a number of different growth rates. Studies have demonstrated that not all phases of the growth cycle expand proportionately with an increase in the generation time. The major expansion occurred during the G_1 period, that is, the unbudded period, and preceded the pheromone–cdc 28 step. When cultures are deprived of several nutrients, cells arrest as unbudded cells. These cells have unduplicated spindle plaques, which is characteristic of cells arrested before the pheromone–cdc 28 step. Thus it appears that the cdc 28 event may actually initiate the "start" signal.

Cell size varies with the growth rate. At faster growth rates (generation times of 2 to 3 hours) the size of the cell at bud initiation varies with the growth rate, but at slower growth rates (generation times greater than 3 hours) the cell size at bud initiation reaches a lower limit and becomes constant and independent of the growth rate. Cells must attain a certain size before a specific cell cycle event can be completed. When a culture is starved (nitrogen limitation, for example) the cells become very small. If these cells are resuspended in a medium that supports a higher growth rate the cells are, for a brief period of time, arrested in the G_1 period. This arrest reflects the additional growth that is required for cells to reach the critical size characteristic of the new medium.

The results of these and other experiments with cdc mutants have shown that the inner pathway associated with nuclear division (see Figure 14-28) and the outer pathway associated with budding are presumably independent of each other but are connected and regulated by the "start" event. Budding and nuclear migration occur in the absence of DNA synthesis and nuclear division, and vice versa. No cell cycle event, however, can occur unless the preceding event in the pathway has been completed.

Septum formation and bud separation

Septum formation is the last event in the yeast cycle and is dependent on DNA synthesis and nuclear division. These associated events ensure that the bud will possess a nucleus. Septum formation and bud separation were discussed in Chapter 6.

FACTORS AFFECTING GROWTH

In addition to their chemical needs, microorganisms are affected by the physical nature of their environment. Environmental factors, such as temperature, pH, osmotic pressure, and radiation, influence what type of microorganism can survive in a particular ecological niche. Habitats that are exposed to extremes of temperature, pH, etc., are colonized by microorganisms that exhibit unusual structural or chemical characteristics.

Temperature

Growth is the culmination of a series of enzymatic reactions, and because of this, temperature plays a very important role in the cell cycle. The optimum* temperature for enzyme activity in microorganisms is dependent on the ecological site of the species under consideration and the enzyme's location (extracellular or intracellular). Certain temperatures are lethal to some microorganisms, but other microorganisms not only survive at these temperatures but are actually capable of growth. Microorganisms can grow over a temperature range of 30° C or more, but the **optimum growth temperature** is within a very narrow range of temperatures (Figure 14-29). Above the optimum range enzymes become denatured, and the rate of growth drops precipitously, producing what is called the **maximum growth temperature**. Below the optimum temperature range the rate of growth decreases slowly until a point is reached, called the **minimum growth temperature**, where growth ceases. Microorganisms have been divided into three groups based on their optimum growth temperature: **psychrophiles** (low temperatures), **mesophiles** (moderate temperatures), and **thermophiles** (high temperatures). How each group has evolved to grow within particular temperature ranges and how each responds to temperatures outside its range is the subject of the following discussion.

Psychrophiles. Generally speaking, psychrophiles exhibit an optimum range of growth between 0° and 20° C. This is not to say that psychrophiles cannot grow above these temperatures. Some can, albeit more slowly than at psychrophilic temperatures. Psychrophiles, as one might expect, are found in great numbers in the Antarctic, where temperatures fall below 0°, but they are also found in cold soil, streams, ocean water below the thermocline, rivers, and lake muds of temperate climates. Consequently, psychrophiles can play a role in biodegradation in cold-water lakes and in refrigerated foods.

One of the most interesting questions confronting scientists concerns the physiological bases of psychrophilic growth. Experiments have demonstrated that temperature may affect a wide variety of different microbial processes. The ribosomes in some psychrophiles become more unstable as the temperature increases above 20° C, and this results in a reduction in the rate of protein synthesis. The ribosomes of some psychrophiles possess a protein factor that enables protein synthesis to occur at 0° C. Removal of the factor causes protein synthesis at 0° C to cease, but at higher temperatures (25° C) protein synthesis is not affected. Many organisms appear to protect themselves from colder temperatures by increasing the amount of unsaturated fatty acids in membranes as the temperature decreases and approaches 0° C. Comparison of mesophilic and psychrophilic strains of the genus *Bacillus* has revealed that the unsaturated fatty acid content is considerably higher in the psychrophilic strains than in the mesophilic strains. Elevated temperatures, such as those in the mesophilic range, affect the cell membrane of psychrophiles and cause the leakage of intracellular contents, which may lead to lysis of the cell. The inability of psychrophiles to grow at elevated temperatures may also be due to heat inactivation of certain enzymes. There are many key enzymes in metabolism, and inactivation of any one of them could inhibit growth—for example, an enzyme in glycolysis or an enzyme required in protein synthesis. The increase in temperature above the psychrophilic range is not always lethal, and a return to psychrophilic temperatures restores the growth of the microorganism.

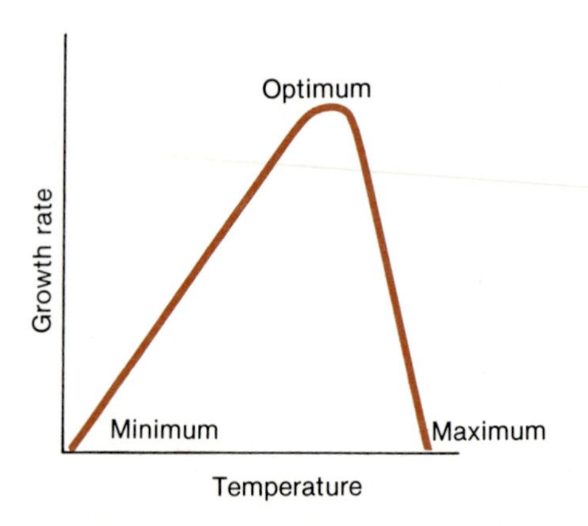

Figure 14-29 *Effect of temperature on growth rate.*

* *Optimum* refers to the temperature at which the growth is maximal or enzymes work the fastest.

Thermophiles. The temperature span for optimum growth of different thermophiles is 40° to 80° C, but few thermophiles can live at temperatures of 90° to 100 ° C (see Point of Interest). Thermophiles are found among many groups of microorganisms, such as the cyanobacteria, the photosynthetic bacteria, gram-positive and gram-negative bacilli, and the protozoa. Most species are found in hot springs in many areas of the world. Thermophiles have ribosomes, enzymes, enzyme systems, metabolic pathways, and regulatory devices that are functionally the same as those found in mesophiles. The ability of thermophiles (but not mesophiles) to survive at elevated temperatures is a characteristic that has not been fully explained. Experiments have demonstrated that the cytoplasmic proteins from thermophiles are resistant to temperatures that denature mesophilic proteins. Many theories have been proposed to explain this thermal resistance, but it may be that the molecular mechanisms vary from one group to another.

One mechanism that has received considerable support is that macromolecules are inherently stabilized. Stability could be the result of the substitution of polar amino acids for nonpolar ones, resulting in an increase in ionic and hydrogen bonding, particularly in the interior of the molecule. One example is the salt bridge resulting from the electrostatic interaction between ionizable groups such as NH_3^+ and COO^- of amino acids. This would alter the conformational properties of the macromolecule and make it resistant to denaturation by heat. A comparison of the lactic dehydrogenases of thermophilic and mesophilic species of bacteria has revealed that this type of change does take place. Stability could also be obtained by a decrease in size of macromolecules.

A second mechanism for stability may be the result of biochemical modification. The transfer RNA of an extreme thermophile *(Thermus thermophilus)* shows modifications not evident in moderately thermophilic microorganisms. The modification involves the replacement of ribothymidine by 5-methyl 2-thiouridine, which apparently stabilizes the molecule at high temperatures. The replacement reaction is activated when the organism is grown at higher temperatures, and the mole percent of the replacement component increases with the increase in temperature.

A third mechanism that may be used for stability in-

Point of Interest

HOT BACTERIA

We have been conditioned to believe that all microorganisms will die when exposed to a temperature of 120° C at a pressure of 15 pounds per square inch for 15 to 30 minutes. It now appears that we may have to revise our thinking. Baross and others have described communities of bacteria from volcanic vents on the ocean floor where temperatures of 360° C are reached. These researchers have isolated some of these bacteria and grown them in enriched seawater enclosed in a titanium syringe surrounded by a chamber heated to 250° C and pressurized at 265 atmospheres (the hydrostatic pressure found at the vents). At such high pressures water does not boil until it reaches 460° C. At a temperature of 250° C and 265 atmospheres some of the bacteria doubled every 40 minutes, a rate of growth comparable to most mesophilic bacteria at ambient temperatures. Most thermophilic organisms studied, before Baross's discovery, had optimum temperatures for growth between 70° and 85° C (an organism also isolated from a continental volcanic area was shown in 1982 to grow optimally at 105° C in the presence of sulfur, hydrogen, and carbon dioxide). Baross' discovery in 1983 now makes us wonder what are the upper temperature limits for optimal growth of bacteria. In addition, we wonder what forces hold the DNA helix together and stabilize proteins and lipids. The extreme thermophilic vent bacteria are strict anaerobes, appear very flat (see figure at right), and probably are members of *Archaebacteria.* They generate hydrogen, carbon dioxide, and large quantities of methane during growth. The fact that these organisms have highly thermostable enzymes makes them very attractive for industrial research purposes. Many industrial processes utilizing bacteria have high temperature requirements.

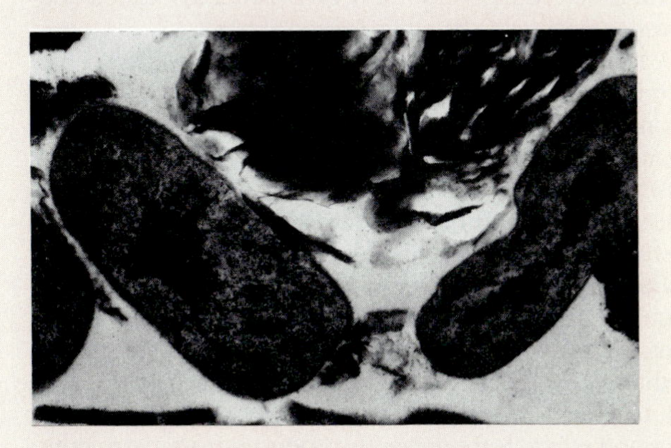

Transmission electron micrograph of ultrathin sections of mircoorganisms taken from sea floor where temperatures reach 350° C. Note the internal membranes (mesosomes?) and thick wall and lack of any definable internal structures.

volves the interaction of macromolecules with cofactors such as divalent cations (Ca^{++}, Mg^{++}, etc.) and polyamines. Thermophiles contain several types of polyamines, some of which are unique. They appear to stabilize the ribosome-tRNA-mRNA complex and ensure its activity. The complex is inactive at higher temperatures in mutants that are unable to synthesize polyamines. Most of the polyamines are related to spermidine, which is found abundantly in nature and has the structure of:

$$\begin{array}{ccccccccccccccccccccccc}
 & & H & & H & & H & & H & & H & & H & & H & & H & & H & & H & & \\
 & & | & & | & & | & & | & & | & & | & & | & & | & & | & & | & & \\
H & — & N^+ & — & C & — & C & — & C & — & N^+ & — & C & — & C & — & C & — & C & — & N^+ & — & H \\
 & & | & & | & & | & & | & & | & & | & & | & & | & & | & & | & & \\
 & & H & & H & & H & & H & & H & & H & & H & & H & & H & & H & &
\end{array}$$

The cell membrane of microorganisms also changes with variations in temperature, and these changes are believed to play a minor role in cellular stability. The melting point of the membrane fatty acids also increases with increases in temperature; thus the fatty acids synthesized at elevated temperatures become more branched, longer, and more saturated.

It has been proposed that most thermophilic groups of microorganisms evolved from mesophily to thermophily, since the prokaryotes are believed to have evolved in a sea that was cool to warm but not hot. The evolution to thermophily may have been one of selective advantage in order to escape predation and eliminate other competitors for nutrients. It has been assumed that this evolutionary process occurred in a stepwise manner through a series of single mutations, with each mutation resulting in an increase in heat stability for only a single protein. Others have suggested, however, that the potentially thermostable components of the cell are controlled by the products of a few genes. The genes could conceivably be plasmid associated and easily transferable to other related species.

Resistance to heat is a property associated with the endospores of such bacteria as the clostridia and is a constant problem in home canning because of the organism's ability to release potent toxins *(Clostridium botulinum* toxin). Some spore-forming clostridia are also thermophilic, and the vegetative cells can grow at high temperatures. Thermophilic clostridia can ferment a wide variety of organic substrates, including the high molecular weight cellulose. They break down cellulose into a number of products, including acetic acid, ethanol, hydrogen gas, carbon dioxide, lactic acid, and butyric acid. These products, particularly the organic acids and ethanol, are presently produced from petroleum and natural gas. The clostridia have been considered for production of these chemicals to conserve petroleum. Thermophilic clostridia that ferment cellulose can be grown in mixed cultures with methane-producing bacteria. The products of clostridial fermentation (acetic acid and hydrogen gas) are used by the methane-producing bacteria to generate methane gas. Methane gas is also a source of fuel for certain industrial operations:

$$\underset{\text{Acetic acid}}{CH_3COOH} + \underset{\text{Hydrogen gas}}{2H_2} \rightarrow \underset{\text{Methane}}{2CH_4}$$

Mesophiles. Mesophilic microorganisms make up the majority of the species known to humans. Some species are indigenous as well as pathogenic to humans and other warm-blooded animals. This explains why indigenous microflora in and on the surface of humans grows best at 37° C (body temperature). The optimum temperature for growth of mesophiles is between 25° and 40° C, but they have minimum and maximum temperatures of 10° and 50° C, respectively. The ability of mesophiles to survive at cold as well as hot temperatures appears to be related to the composition of the fatty acids in the cell membrane. When the membrane fatty acids of mesophiles are compared at psychrophilic and thermophilic temperatures, there is observed to be an increase in the melting point of the fatty acids as the temperature increases. There is also an increase in unsaturated fatty acids as the temperature decreases from 43° to 10° C in *E. coli.* Fatty acid changes in the membrane alter the viscosity, which in turn affects nutrient transport and ultimately growth.

Temperature and mutation. Mutations in the microbial genome can produce temperature-sensitive mutants unable to grow at temperatures that normally support growth, such as 37° C, but able to grow at lower temperatures, such as 25° C. (Temperature-sensitive mutants are discussed in Chapter 10.) Temperature-sensitive mutants have been important tools for the researcher wishing to study cellular processes. Mutations involving certain cellular activities can be assayed because the mutant can express itself at one temperature but not at another. Colonies of cells or liquid cultures can be detected at the permissive temperature and then switched to the nonpermissive temperature to assay for the defect. We have already discussed the use of temperature-sensitive mutants in characterizing the cell cycle of yeasts (see p. 407).

Temperature-sensitive mutants of a large number of viruses have also been studied in animal models to characterize the biochemical, replicative, and genetic events that are characteristic of the virus-infection process. Temperature-sensitive mutants of virulent viruses, such as the influenza virus, cytomegalovirus, and respiratory syncytial virus, are being evaluated for use as vaccines (see Chapter 10).

Temperature has been used as a technique to cure microorganisms of plasmid activity. A microorganism that is virulent because of plasmids that control the synthesis of toxins can be made avirulent by subculturing at higher temperatures. In some instances the avirulent

Figure 14-30 *Effect of high and low salt concentration on halophiles.*

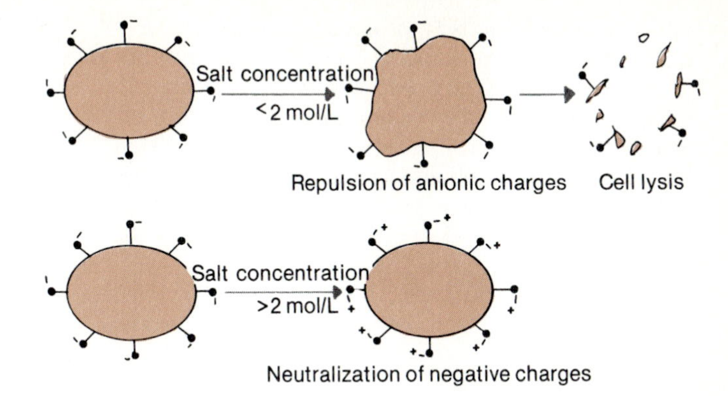

species appears to carry a lower molecular weight plasmid than the virulent species, which suggests that a deletion has taken place.

Concept Check

1. **What techniques are available to the microbiologist for determining microbial growth? What are the advantages and disadvantages of each?**
2. **What are the stages that make up the eukaryotic cell cycle? What factors have been demonstrated to affect this cycle?**
3. **Why are some microorganisms able to survive at low temperatures while others cannot?**
4. **What mechanisms may be operating in the microbial cell that enable it to grow at extremely hot temperatures?**

Osmotic pressure

Microorganisms are often in environments in which the solute concentration is lower (hypotonic) than the cell cytoplasm, but the cell does not lyse because of the nature of the rigid cell wall. (See Chapter 5 for a discussion of osmosis.) There are a few ecological sites, however, in which the environment outside the cell is extremely hypertonic, such as the Dead Sea and inland lakes where evaporation has created a saline environment. The organisms living in these areas not only have evolved mechanisms that enable them to tolerate a saline environment and prevent dehydration, but some are unable to survive unless the concentration of salt is between 3 and 6 mol/L. These organisms are called **obligate** or **extreme halophiles** and are *Archaebacteria* belonging to the genera of bacteria called *Halobacterium* and *Halococcus.* The structure of the cytoplasmic membrane is one factor that enables these organisms to exist in an environment that will not support any other cell type (Figure 14-30). The lipids of halophilic bacteria are highly acidic (negatively charged because they are deprotonated at physiological pH), and when the organisms are placed in a medium in which the salt concentration is below 2 mol/L, the outer portion of the envelope disintegrates. Presumably, there is a repulsion between the anions of the lipid, which makes the envelope unstable, but at elevated salt concentrations this effect is neutralized by the abundant cations.

Proteins are also affected by high concentrations of salt. The nonpolar hydrophobic groups that arrange themselves on the inside of protein molecules are responsible for the conformational properties of proteins and hence their activity. Hydrophobic bonds are weak, and high concentrations of salt increase the strength of these hydrophobic interactions, making the molecule more rigid and less susceptible to unfolding and denaturation. Salts decrease the solubility of the protein, that is, the interaction between charged amino acids and water decreases while protein-protein interactions increase. The ability of the halophiles to survive and prevent dehydration, an osmotic problem, is related to their ability to concentrate certain solutes, for example, polyhydric alcohols such as glycerol by xerotolerant (sugar-tolerant) yeasts and potassium by salt-tolerant *Halobacterium* and *Halococcus.* In the case of *Halobacterium* and *Halococcus,* in vitro experiments have shown that their enzymes have optimum activity when the sodium chloride concentration is between 0.5 and 1.0 mol/L and that concentrations between 2 and 6 mol/L are inhibitory. Potassium, however, at concentrations between 2 and 6 mol/L is a relatively poor inhibitor of enzyme activity; thus halophilic enzymes can maintain maximum activity at high intracellular levels of salt by concentrating potassium.

Halophiles have generated considerable interest and research in the National Aeronautics and Space Administration (NASA). The search for extraterrestrial life on planets where there is low water activity makes halophilic microorganisms of particular interest.

One of the reasons for mixing high concentrations of

salt or sugar with foods is to prevent growth of microorganisms through dehydration. The addition of 12% sodium chloride is a favorite technique for reducing microbial growth, but this technique has no effect on certain halophiles that can spoil many types of dried foods. Halophiles have been implicated in the spoilage of brines used to pickle olives (see Chapter 21).

Hydrostatic pressure

The pressure exerted by a column of water on the bottom of the column is referred to as **hydrostatic pressure.** Hydrostatic pressure is a physical characteristic that affects all life but especially life in the ocean's depths. The ocean has an average depth of 3500 m, but in areas such as the bottom of the Mariana Trench, located near the Philippines, the depth is approximately 10,800 m. Pressure increases 1 atm for every 10 m, and at the bottom of the Mariana Trench there is nearly 1160 atm of pressure. The study of effects of hydrostatic pressure on organisms is called **barobiology.** Microorganisms that tolerate elevated hydrostatic pressure are called **barotolerant**, and those that grow better at elevated hydrostatic pressures than at normal atmospheric pressure are called **barophilic.**

Microorganisms show a wide range of sensitivity to pressure; the growth of some microorganisms is inhibited by 100 atm of pressure, whereas others are able to grow under pressures of 1000 atm. Those microorganisms possessing gas vacuoles are extremely sensitive to changes in hydrostatic pressure because an increase in only a few atmospheres will cause the vacuoles to collapse. Most microorganisms do not contain gas vacuoles, and increases in hydrostatic pressure affect them in different ways. In prokaryotic cells increased hydrostatic pressures affect biochemical processes, whereas in eukaryotic cells major damage is to microtubular structures. Most of the microorganisms isolated from the ocean's depths grow better at ambient pressures than in their normal habitat and are referred to as **facultative barophiles.** Only recently has an obligate barophilic bacterium been isolated from the ocean. Yayanos and others in 1981 isolated an obligate barophilic bacterium from the Mariana Trench. This bacterium, which was recovered from a decomposing crustacean, is unable to grow at 2° C at pressures less than 400 atm but can grow at pressures as high as 1035 atm. Growth of the obligate barophile is extremely slow (30-hour generation time), and this is due in part to the extremely low temperature (approximately 1° to 2° C) that it encounters in its environment.

Microorganisms that are normally found at ambient temperatures have a wide range of response to increases in pressure in accordance with their physiological state. Sensitivity to atmospheric pressures has been studied in the bacterium *Streptococcus faecalis. S. faecalis* is **aciduric** and can tolerate high concentrations of acid produced during growth. Acid toleration is a result of the cell's ability to concentrate potassium ions during the transport of protons out of the cell through the cytoplasmic membrane—via ATPase. Protons released into the medium are therefore prevented from reentry into the cell. When *S. faecalis* is grown at high hydrostatic pressure, ATPase is severely affected, and growth declines precipitously because of the reentry of protons into the cytoplasm of the cell. The growth of many other bacterial species also appears to be inhibited by high pressures due to the accumulation of acids in the cell. *S. faecalis* barotolerance can be shown to be associated with the type of carbohydrate catabolized. When pyruvate is the carbohydrate source, the organism can tolerate only 200 atm of pressure. When ribose is the carbohydrate source, the organism can tolerate 450 atm of pressure, and when the carbohydrate is a hexose such as glucose or galactose, the organism can grow at hydrostatic pressures of 550 atm. Neutralization of acid by various ions is important for barotolerance; for example, ammonia production from the hydrolysis of arginine during glycolysis or the presence of calcium or magnesium in the growth medium enables the microorganism to grow at elevated hydrostatic pressures. It also appears that some microorganisms may reduce their size to become more resistant to pressure. A psychrophilic bacterium isolated from the Antarctic in 1976 responded to starvation by decreasing its cell size and in this state was also found to be more resistant to increases in pressure.

Hydrogen ion concentration (pH)

When cultivated in closed systems, such as a laboratory flask, microorganisms excrete products of metabolism, such as acids, into the surrounding medium. Eventually the pH of the medium drops, and growth is inhibited because enzyme activity is affected by the hydrogen ion concentration. The low pH can also affect the solubility and ionic state of nutrients, making them inaccessible to the organism. Buffer systems are normally added to a culture medium to prevent rapid drops in pH. One of the most effective buffering systems uses a mixture of K_2HPO_4 and KH_2PO_4, which, when added to nutrient media, enables one to produce a large population of microorganisms in a relatively short period of time.*

Microorganisms live in a wide variety of habitats, but the more extreme the pH the fewer the species that will be found there. Most microorganisms flourish at pHs near neutrality (pH 6 to 8), which is the most effective range for phosphate buffering capacity. A few microbial species can flourish at extremes of pH. We may divide microorganisms into three categories based on their

* Complex media at pHs between 6 and 8 have some buffering capacity because of the presence of proteins and amino acids.

Table 14-2 ***Obligately acidophilic bacteria***

Metabolism	Organism	pH optimum
Chemoautotrophic	*Sulfolobus acidocaldarius**	2.0
	*Thiobacillus acidophilus**	3.0-3.5
	Thiobacillus ferrooxidans	1.8-2.4
	*Thiobacillus kalobis**	2.4
	*Thiobacillus organoparus**	3.0
	Thiobacillus thiooxidans	2.5
Heterotrophic	*Acetobacter acidophilus*	3.1
	Bacillus acidocaldarius	3.0
	Bacillus coagulans	3.5-4.0
	Coxiella burnetii	4.5-5.0
	Flavobacterium acidurans	4.1
	Thermoplasma acidophilum	1.8

*Facultatively heterotrophic.

growth properties at different pHs: alkalophiles, neutrophiles, and acidophiles. **Alkalophiles** grow in a pH range of 7 to 11 with an optimum of 10. **Neutrophiles** grow in a pH range of 4 to 9 with an optimum near neutrality. **Acidophiles** live at low pHs and can be divided into facultative and obligate. Facultative acidophiles grow well at acid pHs but can also grow at pH values of 7 or above. Fungi, for example, are facultative acidophiles capable of growth at pHs of 3 or less but are also capable of growth at pHs above 7. Obligate acidophiles (Table 14-2) grow normally at pHs of 3 or less and usually cannot grow above this pH. In fact, pHs near neutrality can lead to disintegration of the organism. The rest of our discussion will center on the acidophiles.

Proteins including enzymes, whether isolated from acidophiles, neutrophiles or alkalophiles, exhibit a narrow range within which they are most active (that is, pH 6.8 to 7.2). Acidophiles must be able to maintain a constant intracellular pH by some homeostatic mechanism when the external pH is very low. In respiring bacteria the pH homeostatic mechanism is dependent on the respiratory system found in the cytoplasmic membrane. pH homeostasis is maintained by the chemiosmotic mechanism, and in acidophiles the internal pH is maintained at approximately 6.5 (the same mechanism works in neutrophiles and alkalophiles). Thus the proton pump of prokaryotic cells has a dual function: energy transduction and pH homeostasis (see Point of Interest).

Resistance of acidophiles to pH was first examined in relationship to structural and compositional peculiarities in the thermoacidophiles *Sulfolobus acidocaldarius* and

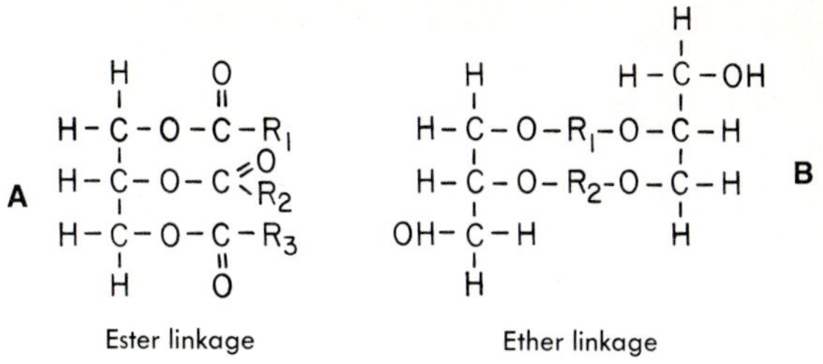

```
                                                  H
                                                  |
     H   O                  H                 H - C - OH
     |   ||                 |                     |
 H - C - O - C - R1     H - C - O - R1 - O - C - H
     |         //O          |                     |
A H - C - O - C         H - C - O - R2 - O - C - H     B
     |         \R2          |                     |
 H - C - O - C - R3    OH - C - H                 H
     |       ||             |
     H       O              H

   Ester linkage                Ether linkage
```

Figure 14-31 *Lipid structure in mesophile and thermophile.* ***A,*** *Ester linkage in mesophile.* ***B,*** *Ether linkage in thermophile.* R *represents fatty acid chains.*

Thermoplasma acidophilum, members of the Archaebacteria. *S. acidocaldarius* is found throughout the world in thermal acidic soils and acidic hot springs, where it oxidizes elemental sulfur to sulfuric acid, creating an environment in which the pH is less than 3.0. The cell wall of this species is unique in that no peptidoglycan is present; however, the cell membrane is covered with a regular polyhexagonal array of protein subunits that provide stability. *T. acidophilum* belongs to the mycoplasmas, which do not possess cell walls. Its habitat is restricted to hot acidic environments associated with heating coal refuse piles, where optimum growth is obtained at pH 2 and 59° C. The internal pH in both bacteria is approximately 6. The mechanism of survival when the external pH is 3 or below has been a subject for considerable speculation. Again, the cell membrane appears to play a vital role. Both acidophilic species possess membrane lipids that contribute to thermophilic survival at high temperatures. The membrane lipids possess ether linkages instead of the usual ester linkages seen in mesophilic fatty acids. Some are diethers and consist of two glycerol moieties joined together by two long-chain alkyl groups (Figure 14-31). These ether linkages are believed to increase the stability of the membrane in acidic environments. An additional characteristic of *T. acidophilum* appears to enable this organism to survive acidic pH. The proteins of the membrane are similar to those of mesophiles except that in *T. acidophilum* the number of free carboxyl groups and amino groups is reduced by one half. This is believed to improve stabilization of the membrane over wider pH ranges. If the mesophilic content of carboxyl groups were present in the acidophile, then ionization at low pHs would occur, resulting in charge repulsion on the membrane and subsequent disintegration.

There is a third type of thermoacidophile, called *Bacillus acidocaldarius,* which differs from the previously mentioned types. This organism is a typical prokaryote having a peptidoglycan-containing cell wall, and it does not contain the tetraether-type fatty acids. It does possess a high content of cyclic fatty acids, which are believed to create a tighter and more stable membrane.

Oxidation-reduction (redox) potential

Redox potential is an environmental factor that is known to affect microbial growth. Only in most recent years has any information been available as to how redox potential might exert its control. We now know that many reactions in the cell are catalyzed by enzymes that contain redox active disulfide groups. Thioredoxin, a hydrogen carrier protein, has been implicated in the regulation of enzyme activity by thioredox control. One of the im-

Point of Interest

A MECHANISM FOR REGULATING THE INTERNAL pH OF THE CELL

Respiring bacteria extrude protons during chemiosmosis, making the inside cytoplasm of the cell relatively basic (negatively charged) as opposed to the acidic (positively charged) exterior of the cell. Yet the internal pH of the cell remains relatively constant over a wide range of external pH values (Table 14-3). The bacterial cell possesses a mechanism for returning protons into the cell and maintaining or regulating the internal pH. This system of proton return is also driven by the proton motive force or one of its carrier proteins and is referred to as a **secondary transport system** (the proton motive force coupled to the synthesis of ATP is a primary transport system). The secondary transport system is also associated with transport of nutrients such as amino acids. Secondary transport can occur by one of three mechanisms:

1. Uniport—only one solute is transferred by a carrier protein
2. Symport—two or more different solutes are transferred by one carrier in the same direction
3. Antiport—two or more different solutes are transferred by one carrier in opposite directions

Most cells use antiport systems for the removal of cations from the cytoplasm. One type of antiport system in bacteria is the K^+/H^+ antiporter, and it works in the following way (see accompanying figure): Protons extruded by chemiosmosis in respiratory bacteria are taken up by the K^+/H^+ antiporter and pass into the cytoplasm while K^+ is removed from the cytoplasm. The K^+ that was extruded returns through an ATPase in which ATP is hydrolyzed and energy is released to maintain the system. This antiport system appears to be one mechanism for maintaining pH homeostasis in the bacterial cell, but other mechanisms may also be involved.

Table 14-3 *Intracellular pH values in various microorganisms at different external pH values*

Microorganism	Optimum pH for growth	External pH	Internal pH
Acidophilic microorganisms			
Bacillus acidocaldarius	3 to 4	3	5.8 to 6.5
Thermoplasma acidophila	2	2 to 6	6.4 to 6.9
Neutrophilic microorganisms			
Streptococcus faecalis	6.5 to 7.2	6	7.2
E. coli K12	6.5 to 7.2	6	8
		7	7.7
		8	7.7
Alkalophilic microorganisms			
Bacillus alcalophilus	10.5	9	7.6
		11.5	9

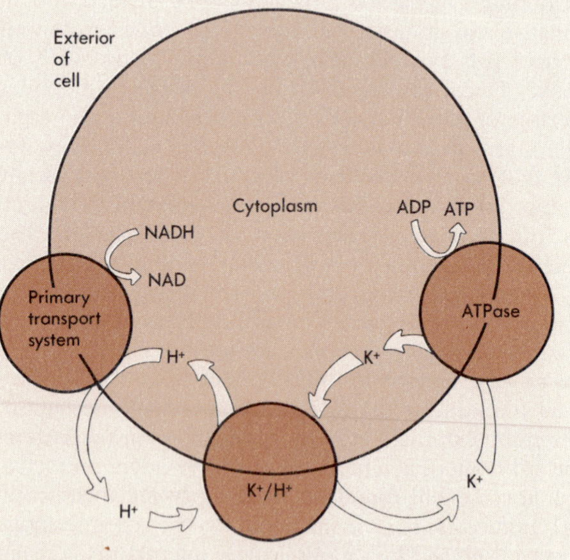

K^+/H^+ antiport system for maintaining internal pH. Protons extruded by primary transport system via chemiosmosis during respiration are taken up by secondary transport system (K^+/H^+). Secondary system carries protons back into cytoplasm accompanied by removal of K^+ from cytoplasm. K^+ removed by secondary system is returned to cytoplasm through ATPase resulting in ATP formation.

portant cellular functions of this protein is as a hydrogen donor for the enzyme ribonucleotide reductase. The latter enzyme is involved in the formation of deoxyribonuclotide precursors of DNA from the corresponding nucleotides:

$$\text{Thioredoxin}\begin{matrix}\text{SH}\\ \text{SH}\end{matrix}\ (\text{reduced}) + \text{Ribonucleotide diphosphates} \xrightarrow[\text{reductase}]{\text{Ribonucleotide}} \text{Thioredoxin}\begin{matrix}\text{S}\\ |\\ \text{S}\end{matrix}\ (\text{oxidized}) + \text{Deoxyribonucleotide diphosphates}$$

Thioredoxin is a protein found in many cellular systems. It contains two catalytically active cysteine residues that exist in a reduced thioredoxin $(SH)_2$ or oxidized (thioredoxin-S_2) form. The hydrogen used to reduce thioredoxin is supplied by NADPH in an NADPH-thioredoxin reductase–catalyzed reaction or by ferredoxin in a ferredoxin-thioredoxin reductase–catalyzed reaction in fermentative and phototrophic bacteria:

$$1.\ \text{Thioredoxin-S}_2 + \text{NADPH} + \text{H}^+ \xrightarrow[\text{reductase}]{\text{Thioredoxin}} \text{Thioredoxin-(SH)}_2 + \text{NADP}^+$$

$$2.\ 4\ \text{Ferredoxin}_{ox} + 2\text{H}_2\text{0} \xrightarrow[\text{thylakoids}]{\text{Light,}} 4\ \text{Ferredoxin}_{red} + \text{0}_2 + 4\text{H}^+$$

$$2\ \text{Ferredoxin}_{red} + \text{Thioredoxin-S}_2 + 2\text{H}^+ \xrightarrow[\text{thioredoxin reductase}]{\text{Ferredoxin}} 2\ \text{Ferredoxin}_{ox} + \text{Thioredoxin-(SH)}_2$$

Another enzyme that appears to be regulated by redox potential is the phosphoenolpyruvate sugar phosphotransferase system. The activity of one of its proteins is coupled to the oxidation-reduction potential. A sulfhydryl group associated with E_{II} (see Chapter 5) of the transport system undergoes reversible oxidation to a disulfide (—S—S—) form. In a reduced form (SH_2) the protein has high affinity for its substrate (certain hexoses), but the oxidized form has a 10^2 to 10^3 times lower affinity for the same substrates. There is evidence that two redox-sensitive disulfide groups are present in the transport protein and that they are located on opposite sides of the membrane. Thus the redox potential in the cytoplasm as well as the proton motive force across the cytoplasmic membrane controls the activity of membrane functions.

Electromagnetic radiation

Electromagnetic radiation consists of discrete packets of light quanta or photons having properties of waves and particles. As a wave, a photon can be characterized by its wavelength, and as a particle, it can be described by its quantum of energy. The different types of radiation in the electromagnetic wave spectrum are illustrated in Figure 3-3. The energy of a photon is inversely related to its wavelength. We have already described the importance of visible light as a source of energy for photosynthetic microorganisms. Microorganisms such as the cyanobacteria and purple and green bacteria possess pigments capable of absorbing light at various wavelengths (see Chapter 12). This light energy is eventually converted to chemical energy in the cell. Ultraviolet radiation, x-rays, and gamma rays have shorter and more energetic wavelengths than visible light and therefore are capable of causing damage to biological molecules. A second class of radiation not part of the normal electromagnetic spectrum, is called particulate and is composed of rapidly moving atomic particles such as electrons and protons. These particles are emitted from radioactive material and from man-made accelerating machines such as cyclotrons and nuclear reactors, and they also damage biological molecules. Radiation based on its effects on biological material can be divided into two types: **ionizing** and **nonionizing**. Ionizing refers to the acquisition of a charge by removal of an electron or electrons resulting in the formation of a highly reactive and positively charged molecule.

Nonionizing radiation. **Ultraviolet light** is the most relevant type of nonionizing radiation. Most of the harmful effects of ultraviolet radiation are due to wavelengths between 240 and 300 nm. The effect of nonionizing radiation on biological molecules is related to the amount of energy absorbed by the molecule. When sufficient energy is absorbed, an outer valence electron of a molecule can be raised to a higher energy level, converting it to an excited state and thus increasing its probability of undergoing a chemical reaction above what it is when the molecule is in the ground or unexcited state. Proteins and nucleic acids are the principal constituents in the cell that absorb ultraviolet light. This absorption is due primarily to the presence of molecules containing conjugated bonds (alternating single and double bonds). In nucleic acids all of the bases (adenine, guanine, thymine, cytosine, and uracil) contain conjugated bonds. In proteins the amino acid cysteine is a major target of irradiation because of the high chemical reactivity of the disulfide bond while aromatic amino acids such as tryptophan, tyrosine, and phenylalanine, which contain conjugated bonds, are also absorbers of ultraviolet radiation. Once the molecule is excited it may emit the energy as light (fluorescence) or the molecule may undergo chemical transformation. Most of the damage incurred from ultraviolet radiation is due to dimer formation involving the DNA molecule (see Figure 10-15). The thymine-thymine dimer is the most frequently formed dimer, although the thymine-cytosine and cytosine-cytosine dimers may also appear after radiation. Pyrimidine-pyrimidine dimers slow down DNA synthesis but do not absolutely block it. Since this type of lesion is repairable (see Chapter 10), it is not the number of

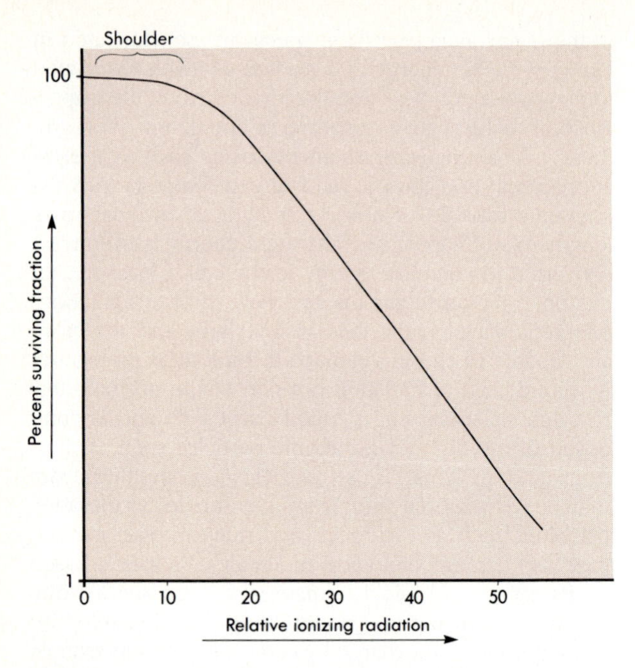

Figure 14-32 *Typical survival curve of a population of bacteria exposed to radiation. Survival is measured as the ability to form a colony. The width of the shoulder in the curve is usually taken as a measure of the amount of damage that can be accumulated before the repair systems become saturated. Once the repair systems become saturated the curve becomes linear.*

lesions in the DNA that cause mutations in the cell or cause cell death but the number and site of the lesions remaining after the repair processes (see dark and light repair processes, Chapter 10) have culminated. When cultures of bacteria (or other microorganisms) are exposed to increasing doses of radiation, a survival curve can be established (Figure 14-32). Usually a shoulder is observed in the curve. Theories to explain this phenomenon include the multitarget theory, which holds that there is more than one target in the cell and each has to be hit to produce an effect. A second theory, the multihit theory, holds that there is only one target but that a certain number of hits are required to produce the desired effect.

Ionizing radiation. The energies transmitted by ionizing radiation (1×10^{-5} to 10^2 nm) cause electrons to be removed from the absorbing molecules. As the electron traverses matter, it loses energy by producing ionizations and leaves behind a positive ion. When its energy is dissipated, the electron is captured by a suitable atom yielding a negative ion. Since cells are composed of 70% to 90% water, most of the energy absorbed and ionizations produced by radiation occur in water. The radiolysis of water releases products such as the hydroxyl radical (OH·), which is the most damaging to the DNA molecule:

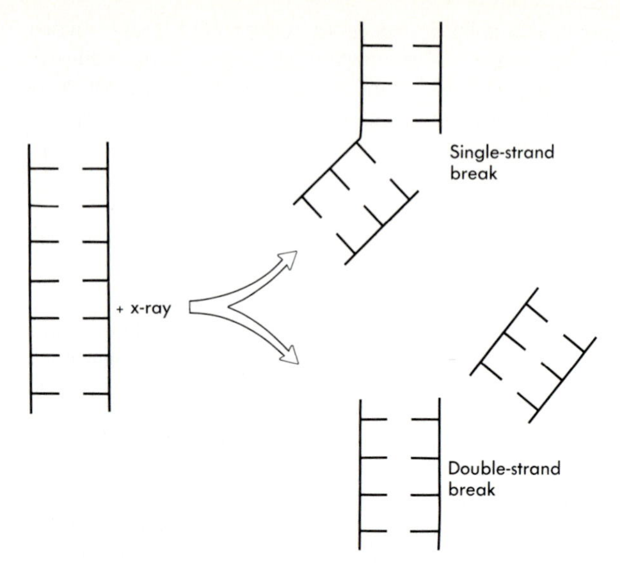

Figure 14-33 *The effect of single- and double-strand breaks in the DNA molecule.*

$$H_2O \xrightarrow{\text{radiation}} H_2O^+ + e^-$$
$$H_2O + H_2O \longrightarrow OH^- + H_3O$$

Hydroxyl radicals interact with DNA bases or deoxyribose. For example, the hydroxyl radical can interact with the hydrogen atoms on any of the five carbon atoms of deoxyribose. This leads to strand breaks and usually involves the C−3′-phosphate ester bond. Double-strand breaks involving sites on each strand that are opposite each other occur in an amount approximately 5% of the single-strand break frequency (Figure 14-33). Single-strand breaks are repairable; however, double-strand breaks have been shown to be repairable in only a few species. The effects of ionizing radiation appear also to be influenced by the presence or absence of ox-

Concept Check

1. What are the mechanisms by which *Halobacteria* prevent dehydration?
2. Why do elevated hydrostatic pressures inhibit growth?
3. How can bacteria maintain an internal pH of 6.5 when they are in an environment in which the pH is less than 3.0?
4. How can the oxidation of sulfhydryl groups on certain enzymes in the cytoplasmic membrane regulate nutrient transport?
5. How are ionizing and nonionizing radiations able to affect the genetic material of the cell?

ygen. When equivalent amounts of radiation are applied to populations of anaerobes and aerobes, radiation damage is more evident in the aerobic culture than in the anaerobic one. It is believed that ionizing radiation converts molecular oxygen to the **superoxide** radical, which can react with DNA and cause single-stranded breaks and probably damage other biological molecules (see Chapter 12).

SUMMARY

1. Because of its size, a single prokaryotic cell cannot be properly evaluated, but if a cell culture is synchronized so that all of the cells are in the same stage of development, the growth data can be extrapolated to single cells.
2. Synchronization of cultures can be obtained by manipulating environmental factors such as temperature, light, and nutrition or by using selection techniques. Two selection techniques include filtration and density centrifugation, in which cell size is the parameter used to separate synchronized from nonsynchronized cells.
3. The bacterial cell cycle, which includes growth, septum formation, and cell division, is controlled by certain factors, only two of which have been amenable to investigation: DNA replication and cell envelope synthesis. DNA synthesis continues throughout the entire cell cycle and is coordinated with cell envelope synthesis and septum formation. This coordination terminates with each cell receiving one or more copies of the circular chromosome.
4. The growth of a population of cells is easily measured in the laboratory using either batch or continuous cultures. In batch cultures the bacterial population goes through a number of different growth phases, which include the lag, log, stationary, and death phases. The growth rate (k) remains constant during the log phase and is expressed as $k = \ln_2/$generation time.
5. Continuous cultures are prepared in devices such as chemostats in which the bacterial population is maintained at a particular density and in a state of logarithmic growth.
6. Microorganisms respond to nutrient limitation by structural as well as metabolic changes. In chemoheterotrophs the mass of cells produced per unit of limiting nutrient is a constant, *Y*, provided the carbon and energy source are the same.
7. Within certain limits there is a linear relationship between the dry weight of organisms produced and the concentration of the energy source—a relationship expressed as Y_{ATP}. The energy required for those processes beyond that needed for growth is called the maintenance energy.
8. The actual measurements of growth, that is, the number of cells present or the rate of growth of a culture sample, can be determined by employing a number of techniques, which include measurements involving direct microscopic count using special engraved slides, colony count, electrical resistance of cells, and cell density, and cell mass. The most frequently used technique measures cell density by employing a spectrophotometric device that measures the light scattered by a bacterial population.
9. Many factors control growth rate but one of the most important is the ribosome content. Shift up and shift down experiments have demonstrated that in slowly growing cultures there is an excess of ribosomes—enough to maintain the slow growth rate and additional ribosomes to support a faster growth rate if the culture is shifted to a rich medium. In rapidly growing cells all ribosomes are used at peak efficiency.
10. The cell cycle in eukaryotes is divided into four phases: G_1, S, G_2, and M. The G_1 period is associated with microtubule formation as well as initiation of DNA synthesis and the formation of a bud. The S period is concerned with the formation of the mitotic spindle. During the G_2 period there is nuclear division and spindle elongation and the nucleus migrates into the neck of the bud. During the M period there is completion of nuclear division and separation of daughter cells.
11. Cell cycle control in the eukaryote *Saccharomyces cerevisiae* revolves around the G_1 period and a single event called "start". Conditions such as adequate nutritional levels and attainment of critical size are required if the cell cycle is to proceed beyond "start" during asexual division. In cells with a sexual cycle a pheromone produced by the two mating types is also required to complete the "start" event. The formation of a bud during the asexual cell cycle occurs at the poles of the mother cell. New cell wall is produced at the site of bud formation until a septum is formed and the bud separates from the mother cell.
12. The growth of microorganisms is affected favorably or adversely by environmental factors. One of the most widely studied factors is temperature. Microorganisms, because of their wide variations in temperature requirements, have been divided into three groups: psychrophiles, mesophiles, and thermophiles. This temperature variation is related to the chemical composition as well as the molecular architecture of the bacterial cell. Other factors that influence the cell are pH, osmotic pressure, hydrostatic pressure, oxidation-reduction potential, and electromagnetic radiation.

Self-study Quiz

Multiple choice

1. What method is the most effective way to determine the viable count of a culture?
   A. Chemostat
   B. Spectrophotometry
   C. Colony count
   D. Coulter count
2. Which of the following would create a synchronous culture of bacteria?
   A. Use of a substance that halted mitotic spindle formation and then removal of cells from the culture containing the reagent to a medium that allowed mitosis
   B. Growth of a culture by repetitively cycling the culture between 42° and 28° C
   C. Lysing of the cells in the culture with an autolysing enzyme and then placement of the autolysate on a growth medium
   D. Taking of cells from the stationary phase of growth and placement in a fresh medium
3. If the generation time for an organism in culture is 30 minutes and there are 1000 cells in the log phase culture at 10 AM, how many cells will there be in the culture at 11 AM?
   A. 2000
   B. 3000
   C. 4000
   D. No way to determine
4. The optimum activity of enzymes isolated from a thermophile would be expected to be in what range?
   A. 0-20° C
   B. 20°-40° C
   C. 40°-60° C
   D. There is no correlation between growth range and enzyme activity
5. Where in nature would you expect to isolate a facultative barophile?
   A. From great ocean depths
   B. From salt marshes
   C. From ice in the Arctic Circle
   D. From the bottom of hot springs
6. Temperature-sensitive mutants are characterized by which of the following?
   A. Able to grow either above or below a temperature for which they are sensitive but not at a specific temperature
   B. Easily killed by temperatures above 10° C
   C. Capable of growth at permissive, lower temperatures, but not at nonpermissive elevated temperatures
   D. Capable of altering the temperature of the environment
7. Approximately what concentration of bacterial cells is the least number of cells that will impart an observable turbidity to a medium?
   A. 1×10^3 cells/ml
   B. 1×10^5 cells/ml
   C. 1×10^7 cells/ml
   D. 1×10^{11} cells/ml
8. After a shift from a nutrient-poor to a nutrient-rich medium, which of the following shows the greatest increase in the bacterial cell?
   A. RNA
   B. DNA
   C. Protein
   D. Lipid
9. Which of the following set of conditions must be met before initiation of the asexual cell cycle in yeast?
   A. Attainment of critical size and DNA content
   B. Presence of adequate nutrients and attainment of critical size
   C. Attainment of critical size and presence of a pheromone
   D. Presence of pheromones from both mating types
10. Which of the following statements concerning pH is *not* true?
    A. Facultative acidophiles live at low pH but can grow at pH 7 or above
    B. The cytoplasm of acidophiles is the same as the pH of their environment
    C. Most microorganisms flourish at pHs between 6 and 8
    D. In respiring bacteria the internal pH of the cell is controlled by chemiosmosis

Matching

1. ________ Petroff Hauser counter
2. ________ Minicell
3. ________ Batch culture
4. ________ Turbidostat
5. ________ Lag phase
6. ________ Log phase
7. ________ Continuous culture
8. ________ Y_{ATP}
9. ________ Pheromone
10. ________ Obligate halophile
11. ________ Antiport
12. ________ Death and decline phase
13. ________ Spectrophotometer
14. ________ Diauxic
15. ________ Barotolerant

A. Type of growth curve obtained when organism grown on two fermentable sugars
B. Requires high concentration of salt to survive
C. Mechanism in which two or more different solutes are transferred by one carrier in opposite directions
D. Compound that increases budding in yeast cells
E. Tolerance to elevated hydrostatic pressures
F. Device used to measure turbidity of a culture
G. Phase of growth in which cell division is curtailed until sufficient enzymes are synthesized
H. A phase of growth in which cells are dying exponentially
I. A small anucleated cell resulting from unequal cell division
J. A device for measuring numbers of cells
K. Phase of growth characterized as "balanced"
L. A continuous culture device in which a photoelectric device measures cell density
M. A closed growth system containing a defined amount of medium
N. Average dry weight of cells produced per mole of ATP used
O. An open culture system in which media are replenished and wastes removed

Thought question

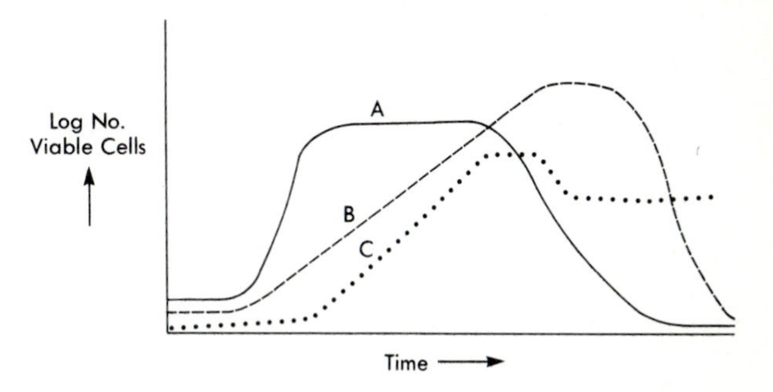

Shown above are the growth curves of three different organisms (A, B, and C) growing together in the same culture flask. Answer the questions concerning the curves.

A. Explain how you might obtain the data to plot each of the curves.
B. Which organism has the shortest generation time during log phase? Explain.
C. Which organism has the longest lag phase? Explain.
D. Which organism reached the heaviest concentration of viable cells? Explain.
E. If one of the organisms is a sporeformer, which organism is it? Explain.
F. Would the three curves be the same as depicted if the three organisms were grown in a chemostat? Explain.
G. Would the curves be identical if you changed the temperature at which you incubated the flasks? Explain.

SELECTED READINGS

Books

Brock, T.D. *Thermophilic microorganisms and life at high temperatures.* Springer Verlag, New York, Inc., New York, 1978.

Dawes, I.W., and Sutherland, I.W. *Microbial physiology.* Halsted Press, New York, 1976.

Doelle, H.W. *Bacterial metabolism* (2nd ed.). Academic Press, Inc., New York, 1975.

Ingraham, J.L., Maaløe, O. and Neidhardt, F.C. *Growth of the bacterial cell,* Sinauer, Sunderland, Me., 1983.

Mandelstam J., and McQuillen, K. (Eds.): *Biochemistry of bacterial growth* (3rd ed.). Halsted Press, New York, 1982.

Moat, A.G. *Microbial physiology.* John Wiley & Sons, Inc., New York, 1979.

Shilo, M. (Ed.). *Strategies of microbial life in extreme environments.* Verlag Chemie International, Inc., New York, 1979.

Tempest, D.W. The continuous cultivation of microorganisms, Part 1, Theory of the chemostat. In Norris, J.R., and Ribbons, D.W. (Ed.). *Methods in microbiology* (Vol. 2). Academic press, Inc., New York, 1976.

Journal articles

Baross, J.A., and Deming, J.W. Growth of "black smoker" bacteria at temperatures of at least 250° C. *Nature* **303:**423, 1983.

Bayley, S.T., and Morton, R.A. Recent developments on the molecular biology of extremely halophilic bacteria. *CRC Crit. Rev. Microbiol.* **6:**151, 1978.

Booth, I.R. Regulation of cytoplasmic pH in bacteria. *Microbiol. Rev.* **49**(1):359, 1985.

Brown, A.D. Microbial water stress. *Microbiol. Rev.* **40:**803, 1976.

Cabib, E., Roberts, R., and Bower, S.B. Synthesis of the yeast cell wall and its regulation. *Annu. Rev. Biochem.* **51:**763, 1982.

Cole, R.M. Symposium on the fine structure and replication of bacteria and their parts. III. Bacterial cell wall replication followed by immunofluorescence. *Microbiol. Rev.* **29:**326, 1963.

Edwards, C. The microbial cell cycle. Aspects of Microbiology(#3). American Society for Microbiology, Washington, D.C., 1981.

Harder, W., and Dijkuizen, L. Physiological responses to nutrient limitation. *Annu. Rev. Microbiol.* **37:**1, 1983.

Harrison, A.P., Jr. The acidophilic thiobacilli and other acidophilic bacteria that share their habitat. *Annu. Rev. Microbiol.* **38:**265, 1984.

Hartwell, L.H. *Saccharomyces cerevisiae* cell cycle. *Microbiol. Rev.* **38:**164, 1974.

Higgins, M.L., and Shockman, G.R. Prokaryotic cell division with respect to wall and membranes. *CRC Crit. Rev. Microbiol.* **1:**29, 1971.

Holmgren, A. Thioredoxin. *Annu. Rev. Biochem.* **54:**237, 1985.

Jannasch, H.W., and Taylor, C.D. Deep sea microbiology. *Annu. Rev. Microbiol.* **38:**487, 1984.

Marquis, R.E. Microbial barobiology. *Bioscience* **32**(4):267, 1982.

Morita, R.Y. Psychrophilic bacteria. *Microbiol. Rev.* **39:**144, 1975.

Rogers, H.J. Bacterial growth and the cell envelope. *Microbiol. Rev.* **36:**347, 1972.

Slater, M., and Schaechter, M. Control of cell division in bacteria. *Microbiol. Rev.* **38:**199, 1974.

Yayanos, A.A., Dietz, A.S., and van Boxtel, R. Obligately barophilic bacterium from the Mariana Trench. *Proc. Natl. Acad. Sci. USA* **78:**5212, 1981.

CHAPTER

15

Control of microbial growth: chemotherapy, sterilization, and disinfection

Chapter Outline

Introduction

Chapter 14 shows that the growth of microorganisms is affected by many environmental factors. This chapter discusses how some of these factors are used under controlled conditions to deliberately affect microbial growth. The deliberate control of microbial growth is important for two reasons. First, the control or destruction of microorganisms is necessary in the treatment of infectious disease. Second, under certain conditions, such as experimental research, it is important that the microbial culture consist of only one species type. Thus the culture should be **pure** or **axenic**. This chapter describes the major antimicrobial agents used to treat infectious disease, discusses those antimicrobial agents used to prevent infections, describes the mechanisms used by microorganisms to resist the action of antimicrobial drugs, describes how the activity of a drug can be tested in the laboratory, and discusses the physical and chemical methods used to destroy microorganisms that contaminate both animate and inanimate objects.

CHEMOTHERAPY

The term **chemotherapy** was introduced into scientific literature by the German chemist Paul Ehrlich in the late 1890s. He defined it as the use of drugs to injure the invading organism without injury to the host. Before 1890 only three agents could be considered chemotherapeutic agents: chinchona, for the treatment of malaria; mercury, for the treatment of syphilis; and ipecacuanha, for the treatment of amebic dysentery. Ehrlich's initial research was concerned with the affinity of chemicals for tissues, and these studies led to significant discoveries in histological staining that are still used today. He was also intrigued by the selective binding occurring between certain molecules, such as proteins, and components of the immune system. This led him to propose that drugs bind to cellular components by bonds. He suggested that chemotherapeutic agents possess two reactive groups: one concerned with fixation or transport to the tissue and the other causing the fatal lesion in the parasite. Ehrlich coined the term *magic bullets* for the ability of some drugs to bind to specific targets in the cell. Ehrlich predicted that the function of the drug would be to upset the parasite's metabolism, after which the host's natural defenses would take over.

The only chemotherapeutic agent of clinical significance discovered by Ehrlich was Salvarsan-606, an arsenical used in the treatment of syphilis. The number 606 refers to the trials employed by Ehrlich's research team to find the correct dosage for treatment of humans. Ehrlich's greatest contributions to chemotherapy were his original ideas, which provided the seed from which germinated new ideas and the inevitable discovery by others of a host of chemotherapeutic agents.

Another milestone in the history of chemotherapy was reached in 1935 by Domagk's discovery of prontosil, a dye used to treat streptococcal infections. Prontosil in the mammalian body is converted to sulfanilamide, which inhibits microbial activity by binding to bacterial enzymes (Ehrlich's receptors). As noted earlier (see Chapter 1), the isolation of penicillin, the antimicrobial substance produced by the fungus *Penicillium,* was a significant achievement in microbiology. Penicillin not only saved many lives in World War II but its discovery encouraged others to search for chemotherapeutic agents produced by other microorganisms.

General concepts and terminology associated with chemotherapeutic agents

Selective toxicity. Ehrlich's definition of chemotherapy implies that the chemical agent is selectively toxic, but this is not always true, since most drugs show some degree of toxicity to the cells of the host. Selective toxicity is best described by some examples in which the chemical agent inhibits microbial activity by interfering with metabolic processes. Sulfanilamide, for example, which interferes with folic acid synthesis in microorganisms, has no effect on humans, because folic acid is not synthesized in the body and must be supplied in the diet. Many chemotherapeutic agents affect protein synthesis because of their ability to bind certain components of the 70s ribosomal complex. The binding of chemotherapeutic agents to microbial ribosomes is a function of differences between the 70S ribosomes of the microorganism and the 80S ribosomes of the mammalian cell. The cell membrane of mammalian cells represents to some degree a permeability barrier to chemotherapeutic agents. Microorganisms that invade mammalian cells (intracellular parasites) are much more difficult to combat with drugs than those microorganisms that do not invade cells (extracellular parasites). Finally, the rate of metabolism and reproduction of microbial cells is much faster than that of mammalian cells; therefore the inhibition of microbial function occurs more rapidly than that of host function. This usually leads to a faster rate of degradation of antimicrobials by microorganisms.

At this point in our discussion a distinction should be made between a chemotherapeutic agent and an antimicrobial agent. All chemotherapeutic agents are antimicrobials, but not all antimicrobials are chemotherapeutic. Antimicrobial agents can inhibit the growth of or kill microorganisms, but they are not necessarily selectively toxic. For example, many antimicrobial agents cannot be administered internally because of toxicity to the host. Sulfuric acid is certainly an antimicrobial, but taken internally it would destroy host tissue.

Sources. Many microorganisms produce compounds that in small quantities inhibit the growth of or kill unrelated species of microorganisms. These biologically produced compounds are called **antibiotics.** Soil microorganisms, such as bacterial genera *Streptomyces* and *Bacillus,* are the principal producers whereas fungi, such as *Penicillium* and *Cephalosporium,* are the source of other antibiotics, for example, penicillin and cephalosporin. Compounds that have antimicrobial activity and chemotherapeutic value can also be synthesized in the laboratory and are called **synthetic** agents, for example, the sulfa drugs. Finally, antibiotics produced by some microorganisms can be modified in the laboratory by adding chemical groups to the existing nucleus, and these are called **semisynthetic** agents (Figure 15-1). Nearly all penicillins and cephalosporins used today are semisynthetic.

How do antibiotic-producing microorganisms protect themselves from the action of these compounds? In some instances, a number of different mechanisms could be working. For example, the antibiotic could be produced within the bacterial cytoplasm in an inactive form. Activation of the antibiotic could occur as it was being exported from the cell. *Streptomyces rimosus* and *S. aureofaciens* produce antibiotics called tetracyclines.

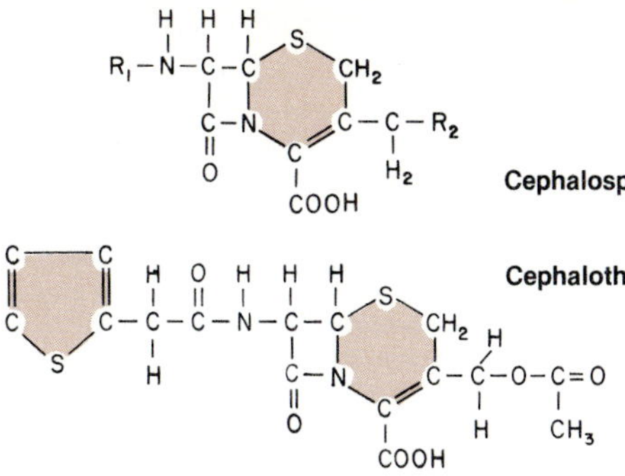

Cephalosporin nucleus

Cephalothin (semisynthetic cephalosporin)

Figure 15-1 *Structure of cephalosporin nucleus as it is produced by microorganisms. R_1 and R_2 are sites on which chemical units can be added. Cephalothin is semisynthetic compound in which two chemical units have been added to R_1 and R_2 of cephalosporin nucleus.*

Tetracyclines affect ribosomes and thus inhibit protein synthesis. Studies have demonstrated that ribosomes isolated from cells not engaged in tetracycline synthesis are more sensitive to inhibition by tetracycline in vitro than those isolated from cells actively synthesizing the antibiotic. *Streptomyces erythreus,* the producer of erythromycin, protects its ribosomes from the inhibitory effects of the drug by modification processes involving methylation. Specifically, the adenine residues in 23S rRNA are dimethylated at the N^6 position. A number of "producer" antibiotic resistance genes have now been characterized. In some instances these natural resistance genes are the same type of resistance genes found in those pathogenic bacteria against which the antibiotic is used and which become resistant to the antibiotic. Further discussion of antibiotic resistance in the clinical situation appears on p. 436.

Activity spectrum. Chemotherapeutic agents do not affect all microbial species in the same way, and this variation can be described as an **activity spectrum.** A **broad-spectrum** chemotherapeutic agent is generally associated with antibacterial agents that are active against gram-positive as well as gram-negative bacterial species. A **narrow-spectrum** drug is active against one or only a few species within either the gram-positive or the gram-negative groups. The activity of a drug is also influenced by the concentration of microorganisms at the site of infection as well as by the type of tissue affected. The number of microorganisms at a body site may be as high as 1×10^9/g of tissue, and greater concentrations of drug may be required to destroy the infectious agents. Some tissues, such as nerve tissue, are relatively impermeable to many antimicrobial agents carried there by the bloodstream, and infection in those sites is very difficult to treat. The concentration of drug at the site of infection often will determine whether or not the organism can be destroyed.

An antimicrobial agent may be described as -cidal or -static. The suffix may be added to the type of microorganism affected by the drug, such as a bactericidal agent, fungicidal agent, etc. A **-cidal** agent kills the microorganism, whereas a **-static** agent inhibits growth. For some drugs it is the concentration that determines what the activity will be. If the concentration of a bactericidal agent is too low to kill microorganisms, it may act as a bacteriostatic agent on a population of microorganisms. One of the disadvantages of bactericidal agents is that gram-negative bacteria release a highly toxic endotoxin when lysed, which can cause serious complications in the host. Second, the effect of a bactericidal agent may be so rapid that it does not allow enough time for the host's immune system to respond and synthesize antibodies. Antibodies are beneficial not only in warding off the initial infection, but in helping to prevent future infections. The advantage of **bacteriostatic** agents is that they allow the immune system of the host to destroy the invading pathogen. One of the disadvantages of bacteriostatic agents is that the growth-inhibiting concentration of the drug must be maintained in the bloodstream for a relatively long period of time in order for it to be effective.

Toxicity. Many chemotherapeutic agents used currently are to some degree toxic to mammalian cells, but fortunately the drugs have a greater affinity for the microbial component than for mammalian cell structures and are thus less toxic to the latter. The degree of tissue toxicity is related to many factors: the type of drug being used, the dosage administered, the length of time of treatment, and the condition of the patient at the time of treatment. Under normal treatment procedures maximum therapeutic levels of chemotherapeutic agents are not injurious to mammalian tissue. Toxicity, when it does occur, may be direct or indirect. Indirectly, the drug may induce an allergic response in the host, resulting in the activation of components of the immune system, which can damage tissue. Direct toxicity may involve destruction or irritation of sensitive tissues such as those of the kidney, liver, blood, nervous tissue, etc. Some drugs, such as the aminoglycosides (gentamicin is an example) are toxic to kidneys (nephrotoxicity) and interfere with the function of the organ by damaging tubular cells. Once the cells have been damaged, the drug begins to accumulate to higher concentrations in the kidney, and this can cause permanent damage, necessitating dialysis for patient survival. There are some drugs that when used even at minimal therapeutic concentrations are very toxic to host tissue. In some instances the toxic drug may be the only one available, and its use may be necessary in life-threatening situations. Often drugs that are very toxic when administered for systemic infections can be used topically to treat infections with no harm to the sensitive internal organs. Bacitracin and neomycin, for example, are so toxic to internal organs that they cannot be administered for systemic infections, but they can be used in topical ointments for superficial skin infections.

A technique of drug delivery to infected tissue, by bypassing healthy tissue, is being currently developed. One approach is to load medication into liposomes (see Point of Interest).

Major chemotherapeutic agents based on mechanism of action

Inhibitors of cell wall synthesis. Agents that inhibit cell wall synthesis are essentially nontoxic* to the host since cell walls are absent in mammalian tissue. The best examples of inhibitors of cell wall synthesis are the **penicillins** and **cephalosporins**, both of which possess a **β-lactam** ring (Figure 15-2). The mechanism of action of penicillin at one time was thought to be relatively simple and involve a single target site. Penicillin, because of its structural resemblance to the D-alanyl-D-alanine portion of the cell wall glycopeptide (Figure 15-3), specifically inhibits the linking of neighboring subunits via transpeptidation. β-lactams bind to the transpeptidase and inactivate it. The transpeptidation reaction lends structural support to the cell wall, but in the presence of penicillin the cell wall is mechanically weak and results in cell lysis. The mechanism of action of penicillin is apparently more involved and appears to vary between groups of bacteria and to be influenced by a variety of other factors. There now appears to be several possible target sites for penicillin, which are located in the cytoplasmic membrane and are referred to as **penicillin-binding proteins (PBPs)**. All penicillin-sensitive microorganisms have PBPs, with as few as three types in *Neisseria gonorrhoeae* and as many as seven types in *Escherichia coli.* The total number of PBP molecules that can be found in a single cell ranges from 1000 to 10,000 molecules. PBPs are enzymes that are responsible for the final stages of peptidoglycan assembly, that is, the incorporation of precursors into the established peptidoglycan layer (Table 15-1). The sensitivities of the PBPs to β-lactams, such as penicillins and cephalosporins, varies from species to species and is related to the type of β-lactam and its concentration. Through the use of mutants, investigators have been able to more precisely de-

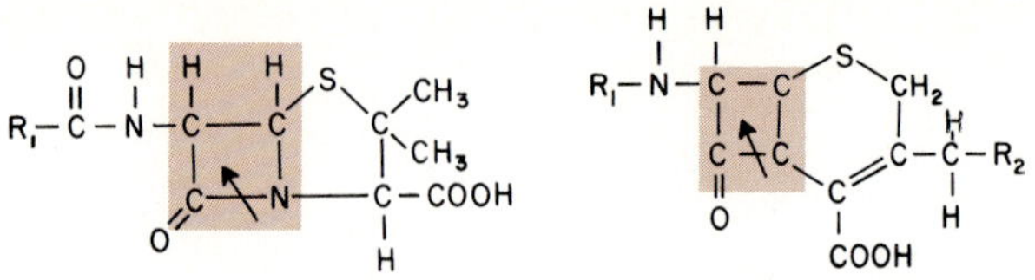

* Although not considered directly toxic to tissue, penicillin can elicit an allergic response in up to 5% of the human population.

Penicillin nucleus

Cephalosporin nucleus

Figure 15-2 *Structure of penicillin and cephalosporin nuclei demonstrating β-lactam rings* (colored squares). *Arrows indicate sites of action of β-lactamase that cleave β-lactam rings.*

Point of Interest

LIPOSOMES: A NEW DRUG DELIVERY SYSTEM

Liposomes are artificial phospholipid vesicles that are similar in properties to biological membranes. They are formed when a high concentration of phospholipid is mixed with water (see Chapter 2). As liposomes form, any water soluble molecules added to the water are incorporated into the aqueous spaces in the interior of the liposome sphere, whereas lipid soluble molecules added to the solvent during vesicle formation are incorporated into the lipid bilayer. Liposomes can therefore be used to transport a variety of chemical agents; one of their potentially commercial values is in the transport of chemotherapeutic drugs to treat disease. Many drugs show great promise because of their ability to destroy diseased cells, as well as kill infectious agents that attack a particular cell or tissue type. Oftentimes these drugs must be administered in small doses because they can randomly attack healthy cells. Liposomes may be used to surmount this obstacle.

In the body of the host, liposomes can absorb to various cells. When they attach to phagocytic cells they become engulfed. Following engulfment the contents of the liposomes are released into the cell. Liposomes are readily removed from the circulation of the host by phagocytic cells of the lymph nodes, spleen, liver, etc. Infectious diseases in which the parasite resides in phagocytic cells have been especially amenable to treatment with drug-bearing liposomes. Liposome therapy has been shown to be particularly effective in experimental treatment of a tropical disease called leishmaniasis (see Chapter 28). Treatment of the disease normally involves the use of highly toxic antimony compounds. The concentration of antimony strong enough to kill the parasite may also kill the patient. Consequently, the concentration of drug that can be safely used is often inadequate to destroy the parasite. When liposomes loaded with antimony are injected intravenously into an animal, the same cells that harbor the parasite also ingest liposomes. The drugs delivered with liposomes were shown to totally destroy the parasite, whereas injection of the drug alone at the same concentration had negligible effects.

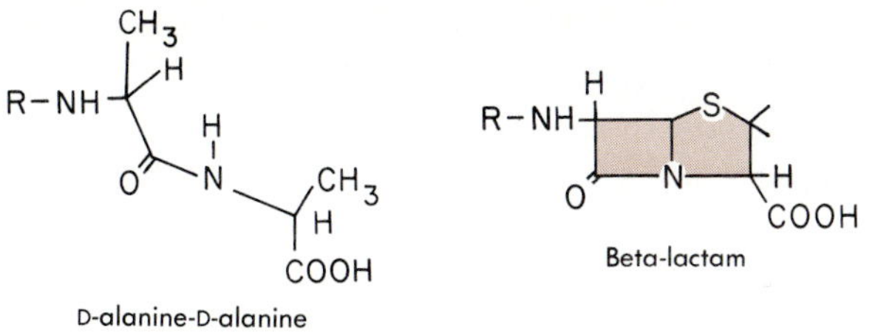

D-alanine-D-alanine

Beta-lactam

Figure 15-3 *Homology between* D*-alanine-*D*-alanine and β-lactam antibiotic, such as penicillin.*

termine the effect of β-lactams on specific PBPs. For example, at low concentrations of antibiotic, different morphological forms of a rod-shaped mircoorganism may appear (Figure 15-4). PBP 1A and B appear to be involved in cell elongation. PBP 2, when bound by certain β-lactams, causes a change in cell shape in which the normal rod shape of the microorganism changes to a stable round form (Figure 15-5). It has been suggested that PBP 2 might play a role in regulating glycan chain length in the cell wall. PBP 3 appears to be associated with cell division, more specifically, septation. Delay of septation can result in the microbial cell developing into long filaments (Figure 15-5). Frequently the PBP that causes the bactericidal response is the one that has the highest affinity for the β-lactam. Although cell lysis is the

Table 15-1 *Functions of the PBPs in* E. coli

PBPs	Function
1A	Peptidoglycan transglycolase* (responsible for the polymerization of glycan chains of peptidoglycan)
	Peptidoglycan transpeptidase (responsible for cross-linking peptide side chains)
1B	Peptidoglycan transglycolase*
	Peptidoglycan transpeptidase
2	Peptidoglycan transpeptidase
3	Peptidoglycan transglycolase*
	Peptidoglycan transpeptidase
4	D-Alanine carboxypeptidase (responsible for cleaving off the terminal D-alanine of the peptide side chain and preventing it from acting as a donor in transpeptidation)
	Peptidoglycan transpeptidase
5 and 6	D-Alanine carboxypeptidase

*Not affected by penicillin.

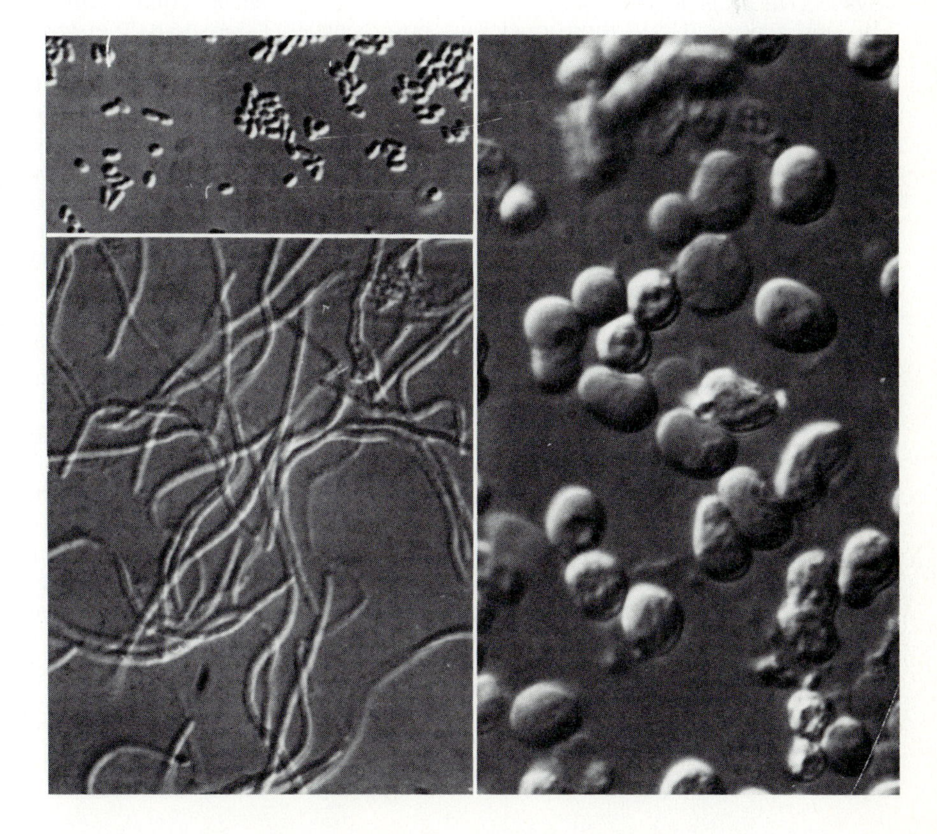

Figure 15-4 *Interference phase-contrast microscopy showing effects of different penicillins on the bacillus* Proteus mirabilis. *(×2500.) Upper left cells are normal and untreated; lower left cells were treated with subinhibitory concentrations of ampicillin and filamentous forms were produced; cells on right were treated with subinhibitory concentrations of mecillinam and round cells were produced.*

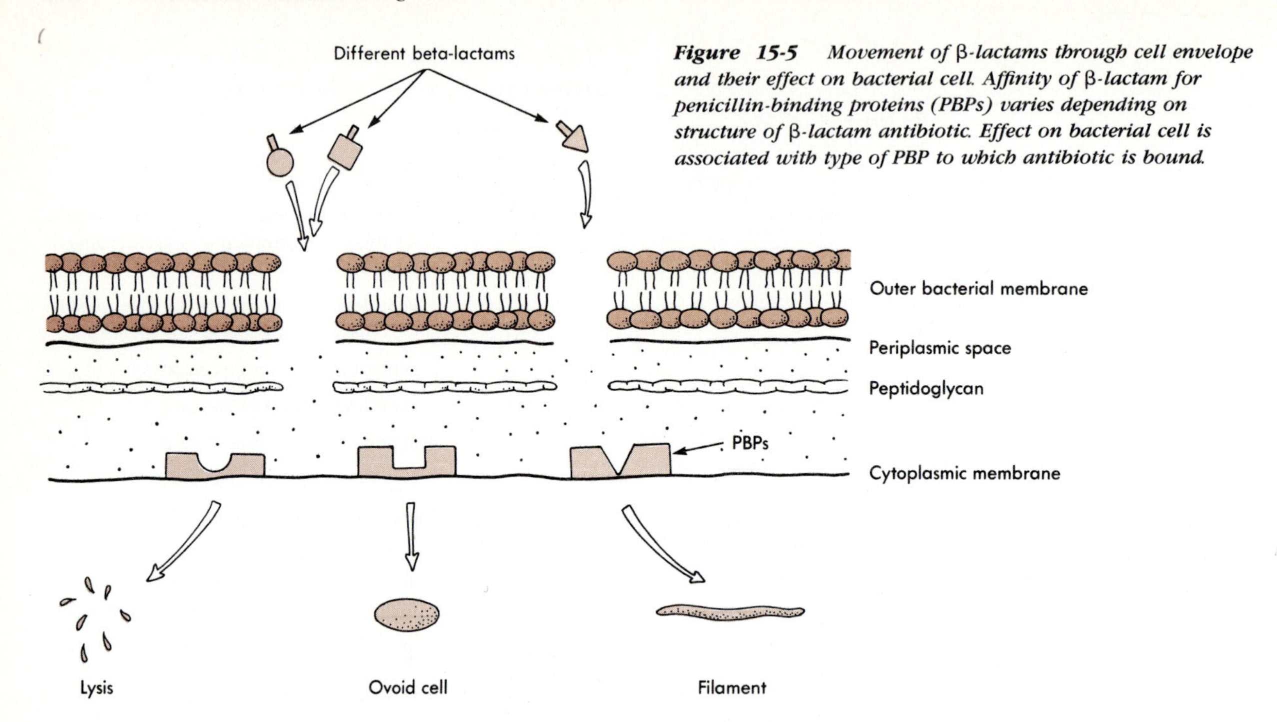

Figure 15-5 *Movement of β-lactams through cell envelope and their effect on bacterial cell. Affinity of β-lactam for penicillin-binding proteins (PBPs) varies depending on structure of β-lactam antibiotic. Effect on bacterial cell is associated with type of PBP to which antibiotic is bound.*

most frequent response to the action of β-lactams, not all bacteria lyse when killed by β-lactams, for example, *Streptococcus pyogenes* does not.

It now appears that in those organisms possessing **murein hydrolases** (autolytic enzymes* that hydrolyze the bonds between the *N*-acetylmuramic acid residues and the L-alanine residues of the peptide chains of the cell wall), two steps may be involved in the irreversible destruction of the cell. First, penicillin inhibits growth by binding to PBPs and interfering with peptidoglycan synthesis. Second, a murein hydrolase is activated to cleave cell wall, resulting in the release of all the peptide units and leaving a structurally weak cell wall. This is followed by exposure of the cytoplasmic membrane to osmotic imbalance and eventual lysis of the cell. Penicillin, which normally lyses *Streptococcus pneumoniae,* will only inhibit growth if the organism possesses a defective murein hydrolase. The relationship between lysis and the loss of viability is not yet clear in other organisms, and other mechanisms for the action of penicillin may exist.

Commonly prescribed drugs that inhibit cell wall synthesis but by different mechanisms than the penicillins and cephalosporins are **bacitracin**, **vancomycin**, and **isoniazid**. Cycloserine also inhibits cell wall synthesis, but it is seldom used as a chemotherapeutic agent (Table 15-2). The stages in cell wall synthesis where the various antimicrobials exert their effects are outlined in Figure 15-6. The synthesis of cell wall in *S. aureus* can be broken down into four stages:

Stage 1. This stage is involved in cell wall–precursor formation and takes place in the cell cytoplasm. The first reaction involves the covalent binding of uridine triphosphate (UTP) to *N*-acetylglucosamine to form UDP-*N*-acetylglucosamine. This latter compound goes through a series of reactions to form the precursor UDP-acetylmuramyl tripeptide, containing the amino acids L-alanine, D-glutamic acid, and L-lysine. (This peptide formation does not involve ribosomes.) An alanine dipeptide is then attached to the tripeptide to form muramyl pentapeptide. The two enzymes involved in these reactions are alanine racemase, which converts L-alanine to D-alanine, and D-alanyl-D-alanine synthetase, which attaches the alanyl-alanine dipeptide to the tripeptide. These enzymatic steps are inhibited by the antibiotic **cycloserine**, which is a structural analogue of D-alanine (Figure 15-7). The muramyl pentapeptide is now ready to take part in the second stage of cell wall synthesis.

Stage 2. This stage takes place on the cytoplasmic membrane. The UDP-acetylmuramyl pentapeptide formed in stage 1 is linked to UDP-*N*-acetylglucosamine to produce a disaccharide. The linkage of these molecules is brought about by a lipid carrier

* Autolytic enzymes are used normally during cell wall growth. They break down specific areas of the cell wall so that new cell wall components can be inserted before cell division.

Table 15-2 *Characteristics of chemotherapeutic agents that inhibit cell wall synthesis**

Agent	Specific mechanism of action	Activity spectrum	Toxicity	Comments
Penicillins				
Penicillin G	Prevents cross-linking of peptidoglycan strands in cell wall	Gram-positive bacteria; bactericidal	Nontoxic	Sensitive to β-lactamase
Ampicillin	Same as above	Gram-positive and gram-negative bacteria; bactericidal	Nontoxic, but rashes are frequent	Same as above
Carbenicillin	Same as above	Gram-negative bacteria; bactericidal	Nontoxic	Same as above
Nafcillin	Same as above	Gram-positive bacteria; bactericidal	Nontoxic	Resistant to β-lactamase
Methicillin	Same as above	Gram-positive bacteria; bactericidal	Nontoxic	Resistant to β-lactamase
Cloxacillin	Same as above	Gram-positive bacteria; bactericidal	Nontoxic	Resistant to β-lactamase and acid resistant
Cephalosporins				
Cephalothin	Same as penicillin	Gram-positive and gram-negative bacteria; bactericidal	Nontoxic	Resistant to β-lactamase
Cephaloridine	Same as above	Gram-positive and gram-negative bacteria; bactericidal	Toxic to kidneys	More active than cephalothin against gram-positive bacteria
Cephalexin	Same as above	Gram-positive and gram-negative bacteria; bactericidal	Nontoxic	Can be taken orally
Cefotaxime	Same as above	Gram-positive and most gram-negative bacteria; bactericidal	Nontoxic	Greater spectrum of activity than all other cephalosporins
Bacitracin	Interferes with transfer of peptidoglycan components in cell wall	Gram-positive bacteria	Toxic to kidneys	Used in topical ointments
Vancomycin	Interferes with lipid carrier of cell wall components	Gram-positive bacteria	Toxic to kidneys and hearing mechanism	Used primarily for staphylococcal infections
Isoniazid	Inhibits synthesis of mycolic acids required in cell wall synthesis of mycobacteria	Mycobacteria	Toxic to liver and peripheral and central nervous system	Used only in treatment of tuberculosis

*The structure of various semisynthetic penicillins can be found in Chapter 22 (Figure 22-1).

molecule (a C_{55} isoprenoid alcohol called **bactoprenol**). This results in the formation of a disaccharide-phospholipid complex. To this molecule is added five glycine residues, which are attached to the ε-amino group of lysine. Glycyl tRNA is the amino acid donor, but no ribosomes are involved in this peptide synthesis. The α-carboxyl group of the glutamic residue in the pentapeptide is amidated. The pentaglycinebridge is characteristic of *Staphylococcus aureus*. Other gram-positive organisms have bridge units consisting of other amino acids. Some species, both gram-positive and gram-negative, have no bridge units.

Stage 3. The phospholipid carrier transports the di-

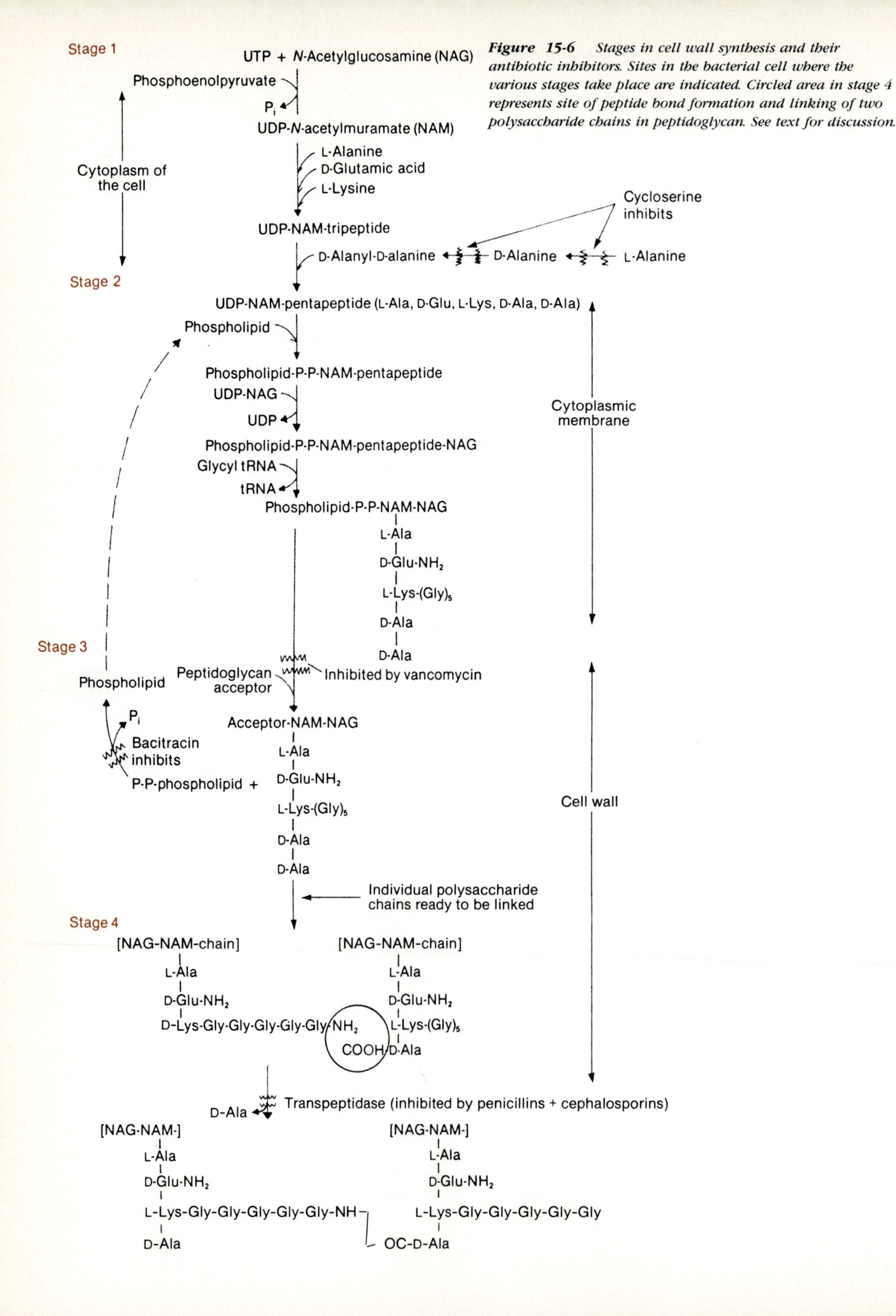

Figure 15-6 *Stages in cell wall synthesis and their antibiotic inhibitors. Sites in the bacterial cell where the various stages take place are indicated. Circled area in stage 4 represents site of peptide bond formation and linking of two polysaccharide chains in peptidoglycan. See text for discussion.*

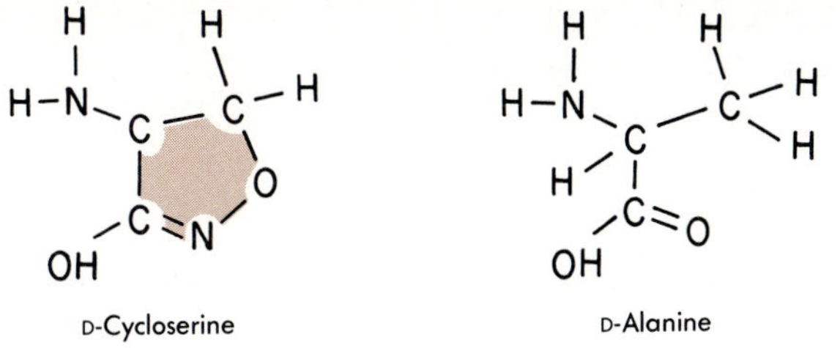

Figure 15-7 *Structural similarities of D-cycloserine and D-alanine.*

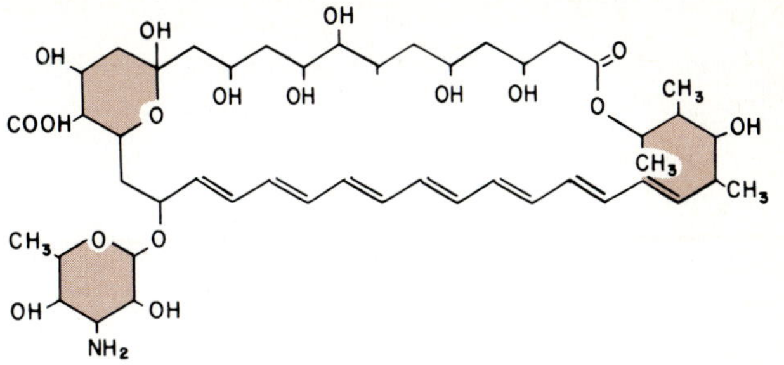

Figure 15-8 *Structure of antifungal agent amphotericin B.*

saccharide across the cytoplasmic membrane to preexisting portions of the cell wall and in the process is cleaved. The transport of the disaccharide across the cytoplasmic membrane is believed to be greatly facilitated because of the hydrophobic nature of the lipid carrier. The phospholipid carrier with its two phosphates is dephosphorylated and can continue with another round of stage 2 cell wall synthesis. Vancomycin inhibits the reaction in which the disaccharide is coupled to preexisting cell wall, and this results in the premature release of the lipid carrier. The enzyme catalyzing the reaction is peptidoglycan synthetase. **Vancomycin** is believed to bind to the acyl D-alanyl-D-alanine portion of the cell wall precursors, and it prevents the latter's binding to peptidoglycan synthetase. **Bacitracin** is believed to form a complex with the lipid carrier and inhibit the dephosphorylation of the lipid, thus preventing its regeneration.

Stage 4. The last stage involved in cell wall synthesis is the cross-linking of peptidoglycan, which takes place outside the cytoplasmic membrane. The cross-linking is catalyzed by a transpeptidase that splits the terminal D-alanine linkage and forms a peptide bond between the terminal glycine of the pentaglycine side chain and the next-to-last (penultimate) alanine of an adjacent chain. The unused terminal D-alanines are removed by a carboxypeptidase. **Penicillins** and **cephalosporins** inhibit the transpeptidation reaction.

Concept Check

1. What is meant by selective toxicity? What are the reasons for selective toxicity?
2. What mechanisms might exist for antibiotic-producing microorganisms to protect themselves from the action of the antibiotic?
3. Why does a single microorganism demonstrate a wide spectrum of sensitivity to penicillins, depending on the penicillin? Is this relationship associated with PBPs?
4. What are the various inhibitors of cell wall synthesis and the specific biochemical reaction they inhibit?

Inhibitors of cell membrane function. Drugs that affect the cell membrane bring about permeability changes in the microbial cell. The principal drugs currently used that affect the cell membrane are **nystatin**, **amphotericin B**, the **polymyxins**, and the **imidazole derivatives.** Amphotericin B and nystatin are **polyene antibiotics** characterized by a conjugated double-bond system (Figure 15-8) and are produced by species of *Streptomyces.* Amphotericin B binds irreversibly to the sterol component of the fungal membrane, ergosterol, but there is a lower degree of binding to the mammalian cell membrane because the principal sterol is cholesterol and not ergosterol. Polyenes, after binding to the fungal plasma membrane, cause it to become leaky to nutrients and ions. Cell death is believed to be a result of the increased permeability of the membrane to protons and abolishment of the proton motive force generated by the fungal plasma membrane ATPase. Thus uptake of nutrients is severely restricted. The permeability changes are believed to be a result of the formation of several polyene-sterol complexes that form large pores in the fungal plasma membrane. Other structures in the cell bounded by membranes are also affected by polyenes.

Polymyxins are polypeptide antibiotics produced by *Bacillus polymyxa.* They act as cationic detergents and interact with phospholipids, such as phosphatidylethanolamine in the bacterial cytoplasmic membrane, thus disrupting it and causing leakage of cellular contents. The precise chemical interaction that leads to cell death is not known. Polymyxins are also toxic to host cell membrane when taken internally. Most preparations containing polymyxin are for skin infections.

The imidazole derivatives include such drugs as **miconazole**, **clotrimazole**, **econazole**, **isoconazole**, and **ke-**

Table 15-3 Characteristics of chemotherapeutic agents that inhibit cell membrane function

Agent	Specific mechanism of action	Activity spectrum	Toxicity	Comments
Nystatin	Binds to sterols in cell membrane	Narrow-spectrum antifungal agent; fungicidal	Cannot be administered IV or IM*; relatively nontoxic taken orally or topically	Particularly useful in treatment of *Candida* infections
Amphotericin B	Binds to sterols in hyphal or cell membrane	Broad-spectrum antifungal agent; fungicidal	Toxic to all tissues; renal toxicity common; thrombophlebitis† with IV injections	Methyl ester of amphotericin B is less toxic
Polymyxin B and polymyxin E (colistin)	Competitively displaces magnesium and calcium from phosphate groups on membrane lipids	Gram-negative bacteria; bactericidal	Depresses muscle response to stimulation; impairs kidney function	Important only as topical agent; rarely used systemically
Imidazole derivatives miconazole, clotrimazole, econazole, isoconazole	Interferes with biosynthesis of lipids in fungal cell, especially sterols (?)	Primarily broad-spectrum antifungal agent, but also effective against some bacteria	Toxic if given systemically; drug interferes with enzyme activity of liver microsomes	Used primarily in topical ointments
Ketoconazole	Same as other imidazole derivatives	Broad-spectrum antifungal agent	Very few side effects when administered orally	Potentially most important antifungal drug; has cured progressive fungal infections that were unresponsive to other drugs

**IV*, intravenous; *IM*, intramuscular.
†Thrombophlebitis is an inflammation of a vein associated with blood clot.

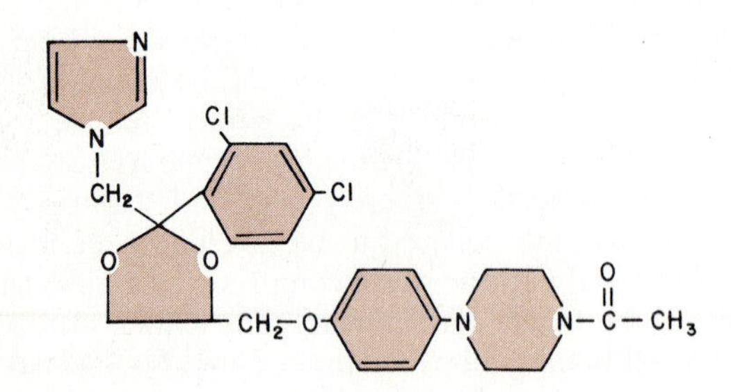

Figure 15-9 *Structure of antifungal agent ketoconazole.*

toconazole. They differ from other cell membrane inhibitors because they have a broad spectrum of activity. Miconazole, econazole, and isoconazole are used primarily as topical agents because they are toxic or because effective blood levels of the drug cannot be maintained when administered systemically. Ketoconazole (Figure 15-9) is one of the most recently developed imidiazole derivatives and offers many advantages over other antifungal agents. The drug, in addition to being broad spectrum, is also important because (1) it is active after oral administration; (2) therapeutic levels of the drug in the bloodstream are obtained rapidly after a single dose and are maintained for several hours; (3) except for some unusual circumstances, such as diabetes, it shows little toxicity when administered systemically; and (4) it has not given rise to fungal resistance.

Table 15-3 outlines the principal characteristics of the cell membrane inhibitors.

Inhibitors of protein synthesis. There are many drugs used today that are classified as inhibitors of protein synthesis, and in most instances they exhibit greater affinity for bacterial ribosomes than for mammalian ribosomes. The major inhibitors of protein synthesis are **chloramphenicol, tetracycline, amikacin, streptomycin, neomycin, kanamycin, gentamicin, clindamycin, lincomycin, tobramycin,** and **spectinomycin**. Many of these inhibitors belong to a group called the **aminoglycosides,**

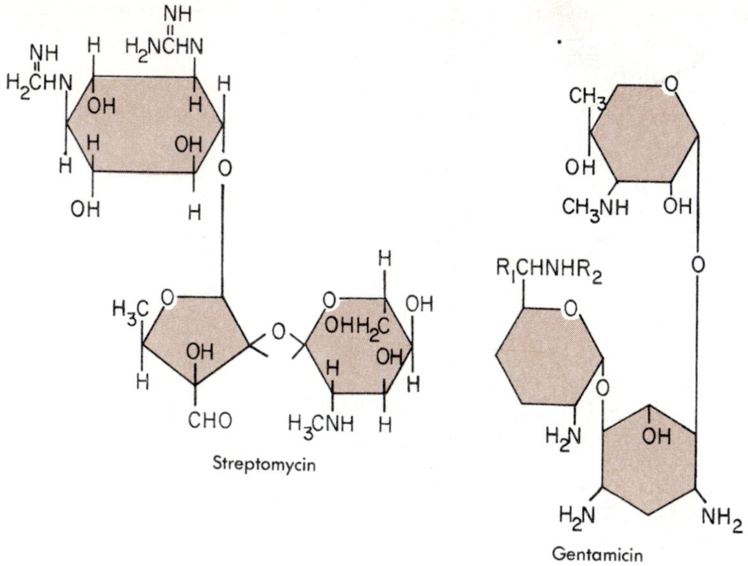

Streptomycin

Gentamicin

Figure 15-10 *Structure of some important aminoglycosides.*

which are structurally similar, have similar spectrums of activity, and exhibit similar toxic reactions in the host.

Streptomycin is the prototype of the aminoglycosides (Figure 15-10). It is believed to produce two effects in the microbial cell. It can bind to free 30S ribosomes before they engage the 50S ribosome to form a complex on the mRNA, and it can also bind to the 30S ribosome while it is engaged in chain elongation. When streptomycin binds to free 30S ribosomes, protein synthesis is completely blocked because chain elongation cannot occur, even though the 30S subunit complexes with the 50S subunit on the mRNA. Chain elongation is prevented because streptomycin prevents binding of the initiating codon, formyl-methionyl tRNA. When streptomycin binds to chain elongating ribosomes, there is ribosome distortion, which causes misreading and slows down protein synthesis. Streptomycin therefore produces a bacteriostatic effect, which in time becomes bactericidal. At high concentrations of streptomycin all the free 30S subunits can be bound with drug, thus preventing their initiation of protein chain elongation. This leads to cell death.

Table 15-4 describes the principal characteristics of the aminoglycosides, and other inhibitors of protein synthesis.

Inhibitors of nucleic acid synthesis. The principal inhibitors of nucleic acid synthesis are **rifampin** and **nalidixic acid**, which are antibacterial agents, and **griseofulvin** and **flucytosine**, which are antifungal agents (Table 15-5). Rifampin (also known as rifampicin) is an inhibitor of DNA-dependent RNA polymerase and prevents RNA synthesis in the cell. It is believed that following the formation of the first phosphodiester bond during RNA synthesis, a rifampin-RNA polymerase complex is formed that prevents further elongation of the RNA molecule.

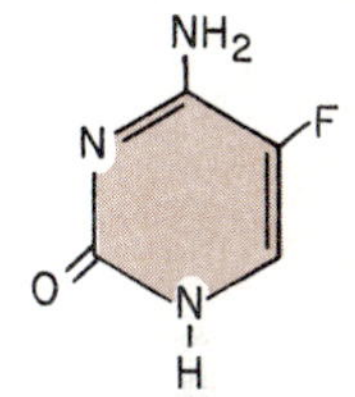

Figure 15-11 *Structure of flucytosine (5-fluorocytosine).*

Griseofulvin is believed to indirectly inhibit the synthesis of the hyphal cell wall of fungi by inhibiting nucleic acid synthesis and mitosis. The drug is effective only on growing cells, and it has been suggested that it interferes with microtubular activity. If this is true, then it might explain the drug's effect on cell wall synthesis, since it has been speculated that microtubules are involved in the transport of components that are used for cell wall synthesis (see Chapter 7).

Flucytosine (5-fluorocytosine) is a fluorinated pyrimidine (Figure 15-11) originally developed as an antimetabolite for cancer therapy. When used in the treatment of fungal infections, it is converted to 5-fluorouracil by organisms possessing the necessary deaminases. It is the latter reaction that makes flucytosine a selective drug. In mammalian cells, cytosine deaminase concentration is low and the reaction is virtually nonexistant. 5-Fluorouracil is converted to 5-fluorouridylic acid, which is incorporated into nucleic acid. 5-Fluorouracil substitutes for cytosine during RNA replication and heavily substituted RNA is translated into abnormal protein in the microor-

Table 15-4 *Characteristics of chemotherapeutic agents that inhibit protein synthesis*

Agent	Specific mechanism of action	Activity spectrum	Toxicity	Comments
Aminoglycosides				
Streptomycin	Binds to 30S ribosomes and causes incorrect codon-anticodon interaction	Bactericidal for gram-positive and gram-negative bacteria	Vestibular (internal ear) and auditory damage (hearing impaired)	Used with other drugs in tuberculosis
Neomycin	Similar to streptomycin	Same as streptomycin	Auditory and kidney damage	Cannot be used systemically
Kanamycin	Similar to streptomycin	Same as streptomycin	Auditory and kidney damage	Less toxic than neomycin
Gentamicin	Similar to streptomycin	Same as streptomycin, but also active against *Pseudomonas aeruginosa* and yeast	Auditory and kidney damage	Related drug called tobramycin is less toxic
Amikacin	Similar to streptomycin	Same as streptomycin, but with wider range; also effective against *Pseudomonas aeruginosa*	Auditory and kidney damage	Is acetylated derivative of kanamycin
Tetracyclines	Binds to 30S ribosome and prevents binding of aminoacyl tRNA to ribosome	Bacteriostatic against gram-positive and gram-negative bacteria; has widest spectrum of activity of any antimicrobial	Causes discoloration of teeth if administered during first 6 years of life	Prolonged use can cause overgrowth of yeasts
Chloramphenicol	Inhibits peptide bond formation	Bacteriostatic against gram-positive and gram-negative bacteria	Prolonged use leads to effects on bone marrow; can cause aplastic anemia	Cannot be used to prevent infection in premature infants, because it is liver toxic and may cause death
Erythromycin	Binds to 50S particle and prevents formation of 70S ribosome; has no effect on active 70S ribosome	Bacteriostatic against gram-positive bacteria or bactericidal, depending on concentration	Relatively nontoxic	A substitute for those sensitive to penicillin
Lincomycin and clindamycin	Acts on 50S ribosome and inhibits initiation of protein synthesis	Similar to erythromycin	Relatively nontoxic	Can cause pseudomembranous colitis* due to destruction of intestinal microflora and elaboration of toxin by clostridia
Spectinomycin	Similar to chloramphenicol	Bacteriostatic against gram-positive bacteria	Relatively nontoxic	Useful in treatment of gonorrhea caused by penicillin-resistant strains

*Colitis is an inflammatory condition of the colon.

ganism. Flucytosine may also affect DNA synthesis (Figure 15-12).

Many nucleic acid inhibitors are antiviral agents, but they also adversely affect mammalian nucleic acid synthesis and cell replication. Most antiviral drugs are nucleoside analogues and include such agents as **adenine arabinoside** (**ara-A**), **cytosine arabinoside** (**ara-C**), **5-iodo-2-deoxyuridine** (**IDU**), and **acyclovir**. IDU, for example, is incorporated into DNA, resulting in a defective DNA that inhibits cellular, as well as viral replication. Many of the antiviral drugs are therefore used as topical agents and are not used systemically.

Table 15-5 *Characteristics of chemotherapeutic agents that inhibit nucleic acid synthesis*

Agent	Specific mechanism of action	Activity spectrum	Toxicity	Comments
Rifampin	Binds to DNA-dependent RNA polymerase and inhibits RNA synthesis	Bacteriostatic; used only in combination with other drugs in treatment of tuberculosis (bactericidal for *Mycobacterium tuberculosis)*	Relatively nontoxic	Drug-resistant forms emerge readily if used alone
Nalidixic acid	Inhibits DNA synthesis, presumably by affecting a DNA unwinding enzyme	Gram-negative bacteria	Skin sensitivity to light; gastrointestinal upset	Used primarily for gram-negative urinary tract infections
Griseofulvin	Exact mechanism unknown	Dermatophytic fungi; fungistatic	Toxicity rare	Administered only orally
5-Iodo-2-deoxyuridine (IUDR)	Interferes with incorporation of thymidine into viral DNA	Antiviral agent used only topically in treatment of keratitis (inflammation of cornea) caused by herpesvirus	Prolonged use can damage corneal epithelium	
Adenine arabinoside (Ara-A)	Exact mechanism unknown, but does affect DNA polymerase	Antiviral agent is used systemically in treatment of some herpes simplex infections, such as encephalitis, varicella-zoster (chickenpox/shingles), and cytomegalovirus infections	Relatively nontoxic but can cause gastrointestinal upset	
Methisazone	Exact mechanism unknown, but blocks formation of viral structural protein	Antiviral agent used only in treatment and prevention of smallpox	Gastrointestinal upset	
Flucytosine (5-fluorocytosine)	Substituted for cytosine in RNA; causes formation of abnormal protein; may also inhibit DNA synthesis	Antifungal agent used for systemic infections, used in combination with amphotericin B	Relatively nontoxic	When used alone resistant strains emerge rapidly
Acyclovir	See discusion in text			

Acyclovir is a guanine derivative (Figure 15-13) that has been approved for use in the treatment of genital herpes. Acyclovir is phosphorylated by cellular kinases to a triphosphate derivative that specifically inhibits DNA polymerase, the enzyme required for DNA synthesis.

The high degree of specificity of acyclovir is related to its mechanism of action, which requires that the compound is first phosphorylated to acyclovir monophosphate. This phosphorylation is catalyzed by a virus-encoded thymidine kinase. Phosphorylation of acyclovir is severely restricted in mammalian cells in the absence of virus-induced kinase. Acyclovir monophosphate is converted to the diphosphates and triphosphates by host-cell kinases. Acyclovir triphosphate is a potent inhibitor of herpesvirus–induced DNA polymerase but has little effect on host-cell DNA polymerase. Acyclovir triphosphate can also serve as a substrate for herpesvirus–induced DNA polymerase, resulting in incorporation of acyclovir into viral DNA with subsequent early chain termination.

A very effective antiviral agent is **interferon**, which is produced by the host in response to viral infection. Characteristics of interferon are discussed in Chapter 17.

Antimetabolites. Drugs that are structurally similar to cellular metabolites and compete with the natural substrate for incorporation into functionally important components of the cell are called **antimetabolites**. The

Figure 15-12 *Mechanism of action of flucytosine (5-fluorocytosine). 5-Fluorocytosine is converted to 5-flourouracil diphosphate (5-FUR-P-P) through series of enzymatic reactions. Depending on presence and amount of specific enzymes present in cell, 5-FUR-P-P may be converted to ribose or deoxyribose derivative and thus inhibit RNA or DNA mechanisms, respectively, in cell.*

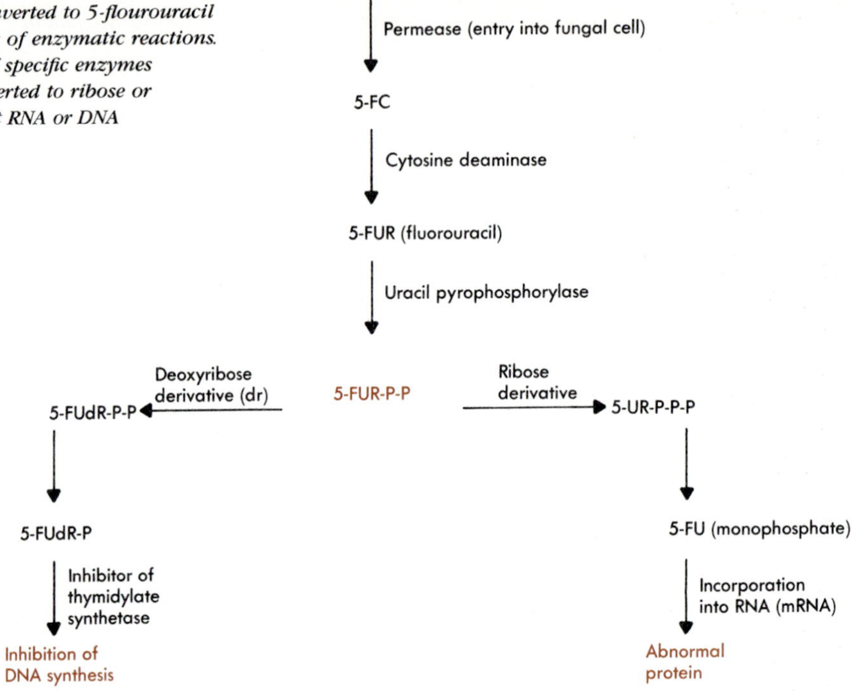

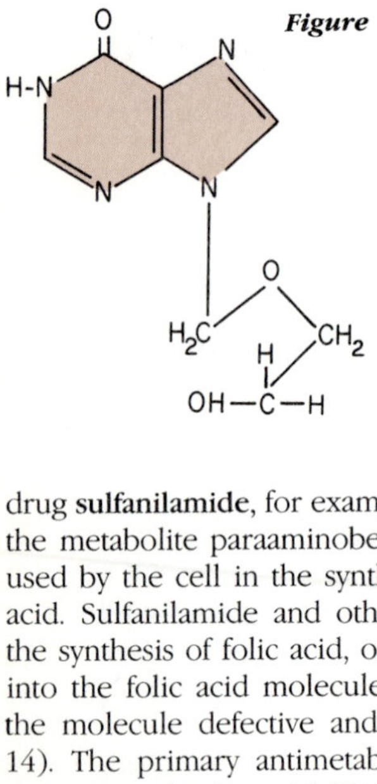

Figure 15-13 *Structure of acyclovir.*

drug **sulfanilamide**, for example, is structurally similar to the metabolite paraaminobenzoic acid (PABA). PABA is used by the cell in the synthesis of the coenzyme folic acid. Sulfanilamide and other sulfa drugs can prevent the synthesis of folic acid, or they may be incorporated into the folic acid molecule in place of PABA, making the molecule defective and nonfunctional (Figure 15-14). The primary antimetabolites are the sulfa drugs, **paraaminosalicylic acid (PAS)**, **trimethoprim**, **ethambutol**, **nitrofurantoin**, and **pyrimethamine**. The characteristics of these agents are outlined in Table 15-6.

Chemotherapeutic agents to prevent disease (prophylaxis)

Chemotherapeutic agents can be used not only to treat disease, but also to prevent disease; that is, they can be used prophylactically. Prophylaxis, if not used with some discretion, can lead to severe problems because it can create an environment in the host in which sensitive microorganisms are removed and only resistant ones remain. Most chemotherapeutic agents are not so exactingly selective that they inhibit or kill only the infectious agents causing the disease. They also affect other microorganisms in the surrounding microflora that, although not considered primary pathogens, can cause serious and even fatal infections in the debilitated and weak patient. One of the most important aspects of prophylaxis is the time of drug administration. If the drug is used over too long a period, drug-resistant microorganisms invariably appear. Some instances in which prophylaxis may be useful include using the drug isoniazid for tuberculosis; using certain drugs before intestinal surgery to prevent infection caused by spillage of intestinal contents; administering drugs to patients with heart abnormalities who are undergoing intestinal, genitourinary tract, or oral surgery—procedures that displace organisms into the bloodstream that could infect the heart; and administering drugs, such as rifampin, to prevent the spread of meningitis caused by *Neisseria meningitidis*.

Chemotherapeutic agents in agriculture

Nutrition experiments on cattle in the 1950s demonstrated that a feed containing mash from antibiotic fermenters stimulated growth. The same results were obtained when antibiotics, such as tetracycline, were

Table 15-6 *Characteristics of chemotherapeutic agents that act as antimetabolites*

Agent	Specific mechanism of action	Activity spectrum	Toxicity	Comments
Sulfonamides	Analogues of paraaminobenzoic acid (PABA); incorporated into folic acid precursors or inhibits incorporation of PABA	Gram-positive and gram-negative bacteria; bacteriostatic; antiprotozoal when used in combination with other drugs	Gastrointestinal; renal obstruction due to cystallization of drug	Many derivatives with variety of useful properties
Paraaminosalicylic acid (PAS)	Same as sulfonamides	Tuberculosis bacillus; bacteriostatic	Gastrointestinal disturbances	Always used in combination with other drugs
Trimethoprim	Analogue of portion of dihydrofolate molecule	Often used in combination with sulfonamides; bacteriostatic for gram-positive and gram-negative bacteria and similar to sulfonamides; also effective against protozoan that causes malaria	Same as sulfonamides	Important in treatment of urinary tract infections
Ethambutol	Unknown	Used only in combination with other agents in treatment of tuberculosis	Relatively nontoxic but can cause optic neuritis (inflammation of optic nerve)	
Nitrofurantoin	Unknown—may affect carbohydrate metabolism	Gram-positive and gram-negative bacteria; bactericidal	Somewhat toxic, causing gastrointestinal disturbances	Often used to treat urinary tract infections
Pyrimethamine	Similar to trimethroprim	Active against protozoan that causes toxoplasmosis	Depression of bone marrow, causing a decrease in blood platelets	Can also be used with sulfonamides as alternate therapy for treatment of malaria and other protozoan infections

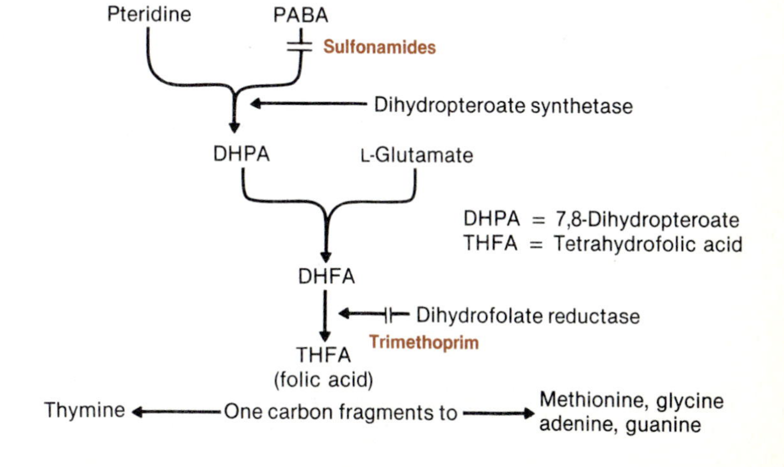

Figure 15-14 *Mechanism of action of sulfonamides and trimethoprim. Figure shows reactions that are involved in synthesis of folic acid (tetrahydrofolic acid, THFA) and sites at which sulfonamides and trimethoprim act as inhibitors. Sulfonamides either prevent formation of dihydropteroate (DHPA) or are incorporated in place of paraaminobenzoic acid (PABA). Trimethoprim inhibits enzyme dihydrofolate reductase, which prevents conversion of dihydrofolic acid (DHFA) to folic acid.*

purposefully added to animal feed. The reasons for this antibiotic effect are not understood, but one half of the antibiotics now produced in this country are used to feed farm animals. The Federal Drug Administration (FDA) estimates that 80% of swine, 60% of cattle, and 30% of chickens are raised with antibiotic-laced feed. Controversy exists concerning the health hazard believed to exist from the use of antibiotics, such as penicillin and tetracycline in animal feeds. Antibiotics in animal feeds are considered potentially hazardous because of the following:

1. Low levels of antibiotics in animal feeds promote the emergence of antibiotic-resistant bacteria in the animal's intestinal tract, particularly members of the Enterobacteriaceae, such as *E. coli*, and *Salmonella* species. When raw or improperly cooked poultry products are consumed by humans, these resistant microorganisms may colonize the human intestine and transfer their plasmid-associated resistance (R) factors to human strains. Many human infections caused by these antibiotic-resistant microorganisms are difficult to treat with our present arsenal of chemotherapeutic agents. This is because the plasmids acquired by a single bacterium may contain genes conferring resistance to as many as eight antibiotics. As long as the antibiotic is present in the feed the antibiotic-sensitive bacterial strains will be killed, while the antibiotic-resistant strains will proliferate.
2. Low levels of antibiotics in feeds may also result in residual levels of the antibiotic in animal tissue. Ingestion of such meat could favor the emergence of resistant microorganisms in the intestinal tract of the meat consumer. In addition, these antibiotic residues could induce an allergic response in the host. Some cases of allergic responses to milk contaminated by penicillin residues have been reported, particularly from foreign countries.

Concept Check

1. Name two chemotherapeutic agents that inhibit cytoplasmic membrane activity, inhibit nucleic acid synthesis, inhibit protein synthesis, and act as antimetabolites.
2. How can streptomycin produce two different effects in sensitive cells?
3. Describe three conditions in which prophylaxis with antimicrobials is recommended. Why is antimicrobial prophylaxis not employed to prevent most other infectious diseases, for example, syphilis?
4. Why is the incorporation of antimicrobials into animal feeds a controversial issue?

Proponents of the use of antibiotics in feed contend that most animal bacteria are poor colonizers of humans and that the exchange of R factor plasmids between animal and human strains of the Enterobacteriaceae is highly unlikely under normal conditions. Recent studies demonstrate that animal-to-human transmission of antimicrobial-resistant bacteria does take place (see Point of Interest), but the significance of this transfer is not always clear.

Microbial resistance to drugs

The ability of microorganisms to develop resistance to drugs is one of the major reasons why new drugs are constantly being sought and evaluated. Resistance to chemotherapeutic agents may be the result of natural resistance, mutation, or plasmid-associated resistance.

Natural resistance. Natural resistance to a drug occurs among a population of microorganisms because of some physical growth or metabolic factor that prevents the action of the drug in a normally sensitive population. Natural resistance can be exemplified in the following ways. Some microorganisms, such as bacteria, are inhibited by certain chemotherapeutic agents only during active growth and cell division. Penicillin, for example, which inhibits cell wall synthesis, is bactericidal *only* when the cell is actively synthesizing cell wall.

Drugs that inhibit metabolic processes within the cell must first penetrate the cell wall and cytoplasmic membrane. The envelope of some microbial cells is impermeable to certain drugs, and this impermeability may be the result of a structural component or a biochemical property. Some bacteria are resistant to the drug tetracycline, which is an inhibitor of protein synthesis. The resistance to tetracycline is not always due to alterations in the protein-synthesizing machinery of the microbial cell, but to changes in the cell envelope. This can be proved by removing the cell envelope of the resistant cells and testing the effects of tetracycline on protein synthesis in cell extracts. Tetracycline inhibits protein synthesis of the cell extract, which shows that resistance of the whole cells is due to a permeability factor. It is assumed that other microbial species to which the resistant species is related are normally sensitive to the drug.

Mutation. Drug resistance in the clinical setting may be the result of a single mutational event or may result from a series of separate mutations. Single-step resistance implies that in a population of cells resistance to a specific concentration of drug has increased significantly with the mutation. For example, the concentration of drug required to inhibit a microbial population may be 100 times the normal concentration. Each mutation in multistep resistance results in a slight increase in resistance. Significant resistance to penicillin, for example, may occur only after four or five single-step mutations in the microorganism. Mutational resistance occurs more frequently in the clinical setting when there is prolonged antimicrobial therapy and a single drug is used. Treatment of tuberculosis in earlier times was confined to the use of one drug, and this invariably led to drug-resistant mutants. Today several drugs are used in combination to treat chronic diseases such as tuberculosis, because if resistance to one drug occurs, the second and third drugs will destroy any resistant mutants.

Plasmid-associated resistance. Clinically, plasmid-associated drug resistance is potentially the most important mechanism of microbial resistance. This type of resistance is usually due to the formation of drug-inactivating or drug-modifying enzymes, but other mechanisms are also involved. As noted in Chapter 8, antibiotic-resistance genes, particularly those existing in transposons, are easily transferred to other cells during conjugation. They can also be transferred intracellularly between the different autonomous genetic units that are present in the cell (plasmid, chromosome, viral genome). In some instances a single plasmid may possess several drug-resistant genes, or the cell may contain several plasmids, with each possessing one or more antibiotic-resistant genes. Drug-resistant plasmids are usually maintained in the microbial population only as long as the drug is present in the patient's tissue. Once the patient is no longer using the drug, the resistant strains are replaced by drug-sensitive ones. Drug-resistant strains are often acquired by patients in the hospital, where they abound and where they can be transferred from nurse to patient, patient to patient, physician to patient, etc. Drug-resistant strains may also be acquired from domestic animals because their feed contains chemotherapeutic agents (see "Chemotherapeutic Agents in Agriculture").

Mechanisms of resistance (Table 15-7)

Altered transport. The ineffectiveness of certain antimicrobials (those that affect certain cytoplasmic properties of the cell) is usually due to alterations within the cell envelope (cytoplasmic membrane and cell wall). Resistance may result from changes in the lipid content of

Point of Interest

TRANSMISSION OF BACTERIA FROM ANIMALS TO HUMANS

Concern about the development of antibiotic-resistant strains of bacteria that are transferable to humans prompted the FDA in 1977 to propose a ban on the use of antibiotics in animal feed. However, no one clearly demonstrated that human illness is related to the use of antibiotics in animal feed, and the FDA proposal stagnated in Congress. In 1983 more solid evidence implicating the link was obtained. Eighteen persons clustered in four Midwestern states were afflicted by severe diarrhea caused by multiply-resistant *Salmonella enteritidis.* Twelve of these individuals had taken antimicrobials for nondiarrheal medical disorders (bronchitis, pharyngitis, etc.). Within 24 to 48 hours of taking the antimicrobials they became ill with salmonellosis. What had occurred was that most of these patients had asymptomatic salmonella infections and the use of antimicrobials, to which the salmonella was resistant, created a selective pressure that allowed growth of *S. enteritidis.* Some of these patients in the outbreak lived together, whereas others represented isolated cases. Epidemiological and laboratory investigations revealed that all patients had eaten hamburger originating from South Dakota cattle that had been fed subtherapeutic doses of chlortetracycline for growth promotion. The *S. enteritidis* isolates were all resistant to ampicillin, carbenicillin, and tetracycline. These *S. enteritidis* strains were different from 91 other samples of *S. enteritidis* taken from infected animals across the United States. The Centers for Disease Control has documented the transmission of salmonella from animal to humans since 1971. Their studies show the impact of infections caused by antimicrobial-resistant strains compared to antimicrobial-sensitive strains. In outbreaks involving antimicrobial-resistant organisms, 4.2% died from salmonellosis (13 deaths among 312 patients), whereas outbreaks caused by antimicrobial-sensitive organisms resulted in only 0.2 % fatalities (4 deaths among 1429 persons).

The study with antimicrobial-fed South Dakota cattle did not address the question as to how much human illness is caused by consumption of meat contaminated by bacteria. In 1984 and 1985 a study by Seattle researchers showed that one half the cases of diarrheal illness among local residents was caused by consumption of chicken contaminated by bacteria. They surveyed chicken sold at grocery stores and monitored the incidence of enteric disease among 300,000 members of a local health maintenance organization for 20 months. The study revealed that individuals who ate chicken had a 20% greater risk of contracting enteritis caused by *Campylobacter jejuni* than those who did not. One half of the victims (112) had eaten chicken before the onset of illness. Bacteria isolated from victims and chickens were similar in various laboratory tests, and both showed the same pattern of antibiotic resistance. Although the Seattle study did not find out whether the implicated chickens had been fed antibiotics to promote growth, it did demonstrate a link between human health problems and bacteria in meat.

Table 15-7 *Major mechanisms of microbial drug resistance*

Mechanism	Drug(s)	Organisms involved
Altered transport system		
Reduced uptake of drug	Tetracyclines	Gram-negative enterobacteria
Membrane not energized	Aminoglycosides	Anaerobes
Enzymatic inactivation		
β-lactamases	Penicillins and cephalosporins	Gram-positive and gram-negative bacteria
Chloramphenicol acetyltransferase	Chloramphenicol	Gram-positive and gram-negative bacteria
Enzyme modification		
Chloramphenicol acetyltransferase	Chloramphenicol	Gram-negative enterobacteria, *Pseudomonas* species
Aminoglycoside modification (phosphotransferase, adenyltransferase, acetyltransferase)	Aminoglycosides	
Alteration of antimicrobial targets		
DNA gyrase	Nalidixic acid	Gram-negative enterobacteria
RNA polymerase	Rifampin	Gram-negative enterobacteria
Penicillin-binding proteins	Penicillin	*Neisseria gonorrhoeae, Streptococcus pneumoniae, Staphylococcus aureus, Escherichia coli*
Methylated 23S RNA	Erythromycin and lincomycin	Staphylococci
30S ribosome	Streptomycin	Gram-negative enterobacteria
Synthesis of resistant metabolic pathways		
Dihydrofolate reductase	Trimethoprim	Gram-negative enterobacteria
Dihydropteroate synthetase	Sulfonamides	Gram-negative enterobacteria, staphylococci

the cell wall, alteration in the porin size of the cell wall of gram-negative bacteria, changes in the cell wall configuration, alteration of a specific transport mechanism, or enzymatic modification of drugs before their entry into the cytoplasm. Let us look at some examples of resistance mechanisms. Many of the enterobacteria carry plasmids that apparently code for cell envelope proteins that prevent penetration of the tetracycline antibiotics. Enterobacteria species also are resistant to β-lactamases because of alteration in the transport channels or porins, located in the outer membrane of the cell wall. The enterobacteria and pseudomonads also carry plasmids that code for aminoglycoside (streptomycin, gentamicin, kanamycin, etc.)-modifying enzymes, which are located in the periplasmic space.

The aminoglycoside modifying enzymes include:

1. Aminoglycoside phosphotransferase (APH), which catalyzes the phosphorylation of the hydroxyl group of the antibiotic:

$$\text{Aminoglycoside-OH} + \text{ATP} \xrightarrow{\text{APH}} \text{Aminoglycoside-O}-\overset{\overset{\displaystyle O}{\|}}{\underset{\underset{\displaystyle OH}{|}}{P}}-\text{OH} + \text{ADP}$$

2. Aminoglycoside adenyltransferase (AAD), which catalyzes the adenylation of the hydroxyl group of the antibiotic:

$$\text{Aminoglycoside-OH} + \text{ATP} \xrightarrow{\text{AAD}} \text{aminoglycoside-O-AMP} + {}_i\text{PP}$$

3. Aminoglycoside acetyltransferase (AAC), which catalyzes the acetylation of a hydroxyl or amino group of the aminoglycoside.

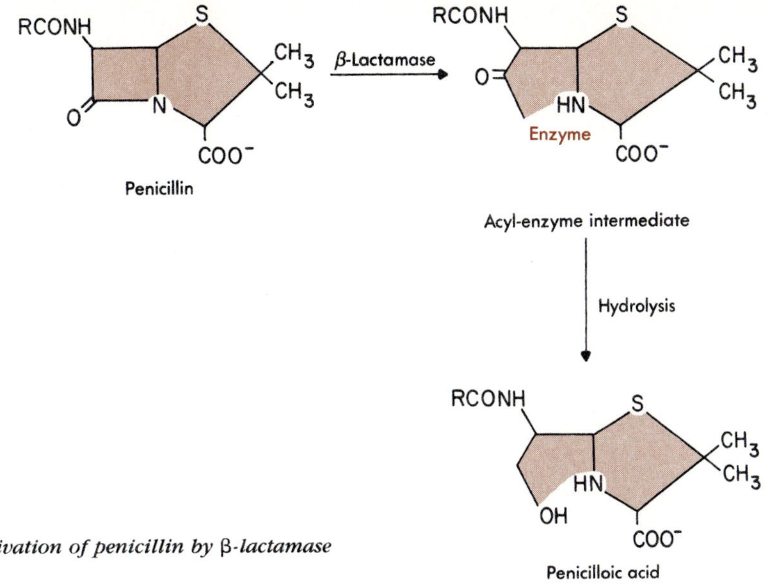

Penicillin —β-Lactamase→ Acyl-enzyme intermediate (Enzyme) —Hydrolysis→ Penicilloic acid

Figure 15-15 *Inactivation of penicillin by β-lactamase (penicillinase).*

$$\text{Aminoglycoside-OH} + \text{acetyl CoA} \xrightarrow{\text{AAC}} \text{aminoglycoside—O—}\overset{\overset{\displaystyle O}{\|}}{C}\text{—CH}_3 + \text{CoA-SH}$$

$$\text{Aminoglycoside-NH}_2 + \text{acetyl CoA} \xrightarrow{\text{AAC}} \text{aminoglycoside-}\underset{\underset{\displaystyle H}{|}}{N}\text{—}\overset{\overset{\displaystyle O}{\|}}{C}\text{—CH}_3 + \text{CoA-SH}$$

Enzymatic inactivation of the drug. The most ubiquitous drug-inactivating enzymes are the β-lactamases (Figure 15-15). There are nearly as many different β-lactamases as there are β-lactam antibiotics, which explains the relentless search for new penicillins. β-lactamases are found in gram-positive and gram-negative bacteria. They may be encoded for by genes located in the chromosome, or they may be found in plasmids, but sometimes they are found in both types of replicons in the same cell. Gram-positive bacteria produce the enzyme extracellularly, whereas gram-negative bacteria retain the enzyme in the periplasmic space. The β-lactamases in gram-negative bacteria are associated with transposons. The indiscriminate transfer of transposons may account for the widespread occurrence of antibiotic-resistance genes in many species of gram-negative bacteria.

A cytoplasmic-containing, drug-inactivating enzyme is chloramphenicol acetyltransferase (Figure 15-16). The enzyme is encoded in a plasmid that is carried by both gram-positive and gram-negative bacteria. There is still some dispute as to the location of some drug-inactivating enzymes. Some may be found in the cytoplasm, as well as the periplasmic space.

Alteration of the target site. Several components involved in replication and protein synthesis are the targets of some antimicrobials, but microorganisms have also evolved mechanisms of resistance. One type of alteration is plasmid associated and is found among the staphylococci. These organisms possess an enzyme that methylates the two adenine residues on the 23S RNA molecules of the 50S subunit of bacterial ribosomes. This type of alteration prevents the binding of erythromycin and lincomycin to the 50S subunit and prevents protein synthesis. Interestingly enough as previously indicated the organism from which erythromycin is produced, *Streptomyces erythreus*, carries out the dimethylation of 23S RNA that protects it from the antibiotic.

Alteration of the 30S ribosome of the gram-negative enterobacteria makes these organisms resistant to streptomycin. A protein in the 30S subunit is the target of the antibiotic, and a single amino acid change in either of two positions on the protein prevents binding of the drug. Enzyme inactivation and not ribosomal alteration, however, is the most frequently encountered type of streptomycin resistance in clinical medicine.

An important mechanism of resistance is alteration of the penicillin-binding proteins in species such as *Neisseria gonorrhoeae, Streptococcus pneumoniae* (South Africa strains), *Streptococcus faecalis*, and *Staphylococcus aureus*. Penicillin binding proteins 1, 2 and 3 appear to be most important in resistance to penicillin in gram-negative bacteria, but this mechanism of alteration is not fully understood. Penicillin may induce a change in the PBPs that interferes with normal cell surface activities,

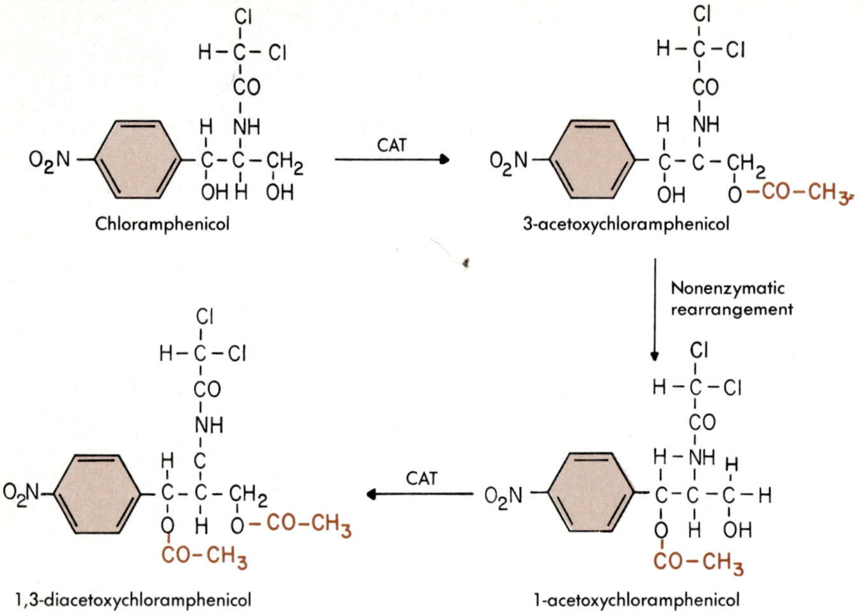

Chloramphenicol

CAT

3-acetoxychloramphenicol

Nonenzymatic rearrangement

1-acetoxychloramphenicol

CAT

1,3-diacetoxychloramphenicol

Figure 15-16 *Inactivation of chloramphenicol by chloramphenicol acetyltransferase (CAT).*

such as cell division and septation. Thus the microbial cells become tolerant to the presence of the antibiotic, resulting in bacteriostasis rather than bactericidal events.

Synthesis of resistant metabolic pathways. Some plasmids found in the enterobacteria and staphylococci code for metabolic enzymes that substitute for the chromosome-mediated enzymes. The site of inhibition of sulfonamides, for example, is the enzyme dihydropteroate synthetase, which is required for dihydrofolate synthesis. The antimicrobial competes with PABA, the normal metabolite of the enzyme (see Figure 15-14) but sulfonamide-resistant bacteria possess plasmids that code for a dihydropteroate synthetase that is resistant to the drug.

The enzyme dihydrofolate reductase, which is also involved in folate synthesis, may also be coded for by a plasmid. The plasmid-associated enzyme is resistant to trimethoprim, which normally inhibits the chromosome-mediated enzyme.

Increased production of a drug-competing metabolite. An infrequent mechanism of resistance involves competition between sulfa drugs and PABA for the active site of the enzyme. Increased resistance to the sulfonamide may occur when a mutation causes an increase in the metabolite (in the case of sulfa drug resistance, PABA). This is a quantitative resistance, because the concentration of drug required to inhibit the microorganism may have to be elevated considerably. In fact, the concentration of drug may be so high that it becomes toxic to human tissue, and a substitute drug must be selected for therapy.

Antibiotic tolerance. A novel response, **antibiotic tolerance**, discovered in 1970 followed the characterization of a mutant of *Streptococcus pneumoniae.* A culture of the wild type of this organism was lysed and killed during exposure to penicillin but the mutant did not lyse and exhibited only a slow loss of viability. Penicillin was therefore bactericidal to the wild type but only bacteriostatic to the mutant. Originally believed to be only laboratory mutants, tolerant species have been found among clinical isolates. The mechanism of tolerance, which so far has been associated with β-lactams, is not understood. A leading hypothesis is that tolerant species have a low level of autolytic enzymes, such as the peptidoglycan hydrolases that hydrolyze covalent bonds in the cell wall (Figure 15-17). Regardless of the mechanism, tolerance improves the chances of bacterial survival during antibiotic treatment and may be regarded as a type of resistance.

Measurement of antimicrobial activity. The activity of a drug can be measured quantitatively and qualitatively, and in each type of test certain standards must be maintained so that appropriate evaluations can be made. The tests are used not only to evaluate recently discovered drugs, but they are also used in the clinical laboratory to determine what chemotherapeutic agents will be useful in the treatment of specific infectious diseases.

Quantitative in vitro susceptibility tests. Quantitative in vitro susceptibility tests are used to determine more exactly the concentration of drug that will either inhibit the growth of microorganisms (the **minimal inhibitory concentration**, or **MIC**) or will kill microorganisms (the **minimal bactericidal concentration**, or **MBC**).

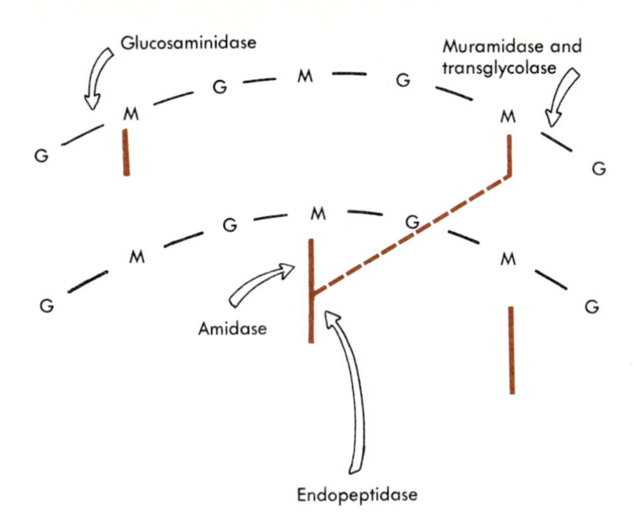

Figure 15-17 *Site of action of various autolytic enzymes found in bacterial cell wall. Two peptidoglycan strands are indicated with stem peptides but only one crosslink peptide (------) is shown.* G, N-*acetylglucosamine;* M, N-*acetylmuramic acid.*

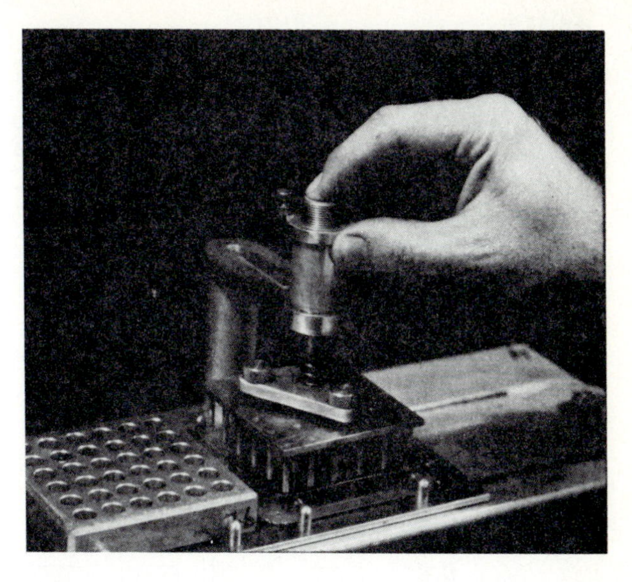

Figure 15-18 *Steers inocula-replicating device. Plate on left contains 36 reservoirs for bacterial inocula. Bottom plate on right contains antibiotic-containing medium on which bacteria will be seeded.*

Quantitative tests are not routinely used in the clinical laboratory but are more appropriate for the experimental laboratory. They have more application in the pharmaceutical laboratory, where exact concentrations of drug must be known in order to establish levels of drug for future animal and human studies. Two in vitro tests are available for determining the MIC of a drug: the agar dilution test and the broth dilution test. In the **agar dilution technique** the antimicrobial is incorporated into a liquefied agar medium (45° to 50° C), which is then mixed and poured into an agar plate. A series of plates are set up so that each will have a different concentration of drug. An inoculating device (Figure 15-18) is used that can seed up to 36 different microbial species on the surface of the agar. The microorganisms selected are commonly rapid-growing pathogens, such as members of the Enterobacteriaceae, *Staphylococcus aureus,* enterococci, *Pseudomonas* species, and other genera with similar growth rates. Slow growers give spurious results that cannot be extrapolated to the in vivo situation. The seeded plates are incubated overnight at 35° to 37° C, and the MIC is read as the lowest concentration of the seeded microorganism. The MIC and MBC can be calculated in the **broth dilution** tests, and the details of the design of this test are outlined in Figure 15-19. MICs obtained from these assays can be used in the clinical laboratory to establish the dosage that will be effective in treatment of human infections (see Interpretation of the Results of Disk Diffusion).

Qualitative antimicrobial sensitivity tests (disk diffussion). The **disk diffusion** test developed by Bauer, Kirby, Sherris, and Turck is an agar diffusion technique recommended by the Food and Drug Administration and is used widely in the hospital clinical laboratory for determining antimicrobial activity. The basic principle of the disk diffusion test is that a standard concentration of microorganisms is seeded on an agar plate, on which is placed a disk containing an antimicrobial of known potency. Those microorganisms inhibited by the antimicrobial, following suitable incubation, will show a zone of inhibition the diameter of which is proportional to the susceptibility of the microorganism, as well as other factors, such as rate of diffusion of antibiotics, temperature of incubation, and physical and chemical characteristics of the agar medium (Figure 15-20).

The agar medium employed in the disk diffusion tests is Mueller-Hinton agar, which contains no inhibitory substances and supports the growth of a wide variety of microorganisms. The depth of the agar in the plastic culture plate must be 4 mm, and any variation in depth will influence the zone of inhibition because of changes in the rate of drug diffusion.

Filter paper disks of specified weight and size and absorbability contain a specific amount of antimicrobial. Once the disk is placed on the agar, the antimicrobial diffuses through the agar, producing a gradient of concentrations, with the highest concentration of drug nearest the disk.

The antimicrobial selected for testing the susceptibility of a microorganism is usually a representative of a family of antimicrobials that have similar modes of action and toxicity. Generally speaking, the type of antimi-

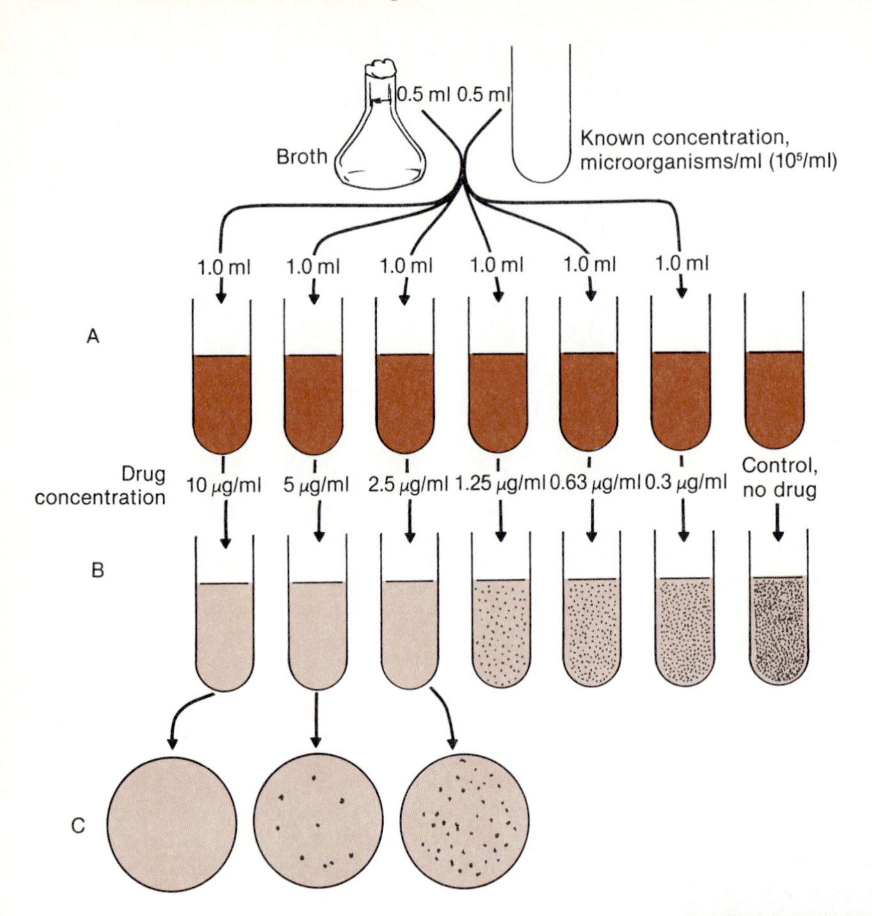

Figure 15-19 *Broth dilution test used to determine minimal inhibitory concentration (MIC) and minimal bactericidal concentration (MBC) of antimicrobial compound.* ***A,*** *Each tube contains 9 ml of sterile broth and different concentrations of antimicrobial. Concentration of organisms in each tube is so small that there is no visible turbidity. Tubes are incubated at 35° C for 12 to 18 hours.* ***B,*** *Tubes are examined for macroscopic growth. Lowest concentration of drug showing no growth is MIC. In this case it is 2.5 μg/ml.* ***C,*** *To determine MBC, loopfuls are removed from tubes in* ***B*** *showing no growth and are streaked on agar plates. Plate showing no growth indicates drug concentration that destroys microorganisms.*

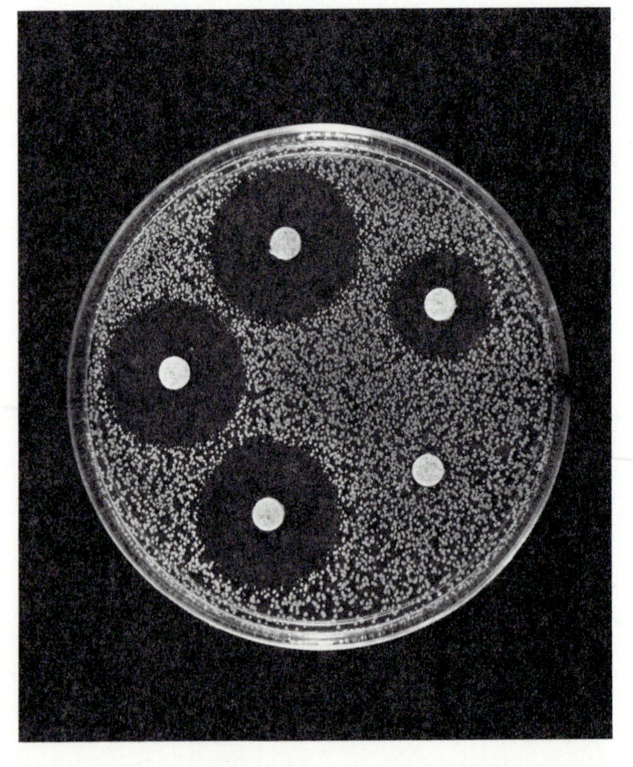

Figure 15-20 *Disk diffusion test. Agar medium is inoculated with microbial species, followed by placing of antibiotic-containing disks on surface of medium. Plates are incubated for 18 to 24 hours and then examined for zones of inhibition. Plate demonstrates that zones of varying diameter surround disks except for disk in lower right. Seeded microorganisms are totally resistant to antimicrobial present in disk at lower right.*

Table 15-8 *Zone-size interpretive chart: a modified Kirby-Bauer table*

Chemotherapeutic agent	Disk potency (μg)	Inhibition zone diameter (MM)		
		Resistant	Intermediate*	Sensitive
Ampicillin (gram-negative bacteria)	10	11 or less	12 to 13	14 or more
Ampicillin (staphylococci and penicillin G–susceptible organisms)	10	20 or less	21 to 28	29 or more
Chloramphenicol	30	12 or less	13 to 17	18 or more
Clindamycin	2	14 or less	15 to 16	17 or more
Nitrofurantoin (urinary tract infections only)	300	14 or less	15 to 16	17 or more
Gentamicin	10	12 or less	13 to 14	15 or more
Tetracycline	30	14 or less	15 to 18	19 or more

*When the zone size falls in the intermediate range, the organism is said to be neither sensitive nor resistant to the drug. The test can be repeated to ensure that all standard conditions have been maintained. More often, another antimicrobial is tested—one that will be more effective against the microorganisms; that is, the zone diameter will be in the sensitive range.

crobial selected for testing is dictated by the site of infection (urinary tract, spinal fluid, etc.), since certain species of microorganisms are frequently associated with such sites of infection in the human body. The potency of the antimicrobial in the disk is selected on the basis of previous determinations of MIC concentrations with over 100 strains of bacteria (recall that the agar dilution and broth dilution techniques were used to obtain these MICs) and by concentrations of the antimicrobial that are achievable in humans without causing severe toxicity.

In the clinical setting the microbe is isolated from the patient and is tested against specific concentrations of the antimicrobial. The concentration of microorganisms used is approximately 3×10^8 cells/ml, which is obtained from an actively growing culture. A swab is used to spread an inoculum across the surface of the plate, and within 15 minutes after the inoculation the antimicrobial disks are placed on the surface of the plate so that they are at least 15 mm from the edge of the plate and at least 15 to 20 mm from each other. This procedure prevents overlapping of zones produced after incubation of the culture plates.

Interpretation of the results of disk diffusion. Inoculated plates used to demonstrate microbial susceptibility are incubated at 35° to 37° C for 18 hours, and any zones of inhibition around the disks are measured with a micrometer (Figure 15-20).

The outer limit of the zone contains a concentration of antimicrobial that is similar to the MIC of that antimicrobial to a specific microorganism. Actually, there is no way to predict in vivo success based on in vitro results, and for this reason interpretive charts, which represent a compilation of clinical studies, have been established. These charts, which have been referred to as **Kirby-Bauer tables**, present a selection of zone diameters based on the response of microorganisms to a specific concentration of drug. The zone diameters are interpreted as *susceptible, intermediate,* and *resistant*. The term *susceptible* implies that an infection caused by the strain tested will respond favorably to the indicated drug. *Resistant* implies that the strain tested is not inhibited by the drug at therapeutic concentrations. *Intermediate* usually implies that an alternative drug should be tested to obtain favorable therapy. An example of a Kirby-Bauer table is Table 15-8. New drugs require publication of updated charts at frequent intervals. The selection of cutoff points for these three categories is finally based on years of clinical studies. As long as certain standards such as agar medium, inoculum size, and disk potency have been adhered to, the clinician can evaluate the potential effectiveness of a drug in the treatment of infections by comparing the zone diameter of his patient's pathogen with the Kirby-Bauer tables.

Commercial testing systems. Commercial antimicrobial susceptibility testing systems are also available for clinical laboratories. There are basically two types of systems. One type uses a broth dilution technique for determining the MIC and MBC. The broth dilution technique, which gives a more definitive determination of MIC and MBC than the agar dilution procedure, is ordinarily very time consuming. A microdilution procedure has been developed commercially in which microtrays are filled with antibiotics added by the manufacturer. Some clinical laboratories use this technique instead of the disk diffusion technique. A second type of test uses growth curve kinetics by measuring the turbidity of antibiotic-treated and untreated cell suspensions. These measurements are evaluated by a built-in computer.

Concept Check

1. What are the three most important ways microorganisms can develop resistance to drugs? Which one is the most important and why?
2. Explain the biochemical mechanism for microbial resistance to antimicrobials. Where are the majority of genetic determinants for these resistance mechanisms located?
3. How does antibiotic tolerance differ from other mechanisms of antimicrobial resistance?
4. What is MBC and MIC? How are they measured in the laboratory?
5. What is the value of the disk diffusion test? What factors influence the results?

STERILIZATION AND DISINFECTION

Terminology

Procedures used to control microbial growth do not always elicit the same results. Some agents destroy all forms of microbes, whereas other agents only inhibit their growth (and as soon as the chemical agent has been sufficiently diluted, the microorganisms resume growth). It is therefore important for the research microbiologist, nurse, physician, laboratory technician, and student laboratory worker to apply the appropriate procedures and to understand the terminology that characterizes these procedures.

Sterility is a term used to denote that the object or solution being treated is devoid of any viable form of life. Some mechanisms used to achieve sterility have a greater probability of occurring per unit time than do others. Methods that achieve sterility in the shortest time possible without destroying the object being treated are the most useful. Because the microbial spore is the most resistant form of life to be encountered, it is used to assess the reliability of the sterilization. A less precise term is **disinfection**, which implies that the object being treated is rendered devoid of infectious microorganisms. For example, an object may be contaminated by several different microbial species: some pathogenic and others nonpathogenic. A disinfectant would at the very least destroy any vegetative forms that could cause disease but not necessarily the nonpathogenic types. Most disinfection processes use chemicals. In general heat is the most widely used sterilizing agent. The term **antiseptic** refers to the application of a nonchemotherapeutic agent to the surface of tissue that minimally would inhibit the growth of potentially pathogenic species. Antiseptics must be of such a nature that they do not harm tissue. **Sepsis** refers to the breakdown of living tissue due to the action of microorganisms, and any procedure that prevents this condition is called **aseptic technique**. Aseptic techniques imply that precautions have been taken to prevent the introduction of microorganisms into the body by eliminating or reducing the number of microorganisms from the environment as well as objects that come in contact with tissue. Germ-free animals (**gnotobiotic** animals), for example, represent a sterile system that can be broken by introducing microorganisms.

The term *aseptic technique* is also used to describe those conditions that the microbiologist strives for in transferring cultures in the laboratory. A researcher studying the metabolic or genetic characteristics of a microbial species cannot afford to have the culture contaminated by other microorganisms.

Physical methods

Heat

Moist heat. Water is important in biochemical reactions because it can cause the hydrolysis of macromolecules such as proteins, nucleic acids, and lipids. Moisture at elevated temperatures causes the coagulation of macromolecules, but the lethal effects of heat are probably due first to the disintegration of cytoplasmic membrane and inactivation of enzymes. Some microbial forms, such as the bacterial spore, are essentially devoid of free water, and this property, as well as the thick refractile outer coat of the spore, provides the cell with relative thermoresistance. Bacterial spores are the most heat resistant form of microbial life, and any procedure that destroys spores will also destroy other microbial life.

Moist heat at temperatures of 60° and 80° C and applied for 30 minutes is sufficient to kill vegetative bacteria, viruses, and fungi, but not bacterial spores. Boiling (100° C) can destroy bacterial spores, but the time required depends on the bacterial species and may range from 30 minutes to 5 to 6 hours. Boiling kills most vegetative forms of life in 3 to 5 minutes, but because of its corrosive effects it is not recommended for sterilization of metallic instruments. **Pasteurization**, which uses moist heat, is employed on palatable liquids such as milk, beer, and wine. Pasteurization destroys vegetative pathogens that may be derived from the soil, cattle, or humans but does not affect many of the microorganisms that spoil milk. There are two techniques for pasteurization of milk: in one a temperature of 62.9° C is applied for 30 minutes to large vats of material, whereas in the other thin films of milk pass over pipes heated to 71.5° C for 15 seconds (see Chapter 21).

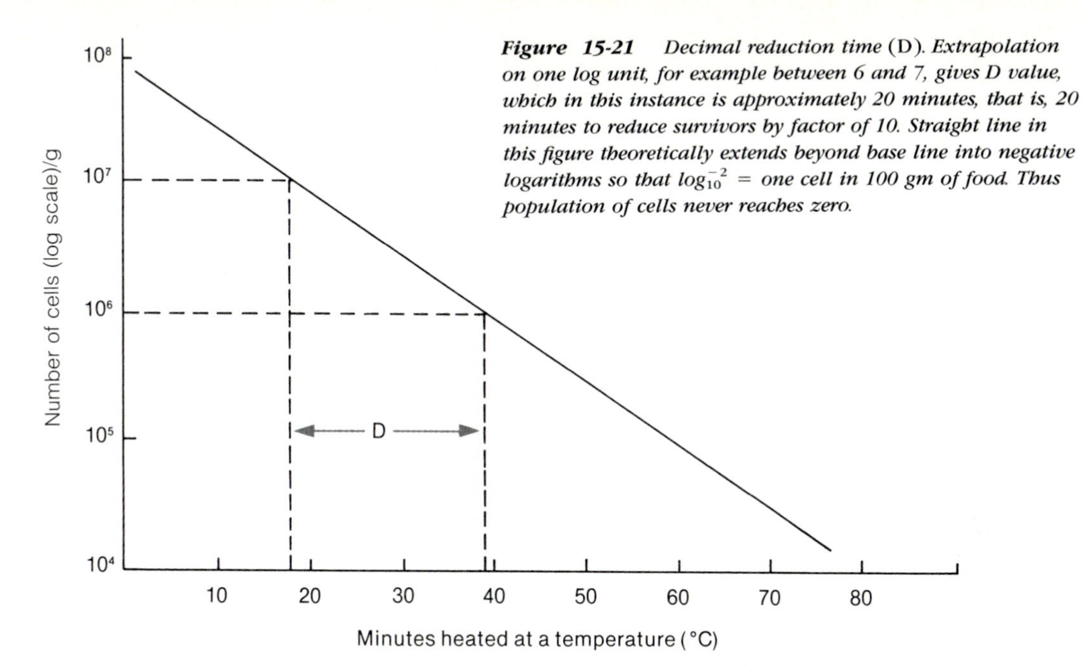

Figure 15-21 *Decimal reduction time* (D). *Extrapolation on one log unit, for example between 6 and 7, gives D value, which in this instance is approximately 20 minutes, that is, 20 minutes to reduce survivors by factor of 10. Straight line in this figure theoretically extends beyond base line into negative logarithms so that* $\log_{10}^{-2}$ *= one cell in 100 gm of food. Thus population of cells never reaches zero.*

The most preferred method of sterilization is steam under pressure, which is called **autoclaving.** The autoclave is a chambered device in which saturated steam is generated and placed under pressure. The temperature and pressure are adjustable, but the most satisfactory conditions for sterilization are 121° C and 15 pounds per square inch (psi) applied for a minimum of 15 minutes. These conditions are known to destroy any microbial life (provided that moisture touches every surface to be sterilized).

When bacterial cells are killed by heat they do so in a logarithmic fashion. A valuable way to characterize heat inactivation of bacteria is to determine the fraction of survivors after a period of time. For example, the time in minutes required for a tenfold (one logarithm) reduction in the population is the **decimal reduction time** or **D value** (Figure 15-21). Increases in temperature result in decreases in the D value. The D value is a measure of heat resistance, which makes it possible to determine beforehand a heating process that will result in a specific level of killing (Figure 15-22). This information is particularly useful for processing canned foods (see Chapter 21). The D value is useful in many other aspects of microbiology, because it can be used to estimate or compare survival in the presence of germicides, acids, or any adverse chemical or physical environment in which microbial death ocurrs.

When D values are plotted against corresponding temperatures, a **thermal death time (TDT)** curve is produced (Figure 15-23). Thermal death time is the time necessary to kill a given number of microorganisms at a specified temperature. Examination of various points on the TDT curve (also a straight line) will determine the resistance of the microorganisms at various tempera-

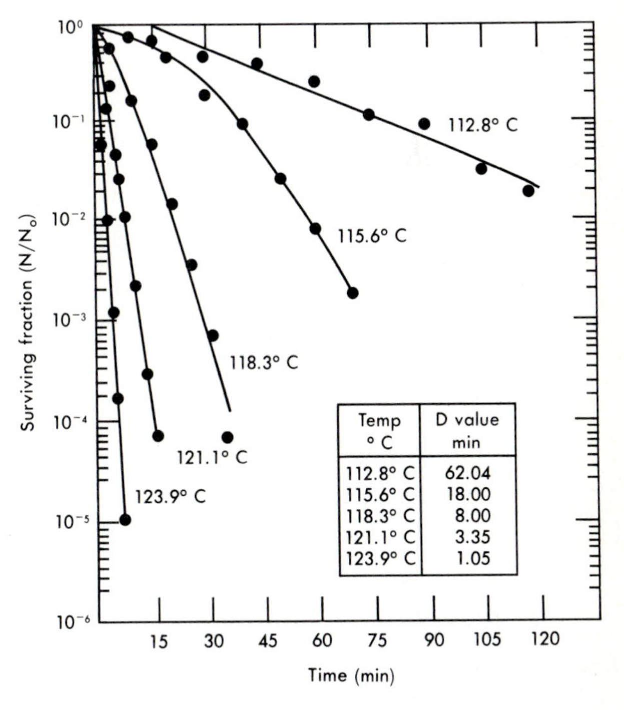

Temp °C	D value min
112.8° C	62.04
115.6° C	18.00
118.3° C	8.00
121.1° C	3.35
123.9° C	1.05

Figure 15-22 *Thermal inactivation of spores of* B. stearothermophilus. *Heat inactivated (100° C for 15 minutes) spores were treated at various temperatures for various times, plated on growth media, and incubated at 55° C for 24 hours. Surviving fraction* (N/N_0) *was number of viable spores at given time divided by number of spores at time zero. Curvlinear portions (112.8°, 115.6,° and 118.3° C) probably a result of some spores being injured but none inactivated. Rapid inactivation and no injury at 121.1° and 123.9° C, thus no curvlinear portion.*

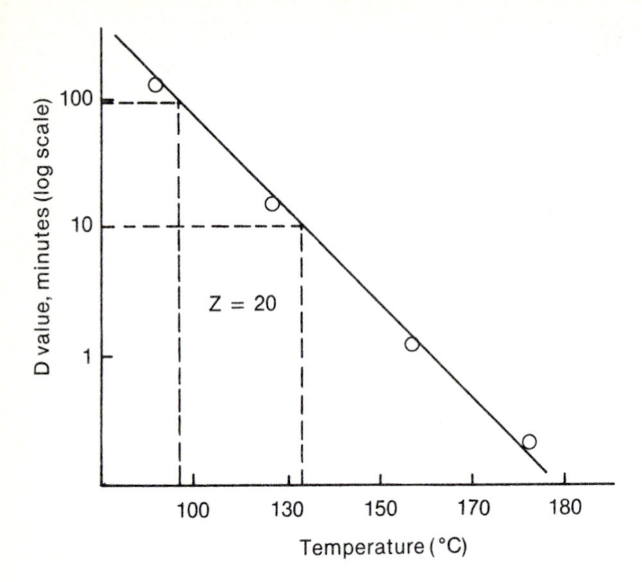

Figure 15-23 *Thermal death time* (TDT) *curve. Four D values* (o) *have been graphed. Z is the degree centigrade required for TDT to traverse one log cycle, here approximately 35 minutes.*

tures. The slope of the curve is expressed by Z, which is the degree Celsius required for the TDT to traverse one log cycle (or tenfold change in the D value).

Dry heat. Dry heat does not inactivate biological molecules as quickly as moist heat. It is a process that dehydrates the cell, causing solute precipitation and oxidation of macromolecules rather than their coagulation. Dry heat is used in the range of 160° to 180° C for 1 to 2 hours and is appropriate wherever moist heat cannot be applied, such as the sterilization of certain glassware, powders, and oils.

Cold. Extremes of cold can destroy microorganisms, but it is not a practical technique for sterilization. The effect of cold is dictated by the extremes of temperature attained. The metabolism and growth of most microorganisms is reduced at refrigeration temperatures (2° to 6° C). This principle is effectively employed in the preservation of foods, biological solutions, and microorganisms. Mesophilic bacteria and fungi reproduce slowly at refrigeration temperatures, and some psychrophiles can divide every 1 to 2 hours. This type of response can be readily observed if food is left in the refrigerator for a few weeks. Microorganisms can be preserved at refrigeration temperatures by cultivating them on agar slants or slants covered with mineral oil. Microbial metabolism occurs rapidly enough on slants without mineral oil that the products of metabolism destroy many viable cells within a few days. The presence of mineral oil over the slant, however, not only reduces microbial metabolism, but also prevents the culture from drying out, and the culture can be preserved for several months.

Freezing at 0° C may result in destruction of vegetative microbial cells due to the formation of large ice crystals, which on warming enlarge and exert enough pressure on cell membranes to rupture them. There is an increase in solute concentration in the cell via dehydration, and it may be that solute precipitation (salting out of proteins) in some way also injures the cell during the freezing process. Many microbial cells can be "quick frozen" at temperatures of −50° to −90° C, even though many cells are killed. This technique is used commonly in the microbiological laboratory to preserve isolated species for long periods of time (months to years). Another technique, which employs liquid nitrogen (−76° C or lower) to freeze and maintain cells at temperatures near absolute zero, has given rise to the science called **cryobiology.** Cryobiology has its greatest application in higher animals, including humans, where the preservation of sperm, embryos, and various organs for future implantation is of genetic and economic significance. Microbial cultures can be maintained indefinitely, with no loss in the population, by using liquid nitrogen, but **lyophilization** (freeze-drying) is a more practical way of preserving microorganisms for storage as well as for shipment to other laboratories anywhere in the world. The basic principle of lyophilization is that the culture broth is dried in a glass vial while in the frozen state by removing the water through a process called **sublimation**; that is, water is removed from the frozen state as a vapor by using a high-vacuum system in which a desiccant absorbs water. The glass vial is sealed under a vacuum and stored. Bacteria can be removed from the vial by removing the tip and adding sterile distilled water to produce a suspension. An aliquot of the suspension can be removed and transferred to a suitable growth medium, and a percentage of the organisms that have survived will replicate.

Filtration. Filtration is a common technique for separating microorganisms of different sizes, but it is also useful for sterilization where other techniques are not appropriate, for example, solutions that would be harmed by heat (toxins, vaccines, and enzymes). Today most microbial filtration processes are performed with membrane filters that are composed of cellulose esters or other polymers. They can be obtained in different pore sizes, ranging from those that retain the smallest viruses to those that retain the largest mammalian cells. The solution (filtrate) is saved from the filtration process, and the filter retaining the contaminating microorganisms is discarded. Filters can also be used to measure the number of microorganisms in a solution, for example, when testing possible microbial contamination of a water supply. The water sample is filtered, and the microbial contaminants are retained on the filter. The membrane filter is then placed on an absorbent pad saturated with culture medium in a Petri dish. The nutrients penetrate the filter, and any deposited bacteria

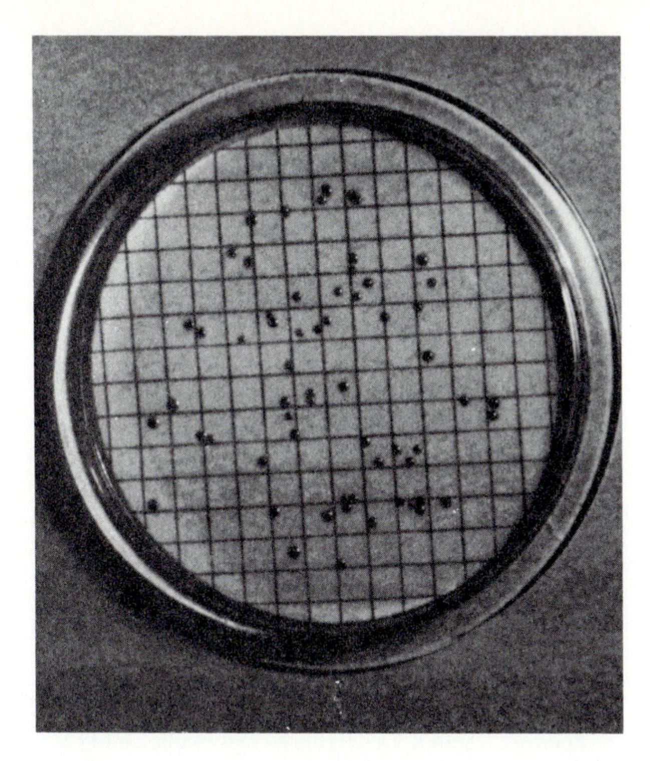

Figure 15-24 *Bacterial colonies on gridded membrane filter.*

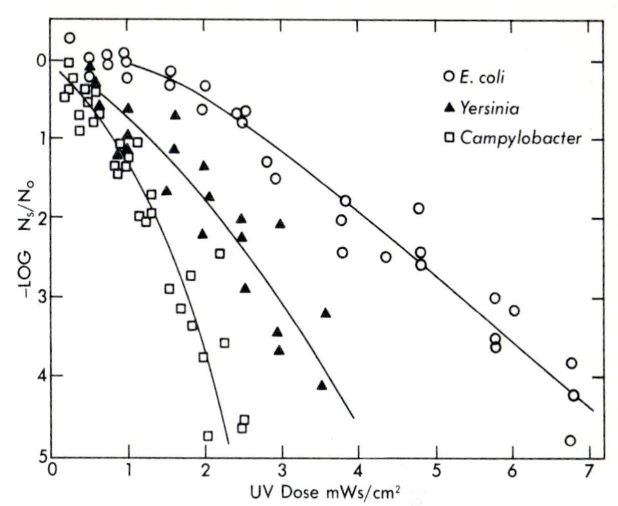

Figure 15-25 *UV inactivation curves of* Campylobacter jejuni, Yersinia enterocolitica, *and* Escherichia coli. *Cultures were in form of suspension at depth of 1 cm in Petri dish at room temperature.* mWs/cm^2, *milliwatt-seconds per square centimeter.*

will reproduce and give rise to discrete colonies (Figure 15-24). The number of colonies can be correlated with the quantity of fluid filtered and is an indicator of the degree of contamination of the original water source.

Radiation. Radiation used for controlling microbial growth can be characterized as either nonionizing, such as ultraviolet (UV) light, or ionizing, such as x-rays and gamma rays. Each type under controlled conditions can be used to produce viable mutants (see Chapter 10) but each can also be microbicidal. UV light is absorbed by DNA at wavelengths between 240 and 280 nm. It has low penetrability and as a microbicidal agent is used primarily for irradiation of air and flat, nonporous surfaces. UV light can be used in hospitals, cafeterias, and meat-packing plants. The UV light fixtures can be so designed that any occupants in the room are protected from the harmful rays. UV light is also used to disinfect drinking water in some European countries. The parameter used for comparing bacterial sensitivity is the 3-log (99.9%) inactivation dose. UV inactivation curves for three bacterial species that contaminate water are illustrated in Figure 15-25. UV light generated by the sun is not a particularly effective microbicidal agent, because the earth's atmosphere absorbs most of the short penetrating rays and allows only the longer rays, such as those between 2850 and 2950 Å, to reach the earth's surface. Any microbicidal effects of the sun's rays are at their peak only on clear days that are dry and devoid of particles in the air. Some microorganisms can repair the effects of UV light, and the molecular mechanisms involved are discussed in Chapter 10.

Ionizing radiations are characterized by their ability to impart their energy to molecules, such as water, and cause the dislodgment of electrons, resulting in the formation of ions that act as oxidizing agents (see Chapter 14 for full discussion of ionizing radiation). Ionizing radiations have been used for sterilization of pharmaceuticals and disposable medical supplies, such as gloves, plastic items, catheters, and sutures.

Concept Check

1. Why is heat the most important sterilizing agent, while chemicals are primarily disinfecting agents?
2. What is the difference between an antiseptic and a disinfectant? Are there situations where one can be substituted for the other?
3. What are the advantages/disadvantages of moist heat vs. dry heat? Why is filtration sometimes necessary for sterilization?
4. How is radiation used as a sterilizing agent?
5. What is meant by decimal reduction time? How is it measured, and what is its value?

Chemical methods

Chemical agents used to control microorganisms on inanimate objects are called disinfectants, whereas those appropriate for mammalian tissue are called antiseptics. Often the two terms are used interchangeably. The response of microbes to chemical compounds is variable, and usually no single agent is equally effective against all microorganisms. The response of the microorganism to the antimicrobial is governed by the following:

1. Characteristics of the microorganism. The chemical makeup of the microorganism can influence the type of disinfectant used. Some microorganisms possess capsules or slime, whereas others may be in the spore state. The agent of tuberculosis *(Mycobacterium tuberculosis)* possesses high concentrations of lipid in its envelope and is consequently resistant to many water-soluble disinfectants. Certain viruses, such as the influenza virus and herpesvirus, possess lipid envelopes that protect them from some chemical agents.
2. Concentration of the disinfectant. Those disinfectants that are -cidal (kill) at high concentrations are often -static (inhibit) at lower concentrations. If the concentration is low enough, some disinfectants may actually stimulate growth. Even if the disinfectant is -cidal at several concentrations, the time required for cell death will sometimes increase by several factors at lower concentrations. For alcohols, however, too high a concentration may be ineffective. Ethyl alcohol at 95% concentration is less effective than at 75%, and this is related to the catalytic activity of water, whose concentration is less in the 95% alcohol solution.
3. pH of the disinfectant. Disinfectants appear to be most active at pHs slightly acidic to neutral. pH affects the degree of ionization, a property that is important for some disinfectants. If the pH of the disinfectant or the material being disinfected is at an extreme of pH, a reduction in activity can be expected. Chlorinated compounds, for example, release their active chlorine atoms at neutral or acidic pHs, but release is prevented at alkaline pHs.
4. Interfering matter. Most disinfectants interact with extraneous organic or inorganic material, as well as with the population of microorganisms. Disinfectants that inactivate enzymes can also be expected to interact with any protein. If, for example, a dissociable ion from a disinfectant is the active agent, it can be neutralized by interaction with organic matter, or it can be chelated by inorganic ions and removed from solution.
5. Time. No time limit can be established for the disinfection of objects, solutions, or body surfaces. Disinfection is controlled by the concentration of the disinfectant, the number of microorganisms present, the presence or absence of interfering organic matter, and other factors. The only way to determine the actual time required for complete disinfection is to perform a death curve in which the variables can be controlled. In such experiments the rate of death, like that of growth, is exponential.

Evaluation of disinfectants. Phenol and phenol derivatives for many years represented the major commercial disinfectants, and because of this, evaluations of other disinfectants based on time and concentration were related to comparisons with phenol. Phenol comparisons now have little meaning, since most of the new disinfectants are not phenols or their derivatives. Other factors play an important role in activity; therefore disinfectants must be tested under actual or simulated conditions for which they are being used.

Classes of disinfectants. There are many hundreds of disinfectants, and not all of them can be described in this book. Instead, they are divided into various classes in which only the most important ones of the group are discussed. The groups include phenols, halogens, alcohols, surfactants, alkylating agents, and heavy metals.

Phenols. Phenol (also called **carboxylic acid**) was first reported as a disinfectant by Joseph Lister in 1867. He applied phenol to surgical instruments and surgical dressings and reported a dramatic decrease in postoperative infections. Phenol is irritating to the skin and is not used today as an antiseptic, but it can be used as a disinfectant for organic matter, such as sputum and feces, that may harbor pathogenic species. Phenols have high lipid solubility and are thus active against the organism causing tuberculosis (the bacterium causing disease has a high concentration of lipids in the cell envelope) and related species.

Phenol at high concentrations appears to act as a cytoplasmic poison by disrupting the cell wall and precipitating cell wall proteins. Once the wall is disrupted, there is leakage of important cellular metabolites. At lower concentrations, phenol inactivates important enzyme systems in the cell. Apparently the free hydroxyl group is the reactive component of the phenol molecule and chemical substitution at this site modifies activity. When two phenols are linked they are called **bisphenols.** Bisphenols show the greatest amount of antimicrobial activity when they contain halogens, and the two phenols are linked in positions adjacent to the hydroxyl groups (Figure 15-26).

At one time one of the most widely used bisphenols was hexachlorophene. It is an antiseptic that was used in many commercial soaps, surgical scrubs, deodorants, and for prophylactic bathing of infants. It is particularly active against gram-positive bacteria, especially *Staphylococcus aureus,* which is a common cause of serious infection in newborns. It was discovered, however, that

Table 15-9 *Killing action of various concentrations of alcohol against* Streptococcus pyogenes

Alcohol		Exposure of test organism to germicide*																
		Seconds					Minutes											
Volume (%)	Weight (%)	10	20	30	40	50	1	1½	2	3	3½	4	5	10	15	30	45	60
100	100	+	+	+	+	+	+	−	−	−	−	−	−	−	−	−		
95	92	−	−	−	−	−	−	−	−	−	−	−	−	−	−	−		
90	85	−	−	−	−	−	−	−	−	−	−	−	−	−	−	−		
80	73	−	−	−	−	−	−	−	−	−	−	−	−	−	−	−		
70	62	−	−	−	−	−	−	−	−	−	−	−	−	−	−	−		
60	52	−	−	−	−	−	−	−	−	−	−	−	−	−	−	−		
50	42	+	+	−	−	−	−	−	−	−	−	−	−	−	−	−		
40	33	+	+	+	+	+	+	+	+	+	+	−	−	−	−	−		
30	24	+	+	+	+	+	+	+	+	+	+	+	+	+	+	+	−	−
25	20	+	+	+	+	+	+	+	+	+	+	+	+	+	+	+	+	−
20	16	+	+	+	+	+	+	+	+	+	+	+	+	+	+	+	+	−

*+, growth; −, no growth.

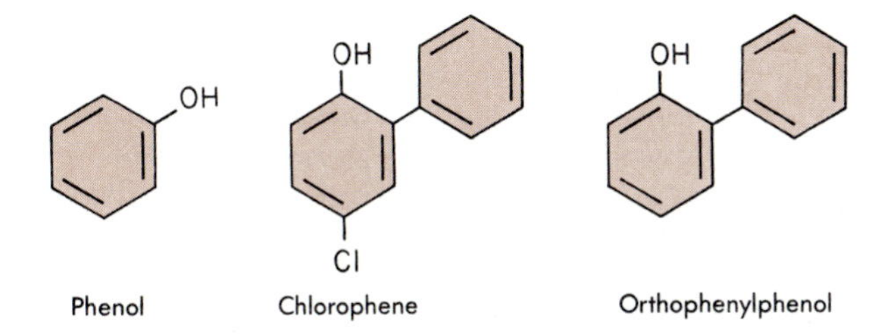

Figure 15-26 *Structure of phenol and two popular derivatives.*

hexachlorophene penetrates the skin and accumulates in brain tissue, causing neurological disorders after repeated use on infants. Hexachlorophene is now a prescription item and is used less widely. Cresol, a phenolic compound found in coal tar derivatives, has been used as a preservative for certain therapeutic preparations. Other uses have been in soaps and detergents.

Alcohols. Alcohols are capable of killing vegetative forms of bacteria and fungi but are not active against spores. The alcohols most widely used as disinfectants are ethyl and isopropyl alcohol. Both are more effective at concentrations between 60% and 90% than at concentrations between 90% and 100% because the increased amount of water is required to denature proteins. Once the concentration of alcohol falls below 50% it is difficult to predict its killing power (Table 15-9). Alcohols also act as lipid solvents and exert their activity on microbial membranes, as well as the lipid envelope of viruses. Ethyl alcohol is used in the gram stain and acid-fast stain and presumably it aids in the differentiation of organisms based on the amount of lipid in their cell envelopes.

Halogens. The halogens—chlorine, bromine, fluorine, and iodine—are very strong oxidizing agents. Some commercial disinfectants contain either chlorine or iodine. They inactivate proteins by oxidizing sulfhydryl groups, resulting in the formation of sulfur-sulfur bonds, which cause changes in protein conformation and hence activity. Chlorine in liquid form is an effective bactericidal agent and is used in water treatment facilities and in swimming pools. Chlorine in water forms hypochlorous acid (HOCl) and hydrochloric acid (HCl):

$$Cl_2 + H_2O \rightleftharpoons HOCl + H^+Cl^-$$

Hypochlorous acid is then subject to additional reactions that can include disinfection, reaction with various inorganic and organic compounds, or dissociation to hydrogen and hypochlorite ions (OCl^-):

$$HOCl \rightleftharpoons H^+ + OCl^-$$

The disinfection capabilities of hypochlorous acid and hypochlorite are nearly equivalent. Chlorinated lime (calcium hypochlorite) in 1% to 5% solutions is an effective disinfectant used in the sterilization of dairy barns and slaughterhouses, but it is also corrosive and has an offensive odor.

Iodine is poorly soluble in water and is found in preparations such as sodium iodide or potassium iodide. The most widely used antiseptic containing iodine is called **tincture** of **iodine** and consists of approximately 2% iodine and 4% sodium iodide in a 50% alcohol solution. The free iodine released in these preparations is toxic to the skin, and the alcohol solution causes a stinging sensation on open wounds. To combat these problems, iodine has been combined with an organic molecule and is called an **iodophor.** Iodophores (for example, povidone-iodine [Betadine]) are among the most frequently used antiseptics in the hospital.

Surface-active agents. **Soaps** and **detergents** belong to that class of agents called surface-active agents or **surfactants.** They have the property of lowering surface tension and increasing the miscibility of molecules. This is possible because they have both hydrophobic and hydrophilic groups, which help to separate molecules that have a tendency to be strongly attracted to each other. Soaps are sodium or potassium salts of fatty acids that function primarily as cleansing agents that increase the miscibility of oils and water. Thus they permit mechanical removal of microbes on the skin's surface. They are microbicidal for only a few species, such as species of *Streptococcus.* Many soaps contain antimicrobial agents in the form of halogenated compounds that are bacteriostatic or bactericidal. These antimicrobials help to prevent skin infections and reduce body odor. Detergents are synthetic surfactants that also possess hydrophobic and hydrophilic groups. They are divided into three types based on polarity, but only those that ionize are effective as disinfectants. **Anionic detergents** possess a hydrophobic hydrocarbon chain and a hydrophilic anion such as a carboxyl or sulfate group, which is the active component (Figure 15-27). They are more active in acidic solutions and affect the lipoprotein component of the cell membrane. **Cationic detergents** are the most microbicidal of the detergents. They are composed of cationic groups that dissociate to yield a positively charged ion, which is the active component and is either ammonium, phosphonium, or sulfonium. The most popular cationic detergents are the **quaternary ammonium** ion compounds also called **quats** (Figure 15-28). Their microbicidal activity is associated with their ability to interact with the bacterial cell membrane. The cationic portion is attracted to the negatively charged phosphate group of the membrane lipid, and the hydrophobic hydrocarbons of both the detergent and the membrane interact. Organic matter such as blood or tissue interferes with the activity of cationic detergents; thus surgical instruments must be thoroughly cleaned before disinfection with quats. Furthermore, quats are neutralized by anionic detergents and soaps. They are most active against gram-positive bacteria and some species of fungi. Today quats are used primarily in sanitizing glasses and flat, nonporous surfaces such as those in restaurants.

A. Soap
$[C_{17}H_{33}COO^-]\ Na^+$
Sodium oleate

B. Synthetic detergent
1. Anionic detergent
$[C_{12}H_{25}OSO_3^-]\ Na^+$
Active anion component — Inactive cation
Sodium lauryl sulfate

2. Cationic detergent 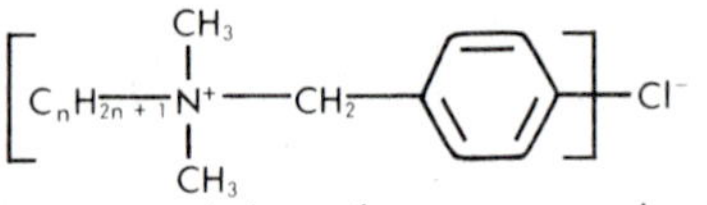
$[C_nH_{2n+1}-N^+(CH_3)_2-CH_2-C_6H_5]\ Cl^-$
Active cation — Inactive anion
Benzalkonium chloride

Figure 15-27 *Comparison of soaps and synthetic detergents.*

Akylating agents. Some of the amino acids on proteins possess R groups, such as SH, COOH, and NH_3, that ionize and release hydrogen. When these hydrogens dissociate from the amino acid, their place can be occupied by alkyl groups. Nucleic acids can also be affected in the same way. The most active alkylating agents are formaldehyde, glutaraldehyde, β-propiolactone, and ethylene oxide (Figure 15-29). **Formaldehyde** is mainly used as a preservative and is a component of embalming fluid. Because of its irritation of tissue and odor, it is seldom used as a disinfectant. **Formalin** is a 37% solution of formaldehyde gas and is used for the preservation and fixation of tissue. Formalin is used in microbiology primarily in the preparation of vaccines, because it inactivates microorganisms but does not destroy their antigenic (antibody-stimulating) properties. **Glutaraldehyde**, a dialdehyde, is one of the preferred disinfectants because it is active against bacteria and viruses in a 2% solution, but spores are destroyed only after 10 hours. It is too toxic for use as an antiseptic. The commercial product, which is called Cidex, is activated by the addition of sodium bicarbonate. **Ethylene oxide** is a gas that when placed under pressure is a most effective disinfectant for linens, sheets and bedding, and any articles or delicate instruments that cannot be sterilized by autoclaving. Its use is limited because of its toxicity if inhaled; explosiveness, if not diluted with inert gases such as carbon dioxide or freon; special equipment needs; and length of time to achieve sterility (4 to 12 hours) at room temperature. Furthermore, because of its toxicity, materials disinfected by this manner must be

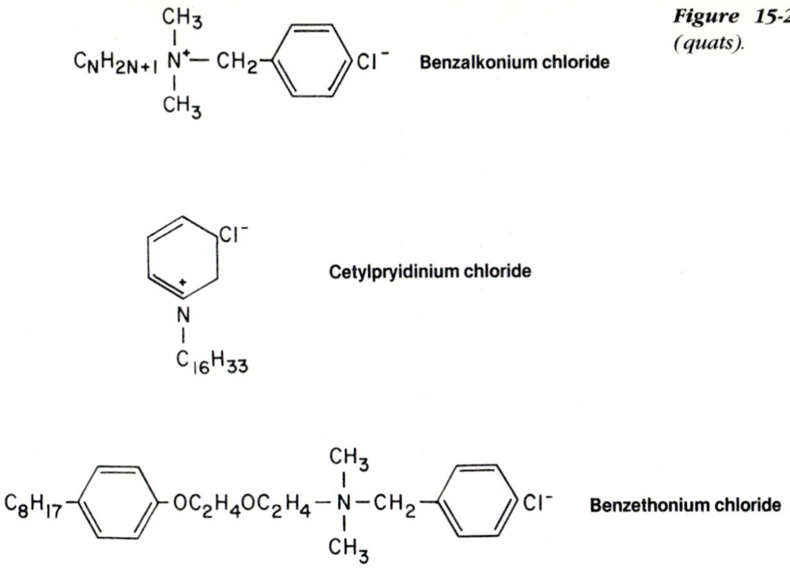

Benzalkonium chloride

Cetylpryidinium chloride

Benzethonium chloride

Figure 15-28 *Three commonly used cationic detergents (quats).*

degassed before use. **β-Propiolactone** is used in the gaseous phase to sterilize liquids. When added to fluids, it sterilizes and then becomes hydrolyzed to an inactive form. It is effective against bacteria, fungi, viruses, and spores but its use has been restricted because of its potential carcinogenic properties.

Heavy metals. Some heavy metals are bacteriostatic because of their ability to bind to proteins and inactivate enzymes by tying up sulfhydryl groups. **Mercury** and **silver** are the only inorganic metal ions that are useful as disinfectants (Table 15-10). Organic mercury compounds, such as thimerosal (Merthiolate) and merbromin (Mercurochrome) were once used in the treatment of superficial wounds, but they have been replaced by organic iodine compounds. **Silver nitrate** is used prophylactically for the prevention of gonococcal conjunctivitis, and many states require that newborns have a 1% solution of silver nitrate or a substitute instilled into the eyes at birth. Erythromycin is frequently used because it is effective for gonococcal conjunctivitis plus *Chlamydia trachomatis,* which is a leading cause of neonatal conjunctivitis.

Resistance to metals is a plasmid-associated property found in both gram-negative and gram-positive bacteria. Mercury resistance is expressed only after exposure to subtoxic levels of Hg^{2+}. The resistance mechanism involves a mercury specific transport system plus a cytoplasmic reductase that converts Hg^{2+} to Hg^0, which is less toxic. Strains that are resistant to organomercurials specify a second cytoplasmic enzyme called organomercurial lyase. This enzyme cleaves the C-Hg bond to release Hg^{2+}, which is volatilized by the reductase. Discontinuation of mercurial disinfectants has resulted in a decreasing incidence of mercury resistance.

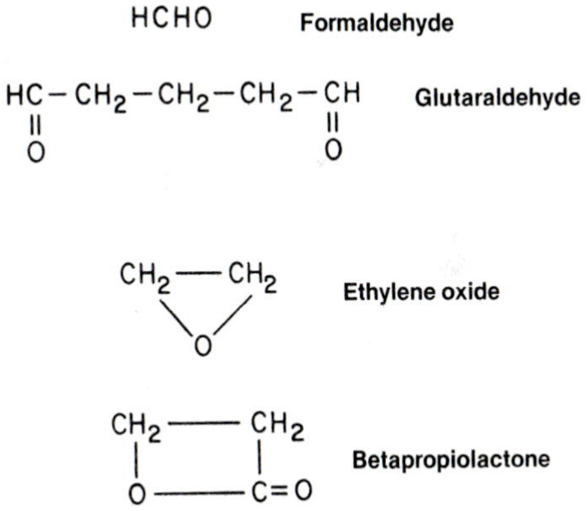

HCHO Formaldehyde

Glutaraldehyde

Ethylene oxide

Betapropiolactone

Figure 15-29 *Structure of alkylating agents, formaldehyde, glutaraldehyde, ethylene oxide, and betapropiolactone.*

Concept Check

1. What factors must be considered before a disinfectant is used?
2. What are the advantages and disadvantages of the following classes of antimicrobials: phenols, chlorine, tincture of iodine, quaternary ammonium compounds, soaps, formaldehyde, glutaraldehyde, and ethylene oxide?
3. What heavy metals are used as antimicrobials? List examples in which they are used.

Table 15-10 *Common mercury and silver compounds used as disinfectants and antiseptics*

Metal and common name	Structural formula	Use
Mercury—inorganic compounds		
Mercuric chloride	$HgCl_2$	As 0.1% aqueous solution or in 70% alcohol solution on skin
Mercuric oxide	HgO	In ointment (1%) for conjunctivitis
Ammoniated mercury	$HgNH_2Cl$	In ointment (3%) for eye infections
Mercury—organic compounds		
Thimerosol (merthiolate)*	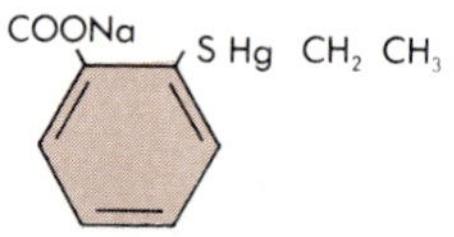 COONa; S Hg CH_2 CH_3	Aqueous solution for eye infections; 0.1% tincture or 0.1% ointment on skin
Nitromersol (metaphen)	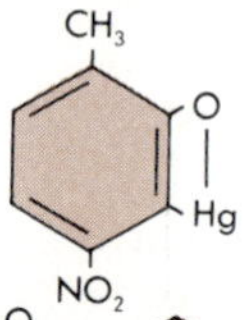 CH_3; O; Hg; NO_2	Aqueous solution or tincture on skin
Merbromin (mercurochrome)	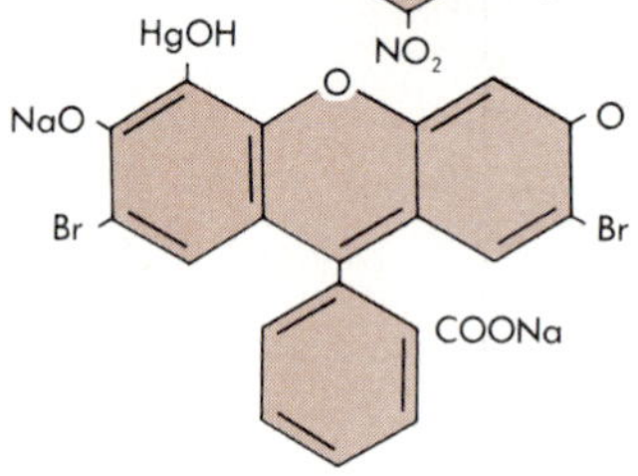 HgOH; NaO; O; O; Br; Br; COONa	2% solution on skin; 1% solution for bladder irrigation
Silver—inorganic compounds		
Silver nitrate	$AgNO_3$	1% solution for eye infections caused by *Neisseria gonorrhoeae,* application to newborns was state requirement but antibiotic replaced it
Silver—organic compounds		
Silver picrate	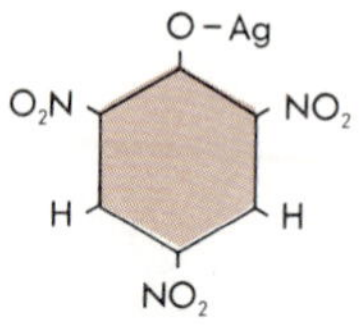 O–Ag; O_2N; NO_2; H; H; NO_2	To treat vaginal fungal infections; also in glycerine suppositories
Silver sulfadiazine	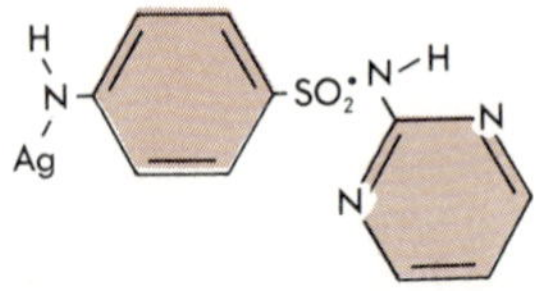 H; N; Ag; SO_2; N–H; N; N	For treatment of burns in form of cream

*Proprietary names in parentheses.

SUMMARY

1. The difference between antimicrobials and chemotherapeutic agents is that chemotherapeutic agents are selectively toxic for the parasite and relatively nontoxic to the host tissue, whereas antimicrobials may or may not possess these attributes. Most of the chemotherapeutic agents used in the treatment of infectious disease are produced by microorganisms; that is, they are antibiotics.
2. The spectrum of activity of chemotherapeutic agents may be narrow or broad. In addition, they may either kill the microorganism or simply inhibit its growth.
3. Chemotherapeutic agents affect microorganisms in a number of ways; they may (1) inhibit cell wall synthesis (penicillins, cephalosporins, etc.), (2) inhibit protein synthesis (chloramphenicol, tetracyclines, aminoglycosides, etc.), (3) inhibit cell membrane function (nystatin, amphotericin B, polymyxin, and imidazole derivatives), (4) inhibit nucleic acid synthesis (nalidixic acid, rifampin, and griseofulvin), and (5) act as antimetabolites (sulfa drugs).
4. Chemotherapeutic agents can be used to prevent disease (prophylaxis), but only under special circumstances, such as to prevent tuberculosis, postoperative infection, and meningococcal meningitis.
5. Antibiotics have been used in animal feed to promote animal growth, but they may be hazardous in that antimicrobial-resistant species, infectious for humans, may be produced.
6. The resistance of microorganisms to antimicrobials may be due to natural phenomena, mutation, or transferable plasmids. Natural resistance is usually associated with the structural or biochemical properties of the organism. Mutation in microorganisms can produce resistance to drugs by (1) affecting the permeability of the cell wall and cytoplasmic membrane, (2) altering the target site of the antimicrobial (such as the ribosome), and (3) increasing the synthesis of drug-inactivating enzymes. Plasmids containing genes that code for many drug-inactivating enzymes may be transferred from one microbial species to another via conjugation.
7. Laboratory measurement of antimicrobial activity can be quantitative or qualitative. Most measurements are qualitative and determine the minimal inhibitory concentration (MIC) of the drug against a microorganism using agar or broth dilution tests. The zone diameters produced on the agar surface can be correlated with interpretive charts and provide the physician with a method for determining what drug and what dosage to use in the treatment of the infection.
8. The process whereby an object or solution is rendered devoid of microorganisms is called sterilization. Agents that kill or control the growth of microorganisms are called antimicrobials and consist of two types: antiseptics and disinfectants. Disinfectants can kill microorganisms but are not necessarily applicable to the body surface. Antiseptics may inhibit the growth of microorganisms or kill them, but they can be applied to the surface of the body.
9. One of the most effective means of killing any type of microorganism is through the application of moist heat under pressure (autoclaving); however, this technique is not always applicable. Other physical methods include filtration and radiation.
10. Chemical agents used as sterilizing agents are subject to certain criteria, including the type of microorganism involved, concentration of the chemical, temperature, pH, and presence or absence of interfering matter. Various classes of disinfectants exist, including phenols, alcohols, halogens, surface-acting agents, alkylating agents, and heavy metals.

Self-study Quiz

Multiple choice

1. The enzyme responsible for the polymerization of glycan chains in the peptidoglycan is:
   A. Peptidoglycan transglycolase
   B. Peptidoglycan transacetylase
   C. Peptidoglycan transpeptidase
   D. Carboxypeptidase
2. Which of the following would *not* be appropriate for treating eye infections?
   A. Mercuric chloride
   B. Silver nitrate
   C. Mercuric oxide
   D. Merthiolate
3. Which of the following is *not* an antibiotic?
   A. Penicillin
   B. Sulfanilamide
   C. Cephalosporin
   D. Erythromycin
4. Which of the following antimicrobials is believed to bind to the cell wall transpeptidase and inactivate it?
   A. Erythromycin
   B. Streptomycin
   C. Tetracycline
   D. Cephalosporin
5. Which of the following statements best characterizes penicillin-binding proteins?
   A. Porin-like molecules in the cell wall
   B. Antibiotic transport proteins located in the periplasm
   C. Enzymes involved in the terminal stages of cell wall synthesis
   D. Cytoplasmic membrane proteins associated with antibiotic transport
6. Acyclovir is a drug that inhibits:
   A. DNA-dependent RNA polymerase
   B. DNA gyrase
   C. DNA-dependent DNA polymerase
   D. Thymidine kinase
7. The use of antibiotics in animal feed is of concern because of the transfer of antibiotic-resistant organisms, such as ____________, from animals to humans.
   A. *Staphylococcus*
   B. *Bacteroides*
   C. *Shigella*
   D. *Salmonella*
8. Which of the following is an aminoglycoside?
   A. Erythromycin
   B. Tetracycline
   C. Streptomycin
   D. Polymyxin
9. The term applied to a chemical agent that inhibits the growth of potentially pathogenic organisms but is not harmful to tissue is:
   A. Sepsis
   B. Chemotherapeutic agent
   C. Disinfectant
   D. Antiseptic
10. The group of chemicals that are used primarily in sanitizing glasses and flat, nonporous surfaces such as in restaurants are:
   A. Alkylating agents
   B. Quaternary ammonium compounds
   C. Heavy metal–containing solutions
   D. Iodophors

Short answer

In a broth dilution antibiotic susceptibility test the following results were obtained (*S. pyogenes* + broth + antibiotic into each tube).

A. What is the MIC indicated?
B. What is the MBC indicated?
C. Would you expect the MIC to be the same for this antibiotic if the organism were *E. coli?* Explain.

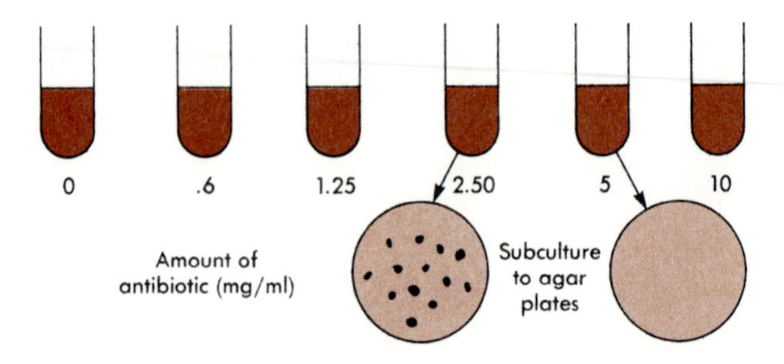

Matching

Match the antibiotics listed below with their basic mechanism of action.

________	1. Cycloserine	A. Cell wall synthesis inhibition
________	2. Nalidixic acid	B. Cell membrane function inhibition
________	3. Nystatin	C. Protein synthesis inhibition
________	4. Cephalexin	D. Nucleic acid synthesis inhibition
________	5. Gentamicin	E. Antimetabolite
________	6. Spectinomycin	
________	7. Ampicillin	
________	8. Tetracycline	
________	9. Sulfanilamide	
________	10. Polymixin B	

Match the physical control method in the left-hand column with its principal mechanism of action.

________	11. Dry heat	A. Coagulation of protein and enzymes
________	12. Moist heat	B. Rupture of cell membrane
________	13. Freezing	C. Formation of hydroxyl and superoxide radicals
________	14. Ionizing radiation	D. Oxidation
________	15. UV radiation	E. Formation of thymine dimers

Match the chemical control agent in the left-hand column with its principal mechanism of action.

________	16. Alcohols	A. Disruption of membrane function
________	17. Iodophors	B. Alkylation of nucleic acids
________	18. Ethylene oxide	C. Oxidation of proteins and enzymes
________	19. Silver nitrate	D. Binding to sulfhydryl group to inhibit enzyme function
________	20. Quaternary ammonium compounds	

Thought questions

1. Antibiotic therapy of infectious disease is not a trivial therapy and is not just a matter of isolating a causative organism, testing its antibiotic sensitivity, and prescribing an antibiotic. Describe other factors that must be considered in the choice and use of antibiotic therapy for a bacterial infection.
2. A significant problem in epidemiology and antibiotic therapy is the development of penicillinase-producing strains of *Neisseria gonorrhoeae* (PPNG) in several areas of the world (especially the Phillippines, Singapore, and West Africa).
   a. One of the contributing factors is the prophylactic use of oral penicillin, especially among prostitutes in the endemic areas. Explain how this contributes to the problem.
   b. In mid-1976 the first isolate of PPNG was made in the United States. Since that time several hundred additional cases have been identified. The PPNG strains are so prevalent in the Los Angeles area that PPNG is considered endemic there. When gonorrhea does not respond to penicillin therapy what should be done?
   c. Spectinomycin is the second antibiotic of choice in treatment of gonorrhea. Would you expect PPNG strains to be sensitive or resistant to spectinomycin? Explain your rationale.
   d. Could strains be resistant to both? If so, what would you recommend for therapy?
3. Probably the most common application of microbiology is in the control of microorganisms (both infectious and noninfectious organisms).
   a. Think back over the last 24 hours and make a list of all the activities and products that you encountered that involve either directly or indirectly the application of microbiological control measures. Your list should have at least 20 items.
4. From the data in Figure 15-22 determine the *Z* value from a TDT curve.

SELECTED READINGS

Books

Barry, A.L. *The antimicrobic susceptibility test: principles and practices.* Lea & Febiger, Philadelphia, 1976.

Block, S.S. (Ed.). *Disinfection, sterilization and preservation* (3rd ed.). Lea & Febiger, Philadelphia, 1985.

Levy, S.B., Clowes, R.C., and Koenig, E.L. (Eds). *Molecular Biology, pathogenicity, and ecology of bacterial plasmids.* Plenum Press, New York, 1981.

Lorian, V. (Ed.). *Antibiotics in laboratory medicine.* The Williams & Wilkins Co., Baltimore, 1980.

Mitsuhashi, S. (Ed.). *Drug resistance in bacteria.* Thieme-Stratton, Inc., New York, 1982.

Russell, A.D., Hugo, W.B., and Ayliffe, G.A.J. (Eds.). *Principles and practice of disinfection, preservation, and sterilization.* The C.V. Mosby Co., St. Louis, 1982.

Journal articles

Abraham, E.P. The beta-lactam antibiotics. *Sci. Am.* **244**:76, 1981.

DeClercq, E. Specific targets for antiviral drugs. *Biochem. J.* **205**:1, 1982.

Demain, A.L. How do antibiotic-producing microorganisms avoid suicide? *Ann. N.Y. Acad. Sci.* **235**:601, 1974.

Dolin, R. Antiviral chemotherapy and chemoprophylaxis. *Science* **227**:1296, 1985.

Foster, T.J. Plasmid determined resistance to antimicrobial drugs and toxic metals in bacteria. *Microbiol. Rev.* **47**:361, 1983.

Guiney, D.G., Jr. Promiscuous transfer of drug resistance in gram-negative bacteria. *Rev. Infect. Dis.* **149(3)**:320, 1984.

Handwerger, S., and Tomasz, A. Antibiotic tolerance among clinical isolates of bacteria. *Rev. Infect. Dis.* **7(3)**:368, 1985.

Holmberg, S.D., et al. Drug-resistant salmonella from animals fed antimicrobials. *N. Engl. J. Med.* **311(10)**:617, 1984.

Koch, A.L. Evolution of antibiotic resistance gene function. *Microbiol. Rev.* **45**:267, 1981.

Lengyel, P. Biochemistry of interferons and their actions. *Annu. Rev. Biochem.* **51**:251, 1982.

Sabath, L.D. Biochemical and physiologic basis for susceptibility and resistance of *Pseudomonas aeruginosa* to antimicrobial agents. *Rev. Infect. Dis.* **6**(suppl.3):643, 1984.

Sanders, C.C., and Sanders, W.E., Jr. Emergence of resistance during therapy with newer beta lactam antibiotics: role of inducible beta lactamases and implications for the future. *Rev. Infect. Dis.* **5**:639, 1983.

Waxman, D.J., and Strominger, J.L. Penicillin-binding proteins and mechanism of action of beta lactam antibiotics. *Annu. Rev. Biochem.* **52**:801, 1983.

PART IV

Viruses

CHAPTER

16

General characteristics of viruses

Chapter Outline

Introduction

Viruses are infectious agents that survive only on a living host and cannot be cultivated on ordinary laboratory media. The infectious nature of what we now call viruses was demonstrated by Iwanowsky and Beijerinck in the 1890s (see Chapter 1 for discussion of virology). The true nature of the virus was not unveiled until after 1935. Between 1935 and 1960 the following events contributed to virus elucidation:

1. The tobacco mosaic virus (TMV) was isolated in the form of crystals from infected tobacco plants by Stanley in 1935. The virus had the ability to reproduce in the plant host and to cause disease before and after repeated crystallization in the laboratory.
2. The study of bacterial viruses (bacteriophages) revealed, through the Hershey Chase experiments (see Chapter 8), that only the viral nucleic acid enters the host cell (this is not true for animal viruses, in which the entire virus particle enters the cell). This discovery demonstrated the importance of the nucleic acid in determining the characteristics of the virus.
3. The development of tissue culture provided a means of growing animal and plant viruses without interference from an immune system.
4. The development of the electron microscope permitted scientists to visualize the virus and to morphologically distinguish one virus from another.

This chapter characterizes the basic morphology of viruses particularly bacterial and animal viruses. The ways in which bacterial, plant, and animal viruses can be

Point of Interest

DISCOVERY OF BACTERIOPHAGES

A subcellular infectious agent in plants (tobacco mosaic virus) had been discovered in 1899. F. W. Twort in England was one of many bacteriologists attempting to grow in vitro these subcellular infectious agents called viruses. In his approach, Twort thought that pathogenic viruses that could not be grown on laboratory media were descendants of nonpathogenic, less fastidious viruses that would grow on simple laboratory media. He inoculated nutrient agar with smallpox vaccine fluid hoping that one of the "nonpathogenic" forms would grow. The colonies that grew on the agar were micrococci that had contaminated the vaccine. However, if he continued incubation of the plates, some of the colonies underwent what he called a "glassy transformation." They became watery looking and then transparent and were no longer viable. If a healthy colony of micrococci were touched with this glassy material, they underwent the same transformation. This glassy material could be diluted and filtered through porcelain filters and still cause transformation. This agent of "disease" could be transferred from one micrococci to another but could not be grown by itself on any medium. It retained its activity for over 6 months but lost activity if heated to 60°C for one hour. Twort published his findings in 1915 with three explanations for the phenomenon. The last explanation was that the agent of disease was a virus that grew on and infected micrococci.

At the same time, d'Herelle observed a similar "disease" of bacteria while studying diarrhea of locusts in Mexico. He isolated pure cultures of coccobacilli from the sick locust. Several times while growing the coccobacilli on agar, he noticed clear, circular spots on the agar. He found that the agent that caused these clear spots could pass through porcelain filters.

He returned to Paris in 1915 and worked on an army epidemic of dysentery. He isolated the dysentery bacilli and kept adding filtrate from the feces of the same sick soldier to the cultured isolates. At first nothing happened, but on the fourth day he placed filtrate in a tube of broth culture that had been turbid the night before. The broth culture became clear and aliquots from it could not be cultivated on nutrient agar. He called these filterable agents bacteriophage (phage meaning devour) and attributed this agent to the sick man's recovery. He published this work in 1917 and drew immediate attention because of the medical potential of these phages.

cultivated and assayed in the laboratory; the effect of chemical and physical agents on virus; and the importance of viruslike agents, such as viroids and satellites, as infectious agents are also discussed. This chapter also points out some of the unusual relationships that exist between fungal and insect viruses and their hosts.

GENERAL CHARACTERISTICS OF BACTERIAL AND ANIMAL VIRUSES

Viruses are given independent status because of several distinguishing features that set them apart from other microbial agents: (1) they are acellular and are not enclosed in a membrane, (2) they replicate by a self-assembly process and not by binary fission, budding, or other mechanisms associated with other microbial agents, (3) they are obligate intracellular parasites that infect virtually all types of cells (plants, animals, and bacteria), (4) they contain either RNA or DNA but not both as their genetic material, (5) they lack an energy-generating system, and (6) they are unable to demonstrate any significant metabolic activity. The relationship between the host and the infecting virus is a specific one, and any given virus may have a restricted host range. The host cell, however, may be infected by different viruses. Certain viruses have been shown to have a broad range and may be able to infect humans, hogs, horses, some birds, and insects. At the other extreme, a virus may be able to infect only one strain of a species; for example, there is one bacterial virus that can infect only one strain of *Escherichia coli.* Since many viruses infect only one strain of bacteria, they are sometimes used to distinguish one bacterial strain from another (the technique is called **typing**).

Size

All viruses, except the smallpox virus, are beyond the limits of resolution with the compound microscope. Since organisms smaller than 0.2 μm cannot be observed with the compound microscope, it follows that most viruses must be less than 0.2 μm. The smallpox virus is 250 nm in diameter, and the foot and mouth disease virus, one of the smallest viruses, is approximately 10 nm in diameter. Even though microbiologists in the early twentieth century were unable to see viral particles, they were still able to isolate and experiment with them, free from contaminating material. They did this by passing virally contaminated culture or tissue extracts through filters that other microbial agents could not penetrate. The comparative size and shape of several

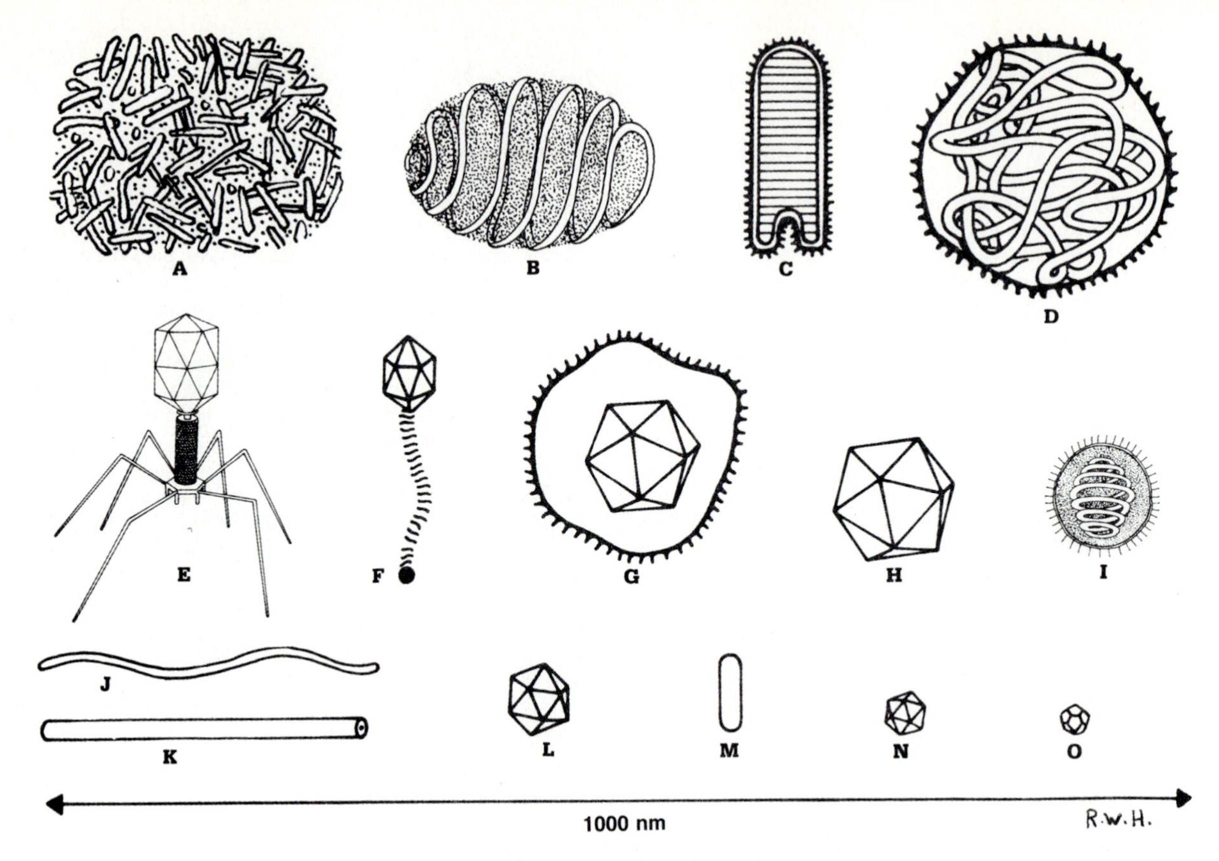

Figure 16-1 *Comparative size and shape of various groups of viruses. Note these groups represent a fraction of many diverse forms of viruses.* ***A,*** *Poxvirus.* ***B,*** *Orf virus.* ***C,*** *Rhabdovirus.* ***D,*** *Parainfluenza virus (mumps).* ***E,*** *T-even bacteriophage.* ***F,*** *Flexuous-tailed bacteriophage.* ***G,*** *Herpesvirus.* ***H,*** *Adenovirus.* ***I,*** *Influenza virus.* ***J,*** *Filamentous flexuous virus.* ***K,*** *Rodlike virus.* ***L,*** *Polyoma/papilloma virus.* ***M,*** *Alfalfa mosaic virus.* ***N,*** *Picornavirus (poliovirus and plant virus).* ***O,*** *Bacteriophage* $\phi\chi174$.

groups of viruses are illustrated in Figure 16-1. The structural components found in a virus particle are demonstrated in Figure 16-2.

Chemical composition

Nucleic acid. Viruses, with only minor modifications, are composed of a central core of nucleic acid surrounded by a protein coat called the **capsid.** A completely assembled and infective virus particle is called a **virion.*** The virus is unique because the genetic information is encoded in either DNA or RNA, but not both, and the nucleic acid may be either single stranded or double stranded. The nucleic acid when isolated from the virus may be a linear or circular molecule, and in certain groups of viruses the nucleic acid may even appear in several distinct segments.

Unlike other microorganisms in which a variety of metabolic enzymes and organelles, such as ribosomes and mitochondria, are present in the cell cytoplasm, the virus contains, besides its nucleic acid only a limited number of macromolecules. There are exceptions, of course; for example, a few viruses contain either a DNA or an RNA polymerase associated with the nucleic acid core. These polymerases are used by the virus for replication during the infection process. One group of viruses, called oncogenic (cancer-causing) RNA viruses, possesses a unique enzyme, reverse transcriptase, which is a DNA polymerase that requires an RNA template. The significance of these enzymes is discussed in Chapter 17.

Protein coat. The protein coat or capsid surrounding the nucleic acid is made up of individual protein subunits called **capsomeres** (Figure 16-3). The capsid subunits are arranged in different types of symmetry, which are discussed later. Only the genera *Rotavirus* and *Reovirus* possess two capsids, both containing capsomeres (Figure 16-4), but this is the exception and not the rule. The capsid with its enclosed nucleic acid is

* There are instances where nucleic acid alone may be infective, a process called **transfection**, and give rise to infectious virus.

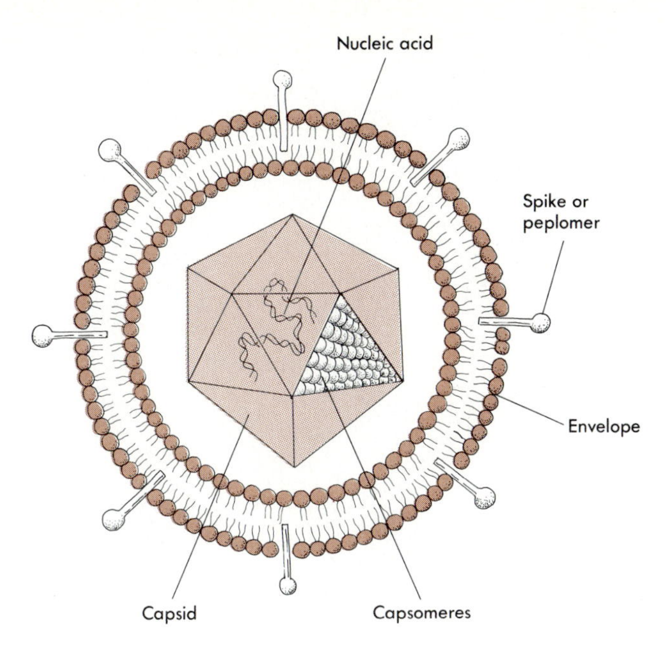

Figure 16-2 *Basic structure of a virus using animal virus as a model.*

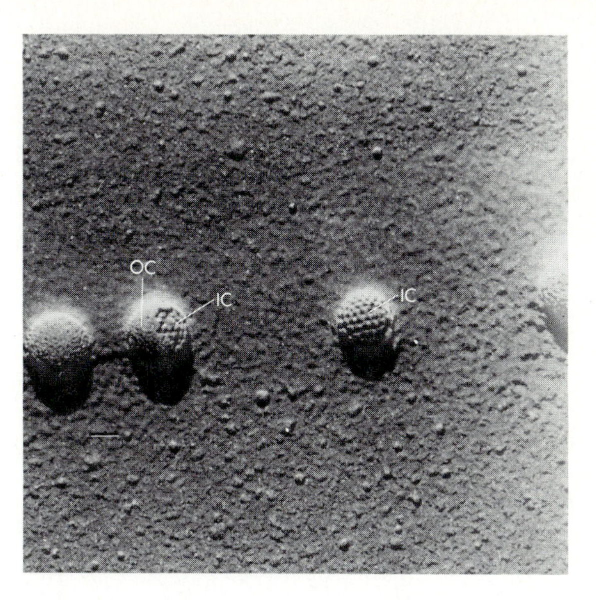

Figure 16-4 *Electron micrograph of* Rotavirus *showing apparent double-capsid symmetry.* IC, *Inner capsid;* OC, *Outer capsid. Bar = 50 nm.*

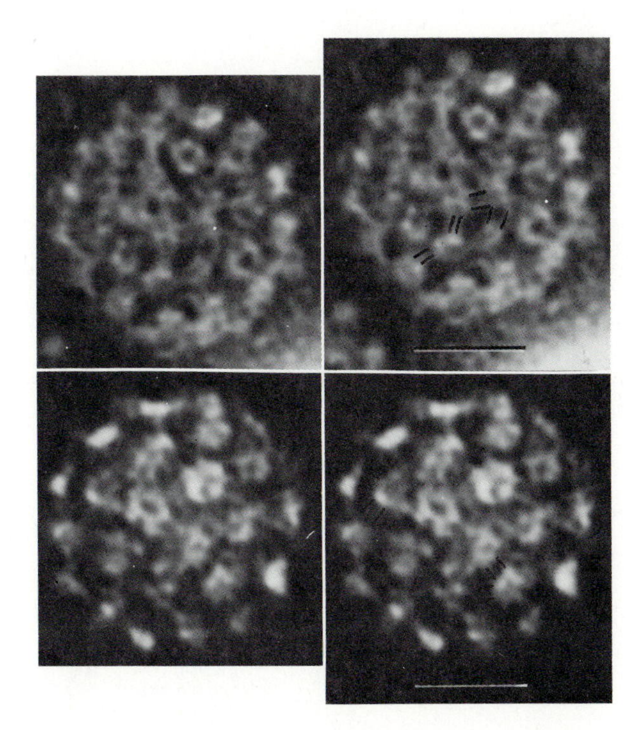

Figure 16-3 *Electron micrograph of papilloma virus demonstrating individual capsomeres that make up capsid. Lower photographs indicate bridges that appear to connect capsomeres in upper micrographs. Bar = 25 nm.*

called a **nucleocapsid**. The nucleocapsid is the virion if no other structures are involved; however, some viruses possess a lipid envelope surrounding the nucleocapsid (see next section). The principal function of the capsid is to protect the nucleic acid from any environmental conditions that might affect it. This protection is due to the manner in which the amino acids of polypeptides are arranged on the external surface of the capsid. It appears that the amino acid R groups on the external capsid surface are made up of nonpolar groups that interact (hydrophobic bonding) to exclude water, thus making the protein unusually resistant to pH, temperature, and radiation. This technique of water exclusion is similar to that found in the spore coats of *Bacillus* and *Clostridium*. The capsid also provides the virus with a means of attachment to host cells and is therefore critical in determining host and tissue specificity.

Lipid envelope. Some viruses contain a membrane or envelope that surrounds the nucleocapsid (see Figure 16-2). The membrane is acquired by the virus, after replication, as it passes through the host's cytoplasmic membrane or nuclear membrane (a process called **budding**) and is composed of lipids and proteins. Glycoproteins make up the outer portion of the envelope, and they often appear as **spikes** (also called **peplomers**) in the electron microscope. The protein components of the glycoprotein are viral in origin, but the carbohydrate components, as well as the lipid moieties, are derived from the host cell. The protein components of the envelope can have a variety of functions. The spikes on the influenza virus, for example, are of two types (Figure 16-

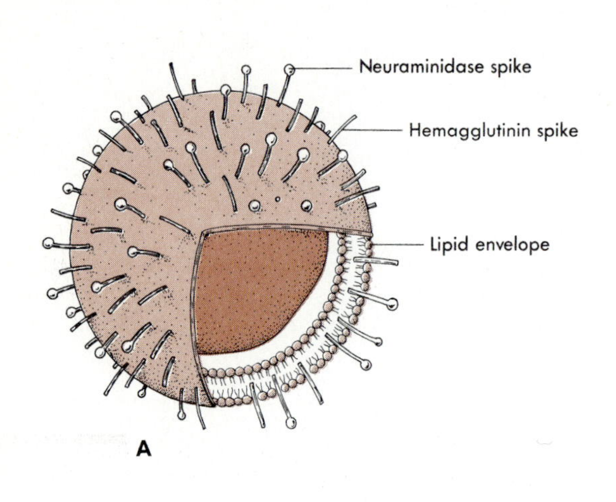

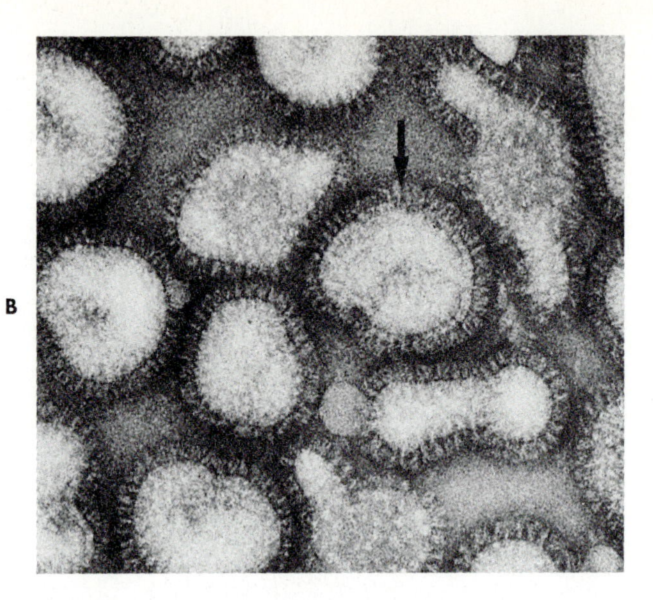

Figure 16-5 *Lipid envelope of viruses.* ***A,*** *Influenza virus containing two types of spikes in its lipid envelope: neuraminidase and hemagglutinin.* ***B,*** *Electron micrograph of influenza virus. (× 180,000.) Spikes are easily observed* (arrow).

5, *A* and *B*): one is an enzyme called **neuraminidase**, and the other is a **hemagglutinin**. Neuraminidase removes *N*-acetylneuraminic acid from the polysaccharide found on the surface of mammalian cells. The enzyme functions not only for attachment and spread of virus on epithelial surfaces but also as a means of intracellular release from already-infected cells. Hemagglutinins bind to red blood cells and cause them to form aggregates; that is, they become agglutinated. Not all enveloped viruses possess spikes, but those that do appear to have a built-in mechanism for attachment to host tissue. Enveloped viruses lose their infectivity once the envelope is destroyed or removed by agents such as bile salts, which are present in the alimentary tract. It is not surprising that enveloped animal viruses do not use the alimentary tract as a means of infecting the host.

Shape (capsid symmetry)

The shape as well as the size of the virus is related to the size of the viral genome and to the nature of the capsid proteins. Some viral genomes are so small that they code for only a few structural proteins. Viruses that exhibit helical capsid symmetry have only one type of capsid protein, whereas viruses with other types of symmetry may have several types of capsid proteins. The interactions of these protein subunits (the bonding between their amino acid R groups) determines the specific geometric patterns of the virus capsid, which may be represented as **helical**, **icosahedral**, or **complex**. Diagrammatic examples of capsid symmetry are illustrated in Figure 16-6.

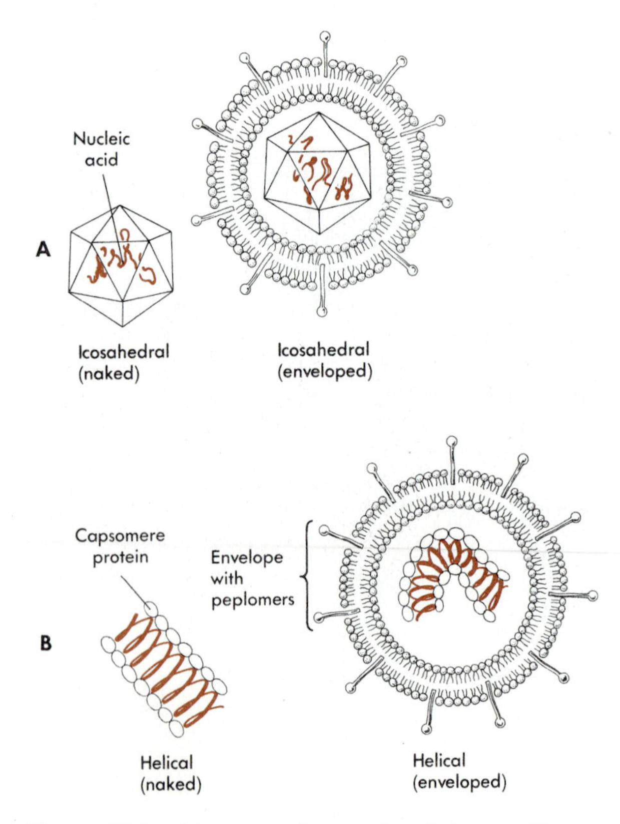

Figure 16-6 *Diagrammatic examples of viral capsid symmetry.* ***A,*** *Icosahedral.* ***B,*** *Helical. Complex virus structure is characteristic of poxviruses and the bacterial virus T4 (see Figure 16-14).*

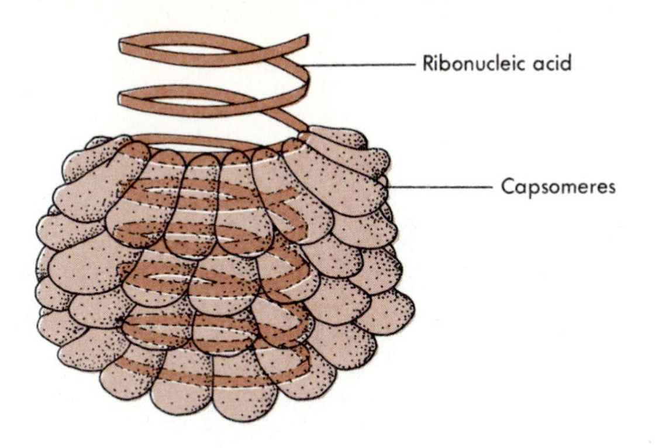

Figure 16-7 *Tobacco mosaic virus.*

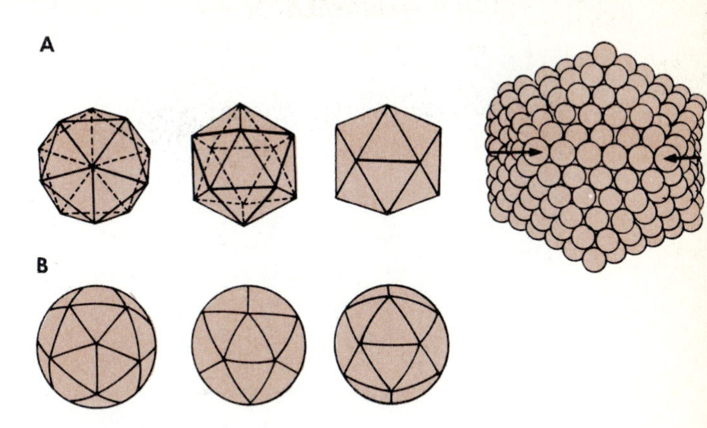

Figure 16-9 *Icosahedron viewed in three positions.* ***A,*** *Icosahedra* (upper left to right) *seen in fivefold, threefold, and twofold axis of rotational symmetry. On far right is model showing fivefold axis* (arrow), *using 252 table tennis balls, of rotational symmetry and capsomere position.* ***B,*** *Spherical models with icosahedral symmetry with same axes as in* ***A.*** *Lower surface not shown in any view.*

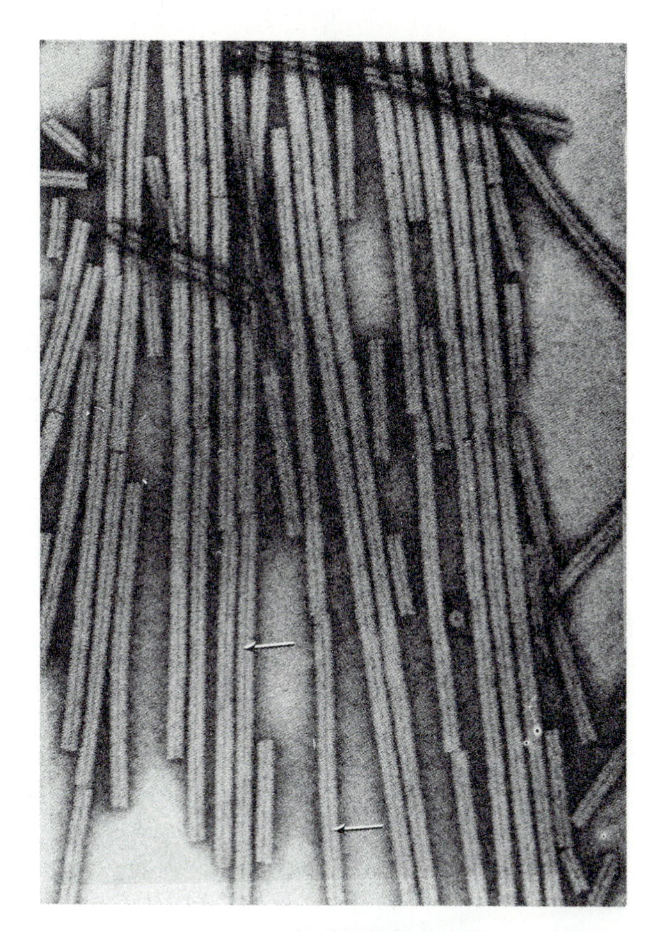

Figure 16-8 *Electron micrograph of TMV particles. Stain has penetrated into axial hole* (arrows), *which has 4 nm diameter.*

Helical capsid symmetry. The identical protein subunits of helical capsids are arranged end to end in the shape of a rod, with each turn of the rod possessing the same number of subunits. Our understanding of this arrangement has come from x-ray diffraction studies of the tobacco mosaic virus (TMV). A model of the virus, illustrated in Figure 16-7, demonstrates that the protein subunits form a spiral-shaped structure. Noncovalent bonds are formed between amino acids on adjacent subunits on the same turn of the spiral, as well as between adjacent turns of the spiral. The vast number of noncovalent bonds gives the capsid tremendous structural stability, similar to the hydrogen bonding between strands of the DNA molecule. A chain of RNA follows the protein helix and is located between the protein subunits. This produces an empty hole extending along the axis of the virus, which can be revealed by electron microscopic examination (Figure 16-8). The TMV is a naked virus (contains no envelope), and its helical nucleocapsid gives it the appearance of a rod. Many helical animal viruses, such as the influenza virus, possess envelopes that give the virus a nearly spherical shape. The nucleocapsid of the influenza virus exhibits helical symmetry, but its rod-shaped spring is flexible (this means that there are fewer forces holding the neighboring coils of the spring together). The nucleocapsid therefore is coiled within the envelope (see Figure 16-5). The flexibility of the spring could be compared to the bending of α-helical proteins due to the positioning of proline or hydroxyproline residues (see Chapter 2).

Icosahedral capsid symmetry. Spherically shaped nucleocapids are constructed according to an icosahedral (20-sided) symmetry (Figure 16-9). Capsomeres that are connected to five neighboring capsomeres are called **pentamers.** Capsomeres that are connected to six neighboring capsomeres are called **hexamers.** A model demonstrating this arrangement in spherical viruses is illustrated in Figure 16-10. The most basic capsid is com-

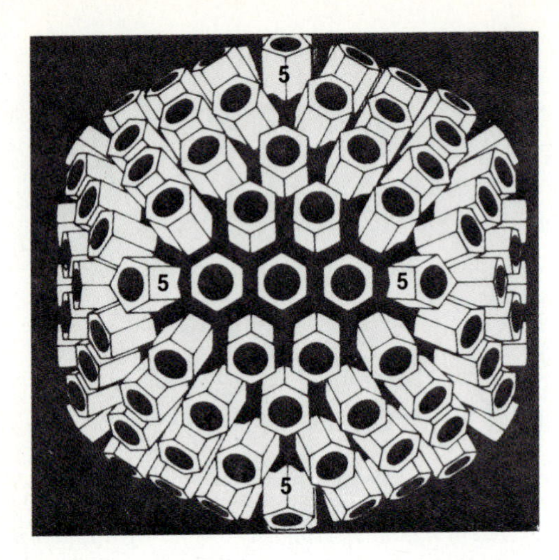

Figure 16-10 *Pentamer location in icosahedron. Only four pentamers from this plane are labeled. Remaining capsomeres are hexamers (six sided).*

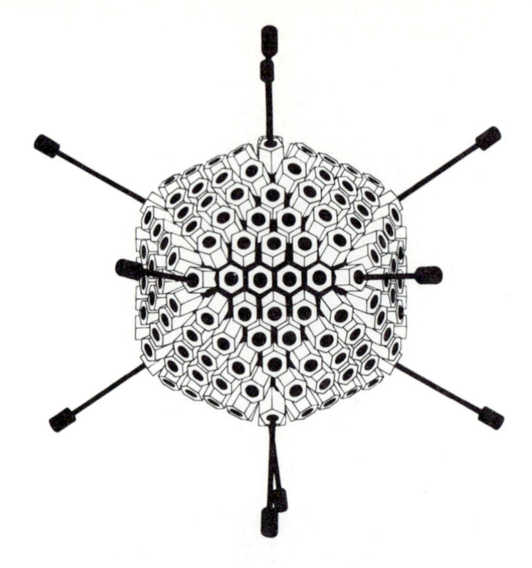

Figure 16-11 *Adenovirus. There are 240 hexamers and 12 pentomers. Antenna-like fibers called penton fibers are attached to penton bases.*

posed of 12 capsomeres. The number of capsomeres increases with larger viruses, but the basic symmetry is generally maintained. To date, the maximum number of virus capsomeres is 252, which can be found in the adenoviruses (Figure 16-11). The arrangement of the nucleic acid in the core of icosahedral viruses is not completely understood. It appears to be one tightly coiled unit in some viruses, whereas in others the nucleic acid is in coiled fragments. Electron micrographs of some of the viruses that possess icosahedral symmetry are presented in Figure 16-12.

Complex or combined symmetries. Some viruses have a symmetric construction that is complex or a combination of helical and icosahedral symmetry. Several viruses fall into this group, including poxviruses and orf virus.* The poxvirus in electron micrographs looks to be brick shaped, with filaments or tubules appearing to be wrapped about the virus particles (Figure 16-13). The best examples demonstrating combined symmetries are the T-even phages, which infect the bacterium *E. coli* and are commonly referred to as **coliphages.** Nearly all phages possess icosahedral symmetry, but T-even phage possesses additional components. The basic features of this virus are illustrated in Figure 16-14. The head of the virus exhibits the typical icosahedral symmetry, with the nucleic acid tightly packed into the capsid. A **collar** structure is located at the base of the head and connects to a **tail** assembly that is in the shape of a helical sheath surrounding an inner hollow core. It is through this core that DNA, located in the head, can be injected into the bacterium during the infection process. See Chapter 17 for details of this process. A hexagonal *baseplate* and

* The *orf virus* is structurally related to the poxviruses. It causes a disease in humans called contagious pustular dermatitis.

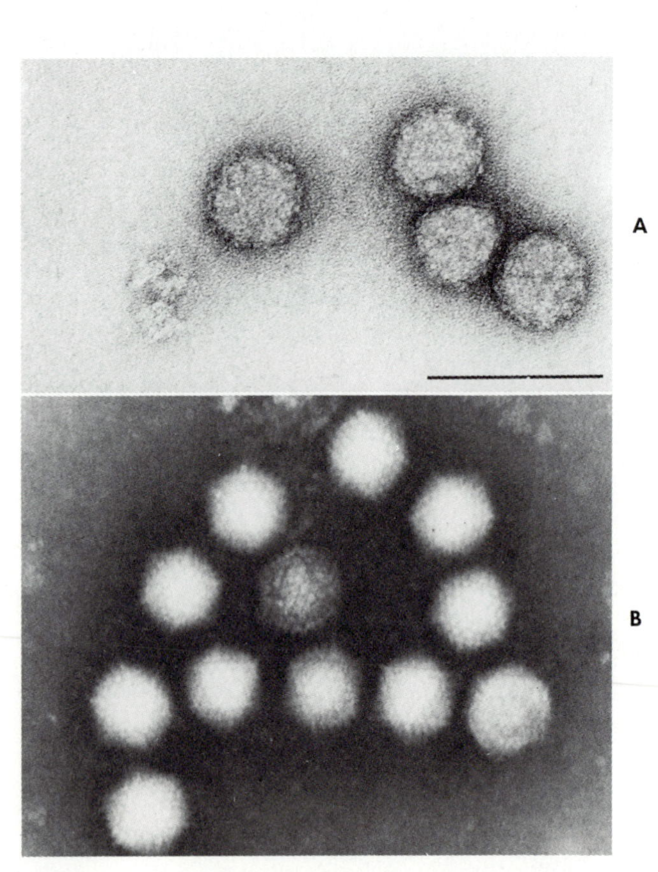

Figure 16-12 *Electron micrographs of two viruses that possess icosahedral symmetry.* **A,** Alphavirus. Alphaviruses *are important causes of encephalitis. B, Adenovirus. Adenoviruses are important causes of respiratory disease. Bar = 100 nm.*

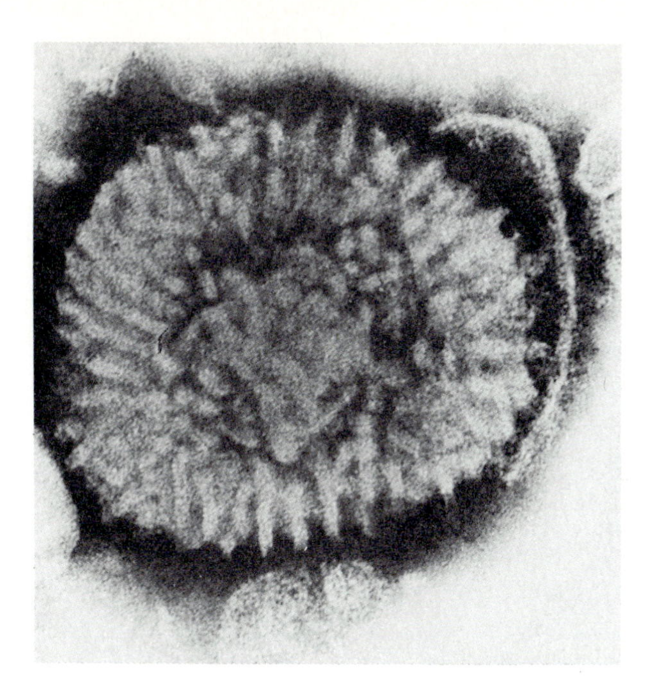

Figure 16-13 *Electron micrograph of poxvirus (vaccinia) demonstrating central core and lateral bodies surrounded by outer membrane. Vaccinia was virus originally used in smallpox vaccine.*

tail fibers, which function in the adsorption of virus to the bacterial cell wall, are located at the proximal end of the sheath. The basic shapes of other phages are illustrated in Figure 16-15.

CULTIVATION

Animal virus cultivation

Viruses cannot be cultivated in nutrient media, but they can be cultivated in their appropriate living hosts, such as bacteria, fungi, plants, and animals. For those viruses that infect higher eukaryotes the simplest technique is to cultivate them in tissue or cell culture. A variety of animal cells may be infected by one animal virus, or the virus may be restricted to growth in one cell type. Human and monkey kidney cells, for example, support the growth of many animal viruses. Cultivation can be used to increase the number of viral particles so that they can be used for biochemical or biophysical studies and in the preparation of vaccines. Growth in tissue or cell culture also permits the isolation of virus from infected patients and thus serves as a diagnostic tool. Cells are first grown in a flask or dish containing the appropriate medium, and as they divide they form a one-cell-thick layer

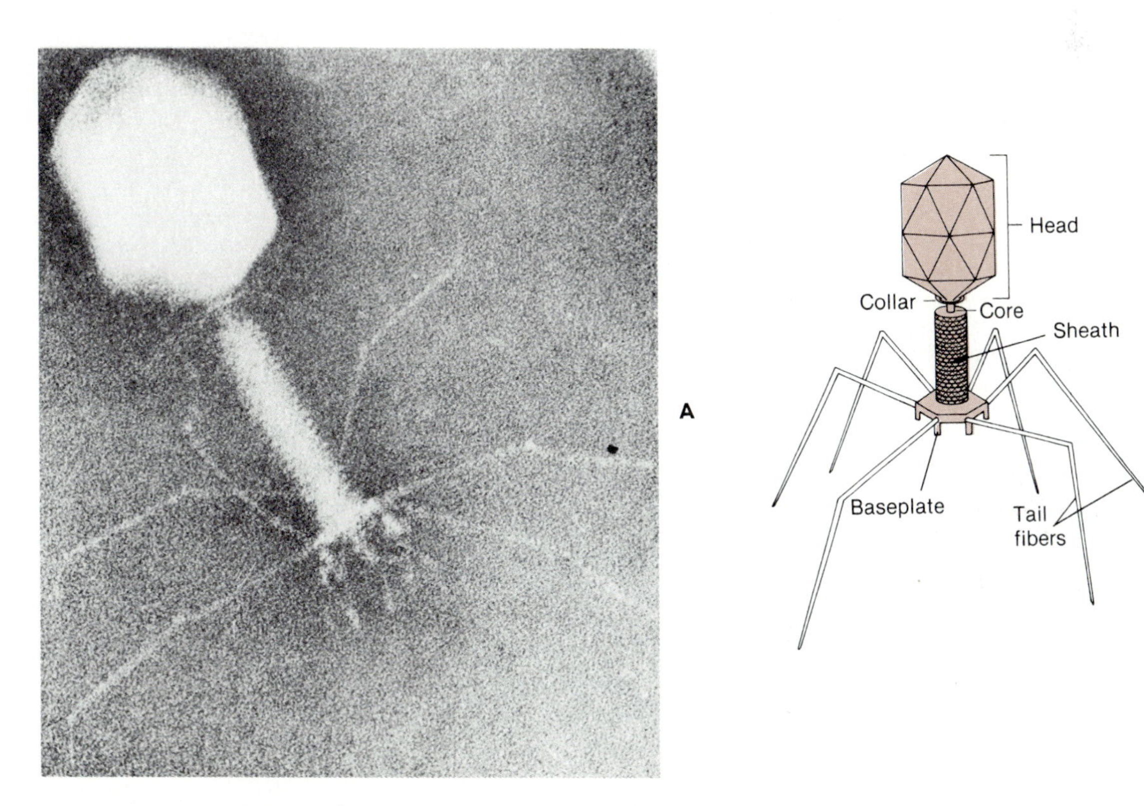

Figure 16-14 ***A,*** *Electron micrograph of T4 bacteriophage.* ***B,*** *T4 bacteriophage model.*

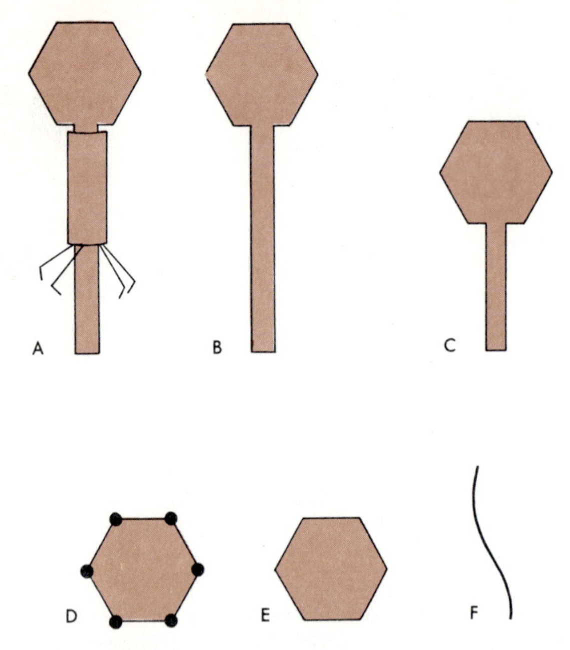

Figure 16-15 *Characteristics of various morphological forms of bacteriophages. Examples are those that infect* Escherichia coli. ***A,*** *Contractile tail, double-stranded DNA: T2 and T4 bacteriophages.* ***B,*** *Long noncontractile tail, double-stranded DNA: T1 bacteriophage and lambda (λ).* ***C,*** *Short noncontractile tail, double-stranded DNA: T3 bacteriophage.* ***D,*** *No tail, large capsomeres, single-stranded DNA: bacteriophage φχ174.* ***E,*** *No tail, small capsomeres, single-stranded RNA: MS2 bacteriophage.* ***F,*** *Filamentous, no head, single-stranded DNA: M13 bacteriophage.*

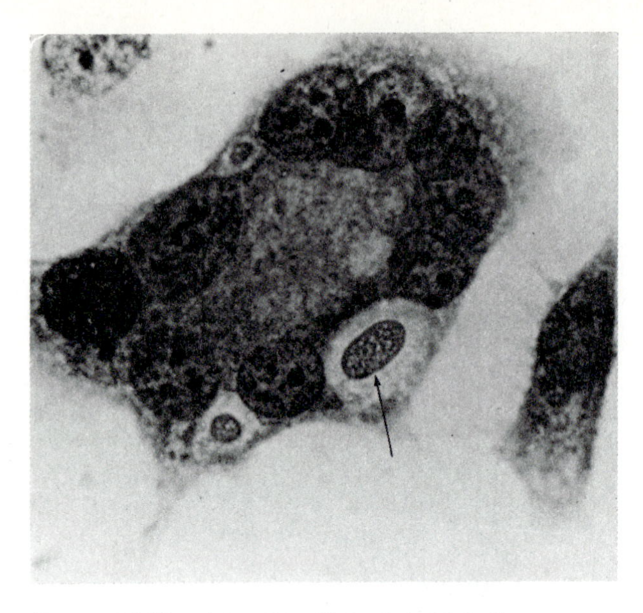

Figure 16-16 *Cytopathic effects of viral infection. Multinucleated giant cells with intranuclear inclusion* (arrow).

(monolayer) over the glass or plastic surface. Virus is then added to this culture, and viral growth is monitored by some change in the host cells called **cytopathic effects**. The most dramatic effect on the host cell is cell death. Cytopathic effects are observable with the compound microscope and sometimes with the naked eye.

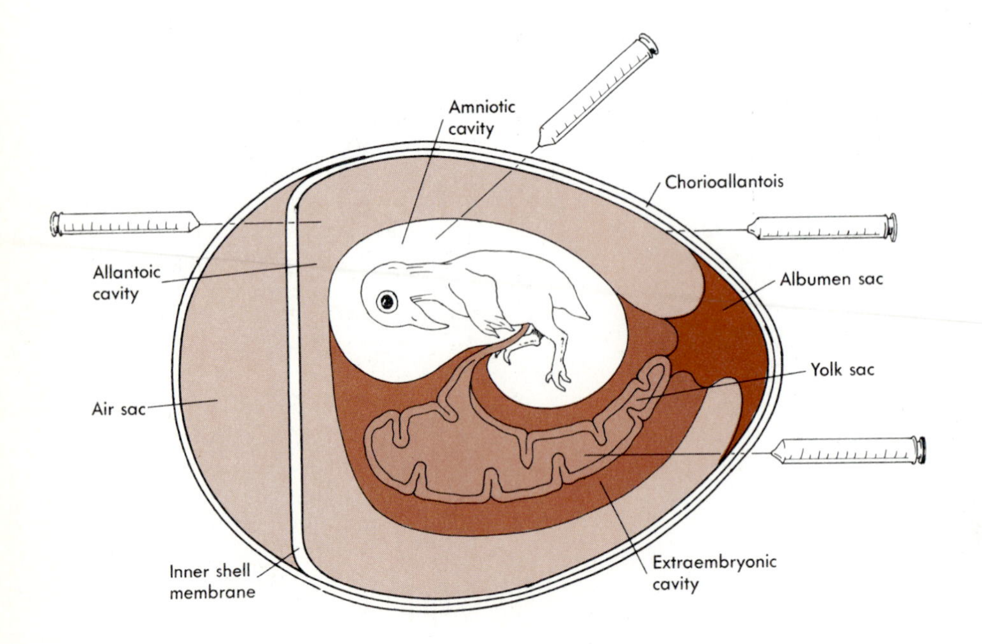

Figure 16-17 *Possible sites of viral inoculation into embryonated egg.*

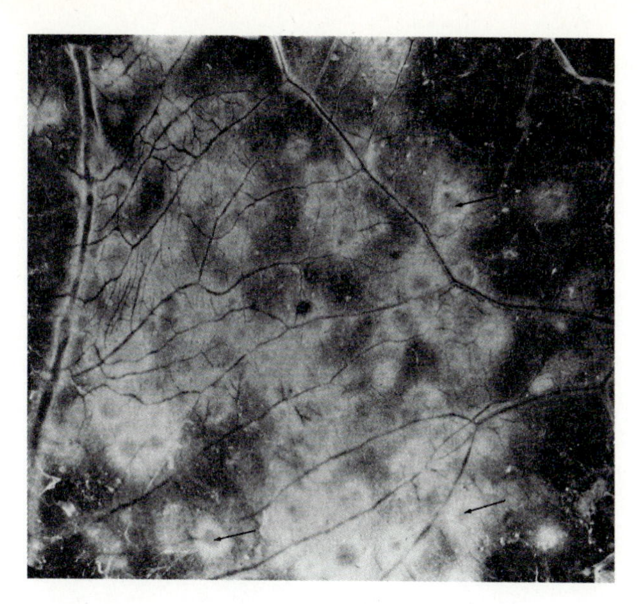

Figure 16-18 *Numerous pocks* (arrows) *produced on chorioallantoic membrane of embryonated hen's egg after 5 days of incubation at 35.5° C.*

They may take the form of cell rounding, fusion of cells into giant cells, or the formation of **inclusion bodies** (Figure 16-16). Inclusion bodies, which are usually observed following some staining procedure, are actually viral proteins, nucleic acids, or some unusual cell product that accumulates in the cell during infection. Some inclusion bodies are so distinctive, for example, in the case of rabies virus infection, that they are of diagnostic significance (**Negri bodies**).

A few viruses that are difficult to grow in cell culture can be cultivated in an embryonated egg. Virus can be introduced into the chorioallantoic membrane, amniotic sac, allantoic sac, or yolk sac (Figure 16-17). Viral multiplication is indicated by death of the embryo, the formation of lesions called **pocks** on the various membranes (Figure 16-18), or by assaying the fluid from infected eggs.

Virus particles can be separated from the host cells and concentrated following the infection process. Separation of virus from extraneous host material can be accomplished using such techniques as chromatography, centrifugation, and electrophoresis. Once the viral particles have been concentrated, it may be necessary to determine those particles that are infectious and those that are not; that is, one must determine the titer (quantity) of the virus suspension (see Figure 16-24). Infectious virus can be treated with inactivating agents, such as phenol or formaldehyde, if it is to be used in certain vaccines.

Cultivation of bacterial viruses (phages)

Phages are easily cultivated in the laboratory by using the following procedure (Figure 16-19):

1. An overnight culture of the indicator bacteria (bacteria to be infected) is prepared.
2. A phage suspension, specific for the host bacteria, is diluted so that the final dilution contains 50 to 100 phage/ml.
3. To 2.5 ml of top agar (soft agar maintained in a water bath at 47° C) is added an aliquot of the phage suspension and one to two drops of the overnight bacterial culture.
4. The top agar suspension is mixed and then spread over the surface of a Petri plate containing bottom agar (hard agar). About 5 to 10 minutes is allowed for the top agar to solidify and then the plate is inverted and incubated at 37° C for 5 hours.
5. Two and one-half hours before the 5-hour period is reached, a broth tube is inoculated with a drop of the overnight bacterial culture.
6. When the 5-hour incubation period has been reached, the infected bacterial culture on the agar in the Petri dish will be examined. The plate will possess a lawn of confluent bacterial growth (uninfected bacterial cells) within which are small clear, nearly circular areas called plaques (Figure 16-20). Plaques represent the sites where phage-infected cells are deposited. The plaques are not produced by the initially infected bacterial cells but by infection of many thousands of adjacent cells on the agar surface. Each initially infected cell releases approximately 75 to 100 phage particles. These particles in turn infect adjacent bacterial cells, which become lysed and produce the typical plaque.
7. One of the plaques is punched out by using a piece of thin wall glass tubing and transferred to the bacterial culture that had been incubating for 2 and one-half hours. Removing only one plaque ensures that the final virus preparation is derived from a single clone.
8. The infected culture, although initially turbid, will subsequently become clear as more and more cells become infected and lyse.
9. Chloroform is added to the tube to enhance lysis of infected cells and to kill any uninfected cells, such as phage-resistant mutants.
10. The chloroformed suspension is passed through a bacteriological filter to remove bacteria and debris, and the filtrate containing only phage can be stored. As many as 10^{11} phage/ml can be obtained by this technique.

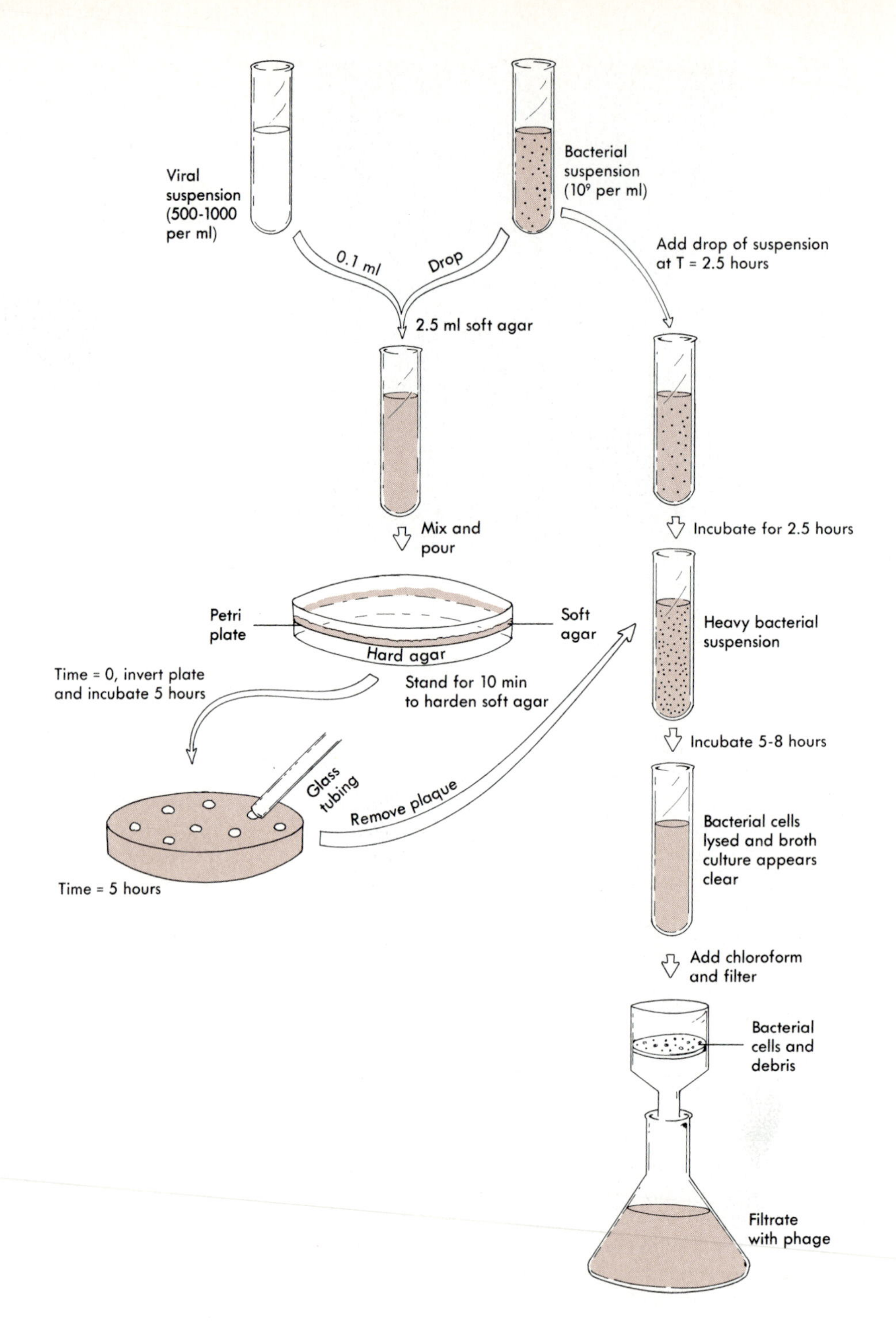

Figure 16-19 *Technique for cultivation of bacterial virus (see text for details).*

ASSAY OF VIRUSES

Animal virus assays

Determining the number of viable or nonviable particles in animal viral suspensions is not as easy as the colony-counting techniques or other measurements used in the enumeration of bacteria. Three of the most useful techniques for counting viruses are:

1. Counting the total number of particles that look like virus with the aid of the electron microscope. Virus particles can be mixed with a known diluted

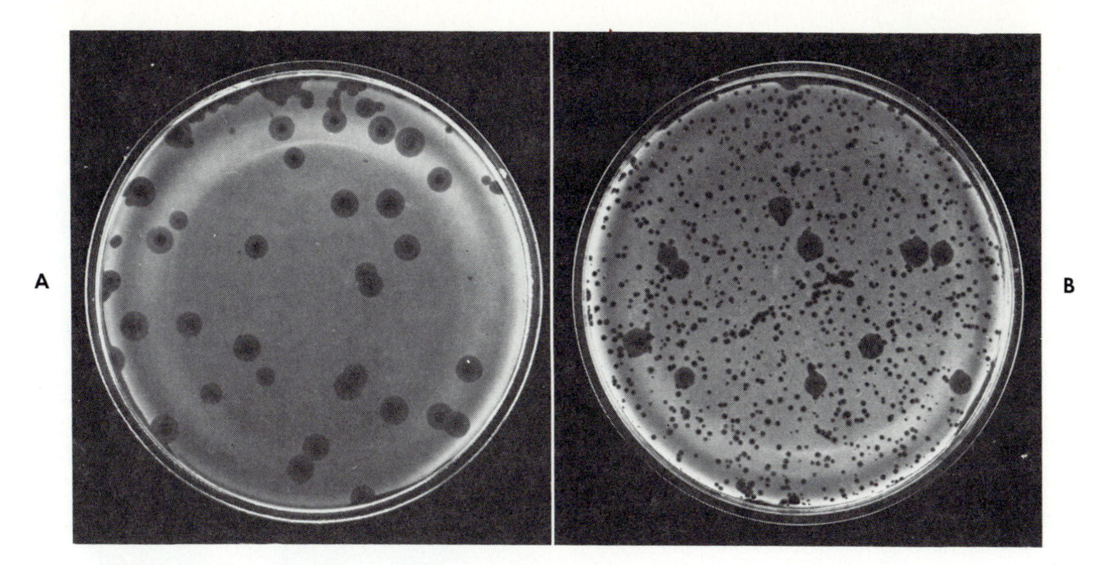

Figure 16-20 *Bacterial plaques. Clear areas represent sites on lawn of bacteria where viral particles have been deposited and have reproduced, resulting in lysis of bacterial cells.* ***A,*** *Uniform plaques produced from single strain of phage.* ***B,*** *Plaques of varying size due to mixture of phage strains.*

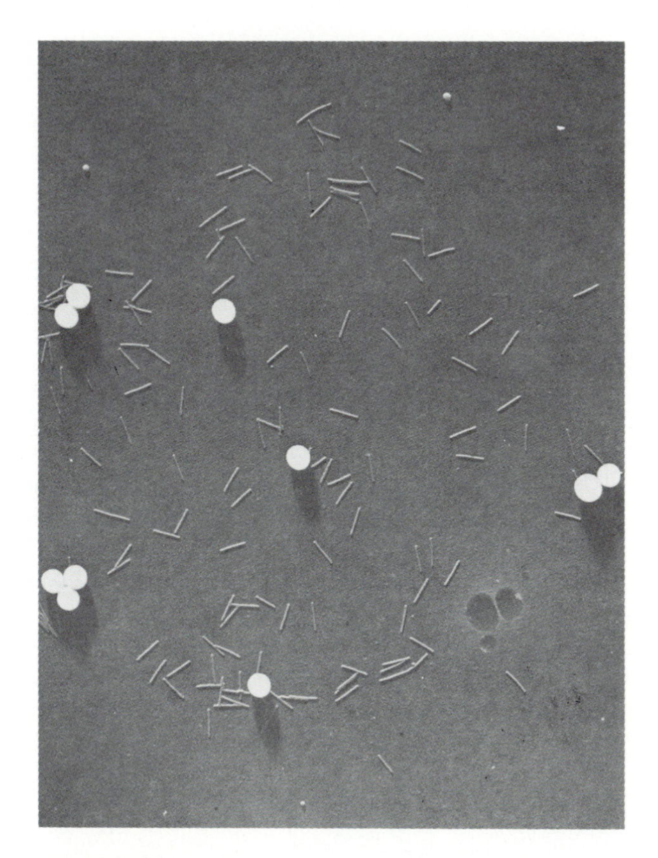

Figure 16-21 *Electron micrograph of rod-shaped TMV particle suspension mixed with spherical latex particles.*

suspension of latex particles and then observed microscopically (Figure 16-21). The number of viral particles observed times the dilution factor for the latex particles provides a measure of the concentration of virus. For example, if there were 10^9 latex spheres per milliliter of suspension and the electron microscope count revealed 20 virus particles for each latex sphere, then the number of viral particles in the original suspension would be 20×10^9/ml or 2×10^{10}/ml. This technique does not differentiate infectious from noninfectious virus.

2. **Agglutination** of red blood cells (hemagglutination) is a characteristic property of the proteins found on some viruses. Hemagglutination is usually due to the intact virus but may also result from proteins released by denaturation of virus or by viral proteins bound to the membrane of virus-infected cells. The hemagglutinin or viral protein acts to join two red blood cells by forming a bridge between them. Large aggregates of red blood cells will form, if sufficient hemagglutinin is present, and fall to the bottom of the tube. The characteristic pattern of hemagglutination readily distinguishes it from nonagglutinated red blood cells. In an actual test virus-containing fluid is first prepared at various dilutions. The highest dilution of virus that can still cause agglutination of red blood cells is called the **hemagglutination titer**. Both infectious and noninfectious particles as well as fragments of infected cells may cause aggluti-

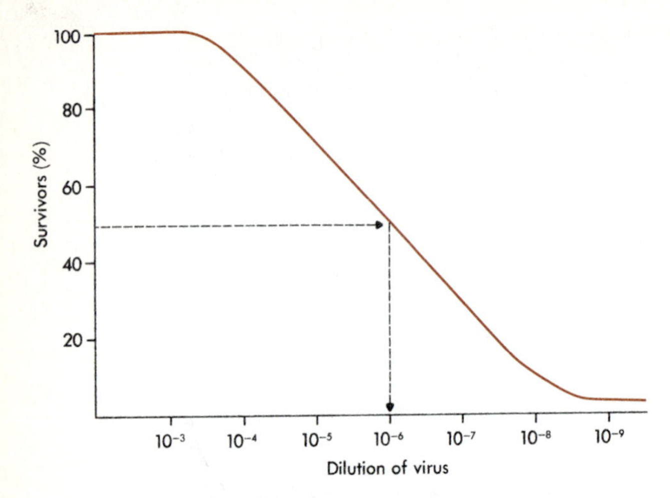

Figure 16-22 *Hypothetical plot determining lethal dose (LD_{50}) for 50% of animals inoculated with dilution of virus. Extrapolation* (dashed line) *of curve shows that at 10^{-6} dilution of virus, 50% of animals were killed.*

Figure 16-23 *Plaque formation. Monolayer of cells was cultured in flat-surfaced bottles. Cell culture on right is uninfected, whereas culture on left was inoculated with a rotavirus. Clear areas are plaques.*

nation of red blood cells, but differentiation between the former two is not possible with this technique.

3. Biological effects such as animal death, cytopathic effects, the number of animals infected, or the number of pocks produced can be quantified. The virus titer is expressed as the **infectious dose (ID_{50})** or **lethal dose (LD_{50})** that causes infection or death, respectively, in 50% of the animals inoculated. For example, suppose we have prepared a virus suspension of unknown concentration. We can prepare dilutions of the virus ranging from 10^{-3} to 10^{-8}; that is, we have diluted the suspension from 1000 to 100,000,000 times. Next, an aliquot of each dilution is used to infect 100 animals. A period of time is allowed for the virus to cause a positive response in the animals. The number of animals showing a positive response at each dilution is plotted (Figure 16-22) and the dilution at which 50% of the animals show a positive response is extrapolated.

The **plaque assay** is one of the most widely used assays for animal virus. Monolayers of cells are inoculated with various dilutions of virus. After several days the cytopathic effects can be macroscopically distinguished from uninfected cells (Figure 16-23). The sites of infection are referred to as plaque-forming units (PFUs). There is a direct proportionality between the number of virus particles and the number of PFUs.

Assay of bacterial viruses

The initial steps used for the cultivation of bacterial viruses can be utilized for assaying unknown phage suspensions (Figure 16-24). Serial hundred-fold dilutions are made of the phage suspension. An aliquot of a diluted phage suspension (0.1 ml) plus a drop or two of the indicator bacteria are added to 2.5 ml of soft agar. This suspension is poured over hard agar in the Petri dish, allowed to harden, and then incubated overnight. This procedure is used for all the phage dilutions. The titer, or number of phage particles in the original phage suspension, is the product of the number of plaques on the plate inoculated with the most favorable dilution (the one that results in giving the closest to 200 plaques per plate—a statistically favorable number) and the total

Concept Check

1. **What are the characteristics of viruses that distinguish them from true bacteria such as *E. coli*?**
2. **Explain the arrangement of capsid subunits in viruses with helical, icosahedral, and complex symmetries.**
3. **Explain the technique used to cultivate bacterial viruses. How can you be sure that cultivated virus was derived from a single clone?**
4. **What are three techniques for counting viral particles in a suspension?**

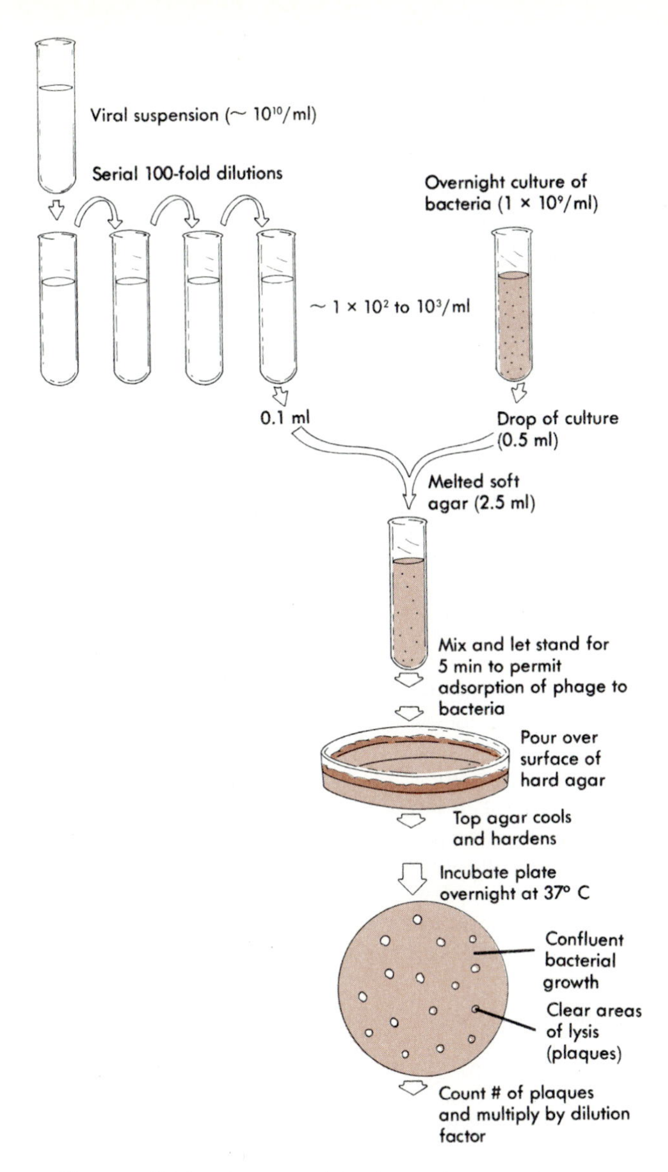

Figure 16-24 *Steps in cultivation and assay of bacterial viruses. For this example we assumed unknown phage suspension to be approximately 10^{10} phage/ml. See text for details of process.*

dilution factor. For example, if 0.1 ml of a 10^7 diluted phage suspension produced 150 plaques then the titer per ml of the original phage suspension is $10 \times 10^7 \times 150 = 1.5 \times 10^{10}$ phage/ml.

EFFECT OF CHEMICAL AND PHYSICAL AGENTS ON VIRUSES

Because of the chemical simplicity of viruses, only a few components are susceptible to the effects of chemical and physical agents, such as the nucleic acid, protein coat, or lipid membrane of enveloped viruses. Some of the chemical and physical agents are of practical value in terms of their ability to inactivate virus and prevent the transmission of virus, as well as their use in the preparation of vaccines. Studies in which the structure and arrangement of various viral components are being explored also require the use of some of these agents. Antiviral agents used in the treatment of human disease are discussed in Chapter 15.

Chemical agents

Many chemical agents can inactivate viruses, but most cannot be used in the treatment of viral infections because of their deleterious effects on host tissue. Some of the more commonly used agents to inactivate viruses are:

1. Lipid solvents. Agents such as ether, chloroform, and anionic detergents solubilize lipid, a component of the membrane of enveloped viruses. Nearly all enveloped viruses are inactivated by lipid solvents.
2. Phenol. Phenol is commonly used to denature proteins and is active against the capsid proteins and protein components of enveloped viruses. The peplomers of the influenza virus, for example, are destroyed by phenol, which is a component of several disinfectants (see Chapter 15).
3. Formaldehyde. Formaldehyde reacts with the amino groups of nucleic acids and proteins. Formaldehyde does not drastically affect the antigenic properties of viruses and for this reason has been used in the preparation of inactivated vaccines.
4. Other agents. Other disinfecting agents that can inactivate viruses include 70% to 90% isopropyl or ethyl alcohol, iodophors, sodium hypochlorite, and 2% glutaraldehyde. These agents are also discussed in Chapter 15.

Photoinactivation. Certain dyes such as neutral red, proflavine, and toluidine blue can penetrate the capsid of some viruses and bind to the nucleic acid. A **photooxidation** process, which renders the nucleic acid inactive, takes place if the virus is subjected to visible light. Neutral red has been used to treat localized herpesvirus (coldsores) infections in humans. Such treatment, however, has been discontinued because of the dye's allergenicity and oncogenic (cancer causing) potential when used in animal studies.

Physical agents

Temperature. Viruses are usually preserved by storage at subfreezing temperatures (−70° C in a mechanical freezer or −196° C in liquid nitrogen) but are inactivated by temperatures of 50° to 70° C for 1 hour. Some

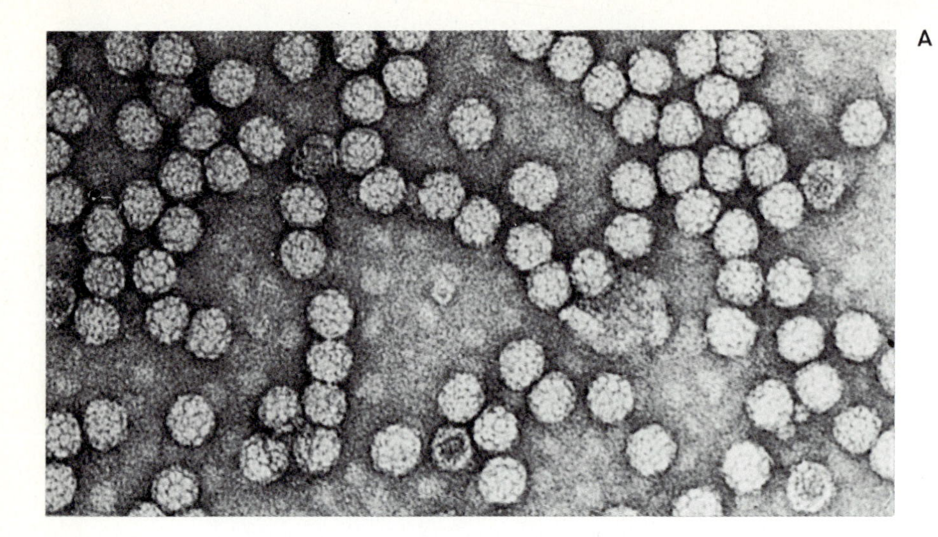

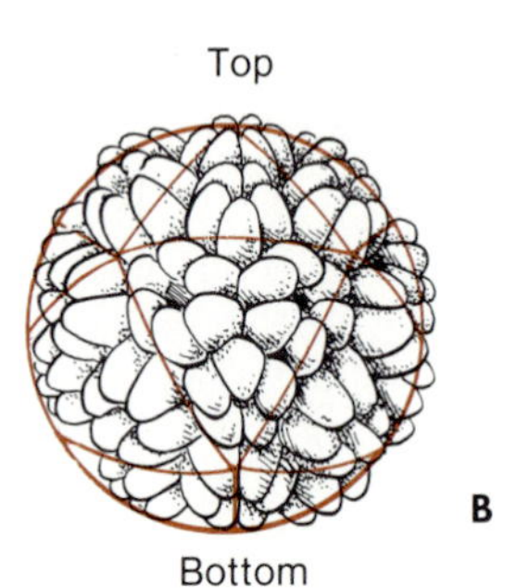

Figure 16-25 *Turnip yellow mosaic virus (TYMV).* ***A,*** *Electron micrograph of TYMV.* ***B,*** *Model of how outer surface of TYMV is believed to appear.*

viruses such as the RNA tumor viruses are inactivated at 37° C. Heat alters the conformation of proteins and nucleic acids by causing them to unfold and become denatured. Sterilization temperatures, such as autoclaving at 121° C for 15 minutes at 15 lb/in^2 pressure or dry air sterilization at 160° to 180° C for 2 to 3 hours, will destroy all viruses.

Radiation. X-rays, ultraviolet (UV) light, and some ionizing particles can inactivate viruses. X-rays impart energy to the medium surrounding the nucleic acid of the virus, which causes the stripping of electrons from some elements and their capture by others. These interactions result in the formation of ions and various radicals that can affect sensitive components of the nucleic acid. In addition to these indirect effects, it is believed that x-rays may also directly affect the nucleic acid by causing breaks in the sugar-phosphate chains. The DNA molecule remains intact if a break occurs in only one strand, but a second break directly opposite the first break or near will result in breakage of the entire molecule. UV light also affects nucleic acids and does so directly. The bases on the nucleic acid have a maximum absorption in the UV spectrum at 2600 Å (260 nm). One of the effects of UV absorption is the formation of dimers, which result from the abnormal bonding between adjacent thymine residues. Dimerization causes an inflexibility within the molecule that twists the molecule out of shape, making replication difficult. See Chapters 10 and 14 for further discussion of the effect of radiation on nucleic acids.

PLANT VIRUSES

There are records of plant diseases as early as the sixteenth century—long before viruses were recognized as infectious agents. Variegation in the color of tulips was observed in 1570 and it was later suggested that this unusual characteristic was due to a transmissible disease. It was not until the twentieth century that tulip variegation was discovered to be due to a virus transmitted by aphids. Despite their early discovery, plant viruses (other than TMV) have not received as much attention as animal or bacterial viruses. They are more difficult to isolate and cultivate than animal and bacterial viruses and, except for a few types, much less is known concerning their morphology and ultrastructure. Viral diseases in plants are discussed in Chapter 20.

Characteristics

Size and shape. Plant viruses, like other viruses, are composed of a protein coat and an inner core of nucleic acid. They are of two types: helical, or rod-shaped particles and icosahedral particles. The size of rod-shaped particles ranges from 15 nm in diameter by 300 nm in length for the TMV to 66 nm in diameter by 227 nm in length for the larger lettuce necrotic virus. Much more is known about the morphology of rod-shaped particles than about the morphology of icosahedral viruses, which range from 30 to 60 nm in diameter and are difficult to distinguish from cellular structures. The basic morphology of rod-shaped viruses is discussed earlier in this chapter under "Helical Capsid Symmetry." Although, unfortunately, less information is available concerning the morphology of icosahedral plant viruses, one of the more widely studied icosahedral viruses is the turnip yellow mosaic virus (Figure 16-25).

Nucleic acid. There are over 300 different plant viruses belonging to approximately 20 taxonomic groups, and 18 of these groups possess an RNA genome that is single stranded. The two groups containing DNA viruses

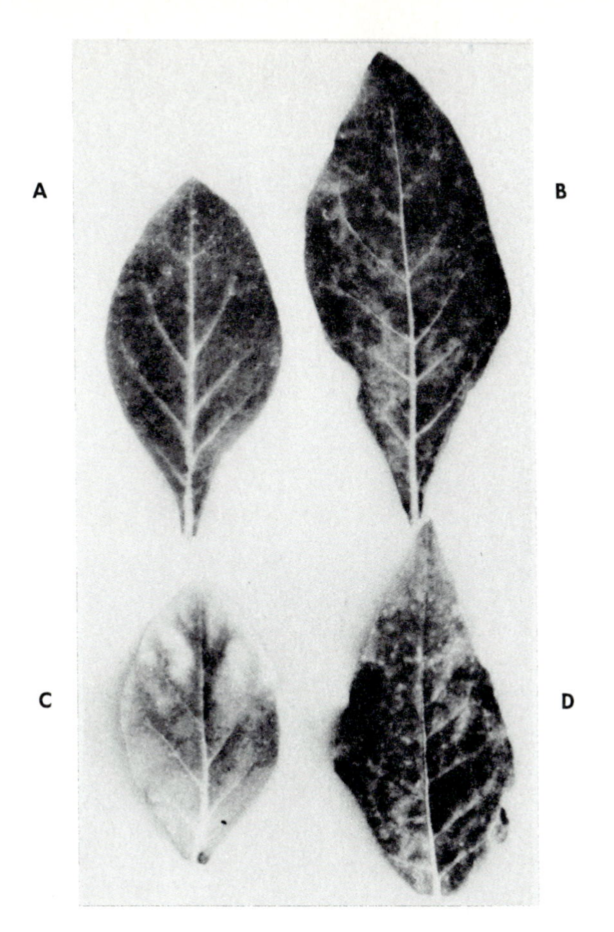

Figure 16-26 *Tobacco leaf used as indicator plant for viral infection.* ***A,*** *Uninfected leaf.* ***B,*** *Infection by lettuce speckle mottle virus.* ***C,*** *Infection by beet western yellow virus.* ***D,*** *Infection by both viruses.*

are the caulimoviruses (double-stranded DNA) and the geminiviruses (single-stranded DNA). The manner in which RNA is associated with the capsid protein is known only for some of the rod-shaped viruses such as the TMV (see Figure 16-7).

It was discovered in 1956 that, unlike many other viruses, the nucleic acid of plant viruses was by itself infectious. Purification of nucleic acid and storage at subfreezing temperatures for over 2 years did not reduce its infectivity. Experiments comparing the infectivity of naked viral nucleic acid and intact virions demonstrated that the lesions, characteristic of infection, appeared more rapidly when naked nucleic acid was used. It is assumed that the time required to uncoat intact viral particles is responsible for the delay in the appearance of cytopathic effects.

Cultivation. Plant viruses can be cultivated by both tissue culture and cell culture. Growth of virus in tissue culture is best obtained if the tissue, for example, root or stem tips, is removed from infected plants and cultivated in nutrient media in the laboratory. Infection of healthy tissue in vitro can be brought about by pricking an inoculum of virus into the tissue by means of pins or through the use of an abrasive material. It is also possible to infect the tissue culture by using the natural vector of the virus. Zoospores of certain fungi, for example, are known to infect plants and transmit virus. Insect vectors can also be used to infect cultures in vitro.

The use of cell culture for cultivation of plant viruses has not reached the state of the art that it has for animal virus cultivation. Isolation of single plant cells for cultivation in the laboratory is more difficult than for animal cells. One technique that has proved valuable is the use of monolayers of insect cells isolated from such vectors as leaf hoppers and aphids for virus cultivation.

Testing for virus. A rapid technique for determining whether the sap of a plant is infected by virus is an important asset in horticulture and has many economic implications. The sap of infected plants often contains only a few viral particles, and microscopic detection, or use of cell or tissue cultures, is futile.* **Indicator plants**, such as tobacco (Figure 16-26) or spinach, have been used successfully to indicate infection by viruses that infect the same or different species of a plant. Indicator plants can be used to determine what percentage of a crop is virus infected; this is especially important for those viruses that cause **latent** infection (they do not initially elicit symptoms in their natural host and require inducement). Often when a latent virus is transferred from its natural host to another plant species, such as an indicator plant, symptoms of infection appear within a few days. This technique has been especially important when grafting procedures are being considered and also in testing potatoes and stone fruits (pears, plums, peaches, etc.) for latent virus. The indicator plant should have large leaves and demonstrate discrete localized lesions following infection. In some instances virus can actually be assayed on the indicator plant if the lesions are discrete and do not coalesce.

Viroids. Viroids are viruslike agents that are smaller than ordinary viruses and do not possess a protein coat. The discovery of viroids, like so many other scientific discoveries, reads like a detective story. Spindle tuber disease in potatoes had been recognized by pathologists since the early 1920s, but no infectious agents could be recovered from diseased tissue. The disease could be transmitted to healthy tissue, even tomatoes, by using extracts of infected tissue. A virus was suspected of causing the disease, and extracts of infected potato tissue were centrifuged, first at low speeds to remove cellular debris, and then at higher speeds to pellet the infectious agent. Infectious activity remained, however, in the supernatant and not in the pellet. The reason for this

* The number of virus particles required to infect plant cells is often very large and ranges from 10^5 to 10^9 particles.

would become clear when later studies revealed that the agent had a molecular weight of 130,000 (most viruses have a molecular weight of 1 million). The infectious supernatant material was insensitive to the chemical agents that solubilize lipid and was also insensitive to phenol. These results indicated that the agent did not possess an envelope or, for that matter, a protein coat. The infectious material was sensitive to RNase but not DNase. Chemical and electron microscopic studies have clearly shown that the viruslike agent is a naked RNA molecule devoid of a protein coat. It is a closed, circular, single-stranded RNA molecule, which exists as a highly base-paired rod-like structure. There are short double-stranded segments interrupted by unpaired regions (Figure 16-27), which form loops. No other agent in nature has been found to possess these characteristics. Viroids are known to cause disease in six other plants, including cucumbers and chrysanthemums. How viroids interfere with the growth process in plants is still a mystery. Viroids have not been isolated in animals. Nonetheless, they are suspected of causing animal diseases, such as arthritis and possibly some brain diseases, although other agents may be involved. Despite their infectious nature, viroids are used in a practical way. In some arid areas of the world, growers actually infect citrus plants with pathogenic viroids. The viroids stunt the trees and increase the yield of fruit while decreasing the tree's water requirements.

Viroids, like other infectious agents, possess nucleic acid but some infectious agents, called **prions**, appear to be devoid of nucleic acid (see Point of Interest). The length of the viroid (240 to 380 nucleotides) is so small that information coding for a viral-replicating enzyme would be impossible. The viroid must apparently rely on the host's enzymatic machinery but the mechanism has not been conclusively determined. In vitro experiments have demonstrated that viroid RNA can be replicated in the presence of host cell RNA–dependent RNA polymerase. At present the only model for replication appears to be the rolling circle (see Chapter 8).

Satellites. Satellites are RNA molecules associated with RNA plant viruses and have no nucleotide sequence homology to either host or viral genomes. They are dependent for their replication on helper virus and in most instances interfere with the synthesis of helper virus. Satellites can be divided into two groups: **satellite viruses** that contain information for synthesis of their own coat protein and **satellite RNAs** that do not have the information for a coat protein and use that of a helper virus. The RNAs of several satellites have been sequenced. The RNA of the tobacco necrosis satellite virus, for example, is 1239 nucleotides in length; 585 nucleotides encode for a single protein, the coat protein, whereas 622 nucleotides represent a noncoding region. The RNA appears to have a high degree of base pairing like a transfer RNA molecule and is thus resistant to nuclease attack.

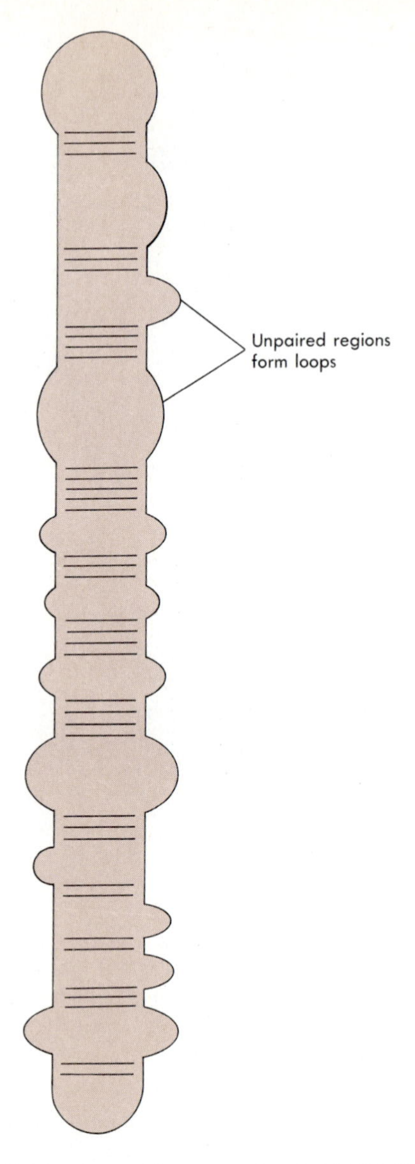

Figure 16-27 *Proposed structure for viroid RNA. Bars equal bonding between adenine and uracil, guanine and cytosine and guanine and uracil. Short base paired regions alternate with shorter unpaired regions resulting in a structure resembling double-stranded RNA.*

The potential significance of satellites in plant disease cannot be underestimated either in their ability to attenuate or modify disease or to exacerbate it. In China, satellite RNA has been used to control disease caused by the cucumber mosaic virus in tomatoes and peppers. Inoculation of the seedlings of these plants with satellite RNA resulted in an increase of 30% more fruit. Similar results were obtained using satellite RNA of TMV. However, these encouraging results must be tempered with

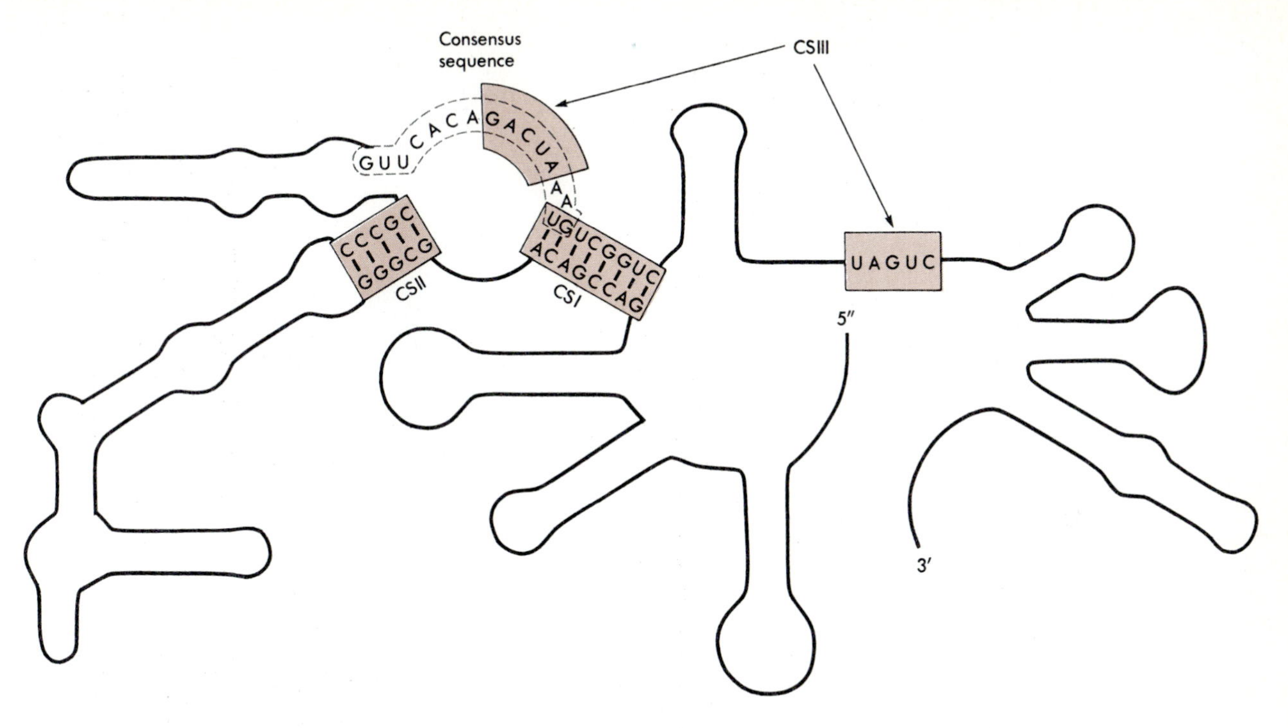

Figure 16-28 *Primary and secondary structure of group I intron. Pairing of sets of complementary sequences* (colored boxes) *in CS 1 and CS 2 provide molecule with secondary structure. Tertiary structure or folding of molecule is acquired by pairing between bases in CS 3. The 16 nucleotide consensus sequence is outlined by dashed lines.*

the realization that satellites may endanger other crops and mutate to a virulent strain.

The origin of satellites is obscure, because many were first observed only in the laboratory, although some have been detected in field plants. There is speculation that satellites have an origin similar to that of viroids because of their similarity in size and structure. Evidence suggests that viroids and plant satellite RNAs may actually be **escaped introns**. The type of introns thought to be involved are called class I introns, and they are found in nuclear rRNA genes, mitochondrial mRNA and rRNA genes, and chloroplast tRNA genes. For example, class I introns of the organism *Tetrahymena thermophila,* a ciliated protozoa, are not only similar in size to the potato spindle tuber viroid (PSTV) but also have structural similarities. A 16 nucleotide phylogenetically conserved sequence, which characterizes all group I introns, is also found in viroids. In addition, group I introns have three sets of complementary sequences that have comparable counterparts in viroids. These complementary sequences or boxes, when paired, provide secondary and tertiary structure to the intron (Figure 16-28). Maintaining the pairing between complementary sequences may require stabilization by protein molecules. Although viroid RNA in vitro appears as a rod-shaped structure, PSTV has been found in vivo to be complexed with a ribonucleoprotein. Because the order in which the conserved regions are arranged is the same in viroids as in introns, the in vivo structure of viroid RNA may be the same as certain introns. The evolutionary relatedness between the two RNA molecules seems to be indisputable.

OTHER VIRUS-HOST ASSOCIATIONS

There is much less information available on the structure, composition, and physiology of infection by viruses that infect or are associated with other organisms such as fungi, protozoa, and some insects. The information that is available is most interesting because it points out some unusual relationships that exist between virus and host.

Fungal viruses

Until 1960 the existence of fungal viruses (mycoviruses) was suspected, but none had been visualized. During the search for an antiviral agent that would combat such diseases as paralytic polio, an unusual chemical was found in the mold *Penicillium.* The chemical was capable of stimulating an antiviral compound called **inter-**

feron in animal cells. Investigators later discovered that the interferon inducer was a double-stranded RNA particle. All fungi examined so far contain double-stranded RNA viruses, most of which are spherical with diameters of 25 to 48 nm (Figure 16-29). The free virus particles isolated from fungi are not infectious, nor are they considered to be defective virus particles. The virus probably exists as a latent or persistent virus in the fungal cell and does not replicate until induced by some environmental factor. The genome of the virus is segmented, and more than one double-stranded RNA species is distributed among the viral particles in the cytoplasm. Genome segmentation is characteristic of all viruses containing double-stranded RNA.

Fungal viruses are usually found in aggregates in older hyphal cells and during vegetative growth are apparently transmitted by cytoplasmic exchange (plasmogamy) with adjacent cells. This type of transmission is an alternative to the usual type of transmission in which attachment to a cell surface is required. Virus can also be found in fungal spores, which represents a means of transmission during the nonvegetative phases of the fungal life cycle.

Fungal viruses are usually considered avirulent, but their presence in tissue causes structural modification and reduced growth of mushrooms. Another example of virus alteration of host phenotype is exemplified in the "**killer**" system of the yeast *Saccharomyces cerevisiae*. Killer strains of yeast secrete a toxin that is lethal to sensitive strains of yeast but to which the killer is immune (Figure 16-30). The killer factor is a double-stranded RNA, probably a fungal virus, and is believed to be controlled by certain chromosomal genes of the host. One set of chromosomal genes maintains the killer characteristic, and a mutation in this set of genes can result in loss of the killer virus, or there can be an inability of the yeast to grow. Another set of genes controls the ability to secrete the toxin as well as maintain resistance to toxin.

Insect viruses. There are over 300 known insect viruses. They are either rod shaped or have icosahedral symmetry. Some have caused epizootics (epidemics among animals) that have ravaged large populations. The majority of insect viruses differ from animal and plant viruses in that individual virus particles are encased either singly or in large numbers in protein crystals that are insoluble in water. The protein-encased viral particles are called **polyhedral inclusion bodies** or **polyhedroses** (Figure 16-31). Polyhedroses may form in the nucleus or cytoplasm of the infected insect. Nuclear polyhedroses multiply primarily in skin, blood, and tracheal tissue, and cytoplasmic polyhedroses multiply in epithelial cells of the insect gut. These protein crystals are produced in large numbers in infected insect larvae and are released when the insect dies and disintegrates. The protein crystals can remain infectious for years in the environment because they are insoluble in water when released from the host.

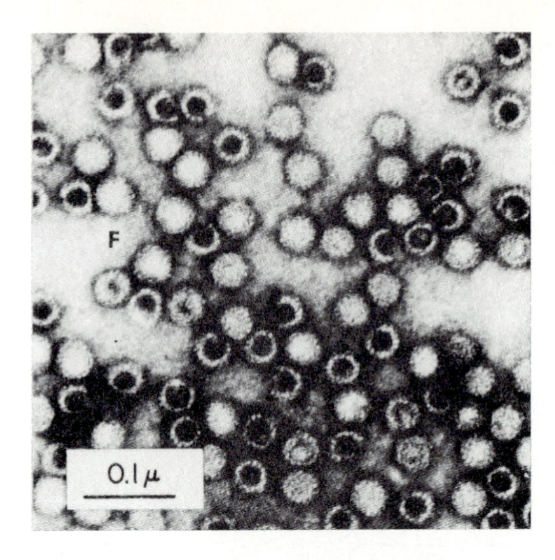

Figure 16-29 *Electron micrograph of virus isolated from fungus* Penicillium stoloniferum. *Scale = 100 nm.*

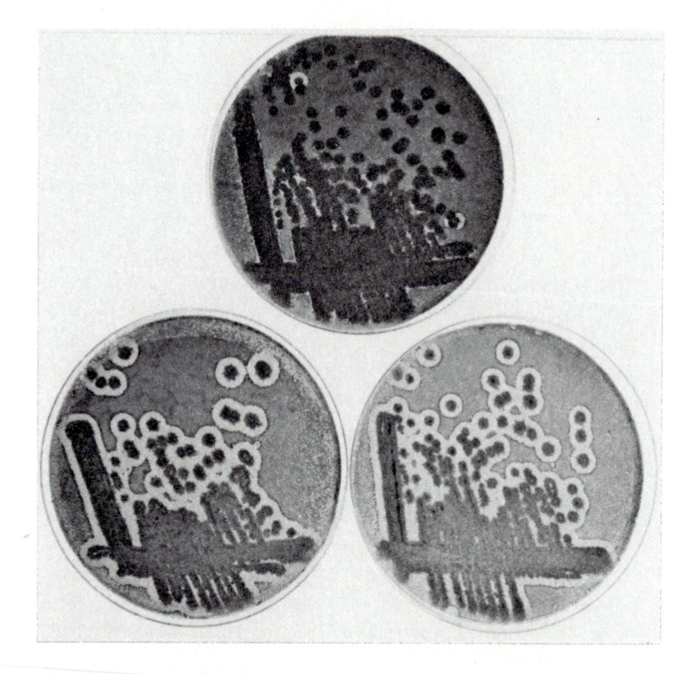

Figure 16-30 *Killer system in yeast. Yeast strain sensitive to "killer" factor is seeded into agar over which is streaked test strains of yeast. Upper photograph shows that killer-sensitive yeast strain, which covers surface of plate, grows up to colonies of test strain. Test strain therefore produces no diffusible "killer" factor. Lower photograph shows clear areas of no growth around test strain, indicating that toxic diffusible factors had been produced.*

Viruses are being used to control certain insect pests, such as the sawfly in Canada. This pest was accidentally

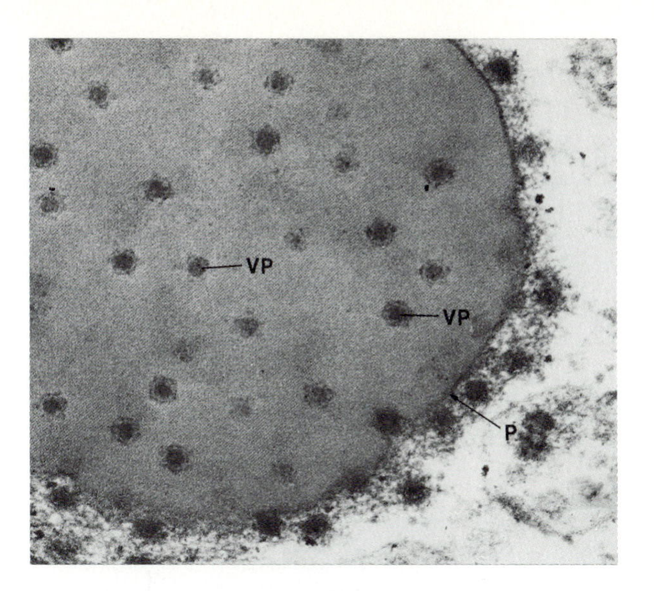

Figure 16-31 *Cytoplasmic polyhedroses (insect virus). Electron micrograph of midgut cell in early stages of infection. (× 150,000.) Polyhedroses* (P) *contain dense bodies that are viral particles* (VP).

introduced into Canada from Europe, and without any natural enemies, it decimated forests. One viral preparation (Virion H) is being used in the United States to control the cotton bollworm and the tobacco budworm. The techniques for propagation of virus for biological control are infection of insect larvae and infection of stable insect tissue culture lines. Once the virus is purified, it can be administered with a spraying machine for fields, or it can be used for individual treatment of trees.

Some insects, such as bloodsucking arthropods (mites, ticks, or mosquitoes, for example), acquire virus by feeding on vertebrate animals. The virus usually multiplies in the salivary glands but does no obvious damage to the arthropod. The virus can be transmitted to vertebrates, including humans. The viruses causing such infections belong to several taxonomic groups, the majority of which are togaviruses.

CLASSIFICATION

Living organisms such as bacteria and fungi are classified based on their many biological properties, with size

Point of Interest

PRIONS: INFECTIOUS AGENTS WITHOUT NUCLEIC ACIDS?

All organisms, including prokaryotic and eukaryotic microorganisms or infectious agents, such as viruses and viroids, possess nucleic acids. Nucleic acids specify the structure of enzymes and other proteins that define an organism's form and development. In addition, nucleic acids are also required for reproduction and perpetuation of the species. However, nucleic acids may not be part of all infectious agents, such as the proteinaceous infectious particle or prion.

Two diseases originally thought to be caused by viruses are now suspected to be caused by prions: scrapie, a neurological disease of sheep, and Creutzfeldt-Jakob disease, a neurological disorder of humans. Kuru, a slow disease involving the nervous system observed in certain tribes of New Guinea, is also believed to be caused by prions. More importantly, prions may cause Alzheimer's disease and other neurological disorders whose cause is not yet known. Alzheimer's disease is the most common form of senility and a leading cause of death in the United States among the aged.

Identifying the prion is painstaking labor for those involved in its investigation. Scrapie can be transmitted to hamsters by taking infected brain tissue and inoculating it into healthy animals, such as hamsters and mice. Extracts of infected brain tissue are subjected to purification steps involving the use of nucleases and proteases. Nucleases digest any nucleic acids present in the extract, whereas proteases digest cellular proteins (the prion is resistant to proteases). The addition of nucleases has no effect on the biological activity of prion material but agents, such as phenol, urea, and sodium dodecyl sulfate, that denature proteins severely reduce the biological activity of prions.

Prions cluster in the brain tissue of sheep with scrapie to form clumps called plaques. Similar plaques have been observed in the brains of victims of Alzheimer's disease. Prion antibodies have been used to determine the presence of prions in the brains of Alzheimer's disease victims but none have so far been detected. Prions are now being sought in the brains of victims of multiple sclerosis, Lou Gehrig's disease (amyotrophic lateral sclerosis), and Parkinson's disease.

Recently researchers located the gene that encodes the major prion protein. Surprisingly the gene was found in normal, as well as infected, cells. How prions replicate and damage cells is the subject of many hypotheses. One theory is that prions are abnormal versions of some ordinary protein found in the brain. When a prion invades a cell, it somehow turns on a gene that makes a normal protein and in the process, alters the protein. The modified protein is slowly churned out as more prions. In time there is an accumulation of prions that destroy the normal cell.

Despite the apparent association between prions and infectious disease, a large body of scientists believe the experimental results are inconclusive. The controversy continues.

Table 16-1 *Scheme for the classification of some bacterial, plant, and animal viruses*

Nucleic acid	Mol. wt. × 10^6	Capsid symmetry	Envelope	No. of capsomeres	Physical type of nucleic acid	No. of genes	Example	Disease or common name
DNA	200	Helical	Present	—*	DS*	400	Poxvirus	Smallpox (animal virus)
	1.6	Icosahedral	Absent	12	SS *	7	φχ174	Bacterial virus (bacteriophage)
	30	Complex (icosahedral head, helical tail)	Absent	32	DS	—	T phage	Bacterial virus (bacteriophage)
	23	Icosahedral	Absent	252	DS	50	Adenovirus	Respiratory disease
RNA	2	Helical	Absent	—	SS	10	Tobacco mosaic virus (TMV)	TMV disease
	4	Helical	Present	—	SS	15	Influenza virus	Influenza
	15	Icosahedral	Absent	—	DS	40	Reovirus	Gastroenteritis
	2	Icosahedral	Absent	—	SS	—	Napovirus	Turnip yellow mosaic virus (TYMV)

**DS,* Double stranded; *SS,* single stranded; (—), unknown.

and shape playing minor roles. Biological properties cannot be used in viral classification schemes, since the virus relies on the host's biological properties for its own development. Logically, size, shape, and structure of viral components would seem to be the most useful characteristics for classification. Advances in microscopy and biochemical techniques have provided a more extensive classification scheme, but they are of recent origin, and many viruses have eluded detailed microscopic or biochemical characterization. The principal viral characteristics that are being used today for classification include type of nucleic acid, capsid symmetry, number of capsomeres, diameter of the virus, molecular weight of nucleic acid, and number of genes. Table 16-1 illustrates a typical scheme for the classification of certain plant, animal, and bacterial viruses. A detailed scheme for the classification of viruses that infect humans is presented in Chapter 27.

Concept Check

1. What are some chemical agents that inactivate viruses, and what are the mechanisms involved?
2. What is the effect of temperature on viruses and the molecular components involved?
3. What characteristics distinguish plant viruses from animal viruses?
4. How do viroids differ from satellites? What is the relationship of viroids to prions?
5. What is the evidence suggesting that viroids are escaped introns?
6. What are the characteristics that are used to differentiate and classify viruses?

SUMMARY

1. Viruses are submicroscopic, obligate, intracellular parasites that rely on their host for energy and nutrients. They range in size from 10 nm to over 250 nm.
2. Chemically, viruses consist of a core of nucleic acid, either RNA or DNA, surrounded by a protein coat. The coat protein, or capsid, is composed of individual protein units called capsomeres. Some viruses also contain a lipid envelope, which may contain spikes or protein called peplomers. The complex infective viral particle is called a virion.
3. The shape of the viral capsid conforms to certain geometric patterns: helical, icosahedral, or complex. Helical capsids are usually associated with rod-shaped viruses, whereas icosahedral capsids are found in spherical viruses. Complex symmetry is usually a combination of helical and icosahedral capsid symmetries.
4. Viruses can be grown in tissue culture or embryonated eggs. Growth often results in cellular degeneration, which can be detected as a cytopathic effect.
5. The number of animal viruses in a suspension can be determined by applying certain techniques: electron microscopy, agglutination of red blood cells, or measurement of cytopathic effects. Phage can be assayed by measuring the number of plaques produced.
6. Viruses can be inactivated by various chemical and physical agents. Chemicals such as formaldehyde, phenol, and lipid solvents can inactivate viruses. Photoinactivation following staining of viruses can also be used. Radiation can inactivate viruses directly via interaction with the viral nucleic acid or indirectly by generating oxidizing agents as it interacts with the cytoplasm of the cell.
7. Plant viruses are rod shaped or spherical. Rod-shaped viruses can be as large as 227 nm. Most plant viruses contain RNA, and the nucleic acid by itself is infectious, a characteristic that distinguishes it from nearly all other viruses.
8. Plant virus can be assayed by determining cytopathic effects on certain indicator plant leaves. Certain viruslike particles are without a capsid but are infectious for certain plants and are called viroids. Satellite RNA molecules are associated with plant viruses but have no nucleotide sequence homology with them. They rely on helper virus for their replication.
9. Fungal viruses all contain double-stranded RNA and are spherical. Nearly all appear to be avirulent. It is believed that they are probably defective and exist in a latent state in the cell.
10. Insect viruses are rod shaped or icosahedral. Viral particles are encased in protein and are referred to as polyhedral inclusion bodies. They may be found in the nucleus or cytoplasm of infected insect cells. Some insect viruses are used to control insect populations that devastate crops or other plants.
11. Viruses are classified according to certain parameters: type of nucleic acid, capsid symmetry, capsomere number, size of virus, and molecular weight of nucleic acid.

Self-study Quiz

Completion

1. An infective viral particle is called a ______________.
2. Capsomeres are protein subunits making up the ________.
3. RNA molecules, with no homology to viral or host genomes, requiring helper virus to replicate are called ______________.
4. The joining or clumping of red blood cells by viruses is called ______________.
5. Viroids are viruslike agents that are smaller than viruses and lack a ______________.
6. Protein-encased viral particles found in the nucleus of insect cells are called ______________.

Short answer

1. In the study of a plant disease researchers obtained the following results. What conclusion can be drawn from each observation?
   a. When a branch from a plant in which disease is suspected but symptoms are not readily observable is grafted onto another closely related plant, the healthy indicator plant begins to show symptoms of disease.
   b. The infectious agent can be isolated from the gut-lining cells of aphids.
   c. The infectious agent is not inactivated by phenol, chloroform, or DNase.
2. Predict the effects of the following treatments of a sample containing an animal virus lacking an envelope. Mark *(X)* if the treatment would inactivate the virus; mark *(O)* if it would have little effect.
   ________ a. Formaldehyde treatment
   ________ b. Chloroform
   ________ c. $-196°$ C
   ________ d. $121°$ C in autoclave for 20 minutes
   ________ e. Ultraviolet light
   ________ f. Drying in the air

Multiple choice

1. Which of the following would be characteristic of animal viruses and not of phage?
   A. Nucleic acid
   B. Envelope
   C. Head/collar/tail
   D. Polyhedroses
2. How can virus infection of an inoculated embryonated egg be detected?
   A. Formation of pocks on membranes
   B. Gas formed inside of shell
   C. Change in consistency of the yolk in yolk sac
   D. Bleeding of chick
3. Identification of a low molecular weight infectious agent that is not inactivated by lipid solvents, phenol, or DNase would lead to the characterization of the agent as what?
   A. Infectious DNA
   B. An RNA virion
   C. A DNA virion
   D. An RNA viroid
4. Which of the following is true concerning fungal viruses?
   A. They are also infectious for humans, as well as fungi.
   B. Free viral particles are not infectious.
   C. Fungal viruses are often transmitted by insects that feed on fungi.
   D. Fungal viruses are responsible for variegation of certain plants and flowers.
5. What type of microscopy is usually necessary to observe viruses?
   A. Darkfield
   B. Phase-contrast
   C. Fluorescent
   D. Electron
6. What virus might be observed with the compound brightfield microscope?
   A. Rabies virus
   B. Poliovirus
   C. Smallpox virus
   D. Foot and mouth disease virus

7. What enzyme is associated with certain RNA, oncogenic viruses?
   A. Photooxidative enzymes
   B. Reverse transcriptase
   C. Neuraminidase
   D. RNA polymerase
8. What part of some animal viruses is derived from the host cell membranes?
   A. DNA core
   B. Capsid
   C. Inclusion body
   D. Envelope
9. During the cultivation of phage, chloroform is used to:
   A. Remove the lipid coat of the virus
   B. Kill any unadsorbed virus
   C. Lyse infected bacterial cells
   D. Prevent adsorption of virus to walls of the container
10. Which of the following sentences is true?
   A. Prions are viruses that are the cause of some neurological disorders.
   B. Prions are proteinaceous infectious particles that are the cause of several plant diseases.
   C. Prions are RNA viruses that are the cause of certain neurological disorders.
   D. Prions are proteinaceous infectious particles that are believed to be the cause of certain neurological disorders.

Thought questions

1. When viruses were first discovered there was debate as to whether one could consider them to be "living" or "nonliving" entities. What characteristics could be cited to support both sides of the argument?
2. Offer explanations as to why viruses exhibit a fairly narrow host range. In other words, what is involved in the virus-host relationship that prevents a virus from entering and replicating in any living cell?

SELECTED READINGS

Books

Dulbecco, R., and Ginsberg, H.S. *Virology.* Harper and Row, New York, 1980.

Fenner, F., and White, D.O. *Medical Virology* (3rd ed.). Academic Press, Inc., New York, 1986.

Fields, B.N. (Ed.). *Virology.* Raven, New York, 1985.

Fraenkel-Conrat, H., *The Viruses.* Plenum Press, New York, 1985.

Harrison, S.C. *Structure of Viruses.* In *The Microbe 1984.* Part 1: Viruses. 36th Symposium Society for General Microbiology, Cambridge University Press, Cambridge, 1984.

Jurnak, F.A., and McPherson, A. (Eds.). *Virus Structures.* John Wiley & Sons, Inc., New York, 1984.

Robertson, H.D., Howell, S.H., Zaitlin, M., and Malmberg, R. (Eds.). *Plant Infectious Agents: Viruses, Viroids, and Satellites.* Cold Spring Harbor Laboratory, Cold Spring Harbor, New York, 1983.

Journal articles

Francki, R.I.B. Plant virus satellites. *Annu. Rev. Microbiol.* **39:**151, 1985.

Matthews, R.E.F. Viral taxonomy for the nonvirologist. *Annu. Rev. Microbiol.* **39:**451, 1985.

McCauley, J.W., and Mahy, B.W. Structure and function of the influenza virus genome. *Biochem. J.* **211:**281, 1983.

Prusiner, S.B. Prions. *Sci. Am.* **251:**50, 1984.

Reisner, D., and Gross, H.J. Viroids. *Annu. Rev. Biochem.* **54:**531, 1985.

CHAPTER 17

Viral replication and tumor development

Chapter Outline

Introduction

Viral replication occurs only in hosts in which metabolites and metabolic machinery can be subverted for viral needs. Although viral replication often causes immediate destruction of the host cell, not all cells infected by virus will die. Relatively recent investigations demonstrated that some viruses persist in cells for an indeterminate period of time. A few viruses also cause certain cancers. To develop effective methods for preventing or controlling viral disease, it is important to understand the mechanisms by which the virus infects the cell.

This chapter discusses the mechanisms used by bacterial, animal, and plant viruses to replicate and cause lysis of the cell or to persist in cells without replication. The chapter also shows the relationship between viruses and cancer, as well as detailing the current theories as to how "cancer" genes work.

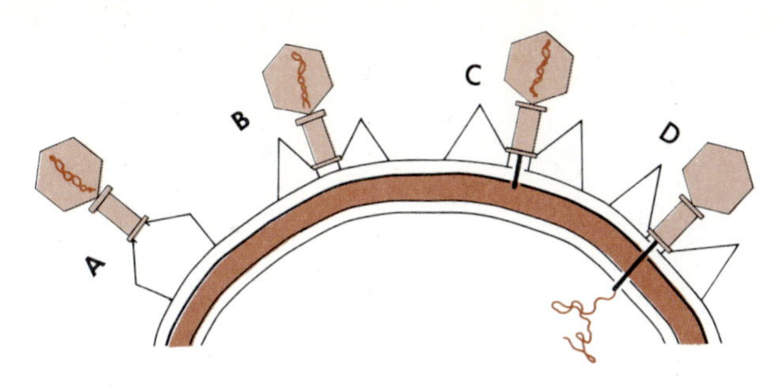

Figure 17-1 *Bacteriophage infection process.* ***A,*** *Attachment of long tail fibers of virus to cell wall.* ***B,*** *Adsorption of tail pins to cell wall.* ***C,*** *Contraction of tail sheath and injection of core into cell wall.* ***D,*** *DNA injection through core into cytoplasm of cell.*

INFECTION BY BACTERIAL VIRUSES (BACTERIOPHAGES)

At first glance the reproduction of a virus within a host might appear to be a relatively simple process, because the virus is composed of only a core of nucleic acid surrounded by a protein coat (plus a lipid envelope in some animal viruses). There are several barriers that the virus must overcome before it can faithfully reproduce itself in the host cytoplasm. The virus must first attach to the surface of the bacterial cell, and this requires some receptor specificity. Second, the virus must penetrate the cell wall and cytoplasmic membrane. Third, once the viral nucleic acid is present in the cytoplasm, it must be protected from degradation. Fourth, viral protein synthesis and viral maturation must compete with host metabolic activities.

The life cycles of phage are divided into two types: the **lytic** and **lysogenic cycle.** The lytic cycle results in the formation of progeny phage and lysis of the bacterial cell. Lysogeny usually results in the viral nucleic acid becoming part of the bacterial chromosome, and no phage progeny are produced. Our discussion first centers on the life cycle of virulent or lytic phage, such as T4.

Lytic life cycle of T4 phage

Adsorption to the bacterial cell. Bacterial viruses, like other viruses, show a high degree of specificity in terms of the bacterial species that they can infect. The interaction between specific adsorption sites on the virus and on the host cell is genetically determined, because mutations in the bacterial or viral genome, which alter surface structures, greatly influence adsorption. The T series of bacteriophages (*T* stands for type) that infect gram-negative bacteria interact with bacterial receptor sites that may include (1) the lipopolysaccharide of the outer membrane of the cell wall, (2) the lipoprotein of the outer membrane of the cell wall, and (3) outermembrane proteins, such as porins, or transport proteins, such as those involved in iron transport. Some filamentous phages, such as MS2 and f^2, bind to the F pilus. Adsorption to pili occurs only if the bacterial cell is "male," and if the pili are mechanically removed, no phage adsorb. The viral receptor sites also vary from one group of viruses to another. The T series possess **tail fibers** (see Figure 16-14) that aid in attachment, but not all bacterial viruses possess such appendages. T4 is believed to bind to a lipopolysaccharide receptor (such as a sugar residue of the O antigen) by a single tail fiber (Figure 17-1). Once the virus is attached to the bacterial surface, interaction between other tail fibers and other lipopolysaccharide receptor sites prevent it from leaving the bacterial surface. Irreversible anchorage occurs by interaction of baseplate pins and the bacterial surface. The specific site-interaction between host and virus in some systems is influenced by certain cations. The addition of specific ions, such as calcium, promotes adsorption, and this may be related to the surfaces of the virus and host, which are naturally electronegative and tend to repel one another. The specificity between host cell wall receptors and viral surface receptors is best demonstrated by the fact that the nucleic acid of some viruses is capable of replication in the cytoplasm of a wide range of different host cells, whereas these same viruses with intact protein coats and/or envelopes can infect only one cell type.

Penetration of the cell wall. A major difference between bacterial and animal virus infection is that only the nucleic acid of bacterial viruses enters the cytoplasm of the cell, although there are exceptions. The filamentous single-stranded DNA phages penetrate the cell with an intact capsid. Once the **baseplate pins** of T4 attach to the bacterial surface, there is contraction of the **tail sheath**. This forces the phage **head** and **collar** to be pulled closer to the bacterial surface and also causes a portion of the **tail tube** to be forced through the cell wall. DNA is released by the capsid and injected through the membrane pore into the cytoplasm (Figure 17-2). The cytoplasmic membrane becomes temporarily leaky following penetration (but later repaired by phage). If many viral particles were to absorb and penetrate the cell wall, the exposed membrane would disintegrate, resulting in lysis of the cell. The latter process is called "lysis from without," that is, the cells lyse before the virus has a chance to replicate in the bacterial cytoplasm and cause "lysis from within."

Protection of viral DNA. The viral nucleic acid is potentially susceptible to hydrolytic enzymes (nucleases) within the cell. Bacterial cells and other organisms protect themselves from invasion by foreign DNA by using restriction **endonucleases** (see Chapter 11) that cleave specific internal areas of the foreign DNA molecule. Endonucleases distinguish bacterial DNA from phage DNA by methyl groups placed on key nucleotides of the bacterial DNA. Methylation appears to be important for the fidelity of bacterial DNA replication, DNA repair activities, and site-specific recombination. Sometimes the bacterial methylating enzymes methylate viral DNA, rendering it resistant to bacterial endonuclease. The DNA sites to which the endonuclease and methylase are specific vary from species to species. The T-even phages contain DNA in which cytosine is replaced by **hydroxymethyl cytosine (HCM)**—a modification process that would appear to protect viral DNA. *Escherichia coli,* however, has evolved nucleases that recognize HMC. T-even phages code for an enzyme that glucosylates HMC and makes phage DNA resistant to *E. coli* nucleases. Thus the long-term virus-host association results in important evolutionary changes. Perhaps in time *E. coli* will evolve a mechanism to degrade DNA, containing glucosylated HMC.

Inhibition of host metabolism. Replication of the viral genome and synthesis of viral proteins require the energy-generating systems of the host. The bacterial ribosomes, as well as many of the bacterial enzymes required for protein synthesis and nucleic acid synthesis, are also used by the virus. Energy-producing systems in the bacterial cell are not significantly altered following viral infection, but the virus will either shut down or modify other host metabolic systems. Some viruses, for example, increase the activity of already-existing enzymes to ensure maximum activity during replication. The mechanism for shutdown of *E. coli* macromolecule synthesis by T4 is not totally understood. Studies show that the adsorption of phage particles lacking DNA (called **ghosts**) is sufficient to turn off host macromolecule synthesis. One of the first viral enzymes synthesized in the infected cell is a DNA nuclease. However, it is synthesized after host macromolecule synthesis ceases and its function appears to provide nucleotides for viral nucleic acid synthesis.

Another mechanism for shutting down host DNA synthesis is **nuclear disruption** in which nuclear areas (nucleoids) located centrally before infection move toward the cytoplasmic membrane after infection. Studies have shown that it takes much longer for host DNA synthesis to be shut off in mutants in which this morphological change does not take place. The following paragraphs explain how the virus alters bacterial transcription.

Viral transcription. Viral proteins are synthesized in a sequential manner and to do this the timing of mRNA synthesis is regulated. T4 controls the timing of protein synthesis by altering the activity of *E. coli* RNA polymerase. Some viruses carry the information for synthesis of their own RNA polymerase but not T4. T4 must control host RNA polymerase activity by using modification techniques. Before describing these modification techniques, it should be mentioned that the mRNAs of T4 can be divided into two basic groups:

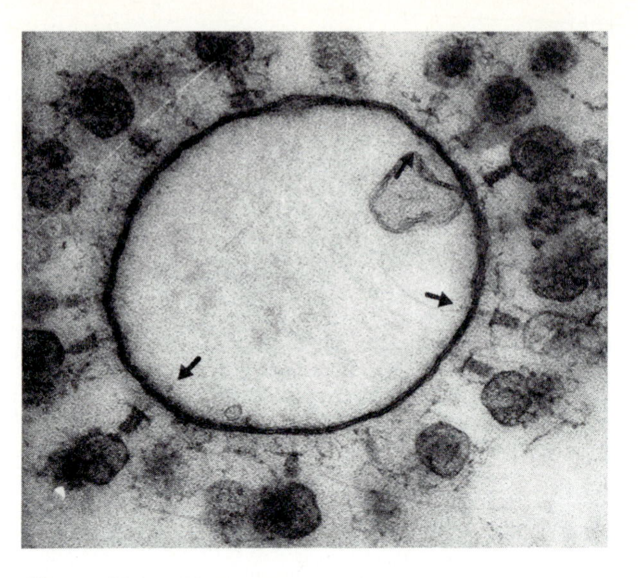

Figure 17-2 *Electron micrograph demonstrating tail tube penetration of cell wall. Several T4 bacteriophage particles can be seen to be adsorbed to reconstituted cell wall. Arrows point to tail tube of virus, which has punctured cell wall. Reconstitution of cell wall makes it appear larger than it does in its natural state.*

1. Early mRNAs are synthesized before viral DNA synthesis occurs. Early mRNAs contain information for the enzymes and proteins associated with DNA synthesis and regulation of mRNA synthesis.
2. Late mRNAs are synthesized after viral DNA synthesis occurs. Late mRNAs contain the information required for synthesis of phage structural proteins and assembly of those proteins into mature virions. In addition, the proteins required for lysis of the bacterial cell are also produced during the late period.

Transcription of some of the early viral mRNAs requires unmodified *E. coli* RNA polymerase. About 30 seconds after infection, *E. coli* RNA polymerase undergoes some changes that alter its specificity. A viral head protein injected with viral DNA during the infection process enzymatically binds ADP-ribose (called **ADP-ribosylation**) to one of the alpha subunits of RNA polymerase. About 2 minutes after infection there is an additional ADP ribosylation of the second alpha subunit. This modification is believed to reduce the recognition between the sigma subunit and the RNA polymerase core, thus host promoters are not recognized. There is also evidence to suggest that translation of host mRNA can be reduced by viral modification of transfer RNA or by alteration of the ribosomes. For example, protein-modified ribosomes are unable to bind to host mRNA, thus preventing the latter's translation.

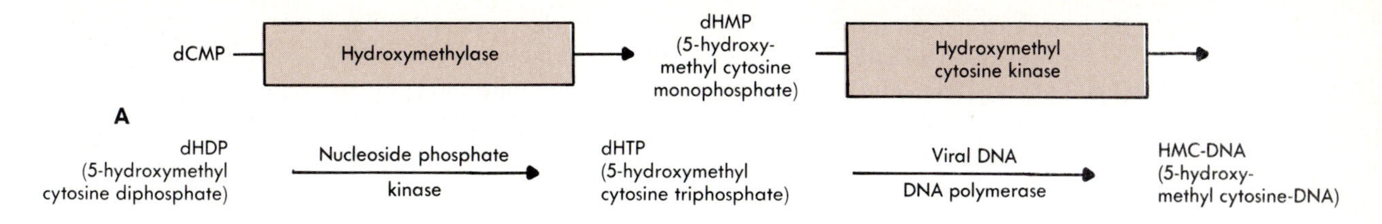

Figure 17-3 *Hydroxymethylcytosine (HMC) metabolism.* ***A,*** *Pathway for synthesis of 5-hydroxymethyl cytosine from deoxycytosine monophosphate (dCMP) and its incorporation into DNA. Phage encoded enzymes are colored.* ***B,*** *Glucosylation of 5-hydroxymethylcytosine.*

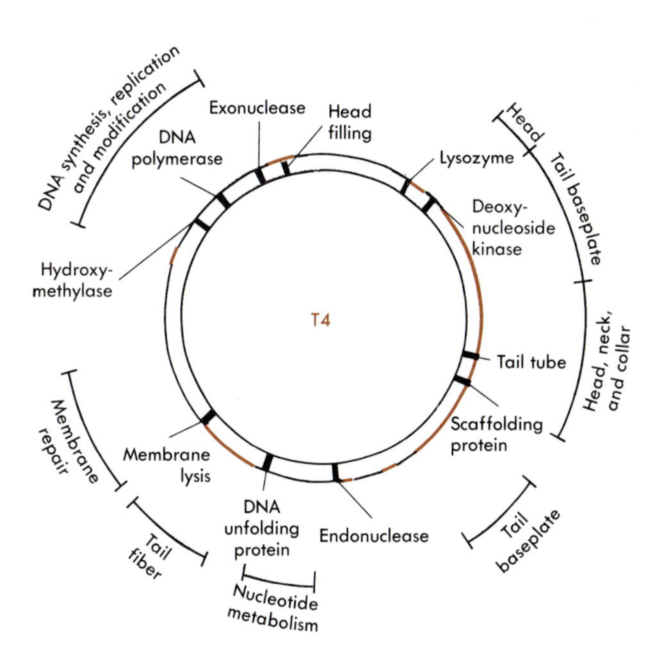

Figure 17-4 *T4 genetic map showing positions of genes involved in DNA synthesis and phage assembly. Inside circle shows relative position of some important genes. Colored portions of circle represent late genes, most of which are located on right side of map.*

Replication of T4 DNA. The deoxyribonucleoside monophosphates required for synthesis of DNA are provided via the degradation of host DNA. **Virally-encoded endonucleases** breakdown host DNA into fragments of double-stranded DNA, which is then degraded by a **virally-encoded exonuclease.** The mononucleotides released by viral degradative enzymes are converted to nucleoside triphosphates by host enzymes but there is an exception. T4 DNA, unlike *E. coli* DNA, contains HMC, therefore phage-encoded enzymes are required. These two viral enzymes are **hydroxymethylase** and **hydroxymethyl cytosine kinase** (Figure 17-3). Glucosylation of HMC occurs after it is incorporated into DNA, thus protecting it from host nucleases.

Several T4 genes (Figure 17-4) code for proteins that are associated with T4 DNA replication and include DNA polymerase and binding proteins, such as helix destabilizing proteins. T4 DNA is a long, linear molecule that is replicated bidirectionally. A major difference between viral DNA synthesis and cellular DNA synthesis is that repeated initiations (there are several origins on the DNA) can take place on the viral chromosome but not on cellular chromosomes. Reinitiations at replication origins begin even before a complete replication cycle has been completed; this leads to the formation of a large number of DNA copies. This is one reason that T4 DNA replication is so complex. Another reason is that T4 DNA undergoes genetic recombination. T4 DNA molecules have single-stranded ends because of the removal of primer RNA and the inability of DNA polymerase to fill in from a 5′ end. Recombination between these DNA molecules results in the formation of molecules of a length greater than that found in the mature phage (Figure 17-5) and are called **concatemers**. The recombination process involves an exchange between redundant ends of the DNA molecule (see Chapter 11).

Structural protein synthesis and assembly. During the synthesis of viral proteins in the cell there is a period called the **eclipse**, in which no mature viral particles can be detected (Figure 17-6). The assembly of phage particles takes place via three independent pathways involving head, tail, and tail fibers. An abbreviated version of these pathways is illustrated in Figure 17-7.

Production of the T4 viral particle, unlike helical or isometric viruses, cannot be characterized as a total self-assembly process. Various proteins and enzymes are required in the assembly process and are not part of the mature virus particle. The capsid is apparently assembled with the help of **scaffolding proteins**, which are removed once the capsid has been completed. The scaffolding protein species, which number in the tens to hundreds, regulate the polymerization of the capsid subunits and in their absence the capsid protein forms disorganized aggregates.

The mechanism by which DNA is packaged into the capsid (**encapsidation**) is not yet totally understood. The

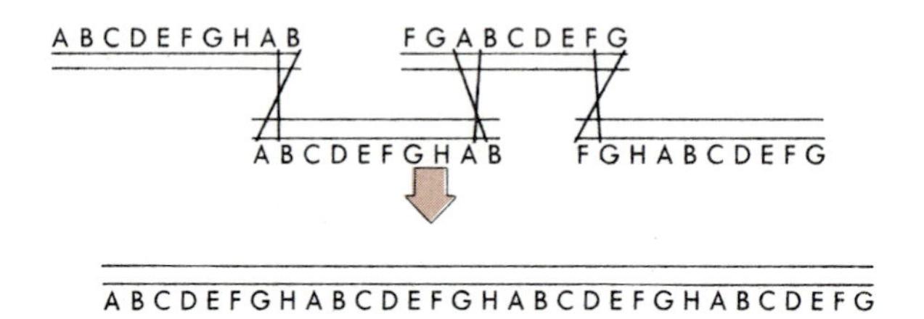

Figure 17-5 *Concatamer formation by recombination.*

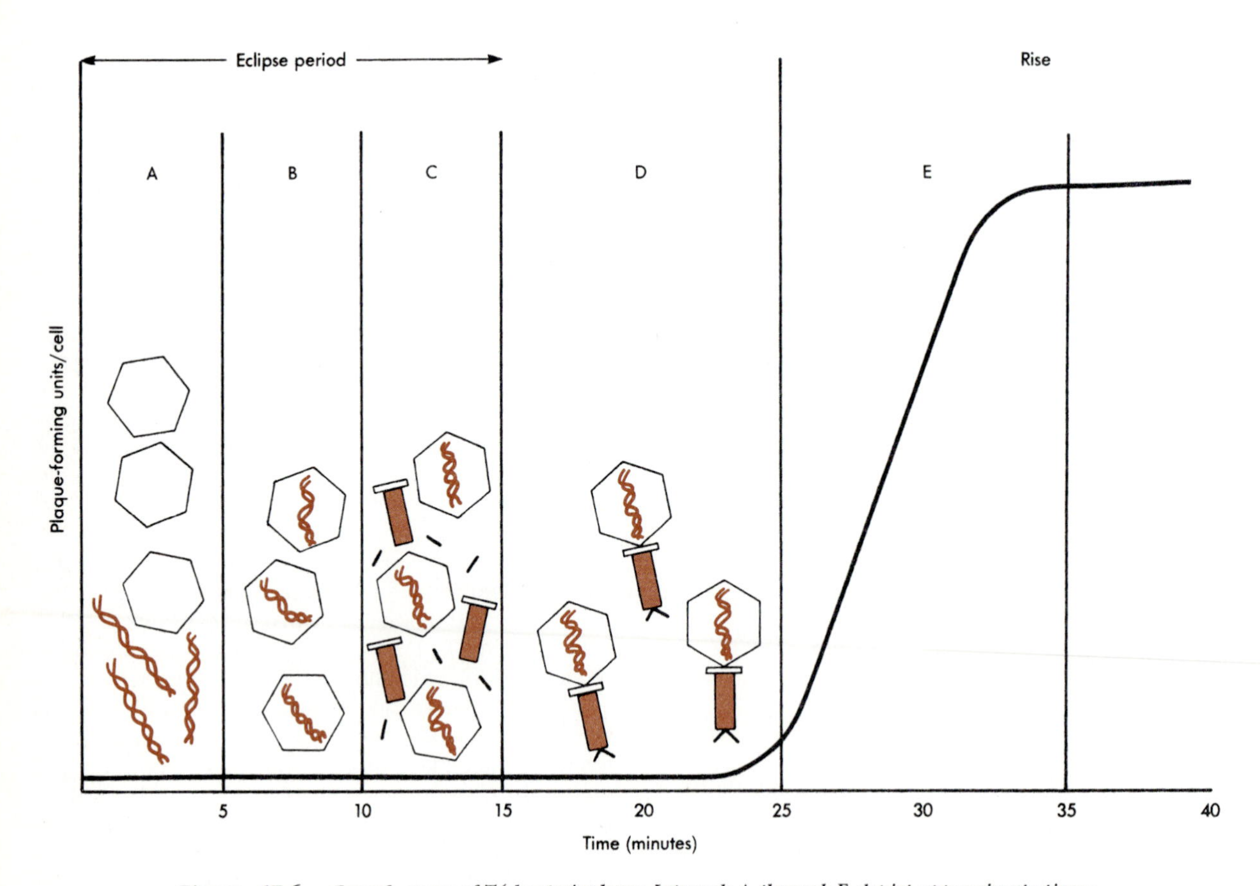

Figure 17-6 *Growth curve of T4 bacteriophage. Intervals* A *through* E *depict approximate times that various viral components are produced and assembled in infected bacterial cell. Mature phage are not assembled until 20 to 25 minutes following infection. Sudden rise in detectable phage particles is due to lysis of bacterial cell(s) and immediate release of all phage. Average number of phage progeny produced per infected cell is called burst size.*

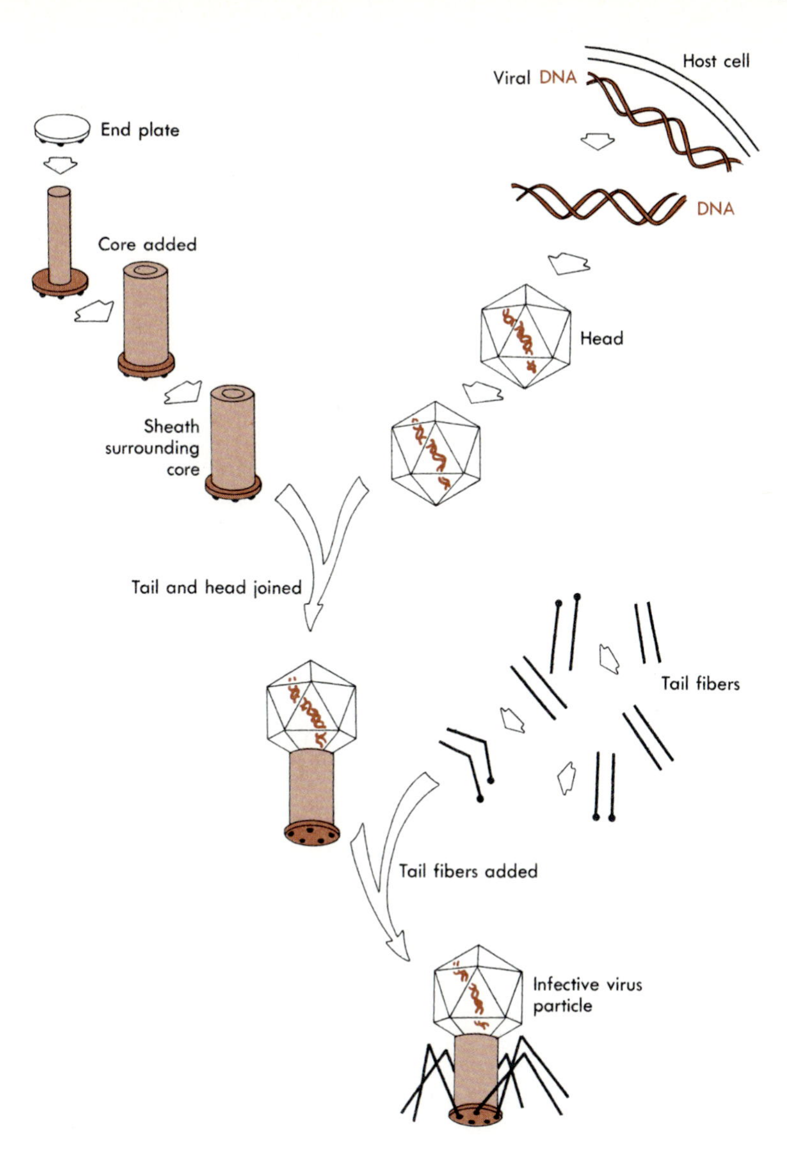

Figure 17-7 *Abbreviated scheme showing assembly pathway of bacteriophage T4. Formation of complete viral particle in T4 bacteriophage is brought about by integration of three independent pathways. Each pathway leads to formation of structural unit: head, tail, and tail fibers, which are later joined to produce infectious particle. If pathway events that lead to formation of head are out of sequence, abortive particles are produced. Prehead particle is believed to be produced and is later filled with DNA. Tail fibers do not attach to free base plates or to fully assembled tail unless tail is already attached to head. Tail has to combine with specific fragments, and head must interact with other fragments before tail and head join firmly.*

Concept Check

1. How are phage able to adsorb to the cell and penetrate into the cytoplasm?
2. How is the bacterial DNA protected from its own nucleases, and how does the bacterial cell protect itself from foreign DNA?
3. What mechanisms operate that enable the virus to shut down host metabolism?
4. What are the major events that occur during early phage transcription? Late phage transcription?
5. How is phage DNA replication different from bacterial DNA replication?

packaging is believed to involve a specificity between one end of the DNA molecule and a site on the capsid. The threading of DNA into the capsid requires phage-specified packaging proteins, as well as energy (Figure 17-8). The substrate for DNA encapsidation is a concatameric DNA (see Chapter 8). The free end of the concatameric molecule enters the head and is apparently wound into concentric coils like a solenoid until the head is filled. The concatemer is cut with enzyme, thus removing excess DNA. The packaging of DNA genomes from concatamers is called **headful**. Because the phage head can hold slightly more than a single genome, T4 chromosomes are terminally redundant and permuted. Thus the final sequence on the DNA that is packaged is a duplicate of the first sequence that entered the head: **redundancy**. However, the initial sequence on the DNA

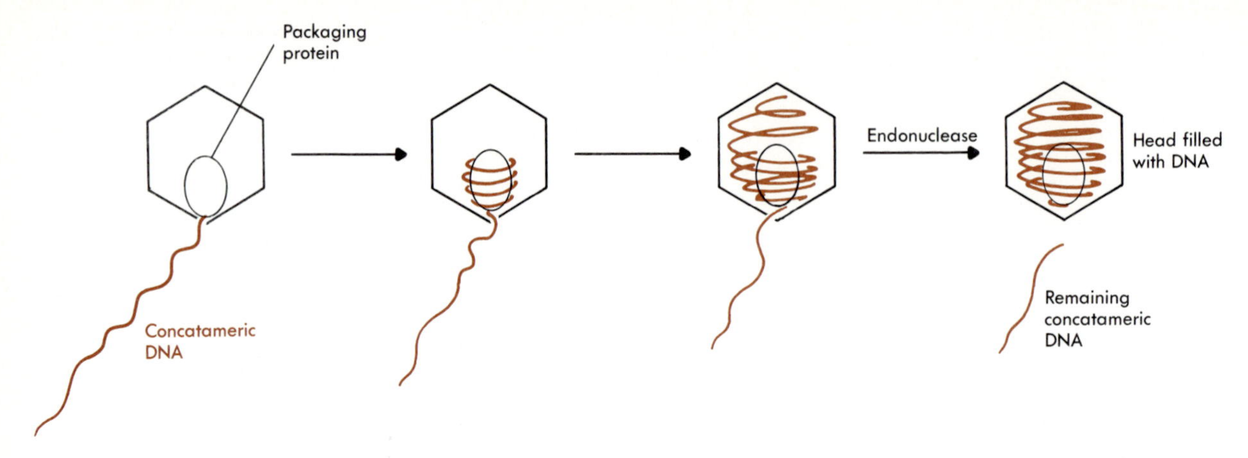

Figure 17-8 *Model for filling phage head with DNA. Packaging protein (also called spool protein) in phage head recognizes concatameric DNA and begins to coil into head. Packaging continues until head is filled with DNA and then endonuclease cleaves terminal end of DNA.*

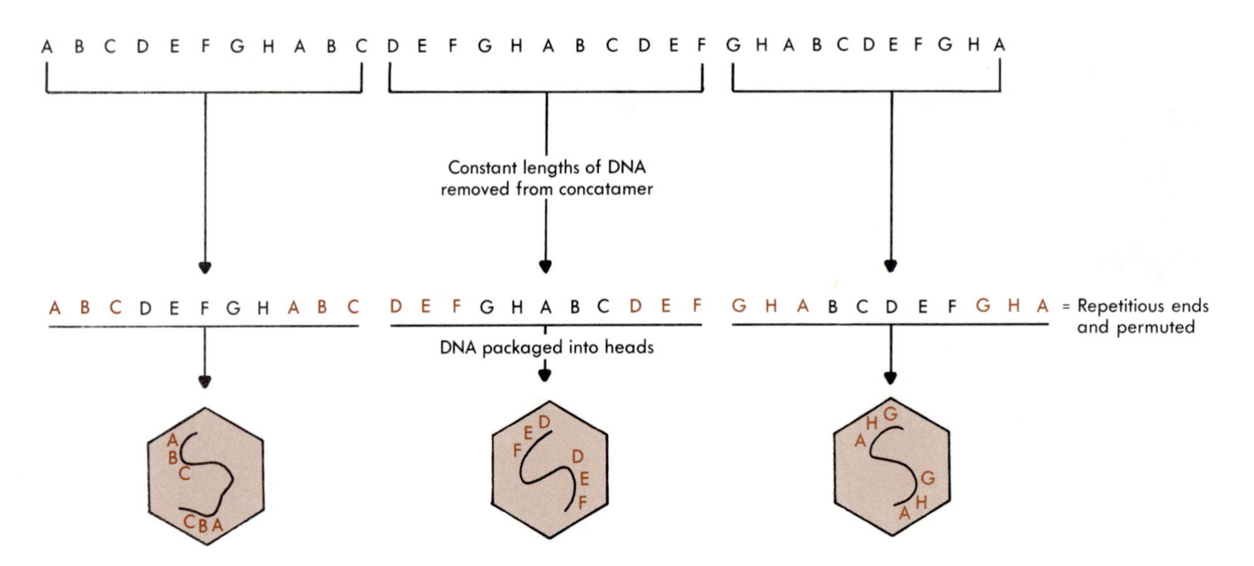

Figure 17-9 *Mechanism for producing phage DNA that is redundant and permuted.*

molecule to be packaged in the second phage particle is different from the initial sequence of the DNA molecule packaged in the first phage: permuted (Figure 17-9).

After encapsidation of T4 DNA the assembly process continues with the addition of the tail and tail fibers (Figure 17-7) until a mature virion is produced.

Release from the cell. Release of newly synthesized T4 phage particles occurs approximately 25 to 30 minutes after infection (when carried out at 37° C). At the end of this period, all phage progeny have been released. The total number of phage released per cell is called the **burst size**, which is from 50 to 100 for T4. Release of virus from the bacterial cell is accomplished by the activity of enzymes produced during the late period of the infectious cycle. Some enzymes dissolve the cytoplasmic membrane, whereas one enzyme, **lysozyme**, causes disruption of the cell wall and breakdown of the cell. There are some filamentous phage that are released continuously from the cell without disrupting the cell wall, but they are the exception.

Lysogenic life cycle of phage lambda

As discussed previously lysogeny is a viral infection process in which the bacterial cell does not lyse but instead

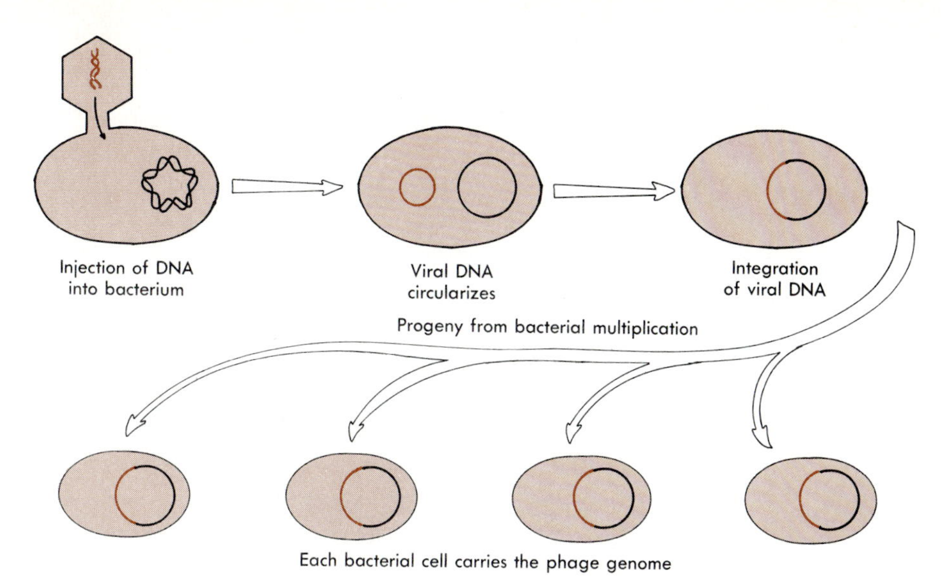

Figure 17-10 *Lysogenic life cycle. Viral DNA is injected into host as linear molecule. Viral DNA circularizes and becomes integrated into bacterial genome. Replication of bacterial cell results in progeny carrying phage genome in integrated state.*

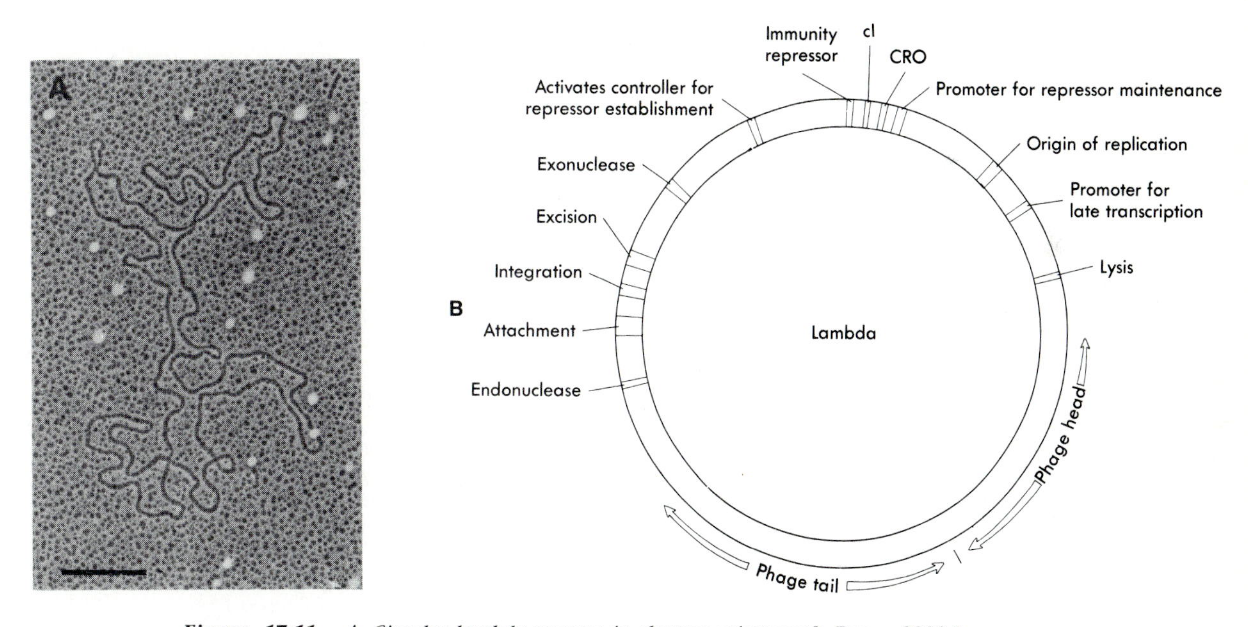

Figure 17-11 ***A,*** *Circular lambda genome in electron micrograph. Bar = 2000 bases.* ***B,*** *Abbreviated and functional map of lambda. Function of* cro *and* cI *are discussed in Figure 17-12.*

usually carries the virus integrated into the host genome.* The lysogenic cycle of lambda, whose host is *E. coli,* can be divided into four phases (Figure 17-10):

1. Lambda DNA (Figure 17-11) is injected into the bacterium.
2. Lambda genes code for a repressor that prevents transcription of the lambda genome except for a few genes that encode the enzyme **integrase.**
3. Integrase catalyzes the insertion of the lambda genome into the *E. coli* genome.
4. The bacterium multiplies with each of the progeny carrying a lambda genome.

Characteristics of the infection process. Lambda and other viruses that engage in lysogeny are called **temperate** viruses. Bacterial cells infected by temperate viruses are **lysogens,** and the process of forming a lysogen is **lysogenization.** Once the temperate virus integrates into the bacterial genome, it is called a **prophage**

* Some phages, such as P1, do not integrate into the host *(E. coli)* genome but instead remain in the cytoplasm as an independently replicating circular DNA molecule.

or **provirus** (the latter term implicates not only bacterial viruses, but also animal viruses that are capable of integrating into the genome of their hosts). Once the bacterial cell has been lysogenized, it cannot be reinfected by a phage of the same type. This resistance to reinfection is called **immunity** and is related to a repressor protein discussed shortly.

Integration and excision (lysogeny vs. lysis). Immediately after injection of lambda DNA into the cytoplasm of *E. coli,* two genetic functions are initiated: synthesis of the enzyme integrase and synthesis of a repressor. The enzyme **integrase** is necessary for insertion of the lambda genome into the *E. coli* genome, whereas **repressor** prevents transcription of lambda genes. The integration of lambda is an example of site-specific recombination, a process that was discussed in Chapter 11 (p. 276).

A lysogen can multiply repeatedly, producing progeny that carry lambda as prophage. There are circumstances, however, in which the lytic state can be induced. These circumstances include damage to the DNA by irradiation with ultraviolet light or exposure to chemical mutagens. Induction of the lytic state begins with excision of the lambda genome. Excision is basically a reversal of the integration process. The control mechanisms that determine whether an infected bacterial cell will become a lysogen or become lysed is a complex process. This mechanism is discussed only in generalities. A more detailed discussion of this process can be found in the selected readings listed at the end of the chapter.

Two regulatory proteins, **repressor protein** and **cro protein**, are associated with control of the lytic and lysogenic state. The lambda repressor protein is coded by the **cI** gene, whereas the cro protein is coded by the **cro** gene. Each one is capable of preventing the synthesis of the other (Figure 17-12). During the lysogenic state the repressor protein predominates. The repressor protein prevents the transcription of any genes required for lysis of the bacterial cell, but it does not repress its own (repressor) synthesis. When the virus enters the lytic state, the cro protein predominates. The viral nucleic acid is released from its integrated state by phage-coded enzymes and circularizes. Viral functions associated with lambda replication and lysis of the bacterial cell are initiated.

The repressor protein is a diffusible product that will bind to the operators of any superinfecting lambda that enters the lysogen. This accounts for the specific immunity of the lysogen. If the repressor protein is dominant early during infection, then the lysogenic response prevails; if cro protein is dominant early during infection, then the lytic response is initiated. What factors control the level of these proteins is not yet known. The physiological state of the host, for example, log phase vs. stationary phase cells, and host regulatory proteins may

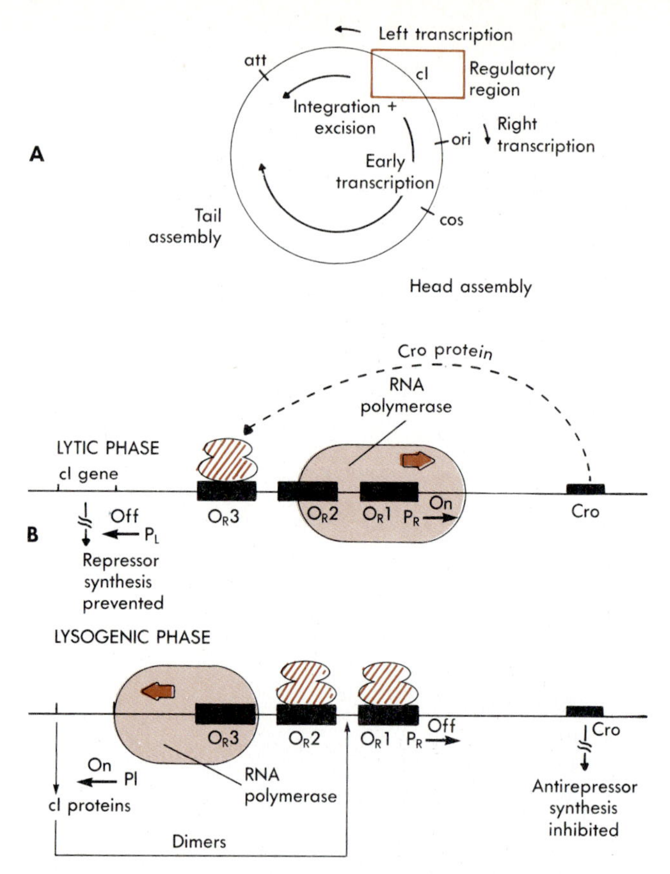

Figure 17-12 *Mechanisms controlling lytic and lysogenic states of bacteriophage lambda (λ).* ***A,*** *Simplified map of lambda genome. Regulatory region is controlled by repressor cI. Arrows inside circle indicate transcription units involved in lytic and lysogenic growth. Note that regulatory region lies between them, therefore if transcription to right of regulatory region is hindered, lysogeny predominates, while hindrance of leftward transcription promotes lytic growth. Boxed regulatory region* (color) *is expanded* ***B. B,*** *Simplified version of regulatory region of lambda. In lytic phase* Cro *gene codes for antirepressor protein (cro protein) that binds to specific site on operator called* O_R3. *This binding prevents RNA polymerase from occupying* O_R3 *and blocks transcription from the left promoter* (P_L). *In addition, binding of cro protein prevents synthesis of cI repressor. Because operator sites* O_R1 *and* O_R2 *are unoccupied, transcription to right is not prevented and RNA synthesis is initiated at right promoter or* P_R. *In lysogenic phase repressor* cI *codes for dimeric repressor protein whose dimers bind cooperatively to operator sites* O_R1 *and* O_R2. *This prevents RNA polymerase from binding to these sites, and transcription from right promoter is turned off. In addition, antirepressor cro synthesis is inhibited. RNA polymerase can bind to* O_R3 *and transcription from the left operator,* P_L, *is turned on.* cI, *Repressor protein;* cro *antirepressor protein;* ori, *origin of replication;* cos, *site where cohesive ends of linear genome close to produce circle;* att, *attachment sites for viral genome to bacterial genome.*

play important roles in this process. We should also mention that the lysogenic state is naturally an unstable one and the repressor of lambda genes is lost with a frequency of 10^{-3} in a population of cells.

Other aspects of the lysogenic process. Besides their importance in understanding molecular processes, phage lambda and other lysogenic viruses have been associated with other phenomena and some have practical applications. Lysogeny has played an important role in human disease. The rash associated with scarlet fever is a result of infection by streptococci that has been lysogenized. The information for production of a toxin that causes the rash is carried in the viral nucleic acid. The toxin that causes the symptoms of diphtheria is also produced from genetic information carried on the virus that lysogenizes the bacterium *Corynebacterium diphtheriae* (see Chapter 23). Nonlysogenic *C. diphtheriae* does not cause severe disease in humans. In the two examples cited, the bacterial cell has acquired new characteristics because of lysogeny by a bacteriophage, a process called **lysogenic conversion.**

Phage conversion changes have also been noted in *Salmonella* outer membrane O antigens. *Salmonella* possesses specific sugar arrangements in the outer membrane that define the organism's O antigens. Infection by specific phages can lead to changes in the arrangement or chemical identity of these sugar components. This type of conversion could be significant if it occurred during infection, because the host's antibodies would not immediately recognize the new antigens. This would provide time for the bacteria to multiply and cause serious damage to the host.

Lambda has also played a role in recombinant DNA technology (see Chapter 11) and some reasons for this are the following:

1. The host of lambda is *E. coli*, which is the organism most frequently used in the host-vector system.
2. Middle segments of the lambda genome are not required for lytic growth, and this allows for the insertion of foreign DNA in its place.
3. Lambda can be used in cosmid production, that is, plasmids containing the cohesive ends of lambda.

Concept Check

1. **Diagram a typical phage "growth" curve indicating what occurs during the early and late stages of the cycle.**
2. **What is the relationship between concatamer formation and packaging of DNA into the phage head?**
3. **What is the process of lysogeny and the mechanism by which the lytic cycle is repressed?**
4. **How is lysogeny related to the infectious disease process in animals and humans?**

ANIMAL VIRUS INFECTION

Adsorption to the animal cell

Infection of a cell by a particular virus requires that some specificity exist between molecules on the cell surface and molecules on the viral surface (Figure 17-13). Viral receptors for enveloped viruses appear to be glycoproteins, such as the hemagglutinin (HA) of the influenza virus. The nature of specific receptors on nonenveloped viruses is largely unknown. The membrane glycoprotein of influenza virus is synthesized as a single polypeptide but during the infection process it is cleaved into two chains, HA_1 and HA_2, which are covalently attached by disulfide bonds (Figure 17-14, *A*). The two-chain monomer makes up a trimer unit on the membrane surface (Figure 17-14, *B*). Six oligosaccharide chains are attached to the lateral surface of HA_1 by asparagine residues, whereas one oligosaccharide is found attached to the HA_2 chain. The *C*-terminal of the HA_2 contains a group of hydrophobic amino acids that anchor the glycoprotein to the viral membrane. An adjacent group of hydrophilic amino acids projects into the cytoplasmic side of the lipid membrane.

The cell receptors that determine the binding capacity of viruses is largely unknown. Many viruses use different receptors, thus a particular cell may be susceptible to infection by one virus and resistant to infection by another virus. For some viruses, receptors for adherence can be found on numerous cell types. For example, the influenza virus binds to a sialic acid–containing receptor on the host surface via a region on the distal tip of the HA glycoprotein (Figure 17-14). The paramyxoviruses (measles and mumps viruses) apparently bind to many cell types via the neuraminic acid-containing glycoproteins or glycolipids of the host cell. For other viruses, host receptors are limited to one or a few cell types, for example, the rabies virus has an affinity for neural cells. The number of host receptor sites may vary from 10^4 to 10^5 per cell.

The adsorption process affixes the virion to the cell membrane and is a prerequisite for entry of virus into the cytoplasm of the cell.

Penetration

There is no single mechanism for virus penetration into the cell. The mechanism used depends on the virus type. Many nonenveloped viruses enter the cell by transfer of the nucleocapsid into the cytoplasm by a process

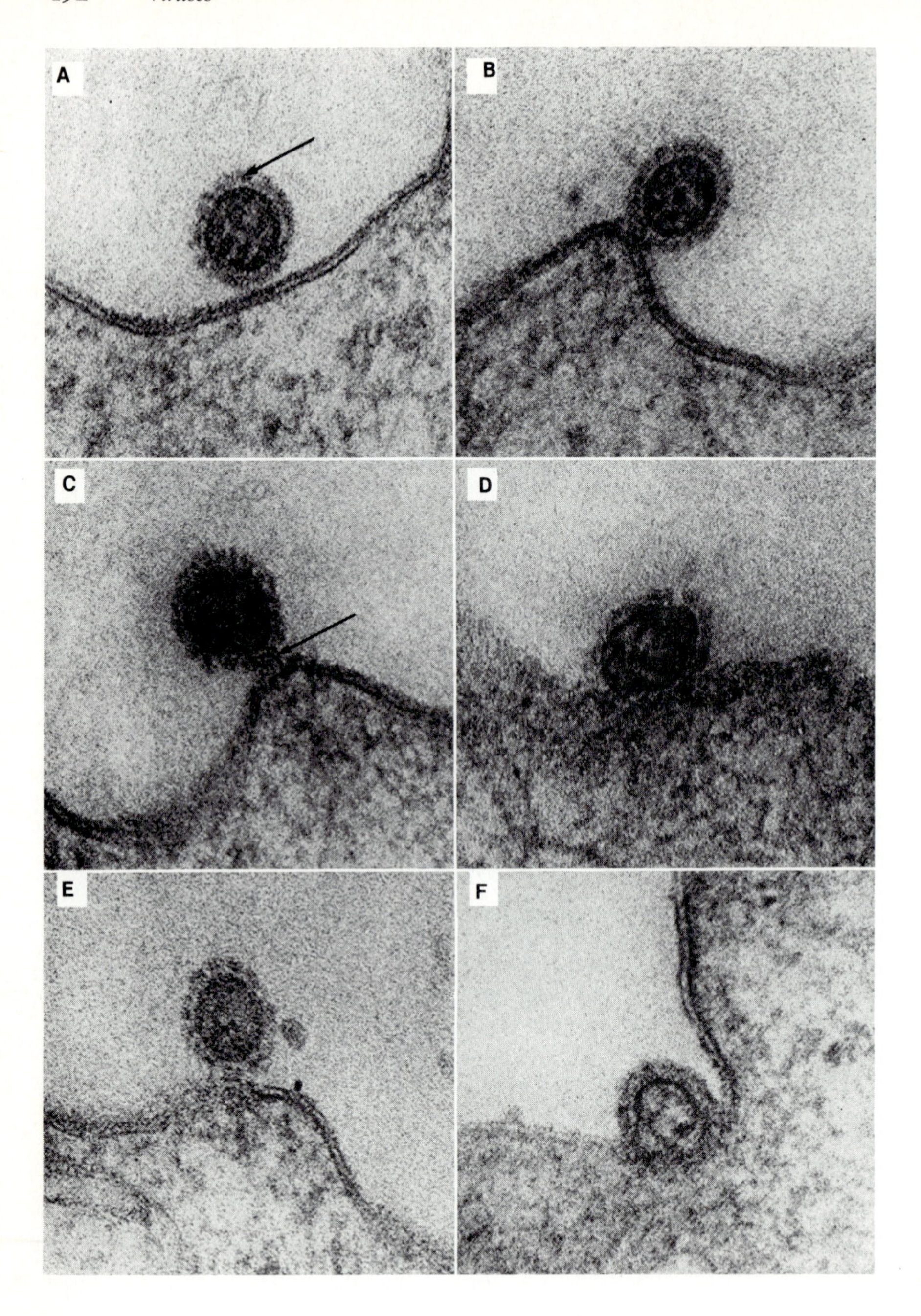

Figure 17-13 *Electron micrograph demonstrating animal virus adsorption to plasma membrane (×200,000).* ***A*** *and* ***B****, Virus has envelope* (arrow) *that makes contact with plasma membrane of cell.* ***C****, Bridge (→) appears to connect virus with plasma membrane.* ***D*** *and* ***E****, Plasma membrane appears indistinct.* ***F****, Virus has merged with cell, and envelope surrounding virus core appears to be disrupted.*

called **adsorptive endocytosis** (Figure 17-15, *A*). Adsorptive endocytosis requires energy, a characteristic that distinguishes it from a similar process called viropexis (see Chapter 7). Some nonenveloped viruses attach to the cytoplasmic membrane and appear to be engulfed by it. The viral capsid undergoes a conformational change that releases the viral nucleic acid into the cytoplasm (Figure 17-15, *A*).

Many enveloped viruses enter the cell by a fusion process involving the cytoplasmic membrane of the host cell and the viral membrane (Figure 17-15, *B*). This process results in the transfer of the nucleocapsid into the cytoplasm with the viral membrane temporarily remaining integrated into the cytoplasmic membrane. The entry of the nucleocapsid into the cell is particularly important for those viruses, such as the retroviruses, that carry their own replicating enzymes within the capsid. Apparently not all enveloped viruses enter the cell by

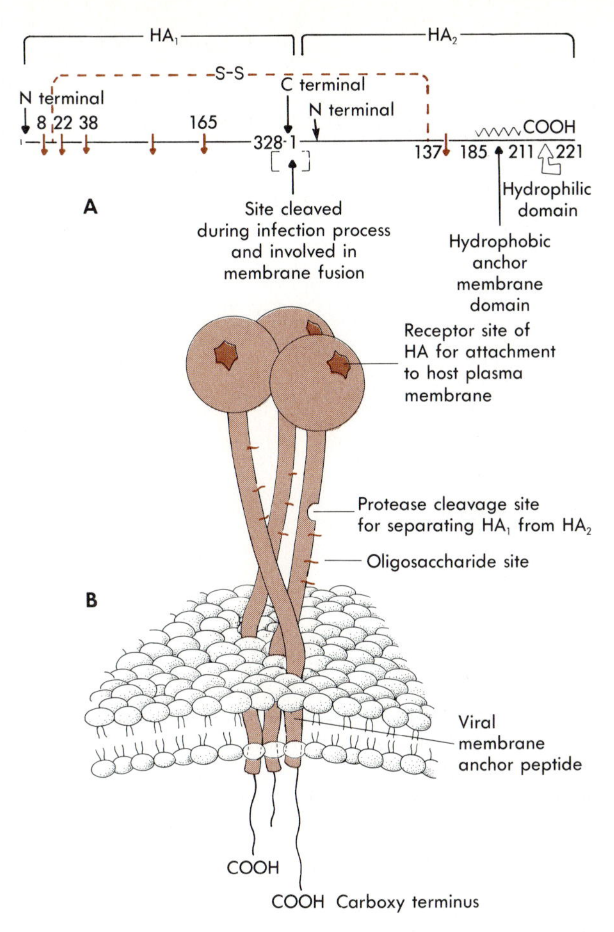

Figure 17-14 *Hemagglutinin structure.* ***A,*** *Schematic drawing of influenza virus hemagglutinin demonstrating important functional sites. Colored arrows point to sites of attachment of oligosaccharides of this glycoprotein. Single polypeptide consists of two chains, HA_1 and HA_2, held together by disulfide bonds (dotted line).* ***B,*** *Model showing relationship of hemagglutinin in lipid membrane of influenza virus. Hemagglutinin forms elongated stalk capped with globular region. Each glycoprotein spike or peplomer (as seen in electron micrographs) is actually a trimer consisting of three hemagglutinin molecules. Oligosaccharides of hemagglutinin are found primarily on stalk region. Globular regions on stalk possess sites for attachment to mammalian plasma membrane. Protease cleavage site represents region where cellular protease cleaves hemagglutinin (see* ***A****) and promotes fusion of virus with plasma membrane of host cell.*

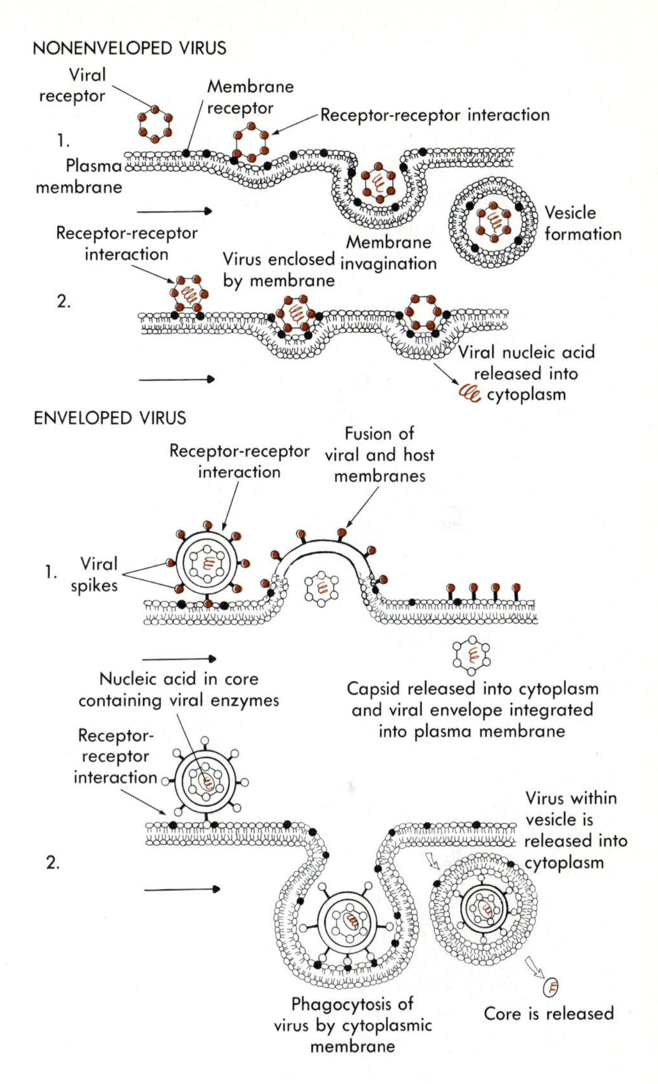

Figure 17-15 *Mechanisms for penetration of animal virus into mammalian cell. Nonenveloped virus:* 1 *represents adsorptive endocytosis.* 2, *Virus surrounded by cytoplasmic membrane, but only viral nucleic acid injected into cell. Enveloped virus:* 1 *represents fusion of cytoplasmic membrane with viral envelope and only nucleocapsid injected into cell.* 2 *represents endocytic process and transfer of whole virion into cytoplasm of cell.*

the fusion process. Some enveloped viruses appear to enter the cell by an endocytic process in which the entire virion enters the cytoplasm followed by degradation of the envelope within the cytoplasmic vacuole thus formed (Figure 17-15, *B*).

There is no guarantee that adsorption will be followed by infection of the cell. As mentioned earlier the HA glycoprotein of influenza is required for adsorption to the cell surface. However, the virus does not become infectious until the HA is cleaved by a **protease** separating HA_1 from HA_2. Such proteases are abundant in the respiratory tract where the influenza virus primarily replicates. This proteolytic cleavage apparently enables certain hydrophobic amino acid residues at the N terminus of HA_2 to penetrate the cytoplasmic membrane of the host and initiate the fusion process (Figure 17-16).

Uncoating

Before the virus can begin to replicate in its host, the coat protein must be removed. Uncoating usually occurs at the same time as transport to the site of replication, which is either the cytoplasm or nucleus. For viruses like the picornaviruses (the poliovirus is an example), removal of the coat protein is associated with membrane-associated host enzymes. This releases the RNA genome directly into the cytoplasm where it becomes associated with ribosomes on the rough endoplasmic reticulum. For other viruses the coat proteins are removed directly in the cytoplasm.

Special barriers are faced by those viral genomes that replicate in the nucleus—a characteristic of most DNA viruses. The nucleus consists of outer and inner membranes punctuated by nuclear pores but the pores are not large enough to permit entry of virus. Adenovirus enters the cytoplasm of the cell as an intact virion and is then transported to the nuclear membrane via microtubules. Adenoviruses attach to microtubules by means of a hexon capsomere. Once at the nuclear membrane, the virion breaches the nuclear membrane much the same way it did the cytoplasmic membrane.

For uncoating of most virions, the uncoating enzymes are supplied by the host cell. This is an invaluable aid to the virus but later these host enzymes must be suppressed if the virus is to be successfully assembled and released from the cell.

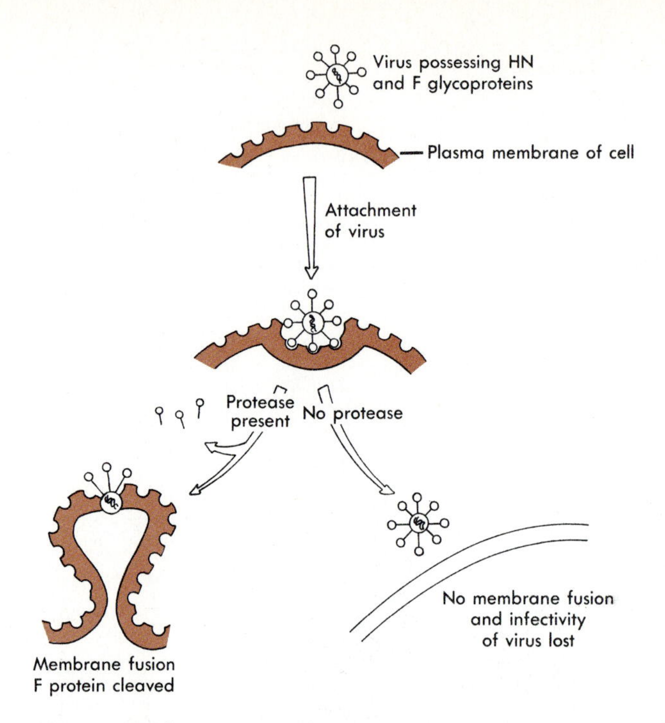

Figure 17-16 *Paramyxovirus penetration of cell and role of HN and F glycoproteins. HN glycoprotein is required for membrane attachment. Once attachment has taken place, F protein is cleaved by protease (cellular), which promotes membrane fusion and infectivity. Cells deficient in protease do not show membrane fusion and infectivity is lost.*

Inhibition of host functions

As mentioned previously, phage usually inhibit host macromolecule synthesis early during the infection process. This also occurs during some animal virus infections but inhibition of host function also occurs much later during infection. The mechanisms of inhibition of host DNA, RNA, and protein synthesis by most animal viruses is poorly understood except for poliovirus infection. Poliovirus inhibits cellular protein synthesis soon after infection. The key to the mechanism of inhibition appears to be the nature of the 5′ end of mRNA. The 5′ end of poliovirus mRNA is not capped as is cellular mRNAs but instead terminates in the nucleotide represented as pUp. The poliovirus encodes a protein that inactivates the *cap-binding complex* required for attachment of "capped" mRNAs to ribosomes (see Chapter 9). The cap-binding complex is made up of three proteins. Early during infection a poliovirus-encoded protease cleaves one of the cap-binding proteins, thus preventing it from attaching to ribosomes.

Inhibition of host protein synthesis in some other picornaviruses appears to be by other mechanisms. For example, as viral mRNAs accumulate in the cytoplasm, they outcompete cellular mRNA for one or more of the factors involved in initiation of protein synthesis.

Inhibition of host protein synthesis by herpesviruses is believed to involve a viral product that dissociates cellular polyribosomes; this is followed by degradation of cellular mRNA. Ribosomes thus become available for virus mRNA and are assembled into virus polyribosomes.

Despite these examples of virtually total shutdown of host macromolecule synthesis, infection by some viruses has practically no effect on cellular macromolecule synthesis. What should be learned from this is that there is a delicate association betwen virus and host. The intracellular environment of the host must be used by the virus but it must not be so seriously damaged that viral replication and assembly are also impaired.

Replication

Animal virus genomes differ considerably in size, gene composition, and gene organization from phage genomes. The animal virus genome may be single or double-stranded RNA or single- or double-stranded DNA. In addition, all the viral genes may be contained in a single chromosome (**monopartite**) or they may be distributed among several chromosomes (**multipartite**). Animal viruses have evolved a multitude of replication techniques to accommodate these genome variations. Viral replication occurs either in the cytoplasm or in the nucleus of

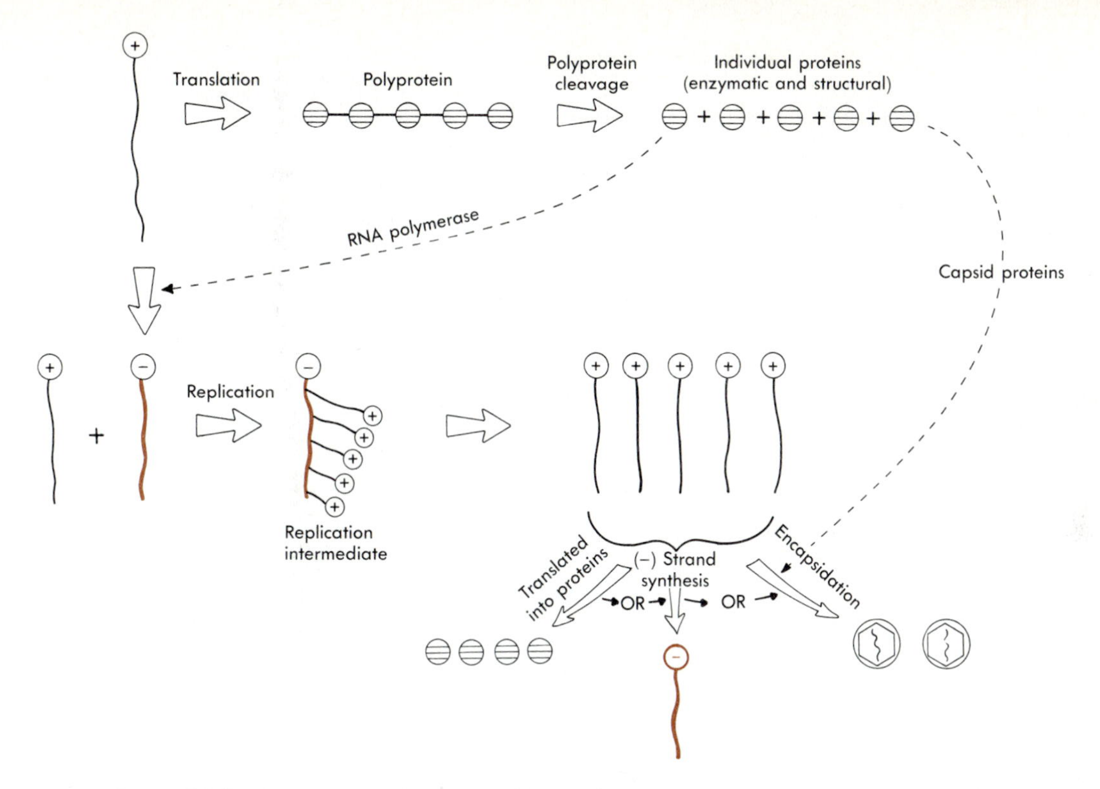

Figure 17-17 *Positive-stranded virus replication (poliovirus, for example). Positive strand of virus is translated in cytoplasm into polyprotein. Cleavage of polyprotein releases individual proteins, one of which is RNA polymerase and some of which are involved in virus capsid formation. RNA polymerase is used to transcribe positive-strand RNA into negative-stranded molecule. Negative strand serves as template for synthesis of positive-stranded RNA. Actually, several positive strands can be found in various stages of replication on negative strand (the complex is called replication intermediate). Positive strand RNA may serve as genome for encapsidation into mature virion, may be translated into protein, or transcribed into negative-stranded RNA.*

the cell. Cellular replication, which involves double-stranded DNA, occurs only in the nucleus; therefore only those viral DNA genomes that are replicated in the nucleus will have access to cellular-replicating enzymes. The cell lacks enzymes for transcribing viral RNA genomes that replicate in the nucleus or cytoplasm or viral DNA genomes that replicate in the cytoplasm. Our discussion begins with those viruses that have the simplest replicating scheme.

Single-stranded RNA viruses. The RNA carried by the virus may be referred to as positive (+) or negative (−) strand. A positive-stranded RNA behaves as an mRNA molecule and is thus translated into protein. A negative-stranded RNA is one that basically does not function as an mRNA but is used as a template for synthesis of positive strands.

The poliovirus is an example of a *positive single–stranded RNA virus*. The events that characterize poliovirus replication are the following (Figure 17-17):

1. Immediately following entry into the cytoplasm the viral genome is bound by ribosomes and is translated in its entirety into a polyprotein. The polyprotein is cleaved into individual proteins that include enzymes and structural proteins.
2. One of the proteins cleaved from the polyprotein is an RNA polymerase (also called **replicase**). Replicase catalyzes the synthesis of negative strand RNA whose function is to act as a template for synthesis of positive-stranded RNA.
3. While the first positive-strand copy is being synthesized on the negative strand, a second copy is synthesized even before completion of the first copy. Actually as many as five positive strands can be simultaneously copied on the negative strand. This multibranched structure is called a **replication intermediate**.
4. The positive strands serve three functions. They can act as an mRNA for synthesis of proteins, be a template for negative-stranded RNA, or serve as the viral genome for progeny poliovirus.

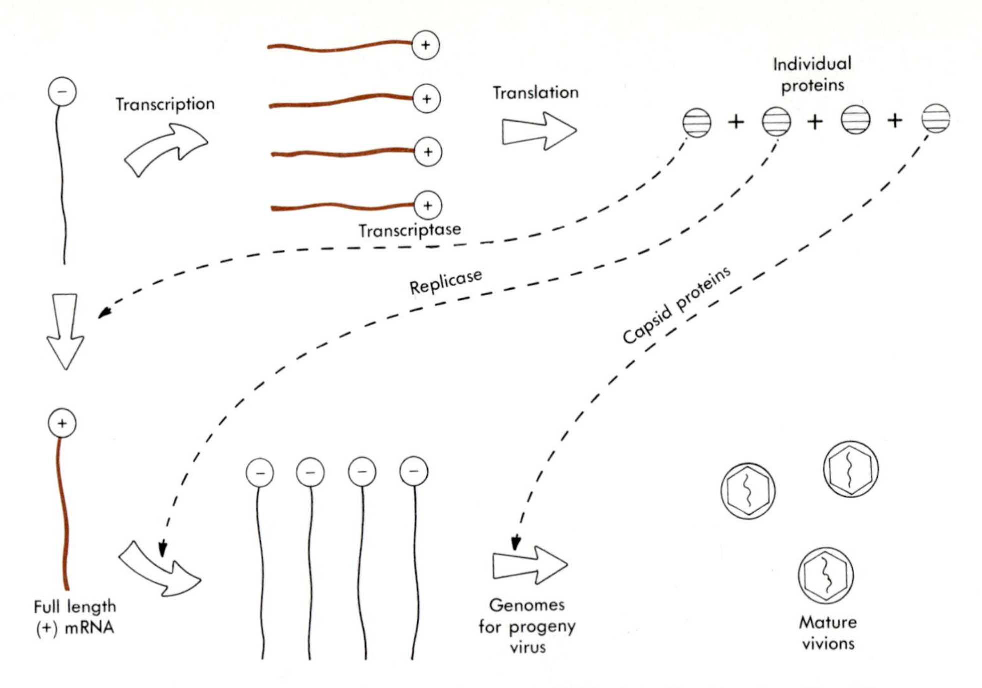

Figure 17-18 *Replication of negative single-stranded RNA virus. Negative strand in cell cytoplasm is transcribed into positive-stranded RNA. Positive strands are translated into individual proteins, such as transcriptase, replicase, and structural capsid proteins. Transcriptase is used to produce positive strand mRNA, whereas the latter uses replicase to produce several negative-stranded RNA molecules. Negative strands are encapsidated by structural proteins into mature virions.*

Negative single–stranded RNA viruses include the orthomyxoviruses (cause of influenza) paramyxoviruses (cause of measles and mumps), and others. Isolated negative-stranded RNA is not infectious because permissive host cells do not possess the necessary enzymes to transcribe the genome into a functional mRNA (positive strand). Consequently the virus must carry within its capsid the appropriate transcriptase.

The initial event following entry of the virus into the cytoplasm is the transcription of each gene on the negative strand into a positive-strand mRNA (Figure 17-18). Each monocistronic mRNA codes for a single protein. These proteins serve distinct functions. First, some proteins are involved in the replication of a full length positive-strand RNA from the negative strand genome. Second, some proteins engage in replication of the full length positive strand into negative-strand RNAs, which serve as the genome for progeny virus. The remaining proteins function as structural proteins during the assembly of mature virions.

The retroviruses are single-stranded RNA viruses that are monopartite but diploid. The RNA is composed of two identical positive strands that are bound noncovalently at their 5′ ends by a specific host tRNA molecule. Retroviruses are an unusual group of viruses because they use their genome to produce a double-stranded DNA molecule. This is accomplished through the activity of an enzyme called **reverse transcriptase** that is carried by the virus. The name of the enzyme indicates that the normal flow of information DNA→RNA is reversed to RNA→DNA. The DNA produced through the activity of reverse transcriptase is capable of inserting into the host genome as a provirus (similar to lysogeny in bacterial viruses).

Concept Check

1. What are the various ways in which animal viruses can penetrate the cell? Why do these variations seem necessary?
2. How are animal viruses able to bind to specific cell types?
3. What enzymes are carried by some animal viruses, and what is their function during infection?
4. What are the ways in which an animal virus may inhibit host protein or nucleic acid synthesis?
5. What is the distinction between positive single–stranded RNA replication and negative single–stranded RNA replication?

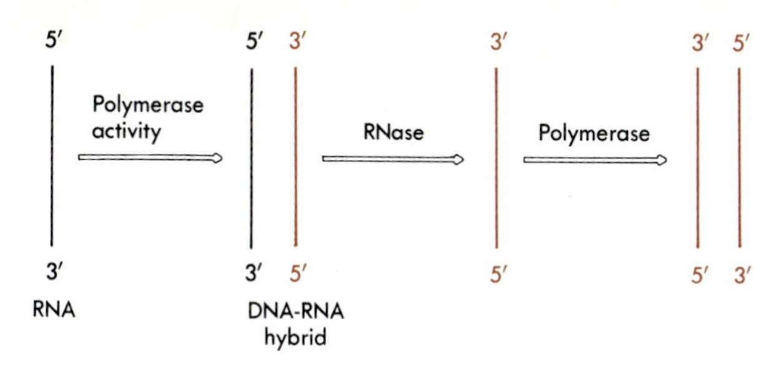

Figure 17-19 *Sequence of events in which reverse transcriptase converts single-stranded RNA into double-stranded DNA. Mechanism of priming for this sequence of events is not known.*

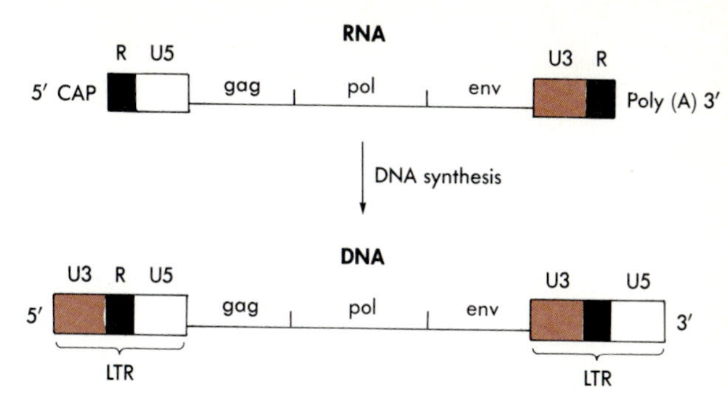

Figure 17-21 *Structure of RNA and DNA molecules of retroviruses. Viral RNA contains 5' cap and 3' polyadenylated cap enclosing three viral genes:* gag, pol, *and* env. *Repeated sequences (*R*) appear at both ends of RNA and at both ends are unique sequences (U3 for 3' end and U5 for 5' end). During synthesis of retroviral DNA from RNA there is duplication of U5 and U3. This results in formation of structure U3-R-U5 at both ends of DNA, and these represent long terminal repeat (LTR) sequences.*

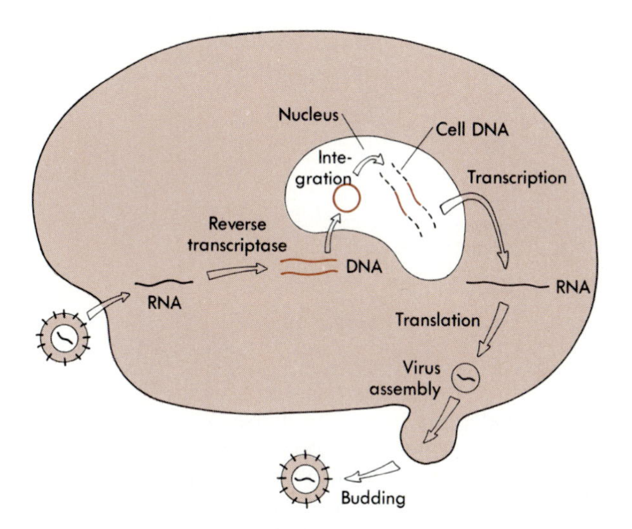

Figure 17-20 *Life cycle of retrovirus. Virus coat removed in cytoplasm of cell where single-stranded viral RNA genome is converted into double-stranded DNA molecule. Viral DNA enters nucleus and converts to circular form that is integrated into host DNA. Transcription of proviral genome results in formation of viral RNA. Viral RNA is carried into cytoplasm where virus acquires protein coat. Budding of virus through plasma membrane results in formation of mature enveloped virion.*

After entry into the cytoplasm of the cell the envelope and capsid of the retrovirus are removed. In the cytoplasm a single RNA strand acts as a template for synthesis of a complementary DNA strand (Figure 17-19). This results in the formation of a DNA-RNA hybrid. The RNA of the hybrid is degraded through the activity of RNase, whereas the DNA copy of RNA serves as a template for synthesis of a complementary DNA strand. Reverse transcriptase functions in all of these reactions. The double-stranded DNA moves to the nucleus where it becomes integrated into the host genome (Figure 17-20). The proviral DNA can be transcribed into mRNA leading to the formation of viral proteins. How virion RNA is synthesized is not understood. Viruses are assembled in the cytoplasm with reverse transcriptase and a host-specified tRNA enclosed in the capsid. An envelope is acquired as the virus buds through the cytoplasmic membrane. Virions may bud over a long period of time without injury to the cell.

The genomic retrovirus RNA resembles eukaryotic mRNA and has a 5' cap and a polyadenylated 3' end. The genome of retroviruses consists of three genes: **gag** (group specific *a*ntigen), which encodes for the internal structural protein; **pol** (polymerase), which encodes for reverse transcriptase; and **env** (envelope), which encodes for the envelope glycoprotein. The genomes of those retroviruses that cause cancer (called oncogenic retroviruses) contain other genes (called **oncogenes**) that are derived from cellular DNA. These oncogenes often replace one of the structural genes and thus the virus becomes replicative-defective and requires helper virus to produce mature virions.

Near the 5' and 3' ends of the retrovirus genome are special regions that are required in replication (Figure 17-21). There are short terminal repeats referred to as (R) at the 5' and 3' ends and adjacent to them are sequences called U5 (at the 5' end) and U3 (at the 3' end). U3 and U5 are copied twice during the synthesis of DNA. This duplication results in the structure U3-R-U5 at both ends and these are **long terminal repeats (LTRs)**. The integrated retroviral DNA therefore consists of three structural genes flanked by an LTR at each end, and this makes it longer than its RNA template. A consequence of the structural gene-LTR arrangement is that the U3 region is placed upstream of the site where viral RNA synthesis starts. The U3 region of the

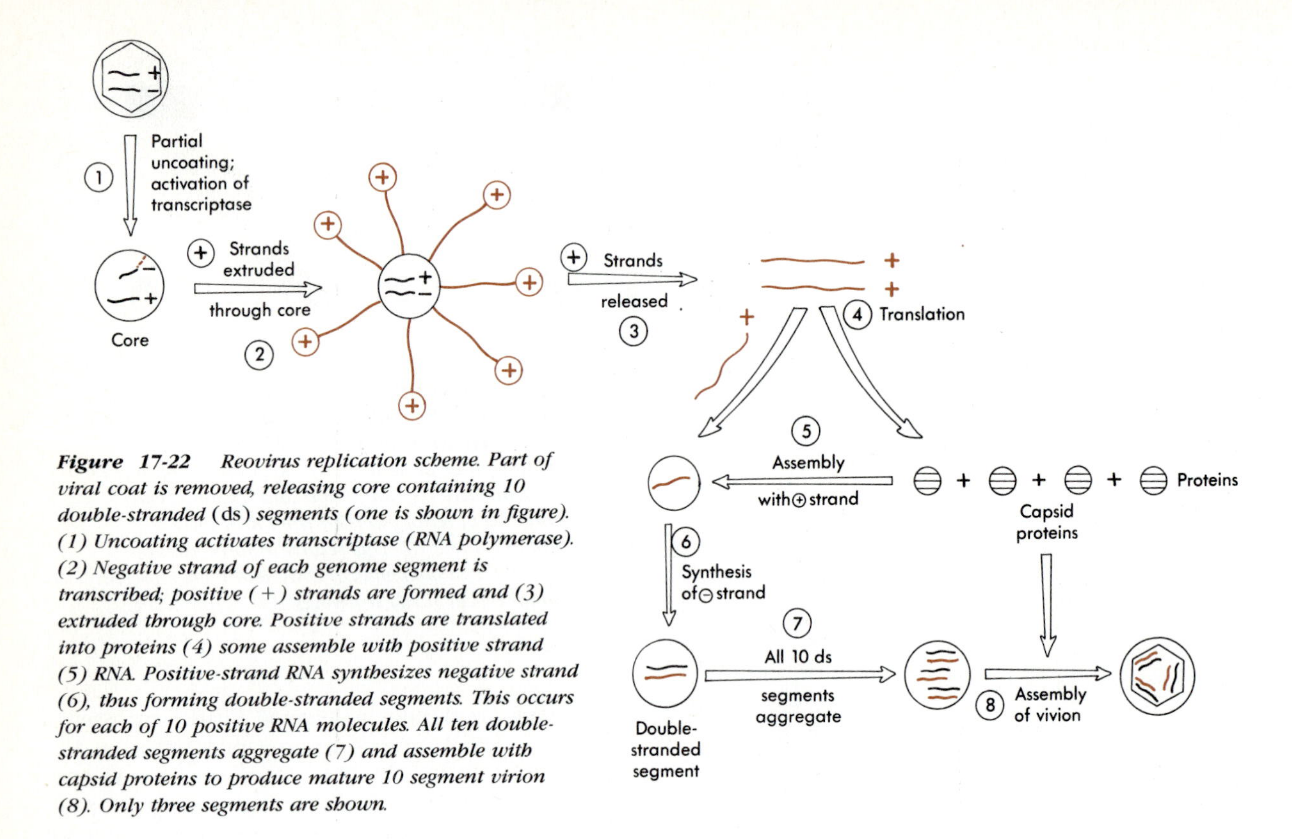

Figure 17-22 *Reovirus replication scheme. Part of viral coat is removed, releasing core containing 10 double-stranded* (ds) *segments (one is shown in figure). (1) Uncoating activates transcriptase (RNA polymerase). (2) Negative strand of each genome segment is transcribed; positive (+) strands are formed and (3) extruded through core. Positive strands are translated into proteins (4) some assemble with positive strand (5) RNA. Positive-strand RNA synthesizes negative strand (6), thus forming double-stranded segments. This occurs for each of 10 positive RNA molecules. All ten double-stranded segments aggregate (7) and assemble with capsid proteins to produce mature 10 segment virion (8). Only three segments are shown.*

DNA serves as an upstream promoter for viral RNA synthesis.

The structural gene-LTR arrangement of the retroviral DNA is similar to transposable elements discussed in Chapter 8. Both repeat sequences on transposable elements, and LTRs on retroviruses mediate their integration into the host genome. Retroviruses transpose genes but they do it through the agency of RNA within viral particles rather than by direct transposition of DNA from one position to another.

Double-stranded RNA viruses. Reovirus is an example of a double-stranded RNA virus. A basic feature of the virus is the presence of a segmented genome. Reovirus contains 10 double-stranded RNA segments, each RNA segment is translated into a unique mRNA, the capsid coat is not completely uncoated and the virion RNA is not released as free RNA molecules, and all of the enzymes required for transcription are carried by the virus.

Once the reovirus has penetrated into the cytoplasm of the cell, uncoating begins (Figure 17-22). Uncoating removes only 50% of the outer capsid leaving the capsid porous particularly at the vertices. Uncoating activates RNA polymerase, carried by the virus, and this results in synthesis of positive strands from each negative strand of the double-stranded RNA segments. Each of the positive-strand mRNAs is capped in the core of the virus by virus-encoded enzymes. The mRNA of this group of viruses does not possess a poly(A) tail—a characteristic that distinguishes it from all other virus mRNAs. The transcribed mRNA molecules are extruded through the exposed capsid vertices but the double-stranded segments of the virus remain in the core.

Each of the positive mRNA strands extruded from the capsid core has two functions. First, they are translated into viral proteins. Second, one RNA of each of the ten species will assemble within one of the newly synthesized viral proteins. Each RNA will serve as a template for synthesis of the complementary strand; thus a double-stranded segment is formed. Each of the polypeptides self-assemble to form the 10 RNA segment core of the virus. This is followed by assembly of the icosahedral shell around the core and formation of the mature virion.

DNA Viruses. The genomes of DNA viruses may be single-stranded or double-stranded. Only a few animal viruses, such as the parvoviruses, possess a single-stranded DNA genome. This section concentrates only on the double-stranded DNA viruses. The life cycles of the DNA viruses, with minor exceptions, are relatively complex. Some of the viruses possess as many as 100 genes, which provides the virus with great versatility but has also made analysis of replication schemes difficult. One of the best understood DNA virus life cycles is the

simian virus 40 (SV-40), a member of the papovaviruses. SV-40 is a tumor-forming virus when it infects some cells but in its natural host can cause lysis of the cell.

The SV-40 genome is a 5243 base-paired, supercoiled DNA that is complexed with host histones, and gives it the appearance of nucleosomes. Virus enters the cytoplasm of the cell and is then transported intact into the nucleus where uncoating and transcription begin. Following uncoating, and before DNA replication, an early 2500 base-pair transcript is produced. Two of the products of expression of this early transcript are designated **T** and **t proteins** or **antigens**. Large T is a multifunctional protein confined to the nucleus, whereas small t, whose function is not understood, is found in both the cytoplasm and nucleus. Large T functions include initiation of viral DNA replication, repression of transcription of the early region of the genome, enhancing expression of the late region, and initiating the transformation process (tumor formation) when transformable cells are infected (transformation is discussed later) and enzymic activities such as ATPase and kinase (protein phosphorylation) activities.

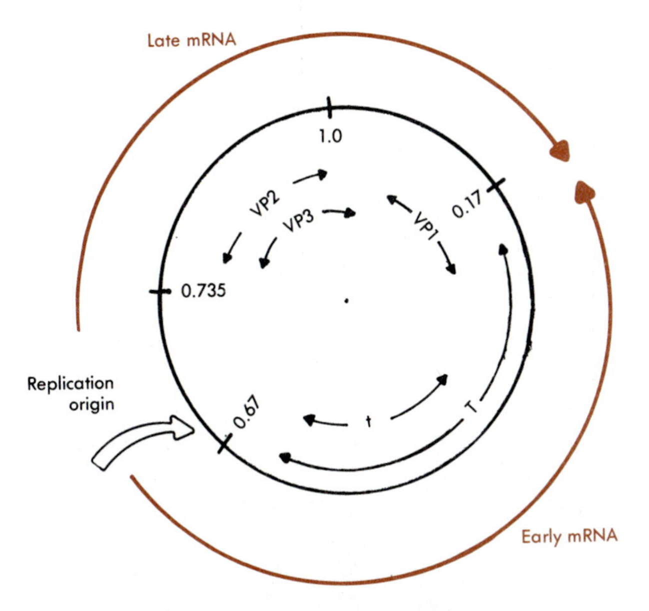

Figure 17-23 *Transcription map of the SV-40 genome. Illustration shows location of genes* T, t, VP1, VP2, and VP3, *replication origin, and early and late mRNA.*

The replication origin is at position 0.67 on the SV-40 transcription map (Figure 17-23). The area around the replication origin consists of inverted repeat sequences plus a 17 base pair AT-rich region (Figure 17-24). The T antigen binds to the replication origin at four points and provides protection from cellular nucleases, as well as regulating early transcription.

Transcription is a fairly complex process and like other viral systems can be divided into early and late stages. The early and late mRNA are transcribed from different strands of the DNA (see Figure 17-23) and in different directions. The mRNAs are transported to the cytoplasm where protein synthesis is initiated. Initiation of synthesis of the early T and t antigens begins at a single-start signal (AUG) but the stop codons for the two proteins are at different sites. The synthesized T and t proteins differ in size and activity because of the way in which introns are removed (Figure 17-25). The late proteins of SV-40 are VP1, VP2, and VP3. They are produced by a splicing technique and use of overlapping reading frames that enable the virus to store a significant amount of information in a small DNA segment.

Viral assembly and release

Shortly after viral nucleic acid replication begins, the synthesis of structural protein, such as capsids, is initiated in the infected cell. Viral maturation involves the formation of nucleocapsids, a process that takes place in the nucleus or cytoplasm of the cell (Figure 17-26).

Assembly of naked viruses is a sequential interaction of polypeptides to form subunits, followed by association with the viral genome. Many viruses use virus-specified polypeptides to bind to progeny genomes at some point in the assembly process, and this prevents further

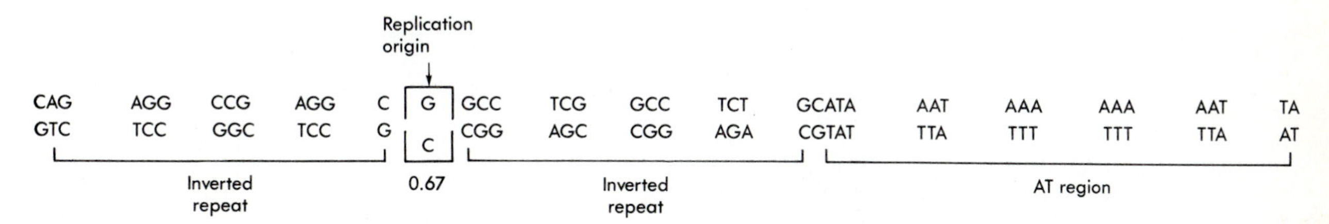

Figure 17-24 *Base sequence at replication origin of SV-40. Inverted repeats appear on either side of base, which is indicated as replication origin. To right of one inverted repeat is 17 base-pair region, rich in AT base pairs.*

transcriptional events. Naked viruses, unlike enveloped viruses, have a tendency to accumulate in the cell and are released only with the death and lysis of the host cell. Considerable evidence indicates that structural proteins of the virus inhibit host metabolism and that they are in some way responsible for lysis of the cell.

The assembly of enveloped viruses (Table 17-1) is a more complex event involving considerable virus-host relationships. To prepare for release from the cell, the enveloped virus interacts with host cell membrane in a process called **budding** (Figure 17-27). During the infection process virus-specific proteins appear throughout the membrane system of the infected cell. These virus proteins have a tendency to aggregate and form patches, and, in the process, cell membrane proteins are displaced. At the same time, viral glycoproteins begin to appear on the surface of the cell membrane. When the viral nucleocapsid is formed, it binds to those virus-specific proteins lining the cytoplasmic side of the cell membrane and becomes wrapped up by the viral patch. This interaction appears to trigger evagination of the cell membrane. As the virus-membrane complex continues to evaginate, the cytoplasmic membrane pinches off and a virus has been formed by budding. As the cytoplasmic membrane pinches off, there are two fusion events. First, the envelope of the virus fuses to establish integrity of the virion. Second, the cytoplasmic membrane fuses to reestablish membrane continuity.

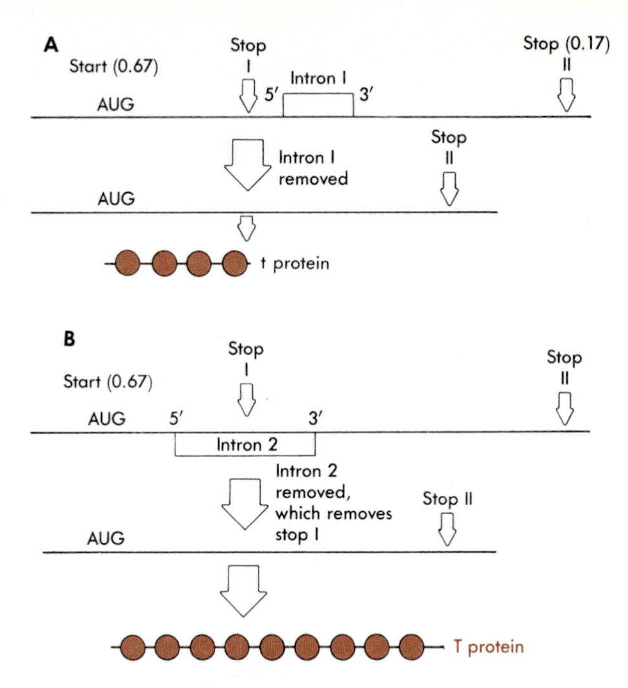

Figure 17-25 *Processing of early mRNA and formation of* T *and* t *proteins.* ***A,*** t *protein is synthesized from mRNA in which intron I has been excised. Translation of mRNA terminates at Stop I and does not involve Stop II.* ***B,*** T *Protein is synthesized from mRNA in which intron 2 is removed. Removal of intron 2 also removes Stop I site, therefore message is translated and does not terminate until Stop II. Note that intron 2 has same 3 terminus as intron 1 but has 5 terminus nearer to 5 of mRNA.*

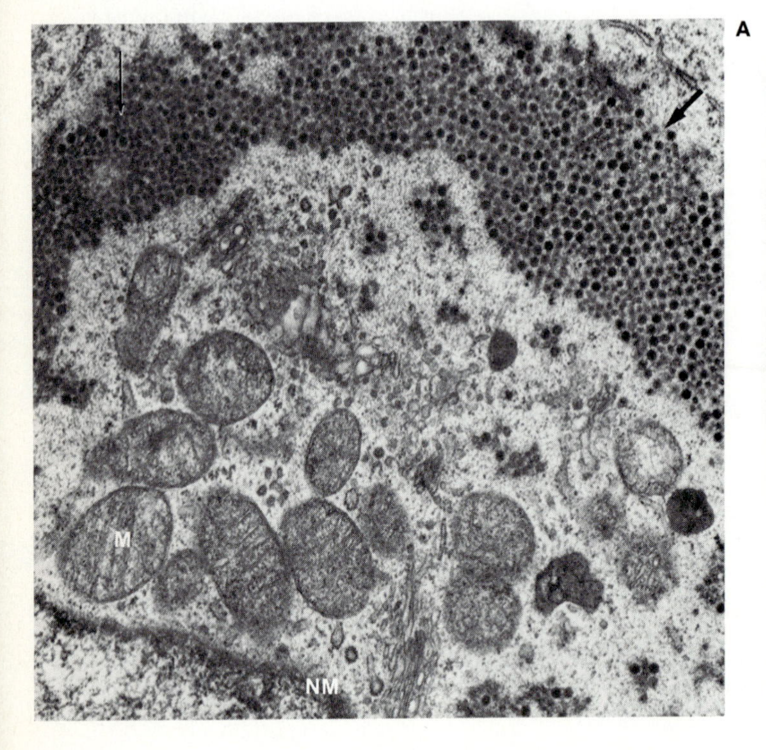

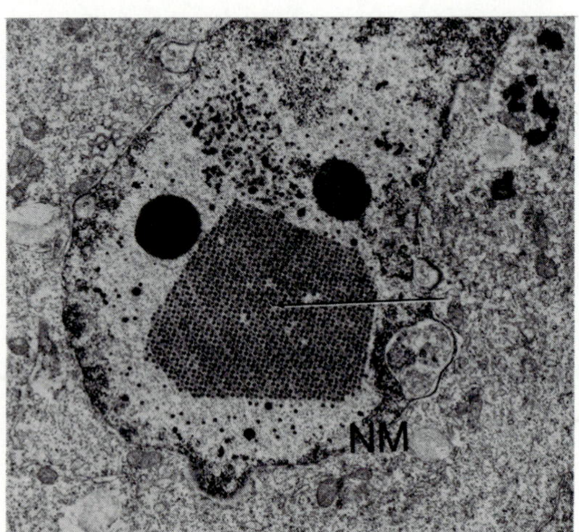

Figure 17-26 *Electron micrographs demonstrating assembled viral particles in cytoplasm or nucleus of cell.* ***A,*** *Cytoplasmic inclusions (reovirus). Arrows point to viral inclusions. (×50,000.)* NM, *Nuclear membrane;* M, *mitochondria.* ***B,*** *Nuclear inclusions (adenovirus). (×19,000.) Viral inclusions* (arrow) *are surrounded by nuclear membrane* (NM).

Table 17-1 Enveloped animal viruses

Viral family	Type of nucleic acid	Representative types	Assembly on cellular membrane
Arenaviridae	RNA; ss*	Lassa fever, lymphocytic choriomeningitis	Yes
Bunyaviridae	RNA; ss	LaCrosse, snowshoe hare	Yes
Coronaviridae	RNA; ss	Murine hepatitis, human respiratory tract infections	Yes
Herpesviridae	DNA; ds*	Herpes simplex, Epstein-Barr, pseudorabies	Yes
Iridoviridae	DNA; ds	African swine fever, Tipula iridescent	No†
Myxoviridae	RNA; ss	Influenza	Yes
Paramyxoviridae	RNA; ss	Parainfluenza, Newcastle disease; measles; mumps; canine distemper	Yes
Poxviridae	DNA; ds	Vaccinia, smallpox, fowlpox	No
Retroviridae	RNA; ss	Avian leukemia, avian sarcoma, murine mammary carcinoma, feline leukemia, primate leukemia, murine leukemia	Yes
Rhabdoviridae	RNA; ss	Vesicular stomatitis, rabies	Yes
Togaviridae	RNA; ss	Semliki Forest, Sindbis, rubella	Yes

**ss*, Single strand; *ds*, double strand.
†The envelopes of these viruses are not derived from the preexisting cytoplasmic membrane but are made de novo in the cytoplasm of the cell.

For viruses, like herpesviruses, that are assembled in the nucleus, membrane envelopment occurs at the nuclear membrane. After the enveloped herpesvirus is released from the nucleus, it appears to be wrapped in a second envelope in the cytoplasm. This new envelope is believed to be derived from the endoplasmic reticulum; thus the virus with its glycoprotein envelope is shielded from contact with the cell cytoplasm and its degradative enzymes.

Viruses that bud from the cell have a variety of effects on the cell. Some are cytolytic, for example, the togaviruses and paramyxoviruses, whereas others like the retroviruses are virtually noncytolytic. Because viral glycoproteins have been inserted into the host cell membrane by enveloped viruses, the cell becomes subject to an immune response.

Effect of viral infection on the cell

The cell is affected in many ways following viral infection. Host metabolism is altered usually in specific ways and tends to correlate with pathways required by the virus. In morphological terms the host cell is significantly affected by viral infection, and these changes, called **cytopathic effects (CPE)**, are recognizable. For example, inclusion bodies discussed in Chapter 16 are a common CPE. Viral infection at the cellular level results either in lysis of the cell and release of virus or the cell remains viable and virus persists in the cell for an indefinite period of time.

Cell lysis. Lysis of the cell is the most common response to viral infection. The mechanisms by which the virus affects the outcome are unknown. Some studies suggest that a structural protein of the virus is involved. One possibility is that lysosomes in the cell become destabilized during infection, allowing enzymes, such as nucleases and proteases, to leak into the cytoplasm. Cell lysis is characteristic of infection by naked (nonenveloped) viruses but some enveloped viruses can also be cytolytic.

Persistence. Under ordinary conditions most viral infections are self-limiting. This means that the host's immune system eventually responds to the presence of the virus, destroys the virus, and the host recovers. All of this occurs within a recognizable time period. Some virus infections, however, may result in virus persisting in the host. Persistent infections can be defined as those in which virus can be continuously recovered from the host well beyond the usual time period of the disease without causing disease. Persistence is often used interchangeably with other virus-host interactions termed *chronic, latent,* and *slow.* The following criteria distinguish them:

1. Latency is a state in which the virus remains silent for long periods of time and cannot be recovered, and no symptoms are detected.

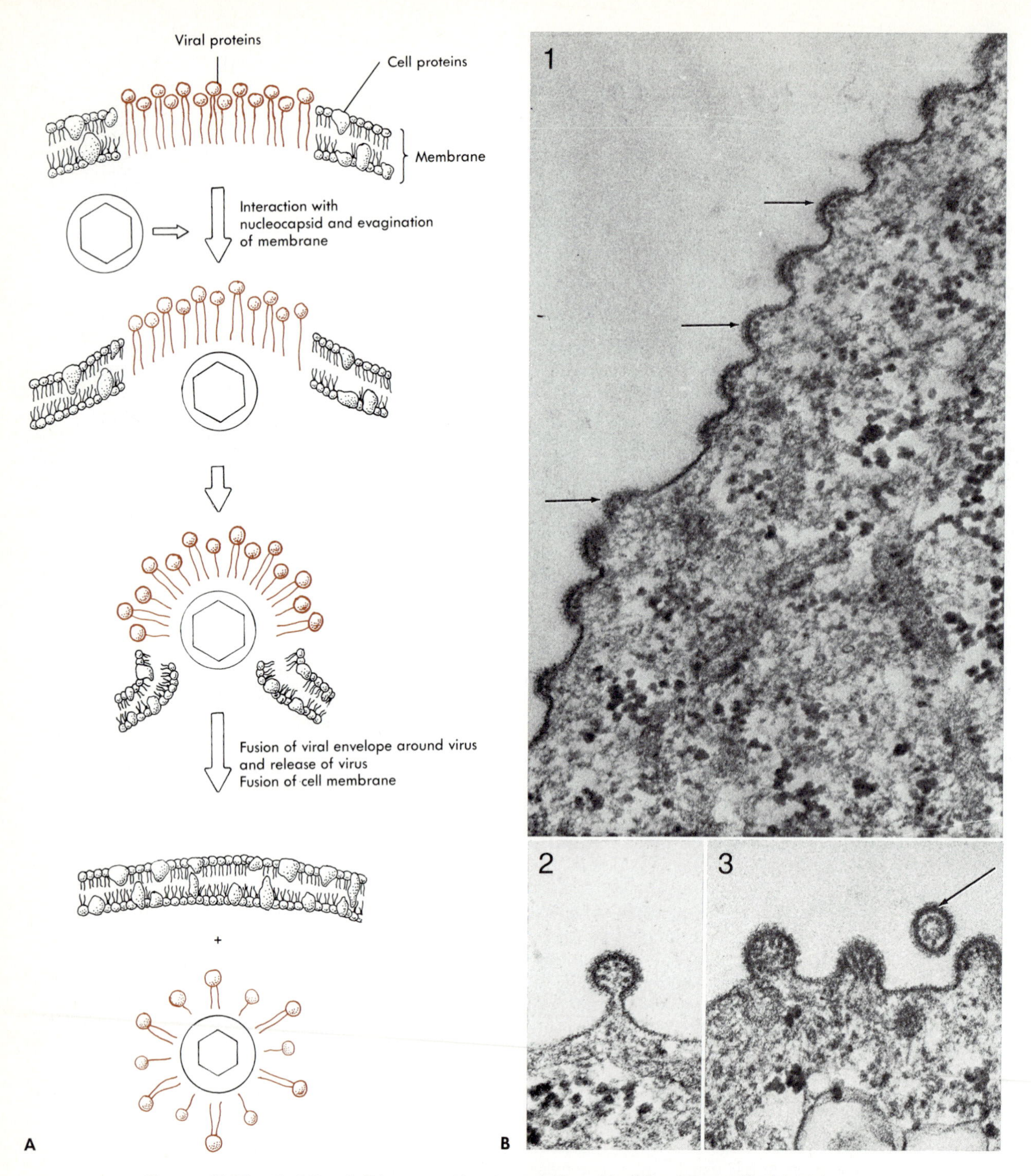

Figure 17-27 *Budding. **A,** Diagrammatic representation of budding process. Viral proteins appear as patches in host's plasma membrane and are inserted there during infection process. When new viral particles are produced in cell, nucleocapsids interact with plasma membrane causing it to evaginate. During evagination, viral envelope proteins engulf nucleocapsid producing mature virion, which is released from plasma membrane into environment. **B,** Electron micrographs demonstrating process of budding in animal virus (respiratory syncytial virus). **1,** Viral particles* (arrows) *appear to bulge at cell membrane. (×75,000.) **2,** Virion just before pinching off from cytoplasmic membrane. (×86,000.) **3,** One virion has budded* (arrow), *and three others are about to bud. (×76,000.)*

2. Chronic is similar to persistent except that disease symptoms are present.
3. Slow virus infections (for example, kuru and Creutzfeldt-Jakob disease) develop slowly, affect the central nervous system, and after many years lead to death. Whether these agents are viruses is still a matter of speculation (see Point of Interest, p. 477).

Further discussion of persistence is also presented in Chapter 26.

Mechanisms of viral persistence. Viral persistence may be caused by a variety of mechanisms and each enables the virus to elude the host's immune system. Some of the more favored hypotheses include the following:

1. Integration of the viral genome into the host genome is a characteristic of those retroviruses that transform cells. These viruses may persist for years in cells and then suddenly cause malignancies (tumors). Transformation is discussed in the next section.
2. Some viruses, such as the rubella virus (German measles), may escape detection by the host's immune system because they are present during a time of immunological immaturity, that is, during fetal development. The rubella virus, which may be acquired during pregnancy, can persist in the newborn for 12 to 18 months after birth. The virus reduces the rate of growth of host cells and is a cause of severe birth defects.

   Conditions associated with suppression of the immune system can also lead to persistence or to reactivation of a latent infection. Immune suppression may result from drugs used before transplantation or they may result from some underlying disease that affects the immune system.
3. Viruses that have affinity for neural tissue may be able to evade the immune system by being retained in the central nervous system. This tissue is nondividing and does not favor viral replication. The slow diseases could be an example of this scenario.
4. Defective interfering viral particles **(DI particles)** are found after most viral infections. These particles have lost part of their genome and require simultaneous infection of host cells by a related helper virus to restore the deleted function. In addition, DI particles inhibit replication of infectious helper virus by diverting gene products from helper virus to DI particles. Many of the oncogenic retroviruses are helper-dependent defective viruses carrying cellular tumor—causing genes recombined with viral gene segments. Thus these viruses can contribute to the persistence of viral infections. Satellite viruses are similar to DI particles (see Chapter 16).
5. Antigenic change on the surface of infected cells may lead to persistence. The virus causing cold sores (herpes simplex) is believed to persist in nervous tissue because of antibodies binding to the surface of the infected cell. This binding inhibits viral replication and allows the virus to persist for indefinite periods of time. Activation of the virus can occur at any time, permitting virus to replicate and migrate to the facial area and produce coldsores.

Interferon production. **Interferon** is a biological regulator of cell function and is produced by a variety of vertebrates, including humans. Its concentration in the host is elevated in response to viral infection but it is produced in incredibly small amounts. In addition to its antiviral activity, interferon also possesses antitumor activity.

Interferon is a glycoprotein of low molecular weight. It is produced not only in whole animals or in animal cell culture in response to viral infection, but can also be produced in response to nonviral agents such as endotoxins, certain bacterial species, and other products. Interferon produced during infection by virus is excreted by the infected cell and interacts with other cells to alter their metabolic and immunological properties. These adjacent cells are not only protected from infection by the original invading virus, but are also protected from infection by unrelated DNA or RNA viruses. The mechanism of action of interferon is not completely understood, but the following sequence of events represents the most currently postulated model (Figure 17-28). The mammalian cell possesses genes that code for interferon production. Interferon, excreted by the originally infected cell, comes in contact with adjacent cells (interferon can also affect to some degree the cell that produced interferon originally) and shuts down viral protein synthesis. The manner in which interferon inhibits protein synthesis has been a subject of speculation for over 20 years, but it is now believed that it induces the synthesis of two enzymes that are initially inactive. The enzymes are activated by a double-stranded RNA molecule produced during viral infection. The two enzymes—**protein kinase (PK)** and **oligoadenylate synthetase** ($2'$-$5'[A]_n$ synthetase)—are the instruments for inhibition of protein synthesis. Protein kinase catalyzes the phosphorylation of a protein (IF-2) that functions as an initiator of protein synthesis. Phosphorylation inactivates this protein; thus protein synthesis is inhibited. Oligoadenylate synthetase catalyzes the synthesis of an unusual nucleotide called $2'5'$-oligoadenylic acid, which appears to function as an activator of a ribonuclease that degrades viral mRNA.

The broad antiviral (RNA and DNA viruses) activity induced by interferon is species specific, and protection against virus occurs only in species in which the interferon is produced. Thus protection of humans from viral infection can occur only if the interferon is produced in human cells or by recombinant techniques in bacteria. One approach to interferon production is to induce it in humans by using a synthetically produced compound. One such inducer is a double-stranded RNA molecule called polyinosinic: polycyticylic acid (poly [I]:poly [C]), which is also very toxic to humans. Many different interferons can be produced from human cells, but only two of them have been purified: one is produced from leukocytes and the other is produced from skin fibroblasts.

Interferon, in addition to its antiviral activity, affects the mammalian cell by alteration of its cell surface, modification of the immune response, and inhibition of cell division. Clinical trials in 1980 with certain human cancers (osteogenic sarcomas and breast cancer) indicated that interferon might eventually play a role in antitumor therapy and prophylaxis. Interferon is now produced by genetic engineering techniques and has been licensed for use in the treatment of hairy cell leukemia. This particular malignancy has responded to treatment with interferon.

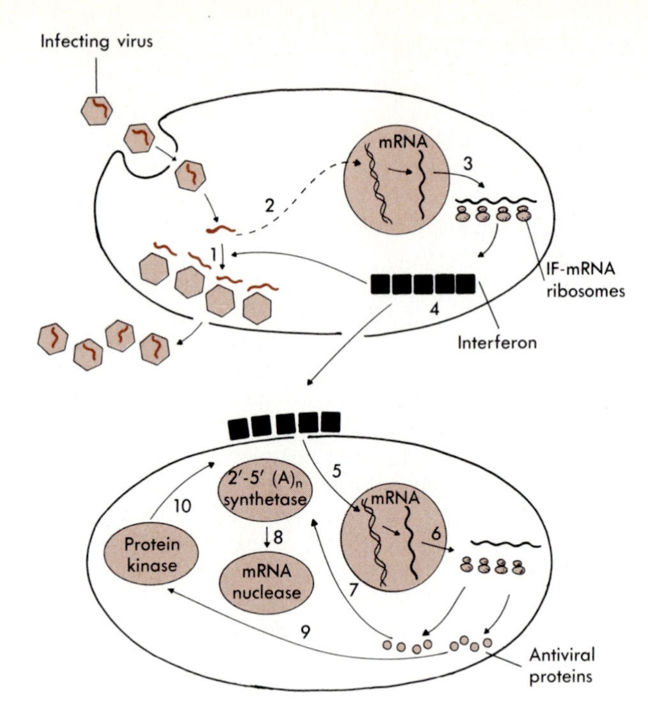

Figure 17-28 *Proposed mechanism of interferon action. Infecting virus* (1) *produces new viral particles but also* (2) *induces cell to produce interferon mRNA (IF-mRNA), which* (3) *is translated into interferon molecules* (4). *Interferon released by cell may inhibit development of virus in same cell, or it may bind to adjacent cells and* (5) *induce cell to produce antiviral proteins* (6), *such as* $2'$-$5'$ $(A)_n$ *synthetase* (7), *which activates* (8) *an mRNA endonuclease that cleaves viral mRNAs. A second enzyme, protein kinase* (9) *phosphorylates and thus inactivates a factor, IF-2, required for protein synthesis.*

Concept Check

1. Describe the infection process of retroviruses. What mediates their integration into the host genome?
2. What characteristics of reoviruses and reovirus replication distinguish them from other viruses and their modes of replication?
3. What characteristics distinguish the replication process of DNA animal viruses from RNA viruses?
4. What is the function of *T* and *t* proteins in SV-40 replication?
5. What is the budding process, and how are viruses released from infected cells?
6. List and explain the effects of viral infection on host cells. What is interferon and what is its mechanism of action?

VIRUSES AND CANCER

Certain cancer-causing agents have been defined and certain types of cancer can be controlled, but we are still unable to determine the mechanism of cancer formation. Cancer may be defined as a malignant growth or tumor resulting from the transformation of cells that demonstrate unrestricted growth. The tumor may remain localized, or it may spread (metastasize) to other sites. Some viruses can cause tumors in animals, and many investigators have suspected that they may also be involved in human malignancies. There are several DNA viruses that cause tumors in animals, including SV-40, adenoviruses, and cytomegaloviruses, but among the RNA viruses only the retroviruses produce tumors. The only viruses known to cause cancer in humans are the human T-lymphotropic viruses (HTLV). HTLV-I and HTLV-II are linked with human leukemias and lymphomas. HTLV-III is the cause of acquired immune deficiency syndrome (AIDS), which predisposes its victims to cancer and other diseases (see Chapter 25). The virus causing AIDS is now referred to as the **human immunodeficiency virus**, or **HIV**.

Transformation of cells by virus

All tumor cells become transformed, that is, they acquire properties not found in uninfected cells. One property that is easily recognizable is the loss of function, or **density-dependent growth** (formerly called **contact inhibition**). Uninfected cells that grow on glass or plastic surfaces in tissue culture stop division once a monolayer has formed. Some of the cells lose the property of density-dependent growth if certain oncogenic viruses are

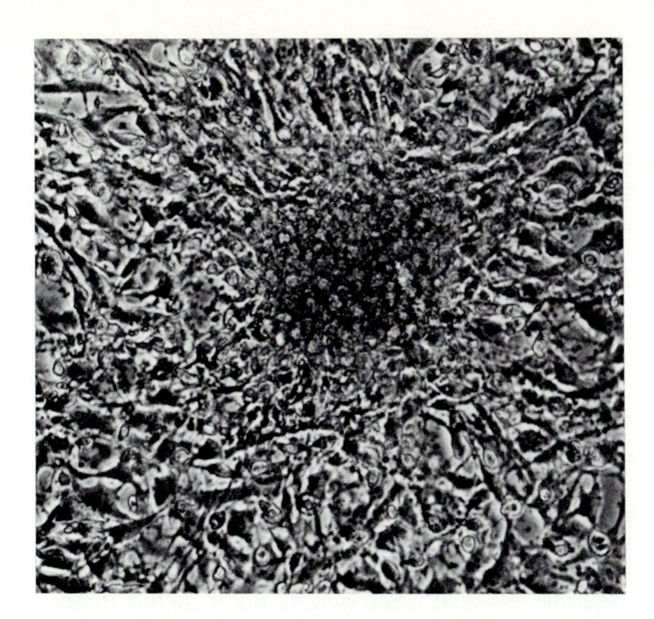

Figure 17-29 *Focus. Focus represents aggregated cells that have been transformed by virus, such as oncogenic virus.*

allowed to infect the monolayer of cells. The infected cells begin to divide, and the progeny cells accumulate at the site of infection producing a multilayered mass called a **focus** (Figure 17-29). The assay used to quantitate a tumor is called a **focus assay**. Other properties of transformed cells are outlined in the box below. The two most important properties of transformed cells are (1) they possess one or more integrated provirus and (2) transformed cells often produce tumors when injected into susceptible animals.

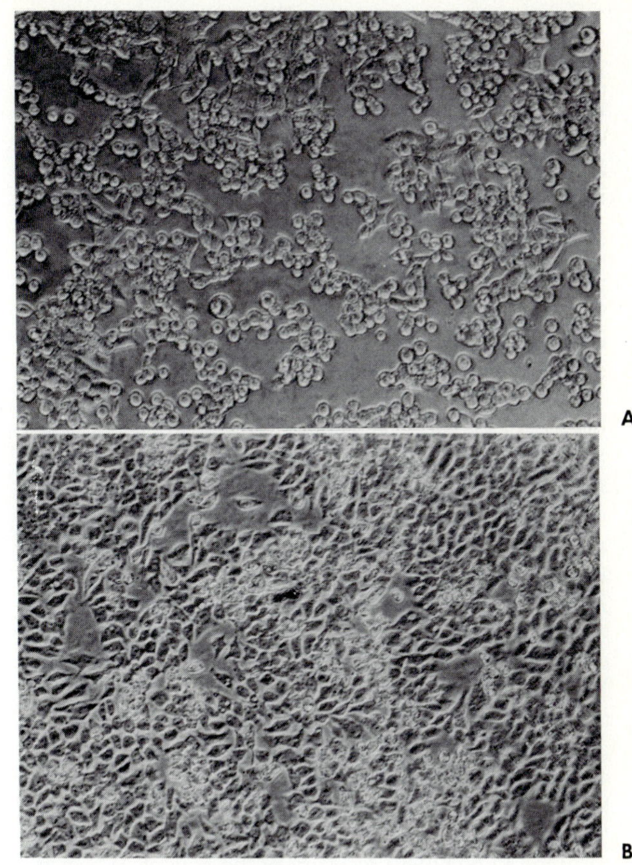

Figure 17-30 *Transformation of cell culture by cytomegalovirus.* ***A,*** *Uninfected control.* ***B,*** *Infected culture. Note rounding up of cells as compared to control.*

SOME IMPORTANT PROPERTIES OF TRANSFORMED CELLS

1. Viral nucleic acid becomes integrated into or associated with cellular DNA.
2. Transformed cells are rounder than normal cells and show irregular patterns of orientation in cell culture (Figure 17-30).
3. Transformed cells lose the property of density-dependent growth and show a loss of growth-inhibition characteristic of normal cells.
4. Some transformed cells produce tumors when injected into susceptible animals. Whether the tumors are benign or malignant is dependent on the transformed cell used and the type of host being challenged.
5. Following transformation, a virus-specific antigen appears on the surface of many tumor cells. It is the tumor-specific transplantation antigen (TSTA). Other antigens may appear on the surface of the host cell that make it recognizable by the immune system.
6. Chromosomal abnormalities, including breakage of some chromosomes and duplication of others, appear after transformation.
7. Changes occur in the cytoplasmic membrane of transformed cells. Transformed cells are more permeable to metabolites, such as simple sugars, than are normal cells. Changes in the glycoprotein, lipid, and carbohydrate composition of the cytoplasmic membrane may occur. The changes vary, depending on the composition of the growth medium and growth rate of the cells.
8. The agglutinability of cells by certain glycoproteins called **lectins** or **agglutinins** is altered in transformed cells. Transformed cells are agglutinated more readily by lectins because of alteration of the lectin-binding sites.

Suspected human oncogenic viruses

Indirect methods must be used in human studies to document viral agents as causes of cancer. The suspected virus cannot be isolated and inoculated into human subjects, as might be done with animals. Indirect methods include electron microscopy, examination of serum for antibodies, and culture and inoculation of experimental animals. Electron microscopy is often handicapped by the fact that few complete viral particles are present in the infected tissue. Examination of serum is handicapped because some tumor-forming viruses in animals do not elicit an antibody response in the host. The absence of antibodies to the virus is therefore not indicative of lack of infection. The use of experimental animals often represents an artificially manipulated condition. Usually the cell line or animal used is from another species, or the virus and/or culture must be preconditioned to produce positive results. Caution must be taken in extrapolating these indirect results to the human condition.

Oncogenic DNA viruses. Members of the herpesviruses appear to be the most likely DNA virus to cause cancer in humans. There is considerable evidence that the Epstein-Barr virus (EBV) may play a role in human cancer. EBV is a cause of the disease **infectious mononucleosis** (the kissing disease) that frequently affects college-age individuals. Infectious mononucleosis is a self-limiting disease that runs a defined time course, usually resulting in complete recovery of the patient. However, EBV may also be involved in the development of Burkitt's lymphoma and nasopharyngeal carcinoma. Burkitt's lymphoma, a type of cancer, is endemic in Africa and affects lymphatic tissue, particularly in the face and neck. Lymphocytes under normal conditions have a short and finite growth period when cultured; however, lymphocytes isolated from patients with Burkitt's lymphoma produce permanent growing cell lines in tissue culture. The EBV genomes are always found in the tumor tissue of patients with Burkitt's lymphoma and nasopharyngeal carcinoma.

Oncogenic RNA viruses. Retroviruses induce a variety of malignancies in animals and are widely distributed in nature. It should be remembered that not all retroviruses are oncogenic or even pathogenic. The most important characteristic of retroviruses, in terms of their ability to cause malignancies, is their noncytocidal nature and their ability to integrate into the host genome. One of the most important divisions of retroviruses suspected of causing human cancers is the **Type C oncovirus.** The oncogenes (*onc* gene) of type C, as well as other oncoviruses, are the only genes known that initiate and maintain cancers.

In 1911 Rous demonstrated that sarcomas (muscle tumors) could be induced in chickens by a virus now referred to as the **Rous sarcoma virus (RSV)**. It was not until the 1970s, however, that the gene responsible for this type of sarcoma was isolated. The RSV oncogene is called the **src** gene. An interesting result from studies on this virus was that the *src* gene is not truly a viral gene. This discovery led to important developments in the understanding of cancer and the possible role that some viruses may play in this disease.

Oncogene vs. proto-oncogene. The *src* gene is nearly an exact copy of a gene found in all chickens. In other words, the viral oncogene (referred to as **v-onc**) represented an altered version of a normal cellular gene. A normal cellular gene that is capable of conversion to an oncogene is called a **proto-oncogene (c-onc).** What is believed to occur is that during infection of chicken cells by retrovirus the proto-oncogene is picked up by the virus via a recombinational event and in the process becomes a cancer gene (Figure 17-31). Typically this results in the onc-specific sequence replacing essential virion genes (only in RSV is the *onc* gene added to the three essential virion genes). Consequently the retrovirus becomes replicative-defective. It should be noted that only a few specific cellular genes are candidates as proto-oncogenes.

Over 40 *c-onc* or *v-onc* genes have been identified and isolated (Table 17-2). DNA sequences homologous to most of them have been identified in a variety of uninfected cells of birds, cats, mice, monkeys, and humans. In each case the oncogene codes for a protein similar to the normal protein of the cell. It appears that proto-oncogenes have been highly conserved during evolution, for example, the proto-*src* gene is found not only in all vertebrates, but in the fruit fly *Drosophila.* The *ras* oncogene and its cellular counterpart differ by only a single nucleotide.

What do oncogenes do? The *v-src* gene found in avian species, the *v-ras* gene found in mice, and the *v-sis* gene found in monkeys have been among the most widely studied oncogenes. Presumably each oncogene is instrumental in generating the tumor from which it was isolated. Human tumors that yield oncogenes are not induced by retroviruses, yet some cellular genes homologous to several retroviral oncogenes have been found in altered form. The *src* gene, for example, codes for a protein called *pp60-src* (*pp* refers to phosphoprotein; *60* represents 60,000, the molecular weight of the protein). This protein phosphorylates the amino acid residues, especially tyrosine, of cellular proteins. The enzyme is called **tyrosine-specific protein kinase.** One of the cell proteins that is believed to be the major target of pp60-src is vinculin. Vinculin is a cytoskeletal protein that is important for the adhesive properties of cells to surfaces and to adjacent cells. It appears that during viral infection there is increased phosphorylation of tyrosine in the vinculin molecule, and this causes the cytoplasmic membrane to lose its adhesive properties and to pile up and metastasize, which is one of the properties of can-

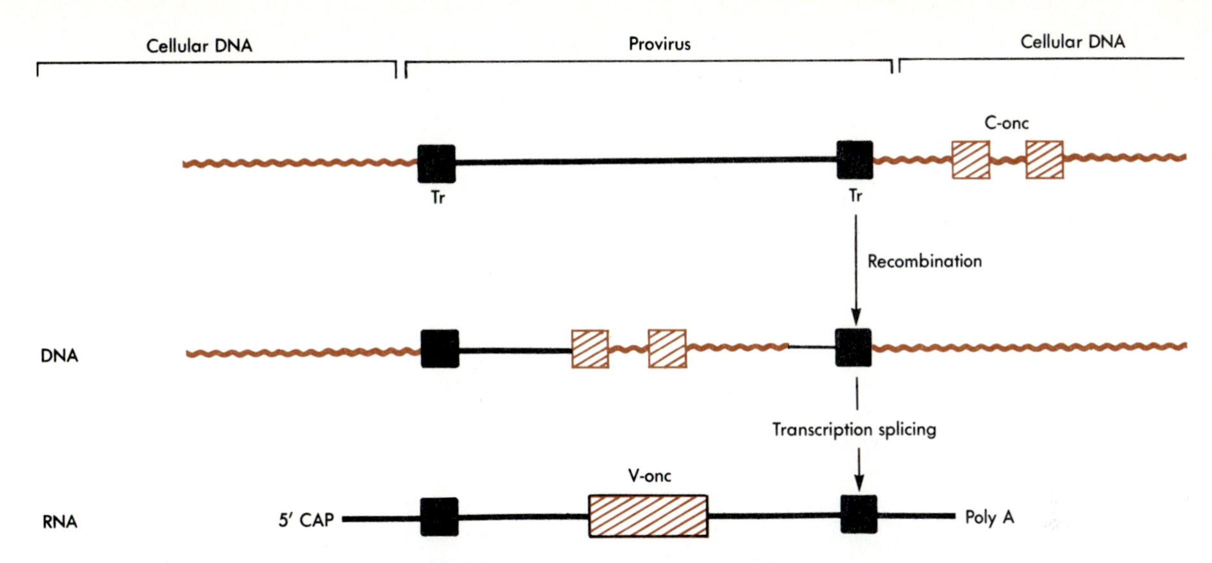

Figure 17-31 *Postulated mechanism for transduction of cellular oncogenes* (c-onc) *by retroviruses. Provirus carrying terminal repeats* (Tr) *is integrated into cellular DNA* (colored) *near a cellular oncogene* (c-onc). *Recombination event between provirus and cellular DNA results in* c-onc *genes being incorporated into proviral DNA. Transcription of proviral DNA and splicing of* c-onc *gene transcripts results in formation of retroviral RNA containing viral oncogene* (v-onc). *Encapsidation of retroviral RNA produces retrovirus capable of transducing oncogene.*

Table 17-2 *Partial list of the known oncogenes, the animals from which they were isolated, and the function of the protein for which they code*

Name of oncogene	Source of retrovirus	Function of protein
src	Chicken sarcoma	Tyrosine-specific protein kinase
yes	Chicken sarcoma	Tyrosine-specific protein kinase
ab1*	Mouse leukemia	Tyrosine-specific protein kinase
sis	Monkey sarcoma	PDGF-like growth factor†
Ha-ras*	Rat sarcoma	GTP binding†
Ki-ras*	Rat sarcoma	GTP binding
myc*	Chicken leukemia	DNA binding
myb*	Chicken leukemia	?

*Oncogenes in which the proto-oncogene has been identified in human tumors.
†*PDGF,* Platelet-derived growth factor; *GTP,* guanosine 5′-triphosphate.

cer cells. Reasons for some of the other properties of tumor cells have so far escaped detection. Since abnormal cell proliferation is characteristic of tumor cells, growth factor proteins are being intensively studied. Two proteins that may be involved are **epidermal growth factor** (EGF) and **platelet-derived growth factor** (PDGF). The *sis* gene appears to code for one of the two protein chains of which PDGF is composed. Excessive production of growth factors could lead to excessive cell proliferation when the cell otherwise would not divide.

The genetic factors responsible for expression of oncogenes are still a mystery. One possibility is that there is overexpression of a proto-oncogene product. This could be the result of attachment of retroviral sequences to specific cellular genes. The LTRs found on retroviruses when inserted upstream of the cellular gene act as enhancers and increase transcription of the proto-oncogene. Some human cancer cells have been shown to carry extra copies of certain oncogenes, a process called **amplification.** Thus an increase in a growth factor could be deleterious to the cell. Another possibility is that a mutation has altered the structure and function of a cellular protein (for example, the *src* gene product discussed earlier). In children with Burkitt's lymphoma, the *c-myc* gene has been translocated from chromosome 8 to three other chromosomes, each associated with immunoglobulin synthesis. Thus the *c-myc* gene is removed from its normal position and is placed under

Table 17-3 *Comparison of plant, bacterial, and animal viruses*

Characteristic	Plant virus	Animal virus	Bacterial virus
Nucleic acid present	Primarily RNA	RNA and DNA	RNA and DNA (predominant type)
Host range	Limited almost exclusively to plants	Animals, but some can replicate in invertebrates, such as arthropods	Limited to bacteria
Method of spread in nature	Some require vectors (insects, fungi, animal parasites); also spread by wind and water	Most do not require vectors; direct or indirect contact is most predominant method	Does not require vectors
Mechanism of infecting host tissue	Requires abraded area on plant surface	Tissue need not be damaged for infection to occur	Infection does not require damage to bacterial cell envelope
Survival in host	Indefinite, since host lacks antibodies	Host can inactivate virus because of antibodies, but some can remain latent for many years	Except for lysogeny, infection results in lysis of host cell
Tissue specificity	Nearly all plant tissue of the infected species (except for certain meristematic tissue) is infected	Most animal viruses are limited to specific tissue	

control of immunoglobulin genes. It is possible that the *c-myc* gene in its new position is transcribed more often than in its normal position. There is also some evidence that the proto-oncogenes require an activator, for example, a chemical carcinogen, before they become oncogenic. A mutated proto-oncogene was observed in the cancerous lung tissue of a patient but the gene from normal tissue in the same patient did not possess the mutation. Thus a mutagenic agent could have caused a mutation in the proto-oncogene, thereby rendering it oncogenic. Finally, there is some circumstantial evidence that two oncogenes must cooperate to transform a cell.

HTLV-I and II do not carry any of the two dozen or so known oncogenes and none of their sequences appear to be related to cellular sequences. How they cause cancer is also a mystery, although one mechanism appears to be a possibility. It is postulated that the LTR sequences, which carry retrovirus promoters, are involved. It is envisaged that the virus produces a protein called **trans-activating protein** that increases transcription from the viral promoter in the LTR. Thus no matter where the virus integrates on the genome the viral protein enhances transcription from the viral promoter and replication of virus ensues. In addition, the viral protein factor would also increase transcription of cellular genes (the trans factor of the protein), such as those that are involved in cell division. The viral genes coding for the protein activator have been identified in the HTLVs, and experiments are underway to determine their regulatory role.

INFECTION BY PLANT VIRUSES

Before discussing the growth of plant viruses it would be informative for the student to first note some of the characteristic similarities and differences between plant, bacterial, and animal viruses. These characteristics are elaborated on in Table 17-3. A more detailed discussion of plant diseases is presented in Chapter 20.

Virus-cell interaction

The usual adsorption step observed in bacterial and animal virus infection processes is not part of the plant virus infection process. The cell wall of plants is much thicker than that of bacteria and fungi and cannot be penetrated by viruses. Plants can be infected when the surface tissue is abraded, for example, when sap from an infected plant comes in contact with an abraded area of a healthy plant. Most plant virus infections, however, are the result of the injection of virus into tissue via plant-sucking and chewing insects (mites, aphids, beetles, and leaf hoppers). Occasionally some soil-inhabiting vectors such as fungi and nematodes (roundworms)

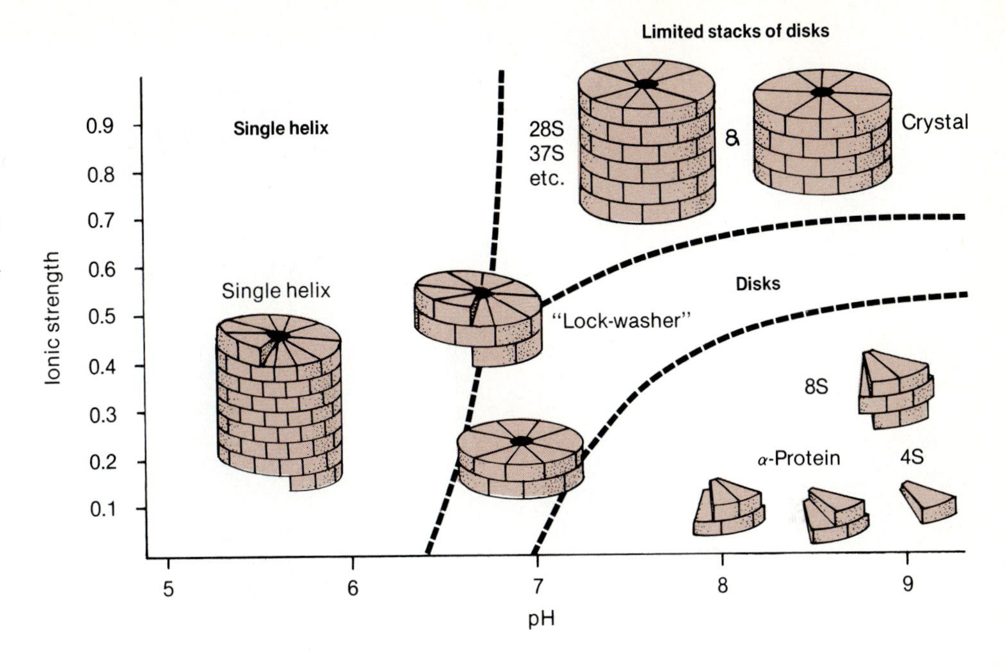

Figure 17-32 *Ranges of pH and ionic strength over which specific reproducible units of tobacco mosaic virus (TMV) protein engage in formation of helical unit. Aggregates (8S and 4S) are formed initially at alkaline pH and low ionic strength. These aggregates form disks (20S and 28S) that polymerize at acidic pH to produce helical structure.*

can damage plant tissue and transmit virus. A large inoculum of virus (up to 1×10^5 viral particles) is required to infect plant tissue (at least in the laboratory). Once the plant cells have been infected, viral particles are disassembled and viral nucleic acid spreads to surrounding cells and replicates in them. Viral titers may continue to increase and remain at high levels if the plant tissue does not die.

Assembly and maturation

Very little is known concerning the biochemistry of replication of plant viruses. The **tobacco mosaic virus** (TMV) is the most widely studied, but many of the subtle viral mechanisms that are known for bacterial viruses remain an enigma for plant virus infections. Nearly all plant viruses have an RNA genome, either single or double stranded. Only single-stranded RNA is potentially capable of being infective in the absence of a protein coat. Single-stranded RNA acts as mRNA and does not require transcription. The host ribosomes are used by plant viruses just as in bacterial and animal virus infections. Although plant organelles such as chloroplasts and mitochondria possess ribosomes, it is the cytoplasmic ribosomes that are used by the plant viruses. Few proteins other than the coat protein have been identified in infections by plant viruses or in plant viruses.

It appears that different viruses use different cellular sites for nucleic acid replication. Virus-associated material in TMV, for example, has been found to pass from the nucleolus through the nuclear membrane and into the cytoplasm. RNA synthesis in the turnip yellow mosaic virus occurs in the chloroplasts.

Concept Check

1. What changes occur in a cell that has been transformed following virus infection?
2. What is the distinction between an oncogene and proto-oncogene and the mechanism that distinguishes them?
3. What are some of the current theories as to how oncogenes cause malignant (cancer) tumors?
4. In what ways does the plant viral infection process differ from other viral infection processes?

All plant viruses are assembled in the cytoplasm. The TMV consists of 2200 identical protein subunits that are arranged in the shape of a helix, with 16⅓ subunits per turn of the helix. RNA is embedded between successive turns of the helix (see Figure 16-7). Laboratory studies have demonstrated that the protein subunits form aggregates of various sizes that can polymerize, depending on the ionic strength and pH of the solution (Figure 17-32). Subunit aggregates are produced at alkaline pH that result in the formation of disks or cylinders. Larger aggregates or cylinders are formed as the pH is decreased, and at acidic pH the aggregates are converted into helical structures (Figure 17-33). Intact disks may then be added to the growing helix. The assembly of the intact

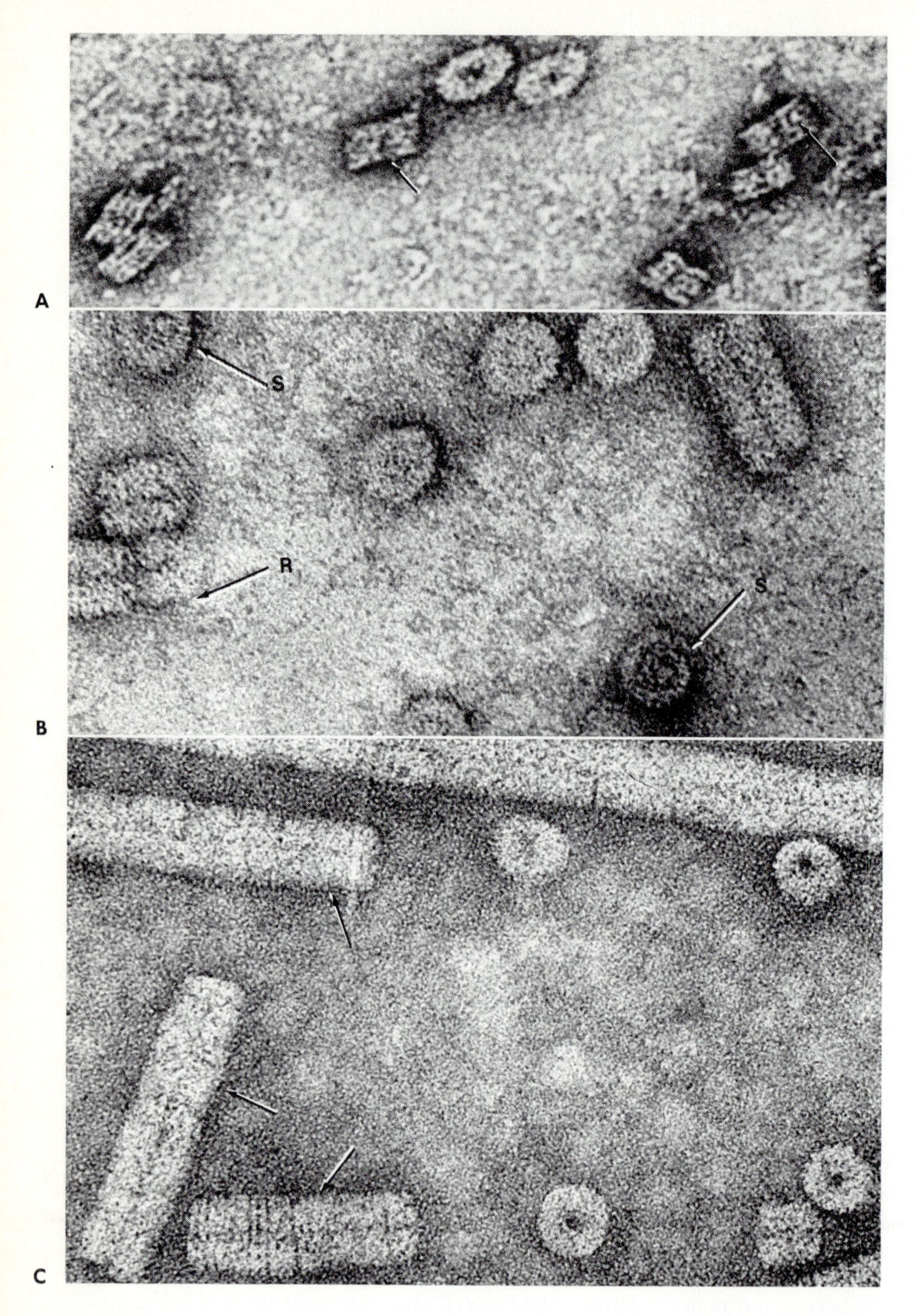

Figure 17-33 *Electron micrographs demonstrating effect of pH on TMV protein aggregation and formation of helical structure.* ***A,*** *Starting material is 28S aggregate formed at pH 8.0. Most of protein is in form of stacks of 4 rings or 2 disks (arrows).* ***B,*** *Protein has been placed in acetate buffer pH 5.0 for 10 minutes. Note that some short rods (R) are visible, but most protein aggregates are short, and some have their centers filled and show spiral (S) appearance or transformation to helix.* ***C,*** *Protein has been in acetate buffer pH 5.0 for 3 hours. Although there are some perfect single helical rods (arrows), there are some short stacks of disks present, and when seen end on, they do not have filled centers or spiral appearance. (×500,000.)*

TMV particles is believed to occur in the same manner as protein assembly in vitro. RNA and protein subunits interact to produce one or two turns of the helix. This then acts as a nucleus for further polymerization and elongation of the helix. The initiation of polymerization is believed to occur at a unique sequence of nucleotides in the TMV RNA.

Mature viral particles can spread from one cell to another through the plasmodesmata (channels connecting adjacent cells) or through the vascular tissues (phloem) until the entire plant is infected.

Satellite virus and viroids

Satellite virus and viroids are associated with plant diseases (see Chapter 16).

SUMMARY

1. Infection by phage follows a pattern of adsorption to the cell surface, penetration of the nucleic acid into the cytoplasm, nucleic acid replication, viral assembly, and finally release from the cell.
2. The phage subverts or breaks down the metabolism of the host, and in so doing, the virus is supplied the necessary nucleotides required for its own nucleic acid synthesis. In addition, modification of the protein-synthesizing system provides a mechanism for recognition of viral mRNA but not bacterial mRNA. Destruction of DNA prevents host cell replication or transcription.
3. The viral reproduction cycle in the cell is initiated by the formation of early and late proteins, but no mature viral particles are yet produced. The viral nucleic acid is then packaged into the phage head protein, and this is followed by the addition of tail and tail fiber proteins. Virally induced lysozyme breaks down the cell wall and causes the release of the mature virus.
4. Some viral infections do not result in lysis, and the virus remains within the host, either integrated into the host's nucleic acid or attached to the cell membrane. Under such circumstances the bacterial cell is said to be lysogenized. Two regulatory proteins, repressor protein and cro protein, are associated with control of the lytic and lysogenic state of the virus.
5. During animal virus infection the entire virus, not just the nucleic acid, is taken into the cell in a process similar to phagocytosis. An eclipse period is also present, during which many viral components are synthesized but no mature virus is formed.
6. The life cycle of animal viruses can be divided into early and late functions. Replication and transcription of the viral genome may occur in the cytoplasm or nucleus, depending on the virus. Replication may involve single- or double-stranded DNA or single- or double-stranded RNA.
7. For RNA viruses the genome may be considered positive or negative depending on whether the RNA acts as an mRNA (positive) or must be transcribed (negative) into a functional mRNA. Most RNA viruses contain either single (positive) or (negative) strand genomes.
8. Reoviruses contain 10 double-stranded segments of RNA. Retroviruses contain two positive RNA strands that are covalently bound by a tRNA into a single-stranded molecule.
9. Most DNA viruses contain many more genes than RNA viruses do and their replication is relatively complex.
10. Structural proteins of viruses are produced in the cytoplasm but viral assembly may occur in the cytoplasm or nucleus depending on the virus. Most naked viruses, after assembly, accumulate in the cell until the cell lyses. Enveloped virus assembly is more complex and requires budding through the nuclear or cytoplasmic membrane.
11. The animal cell is affected by virus infection in two basic ways: the cell will lyse or the virus will persist in the cell. The terms *chronic, slow,* and *latent* virus infection are similar to yet distinct from persistent infections. Persistence is the way of escaping detection by the host's immune system and the mechanisms may be varied. In addition, interferon is an antiviral glycoprotein produced by the vertebrate host in response to viral infection. Interferon is an inhibitor of viral protein synthesis. It has also been demonstrated to reduce malignant growth.
12. Transformation of animal cells by virus can result in tumor formation. These tumor-forming viruses are called oncogenic. Transformation requires incorporation of the viral nucleic acid into the host genome. If the viral nucleic acid is RNA, a special enzyme called reverse transcriptase is carried by the virus to convert RNA into DNA before incorporation into the DNA of the host.
13. Transformed cells exhibit many different properties than uninfected cells: cells pile up, cell membrane antigens are changed, and cell metabolism is altered.
14. There is some evidence that viruses cause cancer in humans. One DNA virus, EBV, has been implicated in lymphomas of African children. Retroviruses, such as HTLV, are RNA viruses linked to human cancers.
15. The oncogenes so far isolated from retroviruses are homologues of certain cellular genes, called proto-oncogenes. The abnormal behavior of the proto-oncogene is believed to be somehow associated with tumor formation.
16. The mechanisms of plant virus replication are not well understood. Plant viruses usually enter the cell via abrasions or by injection via plant-sucking insects. The viral nucleic acid is usually RNA, and replication may occur at different sites within the cell. Following assembly into mature virus, viral particles can spread from one cell to another via the plasmodesmata or through the vascular phloem.

Self-study Quiz

Completion

1. The modified base on the T_4 genome that protects it from cellular nucleases is called ________________.
2. T4 modifies RNA polymerase by covalently adding _____ to the alpha subunits.
3. Concatamers are formed by recombination between ____ ends of DNA molecules.
4. The proteins that help to assemble the phage capsid into an orderly array of subunits are called ________________ proteins.
5. The virally encoded enzyme that helps the phage release through the cell wall is called ________________.
6. The packaging of DNA genomes into the head of the phage is called ________________.
7. The regulatory protein that predominates when the phage enters the lytic cycle is called ________________.
8. The HA glycoprotein, which is part of the influenza envelope, is used for binding to the cell surface but the infection process ceases unless a cellular ____________ ______ cleaves the HA chain.
9. Most naked animal viruses enter the cell by a process called ________________.
10. Viral oncogenes have homologous cellular counterparts called ________________.

Short answer

Describe the function(s) of each of the following enzymes during the viral replication cycle:

1. Endonuclease
2. Exonuclease
3. Reverse transcriptase
4. RNA polymerase
5. DNA polymerase
6. Integrase

Multiple choice

1. Which of the following is an oncogenic virus?
   A. Phage lambda
   B. Poliovirus
   C. TMV
   D. RSV
2. Viral glycoproteins would be found in what structure of the virus and are associated with which phase of viral replication?
   A. Envelope, adsorption
   B. Capsid, uncoating
   C. Nucleic acid, nucleic acid replication
   D. Nucleocapsid, mRNA transcription
3. A temperature bacteriophage is associated with which of the following?
   A. Oncogenesis
   B. Viral budding
   C. Plant viruses
   D. Lysogeny
4. Retroviruses are a diverse group of RNA viruses that have in common which of the following?
   A. They are enveloped and produce a reverse transcriptase.
   B. They are enveloped and produce neuraminidase.
   C. They are single-stranded plant viruses.
   D. They are enveloped and contain as many as 10 genome segments.
5. The virus that has its genome complexed with cellular histones is:
   A. SV-40 virus
   B. Rous sarcoma virus
   C. Poliovirus
   D. Reoviruses
6. Where would you find the prophage in a lysogenized bacterium?
   A. In the bacterial cell genome during normal growth and division
   B. In the cytoplasm of the bacterial cell during normal growth and division
   C. In the cell membrane of a bacterial cell undergoing lysis
   D. Outside the bacterial cell following induction
7. Is the growth curve of a lytic bacteriophage the same as that of a bacterial culture?
   A. Yes, it is identical in all respects.
   B. Yes, but there are many more bacteriophage produced.
   C. No, because from each bacterial cell lysed a large number of viruses are released.
   D. No, because only a few phage are continuously released from the dividing bacterial cell.

PART V

Environmental and Applied Microbiology

CHAPTER

18

Concepts of microbial ecology

Martin Alexander
R. F. Boyd

Chapter Outline

Introduction

Every microbial species has one or more habitats in which it lives. A habitat may be the sea, fresh waters, soils, decaying organic matter, plant or animal surfaces, or the tissues or body fluids of animals. Often for the microbiologist, the microhabitat is more important from a conceptual viewpoint because of the size of individual microbial cells or filaments. Microbial ecology is concerned with the interrelationships between the microorganisms and the environment or habitat in which those organisms live.

Microorganisms are unique tools for ecological studies, and the ability to use microbial mutants deficient in traits of presumed ecological importance and to perform rapidly a variety of studies assessing biochemical interrelationships have made microorganisms especially valuable for defining ecological principles. Many applied problems have appeared in recent years for which information on microbial ecology is of practical importance. For example, the increase in pollution of fresh waters by human activities has led to unwanted and often obnoxious blooms of algae and cyanobacteria, and attempts to control these blooms have fostered interest in the nutrition of the responsible microorganisms. Research on chemical pollution of waters and soil has established that microorganisms are the sole or major agents of biodegradation, and this knowledge has prompted many studies of microbial transformations of polluting chemicals. Moreover, microorganisms have recently been found to produce a variety of hazardous products in waters and soils, and in this way they are themselves responsible for chemical pollution.

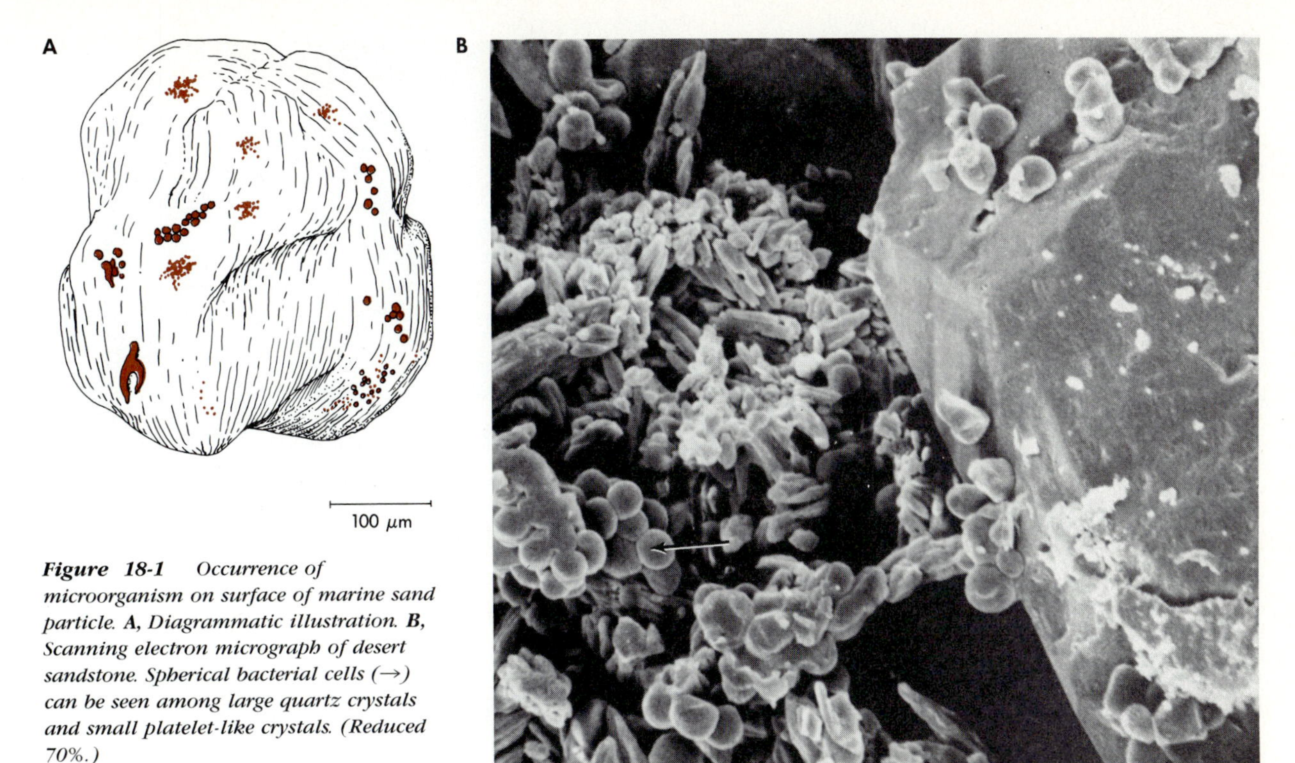

Figure 18-1 *Occurrence of microorganism on surface of marine sand particle.* ***A,*** *Diagrammatic illustration.* ***B,*** *Scanning electron micrograph of desert sandstone. Spherical bacterial cells (→) can be seen among large quartz crystals and small platelet-like crystals. (Reduced 70%.)*

This chapter outlines the basic concepts and terminology of microbial ecology. Also included in the chapter are discussions on the factors that influence a microorganism's ability to colonize a particular habitat and the factors contributing to its persistence in that habitat, how the metabolic activities of microorganisms contribute to the cycling of the major chemical elements, how microbial activities in the environment contribute to pollution, and how microorganisms can be used to extract heavy metals from low-grade ores.

CONCEPTS AND TERMINOLOGY

From the ecological viewpoint, it is important to understand certain terms in what is sometimes called the ecological hierarchy. At the lowest level of that hierarchy is the individual microorganism. This may be the single cell of the unicellular bacterium or protozoan, the resting body of a fungus, or a hyphal strand. All cells or hyphae of a single type that reside in a single habitat, or microhabitat, represent a distinct **population**. Those populations that represent a strain or species of economic or health importance have achieved special prominence. The various populations inhabitating a given site represent the **community**. Considerable information exists on community composition when the environment is the body of the human or higher animal, but less information exists on the communities of marine, freshwater, or terrestrial environments. The various communities in a particular environment together with the nonliving, or **abiotic**, surroundings constitute the **ecosystem**. Each ecosystem has a unique collection of organisms and inorganic and organic components.

Many misconceptions about microbial ecology exist because of the lack of understanding of the microenvironment. At the microscopic level, which is appropriate for considering microorganisms in nature because of their minute size, there may be major gradations in ecologically important variables. In a zone with a diameter of less than 0.1 mm, enormous differences in the amounts of organic substrates, inorganic nutrients, inhibitory substances, and oxygen may prevail. At one site, the level of one or more of these factors may be high, and at a site less than 50 μm away, the concentrations may be low. Thus, two distinctively different populations may reside in what appears, to the macroscopic viewer at least, to be the same site, but the organisms in fact are living and functioning in different microenvironments (Figure 18-1). Analyses of microhabitats are difficult because most of the currently available physical, chemical, and often biological procedures are not useful at the microscopic scale; hence a conclusion based on a property of the macroenvironment and analyses performed by techniques not suitable for the microenvironment frequently is at odds with the properties of the organisms presumed on the basis of tests carried out in laboratory culture. It is critical to bear in mind that the microenvironment is the functional habitat of the microorganism, whether the microorganism is performing its various metabolic activities and reproducing in soils, waters, plant tissues, or different sites within the animal or human body.

The obviously marked differences in species composition of microbial communities are at odds with the oft-stated view that microorganisms are potentially everywhere. This view is derived from observations that microorganisms are easily transported with currents of air and water or on animal and plant surfaces, and they are ingested or otherwise penetrate animal bodies and plant tissues. Furthermore, many microbial species grow rapidly in the laboratory and presumably so in nature. Given that microorganisms are so easily disseminated and grow so rapidly, how can one explain the fact that individual ecosystems have characteristic microbial communities? These characteristic communities are evident in oceans, lakes, soils, plant surfaces, the human mouth, or in stool samples.

The reason is simply that, although microorganisms indeed are potentially everywhere and some can grow rapidly, the environment selects from among the invaders and permits only a few species to become established. Each ecosystem has certain physical, chemical, and biological factors that govern the community composition. These factors may be types and concentrations of organic nutrients, toxins, surfaces receptive to microbial development, temperature, and acidity. These crucial factors determine whether a species that has reached that environment will or will not grow or even survive. The environment sorts out the arrivals with the physiological properties that are attuned to the particular local circumstances. Populations not meeting these environmental stresses are eliminated early in the development of the community. The environment thus selects the fit microorganisms and rejects those that are unfit. The former will survive and multiply, and the latter are eliminated.

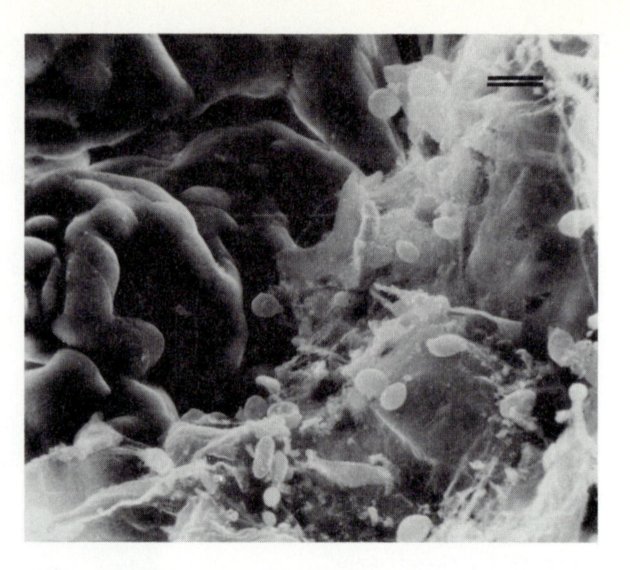

Figure 18-2 *Electron micrograph of yeast cells in mucus covering epithelial cells of the cecum of a mouse. Bar = 10 μm.*

NATURAL SELECTION AND FITNESS

Each microbial species has unique biochemical or morphological properties. These properties permit the development of that organism in some environments but not in others, and it is these traits that determine the geographical patterns of microorganisms or the microenvironments in which they reside. It is on these biochemical and morphological traits that natural selection operates, and the interactions between natural selection and these properties determine the distribution of the species. Based on these traits, some species have a selective advantage in one environment, and others are favored in dissimilar environments. The outcome of the selection is survival or, at some time, active metabolism. To survive the selective processes, the cell or filament must win in the struggle for existence. The unsuited species, in contrast, has a small chance for survival at that site and is usually eliminated.

The identities of most of these ecologically important attributes are difficult to establish because most microbial communities contain many different species, the organisms are extremely small, and the interactions among the populations are numerous. A few of the attributes are simple to establish. Thus, the fit species must be able to tolerate the extremes of temperature, salinity, or osmotic pressure to which the environment is exposed. It may need to endure periods of drying. If it is in the body, it must be able to endure attack by phagocytes. These are all necessary conditions for an organism to exist, but they are not sufficient conditions. Usually, there are many more hidden than visible traits of ecological importance. The lack of a single trait may result in the elimination of a species. However, only the most obvious of these traits are known. Let us examine a few of those properties of microorganisms that determine fitness.

Factors that contribute to the selection process

The obvious factors contributing to selection can be dispensed with quickly: the inhabitant of any environment must be able to tolerate the **ambient moisture** levels, **pH, hydrostatic pressure** if it exists in the water column

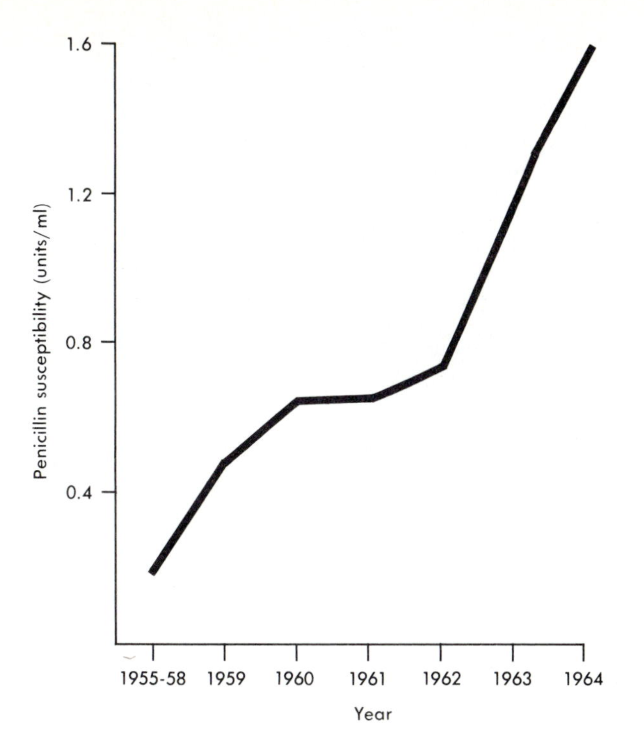

Figure 18-3 *Rise in resistance to penicillin of isolates of* Neisseria gonorrhoeae *not eliminated by concentrations of antibiotic commonly used in therapy.*

of the ocean, **oxygen** levels, **salinities**, and **temperature** of that locality. (These factors were discussed in Chapter 10.) Because many microbial communities are exposed to intense **sunlight**, the ability of an organism to maintain its viability in the presence of lethal light intensities is of crucial importance. Sunlight kills many populations. To cope with the potential harm from sunlight, some species have carotenoid or melanin pigments that are chief fitness traits in sun-drenched environments. Populations possessing these pigments are common on leaf and water surfaces and in salt ponds exposed to high light intensities, and they dominate among the species that are transmitted for long distances through the atmosphere. Many populations are retained on surfaces in regions with moving liquids, for example, microorganisms growing on rocks in streams, bacteria on pipes containing flowing fresh water, protozoa in the intestine, or bacteria on epithelial surfaces of the mouth and intestinal tract (Figure 18-2). These organisms have a characteristic fitness trait: an **attachment** device or other means to allow the organism to be retained on the surface. Because the cell constituents of many microorganisms have high specific gravities, they would usually settle to the bottom of lakes and the ocean, but the organism that is native to aquatic communities must have some special **buoyancy** property that allows it to be maintained in such environments. These populations commonly have cell constituents, such as gas vacuoles (see Chapter 5), that compensate for the otherwise high specific gravity of the cell or filament. For some protozoa, small **size** is an ecologically critical trait—without it, they would be unable to reside in soil or sand, which have very small pores between the soil particles or sand grains. These pores would prevent the active movement and feeding of cells having large sizes. In contrast, large size is important in some aquatic environments because big cells are not easily consumed by small predators, which often feed avidly on minute microorganisms.

Natural selection is also obvious among microorganisms associated with human or animal disease. In these instances, the selective force is often the host. The successful parasite or disease agent makes use of the many nutrients within the host, and it copes with host defenses. In contrast, a nonparasitic species often owes its failings in a host to its inability to become established in the environment, which in this case is the body of the potential host. This lack of establishment may be related to the absence of enzymes allowing the population to make use of host constituents or of structural properties that allow it to ward off host defenses. In these instances, the enzymes or morphological properties are essential **fitness traits.**

Natural selection is also evident in human medicine and in the gain in properties by an invasive microorganism that converts the invader from an avirulent to a virulent strain. Antibiotics are of enormous importance in human medicine, but associated with their use is the appearance of drug-resistant bacteria. Their appearance is a reflection of the forces of natural selection because the drug-resistant variant form of the pathogen has a unique fitness trait that is absent from the original species: it possesses physiological or structural properties that allow it to grow in the presence of a stress not tolerated by the original population. The latter did not have these special traits. This is evident in the increase in resistance among many bacteria to penicillin, an outcome of the widespread use of the antibiotic in chemotherapy (Figure 18-3). The development of many microorganisms in human, animal, and plant hosts also results in a shift in the population from one dominated by avirulent individuals to one dominated by virulent cells. The host is the selective factor for the virulent pathogens. The avirulent cells, because they do not have the requisite fitness traits, do not win the struggle for existence in the environment. Here, the host is the agent of natural selection.

SPECIES DIVERSITY

The number of species in a given habitat or microhabitat is often large, and soils and sewage harbor a multi-

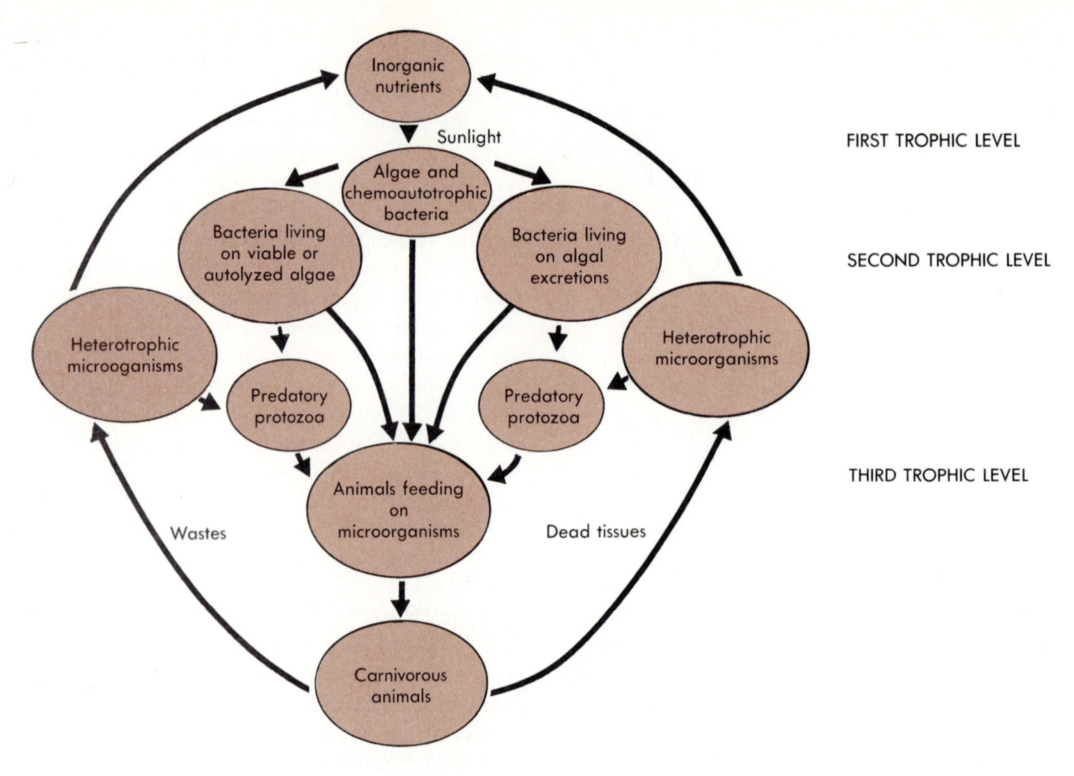

Figure 18-4 *Typical food chain.*

tude of species in a small area. In contrast, some communities have few species, and some may have only one. The reasons for these differences in species diversity and numbers are usually unknown. Few species are present in environments that have a high intensity of some deleterious factor, as in localities with high temperatures, where drought is common, or with high levels of salts or, as in many foods, sugars. Conversely, the presence of many different populations at a particular site is linked with nutrient deficiencies; many algae are present in nutrient-poor waters, and many bacteria are present in soils having little available organic matter. Why so many species coexist at a particular site has long intrigued ecologists. Possibly, the coexistence is attributable to the dissimilar requirements of the various populations, and species possessing the diverse physiological properties are able to live together. In many environments, a species that might be expected to dominate is suppressed by parasites or predators, and thus over-exploitation by a single species is rare. Some environments have mechanical barriers or gradients of stresses or nutrients, so that one species exists at only one level of the gradient or is protected from harmful neighbors by the barriers.

Food chains

Food chains also permit the coexistence at a particular site of many dissimilar species. Growth of the indigenous populations requires sources of energy, carbon, other elements, and growth factors. The energy source for higher plants, algae, cyanobacteria, or photosynthetic bacteria is light; for heterotrophs it is an organic molecule; and for chemoautotrophs it is an inorganic ion or molecule (Figure 18-4). These organisms, which are at the base of food chains (**first trophic level**), proliferate at the expense of their energy source and excrete organic compounds. These excretions, or products released when the cells die, sustain other populations. Moreover, the cells of the populations using the first energy sources serve as nutrients for predators such as protozoa and for a variety of microbial parasites. The organisms at the **second trophic level** live using products or cells of those species using the energy sources that enter the environment from the outside. In turn, these secondary feeders will themselves be nutrients or will excrete nutrients for new feeders (the **third trophic level**), and the latter in their turn provide nutrients for still other organisms. Complex food chains of these sorts are common in soils, natural waters, and sewage.

Hence, even in an environment in which the only energy source is sunlight, a variety of species will coexist.

NUTRITION AND ECOLOGY

The habitat of a species is the source of all of its nutrients, inorganic or organic. From its surroundings, each species must obtain a variety of inorganic substances, an energy and a carbon source, and any ancillary compounds required for multiplication. The absence of a species from an environment could result from the lack of one or more of its essential nutrients. Conversely, if a population grows, all of its needed nutrients must be provided. Chemical analysis of that environment may show none of the nutrient to be present, but this results from the entire supply being assimilated by the proliferating populations.

Not all nutrients are ecologically significant. Should the supply of a particular needed compound or ion be greater than the demand of the entire community, that substance probably is not significant in determining the activity or the occurrence of individual species: all inhabitants get what they require. In contrast, should the supply be less than the demand, the species will interact for the limited supply, and an organism that is unable to obtain what it requires will not be established.

Growth factors

Particular attention has been given to growth factors because many bacteria, protozoa, fungi, and even aquatic algae require them for development. Each of these exacting species is unable to multiply unless its demand for these essential organic compounds is met. The microorganism that requires several growth factors is at a disadvantage in many ecosystems because these factors may not always be present. Nevertheless, microorganisms that need growth factors are widespread, meaning that the species at some time and place must find a source of the essential nutrients. Indeed, many environments that one might not consider to contain growth factors do in fact contain them. A high percentage of the bacteria in soil and marine sediments and algae in the oceans require growth factors, and a supply must thus be present. If the gain of a requirement for growth factors makes a microorganism more versatile physiologically, then a fastidious species may be favored provided its nutritional demands are satisfied. Growth factor–requiring species are often highly active metabolically, so natural selection may favor them over more fastidious organisms in environments that have an adequate supply of the needed substances.

For a species to cause disease, it must find a host, invade it, and multiply. In many instances the potential parasite is unable to proliferate because of defense mechanisms of its host. In some instances, the lack of proliferation is the result of the absence from the site in the host body of a full complement of microbial nutrients. A number of bacteria requiring amino acids for growth fail to cause harm to animal or plant hosts because they cannot obtain needed amino acids and thus cannot reproduce and cause disease.

Nutrient levels in the biosphere. Most of the inhabitable sites of the biosphere contain extremely low levels of available organic matter. About 99% of the volume of biosphere is made up of oceans, and in oceans and lakes the total amount of organic carbon is usually 1 to 25 μg/ml. If this is the total amount of organic matter, the supply of available organic carbon must be much lower. Nevertheless, many microbiologists use extremely rich media in the laboratory, and hence the organisms that they isolate may not represent those that function in the natural nutrient-poor habitat. Based on the concentration of nutrients they need, microorganisms may be divided in two categories: **eutrophs**, or **copiotrophs**, which grow at the high levels of nutrients common in laboratory media but only in certain environments, and **oligotrophs**, which grow at low concentrations of nutrients, especially organic carbon. It is still not clear why the latter do not grow at high nutrient concentrations, and the former do not multiply at low nutrient levels. Oligotrophs are probably important in natural waters because of the low levels of organic matter. Moreover, the widespread concern with chemical pollution of waters and soils is directed at compounds whose concentrations are usually no greater than 100 parts per billion. At these concentrations, only oligotrophs are presumed to grow.

Lakes, rivers, and streams have been receiving an increasing supply of inorganic nutrients. These nutrients come in part from natural sources. In part, they come from the activities of society. Thus, detergent formulations contain phosphorus, fertilizers contribute nitrogen, and city sewage provides both of these elements as well as others, and manure applied to farmland adds essential nutrients to water underlying the soil. The latter nutrients move into surface waters as the groundwaters move laterally. As a result of the increasing amounts of phosphate, nitrate, and sometimes other nutrients in lakes, rivers, and streams, the **biomass** of algae and cyanobacteria increases (Plate 19). These species are commonly not limited by the energy source because they obtain energy from sunlight, and their carbon is obtained from carbon dioxide dissolved in the water. Carbon dioxide is usually not in limited supply in waters. If nitrogen or phosphorus, for example, limit the photosynthetic populations, the introduction of phosphate or nitrate into a lake would stimulate these populations and increase the rate of photosynthesis. In the

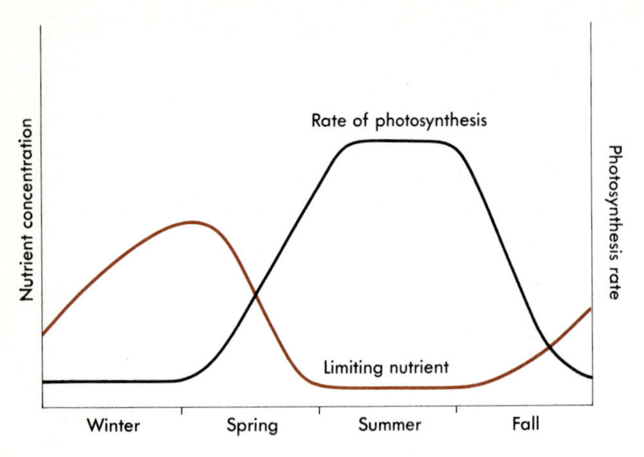

Figure 18-5 *Relation between rate of algal photosynthesis and concentration of limiting nutrients in lake water.*

fall, the organisms die and release the element (nitrate or phosphate) again (Figure 18-5). The excessive growth of algae and cyanobacteria is one of the obvious problems of **eutrophication**, the increasing fertility of waters. Growth of algae, cyanobacteria, or rooted plants in eutrophic waters may be so massive that the movement of boats through waterways is reduced (Plate 17). More often, however, the practical concern with eutrophication arises from dying of the algae and cyanobacteria. During the decline of these populations, bacteria proliferate and produce offensive tastes and odors. The bacterial activity may be so pronounced that oxygen is totally removed from the water. This leads to the death of fish, which are aerobes. In addition, toxins may appear in the water, and these toxins may affect fish as well as human consumers of what was once a supply of potable water. The practical consequences of eutrophication have led to governmental action and also a renewed interest in the nutrition of aquatic populations.

COLONIZATION AND SUCCESSION

Many environments initially have a small biomass and few populations. This is true of lava that has cooled after a volcanic eruption and the surfaces of rocks exposed by a breakage of the rock. Tissues of plants and animals and body fluids are also usually free of microbial life. The newborn infant and the seedling emerging from the seed similarly do not initially contain a microbial community. Fresh fruits that become bruised or canned foods on exposure to the air also are initially free of any forms of life. Surfaces that directly or indirectly make contact with the environment are rapidly invaded, and microbial **colonization** is initiated. **Succession** is the replacement of populations and community types that occurs with time. The initial populations in the previously lifeless regions are the **pioneer species**, but a community dominated by pioneer species is soon replaced through a series of successional events. These events are rarely abrupt, and a slow progression of species changes is characteristic of sites subject to microbial colonization.

The identity of the pioneer is determined by the physical and chemical properties of the environment undergoing colonization. This is evident in the differences in pioneer communities of different types of food products and in the dissimilarities in the bacterial communities of the gastrointestinal tracts of newborn children fed mother's and cow's milk.

In an environment with little organic carbon, as in waters or on the surfaces of rocks, the pioneers are usually algae, cyanobacteria, or lichens. The lichen is a symbiosis of a fungus with an alga or a cyanobacterium (Plates 18 and 19). In aquatic sediments where there is considerable organic matter, the pioneers are different. The succession of bacteria in these environments, where the rate of organic deposition exceeds the supply of dissolved oxygen, is related to depth of sediment and availability of electron acceptors. A thin aerobic zone that appears at the surface is dominated by organisms using aerobic respiration (Table 18-1). The zones below the aerobic zone are anaerobic, and the organisms use terminal electron acceptors other than oxygen, that is, nitrate, sulfate, and carbonate, as well as organic molecules. The metabolically efficient microorganisms appear near the surface, whereas the less efficient ones appear at greater depths in the sediment. In contrast, initially sterile sites that are rich in organic materials have quite different pioneers. In this regard, considerable attention has been given to the colonization of the newborn baby because such colonization is of considerable practical as well as ecological interest, especially if the potential colonist is a pathogen.

Specificity of colonization

Microorganisms are specific for the sites that they colonize. This specificity, in the newborn infant, for exam-

Concept Check

1. Distinguish between the terms *population, community,* and *ecosystem.*
2. How does natural selection operate in microbial communities?
3. What is meant by a food chain? What are the organisms associated with the different levels of the chain?
4. What is the sequence of events that leads to eutrophication?

Table 18-1 *Succession in aquatic anaerobic sediments governed by energy-yielding oxidation-reduction reactions*

Metabolic process	Biogeochemical zone	Prominent bacteria	Chemical equation	ΔG°*
Aerobic respiration	Aerobic zone	Aerobic heterotrophic	$CH_2O + O_2 \rightarrow CO_2 + H_2O$	−686
Anaerobic respiration	Redoxocline†	Nitrate-reducing and denitrifying	$5\ CH_2O + 4\ NO_3^- + 4\ H^+ \rightarrow 2\ N_2 + 5\ CO_2 + 7\ H_2O$	−579
Anaerobic respiration	Sulfate-reducing zone	*Desulfovibrio, Desulfotomaculum*	$2\ CH_2O + SO_4^{2-} \rightarrow S^{2-} + 2\ CO_2 + 2\ H_2O$	−220
Anaerobic respiration	Carbonate-reducing zone	Methanogenic	$2\ CH_2O \rightarrow CH_4 + CO_2$	−99

*Expressed in kilocalories per mole of glucose equivalent oxidized; ΔG°, change in Gibbs free energy.
†Redoxocline is a zone in which the redox range varies from +200 mV to ±50 mV.

Table 18-2 *Percentage of microorganisms isolated from the saliva and gingival crevice*

Organism or group	Saliva	Gingival crevice
Streptococci	40-50	20-30
Diphtheroids	5	15-25
Bacteroides	5	15-18
Fusobacterium	2-3	4-6
Veillonella	14-16	10-12
Neisseria	1-2	1
Spirochetes	None	1-3
Actinomyces	1-2	8-10
Vibro	1	1-3

ple, may be associated with development of certain species only in the mouth.

The oral cavity just before birth is essentially an environment devoid of microorganisms. Microorganisms begin to appear at birth and immediately thereafter because of (1) contamination as the fetus passes through the birth canal, (2) feeding from the mother's breast, and (3) contact with the delivery team or family members. Many of the microorganisms, for example, *E. coli,* that contaminate the oral cavity are transients that are unable to attach to the oral mucosa and become permanent residents. Studies have demonstrated that *E. coli* shows little affinity for epithelial cells in the oral cavity, yet it binds strongly to epithelial cells in the urogenital or intestinal tract. This inability to bind results in the organisms being washed out of the mouth and into the stomach. Four days after birth, the organisms in the oral cavity are species of *Streptococcus, Neisseria, Staphylococcus, Veillonella,* and lactobacilli, with the streptococci predominating. Notably absent are the obligate anaerobes.

The eruption of teeth creates an environment that will support the growth of obligate anaerobes—the gingival crevice, the area between the gingiva (gums) and the tooth surface. What makes it so ideal for anaerobes is its low oxidation-reduction potential and the fact that the gingiva supplies metabolites necessary for the growth of many anaerobes. The obligate anaerobes associated with the gingival crevice include *Bacteroides, Fusobacterium,* and actinomyces. A comparison of the percentage of organisms in the saliva and gingival crevice is illustrated in Table 18-2.

When teeth develop, two new species of *Streptococcus* also appear—*S. mutans* and *S. sanguis.* Both species either directly or indirectly are capable of binding to the enamel on the tooth surface. When adults lose their teeth, these two species disappear as well as obligate anaerobes. If dentures are worn, *S. sanguis* and *S. mutans* reappear, as well as some anaerobes. The mechanism of attachment of *S. mutans* to the tooth surface is discussed in Chapter 23.

Colonization of specific sites is also evident among the algae that are uniquely adapted to grow on rocks on coastlines buffeted by intense wave action. Toxin-producing bacteria associated with diarrhea are also specific for their sites of colonization, and these organisms are specialized for the colonization of surfaces of the intestine. Often, the specificity of microoganisms for individ-

Table 18-3 Barriers to colonization by microorganisms

Environment	Barrier postulated or demonstrated
Skin	Fatty acids
Lungs, stomach	Mucosa
Nose	Lysozyme
Blood	Phagocytes, antibodies
Intestine	Mucosa
Fish tissue	Protamines
Milk	Peroxidase, agglutinins
Roots	Cork layer
Fruits	Cuticle layer
Trees	Gums, resins, tannins
Plant tissue	Phenolic compounds, glycosides
Virus-infected cells	Interferon

ual sites is related to their having a peculiar surface constituent. In turn, the habitat may have specific receptors or other structural features that allow for colonization.

Barriers to colonization

No species will gain entry to every site to which it is transported. The reason is that properties of the environment exclude many arrivals. These properties are often called **barriers to colonization,** that is, conditions restricting population size or activity or preventing the establishment of an arriving species. The barriers are frequently **mechanical.** Mechanical barriers include the skin, mucous membranes, the cuticle of fruits, the bark of trees, and the gum that is exuded by certain plants when they are wounded (Table 18-3). Breaching of these barriers permits colonization, as when the human skin is damaged or plants are bruised.

A particular species may fail to become established because of a chemical barrier in a given environment. Acidity is a common chemical barrier (Figure 18-6) and probably explains the failure of many species to become established in the stomach or in acid foods, such as citrus and tomato products. Organic acids themselves, in addition to the low pH associated with them, are also major barriers to the establishment of microorganisms; witness the long life of vinegar that contains acetic acid or of pickles and sauerkraut, products that contain lactic acid. The inability of many bacteria to colonize the surface of the eyes or the mucosa of the nose has been related to the presence of an enzyme, lysozyme, that lyses many of the potential colonists. The defense mechanisms of humans, animals, and plants also are often chemical barriers, as in the cases of antibodies and the antimicrobial agents produced by plants.

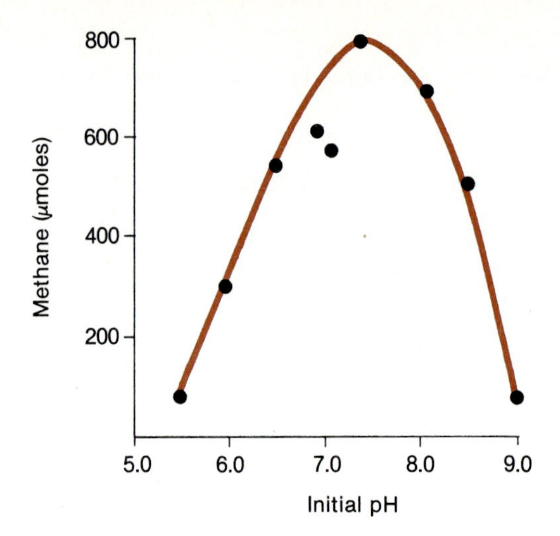

Figure 18-6 *Effect of pH on methane production in thermophilic methanogen cultivated in laboratory.*

Composition of the community

As the pioneers are displaced and new species arise in succession, the community is altered. Some organisms are totally eliminated, and others become relatively rare. At the same time, new organisms come to be dominant. The changes in community composition as a result of succession may arise because the early arrivals produce compounds that inhibit themselves, they produce nutrients useful to the late arrivals, or they expose organic material that was not previously accessible. The composition of the community may also be altered because of modifications not brought about by the indigenous populations, for example, changes in light intensity and temperature with the seasons of the year or the feeding on microorganisms in waters by small animals. In some cases, the pioneers or other species early in succession reduce the supply of inorganic or organic nutrients, and the new arrival is better able to grow at the new nutrient concentrations than are the earlier populations.

Regardless of whether a species is evident during initial colonization, late in succession, or in the finally established community, it has certain features that dictate at which stage it appears. For some organisms, the unique trait in colonization, succession, or community establishment is their rapid growth. For other species, it is their capacity to tolerate nonbiological stresses or, for pathogens to overcome barriers erected by the animal or

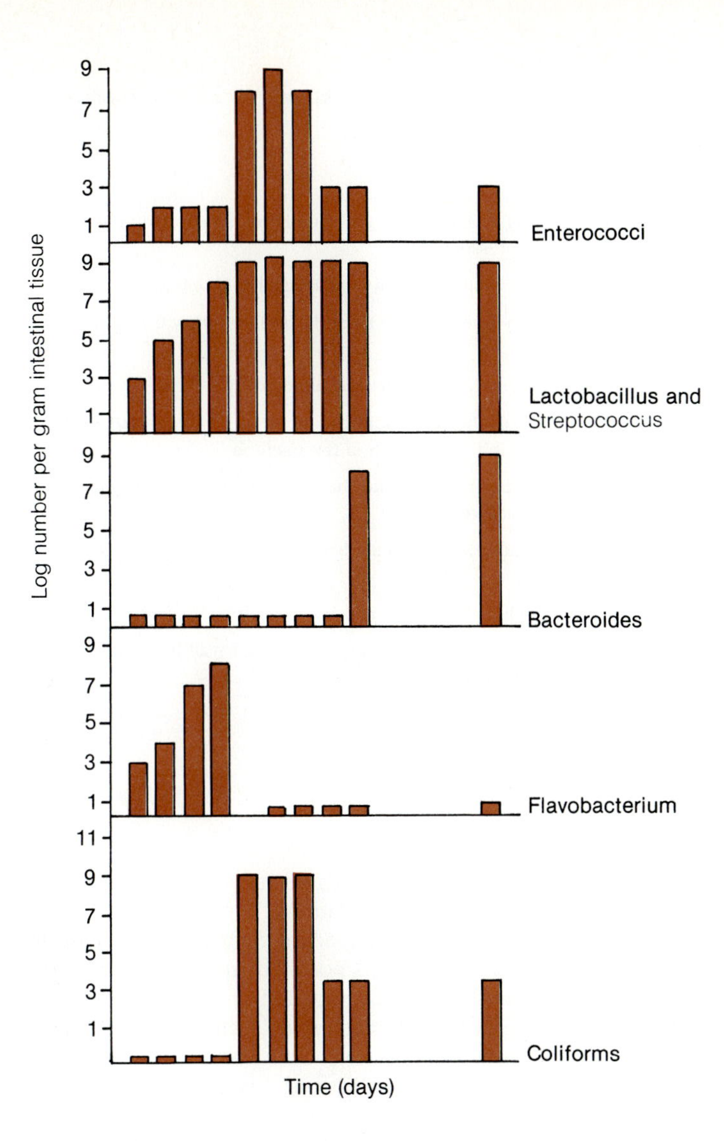

Figure 18-7 *Succession of microbial populations in large intestine of mice following birth. No samples collected after 18 days. Initial colonizers are lactobacilli, enterococci, and flavobacteria. This is followed by facultative anaerobes such as coliforms. Finally, the anaerobic colonizers,* Bacteroides, *predominate.*

plant host. For a population growing on rocks in a stream, its particular position in the colonization-succession sequence may result from its capacity to produce specialized structures allowing it to adhere to the surfaces of the rocks. However, only a few of the distinctive advantages possessed by individual species are presently known.

In most habitats, wave after wave of populations appear and then recede (Figure 18-7). These waves may succeed one another rapidly, or they may proceed slowly. Ultimately, a community is established that is relatively stable. In this community, known as the **climax community**, the dominant organisms and the relative abundance of the species are reasonably constant. The species are in physiological and nutritional equilibrium with their environment, and these are the species that have been selected out among the numerous arrivals for their ability to grow and tolerate all of the biological and nonbiological stresses to which that environment is exposed. To maintain the climax community, there must be a continual inflow of energy. That energy may come from the organic carbon formed by algae in photosynthesis, the excretions of plant roots, food ingested by animals or the organic matter introduced into streams and rivers. It is because of the known composition of many climax communities that microbiologists know where to look for a particular organism they wish to isolate. In the next section we will see how the members of a community interact with each other.

INTERACTION AMONG POPULATIONS

The microorganisms within a community, whether it be the intestinal tract, soil, or water, engage in various types

Table 18-4 *Examples of compounds toxic to one species but removed from the environment by a second species*

Compound	Details of interrelationship
Concentrated sugar solutions	Osmophilic yeasts metabolize the sugar and thereby reduce the osmolarity, so allowing the growth of species that are sensitive to high osmotic pressure.
Oxygen	Aerobic organisms may reduce the oxygen tension, thus allowing anaerobes to grow.
Hydrogen sulfide	Toxic hydrogen sulfide is oxidized by photosynthetic sulfur bacteria, and the growth of other species is possible.
Food preservatives	The growth inhibitors benzoate and sulfur dioxide are destroyed biologically.
Lactic acid	The fungus *Geotrichum candidum* metabolizes the lactic acid produced by *Streptococcus lactis;* the acid would otherwise accumulate and inhibit the growth of the bacteria.
Mercury-containing germicides	*Desulfovibrio* species form hydrogen sulfide from sulfate, and the sulfide combines with mercury-containing germicides and permits bacterial growth.
Antibiotics	Enzymes are produced by some species of bacteria that break down antibiotics; thus the growth of antibiotic-sensitive species is allowed.
Phenols	Some bacteria can oxidize phenols and thereby allow other species to grow.
Trichlorophenol	A number of gram-negative bacteria can absorb trichlorophenol in their cell wall lipids and thereby protect *Staphylococcus aureus* from its action.

of relationships with other microorganisms or with higher organisms. Some of these associations, such as the lichen and rumen symbiosis, were discussed previously (Chapters 6 and 14 respectively). The following discussion will center on the beneficial and antagonistic effects that develop in the microenvironment, particularly the soil.

Beneficial interactions

Three types of beneficial interactions can be distinguished: **commensalism**, **protocooperation**, and **mutualism**.

Commensalism. In commensalism, one of the two interacting species benefits. The second species is not helped or harmed by its associate. Many examples of commensalism have been described. For example, a facultative aerobe will remove oxygen to permit the proliferation of the strict anaerobe. This is not harmful to the first population, and it is beneficial to the second. Such an association may account for the finding of strict anaerobes in aerated soils (Table 18-4). Another example involves the decomposition of an organic compound by one population that converts that substrate to a product used by a second population. If the latter is unable to use the original substrate, the relationship is commensalism. Thus, when cellulose is added to soil, most of the many species appearing are unable to use the added polysaccharide but rather use the products of the cellulose-degrading populations.

The addition of a degradable organic compound to soil contaminated with a poorly degraded organic compound stimulates the degradation of the latter in a nonspecific way. This phenomenon is called **cometabolism** and the organisms do not obtain energy from the reaction or use carbon for biosynthesis. Cometabolism may yield products that are more toxic than the compounds from which they are derived. Some of the products, however, may be available for degradation by other microbial populations and thus may generate a commensalistic interaction. Cometabolism has been shown to be important in the degradation of DDT and other insecticides.

Protocooperation. In protocooperation, both of the interacting species benefit, but the association is transitory and not obligate for the growth of the two species. In protocooperation, there are no fixed partners, but any population having the necessary physiological properties can interact with any other population for which the association is appropriate. A simple example involves a nitrogen-fixing bacterium that lives together with a green alga in environments that are poor in nitrogen and organic carbon. The former organism gets its nitrogen from dinitrogen, and the latter obtains its carbon from carbon dioxide. If the first organism excretes a nitrogen compound that is useful to the second and

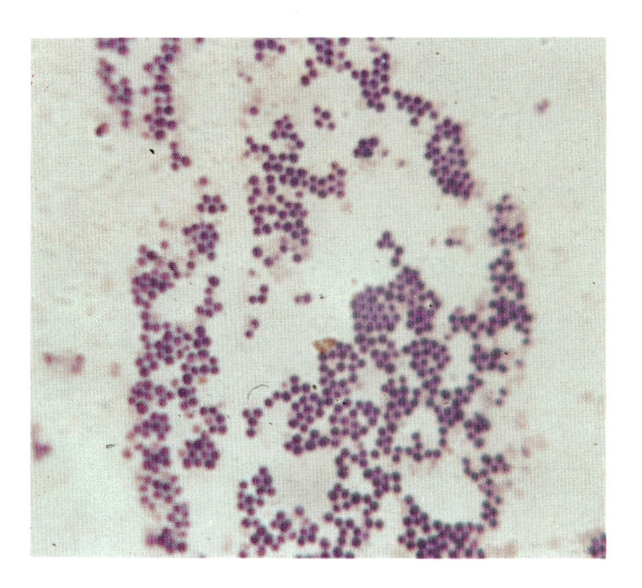

Plate 1 *Gram stain of gram-positive staphylococci. (×4308.)*

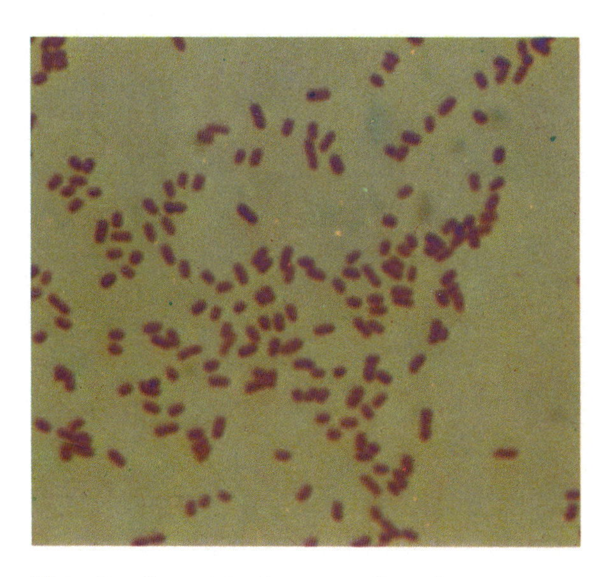

Plate 2 *Gram stain of gram-negative rod* Escherichia coli. *(×4342.)*

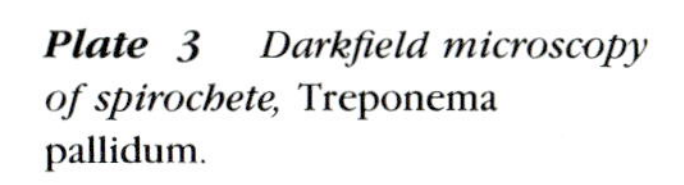

Plate 3 *Darkfield microscopy of spirochete,* Treponema pallidum.

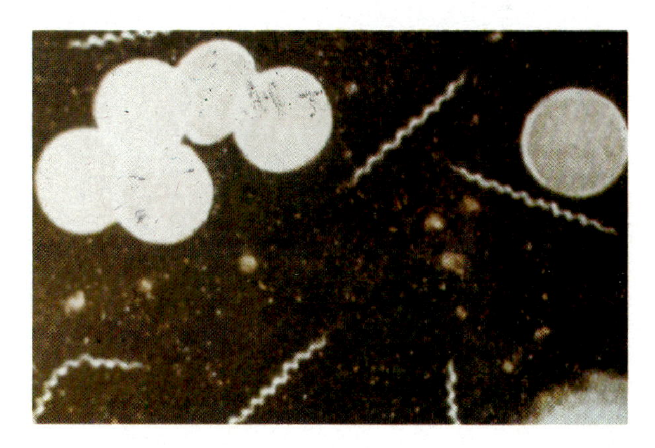

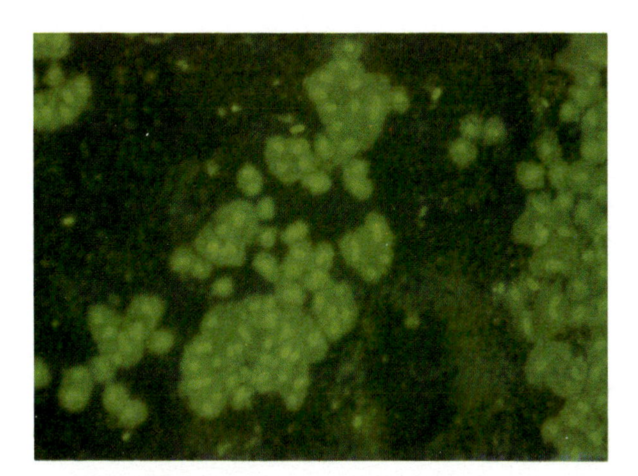

Plate 4 *Capsule stain.* Pasteurella haemolytica *stained with fluorescein isothiocyanate–conjugated antiserum. (×500.) Wide capsules can be observed surrounding cell.*

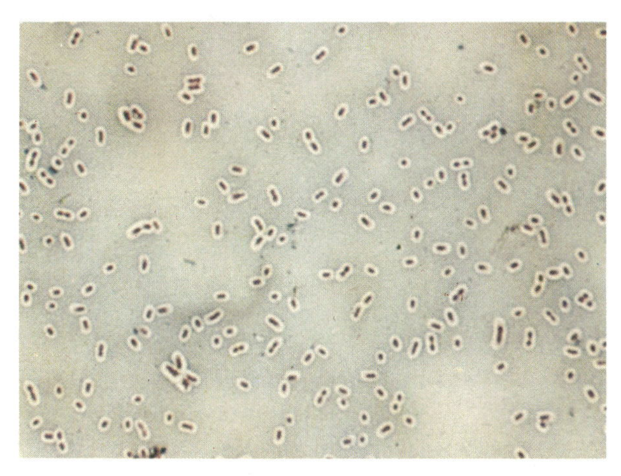

Plate 5 Pasteurella haemolytica *stained with Congo red and counterstained with Maneval stain. Cells appear red surrounded by clear capsular zone.*

Plate 6 *Fungal colony demonstrating fuzzy appearance and presence of aerial spores.*

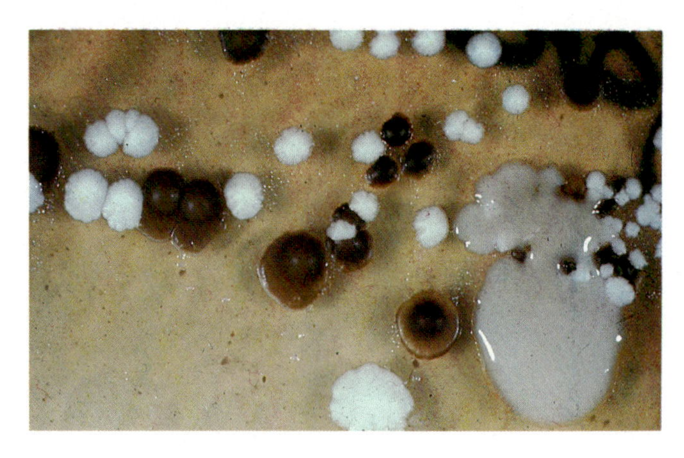

Plate 7 *Fungal colonies on agar demonstrating mucoid appearance and absence of aerial spores.*

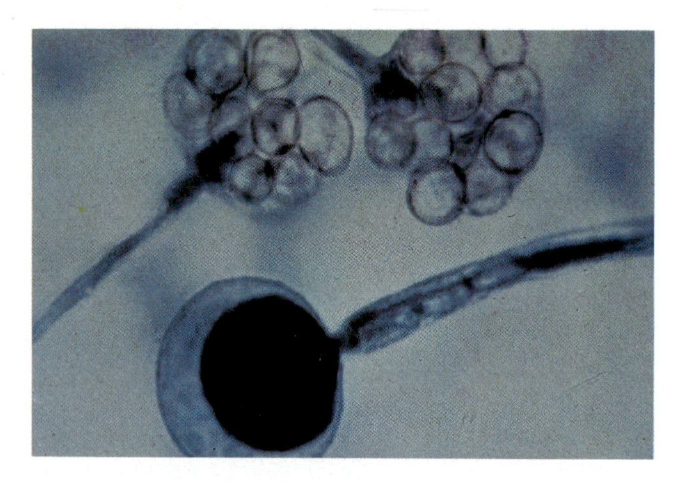

Plate 8 *Species of* Mucor *in which spores can be observed inside sporangium.*

Plate 9 *Peach leaf curl in apricot leaf caused by species of Taphrina.*

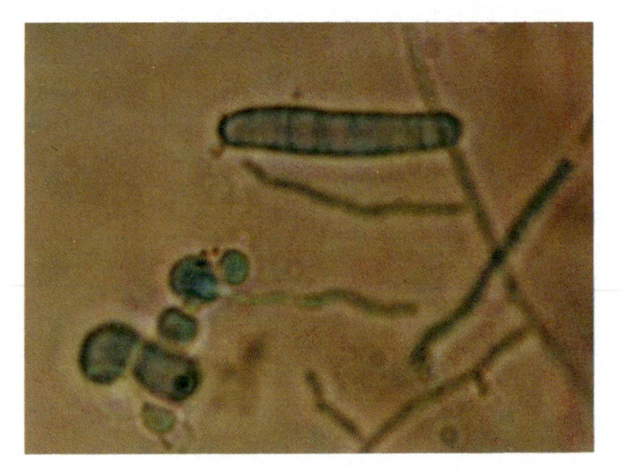

Plate 10 *Preparation of* Trichophyton rubrum, *a cause of ringworm. Microconidia appear tear shaped, whereas macroconidia are long and slender. (×688.)*

Plate 11 *Poisonous mushroom,* Amanita muscaria.

Plate 12 *Poisonous mushroom,* Amanita verna, *also called Angel of Death.*

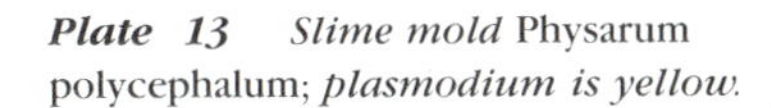

Plate 13 *Slime mold* Physarum polycephalum; *plasmodium is yellow.*

Plate 14 *Slime mold* Ceratiomyxa fruticulosa *showing white fruiting body.*

Plate 15 *Slime mold* Hemitrichia stipitata *showing stalked red fruiting bodies on surface of log.*

Plate 16 *Algal bloom in cyprus swamp.*

Plate 17 *Massive amounts of algae scooped from waterway that has received excessive nutrients.*

Plate 18 Cora pavonia *and* Parmelia enderythrea *lichen.*

Plate 19 Herpothallon sanguineum *lichen growing on tree trunk in Amazon Valley.*

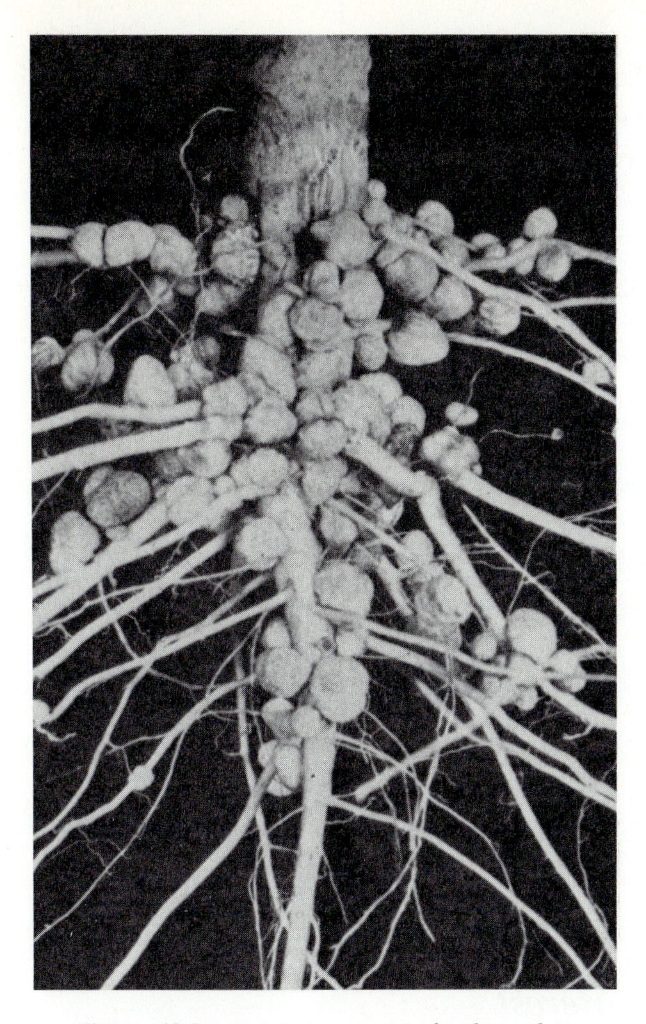

Figure 18-8 *Nodulation on root of soybean plant.*

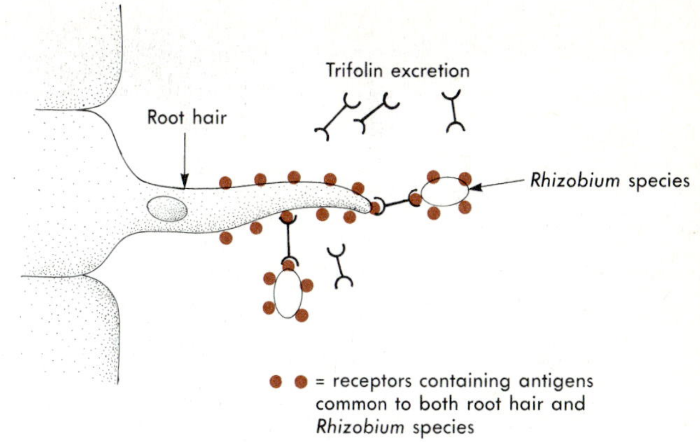

Figure 18-9 *Model for recognition between clover root hair and specific* Rhizobium *species. Root secretes a glycoprotein called trifoliin A. Trifoliin serves as a bridge that binds* Rhizobium *to the surface of root. Receptors on both root surface and bacterium are recognized by trifoliin A.*

the second releases an organic compound that is used for growth of the first population, the two grow well together. Another well-known example involves a species that converts a complex molecule to a product toxic to itself. The toxin producer cannot grow for long on the complex molecule because of the inhibitors it generates, but a second species may live in protocooperation if it cannot use the complex molecule but uses the toxin for growth. In this way, the second species grows and benefits the first, and the first is helpful to the second by giving it a waste product that is also a nutrient.

Protocooperation is a property used in the commercial production of yogurt, a dairy product made by the mixed fermentation of milk using *Streptococcus thermophilus* and *Lactobacillus bulgaricus.* Formic acid, which is a product of lactate metabolism by *S. thermophilus,* is a growth stimulant for *L. bulgaricus. L. bulgaricus* growth on the protein casein results in the release of amino acids, which in turn stimulate growth of *S. thermophilus.*

Mutualism. Mutualism, which is beneficial to both partners, is important in soil and is required either for carrying out some reaction or for the existence of the organisms in significant numbers in nature. In contrast with protocooperation, the partners in mutualism are fixed. Possibly the most important association in terrestrial environments is the one involving *Rhizobium* and *Bradyrhizobium* and leguminous plants. Although legumes may exist apart from these bacteria and the latter may persist in soil in the absence of the plant, dinitrogen is assimilated by neither of the organisms in nature when apart from their partners. It has recently been discovered that the bacteria may fix dinitrogen in culture, and they can use the gaseous form of the element as a nutrient under these conditions, but significant dinitrogen assimilation in nature appears to require that the two organisms live together. The seat of the symbiosis is the nodule, which is a swelling on the root of the host plant (Figure 18-8). The bacterium invades the root, causes root tissue to proliferate, and then begins to assimilate dinitrogen when protected in the nodule microenvironment (see Chapter 13). The interaction between the bacterium and the root is believed to be mediated by chemoattractants. In the association between *Rhizobium* spp. and the roots of white clover, for example, a glycoprotein called trifoliin A is secreted by the plant. Trifoliin A recognizes receptors on the surface of the root hair as well as the receptors on the surface of the *Rhizobium* species (Figure 18-9). Trifoliin A therefore serves as a bridge binding the bacterium to the root surface.

A dinitrogen-fixing mutualism involving nodules on

Table 18-5 Principal economically important nitrogen-fixing systems

Plant	Symbiont	Principal uses
Food and grain legumes (peas, beans, soybeans, chick peas, peanuts)	*Rhizobium* and *Bradyrhizobium* species	Human and animal food and food products (e.g., soya protein)
Forage legumes (alfalfa, clover)	*Rhizobium* species	Livestock fodder, green manure
Sward legumes (clover, subclover, vetch)	*Rhizobium* species	Upgrading grazing land
Leguminous shrubs and flowering plants (lupines, sweet peas)	*Rhizobium* species	Horticulture, landscaping
Azolla (water fern)	*Anabaena*	Green manure
Alder	*Frankia*	Reforestation, interplanting

roots is also prominent among certain nonleguminous plants (Table 18-5 and Point of Interest). In these instances, the microorganism is an actinomycete of the genus *Frankia*. The microorganism is not known to proliferate in soil, or to assimilate dinitrogen, in the absence of the higher plant, and the nonlegumes similarly are unable to make use of dinitrogen in the absence of the nodules engendered by the actinomycetes. The effects of both of these associations, that involve *Rhizobium* and *Frankia,* are evident in the development of plants bearing nodules in nitrogen-poor fields and the consequent increase in the nitrogen level of the ecosystem.

Point of Interest

NITROGEN FIXATION IN NONLEGUMINOUS PLANTS

Nitrogen fixation in leguminous plants is being heavily researched in industrialized and developing countries in hopes of finding and/or developing new strains of *Rhizobium*. The *Rhizobium*-legume symbiosis is the most practical and economical system for producing protein-rich food crops. It is estimated, however, that *Rhizobium*-legume symbiosis is responsible for only 30% of the total nitrogen fixed annually. Nitrogen fixation by nonlegumes is, therefore, on a quantitative basis, a more important process. Nitrogen fixation is particularly important in the association that exists between nonleguminous trees and bushes and certain bacteria. These plants (such as alder or elm trees) possess root nodules infected with the actinomycete *Frankia* (the plants involved in this association are referred to as **actinorhizal**). *Frankia* is a filamentous bacterium that has been isolated and cultivated in the free-living state. The large number of vesicles produced in the bacterium during nitrogen fixation are believed to be the sites of this activity. The vesicles, therefore, are equivalent to the heterocysts of cyanobacteria (see Chapter 13). Hyphae invade the root cortical cells and this leads to nodule formation. Actinorhizal plants are important because they can colonize poor soils, contribute to soil regeneration and fertilization, and play a role in preventing soil erosion. The alder, for example, is found in temperate zones of the northern hemisphere and contributes a significant amount of nitrogen to the soil, which in turn stimulates the growth of other trees.

Nitrogen-fixing cyanobacteria such as species of *Anabaena* are found in association with the water fern *Azolla*. Six species of *Azolla* are known and widely distributed. They develop at the surface of ponds, canals, and rice paddies. The symbiont *Anabaena* can be isolated from the plant, but it cannot be grown in the free-living state. During vegetative growth of the plant, the proportion of heterocysts (sites of nitrogen fixation) increase by 30%. Ammonia produced by the symbiont is assimilated by glutamine synthetase of the host. *Azolla* is routinely inoculated into rice paddies as a source of green manure and is estimated to increase rice productivity by 20% to 40%.

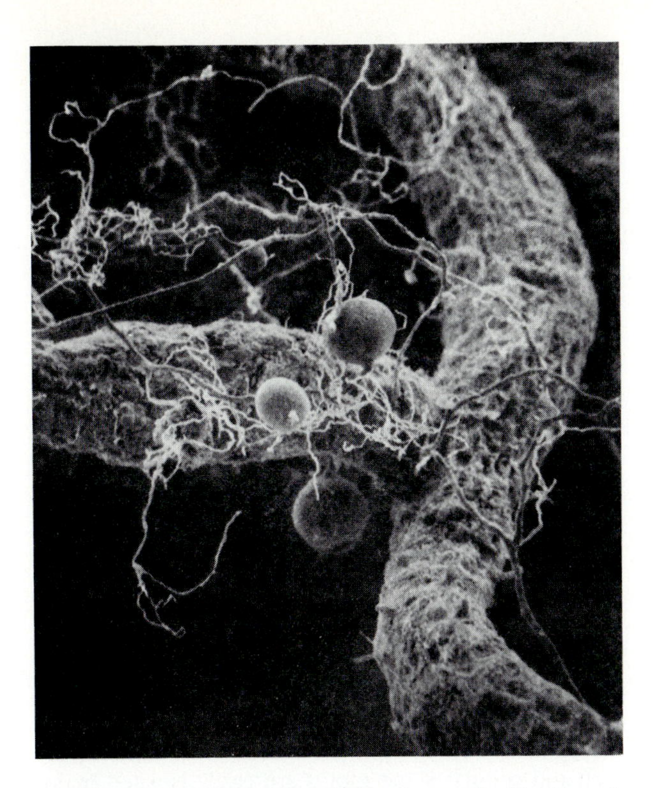

Figure 18-10 *Mycorrhizae. Light micrograph of portion of soybean root infected by fungus* Glomus mosseae. *Many filaments are seen to cover root surface. Spherical objects are chlamydospores of fungus.*

Mycorrhizae are associations of fungi with plant roots (Figure 18-10). The microorganism is known as the mycorrhizal fungus. This association is extremely common, and it has been estimated that 80% of the species of higher plants bear these root structures. As a rule, mycorrhizal fungi are poor competitors and are difficult to find in nature. For active and significant development, they must be in association with the root. Hence, the microorganism is benefited. Considerable evidence exists to show that the higher plant is benefited as well. For example, pine or spruce trees grow well in fertile but poorly in nonfertile areas. However, when the roots of these conifers are infected by appropriate fungi and mycorrhizae appear, the plants are able to grow at low nutrient levels. Particular attention has been given to the usefulness of mycorrhizal fungi in providing phosphorus to plants that are growing in soils that contain phosphorus in forms largely unusable by the fungus-free macrosymbiont.

Several other types of mutualistic associations may be prominent in specific areas. Lichens containing one of several fungal groups and either a green alga or a cyanobacterium are widespread in deserts, on mountainsides in cooled volcanic outflows, and in other areas that have little organic matter and low photosynthetic activity (Plates 18 and 19). Because of their nitrogen-fixing activity, an aquatic fern *(Azolla)* together with the cyanobacterium that the fern contains are prominent in paddy fields in Asia that are planted with rice. Another association that may be significant involves protozoa with their intracellular bacteria. The bacteria are located in the cytoplasm of the protozoan and do not exist apart from the animal. In this way, the bacterium finds its home. The protozoa appear to benefit: they will not grow in laboratory cultures if the bacteria are removed by means of antibacterial agents.

Antagonisms

Antagonisms are easily demonstrated. Thus, many bacteria multiply rapidly when introduced into samples of soil that have been sterilized. On the other hand, the same organisms fail to grow and may not even survive when added to nonsterile soils. Similarly, a pathogenic bacterium or fungus has a devastating effect on higher plants when the two are introduced into samples of sterilized soils, but the same pathogens have little or no effect when introduced together with their host plants in nonsterile samples of the same environment. The antagonism may result from **competition**, **amensalism**, **parasitism**, or **predation**.

Competition. Competition is a rivalry between two species for a limiting factor in the environment. The less active species is suppressed, but the more active competitor is affected to a slight degree. Competition occurs when the demand for an environmental factor exceeds the supply. The major factors for which microorganisms may compete are **carbon source**, **space**, and **oxygen**. Most attention has been given to the competition for carbon because it is so often in limited supply. Although competition for space and oxygen may also be common, little information exists on their significance.

Many species introduced into soil probably are eliminated because they do not compete for the small supply of available carbon. An organism does not die from competition, but rather it is eliminated because of starvation. If other populations residing in the habitat are able to metabolize and use the supply of available carbon before an introduced species can, then the latter will not become a significant member of the community or it will soon perish. It is not certain what makes a microorganism a good competitor, but rapid growth rate, use of a broad range of substrates, and the capacity to move from a depleted supply of organic matter to a site where there are still nutrients are likely to be important in competition (Figure 18-11).

Amensalism. In amensalism, one species is suppressed by the products of another. The toxic product may be a high hydrogen ion concentration, an organic acid that is toxic even apart from its acidity, carbon dioxide, or hydrogen sulfide. Surprisingly, carbon dioxide is ecologically important in some areas because many species are sensitive to the high levels of carbon dioxide that are common in soil. In addition, antibiotics represent another mechanism of amensalism. When tested under laboratory conditions, actinomycetes, fungi, and bacteria of many genera are able to produce antibiotic compounds. Yet, although most antibiotics of practical importance are derived from soil organisms, especially the actinomycetes, there still is not agreement on the significance of **antibiosis** in soil. The arguments against antibiotics being significant include the following: (1) antibiotic producers, especially the actinomycetes, often do not dominate the community and (2) many antibiotic-sensitive species are present and are often quite numerous in soil. Conversely, toxic substances have been obtained from soils, and these may be similar or identical to antibiotics of importance in chemotherapy. Nevertheless, no evidence exists that the toxins that have been obtained from soil that inhibit some fungi and bacteria are similar or identical to antibiotics that are used in medicine. Instead, those compounds that have been characterized as soil toxicants appear to be compounds like ammonia, ethylene, ethyl alcohol, certain simple fatty acids, and a number of high-molecular-weight organic compounds.

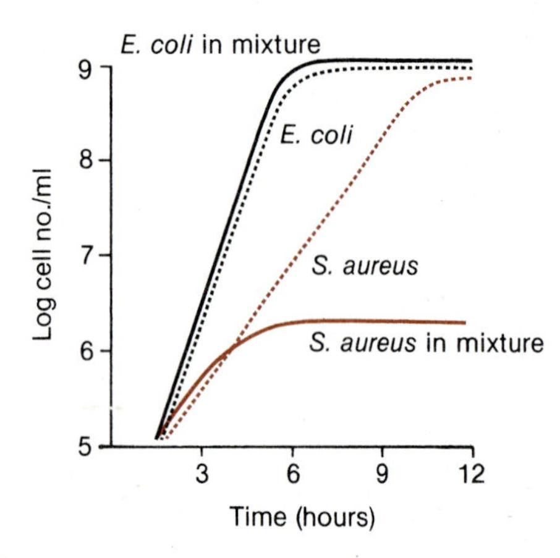

Figure 18-11 *Competition between* Escherichia coli *and* Staphylococcus aureus *when species were grown either alone or as mixture in batch cultures.*

Parasitism or predation. Some of the predatory and parasitic organisms that are found in soil are given in Table 18-6. Viruses can be obtained from soil, but evidence for their significance in community activities is not rigorous. Protozoa, myxobacteria (see Point of Interest), and cellular slime molds are easily isolated by in-

Table 18-6 *Predatory and parasitic relationships*

Feeder	Susceptible host	Mechanism	Source
Virus	Bacteria including actinomycetes, cyanobacteria	Intracellular parasitism	Soil
Protozoa	Bacteria, protozoa	Predation	Soil, water
Myxobacteria and cellular slime molds (Acrasiomycetes)	Bacteria	Lysis and predation	Soil
Fungi	Fungi	Hyphal penetration	Soil
Predaceous fungi	Nematodes and amebas	Hyphal appendages	Soil
Actinomycetes	Fungi	Antibiotic-induced lysis	Soil
Actinomycetes and other bacteria, fungi	Fungi	Enzymatic lysis	Soil
Actinomycetes and other bacteria	Bacteria	Enzymatic lysis	Soil
Bacteria, fungi	Algae	Parasitism	Soil
Bdellovibrio (bacterium)	Bacteria (mostly gram-negative)	Cell penetration (periplasm)	Soil, water
Vampirococcus (bacterium)	Bacteria (phototrophs such as *Chromatium)*	Attaches to cell wall	Fresh water
Daptobacter (bacterium)	Bacteria (phototrophs such as genera of Chromatiaceae)	Cell penetration (cytoplasm)	Fresh water

Point of Interest

THE MYXOBACTERIA

The myxobacteria are found in soil, in decaying vegetation, on the bark of trees, and in animal dung. They are among the several genera referred to as gliders. Gliding allows movement within dense substrates such as soil or rotting wood. Many species appear as brightly colored orange, red, or yellow colonies. They degrade and feed on macromolecules by releasing hydrolytic enzymes and have the capacity to degrade viable as well as dead bacterial cells. This bacteriolytic activity is possible because the extracellular enzymes released from the myxobacteria degrade the cell wall. Other extracellular enzymes—proteases, nucleases, and lipases—are also important for the hydrolysis of proteins, nucleic acids, and lipids, respectively, which are released by lysed bacterial cells.

The myxobacteria are an interesting group not only because of their gliding ability and predatory properties but also because of their differentiation process. During the developmental cycle, individual rod-shaped cells aggregate and form multicellular **fruiting bodies** (figure below), which in some species are very simple. Other species produce a stalk, which arises from an aggregated mass of cells, with an apical bulb (figure at right). Individual vegetative cells of myxobacteria remain dispersed in a liquid nutrient-sufficient medium, but when nutrients become depleted on an agar medium, dispersed cells begin to aggregate and switch from vegetative growth to fruiting body formation. In response to starvation, individual cells in the fruiting bodies of some species enter into a dormant state and are called **myxospores.** The vegetative cells in other species become encased in a protective structure called a **cyst.** Under proper environmental and nutritional conditions, either the myxospores are released to germinate or encysted vegetative cells are released to reproduce immediately. Vegetative cells produced from the fruiting body form swarms, which enter into a feeding process.

The differentiation process, which involves single cells as well as fruiting bodies, suggests that a chemical signal (chemotactic response) initiates cell-to-cell interaction. This process is believed to be similar to one that occurs in the slime molds (see Chapter 6).

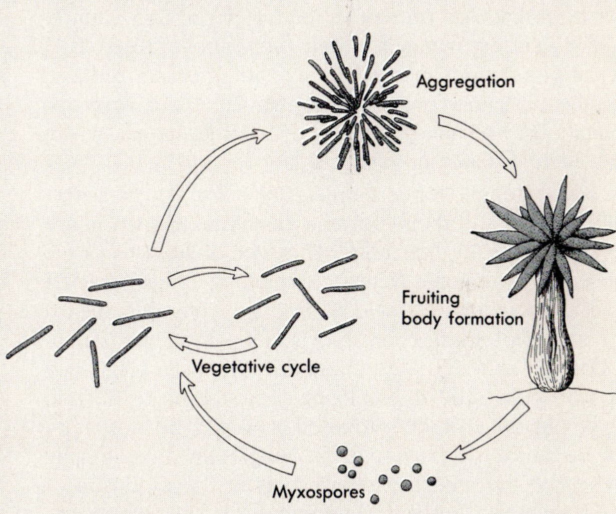

Developmental cycle of myxobacteria. Individual vegetative cells under adverse conditions, such as nutrient depletion, aggregate to form mass of cells that develop into fruiting body with stalk and simple or complex bulb. Many of these fruiting body cells develop spores that, under better environmental conditions, will germinate to produce vegetative cells.

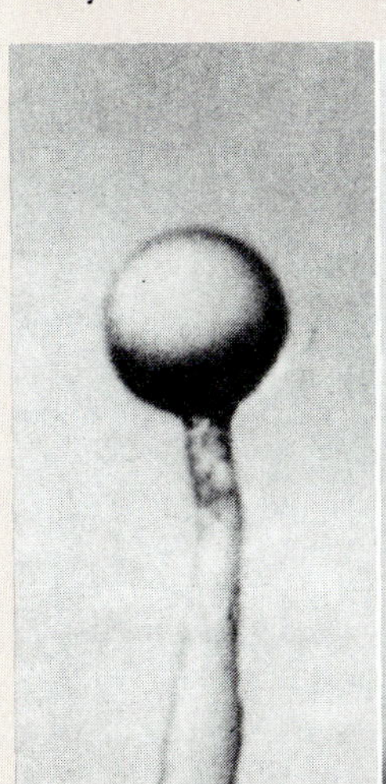

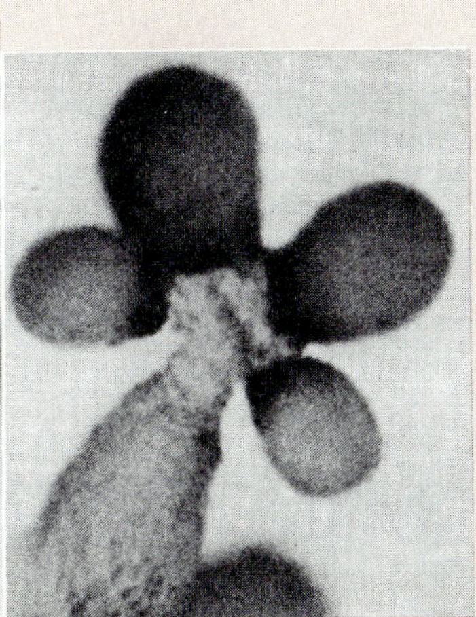

Multicellular fruiting bodies of members of myxobacteria. ***A,*** *Myxococcus.* ***B,*** Stigmatella aurantiaca. ***C,*** Chondromyces apiculatus.

oculating crumbs or dilutions of soil on an agar medium containing bacterial cells as the sole nutrient. It is believed that these predators obligately rely on predation in nature, and hence they must influence bacterial numbers or activities.

Protozoa are the major microbial predators in water, where they feed on bacteria. This predatory characteristic leads to increased conversion of organic nitrogen to inorganic forms of nitrogen and occurs in the following way. Algae that obtain their energy from the sun release carbon that becomes part of the dissolved organic matter pool. Bacteria present in the water can utilize nitrogen-containing matter for their biosynthetic needs. Bacterial populations increase in size but are soon reduced by the predatory protozoa such as flagellates. Protozoa metabolism results in the excretion of ammonia, which can then be taken up by the algae.

Some fungi (for example, *Rhizoctonia*) parasitize other fungi, and their filaments penetrate and do injury to the hyphae of their hosts. This type of attack is rarely observed in nature. Nematode-trapping fungi attracted considerable interest at one time as a possible means for the biological control of plant-pathogenic nematodes, but although these fungi are evident in soils, they do not seem to be able to bring about the major decline in populations of nematodes that would be necessary for practical disease control. Fungi may occasionally penetrate the ameba and bring about the digestion of the protozoan, but this phenomenon is only rarely observed in soils. The production by one species of an antibiotic affecting a second may lead to the lysis, or digestion, of cells or filaments of a second population, but the ecological significance of this type of interaction depends on whether antibiotics are produced in soils; as pointed out above, this issue is still not resolved.

On the other hand, enzymatic lysis is prominent. The organisms that are lysed include fungi and bacteria. Because of their large sizes, lysis of fungi, sometimes termed **mycolysis**, is readily detected by microscopic means. The resting structures of some fungi endure for long periods of time, and these are obviously, therefore, not readily prone to lysis. However, once the hyphae emerge from the resistant structures, enzymatic lysis may take place rapidly. This leads to a decline in the activity of the fungi. The hypha that is susceptible to lysis must be able to develop on some organic material and regenerate the resting structures or penetrate plant cells in order for the lysis-susceptible species to endure.

Algae are also prone to lysis by other microorganisms, and they also may be susceptible to parasitism by bacteria and fungi. Although such lysis and parasitism have been characterized in culture and, to some extent, in aquatic environments, their role in soil is still obscure.

One of the best known examples of a predatory bacterium is *Bdellovibrio,* a common inhabitant of soils and water. This gram-negative organism possesses a polar flagellum and can move at the rate of 100 cell lengths per second, compared to 10 cell lengths per second for its prey, *E. coli.* The predator attaches to the cell surface and after about 10 minutes moves through the outer membrane into the periplasmic space (Figure 18-12) where growth of the predator will occur. The flagellum of the bacterium is detached and remains outside the host cell. Penetration through the layers of the peptidoglycan by glucanases, for example, results in the loss of diaminopimelic acid, and this reduces the cross linking of the peptidoglycan. This alteration is responsible for the conversion of a rod-shaped cell to a spherical one called a bdelloplast. When *Bdellovibrio* attaches to the surface of its prey, the cytoplasmic membrane of the host is damaged, resulting in the flow of metabolites from the cytoplasm into the periplasmic space. *Bdellovibrio* produces several enzymes that can degrade host polymers into their monomeric units, and these are assimilated directly into cell substance. *Bdellovibrio* grows as a filament as long as the food supply lasts. Prior to lysis of the invader's growth chamber, there is resynthesis of flagella and the filament fragments into unit cell lengths, each with a molecule of DNA. The ability of *Bdellovibrio* to assimilate host monomers from degraded proteins, polysaccharides, nucleic acids, and lipids reduces energy expenditures.

Two other gram-negative predatory bacteria, *Vampirococcus* and *Daptobacter,* differ from *Bdellovibrio* in their mechanism of host destruction. *Vampirococcus* attaches to the cell wall of its host, where it reproduces and apparently sucks out the cytoplasmic contents of its victim. *Daptobacter* penetrates the cytoplasm of its victims, where it divides and devours the cytoplasmic contents.

MICROORGANISMS AND BIOGEOCHEMISTRY

Biogeochemistry is concerned with biological activities that affect the chemistry of the earth. Microorganisms are extremely important in this regard because processes they bring about alter the chemistry of the atmosphere, oceanic and inland waters, marine and freshwater sediments, and soils. In some of the geochemical processes in which they are implicated, microorganisms bring about changes that are essential to life on earth. Other reactions are not critical for the maintenance of life but alter the chemistry of the earth.

Microbial activities

Biogeochemical cycles. Particular attention has been given to the role of microbial communities in biogeochemical cycles. In some cycles, the elements are circulated from inorganic to protoplasmic and thus organic forms, and then the elements are converted back to the

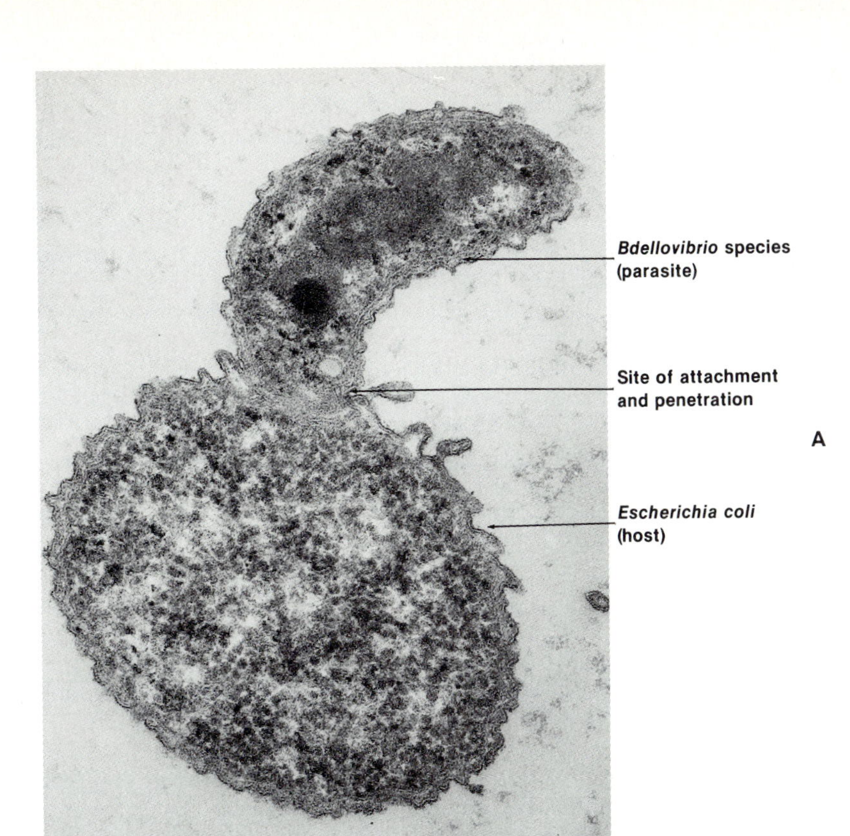

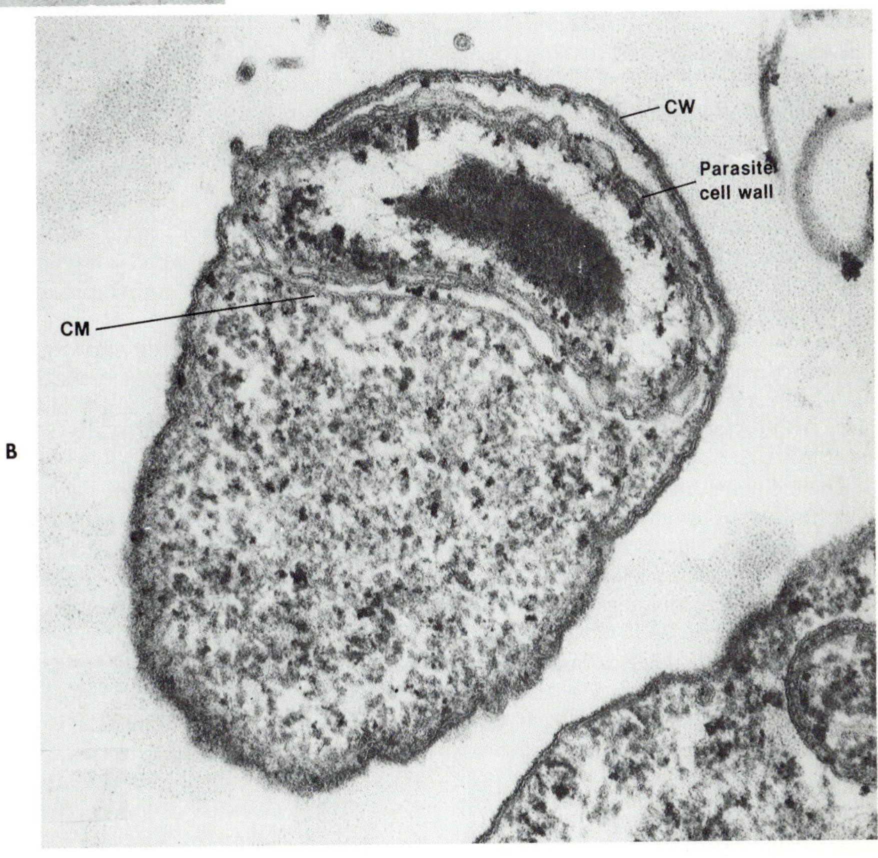

Figure 18-12 *Parasitization of* E. coli *by* Bdellovibrio *species.* ***A,*** *Electron micrograph demonstrating attachment of parasite to host bacterium (×154,000).* ***B,*** *Parasite has not penetrated cytoplasmic membrane (CM) of host but occupies space between it and cell wall (CW) that is periplasm.*

inorganic state. In other cycles, no organic molecules are involved or are significant, but rather microorganisms transform inorganic compounds or ions from one state to another and then back to the original form again. Some of these processes involve an oxidation, and subsequently in the cycle there is a reduction; this is true, for example, of the cycles of nitrogen, sulfur, and iron.

Interest in microbial cycling of the elements has risen dramatically, for several reasons. Microbial ecologists and other environmental scientists are interested in these transformations in order to obtain new information. Also, there is enormous concern with these cycles because they are important, indeed critical, for food production and hence for human life. For example, the rates of certain steps in the nitrogen cycle often limit plant growth and thus, indirectly, the amount of meat that can be produced. In waters, the microbial cycling of phosphorus may limit the development of algae and hence control fish production. Another reason for the interest is to be able to predict the regional or global consequences of new technologies. Some of these technologies are not restricted to small regions and could affect the global cycling of an element that is essential for human health or well-being. It has been suggested that nitrogen-containing pollutants discharged from cities and agriculture or sulfur emitted in the burning of coal and the operation of mines may have catastrophic consequences. This has led many scientists to try to find out whether these pollutants will seriously affect global or regional cycles essential for life.

Another reason to examine the role of microorganisms in biogeochemical cycles is to explain anomalies in the amount of certain pollutants in the atmosphere. Studies designed to explain these anomalies have disclosed that microbial communities are probably major sources of some air pollutants and are active in preventing the accumulation of other pollutants in the atmosphere. This they do by activities not in the air but rather in the underlying soils and oceans. These findings have been made from assessments of the cycling of sulfur, carbon monoxide, and hydrocarbons in the atmosphere.

For elements for which the cycle is complete, there is no long-term upset in the earth's chemistry as a result of the cycling. Thus, many of the steps in the nitrogen cycle result in major chemical changes, but these are balanced by reactions that go in the reverse direction. In contrast, certain elements undergo noncyclic modifications, and these lead to lasting modifications in the distribution of the elements or compounds wherein the element is found. Such modifications take place either on or below the earth's surface, for example, the accumulation of manganese and iron deposits on the ocean floor.

Some of the elements that are transformed microbiologically can be divided into two groups:

1. Elements important in metabolic reactions or cells of microorganisms
   Major: C, H, O, N, P, S, K
   Minor: Fe, Mg, Cu?, Zn?, Co?
   Needed only by some microorganisms: Si, Mn, Ca, Na, Mo, B, V, Cl, Se, Ni
2. Elements not biologically important (in cell structure)
   Field evidence for transformation: Al, As, Hg, Pb
   Only laboratory evidence: U, Bi, Te, Os, Cd, Ag, Au, Sb, Ba

Even though many elements are not biologically important, microbial communities oxidize or reduce elements or form complexes with them that are of geochemical significance.

Mineralization. One of the major microbial processes in geochemistry is mineralization. Mineralization is the conversion of the organic form of an element to the inorganic state. If the element in inorganic form is important as a nutrient (such as ammonium, phosphate, and sulfate), mineralization, by converting the organic complexes to inorganic products, regenerates the nutrients for aquatic or terrestrial plants. Carbon, nitrogen, phosphorus, and sulfur, because they exist in organic complexes in plants, animals, and microorganisms, are thus subject to mineralization. In geochemical terms, mineralization decreases the biochemical complexity of the ecosystem, and it also prevents the accumulation of organic materials in waters and soils. At the same time, the mineralization of carbon provides energy to heterotrophs, and thus carbon mineralization is a driving force for many biogeochemical cycles. Nitrogen mineralization is critical in many soils because it is inorganic nitrogen that is assimilated and used by higher plants. The rate of formation of inorganic nitrogen by microbial mineralization of organic compounds thus ultimately regulates plant growth in natural systems. Similarly, phosphorus mineralization is important in many waters because aquatic algae use phosphorus in the inorganic state, and the phosphorus supply is often less than the algal demand.

Oxidation and reduction. In addition to converting organic to inorganic compounds, microorganisms can alter the oxidation-reduction states of biologically important compounds. Microbial oxidations are also prominent in waters and soils. The elements subject to oxidation include carbon, nitrogen, sulfur, hydrogen, and iron. The energy released in the oxidations may be captured by individual populations. It is, by definition, the heterotroph that is able to capture the energy released by oxidation of carbon compounds. Also by definition, it is the chemoautotrophs that capture energy released by oxidation of other elements. However, the energy re-

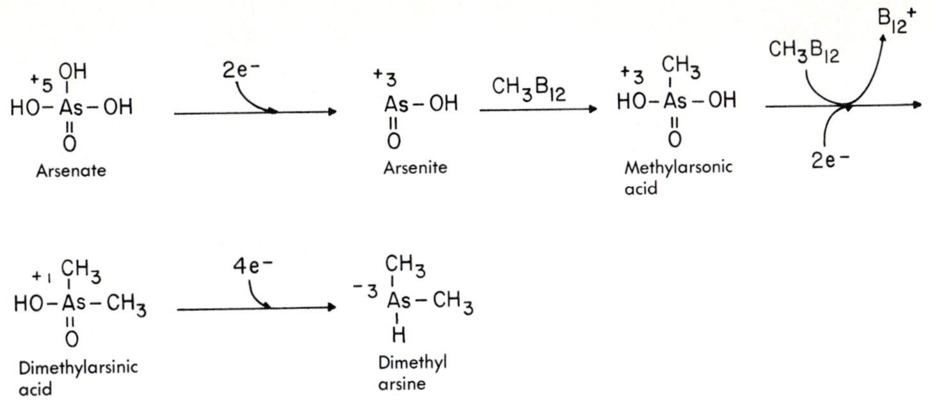

Arsenate ($^{+5}$ $HO{-}As(OH)({=}O){-}OH$) $\xrightarrow{2e^-}$ Arsenite ($^{+3}$ $As{-}OH$, $=O$) $\xrightarrow{CH_3B_{12}}$ Methylarsonic acid ($^{+3}$ $HO{-}As(CH_3)({=}O){-}OH$) $\xrightarrow{CH_3B_{12} \rightarrow B_{12}^+;\ 2e^-}$

Dimethylarsinic acid ($^{+1}$ $HO{-}As(CH_3)({=}O){-}CH_3$) $\xrightarrow{4e^-}$ Dimethyl arsine ($^{-3}$ $H{-}As(CH_3){-}CH_3$)

Figure 18-13 *Methylation of arsenic. Reductive process results in conversion of arsenic from valence of +5 to −3. These reactions carried out using extracts of* Methanobacterium *species under anaerobic conditions in presence of hydrogen gas (H_2) and ATP. Electrons required in reduction process come from oxidation of hydrogen gas, that is, $H_2 \rightarrow 2e^- + 2H^+$. CH_3B_{12} is methylcobalamin—a methylated form of vitamin B_{12}.*

leased in oxidation of some elements may not be captured. In these cases, the energy is lost as heat. In many natural environments, the sulfur and nitrogen in organic or inorganic compounds may be oxidized by microbial activities, yet the responsible populations acquire little or no energy thereby. This contrasts with autotrophs that oxidize inorganic nitrogen or sulfur compounds or ions, getting benefit from the reactions they carry out. With some elements, the oxidation is not the result of an enzymatic process but rather results from a product excreted by a microorganism; this is one way that microorganisms alter iron or manganese ions.

Reductions are also characteristic of microbial activities in natural environments. The reduction may be of an organic substrate, in which case reduced carbon-containing products accumulate. In habitats having oxygen, a consequence of the reducing activity of microbial communities is to convert oxygen to water. This is, of course, the key reaction of aerobic populations. When oxygen is not present but oxidized forms of iron, manganese, sulfur, or nitrate or carbonate ions are, these ions may be reduced. Such reductions are believed to account for significant accumulations of reduced forms of iron and manganese, the appearance of sulfides, the production of dinitrogen and nitrous oxide from soils, and the generation of methane from anaerobic habitats.

Methylation. Microbial methylation may occasionally be ecologically important. In methylation, a methyl group is attached to an element to give methyl derivatives. Depending on the element, one, two, three, or four methyl groups may be attached to mercury, sulfur, tellurium, arsenic, or lead. A few of these methylations may be geochemically prominent; methyl derivatives of sulfur, for example, are important means for the emission from soil of volatile sulfur. Such compounds may be major contributors to the sulfur content of the atmosphere. Great attention has been devoted to methylations that convert an element into a highly toxic form. By adding methyl groups to mercury in sediments, for example, microorganisms create major contamination problems in fresh waters. Methyl forms of arsenic are also highly toxic and are readily produced by microbial action (Figure 18-13).

Carbon cycle

Microbial communities are major participants in the carbon cycle (Figure 18-14). In the oceans, algae concentrate carbon from the atmosphere (where, as carbon dioxide, it is extremely dilute) and convert it to products that can be used by other species. The light energy is thus made available to heterotrophs. In geological terms, the rate of photosynthesis is rapid relative to the amount of carbon available in the atmosphere. Thus, possibly 5% to 10% of the carbon dioxide in the air is converted to organic carbon in algal cells inhabiting marine waters and in higher plants on land each year. To prevent depletion of the atmospheric supply of carbon dioxide, the algal and plant carbon must be returned rapidly to the atmosphere. To explain the almost constant level of carbon dioxide, it is generally assumed that the rate of photosynthesis equals the rate of carbon dioxide formation. The latter, of course, is carbon mineralization. It is widely believed that most of the carbon dioxide that is returned to the atmosphere comes from microbial activity in soils and in waters, although respiration of plants makes a significant contribution to the regeneration of carbon dioxide.

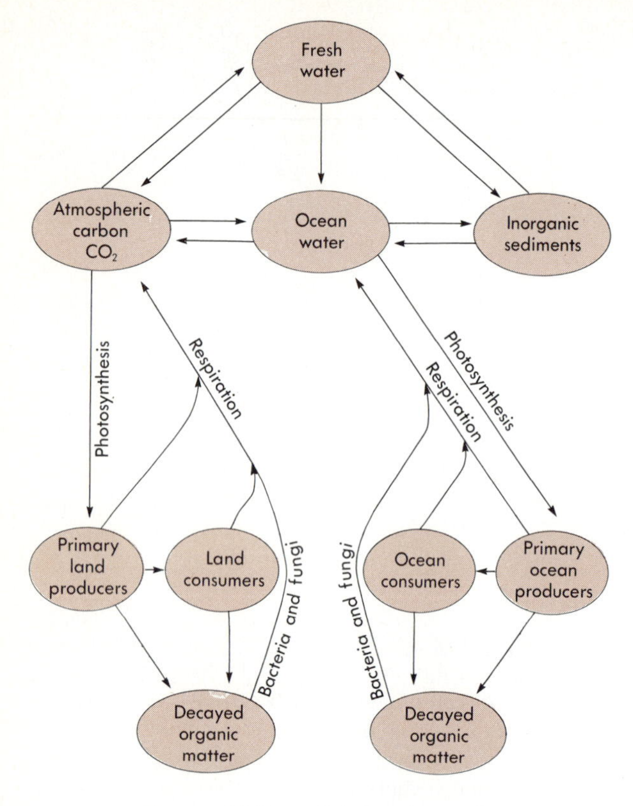

Figure 18-14 *Global carbon cycle.*

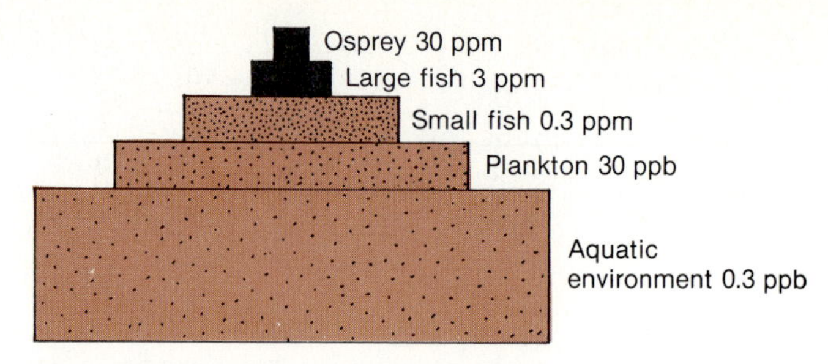

Figure 18-15 *Biomagnification of DDT. Dissolved DDT in aquatic environment moves through plankton and other trophic levels where it is finally concentrated in osprey (fish hawk) tissue. DDT concentrations are idealized and, although typical, do not represent actual analytical data.*

The populations responsible for mineralization are unknown. It is generally assumed that bacteria are particularly important, especially in waters, but that fungi and bacteria dominate in soil. Although the responsible populations have yet to be identified, no doubt exists that the number of different organic compounds that can be mineralized is enormous. Any compound that is biologically formed is mineralizable in environments where oxygen is available. Thus, all of the polysaccharides, lignins, proteins, nucleic acids, and lipids of plants, microorganisms, and animals are transformed, either rapidly or slowly, to the inorganic form of carbon that enters the atmosphere (Table 18-7). Were it not true that every biologically formed organic compound is mineralized, then those resistant compounds would accumulate in nature. Given that the biosphere is more than 1 billion years old, that accumulation would be massive. The only places at the surface of the earth where such accumulations do in fact occur are in the deposits of peat and related organic materials. These in turn may be subjected to nonbiological forces to give rise to coal. The fact that such massive deposits of organic matter exist points to a deficiency of microorganisms, namely, that they do not mineralize rapidly or completely in natural habitats that have little or no oxygen. It is in these anaerobic environments where such accumulations do arise.

Although it is generally accepted that every biologically produced organic compound is mineralizable when oxygen is present, the same is not true for many synthetic organic compounds. Some synthetic organic molecules are resistant to microbial attack, either partially or completely, and thus their release into waters and soils is not a prelude to mineralization or biodegradation. If these resistant organic molecules happen to be toxic to humans, animals, or plants, they are deemed to be pollutants. Such is the case for some of the early insecticides, such as DDT, and it is also true of a variety of pollutants from the chemical industry, such as polychlorinated biphenyls (PCBs), which are no longer allowed to be used in the United States. Although the levels of these pollutants may be relatively small in the aquatic or soil environment, they accumulate in both eukaryotic and prokaryotic cells. By the time the pollutant has reached the higher trophic layers in the food chain, it has increased severalfold, a process called **biomagnification** (Figure 18-15).

Let us use one example in the carbon cycle to show how basic studies of microbial biogeochemistry point to potential problems. The atmosphere contains about 10^{12} kg of CH_4 (methane) but the actual concentration in the air is small. In the past, no one has worried about methane in the atmosphere because it is relatively nonreactive. Geochemists estimate that 80% of the methane that is released to the atmosphere is produced by microorganisms in anaerobic environments. These environments may be marshes, swampy regions, or the paddy fields of Asia, but significant amounts also come from fermentation within the gastrointestinal tract of animals. The quantity of methane thus formed is large, and if there were no means for destroying the methane in the atmosphere, the concentration should increase quite rapidly. Yet, the concentration is essentially constant. This indicates that there is a major means for destroying

Table 18-7 *Polymers released during the decomposition of invertebrates and plants and the microorganisms believed to be associated with their mineralization in the soil and water*

Polymer	Monomer composition	Primary environmental source	Organisms involved
Chitin	*N*-Acetyl-D-glucosamine	Primarily arthropods, crustacea, and fungi	Actinomycetes such as *Streptomyces; Pseudomonas, Bacillus,* and *Clostridium*
Cellulose	Beta-1,4-glucose	Higher plants and algae	Fungi (*Aspergillus, Fusarium, Penicillium*) Bacteria (*Cytophaga, Nocardia,* and *Streptomyces*)
Lignins	*p*-Hydroxyphenyl-propanes?	Higher plants	Fungi (*Agaricus, Fomes*) Bacteria (*Arthrobacter, Flavobacterium, Micrococcus, Pseudomonas*)
Pectin	Mostly polygalacturonic acid	Higher plants	Many bacteria such as *Erwinia* spp.
Hemicellulose	Complex of pentoses, hexoses, and uronic acid such as galacturonic acid	Higher plants	Many fungi and bacteria (many actinomycetes, *Bacillus* spp.)

Table 18-8 *Recalcitrant molecules**

Pesticides and herbicides	Polymers	Other synthetics
DDT (15 years)	Nylon	TNT
Chlordane (15 years)	PVC	Dioxin
Heptachlor (14 years)	Polyethylene	Chlorinated alkane
Aldrin (15 years)	Polypropylene	ABS surfactants
2,4,5-T (190 days)		

*Figures in parentheses represent known times for persistence of the compounds in the environment. Generally speaking, long-chained synthetic polymers are virtually nonbiodegradable. They are, however, susceptible to photochemical damage, which then renders them more susceptible to biodegradation.

the methane in the atmosphere. It is currently believed that one of the chief mechanisms by which this methane is destroyed is by chemical oxidation in the upper atmosphere. However, this gives rise to carbon monoxide (CO), and carbon monoxide in the atmosphere, even at a very low concentration, is a significant pollutant. Such extrapolations lead to a surprising hypothesis: by producing methane, which is then converted in the atmosphere to carbon monoxide, microorganisms in marshes, flooded soils, and the alimentary tract of animals are major sources of a significant air pollutant, indirect though their contribution is.

Recalcitrant molecules. Microbiologists whose focus of interest is solely the laboratory are proud of their successes in obtaining isolates that are able to grow on, and therefore usually mineralize, a variety of organic compounds. As pointed out previously, microbial communities in natural habitats are extremely versatile, and individual species are able to metabolize a wide variety of organic compounds of biosynthetic origin. Summing up the versatilities of all of the representative populations within the community results in a community of impressive mineralizing potential. Based on this versatility, the assumption gained popularity that microbial communities were omnipotent and that they could degrade not only the compounds of natural origin but any organic compound, be it of synthetic or biological origin. Nevertheless, environmental scientists know that this assumption is not true. Many biologically formed products are resistant to microbial attack under certain circumstances, and several synthetic chemicals are resistant under all circumstances to mineralization. These chemicals are **recalcitrant molecules** (Table 18-8).

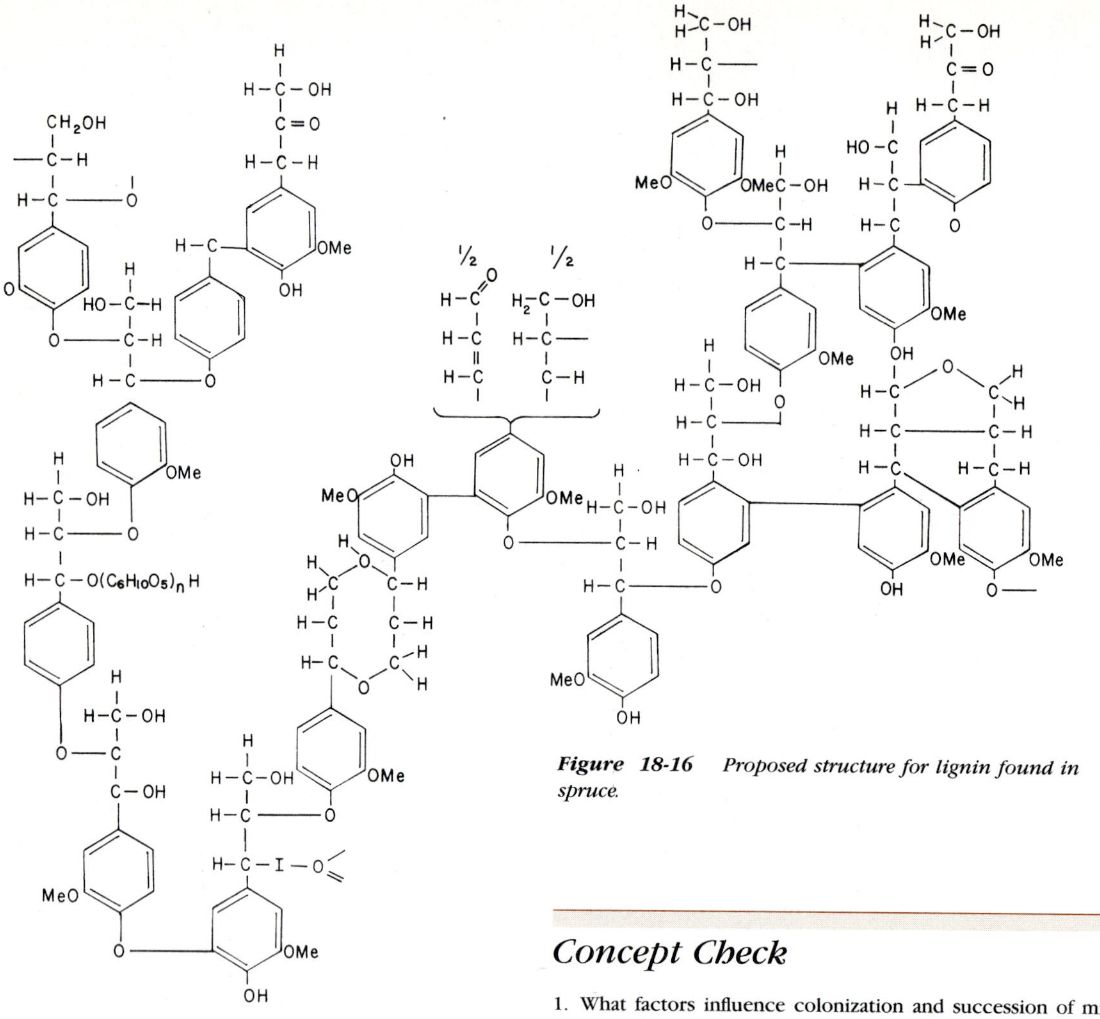

Figure 18-16 *Proposed structure for lignin found in spruce.*

Concept Check

1. What factors influence colonization and succession of microorganisms? What are some of the barriers to colonization?
2. Why is it important to understand the microbial cycling of elements?
3. What is the ecological importance of mierobial mineralization, oxidation, reduction, and methylation?
4. What are some of the natural and synthetic compounds that are resistant to microbial degradation? What is the relationship of recalcitrant synthetic molecules to the process called biomagnification?
5. Discuss the differentiation process of myxobacteria.

Among the biologically formed products that are recalcitrant and persist for long periods in nature are **lignin**, which is a major constituent of higher plants, and the humus of soils (Figure 18-16). In addition, because the spores of bacteria and resting structures of various fungi endure in well-aerated soils, which contain highly diverse and active communities, one must conclude that the organic compounds at the surfaces of these structures are recalcitrant. Were they not resistant to microbial mineralization, the spores and resting structures would be destroyed.

Recalcitrant synthetic compounds have achieved prominence because they are sometimes significant pollutants. In this category are many pesticides, a large number of organic compounds in industrial wastes, synthetic polymers used for fabrics or packing material, and products formed when organic compounds that are in natural waters are chlorinated as a step in municipal water treatment. Many of the recalcitrant molceules are halogenated hydrocarbons such as polychlorinated biphenyls (PCBs), polyvinyl chloride and other synthetic molecules, such as polyethylene and polystyrene (Figure

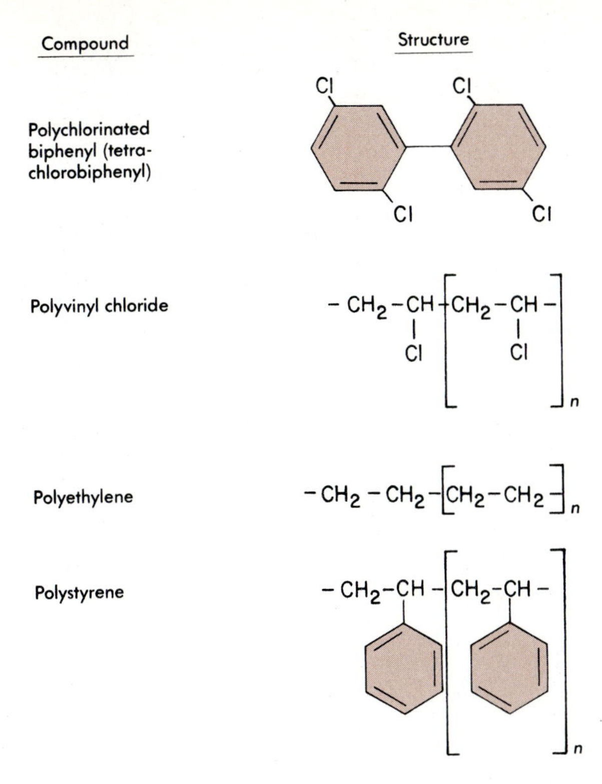

Figure 18-17 *Structure of some synthetic materials that demonstrate various levels of resistance to biodegradation.*

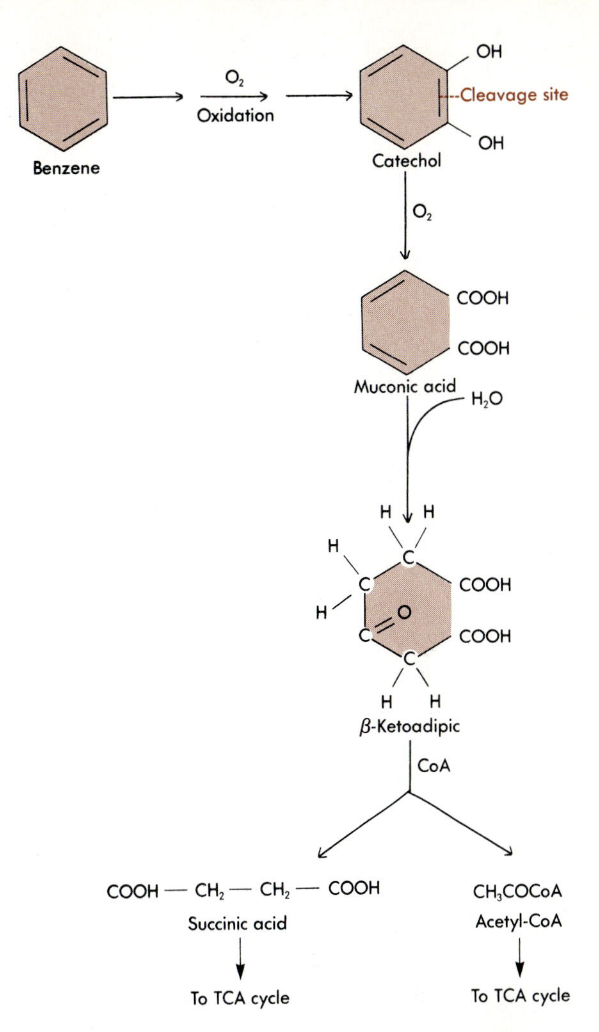

Figure 18-18 *Example of steps involved in microbial degradation of benzene. In example, only one cleavage site is indicated (-----). Hydroxylation and carboxylation of benzene result in its conversion to compounds that can enter tricarboxylic acid cycle.*

18-17). The halogenated compounds are especially resistant to mineralization because the carbon-halogen bond is very stable and considerable energy is required to cleave it. Biphenyl groups, for example, can serve as carbon and energy sources for some microorganisms, since their benzene rings can be hydroxylated and carboxylated (Figure 18-18). The presence of chlorine (or other halogens such as bromine) atoms on the benzene ring is not conducive to hydroxylation. Therefore, the resistance of the molecule to microbial mineralization is often related to the number of halogen atoms on the ring structure. The fact that these compounds are relatively resistant to microbial mineralization illustrates the fact that microbial communities are not infallible. Such a metabolic failing is revealed when pollutants persist in waters and soils. Establishing the reasons why organic compounds resist mineralization and determining the physiological bases for the failings of microorganisms are major areas of current research.

Inoculation of soils with microorganisms having special degradative properties is being used to clean up areas where there have been accidental spills of recalcitrant products or where toxic chemicals are stored and leak into the environment (Figure 18-19). Recombinant

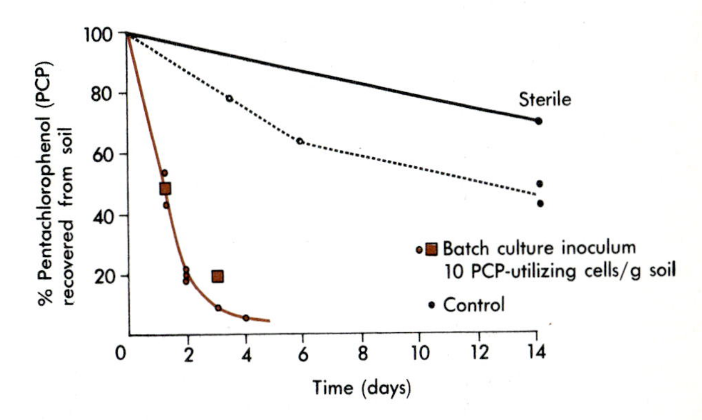

Figure 18-19 *Degradation of pentachlorophenol in soil by indigenous and inoculated bacteria under laboratory conditions at 30° C.*

DNA technology is also being used to construct microorganisms with special degradative properties.

Nitrogen cycle

Many reasons exist for microbiologists to focus on the nitrogen cycle (Figure 18-20). This element is essential for plants and animals. On land and in certain inland and coastal waters, nitrogen frequently limits the growth of higher plants or aquatic algae. The global food shortage is associated with deficiencies of nitrogen more often than any other element; hence, an international effort is underway to increase the ways by which microorganisms make nitrogen available to plants. In addition, excess nitrogen in fresh waters represents a significant form of pollution. Finally, microorganisms make volatile oxides of nitrogen, and these, on entering the air, contribute greatly to atmospheric pollution.

Mineralization. One of the critical steps in the nitrogen cycle is mineralization. As nitrogen is mineralized, it is converted from the organic form, such as proteins and peptides, which higher plants and most algae cannot use, to the inorganic form, such as ammonium (NH_4^+), that is readily assimilated by photosynthetic organisms on land and in waters. Nitrogen, as ammonium, is also oxidized and is thereby converted in aerated soils and oxygenated waters to nitrate. This process is known as **nitrification** and is carried out by nitrifying bacteria such as *Nitrobacter* and *Nitrosomonas.* Nitrate can be used as a nutrient source by microorganisms, but it also may be reduced to gaseous products when oxygen is absent. This reduction, which is known as **denitrification**, removes nitrate from soils and waters and converts it into dinitrogen, nitrous oxide, and other oxides of nitrogen. Dinitrogen (N_2) is used by a variety of heterotrophic bacteria, cyanobacteria, and a few actinomycetes; this process of **nitrogen fixation** (see Chapter 13) returns nitrogen from the gaseous form to compounds that other microorganisms, plants, and animals can assimilate. These processes occur in soils, the oceans, and lakes.

The substrates for mineralization are plants on land, the plankton in waters, the organic materials native to soils and sediments, and other materials that may be introduced above or below the surface of terrestrial and aquatic ecosystems. The microorganisms involved in mineralization are largely unknown, but many bacteria and fungi are able to use organic complexes of this element and convert them to ammonium. However, the product of the reaction is often at indetectable or low levels because algae or higher plants occupying the habitat assimilate the ammonium, or it is used by microorganisms either as a nutrient or as a substrate for nitrification.

Nitrification. Nitrification is an aerobic process in which ammonium is oxidized to nitrate. The process is carried out by autotrophic lithotrophic bacteria and occurs in two steps involving two groups of bacteria. The first reaction is carried out by *Nitrosomonas* species:

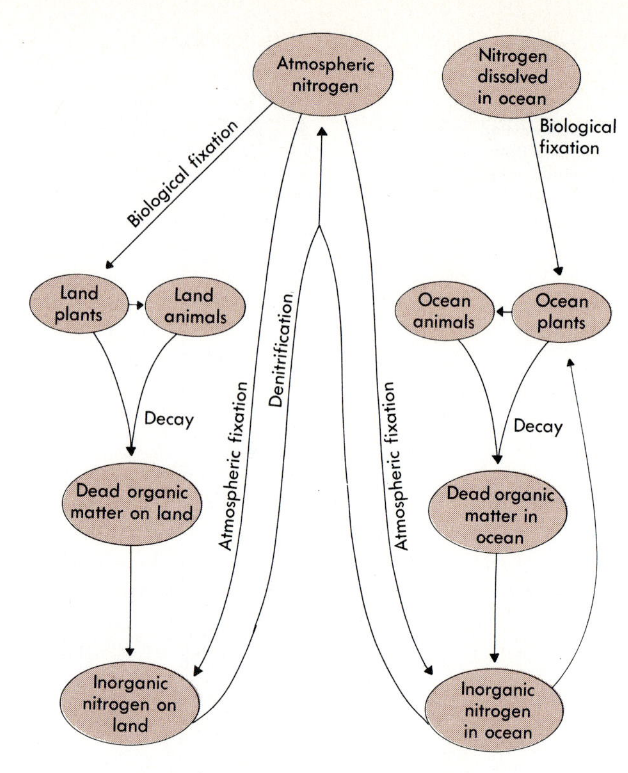

Figure 18-20 *Global nitrogen cycle.*

$$NH_4^+ + 1.5\ O_2 \rightarrow NO_2^- + H_2O + 2H^+ \quad \Delta G^{o\prime} = -66.2\ \text{Kcal/mole}$$

In the second step nitrite (NO_2^-) is oxidized to nitrate by *Nitrobacter* species:

$$NO_2^- + 0.5\ O_2 \rightarrow NO_3^- \quad \Delta G^{o\prime} = -17.4\ \text{Kcal/mole}$$

During this oxidation process nitrogen goes from a valence of $-3(NH_4^+)$ to $+5(NO_3^-)$. The final electron acceptor, which is oxygen, is preceded by a cytochrome system, and ATP is produced. The energy yield from the two oxidation steps is relatively small. Thus, the reduction of carbon dioxide by these bacteria, requires considerable energy and results in small numbers of cells and little contribution to the organic nitrogen pool.

In most environments, nearly all of the nitrate found is formed microbiologically, and only a small amount is returned to waters and soils in the form of precipitation. The nitrate that is generated by the indigenous community is a nutrient for plants or microorganisms, but it can also be utilized by denitrifying bacteria. The nitrate that is formed in soil is important, moreover, not only because it is a nutrient for plants, but also because it is readily leached out of soil and thus enters underground waters, from which it may enter supplies of drinking water. High levels of nitrate are undesirable in drinking water because of deleterious effects on infants.

Several related genera of bacteria common in many soils and waters carry out the same nitrification reactions. However, indirect evidence exists for the functioning, at least in terrestrial environments, of heterotrophs that oxidize nitrogen in organic compounds, but they get little or no energy from the transformation. These organisms require a carbon source to provide energy for biosynthesis. The identity of these heterotrophs has yet to be determined.

The process of nitrification is extremely sensitive to environmental stress. Indeed, it is one of the most sensitive microbial processes in aquatic and terrestrial environments. For this reason, it is often used as a measure of stresses, and a pollutant that does not inhibit nitrification probably will not reduce the rate of any other process brought about by nonphotosynthetic microorganisms. Conversely, an inhibition of the nitrifying populations suggests that other species may be subject to environmental or chemical stress.

Denitrification. When oxygen is absent, nitrate can serve as a final electron acceptor and become reduced. Nitrate reduction is called **denitrification** when the products are molecular nitrogen (N_2), nitrous oxide (N_2O), or nitric oxide (NO). Nitrate reduction takes place as a result of cellular metabolic reactions in which there is the oxidation of reduced substrates such as sugars, H_2S, or H_2. In the process, nitrate is first reduced to nitrite by the enzyme nitrate reductase. Further reduction of nitrite results in the formation of N_2O, NO, and N_2 (see Chapter 13). Reduction does not involve transfer of electrons to cytochrome oxidase (oxygen is the substrate for cytochrome oxidase); thus, only two ATP molecules are produced. Denitrification occurs primarily in habitats where there has been a rapid decomposition of organic matter and anaerobic conditions prevail—deep waters or sediments and water-logged soils. The organisms involved are primarily facultative anaerobes such as *Thiobacillus denitrificans* and species of *Pseudomonas, Bacillus, Micrococcus,* and *Achromobacter.* The oxides released during denitrification are significant pollutants. Nitrous oxide diffuses from soils to the lower atmosphere, and from there to the stratosphere. In the stratosphere, it reacts with ozone (O_3). The ozone is then partially destroyed, and this destruction results in an increased incidence of ultraviolet light reaching the earth. Ultraviolet light is hazardous to many animals and plants and is a cause of skin cancer of humans. Most or nearly all of the nitrous oxide is released as a result of microbial activity, especially in soils, and it has thus been suggested that microorganisms are, indirectly to be sure, causes of skin cancer. The other two oxides of nitrogen, nitric oxide and nitrogen dioxide (NO_2), are themselves toxic to humans, and the calculations of atmospheric scientists suggest that most of these two pollutants are generated not by automobiles and other industrial processes, as is widely believed, but rather result from microbial activity in nature, especially from soils.

Nitrogen fixation. The sole organisms that can bring about nitrogen fixation are microorganisms. The responsible species either live free in soils and waters or form symbiotic associations with higher plants. Some (the cyanobacteria) are symbionts with fungi in lichens. The symbiotic associations with higher plants involve the species *Rhizobium* and *Bradyrhizobium* discussed previously.

The processes involved in nitrogen fixation make the element available to the responsible microorganisms. If the association involves a higher plant, that nitrogen is incorporated into plant tissue. In the free-living bacteria and cyanobacteria, the nitrogen is released as a result of (1) excretion by the active populations, (2) their decomposition by other organisms, or (3) the feeding on these populations by predators. In this way, the nitrogen, which originally was in an unavailable form (dinitrogen), is provided to other organisms inhabiting the same ecosystem. Because nitrogen fixation serves much the same function as chemical fertilizers in agriculture and fertilizers are expensive in developed countries and unavailable in many of the poor countries, considerable research is underway to enhance nitrogen fixation. Particular attention is being given to increasing the activity of the symbiosis between *Rhizobium* and leguminous plants. In addition, attempts are being made to determine how to increase the nitrogen-fixing activity of bacteria around the roots of cereals and of cyanobacteria that inhabit paddy fields, which are important for rice production in much of Asia. Genetic engineers are also attempting to insert the genes encoding nitrogen fixation into nonleguminous plants. This difficult process is examined in Chapter 20.

Sulfur cycle

Sulfur is a ubiquitous element that, like carbon and nitrogen, can exist in several oxidation states (from +6 as in sulfate [SO_4] to −2 as in hydrogen sulfide [H_2S]). In addition there are inorganic intermediates such as sulfite (SO_3^{2-}), thiosulfate ($S_2O_3^{2-}$), polythionate ($S_nO_6^{2-}$), and elemental sulfur (S^0).

Microbial populations at or below the soil surface, in marine and freshwater sediments, and in natural waters are important in the cycling of sulfur (Figure 18-21). Sulfur is present in ores and rocks as sulfides, and this abundant form of sulfur is oxidized when oxygen is introduced into the environment. The responsible populations are members of the genus *Thiobacillus.* These autotrophic bacteria oxidize the sulfides and use them

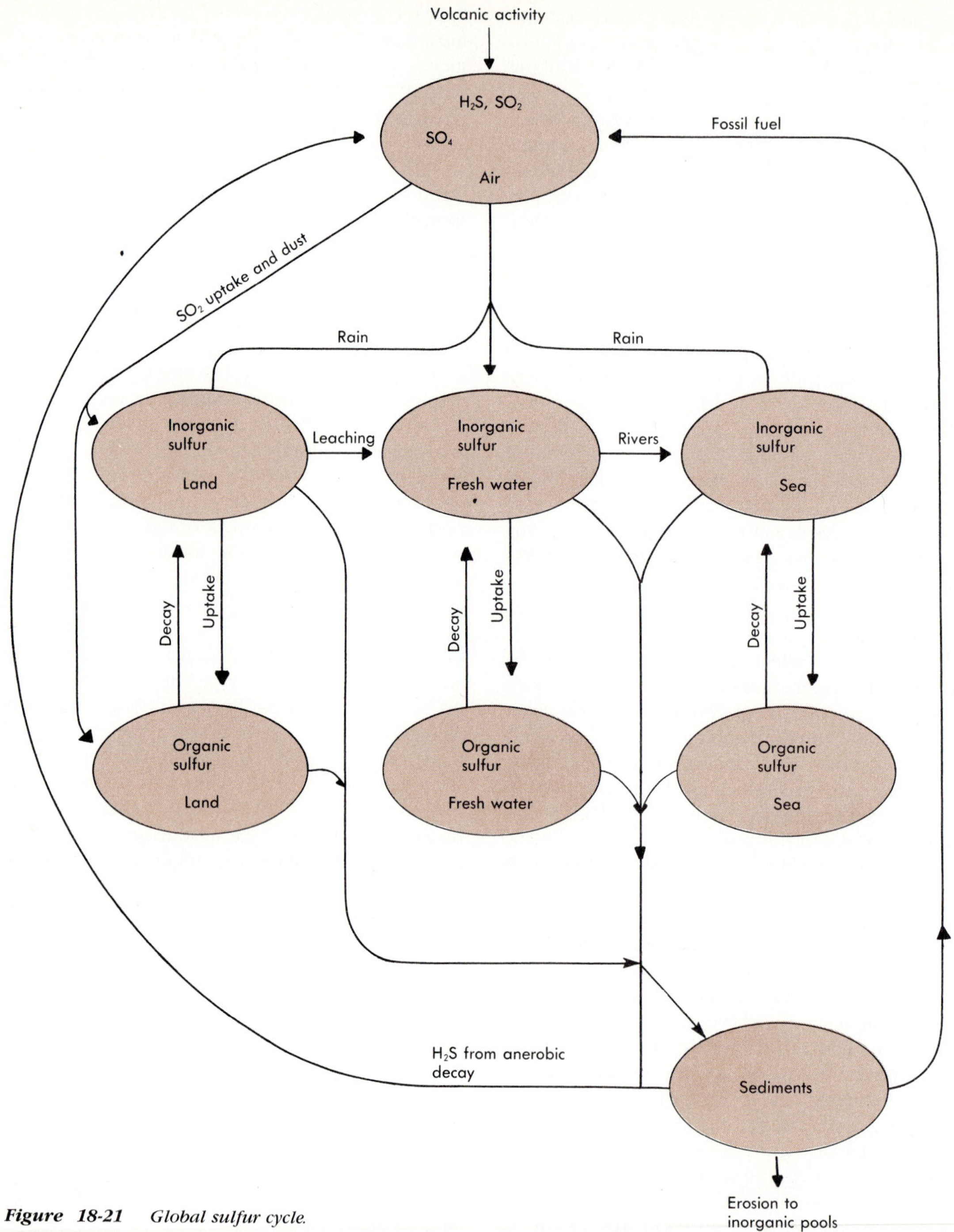

Figure 18-21 *Global sulfur cycle.*

as energy sources for growth. As a result, sulfate is produced, and the acidity of the surroundings increases. The acidity increases because the reaction gives rise not simply to sulfate but rather to sulfuric acid. Oxidations that are carried out by heterotrophs occur in soils and in waters, where the energy sources are organic compounds of sulfur. The sulfur in these compounds is unavailable for higher plants and for many microorganisms. However, the responsible populations mineralize the sulfur and also convert it to sulfate, but in these instances, the populations are heterotrophic. This sulfate is then assimilated by the indigenous populations of plants and microorganisms.

In lakes anaerobic photosynthetic bacteria may be quite important. These organisms use light as their energy source, but at the same time, some oxidize hydro-

Table 18-9 Sulfur transformations and the organisms involved

Process and substrate	Habitat	Product	Organisms involved
Reduction			
Sulfate (SO_4^{2-})	Anaerobic (marshes and sediments)	H_2S	*Desulfovibrio, Desulfobacter, Desulfotomaculum, Desulfobulbus*
Elemental sulfur (S^0)	Anaerobic (marine and fresh water)	H_2S	*Desulfuromonas acetoxidans*
Oxidation			
H_2S or intermediates $S_2O_3^{2-}$ or $S_4O_6^{2-}$	Aerobic	S^0	Colorless sulfur bacteria *(Beggiatoa, Thiothrix, Achromatium,* and *Thiobacillus)*
H_2S	Anaerobic and shallow enough for light	S^0	Phototrophic sulfur bacteria (purple, green, and brown sulfur bacteria, gliding green bacteria)
S^0	Anaerobic and shallow enough for light	H_2SO_4	Phototrophic sulfur bacteria
S^0	Anaerobic	H_2SO_4	*Thiobacillus denitrificans*
S^0	Aerobic (hot springs and hot acid soils)	H_2SO_4	*Sulfolobus*
Metal sulfide (e.g., FeS_2)	Aerobic	$FeSO_4 + H_2SO_4$	*Thiobacillus ferrooxidans*

gen sulfide. The hydrogen sulfide that they use is generated not in the lighted zone of the water, where other photosynthetic species grow, but rather in the underlying anaerobic region.

In marshes, sediments, poorly drained soils, and other anaerobic environments, sulfate is reduced by bacteria that do not use oxygen in their metabolism. These bacteria, a representative of which is *Desulfovibrio,* proliferate and convert sulfate to hydrogen sulfide. For the process, these anaerobes require an energy source, which is often organic, but it may also be hydrogen gas. The hydrogen sulfide that these bacteria excrete may reach high levels and become toxic to fish or to higher plants growing in poorly aerated soils.

Geological microbiologists have been especially interested in the sulfur cycle. They have given particular attention to sulfuric acid formation from ores because of the acidity that is generated. This acid may solubilize many elements, and the elements thus solubilized are mobile and move from place to place. In some instances, the soluble form of the element is toxic. Geological microbiologists have also been interested in sulfate reduction, which, by generating hydrogen sulfide, causes the precipitation of metallic sulfides. Geological deposits of sulfides may be major sources of the metal associated with the sulfur. Organisms involved in the sulfur transformation process are outlined in Table 18-9.

Phosphorus cycle

Phosphorus, unlike the elements discussed so far, is not present in the atmosphere as a gas; it exists primarily in one form—**phosphate**. The major reserve of phosphorus in the biosphere is in its insoluble form in rocks and sediments. Phosphorus is present in organic molecules released during the decomposition of plants, animals, and bacteria in soil and aquatic environments. Incorporation of phosphorus as phosphate into organic matter was discussed on (p. 358). Phosphorus is much less common in aquatic environments or habitats except when pollution occurs from sewage runoff or from farm lands fertilized with nitrogen and phosphorus. The global phosphorus cycle is illustrated in Figure 18-22.

Iron cycle

In Chapter 13, we discussed the importance of iron as a component of cytochromes and its requirement for pathogenicity by some bacteria. Iron is abundant in the earth's crust but is usually unavailable to microorganisms in its normally insoluble form of hydroxides and sulfides or combined with other metals such as magnesium, manganese, or calcium in the form of silicates. Microorganisms interact with iron in three ways: (1) iron scavenging from the environment through the use of chelates such as the siderophores (most bacteria carry out this process or a related one; see Chapter 13), (2) iron oxidation, and (3) iron reduction.

Microorganisms that oxidize iron are acidophilic bacteria such as *Thiobacillus ferrooxidans* and the thermoacidophile *Sulfolobus.* These microorganisms oxidize ferrous iron as well as sulfur compounds. Iron oxidation results in the transformation of ferrous iron (as $FeSO_4$) to ferric iron (as $Fe_2(SO_4)_3$) according to the following reaction:

$$4FeSO_4 + 2H_2SO_4 + O_2 \rightarrow 2Fe_2(SO_4)_3 + 2H_2O$$

It is the oxidation of sulfur to sulfuric acid by the thiobacilli and others that acidifies the environment and provides the stabilization of iron in solution. Otherwise iron will precipitate out at pHs near neutrality. *Gallionella* is a stalked organism similar to *Caulobacter* that is also found in acidic waters containing ferrous iron. It oxidizes ferrous iron to ferric hydroxide, which accumulates in the stalk. Some bacterial species may or may not oxidize iron, but they are still associated with iron deposits. *Leptothrix* species, which are pH neutrophiles, are believed to be associated with the accumulation and precipitation of iron oxides in lake and cave iron deposits (Figure 18-23).

Accumulation of reduced (ferrous) iron occurs in anaerobic environments such as anaerobic marine sediments or other stratified zones in which organic matter accumulates and oxygen is depleted. Iron reduction is carried out by many bacterial genera including *Bacillus, Aerobacter, Pseudomonas, Escherichia, Proteus, Clostridium, Achromobacter,* and *Staphylococcus.* Ferric iron can be reduced by microbial excretion products such as formate or sulfide or by lowering the oxidation-reduction potential and pH of the environment. The exact mechanisms of microbial iron reduction have not been sufficiently determined. It should be noted that some bacteria have intracellular inclusions called **magnetosomes** that contain iron in the form of magnetite (Fe_3O_4). These organisms are routinely found in sedimentary habitats (see Chapter 5).

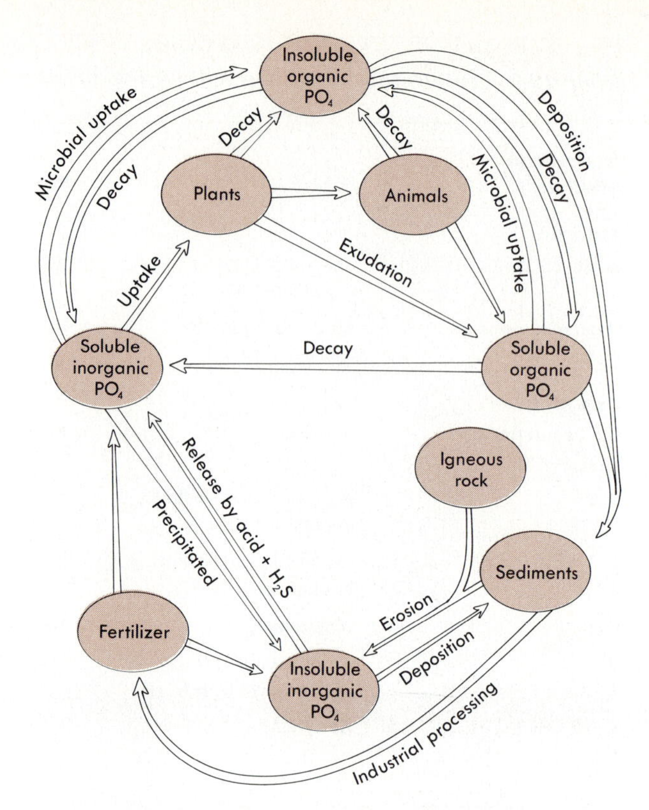

Figure 18-22 *Global phosphorus cycle.*

Concept Check

1. What are the distinctions between denitrification, nitrification, and nitrogen fixation? What groups of organisms are associated with each process? In what part of the environment do they take place?
2. What forms of sulfur exist in the environment, and what are their microbial transformation products?
3. What are the different ways in which microorganisms interact with iron in the environment?
4. What is the consequence of too much phosphate in the environment? How is inorganic phosphorus incorporated into inorganic molecules?

MICROORGANISMS AND POLLUTION

Pollution has a major impact on microbial processes, and microorganisms in turn may have major impacts on pollutants. Classically, the microbial role in pollution was largely assigned to pathogenic species, which themselves are pollutants. However, within the last two decades, it has become evident that microbial processes are affected by chemical pollutants introduced into waters and soils, and that the indigenous communities of these environments are major means both for destroying and for generating chemical pollutants.

One of the most tragic incidents of indirect microbial-associated pollution occurred in Japan in the 1950s. Mercury salts were discharged into Minamata Bay, from which fish and shellfish are harvested and eaten by the local community. Mercury salts, although toxic, are ex-

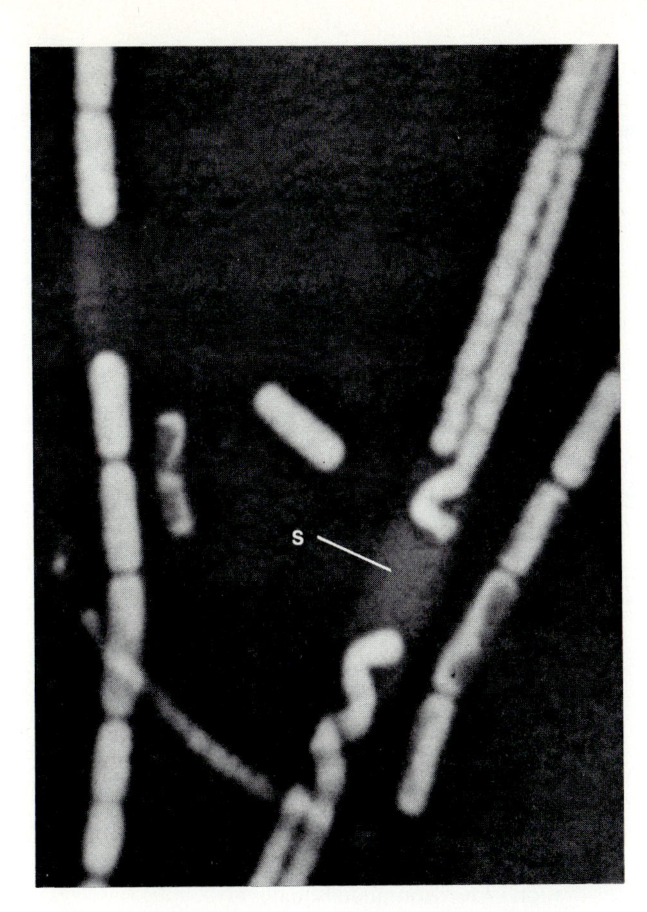

Figure 18-23 *Sheathed bacteria. Negatively stained preparation of species of* Leptothrix *(×4500). Sheath* (S) *is seen to enclose two strands of rod-shaped cells.*

creted rapidly when ingested by humans and other animals. However, when they reach the anaerobic sediments of a bay, they are believed to be converted by a microbial process to methyl mercury, which tends to accumulate in lipid tissue and to be poorly excreted from the body. Once methyl mercury is in the food chain, ingestion of contaminated fish and shellfish by humans results in the buildup of large concentrations. Methyl mercury is a neurotoxin; in this one Japanese village, more than 100 inhabitants were affected by severe and permanent neuromuscular conditions; some even died. Anaerobic methylation of mercury occurs by transfer of methyl groups by a carrier, vitamin B_{12}. The process is carried out by sulfate-reducing bacteria and methanogenic bacteria:

$$Hg^{2+} \xrightarrow{CH_3\text{–vit } B_{12}} HgCH_3 \xrightarrow{CH_3\text{–vit } B_{12}} Hg(CH_3)_2$$

Other metals such as arsenic can also be methylated by microorganisms under anaerobic conditions (see p. 535).

A chemical that enters water or soil as a result of human activities may reduce the rate of a microbial process in a fashion that is deleterious to other species residing in the same environment. In waters, microbial photosynthesis is extremely important. The photosynthetic algae and cyanobacteria are at the base of food chains on which fish and other aquatic animals rely. An inhibition of the rate of microbial photosynthesis in these habitats could result in a diminution in the food supply for, and hence in the yield of, fish or other animals. Some chemicals, it now appears, do cause such inhibitions. A reduction in the rate of the microbial mineralization of an element that is in insufficient supply in a terrestrial or aquatic ecosystem likewise will have a major effect on the function of that ecosystem. For example, should a pesticide or industrial chemical introduced into soil reduce the rate of microbial conversion of organic nitrogen to the inorganic forms used by higher plants, this suppression of microbial activities will be reflected in a diminished rate or extent of plant growth. Similarly, because nitrogen fixation by legumes is important in many agricultural lands as well as in natural ecosystems, an inhibition of *Rhizobium* will reduce the amount of nitrogen that is returned to the environment as a result of the legume-*Rhizobium* symbiosis.

The microbial modification of organic and inorganic compounds or ions may likewise be of profound importance to environmental quality. Thus, microbial populations mineralize many industrial pollutants and pesticides, and these mineralization reactions, by converting the organic chemical to inorganic products, destroy the toxic substances and convert them to harmless substances. In nature, except in areas of high light intensity, the only means to destroy organic pollutants completely and to convert them to inorganic products is by a microbial transformation. These conversions are known as **detoxications** because the processes destroy the inherent toxicity of the molecule. On the other hand, microorganisms also bring about **activation**, a reaction in which a compound of little or no toxicity is converted to a highly toxic molecule. Microorganisms are now recognized as being involved in several activation reactions and in thus generating toxic chemicals. In this way, microorganisms are not solely agents of destroying pollutants, but they also create them. Many examples of microbial detoxication are known, but only a few instances of the formation of toxic chemicals by microbial populations in soils and waters have been described to date.

As pointed out above, microorganisms pollute the air with **nitrogen oxides**. They are also responsible for forming other air pollutants and for destroying certain volatile pollutants. **Carbon monoxide** (CO) is a notable example. This gas is constantly being generated as a result both of combustion and of photochemical reactions in the air. The level of carbon monoxide in the atmosphere is not increasing, because microbial activity, ap-

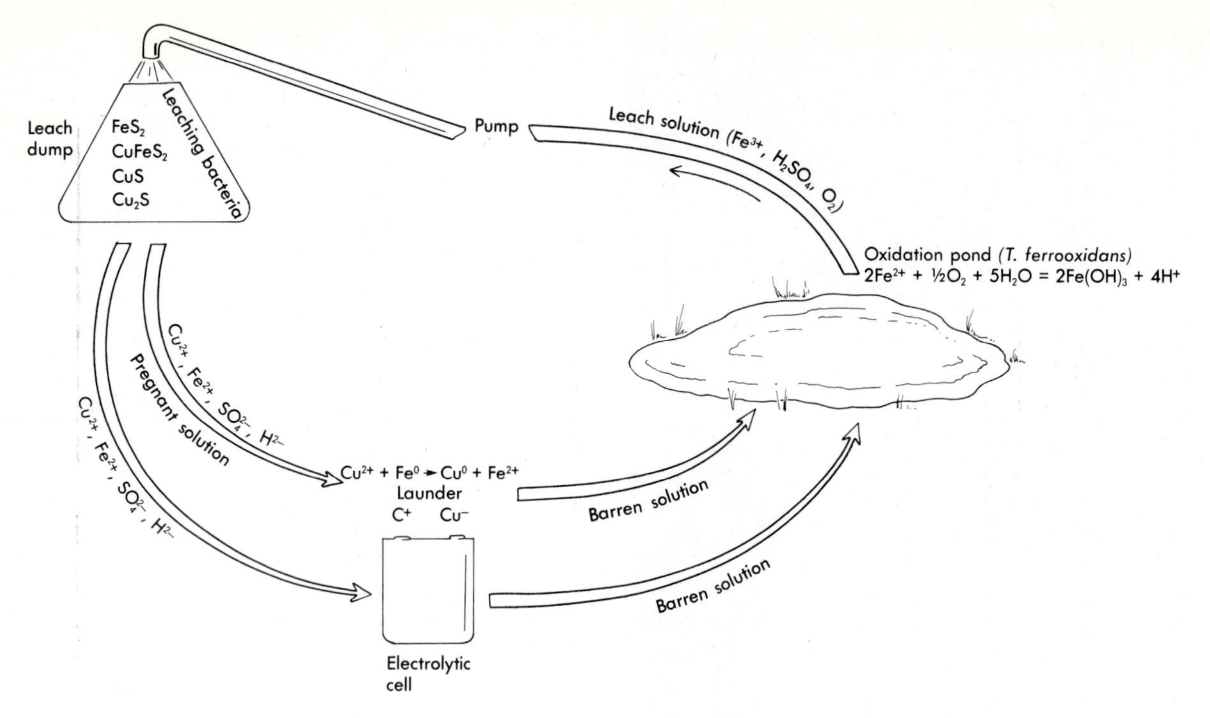

Figure 18-24 *Diagrammatic representation of metal extraction technique for copper sulfide ores.*

parently in soil, converts the carbon monoxide to carbon dioxide. Although this oxidation and detoxication occur readily in soil, the responsible organisms—bacteria or fungi or even the genera involved—are not known as yet. Microorganisms also form carbon monoxide in natural environments, but the net effect appears to be destruction rather than emission of this highly toxic gas.

Volatile hydrocarbons are also produced by microorganisms. Although the concentration of hydrocarbons in the atmosphere is small, natural communities are constantly generating a number of these volatile compounds. Conversely, microorganisms are active in destroying volatile hydrocarbons. Some of this destruction may take place in aquatic environments, but probably most occurs in soil.

Sulfur and other elements are also converted to volatile products. Some of these interest toxicologists, and others interest geochemists and atmospheric scientists. The volatilization is often the consequence of microbial addition of methyl groups to an ionic form of the element. In addition, microorganisms generate other volatile products, for example, carbon disulfide.

This recently recognized role of microorganisms in destroying many synthetic pollutants and in creating new pollutants is serving as impetus for much research. Nevertheless, it still is not clear to what extent microorganisms are major contributors to pollution, but they are unquestionably major agents in preventing the deterioration of environmental quality.

MICROORGANISMS AND METAL EXTRACTION

The ability of chemoautotrophs to obtain their energy from the oxidation of inorganic substances and to fix carbon dioxide into organic matter is being exploited by the metal industry and associated industries. Conventional nonbiological mining procedures and ore processing have resulted in a depletion of high-grade ores, leaving behind low-grade ores whose processing by conventional means is not profitable. In addition, many minerals and fossil fuels (such as coal) exist as sulfides, and their smelting and burning, respectively, create environmental problems. Biological processing of minerals is generally more economical and creates fewer pollution problems than conventional nonbiological processes. Research is now underway to refine techniques for biological mining. To date only copper and uranium are recovered by microbiological processes.

Microbiological mining

Thiobacillus ferrooxidans has been the most widely studied organism for biomining. It can oxidize metal sulfides to sulfuric acid, which solubilizes the metals in the ore and promotes their recovery.

Copper leaching. Low-grade ores are usually transported to a dump site and placed in piles as high as 100 feet. Acidified water is applied to the top of the pile, and when conditions are suitable, *T. ferrooxidans,* which is indigenous in such sulfidic rock, proliferates. *Thiobacillus thiooxidans* is also present in the pile but at lower concentrations. Both of these organisms feed on the copper sulfide (primarily $CuFeS_2$), releasing metals into an effluent called a **pregnant** solution (Figure 18-24). The effluent is collected, and the copper is concentrated by precipitation with scrap iron in a basin called a **launder**. Following recovery of the copper, the solution, now referred to as **barren**, is collected in an oxidation pond where ferrous iron is oxidized to ferric iron by *T. ferrooxidans*. Some of the ferric iron precipitates out, and the remaining solution, which is ferric sulfate, is used as a leaching solution and pumped over the pile as a fine spray. The precipitation of some of the ferric iron in the oxidation pond is important because too much of it in the leaching solution could lead to its precipitation on the metal sulfide crystals in the ore and prevent oxidation. The internal temperature of a leaching pile approaches 70° to 80° C, a temperature not conducive to *T. ferrooxidans* growth, and there is some doubt as to its role in sulfur oxidation when these high temperatures are reached. It is now believed that *Sulfolobus* species, which are acidophilic thermophiles, may play an equally important role in sulfur oxidation.

Uranium leaching. Bacterial leaching of uranium ore can take place naturally as long as pyrite (FeS_2) is present and the ore does not contain high concentrations of carbonates (acid conditions are required in the leaching process and carbonates buffer the pH rather than lower it). Bacteria such as *T. ferrooxidans* play an indirect role in the extraction of uranium from ore. Instead of directly attacking the uranium mineral (uranium oxide), bacteria generate ferric iron from the oxidation of pyrite also present in the ore according to the following equation:

$$2\,Fe^{2+} + 0.5O_2 + 2H^+ \rightarrow 2Fe^{3+} + H_2O$$

The ferric iron hydrolyzes:

$$Fe^{3+} + 3H_2O \rightarrow Fe(OH)_3 + 3H^+$$

The dissolved ferric iron in the form of ferric sulfate oxidizes minerals containing uranium. The uranium is converted from a valence of +4 to a valence of +6 according to the following fraction:

$$UO_2 + Fe_2(SO_4)_3 + 2H_2SO_4 \rightarrow UO_2(SO_4)_3^{4-} + 2FeSO_4 + 4H^+$$

The dissolved uranium can be recovered by concentration through ion exchange.

Desulfurization of fossil fuels such as coal

Burning of fossil fuels, such as coal and petroleum products, results in the contamination of air by sulfur and nitrogen oxides. These air contaminants are believed to be responsible for acid deposition (acid rain). Acid deposition can produce long lasting effects on lakes, agricultural lands, and forest reserves. Sulfur compounds present in petroleum and coal can also lead to the corrosion of pipelines and equipment (see Point of Interest). Nonbiological desulfurization techniques are expensive and time consuming; thus, a biological mechanism is being sought to resolve these problems. A wide range of sulfur-containing compounds are found in coal and petroleum. The major sulfur compound in coal is pyrite (FeS_2) while organic sulfur compounds are more prevalent in petroleum products.

The solubilization of pyrite in coal by *T. ferrooxidans* takes place by direct and indirect means. In the direct method, *T. ferrooxidans* attaches to pyrite crystals in the presence of air and carries out an oxidation in which sulfide is converted to sulfate and ferrous iron is converted to ferric iron:

$$FeS_2 + 3.5O_2 + H_2O \rightarrow FeSO_4 + H_2SO_4$$
$$2FeSO_4 + 0.5O_2 + H_2SO_4 \rightarrow Fe_2(SO_4)_3 + H_2O$$

In the indirect method, the oxidation of pyrite occurs first by nonbiological means:

$$FeS_2 + Fe_2(SO_4)_3 \rightarrow 3FeSO_4 + 2S$$

The generated ferrous sulfate is reoxidized by *T. ferrooxidans* to ferric sulfate as in the direct method. The elemental sulfur generated in the indirect method is oxidized by *T. ferrooxidans* and *T. thiooxidans* to produce more sulfuric acid:

$$2S + 3O_2 + 2H_2O \rightarrow 2H_2SO_4$$

Efforts are now underway to apply the natural phenom-

Concept Check

1. **What is the effect of microbial activity on mercury contamination?**
2. **What are the beneficial as well as possible detrimental effects of microbial activity in the environment?**
3. **Explain the basic process of copper leaching by microorganisms.**
4. **Explain the chemical process of metal corrosion and what microorganisms may be involved.**

Point of Interest

CORROSION AND THE RUST BUGS

Metal corrosion each year in the United States causes approximately $170 billion worth of damage; much of this damage is in the pipelines of oil and gas facilities and in pipelines and industrial storage tanks for hazardous waste. What is this corrosion process, and what is the role of microorganisms?

Corrosion is an electrochemical reaction common to metals in aqueous habitats. It involves the movement of metal from the surface of the metal object into solution as an electrically charged species. Metal removal occurs at anodic sites where the oxygen concentration is low:

$$\text{Metal} \leftrightarrows \text{Metal}^{2+} + 2 \text{ electrons}$$

The electrons that escape during metal dissolution are captured in a separate reaction involving a cathodic site. The reaction at the cathodic site usually involves oxygen and occurs at neutral or alkaline pH:

$$0.5O_2 + H_2O + 2e^- \leftrightarrows 2OH^-$$

In acid solution the electrons that escape interact with the hydrogen ions of acid compounds and form hydrogen gas:

$$2H+ + 2e^- \leftrightarrows H_2$$

The establishment of an anodic site on a metal surface frequently occurs because of imperfections on the surface of the metal or from a film or concentration of cells on the surface **(biofilm).** These surface deposits decrease the amount of oxygen available to the underlying surface (note that oxygen normally binds to a metal surface, forming a metal-oxide skin that acts as a corrosion inhibitor). Thus, different areas of the metal surface develop different electrochemical potentials, and metal dissolution takes place.

Many microorganisms may be involved in the corrosion process, but the most important appear to be the sulfate-reducing bacteria, such as *Desulfotomaculum* and *Desulfovibrio.* The corrosion process is believed to occur in the following way. Aerobic microbial settlers attach to a vulnerable site on the metal surface (figure below). These bacteria often produce extracellular polymers that enhance their attachment to the metal surface and promote the attachment of other bacteria, including the obligately anaerobic sulfate-reducing bacteria. As extracellular polymer is produced, the film on the surface of the metal becomes thicker and an anaerobic environment develops beneath them. Aerobes utilize oxygen at the surface and release byproducts that can be utilized by heterotrophic anaerobes. These anaerobes begin to proliferate, and their fermentation products, such as acetate, support the growth of sulfate reducers that are closest to the metal surface. The corrosion-enhancing material are acids released during anaerobic microbial metabolism.

Under aerobic conditions the growth of *Thiobacillus* spp. results in the formation of sulfuric acid and low pH. This organism is responsible for considerable deterioration of stonework and concrete structures, such as sewage pipes and cooling towers. *T. ferrooxidans* is used in the metal leaching for the mining of low-grade ores and is responsible for the corrosion of iron pipes and pumping equipment machinery used in that process. Organic acids are also responsible for the corrosion of aluminum aircraft fuel tanks. The fungus *Cladosporium resinae* uses the kerosene fuel in the tanks as a carbon and energy source, and the organic acids released during metabolism corrode the metal surface.

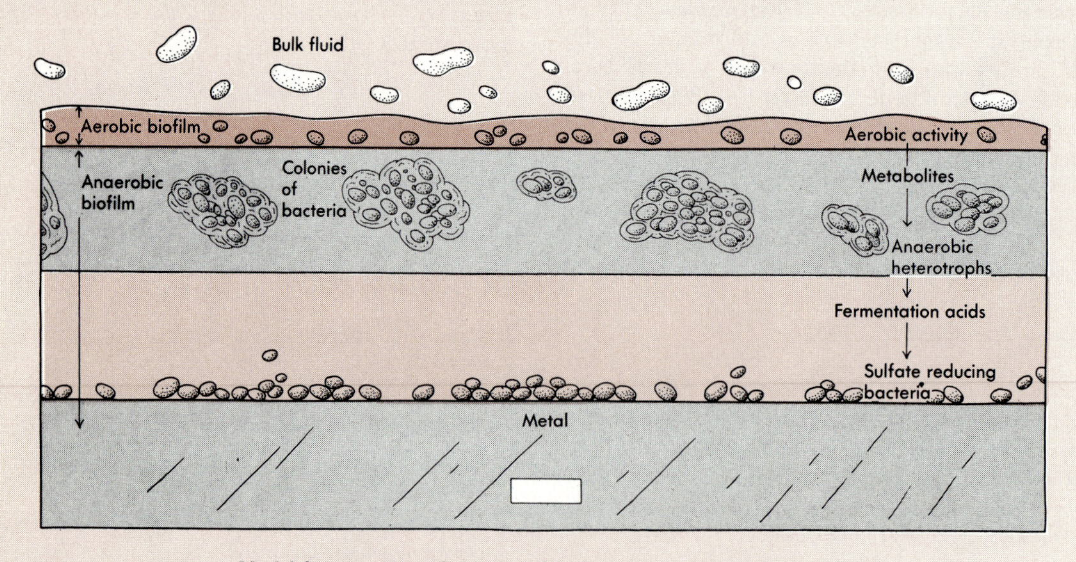

Model for corrosion of metal surface by sulfate-reducing bacteria.

enon of biological oxidation in the marketplace. Mixed cultures of *T. ferrooxidans* and *Sulfolobus acidocaldarius* are being inoculated into coal slurries in desulfurization studies. One of the problems with using *T. ferrooxidans* is that it is sensitive to several heavy metals present in ores. It is hoped that, through genetic manipulation, plasmids possessing metal-resistant genes can be transferred from other organisms to *T. ferrooxidans.*

The reactions that occur naturally when coal is mined and exposed to the air result in acid formation and represent an important pollution problem. Acid drainage from active and abandoned coal mines causes the leaching out of heavy metals at coal storage sites or disposal sites for coal waste or mining products. Large concentrations of heavy metals such as arsenic and cobalt have been found in streams located near the mines. Studies are now underway to find organic compounds that can be sprayed over extracted coal to prevent bacterial oxidation processes.

SUMMARY

1. The composition of the microbial community is determined by the available nutrient supply, certain critical physical and chemical factors, and interactions among those species able to tolerate the abiotic stresses.
2. Natural selection operates on the species that enter an environment, and the species that survive and grow have fitness traits allowing them to persist and multiply in that habitat.
3. The communities within the ecosystem may have a low or high species diversity. These species interact in food chains that permit the coexistence of populations with different nutritional patterns. Some organisms are pioneers and appear early, others appear late in successions, but some microorganisms are typically members only of climax communities.
4. The interactions that may be harmful to individual species in soil are competition, amensalism, parasitism, and predation.
5. Species may benefit from neighboring populations because of commensalism, protocooperation, or mutualism.
6. Microbial transformations in soils and natural waters affect the chemistry of the earth and of the atmosphere, and these processes may also have immediate, practical effects.
7. Microbial transformations may be beneficial because they bring about the cycling of carbon, nitrogen, sulfur, and other elements required by biotic systems. Some transformations, however, can lead to the formation of polluting compounds in natural environments.
8. Many biologically formed products are resistant to mineralization. Some of these products are pollutants such as pesticides or industrial wastes. Halogenated hydrocarbons are among the most resistant to microbial degradation.
9. Organic nitrogen is mineralized to ammonium. Ammonium is oxidized to oxides of nitrogen (nitrite, nitrate) as well as dinitrogen (N_2). Dinitrogen can be fixed by free-living microorganisms and some organisms that infect the roots of leguminous and nonleguminous plants.
10. Sulfur is present in the environment primarily as sulfides that can be used as energy sources by some bacteria. Microorganisms can mineralize sulfur to sulfate—the form of sulfur assimilated by plants and microorganisms.
11. Phosphorous exists in the environment primarily as phosphate and is the principal form assimilated by microorganisms.
12. Iron in the environment is usually in an insoluble form. It can be scavenged by siderophores, or it can be oxidized or reduced. Iron oxidation is a property of the acidophilic bacteria that also oxidize sulfur compounds.
13. Microorganisms can generate pollutants by converting compounds such as mercury salts to derivatives that accumulate in the food chain. Microorganisms can also destroy the toxicity of compounds such as industrial wastes.
14. Autotrophic microorganisms are being used in the metal industry to recover heavy metals from low-grade ores. *Thiobacillus ferrooxidans* is the most widely utilized in biological ore mining because it can oxidize metal sulfides to sulfuric acid, which solubilizes metals and permits their recovery. This organism is also being used to remove sulfur from coal.

Self-study Quiz

Completion

1. The replacement of populations and community types within a given environment over time is called ________________.
2. The increased fertility of water caused by the addition of nutrients is called ________________.
3. The ultimately established, fairly stable community in a given environment is called the ________________ community.
4. The process of concentration of a recalcitrant molecule in each higher trophic level of a food chain is called ________________.
5. The conversion of ammonium to nitrate is called ________________.
6. The environmental product formed from the oxidation of sulfides by *Thiobacillus* is called ________________.
7. The process of conversion of organic pollutants to inorganic compounds is called ________________.
8. Organisms that grow at high nutrient levels are called ________________.
9. The conversion of the organic form of an element to the inorganic state is called ________________.
10. Nitrogen ________________ is associated with the infected roots of leguminous and nonleguminous plants.
11. Classify the following microbial conversions as
    M = Mineralization
    OR = Oxidation-reduction
    CH_3 = Methylation
    a. Protein ⟶ Nitrate
    b. Nitrate ⟶ Ammonium
    c. Mercury ⟶ Methyl mercury
    d. Methane ⟶ CO_2

Multiple choice

1. Which of the following levels of the ecological hierarchy would be inclusive of the other levels?
   A. Population
   B. Ecosystem
   C. Abiotic surroundings
   D. Community
2. What type of environment would favor oligotrophs?
   A. Anaerobic areas of the gastrointestinal tract of animals
   B. Mountain lake with low level of dissolved nutrients
   C. The stratosphere
   D. A river just below the outfall of raw sewage
3. The organism most frequently associated with metal-leaching processes is
   A. *Cladosporium resinae*
   B. *Desulfotomaculum*
   C. *Thiobacillus ferrooxidans*
   D. *Leptothrix*
4. If samples are removed at 1-day intervals from a vat of cabbage that is being ripened to produce sauerkraut, it is observed that first one microbial group predominates, then another, and then still another. This is an example of:
   A. Eutrophication
   B. Nitrogen fixation
   C. Succession
   D. Trophic levels of the food chain
5. Which of the following groups of organisms make their appearance in the oral cavity with the eruption of teeth?
   A. *Staphylococcus* and *Neisseria*
   B. *Spirochetes* and *Staphylococcus*
   C. *E. coli* and *Staphylococcus*
   D. *Streptococcus mutans* and *S. sanguis*
6. What is the significance of mercury methylation?
   A. The toxic compound methyl mercury is introduced into the food chain.
   B. Mercury is effectively and safely detoxified.
   C. The mercury needed by plants is removed from the soil, causing a loss in fertility.
   D. It creates adverse oxidation-reduction potentials favoring growth of anaerobes.

7. If carbon dioxide is rapidly removed from the atmosphere by photosynthetic organisms, by what process is it replaced?
   A. Mineralization steps in the carbon cycle
   B. Methylation reactions of the carbon, nitrogen, and sulfur cycles
   C. Leaching of carbonates from the soil by H_2SO_4, produced in the sulfur cycle
   D. Carbon dioxide fixation steps of the carbon cycle
8. Finding a high level of DDT in the eggs of fish-eating birds is an example of:
   A. Detoxification by microorganisms
   B. Activation by animal cell metabolism
   C. Eutrophication by an oligotroph
   D. Biomagnification of a recalcitrant molecule
9. What seems to be the limiting factor that prevents overwhelming algal and cyanobacterial growth in natural waters?
   A. Availability of nutrients
   B. The presence or absence of a specific growth factor
   C. The amount of direct sunlight
   D. The maintenance of low pH and aerobic conditions
10. Where would you look to isolate bacteria capable of nitrogen fixation?
   A. Water-logged soils of a swamp or marsh
   B. In water droplets on the leaves of plants
   C. In root nodules of leguminous plants
   D. In a nutrient-rich freshwater lake

Identification

Classify the following interactions as beneficial (+) or antagonistic (−), and identify the specific type of interaction (commensalism, predation, competition, etc.)

Interaction	(+) OR (−)	TYPE
a. Nitrogen-fixing bacteria living in the environment of an algae	________	________
b. *Bdellovibrio* penetrating *Erwinia* cells	________	________
c. Two aerobic bacteria occupying the same low-oxygen niche	________	________
d. *Frankia* in plant nodules	________	________
e. Aerobic bacteria and anaerobic bacteria occupying the same low-oxygen niche in sequence	________	________

Thought questions

1. As a chemist for a company producing herbicides, you synthesize a new fluorinated polymerized hydrocarbon molecule that in the laboratory is very selective in its inhibition of crabgrass and nontoxic to all other plants and animals against which you have tested it. When placed in sterilized soil in which crabgrass is growing, it is absorbed through the roots of the plant and kills the crabgrass after 3 days of exposure without harming any other plants. A vice-president for marketing hears of your discovery, recognizes the tremendous potential of such a product, and begins to push for immediate production and marketing. A. You tell him to slow down because there are several tests that you must run. What are those tests? B. Describe what the ideal results for your test would be.
2. Describe in detail how you would assess the biodegradability of a new compound.

SELECTED READINGS

Books

Alexander, M. *Microbial ecology.* John Wiley & Sons, New York, 1971.

Atlas, R.M., and Bartha, R. *Microbial ecology: fundamentals and applications.* Addison-Wesley Publishing Co., Inc., Reading, Mass., 1981.

Bolin, B., and Cook, R.C. (Eds.). *The major biogeochemical cycles and their interactions.* SCOPE Report No. 21, John Wiley & Sons, New York, 1983.

Burns, R.G. Carbon Mineralization by Mixed Cultures. In Bull, A.T., and Slater, J.H. (Eds.). *Microbial interactions and communities.* Academic Press, London, 1982.

Burns, R.G., and Slater, J.H. (Eds.). *Experimental microbial ecology.* Blackwell Scientific Publications, Oxford, 1982.

Cambell, R. *Microbial ecology.* Blackwell Scientific Publications, Oxford, 1977.

Ehrlich, H.L. *Geomicrobiology.* Marcel Dekker, Inc., New York, 1981.

Ellwood, J.N., et al. (Eds). *Contemporary microbial ecology.* Academic Press, Inc., New York, 1980.

Fletcher, M., and McEldowney, S. Microbial attachment to nonbiological surfaces. In Klug, M.J., and Reddy, C.A. (Eds.). *Current perspectives in microbial ecology.* American Society for Microbiology, Washington, D.C., 1984.

Freney, J.R., and Galbally, I.E. (Eds). *Cycling of carbon, nitrogen, sulfur, and phosphorous in terrestrial and aquatic ecosystems.* Springer-Verlag, New York, 1982.

Payne, J.W. (Ed). *Microorganisms and nitrogen sources.* John Wiley & Sons, New York, 1980.

Skinner, F.A., and Carr, J.G. (Eds). *The normal microbial flora of man.* Academic Press, Inc., New York, 1974.

Stevenson, L.H., and Colwell, R.R. (Eds). *Estuarine microbial ecology.* University of South Carolina Press, Columbia, S.C., 1973.

Summers, A.O. Genetic adaptations involving heavy metals. In Klug, M.J.M., and Reddy, C.A. (Eds.). *Current perspectives in microbial ecology.* American Society for Microbiology, Washington, D.C., 1984.

CHAPTER 19

Aquatic microbiology

F.L. Singleton
R.F. Boyd

Chapter Outline

Introduction

The hydrosphere, that is, the earth's aquatic environment, can be divided into two categories on the basis of location: surface waters and sub-surface waters (Figure 19-1). Of the surface waters, two types are common. **Marine waters**, characterized by high salt concentration (high salinity), cover most of the earth's surface. **Freshwater** environments, located inland, contain low concentrations of salts. A third type of aquatic environment shares characteristics of both marine and freshwater environments. These systems, **estuaries**, are located in coastal regions, where the topography permits the entrance of fresh water into a basin with a free exchange of waters with the ocean (Figure 19-1). Estuaries are therefore transition areas between freshwater and marine systems.

The aquatic community is made up of both eukaryotic and prokaryotic organisms. Although many types of aquatic bacteria are involved in the decomposition of organic materials, they contribute other processes vital to the function of this ecosystem. Such processes include cycling and regeneration of nutrients, serving as a food source for higher–trophic level organisms (filter-feeding invertebrates), and contributing to the overall metabolic activities involved in maintaining the internal stability of an ecosystem.

The primary purposes of this chapter are to (1) discuss the techniques used by microbiologists to study bacteria from aquatic environments, (2) discuss how bacteria adapt to an aquatic environment and their relationship to other members of the aquatic community, (3) discuss how microorganisms are utilized in the recycling of wastewater, and (4) explain the procedures used in the treatment of water for human consumption.

METHODS OF STUDYING AQUATIC MICROORGANISMS

The methods employed to study aquatic microorganisms include those routinely used in other disciplines of microbiology as well as some that have been developed for use with aquatic microorganisms. Such methods include procedures to enumerate the organisms and to study their activity.

Direct techniques of enumeration

The size of aquatic microbial populations can be estimated by direct or indirect techniques. The direct technique that has been employed for decades to enumerate aquatic bacteria employs solid culture media and is referred to as the **spread plate technique.** An appropriate dilution of a water sample is spread over the surface of the medium, and after incubation, the colonies are counted. If it is assumed that one colony represents one cell in the dilution, the number of viable organisms can be determined. This technique yields only an estimate of the bacterial community, since more than one cell can give rise to a single colony. More important, no culture medium or laboratory condition provides the environmental conditions necessary for the growth of all the viable organisms in a natural mixed bacterial community, which may include populations of heterotrophs, autotrophs, mesophiles, psychrophiles, and others. The spread plate technique is superior to the **pour plate technique**, since the former does not expose the organisms to temperatures required to keep agar molten (>42° C). In the pour plate procedure a water sample is mixed with molten agar, and the agar is poured into a sterile petri dish. Bacterial colonies develop on and within the agar.

Often as little as 0.01% (or less) of the total community will grow on a culture medium, but the culturable bacteria have been employed to study the bacterial community of aquatic systems. Although the spread plate technique provides a means for enumerating only a small fraction of a bacterial community, it is used today not only as a means of counting bacteria but also for obtaining data that can be compared with previous studies.

Other techniques employed to enumerate aquatic bacteria or specific bacterial populations include extinction dilution, the most probable number (MPN) technique, and membrane filters. All such techniques require the organisms to grow on or in a culture medium. For example, the **MPN technique** provides a statistical estimate of the number of organisms in a water sample when multiple samples of a series of dilutions are inoculated into a culture medium. The sample is first diluted to the estimated point of extinction (no organisms in the final dilution). A number of replicate dilutions (from 3 to 10 replicates per dilution) at the dilution step before the extinction step are used in this technique. The replicate dilutions are scored either positive (growth) or negative (no growth) after they have been placed in suitable growth media and incubated a designated period of time. The number of positives and negatives for each dilution are counted and, with the help of a statistical chart (Table 19-1), a most probable number of organisms per milliliter of the original sample can be determined. The advantage of the MPN technique is that it avoids use of solid media. This is important because many algae do not form colonies on solid media. MPN tests, however, are more laborious and less accurate than plate counts and one cannot identify organisms.

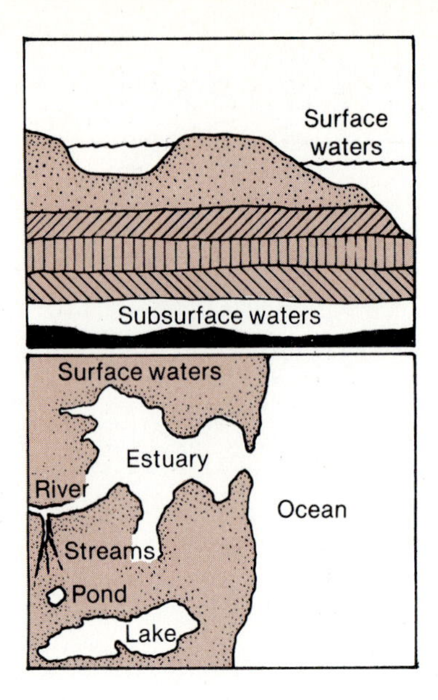

Figure 19-1 *Major types of aquatic systems.*

One of the simplest and quickest methods of bacterial enumeration is the **membrane filter technique**. Membrane filters are employed to collect and enumerate aquatic bacteria by filtering a known volume of water and placing the filter onto an appropriate culture medium. After a suitable incubation period, colonies are counted.

The most direct method to determine the number of individuals in a microbial population or community is

Table 19-1 *Abbreviated most probable number (MPN) table*

Number of positive tubes at the stated dilution			MPN (cells/ml)
10^0	10^{-1}	10^{-2}	
0	1	0	0.18
1	0	0	0.20
1	1	0	0.40
2	0	0	0.45
2	0	1	0.68
2	1	0	0.68
2	2	0	0.93
3	0	0	0.78
3	0	1	1.1
3	1	0	1.1
3	2	0	1.4
4	0	0	1.3
4	0	1	1.7
4	1	0	1.7
4	1	1	2.1
4	2	0	2.2
4	2	1	2.6
4	3	0	2.7

To understand the table, take a hypothetical example in which a wastewater sample is diluted to 10^{-1} and 10^{-2}. We take five 1 ml samples from undiluted wastewater and from each of the dilutions inoculate them into suitable growth media. The dilutions at which growth is observed or not observed are noted. Following incubation we find that four tubes show growth at 10^0, three tubes show growth at 10^{-1}, and no tubes show growth at 10^{-2}. If we look at the chart, we find that there are 2.7 viable cells/ml of wastewater.

direct microscopic observation. Numbers of planktonic organisms are usually determined in this manner. The planktonic community can be divided into three major groups: (1) **phytoplankton** (free-floating algae), (2) **zooplankton** (free-floating animals), and (3) bacteria, which are often referred to as **bacterioplankton.** However, when one is dealing with bacteria, microscopic observation of a water sample is of limited value, since bacteria are transparent and the number of bacteria in most waters (in the range of 1×10^6 per milliliter) is too low to allow for accurate enumeration. Therefore the organisms must either be stained to be visualized or observed by a microscopic technique other than ordinary light microscopy. In either case, when one is working with most natural water sources, the bacteria must be concentrated. This is usually accomplished by filtering a water sample through a membrane filter with a pore size of 0.2 or 0.45 μm.

Figure 19-2 *Bacterial cells (light areas) stained with fluorescent dye and observed with epifluorescent microscope.*

A combination technique of membrane filtration and staining the cells can be of limited value, since it is not always possible to differentiate between cells and small particulate materials. Therefore improvements in methods of microscopic counting of aquatic bacteria include the use of fluorescent dyes (such as acridine orange) that have an affinity for binding to specific cellular components. When used in conjunction with **fluorescent microscopy** (see Chapter 3), such stains allow the investigator to differentiate cells from particulate materials (Figure 19-2).

Another application of fluorescent microscopy to aquatic microbiology includes the use of **fluorescent antibodies** to enumerate specific types of aquatic microorganisms. When it is properly applied to a water sample containing cells with which the antibody is reactive and observed with a fluorescent microscope, only those cells with the antibody attached will be fluorescent.

One of the major drawbacks of microscopic techniques to enumerate bacteria is the inability to differentiate living and dead cells. Microscopic counting of aquatic bacterial populations will yield counts that overestimate the metabolic potential of the populations, since both living and dead cells are counted.

Indirect techniques for determining community size

Techniques have been developed that permit estimating the size of the microbial community indirectly. These are based on **quantifying a specific cellular component** (Table 19-2). In the past, determinations of cellular carbon, nitrogen, phosphorus, or other chemicals were employed as an index of the biomass, that is, the quantity of organisms, in a pure culture. However, such techniques are of little value for natural water samples, since the water may contain noncellular quantities of such materials. Therefore, specific cellular components such as ATP, muramic acid, poly-β-hydroxybutyrate, and lipids have been successfully employed to estimate the microbial biomass in water samples or aquatic sediments. In addition, by differentially filtering water through filters of selected pore sizes, it is possible to separate the various components of the community, that is, separate free-living bacteria from phytoplankton and zooplankton.

The heterotrophic activity of aquatic microbial communities is usually measured with radio-labeled substrates such as ^{14}C glucose, amino acids, or other organic substrates. In heterotrophic activity assays, a water sample is placed in a clean, sterile container and a low concentration of the substrate is added. After appropriate incubation periods, the organisms are killed and acid is added to the samples to drive off any ^{14}C carbon dioxide formed during metabolism of the substrate. The cells and the ^{14}C carbon dioxide are collected, and the amount of radioactivity is measured with a scintillation counter. From this information the rate of uptake and mineralization of the substrate is calculated. These parameters are indicative of the potential activity of the microbial community.

Table 19-2 *Components of microbial cells that have been employed as a means of determining the microbial biomass in aquatic systems*

Component	Comment
Total organic carbon	Assay does not distinguish between living and nonliving
Total organic nitrogen	Assay does not distinguish between living and nonliving
Adenosine triphosphate	Assay measures only living organisms; does not differentiate types of organisms
Lipopolysaccharide	Assay is specific for gram-negative bacteria
Poly-β-hydroxybutyrate	Assay is specific for this bacterial storage product
Lipids Phospholipids Glycolipids	Assays measure total quantity of indicated lipids, both living and nonliving

ADAPTATION TO THE AQUATIC ENVIRONMENT

Organisms inhabiting an aquatic environment are subjected to spatial and temporal fluctuations in physiochemical parameters. For example, in a deep lake the difference in temperature between surface and bottom waters is several degrees. Likewise, the temperature variation between seasons can be as much as 20° to 30° C. Similar variations in dissolved oxygen concentrations, pH, light intensity, and other parameters occur.

Aquatic microorganisms exist in the environment in which the concentration of nutrients is usually low. For example, the concentration of dissolved organic carbon in open ocean waters is in the milligrams-per-liter range. Bacteria, that grow and survive on these low concentrations of nutrients must possess efficient uptake mechanisms and efficiently use nutrients that become available. Such bacteria are referred to as **oligotrophic** (see p. 394 on nutrient limitation). Not only are oligotrophic bacteria adapted for growth on low concentrations of nutrients, but many are inhibited by the high nutrient concentrations characteristic of common laboratory culture media, that is, in the range of grams per liter.

Not all aquatic bacteria are oligotrophic. Indeed, in many nutrient-rich waters a large proportion of the culturable bacteria are **eutrophic (copiotrophic)** and are not inhibited by high nutrient concentrations. These bacteria are sometimes referred to as heterotrophic. However, many microbiologists prefer not to use the term in this way, since it can be confused with the term defining the type of metabolism by which an organism obtains carbon and energy.

The occurrence and distribution of oligotrophic and eutrophic bacteria in an aquatic environment is dependent on the concentration of nutrients. Although both nutritional types occur in most aquatic environments, some investigators consider the oligotrophs to be "true"

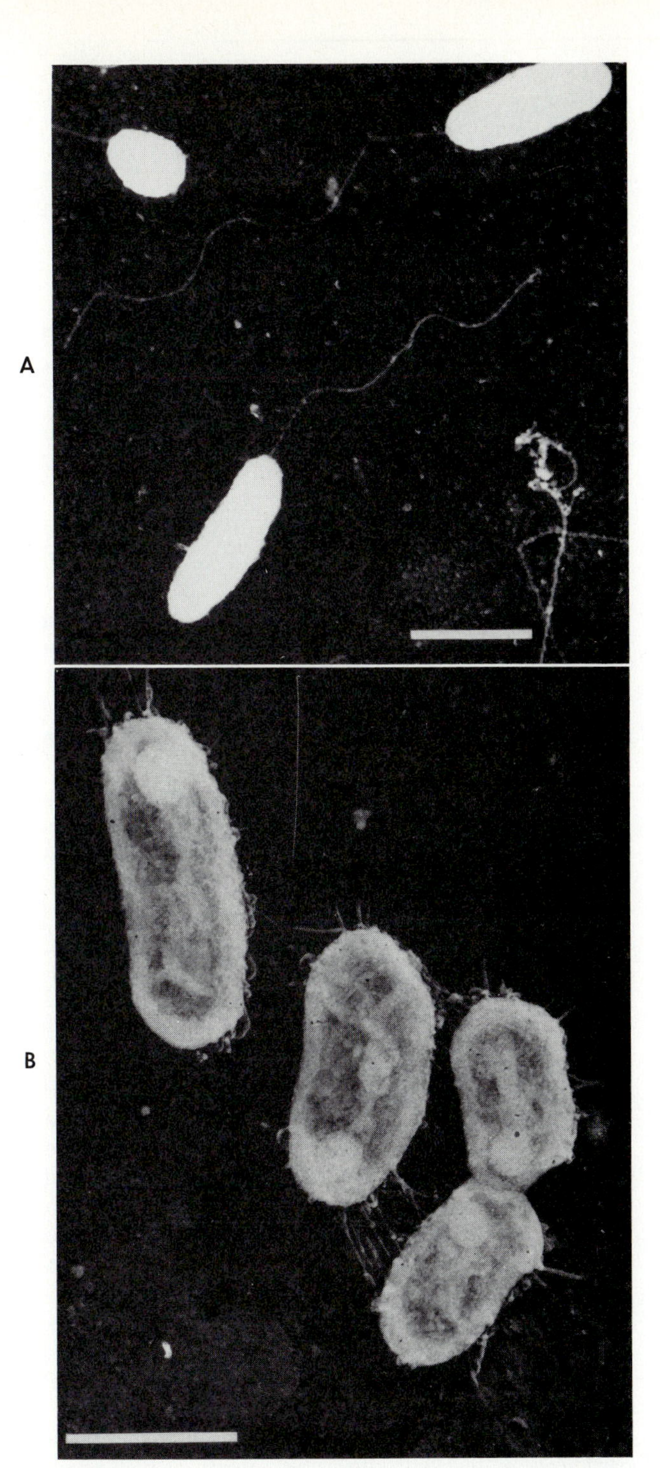

Figure 19-3 *Starvation induced effects on* Vibrio *species. Scanning electron micrograph (bar = 1.0 μm).* ***A,*** *Log phase cells of* Vibrio *washed in artifical seawater. Note polar flagellum.* ***B,*** Vibrio *starved for 5 hours in artificial seawater. Note surface morphology has changed such that large number of filaments have been produced.*

aquatic bacteria, since they possess mechanisms for survival in a nutrient-poor environment. For example, in response to extended periods of limited nutrient supply, some aquatic bacteria decrease their biomass and metabolic activities until nutrients become available. These cells of reduced size and activity survive for weeks or months and, when nutrients become available, resume normal metabolic activities. Occasionally in nutrient-starved bacteria, a change in cell surface morphology enables the organism to attach to areas of increased nutrient concentration (Figure 19-3). The areas of nutrient scavenging may be particulate surfaces or the surfaces of other organisms. Not only do these associations provide a favorable microenvironment for the bacteria, but this is another ecological strategy for several species to coexist within a common environment.

The distribution of algae in aquatic environments is also influenced by the concentration of nutrients in the water. For example, diatoms are more abundant in waters of low nutrient concentration. Since phytoplankton are photosynthetic, their distribution is influenced by the availability of nutrients required for metabolic processes involved in photosynthesis.

Whereas the distribution of bacteria and phytoplankton are directly influenced by the availability of nutrients, zooplankton are seemingly indirectly influenced by this parameter. The major food sources of zooplankton are bacteria and algae. Therefore the availability of a food source determines the ability of a predator species to survive within a system.

MICROBIAL ASSOCIATIONS IN THE AQUATIC ENVIRONMENT

Bacteria and animals

Bacteria that exist in association with zooplankton or other aquatic animals are referred to as **epizooic**. There is doubt as to whether specific associations exist between aquatic animals and specific bacteria species in the sense that a given type of bacteria will associate with a particular type of organism. Rather, associations such as this appear to depend on which bacteria first colonized the host. In such an association the bacteria are provided with a convenient source of nutrients since many bacteria colonize the feeding appendages and the mouth parts (Figure 19-4). It is believed by many that in an association between bacteria and zooplankton the host derives the benefit of a convenient source of vitamins and other growth factors.

Bacteria and plants (algae)

Bacteria that exist in association with plants or algae are referred to as **epiphytic**. Epiphytic bacteria have been the subject of numerous studies designed to determine

Figure 19-4 *Bacteria attached to mouth parts of copepod. Arrow points to chain of spherical cells.*

Table 19-3 *Cyanobacteria common to many aquatic environments*

Genus	Characteristic
Anabaena	Common in blooms
Coelosphaerium	Common in blooms
Lyngbya	Common to most aquatic environments
Oscillatoria	Common to most aquatic environments

Concept Check

1. Describe the techniques of spread plate, pour plate, and most probable number for estimating microbial populations. What indirect techniques are used for estimating microbial populations?
2. How do aquatic microorganisms respond to nutrient starvation?
3. What are the predominant types of associations between microorganisms and other organisms in the aquatic environment? How can such associations benefit the organism? Why aren't microorganisms found on the surface of diatoms?

the exact relationships between these organisms. It is to the benefit of both the algae and the bacteria that such associations occur. Within an aquatic system most of the photosynthesis is carried out by the algae and cyanobacteria (Table 19-3) in fresh water and by algae in oceans. However, of the organic carbon produced during photosynthesis, a large fraction (30% or more) can be lost from some algal cells to the environment. Loss of organic carbon usually is in the form of glycolic acid. Bacteria in the slime layer surrounding the algal cells, as well as planktonic bacteria, use glycolic acid and similar organic compounds as a growth substrate. In turn, the bacteria provide the algae with some vitamins and other growth factors.

Evidence for the association between algae and bacteria has been obtained from studies carried out in many aquatic environments; these studies have demonstrated the distribution patterns of these organisms to be parallel. Also, examination of the slime layer of many algal cells reveals a very heavy bacterial load. However, it is important to note that not all members of the phytoplankton are covered with bacteria. Diatoms are known to be free of bacteria. This is important for the survival of the diatoms, since they exist within delicate exoskeletons containing pores to allow the entrance and exit of water (and nutrients). A heavy bacterial load on the surface of these organisms would interfere with the exchange of water. Diatoms may prevent bacteria from attaching to the exoskeleton by releasing various organic acids that are inhibitory to the bacteria.

Organisms and inanimate objects (surfaces)

Aquatic microorganisms are associated with inanimate particulate materials in the water column as well as on the surface of the water. Such organisms are referred to as **epilithic**. This type of association is apparently more beneficial to bacteria than to other microbes. These bacteria are present at the interface between a particle and the water. The other major interface in the aquatic environment is between the water surface and the atmosphere. Organisms located at interfaces have a greater

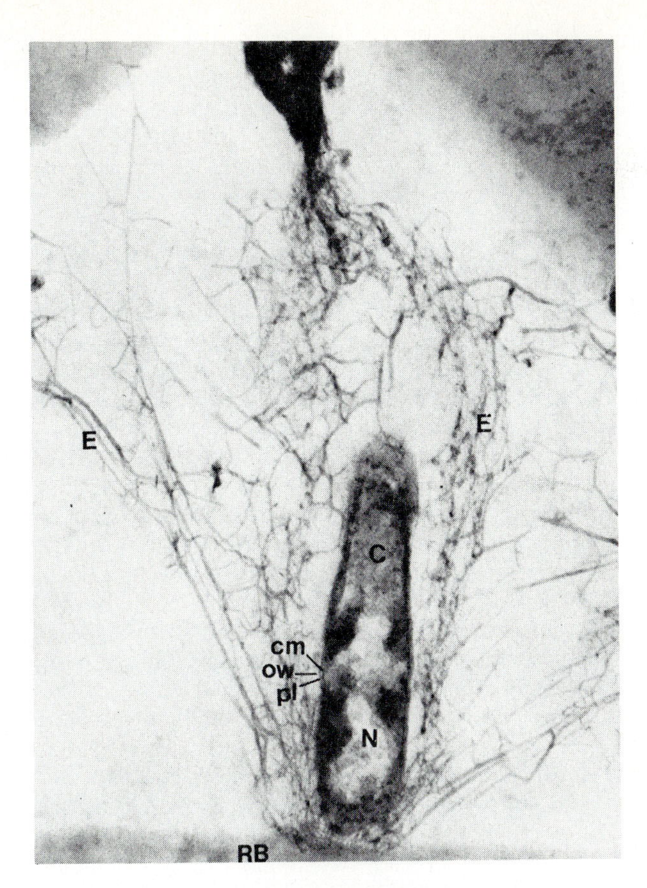

Figure 19-5 *Transmission electron micrograph of thin section of bacterium attached to surface of resin block* (RB). *Exopolymer* (E) *appears as thin fibrils attaching cells to surface. Note internal structure of attached bacterium.* C, *Cytoplasm;* N, *nuclear region;* cm, *cytoplasmic membrane;* pl, *peptidoglycan layer;* ow, *outer wall.*

chance of survival, since the nutrient concentration is many times that found in the water column. Nutrient concentrations at interfaces can support the vigorous growth of bacteria. Therefore the number of bacteria present at an interface is much larger than that in the open water column.

At solid-liquid interfaces, proteins, various other organic molecules, and inorganic molecules are concentrated as a result of electrostatic attraction and other physical forces. At the liquid gas interface on the water's surface, hydrophobic molecules such as lipids and some types of proteins are found in increased concentrations. Regardless of the type of interface, many of the materials present can be used by bacteria as a growth substrate.

Microorganisms become located at interfaces by chemotactic response to the region of increased nutrients (see Chapter 5), or they are transported there by currents. Also, some bacteria become localized at the surface interface by increasing the hydrophobicity of the cells. However, those cells at a liquid-solid interface are initially maintained in position by electrostatic attraction. Many bacteria become fixed in place by producing an extracellular material (usually polysaccharide or protein in composition) that serves to glue the cells to the surface of the particle (Figure 19-5).

Table 19-4 *Concentration of the nine major ions in natural seawater*

Ion	Percent of total (dry weight)
Chlorine	55.0
Sodium	30.6
Sulfate	7.7
Magnesium	3.7
Calcium	1.2
Potassium	1.1
Bicarbonate	0.4
Bromine	0.19
Strontium	0.02

AQUATIC MICROBIAL COMMUNITIES

The presence of free-floating bacteria has been observed in virtually all aquatic systems. However, the numbers and types of planktonic bacteria depend in large part on the characteristics of a system. The existence of a true planktonic community in streams has been questioned, since any planktonic organisms would be moved downstream. Therefore there would have to be a continual input of organisms in the headwaters of those systems. However, in open-water sources such as lakes and the ocean, planktonic bacteria are known to exist.

Marine environment

Open ocean surface waters. The marine environment covers approximately 71% of the earth's surface and has an average depth of approximately 3500 m. Although the majority of the solar radiation reaching the earth enters the marine environment, the average temperature of ocean waters is less than 4° C. Another characteristic of marine waters is a high salt content. The salinity of open ocean waters ranges from 34.0 to 35.8 parts per thousand (ppt), depending on the latitude. Thus marine waters have a relatively constant chemical composition. The typical salt content of an open ocean water sample is given in Table 19-4. Note that the nine ions listed constitute more than 99% of the total amount of salt in the sample. Other elements are present in ocean waters but in very low concentrations. The exact chemical composition of an ocean water sample depends on the physiochemical parameters of a given region.

Marine bacteria. The bacterial community of open ocean waters comprises numerous species. Although those species that grow on culture media have been well characterized, the degree to which they represent the entire community is unknown. The culturable portion of open ocean bacterial communities is usually dominated by **chemoheterotrophic gram-negative bacteria** (80% to 90% of the total), especially members of the genera *Pseudomonas* and *Vibrio.* Other gram-negative genera commonly isolated from marine waters include *Photobacterium, Flavobacterium, Acinetobacter, Altermonas,* and *Alcaligenes.* Species of these genera are associated with marine animals. These latter genera make up a minor fraction of the total as compared with *Pseudomonas* and *Vibrio.* A larger proportion of the culturable bacteria isolated from the surface microlayer, or microfilm, are gram-positive, since these bacteria tend to be more resistant to the effects of solar radiation. *Staphylococcus* and *Micrococcus* species are commonly isolated from the surface microlayer. (Species of these groups are also found on the skin surface of vertebrates.)

Members of the genus *Photobacterium* (and some *Vibrio* species) share a trait that is apparently restricted to a few species of marine bacteria—the ability to emit visible light (Plates 20 and 21). They are widely distributed in the marine environment and may be free living, saprobes, commensalistic, or parasitic. The largest number are found as commensals in the intestinal tract of marine animals. These species hydrolyze **chitin**, a major component of the shells of marine invertebrates. The marine luminous bacteria differ from their terrestrial counterparts in that the former have a specific and relatively high requirement for sodium. Sodium is required for certain enzymatic reactions as well as for maintenance of the integrity of the cell wall. Many species of fish produce their own light, but some have specialized organs in which the luminuous bacteria reside and provide light. This relationship benefits the bacteria, which receive nutrients from the host, while the fish is supplied light, which may be used for attracting prey, as a means of communication, or as a mechanism for avoiding predation. The light-emitting reaction of luminous bacteria is catalyzed by the enzyme **luciferase**. Molecular oxygen is used to oxidize reduced flavin mononucleotide (FMN) and an aliphatic aldehyde according to the following reaction:

$$\mathrm{FMNH_2 + O_2 + RCHO \xrightarrow{luciferase} FMN + H_2O + RCOOH + light}$$

The light emitter is believed to be a luciferase-flavin complex in which the flavin is brought to an excited state and a portion of the excitation energy is released as a photon of light. The function of the reaction for the bacterial cell is not clear, but it may be an alternate route for electron transport and the formation of energy during periods of low oxygen tension, since luciferase is extremely sensitive to even small amounts of oxygen.

One of the characteristics of marine habitats is that they are dominated by photosynthetic microorganisms. Where anaerobic conditions exist and light is present, anoxygenic photosynthetic bacteria and not chemoheterotrophs tend to dominate. Surface waters of the open ocean also contain significant populations of oxygenic cyanobacteria (Table 19-3). In addition to contributing to the total quantity of the biomass, the cyanobacteria are responsible for the addition of significant quantities of nitrogen to the marine ecosystem as a result of their nitrogen-fixing activities (see Chapter 13). In tropical sediments where there are extremes of salinity and high light intensity, stromatolites occur. **Stromatolites** are fossilized limestone mats of cyanobacteria (Plate 22), and some have been found that are over 3 billion years old. Some scientists suggest that part of the reason for the dominance of photosynthesizers is the high ionic composition of seawater. It has been hypothesized that the high amounts of carbon and energy required to maintain osmotic pumps are best fulfilled by photosynthesizers. Mechanisms for maintaining an osmotic balance in the cell may include the accumulation of several types of molecules, both inorganic and organic (also see Osmotic Pressure in Chapter 14). Accumulation of potassium ions (K^+) has been demonstrated to be an important osmotic regulatory device by some chemoheterotrophs, but organic molecules (amino acids, polyols, and disaccharides) may also be important for photosynthetic microorganisms. For example, glycerol, produced from stored starch, is an osmoregulatory solute in certain algae. Glycerol also appears to play a part in protecting enzymes in the cell from the inhibitory effects of high concentrations of inorganic ions.

In general, the number of bacteria in the marine environment is approximately 1×10^6/ml, but this number decreases with increasing distance from land. Increased bacterial loads of near-shore waters are due to increased nutrient concentrations that support growth of the **autochthonous (native)** marine bacteria. Near land masses, nutrients enter the marine environment through surface runoff as well as through rivers and other freshwater systems emptying into the ocean. Not only do nutrients enter marine waters in this manner, but large numbers of **allochthonous (foreign)** bacteria are also contributed. Therefore, with increasing distance from land, nutrients are diluted, allochthonous bacteria die off, and the total number of bacteria decreases.

Many factors are known to influence the dying off of allochthonous bacteria in the marine environment. Among these are inhibition by increased solar radiation exposure, nutrient deprivation, predation by filter feeding invertebrates, high salt concentrations, and heavy-metal toxicity, especially by nickel ions.

In addition to a horizontal distribution pattern, that is, decreasing numbers with increasing distance from land, the bacterial community also has a vertical distribution pattern. Relative to numbers of organisms in the open waters, the largest numbers of bacteria are found on the surface of the ocean. A thin microlayer covers the surface of ocean waters. This microlayer (microfilm) is approximately 30 μm thick and is composed of an assortment of hydrophobic materials, including hydrocarbons, lipids, and various organic and inorganic molecules. Many of the materials making up the surface microlayer can be used by bacteria as a growth substrate. Therefore relatively large numbers of bacteria are associated with this microlayer.

Below the surface microlayer, the number of bacteria decreases significantly. Within the water column, the number of bacteria decreases slightly with increasing depth. This is probably a result of a combination of factors, including decreased temperature (<4° C) and increased hydrostatic pressure. Within the upper levels of the water column, the distribution of bacteria is relatively constant, since wave action and surface currents prevent stratification. This distribution pattern appears to be unchanged in the upper few hundred meters of many regions of the ocean. However, in some regions, a second increase in the bacterial community has been observed to occur in conjunction with increases in the phytoplankton community. In such waters, the phytoplankton are stratified at a certain depth (usually between 20 and 50 m). As the phytoplankton grow and release organics into the water column, the bacterial community increases.

Marine algae. The distribution of marine algae is similar to that of the bacteria—increased numbers of organisms in coastal waters. The composition of the phytoplankton community differs not only with increasing distance from land but also with latitude. For example, this community is dominated by diatoms in the cold Arctic and Antarctic waters. In the warm topical waters the flagellated algae are dominant.

The phytoplankton community of open ocean waters is restricted to the surface waters that receive light—the **photic zone.** The photic zone varies in depth but is usually less than 100 m, depending on the characteristics of the water of a particular region. The depth of the photic zone is determined primarily by the concentration of suspended particulate materials, including planktonic organisms, in the water column. Planktonic organisms absorb and/or scatter (reflect) light in the water column. Those organisms that absorb light do so by virtue of intracellular pigments. Such pigments are employed by photosynthetic bacteria and algae to trap solar radiation for the photosynthetic process. Many marine (and freshwater) nonphotosynthetic bacteria contain pigments that serve to protect the cells from the adverse effects of solar radiation. The presence of pigments in organisms in the water column influences the color of the water column. For example, coastal waters are greenish colored as a result of the presence of yellow pigments in the phytoplankton.

On the average, the photosynthetic activities of the phytoplankton community annually contribute approximately 50 g of carbon per square meter of open ocean surface water. This amounts to a total annual contribution of more than 16 billion tons of carbon to the open waters of the marine environment, or 35% of the world's primary production. Much of this carbon is recycled within the marine system and returned to the atmosphere (see Carbon Cycle).

Marine zooplankton. In waters containing large populations of bacteria and phytoplankton, the zooplankton are also present in increased numbers. These filter-feeding organisms, which serve as the primary food source for many marine organisms of higher tropic levels, consume the smaller phytoplankton and bacteria. The composition of the zooplankton community, like that of the bacteria and phytoplankton, varies within the ocean environment, depending on the physicochemical characteristics of the water and the availability of food. The relationship between the occurrence of zooplankton and their food source is exemplified by the fact that many of the zooplankton carry out diurnal (repeated every day) migration within the water column. Diurnal migration is beneficial to the zooplankton, since during the night they migrate to the photic zone and feed on the phytoplankton. During the daytime they migrate back to the dark, cool waters, where they hide from predators.

Marine zooplankton also demonstrate increased numbers near land masses—a result, no doubt, of the availability of food (bacteria and algae). The species composition of the zooplankton population is dependent on the region of the ocean in consideration.

Marine sediments. The sediments of the oceans contain a large microbial community that is dominated by the bacteria. There are significantly larger numbers of bacteria in the sediment than there are in the overlying waters. Also, the composition of the sediment community differs considerably from that of the water column. The presence of a large community is due to the high concentrations of nutrients in the sediment. The decomposition of chitin, the major structural component of the exoskeleton of invertebrates, and the tissues of other organisms occurs in the sediment. Decomposition processes and their rates are depth dependent, since temperature decreases and hydrostatic pressure increases with increasing depth. Therefore recycling of materials occurs at a faster rate in coastal waters, especially in tropical or subtropical waters, than in the deep ocean. The sediment therefore supports a larger and more active community than the nutrient-poor water.

The number of bacteria as well as the composition of the community changes with depth into the sediment as well as the distance of the sediment from the surface. With increasing depth in the sediment, the total number of bacteria decreases and the percentage of gram-positive bacteria increases.

The proportion of the bacterial community demonstrating a requirement for oxygen also changes with sediment depth. Only the upper most levels of the sediment are aerobic. Therefore, below the aerobic layer of the sediment this environment is anaerobic and is dominated by anaerobes, primarily of the genus *Clostridium.* Other anaerobes include *Desulfovibrio,* which reduces sulfate to H_2S. Below the sulfate area of the sediment reside the anaerobic methanogens. Numerous *Bacillus* species have also been isolated from marine sediments.

The deep ocean. Some trenches in the ocean floor have recorded depths of more than 10,000 m. At one time it was believed that the depths of the oceans were without life; they appeared to be an environment that could not be inhabited. For example, this environment is characterized by low temperature (<4° C), complete absence of light, and extremely high hydrostatic pressures (as high as 1100 atm; see Chapter 14). The microbial community of these waters is dominated by bacteria and zooplankton. Various crustaceans and some fishes also inhabit this seemingly hostile environment.

As early as 1884 the existence of bacteria in water samples collected from depths of 8400 m was demonstrated. Subsequent studies have demonstrated the existence of viable bacteria at depths of more than 10,000 m (see Chapter 14).

Bacteria inhabiting surface waters of the ocean have decreased metabolic activity when exposed to conditions characteristic of the deep ocean. Changes in activity are a result of the combined influence of low temperature and high pressure. Since bacteria have a slower metabolism in a low-temperature or low-temperature and high-pressure environment, the presence of viable bacteria in deep ocean water samples was believed to be a result of cells settling from the overlying surface waters. However, some investigators hypothesized that the deep ocean contained a resident bacterial community that had adapted to conditions of low temperature and high pressure. Such organisms would therefore be **psychrophilic** and **barophilic**. Studies of the activity of the bacterial community of deep ocean waters demonstrated very slow rates of metabolism—so slow, in fact, that an incubation period of many weeks was required to measure them. This led many to believe that there was not a true barophilic community, or if there was one, it was not present in the water column. Subsequent studies of the bacterial flora of the intestinal tract of invertebrates that also inhabit the deep ocean have demonstrated the existence of metabolically active bacteria. Furthermore, some of these bacteria are more metabolically active under high pressure. Recently the existence of bacteria that require high pressure for growth has been reported. Thus the deep ocean contains a barophilic bacterial community that is involved in the recycling of organic materials and that appears to reside primarily in the intestinal tracts of animals inhabiting these waters.

Another deep sea microbial community was discovered in the early 1980s in a systematic search for volcanic activity. Researchers discovered that deep sea thermal vents (produced by volcanic activity) emit sea water that has been transformed into a "hydrothermal" fluid. This fluid is rich in metals, hydrogen sulfide, and molecular hydrogen leached from volcanic rock. The vents appear from 1 to 3 miles down on the ocean floor, depths at which life is usually minimal to nonexistant, but microorganisms were found at three major sites: (1) in emitted warm vent water, (2) in mats associated with the vent system, and (3) in association with vent invertebrates. In each instance, the prokaryote relies on chemosynthesis as its life support system. The vent waters contain reduced sulfur compounds such as hydrogen sulfide, and the prokaryotic organisms predominating there belong to the genus *Thiomicrospira.* On the surfaces surrounding the vents are deposits of manganese and iron on which microbial mats are formed. The bacteria associated with these mats are *Thiomicrospira,* species resembling the stalked *Hyphomicrobium* (see Point of Interest), and the filamentous *Thiothrix.* Thermophilic species of the genus *Methanococcus* have also been isolated in the vents. Bacteria also form a symbiosis with two of the major invertebrates: clams and tube worms (see Fig. 13-9). The tube worms, which are approximately 2.5 m in length and 5 cm in diameter, are mouthless and gutless and possess a body cavity consisting of spongy tissue, the **trophosome**. Most of the trophosome is composed of sulfur-oxiding bacteria that possess the Calvin-Benson cycle enzyme ribulose biphosphate carboxylase and the enzymes for sulfur oxidation. Carbon dioxide is derived from bicarbonate released from basalt rock on the ocean floor. The sulfur-oxiding bacteria form the base of the food chain in the thermal vent and provide food for the animal species. The symbiosis between the bacteria and tube worms (and other animals) is **endosymbiotic** and **mutualistic**. The tube worm receives reduced carbon molecules from the bacteria and in turn provides the bacteria with materials required for its chemoautotrophic existence: carbon dioxide, oxygen, and hydrogen sulfide. Hydrogen sulfide in the vent wa-

Point of Interest

HYPHOMICROBIUM

Bacterium *Hyphomicrobium* belongs to a diverse group of organisms that produce stalks, hyphae, and various appendages during the growth cycle. It can be found in any aquatic habitat, especially those that are nutrient deficient: manganese deposits, water pipes, laboratory water baths, and acid mines. The genus *Caulobacter* is related to *Hyphomicrobium,* and both have been studied for their unusual cell cycle in which there is unequal cell division and the formation of dissimilar daughter cells by budding.

The cell cycle of *Hyphomicrobium* can be divided into three stages: **swarm cell, hyphal development,** and **bud formation** (left figure). Swarm cells are oval to bean shaped and are motile by a polar flagellum. The flagellum not only functions as a mechanism for dispersal in the environment but may also be used for attachment to solid substrates. At the end of the swarm stage, hyphae begin to develop, usually at the poles of the cell (right figure). The hyphae are usually five times longer than the swarm cell, and some species may produce as many as five hyphae. During hyphal development, there is no change in the morphology of the main cell body. Several mechanisms have been suggested for hyphal development, including increasing the surface area for nutrient absorption in nutrient-deficient habitats, preventing the organism from sinking in aquatic habitats, and an alternate site of division for cells that do not divide equally by binary fission.

The bud stage is observed as a swelling at the tip of the hypha. Eventually the bud enlarges to approximately the same size as the mother cell. Other daughter cells can be produced from any one mother cell, provided their development is preceded by hyphal growth. The number of daughter cells is limited to eight in *Hyphomicrobium* because the components of the mother cell are not renewed, as might occur in binary fission. Therefore the mother cell undergoes an aging process. At some time during development, a flagellum is produced by the bud; under microscopic examination it appears that the bud is towing the mother cell. When the bud attains a critical size, it separates from the mother cell to become a swarm cell.

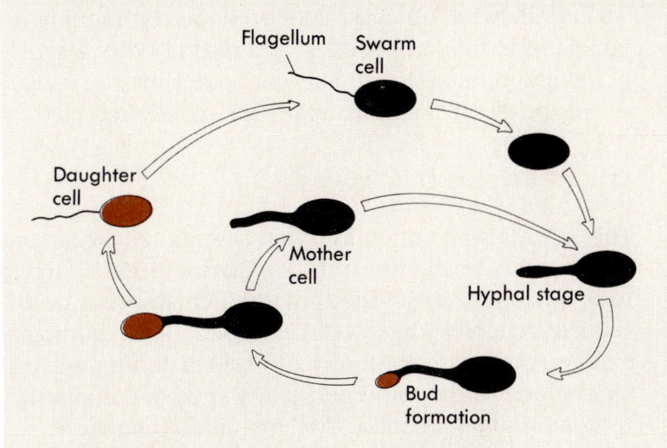

Diagrammatic representation of cell cycle of Hyphomicrobium. *Swarm cell represents motile stage of organism, enabling it to spread in its habitat. Motility is usually produced by single polar flagellum. Later during growth, hyphae are produced at poles* (Hyphomicrobium *usually produces only one hypha, but other budding bacteria may produce hyphae at both poles). Towards end of hyphal stage, there is swelling at tip of hypha, which develops into bud, this is first indicator of daughter cell production. Bud increases in size until it reaches maximum size and then is released from mother cell.*

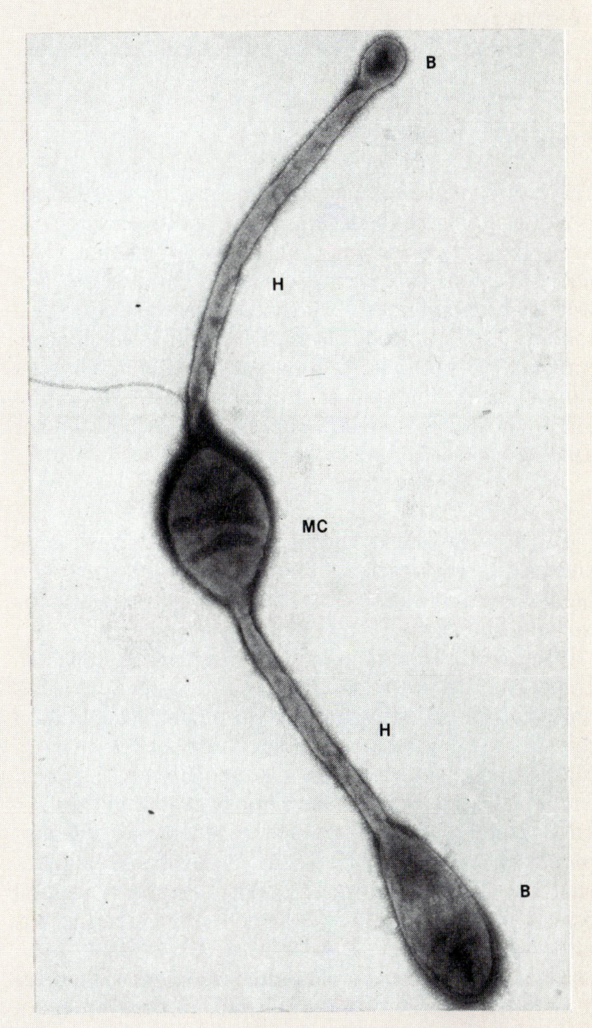

Electron micrograph of Hyphomicrobium. *Mother cell* (MC) *has two hyphae* (H)*, at tips are two budding daughter cells* (B) *in different stages of development.*

ter is absorbed by the tube worm into its circulatory system. Hydrogen sulfide is a very toxic molecule that can block respiration by poisoning cytochrome oxidase and can also bind to hemoglobin thereby blocking the binding sites for oxygen. The tube worms have evolved a mechanism for preventing these toxic reactions. The hemoglobin molecule of tube worms is not transported in red blood cells but instead circulates freely in the serum. It binds hydrogen sulfide at a site distinct from the oxygen-binding site. When complexed to hemoglobin, hydrogen sulfide is not oxidized. The hydrogen sulfide is then transported to the cells of the trophosome, where it is absorbed by the bacterial symbionts. Only in the bacteria is hydrogen sulfide oxidized, and it is this oxidation that drives the synthesis of ATP and fixation of carbon dioxide. In this unusual environment the vent bacteria serve as the "green plants" and their "sunlight" is hydrogen sulfide and reduced carbon compounds.

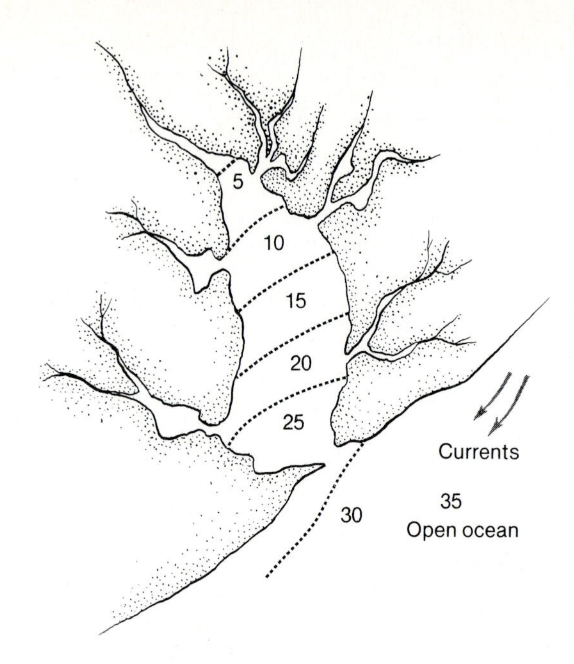

Figure 19-6 *Pattern of salinity in hypothetical estuary. Numbers represent salinity in parts per thousand.*

Estuarine environment

The estuarine environment is unique in the sense that it is a transition zone between marine and freshwater systems. In a typical estuary (Figure 19-6) fresh water enters in regions away from the ocean. Since there is a free exchange of waters with the ocean (tidal action), a salinity gradient extends from the very low salt concentrations (<5 ppt) where fresh water enters to salt concentrations that are typical of ocean water (>30 ppt). At any point in the estuary, bottom waters have a higher salinity than surface waters.

The microbial community of an estuary consists of organisms entering in fresh waters and in marine waters. The estuarine environment supports a very large and active biological community. The estuaries are highly productive and supply humans with tremendous quantities of shellfish and other organisms.

The microbial community of an estuary is similar in composition to that of other aquatic systems. Species characteristic to the marine environment are common to estuary, especially the regions that contain waters of high salinity. Some organisms are apparently restricted to the estuarine environment. For example, in the bacterial community, the dominant genera are *Pseudomonas* and *Vibrio.* Of the *Vibrio* species commonly isolated, *V. parahaemolyticus* appears to be a natural estuarine organism whose distribution is limited to this environment. Another closely related species, *V. cholerae,* the causative agent of epidemic cholera in humans, also appears to be a natural inhabitant of the estuarine environment.

Since estuaries receive inputs of nutrients from freshwater systems as well as surface runoff from the environs, estuarine waters are more densely populated with microorganisms than ocean waters. The bacterial community is one or more orders of magnitude larger than that of ocean waters. The phytoplankton community also contains more individuals than is the case with ocean waters. Likewise, estuaries support large communities of invertebrates and vertebrates. As a result of the presence of increased numbers of different organisms, the estuary is very productive and materials are rapidly recycled.

Freshwater environments

The microbial community of freshwater environments differs considerably from that of marine waters. Organisms inhabiting freshwater environments do not have a requirement for high salt concentrations. Although many organisms inhabiting fresh waters and marine waters belong to the same genus, the species composition of communities of these systems differs markedly. For example, the bacterial community of freshwater environments is made up predominantly of members of the genera *Pseudomonas* and *Vibrio,* as is that of marine waters, but different species are characteristic of each type of system.

The number of organisms in freshwater communities is dependent on the physiochemical characteristics of the system in question. For example, fewer organisms are present in systems containing nutrient-poor waters, such as pristine mountain streams, than those containing nutrient-rich waters, such as may be found in lowland streams or rivers.

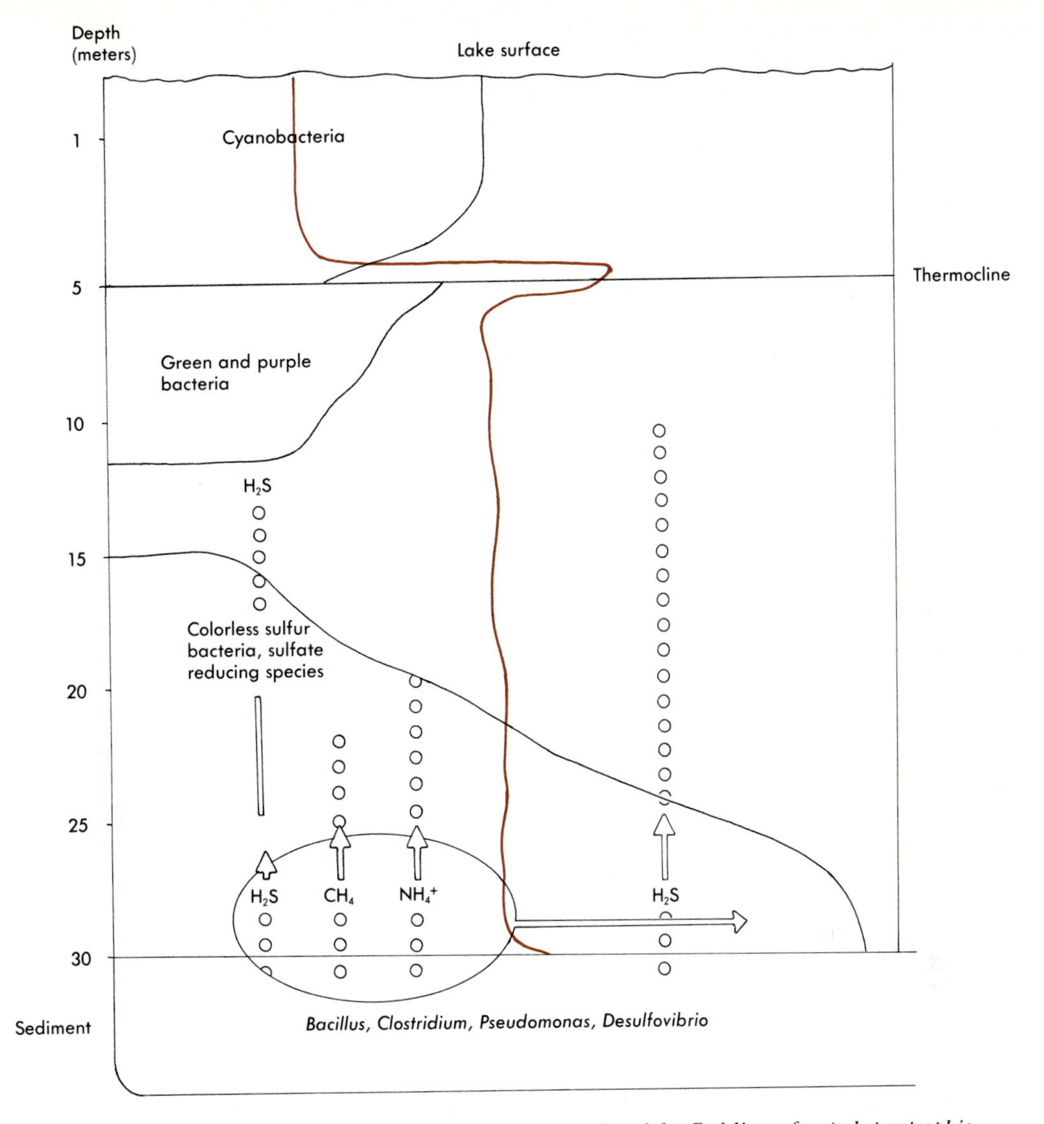

Figure 19-7 *Distribution of major groups of bacteria in a lake. Red line refers to heterotrophic bacteria whose concentration is highest at thermocline and at water-sediment interface. Hydrogen sulfide, ammonium, and methane are seen as bubbles arising from anaerobic zone of sediment.*

Lakes. The occurrence and distribution of organisms within the water column is dependent on the physicochemical nature of the lake as well as the climate of the region in which it is located. As in other environments, the determining factor for the number of organisms in a lake is nutrient availability. Nutrients enter lakes in surface runoff as well as in any other waters that empty into the lake basin. In regions where the area is highly productive, that is, forests or farmlands, which also contain a productive photosynthetic community that contributes to the concentration of nutrients (organic carbon), the concentration of nutrients may be very high. In these types of lakes, termed **eutrophic**, often a large and active biological community is present. Conversely, lakes that receive small inputs of nutrients and that are nutrient poor themselves **(oligotrophic lakes)** support fewer and different types of organisms than eutrophic systems.

A lake is divided into zones based on temperature characteristics. The **thermocline** represents that zone where there is a rapid decline in temperature as opposed to the zone above **(epilimnion)** and the zone below **(hypolimnion)** it (Figure 19-7). A large concentration of bacteria is present in the lake surface microlayer. Within the water column below the microlayer, the concentration of bacteria decreases. Near the surface where oxygen concentrations are high, **cyanobacteria** are represented in high numbers. At depths of 5 to 10 meters where the level of oxygen decreases, hydrogen sulfide is present, and there is still sufficient light, members of the Chlorobiaceae (**green bacteria**) and Chromatiaceae (**purple bacteria**) predominate. Heterotrophic bacteria may also be present in this layer, provided organic matter is present. They are usually located at the thermocline or in the sediments where dead organic matter is deposited. Both green and purple bacteria use hydrogen sulfide as an electron donor, and its oxidation is coupled to the fixation of carbon dioxide. The oxidation of H_2S results in the formation of sulfur granules or globules that may remain inside or outside the cell depending on the species. Sulfur in the cell may be slowly oxidized to sulfate. The reactions that occur during H_2S oxidation may be expressed as follows:

$$2\,H_2S + CO_2 \xrightarrow{\text{light}} 2\,S^0 + H_2O + CH_2O\ \text{(carbohydrate)}$$

$$2S^0 + 3\,CO_2 + 5H_2O \xrightarrow{\text{light}} 2\,H_2SO_4 + 3(CH_2O)$$

The hydrogen sulfide used by the green and purple bacteria is supplied by the activities of microorganisms in the sediment or at the water-sediment interface. The organisms in the sediment are anaerobes, some of which reduce sulfate to hydrogen sulfide. Although hydrogen sulfide is a toxic compound, it is photooxidized by the green and purple bacteria and rendered harmless. Thus, a cycle driven by light energy is maintained between the anaerobic microflora in the sediment and the phototrophic bacteria near or at the surface of the lake.

Within the lake ecosystem, the phytoplankton carry out the majority of the photosynthetic addition of carbon to the environment. However, in many systems a significant fraction of the total primary production comes from photosynthetic bacteria. Zooplankton prey on the phytoplankton and bacteria and themselves serve as a food source for higher-trophic-level organisms.

There is a seasonal pattern in the distribution of many organisms within lakes as a result of the influence of climate. Algae are usually restricted to the photic zone, and their abundance is controlled in large part by water temperature. Algae demonstrate increased metabolic activity during the warmer seasons.

Maximum populations of bacteria are present during the spring and fall. This is especially true for lakes in temperate regions. During winter the water temperature of a temperate region lake is somewhat uniform from the surface to the sediment, especially if the lake is covered with ice. Water has a maximum density of 4° C. At temperatures above or below 4° C, the density decreases. Therefore as ice melts or as waters with a temperature of less than 4° C begin to warm, the density increases, resulting in a circulation pattern being established as the denser water moves to the bottom of the lake. During the fall, when the water temperature begins to decrease, the density increases and the denser water moves to the bottom of the lake, again establishing a circulation pattern. Such circulation patterns result in complete mixing of the water column, and materials on or in the sediments (**benthic matter**) are swept up into the water column. These events, occurring in the spring and fall, are referred to as spring and fall overturn, respectively. When an overturn occurs, the nutrient concentration in the water column increases dramatically. This results in bacterial and phytoplankton blooms, which are followed by blooms in the zooplankton community. Finally, as the zooplankton die from the lack of food, a second bacterial bloom occurs as the zooplankton are decomposed.

Different physiological types of bacteria in the water column also occur on a seasonal basis. For example, during the winter the concentration of dissolved oxygen in a lake is very low. In some lakes, with the exception of the surface layer, the water column may be anaerobic. This is a result of little oxygen being added to the water by photosynthesis and of aerobic organisms using the oxygen that is available. Therefore the water column may contain more anaerobic bacteria than aerobic bacteria during the winter. During the summer, when a lake becomes stratified, the colder waters of hypolimnion (lower region of a stratified lake) are anaerobic.

It is important to note that not all lakes (or ponds) undergo seasonal cycling and stratification as described above. Depending on the depth of the system as well as the climate, a system may never become stratified, as occurs in systems where the water temperature remains considerably more than 4° C. In some regions a system may be stratified during the summer, but only a fall overturn will occur. In any case, when an overturn oc-

Concept Check

1. Describe the most common genera of bacteria in marine surface waters. In freshwaters. In estuaries.
2. What was the significance of the discovery of microbial life as deep as 4 miles in the ocean?
3. Outline the cell cycle of *Hyphomicrobium.*
4. What groups of bacteria are found at various levels of a lake? What gases do they release, and what if any benefit do they have?
5. Describe the symbiosis between vent bacteria and tube worms.

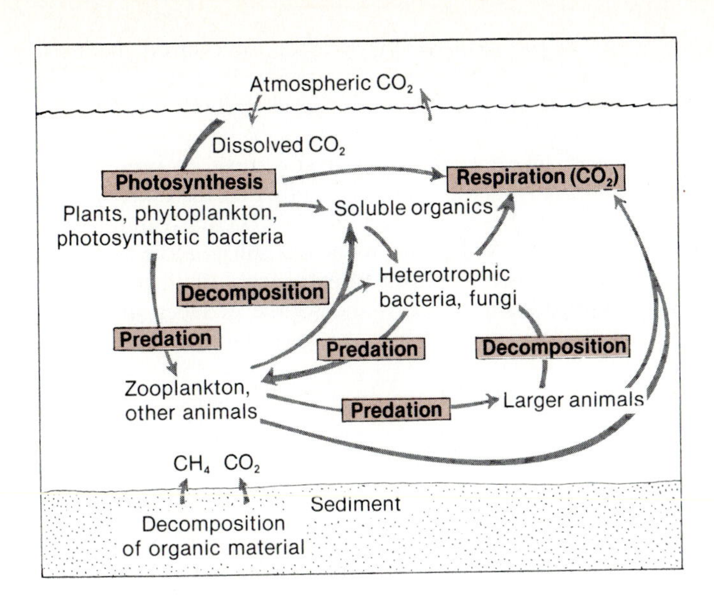

Figure 19-8 *Carbon cycle in aquatic environment.*

curs, the result will be as described above—blooms in the bacterial, phytoplankton, and zooplankton communities. This is a result of the influx of nutrients to the water from the sediment. When an overturn occurs and the water circulation resuspends detritus (disintegrated debris) into the water column, the soluble nutrients are used by the planktonic communities.

Streams and rivers. The microbial community of flowing-water systems is similar in many ways to those of other aquatic systems. However, the majority of the members of communities of flowing-water systems exist in association with objects within these systems. The development of algal mats on the surface of rocks on the bottom of streams is well known. Within mats of attached algae, epiphytic bacteria are common. It is easy to realize the necessity of organisms existing in association with solid surfaces in such systems. In fact, the existence of a true planktonic community in a flowing-water system can be questioned, since such organisms would be moved downstream. For a planktonic community to exist, organisms would have to be continually added to the system. Therefore, the presence of organisms in the water column proper would be a result of attached species becoming dissociated from their substratum. Also, in flowing-water systems, attached species would be favored, since they would only have to remain on a surface and have nutrients transported to them in the water currents.

In flowing-water systems, the smallest numbers of microorganisms are in pristine mountain streams in which the nutrient concentration is very low. As the waters flow toward the low-lying areas, more nutrients enter the water column and the number of organisms increases.

Within the bacterial community the majority of species are gram-negative. The changes in species composition of the culturable proportion of the community depend on several factors, including the availability and types of nutrients and the composition of the land through which the system flows.

Freshwater sediments. Most decomposition of organic materials occurs on or in the sediments. When particulate organic materials such as plant or animal tissues are broken down by bacterial activities, a portion of the material will be converted into a soluble form. Likewise, mineralization of plant and animal tissues releases inorganic nutrients such as nitrogen, phosphorus, and sulfur in soluble form. The bacterial flora of freshwater sediments is similar to that of marine and estuarine sediments and is dominated by the bacteria. The proportion of gram-positive organisms increases with increasing depth into the sediment. This is due primarily to the increasing abundance of *Clostridium* species in the anaerobic zones of the sediment. *Bacillus* and other gram-positive bacteria, including coryneforms and filamentous forms, are also abundant in aquatic sediments. Among the gram-negative bacteria are *Pseudomonas* species, which carry out denitrification, and *Desulfovibrio* species, which produce H_2S.

AQUATIC NUTRIENT CYCLES

Carbon cycle

The cycling of carbon in the aquatic environment is a result of the combined metabolic activities of all the members of the biota (Figure 19-8). The majority of the carbon entering an aquatic system is converted into the biomass of photosynthetic organisms. Some of the photosynthetically fixed carbon is incorporated into the biomass of heterotrophs, which use dissolved organics released into the water column by the photosynthetic organisms. In a simplified model system such as that depicted in Figure 19-8, a large proportion of the biomass of these organisms is converted into the biomass of filter-feeding invertebrates. Likewise, higher-trophic-level organisms, that is, larger invertebrates and vertebrates, prey on the filter-feeders. At each step in the food chain as described above, the biomass of the prey organisms, which is converted into the predator biomass, returns to the atmospheric carbon pool as carbon dioxide or remains in the system in the form of waste products from the predator organisms. Waste products of aquatic organisms decomposed by portions of the microbial community are incorporated into the biomass of the decomposers or are lost from the system as carbon dioxide. Ultimately the biomass of all organisms is acted on by the microbial community involved in the decomposition of organic matter and is either converted into the biomass of the decomposers or lost from the system as carbon dioxide.

The majority of the plant and animal biomass becomes a part of the benthos, where it undergoes decomposition. The microbial communities responsible for the decomposition of the plant and animal biomass differ. The biomolecules of animal tissues (proteins, carbohydrates, lipids, etc.) are easily broken down by heterotrophic bacteria that produce the necessary extracellular enzymes. However, when the plant biomass contains a large proportion of cellulose and lignin, it is initially acted on by members of the fungal community. This is true in both marine and freshwater systems. The principal fungi are the *Phycomycetes,* which resemble some algal groups. The fungi colonize the lignocellulosic materials and, as a result of their activities, break these materials down into smaller particulates that are then colonized by bacteria, and the decomposition process continues. Therefore, the decomposition of plant materials requires the presence of both fungi and bacteria.

The majority of the decomposition of organic materials, of either plant or animal origin, occurs in the sediment or on the surface of the sediment. This environment is often partially or completely anaerobic. This results in populations of anaerobic and aerobic organisms being involved in decomposition processes. Indeed, both metabolic types are necessary for the cycling of organic materials in the aquatic environment.

One of the major products of these decomposition processes in many aquatic sediments is methane (CH_4). Methanogenesis is an important process, not only in carbon recycling in an ecosystem, but also from an economic standpoint, since this is an increasingly attractive alternative energy source. In order for methane to be produced in an aquatic sediment, three different fermentation processes are involved (Figure 19-9). Biologically produced methane either enters the atmosphere or is metabolized by another specialized bacterial group, methane oxidizers (methyltrophs), which use single-carbon compounds as a carbon and energy source. Methanogenesis does not occur in all sediments, since different metabolic groups must be present and interact. Also, the presence of other bacterial groups can inhibit methane production. This occurs commonly in sediments containing an active sulfate-reducing community, since these organisms outrival the methanogens for hydrogen, which is the source of electrons in the reproduction of sulfate to hydrogen sulfide.

Nitrogen, phosphorus, and sulfur cycles

Nitrogen, phosphorus, and sulfur cycles are discussed at length in Chapter 18.

SEWAGE AND WASTEWATER MICROBIOLOGY

The beginning of aquatic microbiology as a discipline is due in part to studies on sewage contamination of water sources. After bacteria were established as causative agents of many diseases, microbiologists noted that diseases such as typhoid fever, cholera, and bacillary dysentery were spread via contaminated water. The ability to detect these organisms in water sources and to treat wastewater to prevent the spread of disease became of primary public health importance. Procedures to monitor aquatic systems to determine to what degree, if any, a system is contaminated by sewage have resulted in dramatic reductions in the occurrence of waterborne diseases.

Sewage (fecal) contamination of water is determined on the basis of the presence of bacteria specific for the intestinal tract of mammals, that is, the **coliforms.*** These organisms are detected with the use of various selective and differential media designed for their isolation. All such media contain lactose as the primary substrate, since lactose fermentation is one of the defining characteristics of enteric bacteria. The coliforms are divided into two groups on the basis of the temperature at which they ferment lactose. Total coliforms ferment lactose at 37° C, whereas fecal coliforms ferment lactose at 44.5° C. Likewise, the fecal streptococci, members of the genus *Streptococcus* indigenous to the intestinal tract of mammals, are also useful in evaluating whether a water source is contaminated by sewage. The dramatic decline in waterborne diseases since the nineteenth century is a result of the development of methods to properly treat sewage and wastewater as well as the use of coliforms as indicators of sewage contamination in domestic water supplies.

* The coliforms are a heterogeneous group of gram-negative rods that may be found in the intestinal tract. They include *E. coli* and species of *Klebsiella, Enterobacter, Serratia,* and others.

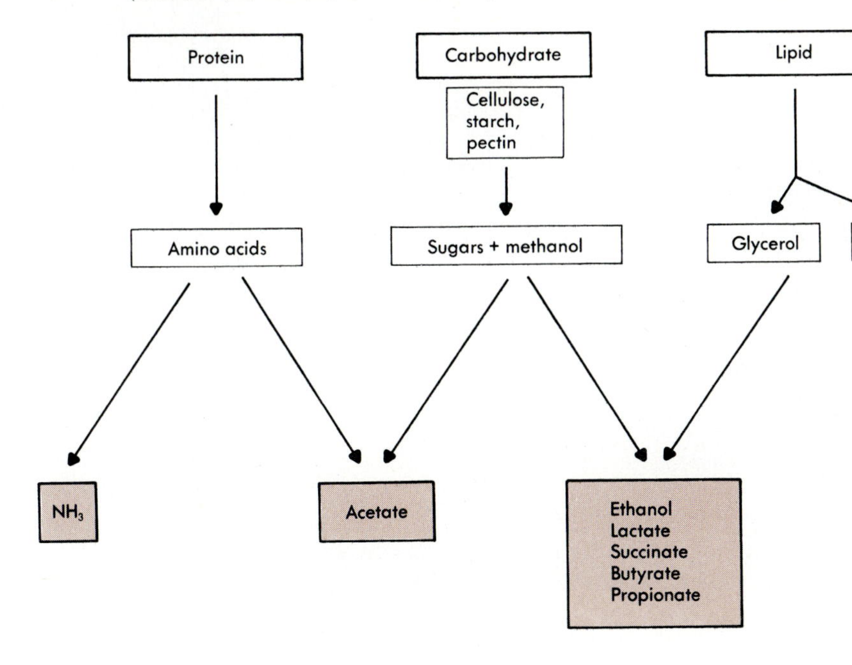

B Secondary fermentation *(Acetobacterium, Clostridium aceticum)* methanogenic bacteria (*Methanobacterium, Methanobacillus,* and other methanogens)

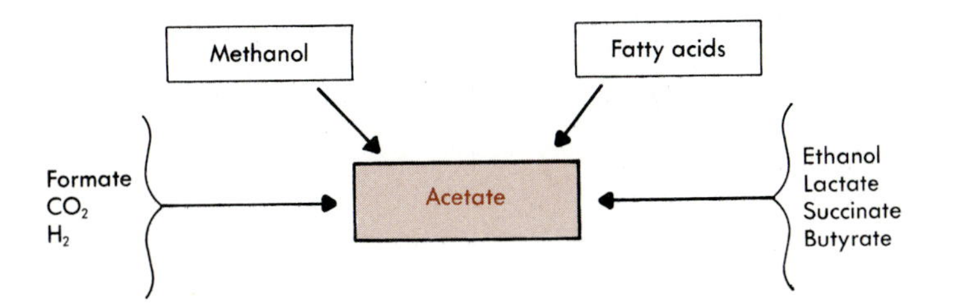

C Tertiary fermentation (same organisms as in secondary fermentation)

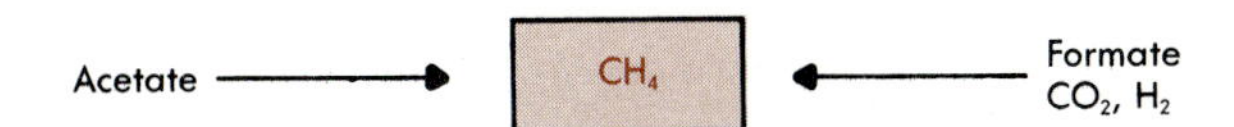

Figure 19-9 *Anaerobic breakdown of polymers into methane. Methane formation may take place in anaerobic environments such as rumen, marine sediments, and anaerobic digesters. Process can be divided into three fermentation processes.* ***A,*** *In primary fermentation macromolecules are fermented by nonmethanogenic bacteria, such as* Clostridium, Bacteroides, *and* Selonomonas, *into their monomeric units. The latter are further fermented to acetate, ethanol, and organic acids such as lactate.* ***B,*** *Secondary fermentation involves degradation of fatty acids, methanol, organic acids, and single carbon compounds to acetate. Principal organisms carrying out secondary fermentation are methanogens.* ***C,*** *Tertiary fermentation process involves conversion of acetate or formate, CO_2, and H_2 to methane by methanogens.*

Sewage treatment

Microorganisms are used in the treatment of sewage and wastewater. It would be impossible to safely recycle water without the activities of these organisms. The three objectives of sewage treatment are (1) to mineralize organic material to reduce the concentration of organic materials; that is, to lower the **biochemical oxygen demand** (BOD; a measure of the amount of oxygen taken up by a water sample during a 5-day incubation period at 20° C, reflecting the quantity of oxidizable organic matter present in a sample); (2) to remove minerals from the water; and (3) to kill pathogenic organisms. These objectives are met by modern treatment processes (Figure 19-10).

Figure 19-10 Aerial view of modern wastewater treatment plant. This treatment plant is located in Williamsburg, Virginia.

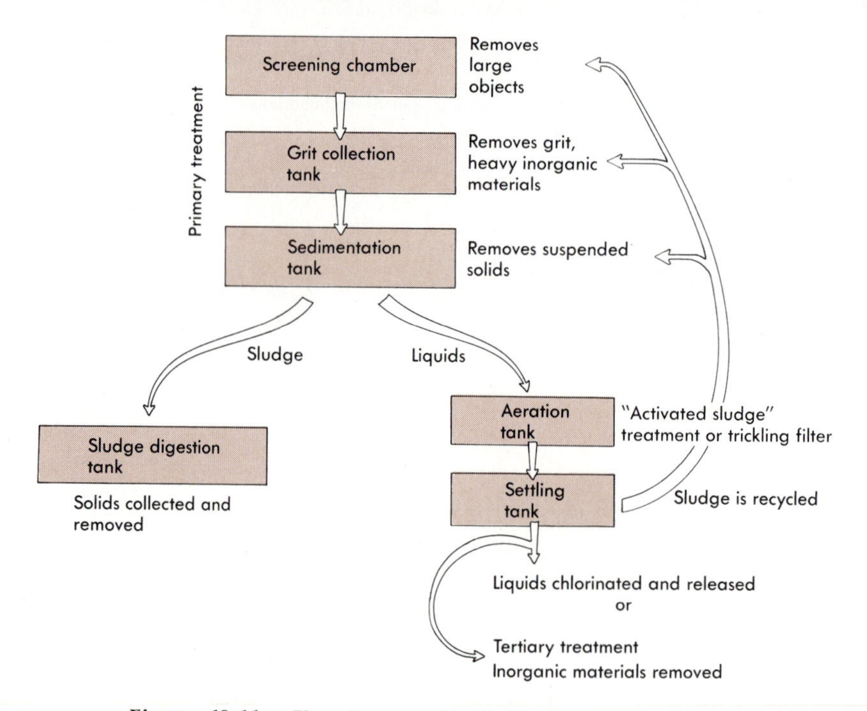

Figure 19-11 *Flow diagram of typical sewage treatment plant.*

The treatment of sewage and wastewater consists of three stages: primary, secondary, and tertiary (Figure 19-11). The objective of **primary treatment** is to remove coarse solid materials, whereas the objective of **secondary treatment** is to remove dissolved organic materials from the water by allowing microorganisms to metabolize (mineralize) them. Secondary treatment may also provide for digestion of the particulate material concentrated during the primary stage (the concentrated particulate is also called **sludge**). Tanks called sludge diges-

Figure 19-12 *Wastewater undergoing treatment in activated-sludge system.*

ters are used for this purpose. The **tertiary stage** of wastewater treatment, which consists of the removal of minerals from the water, is expensive and not practiced in all regions. Mineralization of organic materials without the removal of the minerals is of limited value, since organisms in waters that receive sewage treatment effluents will use the minerals. In such cases algal and bacterial blooms often occur.

In a modern wastewater treatment facility the wastewater is initially collected in a settling tank, where particulate materials are removed by sedimentation. Materials such as oil or grease, which float on the surface, are removed with a skimmer and can be combined with the sediment in sludge digestion. The materials collected at this stage (the **primary sludge**) can be further concentrated by the removal of excess water and at least partially degraded in a digester (see below). The wastewater liquor is then pumped to another system for secondary treatment, during which the particulate and dissolved organic materials are mineralized by microorganisms. In the mineralization process, portions of the organic materials are converted to the microbial biomass and/or removed from the system as gaseous carbon compounds (carbon dioxide, methane, etc.). Primary treatment may remove as much as 80% of the BOD, but in most domestic sewage treatment facilities only 30% to 40% of the BOD is initially removed.

Secondary treatment systems are either aerobic or anaerobic. During aerobic processing, the microorganisms mineralize the organic materials in an aerobic system. Different types of aerobic wastewater treatment systems are employed in different regions. However, of these, the two most common methods of aerobic treatment, and the ones used by most cities, are activated-sludge and trickling-filter systems.

Aerobic secondary treatment. In **activated-sludge systems** (Figure 19-12), the wastewater liquor contains a large bacterial community, most of which is within the "floc" (an organic matrix), which is produced by the bacterium *Zoogloea ramigera.* As *Z. ramigera* grows in the wastewater liquor, it is reported to produce an extracellular organic slime, in which other bacteria become embedded. Organisms within this zoogloeal mass metabolize dissolved organic materials in the wastewater liquor, and the mass increases in size. Within activated-sludge systems, a large protozoan community, composed primarily of ciliates such as *Vorticella,* also thrives. The protozoa, responsible for "grooming" the zoogloeal mass by grazing on it, are consumed by other organisms present in the system. Thus, a food chain is created, and at each stage of the food chain a fraction of the original organic material is removed from the system as carbon dioxide.

After a period sufficient to allow the microbial com-

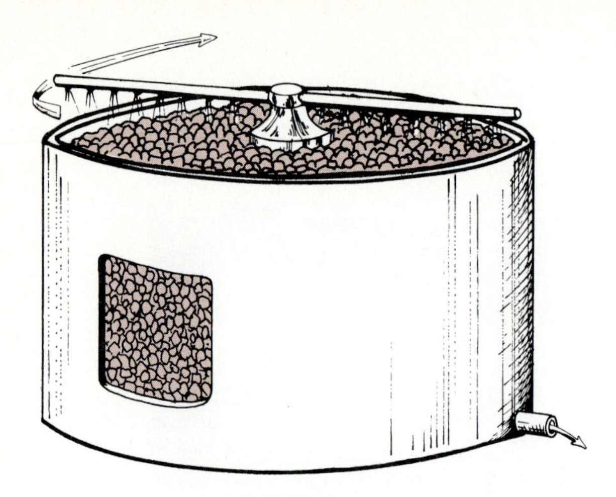

Figure 19-13 *Diagram of trickling-filter wastewater treatment system. See text for details.*

Figure 19-14 *Diagrammatic representation of the biodisc. Biodisc is partially submerged in sludge and rotated to create aerobic conditions necessary for rapid breakdown of matter.*

munity to incorporate the bulk of the organic materials into the biomass (or respire it off as carbon dioxide), aeration is ceased and the floc settles to the bottom of the system. The wastewater is then removed and either is chlorinated to kill any pathogenic microorganisms that survive the activated-sludge process and released into the environment or undergoes tertiary treatment (see below).

The bacterial community of the activated sludge system, in addition to *Z. ramigera,* consists of bacteria typical of an aquatic system, that is, *Pseudomonas, Vibrio, Flavobacterium,* and others present in lower numbers. Bacteria common to the human intestinal tract *(Bacteroides, Fusobacterium, Eubacterium,* and members of the Enterobacteriaceae) do not contribute significantly to the community of the zoogloeal mass.

A second aerobic system that is commonly employed for the treatment of wastewaters in many cities, is the **trickling filter** (Figure 19-13). It is also based on the concept of forming a short food chain within a closed system in which the bacteria use dissolved organic materials and protozoa consume the bacteria. In trickling-filter systems, wastewater liquor is slowly sprayed over a bed of rocks covered by a film of microorganisms. In this system, however, fungi and algae make up the majority of the biomass. The biological film covering the rocks (or other solid support) is approximately 2 mm thick and consists of three layers, each with a characteristic microbial flora. The inner layer is composed of bacteria, algae, and fungi. The middle layer primarily contains fungi and algae. Fungi are the dominant organisms in the outermost layer. As the wastewater slowly trickles over the surface of the biofilm, dissolved organics are used by the microbial flora. Protozoa are present in large numbers and groom the biofilm.

This system is very effective in reducing the concentration of organic material in the wastewater, and the BOD may be reduced by as much as 90%. The role of algae in this system is not well understood. It is believed by many that algae are involved in the removal of some minerals from the water, especially phosphates. Water exiting the trickling filter can be either chlorinated to kill pathogens or processed in tertiary treatment.

Another type of aerobic system is the "biodisc" system. Discs, usually plastic, are closely spaced together and used to rotate through the sewage effluent (Figure 19-14), becoming coated with a microbial slime. The discs are only partially submerged; therefore the microbial community is efficiently aerated. Although highly efficient, the biodisc system is very costly because of the initial investment.

Anaerobic treatment systems. Anaerobic treatment processes can also be employed as a secondary treatment or, as often occurs, in conjunction with aerobic systems. More often than not large-scale anaerobic systems or digesters are used to process sludge obtained from primary and secondary treatments or for treatment of industrial effluents with a high BOD. Anaerobic systems tolerate larger quantities of organic matter than aerobic systems, but they are not as efficient as aerobic systems in lowering the BOD. The simplest and most recognizable anaerobic system is the septic tank, which is found most often in rural areas that do not have sewer systems (Figure 19-15).

Once a system is filled with wastes and becomes anaerobic (in a matter of minutes if the system is closed from the atmosphere), the anaerobic bacteria begin to grow and metabolize organic materials. Anaerobic digestion takes place in two major steps. First, complex organic molecules are depolymerized into acids, carbon dioxide, and hydrogen. Then the products of the first step are the substrates for the methanogenic bacteria according to the following reaction (also see Figure 19-9):

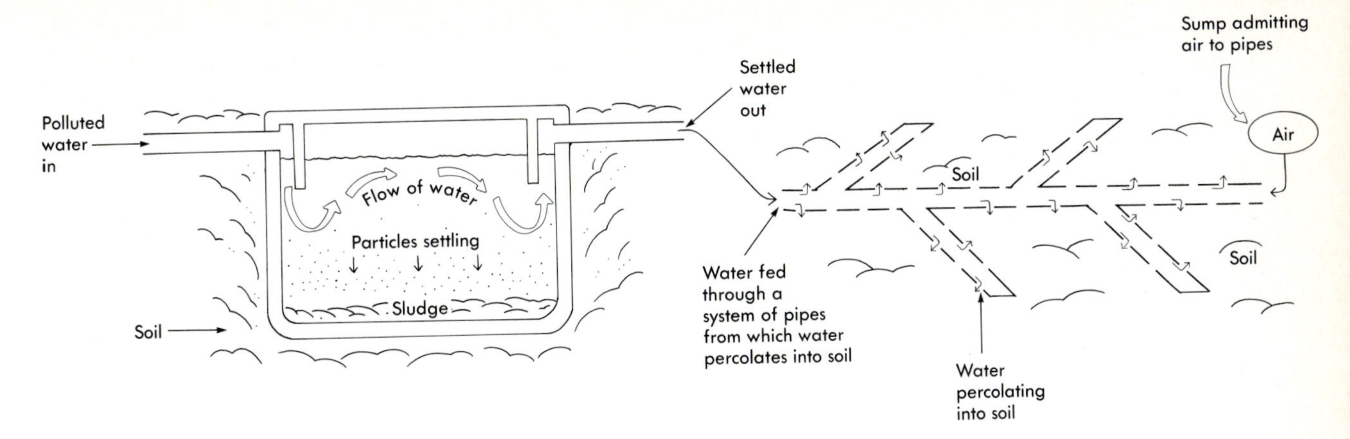

Figure 19-15 *Vertical section through septic tank set in soil. Reduction of flow rate of sewage in tank causes suspended material to settle out as sludge. Sludge is digested anaerobically with release of methane as one major end product. Liquid phase of sewage percolates through soil and undergoes aerobic treatment.*

$$\underset{\text{(carbon source)}}{CO_2} + \underset{\text{(possible electron donors)}}{\text{H}_2\text{, acetate, formate, methanol}} \xrightarrow{\text{methanogenic bacteria}} \underset{\text{(methane)}}{CH_4}$$

The methane produced by this system can be vented and used as a fuel source by the treatment plant or by other facilities. Optimal performance of anaerobic digestion requires continual maintenance of such parameters as temperature, pH, retention time, carbon to nitrogen and carbon to phosphorus ratios, and concentration of heavy metals. In addition to methane gas, small quantities of nitrogen, hydrogen sulfide, and carbon dioxide gases are also products of anaerobic digestion. Sludge is reduced in volume and may be disposed of at restricted landfill sites or may be subjected to aerobic composting. Aerobic composting kills potential pathogens, and the material can then be disposed of at any landfill or used as a soil conditioner. Sewage application to agricultural lands is a common method of waste disposal; however, this practice may result in the accumulation of pollutants contained in the sludge. Heavy metals in sludge are toxic to plants, and some metals can be translocated into vegetable products and subsequently consumed by animals or humans. Heavy metals are known to affect microbial soil processes such as nitrogen fixation. Many scientists believe that sludge should be used as a means of recovery of heavy metals, which are becoming increasingly scarce.

In general, after the secondary treatment process, wastewaters are chlorinated to kill any remaining organisms and are released into the environment (usually into a river or other water source). Any minerals that remain in the water support the growth of organisms in the receiving waters. Thus, to prevent harmful blooms of algae and bacteria, in limited situations the treated waters are processed through tertiary treatment to remove dissolved minerals. The addition of lime ($Ca[OH]_2$) will precipitate phosphates as insoluble calcium phosphate ($Ca_5OH[PO_4]_3$). Also, alum ($Al_2[SO_4]_3$) is used in this manner in some regions. Once a precipitating agent is added, the waters are allowed to stand, and the precipitate settles. The phosphate-free water is then removed from the system. It is important to remove phosphates from wastewater, since an algal bloom can be stimulated by excess phosphates. If a bloom is initiated, it can be perpetuated by nitrogen-fixing bacteria and cyanobacteria as they supply the system with nitrogen to support the growth of other organisms. Toxic substances and salt are also removed by tertiary treatment. Unfortunately,

Concept Check

1. What are the steps involved in the anaerobic breakdown of polymers into methane and the organisms involved?
2. Describe a typical aquatic carbon cycle. Which organisms would be involved in each stage of the cycle?
3. What are the major objectives of sewage treatment?
4. What is the difference between anaerobic and aerobic secondary sewage treatment, and what organisms are involved? What are the advantages and disadvantages of each?

tertiary treatment is rarely practiced, since the expense is prohibitive.

In the next section we will examine the techniques employed to treat water that is to be used for human consumption, that is, potable water.

MICROBIOLOGY OF DOMESTIC WATER

Water treatment

Water treatment is defined as the manipulation of a water source to achieve water quality that meets specific standards set by the community through its regulatory agencies. In earlier times water treatment focused on elimination of waterborne disease such as typhoid fever, cholera, and amebiasis. In the last 20 to 30 years, public health concerns have shifted from acute illness to chronic health effects of trace amounts of organic and inorganic pollutants arising from agricultural or industrial practices. Approximately 35 water quality parameters are included in the Environmental Protection Agencies drinking water standards. It is expected that 50 additional parameters will be required by the 1990s. The following paragraphs will elaborate on those municipal procedures used to achieve water quality.

Components in water supply that affect water quality. Water, such as river water, contains inorganic and organic components as well as microorganisms that require reduction or elimination before it can be considered suitable for human consumption. The predominant inorganic minerals in river water are calcium and bicarbonate (Table 19-5), both of which contribute to water hardness. Inorganic minerals such as iron, nitrate, sulfur, and manganese also contribute to the deterioration of water quality if their concentrations are sufficiently high.

Organic compounds in water are derived from the natural decomposition of plant and animal matter and from industrial, urban, or agricultural pollution. Naturally occurring organics include fulvic and humic acids, which are plant degradation products. Other naturally occuring organics may include geosmin, a product of microbial metabolism, which imparts a musty or earthy odor and taste to water. Synthetic organic materials include pesticides, polychlorinated biphenyls, halogenated methane and ethane derivatives, and others. Microbiologically, water contains bacteria, viruses, algae, protozoa, and some fungi. The agents of waterborne disease include bacterial agents such as *Salmonella, Shigella, Vibrio,* and *Campylobacter;* protozoa such as *Giardia lamblia;* and viruses such as hepatitis A and enteroviruses. Most of these agents cause gastrointestinal symptoms. The epidemiology of these infectious agents is discussed in Chapter 26. Algae present in water are not considered an important cause of disease, but they must be removed to prevent clogging of filter beds during water treatment procedures.

Table 19-5 *Major ions found in an average river, listed in order of decreasing concentration*

Constituent	Effect on water quality
HCO_3	Water hardness
Ca	Water hardness
SiO_2	Concentration in most waters is not sufficient to present problems
SO_4	At very high concentrations, is corrosive to concrete and cement pipes and can impart unpleasant odor; *Desulfovibrio* reduces sulfate to sulfides that are corrosive; *Thiobacillus* can oxidize sulfur and sulfur compounds to the corrosive sulfuric acid (H_2SO_4); concentration in most water is too low to be of concern
Cl	At high concentrations, causes salty taste; corrosive to metals; toxic to plants; concentration in most water supplies is too low to be of concern
Na	Toxic to plants, fish, and wildlife at high concentrations; cause of hypertension in humans; levels in most waters is usually very low
Mg	Water hardness
K	Levels too low to be of concern
NO_3	High concentrations gives bad taste to beer in brewing process; can cause blood disorders in infants; can cause eutrophication in lakes
Fe	At high concentrations, promotes growth of iron bacteria such as *Gallionella, Leptothrix,* and *Sphaerotilus* and ultimately clogging of pipes; stains fixtures and laundry
Mn	Can stain fixtures and laundry; at very high concentrations can cause unpleasant taste to water and can promote growth of microorganisms such as *Hyphomicrobium*

Processes and operations in water treatment. What kind of processes are to be used in water treatment depends on the source of water (ground water or surface water), as well as other factors such as level of human

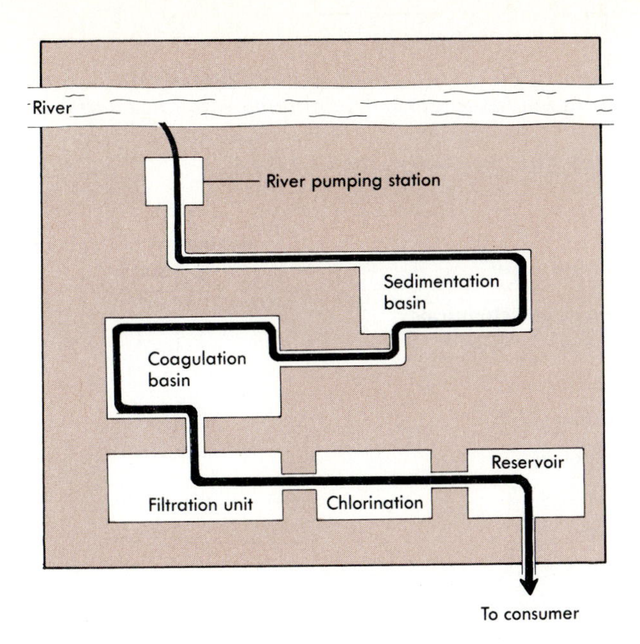

Figure 19-16 *Flow diagram of procedures used in treatment of water for municipal use.*

activity (industrial, agricultural) in the area, and geological factors (soil porosity and mineral content). We will consider the treatment of river water that is relatively poor in quality and highly turbid.

Sedimentation. Sedimentation is not always necessary as an initial step in water purification unless the water contains a high degree of suspended material. Turbid water is pumped from the river (Figure 19-16) and directed to sedimentation basins where gravity separation of suspended material takes place. Matter suspended in water that is heavier than water will settle to the bottom. Algae and bacteria also become part of the sediment, and that can create problems. Algal growth, for example, contributes to the odor and bad taste of water. A new technique to bring algae and organic matter to the surface is being used in some water treatment facilities. Air is dispersed through the water, creating bubbles that attach to the particulate material. This causes the particulate matter to rise to the surface where it can be skimmed off.

Coagulation. If sedimentation basins are not used water can be initially treated in a process called coagulation. Coagulation is used in nearly all water treatment facilities. The basis of the coagulation process is that fine particulate material can be removed from water by the addition of inorganic or organic chemicals. The most frequently used chemicals are metals such as **aluminum** or **iron** which hydrolyze to form insoluble precipitates. Synthetic organic polyelectrolytes such as **polyacrylamides** are also used in some facilities. They rapidly absorb on the surface of particulate matter and accelerate the latter's aggregation. Removal of the particulate matter by coagulation also assists in the removal of toxic contaminants such as heavy metals, pesticides, and viruses known to be associated with organic matter in water.

The aggregates produced by coagulation are removed by physical means such as **gravity sedimentation, flotations,** or **filtration** through screens or granular material such as sand. When sand is used, water is passed through it until the separated particulate matter begins to accumulate on the surface of the sand and clogs it. The top layer of sand is then removed and replaced with fresh sand. This filtration process is the last treatment barrier for removal of undesirable particulate matter. In addition, by the end of the process, from 98% to 99% of the bacteria originally found in the raw water have been removed. The water is now ready for disinfection.

Disinfection. **Chlorination** is the most widely used disinfection process in the United States. Chlorination not only destroys microorganisms but also controls the odor and taste of water. Chlorination may involve the use of free chlorine, which reacts with water to form the active component, hypochlorous acid (HOCl), or combined chlorine, such as chloramines. Monochloramine (NH_2Cl), for example, is formed by the reaction with aqueous ammonia:

$$NH_3 + HOCl \rightleftharpoons NH_2Cl + H_2O$$

One other disinfectant that has widespread use in Europe but limited use in the United States is **ozone**. Ozone (O_3) is a highly reactive gas that is formed by electrical discharge in the presence of oxygen:

$$3O_2 + \text{energy} \rightleftharpoons 2O_3$$

Ozone possesses a high level of energy and is a better oxidizing agent than chlorine. Ozone breaks down to oxygen and leaves no residual in water, whereas chlorine remains at some concentration in treated water. The prohibitive feature of ozone treatment is the cost—three to five times more than chlorination processes.

The disinfection process causes the death of nearly all microorganisms within 30 minutes. The water is not considered microbiologically safe, however, until a measurement of coliforms is performed (see "Drinking Water Standards"). After the disinfection process, the water is pumped to storage tanks from which it flows to the consumer.

It should be noted that some purification processes require additional steps and may include (1) removal of specific minerals that cause hardness, (2) removal of heavy metals that may be characteristic of the water supply, (3) adjusting pH if the water is too acid or alkaline, and (4) addition of fluorine to control tooth decay (dental caries).

Drinking water standards

Standards for water quality were originally a function of the Public Health Service until 1970. With passage of the Safe Drinking Water Act in 1974, the federal government, through the Environmental Protection Agency (EPA), was given the authority to set standards for drinking water quality delivered by public water supplies. The most significant standards change by the EPA was the designation of turbidity as a health-related rather than an esthetic parameter. This change was based on studies showing that the particulate material responsible for turbidity could harbor microorganisms and protect them from the action of disinfecting agents. In the United States there are a number of drinking water regulations, among them the National Interim Primary Drinking Water Standards adopted by the EPA in 1975. It contains the maximum contaminant levels for a number of inorganic and organic chemicals, physical parameters, radioactivity levels, and bacteriological factors (Table 19-6).

Standard tests for the bacteriological quality of water specify that water is safe for human consumption if it does not contain more than an average of 1 coliform per 100 ml. Coliforms such as *E. coli* are nonpathogenic (except for some unusual strains) and are excreted normally in the feces at concentrations of 5×10^7/g. Since waterborne diseases originate from the same source as coliforms, the presence of the latter indicates potential contamination by pathogens such as *Salmonella* and *Shigella*. The tests used to determine the presence of coliforms in potable water systems can be divided into **presumptive**, **confirmed**, and **completed** (Figure 19-17). The presumptive test is based on gas production resulting from the fermentation of lactose. In the test, 10 ml portions of water are transferred to fermentation tubes containing lactose and an inverted vial. Following incubation for 24 hours at 35° C, the tubes are examined for growth and the formation of gas (identified by the presence of bubbles in the inverted vial). Gas formation indicates a positive test—coliforms present. The presumptive test is confirmed by transferring aliquots from the positive tubes to brilliant green lactose broth (BGLB) or to eosin–methylene blue (EMB) agar plates and incubating them for 48 hours at 35° C. If no gas is produced in the BGLB tube or if the colonies on the EMB plate do not have dark centers and a metallic sheen, then the test is negative—no coliforms present. The confirmed test is usually the last test for coliform detection; however, occasionally a completed test is employed. In the completed test, samples from the BGLB tube or EMB plate are transferred to lactose broth and to an agar slant for incubation at 35° C. If gas is produced in the lactose broth, the test is positive—coliforms present. A portion of growth from the agar slant is subjected to a gram stain. Presence of gram-negative rods without spores is considered a positive test—coliforms present.

Table 19-6 *Environmental Protection Agency National Interim Primary Drinking Water Regulations (1975); unless otherwise indicated all contaminant levels are expressed as mg/liter (L)*

Constitutent	Maximum contaminant level
Inorganic chemicals	
Arsenic	0.05
Barium	1
Cadmium	0.010
Chromium	0.05
Fluoride	1.4-2.4
Lead	0.05
Mercury	0.002
Nitrate (as N)	10
Selenium	0.01
Silver	0.05
Organic chemicals	
Chlorinated hydrocarbons	
Endrin	0.0002
Lindane	0.004
Methoxychlor	0.1
Toxaphene	0.005
Chlorophenoxys	
2,4-D	0.1
2,4,5-TP Silvex	0.01
Physical parameters	
Turbidity (turbidity units; TU)	1
Radioactivity	
Gross alpha (pCi*/L)	15
Radium 226 and 228 (pCi/L)	5
Tritium (pCi/L)	20,000
Strontium 90 (pCi/L)	8
Bacteriological factors	
Coliform bacteria (per 100 ml)	1

*pCi (picocurie) = 10^{-12} curie.

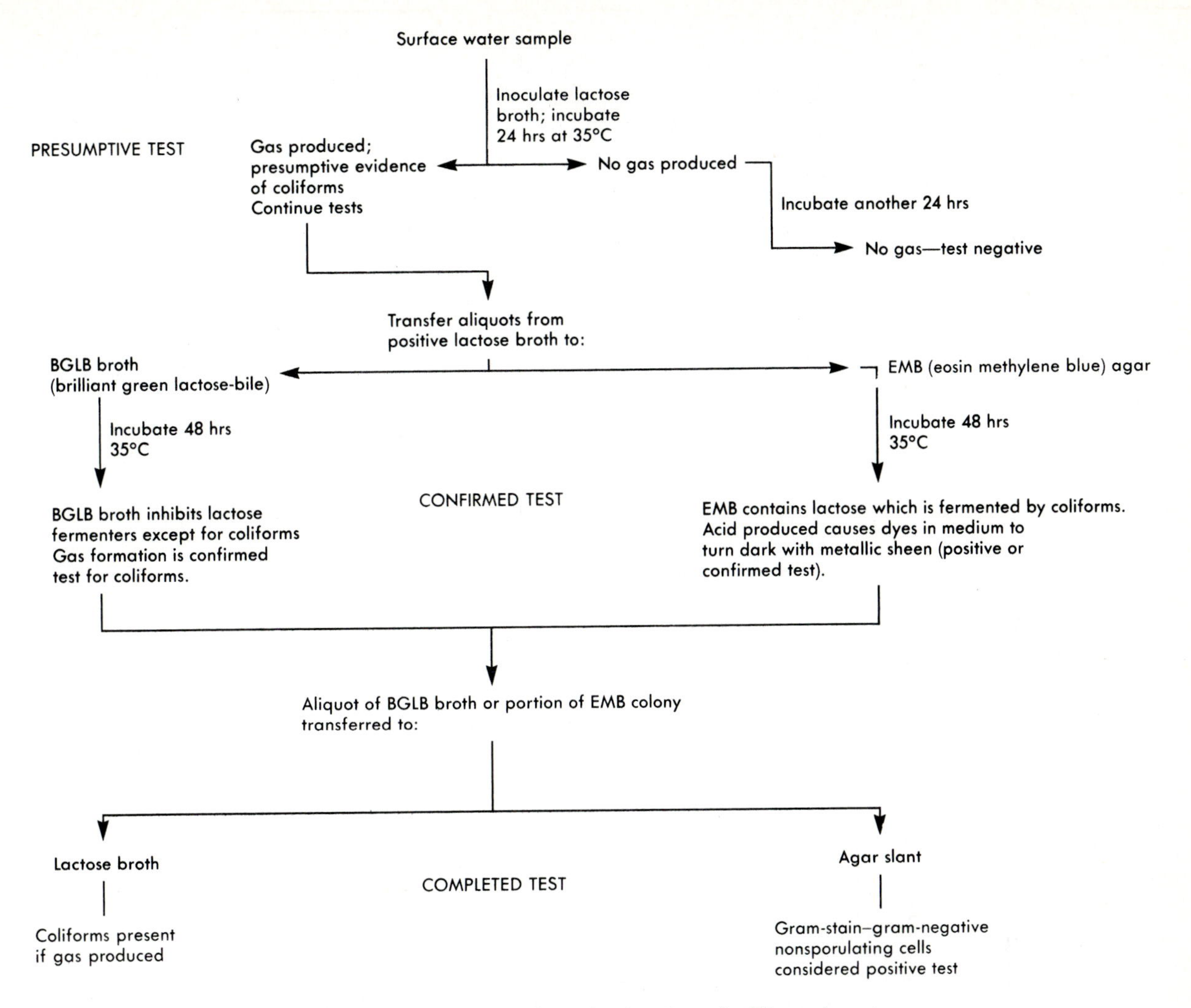

Figure 19-17 *Basic scheme for detection of coliforms in water.*

The water analysis is incomplete unless the number of coliforms is determined as well. A **multiple tube-fermentation technique** can be used in which the results are expressed as the most probable number (MPN; see p. 554 and Figure 19-18). Another technique that can be used to quantify coliforms is the **membrane filter method.** This procedure involves the passing of a measured water sample through a membrane filter to remove bacteria. The filter is then placed on a growth medium (EMB, for example) in a petri dish. The number of coliforms (colonies with dark centers and metallic sheen) are counted and expressed in terms of number per 100 ml of water. The membrane filter technique has several advantages over the MPN technique: (1) a greater volume of water can be examined, (2) the test can be performed quickly, (3) the test is more quantitative, and (4) less laboratory equipment is required.

Concept Check

1. What minerals contribute to the deterioration of water quality? What are the reasons for this, and what microorganisms, if any, may be involved?
2. What microorganisms are associated with waterborne disease?
3. Briefly outline the steps in the purification of water. How does each contribute to improving the quality of water?
4. Outline the steps in determining the presence of coliforms in water.

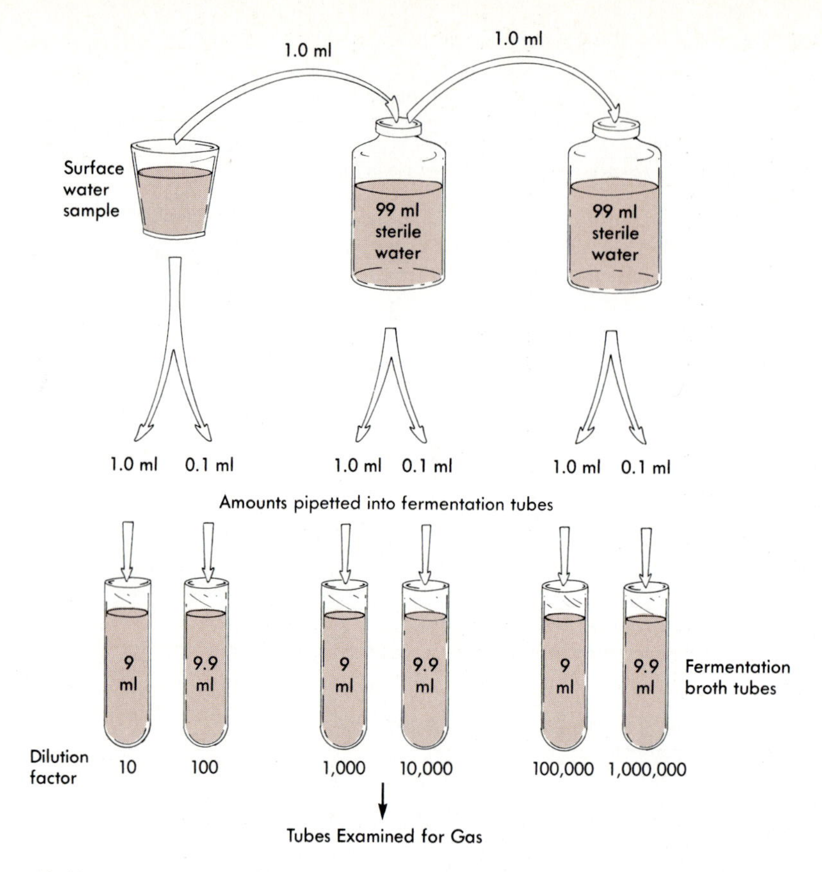

Figure 19-18 *Technique for preparation of multi-tube fermentation method for evaluating most probable number (MPN) of coliform bacteria in water. One milliliter sample of water source is inoculated into 99 ml of sterile water and mixed. This produces a 100-fold dilution of the original water sample. Then 1 ml of the 100-fold dilution mixture is inoculated into another 99 ml bottle. This produces another 100-fold dilution. Then 1.0 and 0.1 ml samples from undiluted water source and dilution bottles are inoculated into lactose fermentation tubes and incubated for 48 hours at 35° C. Number of positive reactions (gas produced) for each of serial dilutions are recorded and MPN determined.*

SUMMARY

1. Homeostasis of aquatic ecosystems is maintained in large part by members of the microbial community. The combined metabolic activities of the various microbial populations function in the cycling and regeneration of materials within these systems.
2. Aquatic bacteria are important not only in the decomposition of organic matter but also as an important member of the biotic community. They are involved in the decomposition and recycling of organic materials as well as serving as a food source for many invertebrates and furnishing organisms of other trophic levels with vitamins and other growth factors.
3. Aquatic populations can be measured by direct and indirect techniques. Direct techniques include spread plate, most probable number, and membrane filter. The most direct technique is microscopy, which can be combined with staining techniques.
4. Indirect measuring techniques include determining the concentration of specific cellular materials such as ATP and poly-beta-hydroxybutyrate. Heterotrophic activity can be determined by the uptake of radioactive carbon sources.
5. Aquatic bacteria are found in the free-floating (planktonic) state, in association with other organisms, or in association with particulate materials. Such associations confer ecological advantages on members of the bacterial community.
6. In associations between bacteria and other organisms, it appears that the bacteria are provided with a favorable microenvironment in which to live, and

the host organism either is not affected by the resident bacterial flora or uses growth factors produced by the bacteria.

7. All natural aquatic systems have a resident microbial flora. Each type of system contains a characteristic community. Waters of the open ocean contain approximately 1×10^6 bacteria per milliliter.
8. The bacterial community exhibits certain distributional patterns. Numbers of bacteria decrease with increasing distance from land masses, a result of decreasing nutrient concentrations. Also, the number of bacteria tends to decrease with increasing depth. Even so, the deep ocean contains a resident microbial population.
9. It appears that much of the recycling of organic matter in the deep ocean (depths greater than 2000 m) occurs in the intestinal tracts of other organisms that also inhabit this environment.
10. The bacterial community of open ocean waters includes members of the genera *Pseudomonas, Vibrio, Photobacterium, Flavobacterium, Acinetobacter* primarily, in addition to lesser numbers of other gram-negative species and some gram-positive species *(Staphylococcus* and *Micrococcus).*
11. The estuary is a transition zone between fresh and marine water. The microbial community is similar in composition to other aquatic environments. Certain *Vibrio* species are restricted to the estuary.
12. The bacterial community of freshwater systems is similar in composition to that of the marine environment. However, members of these bacterial communities do not have a requirement for high concentrations of salt. The composition of these systems is influenced by the type of system, as well as the physicochemical water parameters. Likewise, the size of freshwater bacterial communities depends on the concentration of available nutrients.
13. The microbial community of aquatic systems is essential to the various nutrient cycles within these systems. Each species is involved in some manner in one or more nutrient cycles. The ability of aquatic microorganisms to degrade and recycle organic matter is the basis of some stages of modern wastewater treatment processes.
14. Sewage treatment is divided into three stages: first, to remove large particles; second, to remove dissolved organic matter; and third, to remove minerals and toxic components. Secondary treatment systems are aerobic or anaerobic. Aerobic systems use either activated sludge or trickling filter techniques.
15. Anaerobic waste removal, which is not as efficient as aerobic systems, takes sludge from primary and secondary aerobic treatments and converts its products to methane, which can be used as fuel.
16. Domestic water is processed in a step-wise procedure that removes or reduces particulate matter, chemicals, and microorganisms. These procedures include sedimentation, coagulation and disinfection.
17. Chlorination is the principal method of disinfection in the United States. Water is considered microbiologically safe for human consumption when there is no more than 1 coliform per 100 ml. The routine bacteriological procedures used to determine coliforms involves three tests: presumptive, confirmed, and completed.
18. The number of coliforms in water samples can be determined by one of two types of tests: the most probable number and the membrane filter technique.

Self-study Quiz

Multiple choice

1. A water sample was taken during the treatment of sewage, and its BOD was found to be high. What does this indicate concerning the treatment of the sample?
   A. It has received acceptable treatment and is ready to be released.
   B. It has not received adequate treatment and should not be released until it has been further mineralized.
   C. It is in the anaerobic phase of its treatment and needs only to receive precipitation of the inorganic ions by chlorine to be rendered acceptable.
   D. The inorganic content is still high and must be precipitated out to make it safe for drinking.
2. Which of the following methods of counting the number of microorganisms present in water will overestimate the number of viable cells present?
   A. Pour plate technique
   B. MPN
   C. Direct microscopic count using membrane filter and fluorescent stain
   D. Spread plate technique
3. You would not expect to find living cells of phytoplankton in which of the following aquatic environments?
   A. Estuaries
   B. Freshwater lakes
   C. Deep trenches of the ocean
   D. Photic zones of open marine waters
4. The average expected number of microorganisms in the aquatic environment is in the range of:
   A. 1×10^2 organisms/ml
   B. 1×10^6 organisms/ml
   C. 1×10^8 organisms/ml
   D. 1×10^{12} organisms/ml
5. Where would you expect to isolate oligotrophic microorganisms?
   A. In water of nutrient-poor mountain streams
   B. In freshwater bottom sediments
   C. In upper layers of marine coastal waters
   D. In sediment from an estuary
6. An epiphytic association is
   A. An association between bacteria and zooplankton
   B. An association between bacteria and algae
   C. An association between zoophytic and lithophytic organisms
   D. An association between bacteria and inanimate particles
7. To what are the seasonal differences in the physiological types of bacteria in the water column of freshwater lakes due?
   A. Algal blooms
   B. Seasonal overturning of the layers of the lake
   C. Seasonal activities of freshwater predators
   D. Seasonal runoff of surface waters into the lake
8. What group of microorganisms are methanogenic, and where are they found?
   A. Cyanobacteria in the upper layers of open waters
   B. Nitrogen-fixing algae in the sediments of lakes
   C. Bacteria in aquatic sediments
   D. Zooplankton in marine sediments
9. Which of the following is most often the limiting nutrient for algal growth in a freshwater environment?
   A. Organic carbon
   B. Phosphates
   C. Sulfates
   D. Organic nitrogen
10. What is indicated by the identification of coliform bacteria in water?
    A. The water contains a large number of pathogens.
    B. Fecal contamination of the water has occurred.
    C. The water has a high BOD.
    D. There is a heavy turnover of sediment organisms.
11. The trophosome of tube worms found at ocean depths of 3 miles is composed of which of the following types of microorganisms?
    A. Cyanobacteria and other photosynthesizers
    B. Primarily chemoheterotrophs
    C. Primarily chemoautotrophs such as the sulfur-oxidizing bacteria
    D. Coliforms such as *E. coli*
12. At a depth of 5 to 10 m in a lake where the level of oxygen is low and hydrogen sulfide is present, which of the following organisms predominates?
    A. Cyanobacteria
    B. Methanogens
    C. Green and purple bacteria
    D. *Pseudomonas*
13. Which of the following metals is used in the coagulation process to remove fine particulate material?
    A. Calcium
    B. Zinc
    C. Aluminum
    D. Sodium

14. The sequence of procedures in the purification of water is:
    A. Sedimentation, filtration, coagulation, disinfection
    B. Sedimentation, disinfection, coagulation, filtration
    C. Disinfection, sedimentation, coagulation, filtration
    D. Sedimentation, coagulation, filtration, disinfection
15. The confirmed test to determine coliforms requires which of the following media?
    A. Lactose broth
    B. Nutrient agar slant
    C. Eosin–methylene blue agar
    D. None of the above

Thought questions

1. You are asked to enumerate the bacteria present in three samples of aquatic environment(s): (1) water from a mountain stream, (2) sediment sample from a freshwater lake, and (3) sediment sample from a deep ocean trench.
   a. Describe the methods you would use for each sample.
   b. Which sample would you expect to have the greatest microbial count?
2. Shown below is a map of sampling sites from a stream above and below the outfall of a pipe dumping large amounts of raw sewage into the stream. On the plots below indicate the relative amounts of microorganisms, organic molecules, and inorganic molecules that you expect.

True and false

________1. Members of the aquatic genus *Photobacterium* are noted for being phosphoclastic.

________2. The salt concentration of the estuary exceeds that of both freshwater and marine environments.

________3. Volutin is a polyphosphate storage product found in some bacteria.

________4. Marine zooplankton are filter-feeding organisms that consume phytoplankton and bacteria.

________5. The distribution of algae in aquatic environments is influenced by the concentration of nutrients in the water.

Short answer

1. Identify the stage or process of sewage treatment that is intended to accomplish the following:
   a. Removes suspended solids following biological treatment
   b. Removes inorganics by the addition of lime and alum to the treated wastewater
   c. Uses an organic matrix called a floc in an aerobic method of secondary treatment

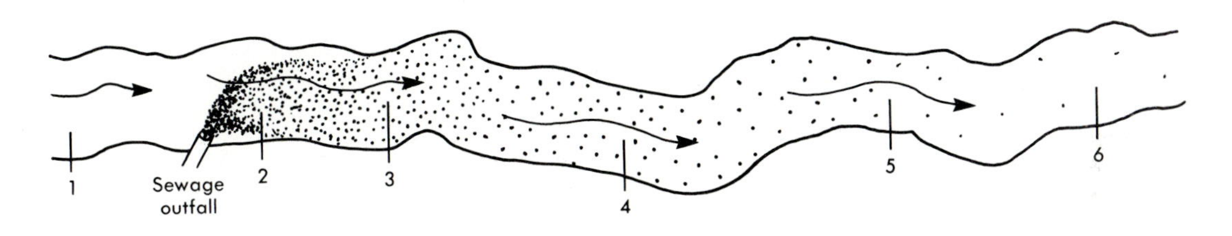

MICROORGANISMS

——Algae——
----Bacteria----
······Protozoa······

ORGANIC MOLECULES

— Organic Carbon —

INORGANIC MOLECULES

——NO_3——
------PO_4------
········NH_4········

SELECTED READINGS

Books

Anderson, J.M., and Macfadyen (Eds.). *The role of terrestrial and aquatic organisms in decomposition processes.* Blackwell Scientific Publications, Oxford, 1976.

Cushings, D.H., and Walsh, J.J. (Eds.). *The ecology of the seas.* W.B. Saunders Co., Philadelphia, 1976.

Litchfield, C.D. (Ed.) *Marine microbiology. Benchmark papers in microbiology 11.* Dowden, Hutchinson and Ross, Inc., Stroudsburg, Pa., 1976. Distributed by Halsted Press, New York.

Montgomery, J.M. *Water treatment: principles and design.* John Wiley & Sons, New York, 1985.

Medwell, D.B., and Brown, C.M. (Eds). *Sediment microbiology.* Academic Press, London, 1982.

Postgate, J.R. *The sulfate reducing bacteria* (2nd ed.), Cambridge University Press, Cambridge, 1984.

Reid, G.K., and Wood, R.D. (Eds.). *Ecology of inland waters and estuaries* (2nd ed.). Van Nostrand Reinhold Co., New York, 1976.

Sieburth, J.M. *Sea microbes.* Oxford University Press, New York, 1979.

Slater, J.H., Whittenbury, R., and Wimpenny, J.W.T. (Eds.). *Microbes in their natural environments.* Cambridge University Press, New York, 1983.

Tait, R.V. *Elements of marine ecology: an introductory course* (3rd ed.). Butterworth Publishers, Inc., Woburn, Mass., 1981.

Journal Articles

Amy, P.S., and Morita, R.Y. Starvation survival pattern of sixteen freshly isolated open ocean bacteria. *Appl. Environ. Microbiol.* **45**:1109, 1983.

Azam, F., and Hodson, R.E. Size distribution and activity of marine microheterotrophs. *Limno. Oceanogr.* **22**:492, 1977.

Azam, F., et al. The ecological role of water column microbes in the sea. *Mar. Ecol. Prog. Ser.* **10**:257, 1983.

Childress, J.J., Felbeck, H., and Somero, N. Symbosis in the deep sea. *Sci. Am.* **256**(5):114, 1987.

Jannasch, H.W., and Taylor, C.D. Deep sea microbiology. *Annu. Rev. Microbiol.* **38**:487, 1984.

Joint, I.R., and Morris, R.J. The role of bacteria in the turnover of organic matter in the sea. *Oceanogr. Mar. Biol. Annu. Rev.* **20**:65, 1982.

LaRiviere, J.W.M. Microbial ecology of liquid waste treatment. *Adv. Microb. Ecol.* **1**:215, 1977.

LeChevallier, M.W., and McFeters, G.A. Recent advances in coliform methodology for water analysis. *J. Environ. Health* **47**:5, 1984.

Meybeck, M. Carbon, nitrogen, and phosphorous transport by world rivers. *Am. J. Sci.* **282**:401, 1982.

Pipes, W.O. Microbiology of waste water treatment. *J. Water Pollut. Control Fed.* **53**:1142, 1981.

Robinson, J.B., and Tuovinen, O.H. Mechanisms of microbial resistance and detoxification of mercury and organomercury compounds: physiological, biochemical and genetic analyses. *Microbiol. Rev.* **48**:95, 1984.

CHAPTER

20

Agricultural microbiology

Martin Alexander
R.F. Boyd

Chapter Outline

Introduction

Agricultural microbiology is concerned with those microorganisms found in the soil and their influence on plants, as well as those microorganisms that cause disease in animals. Soil microorganisms are not only responsible for the cycling of elements important in plant nutrition (see Chapter 18) but some can cause disease in plants, as well as in humans and animals. This chapter outlines the types of microorganisms found in the soil and discusses their influence on the environment; it also discusses the abiotic factors that influence soil populations and the beneficial and antagonistic associations between microorganisms and plants, shows how the decomposition of plants by microorganisms is important both ecologically and economically, and demonstrates how genetic engineering is being used in plant science for human benefits. Farm animals and their influence on humans are also briefly discussed.

SOIL MICROBIOLOGY AND EXPERIMENTAL RESEARCH

Several areas of soil microbiology have been prominent for many years. Microbiologists have long been interested in an ecosystem that contains a vast number of dissimilar species and morphological types and that display a variety of beneficial and detrimental interactions among them. The indigenous communities have also achieved prominence because it has long been evident that they are essential for the nutrition of higher plants and the maintenance of soil structure. Plant pathologists and, to a lesser extent, medical and veterinary scientists have also examined the communities of this habitat because several of the native species cause disease. Furthermore, even the early soil microbiologists realized that the populations they were studying were critical to the maintenance of life and the functioning of the biosphere.

New areas of concern have arisen in recent years. One of these is the role of indigenous heterotrophs in detoxication. Soils receive herbicides, insecticides, fungicides, municipal and industrial wastes, and a variety of other toxic substances, and many of these are destroyed before the concentration of the toxicants rises to a point where they are of ecological concern. By thus acting as agents of biodegradation, microorganisms in the soil are a significant means of ridding natural environments of potentially hazardous chemicals and wastes. It has also been increasingly apparent that the indigenous communities frequently fail to degrade or detoxify many toxic substances. These failings are evident in the increasing pollution of ground and surface waters in many regions and in the persistence of certain types of pesticides, for example, DDT. Furthermore, recent research has established a role for microorganisms in creating pollutants. Thus microorganisms in soil form nitrate, and the downward movement of nitrate from soil to the underlying groundwater leads to levels of this anion that may be toxic to infants consuming the water. If the microbiologically formed nitrate is assimilated by plants to levels that become extremely high, as occurs with certain vegetables, a precursor for human carcinogens may become abundant. The formation of nitrous oxide in soil, the effect of this nitrous oxide on the ozone shield in the atmosphere, and the relationship of these processes to skin cancer in humans are considered in Chapter 18.

BACTERIA

Bacteria are numerically the dominant group of microorganisms in soil. However, because their cells are small, the total biomass of bacteria frequently is less than that of fungi. However, in terms of metabolic activity, certain processes in well-aerated soils are dominated by bacteria, and under anaerobic conditions, bacteria are chiefly responsible for biochemical changes underground.

Table 20-1 ***Typical morphology of bacteria obtained by plate counts of soil***

Morphological group	Bacteria in class (%)
Long rods, non-spore-forming	<1 to 7
Short rods, non-spore-forming	
Gram-positive	8 to 23
Gram-negative	13 to 27
Gram-variable	6 to 12
Rods that become coccoidal	2 to 13
Coccoidal rods	18 to 54
Cocci	<1 to 8
Rods, spore-forming	6 to 10

Because of the wide range of physiological and nutritional types of bacteria, no one culture medium and no single method are considered adequate for defining and estimating the size of the bacterial community of soil. Plate counts on agar media give large numbers of bacteria, frequently from 10^6 to 10^8 or more per gram of soil. The size of the community varies with the location and the environmental conditions that prevail. However, microscopic examination of soils indicates far greater population sizes, and often plate counts give values less than 1% or sometimes as much as 10% of the total count as indicated by microscopic means. Even if many of the cells seen under the microscope are nonviable, it is clear that plating procedures do not estimate the total community. The bacterial biomass even accounts for a significant part of the total soil mass, figures as high as 0.40% being common.

The bacterial community is dominated by 25 genera. Several other genera are present occasionally or in small numbers. The major species tend to be members of the genera *Arthrobacter, Pseudomonas, Agrobacterium, Alcaligenes, Flavobacterium,* and *Bacillus.* Although always present in small numbers, pathogens affecting humans, livestock and other animals, cultivated plants, and wild species of higher plants are present. Among the animal pathogens are species of *Clostridium, Bacillus, Listeria, Erysipelothrix, Coxiella,* and *Streptococcus.* Because roots of higher plants develop underground, the pathogens of higher plants are of particular significance. These include species of *Agrobacterium, Corynebacterium, Erwinia, Pseudomonas,* and *Xantho-*

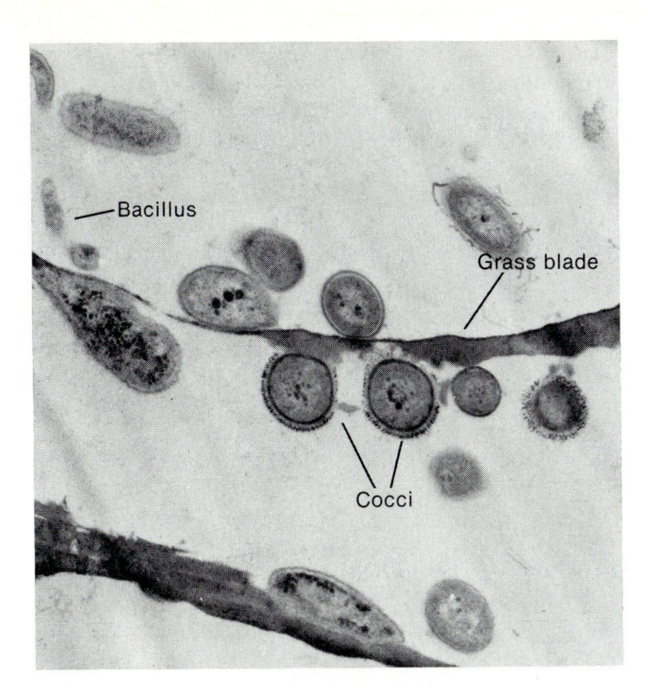

Figure 20-1 *Transmission electron micrograph of encapsulated cocci and irregularly shaped bacilli attaching to and degrading grass. (×18,000.)*

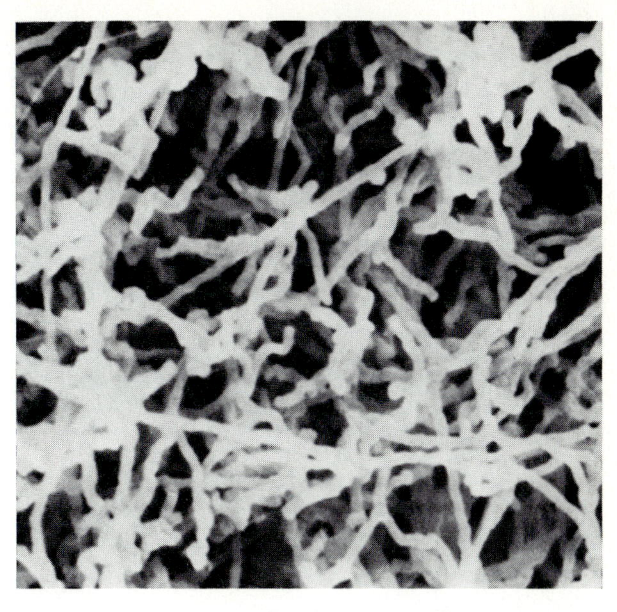

Figure 20-2 *Scanning electron micrograph of filamentous actinomycete* (Nocardia). (×500.)

monas. The percentages of the major soil mórphological types in a representative study are shown in Table 20-1.

It is often assumed that bacteria require no growth factors. This is untrue. A large percentage of the indigenous community is unable to grow in the absence of one or several B vitamins or amino acids. These growth factors are not present in soil in the absence of biological activity, so their presence indicates a role either for the plant in excreting growth factors or, more likely, for microorganisms in excreting the growth factors required by other inhabitants of the ecosystem.

Bacteria stand out because of the many different processes they carry out. Heterotrophic bacteria are able to metabolize and mineralize many low-molecular-weight organic compounds, and they carry out a varied set of biochemical processes and decomposition reactions under a wide range of conditions. They destroy many natural organic materials (Figure 20-1), with the possible notable exception of lignin, and they also decompose many synthetic organic compounds. The bacteria are particularly significant in oxidizing and reducing ions of nitrogen, sulfur, iron, and manganese. They are dominant members of the microflora in the absence of oxygen. Furthermore, many grow particularly quickly, and they thus often assume dominance. However, fast growth rates are only advantageous if the substrate being used is readily metabolized, and compounds that are quickly destroyed are not widely added to or generated in soil, except in the vicinity of plant roots. Microscopic observation of soil shows the presence of small bacteria, often smaller in size than 0.3 μm. These organisms are largely unknown in culture, and hence their role in nature has yet to be established.

FILAMENTOUS BACTERIA AND FUNGI

Bacteria

Actinomycetes are especially common in soil (see Point of Interest). Taxonomically, the actinomycetes are bacteria, but they are often considered separately in soil microbiology because of their morphological characteristics, their slow growth rates, and the limited range of substrates for which they can effectively compete. Because of their slow growth, they often appear on agar media slowly, and the microbiologist who incubates plates for short periods of time frequently misses these organisms because their colonies only appear after 1 or more weeks of incubation. If the plates are incubated sufficiently long, it is evident that the actinomycetes represent the largest group of bacteria.

It is not clear to what extent the actinomycetes exist in soil as hyphae (Fig. 20-2) and to what extent the large numbers arise from the occurrence of underground conidia. However, it is generally believed that the filaments appear only occasionally and that the large numbers of colonies appearing on agar media inoculated with soil dilutions reflect the persistence of the asexual spores. If this view is correct, the actinomycetes are not as significant contributors to many of the biochemical transformations in soil as their numbers would suggest. Their role would largely be linked to the availability of substrates or conditions particularly suited for their proliferation.

FUNGI

Plate counts of fungi are of little value. The reason is that the fungi, as well as actinomycetes, on plates inoculated with soil dilutions may arise from long filaments, a fragment of a hypha, a single conidium, a cluster of conidia, or one of several types of structures produced in soil by fungi. Hence, although plate counts of fungi in soil are almost always low, this does not indicate their true biomass. Measurements of their biomass by several techniques show that they are extremely abundant, and frequently the mass of fungi exceeds that of all other microbial groups in soil. A better appreciation of the fungal mass can be obtained by considering the lengths of their mycelium: as short as 10 meters per gram of surface soil, but sometimes as high as 100 meters per gram, and occasionally in excess of 1000 meters per gram.

Point of Interest

ACTINOMYCETES

Actinomycetes are a large group of bacteria comprising several diverse genera of bacteria. Actinomycetes are usually gram-positive filamentous species that may or may not produce spores. The filaments resemble the hypae of the fungi and are capable of forming mycelia. A mycelium may remain attached to the surface (substrate mycelium), or aerial filaments may arise directly from the substrate mycelium. The individual filaments of hyphae are subdivided into units in which there is an inward growth of the cell wall at regular intervals along the filament (a process called septation). Each septated unit in the filament contains at least one DNA molecule. Some of the different types of filaments produced by actinomycetes are illustrated in figure below right. One of the unusual characteristics of this group is that reproduction often leads to the formation of spores that may appear singly or in chains. The spores produced by most species are not heat resistant as are the endospores of the genera *Bacillus* and *Clostridium.*

Actinomycetes make up approximately 20% to 60% of the total microbial population of the soil. The so-called "earth smell" of soil is due to the production of perpenoids (**geosmins**) and extracellular enzymes produced by actinomycetes. They produce a multitude of extracellular enzymes that are capable of degrading plant and animal organic matter. The genus *Arthrobacter,* related to actinomycetes, makes up a large percentage of soil microorganisms, and many species are capable of slowly degrading recalcitrant substrates, such as petroleum hydrocarbons and insecticides. Some species can emulsify oil when growing on hydrocarbons by producing compounds such as fatty acids. These emulsifying agents help make the recalcitrant molecules more susceptible to biological and abiological degradative processes.

The genus *Streptomyces* is an important actinomycete, and several hundred species can be found in the soil. *Streptomyces* species produce a multitude of extracellular enzymes that enable them to degrade organic molecules resistant to degradation by most other bacteria. *Streptomyces* species produce specialized hyphae from which spores (conidia) are formed. Many secondary metabolites (by-products formed during the synthesis of essential primary metabolites) are synthesized by *Streptomyces* and have industrial and commercial value. One group of secondary metabolites is the antibiotics. Nearly 90% of all the known antibiotics are produced by species of *Streptomyces,* for example, streptomycin, kanamycin, erythromycin, and amphotericin B. There appears to be a close relationship between spore formation and antibiotic production. Cells of *Streptomyces* cease vegetative growth and produce spores when nutrients become limiting. Spore formation is accompanied by antibiotic production. Experiments show that soil-produced antibiotics inhibit the growth of microorganisms that compete with *Streptomyces.* However, only small amounts of antibiotic are produced and it is believed that antibiotics have other functions. For example, one antibiotic produced by a species of *Streptomyces* has been isolated and shown to inhibit the germination process. Other species of *Streptomyces* produce an antibiotic that acts like an alkali-metal transporting agent. Much research must still be performed to evaluate the function of antibiotics in the soil.

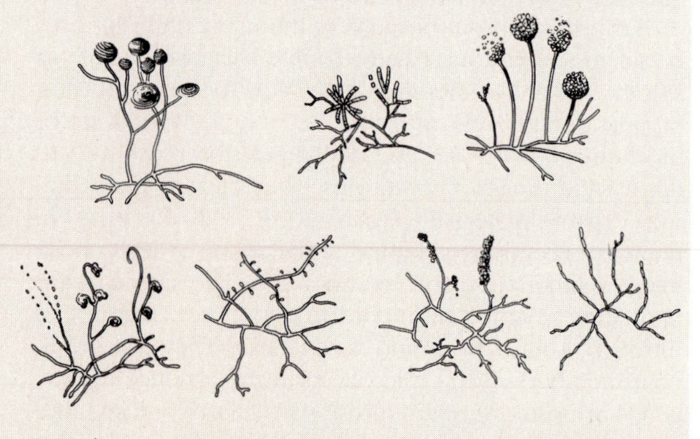

Types of aerial filaments produced by actinomycetes.

Mycologists have devoted much time to identifying the dominant species and genera of fungi. Even now, new species and genera are still being described. The numbers of genera thus are still not known, but the species diversity of filamentous fungi is unquestionably impressive. The fungal flora includes species that lack sexual stages as well as those producing abundant sexual spores. Some species produce spores in asci, and others produce motile zoospores. There are many dissimilar genera of fungi in the soil. The most frequently isolated belong to the Fungi Imperfecti and include species of *Aspergillus, Geotrichum, Trichoderma,* and *Penicillium* (see Chapter 6 for a more complete discussion).

Although the hyphae of some fungi develop quite slowly, and thus the organisms resemble the actinomycetes in not being effective competitors for readily available substrates, the hyphae of other species develop quickly. Organisms possessing these characteristics may dominate when substrates available to them are present. Fungi are particularly active by their use of the two major types of polysaccharides in plants—cellulose and hemicellulose—and they are the chief organisms in decomposing lignin aerobically (there is increasing evidence that lignin may be partially mineralized under anaerobic conditions by bacteria). However, when the nutrient supply falls to low levels or the environment becomes hostile for other reasons, fungi endure because of the production of specialized resting structures. These structures include chlamydospores, oospores, ascospores, and others. Species producing these structures thus have a means for enduring adversity, and the hyphae reemerge from the resistant body once the conditions in soil become again suitable for fungal development.

Fungi are especially important in soil in the decomposition of resistant compounds, especially those that are not readily utilized by bacteria. These filamentous organisms decompose organic nitrogen compounds that are abundant in soil and convert them to ammonium, a process of significance to higher plants, which require inorganic forms of nitrogen. They are also significant both in breaking down and in forming **humus**, the native organic fraction of soils. The filamentous organisms, by virtue of the filaments themselves, are able to bind soil particles together, and the binding of the small particles into larger aggregates improves soil aeration and water movement. These aggregates are helpful in allowing for the room necessary for the growth of roots and their extension through the soil matrix. The mutualistic association between fungi and plant roots, called **mycorrhizae**, was discussed in detail in Chapter 6. A number of fungi are prominent because of their role in disease, often in diseases of higher plants but occasionally in those of both humans and animals.

ALGAE AND CYANOBACTERIA

The presence of algae and cyanobacteria is easily shown. They are able to bring about photosynthesis, and thus their development is readily noted when soils are inoculated into an inorganic nutrient solution that is incubated for several weeks in the light. However, their significance in most soils and many areas of the world is limited because the algae and cyanobacteria seem to become metabolically active only where water is abundant and light is present. Below the soil surface, water is abundant, but the absence of sunlight restricts their capacity to grow in this dark habitat, especially when they must compete with fast-growing bacteria and fungi for the simple organic molecules that some algae and cyanobacteria can metabolize.

On the other hand, algae and cyanobacteria are often noted at the surface. They grow when moisture is abundant, a circumstance that tends to be confined to areas in which the soil is protected from high light intensities or to brief periods immediately following rainfall. Under these conditions, algae and cyanobacteria add organic matter to the soil as a result of photosynthesis. Once the organisms die, the dead cells serve as nutrients for bacteria and fungi, thus making available the organic compounds synthesized from carbon dioxide. Algae and cyanobacteria are important in starting the life cycle on rocks, in volcanic areas, and in other regions largely free of life. The primary colonizers in these areas are free-living algae and cyanobacteria as well as lichens. In such areas, products excreted by the primary colonizers or substances released as they are being decomposed cause a slow but still appreciable destruction of the rock material. This destruction, over considerable periods of time, brings about weathering of the rocks and the initiation of soil formation.

Flooded paddy fields in southeast Asia often contain enormous masses of filamentous algae and cyanobacteria. In such ecosystems, light intensities are often high, the temperature is warm, inorganic nutrients are sometimes abundant, and moisture is not limited. Here, the algae and cyanobacteria have a prominent role in the ecosystem. As they carry out photosynthesis, they add appreciable organic matter to the water. The organic matter in the cells in turn supports the development of many herbivorous animals that grow in the water. The photosynthetic organisms also release oxygen, and this process may improve the aeration status of the otherwise poorly aerated soils and marshes. In addition, certain cyanobacteria are active in nitrogen fixation in many paddy fields. Rice has been grown almost every year for thousands of years in much of Asia, often with no signs of nitrogen deficiencies. The fields in which the rice is grown contain a rich flora of nitrogen-fixing cyanobacteria, and they add sufficient nitrogen to the soil and floodwater to permit the development of a crop of rice.

However, the rice yield associated with reliance solely on cyanobacteria for nitrogen is low. Therefore an international effort is now underway to find means to increase the rate of nitrogen fixation by the cyanobacteria.

PROTOZOA

Among the subterranean protozoa are representatives of many groups of flagellates, amebas, and cilates (Table 20-2). These animals have dissimilar methods of locomotion, namely by flagella, pseudopodia, and cilia, respectively. All three groups may be significant in soil, but their numbers vary appreciably from place to place and time to time. Because protozoa are unicellular organisms, determinations of their numbers have meaning. Such counts indicate that the numbers may be as low as 10,000, but sometimes the values may exceed 100,000 or more per gram of soil. Nevertheless, the counts may be misleading in attempting to assess protozoan contributions to the microbial biomass and activity, because their cells are frequently large and therefore one protozoan cell has a greater biomass than the cells of other unicellular organisms. On the other hand, many protozoa exist at any one time not in the metabolically more active forms but rather as cysts. Cysts have low metabolic activity. As a rule, the flagellates are abundant, and the ciliates are reasonably uncommon.

Certain protozoa grow in the absence of other microorganisms. However, although able to decompose organic materials, these protozoa probably are not significant in this mode of nutrition because they do not compete effectively with bacteria and fungi in the same environment. Nevertheless, protozoa are avid predators (Figure 20-3), and they are able to consume thousands of bacteria per protozoan cell division. The predatory habit of protozoa appears to be more common than the nutritional pattern requiring feeding on nonviable organic materials.

Counts and microscopic examination of soil suggest that these small animals proliferate when the bacterial community becomes large, as occurs around plant roots or following the addition of significant quantities of readily available organic materials. Such proliferation is shown in Figure 20-4. At that time, the protozoan emerges from its cyst and begins to feed voraciously. Their abundance increases markedly, and at the same time, the number of bacteria falls. Nevertheless, the edible bacteria are not totally eliminated by predatory protozoa. Some mechanism exists to allow bacteria to coexist with species that feed on them. This coexistence is evident by the persistence in perpetuity of protozoa and bacteria in soil. A common explanation for the persistence of predator and prey is that the predation rate of protozoa declines with the fall in density of the bacteria they consume, and the two groups of organisms are able to match their growth and predation rates so that neither microbial type is eliminated. Predation may appear to reduce the biochemical activity of the community, but it does allow nutrients retained within the cell of the bacterial prey to be made available to succeeding populations (see Chapter 18). In the absence of any breakdown of the bacterial cells, elements of nutritional importance would not be recirculated, and the rate of biochemical transformations would then decline. In this way, protozoa may allow for a greater activity in soil than would be true in their absence.

Table 20-2 *Major genera of protozoa found in the soil*

Group	Examples (Genera)
Ciliates	*Paramecium*
	Colpoda
	Nassula
	Pleuronema
	Vorticella
	Stylonycha
Flagellates	*Peranema*
	Astasia
	Olicomonas
	Bodo
Rhizopods	*Amoeba*
	Acanthamoeba
	Pelomyxa
	Arcella
	Euglypha

Concept Check

1. List the major genera of bacteria found in the soil.
2. What are the important actinomycetes found in the soil, and what is their ecological and/or commercial value, if any?
3. What are the major genera of fungi found in the soil? What are some of their major benefits to the environment?
4. What contributions do algae, cyanobacteria, and protozoa make to soil ecology?
5. What problems arise in trying to assess quantitatively the abundance of bacteria and actinomycetes in the soil?

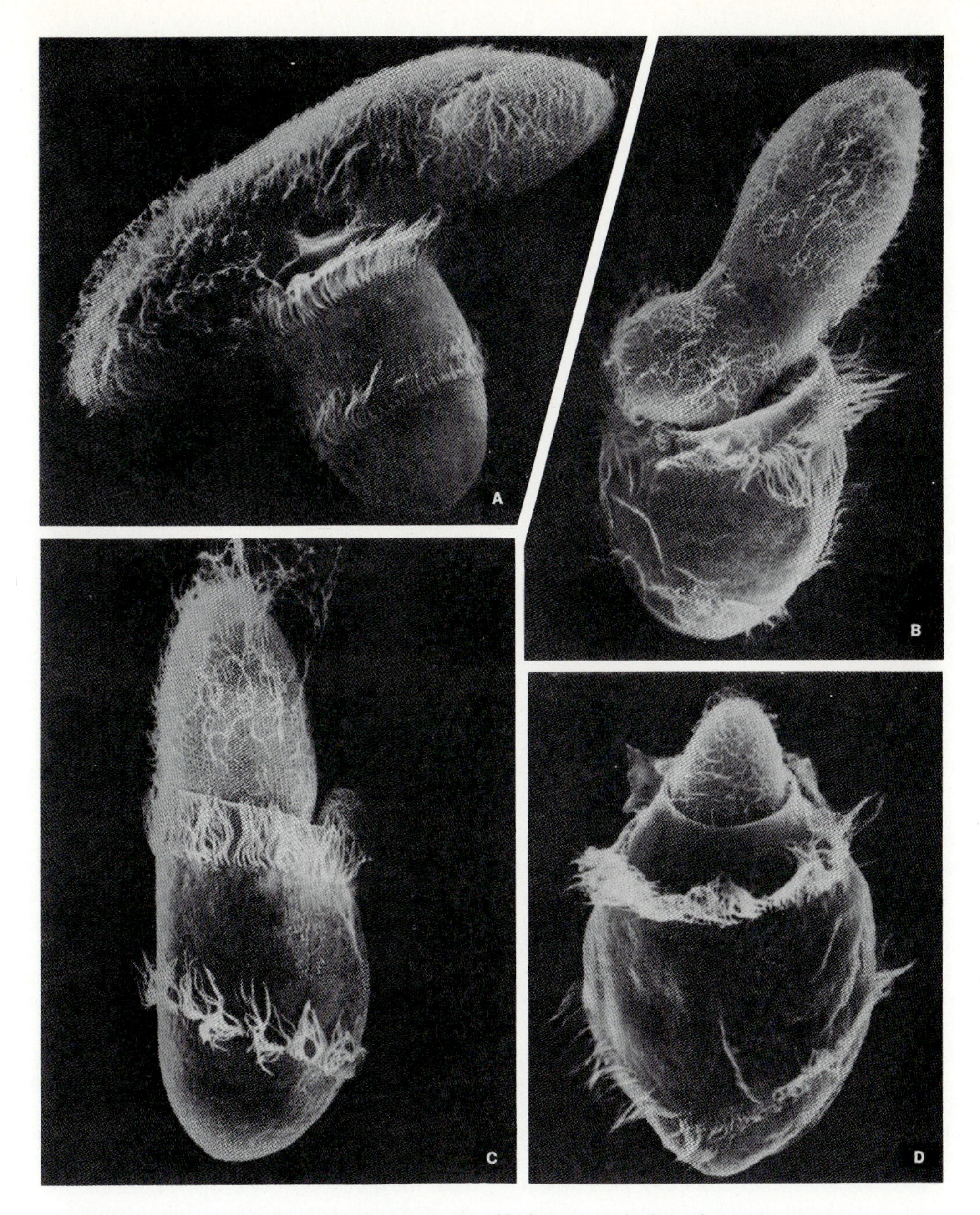

Figure 20-3 *Scanning electron micrographs of* Didinium *species ingesting protozoan* (Paramecium). ***A,*** *Cytosome (mouth-like cavity) of* Didinium *species has expanded.* ***B,*** Paramecium *is being folded as it is ingested.* ***C,*** Paramecium *has been partially engulfed.* ***D,*** *Most of* Paramecium *has been engulfed.*

ECOLOGICAL GROUPS

Indigenous heterotrophs

The terrestrial community contains fast-growing bacteria and fungi that use readily available nutrients in the environment as well as other heterotrophs that metabolize resistant materials in added organic substances. Because they use resistant compounds, the latter populations grow slowly in nature. Other heterotrophs use partially decayed organic matter, humus, microbial products, or microbial cells themselves. Parasites and predators are the inhabitants that live on the cells. The fast-growing

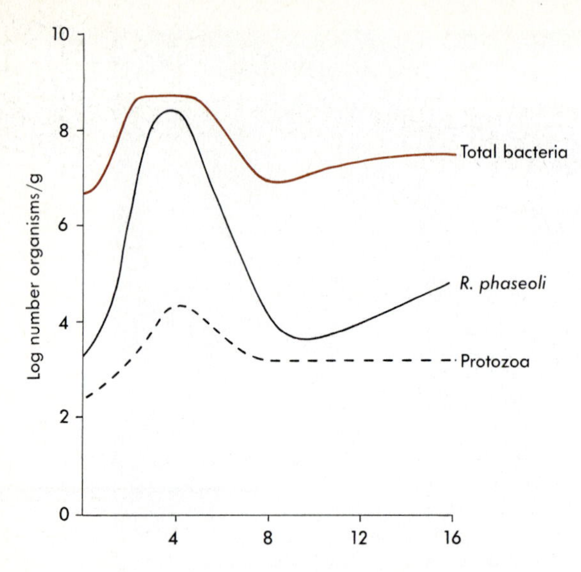

***Figure 20-4** Growth of total bacteria and* Rhizobium phaesoli *and subsequent development of protozoa on bean roots. Seeds were inoculated with* R. phaesolii *and planted in soil on day O.*

heterotrophs respond immediately to nutrient additions, and these bacteria and fungi are active in the initial stages of many biochemical transformations, such as the decomposition of sugars, starch, organic acids, and proteins. The numbers of these fast-growing organisms fluctuate greatly, depending on the nutrient supply. Slow-growing populations using organic materials added to soil and those living on decayed organic matter and humus do not respond too readily to recent nutrient additions, and thus the size of their populations or their biomass is not as variable as that of the fast-growing organisms.

Parasites

Many species are soil invaders. A variety of such transient species enter the environment in diseased plant or animal tissues, animal manure, or sewage sludge applied to the land. The transient behavior of these organisms is evident in the rapid decline of many of the species that are introduced onto or into soil together with sewage sludge. On the other hand, some soil invaders are not transient but still are not native. These are species that persist because they produce a resistant stage. Among this group are animal pathogens such as *Bacillus anthracis* or plant pathogens that survive below ground but do not proliferate. The latter group, which includes many fungi, infects their hosts when it is growing at that soil site. Free-living protozoa, such as ameba, can be found in moist soil. Under adverse environmental conditions these organisms can form cysts. Some amebas are also parasites for vertebrate and invertebrate animals. Some species, such as *Entameba histolytica,* parasitize the human intestine and are released into soil that is fecally contaminated. In the soil, *E. histolytica* also forms cysts that are resistant to levels of chlorine used for water purification.

Root-associated organisms

The true **rhizosphere** inhabitants are not really native to soil and are also not transient. The rhizosphere is the zone immediately surrounding plant roots and is dominated by the activities of the root. The inhabitants of this zone are truly root-associated organisms, although many exist in small numbers in soil apart from the root system. The rhizosphere is discussed at greater length on page 594. In this zone are also found unique fungi that form symbiotic associations with roots of appropriate plant species, that is, **mycorrhizal** fungi.

ABIOTIC FACTORS

Various nonbiological factors affect the composition of the soil community and its activities. The major abiotic determinants are nutrient supply, moisture level, pH, oxygen supply, oxidation-reduction potential, temperature, the presence of inhibitory substances, colloidal surfaces (such as those of clay and humus), and salts.

Nutrients

As a rule, the nutrients affecting the microbial community are almost only organic because few inorganic nutrients stimulate microbial populations and activity in soils (for example, the lithrotrophs derive energy from oxidation of inorganics). A specific type of organic nutrient will favor a specific group of populations that have the capacity to decompose that organic substrate. Thus natural selection favors individual species that are able to metabolize sugars, cellulose, or proteins. In turn, the products of metabolism of these initial populations are used to satisfy the nutritional needs of others. In this way, the addition of one type of organic material leads to profound changes not only in the numbers of microorganisms but in the composition of the community. However, there is no direct correlation between the percent of organic matter found in soil by chemical analysis and the total number of microorganisms; too many other factors control population sizes. The populations that respond following addition of plant residues, animal wastes, or individual organic compounds may therefore live on the introduced nutrients, the excretions of the primary populations, cells or filaments that **autolyze** (that is, the cells or hyphae are digested by their own enzymes), or by utilizing the cells of the ear-

Table 20-3 *Microbial metabolism in anaerobic soils*

	Organic matter decomposition	Metabolism	Eh volts	Organic acids
Stage 1 (rapid)	Aerobic to semianaerobic	O_2 respiration, NO_3^- reduction, Mn^{4+}, Fe^{3+} reduction.	+0.6 to +0.3	Only accumulated if fresh organic matter present
Stage 2 (slow)	Anaerobic	SO_4^{2-} reduction, CH_4 production	0 to −0.22	Rapid accumulation in early stages, rapid decrease in advanced stage

lier populations as prey or parasites. The organisms that are active as predators on microbial cells in soils include protozoa, myxobacteria, and a variety of fungi.

Moisture

The moisture supply affects the community in four ways. First, it provides the water essential for replication. At low water levels, the increase in population sizes and activities with increasing moisture is a result of supplying the water that is needed for microbial development. Second, moisture affects the diffusion of oxygen into the soil. Thus as soil approaches the water-logged condition, its pores become filled with water. Third, the soil solution carries many microbial nutrients, and moisture is therefore important in governing the supply of water-soluble nutrients. Finally, a soil with a high water content is not as subject to rapid temperature changes as a soil with little water.

Hydrogen ion concentration

The pH of soil is important because it alters the composition of the community and also directly influences the populations within the community. In acid soils, fungi are abundant, and bacteria and actinomycetes are less active than in neutral environments. The dominance of fungi under acid conditions is not a result of their being favored directly by the acidity but rather is a result of the elimination of competitors that do not tolerate the lower pH. Certain populations and the processes they carry out are completely inhibited under acid conditions. For example, the formation of nitrate is usually abolished as the pH of soil declines, although certain soils contain nitrate-forming populations that appear to be relatively less sensitive to acidity. The infection of legume roots by its symbionts, *Rhizobium,* is also affected deleteriously by acidity, and, although legumes continue to grow at moderate acidity, infection by the root-nodule bacteria is not initiated, so that nitrogen-fixing nodules are not generated.

Aeration

Aeration affects the microbial community because most of its members are aerobes. Nearly all soil fungi and actinomycetes and most of the bacteria require oxygen for replication. The anaerobic members of the community are chiefly bacteria, which may be either facultatively or strictly anaerobic. The concentration of oxygen in the soil is in a state of flux—the gas coming in from the atmosphere, the microorganisms in the soil matrix using it up in line with their metabolic demand. If the supply of readily available organic nutrients is large, heterotrophs use up the oxygen present, and the demand for the gas exceeds the supply. Under these conditions, the size of the community and its activity are limited by the rate of entry of oxygen from the overlying atmosphere. Surprisingly, in what appears to be a well-aerated soil, anaerobic pockets exist. In these pockets are strictly anaerobic bacteria, so that a well-oxygenated soil still will support processes that are known to be carried out only by obligate anaerobes. Similarly, although denitrification ($NO_3 \rightarrow N_2$ or N_2O) proceeds only in the absence of oxygen, denitrification proceeds in well-aerated soils because the responsible bacteria exist in microenvironments where there is too little or no oxygen for their metabolism and proliferation.

Oxidation-reduction potential

The oxidation-reduction potential is a measure of the electropositivity or electronegativity of the environment. At a positive potential, conditions are oxidizing. At a negative potential, conditions are reduced. The chief oxidant is oxygen, and its presence raises the potential. Reducing agents lower it. Oxidation-reduction potential is ecologically important because aerobes need environments with high potential (that is, oxidized environments), although they lower the potential as a result of their metabolism. Anaerobes, in contrast, need low potentials or reduced environments (Table 20-3). It is difficult to differentiate, however, between the ecological effects of oxygen and those arising from oxidation-re-

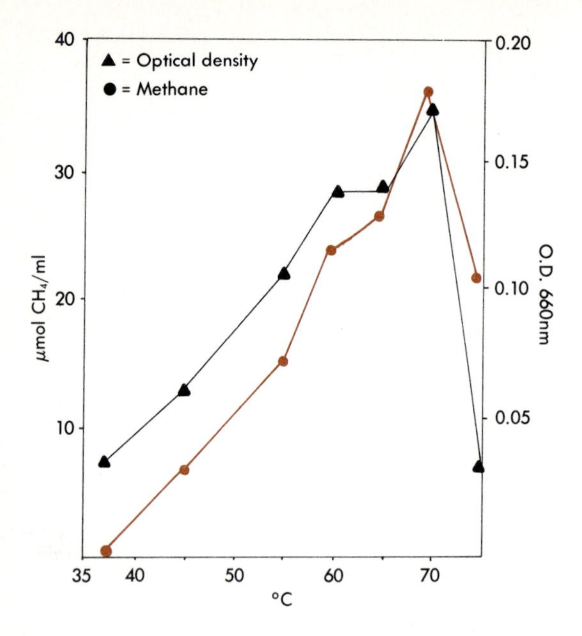

Figure 20-5 *Effect of temperature on methane production by culture of* Methanobacterium *species grown in presence of hydrogen and carbon dioxide gases and incubated for 48 hours at the different temperatures.*

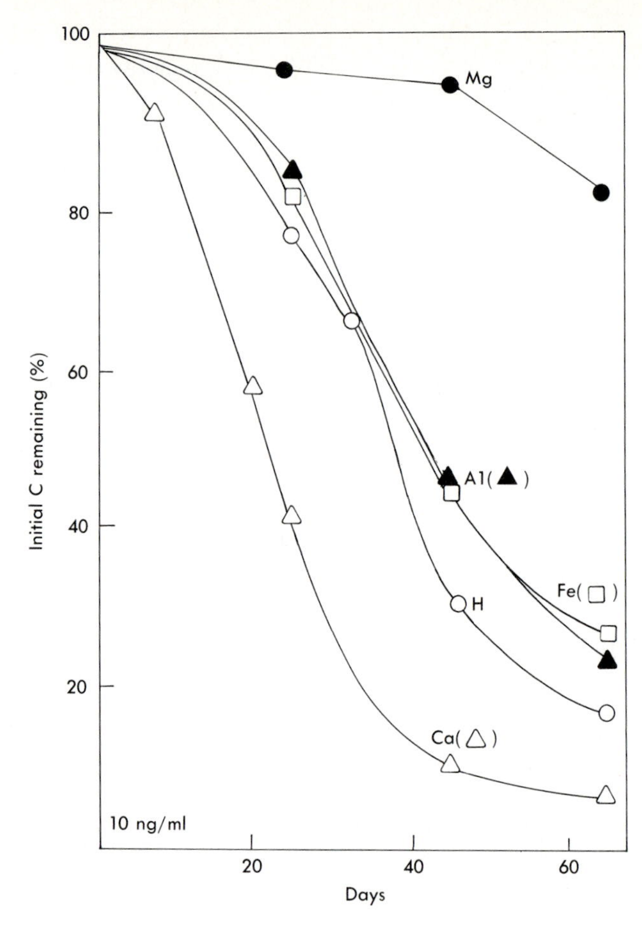

Figure 20-6 *Effect of various metal cations on mineralization of citrate by sewage microorganisms. Citrate forms complexes with various metal cations because of its carboxyl groups that act as chelators (binding agents). At 10 nanograms of carbon per milliliter, rates of mineralization of iron, aluminum, calcium, and hydrogen citrate were not significantly different from one another. However, magnesium citrate was only slowly mineralized.*

duction potential. Aerobes and aerobic processes are dominant when the oxygen level and the potential are high, and anaerobes and anaerobic processes are prominent when the oxygen level is low or none is present and the potential is also low.

Temperature

Temperature affects community composition and its activities in soil, as in other natural environments. Each species has a minimum, an optimum, and a maximum temperature, but each is interacting with other members of the community. As a result, organisms are often not dominant at their optimum temperature but rather are prominent at those temperatures at which their activity is maximum relative to competing populations. Soils contain few strict psychrophiles, even in areas that have long, cold winters, and in these regions, the community is usually dominanted by cold-tolerant mesophiles. When soil is frozen, there is little if any microbial growth, but enzymes continue to function. Hence, low metabolic activity may be evident. Once frozen soil is thawed in the spring, the population size may increase rapidly. This increase is in part a result of the greater warmth, but it may in part be attributed to the breaking up of hyphae, cell clusters, or colonies by the thawing process. With increasing temperature, moreover, activity of the indigenous populations and their growth rates increase, and the increase is proportional to the rise in temperature (Figure 20-5). In periods of the year when the habitat becomes hot, the previously dominant members of the community may be killed, so that areas with high temperatures are dominated by species able to cope with the heat stress. Nevertheless, few soils get sufficiently hot and still contain adequate water to permit the active replication of thermophiles; the high temperatures at the soil surface in the tropics are usually accompanied by drying so that opportunities for proliferation of thermophiles are not common.

Inhibitory substances

Inhibitory substances may be ecologically significant. The inhibitors include inorganic substances that may be effective only at high concentrations, such as the inorganic salts that are abundant in saline soils. Some inorganic inhibitors are ecologically significant at low con-

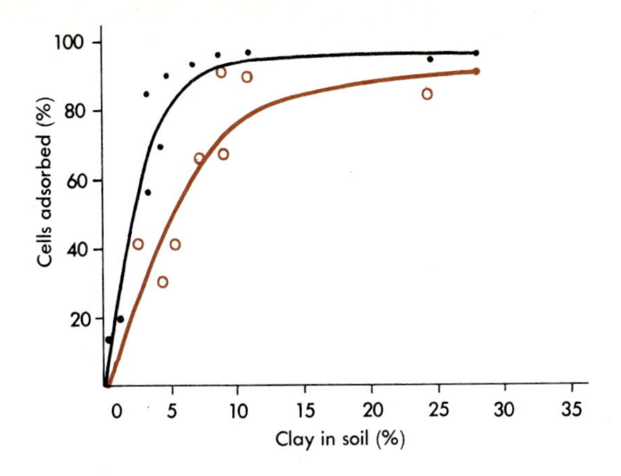

Figure 20-7 *Increase in percent of cells of* Bacillus cereus *variety* mycoides (solid circles) *and* Serratia marcescens *adsorbed in soils with various amounts of clay.*

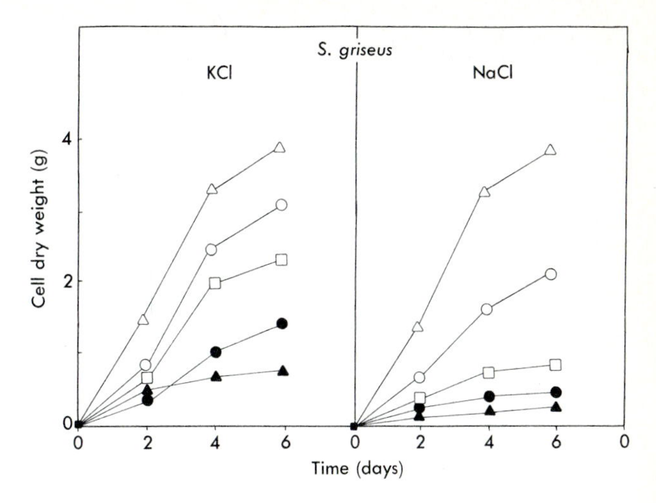

Figure 20-8 *Effect of salt concentration (potassium chloride—KCl) and sodium chloride (NaCl) on the growth of* Streptomyces griseus *in broth.* △, *Basal;* ○, *0.25 M;* □, *0.5 M;* ●, *0.75 M;* ▲, *1.0 M.*

centrations; for example, ionic arsenic, copper, and hydrogen sulfide are sometimes important. Similarly, toxic cations are present in some regions because of the return to soil of sludges that are rich in cationic forms of heavy metals. Many organic compounds form complexes with metal ions and their mineralization may be affected by the type and concentration of metal in the environment. Experiments demonstrating this effect are characterized in Figure 20-6.

In addition to inorganic inhibitors, a number of organic compounds that are produced biologically may be harmful to members of the community. Some of these compounds may be generated in large or small amounts by plant roots or microorganisms. Evidence exists that shows simple organic acids may be responsible for the resistance to decomposition of organic materials in flooded acid peats. Antibiotics may also have an ecological impact, although controversy still rages about their role in nature.

Colloidal substances

Soils are characterized by enormous surface areas. These large surface areas are the result of the presence of small particulate materials—clays and colloidal organic matter. The surface areas are highly reactive, but surface effects have only recently attracted widespread attention. They affect the community because the surfaces of clay and colloidal organic substances react with nutrients, microbial cells, and their extracellular enzymes. These colloidal materials have a charge, and they retain substances on their surfaces in part because of the charge on the colloid and in part by other mechanisms. The relation between the number of bacteria retained by soils and their clay content is shown in Figure 20-7. The activity of the adsorbed microorganism or its enzymes is different from that of the organism or enzyme in the free state. Similarly, adsorbed nutrients are often less available for use than nutrients that exist free in the soil solution. At low nutrient levels, surfaces may stimulate microbial activity because they concentrate the nutrients, which favor biological processes, and colloids may also be beneficial because they retain or adsorb toxins.

Salts

Salinity in irrigated agriculture has become an important problem and has stimulated research on the effects of salt stress in both plants and microorganisms. Inorganic salts are the chief agents of osmotic effects. As discussed in Chapter 14 many prokaryotes respond to salt or osmotic stress by accumulating amino acids, such as proline or glutamate, whereas eukaryotes, such as fungi and algae, accumulate polyols such as glycerol. Salts normally have an inhibitory effect on growth (Figure 20-8). Many land areas are extremely rich in salts, and some saline soils in the tropics are so rich in salts that higher plants are unable to grow. Nevertheless, microorganisms are still able to proliferate because of the presence of salt-tolerant or halophilic species. Among the halophiles in soil are bacteria, particularly actinomycetes.

NUTRIENT CYCLES

Nitrogen and other elements

For plants and for agriculture, the most important of the elements coming from the soil is nitrogen. More often than any other element, it is nitrogen that limits plant growth and food production. Because the supply of this element is regulated almost entirely by microbial activity, soil microbiologists have focused considerable attention on the transformations of the nitrogen cycle. The nitrogen cycle on a global basis was discussed earlier in chapter 18. Nearly every step of the cycle as it occurs on land, except for the uptake of inorganic nitrogen by plants and the consumption of plants by animals, is brought about by microorganisms (Figure 20-9).

The nutrient cycles involving other elements such as carbon, sulfur, phosphorus, and iron are discussed in Chapter 18.

PLANT MICROBIOLOGY

The various interactions occurring between microorganisms and other members of their community are discussed in Chapter 18. This part of the chapter is devoted to the specific relationships between microorganisms and plants. This discussion begins with some of the beneficial and antagonistic associations that exist between plants and microorganisms.

Beneficial associations (the rhizosphere community)

Many microorganisms in the rhizosphere are saprophytic, that is, living off dead plant materials. The nutrition of other microorganisms in the rhizosphere, however, depends on living cells. Mycorrhizal fungi, for example, cause an increase in plant growth and protect the plant against potential pathogens (see Chapter 6). Bacteria, such as *Rhizobium*, are important in nitrogen fixation in leguminous plants (see Chapter 13). These beneficial interactions involve specific species of bacteria and fungi.

The area around a root surface generally contains more microorganisms per gram of soil than a soil area that is root-free. Plant metabolism provides a source of nutrients that are excreted through the roots to the surrounding microbial population. The bacterial community in this restricted area often exceeds 3×10^9/gm of rhizosphere soil. A soil sample at some distance from the rhizosphere might have a maximum of 1×10^8 bacteria/gm of soil (Table 20-4). Most of the rhizospheral bacteria are gram-negative, aerobic, rod-shaped, non-spore-forming organisms, such as *Pseudomonas*, *Azotobacter*, *Azospirillum*, and *Alcaligenes*. Fewer filamentous bacteria, such as actinomyces, or fungi are evident by microscopic examination. The bacterial cells are not distributed uniformly over the surface of roots and root hairs but rather exist as small colonies, presumably at sites of plant excretions. Protozoa also occur in the rhizosphere and most are small amebae or genera, such as *Bodo* and *Olicomonas*. The microbial populations in the rhizosphere are especially active metabolically, thus they probably have a major impact on the physiology of the root system.

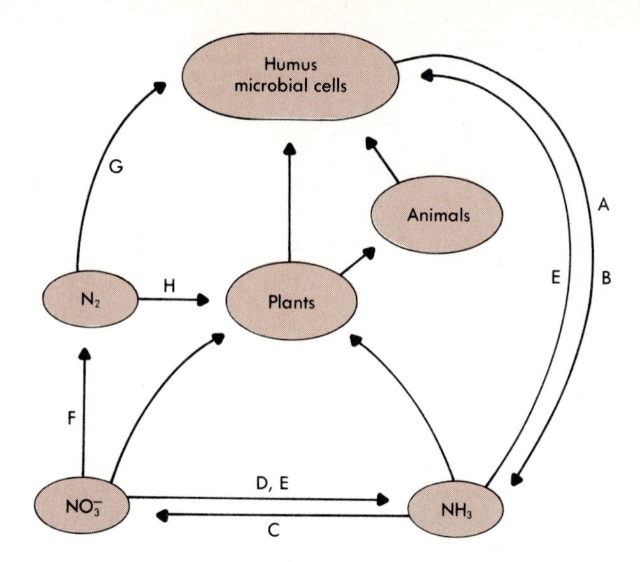

Figure 20-9 *Nitrogen cycle.* ***A,*** *Ammonification;* ***B,*** *mineralization;* ***C,*** *nitrification;* ***D,*** *nitrate reduction;* ***E,*** *immobilization (for example, assimilation by heterotrophs);* ***F,*** *denitrification;* ***G,*** *nitrogen fixation, nonsymbiotic;* ***H,*** *nitrogen fixation, symbiotic.*

Table 20-4 *Colony counts of bacteria in root-free and rhizosphere soil of crop plants and wild species*

	Colony counts of bacteria (10^6/gm soil)		
Species	**Rhizosphere**	**Root-free soil**	**Ratio rhizosphere to root-free soil**
Trifolium pratense (clover)	3260	134	24
Avena sativa (oats)	1090	184	6
Linum usitatissimum (flax)	1015	184	5
Zea mays (corn)	614	184	3
*Atriplex babingtonii**	23.3	0.016	1455
*Ammophila arenaria**	3.58	0.016	233
*Agropyron junceum**	3.56	0.016	222

*Last three species are sand dune plants and their numbers are low.

Table 20-5 *Compounds excreted by plant roots*

	Examples
Sugars and oligosaccharides	
Pentoses (five-carbon sugars)	Xylose, arabinose, ribose, deoxyribose
Hexoses (six-carbon sugars)	Glucose, fructose, galactose, mannose
Disaccharides	Sucrose, maltose
Trisaccharides	Raffinose
Oligosaccharides	
Amino acids	About 20 types
Growth factors	Several B vitamins, choline, inositol
Organic acids	Tartaric, oxalic, citric, malic, acetic, propionic, unsaturated fatty acids, and *p*-hydroxybenzoic acid
Nucleic acid derivatives	Adenine, cytidine, guanine, uridine
Several enzymes	
"Factors" stimulating nematodes or fungi	Nematode-hatching factor, nematode-attracting factor, fungal spore–germination factors, fungal spore–germination inhibitors, fungal zoospore–attracting factors

The chief effect of the plant in the rhizosphere zone probably is providing available carbon for microbial replication. Some of the compounds that are excreted by roots of higher plants are shown in Table 20-5. In young plants, these products are probably true excretions that are liberated by metabolically active roots. In contrast, older plants probably release considerable amounts of tissues and cells that are sloughed off as the root moves through the soil. In the rhizosphere, simple organic molecules, several sugars, and amino acids are released, and their excretion is continuous. Conversely, the substances available to the soil community at a distance from roots are complex in structure; they are sugar polymers (polysaccharides) and amino compounds that are in organic complexes that are not readily available for microbial metabolism. These differences in the identities of the organic materials in soil at large and in the rhizosphere presumably are the chief reasons for the differences between the populations in the soil and the root habitat.

The stimulation in the rhizosphere is selective and affects only certain types of bacteria. The groups that appear to be most benefitted are species that require no growth factors and those that require just a few amino acids. These are characteristically fast-growing species that use simple organic molecules as carbon sources. The amino acids that allow for growth of the populations that need them are probably not derived from root excretions, which contain such compounds, but rather are excreted by the vast numbers of indigenous bacteria. As fast-growing organisms, these bacteria are much more biochemically active on an individual basis than the members of the surrounding soil community. The fungi, which are not especially numerous, may still be particularly important if they are agents of plant disease. The characteristic response in the rhizosphere among the fungi is to favor hyphal forms and to reduce the relative abundance of spores and other stages that are less metabolically active than the hyphae. In addition, mycorrhizal fungi are characteristic of this zone, although they are never abundant.

The rhizosphere community may be beneficial to the root and helpful to the development of the higher plant in several ways. The indigenous species mineralize the organic forms of various elements, and the conversion of organic nitrogen, phosphorus, and sulfur to inorganic products is crucial for plant development because plants cannot use, or use effectively, the organic complexes of the elements that are present in the soil. In the case of nitrogen, for example, microorganisms in the rhizosphere can produce ammonium by degradation of organic material. Ammonium is taken up by plants or microbes as soon as it is released. Ammonium can be oxidized to nitrate in the presence of oxygen by nitrifying bacteria, such as *Nitrosomonas,* but nitrification in the rhizosphere is disputed simply because nitrogen is limiting and free ammonium is unavailable. If ammonium fertilizers are used in areas where nitrification is significant, the nitrate produced by microbial activity could be quickly lost because of leaching. Nitrification-inhibiting chemicals are commercially available to prevent this loss of nitrogen. The bacteria in the rhizosphere may also aid the plant by solubilizing insoluble, inorganic forms of a number of elements. In soils, much of the phosphorus, iron, manganese, and calcium are insoluble, and the indigenous community has the capacity to convert the insoluble—and thus poorly available—forms of these elements to soluble products that plants assimilate readily. The solubilization may be the result of changes in pH, oxidation-reduction potential, or the excretion by bacteria of **chelating agents** that solubilize unavailable nutrients. These chelated nutrients then may be used by the plants. In culture, rhizosphere bacteria synthesize and excrete auxins, gibberellic acid, and other growth regulators (Figure 20-10), and these compounds bring about changes in cell development and root formation and also induce other responses in higher plants. It is believed that these products are also synthesized in the rhizosphere and presumably thus influence plants.

If antibiotics are indeed produced below ground, the likeliest environment for such activities is the rhizosphere because of the large supply of available carbon. These nutrients are needed to sustain the replication of antibiotic producers. If antibiotics are generated in the root zone, they may alter the composition of the community. In this regard, particular interest lies in the possibility that antibiotic producers alter the populations of plant pathogens.

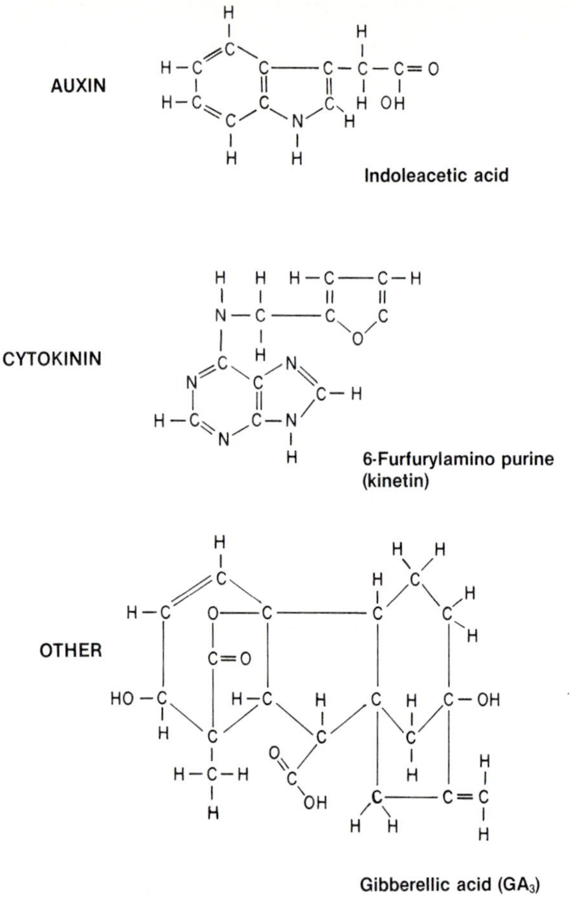

AUXIN

Indoleacetic acid

CYTOKININ

6-Furfurylamino purine (kinetin)

OTHER

Gibberellic acid (GA_3)

Figure 20-10 *Structure of some important plant hormones and regulators.*

Concept Check

1. What are the major abiotic factors that affect the microbial composition of the soil community? List the major bacterial genera or species that are indigenous to soils where these factors are at a maximum or minimum. Example: a soil that has a high pH vs. one that has a low pH.
2. What contributions do bacteria, fungi, and protozoa make to the soil?
3. What is the importance of the rhizosphere community to plant development? What are the major groups of bacteria found in the rhizosphere?
4. Explain the nutrient cycle for nitrogen, explaining the organisms and reactions involved in mineralization, nitrification, and denitrification.

Antagonistic associations (parasitism)

Many microbial species in the soil are capable of causing disease in plants. **Phytopathology** describes plant diseases. Some plant pathogens are primarily saprophytes and cause disease only when the plant host is weakened. For example, *Penicillium* and *Rhizopus* can cause rotting in fruits but only when there is a wound. They will not attack healthy fruits. However, other soil microorganisms appear to spend most of their life cycle as plant pathogens actively attacking plants and killing them. As saprophytes, they have difficulty competing with other microorganisms in the soil. For example, some fungi, such as *Fusarium,* can survive on the dead and decaying plant that it has destroyed, but if the affected crop is not planted the following year the fungus will eventually die out. First, the mechanisms by which plant pathogens are transmitted to plants are examined.

Transmission of plant pathogens. The number of microorganisms or inoculum required to cause infection in plants may vary from a single fungal spore to several million viruses, depending on the site of infection, the virulence of the pathogen, and environmental factors, such as temperature and humidity. Microbial

Table 20-6 *Characteristics of the major genera or groups of phytopathogenic bacteria*

Nonfastidious	
Pseudomonas	Gram-negative, motile, rod-shaped bacterium. About 80 species associated with plant disease. In culture many display fluorescent behavior. Most cause leaf and fruit spotting.
Xanthomonas	All species (about 75) are plant pathogens. Gram-negative, motile rods that exhibit a yellow to orange pigment in culture.
Agrobacterium	Gram-negative, motile, rod-shaped bacterium. About 6 species cause tumors (galls) or hairy root symptoms on plants. One species, *A. tumefaciens* causes crown gall by transferring a plasmid to the plant.
Erwinia	Gram-negative, motile, rod-shaped bacterium that attacks all parts of the plant except roots. About 16 species cause dropping **(wilt)** of plants because of insufficient water, and a softening or discoloration of plant roots and stems (also called **rot**).
Corynebacterium	Gram-positive, irregularly shaped rods that are primarily nonmotile. Twelve to fifteen species are plant pathogens. Many can invade vascular tissue to cause wilt, whereas others cause surface discoloration.
Streptomyces	Filamentous, nonmotile, gram-positive bacteria. Two species occur as causes of warts and scabs on underground plant parts, such as potatoes.
Fastidious	
Spiroplasma	A genus of mycoplasmas (no cell wall) that is pleomorphic (spherical to helical or branched filamentous types). Motile by rotary motion of cell. They are part of the microflora of insects and have an alternate habitat in plants. They cause a disease of vascular tissue-phloem, the nutrient transport vessels. Spiroplasma are transmitted by insect vectors, such as leaf hoppers, and cause dwarfing and yellowing of plants.
Other	Some are rod-shaped and are capable of invading water transport vessels (xylem). They are transmitted by leaf hoppers and cause wilting of plants.

agents capable of infecting the plant can be found in soil, water, infected plants, and various vectors, such as insects. Fungal spores, for example, are found in the soil and may be expelled from spore-bearing structures and disseminated into the air for many miles. Bacteria may be disseminated by air, but this is not how they are usually transmitted. Water is an important method of dissemination of microorganisms, particularly of bacteria and fungi. Water droplets may splash on the ground and disseminate spores, or the pathogen may be carried long distances in ditches or irrigation canals. Many of the obligate parasites, such as viruses, viroids, and mycoplasmas, are disseminated in vectors such as aphids or leaf hoppers.

Characteristics of phytopathogenic bacteria. Phytopathogenic bacteria can be divided into two basic groups: **fastidious** and **nonfastidious.** Most phytopathogenic bacteria are nonfastidious and are easily cultivated in the laboratory. Fastidious bacteria, such as *Spiroplasma,* are restricted to the internal vascular tissue of the plant and are difficult to cultivate free from their plant host. A group of bacteria that has not been cultured or concentrated in a purified form is called **mycoplasma-like organisms (MLOs).** MLOs are pleomorphic, nonhelical, wall-less prokaryotes. Both MLOs and *Spiroplasma* are restricted to the same vascular tissue and are transmitted by the same insect vectors. The most established bacterial pathogens are nonfastidious, belonging to the genera *Pseudomonas, Erwinia, Corynebacterium, Xanthomonas, Agrobacterium,* and *Streptomyces.* Some of the characteristics of these genera are outlined in Table 20-6. We will devote our discussion to one of them, *Agrobacterium,* the agent of crown gall.

Crown gall. Crown gall is a hyperplasia (abnormal multiplication of cells) of dicotyledenous plants caused by the bacterium *Agrobacterium tumefaciens.* The proliferation of plant tissue (tumor formation) is due to a large **Ti** (tumor inducing) plasmid carried by the invading bacterium. A specific part of the Ti plasmid, called **T-DNA,** is transferred to the plant cells. T-DNA encodes for the pathogenic properties of the bacterium, as well as the information for synthesis of amino acid derivatives **(opines)** required for survival of the bacterium (Figure 20-11) (see Point of Interest, p. 599).

Genetics of crown gall. The T-DNA injected into the host integrates into plant nuclear DNA. The mechanism

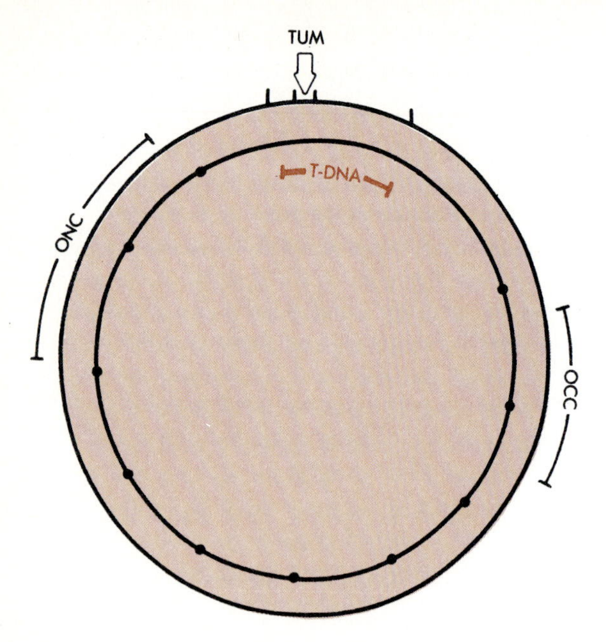

Figure 20-11 ***Map of the T1 plasmid showing T-DNA sequence and location of sites associated with pathogenicity or disease production.*** TUM, ***Site concerned with tumor morphology in plant;*** ONC, ***Sites associated with virulence or disease production;*** OCC, ***Sites associated with octopine (amino acid derivative) catabolism.***

whereby the T-DNA is transferred into the host genome is not known. Ti plasmids are classified on the presence of genes that code for opine synthesis. Opines are amino acid derivatives that are synthesized only in tumor tissue and are usually, but not always, catabolized by the invading bacterium. Some plasmids code for catabolism of octopine and related compounds and are called octopine plasmids; some code for degradation of nopaline and related compounds and are called nopaline plasmids; and others code for degradation of argopine (Figure 20-12). The Ti plasmids have a wide host range and all appear to have a common homologous region that encodes functions that affect tumor morphology. T-DNA appears to have recognition sequences at its borders in the form of repeating base pairs. These signals may have some role in the excision, transfer, and integration of the Ti plasmid. The base pair repeats, flanking the T-DNA, resemble the *att* repeats flanking the prophage form of lambda bacteriophage (see Chapter 17). To date, however, no T-DNA enzymes have been detected that fulfill a function for T-DNA insertion into the plant genome. Apparently the genes encoding these enzymes must lie outside the T-DNA.

The genes that specify octopine and nopaline synthesis are not functional in *A. tumefaciens* because opines are not found in bacteria. The presence of these prokaryotic genes becomes detectable only in the infected host plant. What is interesting about this property is that the genes for opine synthesis are flanked by eukaryotic gene signals, for example, the "TATA" sequence. These signals are found 5′ to the transcription initiation sites of the T-DNA. This suggests that direct control of T-DNA gene expression is regulated by the plant cell.

Once the T-DNA is incorportated into plant nuclear DNA the plant is immune to infection by other *Agrobacterium* isolates. Treatment of infected plants with growth hormones produces a class of regenerants that lose the central region of T-DNA but maintain the border fragments. These plants are completely normal, and this characteristic has provided support for the view that T-DNA can be manipulated to remove the harmful oncogenes and thus be used as a gene vector for normal plants in genetic engineering studies (see p. 604). Investigations have already indicated that when the oncogenes are removed, and the opine genes left intact along with the recognition nucleotide sequences, the T-DNA can insert into plants with no harm to them.

The T-DNA of *A. rhizogenes* is part of the plasmid referred to as **Ri**. *A. rhizogenes* causes a disease called hairy root on nursery stock. Plants infected by this bacterium show a minimal disturbance in plant hormone balance and thus *A. rhizogenes* has appeal as a genetic engineering vector.

Tumor formation. *A. tumefaciens* survives in the soil and attacks many plants. It is especially virulent for stone fruits, grapes, roses, and cane fruits. The organism binds to the cells, roots, or stems of tissue that have been abraded by cultural practices, insects, or grafting. Wounding appears to be necessary for tumor induction. Once the bacteria are present at the wound site, they attach to plant cell walls. This interaction seems to involve the lipopolysaccharide of the bacterium and polysaccharide portion of the plant cell. Plasmid DNA is transferred from the bacterium to the plant cell by passage through the plant cell wall. The bacterium cannot penetrate the plant cell wall and remains outside the plant cell. Only the T-DNA of the Ti plasmid becomes integrated into the plant DNA. T-DNA genes code not only for opine synthesis and catabolism but for synthesis of growth hormones (auxins and cytokinins). Recent evidence demonstrated that a product of one of the T-DNA morphology genes is an enzyme that is involved in the formation of a biologically active cytokinin. Tissue transformed by *A. tumefaciens* grows in the absence of exogenously added plant hormones, such as indole acetic acid. The T-DNA appears to induce an imbalance of plant hormones: auxins, such as indole acetic acid, and cytokinins, such as zeatin, in the infected cell. This leads to the proliferation of cells and the formation of tumors (galls).

Galls may be produced anywhere on the plant, but they occur most often on roots or in the stems near the soil line (Plate 23). The young tumors have peripheral

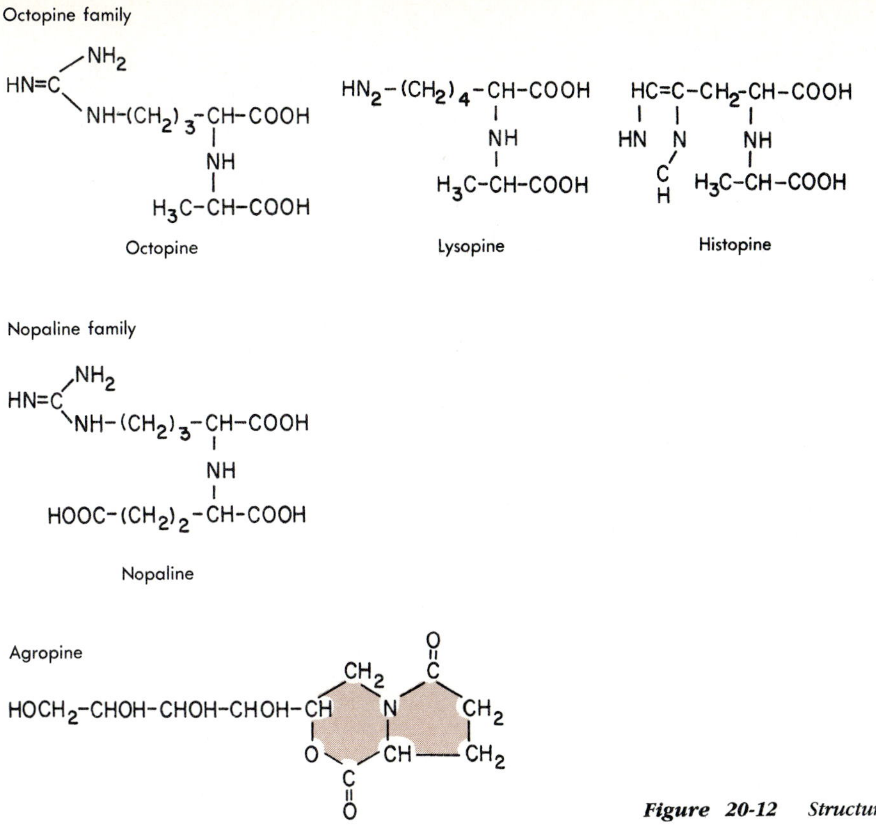

Octopine family

$HN{=}C(NH_2)-NH-(CH_2)_3-CH(COOH)-NH-CH(CH_3)-COOH$

Octopine

$HN_2-(CH_2)_4-CH(COOH)-NH-CH(CH_3)-COOH$

Lysopine

$HC{=}C-CH_2-CH(COOH)-NH-CH(CH_3)-COOH$ (ring: HN, N, C–H)

Histopine

Nopaline family

$HN{=}C(NH_2)-NH-(CH_2)_3-CH(COOH)-NH-CH(COOH)-(CH_2)_2-COOH$

Nopaline

Agropine

$HOCH_2-CHOH-CHOH-CHOH-CH$ (ring: CH_2, N, C=O, CH_2, CH_2, CH, O, C=O)

Figure 20-12 *Structure of some important opines.*

Point of Interest

AGROBACTERIUM AND TRANSFER OF ITS PLASMID

Agrobacterium tumefaciens, the bacterium that causes crown gall disease, affects hundreds of different plants and causes millions of dollars worth of damage every year. Mosses and ferns are resistant probably because the bacterium can not bind to the cell wall of these plants. The bacterium enters a plant through a wound and causes a tumorlike growth called the gall. Galls can be as small as a millimeter or several feet weighing nearly 100 pounds. Cells of *A. tumefaciens* contain 1 or more large plasmids which have about 100 genes on them that code for the compounds causing the tumors. The plasmid (called Ti) from the bacterium can pass through the plant cell wall and part of the plasmid (called T-DNA or transferred DNA) becomes integrated into the plant chromosome. The genes from the bacterium that now lie on the plant chromosome are expressed like any other plant gene. It may be that this plasmid represents ancient plant genes that were originally transferred from plant to *A. tumefaciens.*

Genes (from the bacterial plasmid now on the plant chromosome) code for growth hormones such as auxins and cytokinins, as well as opines. Auxins and cytokinins stimulate the growth of the tumor, while opines provide a food source for the bacterium.

Opines are very special compounds derived from amino acids. They are produced only by plant cells infected with T-DNA. In addition to acting as a food source, opines also increase the rate that Ti plasmids can be transferred to other strains of *A. tumefaciens* that do not have the plasmid. These avirulent strains then become virulent because they acquire the Ti plasmid, which means they can now infect and cause gall-tumors. There are six different strains of *A. tumefaciens* based on their ability to metabolize different opines. For example, the lysopine strains use lysopine, the histopine strains use histopine, and the other strains use other specific opines.

The fact that *A. tumefaciens* can insert part of its plasmid into a plant's chromosome has important applications for genetic engineering. using the bacterial plasmid as a vector (these are discussed later), beneficial genes such as pest resistance could be put on the plasmid. The bacterium could then transfer the genes to the plant. In this way, many new traits could be added to plants. It is believed that such techniques may eventually lead to improved world food production. For example, if the nitrogen-fixing genes (nif) could be transferred to plants that do not have the capability to fix nitrogen, the plant's productivity could be greatly increased.

cells that are unprotected by tough epidermal layers and are attacked by insects and microorganisms. The affected tumor cells decay and discolor to impart a brown or black color to the tumor. Tumors may rot partially or completely, with new tumors appearing at the same or different sites in the next growing season. The size of the tumors ranges from less than 1 cm to 30 cm or more in diameter. Tumors may enlarge to the point where they exert enough pressure on surrounding vascular tissue to reduce the flow of water and photosynthetic products to healthy tissue. This condition ultimately leads to a reduction in plant size and fruit or flower yield.

Crown gall can be prevented by growing uninfected plant materials in fields that are known to be free of the pathogen. Those plants known to be susceptible to crown gall should be handled carefully to avoid damage to roots and crowns. A biological control measure has been developed that involves soaking seedlings or rootstock in a suspension of the bacterium *Agrobacterium radiobacter* strain 84. This organism can inhibit or kill *A. tumefaciens* by liberating a nucleotide antibiotic called **agrocin 84**. The genes for agrocin production are contained on a plasmid in *A. radiobacter;* however, the gene for agrocin sensitivity is located on the Ti plasmid of *A. tumefaciens.* The gene apparently codes for a protein that permits the transport of agrocin across the cytoplasmic membrane.

Characteristics of phytopathogenic fungi. Only 50 out of 100,000 species of fungi can cause disease in humans, whereas some 500 to 600 species are estimated to cause disease in plants and many more are important as wood-rotting organisms or as part of mycorrhizal associations. Many fungi are obligate parasites. Other fungal associations range from very distinctive rot- or canker-causing organisms, which use a variety of extracellular enzymes and toxins to invade tissues, to facultative species, which ordinarily exist on organic debris but given suitable host and environmental conditions are capable of causing plant disease.

Only the Myxomycetes among the slime molds is a significant plant pathogen. Two important genera of slime molds are *Plasmodiophora* and *Polymyxa.* Many genera of the true fungi (Eumycota) are plant pathogens. The most important of the lower Eumycota are the genera *Pythium, Phytophthora, Albugo,* and *Peronospora.* Representatives of the higher Eumycota include members of the Ascomycetes, Basidiomycetes, and Deuteromycetes. Genera in the Ascomycetes include *Taphrina, Erisiphe,* and other powdery mildews, and *Ceratocystis, Endothia,* and *Claviceps.* Genera in the Basidiomycetes include a large group of wood-rotting fungi, such as *Fomes* and *Armillaria,* as well as rusts, such as *Puccinia,* and smuts, such as *Ustilago.* Some of the most common plant pathogenic genera occur in the Deuteromycetes, for example, *Verticillium, Fusarium, Botrytis* (Plate 24), and *Phomopsis.* Dutch elm disease is one of the important fungal diseases associated with trees.

Dutch elm disease. Up until the 1950s many streets of the United States were lined with the graceful and shady American elm tree. Today, except for some areas of the Northeast, the American elm has virtually disappeared. The trees were killed by a fungus, an ascomycete called *Ceratocystis ulmi,* which was transmitted by bark beetles. The disease is presumed to have originated in Asia, where the Chinese elm is resistant to the fungus. During World War I Chinese laborers were brought to Flanders to dig ditches, and they are believed to have brought with them wooden wicker baskets made from the Chinese elm. Apparently, the vector beetles, which were carried in the bark, escaped and proceeded to ravage millions of elms in Europe. The disease was eventually called Dutch elm disease because of its initial appearance in the Netherlands. Furniture manufacturers in the United States, despite importation quarantines, obtained diseased logs from Europe and helped unleash the disease that practically destroyed most American elms (Figure 20-13).

The female bark beetles, which carry spores of the fungus on or in their bodies, deposit eggs in the bark of the stems of weakened or recently killed elm branches. The beetle larvae eat the outside layer of wood and the inner layer of bark, which results in the formation of tunnels. The fungal spores carried by the adult beetle germinate and produce compact or fused conidiophores with clustered sticky spores. When the new adult beetles emerge from the tunnels to fly away, the sticky spores become attached to their bodies. The spore-laden beetles fly away to feed on live trees and in the process of boring through the bark infect the sapwood. Dutch elm disease is a vascular wilt, or upset in water balance, in which infected branches lose turgor (cell rigidity or distension), become limp, and eventually die. Spores can be produced that will be disseminated to other areas of the plant to initiate new sites of infection if the large xylem vessels are reached by the invading mycelium. Rapid death of infected trees is believed to be due to the obstruction of xylem vessels and upsetting the water balance. The vascular tissue appears brown, and the outer layer of wood appears streaked, mottled, or brown to green. The tree may die within a few weeks if there has been a generalized vascular infection.

Dutch elm disease is best prevented by removing recently killed elm wood, since beetles lay their eggs in killed or diseased trees. Recent plantings can be protected by applying insecticide such as methoxychlor early in the spring to destroy newly emerging adult beetles. Control in individual trees has also been obtained by injecting systemic fungicides into the trunks or roots. This practice arrests the disease and prevents further spread.

Characteristics of phytopathogenic viruses and viroids. Of some 500 transmissible agents that cause dis-

Figure 20-13 *Dutch elm disease. **A,** Uninfected tree. **B,** Tree infected by* Ceratocystis ulmi, *the agent of Dutch elm disease.*

ease in plants, over 200 have been identified as viruses. It is likely that most of the other, unknown diseases will prove to have a viral origin as well. A few are suspected to be viroids, whereas others may prove to be plasmids or other infectious entities not yet characterized or known.

Virus-caused diseases rank second only to those induced by fungi in frequency of occurence and economic importance. Since they are submicroscopic entities, viruses are difficult to detect and characterize. They are frequently vectored by insects, nematodes, mites, or fungi and may persist in an infectious state on tools, debris, or even human hands. Many are passed from plant to plant by vegetative propagation and/or seeds and pollen.

Viruses are resistant to most classes of pesticides and therefore cannot be controlled by usual chemical means. Control measures are centered on prevention and early identification, rather than cure of diseased plants.

Plant virus characteristics such as size, morphology, and nucleic acid composition are discussed in Chapter 16.

Less than a dozen diseases are known to be caused by viroids, although several more diseases are suspect. The biology of the viroid is understood primarily in terms of chemical properties, which are discussed in Chapter 16. Most biochemical processes involved in infection, multiplication, and movement within plants are a matter of conjecture. Transmission is via plant propagation or mechanically on tools, but vectors may also be involved.

Tobacco mosaic virus disease. The tobacco mosaic virus (TMV) is the cause of disease in over 15 families of plants, the most important of which are tomato and tobacco. The TMV is an RNA virus (see Chapter 16) that is among the most heat stable of all plant viruses. It is resistant to 90° C for 10 minutes and remains viable in dried plant tissue when heated to 120° C for 30 minutes. The virus has been known to remain viable in dried plant tissue for several years.

TMV is usually transmitted mechanically to susceptible plants by workers whose hands have been contaminated by the virus. Tomato and tobacco leaves have leaf hairs that are easily broken off, and virus brought into contact with such abrasions can cause infection. Transplants can also be infected by virus that has overwintered in crop residues. Transmission by insect vectors, which is the most common method of transmission of all other viruses, is of minor consequence in the transmission of TMV.

TMV moves from cell to cell via openings called plasmodesmata or by connecting cytoplasmic strands. It may also move longer distances in vascular tissues. It is capable of infecting all plant tissues. Viral nucleic acid replication occurs in the nucleus of the cell, but viral assembly occurs in the cytoplasm, the site of protein synthesis. Mature viral particles, however, have been found in vacuoles, in the nucleus, and in chloroplasts. Two types of cytoplasmic inclusions can be observed following infection by TMV: first, an amorphous body consisting of viral particles, mitochondria, and other cellular debris and, second, crystal-containing layers of TMV particles. Symptoms of infection include chlorosis, a yellowing of normally green chlorophyll-containing tissue; mottling; dwarfism; discoloration; and sometimes necrosis. Mottled dark green and light green areas are the most frequent symptoms of infection (Plate 25). The light green areas represent sites where chloroplasts have been damaged. Reduction in photosynthesis leads to loss of carbohydrates and dwarfism. The tobacco plant is not killed, but the quality and quantity of the crop is severely reduced. Infection of tomato plants produces mottling and some reduction in crop yield. If the plant has been infected at a very early stage, there can also be internal discoloration of the fruit.

Sanitation and the use of disease-resistant varieties of plants are the best methods of control. Some disease-resistant varieties, however, do not yield a high-quality crop. Contaminated seedbeds or fields should not be used for planting a tobacco crop for at least 2 years. Workers handling and removing diseased plants should wash their hands with soap and water before handling healthy plants. Since milk inactivates TMV, it is often sprayed over plants during transplanting. Chewing and smoking of tobacco during cultivation practices should be avoided. Workers who tend tomatoes in glass houses are advised to keep a separate set of coveralls for each house to prevent the spread of TMV by their clothes.

Decomposition of plant remains

A major activity of the subterranean community is the decomposition of roots, leaves, stems, and other residues of higher plants. The process results in microbial replication because the heterotrophs are using the organic compounds they mineralize as nutrient sources. The conversions also are critical biogeochemically because they return to the atmosphere the limiting nutrient for photosynthesis, namely carbon dioxide. In the soil, the same processes of decomposition are also critical because the elements contained in the plant remains are transformed back to the inorganic state. Every element that exists in organic complexes in plants is mineralized, and the nutrient elements are thus converted back to forms that are assimilated by root systems. In this way, the indigenous populations are required for the continued growth of higher plants. In the absence of chemical fertilizers, plants rely almost entirely on the microbial decomposition of organic materials to generate the nutrients that the roots must obtain. Only when fertilizers are used in large amounts is there another significant source of these nutrients.

The decomposition of carbonaceous materials is favored by an adequate supply of available nitrogen and phosphorus. Although other nutrient elements are needed by heterotrophic microorganisms for proliferation, their supply is usually greater than the microbial demand for them. As a rule, acidification tends to reduce the rate of decomposition. Good aeration and favorable moisture levels, by favoring aerobiosis, enhance carbon turnover. Furthermore, rapid decomposition is enhanced by a fine state of disintegration of the organic residue, and anything that increases the surface area for microbial attack—as does breaking down the physical structure of plant remains—tends to increase the rate of breakdown. In this connection, earthworms, insects, and other soil animals have a major impact on the degradation of organic substances.

The addition of carbonaceous materials to soil induces an immediate increase in microbial biomass and cell density. This is reflected in the rapid evolution of carbon dioxide and the disappearance of oxygen. Such additions of carbon also change the composition of the microflora, the community is dominated by fast-growing species if the plant remains are readily available and by slow-growing populations if the materials are largely cellulosic or are rich in lignin. After several days, the high initial rate of carbon dioxide evolution diminishes, and the large demand of the microflora for oxygen declines. The phase of rapid carbon dioxide liberation parallels the disappearance of readily available organic substrates. Then, when only the more resistant compounds remain, the rate of carbon dioxide evolution falls off. In the process, the simple, water-soluble compounds (such as sugars and amino acids), which are never abundant, disappear first. Subsequently, the polysaccharides, such as cellulose and several hemicelluloses, disappear more slowly. Lignin is metabolized at a slow rate under all circumstances. Every organic constituent of higher plants, so far as is known, can be decomposed by the microflora of soils, at least under aerobic conditions. Furthermore, regardless of the complexity or uniqueness of the organic constituent, it is metabolized by the same general metabolic pathways. The diversity of substances available to the microflora in this manner is shown in Table 20-7.

The decomposition of plant litter can be used for some practical purposes: garden composts, silage, and industrial processes.

Garden composts. Garden composts are produced from a mixture of nonwoody plant material heaped into large piles. Microorganisms trapped in the pile break down the plant material and heat is released. The temperature in the center of the pile rises high enough to kill weed seeds and undesirable microorganisms. The pile is turned periodically to ensure that all parts of the pile are heated. Lime is added to increase the pH, because plant decomposition (especially leaves such as

Table 20-7 *Major constituents of plants*

Constitutent	Composition	Site in cell
Cellulose	Polymer of glucose: beta 1,4 linkages; straight chain $(C_6H_{10}O_5)_n$*	Cell wall
Hemicelluloses	Polymers of pentoses ($C_5H_{10}O_5$: xylose, arabinose), hexoses ($C_6H_{12}O_6$: mannose, glucose, galactose), or uronic acids ($C_6H_{10}O_7$: glucuronic, galacturonic)	Cell wall
Starch	Polymer of glucose: alpha 1,4 linkages; branched chain $(C_6H_{10}O_5)_n$	Storage carbohydrate
Lignin	Polymer of aromatic nuclei (see Figure 18-16)	Cell wall
Pectic substances	Polymer of galacturonic acid $(C_6H_8O_4)_n$ and other sugars	Cell wall
Insulin	Polymer of fructose	Storage carbohydrate
Hydrocarbons (aliphatic)	Straight chain compounds of carbon and hydrogen	Derived from fats, oils, and waxes

*In the formation of the polysaccharide, one water molecule is lost per repeating sugar unit.

oak) release acids that decrease the pH. Sometimes a nitrogen source, such as commercial fertilizer, is added to reduce the carbon to nitrogen ratio.

Initially the compost heap favors the growth of mesophilic bacteria and fungi but as the temperature of the heap rises, thermophilic fungi begin to dominate. These thermophilic species include *Mucor pusillus, Aspergillus fumigatus, Humicola insolens,* and *Thermomyces lanuginosus.* After 5 to 6 days, when the temperature is above 55° to 60° C, these species are replaced by thermophilic bacteria. Even though temperatures in the center of the pile may kill many beneficial microorganisms, when the pile is turned the microorganisms in the cooler portions of the pile will reinvade. After 4 to 6 weeks much of the carbon in the compost is lost but most of the soluble nitrogen is retained in an immobilized form. There is practically no free ammonium or nitrate after this period. The compost is now ready for use as a soil conditioner and/or source of nutrients for soil.

Silage. Silage, or food used to feed farm animals, is produced from plant materials by a mechanism that is opposite to that used for compost formation. For example, **anaerobic conditions** are maintained in silage formation to prevent heating and to reduce the amount of plant decomposition. The plant material for silage formation should be dry and contain a high percentage of highly fermentable sugars, such as those found in forage legumes. The plant material is chopped to form a compost heap and then put into specially designed silos or other containers to exclude air.

The major microorganisms involved in fermentation of the plant material are *Lactobacillus* and *Streptococcus.* The initial fermentation can be enhanced by adding molasses, because sugars are quickly converted to lactic acid. During the first few days of fermentation the major microbial species are *Lactobacillus curvatus* and *L. plantarum* but species of *Xanthomonas* and *Pseudomonas* may also be present. The pH of the pile becomes acidic after about 7 days, resulting in the proliferation of acid-tolerant lactobacilli, such as *L. buchner* and *L. brevis.* If acid conditions are produced too soon, *Clostridium* species can create a problem by converting nitrogen to toxic amines or by fermenting sugars to butyric acid, which spoils the silage. Once anaerobic and acid conditions are established the silage will keep for long periods of time. Exposure to air, however, will allow the introduction of fungi such as *Aspergillus fumigatus,* which can produce mycotoxins (see Chapter 21).

Industrial processes. Waste products of agriculture are being studied by the industrial sector for the production of single-cell protein and fuels, such as ethanol. These topics are discussed at some length in Chapter 22, Industrial microbiology.

Genetic engineering and plant science

A considerable amount of controversy has been generated about the use of genetically engineered microorganisms in agriculture. No uniform procedures have been agreed on because ecologists fear the unleashing of potentially dangerous organisms into the environment. Molecular biologists argue that genetic engineering has been going on naturally in bacteria and higher organisms for many thousands of years. Ecologists counter that argument by stating that natural genetic engineering occurs sporadically and is in no way similar to the purposeful introduction of huge numbers of microorganisms into the environment. There is some uniform agreement, however, that the best way to proceed is to evaluate the merits and drawbacks of each individual case. Two questions that should be answered before genetically engineered microorganisms are released into the environment are: (1) what is the likelihood of the new genes being transferred to other species and what will be their impact, and (2) will the organisms be spread rapidly in the environment?

Genetic engineering techniques. Genetic engineering techniques involving plants are of two types. One involves gene modification of soil microorganisms for the purpose of plant protection. These microorganisms are manipulated in the laboratory and then released into the appropriate site in the soil community. This technique is being used to prevent frost damage to plants (see Point of Interest). In another example, pesticidal genes are inserted into microorganisms that naturally colonize crops. The delta toxin gene from *Bacillus thuringiensis* has been inserted into the plant bacterium *Pseudomonas fluorescens,* which can colonize roots of plants such as corn. The spliced gene codes for a toxin that kills insects (*B. thuringiensis* has been used as an insecticide for many years, see p. 664). Preliminary tests demonstrated that the engineered bacteria colonize the roots and kill worms, such as black cutworms, that nibble on the roots. When the bacteria are ingested, the toxin in the bacterium is activated and kills the pest. The toxin does not kill beneficial insects and does not persist for any length of time in the environment. The second strategy of genetic engineering techniques introduces prokaryotic genes into plants in an effort to provide them with capabilities beyond those of normal plants. Studies of gene transfer into dicots, such as tomato, potato, and tobacco, is in an advanced state because natural genetic engineering occurs when plants are infected by the bacterium *Agrobacterium,* the agent of crown gall. Molecular geneticists can insert specific genes into *Agrobacterium* Ti plasmid or other plasmids that can be carried by the bacterium.

T-DNA as a vector. T-DNAs of Ti plasmids are ideal vectors because agrobacteria have a broad host range and can infect virtually all dicots. In addition, the integrated T-DNA is transferred to progeny during normal plant cell division. Originally the genetic engineering experiments using T-DNA involved the following:

1. The T-DNA region of the Ti plasmid was cloned into the plasmid pBR 322 (see Chapter 11).
2. The desired genes were inserted into the T-DNA.
3. The hybrid plasmid was reintroduced into agrobacteria harboring an intact TI plasmid.
4. The chimeric plasmid was induced to insert into the intact Ti plasmid.

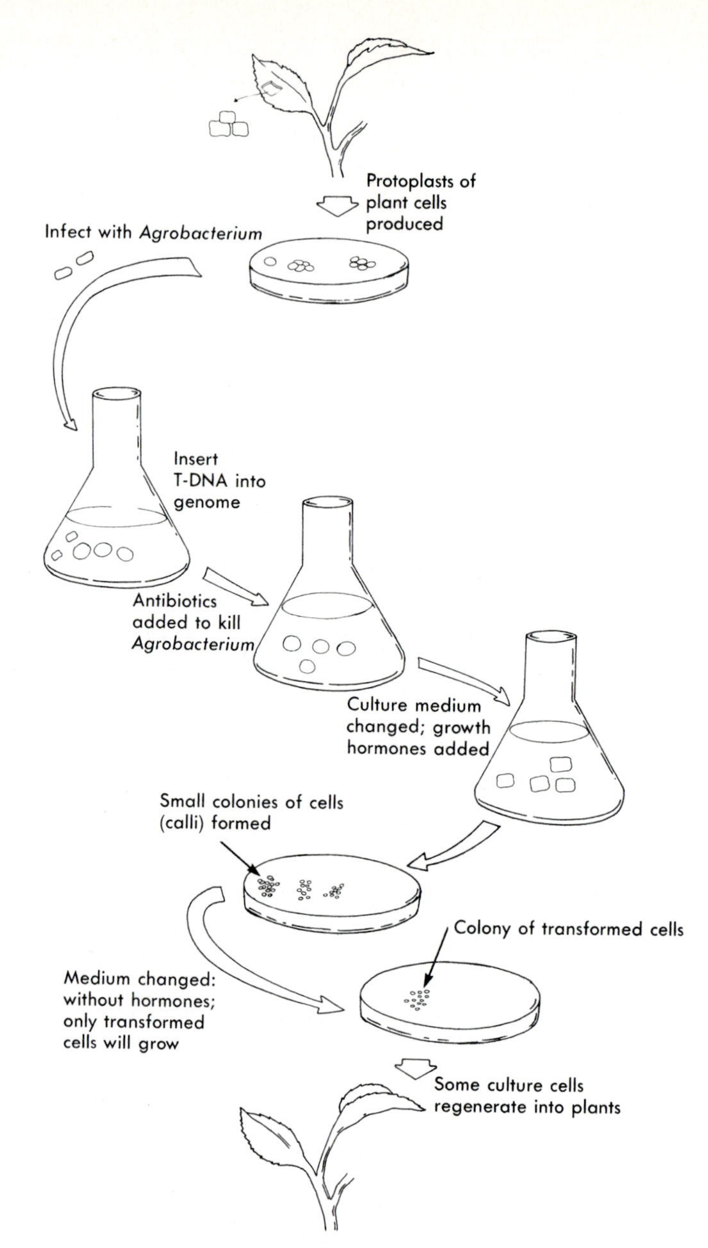

Figure 20-14 *Technique for formation of protoplasts from plant cells and their infection with* Agrobacterium.

This procedure has been simplified because all that is necessary for *Agrobacterium* to infect and transform plant cells is an intact T-DNA region and a part of the region of the Ti plasmid called *vir,* or virulence region. In addition, the T-DNA and *vir* regions do not have to be on the same plasmid. Thus the T-DNA (and whatever other genes have been introduced into it) can be incorporated into the plant genome.

A convenient method for infection and transformation of plant cells is to use protoplasts. Protoplasts of leaf cells are prepared and put into culture (Figure 20-14). When the protoplasts are beginning to form cell walls, they are infected with *Agrobacterium.* Later, antibiotics are added to the medium to kill the bacteria. The cells are grown in a medium containing plant growth hormones for a few weeks until colonies of undifferentiated cells are formed. The culture medium is changed to one lacking plant growth hormones. Thus only the cells transformed by *Agrobacterium* will be able to grow and divide. The transformed cells sometimes spontaneously regenerate into plants carrying T-DNA.

Agrobacterium does not usually infect monocots, such as corn, but monoclonal protoplasts can be transformed by T-DNA. The major problem with monocots, however, is the difficulty in regenerating whole plants from transformed protoplasts. To date there is no evidence that this can be done.

Practical applications of Agrobacterium. Agrobacterium transformation of plant cells has been used to create herbicide resistance in some crop plants. Herbicides, such as glyphosates, kill a variety of weeds but also kill some major crop plants. Application of the herbicide is therefore a tedious and time-consuming process. The inhibitory effects of glyphosate is due to its ability to suppress the activity of an enzyme called **EPSP synthase,** which is required in the synthesis of the amino acids phenylalanine, tyrosine, and tryptophan. A mutant EPSP synthase gene was isolated from a glyphosate-resistant species of *Salmonella.* The protein encoded by the mutant gene differs from the normal gene product by only one amino acid. The mutant gene was incorporated into a plant plasmid, which was then transferred by *Agrobacterium.* In laboratory experiments the herbicide-tolerant gene was successfully introduced into tobacco, soybean, and corn cells. A second method to insure resistance to glyphosate is to remove the plant EPSP gene and attach it to a viral enhancer sequence. This results in plants producing 20 to 50 times the usual amount of enzyme and making the plant tolerant to the herbicide.

Despite some success with genetic engineering experiments in plants there are still two major limitations. First, transformation of monocots is limited to monocot-protoplasts, which to date do not regenerate whole plants. Second, the genes necessary to increasing crop yields have as yet not been identified, cloned, or inserted into a suitable vector, such as T-DNA.

Point of Interest

ICE NUCLEATION

Frost is a serious cause of injury to plants. Ice formed in or on frost-sensitive plants spreads both intracellularly and extracellularly, causing mechanical disruption of cell membranes. Over 1 billion dollars in lost plant production occurs each year in the United States because of frost injury, thus frost has an economic impact.

Water, like many other liquids, does not spontaneously convert from the liquid phase to the solid phase at the melting point of the solid phase. Liquids like water can be supercooled to several degrees centigrade below the melting point of the solid phase and will freeze only if a catalyst is present. Catalysts may be in the form of inorganic salts, dust particles, or bacteria, and they are called **ice nuclei.** The frost sensitivity of most plants can be explained by the fact that they harbor large populations of bacteria on their surfaces. These surface bacteria are an abundant source of ice nuclei. *Pseudomonas syringae, Erwinia herbicola,* and *Pseudomonas fluorescens* are the most common ice-nucleating active bacteria found on plants in the United States. These bacteria catalyze ice formation at temperatures as warm as $-1°$ C. This contrasts with plants cultivated free of bacteria on their surfaces, in which case the plants may be able to supercool and withstand temperatures of $-7°$ to $-14°$ C without damage. Most plants in the field are injured at temperatures above $-5°$ C because bacteria limit the ability of the plant to supercool.

Not every bacterial cell on the surface is active as an ice nucleus at a given time. The fraction of cells that become active increases with a decrease in temperature. In vitro experiments have demonstrated that medium composition, growth temperature, and aeration profoundly affect the ice nucleation efficiency of the bacteria. Live, as well as dead, bacterial cells are active in ice nucleation. A single ice nucleus is thought to be sufficient to initiate ice formation and cause injury to an entire leaf, fruit, or flower. It appears that the bacterial component responsible for ice nucleation is a protein. To prevent frost injury, scientists have looked for methods that would control ice nucleation active bacteria. One promising technique is to use species of bacteria that are antagonistic to the natural ice-nucleating bacteria. Antagonistic bacteria have been selected on their ability to colonize the plant surface and to effectively compete with the native bacterial species. Preliminary tests show that non-ice-nucleating active bacteria applied to the surfaces of young plants reduced the number of ice-nucleating bacteria by 10 to 1000-fold when compared to untreated plants. Scientists have since used the original ice-nucleating bacteria and by recombinant DNA techniques have spliced out the ice-nucleating gene. The new strain is being tested in the field.

Table 20-8 ***Animal diseases transmitted to humans by means other than insect vectors***

Disease	Type of etiological agent and species	Animal reservoir	Method of spread to humans
Tuberculosis	Bacterial: *Mycobacterium bovis*	Cattle	Ingestion of contaminated milk
Brucellosis	Bacterial: *Brucella* species	Goats, cattle, sheep, swine	Ingestion of contaminated milk, meat, or related products
Anthrax	Bacterial: *Bacillus anthracis*	Cattle, horses, sheep, goats, swine, wild animals	Contact with animal hides, wool, etc; ingestion of contaminated meat
Q fever	Bacterial: *Coxiella burnetii*	Cattle	Contact with infectious contaminated parturient (at birth) products, inhalation of contaminated dust
Psittacosis and ornithosis	Bacterial: *Chlamydia psittaci*	Domestic birds	Inhalation of contaminated dust
Leptospirosis	Bacterial: *Leptospira* species	Cattle, dogs, swine, rodents	Contact with infectious urine and contaminated water
Salmonellosis	Bacterial: *Salmonella* species	Cattle, swine, dogs, rodents	Ingestion of contaminated foods
Ringworm	Fungal: species of *Microsporum* and *Trichophyton*	Cattle, horses, cats, dogs	Contact with animal
Rabies	Viral: Lyssavirus	Cattle, dogs, cats, skunks, foxes, bats	Bite of infected animal or inhalation of aerosols in bat caves

INFECTIOUS DISEASE IN FARM ANIMALS

Diseases of farm animals should concern us for two important reasons. First, some diseases of animals are transmissible to humans. These diseases are called **zoonoses.** Second, disease in farm animals used for human consumption can also result in tremendous economical loss and can affect the eating habits of our global population.

Zoonoses

Diseases of lower animals may be transmitted to humans by direct or indirect contact or by insect vectors. The only major disease transmitted by insect vectors from farm animals to humans is **leishmaniasis.** Leishmaniasis is a tropical disease transmitted by a sand fly. The parasite causing leishmaniasis is a protozoan that infects not only cattle but also cats, dogs, sheep, and wild rodents. Several other diseases are transmitted from animals to humans by insect vectors but the animal is usually a domestic pet or a rodent. These diseases include tularemia (deer), Rocky Mountain spotted fever (wild rodents), and plague (rats and ground squirrels). Chapter 27 discusses some of these diseases.

Many diseases of farm animals and domestic animals can be transmitted to humans by direct or indirect contact. These diseases are outlined in Table 20-8. Some of these diseases are discussed in Chapters 27 and 28.

Infectious diseases of farm animals not transmitted to humans

Farm animals are susceptible to infectious disease for two important reasons: (1) they are in contact with the soil and with insect vectors that harbor infectious microorganisms, and (2) they shed or excrete potential pathogens that may contaminate food, water, or the surrounding environment. Infections from contaminated food or water occur less frequently in adult farm animals but young animals are especially prone to infections of this type. One of the most frequent causes of death in farm animals is diarrhea.

Diarrhea in farm animals. Diarrhea, which is responsible for the loss of millions of farm animals each year, is caused by bacterial and viral agents. Bovine virus diarrhea and transmissible gastroenteritis in pigs are caused by a pestivirus and coronavirus, respectively. One of the primary bacterial agents that causes diarrhea in farm animals is *E. coli.* Although *E. coli* is a normal inhabitant of the intestinal tract, some strains, called **enteropathogenic**, cause a diarrheal disease in calves, lambs, and swine referred to as **colibacillosis** or **scours**. The disease-producing ability of this strain of *E. coli* is due to two factors: the ability of the organisms to produce enterotoxin (a toxin that affects the intestinal tract) and the presence of a pilial adhesin designated **K88** for swine and **K99** for calves. Normal *E. coli* strains that inhabit the intestinal tract of healthy animals can be converted to enteropathogenic strains when the genes coding for enterotoxin and pilial adhesin are transferred during conjugation. The transfer of only one factor is not sufficient to make the strain enteropathogenic. The K88 and K99 piliated adhesins have been purified and make up a vaccine called Vicogen.

Concept Check

1. **Discuss the mechanism by which *Agrobacterium* causes crown gall. What is the importance of opines? Why may *Agrobacterium* be important in genetic engineering studies?**
2. **Describe the characteristics of at least three nonfastidious phytopathogenic bacteria. What are the fastidious phytopathogenic bacteria and how do they differ from the nonfastidious types?**
3. **What are the principle plant macromolecules that are broken down by microorganisms?**
4. **Briefly outline the techniques for making compost and silage. What are some of the organisms involved and what are their functions?**
5. **What is meant by ice nucleation and how are genetic engineering techniques involved?**
6. **Describe some diseases transmitted from farm animals to humans. What is the importance of *E. coli* as a normal commensal and disease-producing agent in farm animals?**

SUMMARY

1. The microbial community of soil contains many bacteria, actinomycetes, fungi, algae, and protozoa. No currently available method gives a true estimate of the numbers of viable bacteria. Many of these bacteria require growth factors. Actinomycetes are common in soil, but their role in this environment is uncertain.
2. The biomass of fungi is large, and these organisms are important in the decomposition of plant polysaccharides.
3. Algae and cyanobacteria often have little activity because of their need for sunlight and water, but their activity may be pronounced when these two needs are satisfied. Protozoa assume prominence because of their voracious feeding on bacteria.
4. The populations and activities of the various microorganisms are determined by the nutrient supply, moisture, pH, oxygen level, oxidation-reduction potential, temperature, inhibitors, and salts in the soil.
5. Because of actions among the various species, a microorganism is usually not most abundant or most active at the optimal intensity of these factors as determined in pure culture.
6. The rhizosphere represents a soil-plant root community that is involved in beneficial and antagonistic associations. Nutrients are supplied to bacteria and fungi through plant roots, whereas bacteria and fungi are involved in mineralization of organic compounds and solubilization of inorganic elements required by the plant. Antagonistic associations involve parasitism by bacteria and fungi.
7. Most parasitic microorganisms are disseminated to plants via water, but obligate parasites, such as viruses, viroids, and MLOs, are transmitted primarily by insect vectors.
8. Crown gall caused by *Agrobacterium* is an important plant disease because it represents a natural genetic engineering mechanism. This mechanism also offers hope in laboratory manipulated experiments to reduce the plants sensitivity to frost, herbicides, and parasites.
9. One of the most devastating fungal diseases of plants is Dutch elm disease. Bark beetles carry spores of the fungus and are responsible for transmission of the disease, which has virtually destroyed American elms.
10. TMV disease is an important disease of tobacco and tomato. The virus is transmitted usually by the contaminated hands or tools of farm workers.
11. Soil microorganisms play a vital role in decomposition of plant debris. This process is also used to produce garden composts and silage for farmers. Industry is also interested in plant decomposition for single-cell protein and certain fuels.
12. There are two types of plant genetic engineering techniques found in the laboratory. One involves modification of the genetic material of a soil microorganism for plant protection, and the second involves introduction of prokaryotic genes into plants to provide them with special attributes not found in normal plants. The plasmid of *Agrobacterium* is being used in genetic engineering studies. Transformation of plant cells can also be accomplished using leaf cell protoplasts.
13. Microorganisms in the soil and surrounding environment are capable of causing infectious disease in farm animals. Sometimes the infectious agent found in farm animals can be transmitted to humans. Such diseases are called zoonoses.

Self-study Quiz

Multiple choice

1. The chief effect of the plant in the rhizosphere is the provision of what essential nutrient for microbial growth?
   A. Water
   B. Oxygen
   C. Carbon
   D. Nitrates
2. Which of the following is true concerning the organisms found in a soil sample?
   A. Usually one species predominates in a given sample.
   B. It is easy to isolate all types of soil-inhabiting organisms in a sample by simply adding soil to the isolation medium.
   C. As for the number of independent cells (cell count) in a soil sample, bacteria are found in greatest numbers.
   D. Bacteria provide the greatest biomass to all soil samples.
3. What microbial group is principally responsible for the decomposition of lignin?
   A. Bacteria
   B. Algae
   C. Protozoa
   D. Fungi
4. Which of the following groups is capable of nitrogen-fixation?
   A. Actinomycetes
   B. Protozoa
   C. Algae
   D. Cyanobacteria
5. What group of organisms would be favored in acid soils?
   A. Fungi
   B. Bacteria
   C. Protozoa
   D. Algae
6. Production of antibiotic substances by actinomycetes in the soil, which suppress the growth of competitors in the environment, is an example of what type of interaction?
   A. Commensalism
   B. Parasitism
   C. Predation
   D. Amensalism
7. What characteristic of the genus *Bacillus* would give it a survival potential in the soil?
   A. Conidia formation
   B. Anaerobic growth
   C. Thermophilic growth optima
   D. Endospore formation
8. What limits the distribution of cyanobacteria in soil environments?
   A. Availability of oxygen
   B. Availability of light
   C. Availability of growth factors
   D. Availability of lignin
9. Which of the following is *not* considered to be a prominent member of the rhizosphere?
   A. *Pseudomonas*
   B. *Xanthomonas*
   C. *Actinomyces*
   D. *Azotobacter*
10. Obligate phytopathogenic parasites are transmitted usually by:
    A. Water
    B. Wind
    C. Birds
    D. Insect vectors
11. Crown gall is a disease caused by:
    A. *Agrobacterium,* which carries disease-producing traits on the chromosome.
    B. *Agrobacterium,* which carries disease-producing traits encoded on T-DNA.
    C. *Ceratocystis ulmi,* which carries disease-producing traits on a plasmid.
    D. Tobacco mosaic virus.
12. Which of the following would most likely cause disease by infecting vascular plant tissue such as phloem?
    A. *Pseudomonas*
    B. *Spiroplasma*
    C. *Actinomyces*
    D. *Xanthomonas*
13. Which of the following bacteria are considered to be detrimental to silage formation?
    A. *Pseudomonas* and *Xanthomonas*
    B. *Clostridium* and *Aspergillus fumigatus*
    C. *Lactobacillus brevis* and *L. buchner*
    D. *Azotobacter* and *Azospirillum*

14. *Bacillus thuringiensis* has become important in genetic engineering in plants because it possesses:
    A. A plasmid that can be transferred to plants.
    B. Genes that code for an antibiotic used to treat microbially infected plants.
    C. Genes that code for an enzyme that is used in protoplast formation.
    D. Genes that code for a toxin that kills insects.
15. Which of the following diseases is transmitted from farm animals to humans by an insect vector?
    A. Rabies
    B. Anthrax
    C. Leishmaniasis
    D. Ringworm

Short answer

Given below is a cross-section of fertile soil.

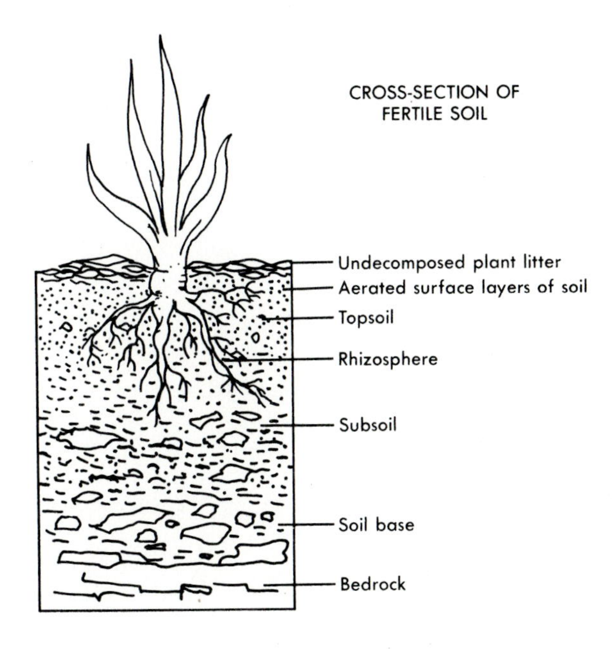

1. Identify the regions in which you would expect to find the following organisms (you may want to list more than one):
    A. *Rhizobium* ________.
    B. Algae ________.
    C. Heterotrophic bacteria ________.
    D. Heterotrophic fungi ________.
    E. Lignin decomposers ________.
    F. Protozoa ________.
2. Rank order the soil areas as to the available organic nutrients.
    A. ________ (least available)
    B. ________
    C. ________
    D. ________
    E. ________
    F. ________
    G. ________ (most available)
3. Where would you expect to find the following processes occurring?
    A. Aerobic processes ________
    B. Anaerobic processes ________
    C. Nitrogen fixation ________
    D. Denitrification ________

Thought questions

1. In the preparation of cultures for gram-staining experiments, a common practice is to use a mixed suspension of *E. coli* (gram-negative rod) and *S. aureus* (gram-positive coccus) as the control smear to validate proper staining procedure. When both cultures are grown independently in nutrient broth, *E. coli* reaches a population of about 1×10^8 cells/ml, but the *S. aureus* culture only reaches 1×10^6 cells/ml. a. How would you explain this? b. What characteristics give *E. coli* an advantage?
2. Most soil microorganisms seem to have a world-wide distribution. a. What data could you gather to determine if the same organisms found in the soil of different continents are truly identical? b. If they are identical, what hypothesis could you offer to explain the phenomenon? c. There are exceptions (that is, organisms limited to certain endemic regions) to this observation. How might they be accounted for?

SELECTED READINGS

Books

Alexander, M. *Microbial ecology.* John Wiley & Sons, New York, 1971.

Atlas, M., and Bartha, R. *Microbial ecology: fundamentals and applications.* Addison-Wesley Publishing Co., Inc., Reading, Mass., 1981.

Fletcher, M., and Floodgate, G. (Eds.). *Bacteria in their natural environment.* Academic Press, Inc., Orlando, Fl., 1985.

Goodfellow, M., Mordarski, M., and Williams, S.T (Eds.). *The biology of the actinomycetes.* Academic Press, Inc., New York, 1984.

Klug, M.J., and Reddy, C.A. (Eds.). *Current perspectives in microbial ecology.* American Society for Microbiology, Washington, D.C., 1984.

Moran, J.M., Morgan, M., and Wiersma, J. *Introduction to environmental science* (2nd ed.). W.H. Freeman, New York, 1985.

Journal articles

Barnhart, C.L., and Vestal, J.R. Effects of environmental toxicants on metabolic activity of natural microbial communities. *Appld. Environ. Microbiol.* **46**:970, 1983.

Bevan, M.W., and Chilton, M. T-DNA of *Agrobacterium* T1 and R1 plasmids. *Annu. Rev. Genet.* **16**:357, 1982.

Chilton, M. A vector for introducing new genes into plants. *Sci. Am.* **248**:50, 1983.

Horvath, R.S. Microbial co-metabolism and degradation of organic compounds in nature. *Microbiol. Rev.* **36**:146, 1972.

Harder, W., and Dijkhuizen, L. Physiological response to nutrient limitation. *Annu. Rev. Microbiol.* **37**:1, 1983.

Lindow, S.E. The role of bacterial ice nucleation in frost injury to plants. *Annu. Rev. Phytopathol.* **21**:363, 1983.

Newman, E.I. Root microorganisms: their significance in the ecosystem. *Microbiol. Rev.* **53**:511, 1978.

Suslow, T.V. Role of root colonizing bacteria in plant growth. In Lacy, G., and Mount, M. (Eds.). *Phytopathogenic prokaryotes,* vol. 1., Academic Press, Inc., New York, 1982.

CHAPTER 21

Food microbiology

Robert Deibel
R.F. Boyd

Chapter Outline

Introduction

According to various estimates, one third of our food is lost to spoilage during growth, harvesting, processing, and distribution. The food microbiologist is primarily associated with the manufacturing segment of the food production industry, whose functions are to ensure a wholesome food supply free from disease and to endeavor to extend the shelf life of foods.

Every so often, previously unrecognized pathogens *(Yersinia, Campylobacter)* or toxins (mycotoxins) are associated with foods, and methods of detection and quantitation as well as procedures to eliminate an offending organism must be investigated. Improved methods of preservation are a neverending quest. Relatively recent technological advances involving anaerobic and controlled-atmosphere packaging, intermediate-moisture foods, and the use of preservatives have extended the shelf life of some foods, but the quest continues.

The purposes of this chapter are to describe how foods are examined for microbial contamination and the sources of microbial contamination, discuss the methods by which processed foods are preserved, discuss the groups of microorganisms associated with specific foods such as meats and dairy products, and outline the various types of microbial or chemical contamination that lead to food poisoning.

MICROBIOLOGICAL EXAMINATION OF FOODS

Sampling

The procedures used in sampling foods for bacteriological analyses are important, because bacteria are not randomly distributed in foods. Measures must be taken to enhance the procurement of a sample that approaches a true microbiological profile of the product. If the product is in discrete units (drums, bags, cartons), the square root of the number of units in a shipment or manufacturing code is sampled and composited. After blending, serial dilutions are made and the analytical sample is withdrawn. Multiple sites are sampled when bulk products are examined, and the individual samples are blended before the analytical sample is withdrawn.

Analyses

For qualitative tests (sometimes referred to as presence/absence tests) a given amount of the analytical unit is introduced into a growth medium. Incubation in the growth medium is usually followed by transfer of an aliquot of the culture to selective-differential media to detect the organism in question. Qualitative tests are frequently used to detect enteric pathogens such as *Salmonella* and *Shigella.* The isolation of *Salmonella,* for example, first involves transferring a food sample to a nonselective enrichment medium such as lactose broth (Figure 21-1). This procedure enhances the isolation of salmonellas, but other microorganisms are also increased. A selective enrichment medium such as tetrathionate brilliant green broth is then used to select for only *Salmonella.* Following incubation in the selective medium an aliquot of the culture is transferred to brilliant green agar where the salmonellas produce a characteristic colonial appearance. The suspected *Salmonella* colony is then subjected to biochemical testing by inoculation into triple sugar iron agar (TSI or lysine agar. These media indicate biochemical reactions typical of the salmonellas. If the biochemical tests are positive for *Salmonella,* confirmation is made by subjecting isolates to serological testing (see Point of Interest, Chapter 25). For quantitative tests such as those employed to enumerate coliform bacteria and *Staphylococcus aureus* and to ascertain the "total count" in a food product, the analytical sample is mechanically blended with a diluent such as dilute phosphate buffer or 0.1% peptone. Many foods are acidic, and after the initial blending the pH value is adjusted to approximately 7.0 to avoid acidification of the growth medium in the lower dilutions. Microbial counts can be determined by transferring aliquots of various dilutions to appropriate growth media employing such techniques as the spread plate and pour plate, or the most probable number can be determined by turbidity evaluations. The enumeration of *S. aureus*

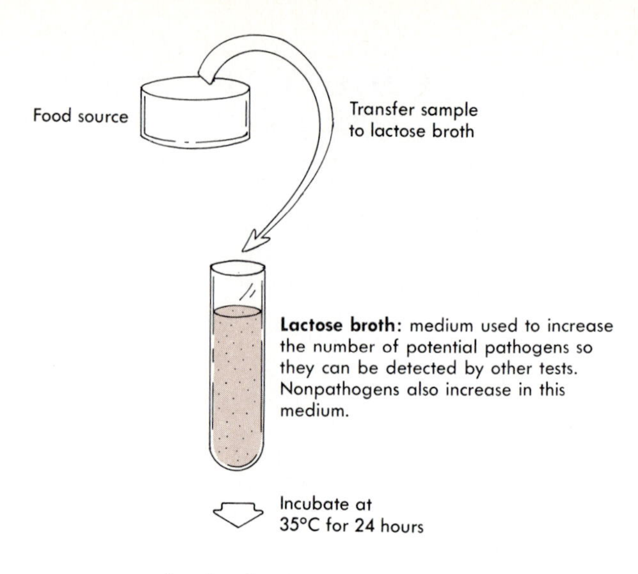
Food source
Transfer sample to lactose broth
Lactose broth: medium used to increase the number of potential pathogens so they can be detected by other tests. Nonpathogens also increase in this medium.
Incubate at 35°C for 24 hours
Transfer aliquot to

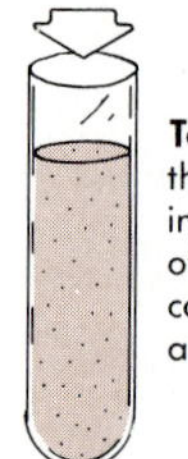
Tetrathionate brilliant green broth: thiosulfate and brilliant green dye in this medium inhibit growth of gram-positive bacteria and coliforms while permitting *salmonella* and *shigella* to grow.
Incubate at 35°C for 24 to 48 hours
Streak an aliquot on triple sugar iron agar and stab to butt of tube
Incubate at 35°C for 24 hours
Triple sugar iron agar: contains the sugars lactose, glucose, and sucrose. Pathogens ferment all three sugars, producing acid that affects pH indicator (phenol red) and changes color of medium from red to yellow. The medium also contains ferrous sulfate, which *salmonella* convert to H_2S, which causes a blackening of the agar.
Other biochemical tests
Serological confirmation

Figure 21-1 *Procedure for determining the presence of* Salmonella *in foods.*

from a diluted food sample first involves growth on a medium that is high in salt or other inhibitor of nonstaphylococcal organisms, for example, mannitol salt agar. Pathogenic *S. aureus* is characterized by its ability to produce the enzyme **coagulase**. A positive coagulase test occurs when a suspension of the organisms is capable of clotting rabbit plasma that contains an anticoagulant.

Specifications and standards

Many manufacturers impose a specific limit on the number of total bacteria or specific organisms that occur in their finished product. These limits, or specifications as they are called in the industry, are also placed on raw incoming products, and often they are included in purchasing contracts. If a regulatory agency such as the Food and Drug Administration or the United States Department of Agriculture sets a limit, it is referred to as a standard (as opposed to a specification), and it is an enforceable law. Products that exceed the standard are subject to seizure and destruction.

Bacterial numbers in foods

The specific organism and its numerical occurrence are factors in determining the wholesomeness of the food. For some specific pathogens, such as *Salmonella,* the limit in a ready-to-eat food product is essentially zero or undetectable in a given analytical unit. On the other hand, the occurrence of 10^3 to 10^4 *S. aureus*/gm in a cheese or sausage product may not be significant. *Salmonella* species are intestinal pathogens capable of invading the intestinal tract, and as few as 17 organisms have been shown to cause disease. Staphylococci are not invasive and are unable to penetrate healthy tissue. Certain strains of *S. aureus* are capable of producing an enterotoxin in food, but they are usually only human strains (for example, a food handler could contaminate food), and at least 10^6 or more organisms are required to cause disease symptoms.

The **shelf life** of a product is often a function of the holding temperature and the number of bacteria after manufacturing or harvesting. Significant effort is expended to keep the number to a minimum to enhance a long shelf life. With perishable products, especially those requiring refrigeration, the lower the total count (or **load**), the longer the shelf life. Marketed products such as pasteurized milk or sausage will often have total counts of 10^3 or 10^4 bacteria/ml or gram. The product is still quite acceptable, and no off flavors or odors can be detected. At a level of 10^7 bacteria/ml or gram the product may be in the incipient stage of spoilage, and counts of 10^8 to 10^{10} usually but not always give rise to souring or off odors. Thus, relatively large numbers of bacteria must be present in a food before detection by the senses (**organoleptically**).

In the manufacture of some foods such as cheese and fermented sausage, a starter culture (pure culture of a specific organism) is employed, and the growth of large numbers of bacteria is encouraged in the manufacturing process. The finished products often contain total counts of 10^8 to 10^{10} bacteria/gm and the absence of high counts may reflect manufacturing failure. Thus the organism, its numbers, and the type of product must be considered when the significance of bacteria in a food product is evaluated.

CONTAMINATION AND MICROBIAL GROWTH IN FOODS

It is virtually impossible to avoid microbial contamination of foods during harvest and subsequent processing. Our only salvation is to keep the contamination to a minimum throughout processing as well as to control the subsequent growth of the contaminating microorganisms. Sanitation and good management help control the level of contamination, and rapid movement of the product through the food chain as well as a variety of processing procedures (pasteurization, canning, refrigeration, salting, drying, and acidification) are used to control the subsequent growth.

Sources of contamination

Many raw food items have a normal flora that is often associated with the immediate environment from which it was harvested (Table 21-1). Subsequent handling and processing allow growth of the normal flora or a portion of it. Quite often the unrestricted growth of the normal flora will develop into the spoilage flora. For example, when harvesting milk, the producer endeavors to reduce the load during harvest by cleansing the udder and by cleaning and sanitizing the milking equipment, pipelines, and holding tanks. The milk in the udder is essentially sterile, and the initial contamination is derived from the teat canal and the milking machine, pipelines, and holding tanks. The milk parlor environment also contributes to the initial load. Producers refrigerate the holding tank and generally keep the milk 1 to 2 days. After pickup on the farm, the milk is transferred to large holding silos at the dairy plant. The sanitation of these silos, as well as the ambient temperature and the length of the holding time, may cause an increase in the load. The milk is pasteurized to kill the pathogens, and some nonpathogens are also killed. Many nonpathogens survive this milk thermal processing, and these organisms constitute the potential spoilage flora (Plate 26).

Processing equipment. The processing environment provides sources of contamination. Good management and proper sanitation help reduce this source. In the past the Food and Drug Administration has recalled or

Table 21-1 *Primary sources of microorganisms that contaminate foods*

Food environment	Most frequently encountered genera in food
Soil and water	Bacteria
	Alcaligenes, Bacillus, Citrobacter, Clostridium, Enterobacter, Micrococcus, Pseudomonas, Serratia
	Fungi
	Aspergillus, Rhizopus, Penicillium, Fusarium
Plants and plant products	Bacteria
	Acetobacter, Pseudomonas, Erwinia, Flavobacterium, Lactobacillus, Leuconostoc, Pediococcus, Streptococcus
	Fungi
	Saccharomyces, Torula
Intestinal tract of infected humans or animals	Bacteria
	Escherichia, Salmonella, Bacteroides, Lactobacillus, Proteus, Vibrio, Shigella, Staphylococcus, Streptococcus, Clostridium, Citrobacter, Pseudomonas
Air and dust	Any of the genera mentioned above but particularly bacteria such as species of *Bacillus* and *Micrococcus* and fungal species such as *Cladosporium, Penicillium, Aspergillus*

seized a product when it was observed that the product was produced in an unsanitary environment. Processing equipment such as machines, cooling tunnels, conveyor belts, and pipes are important sources of contamination. Food manufacturers expend considerable time and money cleaning and sanitizing their manufacturing facilities. Bacteria in the environment or on the machinery inevitably are transferred to the product.

Water and air. The microbiological quality of water and air that comes into direct contact with food must be monitored for quality assurance. Neither water nor air has a normal flora—both become contaminated from the environment before their introduction to the food or food processing plant. Species of *Pseudomonas* are a common water contaminant because of their ability to survive in limited nutrient environments. Water used in foods and for cleaning purposes must meet drinking water standards (500 organisms or less/ml). Quite often, water that is recirculated in a plant is filtered to remove food particles and rechlorinated to reduce the microbial load. Many food processing plants use large quantities of air for drying, in cooling tunnels and in the general food processing environment. Although the air is coarsely filtered (chiefly to remove particulate contamination), spores and vegetative cells readily pass through the filters and contaminate the processing areas.

Food ingredients. The chief sources of microbial contamination of a processed food are the food ingredients. The food processor must therefore maintain a substantial quality assurance program on the raw ingredients. Various microbial specifications are set for each incoming ingredient, and frequently the product is released for manufacture only after laboratory clearance. This is one of the several reasons for the continuing interest in rapid bacteriological methods, because it is costly to warehouse incoming ingredients before laboratory release.

Rework. Another potential source of microbial contamination is **rework**. Rework is partially or completely manufactured product that does not fully comply with various specifications set by the company. For example, candy bars that are not completely enrobed with chocolate, defectively packaged product, and the end portions of bologna and other cured meat loaf products that do not yield a uniform slice are returned to the beginning of the process and incorporated into new product. Just about all segments of the food industry use rework because it would be exorbitantly expensive to discard it. Reworking of a chemically or microbiologically defective product is not condoned because the load of the new product often results in process failure. Generally, product that has not left the manufacturing facility can be reworked; however, product that has been in the distribution chain is destroyed.

Control of contamination and growth in food processing plant

Food manufacturers must be aware of four cardinal considerations associated with the control of microbial contamination and growth in their plants: plant layout, compartmentalization, traffic control, and water control.

Plant layout. This first consideration centers on an orderly sequential processing scheme in which the "flow" of the product originates at one end of the plant and proceeds stepwise throughout, terminating with the finished product at the opposite end. Generally, the raw products and early processing mixtures will contain the highest load, and the finished product will have the lowest load. The decrease in load is usually associated with processing barriers such as heat, drying, and salting.

Compartmentalization. Physical barriers are placed at strategic points between processing areas to preclude cross-contamination between high- and low-load product during manufacture. For example, in the manufacture of chocolate, the raw cocoa bean is received in large jute sacks. The sacks are opened and inspected for discolored beans, twigs, stones and other undesirable matter. The beans are known to have a high microbial load and occasionally to be contaminated with *Salmonella.* The beans are conveyed to a large hopper that feeds the roasting machine. A pipe or chute through the wall that separates the raw and roasting areas radically decreases the potential for cross-contamination of roasted product with raw product. Roasting does not kill all bacteria, and the subsequent dehulling step generates dust and offers potential airborne contamination of the finished product. The dehulled bean or nib is then ground and heated, yielding the chocolate liquor (finished product), which is the basic ingredient for milk chocolate.

Traffic control. Traffic control in compartmented areas is difficult to maintain. Plant personnel may frequently have functions in two or more compartmented areas. Lift or fork trucks and hand carts also can carry contamination from one area to another. Some manufacturers place foot baths containing a bactericidal agent between critical areas to reduce cross-contamination.

Water control. The growth of microorganisms in a food processing plant depends on three conditions: the presence of organic matter (the food being processed), the presence of microorganisms, and water. We cannot remove the food being processed, and we cannot totally exclude microorganisms, but we can control the availability of water in the processing environment. Water is an absolute requirement for all life and by controlling it we can also control microbial growth. Roof leaks, pipe leaks, faulty water pump seals, and the unwarranted use of water (especially for cleaning purposes) are the chief violations of water control. When the above three conditions are present, large microbial populations can be the result. The organisms can be disseminated by foot and vehicular traffic, or the mixture can dry, aerosolize, and be spread by air currents. In many instances water must be used for cleaning; however, its judicious use must be practiced, and drying procedures after cleanup should be instituted. Many manufacturers of dried products (such as candies, flour, starch, and bakery items) use central or portable vacuum cleaners to avoid a wet cleanup. Water control is an important and often overlooked aspect of controlling microbial growth in a food processing plant.

Role of technology in microbial contamination and growth

Whenever technological alterations are made in the processing and packaging of a food product, the potential exists that an alteration of the microbiology of the product can take place. The advent of vacuum packaging, for example, has increased the industry's awareness of the potential for *Clostridium botulinum* growth in many of these anaerobically packaged products. The so-called **controlled atmosphere packaging**, wherein the product is packaged in a gaseous environment other than air, decreases the growth potential for certain groups of spoilage bacteria. Thus the concept that technology dictates microbiology is pervasive in the food industry, and food microbiologists analyze technological changes before, during, and after their innovation to assess the microbiological alterations that may occur.

Investigation of spoilage problems in the manufacture of foods

Several approaches may be employed in investigating a microbial spoilage problem. Assume that the spoilage organism is known or the associated condition reflects the growth of a particular group of bacteria. For example, the souring of sliced, prepackaged ham is associated with the growth of lactic acid bacteria of the genera *Streptococcus, Lactobacillus,* and *Leuconostoc.* If the organism is not known, then it must be identified (usually the predominant organism in the spoiled product). The processing train of the product is charted, and the time and temperature throughout the processing are recorded. At significant (or the so-called critical) points in the processing train, samples are taken. These samples, as well as samples of raw ingredients and finished product, are then analyzed for the offending microorganism.

Assume that a sliced ham has been observed to have a short shelf life, and souring is evident after 1 week in the display case. Investigation at the plant indicates that the raw and pickled ham were within normal limits and the time and temperature of cooking were adequate. Examination of the slicing machine, however, indicates bacterial counts of 10^9/gm. Thus as the pickled ham was being sliced it was also being inoculated with large numbers of microorganisms. A more frequent and careful cleaning of the machine during the manufacturing day would probably solve the problem. If the ham had been insufficiently cooked during the processing, lactobacilli would be present in high numbers on the product. Remedial measures in this instance center on control of established thermal processing procedures.

Concept Check

1. What are the procedures used for the sampling and analysis of food suspected of microbial contamination?
2. What are the various ways in which food is contaminated?
3. Which factors must be considered when manufacturers seek to control microbial contamination?
4. What are the primary sources of microorganisms that contaminate foods, and what are the principal bacteria and fungi associated with each?

FOOD PROCESSING AND METHODS OF PRESERVATION

Shelf life and coding data

The shelf life of a product is defined as the expected time that a product will remain organoleptically acceptable. Sometimes the shelf life of the product is dictated by chemical rather than microbiological considerations.

Many food manufacturers code date their product. The code date (or lot or code number) reflects a given manufacturing lot of product made in a given location during a specified period of time. For example, the code A241B2 could indicate Plant A, the 241st day of the year, B shift, and the second line in the plant. In this manner manufacturers can identify specific manufacturing lots. If a lot is found to be defective, the incriminated lot can be removed easily from the market and the offending area of manufacture identified. In another type of code dating the manufacturer will indicate the expected shelf life of the product (for example, "good until blank date" or "must be sold before blank date").

Thermal processing of foods

Heat is one of the most common methods of food preservation. The thermal processing schedules of foods are generally calculated to kill a given segment of the contaminating microbial flora; in most instances, the food is cooked as well. In certain procedures, such as the blanching of vegetables, the thermal process is calculated only to inactivate the plant's enzymes, which contribute to vegetable ripening and maturation. The heat and the cleansing of the product also help reduce the microbial load.

Table 21-2 *D values of Salmonella senftenberg and S. typhimurium in milk chocolate**

Temperature (°C)	D values (avg of three trials) (min)	
	S. senftenberg	*S. typhimurium*
70	440	816
80	116	222
90	36	75

*These values are extremely high because milk chocolate contains a high fat content and low levels of water. This greatly increases the heat resistance of the organisms present in the product. Milk chocolate has been involved in a few outbreaks of disease by salmonella.

Effect of high temperature. Increases in temperature above the optimum range of growth for the organism result in cell injury and eventually death of the cell. One of the most important considerations in a processing plant is to determine the amount of heat required to produce a commercially sterile product. First, the most heat-resistant mesophilic (thermophiles are found in niches not associated with food and also are not causes of disease) sporeformer that is encountered in the food is isolated. A given number of spores are mixed with a slurry of the food. The slurry is dispensed into glass vials that are sealed and then heated in an oil bath. The latter achieves temperatures above 100° C. The vials are removed from the bath at definitive time intervals and cooled, and each slurry is plated quantitatively to determine the number of survivors. The plot of the logarithm of the number of microbial survivors against time of exposure to a particular temperature yields a "straight line." From this curve the **decimal reduction time (D)**, or time in minutes required to kill 90% of the population, can be determined. The *D* valves of two *Salmonella* species are described in Table 21-2. The information from *D* values is particularly useful for processing canned foods. The D value has use in many other aspects of microbiology, because it can be used to estimate or compare survival in the presence of germicides, acids, or any adverse chemical or physical environment where death is incurred (see Chapter 15 for discussion of D values).

The resistance of microorganisms at various temperatures can be determined by plotting a **thermal death time (TDT)** curve (see chapter 15 for discussion).

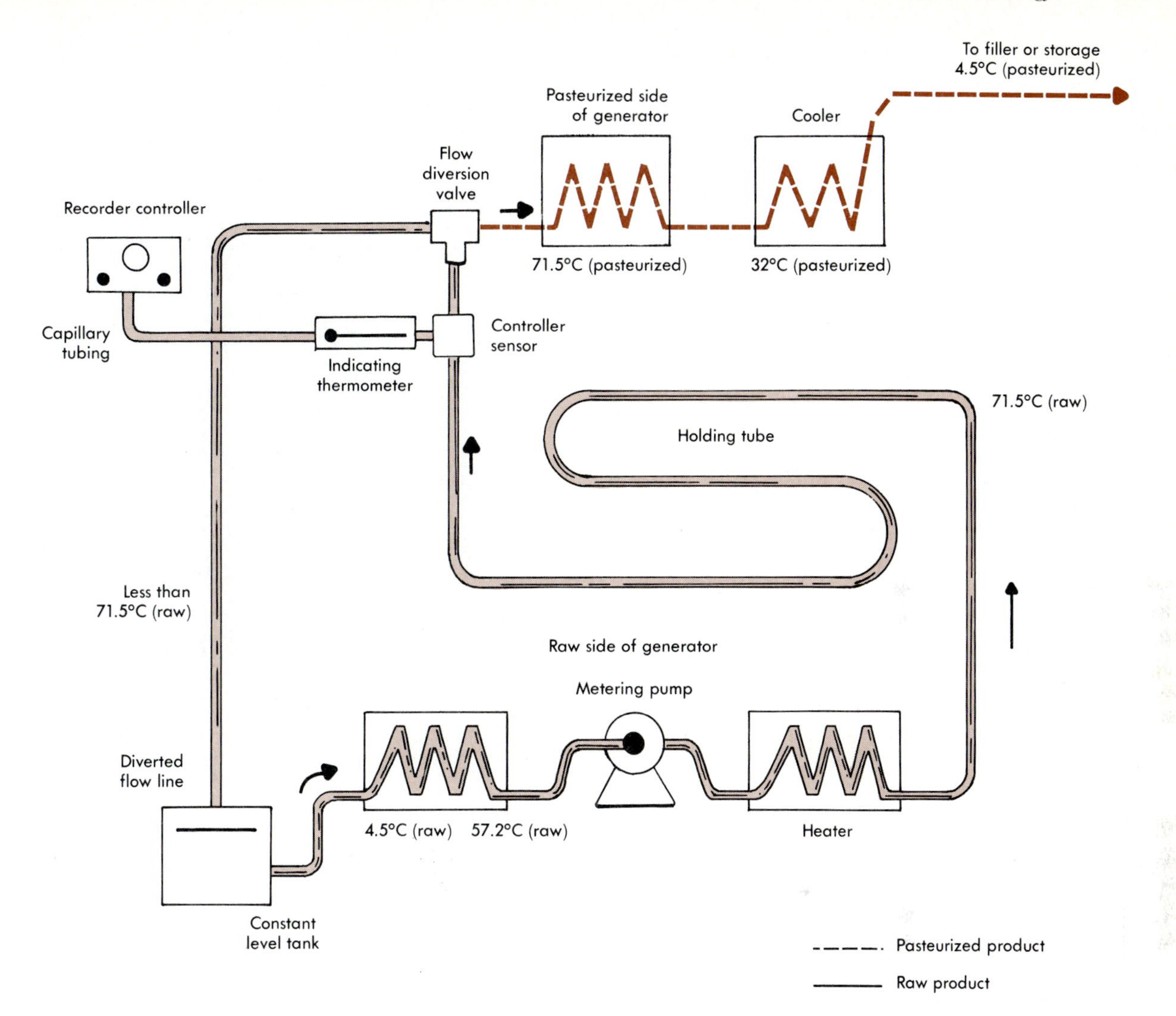

Figure 21-2 *High-temperature short-time (HTST) pasteurization technique.*

12D concept. One of the most important considerations in commercial canning is to ensure the destruction of *Clostridium botulinum* spores. This mesophilic organism produces spores that are comparatively heat sensitive, and the finding of botulinal spores in a retorted product (retorts are devices for heat processing canned foods) indicates gross underprocessing. To be assured that spores of *C. botulinum* are destroyed, a concept was established that would ensure that any population of the most resistant spores would be reduced to 10^{-12} of their original numbers—in other words, a **12 decimal** or **12D concept.** Previous studies have shown that if 10^{12} botulinal spores/gm of food are heated at 121° C, the time required for a 12 decimal reduction is 2.52 minutes. Thus, time and temperature relationships are important in producing sterility in canned food products.

Pasteurization. Any thermal process in the food industry in which the highest temperature reached in the food is under 100° C is considered a pasteurization. Many foods such as dairy products, cured meats (bologna, wieners), fruit juices, bakery goods, vinegar, and wines and beers are pasteurized. Some pasteurization procedures are relatively mild, and many vegetative cells are not killed, but in others only bacterial spores survive. Thus throughout the food industry there is a broad spectrum of pasteurization procedures.

The amount of heat that is used to pasteurize milk is designed to kill the most heat-resistant pathogen (vegetative cell) that occurs in the product. Previously the processes were designed to kill *Mycobacterium tuberculosis,* but in more recent times it has been observed that the rickettsial agent of Q fever, *Coxiella burnetii,* is more heat resistant, thus necessitating a revision of the thermal processing procedure. There are two general procedures that are employed for milk pasteurization. **Low-temperature holding (LTH)** is used for vat-type or batch pasteurization, and milk is heated to 62.9° C for 30 minutes. The previously employed 61.7° C for 30 minutes did not kill *Coxiella burnetii.* The **high-temperature short-time (HTST)** procedure is more amenable to continuous modern processing, and the milk is heated to 71.5° C for 15 seconds (Figure 21-2). This thermal treatment kills *Coxiella burnetii,* and no revision is required. Both procedures ensure a palatable product that is free of all known pathogens. However, many other vegetative bacteria can survive, such as certain nonpathogenic streptococci, lactobacilli, and some micrococci. Eventually, these bacteria will grow and spoil (sour) the milk if it is held beyond its normal shelf life.

Steam-heated retorts. The majority of canned foods are thermally processed in large, steam-heated retorts (Figure 21-3), which operate on the same principles as a pressure cooker or autoclave. The temperature may vary depending on the composition of the product, size of the container, type of retort, and other factors. The amount of heat administered to the product is designed to kill the most heat-resistant mesophilic sporeformers such as *Clostridium botulinum,* the cause of botulism. The temperature for canning high-acid foods may be 100° C and 121° C for low-acid foods. The reason for the difference in temperature is that acids are inhibitory to the growth of *C. botulinum.* The recent development of less acid strains of tomatoes has contributed to more cases of botulism from home canning. Steam-heated retorts are not designed to kill thermophilic sporeformers,

Figure 21-3 *Steam-heated retort for thermal processing of canned foods.*

Figure 21-4 *Spoilage in canned goods.* ***A,*** *Can with leaky seam, subject to microbial spoilage.* ***B,*** *Normal can.* ***C,*** *"Swell" caused by growth of gas-producing bacteria in can or chemical production of hydrogen from acid in fruit reacting with metal container.* ***D,*** *Spoilage of canned fruit caused by gas-producing yeasts (pressure has forced fruit juice through seal of lid).*

Table 21-3 *Some of the major microorganisms involved in the spoilage of canned foods*

Canned food	Spoilage organism(s)	Product defect
Red meats		
Commercially sterile	Sporeformers (species of *Bacillus* and *Clostridium*)	Gas, putrefaction
Semipreserved	*Streptococcus, Bacillus, Clostridium*	Souring, gas discoloration, putrefaction
Vegetables		
Corn, green beans, peas	*Bacillus stearothermophilus, Clostridium sporogenes*	Flat-sour
	C. thermosaccharolyticum	Gas
	Desulfotomaculum nigrificans	Sulfide (rotten egg) smell
Tomatoes	*Bacillus coagulans*	Flat-sour
	Clostridium pasteurianum, C. butyricum	Butyric acid smell
Fruit	*Clostridium butyricum*	Butyric acid smell

because in most instances the product would be overcooked and organoleptically unacceptable. In addition, thermophilic sporeformers are not agents of disease, and the growth of thermophiles does not take place under normal ambient temperatures of holding (less than 50° C). It has been estimated that 10% to 30% of the retorted canned goods on the market are not sterile and contain thermophilic spores. The term **commercial sterility** has been used to designate this situation. One of the most commonly encountered thermophilic sporeformers is *Bacillus stearothermophilus.* Studies have shown that 10^6 spores of this species may survive 100° C for up to 20 hours. They produce flat-sour spoilage of low-acid canned foods.

Microbial spoilage of canned goods. Microbial spoilage of canned goods falls into three categories: preprocessing, underprocessing, and can leakage (Figure 21-4). Preprocessing spoilage occurs before retorting when the food has a high load and it is not thermally processed soon enough to prevent extensive bacterial growth. The condition is detected by observing large numbers of bacteria in stained preparations coupled with the absence of viable bacteria. Heating does not alter the staining propensities of bacteria. Underprocessed products usually contain large numbers of one or two species of viable, mesophilic sporeformers such as *C. botulinum* and *C. sporogenes.* Spoilage caused by can leakage merits special attention. Areas where the body seam intersects with either of the lead seams are the most susceptible to leakage. The leaks often reseal on cooling after heating, but occasional contamination occurs (estimated at about 1/5000 cans). Non-heat-resistant, viable vegetative microbes are encountered in these instances. Swollen cans (caused by microbial gas production) and sour or off tastes are the usual type of defects encountered (Table 21-3).

The vast majority of swollen cans that are seen in the marketplace are not associated with microbial spoilage but with **hydrogen swells.** This condition is associated with acid foods (such as fruits and fruit juices) wherein the acid reacts with the tin of the can producing hydrogen gas (Figure 21-4, *C*), as represented by the equation:

$$Sn^{\circ} + 2\,H^{+} \rightarrow Sn^{++} + H_2$$

This is essentially an electromotive series replacement reaction, and the amount of gas produced depends on the acidity of the food and the composition and construction of the can. Foods containing large amounts of nitrate (spinach, beans) can also cause a hydrogen swell, wherein the nitrate reacts within the tin, producing hydrogen gas. In both types of hydrogen swells there is an absence of viable bacteria and usually a pitting and blackening of the inside of the can. The food microbiologist must be able to differentiate between hydrogen and microbial swelling, because each involves different approaches for rectification. Usually, canned foods have a shelf life of about 14 months, after which the incidence of hydrogen swells increases significantly.

Preservation of foods by refrigeration and freezing

The generation time of microorganisms and the rates of chemical reactions have temperature optimums. Deviations from the optimums decrease the rates, and if the deviations are extreme, zero rates are approached. Decreases in temperature increase the generation time of the microorganisms and extend the shelf life of the

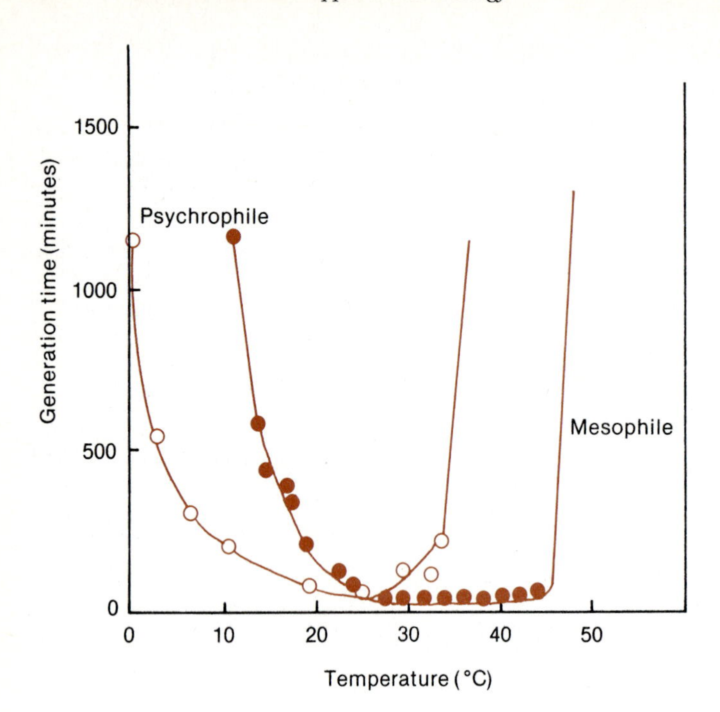

Figure 21-5 *Effect of temperature on generation time of mesophilic (●) and psychrophilic (○) microorganism.*

food. Freezing generally incurs a complete cessation of growth and radically extends the shelf life from a microbiological standpoint.

Chilling temperatures. Chilling may be defined as reducing the temperature of food to a range of 0° to 10° C. Psychrophilic bacteria, especially *Pseudomonas* and allied aerobic gram-negative *Flavobacterium,* can grow in this temperature range and eventually spoil the food. There are also several genera of mesophilic microorganisms that are capable of growth between 0° and 10° C and are referred to as **psychrotrophs**. As a generality, if a food spoils in 20 days at 0° C, it will spoil in 10 days at 5° C and in 5 days at 10° C. Thus relatively small temperature variations markedly influence the growth rate of the psychrophilic bacteria (Figure 21-5). Most pathogenic bacteria do not grow below 10° C, and at 5° C almost all pathogens that occur in foods are inhibited. There is one type of *Clostridium botulinum* (type E) that can grow at 3.3° C, but it takes several months to produce toxin.

Low-temperature control is frequently a problem in industry. Large coolers lack temperature uniformity, and fans must be strategically located to assist in air distribution. Humidity control in these coolers may also be difficult. If the humidity is too high, water may accumulate on the product (such as sides of pork and beef) and accelerate bacterial growth. On the other hand if the humidity is too low, the product is dehydrated, thus incurring financial losses.

As a final word regarding low-temperature control in the food industry, it is axiomatic that more spoilage problems occur in the summer months as compared with the winter months.

Freezing temperatures. The freezing of foods essentially stops microbiological growth. The shelf life is then dictated by chemical (for instance rancidity) and physical (moisture migration) limits. Freezing does not kill all bacteria. Gram-positive bacteria, yeasts, molds, and microbial spores are relatively resistant to killing by freezing, whereas gram-negative bacteria are more susceptible to freezing. Initially there is a large mortality from freezing, but a proportion of the cells survive. For instance, 30% to 70% of *E. coli* and 50% to 80% of *Pseudomonas* populations are killed by freezing. Those viable cells that remain die gradually when stored in the frozen state. Freezing inhibits growth but has no practical killing effect on microorganisms, nor does it affect the activity of any microbial toxin that may be present. Microbiologists use freezing to maintain bacterial cultures (for instance, frozen in milk) for years, especially gram-positive bacteria as well as yeasts and fungi. Some of the effects of freezing on microorganisms follow.

1. Loss of cytoplasmic gases, such as oxygen and carbon dioxide
2. Change of pH in the cell
3. Loss of cellular electrolytes
4. Some denaturation of proteins
5. Metabolic injury

With regard to metabolic injury, on thawing, some microorganisms have increased nutritional requirements, at least for several cell divisions and then revert to their normal state of growth.

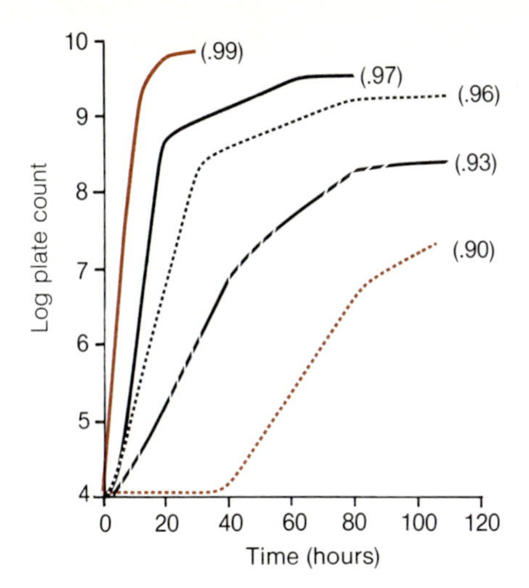

Figure 21-6 *Effect of water activity (a_w) on growth of strain of* Staphylococcus aureus *grown in medium containing protein hydrolysate. The a_w of growth medium is indicated in parentheses.*

Microorganisms can grow during the freezing (and thawing) of a food. Consequently, many procedures and mechanical innovations have been devised to enhance a rapid freezing. Thawing, especially at the consumer level, should be as rapid as possible to preclude microbial multiplication. The refreezing of foods should be approached with caution because of the potential for increasing the load during repeated freezing and thawing.

Preservation by controlling the water content of food

All forms of life require water—some more, some less. Bacteria, yeasts, and some fungi require a definitive aqueous environment, but some fungi can grow on the surface of a substrate lacking an aqueous environment if sufficient humidity is present. The water content of food can be a controlling factor for the inhibition of microbial growth. This is accomplished by water removal (drying) or by chemically binding the water with water-soluble substances (solutes). The amount of water that is available for microbial growth can be measured accurately using the concept of **water activity (a_w).**

When a volume of pure water is placed in a closed container, an equilibrium is established between the water molecules in the air and in the liquid phase. The water in the air creates a vapor pressure (P_w). When a solute, sodium chloride, is added to an identical container with an equal volume of water, again, an equilibrium is established, and the air above the solution will always have a vapor pressure less than that of pure water, that is, $P_s < P_w$. The water activity of the solution is defined by the ratio of the vapor pressures, or $a_w = P_s/P_w$. Pure water has an a_w value of $P_w/P_w = 1.000$, and the water activity values of foods and solutions (or broth media) can never exceed unity.

Figure 21-7 *Freeze dryer. One of many ways of drying foods or microorganisms. It differs from other equipment by reducing moisture content without use of heat. Temperature used is at least −20° C.*

All food has a water phase, and it is in this phase that microbial growth takes place. The solutes in the food are concentrated in the water phase, and the water activity of the food can be measured in a manner analogous to that described above, wherein the food is placed in a closed container, and after equilibrium the vapor pressure is measured, and the a_w is calculated.

Each microbe has an optimum a_w for growth, as well as a maximum and a minimum (Figure 21-6). In food microbiology the minimum a_w values are of paramount concern. Deviations from the optimum increase the generation time, and extreme deviations may cause death. An exception to this generality is the **lyophilization**, or freeze-drying, of microorganisms, which is an excellent procedure to prevent microbial growth (Figure 21-7) and can be used to preserve microorganisms in the laboratory. In this instance the drying is accomplished from

Table 21-4 ***Minimum levels of water activity (a_w) of some microorganisms at temperatures near their optimum***

Microorganism	a_w
Fungi	
Aspergillus candidus	0.75
Mucor plumbeus	0.93
Penicillium chrysogenum	0.79
Chryosporium fastidium	0.69
Yeasts	
Saccharomyces cerevisiae	0.90
S. rouxii	0.62
Bacteria	
Bacillus cereus	0.95
B. subtilis	0.90
Clostridium botulinum type E	0.97
Escherichia coli	0.95
Staphylococcus aureus	0.86
Salmonella species	0.95
Lactobacillus viridescens	0.95

Table 21-5 ***Water activity of some major groups of foods***

Food	a_w
Fresh meat, fish, fruits, and vegetables	0.98 and above
Canned fruit in light syrup	
Canned fruits in heavy syrup	0.93 to 0.97
Evaporated milk	
Fermented sausages	
Processed cheese	
Lightly salted fish, pork, and beef	
Raw ham	0.85 to 0.92
Dried beef	
Dry sausage	
Intermediate-moisture foods	0.60 to 0.84
Dried fruit	
Flour	
Jams and jellies	
Nuts	
Cereals	
Chocolate	Less than 0.60
Honey	
Dried milk and vegetables	

the frozen state (sublimation). The limiting a_w value for certain microbes is presented in Table 21-4 and should be compared with the a_w values of some common foods (Table 21-5). In those foods with a high a_w most microorganisms will grow, whereas in those foods with a low a_w most microorganisms are inhibited.

Two foods can have identical water concentrations, but if one of the foods has a higher content of solute, its a_w will be lower (that is, the greater the solute concentration, the lower the a_w value), and its shelf stability will be greater. In more recent times pet feeds and a limited number of foods (such as toaster pastries) have appeared on the market wherein the a_w of the product is adjusted to about 0.84 or lower. These are referred to as intermediate-moisture foods. On occasion if the a_w is not carefully controlled, strict **osmophilic** (requiring high concentration of solute) fungi and yeasts can grow and spoil the product. The growth of these organisms in the laboratory is relatively easy if 40% sucrose is added to the medium. The organisms will not grow in common laboratory media unless high concentrations of a solute are added. These osmophiles are ubiquitous, and quality control procedures must be instituted to assure their control.

Many food products are dried, both at the producing and manufacturing levels. Cereals, onions, and nuts are examples of products that are solar dried. Manufacturers employ a variety of methods, including spray, tunnel, and drum drying for the production of instant coffee and tea, dried milk, and a spectrum of other products. In all instances the goal is to reduce the free water content of the food, thus increasing the solute concentration. At some point, depending on the food, a sufficiently low a_w value is reached, and the shelf stability of the product is radically increased. It should be remembered that although microorganisms are killed by drying, many, including pathogens, survive. Thus the situation is equivalent to the freezing of foods. One of the most troublesome groups of microorganisms in dried foods is fungi, particularly species of *Aspergillus.*

Preservation by acidification

Many of our food products are acidified to enhance their shelf stability. Some pickles, specialty meat prod-

***Table 21-6** Common preservatives used in the food industry*

Preservative	Microorganism affected	Food
Benzoic acid and benzoates	Yeasts and filamentous fungi	Soft drinks, relishes, pickles, margarine, salad dressings
Sorbic acid and sorbates	Filamentous fungi	Syrups, jellies, cakes, apple cider, semipreserved fish and meat products
Propionic acid and propionates	Filamentous fungi	Bread, cakes
Sulfur dioxide, sulfites	Many microbial types	Wine, dried fruit, dehydrated mashed potatoes
Sodium diacetate	Filamentous fungi	Bread and bakery products
Polymyxin B (antibiotic)	Bacteria	Yeast cultures for beer
Sodium nitrite	Bacteria	Meat curing, smoked salmon
Sodium chloride and sugars	Many microbial types	Sodium chloride for meats, sugars for jellies and preserves
Ethyl formate	Yeasts and filamentous fungi	Dried fruits, nuts

ucts, certain vegetables, as well as mayonnaise and a variety of salad dressings, are chemically **acidulated** (the biologically acidulated or fermented products such as cheese, sausage, and sauerkraut are presented in another section). The most commonly used acidulant is acetic acid or vinegar, although citric acid and adipic and fumaric acids are sometimes employed. The principle involved is to lower the pH value of the food such that microbial growth is inhibited. Lowering the pH can also reduce microbial heat resistance in foods that are to be eventually processed by heat. In general the minimum pH for the growth of bacteria is 4.4, but in yeasts and fungi the minimum pH may be as low as 1.5. On occasion spoilage of acidulated foods occurs if the pH is not controlled; the spoilage organisms involved are generally yeasts and gas-producing lactic acid bacteria.

Preservation by vacuum packaging and modified atmospheres

The advent of vacuum packaging radically enhanced the shelf life of a wide variety of food products. All fungi, some yeasts, and many of the gram-negative, psychrophilic bacteria are strict aerobes, and their growth in vacuum-packaged products is precluded. In addition, the growth rate of facultative organisms is diminished under anaerobic conditions. This technological triumph is not without some shortcomings, because anaerobiosis affords the growth of *Clostridium botulinum*. This is a drawback, and unless other means of precluding growth and toxin formation are incorporated, vacuum packaging cannot be used.

In the immediate past, substantial effort has been expended in investigating the packaging of foods in atmospheres other than air. The most effective gas used for this purpose is carbon dioxide, and commercial application of the technology is found in the packaging of cheese, fresh meats, and seafoods. Here again, oxygen is excluded, but the carbon dioxide itself has an inhibitory effect on many spoilage microorganisms.

Preservation with chemicals

The addition of chemicals to foods is closely regulated by the Food and Drug Administration, as well as individual state regulatory agencies. The FDA publishes and strictly enforces a list of accepted chemicals, called the **GRAS list.** GRAS is an acronym for Generally Recognized As Safe. The list not only specifies what chemicals can be used but also their maximum concentration and the specific foods to which the chemical can be added. To have a new chemical added to the GRAS list, extensive acute and chronic toxicity tests must be performed.

Some of the more common preservatives allowed in foods are listed in Table 21-6. In many instances, without these chemicals the production of the food involved would be impossible (dehydrated mashed potatoes which contain sulfite [see Point of Interest] serve as a good example), or the decreased shelf life would render it economically impossible to market the product.

Nitrite is a preservative that deserves special attention. Nitrite is converted to a variety of compounds as a result of either interaction with components of the meat or microbial metabolism. Nitrite at acid pH ionizes to yield nitrous acid (HNO_2), which is converted to nitric oxide (NO). Nitric oxide reacts with meat pigments such as myoglobin to produce water-insoluble pigments (red nitric oxide-myoglobin). Heating produces a pink nitrosyl hemochrome, which is stabilized by ascorbates (ascorbates are chelating agents). Nitrite also adds flavor to certain meats such as pork, but the chemical basis of this reaction is unknown. Nitrites can react with secondary amines, tertiary amines, and quaternary ammonium compounds to form **nitrosamines** (Figure 21-8). The legal limits of nitrite have been lowered in recent years when some nitrosamines were found to be carcinogenic. The discovery that cooking of cured meats such as bacon resulted in the formation of nitrosamines has prompted a lowering of the legal limits for nitrite in such foods. There have been suggestions that nitrites should be banned from all foods, but to date no such action has been taken. Nitrites have never been shown to cause cancer in humans. One of the most important functions of nitrite, in conjunction with decreased pH, salt content, and temperature, is as an inhibitor of *Clostridium botulinum*. This function is particularly important in anaerobically packaged products such as hams and sausages.

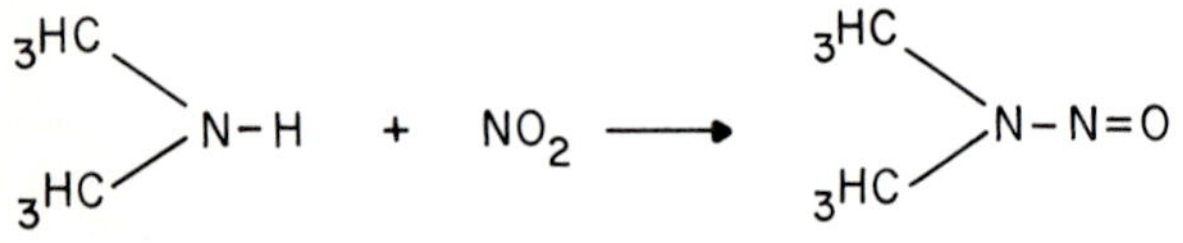

$$(H_3C)_2N{-}H + NO_2 \longrightarrow (H_3C)_2N{-}N{=}O$$

Figure 21-8 *Formation of dimethyl nitrosamine from dimethylamine.*

Barrier concept. In many instances a single preservative or preservative condition (that is, a **barrier**) will not afford the requisite inhibition to provide a reasonable shelf life. Therefore combinations of these preservatives are frequently employed. As an example, consider a 10-pound canned ham (small 1- to 2-pound hams

Concept Check

1. Diagrammatically illustrate a typical survival curve for bacteria exposed to heat. How do you determine the decimal reduction time from the curve?
2. What is meant by thermal death time (TDT), and how is it determined? What is the 12D concept?
3. What are the different techniques for preserving foods? What types of foods are best suited for the technique?
4. What is meant by the water activity of foods, and why is this concept so crucial to the food microbiologist?
5. Name the major chemical compounds used to preserve foods.

Point of Interest

SULFUR DIOXIDE AND SULFITES: SHOULD THEY BE BANNED?

Sulfur dioxide and sulfites have been used for many years as preservatives. Sulfur dioxide is used to control the growth of undesirable organisms in soft fruits, wines, sausages, and fresh shrimp. Sulfur dioxide is also added to a variety of foods as an antioxidant or reducing agent to inhibit (1) certain enzyme-catalyzed reactions such as enzymic browning and (2) non-enzymic browning. In many instances the chemical is added as a sulfite or metabisulfite—compounds that can generate sulfur dioxide. The three major sources of SO_2 are as follows:

Sodium or potassium sulfite—Na_2SO_3, K_2SO_3
Sodium of potassium bisulfite—$NaHSO_3$, $KHSO_3$
Sodium metabisulfite—$Na_2S_2O_5$

Sulfites have been used in many restaurants to preserve the crispness and color of raw fruits and vegetables. Sulfites, however, have been shown to be responsible for allergic reactions, particularly among asthmatics. Thirteen recorded deaths up to 1986 have been attributed to sulfites, and most of these were linked to the use of the chemical in salad bars. The Food and Drug Administration has banned the use of sulfites on fresh fruit and vegetables sold loose or packaged in cellophane. The ban may later be applied to other processed and packaged foods. Sulfites, for example, are used as a bleaching agent in maraschino cherries, and as a dough conditioner in products such as crackers, cookies, and pizza dough. Sulfites are used extensively in the potato industry for canned, frozen, and dehydrated products. Lack of a substitute for sulfite in this industry may have considerable economic implications.

The mechanism of action of sulfites is not known, but they appear to interfere with respiratory processes involving nicotinamide adenine dinucleotide (NAD). Experimental studies have also demonstrated that sulfites may also be mutagenic—causing deaminations such as the conversion of cytosine to uracil.

Figure 21-9 *Milking machine.*

Table 21-7 *Microbiological standards for Grade A dairy products, manufacturing milk, and certified milk*

Product	Maximum number per gram	
	Standard plate count	**Coliform count**
Grade A products		
Raw milk, at pickup	100,000	—
Raw milk, commingled	300,000	—
Pasteurized milk	20,000	10
Pasteurized, condensed milk	30,000	10
Pasteurized, condensed whey	30,000	10
Nonfat dry milk	30,000	10
Dry whey	30,000	10
Milk for manufactured dairy products		
Grade 1	500,000	—
Grade 2	3,000,000	—
Grade 3	Over 3,000,000	—
Certified milk		
Raw certified	10,000	10

—, no standards.

can be heated to commercial sterility because of heat penetration considerations). The ham is cured with sodium chloride and sodium nitrite. The can is evacuated (anaerobiosis), and the ham is heated to 74° C. After heating, the product is kept at refrigerator temperatures until consumed. None of the preservatives or preservative conditions alone would be sufficient to provide shelf stability, but together a relatively stable product with at least a 6-month shelf life is produced.

MICROBIOLOGY OF SPECIFIC FOODS

Dairy products

The microbiology of raw milk in the past 30 years has been radically altered by technological innovations such as mechanically milking the cow, a closed milking system to decrease barn contamination, and on-the-farm refrigeration during holding. Previously, lactic acid bacteria *(Lactobacillus, Leuconostoc, Streptococcus,* and *Pediococcus)* were the predominant flora, but in today's milk supply gram-negative rods are predominant. Milk residues left on equipment that handle milk result in the growth of many gram-negative rods such as *Pseudomonas, Alcaligenes, Flavobacterium,* and *Chromobacterium.*

Milk. Milk is divided into grades on the basis of the milking facilities at the producer level (Figure 21-9). There is no other food that is regulated as closely as fluid milk. Fluid grade A milk for human consumption has the strictest standards (Table 21-7). Most communi-

Table 21-8 *Some of the major bacterial species used as starter cultures in cheese making*

Culture	Foods used in	Function
Streptococcus lactis subspecies *diacetilactis*	Butter, sour cream, cultured buttermilk	Acid and flavor
Streptococcus lactis	Cottage cheese, cultured buttermilk	Acid and flavor
Streptococcus cremoris	Cottage cheese, cultured buttermilk	Acid and flavor
Lactobacillus acidophilus and *Streptococcus thermophilus*	Acidophilus milk, Swiss cheese, yogurt, Emmenthaler cheese	Acid
Leuconostoc cremoris	Cultured buttermilk, cottage cheese, butter	Flavor
Lactobacillus bulgaricus, L. lactis, and *L. helveticus*	Yogurt, kefir, kurniss, Swiss cheese, and related cheeses	Acid and flavor
Propionibacterium freundenreichii	Emmenthaler, Swiss, and related varieties of cheese	Flavor and eye formation

ties throughout the United States permit the sale of only grade A milk for human consumption. Some allow the sale of "certified raw milk"; however, the government is considering banning the sale of it. This type of milk is obtained from herds certified by testing to be free of *Mycobacterium tuberculosis* and *Brucella* species but is not pasteurized. Grades 1, 2, and 3 are used for manufactured dairy products such as cheese.

The enzyme **alkaline phosphatase** normally occurs in cow's milk (but not in all milks—goat's milk is devoid of it), and it has the unique characteristic of just being thermally inactivated by the processes employed for pasteurization. A simple test has been devised such that a manufacturer can rapidly verify the adequacy of the pasteurization process. The pathogens *M. tuberculosis* and *Coxiella burnetii* are destroyed before the enzyme is inactivated, thus providing a margin of safety. On rare occasions some bacteria may grow in the raw milk and produce a thermally stable phosphatase that will give a false-positive result.

Dried milk. A large amount of milk is dried for manufacturing purposes as well as human consumption. Two products, nonfat dry milk and whole dry milk, account for most of the tonnage. They are used in the manufacture of candy (milk chocolate), ice cream, some sausages, and a variety of preblended bakery items such as cake mixes. The milk is preheated and concentrated to half of its original water content. It is then sprayed as a fine mist into a hot air chamber (115° to 125° C), and the dried particles collect at the bottom and are conveyed to the packaging room. Although the incoming air is sufficiently hot to kill vegetative forms, the effect of evaporative cooling on the individual particles precludes a complete killing. In addition, water problems in the driers and the equipment subsequently used sometimes affords bacterial growth or contamination of the product. The isolation of *Salmonella* from these products is not rare.

Milk starter cultures. Many years ago the dairy industry pioneered the use of **starter cultures**, the purposeful addition of bacteria to milk, to effect unique organoleptic characteristics in the product as well as to aid in its preservation. In modern times this has become a science (Table 21-8). Not only do the starter cultures acidify the milk, but also they produce specific flavoring compounds during the fermentation, and the enzymes liberated by the billions of lysed cells during the subsequent ripening period also contribute to flavor. Starter cultures are used in the production of milk products (buttermilk, yogurt), some types of butter, and a variety of cheese, both unripened (cottage, cream) and ripened (cheddar, Swiss). The terms ripening, aging, and curing refer to the holding of a cheese for a period of time (variable with the type) at less than ambient temperature. The exact contribution of the enzymes liberated by lysis during ripening is not understood. Lipolytic and proteolytic activities are readily detected, but the exact subtle flavoring reactions have defied description.

The dominant flavoring compounds produced during the fermentation are **diacetyl** and **acetoin** (also termed acetylmethylcarbinol). The origin of these compounds is the citric acid in the milk, and although the concentration of citrate in milk is only 0.04% to 0.1%, its importance is not commensurate with its concentration. It is against the law to add chemically synthesized diacetyl or acetoin. The pathway from citrate to diacetyl is illustrated in Figure 21-10.

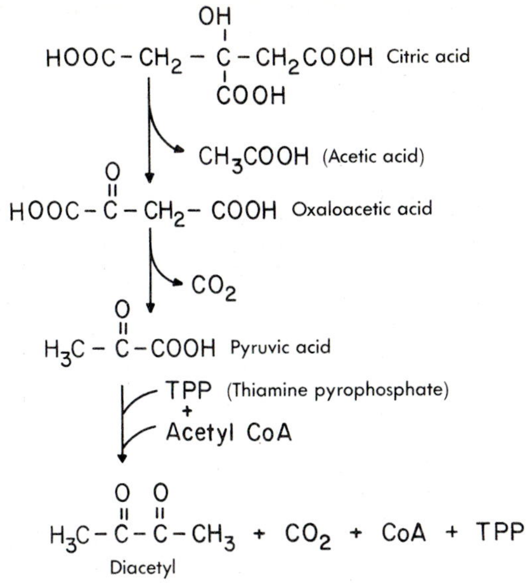

$$HOOC-CH_2-\underset{COOH}{\overset{OH}{C}}-CH_2COOH \quad \text{Citric acid}$$

$$\downarrow \rightarrow CH_3COOH \text{ (Acetic acid)}$$

$$HOOC-\overset{O}{\overset{\|}{C}}-CH_2-COOH \quad \text{Oxaloacetic acid}$$

$$\downarrow \rightarrow CO_2$$

$$H_3C-\overset{O}{\overset{\|}{C}}-COOH \quad \text{Pyruvic acid}$$

$$\downarrow \leftarrow \text{TPP (Thiamine pyrophosphate)} + \text{Acetyl CoA}$$

$$H_3C-\overset{O}{\overset{\|}{C}}-\overset{O}{\overset{\|}{C}}-CH_3 + CO_2 + CoA + TPP$$

Diacetyl

Figure 21-10 *Pathway for formation of diacetyl from citric acid.*

The starter cultures employed in fermented milk and cheese production fall into two categories based on the temperature of fermentation. The low-temperature fermentations are conducted at about 30° to 38° C and employ homofermentative lactic streptococci and heterofermentative *Leuconostoc* organisms (see Point of Interest). The high-temperature fermentations are conducted at 49° to 54° C and generally use lactobacilli and *Streptococcus thermophilus*. Some of the more common starter cultures are listed in Table 21-8. In most instances two strains are used, and the balance, or proportion, of the strains contributes to the flavor of the product. *Lactobacillus bulgaricus* and *Streptococcus thermophilus* are used in the production of yogurt. *S. thermophilus* stimulates *L. bulgaricus* by producing formic acid, and the proteolytic activity of *L. bulgaricus* releases certain amino acids, particularly valine, which stimulates *S. thermophilus*. Thus more acid is produced by the combination of organisms (synergism) than by the two species grown independently of each other.

Bacteriophage and starter cultures. Cheese and fermented milk manufacturers are constantly on the alert for bacteriophage development in the starter cultures during scale up or in the fermentation vat. Bacteriophages are specific for the strain of the starter. For example, if there are 100 strains (each strain constitutes an isolate from a different source) of *S. lactis*, there is an excellent probability that a given strain of bacteriophage will infect and lyse only several strains of *S. lactis*. The host range of the bacteriophage is easily determined in the laboratory, and to circumvent starter failure, the starter strains are constantly rotated to avoid bacteriophage buildup in the plant.

Cheese. All cheeses are produced by lactic fermentation of milk. In cheesemaking the protein-precipitating enzyme mixture, rennet extract, is added shortly after the starter culture decreases the pH value. These enzymes, in conjunction with the development of acid, aid in curd formation; the remaining fluid portion or whey is separated. The subsequent treatment of the curd varies with the type of cheese to be made. Regardless of type, most cheese is ripened for at least 60 days. Considerable microbiological activity takes place during the ripening of some cheeses, whereas in others it is radically diminished. For example, during ripening, gas-producing propionibacteria grow in Swiss cheese producing the "eyes" as well as the characteristic flavor, whereas in cheddar ripening microbiological activity is relatively minimal.

Another example of the use of microorganisms to produce a variety of products is seen in the manufacture of surface-ripened cheese. Previously, examples were given of internally ripened cheese. The surface-ripened varieties (Limburger, brick, Camembert, Brie, and others) usually depend on the growth of aerobic organisms and the diffusion of enzymes or end products of metabolism into the cheese during ripening. In the manufacture of Limburger, which starts with the normal lactic fermentation of the milk, the curd is cut, formed into small rectangles, and rubbed lightly with salt. At the beginning of the ripening period the filamentous fungus *Geotrichum* and aerobic yeasts grow, oxidize the lactic acid, and raise the pH value to approximately 6.0. This affords the subsequent growth of the acid-sensitive, orange-pigmented, salt-tolerant, proteolytic organism *Brevibacterium linens*. As this organism grows, its proteolytic enzymes diffuse into the cheese, giving it the characteristic soft texture, flavor, and odor, and the surface assumes an orange coloration. Brick cheese is essentially a mild form of Limburger. In the surface-ripened varieties of cheese, the pH value must be returned to near neutral values by lactic acid–oxidizing organisms to facilitate the growth of the surface ripening organisms. In the production of Camembert cheese the surface-ripening organisms is *Penicillium camembertii*. The exact flavoring contributing principle is unknown.

Frequently the ripening organisms are sprayed on the curd or into the air of the ripening room. Sometimes the individual pieces of cheese are turned and rotated by hand to ensure surface inoculation and distribution of the ripening organism.

The ripening of blue or Roquetort cheese is peculiar (Plate 27). After the initial lactic fermentation, *Penicillium roquefortii* is mixed with the curd, and small circular "wheels" are formed by compression. Each wheel is multiply pierced with small, solid needles, and the cheese is allowed to ripen for 3 to sometimes 10 months. At first it was thought that the piercing was necessary to afford the entry of oxygen for the fungus

Point of Interest

THE LACTIC ACID BACTERIA

The lactic acid bacteria comprise a group of gram-positive cocci and gram-positive bacilli whose metabolism results in the formation of large quantities of lactic acid. They have been isolated from humans, animals, dairy products, wines, and plants but have not been found outside of these specialized habitats. The principal genera are the rod-shaped *Lactobacillus* and *Bifidobacterium* and the spherically shaped *Streptococcus, Leuconostoc,* and *Pediococcus.* The streptococci are primarily associated with birds, animals, and humans. Among the lactic acid bacteria the streptococci are more capable of causing human infections (strep throat, rheumatic fever, scar-

Glucose
2 ATP
2 ADP
Fructose 1,6-diphosphate
Aldolase
Glyceraldehyde 3-phosphate + Dihydroxyacetone phosphate
NAD
2 ADP
$NADH_2$
2 ATP
2 ADP
2 ATP
Pyruvate
$NADH_2$
NAD
Lactate

Homofermentative species

Homofermentative lactics
Lactobacillus
- *L. bulgaricus*
- *L. casei*
- *L. delbrueckii*
- *L. salivarius*

Pediococcus species
Streptococcus
- *S. bovis*
- *S. cremoris*
- *S. diacetilactis*
- *S. thermophilus*

Glucose
NAD
ATP
$NADH_2$
ADP
6-phosphogluconic acid
NAD
$NADH_2$
CO_2
Xylulose 5-phosphate
Ketolase
Glyceraldehyde 3-phosphate
Acetyl phosphate
NAD
2 ADP
$NADH_2$
2 ATP
$NADH_2$
NAD
Pyruvate
Acetaldehyde
$NADH_2$
NAD
$NADH_2$
NAD
Lactate
Ethanol

Heterofermentative species

Heterofermentative lactics
Lactobacillus
- *L. brevis*
- *L. cellobiosus*
- *L. sanfrancisco*
- *L. viridescens*
- *L. fermentum*
- *L. confusus*

Leuconostoc
- *L. cremoris*
- *L. dextranicum*
- *L. lactis*
- *L. mesenteroides*

Pathways of fermentation for heterofermentative and homofermentative lactic acid bacteria plus other species. Major homofermentative and heterofermentative microorganisms included.

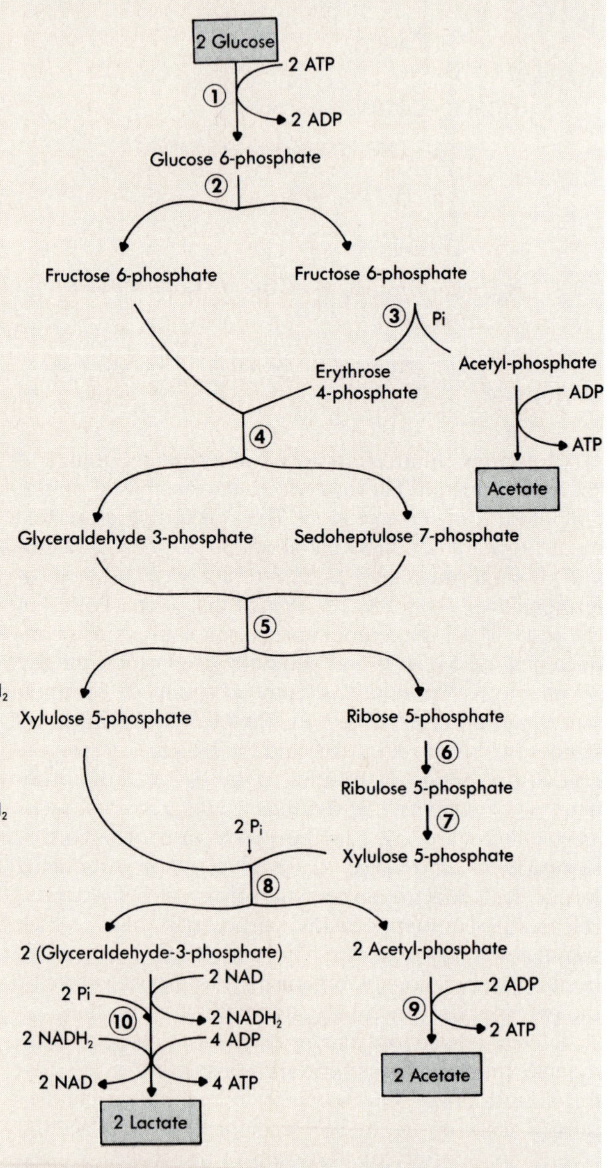

The bifidum pathway in which glucose is converted to lactate and acetate. Enzymes indicated are 1, *hexokinase;* 2, *glucose 6-phosphate isomerase;* 3, *fructose 6-phosphoketolase;* 4, *transaldolase;* 5, *transketolase;* 6, *ribose 5-phosphate isomerase;* 7, *ribulose 5-phosphate 3-epimerase;* 8, *xylulose 5- phosphate phosphoketolase;* 9, *acetate kinase;* 10, *enzymes as in EMP pathway.*

Point of Interest—cont'd.

let fever, and others). More will be said about the streptococci in later chapters. The pediococci and *Leuconostoc* are usually associated with plants or plant products and have some industrial importance (see Chapter 22). *Bifidobacterium* is found in the soil and may also be associated with plants.

The lactic acid bacteria have some common characteristics: (1) complex nutritional requirements, (2) inability to utilize a large complement of organic substrates, (3) lack of functional electron transport system, and (4) anaerobic growth but tolerant of the presence of oxygen. Their energy is obtained via substrate phospyhorylation during carbohydrate fermentation. Based on their fermentation end products the lactic acid bacteria have been divided into three physiological groups: heterofermenters, homofermenters, and those that utilize the *bifidum pathway.* The ***heterofermenters*** dissimilate glucose via the hexose monophosphate shunt (see figure at far left). They lack the enzyme fructose diphosphate aldolase, which splits fructose-1,6-diphosphate into glyceraldehyde-3-phosphate and dihydroxyacetone phosphate. The products from heterofermentation are lactic acid, ethanol, and carbon dioxide. The ***homofermenters*** utilize the EMP pathway, and over 85% of their fermentation product is lactic acid. Those organisms using the bifidum pathway, *Bifidobacterium bifidum,* convert glucose to lactate and acetate in a series of reactions illustrated in figure at left. Overall, 2 moles of glucose are converted to 2 moles of lactate and 3 moles of acetate. The Y_{ATP} yield (moles of ATP produced per mole of substrate used; see Chapter 14) for most fermenters of glucose is near 10.5 when the EMP pathway is used and 8.3 when the Entner-Doudoroff pathway is the major fermentation pathway. The bifidum pathway yields 2.5 to 3.0 ATP molecules per mole of glucose and the Y_{ATP} is 10.3 for those organisms using this pathway.

The lactobacilli not only produce large quantities of lactic acid but are also more tolerant of acid than any other group of microorganisms. Consequently, in environments rich in organic materials where microbial metabolism has produced significant quantities of acid, the lactobacilli can thrive, whereas other microorganisms are incapable of growth. These physiological properties have been exploited for industrial purposes. Lactobacilli have been used in the following processes: (1) the fermentation of cabbage to produce sauerkraut *(Lactobacillus plantarum),* (2) ripening of cheddar cheese from raw milk *(L. casei),* and (3) in brewer's mash *(L. delbrueckii)* to lower the pH and prevent other microbial growth before inoculation with yeast. A more detailed discussion of the industrial importance of the lactobacilli appears in Chapter 22).

Lactobacilli are found in the oral cavity, intestinal tract, and vagina of humans. It is postulated that species of lactobacilli are instrumental in keeping the population of other microorganisms under control because of the acidic nature of their fermentation products. They are not considered to be highly infectious for humans.

---

metabolism, but subsequently it was observed that the escape of metabolic carbon dioxide was equally important. As the fungus grows, it produces blue areas in the cheese and blue "veins" along the pierced channels. During growth the fungus produces lipolytic and proteolytic enzymes that are associated with flavor; it also sporulates. Fungal spores, unlike bacterial spores, are metabolically active (this is why they are heat sensitive). In fact, it is the fungal spore and only the spore that produces the enzyme for the flavoring reaction:

$$\underset{\textbf{Caprylic acid}}{CH_3(CH_2)_6COOH} + O_2 \longrightarrow \underset{\textbf{Methyl-}n\textbf{-amyl ketone}}{CH_3(CH_2)_4COCH_3} + CO_2 + H_2O$$

The caprylic acid is derived from the hydrolysis of fat, and the ketone, derived from its oxidation, gives the characteristic peppery, sharp flavor that is associated with blue cheese.

In many of the fermentations discussed above, there are definitive, stepwise appearances of specific microbial species or groups. These can be referred to as floral succession, which sometimes proceeds without human intervention.

Meat products

The meat industry can be conveniently divided into segments on the basis of their products. Using this scheme, fresh, cured, and fermented meats constitute the major products; and animal by-products such as gelatin, animal feeds, and pharmaceuticals are the minor products. The meat industry is constantly concerned with the perishability and safety of their products because of the inherent susceptibility of the product to microbial contamination and growth. The physical nature of meats does not afford a streamlined processing approach such as that enjoyed in fluid milk processing and other segments of the food industry. Some of the major genera of microorganisms that are associated with meats are outlined in Table 21-9.

Each time meat is cut, the two surfaces are essentially inoculated. Further contamination from cutting boards, chutes, and conveyor belts adds additional bacteria to the product. Temperature is the only barrier that can be imposed on fresh meats. The short shelf life of fresh meats necessitates adequate sanitation, excellent temperature control, and rapid distribution.

Table 21-9 *Some of the major genera of microorganisms found on meats*

Genera	Fresh meats	Processed meats	Vacuum packaged meats	Bacon
Bacteria				
Acinetobacter	++	+	+	+
Aeromonas	++			+
Lactobacillus	+	++	++	
Moraxella	++			+
Pseudomonas	++		+	
Streptococcus	+	++	+	
Pediococcus	+	+	+	
Corynebacterium	+	+	+	+
Filamentous fungi				
Cladosporium	++	+		
Aspergillus	+	++		
Geotrichum	++	+		
Mucor	++	+		
Penicillium	+	++		
Rhizopus	++	+		
Sporotrichum	++			
Yeast				
Candida	+	+		
Rhodotorula	+			
Torulopsis	+	+		
Trichosporon		+		

+, Known to occur; + +, most frequently found; blank spaces indicate absence of the organism.

Fresh meats. The microbiology of fresh meats is similar to that of milk in that psychrophilic species, especially *Pseudomonas* (the others are *Acinetobacter* and *Moraxella*), predominate. After slaughter the sides of pork and beef are chilled to facilitate the subsequent cutting or breakdown of the carcass. Various fluids from the hanging carcasses and the carcasses themselves provide nutrient sources for psychrophiles in the large coolers. Airborne contamination is unavoidable, and the subsequent cutting and sawing of the carcass plus handling and contamination from knives and cutting blocks serve to spread the contamination. The low temperatures of holding enrich for psychrophiles.

The shelf life of fresh meats is directly proportional to the *Pseudomonas* or total count. In fact, many processes assess proper sanitation and holding and handling procedures in a plant with a *Pseudomonas* count. Most, but not all *Pseudomonas* species produce a pigment (fluorescein) that fluoresces under ultraviolet light. *Pseudomonas* colonies can be differentiated using this technique, and an approximate estimate of their numbers can be made. The count is easily determined with a peptone medium that is devoid of carbohydrate and enriched with magnesium and sulfate ions. The medium enhances the production of the pigment, but it offers no selectivity.

The meat sample is quantitatively plated in the same manner as that used for determining a total count. After incubation, only the fluorescing colonies are counted and the number of pseudomonads are estimated.

The shelf life of fresh meats can be as short as 3 and as long as 10 days. It is a function primarily of the initial load and the temperature of holding. Eventually, the meat will spoil regardless of the chilling temperature (Plate 28). An early indication of spoilage is the loss of the red color, or "bloom." The meat appears brown and may assume a surface slime that is chiefly bacteria. People often purchase meat on the basis of its color. This is the paramount reason mitigating against frozen, fresh meat because freezing destroys the bloom. If the bacterial population is sufficiently high, an off taste can be detected in the cooked meat.

Pathogenic bacteria can grow or survive in fresh meats. For example, various surveys have indicated that about 50% of marketed pork and poultry cuts contain *Salmonella*. The incidence of these organisms in beef is somewhat less. This is one reason why consumers and caterers must be careful of cross-contamination between fresh meats and prepared foods. *Salmonella* are ubiquitous, and their elimination from the meat supply is impossible.

Cured meats. Cured meats such as ham, corned beef, and bacon as well as comminuted or blended meats, including weiners, bologna, and sausage-loaf products, are manufactured using nitrite and usually salt. In many products sugar (primarily sucrose) is added for taste. The functions of nitrite have already been discussed.

The thermal process administered to bacon (maximum temperature is about 53° C) during smoking is not sufficient to kill the nematode, *Trichinella spiralis,* the etiological agent of trichinosis (Plate 29). Consequently, this is considered to be a fresh meat, and it must be cooked before consumption. Other cured meats are heated above 59° C, which kills this organism and cooks the meat. In most instances, the thermally processed meats are sliced or prepackaged, and in these operations postprocessing contamination is incurred. The microbial barriers in these products are salt, nitrite, anaerobic conditions, and low temperatures of holding. In many instances shelf lives of 60 days (weiners, sausage) or 6 months (perishable canned products) are readily achievable.

Cured meat spoilage problems usually result from inadequate sanitation or improper temperature control. The chief spoilage organisms are the lactic acid bacteria (streptococci, leuconostocs, and pediococci). These facultatively anaerobic organisms are salt tolerant, some produce gas (heterofermentative), and all produce lactic acid as an end product of carbohydrate fermentations. Spoilage results from souring, gas production in prepackaged items that bloat or puff the package, and slime production. Some species, notably those of *Leuconostoc* and *Lactobacillus,* are capable of synthesizing copious amounts of viscous polysaccharide (both levans and dextrans) when grown with sucrose in the medium. Although the slime is not a health hazard, it is organoleptically unacceptable. The lactic acid bacteria lack the enzyme catalase and accumulate hydrogen peroxide if they grow in the presence of oxygen. Some of the salt-tolerant lactobacilli and leuconostocs can grow on cured meats, and after the consumer opens the vacuum package or if there is a leak, the organisms produce hydrogen peroxide, which oxidizes the cured red meat pigment to a derived green porphyrin. This defect is commonly encountered, and again, although it is not harmful to health, it is unsightly.

Fermented meats. Ancient humans are credited with the discovery that if ground and salted meat with added sugar is held at warm temperatures after a period of time, the meat develops a pleasing tang and its perishability is decreased. This would be called a "wild fermentation" because it reflects the growth of inherent contaminants. Relatively recently the fermented sausage industry used starter cultures, and currently several *Pediococcus* and *Lactobacillus* cultures are commercially available. Their chief function in the manufacturing process is the rapid and characteristic production of lactic acid to produce tang. No other end products of metabolism have been identified as enhancing flavor. Although the pH of the products (summer sausage, thuringer, salami) is in the range of 4.8 to 5.0, like cheese they are still perishable and must be kept under refrigeration to ensure a long shelf life.

Fish

Fish, along with red meats and poultry, constitutes a major source of animal protein in the worldwide human diet. The microbiology of fish is complex and not completely understood. The manner in which microbes penetrate the skin after death and initiate growth in the flesh remains to be elucidated. Data accumulated in the last few decades have firmly established that the growing environment of the fish is the major determinant of the organism's microbiology. Other determinants are the species of fish, the season of harvest, and the salinity of the water (fresh water versus marine water). It is axiomatic that if the growing and harvesting environment of fish is polluted chemically or microbiologically, the fish are polluted. Another dictum is that fish caught close to land masses have a significantly higher microbial load as compared with fish caught a distance from land or in the open waters.

The skin is covered by a mucilaginous slime that is secreted by cells in the skin, and the amount and composition of the slime varies with the fish species. The slime layer affords a luxuriant growth of many different bacteria, and these bacteria, plus organisms from the gut and gill area, constitute significant sources of contamination.

After harvest, the larger fish are dressed aboard the ship and then iced or mechanically chilled. Smaller va-

Table 21-10 *Stages in fish spoilage*

Stage	Tissue changes	Organoleptic changes	Bacterial count
Stage 1 (0 to 5 days in ice)	Rigor mortis; ATP → inosine; slight increase in TMA; changes in bacterial types	Eyes bright, flesh firm, color good, gills bright, odor fresh	10^2 to 10^3/cm^2
Stage II (5 to 10 days in ice)	Bacterial growth becomes apparent; inosine → hypoxanthine; TMAO → TMA; NH_3 increases	Eyes begin to dull, gill color fades, skin color fades, odor neutral to slightly fishy, texture softening	10^3 to 10^6/cm^2
Stage III (10 to 14 days in ice)	Rapid bacterial growth and penetration of tissues; hypoxanthine → xanthine, uric acid, etc.; TMA increases rapidly; TVB and TVA increase	Eyes sunken, gills discolored and slimy, skin bleached, odor sour and fishy, texture soft	10^6 to 10^8/cm^2
Stage IV (over 14 days in ice)	Bacterial numbers stationary; some changes of species; general deterioration of flesh; TVB and TVA increase rapidly; TMA increases or levels off; proteolysis begins; H_2S and other products formed	Eyes opaque and sunken, gills bleached and slimy, skin very slimy, texture very soft, odor offensive	Approximately 10^8/cm^2

ATP, Adenosine triphosphate; *TMA,* trimethylamine; *TVA,* total volatile acid; *TVB,* total volatile base; *TMAO,* trimethylamine oxide

rieties are similarly treated but not dressed. The fish may remain aboard ship for several days to several weeks, and during this time, as well as subsequently, the chilling temperature selects for psychrophiles such as *Pseudomonas, Acinetobacter,* and *Moraxella.* The fish industry employs wooden boxes, bins, and tables for processing and handling, and after several uses the wood surfaces become pitted and uneven. These surfaces are practically impossible to clean and sanitize, and huge bacterial populations develop and serve as important sources of contamination. Only recently (and slowly) has the adoption of stainless steel been instituted. During human handling, depending on the degree and the prevailing sanitary conditions, fish may be significantly contaminated with meosphiles.

Aside from chilling, fish are preserved by salting, drying, heating, and freezing. These methods are employed to facilitate and extend shelf life so that inland markets can be made available. The rapid rate of spoilage of fresh fish precludes their distant distribution without some form of further processing, including freezing.

The spoilage of fish occurs in stages and is associated with proteolysis and strong odors that may include sulfide, fecal, and ammoniacal descriptions (Table 21-10). Fish muscle, unlike mammalian, contains high concentrations of trimethylamine oxide. Some bacteria, such as *Proteus,* possess enzymes to convert the oxide to trimethylamine, which is associated with the characteristic fishy odor. Other spoilage organisms associated with seafoods are outlined in Table 21-11.

Food poisoning associated with fish. There are three types of human food poisoning that are peculiar to fish. *Scombroid food poisoning* is associated with fish of the families Scomberesocidae and Scombridae, which include tuna and mackerel. These fish contain high concentrations of the amino acid histidine in their tissues. The gut and surface of fish, as well as the processing environment, are contaminated with enteric bacteria (especially *Proteus morganii*), and if the organisms grow extensively in or on the fish, the histidine can be decarboxylated to *histamine.* The amine is stable to cooking temperatures and the acidity of the stomach. The ingestion of sufficient amounts of the amine produces characteristic allergic symptoms (facial flushing, tingling of the extremities, and shortness of breath), commonly called scombroid food poisoning.

Vibrio parahaemolyticus, a halophile requiring about 5% sodium chloride in laboratory media, was first isolated and associated with food poisoning in Japan. The

Table 21-11 *Spoilage organisms associated with seafoods*

Seafood	Defect	Microorganism(s)
Fresh fish	Off odors	*Pseudomonas, Acinetobacter, Proteus, Vibrio, Moraxella*
	Fruity	*Pseudomonas*
	Ammoniacal	*Pseudomonas*
	Rotten egg (H_2S)	*Pseudomonas*
Salted fish	Pink	*Halobacterium, Halococcus*
Oysters	Pink	Yeasts
Shrimp	Off odor	*Pseudomonas*
Squid	Yellow discoloration	*Pseudomonas putida*
	Red discoloration	*Serratia marcescens*
Crayfish	Sweet to foul odor	*Pseudomonas, Lactobacillus,* coryneform bacteria

symptoms of vomiting, diarrhea, headache, and fever are common. The vibrio gains entry with the consumption of raw fish. Generally, the afflicted recover within 1 week. The organism and outbreaks of its associated food poisoning have been described throughout the world. Several outbreaks in the United States have occurred as a result of cross-contamination between raw and cooked fish or fish dishes.

Clostridium botulinum (see p. 636) organisms that produce type E toxin appear to be restricted to fresh and marine coastal areas, especially the coastal muds. The organisms are free living and contaminate the slime layer and gut of fish. After heating or smoking of the fish, all vegetative forms of all bacteria are destroyed, but the spores remain viable. If temperature abuse occurs after heating, the normal spoilage flora (*C. botulinum* is not part of normal flora) does not grow and the consumer is not aware (because of lack of odor) of the potentially dangerous situation. If botulinal spores are present, they germinate and the organism grows with the concomitant formation of toxin. The product will appear quite acceptable, but it is lethal. The fish do not have to be vacuum packaged for growth to occur, because penetration of the flesh provides sufficient anaerobiosis for growth. The distribution of this organism is worldwide. Fortunately, the disease is not common.

Vegetables and fruits

The surfaces of fresh fruits and vegetables harbor a predominantly gram-negative flora *(Erwinia* and *Pseudomonas),* which may range from a few thousand to 100,000/cm^2, depending on the product and the conditions of growth. Also encountered with a high incidence but in low numbers are lactic acid bacteria, yeasts, and fungi. Frequently, various members of the resident flora are encountered within the tissues of sound fruits and vegetables. In some manner these microorganisms are incorporated during plant growth; and although they do not multiply extensively, they remain viable. The chief procedures to maintain fruits and vegetables in the fresh state are chilling and the use of controlled atmospheres. During the storage period various bacteria and fungal populations can develop and cause spoilage. Spoilage may consist of acrid and putrid odors, texture defects including various types of "rots", and discolorations (Plate 30).

Some vegetables are fermented to produce characteristic flavors as well as to extend shelf life. Cabbage, cucumbers, and olives undergo a lactic acid fermentation, and the floral succession of these fermentations is similar. In the manufacture of sauerkraut the cabbage is shredded, salted, and added to large tierces, or vats. Oxygen is excluded from the tierce by covering with plastic sheeting and adding water or weights. The salt disrupts the osmotic balance in the shredded cabbage, and nutrient-containing juice is removed from the tissues. Sufficient carbohydrates, vitamins, and other nutrients are in the juice to afford luxurient growth of the nutritionally fastidious lactic acid bacteria. In the floral succession resident coliforms rapidly appear, followed by a burst of *Leuconostoc.* After a brief period the acid-tolerant, homofermentative species such as *Lactobacillus plantarum* predominate. The bursts in the floral succession are controlled by the pH of the juice.

Attempts have been made to employ lactic starter cultures to replace the "wild" vegetable fermentations. In most instances a product with at best equal acceptability was produced and at considerably more expense because of the cost of making the starter. Currently, all fermented vegetables use the wild fermentation procedure for commercial production.

Bread

Breadmaking has been practiced for centuries, and in many areas throughout the world a bread or breadlike product is produced that is peculiar to that area. This is made possible by the many raw ingredients and processing procedures that can be used. Most, but not all, breads are leavened. Leavening is the production of gas in the dough before baking, and this gives the finished product a light, porous texture. Many baked products such as pretzels, doughnuts, crackers, wafers, and bread sticks are leavened.

Two types of leavening are used in industry, and both produce carbon dioxide. Chemical leavening usually involves the addition of a weak acid and sodium bicarbonate to the dough. Fermentative yeasts, principally *Saccharomyces cerevisiae,* are used in biological leavening. The yeasts also produce aroma and flavoring compounds that are unknown and cannot be duplicated chemically. The yeasts are produced by a separate industry. They are grown on molasses from sugar processing, and most commercial yeast used today is dried and packaged as a powder. In breadmaking flour, water, and salt are mixed, and the yeast is added. After an initial fermentation, the dough is cut, placed in pans, and held for the final fermentation. During baking, all vegetative forms are killed, and only bacterial spores survive. The most important of these is *Bacillus subtilis.* This species is a cause of spoilage within the loaf called "rope" in which the bread appears stringy, soft, and brown and has an odor of ripe cantaloupe. The mucoid nature of the contamination is caused by the production of a capsular-like material by the offending microorganism. Most commercial bakeries add propionate to the product to inhibit fungi. Essentially, this adds 1 to 2 days to the shelf-life. Fungi are more evident on the surface of the bread, and the type of defects they cause are outlined in Table 21-12.

Table 21-12 *Microorganisms causing spoilage in bread*

Microorganism	Defect
Bacteria	
Bacillus subtilis	Ropy, slime
Lactic acid species	Sour
Coliforms	Sour
Serratia marcescens	Red
Fungi	
Rhizopus nigricans	Black mold
Penicillium	Blue mold
Neurospora	Pink mold

Sourdough breads are made using a lactic fermentation with an acid-tolerant yeast. Usually a portion of the first fermentation dough is saved, and the following day small amounts of it are used continuously to initiate the fermentation again (termed back inoculation). All sourdoughs are made using a wild fermentation.

The organisms involved in the manufacture of San Francisco sourdough breads have been isolated recently and identified. The lactic acid bacterium is a new species of *Lactobacillus (L. sanfranciscus),* and this heterofermentative organism is peculiar in that it ferments maltose but not glucose or other common sugars. Two acid-tolerant yeasts were identified as *Saccharomyces exiguus* and *S. inusitatus.* Typical baker's yeast *(S. cerevisiae)* was not detected, probably because of its acid sensitivity.

Oriental foods

The foods produced by fermentation in the Orient represent an exceptionally large industry. Products such as soy sauce, tempeh, sufu, and tea fungus have a market value of approximately 40% of the antibiotic industry. The fermented products and the organisms involved are outlined in Table 21-13.

Most of the oriental fermented foods use more than one microorganism in contrast to the single, pure cultures used in the West. The microorganisms may be added in succession as pure cultures but occasionally they are introduced simultaneously. Mixed cultures offer several advantages, some of which are as follows:

1. Increased yield of product. Pure cultures invariably transform substrate less efficiently than mixed cultures. There is a great enzymatic diversity in mixed cultures, and therefore substrates and their metabolically formed products are more easily transformed. In addition, growth of one of the species in the mixed culture is often enhanced.

2. Less contamination. Pure cultures are invariably subject to contamination. Unwanted microbial species find it difficult to contaminate a culture in which the product of metabolism by one species of the mixed culture is utilized by the second species.

3. Phage resistance. In pure cultures the microbial species is subject to attack by phage. Mixed cultures demonstrate greater resistance to attack by one phage.

Table 21-13 *Oriental food fermentations and the organisms involved*

Food	Microorganisms used	Substrate	Food use
Soy sauce	*Aspergillus sojae* *Aspergillus oryzae* *Saccharomyces rouxii* *Candida etchellsii* *Candida versatilis* *Pediococcus halophilus* *Lactobacillus delbrueckii*	Soybeans	Flavoring agent for foods
Miso	*Aspergillus oryzae* *Aspergillus sojae* *Saccharomyces rouxii* *Candida etchellsii* *Pediococcus halophilus*	Rice, barley, or soybean	Soup base or flavoring agent for fish, meats, and vegetables
Tempeh	*Rhizopus oligosporus*	Soybeans or coconuts	Snack
Sufu	*Actinomucor elegans* *Mucor dispersus*	Soybean	Chinese cheese or bean cake
Lao chao	*Amylomyces rouxii* *Rhizopus chinensis* *Rhizopus oryzae* *Saccharomycopsis fibuligera* *Saccharomycopsis malanga*	Rice	Dessert
Tape ketan	*Amylomyces rouxii* *Rhizopus chinensis* *Saccharomycopsis fibuligera* *Saccharomycopsis malanga*	Rice	Dessert or snack

4. Mixed cultures carry out transformations that are often impossible with a single species. For example, a mycelial fungal species producing the enzyme amylase can reduce starch to glucose units, and the latter can be utilized by a second microorganism, such as a yeast, resulting in the production of alcohol or other products. Mixed cultures may also be an alternative to genetic manipulation of a single species to carry out more than one transformation process.

Production of soy sauce (shoyu). Several steps are involved in the production of soy sauce (Figure 21-11). Soy beans are first moistened and cooked with steam under pressure. Wheat kernels are roasted at 170° to 180° C for a few minutes and then crushed. These two components are mixed and inoculated with seed fungus containing the species *Aspergillus oryzae* and *A. sojae.* These species are more tolerant of temperature fluctuations than are other species. They are a source of enzymes that will break down the starch, fats, and proteins of the substrates. Approximately 20% of the wheat starch is used for the growth of the *Aspergillus* species.

The second step in fermentation involves preparation of the mash, which is called **moromi**, and the mixing of salt water. Pure cultures of the yeast *Saccharomyces rouxii* and the bacterium *Pediococcus halophilus* are added to the mash. Other yeast species such as *Candida etchellsi* and *C. versatilis* may also be present in the mash to add flavor. The moromi is transferred to deep fermentation tanks and held there for 4 to 8 months with occasional mixing with compressed air to ensure blending and enhancement of microbial growth. The original fungi have been killed by this time, but the microbial enzymes are present in the mash to hydrolyze the proteins to amino acids and small peptides. The remaining starch is converted to glucose. Half of the glucose is fermented by a member of the lactic acid bacteria *(Pediococcus* or a strain of *Lactobacillus delbrueckii)* to lactic acid and to alcohol by the yeast *Candida.* Both are added periodically during the fermentation process.

After fermentation the liquid is pressed from the moromi, refined, pasteurized, and packaged. Pasteurization kills the yeast population and inactivates any enzymes present.

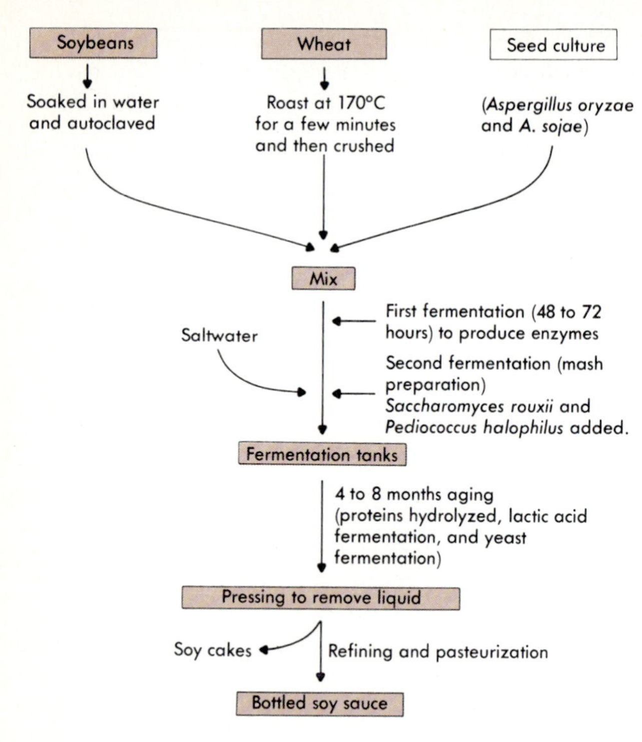

Figure 21-11 *Flow diagram for soy sauce production. Soybeans and wheat following treatment are mixed with seed culture and allowed to ferment for 48 to 72 hours to produce maximum amount of enzymes. Second fermentation involves mixing of salt water with mixture and transferring to deep fermentation tanks. Second fermentation takes 4 to 8 months with occasional mixing with compressed air. During this fermentation starches are converted to sugars and sugars are fermented to lactic acid and alcohol. Proteins are also hydrolyzed to amino acids. Following second fermentation, liquid is pressed from mixture, refined, pasteurized, and packaged.*

Concept Check

1. Name one or two of the principal microorganisms associated with the spoilage of milk, fresh meat, cured meat, fermented meat, fish, vegetables, fruit, and bread.
2. What are the contributions of the lactic acid bacteria to the food industry? What common metabolic characteristics do they demonstrate?
3. How are the cheeses made, and what are the organisms involved?
4. Name some of the major oriental foods produced by microbial activity and the microorganisms involved.

Quality soy sauce will contain 1% to 2% nitrogen (half as free amino acids); 3% to 5% reducing sugar (glucose); 2% to 2.5% ethanol; 1% to 1.5% polyalcohol (glycerol); 1% to 2% organic acid (lactic); and 17% to 18% sodium chloride.

FOOD POISONING

Food poisoning is a general term reflecting the passage of any deleterious chemical substance, microorganism, or microbial toxin by foods. The chemical poisons include those produced by poisonous plants and fish as well as the inadvertent addition of toxic chemicals such as insecticides and germicides. Microbial food poisonings may result from ingestion of preformed toxins (intoxications) or from toxin production by the microorganism in the intestinal tract. *Clostridium botulinum* and *Staphylococcus aureus* produce toxins in foods before ingestion, whereas *Clostridium perfringens* and species of *Salmonella* produce toxins in the intestines.

By far the majority of abuses that cause food poisoning outbreaks occur at the end of the food chain and involve temperature abuse. Most poisonings occur as a result of ignorance or indifference on the part of food handlers. In industry this is generally recognized, and considerable attention has been focused on the education of food handlers, especially restaurant personnel, caterers, and those in the home. The available data indicate that only occasionally does the insult occur at the food manufacturing level.

Bacterial food poisoning (Table 21-14)

Clostridium botulinum. *C. botulinum* is the cause of a foodborne intoxication called botulism. The first recorded outbreak of suspected botulism occurred in Germany in 1793 when 13 people were stricken with a food poisoning that produced a neuromuscular paralysis, and six died. All had eaten blood sausage that was prepared by washing out pig's stomachs, filling with blood and other ingredients, tying up the casings, boiling briefly, and then smoking the product. The sausages were stored at room temperature for several weeks. Uninvestigated outbreaks of sausage-related food poisonings continued until the early 1800s when a medical officer made a detailed study of the disease. He observed that if air pockets remained in the sausage, the latter did not give rise to toxic symptoms. He also noted that boiling and smoking in the manufacturing process led to poisoning of the sausages. These findings, although not obvious to the earlier investigator, in retrospect indicated that, if an organism was involved, it was capable of growing only under anaerobic conditions and was resistant to short periods of boiling and smoking. This type

Table 21-14 *Confirmed foodborne disease outbreaks, cases, and deaths, by etiologic agents, United States, 1982*

Etiologic agent	Outbreaks No.	Outbreaks (%)	Cases No.	Cases (%)	Deaths No.	Deaths (%)
Bacterial						
Bacillus cereus	8	(3.6)	200	(1.8)	0	(0.0)
Brucella	1	(0.5)	3	(<0.1)	0	(0.0)
Campylobacter jejuni	2	(0.9)	31	(0.3)	0	(0.0)
Clostridium botulinum	21	(9.5)	30	(0.3)	5	(20.9)
Clostridium perfringens	22	(10.0)	1,189	(10.8)	0	(0.0)
Escherichia coli	2	(0.9)	47	(0.4)	0	(0.0)
Salmonella	55	(25.0)	2,056	(18.6)	8	(33.3)
Shigella	4	(1.8)	116	(1.1)	0	(0.0)
Staphylococcus aureus	28	(12.7)	669	(6.0)	0	(0.0)
Streptococcus Group A	1	(0.5)	34	(0.3)	0	(0.0)
Vibrio cholerae O1	1	(0.5)	892	(8.0)	11	(45.8)
Vibrio cholerae non-O1	1	(0.5)	7	(0.1)	0	(0.0)
Vibrio parahaemolyticus	3	(1.4)	39	(0.4)	0	(0.0)
Yersinia enterocolitica	2	(0.9)	188	(1.7)	0	(0.0)
TOTAL	151	(68.7)	5,501	(49.9)	24	(100.0)
Chemical						
Ciguatoxin	8	(3.6)	37	(0.3)	0	(0.0)
Heavy metals	5	(2.3)	26	(0.2)	0	(0.0)
Monosodium glutamate	3	(1.4)	10	(0.1)	0	(0.0)
Mushrooms	4	(1.8)	9	(0.1)	0	(0.0)
Scombrotoxin	18	(8.2)	58	(0.5)	0	(0.0)
Shellfish	1	(0.5)	5	(<0.1)	0	(0.0)
Other	8	(3.6)	75	(0.7)	0	(0.0)
TOTAL	47	(21.4)	220	(1.9)	0	(0.0)
Animal Parasite						
Trichinella spiralis	1	(0.5)	4	(<0.1)	0	(0.0)
TOTAL	1	(0.5)	4	(<0.1)	0	(0.0)
Viral						
Hepatitis A	19	(8.5)	325	(2.9)	0	(0.0)
Norwalk virus	2	(0.9)	5,000	(45.2)	0	(0.0)
TOTAL	21	(9.4)	5,325	(48.1)	0	(0.0)
Confirmed total	**220**	**(100.0)**	**11,050**	**(100.0)**	**24**	**(100.0)**

of food poisoning later became known as **botulism**, derived from the Latin *botulus,* a sausage. Outbreaks of botulism were later found to be associated with smoked fish and ham. In the late 1800s an incident of food poisoning involving ham resulted in neuromuscular paralysis and caught the attention of the bacteriologist, Van Ermengem. He procured the contaminated ham, prepared extracts from portions of it, and injected it into laboratory animals who developed a paralytic disease. He later isolated the causative agent by growing it in gelatin (agar had not yet been discovered) under hydrogen gas and called it *Bacillus botulinus.* From these preliminary studies several characteristics of the disease and the organism had been observed.

1. The disease is an intoxication and not an infection.
2. The organisms releasing the toxin grow only under anaerobic conditions.
3. Toxin is resistant to digestive enzymes.
4. The toxin can be inactivated if it is subjected to boiling for more than 5 minutes.
5. In the smoking process for meats if the salt concentration is high enough, the organism causing botulism cannot multiply.

The gram-positive spore-forming organism causing botulism is now classified as *Clostridium botulinum* (Figure 21-12). The heat-resistant spores of *C. botulinum* that contaminate food germinate under anaerobic conditions and during vegetative growth produce a toxin that is released when the bacterial cell lyses. The toxin from some *C. botulinum* strains is activated by enzymes found in the intestinal tract. The toxin, which is a neurotoxin, is one of the most potent toxins known to humans. It is absorbed into the bloodstream and carried to a number of organs and tissues where it affects motor nerves. The initial symptoms of disease include double vision and nausea, which are later followed by difficulty in breathing and finally cardiac arrest.

The botulinal toxins are quite heat labile, and heating a contaminated food to boiling destroys the toxin. Some foods, such as cooked (but not cured) meats and smoked fish, afford the growth of *C. botulinum* in the deep tissue areas. These products need not be vacuum packaged to support growth, and refrigeration must be employed to inhibit growth. Aside from thermal retorting, salt, nitrite, and chilling temperatures (or combination thereof) are used to inhibit the growth of this organism. More detailed explanations of the action of the toxin appear in Chapter 23.

Food-borne botulism in the United States is associated with foods preserved at home as well as with commercial food products (Table 21-15). Home canning of vegetables, fish, or other meats has been most frequently implicated in botulism. The canning process creates anaerobic conditions, and if the sterilization process is inadequate any viable spores can germinate and produce toxin. Botulism can be prevented by careful sterilization of foods to be canned or by adequate cooking of foods before consumption. As a case in point, in 1976 a boy became ill after eating a frozen beef pie that had been heated to 218° C for just 20 minutes and then left overnight in the gas oven. The pie was later eaten without further heating. Apparently the cooking had heat-shocked clostridial spores and driven off any residual oxygen, thus permitting vegetative cells to multiply and produce toxin.

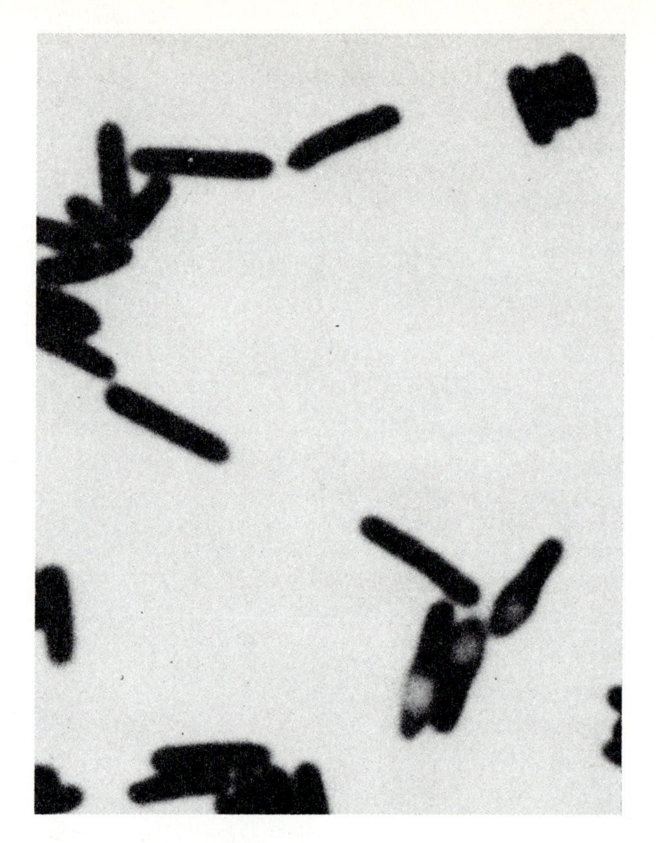

Figure 21-12 *Sporulating cells of* Clostridium botulinum.

Several cases of infant botulism have been recorded in the United States, and these resulted from the ingestion of foods, particularly honey, contaminated with *C. botulinum* endospores. Unlike other types of botulism, infant botulism results from the release of toxins in the intestine when the ingested spores reach the intestinal tract. Why this occurs in some infants and not others is still a mystery. Infant botulism can cause death suddenly in infants, and it is being investigated as one possible cause of *sudden infant death syndrome* (SIDS).

Treatment of botulism is through injection of an antitoxin made from three of the major toxins of *C. botulinum,* toxins of types A, B, and E. The antitoxin has no effect on toxin that is already absorbed to nerve cells.

Staphylococcus aureus. Staphylococcal food poisoning is a major cause of food poisoning, and over 2000

Table 21-15 *Food products causing botulism outbreaks 1899-1977**

Botulinum toxin type	Vegetables	Fish and fish products	Fruits	Condiments†	Beef‡	Milk and milk products	Pork	Poultry	Other	Unknown[2]	Total
A	115	11	22	17	6	3	2	2	8	9	195
B	31	4	7	5	1	2	1	2	3	3	59
E	1	25							3	1	30
F					1						1
A & B	2										2
Unknown§	2	1		1						6	10
TOTAL	151	41	29	23	8	5	3	4	14	19	297

*For period 1899-1973 includes only outbreaks in which toxin type was determined and for period 1974-1977 includes all outbreaks.
†Includes outbreaks traced to tomato relish, chili peppers, chili sauce, and salad dressing.
‡Includes one outbreak of type F in venison, and one outbreak of type A in mutton.
§Categories added for period 1974-1977.

cases are recorded each year in the United States. Intoxication is caused by strains of *Staphylococcus aureus* that produce a heat-labile **enterotoxin.*** The organism is consistently found on the skin and mucous membranes of humans, and avoidance of food contamination is impossible. Toxin is usually produced by organisms that contaminate foods containing protein. Rigid temperature control must be practiced to inhibit microbial growth. The organism produces six serologically different exotoxins, which are referred to as enterotoxin A through F. Generally type A is found in most food poisoning, with ham as the most common vehicle. As little as 0.15 μg of the toxin will elicit symptoms.

Contamination of food is usually due to food handlers who carry the organisms on their skin with or without skin conditions such as pimples or boils. Once the food is contaminated, improper handling temperatures allow multiplication of the staphylococci and subsequent elaboration of toxin. Outbreaks of food poisoning (intoxication) are often associated with local festivities such as weddings and large picnics where food is prepared in advance and held at temperatures that permit bacterial growth. Outbreaks due to commercial procedures also occur, although infrequently. For example, in 1979 salami from a processing plant was implicated in several cases of food poisoning. In the production of fermented sausage, meat is intentionally temperature controlled to permit the growth of lactobacilli that inhibit the growth of other microorganisms. If the procedure is not monitored properly *S. aureus* organisms may grow on the surface of the sausage and produce enterotoxin. The curing procedure, which takes 1 to 2 months, will eventually kill the staphylococci but the enterotoxin remains.

Food poisoning symptoms may appear within 30 minutes following ingestion of enterotoxin but may take as long as 8 hours. The primary symptom is vomiting, but diarrhea may also appear. The symptoms are short lived and seldom fatal.

Clostridium perfringens. *Clostridium perfringens* belongs to a group of bacteria referred to in Bergey's Manual as *endospore-forming gram-positive rods and cocci.* The clostridia are anaerobic sporeformers that inhabit soil, water, and vegetation and play a major role in the decomposition of plant and animal remains. A few species, including *C. perfringens,* parasitize the intestinal tract of humans and various animals. This bacterium, which causes gas gangrene (see Chapter 27), is also a cause of one of the most important types of food poisoning in the United States. *C. perfringens,* because of its ubiquitous nature, is a frequent contaminant of foods containing poultry and other meats. Gravies, for example, provide a relatively anaerobic environment and an excellent growth medium. The contaminated product is usually cooked insufficiently, and spores germinate, releasing vegetative cells that multiply. *C. perfringens* food poisoning differs from other types of food poisoning because there is no ingestion of preformed toxin. Instead,

* Enterotoxins are toxins either ingested or released in the intestine that cause intestinal symptoms such as diarrhea.

vegetative cells are ingested in the food and in the intestinal tract some of them sporulate. It is during the sporulation process that an enterotoxin is produced and gastrointestinal disturbances ensue. Approximately 1×10^8 or more vegetative cells must be ingested before symptoms appear. The typical symptoms, diarrhea and abdominal pain, begin 9 to 17 hours after ingestion of the contaminated food, and recovery is complete within 12 to 24 hours.

Salmonella. Members of the genus *Salmonella* are gram-negative rods that in most instances are peritrichously motile. They are inhabitants of the intestinal tract of various animals and are all capable of causing disease in humans. For example, salmonellas inhabiting the intestinal tracts of imported turtles has caused human infections. The salmonellas are hardy organisms capable of survival for months in moist environments and in the frozen state. All of the salmonellas are closely related by surface antigens, and instead of reporting them as species they are now referred to as serotypes. There are over 1500 serotypes, and proper identification in the laboratory requires serological and biochemical testing.

Diseases caused by Salmonella are called **salmonelloses** and are a result of ingestion of organisms contaminating food, water, or fomites (inanimate objects). Salmonella can cause three types of disease: gastroenteritis, typhoid fever, and bacteremia. Gastroenteritis due to food poisoning is the least severe, although *Salmonella* are the major cause of food poisoning (see Table 21-14). The number of Salmonella organisms required to cause disease depends on the serotype and host factors. For example, gastric acids kill a large percentage of ingested Salmonella, but if an individual has taken antacids, the number of organisms required to cause disease may be very small (less than 100 organisms). Diarrhea, the major symptom, appears 4 to 5 days after ingestion of the bacilli. The cause of diarrhea is similar to that due to the organism causing cholera *(Vibrio cholerae)* and pathogenic *E. coli.* The organism releases factors that stimulate intestinal adenyl cyclase to produce cyclic AMP (cAMP), and this results in an oversecretion of fluids and electrolytes by intestinal tissue. The diarrhea may last from 3 to 5 days.

Other bacterial agents (Table 21-14). Many pathogens that cause foodborne illness escape detection because of late or incomplete laboratory investigation. *Escherichia coli, Bacillus cereus, Yersinia enterocolitica, Vibrio parahaemolyticus,* and *Campylobacter jejuni* cause foodborne illness, but the extent of their role in disease has not been determined. In the United States many laboratories do not look for *Campylobacter* unless they are asked to. In England, for example, *Campylobacter* surveillance since 1977 has revealed that *Campylobacter* is the most common cause of diarrhea and surpasses *Salmonella. Bacillus cereus* is an aerobic sporeformer found in the soil. The primary food vehicle for this organism is cereal dishes. The most severe symptoms have resulted from ingestion of contaminated fried or boiled rice. The incubation period of disease caused by *B. cereus* is between 2 and 5 hours. There are two toxins elaborated by the organism that are important in disease—one associated with vomiting and the other with diarrhea.

Other bacteria suspected of being involved in foodborne disease are Group D *Streptococcus, Citrobacter, Enterobacter, Klebsiella,* and *Pseudomonas.* These genera are infrequently detected because they are not considered during investigation and laboratory diagnosis of foodborne illness. The confirmed cases of foodborne disease caused by bacterial agents in 1982 are listed in Table 20-14. This table confirms the pattern alluded to earlier—the primary offenders are *Salmonella, S. aureus,* and *C. perfringens.*

Animal parasite food poisoning

Trichinella spiralis is a nematode that is associated with a foodborne illness called **trichinosis.** Nematodes are unsegmented round worms that are among the most abundant animals on earth. Many are parasitic, but others are free-living marine, freshwater, and soil species. An acre of fertile soil may contain as many as 9 billion of them. Animals, insects, plants, and humans are parasitized by nematodes. *Trichinella spiralis* is commonly found in carnivorous mammals where it encysts in muscle tissue (Plate 29). The organism is transmitted to humans when the latter ingests undercooked or underfrozen bear, wild pig, walrus, and other meats. The ingested parasite penetrates the intestinal wall and is carried via the hepatic-portal circulation to the liver, heart, lungs, and arterial system. Once in the arterial system the organism is circulated throughout the body, and any organ or tissue may be invaded. The symptoms of infection can occur within 2 days and resemble ordinary food poisoning (gastroenteritis). Once the organs and tissues are invaded, however, severe clinical symptoms may be manifested, such as pneumonia, meningitis, nephritis, muscular pain, difficulty in breathing, or heart damage. This disease left untreated can result in death.

Viral food poisoning (Table 21-14)

The principal viral agent associated with food poisoning is the **hepatitis A virus (HAV).** HAV is a nonenveloped icosahedral virus whose genome is a single-strand RNA molecule. The virus cannot be cultivated in cell culture or tissue culture; therefore isolation is difficult. Virus is isolated and identified by immune electron microscopy. Serum from those who have recovered from the disease is applied to a stool sample of an individual suspected of HAV infection. Since the serum contains antibody specific for the virus, the antibody acts as a binding agent.

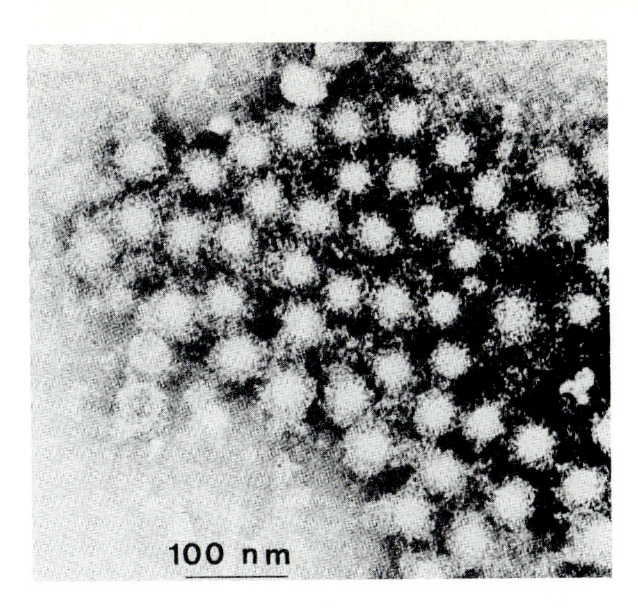

Figure 21-13 *Electron micrograph of hepatitis A virus aggregated by human antibody.*

Large aggregates of virus are produced and can be visualized by electron microscopy (Figure 21-13).

Hepatitis is usually transmitted by the fecal-oral route, that is, fecal contamination of food, water, and hands followed by transfer to the mouth can lead to disease. Outbreaks of disease have been traced to food handlers who contaminate food. Most infections occur where there is poor hygiene: developing countries, mental institutions, day-care centers, and other closed institutions. Since cooking inactivates the virus, outbreaks of foodborne disease almost always involve imporperly cooked food. For example, eating uncooked oysters from an area that has been contaminated with raw sewage has led to outbreaks of HAV infections. Hepatitis A is a relatively mild disease compared to the disease caused by the hepatitis B virus. Hepatitis A symptoms include jaundice, nausea, and vomiting. The disease subsides on its own with no major aftereffects. Most infections can be prevented by simple handwashing and proper cooking of food.

Occasionally a virus called the **Norwalk virus** is responsible for outbreaks of gastroenteritis. Large numbers of cases are often associated with each outbreak.

Fungal food poisoning

Certain fungi produce substances that are toxic to humans and animals. Although these toxins were known to veterinarians for some time, it was not until recently that the potential for involvement in human health was noted. Most of the fungal toxins are called mycotoxins, and one that is commonly encountered is **aflatoxin**, which is produced by strains of *Aspergillus flavus* and

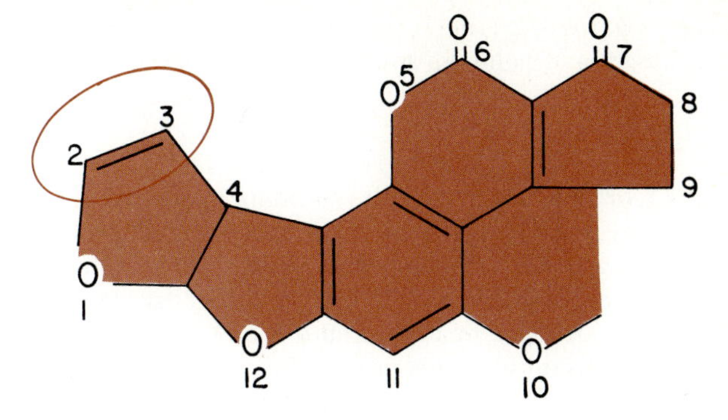

Figure 21-14 *Structure of aflatoxin B. The encircled area represents the site of biological activity of the toxin.*

Penicillium species. In the United States aflatoxin poisoning is more frequently associated with corn and peanuts. There is a measurable amount of aflatoxin in peanut butter, and this represents a potential cancer risk.

Aflatoxins are one of the naturally occurring mutagens. Their mutagenicity occurs when they are metabolically activated by the host system (for example, enzymes in the liver). The biological activity of aflatoxin and related compounds is associated with the C_2-C_3 double bond (see Figure 21-14). If this bond is reduced, there is a 200 to 500-fold decrease in mutagenicity as determined by the Ames test (see p. 268) and a 150-fold decrease in carcinogenicity as determined by experiments in animal cells. It is believed that the C_2-C_3 site undergoes oxidation in the host and that there is an electrophilic attack at the N_7 of guanine and adenine with the subsequent binding of the aflatoxin to DNA. Substitution at the N_7 site produces a positive charge in the imidazole ring of guanine and adenine that can lead to lability of the glycosidic bond. Thus loss of purines from the DNA molecule is possible. Aflatoxins are capable of interaction with other molecules, and therefore processes leading to tumor formation may not depend solely on the genetic activation of aflatoxins. Extremely sensitive methods for the detection and quantitation of aflatoxin in foods have been developed. Currently, the standard for aflatoxin in peanuts, peanut butter, and peanut products is less than 15 parts per billion.

Chemical food poisoning (Table 21-14)

Heavy metals. Various chemicals such as heavy metals (antimony, lead, cadmium, copper, tin, and zinc), can cause gastrointestinal illness when present in excessive amounts in foods.

Monosodium glutamate. Monosodium glutamate is a chemical that is used as a flavor enhancer in various manufactured foods; excessive amounts can cause intoxication. The clinical syndrome includes a burning sensation in the chest, neck, abdomen, and extremities plus a sensation of lightness and pressure over the face.

Fish poisoning

Ciguatera. Ciguatera fish poisoning may result from the ingestion of a variety of marine species such as barracuda, grouper, sea bass, and sharks. These fish become toxic by feeding on herbivorous fish that have ingested toxic algae or toxic matter found in coral reefs. The incubation period is usually 2 to 8 hours, followed by symptoms of gastrointestinal upset and numbing of facial areas and extremities. Death may occur from respiratory paralysis.

Paralytic and neurotoxic shellfish poisoning. This type of poisoning results from the ingestion of shellfish such as oysters, mussels, and clams that have fed on certain dinoflagellates such as *Gonyaulax catenella.* Blooms of these organisms on the ocean produce a characteristic "red tide" (see Chapter 6). The incubation period is 30 minutes to 3 hours. Symptoms of disease are similar to those for ciguatera poisoning. Death results from cardiovascular collapse.

Scombroid poisoning. Scromboid fish poisoning results from ingestion of fish of the order Scrombrodei, discussed earlier.

Mushroom poisoning. A variety of mushrooms are toxic when ingested by humans, and some are particularly lethal. Mushroom poisoning is discussed in Chapter 28.

Concept Check

1. **Name the bacterial agents that are the most frequent cause of food poisoning. How does the food usually become contaminated?**
2. **What are the similarities and differences in the symptoms and treatment of the three major food poisonings caused by bacteria?**
3. **What are the microbial agents that are involved in fungal, viral, and animal parasite food poisoning?**
4. **What are aflatoxins? What is their effect on the host?**

SUMMARY

1. Foods that are processed and sold to the public undergo careful scrutiny before sale. Samples of food are first composited and then subjected to bacteriologic analysis. A variety of tests are used on the sample to detect pathogens as well as to determine the total number of nonpathogenic species that may be characteristically found in the food. The limit of bacteria for some foods is specified by the Federal Drug Administration and is called a standard, which is enforceable by law.
2. The major sources of contamination and microbial growth in foods are many. The first source is the environment from which the food was harvested such as soil or water. Many raw foods have a microbial flora that under proper conditions can act as a cause of spoilage. Further sources of contamination are provided in the processing plant and include the processing equipment, water and air in the plant, the food ingredients being used to produce a processed food, and rework.
3. Control of microbial contamination requires a concerted effort on the part of all divisions in the plant. Plant layout is important to ensure that the product passes without interruption through its various processing phases. Physical barriers (compartmentalization) are placed at specific points in the processing train to prevent cross-contamination. Traffic control is maintained at a minimum to prevent contamination. Last, control of water leaks from the roof, pipes, and water pumps helps prevent one of the major ingredients for growth, water, from reaching the microbial environment.
4. The manufacturer preserves foods in a variety of individual ways but usually a combination of techniques is used. Food can be preserved by heating, which may include employing temperatures as low as 62.9° C (pasteurization) to kill *Mycobacterium tuberculosis* and *Coxiella burnetii* in milk. Canned foods may be heated to 121° C or higher to destroy *Clostridium botulinum,* the organism that produces one of the most lethal toxins known to humans. The heating process that will result in a specific level of killing is determined beforehand by the decimal reduction time. The time required to kill a given number of microorganisms at a specific temperature (thermal death time) is also determined. The minimum heating process (in terms of temperature and time) that will reduce the probability of survival of the most resistant *C. botulinum* spores to 10^{-12} (12D concept) is also selected.
5. Low temperature (below 10° C) inhibits the growth of almost all pathogens but does not inhibit the growth of psychrophilic microorganisms such as *Pseudomonas.* Freezing only inhibits growth of microorganisms but has no practical killing effect on them.
6. Reducing the water content of foods can prevent microbial growth and can be accomplished by drying or by the addition of high concentrations of solutes such as sucrose and salt. The pH of the food can be reduced to prevent microbial growth. Foods that are

acidified do not support the growth of most bacteria, but organisms such as fungi and yeasts can grow at acid pH and cause spoilage.

7. Vacuum packaging prevents growth of aerobic microorganisms but does not inhibit anaerobic microorganisms such as *C. botulinum.* Many foods are treated with a variety of chemicals that are inhibitors of growth. They include benzoates, propionates, nitrites, sulfites, and others.
8. Foods have a typical microbial flora that may be pathogenic for humans or may give rise to spoilage problems. The major pathogens are *M. tuberculosis* and *Coxiella burnetii* (milk and milk products), *Salmonella* (meats and meat-containing products), and *C. botulinum* (meats, fruits, and vegetables). Milk is closely regulated, and specific standards are enforced for raw and pasteurized forms.
9. Many foods are also produced from the action of specific microorganisms called starter cultures. The metabolic products of these organisms are responsible for the formation of processed foods. Milk products (yogurt, cheese, butter), meats (sausage), vegetables (sauerkraut), and bread are examples of foods produced from starter cultures. Several oriental foods are produced from fermentation of soybeans and rice.
10. The microorganisms responsible for most outbreaks of food poisoning in the United States are bacteria and include *S. aureus, C. perfringens,* and *Salmonella.* The only animal parasite of consequence is *Trichinella spiralis,* the agent of trichinosis. Some fungi, such as species of *Aspergillus* and *Penicillium,* produce a toxin, usually in grains, called aflatoxin that can cause serious illness. The major viral agent of consequence is the hepatitis A virus (HAV). Various chemicals can also be a cause of foodborne illness, and these include heavy metals, monosodium glutamate (used as a flavoring agent in many foods), mushroom poisoning, and fish poisoning, resulting from the ingestion of types of fish that themselves have ingested toxic organisms.

Self-study Quiz

Completion

1. The value that describes the number of minutes required to kill 90% of the organisms present in a food sample is called ________________.
2. The heating process designed to kill the most heat-resistant, vegetative cell pathogens present in a food sample is called ________________.
3. The process of freeze-drying a food product is called ________________.
4. The term used to describe the amount of water available in food for microbial growth is ________________.
5. The genus of bacteria whose count is used to measure the shelf life of fresh meat is ________________.
6. The halophilic organism that produces a food poisoning associated with the ingestion of raw fish is called ________________.
7. The majority of canned foods are thermally processed in steam-heated vessels called ________________.
8. The fungal toxin produced by a species of *Aspergillus* and *Penicillium* associated with the growth of the organism in grains and peanuts is ________________.
9. The principal species of microorganisms used to provide biological leavening is ________________.
10. The two principal flavoring compounds produced from citric acid in fermented dairy products such as butter, buttermilk, and cottage cheese are __________ and ________________.

Multiple choice

1. Which of the following would be of greatest concern to the food microbiologist?
   A. Finding a few *Clostridium botulinum* cells in a canned food
   B. Finding a few *Staphylococcus aureus* in raw hamburger
   C. Finding a moderate number of *Pseudomonas* species on the surface of a side of beef
   D. Finding a few *Salmonella* species on the skin of a dressed chicken
2. What does the TDT measure?
   A. The amount of additive in a food
   B. The time required to kill 10^{12} spores of *C. botulinum*
   C. The time required to kill a given number of microorganisms at a specific temperature
   D. The time required to kill 90% of the cells in a food
3. What is accomplished by pasteurization of a product?
   A. Sterilization of the food
   B. Killing of the most heat-resistant pathogen that might be present in the product
   C. Prevention of microbial growth in the food
   D. Killing of the spores of *C. botulinum* in the food
4. Which of the following prevents microbial growth by provision of a low pH in food?
   A. Lyophilization
   B. Canning
   C. 12D concept
   D. Pickling
5. What would be inferred from finding alkaline phosphatase present in a carton of milk on the grocer's shelf?
   A. The milk did not receive adequate pasteurization.
   B. *Mycobacterium tuberculosis* is growing in the milk.
   C. The milk is spoiled.
   D. Bacteriophage have lysed the starter culture.
6. You want to put a certain perservative into a new food product that you are developing; your supervisor tells you to check the GRAS listing. Why?
   A. If the preservative is listed there you can add it in the amounts specified but if it is not listed you must carry out a long, expensive testing to ensure safety.
   B. The listing will tell you what organisms it is effective against.
   C. The listing will tell you how much to add to the type of food product you are developing.
   D. The listing will indicate how stable the agent is so you can judge whether it is compatible with the processing procedure that you intend to use.

7. What produces hydrogen swelling in canned products?
   A. Growth of *S. aureus* in the food
   B. Leakage of oxygen into the can
   C. Reaction of acid in the food with the tin of the can
   D. Microbial spoilage of the food
8. Which of the following toxins is associated with canned or vacuum-packaged foods?
   A. Staphylococcal toxin
   B. Ciguatoxin
   C. Aflatoxin
   D. Botulinum toxin
9. The organisms used in soy sauce fermentation to produce enzymes that degrade proteins, starch, and fats are:
   A. *Pediococcus halophilus* and *Candida versatilis*
   B. *Lactobacillus acidophilus* and *L. delbrueckii*
   C. *Candida etchellsi* and *Saccharomyces rouxii*
   D. *Aspergillus oryzae* and *A. sojae*
10. Eating raw oysters from areas contaminated with raw sewage can lead to foodborne disease caused primarily by
   A. *Salmonella* species
   B. *Trichinella spiralis*
   C. Hepatitis A virus
   D. *E. coli*

Thought questions

1. Your technician has submitted the following report for your approval:

Quantitation of specific organisms in product XYZ

Organisms quantitated	*Count/gm of product*
E. coli (MPN)	15
Coliforms (MPN)	1
Yeast	1500
Filamentous fungi	10
Staphylococcus aureus	520
TOTAL COUNT	1432

   Can you detect at least three problems with this report?
2. A suspension of spores was prepared from a pure culture and mixed with a food product. After 10 minutes of heating at 230° F, the food contained 420 spores/gm. After 15 minutes of heating at 230° F, the food contained 42 spores/gm. What is the D value? How many spores will survive 20 minutes of heating? What was the initial spore count/gm before heating?
3. The kitchens and dormitory rooms of college students are ideal places to find foods that demonstrate the principles of food spoilage. After a recent lecture on food microbiology, the professor asked members of the class to look for items within their living quarters that demonstrated some aspect of the lecture just completed and to bring them to the next class period. He was inundated with some rather vicious-looking items, among which were those listed below. In each case briefly describe the microbiological basis of the problem and comment on the safety of consuming the items.
   A. Two pieces of bologna that were covered with a slimy layer. When the two pieces were pulled apart, long viscous strings hung between the two pieces.
   B. A carton of cottage cheese, with bluish green streaks between the curds, that smelled similar to roquefort cheese.
   C. A carton of milk that had exceeded its consumption date by 3 weeks. The milk looked perfectly normal but had a sharp, bitter taste to it.
   D. A sealed jar of home-canned green beans whose liquid appeared cloudy, with tiny bubbles trapped between the beans.
   E. A fair-sized whole, cleaned perch that was mushy to the touch and fluoresced when put under a black light in the laboratory.
   F. A bottle of wine that was cloudy and smelled like acetic acid.
   G. Two pounds of hamburger meat that smelled sour and rancid.

SELECTED READINGS

Books

Ayres, J.C., Mundt, J.O., and Sandine, W.E., *Microbiology of foods,* W.H. Freeman & Co., San Francisco, 1980.

Banwart, G.J. *Basic food microbiology.* Avi Publishing Co., Westport, Conn., 1975.

Blake, P.A. Prevention of food-borne disease caused by *Vibrio species,* p. 579. In Colwell, R.R. (Ed). *Vibrios in the environment,* John Wiley & Sons, Inc., New York, 1984.

Board, R.G. *A modern introduction to food microbiology,* Blackwell Scientific Publications, Inc. Oxford, 1983.

Brown, M.H. (Ed). *Meat microbiology,* Applied Science Publishers, Ltd., New York, 1982.

Defigueiredo, M.P., and Splittstoesser, D.F. *Food microbiology, public health and spoilage aspects,* AVI Publishing Co., Westport, Conn., 1979.

Fields, M.L. *Fundamentals of food microbiology,* AVI Publishing Co., Westport, Conn., 1979.

Food and Drug Administration. *Bacteriological analytical manual for foods* (5th ed.). Association of Official Analytical Chemists, Washington, D.C., 1978.

Jay, J.N. *Modern food microbiology* (2nd ed). D. Van Nostrand Co., New York, 1978.

Roberts, T.A., and Skinner, F.A. (Eds). *Food microbiology: advances and prospects,* Academic Press, Inc., New York, 1983.

Journal articles

Bryan, F.A. Factors that contribute to outbreaks of foodborne disease. *J. Food Protection* **41**:816, 1978.

Daeschel, M.A., and Fleming, H.P. Selection of lactic acid bacteria for use in vegetable fermentations. *Food Microbiol.* **1**:303, 1984.

Hauschild, A.H.W. Assessment of botulism hazards from cured meat products. *Food Technol.* **36**:95, 1982.

Hesseltine, C.W. Microbiology of oriental fermented foods. *Annu. Rev. Microbiol.* **37**:575, 1983.

Sawyer, C.A., and Pestka, J.J. Foodservice systems: presence of injured bacteria in foods during food product flow. *Annu. Rev. Microbiol.* **39**:51, 1985.

Shandera, W.X., Tacket, C.O., and Blake, P.A. Food poisoning due to *Clostridium perfringens* in the United States. *J. Infect. Dis.* **147**:167, 1983.

Speck, M.L., et al. Overview: use of microbial cultures to increase the safety, shelf life and nutritive value of food products. *Food Technol.* **35**:71, 1981.

CHAPTER 22

Industrial microbiology

Glenn Chamblis
R.F. Boyd

Chapter Outline

Introduction

The use of microorganisms or microbial products to carry out processes of commercial importance has been going on for thousands of years. Of the earliest known such processes are the conversion of grape juice to wine, the production of beer from grain starch, the use of yeast to cause bread to rise, and production of yogurt from milk. Putting microorganisms to work is the job of the industrial microbiologist. Most of the things that microbes make can be produced more cheaply by synthetic means. The range of products that natural microorganisms help to produce is enormous, ranging from plastics to pesticides. The development of genetic engineering techniques may enable scientists to use microorganisms in ways that might have been scoffed at only 20 years ago.

This chapter outlines various commercial products made with the help of microbial activities. How these products function in the marketplace and the steps involved in the commercial production of products, such as antibiotics, amino acids, steroids, single-cell protein, and alcoholic beverages are also described. The techniques used to develop improved strains of yeast for alcoholic beverage production and how genetic engineering is being used to improve microbial strains are also discussed, as well as how this may be applied to many industrial processes in which microorganisms are used.

HISTORY OF INDUSTRIAL MICROBIOLOGY

Most microbial products of commercial importance in earlier times developed empirically without any knowledge that living microorganisms were the causative forces driving the processes. This did not change until the mid 1800s when Pasteur showed that yeast is the causative agent in alcohol fermentation. Because of Pasteur's work it was discovered that many natural processes of industrial importance were caused by microorganisms. As the causative organisms for the various industrial processes were identified, it became common practice to use pure cultures of the organisms to carry out the processes. This resulted in a more uniform product with less batch-to-batch variability. The use of pure culture also opened the possibility that genetic variants of the causative organisms could be isolated—which would be more efficient or would give a better product, or both—than did the original culture.

The knowledge that microorganisms can mediate chemical conversions such as glucose to ethanol prompted microbiologists to search for new and useful microbial processes. For example, at the beginning of the twentieth century a great demand for organic solvents such as acetone, butanol, and ethanol arose, especially with the coming of World War I. The chemical production of these solvents was not efficient and was expensive at that time. Consequently a program was undertaken for screening microorganisms for their ability to produce organic solvents as a result of starch fermentation. Chaim Weizmann isolated a bacterium, today called *Clostridium acetobutylicum,* which did efficiently produce acetone and butanol from starch fermentation. This organism served as the basis for industrial production of acetone and later butanol for several years until it became cheaper to produce them chemically from petrochemicals. At the height of the microbial production of acetone and butanol there were production plants in the U.S., Canada, and England, and as late as the 1940s fermentation provided 65% of the butanol and 10% of the acetone produced. Had it not been for Pasteur's pioneering work, in all likelihood, this industry would never have come into being.

During the same time that advances in the use of intact microorganisms for industrial purposes were being made, it was realized that microbial products, such as enzymes, had industrial potential. One of the earliest pioneers in this area was Jokichi Takamine who received a U.S. patent on the "Process of Making Diastatic Enzyme" in 1894. The enzyme preparation that was obtained by extracting moldy bran with water, precipitating the protein with alcohol and drying. . . "possesses the power of transforming starch into sugar for use in various industries. . ." This early work of Takamine and that of others has led to the development of today's $300,000,000 microbial enzyme industry.

The development of genetic engineering technology during the 1970s was the dawning of a new era in industrial microbiology. In the past to obtain a microorganism to carry out a specific function (for example, for use in cleaning up oil spilled in the ocean), one had to screen naturally occurring species or mutants for that ability, and often a species with the desired characteristics could not be found. With the advent of genetic engineering it has become possible to take desirable genetic characteristics from several organisms and combine them into a single microorganism tailor-made to carry out a specific function. Indeed a patent has been issued on a pseudomonad microorganism that was constructed to clean up oil spills. At this early stage of the game the future potential for tailoring microorganisms for specific functions seems to be limited only by the inventiveness of the industrial microbiologist.

During the past 100 years or so since Pasteur's time industrial microbiology has made tremendous advances. Entire industries, based on the ability of microbes to produce desired products, such as antibiotics, have developed. The rapid developments in the field of enzyme technology and the potential for development in the genetic engineering field promise that the field of industrial microbiology has a bright future.

Industrial microbiologists use the word *fermentation* differently than do microbiologists in general. To the classical microbiologist fermentation is an anaerobic process by which facultative or anaerobic microorganisms obtain energy from an organic compound, often a carbohydrate. Industrial microbiologists apply the term *fermentation* to any process by which a microorganism produces a desired product regardless of whether the process is aerobic or anaerobic. For example, industrial microbiologists refer to the production of penicillin by the aerobic fungus *Penicillium chrysogenum* as the penicillin fermentation.

ANTIBIOTICS

The primary industrial importance of antibiotics is in their use as chemotherapeutic agents in the treatment of diseases caused by microorganisms. Since the discovery of penicillin in 1928 several thousand antibiotics have been isolated, but less than 100 have proved useful as chemotherapeutic agents for human use. The specific effects of antibiotics on microorganisms in the treatment of human disease and their possible usefulness to the organisms producing them are discussed in Chapter 15.

Commercial production of antibiotics

Today penicillin and other antibiotics are produced predominately in huge batch cultures (Table 22-1). The producing organism is grown in **fermenters**, which are

Table 22-1 *1979 U.S. sales of antibiotics produced by American companies*

Category of antibiotic	**Market value (US dollars)**
Penicillins	$220,943,000
Other antibiotics such as cephalosporins, tetracycline, erythromycin, and streptomycin	$638,297,000
Antibiotic-sulfonamide combinations	$ 16,921,000
Topical antibiotics such as bacitracin or neomycin	$ 17,064,000

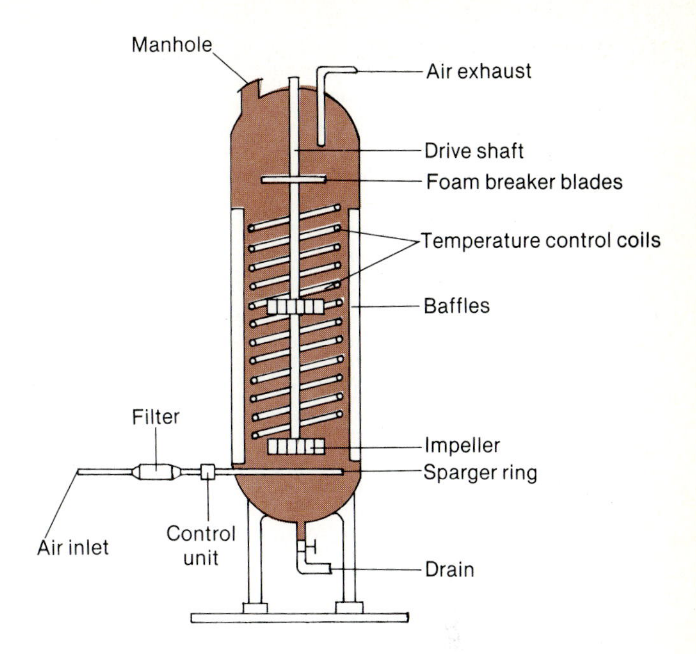

Figure 22-1 *Fermenter used in production of antibiotics.*

cylindrical and have capacities of 35,000 gallons or more (Figure 22-1). Most fermenters are made of stainless steel and contain large impellors for mixing the culture. Antibiotic producers are aerobic, so provision must be made for supplying large amounts of air to the culture. This is one of the biggest engineering problems to overcome because the energy input from stirring and aeration create unwanted heat. The fermentation vessel must contain coils through which a coolant or steam can circulate for temperature control and sterilization. Generally the large fermentations produce so much heat that the culture medium must be cooled to maintain the ideal temperatures for growth and antibiotic production. Often the optimum temperature for antibiotic production is different from the optimum growth temperature. Through the use of computers the temperature of the fermentation can be programmed such that both growth and antibiotic production occur at their optimum temperatures.

Inoculum production. Strains that have been selected for high antibiotic production tend to lose potency when carried for many generations on agar. For that reason stock cultures are made and stored away, either frozen or lyophilized.

In preparation for inoculating a large fermentation vessel, a frozen or lyophilized stock culture is retrieved and spread on solid growth medium. Growth from the solid medium is used to inoculate liquid medium in a shake flask. When the shake flask culture is near maturity, it is used to inoculate a larger culture (10 to 100 times the size of the first shake flask culture). The second culture is in turn incubated and used to inoculate larger and larger cultures until, finally, enough inoculum is available for a large production fermenter. The size and number of cultures needed is dictated by the size of the final fermenter vessel. Usually the volume of each culture is 10 to 100 times greater than that of the previous culture. The latter inoculum cultures are grown in stainless steel stirred-fermenters, similar to the ultimate fermentation vessel except smaller.

Strict sterile conditions and techniques are maintained throughout inoculum buildup to ensure that the culture does not become contaminated. The culture is also checked for contamination at each stage of scale-up. If contamination does occur, the process is usually aborted, since contamination adversely affects antibiotic production.

Culture medium. The composition of the medium can radically affect antibiotic production, and a medium that might be good for the production of one antibiotic may not be at all good for the production of another antibiotic. An important goal of industrial microbiologists is to design the ideal culture medium for the production of each antibiotic; however, the ideal culture medium in economic terms may not be the medium that generates the greatest yield of antibiotic. This is because costs must be considered. The best medium is the one where the cost per unit of antibiotic produced is least. A culture medium for antibiotic production, like any other culture medium must provide a source of carbon, energy, nitrogen, phosphorus, sulfur and trace elements. Most industrial culture media are complex, containing organic components, such as **corn steep liquor**, a byproduct of corn starch manufacture, as a source of growth factors and a source of sugar such as sucrose or a sugar derivative such as molasses as the carbon source. Meals such as soybean meal are often the main nitrogen source. Inorganic nutrients include calcium,

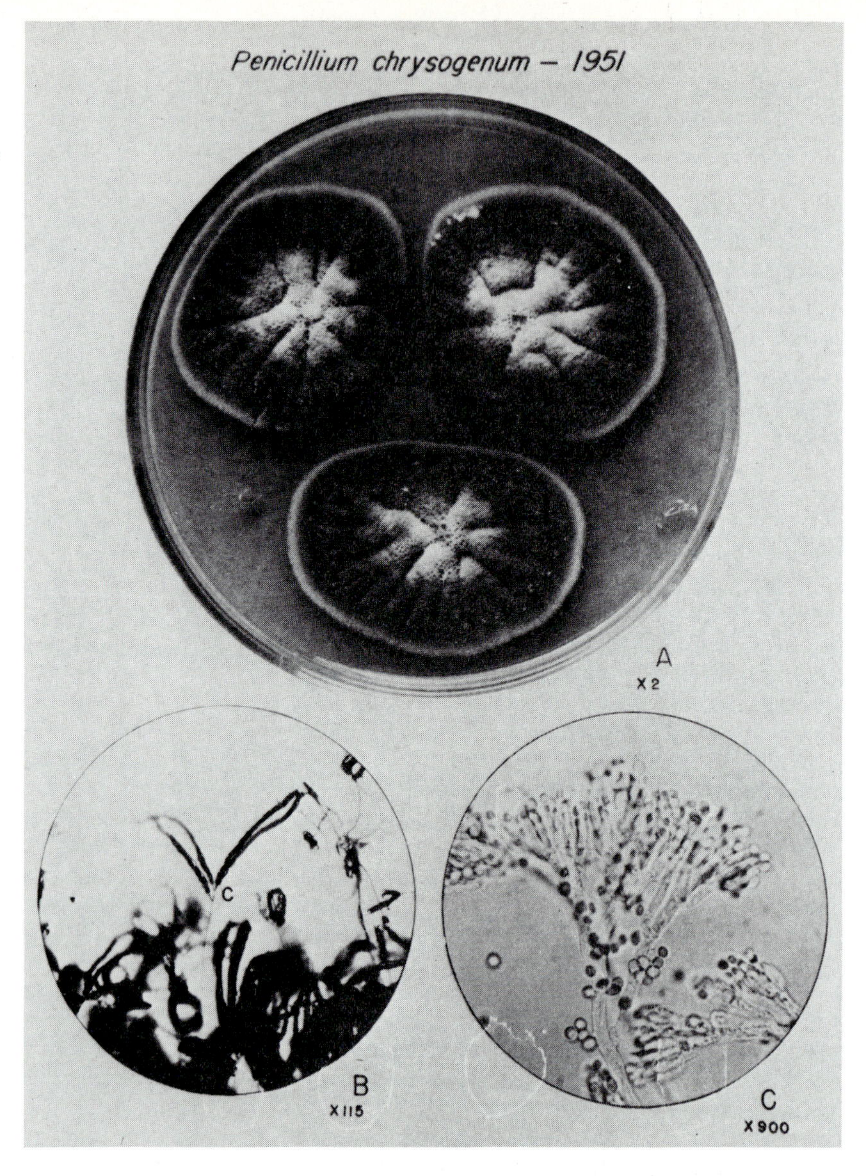

Figure 22-2 Penicillium chrysogenum *Thom NRRL 1951, strain isolated from moldy cantaloupe at Northern Regional Laboratory and parent source of high-yielding strains now used for commercial production of penicillin by submerged culture process.* ***A****, Isolated colonies showing cultural characteristics.* ***B****, Small portion of colony margin in* ***A*** *magnified showing character of single spore-bearing apparatus, or "penicillus."* ***C****, Single penicillus, highly magnified, showing details of structure and manner in which spores are borne.*

magnesium, iron, manganese, phosphate, potassium, and others. Often the medium is mixed in a concentrated form and sterilized. It is then pumped, with dilution, into the presterilized fermenter. During the fermentation process the pH of the culture medium is monitored and controlled by the addition of acid or base.

Fermentation process. Maximum antibiotic production occurs after growth of the producing organism has ceased, which means that antibiotic synthesis occurs in a period when the growth medium is usually depleted of nitrogen sources and limited for carbon and energy sources. Indeed it has been observed that if glucose is fed slowly to penicillin-producing cultures of *Penicillium chrysogenum,* penicillin production greatly increases (Figure 22-2). If glucose is fed to the culture too rapidly, penicillin synthesis is inhibited; therefore, the rate at which glucose is fed to the culture must be closely monitored. One of the important recent advances in fermentation technology is the use of computers to control various parameters of a fermentation such as the rate of glucose feeding, temperature, rate of aeration, and stirring rate.

An antibiotic fermentation typically lasts several days to about a week. Exactly when a fermentation is terminated depends on several variables; however, they are normally terminated when the rate of production of the particular antibiotic under synthesis falls below a certain level. This level is usually determined by a cost-benefit-type analysis. The pathway for penicillin biosynthesis is illustrated in Figure 22-3.

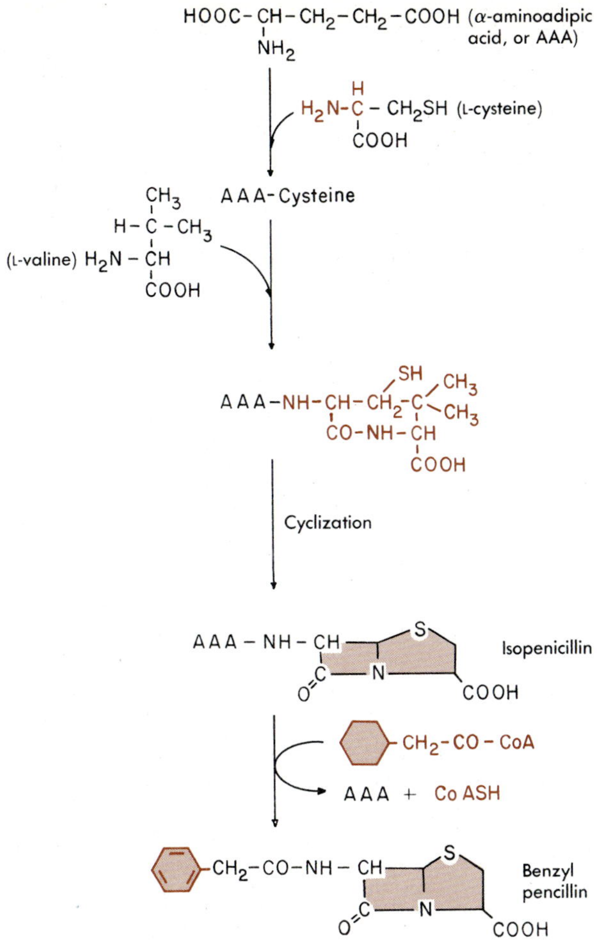

HOOC-CH-CH$_2$-CH$_2$-COOH (α-aminoadipic acid, or AAA)
NH$_2$

H$_2$N-C(H)-CH$_2$SH (L-cysteine)
COOH

AAA-Cysteine

(L-valine) H$_2$N-CH(COOH)-C(H)(CH$_3$)-CH$_3$

AAA-NH-CH-CH$_2$-SH ... C(CH$_3$)(CH$_3$)
CO-NH-CH
COOH

Cyclization

AAA-NH-CH ... S ... N ... C=O ... COOH Isopenicillin

CH$_2$-CO-CoA

AAA + CoASH

CH$_2$-CO-NH-CH ... S ... N ... C=O ... COOH Benzyl pencillin

Figure 22-3 *Pathway for synthesis of penicillin by* Penicillium chrysogenum.

Product recovery. Most antibiotics are excreted into the culture medium by the producing cells. To recover soluble antibiotics from the culture fluids, the cells and other particulate matter are first removed by filtration. The exact method for recovering an antibiotic from the clarified culture fluid varies from antibiotic to antibiotic. Some of the techniques used include solvent extraction, adsorption of the antibiotic to and elution from activated charcoal or diatomaceous earth, and selective precipitation of the antibiotic. Two or more of these and other techniques are usually required to obtain an antibiotic in pure form. The purification procedure is extremely important because antibiotics must contain a mininum of impurities if they are to be used on humans.

Strain improvement

When first isolated from nature, most antibiotic-producing microbes produce little antibiotic, usually only a few micrograms per milliliter of culture fluid. With so little antibiotic being produced, the production costs per unit of antibiotic produced are high. One way to reduce this cost per unit is to develop variants (mutants) of the organisms that produce greater amounts of the antibiotic. Programs for doing this are called strain improvement or strain development programs.

Strain improvement programs are based on improving antibiotic production by genetically altering the producing organism. This is done by subjecting a culture of the organisms to a mutagen such as ultraviolet light, x-rays, or a chemical mutagen so that a certain percentage of the cells in the culture are killed. Among those that survive the mutagenesis, a small percentage will be so mutated that they will produce more of the antibiotic than did the wild type strain, so-called **overproduction.** The difficulty with this is finding the few mutants that overproduce the antibiotic among the millions of survivors that do not. In the first strain improvement programs each survivor of mutagenesis has to be tested. This is done by spraying a suspension of an indicator organism onto agar medium containing isolated colonies of survivors. After growth of the indicator organism, colonies of survivors that overproduce the antibiotic give greater zones of inhibition of the indicator organism than do the colonies of the other survivors. Also, the survivors can be tested by individually culturing each survivor in liquid medium and testing it for amount of antibiotic produced. Often it is necessary to test thousands of survivors of a mutagenesis to find a single overproducing variant. As can be imagined this is laborious and time consuming.

Over the years correlations between certain types of mutations and increased antibiotic production have been found. For example, it has been observed that mutants of *Penicillium chrysogenum* that overproduce penicillin also have **altered colony color or morphology.** Conversely, not all mutants with altered colony color or morphology overproduce the antibiotic, but the percent of them that do is usually much higher than that among total survivors. Thus by testing survivors with altered colony morphology, the chance of finding an overproducer is greatly increased. Another mutant phenotype that has been correlated with antibiotic overproduction is **auxotrophy.** Certain mutations causing amino acid auxotrophy tend to cause antibiotic overproduction. Again not all such mutations affect antibiotic synthesis.

Many antibiotics inhibit the growth of the organism that produces it. It has been found that mutants resistant to the antibiotic they produce tend to overproduce the antibiotic. Today one of the first things that is done in strain improvement programs is to isolate mutants resistant to the antibiotic they produce.

Naturally occurring antibiotic resistance in bacteria is a result of genes encoding proteins that block antibiotic transport, modify the host-target site for the antibiotic, or inactivate the antibiotic (see Chapter 15 for discussion). For example, *Streptomyces erythreus,* the producer of the antibiotic erythromycin, modifies its ribosomes so that the synthesized antibiotic has no affinity for them. These antibiotic resistance genes have been identified and isolated in a number of cases.

In a strain improvement program when an overproducing mutant is isolated, it is subjected to mutagenesis. The survivors of this mutagenesis are tested for variants that produce more antibiotic than did the original mutant. If one is found, it will be subjected to mutagenesis, and the whole process repeated. A strain improvement using *Penicillium chrysogenum* has been carried through 20 rounds of mutagenesis. A variant was isolated from the last mutagenesis that produced more than 1000 times as much penicillin as did the original strain.

Search for new antibiotics

Pathogenic microorganisms continue to develop resistance to currently used antibiotics; thus pharmaceutical companies are constantly searching for new antibiotics. Their search takes at least two directions. There is the never ending search for organisms that produce novel antibiotics. This is done by examining samples, most commonly from soil, for microbes that produce antibiotics. In the initial screening a soil sample is diluted sufficiently to give isolated colonies and spread on solid growth medium. This growth medium is incubated to allow colony formation of the microbes present in the sample of soil. After colonies have formed, the surface of the medium is carefully sprayed with a suspension of an indicator organism such as *Staphylococcus aureus* or *Bacillus subtilis* and reincubated. During this incubation period the indicator organism will form a "lawn" of microbial growth covering the entire surface of the solid medium except around those colonies that produced, during the first incubation, an antibiotic that inhibits growth of the indicator organism. The antibiotic producers are isolated in pure form and tested against a large battery of indicator organisms. By comparing the responses of these indicator strains to the unknown antibiotic with their responses to known antibiotics, one can deduce with accuracy whethei the unknown antibiotic has been previously identified. As more and more antibiotics are identified, it becomes more and more difficult to find new ones. To increase the probability of finding new antibiotics, more exotic samples must be screened and more imaginative new screening methods must be developed. This approach is expensive and other methods are being evaluated.

One mechanism for obtaining therapeutically active antibiotics is to select mutants of the parent antibiotic-producing microorganism. This procedure can result in the production of an antibiotic whose structure is similar to the antibiotic of the parent. Sometimes the derivative is more active than the parent molecule. A technique called **cosynthesis** has also been used to isolate modified antibiotics. In this technique one mutant is selected that accumulates an intermediate of the antibiotic. A second mutant is selected that has a mutation at a step before the one found in the first mutant. When the two mutants are cultured together the intermediate synthesized by the first mutant may be transformed by the second mutant into a new end-product with better therapeutic qualities. An important technique used successfully in the development of new antibiotics is to chemically alter the existing ones. These chemically altered derivatives are called **semisynthetic.** A strain is used that produces the nucleus of the antibiotic. Penicillin G or benzyl penicillin (Figure 22-4), for example, is produced by *Penicillium chrysogenum* when phenylacetic acid ($C_6H_5 \cdot CH_2 \cdot COOH$) is added to the fermentation medium. If phenylacetic acid is withheld from the medium the penicillin nucleus, 6-amino penicillanic acid (6-APA) is produced. Once the 6-APA is isolated, various side chains can be synthetically added (the process is called **acylation**) to the molecule. Most penicillins and cephalosporins used today are semisynthetics. If biological methods were the only way of obtaining antibiotics, the various penicillins and cephalosporins currently available could not have been obtained. Genetic engineering promises to obtain new or modified antibiotics by combining genes of different biosynthetic pathways in one organism. Genetic engineering is discussed in Chapter 11.

Concept Check

1. **Why does the term fermentation have different meanings among microbiologists?**
2. **What are some of the problems that must be addressed to make commercial production of antibiotics economically feasible?**
3. **What are the major schemes being used to develop new antibiotics? Which holds the most promise and why?**

ENZYMES

Only a few of the thousands of enzymes produced by microorganisms have widespread commercial application (Table 22-2). The number of microbial species producing enzymes of commercial importance is limited and includes both fungi and bacteria. From an industrial

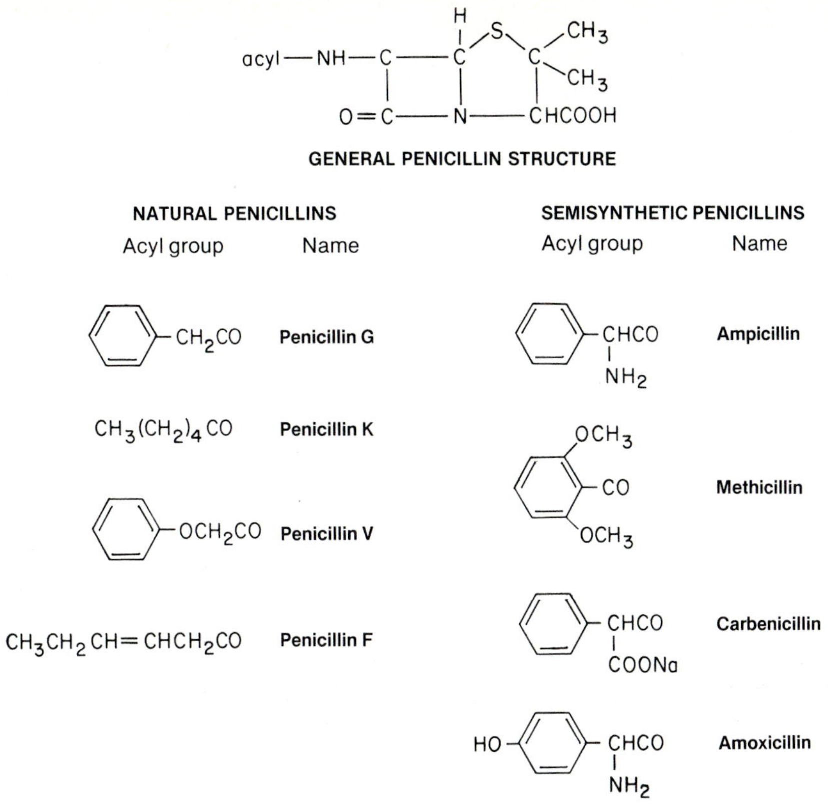

acyl—NH—C——C(H)—S—C(CH3)(CH3)
O=C——N——CHCOOH

GENERAL PENICILLIN STRUCTURE

NATURAL PENICILLINS

Acyl group	Name
C_6H_5-CH_2CO	**Penicillin G**
$CH_3(CH_2)_4CO$	**Penicillin K**
C_6H_5-OCH_2CO	**Penicillin V**
$CH_3CH_2CH{=}CHCH_2CO$	**Penicillin F**

SEMISYNTHETIC PENICILLINS

Acyl group	Name
C_6H_5-$CH(NH_2)CO$	**Ampicillin**
2,6-$(OCH_3)_2C_6H_3$-CO	**Methicillin**
C_6H_5-$CH(COONa)CO$	**Carbenicillin**
HO-C_6H_4-$CH(NH_2)CO$	**Amoxicillin**

Figure 22-4 *Structure of molecules of penicillin and some of its derivatives.*

Table 22-2 *Worldwide sales of enzymes in 1980*

Enzyme	Amount (pounds)	Sales (U.S. dollars)
Bacterial protease	1,060,000	\$66,000,000
Glucamylase	700,000	\$36,000,000
α-Amylase	640,000	\$12,000,000
Glucose isomerase	140,000	\$56,000,000
Rennin	52,000	\$64,000,000

point of view there are distinct advantages to having microbes produce enzymes. First, microbes are generally easy to grow, and with the huge fermenters already in existence, large quantities can be grown. Second, in contrast to plants and animals, microbes are easy to manipulate genetically for strain improvement. Third, some of the enzymes are excreted into the medium where they are easily separated from the cells.

Industrially useful enzymes

α-Amylase. α-Amylase is an enzyme that attacks internal α-1,4-glucosidic links in starch molecules, producing dextrins, oligosaccharides, and maltose as the major products. The principal microorganisms used for the commercial production of α-amylase are the bacterium *Bacillus subtilis* and the fungi *Aspergillus oryzae* and *Aspergillus niger*. The major uses of α-amylase follow.

Breadmaking. Yeast lack amylases for the hydrolysis of starch, therefore, α-amylase is often added to bread dough to promote the breakdown of sufficient starch to provide sugar, which yeast can convert to carbon dioxide and cause the dough to rise. The amylase must be used judiciously to prevent excessive starch breakdown and the consequent reduction in bread quality.

Brewing. Many American beer makers use adjuncts such as corn or rice to augment the starch content of malt. To aid the amylases already present in malt in the breakdown of the adjunct starches, α-amylase is often added. *Bacillus subtilis* α-amylase is frequently used for this purpose because of its great heat stability.

Alcohol production. α-Amylase is used in the distilling industry to reduce the viscosity of the gelatinized starch produced after the cooking of grains used in the production of whiskey and other distilled alcoholic beverages. Here as in the brewing industry heat stable amylases are desired. α-Amylase is also used in the production of gasohol from corn starch. Corn is ground into a meal that is cooked to gelatinize the starch, which is in turn treated with α-amylase to reduce the viscosity of the starch and to provide yeast with fermentable sugars.

Sugar syrups. One of the major uses of α-amylase is in the hydrolysis of starch for the production of sugar syrups. In recent years the demand, most notably by the soft drink industry, for sugar syrups with a high fructose content has increased tremendously. Production of these syrups from corn starch has become a multibillion-dollar industry. The process involves several enzymes including α-amylase, which convert starch to a mixture of oligosaccharides and maltose (Figure 22-5) These are converted to glucose by the enzyme glucamylase. The glucose is isomerized to fructose by glucose isomerase. The last enzyme does not convert all the glucose to fructose, so the final product contains a mixture of glucose and fructose. Since fructose is considerably sweeter than glucose, it is desirable to have as much glucose converted to fructose as possible.

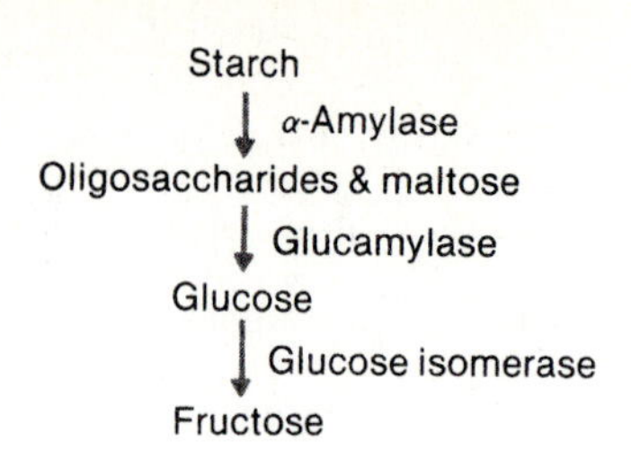

Figure 22-5 *Enzymatic conversion of starch to fructose.*

Additional uses. α-Amylase has been used in spot removers. Instant hot breakfast cereals such as oatmeal are treated with α-amylase, which partially degrades the starch present in the oat flakes and consequently reduces cooking time. α-Amylase has been used to aid in the removal of silver from used photographic film, and it is also used in the desizing of textiles.

Proteases. Proteases are enzymes that hydrolyze the peptide bonds in proteins. Those proteases that only hydrolyze internal peptide bonds are called endoproteases (or endopeptidases), and those that hydrolyse terminal peptide bonds are called exoproteases. Certain proteases are quite specific and only hydrolyze peptide bonds between specific amino acids; consequently, these proteases only partially degrade proteins to short peptides or oligopeptides. Industrially proteases have many uses, several of which follow.

Baking industry. The gluten proteins in wheat flour are important in determining dough mixing time and bread loaf quality; however, the quantity of these proteins varies considerably from one batch of flour to another. Bakers have found that careful addition of protease will reduce dough mixing time and improve loaf quality; however, overaddition of protease will reduce loaf quality. It has been estimated that over half the bread baked in the United States has protease added to the dough. The protease produced by *Aspergillus oryzae* is most commonly used in the baking industry.

Meat tenderizer. Proteases that only partially degrade protein are used to tenderize meat. The tenderizers can be applied to the surface of a piece of meat shortly before it is cooked. Proteases in the tenderizer preparation diffuse into the meat and cleave protein molecules in the muscle tissue. This partial degradation of the muscle protein tenderizes the meat without seriously affecting its quality. There are several tenderizing preparations available to the consumer most of which contain about 2% commercial **papain** (a proteolytic enzyme found in fruit of papaya tree) or 5% fungal protease as the active ingredient. The proteases in the meat are inactivated by heat during cooking and by stomach acids. A second method of tenderization that is reportedly used on a fairly large scale is the direct injection of a proteolytic enzyme solution into the vascular system of animals shortly before slaughter. In this case the proteases are distributed throughout the animal's body, thus giving a more uniform tenderization of large pieces of meat than can be achieved with surface application.

Laundry products. A major use of proteases is in laundry products. The bacterial proteases, which are active in alkaline conditions (so called **alkaline proteases**), are the most widely used, and the most commonly used alkaline protease is produced by *Bacillus subtilis.* The protease attacks proteinaceous material such as milk, egg, or blood that may soil clothes and facilitates its removal. Initially when proteases were incorporated into laundry powders, there were problems especially among workers handling the proteases in the soap factories. Apparently dust containing the proteases was inhaled by the workers. The inhaled proteases irritated the mucous membranes of the respiratory tract and also caused allergic reactions. These problems apparently have been controlled by new formulations of the enzyme preparation that eliminate dust formation.

Additional uses. Proteases are used extensively in the tanning of leather to remove proteinacious material from the fresh skin and in the softening of skins. Proteases are used as digestive aids as well as in the chillproofing of beer. A haze can form in beer on chilling as a result of precipitation of protein. The formation of this haze is prevented by treating the beer with protease to remove the protein before it is chilled.

Rennin. Rennin is a proteolytic enzyme found in the stomach of suckling calves. It causes the casein in milk to form curds, which is one of the first steps in cheese manufacture. Recently the use of bovine rennin in cheese making has been partially replaced by the use of

Plate 20 *Luminous bacteria in broth in room light.*

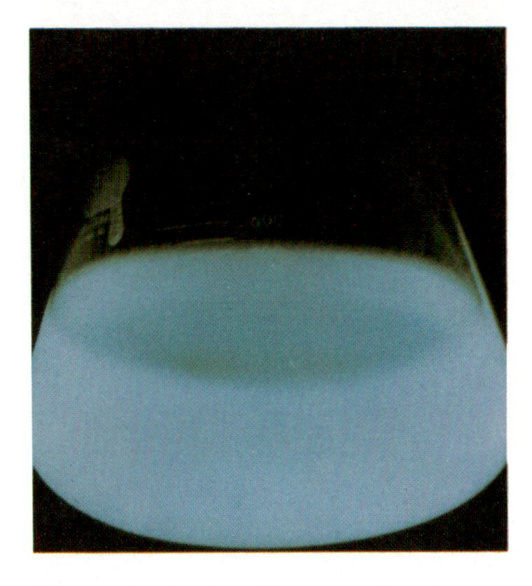

Plate 21 *Luminous bacteria in broth in its own light.*

Plate 22 *Stromatolites in intertidal zone at Shark Bay, Western Australia. Largest structures are about 1.5 meters across. Formed during time of slightly higher sea level, perhaps 1000 to 2000 years ago; now exposed to atmosphere and cemented by form of calcium carbonate ($CaCO_3$) known as aragonite, or rocks.*

Plate 23 *Crown gall caused by* Agrobacterium tumefaciens. *Galls appear at soil line of plant.*

Plate 24 *Pinot Noir with* Botrytis, *also called "noble rot." This fungus causes increase in sugar levels of grapes, facilitating production of sweet dessert wines.*

Plate 25 *Mottled appearance of field tobacco infected by tobacco mosaic virus.*

Plate 26 *Spoiled milk.*

Plate 27 *Blue cheese.*

Plate 28 *Spoiled meat.*

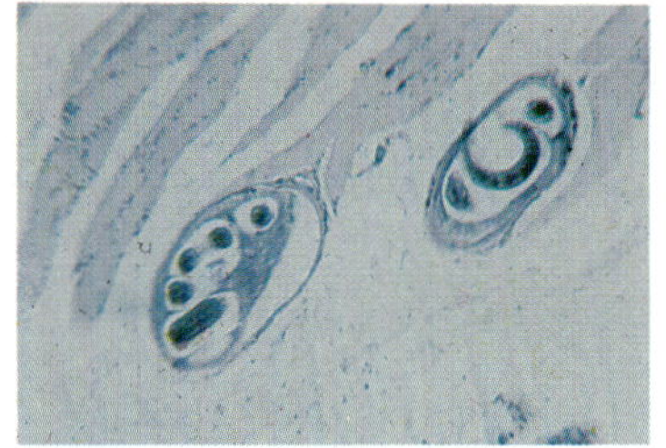

Plate 29 *Trichinella spiralis larvae encysted in muscle.*

Plate 30 *Spoiled fruit and vegetables.*

$$2\ \text{Glucose} + 2O_2 \xrightarrow{\text{Glucose oxidase}} 2\ \text{Gluconic acid} + 2H_2O_2$$

$$2H_2O_2 \xrightarrow{\text{Catalase}} 2H_2O + O_2$$

$$\text{Net:}\quad 2\ \text{Glucose} + O_2 \longrightarrow 2\ \text{Gluconic acid} + 2H_2O$$

Figure 22-6 *Reactions by which glucose oxidase and catalase remove atmospheric oxygen.*

renninlike proteases produced by the fungi *Mucor pusillus* and *Mucor miehei.*

Glucose oxidase. Glucose oxidase is an enzyme that oxidizes glucose to gluconic acid with the concomitant generation of hydrogen peroxide. Its primary uses are in the food industry where it is used to remove from dried eggs and other dried products glucose, which forms undesirable crystals during the drying process. In combination with catalase, glucose oxidase is used to remove oxygen from mayonnaise, fruit juices, and other preserved products to prevent their deterioration and discoloration. Figure 22-6 shows the reactions.

Industrially glucose oxidase is produced by *Aspergillus niger.* It is an intracellular enzyme, which means that the *A. niger* cells must be broken to release the enzyme before it can be purified. Because cell breakage releases all the cellular enzymes, purification of glucose oxidase is a complex process, involving many steps.

Miscellaneous enzymes. **Pectinases** are enzymes that breakdown pectins found in plant tissue. They are used commercially for clarifying fruit juices. The protease **streptokinase** and the nuclease **streptodornase** are both produced by *Streptococcus hemolyticus* and are used medically in the cleansing of wounds. Streptokinase can convert plasminogen, the proenzyme of human serum, to plasmin. Plasmin can digest fibrin, which is important in clot formation. Preventing clots is important particulary during surgeries such as those involving the heart. Streptodornase liquifies the nucleoproteins of dead cells or pus and facilitates the drainage of a traumatized area. Streptodornase has been used in dentistry, sometimes in conjunction with streptokinase. **Invertase** from yeast is used in the confectionary industry to prevent the crystallization of sucrose and to make soft-centered candies. It hydrolyses sucrose to glucose and fructose, which is sweeter than sucrose.

Enzyme production

Today essentially all commercially important bacterial enzymes are produced by **submerged culture** in large fermenters by procedures essentially similar to those used for antibiotic production. Most fungal enzymes are also produced by submerged culture; however, some fungal amylase is produced by the **Koji process** in Japan. In the Koji process for amylase production moist sterile cereal bran (either wheat or rice) is inoculated with spores of *Aspergillus oryzae.* The inoculated bran is either spread in shallow trays or placed in rows in sterile, constant-temperature rooms. The spores germinate and the fungus begins to grow, producing mycelia, which permeate the bran. The fungus produces extracellular amylases that diffuse into the bran, breaking down the starch for use by the mycelia. Every so often the moldy bran is turned to maintain relatively constant conditions throughout the heap. After several days to a week, the enzyme is removed from the moldy bran by water extraction and is further purified or concentrated.

Immobilized enzymes

Until recently it was practical to use enzymes only once because of the difficulty involved in recovering the enzyme from the reaction mixture where it is employed. To reduce the costs connected with their use, ways were sought to make enzymes recyclable. In recent years this has resulted in the development of techniques that allow easy recovery of enzymes from reaction mixtures, thus permitting their reuse. This technology involves affixing enzymes to inert, insoluble matrices such as small polyacrylamide beads, tiny glass beads, or cellulose particles. Enzymes affixed to insoluble supports or encapsulated within a membrane are called **immobilized enzymes** (Figure 22-7). Another method of enzyme immobilization is to mix the purified enzyme with a cross-linking agent such as glutaraldehyde. The cross-linking agent binds enzyme molecules together, producing insoluble aggregates. These aggregates retain enzymatic activity. A third method is to entrap enzyme molecules within the matrix of a gel such as agarose or polyacrylamide.

Currently there are several methods of employing immobilized enzymes (Figure 22-8). The most widely used is the **batch method**, in which an immobilized enzyme is mixed with a batch of substrate. While this mixture is stirred, the enzyme reacts with the substrate, converting it to product. When the reaction is complete, the immobilized enzyme is removed from the mixture by filtration or centrifugation, leaving a clean solution of product that can be further purified if necessary. The immobilized enzyme after filtraton or centrifugation can

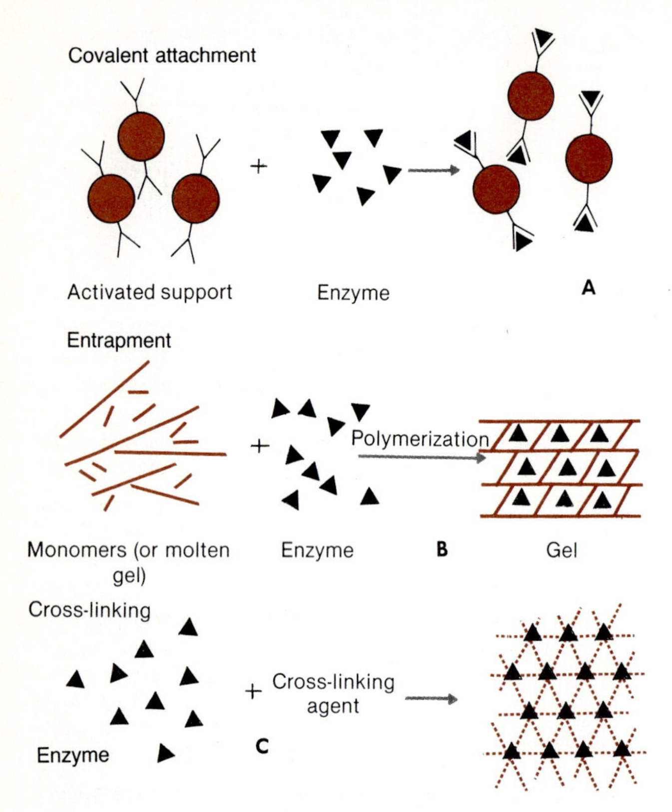

Figure 22-7 *Methods for enzyme immobilization, which still maintain availability of active sites.* ***A,*** *Soluble enzyme molecules are covalently attached to insoluble support, such as tiny glass, polyacrylamide, or cellulose beads or iron filings.* ***B,*** *Soluble enzyme is mixed with acrylamide monomers or molten gel such as starch or agar. When acrylamide is polymerized or gel is set, enzyme becomes entrapped within matrix of resulting gel. For employment, gel is crushed to small particles.* ***C,*** *Soluble enzyme molecules are cross-linked to each other, forming insoluble aggregates that still retain enzymatic activity.*

be washed with buffer and reused. The batch method is relatively inexpensive, efficient, easily scaled up to large batches, and technologically simple.

In the **packed bed method**, a suspension of an immobilized enzyme is placed in a column (usually of glass) that is plugged at each end. The immobilized enzyme is allowed to settle and form a packed bed. Substrate solution is pumped into the column through an inlet pore that is covered with a screen to prevent escape of the immobilized enzyme through the bottom plug. As the substrate percolates up through the immobilized enzyme bed, it encounters enzyme molecules and is converted into product that eventually emerges from the outlet pore, which is also covered with a screen in the upper plug of the column. An advantage of this method is that it can be used as a continuous process in which substrate is continuously pumped into the bottom of the column, and product is continuously exiting at the top of the column. A disadvantage of this method is that the packed bed tends to become clogged, preventing efficient flow of the substrate through the column.

The **fluidized bed method** was developed to overcome the clogging problem of the packed bed method. In this method immobilized enzyme is placed in a cylinder that has a larger diameter at the top than at the bottom. The substrate solution is pumped into the bottom of the cylinder at a sufficient rate to prevent the immobilized enzyme particles from becoming packed. As the diameter of the cylinder increases toward the top, the flow rate of the solution will decrease, allowing the enzyme particles to settle so that no enzyme particles reach the product outlet on the top of the cylinder.

A fourth method is being developed in which enzyme is affixed to the inside of a tube. As a substrate solution is passed through the tube, the enzyme converts the substrate molecules to product so that a solu-

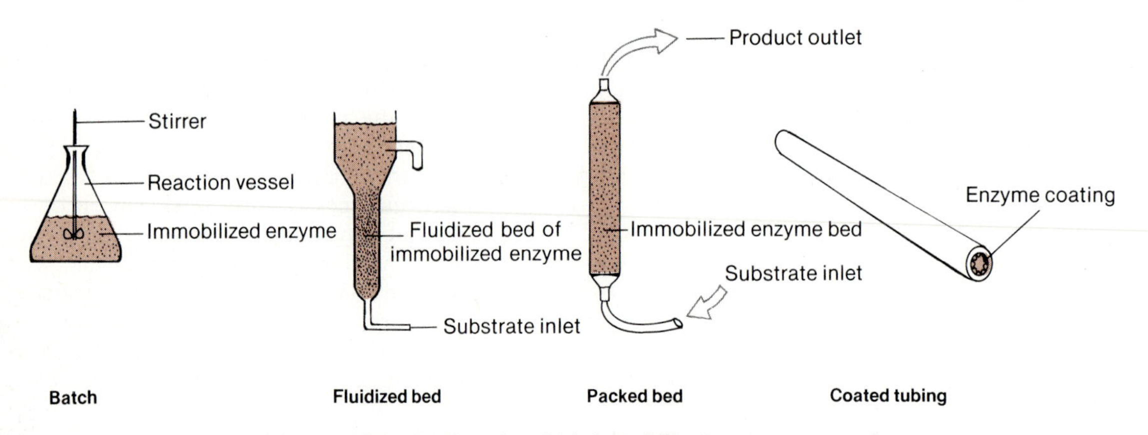

Figure 22-8 *Methods by which immobilized enzymes are used.*

Table 22-3 *Worldwide sales of amino acids in 1980**

Amino acid	Amount (pounds)	Sales (U.S. dollars)
Glutamic acid	600,000,000	$1,080,000,000
Methionine	210,000,000	$ 246,000,000
Lysine	100,000,000	$ 258,000,000

*Industrially glutamic acid is made entirely by fermentation. Approximately 80% of the commercial lysine is made by fermentation. Methionine and the remaining lysine are synthesized chemically.

tion of product emerges from the other end of the tube. Potentially, each of the latter three methods can be used as a continuous process. Such continuous processes have been maintained for several months on an experimental scale.

There are several important advantages to the use of immobilized enzymes. It has been found that the immobilization of certain enzymes also stabilizes them. Since immobilized enzymes are so easy to remove from the reaction mixture, purification of the reaction product is facilitated. The potential of making a continuous process is an important economic advantage in that it provides a continuous supply of product.

AMINO ACID PRODUCTION

The production of amino acids by microbial fermentation has become an important industrial process in recent years. Of the three most important commercial amino acids—glutamic acid, lysine, and methionine—glutamic acid and over 80% of the lysine are produced predominantly by fermentation.

Until the late 1950s most amino acids used commercially were obained by hydrolysis of plant proteins such as wheat gluten and soybean protein. At that time the amino acid in greatest demand was **glutamic acid**, which was used as a flavoring agent in Oriental and to a lesser extent other cuisines. The high cost of glutamic acid production prompted Japanese researchers to search for alternative methods of producing amino acids. In 1957 Kinoshita and his colleagues reported the isolation of a bacterium that excreted glutamic acid in large amounts. The organism was originally called *Micrococcus glutamicus* but has subsequently been renamed *Corynebacterium glutamicum*. Shortly after its discovery it was successfully used to produce glutamic acid on an industrial scale. Today the production of amino acids by microbial fermentation is a billion-dollar industry (Table 22-3), especially in Japan where glutamic acid, lysine, proline, and valine are produced by microbial fermentation.

Glutamic acid: case study

When Kinoshita began his search for microorganisms that excreted glutamic acid, he knew that most bacteria synthesize glutamic acid but that few excrete it in significant amounts. It was therefore important for him to develop a simple method of screening large numbers of bacterial isolates for their ability to excrete glutamic acid. In developing such a screening procedure, he took advantage of the fact that the bacterium *Leuconostoc mesenteroides* must be supplied with glutamic acid to grow. The screening procedure that was developed was beautifully simple. Many bacterial sources were screened in the search for glutamic acid excreters including soil, sewage, water, and strains from stock cultures. The general procedure that was developed was to take a sample such as soil, dilute it, and plate it for isolated colonies on solid medium lacking glutamic acid. The plates were incubated to allow well-defined colonies to form. These colonies were replicated onto a second plate by the replica-plating technique devised by Lederberg, which involves pressing a piece of sterile velveteen cloth onto the surface of the plate containing the colonies to be replicated (see Chapter 11). In the Kinoshita screening procedure, once the original plate has been replicated, the bacteria remaining on it were killed by ultraviolet irradiation. It was then overlaid with cooled, molten agar growth medium that lacked glutamic acid but had been seeded with *Leuconostoc mesenteroides*. After the seeded agar overlay had solidified, the plate was reincubated. If any of the bacteria in the original colonies had excreted glutamic acid, it would diffuse up into the overlay, providing *L. mesenteroides* with glutamic acid and allowing it to grow. After sufficient incubation of the overlay plates there would be growth of the *L. mesenteroides* cells above colonies that had excreted glutamic acid. There would be no growth of the *L. mesenteroides* above the colonies where no glutamic acid was excreted. The extent of *L. mesenteroides* growth was directly proportional to the amount of glutamic acid that had been excreted by the colony below it. Using this screening procedure, not only could Kinoshita identify glutamic acid excreters, but he could get an idea of how much glutamic acid they excreted.

This procedure allowed the screening of thousands of microorganisms from nature and from culture collections. Many of the organisms were found to excrete small amounts of glutamic acid but not enough for industrial purposes. The screening program did, however, turn up *Corynebacterium glutamicum*, which produces large amounts of glutamic acid.

Studies on glutamic acid production by *C. glutamicum* revealed that it required biotin for growth. When

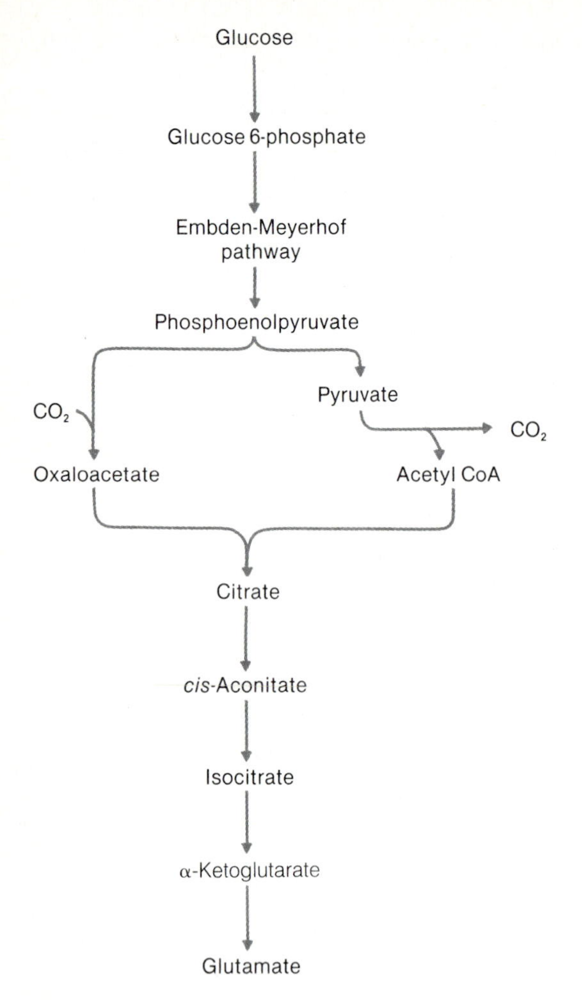

Figure 22-9 *Glutamic acid pathway of* **Corynebacterium glutamicum.**

Figure 22-10 *Outdoor fermentation tanks used by the Kyowa Hakko Kogyo Co. Ltd., of Japan for production of the amino acids, glutamic acid, and lysine. This picture shows seven of 20 such fermentation tanks at this site that are employed by the company in their annual production of at least 20,000 tons of glutamic acid and 10,000 tons of lysine. Most glutamic acid is used for production of monosodium glutamate (MSG), a flavor enhancer.*

there were growth-limiting amounts of biotin in the growth medium, glutamic acid was excreted; however, with excess biotin in the growth medium, little or no glutamic acid would be excreted. This proved to be an important factor in developing the culture medium for glutamic acid production. Glucose is a precursor in the synthesis of glutamic acid (Figure 22-9) and growth substrates containing glucose were preferred for maximum glutamic acid production. Many cheap glucose-containing growth substrates, such as molasses, contained sufficient biotin to prevent glutamic acid excretion. This meant that the cheapest growth substrate could not be used. To control the biotin concentration, more expensive synthetic media had to be developed, consequently adding to the cost of glutamic acid production; yet it was still cheaper to produce it by microbial fermentation than by extraction from wheat or soy beans.

The basis for the biotin effect is not well understood but there is evidence that suggests that limiting biotin affects the chemical composition of the cytoplasmic membrane. Membrane synthesized during biotin limitation permits glutamic acid to be excreted from the cell.

Industrially, glutamic acid fermentation is carried out in submerged culture in large fermenters (Figure 22-10) similar to those used in antibiotic production. During fermentation 60% or more of the glucose or other saccharide growth substrate is converted to glutamic acid. In addition to the biotin concentration in the growth medium, several factors can dramatically affect glutamic

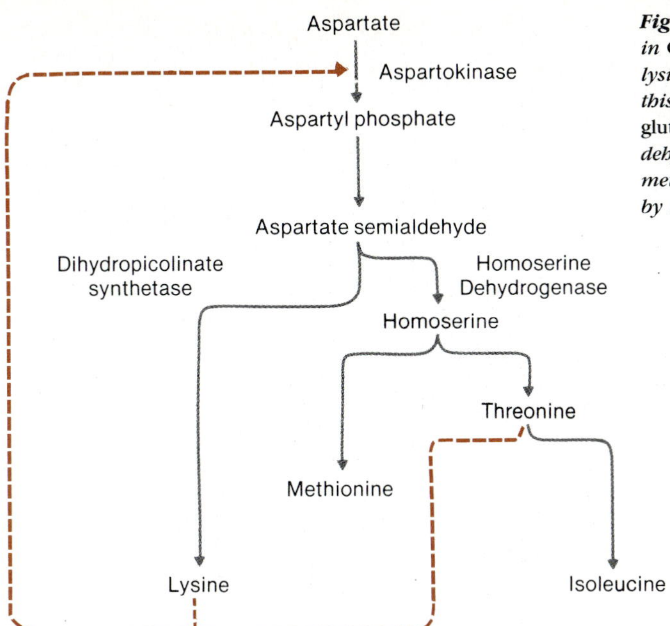

Figure 22-11 *Branched pathway leading to lysine synthesis in* C. glutamicum. *Dashed lines leading from threonine and lysine to aspartokinase depict concerted feedback inhibition of this enzyme by these two amino acids. In mutant of* C. glutamicum *used in lysine production, enzyme homoserine dehydrogenase is missing; thus required amino acids, methionine, threonine, and isoleucine, cannot be synthesized by this mutant.*

acid production. In the course of a fermentation, if the cells do not receive adequate oxygen, little glutamic acid is produced; instead, lactic and succinic acids accumulate. With too much oxygen, α-ketoglutaric acid is produced instead of glutamic acid. The pH of the fermentation must be carefully controlled; if the pH drops too low, glutamic acid production will cease. Another important variable that must be controlled is temperature. The optimum temperature for glutamic acid production ranges from 30° to 35° C.

Lysine

Lysine is an essential amino acid and as such is an important food additive. Bread is fortified with lysine in some Third World countries, and in the United Staes certain breakfast cereals are fortified with this amino acid. Lysine is also used as a supplement in animal food. Because of these uses the demand for lysine is strong; and in excess of 50,000 tons is produced annually, predominantly by Japanese companies.

The microorganism that is used to produce lysine is a mutant of the glutamic acid producer *Corynebacterium glutamicum*. Lysine is a product of a branched metabolic pathway that, on the other branch, also produces methionine, threonine, and isoleucine (Figure 22-11). The concerted action of both lysine and threonine is required to regulate lysine formation in *C. glutamicum*. When both of these amino acids are present in excess, this pathway is shut down and no lysine or threonine is synthesized. When either lysine or threonine is absent or is present in a low amount, the pathway continues to function no matter how much of the other is present. This characteristic was taken advantage of for the development of a strain of *C. glutamicum* that would excrete lysine. To keep the concentration of threonine low, a mutant of *C. glutamicum* was isolated that could not make homoserine from aspartic semialdehyde. This mutant had the ability to make lysine but was unable to make methionine, threonine, and isoleucine; therefore it could not grow unless homoserine or methionine and threonine were added to the culture medium. The isoleucine requirement was met by the conversion of threonine to isoleucine (Figure 22-11). When either homoserine or methionine and threonine were added in large amounts, no lysine was excreted by the mutant; but when the medium contained a growth-limiting amount of threonine and no homoserine even in the presence of excess methionine and isoleucine, the mutant would excrete large amounts of lysine. Under these conditions of limited threonine the regulatory controls of the pathway perceived a need for threonine and continued to function in an effort to supply the needed threonine. The mutation prevented any threonine synthesis and caused all the aspartic semialdehyde to be shunted into lysine synthesis. In this way vast excesses of lysine are produced. Chapter 13 discusses these regulatory mechanisms

The strain of *C. glutamicum* used for lysine production retains the biotin requirement of the parent. Under conditions of biotin excess (30μg/liter) lysine is excreted, but when biotin is limited in the culture medium, glutamic acid is excreted instead of lysine, even by the mutant.

Cane blackstrap molasses is the most commonly used growth substrate in the industrial fermentation of lysine. Yields of 3 g and more of lysine per 10 g of substrate used have been reported.

Concept Check

1. What are the major enzymes commercially produced from microoorganisms and their commercial value?
2. What techniques are used to recycle enzymes used in industrial processes? How does each method work?
3. What method was used by Kinoshita for screening bacterial isolates that excrete glutamic acid?
4. What are the most important industrial amino acids, and what is their commercial value?

ORGANIC ACIDS AND SOLVENTS

In the past, many organic acids and solvents were produced commercially by microbial fermentation. In the early part of this century **butanol** and **acetone** produced in the United States was by microbial fermentation. Although butanol continued to be produced for several years by microbial fermentation, chemical synthesis eventually replaced fermentation.

The early production of butanol and acetone by microbial fermentation led to the development of techniques for production of several organic acids by microbial fermentation, some of which continue to be so produced today.

Citric acid

Citric acid (Figure 22-12) is used in large volumes as an industrial chemical. In 1980 in the United States alone 180,000 metric tons of citric acid were produced by two companies, Miles Laboratories and Pfizer. It is produced by cultures of *Aspergillus niger*. At 1980 prices the value of the citric acid produced in the United States was approximately $100,000,000.

Citric acid is a strong **chelating agent** of cations in addition to being a strong organic acid. Based on these two properties it has many applications. In the food industry it is used as an **acidulant** in soft drinks, jams, jellies, candies, desserts, frozen fruits, and wines. In the pharmaceutical industry it is used in effervescent products such as Alka-Selzer. Another reason it is so widely used in both the food and pharmaceutical industries is that it is rapidly and completely metabolized in the body.

There are several uses of citric acid based on its chelating properties, including electroplating of metals, leather tanning, and reactivation of clogged oil wells. One reason oil wells become unproductive is that iron deposits clog the sands around the well mouth, preventing passage of the oil. The iron deposits can be removed by pumping into the well a solution of citric acid, which sequesters iron ions causing solubilization of the deposits.

Historically citric acid was produced from citrus fruits such as lemons. In the late nineteenth and early twentieth century, Italy, because of its capacity to produce lemons, had a virtual monopoly on citric acid production and consequently demanded an artificially high price for it. This monopoly was broken in 1923 when a microbial process for citric production was developed in the United States. Since that time the microbial process has become the dominant method for citric acid production. *A. niger,* which is the producing organism, is a highly aerobic filamentous fungus and must be continually supplied with air (oxygen). Consequently the early production methods relied on growing *A. niger* in shallow pans containing liquid growth medium. The *A. niger* would form floating mats on the surface of the liquid, and as the fungus extracted nutrients from the growth medium, it would excrete citric acid into it. Glucose, the preferred substrate for citric acid production, was provided by adding molasses to the growth medium. This method is still used by Pfizer today.

Citric acid is also produced by submerged culture in large fermenters of up to 25,000 liters capacity. The carbohydrate sources for submerged fermentation are molasses and hydrolyzed grain (especially corn) starches. These are treated with ion exchange resins to remove iron and manganese ions, which inhibit citric acid production. The culture is aerated by pumping air into the bottom of the fermenter and allowing it to bubble up through the stirred culture. The pH of the culture must be maintained at 3.5 or less to prevent the formation of oxalic and gluconic acids at the expense of citric acid. The length of the fermentation can vary from as few as 4 to as many as 15 days with as much as 95% of the glucose in the substrate being converted to citric acid.

The exact metabolic pathway by which *A. niger* produces citric acid has not been elucidated, but it is assumed that glucose is converted to pyruvic acid by glycolysis. The pyruvate is converted to acetyl CoA, which is condensed with oxaloacetate by the first enzyme of the Kreb's cycle to give citric acid, which is then excreted. Why citric acid should be excreted rather than being further metabolized via the Kreb's cycle is as yet unknown.

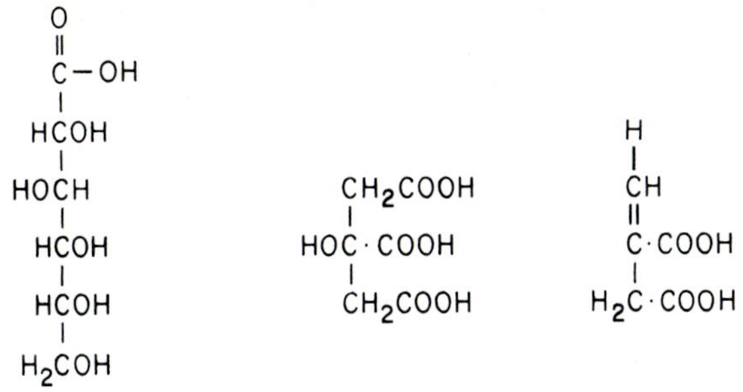

```
    O
    ||
    C—OH
    |
   HCOH                                         H
    |                                           |
  HOCH              CH2COOH                     CH
    |                 |                         ||
   HCOH            HOC·COOH                     C·COOH
    |                 |                         |
   HCOH             CH2COOH                   H2C·COOH
    |
   H2COH
```

Gluconic acid **Citric acid** **Itaconic acid**

Figure 22-12 *Structures of citric acid, gluconic acid, and itaconic acid.*

Gluconic acid

Another organic acid produced by a microbial process that is of commercial importance is gluconic acid (Figure 22-12). Although precise figures are not available, it has been estimated that annual worldwide production exceeds 30,000 metric tons. The price of gluconic acid is in the neighborhood of $1.00/kg, making its production a $30,000,000 a year industry.

Gluconic acid is also a chelating agent, and because of that it is widely used in cleaning products. It is incorporated into detergents used in cleaning soft drink, beer, and milk bottles. It is also used in dishwashing soaps for domestic dishwashers to prevent mineral deposits on glassware. Incorporation of gluconic acid into soap helps prevent the formation of "bathtub ring" in hard water. Gluconic acid is an excellent metal cleaner because it aids in rust removal without adversely affecting the integrity of the unaffected metal. In electroplating, the addition of gluconates to the baths enhances the gloss of the plated metal.

In the food industry gluconic acid is used as an acidulant in baking powders, breads, sausages, cheeses, and other foods. In the pharmaceutical industry, gluconic acid has many uses. Calcium gluconate is used to treat or prevent calcium deficiency especially during pregnancy. Antibiotics such as tetracycline are formulated with gluconates to extend their stability and reduce toxicity.

A potentially large market for gluconic acid is as an additive for concrete. It has been reported that the addition of salts of gluconic acid to cement favorably affects several parameters including setting time, strength, volume, and water resistance.

Gluconic acid is produced in submerged culture under increased atmospheric pressure by a strain of *Aspergillus niger*. The organism contains the enzyme glucose oxidase, which converts glucose directly to gluconic acid. Molecular oxygen is required not only for growth of the fungus but also for gluconic acid synthesis. Inadequate provision of oxygen can limit gluconic acid production; thus copious quantities of air are pumped through the culture in the form of small bubbles that facilitate dissolution of oxygen into the culture medium. Glucose is the preferred substrate and is provided to the culture either as glucose crystals or as a syrup produced by treatment of corn starch with α-amylase and glucamylase, which together hydrolyse starch to glucose. It has been reported that as much as 95% of substrate glucose is converted to gluconic acid.

Itaconic acid

Itaconic acid (Figure 22-12) is another organic acid produced by microbial culture that is of commercial importance. It is a "specialty item" and consequently commands a premium price. Its primary use is as a comonomer in emulsion latexes, which are used in paper adhesives, caulking compounds and putty and in the manufacture of carpet backing.

Itaconic acid is produced by submerged culture of a strain of *Aspergillus terreus* isolated from a sample of Texas soil by Dr. K.B. Raper (Figure 22-13). The preferred substrate for itaconic acid production is molasses, which yields 50% to 60% of the usable sugar being converted to itaconic acid. Control of the pH of the culture is important during production. When the pH is kept below neutral, itaconic acid accumulates in the culture medium; however, if it rises above neutral, itaconic acid becomes a choice growth substrate for *A. terreus* and is quickly consumed.

Lactic acid

Lactic acid is a large-volume, specialty chemical that is produced in part by microbial fermentation. It is also synthesized chemically from lactonitrile and acetaldehyde by the Monsanto Chemical Corporation. Lactic acid is used in the manufacture of plastics and in the food industry as an acidulant and a flavoring agent. It is also used in the pharmaceutical industry and in leather tanning.

Industrially *Lactobacillus delbrueckii* is used for the microbial production of lactic acid (Figure 22-14). During fermentation, glucose is converted to two molecules of lactic acid. Since this is a true microbial fermentation, oxygen is not required; thus there is no need to pump air into the culture. The optimum temperature of the lactic acid fermentation is 49° C, which inhibits the growth of most other organisms. For this reason the maintenance of strict sterile conditions is not as important as with most other fermentations. Substrates that can be fermented for the production of lactic acid in-

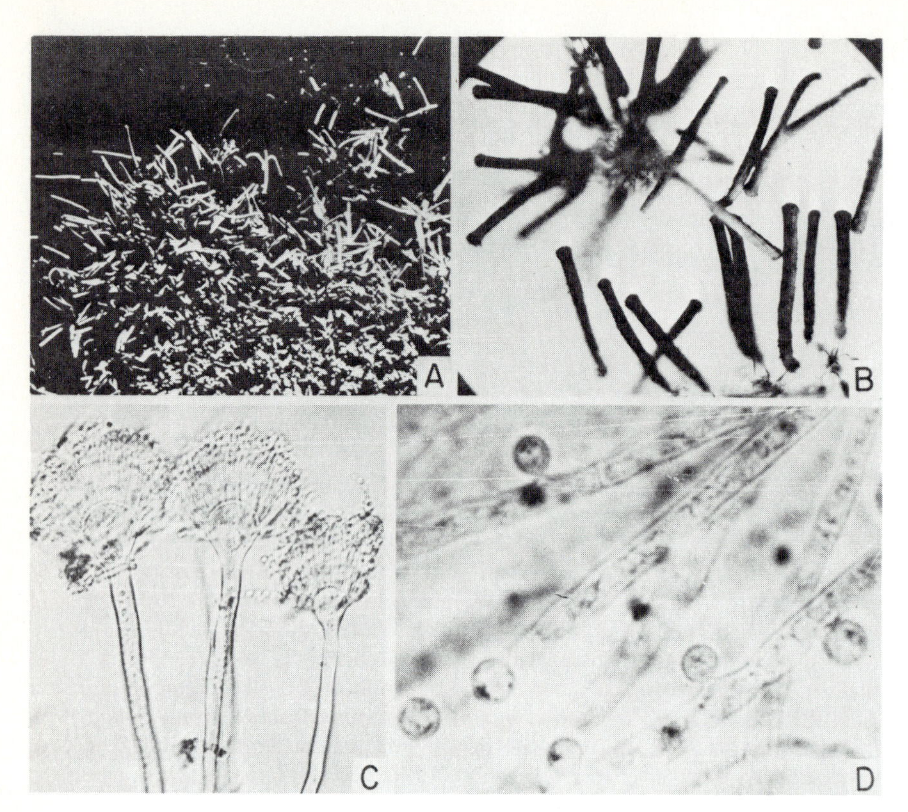

Figure 22-13 Aspergillus terreus, *isolated from Texas soil by Dr. Kenneth Raper, is used for production of itaconic acid.*

clude molasses, whey, and glucose. Yields of lactic acid equalling 85% of the carbohydrate in the substrate have been obtained.

Ethanol

There was great interest, especially in the Midwest, in producing ethanol from grain starch by microbial fermentation. This ethanol would be blended with gasoline in a ratio of one part ethanol to nine parts gasoline and be used to fuel motor vehicles. At the height of interest in this process there was a bumper corn crop and fuel prices were increasing rapidly. The rationale for producing the ethanol-gasoline mixture called **gasohol** was to create a market for the excess corn and perhaps help reduce or stabilize fuel prices. The production of gasohol was not successful from an economic point of view because the cost of ethanol production was more than that of gasoline, so gasohol was more expensive than straight gasoline. Gasohol, however, continues to be sold in isolated areas of the Midwest. There has been renewed interest in ethanol because of its ability to substitute for lead, which is an octane booster and antiknock compound. The Environmental Protection Agency. (EPA) recently mandated the gradual phasing out of tetraethyl lead in liquid fuels.

Ethanol production for gasohol is based primarily on the fermentation of glucose from corn starch by the yeast *Saccharomyces cerevisiae.* Since this yeast lacks amylases, corn starch must first be treated with amylase to provide fermentable substrates. The substrates are fermented to two ethanol and two carbon dioxide molecules per glucose equivalent of substrate used. The fermentation stops when the concentration of ethanol in the culture liquid approaches about 15%; thus the culture fluid must be distilled to obtain the pure alcohol that must be used to blend with gasoline. The distillation step is expensive in terms of energy required and is the primary obstacle in preventing mass production of ethanol for fuel purposes by fermentation. Dried or wet stillage which is a byproduct of distillation is used as protein-rich livestock ration. The CO_2 produced during fermentation can be trapped and used in greenhouses or for dry ice.

The production of ethanol from cellulose, a component of plant material, is being seriously researched to-

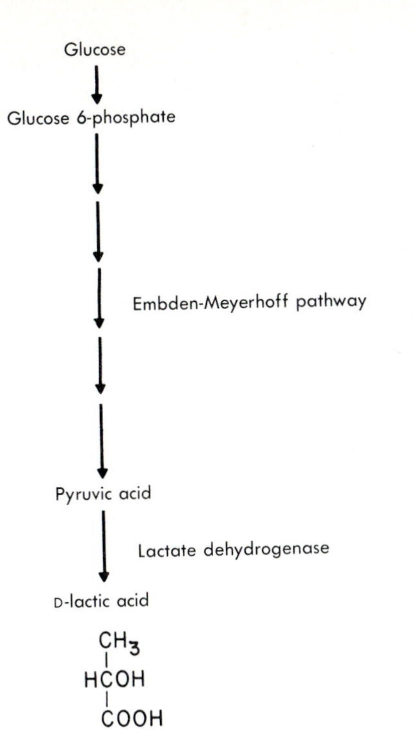

Figure 22-14 *Lactic acid pathway of* Lactobacillus delbrueckii.

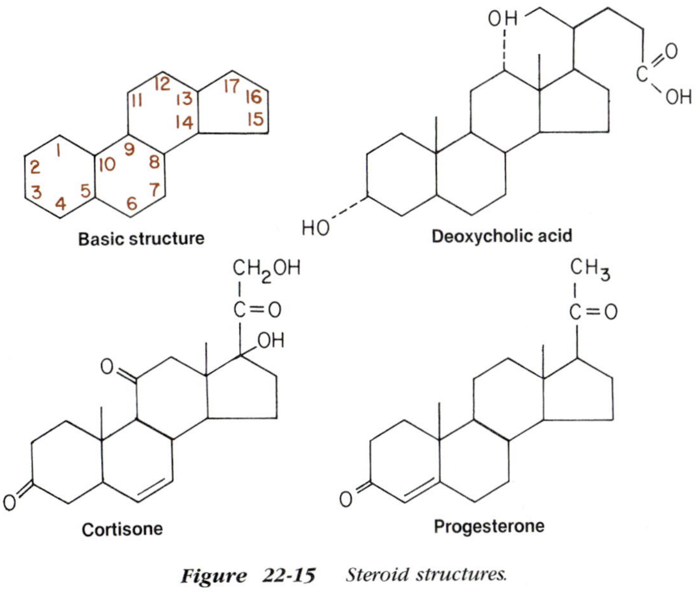

day, but a major problem with this is the conversion of cellulose to a fermentable substrate. Many organisms have been described that produce enzymes that attack cellulose, but so far no enzymatic process for the conversion of cellulose to glucose has been developed on a commercial scale. The most promising at this point is based on the complex of celluloytic enzymes produced by the fungus *Trichoderma reesei* (formerly *T. viride*).

MICROBIAL TRANSFORMATION OF STEROIDS

The demonstration by Hench and coworkers in 1949 and 1950 that the cortical hormones, cortisone and cortisol, isolated from adrenal glands are effective in the treatment of rheumatoid arthritis created a tremendous demand for these steroids. Initially most of the steroid hormones were obtained from animal tissue, but a chemical synthesis was developed by Merck and Company in 1949. The starting material for this synthesis was deoxycholic acid, and it took 30 steps to synthesize cortisone (Figure 22-15). The most difficult portion of the synthesis was the attachment of an oxygen atom to carbon atom 11, which took 10 to 12 steps. The yield of cortisone was initially very low. In 1952 workers at the

Figure 22-15 *Steroid structures.*

Upjohn Company reported that *Rhizopus nigricans* would specifically hydroxylate carbon 11 of progesterone to yield α-hydroxyprogesterone. This introduced the use of microorganisms as catalysts to carry out specific transformations of steroid compounds. It also initiated a large-scale search for additional organisms to carry out other steroid transformations. That search has been fruitful, and today there are hundreds of organisms known to carry out specific modifications of steroids.

There are important advantages to using microbes to carry out chemical reactions in the synthesis of steroid hormones. The transformations are extremely specific, with yields approaching 100% of the substrate being converted to product. Additionally the transformations are relatively rapid and easy to carry out on an industrial scale. These advantages were immediately recognized by industry; thus microbial transformations became an integral part of steroid hormone synthesis. Examples of microbial transformations that are of industrial importance are the C-11 oxygenation by *Rhizopus nigricans* and other fungi, the C-16 α-hydroxylation by streptomycetes, the dehydrogenation of the A ring of glucocorticoids and analogs by *Arthrobacter simplex* and others, and the 3-hydroxydehydrogenation by many microorganisms.

In general the transformations are carried out by growing the transforming organism in cultures of up to 30,000/gallon in volume. When the culture becomes dense, near the end of exponential growth, the steroid compound to be transformed is added to the culture. After the transformation is complete, the steroid can be recovered from the culture medium by selective extraction with organic solvents. Since steroids are only slightly soluble in water, they are generally dissolved in water-miscible solvents before addition to the culture. It has been found that solid steroids in the form of fine powders can be added directly to the culture, and transformation will occur almost quantitatively.

Microbial transformation continues to play an important role not only in the manufacture of cortisone, cortisol, and their derivatives but also in the manufacture of steroid hormones used as antifertility agents.

MICROBIAL INSECTICIDES

Insects have been a serious threat to food production and storage since the earliest days of agriculture. They are also important vectors in the transmission of disease such as malaria, encephalitis, yellow fever, and bubonic plague. The development of chemical insecticides such as DDT provided the first major hope that insects could be eliminated as an important threat to the wellbeing of humankind. This hope has not been entirely realized for several reasons. One important reason is that the target insects developed resistance to the insecticides. Second, the insecticides lacked specificity; that is, they affected a wide variety of insects, both harmful and helpful species, as well as other animals. Massive application of the long-lived chemical insectides has caused their accumulation in the food chain. As a result of this, populations of several predator animals were seriously depleted to the verge of extinction. Consequently, the use of some of the most effective insecticides, such as DDT, was banned in the United States. These and other reasons have prompted a search for effective alternatives to chemical insecticides.

Several microbes are either insect pathogens or they produce toxins that affect insects and consequently are being considered or are used as insecticides. Microbial insecticides offer several potential advantages. In general they are specific, affecting only a limited group of insects. They are relatively short-lived in the environment, which is an advantage because they do not accumulate, but a disadvantage in that their insecticidal action is of short duration. Perhaps most important, many potent microbial insecticides appear to affect only insects and have no harmful effect on a variety of test animals, including both mammals and birds.

Probably the most widely used microbial insecticide is the bacterium *Bacillus thuringiensis,* although viruses, fungi, and other bacteria have potential for widespread use. *B. thuringiensis* is prepared by several companies and sold under trade names such as Thuricide. It appears that the insecticidal agent is not the bacterium itself but a protein toxin that it produces. *B. thuringiensis* produces the toxin in the form of a crystalline protoxin during the sporulation process. In the susceptible insect the alkaline pH of the midgut and its proteases convert the protoxin to an active toxin. In the suscpeptible insect the toxin disrupts the gut cells and induces paralysis. It is effective against more than 100 caterpillar pests including the cabbage worm, cabbage looper, alfalfa caterpillar, gypsy moth, and tobacco budworm. A serovar (serological variety) of *B. thuringiensis (B. thuringiensis* ssp. *israelensis)* produces a toxin that is toxic to mosquito and blackfly larvae but ineffective against caterpillar pests. The larvicide is sold under the trade name *Teknar* and *Bactimos.*

Other *Bacillus* species are also effective insecticides. *B. popilliae* is effective in the control of the Japanese beetle and has been produced commercially for several years for this purpose. This bacterium apparently invades the Japanese beetle grub and multiplies profusely, causing what has been called "milky disease," which is generally fatal. Experimentally *B. sphaericus* has been shown to be toxic to several species of mosquitoes.

Viruses have great potential as agents of insect control but their use is limited by concerns of safety for mammalian species. The viruses considered for commercial use belong to the family of insect viruses called

baculoviruses or Baculoviridae. In its host the virus is enclosed in a proteinaceous shell called an **inclusion body** (see Chapter 16). The most successful applications of viruses controlled sawflies in Canadian forests in the 1930s and 1940s. The first commercial viral pesticide for use in agricultural crops was marketed in the United States in 1975 under the trade name *Elcar*. The virus was used to control the cotton bollworm. One of the disadvantages of virus use is the inconvenience and expense of cultivating virus in the laboratory. In addition, viruses can be rapidly inactivated by ultraviolet light when applied to the plant surface. One advantage is that in the infected host the viruses are released as inclusion bodies resistant to lytic bacterial enzymes and can persist in soil for long periods of time. These characteristics give viruses the potential to bring about long-lasting control of some pests.

Fungi have been used to control pests but their use has been restricted to Soviet bloc countries. A major problem with their use is that fungal spores are subject to desiccation that limits their use as an inoculum for control programs. In addition, a high relative humidity is required during the initial stages of infection.

Growth of microbes for the preparation of commercial insecticides varies from organism to organism. Certain microbes such as *B. thuringiensis* will grow on artificial media and can be grown in large fermenters in submerged culture. When the culture reaches the proper stage, the cells or active agent are harvested and used to compound the insecticide. Other microbes, such as viruses, some fungi, and *B. popilliae,* must be grown in living insects, which is technologically much more difficult than growing cells in submerged culture. A colony of the host insect must be maintained. Insects from the colony are infected with the microbial agent, which multiplies in the host insects. The microbial material used in compounding the insecticides is extracted from the infected insects. In the case of *B. popilliae* the material extracted from the infected insects is composed of endospores, whereas with viruses it is the infectious viral particles themselves.

Concept Check

1. What are the major organic acids produced by microbial activity? What is their commercial value?
2. Describe the metabolic process that results in the formation of ethanol.
3. How are microorganisms used to transform steroids?
4. What are the advantages of microbial insecticides over chemical insecticides such as DDT? Are there any disadvantages?

With more and more attention being given to natural methods to control insect populations, more and more interest is being placed on microbial insecticides. As research continues and expands, it is likely that many important new microbial insecticides will be developed.

MICROBIAL POLYSACCHARIDES

Polysaccharides are produced by a wide variety of microorganisms. Those of primary industrial interest are found on the outside of the microbial cell wall and membrane and are called **exopolysaccharides.** They occur in two distinct forms: attached to the microbial cells that synthesize them as discrete physical structures termed capsules; or secreted from the cell into the surrounding environment in the form of soluble slime.

The chemical structures of the microbial polysaccharides fall into two classes: those composed of a single repeating structural unit, **homopolysaccharides;** and those constructed of two or more monomers, **heteropolysaccharides.** Homopolysaccharides of microbial origin are less abundant than heteropolysaccharides. Dextrans that are polymers of glucose are essentially the only group of microbial homopolysaccharides to have significant industrial applications at present. Heteropolysaccharides, which are composed of neutral sugars and, commonly, uronic acids, have numerous industrial uses.

The industrial uses of polysaccharides are based primarily on their ability to alter the rheological (flow) properties of aqueous solutions, either by gel formation or by alteration of flow characteristics. Polysaccharides are widely used by the food industry as suspending agents, as thickeners and as inducers of gel formation. The annual consumption of polysaccharides by this industry in the U.S. alone exceeds 36,000 tons. Generally, polysaccharides are used in instant foods, salad dressings, sauces, whips, toppings, processed cheeses, and dairy products.

An important potential use of microbial polysaccharides is in the recovery of tertiary oil by the petroleum industry. Pumping an aqueous polysaccharide solution into an oil well that has had most of the readily accessible oil removed significantly enhances the production of the well. Another use of polysaccharides by the oil industry is in drilling muds. The size of this latter market was recently quoted as 1830 tons annually.

The two microbial polysaccharides that are most widely used by industry today are **dextran** and **xanthan.** Other polysaccharides that are of potential use are curdlan, pullulan, and microbial alginate.

Dextrans are α-linked polymers of glucose produced by several bacterial species including *Klebsiella* spp., *Acetobacter* spp., streptococci, and *Leuconostoc* spp. Most dextrans for industrial purposes are produced by

Leuconostoc mesenteroides. Dextrans differ in molecular weight, ranging from 1×10^7 to 3×10^8, and in the varying amounts of different α-linkage (e.g., α1-3- α1-4-, or α1,6-). The two major uses of dextrans are as blood expanders and as the basis for a wide range of absorbents for use in the biochemical and pharmaceutical industries as well as in the research laboratories.

Xanthan is produced by *Xanthomonas campestris.* It is a complex heteropolysaccharide consisting of a substituted cellulose molecule with trisaccharide side-chains on alternate glucose residues. Xanthan is widely used for both food and nonfood uses. The major food use of xanthan is as a stabilizer in French dressing, fruit-flavored beverages, processed cheese, and other dairy products. It is also used in instant desserts, toppings and whips, and frozen and canned foods. Nonfood uses include as a stabilizer of paints and a suspending agent for laundry starch and in metal pickling baths and for paper finishing.

Alginates are currently produced commercially from brown algae. They consist of alternating D-mannuronic acid and L-glucuronic acid units. Recently it was discovered that *Azotobacter vinelandii* and *Pseudomonas aeruginosa* each produce alginates. There are several potential advantages to producing alginates by microbial fermentation. There would be no seasonal variations in production as occur with harvesting marine algae. Alginates produced by microbial fermentation would be free from the pollution now found in algal beds, and yields of known composition could be better assured.

Alginates have a wide range of applications with about 50% of that produced being used in the food industry where it serves as a stablizer for ice creams, instant desserts, frozen custards, and cake mixes. It is also used in the brewing of beer for its ability to enhance "head" or foam formation. Alginates also have numersous applications in the pharmaceutical industry, primarily because of their value in forming stable emulsions.

Curdlan is a homopolysaccharide compound of β-1-3-linked glucose units that is produced by *Alcaligenes faecalis.* It has been proposed for a number of applications including a gelling agent, a thickener, and a stabilizer in foods. Because it is not degraded in the human body, it has been suggested as an ingredient in low-calorie foods.

Pullulan is a homopolysaccharide composed of maltotriose units linked by β1-6-bonds. It is produced by the fungus *Aureobasidium pullulans.* Films made from pullulan are claimed to have antioxidant properties and have been proposed as a biodegradable material for food packaging and coating.

Although the present production of microbial polysaccharides is relatively small, the potential market is large, and the new polysaccharides are continuously being developed.

PRODUCTION OF SINGLE-CELL PROTEIN

For many years scientists have toyed with the idea of producing edible protein from microbes. There are several outstanding advantages to doing so. In contrast to conventional sources of edible protein, such as poultry, pork, or beef, microbes double in weight in a matter of hours or less, whereas a baby pig or chicken may take weeks to do so. The facilities to produce large amounts of microbial or single-cell protein (SCP) require very little land, at most only a few acres. Microorganisms are eaten in small quantities by humans in cheese and yeast. Populations living around Lake Chad in Africa have used a filamentous cyanobacterium called *Spirulina* as food for many centuries (see Point of Interest).

Because the production of single-cell protein is capital intensive, it is normally uncompetitive pricewise with more conventional forms of protein such as soya and fish meal protein. The climatic changes in the 1960s that limited soybean and anchovy production caused the prices of these more conventional forms of protein to escalate dramatically. This opened the door for the production of SCP. Several projects were started primarily in Western Europe where the protein shortage was especially critical. These projects were based on growing microbes in large fermenters using a variety of growth substrates including starch, methanol, and hydrocarbons derived from petroleum. Technologically these projects were feasible, but the climatic changes of the 1960s reversed themselves, allowing increased production of soya protein and fish meal, driving down their prices. At about the same time the price of oil rose sharply. These factors combined to make the cost of SCP production uncompetitive again. In addition to these factors there were problems with contamination of the final protein product in the processes that rely on the hydrocarbons as growth substrate.

One industrial process for making single-cell protein uses methanol fermentation by methylotrophic bacteria and fungi. In methanol fermentation, bacterial methylotrophs convert methanol to carbon dioxide according to the following equations:

$$\underset{\textbf{Methanol}}{CH_3OH} + \xrightarrow[Q \;\rightarrow\; QH_2]{\text{Methanol dehydrogenase}} \underset{\textbf{Formaldehyde}}{HCHO}$$

where Q is a quinone coenzyme called methoxatin.

$$HCHO \xrightarrow[NAD \;\rightarrow\; NADH_2]{H_2O \quad \text{Formaldehyde dehydrogenase}} \underset{\textbf{Formic acid}}{HCOOH}$$

$$HCOOH \xrightarrow[\text{Formate dehydrogenase}]{NAD \;\rightarrow\; NADH_2} CO_2$$

Methylotrophs growing on methanol may assimilate carbon dioxide via the Calvin-Benson cycle (ribulose diphosphate cycle), by condensing formaldehyde with ribulose monophosphate (the Quale cycle) (Figure 22-16) or by the serine pathway in which formaldehyde is condensed with glycine to produce serine. Regardless of the cycle, some of the major products of single-carbon assimilation are pyruvate and acetyl-CoA, which can feed into primary metabolic pathways.

Today much of the SCP being produced commercially is done so on a small basis and for the most part uses waste materials as the growth substrate. In one pro-

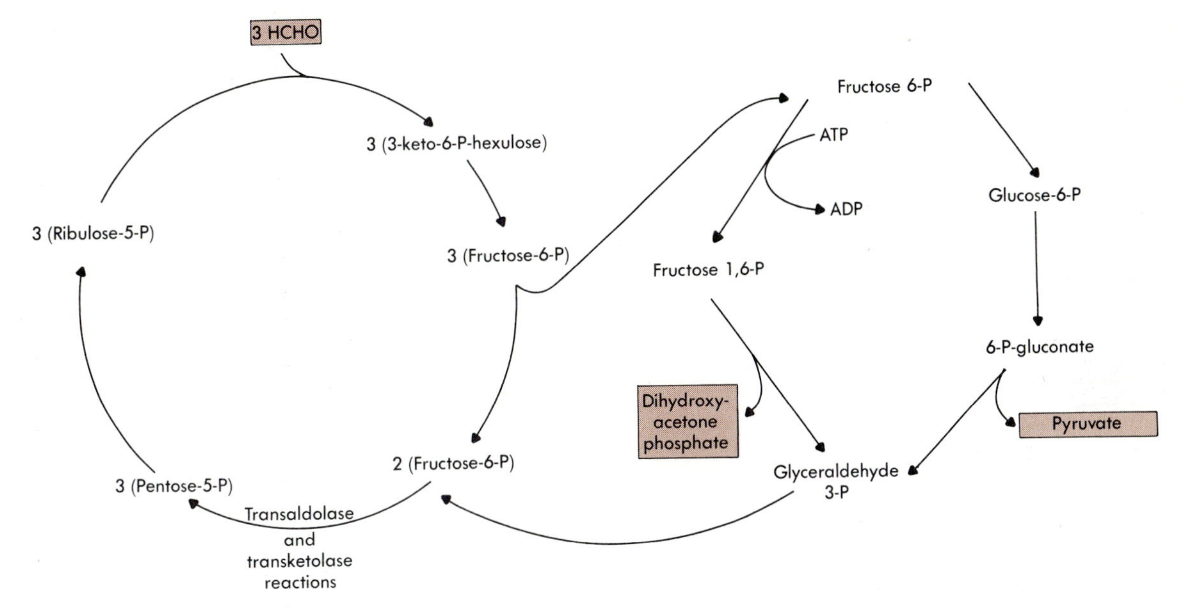

Figure 22-16 *The Quale cycle (ribulose monophosphate pathway). Carbon is assimilated by the condensation of formaldehyde (HCHO) with ribulose 5-phosphate. Major products are dihydroxyacetone phosphate and pyruvate.*

Point of Interest

CHICKEN BREASTS FROM FUNGI

Commercial production of microorganisms as a food source for humans may soon become a reality. Some British biotechnology firms are currently involved in producing meatlike products from the fungus *Fusarium graminearum.* The organism is microscopic and belongs to the same family as truffles. The fungus produces tiny filaments or hyphae that resemble in strength and size the fibers found in meat. The fungus can be formulated into meatlike analogs, which are chewy and have the texture of meat.

The organism is grown at 30° C and pH 6.0 in a medium containing glucose syrup, mineral salts, trace metals, choline, and biotin. After fermentation the fungus is subjected to thermal shock that kills the organism and activates enzymes that degrade RNA. The ribonucleotides pass through the cell wall into the surrounding medium. Removal of RNA is required by the World Health Organization because large amounts of dietary RNA can induce gout and other related diseases.

The fungus is harvested by vacuum filtration and forms a mat of intertwined hyphae on the filter. At this point the harvested fungus is bland in color, taste, and aroma. A certain amount of water is retained in the filtered fungus so that it retains its texture and can be molded into specific products.

The filter cake is mixed with color, flavorings, and egg albumin and then steamed. If the filter cake is to be used as "chicken breasts," it is further processed by putting it through a rolling machine that aligns the matted fibers and produces thin layers that give it a meatlike texture. The filter cakes can be broken into small bite-sized pieces to be used as snacks or they can be molded into "fish cakes" or "chicken and ham patties." The filter cakes can be rehydrated in less than 20 seconds and when cooked do not shrink. The cakes can be canned or frozen for up to 3 years with no deterioration.

The myco-protein venture is an attempt to provide a source of protein for the millions of starving people in the world. The protein content of the myco-protein is 44% as compared to 57% for lean steak. The lipid content of myco-protein is less than half of that found in lean steak, and myco-protein contains no cholesterol. Myco-protein products are now being test marketed and scale-up procedures are being evaluated.

cess waste water from potato processing, which contains as high as 3% solids, much of which is starch, is used to grow a mixed culture of the yeasts *Candida utilis* and *Endomycopsis fibuliger*. The reason for the mixed culture is that *E. fibuliger* produces extracellular amylases, which break down the starch, allowing *C. utilis*, which produces no amylase, to grow. The *C. utilis* is the more desirable protein source, and it grows much faster than *E. fibuliger*, so that in the end it acccounts for about 98% of the microbial mass in the mixed culture. The final product in which the biomass from the fermentation has been concentrated and dried is used primarily as an animal feed supplement.

The future of large-scale SCP production will depend on the price of competing protein sources, its acceptance as a human food, and the availability of inexpensive growth substrates.

PRODUCTION OF ALCOHOLIC BEVERAGES

Brewing

Humans have been making alcoholic beverages, including beer, for thousands of years, perhaps since the dawn of civilization. Today beer is consumed in vast quantities. In 1977 the annual per capita consumption of beer in the United States was 22.7 gallons for the population as a whole and 35.1 gallons when only persons over 21 years of age were considered. In terms of 12-ounce bottles, each adult in America consumed, on the average, about 370 bottles of beer in 1977, or slightly more than a bottle of beer every day.

Beer ingredients. Historically, beer was made from malt, hops, water, and yeast. Today many American breweries substitute adjuncts such as corn or rice starch for a portion of the malt.

Malt. Malt is made from germinated barley grains in facilities called malt houses. In Europe, most breweries have their own malt houses where they make malt by their own recipes. For the most part American breweries buy their malt from companies that specialize in malt production; however, some breweries do continue to make their own malt.

Malt is important in determining the character of beer. It contains enzymes, α and β amylases, that convert malt and adjunct starch into maltose and dextrins. The maltose is fermented to ethanol and carbon dioxide by the yeast. The **dextrins**, which are not fermented, contribute to the "body" of the beer; the more dextrins, the "heavier" the beer. Dextrins are also a major source of calories in beer. In the low-calorie beers, which have become popular in recent years, the dextrins have been removed by treatment with special enzymes, **glucamylases**, that convert the dextrins to sugars, which are fermented by the yeast. The low-calorie beers are often called "lite" or "light", not only because they are low in calories but also because they are light bodied because of their lack of dextrins.

In addition to the amylase enzymes, malt contains proteins that are important in "head" formation and contribute to beer flavor. Beer color is also determined by the malt used. All beer, even the dark beers, is made primarily from light malt, which gives beer a yellow color. The darker beers are made by mixing darker malts with the light malt.

In the malting process barley grains are cleaned of extraneous material and placed in a steeptank where they are soaked with water for 2 days, during which time the grains swell. Afterward excess water is·removed, and the swollen barley is transferred to the malting chamber for 5 to 7 days. The barley grains germinate in the malting chamber where high humidity and constant temperature are maintained. When the embryo reaches the proper size, the germinated barley is transferred to a kiln where the embryo is heat killed and the germinated grains, now called **malt**, are dried. The type of malt produced is determined by the amouont of heat applied in the kiln. Light malt is heated to 80° C and dark malt to 105° C. Very dark malt is heated to over 200° C.

During the malting process the starch in the barley seeds is partially broken down into maltose and dextrins by the amylases, which are activated when the seeds germinate. On the average light malt contains 50% starch and 10% sugars. During light malt production the heating in the kiln does not totally inactivate the malt amylases; therefore when the malt is later mixed with water in the mash tub they continue to degrade the malt starch into dextrins and fermentable sugars. Dark malts provide mainly color and taste to beer and some starch.

Hops. Hops are added to beer primarily as a flavoring agent. They give beer its characteristic bitter taste and its aroma. The bitter taste is caused by acids and resins contained in the hops. The essential oils in hops contribute to the development of beer aroma and to some extent to the beer flavor. Hops also contain **tannins**, which contribute some bitterness to beer, but which serve mainly to cause coagulation of undesirable malt proteins, forming a sediment called **break**. Hops are the female flower of the hop plant, *Humulus lupulus*.

Water. Beer is 85% to 92% water, which consequently is a very important factor in determining beer quality. Generally hard water makes the best beer. Water hardness is generally expressed as the concentration of calcium salts it contains. In Germany most beer is brewed with water with a hardness of 600 ppm, whereas the water used in English ales has a hardness of around 1800 ppm. In America most breweries adjust the water hardness to about 350 to 500 ppm. In addition to the proper hardness, the water used for beer should be unpolluted and pleasant to the taste. Because of this many breweries advertise that they use water from virgin springs or some other pristine source.

Yeast. Another critical element in determining the final quality of beer is the yeast used to ferment the sugar from the malt or adjunct to carbon dioxide and ethanol. Two general types of yeast are used: **bottom-fermenting yeast** and **top-fermenting yeast**. For the production of lager beers a bottom-fermenting strain of yeast is used. One such strain, *Saccharomyces carlsbergensis*, was discovered by workers at the Carlsberg Laboratories in Copenhagen, Denmark. Also strains of *Saccharomyces cerevisiae* are used. During fermentation the yeast cells settle to the bottom of the fermentation vessel, where they continue fermentation until the fermentable sugars are exhausted from the medium. Top-fermenting yeast strains are used in the production of ales. Generally speaking top-fermenting yeasts work better at warmer temperatures (15° to 18° C) than do the bottom-fermenting strains (10° to 15° C) and are hardier. After a batch of beer has been fermented, the yeast are harvested and reused to inoculate fresh batches of beer. The harvested yeast are monitored carefully for vigor and contamination, and as long as they remain vigorous and uncontaminated, they may be reused many times.

Adjuncts. Adjuncts are added during the brewing process to increase the alcohol content of the beer. They are starchy materials that are enzymatically converted to fermentable sugars. Adjuncts add little or nothing to the taste, aroma, or color of beer. In Europe adjuncts are not used in brewing; hence malt is the only source of fermentable sugars. The all-malt beers have an intense yellow color and are heavy bodied. Americans seem to prefer beer that is pale yellow in color and lighter in body than European beer. These qualities can only be obtained if less malt is used. Because malt is the source of fermentable sugars a reduction in malt content would reduce the amount of sugar added and consequently would reduce the alcohol content of the beer if adjuncts were not added to compensate for the lost sugar. The most commonly used adjuncts are corn or rice starch; however, other products such as potatoes, cassava, soybeans, and sorghum have been used. Generally about 1.5 pounds of adjunct starch is added for each 3.5 pounds of light malt.

Brewing process. At the brewery malt is cleaned, milled to a powder, weighed, and then placed in the mash tub where it is mixed with water. Independently the adjunct, either ground corn or rice, is placed in a cooker with water and a small amount of milled malt. The temperature of the cooker is slowly brought to boiling. During the slow temperature rise amylases and proteases from the malt act, to some extent, on the adjunct starch and protein until these enzymes are inactivated when the temperature rises about 72° C. Boiling gelatinizes the adjunct starch, making it more amenable to breakdown by the malt amylases. After boiling, it is mixed with the wetted malt in the mash tub. The temperature is raised in the mash tub to stimulate starch breakdown by the malt amylases while proteases from the malt help breakdown undesirable malt proteins. When all the starch from the malt and the adjunct has been converted to dextrins and maltose, the temperature is raised to 72° C to inactivate the amylases. This temperature also causes some protein precipitation.

The solids in the mash are removed by transferring the mash to a **Lauter Tub**, where the solids are allowed to settle and the liquid is drawn off. The liquid, which is now called **wort**, is transferred to the brew kettle (Figure 22-17, *A*), usually made of copper, where hops are added and the mixture is boiled for about 1½ hours. The boiling serves several purposes. It promotes the extraction of the acids, resins, and tannins from the hops; it causes coagulation of unwanted proteins; it concentrates the wort; and finally it sterilizes the wort. The hot wort from the brew kettle is passed through a strainer to remove the hop remnants and into a settling tank to allow the coagulated protein and other insoluble material to settle out. The sediment, known as **trub**, is discarded. The clarified wort is passed through coolers to bring the temperature to about 20° C and is then passed to the fermenter (Figure 22-17, *B*), the yeast are added and fermentation takes place. Fermentation using bottom-fermenting yeast is carried out at 6 to 12° C and normally takes 8 to 10 days for completion. With top-fermenting yeast the temperature is higher (14 to 23° C), and the fermentation takes less time (5 to 7 days). When the fermentation is complete, the **green beer**, as it is now called, is carefully removed from the settled yeast in the fermenter to a storage tank, which is refrigerated to 0° C. The beer is stored here for a few weeks to several months, during which time unstable proteins, yeast, resins, and other undesirable substances precipitate from the beer. Esters are produced during this storage period, and the harsh taste of the green beer disappears.

Concept Check

1. What are some important microbial polysaccharides used in industry? What are their functions?
2. Describe the process used in Britain to use fungi as a source of food (myco-protein).
3. What metabolic reactions are being exploited for making single-cell protein?
4. What is the beer-making process? Why are hops added to beer? What are adjuncts and why are they added to beer? In the making of malt, why are the barley seeds allowed to germinate?

A

B

Figure 22-17 ***A,*** *Brew kettles. Boiling of wort with hops in brew kettles is important in determining beer flavor. Additionally, it causes inactivation and precipitation of enzymes and protein in wort.* ***B,*** *Primary fermentation tanks. After being boiled in brew kettle, wort is cooled, filtered, and put in primary fermentation tanks where it is pitched, or seeded, with yeast. In these tanks, yeast consume fermentable sugars and produce carbon dioxide and ethanol and more yeast cells.*

After aging the mature beer is filtered, carbonated under pressure, pasteurized by heating for 5 minutes or more at 60° C, and canned or bottled. Some beer is not pasteurized but is filter-sterilized before being canned or bottled.

Types of beer. The most common type of beer made in the United States is **lager beer**. The word *lager* comes from the German verb *lagern*, which means to store. Lager beer quite literally is stored beer. It is produced by the process described above in which the green beer is stored in a refrigerated tank to allow it to mellow and mature. American lager beer is made from light malt and adjuncts so it has a pale yellow color and is light bodied. **Bock beer** is a lager beer that is made from a mixture of light and dark malts with little or no adjunct added. Generally bock beer is darker colored and heavier bodied than regular lager beer. European lager beers are generally made from light malt with little or no adjunct added. **Pilsener beer** is a type of lager beer that is made in Pilsen, Czechoslovakia. It is light in color, not sweet, with good hop aroma.

Ale, porter, and **stout** are all malt beverages made with top fermenting yeast. Generally they are darker in color, heavier bodied, and hopped at a higher rate than is lager beer.

The low calorie beer being sold today is lager beer that has been treated to remove dextrins.

Malt liquor is brewed from malt without added adjunct and usually has a higher alcohol content than lager beer.

Wine making

Wine, like beer, has been produced for thousands of years and is produced by the fermentation of sugars to ethanol by yeast. The wine production process is much simpler than that for the production of beer because wine is made from fruit juices, which contain the fermentable sugars glucose and fructose and little or no starch, whereas beer requires that grain starch be converted to fermentable sugar.

The most important wine-producing regions of the world include Western Europe, the United States, and several South American countries. The major producing countries in Western Europe are Italy, France, West Germany, and Spain. In the United States, California produces by far the most wine, with New York state being the second most important wine-producing state. Both Chile and Argentina are developing into important wine producers in South America. They produce some very fine quality wines.

Production of grape wine. Grape wine is produced from selected varieties of grapes. White wines are generally produced from white grapes such as the **Chardonnay** variety, whereas red wine is made from red grapes such as **Cabernet Sauvignon**. The following outlines the production of red table wine.

Grapes are harvested when they reach the proper degree of ripeness. The sugar-to-acid ratio is an important determinant of the degree of ripeness. If the grapes are too ripe, there will be too much sugar and too little acid so that wine produced from such grapes will have a high alcohol content and will possess an insipid (no distinctive qualities) taste due to insufficient acid. Wines made from underripe grapes have reduced alcohol content and excess acid and consequently are not very desirable. Grapes at the optimum degree of ripeness usually give wine with an alcohol content of about 12.5% by volume.

The harvested grapes are crushed carefully so that only the skins are broken and not the stems and seeds which contain bitter tannins. The stems are removed from the broken grapes, now called **must**. Sulfite is added to the must to about 100 ppm to inhibit the growth of bacteria, undesirable yeast, and fungi that may have been on the grapes when they were crushed. *Saccharomyces cerevisiae* var. *ellipsoideus*, which is the strain of yeast used to ferment the sugars in the grape juice to ethanol and carbon dioxides, is not inhibited by this concentration of sulfite.

Approximately 6 hours after sulfite treatment the crushed grapes in the fermentation tank are inoculated with a pure culture of the yeast. The contents of the fermentation tank are mixed, or stirred, twice a day during the early stages of fermentation to facilitate temperature equalization, the extraction of color and tannin from the grape skins, and the aeration of the must which is necessary for rapid yeast multiplication. Stirring is ceased when the right density of yeast has been reached, and the fermentation proceeds anaerobically.

The temperature of the fermentation is carefully controlled. The optimum temperature range for bouquet and aroma development is from 21° to 24° C. The fermentation proceeds for 3 to 5 days until the proper amount of color and tannin has been extracted from the grape skins. The winemaker decides when that point has been reached, and the wine is carefully drawn away from the skins and seeds and transferred to a second tank. The fermentation continues anaerobically in the second tank until the fermentable sugars are exhausted, usually within 11 days at a temperature of 21° C. As the alcohol content of the wine increases various components such as yeast cells, tannins, proteins, and pectins form a sediment on the floor of the tank. After the fermentation is complete, the wine is carefully drained from the tank without disturbing the sediment and is transferred to **aging casks** where it is aged for several months to several years. The aging casks are generally made of white oak or sometimes of redwood. The bouquet of wine develops through the formation of esters from alcohols and acids during the aging process. The flavor, taste, and color of the wine are also affected by the aging process which for high quality red wines may be as long as 5 years.

After the wine has been aged, it is filtered, bottled, and sold. Good wines continue to age for several years in the bottle.

Strain improvement in yeasts. Strain improvement for substrate use, product tolerance, and reduction of unwanted secondary metabolites are a few of the characteristics that geneticists seek in yeast strains, particularly those used in production of alcohol.

Brewing strains of yeast are often polyploid (having more than two haploid sets of chromosomes) or aneuploid (having a chromosome number that is not an integral number of a normal genome). Polyploid yeasts are genetically more stable and less susceptible to mutation than their haploid or diploid counterparts. Because most of the important characteristics of industrial yeasts are under the control of a single gene or limited number of genes, they are still amenable to genetic analysis. The following techniques are available for strain improvement.

Selection. New substrains of industrial yeasts arise through mitotic recombination during vegetative growth. Mitotic recombination occurs infrequently and is not the most suitable method for strain improvement. However, selection following use of a mutagenic agent is used to a considerable extent for improvement in wine, brewing, and distiller's yeasts. Low doses of mutagen increase the rate of mitotic recombination and provide an excellent means of obtaining the expression of recessive mutations. In wine yeasts the characteristics most selected for are increased ethanol tolerance, decreased foaming, and increased flocculation to improve sedimentation. In brewing yeasts, mutagen-induced mutations have been used to select auxotrophs for isoleucine and valine so that levels of undesirable products, such as diacetyl and acetoin, which give a displeasing taste, can be reduced in the beer.

Cross-breeding or hybridization. Cross-breeding of yeasts is complicated by the lack of mating type, poor sporulation, and poor spore viability in polyploid industrial strains. Although conventional cross-breeding techniques are possible, hybrids with more favorable attributes than their parents have been few. A greater understanding of yeast recombination and fine structure of yeast chromosomes is needed before cross breeding experiments can be used with any success.

Cytoinduction. The rare mating of polyploid industrial strains has been exploited to introduce cytoplasmic elements into brewing strains without transfer of nuclear genetic elements from a nonbrewing parent. Thus the progeny have the nuclear genotype of one parent and the cytoplasmic genotype of the other parent. This process is called **cytoinduction** and requires that the nonbrewing parent possesses a mutation called *kar I*—a mutation that impairs nuclear fusion (Figure 22-18). Cytoinduction has been used to do the following:

1. Produce hybrids containing the mitochondrial genome from a selected parent. The specific effect of the mitochondrial genes on the cell can be determined using this technique.
2. Transfer double-stranded RNAs that encode toxin production, for example, the killer factor (see Chapter 16). Killer factors are a nuisance to the brewing industry because invasion by a killer strain can halt the brewing operation by destroying the producing strain of yeast. In addition, the killer strain gives the beer a foul flavor. The solution was to breed strains that lacked the genes encoding the off flavor but carried killer particles. Carriage of the killer particles makes the yeast strain immune to attack by other killers.
3. Transfer plasmids, which contain specific genetic determinants that have been prepared by genetic engineering techniques.

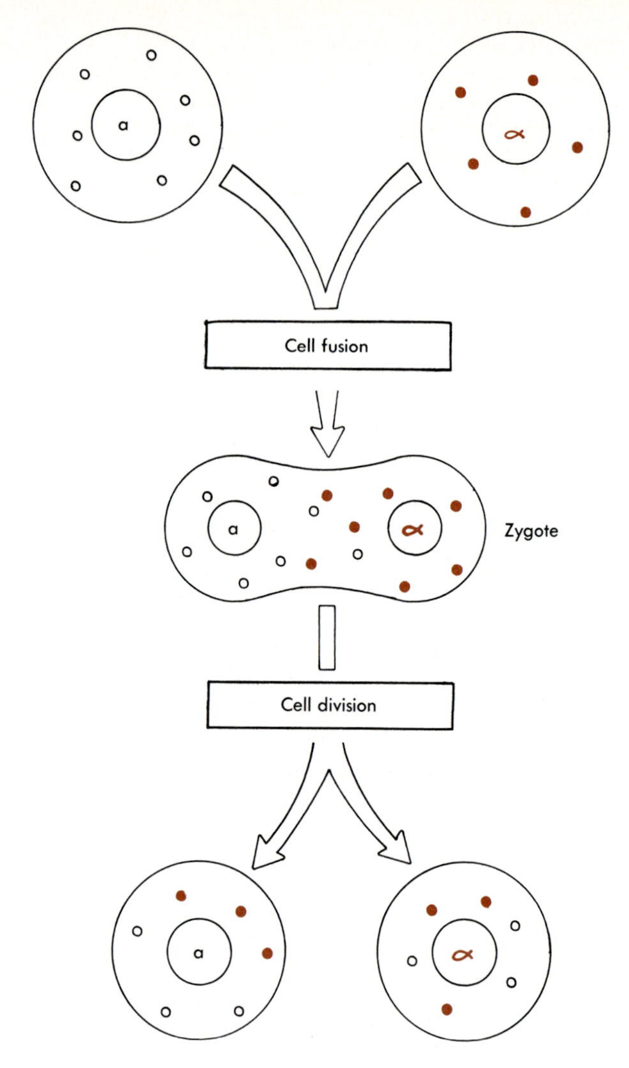

Figure 22-18 *Cytoinduction. Mating types of* Saccharomyces *fuse to form zygote. Because of mutation, nuclear fusion does not take place in zygote. In subsequent division zygote divides producing haploid cells with cytoplasm of both parents. Colored components in this example represent mitochondria.*

Protoplast fusion. Yeast protoplasts can be prepared by treating them with enzymes that remove the cell wall and leave only the cytoplasmic membrane surrounding the cytoplasmic contents (Figure 22-19). When protoplasts are cultivated in the presence of polyethylene glycol, fusion of the protoplasts can be obtained. The fused protoplasts are induced to regenerate their cell walls and to initiate cell division by transferring them to solid media containing 3% agar and sorbitol. Protoplast fusion is especially useful in producing hybrids that are impaired in mating and spore formation or lack a sexual cycle. Thus conditions that do not favor

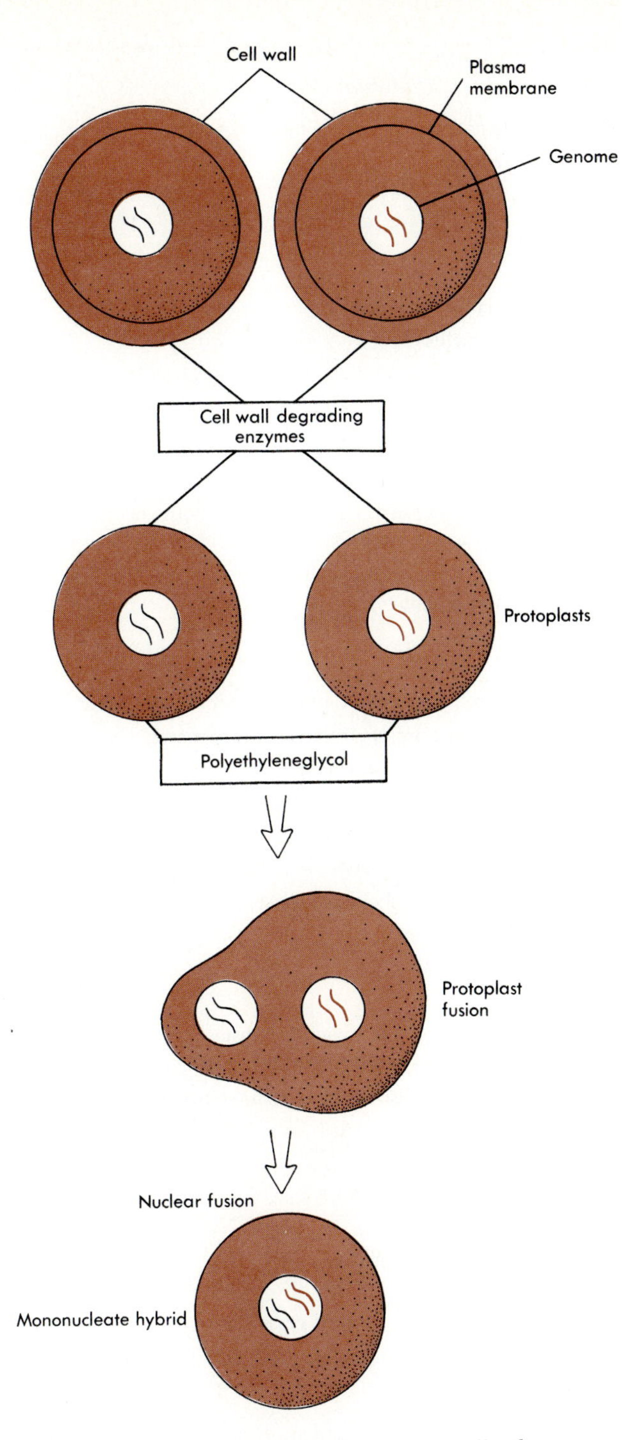

Figure 22-19 *Yeast protoplast fusion. Two cells of differing genotype are subjected to wall-degrading enzymes. Resulting protoplasts are mixed in medium containing polyethylene glycol, which promotes fusion. Nuclear fusion may occur before, during, or after mitosis to produce mononucleate hybrid. Recombination can occur between parent chromosomes, and portions of either parent chromosomes can be eliminated.*

microbial diversity can be overcome. Unfortunately, to date protoplast fusion is not specific enough for genetic modification, and the hybrid is usually very different from either fusion partner.

Transformation. Successful transformation of yeasts has been obtained using various plasmid vectors. A normal plasmid of *Saccharomyces* strains is called **two-micron DNA plasmid**. Various genes, both bacterial and yeast, have been incorporated into the plasmid for transformation of various yeasts. The use of plasmid vector systems with recombinant DNA techniques offers great promise for not only improving brewing yeast strains but for producing foreign proteins of commercial significance by yeasts. There exists methods for isolating and amplifying yeast genes by using *E. coli* in the engineering procedure. Potentially, any cloned gene can then be used to transform yeast.

OTHER INDUSTRIAL PRODUCTS

It would be impossible to discuss at length all of the industrial products that are produced by microorganisms. Some of these products have been discussed in Chapter 21, Food Microbiology. To make our list more complete, Table 22-4 outlines the various microbial species and products with which they are associated in industrial processes.

GENETIC ENGINEERING AND INDUSTRIAL MICROBIOLOGY

The industrial production of compounds may be accomplished in one of three ways: chemical synthesis, biological synthesis, or a combination of the two. The type of process to be used, if all three are available, usually is governed by the cost factor. The rising costs of raw materials, such as petroleum for chemical synthesis, and the advent of recombinant DNA technology (see Chapter 11) have made biological synthesis more attractive. In the near future biological synthesis will replace many of the current synthetic chemical techniques.

The basic advantage of recombinant DNA technology is that the development of superior strains of an organism need not be based on the randomness of spontaneous mutation or even induced mutations, which are time consuming and costly. Genetic engineering offers a technique by which genes can be exchanged among unrelated organisms, making possible the replacement of an existing metabolic pathway with one that will be more efficient in synthesizing a product for industrial use.

It is unwise to think that recombinant DNA technology can be applied to all industrial processes in which microorganisms might be used. Two of the major re-

Table 22-4 *Some important industrial products obtained with the aid of microbial activity*

Organism	Product
Pharmaceuticals	
Penicillium chrysogenum (fungus)	Penicillins
Cephalosporium acremonium (fungus)	Cephalosporins
Streptomyces species (bacterium)	Streptomycin, tetracyclines, kanamycin and others
Bacillus brevis (bacterium)	Gramicidin S
Bacillus subtilis (bacterium)	Bacitracin
Bacillus polymyxa (bacterium)	Polymyxin B
Arthrobacter simplex (bacterium)	Steroid transformation
Rhizopus nigricans (fungus)	Steroid transformation
Escherichia coli (bacterium, genetically engineered)	Insulin, interferon, human growth hormone, somatostatin, various vaccines
Polysaccharides	
Xanthomonas campestris (bacterium)	Xanthan gum
Leuconostoc mesenteroides (bacterium)	Dextran
Single-cell proteins	
Saccharomycopsis lipolytica (fungus-yeast)	Microbial protein from petroleum alkanes
Candida utilis (fungus-yeast)	Microbial protein from paper-pulp waste
Methylophilus methylotrophus (bacterium)	Microbial protein from growth on methane or methanol
Organic acids and solvents	
Saccharomyces cerevisiae (fungus-yeast)	Ethanol from glucose
Clostridium acetobutylicum (bacterium)	Acetone and butanol
Aspergillus niger (fungus)	Citric acid and gluconic acid
Aspergillus terreus (fungus)	Itaconic acid
Lactobacillus delbrueckii (bacterium)	Lactic acid
Insecticides	
Bacillus thuringiensis (bacterium)	Bacterial protein toxin used against cabbage worm, cabbage looper, gypsy moth, and others
Bacillus popilliae (bacterium)	Endospores used against Japanese beetle
Vitamins	
Eremothecium ashbyi (fungus-yeast)	Riboflavin
Pseudomonas denitrificans (bacterium)	Vitamin B_{12}
Amino acids	
Corynebacterium glutamicum (bacterium)	L-Lysine and L-glutamic acid
Food and beverages	
Saccharomyces cerevisiae (fungus-yeast)	Wine, ale, sake, baker's yeast
Saccharomyces carlsbergensis (fungus-yeast)	Lager beer
Saccharomyces rouxii (fungus-yeast)	Soy sauce
Penicillium roquefortii (fungus)	Blue cheeses
Penicillium camembertii (fungus)	Camembert and brie cheeses
Streptococcus thermophilus (bacterium)	Yogurt
Lactobacillus bulgaricus (bacterium)	Yogurt
Propionibacterium shermanii (bacterium)	Swiss cheese
Gluconobacter oxydans subsp. *suboxydans* (bacterium)	Vinegar
Lactobacillus sanfranciscus (bacterium)	Sour french bread

strictions are vector selection and insufficient information about the location and number of industrially useful genes in certain microorganisms. Vectors for the genes to be cloned should supply efficient transcriptional and translational signals. In addition, many products that are industrially useful are encoded by a single gene but products, such as antibiotics, are the consequence of a synthetic process that may encompass up to 30 different genes. At this time the biochemistry of all the reactions encoded by the genes is not understood. If the genes are widely dispersed on the chromosome, their manipulation becomes increasingly difficult.

The improvement of strains by genetic engineering may take different forms and includes the following.

Gene amplification

Enzyme production may be increased by increasing the number of copies of the gene (gene amplification) that codes for the enzyme. A microorganism containing one copy of a gene will produce 1X amount of the enzyme, but if the number of copies of the gene is increased to 20, the amount of enzyme produced will be 20X. A consequence of this would be a twentyfold reduction in the unit cost of the product of the enzyme. Recently the gene for α-amylase was cloned in *Bacillus subtilis*. The hybrid plasmid containing the cloned α-amylase gene was maintained at 50 copies per cell. The production of α-amylase by the strain carrying the cloned gene was 50 times that of the same strain lacking the cloned gene.

Vaccine production

Many disease-causing organisms and viruses are difficult to grow and often extremely dangerous to manipulate. The production of vaccines, on an industrial scale, from these organisms is often expensive and hazardous if possible at all. Removing antigen-coding gene(s) from the disease-causing organism and placing them in nonpathogenic bacteria will permit less expensive and less hazardous vaccine production. The production of a vaccine for hoof-and-mouth disease in cattle has been accomplished through the use of genetic engineering and is commercially available, as is a vaccine against swine diarrhea.

Fermentation potential

There is a widespread demand for light beers that have reduced carbohydrate content. Malt wort is made up of approximately 22% dextrins, which are not fermented by naturally occurring strains of *Saccharomyces cerevisiae*. Efforts are now underway to introduce genetic material for dextrin fermentation into *S. cerevisiae*.

Many of the fungi that carry out industrial fermentations are filamentous and form mats. Fungal mats become so intertwined in the large fermenter that efficient nutrient absorption is lost. Recombinant DNA technology may be applied to the development of single-cell fungi, which would make fermentation more efficient.

Product tolerance

Organisms that are selected because of their ability to produce copious amounts of a product are sometimes destroyed by the same product, for example, butanol. Recombinant DNA technology is being used to develop product-tolerant strains.

Product excretion

Some microorganisms do not excrete products, such as amino acids, in large quantities. It appears possible that through genetic engineering microbial strains with leaky membranes for product excretion can be developed.

The inability to excrete products for example as enzymes is associated with gram-negative bacteria, such as *Escherichia coli*, which is the host for several industrial products. Generally *E. coli* retains extracellular enzymes in the periplasm, and the cells must be treated to release them. This treatment also releases other soluble proteins present in *E. coli* and makes purification more difficult and expensive. *E. coli* also produces endotoxin, which contaminates the desired product and can be even more difficult to remove. Serious efforts are underway to develop host-vector systems using gram-positive bacteria that would excrete enzymes into the culture medium and would not produce endotoxins or other toxic substances.

Product purification

Pharmaceuticals, such as peptides or proteins (for example, interferon), when produced by genetic engineering, must be purified. Monoclonal antibodies will probably become the means for this purification. One technique called **affinity purification** uses a solid matrix to which adhere monoclonal antibodies specific for the protein to be purified (Figure 22-20). When microbial products containing wanted and unwanted proteins are passed through a column to which are attached monoclonal antibodies, only the protein specific for the antibodies will bind and everything else passes through. Mild chemical treatment of the column breaks the antibody-antigen bond, and the product can be collected at the bottom of the column. This technique isolates viral, microbial, and parasitic antigens for vaccine production, as well as hormones, enzymes growth factors, and other proteins.

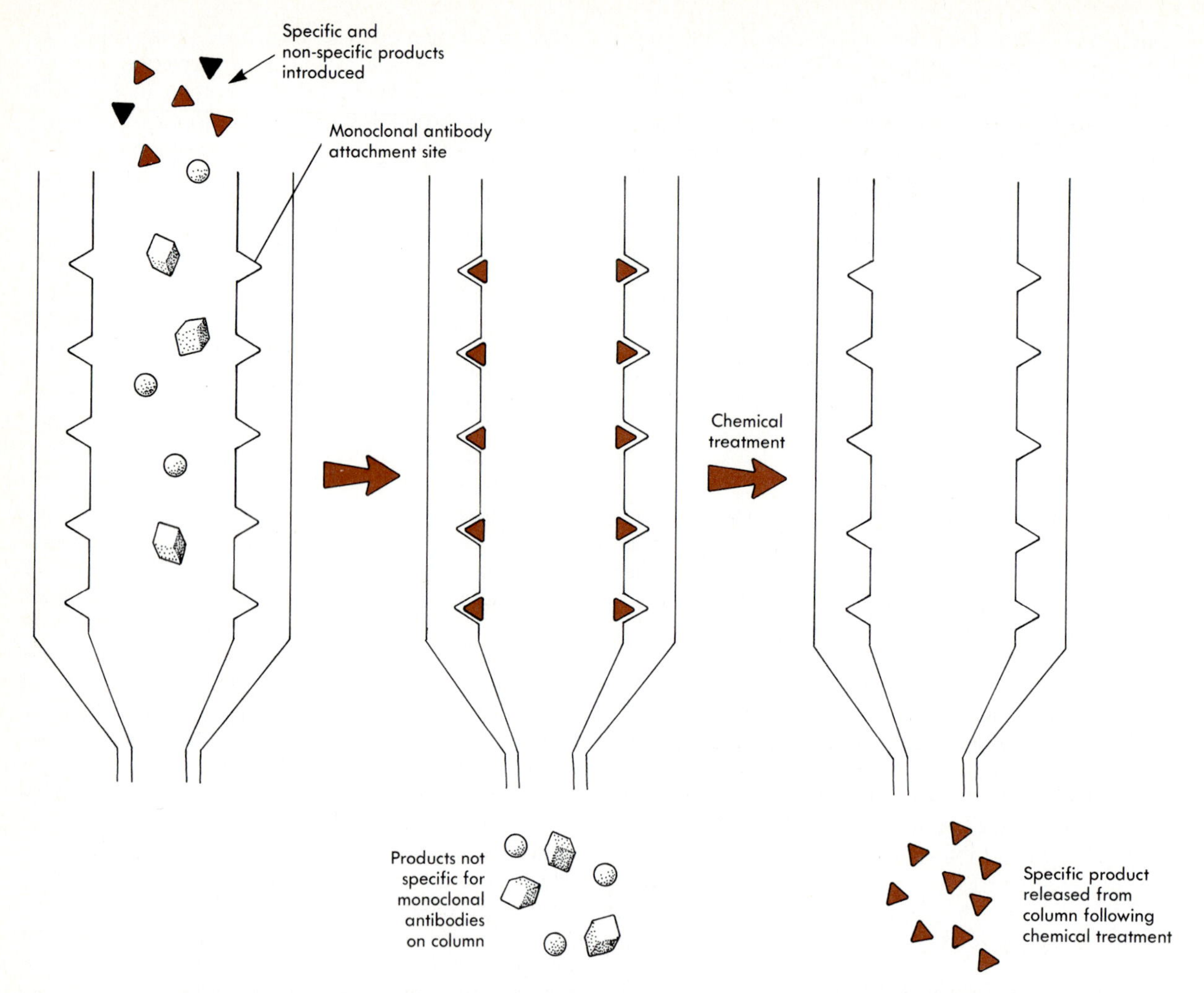

Figure 22-20 *Affinity purification. Column to be used for separation contains solid matrix to which are attached monoclonal antibodies specific for product (▲) to be purified. Wanted product is part of mixture containing unwanted components (○, □). Product to be purified attaches to monoclonal antibodies via antigen-antibody interaction, whereas unwanted components pass through column. To remove product from column, latter is treated with chemical that breaks antigen-antibody bond.*

Microbial production of nonmicrobial products

One of the most important potential uses of genetic engineering is to have microbes serve as factories for the production of specific products not normally produced by microbes. Many proteins produced in the human body have great potential in the treatment of human disease. For example, insulin is the only treatment for certain types of diabetes, and interferon has potential in the treatment of some forms of cancer. The technique originally used to clone rat insulin genes is illustrated in Figure 22-21. Fortunately insulin can be obtained from other than human sources, but interferon active in humans must be obtained from humans. To obtain interferon in sufficient amounts for evaluation, human interferon genes have been cloned into bacteria where inteferon can now be produced in large amounts. This increased production makes it possible to investigate its potential as a therapeutic agent. Other nonmicrobial proteins of interest are human peptide hormones, such as growth hormone, which are difficult to isolate in sufficient quantity to study fully their effects.

Another nonmicrobial product now available via genetic engineering is rennin (also called chymosin). Rennin is an enzyme used in cheese making that is ordinarily obtained from the stomach lining of calves. Recently a species of *Aspergillus* has been engineered to secrete

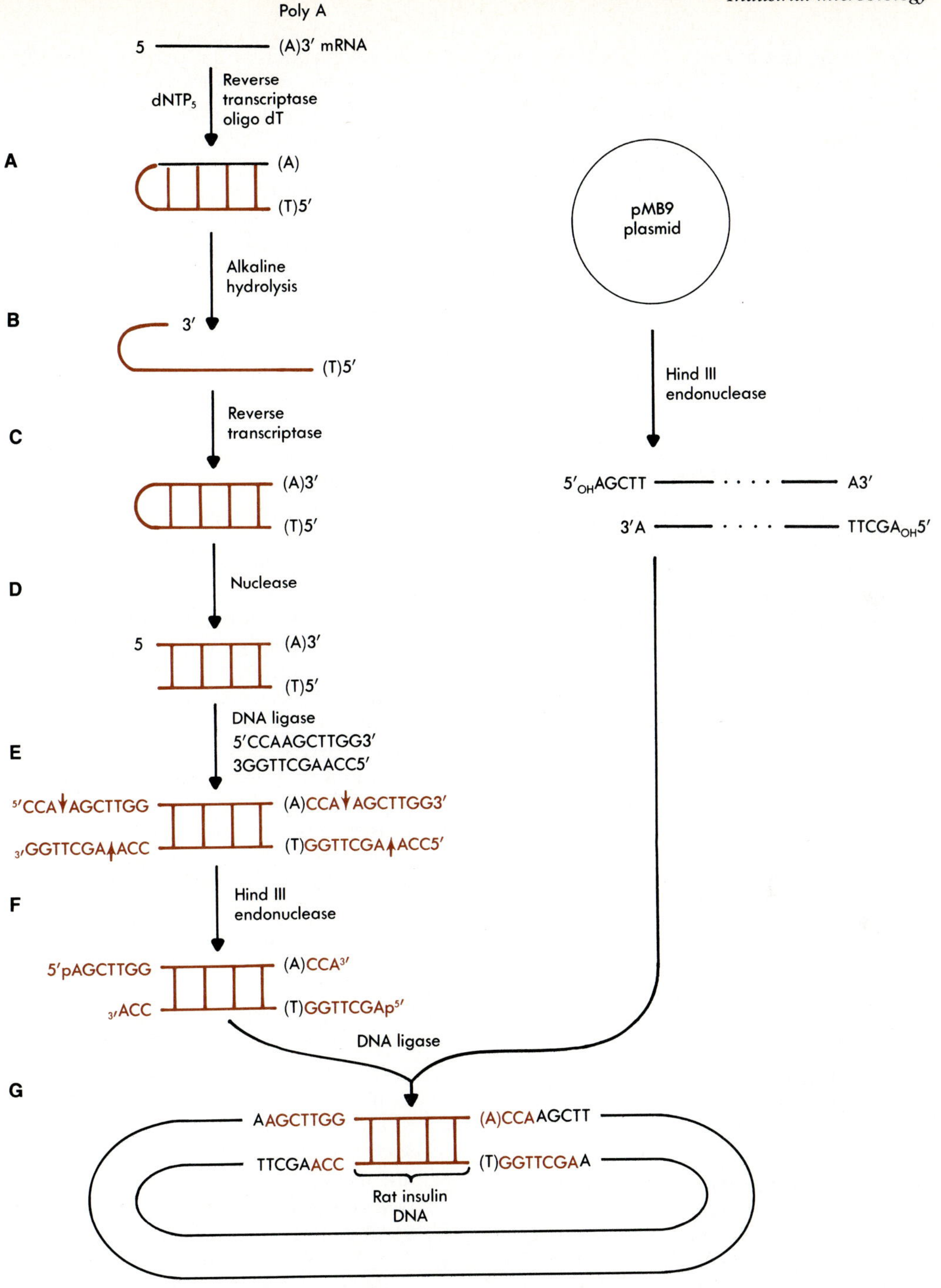

Figure 22-21 *Cloning rat insulin gene.* ***A,*** *Rat insulin mRNA is isolated. Insulin mRNA contains poly A tail. In presence of reverse transcriptase plus oligodeoxythymidine primer and nucleoside triphosphates (NTP_5) mRNA is transcribed into complementary DNA (cDNA, in color).* ***B,*** *RNA removed by alkaline hydrolysis.* ***C,*** *Single-stranded DNA becomes double-stranded via reverse transcriptase producing hairpin structure.* ***D,*** *Hairpin region is removed by nuclease.* ***E,*** *By means of action of DNA ligase, chemically synthesized nucleotide sequences, which are recognized by Hind III endonuclease, are attached at both ends of cDNA (*arrows *point to sites attacked by endonuclease).* ***F,*** *Endonuclease produces sticky ends. Plasmid (pM B9) is similarly treated with endonuclease.* ***G,*** *Rat insulin cDNA is inserted into plasmid via DNA ligase. Plasmid will be used to transform* E. coli.

the protein in its active form. Earlier experiments with yeasts and *E. coli* resulted in the enzyme not being secreted and even when isolated to still be in an inactive form (prochymosin). This problem is overcome by fusing the gene for rennin synthesis to that of another yeast enzyme that is ordinarily secreted. This resulted in the secretion of the enzyme in an active form. Even though yeast is not used for the commercial production of rennin, the results of these studies will be important when other nonmicrobial proteins are cloned.

Concept Check

1. What are the steps followed in wine making, and what precautions must be taken to produce an acceptable product?
2. What techniques are being currently exploited in an effort to improve yeast strains? Which shows the most promise?
3. What are major stumbling blocks in the use of recombinant DNA technology in industrial microbiology?
4. What aspects of industrial microbiology will be improved with the use of genetic engineering?

SUMMARY

1. The products of industrial microbiology can be divided into the following groups:
   a. Pharmaceuticals, which include antibiotics, steroids, insulin, human growth hormone, somatostatin, and interferon
   b. Bioinsecticides
   c. Polysaccharides, such as dextran
   d. Vitamins, such as riboflavin
   e. Single-cell proteins
   f. Industrial chemicals, such as ethanol and citric acid
   g. Food and beverages, such as cheeses, alcoholic beverages, yogurt, and vinegar
   h. Enzymes, such as alpha amylase and proteases
   i. Amino acids, such as lysine and methionine
2. Efficient antibiotic production depends on factors controlling inoculum production, culture medium, the fermentation process, and product recovery. Improvement of antibiotic-producing strains relies on genetic alterations that increase the amount of antibiotic produced, as well as increase the resistance to the antibiotic produced. Antibiotic derivatives can be obtained by mutations involving the parent strain. However, an important technique for obtaining new antibiotic derivatives is to chemically alter the nucleus of the preexisting antibiotic.
3. Important industrial enzymes produced by microorganisms include α-amylase, which is used in bread making, brewing, alcohol production, sugar syrups, spot remover, and proteases, which are used in the baking industry as meat tenderizers and in laundry products. Most enzymes are produced by procedures similar to those for antibiotic production.
4. Amino acids, such as glutamic acid, lysine, and methionine, are produced by fermentation processes similar to those for antibiotic production. Amino acids are frequently used to fortify foods. Organic acids and solvents, such as ethanol, citric acid, gluconic acid, itaconic acid, and lactic acid are all produced by microbial fermentation. Ethanol is being used as a fuel additive; citric acid is used as an acidulant in many foods; gluconic acid is used as a cleaning agent and acidulant in foods; itaconic acid is used in paper adhesives; and lactic acid is used as an acidulant and flavoring agent in foods.
5. Polysaccharides, such as dextrans, alginates, and xanthan, are produced by microorganisms and are used in the food industry as thickeners and suspending agents. Dextrans are used as a blood expander.
6. Microbial cells are used in certain industrial processes for the production of consumer products of nonmicrobial origin. An example of this is the use of microbial cells to carry out specific transformation reactions in the synthesis of steroid hormones such as cortisone. In some instances, microbial cells are used as consumer products. Microbial insecticides often are based on viable preparations of bacterial endospores.

7. Microorganisms are being considered as sources of food particularly for animals. The growth medium is usually a waste product from some other commercial process. Small pilot projects are engaged in the use of fungi as a source of human foods.
8. The brewing industry relies on yeasts such as *Saccharomyces cerevisiae* to convert sugars to ethanol and carbon dioxide. During the brewing process, malt and adjunct are broken down to maltose and dextrins by amylase present in the malt. The liquid is separated and transferred to a kettle where hops are added and the mixture is boiled. Yeast is later added to the clarified material and fermentation results in the conversion of maltose to carbon dioxide and ethanol.
9. In wine making, ripened grapes are crushed and stems are removed. Sulfite is added to inhibit growth of undesirable microorganisms. Later the crushed grapes are inoculated with yeast and fermentation proceeds for 3 to 5 days. The wine is separated from skins and seeds, and fermentation continues for 11 days. The wine is separated from sediment and placed in aging casks where it ages for several months to years.
10. Strain improvement of yeasts for brewing industry and other enterprises can be accomplished through selection of mutants, cross-breeding, cytoinduction, protoplast fusion, and transformation. Genetic engineering offers a means of improving strains for industrial purposes but restrictions are vector selection and insufficient knowledge of the location and number of genes involved in product formation. Strain improvement by genetic engineering may take the following forms: gene amplification, vaccine production, increased fermentation, product tolerance, product excretion, and product purification.
11. Genetic engineering offers a potential for production of nonmicrobial products. Insertion of animal genes into microorganisms has resulted in production of human and animal insulin, interferon, and growth hormones.

Self-study Quiz

Completion

1. The enzyme used to partially break down grain starches to produce instant breakfast cereals is called ________.
2. The sediment of malt proteins precipitated by tannins of hops is called ________.
3. The enzyme used to form curds from milk in cheese making is called ________.
4. The mating process that results in progeny having the nuclear genotype of one parent and the cytoplasmic genotype of the other parent is called ________.
5. The agent added to grape juice to prevent growth of bacteria and fungi during fermentation to make wine is called ________.
6. The type of yeast used to produce ales is called ________.
7. The term used by industrial microbiologists to designate any process by which a microorganisms produces a product is called ________.
8. The bifidum pathway used by lactic acid bacteria results in the formation of ________ and ________.

Short answer

1. What is the basic approach to antibiotic-producing strain improvement? Briefly describe how this is done.
2. Describe how cornstarch is converted to sugar syrups.
3. What is the most significant advantage in using immobilized enzymes?
4. How is glutamic acid produced?
5. Identify the contribution of each of the ingredients of beer listed below:
   a. Malt
   b. Hops
   c. Yeast
   d. Adjuncts
6. List several potential areas in which recombinant DNA technology can be expected to improve industrial microbiology.

Multiple choice

1. Which of the following organisms would you add to grape juice to produce wine?
   A. *Aspergillus niger*
   B. *Gluconobacter oxydans*
   C. *Saccharomyces cerevisiae var. ellipsoideus*
   D. *Saccharomyces carlsbergensis*
2. What determines the color of beer?
   A. The color of the hops flower
   B. The color of the malt
   C. The type of yeast used to ferment the malt
   D. The color of the adjuncts added
3. Meat tenderizer is what type of product?
   A. Fungal protease
   B. Bacterial α-amylase
   C. Glutamic acid
   D. Glucamylase
4. Which of the following conditions are used in the production of antibiotics?
   A. Cultures are grown under aerobic conditions in larger fermenters.
   B. Cultures are grown in huge shallow pans containing media to reduce the buildup of metabolic heat.
   C. The media is trickled over a large column of glass beads on which the organisms grow.
   D. Cultures are grown in huge anaerobic vats.
5. The hydrolysis of sucrose to glucose and fructose is carried out by which enzyme?
   A. Pectinase
   B. Gluconase
   C. Invertase
   D. Rennin
6. Why is the microbial transformation of steroids often preferred over chemical modification?
   A. The microbially transformed steroids are more active than the chemically transformed product.
   B. Other parts of the steroid ring are likely to be altered by chemical conversion.
   C. Microbial conversions are extremely specific, with about 100% transformation of substrate to product.
   D. Microbial conversions are carried out anaerobically without heat, thereby conserving energy costs.

7. For what are alginates used?
   A. Blood extenders
   B. Food and pharmaceutical stabilizers
   C. Single-cell protein foods
   D. Insecticides
8. To produce a light beer, which of the following would be removed?
   A. Hops
   B. Tannins
   C. Dextrins
   D. Wort
9. Which of the following cannot be expected from application of recombinant DNA technology to industrial microbiology?
   A. Development of bacterial strains that can produce five or six different antibiotics in culture at the same time
   B. Production of more specific and safer vaccines
   C. Increased production of a microbial product from a single cell by increasing the number of copies of the gene in the cell
   D. Development of product tolerant strains
10. What enzyme would you use to remove oxygen that can cause unwanted oxidation reactions in foods such as fruit juices and mayonnaise?
    A. Invertase
    B. Glucose oxidase
    C. Alkaline protease
    D. Streptodornase

Thought questions

1. Choose three processed foods (for example, sandwich spread, canned chili, instant oatmeal, ice cream, or instant pudding). Read the label of each and list the ingredients that were produced by a microbial process.
2. Look back through the chapter and note the general principles involved in brewing a fermented beverage product from grains.
   A. List those principles
   B. Using the principles listed above, describe how you would make a fermented beverage from rice (known as *sake* in Japan)
3. You and six other people are stranded on a deserted atoll in the South Pacific. Shortly after landing, one of your colleagues develops a nasty infection from a cut. Using only the primitive materials at hand, how would you develop a primitive antibiotic therapy?

SELECTED READINGS

Books

Amerine, M.A., Berg, H.W., and Cruess, W.V. *The technology of wine making (4th ed.)*. AVI Publishing Co., Inc., Westport, Conn., 1980.

Biomass Process Handbook. Technical Insights, Inc., Fort Lee, N.J., 1981.

Bull, A.T., Ellwood, D.C., and Ratledge, C. *Microbial technology: current state, future prospects*. Cambridge University Press, Cambridge, 1979.

Demain, A.L., and Solomon, N. *Biology of industrial microorganisms*. Benjamin/Cummings, Menlo Park, Calif., 1985.

Hugo, W.B., and Russell, A.D. *Pharmaceutical microbiology*. Blackwell Scientific Publications, Oxford, 1977.

Lafon-Lafouracade, S. Wine and brandy. In Rehra, H., and Reed, G. (Eds.). *Biotechnology,* vol. 5, Verlag Chemie, Wenhein, 1983.

Miller, B.M., and Litsky, W. *Industrial microbiology*. McGraw-Hill Book Co., New York, 1976.

Norris, J.R., and Richmond, M.H. *Essays in applied microbiology*, John Wiley & Sons, Ltd., Chichester, England, 1981.

Peppler, H.J., and Perlman, D. *Microbial technology.* Vol. 1, *Microbial processes*. Vol. 2, *Fermentation technology.* Academic Press, Inc., New York, 1979.

Rose, A.H. (Ed.). Alcoholic beverages. *In Economic Microbiology.* Vol. I, Academic Press, Inc., London, 1977.

Smith, J.E., and Berry, D.R. The filamentous fungi. Vol. 1. *Industrial mycology*. Halsted Press, New York, 1975.

Smith, J.E. *Biotechnology principles in aspects of microbiology* (No. 11). American Society for Microbiology, Washington, D.C., 1985.

Wang, D.I-C., et al. *Fermentation and enzyme technology*. John Wiley & Sons, Inc., New York, 1979.

Wiseman, A. (Ed.). *Principles of biotechnology.* Surrey University Press, Glasgow, 1983.

Yamada, K., Kinoshita, S., Tsunoda, T., and Aida, K. *The microbial production of amino acids*. Halsted Press, New York, 1972.

Journal Articles

Detroy, R.W., and Julian, G. Biomass conversion: fermentation, chemicals, and fuels. *Crit. Rev. Microbiol.* **10**(3):203, 1984.

Industrial microbiology, *Sci. Am.* **245**:Entire Issue, 1981.

Spencer, J.F.T., and Spencer, D.M. Genetic improvement in industrial yeasts. *Annu. Rev. Microbiol.* **37**:121, 1983.

PART VI

Infection and immunity

CHAPTER

23

Host-parasite interactions

Chapter Outline

Introduction

A silent but constant "war" takes place on and within the human body. Our bodies are colonized and can be colonized by microorganisms whose activities, if not controlled by certain natural host defense mechanisms, can lead to disease. This is a concept that is sometimes difficult to imagine for the healthy individual because the "silent war" produces no overt body responses or damage to the host. Many microorganisms are equipped to cause disease, but some are better equipped than others. Similarly, the host has several physical and chemical attributes that can prevent any harmful effects from microbial colonization. This host resistance varies from individual to individual, depending on genetic (innate) and environmental factors. The relationship between host and parasite includes not only humans but also other animals and even plants. Some of these relationships are discussed in earlier chapters.

This chapter discusses the relationship that exists between parasite and host, as well as indigenous microorganisms and the host; the mechanisms used by the parasite to invade the host at various body sites; the host's first and second lines of defense against an invading parasite; the mechanisms of phagocytosis as a deterrent to microbial invasion and how some microorganisms evade this process; and the mechanisms by which microorganisms can damage host tissue.

CONCEPTS OF HOST-PARASITE RELATIONSHIPS

Commensals

Areas of the human body, such as the skin, mouth, gastrointestinal tract, nose, throat, and genitalia, are the abode for a large number of microorganisms, especially the bacteria. Very few species of fungi or protozoa are indigenous to humans.* Organisms indigenous to the host are called **commensals** (often referred to as "normal flora"), and the relationship is one in which the commensal benefits and the host is neither helped nor harmed. The kind of microorganisms that can inhabit each body site is influenced by the chemical and physical characteristics of the microorganism, the chemical and physical characteristics of the host habitat, and other microorganisms already present in the habitat.

The total number of commensal species, within certain limits, remains relatively constant at specific sites in the host. This constancy is maintained by the limitation of nutrients produced in the host or by metabolic products excreted by neighboring commensals. Some examples should make this clearer. (1) The bacterium, *Bacteroides melaninogenicus,* which inhabits the area between the gums (gingiva) and tooth surface, requires vitamin K for growth. Vitamin K is excreted by other resident microorganisms, such as species of *Veillonella.* (2) Many commensals ferment sugars, which results in the production of acids that reduce the pH of the environment to a level that inhibits the growth of some microorganisms. (3) Commensals in the intestinal tract produce fatty acids that are lethal to some microorganisms attempting to colonize intestinal surfaces. (4) Usually, gram-positive bacteria control the number of gram-negative species and vice versa, and the presence of both groups influences the concentration of yeasts. How this system works can be best explained by an example. Certain chemotherapeutic drugs, such as antibiotics, can severely reduce the population of gram-positive and gram-negative bacteria in many body sites, such as the mouth. Yeast cells, without the controlling influences of gram-positive and gram-negative bacteria, exhibit unrestricted growth and create a disease condition.

Parasites

A parasitic relationship is one in which the microorganism benefits but the host is harmed. Most parasitic relationships occur when nonindigenous microorganisms colonize the host, but occasionally commensals can also become parasites if the host has some underlying condition or disease that impairs normal defense mechanisms. Microorganisms that cause disease are called **pathogens**, but some are more pathogenic than others.

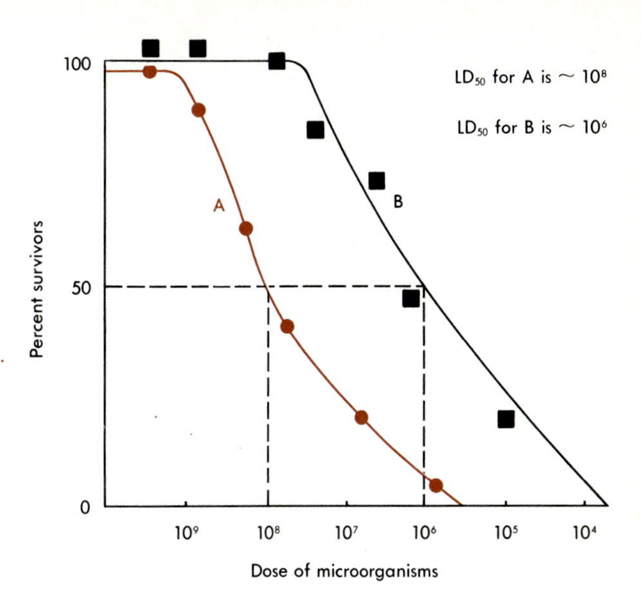

Figure 23-1 *A hypothetical plot measuring the dose of microorganisms required to kill 50% of the test animals. Two different microbial strains of same species are indicated, A and B.*

The ability to become a pathogen is often related to many factors of either host or microbial origin. Certain strains of *Escherichia coli,* for example, that colonize the intestinal tract owe their pathogenicity not only to their ability to adhere to intestinal epithelium but also to the production of a toxic substance. The absence of one of these properties usually renders the microorganism harmless. Pathogenicity is a qualitative phenomenon; however, the measure of an organism's disease-producing traits can be quantitated. The term **virulence** expresses this quantitative measurement. For example, the dose of microorganisms or amount of toxin required to produce an effect in the host can be measured in the laboratory. The terms LD_{50} and ID_{50} refer to the dose of microorganisms that is lethal and infectious, respectively, for 50% of the animals or other hosts used in the experiment. In the example in Figure 23-1, organism B is more virulent than organism A because fewer B organisms are needed to kill 50% of the animals. The virulence of an organism can be maintained if the microorganism replicates in the host but it may be lost if the organism is cultured outside the host. Bacteria can lose their virulence by repeated passage on laboratory media, and the virulence of viruses is affected by repeated passage in cell culture. Some bacteria owe their virulence to the production of a capsule or slime layer that prevents digestion by certain cellular elements in the blood called phagocytes. Repeated passage on laboratory media can result in the formation of mutants that lose the property of capsule production and are better suited to growth on artificial media than in the host.

* Those microorganisms that are part of the normal microflora of humans are described in Chapter 26, p. 788.

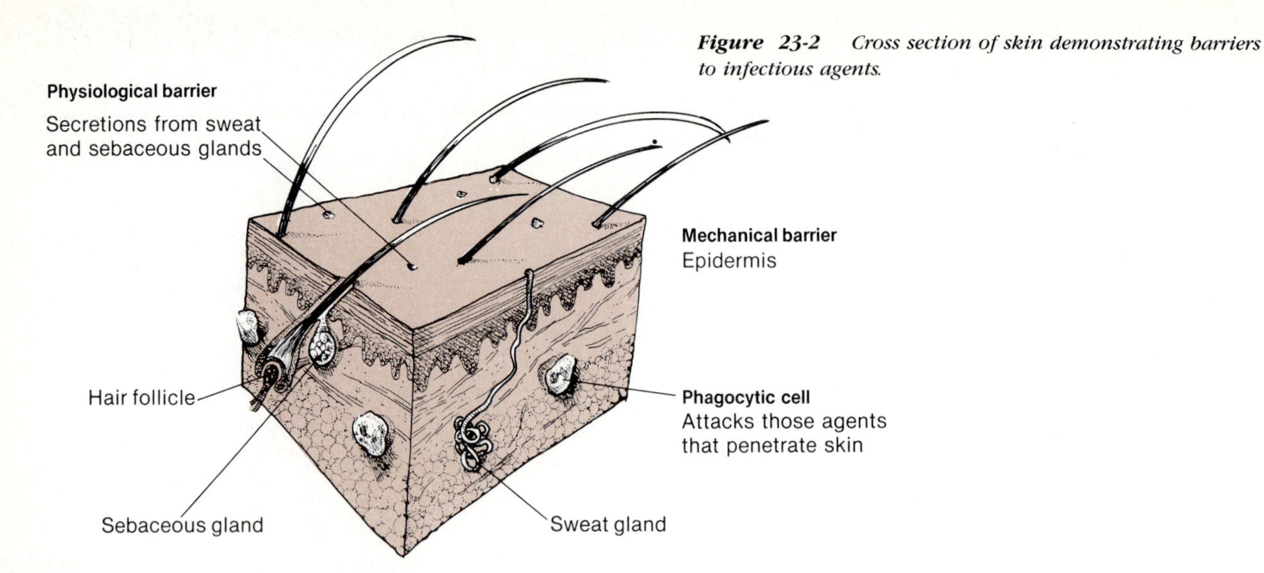

Figure 23-2 *Cross section of skin demonstrating barriers to infectious agents.*

This reduction or loss of virulence is called **attenuation.** Attenuated microorganisms exhibit nearly all of the other properties of the virulent strain except they are unable to cause disease. Attenuation is a technique used in a production of vaccines, which are microorganisms or material derived from microorganisms or other parasites such as viruses, that induce the host's immune system to resist disease. Vaccines have reduced the incidence of infectious diseases and prevented early deaths from afflictions such as smallpox, polio, and diphtheria. Vaccines are discussed in Chapter 24.

Two terms that are sometimes misinterpreted when considering parasitic interactions are infection and disease. **Disease** is defined in general terms as any disturbance of the structure or function of an organism in which there are characteristic signs or symptoms. Diseases caused by microbial agents are called **infectious.** When you go to a doctor to find out the cause of your runny nose or sore throat, he invariably will say that you have a viral or bacterial infection. An infection, however, is not the same as a disease, even though the two terms are used interchangeably. For example, a considerable percentage of the population carry staphylococci in the anterior nares (nose). They may carry this organism for a few days, weeks, or even months. If the organisms do not cause the formation of antibodies or elicit a hypersensitive response in the host, they are merely **colonizers.** If the colonizers cause the host to respond by producing antibodies or elicit a hypersensitive response, the organisms have caused an **infection.** Organisms that colonize the host and initiate overt manifestations such as fever, swelling, abscess formation, and other responses have produced an infectious disease. The distinction between disease and infection is important because treatment may be required for the former and not the latter. Unnecessary treatment using toxic drugs or drugs that may give rise to secondary infections is not in the best interest of the patient. The distinction between infection and disease is not always clearcut. The distinction may depend on the microbial species and condition of the host.

MECHANISMS OF PATHOGENESIS AND HOST RESISTANCE

There are four conditions that must be satisfied before an infectious agent can cause disease: it must (1) gain entry into the host, (2) colonize host tissue, (3) resist the defenses of the host, and (4) damage host tissue.

Entry of microorganisms into the host

There are several sites for the entry of microorganisms into the host: the skin, respiratory tract, genitourinary tract, and intestinal tract.

Skin and mucous membranes. The host's first line of defense is the intact skin (Figure 23-2) and mucous membranes. The skin, because of its thick keratinized layer, is the most effective barrier to entry by foreign agents. Microorganisms that do adhere to the skin are easily removed by washing or by the natural shedding (desquamation) of epithelial cells. Secretions of the sebaceous glands and sweat glands contain antimicrobial factors, in the form of fatty acids or other organic acids, that reduce the pH. Acidic pH is an inhibitor to the growth of many microorganisms, particularly bacteria. The continuity of the skin is broken in such areas as the eye, genitourinary tract, and mouth, but these areas possess other barriers to infection. The conjuctiva is a transparent layer that protects the eye from invasion by microbial agents by the flushing action of tears. Tears, as

Figure 23-3 *Barriers to infectious agents in the respiratory tract.*

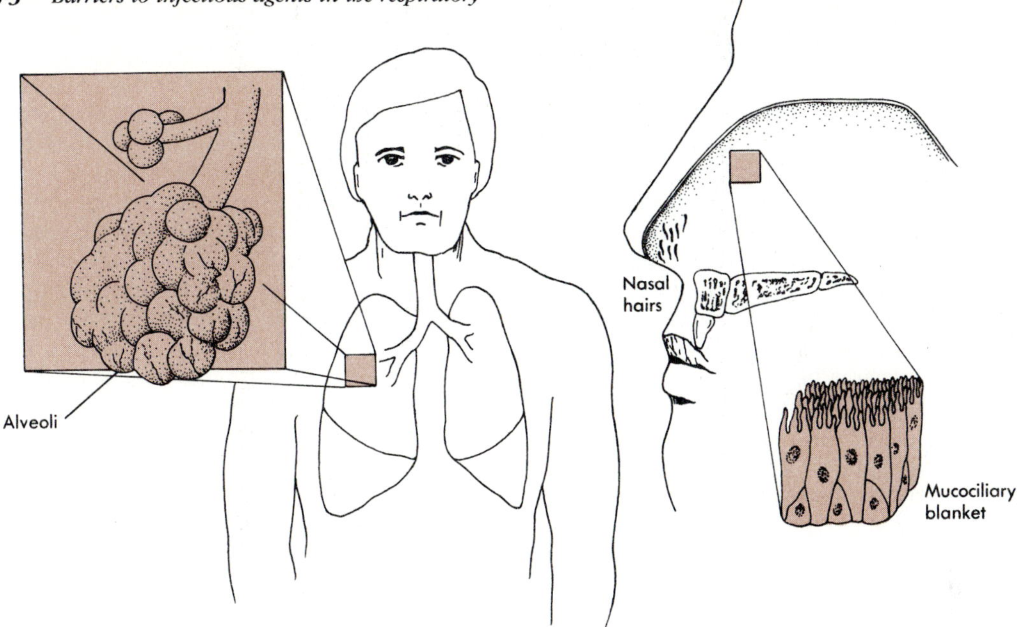

well as other body fluids, also contain **lysozyme**, an enzyme capable of breaking down the peptidoglycan layer of bacterial cell walls (see Figure 5-6, p. 107). The mucous membranes of the respiratory tract, genitourinary tract, and intestinal tract are barriers to microbial invasion and are discussed later.

The hard epithelial layer of the skin can be penetrated when there are abrasions or wounds, or when a skin condition, such as eczema, alters the continuity of the skin and makes it susceptible to infection. Most infections of the eye are also due to abrasions of the conjunctiva followed by touching the eye with contaminated fingers or objects. The epithelial barrier can also be broken by biting insects or animals. Fleas, ticks, mosquitoes, and lice are arthropods that can penetrate the skin and inject the agents of diseases such as malaria, plague, and encephalitis into the bloodstream. Animal bites or scratches are responsible for diseases such as rabies and cat scratch fever. Indigenous species of microorganisms on the skin *(Staphylococcus, Propionibacterium)* can cause infections such as acne, but only under abnormal physiological conditions in the host, such as sebaceous glands clogged by secretions.

Respiratory tract. A very large proportion of infectious diseases seen in the world today are respiratory diseases. The respiratory tract possesses several innate mechanisms for handling the presence of foreign agents, including microorganisms suspended in the air (Figure 23-3). A **mucociliary blanket** covers much of the respiratory tract and acts as a trap for foreign matter. It consists of mucus-secreting cells that contain **lysozyme**, and **cilia**, which are projections of epithelial cells. The mucociliary blanket traps particles up to 10 μm in diameter and expels them by using the sweeping motion of the cilia. Particles larger than 10 μm are usually trapped by hairs lining the nasal cavity. Particles in the lower respiratory tract are swept up to the throat; particles in the upper respiratory tract are swept to the back of the throat and then swallowed. The **alveoli**, which are the terminal portions of the lower respiratory tract, do not possess a mucociliary blanket, but they are lined with **phagocytic cells (macrophages)** that ingest foreign matter.

The successful colonization of the respiratory tract by microorganisms is often related to abnormal physiological conditions in the host. Disturbance or removal of the mucociliary blanket by prior conditions, such as bronchitis or viral influenza, can lead to microbial colonization of the unprotected epithelial cell layer.

Intestinal tract. Infections of the intestinal tract occur most often from the ingestion of contaminated food or water. Any exogenous (noncommensal) microorganisms that enter the intestinal tract in this manner will encounter many obstacles before they can establish a site for colonization. Epithelial surfaces, such as gingiva (gums) and the hard palate are **keratinized** and offer a mechanical barrier to infectious agents. The **mucus** coating the oral cavity, including the teeth, helps to trap microorganisms, and the saliva that is formed carries microorganisms to the back of the throat, where they are swallowed. Once microorgamisms are swallowed, many are rapidly destroyed by **stomach acids**. One microorganism that is able to somehow evade the acidic environment and penetrate the stomach mucosa is *Campy-*

lobacter pyloridis. This motile organism appears to attach to epithelial cells of the mucosa where it multiplies and releases inflammatory products. *C. pyloridis* is a major cause of gastritis. Some microorganisms survive stomach acidity and are transported to the intestinal tract, where **bile salts** and **intestinal enzymes** can attack them. Even if the infectious agent survives these barriers, other host factors can contribute to its demise. Mucus lines the epithelium of the intestinal tract and acts as an obstruction to microbial adherence and penetration. Part of the intestinal secretion contains a special class of antibodies called **secretory IgA (sIgA)** that protects the immune individual against infection. (See Chapter 24 for a discussion of IgA.) These factors, combined with the **peristaltic action** of the intestine, which moves intestinal contents from the upper to the lower intestine, are major obstacles to colonization by infectious agents. Indigenous microorganisms can also be important in protecting the host from invasion by noncommensal species. The number of commensals that inhabit the intestinal tract is staggering, and some groups, such as the *Bacteroides,* can be found in concentrations as high as 1×10^{10}/gm of feces. These species release products of metabolism, such as fatty acids, that prevent "outsiders" from colonizing the intestinal surface.

The elimination of commensal species or the reduction in their number by certain antimicrobial agents can reduce the intestinal population to a point at which the commensal species exert no effect on ingested microorganisms.

Microorganisms can infect the intestinal tract if they possess special properties that enable them to resist host defense mechanisms, or if the natural barriers of the host have been altered by an underlying disease or condition (see Point of Interest). Many viruses, such as the enteroviruses (the poliovirus is an enterovirus), are resistant to acid, bile salts, and the action of intestinal enzymes. There are conditions in which the acid content of the stomach may be reduced, thereby permitting the entrance of acid-sensitive microorganisms. Microorganisms sensitive to acid can also be protected by entrapment in large particles of food that escape the effects of acid in the stomach. Infection may also be influenced by the number of pathogens ingested, as in the case of salmonella food poisoning in which the symptoms appear only after ingestion of microorganisms above a minimal number.

Genitourinary tract. The urinary tract and bladder are essentially sterile environments that, in the healthy individual, are not invaded by microorganisms. The absence of microorganisms in these environments is primarily a result of the flushing action of urine. The epithelial cells of the vagina, between puberty and menopause, contain large amounts of glycogen, which is metabolized to acids that prevent colonization by microorganisms other than commensals. Urinary tract infections are still more prevalent in females because of the shortness of the urethra and the proximity of the urethra to the anus, where gram-negative bacterial species are abundant. The vaginal secretions before puberty and after menopause are slightly alkaline, which also pro-

Point of Interest

HOSPITAL ACQUIRED INFECTIONS

Many patients enter hospitals for treatment for a disease, surgery, or injury and end up acquiring another new infection during their hospital stay. Such infections are called **nosocomial** infections. Many of the organisms that produce these hospital infections are common bacteria that do not cause overt disease under normal conditions. These bacteria are opportunists that can cause infections if the body's normal outer defense mechanisms are bypassed. The skin is broken as a defense barrier in surgery, injuries, or burns. Hospital procedures can introduce bacteria into the urinary tract by catheterization, into the respiratory tract from respirators or aerosol medications, or into the bloodstream by intravenous feeding. In addition, hospital cancer and transplant patients have suppressed immune systems due to radiation or drug therapy.

The bacteria involved in most hospital acquired infections are *E. coli, S. aureus,* and *P. aeruginosa.* These are all bacteria that are not harmful under normal conditions and are actually a part of the normal human flora or common in the environment. However, all there bacteria produce either harmful extracellular enzymes, endotoxins, or exotoxins or have a means to resist phagocytosis, and thus can cause disease if the body's outer defense mechanisms are overcome or if the immune system is suppressed. In addition, *S. aureus* and *P. aeruginosa* are resistant to many disinfectants used in hospitals and thus remain in the environment. A large number of hospital personnel carry *S. aureus* as a part of the normal nasopharynx flora and it is commonly found on the skin. *E. coli* is a member of the normal intestinal fora and is especially important in infection when perforations of the intestine occur. *P. aeruginosa* is found in water and is very common in the environment.

Because of the extensive use of antimicrobial medications in hospitals, a large number of the bacteria found in hospitals are resistant to many commonly used antibiotics. Thus in order to cure diseases caused by these hospital acquired, resistant bacteria, antibiotics with an increased number of side effects may need to be used that may further harm the patient.

Table 23-1 *Host sites that are barriers to infection and nonmicrobial factors that favor microbial entry into the host*

Potential site of entry	Factors that prevent microbial entry	Factors favoring microbial entry
Skin and mucus membrane	Flushing action of tears; lysozyme in body fluids; secretion of sebaceous glands	Abrasions or wounds due to injury or animal and insect bites
Respiratory tract	Mucociliary blanket; phagocytic cells in alveoli; hairs lining nasal cavity	Abnormal physiology of host; loss of mucociliary blanket due to infectious or noninfectious agents
Intestinal tract	Keratinized epithelial surfaces; mucus coating oral cavity; stomach acids, bile salts, and intestinal enzymes; peristaltic action of intestine; secretory IgA in intestinal secretions; normal microbial flora	Conditions that reduce the microbial flora (antimicrobial therapy); conditions that result in elevation of pH
Genitourinary tract	Flushing action of urine; acidic conditions in vagina	Obstruction of urinary tract; abrasions from catheter insertion or sexual intercourse

vides an environment for the growth of unwanted microbial species. Obstruction of the urinary tract, which prevents the free flushing action of urine, is an important reason for the appearance of urinary tract diseases. Abrasions resulting from the insertion of catheters and other instruments as well as from sexual intercourse can also lead to the introduction of potentially pathogenic microorganisms and their colonization of the genitourinary tract.

Table 23-1 outlines host factors that are barriers to microbial invasion and those nonmicrobial factors that interfere with these barriers.

Colonization and multiplication of pathogens in the host

Growth on epithelial surfaces. Both the indigenous microbial population of the host and nonindigenous parasites colonize epithelial surfaces by highly specific interactions. These interactions involve microbial surface components referred to as **adhesins**. The adhesin may be a capsular sugar complex, pili, or other components. Little is known of the molecular architecture at the site of microbial adherence, and only now are microbial adhesins being isolated, purified, and characterized as to their role in the infectious disease process. Discussions of some of the parasites in which surface components have been implicated in the attachment process follow.

Escherichia coli. The bacterium *E. coli* is normally a commensal in the intestinal tract, but pathogenic strains also exist. Both pathogenic and nonpathogenic strains produce chromosome and plasmid-mediated **pilial adhesins** (Figure 23-4). All the pilial adhesins studied to date have an affinity for carbohydrate receptors on the epithelial surface. Many receptors for the pilial adhesion contain the sugar mannose, whereas others do not; thus adhesins are referred to as **mannose-sensitive** and **mannose-resistant**. This is based on the ability of D-mannose and its derivatives to bind to and block the binding sites of the pili, which otherwise would bind to cell recep-

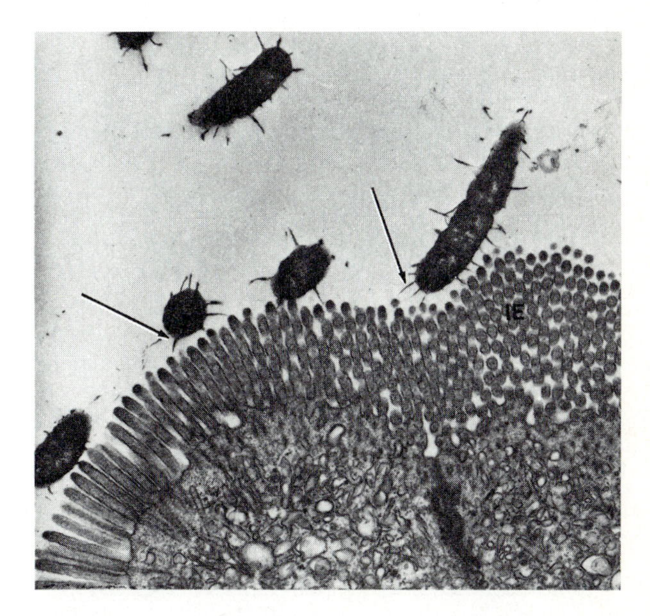

Figure 23-4 *Electron micrograph demonstrating attachment of* E. coli *to intestinal epithelium* (IE). *Arrows point to attachment-pili of microorganism.*

tors. Many epithelial surfaces contain mannose receptors, including phagocytes, the immune cells that aid in destruction of microbial parasites. This implies that although some strains of *E. coli* can initiate disease by attaching to the intestinal epithelium if they penetrate deeper into tissue where phagocytes reside, they could be destroyed because of binding to phagocytes.

Influenza virus. The influenza virus, as well as those viruses that cause the common cold (rhinoviruses and coronaviruses), confines its activities to the epithelial surface. The influenza virus and other enveloped viruses possess glycoproteins projecting from the envelope surface (see Figure 16-5). These **glycoproteins** are important for adsorption and appear to play a role in penetration and virulence. The hemagglutinin glycoprotein spike of the influenza virus adheres to the host cell by binding to the neuraminic acid component of the cytoplasmic membrane.

Streptococcus pyogenes. The bacterium *S. pyogenes* is a cause of streptococcal sore throat and one of its consequences, rheumatic fever. The ability of the microorganism to adhere to the respiratory epithelium is correlated with the presence of a protein called the **M protein** (Figure 23-5) and **lipoteichoic acid** on the cell envelope. These two components form a fibrillar network on the surface of the streptococcus by virtue of the interaction of the polyanionic backbone of the lipoteichoic acid and the positively charged M protein. This interaction permits exposure of the lipid ends of lipoteichoic acid and promotes attachment to receptor cells on the host cell surface (Figure 23-6). The receptor on the host cell is believed to be an albumin-like protein.

Streptococcus mutans. The bacterium *S. mutans* is the agent of dental decay (caries). The ability of some strains to adhere to the enamel of the tooth appears to be related to several surface components. *S. mutans* is known to interact with several types of glycoproteins found in whole saliva and secretions from the parotid gland. In addition, the microorganism produces an insoluble glucose polymer from sucrose called **dextran**. In vitro experiments have demonstrated the importance of dextran for attachment to a variety of surfaces. Those strains of *S. mutans* that produce **levan** (fructose-containing polymer) adhere poorly to similar surfaces. It is believed that the enzyme **glucosyltransferase**, which catalyzes the formation of dextran, forms a complex with the dextran and the microbial surface and promotes attachment to the tooth surface (Figure 23-7).

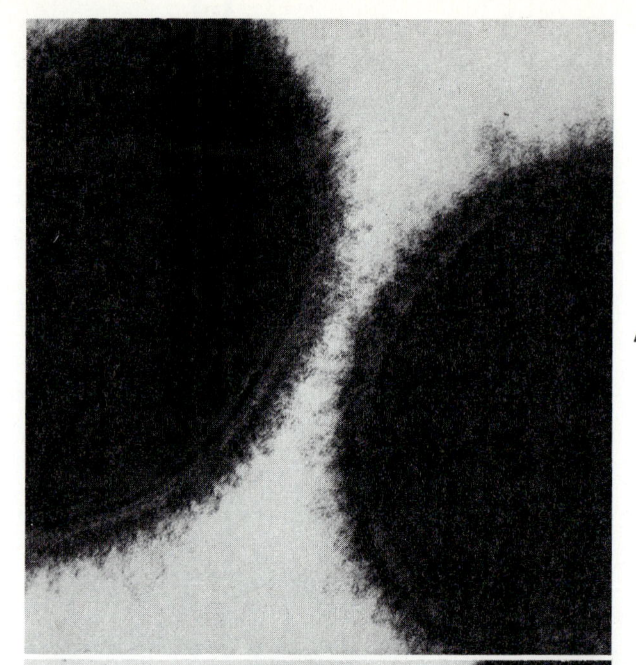

A

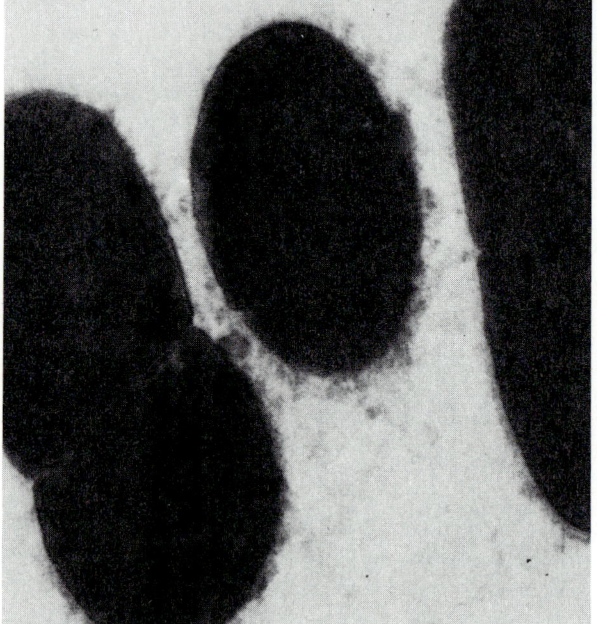

B

Figure 23-5 *Electron micrographs of cross sections of* Streptococcus pyogenes *with M protein,* ***A****, and without M protein,* ***B****. (×84,000.)*

Concept Check

1. How are commensals in the body maintained at relatively constant numbers?
2. What is the difference between the concepts of infection vs. disease and pathogenicity vs. virulence?
3. How can each of the potential sites of microbial entry into the host act as barriers to microbial penetration? How does the microorganism overcome these obstacles?
4. What are microbial adhesins or surface components that enable the pathogen to colonize a surface?

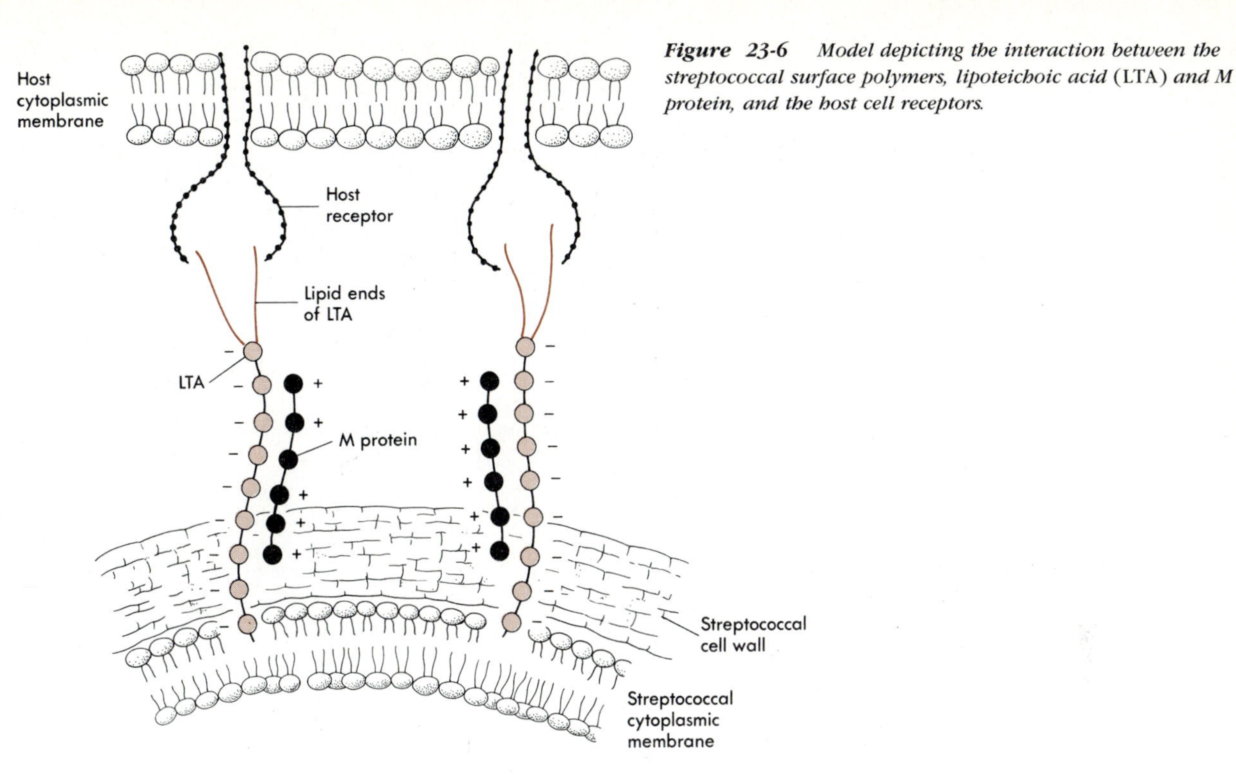

Figure 23-6 *Model depicting the interaction between the streptococcal surface polymers, lipoteichoic acid* (LTA) *and M protein, and the host cell receptors.*

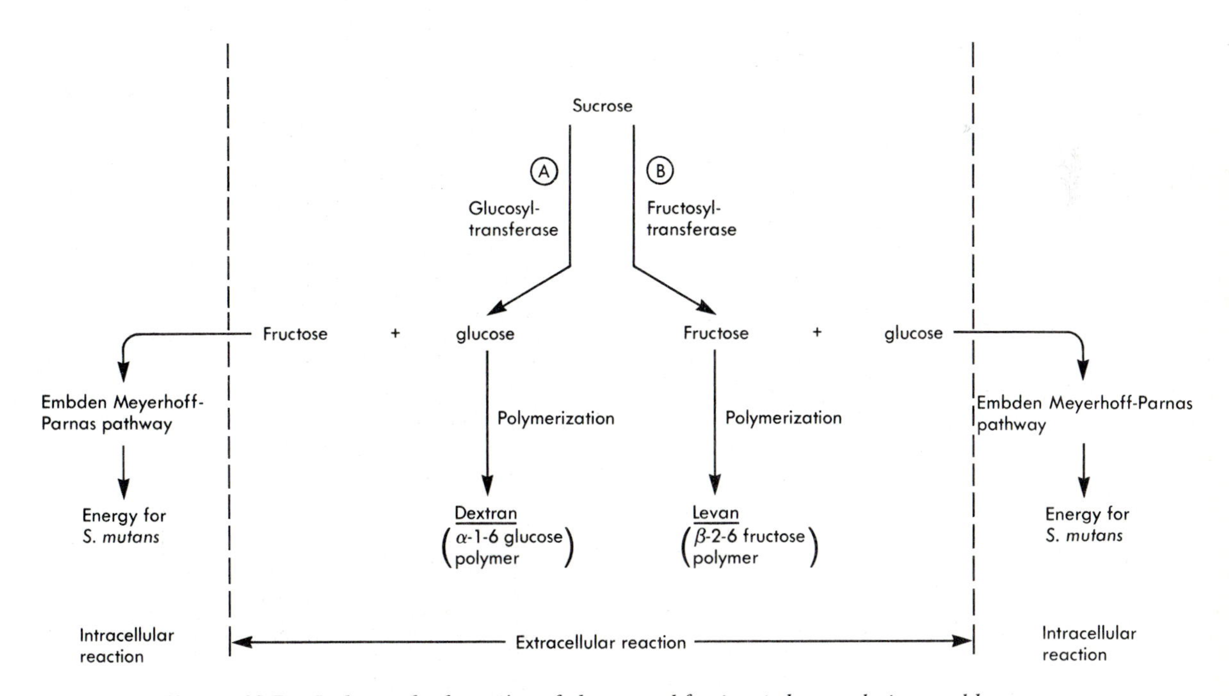

Figure 23-7 *Pathways for formation of glucose and fructose polymers, dextran and levan, respectively, from the disaccharide, sucrose. Two strains of* Streptococcus mutans *are illustrated. Strain A uses enzyme glucosyltransferase to cleave sucrose to glucose and fructose. Same enzyme is involved in formation of the glucan polymer, dextran. Fructose moiety enters the cell and serves as energy source for* S. mutans. *Strain B uses enzyme fructosyltransferase to cleave sucrose to fructose and glucose. Same enzyme is used to form fructose polymer, levan. Glucose moiety enters bacterial cell and serves as energy source.*

Table 23-2 *Bacterial products believed to be important in the spread of microorganisms in tissue*

Microbial product	Microorganism(s) involved	Activity of microbial product
Hyaluronidase	Streptococci, staphylococci, and clostridia	Degrades hyaluronic acid, a component of connective tissue
Coagulase	Staphylococci	Acts on clotting blood, which results in formation of fibrin; is believed to prevent phagocytosis
Collagenase	Species of *Clostridium*	Digests collagen, which is a component of muscle
Hemolysins	Many bacterial species, especially streptococci and staphylococci	Destroys red blood cells; importance is uncertain
Fibrinolysin (streptokinase)	Streptococci	Converts plasminogen in human sera to plasmin, an enzyme that digests fibrin; microorganisms trapped in fibrin clot are free to spread
Lecithinase	Species of *Clostridium*	Destroys lecithin, a component of cell membranes
DNase	Staphylococci	Causes lowering of viscosity of exudates (pus), giving microorganism greater mobility

Neisseria gonorrhoeae. Substantial evidence indicates that piliated *N. gonorrhoeae* adheres more strongly to mucosal surfaces of the urogenital tract than nonpiliated strains. Adhesion is an important factor in the pathogenesis of the gonorrhea-producing organism. Adhesion permits the bacterium to more effectively release toxic secretions, such as lipopolysaccharide, to host cells and also facilitates its invasion of epithelial cells. The receptors on the host cell are believed to be galactose β 1-3 *N*-acetylgalactosamine β 1-4 galactose.

Penetration and spread of microorganisms in the host. Microorganisms that are confined to epithelial surfaces can be spread to other epithelial surfaces by a variety of means (Table 23-2). The table shows that the spreading factors are enzymes that degrade host cellular components. Most studies have shown, however, that many of these enzymes probably play a minor role in virulence. Most of the enzymes play a role in nutrition and only when the bacterial species has multiplied sufficiently at the site of infection do they promote spread.

Surfaces covered with fluid, such as those in the respiratory and intestinal tracts, allow free diffusion of microorganisms, especially if there are excess secretions due to infection. Cilia, if they are not damaged, can sweep fluids to other parts of the respiratory tract, and coughing and sneezing also permit spread to uninfected areas. Microorganisms on the skin are usually spread by scratching and touching uninfected areas of the body with contaminated fingers. Microorganisms that remain on epithelial surfaces can produce symptoms in the host long before antimicrobial factors come into play. The immune system is slow to respond to any microbial agent that remains at the epithelial surface and is effective only during the later stages of infection.

Some microorganisms are virulent because of their ability to penetrate as well as adhere to epithelial tissue, but the microbial factors responsible for penetration are not entirely understood. Experiments with avirulent and virulent strains of some intestinal parasites have demonstrated that a component of the cell envelope, such as a sugar moiety, may be one type of determinant. Certain protozoa can penetrate host cells through the action of **lysosomal** enzymes located at the end of the parasite that is in contact with the host cell (Figure 23-8).

Microorganisms that are able to penetrate the epithelial surface are subject to many host defense measures that could destroy them. The microbial agent must be equipped to resist these host factors and must find susceptible cells on which or in which it can replicate. Once the epithelial surface has been penetrated, the parasite can enter the **lymphatics** and conceivably infect any part of the host. The lymphatic system is a series of vessels and nodes that carry fluid into the circulatory system (Figure 23-9). The vessels of the lymph system are in close proximity to blood vessels, and at very strategic points along their path they become associated with large nodes called **lymph nodes**. Lymph nodes are

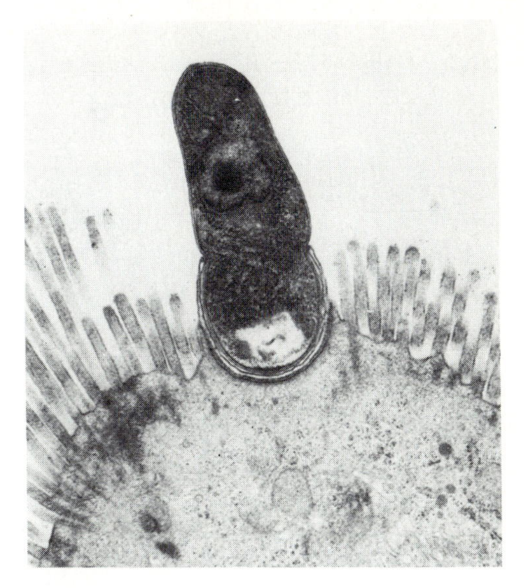

Figure 23-8 *Electron micrograph of protozoan penetrating mucosal surface of guinea pig mucosal cell.*

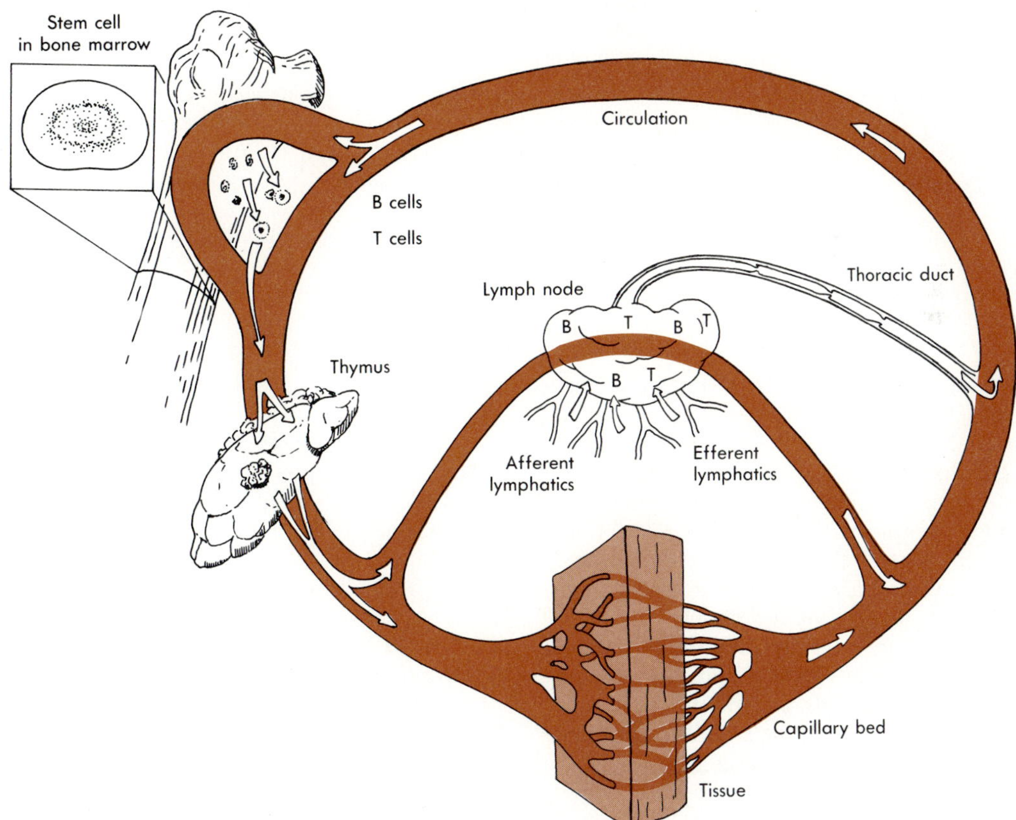

Figure 23-9 *Flow of lymph in lymphatic system of humans. Interstitial fluid (lymph) collects in capillary beds and is transported into afferent lymphatics. Lymph is filtered in nodes and then passed into efferent vessels and finally thoracic duct, which empties into venous circulation. B and T lymphocytes produced in bone marrow enter circulation, and T lymphocytes are further processed in thymus. Lymphocytes enter lymph nodes through circulation and can be released again into circulation. B lymphocytes are associated with formation of antibodies, whereas T lymphocytes are associated with cellular components of the immune system that can destroy foreign matter, including organisms.*

Table 23-3 *Characteristics of the various leukocytes found in humans*

Cell type*	Size	Morphology	Function
Lymphocytes (32% to 35%)	7 to 8 μm	Has very little cytoplasm and is composed primarily of nucleus	Has receptors for interaction with antigen; principal types are T and B cells, which increase in size after interaction with antigen; B cells give rise to plasma cells, which are the actual producers of antibody. T cells release substances that activate other immune cells, such as the macrophage (see below)
Macrophage (3% to 4%)	12 to 20 μm	Contains bilobed nucleus; cytoplasm contains large number of organelles; tissue macrophage or histiocyte is larger than blood macrophage	Primary function is phagocytosis; degradation of microorganisms releases antigens that can be processed for reactions involved in specific immune resistance
Polymorphonuclear leukocytes			
Neutrophil (60% to 65%)	9 to 12 μm	Contains multilobed nucleus; granules in cytoplasm do not stain strongly acidic or basic	Engages in phagocytosis; cells rapidly migrate to areas of acute inflammation
Basophil (0.1% to 0.3%)	18 to 24 μm	Lobulated nucleus; large number of granules stain basic (blue)	Granules contain pharmacologically active compounds, such as histamine, that may be involved in allergic reactions
Eosinophil (2% to 3%)	10 to 14 μm	Lobulated nucleus; cytoplasmic granules stain acidic (red)	Appears to aid in destruction of metozoan parasites, such as parasitic worms (helminths)

*The percentage of each in the blood is in parentheses.

particularly numerous at body joints and in areas where the arms and legs join the body. The lymph nodes are reticulated structures that serve to filter lymph passing through them. Specific areas within the nodes serve as repositories for special cellular elements of the blood called **T** and **B lymphocytes.** Lymph leaves the nodes through efferent vessels that eventually drain into a large lymphatic vessel called the **thoracic duct**. The latter empties directly into the venous system.

Microorganisms can be transported to a variety of body sites by entering the blood via direct or indirect means. Some microorganisms are transmitted directly into the bloodstream following the bite of some arthropod, but most microorganisms find their way into the bloodstream via the lymphatics. During their journey through the bloodstream, microorganisms may remain free or be carried within certain cellular elements of the blood, particularly macrophages (monocytes) and, infrequently, red blood cells. Once in the bloodstream, microorganisms may be disseminated to localize in various organs, the central nervous system, and the skin. For example, measles is caused by a virus that initially infects the respiratory tract but later penetrates into the bloodstream and is then disseminated to cells of the skin, where it multiplies and produces a rash.

Host's second line of defense

Once parasites penetrate the subepithelial layer, they become prey to the host's second line of defense, which includes cellular elements of the blood, the lymphatic system, and other nonspecific resistance factors.

Cellular elements of the blood

Red blood cells (erythrocytes). Red blood cells carry oxygen to tissues and remove carbon dioxide, but they are in no way engaged in defending the host against infection. Very few microorganisms actually use the erythrocyte as a vehicle for dissemination to various parts of the body. The red blood cell is most important for the parasite causing malaria, because it uses the cell not only as a site for growth and replication but also as a "hideaway" from the phagocytic cells of the blood.

White blood cells (leukocytes). The term *white* refers to the sedimentation characteristics of unclotted blood, in which leukocytes appear as a white layer between the red blood cells and the fluid portion of the blood (plasma). Leukocytes (Table 23-3) are found in both blood and tissue and are divided into two major groups: granulocytes and agranulocytes.

Granulocytes are so called because of their microscopic appearance due to the presence of aggregates of hydrolytic enzymes, which are involved in the phagocytic process (see The Phagocytic Process). Granulocytes are thought to have a life span of 14 hours or less. They possess a multilobed nucleus and are frequently referred to as **polymorphonuclear leukocytes** (PMNs). Granulocytes are divided into three types based on their staining characteristics: neutrophils, eosinophils, and basophils (Plates 31 to 33). **Neutrophils**, like macrophages (discussed later), are the most active phagocytic cells and are therefore important in attacking the infectious agent, as well as being responsible for the pathological features of inflammation.

Agranulocytes are divided into two groups: lymphocytes and monocytes. Lymphocytes are a population of cells with different functions. They are composed of T and B cells, which have their origin in the bone marrow and fetal liver (Figure 23-10). They are the principal cells responsible for the immune response of the host (Table 23-3) and are discussed at length in Chapter 24. B cells have a short life span of 5 to 7 days, whereas T cells have life spans measured in months or even years. **Monocytes** and macrophages (Figure 23-11) are the largest of the lymphoid cells and are found in the blood as well as in tissue. The life span of monocytes is uncertain. Monocytes are part of the mononuclear phagocytic system, which also includes tissue phagocytic cells called **macrophages** (Figure 23-11). Cells of the monocyte/macrophage system are derived from white blood cells (monocytes) manufactured in the bone marrow. Monocytes pass out of the blood vessels and become macrophages in tissue. Monocyte/macrophage cells phagocytize damaged cellular material and parasitic microorganisms that appear in the circulation or at the site of infection. Their role in infection is discussed later.

Platelets. Platelets, which are nonnucleated cells, are derived from cells in the bone marrow and vascular sinuses. They have a life span of approximately 10 days and produce a component that is involved in the clotting of blood. They are also involved in reactions that are unfavorable to the host, such as hypersensitivity reactions. **β-Lysin** is a protein released from platelets that is believed to have antimicrobial properties, particularly against gram-positive bacteria.

Fluid elements of the blood. The fluid portion of the blood is called **plasma** and is over 90% water. Proteins, which are produced in the liver, make up 6% to 8% of the plasma and include such molecules as albumin, globulin, prothrombin, fibrinogen, and complement. Globulins play the most important role in immunity, particularly **immunoglobulins (antibodies)**, which are produced by B lymphocytes in response to antigenic stimulation.

Complement is a group of glycoproteins that can be activated in a manner similar to that of the clotting factors in blood. Complement is involved in specific immune reactions involving antibody but can also take part in nonspecific immune reactions. Some complement proteins cause the release of vasoactive compounds, which dilate blood vessels and contract smooth muscle. Complement can also generate chemotactic factors that attract PMNs, and, finally, complement can bind to the bacterial surface and provide help in the phagocytic process. Complement is discussed at more length in Chapter 24.

Lymphatic system. Microorganisms that cannot be walled off and are not killed by leukocytes at the site of infection proliferate and pass into regional lymph nodes. The lymph nodes try to filter out the microorganisms with the help of resident macrophages and circulating neutrophils. If the number of microorganisms entering the node is very large, the interaction between parasite and leukocyte becomes so intense that the nodes become inflamed, swollen, and tender (lymphadenitis). Sometimes the nodes become greatly enlarged and necrotic and are of diagnostic significance. These enlarged nodes are called **buboes** and are characteristic of diseases such as bubonic plague. Localized infections may be terminated if the macrophages in the lymph nodes are capable of destroying the parasite, but some parasites survive and even multiply within the macrophage. The macrophage has a considerably longer life span than the neutrophil and provides the parasite with a vehicle for transport and protection in the circulatory system of the host. Some of the infectious agents that are known to multiply in macrophages are the poxviruses, herpesviruses, and bacterial agents that cause diseases such as tuberculosis, leprosy, and Rocky Mountain spotted fever. Once the macrophage reaches the bloodstream, it proceeds to localize in one or more organs and then engages in reseeding progeny microorganisms into the bloodstream to infect other organs of the body.

Phagocytosis

The phagocytic process. As already discussed, one of the functions of phagocytes, the cellular elements of the blood, is to engulf and destroy microbial parasites. We will now examine how phagocytes are brought to the site of infection and the biochemical events that are associated with digestion of the microbial agent.

Inflammation is the body's response to invasion by a foreign agent, such as an infectious microorganism. Three major events occur during this response:

1. An increase in blood supply to the affected area.
2. An increase in the permeability of capillaries as a result of release of compounds, such as histamines. This permits soluble mediators of immunity, for example, complement, to reach the infection site.
3. Phagocytic cells migrate from the blood into tissue surrounding the infectious agent.

A

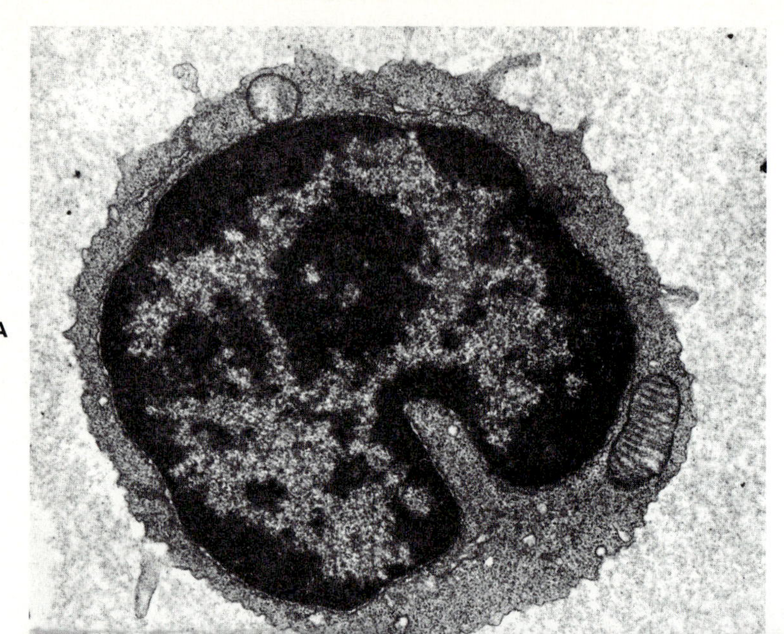

B

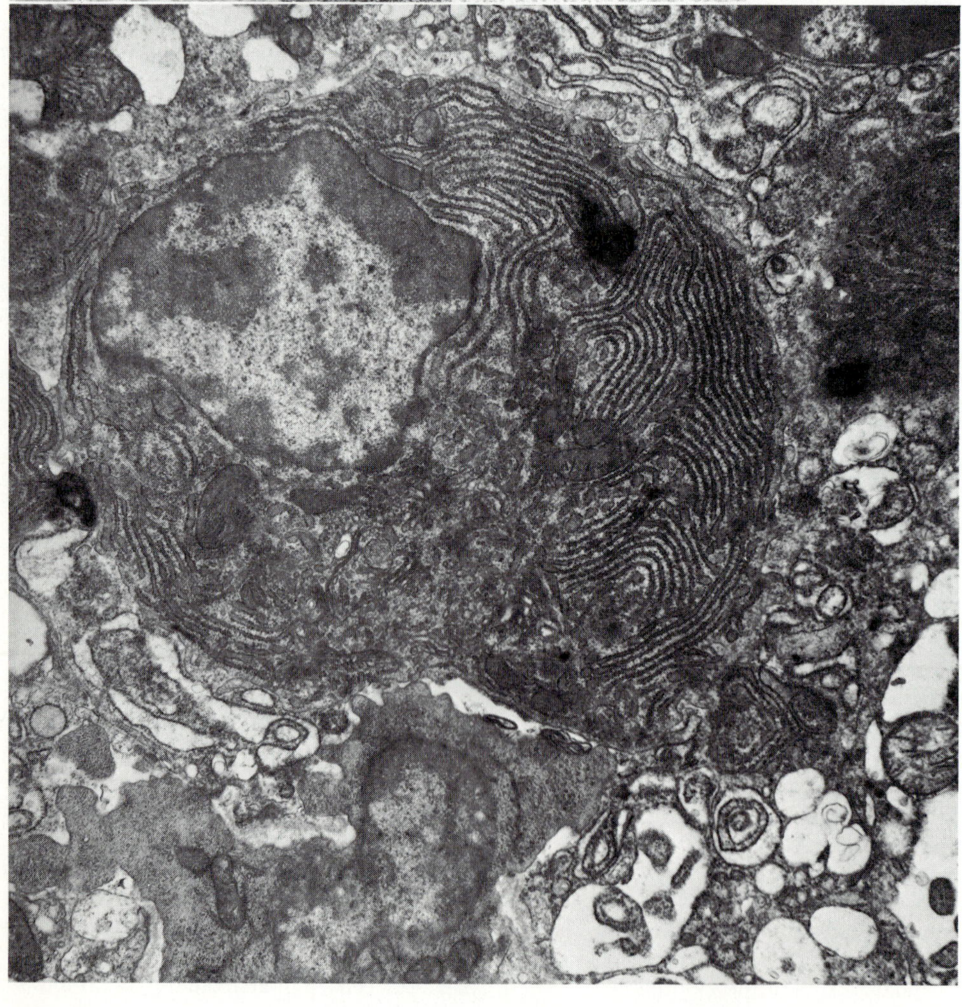

Figure 23-10 *Electron micrographs of lymphocytes.* ***A,*** *T cell lymphocyte in which there is very little cytoplasm and indented nucleus. (×13,000.)* ***B,*** *Plasma cell derived from stimulated B lymphocyte. Cytoplasm is filled with endoplasmic reticulum (curving lines), which is site of immunoglobulin (antibody) synthesis.*

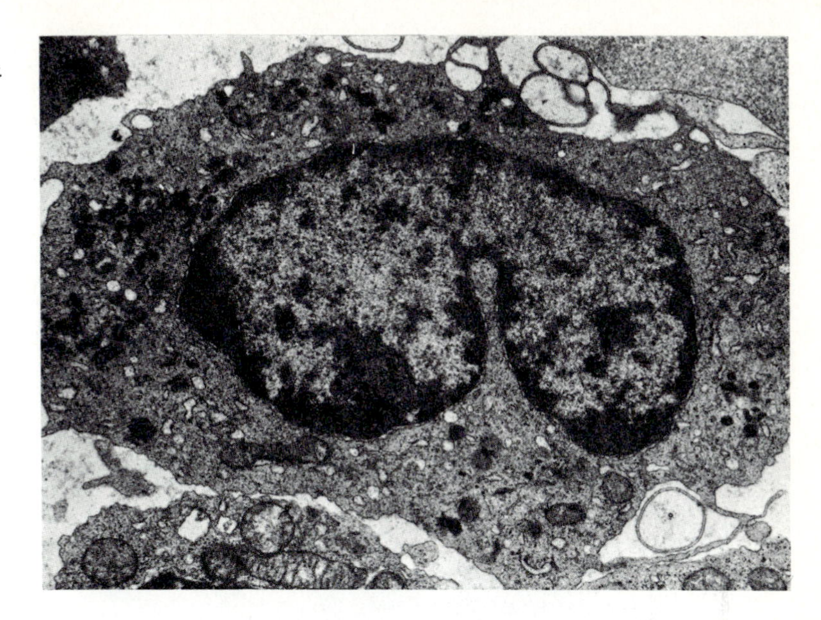

Figure 23-11 *Electron micrograph of macrophage demonstrating horseshoe nucleus. (× 10,000.)*

The migration of phagocytic leukocytes to the site of infection results from the release of chemotactic factors by the infectious microorganism. Phagocytes possess receptors on their surface that allow them to attach nonspecifically to a variety of microorganisms. This attachment is enhanced by complement and antibodies called **opsonins** (the molecular mechanism of this process is discussed in Chapter 24). Once the phagocyte makes contact with the intact microorganism or with cellular debris, a series of events are set in motion whose purpose is to engulf and degrade the target material. These events can be divided into two basic processes: **degranulation** and **respiratory burst activity**.

The phagocytic membrane engulfs the target (Figure 23-12, *A*) and is pinched off inside the cell into a vesicle called the **phagosome** (Figure 23-12, *B*). the phagosome fuses with many **lysosome** granules, which contain hydrolytic enzymes, to produce a **phagolysosome** (Figure 23-12, *C* and *D*). Once fusion has taken place, the hydrolytic enzymes are discharged (a process called degranulation) into the vesicle containing the material targeted for destruction. The hydrolytic enzymes in conjunction with certain microbicidal agents released during metabolism of the leukocyte are used to digest the cellular debris or kill the invading parasite.

Energy is required by the phagocyte to carry out the ingestion process and the subsequent degradative processes. After membrane contact of the phagocyte with the targeted material, there is a respiratory burst of activity in which oxygen consumption is increased significantly. During this respiratory burst certain metabolic reactions take place that result in the formation of microbicidal agents. For example, glucose oxidation in the phagocyte occurs via the hexose monophosphate shunt, resulting in the formation of NADH and NADPH. It is currently believed that an enzyme, NADPH oxidase, catalyzes the one electron reduction of oxygen to superoxide (see Chapter 12), using NADPH as the electron donor. In a subsequent reaction 80% of the superoxide is converted to hydrogen peroxide according to the following reactions:

$$2\ O_2 + NADPH \longrightarrow \underset{\textbf{Superoxide}}{2\ O_2^-} + NADP^+ + H^+$$

$$2\ O_2^- + 2\ H^+ \longrightarrow O_2 + \underset{\textbf{Hydrogen peroxide}}{H_2O_2}$$

Persons with chronic granulomatous disease* possess leukocytes with defects in these reactions, thus indicating the importance of phagocytes in resistance to infection. The superoxide ion is toxic, but it is not believed to be the major microbicidal agent. Microbicidal activity is believed to be associated with hydrogen peroxide formation and the enzyme **myeloperoxidase**, since defects in their formation often lead to recurrent infections in the host. The myeloperoxidase system catalyzes the oxidation of halide ions (chlorine, iodine, and bromine) to **hypohalite ions** in the presence of hydrogen peroxide. For example, chlorine, which is the most predominant halide in the cell, takes part in the following reaction:

$$Cl^- + H_2O_2 \longrightarrow \underset{\textbf{Halite ion}}{ClO^-} + H_2O$$

Experimental evidence indicates that the myeloperoxi-

* Chronic granulomatous disease is a sex-linked recessive disease of males in which infections from pathogens of low virulence recur.

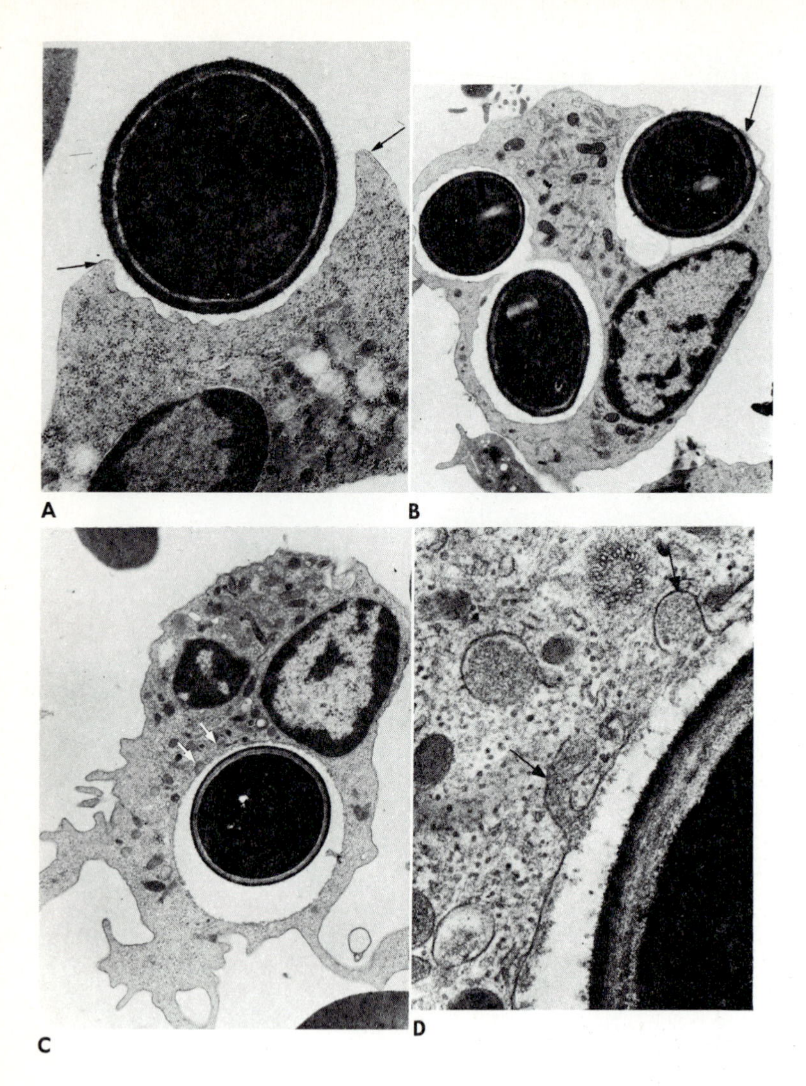

Figure 23-12 *Electron micrograph demonstrating phagocytosis of yeast* Candida albicans. ***A,*** *Organism is being surrounded by cytoplasmic arms* (arrow). *(×15,000.)* ***B,*** *Three organisms are phagocytized; one* (arrow) *shows nearly complete phagosome formation, and other two are completely ingested. (×5000.)* ***C,*** *Thirty minutes after ingestion two lysosomal granules* (arrows) *are fusing with phagocytic vacuole. (×5000.)* ***D,*** *Higher magnification of* ***C*** *showing fusing granules discharging their contents into phagocytic vacuole* (arrow). *(×33,000.)*

dase system may contribute to microbicidal activity in the following ways:

1. Hypohalite ions interact with the bacterial cell wall, causing it to lose its integrity and eventually lyse.
2. The system causes the decarboxylation of amino acids to aldehydes, which are toxic and could kill the cell.
3. The hypohalite produced in the system can react with hydrogen peroxide to produce **singlet oxygen** ($^1O_2$), which is simply a higher energy form of oxygen. Singlet oxygen has the same molecular formula as atmospheric oxygen ($^3O_2$) but differs in the distribution of the electrons around the two oxygen atoms. Singlet oxygen is a highly reactive product and is capable of breaking double bonds that are normally unreactive to atmospheric oxygen. It may be produced in the following way:

$$ClO^- + H_2O_2 \longrightarrow {}^1O_2 + Cl^- + H_2O$$

4. The superoxide produced in the myeloperoxidase system reacts with hydrogen peroxide to produce the **hydroxyl radical** ($OH\cdot$) a highly unstable but extremely toxic product:

$$O_2^- + H_2O_2 \longrightarrow (OH\cdot) + OH^- + O_2$$

The molecular events occurring within the phagocyte that are necessary for microbicidal activity have been identified because of investigations of cells from patients with congenital disorders of phagocytes. The ability of the phagocyte to kill the parasite is largely dependent on microbial surface structures. Gram-positive bacteria, because of their simple cell wall structure, are more easily digested than gram-negative bacteria, and certain microbial species seem to be more sensitive to one phagocytic activity than to another. It is hoped that as a consequence of these investigations it will be possible in the near future to stimulate the phagocytes of patients who have increased susceptibility to infection.

The sequel to phagocytosis. Phagocytosis represents a nonspecific immune response by the host to infectious microorganisms but there is more to come. The phagocytic process is followed by events in which some digested components (antigens) of the microorganism are presented by macrophages to immune cells called T and B cells. T cells respond to microbial antigens by giving rise to other cells that help combat the infectious microorganisms. These cells make up the **cellular immune response.** B cells respond to antigen by synthesizing and releasing noncellular molecules called **antibodies,** which also help ward off infection. B cells are part of the immune response referred to as **humoral immunity.** T and B cells represent components of the specific immune response and are discussed in Chapter 24.

One of the normal host responses to phagocytosis at the site of infection is **fever,** which is a rise in body temperature above 98.6° F. Fever is mediated by proteins, called **endogenous pyrogens,** that are produced in phagocytic cells, such as PMNs and macrophages. Pyrogens act on the temperature-regulating centers of the brain in the anterior hypothalamus. The phagocytic cells releasing the pyrogens can be activated by a number of microbial products, as well as some nonmicrobial components. The suppression of fever is considered an important aspect in the treatment of infectious disease, but the height of the fever in most patients is not sufficient to be detrimental. Prolonged fever may be detrimental to the host by causing an increased heart rate or by causing convulsions, particularly in children. Fever of short duration may actually be beneficial to the host. The use of drugs (antipyretics) to reduce fever in some infections may be detrimental. Recent observations suggest that aspirin (acetylsalicylic acid) used to reduce fever following certain infections may depress the immune response and make the patient more susceptible to complications such as Reye's syndrome (see Chapter 27).

Many infections in the preantibiotic era were treated by inducing hyperthermia in humans. Treatment of gonococcal endocarditis, syphilis, and pneumococcal meningitis for example, included methods for inducing fever. Those earlier studies suggested that fever contributed greatly to the recovery from infection. Many of the investigators in those studies suggested that hyperthermia stimulated host defense mechanisms and did not directly affect the microorganism causing infection. However, not enough controlled studies were developed to provide specific answers.

$CH_2O-CO-CH(C_{24}H_{49})-CH(OH)-C_{60}H_{120}$

$OC-CH(C_{24}H_{49})-CH(OH)-C_{60}H_{120}$

CORD FACTOR

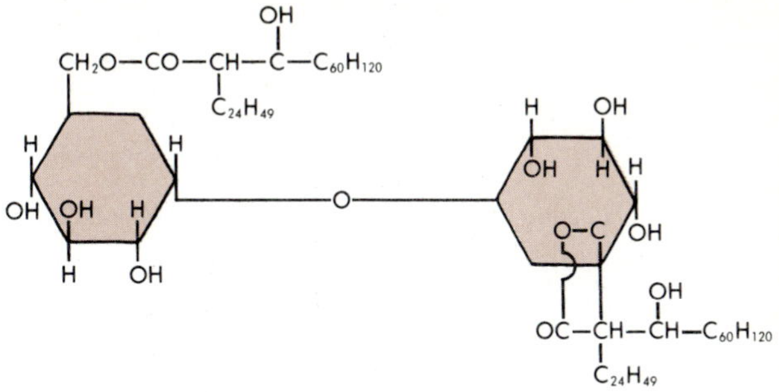

Figure 23-13 *Structure of the cord factor.*

Resistance of the parasite to phagocytosis

Interference with chemotaxis. Phagocytes are attracted to microbial agents by chemotactic substances produced by metabolizing microorganisms. Some microorganisms, such as the agent of tuberculosis, are capable of interfering with this process. The exact mechanism is not known, but a surface lipid called the "cord factor" is believed to be the interfering agent (Figure 23-13). A similar interference mechanism has been observed with *Neisseria gonorrhoeae,* which is the causative agent of gonorrhea. The infectious agent in most patients remains localized in the genitourinary tract, but in a very few instances a strain of this agent can break through the local barriers and enter the bloodstream to be disseminated throughout the body. Comparison of the strains from localized and disseminated infections has revealed that disseminated strains are poor stimulators of chemotaxis as compared with active stimulation by strains causing localized infections.

Interference with ingestion by phagocytes. The ingestion of microorganisms into the phagocytic cell can be prevented by some bacteria. These inhibitors of phagocytosis are surface components, usually capsules belonging to agents described in Table 23-4. The manner in which these components interfere with phagocytosis is not known, but they may prevent adsorption of opsonins, the immune components that are known to aid in phagocytosis. It is important to remember that not all capsule-bearing microorganisms are resistant to phagocytosis.

Concept Check

1. How are microorganisms able to spread from their initial site of colonization? Are there any barriers?
2. How are microorganisms exposed to danger when they enter the blood or lymph systems?
3. Briefly explain the process of phagocytosis and the sequel to this process.
4. What happens chemically when a microorganism is degraded in the phagocyte?

Table 23-4 *Microorganisms in which surface components interfere with digestion by phagocytes*

Microorganism	Surface component	Disease in which organism is involved
Cryptococcus neoformans	Capsule (polysaccharide)	Cryptococcosis, a systemic fungal disease
Streptococcus pyogenes	M protein	Rheumatic fever
Streptococcus pneumoniae	Capsule (polysaccharide)	Bacterial pneumonia
Bacillus anthracis	Capsule (polypeptide)	Anthrax
Salmonella typhi	Vi antigen (amino sugar)	Typhoid fever
Escherichia coli	K antigen (polysaccharide)	Diarrhea
Pseudomonas aeruginosa	Slime layer (polysaccharide)	Infection as cause of death in burn patients
Yersinia pestis	Polysaccharide-protein complex	Bubonic plague
Treponema pallidum	? Hylauronic acid in cell envelope	Syphilis
Staphylococcus aureus	Protein A	Skin infections, pneumonia

Some infectious agents acquire their capsules only after infection. Plague, for example, is a disease in which the infectious agent (*Yersinia pestis*) is harbored by fleas that infect rats and other rodents. The infectious agent does not produce a capsule while harbored in the flea, but when the flea bites a human, some of the transferred organisms are ingested by PMNs and killed, whereas those ingested by macrophages produce a capsule and multiply.

Interference with intracellular digestion by phagocytes. Microbial surface components are apparently involved in the prevention of phagocytic digestion. Microorganisms surviving in the short-lived PMN (Figure 23-14) can be disseminated to other parts of the body. Microorganisms surviving in the long-lived macrophage multiply and are transported to organs where they are responsible for chronic infections such as leprosy, tuberculosis, and brucellosis. It appears that the principal technique for avoiding digestion in the macrophage is to prevent the release of lysosomal enzymes into the phagocytic vacuole (phagosome) that contains the intact microorganisms.

Destruction of the phagocyte. Some microorganisms have the ability to cause the premature death of phagocytes. The streptococci and staphylococci release soluble factors that bring about the destruction of phagocytes (Figure 23-15). The streptococci produce a hemolysin called **streptolysin**, and the staphylococci produce a compound called **leukocidin**. These bacterial products cause the release of lysosomal enzymes into the cell cytoplasm of the phagocyte and bring about its destruction.

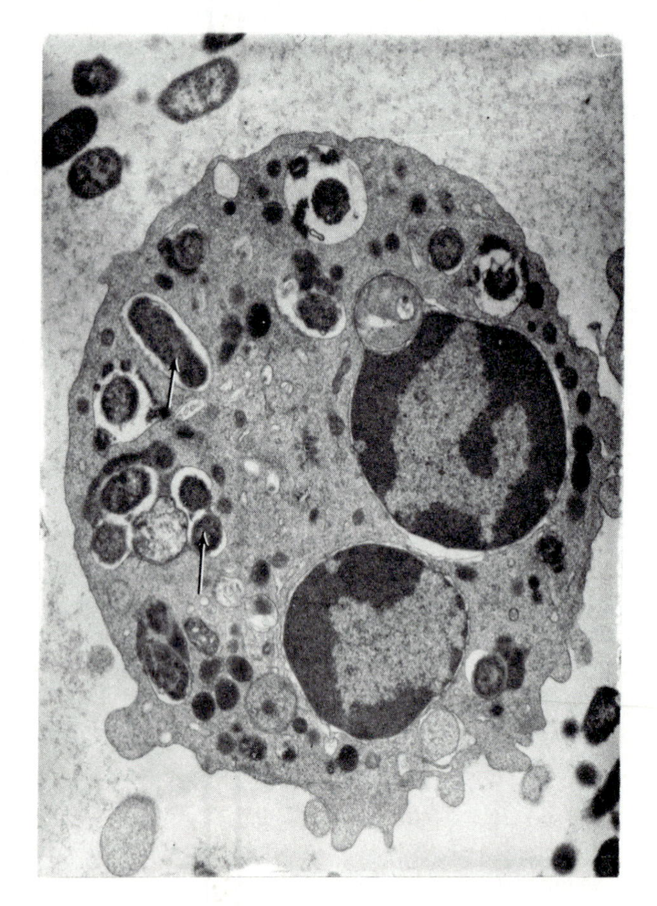

Figure 23-14 *Electron micrograph (×6,600) of human PMN showing presence of organism* Brucella abortus (arrow).

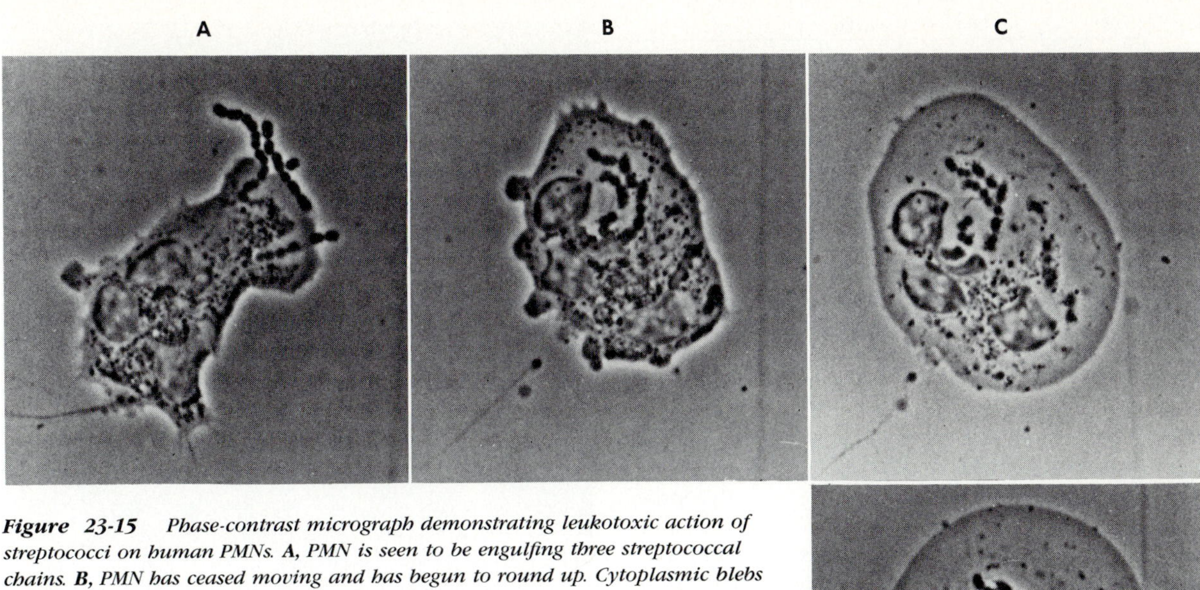

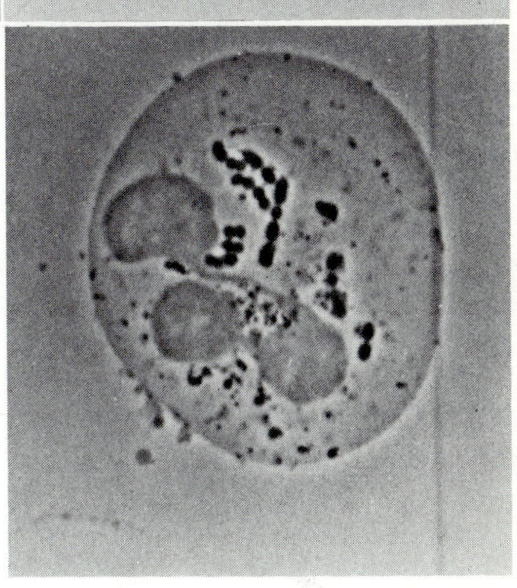

Figure 23-15 *Phase-contrast micrograph demonstrating leukotoxic action of streptococci on human PMNs.* ***A,*** *PMN is seen to be engulfing three streptococcal chains.* ***B,*** *PMN has ceased moving and has begun to round up. Cytoplasmic blebs are seen to emanate from surface of PMN.* ***C,*** *PMN is swollen and immobile.* ***D,*** *Most internal structures of PMN have disappeared, and nucleus appears swollen.*

Defective phagocytic mechanisms. The ability of the microorganisms to withstand phagocytosis is greatly increased if there are inherited abnormalities in the killing mechanism of the phagocyte. Conditions such as myeloperoxidase deficiency and chronic granulomatous disease result in increased susceptibility to bacterial infections because of the inability of the phagocyte to generate superoxide and hydrogen peroxide.

OTHER MECHANISMS OF MICROBIAL RESISTANCE

Not all microbial resistance mechanisms are directly associated with the phagocytic cell. Some microorganisms are able to alter the chemical nature of their surface components so they can evade the host's immune system.

Antigenic diversity. The protozoan *Trypanosoma* evolved a mechanism for evading the host's immune system and causing disease. *Trypanosoma* species are transmitted to humans by the tsetse fly and are the cause of a fatal neurological disease called **sleeping sickness**. Some *Trypanosoma* species infect livestock, and nearly 4 million square miles of Africa are uninhabitable for most dairy and beef cattle. Over 20,000 new cases of human disease are reported each year. In most infectious diseases the infectious agent possesses antigens that stimulate the host to produce antibodies that help kill the invading organism. In addition, vaccines containing the specific antigen can be used to prevent infection by simulating the disease. These two mechanisms of host resistance are of no use in sleeping sickness because the trypanosomes continually change their surface antigens. As soon as the host's immune system is alerted, antibodies are produced in response to the antigens of the infectious agent. About 99% of the parasites are killed by the immune system but 1% survive and change their antigenic coat. This process is repeated until eventually the parasite overwhelms the host by taxing the immune system.

The surface coat of *Trypanosoma* consists of glyco-

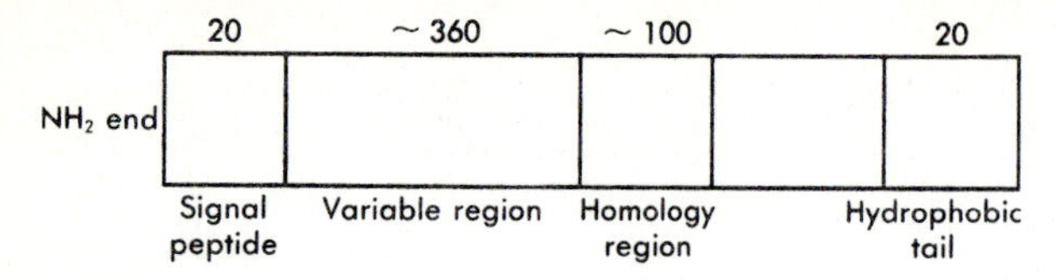

Figure 23-16 ***Protein chain of variable surface glycoprotein (VSG). Protein is composed of approximately 500 amino acids. Last 20 amino acids of carboxyl terminal end are replaced by molecule that anchors VSG into plasma membrane.***

protein molecules that are the same for all individuals of a single clone but different for different clones. The amino acid sequence at the beginning of the glycoprotein of different clones is also different (Figure 23-16). These glycoprotein antigens are referred to as **variable surface glycoproteins (VSGs).** It is now estimated that a single parasite has as many as 1000 or more VSG genes; that is, approximately 10% of its genetic capacity is devoted to immune evasion. The VSG protein chain contains an *N*-terminal signal sequence whose function is to help transport the protein across the plasma membrane of the parasite. This is followed by a region of amino acids that differ in most VSGs and is called the **variable region.** The last region of amino acids is similar in nearly all VSGs and is called the **homology region.** However, the homology region cannot serve as the basis of a vaccine because the folding of the VSG on the surface coat of the parasite is such that the region of homology is not exposed. Only the variable region is exposed to the immune components of blood and tissue.

Molecular mimicry. Some microorganisms can evade the host's immune system by synthesizing or acquiring hostlike antigens. One infectious agent that exhibits this property is the parasite *Schistosoma* whose species parasitize humans and other mammals. Over 300 million people in Africa, the Middle East, and Central and South America are infected by this parasitic worm. The organism is acquired by humans whose feet are exposed to contaminated water, for example, workers in rice paddies. The organism, after penetrating the host, matures and takes up residence in the veins of the intestine or bladder. These organisms are continually exposed to the host's immune system yet they are unaffected by it. *Schistosoma* cells are able to coat themselves with host antigens, such as blood group antigens (A, B, and O) and the histocompatibility antigens, which the immune system uses to distinguish self from nonself or foreign antigens. The immune system of the host is actually responsible for the disease. By acquiring blood group antigens the parasite can infect any human host and go unrecognized. However, once the parasite acquires the histocompatability antigens any other schistosome that infects the host will be recognized as foreign and then destroyed.

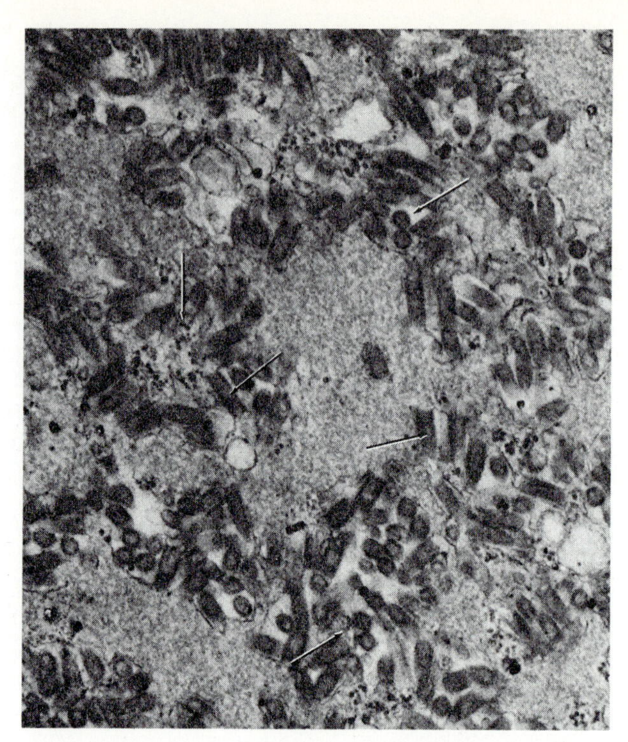

Figure 23-17 ***Viral inclusion bodies. Electron micrograph of brain neuron in suckling mouse demonstrating rabies virus inclusion body within cytoplasm. (×49,000.) Many viral particles*** (arrows) ***can be seen lengthwise and in cross section.***

Damage to host tissue

The symptoms of a disease or the damage to host tissue in most cases is in part related to the rate at which the microorganisms divide or multiply in a given tissue. Some members of the Enterobacteriaceae, which cause a form of bacterial gastroenteritis, produce symptoms within 18 to 24 hours following infection and have division times of 30 minutes or less. The agent of leprosy make take 2 to 3 weeks to divide, and many months are required before disease symptoms are visible. Damage to host tissue may be due to one or more of the following conditions: (1) direct damage by microbial multiplication, (2) microbial toxin production, or (3) immunopathological reactions.

Direct damage by microbial multiplication. Many viruses damage host tissue as a result of replication and the inhibition of host protein, DNA, and RNA synthesis. Inclusion bodies, which represent an accumulation of virally associated components, can often be seen microscopically (Figure 23-17). The site of viral multiplication will determine the type of tissue damage. When the po-

Table 23-5 *Characteristics that differentiate exotoxins from endotoxin*

Exotoxins	Endotoxin
Excreted by living cells	Part of cell wall of gram-negative bacteria and usually released on lysis of cell
Most are polypeptides with molecular weights between 1×10^4 to 9×10^5	Lipopolysaccharide (LPS) complex; lipid A is believed to be toxic factor; low molecular weight
Relatively unstable to temperatures above 60° C	Relatively stable to temperatures above 60° C for several hours with no loss of toxicity
Very antigenic and stimulate formation of antitoxic antibodies	Does not stimulate formation of antitoxin but does stimulate formation of antibodies to polysaccharide of LPS
Can be converted to toxoid* by treatment with formalin, acids, heat or other agents	Cannot be converted to toxoid
Do not produce fever in host	Does produce fever in host
Very toxic in small microgram quantities to laboratory animals	Weakly toxic; lethal for laboratory animals only in hundreds of micrograms

*A toxoid is a toxin that has been treated to remove its toxicity but not its antibody-stimulating capacity.

liovirus, for example, multiplies in the anterior horn cells of the spinal cord, it damages nerve tissue, resulting in a loss of function to the muscles innervated by them. Bacteria may multiply within cells, as well as in tissues and tissue fluids. Some bacteria have the ability to destroy the phagocytes in which they find themselves, but damage from most bacterial infections is due to toxic products, such as microbial toxins.

Microbial toxins. Many microorganisms, especially bacteria, produce toxins. Bacterial toxins are of two types: **exotoxins**, which are produced by the cell and released into the surrounding tissues, and **endotoxin**, which is part of the cell wall of gram-negative bacteria and is released during growth or when the bacterial cell lyses (Table 23-5). Toxins are considered virulence determinants, but only in a few instances has it been unequivocally established that toxins are the sole or major determinant of the disease process. Often there are a number of other factors that together with the toxic factor play a role in the disease process. Viruses do not produce toxins per se, although some of their surface components may be cytotoxic (toxic to mammalian cells). For example, the adenovirus possesses a surface projection, called a penton fiber (Figure 23-18), that may be considered cytotoxic, but other viral components with cytotoxic properties have not been identified.

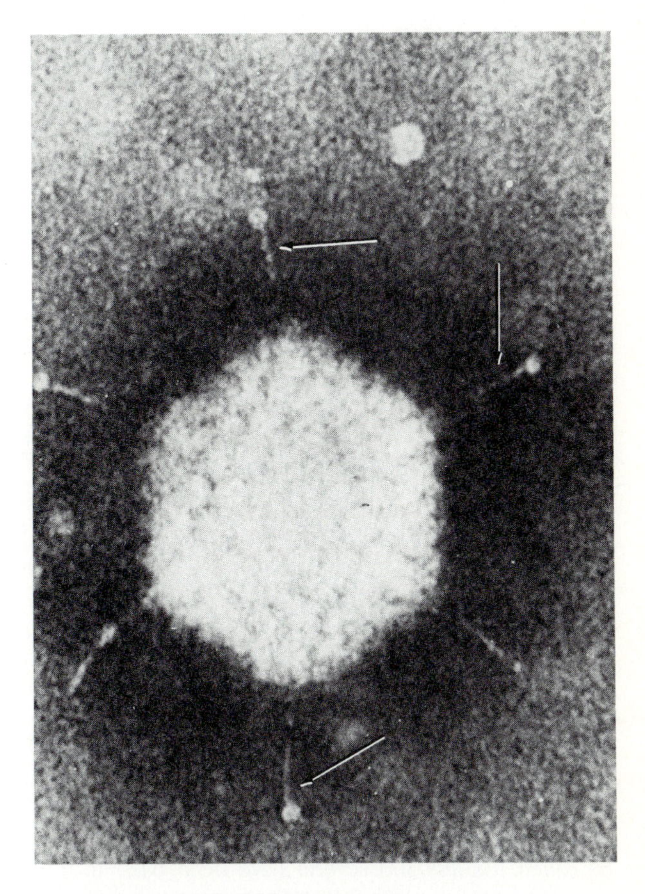

Figure 23-18 *Electron micrograph of adenovirus demonstrating penton fibers* (arrows) *considered to be cytotoxic.*

Exotoxins. Exotoxins are proteins that are produced by both gram-negative and gram-positive bacteria (Table 23-6). They are highly antigenic (stimulate the formation of antibodies) and can be treated with chemical agents that destroy their toxicity but do not affect their capacity to stimulate antibody formation. These inactivated toxins

Table 23-6 *Bacterial exotoxins*

Microorganism	Name of disease and/or toxin	Action	Symptoms produced in host	Toxicity for mice*
Clostridium botulinum	Botulism	A neurotoxin that blocks release of acetylcholine at nerve synapses	Double vision (diplopia), respiratory paralysis	1.2 ng IP for type A; same for humans
Clostridium perfringens	Gas gangrene, alpha toxin	A lecithinase that splits lecithin in cell membranes	Hemolysis, toxemia as cause of death	3.0 μg IV
	Food poisoning, enterotoxin	Stimulates adenyl cyclase and increases AMP, resulting in hypersecretion of water	Diarrhea with loss of water and monovalent ions	
Clostridium tetani	Tetanus (lockjaw)	Inhibits antagonists of motor neurons in spinal cord and brain	Hyperreflex action of muscles and violent spasms of skeletal muscles; respiratory failure as cause of death	1.0 ng; 2.5 ng IP for humans
Corynebacterium diphtheriae	Diphtheria	Inhibits protein synthesis	Heart damage as most common cause of death	1.6 mg SC
Staphylococcus aureus	Food poisoning	A neurotoxin that affects brain center that induces vomiting	Vomiting	40-60 ng IV
	Scalded skin syndrome, exfoliation	Causes intradermal separation	Erythema (reddening) and exfoliation (peeling) of skin	ND
Streptococcus pyogenes	Scarlet fever, erythrogenic toxin	Causes vasodilatation	Skin rash	3-6 mg IV
Bacillus anthracis	Anthrax; toxin is a three-component complex	Causes increase in vascular permeability	Pulmonary edema and hemorrhage	114 μg IV
Vibrio cholerae	Cholera, enterotoxin (choleragen)	Stimulates adenyl cyclase and increases cAMP in gut	Diarrhea with massive losses of water and ions	250 μg IV
Escherichia coli	Traveler's diarrhea, (enterotoxin)	Stimulates adenyl cyclase and increases cAMP in gut	Diarrhea	250 μg IV
Pseudomonas aeruginosa	Exotoxin A	Similar to diphtheria toxin	Not known	3.0 μg IV
Bacillus cereus	Enterotoxin	Similar to cholera toxin	Diarrhea	15 mg IV
Shigella dysenteriae	Shigella toxin	Neurotoxic, cytotoxic, and enterotoxic activity	Dysentery	450 ng IV
Bordetella pertussis	Pertussigen	Stimulates adenyl cyclase and increases cAMP	Whooping cough	15 μg IP

*Quantity/kg body weight.
IP, Intraperitoneal; *IV,* intravenous; *SC,* subcutaneous; ND, no data available.

are called **toxoids** and several have been used in the prevention of disease (for example, tetanus and diphtheria). Many exotoxins are enzymes that cause damage to a number of different components of the cell, and during infection they spread into surrounding tissue or into the bloodstream. For example, some toxins are **phospholipases** that enzymatically break down the plasma membrane of eukaryotic cells. Most bacterial exotoxins are proteins that affect internal targets in eukaryotic cells. They are synthesized in an active or inactive form.

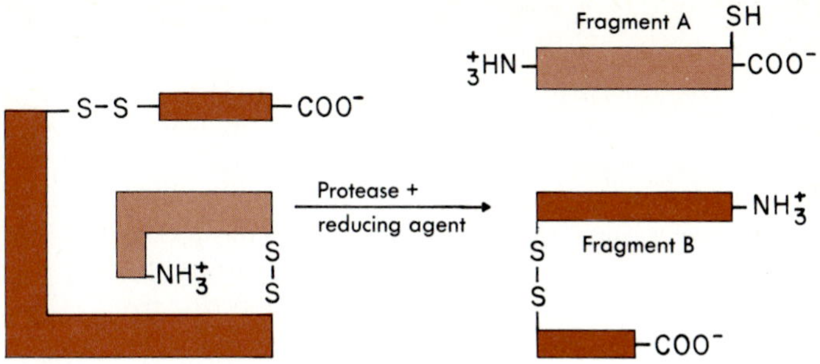

Figure 23-19 *Proposed action of cellular protease on polypeptide toxin of* Corynebacterium diphtheriae.

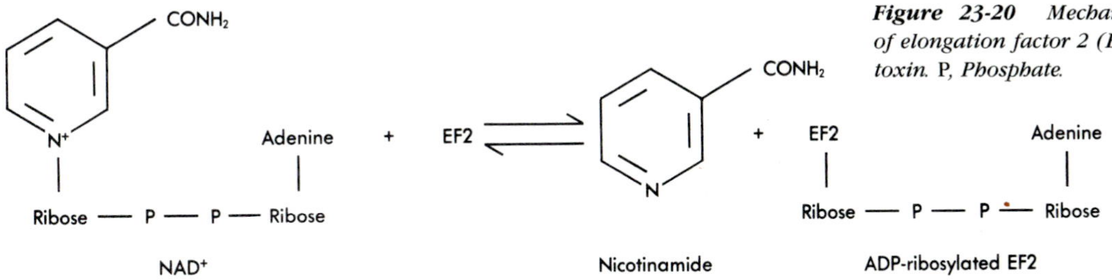

Figure 23-20 *Mechanism of inactivation of elongation factor 2 (EF 2) by diphtheria toxin. P, Phosphate.*

To reach their targets, toxins must bind to receptors on the plasma membrane of the eukaryotic cell and then be transported into the cytosol. The plasma membrane receptors for bacterial toxins are glycolipids or glycoproteins. Studies so far have indicated that the carbohydrate moiety is the component most directly involved in toxin binding. The transport of many toxins across the plasma membrane is similar to the way in which some viruses, hormones, and other biologically active molecules are transported in higher eukaryotes. The transport process is called **receptor mediated endocytosis** and is associated with specialized indented segments on the plasma membrane called **coated pits.** The coating on these pits is a protein called **clathrin.** Once contact has been made between toxin and receptor, the pits invaginate to form intracellular-coated vesicles, which pass into the cytoplasm of the cell. Some vesicles probably fuse directly with lysosomes, whereas others are processed by the Golgi body. Not all toxins are internalized this way and presumably do not require coated pits.

Some of the protein exotoxins have similar characteristics (Table 23-6). For example, the toxins associated with diseases, such as cholera, diphtheria, tetanus, botulism, and others, are polypeptides that have two chains, A and B, linked by disulfide bridges (Figure 23-19). The **B chain** is involved in receptor binding, whereas the **A chain** possesses enzymatic activity. For those toxins synthesized in an inactive form (diphtheria toxin, for example) an additional biochemical step is required before enzymatic (toxic) activity can be realized. In the case of diphtheria toxin, activity is not exhibited until the A chain is separated from the B chain. This separation is the result of protease activity and probably occurs in the coated vesicles following endocytosis. The activity of the A chain may be directed at target sites either in the cytoplasm or on the plasma membrane. Those toxins that act in the cytoplasm, such as diphtheria, pseudomonas, and shigella toxins, inhibit protein synthesis. Fragment A of these toxins catalyzes the ADP-ribosylation of elongation factor 2 (EF 2), and this prevents the transfer of amino acids to the growing polypeptide chain (Figure 23-20). ADP-ribosylation reactions use NAD as the donor of the ADP-ribosyl group, and this type of activity is also associated with other toxins (see below). Those toxins that act at the inner surface of the plasma membrane affect plasma membrane enzymes. For example, cholera toxin (and others) cause increases in adenylate cyclase activity in cells of the intestinal tract. The increase in adenylate cyclase activity is due to the ADP-ribosylation of a regulatory subunit of the enzyme by the toxin. Adenylate cyclase catalyzes the conversion of adenosine 5-triphosphate to cyclic adenosine monophosphate (cAMP). Increased levels of cAMP cause a change in the electrical potential of the plasma membrane, leading to the loss of fluids and electrolytes. Twenty liters of fluid may be lost per day in diarrhea associated with cholera.

The relationship between toxin production and disease is not always simple or direct. The mechanism of exotoxin production by the agent of diphtheria *(Corynebacterium diphtheriae)* is multifaceted and associated with a temperate virus. Toxin is synthesized only by strains of *C. diphtheriae,* lysogenic for bacteriophage carrying the *tox* gene. The level of inorganic iron in the cell controls the production of exotoxin, and as the level of iron increases, the synthesis of exotoxin is decreased.

The link between metal ion and toxin production

may also prove to be a factor in the disease **toxic shock syndrome (TSS)**. TSS is caused by *Staphylococcus aureus* and was originally discovered in women who used tampons. Experiments in the laboratory have shown that toxic shock protein, believed to be one of the factors associated with disease, is controlled by the level of magnesium. Low levels of magnesium stop the growth of *S. aureus* and trigger the release of several proteins, including the toxic shock protein. It is hypothesized that tampons act like ion exchange resins and bind magnesium ions from menstrual fluids. Thus the commensal population of *S. aureus* in the vagina is deprived of magnesium and release toxin.

Exotoxin production by microorganisms is not always associated with infection in the host. Some toxins are produced in foods that have been contaminated by microorganisms. Ingestion of preformed toxins is responsible for the disease symptoms associated with food poisoning.

The exotoxin of *Clostridium botulinum* is a neurotoxin and one of the most potent toxins known to humankind. Less than 0.1×10^{-3} μg can kill a mouse. The disease botulism is acquired by ingestion of foods contaminated with toxin excreted by the organisms. The toxin blocks the passage of nerve stimuli to nerve endings and causes paralysis, which frequently ends in death due to respiratory failure. There are seven types of botulinum toxins. Of these, types C_1 and D are controlled by temperate bacteriophages, and type A is the most frequently incriminated in fatal episodes. The toxin has been recovered in crystalline form and is a complex consisting of a **toxic fraction** and a **hemagglutinin fraction**. The toxin complex (most types) is inactive until activated by proteases (such as trypsin) produced either by the bacterial cell or by a combination of bacterial and host enzymes. Activation takes place when the polypeptide chains of the toxin assume a particular configuration. The toxin blocks cholinergic (mediated by acetylcholine) innervations but not adrenergic (mediated by norepinephrine) innervations. It acts at the junction formed by the nerve and the muscle or structure it innervates but not the affected structure *per se*. The inability of toxin-affected muscle to respond to stimuli is believed to be due to either the inhibition of acetylcholine synthesis (perhaps by preventing uptake of choline) or prevention of its release from motor nerves. Calcium ions appear to play an important role, because they are normally required for acetylcholine release. Studies with intact animals have demonstrated that bathing poisoned muscle in solutions containing high concentrations of calcium ions improves the release of acetylcholine.

The fungi also produce exotoxins, the most active of which is called **aflatoxin**, produced by certain species that contaminate grains. Aflatoxins are discussed in Chapter 21.

Concept Check

1. **What mechanisms are available to microorganisms for evading phagocytosis? What surface components are known to interfere with digestion?**
2. **What characteristics distinguish exotoxins from endotoxins?**
3. **Describe the basic structure and function of exotoxins, such as those that cause diphtheria, botulism, and cholera.**
4. **How does the mechanism of antigenic diversity and molecular mimicry enable some microorganisms to evade the immune system?**

Endotoxins. Endotoxins are the lipopolysaccharide component of the cell wall of gram-negative bacteria. The lipopolysaccharide (LPS), is composed of three components: a core of polysaccharide whose composition in species of the same genus may be the same but different in other genera; the O-specific polysaccharide region, which gives serological specificity to the species and whose alteration may change a virulent species to a nonvirulent one; and the lipid component. The lipid component called lipid A is the actual toxic element of the LPS complex. (See Chapter 25 for a discussion of serological specificity of the LPS complex.)

Endotoxin has been studied extensively, but no direct role in the pathogenesis of any human diseases has been established for it. Studies comparing the symptoms of gram-negative infection in humans with such symptoms in laboratory animals indicate that endotoxin may play an important role in fever production and vascular change. Endotoxin injected directly into animals causes a release of fever-inducing agents from macrophages 60 to 90 minutes after injection. Injection of endotoxin can initiate a series of changes in the circulation: vascular permeability, decrease in blood flow, and shock. Disseminated intravascular coagulation, in which blood elements are clotted, leading to occlusion of the small peripheral blood vessels, can also be a complication of gram-negative infections. Shock and disseminated intravascular coagulation can lead to death, and both phenomena have been observed in gram-negative infections in humans.

Not all gram-negative bacteria possess endotoxin of equal potency, and their effect on the host is variable.

Damage via the immune response (immunopathology). One of the first events that normally occurs in the host after contact with an infectious agent is inflammation. Inflammatory materials are released by both host and microbe, and in most instances the response does not cause major damage to the host. A few infectious diseases are characterized by pathological changes due to immune responses. Rheumatic fever, one consequence of infection by *Streptococcus pyogenes,* is an example of tissue damage via the immune response.

RECOVERY FROM INFECTION

Recovery from infection is hastened by nonspecific immune mechanisms discussed in this chapter and specific immune mechanisms discussed in Chapter 24. Despite all the efforts of the host in defending against infection, there are some infections in which the microorganism is not eliminated but instead persists in the host. **Persistent infections** usually cause only minor illness, but their most important contribution to the host-parasite relationship is that they help the infectious agent to remain in the community and to be spread to others. Transmission of microorganisms that cause persistent infections is discussed in detail in Chapter 26.

Concept Check

1. **Describe the basic structure of endotoxin and which components are immunogenic (stimulate the formation of antibodies).**
2. **What are some of the basic symptoms associated with endotoxin release in the host?**
3. **How can the host's own immune system cause damage during or after infection by a microbial agent?**

SUMMARY

1. In the vertebrate host, microorganisms exist as commensals or parasites. Many factors influence an organism's ability to act as a commensal, including the presence or absence of oxygen at the site of colonization, nutritional requirements of the organism, pH of the environment, and the effect of other microorganisms inhabiting the colonization site. Commensals, by their sheer numbers or the products they elaborate, benefit the host by preventing the colonization of pathogens.
2. Most parasites are nonindigenous species; however, under certain conditions in which the host's defense mechanisms have been reduced, commensals may also become parasitic and harm the host. The ability to harm the host is related to the virulence of the microorganism.
3. Parasites can cause disease when they gain entry into the host, colonize the host tissue and multiply, resist host defense mechanisms, and damage host tissue.
4. The first line of host defense to microorganisms is composed of the natural physical barriers, such as skin and mucous membranes. Host barriers to microbial entry and the nonmicrobial factors that favor colonization are outlined in Table 23-1.
5. The ability to colonize and multiply in the host is often related to the nature of microbial components called adhesins and to the nature of the surface components on the tissue being colonized.
6. The ability to penetrate the host cell or spread to other body sites may be a result of toxic products or enzymes released by the microorganisms that destroy the epithelial surface. Sometimes, microorganisms can be taken up by epithelial cells in a process similar to phagocytosis. Once the epithelial surface is penetrated, microorganisms have a chance to enter the bloodstream.
7. The host's second line of defense includes cellular elements of the blood, such as red and white blood cells and platelets, as well as fluid elements, such as antibodies. In addition the lymphatic system and various phagocytic cells throughout the body are prepared to destroy the invading microorganisms.
8. The parasite can resist host defense mechanisms in different ways. It can interfere with nonspecific host responses, such as leukocyte chemotaxis, the inflammatory response, ingestion by phagocytes, intracellular digestion by phagocytes, and phagocytic survival. The ability of the parasite to interfere with these defense mechanisms is related to the surface components or products released by the parasite.
9. Some parasites evade the immune system by changing the antigenic character of their surface components, such as glycoproteins. Other parasites coat themselves with host antigens and thus go undetected during infection.
10. The parasite damages tissue through the act of multiplication and through the release of toxins—both exotoxins and endotoxins. Toxins are released primarily by bacteria. Vaccines have been prepared against some of the most potent toxins, such as diphtheria and tetanus. Damage to host tissue can also result from inflammatory events (immunopathology) that occur in response to diseases such as rheumatic fever.

Self-study Quiz

Short answer

Next to each system listed describe at least one resistance factor serving as a barrier to microbial entry into the host.

A. Skin and mucous membranes ______________
B. Respiratory tract ______________
C. Intestinal tract ______________
D. Genitourinary tract ______________

Matching

Match the microbial structure or product in the left column with its specific activity or function listed in the right column.

A. Pili ________
B. Capsule ________
C. Diphtheria toxin ________
D. Dextran ________
E. Lecithinase ________
F. Cholera toxin ________
G. Endotoxin ________
H. Botulinal toxin ________

1. Inhibits protein synthesis
2. Produces fever and initiates changes in circulation
3. Can degrade components in membrane
4. Simulates adenyl cyclase in gut
5. Promotes attachment to epithelial surfaces
6. Promotes formation of fibrin in blood
7. Blocks release of acetylcholine at nerve synapsis
8. Promotes binding to oral surfaces
9. Degrades connective tissue component
10. Inhibits phagocytosis

Multiple choice

1. Which of the following represents a microbial technique for evading the specific immune system of the host?
   A. Interference with phagocytosis
   B. Changing surface antigens during the infection process
   C. Production of endotoxins
   D. Destruction of leukocytes
2. An organism that is indigenous to the host and produces no ill effects in the healthy host is called a (an):
   A. Attenuated organism
   B. Pathogen
   C. Commensal
   D. Parasite
3. Secretory IgA is a primary barrier to infection in the:
   A. Genitourinary tract
   B. Respiratory tract
   C. Eye
   D. Intestinal tract
4. Persons at risk for the development of frequent bacterial infections are born with a metabolic deficiency in the ability to produce which of the following enzymes in their leukocytes?
   A. Myeloperoxidase
   B. Leukocidin
   C. Hyaluronidase
   D. Interferon
5. A toxoid is made from which of the following?
   A. Bacterial capsule material
   B. Adhesins
   C. Endotoxin
   D. Exotoxins
6. The enzyme found in tears that is capable of degrading the cell wall material of bacteria is:
   A. Lysozyme
   B. Hyaluronidase
   C. Coagulase
   D. Lecithinase

7. Which of the following terms have no relationship and should not be matched?
   A. Lymphocyte and B cell
   B. Macrophage and monocyte
   C. Endotoxin and gram-negative cell wall
   D. ADP-ribosylation and botulinal toxin
8. Which of the following would be found in the blood of a person suffering from fever?
   A. Granulomas
   B. Endogenous pyrogens
   C. Diphtheria toxin
   D. Superoxides

Thought questions

1. Using the species *E. coli* as an example, explain the conditions under which it would be considered: (1) a commensal, (2) a pathogen, (3) a colonizer, (4) an infectious agent, (5) a virulent organism, (6) and an endotoxin producer.
2. An often encountered sequence of events during the hospital stay of a patient who requires a urinary catheter for any length of time is the development of urinary tract infections followed by gram-negative septicemia (multiplication of the microorganism in the blood). This may be followed by gram-negative pneumonia or the development of septic shock, which may be fatal. Using your knowledge of the host-parasite relationship, trace the sequence of events and discuss the parasite virulence factors, host resistance factors, and host response at each stage of the progressing disease.

SELECTED READINGS

Books

Burnet, M., and White, D.O. *Natural history of infectious disease* (4th ed.). Cambridge University Press, New York, 1972.

Merhagen, S.E. (Ed.). *Molecular basis of oral adhesion,* American Society for Microbiology, Washington, D.C. 1985.

Mims, C. *The pathogenesis of infectious disease* (2nd ed.). Gune and Stratton, Inc., New York, 1982.

Savage, D.C. and Fletcher, M.M. *Bacterial adhesion,* Plenum Press, New York, 1985.

Journal articles

Beaman, L., and Beaman, B.L. The role of oxygen and its derivatives in microbial pathogenesis and host defenses. *Annu. Rev. Microbiol.* **38:**27, 1984.

Boothroyd, J. C. Antigenic variation in African trypanosomes. *Annu. Rev. Microbiol.* **39:**475, 1985.

Brubaker, R.R. Mechanisms of bacterial virulence. *Annu. Rev. Microbiol.* **39:**21, 1985.

Chesney, P.J., et al. The disease spectrum, epidemiology and etiology of toxic shock syndrome. *Annu. Rev. Microbiol.* **38:**315, 1984.

Costerton, J.W., Geesey, G.G., and Cheng, K.J. How bacteria stick. *Sci. Am.* **238:**86, 1978.

Crosa, J.H. The relationship of plasmid mediated iron transport and bacterial virulence. *Annu. Rev. Microbiol.* **38:**69, 1984.

Densen, L.P., and Mandell, G. Phagocytic strategy vs. microbial tactics. *Rev. Infect. Dis.* **2**(5):817, 1980.

Elsbach, P. Degradation of microorganisms by phagocytic cells. *Rev. Infect. Dis.* **2**(1):106, 1980.

Gallin, J.I. Disorders of phagocyte chemotaxis. *Ann. Intern. Med.* **92:**520, 1980.

McNabb, P.C., and Tomasi, T.B. Host defense mechanisms at mucosal surfaces. *Annu. Rev. Microbiol.* **35:**453, 1981.

Middlebrook, J.L., and Dorland, R.B. Bacterial toxins: cellular mechanisms of action. *Microbiol. Rev.* **43**(3):199, 1984.

Mills, E.L., and Quie, P.G. Congenital disorders of the functions of polymorphonuclear neutrophils. Rev. Infect. Dis. **2**(3):505, 1980.

Mogensen, S.C. Role of macrophages in natural resistance to virus infection. *Microbiol. Rev.* **43:**1, 1979.

Morrison, D.C. Bacterial endotoxins and pathogenesis. *Rev. Infect. Dis.* **5**(suppl. 4):733, 1983.

Moulder, J.W. Comparative biology of intracellular parasites. *Microbiol. Rev.* **49**(3):298, 1985.

Peterson, P.K., and Quie, P.G. Bacterial surface components and the pathogenesis of infectious disease. *Annu. Rev. Microbiol.* **32:**29, 1981.

Plaut, A.G. The IgAl proteases of pathogenic bacteria. *Annu. Rev. Microbiol.* **37:**603, 1983.

Roberts, N.J. Temperature and host defense. *Microbiol. Rev.* **43:**241, 1979.

Root, R.K., and Cohen, M.S. The microbicidal mechanisms of the human neutrophils and eosinophils. *Rev. Infect. Dis.* **3**(3):565, 1981.

Sparling, P.F. Bacterial virulence and pathogenesis: an overview. *Rev. Infect. Dis.* **5**(suppl. 4):637, 1983.

CHAPTER 24

Immunology: factors and functions of the immune system

Chapter Outline

Introduction

The previous chapter focused on the nonspecific immune mechanisms used by the host to destroy pathogenic microorganisms. This chapter will examine the body's specific immune devices for destroying infectious agents and other substances that are foreign to the body. The immune system has its origin in lymphoid tissue, specifically the lymph nodes, bone marrow, thymus, and spleen. Lymphoid tissue and the network of vessels associated with it are concerned with (1) the production and circulation of **lymphocytes**, the primary cell type of the immune system; (2) the capture of foreign substances; and (3) the production of specific immune substances by lymphocytes. The products of the immune system specifically react with foreign substances we call **antigens**. The result of this interaction may range from a lifesaving reaction to one which is life-threatening. Fortunately, most of these interactions are lifesaving.

The primary functions of this chapter are to (1) describe how T cells and B cells are involved with the immune response, (2) discuss in detail the structure and function of the immunoglobulin (antibody) molecules, (3) discuss the mechanism used by the immune system to respond to the vast number of antigens, (4) discuss how serum proteins called complement are integrated into the functions of the immune system, (5) describe the various states of immunity, (6) discuss the various types of vaccines that can be produced to prevent disease, and (7) describe the laboratory tests based on immunologic reactions.

WHY AN IMMUNE SYSTEM?

Nonspecific immune mechanisms are often inadequate to cope with foreign agents or substances, especially when a particularly virulent microorganism is involved and is able to evade phagocytic mechanisms. For individuals with defective phagocytic mechanisms even a microorganism that is considered nonpathogenic can cause serious disease. The **specific immune system** of the vertebrate host is endowed with three characteristics that may be expressed when it encounters a foreign agent: **specificity**, **memory**, and **recognition** of self-antigens.

1. Specificity. The immune system specifically recognizes foreign substances, that is, antigens or substances that can stimulate the formation of antibodies. Once contact is made, products of the immune system are elaborated and interact with the antigen.
2. Memory. During responses to foreign substances some lymphocytes give rise to memory cells. Memory cells are primed to act with speed and vigor the next time the same foreign agent is encountered. The existence of this memory capacity makes feasible the process of vaccination. The first time a foreign agent is encountered there is a short lag time before the immune system can generate enough immune products to overcome, for example, an infectious agent. There is such a multitude of antigens in nature that the body could not possibly contain enough cells of each specificity to respond speedily on the first encounter with each and every foreign agent. Memory cells make it possible for the immune system to remember and to be alerted to specific foreign agents that have been previously encountered, for example, through infection or vaccination.
3. Self-recognition. Many of the foreign substances with which the immune system interacts are similar to the body's own molecules, cells, or tissues. The immune system, however, can discriminate between the foreign substances and self substances. It does not normally form products against its own tissues, but there are exceptions (see Chapter 25).

The specificity of responses by the immune system makes possible swift and efficient actions that nonspecific resistance responses are incapable of or that they perform inadequately. Immune system products specifically facilitate the phagocytosis of microorganisms, the destruction of host cells and the microorganisms they contain, and the production of other beneficial actions. We must not forget, however, that some interactions, such as allergies, can be harmful to the host.

DEFINITIONS AND EXPLANATIONS OF FREQUENTLY USED TERMS AND CONCEPTS

antigen (Ag): A substance (specifically proteins, polysaccharides, or nucleic acids that are either soluble or particulate) that incites the immune system to form immune products specific for that substance. Antigens not only cause the formation of immune products but interact with them as well. Antigens are sometimes called **immunogens** when the emphasis is on their ability to incite the formation of immune products (**immunogenicity**). We will, for clarity's sake, use only the term antigen.

antigenic determinant: Small chemical groups on the surface of antigens with which immune products interact.

immune product: These are the products of lymphocytes that have been stimulated by appropriate antigens. The immune products of B lymphocytes are called **antibodies**. T lymphocytes, when stimulated by antigens, proliferate and differentiate into effector T cells. In a general sense, the effector T cells can be regarded as the immune products of a T-cell response.

immunoglobulin (Ig): A glycoprotein composed of four polypeptide chains that functions as the antibody molecule. Functionally, there are two types of immunoglobulins: **surface immunoglobulins**, which are present on the surface of lymphocytes where they act as specific receptors (recognition molecules) for the antigen, and **secreted immunoglobulins**, which are the products of B lymphocytes and appear in the body fluids (humors) as antibodies.

antibody (Ab): An immunoglobulin that has a known specificity for a given antigen. Each antibody has binding sites for an identified antigen.

B lymphocytes (B cells): A major class of lymphocytes that produce immunoglobulins and are primarily involved in humoral immunity, that is, the production of antibodies.

T lymphocytes (T cells): A major class of lymphocytes that are thymus dependent and form effector T lymphocytes on stimulation by antigens. Effector T lymphocytes destroy cells that have specific surface antigens. They also produce **lymphokines**, which are nonantibody mediators associated with inflammatory events and also associated with immunoregulation.

humoral immunity: Immunity provided by antibodies; B cell immunity; antibody-mediated immunity. Examples include toxin neutralization and the increased intensity of phagocytosis.

cell-mediated immunity (CMI): Immunity provided by effector T cells: T-cell immunity. Examples include intracellular parasite elimination, graft rejection, and protection from cancer.

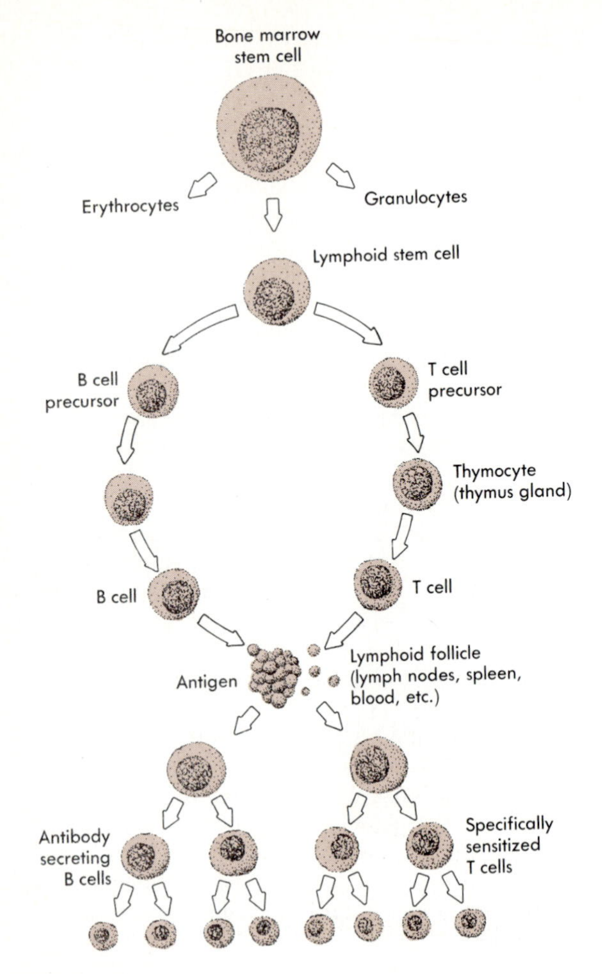

Figure 24-1 *Development of T and B cells. Both cell types have their origin in the bone marrow. T cells migrate to the thymus, where they begin their differentiation. B cells not only originate in the bone marrow but also apparently differentiate there. T and B cells migrate to lymphoid tissue, where they await stimulation by antigens. Stimulated B cells give rise to antibody-producing cells, while T cells become sensitized and divide into a population of cells with different functions.*

T CELLS, B CELLS, AND THE IMMUNE RESPONSE

Derivation and characteristics of T cells and B cells

T cells and B cells represent the two major populations of lymphocytes. They differ not only in the manner in which they are developed but also in distribution and function. Both cell types are derived from stem cells located in the bone marrow, the same site from which red blood cells, monocytes, granulocytes, and platelets are also derived (Figure 24-1).

The *T* of the T cell signifies **thymus dependency**, since the thymus gland at one stage is critical to the development of this population of cells. The thymus, in this role of T cell processing, is called a primary lymphoid organ. The majority of T cells, while in the thymus, are not fully competent to perform their immunological assignments. They travel primarily to lymph nodes and the spleen, which represent secondary lymphoid organs. It is primarily in the spleen and lymph nodes that T cells await activation by antigens and differentiation into specialized cell types. T cells constitute the greater percentage of circulating small lymphocytes (Table 24-1). T cells usually have a life span of months to years. Functionally T cells produce **lymphokines**—substances that exhibit specific biological activities (see p. 715 for further discussion).

Table 24-1 *Percentage of B and T lymphocytes in various human lymphoid depots*

Source	B cells	T cells
Peripheral blood	1 to 15	70 to 75
Thymus	1	99
Spleen	40 to 45	40 to 45
Lymph node	15 to 20	65 to 70
Gut-associated lymphoid tissue	60	20 to 25

The B cells are symbolized with a letter *B* because in early experiments it was discovered that a cloaca-associated lymphoid organ in chickens, called the **bursa of Fabricius**, was critical to its development and competence in the immune response. The organ for differentiation of mammalian B cells is believed to be the fetal liver. From this site B cells migrate to peripheral lymphoid tissue such as lymph nodes and spleen, where they await activation. B cells usually have a life span of days to weeks. Functionally they produce antibodies. In the following paragraphs we will discuss how T and B cells are specifically involved in the immune response.

T cells and cell-mediated immunity. The immune system of the host is stimulated by the presence of a foreign antigen. The host response is initiated by the capture of the antigen by macrophages located at sites such as the skin or lymph nodes (Figure 24-2). Macrophages process the antigen into fragments, which then become associated with the macrophage surface (see box on p. 714 for other functions of macrophages). Once this process takes place, the macrophage behaves as an **antigen-presenting cell**; that is, it is ready to present the antigen to a T cell. This interaction between macrophage and T cell causes the T cell to become **sensitized** or **activated** and this is followed by two major events. First, some T cells become **memory cells** and

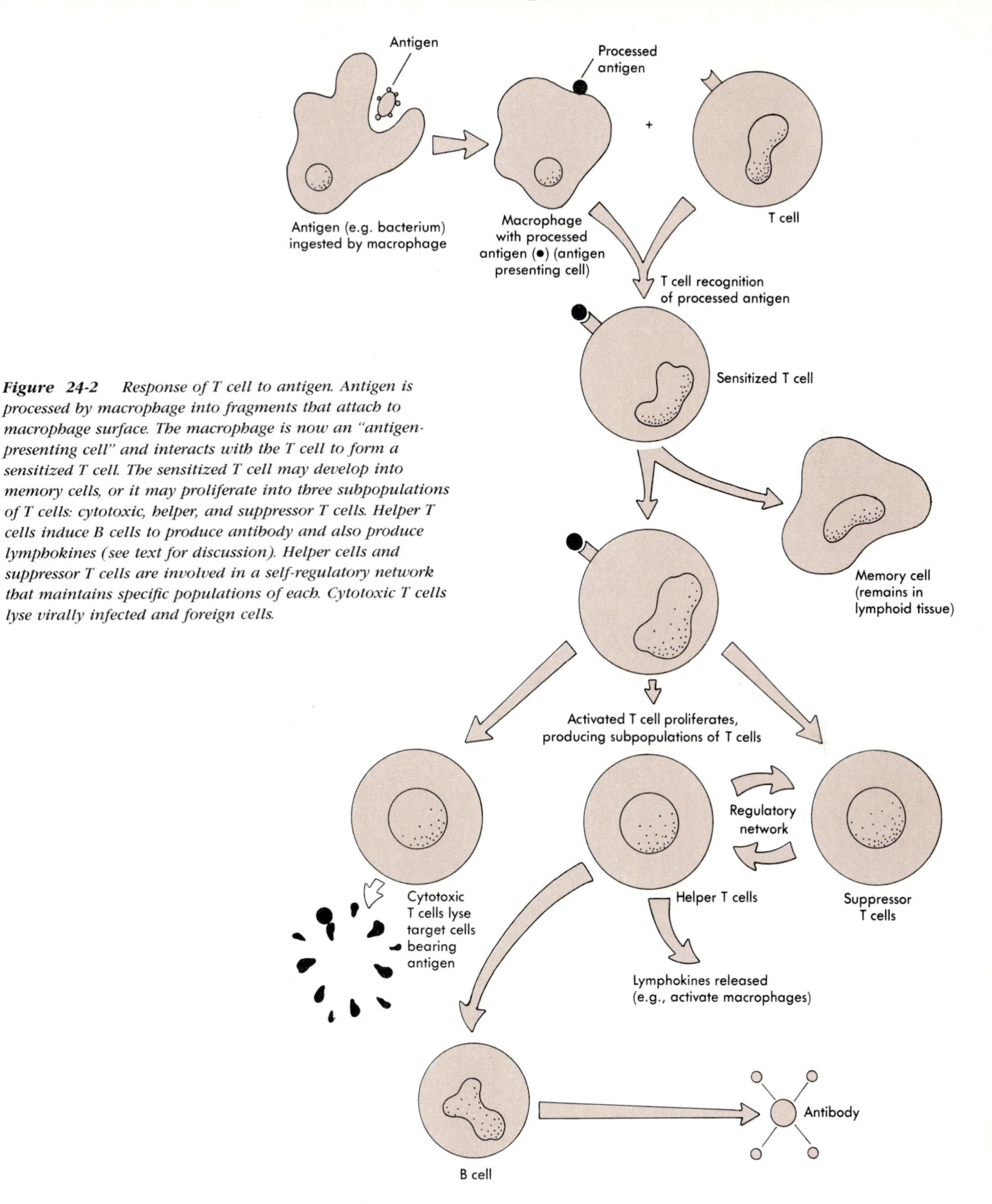

Figure 24-2 *Response of T cell to antigen. Antigen is processed by macrophage into fragments that attach to macrophage surface. The macrophage is now an "antigen-presenting cell" and interacts with the T cell to form a sensitized T cell. The sensitized T cell may develop into memory cells, or it may proliferate into three subpopulations of T cells: cytotoxic, helper, and suppressor T cells. Helper T cells induce B cells to produce antibody and also produce lymphokines (see text for discussion). Helper cells and suppressor T cells are involved in a self-regulatory network that maintains specific populations of each. Cytotoxic T cells lyse virally infected and foreign cells.*

remain in lymphoid tissue. If the same antigen is present in the host at some later time, these cells can respond immediately to the intruder. Second, the T cell can produce a clone of cells that differentiate into populations of activated cells with specific functions and include:

1. **Cytotoxic T cells (Tk):** These cells kill foreign or virus-infected vertebrate cells. They respond to

MACROPHAGE FUNCTIONS

1. Interact with many extracellular molecules, such as proteins and polysaccharides, and process them. These molecules may be free in solution or part of microorganisms.
2. Secrete several products, such as proteases, complement proteins, and growth regulatory factors such as interleukin I(IL-1).
3. Interact with B cells and T cells and become part of immunological response.
4. Bind lymphokines, which are regulatory proteins released by lymphocytes, and become activated. Activated macrophages are more effective in destroying microorganisms and tumor cells.

viral antigens expressed on the surface of the virally-infected cell and kill the cell before virus can replicate.

2. **Helper T cells (Th)**: Helper T cells help B and T cells respond to antigen and can also activate macrophages. Some helper T cells respond to antigen by releasing hormone-like substances called **lymphokines**, some of which are **interleukins** (which means between lymphocytes). Lymphokines can interact with macrophages as well as lymphocytes and their activities will be discussed shortly.
3. **Suppressor T cells (Ts)**: Suppressor cells suppress the responses of specific T and B lymphocytes to antigens. Like helper cells they apparently recognize antigens on the B and T cell surface. Suppressor cells are only active when stimulated by helper cells and the helper cell is itself inhibited by the suppressor cell. These reactions result in a self-regulating network that prevents the T and B cells from continual response to just one antigen. Imbalances in this regulatory role are observed in such diseases as acquired immune deficiency syndrome (AIDS) where suppressor cells outnumber helper cells. T cells perform several protective responses in the cell and include (1) immunity against those infectious agents that have an intracellular habitat (viruses and certain bacteria such as the agent of tuberculosis and those agents that cause chronic infections (fungi); (2) recognition and elimination of aberrant "self" cells, namely cancer cells; and (3) recognition and elimination of foreign cells and tissues as occurs in graft rejection.

The complexity of the T-cell response. The student should be aware that Figure 24-2 is a simplified scheme for the events involved in cell-mediated immunity. The understanding of T cell responses has been hampered by the inability of scientists to identify and characterize the T cell receptor that recognizes the antigen. The question that has been raised for many decades is "How is the T cell able to recognize a large number of foreign antigens?" In other words the T cell faces the same challenge as antibodies in recognizing specific antigens. In 1984 the T cell receptor for antigen was isolated and identified. It is a protein consisting of two chains, alpha and beta, and belongs to a family of molecules that includes antibody and histocompatibility molecules. The T cell receptor gene, like antibody genes, assembles complete genes from smaller coding segments. The beta chain of the T cell receptor, like the heavier of the two chains of which antibodies are composed, is assembled from four coding segments designated **V (variable)**, **D (diversity)**, **J (joining)**, and **C (constant)**. The alpha chain at this time appears to be assembled only from V, J, and C segments. During T cell development the four coding segments are separated but they later rearrange to bring together the V, D, and J regions (see Figure 24-3). The greater the number of different V, D, and J segments the greater the number of T cell receptor proteins that can be made from them. The number of these segments for the alpha and beta chains is somewhere between 15 and 50, which is considerably less than the comparable segments for antibody heavy and light chains. However, another molecule is also involved in the recognition process.

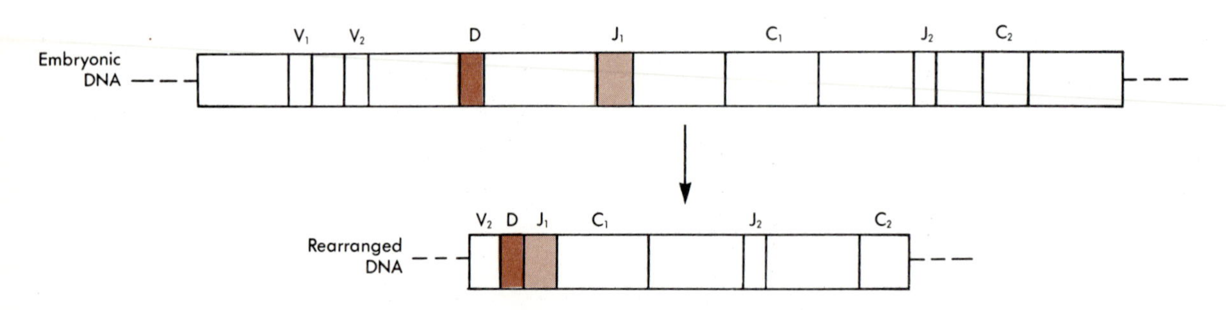

Figure 24-3 *Mechanism by which the dispersed V,D,J, and C segments on embryonic DNA of the T cell receptor gene are rearranged to form contiguous segments.*

Scientists have learned that the recognition between T cell and antigen occurs only with the appropriate **histocompatibility molecules**, located on the macrophage. Histocompatibility molecules are glycoproteins controlled by specific genes and, as indicated previously, belong to the family of molecules like that of the T cell receptor proteins. One class of histocompatibility molecules (class II) is found primarily on macrophages and B cells, while the other class (class I), is found on nearly all body tissues. The histocompatibility molecule is a marker that distinguishes "self" from "nonself." For example, the histocompatibility molecules found on the majority of tissues is responsible for the rejection of transplanted tissue from unrelated individuals. In the special case of interaction between T cells and antigen the histocompatibility markers on the macrophage also impose restrictions on recognition. The allele (mutational state of the gene) of the macrophage histocompatibility molecule determines which antigenic determinant is recognized. Because this class of histocompatibility molecules is expressed only on macrophages and B cells, recognition of some proteins, including "self" proteins, will not occur.

The success of the T cell in the immunological response is also influenced by a class of molecules called lymphokines that stimulate T-cell proliferation and are discussed in the next section.

Lymphokines. Lymphokines, which are produced in minute quantities, act locally in the host. They have been difficult to purify and identify chemically and have therefore been named according to their biological activity. Anywhere from 50 to 100 have been named, and each has several biological activities. Some of the lymphokines, called **interleukins**, are involved in the T cell response and will be discussed with other lymphokines.

INTERLEUKINS. **Interleukin 1(IL-1)** is one of the major molecules released into the circulation after infection and responsible for fever. IL-1 is produced by antigen presenting cells in response to an as yet unknown factor released by the bound T cell. IL-1 acts on T cells by inducing the formation of receptors for interleukin 2(IL-2). T cells that are not activated by antigen do not express IL-2 receptors. **Interleukin 2**, also called **T-cell growth factor**, is released by T cells and binds to receptors on the antigen-stimulated T cells. The binding of IL-2 to the antigen-stimulated T cell causes it to proliferate and enlarge its population of cells (Figure 24-4). Although the interaction of the antigen-presenting cell with its specific antigen receptor on the T cell confers specificity for a given immune response, the interaction of IL-2 with IL-2 receptors determines the magnitude and duration of the immune response. Studies on the control of IL-2 receptor should help us to understand certain diseases in which there is uncontrolled division of T cells, for example, human T cell leukemia virus infection and AIDS.

GAMMA (γ) INTERFERON. The expression of histocompatibility molecules on the macrophage is under regulatory control and involves γ-interferon. For example, when the host is infected the foreign agent is processed by the macrophage and presented to the T cell. The activated T cell responds by producing γ-interferon. Gamma interferon binds to macrophages and induces new expression of mRNA for histocompatibility molecules. An inflammatory reaction ensues in which the activated macrophage is the predominant cell type. When the foreign agent has been processed and eliminated the inflammatory process subsides, activated macrophages with the specific histocompatibility marker are no longer present, and T cell activation is no longer pos-

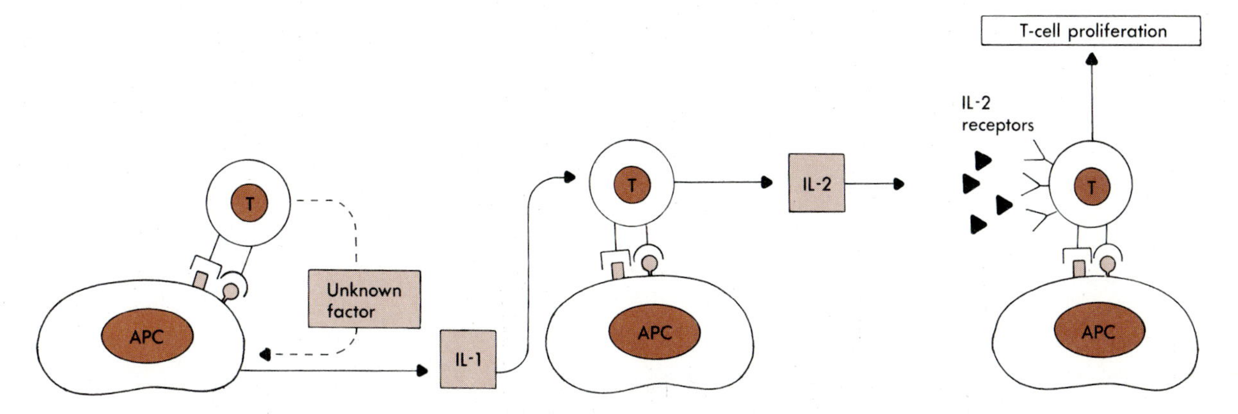

Figure 24-4 *Activation of T cells and factors causing their proliferation. Antigen presenting cells (APC) which contains histocompatibility determinant (■) and antigen determinant (♦) binds to T-cell receptors. This interaction causes T cell to release an unknown factor that induces APC to release interleukin I(IL-I). IL-I acts on T-cell-APC complex and induces interleukin receptors (γ) on T cell and stimulates T cell to release interleukin 2 (IL-2). IL-2 induces antigen-activated T cell into proliferation.*

sible. Only by reexposure to antigen will the T cell again become activated.

MACROPHAGE-ACTIVATING FACTOR. Macrophages activated by this lymphokine are larger, have more lysosomes than nonactivated macrophages, secrete enzymes involved in inflammation, and have increased phagocytic activity. They have increased killing capacities over resting cells because they release relatively large amounts of hydrogen peroxide and superoxide ion, both of which help to destroy microorganisms and possibly tumor cells.

B cells and humoral immunity. The B cell is activated not only by antigen, but by some T cells and their products. Many thousands of B cells are present in the host, and each has a different antigen-binding specificity. Antigen will bind only to those B cells with the appropriate surface receptors. It is important to point out that although differentiated B cells produce antibody endogenously, antibody molecules (such as IgM and IgD) are also located on the surface of precursor B cells. It is these surface antibodies that recognize specific antigens. Antigen-stimulated precursor B-cells differentiate into antibody-producing cells called **plasma cells** and the longer-lived **memory cells** (Figure 24-5), each with the same antigen-binding specificity. Should the same antigen appear at some later time, the memory cells will be ready to respond quickly and with more intensity than at the initial encounter.

Antibody produced by plasma cells can be produced at the rate of 1000 to 2000 molecules per second, and this rate is maintained for a period of 3 to 5 days until the plasma cell dies. Antibodies produced endogenously by plasma cells can specifically interact with molecular groups, called **antigenic determinants**, which are located on the surface of the antigen. The antibody may have a direct effect on the antigen. For example, if the antigen is a toxin, antibody may interact with the toxin and neutralize it. In many instances, however, antibody is not the sole factor in the antigen-antibody response. For example, in the case of an infectious microorganism that possesses a capsule, antibody binds to the capsule of the organism but the organism is not killed. The antibody merely serves to bind the bacterium to a phagocyte, which determines the outcome (destruction) of the bacterium.

Only B cells can produce antibody, but the B cells are subordinate to the T cells, which can either enhance or suppress this capacity. Once the antigen and antibody complex is formed, other factors may be involved, such as complement, which intensifies the phagocytic process; this is discussed later (see p. 727).

Concept Check

1. What three factors make the immune system an improvement over nonspecific mechanisms?
2. What are the differences between T cells and B cells?
3. What is the sequence of events that takes place when a foreign substance enters the tissue of the host?
4. Describe the four populations of cells produced from T cells activated by an antigen.
5. How are interleukins associated with T-cell proliferation?

ANTIGENS

Characteristics

Foreignness. The immune system's ability to discriminate between self- and nonself-antigens occurs during development. Most antigens are foreign and include such substances as bacteria, viruses, fungi, toxins, red blood cells, serum, and vaccines.

There are exceptions to the rule that antigens are foreign. There are human and animal diseases, for instance, that are called autoallergies or autoimmune diseases. In autoimmune diseases products are directed against the person's own molecules, tissues, or cells that are damaging to the self-antigen.

Determinant groups. Specific immune products interact only with specific portions of the antigen molecule (Figure 24-6). An antigen is composed of two parts: a **carrier portion** that makes up the bulk of the molecule and chemical groups called **antigenic determinants**. Functional antigenic determinants (those that interact with antibody) project from the surface of the molecule (Figure 24-7). Chemical groups that are internal (hidden) can come to serve as determinants if the antigen is structurally altered, for example, through enzyme action. The carrier portion also has determinants to which lymphocytes respond. Antigens contain multiple determinants (multivalent), but only those determinants that are on the surface are functional. Antigens may have repeating copies of the same determinant, or there may be distinct determinants, each of which may be present in varying numbers. Determinants are no larger than 5 to 7 amino acids of a protein or 5 to 7 sugar residues of a polysaccharide. The antigen-binding region (**Fab**) of the antibody molecule binds to the antigen determinant group and because of its small size cannot accommodate structures larger than its own binding sites. The antibodies formed are complementary for a given antigenic determinant, and this accounts for specificity, but specificity is not absolute. It is possible for the Fab region to bind to a determinant group that is chemically similar to but not exactly like the specific determinant group. In other words, different antigens A and B may bind the same antibody, a property

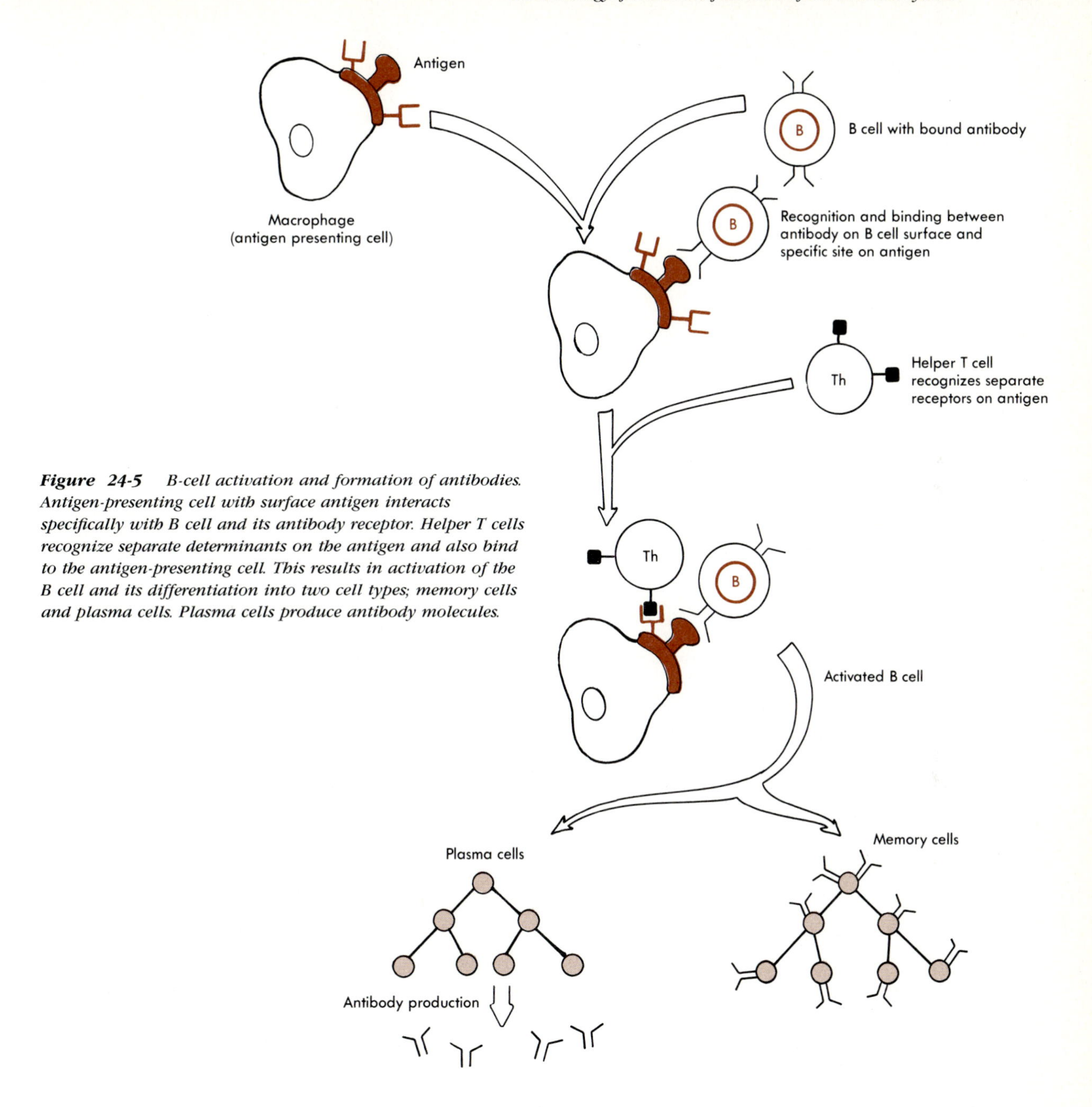

Figure 24-5 *B-cell activation and formation of antibodies. Antigen-presenting cell with surface antigen interacts specifically with B cell and its antibody receptor. Helper T cells recognize separate determinants on the antigen and also bind to the antigen-presenting cell. This results in activation of the B cell and its differentiation into two cell types; memory cells and plasma cells. Plasma cells produce antibody molecules.*

referred to as **cross-reactivity**. A large degree of cross-reactivity indicates a structural similarity between the two antigens.

Size. The size of the entire antigen molecule influences its ability to incite an antigenic response. Usually, the larger the molecule, the greater the antigenic potential. Apparently, the larger the molecule, the greater the number of available determinant groups. Generally, an antigen must have a molecular weight of at least 1000 to stimulate the immune-response mechanism, but more typically, 10,000 is the lower limit.

Chemical composition and complexity. The greater the molecular complexity of the macromolecule, the greater its antigenicity. Most natural substances are constructed of a large number of low molecular weight constituents rather than one or two repetitive determinants. The majority of antigens are natural proteins, such as the proteins found in serum, cells and tissues, enzymes, protein components of microorganisms, and microbial products. Polysaccharides such as those found in microbial capsules are antigenic, but lipids are not antigenic. Organic molecules composed of combinations of pro-

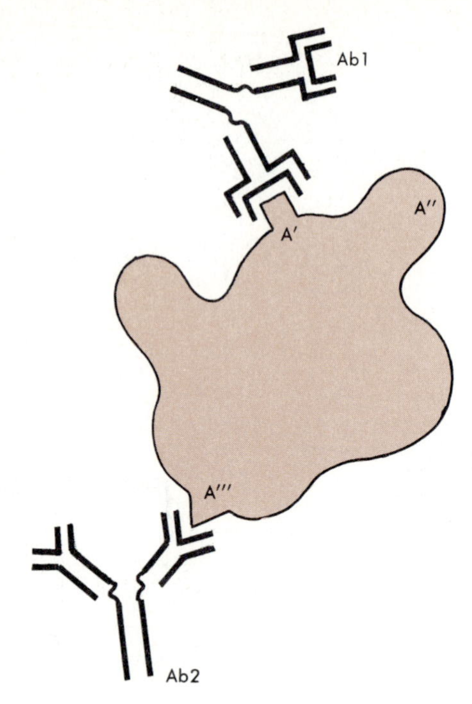

Figure 24-6 *Antigenic determinants. Part of antigen bound by antibody is called antigenic determinant (A', A'', A'''). Most antigens possess many determinants that are bound by different antibody molecules. Ab1 is specific for A'; whereas Ab2 is specific for determinant A'''.*

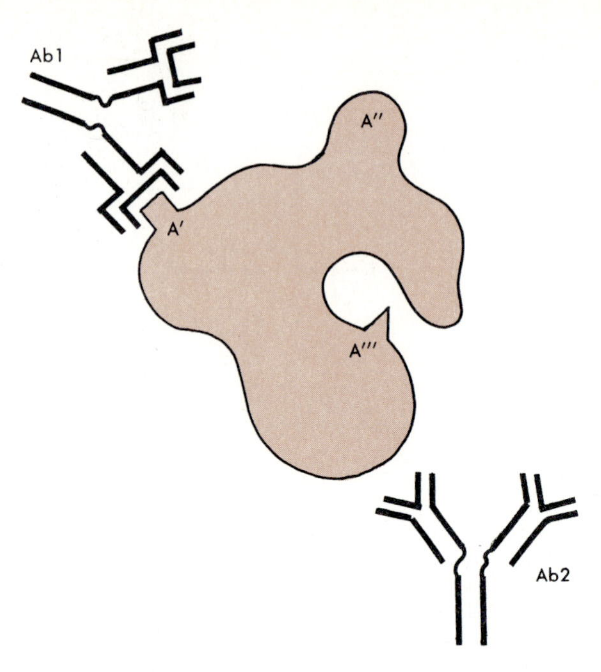

Figure 24-7 *"Hidden" determinants. Only determinants expressed on surface of antigen can be bound by antibody. Thus antigen in figure has in effect only two determinants, and Ab2 does not have access to "hidden" determinant A'''.*

teins, carbohydrates, and lipids such as lipoproteins (cell membrane), glycoproteins (the A and B blood cell group substance), polypeptides (insulin), lipopolysaccharides (endotoxins), and nucleic acids are antigenic.

Haptens. Certain molecules, because of their small size, are not antigenic. When coupled with a carrier molecule (usually a protein), they can be antigenic and are referred to as **haptens**. Antibodies or sensitized lymphocytes formed against the haptens will give a visible reaction when they interact with hapten-carrier conjugates, but not if they are allowed to interact with unconjugated hapten molecules. Haptens are sometimes referred to as partial or **incomplete antigens**. Some of the naturally occurring substances involved in allergic contact dermatitis, such as the catechols of poison ivy, are haptens (see Chapter 25) as well as the fungus-produced antibiotic, penicillin.

IMMUNOGLOBULINS

Immunoglobulins are glycoproteins that make up about 20% to 25% of serum proteins. They are found primarily in the gamma globulin fraction of the serum and on the surfaces of B lymphocytes. Immunoglobulins function in the immune response mechanism in two ways: as surface receptors and as immune products. **Surface immunoglobulins** located on B lymphocytes function as specific receptors for the determinant groups of antigens. As we have already mentioned, plasma cells are the producers of immunoglobulins. Immunoglobulins can be found in body fluids and on mucous surfaces, but most are found in the serum. An immunoglobulin is called an **antibody** when the antigen, for which the immunoglobulin is specific, is known. Antibodies may be more specifically named, for example, **agglutinins**, **precipitins**, **antitoxins**, and **opsonins**, depending on the effect they have on the antigen. Many of these antigen-antibody effects are the basis of tests often employed in the diagnosis of infection and are discussed later in this chapter.

Physicochemical properties

Lines of investigation. Much of the understanding of the nature of the immunoglobulin molecule stemmed from four lines of investigation:

1. The experiments of Edelman and others culminated in 1959 with the revelation that the basic immunoglobulin molecule is composed of two heavy (H) and two light (L) polypeptide chains.

2. Porter in 1959, using controlled enzymatic degradation of the immunoglobulin molecule, obtained two major fragments: the *Fab,* or *a*ntigen-*b*inding *f*ragment,

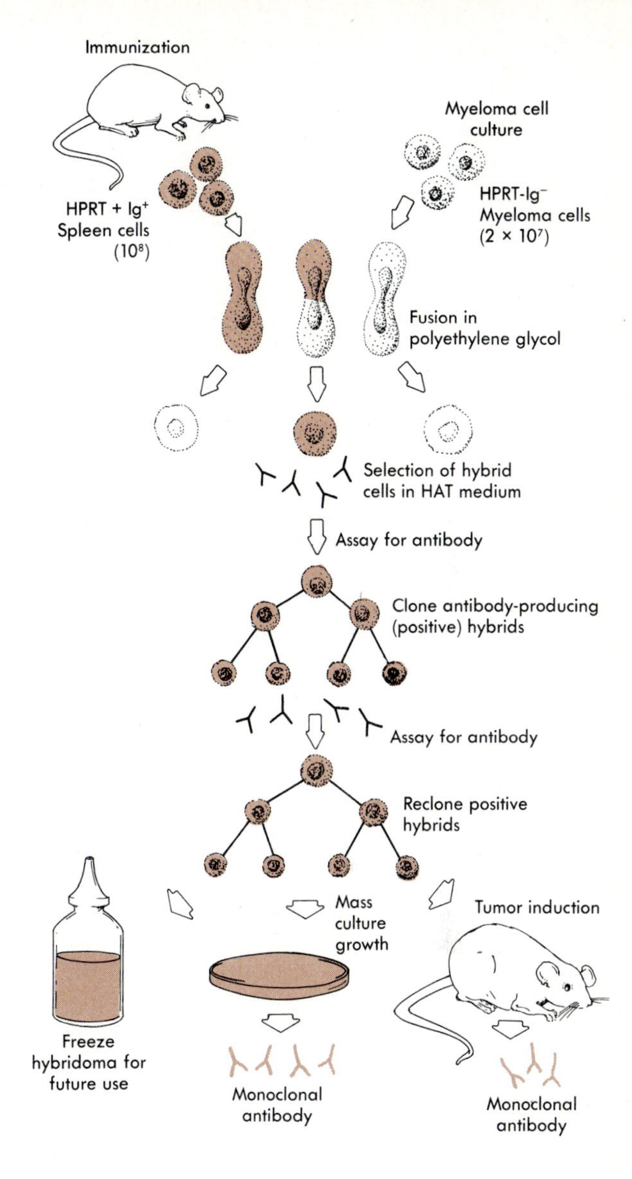

Figure 24-8 *Hybridoma technique. Hybridoma production begins by immunizing mouse with selected antigen. Animal's spleen cells are harvested and mixed with mouse myeloma cells and briefly incubated with polyethylene glycol to promote fusion. Cells are transferred to growth medium containing hypoxanthine, aminopterin, and thymidine (HAT). Unfused myeloma cells and myeloma-myeloma hybrids, which lack enzyme HPRT (hypoxanthine phosphoribosyl transferase), cannot survive in HAT medium. Unfused spleen cells and spleen-spleen hybrids die naturally after a few replications. Remaining spleen cell-myeloma hybrids are assayed for antibody production against antigen, and positive hybrids are cloned. Later they are again assayed for antibody production, and the positive hybrids recloned. At this point, hybridoma can be frozen and stored. To amplify antibody production, hybridomas are injected into mice, in which they produce ascites tumors that generate large amounts of monoclonal antibody. In human spleen cell-human myeloma hybridomas, amplification would be done by growing cells in mass culture or in immunosuppressed animals. Myeloma cells that are used in these experiments have a loss mutation for HPRT. This is important in selecting out myeloma cells that have hybridized with antibody-producing spleen cells. Only myeloma cells that have fused to spleen cells containing HPRT are able to use the hypoxanthine and thymidine in the HAT medium and therefore survive.*

and the *Fc,* or *c*rystalline *f*ragment. Functional activities are related to those fragments.

3. Detailed information about immunglobulin composition was obtained through a third line of investigation, the examination of myeloma proteins. The cells that produce immunoglobulins sometimes become malignant and form clones of cells (plasma cells) that yield large amounts of homogenous (monoclonal) immunoglobulins called myeloma proteins. These monoclonal immunoglobulins are specific for one determinant site but are nonprotective, since they suppress protective antibody formation. Portions of the immunoglobulin molecule called the *L chain* may appear in the urine. These are known as *Bence-Jones proteins.* There are five major classes of immunoglobulins, all of which appear in serum in varying quantities. Investigations of myeloma proteins have revealed how the five classes differ from each other, their amino acid sequences, and other feature of the immunoglobulin molecule.

A recent mode of investigation of immunoglobulins, called cell fusion or the *hybridoma technique,* is an important research tool. It is expected to further illuminate the composition and the genetics of immunoglobulins and to yield important clinical applications. During cell fusion (Figure 24-8) the nucleus of a spleen cell, which is producing antibodies specific for a particular antigen, fuses in vitro with the nucleus of a *myeloma tumor cell.* The tumor cell multiplies and produces antibodies that are specific for a given antigenic determinant when the fused cell is transferred into a suitable experimental animal or is mass cultured. Two qualities are combined in the hybridoma: virtually unrestricted multiplication of

> **APPLICATIONS OF MONOCLONAL ANTIBODIES**
>
> **REPORTED TO DATE**
>
> Routine diagnostic and investigative serology and tissue typing
> Identification and epidemiology of infectious agents
> - Viruses
>   - Influenza: Monoclonal antibodies create genetic changes in vitro; map antigenic domains on hemagglutinin; compare in vitro mutants and in vivo variants; refine epidemiology
>   - Rabies: Monoclonal antibodies reveal antigenic instability; reveal serological subtypes in virus; refine epidemiology
> - Bacteria
> - Parasites
>
> Identification of tumor antigens; classification of leukemias and lymphomas
> Identification of functional subpopulations of lymphoid cells
>
> **ANTICIPATED IN HUMANS**
>
> Passive immunization against
> - Infectious agents
> - Drug toxicity
>
> Provision of graft protection
> Potentiation of tumor rejection
> Manipulation of the immune response
> Targeting of diagnostic or therapeutic agents in vivo for
> - Detection of metastases (spread of disease)
> - Delivery of cytotoxic agents in tumor cells
>
> Vaccine preparation

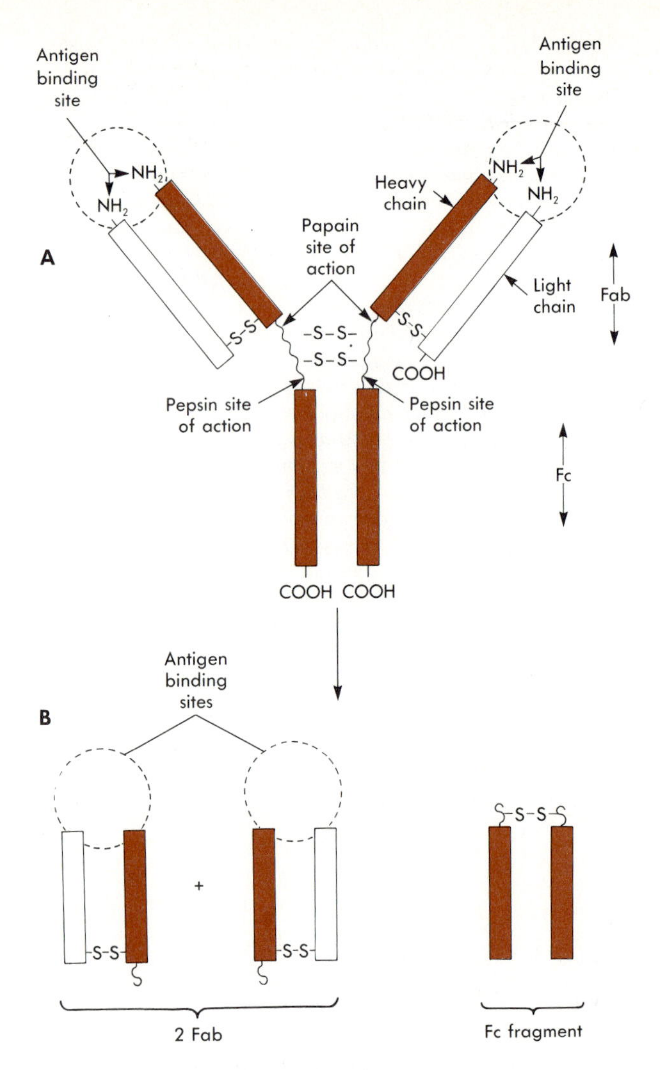

Figure 24-9 *Immunoglobulin molecule.* ***A,*** *Basic structure. Immunoglobulin molecule is composed of two identical heavy (H) chains and two identical light (L) chains. There are two sites for binding antigen.* ***B,*** *Effect of enzymatic activity of pepsin and papain on the immunoglobulin molecule is the release of fragments: two Fab and one Fc fragment.*

immunoglobulin-producing cells and large yields of a monospecific **monoclonal antibody**. Stated another way, the hybridoma technique makes possible the mass production of selected antibodies of a single specificity. The importance of this technique cannot be overstated, both for the potential it offers for improving the understanding of underlying immunological and genetic phenomena and for the potential for clinical applications (see box). As an example of a possible application of cell fusion, anticancer antibodies produced by this technique could have valuable diagnostic and immunotherapeutic applications.

Overall structure of the immunoglobulin molecule. The basic immunoglobulin molecule has four polypeptide chains, two of which are designated as heavy (**H**) chains and two as light (**L**) chains (Figure 24-9). The *H* chains of any particular immunoglobulin molecule are identical to each other, as are the *L* chains idential to each other. The four chains are held together by covalent disulfide linkages as well as noncovalent forces. The **hinge region** is flexible enough to permit bending and thus vary the distance between the antigen-binding sites.

The characteristics and activity of components of the antibody molecule were determined by splitting it with proteolytic enzymes, such as papain and pepsin. Treatment with papain, for example, produces two identical Fab fragments and one Fc fragment (Figure 24-9, *B*). Although the Fab fragments are engaged in antigen binding, it is the Fc fragment that is associated with biological activity. Three of these activities are as follows:

1. The Fc fragment promotes attachment of antibodies to phagocytes and mast cells. This increases the efficiency with which the phagocytic cell can in-

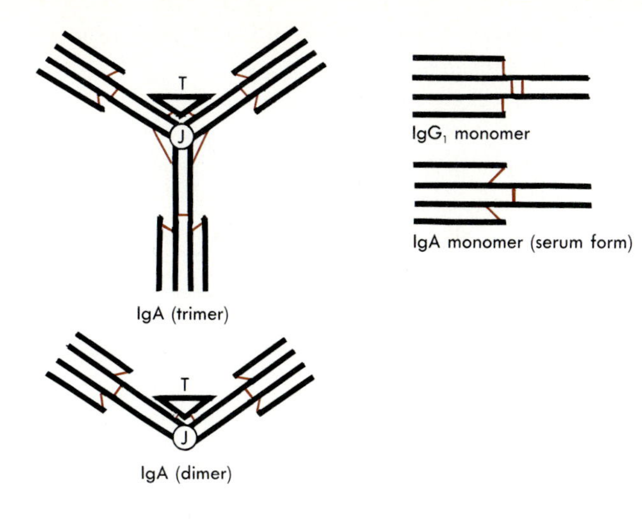

Figure 24-10
Representative structures for the major human immunoglobulin classes: IgA, IgM and IgG (IgD and IgE are not shown). Each contains basic four-chain polypeptide unit and some may exist as polymers (for example, IgM and IgA). Accessory polypeptides J (joining piece) and T (secretory piece) may also be present in polymeric forms.

gest and subsequently destroy microorganisms.

2. The Fc portion is the attachment site for the immunoglobulins IgG and IgM that bind complement (see discussion of complement, p. 727).
3. The placenta possesses receptor sites for binding the Fc portion of IgG, and this promotes transfer of the immunoglobulin across the placenta into the fetal circulation. Other antibodies do not bind to these receptors.

Concept Check

1. **What is an antigenic determinant, and how does it relate to an antigen? To a hapten?**
2. **What are the chemical and functional groups on the antigen?**
3. **Diagrammatically illustrate the overall structure of the immunoglobulin molecule.**
4. **What are monoclonal antibodies, and how are they produced? What is their value to the microbiologist?**

Immunoglobulin classes

Five major classes of immunoglobulins have been described in humans: IgG, IgM, IgA, IgE, and IgD. Each class of antibody has its own distinctive type of H chain. The heavy chains are designated as follows: γ for IgG, μ for IgM, α for IgA, ϵ for IgE, and δ for IgD. The H chain structure influences the conformation of the tail portion of the antibody molecule. It is this conformational difference that provides the antibody molecule with its distinct biological properties. In addition, the five classes of antibody have two types of L chains: kappa (κ) and lambda (λ). An individual antibody molecule will possess either κ or λ, but never both.

All but class IgD are known to exist in other mammals. All five classes are found in humans. IgE and IgD are present at very low levels. The classes differ chemically (in their amino acid content and sequences), structurally (Figure 24-10), and in biological functions (Table 24-2).

IgG. IgG is present in the serum in the greatest concentration of the five classes and accounts for over 70% of serum immunoglobulins. Its structure is that described previously for the basic immunoglobulin molecule.

Functionally IgG antibodies agglutinate and precipitate antigens, and form links between microorganisms and phagocytes, thereby facilitating endocytosis (Figure 24-11). Agglutinations represent reactions involving a *cell-bound* antigen and antibody, whereas precipitations involve *soluble* antigens and antibody. The Fc portion of IgG binds and activates complement, thus inducing an inflammatory response, and it attaches to macrophages, thus facilitating antigen uptake. IgG, which is the only immunoglobulin that crosses the placenta to the fetus, provides the newborn with a passive immunity to some of the diseases against which the mother has immunity.

IgM. IgM is the largest of the immunoglobulin molecules, consisting of five basic immunoglobulin monomers. There are 10 heavy chains, 10 light chains, 10 identical antigen-binding sites, and 5 Fc regions. IgM has an added chain, called the *J chain,* that is believed to be instrumental in joining together the five-unit structure, which is referred to as a pentamer. Disulfide bonds also serve in binding the basic units together. Even though there are 10 antigen-binding sites, experiments indicate that all 10 sites rarely bind to 10 antigenic determinants at one time. Many antigens are too large or complex to fit into all 10 antigen-binding sites at one time. IgM,

Table 24-2 *Selected physiochemical and biological properties of immunoglobulins*

	IgG	IgM	IgA	IgE	IgD
Physicochemical properties					
Sedimentation	7S	19S	7S to 11S	8S	7S
Molecular weight	148,000	900,000	170,000 400,000	200,000	200,000
Serum concentration (mg/100 ml)	1000 to 1500	60 to 180	100 to 400	0.01	3 to 5
Half-life in serum (in days)	18 to 23	5	6	2.5	3
Number of four-chain units	1	5	1 to 3	1	1
Effective combining sites (valences)	2	5	2 to 3	2	?
Biological properties					
Types of antibody	Agglutinin Precipitin Opsonin Neutralizing	Agglutinin Precipitin Opsonin Neutralizing	Secretory antibody on mucous surfaces	Reagin; skin sensitizing	?
Fixes complement	+	+	−*	−*	−
Crosses placenta	+	−	−	−	−
Other distinctive properties	Blocking antibody of atopic allergy hyposensitization	First Ig formed; ABO antibodies, receptor on B cell membrane	Found in serum and external secretions; surface protection	Reaginic antibody of anaphylaxis and atopic allergies; attaches to mast cells	Receptor on B cell membrane

*Activates complement through the alternative pathway.

which is found primarily in the blood, accounts for 5% to 10% of the immunoglobulin in serum.

Functionally, IgM is usually the first immunoglobulin produced in response to antigen stimulation, but the synthesis is not prolonged. The human fetus is capable of forming IgM, so that its presence in fetal serum indicates infection or exposure to antigens. IgM antibodies are very efficient in bringing about agglutination and in activating complement (see discussion of complement system pathways). IgM antibodies figure prominently in blood transfusion reactions because they can destroy the improperly transfused red blood cells of the donor. These agglutinating IgM antibodies are naturally occurring, since they develop spontaneously in humans without a known stimulus.

IgM is present as a surface immunoglobulin on most B lymphocytes that are the precursors of antibody-secreting cells. IgM as a surface immunoglobulin differs from secreted (serum) IgM. It is a monomer instead of a pentamer, and the Fc portion is longer than it is in the secreted IgM molecule. As a surface immunoglobulin IgM provides an antigen-binding receptor that recognizes a specific antigen on contact.

IgA. IgA appears regularly in serum and in body secretions. It is structurally similar to IgG when it appears in the serum, consisting of the typical four-chain immunoglobulin unit. IgA in body secretions such as the gastrointestinal tract, saliva, colostrum, tears, and urine is called **secretory IgA** or **sIgA**. Secretory IgA consists of two and sometimes three basic immunoglobulin units joined together by a polypeptide *J* chain and one other polypeptide, called the secretory piece. IgA is formed in plasma cells located beneath the epithelial lining of mucosal surfaces. The secretory piece, which is a transmembrane glycoprotein receptor, is added to the IgA molecule as it is transported across the mucous mem-

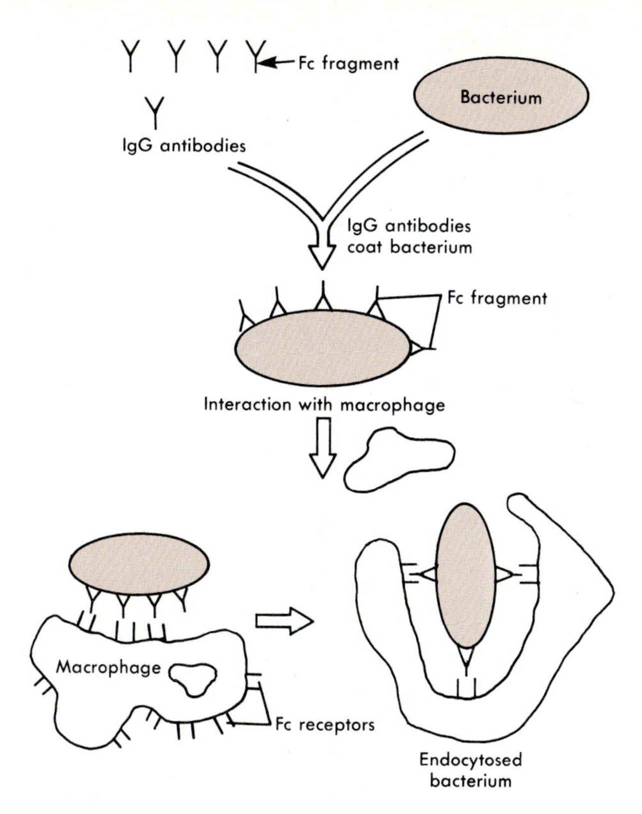

Figure 24-11 *Diagrammatic illustration of how IgG antibody promotes ingestion of an antigen (in this example a bacterium). IgG antibodies bind to bacterium with Fc fragments exposed. Fc receptors on the macrophage interact with Fc fragments of the antibody bound to the bacterium. This interaction promotes ingestion (endocytosis) of bacterium.*

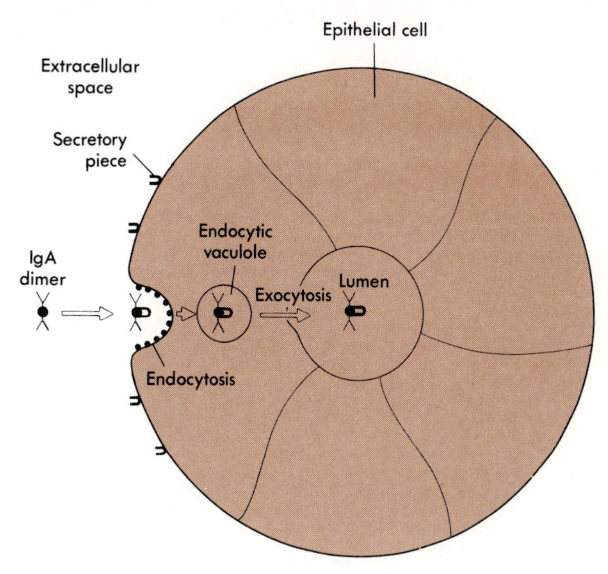

Figure 24-12 *Steps in the addition of the secretory piece to the IgA dimer in the formation of secretory IgA (sIgA). IgA dimer interacts with the secretory piece that is bound to the epithelial surface forming sIgA. sIgA is transported across the cell membrane, forming a vesicle that is then transported into the cell lumen.*

brane (Figure 24-12). Something about sIgA's structure prevents enzymes on the mucosal surface from degrading it.

IgA antibodies in the serum appear to have only a minor protective function. Individuals with serum IgA deficiencies appear not to suffer an impairment of their immunity. Secretory IgA on the mucosal surfaces, however, is immensely important as the first line of specific defense. A major defensive role it performs on the mucosal surface is to prevent the attachment of microorganisms. Attachment of an invading microorganism is a prelude to its penetrating the mucosa and subsequently entering the tissue. Some microorganisms, not to be outdone, produce proteases that destroy secretory IgA. Secretory IgA is regarded as one of the major factors in topical or local immunity.

IgE. The overall structure of IgE is similar to that of IgA with the exception that the heavy chains are larger and they have an additional domain. IgE antibodies are best known for their role in acute allergic disease. They are variously named as **reagin** or **atopic antibody**. In humans they are involved in certain allergies (hypersensitivities) of the immediate type that are referred to as atopic allergies (see Chapter 25). IgE antibodies have a distinctive quality: the predilection to attach to **mast cells** and **basophils** that are rich in pharmacologically active mediators such as histamine, serotonin, and heparin. In an allergic individual IgE antibodies that have been induced as a result of previous antigen (the antigen may be more specifically called an **allergen**) exposure are attached to mast cells in the tissue and to basophils in the blood. The allergen combines with the IgE antibodies that are located on the mast cell surface when the allergic individual is exposed to more of the same allergen (Figure 24-13). This antigen-antibody union initiates the release of the pharmacologically active mediators. The events resulting from the release of these secretory granules include smooth muscle spasms, dilatation and increased permeability of blood vessels, secretion of mucus by exocrine glands, attraction (chemotaxis) of neutrophils and eosinophils, and anticoagulation. The seriousness of the allergic reaction caused through these effects is determined by the amount of the substance released, as well as the suddenness and location of their release.

The beneficial functions of IgE antibodies are not entirely clear. Near the body surfaces IgE antibodies initiate a beneficial inflammatory response when irritants

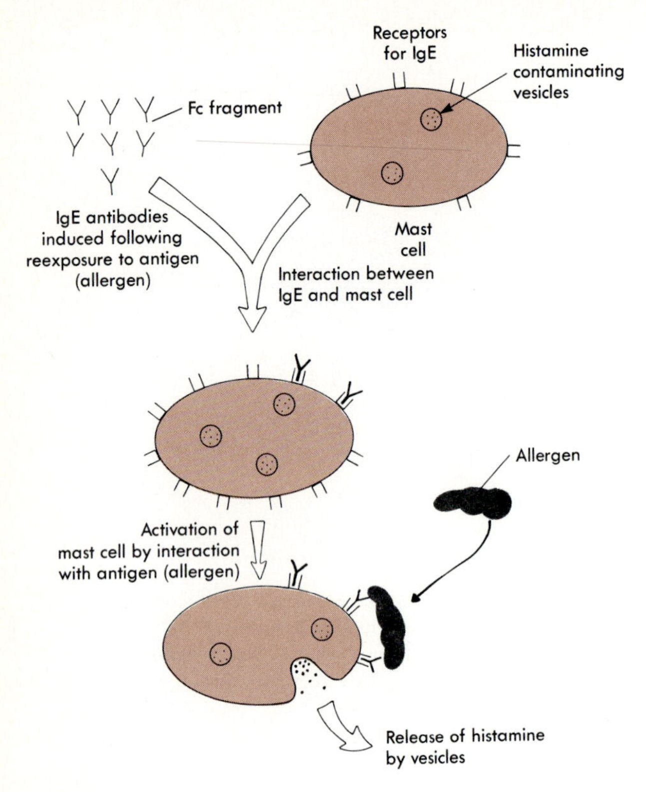

Figure 24-13 ***Diagrammatic illustration of the interaction between IgE and mast cells and the release of histamine, an allergic response.***

penetrate the integument. Many of the allergens that elicit an overexuberant response, such as in anaphylactic shock,* enter the host via injection (antibiotics, antisera), an unnatural route that evolution could not adapt to and guard against. IgE-mediated reactions possibly provide protection against intestinal parasites, since persons with chronic roundworm infections of the intestines have comparatively high levels of IgE in their serum. The IgE, however, does not appear to be worm specific.

IgD. IgD is present in serum in trace amounts and does not have a known protective antibody function. It acts as a surface immunoglobulin on B lymphocytes, just as surface IgM does.

ANTIBODY DIVERSITY

One of the most frequently asked questions by those first studying the immune response is "How does an animal produce antibodies that are specific for the enormous number of antigens, to which it is exposed during a lifetime, without also requiring a vast number of genes?" The number of specific antibody molecules that an animal can generate is believed to be on the order of 10^6 or more. This would mean that 1 million or more animal genes would have to be devoted to just antibody production. Since amino acids, which make up the antigen-binding sites, are encoded in the DNA molecule, the problem of antibody diversity is primarily genetic. To better understand antibody diversity, we will first look closely at the fine structure of the antibody molecule.

Antibody fine structure (constant and variable regions)

Examination of myeloma proteins has demonstrated that both heavy and light chains can be divided into **constant** and **variable** regions. The amino ends of these chains correspond to the variable region, while the carboxyl ends correspond to the constant region (Figure 24-14, *A*). The variable region of both heavy and light chains are made up of approximately 110 amino acids each. However, the constant region in the heavy chain is three times larger than that of the light chain. The differences in amino acid sequences at the variable end of the molecule will contribute to antibody diversity, since it represents the site of antigen binding. Only three portions of the variable region of both light and heavy chains display frequent variation. These areas of diversity are called **hypervariable** regions (Figure 24-14, *B*) and consist of approximately 10 amino acids each. They represent the actual binding sites for antigen. The amino acid sequences in the remaining portions of the variable region remain relatively stable.

Further sequence analysis studies of the H and L chains revealed that the constant region of the heavy chain is more complex than the constant region of the light chain. There are intrachain disulfide bridges (in contrast to interchain disulfide bridges) in the constant region of the heavy chain that cause the formation of three loops (Figure 24-15). Each of the loops in the constant region of the light chain has a considerable amount of sequence homology with each of the other loops. The same type of sequence homology also exists between the variable regions of heavy and light chains. These results showed that there are repeating segments of amino acids (also called **domains**), each of which contains an intrachain disulfide bridge. The heavy chain consists of three domains (**C_H1**, **C_H2**, and **C_H3**), whereas light chains consist of one variable (**V_L**) and one constant (**C_L**) domain.

Generation of antibody diversity

The fact that the antibody molecule consists of distinct variable and constant regions led many investigators to

* Anaphylactic shock is a hypersensitivity reaction resulting from a second exposure to a substance. The affected individual exhibits smooth muscle contraction, dilation of the vascular system, and edema. Hypersensitivity to penicillin, for example, results in over 200 deaths per year.

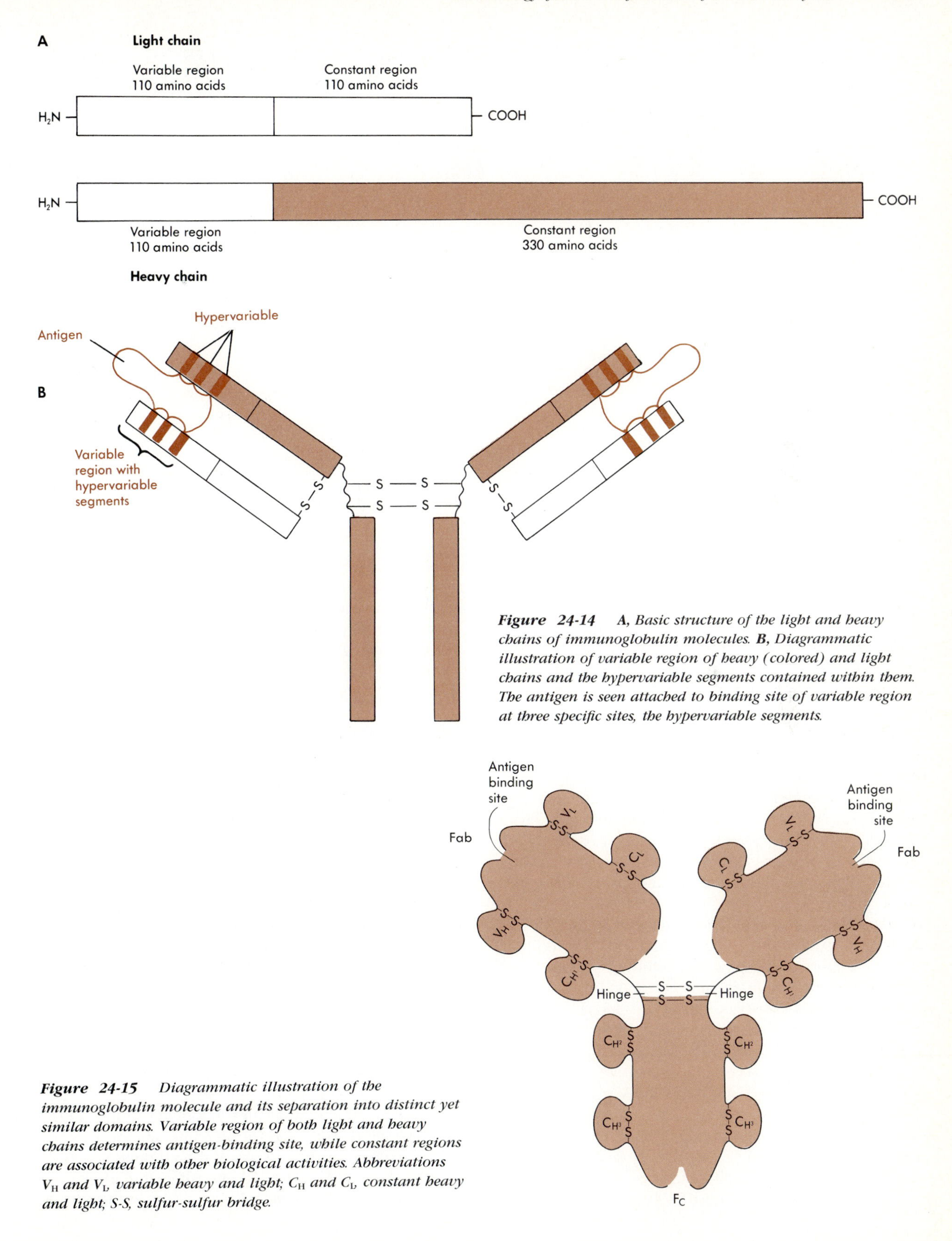

Figure 24-14 *A, Basic structure of the light and heavy chains of immunoglobulin molecules.* ***B,*** *Diagrammatic illustration of variable region of heavy (colored) and light chains and the hypervariable segments contained within them. The antigen is seen attached to binding site of variable region at three specific sites, the hypervariable segments.*

Figure 24-15 *Diagrammatic illustration of the immunoglobulin molecule and its separation into distinct yet similar domains. Variable region of both light and heavy chains determines antigen-binding site, while constant regions are associated with other biological activities. Abbreviations V_H and V_L, variable heavy and light; C_H and C_L, constant heavy and light; S-S, sulfur-sulfur bridge.*

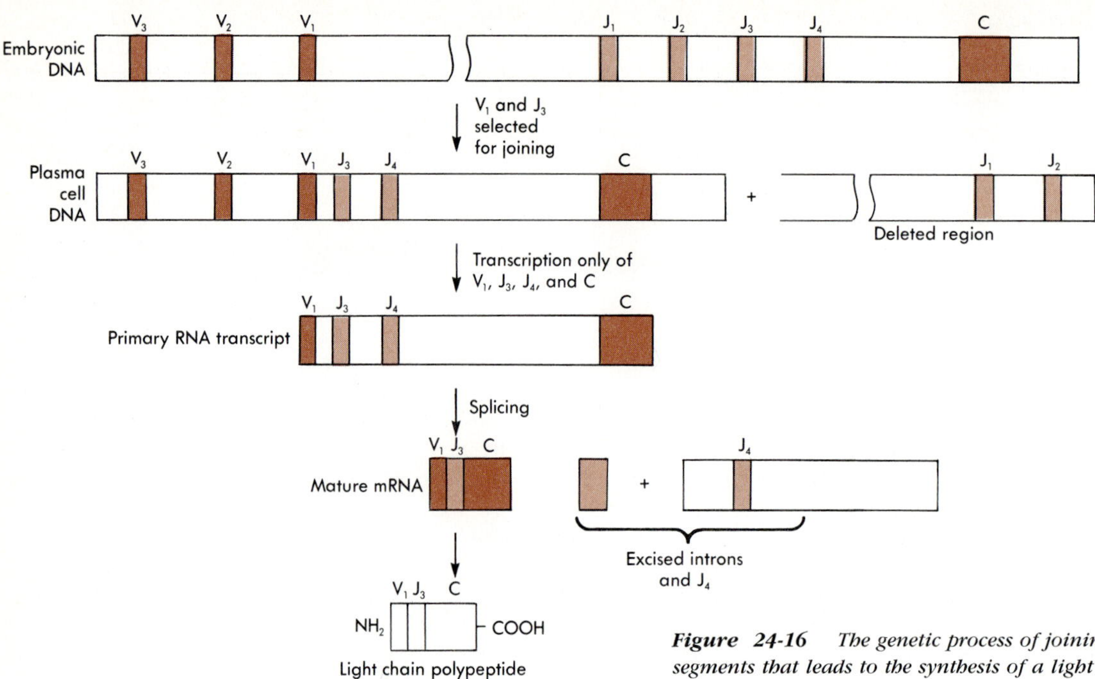

Figure 24-16 *The genetic process of joining V, J, and C segments that leads to the synthesis of a light chain.*

believe that each chain of the V and C regions was encoded by two separate genes that later joined together before being translated into polypeptides. In other words a two-gene, one-polypeptide hypothesis had been proposed. Experiments later showed this hypothesis to be true. What experimenters discovered was that in the embryonic cell, in which the immunoglobulin is not assembled, the genetic sequences encoding the V and C regions (that is, the V and C genes) are not contiguous but instead are located on different parts of the genome. During differentiation of the embryonic cell into a plasma cell the genetic sequences are brought together. We now know that for each immunoglobulin light chain (both kappa and lambda), as well as heavy chains, there exists a pool of genes that encode the information for a single polypeptide. Each gene pool consists of V genes and C genes.

Light chain diversity. Each V genetic region consists of at least two or three V gene segments. In addition, there is a short DNA segment called the **J** region downstream from the V region that is necessary before a complete light chain transcript can be produced. The J region consists of segments interspersed with introns and separated from the C region by an intron and separated from the V region by several thousand bases (Figure 24-16). During differentiation one of the V gene segments will be selected to be placed next to a specific J gene segment. Thus the differences in V and J segments provide a source of diversity. This type of diversity was described earlier with T cell receptors, which exhibit a similar mechanism (see Figure 24-3).

Heavy chain diversity. In addition to the V, J, and C regions, a segment called **D** (diversity) is associated with heavy chain synthesis. The number of D segments is not known, but they can be joined to any $\mathbf{V_H}$ or $\mathbf{J_H}$ segment to create even greater diversity.

Increasing antibody diversity

Up to this point we have seen that the cell is able to recombine selected V, D, and J segments into informational molecules coding for V_H and V_L regions. In the mouse there are approximately 300 gene segments for light chains and 100 V gene segments for heavy chains. By variously combining these gene segments, several thousand different heavy and light chains can be produced. Once the chains are produced, diversity can also be increased to over a million by combining different light and heavy chains as the immunoglobulin is being assembled. Thus we see that even though the cell is capable of responding to a million or more antigens, it is not necessary to possess an equivalent number of genes.

Self vs. nonself-discrimination

Broadly speaking, the immune system regards molecules not belonging to the individual as "nonself" and reacts against them. Molecules belonging to individuals are considered "self" and the immune system does not react against them. The failure of the immune system to react against "self" molecules is called **immunological tolerance**. Immunological tolerance can be induced in experimental animals and has been shown to be related primarily to the following factors:

1. The immunological maturity of the animal. Antigens are more readily accepted by the host if they are present during the maturation of the immune system, for example, in the newborn. Such tolerance, for example, results in the acceptance of tissue or organ grafting between members of the same species that are not genetically identical.
2. The inherent immunogenicity of the antigen. Some antigens are poor stimulators of an immune response, while others induce elevated immune products in the host. A poor stimulator of an immune response is foreign antibodies. For example, antibodies are taken from individuals who have recovered from hepatitis B. These antibodies are used for injection into individuals who have been exposed to the hepatitis B virus, such as oral surgeons, to prevent disease.
3. The dose of the antigen. Most antigens induce an immune response, yet, if the antigen is injected in large doses (50 to 100 times the normal immunizing dose) for a long period of time, tolerance can be achieved.

The discrimination between "self" and "nonself" apparently requires the integration of many signals. At this time we do not know enough about the immunological pathways to describe this method of discrimination.

Concept Check

1. **List the five major classes of immunoglobulins, and explain their principal biologic properties.**
2. **Explain the effect of an IgE-mediated response in an allergic host.**
3. **What is the basis of antibody diversity, and how it is generated?**
4. **What experimental studies have demonstrated mechanisms for immunological tolerance?**

COMPLEMENT SYSTEM

At the beginning of the 20th century it was recognized that the bactericidal activity of serum was due not only to the presence of antibody but also to another serum protein called **complement**. There are at least 20 discrete complement proteins (many of which are proteinases); 13 are activated in a sequential manner. The other complement proteins are inhibitors or inactivators that play a regulatory role.

Complement system pathways

The complement proteins form two interrelated enzyme cascades called the **classical** and **alternative** pathways. The numerical sequence of participation of the complement components is C1,C4,C2,C3,C5,C6,C7,C8, and C9. The two pathways provide two avenues to the cleavage of the complement protein **C3** to **C3b**, the main event in the complement system (Figure 24-17). In some of the steps one of the complement proteins cleaves the protein that is next in the sequence. Some of the smaller fragments that result from cleavage have various biological effects, which are discussed later.

Classic pathway. The classic pathway is activated by **antigen-antibody complexes** in which the antibody is **IgG** or **IgM.** Activation of complement protein C1 leads to the sequential formation of C3 and its cleavage to C3b. **C3b** coats target particles, such as microorganisms, and this is a major biological function of the pathway. C3b coating promotes strong adherence to phagocytic cells. The role of C3b in the induction of phagocytosis is not always clear, but it does appear to enhance ingestion of bacteria of low virulence and to facilitate ingestion of bacteria by activated macrophages.

In the next sequence of events C3b initiates the activation of other complement proteins: C5,C6,C7,C8, and C9. The C5 to C9 fragments form a unit called the **membrane attack complex (MAC).** The biological role of MAC along with other complement proteins is discussed subsequently (also see Figure 24-17).

Alternative pathway. The activation of the alternative pathway does not depend on antigen-antibody complexes. It can be activated by **microbial polysaccharides** and **bacterial lipopolysaccharides.** The alternative pathway is entered at the C3 level, and C1, C4, and C2 are bypassed. Like the classic pathway C3 is converted to C3b, which attaches to the microbial cell. Thus the alternative pathway provides a mechanism for fixation of C3b whenever sufficient quantities of antibody have not had time to develop (during the early stages of infection).

Biological activities of complement

We have seen that a variety of complement fragments are produced during activation of the complement systems. The biological activity that is promoted depends on what fragment is attached to the cell and what cell type to which it is attached. The capacity of complement fragments to attach to cells is important for several reasons: for marking and preparing particles for phagocytosis, for causing the release of pharmacologically active substances, and for causing lysis of cells.

The major biological activities of complement are **chemotaxis, opsonization, anaphylatoxin activity, membrane damage,** and **cytolysis** (Figure 24-17).

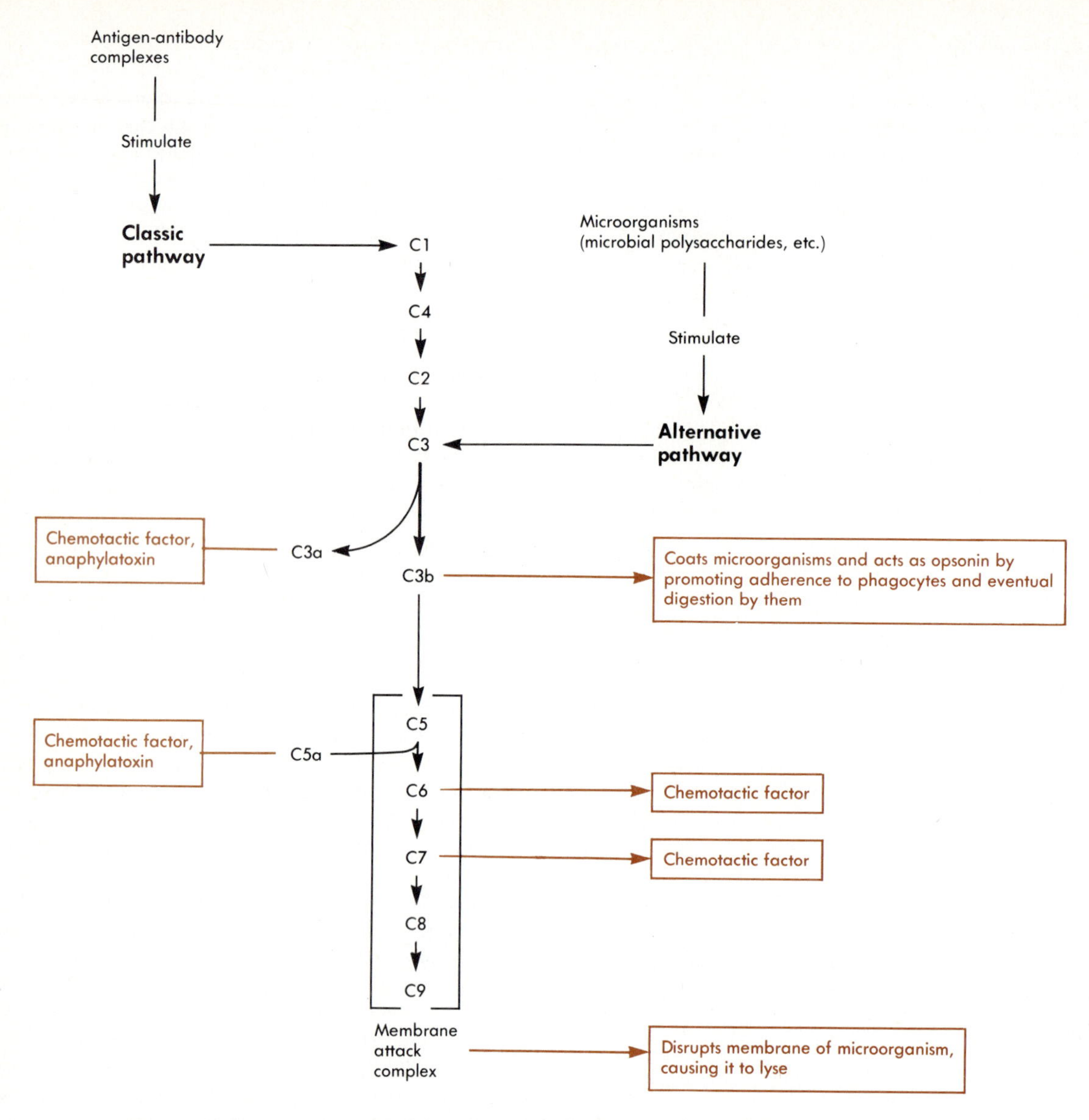

Figure 24-17 *Comparison of the classic and alternative pathways in the complement system. The first part of the pathway is devoted to the production of C3b. The second series of reactions involve cleavage of complement components C5,C6,C7,C8, and C9, which ultimately form a complex called the membrane attack complex. The functions of the various fragments are in colored boxes. A discussion of these functions is detailed in the text.*

Chemotaxis. Complement fragments that attach to bacterial cells and other antigens are the chemical attractants that "mark" the antigen and attract migrating phagocytic cells. The complement system is a source of the principal chemotactic factor **C5a**, which is a cleavage product of complement factor C5.

Opsonization. Some microorganisms are able to exercise their virulence because of their ability to escape ingestion by phagocytes alone. This evasion mechanism can be overcome by a process called opsonization. Opsonization is the process in which certain plasma proteins promote the binding of microorganisms to the phagocytic surface. The plasma proteins that facilitate this process, and thus enhance phagocytosis, are called **opsonins**; examples are **antibodies** and complement fragment C3b. Antibody molecules that act as opsonins are of the **IgG** and **IgM** class and attach to the phagocyte by means of their Fc portion. Phagocytes possess recep-

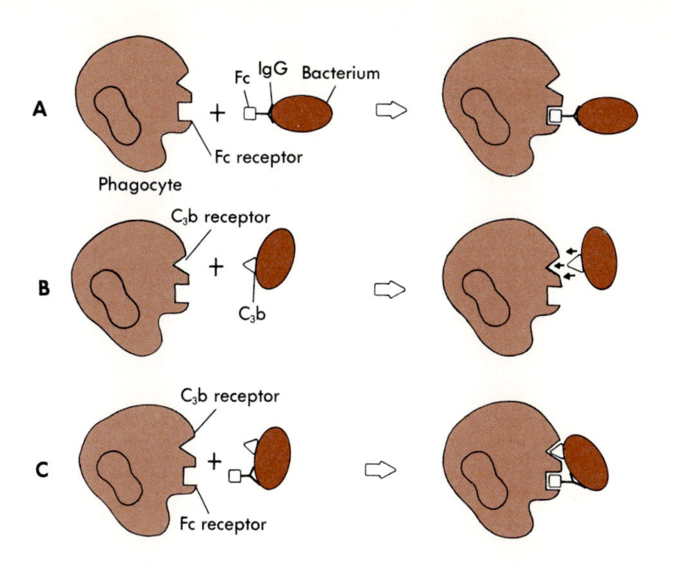

Figure 24-18 *Opsonization process associated with antibody and/or complement C3b.* ***A,*** *The bacterium with only antibody attached to its surface attaches very loosely to the phagocyte receptor site.* ***B,*** *The bacterium only has complement C3b attached and it too binds loosely to the phagocyte receptor site.* ***C,*** *A bacterium with both antibody and complement C3b attached to it surface binds very tightly to the phagocyte, and this provides a greater opportunity for the phagocyte to ingest the bacterial cell.*

tors on their membrane surface for C3b, which forms a bridge between the bacterium and the phagocyte and thus facilitates phagocytosis (Figure 24-18).

Anaphylatoxin activity. Anaphylatoxins are complement peptides (**C3a** and **C5a**) produced in the complement cascade that adhere to the membrane of mast cells and platelets, causing them to release histamine and other pharmacologically active substances. A similar response occurs during anaphylactic reactions, such as penicillin sensitivity. The pharmacologically active substances released from mast cells produce among their more prominent effects vascular permeability and smooth muscle contraction. The activities induced by these substances in modest amounts are beneficial. They allow phagocytes and serum proteins, including antibodies, to pour into the surrounding tissue of infected body sites.

Membrane damage and cytolysis. The membrane attack complex inserts itself into the lipid bilayer of the target cell. This disrupts the membrane and permits free exchange of solutes leading to the net influx of sodium ion and water and finally the lysis of the cell. When antibodies are involved in the process, it is called **immune cytolysis.** A well-known test, the complement fixation test, is based on the involvement of complement in immune cytolysis. The test is discussed later in this chapter.

Other activities. Complement-associated reactions can lead to host cell destruction and tissue damage, largely through intense inflammatory responses. Immune complex disease and some of the blood abnormalities in which there is destruction of red blood cells and platelets are examples of complement-mediated diseases (see Chapter 25).

IMMUNITY

States of immunity

The principal function of the immune system is to provide protection against foreign substances and, apparently, against nonnormal self-materials such as cancer cells. There are terms that are employed to describe an individual's state and type of protection. These especially pertain to protection against infectious agents. The two major categories of immune status are **innate immunity** and **acquired immunity.**

Innate immunity. Innate immunity is also called *natural immunity* or *genetic immunity.* Immune system products, such as antibodies and sensitized T lymphocytes, are not involved. Permanent, genetically endowed physiological, anatomical, and biochemical differences in individuals and between species account for innate resistance. These differences may be due to the lack of suitable receptor sites for the attachment of infectious agents, to differences in body temperature, to oxidation-reduction conditions of the tissues, to toxicity and pH of the body fluid, and to the availability of certain nutritional requirements. For example, resistance of chickens to anthrax infection is based on body temperature. If the temperature is lowered from the normal 39° C to 37° C, chickens become susceptible to anthrax. The bacterium that causes contagious abortion in cattle, *Brucella abortus,* requires erythritol, a carbohydrate, which is available in sufficient quantities in cattle placenta but not in human placenta. Contagious abortion does not occur in humans.

Acquired immunity. Acquired immunity differs from innate immunity in that it involves specific immune products, **antibodies,** and **sensitized T lymphocytes** and can be divided into active and passive forms.

Actively acquired immunity (Figure 24-19). During an infection the antigens of a microorganism stimulate the formation of antibodies. Recovery from infection usually leaves the host with a sufficient number of antibodies to resist infection if there is a subsequent visit by the pathogen. This type of immunity is effective, provided the original infection is survived. Today we are able to use microbial antigens in the form of killed or weakened organisms as well as detoxified bacterial toxins to stimulate antibody formation in the host without the deleterious effects of infection. In addition,

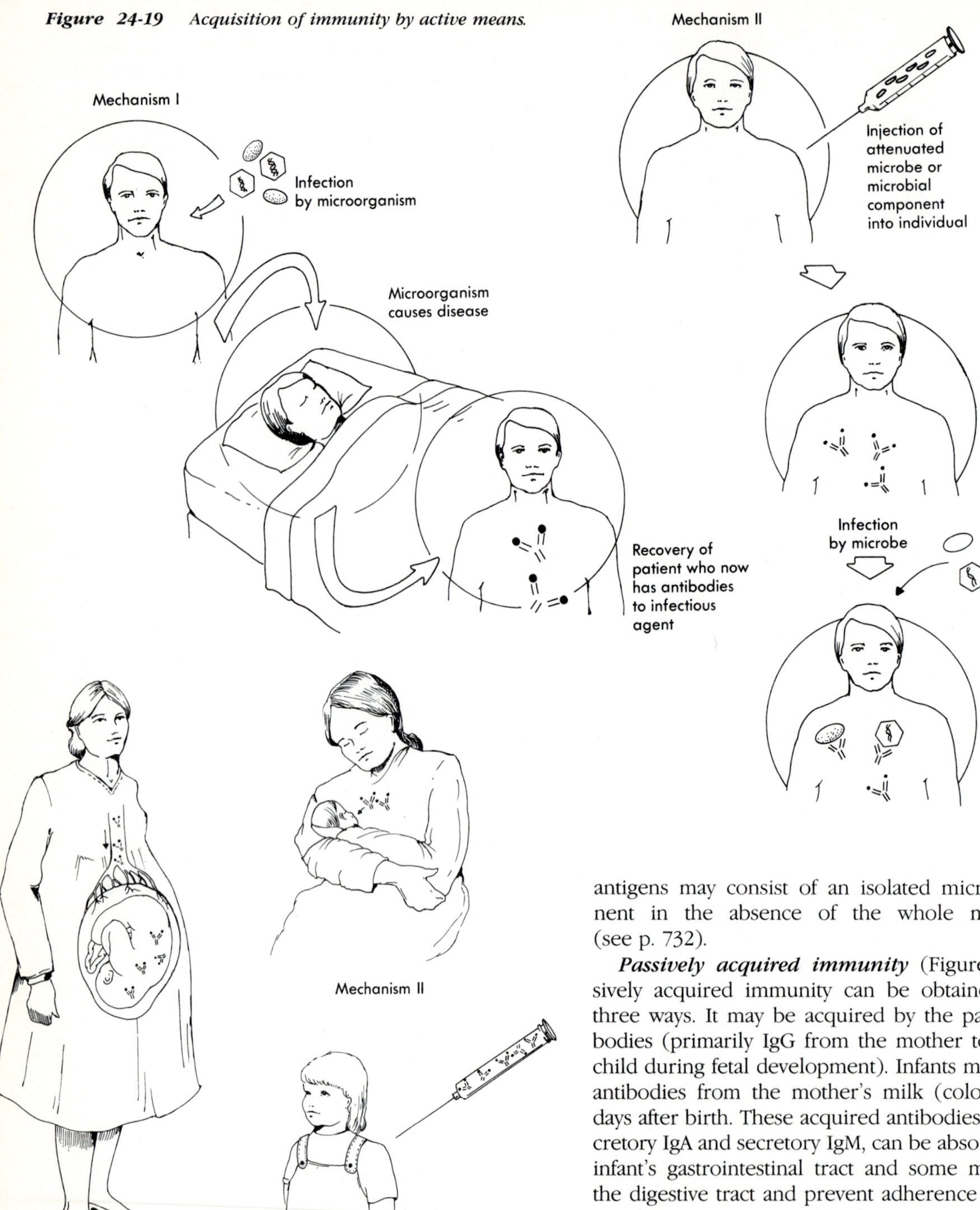

Figure 24-19 *Acquisition of immunity by active means.*

Figure 24-20 *Acquisition of immunity by passive means. Mechanism I, antibodies passed from mother to fetus; mechanism II, infant acquires maternal antibodies by nursing; mechanism III, injection of antibodies (antiserum) into individual.*

antigens may consist of an isolated microbial component in the absence of the whole microorganism (see p. 732).

Passively acquired immunity (Figure 24-20). Passively acquired immunity can be obtained in one of three ways. It may be acquired by the passage of antibodies (primarily IgG from the mother to her unborn child during fetal development). Infants may also obtain antibodies from the mother's milk (colostrum) a few days after birth. These acquired antibodies, primarily secretory IgA and secretory IgM, can be absorbed from the infant's gastrointestinal tract and some may even coat the digestive tract and prevent adherence of any pathogenic microorganism that happens to make contact with it.

A person may also acquire antibodies that have been produced in another human or lower animal. Injection of antisera or gamma globulin pooled from the sera of humans who have recovered from a disease (for example, hepatitis) or sera from animals, such as horses, who have been injected with the antigen of the infectious agent characterize this type of immunity.

Table 24-3 *Comparison of the maximum number of cases of the major human diseases in the prevaccination and postvaccination eras*

Disease	Maximal cases in prevaccination era	Maximal cases in postvaccination era (1983)	Percent change or reduction in cases in postvaccination era
Diphtheria	206,939 (1921)	5	−99.9
Measles	894,134 (1941)	1497	−99.84
Mumps	152,209 (1968)	3355	−97.80
Whooping cough (pertussis)	265,269 (1934)	2463	−99.08
Poliomyelitis	21,269 (1952)	15	−99.03
Rubella (congenital)	20,000 (1964-1965)	22	−99.89

Immunization practices

Active immunization (vaccines and toxoids). Active immunity can be induced by antigenic material that is secreted by the microorganism or that is a part of the microorganism. More specifically, the inducing material is called a **toxoid** if it is a modified toxin and a **vaccine** if it contains living or inactivated microorganisms. Bacterial exotoxins, such as the tetanus and diphtheria toxins, are converted to toxoids by treatment with formalin. To this definition of vaccines we must also add that the inducing material can be a specific component of the microorganism. Vaccines and toxoids work by stimulating the recipient's immune system to produce antibodies against what it considers to be a foreign substance. T cells interact with a microorganism, for example, and induce B cells to produce antibodies that bind to it and help destroy it. The antibodies survive only a few weeks in the recipient, but the B cells are permanently programmed to recognize the microorganism and produce antibodies to destroy it before it can cause permanent damage.

The level of antibody that is obtained from immunization and the persistence of antibody can be increased with some antigens by the addition of **adjuvants** (helping materials) to the antigen. Adjuvants help prolong the antigenic stimulus, or they foster the involvement of increased numbers of immune cells, thus creating conditions that lead to elevated immunity. Toxoids that are alum precipitated, for example, produce a better immunity than toxoids alone.

Most vaccines today are made from harmless (nonvirulent) forms of a pathogenic microorganism. The microbe is either chemically inactivated or killed, as in the Salk polio vaccine, or weakened **(attenuated)** by genetic changes, as in the Sabin polio vaccine. A rare but potential hazard is that not all of the microorganisms in an inactivated vaccine have been killed. The release of one live organism could cause disease. In addition, attenuated organisms by another genetic change could revert to a virulent form. Vaccines are also hard to make and mass-produce. It is extremely difficult to kill or attenuate an organism without also modifying or destroying the important antigens on its surface. Vaccines, such as the pertussis vaccine, contain killed microorganisms, but some of the antigens are believed to be responsible for the occasional inflammatory response that leads to brain damage. In addition, virus from attenuated preparations can be passed from person to person. A vaccine-related disease may not appear in the immunized person but can appear in a close contact such as a family relative. The serious side effects of such a vaccine results from abnormal immunological responses. Although the instances of vaccine-associated disease may be one in several hundred thousand, a totally safe vaccine is the ultimate goal. We should realize that, despite inherent problems in vaccine preparation and delivery, vaccines have significantly reduced mortality for most of the major diseases. When the number of cases of the major human diseases is compared before and after the development of vaccines, we observe nearly a 98% or greater reduction following mass vaccination (Table 24-3).

Genetic engineering techniques will help to develop a second generation of attenuated vaccines that will be safer than previous preparations. We will be able to remove those portions of the infectious microorganism's DNA that are responsible for the disease. For example, an oral typhoid vaccine has been developed in which a genetic change causes the typhoid bacterium to self-destruct after two generations. The mutant absorbs larger than normal quantities of the sugar galactose, most of which it cannot metabolize. Once the vaccine reaches the intestinal tract, it begins to grow; this stimulates an immune response. But before the organism can produce toxins it dies of galactose poisoning (this vaccine is awaiting approval in the United States).

Concept Check

1. Why is the complement system, which is not an integral part of the immune system, always described in association with it?
2. What are some of the biological activities of complement?
3. What is the difference between active and passive immunity? Give some examples.
4. What is the difference between a vaccine and toxoid? What is their function, and what happens to them after they are injected into the body?

Isolated antigens as vaccines (subunit vaccines). One of the drawbacks to the use of live vaccines is their instability and need for refrigeration. This is a serious limitation, particularly for Third World countries where they are needed most. One way to circumvent this problem is to use only the specific surface antigen, or subunit of the microorganism, that is associated with disease and not the whole organism. Some current vaccines employ this technique. For example, the polysaccharides that make up the capsule of the organism *Streptococcus pneumoniae* (one of the important causes of pneumonia) are used in a vaccine. Different strains of *S. pneumoniae* produce slightly different capsules; therefore the vaccine must contain antigens from those strains that are most frequently associated with disease. This type of vaccine is called **polyvalent**, and in the case of *S. pneumoniae,* it contains polysaccharide derived from the capsules of 14 different strains. Another surface component that may prove to be important in vaccines is the pilus. Pili are used by many microorganism as a means of attachment to epithelial surfaces—the essential stage in most infections. However, the problems of strain variation and harvesting techniques represent a hinderance to production of an effective vaccine. A vaccine for gonorrhea using pili is under trial. This vaccine works only if the individual is infected by the same strain of the pathogen *(Neisseria gonorrhoeae),* and there are over 100 different strains. Until recently, the virus of human hepatitis B could not be grown in cell culture, and thus there was no conventional source of this virus. However, human carriers of the hepatitis B virus produce and have circulating in the bloodstream large quantities of the hepatitis B surface antigen. Vaccines against hepatitis B were developed by using surface antigen purified from human plasma. Production of vaccine was limited by the availability of infected plasma. Biotechnology has now circumvented this problem (see below).

The new biotechnology offers considerable promise for the development of safe and effective vaccines. Recombinant DNA technology offers a technique for removing the genes that code for important polypeptides of bacteria and viruses and inserting them into hosts such as bacteria, yeast, or animal cells, permitting production of specific proteins and polypeptides. The surface antigen gene of the hepatitis B virus, for example, has been cloned in yeast. The immunogenicity of the yeast-derived vaccine is the same or greater than that of the human plasma vaccine. Monoclonal antibody production will also be important in identification of those antigens that are relevant to immunity. The value of monoclonal antibodies lies in their ability to define very short chain lengths of polypeptides that carry the restricted sites (**epitopes**) that relate to specific immunity. Identification of these restricted sites, which consist of only a few amino acids, will probably lead to eventual chemical synthesis. This technique is being applied to the development of a vaccine against malaria. Malaria is caused by a protozoan whose life cycle consists of different stages, each stage exhibiting different types of surface antigens. Monoclonal antibodies have been prepared against surface antigens of one stage of this parasite. The gene coding for the surface antigen has been identified, and it will be possible to clone the gene in *E. coli* and produce large quantities of the antigen for use as a vaccine (see Chapter 28 for discussion of malaria vaccine).

Identification of subunits on viruses is being actively pursued and applied to the production of **synthetic vaccines**. One technique for which a patent has been applied involves determining the composition of the different protein molecules residing on the viral surface. This information is then fed into a computer that takes into account various molecular interactions that occur between protein molecules. The computer then selects which of the protein subunits is most likely to protrude from the viral surface. It is usually the protruding component on the microbial surface that is most likely to be recognized by the immune system. The number of subunit possibilities can be reduced by using mutant viruses that are not destroyed by specific antibodies. The subunit regions of normal and mutant virus are compared, and the subunit recognized by the immune system can be identified. These subunits are called **neutralizing epitopes** because they stimulate antibodies but cause no disease. The synthesis of the protein subunit in the laboratory is sometimes difficult, but it is hoped that the use of recombinant DNA technology will resolve this problem. In other words, the gene coding for the subunit could be isolated and inserted into a microbial cell for unrestricted subunit production.

Table 24-4 *Principal bacterial and viral vaccines in current use*

Disease	Vaccine type	Comment
Bacterial		
Pneumonia caused by *Streptococcus pneumoniae*	Mixture of 14 different capsular polysaccharide types	For older adults compromised by other illnesses
Whooping cough (pertussis)	Killed *Bordetella pertussis*	Begun at 3 months of age, followed by boosters
Tuberculosis	Attenuated *Mycobacterium tuberculosis*	Used primarily in Europe
Meningitis caused by *Neisseria meningitidis*	Purified polysaccharide from types A and C	Used primarily for servicemen
Cholera	Crude fraction of *Vibrio cholerae*	For those in areas endemic for the disease
Typhoid fever	Killed *Salmonella typhi*	For those in areas endemic for the disease
Plague	Killed *Yersinia pestis*	For those in areas endemic for the disease
Meningitis caused by *Haemophilus influenzae*	Polysaccharide	For children 24 months of age
Viral		
Rabies	Inactivated virus	For those working in areas where rabies is endemic in wildlife
Smallpox	Attenuated virus	Disease has been eradicated, but vaccine is recommended for laboratory personnel who cultivate virus
Poliomyelitis	Attenuated or inactivated virus	Oral attenuated virus is most widely used; begun at 4 to 5 months of age plus boosters
Mumps	Attenuated virus	Given to infants 15 to 19 months of age
Measles	Attenuated virus	Given to infants 15 to 19 months of age
Influenza	Inactivated or live attenuated virus	Inactivated virus is used in United States; attenuated virus is used in Europe
Rubella	Attenuated virus	Given to infants 15 to 19 months of age and to women of childbearing age who are not pregnant
Yellow fever	Attenuated virus	For those traveling to areas endemic for the disease
Hepatitis B	Subunit vaccine	Ideally will reduce the large number of hepatitis carriers

One of the major deterrents to progress in the development of nonliving vaccines is the limited antigenicity of simple antigens as compared with whole-pathogen vaccines. Some of the very simple synthetic antigens require coupling with longer chain carrier proteins that will fold appropriately to expose the specific epitope determinants. These coupled antigens must also be incorporated into a safe adjuvant such as aluminum hydroxide or phosphate adjuvants. Immunological enhancement of single polypeptide chains can be obtained by incorporating them into synthetic bubbles called liposomes (see Point of Interest, Chapter 15). The disguised liposome is more antigenic than the solitary subunit and does not give rise to side effects in the host.

A list of some of the major bacterial and viral vaccines can be found in Table 24-4, and the major techniques for vaccine production are illustrated in Figure 24-21.

Passive immunization. It is sometimes necessary to artificially provide a passive immunity when a nonim-

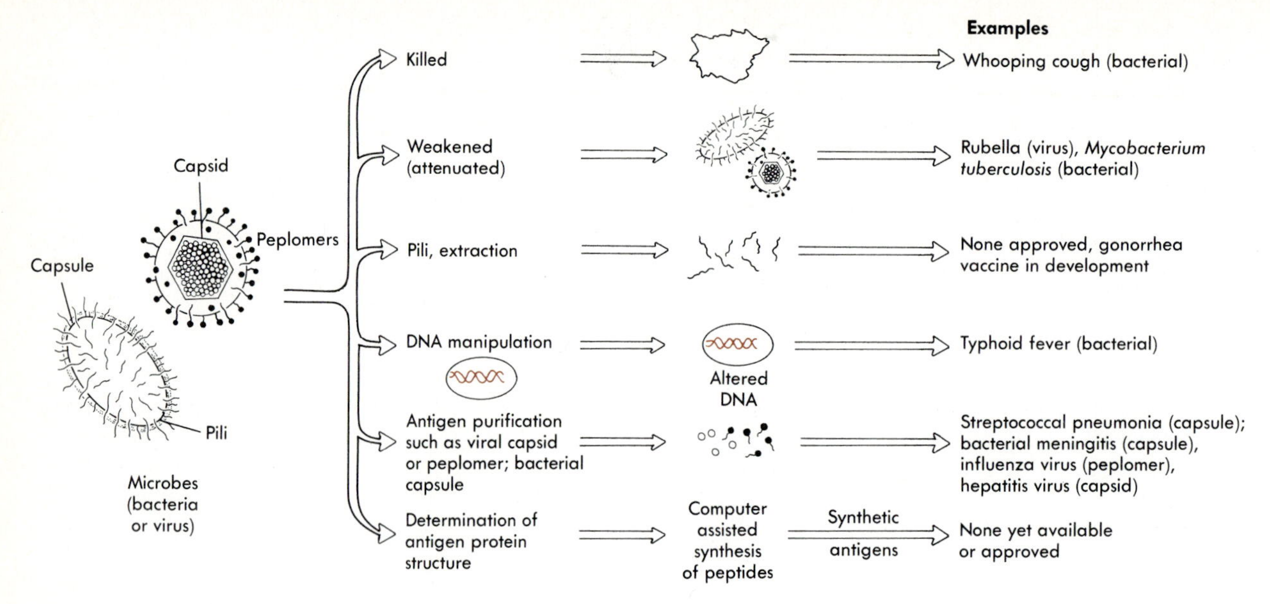

Figure 24-21 *Possible ways in which a vaccine can be prepared from a bacterium or virus.*

mune individual has suffered an exposure to an infectious agent or to a microbial product. The products used for passive immunization usually contain high levels of specific antibodies. Passive immunization is also referred to as **serotherapy** or **immunotherapy**. On a related note, it should be realized that there are foreign proteins present in an antiserum that are developed in other vertebrate species. The antiserum may temporarily protect the recipient against a given harmful agent, but at the same time it may sensitize the recipient to the serum proteins. Thus equine tetanus antitoxin (serum from a horse that has been actively immunized against tetanus toxin) administered to a human may temporarily protect the human against tetanus, but it also sensitizes him or her to horse serum proteins. A later injection of horse serum into the same individual could very well lead to an allergic reaction of as serious a consequence as death caused by an anaphylactic reaction.

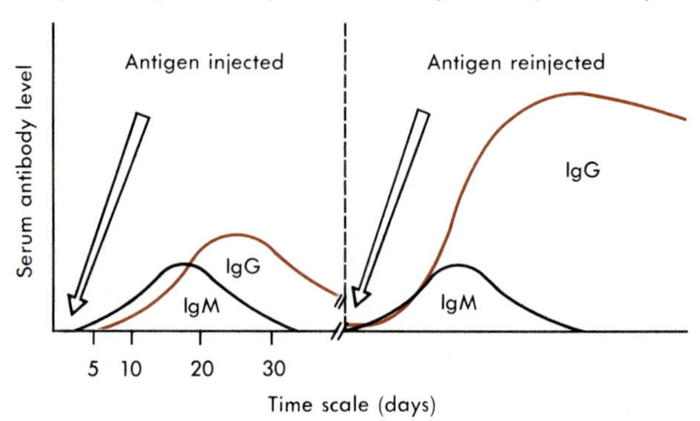

Figure 24-22 *Dynamics of immune (immunoglobulin) response. In primary immune response, IgM is first to appear. It appears on fifth to seventh day and peaks at about 14 days. IgG appears on about tenth day and peaks several days later. In secondary immune response, both IgM and IgG appear or increase within 2 to 3 days. Lag period is shorter, total antibody level far surpasses that of primary immune response, antibody is formed over longer period of time, and much more IgG is formed than IgM.*

Dynamics of the immune response

Primary immune response. The primary immune response occurs the first time the immune system is incited to respond to a given antigen (Figure 24-22). The response varies according to the nature of the antigen, the competence of the host's immune system, the size of the dose, and the route of administration of the antigen. Despite these variations, there is a fairly characteristic pattern of antibody production following the initial exposure to the antigen. Readily detectable antibody does not appear for about 5 days. This is the lag or induction period. The first antibody to appear belongs to immunoglobulin class IgM. The IgM level rises exponentially for several days and then begins to fall until very little residual IgM remains after several weeks. IgG antibody has a longer induction period, and detectable levels start about the time that the IgM level peaks. IgG antibody levels peak later and may remain elevated for a long period of time, or the IgG antibody may disappear.

Some of the proliferating lymphocytes produced on antigen stimulation are set aside as memory cells, an important consideration in the secondary immune response. Because of the lag period a protective level of antibody does not develop soon enough to prevent a natural infection from occurring. The significance of artificially inducing an active immunity whenever practical should be evident.

Secondary immune response. A secondary immune response occurs when there are exposures to the same antigen subsequent to the first exposure. Natural exposure to an infectious agent or subsequent administration of the same vaccine material can serve to incite the secondary immune response. The ensuing doses that are used in artificial immunizations are referred to as *booster doses.* There are marked differences between the primary and the secondary immune responses. Reexposure to the same antigen leads to (1) a shorter lag period for antibodies to appear, (2) a higher and more persistent level of antibodies due to increased number of responding cells produced in the primary response, (3) a predominance of IgG antibodies, and (4) an increased affinity of the antibodies for the antigen.

The prompt response to repeated antigenic stimuli described previously is due to memory cells. Memory cells that are preprimed with the first antigen exposure respond promptly and efficiently to subsequent exposures. This phenomenon is referred to as the **anamnestic response** or the recall phenomenon. Memory cells may persist for many years. It has been shown in humans that tetanus toxoid administered 20 years after primary immunization continues to elicit a prompt and vigorous antitoxin response.

COMPARATIVE IMMUNOLOGY

Most of the immunology in this textbook pertains to the human immune system. When examining immunity in other life forms it is natural to make comparisons with the human immune system. We use as the basis for comparison the major components of the human immune system and its cardinal features: specificity, self-recognition, and memory.

Plant immunity

Plants are well adapted to survive in their environments. This implies that there are anatomical, biochemical, and immunological bases for their resistance to certain diseases. They can to some degree recognize foreign invaders and defend against them. Their immunity falls mostly into the category of natural or genetic immunity rather than acquired immunity. Plant cells possess surface receptors that can be viewed as one mode of natural immunity. Plants can also be bred for disease resistance. What the basis for that resistance might be is not known, but it can be regarded as a form of natural immunity. There is no evidence that plants have anything that compares with the specificity and memory aspects that are such integral features of the vertebrate animal immune system. There are no circulating cells in plants of a type that play such key roles in both natural and acquired immunity in animals.

Plants really do not require an elegant immune system as do the multiorganed species of the higher vertebrates, with their highly specialized tissues and organs. Plants recover more readily from injury because of great regenerative powers and relatively unlimited growth, and because their somatic tissues can differentiate on a broad scale.

Some products and reactions of plants can be interpreted as manifestations of an immune response. Plants are known to produce **phytoalexin** that is nonspecifically toxic to invading fungi. Grafting experiments reveal that **autografts**, **allografts**, and some **xenografts** are tolerated. Autografts are self-grafts; allografts are grafts between nonidentical members of the same species; and xenografts are grafts between members of different species. Graft rejection is regarded as a prime indicator of cellular immunity.

Invertebrate immunity

Invertebrates, because of their high reproductive capacity, short life span, and less-complex anatomical makeup, probably require a less-elegant immune system than more highly developed vertebrate forms. All animals, including invertebrates, exhibit some process that enables them to discriminate self from nonself. There is evidence for immunological memory in some invertebrates, especially in respect to transplantation immunity. In regard to specificity, invertebrates do not have immunoglobulins, and there is no evidence of an inducible humoral immunity. It is possible that another type of molecule might be discovered that has specific immune function. Graft rejection is observed in most of the invertebrate phyla. The second time the recipient is exposed to a graft from the same source, there is often an infiltration of the graft rejection site by cells that have a morphological resemblance to lymphocytes.

Invertebrates protect themselves essentially by three mechanisms: phagocytosis, encapsulation, and nonspecific humoral substances. Phagocytosis is universal among invertebrates. **Archeocytes**, **amebocytes**, **plasmatocytes**, and **granulocytes** are examples of phagocytic cells among the invertebrate phyla. Foreign objects that are too large to be endocytosed are encapsulated. Amebocytes of mollusks and earthworms and hemocytes of insects, for example, collect around the involved area, and a capsule, whose description varies according to the species involved, comes to surround the foreign object. A number of invertebrates have a "solid body," and there are no body fluids (humors) or circulatory system. Their defense depends on mobile leukocytes. Humoral components do appear in invertebrates that have a body cavity (coelomates). Nonspecific humoral substances that are not well characterized are found among the coelomates. The substances include lysozyme and other types of bacteriocidins, agglutinins for vertebrate red blood cells and some bacteria, and opsonin-like molecules in mollusks.

Vertebrate immunity

All vertebrates possess cell-mediated immunity that is highly developed and specific. They also produce at least IgM, and they usually possess immunological memory. All have a thymus and spleen, although of a primitive form in the cyclostomes (hagfish and lamprey), and all have lymphocytes. Vertebrates reject allografts and xenografts, but in the more primitive vertebrates the rejection of the first-set graft (the first time the recipient is exposed to a certain graft) is much slower than in the higher vetebrate forms. This difference is probably related to weaker transplantation (**histocompatability**) antigens among the lower forms. Graft rejection is also slower among the cold-blooded animals; therefore the speed of rejection may be related to body temperature. Second-set grafts are typically rejected in an accelerated manner.

Immunoglobulins with the specificity and inducibility characteristics of mammallian antibodies are found in all the classes of vertebrates. The general structure of the immunoglobulin is comparable to that of mammals. They consist of the typical heavy chains and light chains. IgM, which is the only class of immunoglobulins that appears in all vertebrates, is a pentamer, just as in mammals. In a few species, however, IgM has a tetrameric or hexameric configuration. Membrane (receptor) IgM is found in B cells (or their presumed equivalents) of all vertebrates. The number of immunoglobulin classes increases as one proceeds up the phylogenetic scale (Cyclostomes—Ichthyes—Amphibia—Reptilia—Aves—Mammalia). Fishes and urodelian amphibians (salamanders, newts) have IgM only. Vertebrates from anuran amphibians (frogs, toads) to mammals have two to three classes of immunoglobulins. Birds produce a third class that is homologous to mammalian IgA.

From a phylogenetic viewpoint it would appear that higher forms built on the nonspecific resistance mechanisms of phagocytosis and inflammation that existed in the invertebrate forms. The higher forms added specific humoral immune mechanisms and the amplification systems of coagulation and complement.

Concept Check

1. **What are the advantages of live vs. killed or attenuated vaccines? Disadvantages?**
2. **How have genetic engineering and monoclonal antibody contributed to vaccine development?**
3. **What are the major bacterial and viral vaccines used today? Which are live vaccines and which are not?**
4. **What are the distinctions between a primary and a secondary immune response?**
5. **What are the similarities and differences between the human immune response and the immune response of plants, invertebrates, and vertebrates?**

SEROLOGY

Antibodies and sensitized lymphocytes are used extensively in identification and in monitoring of test procedures. The facts that immune products have high specificity for antigens and that their reactions with antigens are usually readily detectable make immune product–antigen interactions highly useful for identification and test monitoring. The techniques described in this chapter are mostly in vitro procedures that are encompassed under the heading **serology**. This term implies a study of serum, but serological tests also make use of material containing immune products other than serum.

Before venturing any further, some explanation of the term serum is needed. About 60% of blood consists of a liquid portion called **plasma**. Plasma consists of 91% to 92% water, and most of the remaining 8% is protein. There are three major types of protein in the plasma: **albumins**, which maintain osmotic properties; **immunoglobulins** or antibodies, which defend the organism against foreign invaders; and **fibrinogen**, which is responsible for clotting. When fibrinogen and cells have been removed from the plasma by clotting, the resulting fluid is called **serum**. It is the detection and quantitation of serum antibodies that form the basis of serological tests.

Application of in vitro testing

There is a broad range of procedures and applications in serological testing. Most infectious agents and other antigenic substances induce the production of immune products in animals. If the induction period is long enough, suitable amounts of the immune products can be obtained from the host and appropriate tests conducted. There are situations in which antigen and not the immune product is recovered from the host. Whenever a serological test is being performed, either the antigen or the immune product is known. Serological tests are used for the following purposes:

1. Confirmation of identifications made by other clin-

ical or laboratory procedures. An infecting organism, for example, may be tentatively identified by biochemical tests, but a question may remain as to its full identity. Serological tests are sometimes used to substantiate identifications made by other means.

2. Immunodiagnosis. Sometimes it is not possible to isolate the antigen, such as the infectious agent, from the patient or it is impractical to do so. The antigen, however, leaves its mark in the form of an immune product. Thus a child may have had an infection that clinically appears to be mumps. Many diagnostic laboratories are not equipped to isolate the mumps virus from the patient, but they can recover antibodies. After the child's immune system has had time to produce a detectable level of antibodies, blood is drawn and the serum is harvested. The serum is tested for the presence of antimumps antibodies.

3. Direct examination of clinical material. A preliminary presumptive diagnosis is made by performing an immunodiagnostic test directly on a clinical specimen such as a body tissue. Sometimes this is the only option available. Immunofluorescence tests and other immunohistochemical tests, for example, are versatile tools for locating "unknown" antigens on cells and tissues and other clinical materials. Parts of the specimen that are used for immunodiagnosis may be submitted for culturing and the cultures used for subsequent routine identification procedures.

4. Following the course of a disease. It is good clinical management in some diseases to obtain serum specimens periodically and to examine the serum for changes in **titer** (amount of antibody). A stabilization of or decrease in titer may indicate that the disease has been overcome or is waning, whereas an increase in titer usually indicates that the disease is continuing. The determination of antistreptococcal antibodies during streptococcal disease is an instance of this application.

5. Determining the serotype or serogroup of microorganisms. In some cases the identification of a pathogen has been made to the species level, and that information is sufficient to determine the treatment for a given patient. Strains of a species of microorganisms sometimes develop significantly different characterstics because of mutation and natural selection. The identification of such strains through serological testing may be useful information for epidemiological purposes and for determining the control mode to be used against the strain.

6. Determining immune status. It is sometimes useful to know if a person has antibodies to a given microorganism or to a given microbial product. A woman of childbearing age, for example, may want to know if she has antibodies to the German measles virus. A blood sample is obtained and a suitable test conducted for serum antibodies. The serological tests may also be used on populations to determine their past or present exposures to infectious agents.

7. Although serology is primarily used in association with the infectious disease process, this is not its only function. Serology is an important tool for investigating the molecular architecture of the microbial cell. For example, it is useful in bacterial classification. Detecting differences in cell surface structures (pili, flagella, cell wall, capsules, and slime layers) can be helpful in differentiating bacteria. For example, serological studies of the streptococci by Lancefield in the 1930s resulted in the division of the genus into different serological groups. Antisera prepared against purified proteins can be used to detect serological cross-reactions of homologous proteins from other bacteria. The immunological relationships among enzymes from related bacteria has been used to verify results obtained from other studies, such as rRNA/DNA homology experiments. Such serological studies have also shed some light on phylogenetic relationships.

General principles of serological testing

Most serological tests involve serum antibodies; thus most of the following discussion pertains to tests for antibodies rather than tests for sensitized lymphocytes. The fact that antibodies specifically interact with a given antigen (specificity is not necessarily absolute) makes it possible to use in vitro antigen-antibody reactions for identification.

The components of a serological test consist basically of an **antigen** and an **antiserum** or serum sample. One of the components of a serological test, either the antigen or the antibody, usually must be known (identified). The antigen of the test might be a microorganism whose identity is unknown or whose identity, based on other laboratory tests, is to be confirmed. This "unknown" microorganism is tested against known antisera. The specificity and antibody content of the antiserum is known because it was obtained from animals that were deliberately immunized with a known antigen. (Most antisera and antigens remain serologically active for long periods of time when properly stored.) Thus a suspected typhoid bacterium isolated from a patient can be tested against various antibacterial antisera. Suspensions of the isolated typhoid (*Salmonella*) bacteria are tested against antisera specific for suspected microorganisms, including an antisalmonella serum (antiserum). A visible reaction occurs only if the specimen containing typhoid organisms is mixed with the antityphoid serum; thus the causative agent of infection is identified.

The antibody may be the unknown and unidentified component in a serological test. It is possible in many instances to determine what caused the patient's infection by testing his or her serum against known antigens. Thus the serum from a patient who had typhoid fever (assume that the causative agent was not isolated and therefore not identified) is checked against known sus-

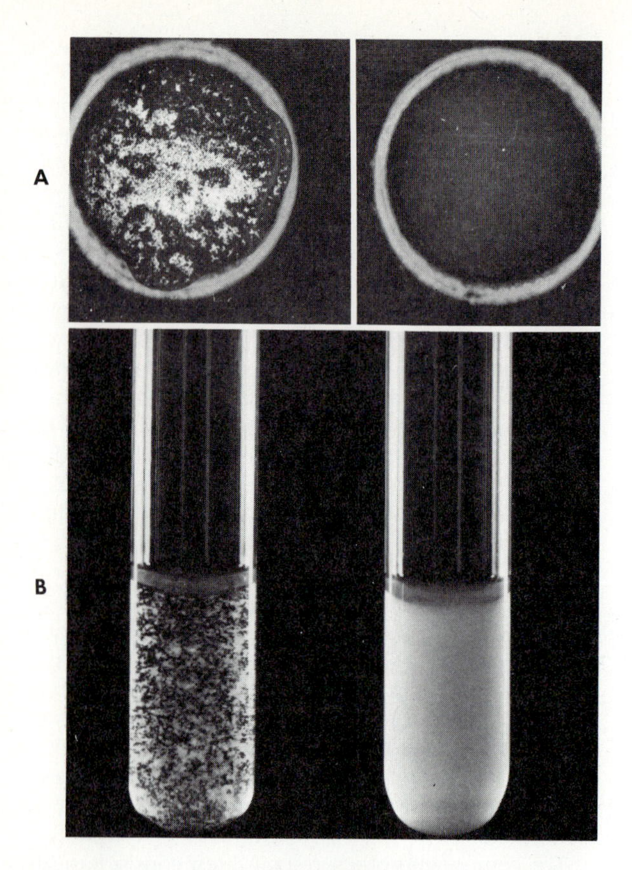

Figure 24-23 ***A,*** *Slide agglutination of bacteria is a rapid method of identifying unknown bacteria with known antisera or the reverse.* ***B,*** *Tube agglutination can be used for same purpose. In each case positive test is seen at left and negative control at right.*

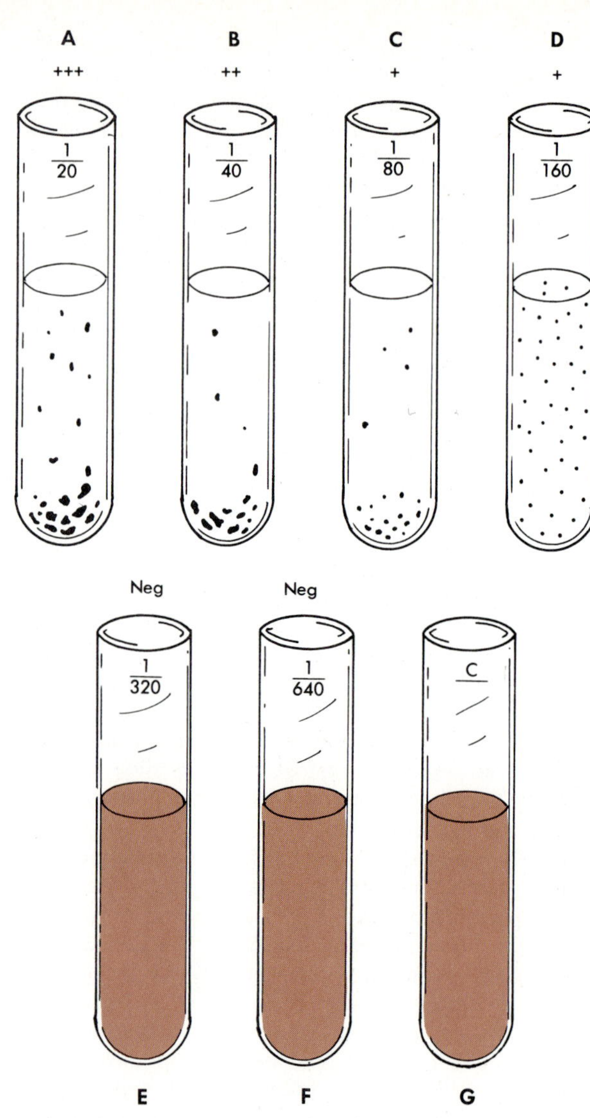

Figure 24-24 *Bacterial agglutination, test tube method.* ***A-F,*** *Dilute 0.2 ml of patient's serum serially in isotonic saline through first six tubes. Add 1 ml of suspension of bacteria to each tube through all seven. Final dilutions are indicated by numbers on each tube (such as 1 part serum in 20 parts, etc., suspension).* ***G,*** *Control tube contains only bacterial suspension and saline. Note maximal agglutination of bacteria* (large masses) *in* ***A, B,*** *and* ***C,*** *some in* ***D,*** *and none in* ***E,*** *and* ***F.*** *Agglutination test is positive; titer of serum is 160.*

pensions of different bacteria, including *Salmonella* (typhoid) bacteria. Again, because antigen-antibody reactions are specific, the serum of the patient gives a visible reaction with the suspension of *Salmonella* bacteria but not with the other test antigens.

Quantitation in serological tests. Serological tests may be as simple as placing an antigen and some serum onto a slide or into a tube and observing whether or not a reaction occurs (Figure 24-23). The testing of just one concentration of both components, however, is not always satisfactory. Single concentration reactions may represent, for example, cross-reactivity or former disease. It is important in some instances to show a change in the antibody level against a given antigen. Serological tests therefore, for these and other reasons, usually have a quantitative aspect. Serial dilutions of either the serum or the antigen are made with the purpose of determining a **titer**. Measured dilutions (usually of the serum) are made, and the highest dilution that gives a detectable reaction is recorded as the titer. Suppose that in a six-tube series of doubling dilutions of serum (1:20, 1:40, 1:80, 1:160, etc.), tubes 1, 2, 3, and 4 gave a detectable reaction (were positive). The titer of the serum would be reported as 160 (Figure 24-24). (The titer is usually expressed as the reciprocal of the highest dilution that gives a positive reaction.) The titer level is in many instances diagnostic, particularly when a significant change in titer can be shown between a serum sam-

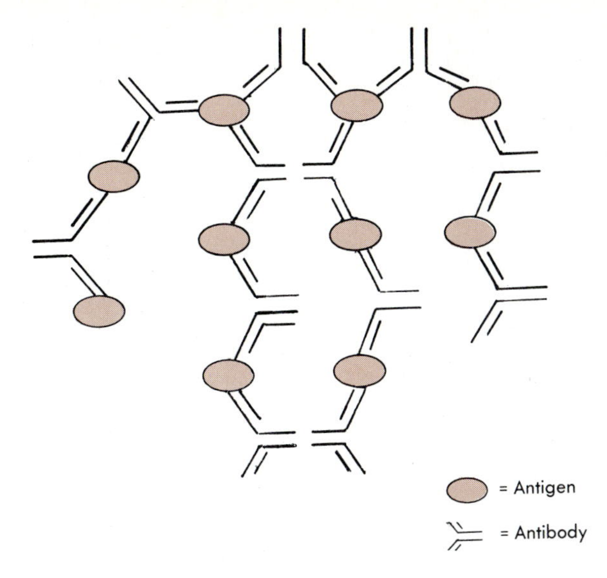

Figure 24-25 *Lattice formation from the interaction of a bivalent antibody and multivalent antigen.*

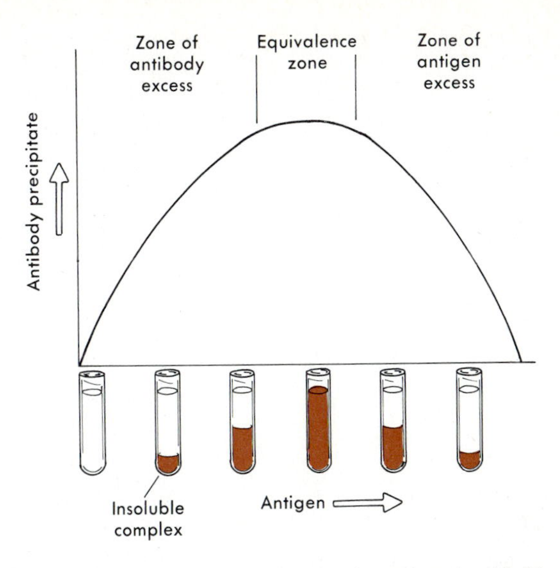

Figure 24-26 *The precipitin curve. As antigen is added to antiserum, an insoluble complex is formed and increases to its maximum at the equivalence zone. Addition of antigen past equivalence zone results in formation of small soluble complexes and a decrease of insoluble complexes. Tubes demonstrate relative amount of insoluble complex (in red) at specific concentrations of antigen.*

ple that is collected early (an acute-phase specimen) in the course of an illness and a sample collected late in the illness (a convalescent-phase specimen). There is enough experience with most serological test types to interpret the significance of a titer for a given antigen-antibody system. Titers can also signify levels of protection, and they may be indicative of the severity and of the progress of a disease.

Types of serological tests

The union of antigen and antibody takes place under a variety of conditions, and the reaction may be measured in different ways. Longstanding serological tests are **agglutination, precipitation, immune cytolysis, virus neutralization, toxin neutralization, complement fixation,** and **capsular swelling (quellung).** These are what might be called "classical" tests and have been used for many years. The test names are mostly descriptive of what happens to the antigen. More recently designed tests are named according to the nature of the technique that is involved: **fluorescent antibody, immunodiffusion, immunoelectrophoresis, immunoperoxidase,** and **radioimmunoassay,** to name the more important ones.

The test antigen from the standpoint of its physical makeup is described as either **particulate,** such as a bacterium or red blood cell, or **soluble,** such as molecules in solution. Most antigens (polysaccharides, proteins, etc.) possess more than one antigenic determinant, that is, they are **multivalent.** If bacterial cells are used as the test antigen, they contain several different antigens (flagella, pili, capsular polysaccharide, etc.), with each possessing several determinants. Antibodies that are at the very least bivalent can interact with multivalent antigens to produce lattices that develop into larger complexes, which may be seen in the test tube or under the microscope (Figure 24-25). Three tests that are based on this multivalent principle are **precipitation, agglutination,** and **complement fixation.** These tests and others are explained in the following sections.

Precipitation. When small amounts of soluble antigen are added to a solution of antibody, the binding sites on the antigen become filled. This results in the formation of insoluble complexes that precipitate from solution and can be measured (Figure 24-26). The antibodies causing the precipitation are called **precipitins.** The size of the insoluble complex depends on the number of binding sites on the antigen. The initial reaction with small amounts of antigen is called the stage of **antibody excess.** As more antigen is added, lattice formation becomes predominant, that is, the antibody molecule, because it will have at least two binding sites, can cross-link two separate antigen molecules. Eventually a point is reached, following addition of antigen, where no more free antibody is present, and the largest antigen-antibody complexes are formed. This point is called **equivalence.** As more antigen is added, smaller antigen-antibody complexes are formed because it is unlikely that a single antigen can be bound by more than one antibody molecule. This is the region of **antigen excess.**

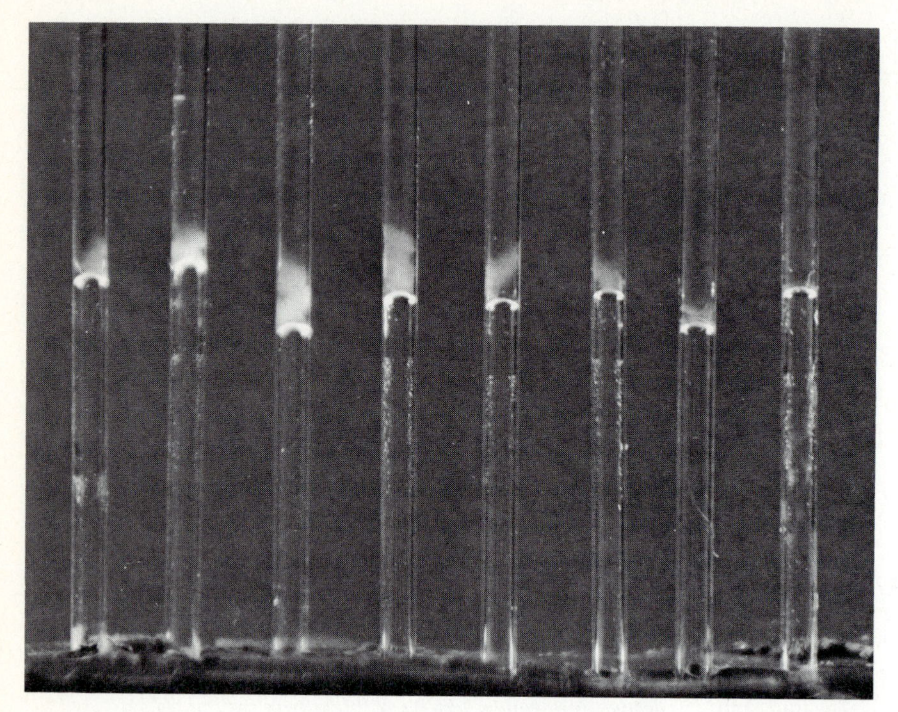

Figure 24-27 *Capillary tube precipitin test. Dilutions of soluble antigen and constant amount of specific antiserum were drawn into capillary tubes. The precipitate form falls out of solution and accumulates at bottom of the fluid column. Tubes 3 and 4 show greatest amount of precipitate. White semicircle in each tube is caused by light reflection from interface where antiserum and antigen dilution meet. Final tube is negative control. The height of the precipitate in each tube can be measured and used as estimate of antibody present.*

The complexes formed in this region of antigen excess are soluble.

The in vitro precipitation test just described is equivalent to the in vivo situation where antigen-antibody complexes are formed. For example, early during infection the antigen is in excess, and later less soluble antigen-antibody complexes are formed in the region of equivalence and antibody excess. Antigen-antibody complexes are normally cleared by phagocytic mechanisms. Some antigen-antibody complexes, because of several factors, including solubility and size, are not cleared and instead gain entry into body tissues and cause immune complex disease, such as rheumatic fever. The complexes formed in antibody excess are cleared very rapidly from the circulation, while the small soluble complexes formed in antigen excess are cleared least well by phagocytic mechanisms. Studies have demonstrated that macrophages do not bind very efficiently to antigen-antibody complexes that have been formed at extremes of antigen excess.

Various techniques are used to demonstrate precipitation reactions, such as the **ring test**, the **capillary test**, and **diffusion tests**. In the ring test a layer of antiserum is placed into the bottom half of a small-diameter tube. A layer of the antigen solution is carefully placed over the serum sample. A positive test is characterized by a region of precipitate developing at the interface (where the two layers meet). The capillary test is performed by sequentially drawing, by capillary action, antiserum and antigen solutions into the tube. A precipitate usually develops in minutes in a positive test (Figure 24-27).

Immunodiffusion. Many serological precipitation tests are now conducted in gels or other support media. Distinct lines of precipitate form in gel where optimal portions of diffusing antigen and antibody molecules meet.

The gels used today include **agarose**, a neutral polysaccharide that is extracted from agar; **cellulose acetate film**; and **polyacrylamide gels**. These are usually placed in Petri dishes or as a thin film on a slide or other support. Wells or troughs are usually made in the gel and filled with the antigen and antibody solutions. The number of wells used, the location and spacing of the wells, and the configuration of the wells (circular, rectangular, etc.) are dictated by the type of reagents and the information being sought and to some degree by personal preferences. Immunodiffusion tests can be varied as to whether one or both of the reagents diffuse **(single diffusion, double diffusion)** and as to the directions of the diffusing reagents **(radial diffusion, one-dimensional diffusion)**. Single diffusion means that only one of the reagents diffuses from the deposit site. The other reagent is uniformly distributed throughout the gel because it was incorporated into the molten gel before the gel hardened. Double diffusion requires that both antigen and antiserum are placed in separate wells or troughs, resulting in both reagents diffusing through the gel. In radial diffusion the reagent diffuses in all directions (360 degrees) from the deposit site, whereas in one-dimensional diffusion the reagent diffuses in only one direction.

In a double diffusion test two wells are made in the

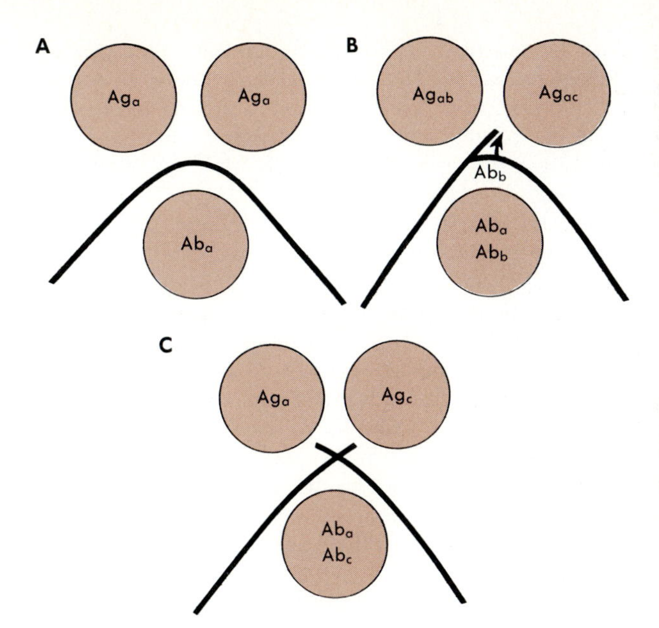

Figure 24-28 *Immunodiffusion.* ***A,*** *Line of confluence obtained with two antigens that cannot be distinguished by antiserum used.* ***B,*** *Spur formation by partially related antigens having common determinant a but individual determinants b and c reacting with mixture of antibodies directed against a and b. Antigen with determinants a and c can only precipitate antibodies directed to a. Remaining antibodies (Ab_b) cross precipitin line to react with antigen from adjacent well, which has determinant b, giving rise to "spur" over precipitin line.* ***C,*** *Crossing lines formed with unrelated antigens*

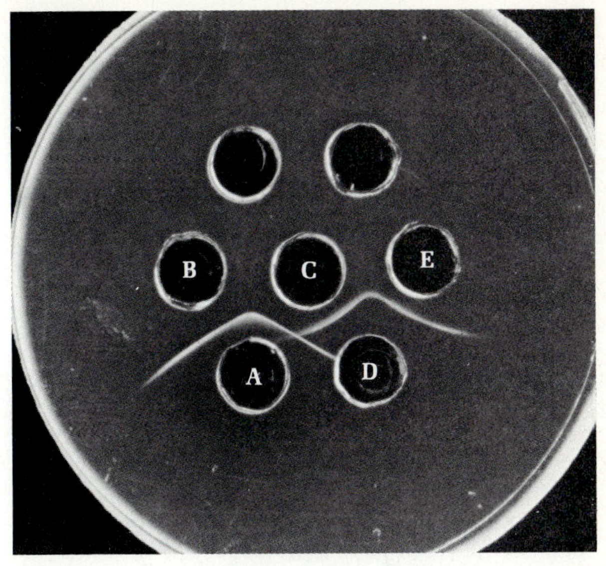

Figure 24-29 *Double diffusion (Ouchterlony) test shows two reactions of serologic identity and a reaction of nonidentity. Well A contains an antiserum to an* E. coli *toxin. The toxin was placed in wells B and C. Well D contains antiserum to human lactoferrin. Lactoferrin was placed in wells C and E.*

gel. Antiserum is placed in one well and antigen in the other. The molecules from each of the wells diffuse in all directions (radial) in ever-widening circles. Some of the molecules meet someplace between the two wells. A line (band) of precipitate forms if the serum contains antibodies specific for the antigen. If the antigen preparation contains three different antigens and if the antiserum has antibodies against the three antigens, theoretically three distinct lines of precipitate will develop between the two wells.

Another basic immunodiffusion setup consists of three equidistantly spaced wells, one well containing antiserum and the other two containing antigen. Three principal types of reaction (precipitation patterns) may be detected. Assume that in this example each antigen-containing well contains only a single antigen. The reactions show whether the antigens in the two wells are serologically **nonidentical**, **identical**, **indistinguishable**, or **partially identical** (Figures 24-28 and 24-29). If the antigens are nonidentical, two separate lines of precipitate form, and the lines merely cross in the common area where the lines of precipitate intersect. If the two antigens are serologically identical, a single line of precipitate develops that is curved in the common area where the two lines of precipitate meet. In the reaction of partial identity a single curved line develops, but a spur also occurs in the common area. The single curved line indicates that the two antigens share certain determinants. The spur indicates that the antigens are nonidentical.

In a typical single radial immundiffusion test, antiserum at relatively low concentration is incorporated into the agar. A single well is made in the agar into which the antigen is placed. When antigen diffuses from the well it is initially at a very high concentration and forms soluble complexes. As the antigen diffuses further its concentration decreases and reaches a point of optimal concentration with the antiserum. At this point a ring of precipitate is formed. The greater the concentration of antigen, the greater the diameter of the ring (Figure 24-30). By incorporating antigen of known concentration into different wells a calibration can be obtained. Once the curve has been established the amount of antigen in unknown samples can be tested and its concentration determined. Single radial immunodiffusion, therefore, offers a method of quantitating antigen, whereas the other immunodiffusion tests are for identification.

Agglutination. Agglutination, unlike precipitation,

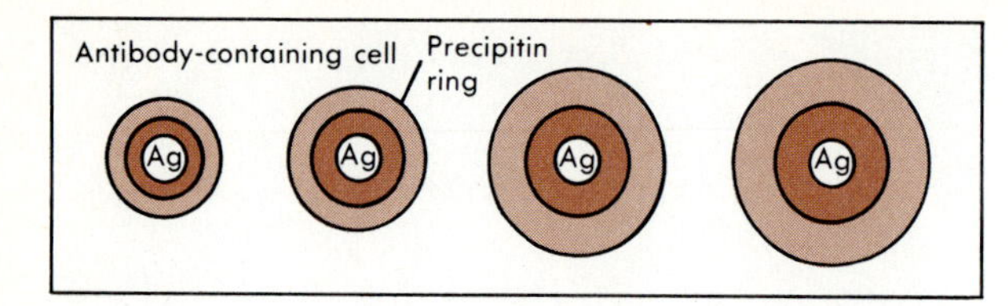

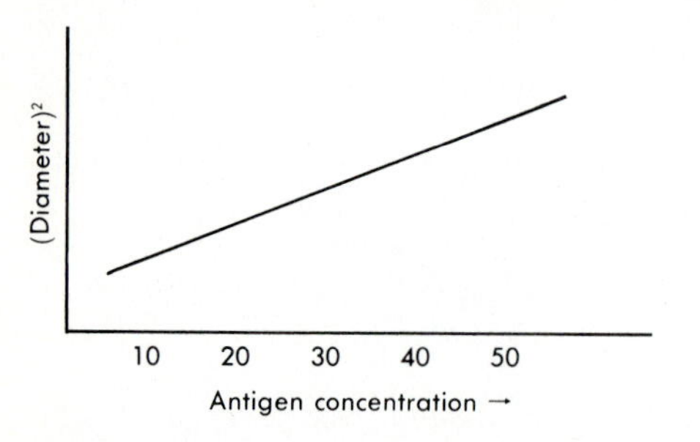

***Figure 24-30** Single radial immunodiffusion. Antibody is added to an agar gel (45° C) which is then poured on slides and allowed to set (20° C). Wells are punched in agar, and standard volume of test antigen at different concentrations are placed in wells. Plates are left for 24 hours or more, during which time antigen diffuses out to form soluble complexes (antigen excess) with the antibody. These continue to diffuse outward binding more antibody until an equivalence is reached and a precipitate or insoluble complex, is formed. Area within precipitin ring, measured as ring diameter squared, is proportional to antigen concentration. Once the curve has been established, unknowns can be derived by interpolation from the standard curve. The process can be reversed to determine the concentration of unknown antibody.*

involves antibodies that aggregate cellular or insoluble antigens. The antibodies that bring about agglutination are called **agglutinins.** The insoluble antigen may be a component on the cell wall of a bacterium or a component on the surface of a red blood cell. Addition of antibodies (antiserum) to a solution containing cellular antigen will cause the formation of crosslinks between antigenic determinants on the cellular particles. During the period of incubation the particles clump **(agglutinate)** and settle out of suspension to the bottom of the tube or the test plate (see Figures 24-23 and 24-24). The reaction can be affected by the class of antibody involved. IgM antibodies, which have five antigen-combining sites, are more effective in agglutinating particles than are IgG antibody molecules, which possess two antigen-combining sites. When the particles are red blood cells the process is called **hemagglutination.**

A normally soluble antigen can be combined with a number of insoluble particles, such as red blood cells or latex beads, thereby rendering them insoluble. Soluble antigen can be attached to red blood cells by treating the latter with tannic acid, which alters the cell membrane surface. Agglutination involving antigenic determinants that are not an intrinsic component of the particle (red blood cell or latex bead) is called **passive hemagglutination.** The agglutination reaction is widely used to determine the presence of antibody and to estimate the relative amount of antibody in an antiserum. The test has been used to diagnose bacterial, viral, and protozoal diseases.

Several viruses, such as the measles and influenza virus, have the ability to agglutinate red blood cells. Although this is a nonimmunological phenomenon it serves as the basis of a diagnostic test called **hemagglutination inhibition (HI).** For example, we may wish to determine if a patient suspected of having influenza has antibodies to the virus in the serum. In the test we mix the patient's serum with known concentrations of influenza virus antigen and red blood cells. If the patient's serum contains antibody to the influenza virus, hemagglutination will be inhibited. If no virus antibody is present, hemagglutination will occur.

Virus + RBCs → Hemagglutination

Serum antibody + Virus + RBCs → No hemagglutination

The test is carried out in the laboratory by preparing dilutions of the patient's serum and adding virus antigen of known hemagglutinating activity. The mixture is incubated for 1 hour at 37° C, and then red blood cells are added. The highest dilution of serum that gives complete inhibition of hemagglutination is considered the end point of the test.

Concept Check

1. What features of an in vitro immune product–antigen interaction make it useful for identification and monitoring procedures?
2. What are the various applications of serological tests?
3. Why is it usually necessary to establish a titer? Why not use a single dilution and determine whether the reaction is positive or negative?
4. What is the basic principle behind the following serological tests: agglutination, precipitation, immunodiffusion?
5. What is meant by the equivalence point in the precipitation test?

Complement fixation. The complement fixation test was devised before the true physicochemical and biological nature of the complement system was discovered. Complement, when used in the context of complement fixation test, refers to a thermolabile component that is present in fresh serum. The feature of comple-

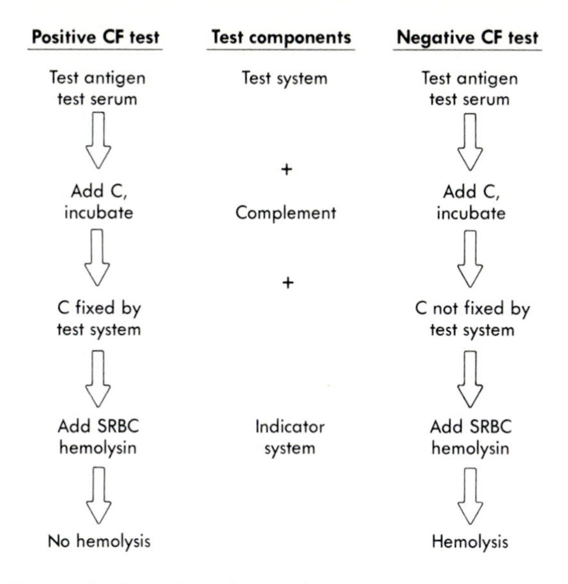

Figure 24-31 *Complement fixation test.* C, *Complement;* SRBC, *sheep red blood cells.*

ment that is used in the test is its ability to attach to IgG and IgM antibodies which have interacted with their specific antigens. If the antigen is a cell, the final result of the antigen-antibody-complement union is lysis of the cell, which is a visible reaction. The test consists of an **indicator system** and a **test system** (Figure 24-31). The complement fixation test is especially useful when an antigen-antibody union by itself does not give a visible reaction. A typical situation might be in a diagnostic laboratory where the question is, "Does the patient have antibodies against the mumps virus?" When the patient's serum is mixed with the mumps virus test antigen, a visible reaction does not occur, even though there are antimumps antibodies in the serum. The indicator system consists of sheep red blood cells (SRBCs) as the antigen of the indicator system and an antiserum that contains antibodies against the SRBCs. The antiserum is called **hemolysin** or **amboceptor**. When SRBCs and the hemolysin are allowed to interact in the presence of complement, the SRBCs lyse. The lysis is a visible reaction. The complete test is set up in this sequence: the patient's serum and the test antigen are mixed. A measured amount of complement (commercially available lyophilized guinea pig serum) is added to the test system (patient's serum and the known or suspected antigen). The indicator system (SRBCs and hemolysin) is added to the tube containing the test system and complement. Either the test system or the indicator system binds the complement. In a positive test there are antibodies in the patient's serum specific for the test antigen. The complement in the tube is fixed to the antigen-antibody complex and is not available to the indicator system when the latter is added to the tube. Hemolysis

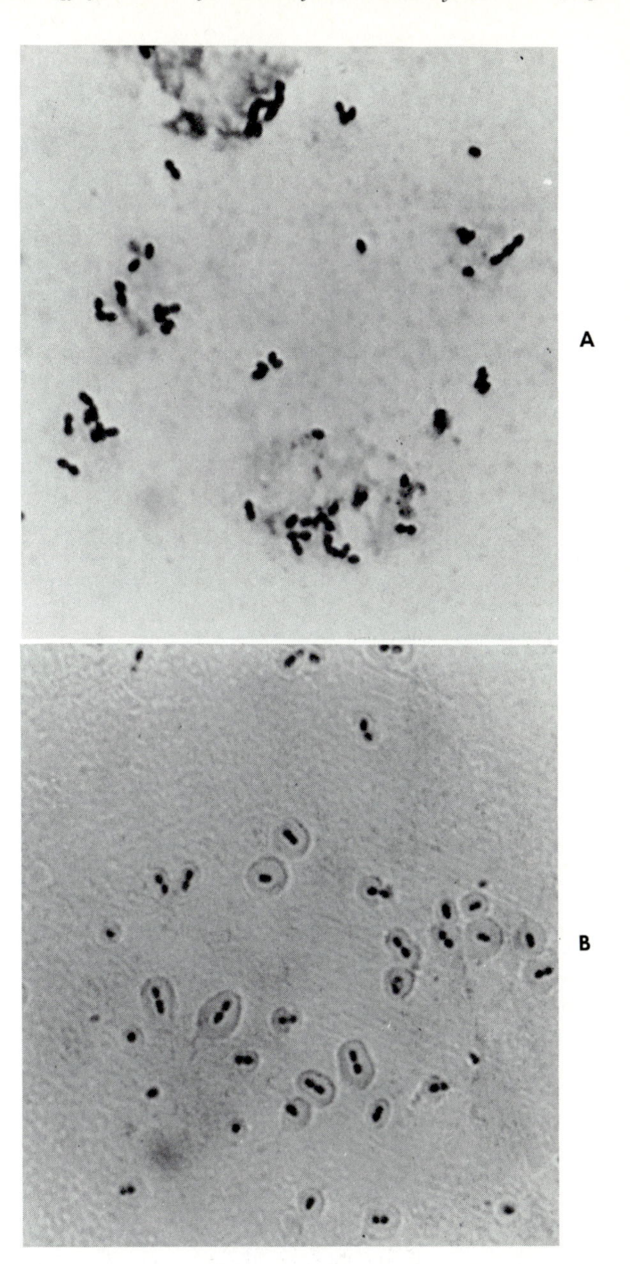

Figure 24-32 *Capsular swelling (quellung) of bacteria.* ***A,*** *Before application of antiserum. Note very small halo around cells indicating capsule.* ***B,*** *After application of antiserum. Note enlarged capsule.*

therefore does not occur in a positive test. In a negative test, because the patient's serum does not contain antibodies for the test antigen, the complement is not fixed. The complement can therefore interact with the indicator system and produce hemolysis.

Capsular swelling (quellung). Antibodies against the capsular material of bacteria can bring about an apparent enlargement of the capsule (Figure 24-32), a reaction that is visible under the microscope. This phenomenon is useful in identification procedures and in serotyping. **Serotyping** refers to serological procedures that are used to differentiate strains of a species that have differences in the antigenic composition of a structure, product, etc. (see Point of Interest). There are, for example, over 80 strains of the bacterial species *Streptococcus pneumoniae,* which is a leading cause of pneumonia. The strains or stereotypes in this instance are based on differences in the antigenic composition of the capsular material of each strain. These differences can be detected by capsular swelling reactions using antisera that are specific for the capsular types. Serotyping is also possible with other serological techniques.

Neutralization tests. Neutralization tests are of a type that determine whether the activity of a toxin or of an infectious agent has been neutralized by an antibody. It is sometimes impossible or impractical to detect serum antibodies to a bacterial exotoxin or to a given virus in vitro. Animals or tissue culture cells are used as the "indicator" system in neutralization tests. The toxin or the virus (the antigens of the test) have a known effect on an animal or on tissue culture cells. The effect in the indicator-animal might be death, paralysis, or skin lesions. The exotoxin of *Clostridium botulinum,* for example, causes an often-fatal type of food intoxication. Diagnosis is made by collecting filtrates of the suspect food or of the serum, stool, or vomitus of the patient and then injecting the filtrates into mice, some of which have received botulinum antitoxin. If there is toxin in

Point of Interest

SEROTYPING OF *SALMONELLA*

Salmonella is but one of a group of gram-negative bacteria called enterics or those species that inhabit the intestinal tract. Other enteric genera include *Citrobacter, Escherichia, Enterobacter,* and *Yersinia.* With certain enterics such as *Salmonella* serotyping has been an important tool in epidemiology and classification. *Salmonella* is routinely divided into three species; *S. typhi, S. choleraesuis,* and *S. enteritidis,* with various serotypes. Since all *Salmonella* are so closely related, they are now reported as serotypes.

Enterics like *Salmonella* possess three kinds of antigens that can be used in the identification process: *H* or flagellar antigen; *O* or somatic antigen, which refers to the lipopolysaccharide component of the cell wall; and *K* or capsular antigen. Most *Salmonella* do not possess a capsule, and serotyping is dependent on identification of somatic (whole cell) and flagellar antigens.

The lipopolysaccharide consists of three regions (see figure below). Region I is a polymer of repeating oligosaccharide units that represent the outermost surface of the cell. It is called the O specific polysaccharide region, and it is this region in which there is antigenic diversity due to the character and arrangement of the sugar units. Identification of this region permits the placement of the *Salmonella* into **serogroups,** of which there are over 30 (for example, A,B,C_1C_2,D, etc.). The **serotype** is determined by using *H* typing antisera for flagellar antigens (for example, a,b,c,d,1,2,3, etc.). Most motile *Salmonella* organisms possess two antigenic flagellar forms referred to as **phases.*** These phases share the same O antigen but possess different H antigens and serotype identification requires identity of *H* antigens in both phases.

Region II is the core polysaccharide, which is constant for each genus but different between genera.

Region III is lipid A, the toxic portion of the cell surface, and is virtually nonantigenic.

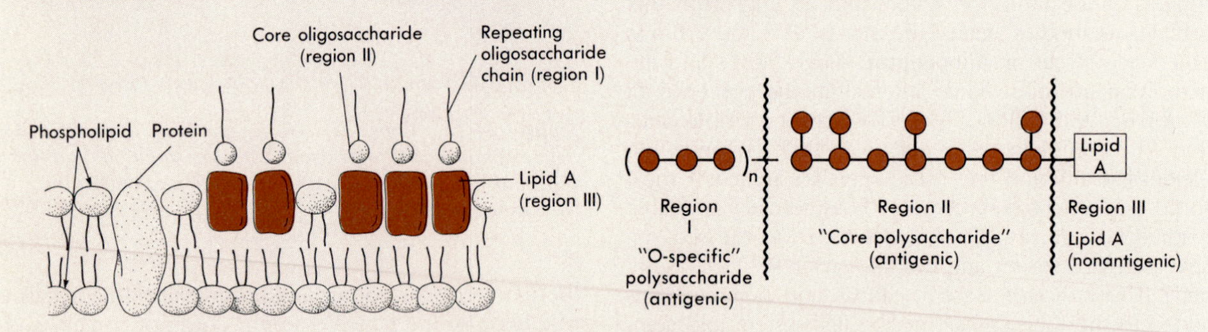

Structural relationship of the lipopolysaccharide of gram-negative bacteria. Illustration at right shows relative position of components of the outer membrane of gram-negative bacteria. Illustration at left shows how polysaccharides are associated with lipid A. Color circles represent individual sugar units.

*The genetics of "phasing" is discussed in Point of Interest in Chapter 9.

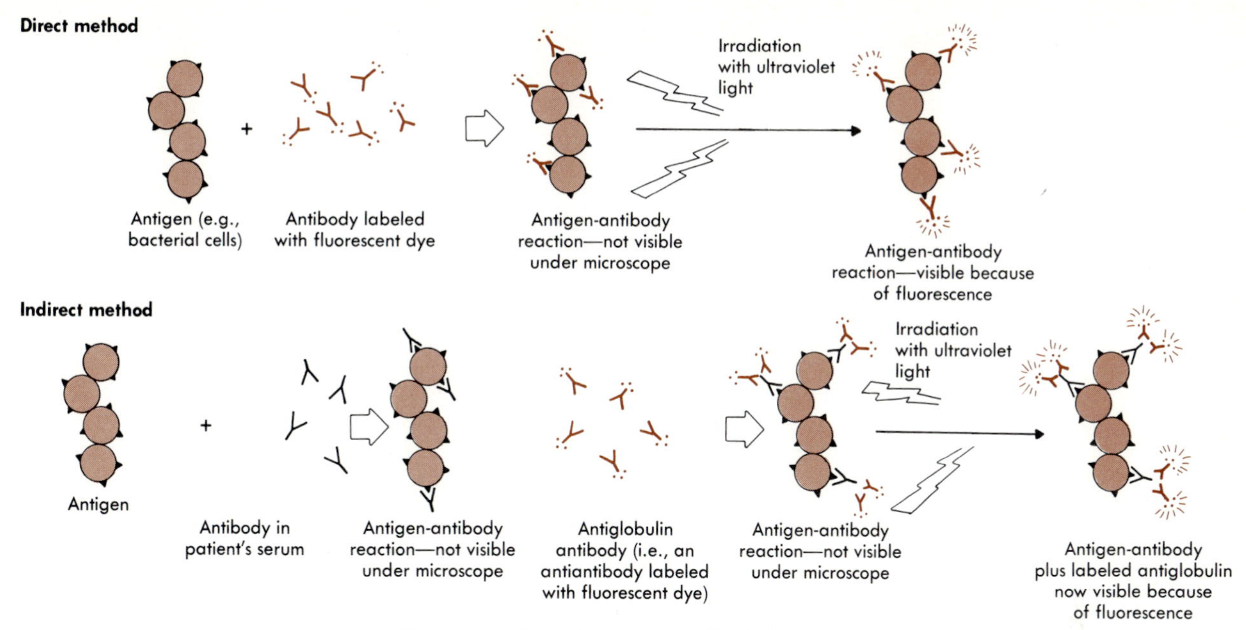

Figure 24-33 *Fluorescent antibody techniques.*

the filtrate, all the mice except those that received antitoxin will be killed. The objective of a neutralization test, rather than testing for a toxin or a virus, might be to determine whether a patient's serum has antibodies to a given toxin or virus. The patient's serum and a measured amount of toxin or virus are mixed in the test tube and injected into a suitable laboratory animal that gives a known reaction to the toxin or virus. If the patient's serum contains antibodies specific for the toxin or virus, the toxin or virus is no longer free to bring about the effect it would typically produce if it had not been neutralized by antibodies.

Tests that incorporate chemical markers. Antigen-antibody interactions are also detectable in immunological tests in which chemical markers have been conjugated to antibodies or antigen. These tests are useful in cases where (1) an antigen-antibody union by itself is not readily detectable, (2) antibodies or antigens are more easily detected directly in clinical material, and (3) greater sensitivity and/or faster results are attainable. Three very important tests in this category are **immunofluorescence (fluorescent antibody technique), enzyme-linked immunosorbent assay (ELISA),** and **radioimmunoassay (RIA).**

Immunofluorescence. Certain dyes, called **fluorochromes,** when irradiated with ultraviolet rays emit visible light. The dyes can be conjugated (coupled) to antibody molecules without changing the antibody's capacity to bind to the antigen, and they can also be attached to antigens. Immunofluorescence can be applied in a direct or indirect way (see Figure 24-33). One application of this technique is in determining whether or not an animal had rabies. Laboratory diagnosis of rabies includes the examination of brain tissue of the rabies-suspected animal for **Negri bodies** and the injection of homogenized brain tissue into mice. Negri bodies, which are viral inclusion bodies elicited by the rabies virus, are not always visible in the brain tissue of the infected animal. In addition, with injections of homogenized brain tissue into mice, 3 weeks must elapse before it can be determined whether or not the mice have acquired rabies. A diagnosis, however, can be made within hours-by-using the fluorescent antibody technique. Sections or impression smears of the animal's brain tissue are exposed to an antiserum that contains conjugated antibodies specific for rabies virus. Antibodies will attach to the virus if rabies virus is present in the tissue. When the preparation is irradiated with ultraviolet light rays, the fluorochrome attached to the antibodies emits light that is readily visible when the preparation is examined with a fluorescent microscope.

ELISA and RIA tests. The ELISA and RIA tests are similar in principle to immunofluorescence, but they are easier to interpret. Enzymes, such as **peroxidase** and **alkaline phosphatase,** are conjugated to antibodies or to antigens in the ELISA tests (Figure 24-34). The enzymes are detected by adding a specific chromogenic enzyme substrate to the preparation. This produces a colored product that can be read visually or measured spectrophotometrically. ELISA tests are less expensive and not as laborious as immunofluorescence tests. The RIA technique uses antibodies or antigens that have been la-

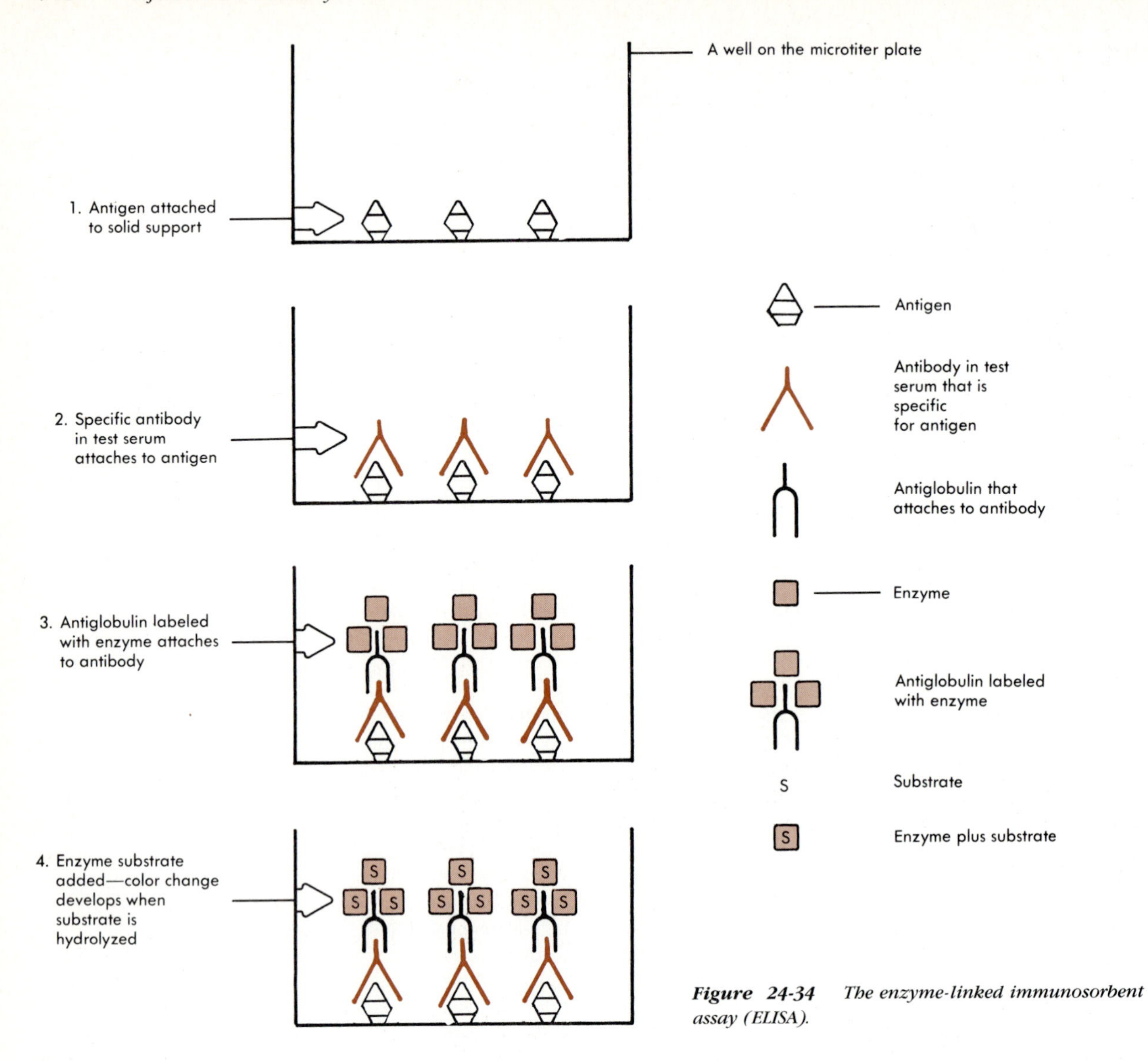

Figure 24-34 *The enzyme-linked immunosorbent assay (ELISA).*

beled with radioactive isotopes. Radioactivity is detected by such means as radioisotope analyzers and by autoradiography (photographic emulsions that show up areas of radioactivity). RIA tests are easier to interpret than immunofluorescence, but they are expensive, and there is some risk because of radioactive material.

Immunoelectrophoresis. Charged molecules, as found, for instance, in antigen preparation and in serum, when subjected to direct current in a gel matrix, are separated because their **electrophoretic mobilities** differ. It is very unlikely that different molecules have the same electrophoretic mobility. The proteins that are present in a solution or in serum will be deposited in certain sites based on each protein's electrophoretic mobility. The electrophoresed protein can be stained and the location noted on the support medium. The electrophoretic pattern of the proteins of normal serum is well known. Specific bands or peaks are associated with specific serum proteins. Electrophoretic analyses may reveal significant departures from the normal serum pattern, and certain diseases can be linked to the absence or change in specific peaks (Figure 24-35).

The specificity that exists between an antigen and an antibody can be applied to the electrophoretic characteristics to make the reaction immunoelectrophoretic. Immunoelectrophoretic techniques offer certain advantages over immunodiffusion methods because lines of precipitate can develop sooner, and diagnosis or identification can be speeded up. Better separations can be obtained by this method when there are multiple antigens in the antigen preparation. Two antigens may have mobilities that are not easily differentiated by diffusion. Electrophoresed molecules can in many instances be made to move in a specific direction. This latter feature

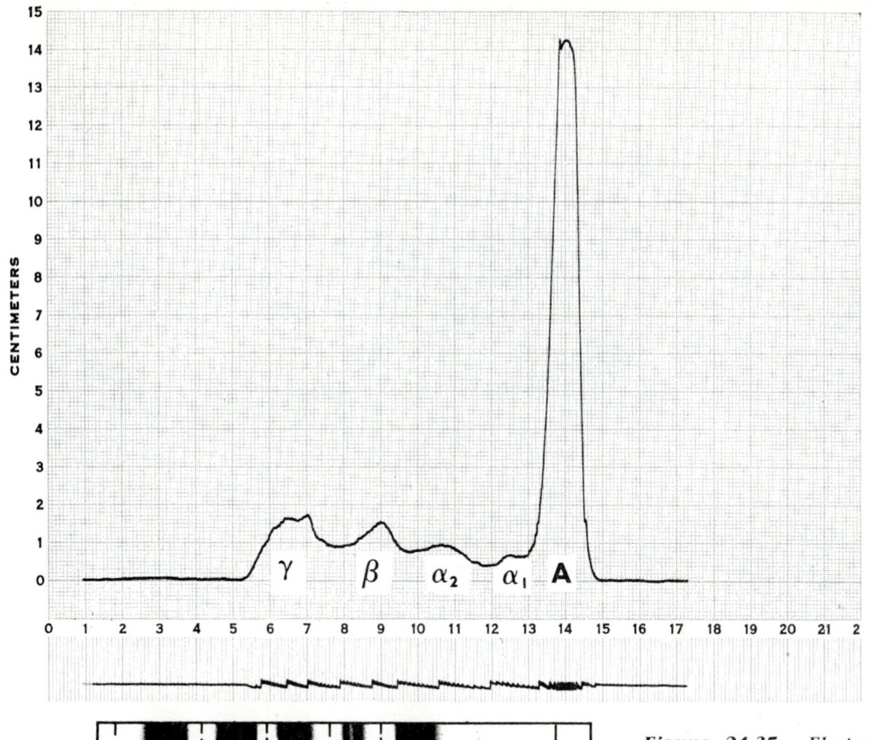

Figure 24-35 *Electrophoresis of normal human serum. Human serum was electrophoresed on cellulose acetate strip* (lower part of illustration). *Recording densitometer was then used to scan strip, and curve was obtained, as shown on graph. A, Albumin; α_1 and α_2, α-globulins; β, β-globulins; γ, γ-globulins.*

is important when only small quantities of the reactants are available. The test is more sensitive because the available molecules are concentrated in one area, whereas in radial diffusion the molecules diffuse in all directions.

In a typical test multiple antigens are separated by electrophoresis (Figure 24-36), but the separated antigens are not visible at this point. After the electrophoretic separation a trough, which parallels the line along which the antigen molecules were deposited, is filled with antiserum. The antigen molecules that diffuse through the gel in a direction that is perpendicular to the antigen line of electrophoretic migration make contact with antibody molecules. Lines of precipitate develop at the contact point where specific antibody and antigen meet. There are many modifications to the immunoelectrophoretic technique: **counterimmuno-electrophoresis (CIE), one-dimensional rocket immunoelectrophoresis**, and **two-dimensional immunoelectrophoresis.** Only counterimmunoelectrophoresis is described here.

The antigens and antibodies in counterimmunoelectrophoresis are made to flow in opposing directions toward each other and to form precipitate bands where their paths meet (Figure 24-37). The antigen and antibody molecules must have different **isoelectric*** pH points, and the isoelectric pH points need to be known or empirically determined for this sensitive and rapid technique to be applicable. The gel is buffered to a pH that is intermediate to the isoelectric pHs of the two reactants. Under these conditions the antigen molecules have a net charge that is opposite that of the antibody molecules, and hence they migrate in opposing directions toward each other: one to the positive pole (anode [+]) and the other to the negative pole (cathode [−]).

In vivo tests

Some immunological tests, usually in the form of a diagnostic skin test, are conducted in patients. Measured amounts of antigens are injected and the reaction noted. In some tests there is no detectable reaction, because the patient has antibodies specific for the antigen. For

* The isoelectric pH point is the pH at which electrophoretic migration does not occur, that is, the net charge on the molecule is zero.

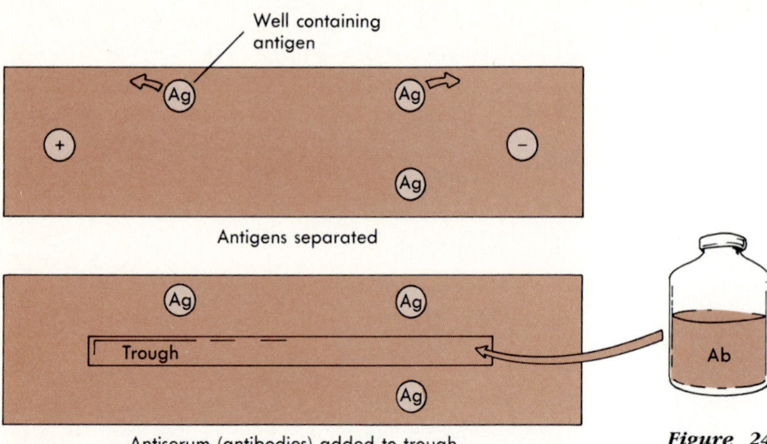

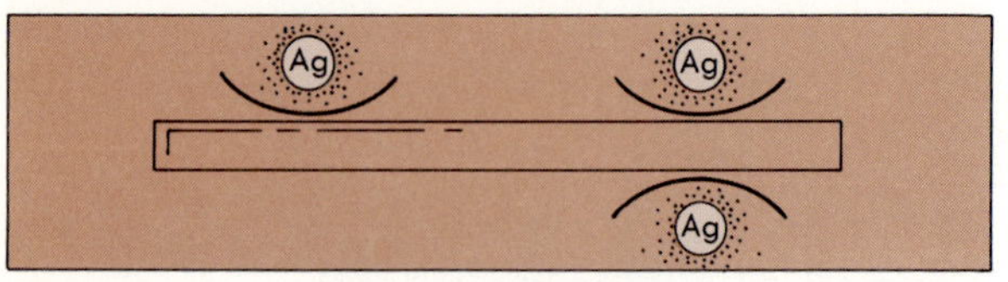

Figure 24-36 *Immunoelectrophoresis. Antigens separated in agar gel by placing electric charge across it. A pH is selected, which separates antigens so that positively charged proteins move to negative electrode and negatively charged particles move to positive electrode. Antiserum containing antibodies (Ab) for the antigens are placed in trough cut between the antigen wells. The antigen and antibody diffuse into the agar, forming precipitin arcs where they interact.*

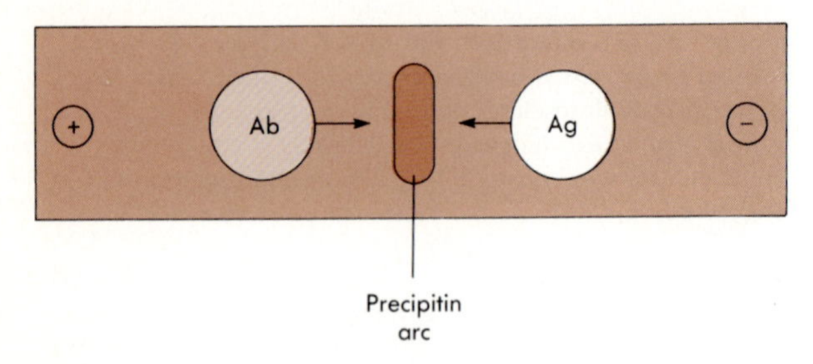

Figure 24-37 *Countercurrent electrophoresis. Antigen and antibody are placed in wells in an agar gel in which the pH renders the antibody positively charged and antigen negatively charged. When a voltage is applied across the gel, the antigen and antibody move toward each other and precipitate.*

Concept Check

1. Why is complement added to the complement test system before the indicator system is added to the same tube?
2. Why is serotyping of bacteria important? How is it applied to *Salmonella?*
3. What three tests incorporate chemical markers conjugated to antibodies? What is the nature of the chemicals, and how do they work? What are their advantages over tests such as agglutination?
4. What is the value of in vivo serological tests?

example, in the Schick test, which is used to determine if a person has antibodies to diphtheria toxin, a small amount of toxin is injected into the skin. There is no reaction in the patient who has sufficient antibodies to diphtheria toxin, but the patient who lacks antibodies to diphtheria toxin develops a skin reaction that is typically caused by the unneutralized toxin. In other tests the patient develops a reaction if he has formed immune products against the test antigen. This is especially the case when cell-mediated immunity is involved, as in **tuberculin testing** (Figure 24-38) and in testing for fungus infections. In vivo tests are usually named after the individual who devised the test or after the test material that is extracted from the microorganism: **Schick test**, **Mantoux test**, **Prausnitz-Kuestner test**, and **tuberculin test**, to name a few. In vivo tests are usually adjudged positive or negative. They do not lend themselves to determining a titer.

SUMMARY

1. The immune system specifically interacts with foreign substances (antigens) that have certain characteristics. Memory cells of the immune system are prealerted to that same substance for future encounters. These features, specificity and memory, enable the body to cope with foreign substances much

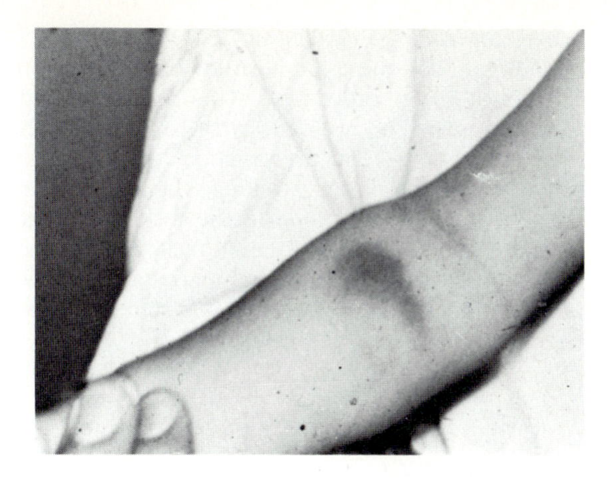

Figure 24-38 *Positive tuberculin skin test.*

more effectively than can be done by nonspecific mechanisms. This system, with its ability to recognize specific chemical groupings, called antigenic determinants, on the surfaces of molecules must of necessity be able to recognize self-substances and refrain from attacking them.

2. Two major classes of lymphocytes, B cells and T cells, specifically recognize antigens and become sensitized lymphocytes, which form immune products called immunoglobulins and lymphokines, respectively. The products specifically interact with antigens or effector cells.
3. B cells in mammals develop in the bone marrow. Many of them emigrate to peripheral lymphoid organs, especially to the spleen and to the lymph nodes. B cells have receptor immunoglobulins on their membranes that are specific for given antigenic determinants. Contact with the antigenic determinants causes the B cell to multiply and differentiate to a cell type called a plasma cell. The plasma cell secretes immune products called immunoglobulins or antibodies that specifically interact with the antigens that elicited their production.
4. T cells derive from bone marrow stem cells, from where they journey to the thymus. The journey through the thymus conditions them to function as T cells. They travel to peripheral lymphoid organs where, through specific receptors, they are activated by antigens. Antigen contact causes a T lymphocyte to proliferate and form sensitized lymphocytes. Sensitized lymphocytes may directly destroy the antigen. They secrete nonspecific products called lymphokines. T cells also have immunoregulatory roles as helpers, suppressors, and regulator cells.
5. The basic immunoglobulin molecule consists of four polypeptide chains: two heavy chains and two light chains. The chains and domains within the molecules are linked by sulfhydryl bonds. One end of the basic molecule has two antigen-binding sites (Fab), each of which is made up of the variable regions of a light chain and a heavy chain. The biological differences in the immunoglobulins are due to differences in the heavy chain. The other end of the Ig molecule, depending on the Ig class, called the Fc fragment, has various biological functions: attachment to phagocytes and mast cells, binding site for complement, and passage through the placenta.
6. There are five Ig classes based on physiochemical and biological activity differences. Antibodies are immunoglobulins with a known antigenic specificity; that is, the substance with which they interact is known. The immune activities of antibodies include neutralization of toxins and viruses, promotion of phagocytosis of antigens, and lysing of cells.
7. Antibodies can be produced against a multitude of different antigens because the genes coding for heavy and light chains contain a variety of segments that can be rearranged during lymphocyte differentiation.
8. Immune products very often are not the final effectors on an antigen. The union of an antigen and the immune product frequently initiates the participation of other body elements such as macrophages and the complement system. It is these elements that determine the ultimate fate of the antigen.
9. The complement system consists of a number of serum proteins that exhibit a major beneficial (and harmful) role, both in nonspecific resistance and in immunity. Complement's activities include cell lysis, phagocytosis, chemotaxis, and stimulation of inflammatory responses.
10. The prime function of the immune system is to provide specific protection (immunity). Terminology has been developed that signifies the various immune states: innate vs. acquired, actively acquired vs. passively acquired. The development of the active immune state and the level of immunity attained are related to such considerations as the use of adjuvants; the type of antigen, and the frequency and spacing of the exposure to the antigen; types and order of appearance of immune products during the first antigen exposure (primary immune response) and during subsequent exposures (secondary immune response); and the role of memory cells.
11. The union of antigen and immune products in vitro, is generally a visible reaction, or it can be made visible by a variety of test techniques. Since there is a high degree of specificity of an immune product for its corresponding antigen, antigen-immune product reactions are a very sensitive means of identification and monitoring.

12. In microbiology serological tests are used not only to identify microorganisms and their products or the antibodies they evoke, but also to (a) confirm a diagnosis made by other laboratory procedures, (b) follow the course of a disease, (c) determine the serotype and serogroup, and (d) determine the level of protection from disease.
13. There is a quantitative aspect to many serological tests. Dilutions of either the antigen preparation or of the test serum make possible the establishment of a titer. Not only can it be determined that a given serum, for example, contains antibodies for a specific antigen, but it is possible to determine the level of antibody present. The antibody level (titer) in turn may be significant for diagnosis, for determining the level of protection, or for following the course of infection.
14. There are a large number of basic serological test techniques, and there are a variety of modifications of the basic test procedures. The older "classical" tests are named according to what happened to the antigen: agglutination, precipitation, cytolysis, neutralization, and capsular swelling. The more recent tests are named according to the technique used: immunofluorescence, immunodiffusion, immunoelectrophoresis, and radioimmunoassay. In vivo tests, which are largely performed as skin tests, yield significant diagnostic information.

Self-study Quiz

Multiple choice

1. Which of the following classes of immunoglobulins is most often associated with immediate type (type 1) allergic reactions?
   A. IgA
   B. IgE
   C. IgM
   D. IgG
2. The anamnestic response refers to the
   A. Anaphylactic reaction to injected antigen
   B. Complement reactions to antigen in the bloodstream
   C. More rapid and efficient response to a second or subsequent dose of antigen
   D. Response to passively transferred antibodies
3. If the bursa equivalent does not develop properly, which of the following types of cells would be lacking?
   A. Macrophages
   B. B-cell lymphocytes
   C. Stem cells
   D. T-cell lymphocytes
4. Which of the following is responsible for attracting circulating PMN leukocytes to the inflammatory site?
   A. Interferon
   B. Lysozyme
   C. Chemotactic substances
   D. Macrophages
5. Which of the following would be most directly responsible for the synthesis of lymphokines?
   A. Thymus
   B. Macrophage
   C. Activated T cells
   D. Mast cells
6. If you separated the peptide chains of an immunoglobulin monomer by reducing all the disulfide bonds, which of the following would be true?
   A. The Ig could bind only one antigen molecule.
   B. The complement cascade would be activated.
   C. The normally Y-shaped molecule takes on a T shape.
   D. The Ig could not bind its antigen.
7. Protective antibodies derived from recovery from a childhood disease represents what type of immunity?
   A. Active, natural
   B. Active, artificial
   C. Passive, natural
   D. Passive, artificial
8. What is the end product of the stimulation of the immune response system by an antigen that activates the cell-mediated immune response system?
   A. Complement
   B. Circulating antibodies against the antigen
   C. Activated T cells
   D. IgG

9. What is the functional (actual) valence of IgM antibody?
   A. 2
   B. 5
   C. 6
   D. 10
10. Which of the following is not a specific protective function of the cell-mediated immune response system?
    A. Immunity against intracellular infectious agents
    B. Recognition and elimination of aberrant "self" cells
    C. Inactivation of microbial toxins in circulation
    D. Recognition and inactivation of foreign cells

Matching

Match the immunoglobulins and cell types in the left column with their function or characteristic listed in the right column.

A. IgA ________
B. IgG ________
C. IgM ________
D. IgD ________
E. IgE ________
F. Helper T cell ________
G. Killer T cell ________
H. Suppressor T cell ________
I. Macrophage ________
J. B cell ________
K. Memory cell ________

1. Capable of binding complement
2. Accounts for over 70% of serum immunoglobulins
3. Attaches to mast cells
4. No known protective immune function
5. Can cross the placenta
6. Secretory antibody on mucosal surface
7. Found as a pentamer in serum
8. Has the longest half-life in serum
9. Aids antigen-stimulated development of B-cell function
10. Responsible for anamnestic response
11. Directs cellular attack of antigen-labeled cells
12. Blocks T cell and B cell response
13. Controls maturation and function of T_h and T_s cells
14. Thymus independent lymphocytes

Completion

1. The immunological test used to differentiate strains on the basis of antigenic composition of the cell is called ________________.
2. The amount of antibody in a serum sample is called ________________.
3. A neutral polysaccharide used to form gels for diffusion of antigen and antibody is called ________________.
4. The most common type of in vivo immunological test is called ________________.
5. The immunological test that employs radioactive labeling is called ________________.
6. The immunological test that takes advantage of the different size and charge of the antigens is called ____________.
7. An antiserum that produces the lysis of red blood cells in the presence of complement is called ________________.

Short answer

1. List the materials that would be needed to assess serologically the following situations or to conduct the following tests:
   a. A complement fixation test to identify the presence of mumps virus antibodies in a child's serum
   b. Whether a protein isolated from serum is identical to a known serum protein using the immunodiffusion technique
   c. The antibody titer of serum using the latex particle agglutination test
   d. The type of human *C. botulinum* toxin found in a culture medium
2. On the figures below draw in the precipitation lines depicting each of the reactions listed:
   a. Identical antigens in wells 1 and 2 and antibody in well 3:

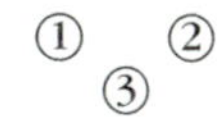

   b. Nonidentical antigens in wells 1 and 2 and antibody against both in well 3:

① ②
③

3. Interpret the results of the following serological tests.
   a. Unknown serum + latex particles with Agx attached—agglutination of latex particles.
      Did the serum contain antibodies against Agx?
   b. Unknown antigen + Anti-X antiserum + complement + sheep red blood cells—hemolysis.
      Was antigen-X the unknown antigen?

Thought questions

1. The human immune response system is a complex, integrated mechanism that is dependent on (a) the proper development of tissues and cells, (b) the proper programming of specific cells, and (c) careful control of the functioning of what are potentially self-destructive reactions within the human body. From what you know of the factors involved and functions of the immune response system, in the following situations predict how the immune response system will be affected and what problems would be encountered by a person suffering from the defect.
   a. The thymus gland does not develop during fetal growth.
   b. A person inherits a metabolic defect so that he cannot synthesize the C_3 complement serum protein.
   c. A person cannot synthesize the secretory piece of IgA immunoglobulin.
   d. A clone of mutant T cells arises that is no longer capable of distinguishing self from nonself tissues.
   e. A virus infects and destroys the helper T cell (T_h) lymphocyte population.

2. Currently, extensive research is being conducted in an attempt to develop a vaccine against the organism that contributes to the development of dental caries (tooth decay).
   a. You would expect the best protection by stimulating the formation of specific antibodies of what class?
   b. How might you immunize a person?

SELECTED READINGS

Books

Barrett, J.T. *Textbook of immunology* (5th ed.). The C.V. Mosby Co., St. Louis, 1986.

Bellanti, J.A. *Immunology III.* W.B. Saunders Co., Philadelphia, 1985.

Bier, O.G., and others. *Fundamentals of immunology.* Springer-Verlag New York, Inc., New York, 1981.

King, L.S. (Ed.) *Immunology: readings from Scientific American.* W.H. Freeman & Co., Publisher, San Francisco, 1976.

Myrvik, Q., and Weiser, R.S. *Fundamentals of immunology* (2nd ed.). Lea & Febiger, Philadelphia, 1984.

Robbins, J.B., Hill, J.C., and Sadoff, J.C. (Eds.). *Seminars in infectious diseases, Vol. 4. Bacterial vaccines.* Thieme-Stratton, Inc., New York, 1982.

Roitt, I. *Essential immunology* (5th ed.). Blackwell Scientific Publications, Ltd., Oxford, 1984.

Stites, D.P., et al. *Basic and clinical immunology* (6th ed.). Lange Medical Publications, Los Altos, Calif., 1986.

Journal articles

Centers for Disease Control. General recommendations on immunization. *MMWR* **32:**1, 1983.

Fearon, D.T., and Wong, W.W. Complement ligand interactions that mediate biological responses. *Annu. Rev. Immunol.* **1:**243, 1983.

Hanson, L.A., et al. Human milk: defense against infection. *Prog. Clin. Biol. Res.* **61:**147, 1981.

Hayward, A.R. Development of lymphocyte responses and interactions in the human fetus and new born. *Immunol. Rev.* **57:**39, 1981.

Hilleman, M.R. Newer directions in vaccine development and utilization. *J. Infect. Dis.* **151:**407, 1985.

Honju, T., and Habu, S. Origin of immune diversity: genetic variation and selection. *Annu. Rev. Biochem.* **54:**803, 1985.

Leder, P. The genetics of antibody diversity. *Sci. Am.* **246:**102, 1982.

Lerner, R.A. Synthetic vaccines. *Sci. Am.* **248**(2):212, 1983.

Milstein, C. Monoclonal antibodies. *Cancer* **49:**1953, 1982.

Muller-Eberhard, H.J. Chemistry and function of the complement system. *Hosp. Pract.* **12:**33, 1977.

Nossaman, J. Immunoglobulins: structure, genetics, diversity. *Lab. Med.* **12:**284, 1981.

Pollock, R.R., Teillaud, J.L., and Scharff, M.D. Monoclonal antibodies: a powerful tool for selecting and analyzing mutations in antigens and antibodies. *Annu. Rev. Microbiol.* **38:**389, 1984.

Reid, K.B.M., and Porter, R.R. The proteolytic activation system of complement. *Annu. Rev. Biochem.* **50:**435, 1981.

Rocklin, R.E., Bendtzen, K., and Greineder, D. Mediators of immunity: lymphokines and monokines. *Adv. Immunol.* **29:**55, 1980.

Talmadge, D.W. Recognition and memory in the cells of the immune system. *Am. Sci.* **67:**73, 1979.

Tonegawa, S. The molecules of the immune system. *Sci. Am.* **25**(3):122, 1985.

Tonegawa, S. Somatic generation of antibody diversity. *Nature* **302:**575, 1983.

Winkelstein, J.A. Complement and natural immunity. *Clin. Immunol. Allergy* **3:**421, 1983.

CHAPTER

25

Immune disorders: immunodeficiencies and hypersensitivities

Chapter Outline

Introduction

The term **immune** implies protection, which is indeed the prime and expected function of the immune system. However, there are genetic, developmental, and acquired defects that cause the immune system to overreact or underreact and they can lead to immune disorders. Immune disorders can be placed into two broad categories: immunodeficiencies and hypersensitivities.

This chapter describes the various immunodeficiency diseases and how they develop from defects in the immune system, the characteristics of one of the most devastating immunodeficiencies, acquired immune deficiency syndrome (AIDS); graft rejection and the characteristics of the various types of hypersensitivities; and in brief the relationship of cancer to immune disorders. Graft rejection is customarily discussed with allergies because the immune system damages the transplanted tissue that is intended to become part of the graft recipient's tissues. Cancer is discussed because the development of cancer is probably caused by proliferation of aberrant cells that have escaped the postulated cancer surveillance action of the immune system.

Table 25-1 *Infection as a clue to immune dysfunction*

Type of infection	Possible dysfunction
Severe and/or recurrent infection with pyogenic (pus-forming) bacteria such as species of *Streptococcus* and *Staphylococcus*	B cells (except in IgA deficiency); phagocytic system (PMN-killing* defect, splenic dysfunction); complement (deficiency of C1, C2, C3, C5, C6, C7, or C8)
Severe infection with herpesvirus, cytomegalovirus, chickenpox virus, live vaccine	T cells (SCID† treated malignancy)
Severe infection with hepatitis virus, echovirus, vaccine-strain poliomyelitis virus	B cells (deficiency of IgG, IgM, or IgA)
Resistant superficial candidiasis (a fungal infection caused by species of *Candida)*	T cells (SCID, thymic hypoplasia, chronic mucocutaneous candidiasis, steroid therapy)
Systemic infection with opportunistic fungi such as species of *Nocardia, Candida,* and *Aspergillus*	T cells (Hodgkin's disease); phagocytic system (PMN-killing defect)
Pneumonia caused by *Pneumocystis carinii*, an important pathogen in AIDS patients	T cells (SCID, treated leukemia); B cells (very rare)
Giardiasis (an infection by the animal parasite *Giardia lamblia*)	B cells (deficiency of IgG, IgM, or IgA)
Sudden severe sepsis	B cells (deficiency of IgM); spleen; complement

**PMN*, Polymorphonuclear neutrophil leukocyte.

†*SCID,* severe, combined immunodeficiency disease. SCID is an inherited disease that is sex linked and involves both the humoral and cellular arms of the immune system. Children often die before 1 year of age as a result of microbial infections.

IMMUNODEFICIENCY DISEASES

Deficiency diseases involving the immune system proper can be subdivided into two categories: primary immunodeficiencies that are of genetic origin and secondary immunodeficiencies that are due to underlying nonspecific causes, including malnutrition, drugs, x-rays, and cancer.

Primary immunodeficiencies

Genetic and developmental disorders involving lymphoid tissue may affect the B cell component, the T cell component, or both. Combined B and T cell disorders affect the lymphoid stem cell population of the bone marrow. Primary immunodeficiencies often make the individual easy prey for infectious agents, and the type of infection can be a clue to the type of immune dysfunction (Table 25-1). Primary immunodeficiencies are usually detected within the first 6 to 8 months of life because of recurrent infections or because of disorders of the skin, intestines, connective tissue, or vascular tissue in the newborn child (see box). The naturally acquired passive immunity obtained through the transplacental transfer of maternal IgG antibodies has diminished appreciably by 6 to 8 months following birth.

Defects in immune system accessory elements. Strictly speaking, some of the diseases that are categorized as immune disorders do not belong to the elements of the immune system proper, but to cells and systems that interact with the immune system.

Defects in phagocytes. Phagocytic cells may be defective at some point in the chain of chemical events that lead to intracellular killing. As discussed in Chapter 24, children who have chronic granulomatous disease (CGD) have polymorphonuclear leukocytes (PMNs) that are morphologically normal, but they are unable to gen-

MANIFESTATIONS OF PRIMARY IMMUNODEFICIENCIES

- Infection (often by microorganisms that are weakly pathogenic)
- Failure to thrive (those suffering from SCID)
- Intestinal disorders: chronic diarrhea, malabsorption,* giardiasis (diarrhea caused by protozoa, *Giardi lamblia*)
- Skin disorders: eczema, seborrhea,† warts (IgM deficiency)
- Collagen-vascular disorders: vasculitis‡—disease resembling systemic lupus erythematosus (see "Defects in the Complement System")

*Malabsorption is the inability to absorb nutrients from the intestine.

†Seborrhea is a disease of the sebaceous glands in which secretions (sebum) collect on the skin and form an oily coating or scales.

‡Vasculitis is an inflammation of blood vessels.

erate the superoxide ion that is necessary for the production of an intracellular bactericidal substance.

Defects in the complement system. Genetic defects are known for most of the complement factors. Such defects are associated with recurrent and disseminated bacterial infections, edema, and possibly lupus erythematosus. **Lupus erythematosus** (Latin: *red wolf*) is an immune complex disease that may affect only the skin or the skin and internal organs. The disease gets its name from the characteristic red rash that appears across the nose and upper checks. One of the complications of systemic lupus erythematosus is the formation of antibodies against various cellular components, especially of the kidneys (glomerulonephritis).

B cell immunodeficiency. Sex-linked agammaglobulinemia, or **Bruton's disease**, is a congenital immunodeficiency involving B lymphocytes and is seen almost exclusively in males. Once a child with this immunodeficiency lives beyond 6 months of age (the age when maternal antibodies become depleted), he is subject to a succession of infections by pyogenic (pus-forming) cocci, such as streptococci and staphylococci, as well as *Haemophilus influenzae* and *Pseudomonas aeruginosa.* Victims are able to reject grafts, however, because the T cell system is fully functional, and they can cope with tuberculosis, as well as fungal and viral infections that the T cell system normally controls.

T cell immunodeficiency. The prototype of this deficiency is **DiGeorge's syndrome**, in which the thymus gland fails to develop properly. Spasms (tetany), due to low calcium levels, are an early indication of this disease because the parathyroid gland, which controls calcium metabolism, also develops from tissue in this same area. The immune system deficiency is characterized by a poorly developed thymus gland and an absence or diminution in numbers of T cells in the circulation and in the thymus-dependent areas of the peripheral lymphoid organs. Serum Ig levels may be normal, but antibody-mediated reactions are affected because of the T cell–B cell interactions that are required in order for B cells to respond to T cell–dependent antigens. T cell–mediated immune functions, such as graft rejection and the development of delayed hypersensitivity reactions to skin test antigens, are markedly impaired. Infants with this defect are especially vulnerable to viral and fungal infections. Extensive infections of the skin and mucous membranes with the fungus *Candida* (mucocutaneous candidiasis) often occurs (see Figure 28-8). (NOTE: An extract called transfer factor, obtained from human T lymphocytes, is sometimes used to treat this condition.) Serious respiratory tract infections caused by *Pneumocystis carinii* are not uncommon and are generally an indication of an immunodeficiency, because this organism is ordinarily nonpathogenic in individuals with a normal immune system. Treatment of the immunodeficiency aspect of the disease involves the transplantation of thymus tissue obtained from fetuses.

B and T cell immunodeficiencies. Defects in the stem cell population of B and T lymphoid cells result in a condition referred to as a **severe, combined immunodeficiency disease (SCID).** Swiss-type agammaglobulinemia is one type of SCID. A child suffering from this type of immunodeficiency usually succumbs within the first 2 years to any one of a variety of infectious agents, including microorgansims of low pathogenicity. Some patients can be successfully reconstituted immunologically through the administration of compatible bone marrow cells from a sibling with well-matched histocompatibility antigens.

Secondary immunodeficiencies

Defects in what was originally a normal immune system may occur secondarily as a result of malignancies, infections, drugs, malnutrition, and therapies that affect the lymphoid system (Table 25-2). These deficiencies usually occur later in life and are not ordinarily as severe as the primary immunodeficiences. Graft recipients have secondary immunodeficiency when they are immunosuppressed. Immunosuppressive measures are employed in graft recipients so that their immune system is less able to reject (immunoreject) the graft, but this also makes the immunosuppressed graft recipient vulnerable to infections.

The disease known as **Acquired Immune Deficiency Syndrome (AIDS)** was first recognized in 1981 among the male homosexual community. Affected individuals would initially show weight loss, fatigue, swelling of lymph nodes, and diarrhea. Six to eight months later the immune system would fail, and the patient would fall prey to severe disease such as pneumonia caused by a rare infectious agent, *Pneumocystis carinii* (a protozoan), or Kaposi's sarcoma, is a rare neoplastic malignancy seen primarily in elderly men and is rarely fatal. However, AIDS victims are generally young, and case fatality ratios are as high as 60% for cases first diagnosed a year previously.

The type of immunosuppression caused by the agent of AIDS is unique. Ordinarily, healthy individuals have twice as many helper T cells as suppressor T cells, but in AIDS victims the ratio is reversed, and the helper T cells, which promote antibody formation, are depleted. Thus the patient is subject to infection by a multitude of opportunistic microorganisms that ordinarily do not harm the healthy individual.

Epidemiological studies now indicate that AIDS victims fall into four general categories:

1. Seventy-three to seventy-five percent are homosexual or bisexual men in their 30s or 40s. They have had prior infections with agents of sexually transmitted disease (*chlamydia, Neisseria gonorrhoeae* or cytomegalovirus) and have had many sexual partners.
2. Fifteen to twenty percent are intravenous drug

Table 25-2 *Some factors predisposing to secondary immunodeficiency*

Factor	Impaired immune component
Age	
Newborn (especially if premature)	B and T cells; phagocytic system; complement (severe bacterial and viral infections)
Elderly	B and T cells (anergy,* autoimmunity)
Lack of breast-feeding	IgA (decreased mucosal protection in intestines, increased frequency of respiratory and gastrointestinal tract infections)
Immunosuppressive therapy	
Corticosteroids	B and T cells; phagocytic system (decreased PMN chemotaxis)
Cyclophosphamide and 6-mercaptopurine	B and T cells; phagocytic system (decreased number of PMNs)
Irradiation	T cells; phagocytic system (decreased number of PMNs)
Diseases involving lymphoid system	
Hodgkin's disease	T cells
Lymphoma, chronic lymphatic leukemia	B cells
Multiple myeloma	B cells
Thymoma	B cell (hypogammaglobulinemia)
Splenectomy (anatomical or functional as in sickle cell disease)	Phagocytic system (decreased clearance of pneumococci and other bacteria by spleen and macrophages)
Other	
Severe infections	T cells (decreased delayed hypersensitivity); phagocytic system (PMN-killing defect following severe bacterial infections)
Burns	B and T cells
Malnutrition	T cells
Malignancy	T cells
Diabetes mellitus	Phagocytic system (decreased PMN chemotaxis)
Systemic lupus erythematosus	T cells; complement

*Anergy: lack of immunity to an antigen.

abusers who are heterosexual men and women.

3. Three percent are infected during blood transfusions or from coagulation products used to combat hemophilia.
4. One percent are predominantly women who apparently acquire the disease through heterosexual contact with an infected partner.

Transmission of the AIDS virus requires direct insertion of the virus into the bloodstream, for example, through the use of contaminated hypodermic needles, blood transfusions, or some sexual practice in which the mucosa is injured and virus can enter the bloodstream. There is no evidence to date that the virus is spread in any other way. There is also good evidence that the virus is not spread by prolonged exposure to AIDS patients.

By late February of 1987 just over 31,000 cases of AIDS were reported in the United States. It is believed that over 1 million people in the United States have been exposed to the AIDS virus and carry antibodies to it in their blood. Since the incubation period for the disease is at least from 2 to 5 years, it may be several years before we fully understand the epidemiology of the disease. A few long-term studies have shown that about 65% of those with antibodies to the AIDS virus will remain free of symptoms for at least several years. Some of the remainder will show AIDS-related conditions, such as swollen lymph nodes, and remain stable for some time, whereas the rest will experience degeneration of the immune system.

It is important to understand that AIDS is no longer a disease that affects only homosexual men. Heterosexual men have been shown to acquire AIDS from prostitutes and to transmit the virus to other female sexual partners. All sexually active persons—whether homosexual, heterosexual, or bisexual—must realize that the greater the number of sexual contacts, the greater the risk of acquiring AIDS. Most cases of AIDS among children are a result of transmission from the mother in the perinatal (at or around birth) period.

The agent of AIDS is a human retrovirus similar in morphology to the human T cell leukemic virus (HTLV-I) and is called the human immunodeficiency virus, or **HIV**. Both HTLV-I and HIV are tropic for human cells. HTLV-I has the capacity to transform T cells and make them proliferate, but HIV is cytolytic and destroys T cells. Although the AIDS virus infects T-lymphocytes, it is not fastidious about the cells it infects. The AIDS virus has been isolated from the brain, as well as cerebrospinal fluid, of some of its victims. In addition, several types of cancer, such as Hodgkins disease, have appeared in individuals infected with the AIDS virus. It is still a mystery as to which molecules on the surface of cells constitute the receptor (see Point of Interest).

The nucleotide sequence of the AIDS virus has been determined, and there is considerable variation from one isolate to the next. The nucleotide differences appear to be concentrated in the *env* gene, which codes for the major capsid protein of the virus. This surface protein is the likely target of protective antibodies. If the nucleotide variation is too great in the *env* gene, then a single vaccine might not be able to generate immunity against all strains. In addition, the variation could affect the reliability of diagnostic tests for the presence of the AIDS virus. Diagnostic test kits currently use viral protein to identify antibodies to the virus in blood samples from infected individuals. The tests are currently used at blood banks, plasma collection centers, health departments, and selected clinical centers in the United States. The tests have been highly useful for screening blood and have been responsible for detecting several thousand units of potentially infectious blood in the U.S. blood supply.

There is no successful treatment of AIDS; however, some drugs have shown promise in preventing replication of the virus in infected cells. One of these is azidothymidine (AZT), which belongs to a broad family of 2′3′-dideoxynucleoside analogues. These drugs appear to be phosphorylated in the cytoplasm to yield 2′3′-dideoxynucleoside triphosphates that are natural substrates for cellular DNA polymerases and reverse transcriptase. These analogues compete with the binding of normal nucleotides to DNA polymerase or can be incorporated into the DNA chain and bring about chain termination. This is possible because the 3′ sugar sites on the analogue are occupied by a chemical group other than a hydroxyl (Figure 25-1). Consequently, it is not possible to form a normal 5′-3′ phosphodiester linkage that is necessary for chain elongation. AZT is extremely toxic.

No chain elongation

X = N_3, H, etc.

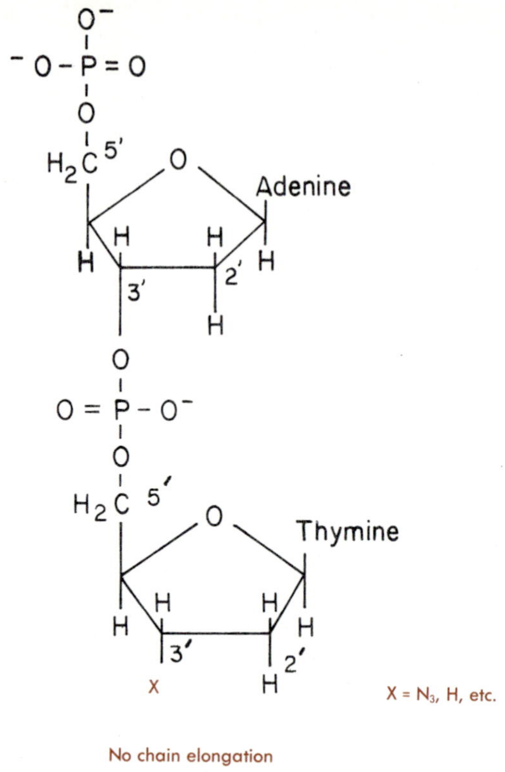

Figure 25-1 *One mechanism for activity of dideoxynucleosides against HIV. When 3′ carbon on sugar is modified by substitution of groups, such as N_3 or H for normal hydroxyl (OH) group no 3′ to 5′ phosphodiester bond can be formed between thymidine and next incoming nucleotide. This prevents elongation of chain as indicated in illustration.*

HYPERSENSITIVITIES (ALLERGIES)

The in vivo reactions of immune products with their homologous antigens, are beneficial to the host by virtue of their effect on the foreign agent (neutralization, cytolysis, phagocytosis). However, sometimes in vivo immune product-antigen reactions damage the host's tissues, cells, or systems. Hypersensitivities (or allergies—the terms are usually used interchangeably) are characteristic of an immune system that responds in a manner outside its normal response. Most allergic reactions are responses of a normal immune system. Whether or not an overresponse (allergic reaction) is going to occur and the nature of the response are usually determined by a combination of conditions, including (1) the nature of the antigen, (2) the amount of antigen that is introduced, (3) the route by which the antigen exposure occurs, (4) the chronological spacings and frequency of antigen exposure, and (5) the immune responsiveness of the host. The injection into an animal of the same antigen under the varied conditions cited above can lead to different allergic responses, among which are **anaphylactic shock**, the **Arthus reaction**, and **serum sickness**. A fatal anaphylactic shock reaction can be elicited by first sensitizing the animal with a small amount of ovalbumin administered by injection, waiting 10 to 21 days for antibodies to be induced, and then challenging the sensitized (to ovalbumin) animal with a larger amount of the same antigen. An Arthus reaction, which is characterized by an intense inflammatory reaction of the skin that leads to localized tissue necrosis can be elicited by giving a series of 3 to 5 ovalbumin shots about every 5 to 7 days into the skin at approximately the same site. A serum sickness type of reaction can be elicited by the injection of a single, but large, dose of ovalbumin. The symptoms of serum sickness include el-

Point of Interest

AIDS VIRUS: POSSIBLE MODE OF PATHOGENESIS

The AIDS virus is a latent retrovirus that may remain silent for years in the host. It binds to a selective subset of T cells called T4 helper inducer cells. These cells are defined by the presence of surface molecules called CD-4. CD-4 molecules in conjunction with the histocompatibility complex appear to be the receptor for virus. After binding to a cell, HIV enters the target cell probably by fusion with the cytoplasmic membrane. The virus is then uncoated, and RNA is released into the cytoplasm. Uncoated RNA is used as a template for production of DNA by reverse transcriptase. The DNA may be circularized and can apparently remain in an unintegrated or integrated form in the cell. Later, when activated, viral DNA is transcribed into mRNA and viral genomic RNA, and this is followed by production of viral proteins. The trigger that results in reproduction of the virus appears to be the normal antigenic activation of T4 infected lymphocytes. It is estimated that only 10% of T4 infected cells of infected individuals contain HIV, but when these respond to a foreign antigen the virus will reproduce, leading to cell death and spread of virus to new cells. The link between T cell activation and viral replication is believed to be associated with a control region located in the long terminal repeats (LTR) at the ends of the viral genome. One of the roles of T cells is the release of lymphokines. The region on the T cell genome that controls the release of lymphokines is similar in sequence homology to the LTR control region of HIV. Thus the factors that turn on lymphokine production may also activate HIV genomes.

HTLVs and HIV also contain genes that code for proteins that stimulate the expression of the viral genome after activation has occurred. These genes, designated trans-activation of transcription (TAT) enable rapid replication of virus (see figure below). The principal element of this autostimulatory pathway is an effector protein called the trans-activator protein. It is believed that the trans-activator protein interacts with sequences at the 5′ terminus of mRNA species and thereby increases translational efficiency (a positive regulatory device). The trans-activator protein possesses a cluster of positively charged amino acids that are believed to interact with a specific region on the 5′ mRNA terminus. This region on the mRNA corresponds to a special region in the LTR of the viral genome. The TAT gene product also enhances host regulatory genes. One of the host genes that is activated contains the information for the receptor protein used by the virus to enter the cell. It has been postulated that dying T cells release the activator protein, which helps the virus to infect other T cells.

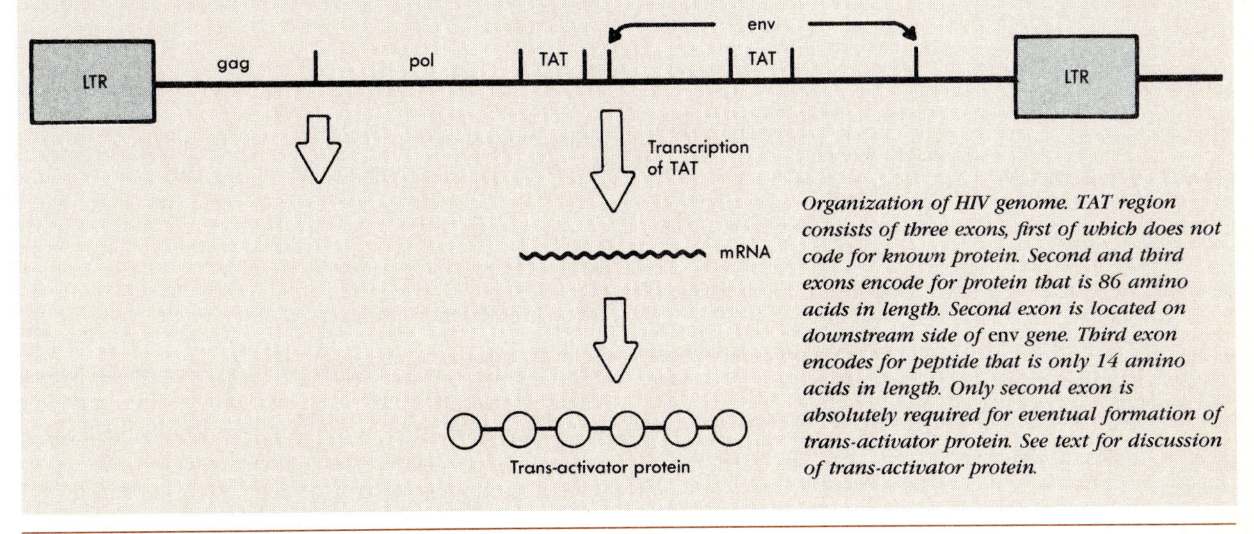

Organization of HIV genome. TAT region consists of three exons, first of which does not code for known protein. Second and third exons encode for protein that is 86 amino acids in length. Second exon is located on downstream side of env *gene. Third exon encodes for peptide that is only 14 amino acids in length. Only second exon is absolutely required for eventual formation of trans-activator protein. See text for discussion of trans-activator protein.*

Table 25-3 *Principal characteristics of immediate and delayed hypersensitivities*

	Type of allergy	Examples of antigens	Coupler: IG class or T cells	Principle effector cells and their mediators	Principal pathophysiological effect
Immediate: humoral hypersensitivity; antibody mediated; passively transferable via serum	Systemic anaphylaxis	Penicillin, hymenoptera* stings, foreign serum	IgE	Mast cells: Histamine Kinins	Capillary permeability, smooth muscle contraction, exocrine secretions (e.g. liver)
	Atopic allergies (localized anaphylaxis)	Pollens, house dust, animal scales or hair	IgE	Mast cells: Histamine Kinins	Exocrine secretions
	Arthus reaction	Ovalbumin, foreign serum	IgG and/or IgM plus complement	Cells of acute inflammation	Induration and dermonecrosis due to localized destruction of blood vessels
	Serum sickness	Foreign serum, penicillin	IgG and/or IgM plus complement	Cells of acute inflammation; mast cells	Vasoactive, edema, inflammation; hives, arthralgia,† lymph node enlargement
	Immune (toxic) complex	Soluble antigens, viruses	IgG or IgM plus complement	Cells of acute inflammation	Kidney (basement membrane) damage; damage to other organs and tissues; vasculitis
Delayed: cell-mediated hypersensitivity; sensitized lymphocyte mediated; passively transferable via lymphocytes and transfer factor	Tuberculin (infectious allergy)	Tuberculin, fungi	T cells	Lymphocytes: Lymphokines Macrophages	Antigen lysis or isolation; Destruction of host cells and tissues that surround antigen
	Allergic contact dermatitis	Poison ivy, industrial chemicals, cosmetics	↓	↓	↓
	Graft rejection	Foreign tissues, organs, cells	↓	↓	↓
	Cancer surveillance	Cancer cells	↓	↓	↓
	Autoimmune disease	Self-antigens	T cells and antibodies	Lymphokines and immune complexes	↓

*Hymenoptera are bees, wasps, hornets, etc.
†Arthralgia is joint pain.

Concept Check

1. What are the differences between primary and secondary immunodeficiencies?
2. What are the differences in host response to B cell immunodeficiencies? To T cell immunodeficiencies?
3. Describe the pathogenic mechanism of the virus causing AIDS. What is the role of the trans-activator protein?
4. How does an immunodeficiency differ from an allergy?
5. What conditions determine whether an allergic reaction will occur, and what is the nature of the response?

evated temperature, lameness due to inflamed and enlarged joints, and enlarged lymph nodes.

Categorization

Immediate and delayed hypersensitivities. The terms **immediate** and **delayed** refer to the time of appearance of skin lesions when a sensitized individual is skin challenged with the antigen that induced the hypersensitive state. If the reaction takes place within minutes of the administration of the skin test dose, the allergy is categorized as being the immediate type; if it is hours to days, it is the delayed type. Immediate hypersensitivities are triggered by antibodies interacting with their appropriate antigens; thus immediate hypersensitivities are

Table 25-4 *Classification of hypersensitive reactions*

	Anaphylactic	Cytotoxic	Immune complex	T-cell dependent
Immunoglobulin	IgE	IgG, possibly other	IgG, IgM, etc.	None
Antigens involved	Heterologous	Autologous or hapten modified	Autologous or heterologous	Autologous or heterologous
Complement involved	No	Yes	Yes	No
Cellular involvement	Mast cells and basophils	Red and white blood cells, platelets, etc., as targets	Host tissue cells	Host tissue cells
Chemical mechanism	Mast cell products and others	Complement-dependent cytolysis	Complement-dependent reactions	Lymphokines
Examples	Anaphylaxis, hay fever, food allergy	Transfusion reactions, Rh disease	Arthus reaction, serum sickness, pneumonitis	Allergy of infection, contact dermatitis

also termed **humoral hypersensitivities.** Delayed hypersensitivities involve sensitized T lymphocytes and are termed **cell-mediated hypersensitivities.** Immediate hypersensitivities are transferable via serum, and delayed hypersensitivities are transferable via intact T lymphocytes or via an extract of T lymphocytes called **transfer factor** (Table 25-3).

Gell and Coombs classification. Another commonly used classification is reflective of the underlying mechanism of the hypersensitivity reaction. Allergic disorders are divided into four types, which are modifications of the original proposal by Gell and Coombs (Table 25-4).

Type I: anaphylactic, reagin-dependent reactions
Type II: cytotoxic and cytolytic reactions
Type III: immune or toxic complex reactions
Type IV: delayed hypersensitivity, T cell–mediated reactions

Descriptions in the literature commonly refer to allergic reactions as a type I reaction, a type II reaction, and so on.

Type I: anaphylactic, reagin-dependent reactions

Antibodies of class IgE (see Chapter 24) are involved in type I allergic reactions that are the immediate type. Recall that IgE antibody is cytotropic for mast cells and basophils; that is, it preferentially binds to those cell types. IgE is also called **reagin** or **reaginic antibody**. There is a hereditary predisposition to produce antibodies of this class. A positive history of some clinical allergy usually can be found among close relatives of an individual with reaginic types of allergy. It is estimated that 10% of the population have overt atopic allergies (common, clinical allergies) and that as high as 40% to 50% have minor, undetected allergies of this type.

Type I reactions are the result of the patient at some time becoming sensitized to an antigen (allergen). An allergic reaction may occur on a subsequent exposure to the same allergen. The combination of the allergen and the antibody on the mast cell surface causes the mast cell to release (degranulate) the pharmacologically active substances that cause capillary permeability, vessel dilation, smooth muscle contraction, and mucous membrane responses (Figure 25-2). These effects may remain essentially localized, as in the common clinical (atopic) allergies, or they may occur systemically on a grander scale, as in an anaphylactic shock reaction.

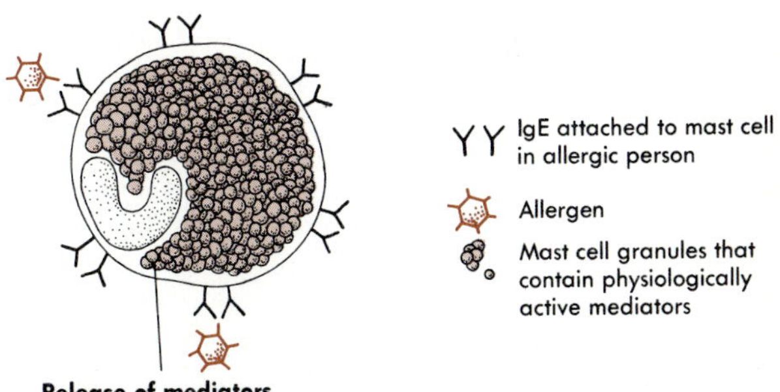

Release of mediators
Histamine
SRS-A (slow-reacting substance of anaphylaxis; also called leukotriene)
PAF (platelet activation factor)
ECF-A (eosinophil chemotactic factor of anaphylaxis)

Effects of mediators
Smooth muscle contraction
Vasodilatation
Increased vascular permeability
Mucous secretions

Diseases
Systemic anaphylaxis
Atopic allergies (local anaphylaxis)
- Rhinoconjunctivitis (e.g., hayfever)
- Asthma
- Atopic dermatitis (e.g., hives)
- Gastrointestinal allergies (e.g., food allergies)

Figure 25-2 *Type I (anaphylactic) hypersensitivity.*

Table 25-5 *Clinical data from four cases of human anaphylaxis*

Case no	Sex	Age (yr)	Agent	Dose (ML)	Route of administration	Estimated time from challenge to death (min)	Symptoms and signs	Known prior exposure	"Allergic history"
1	F	39	Penicillin	1.5	Intramuscular	60	Generalized warmth; tightness of throat; respiratory distress; cyanosis;* convulsion; respiratory failure	Yes	"Hives" 2 weeks before death; allergen unknown
2	F	21	Guinea pig hemoglobin	0.2	Subcutaneous	16	Headache; wheezing; cyanosis	No	"Asthma"; skin sensitivity to dog hair, shell fish, ragweed, timothy, orris root, and house dust
3	M	52	Bee venom	—	Subcutaneous	20	Unknown	Yes	Severe local reaction to bee sting 20 years previously
4	M	56	Hay fever desensitization vaccine	1/16	Subcutaneous	45	Difficulty breathing	Yes	"Hay fever" injection was eleventh in series of weekly desensitization injections

*Cyanosis is blueness of skin caused by reduced oxygen content.

Anaphylaxis. The word **anaphylaxis** means "unprotected" or "unguarded." The sensitized individual is in an unguarded state in respect to a reaction to a given allergen. Systemic anaphylactic reactions usually develop with suddenness and often result from the injection of drugs and antisera and from stings of *Hymenoptera* insects, such as wasps, hornets, and bees. The reactions include combinations of reddening (erythema) of areas of the skin, hives (urticaria), and itching; severe respiratory difficulties due to the accumulation of fluids and cells in lung tissue and constriction of the respiratory bronchioles; airway obstruction from laryngeal edema; hypotension (shock resulting from vascular permeability and collapse); and abdominal cramps, vomiting, and diarrhea (Table 25-5). A systemic anaphylactic reaction can be fatal in minutes. Countermeasures to be used in systemic anaphylactic reactions include (1) administration of epinephrine to counteract the effects of histamine and other vasoactive mediators, (2) maintenance of an airway, and (3) drugs to maintain adequate blood pressure whenever shock occurs.

Atopic allergies (local anaphylaxis). Atopic allergies are "out-of-place" reactions, as the word **atopy** signifies, and their underlying immune mechanism is the same as for systemic anaphylaxis. Atopic allergies in humans are IgE mediated; therefore there is a hereditary predilection toward forming these allergies. Atopic allergies are the common, spontaneously occurring, clinical allergies such as hay fever, hives, asthma, and food allergies (see box).

The management of atopic allergies includes appropriate testing (Figure 25-3), avoidance of the identified allergen whenever possible, hyposensitization and/or desensitization, and the use of drugs appropriate to the allergic condition. Hyposensitization/desensitization procedures ("allergy shot") attempt to decrease (hyposensitize) or eliminate (desensitize) the allergic state through the periodic administration of the same allergenic substance that caused the allergy. One might expect such a procedure to increase the patient's allergic reactivity, but in certain atopic allergies, notably in ragweed pollen hay fever, the allergic reactivity is greatly reduced. Administration of the allergen by injection evokes the formation of antibodies of class IgG, which are referred to as **blocking antibodies.** There are several hypotheses as to how hyposensitization and/or desensitization works. IgG antibodies are believed to have a greater affinity for the allergen, thereby preventing the

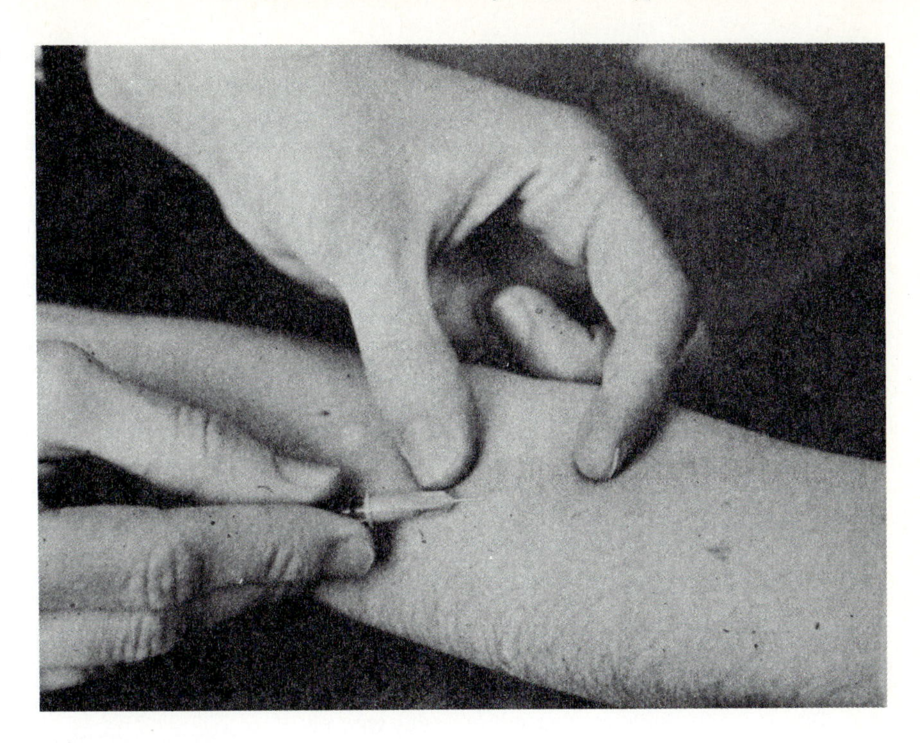

Figure 25-3 *Skin test for atopic allergy.*

allergen from uniting with the mast cell–bound IgE antibodies. The IgG antibodies may prevent the induction of additional IgE antibodies specific for the same allergen. IgG antibodies may not be involved at all, and suppressor T cell activity may cause a decrease in IgE synthesis.

The drugs that are used to manage or give relief in atopic allergies include antihistamines, bronchodilators, decongestants, expectorants, and antiinflammatory agents. Antihistamines, incidentally, are of little value in systemic anaphylactic reactions because of the extent and the explosiveness of the reaction, but symptoms can be eliminated with norepinephrine if time permits.

Type II: cytotoxic and cytolytic reactions

Cytotoxic and cytolytic reactions are cell-damaging and cell-destroying reactions, respectively, in which essentially an antigen-antibody union occurs on the cell surface. The cell can be destroyed by phagocytosis or lysis. The complement system is frequently involved in these reactions.

Cytolytic reactions take place under a variety of conditions: (1) the antigen may be a structural component of the cell membrane, or it may be an extrinsic antigen that has attached to the cell; (2) the antigen that evoked the production of the involved antibody may be a cross-reacting antigen; that is, antigenic determinants on the host's own cells may be identical or sufficiently similar to antigenic determinants of foreign antigens to cross-react with the involved antibody; (3) drug-induced

COMMON CLINICAL ALLERGIES

Allergic respiratory disease (ARD)

- Pollinosis—seasonal allergic rhinitis.* Pollens are from trees and grasses in spring and early summer, weeds in late summer. Pollens are mostly wind borne but also insect borne.
- Perennial allergic rhinitis (PAR)—year-round. Common allergens are house dust and animal hair, feathers and dander. Among the many allergens of house dust are mold spores, animal hair, and mites.
- Bronchial asthma. Common allergens are those cited under ARD and PAR. The air sacs are overdistended, plugs of mucus fill bronchial passages, and smooth muscles enlarge (hypertrophy) to thicken and narrow the walls of the bronchi. Symptomatic relief is obtained by bronchodilators that relax the muscles of the bronchi, and by expectorants and liquefacients that dissolve and expel the mucus plugs and the other accumulations.
- Gastrointestinal allergies—colic and possibly ulcerative colitis.†
- Allergic skin disorders
  - Atopic dermatitis. This can be divided into stages: infant, childhood, and adolescent-adult.
  - Allergic contact dermatitis (ACD). A variety of allergens, many serving as haptens, elicit nonatopic skin allergies via the delayed hypersensitivity mechanism.

*Rhinitis is inflammation of nose or nasal membranes.
†Colitis is inflammation of the colon.

changes may cause the appearance of new antigens on the cell surface; and (4) an antigen may combine with an antibody in circulation, and this complex attaches to red blood cells, which in turn leads to red blood cell damage.

Most of the diseases resulting from cytolytic reactions fall into the category of hematological cytolytic diseases. The cell types primarily damaged or destroyed are red blood cells granulocytic white blood cells, platelets (thrombocytes), and vascular endothelial cells (cells that form the interior layer of vessels). The cells that are destroyed are usually the patient's cells, but they may be transfused or translocated cells, as in red blood cell transfusion reactions or Rh incompatability disease. The destruction of red blood cells of the patient leads to hemolytic anemia; that is, the patient has an anemia due to the destruction of red blood cells.

Type III: immune or toxic complex reactions

Type III reactions occur when soluble antigen and the corresponding antibody unite and activate complement. This combination can lead to intense inflammation and tissue destruction. Conditions that promote this type of reaction include the slight excess of soluble antigen to antibody and the presence of IgG and IgM antibodies. If either antigen or antibody are present in considerable excess of each other, the complexes that form are eliminated uneventfully. Complement fragments C3a and C5a, as well as the trimolecular complement complex C5, 6, 7, cause PMNs to accumulate in the affected areas. The lysosomal enzymes of the PMNs cause various types of damage, especially to blood vessels, basement membranes, and joints. (Figure 25-4).

Serum sickness. Serum sickness has long been known as an immune disease, but the underlying immune complex cause was not fully appreciated for many years. Historically, serum sickness was observed in patients who received large single amounts of serum for the treatment or prophylaxis of such diseases as diphtheria, tetanus, and pneumococcal pneumonia. The serum, more specifically named antiserum or antitoxin, was administered to provide passive immunity. The serum was whole-animal serum, which naturally contains foreign antigenic proteins. Manifestations of serum sickness would appear about 7 or more days after the injection. Sufficient antibody would have been produced by this time to form complexes with antigen that remain in the tissues. Small amounts of protein would ordinarily have been eliminated by this time, and thus no antigen would have remained to interact with the antibody. Clinical manifestations would include arthralgia (painful and enlarged joints), fever, urticaria (hives), and lymphadenopathy (enlargement of lymphoid tissue). A distinctive feature of serum sickness is that only a single large or persistent dose of antigen would serve both to "sensitize" the patient and later to trip off the allergic reactions. Serotherapy is no longer widely used.

Immune complex disease. The cause of a number of human and animal diseases involves immune complex formation. The conditions are as previously described in the introduction to type III reactions. The antigens include soluble proteins, nucleic acids, viruses, and microbial proteins. Immune complex formation is at least partially involved in the pathogenesis of poststreptococcal kidney tubular inflammation (glomerulonephritis), lupus erythematosus, pneumonitis (farmer's lung, pigeon breeder's lung), rheumatic fever, rheumatoid arthritis, a number of viral diseases (for example, hepatitis B virus disease), and other diseases. The antigens in some of these diseases are derived from the patient, and they are additionally categorized as autoimmune disease.

Concept Check

1. **What is the basis for categorizing hypersensitivities as immediate or delayed? What is the rationale for the Gell and Coombs classification scheme?**
2. **What are the effects of IgE antibodies in type I allergic reactions?**
3. **Why are antihistamines used in some atopic allergies but not in systemic anaphylaxis?**
4. **Describe the effects of type III or immune complex reactions.**
5. **Under what conditions does immune complex disease develop?**

Type IV: delayed hypersensitivity, T cell–mediated reactions

T lymphocytes are the recognition agent, coupler, or trigger for delayed hypersensitivity reactions such as the tuberculosis skin test. Delayed hypersensitivities can be passively transferred via intact sensitized T cells or in humans via an extract of sensitized T cells. The extract is called **transfer factor** and has been used clinically with some success against conditions such as mucocutaneous candidiasis (see Chapter 28).

Exactly how immunological injury occurs in the host tissues because of a delayed hypersensitivity response is not known with certainty. One can speculate that in T cell responses the nature of the antigen and the tissue locations of the antigen (intracellular, chronic, not easily lysed) are of such a type that the immune system must destroy some of the body's own cells and surrounding tissue to eliminate the antigen. The destruction of tissues may be sufficiently extensive to create a pathologi-

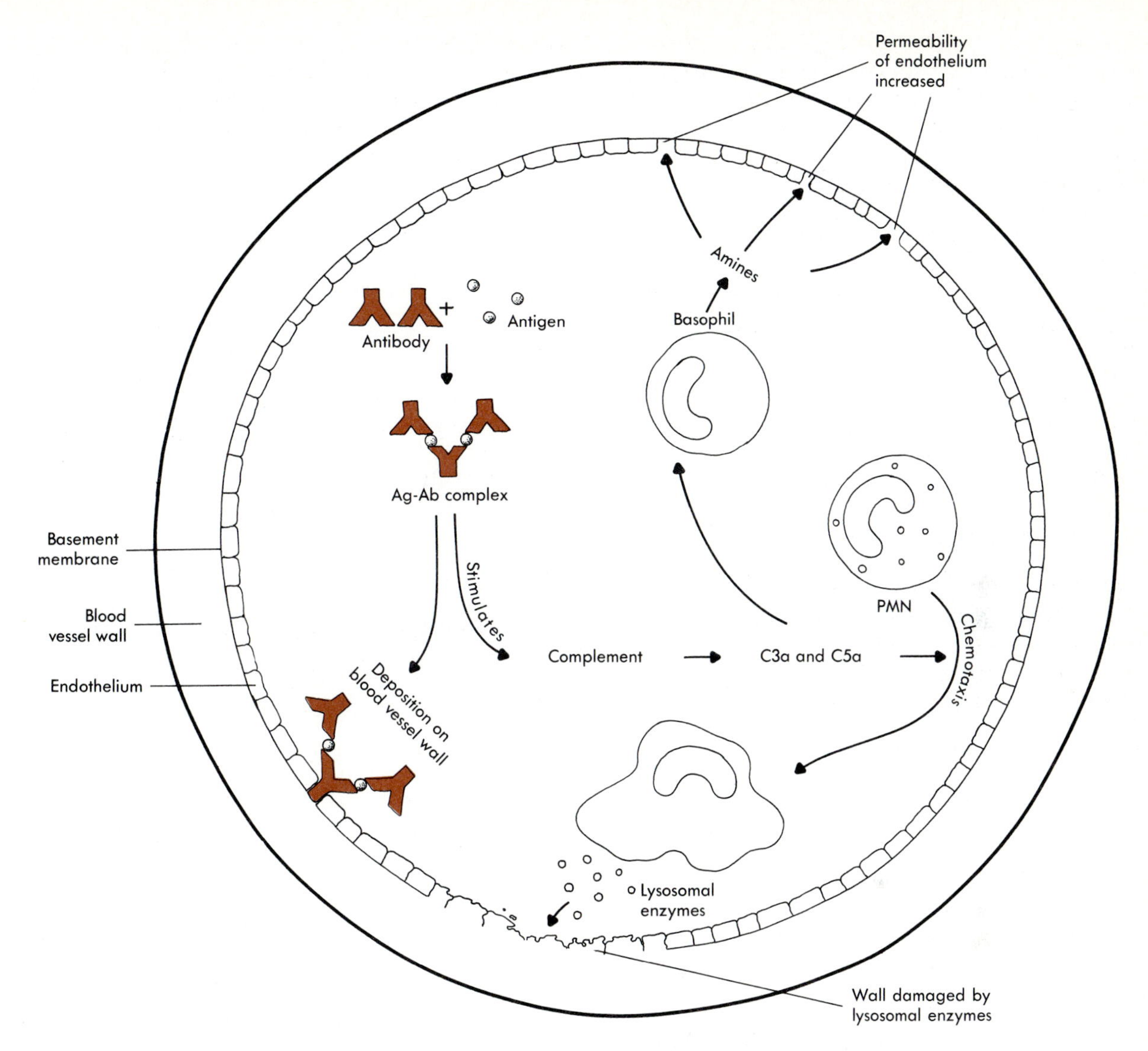

Figure 25-4 *Deposition of immune complexes in blood vessel walls. Antibody and antigen combine to form immune complexes. Complexes act on complement, which in turn acts on basophils to release histamine, which causes increased vascular permeability. With increased vascular permeability, immune complexes deposit on vessel wall. PMNs attracted to site by chemotactic complement peptides cannot phagocytose complexes and so release their lysosomal enzymes to exterior of cell, causing damage to the vessel wall.*

cal clinical condition that falls into the category of delayed hypersensitivity.

Descriptions of delayed hypersensitivities classically include tuberculin hypersensitivity, allergic contact dermatitis (ACD), autoimmune diseases, graft rejection, and cancer surveillance. The student should note that graft rejection and cancer surveillance are not adverse reactions, as their categorization, type IV reactions would imply. The rejection of foreign tissue cells, despite our great desire to replace diseased organs and tissues with transplanted healthy foreign tissue, is basically a beneficial action on the part of the immune system. If it were not for immune rejection of foreign tissue cells, any viable foreign cell, no matter what the species, if it made appropriate contact, theoretically could survive and multipy, It should also be understood that the immunological events and the pathogenesis of a given allergic disease are not necessarily attributable to just a single underlying mechanism, as the categorization into the four types might imply. Combinations of types II, III, and IV reactions contribute to the pathogenesis of autoimmune diseases.

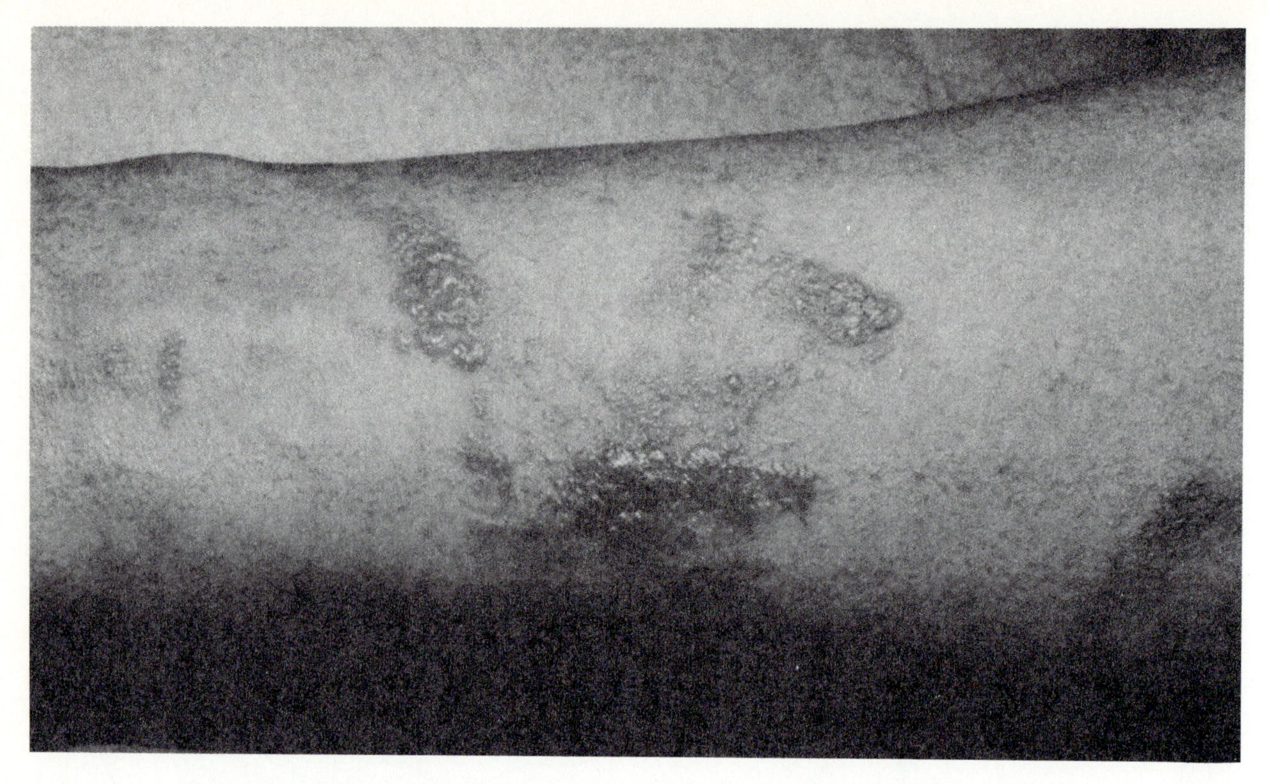

Figure 25-5 *Poison oak contact dermatitis of forearm. Streaks and patches of acute dermatitis strongly suggest plant dermatitis.*

Tuberculin hypersensitivity. Many of the early studies of delayed hypersensitivity were conducted on responses to the bacilli that causes tuberculosis *(Mycobacterium tuberculosis)* and on extracts of tubercle bacilli. The extracts are called **tuberculin** or **tuberculoprotein**. The tissue damage that occurs in tuberculosis is primarily attributable to the reaction triggered by T cells that are sensitized to tuberculoprotein. Chronic infections and intracellular infections tend to evoke type IV reactions.

Diagnostic skin tests. There are useful diagnostic skin tests that are based on a person's delayed hypersensitivity reaction to the test microorganism or to an extract of the microorganism. The best known is the TB skin test (see Figure 24-38), in which tuberculin is introduced intradermally. The response in a tuberculin-positive individual begins in about 8 hours as an erythematous (reddened) area that becomes indurated (firm and hardened) in 12 to 24 hours. The reaction reaches its peak by 48 hours and then subsides. The size and the intensity of the reaction are related to the amount of antigen that was introduced and to the degree of hypersensitivity of the tested individual. Other microbial products that are used in delayed hypersensitivity skin testing are histoplasmin and coccidioidin for the fungal diseases histoplasmosis and coccidioidomycosis, respectively, and lepromin and brucellergen for the bacterial diseases leprosy and brucellosis, respectively.

Allergic contact dermatitis. Skin contact with a variety of simple chemicals can cause allergic contact dermatitis (ACD) (Figure 25-5). Allergens are found in cosmetics, industrial chemicals, dyes, ointments, plant materials, and topically applied chemotherapeutic agents, to name a few. Clinical allergists call ACD allergies **nonatopic allergies** to distinguish them from the immediate-type, hereditarily predisposed atopic allergies.

The classic contact allergy is poison ivy hypersensitivity. The typical reaction develops 18 to 24 hours after a sensitized person contacts the allergen. Sharply defined erythematous areas of the skin are followed by induration and vesiculation (blistering). The blistering may be extensive, and the lesions are then called **bullae**. Oozing, crusting, intense itching, and occasionally infection are typical findings.

Antigens are usually large molecules. How then, can the chemically simple allergens of ACD incite the immune system to develop a delayed hypersensitivity reaction? Many of the allergens of ACD are believed to be haptens that combine with the patient's skin proteins to form the allergen. The haptens are the antigenic determinants, and the proteins of the skin are the carrier molecules for the haptens.

Table 25-6 *Autoimmune phenomena in human disease*

Disease	Specificity of autoantibody or autoreactive lymphocytes	Pathological effect of autoimmune reaction
Autoimmune hemolytic anemia	Anti-RBC	Red cell lysis; anemia
Pernicious anemia	Anti-intrinsic factor	Prevents absorption of vitamin B_{12} by ileum
Addison's disease	Anti-adrenal cortex	Adrenalitis
Myasthenia gravis	Anti-acetylcholine receptor	Impairs neuromuscular transmission
Systemic lupus erythematosus	Anti-DNA	Multiple
Rheumatoid arthritis	Anti-immunoglobulin	Inflammation of synovial (joint) membranes
Glomerulonephritis	Various	Inflammation of glomerular basement membrane (kidney)

Autoallergies or autoimmune diseases. One of the essential features of the immune system is recognition of self-antigens. An immune system attack against the body's own antigens ordinarily does not occur. The immune system, however, does sometimes form immune products against self (auto) antigens, and some grave diseases result.

What are some conditions that would lead to self-attack? Some possibilities are: body proteins are altered so that they are similar to, or share antigen determinants with, foreign antigens; self-antigens were never "marked" or "recognized" as self during fetal development; antigens that are normally never exposed to the immune system (sequestered antigens) become exposed (for example, through injury or surgery); the immune system itself is in some way defective and responds inappropriately; and the immune system recognizes tissues that are diseased and eliminates them.

There is a lengthy list of human diseases, many of them of a less well known variety, whose cause at least includes autoimmune features: Hashimotos's thyroiditis, myasthenia gravis, scleroderma, Sjögren's syndrome, lupus erythematosus, and autoimmune hemolytic anemia, are just a few of these diseases (Table 25-6).

Transplantation rejection. The elimination of grafted tissue is due to immunorejection that is primarily T cell mediated. The terminology that is associated with grafting is outlined in Table 25-7. Autografts and isografts are not immunorejected, because the antigens of the transplant and of the recipient are identical. Most transplants in humans are allografts, which are grafts between individuals of the same species who are not genetically identical. Tissues contain multiple transplantation antigens whose composition is controlled by histocompatability genes. The graft recipient's immune system recognizes the foreign transplantation antigens of allografts and xenografts and immunorejects them. Some histocompatability antigens induce a stronger antigenic response than others and are referred to as major histocompatability antigens. A vigorous immunorejection response occurs when the donor's antigens possess major histocompatability antigens that are lacking in the recipient.

Maneuvers that are undertaken to prevent immunorejection include tissue typing and the use of immunosuppressive agents. Tissue typing is performed in the laboratory to identify the donor's transplantation antigens and to detect the recipient's immunoreactivity to the donor's antigens. Immunosuppressive measures are used to suppress the graft recipient's immune system's responsiveness, with the purpose of forestalling immunorejection. Immunosuppressive agents include drugs, irradiation, and antilymphocytic serum (ALS). Infections, including those by microorganisms that are ordinarily of low pathogenicity, are a constant threat in the immunosuppressed.

Fetal-maternal immunorelationship and graft-versus-host reaction. Two interesting phenomena are related to grafting: the fetal-maternal immunorelationship and the graft versus host reaction (GVHR). The fetus is not immunorejected by the mother's immune system despite the fact that half of the fetal tissue antigens are foreign to the mother. Why the fetus is not immunorejected is not well understood. Possible explanations are as follows: the placenta acts in part as a mechanical barrier to maternal lymphocytes; cytotoxic maternal lymphocytes bind to the placenta and therefore do not reach the fetus; the fetal immune system eliminates maternal lymphocytes that cross the placental barrier; and

Table 25-7 *Transplantation terminology*

Prefix	Meaning	Combining suffixes	Transplantation examples
Auto-	Self	-graft -genetic -antigen -antibody	An autograft is a self-graft (for example, skin from one site on the patient's body moved to another site on the patient's body)
Iso-	Equal; identical with another	-graft -geneic -antigen -antibody	An isograft is a graft between isogeneic individuals (that is, between genetically identical individuals, such as identical [uniovular] twins)
Allo- (homo-)	Similar; like another	-graft -geneic -antigen -antibody	An allograft is a graft between allogeneic individuals (that is, between nonidentical members of the same species)
Xeno- (hetero-)	Dissimilar; unlike another; foreign	-graft -geneic -antigen -antibody	A xenograft is a graft between xenogeneics (that is, between members of different species, such as a graft from ape to human)

maternal antibodies attach to fetal membranes and block the access of maternal lymphocytes to fetal tissue. The last possibility, the inaccessibility of cells to lymphocytes because of the blocking effect of attached antibodies, is known to occur in at least one cancer, neuroblastoma. In neuroblastoma cancer patients the antibodies that are formed to the cancer cell antigens prevent the T cells from destroying the neuroblastoma cancer cells.

GVHR sometimes occurs when immunocompetent tissue from the donor (for example, bone marrow) is transplanted into an immunologically depleted or defective recipient. The transplanted immunocompetent tissue forms immune products against the recipient's tissues. GVHR is sometimes severe enough to cause death.

Tumor immunology. Cancer cells are believed to develop regularly in everyone. The immune system apparently recognizes these cells as being different from normal body cells and eliminates them. The immune system, however, is unable to eliminate rapidly proliferating tumor cells. The importance of the immune system in tumor surveillance is supported by the observation that cancer incidence of the lymphoreticular system in individuals with primary immunodeficiency disease is considerably higher than that in the normal population. Various experimental animal cancers and some human cancers have been shown to have **tumor-specific antigens (TSA)**, which make them different immunologically from normal cells. For example, viral oncogenic genes code for tumor-specific antigens that can be found in the cytoplasmic membrane of host cells (see Chapter 17).

The existence of tumor-associated antigens makes it possible to detect some cancers before they become clinically evident and to more precisely identify the cancer type. The presence of tumor-associated antigens also creates the possibility of controlling cancer by the administration of exogenous immune products and by manipulating the patient's immune system. These avenues continue to be explored with increasingly encouraging results.

Concept Check

1. **Explain the basic mechanism of delayed hypersensitivity. What are some examples of this type of reaction?**
2. **How does allergic contact dermatitis differ from an atopic allergy?**
3. **What are some explanations that have been proposed for the development of autoimmune disease?**
4. **Why does the existence of tumor-specific antigens raise hope for controlling cancer by immunological methods? How might hybridomas fit into the picture?**

SUMMARY

1. The immune system's primary function is to provide specific and speedy protection against microorganisms, against foreign substances, and against the development of cancer. Problems associated with the immune system occur, however, when it becomes defective (immunodeficiencies) or when the system overreacts (hypersensitivities). The defects that lead to immunodeficiencies are hereditary (primary immunodeficiencies), or they are acquired (secondary immunodeficiencies) through malignancy, drug therapy, and infection. Individuals with defects in the immune system suffer a variety of distressing medical problems, many of which are due to infections. One of the most important immunodeficiency diseases is AIDS which is caused by a virus. The disease results from infection of T4 lymphocytes and the decrease in helper T cells.
2. Overreactions (hypersensitivities) occur under circumstances that include the nature of the immunogen and the amount, route, and spacing of immunogen exposure. Hypersensitivities are either immediate or delayed, depending on how soon a sensitized (allergic) individual develops a lesion after skin challenge with the allergen that is reponsible for the hypersensitivity. Immediate hypersensitivities are of the antibody-mediated type. Delayed hypersensitivities are of the cell-mediated type.
3. There are four basic immune mechanisms that account for the tissue damage that occurs in hypersensitivity reactions. In type I reactions (the anaphylactic type) antibodies (IgE in humans) located on the mast cells of sensitized individuals cause the mast cells to degranulate when antigen combines with the antibody. The physiologically active substances contained in the granules produce effects especially on blood vessels, smooth muscle, and exocrine glands. The seriousness of these effects ranges from the common clinical (atopic) allergies to death in a systemic anaphylactic reaction. There is a hereditary predisposition in humans to form immunoglobulins of class IgE.
4. In type II hypersensitivities, cells are damaged or destroyed when antigens and immune product react at the cell surface. Blood cells are the cell type frequently involved in these immune cytolytic reactions.
5. In type III hypersensitivities (the immune complex type), the formation of complexes consisting of soluble antigen, antibodies of class IG or IgM, and complement initiate an intense, tissue-injuring inflammatory reaction.
6. In type IV hypersensitivities (the delayed or cell-mediated type), tissue damage results from the activities of sensitized T-cells or is mediated by lymphokines. Type IV hypersensitivities include the tuberculin type, allergic contact dermatitis, and autoimmune diseases. Graft rejections, fetal-maternal relationship, and cancer surveillance are frequently included in descriptions of type IV hypersensitivities. A distinctive feature of allergic contact dermatitis is that many of the allergens are haptens. In autoimmune diseases immune products are formed against the patient's own cells and tissues.

Self-study Quiz

Multiple choice

1. The genes of HIV that increase translational efficiency are:
   A. env
   B. pol
   C. tat
   D. gag
2. What is the basic developmental defect responsible for DiGeorge's syndrome?
   A. Overproduction of suppressor T cells
   B. Improper formation of bone marrow
   C. Defect in undifferentiated stem cell
   D. Failure in the development of the thymus gland
3. Atopic allergies are associated with production of which of the following?
   A. IgE
   B. Sensitized T cells
   C. Immune complexes
   D. Complement
4. Which of the following diseases is caused by the development of antibodies against certain of one's own immunoglobulins?
   A. Rheumatoid arthritis
   B. Rheumatic heart disease
   C. DiGeorge's syndrome
   C. Acute lymphocytic leukemia
5. The AIDS virus infects which of the following cell types?
   A. B cells
   B. T_h cells
   C. T_k cells
   D. Red blood cells
6. Graft versus host reactions would occur in which of the following situations?
   A. Injection of interferon into a virally infected host
   B. Injection of immunocompetent lymphocytes into an immunosuppressed host
   C. Injection of lymphocytes from an immunodeficient child into one of its parents
   D. Injection of toxoid into a hypersensitive person
7. How could you transfer type IV hypersensitivity from an allergic person to a nonallergic person?
   A. By long-term injection of small doses of the allergen
   B. By transfer of serum
   C. By transfer of the IgA fraction
   D. By transfer of T cells of lymphoid tissue
8. Histamines, when activated, would be responsible for what? Where would they most likely be found before activation?
   A. Chemotaxis—B cells
   B. Contraction of smooth muscle—mast cells
   C. Activation of macrophages—PMN leukocytes
   D. Creating holes in cell membranes—serum
9. Transplants of foreign organs are often rejected by much the same mechanisms that act against cancer cells, in particular:
   A. The interaction of humoral antibodies and soluble antigens
   B. The interaction of humoral antibodies and antigens inside of cells
   C. The interaction of immune cells and foreign antigens on cell surfaces
   D. The interaction of immune cells and antigens of one's self
10. Tissues transplanted between identical twins would be an example of what?
   A. Autograft
   B. Isograft
   C. Allograft
   D. Xenograft

Identification

Identify the type of hypersensitivity associated with the following conditions:

A. Development of hives after the consumption of strawberries
B. An intense inflammatory reaction on the skin at the site of injection of the eighth dose of antilymphocytic serum in a renal transplant patient
C. The transfusion crisis reaction created by the accidental introduction of type A blood into a type O recipient
D. The development of a poststreptococcal glomerulonephritis
E. A positive reaction to a tuberculosis test

Thought question

1. A 17-year-old woman is admitted to the hospital in January with what appears to be severe bacterial pneumonia. Her history is checked, and it is found that she was treated in the hospital's Sexually Transmitted Disease Clinic in October for gonorrhea. Her treatment at that time was a single high-dose injection of penicillin, and no ill effects were noted. Thus it is decided to initiate penicillin therapy for the bacterial pneumonia. Within 15 minutes of the first penicillin injection, the patient becomes restless, short of breath, and develops hives on her face, shoulders, arms, and chest. What appears to be happening? What steps should be taken? If no therapy is given to the patient, what might be the most serious outcome expected? How could she have escaped serious complications in October and show such a violent reaction in January?

SELECTED READINGS

Journal articles

Bach, M.K.: Mediators of anaphylaxis and inflammation. *Annu. Rev. Microbiol.* **36**:371, 1982.

Billingham, R.E.: Immunobiology of the maternal-fetal relationship. *Prog. Clin. Biol. Res.* **70**:339, 1981.

Brodt, P.: Tumor immunology—three decades in review. *Annu. Rev. Microbiol.* **37**:447, 1983.

Fauci, A.S.: AIDS-pathogenic mechanisms and research strategies. *ASM News* **53**(5):263, 1987.

Georgitis, J.W.: Insect stings: responding to the gamut of allergic reactions. *Mod. Med.* **50**:106, 1982.

Glasser, R.J.: How the body works against itself: autoimmune diseases. *Nursing* **7**:38, 1977.

Halpern, S.R., et al. Childhood allergy. *South. Med. J.* **73**:1601, 1980.

Hypersensitivity pneumonitis. In *Microbiology* 1985, Leive, L. (Ed.), pages 168-87. American Society for Microbiology, Washington, D.C., 1985.

Rubin, J.M. Food allergies. *Hosp. Med.* **18**:86, 1982.

Solley, G.O.: Present status of immunotherapy (desensitization) for allergic disorders. *South. Med. J.* **72**:183, 1979.

Thomas, E.D.: The role of marrow transplantation in the eradication of malignant disease. *Cancer* **49**:1963, 1982.

PART

VII

Microorganisms and infectious disease

CHAPTER

26

Epidemiology of community and hospital-associated diseases

Chapter Outline

Introduction

The epidemiology of infectious disease is concerned with determining what microbial agent has caused the disease (the etiology of the disease), the source of the disease, how it was transmitted, and what host and environmental factors could have caused the disease to develop. All these factors are used to determine the natural history of a particular infectious disease. Epidemiological studies are important because from them we have been able to design measures that have been useful in the control, prevention, and eradication of disease. The best example of the value of epidemiological studies is the case of smallpox. Epidemiological studies of smallpox over the years have revealed that (1) there is no evidence of an animal reservoir for the infectious agent; (2) infected individuals who recover do not act as carriers after the infection has subsided; and (3) transmission is by close contact with the microbial agent from highly visible lesions. These characteristics permitted rapid identification of infected persons and their contacts. Immunization of contacts interrupted the cycle of transmission and prevented further spread of the infectious agent. In 1977, after intensive worldwide immunization programs, smallpox was declared eradicated.

This chapter shows how the host-parasite relationship evolved; describes the infectious disease cycle in terms of host and microbial factors that lead to infection, the sources and reservoirs of infectious agents, and how the infectious agent can be transmitted to humans; and discusses the host response to infectious disease and the stages that are involved.

The first and major portion of the chapter is devoted

to the epidemiology of infections that affect the general community. The last section of the chapter discusses the epidemiology of hospital-associated infections, one of our more serious health problems.

EVOLUTION OF THE HOST-PARASITE RELATIONSHIP

The relationship between microorganisms and their hosts has evolved over many centuries. Microorganisms probably first existed as free-living saprophytes, but some of these habits changed with the emergence of plants, animals, and eventually humans. The various types of relationships—such as commensalism and mutualism—that evolved between microorganisms and other organisms are discussed in Chapter 18. Some microorganisms initiated a parasitic existence in humans or animals, thereby causing severe disease and even death. However, a parasitic existence in which the host is always killed is not advantageous to the parasite. Some hosts survived their first encounter with the parasite and developed an immunity to the organism or even entered into a commensal relationship with the microorganism.

History has demonstrated that during the evolution of the host-parasite relationship genetic changes occurred in both groups. Today's benign commensal organism could become tomorrow's parasite and vice versa, depending on the host's resistance and the parasite's potential for causing disease. Modern technology and microbiological research have provided a means to interrupt the evolutionary process and stop parasites before they are able to decimate populations. By using antibiotics and immunization procedures we have been able to prevent major global epidemics, which in earlier times killed as much as 25% of the world's population.

INFECTIOUS DISEASE CYCLE

All the major groups of microbial agents (such as bacteria and viruses) have members that are capable of causing infection in suitable hosts. This potential is dependent on factors such as (1) microbial characteristics, (2) the source, or reservoir, of the microbial agent, (3) the transmission of the microbial agent, (4) the susceptibility of the host, and (5) the mechanisms of exit of the infectious agent from the body and its dissemination to a new host.

Microbial characteristics

The ability of a parasite to cause disease is related to two principal factors: (1) the number of microorganisms, or dose, and (2) the virulence factors of the parasite. Both of these factors in turn are influenced by the susceptibility of the host.

Dose. The number, or **dose**, of microorganisms required to cause disease varies from one microbial species to another and is influenced by the susceptibility of the host and the route of infection. Some clear-cut examples of the effect of dosage have been demonstrated with enteric pathogens, such as *Salmonella* and *Shigella* species. Human volunteers administered an inoculum of 10^3 *Salmonella typhi* developed no disease symptoms, but when the dose was increased to 10^7 or higher, at least 50% of the volunteers developed symptoms. Other studies have also demonstrated that the minimal infecting dose of *S. typhi* can be reduced considerably if bicarbonate is also administered with the bacteria. Many bacteria are destroyed by acidic conditions in the stomach, and any host condition (achlorhydria) or treatment (bicarbonate) that reduces stomach acidity favors microbial resistance. The dose is also affected by the route of infection. Studies of the virus causing the common cold (rhinoviruses) have shown that the dosage required to cause disease may vary by a factor of 200, depending on the site of inoculation in the respiratory tract. The nature of the epithelial surface, the flow of mucus, and the presence or absence of chemical and immune factors may be barriers to the adherence and potential virulence of microorganisms.

Virulence factors. Virulence factors, such as toxin formation, adhesins, capsule formation, that enable the parasite to cause disease in a host have already been discussed in Chapter 23.

Sources and reservoirs of microbial agents

Living agents or inanimate objects that serve as the bridge for contact between the host and the microorganism are referred to as **sources of infection**. Animate or inanimate objects that are necessary to support growth of microorganisms are termed **reservoirs of infection**. The reservoir and source may be the same for one microorganism, or they may be different. Defining the reservoirs and/or sources of microbial agents is an important aspect of epidemiological studies, because if they can be eliminated, the chain of events leading to the infectious state can be broken and disease transmission prevented.

Animate reservoirs. Humans represent the most important reservoirs of infectious agents. The microorganisms found on the host may produce recognizable symptoms that are characteristic of the disease, or the infected person may be a **carrier** of the infectious agent. Carriers may be categorized as **healthy** if they harbor the microorganism but do not develop characteristic disease symptoms at any time. A second type of carrier is the **convalescent** carrier, who has recovered from infection but continues to harbor large numbers of potentially infectious microorganisms. One of the most publicized examples of a convalescent carrier is the case of Typhoid Mary (see Point of Interest).

The most potentially dangerous carriers are those who shed microorganisms via the respiratory or intestinal tract. Coughing or sneezing within 3 feet of a potential host is sufficient to transmit an infectious agent. Food, water, or **fomites** (inanimate objects that harbor microbes) may also be contaminated by the secretions or excretions of carriers and thus act as indirect sources of infection to others.

Some carriers harbor the infectious agents in the bloodstream, and bloodsucking insects, such as mosquitoes, may become infected by biting the host. The infected mosquito in turn can support the growth of the infectious agent and transmit it to susceptible hosts.

Insects belong to the phylum Arthropoda, of which there are two important classes: Insecta and Arachnida. Members of the class Insecta include flies, mosquitoes, fleas, and lice; those of the class Arachnida include ticks and mites (Figure 26-1). Insects that transfer infectious agents in a passive way from one host to another are called **mechanical vectors.** Flies, for example, can act as passive vectors because, whereas many of them do not parasitize humans or animals, they do come in contact with them and their waste products, which may be contaminated. The infectious agents are usually carried on the insect's body parts. Some insects prey on animals or humans, and occasionally a part of the cell cycle of the microorganism develops within the insect. Insects that serve as a host and reservoir for infectious agents are called **biological vectors**, such as the *Anopheles* mosquito, which carries the agent of malaria.

As discussed in Chapter 20, animals play an important role as reservoirs of infectious agents. Infections may be acquired directly by touching infected animals or eating contaminated meat or indirectly by insect vectors. Once humans are infected directly or indirectly by animals, they are usually unable to transmit the infectious agent to others, but there are exceptions. For example, flies may transmit the agent of the intestinal disease salmonellosis to humans. Infected humans, during the course of the disease, may transmit the infectious agent to others by way of contaminated hands.

Inanimate reservoirs. The primary **inanimate reservoirs** for infectious agents are **soil**, **water**, and **food.** Air is not considered a true inanimate reservoir, because it does not provide the organism with the nutrients required for growth or maintenance. Air can be contaminated by organisms that are derived from other sources such as soil, water, or animate hosts and does serve in the transmission of disease (discussed later).

Most saprophytic microorganisms whose primary habitat is the soil are not pathogenic for humans, but there are exceptions, such as some fungal agents. Most of the microorganisms found in the soil that cause disease in humans are inhabitants of plants and animals and are shed or excreted into the soil. Many of these organisms remain in the soil as spores and await transfer to a viable host before germinating. For example, anthrax is a bacterial disease of cattle and various her-

Point of Interest

TYPHOID MARY

A cook named Mary Mallon, who had served a family in New York for nearly 3 years, developed typhoid fever and recovered without incident. Shortly thereafter a visitor to the family developed typhoid fever and the family laundress was also stricken with the disease. In 1902 Mary Mallon went to work as a cook for a new family. Less than 2 weeks after her arrival the family laundress was taken ill with typhoid fever. Soon thereafter seven members of the household also became ill with the disease. In 1904 Mary became a cook for a family on Long Island and within weeks of her arrival four of the seven servants in the household became ill with typhoid fever. In 1906 Mary was engaged as a cook for a new family and six of the eleven members of that family became ill with the disease. Again Mary sought a new job as a cook for a family but this time one of the stricken members died from the disease. Mary was finally suspected as a carrier and was virtually imprisoned in a hospital by the New York Department of Health. Daily stool samples from Mary revealed that on some days there was an enormous number of typhoid bacilli, but on other days no organisms could be found. It was during her imprisonment that Mary was called "typhoid Mary." Mary was not heard from until 1914, when she was engaged as a cook in the Sloane Hospital for Women. Early in 1915 25 cases of typhoid fever occurred at the hospital involving doctors, nurses, and other hospital personnel. The cook was suspected, but Mary quickly left the premises before she could be apprehended. Later, however, she was located by the Department of Health even though living under an assumed name.

A subsequent study of her career showed that Mary Mellon had infected still other individuals than those already mentioned. She was strongly suspected of causing waterborne outbreaks of typhoid in Ithaca, New York in which over 1300 cases had been reported. It was established that a person by the name of Mary Mallon had been employed as a cook in the vicinity of the place where the first cases appeared and from which contamination of the water supply occurred.

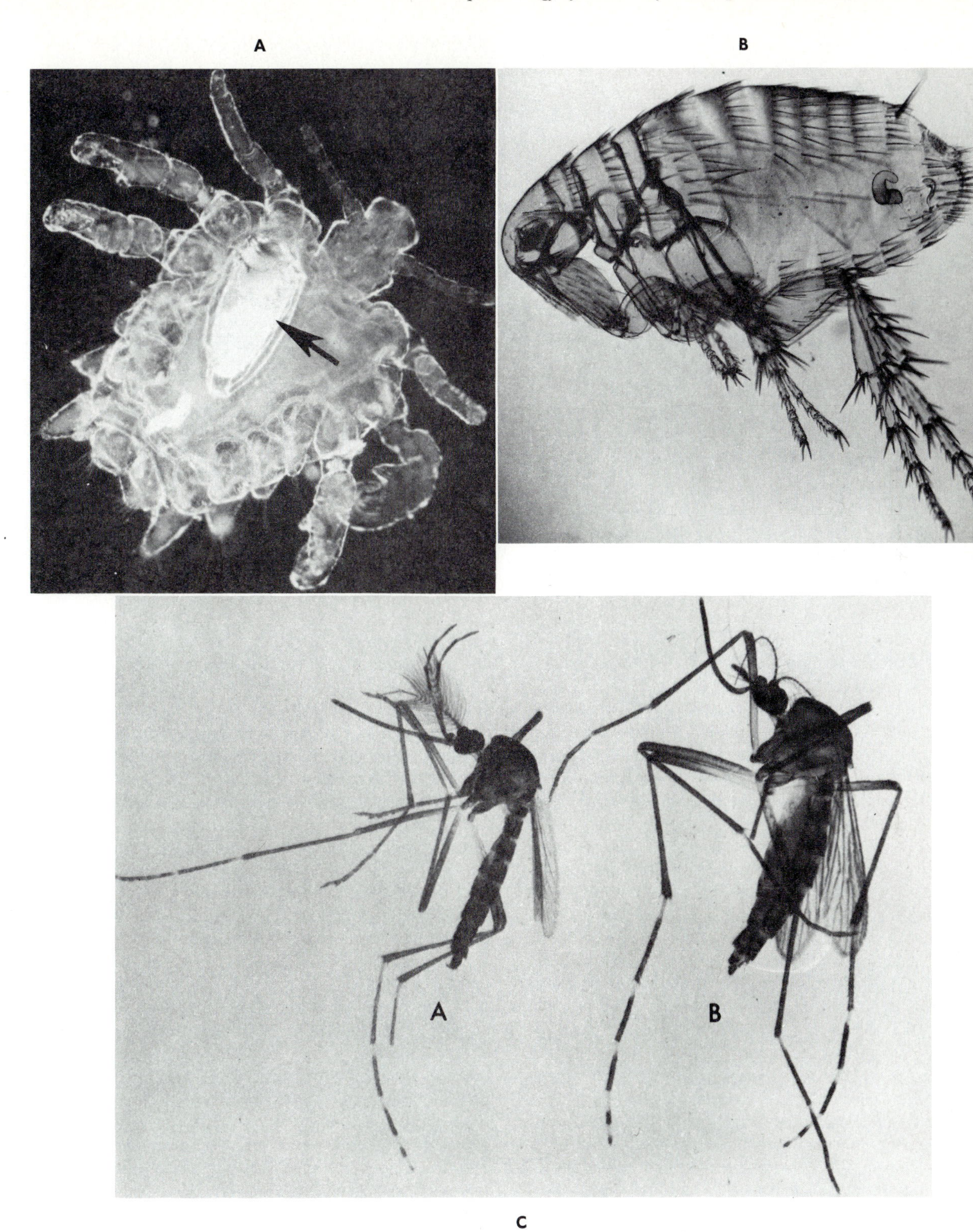

Figure 26-1 *Some important arthropod vectors.* ***A,*** *Crab louse* (Phthirus pubis). *Arrow points to developing egg.* ***B,*** *Rat flea* (Xenopsylla). *Most common vector of plague and murine typhus.* ***C,*** Aedes aegypti *mosquito, agent of yellow fever. In* **C:** A, *male;* B, *female.*

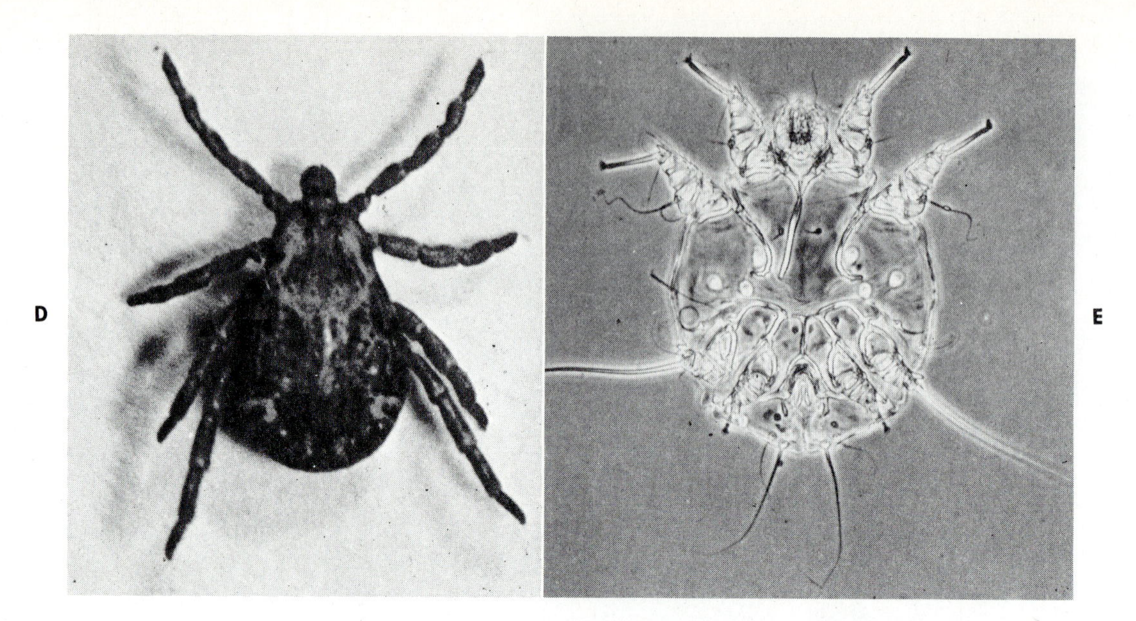

Figure 26-1, cont'd. ***D, Wood tick*** (Dermacentor). *One species of this genus transmits agents of Rocky Mountain spotted fever and tularemia in humans and tick paralysis in pets.* ***E, Itch mite*** (Sarcoptes scabiei).

bivorous animals. The spores of the microorganism causing this disease are found in the soil and are ingested by grazing animals. The spores germinate in the animal host and release vegetative cells that multiply and produce disease. If the diseased cattle are destroyed and only buried, the vegetative cells sporulate and are returned to the soil as resistant spores. It is therefore necessary to either cremate the animals or bury them in lime to destroy the vegetative cells and prevent sporulation.

Foods are important reservoirs of microorganisms because they contain nutrients that support the growth and maintenance of microorganisms. Food may be contaminated if it is obtained from infected animals, or it may be contaminated during or after processing (see Chapter 21). Food-borne disease results from the ingestion of foods, including milk, contaminated by microorganisms that either directly or indirectly cause gastrointestinal symptoms. Indirectly, some contaminating microorganisms release potent toxins into the food, which when ingested cause intestinal symptoms. This is not a true infection but an **intoxication** (as occurs with the organisms causing botulism). Some microorganisms, such as *Shigella,* that contaminate food are ingested, replicate, and invade the intestine to cause disease—a direct course of events. Improper cooking or preservation of foods is the cause of infection or intoxication in nearly all food-borne disease. The vehicles for transmission, by order of importance, are meat and poultry; salads containing chicken, turkey, potatoes, or eggs (particularly mayonnaise); shellfish; dairy products, and fruits and vegetables. Many animals harbor potentially infectious microbial species in their intestinal tract, such as *Salmonella* species, that can infect humans. Meat and particularly poultry contaminated by these species must be thoroughly cooked to prevent their multiplication. *Staphylococcus aureus,* which can be carried in the nasal cavity and can be a source of food contamination by food handlers and others, is a source of food intoxication. The microorganisms contaminating a salad, for example, will multiply and elaborate toxin, which may or may not be heat stable. Although not appropriate for salad, heating foods may be necessary to destroy the toxin and prevent gastrointestinal symptoms. The microbial agents involved in food-borne disease are outlined in Table 26-1.

Water, which is seldom distilled, usually contains inorganic ions and often is contaminated with organic material derived from the soil, air, animals, or plants. Saprophytes inhabiting the water are seldom pathogenic for humans, but water contaminated with organisms derived from fecal material may become an important source of pathogens for humans. Waterborne disease results from consumption of contaminated water that has not been properly prepared for drinking. Most outbreaks involve semipublic and municipal water systems, including recreational areas. Several cases of severe gastrointestinal symptoms in a town in Colorado were traced to a sewer line that had become obstructed, resulting in leakage of sewage into a creek supplying wa-

Table 26-1 *Microbial agents involved in food-borne disease*

Microbial agent	Clinical syndrome	No. cases/year in humans in the United States*
Bacterial		
Bacillus cereus	Gastrointestinal symptoms	70
Brucella species	Fever; enlargement of lymph nodes, spleen, and liver	120
Clostridium botulinum	Intoxication causing muscle paralysis, leading to double vision, difficulty in breathing, and death	110 (includes infant botulism)
Clostridium perfringens	Primarily diarrhea	1160
Escherichia coli	Primarily diarrhea	1
Salmonella species	Primarily diarrhea	2500
Shigella species	Primarily diarrhea	345
Staphylococcus aureus	Intoxication causing gastrointestinal symptoms and vomiting	2900
Group A streptococci	Fever, sore throat, other upper respiratory tract symptoms	300
Vibrio cholerae	Diarrhea	4
Vibrio parahaemolyticus	Diarrhea	12
Yersinia enterocolitica	Diarrhea	300
Campylobacter jejuni	Diarrhea	ND†
Viral		
Hepatitis A	Jaundice and gastrointestinal symptoms	130
Parasitic (helminths)		
Trichinella spiralis	Fever, orbital (eye) edema, muscle soreness	61

*Numbers represent only confirmed cases of foodborne disease and are based on an average for the years 1982-1984. Some microbial agents are transmitted by means other than food, for example, *Salmonella, Shigella, Campylobacter,* and hepatitis A, are also transmitted by water and other means.
†*ND;* no accurate data available.

Point of Interest

UNUSUAL SOURCES OF INFECTION

Sources of infection sometimes go unrecognized because they are uncommon. Plague, for example, is normally transmitted to humans by the bite of infected fleas harbored by rats. Two cases of human plague associated with domestic cats were reported from the west coast of the United States in 1981. One of the patients became ill one morning, complaining of shortness of breath, vomiting, and diarrhea. The following day a dry, nonproductive cough developed, and the patient decided to go to a physician. The physician prescribed antibiotics for a presumed urinary tract infection. After 2 days the patient's condition worsened, the cough became productive, and the patient was admitted to a hospital. The patient died 4 hours after admission. Sputum samples revealed *Yersinia pestis,* the etiological agent of plague. Later investigations revealed that 8 days before the patient died, her cat had brought home a dead chipmunk. The cat had become ill with coughing and bloody nasal discharge and died 5 days later. The cat was exhumed and demonstrated to be infected with *Y. pestis.* Eating or mouthing of infected rodents (in this case, the chipmunk) by the cat was the most probable mode of infection. Contact with the nasal secretions from the cat was the source of infection for the patient.

Table 26-2 *Microbial agents involved in waterborne disease*

Microbial agent	**Clinical syndrome**	**No. cases/year in humans in the United States***
Bacterial		
Escherichia coli	Diarrhea and fever	ND†
Salmonella species	Diarrhea and fever	1150
Shigella species	Diarrhea and fever	15
Vibrio cholerae	Diarrhea and vomiting	ND
Campylobacter jejuni	Diarrhea	875
Yersinia enterocolitica	Diarrhea	ND
Leptospira species	Headache, conjunctivitis, meningitis	75
Parasitic (protozoa)		
Giardia lamblia	Chronic diarrhea, fatigue, weight loss, abdominal cramps	2205
Entamoeba histolytica	Dysentery (severe diarrhea), fever, chills, or mild diarrhea	ND
Viral		
Hepatitis A	Jaundice, gastrointestinal symptoms	160
Parvovirus-like agents	Vomiting, watery diarrhea, abdominal cramps	ND
Rotavirus	Vomiting, watery diarrhea, abdominal cramps	ND
Enterovirus (poliovirus is one example)	Variable depending on virus but may be paralytic polio to aseptic meningitis to common cold	ND

*Numbers represent only confirmed cases of waterborne disease and are based on an average for the years 1982-1984 in the United States. Some microbial agents such as *Salmonella, Shigella, Campylobacter,* and hepatitis A are also transmitted via food and other means. (See Table 26-1.)
†*ND,* No accurate data available.

ter to the city. A protozoan called *Giardia lamblia* found contaminating the water was the cause of the outbreak of giardiasis. Other large outbreaks also result from leakage of sewage into private wells, with the latter acting as a source of waterborne disease for those served by the well. Nearly all of the agents of waterborne disease cause gastrointestinal symptoms and are outlined in Table 26-2. Microbial contamination of water supplies is a very important aspect of hospital-associated infections (see later in chapter).

Uncommon sources. The sources of infection that have been discussed so far are relatively common. Occasionally, the sources are unusual and require prolonged investigation by the epidemiologist (see Point of Interest).

Transmission

Microorganisms are transmitted to susceptible hosts by four routes: **contact**, **air**, **common vehicle**, and **vector borne**. Vector borne disease was discussed previously. Transmission may involve movement of microorganisms from animate or inanimate reservoirs to humans.

Contact

Direct contact. Direct contact implies actual physical interaction with the source of the microorganisms (Figure 26-2). Human-to-human transmission occurs most often through touching, kissing, or sexual intercourse. Contact with oral secretions or body lesions is one of the most common mechanisms for the spread of infectious agents. Many viral respiratory tract diseases are transmitted by touching the mucous membranes of the mouth, nose, or eyes and then touching another potentially susceptible host in these same areas. Enteric diseases, such as salmonellosis and hepatitis A, can also be transferred by the hands, but transfer is primarily by the fecal-oral route. Sexually transmitted diseases are transmitted primarily by sexual contact, but infected nursing mothers may also transmit the microbial agent to their offspring by direct contact. The agent of syphilis, as well

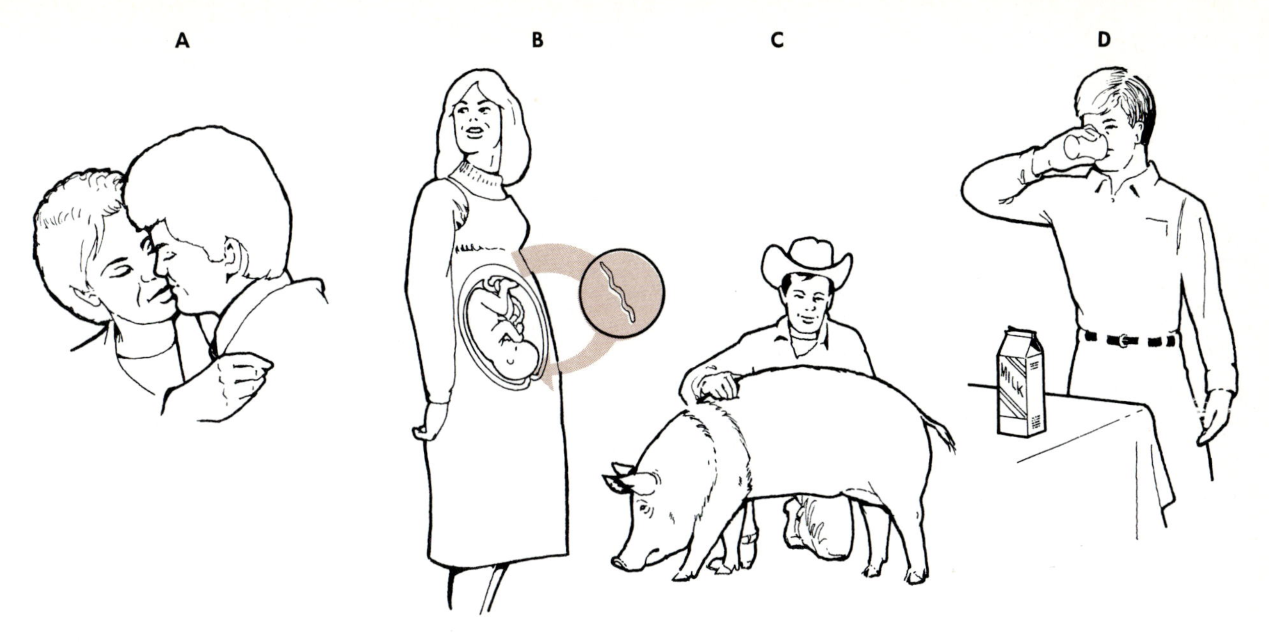

Figure 26-2 *Examples of how microorganisms are transmitted by direct contact.* ***A,*** *Touching or kissing.* ***B,*** *Congenital transfer.* ***C,*** *Contact with animals.* ***D,*** *Ingestion of contaminated food or food particles (milk).*

as some other microorganisms, can be transferred across the placental barrier to the developing fetus (in utero transfer). The agent of gonorrhea cannot be transferred across the placenta, but it can contaminate the conjunctiva of the newborn during passage through the birth canal. The agents that are transferred congenitally are primarily viruses (see Table 27-8).

Microorganisms may also be transmitted by direct contact with animals or animal products. The majority of infections in this category occur among those whose occupations place them in contact with animals. Farmers, dairymen, abattoir workers,* veterinarians, and hide and leather workers come in contact with infected animals or contaminated animal products. Ingestion of contaminated meat or animal products such as milk or cheese occurs more frequently in underdeveloped countries where domestic animals are not properly cared for or vaccinated. Scrofula caused by ingestion of milk contaminated with *Mycobacterium bovis* is almost unheard of in the United States because of pasteurization and dairy control, but in underdeveloped countries it is a frequent cause of infection because many persons drink raw milk. Raw milk, even when strictly controlled, may be contaminated with microorganisms indigenous to the cows. *Campylobacter jejuni* can be excreted directly into milk by cows who show no apparent illness. Outbreaks of gastrointestinal disease due to *C. jejuni* were recorded in 1985 among schoolchildren who sampled raw milk on dairy tours. The FDA in late 1985 issued a "milk advisory" to all states recommending that children not be permitted to sample raw milk.

Indirect contact. Indirect contact implies that the infectious agent persists on an object (usually inanimate) from which there is transfer to a susceptible host (Figure 26-3). The intermediary object may be contaminated from an animate or an inanimate source. A variety of objects, such as clothes, eating utensils, hospital thermometers, drinking glasses, and bedding, may be the intermediary. Most infections acquired in this way usually involve a hand-to-mouth transfer of the infectious agent.

Air. Infectious microorganisms can reach the air in a number of different ways. They may be dispersed from the respiratory tract by sneezing or coughing or other related activities. These air-borne mucous secretions are sometimes so small that they evaporate and remain suspended in the air for prolonged periods of time and are called **droplet nuclei**. Some microorganisms that are known to be transmitted by droplet nuclei are the viral agents causing smallpox and measles and the bacterial agent causing tuberculosis. The very large secretions expelled from the respiratory tract can contaminate clothing and various inanimate surfaces. These particles dry and become **dust particles**, and if the infectious agent is resistant to drying (smallpox), it may remain infectious for varying lengths of time. Dust particles may become airborne by dispersal from bedsheets or clothing, or by

* An abattoir is a slaughterhouse.

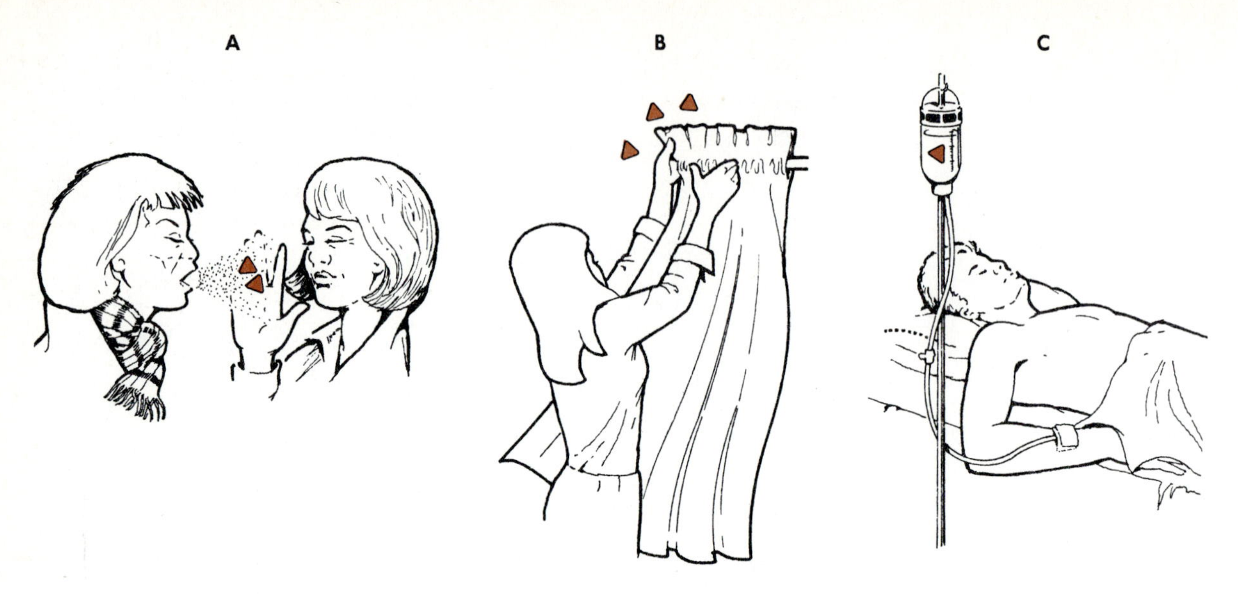

Figure 26-3 *Examples of how microorganisms are transmitted by indirect contact.* ***A,*** *Sneezing or coughing.* ***B,*** *Inhalation or contact with contaminated dust particles.* ***C,*** *Contaminated animate object or solution is common vehicle of infection (contaminated infusion fluid). Microbial contaminants (▲) are enlarged for effect.*

currents of air. For example, Q fever is a respiratory tract infection in humans caused by contact with dried parturition (during birth) products of cattle. These particles remain on the floor of the barn, where they can be disseminated by air currents and inhaled. Infrequently, particles from the skin, urinary tract, or genitourinary tract may also be discharged into the environment to contaminate soil or inanimate objects. Once the particles dry, they may be carried by air currents to infect susceptible hosts.

Many soil microorganisms produce spores, some of which are easily disseminated in the air. In the United States fungal spores of *Coccidioides immitis* are spread over the desert floor in the Western United States after winter and spring rains. During the hot, dry summers violent dust storms spread the spores in the air, and persons traveling through the area during the storms are exposed to them. Inhalation of large numbers of spores can cause serious disease or allergy, even in the healthy host. Fortunately, most infections do not result in serious disease.

Common vehicles. Common-vehicle transmission refers to the acquisition of microorganisms by a large group of individuals from a common inanimate vehicle or source. This type of transmission is frequently associated with hospital-associated infections (see pg. 790).

Concept Check

1. What microbial factors influence the events in an infectious disease cycle?
2. What is the difference between healthy and convalescent carriers, and what are examples of each?
3. What are the primary inanimate reservoirs for infectious agents? Name a disease that is associated with each reservoir.
4. By which routes are microorganisms transmitted to susceptible hosts? Name a disease transmitted by each route.

Host susceptibility

A multitude of factors can influence the outcome of an interaction between microorganism and host, and one of the most important is the genetic constitution of the host. For example, certain groups of individuals are resistant to tuberculosis, whereas others, such as blacks, are highly susceptible to the disease. Genetic factors probably operate at the level of the immune response. For example, aggressive macrophage function could conceivably be responsible for destruction of intracellular parasites such as the agents of tuberculosis and brucellosis. Individuals with a heterozygous form of sickle cell anemia, a genetic disease, are resistant to malaria. Malaria is caused by a protozoan that invades and multiplies in the normal red blood cell; in a sickle cell, however, the microbe is destroyed by the macrophage phagocytic system.

The **very young** and the **aged** are more suceptible to infection than other age groups. The immune response is greatly reduced in both groups, and some infections that would produce mild symptoms in the adult can cause death in these more susceptible groups. The mechanical and physiological barriers to infection are not developed in the very young, and in the very old these barriers have been compromised by the aging process and by any underlying conditions or diseases past or present. However, many infants are less susceptible to certain infections than the adult because of maternal antibodies that are transferred to the fetus. This protection seldom exceeds 6 to 10 months following birth and is available only if the mother has come in contact with the infectious agent prior to the delivery of the infant. Transfer of both humoral and cellular factors can also occur via breast milk.

Other factors that may in part influence susceptibility to infection include hormones, nutrition, social customs, personal hygiene, and various environmental factors, such as temperature, humidity, and air quality. Underlying conditions or disease also predispose the host to infection and are discussed later.

Release of microorganisms from the host

The successful infectious agent finds a host that will support its growth and development, but to maintain itself in the environment, the microbe must be retained in the originally infected host or transferred to another susceptible host. Most microorganisms that are transferred from animals or arthropods to humans reach a dead end because they cannot be transmitted to another human host. Diseases such as anthrax or brucellosis, which are obtained from cattle, are maintained in nature because of animal reservoirs, but once a human is infected, there is no human-to-human transfer. Pathogenic microorganisms must be spread to other individuals if they are to be maintained in nature, and this spread may take one of two forms: horizontal or vertical. **Horizontal spread**, which is the most common mechanism, implies that one individual infects another individual by contact, through insect vectors, or by contaminated water, air, food, or fomites. For example food handlers carrying infectious microorganisms (e.g., *Salmonella, Shigella, Staphylococcus*) on their hands can contaminate food and indirectly cause infections in those who eat the contaminated food. **Vertical spread** implies that the infectious agent is transferred from parent to offspring via ovum, sperm, or by contact between placenta and offspring. For example, a pregnant mother who has syphilis can transmit the infectious microorganism across the placenta into the developing fetus. The spread of microorganisms from one individual to another usually takes place within a few days during the infection, and before the immune system has a chance to destroy the microorganisms; however, there are exceptions. The immune system of some individuals is not sufficient to destroy the pathogen, and the microorganism persists in the host, sometimes for the lifetime of the host.

Persistence. As mentioned earlier, some individuals are carriers of infectious microorganisms. This type of microbial persistence, due to the inability of the host to eliminate the parasite, can give rise to different disease conditions referred to as chronic, slow, and latent. **Chronic diseases**, such as tuberculosis or hepatitis B, are characterized by the presence of disease symptoms over a long period. Chronic diseases can be the result of defective phagocytic activity or may be caused by a lack of specific antibodies to aid in the phagocytosis of the pathogen.

Slow diseases are characterized by the presence of the microbial agent for several years in the host without noticeable symptoms. Suddenly the individual develops symptoms that progress rapidly, usually ending in death in just a few weeks or months. **Kuru**, for example, is a slow disease believed to be caused by a prion (see Point of Interest, Chapter 16). The disease affects a tribe (Fore) in New Guinea that was once suspected of practicing cannibalism. It was reported that the infected brains from dead tribal enemies were handled by women of the village and served uncooked as a ceremonial food. The infectious agent is believed to be transmitted through abraded areas of the skin during the handling procedures. The first appearance of symptoms are from 3 to 18 years after contact with the infectious agent, and during this period the infectious agent progressively destroys nervous tissue. Once symptoms have appeared, death occurs in 3 to 9 months.

The infectious agent in some persistent infections may enter a **latent** state in which there is no shedding and the symptoms of disease are not present. This latency may be of the intermittent type, for example, the herpesvirus that is the causative agent of coldsores. The symptoms subside following the initial infection, but the virus travels to local neurons, where it remains concealed from the immune system. The virus can be intermittently activated by such factors as sunlight, hormonal changes (menstruation), and anxiety. The virus then migrates down the nerve fiber to the skin surface, where it multiplies and causes a recurrence of coldsores. Another type of latent infection is one in which the agent remains quiescent for a number of years and then becomes activated. The disease chickenpox, for example, is a childhood infection in which the viral agent remains latent for years until adulthood. At this time, under certain conditions, it may erupt into a disease called shingles.

A brief synopsis of the factors and events associated with the infectious disease cycle is presented in Table 26-3.

Table 26-3 *Outline of factors and events in the infectious disease cycle*

Factor	Comment
Microbial	
Dose	The number of microorganisms that cause disease affected by the susceptibility of host and initial site of entry.
Virulence traits	Related primarily to microbial surface components and host receptors that permit colonization and the ability of microorganisms to produce toxins.
Sources and reservoirs	
Animate (human)	Humans may be carriers who can transmit infectious agents directly or indirectly by contaminated food or water.
Animate (animal)	Infected animals make contact with humans. Insects may transmit infectious agent from infected animal to human.
Inanimate	Soil, food, or water contaminated by infectious agents can be ingested by humans. Disease may result from multiplication of the infectious agent in the host or from ingestion of preformed toxins (food poisoning).
Transmission	
Direct contact	Human-to-human contact such as touching, kissing, or sexual intercourse. In utero transfer from mother to fetus. Contact with infected animals or contaminated animal products.
Indirect contact	Usually inanimate articles are contaminated by humans or animals. Air contaminated by microorganisms or their spores can be inhaled.
Common vehicle	A single object (usually inanimate) is contaminated and a source of infection to many individuals.
Arthropod-borne	Infectious microorganism has one stage of development in arthropod before transmission to humans.
Host susceptibility	Genetic constitution of host influences immune response and host defense mechanisms. Very young and old are more susceptible to infection.
Release of microorganisms	To be successful, infectious microorganism must be maintained in the host and spread to others. Spread may be horizontal through contact or vertical via parent to offspring. Persistence of organism in host may be a chronic type producing various grades of symptoms for a long period of time or it may latent, in which case microbial agent is intermittently activated to produce symptoms.

CHARACTERISTICS OF INFECTION IN THE HOST

Clinical stages of infectious disease

Acute vs. chronic diseases. The clinical response of the host to an infectious disease agent may be characterized as acute or chronic. **Acute** diseases are those in which symptoms are manifested for a short period but then subside when the host's immune system has responded. **Chronic** diseases are characterized by the appearance of symptoms over an extended period of time due to an inefficient immune response.

Acute disease may be divided into four distinct periods: the incubation period, the prodromal period, the acute stage, and the convalescent period. The **incubation period** is that time between contact with the infectious agent and the appearance of the first symptoms. It is during this period that the infectious agent begins to multiply at the initial site of infection. Sometimes the initial site of infection acts as a source of dissemination of either microbial products or microbial cells to other parts of the body. Diseases in which organisms are spread throughout the body to infect organs or tissues are referred to as **systemic**. The length of the incubation period may be influenced by the initial dose of microorganisms, the rate of multiplication of the infectious agent, the immune response of the host, and the time for pathogenic factors, such as bacterial toxins, to reach a susceptible organ or tissue. The incubation period may be as short as a few hours, such as in food poisoning, or as long as several years, such as in leprosy.

The first sign of symptoms is called the **prodromal period** and usually lasts only 1 or 2 days. The symptoms might include a headache or upset stomach, but usually consist of just a general feeling of not being well.

The **acute period** represents the peak of microbial activity in the host and therefore the peak expression of host symptoms. **Fever**, which is characteristic of the acute stage in most infectious diseases, is often an indicator of the nature of the disease. Fluctuations in fever during the course of a disease are often associated with an increase of microorganisms in the bloodstream. The suffix *emia* denotes organisms in the bloodstream (for example, bacteremia, fungemia, or viremia), but if the bloodstream serves as a site for microbial multiplication, the term *septicemia* is applied.

The **convalescent period** is a stage of recovery from disease and is characterized by a rapid decline in fever and/other symptoms because of an effective response to the infection.

Not all infections by a microorganism result in acute disease manifestations. Sometimes the host may demonstrate no symptoms or minor symptoms, such as a slight fever or cough. The immune system of the host, probably because of previous encounters with the microbe, responds rapidly, and the infection ceases. Such diseases are called **subclinical** or **inapparent**.

Some diseases reach a dead end in the host because they are not transmissible from one host to another. Such diseases are called **noncommunicable.** For example, humans acquire tetanus by puncture wounds contaminated with spores of *Clostridium tetani* but the disease cannot be transmitted from the infected individual to another host. Diseases that are transmissible from one individual to another, (for example, measles) are called **communicable**.

CHARACTERISTICS OF INFECTIOUS DISEASE IN A POPULATION

Types of infections

Epidemiological investigations may be concerned with diseases that are divided into three types based on their occurrence in a population: epidemic, endemic, and pandemic. **Epidemics** are defined as either small or large numbers of cases of disease that are always in excess of what is expected or what is normal for that population. **Endemic** refers to disease that is continuously present in the population, and the frequency of disease remains at predictable levels (endemic levels) within the community. **Pandemic** implies overwhelming outbreaks of disease that involve countries or are worldwide. Influenza, for example, has been implicated in several pandemics.

Patterns of infectious disease

The distribution pattern of infectious disease in a community may be influenced by many factors, including the number of susceptible hosts, the population density (the frequency and closeness of contacts), the communicability of the agent, and, if relevant, the density of the vector, for example, mosquitoes.

The introduction of a new infectious agent into a community can result in a very explosive epidemic because few if any individuals will show resistance. Over a period of time, as the outbreak continues and people recover from the infection, a sufficient number of immune persons will be present to act as a barrier to the continued rapid spread of the disease among susceptible persons. Furthermore, if the same disease reappears, the immune individuals will prevent rapid spread of the disease, and only isolated cases among the susceptible hosts may appear. This type of immunity, which can protect susceptible persons, is called **herd immunity**. The protection is lost, however, if the infectious agent is highly communicable or if the community is in very close contact.

The frequency with which epidemics occur is a reflection of the immune status of the population. Influenza is cyclic and appears in epidemic proportions every 2 to 4 years. The reason for this is that infection by one strain of virus confers immunity against that strain for 1 to 2 years, but as new strains appear, the population as a whole may show little resistance to the new strain.

How quickly an infectious disease spreads is usually related to the source of infection, the method of transmission, and the incubation period of the microbial

Concept Check

1. What factors influence the susceptibility of a host to an infectious agent?
2. What is meant by persistence (also see Chapter 16) and the relationship between chronic and latent disease?
3. Outline the clinical stages of infectious disease.
4. How can epidemic diseases become endemic and vice versa? What are some reasons for these reversals?
5. What factors influence how rapidly an infectious disease spreads in a community?

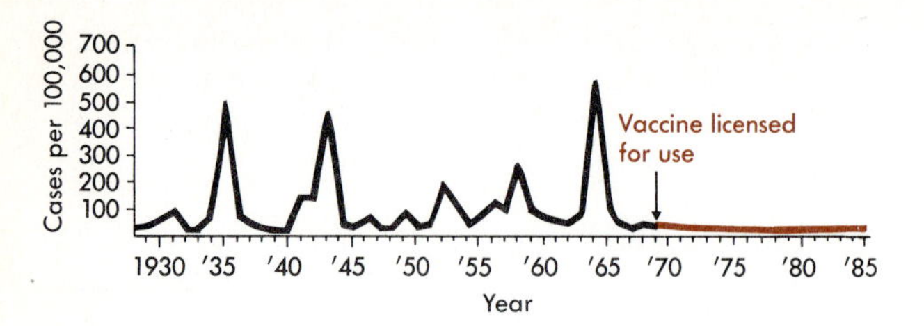

Figure 26-4 ***Rubella incidence in 10 selected areas of United States, 1928-1985. Peaks represent epidemic outbreaks, which appear in 6- to 9-year cycles. Last epidemic outbreak occurred in 1964; no further outbreaks were seen, because rubella vaccine was licensed in 1969.***

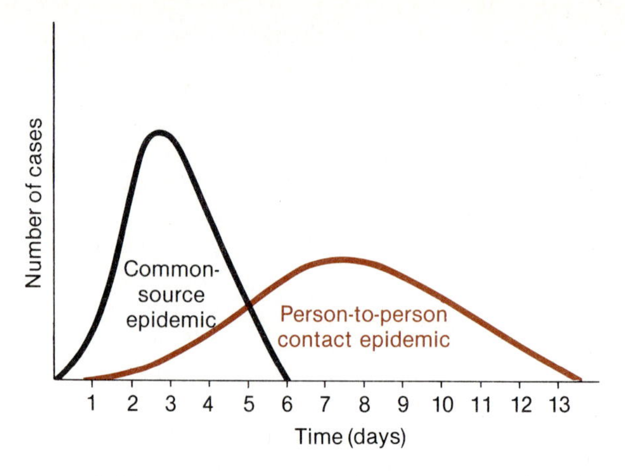

Figure 26-5 ***Comparison of epidemic curves in which there is common source and person-to-person transmission. Common source has rapid upswing in cases as compared with gradual upswing in epidemic caused by person-to-person transmission. Downswing of both curves is more gradual than upswing but is steeper in common-source epidemics.***

agent. Diseases that have a common source, such as a contaminated water or food supply, and a short incubation period will appear quickly and involve large groups of individuals. Common-source outbreaks of food poisoning caused by bacteria, such as *Staphylococcus* and *Salmonella,* result in symptoms within hours of ingestion of the contaminated food or water and frequently involve large groups of people (such as people at picnics and hospitalized patients).

Epidemiological methods

In infectious disease the causative agent, also called the **etiological agent,** is only one factor in determining the outcome of infection. Virtually every disease or condition is the consequence of several factors. Therefore the epidemiologist is concerned with analyzing these factors in a biostatistical manner. There are three main types of epidemiological investigations: descriptive, analytical, and experimental. We will discuss only the descriptive approach.

Descriptive epidemiology. Descriptive epidemiology, which is the most frequent investigation used, examines the occurrence of disease in terms of persons affected, the place at which infection occurred, and the time period over which the disease appears.

Person. Any factor that can influence the development of disease in an individual must be considered and evaluated by the epidemiologist. One of the most important factors is age, because the very young and the very old are more susceptible to infection than other age groups. Other factors that are important are sex, race, socioeconomic status, marital status, and underlying disease or conditions.

Place. The number of specific diseases that occur within a population is influenced by the geographic area where contact was made between the host and the infectious agent. Diseases acquired from animals would be more prevalent in individuals such as farmers, abattoir workers, and veterinarians than in those individuals whose occupations do not bring them in contact with animals. Determining the place of contact between host and infectious agent is not always sufficient in epidemiological investigations. A hospital-acquired disease, for example, may result from receiving contaminated medication, and the site of the contamination (factory) is more important than host contact.

Time. Time may be considered by the epidemiologist in different ways. A seasonal trend may be examined in which variations in disease occur from season to season. Respiratory tract infections occur more frequently in the winter months, when conditions such as confinement and close proximity to other individuals favor droplet transmission. Variations in disease may also be related to changes that occur over a period of years. A disease such as polio may be examined over a period of several years to determine what effect hygienic practices or immunization practices have had on the course of the disease in a population.

Epidemics are graphically represented as the number of cases of disease plotted against time (Figure 26-4). The shape of the curve is influenced by many factors, including the characteristics of the infectious disease agent, the incubation period, the susceptibility of the host, environmental factors, and the method of transmission. The first case in the epidemic is called the **index case.** A sharp increase in the number of cases occurs with diseases that have a short incubation period, and for which there is a common source, such as food contaminated by *Salmonella* organisms. The downslope of such a curve is relatively steep, but more gradual than the upswing (Figure 26-5). In diseases transmitted by person-to-person contact, such as many respiratory tract infections, the curve is characterized by a gradual up-

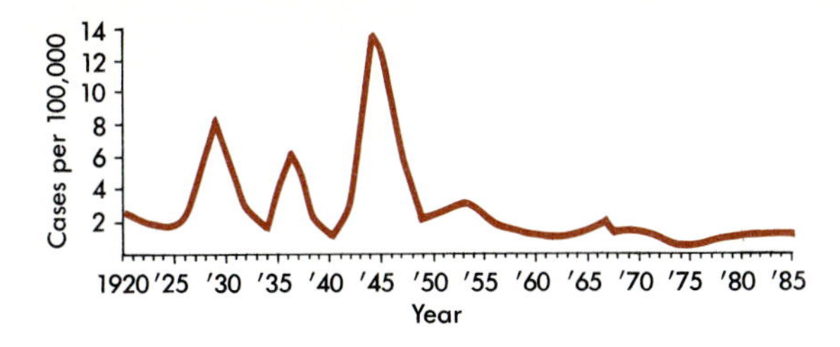

Figure 26-6 *Reported meningococcal infection rates by year, United States, 1920-1985.*

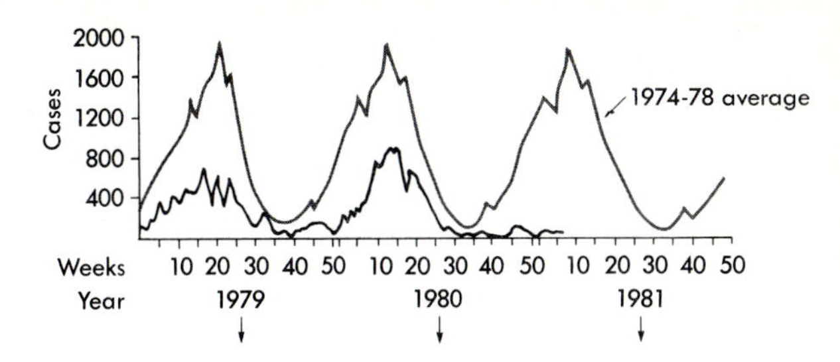

Figure 26-7 *Measles incidence, United States, 1979-1981 and 1974-1978 average.*

swing, a flatter peak, and a very gradual downswing (Figure 26-5). The downslopes of both curves represent cases with longer incubation periods and a decreasing number of susceptible persons. In diseases such as hepatitis, in which the incubation period is several weeks, the number of diagnosed cases of disease may be so spread out that it gives the impression that no epidemic has taken place.

Measurements in epidemiology

Once all of the data have been collected and tabulated, an **infection rate** can be determined. Such a rate expresses the probability of occurrence of an event and is correlated with a specific time interval and population (Figure 26-6). The rate is expressed as $k(x/y)$, where x is the number of times an event such as an infection has taken place during a specified time interval; y is the population from which the infectious disease is measured; and k is a multiplication factor based on the population measured (100, 1000, 10,000, etc.) and is used to describe the rate as a whole number rather than as a fraction. Sometimes the term **ratio** is used instead of rate to compare the frequency of one occurrence with another. For example, a **case-fatality ratio** compares the number of fatalities that has occurred in a specified number of cases. There are two types of rates: prevalence and incidence. The number of new cases that appear in a certain population over a specified period is referred to as **incidence** (Figure 26-7). The number of new cases in addition to the old cases that are recorded over a period is referred to as **prevalence**. Prevalence takes into account all of the new and old cases of disease and is influenced by the duration of infection. A patient with an infection that lasted 6 weeks would be counted only once in an incidence survey, but if a prevalence survey were taken twice during the 6-week period, the infection would be counted twice.

Concept Check

1. What factors must an epidemiologist be concerned with during an investigation of an epidemic?
2. What is meant by an infection rate? Can you give an example? What is the difference between the incidence of a disease and its prevalence in a population?
3. Explain by a curve the distinction between a disease caused by a common source and one caused by person-to-person contact.

EPIDEMIOLOGY OF HOSPITAL-ASSOCIATED INFECTIOUS DISEASES

Hospital-associated diseases are those that develop in the patient during his stay in the hospital and that were not present at the time of admission. Hospital-associated diseases are also called **nosocomial**, which is derived from the Greek; *nosos,* disease; *komeion,* to take care of. A recent study indicates that from 5% to 10% of the patients in acute-care hospitals will develop a nosocomial disease. Nosocomial diseases are an important health problem, not only because of the increased risk on the patient's life but also for economic reasons.* One of the ironies is that the improvements and advances in hospital medical care are also responsible for many nosocomial diseases. Increased antibiotic therapy, the use of therapeutic and diagnostic equipment (such as intravenous and urinary catheters), surgical procedures, and transplantation have increased the risk of nosocomial disease. However, it is important to understand that the percentage of patients who might die without these therapies and procedures is much greater than the percentage of patients who acquire diseases from their use.

* Approximately 34 million people are admitted to acute-care hospitals each year. If 5% acquire nosocomial diseases and require the average 7 days of additional care, the cost of the excess hospitalization is approximately 3.4 billion dollars per year.

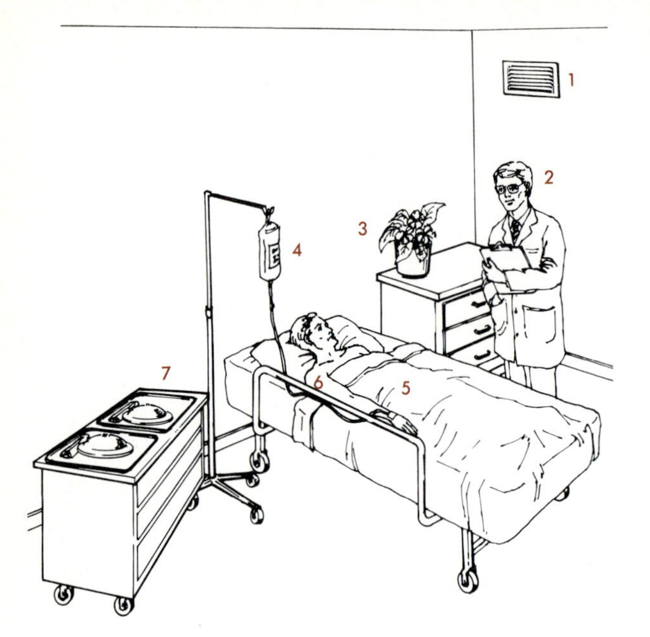

Figure 26-8 *Some possible sources of infection to hospitalized patient.* ***1,*** *Contaminated air flowing from airducts.* ***2,*** *Physician or others who carry microorganisms.* ***3,*** *Microorganisms contaminating plants.* ***4,*** *Contaminated intravenous fluids.* ***5,*** *Contaminated intravenous catheter.* ***6,*** *Patient's own microbial flora.* ***7,*** *Food contaminated by food handlers.*

PRINCIPAL MICROORGANISMS INDIGENOUS TO THE ORAL CAVITY AND RESPIRATORY TRACT

BACTERIA

Actinomyces species
Acinetobacter species
Corynebacterium species
Branhamella species
Fusobacterium species
Lactobacillus species
Moraxella species
Mycoplasma species
Peptostreptococcus species
Spirochetes
Streptococci such as enterococci and *S. pneumoniae*
Veillonella species
Spirochetes
Streptococci such as enterococci and *S. pneumoniae*
Veillonella species
Viridans streptococci

FUNGI

Candida albicans and other yeasts

PROTOZOA

Entamoeba gingivalis
Trichomonas tenax

Sources of infection

The organisms causing disease in the hospital may be from endogenous or exogenous sources (Figure 26-8). **Endogenous diseases** are those caused by the patient's own flora, and **exogenous diseases** are those that arise from sources other than the patient. The microbial flora (see boxes) of the patient, which is made up of microorganisms that are considered "nonvirulent," can cause endogenous diseases under circumstances in which (1) the organism is displaced from its normal habitat (under such circumstances the new habitat often promotes the rapid multiplication of the organism); (2) conditions alter the environment of the organism, such as an irritant (foreign object) or alteration of the blood supply to a body site; (3) multiple or broad-spectrum antibiotics reduce the number of competing microorganisms; and (4) the immunological defenses of the host have been suppressed by drugs, such as steroids, or by radiation therapy.

Exogenous sources of infection may be classified into animate and inanimate. **Animate** sources of infection include (1) other patients in the hospital with infectious disease who may represent a risk to other patients as well as hospital staff, (2) hospital staff who serve as a

PRINCIPAL MICROORGANISMS INDIGENOUS TO THE INTESTINAL TRACT

BACTERIA

Achromobacter species
Aeromonas species
Alcaligenes faecalis
Bacteroides species
Clostridium species
Corynebacterium species
Gram-negative enteric bacilli such as species of *Proteus* and *Enterobacter*
Enterococci
Flavobacterium species
Fusobacterium species
Lactobacillus species
Mycoplasma species
Peptostreptococcus species
Pseudomonas aeruginosa
Staphylococcus aureus
Viridans streptococci

FUNGI

Candida albicans and other yeasts

Table 26-4 *Types of microorganisms associated with certain exogenous sources*

Source	**Common types of microorganisms**
Outside air	Fungi, bacterial spores, soil-associated bacteria and fungi
Urinary catheters	Bacteria associated with contaminated antiseptics, instruments, drainage tubes, and/or collecting bags; *E. coli,* enterococci, and species of *Proteus, Enterobacter, Klebsiella, Serratia, Pseudomonas,* and *Candida*
Intravenous therapy equipment	Gram-negative bacteria, species of *Enterobacter, Flavobacterium, Pseudomonas, E. coli, S. aureus,* and *S. epidermidis*. Species of *Candida,* and the hepatitis B virus
Respiratory therapy equipment	*S. aureus, Streptococcus pneumoniae,* and *Pseudomonas*
Water systems (such as water softeners, deionization units, hemodialysis machines, and hydrotherapy equipment)	Gram-negative bacteria and associated bacterial endotoxins, species of *Acinetobacter, Serratia, Aeromonas, Klebsiella, Escherichia,* and the hepatitis B virus
Antiseptics	Gram-negative bacteria, species of *Pseudomonas, Flavobacterium, Acinetobacter, Klebsiella,* and *Serratia*; *Candida*

PRINCIPAL MICROORGANISMS INDIGENOUS TO THE GENITOURINARY TRACT*

BACTERIA

Acinetobacter species (V)
Bacteroides species (EG)
Clostridium species (V)
Corynebacterium species (V, EG, AU)
Gram-negative bacilli such as species of *Proteus, Escherichia, Serratia* (V, EG, AU)
Enterococcus (V, EG, AU)
Haemophilus vaginalis (V, AU)
Mycobacterium species (V, AU, EG)
Neisseria species (V, EG, AU)
Peptostreptococcus species (V, EG)
Staphylococcus aureus (V, EG)
Staphylococcus epidermidis (V, EG)
Streptococcus agalactiae (V)
Viridans streptococci (V, EG)

FUNGI

Candida albicans and other yeasts (V, EG, AU)

PROTOZOA

Trichomonas vaginalis (V, AU)

*Specific sites are indicated as follows: *V,* vagina; *EG,* external genitalia; *AU,* anterior urethra.

PRINCIPAL MICROORGANISMS INDIGENOUS TO THE SKIN

BACTERIA

Corynebacterium species
Gram-negative enteric bacilli
Mycobacterium species
Propionibacterium acnes
Staphylococcus aureus
Staphylococcus epidermidis
Viridans streptococci

FUNGI

Candida albicans and other yeasts
Trichophyton species

reservoir of potentially infectious microorganisms, and (3) visitors who not only may serve as a source of infection but also may themselves be at risk of acquiring an infection from the patient.

There are many **inanimate** objects within the hospital environment that are sources of infection to the patient (Table 26-4). These sources may also represent reservoirs where microorganisms, particularly bacteria, replicate and maintain themselves. Bacteria can grow in a variety of moist environments, including water softeners, whirlpools, sinks, soap dispensers, hemodialysis machines, and humidifers. Moist environments support the growth of many gram-negative bacteria, such as species of *Pseudomonas, Klebsiella, Serratia,* and *Flavobacterium,* which can survive in water for up to 1 year. These organisms are especially troublesome because they are able to resist many disinfectants, such as chlorine and quaternary ammonium compounds, which are used as antimicrobials in the hospital. One of the organisms, *Pseudomonas aeruginosa,* deserves special mention. This organism can be regarded as a water organism that has become one of the most important causes of nosocomial infections (see Point of Interest).

In addition to the usual patient-care supplies and equipment, food may be a direct or indirect source of infection. Food may be contaminated before coming to the hospital and, when improperly cooked, can be a source of enteric infections from such bacterial agents as *Salmonella* and *Shigella.* Food may also be contaminated in the hospital by food handlers who may be carriers of infectious agents. Nosocomial disease from contaminated food occurs infrequently.

Transmission of infectious agents

Microorganisms can be transmitted within the hospital environment by any of four avenues associated with infections outside the hospital. Some microorganisms may be transmitted by more than one route; therefore the epidemiology of nosocomial infections may vary from one epidemic to another.

Contact transmission. The hospitalized patient can be infected by direct or indirect contact with microorganisms. The hands of the physician or nurse may be inadequately washed after contact with a contaminated environment, and microorganisms may be transferred directly to the patient. Indirect contact implies that the transfer of microorganisms to the patient takes place by means of an intermediary object. Any type of patient-care supply or therapeutic and diagnostic device may be contaminated by microorganisms. These sources include objects such as needles, syringes, bedding, catheters, cystoscopes, rectal thermometers, bedpans, antiseptic solutions, medicines, and shaving brushes.

Common-vehicle transmission. Common-vehicle transmission refers to the transmission of multiple cases of a disease in which the source of infection is usually contaminated food or water, or medicines or blood. One of the severest nosocomial epidemics in the United States occurred in 1970 and 1971 when 378 cases of septicemia resulted in 40 deaths. The common vehicle was contaminated infusion fluid in which a screw-cap closure was improperly sterilized. Blood and blood products can also be contaminated with microorganisms directly or indirectly and when used for transfusions are a frequent cause of fatal infections. Bacterial species such as species of *Staphylococcus* and *Pseudomonas* and viral agents such as the hepatitis B virus are the most common cause of diseases resulting from contaminated blood or blood products.

Air transmission. The air may serve to transmit any infectious agent that can survive outside the host for at least a short interval of time. Many fungal and bacterial spores are found naturally in outside air and can be spread in the hospital by improperly designed ventilating systems. Other microorganisms can be suspended in droplet nuclei or dust particles and, depending on air currents, may be transmitted for long distances. Respiratory agents, such as those causing tuberculosis, are spread by coughing or sneezing. Staphylococci, which are found on the skin, can be spread following desquamation (normal peeling of the skin), particularly in patients with eczema and related skin conditions. Patients with open wounds, such as burn patients or surgical patients, are the most vulnerable to infection by the airborne route.

Vector-borne transmission. Vector-borne transmission is unreported in the United States but is important in many underdeveloped countries of the world. *Salmonella* and *Shigella* can be carried on the body parts of flies, and some vectors can transmit disease agents, such as those causing malaria or plague, via the bite of the arthropod.

Microorganisms involved

Bacteria are the leading cause of nosocomial diseases, and viruses are a distant second (Figure 26-9). Occasionally fungi cause disease, but rarely are protozoa involved. One of the factors that make the epidemiology of nosocomial diseases so much different from community-associated diseases is that the majority of microorganisms involved are considered noninvasive. Diseases can be caused by microorganisms that do not elaborate potent toxins or enzymes that permit invasion of epithelial tissue or spread within tissue. Most nosocomial diseases in the United States are caused by **gram-negative bacilli**, such as *Escherichia coli,* which under ordinary circumstances are harmless commensals in the intestinal tract. Disease is also related to the dose, or number, of microorganisms that are initially involved at the site of infection. The dose necessary to cause nosocomial disease is often much smaller than the dose required to cause community-associated diseases, but this is also influenced by other factors, such as the mode of transmission, host susceptibility, and site of infection. Organisms that are naturally invasive and virulent will cause diseases whose symptoms appear rapidly and that often terminate in a fatal response. The specific microorganisms associated with various sites of infection are discussed later.

Point of Interest

PSEUDOMONAS AERUGINOSA

Pseudomonas aeruginosa is a persister in the environment and can be found in water, on vegetables, in the soil, and in the intestinal tract of humans. Its survival capacity in the hospital is remarkable and it is one of the factors responsible for its association with disease. It is also a highly virulent organism for patients whose immune systems have been suppressed either by underlying disease, by drugs used to treat neoplasias, or by drugs used prior to transplantation procedures. The organism is also a major pathogen for patients with cystic fibrosis and for burn patients.

The widespread distribution of *P. aeruginosa* is due to its ability to use a wide variety of organic nutrients, to adhere to solid surfaces in aqueous environments (biofilms), and its survival in high concentrations of salt. Adherence is due to the presence of an extracellular polysaccharide composed of alginate (polymer of glucuronic and mannuronic acids). The surface polysaccharide is complex and renders the organism resistant to enzymatic degradation, phage attack, and antibiotic penetration. Once the organism attaches to a surface it divides and produces microcolonies (biofilms) encased by the secreted polysaccharide. The microcolonies continuously release swarmer cells to colonize new surfaces (see figure below). In individuals with normal immune systems, swarm cells are rapidly killed. The protected microcolony (for example, in the intestine) remains intact but other microcolonies are not produced. When immune defenses are depressed, swarm cells can attach to various organs and tissue to produce other microcolonies. Most antibiotic treatment can kill the swarm cells but the organisms within the microcolony are often protected. One of the challenges to the pharmaceutical industry is to develop antibiotics that will penetrate the microcolony or biofilm.

P. aeruginosa has several potential virulence factors, none of which by themselves is preeminent in disease. Microcolonies contain large numbers of microorganisms and thus a considerable amount of each virulence factor will be produced. The major virulence factors are:

1. Extracellular polysaccharide, which is antiphagocytic
2. Outer membrane lipopolysaccharide, which is resistant to the bactericidal action of serum components
3. Extracellular proteases, which can cleave immunoglobulins and which may also cause necrosis of local tissue
4. Exotoxin A, a protein which is toxic to macrophages and inhibits protein synthesis in a manner similar to diphtheria toxin (catalyzes the transfer of ADP-ribose from NAD to eukaryotic proteins)

Both mucoid and nonmucoid strains of *P. aeruginosa* can cause disease. Invariably, mucoid strains are isolated from patients with cystic fibrosis. Patients with this disease appear to be first colonized by nonmucoid strains that later are replaced by mucoid strains. Cystic fibrosis is an inherited disease characterized by abnormal secretion of exocrine glands. When the lung is affected, a thick mucus is secreted that can block bronchioles. Secondary infections caused by *P. aeruginosa* can exacerbate this problem and cause permanent changes in the lung.

P. aeruginosa, because of its resistance to many antimicrobials, was at one time an important cause of death in burn patients. The administration of a vaccine called *Pseudogen* combined with hyperimmune globulin has caused a 60% decrease in Pseudomonas septicemia. Mortality from infection by *P. aeruginosa* has been virtually eliminated in this group of patients.

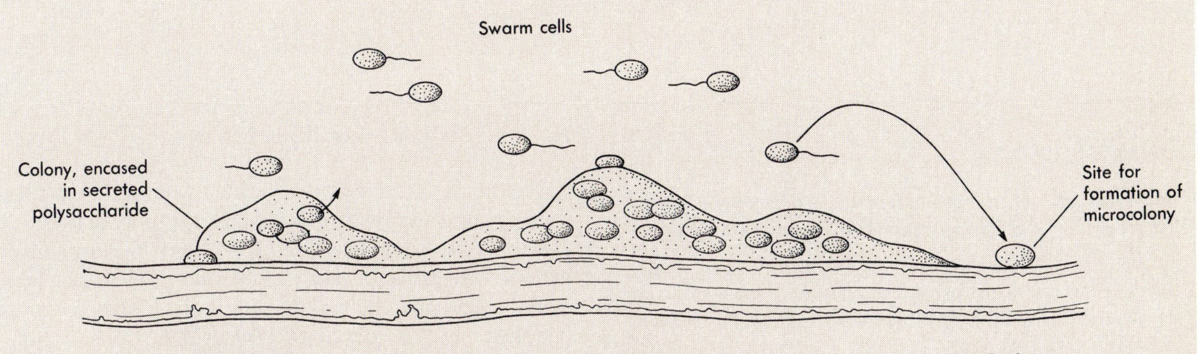

Diagrammatic representation of microcolony or biofilm of Pseudomonas aeruginosa *attached to intestinal wall.*

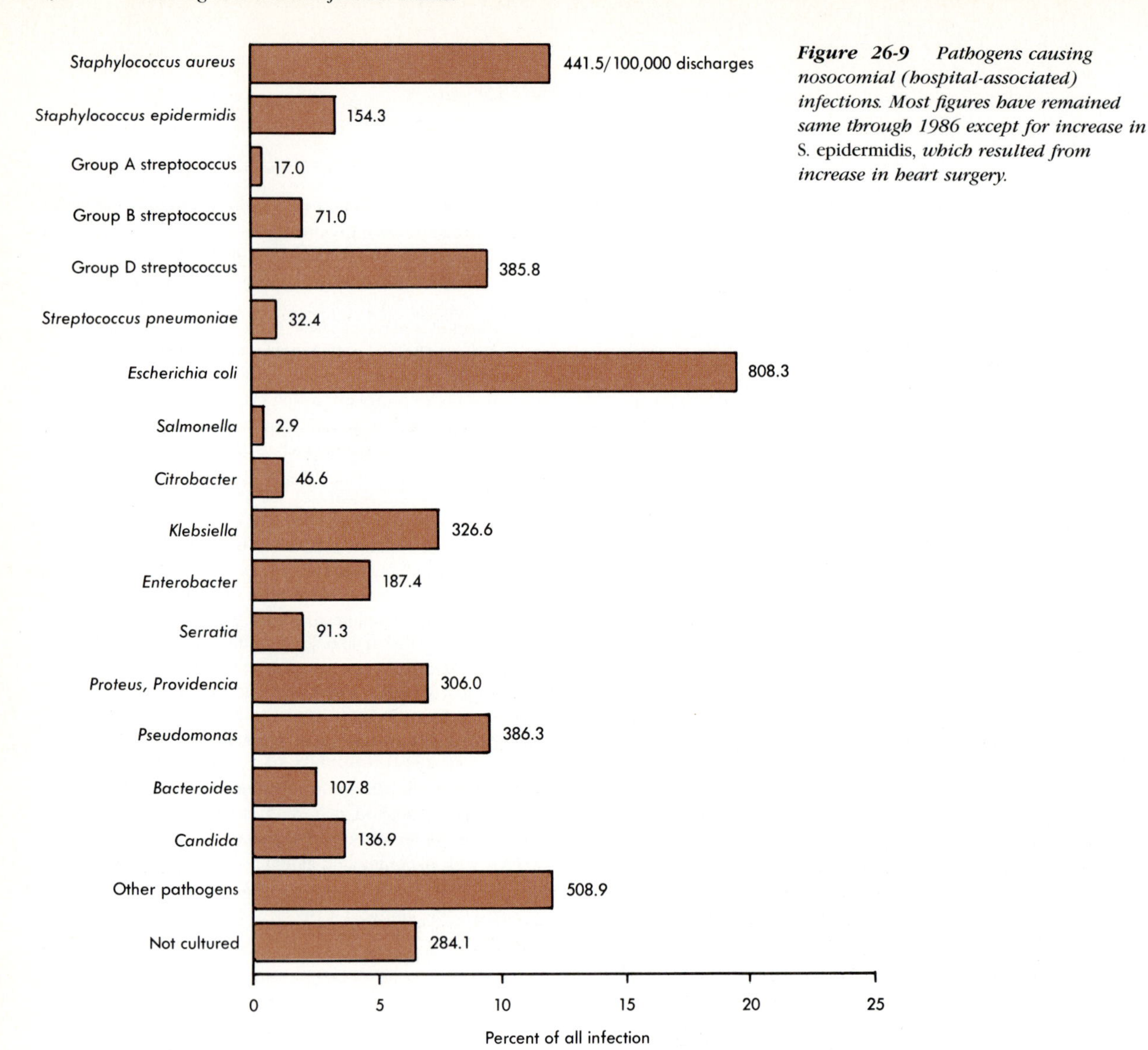

Figure 26-9 *Pathogens causing nosocomial (hospital-associated) infections. Most figures have remained same through 1986 except for increase in* S. epidermidis, *which resulted from increase in heart surgery.*

Concept Check

1. What are some of the most important endogenous species of bacteria and fungi that are a cause of hospital-associated disease?
2. What are the principal exogenous sources of infection to the hospitalized patient? What group of microorganisms is associated with each source?
3. What are the characteristics of *Pseudomonas* species that make them troublesome for hospitalized patients?
4. Which group of bacteria are the most common cause of hospital-associated disease? Why?

Host factors contributing to disease

The ultimate factor that determines whether a microorganism can cause disease is the susceptibility of the host. Susceptibility to infection is increased in hospitalized patients primarily because of (1) age, (2) presence of some underlying chronic disease such as diabetes or alcoholism, and (3) conditions (transplantations, tumors) that require the use of drugs that suppress the immune system.

Hospital procedures contributing to disease

There has been an increase in recent years in the number of mechanical devices that are inserted into or come

Table 26-5 *Hospital devices and procedures leading to disease in the patient*

Device or procedure	Function	Organisms most frequently involved
Catheter	To relieve temporary obstruction or obtain urine specimen	Nearly 70% caused by gram-negative bacteria; over 30% caused by *E. coli*
Nebulizers	Used to saturate gases with water vapor for dispensing medication in aerosol form	Gram-negative bacteria and staphylococci major cause of infections; most infections result from contamination of equipment by hospital personnel
Prostheses	Artificial devices such as heart valves, and hip, knee, and finger joints inserted to correct body defects	Totally 80% of infections caused by staphylococci
Hemodialysis units	Devices for removing waste products such as those accumulating from kidney disease	Bacteria such as staphylococci and species of *Pseudomonas* most frequent cause of infection; hepatitis B virus also common cause of disease
Endoscopes	Instruments used to examine mucosal areas of various body orifices such as colon and bladder	Enterococci from patient's flora most frequent cause of infection; *Pseudomonas aeruginosa* is cause when endoscope contaminated by physician or nurse
Surgery	Removal or repair of various body tissues or organs	*Staphylococcus aureus* and *E. coli* together account for over 30% of infections

Figure 26-10 *Scanning electron micrograph of catheter removed from patient. Spherical-shaped cells (yeast—*Candida albicans*) adhere to catheter's surface.*

in contact with the hospitalized patient. An abundance of drugs and solutions have also been developed that keep people alive who otherwise would have died from illness, injury, or old age. The single most important hospital procedure outside of surgery that can make the host prone to infection is **catheterization. Catheters** are flexible tubular devices used to withdraw fluid such as urine from the urinary bladder. Catheters are also used to introduce fluids into the body such as in the intravenous feeding (parenteral nutrition) of a patient or the administration of medications. Urinary catheters are responsible for over 75% of hospital-associated urinary tract diseases. Bacteria and fungi are the major cause of catheter-associated diseases and are frequently derived from the patient's own microbial flora (Figure 26-10).

A second important hospital procedure leading to hospital-associated infections is antibiotic administration. One cannot overestimate the value of antibiotics and related drugs in the treatment of disease. However, the overuse or inappropriate administration of antibiotics can compromise the patient and place him or her in a life-threatening situation. Keeping the patient on antimicrobial therapy for too long a time creates an imbalance in the host's microbial flora. Prolonged use of drugs such as aminoglycosides or broad-spectrum antibiotics reduces the microbial population and permits colonization and subsequent infection by multiple drug-resistant organisms (see Chapter 15).

Those devices and procedures that most frequently lead to hospital-associated infections and the microorganisms involved are outlined in Table 26-5.

Table 26-6 *Sites of infection in the hospitalized patient and organisms involved**

Site	Principal microorganisms	Leading cause of infection or underlying condition
Urinary tract	Gram-negative bacteria such as *E. coli, Proteus* species, *Klebsiella* species, and *Pseudomonas* species	Catheterization
Lower respiratory tract	Gram-negative bacteria such as *Klebsiella* species, *Pseudomonas* species, and *E. coli*, Gram-positive bacteria such as staphylococci	Inhalation of aerosols generated by coughing or sneezing or by nebulizing equipment; aspiration of organisms from oropharynx; invasion of lung via bloodstream
Surgical wound	*Staphylococcus aureus* and *E. coli* equally important	Displacement of patient's own microbial flora into the wound; punctured surgical gloves of physician; contaminated instruments
Blood (bacteremia)	Gram-negative bacteria such as *E. coli, Klebsiella* species, and *Pseudomonas* species	Hospital procedures such as intravenous infusion, catheterization, hemodialysis, and surgical procedures
Burn wounds	*Pseudomonas aeruginosa* and *Staphylococcus aureus*	Degree of burn (first or second); burns cause depression of immune system; burn tissue often infected because of transmission from hospital personnel
Skin and eyes	*Staphylococcus aureus* in neonates and gram-negative bacilli in adults	Infection primarily in neonates; hospital personnel transmit infectious agents to patient via hands; infections may also involve intravenous infusion sites

*Sites of infection and organisms involved are listed in order of importance.

Principal site of infection and the organisms involved

Any organ or tissue of the body may be affected by infectious agents. In the hospital there are specific sites that become infected because of a hospital procedure or because of the underlying condition of the patient. The urinary tract is the most common site of infection during hospitalization. The principal sites of infection and the organisms involved are outlined in Table 26-6.

Routine techniques used to trace the source of infection

Precise identification of microorganisms during epidemiologic studies is an important aspect in the control of hospital-associated infections. Its purpose is to establish differences and similarities between the organism causing disease and organisms isolated from suspected sources. For example, if a cluster of patients in one or more wards of a hospital become ill from infection by a bacterial species, it is important to determine if there is a common source for the transmission of the infectious agent. Specimens are taken from the patients and from attending physicians and nurses and others coming in contact with the affected patients, and samples are also taken from various devices, solutions, and other inanimate objects with which the patient may come into direct or indirect contact. The organisms causing disease in the patient are isolated and identified. The same organism is sought from the other sources and is then subjected to a number of different tests to determine if there is a common source for this hospital epidemic. A variety of different tests can be used to "type" the organism; that is, produce a strain profile. Not all the tests are applicable for identification of all species. One can identify strains by (1) **biotyping**, or determining the pattern of biochemical activity of the organism; (2) **antibiograms**, or determining antibiotic susceptibility patterns; (3) **serotyping**, or identification of strain-specific antigens such as those on the cell surface; (4) **bacteriophage typing**, or determining a bacterial strain's sensitivity or resistance to bacteriophage; and (5) **bacteriocin typing**, or determining the ability of strains to produce specific bacteriocins.

Bacteriophage typing has been used with much success in the identification of strains of *Staphylococcus aureus*. Bacteriophage typing is carried out by taking an inoculum from the culture of the staphylococcal strain to be tested and spreading it on an agar surface so that at the end of an 18-hour incubation period there will be confluent growth on the agar surface. Before the incubation period, however, a drop of different staphylophage is placed at properly identified and demarcated

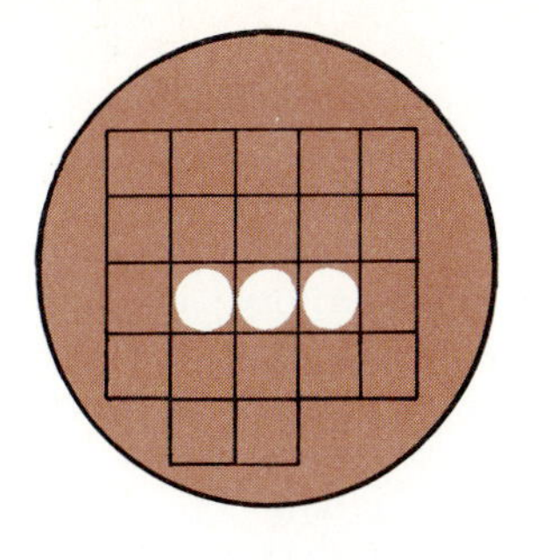

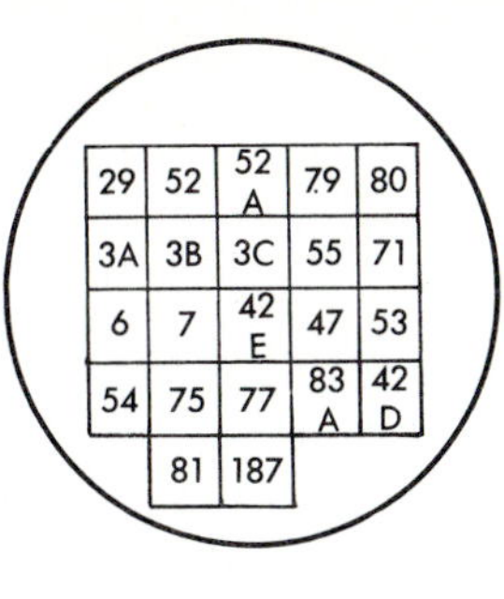

Figure 26-11 *Phage typing of staphylococci. Agar plate is marked off in squares. Plate is then streaked with* S. aureus *and allowed to dry. Specific phage suspensions, corresponding to strains indicated on right, are placed in appropriate squares. After being allowed to dry again, plate is incubated for 12 to 18 hours. Plate on left shows confluent growth except where plaques appear in squares corresponding to phage preparations 7, 42E, and 47.*

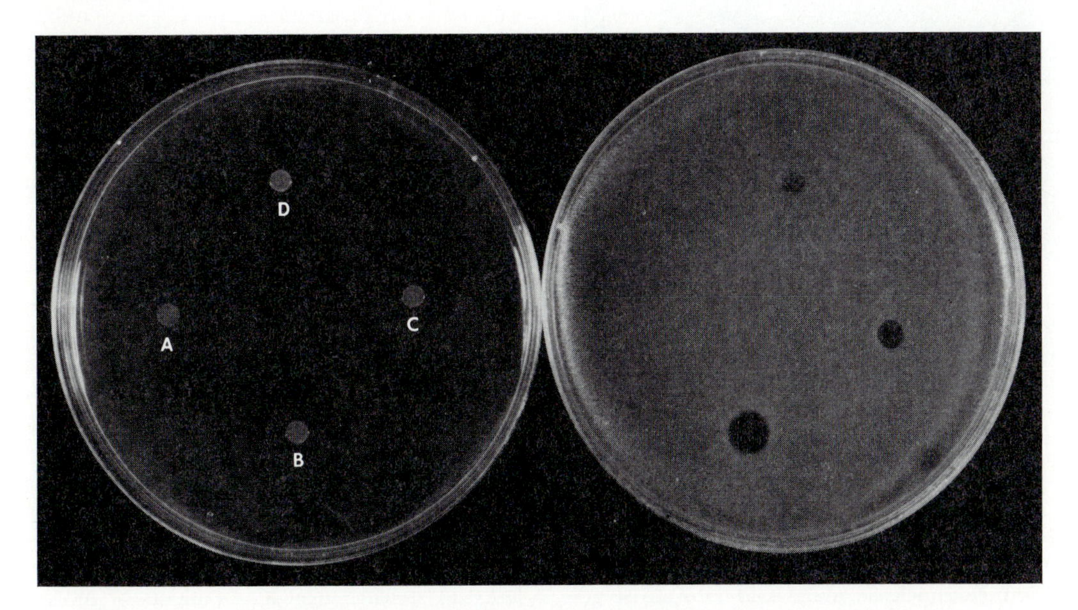

Figure 26-12 *Bacteriocin typing. Examination of pyocin activity against indicator strain in four strains of* Pseudomonas aeruginosa. *Left plate shows four test strains after 6 hours incubation at 30° C and before addition of indicator strain. Right plate shows inhibition zones produced by four strains after addition of indicator strain in agar overlay. Strain A shows no activity; strain B shows pyocin activity extended beyond area of original growth; strain C shows restricted inhibition zone and strain D shows mottled inhibition zone indicative of bacteriophage activity.*

areas on the agar surface. The plate is examined after the incubation period. Areas of confluent growth at the site of bacteriophage inoculation indicate the bacterial strain's resistance to infection by one type of staphylophage, whereas areas of clearing represent susceptibility to the staphylophage strain (Figure 26-11).

Bacteriocin typing is used to differentiate strains of *Pseudomonas aeruginosa, Serratia marcescens, E. coli, Shigella* species and *Proteus mirabilis.* Bacteriocins, which are antimicrobial-like proteins produced primarily by gram-negative bacilli, are active against strains of the same or closely related species. The test can be performed by inoculating a known bacteriocin-producing strain on an agar plate. Next clinical isolates (strains isolated from various sites or individuals suspected of being the source of an epidemic) are inoculated on the plate for overnight incubation, and during the period bacteriocin diffuses into the agar. The plate is read for growth inhibition (Figure 26-12).

Other techniques used to trace the source of infection

Many times the routine typing systems discussed are not applicable to the identification of some microbial species. The hospital epidemiologist must then resort to other techniques. One of the newly developed techniques is plasmid characterization, a technique that has application in basic research. Techniques involving plasmids fall into two categories, direct and indirect.

Direct plasmid analysis involves the hybridization of plasmid DNA from different strains which allows the

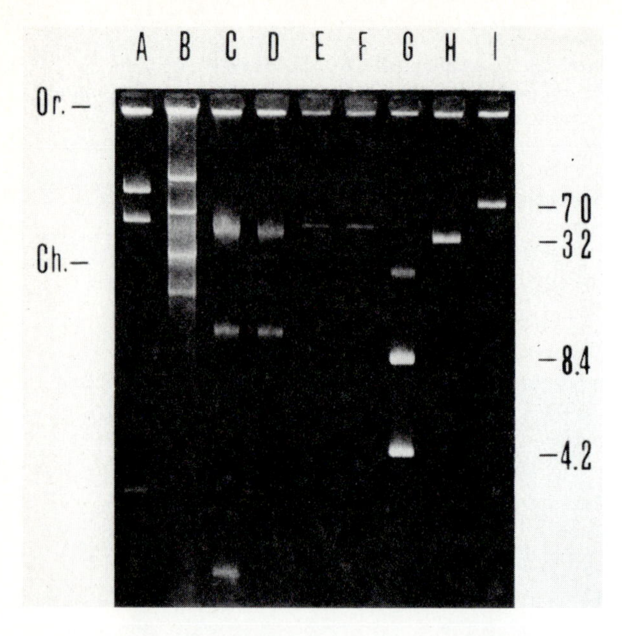

Figure 26-13 *Agarose gel electrophoresis of plasmids extracted from various species of clinical isolates. Lanes A and B,* Pseudomonas aeruginosa *strains; C and D,* Staphylococcus aureus *strains; E and F,* Haemophilus influenzae, *strains. Lane G represents three different plasmids used as molecular weight standards. Numbers indicate molecular weights ($\times 10^6$).* Or, *Origin of electrophoresis;* Ch, *chromosomal DNA.*

quantitative assessment of base sequence homology. Indirect techniques include agarose gel electrophoresis and restriction endonuclease analysis. In the agarose gel technique the clinical isolates are cultivated and lysed and then the supernatant is electrophoresed on agarose gel. The gel is washed and stained with ethidium bromide, transluminated with ultraviolet light, and photographed. Plasmid size is determined by comparing gel migration distances (Figure 26-13) of unknown plasmids with that of plasmids of known molecular size. Plasmids of the same size may have entirely different base sequences, and these differences can be detected by using restriction endonucleases. Restriction endonucleases cleave double-stranded DNA at specific recognition sites and produce a number of plasmid fragments. If two plasmids are of the same size and yield identical patterns of fragments on restriction endonuclease analysis, they are assumed to be identical or nearly so. If the plasmids of equal size have different sequences, then different fragments will be produced from endonuclease digestion.

Individually many of the techniques for differentiating strains of a particular species are not sufficient to provide an absolute differentiating procedure. Usually a combination of two or more procedures is used in epidemiologic studies. In addition a procedure such as antibiograms is influenced by the antibiotics used in the hospital. This method of identification is most likely to be useful only within a localized environment.

Prevention

Accredited hospitals are now required to have an infection-control program that includes (1) surveillance of nosocomial diseases, (2) employee health surveillance (3) an isolation policy for certain diseased patients, (4) education of hospital personnel concerning nosocomial diseases, (5) a microbiology laboratory, (6) environmental sanitation procedures, and (7) accepted techniques for certain hospital procedures, such as catheterization. A discussion of each of these aspects is beyond the scope of this book, but some general procedures that are being used to identify and control nosocomial diseases are highlighted below.

Surveillance. Hospitals employ infection-control practitioners, such as nurses, physicians, or public health officials, to perform surveillance tasks. Surveillance may use different sources to collect data that are processed and disseminated to an **infection-control committee.** The sources of nosocomial information should include (1) reports from the microbiology laboratory; (2) antimicrobial data from the pharmacy; (3) autopsy reports; (4) ward rounds to determine the presence of fevers, types of hospital procedures being performed, antibiotic regimens, and underlying disease of the patients; (5) the employee health clinic; and (6) the outpatient department. Special surveillance activities may also include determining the spread of antibiotic resistance in specific groups of patients or monitoring the occurrence of a viral infection, such as varicella, in a specific group. Once the data have been tabulated and assessed, appropriate measures can be taken to educate hospital personnel and design methods of prevention.

Techniques to prevent nosocomial infections

Handwashing. The single most important procedure that can aid in the prevention of hospital-associated diseases is the practice of handwashing. Nurses, physicians, and other hospital personnel are frequently in contact with patients and thus serve as vectors in the transmission of infectious agents from one patient to another. The handwashing procedures vary, depending on the task to be performed. Handwashing with soap and water is sufficient when the nurse is going to bathe the patient or merely change the bed linen. Handwashing with antiseptic soaps that contain either 3% hexachlorophene, an iodophor, or 4% chlorhexidene should be used before procedures such as surgery or the manipulation of an invasive device such as a catheter. Contact with a lesion or a draining skin infection requires the use of gloves. For special cases, such as patients with hepatitis, gloves are also recommended for persons performing such duties as drawing blood, dismantling dialysis machinery, or changing bedpans.

Disinfection and antiseptic procedures. Equipment and devices used in the diagnosis of patients require decontamination before use, but many of these cannot be autoclaved and require special disinfection techniques. Respiratory therapy equipment can be sterilized by soaking in glutaraldehyde or can be gas sterilized with ethylene oxide. Nebulizers, following an initial sterilization, can be disinfected daily by separating the unit from the patient and nebulizing with 0.25% acetic acid. Depending on the item used in hemodialysis, sterilization by heat, ethylene oxide, or chemical disinfection can be used, but because of the presence of hepatitis B virus, special disinfection techniques may be employed, such as immersion in (1) sodium hypochlorite (0.5% to 1.0%) for 30 minutes, (2) 20% formalin in 70% alcohol for 18 hours, or (3) 2% aqueous alkalinized glutaraldehyde.

The insertion of devices such as cannulas or catheters can result in infection at the skin-cannula or skin-catheter junction. Antibiotic-containing ointments are used to protect against cannula-associated sepsis and are particularly effective against infections caused by staphylococci. The antimicrobials used in the ointments include neomycin, bacitracin, polymyxin, and nitrofurazone, used singly or in combination. Urinary catheters are often impregnated with antimicrobials, and sometimes antibacterial agents are instilled into the bladder to prevent disease.

Antimicrobial surveillance. There is no question that antimicrobials have had a profound influence on the character of hospital-associated disease. Antimicrobials are administered to between 25% and 38% of hospitalized patients and account for up to 35% of the total pharmaceutical cost. The kind and amount of drugs used in different hospitals may vary considerably, but nearly one half of the antimicrobials used are either unnecessary or are used inappropriately. The surveillance of antimicrobials in the hospital is a very important task, because overuse of any drug results in the selection of organisms that are resistant to it. Not only will future diseases be difficult to treat with the drug, but the patient acts as a reservoir for the drug-resistant pathogen, which can be transmitted by hospital personnel to other patients. Fortunately, the resistant microorganisms in the host are at a disadvantage in terms of their ability to compete with other microorganisms unless the antimicrobial is present. Therefore removal of the antimicrobial for a period of time from the therapeutic regimen usually results in the selection of drug-sensitive strains in the patient.

The administration of antibiotics to prevent disease, under certain circumstances, is a worthwhile and often necessary procedure. Many operations and techniques involve areas of the body where the resident flora are in very high numbers, and even minor spillage following surgery is difficult to avoid. Antimicrobial prophylaxis reduces the resident flora, and any organisms that do gain entry to normally sterile sites are destroyed by the drug present in the tissue. The situations in which antimicrobial prophylaxis is justified are indicated in Table 26-7.

Table 26-7 *Situations in which antimicrobial prophylaxis is justified and the antimicrobials used*

Procedure or condition	Antimicrobial used
Vaginal hysterectomies	Cephalosporin or penicillin
Total abdominal hysterectomies	Cephalosporin
High-risk cesarean sections	Cephalosporin
Colorectal surgery	Oral erythromycin-neomycin, kanamycin-metronidazole, or doxycycline
Vascular grafts of the abdominal aorta or lower extremity vasculature	Cephalosporin
Total hip replacement	Cephalosporin or penicillinase-resistant penicillin
Head and neck cancer surgery	Cephalosporin
Traveler's diarrhea	Doxycycline
Prevention of pneumonia caused by *Pneumocystis carinii* in susceptible cancer patients	Trimethoprim-sulfamethoxazole
Recurrent urinary tract infections in female patients	Trimethoprim-sulfamethoxazole

Concept Check

1. What influences the susceptibility of a hospitalized patient to infectious disease? What hospital procedures contribute to his or her condition?
2. What are the principal sites of infection in the hospitalized patient and the microorganism most likely to be involved?
3. What techniques are used to identify bacterial strains associated with an epidemic? What molecular approaches are being used?
4. What three major procedures are used in the hospital to prevent hospital-associated disease?

Each individual entering the hospital is equipped with physical and chemical barriers to infection and responds differently to the presence of microorganisms. Until we are able to define all the factors that contribute to the virulence of the microorganism as well as the resistance of the host, hospital-associated diseases will be inevitable. All we can do at this time is be aware of the factors responsible for such diseases and use all the measures available to us for their prevention.

SUMMARY

1. Epidemiology is the science that deals with factors associated with the frequency of diseae and how disease is transmitted. These factors are rooted in the evolutionary relationship of humans and parasites. Microorganisms are believed to have first existed as free saprophytes that, after contact with humans, developed certain relationships that were either mutualistic, commensalistic, or parasitic. Many parasites, after their initial contact with higher organisms, evolved with their host into a relationship that would enable the host to survive and permit the parasite's continued existence.
2. The development of an infectious disease is dependent on:
   a. The characteristics of the microbial agent, such as dose and virulence.
   b. The source of the microorganisms, such as animate or inanimate sources. Among animate sources, human carriers are the most important means of transmission of infectious agents.
   c. How the microorganisms are transmitted. Microorganisms can be transmitted by direct contact or indirect contact.
   d. The susceptibility of the host. The age and the genetic constitution of the individual often determine susceptibility to infection.
   e. How the microorganisms are disseminated from an infected to an uninfected host.
3. Disease in the host is characterized by various stages. Acute disease that terminates in recovery is divided into four periods: incubation, prodromal, acute, and convalescent or recovery period. Chronic diseases are characterized by the appearance of symptoms over extended periods of time, and subclinical infections are those in which the host is infected but disease symptoms do not appear because of resistance of the host.
4. Diseases in a population are divided into three groups based on their occurrence: epidemic, endemic, and pandemic. Epidemiological methods are used to determine the how, when, where, and why of an infectious disease so that techniques can be developed to prevent its spread. Descriptive epidemiology, which is the most frequently used epidemiological method, measures the occurrence of disease in terms of persons affected, the place at which infection occurred, and the time over which the disease appears.
5. Diseases acquired in the hospital are referred to as nosocomial diseases. Bacteria are the most frequent cause of hospital-associated diseases, but other microorganisms may also be involved. The gram-negative facultatively anaerobic bacilli, such as *E. coli,* are the most common cause of bacterial diseases.
6. The source of disease may be exogenous (outside the host) or endogenous (the host's own microbial flora). Transmission of infectious agents is the same in nosocomial diseases as it is in community-associated diseases. Nurses and physicians are the most frequent carriers of infectious agents.
7. Susceptibility to nosocomial infection is influenced primarily by the immunological state of the host. The immunological state can be depressed by some underlying condition or disease or by hospital-associated procedures. Catheterization and the use of intravenous devices have contributed to the majority of nosocomial diseases. In addition the improper use of antimicrobial drugs that suppress the immune response can lead to infection by drug-resistant species.

8. The principal sites of infection and the microorganisms involved are the urinary tract—*Escherichia coli* and other gram-negative facultatively anaerobic bacilli; the respiratory tract—gram-negative aerobic bacteria and staphylococci; surgical wounds—staphylococci; skin and eyes—staphylococci and gram-negative aerobic bacilli; burn wounds—*Pseudomonas aeruginosa* and *Staphylococcus aureus;* and bacteremia—*E. coli* and other gram-negative facultatively anaerobic bacilli.
9. The most important method for preventing hospital-associated diseases is handwashing by hospital staff. Because many hospitalized patients are subjected to hypodermic injection, disinfection of the site of injection is also an important means of preventing infection. In most hospitals an infection-control committee is in charge of surveying the number, types, and sites of infection in the hospitalized patients and distributing this information to hospital personnel. It is also the purpose of the committee to suggest methods for reducing hospital-associated disease. Such proposals may include reducing the use of certain antimicrobials for treatment or suggesting the use of new disinfectants.

Self-study Quiz

Multiple choice

1. Which of the following epidemiological incidence curves is most consistent with a common-source epidemic?
   A. A
   B. B
   C. C
   D. Both A and B

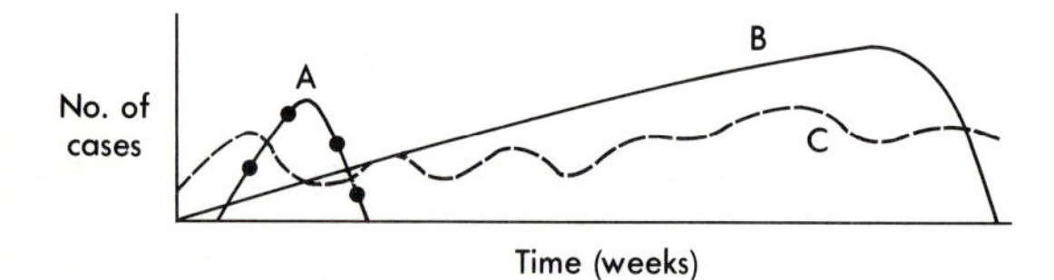

2. In the curves illustrated in question 1, what is the most reasonable explanation of the sudden decline in curve B?
   A. Interferon production
   B. Phenomenon of herd immunity
   C. Removal of the common source
   D. Development of antibiotic susceptibility
3. Which of the following is used to denote the number of new cases of a disease in a certain population over a specified period of time?
   A. Prevalence
   B. Incidence
   C. Fatality rate
   D. Latency
4. A child born with congenital syphilis is an example of which of the following?
   A. Convalescent carrier state
   B. Transmission of a noncommunicable disease
   C. Vertical transmission
   D. Horizontal transmission
5. To be classified as a reservoir, the animate or inanimate object must be:
   A. Able to support the growth of the infectious agent
   B. A biological vector
   C. A water-borne source
   D. Capable of vertical transmission
6. Which of the following is not part of the complex of virulence factors of *Pseudomonas aeruginosa?*
   A. Exotoxin A
   B. Pili
   C. Capsule (alginate)
   D. Extracellular proteases
7. What term would be most applicable to the worldwide spread of influenza during the 1918-1919 outbreak?
   A. Endemic
   B. Latency
   C. Pandemic
   D. Epidemic
8. A disease of animals that is sometimes spread to human populations is known as what?
   A. Persistent infection
   B. Fomite
   C. Intoxication
   D. Zoonosis
9. What is the most common source of hospital-acquired infections?
   A. Patient's own microbial flora
   B. Contaminated food or water
   C. The air
   D. Health care personnel's microbial flora
10. Which of the following conditions does not warrant the use of prophylactic antibiotic therapy?
    A. Vaginal hysterectomy
    B. Childbirth
    C. Total hip replacement
    D. Surgery of the colon

Completion

1. The microbial agent that causes the disease is called the ________________ agent.
2. The stage of an acute disease in which no symptoms are manifested is called the ________________ period.
3. The epidemiological term used to describe the number of deaths in a specified number of cases is called the ________________.
4. The endogenous species of bacteria that is a major cause of hospital-associated urinary tract infection is ________________.
5. Foodborne diseases caused by ingestion of preformed microbial toxins are called ________________.
6. A persistent infection in which the organism is not shed and symptoms are not present the majority of time is called ________________.
7. Patients with indwelling catheters are at most risk of infection by ________________ bacteria.
8. The least likely mode of transmission of hospital-acquired infections is ________________.
9. The principal means of typing bacteria is biotyping, antibiograms, serotyping, bacteriocin typing, and ________________.
10. The single most important procedure known to reduce hospital-acquired infections is ________________.

Thought questions

1. Recently in a Pennsylvania steel town of 45,000 people in the financially depressed Monongahela River Valley, 112 cases of giardiasis were diagnosed in a 30-day period.
   a. Outline the kind of data that an epidemiologist would gather in an attempt to define the epidemiology of the problem.
   b. What steps should immediately be taken to protect the uninfected population?
   c. After the identification of the municipal water supply as the source of the problem, people were told to drink only bottled water or boiled water; however, the number of cases continued to rise for another 15 days following enactment of these measures to reach a total of 227 diagnosed cases. How would you explain this continued rise? (Hint: the incubation period of giardiasis is about 20 days.)
   d. Do these data support a common-source epidemiological pattern or one of person-to-person contact? Explain.
   e. What is the infection rate for this particular outbreak?
2. It has been stated that the two most dangerous microbiological environments that exist in a community are the sewage treatment facility and the hospital. Defend or refute this statement.
3. Burn wards in hospitals do not allow fresh-cut flowers to be brought into the patient's room. Can you offer an explanation for this practice?

SELECTED READINGS

Books

Castle, M. *Hospital Infection Control,* John Wiley & Sons, Inc. New York, 1980.

Centers for Disease Control. *Morbidity and Mortality Reports.* Atlanta (A weekly report that discusses primarily infectious diseases, especially in the United States).

Centers for Disease Control. *Surveillance.* (Annual summaries of specific infectious diseases such as foodborne and waterborne diseases). Centers for Disease Control, Atlanta.

Centers for Disease Control, National Nosocomial Infections Study Report: Annual summary 1978. Centers for Disease Control, Atlanta, 1981.

Evans, A.S. *Viral Infections of Humans: Epidemiology and Control (2nd ed).* Plenum Medical Book Co., New York, 1982.

Mausner, J.S., and Bahn, A.K. *Epidemiology.* W.B. Saunders Co., Philadelphia, 1974.

Mims, C. *The Pathogenesis of Infectious Disease (2nd ed.).* Grune & Stratton, Inc., New York, 1982.

Smith, C.E.G. *Epidemiology and Infections.* Meadowfield Press Ltd., Shildon, England, 1976.

Journal articles

Aber, R.C., and Mackel, D.C. Epidemiologic typing of nosocomial microorganisms. *Am. J. Med.* **70**:899, 1981.

Blair, J.E., and Williams, R.E.O. Phage typing of staphylococci. *Bull. WHO* **24**:771, 1961.

Farmer, J.J., III. Epidemiological differentiation of *Serratia marcescens. Appld. Microbiol.* **23**:226, 1972.

Farrar, W.E., Jr. Molecular analysis of plasmids in epidemiologic investigations. *J. Infect. Dis.* **148**:1, 1983.

Huang, A.S. Persistent viral infections: Introduction. *Rev. Infect. Dis.* **6**(5):649, 1984.

Lillienfield, A.M. Epidemiology of infectious and noninfectious disease: Some comparisons. *Am. J. Epidemiol.* **97**:135, 1973.

Mackowiak, P.A. Microbial latency. *Rev. Infect. Dis.* **6**(5):649, 1984.

Mayer, K.H., and Zinner, S.H. Bacterial pathogens of increasing significance in hospital-acquired infections. *Rev. Infect. Dis.* **7**(suppl 3):371, 1985.

McGowan, J.E., Jr. Changing etiology of nosocomial bacteremia and fungemia and other hospital-acquired infections. *Rev. Infect. Dis.* **7**(suppl. 3):357, 1985.

Steere, A.C., and Mallison, G.F. Handwashing practices for the prevention of nosocomial infections. *Ann. Intern. Med.* **83**:683, 1975.

CHAPTER 27

Human diseases caused by bacteria and viruses

Chapter Outline

Introduction

This chapter is the first of two chapters dealing specifically with infectious disease in humans. It is not important at this stage in your study of microbiology to understand all of the symptoms or clinical aspects of each infectious disease. However, an understanding of epidemiology, host-parasite relationships, and immunology is necessary to understand the basics of all infectious diseases. In this chapter a brief overview of the infectious disease process involving bacteria and viruses is presented. In addition, some of the more important genera and species are described, as well as the special characteristics that enable them to cause infections in humans.

Table 27-1 *Major groups of bacteria that cause infectious disease in humans*

Group	Genus	Species	Disease
Gram positive			
Cocci	*Staphylococcus*	*S. aureus*	Boils, food poisoning
	Streptococcus	*S. pneumoniae*	Pneumonia
		S. pyogenes	Rheumatic fever, scarlet fever
Bacilli			
Aerobic	*Bacillus*	*B. anthracis*	Anthrax
	Mycobacterium	*M. tuberculosis*	Tuberculosis
	Nocardia	*N. asteroides*	Nocardiosis
Anaerobic	*Clostridium*	*C. perfringens*	Gas gangrene
		C. botulinum	Botulism
		C. tetani	Tetanus
	Actinomyces	*A. israelii*	Actinomycosis
Facultative anaerobe	*Listeria*	*L. monocytogenes*	Listeriosis
	Corynebacterium	*C. diphtheriae*	Diphtheria

Continued.

GENERAL CHARACTERISTICS OF BACTERIAL DISEASES

Although characteristics of the infectious disease process are described in previous chapters, the pathogenic process as it applies to infectious bacteria should be briefly summarized. Bacterial infectious agents are transmitted to humans via the respiratory tract; by insect vectors; by contaminated food, water, and fomites; and by sexual contact. One of the first steps in the infection process is the attachment of the bacterium to a host cell by bacterial surface structures called **adhesins**. Even though attachment is necessary, the pathogen must survive and replicate at the attachment site. Replication and survival on host cells is accomplished first by overcoming host defense mechanisms. Microbial defense mechanisms may take the form of proteases that degrade secretory IgA at mucosal surfaces, or they may be in the form of capsular polysaccharides that inhibit phagocytosis by professional phagocytes. For those intracellular bacterial parasites, such as the chlamydia, defense usually involves resistance to degradation by phagocytic enzymes.

Most clinically important bacterial diseases are characterized by tissue damage from the release of toxins (exotoxins or endotoxins). In addition, some bacterial antigens trigger an immunological response by the host that can also lead to severe tissue damage. Still other bacteria, such as *Shigella,* cause tissue damage by their ability to invade epithelial cells, to multiply in the host cell, and to invade adjacent cells. Dissemination from local sites of infection are clearly important in the pathogenesis of many bacterial diseases. In some cases, however, such as diphtheria, the pathogen produces an exotoxin that is transported into the bloodstream, while the organisms remain localized in the respiratory tract, the initial site of infection. Most microorganisms are disseminated from their initial site of infection to various body sites via the bloodstream.

For most students it is difficult to remember what organism causes what disease. To make this task easier, Table 27-1 provides a brief classification scheme of bacteria and the diseases of major clinical significance to humans.

Diseases acquired through the respiratory tract

Bacterial infections involving the respiratory tract are generally acquired by inhalation or ingestion of microorganisms from the air (see Point of Interest, p. 808) or by direct or indirect contact with respiratory secretions. Diseases involving an infection of the respiratory tissue include pneumonia, whooping cough, diphtheria, tuberculosis, and pharyngitis. Meningitis, although it involves the meninges (the membranes surrounding the brain or spinal cord), is caused by a bacterium that initiates infection in the upper respiratory tract but disseminates to nervous tis-

Table 27-1 *Major groups of bacteria that cause infectious disease in humans—cont'd*

Group	Genus	Species	Disease
Gram negative			
Cocci	*Neisseria*	*N. meningitidis*	Meningitis
		N. gonorrhoeae	Gonorrhea
Bacilli			
Aerobic	*Pseudomonas*	*P. aeruginosa*	Burn wounds
	Brucella	*B. abortus*	Brucellosis
	Bordetella	*B. pertussis*	Whooping cough
	Rickettsia	*R. rickettsii*	Rocky Mountain spotted fever
		R. prowazekii	Louse-borne typhus
	Coxiella	*C. burnetii*	Q-fever
	Chlamydia	*C. trachomatis*	Trachoma
Anaerobic	*Bacteroides*	*B. fragilis*	Anaerobic infections
	Fusobacterium	*F. nucleatum*	Fusospirochetal disease
Facultative anaerobic	*Escherichia*	*E. coli*	Enteritis
	Klebsiella	*K. pneumoniae*	Pneumonia
	Salmonella	*S. typhi*	Typhoid fever
	Shigella	*S. flexneri*	Dysentery
	Campylobacter	*C. fetus*	Enteritis
	Yersinia	*Y. pestis*	Plague
	Vibrio	*V. cholerae*	Cholera
	Haemophilus	*H. influenzae*	Meningitis
Spiral	*Treponema*	*T. pallidum*	Syphilis
	Borrelia	*B. recurrentis*	Relapsing fever
	Leptospira	*L. icterohemorrhagiae*	Leptospirosis
Wall-less	*Mycoplasma*	*M. pneumoniae*	Atypical pneumonia

sue. The incidence of infections caused by bacteria acquired by the respiratory tract are outlined in Table 27-2. We will discuss only the most important of these.

Pneumonia. Pneumonia is a condition in which the lungs are inflamed. Pneumonia, caused by microorganisms, may result from inhalation of contaminated dust or respiratory secretions, or it may be caused by aspiration of microorganisms that are normally present in the respiratory tract. Most pneumonias are secondary to conditions or diseases that exist in the patient, for example, bronchitis, viral infection, and alcoholism. An individual suffering from a viral disease of the upper respiratory tract will have a damaged respiratory epithelium, and resistance to infection will be reduced. As a consequence, certain microorganisms, especially bacteria, are capable of unrestricted growth and initiation of disease. The principal bacterial agent causing pneumonia is *Streptococcus pneumoniae; Klebsiella pneumoniae* and a host of other agents can also cause pneumonia (Table 27-3).

Streptococcus pneumoniae, also called the **pneumococcus,** is a natural inhabitant of the upper respiratory tract of humans. It is the primary cause of bacterial pneumonia and is involved in over 50% of the cases of pneumonia in the United States. The organism is a gram-positive coccus that appears in pairs or chains (Figure 27-1). Virulent strains produce a capsule that enables them to resist phagocytosis.

The capsule of *S. pneumoniae* consists of a polysaccharide whose composition varies among different

Table 27-2 *Bacterial diseases acquired through the respiratory tract*

Type of disease	Organism(s) involved	Average no. cases/ year (U.S.)	Number of deaths	Vaccine
Pneumonia*	*Streptococcus pneumoniae* and many others (see Table 27-3)	500,000-1,000,000	50,000-100,000	Yes
Whooping cough	*Bordetella pertussis*	1500-2400	6-10	Yes
Diphtheria	*Corynebacterium diphtheriae*	2-6	1	Yes
Tuberculosis	*Mycobacterium tuberculosis*	23,000-28,000	1800-2100	Yes, but primarily Europe
Pharyngitis,* and its complications; rheumatic fever, scarlet fever	Group A β-hemolytic streptococci	7-15 for rheumatic fever; 200,000-225,000 for streptococcal pharyngitis and scarlet fever	?	No

*Not reportable to the Centers for Disease Control.

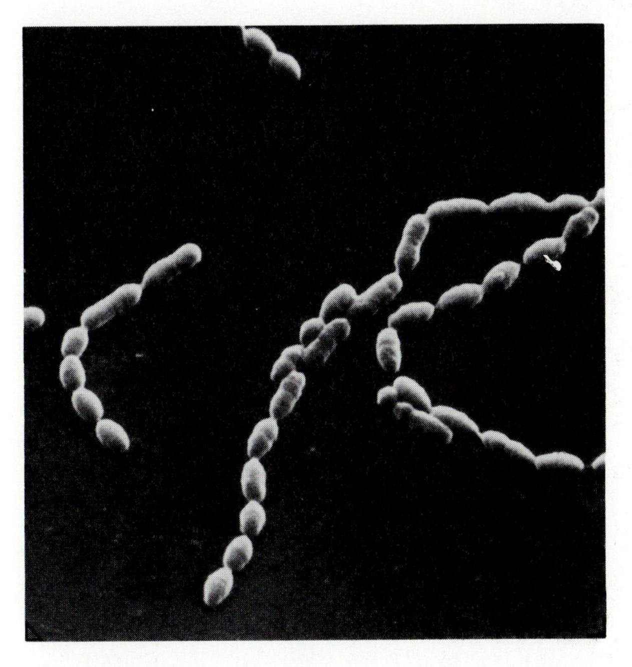

Figure 27-1 *Scanning electron micrograph of a streptococcal species. (×6,800.)*

strains There are 82 capsular types of *S. pneumoniae*. It would be impractical to produce a vaccine against the 82 capsular types, so epidemiologists have determined which types are the most frequent cause of disease. Currently we have a vaccine that contains 14 type-specific antigens. The principal recipients of the vaccine are the elderly with chronic ailments. Over 70% of these individuals carry *S. pneumoniae* in their respiratory tracts. When they become ill or debilitated their immunological defenses are lowered, and the pneumococcus can cause disease. Pneumococcal pneumonia is one of the most frequent causes of death in the elderly.

Whooping cough. Whooping cough, also called **pertussis**, is caused by a gram-negative bacillus called *Bordetella pertussis* that inhabits the upper respiratory tract of humans. The disease is highly contagious and affects primarily infants and children. In the United States, between 1940 and 1948, whooping cough killed nearly three times as many infants under 1 year of age as measles, mumps, chickenpox, rubella, scarlet fever, diphtheria, poliomyelitis, and meningitis combined.

Whooping cough is transmitted by contact with respiratory secretions that contaminate the air or surrounding objects. The first step in the pathogenic process of *B. pertussis* is attachment via adhesins to cilia projecting from the mucosal surface of the respiratory epithelium. The major adhesin appears to be a filamentous protein that causes hemagglutination and is called **filamentous hemagglutinin**, or **FHA**. During infection the microorganism erodes the mucosal lining of the respiratory epithelium, causing repeated coughing. Apparently a toxin, **tracheal cytotoxin**, produced by the microorganism damages the cilia of the respiratory epithelium (ciliary movement is a normal mechanism for clearing foreign material and mucus) and causes this symptom. Later, mucus plugs the air passages, and inhalation of air, following a series of coughs, produces a characteristic "whoop." The accumulation of mucus and respiratory secretions is believed to be a result of microbial **adenyl cyclase** that converts ATP to cyclic AMP (cAMP) and af-

***Table 27-3** Characteristics of bacterial agents that cause pneumonia**

Bacterial agent	Morphology	Population most likely affected	Chemotherapy	Immunization	Comments
Klebsiella pneumoniae	Gram-negative bacillus, capsule produced	Those with underlying respiratory conditions	Cephalosporins, but multiple drug resistance common	None	Causes necrosis of lung tissue
Staphylococcus aureus	Gram-positive coccus	Those infected with influenza virus, or other respiratory viruses	Multiple drug resistance; susceptibility tests required	None	Organism produces coagulase, important in identification
Haemophilus influenzae	Gram-negative coccobacillus	Those with underlying respiratory conditions	Ampicillin or chloramphenicol	None	Ampicillin-resistant strains common
Legionella pneumophila (Legionnaires disease)	Gram-negative bacillus	Those with underlying respiratory conditions	Erythromycin	None	Transmission perhaps by contaminated aerosols
Yersinia pestis (plague)	Gram-negative bacillus	Contact with respiratory secretions of those with pneumonic variety of plague	Streptomycin or tetracycline	Yes, but of limited value	See p. 818
Mycoplasma pneumoniae	Devoid of cell wall	Those in crowded quarters (such as barracks)	Erythromycin or tetracycline	Experimental	Colonies look like fried egg on agar
Coxiella burnetii	Gram-negative bacillus	Those inhaling dried parturition products of barn animals	Tetracycline	None	Belongs to group Rickettsia (see p. 819)
Chlamydia psittaci	Gram-negative, exists in two morphological states: infectious elementary body and noninfectious reticulate body	Primarily infects birds and secondarily infects humans (psittacosis); those inhaling particles contaminated with bird droppings	Tetracycline	None	Pneumonia an infrequent aspect of infection
Streptococcus pneumoniae	Gram-positive diplococcus	The elderly with underlying disease or poor health	Penicillin	Polyvalent vaccine available	Most common cause of bacterial pneumonia

*Many other microbial agents can produce pneumonia such as *Streptococcus pyogenes* and many gram-negative enteric species, but only the primary agents of pneumonia are listed in the table.

fects cellular membrane activity. Whooping cough is considered self-limiting, but the mortality rate may be high as a result of secondary bacterial or viral infections.

Whooping cough does not respond well to antibiotics treatment, therefore immunization is the only control. Before mass immunization in the 1950s, over 100,000 cases of whooping cough occurred in the United States but today as few as 1500 are recorded per year. The disease is still prevalent in most underdeveloped countries where few children have been vaccinated. The vaccine used in the United Staes is part of a mixture called **DTP** (*D,* diphtheria toxoid; *T,* tetanus toxoid; *P,* pertussis vaccine). The vaccine-toxoid is usually administered as three doses at monthly intervals beginning 2 months after birth. Booster doses at age 1 and 2 years and 5 to 6 years are also usual. The pertussis vaccine is not without its risks, particularly for children with some underlying immunological problem. In these children a vaccine side effect is encephalitis (inflammation of the brain). The Japanese use only the filamentous hemagglutinin of *B. pertussis* and not whole cells for pertussis vaccine. The Japanese vaccine does not produce side effects, but scientists in the United States are waiting to see if the vaccine induces a long-

Table 27-4 *Bacterial agents that are a primary cause of meningitis**

Bacterial agent	Morphology	Population most likely affected	Chemotherapy	Immunization	Comments	Estimated percentage of meningitis cases in U.S. caused by specific organism
Streptococcus agalactiae (group B streptococci)	Gram-positive streptococci	Newborns during passage through birth canal	Penicillin or ampicillin plus gentamicin	None	Second only to *E. coli* as cause of neonatal meningitis	3
Haemophilus influenzae type B	Gram-negative coccobacillus, produces capsule	Children under 4 years of age	Ampicillin or chloramphenicol	Vaccine for children 18-24 months of age	Ampicillin-resistant varieties are arising	40
Escherichia coli	Gram-negative bacillus	Neonates and those debilitated as a result of surgery and catheter manipulation	Multiple antibiotic resistance; susceptibility tests required	None	Most frequent cause of neonatal meningitis	5
Streptococcus pneumoniae	Gram-positive streptococcus	Severely debilitated patients	Penicillin	Polyvalent vaccine containing 14 serotypes	Often follows lung infection	22
Neisseria meningitidis	Gram-negative diplococcus	Children less than 5 years of age	Penicillin	Polysaccharide vaccine against group A and C but not B serotypes	Vaccine recommended for contacts at day care centers	25

*Many other microbial agents can cause meningitis, including *Listeria monocytogenes, Flavobacterium meningosepticum,* many gram-negative enteric species, as well as staphylococci and streptococci. Only the primary agents of meningitis are listed in the table.

term immune response before using it in this country.

Meningitis. Meningitis is an inflammatory disease that involves the protective layers of tissue surrounding the spinal cord and brain, called the meninges. If the infection results in the formation of a purulent (pus-containing) discharge, it is called septic meningitis, and aseptic if nonpurulent. One of the major problems in the treatment of meningitis is the relatively poor rate of diffusion of antibiotics from the bloodstream into the central nervous system. Meningitis is therefore a particularly severe disease that can result in a high percentage of fatalities. The two most important agents of bacterial meningitis are *Neisseria meningitidis* and *Haemophilius influenzae*. A wide variety of bacterial agents that cause primary diseases elsewhere in the body may also cause meningitis; they are outlined in Table 27-4.

Neisseria meningitidis (also called **meningococcus**) (Figure 27-2) resides in the upper respiratory tract. The meningococcus is the only bacterium that causes epidemic meningitis. Contact with respiratory secretions from carriers or active cases is the most important ave-

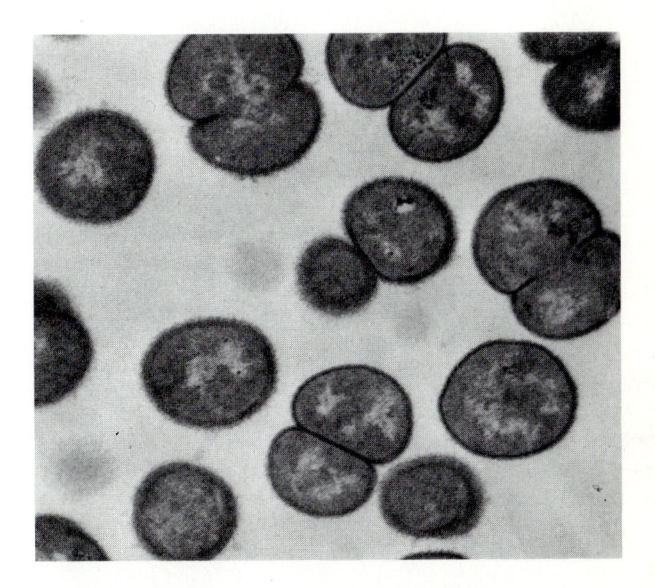

Figure 27-2 *Electron micrograph of a* Neisseria *species demonstrating its diplococcal nature. (×27,000.)*

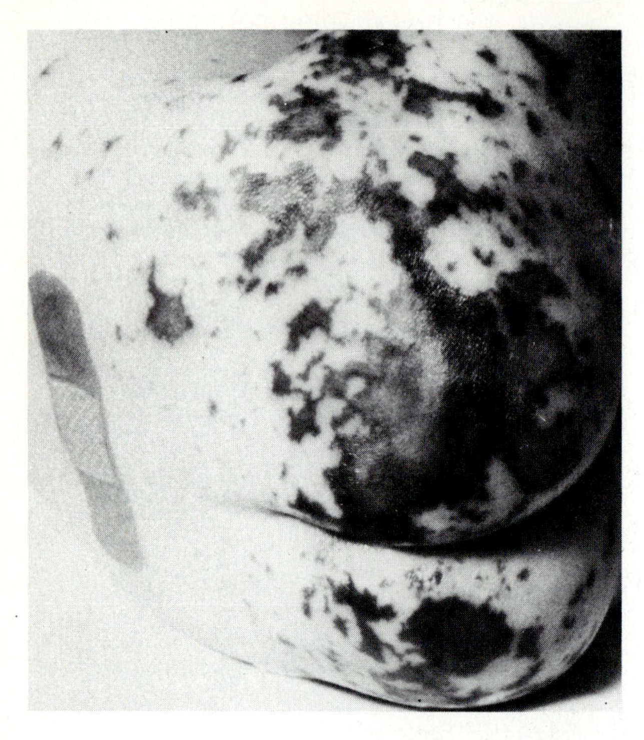

Figure 27-3 *Fulminating meningococcemia. Cutaneous hemorrhages are evident before meningitis is recognized.*

nue of transmission. Infections usually begin with mild nasopharyngitis, which may progress into septicemia (meningococcemia) that may be severe enough to cause death from hemorrhage (Figure 27-3). Septic meningitis is a consequence of the dissemination of the meningococcus from the bloodstream to the central nervous system. The most common symptoms of disseminated disease are stiff neck or back, headache, and nausea.

One of the most important aspects of meningococcal disease is the carrier state. The organism colonizes the upper respiratory tract of carriers and stimulates the formation of antibodies that do not eliminate the organism but do prevent it from multiplying and disseminating to other tissues. It appears that epidemics occur only when the carrier rate exceeds 20%; no major epidemic has occurred in the United States since 1945. Approximately 2600 cases of meningococcal disease were reported in 1986, and most of them were among children less than 4 years of age. Transmission is accelerated when there are crowded conditions, such as in day care centers, military camps, and hospitals.

N. meningitidis is a gram-negative aerobic to facultatively anaerobic coccal bacterium. When observed under the microscope the cocci appear in pairs or tetrads that have a kidney-bean shape. The meningococci have been divided into several serogroups based on the characteristics of a capsular polysaccharide. Groups A, B, and C are most frequently associated with disease. Vaccines prepared from strains of groups A and C are available but not for group B, which is the most prevalent type in

Point of Interest

THE STORY OF LEGIONNAIRE'S DISEASE

Microbiology students many times believe that all the exciting and disease-causing microorganisms have already been discovered, but recent experience with Legionnaire's disease and AIDS indicate that this is not the case.

In 1976 over 200 cases of pneumonia developed during an American Legion Convention in Philadelphia. Over 30 deaths resulted, and no obvious cause for the pneumonia could be detected. The only thing all the victims had in common was that they had all stayed in the same hotel.

An investigation to identify the disease agent was started by the Centers for Disease Control in Atlanta, Georgia. Tissue, blood, and sputum samples were collected from infected and normal individuals. These samples were inoculated into a variety of media under a variety of conditions. A rickettsial or viral agent was suspected. After 5 months involving over 90,000 man-hours and $2 million, a previously unrecognized bacterium was isolated. The bacterium was called *Legionella pneumophilia.* It was a small, gram-negative rod that had complex nutritional requirements and an usually high iron requirement. It had been overlooked on microscopic investigations of the tissues because it did not stain readily.

Although the bacterium has only recently been isolated, it appears to be quite common in water and was found to be present in the cooling towers of the air conditioning unit of the Philadelphia hotel. Thus infectious *Legionella* aerosols were spread via the air ducts of the hotel's central air conditioning system. Although the disease is spread by the airborne route, there is no evidence for person-to-person transmission.

Most cases of *Legionella* are asymptomatic to mild. In all cases in which pneumonia develops, the patient has a deficient immune system or some other predisposing disease. The reason for the large number of pneumonia cases in Philadelphia in 1976 was the presence of a large number of elderly, frail conventioneers. Luckily the disease has been shown to respond well to treatment with erythromycin and can now be easily detected by a serological test using fluorescent antibodies.

the United States. The group B polysaccharide is poorly antigenic.

H. influenzae is a parasite that is part of the normal microbial flora of the respiratory tract of humans and animals. Two strains may be present: **encapsulated** and **nonencapsulated**. The encapsulated strain, called *H. influenzae* type b, is more virulent and more predominant in children under 4 years of age. Older children and adults rarely carry the type b strain. Type b is one of the major causes of meningitis in children under 4 years of age, but the disease in adults is rare. The species name, *H. influenzae,* is a misnomer, because it was mistakenly thought to be the cause of influenza when first isolated in 1892. *H. influenzae* is a gram-negative organism that appears as a coccobacillus or filamentous rod. One of the characteristics of *Haemophilus* species is their requirement of the growth factors called *X factor* (hemin) and *V factor* (nicotinamide adenine dinucleotide, or NAD). Some species, such as *H. influenzae,* require both factors, whereas other species require one or the other. The importance of the V factor was accidentally discovered when a throat culture containing *Haemophilus* species and staphylococci were streaked on agar media. *Haemophilus* colonies are considerably larger when they are in close proximity to a staphylococcal colony because staphylococci are known to excrete NAD into the medium (Plate 34).

The capsule of *H. influenzae* type b, which is a polyribitol phosphate, is the basis of a vaccine licensed in the United States. The vaccine is used primarily for immunization of children between the ages of 18 and 24 months.

Tuberculosis. Tuberculosis has been the major cause of death by infectious agents in the Western world for over 200 years. In the late 1800s villages in the eastern parts of the United States were decimated by tuberculosis, and as many as 30% of the deaths each year were caused by this disease. Even today nearly 4 million people throughout the world die each year from tuberculosis. Tuberculosis has declined in developed countries of the world because of the following factors:

1. Pasteurization of milk, which destroys the bovine organism, originally responsible for many cases of human disease.
2. Development of sanitariums where isolation, rest, and good diet contribute to a decline in transmission of the microorganisms.
3. The discovery of streptomycin (by Waskman in 1943) and other drugs that were effective in treatment and caused an immediate decline in the death rate.

The success of all of these programs is evident from the statistics. Before 1850 the death rate for tuberculosis was 400/100,000 population. Today, the death rate in the United States is less than 2/100,000 population.

Tuberculosis is most frequently acquired by contact with respiratory secretions of patients infected with *Mycobacterium tuberculosis*. Ninety percent of the cases in the United States are of the pulmonary type, but any organ or tissue may be involved. Ingestion of milk contaminated with *M. bovis* is still a major cause of disease in countries where pasteurization is not a routine practice. The first site of involvement, following infection, is the lung, where the lesion is called a **tubercle** (Figure 27-4). The course of the disease depends on many factors but ultimately relates to the cell-mediated immunity of the host. *M. bovis* or *M. tuberculosis* organisms are phagocytized but not killed by alveolar macrophages, where they multiply for several weeks. During this period the microorganisms may be disseminated to every major organ of the body. A hypersensitivity to the bacterial components in the macrophage develops primarily in the lung and results in necrosis around the tuberculous lesion. The necrotic lesion is called **caseation necrosis** (caseation refers to a cheeselike consistency). Hypersensitivity to bacterial components is the basis of the tuberculin skin test (discussed later). The tubercles formed during infection will eventually heal to form calcified nodules, but may continue to harbor viable bacilli. These organisms can be reactivated—if at some later time the host's resistance is lowered—and produce disease a second time.

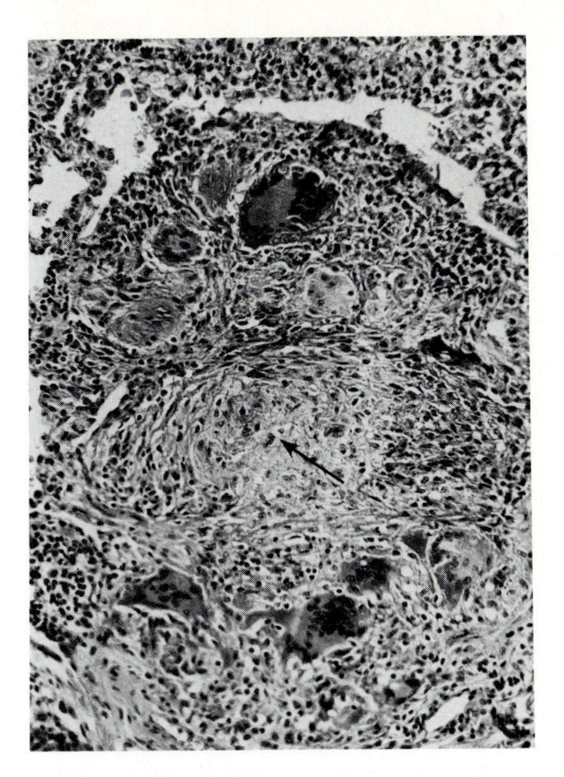

Figure 27-4 *Young "hard" tubercle* (arrow) *with prominent epithelial cells. (×120.)*

The symptoms of tuberculosis, which may occur sev-

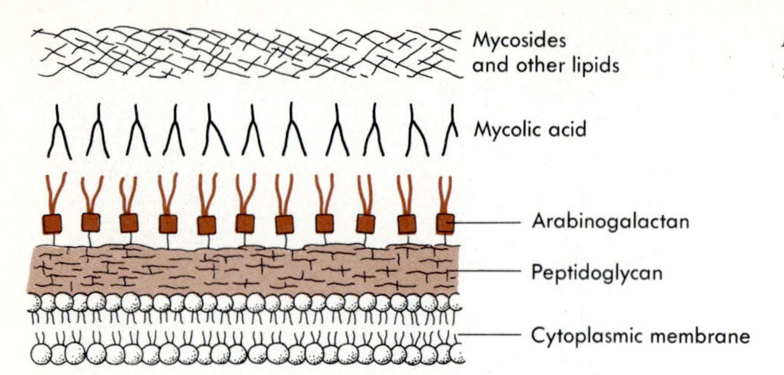

Figure 27-5 *Diagrammatic illustration of the cell wall of mycobacteria.*

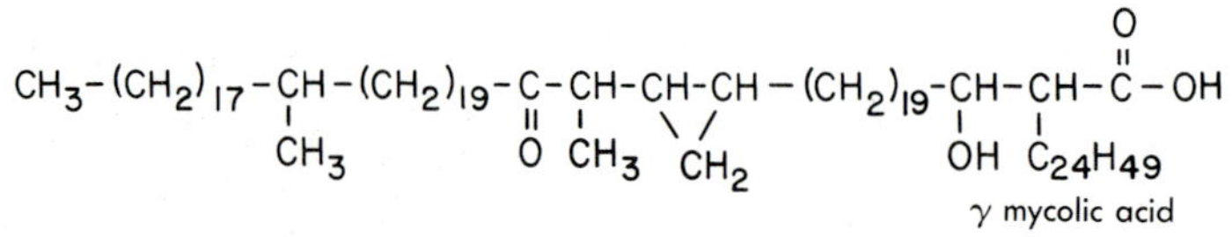

$$CH_3-(CH_2)_{17}-\underset{\displaystyle CH_3}{CH}-(CH_2)_{19}-\underset{\displaystyle O}{\overset{}{C}}-\underset{\displaystyle CH_3}{CH}-CH-\underset{\displaystyle CH_2}{CH}-(CH_2)_{19}-\underset{\displaystyle OH}{CH}-\underset{\displaystyle C_{24}H_{49}}{CH}-\overset{\displaystyle O}{C}-OH$$

γ mycolic acid

Figure 27-6 *Structure of mycolic acid.*

eral weeks or months after infection, are fever, fatigue, and loss of weight. The cough, which is characteristic of pulmonary involvement, may result in expectoration of bloody sputum.

M. tuberculosis, like other mycobacteria, exhibit characteristics that are not associated with most other pathogenic bacteria and include:

1. The cell wall of the mycobacteria contains peptidoglycan; however, the organism cannot be stained with ordinary gram-stain reagents. The peptidoglycan layer is linked to a polymer of arabinose and galactose (Figure 27-5). This arabinogalactan polymer forms a bridge between the peptidoglycan and an outer **mycolic acid** fraction. Outside of this triple layer are glycolipids and peptidoglycolipids called **mycosides** that make up the outer envelope. Over 50% of the lipid in the surface components are in the form of mycolic acids that have chain lengths of up to 91 carbons (Figure 27-6). Much of the remaining lipid is in the form of normal fatty acids. The mycobacteria can be stained in a procedure called the acid-fast stain (see Chapter 3).
2. Those glycolipids that lie outside the cell wall and are not bound to it contain a mycolic acid called the **cord factor** (cord refers to the serpentine cords of the colonies of mycobacterial pathogens that were once believed to be a result of this mycolic acid). The cord factor is covalently linked to the carbohydrate trehalose. Cord factor is toxic for mice and may be associated with virulence in humans.
3. The long incubation period before appearance of symptoms is related to the naturally slow growth of most mycobacteria. Species, such as *M. tuberculosis,* do not show visible colony growth in the laboratory for 2 to 5 weeks. Their doubling time is somewhere between 18 to 24 hours as compared to the 20 to 30 minutes for *E. coli.*
4. Mycobacteria are characterized by their ability to produce carotenoid pigments. In some strains, pigments are constitutive (**scotochromogens**) but in others the pigments are induced by light (**photochromogens**—see Plate 35). The significance of these pigments appears to be a protective mechanism against photodynamic activity and is similar to that associated with photosynthetic microorganisms.

Disease caused by *M. tuberculosis* sensitizes the in-

Concept Check

1. **Which bacterial diseases are acquired through the respiratory tract? Name an example of the bacterial species involved. For which diseases is a vaccine available?**
2. **What are the most significant components of *Bordetella pertussis* that are believed to be associated with disease? How do these components affect the mammalian cell?**
3. **Why is *H. influenzae* so devastating to children between 18 and 24 months but not to children over 4 years of age?**
4. **What is the sequence of events from inhalation of the TB bacillus to the formation of tubercles? Are there any structural or chemical characteristics that would provide an answer to its pathogenicity?**
5. **What characteristics of *M. tuberculosis* make it so distinctive from other bacteria?**

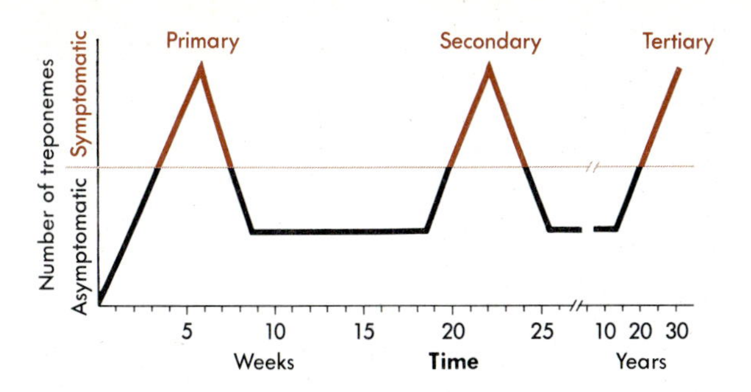

Figure 27-7 *Average time intervals for stages of syphilis. Disease fluctuates between symptomatic and asymptomatic periods. Clinical manifestations of symptomatic syphilis results from multiplication of treponemes within various tissues. Effective host defenses then evolve, decreasing number of treponemes, and resulting in asymptomatic periods.*

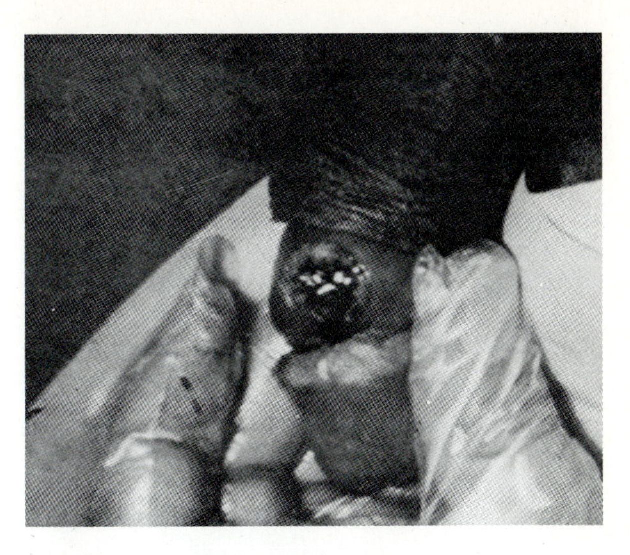

Figure 27-8 *Chancre on penis during primary stage of syphilis.*

dividual to the protein **(tuberculin)** of the bacillus, and this hypersensitivity can be detected by tuberculin skin tests. These tests are an important method for detecting tuberculosis in younger persons in whom the disease may not be recognized.

A vaccine, **BCG** (Bacille-Calmette-Guerin), is available and used in countries where tuberculosis is more common. In the United States, where the disease is less common, it is cheaper to treat individuals with drugs such as **isoniazid**. Isoniazid is sometimes used in combination with other drugs, such as streptomycin, ethambutol, and rifampin, to prevent the emergence of bacilli that develop resistance to one of the drugs.

Sexually transmitted diseases and other diseases acquired through the skin including the eye

Sexually transmitted diseases. Antibiotics have been used successfully for many years to treat sexually transmitted diseases (STDs), such as syphilis and gonorrhea. The relatively recent appearance of untreatable acquired immune deficiency syndrome (AIDS) has made STDs an item of scientific and public concern. In addition to AIDS, some STDs do not always cause overt symptoms in sexual partners and can be a primary cause of infertility, abortion, or fetal damage.

Syphilis. Syphilis is caused by *Treponema pallidum,* an organism belonging to the group called spirochetes. Syphilis is ordinarily transmitted by sexual contact but 10% of the cases are contacted by nonsexual means, because the parasite may be present in lesions in the oral cavity or other parts of the body. Contact with such lesions by kissing, for example, can result in transmission of the parasite. *T. pallidum* can also be transferred during pregnancy to the developing fetus.

Syphilis can be divided into four stages: **primary, secondary, latent,** and **tertiary** (Figure 27-7). The primary stage is characterized by a lesion, called a **chancre,** at the site of entry of the parasite into the host (Figure 27-8). The lesion is filled with organisms (also called **treponemes**), and this represents a highly infectious stage. The chancre heals (end of primary stage) and the organisms are destroyed by the host's immune system (70% of the time), or the organisms invade the vascular system and are potentially distributed to any body tissue. The presence of the organisms in the skin results in a rash characteristic of the secondary stage. The rash is primarily on the trunk, face, palms, or soles. Secondary lesions may also appear in the oral cavity or genitalia. All secondary lesions contain large numbers of infectious treponemes. Syphilis may terminate at the secondary stage, but in one third of the cases, a latent period occurs. Latency is characterized by no symptoms and may last for several weeks or several years. Latency may develop into secondary symptoms or evolve into the tertiary stage. The tertiary stage is brought about apparently by an immunological response by the host, resulting in lesions called **gummas** (Figure 27-9). The tertiary lesions can cause blindness, loss of muscular control, aneurysms of blood vessels, and even death.

The lesions of congenital syphilis can be more damaging than the sexually transmitted variety to the developing fetal tissue. One half of the fetuses will be dead at birth if chemotherapy is not initiated during pregnancy. Fetuses that survive develop lesions manifested in later years by perforations of the soft and hard palates (Figure 27-10), impaired vision, bowing of legs, and other conditions.

T. pallidum is a helically shaped bacterium that is 6 to 14 μm in length. It is best observed by darkfield microscopy, because its width (0.2 μm) makes it barely visible by brightfield microscopy of stained preparations. The coils of the bacterium are evenly spaced 1 μm from each other (Figure 27-11). Laboratory identification is based on serological and direct microscopic examination of lesion material. One of the serological tests, called the **Venereal Disease Research Laboratory (VDRL)**, is a screening test. It detects a nonspecific serum component called **reagin**, which is produced during syphilis and some other diseases and conditions. The **fluorescent antibody absorption (FTA-ABS) test** is used to confirm tests such as the VDRL test. FTA-ABS test detects specific antibodies to *T. pallidum* in the patient's serum (Figure 27-12).

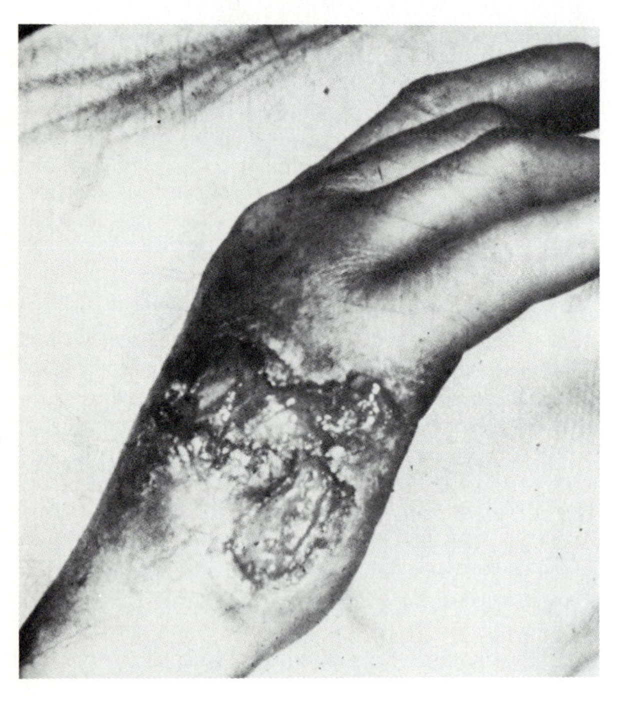

Figure 27-9 *Gumma of hand observed during tertiary stage of syphilis.*

Penicillin is the drug of choice in treatment of syphilis. No vaccine is yet available; this is due in part to our inability to consistently cultivate *T. pallidum* on laboratory media or to identify the virulence factors. *T. pallidum* is ordinarily cultivated in rabbit testicular tissue.

Gonorrhea. The agent of gonorrhea is a gram-negative diplococcus called *Neisseria gonorrhoeae* (Plate 36). The natural habitat of this organism is the mucous membranes of warm-blooded animals. This organism is often referred to as the **gonococcus**.

Gonorrhea is the most frequently reported communicable disease in the United States. In 1986 there were 896,330 cases of gonorrhea reported to the Centers for Disease Control (CDC), which is up from the 883,826 of 1985 but less than the 1,042,900 cases in 1980.

Gonorrheal disease is not limited to the genitalia and can occur in the anal and oral cavities, depending on sexual practices and other conditions. Gonococcal disease may also occur by nonsexual means. The newborn may acquire gonococcal conjunctivitis **(ophthalmia neonatorum)** during passage through an infected birth canal. Untreated, this condition can lead to blindness. Gonococcal conjunctivitis can occur during the first year of life by accidental contamination from a parent or relative, although the same infection in older children is frequently acquired by sexual molestation or premature sexual activities.

The gonococcus lives only a few seconds on the skin, and transmission by doorknobs, towels, or toilet seats is unlikely.

The gonococcus adheres to mucosal tissue by adhesins. In the genital tract the organisms attach to the microvilli that entraps them (Figure 27-13). The microvilli retract, and this places the gonococci at the surface of the mucosal cell proper. The gonococci are then surrounded by the cell membrane of mucous-secreting

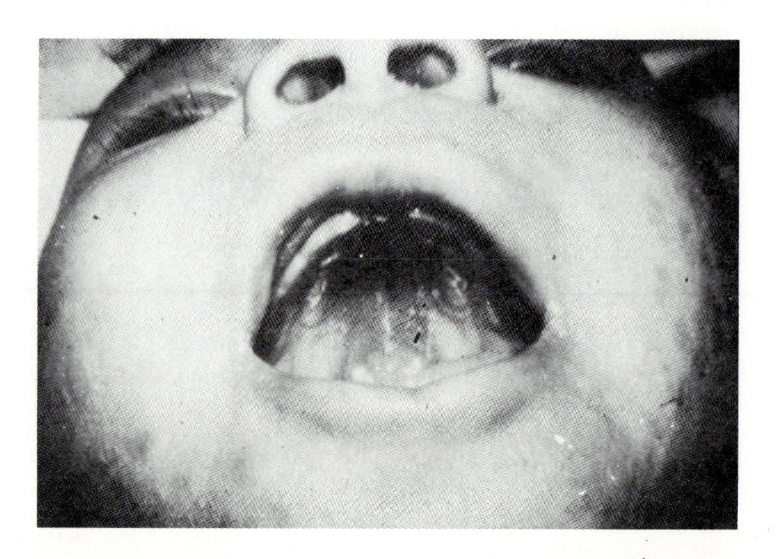

Figure 27-10 *Perforated palate resulting from congenital syphilis.*

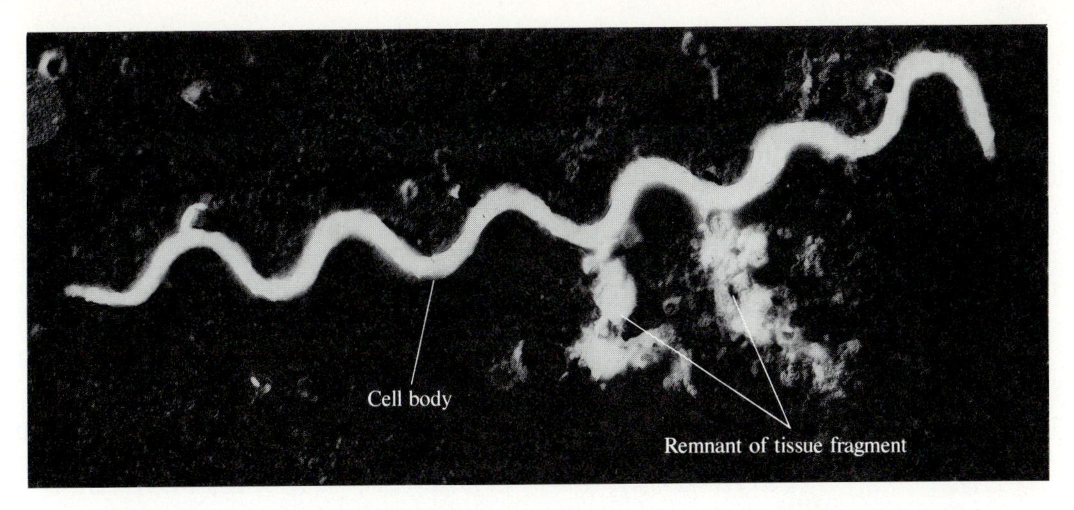

Figure 27-11 *Electron micrograph (×23,000) of* Treponema pallidum *obtained from rabbit testes.*

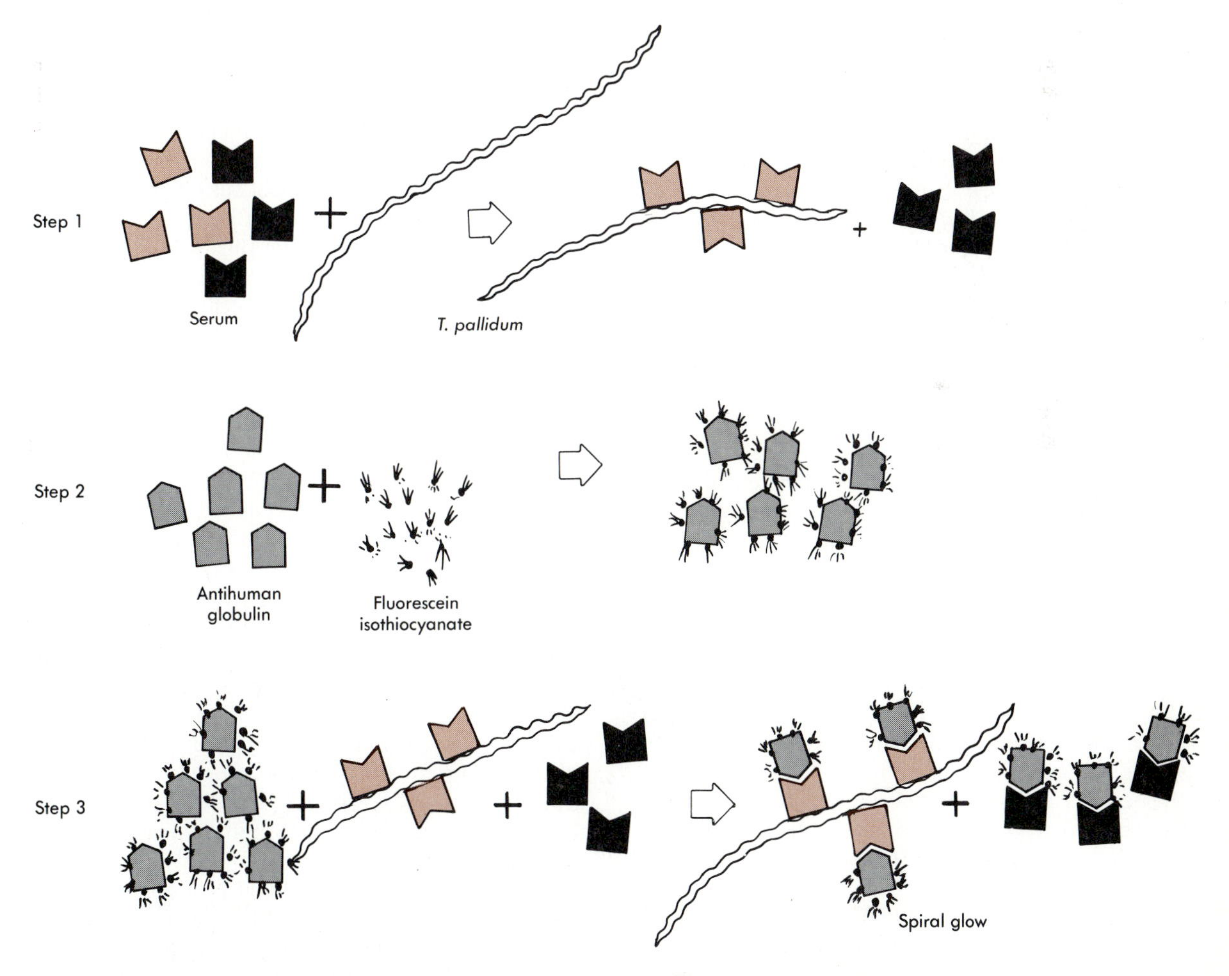

Figure 27-12 *Fluorescent treponemal absorption (FTA-ABS) test. Step 1: patient's serum may contain specific antitreponemal antibodies (■) or nonantitreponemal antibodies (■). Serum is reacted with* Treponema pallidum. *Step 2: antihuman globulin is conjugated with fluorescein isothiocyanate. Step 3: conjugated antihuman globulin will coat any human antibodies, but treponemes will fluoresce only if patient's serum contains antitreponemal antibodies.*

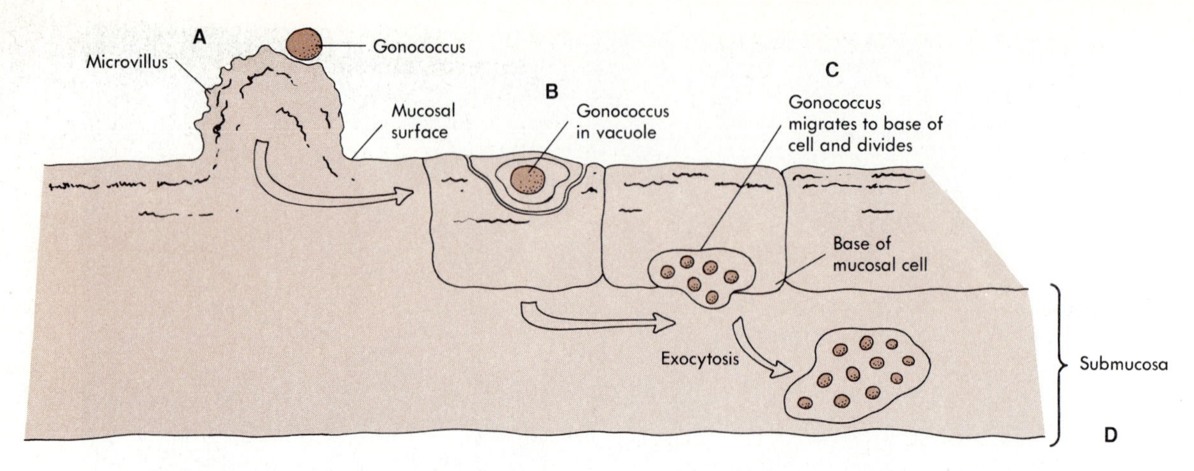

Figure 27-13 *Process of interaction of gonococci with the genital mucosa.* ***A,*** *Gonococcus binds to tip of microvillus on mucosal surface.* ***B,*** *Microvillus retracts and gonococcus in contact with mucosal surface is surrounded by host membrane to form vacuole.* ***C,*** *Vacuole containing gonococcus migrates to base of mucosal cell where gonococcus multiplies.* ***D,*** *Vacuole-containing gonococci are excytosed into submucosa from where they can penetrate into the bloodstream.*

cells in a process similar to phagocytosis. The membrane-bound vacuole containing the gonococci are transported to the base of the mucosal cell, where the organisms multiply. They remain there for about 96 hours and then are exocytosed into the submucosa. It is the penetration of the mucosal surfaces that elicits an inflammatory response and aids in the spread of the parasites. The inflammatory response in humans is a yellowish-white discharge, resulting from inflammation of the urethra (urethritis). In up to 40% of the cases the disease may be asymptomatic in men. The disease may spread to the prostate or testes, which may result in sterility. In women the endocervix is the genital tissue most often infected, and this results in a vaginal discharge. Many infected women, however, are asymptomatic; this represents a reservoir of infection for their sexual partners. In about 15% of women, infection may spread to the fallopian tubes, ovaries, and pelvic peritoneum, giving rise to **pelvic inflammatory disease** (PID). PID can be caused by a variety of microorganisms, including chlamydia, which is discussed later. PID causes necrosis and scarring of the fallopian tubes, which can lead to obstruction and infertility.

Anorectal and oropharyngeal gonorrhea occurs in both men and women who practice rectal intercourse and fellatio, but appears more frequently in homosexual men. The infections are usually asymptomatic, although pharyngitis can occur.

During infection the gonococci do not always remain in the submucosal layer, and in 1% to 3% of the patients the parasite penetrates into the bloodstream. The organism eventually settles in the joints and can cause arthritis.

Penicillin has been the most effective drug in the treatment of uncomplicated gonorrhea, but since 1976 a **penicillinase-producing gonococcus (PPNG)** has shown an unusually high resistance to the drug. The gonococcal penicillinase, whose determinant is located on a plasmid, is retained in the periplasmic space. Even if some bacteria are killed by penicillin, penicillinase is released from the dead cells and may inactivate any residual penicillin. The appearance of PPNG and other penicillin-resistant strains, as well as the coexistence of chlamydia in many gonococcal infections, has led to changes in the chemotherapeutic regimen. Tetracycline to kill chlamydia and spectinomycin to kill PPNG have been added to the chemotherapeutic arsenal for complicated gonorrhea. Ophthalmia neonatorum can be prevented by applying ointment or drops containing tetracycline, erythromycin, or a 1% solution of silver nitrate to the eyes of the infant following birth. Most states require that this procedure be performed.

A vaccine for gonorrhea has not been forthcoming because there are many distinct gonococcal serotypes, and all the antigens associated with inducing a protective immunity have not been identified. Thus protection against one serotype would not prevent repeated infections from other serotypes. Control measures can reduce the incidence of disease if (1) the asymptomatic patient is identified, (2) condoms are used, and (3) sexual partners of an individual who has already been diagnosed as having gonorrhea are treated.

Nongonococcal urethritis. *Chlamydia trachomatis* is the major cause of nongonococcal urethritis (NGU), which is a nonreportable disease in the United States. The estimates of NGU in the United States each year range from 4 to 5 million. NGU is usually not fatal, but is still a major health problem.

The most common manifestation of urethritis is slight pain and a burning sensation following urination. Genital chlamydia have been incriminated in other syndromes, including:

1. Cervicitis. As many as 50% of the women with inflammation of the cervix have been shown to have positive chlamydial infection.
2. Pelvic inflammatory disease (PID). The wearing of intrauterine devices appears to increase the risk of infection by chlamydia. PID can lead not only to infertility but to potential fatal ectopic or tubal pregnancies.
3. Neonatal infection. Approximately 11% of the general population of pregnant women have chlamydiae in the cervix, which can infect the neonate during passage through the birth canal. Conjunctivitis (**inclusion blennorrhea**) acquired during delivery is usually a benign infection. It has been documented that *C. trachomatis* is also the cause of neonatal pneumonia. This organism is believed to account for 30% of all cases of pneumonia in hospitalized infants less than 6 months of age.

C. trachomatis belongs to a genus of obligate parasites that also includes the species *C. psittaci*. *C. psittaci* is the agent of a disease in humans called **psittacosis**, which is acquired by contact with infected birds. *C. trachomatis* is also associated with another sexually transmitted disease called **lymphogranuloma venereum**.

The chlamydiae are approximately 0.2 to 1 μm in diameter, depending on their state of development in host cells (Figure 27-14). They do not possess a peptidoglycan layer, but the cell wall still resembles gram-negative bacteria. There is an outer membrane lying adjacent to a periplasmic space. The rigidity of the cell wall appears to be the result of a single protein located in the outer membrane, which is bound to other proteins by disulfide bridges. This is in contrast to the covalent binding of the outer membrane to the peptidoglycan in gram-negative bacteria. The outer membrane contains lipopolysaccharide (LPS) similar to the LPS of gram-negative bacteria. The species-specific antigens that differentiate *C. trachomatis* from *C. psittaci* are contained in proteins that reside in the outer membrane.

Chlamydiae rely on the host for their source of energy, because they are unable to produce ATP. Nevertheless, chlamydiae are capable of producing their own RNA, DNA, and protein. One of the unique characteristics of the chlamydiae that distinguishes them from all other microorganisms is their developmental cycle in which two forms of the organism exist: the **elementary body (EB)** and the **reticulate body (RB)**. Both forms possess DNA, RNA, and ribosomes and are limited by a cytoplasmic membrane. The infectious cycle of chlamydiae occurs in the following way:

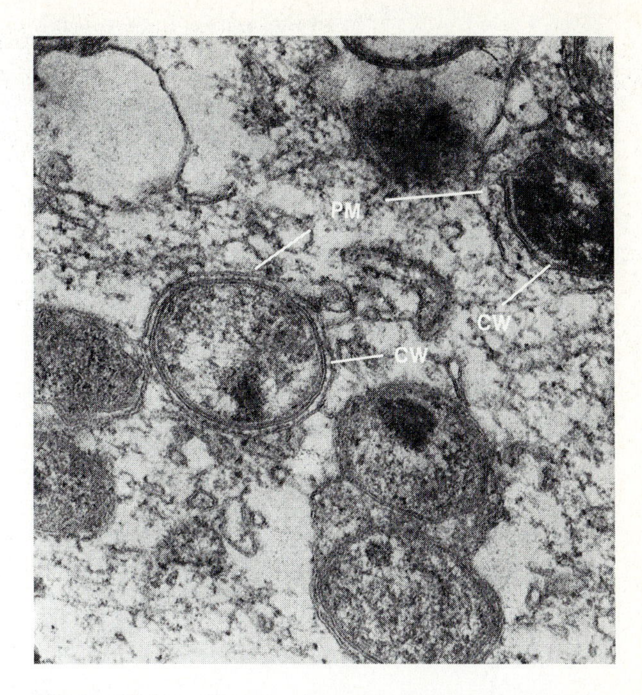

Figure 27-14 *Electron micrograph of infected mammalian cell containing* Chlamydia *organisms. Chlamydial cell can be seen with its cell wall* (CW) *surrounded by cytoplasmic membrane unit called phagosome membrane* (PM).

1. The EB, which is small (350 nm in diameter), is the infectious form of the microorganism that binds to receptors on the host cell (Figure 27-15). The EB is relatively resistant to environmental factors and is not metabolically active. The EB is engulfed by the plasma membrane of the host cell to form a phagosome that is transported into the cytoplasm. Chlamydiae inhibit the fusion of lysosomes to the phagosome, thus ensuring their resistance to the host's most important intracellular defense mechanism.
2. The EB engages in protein, DNA, and RNA synthesis and is converted to a larger (1.0 μm in diameter) reticulate body. The RB multiplies by binary fission, giving rise to more reticulate bodies, which is followed by enlargement of the phagosome. The enlarged phagosome with its numerous reticulate bodies is now called an **inclusion body** (Figure 27-16).
3. Not all of the reticulate bodies multiply and some undergo a reorganization and condensation into infectious elementary bodies.
4. Elementary bodies are released from the inclusion body and the host cell to infect adjacent cells; thus the cycle is repeated.

Other sexually transmitted diseases caused by bacteria. Several other bacterial species also cause sexually transmitted disease. Some of these are the result of ho-

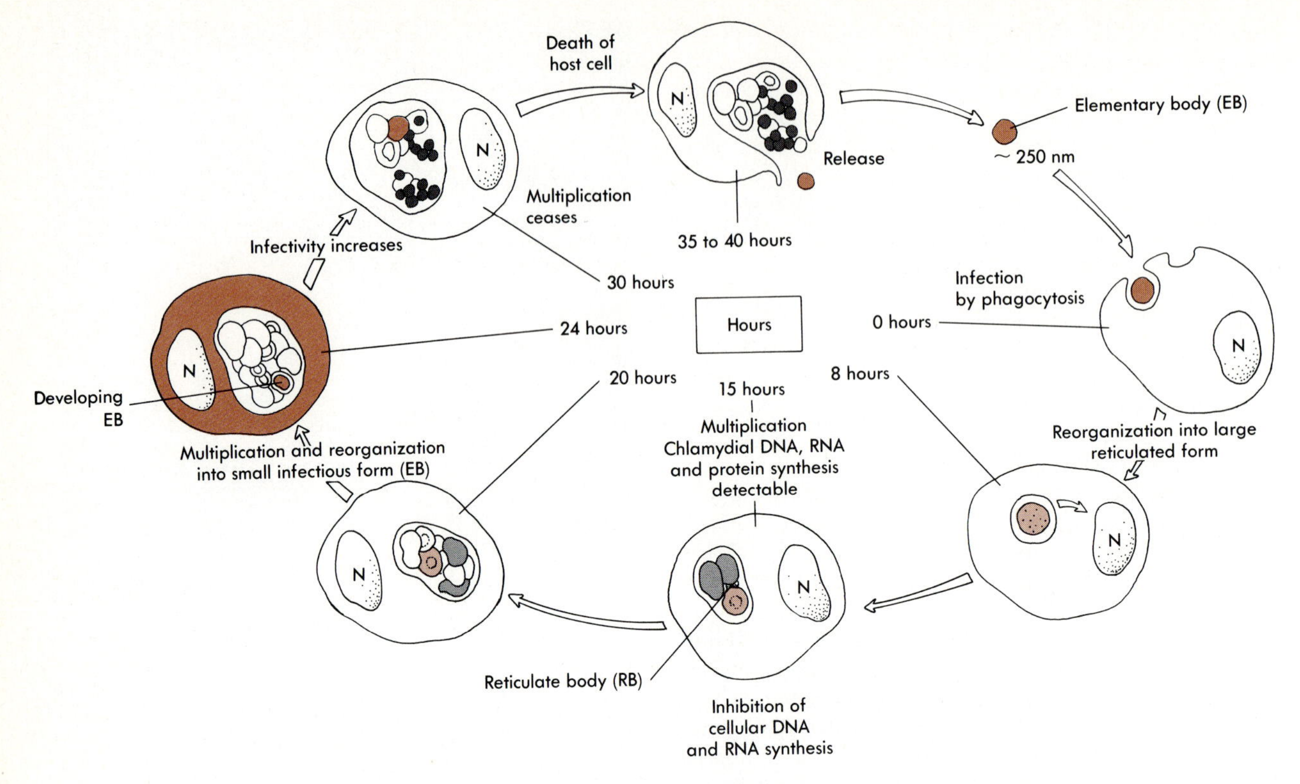

Figure 27-15 *Schematic representation of developmental cycle of* Chlamydia. N, *Nucleus;* EB, *elementary body;* RB, *reticulate body.*

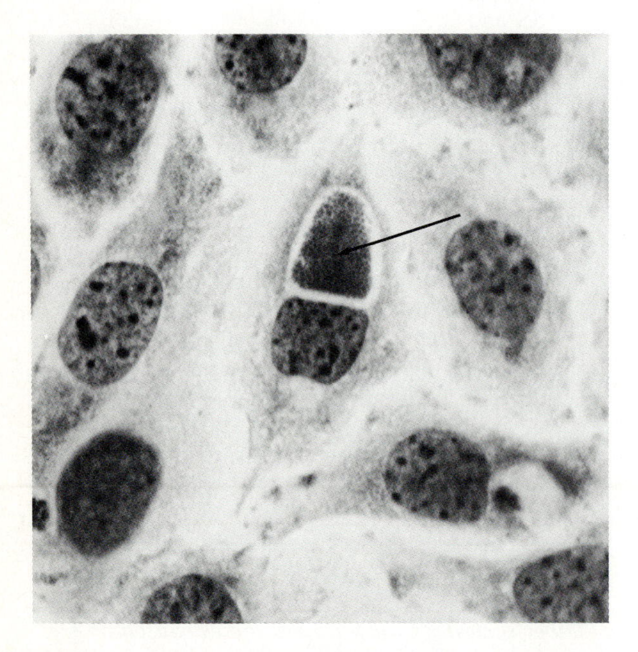

Figure 27-16 *Light micrograph of stained* Chlamydia-*infected culture cells.* Arrow *points to inclusion body.*

mosexual behavior or practices involving oral-genital or oral-anal sex. Therefore microorganisms associated with infections of the intestinal tract can also be considered sexually transmitted diseases. Table 27-5 outlines the major and minor sexually transmitted diseases caused by bacteria.

Diseases involving the skin, mucosa, and eye. Infectious diseases involving the skin, mucosa, and eye include yaws, leprosy, staphylococcal and streptococcal infections, and conjunctivitis. Our discussion is limited to the most frequent cause of skin infections, the staphylococci. Table 27-6 outlines the other microbial agents involved in these infections.

Staphylococcal skin disease. The staphylococci are capable of infecting any tissue but they are best known for food poisoning (discussed in Chapter 21) and for those diseases involving the skin, such as pimples and boils. The species associated with these infections is *Staphylococcus aureus,* which is present on the skin and in the nose. It inhabits the skin with another indigenous species *S. epidermidis. S. epidermidis* causes disease primarily in individuals highly compromised by procedures such as heart transplantation.

The ability of *S. aureus* to cause pimples, boils, etc. is associated with production of an extracellular enzyme, **lipase**. Lipase can degrade the fatty acid secretions

Table 27-5 *Major and minor sexually transmitted diseases caused by bacteria*

Disease	Etiological agent	Cause of infertility, abortion, or infection to fetus	Average no. cases reported between 1983-1986	Estimated no. of infections	Principal chemotherapy
Gonorrhea	*Neisseria gonorrhoeae*	Yes	850,000	2-3 million	Penicillin
Syphilis	*Treponema pallidum*	Yes	27,500	30-50,000	Penicillin
Nongonococcal urethritis	*Chlamydia trachomatis*	No*	NR†	3-4 million	Tetracycline
Lymphogranuloma venereum	*C. trachomatis*	No	250	NA†	Tetracycline
Soft chancre (chancroid)	*Haemophilus ducreyi*	No	751	NA	Erythromycin
Granuloma inguinale	*Calymmatobacterium granulomatis*	No	51	NA	Streptomycin, tetracycline
Enteric disease	*Salmonella* species	No	NR	NA	Usually not warranted
	Shigella species	No	NR	NA	Usually not warranted
	Campylobacter	Yes	NR	NA	Erythromycin, tetracycline

*Pregnant women with *C. trachomatis* infection of the cervix are a source of infection (conjunctivitis, pneumonia) to the newborn.
†*NR,* Not reportable to the CDC. *NA,* Not available.

Table 27-6 *Characteristics of bacterial diseases that affect the skin, mucosa, and eye*

Organism and disease	Morphology of microbial agent	Characteristics of the disease	Average no. cases/year (U.S.)	Vaccine	Chemotherapy
Staphylococcus aureus (pimples, boils, scalded skin syndrome)*	Gram-positive coccus in clusters	Infection of hair follicles and subcutaneous tissue; peeling skin	†	None	Not usually required
Mycobacterium leprae (leprosy)	Acid-fast bacillus	Disfiguring lesions over entire body	220-260	None	Dapsone, a sulfa drug
Treponema pertenue (Yaws)	Gram-negative spirochete	Large elevated lesions over entire body	Not in United States	None	Penicillin
Streptococcus pyogenes (impetigo)*	Gram-positive streptococcus	Contagious skin disease	?	None	Penicillin or erythromycin in extreme cases
Haemophilus aegyptius (conjunctivitis)*	Gram-negative coccobacillus	Inflammation of conjunctiva (pink eye)	?	None	Sulfonamides
Chlamydia trachomatis (trachoma in adults; inclusion conjunctivitis in children)	Gram-negative bacillus	Scarring of eye tissue in trachoma leading to blindness	1500 but 500 million worldwide; 7-9 million are blind	None	Tetracyclines

*Not reportable to CDC.
† No estimates available.

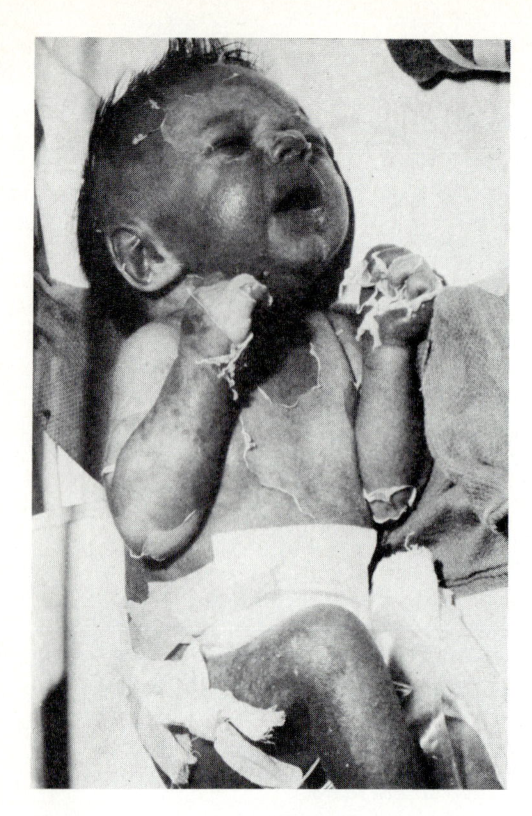

Figure 27-17 *Scalded skin syndrome caused by* Staphylococcus aureus.

found on the surface of the skin. This property apparently enables the organism to colonize and infect the sebaceous glands of the skin. The organism can also thrive in areas of high and low oxidation–reduction potential, therefore *S. aureus* survives in deep tissue.

Another skin condition called **scalded skin syndrome** is caused by a specific strain of *S. aureus* that carries a plasmid-controlled toxin called **exfoliative toxin**. This disease is characterized by a peeling off of the epidermis, revealing a red area underneath, thus the name of the disease (Figure 27-17).

Differentiation of *S. aureus* from *S. epidermidis* is critical to establishing the pathogenicity of the agent of infection. Tentative identification can be made, because most *S. aureus* strains produce a yellow-gold pigment during colonial growth, whereas *S. epidermidis* produces a white pigment. The coagulase test is an important definitive test. Pathogenic strains of *S. aureus* produce coagulase, but *S. epidermidis* does not. In the test, an inoculum of the suspected strain is suspended in plasma in which the clotting factors have been inactivated. If the bacterial strain produces coagulase, the plasma clumps; this can be detected with the naked eye. How coagulase acts as a virulence factor is not understood.

Concept Check

1. What are the symptoms associated with the various stages of syphilis? Which stages are contagious?
2. Describe the most widely used tests for determining the presence of the syphilis organism or its antibodies.
3. What is the most severe aspect of gonorrhea in men? In women? In the neonate?
4. Outline the life cycle of the chlamydiae beginning with the elementary body.
5. What characteristics differentiate *S. aureus* from *S. epidermidis*? What factors enable *S. aureus* to be a common cause of skin infections?

Diseases from animal contact or contact with arthropod vectors

In Chapter 26, several bacterial diseases resulting from contact with animals or arthropod vectors, such as tularemia, anthrax, plague, leptospirosis, undulant fever, and rickettsial diseases such as Rocky Mountain spotted fever were discussed. The following discussion focuses on plague and rickettsial disease.

Plague. Plague, a disease of ancient origin, caused pandemics that resulted in up to 100 million deaths. The last major pandemic of plague was in 1885, and it caused over 12 million deaths. Plague is also called the **black death** because subcutaneous hemorrhages darken the skin during the course of infection.

Many in the Middle Ages thought plague was a contagious disease, and in the fourteenth century, generals catapulted plague victims into each other's cities, hoping to spread the disease. The organism causing plague is a bacillus that was discovered in 1894 by Yersin and Kitasato. They observed that the organism was transmitted by the rat flea, and only when the rat died did the fleas seek other habitats, such as humans. The flea's stomach becomes blocked, and normal feeding is prevented after feeding on blood infected with plague bacilli. In an effort to feed normally the flea regurgitates the bacilli-laden stomach contents and passes them into the victim's punctured skin.

In the United States, wild rodents, such as ground squirrels, marmots, prairie dogs, wood rats, and rabbits, carry the plague bacillus and their attendant fleas. Disease in these animals is called **sylvatic plague**. In addition to becoming infected from the bite of fleas, hunters have been shown to also acquire the disease by contact with infected carcasses during the skinning of rabbits or prairie dogs. Plague is not a contagious disease unless the lungs of the victims are infected **(pneumonic**

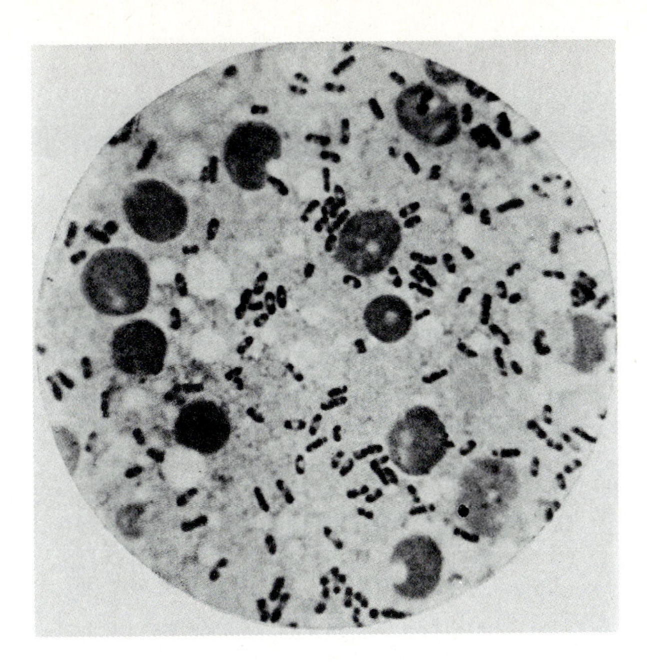

Figure 27-18 Yersinia pestis *isolated from lymph gland and stained. Note bipolar staining.*

plague), and bacilli, present in the saliva, can be transmitted by droplet spread. Symptoms of plague include fever and the appearance of swollen nodes (**buboes**) under the armpits, in the groin, and lymph glands that drain the areas where the bite occurred. The bacilli migrate to the bloodstream to infect all organs and the skin. Death is usually caused by toxemia, resulting from the heavy concentrations of bacilli in the blood.

The etiological agent of plague is *Yersinia pestis,* a gram-negative bacillus that exhibits bipolar staining (Figure 27-18). The organism appears to possess several virulence factors that enable the organism to avoid host defense and cause disease. The major virulence antigens are called **V** and **W** and help the organism to avoid digestion by phagocytic enzymes and to enhance multiplication in the phagocyte.

Streptomycin and tetracycline are effective drugs for treatment. Vaccines have limited success. A formalin-killed vaccine is currently used by employees in areas where plague is endemic and in laboratories that handle plague-infected rodents.

Rickettsial diseases. The genus *Rickettsia* belongs to a group of bacteria (rickettsiae) that also includes the genera *Rochalimaea* and *Coxiella*. Except for *Coxiella burnetii,* all rickettsial diseases are transmitted by arthropod vectors, such as lice and ticks. Nearly all rickettsial diseases cause high fever, rash, and enlargement of the spleen and liver. Most infections are treatable with tetracycline. Two rickettsial diseases are particularly important: typhus fever and Rocky Mountain spotted fever. Both diseases can be fatal, if untreated.

Typhus is transmitted by lice, and lice prevail where humans congregate and filth prevails. The human louse can live only on humans, and as long as clothes are changed infrequently or are not laundered, the louse has a chance to multiply and infect others. The louse is also affected by the typhus organisms and lives only long enough to contaminate neighboring lice and to bite its human host. Lice will leave the dying patient when the fever has become unbearable for both louse and host. Epidemics of typhus still occur, but are seldom found in the United States.

The most common rickettsial disease in the United States is Rocky Mountain spotted fever, which is caused by *Rickettsia rickettsii*. Most cases are observed in the Southeastern and Western United States. The primary vector in the Southeast is the dog tick, which is harbored by the rabbit, fox, woodchuck, deer, and squirrel. The wood tick is the most common vector in the Western United States. The disease is often misdiagnosed because of its similarity to other diseases in which a rash and fever appear. The organism is disseminated from the site of the vector bite to all organs of the body and can cause intravascular coagulation, which is a frequent cause of death. Other human rickettsial diseases are outlined in Table 27-7.

The rickettsiae are obligate intracellular parasites but unlike the chlamydiae are not parasitic for ATP. They possess a tricarboxylic acid (TCA) cycle and have an ATPase that couples respiration to ATP synthesis. They are not deficient in biosynthetic capacities, and yet they can survive only intracellularly. They are, however, extremely sensitive to most environmental conditions whether chemical or physical. Rickettsia are taken into host cells by a phagocytic process. Lysosomes in the cell make contact with the phagocytic-vacuole containing rickettsia to produce a phagolysosome. Before the lysosomal enzymes can attack the rickettsia, the latter produce a **phospholipase A** that breaks down the membrane of the phagosome. The rickettsia are released into the cytoplasm of the cell where they divide rapidly. This resistance mechanism differs from *Coxiella burnetii,* which can remain in the phagosome because it is resistant to acid hydrolases of the lysosome. In fact, the acid environment of the lysosome (pH 4 to 5) is required to activate the nutrient transport mechanisms of the microorganism. These resistance mechanisms are compared with another intracellular parasite, the chlamydia, which inhibits the fusion of lysosomes with phagocytic vacuoles (Figure 27-19).

Diseases resulting from major trauma

Some microbial agents are notorious for their ability to cause severe disease when they are introduced into

Table 27-7 *Human rickettsial diseases*

Disease	Causative agent	Mode of transmission	Geographic distribution	Reservoir of agent
Typhus group				
Epidemic typhus	*R. prowazekii*	Infected louse feces rubbed into abrasion	Worldwide	Humans
Murine typhus	*R. mooseri (typhi)*	Infected flea feces rubbed into abrasion	Worldwide	Rats and field mice
Spotted fever group				
Rocky Mountain spotted fever	*R. rickettsii*	Bite of infected tick	Western hemisphere	Rodents
Asian tick typhus	*R. siberica*	Bite of infected tick	Asia	Rodents and dogs
African tick typhus	*R. conori*	Bite of infected tick	Africa	Rodents and dogs
Rickettsial pox	*R. australis*	Bite of infected mite	Worldwide	Marsupials or house mice
Scrub typhus	*R. tsutsugamushi*	Bite of infected mite	Japan, Southeast Asia, and Pacific islands	Rodents
Q fever	*Coxiella burnetii*	Inhalation of or contact with dust particles contaminated with parturition products of cattle	Worldwide	Cattle, sheep, or goats
Trench fever	*R. quintana*	Bite of human body louse	Worldwide	Humans

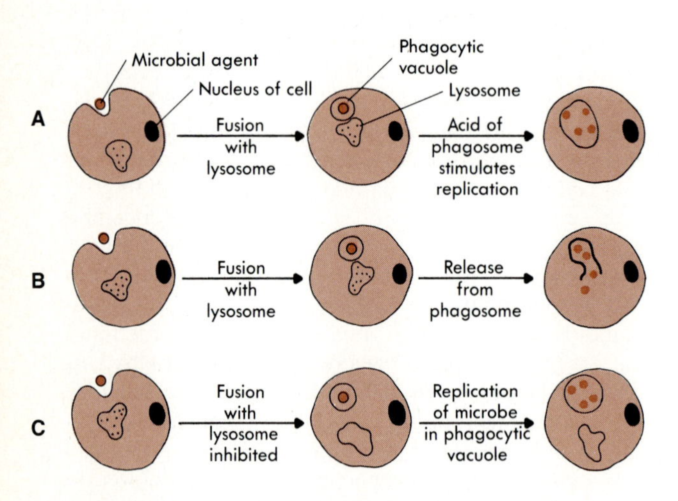

Figure 27-19 *Mechanisms of resistance to phagocytic digestion in three different bacterial species.* ***A,*** Coxiella; ***B,*** Rickettsia; ***C,*** Chlamydia.

deeper tissues by penetrating wounds, burns, or accidental injuries. The organisms that cause these diseases are found in the soil or water, sites from which they are accidentally introduced into the body. Three of the most important bacterial agents found in the soil are species of *Clostridium,* which are gram-positive, endospore-forming anaerobes.

Clostridial diseases. *Clostridium botulinum,* the cause of food poisoning, is infrequently involved in wound infections called **wound botulism**. *C. perfringens,* also a cause of food poisoning, causes a wound infection called **gas gangrene**. Finally, *C. tetani* is responsible for the disease, **tetanus**, which often results from puncture wounds.

Human infections result from the contamination of a wound by spores of the clostridial species. Deep puncture wounds or areas of dead tissue are ideal environments for germination of spores because they have reduced oxygen tension. The virulence of the clostridia results from the production of a variety of enzymes and potent exotoxins.

The toxins of *C. botulinum* and *C. tetani* are the most potent exotoxins known to humans. The toxin of *C. botulinum* was discussed in Chapter 23; it is similar in structure to the tetanus toxin. Botulinum toxin acts primarily on the neuromuscular junction, blocking the stimulation of muscle contraction and thereby causing a muscular paralysis. Tetanus toxin acts on the central reflex apparatus in the spinal cord, causing continual excitability of motor neurons and preventing relaxation of muscles. This is manifested as muscle spasms or tetany.

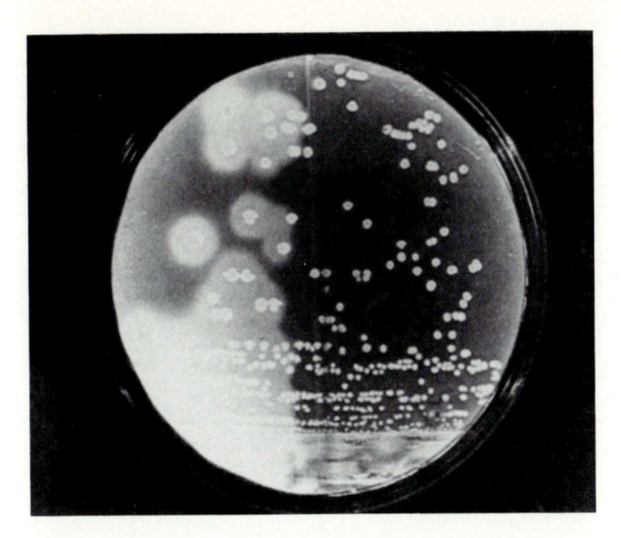

Figure 27-20 *Lecithinase detection in identification of* Clostridium perfringens. *Egg yolk agar plate is cultivated with* C. perfringens *type A. On left half of plate each colony is surrounded by opalescent zone of precipitated lipid released from lecithin in egg yolk. Right half of plate was previously treated with* C. perfringens *antitoxin (anti-lecithinase) and no zone of precipitated lipid appears.*

Convulsive contractions of the jaw, called **lockjaw**, are characteristic of the disease but other muscles are also involved. Few cases of tetanus are observed in the United States because of current immunization practices in which tetanus toxin is part of the DTP vaccine complex.

C. perfringens produces several toxins and the most important is **type A alpha toxin**. Alpha toxin acts as a **lecithinase** and splits lecithin, a component of eukaryotic membranes, into phosphorylcholine and diglyceride. Lecithinase production can be detected when organisms are cultivated on egg yolk agar. One half of the agar plate is swabbed with *C. perfringens* type A antitoxin. The culture is streaked across both halves of the plate and then incubated for 24 to 48 hours. On the half of the plate containing type A antitoxin the areas around the clostridial colonies appear unaffected. On the half of the plate without antitoxin, lecithinase production results in the formation of opaque zones around the bacterial colonies (Figure 27-20). **Collagenase** and **hyaluronidase** are two other enzymes produced by *C. perfringens* that degrade collagen and hyaluronic acid, respectively, of tissue. Tissue degradation leads to necrosis and a reduction in the blood supply, which creates anaerobic conditions and helps promote the spread of infection. During tissue degradation, carbon dioxide and hydrogen gases are some of the metabolic products released that cause swelling of the tissue. Death from gas gangrene is caused by severe toxemia due to alpha toxin.

The treatment of all three clostridial diseases depends on neutralization of any toxin that is not adsorbed to host cells, destruction of vegetative clostridia by antibiotics to prevent further toxin production, and in some cases removing dead tissue that promotes anaerobic conditions.

Diseases acquired through the alimentary tract

Microorganisms are ingested daily, and most are destroyed by enzymes or acids encountered in the stomach and intestinal tract or eliminated in the feces. Ingested bacterial pathogens that survive the intestinal tract may cause a variety of diseases. Some remain localized in the intestinal tract and cause the symptoms of **gastroenteritis**—diarrhea, cramps, nausea, and vomiting. Some microorganisms can penetrate the epithelial barrier of the intestinal mucosa and invade the bloodstream to cause **systemic disease.** A few microorganisms release toxins in food and when ingested cause a disease condition called **intoxication** (discussed in Chapter 21). Gastrointestinal disease can be treated simply by replenishing fluids and electrolytes. Antibiotics are required only when the microorganisms have penetrated the bloodstream.

A variety of bacterial agents are involved in gastroenteritis caused by the ingestion of contaminated food or water; these are outlined in Chapter 26. Our discussion is limited to a specific salmonella infection called **typhoid fever** and to gastroenteritis caused by the *Campylobacter* species.

Typhoid fever. In 1860 Dr. William Budd discovered that contamination of water with human excreta was responsible for the transmission of typhoid fever. Today, chlorination of water has practically eliminated typhoid fever from developed countries of the world. In the United States, 450 to 500 cases of typhoid are reported each year, but worldwide there are approximately 12 to 15 million cases each year. Travelers to most underdeveloped countries are at a high risk for acquiring typhoid.

Unlike most salmonella infections that result only in gastroenteritis, the agent of typhoid, *Salmonella typhi,* is capable of penetrating the intestinal mucosa and advancing into the bloodstream. Salmonella organisms enter lymph nodes where they invade and colonize the macrophage. Some salmonella infect the gallbladder where they multiply and provide a focal point for shedding of microorganisms into the intestinal tract. This carrier state condition can become chronic in the elderly. The classic carrier state, Typhoid Mary, was discussed in Chapter 26.

Typhoid fever is characterized by a persistent fever and abdominal pain. Later a rash **(rose spots)** appears on the upper abdomen. Two of the most important complications from penetration of the intestinal mucosa

Figure 27-21 *Typhoid ulceration of ileum. Note large number of ulcerations* (arrows).

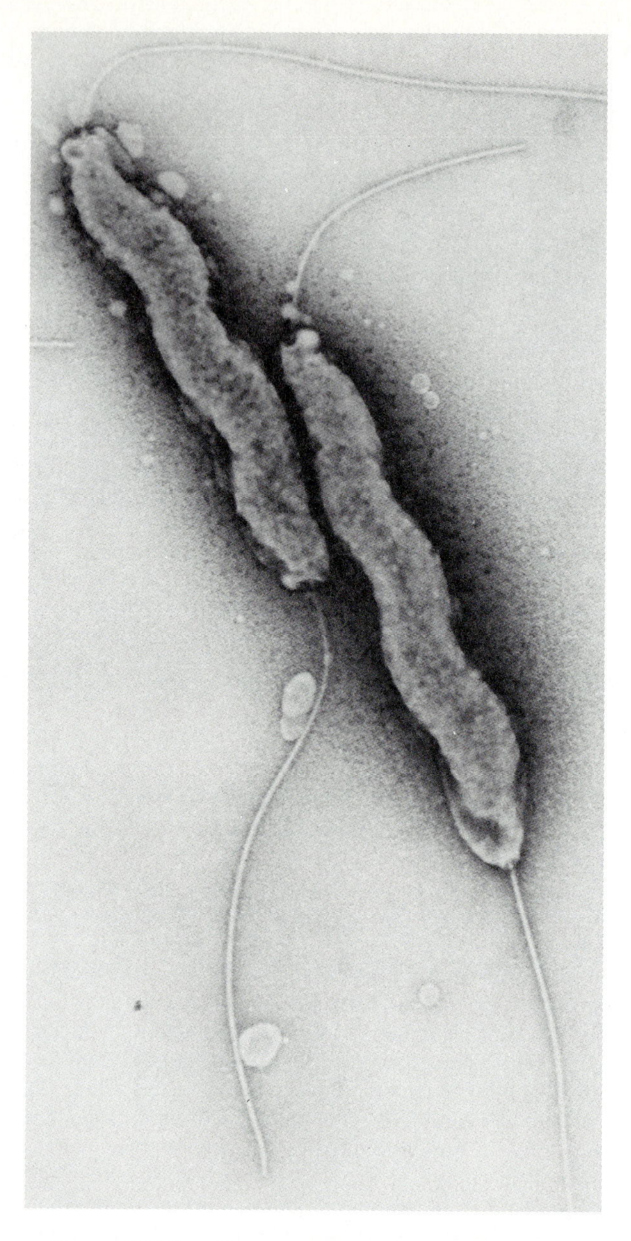

Figure 27-22 *Electron micrograph (×33,740) of* Campylobacter jejuni. *Note curved shape of organisms and polar flagella.*

are intestinal hemorrhage and perforation (Figure 27-21). The major virulence factor associated with the symptoms of disease is the endotoxin of the cell wall of this gram-negative pathogen. The endotoxin causes local inflammatory conditions at the tissue sites of salmonella multiplication. The remaining manifestations of disease are believed to be the result of acid metabolites, free oxygen radicals, and other products released by infected macrophages. Pathogenic strains of *S. typhi* are also surrounded by a capsular polysaccharide called the **Vi antigen.** The Vi antigen is composed of a linear polymer of galacturonic acid. The Vi antigen is believed to exert a pathogenic role by shielding *S. typhi* from the action of complement.

Other salmonella, such as *S. paratyphi A, S. paratyphi B,* and *S. typhimurium* are also capable of causing enteric fever similar to that caused by *S. typhi.* Chloramphenicol or ampicillin is used in treatment of typhoid fever. Vaccines are available for military personnel and those traveling to other parts of the world. Vaccines are also available to those living in areas where typhoid fever is endemic.

Campylobacter gastroenteritis. Campylobacter species are recognized as important causes of gastroenteritis. The most important species and the one most studied is *Campylobacter jejuni,* but *C. coli* is also associated with gastroenteritis.

Campylobacter is a gram-negative, comma-shaped rod, possessing a single unipolar or bipolar flagellum (Figure 27-22). *C. jejuni* is rapidly killed by stomach ac-

Table 27-8 *Viral agents that cause congenital infections*

Agent (common species name)	Consequence of fetal infection
Cytomegalovirus	Hepatitis and jaundice; congenital heart disease; mental retardation
Rubella virus	Abortion, congenital malformation, deafness, cataracts, heart defects, encephalitis
Small pox virus (variola)	Abortion, stillbirth, congenital smallpox
Varicella-zoster virus (chickenpox-shingles)	Low birth weight, bilateral cataracts, atrophic limbs, mental retardation, congenital varicella
Vaccinia virus	Abortion, generalized vaccinia (skin lesions)
Poliovirus	Congenital poliomyelitis
Measles virus	Stillbirth, congenital measles
Mumps virus	Abortion, heart damage
Herpes simplex virus (type II)	Abortion, excessive brain damage, hepatoadrenal necrosis with jaundice
Hepatitis B virus	Neonatal hepatitis, stillbirth, abortion
Echovirus	Hydrocephalus and neurologic consequences, jaundice; most infections are mild or asymptomatic
Coxsackievirus B	Myocarditis and central nervous system involvement

ids but when milk is the vehicle of transmission, the fats in the milk significantly protect the organism during transport through the stomach. Once the organism reaches the intestinal tract it colonizes the mucosa by apparently using the flagella as adhesins. Other microbial surface adhesins may be involved. The watery diarrhea associated with infection is believed to be the result of the release of an enterotoxin similar in activity to the cholera toxin. In other words there is adenyl cyclase activity in the intestinal mucosa that causes abnormal secretory activity.

Concept Check

1. What distinguishes bubonic plague from pneumonic plague? What are the virulence factors of the plague bacillus?
2. What characteristics of most *Rickettsia* distinguish them from other bacteria?
3. How are the *Rickettsia* able to survive in macrophages? How does this differ from other intracellular bacterial parasites?
4. Describe the characteristics of the three major species of *Clostridium,* including the diseases they cause and the virulence factors associated with each.
5. What virulence factors enable *S. typhi* and *C. jejuni* to cause disease symptoms or evade the host's immune system?

GENERAL CHARACTERISTICS OF VIRAL DISEASES

Viruses have been found that are capable of infecting practically every cell type. Their long-term survival within a population of humans depends on their transmissibility from cell to cell or from individual to individual. Unlike some bacteria and fungi that can survive in a resistant form, such as spores, most viruses cannot exist for long periods of time outside a living host and are susceptible to environmental factors. Slight variations in temperature, irradiation, and humidity can destroy most viral agents. Only a few viruses, such as the hepatitis viruses and the poliovirus, are unusually resistant to environmental factors, including many chemical agents that would ordinarily destroy other microbial forms.

Viruses can be transmitted to humans by the same routes as bacteria. The most common method of viral spread is via the respiratory tract. Unlike many bacterial pathogens, viruses do not always remain at the site of entry into the host and instead are disseminated to a variety of organs and tissues. It is usual to separate many viral diseases based on the type of tissue affected rather than site of entry. Many viral agents are able to cross the placenta or to be transferred in utero to the developing fetus. These are listed in Table 27-8.

The effect that viral infection has on human cells is variable. Most damage occurs when the rate of viral multiplication in the cell is high, and this is related to disruption of normal cellular metabolism. The changes that

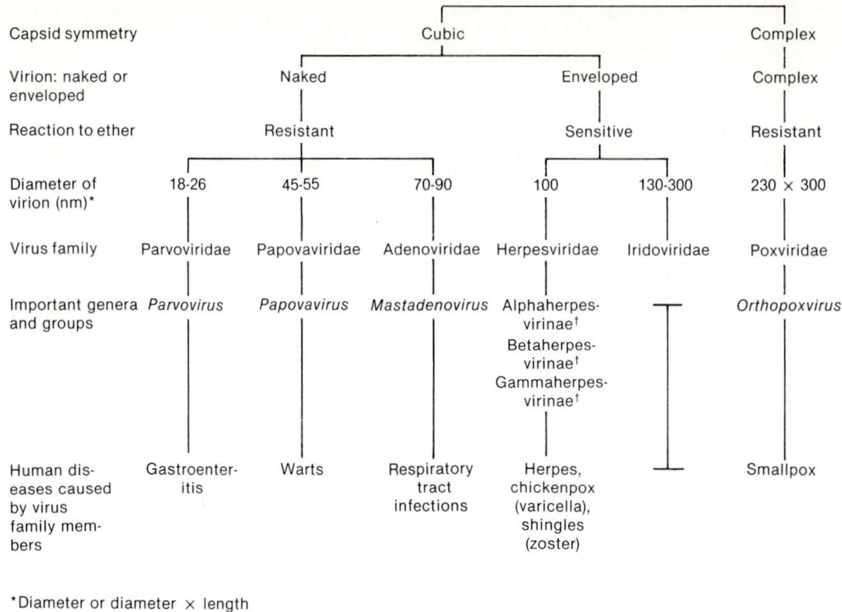

Figure 27-23 *Classification of DNA-containing viruses of vertebrates.*

occur in the cell or tissue following viral disease (cytopathic effects) may be observed microscopically and may take a number of forms (see Chapter 17).

Some viruses that possess envelopes incorporate host cell membrane components into their own structure. Thus antibodies against host-virus complexes may be generated during infection, resulting in damage to the host cell (immunopathology). Most viral diseases are acute, in which there is a short incubation period followed by rapid recovery. Those viral agents that escape the immune system of the host may cause delayed manifestations, or they may be reactivated at a later time to cause disease. Some viruses escape detection by having their nucleic acid incorporated into the host DNA. Thus without a viral protein, the virus nucleic acid, which is not antigenic inside the cell, is not recognized by the host's immune system.

The host, following viral disease, may develop resistance to further infection by the same virus by using specific and nonspecific defense mechanisms. Specific resistance to viral infection can be obtained by antibodies at the site of infection in the bloodstream. Local antibodies, such as secretory IgA, are particularly important in providing protection in the respiratory and gastrointestinal tracts. Cell-mediated immunity also plays a role in resistance to infection and appears to be more important than humoral antibodies. Individuals lacking immunoglobulins but possessing cell-mediated responses recover more rapidly from viral diseases than those possessing immunoglobulins but no cell-mediated response. In some viral diseases, such as AIDS, the helper T cells that are involved in antibody formation are destroyed. This leads to secondary infections by other viruses, bacteria, fungi, and animal parasites. Even when the immune response is functioning it may be detrimental. Immune complexes may develop that evoke an inflammatory reaction causing tissue destruction (see Chapter 25). The single most important nonspecific factor in resistance to viral infection is interferon, which is the protein produced in host cells in response to the presence of the virus (See Chapter 17).

Treatment of viral diseases is difficult, because viral duplication is dependent on host metabolic pathways, and only a few viral processes—some enzymes involved in viral RNA and DNA synthesis—are independent of the host. Most antiviral agents are seldom used in systemic diseases. Treatment of viral disease is concerned primarily with alleviation of symptoms and permitting the host's immune system to eliminate the virus, or an antiserum (hyperimmune globulin obtained from humans who have a high level of antibodies to the virus) may be administered. The most effective method for controlling viral disease is prevention through the use of vaccines. Today, many of the childhood diseases, such as measles, mumps, polio, and rubella, are effectively controlled because of immunization practices.

The viruses that cause disease in vertebrates belong to a number of different families whose separation has been based on the type of nucleic acid, symmetry, and other factors. Figures 27-23 and 27-24 represent abbreviated classification schemes for the major RNA and DNA viruses that infect vertebrates.

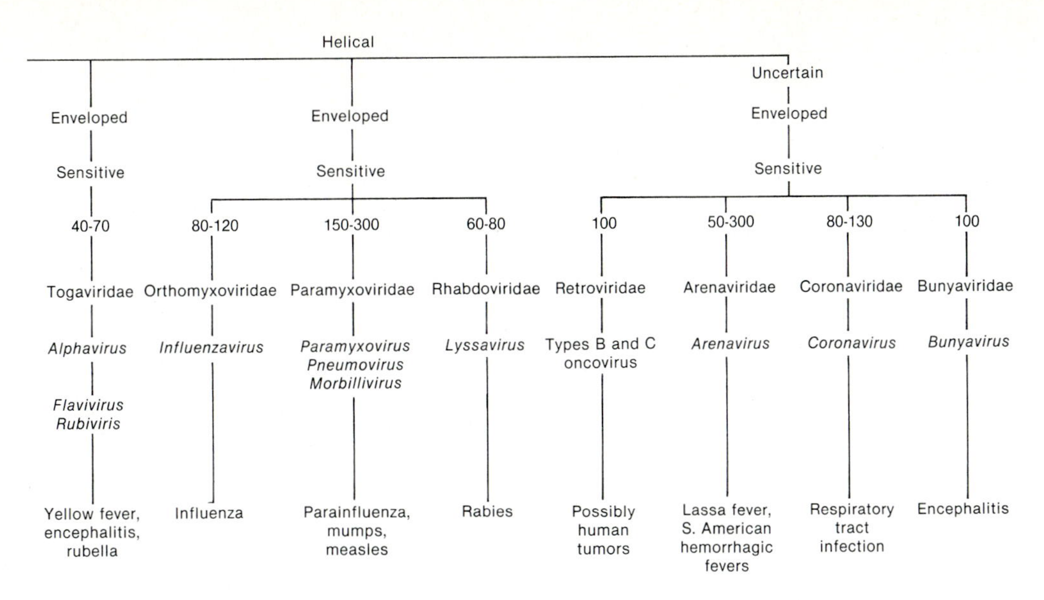

Figure 27-24 *Classification of RNA-containing viruses of vertebrates.*

Viruses that affect nervous tissue

Many viruses are capable of replicating in nervous tissue, but only a few find their way to the nervous system after naturally occurring infections. Viruses may reach the nervous system by three routes: **olfactory** (nasal mucosa), **neural** (nerve), or **hematogenous** (bloodstream). Spread via the nasal mucosa appears to be of minimal importance. A few viruses are spread by the neural route in which they travel along the axons of peripheral nerves and then multiply in the nucleus of the nerve cells. Animal studies demonstrated that the poliovirus, rabies virus, and herpes simplex virus are capable of neural spread to the central nervous system. The most common method of spread to nervous tissue in naturally occurring infections is via the hematogenous route, where the virus multiples outside nervous tissue, as in the epithelium of blood vessels or in lymphatics. This maintains a high concentration of virus in the blood (viremia) and is followed by the passage of virus into the brain or cerebral spinal fluid. Those viruses that use the hematogenous route to infect nervous tissue are the rabies virus, poliovirus, echoviruses, mumps virus, herpes simplex virus, and cytomegalovirus. Viruses are spread more readily by the hematogenous route, but the other avenues can also be used, depending on the initial site of inoculation. Characteristics of viral infections affecting nervous tissue are outlined in Table 27-9. We will briefly discuss two of them: rabies and polio.

Rabies. Rabies is a disease of certain wild animals, such as skunks, bats, racoons, and foxes, and domestic animals, particularly dogs, that can be transmitted to humans. The rabies virus is transmitted to humans via a bite or contact of an abraded area of the skin with the saliva from an infected animal. Occasionally inhalation of aerosolized virus in caves haboring large numbers of bats can be a means of transmission to humans. There have been recorded instances in which rabies was transmitted during corneal transplantation in which the donor died from a disease not recognized as rabies. The virus multiplies at the site of infection for several days and then travels along the peripheral nerves to the central nervous system, where it localizes and initiates an inflammatory response (**encephalitis**) that is usually fatal. Degeneration of nervous tissue leads to the symptoms and manifestations of infection. The patient has difficulty swallowing, and the sight of liquids can induce painful contractions of the muscles involved in swallowing—a condition referred to as **hydrophobia** (fear of water). Death results from respiratory paralysis.

The incubation period for rabies is 2 to 16 weeks, and by the time the symptoms are recognized, it is too late to save the patient. Therefore prompt treatment following the bite of a potentially infected animal is important. The brain of the infected animal should be examined for the presence of virus. Infected nerve cells observed by light microscopy show the presence of Negri bodies, and virus can be detected by electron micrography (Figure 27-25). Smears of brain tissue can also be identified by fluorescent microscopy, which today is the principal method of identification.

There are only three recorded cases of survival from clinically apparent rabies; however, postexposure treatment can be effective if instituted immediately after the bite of an infected animal. Antirabies serum is applied to the wound, followed by active immunization with a recently developed rabies vaccine that is prepared from

Table 27-9 *Viral infections that affect nervous tissue*

Viral agent*	Clinical condition	Mode of transmission	Geographic distribution	Vaccine	Treatment
Poliovirus *(Enterovirus)*	1% to 2% of those infected show paralysis of limbs	Fecal-oral route	Worldwide	Yes	None specific
Rabies virus *(Lyssavirus)*	Encephalitis	Contact of abraded skin with saliva from infected animal or animal bite	Worldwide	Yes (humans and animals)	Antirabies serum followed by vaccination
Western equine encephalitis virus *(Alphavirus)*	Encephalitis	Bite of mosquito	North and South America	None	None specific
Eastern equine encephalitis virus *(Alphavirus)*	Encephalitis	Bite of mosquito	Eastern and Northern United States	None	None specific
St. Louis encephalitis virus *(Flavivirus)*	Encephalitis	Bite of mosquito	Central United States	None	None specific
Venezuela equine encephalitis virus *(Alphavirus)*	Encephalitis	Bite of mosquito	South and Central America and United States	None	None specific
California encephalitis virus *(Bunyavirus)*	Encephalitis	Bite of mosquito	United States	None	None specific
Mumps virus† *(Paramyxovirus)*	Encephalitis	Respiratory secretions	Worldwide	Yes	None specific
Varicella (chickenpox) virus† (subfamily: Alphaherpesvirinae)	Encephalitis	Respiratory secretions	Worldwide	Yes	See Table 27-11
Measles (rubeola) virus† *(Morbillivirus)*	Encephalitis	Respiratory secretions	Worldwide	Yes	Gamma globulin can modify disease
Herpes simplex virus† (subfamily: Alphaherpesvirinae)	Encephalitis	Respiratory secretions	Worldwide	None	None specific
Echoviruses†‡ *(Enterovirus)*	Meningitis	Fecal-oral route	Worldwide	None	None specific
Coxsackieviruses *(Enterovirus)*	Meningitis	Fecal-oral and respiratory secretions	Worldwide	None	None specific
Mumps virus *(Paramyxovirus)*	Meningitis	Respiratory secretions	Worldwide	Yes	None specific

*The common species name of the viral agent is used with the genus in parentheses.
†The condition caused by the agent is a complication of infection and not the primary manifestation of infection.
‡Echoviruses: *e*nteric *c*ytopathogenic *h*uman *o*rphan viruses.

human fibroblast cultures. The vaccine, which requires only 5 or 6 doses (the old vaccine required 23 doses), has proved effective in Europe and in 1980 was made available for general use in the United States. Rabies can be prevented in most human populations by vaccination of domestic animals, such as dogs. Those individuals, such as veterinarians, who are exposed to wild animals can also be vaccinated.

Wild animals now constitute the most important source of infection for humans and domestic animals. Skunks, in particular, represent a threat to humans because of their contact with domestic animals. Rabid dogs

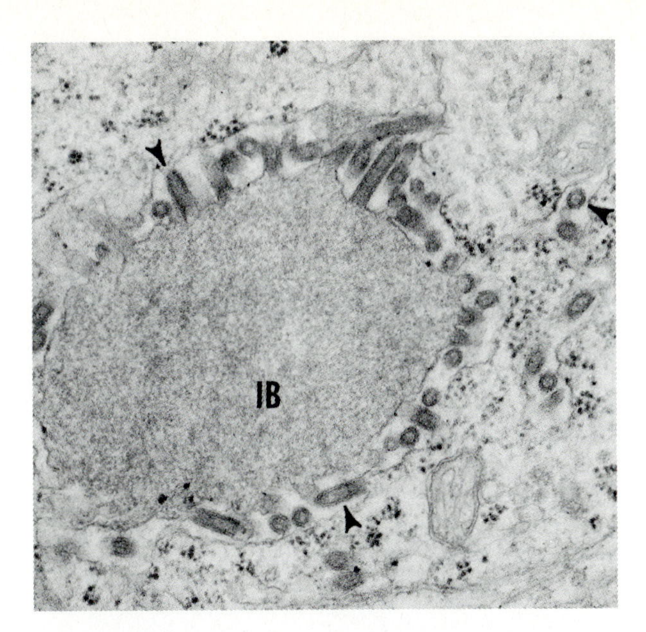

Figure 27-25 *Rabies inclusion bodies (Negri bodies); electron micrograph of rabies virus infection in hamster brain neuron (×50,000). Virus* (arrows) *can be seen budding from intracytoplasmic membrane in association with inclusion body* (IB).

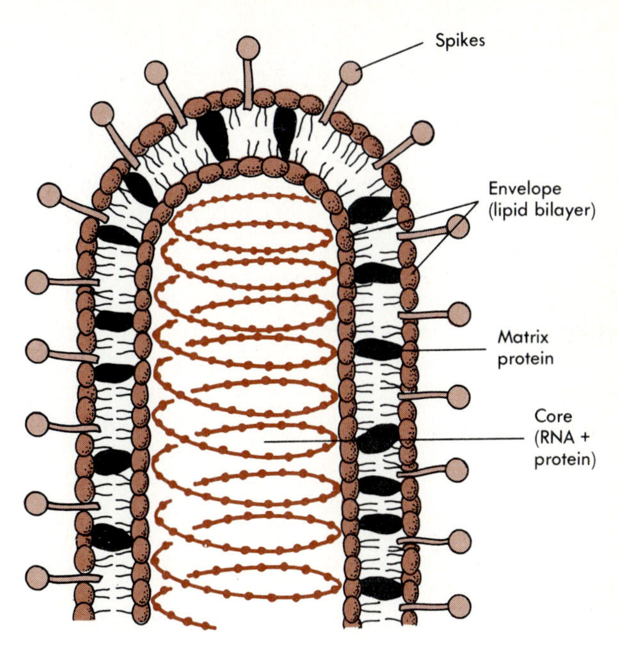

Figure 27-26 *Diagrammatic illustration of basic morphology of rabies virus.*

still present a great health hazard because of their close association with humans. In 1980, for example, over 103 children from an elementary school and a junior high school in Illinois came in contact with a dog with blood on its side, one half of its tail missing, and a limp in its left leg. Twenty students had significant exposure in which some were bitten, scratched, or had open wounds that had been licked by the dog. All 20 children received postexposure prophylaxis and none developed rabies, but the dog died of the disease shortly thereafter.

The rabies virus is cylindrical or bullet shaped (Figure 27-26) and possesses a single strand of RNA. The RNA genome is negative stranded and is not infectious like the poliovirus. This is because replication of virus in the host requires a virus-associated RNA polymerase that is part of the nucleocapsid. The surface of the virus consists of an array of knobbed glycoprotein projections protruding from a lipid bilayer envelope that surrounds the capsid. The glycoprotein projections serve as attachment receptors for the virus to host cells.

Polio. The disease now called poliomyelitis or polio probably originated in ancient times. Various Egyptian inscriptions depict individuals with withered limbs that many scientists believe to have been victims of poliomyelitis. The disease was not recognized as a distinct clinical entity until 1784.

A great deal of public interest in polio resulted from the 1932 presidential election of Franklin D. Roosevelt, a victim of polio. A March of Dimes campaign was originated in 1938 whose sole purpose was to collect money to fight infantile paralysis (the common name used at the time, since the disease affected primarily the very young). John Enders, Thomas Weller, and Frederick Robbins in 1949 succeeded in culturing the poliovirus in human embryonic tissue—a discovery that heralded the eventual development of viral vaccines. Jonas Salk in 1953 produced a formalin-inactivated vaccine that was licensed for use in 1958. Problems with quality control for the inactivated vaccine resulted in over 200 cases of vaccine-associated polio, but these problems were resolved. Most countries today use the oral-attenuated poliovirus vaccine developed by Albert Sabin. Vaccination resulted in a drop in the number of cases of polio in the United States from 340,000 in 1955 to 5 in 1985.

The epidemiology of polio is interesting when urban conditions in which sanitation is at a high or low level are compared. The poliovirus is easily transmitted among infants and children under 4 years of age when hygienic conditions are lacking. The virus can be found in untreated sewage and water supplies, and ingestion of virus is followed by viral multiplication in the oropharynx or intestine. The disease is relatively mild with only minor symptoms and is rarely paralytic. Carriers represent a large pool of infectious virus, and large amounts of virus are shed from the intestinal tract of asymptomatic and symptomatic carriers. Children therefore become immune very early in life because of their contact with carriers. Improvement in sanitation (improved sewage disposal, rigid water treatment, and improved individual hygiene) brings about a change in the epidemiology of poliovirus infection, because few carriers are present and fewer children under 4 years of age become infected. Thus a much larger group of susceptibles over 5 years of age develops in the population.

Paralytic polio, which occurs infrequently, results from invasion by the virus into the bloodstream and then into the central nervous system. Viral multiplication in the central nervous system, particularly in the spinal cord and brain, results in nerve cell damage that eventually affects muscle function. Paralysis may be temporary or permanent, but less than 2% of those infected with poliovirus show signs of paralysis.

Vaccination can reduce the number of cases of paralytic polio. The oral vaccine offers the following advantages: (1) there is no need for subcutaneous injection; (2) there is an induction of intestinal immunity that prevents reinfection from wild* virus; and (3) there is no need for repeated booster doses. Probably the most important reason for use of the live vaccine is the fact that it protects the individual from intestinal infection from wild virus. An individual receiving the inactivated vaccine can later be infected by a carrier of active poliovirus and can remain a carrier and a source of infection to others. A case of vaccine-associated paralysis occurs with the live vaccine in rare instances (once for every 3 million doses of vaccine administered) and is frequently associated with immunodeficient or immunosuppressed patients. It has been estimated that the poliovirus vaccines have prevented over 5 million cases of paralytic polio in the past 20 years.

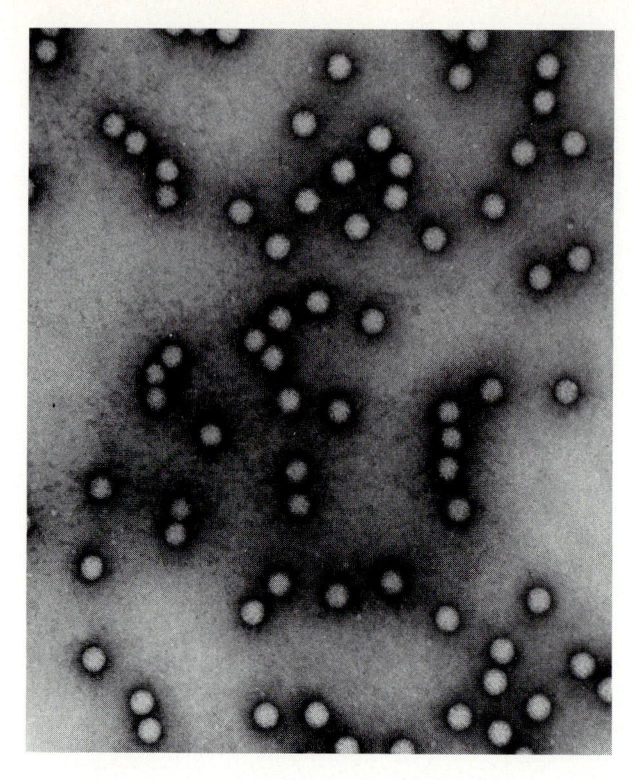

Figure 27-27 *Electron micrograph of poliovirus. (×29,000.)*

Polio is a disease caused by an RNA virus (Figure 27-27). The poliovirus belongs to a group of viruses called the enteroviruses, which are all included in the family Picornaviridae. They are small (20 to 30 nm) icosahedral RNA-containing viruses that are stable to pH 3 to 5. This acid stability allows them to be ingested and reach the intestinal tract unharmed. The genome is a single plus-stranded RNA whose replication is discussed in Chapter 17. Intact virus is infectious only for cultured primate cells, whereas the naked RNA can infect any cell type. This characteristic indicates the importance of recognition between the intact virion and receptors on the cell membrane surface.

Viruses that affect respiratory tissue

The site where virus lodges in the respiratory tract is related to the size of the material in which the virus is lodged. If the virus is present in a respiratory secretion, the larger droplets remain entrapped in the upper respiratory tract, whereas smaller droplets evade entrapment and are carried to the lower respiratory tract. The viscosity of the mucus and its rate of flow influence viral attachment to epithelial tissue. The lower the viscosity and the slower the rate of flow of mucus, the greater the chance for virus to attach to the epithelium. Several viral agents have an affinity for respiratory tissue and include the influenza viruses, the rhinoviruses, respiratory syncytial virus, parainfluenza viruses, and coronavirus (Table 27-10). Discussion of the most important of this group, the influenza viruses, follows.

Influenza. The word *influenza* is derived from the Italian, meaning influence and in the thirteenth century was used to describe a disease that appeared during the winter months. The winter-occurring disease was believed to be caused by the influence of the stars, and its characteristic recurrence during the winter months led to the permanent use of the world **influenza**. Influenza, unlike any other disease, is global and is not endemic to any one area. Most populations have been visited by the disease at one time or another. The first recorded pandemic in 1580 was followed by major outbreaks in 1645, 1743, 1782, 1830, 1837, and 1847. The most severe pandemic appeared in 1918 when nearly 20 million lives were lost, mostly to a major complication, bacterial pneumonia. Large scale epidemics still occur periodically, but none has ever matched the geographic scope and morbidity of the 1918 pandemic.

The influenza virus was first isolated in 1933 and was

* *Wild* refers to virus that exists naturally in the environment and is not a laboratory strain that might be used in a vaccine.

Table 27-10 *Viruses that affect respiratory tissue*

Viral agent*	**Clinical condition**	**Method of transmission**	**Vaccine**	**Treatment**
Rhinovirus types 1A to 114 *(Rhinovirus)*	Common cold observed in all age groups	Close contact with respiratory secretions	None	None specific
Parainfluenza virus *(Paramyxovirus)*	Common cold in older children and adults; in infants more serious illness, such as bronchitis and croup, requiring hospitalization	Close contact with respiratory secretions	Under investigation for infants	None specific
Respiratory syncytial virus *(Pneumovirus)*	Most frequent cause of infant pneumonia and bronchiolitis; infection in older children and adults is milder	Close contact with respiratory secretions	None	None specific
Human coronavirus *(Coronavirus)*	Colds; most infections are asymptomatic; no lower respiratory tract involvement	Close contact with respiratory secretions	None	None specific
Influenza virus types A, B, and C *(Influenzavirus)*	Mild symptoms—fever, headache, sore throat, and conjunctivitis; 2% develop pneumonia	Close contact with respiratory secretions	Yes	Amantadine has been used as prophylactic agent; therapeutically it reduces recovery time

*The common species name is used with the genus in parentheses.

called influenza virus type A. Type A, is the most frequent cause of disease, but type B can also cause infection. Influenza viruses are pleomorphic and may exist as spheres or filaments, although most are spherical. The capsid is surrounded by a lipid coat containing spikes of hemagglutinin and neuraminidase (Figure 27-28). A virus-associated, RNA-dependent RNA polymerase is part of the internal nucleoprotein, but the most abundant protein is the matrix protein (M protein) that lies just beneath the inner surface of the lipid envelope. The viral genome is a single negative–stranded RNA that appears in the nucleocapsid as eight distinct segments. Influenza A, which is responsible for the major pandemics, has undergone considerable antigenic variations since it was first discovered. For this reason the new variants are described by their subtypes. The subtyping is based on the antigenic variation of the hemagglutinin (H) and neuraminidase (N) spikes. There are four hemagglutinin types (H_0,H_1,H_2,H_3) and two types of neuraminidase (N_1,N_2). The source of isolate, geographical location or origin of isolate, strain number, and year isolated is also included. For example, influenza A could be described as: A/swine/USSR/5/85/(H_2N_2).

In the infection process the virus adheres to the respiratory epithelium by the **hemagglutinin** spikes while the **neuraminidase** spikes make contact with their substrate, neuraminic acid, which is part of the mucosa. Viral multiplication leads to the hydrolysis of neuraminic acid and liquification of the mucosa, leaving an underlying epithelial layer unprotected from infection by other microbial agents and promoting the spread of the virus to surrounding tissue. Unabated, the disease process may involve the bronchioles. The incubation period for influenza virus disease is 24 to 48 hours, with most infections showing no symptoms or mild symptoms, such as fever, headache, sore throat, and conjunctivitis. Two percent of the cases are complicated by pneumonia, and most of these are associated with individuals having chronic lung or heart disease. Pneumonia caused by *Staphylococcus aureus* or *Streptococcus pneumoniae,* is the most frequent cause of death after infection by the virus.

It is the genetic instability of the virus that is most responsible for the large-scale pandemics that occurred in 1918-1919, 1957-1958, and 1968-1969, as well as the more periodic epidemics. Changes in the influenza virus nucleic acid result in alterations in the chemical structure of the neuraminidase and hemagglutinin spikes that will affect the outcome of future infections. Antibodies to the hemagglutinin spikes interfere with viral infectivity, whereas antibodies to neuraminidase prevent spread of virus from cell to cell. What this means is that the antibodies produced during infection by one strain of influenza A virus may be unable to ward off future infec-

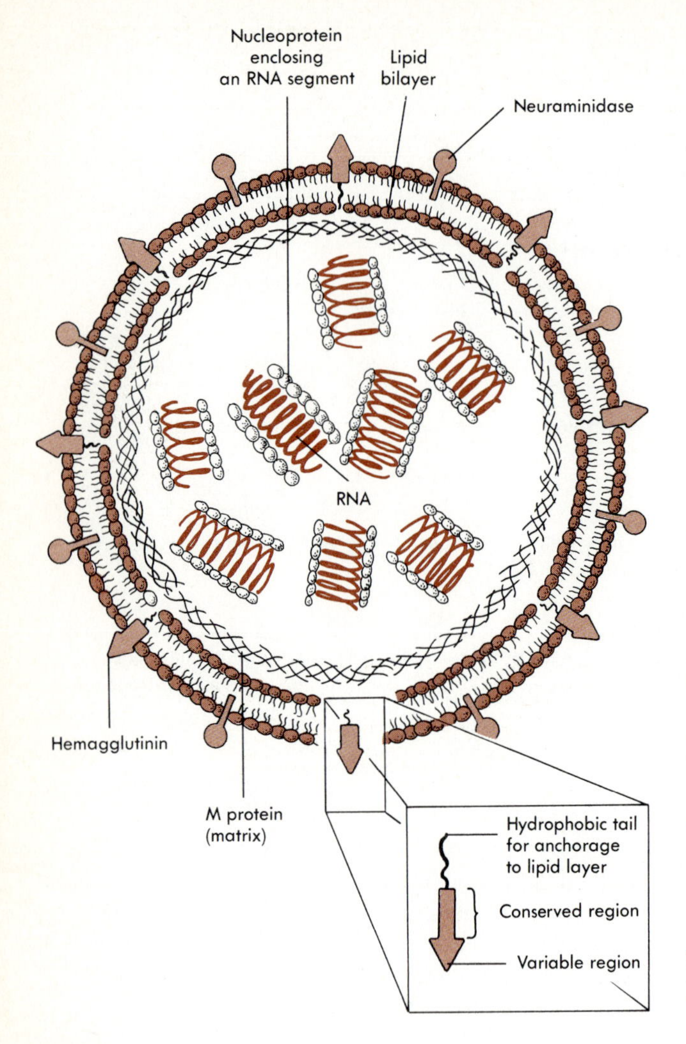

Figure 27-28 *Characteristic morphology of influenza virus that contains segmented genome (8 segments).*

tions by influenza A virus in which a genetic change has taken place. These genetic changes result in variations that may be classified as either minor or major. To understand this, one must first consider that, following infection, antibodies to one virus protect the individual for approximately 2 to 3 years from infection by the same strain of virus. When a minor change, called **antigenic drift**, occurs in the viral spikes, the antibodies that were produced in response to the parent virus do not interact to the same degree with the mutant virus. When a major change, called **antigenic shift**, occurs in the viral spikes, the antibodies produced not only show a reduced affinity for the parent virus, but also do not recognize the mutant's altered spikes. The result is that the individual or population originally infected by the parent influenza virus is unprotected from the infection by the mutant virus in which the antigenic shift has occurred. Thus vacines produced against influenza virus type A of one type might be only 60% effective against a virus that has undergone antigenic shift.

Epidemics of influenza A occur every 2 to 4 years, whereas pandemics occur every 10 years. Considerable genetic change can occur in the virus between epidemics and pandemics, and immunity to reinfection can be related to the genetic character of the virus and the dose of infecting virus. Influenza vaccines are recommended only for high-risk groups, such as the elderly over 65 years of age and those with chronic cardiovascular and pulmonary disorders. In recent years an antiviral drug called **amantadine** has been used with some success as a prophylactic agent in preventing influenza and as a therapeutic agent in decreasing the length of time for recovery from disease.

Influenza virus type B does not show the antigenic variation of its sister virus, influenza A. Influenza B seldom occurs in adults, but in children influenza B virus, influenza A, and some other viruses are believed to be associated with a complication called **Reye's syndrome.** Reye's syndrome is characterized by encephalopathy and fatty degeneration of the viscera. The disease is fatal to approximately 25% of those affected. Nationwide outbreaks of Reye's syndrome (350 to 400 cases per year) have been associated with influenza B outbreaks during the 1973-1974 and 1976-1977 influenza seasons and with influenza A in the 1978-1979 season. In 1985 only 91 cases were reported. It is suspected, but has not yet been proved, that salicylates, such as aspirin, administered for influenza infection contribute to the pathogenesis of Reye's syndrome.

Concept Check

1. Why is the treatment of viral diseases difficult compared to the treatment of bacterial diseases?
2. Which viruses affect nervous tissue? What is the most common avenue that a virus takes to infect nervous tissue?
3. Why would someone argue that factors others than vaccines are responsible for the decline in poliomyelitis?
4. Explain the pathogenic mechanism of the influenza virus on respiratory tissue. What is the relationship of influenza to pneumonia?
5. Why is it so difficult to develop influenza A virus vaccines? What is meant by antigenic shift? Antigenic drift?
6. What is Reye's syndrome and what is its relationship to influenza B infection?

Viruses that affect the skin including sexually transmitted diseases

The intact skin or mucosal surfaces are effective barriers against penetration of viruses. Viruses can penetrate

Table 27-11 *Characteristics of viruses that affect the skin*

Viral agent*	Clinical condition	Method of transmission	Vaccine	Treatment
Rubella virus *(Rubivirus)*	German measles—mild infection with rash when acquired postnatally; when acquired by pregnant woman, virus can be transferred to fetus and cause fetal abnormalities or death	Contact with respiratory secretions or congenital transfer through placenta	Yes, for children over 12 months of age	None specific
Smallpox virus *(Orthopoxvirus)*	Smallpox—development of large pustules on skin; as of 1978 disease has been declared eradicated	Contact with pustules or various fomites contamined by fluid from pustules	Yes, but is no longer recommended in United States; only those who may be exposed to virus (lab workers) receive vaccine	None specific
Varicella-zoster virus (subfamily; Alphaherpesvirinae)	Varicella (chickenpox)—highly communicable disease of children, producing rash; zoster (shingles) causes pain in skin area where nerves have been affected	Varicella by contact with fluid in skin lesions or contact with respiratory secretions; zoster is reactivation of latent virus from primary chickenpox infection	None	Zoster immune globulin is available for passive protection against varicella in high-risk adolescents exposed to chickenpox
Papilloma virus *(Papovavirus A)*	Warts	Direct contact or indirectly by contact with fomites; can be transferred from one part of body to another by autoinoculation	None	Not required, but surgery has been used
Dengue virus *(Flavivirus)*	Dengue fever; rash with joint pain; severe form of disease causes extreme vascular hemorrhage; prevalent in Carribbean	Bite of mosquito	Vaccine not yet available for commercial use	None specific
Measle virus *(Morbillivirus)*	Measles (rubeola); usually mild with skin lesions, as well as oral lesions called Koplik's spots	Inhalation of or oral contact with respiratory secretions	Yes, given 15 months after birth	No specific therapy
Herpes simplex virus (family: Alphaherpesvirinae)	Type I—facial lesions Type II—genital lesions	Type I—contact with oral secretions or facial lesions; Type II—sexual contact	None	Acyclovir is useful for treatment of primary type II disease

*The common species name of the viral agent is used with the genus in parentheses.

these barriers when (1) the skin or mucosa is abraded, such as during sexual contact; (2) the skin is punctured by needles or by animal bites; (3) there is an underlying disease (eczema); or (4) the skin is bitten by arthropod vectors, such as mosquitoes. The above avenues, except for sexual contact, are not associated with skin lesions. For example some viruses enter the host through the respiratory tract and later find their way to the skin, whereas others, such as those associated with sexual contact, remain at or near the initial site of infection. The viruses discussed in this section are those that specifically cause skin lesions. They include two sexually transmitted diseases, genital herpes and venereal warts, and measles, which is not a sexually transmitted disease. AIDS, also a sexually transmitted disease that affects lymphocytes, is discussed in Chapter 25. The characteristics of viruses that affect the skin are outlined in Table 27-11.

Genital herpes. Genital herpes does not cause a life-threatening condition, but it does cause emotional

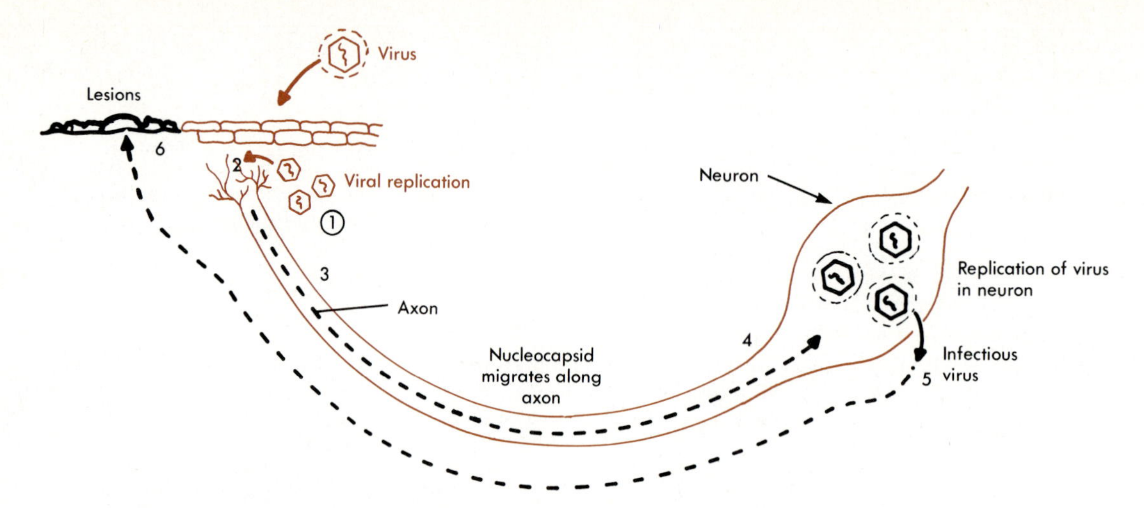

Figure 27-29 *Postulated mechanism by which herpes simplex virus causes infection.* ***1,*** *Virus initially infects skin and replicates at site of inoculation;* ***2,*** *virus is taken up by nerve endings and envelope is removed;* ***3,*** *viral nucleocapsid migrates along axon;* ***4,*** *nucleocapsid in neuron replicates and infectious virus formed;* ***5,*** *infectious virus becomes activated to migrate in axon back to area of initial site of inoculation;* ***6,*** *virus causes lesions on skin.*

trauma because it can recur over and over again. Between 2 and 3 million persons in the United States are believed to be affected by genital herpes.

Genital herpes is caused by the **herpes simplex virus type II (HSV II)**. (HSV I is associated with infections of the facial area, such as coldsores.) The virus replicates in host cells at the site of infection and then spills out to infect adjacent cells, including nerve endings (Figure 27-29). In the nerves the virus is uncoated, releasing a noninfectious nucleocapsid. The nucleocapsid travels toward the neurons. Once inside the neurons the noninfectious nucleocapsid undergoes at least one cycle of replication, resulting in the formation of infectious virus. Infectious virus migrates, from the collection of neurons (ganglia) to the inoculation site where viral replication creates lesions characteristic of disease. Thus the virus makes a round-trip before disease symptoms appear. Herpesviruses are notorious for causing latent infections with recurrences presumably brought about by such factors as stress, hormonal changes, abrasions, and exposure to sunlight. It is believed that the virus remains dormant in sensory or spinal ganglia awaiting reactivation.

Infection in men usually involves lesions on the penis. Penile herpes is extremely painful; however, the condition resolves on its own. Lesions may also appear near the anus or oral cavity, depending on sexual practices. Most cases of genital herpes in women involves the cervix. This condition is often asymptomatic and characterized only by inflammation. Women are more likely to engage in sexual intercourse when infected because of the asymptomatic nature of the disease and therefore are a more frequent cause of herpes transmission to others. Women who acquire primary genital herpes infection during pregnancy excrete large amounts of virus over a 3 to 4 week period, and cesarean delivery may be required to prevent disease and death of the fetus. Viral excretion during recurring genital herpes infection is brief, and HSV antibody is present in the serum of the mother, as well as in that of the fetus. The fetus is therefore protected and cesarean delivery is not required. Herpes can result in abortion and at least one half of those fetuses who survive will be afflicted with some neurological or ocular disorder. Systemic disease may also involve the viscera and central nervous system.

There have been many studies employing various treatments for genital herpes, but few have proved effective. One of the most promising drugs is **acyclovir**, which is specific for viral enzymes and is not metabolically degraded following systemic administration. Acyclovir as a topical ointment is effective for first episodes of genital herpes but has no apparent usefulness for recurrent herpes.

The herpesviruses (Figure 27-30) are icosahedral viruses whose nucleocapsid contains a double-stranded DNA genome. They can be found in a variety of vertebrates, and five are known to infect humans, including **varicella-zoster virus**, the causative agent of two clinical entities: **chickenpox** (varicella) in children and **shingles** (zoster) in adults; **cytomegalovirus**, which causes an important congenital disease, as well as a potentially fatal disease in immunosuppressed adults; the **Epstein-Barr virus**, which is the causative agent of **infectious mononucleosis**, as well as **Burkitt's lymphoma** and nasopharyngeal carcinoma; and HSV I and II.

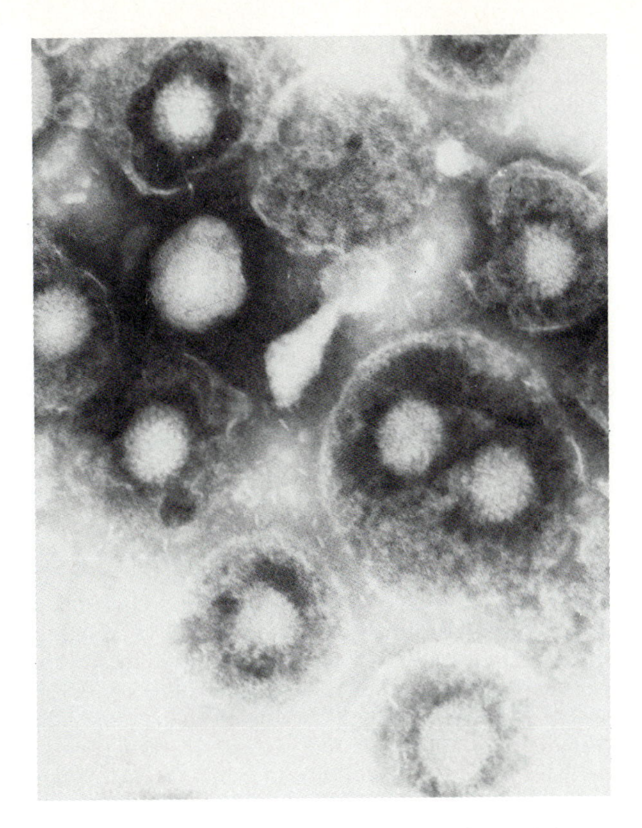

Figure 27-30 *Herpesvirus surrounded by envelope.*

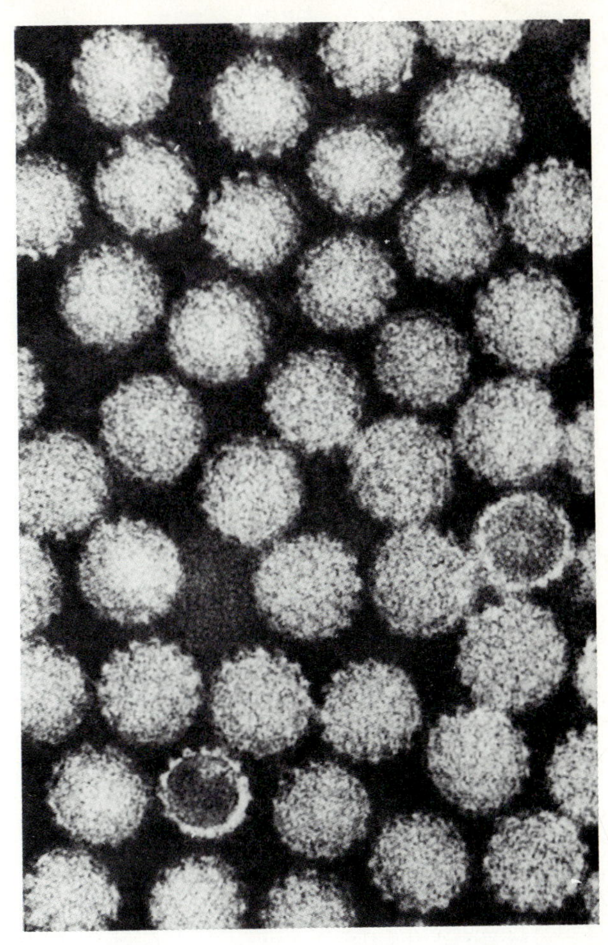

Figure 27-31 *Electron micrograph of human papilloma virus.*

Venereal warts. Venereal warts (**condylomata acuminata**) are soft pink cauliflower-like growths that appear singly or in clusters on the external genitalia and rectum. They are caused by a human papilloma virus (Figure 27-31), and as many as five distinct papilloma viruses may be involved in wart formation. This disease is one of the most common sexually transmitted diseases in the United States, and complications can be severe. The estimated number of cases of genital warts in the United States is 1 million. The smaller growths usually make intercourse difficult or cause rectal pain. Larger growths can make urinating difficult and in women may result in transmission to neonates during childbirth. Neonatal transmission is thought to cause childhood laryngeal papillomatosis. There is some evidence that cervical neoplasia and perianal neoplasia occur in patients with a history of anogenital warts. Spontaneous regression can occur and is most often observed in women after childbirth.

Treatment involves the use of topically administered pharmacological agents, such as podophyllin, dichloracetic acid, colchicine, idoxuridine, and other antiviral agents. If these are unsuccessful, surgery is recommended.

Measles (rubeola). Measles is an acute infectious disease associated primarily with children. It is highly communicable, and before the advent of a measles vaccine in 1967 over 500,000 cases were reported annually in the United States. Infection is acquired by inhalation or oral contact with respiratory secretions. The virus may also enter the body through the conjunctiva. Initially the virus multiples in the respiratory mucosa and regional lymph nodes before it is disseminated through the bloodstream. A rash appears on the skin (Plate 37) about 2 weeks after contact with the virus, and lesions called **Koplik's spots** appear on the oral mucosa. A fever, which may reach as high as 105° F, precedes and accompanies the rash. Measles, although considered a relatively mild disease, can be fatal to those who are immunologically deficient. Most deaths are the result of complications following degeneration of tissue cells by the virus and include bacterial pneumonia and encephalitis. A very rare complication, following measles is called **subacute sclerosing panencephalitis (SSPE)**. SSPE causes a progressive mental deterioration that does not

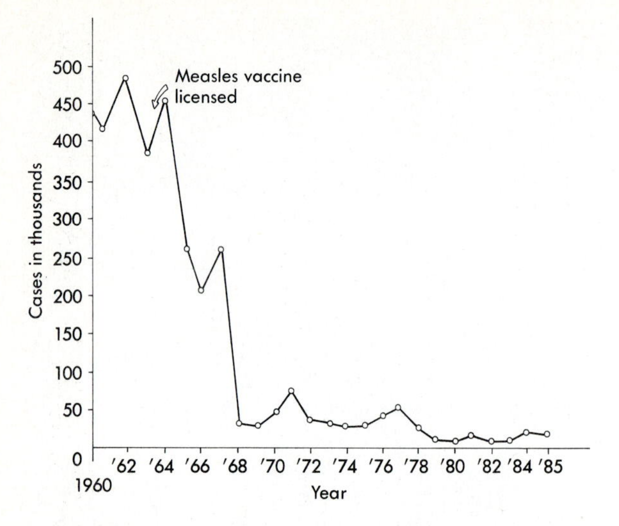

Figure 27-32 *Number of reported cases of measles in the United States, 1960-1985.*

appear until 5 to 10 years after apparent recovery from measles. It has not been determined whether SSPE is caused by the original measles virus or a variant of it.

Infants are protected from the measles virus for at least 6 months because of antibodies acquired from the mother. The measles vaccine, which is administered 15 months after birth, has reduced the incidence of measles from over 450,000 cases in 1964 to 22,000 in 1968, but nearly 60,000 cases were reported in 1977 (Figure 27-32). Some of the rise was due to a lack of parental concern in obtaining the vaccine for their children and to the fact that the original vaccine used between 1963 and 1967 was an inactivated one and was poorly antigenic. Children vaccinated between 1963 and 1967, when later exposed to the measles virus, have manifested a disease that frequently terminates in severe and often fatal forms of pneumonia. A vigorous immunization campaign has resulted in the reduction of the number of cases of measles to approximately 6000 cases in 1986.

Viruses that affect the digestive glands

Viral agents affect various digestive glands, particularly the liver. The following section discusses hepatitis viruses, which are the major viruses that affect the liver. Additional viral agents that infect digestive glands are outlined in Table 27-12.

Hepatitis viruses. The term **hepatitis** means inflammation of the liver. Several viruses, including the agents of infectious mononucleosis and yellow fever, infect the liver, but the most important are the hepatitis viruses.

In the early 1900s it was suspected that the agent of hepatitis was smaller than a bacterium. Because no animal could be infected by the hepatitis agent, the cause of the disease remained a mystery until 1939 when a clue to its identity was revealed from a totally unrelated incident. A large number of individuals were administered a yellow fever vaccine, but shorterly thereafter 27% of the vaccinees developed hepatitis. This result suggested that a filterable agent had contaminated the vaccine, since human serum was used in the viral culture medium and filtration would have removed a bacterial agent from the medium. The use of human serum for other viral cultivations or for transfusions also resulted in outbreaks of hepatitis and thus confirmed the earlier findings on the epidemiology of the disease. It became apparent that hepatitis could be transmitted by different routes.

There are three hepatitis viruses: **hepatitis A virus (HAV)**, **hepatitis B virus (HBV)**, and **non-A, non-B (NANB)**. Until 1986 when HBV was first cultured in the laboratory, little success was obtained in growing the viruses in culture. Most of our information concerning these viruses has come from animal studies or from human infections. All three hepatitis viruses clinically appear the same, that is, there is degeneration of liver tissue and accompanying yellow discoloration of the skin (jaundice). HAV infection is a disease occurring primarily from the ingestion of contaminated food or water (see Chapter 21). We will center our attention on HBV and NANB.

Hepatitis B. HBV for many years was transmitted primarily by contact with the infected blood of patients with active or chronic disease. Today the chief method of spread is by sexual means. The incubation period for the disease is from 50 to 180 days. Up to 30% of those infected with HBV become chronic carriers, and this represents an important health hazard. The world reservoir of hepatitis B carriers is estimated to be between 200 to 300 million. A genetically engineered vaccine was approved for use in 1986, and it is hoped this will circumvent the health problem.

The HBV is a double-shelled virion 40 to 50 nm in diameter with a core 27 nm in diameter. Within the core is a partially double-stranded DNA and DNA polymerase. Two other particle forms of the virus also exist (Figure 27-33): a tubular or filamentous form and a 22 nm spherical particle. During the course of HBV infection several antigenic forms of the virus exist, which include:

Hepatitis B virus: the 40 to 50 nm particle and its core or the infectious form of the virus

Hepatitis B surface antigen: HB_sAg found on the surface of the HBV, tubular form and the 22 nm particle (Figure 27-34).

Hepatitis B core antigen: HB_c is found within the core of HBV

Hepatitis B_e antigen: HB_eAg is closely associated with nucleocapsid of the HBV

The most important of these is the surface antigen,

Table 27-12 *Characteristics of viral agents that affect digestive glands*

Viral agent*	**Clinical condition**	**Mode of transmission**	**Vaccine**	**Treatment**
Yellow fever virus *(Flavivirus)*	Yellow fever—virus multiples in lymph nodes and later infects liver, spleen, and kidney; degeneration of liver or kidney is cause of death	Bite of mosquito *(Aedes aegyptii)*	Yes (attenuated vaccine); immunity lasts at least 10 years	None specific
Mumps virus *(Paramyxovirus)*	Mumps virus multiples in parotid glands, causing their enlargement (parotitis), and later disseminates to testes, ovaries, pancreas, and brain	Direct contact with contaminated saliva or indirectly by contact with contaminated fomities	Yes (attenuated vaccine); given with measles and rubella vaccines	None specific
Cytomegalovirus (subfamily: Betaherpesvirinae)	Cytomegalic inclusion disease virus forms large intranuclear inclusions in cells of liver, salivary glands, kidneys, pancreas, and endocrine glands; fatalities are high in children under 2 years of age; fetal death or cytomegalic disease in infants	Unknown in general population; however, congenital transfer from infected mother to fetus during delivery; venereal transmission possible	None available in United States	None specific
Epstein-Barr virus (subfamily: Gammaherpesvirinae)	Infectious mononucleosis—occurs primarily in children and young adults; lymph node involvement and enlargement of spleen	Not entirely understood; close contact probably required	None available	None specific
Hepatitis viruses	Hepatitis—degeneration of liver with accompanying jaundice	Hepatitis B by contact with blood or blood products and sexually; hepatitis A by contaminated food; non-A, non-B by contact with blood or blood products	Vaccine available for hepatitis B	Administration of gamma globulin can prevent jaundice after virus infection

*The common species name of the viral agent is used with the genus in parentheses.

HB_sAg, which in itself is not infectious, but whose presence in the blood is an indicator of present or past hepatitis B infection. Some hepatitis B carriers also carry a very unusual antigen (see Point of Interest).

HBV, like other hepatitis viruses, is unusually resistant to disinfectants, heat, and other chemical and physical agents that are normally detrimental to other viruses. However, autoclaving is effective in destroying the virus. Disinfection of materials to be used in hospital procedures, involving blood or blood products, and use of disposable items, such as needles and syringes can prevent infection.

NANB hepatitis infection has been recognized only by exclusion of HAV and HBV infection and not by electron microscopy or serological results. NANB virus disease is now believed to be the single most common cause of hepatitis following blood transfusion, because hepatitis B antigen can now be successfully detected in blood donors. Tests are currently being evaluated in humans and chimpanzees to fully determine the nature of infection by NANB.

Viruses that cause diarrhea

The rotaviruses. Diarrhea, especially in children, is a major cause of morbidity and mortality in the devel-

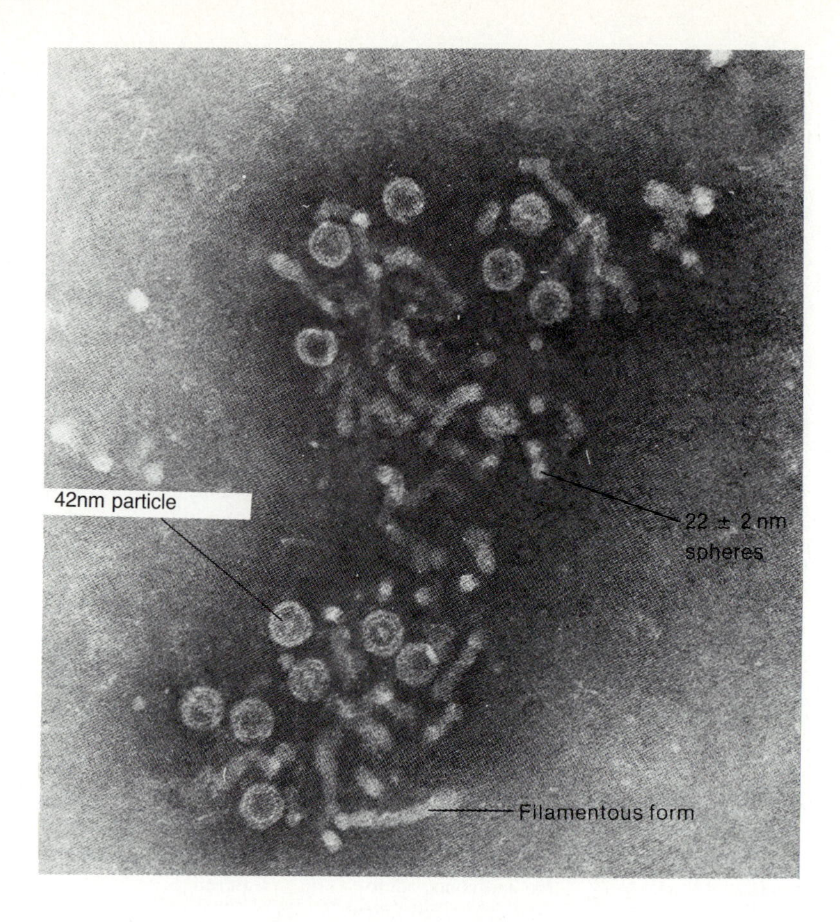

Figure 27-33 *Electron micrograph (×210,000) of HBV. Three morphological forms are observed: spheres 22 ± 2 nm in diameter; filamentous forms approximately 22 nm in diameter and varying length; and 42 nm particles.*

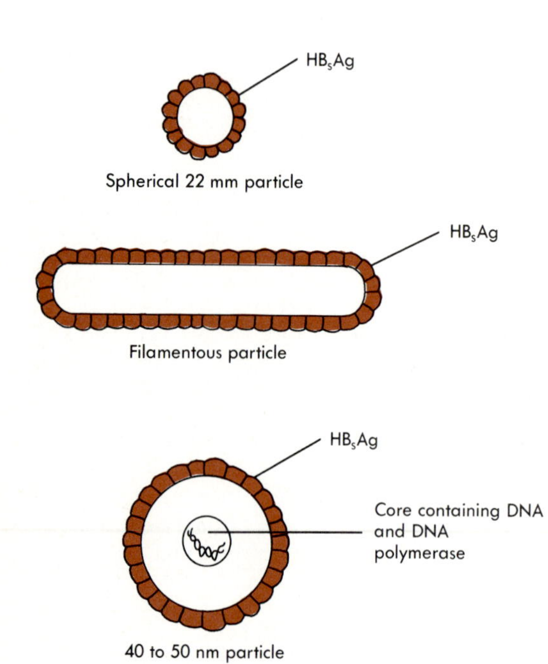

Figure 27-34 *Morphological forms of HBV. Each form contains surface antigen, HB_sAg* (color).

oping countries of the world. The rotaviruses are among the most frequently isolated microbial agents causing diarrhea.

The rotaviruses are double-stranded RNA viruses that have icosahedral symmetry, and in electron micrographs (Figure 27-35) give the appearance of a spokelike pattern (*rota* means wheel in Latin). The rotaviruses that cause human disease are morphologically indistinct from the agents of calf diarrhea, diarrhea in mice, and agents isolated from monkeys, sheep, and foals. Calf diarrhea is of economic importance because it may be a major cause of mortality.

There are two antigenically distinct types of human rotaviruses, with type II being the major cause of disease during the winter months in the Northern Hemisphere. Disease is seen more frequently in infants aged 1 to 11 months and is rarely observed in those over 6 years of age. Infections are acquired by the fecal-oral route, with symptoms appearing within 48 hours. Vomiting and diarrhea, which are the principal symptoms, may lead to severe dehydration. The virus attacks the upper intestine and causes epithelial disturbances that lead to impairment of sodium transport and deficiency in enzymes, such as maltase, sucrase, and lactase. These

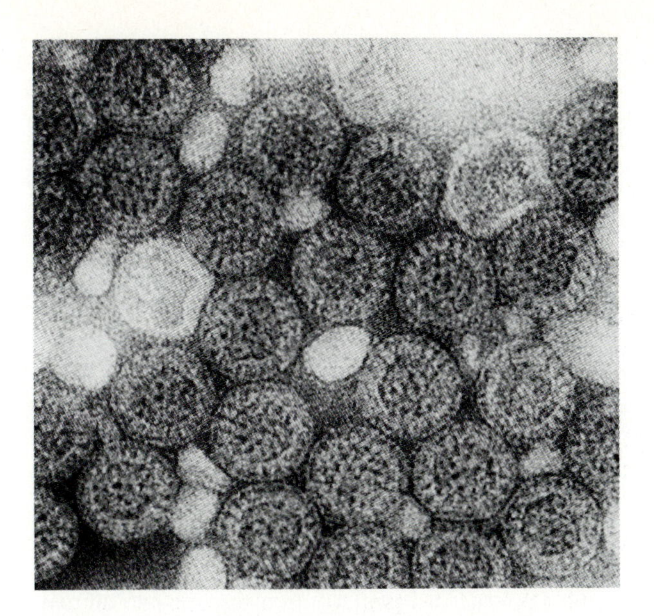

Figure 27-35 *Electron micrograph of human rotavirus. Note spokelike appearance of capsid.*

enzymes break down the most common disaccharides found in the diet; maltose, sucrose, and lactose, respectively.

There are no antimicrobial agents that are effective in therapy. Parenteral rehydration using solutions containing sodium appears to be the most successful method for replenishing fluids and preventing mortality.

Concept Check

1. Describe the various antigenic forms of HBV. Which is the most important indicator of infection?
2. Describe the sequence of events associated with infection by herpes simplex virus.
3. What is the "delta agent" and what characteristics does it share with viroids and satellites?
4. What are venereal warts and why may they be an important health problem?
5. What are the various clinical conditions caused by herpesviruses?
6. Explain the effect of venereal herpes on men. On women. On neonates. What drug is most effective in treatment?

SUMMARY

1. Bacterial diseases may involve any part of the body, but most are acquired through the respiratory tract. The principal respiratory diseases include pneumonia, whooping cough, meningitis, and tuberculosis.
2. Pneumonia is caused primarily by *Streptococcus pneumoniae.* A vaccine containing 14 pneumococcal types has been developed. Whooping cough is caused by *Bordetella pertussis.* The organisms are part of the DTP vaccine that induces

Point of Interest

HEPATITIS: DELTA AGENT, VIROIDS, AND INTRONS

In 1976, certain Italian carriers of HBV were discovered to carry a specific antigen now called the **delta antigen.** The delta antigen occurred only in those already infected with HBV. These patients were more likely to die than those infected with HBV alone. The delta agent has since been found worldwide but incidence of infection varies greatly from one geographic area to another. A bigger surprise was uncovered 10 years later.

In 1986, the delta agent was identified as a viroidlike agent. The delta agent is a single-stranded circular RNA molecule, like viroids, that uses the "rolling circle" mode of replication. The delta agent RNA, however, contains 1678 bases that make it four times larger than typical viroids. The delta agent also contains a protein coat, which is unlike viroids, and this makes it more similar to satellite viruses. Satellite viruses require helper virus for their replication, which is unlike the delta agent, which does not require a helper virus for replication. The delta agent does require helper virus, HBV, for transmission to a susceptible host. Some other similarities also exist between the delta agent and viroids.

Located within the delta agent genome are two short sequences that also exist in virtually all viroids. In addition, there are homologies with group I introns (see Chapter 11) that now suggest a link between viroids, the delta agent, and group I introns. These findings make the hypothesis that viroids and the delta agent are escaped introns even more appealing. There is some speculation that all three classes of RNA molecule may have derived from an as yet unidentified ancestor molecule. More detailed sequence comparisons are needed to solve the puzzle.

immunity to whooping cough, as well as tetanus and diphtheria.

3. Meningitis is a disease of the nervous system, that is, the meninges, which initially begins as an upper respiratory infection. *Neisseria meningitidis* is an important cause of meningitis. *Haemophilus influenzae* type b is an important cause of meningitis in children between the ages of 18 and 24 months.
4. Tuberculosis is the most important cause of death world wide and is caused by the organism *Mycobacterium tuberculosis.* The disease is characterized by the formation of swellings (tubercles) in lung tissue as well as other organs.
5. Those bacterial diseases that affect the skin and eye can be acquired by abrasions, sexual contact, bites of insects, or hospital procedures. Most skin diseases are caused by *Staphylococcus aureus,* which is harbored in the nasal passages by carriers.
6. The possibilities of deformed infants, infertility, and abortion resulting from sexually transmitted diseases have created concern among the medical community as well as the general public. Syphilis and gonorrhea are the most prevalent sexually transmitted diseases, but some other sexually transmitted diseases, although occurring less frequently, can also be devastating.
7. Venereal syphilis is caused by the bacterium *Treponema pallidum.* Disease is characterized by three stages: a primary stage in which a lesion called a chancre predominates, a secondary stage characterized by a rash, and a tertiary stage characterized by lesions called gummas. *Treponema pallidum* can penetrate the placenta and can infect the fetus, causing death or deformities in the infant.

   The most widely used test for diagnosis is the Venereal Disease Research Laboratory (VDRL) test. More specific tests include the fluorescent treponemal antibody absorption (FTA-ABS) test.
8. Gonorrhea is caused by the bacterium *Neisseria gonorrhoeae.* Nearly 1 million cases are recorded each year. During pregnancy the infectious agent can infect the conjunctiva of the fetus as it passes through the birth canal. Disease in women often goes unnoticed and can spread to the fallopian tubes or ovaries and cause sterility. Infrequently, *N. gonorrhoeae* penetrates the bloodstream and causes disseminated disease that results in polyarthritis.
9. Nongonococcal urethritis is caused primarily by *Chlamydia trachomatis.* The disease is a frequent cause of pelvic inflammatory disease, which can lead to infertility. Chlamydiae have an unusual infectious cell cycle that involves two forms of the organism called the elementary body and the reticulate body.
10. Diseases resulting from animal contact or contact with arthropod vectors include plague and rickettsial diseases. Plague is caused by *Yersinia pestis* and most cases are the result of contact with infected fleas. The disease is characterized by buboes that are enlarged lymph nodes. Rickettsial agents infectious for humans are transmitted by arthropod vectors, such as ticks and fleas. Disease is characterized by a rash and fever. Rocky Mountain spotted fever is the most common rickettsial disease in the United States.
11. Bacterial diseases resulting from major trauma include wound botulism, tetanus, and gas gangrene. The organisms causing these diseases are *C. botulinum, C. tetani,* and *C. perfringens,* respectively. All are found in the soil and all produce potent toxins that can cause death.
12. Bacterial diseases acquired through the alimentary tract are usually the result of ingestion of food or water contaminated by organisms, such as *Salmonella* and *Campylobacter.* Most infections are characterized by diarrhea but some salmonella, such as *S. typhi* are capable of penetrating the bloodstream and causing a potentially fatal disease.
13. Viral infection is usually associated with some cytopathic effect, such as the presence of inclusion bodies or other characteristics that can be observed with the unaided eye or by microscopic means. Most viral diseases are spread by the respiratory route.
14. Viruses that affect nervous tissue are usually carried to the site of infection via the bloodstream. The viruses causing such diseases are the poliovirus, echoviruses, mumps virus, herpes simplex virus, and cytomegalovirus. The rabies virus is a bullet-shaped virus. Humans are usually infected by the bite of infected animals such as dogs. The disease is invariably fatal if left untreated. Wild animals, such as skunks and foxes, are the most important sources of infection for domestic animals. Polio is a paralytic disease seldom seen today because of the development of vaccines by Sabin and Salk in the 1950s.
15. Several viruses affect the respiratory tract, but the most important are the influenza viruses, types A and B. Influenza A pandemics, recorded throughout history, have caused the deaths of millions of humans. The most important characteristic of the virus is its ability to mutate, resulting in new strains for which vaccines are totally or partially ineffective. Influenza B disease occurs primarily in children and is not as virulent as the influenza A virus.

16. Several viruses affect the skin including those that are sexually transmitted. Measles is a common childhood disease that is manifested by a bright rash on the body and Koplik's spots on the oral mucosa. The most important sexually transmitted diseases are genital herpes caused by the herpes simplex virus II and venereal warts caused by a papilloma virus. Herpes is characterized by recurrent lesions on the genitalia and may cause fetal death if incurred from the mother during pregnancy. Venereal warts is now being recognized as a possible cause of cervical cancer.
17. The hepatitis viruses are the most important viral agents affecting the digestive glands. They include hepatitis A, hepatitis B, and non-A, non-B virus. All three viruses affect the liver. Hepatitis B is associated with the carrier state and is usually transmitted by contaminated blood or blood products. The mortality in hepatitis B disease is relatively high. Non-A, non-B virus is believed to be the most common cause of hepatitis following blood transfusions and is also associated with the carrier state.
18. The rotaviruses are one of the major causes of viral diarrhea in developing countries. They are frequently associated with infant mortality in these countries.

Self-study Quiz

Completion

1. Scalded skin syndrome is the result of an exfoliative toxin produced by ________________.
2. The bacterial disease considered to be the most important cause of death worldwide is ________________.
3. The acid-fast staining technique is utilized in the identification of ________________.
4. The *Rickettsia* produce an enzyme called ____________ that degrades the phagosome membrane before the organism can be digested by lysosomal enzymes.
5. The coagulase test is used to distinguish pathogenic from nonpathogenic ________________.
6. The organism causing plague is transmitted to humans by the bite of a ________________.
7. The Vi antigen is associated with virulence by ________.
8. The filamentous hemagglutinin is a major adhesin of ________________.
9. The most important cause of meningitis in children under 4 years of age is ________________.
10. The drug used for prophylaxis and treatment of tuberculosis is ________________.
11. Botulinum toxin causes ________________ of muscles while tetanus toxin prevents ________________ of muscles.
12. The letters HN in influenza terminology refer to the ________________ and ________________ of the virus.

Multiple choice

1. Most bacterial infectious agents are transmitted to humans via the:
   A. Respiratory tract
   B. Skin
   C. Genitourinary tract
   D. Intestinal tract
2. Which of the following is considered a major virulence factor for *Streptococcus penumoniae?*
   A. Exotoxin
   B. Lecithinase
   C. Capsule
   D. Endotoxin
3. For which of the following diseases does vaccination occasionally result in encephalitis?
   A. Tetanus
   B. Diphtheria
   C. Whooping cough
   D. Leptospirosis
4. Which of the following is most descriptive of *Neisseria meningitidis?*
   A. Gram-negative, nonencapsulated short rod most often associated with an urogenital tract infection but also producing meningitis
   B. Gram-negative, encapsulated streptococci causing an upper respiratory infection developing into pneumonia
   C. Gram-positive, encapsulated short-rod producing upper respiratory tract infections and bacteremia.
   D. Gram-negative, encapsulated diplococci sometimes producing epidemic meningitis
5. What is PPD?
   A. *Pasteurella pestis* type D cell wall fraction, which is used to immunize against plague
   B. Polyvalent pneumonia derivative, which protects against the most common types of streptococcal pneumonia
   C. Purified protein derivative, which is used for the tuberculin skin test
   D. Polycystic pharyngitis diphtheria, which is one of the complications of untreated diphtheria
6. Which of the following would be associated with the diagnosis of plague?
   A. Rice water stool
   B. Rose spot rash, fever, and abdominal pain
   C. Formation of buboes and high fever
   D. High fever recurring in unpredictable cycles with the sloughing of epidermis
7. What do gas gangrene, tetanus, and botulism all have in common?
   A. Each disease is produced by a species of *Clostridium*
   B. They each produce direct tissue damage from the action of proteinases
   C. They all are produced by enterotoxins
   D. Each is accompanied by headache, high fever, and convulsions

8. Gummas are associated with which of the following?
   A. Primary stage of syphilis
   B. Lymphogranuloma inguinale
   C. Tertiary stage of syphilis
   D. Complications of gonorrhea
9. PID is a common complication of which of the following?
   A. Syphilis
   B. NGU
   C. Hepatitis
   D. Tuberculosis
10. The only sexually transmitted disease for which there is an effective vaccine is:
    A. Syphilis
    B. Gonorrhea
    C. Hepatitis B
    D. Nongonococcal urethritis
11. Acyclovir would be used to treat:
    A. Candidiasis
    B. Genital herpes infection
    C. Secondary syphilis
    D. Chancroid
12. Which of the following would be considered to be noninfectious?
    A. Asymptomatic carrier of *N. gonorrhoeae*
    B. Middle-aged man with genital herpes lesion
    C. 60-year-old man with tertiary syphilis
    D. Child suffering from primary congenital syphilis
13. Which of the following can cause infertility?
    A. *N. gonorrhoeae*
    B. Hepatitis B virus
    C. *Haemophilus ducreyi*
    D. Cytomegalovirus
14. The infectious component of the *Chlamydia* is a (an):
    A. Reticulate body
    B. Inclusion body
    C. Elementary body
    D. Endospore

Matching

Match the disease with the causative agent.

________	1. Sexually transmitted hepatitis	a.	Rotaviruses
________	2. Infant diarrhea	b.	Influenza virus and salicylates
________	3. Common cold	c.	Togaviruses
________	4. Reye's syndrome	d.	Varicella-zoster virus
________	5. Shingles	e.	Hepatitis A virus
________	6. Equine encephalitis	f.	Rubella virus
________	7. German measles	g.	Measles virus
________	8. Infectious mononucleosis	h.	Rabies virus
		i.	Rhinoviruses
		j.	Hepatitis B virus
		k.	Herpes type 1 virus
		l.	Epstein-Barr virus
		m.	Mumps virus
		n.	Poliovirus

Match the disease with primary method(s) of transmission.

________	9. Influenza	a.	Inhalation of respiratory secretions
________	10. Rabies	b.	Contact with saliva of infected animal
________	11. Equine encephalitis	c.	Water or food contaminated with virus
________	12. Herpes type II	d.	Direct contact with lesions
________	13. Measles	e.	Mosquito bite
________	14. Polio	f.	Blood or blood products contaminated with virus
________	15. Hepatitis A	g.	Human body louse
________	16. Hepatitis B	h.	Sexual transmission
________	17. Common cold		
________	18. Infectious diarrhea		

Thought questions

1. Tuberculosis is extremely difficult to treat with antibiotic therapy and requires long-term antibiotic exposure. Knowing the nature of the causative organism, the pathogenicity of the disease, and the human body's response to the agent, what explanations can you propose for this difficulty and length of treatment?
2. Take any of the major diseases discussed in this chapter and follow through the general scheme outlined under "General Characteristics" to illustrate how a bacterial disease is suspected, isolated, and identified, and how decisions about treatment are made. You may need to refer to a diagnostic or clinical microbiology book to help with the specifics.
   Disease____
   a. Relevant clinical history
   b. What specimens should be taken
   c. Relevant microscopic data
   d. Isolation and growth of organisms in the laboratory (media, environmental conditions, etc.)
   e. Cultural characteristics
   f. Biochemical characterization
   g. Choosing a therapy
3. Maternal infections by viral disease agents during pregnancy often have devastating consequences for the developing fetus (see Table 27-8).
   a. Explain why you think that viral infections of the fetus are more common than bacterial or fungal infections.
   b. Explain why these congenital infections of the fetus have such devastating effects.
   c. What implications does this have for the advisability of immunization of pregnant women with live viral vaccines?

2. In January, 1976, a young army recruit died at Fort Dix in New Jersey of an influenza infection. It was assumed to be a Type A strain, but when immunologically tested, was typed as a swine flu strain with many of the same characteristics as the influenza strain involved in the 1918-1919 pandemic. This finding initiated the national Swine Flu immunization program of 1976. The decision to initiate the mass immunization program had significant sociopolitical and scientific implications.
   a. Go to a library and using the lay press as a source (national news magazines or newspaper accounts), trace the chronological events in that program. Note the controversies that arose.
   b. It has been pointed out that the initiation of the program was based more on politics than on scientific merit. What facts seem to support that claim?
   c. In retrospect, how would you assess the success of the program?
   d. Did you receive the swine flu immunization? If so, do you remember why you did? If not, do you remember why you did not?
   e. If you knew then what you know now about influenza, what would you have done? Explain.
   f. What were the positive and negative outcomes of that program?

SELECTED READINGS

Books

Boyd, R.F., and Hoerl, B.H. *Basic medical microbiology* (3rd ed.). Little, Brown & Co., Boston, 1986.

Evans, A.S. *Viral infections of humans* (3rd ed.). Plenum Publishing Co., New York, 1986.

Freeman, B.A. *Burrows textbook of microbiology* (22nd ed.). W.B. Saunders, Philadelphia, 1983.

Hoeprich, P.D. (Ed.). *Infectious diseases* (3rd ed.). Harper & Row, Publishers, Inc., New York, 1984.

Holmes, K.K., Mardh, P.A., Sparling, P.F., and Wiesner, P.J. (Eds.). *Sexually transmitted diseases.* McGraw-Hill, New York, 1984.

Kucera, L.S., and Myrvik, Q.N. *Fundamentals of medical virology* (2nd ed.). Lea and Febiger, Philadelphia, 1985.

Schoolnik, G.K. (Ed.). *The pathogenic neisseriae.* American Society for Microbiology, Washington, D.C., 1984.

Journal articles

Bacterial virulence and pathogenicity. *Rev. Infect. Dis.* **5**(suppl. 4), 1983.

CDC. Diphtheria, tetanus and pertussis guidelines for vaccine prophylaxis and other preventive measures. *Morbid. Mortal. Week. Rep.* **34**:405, 1985.

CDC. 1985 Sexually transmitted disease treatment guidelines. *Morbid. Mortal. Week. Rep.* **34**:No.4s, 1985.

Couch, R.B. The effect of influenza on host defenses. *J. Infect. Dis.* **144**:284, 1981.

Fitzgerald, T.J. Pathogenesis and immunobiology of *Treponema pallidum. Annu. Rev. Microbiol.* **35**:29, 1981.

McCutchan, J.A. Epidemiology of venereal urethritis: comparison of gonorrhea and nongonococcal urethritis. *Rev. Infect. Dis.* **6(5)**:699, 1984.

Middlebrook, J.L., and Dorland, R.B. Bacterial toxins:cellular mechanisms of action. *Microbiol. Rev.* **48**:199, 1984.

Nahmias, A.J., and Norrild, B. Herpes simplex viruses 1 and 2—basic and clinical aspects. *Dis. Month* **25**(10):Entire issue, 1979.

Schachter, J., and Grossman, M. Chlamydial infections. *Annu. Rev. Med.* **32**:45, 1981.

The pneumococcus. *Rev. Infect. Dis.* **3**(2):1981.

Weller, T.H. Varicella and herpes zoster: changing concepts of the natural history, control and importance of a not-so-benign virus. *N. Engl. J. Med.* **309**:1362, 1983.

Woodward, T.C. Rocky Mountain spotted fever. *J. Infect. Dis.* **150**:465, 1984.

CHAPTER

28

Fungal and animal parasite diseases of humans

Chapter Outline

Introduction

Fungal and animal parasite diseases are the least occurring infectious diseases in developed countries, but they occur with great frequency in developing countries. Protozoa, such as the agent of malaria, causes considerable morbidity and mortality in Third World countries. Such diseases concern all of us because immigrants to developed countries can transmit the diseases and because travel to developing countries is a means of acquiring these "exotic" diseases.

This chapter outlines some of the basic concepts of fungal and protozoal diseases in humans and discusses the more important diseases associated with these parasites.

FUNGAL DISEASES

Approximately 100 fungal agents of more than 100,000 or more species are pathogenic to humans. About 20 fungal agents can cause life-threatening infections. The rest of the fungi normally develop in the soil and humans are only an accidental host. Fungi usually exist as spores in the soil, and it is primarily the fungal spore that is transmitted to humans. Diseases caused by fungi are **mycoses.** They are divided into four major subdivisions, based on the site of infection: the **superficial**, the **cutaneous**, the **subcutaneous**, and the **systemic** mycoses. A fifth group, the **opportunistic** mycoses, occur in compromised hosts.

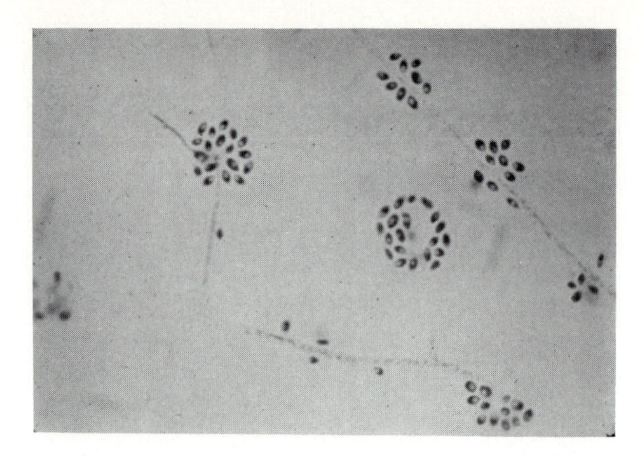

Figure 28-1 Sporothrix schenckii. *Culture demonstrates flowerette arrangement of microconidia along the sides and at the tips of conidiophore.*

General characteristics of fungal diseases

Epidemiology

Superficial and cutaneous mycoses. Superficial mycoses are those that affect only the thick, keratinized layers of epidermis and the outer portions of hair follicles. Superficial mycoses are very rare in the United States but, when occurring, are found primarily in the southern United States. Cutaneous mycoses are those in which the parasite penetrates the keratinized layers of skin but cannot invade deeper tissue, thus remaining just beneath the hard surface. The fungal agents responsible for cutaneous infection are **dermatophytes**, and the diseases they cause are **dermatophytoses, ringworm**, or **tineas.** Some dermatophytes parasitize animals (Plate 38), others infect only humans, and still others are exclusively soil saprophytes. In the host, dermatophytes subsist on the keratin protein in skin, hair, and nails. Most human infections are caused by direct or indirect human-to-human transmission; for example, exchange of bath towels or contact with contaminated inanimate articles. Human infection can also occur by contact with the infected hair or epithelial cells of animals such as horses or dogs. Dermatophytoses, which are the only contagious fungal infections known to occur in humans, are relatively common but seldom fatal.

Subcutaneous mycoses. Subcutaneous mycoses result from parasites penetrating the epithelial layers and reaching the subcutaneous tissue. The fungal agent remains localized in the subcutaneous tissue and does not cause fatal infections. Infections result from puncturing the skin with sharp objects contaminated by soil, such as wood splinters from rose bushes or timber in mines that are contaminated by fungi. Infections in the United States are primarily of an occupational nature; for example, several instances have been reported where infection occurred among nursery workers handling peat moss. The most frequently occurring subcutaneous disease in the United States is called **sporotrichosis**, which is caused by *Sporothrix schenckii* (Figure 28-1). Other subcutaneous diseases are outlined in Table 28-1.

Systemic mycoses. Systemic mycoses are caused by fungi that penetrate epithelial tissue and invade the bloodstream. They are frequently characterized by their effect on respiratory tissue and many times their infections are misdiagnosed as tuberculosis. One of the most common systemic mycoses is **histoplasmosis**, which is endemic in the Mississippi River Valley. The disease agent is found in soil worldwide but is localized in areas that have been enriched with bird excreta, particularly from birds such as starlings, chickens, and crows. The congregation of hundreds of thousands of starlings in small areas of towns in the United States has been associated with outbreaks of histoplasmosis. In some instances the bird excreta was inches thick on the soil. Bird excreta provides a selective medium for the germination of spores of the disease agent, *Histoplasma capsulatum* (Figure 28-2). After germination the vegetative cells produce large numbers of spores, which are disseminated in the air. Inhaling spores results in a pulmonary infection that can spread to other organs and can also be fatal. Other systemic mycoses are also acquired in this manner (Table 28-2).

Opportunistic mycoses. Advances in medicine have enabled us to keep alive patients who, just a few years ago, might otherwise have died. However, new drugs, chemicals, and physical processes not only keep such patients alive but also create compromising conditions in which the patients become susceptible to disease caused by exogenous and opportunistic endogenous microbial agents. The principal endogenous fungal species responsible for opportunistic infections is *Candida albicans.* The major compromising conditions that lead to infection are:

1. Diabetes. One of the side effects of diabetes, especially uncontrolled diabetes, is the formation of ketones, acids, and excess glucose in the bloodstream. These are conditions that favor tissue invasion by fungi, and fungal (as well as bacterial)

Table 28-1 *Characteristics of the subcutaneous mycoses*

Disease	Most common etiological agent	Geographic location	Transmission	Characteristics of disease
Sporotrichosis	*Sporothrix schenckii*	Worldwide, but highest in United States, Mexico, South America, and Africa	Most infections arise from contact with plants	Subcutaneous abscesses in local lymphatics
Lobomycosis	*Loboa loboi*	Northern South America	Contact with plants or soil	Chronic infection; fibrous tumors appear on feet, legs, and face
Mycetoma (maduromycosis)	*Petrillidium boydii* and other species	Worldwide, especially the tropics	Wound contaminated by soil	Abscess formation usually of foot; chronic
Chromomycosis	*Phialophora verrucosa* and other dematiaceous fungi	Worldwide, especially the tropics	Wound contaminated by soil	Most infections involve legs and feet; disease develops slowly; nodules appear that later enlarge
Rhinosporidiosis	*Rhinosporidium seeberi*	India, South America, and United States	Associated with stagnant water	Polyps appear on mucosal surfaces, particularly nose and soft palate

Table 28-2 *Characteristics of the systemic mycoses*

Disease	Etiological agent	Transmission	Primary site of infection	Characteristics of the agent
Blastomycosis	*Blastomyces dermatitidis*	Inhalation of spores from soil	Lung, but disseminated with manifestations in skin and subcutaneous tissue (see Plate 39)	In tissue yeasts have thick wall
Paracoccidioidomycosis	*Paracoccidioides brasiliensis*	Inhalation of spores	Lung, but lesions more prominent in oral cavity and nasal mucosa; systemic spread involves spleen, intestines, and adrenal glands	Yeasts show multiple budding in tissue and may be 30 to 60 μm in diameter
Coccidioidomycosis	*Coccidioides immitis*	Inhalation of only a few spores will produce disease; dark-skinned people more susceptible	Lung; dissemination likely if many spores are inhaled, with lesions in many organs; meningitis fatal in dark-skinned people	Most virulent of systemic agents. Extremely hazardous in laboratory because arthrospores are easily spread; yeast phase is a spherule containing endospores
Cryptococcosis	*Cryptococcus neoformans*	Inhalation of spores from soil or dust contaminated with pigeon droppings	Lung; dissemination to other tissues such as skin, bones, viscera, and central nervous system	Is not dimorphic, only yeast phase present; produces a capsule
Histoplasmosis	*Histoplasma capsulatum*	Inhalation of spores associated with bird dung, especially starlings, chickens, and crows	Lung; dissemination to meninges, adrenal glands, heart, and bone	Mycelial stage shows tuberculate spores

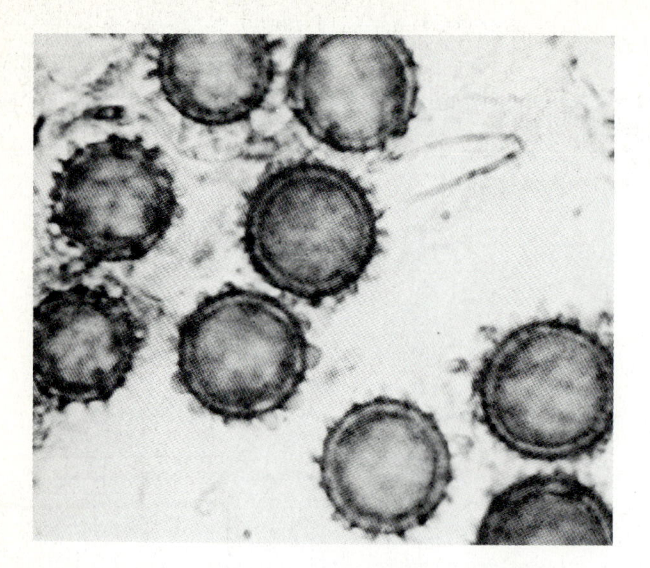

Figure 28-2 Histoplasma capsulatum. *Microscopic appearance of mycelial phase of* H. capsulatum *demonstrating large macroconidia, which are round, heavy-walled spores.*

diseases often appear in the extremities, where circulatory problems exist because of reduced oxygen tension. Fungal species of the order *Mucorales* are responsible for disease in diabetics.

2. Immunodeficiencies. The principal immunological defense mechanism of humans to fungal disease is the cell-mediated response. Any disease, therapy, or condition that imperils cell-mediated immunity, such as steroid use or radiation treatment for cancer, predisposes the patient to fungal infections. Many fatal infections result from tissue invasion by indigenous fungal species such as *Candida.*
3. Antibiotic therapy. Antibiotics such as tetracycline reduce the gram-negative or gram-positive flora of the host. The bacteria are natural competitors with fungi in the oral cavity, intestinal tract, and genitourinary tract, and without bacterial competition fungal species soon take over the habitat.

Pathogenicity. Fungi, which cause human disease, must be able to reproduce at the temperature of the body, 37° C, and must be able to survive in tissue where the oxidation-reduction potential of the tissue may vary from 0 to + 300 millivolts. Most fungi have diverse enzyme systems and can metabolize many different substrates, but some pathogenic species have special requirements. For example, the dermatophytes can infect only keratinized tissue and prefer a temperature lower than 37° C. Consequently, they do not survive in other tissue. Such specialization in most pathogenic fungi is not evident, although there are certain environmental conditions that favor the growth of one species over another. The nutritional and physical requirements of the fungi are supplied by the host at the site of infection. Most fungi grow best in warm, humid conditions, but they are able to adjust to different conditions.

What should not be overlooked, and must be emphasized, is that fungal diseases are the exception and not the rule. Unlike bacterial diseases, in which the severity of disease is determined by virulence factors such as toxins or enzymes, fungal diseases are usually not caused by such factors. Most of the time the fungal agent, which is acquired by either inhalation or puncture wound, induces a cell-mediated immune response in which inflammation and walling off of the fungal agent in a fibrous calcified deposit both occur. Skin tests to determine hypersensitivity to a fungal agent have demonstrated that several million people have had contact with a fungal agent at sometime in their lives, but no disease symptoms had ever become apparent. What, then, is responsible for fungal disease?

The two most important factors that determine a host's susceptibility to fungal disease are the number of organisms to which the host is exposed and the state of health of the host at the time of exposure. Inhalation of a large number of spores can occur in a laboratory where fungal agents are processed and stored. Thousands of spores are produced by the fungi when cultivated in Petri dishes, and air currents passing over an exposed plate can disseminate them. Disease can also occur when a fungus is concentrated in nature and when there is continual human contact, such as diseases occurring as a result of the host's occupation. Fungal diseases, like some bacterial diseases (such as tuberculosis, where the cell-mediated immune response is critical for recovery), are often chronic and may persist in the host for years.

Several of the fungi, such as the dermatophytes and some species of *Candida,* act as allergens and cause allergic responses that are called **ids.** The fungal allergies manifest as skin lesions at sites distant from the initial infection site. Id is an abbreviation of **dermatophytid,** a term used to denote lesions appearing on the fingers from infection by a dermatophyte that initially infected the feet. Other mycotic agents that are inhaled, such as species of *Aspergillus,* can cause asthmalike responses that are also believed to be allergic responses. The allergic response in some individuals is not surprising because humans frequently inhale spores.

Immunity. Normal host defense mechanisms are usually adequate to prevent fungal infections. The cell-mediated immune response appears to play the most important role in preventing disease. Conditions or agents that cause T lymphocyte defects—such as radiation, steroid therapy, and immunosuppressive drugs—predispose the patient to fungal infection. Acquired immunodeficiency syndrome (AIDS) patients, for example, are especially susceptible to infections by fungal agents. The fungi have surface antigens that elicit a poor antibody response, and antibodies that are produced do not

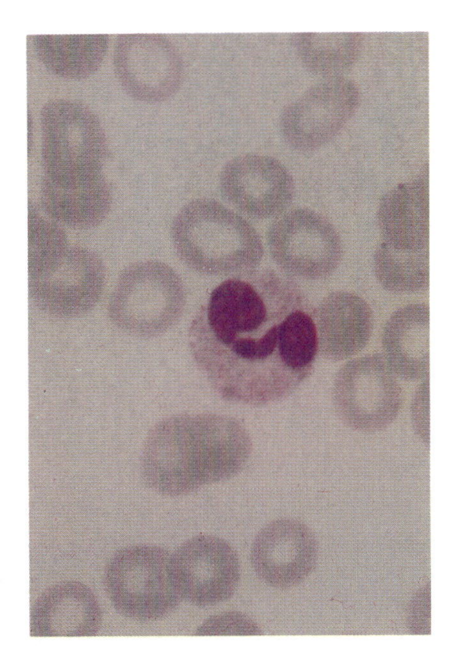

Plate 31 *Band neutrophil.*

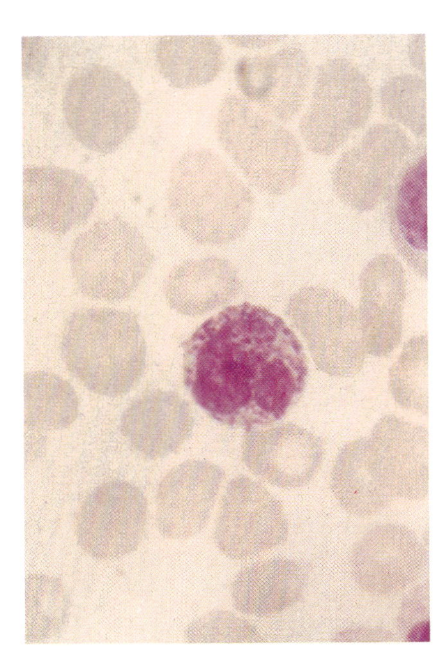

Plate 32 *Typical basophil.*

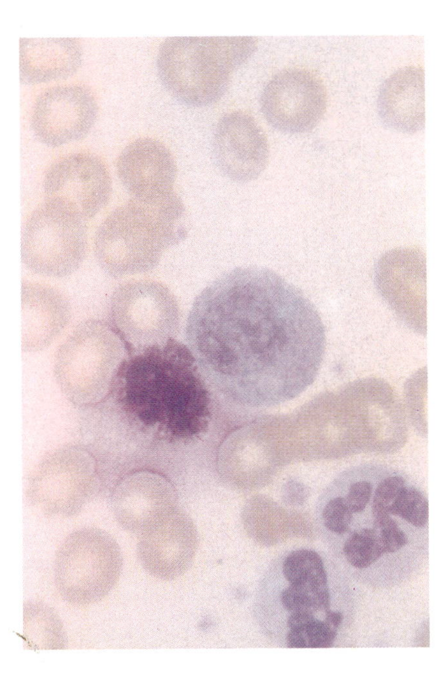

Plate 33 *Basophil (left center), eosinophil (right center), segmented neutrophil (4 o'clock), and band neutrophil (5 o'clock). Note that eosinophil granules are uniform in size and have pale centers.*

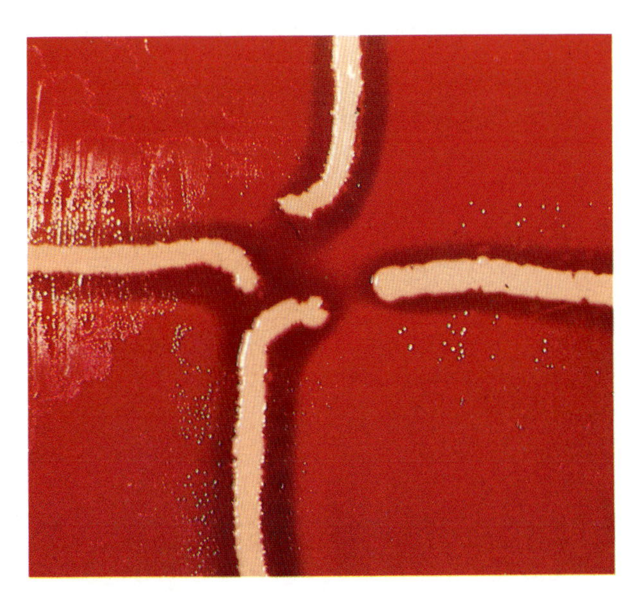

Plate 34 *Satellite growth of* Haemophilus influenzae *around* Staphylococcus aureus *on blood agar. (X1.)*

Plate 35 Mycobacterium kansasii *colonies exposed to light.*

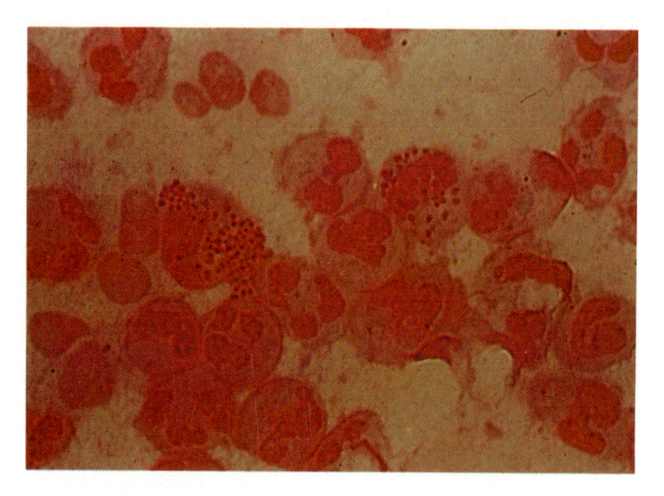

Plate 36 *Urethral smear, gonorrhea; gram-negative intracellular diplococci.*

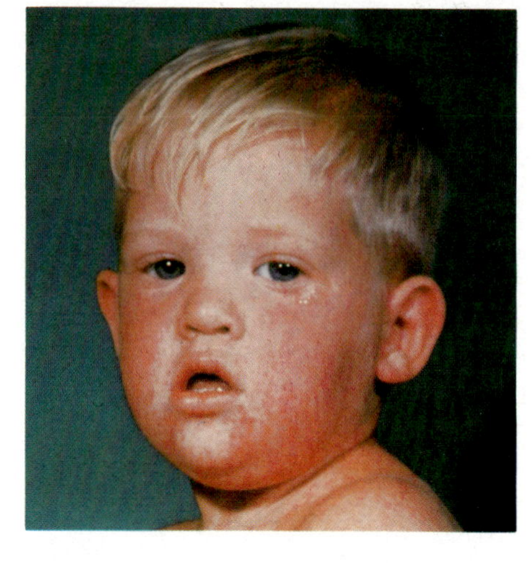

Plate 37 *Measles rash.*

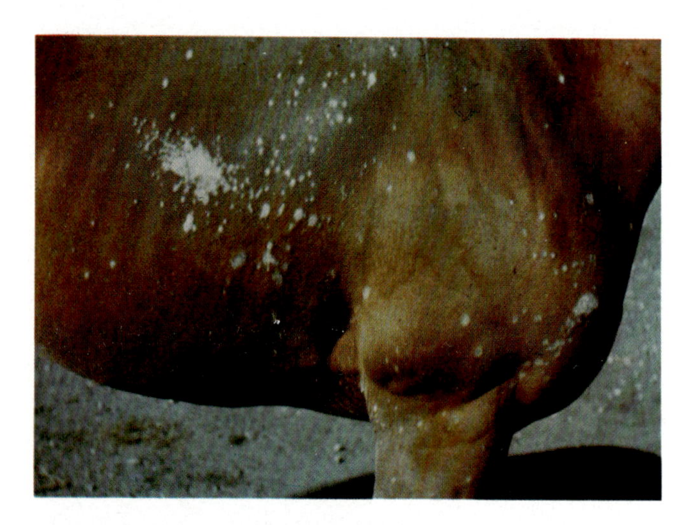

Plate 38 *Ringworm affecting horse.*

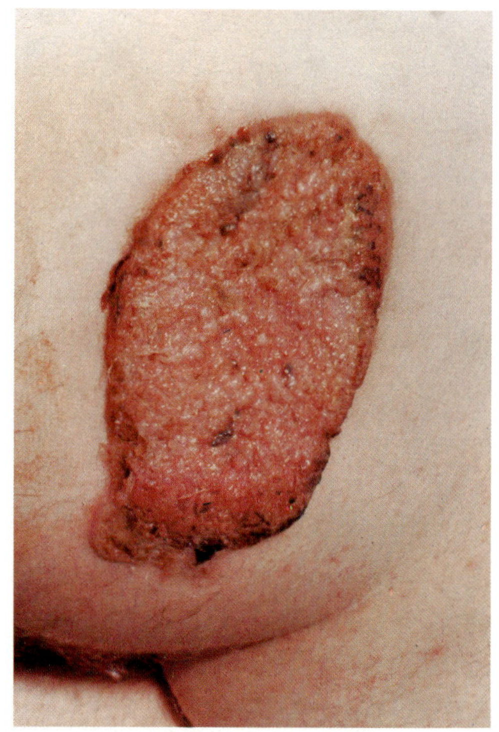

Plate 39 *Cutaneous blastomycosis on buttock.*

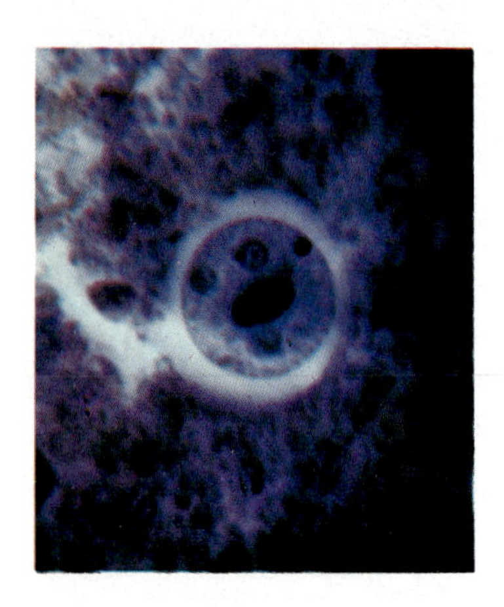

Plate 40 Entamoeba histolytica *cyst. (× 4342.) Iron-hematoxylin stain.*

Table 28-3 *Characteristics of cutaneous mycoses*

Skin disease	Primary etiological agent(s)	Site and/or characteristic of lesion
Ringworm of the body (tinea corporis)	*Trichophyton rubrum, Microsporum canis, T. mentagrophytes*	Raised, circular lesions on face, arms, and other exposed nonhairy body surfaces
Ringworm of the nails (tinea unguium)	*T. rubrum, T. mentagrophytes*	Nails become thickened and discolored
Ringworm of the beard (tinea barbae)	*T. rubrum, T. verrucosum*	Hair follicles on beard filled with fluid; lesions appear reddish
Ringworm of the groin (tinea cruris)	*T. rubrum, T. mentagrophytes, Epidermophyton floccosum*	Also called "jock itch"; red, scaling, itchy lesion
Ringworm of the scalp (tinea capitis)	*T. soudanense, T. tonsurans, M. canis*	Circular bald patches on scalp; broken hair within follicles
Ringworm of the feet (tinea pedis)	*T. rubrum, T. mentagrophytes, E. floccosum*	Spaces between toes contain itchy, red, blisterlike lesions
Dermatophytid (id reaction)	No fungi present in lesion but lesion may become secondarily infected with bacteria	Itchy blisterlike lesions, usually associated with ringworm of the feet

appear to protect the host or aid in recovery from disease. Yet, in some systemic mycoses, the severity of disease seems greatest in those patients with defective antibody responses.

Diagnosis and treatment. Superficial, cutaneous, and subcutaneous disease are easily recognized by their symptoms, which constitute specific lesions on the body surface. Special laboratory techniques are often required for systemic disease.

Chemotherapeutic agents such as nystatin, amphotericin B, 5-fluorocytosine, and ketoconazole are useful in treating fungal infections (see Chapter 15). Vaccines are not available, partly because of the poor antigenicity of fungal antigens and partly because scientists have been unable to identify the major virulence factors.

Concept Check

1. Why are most fungal diseases not contagious? Which ones are contagious?
2. What are some of the characteristics of systemic mycoses?
3. Describe those conditions that frequently lead to fungal infections.
4. How are most fungal infections acquired? Does the host's occupation play a role? What two factors determine the host's susceptibility to disease?
5. Describe the relationship of certain birds to histoplasmosis.

Cutaneous mycoses

Cutaneous mycoses are the most common fungal diseases in humans. They constitute particularly important health problems in tropical countries, where overcrowding occurs and a lack of hygienic practices exist. Three genera of dermatophytes cause infection: *Trichophyton, Microsporum*, and *Epidermophyton.* All three are treatable with griseofulvin, a drug that has an affinity for keratinized epithelial tissue. Topical ointments such as tolnaftate and miconazole are also effective but necessitate longer treatment.

The characteristics of disease caused by dermatophytes depends on the body site because some fungal species may be able to colonize more than one area or type of keratinized tissue. A typical lesion that exemplifies the characteristic ringworm appearance is illustrated in Figure 28-3. The features of the various ringworms, or tineas, are outlined in Table 28-3.

Opportunistic mycoses

Some of the most frequent and often fatal fungal infections are those that occur in compromised patients. Infection by *Candida* species (such infections are also called **candidiasis**) is the most frequently reported fungal disease that is a cause of death. Another less frequently reported, but equally important, opportunistic fungal infection is **aspergillosis**. Aspergillosis is caused by species of *Aspergillus,* whose distinguishing characteristic is the swollen conidiophore (Figure 28-4).

The infectious disease process involving species of *Aspergillus* is a good example of how a compromised

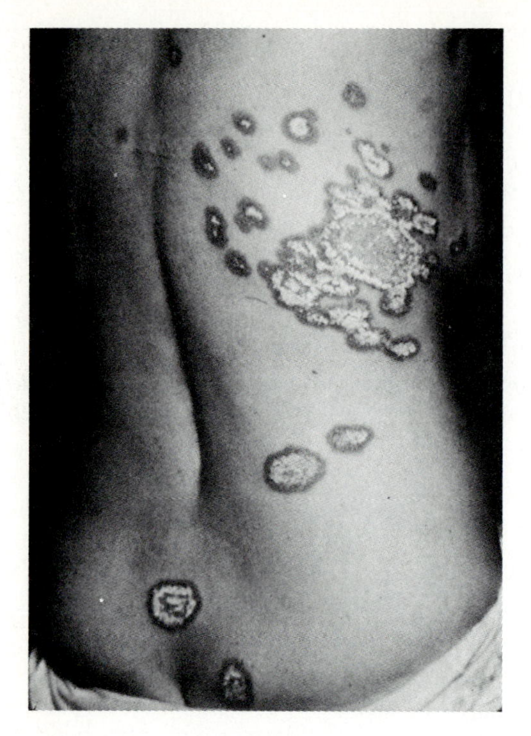

Figure 28-3 *Tinea corporis (ringworm of the body).*

Figure 28-4 *Scanning electron micrograph of* Aspergillus *conidiophore (→) and teminal swelling to which are attached nearly spherical conidia.*

host is subject to infection. People regularly inhale or ingest many aspergillus spores with little or no effect on their health. The host's first line of defense against the spore seems to be the macrophage. Spores are ingested by macrophages, which prevent them from germinating. Even if spores germinate and hyphae are produced, the host's second line of defense, the neutrophil, is ready to ward off attack. Neutrophils are the principal line of defense against hyphal penetration of respiratory tissue. However, when individuals are treated with corticosteroids, macrophages are unable to inhibit germination of spores because of the immunosuppressive effects of the drug. Infection results as a consequence. Corticosteroids are used during renal transplantation procedures to prevent organ rejection. Studies have demonstrated that when the concentration of corticosteroids is reduced, the number of aspergillus infections following transplantation is significantly reduced.

Let us now examine the most frequent cause of opportunistic mycoses, *Candida albicans.*

Candida albicans. *C. albicans* (Figure 28-5) and other species of *Candida* can be found as commensals on the skin, in the oral cavity, in the intestinal tract, and in the genitourinary tract of humans. *C. albicans* is a dimorphic fungus that can grow efficiently at temperatures between 20° to 40° C and at pH levels between 2 and 8. Body sites such as the vagina, where the pH may become very acid, can support luxuriant growth of *C. albicans.* Although *C. albicans* is the most frequently isolated species, other species, such as *C. tropicalis, C. pseudotropicalis,* and *C. parapsilosis,* are also potential pathogens.

Adhesion is necessary for initiation of infection by *C. albicans,* and the most likely candidate for this task is the mannoprotein that makes up the outermost layer of *Candida* organisms. Studies have shown that adhesion can be influenced by the bacterial flora. Bacteria can inhibit extensive colonization of mucosal receptor sites simply by outcompeting *Candida* organisms for the sites or by secreting metabolic products, such as fatty acids, that inhibit fungal activity. If *C. albicans* reaches the bloodstream, it adheres to the endothelial cells lining the walls of blood vessels. The organism evades the circulating phagocytes and then migrates through the endothelial cells into organ tissue (Figure 28-6).

The significance of dimorphism in the disease process is not clear. The yeast phase is considered the pathogenic form in most pathogenic fungi; however, the filamentous form may be favored in *C. albicans* infections. The increased surface area of the filamentous form may be one reason for the increased adherence properties of *C. albicans.* Enzymes, endotoxins, or secreted metabolic products may also contribute to pathogenesis. Proteinase, an enzyme, is presently considered the most likely candidate. Proteinase-deficient strains of *C. albicans* are considerably less virulent than proteinase-producing strains.

One of the interesting observations made during virulence studies is the possible role of steroids. For example, it has been noted that *C. albicans* has a predilection (preexisting tendency) for patients receiving

Figure 28-5 *Electron micrograph of* Candida albicans.

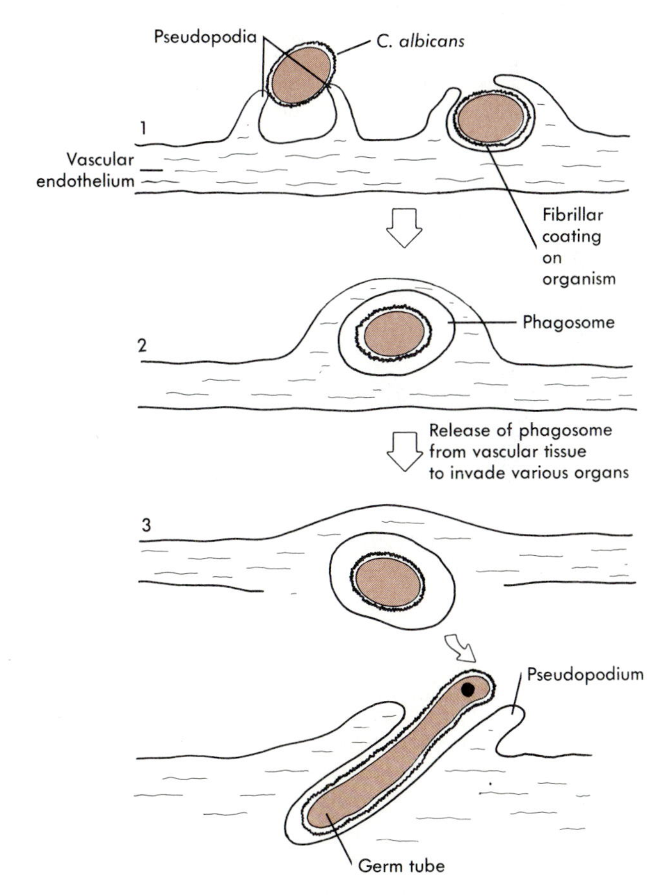

Figure 28-6 *Mechanism of penetration of* Candida albicans *through vascular tissue.* ***A,*** *Penetration of yeast-phase cells.* 1, *Pseudopodia arising from vascular endothelial cells bind to* C. albicans. *Pseudopodia are in contact with fibrillar material on surface of microorganism.* 2, *Fusion of pseudopodia around yeast-phase cell.* 3, C. albicans *within endothelial cell as phagosome. Microorganism or phagosome can be released from endothelium to invade various organs.* ***B,*** *Penetration of filamentous form of* C. albicans. *Microorganism produces a germ tube on endothelial surface. Germ tube is phagocytosed just as yeast-phase cells in* ***A.***

glucocorticoid therapy. *C. albicans* possesses a protein that has a high affinity and selectivity for vertebrate corticosteroids. This raises the possibility that steroid hormones may affect the behavior of yeasts. A similar hormone receptor has been isolated from the fungus, *Paracoccidioides brasiliensis.* The mycelial-to-yeast transition, which is important in the pathogenesis of *P. brasiliensis,* is inhibited by estradiol, the female estrogen hormone, so this yeast shows a marked preference for males. Thus it appears that steroid hormones in the host somehow contribute to altering the pathogenicity of the parasite, either positively or negatively.

The most common manifestation of candidiasis is superficial lesions on the mucosal surface. In very serious superficial forms of the disease the yeast can gain entry into the bloodstream and disseminate to internal organs. However, most disseminated forms of the disease are the direct result of entry into the bloodstream during intravenous feeding. Surgical heart valve implantation and drug therapy for cancer are also procedures that lead to invasion of the bloodstream. Candidiasis takes several forms, which depend on the circumstances that have compromised the patient and include those listed below.

Cutaneous candidiasis. Cutaneous candidal diseases include oral thrush, onychia and paronychia, intertriginous candidiasis, and vulvovaginitis. **Oral thrush** is characterized by the formation of gray-white patches, or pseudomembrane, that may cover the tongue (Figure 28-7), soft palate, or other oral surfaces. The pseudomembrane is composed of spherical yeast cells, hyphal filaments (pseudohyphae) of *C. albicans,* and cellular debris, and may become so large that it obstructs breathing. Oral thrush is most often observed in newborns. They acquire the disease during passage through the birth canal of their mothers, who have vulvovaginal candidiasis. Oral thrush is also frequently observed in the debilitated and aged patient. Other predisposing factors that lead to infection include low pH in the oral cavity of newborns, neoplasias, diabetes, and broad-

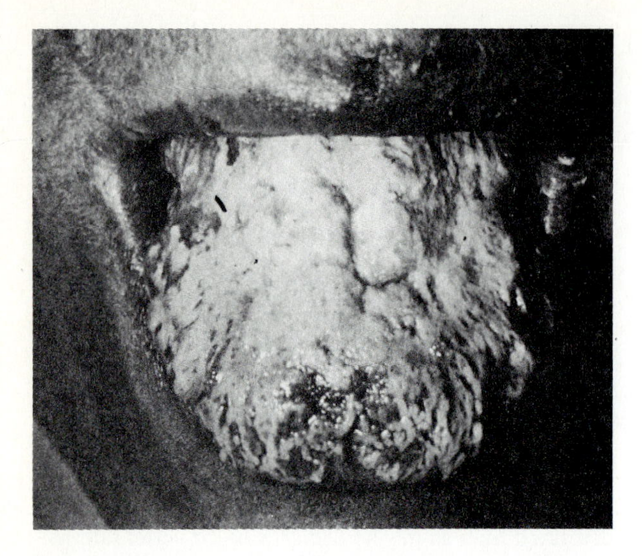

Figure 28-7 *Oral thrush, demonstrating white patches on surface of tongue.*

spectrum antibiotic and steroid administration in older children and adults.

Onychia and **paronychia** are infections of the subcutaneous tissue at the base of the nails of fingers or toes (Figure 28-8). Infections are usually the result of continued immersion of the hands or feet in water and are characterized by swelling, redness, and pain in the affected area.

Intertriginous candidiasis involves the moist areas of the body, such as the axillae, groin, intermammary folds, interdigital spaces, and intergluteal folds. Lesions are red, scaling or moist, and may be painful, depending on their location. Diabetic, obese, and alcoholic patients are more prone to infection, but those with occupations in which hands are frequently in water such as barbers, bartenders, fruit canners, and dishwashers are also susceptible to infection.

Vulvovaginal candidiasis occurs primarily in pregnant women, diabetics, and patients receiving antimicrobial therapy or hormonal contraceptives. Vaginitis caused by *C. albicans* is the second most common form of vaginal infection. The incidence of infection in pregnant women is highest during the third trimester, when the pH of the vagina is very acid. Lactobacilli, which are capable of controlling the population of *Candida* organisms in the vagina, are reduced in number or eliminated by the previously described predisposing factors. *Candida* organisms without competition proliferate and cause disease characterized by a yellow-white discharge and patches of pseudomembrane on the vaginal mucosa. The disease can also be transmitted to the sexual partner during intercourse.

Mucocutaneous candidiasis. Mucocutaneous candidiasis is a superficial disease that includes several clinical varieties of candidiasis such as onychia and paronychia, and vaginal, oral, respiratory, and gastrointestinal candidiasis (Figure 28-8). Disease is observed primarily in children who have defects in cell-mediated immunity, or occasionally in those with endocrine defects or leukemia. Treatment of superficial disease includes using drugs such as clotrimazole, miconazole, or nystatin.

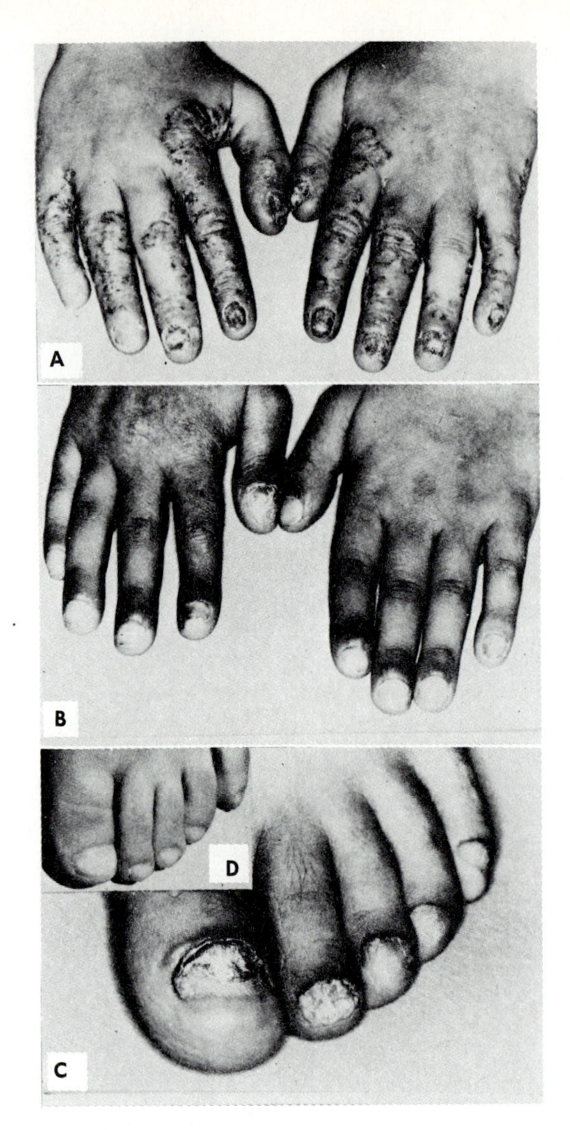

Figure 28-8 *Some forms of chronic mucocutaneous candidiasis in children.* ***A,*** *Skin and nail lesions.* ***B,*** *Paronchyia.* ***C,*** *Onychia.* ***D,*** *Toes of patient in* ***C*** *after treatment.*

Systemic candidiasis. Systemic candidiasis is the most frequent cause of death by any fungal agent. The prognosis of patients with systemic candidiasis is poor, and the death rate may be as high as 85% for those who are immunologically compromised. Amphotericin B and 5-fluorocytosine are used in treatment.

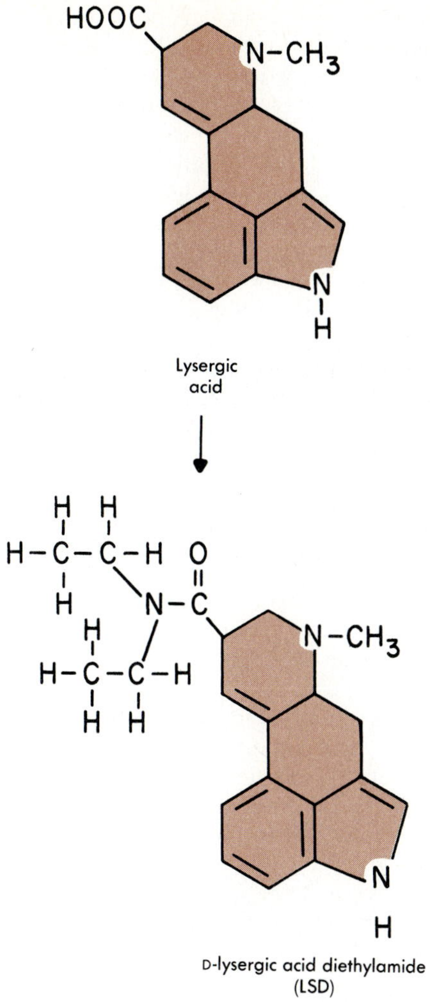

Lysergic acid

D-lysergic acid diethylamide (LSD)

Figure 28-9 *Conversion of lysergic acid to the hallucinogenic drug D-lysergic acid diethylamide (LSD).*

Mycotoxicosis, ergotism, mycetismus

Ingesting toxins of fungal origin cause conditions referred to as mycotoxicosis and mycetismus. **Mycotoxicosis** results from the ingestion of toxins (for example, aflatoxin) formed in foods, such as grains, that have been damaged because of the growth of fungi (also see Chapter 21.)

Ergotism is a disease of considerable historical importance and is occasionally observed in some areas of the world where home milling (processing of grain) is still practiced. Ergotism is caused by ingesting hyphae of the ascomycete, *Claviceps purpurea.* The fungus infects rye and forms dark hyphal masses, **ergots**, that replace the ovaries of rye flowers. The ergots may go unnoticed when the rye is harvested and made into flour for bread. Several armies in the past, including those of Napoleon, were stricken with ergotism, and many soliders died. Ergotism contributed to Napoleon's disastrous retreat from Moscow and his ultimate defeat. Ergots contain alkaloids that are related to lysergic acid, a precursor of the hallucinogenic drug, LSD (lysergic acid diethylamide—see Figure 28-9). Modern baking procedures have virtually eliminated ergotism.

Mycetismus is known as mushroom poisoning. Most of the toxic species belong to the genus *Amanita,* one species of which is called the "angel of death" (Plate 12). The symptoms of toxicosis include nausea, vomiting, and other gastrointestinal disturbances. Some species affect the nervous system, causing confusion, delirium, and visual disturbances, and can cause death. Some mushrooms produce hallucinogenic effects and were used in religious ceremonies for hundreds of years by certain Indian tribes of Mexico and South America. Treatment should be directed toward removing the fungus from the gastrointestinal tract by emetics or cathartics.

Concept Check

1. **List those fungal genera which are causes of cutaneous and superficial mycoses.**
2. **Describe the various conditions caused by *Candida albicans.* What is believed to be the relationship between virulence and steroids?**
3. **What conditions lead to infection by *Aspergillus* species?**
4. **Describe the distinction between mycotoxicosis and mycetismus. What is ergotism?**

ANIMAL PARASITE DISEASES

The parasites that have been discussed so far are members of three of the four major groups of microbial agents, the three groups being the bacteria, viruses, and fungi. However, there are parasites that belong to the animal kingdom. Animal parasites exhibit biochemical and morphological differences from the other three microbial groups. The animal parasites include arthropods, such as ticks and mites, that normally live on the body surface and are **ectoparasites.** They were discussed in Chapter 26. There are also the protozoa, which are single-celled members of the animal kingdom. Protozoa live in their hosts and are **endoparasites.**

General characteristics of protozoal diseases

Life cycles. During a protozoal infection the parasite becomes adapted to its host and often goes through a

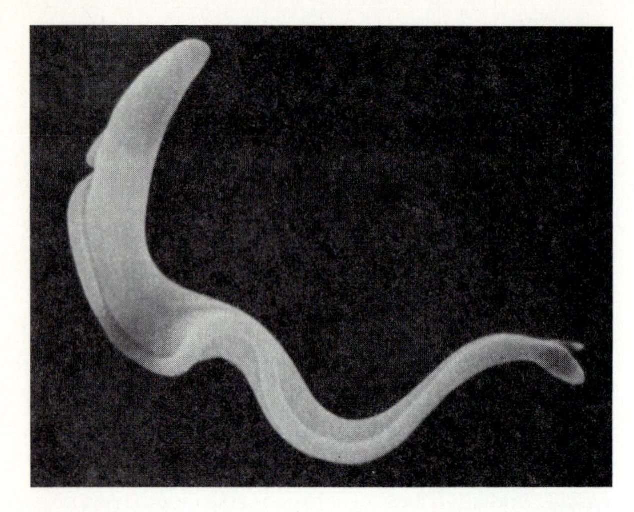

Figure 28-10 *Scanning electron micrograph of trophozoite stage of animal parasite* Trypanosoma brucei.

complicated course of development. The stages in animal parasite development constitute the **life cycle.** Many parasites colonize a number of different hosts and during their evolution have developed varied mechanisms of reproduction to suit the different host environments. The **definitive host** is the host in which the parasite undergoes sexual reproduction. If the parasite does not undergo sexual reproduction, the definitive host is the one believed to be the most important to parasite survival. An **intermediate host** is one in which the parasite does not reach a mature state.

The life cycle may be relatively simple. Some protozoa have only two stages: the motile, parasitic stage, or **trophozoite** (Figure 28-10), and the inactive, resistant form called the **cyst** (Plate 40). The cyst is transmitted to susceptible hosts, but in nearly all cases the trophozoite causes the symptoms of disease. Some protozoa may have stages involving both invertebrate and vertebrate hosts. For example, the agent of malaria develops in the mosquito as well as in humans. Scientists must understand the stages of the life cycle of protozoa because there are changes in surface antigens during those stages. This antigenic diversity makes identification of parasites difficult, and scientists must know which antigens are capable of eliciting a protective response. Vaccine development is impossible without this knowledge.

Epidemiology. Most animal parasite diseases are a consequence of poor hygiene. In addition, factors such as humidity and temperature also exert an influence on the probability of maintaining the parasite in the environment and acquiring the disease. Therefore more diseases are found in tropical and subtropical regions. In developed countries, most of the severe parasitic diseases, such as malaria, have been virtually eliminated.

Most serious diseases caused by protozoa are acquired by ingestion of contaminated food or water or by the bite of an arthropod such as a mosquito. Water may also serve as a vehicle for some unusual types of infections (see point of interest). Protozoa may be accidentally introduced into the body by hypodermic injection of contaminated fluids. For example, red blood cell–containing products may contain the malaria parasite, which then can be transmitted to others by transfusion.

Sexual transmission of the protozoa *Trichomonas vaginalis* is responsible for vaginitis (**trichomoniasis**) in women and urethritis in men. A high prevalence of colonization by *T. vaginalis* occurs in the United States, and the protozoa has been estimated to exist in more than

Point of Interest

A PROTOZOA CAUSING EYE INFECTIONS IN CONTACT LENS WEARERS

Eye infections associated with wearing contact lenses have caused some difficult problems in treatment. Many of the infections were originally diagnosed as being caused by herpes simplex virus. When treatment failed and cultures were taken from the cornea, the agent of disease was discovered to be an amoeba, *Acanthamoeba.*

Members of the genus *Acanthamoeba* are the most common free-living amebas in fresh water and soil. They are found in both air-borne dust and various types of water supplies and are commonly inhaled. *Acanthamoeba* are resistant to freezing, dessication, and the levels of chlorine that are used to disinfect municipal water supplies, swimming pools, and hot tubs.

It now appears that *Acanthamoeba* may contaminate contact lenses or lens-cleaning and lens-soaking solutions. In many described cases the contact lenses were cleaned with home-made salt solutions that were not sterilized. Thermal disinfection can prevent growth of the amoebae.

Treatment of eye infections caused by *Acanthamoeba* is difficult because of the resistance of these organisms to most of the commonly used antimicrobial agents. In some patients the disease was severe enough to necessitate corneal transplantation or removal of the eye. Twenty-four cases of eye infections were reported to the CDC in 1985.

30% of the population. Vaginitis is associated with a decreased vaginal acidity and this may be due to the ingestion by protozoa of bacteria such as lactobacilli that maintain a lower vaginal pH (3-4.5).

A protozoa that can cross the placenta is *Toxoplasma gondii,* the agent of toxoplasmosis. *T. gondii* has worldwide distribution and is believed to exist in over 50% of the population in the United States. Congenital disease is rarely serious, but serious disease can cause visual impairment and mental retardation. The life cycle of *T. gondii* is carried out in cats and other animals. Cats shed cysts in the feces, and these can be acquired by children and adults.

Pathogenicity. Colonization by protozoa seldom results in overt symptoms of disease. Many intestinal protozoa, such as amoebas, are harmless commensals, but occasionally a virulent form can cause intestinal disease. *Giardia lamblia* is an intestinal protozoan found in human populations of the world. Diarrhea caused by *G. lamblia* (**giardiasis**) frequently occurs during or after trips to foreign countries but may also occur among people in their own communities. In 1978, outbreaks of giardiasis occurred among residents of some Colorado ski villages and were linked to contaminated water supplies. Outbreaks of giardiasis continue to occur; most are caused by contamination of water supplies (see Point of Interest).

Most protozoal diseases are initiated by attachment of the parasite to mucosal surfaces. Very little is known of the types of adhesins used for this attachment process. One form of the malaria parasite is believed to possess surface fibrils that recognize receptors on the red blood cell surface. Black people apparently have red blood cells deficient in these receptors and thus are more resistant to malaria than are white people.

Toxins do not appear to play a significant role in the disease process of protozoa, unlike their role in bacteria. However, a protein toxin has been isolated from a pathogenic strain of *Entamoeba histolytica* that has been shown to be cytotoxic to hamster kidney cells. Toxinlike substances have also been isolated from the agents of African sleeping sickness (*Trypanosoma* species) and Chaga's disease (also caused by *Trypanosoma*), but they have not been purified.

Immunity. Little is known concerning the immune response of the host to protozoa. Some of the parasites possess a multitude of surface antigens, and many parasites exist in a variety of stages. The protozoa do stimulate the humoral and cell-mediated immune responses

Point of Interest

GIARDIA AND OUR WATER SUPPLY

Of all the waterborne outbreaks in the United States from the mid 1940s to 1980, *Giardia lamblia* is the leading cause of gastroenteritis, involving nearly 20 thousand cases. *Salmonella* causes approximately 18 thousand cases. It has been estimated that 4% (but perhaps as high as 20%) of the population in this country is infected with this protozoan. In the last few years, the number of cases of giardiasis has increased and now averages around 11,000 cases per year.

Although the classical symptoms of giardiasis are diarrhea, cramps, nausea, vomiting and greasy stool, many people carry *Giardia* and are asymptomatic. The active stage, called the trophozoite, attaches firmly to the intestinal wall by means of a suction disc. The parasite rarely invades the tissue but can cause damage to the microvilli. If the infection is serious, the entire surface of the upper intestine may be covered, and the protozoan may prevent absorption of nutrients, especially fats and vitamin B. It also causes electrolyte imbalance which results in the diarrhea. As the organisms reproduces and passes into the colon, it encysts and is released in the feces.

Giardia cysts are very resistant to chlorine so the chlorination of drinking water alone is inadequate to kill the cysts. Although appropriate sedimentation, flocculation, and filtration remove the cysts, many cities do not carry out these procedures. Because of this, outbreaks have been reported worldwide in large cities such as New York, Rome, and Leningrad.

In 1980, nearly 800 people in Red Lodge, Montana suffered from an outbreak of giardiasis. Ashfall from the Mount St. Helens volcanic eruption and warm sunny weather resulted in unusually heavy snow melts and water run-off. The city water system became overloaded, and the drinking water was not properly treated.

In addition to cities, giardiasis has been reported in campers in the Rocky Mountains and the Adirondack Mountains. Apparently, beavers and other wild animals carry the organism and contaminate the mountain streams. It is also speculated that other campers may have originally contaminated the waters. Drinking from these clear, "pristine" streams has resulted in the transmission of the infection to many backpackers and caused the uncomfortable condition known as "beaver's revenge."

Because the drug metronidazole, which is used to treat giardiasis, is also effective against anaerobic bacteria, it is believed that the disease may actually be the combined action of *Giardia* sp. causing the primary infection and anaerobic bacteria causing a secondary infection. It has also been observed that an antimicrobial substance in mother's milk kills *Giardia*. This is particularly fortunate because infants are especially sensitive to giardiasis, and when they become infected they often die.

Table 28-4 *Characteristics of drugs used in the treatment of animal parasite infections*

Drugs	Parasite affected	Toxicity	Mechanism of action
Thiabendazole	Antihelminthic* drug	May cause vomiting	Blocks production of ATP by inhibiting fumaric reductase, an enzyme involved in electron transport
Mebendazole	Antihelminthic drug (nematodes)	May cause nausea but less toxic than thiabenazole	Prevents uptake of glucose by parasite, which leads to glycogen depletion and eventually inability to produce energy
Metronidazole	Antiprotozoal drug (amebic dysentery, trichomoniasis, and giardiasis)	Gastrointestinal symptoms; possible carcinogen, not used during pregnancy or in asymptomatic infections	Mechanism of action unknown, but it is believed that the drug is reduced in the parasite to a compound that interferes with DNA synthesis
Quinacrine	Antiprotozoal drug (giardiasis)	Vomiting and dizziness	Mechanism of action unknown
Chloroquine	Antiprotozoal drug (malaria and amoebiasis)	Occular toxicity with corneal and retinal changes that may be permanent; also crosses placenta and can cause fetal damage	Drug forms complex with DNA and prevents DNA replication as well as transcription; exact mechanism is not known
Diethylcarbamazine	Antihelminthic drug (filariasis)	Relatively nontoxic but can cause allergic reactions	Mode of action is not known

*Helminths are worms.

of the host, but permanent immunity, with one or two exceptions, does not occur. Developing antibodies is the most important of the immune responses against those parasites that live in the bloodstream (for example, the agent of malaria). Parasites infect red blood cells and spread to new red blood cells by specific receptor attachment. Blocking the parasite binding sites with antibodies prevents attachment to receptors on the red blood cell. This prevents further parasite multiplication.

Cell-mediated immunity is more active against those parasites that live in tissues. For example, the agent of toxoplasmosis is an intracellular parasite that travels through the circulation in monocytes until they encyst in various tissues of the body. In most cases, *Toxoplasma* infections are asymptomatic, and immunity is believed to be related to the cell-mediated response. Humoral antibodies do not appear to play any role in immunity by the host.

The immune response to certain parasites is believed to be responsible for certain pathological conditions associated with the disease. In patients with chronic infections caused by the protozoa, *Trypanosoma cruzi,* antibodies are produced that cross-react with antigens common to the parasite and the heart and vascular tissue of the host. The heart damage in these patients is believed to be caused by this immunological response.

Diagnosis and treatment. In many human diseases, protozoa are shed in the feces in some morphological form. Direct microscopic examination of stained specimens is sufficient for identification of trophozoites and cysts. Less frequently, urine and blood can also be examined for parasites such as *T. vaginalis* and the agent of malaria, *Plasmodium* species.

Not all protozoa elicit an appreciable antibody response, especially those that infect only the intestinal tract, but some serological tests are available for those protozoa that do elicit such a response.

Diseases caused by animal parasites can be treated with a variety of chemotherapeutic agents (Table 28-4), but most agents are toxic to the host as well as the parasite, and drug levels must be carefully controlled. Many of the drugs are effective on certain trophozoites but have no effect on cysts. Cysts are unusually resistant to chemicals at even high concentrations. In fact, some cysts that are to be identified in the laboratory can be stored in a solution of sulfuric acid. The most widely used drugs for protozoal diseases are quinacrine, antimony compounds, and metronidazole.

No vaccines are available for preventing protozoal diseases in humans, primarily because of our difficulty in defining which antigens in the various stages of the cell cycle are involved in antibody formation. Researchers are now using nonvirulent variants of virulent parasites to stimulate antibodies and protect against infection by virulent strains. This is the same principle used in the eighteenth century, by Edward Jenner who developed the cowpox virus to prevent smallpox. Genetic engineering techniques are also being applied to the isolation and identification of antigens for vaccine development.

Table 28-5 *Characteristics of protozoal diseases*

Protozoal agent	Clinical condition	Method of transmission	Chemotherapy	Characteristics of life cycle
Naegleria fowleri	Primary amoebic meningoencephalitis; symptoms resemble meningitis; not common but almost invariably fatal	Through nose following swimming	Amphotericin B	Infects humans and is free-living outside host
Dientamoeba fragilis	Diarrhea in children; no fatalities	May be transmitted in pinworm eggs	Metronidazole	Commensal in humans
Giardia lamblia	Diarrhea	Transmission through contaminated water supplies	Quinacrine and metronidazole	Humans are carriers; beavers, dogs, and other mammals are reservoirs
Trichomonas vaginalis	Symptomatic infection only in females	Venereal	Metronidazole	Found in genitourinary tract
Leishmania species	Leishmaniasis	Bite of sand flies	Antimony gluconate	Dogs are the only important reservoir
L. tropica	Cutaneous ulcers	Same as above	Same as above	Dogs are reservoirs
L. braziliensis	Cutaneous lesions with parasites attacking cartilage of nasal septum	Same as above	Same as above	Sloths, anteaters, arboreal (tree) rodents, and marsupials are important reservoirs

Continued.

Concept Check

1. What are the characteristics of the trophozoite and cyst stages of protozoa?
2. What are believed to be some of the virulence factors for protozoa?
3. What is the relationship of the humoral and cell-mediated immune response of the host to infections by protozoa?

Diseases caused by protozoa

Several protozoa cause disease in humans, and some have been alluded to in the previous discussion. We will discuss two diseases; amoebic dysentery, which affects primarily the intestinal tract, and malaria, a disease of the red blood cells. The other diseases caused by protozoa are outlined in Table 28-5.

Amoebic dysentery. Several amoebas are commensals in the intestinal tract of humans, but only *Entamoeba histolytica* (Figure 28-11) is a pathogen of major

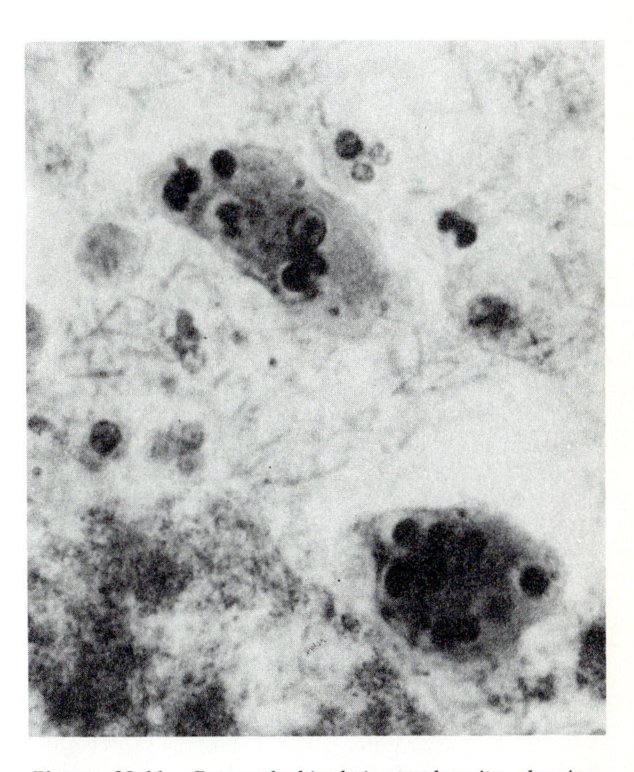

Figure 28-11 Entamoeba histolytica *trophozoites, showing ingested red blood cells and directional motility.*

***Table 28-5** Characteristics of protozoal diseases—cont'd*

Protozoal agent	Clinical condition	Method of transmission	Chemotherapy	Characteristics of life cycle
Leishmania species—cont'd.				
L. donovani	Visceral involvement (kala azar); enlargement of spleen and liver; high mortality	Same as above	Same as above	Dogs and foxes are important reservoirs
Balantidium coli	Common intestinal infection occurring in debilitated patients; severe dysentery	Ingestion of cysts from feces of humans and swine	Oxytetracycline	Infects humans and many lower primates; pigs most important reservoir
Entamoeba histolytica	Amoebiasis; intestinal infection most common with severe dysentery; hepatic infection causes abscesses and pain over liver and may be fatal	Ingestion of food or water contaminated with cysts	Metronidazole; furamide for cyst passers	Trophozoite and cyst stage occur only in single host; humans are important reservoirs, but dogs, pigs, and monkeys also implicated
African *Trypanosoma* species *T. brucei gambiense* *T. brucei rhodesiense*	African sleeping sickness; central nervous system involvement; heart failure, pneumonia, or encephalitis frequent cause of death	Bite of tsetse fly	Suramin and pentamidine isethionate	Humans only reservoir of *T. gambiense;* cattle, hartebeest, bushbucks, and other mammals are reservoirs
American *Trypanosoma cruzi*	Chaga's disease; visceral organs affected, including heart; acute disease in all age groups; chronic disease in adults with cardiac dysfunction	Feces of reduviid bugs	No satisfactory treatment available	Humans and domestic animals serve as hosts; cats and wild animals such as armadillos and oppossums are important reservoirs
Plasmodium species *P. vivax* *P. falciparum* *P. malariae* *P. ovale*	Malaria; chills and fever; anemia and enlargement of spleen and liver in severe cases; fever related to release of parasites from infected red blood cells	Bite of female *Anopheles* mosquito	Chloroquine, primaquine	Life cycle takes place in two hosts: a vertebrate such as humans and an invertebrate, the mosquito
Babesia microti	Nantucket fever (babesiosis); symptoms similar to malaria	Transmitted by ticks	Chloroquine	Life cycle similar to *Plasmodium;* ticks are vectors and intermediate hosts
Toxoplasma gonodii	Toxoplasmosis; congenital infection most severe, involving eye and central nervous system, sometimes resulting in mental retardation and other clinical conditions; adult infection usually asymptomatic	Ingestion of cysts from cat feces; ingestion of uncooked contaminated meat; transfer across placenta to fetus from infected mother	Sulfadiazone and pyrimethamine	Cats are final host, and are infected by eating cysts or infected rodents; parasite infects a wide variety of birds and animals; humans and wild and domestic animals are intermediate hosts

significance. Humans are the most important reservoirs of the infectious agent, but dogs, pigs, and monkeys can also be implicated. Most infections are the result of ingestion of food or water contaminated by cysts (Figure 28-12). Approximately 10% of the world population is affected by this disease; most cases occur in tropical or subtropical areas. In the United States between 3000 to 4000 cases are reported each year. The majority of infections result from poor sanitation and improper disposal of human wastes. Flies and cockroaches can carry cysts

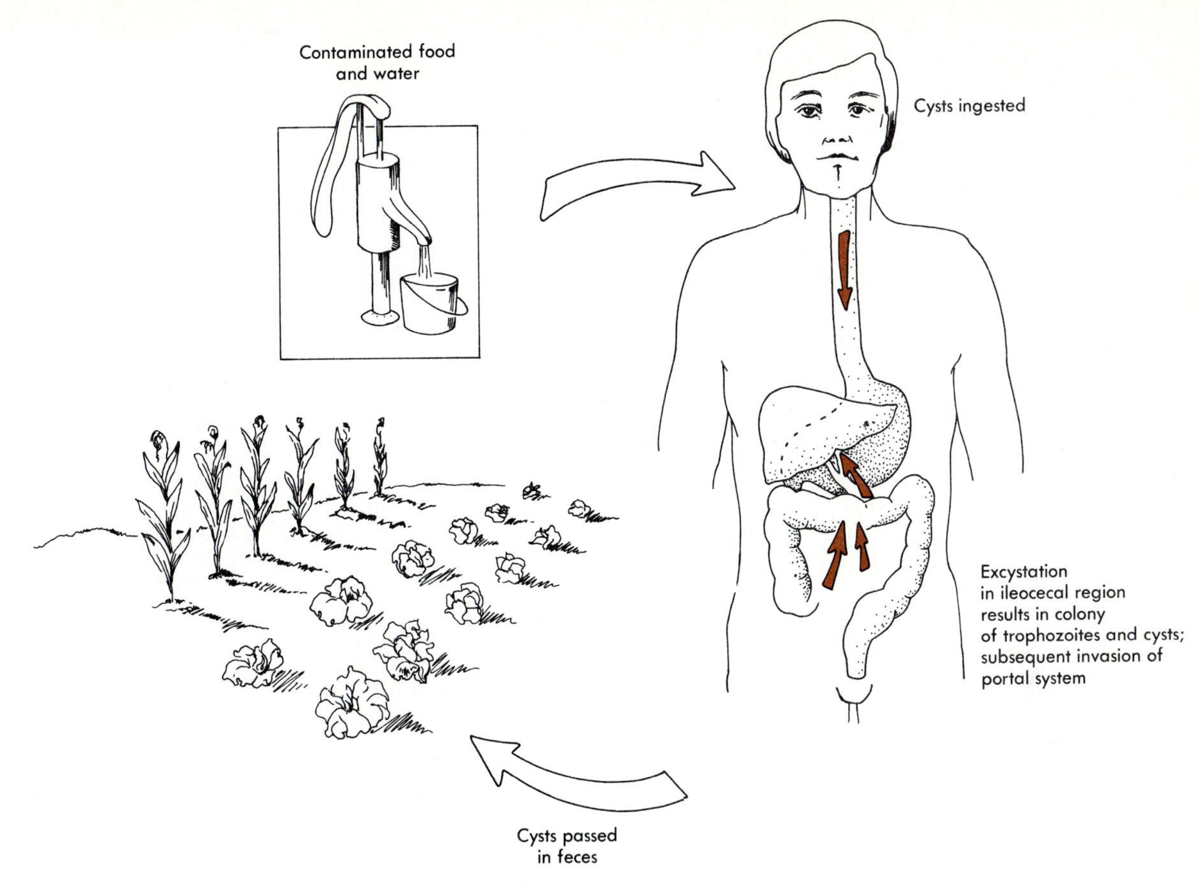

Figure 28-12 *Life cycle of* Entamoeba histolytica. *Food and water contaminated with infective cysts are ingested. Excystation in ileocecal regions results in colony of trophozoites and cysts becoming established in cecum and colon. Penetration of mucosa by trophozoites with subsequent invasion of portal circulation may result in extraintestinal amebiasis. Cysts passed in feces contaminate food and water, which in turn are ingested by humans.*

from contaminated material and are important vectors in transmission to humans. In areas where human feces is used as a fertilizer, infections among the human population are common. Asymptomatic human cyst shedders are prevalent and provide a ready avenue for transmission to family members.

Ingestion of food or water contaminated by cysts initiates the life cycle (Figure 28-12). Excystation occurs in the ileocecal region of the intestine, and both cysts and trophozoites can be found there. Trophozoites that penetrate the intestinal wall produce an ulcer that is usually located in the appendix, cecum, or colon. Death can result from a perforated colon and subsequent peritonitis. Trophozoites that invade the bloodstream can cause secondary disease, especially in the liver. Abscesses that occur in the liver sometimes rupture and pass parasites into the body cavity to infect other organs.

Most disease develops slowly with slight nausea and mild gastrointestinal disturbances. Tissue invasion results in sharp abdominal pain, sometimes accompanied by vomiting and constipation, with or without diarrhea. The disease may progress to the point that dysentery occurs, with as many as 10 to 20 or more bloody stools occurring per day. It is during the latter period that extensive ulceration of the intestine takes place, sometimes followed by secondary bacterial diseases.

The virulent strains of *E. histolytica* are cytolytic and invasive. Pathogenicity is initiated by the adherence of the parasite to host cells by means of an adhesin. The adhesin, which is inhibited by N-acetyl-D-galactosamine, may be associated with microfilaments on the amoebic surface. The microfilaments are believed to make contact with the host cell. Immediately following adhesion, lysis of the host cell occurs; apparently this is associated with a calcium-dependent phospholipase A present in the parasite. Antagonists of calcium-dependent phospholipase A have been shown to prevent the cytolytic activity of the parasite. The mechanism by which cytolysis occurs is still not known. Amoebic phospholipase may act on the plasma membrane of the host cell, causing

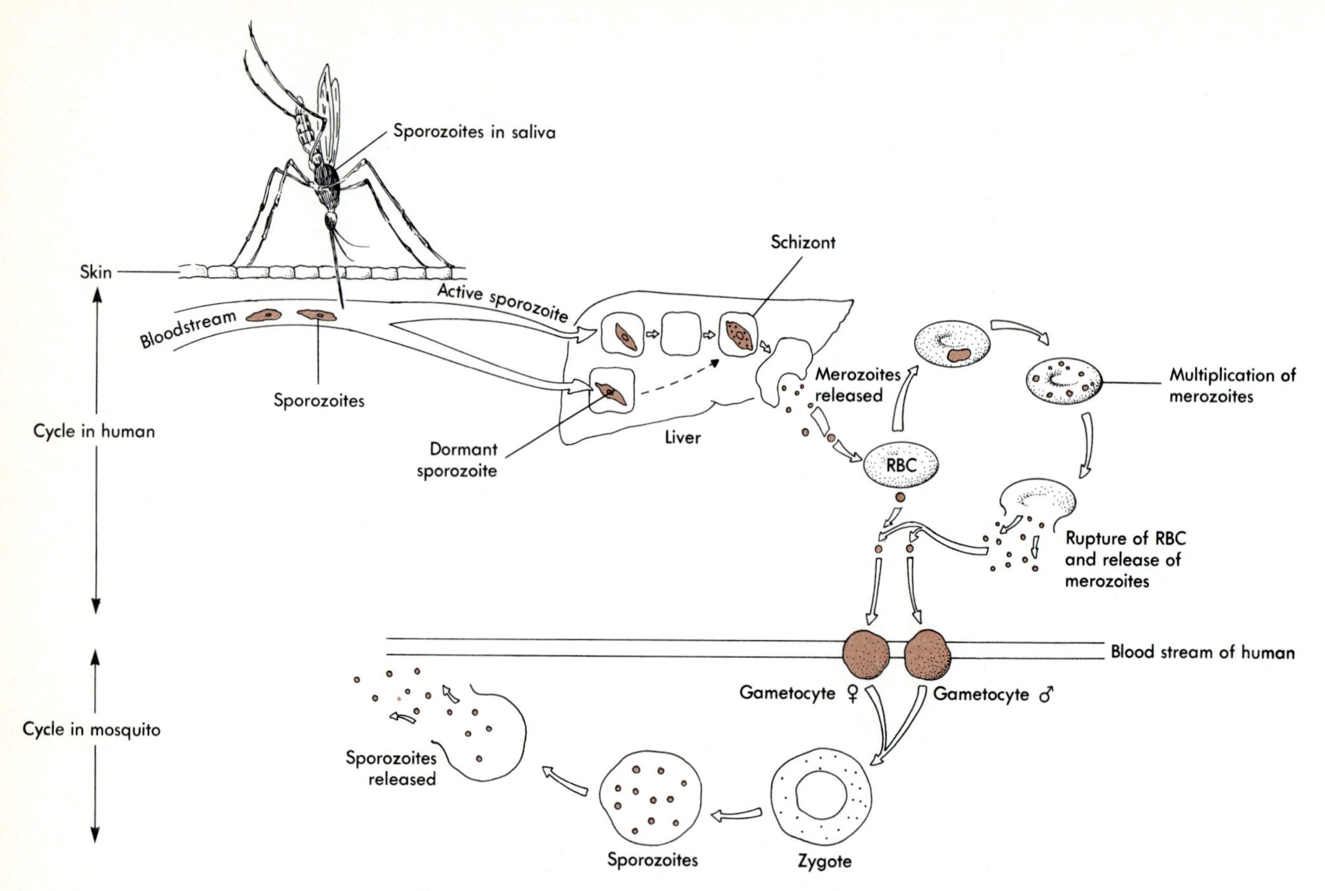

Figure 28-13 *Life cycle of malaria parasite. Mosquito injects sporozoites into bloodstream of human. Sporozoites go to liver, where some invade cells and remain dormant while others are active and develop into structure called* shizont. *Schizont contains thousands of merozoites, a second form of the parasite. Merozoites released from liver invade red blood cells, in which more merozoites are produced to infect other red blood cells. Some merozoites in red blood cells develop into gametocytes, which are ingested by mosquito feeding on blood of human. Gametocyte union produces zygote, which in several stages develops into sporozoites that travel to salivary gland of mosquito. Sporozites are transmitted to humans when mosquito feeds on blood.*

lethal permeability changes, or it may activate host cell phospholipases that degrade the membrane of lysosomes. Release of lysosomal enzymes might result in an autolytic process in the host cell.

Metronidazole is the drug of choice in treatment. Furamide is used almost exclusively for asymptomatic cyst shedders.

Malaria. Malaria is caused by protozoa belonging to the genus, *Plasmodium.* Four species of *Plasmodium* are recognized as agents of malaria, but most severe malaria infections are caused by *P. falciparum.*

The life cycle of the malaria parasite involves three antigenically distinct forms—sporozoite, merozoite, and gametocyte. The cycle begins when the female *Anopheles* mosquito feeds on a human (Figure 28-13). The form of the parasite in the mosquito is called the **sporozoite**, which enters the human bloodstream from the mosquito's saliva. Each sporozoite invades a liver cell and begins an asexual phase called **schizogony** (some of the sporozoites form a dormant body that does not immediately replicate). Each active sporozoite divides, forming a structure called a **schizont**. Each schizont contains thousands of a second form of the parasite, **merozoites**. The schizont ruptures, spilling the merozoites into the bloodstream. Merozoites rapidly invade red blood cells and multiply. The red blood cells burst and release the merozoites, which invade other red blood cells and multiply. This cycle continues until the person dies of anemia, kidney failure, or brain damage, or until the disease is brought under control by the host's immune system or by drugs.

Some of the merozoites in red blood cells develop into sexual forms, **gametocytes**. When a mosquito feeds on infected humans, red blood cells containing gametocytes are ingested. The gametocytes are released from the red blood cells in the stomach of the mosquito, undergo zygote formation, and finally develop into thousands of sporozoites. The sporozoites move from the stomach to the salivary glands, where they remain ready to enter a new host when the mosquito next feeds.

The rupturing of red blood cells causes the typical malaria symptoms of chills and fever; however, the exact mechanism of symptoms is not known. Disease relapses may occur for many years after the initial attack and are caused by the dormant sporozoites in liver cells becoming active and replicating. Malaria usually kills children under age of 5 and is found primarily in tropical and subtropical areas. Thus one third of the world's population is threatened. Currently it is estimated that 60 to 100 million people a year are affected by the disease and from 2 to 4 million deaths occur each year.

People over the age of 5 who survive malaria develop immunity. An immune adult may have a low fever once or twice a year, but the symptoms subside quickly. Immune people in areas where malaria is endemic often carry small numbers of *Plasmodium* organisms in the blood and thus serve as a source of disease to others. Those who carry the parasite may not develop malaria but are susceptible to other diseases because of a depressed immune system. Resistance to malaria in Africa and the Mediterranean has been known for many years.

Genetic resistance among black people is the result of certain heritable diseases such as sickle cell anemia, thalassemia, and glucose 6-phosphate dehydrogenase (G6PD) deficiency, or **favism**. Each of these genetic defects is in some manner associated with the red blood cell. Resistance by individuals with favism is understandable because the malaria agent depends totally on host G6PD for the conversion of hexoses to pentoses in the hexose monophosphate shunt. Why those with sickle cell traits or thalassemia are resistant is not totally clear, but it appears that a red blood cell component plays an important role. A sugar-containing component is absent from the red blood cells of resistant individuals but is present in those susceptible to malaria.

The two principal drugs used in treating malaria are

Point of Interest

MALARIA VACCINES

A vaccine directed against an animal parasite does not yet exist, but the first one will probably be against the agent of malaria. Developing such a vaccine is complex for two reasons: the malaria parasite is so much larger than other infective agents, and it has a life cycle in which there are three antigenically distinct phases—sporozoite, merozoite, and gametocyte. Eventually a vaccine will be developed that will immunize against all three forms of the parasite, but most work at present is on developing a sporozoite vaccine. A single-coat protein covers the surface of the sporozoite, which is the form of the parasite injected into the bloodstream by the mosquito. A sporozoite vaccine would prevent malaria and block its spread.

Until recently there was great difficulty in obtaining sporozoites because they had to be extracted from the salivary glands of mosquitos. Scarcity of sporozoites was also a problem for those investigators wishing to clone the genes that code for the major surface protein, the circumsporozoite (CS) protein. Mosquitos had to be fed the blood of patients in the infective stages; it took about two years to accumulate enough material to do one cloning experiment. A newer technique for obtaining the gene coding for CS protein involves removing the gene directly from the DNA of a parasitic form that can be grown in culture. In this method a nuclease enzyme from the mung bean plant is used to cut the DNA. This nuclease is remarkable in that it happens to clip the DNA just before the beginning and just after the end of the genes of this parasite. The major portion of the CS protein has been shown to consist of 41 tandemly linked tetrapeptides, 37 of which are identical. The tetrapeptide for *P. falciparum* is:

Asparagine(asn)-alanine(ala)-asparagine-proline(pro)

No major variation appears to occur in these particular tandem repeats within the species; a single monoclonal antibody recognizes the same repeating structure in dozens of *P. falciparum* isolates from diverse geographical areas around the world; however, variation exists in other *Plasmodium* species).

The question now is if a vaccine directed against CS protein can protect humans against malaria. Experiments have already demonstrated that antibodies to the protein can protect animals such as mice against malaria. In 1985, a series of tetrapeptides were synthesized and used to determine the dominant determinants that are likely to be involved in the initial interaction between the parasitic surface and the host target cell. The dominant determinant was found to be a dodecapeptide:

Asn-ala-asn-pro-asn-ala-asn-pro-asn-ala-asn-pro

Polyclonal antibodies raised against the synthetic peptide react with the surface of the parasite and neutralize its infectivity. It now appears that the determinant will be the target for vaccine development.

chloroquine, which is used to kill the parasite in the red blood cell, and primaquine, which is used to kill the parasite in the tissue phase (liver cells). If the dormant sporozoites are not eradicated from the liver the *Plasmodium* organisms can reestablish symptoms many years after the initial attack. Chloroquine, which is administered with or without primaquine, is a very effective and inexpensive drug, but *P. falciparum* strains resistant to the drug are becoming more prevalent. Alternative drugs are more expensive and not as effective as chloroquine. Considerable research effort is being directed at developing a vaccine to overcome the current problems associated with chemotherapy and to control this devastating disease (see Point of Interest).

Concept Check

1. Describe the life cycle of *Entamoeba histolytica.*
2. What is the relationship between phospholipase A and *E. histolytica?*
3. List one protozoal disease that is transmitted in the following way: ingestion of contaminated water or food, sexual transfer, congenital transfer, and mosquito bite.
4. What experiments with the agent of malaria make it appear that a vaccine may be available in the near future?
5. Describe the three forms of the malaria parasite.

SUMMARY

1. Fungal diseases are called mycoses and are divided into the following types: systemic, superficial, cutaneous, subcutaneous, and opportunistic.
2. Nearly all fungal diseases are acquired by contact with spores in the soil or by inhalation of spores dispersed in the air. The ability of the fungi to cause disease is related to the number of spores to which the host is exposed and the state of health of the individual. Fungal diseases in the healthy person are rare.
3. Systemic mycoses are caused by organisms that can penetrate epithelial tissue and can be disseminated throughout the body. Most infections are acquired by inhalation of spores resulting initially in respiratory symptoms that mimic tuberculosis. Histoplasmosis is the most common systemic disease in the United States and is acquired by contact with soil enriched with excreta from birds such as chickens. Most diseases are acute and seldom fatal.
4. Superficial and cutaneous mycoses, such as ringworm (tineas), are caused by agents called dermatophytes. The microbial agents colonize keratinized tissue on any body surface (scalp, beard, groin, feet). Many of the microorganisms parasitize animals, and humans are secondarily infected. The most common source of infections for humans is via human-to-human spread. Some of the dermatophytes can cause allergic responses (ids).

5. Subcutaneous mycoses are often the result of puncture wounds. Sporotrichosis is the most common disease and is characterized by lymphatic nodule enlargement near the site of infection.
6. Opportunistic mycoses occur when the host has been compromised by diabetes, immunodeficiencies, antibiotic therapy, or underlying disease. Diseases result from the overgrowth of indigenous species, particularly species of *Candida,* which are commensals in the oral cavity, intestinal tract, vagina, and skin. Diseases involve moist areas of the body, including the mucous membranes and nails.
7. Treatment of fungal diseases is divided between two types of agents: topical and systemic. Topical agents include nystatin, tolnaftate, and miconazole, and systemic agents include amphotericin B and 5-fluorocytosine.
8. Fungal agents can be toxic when ingested (for example, mushrooms), and the resulting condition is called *mycetismus.* Other fungal agents can produce toxic products when they grow in various stored grain crops. Ingestion of grains containing these toxic products, such as aflatoxin, can cause serious disease (mycotoxicosis).
9. Animal parasites, such as the protozoa, have life cycles that are either simple or complex, depending on the number of hosts involved. In many protozoal diseases, only the resistant cyst stage is transmissible to humans.
10. Most infections by protozoa occur in tropical or subtropical regions of the world. The parasite is usually transmitted by ingestion of contaminated water or food, but transmission may also be through insect bites and by congenital and sexual transfer.
11. Whether one becomes infected is related to the number of parasites ingested or number of exposures to the parasite and the ability of the parasite to penetrate epithelial barriers.
12. Unlike other microbial agents, protozoa do not confer a permanent immunity following disease. Antibodies are usually protective for those infections involving the bloodstream, whereas cell-mediated immunity prevails for tissue infections.
13. Many, but not all, protozoal diseases can be diagnosed by examination of cysts in the feces or of trophozoites in the urine or blood. Serological tests can also be used if the parasite is known to induce sufficient antibodies during the course of the disease. The most important chemotherapeutic agents for infections by protozoa are antimony compounds, quinacrine, and metronidazole.
14. Two important parasitic protozoa are *Entamoeba histolytica,* the agent of amebic dysentery, and *Plasmodium* species, the agents of malaria.

Self-study Quiz

Multiple choice

1. Which of the following statements best reflects the pathogenicity of the fungi as a group?
   A. Few fungal pathogens exist, but those species which are pathogenic often produce fatal infections.
   B. A large number of fungal pathogens exist, but they usually produce only mild infections.
   C. Few species of fungi that are pathogenic exist, and they usually become established only in hosts whose resistance is low.
   D. Even though many fungal pathogens exist, infections by fungi are quickly and easily treated and so are of little pathogenic consequence.
2. Which of the following genera would most likely be isolated from an athlete's foot lesion?
   A. *Sporothrix*
   B. *Trichophyton*
   C. Cladosporium
   D. *Candida*
3. What aspects of diabetes mellitus make the physiological condition selective for fungal infections?
   A. Acidosis and elevated blood glucose
   B. Immunodeficiency in circulating antibodies
   C. Increased levels of oxygen in the blood
   D. Lack of insulin in the blood
4. Which of the following would not be effective in treatment of fungal infections?
   A. Nystatin
   B. Amphotericin B
   C. Ketoconazole
   D. Sulfa drugs
5. Which of the following is associated with mushroom poisoning?
   A. Mucormycosis
   B. Mycetismus
   C. Mycotoxicosis
   D. Aspergillosis
6. Ergotism refers to:
   A. A condition caused by the organism *Aspergillus ergota*
   B. A disease caused by a fungal parasite that replaces the ovaries of rye flowers with masses of hyphae
   C. An intoxication caused by ingestion of poisonous mushrooms
   D. A disease caused by *Trichophyton rubrum*
7. What is tinea pedis?
   A. Ringworm of the nails
   B. Ringworm of the foot
   C. Ringworm of the lung
   D. Ringworm of the scalp and torso
8. What aspects of the immune response system are most responsible for protecting against fungal infections?
   A. Humoral antibodies
   B. Wandering macrophage engulfment
   C. The complement system
   D. Cell-mediated immune response
9. Oral thrush is produced by what genera?
   A. *Candida*
   B. *Cryptococcus*
   C. *Aspergillus*
   D. *Histoplasma*
10. What is the primary source of most mycotic infectious organisms producing cutaneous infections?
    A. The soil
    B. Direct or indirect contact with other humans suffering from the infection
    C. Bird feces
    D. Puncture wounds by plants harboring the agent
11. The form of the malaria parasite in the mosquito that is transmitted to humans is called:
    A. Schizont
    B. Merozoite
    C. Sporozoite
    D. Gametocyte
12. Which of the following protozoal diseases would most likely be affected by the humoral immune response?
    A. Giardiasis
    B. Dysentery caused by *Entamoeba histolytica*
    C. Trichomoniasis
    D. Malaria

Matching

Match the disease with the method of transmission

________1. Malaria	a. Sexual
________2. Giardiasis	b. Fly bite
________3. Toxoplasmosis	c. Contaminated water or food
________4. Trichomoniasis	d. Mosquito bite
________5. African sleeping sickness	
________6. Leishmaniasis	

Thought questions

1. In the 1960s and 1970s a dramatic increase in the number of cases of vulvovaginitis caused by *Candida albicans* was reported. This increased incidence has been attributed to the widespread use of oral contraceptives and the introduction of panty hose. Explain the connections.
2. Go to a drug store and find two over-the-counter preparations that are used to treat fungal infections. List the antifungal agent(s) in each.
3. The number of malaria cases in the United States at specific times can be traced to specific programs and worldwide events. Try to identify some of the events affecting these trends.
   a. 1930-1940: Drop in cases from 180,000 to 50,000 per year
   b. 1941-1945: Rise from 50,000 to 60,000 cases per year
   c. 1946-1950: Drop from 60,000 to 2000 cases per year
   d. 1951-1953: Rise from 2000 to 6000 cases per year
   e. 1954-1960: Drop from 6000 to 75 cases per year
   f. 1961-1970: Rise from 75 to 4000 cases per year
   g. 1971-1973: Drop from 4000 to 300 cases per year
   h. 1974-1980: Rise from 300 to 2000 cases per year

SELECTED READINGS

Books

Beck, J.W., and Davies, J.E. *Medical parasitology* (3rd ed.). The C.V. Mosby Co., St. Louis, 1981.

Katz, M., Despommier, D., and Gwadz, R. *Parasitic diseases.* Springer-Verlag New York, Inc., New York, 1983.

Levinthal, R., and Cheadle, R. *Medical parasitology.* F.A. Davis Co., Philadelphia, 1985.

Rippon, J.W. *Medical mycology* (2nd ed.). W.B. Saunders Co., Philadelphia, 1982.

Rose, A.H., and Harrison, J.H. (Eds.). *The yeasts* (2nd ed.). Academic Press Inc., New York, 1984.

Uraguchi, K., and Yamazaki, M. (Eds.). *Toxicology, biochemistry, and pathology of mycotoxins.* John Wiley & Sons, Inc., New York, 1980.

Journal articles

Epstein, J.B., Truelove, E.L., and Izutzu, K.T. Oral candidiasis: pathogenesis and host defense. *Rev. Infect. Dis.* **6**(1):96, 1984.

Gold, J.W.M. Opportunistic fungal infections in patients with neoplastic disease. *Am. J. Med.* **76**:458, 1984.

Goodson, G.N. Molecular approaches to malaria vaccine. *Sci. Am.* **252**(5):52, 1985.

Ravdin, J.I., Murphy, C.F., Guerrant, R.L., and Cong-Krug, S.A. Effect of antagonists on calcium and phospholipase A on the cytopathogenicity of *Entamoeba histolytica. J. Infect. Dis.* **152**:542, 1985.

Rostrosen, D. Adherence of *Candida* species to host tissues and plastic surfaces. *Rev. Infect. Dis.* **8**(1):73, 1986.

Rotrosen, D., et al. Adherence of *Candida* to cultured vascular endothelial cells: mechanisms of attachment and endothelial cell penetration. *J. Infect. Dis.* **152**(6):1264, 1985.

Shepherd, M.G., Poulter, R.T.M., and Sullivan, P.A. *Candida albicans:* biology, genetics, and pathogenicity. *Annu. Rev. Microbiol.* **39**:579, 1985.

Stevens, D.P. Strategies for control of disease in the developing world. XIX. Giardiasis. *Rev. Infect. Dis.* **7**(4):530, 1985.

Werk, R. How does Toxoplasma gondii enter host cells. *Rev. Infect. Dis.* **7**(4):449, 1985.

Zavala, F., et al. Rationale for development of a synthetic vaccine against *Plasmodium falciparum* malaria. *Science* **228**:1436, 1985.

APPENDIX

Scientific measurements and mathematics

TERMINOLOGY

BASIC UNITS

Property	Basic Unit	Symbol
Length	Meter	m
Mass	Kilogram	kg
Time	Second	s
Amount of substance	Mole	mol
Electric current	Ampere	A
Temperature	Kelvin	K
Luminous intensity	Candela	cd
Energy	Joule	J
Pressure	Pascal	Pa
Force	Newton	N

STANDARD PREFIXES

Prefix	Multiplication Factor	Symbol
pico	10^{-12}	p
nano	10^{-9}	n
micro	10^{-6}	μ
milli	10^{-3}	m
centi	10^{-2}	c
deci	10^{-1}	d
deca	10	da
hecto	10^{2}	h
kilo	10^{3}	k
mega	10^{6}	M

METRIC SYSTEM

Length

1 meter (m) = 10 decimeters (dm) =
100 centimeters (cm) = 1000 millimeters (mm) =
1,000,000 micrometers (μm) =
1,000,000,000 nanometers (nm)

1 angstrom (Å) = 0.1 nm = 0.0001 μm

1 kilometer (km) = 1000 m

Examples of interconversions:

1 cm = 10 mm; 1 mm = 0.1 cm; 1 mm = 1000 μm; 1 μm = 0.001 mm; 1 μm = 1000 nm; 1 nm = 0.001 μm; 1 nm = 10 Å

Volume

1 liter (L) = 1000 milliliters (ml) =
1000 cubic centimeters (cc or cm^3)

Mass

1 kilogram (kg) = 1000 grams (g) =
1,000,000 milligrams (mg)

1 mg = 1000 micrograms (μg)

ENGLISH EQUIVALENTS TO THE METRIC SYSTEM

Length

1 inch = 2.540 centimeters

Volume

1 liter = 1.056 quarts

1 cubic inch = 16.38 cubic centimeters

Mass

2.2 pounds = 1 kilogram

1 pound = 453 grams

1 ounce = 28.3 grams

Temperature Equivalents

$$\text{Degrees centigrade (°C)} = \frac{5}{9}(°F - 32)$$

$$\text{Degrees Fahrenheit (°F} = \frac{9}{5}°C + 32$$

$$0° C = 32° F$$

$$100° C = 212° F$$

$$\text{Absolute temperature} = \text{degrees Kelvin (°K)} = 273 + °C$$

MATHEMATICAL OPERATIONS: EXPONENTS

An exponent is a number that indicates how many times the base appears as a factor. In most mathematical operations 10 is the most convenient base. The exponent appears as a superscript on the base. For example:

$$10^3 = 1 \times 10^3 = 1000$$

Therefore the exponent of 10^3 is 3, and 10^3 means $10 \times 10 \times 10$. The exponent of the base 10 determines where the decimal point must be shifted. If the shift is to the right, the exponent is positive:

$$1 \times 10^3 = 1000$$

If the shift is to the left, the exponent is negative:

$$1 \times 10^{-3} = 0.001 = \frac{1}{1000}$$

Multiplication of exponents

During multiplication exponents are added:

$$10^m \times 10^n = 10^{m+n}$$

Some examples are:

$$10^4 \times 10^2 = 10^6 = 1{,}000{,}000$$

$$5.3 \times 10^2 \times 10^3 = 5.3 \times 10^5 = 530{,}000$$

Division of exponents

During division exponents are subtracted:

$$\frac{10^m}{10^n} = 10^{m-n}$$

Some examples are:

$$\frac{2 \times 10^5}{1 \times 10^2} = 2 \times 10^{5-2} = 2 \times 10^3 = 2000$$

$$\frac{5 \times 10^4}{2 \times 10^6} = 2.5 \times 10^{4-6} = 2.5 \times 10^{-2} = 0.025$$

$$\frac{5 \times 10^{-2}}{2 \times 10^3} = 2.5 \times 10^{-2+-3}$$

$$= 2.5 \times 10^{-5} = 0.000025$$

Addition and subtraction of exponents

During addition and subtraction of exponents the terms must have the same exponent; the terms are then either added or subtracted. Some examples are:

$$5 \times 10^{-2} + 3 \times 10^{-2} = 8 \times 10^{-2}$$

$$4.2 \times 10^2 + 2 \times 10^2 = 6.2 \times 10^2$$

$$5 \times 10^{-2} + 2 \times 10^{-3} = 5 \times 10^{-2} + 0.2 \times 10^{-2} = 5.2 \times 10^{-2}$$

$$4 \times 10^3 + 5 \times 10^4 = 0.4 \times 10^4 + 5 \times 10^4 = 5.4 \times 10^4$$

$$5 \times 10^{-2} - 2 \times 10^{-3} = 5 \times 10^{-2} - 0.2 \times 10^{-2} = 4.8 \times 10^{-2}$$

Exponents of exponents

The exponent of an exponent is multiplied by the exponent. For example:

$$(10^3)^2 = 10^6$$

LOGARITHMS (TO BASE 10)

The logarithm of a number is the power to which the base must be raised to equal the number. For our purposes, the base number is 10. For example:

$$\log 10 = 1 \text{ or } 10^1$$

$$\log 100 = 2 \text{ or } 10^2$$

$$\log 1000 = 3 \text{ or } 10^3$$

$$\log 0.1 = -1 \text{ or } 10^{-1}$$

$$\log 0.01 = -2 \text{ or } 10^{-2}$$

$$\log 0.001 = -3 \text{ or } 10^{-3}$$

Multiplication of logarithms

During the multiplication of logarithms the number is added. For example:

$$\log(1000 \times 1000) = \log(10^3 \times 10^3) = \log(10^3) + \log(10^3) = 3 + 3 = 6$$

Division of logarithms

During division logarithms are subtracted. For example:

$$\log(1000 \div 100) = \log(1000) - \log(100) = 3 - 2 = 1$$

APPENDIX B

Bacterial taxonomy

In Chapter 4 you were introduced to bacterial taxonomy, or how bacteria can be arranged into specific groups and how names are given to these groups. The criteria for classification was not discussed in Chapter 4 but is discussed in the following paragraphs.

NUMERICAL TAXONOMY

Numerical taxonomy is a technique devised in 1950 to assist in the arrangement of species into genera based on morphological, biochemical, serological, and other properties. The technique was developed in parallel with the advent of computers to reduce the burden of analyzing by eye the multitude of taxonomic characteristics. Numerical taxonomy is carried out in the following way (see figure on p. A4, at left):

1. Several bacterial strains are examined for a number of relevant properties such as fermentation, pigment production, etc. At least 50 or more properties should be examined.
2. Each property that is measured is given some value. Usually each property is given equal weight. The data obtained from the measurements is scaled, that is, assigned a plus or minus value depending on whether the property is present or absent, respectively.
3. The similarity between the strains is calculated using a computer. The similarity values, based on the chosen set of measured properties between two strains, is plotted and yields a table of similarities (squares) known as a similarity matrix (figure on p. A4, at right).
4. Each of the squares in the checkerboard represents a percentage, with 100% indicating similarity in all the properties measured and 0% indicating complete disimilarity. Thus each strain is compared with every other strain and similarities are scored.
5. The similarities of the organisms are analyzed. The greater the number of characteristics shared by two organisms, the closer their taxonomic relationship. Thus clusters of closely related strains can be arranged into some taxonomic groups. If two organisms show an 80% to 90% similarity profile they are often considered to belong to the same species.

CLASSIFICATION BASED ON PHENOTYPE

Until very recently nearly all classification schemes were based solely on the biochemical and physiological properties of the cell. Many times the characteristics were inappropriately used to define a species. Bacteria possess a diverse number of enzymes that catalyze reactions in metabolic pathways. The products of these pathways as well as the pathway itself can be detected directly or indirectly. Usually a specific substrate is added to a growth medium to detect any biochemical changes that have taken place in the substrate during bacterial growth. The most commonly measured biochemical property is fermentation. For example, one test used to distinguish among *Neisseria* species is based on fermentation of specific carbohydrates. When fermentation occurs, acid is produced, and an acid-base indicator in the

Strains (columns) / Tests (rows)

Tests	A	B	C	D	E	F	G	H	I	J
1	+	+	+	+	+	+	+	+	+	+
2	+		+	+		+	+	+	+	+
3	+				+		+	+	+	
4	+		+	+	+	+	+	+	+	+
5	+	+	+	+	+	+	+	+	+	+
6	+				+			+	+	
7	+	+				+	+	+	+	
8	+	+	+	+		+	+	+	+	+
9	+				+			+	+	
10	+							+	+	

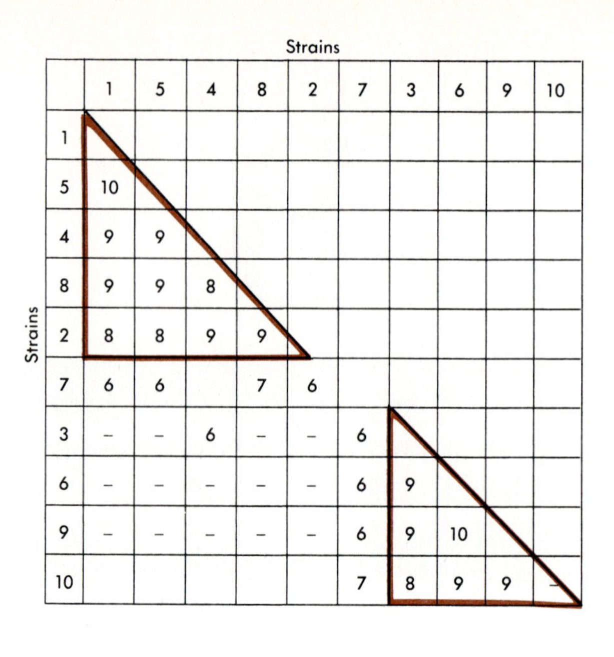

Strains
Strains
1 5 4 8 2 7 3 6 9 10
1
5 10
4 9 9
8 9 9 8
2 8 8 9 9
7 6 6 7 6
3 – – 6 – – 6
6 – – – – – 6 9
9 – – – – – 6 9 10
10 7 8 9 9

Simplified numerical taxonomy scheme. Left, 10 strains listed 1 to 10 and compared using 10 different tests (A to J, top). +, positive for test indicated. Right, strains arranged in sequence so that each strain is closest to strains sharing greatest number of features and farthest from those sharing fewest number of features. Strain 1 is next to strain 5 and strain 10 is farthest from it. For example, if shared features were less than 6, they were left blank. Two distinct clusters in which 8 or more features are shared can be seen; strains 1, 5, 4, 8, and 2 form one cluster and strains 3, 6, 9, and 10 form the other. Strain 7 cannot be grouped, because it does not match any of its members 80% or above.

growth medium can detect such changes. Physiological properties such as oxygen requirement, pH, and temperature requirements can also be used to identify and classify bacteria. Some of these characteristics have been used to group bacteria into sections in Bergey's Manual, for example, anaerobic gram-negative cocci. Although classification of bacteria into species should be based on evolutionary relationships, phenotypic grouping based on biochemical tests is the first step in classification.

CLASSIFICATION BASED ON NUCLEIC ACID STUDIES

The ultimate technique for classifying bacteria or any other organism is to compare the nucleic acid sequences of their genomes. Unfortunately this is not yet feasible for most organisms. However, genetic relatedness of bacteria can be compared by subtler and less demanding techniques.

DNA base composition

The DNA base composition is expressed in terms of the guanine cytosine (G-C) content. The ratio of G:C and A:T in the DNA molecule is one because guanine pairs with cytosine and adenine with thymine. However, the molar ratio of G:C base pairs to A:T base pairs will vary depending on the species of microorganism examined. The G-C content of bacteria ranges from 22 percent to 75 percent but remains the same for strains of a given species. The G:C content is determined by finding the temperature at which the DNA melts, that is, the temperature at which the double stranded DNA unwinds and strands separate (this is approximately 90° C). Denaturation is easy to measure because there is a gradual increase in absorbency at wavelengths of 260 nm as the DNA slowly unwinds. The denaturation process or breaking of hydrogen bonds is directly related to the G:C content because of the extra bond that exists between these nucleotide pairs as compared with A:T base pairs. Even if the size of the genome of two organisms may be the same, the number of G-C base pairs may vary tremendously from one species to another. Likewise the genome of two organisms may differ considerably in size yet have the same G-C(%) content. What then is the value of knowing the G-C content? Suppose we have two organisms that we think should be in the same species based on metabolic characteristics. When we examine the G-C content of these organisms we find that one has a G-C content of 75% while the other has a G-C content of 30%. Such a variation indicates that these two organisms should be placed in different species but not different families. If the G-C content of the organisms were 75% and 65% we could conclude that there might be similarity between them but other tests would have to be carried out to confirm such a conclusion.

Hybridization

The property of strand separation when DNA melts can be used to compare the relatedness of bacteria. If the single strands of melted DNA are incubated at a temperature below the melting point of DNA by slow cooling (for example 25° C below the melting point) the complementary strands of DNA reassociate or renature to form double-stranded duplexes that are identical to the native DNA. This property can be used to compare the DNA from two different organisms. The experiment is often carried out by labeling the DNA of one of the two organisms with an isotope, for example, ^{3}H, ^{14}C, or ^{32}P (figure below). The DNA is extracted from both strains. Strain I contains labeled DNA, whereas the DNA from strain II is unlabeled and present in excess. The two DNA samples are mixed, denatured, and then reassociated at a temperature below the melting point. The greater the similarity of the DNA of the two organisms the greater the degree of hybridization. The specificity of hybridization can be best determined by performing the experiment at different hybridization temperatures. When the hybridization temperature is set to only 10° to 15° C below the melting temperature, only those duplexes that have a high degree of complementarity will persist. When the hybridization temperature is less stringent, for example, 30° to 50° C below the melting temperature, there is association of less related strands of DNA. Strains from the same species typically show 60% to 100% relatedness at higher hybridization temperatures but strains from different species show less than 50% relatedness (Table B-1). DNA-DNA homology studies are useful primarily for determining associations between closely related organisms.

Table B-1 *Effect of temperature on the hybridization of DNA from different strains**

Species	Percent of Hybridization at 60° C†	Percent of Hybridization at 75° C‡
A/A	100	100
A/B	75	74
A/C	72	40
A/D	25	40

*This hypothetical experiment was designed to compare species A with three other organisms: B, C, and D. A/A is a control in which two strains of A have been normalized to 100%. From the data it appears that at an incubation temperature of 60° C, organisms B and C are related, but when the values are compared at higher temperatures there is considerably less homology between them. Organisms A and B, although not the same species, show a degree of relatedness. Organism D is unrelated to A, B, or C.
†This value is 30° C below the melting temperature for DNA.
‡This value is 15° C below the melting temperature for DNA.

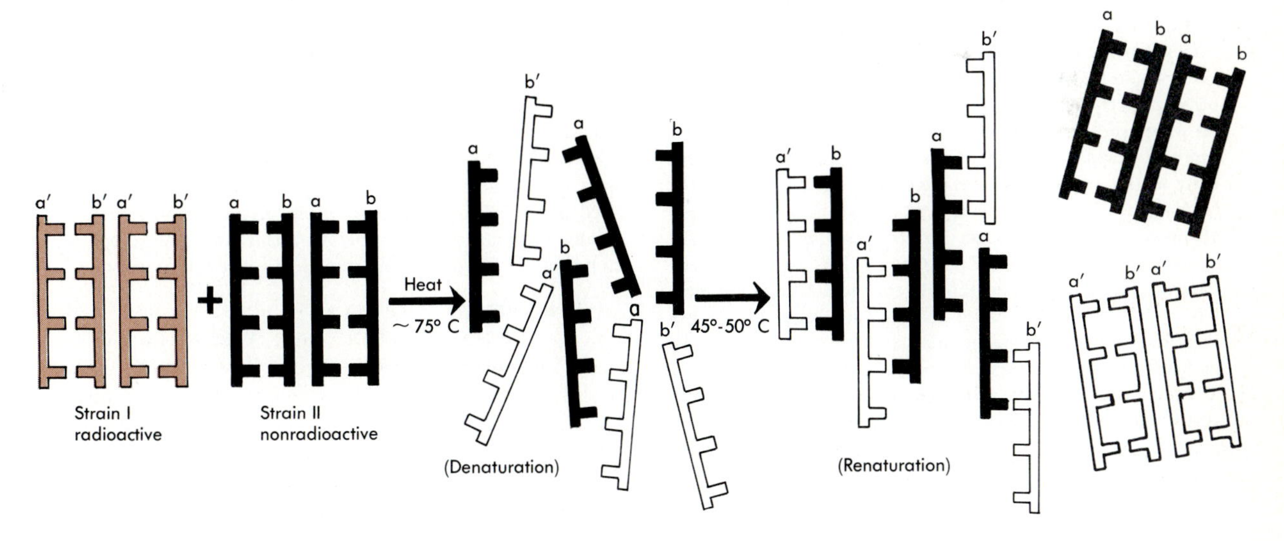

Hybridization experiment. DNA is isolated from strain I, which was grown in radioactive isotope (color), *and from strain II, which was grown in nonradioactive medium* (black). *DNA from the two strains is mixed and then heated at 75° C to denature molecules and produce single strands. Temperature is then reduced 25° to 30° C below denaturation temperature, permitting homologous areas from different strands to bind, for example,* a′ *from strain I binding to* b *from strain II and* b′ *from strain I binding to a from strain II. In addition, complementary strands of the original strains will also bond, that is,* a′ *to* b′ *and* a *to* b.

Only a few cistrons in the bacterial genome are transcribed into ribosomal RNA therefore homology studies involving rRNA and DNA can compare specific portions of the genome. Single-stranded DNA is mixed with labeled rRNA and incubated. After hybridization, RNase is used to remove unpaired segments of RNA from hybrids. Radioactivity is measured, and rRNA binding values can be compared from various organisms. The rRNA-DNA hybrid can also be treated in another way. After hybridization, the thermal stability of the rRNA-DNA duplex can be compared from different organisms. rRNA homology studies are useful for determining the similarity between more distantly related organisms (above the species level).

Sequence analysis

Macromolecules are documents of evolutionary history and can be regarded as fossil records. Sequence analysis of informational molecules such as nucleic acids and proteins provides a history of the relatedness between microorganisms. Proteins such as cytochrome c and ferredoxin, which are involved in electron transport, have been sequenced in various organisms. This technique has been invaluable in speciating some groups of bacteria such as the purple nonsulfur bacteria. The same proteins are not always present in the different organisms examined, and even when they are, are not always of similar structure. Protein analysis therefore has its limitations. The ribosome, which is involved in protein synthesis and is present in all organisms, contains a ribonucleic acid component, 16SrRNA, which can be used to show the relatedness of various species (See Point of Interest, below). More than 200 individual species have been characterized using 16SrRNA sequencing. The results of these studies have demonstrated that the older traditional methods used to define taxonomic levels may not always be reliable. Sequence analysis is expensive and time-consuming and is used primarily as an adjunct to other studies.

Point of Interest

RIBOSOMAL RNA ANALYSIS

The past record of a bacterial cell can be determined by its genetic sequences. Some genes that are present today in the bacterial cell represent genes that presumably existed many eons ago. During this time mutations occurred in the gene but if the basic function of the gene remained the same, then some of the nucleotides sequences of the "old" and "new" gene would be similar or homologous. Protein synthesis is a basic property of all cells and ribosomes are the workbenches on which proteins are synthesized. Ribosomes, therefore represent a cellular component that must have appeared very early in the evolution of the microbial cell. Ribosomal RNA (rRNA), a component of the ribosome, appears to have changed little with respect to its function since it first evolved. In addition, some portions of the rRNA sequences have changed slowly enough to detect ancestral relationships. One rRNA component, called 16S RNA, has proved to be the ideal molecule for sequence analysis. The 16S RNA is 1540 nucleotides long, contains regions of extreme conservatism and regions of extreme variability, and is easy to purify. Analysis is carried out by treating the RNA molecule with particular ribonucleases. This enzyme cuts the RNA molecule at specific sites so that short fragments called oligonucleotides are produced with each ending in a nucleotide containing guanine(G), for example, fragments containing 2, 3 4, or 5 nucleotides, are of little value because they will recur many times within the same RNA molecule. Once the fragment becomes 6 nucleotides or more long it is unlikely that the same oligonucleotide sequence would appear more than once in the same digested molecule. Therefore if we compare the 16S RNA from two different organisms and both possess at least one 6 letter sequence that is the same, then it reflects an ancestral similarity between the two organisms. After the RNA molecule is digested with ribonuclease the various lengths of oligonucelotides are separated. A catalog of oligonucleotides varying in length from 6 to 18 nucleotides is prepared from the organisms being examined. Once the exact sequence of each of the oligonucleotides is known the number of oligonucleotides common to each of the organisms is determined and their degree of relatedness is evaluated. This information along with biochemical, physiological, genetic, serological data is used to establish ancestral relationships. In the very abbreviated catalog of nucleotides seen below organisms 1 and 2 appear to be related but neither appears to be related to organism 3.

NUMBER AND SEQUENCE OF NUCLEOTIDES IN OLIGONUCLEOTIDE	PRESENT IN ORGANISM #
6	
ACACCG	1, 2
AAAUCG	1, 2
CUCUAG	3
7	
CCAACAG	1, 2
UAACACG	1, 2
UCACACG	3
8	
CCCACAUG	1, 2
ACCUAAGA	3

Other techniques

Many other techniques, some of which have been used primarily for identification purposes, can be used to supplement the results obtained from the more conventional classification procedures. Genetic procedures such as transformation experiments have proved useful in determining the degree of relatedness between different taxonomic groups.

Serological procedures also prove useful in determining relateness between organisms. The chemical components of bacteria such as proteins act as antigens and elicit the production of antibodies. When bacterial components such as pili, flagella, and enzymes are injected into a mammal, antibodies to these components are produced in the serum; this serum is called antiserum. The antibodies in this serum will therefore bind to those components that induce their formation. The structural similarity of homologous proteins of different organisms provides evidence of relatedness. Both monoclonal and polyvalent antibodies are important in studying relatedness.

The classification of bacteria has been greatly aided by comparing the chemical composition of bacterial structures. For example, lipid composition and distribution have been used to distinguish the archaebacteria from other prokaryotes. Some lipids, termed *isoprenoid quinones,* play a role in electron transport. Two structural groups of isoprenoids have been recognized and have been used to separate bacterial groups. Cytochrome patterns have been used to separate members of certain taxa. For example, cytochromes c and d, which are present in *Micrococcus,* are virtually absent in *Staphylococcus.* The two genera can be distinguished by a cytochrome oxidase test. Cell wall components such as peptidoglycan, teichoic acid, and lipopolysaccharide are sufficiently diverse in some organisms to permit differentiation between groups and species of bacteria. The side chain of the lipopolysaccharide is used in the serological classification of the Enterobacteriaceae.

In summary, classification of bacteria should be followed in three steps. Examination of phenotypic characteristics such as fermentation reactions and other biochemical reactions is the first step. Second, the organisms should be tested for DNA relatedness. The results of these studies will determine whether the biochemical results reflect genetic relatedness. Third, biochemical, serological, or genetic tests of DNA-related groups permit further differentiation.

Despite the significance of phylogenetic information, most classification schemes have been devised to aid the microbiologist and not the taxonomist. *Bergey's Manual* is recognized as the most practical, yet artifical, classification scheme. The most recent edition (9th) is still under revision. The manual will appear in four volumes, and its divisions are outlined below. Discussions of the specific groups or genera that appear in the Bergey outline are scattered throughout the book. As an aid to the student we have listed the text page number where some of these groups or genera are discussed in some detail.

KINGDOM *PROCARYOTAE*

Division 1. **Gracilicutes**
Division 2. **Firmicutes**
Division 3. **Tenericutes**
Division 4. **Mendosicutes**

SECTION 1.

The spirochetes

Order 1. *Spirochaetales*
Family 1. *Spirochaetaceae*
Genus 1. *Spirochaeta*
Genus 2. *Cristispira*
Genus 3. *Treponema*—p. 811, 812
Genus 4. *Borrelia*
Family 2. *Leptospiraceae*
Genus 1. *Leptospira*

Other Organisms

Hindgut Spirochetes of Termites and *Cryptocercus punctulatus*

SECTION 2.

Aerobic/microaerophilic, motile, helical/vibrioid gram-negative bacteria

Genus *Aquaspirillum*
Genus *Spirillum*
Genus *Azospirillum*
Genus *Oceanospirillum*
Genus *Campylobacter*—p. 640, 822
Genus *Bdellovibrio*—p. 532
Genus *Vampirovibrio*

SECTION 3.

Nonmotile (or rarely motile), gram-negative curved bacteria

Family 1. *Spiromaceae*
Genus 2. *Spirosoma*
Genus 2. *Runella*
Genus 3. *Flectobacillus*

Other genera

Genus *Microcyclus*
Genus *Meniscus*
Genus *Brachyarcus*
Genus *Pelosigma*

SECTION 4.

Gram-negative aerobic rods and cocci

Family 1. *Pseudomonadaceae*
Genus 1. *Pseudomonas*—p. 605, 630, 791
Genus 2. *Xanthomonas*
Genus 3. *Frateuria*
Genus 4. *Zoogloea*—p. 571
Family 2. *Azotobacteraceae*
Genus 1. *Azotobacter*—p. 353, 354
Genus 2. *Azomonas*
Family 3. *Rhizobiaceae*
Genus 1. *Rhizobium*—p. 355, 527
Genus 2. *Bradyrhizobium*
Genus 3. *Agrobacterium*—p. 597-599
Genus 4. *Phyllobacterium*
Family 4. *Methylococcaceae*
Genus 1. *Methylococcus*
Genus *Methylomonas*
Family 5. *Halobacteriaceae*
Genus 1. *Halobacterium*
Genus 2. *Halococcus*—p. 109
Family 6. *Acetobacteraceae*
Genus 1. *Acetobacter*
Genus 2. *Gluconobacter*
Family 7. *Legionellaceae*
Genus 1. *Legionella*—p. 806, 808
Family 8. *Neisseriaceae*
Genus 1. *Neisseria*—p. 368, 692, 807, 808. 812, 814
Genus 2. *Moraxella*
Genus 3. *Acinetobacter*
Genus 4. *Kingella*

Other genera

Genus *Beijerinckia*
Genus *Derxia*
Genus *Xanthobacter*
Genus *Thermus*
Genus *Thermomicrobium*
Genus *Halomonas*
Genus *Alteromonas*
Genus *Flavobacterium*
Genus *Alcaligenes*
Genus *Serpens*
Genus *Janthinobacterium*
Genus *Brucella*
Genus *Bordetella*—p. 805
Genus *Francisella*
Genus *Paracoccus*
Genus *Lampropedia*

SECTION 5.

Facultatively anaerobic gram-negative rods

Family 1. *Enterobacteriaceae*
Genus 1. *Escherichia*—p. 606, 670, 689
Genus 2. *Shigella*—p. 778
Genus 3. *Salmonella*—p. 197, 437, 612, 640, 744, 776, 821
Genus 4. *Citrobacter*
Genus 5. *Klebsiella*—p. 804, 806
Genus 6. *Enterobacter*
Genus 7. *Erwinia*
Genus 8. *Serratia*
Genus 9. *Hafnia*
Genus 10. *Edwardsiella*
Genus 11. *Proteus*—p. 123, 124
Genus 12. *Providencia*
Genus 13. *Morganella*
Genus 14. *Yersina*—p. 779, 818, 819
Other Genera of the Family *Enterobacteriaceae*
Genus *Obesumbacterium*
Genus *Xenorhabdus*
Genus *Kluyvera*
Genus *Rahnella*
Genus *Cedecea*
Genus *Tatumella*
Family 2. *Vibrionaceae*
Genus 1. *Vibrio*—p. 632
Genus 2. *Photobacterium*—p. 560
Genus 3. *Aeromonas*
Genus 4. *Plesiomonas*
Family 3. *Pasteurellaceae*
Genus 1. *Pasteurella*
Genus 2. *Haemophilus*—p. 806, 809
Genus 3. *Actinobacillus*

Other genera

Genus *Zymomonas*
Genus *Chromobacterium*
Genus *Cardiobacterium*
Genus *Calymmatobacterium*—p. 817
Genus *Gardnerella*
Genus *Eikenella*
Genus *Streptobacillus*

SECTION 6.

Anaerobic gram-negative straight, curved, and helical rods

Family 1. *Bacteroidaceae*
Genus 1. *Bacteroides*
Genus 2. *Fusobacterium*
Genus 3. *Leptotrichia*
Genus 4. *Butyrivibrio*
Genus 5. *Succinimonas*
Genus 6. *Succinivibrio*
Genus 7. *Anaerobiospirillum*
Genus 8. *Wolinella*
Genus 9. *Selenomonas*
Genus 10. *Anaerovibrio*
Genus 11. *Pectinatus*
Genus 12. *Acetivibrio*
Genus 13. *Lachnospira*

SECTION 7.

Dissimilatory sulfate- or sulfur-reducing bacteria—p. 543

Genus *Desulfuromonas*
Genus *Desulfovibrio*—p. 548
Genus *Desulfomonas*
Genus *Desulfococcus*
Genus *Desulfobacter*
Genus *Desulfobulbus*
Genus *Desulfosarcina*

SECTION 8.

Anaerobic Gram-Negative Cocci

Family 1. *Veillonellaceae*
Genus 1. *Veillonella*
Genus 2. *Acidaminococcus*
Genus 3. *Megasphaera*

SECTION 9.

The rickettsias and chlamydias

Order 1. *Rickettsiales*
Family 1. *Rickettsiaceae*
Tribe 1. *Rickettsieae*
Genus 1. *Rickettsia*—p. 819
Genus 2. *Rochalimaea*
Genus 3. *Coxiella*—p. 806, 819
Tribe 2. *Ehrlichieae*
Genus 4. *Ehrlichia*
Genus 5. *Cowdria*
Genus 6. *Neorickettsia*
Tribe 3. *Wolbachieae*
Genus 7. *Wolbachia*
Genus 8. *Rickettsiella*
Family 2. *Bartonellaceae*
Genus 1. *Bartonella*
Genus 2. *Grahamella*
Family 3. *Anaplasmataceae*
Genus 1. *Anaplasma*
Genus 2. *Aegyptianella*
Genus 3. *Haemobartonella*
Genus 4. *Eperythtrozoon*
Order 2. *Chlamydiales*
Family 1. *Chlamydiaceae*
Genus 1. *Chlamydia*—p. 806, 814, 815

SECTION 10.

The mycoplasmas

Class 1. *Mollicutes*
Order 1. *Mycoplasmatales*
Family 1. *Mycoplasmataceae*
Genus 1. *Mycoplasma*—p. 111, 806
Genus 2. *Ureaplasma*—p. 111
Family 2. *Acholeplasmataceae*
Genus 1. *Acholeplasma*—p. 111
Family 3. *Spiroplasmataceae*
Genus 1. *Spiroplasma*—p. 111

Other genera

Genus *Anaeroplasma*
Genus *Thermoplasma*—p. 111
Mycoplasma-like Organisms of Plants and Invertebrates—p. 597

SECTION 11.

Endosymbionts

A. Endosymbionts of Protozoa
Endosymbionts of ciliates
Endosymbionts of flagellates
Endosymbionts of amoebas
Taxa of endosymbionts:
Genus 1. *Holospora*
Genus 2. *Caedibacter*
Genus 3. *Pseudocaedibacter*
Genus 4. *Lyticum*
Genus 5. *Tectibacter*

B. Endosymbionts of Insects
 Blood-sucking insects
 Plant sap-sucking insects
 Cellulose and stored grain feeders
 Insects feeding on complex diets
 Taxon of endosymbionts:
 Genus *Blattabacterium*

C. Endosymbionts of Fungi and Invertebrates other than Arthropods
 Fungi
 Sponges
 Coelenterates
 Helminthes
 Annelids
 Marine worms and mollusks

SECTION 12.

Gram-positive cocci

Family 1. *Micrococcaceae*
 Genus 1. *Micrococcus*
 Genus 2. *Stomatococcus*
 Genus 3. *Planococcus*
 Genus 4. *Staphylococcus*—p. 638, 639, 706, 794, 816
Family 2. *Deinococcaceae*
 Genus 1. *Deinococcus*

Other organisms.

"Pyogenic" streptococci—p. 690, 804
"Oral" streptococci—p. 118, 119
"Lactic" streptococci and enterococci
Leuconostoc—p. 628, 629, 657, 666
Pediococcus—p. 635
Aerococcus
Gemella
Peptococcus
Peptostreptococcus
Ruminococcus
Coprococcus
Sarcina

SECTION 13.

Endospore-forming gram-positive rods and cocci

Genus 1. *Bacillus*—p. 604, 634, 665
Genus 2. *Sporolactobacillus*
Genus 3. *Clostridium*—p. 617, 633, 636, 637, 639, 820, 821
Genus 4. *Desulfotomaculum*—p. 548
Genus 5. *Sporosarcina*
Genus 6. *Oscillospira*

SECTION 14.

Regular, non-sporing, gram-positive rods

Genus 1. *Lactobacillus*—p. 603, 626, 628
Genus 2. *Listeria*
Genus 3. *Erysipelothrix*
Genus 4. *Brochothrix*
Genus 5. *Renibacterium*
Genus 6. *Kurthia*
Genus 7. *Caryophanon*

SECTION 15.

Irregular, non-sporing gram-positive rods

Animal and saprophytic corynebacteria—p. 657, 659
Plant corynebacteria
 Genus 1. *Gardnerella*
 Genus 2. *Arcanobacterium*
 Genus 3. *Arthrobacter*
 Genus 4. *Brevibacterium*
 Genus 5. *Curtobacterium*
 Genus 6. *Caseobacter*
 Genus 7. *Microbacterium*
 Genus 8. *Aureobacterium*
 Genus 9. *Cellulomonas*
 Genus 10. *Agromyces*
 Genus 11. *Arachnia*
 Genus 12. *Rothia*
 Genus 13. *Propionibacterium*
 Genus 14. *Eubacterium*
 Genus 15. *Acetobacterium*
 Genus 16. *Lachnospira*
 Genus 17. *Butyrivibrio*
 Genus 18. *Thermoanaerobacter*
 Genus 19. *Actinomyces*
 Genus 20. *Bifidobacterium*—p. 629

SECTION 16.

Mycobacteria

Family 1. *Mycobacteriaceae*
 Genus *Mycobacteria*—p. 617, 618, 809, 810

SECTION 17.

Nocardioforms

Genus 1. *Nocardia*
Genus 2. *Rhodococcus*
Genus 3. *Nocardioides*
Genus 4. *Pseudonocardia*

Genus 5. *Oerskovia*
Genus 6. *Saccharopolyspora*
Genus 7. *Micropolyspora*
Genus 8. *Promicromonospora*
Genus 9. *Intrasporangium*

SECTION 18.

Gliding, non-fruiting bacteria

Order 1. *Cytophagales*
Genus 1. *Cytophaga*
Genus 2. *Sporocytophaga*
Genus 3. *Capnocytophaga*
Genus 4. *Flexithrix*
Genus 5. *Flexibacter*
Genus 6. *Microscilla*
Genus 7. *Saprospira*
Genus 8. *Herpetosiphon*
Order 2. *Lysobacterales*
Family 1. *Lysobacteriaceae*
Genus 1. *Lysobacter*
Order 3. *Beggiatoaceae*
Genus 1. *Beggiatoa*—p. 128
Genus 2. *Thioploca*
Genus 3. *Thiospirillopsis*
Genus 4. *Thiothrix*
Genus 5. *Achromatium*
Family 2. *Simonsiellaceae*
Genus 1. *Simonsiella*
Genus 2. *Alysiella*
Family 3. *Leucotrichaceae*
Genus 1. *Leucothrix*
Family 4. *Pelonemataceae*
Genus 1. *Pelonema*
Genus 2. *Achroonema*
Genus 3. *Peloploca*
Genus 4. *Desmanthus*
Families and Genera incertae sedis
Genus 1. *Toxothrix*
Genus 2. *Vitreoscilla*
Genus 3. *Chitinophagen*
Genus 4. *Desulfonema*

SECTION 19.

Anoxygenic phototrophic bacteria

Purple Bacteria
Family 1. *Chromatiaceae*
Genus 1. *Chromatium*
Genus 2. *Thiocystis*
Genus 3. *Thiospirillum*
Genus 4. *Thiocapsa*
Genus 5. *Amoebobacter*
Genus 6. *Lamprobacter*
Genus 7. *Lamprocystis*
Genus 8. *Thiodictyon*
Genus 9. *Thiopedia*
Family 2. *Ectothiorhodospira*
Genus 1. *Ectothiorhodospira*
Purple nonsulfur bacteria
Genus 1. *Rhodospirillum*
Genus 2. *Rhodopseudomonas*
Genus 3. *Rhodobacter*
Genus 4. *Rhodomicrobium*
Genus 5. *Rhodopila*
Genus 6. *Rhodocyclus*
Green bacteria
Green sulfur bacteria
Genus 1. *Chlorobium*
Genus 2. *Prosthecochloris*
Genus 3. *Ancalochlorus*
Genus 4. *Pelodictyon*
Genus 5. *Chloroherpeton*
Symbiotic consortia
Multicellular filamentous green bacteria
Genus 1. *Chloroflexus*
Genus 2. *Heliothrix*
Genus 3. *Oscillochloris*
Genus 4. *Chloronema*
Genera Incertae Sedis
Genus 1. *Heliobacterium*—p. 73, 74
Genus 2. *Erythrobacter*

SECTION 20.

Budding and/or appendaged bacteria

Prosthecate bacteria

Budding bacteria
Genus 1. *Hyphomicrobium*—p. 563
Genus 2. *Hyphomonas*
Genus 3. *Pediomicrobium*
Genus 4. *Filomicrobium*
Genus 5. *Dicotomicrobium*
Genus 6. *Tetramicrobium*
Genus 7. *Stella*
Genus 8. *Ancalomicrobium*
Genus 9. *Prosthecomicrobium*

Non-prosthecate bacteria

Budding bacteria
Genus 1. *Planctomyces*
Genus 2. *Pasteuria*
Genus 3. *Blastobacter*
Genus 4. *Angulomicrobium*
Genus 5. *Gemmiger*
Genus 6. *Ensifer*
Genus 7. *Isophaera*

Non-budding stalked bacteria
Genus 1. *Gallionella*—p. 544
Genus 2. *Nevskia*
Morphologically unusual budding bacteria involved in iron and manganese deposition
Genus 1. *Seliberia*
Genus 2. *Metallogenium*
Genus 3. *Caulococcus*
Genus 4. *Kuznezovia*
Others
Spinate bacteria

SECTION 21.

Archaeobacteria—p. 73, 109

Methanogenic bacteria—p. 330, 569

Genus 1. *Methanobacterium*—p. 109, 330, 569
Genus 2. *Methanobrevibacter*
Genus 3. *Methanococcus*—p. 109
Genus 4. *Methanomicrobium*
Genus 5. *Methanospirillum*
Genus 6. *Methanosarcina*
Genus 7. *Methanococcoides*
Genus 8. *Methanothermus*
Genus 9. *Methanolobus*
Genus 10. *Methanoplanus*
Genus 11. *Methanogenium*
Genus 12. *Methanothrix*

Extreme halophilic bacteria

Genus 1. *Halobacterium*—p. 114, 321, 322, 411
Genus 2. *Halococcus*—p. 411

Extreme thermophilic bacteria

Genus 1. *Thermoplasma*—p. 413
Genus 2. *Sulfolobus*—p. 413, 544, 547
Genus 3. *Thermoproteus*
Genus 4. *Thermofilum*
Genus 5. *Thermococcus*
Genus 6. *Desulfurococcus*
Genus 7. *Thermodiscus*
Genus 8. *Pyrodictium*

SECTION 22.

Sheathed bacteria

Genus 1. *Sphaerotilus*—p. 360
Genus 2. *Leptothrix*—p. 360, 544
Genus 3. *Haliscominobacter*
Genus 4. *Lieskeella*
Genus 5. *Phragmidiothrix*
Genus 6. *Crenothrix*
Genus 7. *Clonothrix*

SECTION 23.

Gliding, fruiting bacteria

Order 1. *Myxobacteriales*
Family 1. *Myxococcaceae*—p. 531
Genus 1. *Myxococcus*
Family 2. *Archangiaceae*
Genus 1. *Archangium*
Family 3. *Cystobacteraceae*
Genus 1. *Cystobacter*
Genus 2. *Melittangium*
Genus 3. *Stigmatella*
Family 4. *Polyangiaceae*
Genus 1. *Polyangium*
Genus 2. *Nannocystis*
Genus 3. *Chondromyces*
Genus incerta sedis
Genus 1. *Angiococcus*

SECTION 24.

Chemolithotrophic bacteria

Nitrifiers

Family 1. *Nitrobacteraceae*
Genus 1. *Nitrobacter*
Genus 2. *Nitrospina*
Genus 3. *Nitrococcus*
Genus 4. *Nitrosomonas*—p. 545
Genus 5. *Nitrosospira*
Genus 6. *Nitrosococcus*
Genus 7. *Nitrosolobus*

Sulfur oxidizers

Genus 1. *Thiobacillus*—p. 340, 545-548
Genus 2. *Thiomicrospira*—p. 562
Genus 3. *Thiobacterium*
Genus 4. *Thiospira*
Genus 5. *Macrommonas*

Obligate hydrogen oxidizers

Genus 1. *Hydrogenbacter*

Metal oxidizers and depositers

Family 1. *Siderocapsaceae*
Genus 1. *Siderocapsa*
Genus 2. *Naumaniella*
Genus 3. *Ochrobium*
Genus 4. *Siderococcus*

Other magnetotactic bacteria

SECTION 25.

Cyanobacteria—p. 73-76, 587

Others

Order 1. *Prochlorales*
Family 1. *Prochloraceae*
Genus 1. *Prochloron*—p. 73, 74

SECTION 26.

Actinomycetes that divide in more than one plane

Genus 1. *Geodermoatophilus*
Genus 2. *Dermatophilus*
Genus 3. *Frankia*—p. 528
Genus 4. *Tonsilophilus*

SECTION 27.

Sporangiate actinomycetes

Genus 1. *Actinoplanes* (including Amorphosporangium)
Genus 2. *Streptosporangium*
Genus 3. *Ampullariella*
Genus 4. *Spirillospora*
Genus 5. *Pilimelia*
Genus 6. *Dactylosporangium*
Genus 7. *Planomonospora*
Genus 8. *Planobispora*

SECTION 28.

Streptomycetes and their allies

Genus 1. *Streptomyces*—p. 586
Genus 2. *Streptoverticillium*
Genus 3. *Actinopycnidium*
Genus 4. *Actinosporangium*
Genus 5. *Chainia*
Genus 6. *Elytrosporangium*
Genus 7. *Microellobosporia*

SECTION 29.

Other conidiate genera

Genus 1. *Actinopolyspora*
Genus 2. *Actinosynnema*
Genus 3. *Kineospora*
Genus 4. *Kitasatosporia*
Genus 5. *Microbispora*
Genus 6. *Micromonospora*
Genus 7. *Microtetrospora*
Genus 8. *Saccharomonospora*
Genus 9. *Sporichthya*
Genus 10. *Streptoalloteichus*
Genus 11. *Thermomonospora*
Genus 12. *Actinomadura*
Genus 13. *Nocardiopsis*
Genus 14. *Excellospora*
Genus 15. *Thermoactinomyces*

Glossary

aberation disturbance of the rays of light so they are unable to be brought to a sharp focus.

acidophile acid-loving microorganism.

acidulent a substance used to increase acidity.

acquired immunity immunity that is based on prior exposure and the subsequent interaction of specific immune system products with antigens.

active immunity acquired immunity resulting from the presence of antibody or immune lymphoid cells produced in response to an antigenic stimulus.

active site a site on the enzyme for binding of substrate and promoting catalytic activity.

active transport transfer of molecules across a membrane in the direction of increasing concentration by an expenditure of energy.

activated sludge method of sewage treatment where flocs absorb organic material.

acute having a short and relatively severe time course.

adhesin adherence polymers found on the surface of microorganisms.

adjuvant a substance that can be added to a vaccine to increase its effectiveness.

alkalophile microorganism having optimum growth at alkaline pH.

aerobe organism requiring oxygen for growth.

aerosol a suspension of particles in air.

aging cask a barrel-shaped vessel of staves, headings, and hoops usually of oak, used to store wine for a period during which taste and aroma develop before it is bottled.

akinete dormant thick-walled reproductive cell of some cyanobacteria.

ale beer brewed with top-fermenting yeast.

allele one of the alternative forms of a gene that may occupy a particular site on the chromosome.

allergen a more specific term for the antigens that are involved in allergic or hypersensitivity reactions.

allergy an immediate or delayed type of hypersensitivity that follows exposure to an allergen.

allochthonous referring to organisms not indigenous to a given environment.

allograft (homograft) a tissue or organ graft between members of the same species who are not genetically identical.

allosteric (enzyme) a regulatory enzyme whose catalytic activity is affected by the noncovalent binding of a specific metabolite to a site other than the active site.

alternative (alternate) complement pathway pathway in which the complement cascade is activated at the C3 level; immune system products are not required for the activation of this pathway.

ameboid (movement) a type of movement in which the protoplasm of the cell changes shape to produce feet (pseudopodia) that help the cell move.

Ames test a procedure using bacteria to determine the potential carcinogenicity of a substance.

amylase an enzyme that hydrolyses the $\alpha - 1,4$ glucosidic bonds of starch, yielding dextrins and maltose.

anabolism metabolic process involved in the synthesis of cell material.

anaerobe organism that lives in the absence of air (oxygen).

anamnestic response the recall or the "remembering" by the immune system of a prior response to a given antigen (see memory cell).

anaphylatoxin the activated C3 and C5 components of complement that cause mast cells to release histamine and other physiologically active amines that contribute to intense inflammatory reactions.

anaphylaxis a form of immediate hypersensitivity that can occur in anyone.

anaplerotic a reaction or series of reactions that replenishes intermediates used in biosynthetic reactions.

anion a negatively charged ion.

anoxygenic refers primarily to photosynthetic processes in which oxygen is not present or produced.

antheridium structure responsible for production of male gametes.

antibiogram test for determining antibiotic susceptibility pattern of microorganisms.

antibiosis the production by one species of an organic compound that is toxic at low concentrations to another species.

antibiotic a chemical compound produced by microorganisms that can inhibit or kill other microorganisms.

antibody a glycoprotein of the Ig type that is produced by a plasma cell and specifically reacts with the known antigen that elicited its production.

anticodon the triplet of nucleotides in a transfer RNA molecule that complements a specific codon on the messenger RNA.

antigen a foreign substance capable of inducing an immunological response.

antigen-binding site the variable Fab portion of the antibody molecule that binds a specific or closely related antigenic determinant.

antigenic capable of stimulating the formation of antibodies.

antigenic determinants small chemical groups on antigens that elicit the formation of specific antibodies and sensitized lymphocytes.

antimicrobial an agent (chemical or physical) that inhibits growth of or kills microorganisms.

antiseptic a chemical compound that can be used on the surface of living tissue and inhibits bacterial growth.

antitoxins antibodies that are formed in response to bacterial toxins or toxoids and neutralize or inactivate those toxins.

apoenzyme the protein portion of enzyme and requiring coenzyme before becoming catalytically active.

arthropod invertebrate with jointed limbs.

ascocarp sexual fruiting structure of higher ascomycetes.

ascogonium female reproductive structure of an ascomycete.

ascospores a sexual spore of the Ascomycetes.

ascus a saclike structure that holds spores (ascospores) produced by species of fungi.

aseptic a technique in which precautions are taken against microbial contamination.

assimilation the absorption of food and its conversion into cells or tissue.

atopic hypersensitivity immediate-type, IgE-mediated common clinical allergies for which there is a genetically determined predisposition for formation.

atopy a form of immediate hypersensitivity that tends to occur in those with a hereditary predisposition.

attenuation lessening; reduction of the virulence of microorganisms as in a vaccine.

attenuator a transcription termination site located in the interior of an operon.

autoallergy immune response against an individual's own antigens.

autochthonous referring to the indigenous members of an ecological site.

autoclave a chamber in which steam under pressure is used to sterilize objects and solutions.

autogamy union of gametes produced by one individual.

autoimmune disease a disease in which the host system destroys its own tissue.

autolysis digestion of a cell by enzymes produced by that cell at time of death.

autotrophic the ability to use carbon dioxide or bicarbonate as the sole source of carbon.

auxin plant hormones that regulate growth.

auxochrome the dissociable group of a dye that intensifies color.

auxospore the reproductive cell of diatoms.

auxotroph a mutant microorganism that will grow only on minimal media supplemented with a growth factor or factors not required by the normal parent.

axenic pure or uncontaminated culture.

bacillus an organism of the genus *Bacillus;* also used to designate rod-shaped bacteria.

bacteriocin antimicrobial substance produced by bacteria that kills sensitive members of related strains.

bacteriophage a bacterial virus, sometimes referred to as phage.

bacteroid irregular shape of bacterium *(Rhizobium)* associated with symbiotic nitrogen fixation in roots of plant.

baeocyte reproductive cells produced by species of cyanobacteria.

basal body structure at the base of flagella and cilia.

basidiospore a sexual spore formed in the fungal group Basidiomycotina.

basidium a specialized cell of the Basidiomycetes in which the products of meiosis are formed.

basophil a blood leukocyte that is rich in vasoactive substances; the blood counterpart of mast cells. Stains with basic dyes.

B cell (B lymphocytes) a major population of lymphocytes; the B cell stem cells are found in bone marrow (in the bursa of Fabricius in avian species); mature B lymphocytes have cell-surface Ig receptors for specific antigenic determinants; antigen-stimulated B cells proliferate and differentiate into antibody-producing plasma cells.

benthos sediment, or pertaining to life within the sediment.

beta-lactam the ring structure associated with compounds such as the antibiotics penicillin and cephalosporin.

beta-oxidation catabolic process in which two carbon units of a long chain fatty acid are removed to form acetyl CoA.

binary fission an asexual process in which a cell splits into two cells of equivalent size.

binomial two names.

biodegradation the biological destruction of organic compounds.

biofilm a layer of microorganisms attached to a substrate.

biogeochemistry the science that deals with the role of living organisms in affecting the chemistry of the earth and its atmosphere.

bioluminescence production of light by microorganisms.

biomass all of the living organisms within a particular region.

biosynthesis the building up of biochemical compounds.

blast cell a leukocyte that is an intermediate cell type of B and T lymphocytes.

bock beer a dark-colored lager beer brewed with more malt and hops than light-colored lager beers.

BOD biochemical oxygen demand; amount of oxygen required to convert organic matter to carbon dioxide and water under prescribed conditions.

boil a localized abscess resulting from infection of a hair follicle; also called a furuncle.

Brownian movement movement of particles in a suspension due to bombardment by water molecules.

bubo a lymph node that has become enlarged.

budding an asexual reproductive process in which the parent cell retains its identity while the offspring (bud) is produced with a newly synthesized cell wall attached to parent cell; also a process in which an enveloped virus is released through the plasma membrane of a vertebrate cell.

bulk starter method starting with a purchased mother culture, the progressive inoculation and growth of starter cultures in increasing volumes of milk culminating with the inoculating of the cheese fermentation vessel.

bursa of fabricius a cloaca-associated central lymphoid organ in avian species where B stem cells originate.

Cabernet Sauvignon a variety of purple grape used in the production of high-quality red wines.

canker a disease that kills definite and relatively localized areas of bark on branches or trunks of trees.

cannula an artificial tube used for insertion into the body, for example, an artery.

capsid protein coat of a virus.

capsomere an aggregate of polypeptides forming a unit of the viral capsid.

carcinogen an agent (chemical, physical, or biological) that can cause malignant tumor formation.

caries a disease of the calcified tissue of teeth; tooth decay.

carrier a host that harbors infectious microorganisms and can transmit them to others but shows no disease symptoms.

caseation necrosis a type of necrosis in which the tissue resembles an amorphous mass of cheese.

cassava any of several plants of the spurge family grown in the tropics for their fleshy edible rootstocks, which yield a nutritious starch.

catabolism a metabolic process in which foodstuffs are broken down to release energy.

catheter a tubular device used to withdraw fluids from, or introduce them into, body cavities.

cell-mediated immunity (CMI) T cell-dependent immunity.

cellulitis inflammatory condition of the subcutaneous tissue.

central lymphoid organs lymphatic organs where lymphoid cell subpopulations become committed to specific functions; the thymus and bone marrow for T and B cells, respectively; also called primary lymphatic organ.

centriole structure associated with nuclear division and involved in spindle organization.

centromere the point on a chromosome that segregates during the first division of meiosis.

chancre the primary lesion of syphilis.

chardonnay a variety of green grape used in the production of white wine.

chelate process of combining with a metal ion to keep in solution.

chemiosmotic hypothesis the theory that states the usage and synthesis of energy in a cell is the result of a proton electrical gradient across the cell membrane, mitochondrial membrane or chloroplast membrane.

chemoautotroph an organism that derives its energy from the oxidation of reduced inorganic chemicals and its carbon from CO_2 or bicarbonate.

chemoheterotroph an organism that derives its energy from the oxidation of preformed organic chemicals and its carbon from organic substances.

chemolithotroph same as chemoautotroph.

chemostat an apparatus used to maintain bacterial cultures in a state of continuous division.

chemotaxis phenomenon of movement toward or away from a chemical stimulus.

chemotherapeutic chemical agents used in treatment of disease.

chiasma crosslike configuration in a tetrad resulting from a genetic exchange between nonsister chromatides.

chimeric in genetics refers to a DNA molecule derived from different sources.

chitin a polysaccharide found in the shells of invertebrates such as lobsters and makes up the exoskeleton of insects; found in walls of many fungi.

chloroplast an organelle found only in photosynthetic organisms and carrying chlorophyll—responsible for photosynthesis.

chlorosis yellowing of green leaf tissue through lack of green pigment.

chromatid one of two chromosome strands visibly present during mitosis or meiosis.

chromatin (body) a network of deoxyribonucleic acid fibers and histones giving the appearance of a distinct body in the cell.

chronic over a long period of time.

-cidal killing.

cilia small hairlike projections found on the surface of some microorganisms and used for movement; found on respiratory epithelium, cilia are used to direct flow of various secretions and particles. Cilia have same internal structure as eukaryotic flagellum but are shorter.

cirri tuft of cilia used as organelle for motility and feeding in certain protozoa.

cisternae stacked membranes associated with the golgi complex.

cistron a segment of the chromosome used as a pseudonym for gene; it codes for a single polypeptide.

climax community the stable community characteristic of a site.

clone population of cells identical to the parent.

coagulase bacterial enzyme that causes blood plasma to clot.

coccus a spherical bacterial cell.

code date a mixture of letters or numbers that signify the manufacturing place and date of a product; also called a manufacturing lot number.

codon a nucleotide triplet in the messenger RNA that specifies for a particular amino acid or for termination of protein synthesis.

coenocytic a filamentous cell or mycelium not divided by complete septa and containing more than one nucleus; also occurs in slime molds where there is no cell wall or mycelium.

coenzyme an organic molecule that is loosely bound to the enzyme and necessary for the activity of the enzyme.

cofactor an inorganic molecule, such as a metal ion, that increases enzyme activity.

colicin a protein secreted by certain strains of *Escherichia coli* and lethal to other strains of the same species.

colitis inflammation of the colon.

colony a uniform mass of cells derived from a single cell growing on a solid surface.

colostrum the milk secreted a few days before and after childbirth.

cometabolism metabolism of substrates incapable of being used as a carbon or energy source in the presence of alternative, usable substrates.

commensal a relationship between organisms in which one benefits but the other neither benefits nor is harmed.

commercial sterility the thermal processing of canned foods such that only thermophilic spores survive.

communicable easily transmitted.

compensation depth point where photosynthesis and respiration balance in an aquatic environment.

competition a rivalry between two species for a limiting factor in the environment.

competitive inhibitor a molecule that competes with the substrate for attachment to the active site of an enzyme.

complement a complex system of serum proteins that are activated in a sequential fashion (in the complement cascade) by immune complexes or by other substances such as bacterial endotoxins.

concatemer a DNA molecule consisting of two or more separate genomes linked end to end to produce a linear structure.

congenital present at birth.

conidia asexual fungal spores produced on specialized mycelial branches called conidiophores.

conidiophore a hyphal filament on which spores, called conidia, are borne.

conjugation the act of joining together; in bacteria a process in which genetic information is transferred from one cell to another by direct cell contact.

conjunctivitis inflammation of the conjunctiva of the eye.

constitutive enzyme an enzyme produced by a cell under any environmental conditions.

consumer organisms feeding on preformed organic material.

convalescent state of recovery from disease.

co-repressor the end product of a biosynthetic pathway that combines with the aporepressor (a protein) to control an operon.

corn steep liquor a waste product of the corn starch industry; it contains both carbohydrates and nitrogenous compounds and can serve as an inexpensive source of each in the cultivation of microorganisms.

cortex a layer found inside the coat layer of the endospore.

counterimmunoelectrophoresis (CIE) electrophoretically directed movement of antigen and antibody toward each other through a support medium.

cristae foldings of the inner membrane of mitochondria.

crown upper dome of tree, bearing leaves, flowers, and fruits.

cultivar cultivated variety (often not a valid taxon) of plant.

culture a particular strain of microorganism growing in a medium.

cyst any closed cavity or sac, or a type of resting cell in prokaryotes and eukaryotes.

cystoscope an instrument for examining the interior of the urinary bladder.

cytoinduction a mating process resulting in the progeny having the nuclear genotype of one parent and the cytoplasmic genotype of the other parent.

cytokinesis cytoplasmic division into daughter cells following nuclear division in eukaryotes.

cytolytic causing lysis and death of the cell.

cytopathic characterized by pathological change in a cell.

cytophilic antibodies antibodies that bind to cell surface receptors through the Fc portion of the Ig molecule.

cytoplasmic streaming the flow of the cytoplasm in eukaryotic stationary cells that is used to distribute nutrients to all parts of the cell.

cytosol the cytoplasm of the eukaryotic cell in which organelles are suspended.

cytotoxic capable of killing cells.

D value (decimal reduction time) time required to destroy 90% of a population of microorganisms at a given temperature.

damping-off collapse of seedling near soil line from necrotic infection induced by water and fungus.

deamination removal of an amino group.

debilitation loss of normal function.

decomposer organism converts organic material into its inorganic constituents.

degeneracy in genetics a term used to denote that an amino acid has more than one triplet code word.

dehydration removal of water.

dehydrogenation the removal of hydrogens from a compound.

delayed hypersensitivity T cell-mediated hypersensitivity in which the reaction reaches the maximum in about 24 to 48 hours after exposure to the antigen that elicited the original hypersensitive state.

denature to lose the natural configuration, for example, heating an enzyme in which there is change in conformation and usually biologic activity; also stripping a part of complementary strands of DNA.

denitrification microbial reduction of nitrate to N_2, N_2O, and NO.

depurination removal of purine bases.

dermatophyte a fungus that affects the skin.

desiccate to dry.

desquamation shedding of the superficial layers of skin.

detergent synthetic substance that reduces surface tension of water.

detoxication biological destruction of a toxic chemical.

dextrin a soluble polymer of glucose usually derived from starch.

dextrorotatory capable of rotating plane-polarized light to the right.

diarrhea abnormal fluidity of fecal discharge.

dikaryon a mycelium consisting of hyphal elements, each containing two nuclei of unlike mating type.

dimorphism the ability to exist in one of two morphological states; For example, some fungi may exist in filamentous or yeastlike state.

diploid containing twice the haploid chromosome number.

disaccharide a sugar consisting of two monosaccharides connected by glycolytic bond.

disease abnormal condition of the body, usually having characteristic symptoms.

disinfectant a substance that destroys vegetative pathogens.

dissimilation the act of breaking down a component into smaller parts.

domain in immunoglobulin chemistry the folded regions or segments of H and L immunoglobulin chains that are stabilized with disulfide bonds. L chains have one domain in the variable region and one in the constant region. H chains have one domain in the variable region and three or four domains in the constant region.

dynein protein linkages found in the flagella of eukaryotes.

dysentery an intense diarrhea characterized by mucus and blood in the stool.

ecosystem the ecological site where an interaction between living and nonliving components exists.

ectosymbiosis a symbiotic state in which one organism attaches to the surface of another.

eczema inflammatory skin disease characterized by scales and crusts.

edema abnormal accumulation of fluid in tissues causing swelling.

effector T cells T cells that have an effect on specific antigens (such as cytotoxic cells) as opposed to regulatory T cells that have regulatory effects on other lymphocytes.

elephantiasis a condition in which fluids accumulate in various tissues, causing grotesque features.

encephalitis inflammation of brain.

endemic continually present in a community.

endergonic a reaction in which energy is used or absorbed in the system.

endocytosis the process in which soluble or insoluble materials are engulfed by ameboid cells.

endogenous produced within the cell.

endoplasmic reticulum a protoplasmic network in cells of higher animals and plants consisting of a continuous tubular membrane system that courses through the cytoplasm.

endospore a heat-resistant spore occurring within the sporangium of *Bacillus, Clostridium,* and *Sporosarcina.*

endosymbiosis the state in which one cell lives within another, for example, a prokaryotic bacterium living in a eukaryotic protozoan.

endotoxin a toxin derived from the cell wall of gram-negative bacteria, consisting of lipopolysaccharide.

enterics a common name for the family of gram-negative bacteria called the Enterobacteriaceae.

enterotoxin a toxin that gives rise to gastrointestinal symptoms when ingested.

entropy the nonusable energy obtained from chemical reactions and energy yielding processes.

enzyme a protein catalyst that can engage in specific reactions in which a product or products are formed.

eosinophil a type of leukocyte containing granules that have an affinity for acid dyes.

epidemic an outbreak of a disease affecting a large number of individuals in a community.

epidemiology the science that deals with the incidence, transmission, and prevention of disease.

epilimnion surface waters of oceans or lakes above the thermocline.

epilithic association of organism with inanimate materials.

epiphyte microorganism living attached to phytoplankton.

epiphytic living in association with plants or algae.

epitope major antigenic site on antigen that relates specifically to immunity.

epizootic outbreak of disease affecting a large number of animals.

ergot fungal mass of *Claviceps purpurea* found on grasses such as rye; disease of rye and other grasses.

erythema redness of skin.

eschar a dry mass of necrotic (dead) tissue.

etiology study of the cause of disease.

eucarpic type of fungal hyphae in which part of the thallus contains rhizoids.

euchromatin diffuse chromatin that appears during interphase.

eukaryotic of a cell type in which the nuclear material is bounded by a membrane and more complex than prokaryotic cell.

eutroph a species that grows at high nutrient concentrations.

eutrophication biological enrichment of a lake with subsequent depletion of oxygen.

exergonic a reaction in which energy is released.

exfoliation dropping off of scales or layers, for example, skin.

exocytosis extrusion of materials from the cell into the environment via vacuoles.

exon that part of a eukaryotic gene that is transcribed into messenger RNA and codes for a specific protein.

exotoxin toxin produced in the cell and released into the environment.

exudate material (fluids or cells) that has escaped from blood vessels.

fab fragment the fragment of the enzyme-digested antibody molecule that binds antigen, consisting of the combined amino terminals of the variable regions of an H and an L chain.

Fc fragment the crystalizable fragment of the enzyme-digested antibody molecule. Consists of the carboxy terminal half of the two H chains of an antibody molecule. Biological functions include attachment to appropriate cell-surface receptors, complement binding, placental passage.

facultative anaerobes organisms that grow under aerobic or anaerobic conditions.

feeder root system fine root hairs of a tree that absorb nutrients.

fermentation an anaerobic metabolic process that yields energy and uses an organic compound as the final electron acceptor. In industrial microbiology refers to any aerobic or anaerobic process in which a product is produced.

fermenter a large cylindrical vessel, usually made of stainless steel, used to grow microbes on an industrial scale.

fertility (F) factor a genetic factor (circular double-stranded DNA) carried by some microorganisms that permits conjugation to occur with the same or related species.

fibrinogen a plasma protein involved in clotting.

fimbria fringe; in microbial genetics, a small nonconjugative projection on the surface of bacteria; term *pilus* has replaced fimbria.

flagellin the protein making up the flagellum.

flagellum organ of motility in microorganisms.

flavoprotein a protein component of the electron transport system.

flora resident organisms in an area.

fluorescein a fluorescing pigment produced by many pseudomonads and observed by illumination with ultraviolet light.

fluorescent capable of emitting light of one wavelength when exposed to light of another wavelength.

fluorescent antibody (FA) an antibody conjugated with a fluorescent dye that fluoresces on ultraviolet irradiation; used in an immunoidentification procedure; may be visualized by UV or fluorescent microscopy.

fluorochrome a compound that releases its energy following excitation in the form of light.

food chain a group of organisms and their interrelationships based on exchange of nutrients.

folliculitis inflammation of hair follicle.

fomite an inanimate object that may be involved in disease transmission.

fruiting body a specialized structure in some bacteria (Myxobacteria) and some fungi that produce spores.

frustule another term for the cell wall of diatoms.

furuncle a boil.

gall gross enlargement and deformity, usually globose, of plant tissue; caused by *Agrobacterium tumefaciens.*

gamete a sex cell (either male or female) that unites with gamete of opposite sex to produce a zygote.

gamma globulin a fraction of globulin (a plasma protein); identified by electrophoresis; contains the majority of immunoglobulins (antibodies); also signifies the antibody-containing fraction of plasma that is used for prophylaxis or therapy.

gas vesicle a structure found in certain prokaryotes that contains gas and enables the organism to float or maintain certain levels in aqueous environment.

gastroenteritis inflammation of stomach and intestine.

gene a particular segment of a nucleic acid (DNA usually) that codes for a single polypeptide or protein.

generation time time required for a population of cells to double in number.

genome complete set of hereditary characteristics for an organism.

genotype the genetic constitution of an organism.

geosmin compounds produced by some microorganisms that contributes to "earthy" smell of soil.

germination the process of vegetative cell formation from spore following a period of dormancy.

germ tube a filamentous protuberance produced by yeast cells.

gingiva (gums) soft tissue surrounding teeth.

globulins class of proteins found in blood.

glomerulonephritis inflammation of tubules (glomeruli) in the kidney.

glucan a general term given to polysaccharides in which glucose is a component.

glucoamylase an enzyme that removes successive glucose units from the nonreducing ends of dextrin chains.

gluconeogenesis formation of glucose from noncarbohydrate precursors.

glycolysis the anaerobic energy yielding process in which carbohydrates such as glucose are oxidized to pyruvic or lactic acid.

glycoprotein proteins containing sugar sequences.

Goldberg-Hogness box nucleotide sequences in eukaryotic promoter genes that bind RNA polymerase.

Golgi complex membranous complex found in eukaryotes that packages proteins and lipids, which are transferred to selected areas of cell.

graft a portion of tissue used to replace a defective one in the body.

grana condensed bodies of chlorophyll within plant cell's chloroplasts.

granulocyte a type of white blood cell such as neutrophils, eosinophils and blasophils containing granules of hydrolytic enzymes involved in phagocytic digestion.

granuloma a nodule formed in tissue as a result of the accumulation of fused macrophages, which try to destroy infectious agents.

GRAS list an acronym for Generally Recognized as Safe. A list of chemicals that are allowed by law to be added to foods.

green beer freshly fermented beer before it has been aged.

growth factor an organic compound (vitamin, amino acid, purine, or pyrimidine) required in small amounts to support growth of an organism.

gumma granuloma found in tertiary stage of syphilis.

gyrase enzyme capable of catalyzing the interconversion of superhelical (positive and negative) and relaxed forms of DNA.

H chain (heavy chain) the larger of the two types of polypeptide chains of the basic Ig molecule; there are two identical H chains per basic Ig molecule; H chain composition determines the class of the Ig molecule.

habitat physical environment of an organism.

halophile an osmophile that has a specific requirement for increased amounts of an inorganic salt (usually NaCl) for growth.

haploid containing but one set of chromosomes.

hapten a substance that by itself does not elicit an immune response unless conjugated with an immunogenic carrier molecule; may react specifically with immune products in the absence of a carrier molecule.

haustoria hyphae that penetrate host cells for the purpose of obtaining nutrients.

heartwood central cylinder of nonfunctional xylem in woody stem.

helminth parasitic worm.

helper T cells (T_h) a subset of regulatory T cells that facilitate the response of B and other T cells to antigens.

hemagglutination clumping of red blood cells.

hemolysins substances, including antibodies (immune hemolysins), that lyse red blood cells.

hepatitis inflammation of the liver.

heterocyst specialized cell found in some cyanobacteria that carries out nitrogen fixation and lacks O_2 producing photophosphorylation capability.

heteroduplex DNA molecule in which one strand is from one source while the other strand is from another source.

heterokaryon a cell, usually a mycelium, containing more than one genetically distinct type of nucleus in a common cytoplasm.

heterotrophic requiring preformed organic compounds as a source of carbon.

histamine a vasoactive amine released (especially in type I IgE-mediated hypersensitivity reactions) from mast cells and basophils; causes smooth muscle contraction, increased capillary permeability, and increased glandular secretion.

histocompatability antigens tissue antigens whose similarities or dissimilarities determine graft acceptance or graft rejection. Also called transplantation antigens.

histone a basic protein associated with eukaryotic nuclear DNA in chromatin.

holdfast an adhesive-containing structure that enables an organism to attach to a substrate.

holocarpic a type of fungal species in which the thallus has no reproductive structures.

hops female flower of the moraceous vine *Humulus lupulus* and used in making beer.

horizons the layers making up the vertical soil profile.

horizontal spread the spread of infectious agents from one person to another through contact, insect vectors, or by contaminated food or fomites.

host the organism that is susceptible to infection by parasites.

humoral immunity antibody-mediated (B cell) immunity; antibodies are primarily located and transferable in body fluids (humors) such as serum.

humus organic fraction of the soil that is relatively stable.

hydrogen swell the nonmicrobial formation of hydrogen gas in a canned food product, usually resulting from a reduction of an acid food with the tin lining of the can.

hydrolysis breakdown of a polymer into individual monomers by the addition of water.

hydrophilic readily attracted to water.

hydrophobia fear of water.

hydrophobic not attracted to water and unable to mix with it.

hydrosphere earth's aquatic environment.

hymenium spore-bearing (sporogenous) layer of a fungal fruiting body.

hyperplasia increase in the number of cells in a tissue.

hypersensitivity a state of overresponsiveness by the immune system to an antigen.

hypertonic a solution in which there is a higher concentration of solute relative to that found in the interior of a cell (isotonic).

hypha (pl. hyphae) one of the filaments that make up a fungal mycelium.

hypolimnion deep waters of oceans or lakes below the thermocline.

hypotonic a solution having a lower solute concentration than an isotonic solution.

iatrogenic a disorder or disease caused by the diagnosis or treatment by a physician.

icosahedron twenty-sided object.

icterus jaundice; yellow discoloration of skin due to bile pigments.

id an allergic reaction to fungi or fungal products.

idiotype antigenic determinant associated with the antigen-binding site of an antibody molecule.

immediate hypersensitivity antibody-mediated (humoral) hypersensitivity reactions that occur within minutes after antigen is introduced into an individual who has been previously sensitized (i.e., has antibodies) to the antigen.

immobilized enzyme an enzyme that has been attached to or entrapped within an inert matrix.

immunity the state of being protected or resistant, especially to microorganisms and their products.

immunofluorescence fluorescence resulting from the interaction of an immunoglobulin, to which a fluorescent dye has been conjugated, and a specific antigen.

immunogen a substance capable of inducing an immune response.

immunoglobulins (Igs) plasma proteins of the gamma globulin class that are produced by or associated with B lymphocytes; the basic Ig molecule consists of two H chains and two L chains; antibodies are Igs.

immunosuppression the reduction in the capability of the immune system to respond to an antigen; obtained through immunosuppressive drugs, irradiation, and antilymphocytic serum (ALS).

impetigo infection of the skin usually caused by streptococci.

IMVC biochemical tests (indole, methyl red, Voges-Proskauer, and citrate) used in identification of enterics.

incidence the number of new cases of disease in a particular population within a specified time period.

inclusion bodies in virus infections, the highly stainable components, usually virus, found in the cytoplasm or nucleus of the infected cell.

index case the first case observed or detected in an epidemic.

indigenous native to a particular place.

inducer a molecule capable of stimulating the formation of compounds such as enzymes involved in cellular metabolism.

induration a hardened area or lesion.

infectious able to cause disease.

infestation attachment of parasites to superficial layers of the skin.

inflammation nonspecific response to irritants characterized by pain, heat, redness, and swelling.

inoculum a substance (such as microorganisms or serum) introduced into the tissues or culture media.

insertion sequence in genetics, the addition of gene or clustered genes to the chromosome and related to transposons.

in situ the natural environment in which a microorganism resides.

intercalary inserted between components.

interferon a class of proteins produced by vertebrate cells in response to viruses, endotoxins, and certain chemicals; associated primarily with antiviral activity.

intoxication state of being poisoned.

intron a sequence in a eukaryotic gene that is transcribed but is excised from RNA before translation.

in utero within the uterus.

invasion able to invade or penetrate the body.

invertase an enzyme that converts sucrose to a mixture of glucose and fructose.

in vitro outside of the body or performed in artificial environments.

in vivo within the body or within a living organism.

iodophors disinfectants consisting of iodine combined with a carrier molecule.

ionization the separation of a molecule into atoms possessing a charge.

isomer a molecule having the same atoms and groups as another molecule but having a different arrangement.

J chain (joining chain) glycoprotein whose function is believed to be to join together basic IgM molecules to each other, and basic IgA molecules to each other, and thus form polymers (IgM pentamer and IgA dimer and trimer).

jaundice see icterus.

karyogamy fusion of nuclei of cells.

keratin the protein that is part of epidermis, hair, and nails.

killer T cell (T_k) sensitized effector T cells that are cytolytic for corresponding target cells.

kinetochore see centromere.

kinetosome see basal body.

kinins a group of inactive peptides in the blood and tissue that, on activation, are involved in blood clotting and inflammatory processes.

Koji process a process by which molds are cultured on the surface of moist bran.

Koplik's spots small bluish red lesions appearing on the oral mucosa during measles infection.

kuru chronic progressive degenerative disorder of the central nervous system caused by a virus and found among New Guinea natives.

L-form a wall-deficient form of bacteria.

lactoferrin a protein found in external secretions of vertebrates that binds (chelates) iron.

lager beer beer that has been aged to develop its characteristic taste.

lambda (virus) a bacterial virus of *E. coli* noted for its ability to integrate into bacterial chromosome.

latency a state of inactivity.

lauter tub in the brewing process it is the tank where insoluble spent brewers grains are removed after mashing is completed.

leafhopper delicate, leaf-sucking, homopterous insect; often a disease vector.

leavening the production of gas in the manufacture of a baked food product.

lectins glycoprotein found on the surface of plant cells.

legume a group of plants such as peas, beans, and clover, in which nodules form on the roots and contain bacteria that fix nitrogen.

leukocidin a substance produced by some pathogenic bacteria and toxic to white blood cells.

leukocyte white blood cell.

leukopenia a smaller than normal number of circulating leukocytes.

lichen an association of a fungus with an algae, resulting in the formation of a biological unit morphologically distinct from its partners.

ligase an enzyme that catalyzes the covalent union of two pieces of DNA.

light chain (L chain) the smaller of the two types of polypeptide changes (the other is the H chain) that compose the basic Ig molecule; each basic Ig molecule contains two identical light chains.

lignification hardening of tissue from wall deposition of lignins.

lignin complex organic material that in part constitutes wood.

lipid a molecule generally composed of glycerol esterified to fatty acids.

lipopolysaccharide (LPS) the lipid and sugar complex that makes up part of the outer membrane of the gram-negative cell wall.

liposomes synthetic phospholipid sacs (vesicles) having properties of biological membranes.

lithotroph a microorganism that obtains its energy from the oxidation of reduced inorganic compounds.

load total count of bacteria in a consumable product.

local anaphylaxis a type I hypersensitivity reaction that primarily affects a given organ or tissue (such as skin or nasal mucosa).

lophotrichous having a tuft of flagella at a pole.

lymph a fluid derived from tissues that contains white blood cells and is organized into vessels that drain into the blood stream.

lymphadenitis inflammation of a lymph node.

lymphocyte a white blood cell devoid of cytoplasmic granules, associated with immune response and chronic inflammation.

lymphokines soluble factors released by sensitized lymphoid cells (classically T cells) in response to antigenic contact; such factors modulate the biological activity of other cell populations, especially macrophage cells.

lymphoma any neoplasia (abnormal growth) associated with lymphatic system.

lyophilization process of freeze-drying.

lysis rupture or dissolution of a cell.

lysogenic conversion the change in properties of a bacterium resulting from the carriage of a prophage.

lysogeny the state of a bacterial cell in which a bacteriophage genome is in association with the bacterial genome.

lysosome a eukaryotic cell organelle containing hydrolytic enzymes collectively termed acid hydrolases.

macrolide an antibiotic inhibiting protein synthesis, for example erythromycin.

macrophage a large mononuclear cell derived from the mononuclear phagocytic system and having phagocytic activity.

macrophage activation factor (MAF) a lymphokine that increases the capacity of macrophage cells to destroy antigens.

mast cell cell found in connective tissue that stores heparin and histamine.

matrix the ground substance of a component.

meiosis a eukaryotic reproductive process in which two mitotic divisions result in the formation of daughter cells (usually sex cells) receiving the haploid chromosome number.

memory cell a lymphocyte of either B or T cell lineage that was formed during a primary immune response and that is capable of synthesizing specific immune products at an increased rate during an anamnestic response.

meningitis inflammation of the meninges (membranes that surround the brain and spinal cord).

merozoite a plasmodium trophozoite resulting from multiple fission (for example, malaria).

mesophile a microorganism that grows best at temperatures between 20° and 45° C.

mesosome involuted membrane of the bacterial cell sometimes associated with cell septum and nucleus.

metabolism the sum total of all the physical and chemical changes that take place in a cell and maintain the cell's integrity.

metabolite product of a metabolic pathway.

metachromatic granules granules of certain bacteria that often contain phosphate chains, also called volutin.

methylotroph organisms that gain their energy from the oxidation of certain one-carbon compounds other than carbon dioxide.

Michaelis-Menten constant a constant that specifies the quantitative relationship between substrate concentration and maximum velocity for different enzymes.

microaerophile an organism that grows in air but at a concentration of oxygen less than our atmosphere.

microhabitat the microscopic environment that is important to a population or community in nature.

microphage a granulocytic phagocyte.

microtubules tubes that are the structural basis of eukaryotic flagella and are also part of the spindle organization during mitosis of eukaryotes.

mineralization conversion of the organic form of an element to the inorganic state.

minicell the small anucleate daughter cell, resulting from unequal cell division.

mitosis the process in which genetic material is duplicated and distributed equally between daughter cells.

mixotroph organism that can obtain energy from inorganic sources but requires organic carbon.

mold another name for fungi that exhibit branching or mycelium formation.

Monera the kingdom to which all prokaryotes belong.

monocyte see macrophage.

monomer the simplest molecule of a compound.

morbidity sickness; the ratio of sick to well individuals in a community.

mordant a substance added during staining procedures to intensify the staining.

morphogenesis the change in structure as an organism develops.

mortality fatality; the ratio of the number of deaths to a given population in a defined situation.

mosaic disease symptom of leaves with mixed green, light green, and yellow patches.

most probable number (MPN) a statistical expression of estimating the cell number in a culture.

murein see peptidoglycan.

mutagen a substance that increases the mutation rate of an organism.

mutation a process in which a gene undergoes a structural change, which is inherited by offspring.
mutualism a relationship between organisms in which both benefit.
mycelium a matlike organization of fungal hyphae.
mycetismus mushroom poisoning.
mycology the study of fungi.
mycolysis the enzymatic lysis of fungi.
Mycoplasma a taxonomic group of bacteria without cell walls.
mycorrhiza an association of a fungus with the roots of a plant.
mycoses diseases caused by fungi.
mycotoxicosis poisoning of humans and animals following the ingestion of foods containing toxins produced by fungi.
myeloma proteins Igs or portions of Igs produced in large quantities by neoplastic plasma cells (plasmacytoma).
myiasis infestation caused by maggots (fly larvae).
nebulizer a device used to produce fine sprays from liquids.
necrosis death of tissue or cells.
Negri bodies inclusion bodies found in brain cells of those animals infected with rabies virus.
nematode roundworm.
neonate a newborn infant (up to 4 weeks old).
neoplastic a disease.
nephrotoxic toxic to the kidney.
neutrophil a leukocyte in which the granules do not have an apparent affinity for acid or basic dyes.
neutrophile microorganism having optimum growth at neutral pH.
niche an organism's function within a habitat.
nitrification microbial formation of nitrate from ammonia.
nitrogen fixation process in which nitrogen gas (N_2) is converted into nitrogen compounds.
nosocomial pertaining to the hospital.
nucleocapsid the capsid protein of a virus with its enclosed nucleic acid.
nucleoid a region in prokaryotic cells in which there is a high concentration of DNA.
nucleolus a small body residing in the eukaryotic nucleus that contains a high concentration of ribonucleic acid.
nucleosome a structural arrangement on the eukaryotic chromosome in which DNA of 130 to 140 base pairs is associated with eight histone molecules.
nucleotide a compound consisting of a purine or pyrimidine base, a five carbon sugar, and phosphate.
nutrient substances extracted from the environment by cells and used in metabolic processes.
obligate necessary or required.
Okazaki fragments short pieces of single-stranded DNA that are synthesized during replication and later polymerized.
oligotroph a species that grows at low nutrient concentrations.
oncogenic having the ability to cause tumors.
operon a unit of the chromosome consisting of adjacent genes usually involved in similar function and controlled by an operator.
ophthalmic pertaining to the eye.
opine amino acid derivative synthesized only in plant tumors caused by *Agrobacterium tumefaciens*.
opsonin a substance (usually an antibody or complement fragment C3b) that adheres to bacteria, viruses, and antigen-antibody aggregated molecules and makes those particles more susceptible to ingestion by phagocytes.
organoleptic detected by the senses.
osmophile a microorganism that requires a high solute concentration in the medium for growth.
osmosis passage of fluids through a semipermeable membrane.
oxidation the process of adding oxygen to a molecule or the removal of electrons from a molecule.
oxidation-reduction (redox) reaction a reaction in which electrons are transferred from a donor molecule to an acceptor molecule.
oxidative phosphorylation the process of energy (ATP) formation coupled to electron transport with oxygen as the final electron acceptor.
oxygenic the ability to use a photosynthetic process in which oxygen is a byproduct.
PMN or PMNL polymorphonuclear leukocyte of a type called a neutrophil; it is a white blood cell with a multilobed nucleus that is highly phagocytic and the predominant cell type in acute inflammation.
palindromic sequences the same DNA nucleotide sequences that appear on complimentary strands but in opposite directions.
pandemic a worldwide epidemic.
parasite an organism that survives on or at the expense of another living host.
parenteral injection by a route other than oral, that is, intravenous, intramuscular, subcutaneous.
passive immunity immunity in which preformed immune products are acquired by an individual from an external source either naturally (via colostrum or transplacental passage) or artificially (via injection).
passive transport transfer of molecules without the need for energy.
pasteurization a process in which fluids are heated at temperatures below boiling to kill pathogenic microorganisms in the vegetative state.
pathogen an organism capable of producing disease.
pathovar a pathological variety of a species of bacteria.
pellicle thick covering over the cytoplasmic membrane.
peplomer a projection (spike) extending from the outer surface of a virus envelope.
peptidoglycan the rigid structural component of the bacterial cell wall consisting of *N*-acetylmuramic acid and *N*-acetylglucosamine.
periplasm contents of a space (periplasmic) that lies between the outer membrane and cytoplasmic membrane of the gram-negative cell.
permease a transport protein located in the cytoplasmic membrane.
pertussis whooping cough caused by the bacterium *Bordetella pertussis*.
pH symbol for hydrogen ion concentration and a measure of the acidity or alkalinity of a solution; $\log_{10}$ of the reciprocal of the hydrogen ion concentration in molarity.
phage an abbreviated term for bacterial virus (bacteriophage).
phagocytes cells (especially PMNs and macrophages) that are capable of ingesting (endocytosing) substances and subsequently digesting them.
phagocytosis a process in which a cell (phagocyte) engulfs other cells or particles.
phagosome a cell vacuole resulting from phagocytosis of particulate materials.
phagotroph organism that ingests living material as a source of food.

phenotype the genetic makeup of an organism; its observable properties.
pheromone hormonelike peptide produced by eukaryotic microorganisms and associated with mating type.
phlebitis inflammation of blood vessels.
phloem food-conducting tissues in trees.
photoheterotroph an organism that uses light for energy and organic molecules as a source of carbon.
photon a particle of radiant energy.
photosynthesis a process in plants and some microorganisms in which light as an energy source is used to reduce carbon dioxide to organic material.
phototaxis movement toward light.
phycology study of algae.
phylogeny the evolutionary history of a group of organisms.
phytopathogen microorganism pathogenic for plants.
phytoplankton plantlike organisms that are floating or free swimming.
pili small protein projections on the bacterial cell involved in conjugation process, but there may be other functions.
pilsener beer lager beer made according to the beer prevailing in Pilsen, Czechoslovakia.
pinocytosis the engulfment of liquid droplets by a cell.
plaque the structureless debris found on the surface of teeth; also the clear area on an opaque lawn of bacteria or on a monolayer of tissue culture cells due to the lytic activity of viruses.
plasma fluid portion of the blood.
plasma cell cell type regarded as the principal producer of antibodies.
plasmid an extrachromosomal piece of circular DNA.
plasmodium a cellular mass of protoplasm associated with the slime molds.
plasmogamy fusion of protoplasts (such as gametes), often followed by nuclear fusion.
plasmolysis shrinkage of the cell caused by osmotic removal of water.
plasmoptysis swelling or bursting of a cell due to osmotic inflow of water.
platelet small blood components important in blood coagulation.
pleomorphism the state of having more than one form or shape.
pock a pit, spot, or pustule.
polycistronic a messenger RNA in which several coded genes are part of the molecule.
polyhedral inclusion body protein-encased insect virus complex.
polyhedral many-sided.
polymerase an enzyme that engages in the polymerization of monomeric units.
polyploid having a chromosome number more than twice the haploid number.
porin protein pores found in the outer membrane of the gram-negative cell wall.
porter a weak stout that is rich in saccharine matter and contains about 4% alcohol.
Prausnitz-Küstner Test (P-K test) a passive transfer test that is used in atopic allergy diagnosis; Serum from the allergic individual is deposited in the skin of a volunteer and the suspected allergen is injected later into the same site.
precipitin antibody that forms precipitating lattice upon interaction with its homologous soluble antigen.
predation the relationship in which one animal feeds or preys on another.
Pribnow box a nucleotide sequence in the prokaryotic promoter gene that is the site of RNA polymerase binding.
prion proteinaceous infectious particle believed to be responsible for certain neurological disorders of sheep and humans.
proctitis inflammation of the rectum.
prodromal period a stage in infectious disease process in which the first symptoms appear.
producer organism that synthesizes organic material.
profile vertical cross section of soil.
prokaryote a microorganism characterized by the absence of a nuclear membrane; the absence of cytoplasmic organelles and less complex organization than eukaryote.
promoter region the area on a gene or operon that binds RNA polymerase and allows transcription initiation.
properdin a bactericidal protein component of the blood.
prophage the state of a bacteriophage in which the viral genome is integrated into the bacterial genome and is duplicated each cell generation.
prophylaxis protection, for example, against disease.
prostheca a cytoplasmic filled cell wall protrusion from a cell.
prosthesis artificial component substituting for a defective body part.
protease an enzyme that hydrolyzes peptide bonds, thus converting proteins into polypeptides and eventually amino acids.
Protista the kingdom that includes fungi, protozoa, and primitive algae.
protocooperation a transitory association between two species in which both benefit.
proton-motive force the force tending to pull protons across the cell membrane because of the electrochemical potential generated across the membrane.
proto-oncogene a normal cellular gene capable of conversion to an oncogene (capable of causing cancer).
protoplast a cell in which the cell wall has been removed.
prototroph an organism that has no growth factor requirements as compared with an auxotroph, which does have growth factor requirements.
protozoa eukaryotic unicellular microorganisms with animal characteristics.
provirus virus integrated into host animal or human chromosome and transmitted from one generation to another.
pseudopodium a "foot" or temporary extension of the cell membrane of ameboid cells and used for movement.
pseudowild the partial restoration of a function to an organism.
psychrophile microorganisms that grow best at low temperature (0° to 20° C).
psychrotroph mesophilic organism that can grow at psychrophilic temperatures (0° to 10° C).
psittacosis a bird disease transmissible to humans.
pure culture a culture containing only one kind of organism.
purulent associated with the formation of pus.
pus a yellow white substance found in abscesses containing primarily living, dying, and dead leukocytes.
pyrogen an agent that induces fever.
quarantine to detain or isolate individuals because of suspicion of infection.
quaternary ammonium compound a cationic detergent with a central nitrogen atom to which four groups are attached.
quellung reaction a swelling of encapsulated cells resulting from the interaction of specific immune serum with capsular antigens.

R-factor a transferrable plasmid that carries the information for resistance to antimicrobials or other agents.

race group of individuals within a cultivar or species distinguished by pathogenic behavior but not by morphology.

radioimmunoassay (RIA) an immunoassay technique in which one of the test components (antibody or antigen) is labeled with a radioactive isotope; radioactivity is used to indicate that an antigen-antibody reaction has occurred.

raphe a linear slit on the valves of diatoms.

reagin the skin-sensitizing antibody IgE; in syphilis the nonprotective antibody that has specificity for cardiolipin antigens.

recalcitrant molecule a compound that is slowly mineralized or that is not metabolized by microorganisms in nature.

receptor-mediated endocytosis a phagocytic process in which specialized segments on the plasma membrane, called coated pits, act as receptors for substance to be transported.

recombination a process in which genetic information is exchanged between two organisms or in the case of bacteria from a donor to a recipient.

reduction the addition of electrons to a molecule or the gain of hydrogen atoms by a molecule.

regulon a system of more than one operon controlled by a single regulatory protein.

rennin a proteolytic enzyme obtained from calf's stomach which coagulates milk protein.

repression a process occurring at the gene level in which the synthesis of an inducible enzyme is inhibited by a repressor molecule.

reservoir of infection animate or inanimate objects that support the growth of microorganisms.

resistance transfer factor (RTF) a group of genes that permits the transfer of resistance factors during the conjugation process.

resolving power the ability to distinguish two close objects separately when observed with the unaided eye or with a microscope.

respiration an oxidative process in which energy is released from foodstuffs.

reticuloendothelial system (RES) a system of fixed macrophages in spleen, liver, lymph nodes, and bone marrow.

retrovirus a group of RNA viruses that possess reverse transcriptase for transcribing the RNA genome into a DNA intermediate. The agent of some cancers and AIDS.

reverse transcriptase an enzyme (RNA-dependent DNA polymerase) that synthesizes a complementary DNA from an RNA template.

rhinitis inflammation of the nose.

rhizoid filamentous appendages by which organisms such as algae and plants attach to the substratum.

rhizoplane the root surface and its adhering soil.

rhizosphere microenvironment in soil immediately around plant roots.

ribosome a ribonucleoprotein particle found in the cell cytoplasm and involved in protein synthesis.

ringworm fungal infection caused by a group called dermatophytes.

ripening the aging of a food product to enhance the chemical or microbiological development of specific flavors and textures.

rubella German measles.

rubeola red measles.

saprobe an organism that derives its nourishment from dead or decaying material.

sapwood physiologically active zone of wood contiguous to cambium.

sarcoma a solid tumor growing from derivatives of tissue such as embryonal connective tissue, bone, muscle, and fat.

schizogony asexual reproduction by multiple fission of a trophozoite.

secretory IgA (sIgA) composed of two (dimer) or three (trimer) basic IgA molecules, a secretory piece, and a J chain; provides topical or local immunity via sIgA-containing secretions (mucus secretions, colostrum, saliva).

sepsis a toxic condition resulting from the presence of microbes or microbial products in the body.

septicemia condition resulting from multiplication oif microorganisms in the bloodstream.

septum a wall dividing biological units, such as cells in a fungal filament.

sequestering agent an agent that will hold metallic ions in solution usually by inclusion of the ions in an appropriate coordination complex.

serology the study of antigen-antibody reactions in vitro.

serotype subtype of a species that is based on antigenic differences.

serum the clear portion of the blood minus the factors necessary for clotting.

serum sickness a systemic, type III hypersensitivity that develops when soluble antigen that was administered in large quantities remains in the tissues and forms immune complexes with newly formed antibody.

sex factor (F factor) a genetic unit that may be part of the chromosome or may be extrachromosomal; it functions in the transfer of genes during conjugation.

shingles a viral disease of adults resulting from activation of latent measles virus.

siderophore metal-binding molecule.

sieve tube specialized phloem structure that conducts food in plants.

sludge digestion anaerobic treatment of sewage sludge.

somatic relating to the cell.

sorghum an Old World tropical grass similar to Indian corn in habit.

source of infection animate or inanimate objects that serve as a contact between host and microorganism.

species a taxonomic group composed of individuals having closely related characteristics.

specification accepted value in the manufacture of a product.

spheroplast wall-deficient bacterium that retains part of the wall.

spirochete a corkscrew-shaped bacterium.

sporangium a structure that holds asexual spores.

spore the resistant form of a bacterium encospore derived from the vegetative cell; the asexual or sexual reproductive cell of certain organisms.

sporozoite a trophozoire of *Plasmodium* found in mosquitoes.

sporulation the process in which a spore is formed.

springwood early part of the yearly growth ring.

standard a law stating accepted value(s) for a manufactured product.

starter culture the inoculation of a food or beverage with an organism to ensure the development of a specific product.

static growth-inhibiting.

steady-state open system in which raw materials are converted into biomass, also equilibrium in continuous cultures where there is no change in cell mass.

sterilization the process that destroys all living microorganisms.

stoma regulated pore opening in the leaf epidermis for passage of gases and water vapor.

stout a heavy-bodied brew that is darker and sweeter than porter and is made with roasted malt and a relatively high percentage of hops.

stroma the supporting substance of an organelle or structure in which other components may be suspended.

stromatolite fossilized mats of cyanobacteria.

stylet stiff, slender, hollow feeding organ of plant-parasitic nematodes.

subclinical without clinical manifestations of the disease.

submerged culture a technique for culturing molds and other microorganisms in which the microbes are distributed through the culture fluid by vigorous mixing. It is in contrast to surface culture where the organism grows on the upper surface of the culture fluid.

substrate the compound with which an enzyme reacts.

succession replacement of types of populations and communities with time.

sufu oriental food made by fermentation of soybean.

suppressor T cells (Ts) a subset of regulatory T cells that suppresses the activity of B cells and other T cells.

surfactant substance that reduces surface tension; a wetting agent such as detergents and quaternary ammonium compounds.

sylvatic occurring in or affecting wild rodent populations.

symbiosis an association or relationship between two organisms.

synapse the junction between two processes such as two nerve cells.

syndrome a group of symptoms that characterize a disease.

syngamy union of gametes.

systemic not localized in any one part of the body but spread throughout the body.

T cell thymus-dependent lymphocytes involved in cell-mediated immunity.

tannin any of several soluble astringent complex phenolic substances of plant origin.

taxis movement of an organism in a particular direction in response to an external stimulus.

taxon a taxonomic group such as species and genus.

teichoic acid polysaccharide containing ribitol or glycerol phosphates found in gram-positive bacteria.

temperate virus lysogenic virus in which genome is integrated or associated with bacterial genome.

tempeh oriental food made by fermentation of soybean or coconut.

test shell or capsule secreted by a protozoan.

tetany spasms.

thallus the vegetative unit of a fungus.

thermal death point (TDP) that temperature at which all microorganisms in a culture at a pH of 7.0 are killed in 10 minutes.

thermal death time (TDT) the time required to kill all microorganisms of a known population at a predetermined temperature under specified conditions.

thermocline temperature boundary in aquatic environments.

thermophiles microorganisms growing best at temperatures between 45° and 70° C.

thrush a condition affecting the oral mucous membranes caused by the fungus *Candida*.

thylakoids photosynthetic membrane found in eukaryotes and some prokaryotes, such as cyanobacteria, that contains pigments.

tinea a name applied to fungal infections of the skin.

titer the quantity of a substance required to produce a particular reaction.

topoisomerase enzyme that converts DNA into different topological forms.

toxoid an inactivated toxin that has lost its toxigenicity but retains its immunogenicity and thus can be used for active immunization.

tracheid thick-walled cell of xylem tissue that is capable of water transport.

transamination transfer of an amino group from an amino acid to a keto group of metabolite and the formation of a new amino acid.

transcription the process by which the nucleic acid (usually one strand of DNA) acting as a template, is reproduced into a complementary molecule of RNA.

transduction the transfer of bacterial genetic information from one cell to another by a virus (bacteriophage).

transfection the infection of host cells with naked viral DNA.

transformation in bacterial systems the uptake of free nucleic acid (DNA) by a cell and its acquisition of new traits; in virology a process in which virus infection alters the eukaryotic cell, making it malignant.

transition a mutation caused by substitution of one purine for another or one pyrimidine for another.

translation the conversion of the mRNA into a protein product in the process of protein synthesis.

transplantation antigens cell surface antigens that induce an immunorejection response when the transplant recipient is not genetically identical to the transplant donor.

transpiration loss of water vapor through the leaves.

transposition the shift of a gene or chromosome segment to a new position on the chromosome.

transposon a genetic element that can be duplicated and moved to other sites on the same or different genome.

transversion a mutation in which there is a substitution of a purine for a pyrimidine or vice versa.

treponematoses all of the diseases caused by species of *Treponema*.

trichocyst an organelle found in certain eukaryotic microorganisms that ejects a filament through the cytoplasmic membrane.

trichome filamentous structure.

triplet code three nucleotide code specifying for one amino acid.

trophic level organisms in a food chain that have similar relationships to the source of their nutrients.

trophozoite the active vegetative stage of a protozoan.

tropism a growth response of an organism following an external stimulus.

trub the coagulate protein that settles out of freshly boiled wort.

tubercle a nodule.

tuberculin a bacteria-free protein fraction extracted from the bacilli causing tuberculosis (*Mycobacterium tuberculosis*).

tubulin the protein subunit of microtubles.

tumor mass of tissue resulting from uncontrolled growth.

turbidometry measurement of the turbidity of a solution.

ulcer circumscribed area of inflammation found in epithelial lincontiguous vessels.

unsaturated presence of double or triple covalent bonds in a molecule.

urethritis inflammation of the urethra.

urticaria inflammatory reaction characterized by eruptions that itch.
vaccine usually a suspension of organisms killed or attenuated and used for immunization; some vaccines are macromolecules such as proteins or polysaccharides derived from a microbial agent.
vacuole space or cavity in the cell usually filled with some substance.
varicella chickenpox.
vascular containing blood vessels.
vasoactive affecting vessels (especially blood vessels).
vector a carrier of pathogenic microorganisms, in gene engineering a plasmid or virus genome used to carry genes.
vegetative concerned with the growing stage of a microorganism as opposed to the spore state.
venereal transmitted by sexual contact.
vertical spread transfer of infectious agent from parent to offspring via ovum or sperm or by contact between placenta and offspring.
vesicle blister; sac surrounded by unit membrane.
virion a complete virus particle consisting of a core of nucleic acid and a protein capsid.
viroid an infectious subviral particle consisting of nucleic acid without a protein capsid.
viropexis a type of pinocytosis in which a virus enters the cell within a closed membrane.
virulence the relative ability of an organism to cause disease.
walling-off separation of diseased from healthy tissue by barrier tissues produced by diseased plant.
water activity (a_w) the ratio of existing vapor pressure to pure water vapor pressure.
wilt collapse of foliage from water deficiency; may be temporary or permanent.
witches'-broom broomlike proliferation of woody stems originating from closely spread nodes.
wort a dilute solution of sugars obtained from malt by infusion and fermented to form beer.
xenograft a graft between individuals of different species.
xylem water-conducting, woody tissue produced by the cambium.
xylem parenchyma living parenchyma cells associated with the xylem.
yeast a unicellular fungus that does not give rise to mycelial growth.
Z value the degrees Fahrenheit required to reduce the thermal death time by 90% or one log value.
zoa pertaining to animal.
zoonoses diseases of animals.
zooplankton a collective term for nonphotosynthetic microorganisms found in flora and fauna of a body of water.
zoospore a motile reproductive spore common to aquatic fungi.
zygote the diploid cell in eukaryotes resulting from union of male and female gametes.
zymogenous soil microorganisms, primarily rods and spore formers that readily decompose (ferment) organic matter.

Answers to self-study quizzes

CHAPTER 1 HISTORY OF MICROBIOLOGY

Multiple choice

1. E; 2. C; 3. B; 4. B; 5. A; 6. C

Matching

1. I; 2. G; 3. C; 4. F; 5. H; 6. A; 7. J; 8. D; 9. E

CHAPTER 2 BIOCHEMICAL BACKGROUND FOR MICROBIOLOGY

Matching

1. G; 2. D; 3. E; 4. C; 5. F; 6. H; 7. A; 8. B; 9. J; 10. I

Multiple choice

1. C; 2. D; 3. A; 4. C; 5. A; 6. C

Completion

1. Km; 2. competitive; 3. allosteric or regulatory; 4. denaturation; 5. glycine

CHAPTER 3 INTRODUCTION TO MICROSCOPIC TECHNIQUES

Multiple choice

1. A; 2. B; 3. C; 4. B

Matching

1. F; 2. B; 3. E; 4. A; 5. C; 6. D

CHAPTER 4 PROKARYOTES AND EUKARYOTES: A COMPARISON

Identification

1. P and E; 2. P and E; 3. P; 4. P; 5. E; 6. E; 7. P; 8. E; 9. E; 10. E

Multiple choice

1. C; 2. C; 3. D; 4. E; 5. C; 6. E; 7. B; 8. D, 9. B

CHAPTER 5 PROKARYOTIC ULTRASTRUCTURE AND FUNCTION

Matching

1. A, B, D; 2. G; 3. K, E; 4. F; 5. A; 6. H, J; 7. H, D; 8. I; 9. C; 10. A

Multiple choice

1. C; 2. D; 3. A; 4. B; 5. C; 6. C; 7. D; 8. B; 9. C; 10. D

Completion

1. cell wall; 2. passive diffusion; 3. active transport; 4. hydrophobic; 5. polysaccharide; 6. counterclockwise; 7. sex pilus; 8. beta-polyhydroxybutyrate; 9. cortex; 10. gas vacuole

CHAPTER 6 CHARACTERISTICS OF EUKARYOTES

Completion

1. Deuteromycotina; 2. haustoria; 3. thecal plate; 4. dimorphism; 5. schizony or multiple fission; 6. saprobes or saprophytes

Multiple choice

1. D; 2. C; 3. C; 4. C; 5. D; 6. A

Identification

1. F; 2. F; 3. P; 4. A; 5. P, A; 6. F; 7. P; 8. F; 9. F; 10. F; 11. F; 12. P; 13. F

CHAPTER 7 EUKARYOTIC ULTRASTRUCTURE AND FUNCTION

Completion

1. mitochondrion; 2. cell wall; 3. microtubules; 4. ribosome; 5. cytoplasmic membrane; 6. chloroplast; 7. nuclear DNA; 8. mitochondrion; 9. endoplasmic reticulum; 10. nuclear membrane

Multiple choice

1. C; 2. D; 3. D; 4. B; 5. A; 6. B; 7. A; 8. B; 9. B; 10. C

CHAPTER 8 DNA REPLICATION, PLASMIDS, AND OTHER DNA ELEMENTS

Identification

A. DNA gyrase; B. RNA primase; C. DNA ligase; D. DNA polymerase III; E. helicase; F. helix destabilizing proteins; G. DNA polymerase I

Multiple choice

1. B; 2. B; 3. D; 4. C; 5. D; 6. C; 7. B; 8. D; 9. B; 10. C

CHAPTER 9 GENE EXPRESSION AND ITS CONTROL

Multiple choice

1. D; 2. C; 3. B; 4. A; 5. B; 6. A; 7. B; 8. D; 9. B; 10. C

Completion

1. codon; 2. nonsense; 3. transcription; 4. translocation; 5. polysome; 6. catabolite repression (glucose effect); 7. constitutive; 8. sigma

Matching

1. I; 2. H; 3. E; 4. F; 5. C; 6. D; 7. G; 8. A

CHAPTER 10 MUTATION

Multiple choice

1. A; 2. D; 3. B; 4. D; 5. A; 6. B; 7. C; 8. B; 9. A; 10. A

Completion

1. hot spots; 2. tautomers; 3. transition; 4. uracil, adenine; 5. alkylating; 6. frameshift; 7. missense; 8. amber; 9. SOS; 10. transformation

CHAPTER 11 RECOMBINATION, GENE TRANSFER GENETIC ENGINEERING, AND RECOMBINANT DNA

Multiple choice

1. C; 2. B; 3. C; 4. B; 5. A; 6. A; 7. D; 8. A; 9. A; 10. D

Completion

1. vector; 2. cosmids; 3. recA; 4. F-factor genes; 5. lysogeny; 6. synthetic linkers; 7. Hfr; 8. transformation; 9. conjugation; 10. specialized transduction

CHAPTER 12 ENERGY PRODUCTION IN MICROORGANISMS

Completion

1. free energy; 2. dehydrogenases; 3. isoprenoids; 4. phycobiliproteins; 5. Embden-Meyerhof-Parnas pathway; 6. pyruvate and glyceraldehyde 3-phosphate; 7. catalase

Multiple choice

1. A; 2. B; 3. C; 4. C; 5. C; 6. B; 7. C; 8. D; 9. A; 10. B

Matching

1. D; 2. M; 3. K; 4. J; 5. L; 6. I; 7. F; 8. C; 9. B

CHAPTER 13 NUTRITION, BIOSYNTHESIS, AND METABOLIC CONTROL

Multiple choice

1. D; 2. A; 3. D; 4. B; 5. C; 6. C; 7. B

Completion

1. isoenzymes; 2. multivalent inhibition; 3. racemase; 4. nitrogen fixation; 5. siderophores

Matching

1. E; 2. D; 3. J; 4. A; 5. C; 6. B; 7. G

CHAPTER 14 MICROBIAL GROWTH AND FACTORS AFFECTING GROWTH

1. C; 2. B; 3. C; 4. C; 5. A; 6. C; 7. C; 8. A; 9. B; 10. B

Matching

1. J; 2. I; 3. M; 4. F; 5. G; 6. K; 7. O; 8. N; 9. D; 10. B; 11. C; 12. H; 13. L; 14. A; 15. E

CHAPTER 15 CONTROL OF MICROBIAL GROWTH: CHEMOTHERAPY, STERILIZATION, AND DISINFECTION

Multiple choice

1. A; 2. D; 3. B; 4. D; 5. D; 6. C; 7. D; 8. C; 9. D; 10. B

Short answer

a. 2.50 mg/ml; b. 5.0 mg/ml; c. no, because the MIC is specific due to the organism tested and in this case *S. pyogenes* is gram-positive and *E. coli* is gram-negative

Matching

1. A; 2. D; 3. B; 4. A; 5. C; 6. C; 7. A; 8. C; 9. E; 10. B; 11. D; 12. A; 13. B; 14. C; 15. E; 16. A; 17. C; 18. B; 19. D; 20. A

CHAPTER 16 CHARACTERISTICS OF VIRUSES

Completion

1. virion; 2. capsid; 3. satellite RNA; 4. hemagglutination; 5. protein coat or capsid; 6. polyhedroses

Short answer

1. a. the suspected plant is harboring a latent virus; b. the virus is probably transmitted by insects; c. the agent is a viroid
2. a. X; b. O; c. O; d. X; e. X; f. O.

Multiple choice

1. B; 2. A; 3. D; 4. B; 5. D; 6. C; 7. B; 8. D; 9. C; 10. D

CHAPTER 17 VIRAL REPLICATION AND TUMOR DEVELOPMENT

Completion

1. 5-hydroxymethyl cytosine; 2. ADP-ribose; 3. redundant; 4. scaffolding proteins; 5. lysozyme; 6. headful; 7. cro; 8. protease; 9. adsorptive endocytosis; 10. proto-oncogenes

Short answer

1. degradation of host DNA into fragments; 2. degrades DNA fragments produced by endonuclease; 3. converts RNA into DNA; 4. synthesis of RNA; 5. synthesis of DNA; 6. integration of lambda into host DNA

Multiple choice

1. D; 2. A; 3. D; 4. A; 5. A; 6. A; 7. C; 8. D; 9. B; 10. A; 11. C

CHAPTER 18 ECOLOGY

Completion

1. succession; 2. eutrophism; 3. climax community; 4. biomagnification; 5. nitrification; 6. $SO_4^{++}(H_2SO_4)$; 8. eutrophs; 9. mineralization; 10. fixation

Classify

a. M; b. OR; c. CH_3; d. M

Multiple choice

1. B; 2. B; 3. C; 4. C; 5. D; 6. A; 7. A; 8. D; 9. A; 10. C

Classify (short answer)

a. (+) protocooperation; b. (−) parasitism; c. (−) competition; d. (+) mutualism; e. (+) commensalism

CHAPTER 19 AQUATIC MICROBIOLOGY

Multiple choice

1. B; 2. C; 3. C; 4. B; 5. A; 6. B; 7. B; 8. C; 9. B; 10. B; 11. C; 12. C; 13. C; 14. D; 15. C

Short answer

a. secondary treatment; b. tertiary treatment; c. activated sludge

True and False

1. False; 2. False; 3. True; 4. True; 5. True.

CHAPTER 20 AGRICULTURAL MICROBIOLOGY

Short answer

1. (1) D; (2) A; (3) A-E; (4) A-D; (5) A-B; (6) A-D
2. (1) G; (2) F; (3) E; (4) C; (5) D; (6) A or B; (7) A or B
3. (1) A-C; (2) C-E; (3) C-D; (4) E

Multiple choice

1. C; 2. C; 3. D; 4. D; 5. A; 6. D; 7. D; 8. B; 9. C; 10. D; 11. B; 12. B; 13. B; 14. D; 15. C

CHAPTER 21 FOOD MICROBIOLOGY

Completion

1. D value; 2. pasteurization; 3. lyophilization; 4. water activity; 5. *Pseudomonas;* 6. *Vibrio parahaemolyticus;* 7. retorts; 8. aflatoxin; 9. *Saccharomyces cerevisiae;* 10. diacetyl, acetoin

Multiple choice

1. A; 2. C; 3. B; 4. D; 5. A; 6. A; 7. C; 8. D; 9. D; 10. C

CHAPTER 22 INDUSTRIAL MICROBIOLOGY

Completion

1. alpha amylase; 2. break; 3. rennin; 4. cytoinduction; 5. sulfite; 6. top fermenting; 7. fermentation; 8. lactate and acetate

Short answer

1. The basic approach is isolation of strains of the antibiotic-producing organisms that produce significantly greater amounts of the antibiotic than do wild-type strains. This is accomplished by subjecting the antibiotic-producing strains to a mutagen, selecting overproducing survivors, and recycling the overproducing strains to mutagenesis and selection.
2. Starch $\xrightarrow{\text{alpha amylase}}$ Oligosaccharides + maltose
Fructose $\xrightarrow{\text{glucose isomerase}}$ Glucose
(sugar syrup)
3. It allows the recovery of enzymes following their catalytic action for reuse.
4. Growth of *Corynebacterium glutamicum* in submerged cultures in large fermenters with careful control of biotin, oxygen, pH, and temperature.
5. a. Malt: contains alpha- and beta-amylases that convert malt and adjunct starch into maltose and dextrins. Dextrins contribute to the body of the beer. Malt proteins contribute to the head, and malts determine the color of the beer.
b. Hops: primarily a flavoring agent; removes undesirable malt proteins
c. Yeast: fermentation of maltose to ethanol and CO_2
d. Adjuncts: increase alcohol content
6. Gene amplification, vaccine production, single-cell fungi for improved fermentation reactions, production of product-tolerant strains, improved product excretion, and the production of nonmicrobial products by microorganisms.

Multiple choice

1. C; 2. B; 3. A; 4. A; 5. C; 6. C; 7. B; 8. C; 9. A; 10. B.

CHAPTER 23 HOST-PARASITE INTERACTIONS

Short answer

a. keratinized epithelium, sebaceous gland secretions (fatty acids), tears, and lysozyme; b. mucociliary blanket, lysozyme, macrophages; c. mucus, stomach acids, fatty acids, secretory IgA, peristalsis, microbial flora, and bile salts; d. flushing action of urine, acids

Matching

A. 5; B. 10; C. 1; D. 8; E. 3; F. 4; G. 2; H. 7

Multiple choice

1. B; 2. C; 3. D; 4. A; 5. D; 6. A; 7. D; 8. B

CHAPTER 24 IMMUNOLOGY

Multiple choice

1. B; 2. C; 3. B; 4. C; 5. C; 6. D; 7. A; 8. C; 9. B; 10. C

Matching

A. 6; B. 1, 2, 5, 8; C. 1, 7; D. 4, 5; E. 3; F. 9; G. 11; H. 12; I. 11; J. 14; K. 10

Completion

1. serotyping; 2. titer; 3. agarose; 4. diagnostic skin tests; 5. radioimmunoassay 6. immunoelectrophoresis; 7. hemolysin

Short answer

1. a. Test serum, mumps virus antigen, guinea pig complement, sheep red blood cells; b. serum protein, serum protein antibodies, agarose gel slide or plate; c. serial dilution of the test serum, latex particles with antigen attached; d. toxin in culture media, antibody preparations against each of the four human toxin types (A,B,E,F), a series of test animals
2. a.

b.

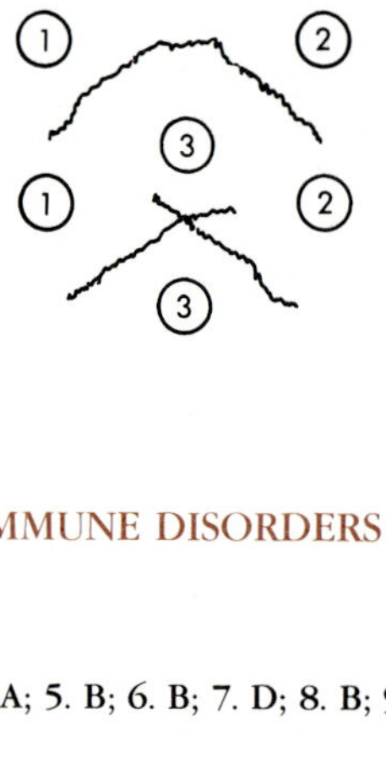

3. a. yes; b. no

CHAPTER 25 IMMUNE DISORDERS

Multiple choice

1. C; 2. D; 3. A; 4. A; 5. B; 6. B; 7. D; 8. B; 9. C; 10. B

Identification

a. type I; b. type III; c. type II; D. type III; E. type IV

CHAPTER 26 EPIDEMIOLOGY OF COMMUNITY AND HOSPITAL-ASSOCIATED DISEASES

Multiple choice

1. A; 2. B; 3. B; 4. C; 5. A; 6. B; 7. C; 8. D; 9. A; 10. B

Completion

1. etiological; 2. incubation; 3. case-fatality ratio; 4. *E. coli* 5. intoxications; 6. chronic; 7. gram-negative; 8. insect vectors; 9. bacteriophage typing; 10. handwashing

CHAPTER 27 HUMAN DISEASES CAUSED BY BACTERIA AND VIRUSES

Completion

1. *Staphylococcus aureus;* 2. tuberculosis; 3. *Mycobacterium;* 4. phospholipase A; 5. *S. aureus;* 6. flea; 7. *Salmonella typhi;* 8. *Bordetella pertussis;* 9. *Haemophilus influenzae;* 10. isoniazid; 11. paralysis of muscles; relaxation; 12. hemagglutining and neuraminidase

Multiple choice

1. A; 2. C; 3. C; 4. D; 5. C; 6. C; 7. A; 8. C; 9. B; 10. C; 11. B; 12. C; 13. A; 14. C

Matching

1. j; 2. a; 3. i; 4. b; 5. d; 6. c; 7. f; 8. l; 9. a; 10. b; 11. e; 12. h; 13. a; 14. c; 15. c; 16. h; 17. a; 18. c

CHAPTER 28 FUNGAL AND ANIMAL PARASITE DISEASES OF HUMANS

Multiple choice

1. C; 2. B; 3. A; 4. D; 5. B; 6. B; 7. B; 8. D; 9. A; 10. B; 11. C; 12. D

Matching

1. d; 2. c; 3. c; 4. a; 5. b; 6. b

Index

D

F

J

K

L

Q

R

Credits

Chapter 1

Figure 1-1 Anderson, D.A., and Sobieski, R.J. *Introduction to microbiology* (2nd ed.), The C.V. Mosby Co., St. Louis, 1980. **Figure 1-5** Historical Picture Service Inc., Chicago. **Figure 1-6** Anderson, D.A., and Sobieski, R.J. *Introduction to microbiology* (2nd ed.), The C.V. Mosby Co., St. Louis, 1980. **Figure 1-8** Historical Picture Service, Inc., Chicago. **Figure 1-9** Air Force Institute of Pathology, AFIP 28-59. **Figure 1-11** Stanley O. Foster, M.D. **Figure 1-13** R.C. Williams. **Figure 1-14** Mizushima, S. *J. Bacteriol.* **140**:1071, 1979.

Chapter 3

Figure 3-1 A, Baush and Lomb, Rochester, N.Y. **B,** Modified from American Optical Corp., Buffalo, N.Y. **Figure 3-5** Corstvet, R.E., et al. *J. Clin. Microbiol.* **76**:1123, 1982. **Figure 3-6** Stevens, A.R., et al. *J. Infect. Dis.* **143**:193, 1981. **Figure 3-9** Nikon, Inc., Instrument Division, Garden City, N.Y. **Figure 3-10 A,** Stahl, M.L., and Williams, F.D. *Appl. Environ. Microbiol.* **41**:801, 1981. **B,** Rosoff, J., and Stibbs, H. *J. Clin. Microbiol.* **23**:905, 1986. **Figure 3-11** Carl Zeiss, Inc., New York. **Figure 3-15** Costerton, J.W. *Annu. Rev. Microbiol.* **33**:459, 1979. **Figure 3-16** Qualls, G.T., et al. *Science* **201**:444, 1978. Copyright 1978 by the American Association for the Advancement of Science.

Chapter 4

Figure 4-2 Redrawn from Fox, G.E., et al. *Science* **209**:457, 1980. Copyright 1980 by the American Association for the Advancement of Science. **Figure 4-3** Newcomb, E.H. and Pugh, T.D. From Raven, P.H., and Johnson, G.B. *Biology.* Time Mirror/Mosby College Publishing, St. Louis, 1986. **Figure 4-4** Photo by F.R. Turner. Courtesy Howard Gest. From Raven, P.H., and Johnson, G.B. *Biology.* Times Mirror/Mosby College Publishing, St. Louis, 1986. **Figure 4-6** Boyd, R.F., and Marr, J.J. *Medical Microbiology.* Little, Brown & Co., Boston, 1980. **Figure 4-7** Carolina Biological Supply Co. **Figure 4-9 A,** Micrograph from Kayama, T., et al. *J. Bacteriol.* **129**:1518, 1977. **B,** Micrograph from P.D. Walker. **C,** Micrograph from Ziolecki, A. *Appl. Environ. Microbiol.* **37**:131, 1979. **D,** Micrograph from Stoeckenius, D. *J. Bacteriol.* **148**:352, 1981. **Figure 4-13** E. Boatman. **Figure 4-14** Zak, O., and Kradolfer, F. *Rev. Infect. Dis.* **1**(5):862, 1979. **Figures 4-15** and **4-16** Raven, P.H., and Johnson, G.B. *Biology.* Times Mirror/Mosby College Publishing, St. Louis, 1986. Photo, D.A. Phillips/Visuals Unlimited. **Figure 4-21** A. Ryter, Institute Pasteur. **Figure 4-22** C.F. Robinow. **Figure 4-23** Raven, P.H., and Johnson, G.B. *Biology.* Times Mirror/Mosby College Publishing, St. Louis, 1986. Courtesy W. Ober. **Figure 4-25** Burdett, I.D.J., and Murray, R.G.E. *J. Bacteriol.* **199**:1039, 1974. **Figure 4-28** Preer, J.R., et al. *Mircobiol. Rev.* **38**:113, 1974. **Figure 4-30 B** and **C,** W. Hodgkiss. **D,** Hoeniger, J.F. *J. Gen. Microbiol.* **40**:29, 1965. **Figure 4-31** Holt, S. *Microbiol. Rev.* **42**:114, 1978. **Figure 4-32** Barlow, D.I., and Sleigh, M.A. *J. Microscopy* **115**:81, 1979. **Figure,** p. 76 Haride, L.P., Balkwill, D.L., and I.E. Stevens, Jr. *Appl. Environ. Microbiol.* **45**:1007, 1985. **Figure,** bottom, p. 76 N.G. Carr.

Chapter 5

Figure 5-1 Ellar, D.J., and Slepecky, R.A. *J. Bacteriol.* **94**:1189, 1967. **Figure 5-16** Sletyr, U.B., and Messner, P. Reproduced with permission from *Annu. Rev. Microbiol.* **37**:311, copyright © 1983 by Annual Reviews, Inc. **Figure 5-17** Babb, J.L. *Infect. Immun.* **19**:1088, 1978. **Figure 5-19** Dazzo, F.B., and Brill, W.J. *J. Bacteriol.* **137**:1362, 1969. **Figure 5-25** Williams, F.D., and Schwarzhoff, R.H. *Annual Review of Microbiology,* Volume 32. Copyright 1978 by Annual Reviews, Inc. **Figure 5-26** Williams, F.D., and Schwarzhoff, R.H. *Annual Review of Microbiology,* Volume 32. Copyright 1978 by Annual Reviews, Inc. **Figure 5-27** A. Ryter, Institute Pasteur. **Figure 5-28** Roger M. Cole. **Figure 5-29** J. Brinton. **Figure 5-31** Laishley, E.J. *Can. J. Microbiol.* **19**:991, 1973. **Figure 5-32** Baxter, M., Sieburth, J. *Appl. Environ. Microbiol.* **47**:31, 1984. **Figure 5-33** Hageage, G. Jr., Eanes, E., and Gherna, R.L. *J. Bacteriol.* **101**:*464, 1970.* **Figure 5-34** Balkwill, D.L., and Blakemore, R.P. *J. Bacteriol.* **141**:*1399, 1980.* **Figure 5-36** Jensen, T.E., and Sicko-Goad, L. *J. Bacteriol.* **106**:683, 1971. **Figure 5-38** P.D. Walker. **Figure 5-39** Z. Yoshii. **Figure, left, p. 111** Razin, S. et al. *Infect. Immun.* **30**:538, 1980. **Figure, right, p. 111** Roger M. Cole.

Chapter 6

Figure 6-1 Coffey, M. *Can. J. Botany* **53**:1285, 1975. **Figure 6-2** Johnson, B.F., and McDonald, I.J. *J. Gen. Microbiol.* **129**:3411, 1983. **Figure 6-3** M.R. McGinnis. **Figure 6-4** Rogers, A.L. in *Manual of clinical microbiology* (4th ed.). American Society for Microbiology, Washington, D.C., 1985. **Figure 6-5 A,** Glen Roberts. **B,** Rogers, A.L. In *Manual of clinical microbiology* (4th ed.). American Society for Microbiology, Washington, D.C., 1980. **C,** Larsh, H.W., and Goodman, N.L. In *Manual of clinical microbiology* (4th ed.). American Society for Clinical Microbiology, Washington, D.C., 1985. **Figure 6-6** Perry, D.F. *Mycologia* **74**:549, copyright © 1980, D.F. Perry and the New York Botanical Garden. **Figure 6-8** From Baker, R.D. In Anderson, W.A.D., and Kissane, J.M. (eds.). *Pathology* (7th ed.). The C.V. Mosby Co., St. Louis, 1977. **Figure 6-10** *Mycologia* **71**:379. Copyright 1979. Dowsett, J.A., and Reid, J.J., and The New York Botanical Garden. **Figure 6-11** *Mycologia* **69**:628. Copyright 1977. Uecker, F.A., and Burdsall, H.H., Jr., and The New York Botanical Garden. **Figure 6-13** Bulmer, G.S., et al. *Infect. Immun.* **20**:262, 1978. **Figure 6-15** Dolan, C.T., et al. *Atlas of clinical mycology.* American Society of Clinical Pathologists, Chicago, 1975. **Figure 6-16** Carolina Biological Supply Co. **Figure 6-17** Mahoney, D.P. and LaFaure, L. S. *Mycologia* **73**:931, copyright 1981, The New York Botanical Garden. **Figure 6-20** Mycologia **71**:445. Copyright 1979. Perry, J.W., Fisher, D.G., and Kuter, G.A., and The New York Botanical Garden. **Figure 6-23** P. Brachet. **Figure 6-24** Carolina Biological Supply Co. **Figure 6-25** From Biology of the Fungi by Ross, I.K. Copyright © 1979 McGraw Hill Book Co. Used with the permission of McGraw-Hill Book Co. **Figure 6-29** Wunderlich, F. *J. Protozool.* **29**:49, 1982. **Figure 6-30** Carolina Biological Supply Co. **Figure 6-31** Beck, J.W., and Davies, J.E. Medical parasitology (ed. 3). The C.V. Mosby Co., St. Louis, 1981. **Figure 6-35** Beck, J.W., and Davies, J.E. Medical Parasitology (ed. 3). The C.V. Mosby Co., St. Louis, 1981. **Figure 6-37** Schmidt, G.D., and Roberts, L.S. *Foundations of parasitology* (2nd ed.). The C.V. Mosby Co., St. Louis, 1981. **Figure 6-41** Barry Rosen, The Ohio State University. **Figure 6-42** Hoops, H.J., and Floyd, G.I. *Phycologia* **18**:424, 1979. **Figure 6-43** Hanic, L., *J. Phycol.* **13**:174, 1979. **Figure 6-45** Pritchard, H.N., and Bradt, P.T. *Biology of nonvascular plants.* The C.V. Mosby Co., St. Louis, 1984. **Figure 6-46** Berdach, J. *J. Phycol.* **13**:243, 1977. **Figure,** p. 152 Beran, K., and Streiblova, E. *Adv. Microb. Physiol.* **2**:143, 1968. Copyright Academic Press, Inc. (London) Ltd.

Chapter 7

Figure 7-1 Howard, R.J., and Aist, J.R. *J. Ultrastruct. Res.* **66**:224, 1979. **Figure 7-2** Hobot, J.A. and Gull, K. *Symp. Soc. Gen. Microbiol.* **30**:207, 1980. Cambridge University Press. **Figure 7-8** Browitzka, M.A. *J. Physiol.* **13**:6, 1977. **Figure 7-10** Roland, J., and Szölliosi, D. *Atlas of cell biology.* Little, Brown & Co., Boston, 1977. **Figure 7-11** A. Stetler, Virginia Polytechnic Institute and State University. **Figure 7-13** Roland, J., Szollosi, A., and Szollosi, D. *Atlas of cell biology.* Little, Brown & Co., Boston, 1977. **Figure 7-14** Fawcett, D.W. *Cell biology.* W.B. Saunders, Philadelphia, 1977. **Figure 7-15**

Olins, A.L., and Olins, D.E. *Am. Sci.* **66**:704, 1977. American Scientist, Journal of Sigma XI, The Scientific Research Society, Inc. **Figure 7-19** Weber, K. *Proc. Natl. Acad. Sci.* (USA) **75**:1820, 1978.

Chapter 8

Figure 8-8 Cairns, J. *Cold Spring Harbor Symp. Quant. Biol.* **28**:43, 1963. **Figure 8-17** R.C. Lambe, Virginia Polytechnic Institute and State University.

Chapter 9

Figure 9-11 Redrawn from Kim, S.H., et al. *Science* **185**:435, 1974. Copyright 1974 by the American Association for the Advancement of Science. **Figure 9-14** Miller, O.L., Jr. *Cold Spring Harbor Symp. Quant. Biol.* **35**:505, 1970.

Chapter 10

Figure 10-11 Redrawn from William Hexter/Henry T. Yost, Jr. *The science of genetics.* Copyright 1976, p. 426. Prentice-Hall, Inc., Englewood Cliffs, N.J. **Figure 10-19** Witkin, E.M. *Microbiol. Rev.* **40**:869, 1976. **Figure 10-20** Redrawn from Witkin, E.M. *Microbiol. Rev.* **40**:869, 1976. **Figure 10-22** Redrawn from Witkin, E.M. *Microbiol. Rev.* **40**:869, 1976.

Chapter 11

Figure 11-15 Redrawn from Jacob, F., and Wollman, E.L. *Competes Rendus Acad. Sci.* **240**:2450, 1965. **Figure 11-16** Redrawn from Taylor, A.L., and Trotter, C.L. *Microbiol. Rev.* **36**:504, 1976.

Chapter 12

Figure, p. 330 Mueller, R.E., et al. *Appl. Environ. Microbiol.* **47:715, 1984. Table 12-1** Adapted from Lehninger, A.L. *Bioenergetics.* W.A. Benjamin Inc., Menlo Park, Calif., 1965.

Chapter 13

Figure 13-9 Jannasch, H.W., and Wirsen, C.O. *Appl. Environ. Microbiol.* **41**:528, 1981

Chapter 14

Figure 14-3 Edwards, C. *Aspects of microbiology* (3). American Society for Microbiology, Washington, D.C., 1981. **Figure 14-6** Redrawn from Cole, R.M. *Bacteriol. Rev.* **29**:326, 1965. **Figure 14-7** Based on Higgins, M.L., and Shockman, G.D. *CRC Crit. Rev. Microbiol.* **1**:29, 1971. Copyright. Chemical Rubber Co., CRC Press, Inc. **Figure 14-15** Kjelleberg, B.A., Humphrey, B.A., and Marshall, K.C. *Appl. Environ. Microbiol.* **46**:978, 1983. **Figure 14-19** Baker, R.M., Singleton, F.L., and Hood, M.A. *Appl. Environ. Microbiol.* **46**:930, 1983. **Figure 14-27** Redrawn from Hartwell, L.H. *Bacteriol. Rev.* **38**:164, 1974. **Figure 14-28** Hartwell, L.H. *Bacteriol. Rev.* **38**:164, 1974. **Figure,** p. 409 Baross, J. and Deming, J.W. Reprinted by permission from *Nature* **303**:423, copyright © 1983, Macmillan Journals Ltd. **Table 14-3** Horikoshi, K., and Teruhiko, A. *Alkalophilic microorganisms.* Springer-Verlag, New York, 1982.

Chapter 15

Figure 15-4 Lorian, V. and Atkinson, B.A. *Rev. Infect. Dis.* **1**(5):797, 1979. **Figure 15-18** Washington, J.A., II, *Mayo Clin. Proc.* **44**:811, 1969. **Figure 15-22** Feeherry, F., Munsey, D., and Rowley D. *Appl. Environ. Microbiol.* **53**:365, 1987. **Figure 15-24** Gelman Sciences, Ann Arbor, Mich. **Figure 15-25** Butler, R., Lund, V., and Carlson, D. *Appl. Environ. Microbiol.* **53**:375, 1987. **Table 15-9** Morton, H.E. *Ann. NY Acad. Sci.* **53**:191, 1950.

Chapter 16

Figure 16-1 Horne, R.M. *The structure and function of viruses.* Copyright 1963 by Scientific American Inc. All rights reserved. **Figure 16-3** Yabe, Y., et al. *Virology* **96**:547, 1979. **Figure 16-4** Roseto, A., et al. *Virology* **98**:471, 1979. **Figure 16-5 B,** Courtesy Mrs. J.D. Almeida and F. Fenner. **C,** From O'Calleghan. **D,** *Prog. Med. Virol.* **22**:152, 1976. S. Karger A.G., Basel. **Figure 16-9** Mattern, C. *Symmetry in virions.* In Nayak, D. (ed.). *Molecular biology of animal virus.* Marcel Dekker, Inc., New York, 1979. Reprinted by courtesy of Marcel Dekker, Inc. **Figure 16-10** Horne, R.W., *Virology,* **15**:348, 1968. **Figure 16-11** R.W. Horne. **Figure 16-12 A,** Enzmann, P.J. *Virology* **95**:1979. **B,** H.D. Mayor. **Figure 16-13** Dales, S. *Am. J. Med.* **38**:699, 1965. **Figure 16-14 B,** Williams, R.C., and Fisher, H.W. *An electron micrograph atlas of viruses.* Charles C Thomas Publisher, Springfield, Ill., 1974. **Figure 16-15** Redrawn from Bradley, D.E. *Microbiol. Rev.* **31**:230, 1967. **Figure 16-16** Yamashiroya, H.M. In Boyd, R.F., and Marr, J. (eds.). *Medical microbiology.* Little, Brown & Co., Boston, 1980. **Figure 16-17** Audiovisual Service Department of Pathology, University of Illinois Medical Center, Chicago. **Figure 16-20** Hawker, L.E. and Linton, A.H. *Microorganisms: function, form and environment* (2nd ed.). Edward Arnold Publishers, Ltd., London, 1979. **Figure 16-21** Symington, J., Commoner, B., and Yamada, M. *Natl. Acad. Sci. USA* **48**:1675, 1962. **Figure 16-23** Satter, S.A., and Ramion, S. *J. Clin. Microbiol.* **10**:609, 1980. **Figure 16-25 A,** Courtesy L. Amos. **B,** Based on model by Finch, J.T., and Klug, A. *J. Mol. Biol.* **15**:315, 1966. Copyright Academic Press, Inc. (London), Ltd. **Figure 16-26** Falk, B.W. *Phytopathology* **69**:612, 1979. **Figure 16-29** Bozarth, R.F., et al. *Virology* **45**:516, 1971. **Figure 16-30** J.R. Preer, Jr. **Figure 16-31** Arnott, H.J. *J. Ultrastruct. Res.* **24**:479, 1968.

Chapter 17

Figure 17-2 Mizushima, S.J. *J. Bacteriol.* **140**:1071, 1979. **Figure 17-10** Gold, M., et al. *Virology* **119**:35, 1982. **Figure 17-13** Walter Brandt. **Figure 17-26 A,** From Karpas, A., et al. *J. Gen. Virol.* **35**:191, 1977. **B,** From Takeuchi, A., and Hashimoto, K. *Infect. Immun.* **13**:569, 1976. **Figure 17-27** Norby, E., et al. *J. Virol.* **6**:237, 1970. **Figure 17-29** Hamada, K. et al. *J. Virol.* **38**:327, 1981. **Figure 17-30** Baskar, J.F., Stanat, S.C., and Huang, E. *Infect. Immun.* **40**:726, 1983. **Figure 17-32** Redrawn from Durham, A.C.H., Finch, J.T., and Klug, A. *Nature* **229**:47, 1971. **Figure 17-33** Durham, A.C.H., et al. with permission from *J. Mol. Biol.* **67**:307, 1972. Copyright Academic Press, Inc. (London) Ltd.

Chapter 18

Figure 18-1 A, Redrawn from Meadows, P.S., and Anderson, J.G. *J. Marine Biol. Assoc. U.K.* **48**:161, 1968. **B,** E.I. Friedman. **Figure 18-3** Modified from Thayer, J.D., et al. In *Antimicrobial agents and chemotherapy—1964.* American Society for Microgiology, Ann Arbor, Michigan, 1964. **Figure 18-4** Modified from Alexander, M.A. *Microbial ecology.* John Wiley & Sons, Inc., New York, 1971. **Figure 18-6** Redrawn from Ferguson, T.J., and Mah, R.A. *Appl. Environ. Microbiol.* **45**:265, 1983. **Figure 18-7** Redrawn from Schaedler, R.W., et al. *J. Exp. Med.* **122**:59, 1965. **Figure 18-8** Raymond C. Valentine, University of California, Davis. **Figure 18-10** Merton F. Brown. **Figure 18-11** Oberhofer, T.R., and Frazier, W.C. *J. Milk Food Technol.* **24**:172, 1961. **Figure 18-12** Burnham, G.F., et al. *J. Bacteriol.* **96**:1366, 1968. **Figure 18-14** Redrawn from Campbell, R. *Microbial ecology.* Blackwell Scientific Publications, Oxford, 1977. **Figure 18-15** Redrawn from Atlas, R.M., and Bartha, R. *Microbial ecology: fundamentals and applications.* Addison-Wesley Publishing Co., Inc., Reading, Mass., 1981. **Figure 18-16** Redrawn from Freudenberg, K. *Science* **148**:595, 1965. **Figure 18-19** Redrawn from Edgehill, R.U., and Finn, R.K. *Appl. Environ. Microbiol.* **45**:1122, 1983. **Figure 18-20** Redrawn from Campbell, R. *Microbial ecology,* Blackwell Scientific Publications, Oxford, 1977. **Figure 18-21** Redrawn from Campbell, R. *Microbial ecology,* Blackwell Scientific Publications, Oxford, 1977. **Figure 18-22** Redrawn from Campbell, R. *Microbial ecology,* Blackwell Scientific Publications, Oxford, 1977. **Figure 18-23** Stokes, J.L., *J. Bacteriol.* **67**:278, 1954. **Figure, bottom, p. 531** Dworkin, M., and Reichenback, H. *Symp. Soc. Gen. Microbiol.* **23**:125, 1973. **Table 18-1** Jones, G.E. Biogeochemical succession of bacterial activities in aquatic sediments. In *Microbiology 1980.* American Society for Microbiology. Washington, D.C., 1980. **Table 18-2** Alexander, M. *Microbial ecology.* John Wiley & Sons, Inc., New York, 1971. **Table 18-3** Alexander, M. *Microbial ecology.* John Wiley & Sons, Inc., New York, 1971. **Table 18-4** Meers, J.L. Growth of bacteria in mixed culture. In Laskin, A., and Lechevalier, H. (eds.). Reprinted with permission from *Microbial ecology.* Copyright CRC Press, Inc., Boca Raton, Florida, 1974. **Table 18-5** Postgate, J.R., and Hill, S. Economic microbial ecology. In Lynch, J.M., and Poole. N.J. (eds.). *Microbial ecology: a concept approach.* Blackwell Scientific Publications, Oxford, 1979.

Chapter 19

Figure 19-2 R.M. Baker and M.A. Hood. **Figure 19-3** Dawson, M.P., Humphrey, B., and Marshall, K.C. *Curr. Microbiol.* 6:195, 1981. **Figure 19-4** A. Huq and R.R. Colwell. **Figure 19-5** C.A. Liebert and M.A. Hood. **Figure 19-10** Hampton Roads Sanitation District. **Figure, p. 563** Hirsch, P. In Holt, J.G. (ed.). *Bergey's manual of systematic bacteriology* (9th ed.). Williams & Wilkins, Baltimore, 1985.

Chapter 20

Figure 20-1 Akin, D.E. *Appl. Environ. Microbiol.* **39**:242, 1980. **Figure 20-2** Yasuko Rikihisa. **Figure 20-3** Wessenber, H., and Antipa, G.J. *J. Protozool.* **17**:250, 1970. **Figure 20-4** Redrawn from Lennox, L.B., and Alexander, M. *Appl. Environ. Microbiol.* **41**:404, 1981. **Figure 20-5** Zinder, S., Anguish, T., and Cardwell, S. *Appl. Environ. Microbiol.* **47**:796, 1984. **Figure 20-6** Madsen, E.L., and Alexander, M. *Appl. Environ. Microbiol.* **50**:342, 1985. **Figure 20-7** Redrawn from Marshall, K.C. *Interfaces in microbiology ecology.* Harvard University Press, Cambridge, Mass., 1976. **Figure 20-8** Killham, K., and Firestone, M.,. *Appl. Environ. Microbiol.* **47**:301, 1984. **Figure 20-9** Redrawn from Alexander, M. *Introduction to soil microbiology* (2nd ed.) John Wiley & Sons, Inc., New York, 1977. **Figure 20-13** R. Jay Stipes, Virginia Polytechnic Institute and State University. **Table 20-1** After Lochhead, A.G. *Can. J. Res.* **C18**:42, 1940. **Table 20-3** Takia, Y. *Folia Microbiol.* **11**:304, 1966. **Table 20-4** Woldendorp, J.W. *Structure and functioning of plant populations.* Verhandeligen der Koninklijke, Nederlandse Akademie van Wetenschappen, Afdeling Naturukunde, Twede Riiks, deel 70, 1978.

Chapter 21

Figure 21-3 Malo, Inc., Tulsa, Okla. **Figure 21-4** Anderson, D.A., and Sobieski, R.J. *Introduction to microbiology* (2nd ed.). St. Louis, The C.V. Mosby Co., 1980. **Figure 21-5** Ingraham, J.L. *Proceedings low temperature microbiology symposium.* Campbell Soup Company, Camden, N.Y., 1961. **Figure 21-6** Troller, J.A. *Appl. Microbiol.* **21**:435, 1971. **Figure 21-7** Karen Siebert, Ralston Purina Co. **Figure 21-9** Ralston Purina Co. **Figure 21-12** P.D. Walker, Wellcome Research Laboratories, Kent, England. **Figure 21-13** Centers for Disease Control, Bureau of Epidemiology, Hepatitis Laboratories, Phoenix, Arizona. **Table 21-2** Goepfert, J.M., and Biggie, R.A. *Appl. Microbiol.* **16**:1939, 1968. **Table 21-4** Adapted from Troller, J.A., and Christian, J.H.B. *Water activity and food.* Academic Press, Inc., New York, 1978. **Table 21-9** Adapted from Jay, J.M. *Modern food microbiology* (2nd ed.). D. Van Nostrand, New York, 1978. **Table 21-10** Adapted from Liston, J. In *Microbial ecology of foods* (vol. 2). The International Commission on Microbiological Specifications for Foods, Academic Press, Inc., New York, 1980. **Table 21-11** Adapted from Banwart, G.J. *Basic food microbiology.* AVI Publishing, Westport, Conn., 1979. **Table 21-15** Centers for Disease Control, Atlanta, Georgia.

Chapter 22

Figure 22-2 K. Raper. **Figure 22-10** H. Samejima, Kyowa Hakko Kogyo Co., Ltd. **Figure 22-13** K. Raper.

Chapter 23

Figure 23-4 Moon, H.J. *J. Infect. Dis.* **136**:S124, 1977 **Figure 23-5** Beachey, E.H., and Stollerman, G.H. In Dumonde, D.C. (ed.). *Infection and immunology in the rheumatic diseases.* Blackwell Scientific Publications, Ltd., Oxford, 1976. **Figure 23-8** Vetterling, J.M., et al. *J. Protozool.* **18**:255, 1971. The Society of Protozoologists. **Figure 23-10 A,** Roitt, I. *Essential immunology* (4th ed.). Blackwell Scientific Publications, Ltd., Oxford, 1980. **B,** Adelstein, E., Barnett, J.T. *Textbook of immunology* (4th ed.). The C.V. Mosby Co., St. Louis, 1983. **Figure 23-11 A,** Roitt, I. *Essential immunology* (4th ed.). Blackwell Scientific Publications, Ltd., Oxford, 1980. **B,** Horowitz, M.A. in Schlessinger, D. (ed.) *Microbiology—1982.* American Society for Microbiology, Washington, D.C., 1982. **Figure 23-12** H. Valdimarsson. In Roitt, I. *Essential immunology* (4th ed.). Blackwell Scientific Publications, Ltd., Oxford, 1980. **Figure 23-14** Robertson, D.C., et al. In Schlessinger, D. (ed.) *Microbiology—1979.* American Society for Microbiology, Washington, D.C., 1979. **Figure 23-15** Sullivan, G.W., and Mandell, G.L. *Infect. Immunol.* **30**:272, 1980. **Figure 23-17** Alyne K. Harrison. **Figure 23-18** Valentine, R.C., and Pereira, H.G., with permission from *J. Mol. Biol.* **13**:13, 1965. Copyright Academic Press, Inc. (London) Ltd.

Chapter 24

Figure 24-6 Redrawn from Thaler, M.S., Klausner, R.D., and Cohen, H.J. *Medical immunology.* J.B. Lippincott Co., Philadelphia, 1977. **Figure 24-8** Redrawn from Scharff, M.D., Roberts, S., and Thammena, P. *Hosp. Pract.* **16**:64, 1981. **Figure 24-23** Barrett, J.T. *Textbook of immunology* (4th ed.). The C.V. Mosby Co., St. Louis, 1983. **Figure 24-24** Smith, A.L. *Microbiology and pathology* (12th ed.). The C.V. Mosby Co., St. Louis, 1980. **Figure 24-27** William Krass. From Barret, J.T., *Textbook of immunology* (4th ed.). The C.V. Mosby Co., St. Louis, 1983. **Figure 24-28** Redrawn from Roitt, I. *Essential immunology* (2nd ed.). Blackwell Scientific Publications, Ltd., Oxford, 1974. **Figure 24-32** R. Austrian. **Figure 24-35** Bauer, J.D., *Clinical laboratory methods* (9th ed.). The C.V. Mosby Co., St. Louis, 1974. **Figure 24-38** Nakamura, R.M. *Immunopathology: clinical laboratory concepts and methods.* Little, Brown & Co., Boston, 1974. **Table 24-1** Moticka, E.J. In Boyd, R.F., and Marr, J.J. (eds.). *Medical microbiology.* Little, Brown & Co., New York, 1980. **Table 24-2** Boyd, R.F., and Hoerl, B.G. *Basic medical microbiology* (3rd ed.) Little, Brown & Co., Boston, 1986.

Chapter 25

Figure 25-3 Midleton, E., Jr., Reed, C.E., and Ellis, E.F. (eds.) *Allergy: principles and practice* (vol. 2.). The C.V. Mosby Co., St. Louis, 1978. **Figure 25-5** Middleton, E., Jr., Reed, C.E., and Ellis, E.F. *Allergy: principles and practice* (vol. 2.). The C.V. Mosby Co., St. Louis, 1978. **Table 25-1** Modified from Meuwissen, H. *Postgrad. Med.* **66**(5):116, 1979. **Table 25-2** Meuwissen, H. *Postgrad. Med.* **66**(5):116, 1979. **Table 25-3** Boyd, R.F., and Hoerl, B.G. *Basic medical microbiology* (3rd ed.). Little, Brown & Co., Boston, 1986. **Table 25-4** Barrett, J.T. *Textbook of immunology* (4th ed.). The C.V. Mosby Co., St. Louis, 1983. **Table 25-5** James, L.P., and Austen, K.F. Reprinted by permission of the New England Journal of Medicine **270**:597, 1964. **Table 25-6** Boyd, R.F., and Marr, J.J. *Medical microbiology.* Little, Brown & Co., Boston, 1980.

Chapter 26

Figure 26-1 A, Schmidt, G.D., and Roberts, L.S. *Foundations of parasitology* (3rd ed.). The C.V. Mosby Company, St. Louis, 1986. Photograph by Warren Buss. **B,** Schmidt, G.D. and Roberts, L.S. *Foundations of parasitology* (3rd ed.). The C.V. Mosby Company, St. Louis, 1986. Photograph by Jay George. **C,** Schmidt, G.D. and Roberts, L.S. *Foundations of parasitology* (3rd ed.). The C.V. Mosby Co., St. Louis, 1986. Photograph by Warren Buss. **D,** Hickman, C.P., et al. *Integrated principles of zoology* (7th ed.). The C.V. Mosby Co., St. Louis, 1981. **Figure 26-6** Redrawn from *MMWR* **30**:*114,1981.* **Figure 26-7** Redrawn from *MMWR* **30**:183, 1981. **Figure 26-9** Redrawn from National Nosocomial Infection Study, 1978. **Figure 26-10** Marrie, T.J., et al. *J. Clin Microbiol.* **18**:1388, 1983. **Figure 26-13** Takahashi, S., and Nagano, Y. *J. Clin. Microbiol.* **20**:608, 1984. **Table 26-4** Favero, M.S. In Fahlberg, W.J., and Gröschel, D. (eds.). Occurence, diagnosis, and sources of hospital-associated infections. Marcel Dekker, Inc., New York, 1980.

Chapter 27

Figure 27-1 Hamada, S. *Microbiol. Rev.* **44**:331, 1980. **Figure 27-2** Dajani, A.S. *Infect. Immun.* **14**:776, 1986. **Figure 27-3** Anderson, W.A.D., and Kissane, J.M. (eds.). *Pathology* (vol 2.). The C.V. Mosby Co., St. Louis, 1977. **Figure 27-7** Redrawn from Fitzgerald, T.J. *Annu. Rev. Microbiol.* **35**:29, 1981. **Figure 27-8** Centers for Disease Control, Atlanta, Georgia. **Figure 27-9** Centers for Disease Control, Atlanta, Georgia. **Figure 27-10** Centers for Disease Control, Atlanta, Georgia. **Figure 27-11** Z. Yoshii. **Figure 27-14** Wyrick, P. and Brownridge, E.A. *Infect. Immun.* **19**:1054, 1978. **Figure 27-16** P. Spears. **Figure 27-17** Glasgow, L.A. *N. Engl. J. Med.* **282**:114, 1970. **Figure 27-18** Gillies, R.R., and Dodds, T.C. *Bacteriology illustrated.* Churchill-Livingstone, Edinburgh, 1976. **Figure 27-20** Olds, R.J. *Color atlas of microbiology.* Year Book Medical Publishers, Chicago, 1975. **Figure 27-21** Air Force Institute of Pathology, AFIP 2803. **Figure 27-22** Pead, P.J.J. *Med. Mi-*

crobiol. **12**:383, 1979. **Figure 27-25** Fred A. Murphy. **Figure 27-27** R.C. Williams. **Figure 27-30** Dr. Erskine L. Palmer, Centers for Disease Control, Atlanta, Georgia. **Figure 27-31** Viac, J. *Br. J. Vener. Dis.* **54**:172, 1978. **Figure 27-32** Centers for Disease Control, Atlanta, Georgia. **Figure 27-33** G.A. Cabral and M. Patterson. **Figure 27-35** R. Kogasaka.

Chapter 28

Figure 28-1 Dolan, C.T., et al. *Atlas of clinical mycology.* American Society of Clinical Pathologists, Chicago, 1975. **Figure 28-2** Dolan, C.T., et al. *Atlas of clinical mycology.* American Society of Clinical Pathologists, Chicago, 1975. **Figure 28-3** Dolan, C.T., et al. *Atlas of clinical mycology.* American Society of Clinical Pathologists, Chicago, 1975. **Figure 28-4** King, E.U., and Brown, M.F. *Can. J. Microbiol.* **29**:653, 1983. **Figure 28-5** King, R. *Infect. Immun.* **27**:667, 1980. **Figure 28-7** Dolan, C.T., et al. *Atlas of clinical mycology.* American Society of Clinical Pathologists, Chicago, 1975. **Figure 28-8** Drouhet, E., and Dupont, C. *Rev. Infect. Dis.* **2(4)**:606, 1980. **Figure 28-10** John E. Donelson. **Figure 28-11** Beck, J.W., and Davies, J.E. *Medical parasitology* (3rd ed.). The C.V. Mosby Co., St. Louis, 1981. **Figure 28-12** Modified from Beck, J.W. and Davies, J.E. *Medical parasitology* (3rd ed.). The C.V. Mosby Co., St. Louis, 1981.

Color plates

Plates 2, 10, and 40 *Stratford B.C.* An atlas of medical microbiology. *Blackwell Scientific Publications, Oxford, 1977.* **Plates 3 and 37** *The Centers for Disease Control, Atlanta.* **Plates 4 and 5** *Corstvet, R.E. et al.* J. Clin Microbiol. **16**:*123, 1982.* **Plates 6 to 8, 29, and 36** *Finegold, S.M., and Martin, W.J.* Diagnostic microbiology *(6th ed.). The C.V. Mosby Co., St. Louis, 1982.* **Plate 9** *Robert Wick, Virginia Polytechnic Institute and State University.* **Plates 11 to 15** *Stanley Flegler, Michigan State University.* **Plates 16 and 17** *William R. Walker, Virginia Polytechnic Institute and State University.* **Plates 18 and 19** *Kjell Sandved.* **Plates 20 and 21** *J.W. Hastings, Harvard University.* **Plate 22** *Raven, P.H., and Johnson, G.B.* Biology. *Times Mirror/Mosby College Publishing, 1986.* **Plate 23** *George H. Lacy, Virginia Polytechnic Institute and State University.* **Plate 24** *Photo R.G. Peterson, Courtesy Monterey Vineyards, Gonzales, Calif.* **Plate 25** *Sue A. Tobin, Virginia Polytechnic Institute and State University.* **Plates 26, 28, 30** *Karen Siebert.* **Plate 27** *Nauvoo Cheese Co.* **Plates 31 to 33** *Miale, J.B.* Laboratory medicine: hematology *(6th ed.). The C.V. Mosby Co., St. Louis, 1982.* **Plate 34** *L.J. Le Beau, Department of Biocommunication Arts, University of Illinois at Chicago.* **Plate 35** *Finegold, S.M., and Martin, W.J.* Diagnostic microbiology *(7th ed.). The C.V. Mosby Co., St. Louis, 1986.* **Plate 38** *Gordon Carter, Virginia Polytechnic Institute and State University.* **Plate 39** *John Utz, M.D.*